Unit Conversions (Equivalents)

Length

1 in. = 2.54 cm
1 cm = 0.394 in.
1 ft = 30.5 cm
1 m = 39.37 in. = 3.28 ft
1 mi = 5280 ft = 1.61 km
1 km = 0.621 mi
1 nautical mile (U.S.) = 1.15 mi = 6076 ft = 1.852 km
1 fermi = 1 femtometer (fm) = 10^{-15} m
1 angstrom (Å) = 10^{-10} m
1 light-year (ly) = 9.46×10^{15} m
1 parsec = 3.26 ly = 3.09×10^{16} m

Volume

1 liter (L) = 1000 mL = 1000 cm^3 = 1.0×10^{-3} m^3 =
   1.057 quart (U.S.) = 54.6 in.^{3}
1 gallon (U.S.) = 4 qt (U.S.) = 231 in.^{3} = 3.78 L =
   0.83 gal (Imperial)
1 m^3 = 35.31 ft^3

Speed

1 mi/h = 1.47 ft/s = 1.609 km/h = 0.447 m/s
1 km/h = 0.278 m/s = 0.621 mi/h
1 ft/s = 0.305 m/s = 0.682 mi/h
1 m/s = 3.28 ft/s = 3.60 km/h
1 knot = 1.151 mi/h = 0.5144 m/s

Angle

1 radian (rad) = 57.30° = 57°18'
1° = 0.01745 rad
1 rev/min (rpm) = 0.1047 rad/s

Time

1 day = 8.64×10^4 s
1 year = 3.156×10^7 s

Mass

1 atomic mass unit (u) = 1.6605×10^{-27} kg
1 kg = 0.0685 slug
[1 kg has a weight of 2.20 lb where g = 9.81 m/s^2.]

Force

1 lb = 4.45 N
1 N = 10^5 dyne = 0.225 lb

Energy and Work

1 J = 10^7 ergs = 0.738 ft·lb
1 ft·lb = 1.36 J = 1.29×10^{-3} Btu = 3.24×10^{-4} kcal
1 kcal = 4.18×10^3 J = 3.97 Btu
1 eV = 1.602×10^{-19} J
1 kWh = 3.60×10^6 J = 860 kcal

Power

1 W = 1 J/s = 0.738 ft·lb/s = 3.42 Btu/h
1 hp = 550 ft·lb/s = 746 W

Pressure

1 atm = 1.013 bar = 1.013×10^5 N/m^2
   = 14.7 lb/in.^{2} = 760 torr
1 lb/in.^{2} = 6.90×10^3 N/m^2
1 Pa = 1 N/m^2 = 1.45×10^{-4} lb/in.^{2}

SI Derived Units and Their Abbreviations

Quantity	Unit	Abbreviation	In Terms of Base Units†
Force	newton	N	kg·m/s^2
Energy and work	joule	J	kg·m^2/s^2
Power	watt	W	kg·m^2/s^3
Pressure	pascal	Pa	kg/(m·s^2)
Frequency	hertz	Hz	s^{-1}
Electric charge	coulomb	C	A·s
Electric potential	volt	V	kg·m^2/(A·s^3)
Electric resistance	ohm	Ω	kg·m^2/(A^2·s^3)
Capacitance	farad	F	A^2·s^4/(kg·m^2)
Magnetic field	tesla	T	kg/(A·s^2)
Magnetic flux	weber	Wb	kg·m^2/(A·s^2)
Inductance	henry	H	kg·m^2/(s^2·A^2)

†kg = kilogram (mass), m = meter (length), s = second (time), A = ampere (electric current).

Metric (SI) Multipliers

Prefix	Abbreviation	Value
exa	E	10^{18}
peta	P	10^{15}
tera	T	10^{12}
giga	G	10^9
mega	M	10^6
kilo	k	10^3
hecto	h	10^2
deka	da	10^1
deci	d	10^{-1}
centi	c	10^{-2}
milli	m	10^{-3}
micro	μ	10^{-6}
nano	n	10^{-9}
pico	p	10^{-12}
femto	f	10^{-15}
atto	a	10^{-18}

PHYSICS

for

SCIENTISTS & ENGINEERS

PHYSICS

for

SCIENTISTS & ENGINEERS

with Modern Physics

Third Edition

DOUGLAS C. GIANCOLI

PRENTICE HALL
Upper Saddle River, New Jersey 07458

Editor-in-Chief: Paul F. Corey
Production Editor: Susan Fisher
Executive Editor: Alison Reeves
Development Editor: David Chelton
Director of Marketing: John Tweedale
Senior Marketing Manager: Erik Fahlgren
Assistant Vice President of Production and Manufacturing: David W. Riccardi
Executive Managing Editor: Kathleen Schiaparelli
Manufacturing Manager: Trudy Pisciotti
Art Manager: Gus Vibal
Director of Creative Services: Paul Belfanti
Advertising and Promotions Manager: Elise Schneider
Editor in Chief of Development: Ray Mullaney
Project Manager: Elizabeth Kell
Photo Research: Mary Teresa Giancoli
Photo Research Administrator: Melinda Reo
Copy Editor: Jocelyn Phillips
Editorial Assistant: Marilyn Coco
Cover photo: Onne van der Wal/Young America
Composition: Emilcomp srl / Preparé Inc.

© 2000, 1988, 1984 by Douglas C. Giancoli
Published by Prentice Hall
Upper Saddle River, NJ, 07458

Photo credits appear at the end of the book, and
constitute a continuation of the copyright page.

Printed in the United States of America

10 9 8 7 6 5 4 3 2 1

ISBN 0-13-021517-1

Prentice-Hall International (UK) Limited, *London*
Prentice-Hall of Australia Pty. Limited, *Sydney*
Prentice-Hall Canada Inc., *Toronto*
Prentice-Hall Hispanoamericana, S.A., *Mexico City*
Prentice-Hall of India Private Limited, *New Delhi*
Prentice-Hall of Japan, Inc., *Tokyo*
Prentice-Hall (*Singapore*) Pte. Ltd.
Editora Prentice-Hall do Brasil, Ltda., *Rio de Janeiro*

CONTENTS

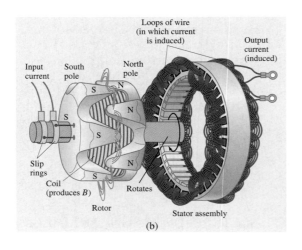

Loops of wire (in which current is induced)

Output current (induced)

Input current

South pole

North pole

Slip rings

Coil (produces B)

Rotates

Rotor

Stator assembly

(b)

APPENDICES

PREFACE

A Brand New Third Edition

It has been more than ten years since the second edition of this calculus-based introductory physics textbook was published. A lot has changed since then, not only in physics itself, but also in how physics is presented. Research in how students learn has provided textbook authors new opportunities to help students learn physics and learn it well.

This third edition comes in two versions. The standard version covers all of classical physics plus a chapter on special relativity and one on the early quantum theory. The extended version, with modern physics, contains a total of nine detailed chapters on modern physics, ending with astrophysics and cosmology. This book retains the original approach: in-depth physics, concrete and nondogmatic, readable.

This new third edition has many improvements in the physics and its applications. Before discussing those changes in detail, here is a list of some of the overall changes that will catch the eye immediately.

Full color throughout is not just cosmetic, although fine color photographs do help to attract the student readers. More important, full color diagrams allow the physics to be displayed with much greater clarity. We have not stopped at a 4-color process; this book has actually been printed in 5 pure colors (5 passes through the presses) to provide better variety and definition for illustrating vectors and other physics concepts such as rays and fields. I want to emphasize that color is used pedagogically to bring out the physics. For example, different types of vectors are given different colors—see the chart on page xxxi.

Many more diagrams, almost double the number in the previous edition, have all been done or redone carefully using full color; there are many more graphs and many more photographs throughout. See for example in optics where new photographs show lenses and the images they make.

Marginal notes have been added as an aid to students to (i) point out what is truly important, (ii) serve as a sort of outline, and (iii) help students find details about something referred to later that they may not remember so well. Besides such "normal" marginal notes, there are also marginal notes that point out brief *problem solving* hints, and others that point out interesting *applications*.

The great laws of physics are emphasized by giving them a marginal note all in capital letters and enclosed in a rectangle. The most important equations, especially those expressing the great laws, are further emphasized by a tan-colored screen behind them.

Chapter opening photographs have been chosen to illustrate aspects of each chapter. Each was chosen with an eye to writing a caption which could serve as a kind of summary of what is in that chapter, and sometimes offer a challenge. Some chapter-opening photos have vectors or other analysis superimposed on them.

Page layout: complete derivations. Serious attention has been paid to how each page was formatted, especially for page turns. Great effort has been made to keep important derivations and arguments on facing pages. Students then don't have to turn back to check. More important, readers repeatedly see before them, on two facing pages, an important slice of physics.

Two new kinds of Examples: Conceptual Examples and **Estimates.**

New Physics

The whole idea of a new edition is to improve, to bring in new material, and to delete material that is verbose and only makes the book longer or is perhaps too advanced and not so useful. Here is a brief summary of a few of the changes involving the physics iself. These lists are selections, not complete lists.

New discoveries:
- planets revolving around distant stars
- Hubble Space Telescope
- updates in particle physics and cosmology, such as inflation and the age of the universe

New physics topics added:
- new treatment of how to make estimates (Chapter 1), including new Estimating Examples throughout (in Chapter 1, estimating the volume of a lake, and the radius of the Earth)
- symmetry used much more, including for solving problems
- new Tables illustrating the great range of lengths, time intervals, masses, voltages
- gravitation as curvature of space, and black holes (Chapter 6)
- engine efficiency (Chapter 8 as well as Chapter 20)
- rolling with and without slipping, and other useful details of rotational motion (Chapter 10)
- forces in structures including trusses, bridges, arches, and domes (Chapter 12)
- square wave (Chapter 15)
- using the Maxwell distribution (Chapter 18)
- Otto cycle (Chapter 20)
- statistical calculation of entropy change in free expansion (Chapter 20)
- effects of dielectrics on capacitor connected and not (Chapter 24)
- grounding to avoid electric hazards (Chapter 25)
- three phase ac (Chapter 31)
- equal energy in E and B of EM wave (Chapter 32)
- radiation pressure, EM wave (Chapter 32)
- photos of lenses and mirrors with their images (Chapter 33)
- detailed outlines for ray tracing with mirrors and lenses (Chapters 33, 34)
- lens combinations (Chapter 34)
- new radiation standards (Chapter 43)
- Higgs boson, supersymmetry (Chapter 44)

Modern physics. A number of modern physics topics are discussed in the framework of classical physics. Here are some highlights:
- gravitation as curvature of space, and black holes (Chapter 6)
- planets revolving around distant stars (Chapter 6)
- kinetic energy at relativistic speeds (Chapter 7)
- nuclear collisions (Chapter 9)
- star collapse (Chapter 10)
- galaxy red shift, Doppler (Chapter 16)
- atoms, theory of (Chapters 17, 18, 21)
- atomic theory of thermal expansion (Chapter 17)
- mass of hydrogen atom (Chapter 17)
- atoms and molecules in gases (Chapters 17, 18)
- molecular speeds (Chapter 18)
- equipartition of energy; molar specific heats (Chapter 19)
- star size (Chapter 19)
- molecular dipoles (Chapters 21, 23)
- cathode ray tube (Chapters 23, 27)
- electrons in a wire (Chapter 25)
- superconductivity (Chapter 25)
- discovery and properties of the electron, e/m, oil drop experiment (Chapter 27)
- Hall effect (Chapter 27)

- magnetic moment of electrons (Chapter 27)
- mass spectrometer (Chapter 27)
- velocity selector (Chapter 27)
- electron spin in magnetic materials (Chapter 28)
- light and EM wave emission (Chapter 32)
- spectroscopy (Chapter 36)

Many other examples of modern physics are found as Problems, even in early chapters. Chapters 37 and 38 contain the modern physics topics of Special Relativity, and an introduction to Quantum Theory and Models of the Atom. The longer version of this text, "with Modern Physics," contains an additional seven chapters (for a total of nine) which present a detailed and extremely up-to-date treatment of modern physics: Quantum Mechanics of Atoms (Chapters 38 to 40); Molecules and Condensed Matter (Chapter 41); Nuclear Physics (Chapter 42 and 43); Elementary Particles (Chapter 44); and finally Astrophysics, General Relativity, and Cosmology (Chapter 45).

Revised physics and reorganizations. First of all, a major effort has been made to not throw everything at the students in the first few chapters. The basics have to be learned first; many aspects can come later, when the students are more prepared. Secondly, a great part of this book has been rewritten to make it clearer and more understandable to students. Clearer does not always mean simpler or easier. Sometimes making it "easier" actually makes it harder to understand. Often a little more detail, without being verbose, can make an explanation clearer. Here are a few of the changes, big and small:

- new graphs and diagrams to clarify velocity and acceleration; deceleration carefully treated.
- unit conversion now a new Section in Chapter 1, instead of interrupting kinematics.
- circular motion: Chapter 3 now gives only the basics, with more complicated treatment coming later: non-uniform circular motion in Chapter 5, angular variables in Chapter 10.
- Newton's second law now written throughout as $m\mathbf{a} = \Sigma\mathbf{F}$, to emphasize inclusion of all forces acting on a body.
- Newton's third law follows the second directly, with inertial reference frames placed earlier. New careful discussions to head off confusion when using Newton's third law.
- careful rewriting of chapters on Work and Energy, especially potential energy, conservative and nonconservative forces, and the conservation of energy.
- renewed emphasis that $\Sigma\tau = I\alpha$ is not always valid: only for an axis fixed in an inertial frame or if axis is through the CM (Chapters 10 and 11).
- rolling motion introduced early in Chapter 10, with more details later, including rolling with and without slipping.
- rotating frames of reference, and Coriolis, moved later, to Chapter 11, shortened, optional, but still including why an object does not fall straight down on Earth.
- fluids reduced to a single chapter (13); some topics and details dropped or greatly shortened.
- clearer details on how an object floats (Chapter 13).
- distinction between wave interference in space, and in time (beats) (Chapter 16).
- thermodynamics reduced to four chapters; the old chapters on Heat and on the First Law of Thermodynamics have been combined into one (19), with some topics shortened and a more rational sequence of topics achieved.
- heat transfer now follows the first law of thermodynamics (Chapter 19).
- electric potential carefully rewritten for accuracy (Chapter 23).
- CRT, computer monitors, TV, treated earlier (Chapter 23).
- use of Q_{encl} and I_{encl} for Gauss's and Ampère's laws, with subscripts meaning "enclosed".
- Ohm's law and definition of resistance carefully redone (Chapter 25).
- sources of magnetic field, Chapter 28, reorganized for ease of understanding, with some new material, and deletion of the advanced topic on magnetization vector.
- circuits with L, C, and/or R now introduced via Kirchhoff's loop rule, and clarified in other ways too (Chapters 30, 31).
- streamlined Maxwell's equations, with displacement current downplayed (Chapter 32).
- optics reduced to four chapters; polarization is now placed in the same chapter as diffraction.

New Pedagogy

All of the above mentioned revisions, rewritings, and reorganizations are intended to help students learn physics better. They were done in response to contemporary research in how students learn, as well as to kind and generous input from professors who have read, reviewed, or used the previous editions. This new edition also contains some new elements, especially an increased emphasis on conceptual development:

Conceptual Examples, typically 1 or 2 per chapter, sometimes more, are each a sort of brief Socratic question and answer. It is intended that students will be stimulated by the question to think, or reflect, and come up with a response—before reading the Response given. Here are a few:

- using symmetry (Chapters 1, 44, and elsewhere)
- ball moving upward: misconceptions (Chapter 2)
- reference frames and projectile motion: where does the apple land? (Chapter 3)
- what exerts the force that makes a car move? (Chapter 4)
- Newton's third law clarification: pulling a sled (Chapter 4)
- free-body diagram for a hockey puck (Chapter 4)
- advantage of a pulley (Chapter 4), and of a lever (Chapter 12)
- to push or to pull a sled (Chapter 5)
- which object rolls down a hill faster? (Chapter 10)
- moving the axis of a spinning wheel (Chapter 11)
- tragic collapse (Chapter 12)
- finger at top of a full straw (Chapter 13)
- suction cups on a spacecraft (Chapter 13)
- doubling amplitude of SHM (Chapter 14)
- do holes expand thermally? (Chapter 17)
- simple adiabatic process: stretching a rubber band (Chapter 19)
- charge inside a conductor's cavity (Chapter 22)
- how stretching a wire changes its resistance (Chapter 25)
- series or parallel (Chapter 26)
- bulb brightness (Chapter 26)
- spiral path in magnetic field (Ch. 27)
- practice with Lenz's law (Chapter 29)
- motor overload (Chapter 29)
- emf direction in inductor (Chapter 30)
- photo with reflection—is it upside down? (Chapter 33)
- reversible light rays (Chapter 33)
- how tall must a full-length mirror be? (Chapter 33)
- diffraction spreading (Chapter 36)

Estimating Examples, roughly 10% of all Examples, also a new feature of this edition, are intended to develop the skills for making order-of-magnitude estimates, even when the data are scarce, and even when you might never have guessed that any result was possible at all. See, for example, Section 1–6, Examples 1–5 to 1–8.

Problem Solving, with New and Improved Approaches

Learning how to approach and solve problems is a basic part of any physics course. It is a highly useful skill in itself, but is also important because the process helps bring understanding of the physics. Problem solving in this new edition has a significantly increased emphasis, including some new features.

Problem-solving boxes, about 20 of them, are new to this edition. They are more concentrated in the early chapters, but are found throughout the book. They each outline a step-by-step approach to solving problems in general, and/or specifically for the material being covered. The best students may find these separate "boxes" unnecessary (they can skip them), but many students will find it helpful to be reminded of the general approach and of steps they can take to get started; and, I think, they help to build confidence. The general problem solving box in Section 4–8 is placed there, after students have had some experience wrestling with problems, and so may be strongly motivated to read it with close attention. Section 4–8 can, of course, be covered earlier if desired.

Problem-solving Sections occur in many chapters, and are intended to provide extra drill in areas where solving problems is especially important or detailed.

Examples. This new edition has many more worked-out Examples, and they all now have titles for interest and for easy reference. There are even two new categories of Example: Conceptual, and Estimates, as described above. Regular Examples serve as "practice problems". Many new ones have been added, some of the old ones have been dropped, and many have been reworked to provide greater clarity and detail: more steps are spelled out, more of "why we do it this way", and more discussion of the reasoning and approach. In sum, the idea is "to think aloud with the students", leading them to develop insight. The total number of worked-out Examples is about 30% greater than in the previous edition, for an average of 12 to 15 per chapter. There is a significantly higher concentration of Examples in the early chapters, where drill is especially important for developing skills and a variety of approaches. The level of the worked-out Examples for most topics increases gradually, with the more complicated ones being on a par with the most difficult Problems at the end of each chapter, so that students can see how to approach complex problems. Many of the new Examples, and improvements to old ones, provide relevant applications to engineering, other related fields, and to everyday life.

Problems at the end of each chapter have been greatly increased in quality and quantity. There are over 30% more Problems than in the second edition. Many of the old ones have been replaced, or rewritten to make them clearer, and/or have had their numerical values changed. Each chapter contains a large group of Problems arranged by Section and graded according to difficulty: level I Problems are simple, designed to give students confidence; level II are "normal" Problems, providing more of a challenge and often the combination of two different concepts; level III are the most complex, typically combining different issues, and will challenge even superior students. The arrangement by Section number means only that those Problems depend on material up to and including that Section: earlier material may also be relied upon. The ranking of Problems by difficulty (I, II, III) is intended only as a guide.

General Problems. About 70% of Problems are ranked by level of difficulty (I, II, III) and arranged by Section. New to this edition are General Problems that are unranked and grouped together at the end of each chapter, and account for about 30% of all problems. The average total number of Problems per chapter is about 90. Answers to odd-numbered Problems are given at the back of the book.

Complete Physics Coverage, with Options

This book is intended to give students the opportunity to obtain a thorough background in all areas of basic physics. There is great flexibility in choice of topics so that instructors can choose which topics they cover and which they omit. Sections marked with an asterisk can be considered optional, as discussed more fully on p. xxv. Here I want to emphasize that topics not covered in class can still be read by serious students for their own enrichment, either immediately or later. Here is a partial list of physics topics, not the standard ones, but topics that might not usually be covered, and that represent how thorough this book is in its coverage of basic physics. Section numbers are given in parentheses.

- use of calculus; variable acceleration (2–8)
- nonuniform circular motion (5–4)
- velocity-dependent forces (5–5)
- gravitational versus inertial mass; principle of equivalence (6–8)
- gravitation as curvature of space; black holes (6–9)
- kinetic energy at very high speed (7–5)
- potential energy diagrams (8–9)
- systems of variable mass (9–10)
- rotational plus translational motion (10–11)
- using $\Sigma \tau_{CM} = I_{CM} \alpha_{CM}$ (10–11)
- derivation of $K = K_{CM} + K_{rot}$ (10–11)
- why does a rolling sphere slow down? (10–12)
- angular momentum and torque for a system (11–4)
- derivation of $d\mathbf{L}_{CM}/dt = \Sigma \boldsymbol{\tau}_{CM}$ (11–4)
- rotational imbalance (11–6)
- the spinning top (11–8)
- rotating reference frames; inertial forces (11–9)
- coriolis effect (11–10)
- trusses (12–7)
- flow in tubes: Poiseuille's equation (13–11)
- surface tension and capillarity (13–12)
- physical pendulum; torsion pendulum (14–6)
- damped harmonic motion: finding the solution (14–7)
- forced vibrations; equation of motion and its solution; Q-value (14–8)
- the wave equation (15–5)
- mathematical representation of waves; pressure wave derivation (16–2)
- intensity of sound related to amplitude (16–3)
- interference in space and in time (16–6)
- atomic theory of expansion (17–4)
- thermal stresses (17–5)
- ideal gas temperature scale (17–10)
- calculations using the Maxwell distribution of molecular speeds (18–2)
- real gases (18–3)
- vapor pressure and humidity (18–4)
- van der Waals equation of state (18–5)
- mean free path (18–6)
- diffusion (18–7)
- equipartition of energy (19–8)
- energy availability; heat death (20–8)
- statistical interpretation of entropy and the second law (20–9)
- thermodynamic temperature scale; absolute zero and the third law (20–10)
- electric dipoles (21–11, 23–6)
- experimental basis of Gauss's and Coulomb's laws (22–4)
- general relation between electric potential and electric field (23–2, 23–8)

- electric fields in dielectrics (24–5)
- molecular description of dielectrics (24–6)
- current density and drift velocity (25–8)
- superconductivity (25–9)
- RC circuits (26–4)
- use of voltmeters and ammeters; effects of meter resistance (26–5)
- transducers (26–6)
- magnetic dipole moment (27–5)
- Hall effect (27–8)
- operational definition of the ampere and coulomb (28–3)
- magnetic materials—ferromagnetism (28–7)
- electromagnets and solenoids (28–8)
- hysteresis (28–9)
- paramagnetism and diamagnetism (28–10)
- counter emf and torque; eddy currents (29–5)
- Faraday's law—general form (29–7)
- force due to changing $\mathbf{B}$ is nonconservative (29–7)
- LC circuits and EM oscillations (30–5)
- AC resonance; oscillators (31–6)
- impedance matching (31–7)
- three phase AC (31–8)
- changing electric fields produce magnetic fields (32–1)
- speed of light from Maxwell's equations (32–5)
- radiation pressure (32–8)
- fiber optics (33–7)
- lens combinations (34–3)
- aberrations of lenses and mirrors (34–10)
- coherence (35–4)
- intensity in double-slit pattern (35–5)
- luminous intensity (35–8)
- intensity for single-slit (36–2)
- diffraction for double–slit (36–3)
- limits of resolution, the λ limit (36–4, 36–5)
- resolution of the human eye and useful magnification (36–6)
- spectroscopy (36–8)
- peak widths and resolving power for a diffraction grating (36–9)
- x-rays and x-ray diffraction (36–10)
- scattering of light by the atmosphere (36–12)
- time–dependent Schrödinger equation (39–6)
- wave packets (39–7)
- tunneling through a barrier (39–9)
- free-electron theory of metals (41–6)
- semiconductor electronics (41–9)
- standard model, symmetry, QCD, GUT (44–9, 44–10)
- astrophysics, cosmology (Ch. 45)

New Applications

Relevant applications to everyday life, to engineering, and to other fields such as geology and medicine, provide students with motivation and offer the instructor the opportunity to show the relevance of physics. Applications are a good response to students who ask "Why study physics?" Many new applications have been added in this edition. Here are some highlights:

- airbags (Chapter 2)
- elevator and counterweight (Chapter 4)
- antilock brakes and skidding (Chapter 5)
- geosynchronous satellites (Chapter 6)
- hard drive and bit speed (Chapter 10)
- star collapse (Chapter 10)
- forces within trusses, bridges, arches, domes (Chapter 12)
- the Titanic (Chapter 12)
- Bernoulli's principle: wings, sailboats, TIA, plumbing traps and bypasses (Chapter 13)
- pumps (Chapter 13)
- car springs, shock absorbers, building dampers for earthquakes (Chapter 14)
- loudspeakers (Chapters 14, 16, 27)
- autofocusing cameras (Chapter 16)
- sonar (Chapter 16)
- ultrasound imaging (Chapter 16)
- thermal stresses (Chapter 17)
- R-values, thermal insulation (Ch. 19)
- engines (Chapter 20)
- heat pumps, refrigerators, AC; coefficient of performance (Chapter 20)
- thermal pollution (Chapter 20)
- electric shielding (Chapters 21, 28)
- photocopier (Chapter 21)
- superconducting cables (Chapter 25)
- jump starting a car (Chapter 26)
- aurora borealis (Chapter 27)
- solenoids and electromagnetics (Ch. 28)
- computer memory and digital information (Chapter 29)
- seismograph (Chapter 29)

- tape recording (Chapter 29)
- loudspeaker cross-over network (Ch. 31)
- antennas, for **E** or **B** (Chapter 32)
- TV and radio; AM and FM (Chapter 32)
- eye and corrective lenses (Chapter 34)
- mirages (Chapter 35)
- liquid crystal displays (Chapter 36)
- CAT scans, PET, MRI (Chapter 43)

Some old favorites retained (and improved):

- pressure gauges (Chapter 13)
- musical instruments (Chapter 16)
- humidity (Chapter 18)
- CRT, TV, computer monitors (Ch. 23, 27)
- electric hazards (Chapter 25)
- power in household circuits (Chapter 25)
- ammeters and voltmeters (Chapter 26)
- microphones (Chapters 26, 29)
- transducers (Chapter 26, and elsewhere)
- electric motors (Chapter 27)
- car alternator (Chapter 29)
- electric power transmission (Chapter 29)
- capacitors as filters (Chapter 31)
- impedance matching (Chapter 31)
- fiber optics (Chapter 33)
- cameras, telescopes, microscopes, other optical instruments (Chapter 34)
- lens coatings (Chapter 35)
- spectroscopy (Chapter 36)
- electron microscopes (Chapter 38)
- lasers, holography, CD players (Ch. 40)
- semiconductor electronics (Chapter 41)
- radioactivity (Chapters 42 and 43)

Deletions

Something had to go, or the book would have been too long. Lots of subjects were shortened—the detail simply isn't necessary at this level. Some topics were dropped entirely: polar coordinates; center-of-momentum reference frame; Reynolds number (now a Problem); object moving in a fluid and sedimentation; derivation of Poiseuille's equation; Stoke's equation; waveguide and transmission line analysis; electric polarization and electric displacement vectors; potentiometer (now a Problem); negative pressure; combinations of two harmonic motions; adiabatic character of sound waves; central forces.

Many topics have been shortened, often a lot, such as: velocity-dependent forces; variable acceleration; instantaneous axis; surface tension and capillarity; optics topics such as some aspects of light polarizarion. Many of the brief historical and philosophical issues have been shortened as well.

General Approach

This book offers an in-depth presentation of physics, and retains the basic approach of the earlier editions. Rather than using the common, dry, dogmatic approach of treating topics formally and abstractly first, and only later relating the material to the students' own experience, my approach is to recognize that physics is a description of reality and thus to start each topic with concrete observations and experiences that students can directly relate to. Then we move on to the generalizations and more formal treatment of the topic. Not only does this make the material more interesting and easier to understand, but it is closer to the way physics is actually practiced.

This new edition, even more than previous editions, aims to explain the physics in a readable and interesting manner that is accessible and clear. It aims to teach students by anticipating their needs and difficulties, but without oversimplifying. Physics is all about us. Indeed, it is the goal of this book to help students "see the world through eyes that know physics."

As mentioned above, this book includes of a wide range of Examples and applications from technology, engineering, architecture, earth sciences, the environment, biology, medicine, and daily life. Some applications serve only as examples of physical principles. Others are treated in depth. But applications do not dominate the text—this is, after all, a physics book. They have been carefully chosen and integrated into the text so as not to interfere with the development of the physics but rather to illuminate it. You won't find essay sidebars here. The applications are integrated right into the physics. To make it easy to spot the applications, a new *Physics Applied* marginal note is placed in the margin (except where diagrams in the margin prevent it).

It is assumed that students have started calculus or are taking it concurrently. Calculus is treated gently at first, usually in an optional Section so as not to burden students taking calculus concurrently. For example, using the integral in kinematics, Chapter 2, is an optional Section. But in Chapter 7, on work, the integral is discussed fully for all readers.

Throughout the text, *Système International* (SI) units are used. Other metric and British units are defined for informational purposes. Careful attention is paid to significant figures. When a certain value is given as, say, 3, with its units, it is meant to be 3, not assumed to be 3.0 or 3.00. When we mean 3.00 we write 3.00. It is important for students to be aware of the uncertainty in any measured value, and not to overestimate the precision of a numerical result.

Rather than start this physics book with a chapter on mathematics, I have instead incorporated many mathematical tools, such as vector addition and multiplication, directly in the text where first needed. In addition, the Appendices contain a review of many mathematical topics such as trigonometric identities, integrals, and the binomial (and other) expansions. One advanced topic is also given an Appendix: integrating to get the gravitational force due to a spherical mass distribution.

It is necessary, I feel, to pay careful attention to detail, especially when deriving an important result. I have aimed at including all steps in a derivation, and have tried to make clear which equations are general, and which are not, by explicitly stating the limitations of important equations in brackets next to the equation, such as

$$x = x_0 + v_0 t + \tfrac{1}{2}at^2. \qquad \text{[constant acceleration]}$$

The more detailed introduction to Newton's laws and their use is of crucial pedagogic importance. The many new worked-out Examples include initially fairly simple ones that provide careful step-by-step analysis of how to proceed in solving dynamics problems. Each succeeding Example adds a new element or a new twist that introduces greater complexity. It is hoped that this strategy will enable even less-well-prepared students to acquire the tools for using Newton's laws correctly. If students don't surmount this crucial hurdle, the rest of physics may remain forever beyond their grasp.

Rotational motion is difficult for most students. As an example of attention to detail (although this is not really a "detail"), I have carefully distinguished the position vector (**r**) of a point and the perpendicular distance of that point from an axis, which is called R in this book (see Fig. 10–2). This distinction, which enters particularly in connection with torque, moment of inertia, and angular momentum, is often not made clear—it is a disservice to students to use **r** or r for both without distinguishing. Also, I have made clear that it is not always true that $\Sigma\tau = I\alpha$. It depends on the axis chosen (valid if axis is fixed in an inertial reference frame, or through the CM). To not tell this to students can get them into serious trouble. (See pp. 250, 283, 284.) I have treated rotational motion by starting with the simple instance of rotation about an axis (Chapter 10), including the concepts of angular momentum and rotational kinetic energy. Only in Chapter 11 is the more general case of rotation about a point dealt with, and this slightly more advanced material can be omitted if desired (except for Sections 11–1 and 11–2 on the vector product and the torque vector). The end of Chapter 10 has an optional subsection containing three slightly more advanced Examples, using $\Sigma\tau_{CM} = I_{CM}\alpha_{CM}$: car braking distribution, a falling yo-yo, and a sphere rolling with and without slipping.

Among other special treatments is Chapter 28, Sources of Magnetic Field: here, in one chapter, are discussed the magnetic field due to currents (including Ampère's law and the law of Biot-Savart) as well as magnetic materials, ferromagnetism, paramagnetism, and diamagnetism. This presentation is clearer, briefer, and more of a whole, and all the content is there.

Organization

The general outline of this new edition retains a traditional order of topics: mechanics (Chapters 1 to 12); fluids, vibrations, waves, and sound (Chapter 13 to 16); kinetic theory and thermodynamics (Chapters 17 to 20). In the two-volume version of this text, volume I ends here, after Chapter 20. The text continues with electricity and magnetism (Chapters 21 to 32), light (Chapters 33 to 36), and modern physics (Chapters 37 and 38 in the short version, Chapters 37 to 45 in the extended version "with Modern Physics"). Nearly all topics customarily taught in introductory physics courses are included. A number of topics from modern physics are included with the classical physics chapters as discussed earlier.

The tradition of beginning with mechanics is sensible, I believe, because it was developed first, historically, and because so much else in physics depends on it. Within mechanics, there are various ways to order topics, and this book allows for considerable flexibility. I prefer, for example, to cover statics after dynamics, partly because many students have trouble working with forces without motion. Besides, statics is a special case of dynamics—we study statics so that we can prevent structures from becoming dynamic (falling down)—and that sense of being at the limit of dynamics is intuitively helpful. Nonetheless statics (Chapter 12) can be covered earlier, if desired, before dynamics, after a brief introduction to vector addition. Another option is light, which I have placed after electricity and magnetism and EM waves. But light could be treated immediately after the chapters on waves (Chapters 15 and 16). Special relativity is Chapter 37, but could instead be treated along with mechanics—say, after Chapter 9.

Not every chapter need be given equal weight. Whereas Chapter 4 might require $1\frac{1}{2}$ to 2 weeks of coverage, Chapter 16 or 22 may need only $\frac{1}{2}$ week.

Some instructors may find that this book contains more material than can be covered completely in their courses. But the text offers great flexibility in choice of topics. Sections marked with a star (asterisk) are considered optional. These Sections contain slightly more advanced physics material, or material not usually covered in typical courses, and/or interesting applications. They contain no material needed in later chapters (except perhaps in later optional Sections). This does not imply that all nonstarred Sections must be covered: there still remains considerable flexibility in the choice of material. For a brief course, all optional material could be dropped as well as major parts of Chapters 11, 13, 16, 26, 30, 31, and 36 as well as selected parts of Chapters 9, 12, 19, 20, 32, 34, and the modern physics chapters. Topics not covered in class can be a valuable resource for later study; indeed, this text can serve as a useful reference for students for years because of its wide range of coverage.

Thanks

Some 60 physics professors provided input or direct feedback on every aspect of this textbook. The reviewers and contributors to this third edition are listed below. I owe each a debt of gratitude.

Ralph Alexander, University of Missouri at Rolla

Zaven Altounian, McGill University

Charles R. Bacon, Ferris State University

Bruce Birkett, University of California, Berkeley

Art Braundmeier, Southern Illinois University at Edwardsville

Wayne Carr, Stevens Institute of Technology

Edward Chang, University of Massachusetts, Amherst

Charles Chiu, University of Texas at Austin

Lucien Crimaldi, University of Mississippi

Robert Creel, University of Akron

Alexandra Cowley, Community College of Philadelphia

Timir Datta, University of South Carolina

Gary DeLeo, Lehigh University

John Dinardo, Drexel University

Paul Draper, University of Texas, Arlington

Alex Dzierba, Indiana University

William Fickinger, Case Western University

Jerome Finkelstein, San Jose State University

Donald Foster, Wichita State University

Gregory E. Frances, Montana State University

Lothar Frommhold, University of Texas at Austin

Thomas Furtak, Colorado School of Mines

Edward Gibson, California State University, Sacramento

Christopher Gould, University of Southern California

John Gruber, San Jose State University

Martin den Boer, Hunter College

Greg Hassold, General Motors Institute

Joseph Hemsky, Wright State University

Laurent Hodges, Iowa State University

Mark Holtz, Texas Tech University

James P. Jacobs, University of Montana

James Kettler, Ohio University Eastern Campus

Jean Krisch, University of Michigan

Mark Lindsay, University of Louisville

Eugene Livingston, University of Notre Dame

Bryan Long, Columbia State Community College

Daniel Mavlow, Princeton University

Pete Markowitz, Florida International University

John McCullen, University of Arizona, Tucson

Peter Nemeth, New York University

Hon-Kie Ng, Florida State University

Eugene Patroni, Georgia Institute of Technology

Robert Pelcovits, Brown University

William Pollard, Valdosta State University

Joseph Priest, Miami University

Carl Rotter, West Virginia University

Lawrence Rees, Brigham Young University

Peter Riley, University of Texas at Austin

Roy Rubins, University of Texas at Arlington

Mark Semon, Bates College

Robert Simpson, University of New Hampshire

Mano Singham, Case Western University

Harold Slusher, University of Texas at El Paso

Don Sparks, Los Angeles Pierce Community College

Michael Strauss, University of Oklahoma

Joseph Strecker, Wichita State University

William Sturrus, Youngstown State University

Arthur Swift, University of Massachusetts, Amherst

Leo Takahasi, The Pennsylvania State University

Edward Thomas, Georgia Institute of Technology

Som Tyagi, Drexel University

John Wahr, University of Colorado

Robert Webb, Texas A & M University

James Whitmore, The Pennsylvania State University

W. Steve Quon, Ventura College

I owe special thanks to Irv Miller, not only for many helpful physics discussions, but for having worked out all the Problems and managed the team that also worked out the Problems, each checking the other, and finally for producing the Solutions Manual and all the answers to the odd-numbered Problems at the end of this book. He was ably assisted by Zaven Altounian and Anand Batra.

I am particularly grateful to Robert Pelcovits and Peter Riley, as well as to Paul Draper and James Jacobs, who inspired many of the new Examples, Conceptual Examples, and Problems.

Crucial for rooting out errors, as well as providing excellent suggestions, were the perspicacious Edward Gibson and Michael Strauss, both of whom carefully checked all aspects of the physics in page proof.

Special thanks to Bruce Birkett for input of every kind, from illuminating discussions on pedagogy to a careful checking of details in many sections of this book. I wish also to thank Professors Howard Shugart, Joe Cerny, Roger Falcone and Buford Price for helpful discussions, and for hospitality at the University of California, Berkeley. Many thanks also to Prof. Tito Arecchi at the Istituto Nazionale di Ottica, Florence, Italy, and to the staff of the Institute and Museum for the History of Science, Florence, for their hospitality.

Finally, I wish to thank the superb editorial and production work provided by all those with whom I worked directly at Prentice Hall: Susan Fisher, Marilyn Coco, David Chelton, Kathleen Schiaparelli, Trudy Pisciotti, Gus Vibal, Mary Teresa Giancoli, and Jocelyn Phillips.

The biggest thanks of all goes to Paul Corey, whose constant encouragement and astute ability to get things done, provided the single strongest catalyst.

The final responsibility for all errors lies with me, of course. I welcome comments and corrections.

D.C.G.

AVAILABLE SUPPLEMENTS

For the Student

Student Study Guide and Solutions Manual
Douglas Brandt, Eastern Illinois University. (0-13-021475-2)
Contains chapter objectives, summaries with additional examples, self-study quizzes, key mathematical equations, and complete worked-out solutions to alternate odd problems in the text.

Doing Physics with Spreadsheets: A Workbook
Gordon Aubrecht, T. Kenneth Bolland, and Michael Ziegler, all of The Ohio State University.
(0-13-021474-4)
Designed to introduce students to the use of spreadsheets for solving simple and complex physics problems. Students are either provided with spreadsheets or must construct their own, then use the model to most closely approximate natural behavior. The amount of spreadsheet construction and the complexity of the spreadsheet increases as the student gains experience.

Science on the Internet: A Student's Guide, 1999
Andrew Stull and Carl Adler (0-13-021308-X)
The perfect tool to help students take advantage of the *Physics for Scientists and Engineers, Third Edition* Web page. This useful resource gives clear steps to access Prentice Hall's regularly updated physics resources, along with an overview of general World Wide Web navigation strategies. Available FREE for students when packaged with the text.

Prentice Hall/*New York Times* Themes of the Times — Physics
This unique newspaper supplement brings together a collection of the latest physics-related articles from the pages of *The New York Times*. Updated twice per year and available FREE to students when packaged with the text.

For the Instructor

Instructor's Solutions Manual
Irvin A. Miller, Drexel University.
Print version (0-13-021381-0); Electronic (CD-ROM) version (0-13-021481-7)
Contains detailed worked solutions to every problem in the text. Electronic versions are available in CD-ROM (dual platform for both Windows and Macintosh systems) for instructors with Microsoft Word or Word-compatible software.

Test Item File
Robert Pelcovits, Brown University; David Curott, University of North Alabama; and Edward Oberhofer, University of North Carolina at Charlotte (0-13-021482-5)
Contains over 2200 multiple choice questions, about 25% conceptual in nature. All are referenced to the corresponding Section in the text and ranked by difficulty.

Prentice Hall Custom Test Windows (0-13-021477-9); Macintosh (0-13-021476-0)
Based on the powerful testing technology developed by Engineering Software Associates, Inc. (ESA), Prentice Hall Custom Test includes all questions from the Test Item File and allows instructors to create and tailor exams to their own needs. With the Online Testing Program, exams can also be administered on line and data can then be automatically transferred for evaluation. A comprehensive desk reference guide is included along with online assistance.

Transparency Pack (0-13-021470-1)
Includes approximately 400 full color transparencies of images from the text.

Media Supplements

Physics for Scientists and Engineers Web Site www.prenhall.com/giancoli
A FREE innovative online resource that provides students with a wealth of activities and exercises for each text chapter. Features on the site include:

- Practice Questions, Destinations (links to related sites), NetSearch keywords and algorithmically generated numeric Practice Problems by Carl Adler of East Carolina University.
- Physlet Problems (Java-applet simulations) by Wolfgang Christian of Davidson College.
- Warmups and Puzzles essay questions and Applications from Gregor Novak and Andrew Gavrin at Indiana University-Purdue University, Indianapolis.
- Ranking Task Exercises edited by Tom O'Kuma of Lee College, Curtis Hieggelke of Joliet Junior College and David Maloney of Indiana University-Purdue University, Fort Wayne.

Using Prentice Hall CW '99 technology, the website grades and scores all objective questions, and results can be automatically e-mailed directly to the instructors if so desired. Instructors can also create customized syllabi online and link directly to activities on the Giancoli website.

Presentation Manager CD-ROM
Dual Platform (Windows/Macintosh; 0-13-214479-5)
This CD-ROM enables instructors to build custom sequences of Giancoli text images and Prentice Hall digital media for playback in lecture presentations. The CD-ROM contains all text illustrations, digitized segments from the Prentice Hall *Physics You Can See* videotape as well as additional lab and demonstration videos and animations from the Prentice Hall *Interactive Journey Through Physics* CD-ROM. Easy to navigate with Prentice Hall Presentation Manager software, instructors can preview, sequence, and play back images, as well as perform keyword searches, add lecture notes, and incorporate their own digital resources.

Physics You Can See *Video*
(0-205-12393-7)
Contains eleven two- to five-minute demonstrations of classical physics experiments. It includes segments such as "Coin and Feather" (acceleration due to gravity), "Monkey and Gun" (projectile motion), "Swivel Hips" (force pairs), and "Collapse a Can" (atmospheric pressure).

CAPA: A Computer-Assisted Personalized Approach to Assignments, Quizzes, and Exams
CAPA is an on-line homework system developed at Michigan State University that instructors can use to deliver problem sets with randomized variables for each student. The system gives students immediate feedback on their answers to problems, and records their participation and performance. Prentice Hall has arranged to have half of the even-numbered problems of Giancoli, *Physics for Scientists and Engineers, Third Edition*, coded for use with the CAPA system. For additional information about the CAPA system, please visit the web site at http://www.pa.msu.edu/educ/CAPA/.

WebAssign
WebAssign is a web-based homework delivery, collection, grading, and recording service developed and hosted by North Carolina State University. Prentice Hall will arrange for end-of-chapter problems from Giancoli, *Physics for Scientists and Engineers, Third Edition* to be coded for use with the *WebAssign* system for instructors who wish to take advantage of this service. For more information on the *WebAssign* system and its features, please visit http://webassign.net/info or e-mail webassign@ncsu.edu.

NOTES TO STUDENTS AND INSTRUCTORS ON THE FORMAT

1. Sections marked with a star (*) are considered optional. They can be omitted without interrupting the main flow of topics. No later material depends on them except possibly later starred sections. They may be fun to read though.

2. The customary conventions are used: symbols for quantities (such as m for mass) are italicized, whereas units (such as m for meter) are not italicized. Boldface (**F**) is used for vectors.

3. Few equations are valid in all situations. Where practical, the limitations of important equations are stated in square brackets next to the equation. The equations that represent the great laws of physics are displayed with a tan background, as are a few other equations that are so useful that they are indispensable.

4. The number of significant figures (see Section 1–3) should not be assumed to be greater than given: if a number is stated as (say) 6, with its units, it is meant to be 6 and not 6.0 or 6.00.

5. At the end of each chapter is a set of Questions that students should attempt to answer (to themselves at least). These are followed by Problems which are ranked as level I, II, or III, according to estimated difficulty, with level I Problems being easiest. These Problems are arranged by Section, but Problems for a given Section may depend on earlier material as well. There follows a group of General Problems, which are not arranged by Section nor ranked as to difficulty. Questions and Problems that relate to optional Sections are starred.

6. Being able to solve problems is a crucial part of learning physics, and provides a powerful means for understanding the concepts and principles. This book contains many aids to problem solving: (a) worked-out Examples and their solutions in the text, which are set off with a vertical blue line in the margin, and should be studied as an integral part of the text; (b) special "Problem-solving boxes" placed throughout the text to suggest ways to approach problem solving for a particular topic—but don't get the idea that every topic has its own "techniques," because the basics remain the same; (c) special problem-solving Sections (marked in blue in the Table of Contents); (d) "Problem solving" marginal notes (see point 8 below) which refer to hints for solving problems within the text; (e) some of the worked-out Examples are Estimation Examples, which show how rough or approximate results can be obtained even if the given data are sparse (see Section 1–6); and finally (f) the Problems themselves at the end of each chapter (point 5 above).

7. Conceptual Examples look like ordinary Examples but are conceptual rather than numerical. Each proposes a question or two, which hopefully starts you to think and come up with a response. Give yourself a little time to come up with your own response before reading the Response given.

8. Marginal notes: brief notes in the margin of almost every page are printed in blue and are of four types: (a) ordinary notes (the majority) that serve as a sort of outline of the text and can help you later locate important concepts and equations; (b) notes that refer to the great laws and principles of physics, and these are in capital letters and in a box for emphasis; (c) notes that refer to a problem-solving hint or technique treated in the text, and these say "Problem Solving"; (d) notes that refer to an application of physics, in the text or an Example, and these say "Physics Applied."

9. This book is printed in full color. But not simply to make it more attractive. The color is used above all in the Figures, to give them greater clarity for our analysis, and to provide easier learning of the physical principles involved. The Table on the next page is a summary of how color is used in the Figures, and shows which colors are used for the different kinds of vectors, for field lines, and for other symbols and objects. These colors are used consistently throughout the book.

10. Appendices include useful mathematical formulas (such as derivatives and integrals, trigonometric identities, areas and volumes, expansions), and a table of isotopes with atomic masses and other data. Tables of useful data are located inside the front and back covers.

USE OF COLOR

Vectors

A general vector	⟶
resultant vector (sum) is slightly thicker	⟹
components of any vector are dashed	⇢
Displacement ($\mathbf{D}$, $\mathbf{r}$)	⟶
Velocity ($\mathbf{v}$)	⟶
Acceleration ($\mathbf{a}$)	⟶
Force ($\mathbf{F}$)	⟶
Force on second or	⟶
third object in same figure	⟶
Momentum ($\mathbf{p}$ or $m\mathbf{v}$)	⟶
Angular momentum ($\mathbf{L}$)	⟶
Angular velocity (ω)	⟶
Torque (τ)	⟶
Electric field ($\mathbf{E}$)	⟶
Magnetic field ($\mathbf{B}$)	⟶

Electricity and magnetism

Electric field lines	⟶
Equipotential lines	⇢
Magnetic field lines	⟶
Electric charge (+)	⊕ or • +
Electric charge (−)	⊖ or • −

Electric circuit symbols

Wire	—
Resistor	–⋀⋁⋀–
Capacitor	–⊣⊢–
Inductor	–⁃⁃⁃⁃–
Battery	–⊣⊢–

Optics

Light rays	⟶
Object	↑
Real image (dashed)	↑
Virtual image (dashed and paler)	↑

Other

Energy level (atom, etc.)	—
Measurement lines	⊢—1.0 m—⊣
Path of a moving object	⇢
Direction of motion or current	⟶

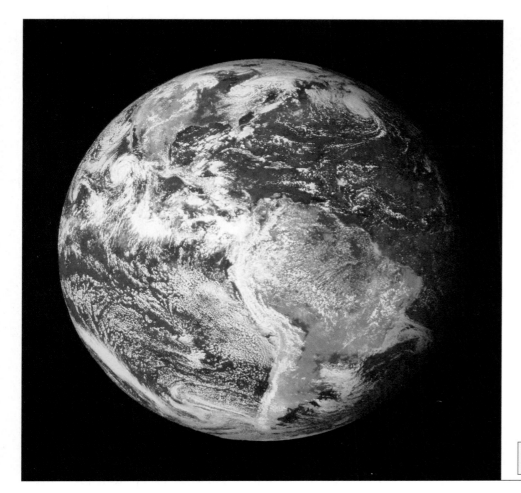

Besides learning some basics about science and its theories, and about measurement and units, you will also learn in this Chapter how to make quick estimates—such as estimating the radius of the Earth using everyday observations. The photo shown here of Earth was taken from about 36,000 km away. North and South America are clearly visible below the clouds in this computer-enhanced photo.

CHAPTER 1

Introduction, Measurement, Estimating

Physics is the most basic of the sciences. It deals with the behavior and structure of matter. The field of physics is usually divided into the areas of motion, fluids, heat, sound, light, electricity and magnetism, which is referred to as classical physics; plus the modern physics topics of relativity, atomic structure, condensed matter, nuclear physics, elementary particles, and astrophysics.

The fundamentals of physics need to be understood by anyone who hopes to make a career in the sciences or technology: physicists, engineers, chemists, astronomers, mathematicians, geologists, biologists. All the sciences use physics as a fundamental basis, and so does engineering. Engineers, for example, must understand how to use the laws of thermodynamics to design heating systems; they must understand optics and electromagnetism to design medical imaging systems; and they must know how to calculate the forces within a structure in order to design a structure that will remain standing (Fig. 1–1). Indeed, in Chapter 12 we will see a worked-out example of how a simple physics calculation—or even intuition based on understanding the physics of forces—would have saved the lives of hundreds of people. We will see many examples in this book of how physics is useful in other fields, and in everyday life.

(a)

(b)

(c)

FIGURE 1–1 (a) This Roman aqueduct was built 2000 years ago and still stands. (b) So does the Golden Gate Bridge, built in 1937. (c) Collapse of the Hartford Civic Center in 1978, just two years after it was built.

The principal aim of all sciences, including physics, is generally considered to be the search for order in our observations of the world around us. Many people think that science is a mechanical process of collecting facts and devising theories. But it is not so simple. Science is a creative activity that in many respects resembles other creative activities of the human mind.

The Nature of Science

Observation

One important aspect of science is **observation** of events, which includes the design and carrying out of experiments. But observation and experiment require imagination, for scientists can never include everything in a description of what they observe. Hence, scientists must make judgments about what is relevant in their observations and experiments. Consider, for example, how two great minds, Aristotle (384–322 B.C.) and Galileo (1564–1642), interpreted motion along a horizontal surface. Aristotle noted that objects given an initial push along the ground (or on a tabletop) always slow down and stop. Consequently, Aristotle argued that the natural state of an object is to be at rest. Galileo, in his reexamination of horizontal motion almost 2000 years later, chose rather to study the idealized case of motion free from resistance. Galileo imagined that if friction could be eliminated, an object given an initial push along a horizontal surface would continue to move indefinitely without stopping. He concluded that for an object to be in motion was just as natural as for it to be at rest. By inventing a new approach, Galileo founded our modern view of motion (Chapters 2, 3, and 4), and he did so with a leap of the imagination. Galileo made this leap conceptually, without actually eliminating friction.

Observation, and careful experimentation and measurement, are one side of the scientific process. The other side is the invention or creation of **theories** to explain and order the observations. Theories, it must be emphasized, are not derived directly from observations. Observations may help inspire a theory, and theories are accepted or rejected based on observation and experiment. *Theories*

The great theories of science may be compared, as creative achievements, with great works of art or literature. But how does science differ from these other creative activities? One important difference is that science requires **testing** of its ideas or theories to see if their predictions are borne out by experiment. *Testing (which can never be exhaustive)*

Although the testing of theories distinguishes science from other creative fields, it should not be assumed that a theory is "proved" by testing. First of all, no measuring instrument is perfect, so exact confirmation cannot be possible. Furthermore, it is not possible to test a theory in every single possible circumstance. Hence a theory cannot be absolutely verified. Indeed, the history of science tells us that long-held theories can be replaced by new ones.

1–2 | Models, Theories, and Laws

When scientists are trying to understand a particular set of phenomena, they often make use of a **model**. A model, in the scientist's sense, is a kind of analogy or mental image of the phenomena in terms of something we are familiar with. One example is the wave model of light. We cannot see waves of light as we can water waves; but it is valuable to think of light as if it were made up of waves because experiments indicate that light behaves in many respects as water waves do. *Models*

The purpose of a model is to give us an approximate mental or visual picture—something to hold on to—when we cannot see what actually is happening. Models often give us a deeper understanding: the analogy to a known system (for instance, water waves in the above example) can suggest new experiments to perform and can provide ideas about what other related phenomena might occur.

You may wonder what the difference is between a theory and a model. Sometimes the words are used interchangeably. Usually, however, a model is relatively simple and provides a structural similarity to the phenomena being studied. A **theory**, on the other hand, is broader, more detailed, and attempts to solve a set of problems, often with great precision. Sometimes, as a model is developed and modified and corresponds more closely to experiment over a wide range of phenomena, it may come to be referred to as a theory. The atomic theory of matter is an example, as is the wave theory of light. *Theories (vs. models)*

Models can be very helpful, and they often lead to important theories. But it is important not to confuse a model, or a theory, with the real system or the phenomena themselves.

Scientists give the title **law** to certain concise but general statements about how nature behaves (that energy is conserved, for example). Sometimes the statement takes the form of a relationship or equation between quantities (such as Newton's second law, $F = ma$.) *Laws*

Scientific laws are different from political laws in that the latter are *prescriptive*: they tell us how we ought to behave. Scientific laws are *descriptive*: they do not say how nature *should* behave, but rather are meant to describe how nature *does* behave. As with theories, laws cannot be tested in the infinite variety of cases possible. So we cannot be sure that any law is absolutely true. We use the term "law" when its validity has been tested over a wide range of cases, and when any limitations and the range of validity are clearly understood. Even then, as new information comes in, certain laws may have to be modified or discarded.

Scientists normally do their work as if the accepted laws and theories were true. But they are obliged to keep an open mind in case new information should alter the validity of any given law or theory.

1–3 Measurement and Uncertainty; Significant Figures

In the quest to understand the world around us, scientists seek to find relationships among physical quantities that can be measured.

Uncertainty

Every measurement has an uncertainty.

Accurate measurements are an important part of physics. But no measurement is absolutely precise. There is an uncertainty associated with every measurement. Uncertainty arises from different sources. Among the most important, other than blunders, are the limited accuracy of every measuring instrument and the inability to read an instrument beyond some fraction of the smallest division shown. For example, if you were to use a centimeter ruler to measure the width of a board (Fig. 1–2), the result could be claimed to be precise to about 0.1 cm, the smallest division on the ruler, although half of this value might be a valid claim as well. The reason for this is that it is difficult for the observer to interpolate between the smallest divisions. Furthermore, the ruler itself may not have been manufactured to an accuracy very much better than this.[†]

FIGURE 1–2 Measuring the width of a board with a centimeter ruler. Accuracy is about ±1 mm.

When giving the result of a measurement, it is important to state the precision, or **estimated uncertainty**, in the measurement. For example, the width of a board might be written as 8.8 ± 0.1 cm. The ± 0.1 cm ("plus or minus 0.1 cm") represents the estimated uncertainty in the measurement, so that the actual width most likely lies between 8.7 and 8.9 cm. The **percent uncertainty** is simply the ratio of the uncertainty to the measured value, multiplied by 100. For example, if the measurement is 8.8 and the uncertainty about 0.1 cm, the percent uncertainty is

$$\frac{0.1}{8.8} \times 100\% \approx 1\%$$

where ≈ means "is roughly equal to."

Assumed uncertainty

Often the uncertainty in a measured value is not specified explicitly. In such cases, the uncertainty is generally assumed to be one or two (or even three) units in the last digit specified. For example, if a length is given as 8.8 cm, the uncertainty is assumed to be about 0.1 cm (or perhaps 0.2 cm). It is important in this case that you do not write 8.80 cm, for this implies an uncertainty on the order of 0.01 cm; it assumes that the length is probably between 8.79 cm and 8.81 cm, when actually you believe it is between 8.7 and 8.9 cm.

Significant Figures

Which digits are significant?

The number of reliably known digits in a number is called the number of **significant figures**. Thus there are four significant figures in the number 23.21 cm and two in the number 0.062 cm (the zeros in the latter are merely place holders that show where the decimal point goes). The number of significant figures may

[†] There is a technical difference between "precision" and "accuracy." Precision in a strict sense refers to the repeatability of the measurement using a given instrument. For example, if you measure the width of a board many times, getting results like 8.81 cm, 8.85 cm, 8.78 cm, 8.82 cm (interpolating between marks as a best estimate each time), you could say the measurements give a *precision* a bit better than 0.1 cm. *Accuracy* refers to how close a measurement is to the true value. For example, if the ruler shown in Fig. 1–2 was manufactured with a 2% error, the accuracy of its measurement of the board's width (about 8.8 cm) would be about 2% of 8.8 cm or about ± 0.2 cm. Estimated uncertainty is meant to take both accuracy and precision into account.

not always be clear. Take, for example, the number 80. Are there one or two significant figures? If we say it is *about* 80 km between two cities, there is only one significant figure (the 8) since the zero is merely a place holder. If it is *exactly* 80 km within an accuracy of 1 or 2 km, then the 80 has two significant figures.[†] If it is precisely 80 km, to within ± 0.1 km, then we write 80.0 km.

When making measurements, or when doing calculations, you should avoid the temptation to keep more digits in the final answer than is justified. For example, to calculate the area of a rectangle 11.3 cm by 6.8 cm, the result of multiplication would be 76.84 cm^2. But this answer is clearly not accurate to 0.01 cm^2, since (using the outer limits of the assumed uncertainty for each measurement) the result could be between $11.2 \times 6.7 = 75.04$ cm^2 and $11.4 \times 6.9 = 78.66$ cm^2. At best, we can quote the answer as 77 cm^2, which implies an uncertainty of about 1 or 2 cm^2. The other two digits (in the number 76.84 cm^2) must be dropped since they are not significant. As a rough general rule, (i.e., in the absence of a detailed consideration of uncertainties), we can say that: *the final result of a multiplication or division should have only as many digits as the number with the least number of significant figures used in the calculation*. In our example, 6.8 cm has the least number of significant figures, namely two. Thus the result 76.84 cm^2 needs to be rounded off to 77 cm^2.

Similarly, when adding or subtracting numbers, the final result is no more accurate than the least accurate number used. For example, the result of subtracting 0.57 from 3.6 is 3.0 (and not 3.03).

Keep in mind when you use a calculator that all the digits it produces may not be significant. When you divide 2.0 by 3.0, the proper answer is 0.67, and not some such thing as 0.66666666. Digits should not be quoted (or written down) in a result, unless they are truly significant figures. However, to obtain the most accurate result, you should normally keep an extra significant figure or two throughout a calculation, and round off only in the final result. Note also that calculators sometimes give too few significant figures. For example, when you multiply 2.5×3.2, a calculator may give the answer as simply 8. But the answer is good to two significant figures, so the proper answer is 8.0.

Scientific Notation

We commonly write numbers in "powers of ten," or "scientific" notation—for instance 36,900 as 3.69×10^4, or 0.0021 as 2.1×10^{-3}. One advantage of scientific notation is that it allows the number of significant figures to be clearly expressed. For example, it is not clear whether 36,900 has three, four, or five significant figures. With powers of ten notation the ambiguity can be avoided: if the number is known to an accuracy of three significant figures, we write 3.69×10^4, but if it is known to four, we write 3.690×10^4.

| **CONCEPTUAL EXAMPLE 1–1** | **Is the diamond yours?** |

CONCEPTUAL EXAMPLE 1–1 **Is the diamond yours?** A friend asks to borrow your precious diamond for a day to show her family. You are a bit worried, so you carefully have your diamond weighed on a scale which reads 8.17 grams. The scale's accuracy is claimed to be ± 0.05 grams. The next day you weigh the returned diamond again, getting 8.09 grams. Is this your diamond?

RESPONSE The scale readings are measurements and do not necessarily give the "true" value of the mass. Each measurement could have been high or low by up to 0.05 gram or so. The actual mass of your diamond lies most likely between 8.12 grams and 8.22 grams. The actual mass of the returned diamond is most likely between 8.04 grams and 8.14 grams. These two ranges overlap, so there is not a strong reason to doubt that the returned diamond is yours, at least based on the scale readings.

[†] If the 80 has two significant figures, some people prefer to write it 80., with a decimal point. This is not always (or even usually) done, so the number of significant figures in 80 can be ambiguous unless something is said about it, such as "about" or "precisely."

1–4 | Units, Standards, and the SI System

The measurement of any quantity is made relative to a particular standard or **unit**, and this unit must be specified along with the numerical value of the quantity. For example, we can measure length in units such as inches, feet, or miles, or in the metric system in centimeters, meters, or kilometers. To specify that the length of a particular object is 18.6 is meaningless. The unit *must* be given; for clearly, 18.6 meters is very different from 18.6 inches or 18.6 millimeters.

Length

The first real international standard was the **meter** (abbreviated m) established as the standard of **length** by the French Academy of Sciences in the 1790s. In a spirit of rationality, the standard meter was originally chosen to be one ten-millionth of the distance from the Earth's equator to either pole,[†] and a platinum rod to represent this length was made. (This turns out to be, very roughly, the distance from the tip of your nose to the tip of your finger, with arm and hand stretched out to the side.) In 1889, the meter was defined more precisely as the distance between two finely engraved marks on a particular bar of platinum–iridium alloy. In 1960, to provide greater precision and reproducibility, the meter was redefined as 1,650,763.73 wavelengths of a particular orange light emitted by the gas krypton 86. In 1983 the meter was again redefined, this time in terms of the speed of light (whose best measured value in terms of the older definition of the meter was 299,792,458 m/s, with an uncertainty of 1 m/s). The new definition reads: "The meter is the length of path traveled by light in vacuum during a time interval of 1/299,792,458 of a second."[‡]

British units of length (inch, foot, mile) are now defined in terms of the meter. The inch (in.) is defined as precisely 2.54 centimeters (cm; 1 cm = 0.01 m). Other conversion factors are given in the table on the inside of the front cover of this book. Table 1–1 presents some characteristic lengths, from very small to very large. See also Fig. 1–3.

FIGURE 1–3 Some lengths: (a) viruses (about 10^{-7} m long) attacking a cell; (b) Mt. Everest's height is on the order of 10^4 m (8848 m to be precise).

(a)

(b)

TABLE 1–1 Some typical Lengths or Distances (order of magnitude)

Length (or distance)	Meters (approximate)
Neutron or proton (radius)	10^{-15} m
Atom	10^{-10} m
Virus [see Fig. 1–3]	10^{-7} m
Sheet of paper (thickness)	10^{-4} m
Finger width	10^{-2} m
Football field length	10^2 m
Mt. Everest height [see Fig. 1–3]	10^4 m
Earth diameter	10^7 m
Earth to Sun	10^{11} m
Nearest star, distance	10^{16} m
Nearest galaxy	10^{22} m
Farthest galaxy visible	10^{26} m

[†] Modern measurements of the Earth's circumference reveal that the intended length is off by about one-fiftieth of 1 percent. Not bad!

[‡] The new definition of the meter has the effect of giving the speed of light the exact value of 299,792,458 m/s.

TABLE 1–2 Some typical Time Intervals	
Time interval	**Seconds (approximate)**
Lifetime of very unstable subatomic particle	10^{-23} s
Lifetime of radioactive elements	10^{-22} s to 10^{28} s
Lifetime of muon	10^{-6} s
Time between human heartbeats	10^{0} s $(= 1$ s$)$
One day	10^{5} s
One year	3×10^{7} s
Human life span	2×10^{9} s
Length of recorded history	10^{11} s
Humans on Earth	10^{14} s
Life on Earth	10^{17} s
Age of Universe	10^{18} s

TABLE 1–3 Some Masses	
Object	**Kilograms (approx.)**
Electron	10^{-30} kg
Proton, neutron	10^{-27} kg
DNA molecule	10^{-17} kg
Bacterium	10^{-15} kg
Mosquito	10^{-5} kg
Plum	10^{-1} kg
Person	10^{2} kg
Ship	10^{8} kg
Earth	6×10^{24} kg
Sun	2×10^{30} kg
Galaxy	10^{41} kg

Time

The standard unit of **time** is the **second** (s). For many years, the second was defined as 1/86,400 of a mean solar day. The standard second is now defined more precisely in terms of the frequency of radiation emitted by cesium atoms when they pass between two particular states. [Specifically, one second is defined as the time required for 9,192,631,770 periods of this radiation.] There are, by definition, 60 s in one minute (min) and 60 minutes in one hour (h). Table 1–2 presents a range of measured time intervals.

Mass

The standard unit of **mass** is the **kilogram** (kg). The standard mass is a particular platinum-iridium cylinder (Fig. 1–4), kept at the International Bureau of Weights and Measures near Paris, France, whose mass is defined as exactly 1 kg. A range of masses is presented in Table 1–3. [For practical purposes, 1 kg weighs about 2.2 pounds on Earth.]

When dealing with atoms and molecules, the **unified atomic mass unit** (u) is usually used. In terms of the kilogram

$$1 \text{ u} = 1.6605 \times 10^{-27} \text{ kg}.$$

The definitions of other standard units for other quantities will be given as we encounter them in later chapters.

Unit Prefixes

In the metric system, the larger and smaller units are defined in multiples of 10 from the standard unit, and this makes calculation particularly easy. Thus 1 kilometer (km) is 1000 m, 1 centimeter is $\frac{1}{100}$ m, 1 millimeter (mm) is $\frac{1}{1000}$ m or $\frac{1}{10}$ cm, and so on. The prefixes "centi-," "kilo-," and others are listed in Table 1–4 and can be applied not only to units of length, but to units of volume, mass, or any other metric unit. For example, a centiliter (cL) is $\frac{1}{100}$ liter (L) and a kilogram (kg) is 1000 grams (g).

Systems of Units

When dealing with the laws and equations of physics it is very important to use a consistent set of units. Several systems of units have been in use over the years. Today the most important is the **Système International** (French for International System), which is abbreviated SI. In SI units, the standard of length is the meter, the standard for time is the second, and the standard for mass is the kilogram. This system used to be called the MKS (meter-kilogram-second) system.

FIGURE 1–4   The standard kilogram.

TABLE 1–4 Metric (SI) Prefixes		
Prefix	**Abbreviation**	**Value**
exa	E	10^{18}
peta	P	10^{15}
tera	T	10^{12}
giga	G	10^{9}
mega	M	10^{6}
kilo	k	10^{3}
hecto	h	10^{2}
deka	da	10^{1}
deci	d	10^{-1}
centi	c	10^{-2}
milli	m	10^{-3}
micro[†]	μ	10^{-6}
nano	n	10^{-9}
pico	p	10^{-12}
femto	f	10^{-15}
atto	a	10^{-18}

[†] μ is the Greek letter "mu."

TABLE 1–5
SI Base Quantities and Units

Quantity	Unit	Unit Abbreviation
Length	meter	m
Time	second	s
Mass	kilogram	kg
Electric current	ampere	A
Temperature	kelvin	K
Amount of substance	mole	mol
Luminous intensity	candela	cd

A second metric system is the **cgs system**, in which the centimeter, gram, and second are the standard units of length, mass, and time, as abbreviated in the title. The **British engineering system** takes as its standards the foot for length, the pound for force, and the second for time.

SI units are the principal ones used today in scientific work. We will therefore use SI units almost exclusively in this book, although we will give the cgs and British units for various quantities when introduced.

* Base vs. Derived Units (optional)

Physical quantities can be divided into two categories: *base quantities* and *derived quantities*. The corresponding units for these quantities are called *base units* and *derived units*. A **base quantity** must be defined in terms of a standard. Scientists, in the interest of simplicity, want the smallest number of base quantities possible consistent with a full description of the physical world. This number turns out to be seven, and those used in the SI are given in Table 1–5. All other quantities can be defined in terms of these seven base quantities,[†] and hence are referred to as **derived quantities**. An example of a derived quantity is speed, which is defined as distance traveled divided by the time it takes to travel that distance. To define any quantity, whether base or derived, we can specify a rule or procedure, and this is called an **operational definition**.

1–5 | Converting Units

Any quantity we measure, such as a length, a speed, or an electric current, consists of a number *and* a unit. Often we are given a quantity in one set of units, but we want it expressed in another set of units. For example, suppose we measure that a table is 21.5 inches wide, and we want to express this in centimeters. We must use a **conversion factor** which in this case is

$$1 \text{ in.} = 2.54 \text{ cm}$$

or, written another way,

$$1 = 2.54 \text{ cm/in.}$$

Since multiplying by one does not change anything, the width of our table, in cm, is

$$21.5 \text{ inches} = (21.5 \text{ in.}) \times \left(2.54 \frac{\text{cm}}{\text{in.}}\right) = 54.6 \text{ cm}.$$

Note how the units (inches in this case) cancelled out. A table containing many unit conversions is found inside the front cover of this book. Let's take some Examples.

⇒ P H Y S I C S   A P P L I E D

How many yards in the 100-m dash

EXAMPLE 1–2 **The 100-m dash.** What is the length of the 100-m dash expressed in yards?

SOLUTION Let us assume the distance is accurately known to four significant figures, 100.0 m. One yard (yd) is precisely 3 feet (36 inches), so we can write

$$1 \text{ yd} = 3 \text{ ft} = 36 \text{ in.} = (36 \text{ in.})\left(2.54 \frac{\text{cm}}{\text{in.}}\right) = 91.44 \text{ cm} = 0.9144 \text{ m}.$$

We can rewrite this result as

$$1 \text{ m} = \frac{1 \text{ yd}}{0.9144} = 1.094 \text{ yd}.$$

Then

$$100 \text{ m} = (100 \text{ m})\left(1.094 \frac{\text{yd}}{\text{m}}\right) = 109.4 \text{ yd},$$

so a 100-m dash is 9.4 yards longer than a 100-yard dash.

[†] The only exceptions are for angle (radians—see Chapter 10) and solid angle (steradian). No general agreement has been reached as to whether these are base or derived quantities.

We could have done this conversion all in one line:

$$100.0 \text{ m} = (100.0 \text{ m})\left(\frac{100 \text{ cm}}{1 \text{ m}}\right)\left(\frac{1 \text{ in.}}{2.54 \text{ cm}}\right)\left(\frac{1 \text{ yd}}{36 \text{ in.}}\right) = 109.4 \text{ yd.}$$

EXAMPLE 1–3 **Area of a semiconductor chip.** A silicon chip has an area of 1.25 square inches. Express this in square centimeters.

SOLUTION Because 1 in. = 2.54 cm, then 1 in.^{2} = (2.54 cm)^2 = 6.45 cm^2. So

$$1.25 \text{ in.}^2 = (1.25 \text{ in.}^2)\left(2.54 \frac{\text{cm}}{\text{in.}}\right)^2 = (1.25 \text{ in.}^2)\left(6.45 \frac{\text{cm}^2}{\text{in.}^2}\right) = 8.06 \text{ cm}^2.$$

EXAMPLE 1–4 **Speeds.** Where the posted speed limit is 55 miles per hour (mi/h or mph), what is this speed (a) in meters per second (m/s) and (b) in kilometers per hour (km/h)?

SOLUTION (a) We can write 1 mile as

$$1 \text{ mi} = (5280 \text{ ft})\left(12 \frac{\text{in.}}{\text{ft}}\right)\left(2.54 \frac{\text{cm}}{\text{in.}}\right)\left(\frac{1 \text{ m}}{100 \text{ cm}}\right) = 1609 \text{ m.}$$

Note that each conversion factor is equal to one. We also know that 1 hour equals (60 min/h) × (60 s/min) = 3600 s/h, so

Conversion factors = 1

$$55 \frac{\text{mi}}{\text{h}} = \left(55 \frac{\text{mi}}{\text{h}}\right)\left(1609 \frac{\text{m}}{\text{mi}}\right)\left(\frac{1 \text{ h}}{3600 \text{ s}}\right) = 25 \frac{\text{m}}{\text{s}}.$$

(b) Now we use 1 mi = 1609 m = 1.609 km; then

$$55 \frac{\text{mi}}{\text{h}} = \left(55 \frac{\text{mi}}{\text{h}}\right)\left(1.609 \frac{\text{km}}{\text{mi}}\right) = 88 \frac{\text{km}}{\text{h}}.$$

When changing units, you can avoid making an error in the use of conversion factors by checking that units cancel out properly. For example, in our conversion of 1 mi to 1609 m in Example 1–4(a), if we had incorrectly used the factor $\left(\frac{100 \text{ cm}}{1 \text{ m}}\right)$ instead of $\left(\frac{1 \text{ m}}{100 \text{ cm}}\right)$, the meter units would not have cancelled out; we would not have ended up with meters.

➡ **PROBLEM SOLVING**

Unit conversion is wrong
if units do not cancel

1–6 | Order of Magnitude: Rapid Estimating

We are sometimes interested only in an approximate value for a quantity. This might be because an accurate calculation would take more time than it is worth or would require additional data that are not available. In other cases, we may want to make a rough estimate in order to check an accurate calculation made on a calculator, to make sure that no blunders were made when entering the numbers.

A rough estimate is made by rounding off all numbers to one significant figure and its power of 10, and after the calculation is made, again only one significant figure is kept. Such an estimate is called an **order-of-magnitude estimate** and can be accurate within a factor of 10, and often better. In fact, the phrase "order of magnitude" is sometimes used to refer simply to the power of 10.

To give you some idea of how useful and powerful rough estimates can be, let us do a few "worked-out Examples."

➡ **PROBLEM SOLVING**

How to make a rough estimate

(a)

$r = 500$ m

10 m

(b)

FIGURE 1–5 Example 1–5. (a) How much water is in this lake? (Photo is of one of the Rae Lakes in the Sierra Nevada of California.) (b) Model of the lake as a cylinder. [We could go one step further and estimate the mass or weight of this lake. We will see later that water has a density of 1000 kg/m^3, so this lake has a mass of about $(10^3 \text{ kg/m}^3)(10^7 \text{ m}^3) \approx 10^{10}$ kg, which is about 10 billion kg or 10 million metric tons. (A metric ton is 1000 kg, about 2200 lbs, slightly larger than a British ton, 2000 lbs.)]

➡ **PHYSICS APPLIED**

Estimating the volume (or mass) of a lake; see also Fig. 1–5

EXAMPLE 1–5 **ESTIMATE** **Volume of a lake.** Estimate how much water there is in a particular lake, Fig. 1–5a, which is roughly circular, about 1 km across, and you guess it to have an average depth of about 10 m.

SOLUTION No lake is a perfect circle, nor can lakes be expected to have a perfectly flat bottom. We are only estimating here. To estimate the volume, we use a simple model of the lake as a cylinder: we multiply the average depth of the lake times its roughly circular surface area, as if the lake were a cylinder (Fig. 1–5b). The volume V of a cylinder is the product of its height h times the area of its base: $V = h\pi r^2$, where r is the radius of the circular base. The radius r is $\frac{1}{2}$ km $= 500$ m, so the volume is approximately

$$V = h\pi r^2 \approx (10 \text{ m}) \times (3) \times (5 \times 10^2 \text{ m})^2 \approx 8 \times 10^6 \text{ m}^3 \approx 10^7 \text{ m}^3,$$

where π was rounded off to 3. So the volume is on the order of 10^7 m^3, ten million cubic meters. Because of all the estimates that went into this calculation, the order-of-magnitude estimate (10^7 m^3) is probably better to quote than the 8×10^6 m^3 figure.

EXAMPLE 1–6 **ESTIMATE** **Thickness of a page.** Estimate the thickness of a page of this book.

➡ **PROBLEM SOLVING**

Use symmetry when possible

FIGURE 1–6 A micrometer, which is used for measuring small thicknesses.

SOLUTION At first you might think that a special measuring device, a micrometer (Fig. 1–6), is needed to measure the thickness of one page since an ordinary ruler clearly won't do. But we can use a trick or, to put it in physics terms, make use of a *symmetry*: we can make the reasonable assumption that all the pages of this book are equal in thickness. Thus, we can use a ruler to measure hundreds of pages at once. If you measure the thickness of the first 500 pages of this book (page 1 to page 500) you might get something like 1.5 cm. Note that 500 pages counted front and back is 250 separate pieces of paper. So one page must have a thickness of about

$$\frac{1.5 \text{ cm}}{250 \text{ pages}} \approx 6 \times 10^{-3} \text{ cm} = 6 \times 10^{-2} \text{ mm}$$

or less than a tenth of a millimeter (0.1 mm).

Now let's take a simple Example of how a diagram can be useful for making an estimate. It cannot be emphasized enough how important it is to draw a diagram when trying to solve a physics problem.

EXAMPLE 1–7 ESTIMATE **Height by triangulation.** Estimate the height of the building shown in Fig. 1–7a, by "triangulation," with the help of a bus-stop pole and a friend.

SOLUTION By standing your friend next to the pole, you estimate the height of the pole to be 3 m. You next step away from the pole until the top of the pole is in line with the top of the building, Fig. 1–7a. You are 5 ft 6 in. tall, so your eyes are about 1.5 m above the ground. Your friend is taller, and when she stretches out her arms, one hand touches you, and the other touches the pole, so you estimate that distance as 2 m (Fig. 1–7a). You then pace off the distance from the pole to the base of the building with big, 1-m-long, steps, and you get a total of 16 steps or 16 m. Now you draw, to scale, the diagram shown in Fig. 1–7b using these measurements. You can measure, right on the diagram, the last side of the triangle to be about $x = 13$ m. Alternatively, you can use similar triangles to obtain the height x:

$$\frac{1.5\,\text{m}}{2\,\text{m}} = \frac{x}{18\,\text{m}}, \qquad \text{so} \qquad x \approx 13\tfrac{1}{2}\,\text{m}.$$

Finally you add in your eye height of 1.5 m above the ground to get your final result: the building is about 15 m tall.

EXAMPLE 1–8 ESTIMATE **Estimating the radius of the Earth.** An easy way to convince yourself that the Earth is round is to watch a ship at sea disappear over the horizon on a calm day. In fact, believe it or not, you can estimate the radius of the Earth without having to go into space (see chapter opening photo). To see how, consider the following data. Suppose you measure the deck of a large sailboat moored on a lake or bay to be 2.0 m above water level. Then you go to the far side of the lake, where you are 4.4 km from the sailboat. Now you lie down right at the water's edge and you estimate that you can see only the upper $\tfrac{1}{4}$ of the sailboat's hull—that is, $\tfrac{3}{4}$ of the hull, or 1.5 m, is below the horizon (hidden behind water). Using Fig. 1–8, where $h = 1.5$ m, estimate the radius of the Earth.

SOLUTION We can use the Pythagorean theorem on the large right triangle in Fig. 1–8, where R is the radius of the Earth, $h + R$ is approximately the hypotenuse, $d = 4.4$ km, and $h = 1.5$ m:

$$R^2 + d^2 \approx (R + h)^2$$
$$\approx R^2 + 2hR + h^2$$

so

$$R \approx \frac{d^2 - h^2}{2h} = \frac{(4400\,\text{m})^2 - (1.5\,\text{m})^2}{3.0\,\text{m}} \approx 6500\,\text{km}.$$

Precise measurements give 6380 km. But look at your achievement! With a few simple rough measurements, and simple geometry, you can make a good estimate of the Earth's radius. You didn't need to go out in space, nor did you need a very long measuring stick.

Another technique for estimating, this one made famous by Enrico Fermi to his physics students, is to estimate the number of piano tuners in a city, say, Chicago or San Francisco. To get a rough order-of-magnitude estimate of the number of piano tuners today in San Francisco, a city of about 700,000 inhabitants, we can proceed by estimating the number of functioning pianos, how often each piano is

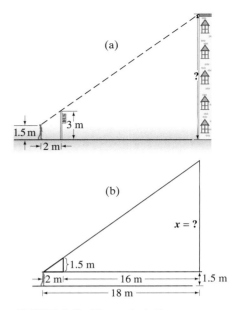

(a)

(b)

$x = ?$

1.5 m

2 m ⟷ 16 m ⟷ 1.5 m

18 m

FIGURE 1–7 Example 1–7. Diagrams are really useful!

FIGURE 1–8 Boat disappearing over the horizon, Example 1–8 (not to scale). R is the radius of the Earth. You are a distance $d = 4.4$ km from the sailboat when you can see only $\tfrac{1}{4}$ of its hull. Because of the curvature of the Earth, the water "bulges out" between you and the boat.

tuned, and how many pianos each tuner can tune. To estimate the number of pianos in San Francisco, we note that certainly not everyone has a piano. A guess of 1 family in 3 having a piano would correspond to 1 piano per 12 persons, assuming an average family of 4 persons. As an order of magnitude, let's say 1 piano per 10 people. This is certainly more reasonable than 1 per 100 people, or 1 per every person, so let's proceed with the estimate that 1 person in 10 has a piano, or about 70,000 pianos in San Francisco. Now a piano tuner needs an hour or two to tune a piano. So let's estimate that a tuner can tune 4 or 5 pianos a day. A piano ought to be tuned every 6 months or a year—let's say once each year. A piano tuner tuning 4 pianos a day, 5 days a week, 50 weeks a year can tune about 1000 pianos a year. So San Francisco, with its (very) roughly 70,000 pianos needs about 70 piano tuners. This is, of course, only a rough estimate.[†] It tells us that there must be many more than 10 piano tuners, and surely not as many as 1000. If you were estimating the number of car mechanics, on the other hand, your estimate would be rather different!

** $\boxed{1-7}$ Dimensions and Dimensional Analysis

When we speak of the **dimensions** of a quantity, we are referring to the type of units or base quantities that make it up. The dimensions of area, for example, are always length squared, abbreviated $[L^2]$, using square brackets; the units can be square meters, square feet, and so on. Velocity, on the other hand, can be measured in units of km/h, m/s and mi/h, but the dimensions are always a length $[L]$ divided by a time $[T]$; that is, $[L/T]$. The formula for a quantity may be different in different cases, but the dimensions remain the same. For example, the area of a triangle of base b and height h is $A = \frac{1}{2}bh$, whereas the area of a circle of radius r is $A = \pi r^2$. The formulas are different in the two cases, but the dimensions in both cases are the same: $[L^2]$.

When we specify the dimensions of a quantity, we usually do so in terms of base quantities, not derived quantities. For example, force, which we will see later has the same units as mass $[M]$ times acceleration $[L/T^2]$, has dimensions of $[ML/T^2]$.

Dimensional analysis

Dimensions can be used as a help in working out relationships, and such a procedure is referred to as **dimensional analysis**.[‡] One useful technique is the use of dimensions to check if a relationship is *incorrect*. A simple rule applies here: we add or subtract quantities only if they have the same dimensions (we don't add centimeters and pounds). This implies that the quantities on each side of an equals sign must have the same dimensions.

For example, suppose you derived the equation $v = v_0 + \frac{1}{2}at^2$, where v is the speed of an object after a time t, when it starts with an initial speed v_0 and undergoes an acceleration a. Let's do a dimensional check to see if this equation is correct. We write a dimensional equation as follows, remembering that the dimensions of speed are $[L/T]$ and (as we shall see in Chapter 2) the dimensions of acceleration are $[L/T^2]$:

$$\left[\frac{L}{T}\right] \overset{?}{=} \left[\frac{L}{T}\right] + \left[\frac{L}{T^2}\right][T^2]$$

$$\overset{?}{=} \left[\frac{L}{T}\right] + [L].$$

The dimensions are incorrect: on the right side, we have the sum of quantities whose dimensions are not the same. Thus we conclude that an error was made in the derivation of the original equation.

[†] A check of the San Francisco Yellow Pages (done after this calculation) reveals about 50 listings. Each of these listings may employ more than one tuner, but on the other hand, each may also do repairs as well as tuning. In any case, our estimate is reasonable.

[‡] The techniques described in the next few paragraphs may seem more meaningful after you have studied a few chapters of this book. Reading this Section now will give you an overview of the subject, and you can then return to it later as needed.

If such a dimensional check does come out correct, it does not prove that the equation is correct. For example, a dimensionless numerical factor (such as $\frac{1}{2}$ or 2π) could be wrong. Thus a dimensional check can only tell you when a relationship is wrong. It can't tell you if it is completely right.

Dimensional analysis can also be used as a quick check on an equation you are not sure about. For example, suppose that you can't remember whether the equation for the period T (the time to make one back-and-forth swing) of a simple pendulum of length l is $T = 2\pi \sqrt{l/g}$ or $T = 2\pi \sqrt{g/l}$, where g is the acceleration due to gravity and, like all accelerations, has dimensions $[L/T^2]$. (Do not worry about these formulas—the correct one will be derived in Chapter 14; what we are concerned about here is a person's forgetting whether it contains l/g or g/l.) A dimensional check shows that the former is correct:

$$[T] = \sqrt{\frac{[L]}{[L/T^2]}} = \sqrt{[T^2]} = [T],$$

whereas the latter is not:

$$[T] \neq \sqrt{\frac{[L/T^2]}{[L]}} = \sqrt{\frac{1}{[T^2]}} = \frac{1}{[T]}.$$

Note that the constant 2π has no dimensions and so can't be checked using dimensions.

Summary

Scientists often devise models of physical phenomena. A **model** is a kind of picture or analogy that seems to explain the phenomena. A **theory**, often developed from a model, is usually deeper and more complex than a simple model.

A scientific **law** is a concise statement, often expressed in the form of an equation, which quantitatively describes a particular range of phenomena over a wide range of cases.

Measurements play a crucial role in physics, but can never be perfectly precise. It is important to specify the **uncertainty** of a measurement either by stating it directly using the ± notation, and/or by keeping only the correct number of **significant figures**.

Physical quantities are always specified relative to a particular standard or **unit**, and the unit used should always be stated. The commonly accepted set of units today is the **Système International** (SI), in which the standard units of length, mass, and time are the **meter**, **kilogram**, and **second**.

When converting units, check all **conversion factors** for correct cancellation of units.

Making rough, **order-of-magnitude estimates** is a very useful technique in science as well as in everyday life.

The **dimensions** of a quantity refer to the combination of base quantities that comprise it. Velocity, for example, has dimensions of [length/time] or $[L/T]$. Working with only the dimensions of the various quantities in a given relationship (this technique is called **dimensional analysis**) it is possible to check a relationship for correct form.

Questions

1. It is advantageous that fundamental standards, such as those for length and time, be accessible (easy to compare to), invariable (do not change), indestructible, and reproducible. Discuss why these are advantages and whether any of these criteria can be incompatible with others.

2. What are the merits and drawbacks of using a person's foot as a standard? Discuss in terms of the criteria mentioned in Question 1. Consider both (a) a particular person's foot, and (b) any person's foot.

3. When traveling a highway in the mountains, you may see elevation signs that read "914 m (3000 ft)." Critics of the metric system claim that such numbers show the metric system is more complicated. How would you alter such signs to be more consistent with a switch to the metric system?

4. Suggest a way to measure the distance from Earth to the Sun.

5. What is wrong with this road sign: Memphis 7 mi (11.263 km)?

*6. Can you set up a complete set of base quantities, as in Table 1–5, that does not include length as one of them?

7. List assumptions useful to estimate the number of car mechanics in (a) San Francisco, (b) your hometown, and then make the estimates.

8. Estimate the number of hours you have spent in school thus far in your life.

9. Discuss how the notion of symmetry could be used to estimate the number of marbles in a one-liter jar.

10. You measure the radius of a wheel to be 4.16 cm. If you multiply by 2 to get the diameter, should you write the result as 8 cm or as 8.32 cm? Justify your answer.

Problems

[The problems at the end of each chapter are ranked I, II, or III according to estimated difficulty, with I problems being easiest. The problems are arranged by Sections, meaning that the reader should have read up to and including that Section, but not only that Section—problems often depend on earlier material. Each chapter also has a group of General Problems that are not arranged by Section and not ranked.]

Section 1–3

1. (I) The age of the universe is thought to be somewhere around 10 billion years. Assuming one significant figure, write this in powers of ten in (a) years, (b) seconds.

2. (I) How many significant figures do each of the following numbers have: (a) 2142, (b) 81.60, (c) 7.63, (d) 0.03, (e) 0.0086, (f) 3236, and (g) 8700?

3. (I) Write the following numbers in powers of ten notation: (a) 1,156, (b) 21.8, (c) 0.0068, (d) 27.635, (e) 0.219, and (f) 22.

4. (I) Write out the following numbers in full with a decimal point and correct number of zeros: (a) 8.69×10^4, (b) 7.1×10^3, (c) 6.6×10^{-1}, (d) 8.76×10^2, and (e) 8.62×10^{-5}.

5. (I) What is the percent uncertainty in the measurement 3.26 ± 0.25 m?

6. (I) What, approximately, is the percent uncertainty for the measurement given as 1.28 m?

7. (I) Time intervals measured with a stopwatch typically have an uncertainty of about 0.2 s, due to human reaction time at the start and stop moments. What is the percent uncertainty of a handtimed measurement of (a) 5 s, (b) 50 s, (c) 5 min?

8. (II) Multiply 2.079×10^2 m by 0.072×10^{-1}, taking into account significant figures.

9. (II) Add 9.2×10^3 s + 8.3×10^4 s + 0.008×10^6 s.

10. (II) What is the area, and its approximate uncertainty, of a circle of radius 3.8×10^4 cm?

11. (II) What is the percent uncertainty in the volume of a spherical beach ball whose radius is $r = 2.86 \pm 0.08$ m?

Sections 1–4 and 1–5

12. (I) Express the following using the prefixes of Table 1–4: (a) 10^6 volts, (b) 10^{-6} meters, (c) 6×10^3 days, (d) 18×10^2 bucks, and (e) 8×10^{-9} pieces.

13. (I) Write the following as full (decimal) numbers with standard units: (a) 286.6 mm, (b) 85 μV, (c) 760 mg, (d) 60.0 picoseconds, (e) 22.5 femtometers (f) 2.50 gigavolts.

14. (I) How many cars is 50 hectocars? What would you be if you earned a megabuck a year?

15. (I) Determine your height in meters.

16. (I) The Sun, on average, is 93 million miles from the Earth. How many meters is this? Express (a) using powers of ten, and (b) using a metric prefix.

17. (I) What is the conversion factor between (a) ft^2 and yd^2, (b) m^2 and ft^2?

18. (II) The Concorde airplane travels at about 2300 km/h. How long does it take to travel 1.0 mile?

19. (II) A typical atom has a diameter of about 1.0×10^{-10} m. (a) What is this in inches? (b) Approximately how many atoms are there along a 1.0-cm line?

20. (II) Express the following sum with the correct number of significant figures: 2.00 m + 142.5 cm + 7.24×10^5 μm.

21. (II) Determine the conversion factor between (a) km/h and mi/h, (b) m/s and ft/s, and (c) km/h and m/s.

22. (II) How much longer (percentage) is a one-mile race than a 1500-m race ("the metric mile")?

23. (II) A *light-year* is the distance light (speed = 2.998×10^8 m/s) travels in one year. (a) How many meters are there in 1.00 light-year? (b) An astronomical unit (AU) is the average distance from the Sun to Earth, 1.50×10^8 km. How many AU are there in 1.00 light-year? (c) What is the speed of light in AU/h?

24. (II) The diameter of the moon is 3480 km. What is the surface area, and how does it compare to the surface area of the Earth?

Section 1–6

(*Note*: Remember that for rough estimates, only round numbers are needed both as input to calculations and as final results.)

25. (I) Estimate the order of magnitude (power of ten) of: (a) 2800, (b) 86.30×10^2, (c) 0.0076, and (d) 15.0×10^8.

26. (II) Estimate how long it would take a good runner to run across the United States from New York to California.

27. (II) Make a rough estimate of the percentage of a house's outside wall area that consists of window area.

28. (II) Estimate the number of times a human heart beats in a lifetime.

29. (II) Make a rough estimate of the volume of your body (in cm^3).

30. (II) Estimate the time to drive from Beijing (Peking) to Paris (a) today, and (b) in 1906 when a great car race was run between those two cities.

31. (II) Estimate the number of dentists (a) in San Francisco and (b) in your town or city.

32. (II) Estimate how long it would take one person to mow a football field using an ordinary home lawn mower.

33. (II) The rubber worn from tires mostly enters the atmosphere as particulate pollution. Estimate how much rubber (in kg) is put into the air in the United States every year. To get you started, a good estimate for a tire tread's depth is 1 cm when new, and the density of rubber is about 1200 kg/m^3.

34. (II) Estimate how many books can be shelved in a college library with 1500 square meters of floor space.

35. (III) You are in a hot air balloon, 200 m above the flat Texas plains. You look out toward the horizon. How far out can you see—that is, how far is your horizon? The Earth's radius is about 6400 km.

* Section 1–7

* 36. (I) The speed, v, of a body is given by the equation $v = At^3 - Bt$, where t refers to time. What are the dimensions of A and B?

* 37. (I) What are the SI units for the constants A and B in Problem 36?

* 38. (II) Three students derive the following equations in which x refers to distance traveled, v the speed, a the acceleration (m/s^2), t the time, and the subscript $(_0)$ means a quantity at time $t = 0$: (a) $x = vt^2 + 2at$, (b) $x = v_0t + \frac{1}{2}at^2$, and (c) $x = v_0t + 2at^2$. Which of these could possibly be correct according to a dimensional check?

General Problems

39. An angstrom (symbol Å) is an older unit of length, defined as 10^{-10} m. (a) How many nanometers are in 1.0 angstrom? (b) How many femtometers or fermis (the common unit of length in nuclear physics) are in 1.0 angstrom? (c) How many angstroms are in 1.0 meter? (d) How many angstroms are in 1.0 light-year (see Problem 23)?

40. Use Table 1–3 to estimate the total number of protons or neutrons in (a) a bacterium, (b) a DNA molecule, (c) the human body, (d) our Galaxy.

41. (a) How many seconds are there in 1.00 year? (b) How many nanoseconds are there in 1.00 year? (c) How many years are there in 1.00 second?

42. One hectare is defined as 10^4 m^2. One acre is 4×10^4 ft^2. How many acres are in one hectare?

43. Estimate the number of bus drivers (a) in Washington, D.C., and (b) in your town.

44. Computer chips (Fig. 1–9) are etched on circular silicon wafers of thickness 0.60 mm that are sliced from a solid cylindrical silicon crystal of length 30 cm. If each wafer can hold 100 chips, what is the maximum number of chips that can be produced from one entire cylinder?

45. Estimate the number of gallons of gasoline consumed by the total of all automobile drivers in the United States, per year.

46. Estimate the number of gumballs in the machine of Fig. 1–10.

FIGURE 1–9 Problem 44. The wafer held by the hand (above) is shown below, enlarged and illuminated by colored light. Visible are rows of integrated circuits (chips).

FIGURE 1–10 Problem 46. Estimate the number of gumballs in the machine.

47. An average family of four uses roughly 1200 liters (about 300 gallons) of water per day. (One liter = 1000 cm^3.) How much depth would a lake lose per year if it uniformly covered an area of 50 square kilometers and supplied a local town with a population of 40,000 people? Consider only population uses, and neglect evaporation and so on.

48. How big is a ton? That is, what is the volume of something that weighs a ton? To be specific, estimate the diameter of a 1-ton rock, but first make a wild guess: will it be 1 ft across, 3 ft, or the size of a car? [Hint: Rock has mass per volume about 3 times that of water, which is 1 kg per liter (10^3 cm^3) or 62 lb per cubic foot.]

49. A heavy rainstorm dumps 1.0 cm of rain on a city 5 km wide and 8 km long in a 2-h period. How many metric tons (1 ton = 10^3 kg) of water fell on the city? [1 cm^3 of water has a mass of 1 gram = 10^{-3} kg.]

50. Hold a pencil in front of your eye at a position where its blunt end just blocks out the Moon (Fig. 1–11). Make appropriate measurements to estimate the diameter of the Moon, given that the Earth–Moon distance is 3.8×10^5 km.

FIGURE 1–11 Problem 50. How big is the Moon?

51. Estimate how long it would take to walk around the world.

52. Noah's ark was ordered to be 300 cubits long, 50 cubits wide, and 30 cubits high. The cubit was a unit of measure equal to the length of a human forearm, elbow to the tip of the longest finger. Express the dimensions of Noah's ark in meters.

53. Jean camps beside a wide river and wonders how wide it is. She spots a large rock on the bank directly across from her. She then walks upstream until she judges that the angle between her and the rock, which she can still see clearly, is now at an angle of 30° downstream (Fig. 1–12). Jean measures her stride to be about one yard long. The distance back to her camp is 120 strides. About how far across, both in yards and in meters, is the river?

FIGURE 1–12 Problem 53.

54. One liter (1000 cm^3) of oil is spilled onto a smooth lake. If the oil spreads out uniformly until it makes an oil slick just one molecule thick, with adjacent molecules just touching, estimate the diameter of the oil slick. Assume the oil molecules have a diameter of 2×10^{-10} m.

55. Compare the percent uncertainty in θ and in $\sin\theta$ when (a) $\theta = 15.0° \pm 0.5°$, (b) $\theta = 75.0° \pm 0.5°$.

56. You are lying on the sand at the edge of the sea, watching a sailboat. If you know (or measured) that the distance from the water to the top of the boat hull is 2.5 m, estimate how far away the boat is when (using binoculars) you can no longer see the hull. The radius of the Earth is 6.38×10^6 m.

A high speed car has released a parachute to reduce its speed quickly. The directions of the car's velocity and acceleration are shown by the green (**v**) and gold (**a**) arrows. Note that **v** and **a** point in opposite directions. Motion is described using the concepts of velocity and acceleration. We see here that the acceleration **a** can sometimes be in the opposite direction from the velocity **v**. We will also examine in detail motion with constant acceleration, including the vertical motion of objects falling under gravity.

CHAPTER 2

Describing Motion: Kinematics in One Dimension

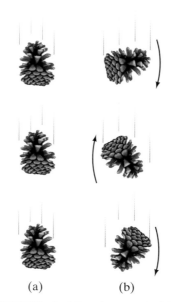

(a)   (b)

FIGURE 2–1 The pinecone in (a) undergoes pure translation as it falls, whereas in (b) it is rotating as well as translating.

The motion of objects—baseballs, automobiles, joggers, and even the Sun and Moon—is an obvious part of everyday life. It was not until the sixteenth and seventeenth centuries that our modern understanding of motion was established. Many contributed to this understanding, particularly Galileo Galilei (1564–1642) and Isaac Newton (1642–1727).

The study of the motion of objects, and the related concepts of force and energy, form the field called **mechanics**. Mechanics is customarily divided into two parts: **kinematics**, which is the description of how objects move, and **dynamics**, which deals with force and why objects move as they do. This chapter and the next deal with kinematics.

We start by discussing objects that move without rotating (Fig. 2–1a). Such motion is called **translational motion**. In the present chapter we will be concerned with describing an object that moves along a straight-line path, which is one-dimensional motion. In Chapter 3 we will study how to describe translational motion in two (or three) dimensions.

We will often make use of the concept, or model, of an idealized **particle** which is considered to be a mathematical point and to have no spatial extent (no size). A particle can undergo only translational motion. The particle model is useful in many real situations where we are interested only in translational motion and the object's size is not significant. For example, we might consider a billiard ball, or even a spacecraft traveling toward the moon, as a particle for many purposes.

2–1 Reference Frames and Displacement

Any measurement of position, distance, or speed must be made with respect to a **frame of reference**. For example, while you are on a train traveling at 80 km/h, you might notice a person who walks past you toward the front of the train at a speed of, say, 5 km/h (Fig. 2–2). Of course this is the person's speed with respect to the train as frame of reference. With respect to the ground that person is moving at a speed of 80 km/h + 5 km/h = 85 km/h. It is always important to specify the frame of reference when stating a speed. In everyday life, we usually mean "with respect to the Earth" without even thinking about it, but the reference frame must be specified whenever there might be confusion.

All measurements are made relative to a frame of reference

FIGURE 2–2 A person walks toward the front of a train at 5 km/h. The train is moving 80 km/h with respect to the ground, so the walking person's speed, relative to the ground, is 85 km/h.

When specifying the motion of an object, it is important to specify not only the speed but also the direction of motion. Often we can specify a direction by using the cardinal points, north, east, south, and west, and by "up" and "down." In physics, we often draw a set of **coordinate axes**, as shown in Fig. 2–3, to represent a frame of reference. We can always place the origin 0, and the directions of the x and y axes, as we like for convenience. The x and y axes are always perpendicular to each other. Objects positioned to the right of the origin of coordinates (0) on the x axis have an x coordinate which we usually choose to be positive; then points to the left of 0 have a negative x coordinate. The position along the y axis is usually considered positive when above 0, and negative when below 0, although the reverse convention can be used if convenient. Any point on the plane can be specified by giving its x and y coordinates. In three dimensions, a z axis perpendicular to the x and y axes is added.

For one-dimensional motion, we often choose the x axis as the line along which the motion takes place. Then the **position** of an object at any moment is given by its x coordinate.

We need to make a distinction between the *distance* an object has traveled, and its **displacement**, which is defined as the *change in position* of the object. That is, displacement is how far the object is from its starting point. To see the distinction between total distance and displacement, imagine a person walking 70 m to the east and then turning around and walking back (west) a distance of 30 m (see Fig. 2–4). The total *distance* traveled is 100 m, but the *displacement* is only 40 m since the person is now only 40 m from the starting point.

Displacement is a quantity that has both magnitude and direction. Such quantities are called **vectors**, and are represented by arrows in diagrams. For example, in Fig. 2–4, the blue arrow represents the displacement whose magnitude is 40 m and whose direction is to the right.

We will deal with vectors more fully in Chapter 3. For now, we deal only with motion in one dimension, along a line, and in this case vectors which point in one direction will have a positive sign, whereas vectors that point in the opposite direction will have a negative sign.

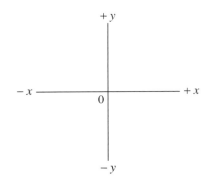

FIGURE 2–3 Standard set of xy coordinate axes.

Displacement

FIGURE 2–4 A person walks 70 m east, then 30 m west. The total distance traveled is 100 m (path is shown in black); but the displacement, shown as a blue arrow, is 40 m to the east.

FIGURE 2–5 The arrow represents the displacement $x_2 - x_1$. Distances are in meters.

FIGURE 2–6 For the displacement $\Delta x = x_2 - x_1 = 10.0\,\text{m} - 30.0\,\text{m}$, the displacement vector points to the left.

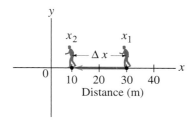

Consider the motion of an object over a particular time interval. Suppose that at some initial moment in time, call it t_1, the object is on the x axis at the point x_1 in the coordinate system shown in Fig. 2–5. At some later time, t_2, suppose the object is at point x_2. The displacement of our object is $x_2 - x_1$, and is represented by the arrow pointing to the right in Fig. 2–5. It is convenient to write

$$\Delta x = x_2 - x_1$$

where the symbol Δ (Greek letter delta) means "change in." Then Δx means "the change in x," which is the displacement. Note that the "change in" any quantity means the final value of that quantity, minus the initial value.

To be concrete, suppose $x_1 = 10.0\,\text{m}$ and $x_2 = 30.0\,\text{m}$. Then

$$\Delta x = x_2 - x_1 = 30.0\,\text{m} - 10.0\,\text{m} = 20.0\,\text{m}.$$

See Fig. 2–5.

Now consider an object moving to the left as shown in Fig. 2–6. Here the object, say a person, starts at $x_1 = 30.0\,\text{m}$ and walks to the left to the point $x_2 = 10.0\,\text{m}$. In this case

$$\Delta x = x_2 - x_1 = 10.0\,\text{m} - 30.0\,\text{m} = -20.0\,\text{m}$$

and the blue arrow representing the vector displacement points to the left. This example illustrates that when dealing with one-dimensional motion, a vector pointing to the right has a positive value, whereas one pointing to the left has a negative value.

2–2 Average Velocity

The most obvious aspect of the motion of a moving object is how fast it is moving—its speed or velocity.

The term "speed" refers to how far an object travels in a given time interval, regardless of direction. If a car travels 240 kilometers (km) in 3 hours, we say its average speed was 80 km/h. In general, the **average speed** of an object is defined as *the total distance traveled along its path divided by the time it takes to travel this distance*:

Average speed

$$\text{average speed} = \frac{\text{distance traveled}}{\text{time elapsed}}. \tag{2–1}$$

The terms velocity and speed are often used interchangeably in ordinary language. But in physics we make a distinction between the two. Speed is simply a positive number, with units. **Velocity**, on the other hand, is used to signify both the *magnitude* (numerical value) of how fast an object is moving and the *direction* in which it is moving. (Velocity is therefore a vector.) There is a second difference between speed and velocity: namely, the **average velocity** is defined in terms of *displacement*, rather than total distance traveled:

Velocity

$$\text{average velocity} = \frac{\text{displacement}}{\text{time elapsed}} = \frac{\text{final position} - \text{initial position}}{\text{time elapsed}}.$$

Average speed and average velocity have the same magnitude when the motion is all in one direction. In other cases, they may differ: recall the walk we described earlier, in Fig. 2–4, where a person walked 70 m east and then 30 m west.

The total distance traveled was 70 m + 30 m = 100 m, but the displacement was 40 m. Suppose this walk took 70 s to complete. Then the average speed was:

$$\frac{\text{distance}}{\text{time}} = \frac{100 \, \text{m}}{70 \, \text{s}} = 1.4 \, \text{m/s.}$$

The magnitude of the average velocity, on the other hand, was:

$$\frac{\text{displacement}}{\text{time}} = \frac{40 \, \text{m}}{70 \, \text{s}} = 0.57 \, \text{m/s.}$$

This difference between the speed and the magnitude of the velocity occurs in some cases, but only for the *average* values.

To discuss one-dimensional motion of an object in general, suppose that at some moment in time, call it t_1, the object is on the x axis at point x_1 in a coordinate system, and at some later time, t_2, suppose it is at point x_2. The elapsed time is $t_2 - t_1$, and during this time interval the displacement of our object was $\Delta x = x_2 - x_1$. Then the average velocity, defined as *the displacement divided by the elapsed time*, can be written

$$\bar{v} = \frac{x_2 - x_1}{t_2 - t_1} = \frac{\Delta x}{\Delta t}, \tag{2–2}$$

Average velocity

where v stands for velocity and the bar ($^-$) over the v is a standard symbol meaning "average."

For the usual case of the $+x$ axis to the right, note that if x_2 is less than x_1, the object is moving to the left, and then $\Delta x = x_2 - x_1$ is less than zero. The sign of the displacement, and thus of the velocity, indicates the direction: the average velocity is positive for an object moving to the right along the $+x$ axis and negative when the object moves to the left. The direction of the average velocity is always the same as the direction of the displacement.

➡ **PROBLEM SOLVING**

+ or − sign can signify the direction for linear motion

EXAMPLE 2–1 **Runner's average velocity.** The position of a runner as a function of time is plotted as moving along the x axis of a coordinate system. During a 3.00-s time interval, the runner's position changes from $x_1 = 50.0$ m to $x_2 = 30.5$ m, as shown in Fig. 2–7. What was the runner's average velocity?

SOLUTION Average velocity is the displacement divided by the elapsed time. The displacement is $\Delta x = x_2 - x_1 = 30.5 \, \text{m} - 50.0 \, \text{m} = -19.5$ m. The time interval is $\Delta t = 3.00$ s. Therefore the average velocity is

$$\bar{v} = \frac{\Delta x}{\Delta t} = \frac{-19.5 \, \text{m}}{3.00 \, \text{s}} = -6.50 \, \text{m/s.}$$

The displacement and average velocity are negative, which tells us (if we didn't already know it) that the runner is moving to the left along the x axis, as indicated by the arrow in Fig. 2–7. Thus we can say that the runner's average velocity is 6.50 m/s to the left.

FIGURE 2–7 Example 2–1. A person runs from $x_1 = 50.0$ m to $x_2 = 30.5$ m. The displacement is −19.5 m.

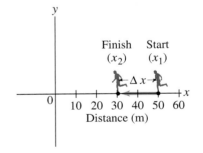

EXAMPLE 2–2 **Distance a cyclist travels.** How far can a cyclist travel in 2.5 h along a straight road if her average speed is 18 km/h?

SOLUTION We want to find the distance traveled, so we use Eq. 2–2 letting Δx be the distance and $\bar{v}$ be the average speed, and then rewrite it as

$$\Delta x = \bar{v} \, \Delta t = (18 \, \text{km/h})(2.5 \, \text{h}) = 45 \, \text{km.}$$

2–3 | Instantaneous Velocity

If you drive a car along a straight road for 150 km in 2.0 h, the magnitude of your average velocity is 75 km/h. It is unlikely, though, that you were moving at precisely 75 km/h at every instant. To describe this situation we need the concept of *instantaneous velocity*, which is the velocity at any instant of time. (This is the magnitude that a speedometer is supposed to indicate.) More precisely, the **instantaneous velocity** at any moment is defined as *the average velocity over an infinitesimally short time interval*. That is, Eq. 2–2 is to be evaluated in the limit of Δt becoming extremely small, approaching zero. We can write the definition of instantaneous velocity, v, for one-dimensional motion as

Instantaneous velocity

$$v = \lim_{\Delta t \to 0} \frac{\Delta x}{\Delta t}.$$

(2–3)

The notation $\lim_{\Delta t \to 0}$ means the ratio $\Delta x / \Delta t$ is to be evaluated in the limit of Δt approaching zero. But we do not simply set $\Delta t = 0$ in this definition, for then Δx would also be zero, and we would have an undefined number. Rather, we are considering the *ratio* $\Delta x / \Delta t$, as a whole. As we let Δt approach zero, Δx approaches zero as well. But the ratio $\Delta x / \Delta t$ approaches some definite value, which is the instantaneous velocity at a given instant.

For instantaneous velocity we use the symbol v, whereas for average velocity we use $\bar{v}$, with a bar. In the rest of this book, when we use the term "velocity" it will refer to instantaneous velocity. When we want to speak of the average velocity, we will make this clear by including the word "average."

Note that the *instantaneous* speed always equals the magnitude of the instantaneous velocity. Why? Because distance and displacement become the same when they become infinitesimally small.

If an object moves at a uniform (that is, constant) velocity over a particular time interval, then its instantaneous velocity at any instant is the same as its average velocity (see Fig. 2–8a). But in many situations this is not the case. For example, a car may start from rest, speed up to 50 km/h, remain at that velocity for a time, then slow down to 20 km/h in a traffic jam, and finally stop at its destination after traveling a total of 15 km in 30 min. This trip is plotted on the graph of Fig. 2–8b. Also shown on the graph is the average velocity (dashed line), which is $\bar{v} = \Delta x / \Delta t = 15 \text{ km}/0.50 \text{ h} = 30 \text{ km/h}$.

To better understand instantaneous velocity, let us consider a graph of the position of a particular particle versus time (x vs. t), as shown in Fig. 2–9. (Note that this is different from showing the "path" of a particle on an x vs. y plot.) The particle is at position x_1 at a time t_1, and at position x_2 at time t_2. P_1 and P_2 represent these two points on the graph. A straight line drawn from point $P_1(x_1, t_1)$ to

FIGURE 2–8 Velocity of a car as a function of time: (a) at constant velocity; (b) with varying velocity.

FIGURE 2–9
Graph of a particle's position x vs. time t. The slope of the straight line $P_1 P_2$ represents the average velocity of the particle during the time interval $\Delta t = t_2 - t_1$.

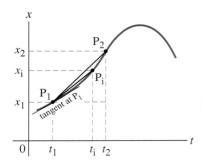

FIGURE 2–10 Same position vs. time curve as in Fig. 2–9, but note that the average velocity over the time interval $t_i - t_1$ (which is the slope of $P_1 P_i$) is less than the average velocity over the time interval $t_2 - t_1$. The slope of the thin line tangent to the curve at point P_1 equals the instantaneous velocity at time t_1.

point $P_2(x_2, t_2)$ forms the hypotenuse of a right triangle whose sides are Δx and Δt. The ratio $\Delta x/\Delta t$ is the **slope** of the straight line $P_1 P_2$. But $\Delta x/\Delta t$ is also the average velocity of the particle during the time interval $\Delta t = t_2 - t_1$. Therefore, we conclude that the average velocity of an object during any time interval $\Delta t = t_2 - t_1$ is equal to the slope of the straight line (or *chord*) connecting the two points (x_1, t_1) and (x_2, t_2) on an x vs. t graph.

Slope of chord joining 2 points on x vs. t graph equals average velocity

Consider now a time t_i, intermediate between t_1 and t_2, at which time the particle is at x_i (Fig. 2–10). The slope of the straight line $P_1 P_i$ is less than the slope of $P_1 P_2$ in this case. Thus the average velocity during the time interval $t_i - t_1$ is less than that during the time interval $t_2 - t_1$.

Now let us imagine that we take the point P_i in Fig. 2–10 to be closer and closer to point P_1. That is, we let the internal $t_i - t_1$, which we now call Δt, to become smaller and smaller. The slope of the line connecting the two points becomes closer and closer to the slope of a line tangent to the curve at point P_1. As we take Δt smaller and smaller, the average velocity (equal to the slope of the chord) approaches the slope of the tangent at point P_1. The definition of the instantaneous velocity (Eq. 2–3) is the limiting value of the average velocity as Δt approaches zero. Thus the *instantaneous velocity equals the slope of the tangent to the curve* at that point (which we can simply call "the slope of the curve" at that point).

Slope of tangent to x vs. t curve equals instantaneous velocity

In Eq. 2–3, the limit as $\Delta t \to 0$ is written in calculus notation as dx/dt and is called the *derivative* of x with respect to t. Thus we can write Eq. 2–3 in calculus notation as:

$$v = \lim_{\Delta t \to 0} \frac{\Delta x}{\Delta t} = \frac{dx}{dt}. \qquad \textbf{(2–4)}$$

Instantaneous velocity

This equation is the definition of instantaneous velocity for one-dimensional motion.

Because the velocity at any instant equals the slope of the tangent to the x vs. t graph at that instant, we can obtain the velocity at any instant from such a graph. For example, in Fig. 2–11 (which shows the same curve as in Figs. 2–9 and 2–10), as our object moves from x_1 to x_2, the slope continually increases, so the velocity is increasing. For times after t_2, however, the slope begins to decrease and in fact reaches zero (so $v = 0$) where x has its maximum value, at point P_3 in Fig. 2–11. Beyond this point, the slope is negative, as for point P_4. The velocity is therefore negative, which makes sense since x is now decreasing—the particle is moving toward decreasing values of x, to the left on a standard xy plot.

If an object moves with constant velocity over a particular time interval, its instantaneous velocity is equal to its average velocity. The graph of x vs. t in this case will be a straight line whose slope equals the velocity. The curve of Fig. 2–9 has no straight sections, so there are no time intervals when the velocity is constant.

FIGURE 2–11 Same x vs. t curve as in Figs. 2–9 and 2–10, but here showing the slope at four different points: At P_3, the slope is zero, so $v = 0$. At P_4 the slope is negative, so $v < 0$.

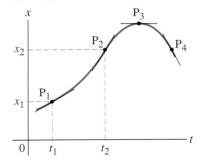

EXAMPLE 2–3 **Given x as a function of t.** A jet engine moves along an experimental track (which we call the x-axis) as shown in Fig. 2–12a. We will treat the engine as if it were a particle. Its position as a function of time is given by the equation $x = At^2 + B$, where $A = 2.10 \text{ m/s}^2$ and $B = 2.80 \text{ m}$, and this equation is plotted in Fig. 2–12b. (a) Determine the displacement of the engine during the time interval from $t_1 = 3.00 \text{ s}$ to $t_2 = 5.00 \text{ s}$. (b) Determine the average velocity during this time interval. (c) Determine the magnitude of the instantaneous velocity at $t = 5.00 \text{ s}$.

SOLUTION (a) At $t_1 = 3.00 \text{ s}$, the position (point P_1 in Fig. 2–12b) is

$$x_1 = At_1^2 + B = (2.10 \text{ m/s}^2)(3.00 \text{ s})^2 + 2.80 \text{ m} = 21.7 \text{ m}.$$

At $t_2 = 5.00 \text{ s}$, the position (P_2 in Fig. 2–12b) is

$$x_2 = (2.10 \text{ m/s}^2)(5.00 \text{ s})^2 + 2.80 \text{ m} = 55.3 \text{ m}.$$

The displacement is thus

$$x_2 - x_1 = 55.3 \text{ m} - 21.7 \text{ m} = 33.6 \text{ m}.$$

(b) The magnitude of the average velocity can then be calculated as

$$\bar{v} = \frac{x_2 - x_1}{t_2 - t_1} = \frac{33.6 \text{ m}}{2.00 \text{ s}} = 16.8 \text{ m/s}.$$

This equals the slope of the straight line joining points P_1 and P_2 shown in Fig. 2–12b.

(c) The instantaneous velocity at $t = t_2 = 5.00 \text{ s}$ equals the slope of the tangent to the curve at point P_2 shown in Fig. 2–12b, and we could measure this slope off the graph to obtain v_2. We can calculate v more precisely, and for any time t, using the given formula

$$x = At^2 + B,$$

which is the engine's position x at time t. Using the calculus formulas for derivatives

$$\frac{d}{dt}(Ct^n) = nCt^{n-1} \quad \text{and} \quad \frac{dC}{dt} = 0,$$

where C is any constant, then

$$v = \frac{dx}{dt} = \frac{d}{dt}(At^2 + B) = 2At.$$

We are given $A = 2.10 \text{ m/s}^2$, so for $t = t_2 = 5.00 \text{ s}$,

$$v_2 = 2At = 2(2.10 \text{ m/s}^2)(5.00 \text{ s}) = 21.0 \text{ m/s}.$$

FIGURE 2–12 Example 2–3. (a) Engine traveling on a straight track. (b) Graph of x vs. t: $x = At^2 + B$.

(a)

(b)

2–4 Acceleration

An object whose velocity is changing is said to be accelerating. A car whose velocity increases in magnitude from zero to 80 km/h is accelerating. That is, acceleration specifies how rapidly the velocity of an object is changing.

Average Acceleration

Average acceleration is defined as the change in velocity divided by the time taken to make this change:

$$\text{average acceleration} = \frac{\text{change of velocity}}{\text{time elapsed}}.$$

In symbols, the average acceleration $\bar{a}$, over a time interval $\Delta t = t_2 - t_1$ during which the velocity changes by $\Delta v = v_2 - v_1$, is defined as

$$\bar{a} = \frac{v_2 - v_1}{t_2 - t_1} = \frac{\Delta v}{\Delta t}. \tag{2–5}$$

Average acceleration

Acceleration is also a vector, but for one-dimensional motion, we need only use a plus or minus sign to indicate direction relative to a chosen coordinate system.

EXAMPLE 2–4 **Average acceleration.** A car accelerates along a straight road from rest to 75 km/h in 5.0 s, Fig. 2–13. What is the magnitude of its average acceleration?

SOLUTION The car starts from rest, so $v_1 = 0$. The final velocity is $v_2 = 75$ km/h. Then from Eq. 2–5, the average acceleration is

$$\bar{a} = \frac{75 \text{ km/h} - 0 \text{ km/h}}{5.0 \text{ s}} = 15 \frac{\text{km/h}}{\text{s}}.$$

This is read as "fifteen kilometers per hour per second" and means that, on average, the velocity changed by 15 km/h during each second. That is, assuming the acceleration was constant, during the first second the car's velocity increased from zero to 15 km/h. During the next second its velocity increased by another 15 km/h up to 30 km/h, and so on, Fig. 2–13. (Of course, if the instantaneous acceleration was not constant, these numbers could be different.)

$t_1 = 0$
$v_1 = 0$

Acceleration
$= 15 \frac{\text{km/h}}{\text{s}}$

at $t = 1.0$ s
$v = 15$ km/h

at $t = 2.0$ s
$v = 30$ km/h

at $t = t_2 = 5.0$ s
$v = v_2 = 75$ km/h

FIGURE 2–13 Example 2–4. The car is shown at the start with $v_1 = 0$ at $t_1 = 0$. It is shown three more times, at $t = 1.0$ s, at $t = 2.0$ s, and at $t = t_2 = 5.0$ s. We assume the acceleration is constant and equals 15 km/h/s. The green arrows represent the velocity vectors; the length of each represents the magnitude of the velocity at that moment. The acceleration vector is the orange arrow. Distances are not to scale.

Careful:
Do not confuse
velocity with acceleration

Note carefully that *acceleration tells us how fast the velocity changes,* whereas *velocity tells us how fast the position changes.* In this last Example, the calculated acceleration contained two different time units: hours and seconds. We usually prefer to use only seconds. To do so we can change km/h to m/s (see Section 1–5, and Example 1–4):

$$75 \text{ km/h} = \left(75 \frac{\cancel{\text{km}}}{\cancel{\text{h}}} \right) \left(\frac{1000 \text{ m}}{1 \cancel{\text{km}}} \right) \left(\frac{1 \cancel{\text{h}}}{3600 \text{ s}} \right) = 21 \text{ m/s}.$$

Then we get

$$\bar{a} = \frac{21 \text{ m/s} - 0.0 \text{ m/s}}{5.0 \text{ s}} = 4.2 \frac{\text{m/s}}{\text{s}} = 4.2 \frac{\text{m}}{\text{s}^2}.$$

We almost always write these units as m/s^2 (meters per second squared). According to the above calculation, the velocity in Example 2–4 (Fig. 2–13) changed on the average by 4.2 m/s during each second, for a total change of 21 m/s over the 5.0 s.

CONCEPTUAL EXAMPLE 2–5 | **Velocity and acceleration.** (*a*) If the velocity of an object is zero, does it mean that the acceleration is zero? (*b*) If the acceleration is zero, does it mean that the velocity is zero?

RESPONSE A zero velocity does not necessarily mean that the acceleration is zero, nor does a zero acceleration mean that the velocity is zero. (*a*) For example, when you put your foot on the gas pedal of your car which is at rest, the velocity starts from zero but the acceleration is not zero since the velocity of the car changes. (How else could your car start forward if its velocity weren't changing— that is, if the acceleration were zero?) (*b*) As you cruise along a straight highway at a constant velocity of 100 km/h, your acceleration is zero.

---

at $t_1 = 0$
$v_1 = 15.0$ m/s

Acceleration
$= -2.0$ m/s^2

at $t_2 = 5.0$ s
$v_2 = 5.0$ m/s

FIGURE 2–14 Example 2–6, showing the position of the car at times t_1 and t_2, as well as the car's velocity represented by the green arrows. The acceleration vector (orange) points to the left.

FIGURE 2–15 The same car as in Example 2–6, but now moving to the left and decelerating. The acceleration is

$$a = \frac{v_2 - v_1}{\Delta t}$$

$$= \frac{-5.0 \text{ m/s} - (-15.0 \text{ m/s})}{5.0 \text{ s}}$$

$$= \frac{-5.0 \text{ m/s} + 15.0 \text{ m/s}}{5.0 \text{ s}} = +2.0 \text{ m/s}.$$

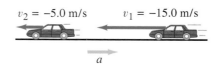

$v_2 = -5.0$ m/s $v_1 = -15.0$ m/s

a

EXAMPLE 2–6 **Car slowing down.** An automobile is moving to the right along a straight highway, which we choose to be the positive *x* axis (Fig. 2–14), and the driver puts on the brakes. If the initial velocity is $v_1 = 15.0$ m/s and it takes 5.0 s to slow down to $v_2 = 5.0$ m/s, what was the car's average acceleration?

SOLUTION The average acceleration is equal to the change in velocity divided by the elapsed time, Eq. 2–5. Let us call the initial time $t_1 = 0$; then $t_2 = 5.0$ s. (Note that our choice of $t_1 = 0$ doesn't affect the calculation of $\bar{a}$ because only $\Delta t = t_2 - t_1$ appears in Eq. 2–5.) Then

$$\bar{a} = \frac{5.0 \text{ m/s} - 15.0 \text{ m/s}}{5.0 \text{ s}} = -2.0 \text{ m/s}^2.$$

The negative sign appears because the final velocity is less than the initial velocity. In this case the direction of the acceleration is to the left (in the negative *x* direction)—even though the velocity is always pointing to the right. We say that the acceleration is 2.0 m/s^2 to the left, and it is shown in Fig. 2–14 as an orange arrow.

When an object is slowing down, we sometimes say it is decelerating. But be careful: deceleration does *not* mean that the acceleration is necessarily negative. For an object moving to the right along the positive *x* axis and slowing down (as in Fig. 2–14), the acceleration *is* negative. But the same car moving to the left (decreasing *x*) and slowing down has positive acceleration that points to the right, as shown in Fig. 2–15. We have a deceleration whenever the velocity and acceleration point in opposite directions.

Instantaneous Acceleration

The **instantaneous acceleration**, a, is defined as the *limiting value of the average acceleration as we let Δt approach zero*:

$$a = \lim_{\Delta t \to 0} \frac{\Delta v}{\Delta t} = \frac{dv}{dt}. \qquad \text{(2–6)}$$

This limit, dv/dt, is the derivative of v with respect to t. We will use the term "acceleration" to refer to the instantaneous value. If we want to discuss the average acceleration, we will always include the word "average."

If we draw a graph of the velocity, v, vs. time, t, as shown in Fig. 2–16, then the average acceleration over a time interval $\Delta t = t_2 - t_1$ is represented by the slope of the straight line connecting the two points P_1 and P_2 as shown. [Compare this to the position vs. time graph of Fig. 2–9 for which the slope of the straight line represents the average velocity.] The instantaneous acceleration at any time, say t_1, is the slope of the tangent to the v vs. t curve at that time, which is also shown in Fig. 2–16. Let us use this fact for the situation graphed in Fig. 2–16; as we go from time t_1 to time t_2 the velocity continually increases but the acceleration (the rate at which the velocity changes) is decreasing since the slope of the curve is decreasing.

FIGURE 2–16 A graph of velocity v vs. time t. The average acceleration over a time interval $\Delta t = t_2 - t_1$ is the slope of the straight line P_1P_2: $\bar{a} = \Delta v/\Delta t$. The instantaneous acceleration at time t_1 is the slope of the v vs. t curve at that instant.

FIGURE 2–17 Graphs of (a) x vs. t, (b) v vs. t, and (c) a vs. t for the motion $x = At^2 + B$. Note that v increases linearly with t and that the acceleration a is constant. Also, v is the slope of the x vs. t curve, whereas a is the slope of the v vs. t curve.

EXAMPLE 2–7 Acceleration given $x(t)$. A particle is moving in a straight line so that its position is given by the relation $x = (2.10 \text{ m/s}^2)t^2 + (2.80 \text{ m})$, as in Example 2–3. Calculate (a) its average acceleration during the time interval from $t_1 = 3.00 \text{ s}$ to $t_2 = 5.00 \text{ s}$, and (b) its instantaneous acceleration as a function of time.

SOLUTION (a) We saw in Example 2–3c that the velocity at any time t is $v = dx/dt = (4.20 \text{ m/s}^2)t$. Therefore, at $t_1 = 3.00 \text{ s}$, $v_1 = (4.20 \text{ m/s}^2)(3.00 \text{ s}) = 12.6 \text{ m/s}$ and at $t_2 = 5.00 \text{ s}$, $v_2 = 21.0 \text{ m/s}$. Therefore,

$$\bar{a} = \frac{21.0 \text{ m/s} - 12.6 \text{ m/s}}{5.00 \text{ s} - 3.00 \text{ s}} = 4.20 \text{ m/s}^2.$$

(b) With $v = (4.20 \text{ m/s}^2)t$, the instantaneous acceleration is

$$a = \frac{dv}{dt} = \frac{d}{dt}\left[(4.20 \text{ m/s}^2)t\right] = 4.20 \text{ m/s}^2.$$

The acceleration in this case is constant; it does not depend on time. Figure 2–17 shows graphs of (a) x vs. t (the same as Fig. 2–12b), (b) v vs. t, which is linearly increasing as calculated above, and (c) a vs. t, which is a horizontal straight line because $a = $ constant.

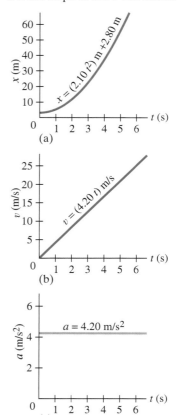

Like velocity, acceleration is a rate. The velocity of an object is the rate at which its displacement changes with time; its acceleration, on the other hand, is the rate at which its velocity changes with time. In a sense, acceleration is a "rate of a rate." This can be expressed in equation form as follows: since $a = dv/dt$ and $v = dx/dt$, then

$$a = \frac{dv}{dt} = \frac{d}{dt}\left(\frac{dx}{dt}\right) = \frac{d^2x}{dt^2}.$$

Here d^2x/dt^2 is the *second derivative* of x with respect to time: we first take the derivative of x with respect to time (dx/dt), and then we again take the derivative with respect to time, $(d/dt)(dx/dt)$, to get the acceleration.

2–5 | Motion at Constant Acceleration

Many practical situations occur in which the acceleration is constant or close enough that we can assume it is constant. The acceleration due to gravity near the Earth's surface is an example. We now treat this situation when the magnitude of the acceleration is constant and the motion is in a straight line. In this case, the instantaneous and average accelerations are equal.

Let a = constant

To simplify our notation, let us take the initial time in any discussion to be zero: $t_1 = 0$. (This is effectively starting a stopwatch at t_1.) We can then let $t_2 = t$ be the elapsed time. The initial position (x_1) and initial velocity (v_1) of an object will now be represented by x_0 and v_0; and at time t the position and velocity will be called x and v (rather than x_2 and v_2). The average velocity during the time t will be (from Eq. 2–2)

x(t = 0) = x₀
v(t = 0) = v₀

$$\overline{v} = \frac{x - x_0}{t - t_0} = \frac{x - x_0}{t}$$

since $t_0 = 0$. And the acceleration, which is assumed constant in time, will be (from Eq. 2–5)

$$a = \frac{v - v_0}{t}.$$

A common problem is to determine the velocity of an object after a certain time, given its acceleration. We can solve such problems by solving for v in the last equation to obtain:

v related to a and t
(a = constant)

$$v = v_0 + at. \qquad \text{[constant acceleration]} \quad (2\text{–}7)$$

For example, it may be known that the acceleration of a particular motorcycle is $4.0 \, \text{m/s}^2$, and we wish to determine how fast it will be going after, say, 6.0 s. Assuming it starts from rest $(v_0 = 0)$, after 6.0 s the velocity will be $v = at = (4.0 \, \text{m/s}^2)(6.0 \, \text{s}) = 24 \, \text{m/s}$.

Next, let us see how to calculate the position of an object after a time t when it is undergoing constant acceleration. The definition of average velocity (Eq. 2–2) is $\overline{v} = (x - x_0)/t$, which we can rewrite as

$$x = x_0 + \overline{v}t. \qquad (2\text{–}8)$$

Because the velocity increases at a uniform rate, the average velocity, $\overline{v}$, will be midway between the initial and final velocities:

Average velocity
(when acceleration is constant)

$$\overline{v} = \frac{v_0 + v}{2}. \qquad \text{[constant acceleration]} \quad (2\text{–}9)$$

(Careful: Eq. 2–9 is not necessarily valid if the acceleration is not constant.) We combine the last two equations with Eq. 2–7 and find

$$x = x_0 + \overline{v}t$$

$$= x_0 + \left(\frac{v_0 + v}{2}\right)t$$

$$= x_0 + \left(\frac{v_0 + v_0 + at}{2}\right)t$$

or

x related to a and t
(a = constant)

$$x = x_0 + v_0 t + \tfrac{1}{2}at^2. \qquad \text{[constant acceleration]} \quad (2\text{–}10)$$

Equations 2–7, 2–9, and 2–10 are three of the four most useful equations for motion at constant acceleration. We now derive the fourth equation, which is useful in situations where the time t is not known. We begin with Eq. 2–8 and substitute in Eq. 2–9:

$$x = x_0 + \bar{v}t = x_0 + \left(\frac{v + v_0}{2}\right)t.$$

Next we solve Eq. 2–7 for t, obtaining

$$t = \frac{v - v_0}{a},$$

and substituting this into the equation above we have

$$x = x_0 + \left(\frac{v + v_0}{2}\right)\left(\frac{v - v_0}{a}\right) = x_0 + \frac{v^2 - v_0^2}{2a}.$$

We solve this for v^2 and obtain

$$v^2 = v_0^2 + 2a(x - x_0), \qquad \text{[constant acceleration]} \quad \textbf{(2–11)}$$

v related to a and x
(a = constant)

which is the useful equation we sought.

We now have four equations relating position, velocity, acceleration, and time, when the acceleration a is constant. We collect them here in one place for further reference (the tan background screen is to emphasize their usefulness):

$$v = v_0 + at \qquad\qquad \text{[a = constant]} \quad \textbf{(2–12a)}$$
$$x = x_0 + v_0 t + \tfrac{1}{2}at^2 \qquad \text{[a = constant]} \quad \textbf{(2–12b)}$$
$$v^2 = v_0^2 + 2a(x - x_0) \qquad \text{[a = constant]} \quad \textbf{(2–12c)}$$
$$\bar{v} = \frac{v + v_0}{2}. \qquad\qquad \text{[a = constant]} \quad \textbf{(2–12d)}$$

Kinematic equations

for constant acceleration

(we'll use them a lot)

These useful equations are not valid unless a is a constant. In many cases we can set $x_0 = 0$, and this simplifies the above equations a bit. Note that x represents position, not distance, and $x - x_0$ is the displacement.

EXAMPLE 2–8 **Runway design.** You are designing an airport for small planes. One kind of airplane that might use this airfield must reach a speed before takeoff of at least 27.8 m/s (100 km/h), and can accelerate at 2.00 m/s². (*a*) If the runway is 150 m long, can this airplane reach the proper speed to take off? (*b*) If not, what minimum length must the runway have?

➡ **PHYSICS APPLIED**
Engineering design

SOLUTION (*a*) We are given the airplane's acceleration ($a = 2.00$ m/s²), and we know the plane can travel a distance of 150 m. We want to find its velocity, to determine if it will be at least 27.8 m/s. We want to find v when we are given:

Known	Wanted
$x_0 = 0$	v
$v_0 = 0$	
$x = 150$ m	
$a = 2.00$ m/s²	

Of the above four equations, Eq. 2–12c will give us v when we know v_0, a, x, and x_0:

$$v^2 = v_0^2 + 2a(x - x_0)$$
$$= 0 + 2(2.0 \text{ m/s}^2)(150 \text{ m}) = 600 \text{ m}^2/\text{s}^2$$
$$v = \sqrt{600 \text{ m}^2/\text{s}^2} = 24.5 \text{ m/s}.$$

➡ **PROBLEM SOLVING**
Equations 2–12 are valid only when the acceleration is constant, which we assume in this Example

This runway length is *not* sufficient.
(*b*) Now we want $(x - x_0)$ given $v = 27.8$ m/s and $a = 2.00$ m/s². So we use Eq. 2–12c, rewritten as

$$(x - x_0) = \frac{v^2 - v_0^2}{2a} = \frac{(27.8 \text{ m/s})^2 - 0}{2(2.0 \text{ m/s}^2)} = 193 \text{ m}.$$

2-6 Solving Problems

In this Section we do several more worked-out Examples of objects moving with constant acceleration. First let us consider generally how to approach problem solving. It is important to note that physics is *not* a collection of equations to be memorized. (In fact, rather than memorizing the very useful Eqs. 2–12, it is better to understand how to derive them from the definitions of velocity and acceleration as we did above.) Simply searching for an equation that might work can be disastrous and can lead you to a wrong result, and will surely not help you understand physics. A better approach to problem solving is to use the following (rough) procedure, which we put in a special "box":

PROBLEM SOLVING

1. **Read** *and reread* the whole problem carefully before trying to solve it.
2. **Draw** a **diagram** or picture of the situation, with coordinate axes wherever applicable. [You can choose to place the origin of coordinates and the axes wherever you like, so as to make your calculations easier. You also choose which direction is positive and which is negative. Usually we choose the x axis to the right as positive, but you could choose positive to the left.]
3. **Write down** what quantities are "known" or "given," and then what you *want* to know.
4. Think about which principles of physics apply in this problem. Then plan an approach:
5. Consider which equations (and/or definitions) relate the quantities involved. Before using equations, be sure their **range of validity** includes your problem (for example, Eqs. 2–12 are valid only when the acceleration is constant). If you find an applicable equation that involves only known quantities and one desired unknown, **solve** the equation algebraically for the unknown. In many instances several sequential calcu-

lations, or a combination of equations, may be needed. It is often advantageous to solve algebraically for the desired unknown before putting in numerical values.
6. Carry out the **calculation** if it is a numerical problem. Keep one or two extra digits during the calculations, but round off the final answers to the correct number of significant figures (Section 1–3).
7. Think carefully about the result you obtain: Is it **reasonable**? Does it make sense according to your own intuition and experience? A good check is to do a rough **estimate** using only powers of ten, as discussed in Section 1–6. Often it is preferable to do a rough estimate at the *start* of a numerical problem because it can help you focus your attention on finding a path toward a solution.
8. A very important aspect of doing problems is keeping track of **units**. An equals sign implies the units on each side must be the same, just as the numbers must. If the units do not balance, a mistake has been made. This can serve as a **check** on your solution (but it only tells you if you're wrong, not if you're right). And: always use a consistent set of units.

$a = 2.00\ \text{m/s}^2$ $a = 2.00\ \text{m/s}^2$

$x_0 = 0$ $x = 30\ \text{m}$
$v_0 = 0$

FIGURE 2–18 Example 2–9.

Known	Wanted
$x_0 = 0$	t
$x = 30.0\ \text{m}$	
$a = 2.00\ \text{m/s}^2$	
$v_0 = 0$	

EXAMPLE 2–9 **Acceleration of car.** How long does it take a car to cross a 30.0-m-wide intersection after the light turns green, if it accelerates from rest at a constant $2.00\ \text{m/s}^2$?

SOLUTION First we make a sketch, Fig. 2–18. Next we make a table, shown in the margin, choosing $x_0 = 0$ and assume the car moves to the right along the positive x axis, and noting that "starting from rest" means $v = 0$ at $t = 0$; that is, $v_0 = 0$. Since a is constant, we can use Eqs. 2–12. Equation 2–12b is perfect since the only unknown quantity is t, which is what we are seeking. Setting $v_0 = 0$ and $x_0 = 0$, we can solve Eq. 2–12b, $x = \frac{1}{2}at^2$, for t:

$$t = \sqrt{\frac{2x}{a}} = \sqrt{\frac{2(30.0\ \text{m})}{2.00\ \text{m/s}^2}} = 5.48\ \text{s}.$$

We can check the reasonableness of our answer by calculating the final velocity v: $v = at = (2.00\ \text{m/s}^2)(5.48\ \text{s}) = 10.96\ \text{m/s}$, and then finding $x = x_0 + \bar{v}t = 0 + \frac{1}{2}(10.96\ \text{m/s} + 0)(5.48\ \text{s}) = 30.0\ \text{m}$, which is our given distance.

EXAMPLE 2–10 **ESTIMATE** **Braking distances.** Estimate the minimum stopping distances for a car, which are important for traffic safety and traffic design. The problem is best dealt with in two parts: (1) the time between the decision to apply the brakes and their actual application (the "reaction time"), during which we assume $a = 0$; and (2) the actual braking period when the vehicle slows down ($a \neq 0$). The stopping distance depends on the reaction time of the driver, the initial speed of the car (the final speed is zero), and the acceleration of the car. For a dry road and good tires, good brakes can decelerate a car at a rate of about 5 m/s^2 to 8 m/s^2. Calculate the total stopping distance for an initial velocity of 50 km/h ($14 \text{ m/s} \approx 31 \text{ mph}$) and assume the acceleration of the car is -6.0 m/s^2 (the minus sign appears because the velocity is taken to be in the positive x direction and its magnitude is decreasing). Reaction time for normal drivers varies from perhaps 0.3 s to about 1.0 s; take it to be 0.50 s.

SOLUTION The car is moving to the right in the positive x direction. We take $x_0 = 0$ for the first part of the problem, in which the car travels at a constant speed of 14 m/s during the time the driver is reacting (0.50 s). See Fig. 2–19 and the Table in the margin. To find x we can use Eq. 2–12b:

$$x = v_0 t + 0 = (14 \text{ m/s})(0.50 \text{ s}) = 7.0 \text{ m}.$$

The car travels 7.0 m during the driver's reaction time, until the moment the brakes are applied.

Now for the second part, during which the brakes are applied and the car is brought to rest. We now take $x_0 = 7.0 \text{ m}$ (result of the first part): see Table to the right. Equation 2–12a doesn't contain x; Eq. 2–12b contains x but also the unknown t. Equation 2–12c, $v^2 - v_0^2 = 2a(x - x_0)$, is what we want; we solve for x, after setting $x_0 = 7.0 \text{ m}$:

$$x = x_0 + \frac{v^2 - v_0^2}{2a}$$

$$= 7.0 \text{ m} + \frac{0 - (14 \text{ m/s})^2}{2(-6.0 \text{ m/s}^2)} = 7.0 \text{ m} + \frac{-196 \text{ m}^2/\text{s}^2}{-12 \text{ m/s}^2}$$

$$= 7.0 \text{ m} + 16 \text{ m} = 23 \text{ m}.$$

Part 1: Reaction time

Known	Wanted
$t = 0.50 \text{ s}$	x
$v_0 = 14 \text{ m/s}$	
$v = 14 \text{ m/s}$	
$a = 0$	
$x_0 = 0$	

Part 2: Braking

Known	Wanted
$x_0 = 7.0 \text{ m}$	x
$v_0 = 14 \text{ m/s}$	
$v = 0$	
$a = -6.0 \text{ m/s}^2$	

The car traveled 7.0 m while the driver was reacting and another 16 m during the braking period before coming to a stop. The total distance traveled was then 23 m. See Fig. 2–20 for graphs of v vs. t and x vs. t.

Under wet or icy conditions, the value of a may be only one third the value for a dry road since the brakes cannot be applied as hard without skidding, and hence stopping distances are much greater. Note also that the stopping distance after you hit the brakes increases with the *square* of the speed, not just linearly with speed: If you are traveling twice as fast, it takes four times the distance.

FIGURE 2–20 Graphs of v vs. t and x vs. t for Example 2–10.

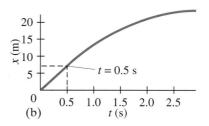

FIGURE 2–19 Example 2–10: stopping distance for a braking car.

EXAMPLE 2–11 **ESTIMATE** **Air bags.** Suppose you want to design an air-bag system that can protect the driver in a head-on collision at a speed of 100 km/h (60 mph). Estimate how fast the air bag must inflate to effectively protect the driver. Assume the car crumples upon impact over a distance of about 1 m. How does the use of a seat belt help the driver?

SOLUTION The car decelerates from 100 km/h to zero in a very short time and a very short distance (1 m). Noting that 100 km/h = 100×10^3 m/3600 s = 28 m/s, we can get the acceleration from Eq. 2–12c:

$$a = -\frac{v_0^2}{2x} = -\frac{(28 \text{ m/s})^2}{2.0 \text{ m}} = -390 \text{ m/s}^2.$$

This enormous acceleration takes place in a time given by (Eq. 2–12a):

$$t = \frac{v - v_0}{a} = \frac{0 - 28 \text{ m/s}}{-390 \text{ m/s}^2} = 0.07 \text{ s}.$$

To be effective, the air bag would need to inflate faster than this.

What does the air bag do? First, it spreads the force over a larger area of the chest. This is better than being punctured by the steering column. Also, the pressure in the bag is controlled to minimize the head's maximum deceleration. The seat belt keeps the person in the correct position against the expanding air bag.

* Two Moving Objects

We now do an Example that is a little more involved, and put it in this separate subsection marked with a * to suggest that it could be considered optional, especially on a first reading.

EXAMPLE 2–12 **ESTIMATE** **Catching a speeder.** A car speeding at 80 mi/h passes a still police car which immediately takes off in hot pursuit. Using simple assumptions, such as that the speeder continues at constant speed, estimate how long it takes the police car to overtake the speeder. Then estimate the police car's speed at that moment and decide if the assumptions were reasonable.

SOLUTION When the police car takes off, it accelerates, and the simplest assumption is that its acceleration is constant. This may not be reasonable, but let's see what happens. We can estimate the acceleration if we have noticed automobile ads, which claim cars can accelerate from rest to 60 mi/h in 6 seconds. So the average acceleration of the police car could be approximately

$$a_P = \frac{60 \text{ mi/h}}{6 \text{ s}} = 10 \frac{\text{mi}}{\text{h} \cdot \text{s}}.$$

We should perhaps change units to proper SI units, but let's try to save time and work with these mixed units. We need to set up the kinematic equations to determine the unknown quantities, and since there are two moving objects, we need two separate sets of equations. We denote the speeding car's position by x_S and the police car's position by x_P. Because we are interested in solving for the time when the two vehicles arrive at the same position on the road, we use Eq. 2–12b for each car:

$$x_S = v_{0S}t = \left(80 \frac{\text{mi}}{\text{h}}\right)t$$

$$x_P = v_{0P}t + \tfrac{1}{2}a_P t^2 = \tfrac{1}{2}\left(10 \frac{\text{mi}}{\text{h} \cdot \text{s}}\right)t^2$$

where we have set $v_{0P} = 0$ and $a_S = 0$ (speeder assumed to move at constant speed). We want the time when the cars meet, so we set $x_S = x_P$ and solve for t:

$$\left(80\,\frac{\text{mi}}{\text{h}}\right)t = \left(5\,\frac{\text{mi}}{\text{h}\cdot\text{s}}\right)t^2.$$

The solutions are

$$t = 0 \quad\text{and}\quad t = \frac{80\,\dfrac{\text{mi}}{\text{h}}}{5\,\dfrac{\text{mi}}{\text{h}\cdot\text{s}}} = 16\,\text{s}.$$

The first solution corresponds to the moment the speeder passed the police car. The second solution tells us when the police car catches up to the speeder, 16 s later. This is our answer, but is it reasonable? The police car's speed at $t = 16\,\text{s}$ is

$$v_P = v_{0P} + a_P t = 0 + \left(10\,\frac{\text{mi}}{\text{h}\cdot\text{s}}\right)(16\,\text{s}) = 160\,\text{mi/h}.$$

Not reasonable, and highly dangerous. More reasonable is to give up the assumption of constant acceleration. The police car surely cannot maintain constant acceleration at those speeds. Also, the speeder, if a reasonable person, would slow down upon hearing the police siren. Figure 2–21 shows (*a*) *x* vs. *t* and (*b*) *v* vs. *t* graphs, based on the original assumption of $a_P = $ constant, whereas (*c*) shows *v* vs. *t* for more reasonable assumptions.

Careful:
Initial assumptions need to be checked out for reasonableness

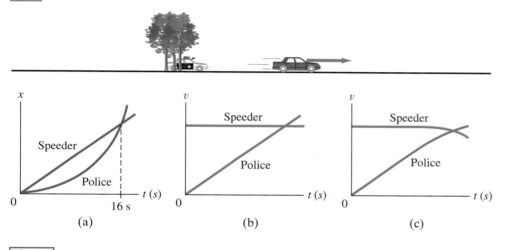

FIGURE 2–21   Example 2–12.

(a)   (b)   (c)

2–7   Falling Objects

One of the most common examples of uniformly accelerated motion is that of an object allowed to fall freely near the Earth's surface. That a falling object is accelerating may not be obvious at first. And beware of thinking, as was widely believed until the time of Galileo (Fig. 2–22), that heavier objects fall faster than lighter objects and that the speed of fall is proportional to how heavy the object is.

Galileo's analysis made use of his new and creative technique of imagining what would happen in idealized (simplified) cases. For free fall, he postulated that all objects would fall with the *same constant acceleration* in the absence of air or other resistance. He showed that this postulate predicts that for an object falling from rest, the distance traveled will be proportional to the square of the time (Fig. 2–23); that is, $d \propto t^2$. We can see this from Eq. 2–12b, but Galileo was the first to derive this mathematical relation.

To support his claim that the speed of falling objects increases as they fall, Galileo made use of a clever argument: a heavy stone dropped from a height of

FIGURE 2–22
Galileo Galilei (1564–1642).

FIGURE 2–23 Multiflash photograph of a falling apple, photographed at equal time intervals. Note that the apple falls farther during each successive time interval, which means it is accelerating.

FIGURE 2–25 A rock and a feather are dropped simultaneously (a) in air, (b) in a vacuum.

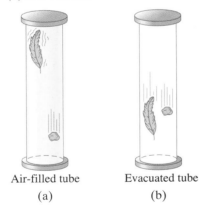

Air-filled tube   Evacuated tube
     (a)   (b)

Acceleration due to gravity

FIGURE 2–24 (a) A ball and a light piece of paper are dropped at the same time. (b) Repeated, with the paper wadded up.

(a)   (b)

2 m will drive a stake into the ground much further than will the same stone dropped from a height of only 0.2 m. Clearly, the stone must be moving faster when it has fallen from a greater height.

As we saw, Galileo also claimed that *all* objects, light or heavy, fall with the *same* acceleration, at least in the absence of air. If you hold a piece of paper horizontally in one hand and a heavier object—say, a baseball—in the other, and release them at the same time as in Fig. 2–24a, the heavier object will reach the ground first. But if you repeat the experiment, this time crumpling the paper into a small wad (see Fig. 2–24b), you will find that the two objects reach the floor at nearly the same time.

Galileo was sure that air acts as a resistance to very light objects that have a large surface area. But in many ordinary circumstances this air resistance is negligible. In a chamber from which the air has been removed, even light objects like a feather or a horizontally held piece of paper will fall with the same acceleration as any other object (see Fig. 2–25). Such a demonstration in vacuum was of course not possible in Galileo's time, which makes Galileo's achievement all the greater. Galileo is often called the "father of modern science," not only for the content of his science (astronomical discoveries, inertia, free fall), but also for his approach to science (idealization and simplification, mathematization of theory, theories that have testable consequences, experiments to test theoretical predictions).

Galileo's specific contribution to our understanding of the motion of falling objects can be summarized as follows:

at a given location on the Earth and in the absence of air resistance, all objects fall with the same constant acceleration.

We call this acceleration the **acceleration due to gravity** on the Earth, and we give it the symbol g. Its magnitude is approximately

$$g = 9.80 \text{ m/s}^2.$$

In British units g is about 32 ft/s². Actually, g varies slightly according to latitude and elevation, but these variations are so small that we will ignore them for most purposes. The effects of air resistance are often small, and we will neglect them for the most part. However, air resistance will be noticeable even on a reasonably heavy object if the velocity becomes large.[†] Acceleration due to gravity is a vector, as is any acceleration, and its direction is downward, toward the center of the Earth.

[†] The speed of an object falling in air (or other fluid) does not increase indefinitely. If the object falls far enough, it will reach a maximum velocity called the **terminal velocity**.

When dealing with freely falling objects we can make use of Eqs. 2–12, where for a we use the value of g given above. Also, since the motion is vertical we will substitute y in place of x, and y_0 in place of x_0. We take $y_0 = 0$ unless otherwise specified. *It is arbitrary whether we choose y to be positive in the upward direction or in the downward direction; but we must be consistent about it throughout a problem's solution.*

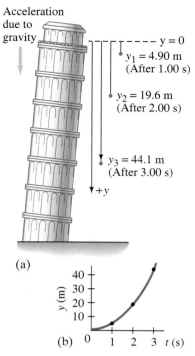

Acceleration due to gravity

$y = 0$

$y_1 = 4.90$ m (After 1.00 s)

$y_2 = 19.6$ m (After 2.00 s)

$y_3 = 44.1$ m (After 3.00 s)

$+y$

(a)

(b)

FIGURE 2–26 Example 2–13. (a) An object dropped from a tower falls with progressively greater speed and covers greater distance with each successive second. (See also Fig. 2–23.) (b) Graph of y vs. t.

EXAMPLE 2–13 **Falling from a tower.** Suppose that a ball is dropped from a tower 70.0 m high. How far will it have fallen after 1.00 s, 2.00 s, and 3.00 s? Assume y is positive downward. Neglect air resistance.

SOLUTION We are given the acceleration, $a = g = +9.80$ m/s^2 which is positive because we have chosen downward as positive. Since we want to find the distance fallen given the time, t, Eq. 2–12b is the appropriate one, with $v_0 = 0$ and $y_0 = 0$. Then, after 1.00 s, the position of the ball is

$$y_1 = \tfrac{1}{2} at^2 = \tfrac{1}{2}(9.80 \text{ m/s}^2)(1.00 \text{ s})^2 = 4.90 \text{ m},$$

so the ball has fallen a distance of 4.90 m after 1.00 s. Similarly, after 2.00 s,

$$y_2 = \tfrac{1}{2} at^2 = \tfrac{1}{2}(9.80 \text{ m/s}^2)(2.00 \text{ s})^2 = 19.6 \text{ m},$$

and after 3.00 s,

$$y_3 = \tfrac{1}{2} at^2 = \tfrac{1}{2}(9.80 \text{ m/s}^2)(3.00 \text{ s})^2 = 44.1 \text{ m}.$$

See Fig. 2–26.

EXAMPLE 2–14 **Thrown down from a tower.** Suppose the ball in Example 2–13 is *thrown* downward with an initial velocity of 3.00 m/s, instead of being dropped. (*a*) What then would be its position after 1.00 s and 2.00 s? (*b*) What would its speed be after 1.00 s and 2.00 s? Compare to the speeds of a dropped ball.

SOLUTION (*a*) We can approach this in the same way as Example 2–13, using Eq. 2–12b, but this time v_0 is not zero but is $v_0 = 3.00$ m/s. Thus, at $t = 1.00$ s the position of the ball is

$$y = v_0 t + \tfrac{1}{2} at^2 = (3.00 \text{ m/s})(1.00 \text{ s}) + \tfrac{1}{2}(9.80 \text{ m/s}^2)(1.00 \text{ s})^2 = 7.90 \text{ m},$$

and at $t = 2.00$ s

$$y = v_0 t + \tfrac{1}{2} at^2 = (3.00 \text{ m/s})(2.00 \text{ s}) + \tfrac{1}{2}(9.80 \text{ m/s}^2)(2.00 \text{ s})^2 = 25.6 \text{ m}.$$

As expected, the ball falls farther each second than if it were dropped with $v_0 = 0$. (*b*) The velocity is readily obtained from Eq. 2–12a:

$$v = v_0 + at$$
$$= 3.00 \text{ m/s} + (9.80 \text{ m/s}^2)(1.00 \text{ s}) = 12.8 \text{ m/s} \quad [\text{at } t = 1.00 \text{ s}]$$
$$= 3.00 \text{ m/s} + (9.80 \text{ m/s}^2)(2.00 \text{ s}) = 22.6 \text{ m/s}. \quad [\text{at } t = 2.00 \text{ s}]$$

When the ball is dropped $(v_0 = 0)$, the first term in the above equations is zero, so

$$v = 0 + at$$
$$= (9.80 \text{ m/s}^2)(1.00 \text{ s}) = 9.80 \text{ m/s} \quad [\text{at } t = 1.00 \text{ s}]$$
$$= (9.80 \text{ m/s}^2)(2.00 \text{ s}) = 19.6 \text{ m/s}. \quad [\text{at } t = 2.00 \text{ s}]$$

We see that the speed of a dropped ball increases linearly in time. (In Example 2–13 we saw that the distance fallen increases as the *square* of the time.) The speed of the downwardly thrown ball also increases linearly ($\Delta v = 9.80$ m/s each second), but its speed at any moment is always higher than that of a falling ball by 3.0 m/s (its initial speed).

FIGURE 2–27 An object thrown into the air leaves the thrower's hand at A, reaches its maximum height at B, and returns to the original height at C. Examples 2–15, 2–16, and 2–17.

Careful:
Velocity and acceleration are not always in the same direction

Careful:
$a \neq 0$ even at the highest point of a trajectory

EXAMPLE 2–15 **Ball thrown upward.** A person throws a ball *upward* into the air with an initial velocity of 15.0 m/s. Calculate (*a*) how high it goes, and (*b*) how long the ball is in the air before it comes back to his hand. We are not concerned here with the throwing action, but only with the motion of the ball *after* it leaves the thrower's hand (Fig. 2–27).

SOLUTION Let us choose *y* to be positive in the upward direction and negative in the downward direction. (Note: This is a different convention from that used in Examples 2–13 and 2–14.) Then the acceleration due to gravity will have a negative sign, $a = -9.80 \text{ m/s}^2$. Note that as the ball rises, its speed decreases until it reaches the highest point (B in Fig. 2–27), where its speed is zero for an instant; then it descends with increasing speed.

(*a*) To determine the maximum height, we calculate the position of the ball when its velocity equals zero ($v = 0$ at the highest point). At $t = 0$ (point A in Fig. 2–27) we have $y_0 = 0$, $v_0 = 15.0 \text{ m/s}$, and $a = -9.80 \text{ m/s}^2$. At time *t* (maximum height), $v = 0$, $a = -9.80 \text{ m/s}^2$, and we wish to find *y*. We use Eq. 2–12c (replacing *x* with *y*) and solve for *y*:

$$v^2 = v_0^2 + 2ay$$

$$y = \frac{v^2 - v_0^2}{2a} = \frac{0 - (15.0 \text{ m/s})^2}{2(-9.80 \text{ m/s}^2)} = 11.5 \text{ m}.$$

The ball reaches a height of 11.5 m above the hand.

(*b*) Now we need to calculate how long the ball is in the air before it returns to his hand. We could do this calculation in two parts by first determining the time required for the ball to reach its highest point, and then determining the time it takes to fall back down. However, it is simpler to consider the motion from A to B to C (Fig. 2–27) in one step and use Eq. 2–12b. We can do this because *y* (or *x*) represents position or displacement, and not the total distance traveled. Thus, at both points A and C, $y = 0$. We use Eq. 2–12b with $a = -9.80 \text{ m/s}^2$ and find

$$y = v_0 t + \tfrac{1}{2} a t^2$$

$$0 = (15.0 \text{ m/s})t + \tfrac{1}{2}(-9.80 \text{ m/s}^2)t^2.$$

In this equation we can factor out one *t* to obtain

$$\left(15.0 \text{ m/s} - 4.90 \text{ m/s}^2 t\right)t = 0.$$

There are two solutions:

$$t = 0, \qquad \text{and} \qquad t = \frac{15.0 \text{ m/s}}{4.90 \text{ m/s}^2} = 3.06 \text{ s}.$$

The first solution ($t = 0$) corresponds to the initial point (A) in Fig. 2–27, when the ball was first thrown and was also at $y = 0$. The second solution, $t = 3.06$ s, corresponds to point C, when the ball has returned to $y = 0$. Thus the ball is in the air for 3.06 s.

CONCEPTUAL EXAMPLE 2–16 **Two common misconceptions.** Explain the error in these two common misconceptions: (1) that acceleration and velocity are always in the same direction, and (2) that an object thrown upward has zero acceleration at the highest point (B in Fig. 2–27.)

RESPONSE Both are wrong. (1) Velocity and acceleration are *not* necessarily in the same direction. When a ball is falling downward, its velocity and acceleration are in the same direction. But when a ball is moving upward, as in Example 2–15, its velocity is upward, whereas the acceleration is downward, in the opposite direction. (2) At the highest point (B in Fig. 2–27), the ball has zero velocity for an instant. Is the acceleration also zero at this point? No. Gravity does not stop acting, so $a = -g = -9.80 \text{ m/s}^2$ even there. Thinking that $a = 0$ at point B would lead to the conclusion that upon reaching point B, the ball would hover there. For if the acceleration (= rate of change of velocity) were zero, the velocity would remain zero, and the ball could stay up there without falling.

EXAMPLE 2–17 **Ball thrown upward, II.** Let us consider again the ball thrown upward in Example 2–15, and make three more calculations. Calculate (a) how much time it takes for the ball to reach the maximum height (point B in Fig. 2–27), (b) the velocity of the ball when it returns to the thrower's hand (point C), and (c) at what time t the ball passes a point 8.00 m above the person's hand.

SOLUTION Again we take y as positive upward. (a) Both Eqs. 2–12a and 2–12b contain the time t with other quantities known. Let us use Eq. 2–12a with $a = -9.80 \text{ m/s}^2$, $v_0 = 15.0 \text{ m/s}$, and $v = 0$:

$$v = v_0 + at,$$

so

$$t = -\frac{v_0}{a} = -\frac{15.0 \text{ m/s}}{-9.80 \text{ m/s}^2} = 1.53 \text{ s}.$$

This is just half the time it takes the ball to go up and fall back to its original position [3.06 s, calculated in part (b) of Example 2–15]. Thus it takes the same time to reach the maximum height as to fall back to the starting point.
(b) We use Eq. 2–12a with $v_0 = 15.0 \text{ m/s}$ and $t = 3.06 \text{ s}$ (the time calculated in Example 2–15 for the ball to come back to the hand):

$$v = v_0 + at = 15.0 \text{ m/s} - (9.80 \text{ m/s}^2)(3.06 \text{ s}) = -15.0 \text{ m/s}.$$

The ball has the same magnitude of velocity when it returns to the starting point as it did initially, but in the opposite direction (this is the meaning of the negative sign). Thus, as we gathered from part (a), we see that the motion is symmetrical about the maximum height.
(c) We want t, given that $y = 8.00 \text{ m}$, $y_0 = 0$, $v_0 = 15.0 \text{ m/s}$, and $a = -9.80 \text{ m/s}^2$. We use Eq. 2–12b:

Note the symmetry: the speed at any height is the same when going up as coming down

$$y = y_0 + v_0 t + \tfrac{1}{2}at^2$$

$$8.00 \text{ m} = 0 + (15.0 \text{ m/s})t + \tfrac{1}{2}(-9.80 \text{ m/s}^2)t^2.$$

To solve any quadratic equation of the form $at^2 + bt + c = 0$, where a, b, and c are constants, the **quadratic formula** gives the solutions

$$t = \frac{-b \pm \sqrt{b^2 - 4ac}}{2a}.$$

We rewrite our equation in the standard form:

$$(4.90 \text{ m/s}^2)t^2 - (15.0 \text{ m/s})t + (8.00 \text{ m}) = 0.$$

So the coefficient a is 4.90 m/s^2, b is -15.0 m/s, and c is 8.00 m. Putting these into the quadratic formula, we obtain

$$t = \frac{15.0 \text{ m/s} \pm \sqrt{(15.0 \text{ m/s})^2 - 4(4.90 \text{ m/s}^2)(8.00 \text{ m})}}{2(4.90 \text{ m/s}^2)},$$

so $t = 0.69 \text{ s}$ and $t = 2.37 \text{ s}$. Why are there two solutions? Are they both valid? Yes, because the ball passes $y = 8.00 \text{ m}$ both when it goes up and when it comes down.

Figure 2–28 shows graphs of y vs. t and v vs. t for the ball thrown upward in Fig. 2–27, incorporating the results of Examples 2–15 and 2–17.

FIGURE 2–28 Graphs of (a) y vs. t, (b) v vs. t for a ball thrown upward, Examples 2–15 and 2–17.

(a)

(b)

FIGURE 2–29 Example 2–18: the person in Fig. 2–27 stands on the edge of a cliff. The ball falls to the base of the cliff.

FIGURE 2–30
Example 2–18, the y vs. t graph.

Deriving the kinematic equations using calculus

EXAMPLE 2–18 Ball thrown upward at edge of cliff. Suppose that the person of Examples 2–15 and 2–17 is standing on the edge of a 50.0-m-high cliff, so that the ball can fall to the base of the cliff as in Fig. 2–29. (*a*) How long does it take the ball to reach the base of the cliff? (*b*) What is the total distance traveled by the ball?

SOLUTION (*a*) Again we use Eq. 2–12b with $a = -9.80 \, \text{m/s}^2$, $v_0 = 15.0 \, \text{m/s}$, and $y_0 = 0$. But this time we set $y = -50.0 \, \text{m}$, the bottom of the cliff, 50.0 m below the initial position ($y = 0$):

$$y = y_0 + v_0 t + \tfrac{1}{2}at^2$$

$$-50.0 \, \text{m} = 0 + (15.0 \, \text{m/s})t - \tfrac{1}{2}(9.80 \, \text{m/s}^2)t^2.$$

Rewriting in the standard form we have

$$(4.90 \, \text{m/s}^2)t^2 - (15.0 \, \text{m/s})t - (50.0 \, \text{m}) = 0.$$

Using the quadratic formula, we find as solutions $t = 5.07 \, \text{s}$ and $t = -2.01 \, \text{s}$. The first solution, $t = 5.07 \, \text{s}$, is the answer to our problem. This is the time it takes the ball to rise to its highest point and then fall to the base of the cliff. To rise and fall back to the top of the cliff took 3.06 s (Example 2–15); then it took an additional 2.01 s to fall to the base. But what is the meaning of the other solution, $t = -2.01 \, \text{s}$? This is a time before our problem even begins. So it isn't relevant to our problem.[†]

(*b*) From Example 2–15, the ball moves up 11.5 m, falls 11.5 m back to the top of the cliff, and then back another 50.0 m to the base of the cliff, for a total distance traveled of 73.0 m. Note that the *displacement*, however, was −50.0 m. Figure 2–30 shows the y vs. t graph for this situation.

The acceleration of an object, particularly rockets and fast airplanes, is often given as a multiple of $g = 9.80 \, \text{m/s}^2$. For example, a plane pulling out of a dive and undergoing 3.00 g's or "gees" would have an acceleration of $(3.00)(9.80 \, \text{m/s}^2) = 29.4 \, \text{m/s}^2$. And the acceleration calculated in the collision of Example 2–11 could be expressed as $(390 \, \text{m/s}^2)/(9.8 \, \text{m/s}^2) = 40 \, g$.

* 2–8 Use of Calculus; Variable Acceleration

In this brief optional Section we use integral calculus to derive the kinematic equations for constant acceleration, Eqs. 2–12a and b. We also show how calculus can be used when the acceleration is not constant. If you have not yet studied simple integration in your calculus course, you may want to postpone reading this Section until you have.

First we derive Eq. 2–12a, starting with the definition of instantaneous acceleration, $a = dv/dt$, which we rewrite as

$$dv = a \, dt.$$

We take the definite integral of both sides of this equation, using the same nota-

---

[†] The solution $t = -2.01$ s could be meaningful in a different physical situation. Suppose that a person standing on top of a 50.0-m-high cliff sees a rock pass by him at $t = 0$ moving upward at 15.0 m/s; at what time did the rock leave the base of the cliff, and when did it arrive back at the base of the cliff? The equations will be precisely the same as for our original problem, and the answers $t = -2.01$ s and $t = 5.07$ s will be the correct answers. Note that we cannot put all the information for a problem into the mathematics, so we have to use common sense in interpreting results.

tion we did in Section 2–5 ($v = v_0$ at $t = 0$):

$$\int_{v=v_0}^{v} dv = \int_{t=0}^{t} a\,dt$$

which gives, since $a = $ constant,

$$v - v_0 = at.$$

Eq. 2–12a

This is Eq. 2–12a, $v = v_0 + at$.

Next we derive Eq. 2–12b starting with the definition of instantaneous velocity, Eq. 2–4, $v = dx/dt$. We rewrite this as

$$dx = v\,dt.$$

We substitute Eq. 2–12a above, $v = v_0 + at$, and integrate:

$$\int_{x=x_0}^{x} dx = \int_{t=0}^{t} (v_0 + at)\,dt$$

$$x - x_0 = \int_{t=0}^{t} v_0\,dt + \int_{t=0}^{t} at\,dt$$

$$x - x_0 = v_0 t + \tfrac{1}{2}at^2$$

Eq. 2–12b

since v_0 and a are constants. This result is just Eq. 2–12b, $x = x_0 + v_0 t + \tfrac{1}{2}at^2$.

Finally let us use calculus to find velocity and displacement, given an acceleration that is not constant but varies in time.

EXAMPLE 2–19 **Integrating a time-varying acceleration.** An experimental vehicle starts from rest $(v_0 = 0)$ at $t = 0$ and accelerates at a rate given by $a = (7.00\ \text{m/s}^3)t$. What is ($a$) its velocity and ($b$) its displacement 2.00 s later?

SOLUTION We cannot use Eqs. 2–12 because a is not constant. Instead, we determine v as a function of t using calculus. From the definition of acceleration, $a = dv/dt$, we have

$$dv = a\,dt.$$

We take the integral of both sides from $v = 0$ at $t = 0$ to velocity v at an arbitrary time t:

$$\int_{0}^{v} dv = \int_{0}^{t} a\,dt$$

$$v = \int_{0}^{t} (7.00\ \text{m/s}^3)t\,dt$$

$$= (7.00\ \text{m/s}^3)\left(\frac{t^2}{2}\right)\Big|_{0}^{t} = (7.00\ \text{m/s}^3)\left(\frac{t^2}{2} - 0\right) = (3.50\ \text{m/s}^3)t^2.$$

At $t = 2.00$ s, $v = (3.50\ \text{m/s}^3)(2.00\ \text{s})^2 = 14.0\ \text{m/s}$.
(b) To get the displacement, we assume $x_0 = 0$ and start with $v = dx/dt$ which we rewrite as $dx = v\,dt$. Then we integrate from $x = 0$ at $t = 0$ to position x at time t:

$$\int_{0}^{x} dx = \int_{0}^{t} v\,dt$$

$$x = \int_{0}^{2.00\ \text{s}} (3.50\ \text{m/s}^3)t^2\,dt = (3.50\ \text{m/s}^3)\frac{t^3}{3}\Big|_{0}^{2.00\ \text{s}} = 9.33\ \text{m}.$$

In sum, at $t = 2.00$ s, $v = 14.0\ \text{m/s}$ and $x = 9.33$ m.

Summary

[The Summary that appears at the end of each chapter in this book gives a brief overview of the main ideas of the chapter. The Summary *cannot* serve to give an understanding of the material, which can be accomplished only by a detailed reading of the chapter.]

Kinematics deals with the description of how objects move. The description of the motion of any object must always be given relative to some particular **reference frame**.

The **displacement** of an object is the change in position of the object.

Average speed is the distance traveled divided by the elapsed time. An object's **average velocity** over a particular time interval Δt is the displacement Δx divided by Δt:

$$\bar{v} = \frac{\Delta x}{\Delta t}.$$

The **instantaneous velocity**, whose magnitude is the same as the *instantaneous speed*, is the average velocity taken over an infinitesimally short time interval (Δt allowed to approach zero):

$$v = \lim_{\Delta t \to 0} \frac{\Delta x}{\Delta t} = \frac{dx}{dt},$$

where dx/dt is the derivative of x with respect to t.

On a graph of position vs. time, the *slope* is equal to the instantaneous velocity.

Acceleration is the change of velocity per unit time. An object's **average acceleration** over a time interval Δt is

$$\bar{a} = \frac{\Delta v}{\Delta t},$$

where Δv is the change of velocity during the time interval Δt.

Instantaneous acceleration is the average acceleration taken over an infinitesimally short time interval:

$$a = \lim_{\Delta t \to 0} \frac{\Delta v}{\Delta t} = \frac{dv}{dt}.$$

If an object moves in a straight line with constant acceleration, the velocity v and position x are related to the acceleration a, the elapsed time t, and the initial position x_0 and initial velocity v_0, by Eqs. 2–12:

$$v = v_0 + at, \qquad x = x_0 + v_0 t + \tfrac{1}{2}at^2,$$

$$v^2 = v_0^2 + 2a(x - x_0), \qquad \bar{v} = \frac{v + v_0}{2}.$$

Objects that move vertically near the surface of the Earth, either falling or having been projected vertically up or down, move with the constant downward **acceleration due to gravity** with magnitude of about $g = 9.80 \text{ m/s}^2$, if air resistance can be ignored.

Questions

1. Does a car speedometer measure speed, velocity, or both?
2. Can an object have a varying speed if its velocity is constant? If yes, give examples.
3. When an object moves with constant velocity, does its average velocity during any time interval differ from its instantaneous velocity at any instant?
4. In drag racing, is it possible for the car with the greatest speed crossing the finish line to lose the race? Explain.
5. If one object has a greater speed than a second object, does the first necessarily have a greater acceleration? Explain, using examples.
6. Compare the acceleration of a motorcycle that accelerates from 80 km/h to 90 km/h with the acceleration of a bicycle that accelerates from rest to 10 km/h in the same time.
7. Can an object have a northward velocity and a southward acceleration? Explain.
8. Can the velocity of an object be negative when its acceleration is positive? What about vice versa?
9. Give an example where both the velocity and acceleration are negative.
10. Two cars emerge side by side from a tunnel. Car A is traveling with a speed of 60 km/h and has an acceleration of 40 km/h/min. Car B has a speed of 40 km/h and has an acceleration of 60 km/h/min. Which car is passing the other as they come out of the tunnel? Explain your reasoning.
11. Can an object be increasing in speed as its acceleration decreases? If so, give an example. If not, explain.
12. An object that is thrown vertically upward will return to its original position with the same speed as it had initially if air resistance is negligible. If air resistance is appreciable, will this result be altered, and if so, how? [*Hint*: The acceleration due to air resistance is always in a direction opposite to the motion.]
13. As a freely falling object speeds up, what is happening to its acceleration due to gravity? Does it increase, decrease, or stay the same?
14. How would you estimate the maximum height you could throw a ball vertically upward? How would you estimate the maximum speed you could give it?
15. A rock is thrown upward with speed v from the edge of a cliff. A second rock is thrown vertically downward with the same initial speed. Which rock has greater speed when it reaches the bottom of the cliff? Ignore the effect of air resistance.
16. You travel from point A to point B in a car moving at a constant speed of 70 km/h. Then you travel the same distance to another point C, moving at a constant speed of 90 km/h. Is your average speed for the entire trip from A to C 80 km/h? Explain why or why not.

17. Describe in words the motion plotted in Fig. 2–31 in terms of v, a, etc. [*Hint*: First try to duplicate the motion plotted by walking or moving your hand.]

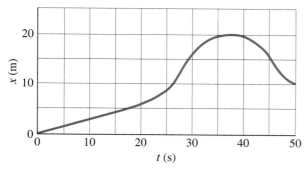

FIGURE 2–31 Question 17, Problems 11, 12, and 84.

18. Describe in words the motion of the object graphed in Fig. 2–32.

FIGURE 2–32 Question 18 and Problem 22.

Problems

[The problems at the end of each chapter are ranked I, II, or III according to estimated difficulty, with I problems being easiest. The problems are arranged by Section, meaning that the reader should have read up to and including that Section, but not only that Section—problems often depend on earlier material. Finally, there is a set of unranked "General Problems" not arranged by Section number.]

Sections 2–1 to 2–3

1. (I) A bird can fly 15 km/h. How long does it take to fly 75 km?

2. (I) What must your car's average speed be in order to travel 280 km in 3.2 h?

3. (I) If you are driving 110 km/h along a straight road and you look to the side for 2.0 s, how far do you travel during this inattentive period?

4. (I) A rolling ball moves from $x_1 = 3.4$ cm to $x_2 = -4.2$ cm during the time from $t_1 = 3.0$ s to $t_2 = 6.1$ s. What is its average velocity?

5. (I) A particle at $t_1 = -2.0$ s is at $x_1 = 3.4$ cm and at $t_2 = 4.5$ s is at $x_2 = 8.5$ cm. What is its average velocity? Can you calculate its average speed from these data?

6. (II) You are driving home from school steadily at 65 mph for 130 miles. It then begins to rain and you slow to 55 mph. You arrive home after driving 3 hours and 20 minutes. (a) How far is your hometown from school? (b) What was your average speed?

7. (II) According to a rule-of-thumb, every five seconds between a lightning flash and the following thunder gives the distance of the storm in miles. Assuming that the flash of light arrives in essentially no time at all, estimate the speed of sound in m/s from this rule.

8. (II) A person runs eight complete laps around a quarter-mile track in a total time of 12.5 min. Calculate (a) the average speed and (b) the average velocity, in m/s.

9. (II) A horse canters away from its trainer in a straight line, moving 160 m away in 17.0 s. It then turns abruptly and gallops halfway back in 6.8 s. Calculate (a) its average speed and (b) its average velocity for the entire trip, using "away from the trainer" as the positive direction.

10. (II) Two locomotives approach each other on parallel tracks. Each has a speed of 95 km/h with respect to the ground. If they are initially 8.5 km apart, how long will it be before they reach each other? (See Fig. 2–33.)

FIGURE 2–33 Problem 10.

11. (II) The position of a rabbit along a straight tunnel as a function of time is plotted in Fig. 2–31. What is its instantaneous velocity (a) at $t = 10.0$ s and (b) at $t = 30.0$ s? What is its average velocity (c) between $t = 0$ and $t = 5.0$ s, (d) between $t = 25.0$ s and $t = 30.0$ s, and (e) between $t = 40.0$ s and $t = 50.0$ s?

12. (II) In Fig. 2–31, (a) during what time periods, if any, is the rabbit's velocity constant? (b) At what time is its velocity the greatest? (c) At what time, if any, is the velocity zero? (d) Does the rabbit run in one direction or in both along its tunnel during the time shown?

13. (II) A dog runs 100 m away from its master in a straight line in 8.4 s, and then runs halfway back in one-third the time. Calculate (a) its average speed and (b) its average velocity.

14. (II) The position of a ball rolling in a straight line is given by $x = 2.0 - 4.6t + 1.1t^2$, where x is in meters and t in seconds. (a) Determine the position of the ball at $t = 1.0$, 2.0, and 3.0 s. (b) What is the average velocity over the interval $t = 1.0$ s to $t = 3.0$ s? (c) What is its instantaneous velocity at $t = 2.0$ s and at $t = 3.0$ s?

15. (II) A car traveling 90 km/h is 100 m behind a truck traveling 75 km/h. How long will it take the car to reach the truck?

16. (II) An airplane travels 2100 km at a speed of 800 km/h, and then encounters a tailwind that boosts its speed to 1000 km/h for the next 1800 km. What was the total time for the trip? What was the average speed of the plane for this trip? [*Hint*: Think carefully before using Eq. 2–12d.]

17. (II) Calculate the average speed and average velocity of a complete round-trip in which the outgoing 200 km is covered at 90 km/h, followed by a one-hour lunch break, and the return 200 km is covered at 50 km/h.

18. (II) An automobile traveling 90 km/h overtakes a 1.10-km-long train traveling in the same direction on a track parallel to the road. If the train's speed is 80 km/h, how long does it take the car to pass it, and how far will the car have traveled in this time? See Fig. 2–34. What are the results if the car and train are traveling in opposite directions?

FIGURE 2–34   Problem 18.

19. (II) A bowling ball traveling with constant speed hits the pins at the end of a bowling lane 16.5 m long. The bowler hears the sound of the ball hitting the pins 2.50 s after the ball is released from his hands. What is the speed of the ball? The speed of sound is 340 m/s.

Section 2–4

20. (I) A sports car accelerates from rest to 95 km/h in 6.2 s. What is its average acceleration in m/s^2?

21. (I) At highway speeds, a particular automobile is capable of an acceleration of about 1.6 m/s^2. At this rate, how long does it take to accelerate from 80 km/h to 110 km/h?

22. (I) Figure 2–32 shows the velocity of a train as a function of time. (*a*) At what time was its velocity greatest? (*b*) During what periods, if any, was the velocity constant? (*c*) During what periods, if any, was the acceleration constant? (*d*) When was the magnitude of the acceleration greatest?

23. (II) A sports car is advertised to be able to stop in a distance of 55 m from a speed of 100 km/h. What is its acceleration in m/s^2? How many g's is this $(g = 9.80 \text{ m/s}^2)$?

24. (II) A particular automobile can accelerate approximately as shown in the velocity–time graph of Fig. 2–35. (The short flat spots in the curve represent shifting of the gears.) Estimate the average acceleration of the car in second gear and in fourth gear.

FIGURE 2–35   The velocity of a high-performance automobile as a function of time, starting from a dead stop. The flat spots in the curve represent gear shifts. Problems 24 and 25.

25. (II) Estimate the average acceleration of the car in the previous problem (Fig. 2–35) when it is in (*a*) first, (*b*) third, and (*c*) fifth gear. (d) What is its average acceleration through the first four gears?

26. (II) The position of a racing car, which starts from rest at $t = 0$ and moves in a straight line, has been measured as a function of time, as given in the following table. Estimate (*a*) its velocity and (*b*) its acceleration as a function of time. Display each in a table and on a graph.

t(s)	0	0.25	0.50	0.75	1.00	1.50	2.00	2.50
x(m)	0	0.11	0.46	1.06	1.94	4.62	8.55	13.79

t(s)	3.00	3.50	4.00	4.50	5.00	5.50	6.00
x(m)	20.36	28.31	37.65	48.37	60.30	73.26	87.16

27. (II) A particle moves along the x axis. Its position as a function of time is given by $x = 6.0t + 8.5t^2$, where t is in seconds and x is in meters. What is the acceleration as a function of time?

28. (II) The position of a body is given by $x = At + 6Bt^3$, where x is in meters and t is in seconds. (*a*) What are the units of A and B? (*b*) What is the acceleration as a function of time? (*c*) What is the velocity and acceleration at $t = 5.0$ s? (*d*) What is the velocity as a function of time if $x = At + Bt^{-3}$?

Sections 2–5 and 2–6

29. (I) A car accelerates from 12 m/s to 21 m/s in 6.0 s. What was its acceleration? How far did it travel in this time? Assume constant acceleration.

30. (I) A car slows down from 25 m/s to rest in a distance of 75 m. What was its acceleration, assumed constant?

31. (I) A light plane must reach a speed of 32 m/s for takeoff. How long a runway is needed if the (constant) acceleration is 3.0 m/s^2?

32. (II) A baseball pitcher throws a baseball with a speed of 44 m/s. Estimate the average acceleration of the ball during the throwing motion. It is observed that in throwing the baseball, the pitcher accelerates the ball through a displacement of about 3.5 m, from behind the body to the point where it is released (Fig. 2–36).

FIGURE 2–36
Problem 32.

33. (II) A world-class sprinter can reach a top speed (of about 11.5 m/s) in the first 15.0 m of a race. What is the average acceleration of this sprinter and how long does it take her to reach that speed?

34. (II) Show that $\bar{v} = (v + v_0)/2$ (see Eq. 2–12d) is not valid for the case when the acceleration $a = A + Bt$, where A and B are constants.

35. (II) A car slows down uniformly from a speed of 22.0 m/s to rest in 5.00 s. How far did it travel in that time?

36. (II) In coming to a stop, a car leaves skid marks 75 m long on the highway. Assuming a deceleration of 7.00 m/s^2, estimate the speed of the car just before braking.

37. (II) A car traveling 55 km/h slows down at a constant 0.50 m/s^2 just by "letting up on the gas." Calculate (a) the distance the car coasts before it stops, (b) the time it takes to stop, and (c) the distance it travels during the first and fifth seconds.

38. (II) A car traveling at 95 km/h strikes a tree. The front end of the car compresses and the driver comes to rest after traveling 0.80 m. What was the average acceleration of the driver during the collision? Express the answer in terms of "g's," where 1.00 g = 9.80 m/s^2.

39. (II) Determine the stopping distances for an automobile with an initial speed of 90 km/h and human reaction time of 1.0 s: (a) for an acceleration $a = -4.0$ m/s^2; (b) for $a = -8.0$ m/s^2.

40. (II) A space vehicle accelerates uniformly from 65 m/s at $t = 0$ to 162 m/s at $t = 10.0$ s. How far did it move between $t = 2.0$ s and $t = 6.0$ s?

41. (II) A 75-m-long train accelerates uniformly from rest. If the front of the train passes a railway worker 140 m down the track at a speed of 25 m/s, what will be the speed of the last car as it passes the worker?

42. (II) Show that the equation for the stopping distance of a car is $d_S = v_0 t_R - v_0^2/(2a)$, where v_0 is the initial speed of the car, t_R is the driver's reaction time, and a is the constant acceleration (and is negative).

43. (II) In designing traffic signals, it is necessary to allow the yellow light to remain on long enough that a driver can either stop or pass completely through the intersection. Thus, if a driver is less than the stopping distance, d_S (calculated in Problem 42), from the intersection, the light must remain on long enough for her to travel this distance plus the width of the intersection, d_I. (a) Show that the yellow light should remain on for a time $t = t_R - v_0/(2a) + d_I/v_0$, where v_0 is a typical expected speed of a car approaching the intersection, and a and t_R are as defined in Problem 42. (b) A traffic engineer expects cars to approach a 14.4-m-wide intersection at speeds between 30.0 and 60.0 km/h. To be safe, he calculates the time for both speeds, assuming $t_R = 0.500$ s and $a = -4.00$ m/s^2, and chooses the longest time to be safe. What is his result?

44. (II) An unmarked police car traveling a constant 95 km/h is passed by a speeder traveling 140 km/h. Precisely 1.00 s after the speeder passes, the policeman steps on the accelerator; if the police car's acceleration is 2.00 m/s^2, how much time passes before the police car overtakes the speeder (assumed moving at constant speed)?

45. (III) Assume in Problem 44 that the speeder's speed is not known. If the police car accelerates uniformly as given above and overtakes the speeder after accelerating for 6.0 s, what was the speeder's speed?

46. (III) A runner hopes to complete the 10,000-m run in less than 30.0 min. After running at constant speed for exactly 27.0 min, there are still 1100 m to go. The runner must then accelerate at 0.20 m/s^2 for how many seconds in order to achieve the desired time?

Section 2–7 [neglect air resistance]

47. (I) If a car rolls gently ($v_0 = 0$) off a vertical cliff, how long does it take it to reach 100 km/h?

48. (I) A stone is dropped from the top of a cliff. It is seen to hit the ground below after 2.75 s. How high is the cliff?

49. (I) Calculate (a) how long it took King Kong to fall straight down from the top of the Empire State Building (380 m high), and (b) his velocity just before "landing."

50. (II) A baseball is hit almost straight up into the air with a speed of about 20 m/s. (a) How high does it go? (b) How long is it in the air?

51. (II) A kangaroo jumps to a vertical height of 2.55 m. How long was it in the air before returning to Earth?

52. (II) A ballplayer catches a ball 3.1 s after throwing it vertically upward. With what speed did he throw it, and what height did it reach?

53. (II) Estimate the maximum speed with which you can throw an object straight up in the air. Describe your method of reaching the estimate.

54. (II) The best rebounders in basketball have a vertical leap (that is, the vertical movement of a fixed point on their body) of about 120 cm. (a) What is their initial "launch" speed off the ground? (b) How long are they in the air?

55. (II) A helicopter is ascending vertically with a speed of 5.60 m/s. At a height of 115 m above the Earth, a package is dropped from a window. How much time does it take for the package to reach the ground?

56. (II) For an object falling freely from rest, show that the distance traveled during each successive second increases in the ratio of successive odd integers (1, 3, 5, etc.). (This was first shown by Galileo.) See Figs. 2–23 and 2–26.

57. (II) If air resistance is neglected, show (algebraically) that a ball thrown vertically upward with a speed v_0 will have the same speed, v_0, when it comes back down to the starting point.

58. (II) A stone is thrown vertically upward with a speed of 23.0 m/s. (a) How fast is it moving when it reaches a height of 12.0 m? (b) How much time is required to reach this height? (c) Why are there two answers to (b)?

59. (II) Estimate the time between each photoflash of the apple in Fig. 2–23 (or number of photoflashes per second). Assume the apple is about 10 cm in diameter.

60. (II) A rocket rises vertically, from rest, with an acceleration of 3.2 m/s^2 until it runs out of fuel at an altitude of 1200 m. After this point, its acceleration is that of gravity, downward. (a) What is the velocity of the rocket when it runs out of fuel? (b) How long does it take to reach this point? (c) What maximum altitude does the rocket reach? (d) How much time (total) does it take to reach maximum altitude? (e) With what velocity does it strike the earth? (f) How long (total) is it in the air?

61. (II) A falling stone takes 0.30 s to travel past a window 2.2 m tall (Fig. 2–37). From what height above the top of the window did the stone fall?

2.2 m

To travel this distance took 0.30 s

FIGURE 2–37
Problem 61.

62. (II) Suppose you adjust your garden hose nozzle for a hard stream of water. You point the nozzle vertically upward at a height of 1.5 m above the ground (Fig. 2–38). When you quickly move the nozzle away from the vertical, you hear the water striking the ground next to you for another 2.0 s. What is the water speed as it leaves the nozzle?

1.5 m

FIGURE 2–38
Problem 62.

63. (III) A rock is dropped from a sea cliff and the sound of it striking the ocean is heard 3.4 s later. If the speed of sound is 340 m/s, how high is the cliff?

64. (III) A rock is thrown vertically upward with a speed of 12.0 m/s. Exactly 1.00 s later, a ball is thrown up vertically along the same path with a speed of 20.0 m/s. (a) At what time will they strike each other? (b) At what height will the collision occur? (c) Answer (a) and (b) assuming that the order is reversed: the ball is thrown 1.00 s before the rock.

65. (III) A toy rocket passes by a 2.0-m-high window whose sill is 10.0 m above the ground. The rocket takes 0.15 s to travel the 2.0 m height of the window. What was the launch speed of the rocket, and how high will it go? Assume the propellant is burned very quickly at blastoff.

* Section 2–8

* 66. (II) Given $v(t) = 25 + 18t$, where v is in m/s and t is in s, use calculus to determine the total displacement from $t_1 = 1.5$ s to $t_2 = 3.5$ s.

* 67. (III) Air resistance acting on a falling body can be taken into account by the approximate relation for the acceleration:

$$a = \frac{dv}{dt} = g - kv,$$

where k is a constant. (a) Derive a formula for the velocity of the body as a function of time assuming it starts from rest ($v = 0$ at $t = 0$). [*Hint*: Change variables by setting $u = g - kv$.] (b) Determine an expression for the terminal velocity, which is the maximum value the velocity reaches.

* 68. (III) The acceleration of a particle is given by $a = A\sqrt{t}$ where $A = 2.0$ m/s$^{5/2}$. At $t = 0$, $v = 10$ m/s and $x = 0$. (a) What is the speed as a function of time? (b) What is the displacement as a function of time? (c) What are the acceleration, speed and displacement at $t = 5.0$ s?

General Problems

69. The acceleration due to gravity on the Moon is about one sixth what it is on Earth. If an object is thrown vertically upward on the Moon, how many times higher will it go than it would on Earth, assuming the same initial velocity?

70. A person jumps from a fourth-story window 15.0 m above a firefighter's safety net. The survivor stretches the net 1.0 m before coming to rest, Fig. 2–39. (a) What was the average deceleration experienced by the survivor when slowed to rest by the net? (b) What would you do to make it "safer" (that is, generate a smaller deceleration): would you stiffen or loosen the net? Explain.

71. A person who is properly constrained by an over-the-shoulder seat belt has a good chance of surviving a car collision if the deceleration does not exceed 30 "g's" $(1.00\,g = 9.80$ m/s$^2)$. Assuming uniform deceleration of this value, calculate the distance over which the front end of the car must be designed to collapse if a crash brings the car to rest from 100 km/h.

15.0 m

1.0 m

FIGURE 2–39
Problem 70.

72. A race car driver must average 200.0 km/h over the course of a time trial lasting ten laps. If the first nine laps were done at 199.0 km/h, what average speed must be maintained for the last lap?

73. A car manufacturer tests its cars for front-end collisions by hauling them up on a crane and dropping them from a certain height. (*a*) Show that the speed just before a car hits the ground, after falling from rest a vertical distance H, is given by $\sqrt{2gH}$. What height corresponds to a collision at (*b*) 50 km/h? (*c*) 100 km/h?

74. Figure 2–40 is a position versus time graph for the motion of an object along the *x* axis. As the object moves from A to B: (*a*) Is the object moving in the positive or negative direction? (*b*) Is the object speeding up or slowing down? (*c*) Is the acceleration of the object positive or negative? Next, for the time interval from D to E: (*d*) Is the object moving in the positive or negative direction? (*e*) Is the object speeding up or slowing down? (*f*) Is the acceleration of the object positive or negative? (*g*) Finally, answer these same three questions for the time interval from C to D.

FIGURE 2–40 Problem 74.

75. Two children are playing on two trampolines. The first child can bounce up to one-and-a-half times higher than the second child. The initial speed up of the second child is 5.0 m/s. (*a*) Find the maximum height the second child reaches. (*b*) What is the initial speed of the first child? (*c*) How long was the first child in the air?

76. A 90-m-long train begins uniform acceleration from rest. The front of the train has a speed of 20 m/s when it passes a railway worker who is standing 180 m from where the front of the train started. What will be the speed of the last car as it passes the worker? (See Fig. 2–41.)

FIGURE 2–41 Problem 76.

77. A first stone is dropped from the roof of a building. 2.00 s after that, a second stone is thrown straight down with an initial speed of 30.0 m/s, and it is observed that the two stones land at the same time. (*a*) How long did it take the first stone to reach the ground? (*b*) How high is the building? (*c*) What are the speeds of the two stones just before they hit the ground?

78. A police car at rest, passed by a speeder traveling at a constant 110 km/h, takes off in hot pursuit. The police officer catches up to the speeder in 700 m, maintaining a constant acceleration. (*a*) Qualitatively plot the position versus time graph for both cars from the police car's start to the catch-up point. (*b*) Calculate how long it took the police officer to overtake the speeder, (*c*) calculate the required police car acceleration, and (*d*) calculate the speed of the police car at the overtaking point.

79. In the design of a rapid transit system, it is necessary to balance out the average speed of a train against the distance between stops. The more stops there are, the slower the train's average speed. To get an idea of this problem, calculate the time it takes a train to make a 36-km trip in two situations: (*a*) the stations at which the trains must stop are 0.80 km apart; and (*b*) the stations are 3.0 km apart. Assume that at each station the train accelerates at a rate of 1.1 m/s² until it reaches 90 km/h, then stays at this speed until its brakes are applied for arrival at the next station, at which time it decelerates at −2.0 m/s². Assume it stops at each intermediate station for 20 s.

80. Consider the street pattern as shown in Fig. 2–42. Each intersection has a traffic signal and the speed limit is 50 km/h. Suppose you are coming from the West at the speed limit and when you are 10 m from the first intersection, all the lights turn green. The lights are green for 13 s each. (*a*) Can you make it through all the lights without stopping? (*b*) Another car was stopped at the first light when all the lights turned green. It can accelerate at the rate of 2.0 m/s² to the speed limit. Can the second car make it through all the lights without stopping?

FIGURE 2–42 Problem 80.

81. A baseball is seen to pass upward by a window 25 m above the street with a vertical speed of 14 m/s. If the ball was thrown from the street, (*a*) what was its initial speed, (*b*) what altitude does it reach, (*c*) when was it thrown, and (*d*) when does it reach the street again?

82. A fugitive tries to catch a freight train traveling at a constant speed of 6.0 m/s. Just as an empty box car passes him, the fugitive starts from rest and accelerates at $a = 4.0$ m/s² to his maximum speed of 8.0 m/s. (*a*) How long does it take him to reach the empty box car? (*b*) What is the distance traveled to reach the box car?

83. A stone is thrown vertically upward with a speed of 10.0 m/s from the edge of a cliff 65.0 m high (Fig. 2–43). (a) How much later does it reach the bottom of the cliff? (b) What is its speed just before hitting? (c) What total distance did it travel?

FIGURE 2–43
Problem 83.

84. Sketch the v vs. t graph for the object whose displacement as a function of time is given by Fig. 2–31.

85. A person driving her car at 50 km/h approaches an intersection just as the traffic light turns yellow. She knows that the yellow light lasts only 2.0 s before turning to red, and she is 30 m away from the near side of the intersection (Fig. 2–44). Should she try to stop, or should she make a run for it? The intersection is 15 m wide. Her car's maximum deceleration is −6.0 m/s², whereas it can accelerate from 50 km/h to 70 km/h in 6.0 s. Ignore the length of her car and her reaction time.

FIGURE 2–44   Problem 85.

86. Pelicans tuck their wings and free fall straight down when diving for fish. Suppose a pelican starts its dive from a height of 16.0 m and cannot change its path once committed. If it takes a fish 0.20 s to perform evasive action, at what minimum height must it spot the pelican to escape? Assume the fish is at the surface of the water.

87. In putting, the force with which a golfer strikes a ball is planned so that the ball will stop within some small distance of the cup, say 1.0 m long or short, in case the putt is missed. Accomplishing this from an uphill lie (that is, putting downhill, see Fig. 2–45) is more difficult than from a downhill lie. To see why, assume that on a particular green the ball decelerates constantly at 2.0 m/s² going downhill, and constantly at 3.0 m/s² going uphill. Suppose we have an uphill lie 7.0 m from the cup. Calculate the allowable range of initial velocities we may impart to the ball so that it stops in the range 1.0 m short to 1.0 m long of the cup. Do the same for a downhill lie 7.0 m from the cup. What in your results suggests that the downhill putt is more difficult?

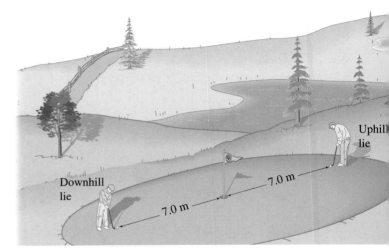

FIGURE 2–45   Problem 87. Golf on Wednesday morning.

88. A car is behind a truck going 25 m/s on the highway. The driver looks for an opportunity to pass, guessing that his car can accelerate at 1.0 m/s², and he gauges that he has to cover the 20-m length of the truck, plus 10 m clear room at the rear of the truck and 10 m more at the front of it. In the oncoming lane, he sees a car approaching, probably also traveling at 25 m/s. He estimates that the car is about 400 m away. Should he attempt the pass? Give details.

89. A stone is dropped from the roof of a high building. A second stone is dropped 1.50 s later. How far apart are the stones when the second one has reached a speed of 12.0 m/s?

90. Bond is standing on a bridge, 10 m above the road below, and his pursuers are getting too close for comfort. He spots a flatbed truck loaded with mattresses approaching at 30 m/s, which he measures by knowing that the telephone poles the truck is passing are 20 m apart in this country. The bed of the truck is 1.5 m above the road, and Bond quickly calculates how many poles away the truck should be when he jumps down from the bridge onto the truck, making his getaway. How many poles is it?

This multiflash photograph of a ping pong ball shows examples of motion in two dimensions. The arcs of the ping pong ball are parabolas that represent "projectile motion." Galileo analyzed projectile motion into its horizontal and vertical components, under the action of gravity (the gold arrow represents the downward acceleration of gravity, **g**). We will discuss how to manipulate vectors and how to add them. Besides analyzing projectile motion, we will also analyze uniform circular motion and how to work with relative velocity.

CHAPTER 3

Kinematics In Two Dimensions; Vectors

In Chapter 2 we dealt with motion along a straight line. We now consider the description of the motion of objects that move in paths in two (or three) dimensions. To do so we first need to discuss vectors and how they are added. Then we will examine the description of motion in general, followed by some interesting special cases including the motion of projectiles near the Earth's surface, and objects that are constrained to move in a circle.

3–1 Vectors and Scalars

We mentioned in Chapter 2 that the term *velocity* refers not only to how fast something is moving but also to its direction. A quantity such as velocity, which has *direction* as well as *magnitude*, is a **vector** quantity. Other quantities that are also vectors are displacement, force, and momentum. However, many quantities such as mass, time, and temperature, have no direction associated with them. They are specified completely by giving a number and units. Such quantities are called **scalars**.

Drawing a diagram of a particular physical situation is always helpful in physics, and this is especially true when dealing with vectors. On a diagram, each vector is represented by an arrow. The arrow is always drawn so that it points in the direction of the vector quantity it represents. The length of the arrow is drawn proportional to the magnitude of the vector quantity. For example, in Fig. 3–1, arrows have been drawn representing the velocity of a car at various places as it rounds a curve. The magnitude of the velocity at each point can be read off this figure by measuring the length of the corresponding arrow and using the scale shown (1 cm = 90 km/h).

FIGURE 3–1 Car traveling on a road. The green arrows represent the velocity vector at each position.

Scale for velocity:
1 cm = 90 km/h

Boldface symbols for vectors

When we write the symbol for a vector, we will always use boldface type. Thus for velocity we write **v**. (In handwritten work, the symbol for a vector can be indicated by putting an arrow over it, a $\vec{v}$ for velocity.) If we are concerned only with the magnitude of the vector, we will write simply v, in italics.

3-2 | Addition of Vectors—Graphical Methods

Because vectors are quantities that have direction as well as magnitude, they must be added in a special way. In this chapter, we will deal mainly with displacement vectors, for which we now use the symbol **D**, and velocity vectors, **v**. But the results will apply for other vectors we encounter later.

We use simple arithmetic for adding scalars. Simple arithmetic can also be used for adding vectors if they are in the same direction. For example, if a person walks 8 km east one day, and 6 km east the next day, the person will be 8 km + 6 km = 14 km east of the point of origin. We say that the *net* or *resultant* displacement is 14 km to the east (Fig. 3–2a). If, on the other hand, the person walks 8 km east on the first day, and 6 km west (in the reverse direction) on the second day, then the person will end up 2 km from the origin (Fig. 3–2b), so the resultant displacement is 2 km to the east. In this case, the resultant displacement is obtained by subtraction: 8 km − 6 km = 2 km.

But simple arithmetic cannot be used if the two vectors are not along the same line. For example, suppose a person walks 10.0 km east and then walks 5.0 km north. These displacements can be represented on a graph in which the positive y axis points north and the positive x axis points east, Fig. 3–3. On this graph, we draw an arrow, labeled $\mathbf{D}_1$, to represent the displacement vector of the 10.0-km displacement to the east. Then we draw a second arrow, $\mathbf{D}_2$, to represent the 5.0-km displacement to the north. Both vectors are drawn to scale, as in Fig. 3–3.

After taking this walk, the person is now 10.0 km east and 5.0 km north of the point of origin. The **resultant displacement** is represented by the arrow labeled $\mathbf{D}_R$ in Fig. 3–3. Using a ruler and a protractor, you can measure on this diagram that the person is 11.2 km from the origin at an angle of 27° north of east. In other words, the resultant displacement vector has a magnitude of 11.2 km and makes an angle $\theta = 27°$ with the positive x axis. The magnitude (length) of $\mathbf{D}_R$ can also be obtained using the theorem of Pythagoras in this case, since D_1, D_2, and D_R form a right triangle with D_R as the hypotenuse. Thus

$$D_R = \sqrt{D_1^2 + D_2^2} = \sqrt{(10.0\text{ km})^2 + (5.0\text{ km})^2} = \sqrt{125\text{ km}^2} = 11.2\text{ km}.$$

You can use the Pythagorean theorem, of course, only when the vectors are *perpendicular* to each other.

The resultant displacement vector, $\mathbf{D}_R$, is the sum of the vectors $\mathbf{D}_1$ and $\mathbf{D}_2$. That is,

$$\mathbf{D}_R = \mathbf{D}_1 + \mathbf{D}_2.$$

This is a *vector* equation. An important feature of adding two vectors that are not along the same line is that the magnitude of the resultant vector is not equal to the sum of the magnitudes of the two separate vectors, but is smaller than their sum:

$$D_R < D_1 + D_2. \qquad \text{[Vectors not along the same line]}$$

In our example (Fig. 3–3), $D_R = 11.2$ km, whereas $D_1 + D_2$ equals 15 km. We generally are not interested in $D_1 + D_2$; rather we are interested in the *vector* sum of the two vectors and its magnitude, D_R. Note also that we cannot set $\mathbf{D}_R$ equal to 11.2 km, because we have a vector equation and 11.2 km is only a part of the resultant vector, its magnitude. We could write something like this, though: $\mathbf{D}_R = \mathbf{D}_1 + \mathbf{D}_2 = (11.2\text{ km}, 27°\text{ N of E})$.

FIGURE 3–2 Combining vectors in one dimension.

FIGURE 3–3 A person walks 10.0 km east and then 5.0 km north. These two displacements are represented by the vectors $\mathbf{D}_1$ and $\mathbf{D}_2$, which are shown as arrows. The resultant displacement vector, $\mathbf{D}_R$, which is the vector sum of $\mathbf{D}_1$ and $\mathbf{D}_2$, is also shown. Measurement on the graph with ruler and protractor shows that $\mathbf{D}_R$ has a magnitude of 11.2 km and points at an angle $\theta = 27°$ north of east.

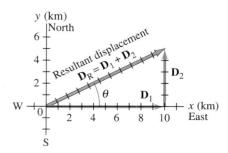

Figure 3–3 illustrates the general rules for graphically adding two vectors together, no matter what angles they make, to get their sum. The rules are as follows:

1. On a diagram, draw one of the vectors—call it $\mathbf{D}_1$—to scale.
2. Next draw the second vector, $\mathbf{D}_2$, to scale, placing its tail at the tip of the first vector and being sure its direction is correct.
3. The arrow drawn from the tail of the first vector to the tip of the second represents the *sum*, or **resultant**, of the two vectors.

Note that vectors can be translated parallel to themselves to accomplish these manipulations. The length of the resultant can be measured with a ruler and compared to the scale. Angles can be measured with a protractor. This method is known as the **tail-to-tip method of adding vectors**.

Note that it is not important in which order the vectors are added. For example, a displacement of 5.0 km north, to which is added a displacement of 10.0 km east, yields a resultant of 11.2 km and angle $\theta = 27°$ (see Fig. 3–4), the same as when they were added in reverse order (Fig. 3–3). That is,

$$\mathbf{V}_1 + \mathbf{V}_2 = \mathbf{V}_2 + \mathbf{V}_1. \qquad \text{[commutative law]} \quad \textbf{(3–1a)}$$

The tail-to-tip method can be extended to three or more vectors (Fig. 3–5) and, as indicated in that figure,

$$\mathbf{V}_1 + (\mathbf{V}_2 + \mathbf{V}_3) = (\mathbf{V}_1 + \mathbf{V}_2) + \mathbf{V}_3. \qquad \text{[associative law]} \quad \textbf{(3–1b)}$$

The left side of this equation means we first add $\mathbf{V}_2$ and $\mathbf{V}_3$ together and then add $\mathbf{V}_1$ to that sum to get the total sum. On the right side, $\mathbf{V}_1$ is added to $\mathbf{V}_2$, and this sum is added to $\mathbf{V}_3$. We see that the order in which two or more vectors are added doesn't affect the result.

A second way to add two vectors is the **parallelogram method**. It is fully equivalent to the tail-to-tip method. In this method, the two vectors are drawn starting from a common origin, and a parallelogram is constructed using these two vectors as adjacent sides as shown in Fig. 3–6b. The resultant is the diagonal drawn from the common origin. In Fig. 3–6a, the tail-to-tip method is shown, and it is clear that both methods yield the same result. It is a common error to draw the sum vector as the diagonal running between the tips of the two vectors, as in Fig. 3–6c. *This is incorrect*: it does not represent the sum of the two vectors. (In fact, it represents their difference, $\mathbf{V}_2 - \mathbf{V}_1$, as we will see in the next Section.)

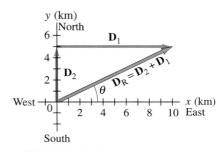

FIGURE 3–4 If the vectors are added in reverse order, the resultant is the same. (Compare Fig. 3–3.)

FIGURE 3–5 The three vectors in (a) can be added in any order with the same result, $\mathbf{V}_1 + \mathbf{V}_2 + \mathbf{V}_3$. In (b) $\mathbf{V}_R = (\mathbf{V}_1 + \mathbf{V}_2) + \mathbf{V}_3$, which clearly gives the same result as in (c) where $\mathbf{V}_R = \mathbf{V}_1 + (\mathbf{V}_2 + \mathbf{V}_3)$, which we simply write now as $\mathbf{V}_R = \mathbf{V}_1 + \mathbf{V}_2 + \mathbf{V}_3$ without parentheses.

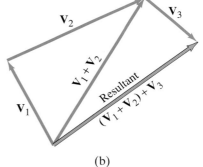

FIGURE 3–6
Vector addition by two different methods, (a) and (b). Part (c) is incorrect.

FIGURE 3–7 The negative of a vector is a vector having the same length but opposite direction.

3–3 Subtraction of Vectors, and Multiplication of a Vector by a Scalar

Given a vector **V**, we define the *negative* of this vector (−**V**) to be a vector with the same magnitude as **V** but opposite in direction, Fig. 3–7. Note, however, that no vector is ever negative in the sense of its magnitude: the magnitude of every vector is positive. A minus sign tells us about its direction.

We can now define the subtraction of one vector from another: the difference between two vectors, **V**$_2$ − **V**$_1$ is defined as

$$\mathbf{V}_2 - \mathbf{V}_1 = \mathbf{V}_2 + (-\mathbf{V}_1).$$

That is, the difference between two vectors is equal to the sum of the first plus the negative of the second. Thus our rules for addition of vectors can be applied as shown in Fig. 3–8 using the tail-to-tip method.

FIGURE 3–8 Subtracting two vectors: **V**$_2$ − **V**$_1$.

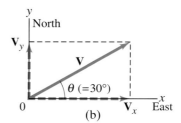

FIGURE 3–9 Multiplying a vector **V** by a scalar *c* gives a vector whose magnitude is *c* times greater and in the same direction as **V** (or opposite direction if *c* is negative).

A vector **V** can be multiplied by a scalar *c*. We define this product so that *c***V** has the same direction as **V** and has magnitude *cV*. That is, multiplication of a vector by a positive scalar *c* changes the magnitude of the vector by a factor *c* but doesn't alter the direction. If *c* is a negative scalar, the magnitude of the product *c***V** is still *cV* (without the minus sign), but the direction is precisely opposite to that of **V**. See Fig. 3–9.

3–4 Adding Vectors by Components

Adding vectors graphically using a ruler and protractor is often not sufficiently accurate and is not useful for vectors in three dimensions. We discuss now a more powerful and precise method for adding vectors.

Consider first a vector **V** that lies in a particular plane. It can be expressed as the sum of two other vectors, called the **components** of the original vector. The components are usually chosen to be along two perpendicular directions. The process of finding the components is known as **resolving the vector into its components**. An example is shown in Fig. 3–10; the vector **V** could be a displacement vector that points at an angle $\theta = 30°$ north of east, where we have chosen the

Resolving a vector into components

FIGURE 3–10 Resolving a vector **V** into its components along an arbitrarily chosen set of *x* and *y* axes. Note that the components, once found, themselves represent the vector. That is, the components contain as much information as the vector itself.

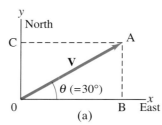

positive x axis to be to the east and the positive y axis north. This vector $\mathbf{V}$ is resolved into its x and y components by drawing dashed lines from the tip (A) of the vector and drawing these lines perpendicular to the x and y axes (lines AB and AC). Then the lines OB and OC represent the x and y components of $\mathbf{V}$, respectively, as shown in Fig. 3–10b. These *vector components* are written $\mathbf{V}_x$ and $\mathbf{V}_y$. We generally show vector components as arrows, like vectors, but dashed. The *scalar components*, V_x and V_y, are numbers, with units, that are given a positive or negative sign depending on whether they point along the positive or negative x or y axis. As can be seen in Fig. 3–10, $\mathbf{V}_x + \mathbf{V}_y = \mathbf{V}$ by the parallelogram method of adding vectors.

Space is made up of three dimensions, and sometimes it is necessary to resolve a vector into components along three mutually perpendicular directions. In rectangular coordinates the components are $\mathbf{V}_x$, $\mathbf{V}_y$, and $\mathbf{V}_z$, and

$$\mathbf{V} = \mathbf{V}_x + \mathbf{V}_y + \mathbf{V}_z.$$

Resolution of a vector in three dimensions is merely an extension of the above technique. We will mostly be concerned with situations in which the vectors are in a plane and two components are all that are necessary.

The use of trigonometric functions for finding the components of a vector is illustrated in Fig. 3–11, where it is seen that a vector and its two components can be thought of as making up a right triangle. We then see that the sine, cosine, and tangent are as given in the figure. Hence the x and y components of a vector V are

$$V_y = V \sin \theta \qquad\qquad \textbf{(3–2a)}$$
$$V_x = V \cos \theta. \qquad\qquad \textbf{(3–2b)}$$

Note that θ is chosen (by convention) to be the angle that the vector makes with the positive x axis.

The components of a given vector will be different for different choices of coordinate axes. It is therefore crucial to specify the choice of coordinate system when giving the components.

Note that there are two ways to specify a vector in a given coordinate system:

1. We can give its components, V_x and V_y.
2. We can give its magnitude V and the angle θ it makes with the positive x axis.

We can shift from one description to the other using Eqs. 3–2, and, for the reverse, by using the theorem of Pythagoras[†] and the definition of tangent (see Fig. 3–11):

$$V = \sqrt{V_x^2 + V_y^2} \qquad\qquad \textbf{(3–3a)}$$
$$\tan \theta = \frac{V_y}{V_x}. \qquad\qquad \textbf{(3–3b)}$$

[†] In three dimensions, the theorem of Pythagoras becomes $V = \sqrt{V_x^2 + V_y^2 + V_z^2}$, where V_z is the component along the third, or z, axis.

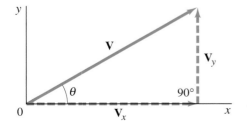

$$\sin \theta = \frac{V_y}{V}$$
$$\cos \theta = \frac{V_x}{V}$$
$$\tan \theta = \frac{V_y}{V_x}$$
$$V^2 = V_x^2 + V_y^2$$

FIGURE 3–11 Finding the components of a vector using trigonometric functions, where θ is the angle with the x axis.

FIGURE 3–12
The components of
$\mathbf{V} = \mathbf{V}_1 + \mathbf{V}_2$ are
$V_x = V_{1x} + V_{2x}$ and
$V_y = V_{1y} + V_{2y}$.

We can now discuss how to add vectors using components. The first step is to resolve each vector into its components. Next we can see, using Fig. 3–12, that the addition of any two vectors $\mathbf{V}_1$ and $\mathbf{V}_2$ to give a resultant, $\mathbf{V} = \mathbf{V}_1 + \mathbf{V}_2$, implies that

Adding vectors analytically (by components)

$$V_x = V_{1x} + V_{2x}$$
$$V_y = V_{1y} + V_{2y}.$$

(3–4)

That is, the sum of the x components equals the x component of the resultant, and similarly for y. That this is valid can be verified by a careful examination of Fig. 3–12. But note that we add all the x components together to get the x component of the resultant; and we add all the y components together to get the y component of the resultant. We do *not* add x components to y components.

If the magnitude and direction of the resultant vector are desired, they can be obtained using Eqs. 3–3.

Choice of axes can simplify effort needed

The choice of coordinate axes is, of course, always arbitrary. You can often reduce the work involved in adding vectors by a good choice of axes—for example, by choosing one of the axes to be in the same direction as one of the vectors. Then that vector will have only one nonzero component.

EXAMPLE 3–1 **Mail carrier's displacement.** A rural mail carrier leaves the post office and drives 22.0 km in a northerly direction to the next town. She then drives in a direction 60.0° south of east for 47.0 km (Fig. 3–13a) to another town. What is her displacement from the post office?

SOLUTION We want to find her resultant displacement from the origin. We choose the positive x axis to be east and the positive y axis north, and resolve each displacement vector into its components (Fig. 3–13b). Since $\mathbf{D}_1$ has

FIGURE 3–13 Example 3–1.

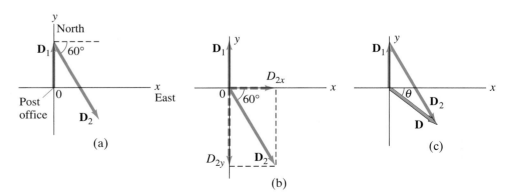

(a)

(b)

(c)

magnitude 22.0 km and points north, it has only a y component:

$$D_{1x} = 0, \quad D_{1y} = 22.0 \text{ km}$$

whereas $\mathbf{D}_2$ has both x and y components:

$$D_{2x} = +(47.0 \text{ km})(\cos 60°) = +(47.0 \text{ km})(0.500) = +23.5 \text{ km}$$
$$D_{2y} = -(47.0 \text{ km})(\sin 60°) = -(47.0 \text{ km})(0.866) = -40.7 \text{ km}.$$

Notice that D_{2y} is negative because this vector component points along the negative y axis. The resultant vector, $\mathbf{D}$, has components:

$$D_x = D_{1x} + D_{2x} = \quad 0 \text{ km} + \quad 23.5 \text{ km} \ = +23.5 \text{ km}$$
$$D_y = D_{1y} + D_{2y} = 22.0 \text{ km} + (-40.7 \text{ km}) = -18.7 \text{ km}.$$

This specifies the resultant vector completely:

$$D_x = 23.5 \text{ km}, \quad D_y = -18.7 \text{ km}.$$

We can also specify the resultant vector by giving its magnitude and angle using Eqs. 3–3:

$$D = \sqrt{D_x^2 + D_y^2} = \sqrt{(23.5 \text{ km})^2 + (-18.7 \text{ km})^2} = 30.0 \text{ km}$$

$$\tan \theta = \frac{D_y}{D_x} = \frac{-18.7 \text{ km}}{23.5 \text{ km}} = -0.796.$$

A calculator with an INV TAN or TAN^{-1} key gives $\theta = \tan^{-1}(-0.796) = -38.5°$. The negative sign means $\theta = 38.5°$ below the x axis, Fig. 3–13c.

The signs of trigonometric functions depend on which "quadrant" the angle falls in: for example, the tangent is positive in the first and third quadrants (from 0° to 90°, and 180° to 270°), but negative in the second and fourth quadrants; see Appendix A. The best way to keep track of angles, and to check any vector result, is always to draw a vector diagram. A vector diagram gives you something tangible to look at when analyzing a problem, and provides a check on the results.

➡ **PROBLEM SOLVING**

Check the quadrant

PROBLEM SOLVING

Adding Vectors by Components

Here is a brief summary of how to add two or more vectors using components:

1. Draw a diagram, adding the vectors graphically.
2. Choose x and y axes. Choose them in a way, if possible, that will make your work easier. (For example, choose one axis along the direction of one of the vectors so that vector will have only one component.)
3. Resolve each vector into its x and y components, showing each component along its appropriate (x or y) axis as a (dashed) arrow.
4. Calculate each component (when not given) using sines and cosines. If θ_1 is the angle vector $\mathbf{V}_1$ makes with the x axis, then:

$$V_{1x} = V_1 \cos \theta_1, \qquad V_{1y} = V_1 \sin \theta_1.$$

Pay careful attention to signs: any component that points along the negative x or y axis gets a negative sign.

5. Add the x components together to get the x component of the resultant. Similarly for y:

$$V_x = V_{1x} + V_{2x} + \text{any others.}$$
$$V_y = V_{1y} + V_{2y} + \text{any others.}$$

This is the answer: the components of the resultant vector.

6. If you want to know the magnitude and direction of the resultant vector, use Eqs. 3–3:

$$V = \sqrt{V_x^2 + V_y^2}, \qquad \tan \theta = \frac{V_y}{V_x}.$$

The vector diagram you already drew helps to obtain the correct position (quadrant) of the angle θ.

(a)

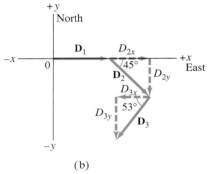

(b)

FIGURE 3–14 Example 3–2.

Vector	Components	
	x (km)	*y* (km)
$\mathbf{D}_1$	620	0
$\mathbf{D}_2$	311	−311
$\mathbf{D}_3$	−331	−439
$\mathbf{D}_R$	600	−750

FIGURE 3–15
Unit vectors **i**, **j**, and **k** along the *x*, *y*, and *z* axes.

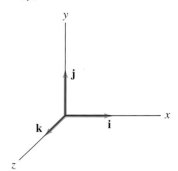

EXAMPLE 3–2 **Three short trips.** An airplane trip involves three legs, with two stopovers, as shown in Fig. 3–14a. The first leg is due east for 620 km; the second leg is southeast (45°) for 440 km; and the third leg is at 53° south of west, for 550 km, as shown. What is the plane's total displacement?

SOLUTION We follow the steps in the above Problem Solving box:
(1) and (2): Already shown in Fig. 3–14a, where we have taken the *x* axis as east (then $\mathbf{D}_1$ has only an *x* component).
(3): It is imperative to draw a good figure. The components are shown in Fig. 3–14b. Notice that instead of drawing all the vectors starting from a common origin, as we did in Fig. 3–13b, here we have drawn them "tail-to-tip" style, which is just as valid and may make it easier to see.
(4): Now we calculate the components:

$$\mathbf{D}_1: \quad D_{1x} = +D_1 \cos 0° = D_1 = 620 \text{ km}$$
$$D_{1y} = +D_1 \sin 0° = 0 \text{ km}$$

$$\mathbf{D}_2: \quad D_{2x} = +D_2 \cos 45° = +(440 \text{ km})(0.707) = +311 \text{ km}$$
$$D_{2y} = -D_2 \sin 45° = -(440 \text{ km})(0.707) = -311 \text{ km}$$

$$\mathbf{D}_3: \quad D_{3x} = -D_3 \cos 53° = -(550 \text{ km})(0.602) = -331 \text{ km}$$
$$D_{3y} = -D_3 \sin 53° = -(550 \text{ km})(0.799) = -439 \text{ km}.$$

Note carefully that we have given a minus sign to each component that in Fig. 3–14b points in the negative *x* or negative *y* direction. We see why a good drawing is so important. We summarize the components in the Table in the margin.
(5): This is easy:

$$D_x = D_{1x} + D_{2x} + D_{3x} = 620 \text{ km} + 311 \text{ km} - 331 \text{ km} = 600 \text{ km}$$
$$D_y = D_{1y} + D_{2y} + D_{3y} = 0 \text{ km} - 311 \text{ km} - 439 \text{ km} = -750 \text{ km}.$$

The *x* and *y* components are 600 km and −750 km, and point respectively to the east and south. This is one way to give the answer.
(6): We can also give the answer as

$$D_R = \sqrt{D_x^2 + D_y^2} = \sqrt{(600)^2 + (-750)^2} \text{ km} = 960 \text{ km}$$

$$\tan \theta = \frac{D_y}{D_x} = \frac{-750 \text{ km}}{600 \text{ km}} = -1.25, \qquad \text{so } \theta = -51°,$$

where we assume only two significant figures. Thus, the total displacement has magnitude 960 km and points 51° below the *x* axis (south of east), as was shown in our original sketch, Fig. 3–14a.

3–5 Unit Vectors

Vectors can be conveniently written in terms of *unit vectors*. A **unit vector** is defined to have a magnitude exactly equal to one (1). It is useful to define unit vectors that point along coordinate axes, and in a rectangular coordinate system these unit vectors are called **i**, **j**, and **k**. They point, respectively, along the positive *x*, *y*, and *z* axes as shown in Fig. 3–15. Like other vectors, **i**, **j**, and **k** do not have to be placed at the origin, but can be placed elsewhere as long as the direction and unit length remain unchanged. Sometimes you may see unit vectors written with a "hat": $\hat{\mathbf{i}}$, $\hat{\mathbf{j}}$, $\hat{\mathbf{k}}$.

Because of the definition of multiplication of a vector by a scalar (Section 3–3), the components of a vector **V** can be written $\mathbf{V}_x = V_x \mathbf{i}$, $\mathbf{V}_y = V_y \mathbf{j}$, and $\mathbf{V}_z = V_z\mathbf{k}$. Hence any vector $\mathbf{V}$ can be written in terms of its components as

$$\mathbf{V} = V_x\mathbf{i} + V_y\mathbf{j} + V_z\mathbf{k}. \tag{3-5}$$

Unit vectors are helpful when adding vectors analytically by components. For example, Eq. 3–4 can be seen to be true by using unit vector notation for each vector (which we write for the two-dimensional case, but the extension to three dimensions is straightforward):

$$\begin{aligned}
\mathbf{V} &= (V_x)\mathbf{i} + (V_y)\mathbf{j} \\
&= \mathbf{V}_1 + \mathbf{V}_2 \\
&= (V_{1x}\mathbf{i} + V_{1y}\mathbf{j}) + (V_{2x}\mathbf{i} + V_{2y}\mathbf{j}) \\
&= (V_{1x} + V_{2x})\mathbf{i} + (V_{1y} + V_{2y})\mathbf{j}.
\end{aligned}$$

Comparing the first line to the fourth line, we get Eq. 3–4.

EXAMPLE 3–3 **Using unit vectors.** Write the vectors of Example 3–1 in unit vector notation, and perform the addition.

SOLUTION In Example 3–1 we resolved $\mathbf{D}_1$ and $\mathbf{D}_2$ into components and found

$$D_{1x} = 0, D_{1y} = 22.0 \text{ km, and } D_{2x} = 23.5 \text{ km and } D_{2y} = -40.7 \text{ km.}$$

Thus we have

$$\mathbf{D}_1 = 0\mathbf{i} + 22.0 \text{ km } \mathbf{j}$$
$$\mathbf{D}_2 = 23.5 \text{ km } \mathbf{i} - 40.7 \text{ km } \mathbf{j}.$$

Then

$$\begin{aligned}
\mathbf{D}_1 + \mathbf{D}_2 &= (0 + 23.5) \text{ km } \mathbf{i} + (22.0 - 40.7) \text{ km } \mathbf{j} \\
&= 23.5 \text{ km } \mathbf{i} - 18.7 \text{ km } \mathbf{j}.
\end{aligned}$$

The components of the resultant displacement, $\mathbf{D}$, are $D_x = 23.5$ km and $D_y = -18.7$ km.

3–6 | Vector Kinematics

We can now extend our definitions of velocity and acceleration in a formal way to two- and three-dimensional motion. Suppose a particle follows a path in the xy plane as shown in Fig. 3–16. At time t_1, the particle is at point P_1, and at time t_2, it is at point P_2. The vector $\mathbf{r}_1$ is the position vector of the particle at time t_1 (it represents the displacement of the particle from the origin of the coordinate system). And $\mathbf{r}_2$ is the position vector at time t_2.

In one dimension, we defined displacement as the *change in position* of the particle. In the more general case of two or three dimensions, the **displacement vector** is defined as the vector representing change in position. We call it $\Delta\mathbf{r}$,[†] where

$$\Delta\mathbf{r} = \mathbf{r}_2 - \mathbf{r}_1.$$

This represents the displacement during the time interval $\Delta t = t_2 - t_1$. In unit vector notation, we can write

$$\mathbf{r}_1 = x_1\mathbf{i} + y_1\mathbf{j} + z_1\mathbf{k}, \tag{3-6a}$$

where x_1, y_1, and z_1 are the coordinates of point P_1 (Fig. 3–16). Similarly,

$$\mathbf{r}_2 = x_2\mathbf{i} + y_2\mathbf{j} + z_2\mathbf{k}.$$

Hence

$$\Delta\mathbf{r} = (x_2 - x_1)\mathbf{i} + (y_2 - y_1)\mathbf{j} + (z_2 - z_1)\mathbf{k}. \tag{3-6b}$$

If the motion is along the x axis only, then $y_2 - y_1 = 0$, $z_2 - z_1 = 0$, and the magnitude of the displacement is $\Delta r = x_2 - x_1$, which is consistent with our earlier one-dimensional equation (Section 2–1). Even in one dimension, displacement is a vector, as are velocity and acceleration.

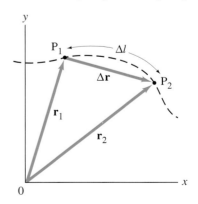

FIGURE 3–16 Path of a particle in the xy plane. At time t_1 the particle is at point P_1 given by the position vector $\mathbf{r}_1$; at t_2 the particle is at point P_2 given by the position vector $\mathbf{r}_2$. The displacement vector for the time interval $t_2 - t_1$ is $\Delta\mathbf{r} = \mathbf{r}_2 - \mathbf{r}_1$.

[†] We used $\mathbf{D}$ for the displacement vector earlier in the chapter for illustrating vector addition. The new notation here, $\Delta\mathbf{r}$, emphasizes that it is the difference between two position vectors.

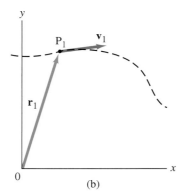

FIGURE 3–17 As we take Δt and $\Delta \mathbf{r}$ smaller and smaller [compare Fig. 3–16 and part (a) of this figure] we see that the direction of $\Delta \mathbf{r}$ and of the instantaneous velocity ($\Delta \mathbf{r}/\Delta t$, where $\Delta t \to 0$) is tangent to the curve at P_1 (part b).

FIGURE 3–18 (a) Velocity vectors $\mathbf{v}_1$ and $\mathbf{v}_2$ at instants t_1 and t_2 for the particle of Fig. 3–16. (b) Direction of the average acceleration in this case, $\bar{\mathbf{a}} = \Delta \mathbf{v}/\Delta t$.

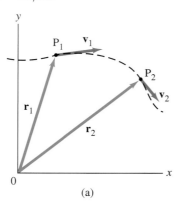

The **average velocity vector**, $\bar{\mathbf{v}}$, over the time interval $\Delta t = t_2 - t_1$ is defined as

$$\bar{\mathbf{v}} = \frac{\Delta \mathbf{r}}{\Delta t}. \qquad (3\text{–}7)$$

Since $\mathbf{v}$ is a product of the vector $\Delta \mathbf{r}$ times a scalar $(1/\Delta t)$, the direction of $\mathbf{v}$ is the same as that of $\Delta \mathbf{r}$; and its magnitude is $\Delta r/\Delta t$.

We next consider shorter and shorter time intervals—that is, we let Δt approach zero so that the distance between points P_2 and P_1 also approaches zero. We define the **instantaneous velocity vector** as the limit of the average velocity as Δt approaches zero:

$$\mathbf{v} = \lim_{\Delta t \to 0} \frac{\Delta \mathbf{r}}{\Delta t} = \frac{d\mathbf{r}}{dt}. \qquad (3\text{–}8)$$

The direction of $\mathbf{v}$ at any moment is along the line tangent to the path at that moment (Fig. 3–17).

Note that the magnitude of the average velocity in Fig. 3–16 is not equal to the average speed, which is the actual distance traveled, Δl, divided by Δt. In some special cases, the average speed and average velocity are equal (such as motion along a straight line in one direction), but in general they are not. However, in the limit $\Delta t \to 0$, Δr always approaches Δl, so the instantaneous speed *always* equals the magnitude of the instantaneous velocity at any time.

The instantaneous velocity (Eq. 3–8) is equal to the derivative of the position vector with respect to time. Equation 3–8 can be written in terms of components starting with Eq. 3–6a as:

$$\mathbf{v} = \frac{d\mathbf{r}}{dt} = \frac{dx}{dt}\mathbf{i} + \frac{dy}{dt}\mathbf{j} + \frac{dz}{dt}\mathbf{k}$$
$$= v_x \mathbf{i} + v_y \mathbf{j} + v_z \mathbf{k}, \qquad (3\text{–}9)$$

where $v_x = dx/dt$, $v_y = dy/dt$, $v_z = dz/dt$ are the x, y, and z components of the velocity. Note that $d\mathbf{i}/dt = d\mathbf{j}/dt = d\mathbf{k}/dt = 0$ since these unit vectors are constant in both magnitude and direction.

Acceleration in two or three dimensions is treated in a similar way. The **average acceleration vector**, $\bar{\mathbf{a}}$, over a time interval $\Delta t = t_2 - t_1$ is defined as

$$\bar{\mathbf{a}} = \frac{\Delta \mathbf{v}}{\Delta t} = \frac{\mathbf{v}_2 - \mathbf{v}_1}{t_2 - t_1}, \qquad (3\text{–}10)$$

where $\Delta \mathbf{v}$ is the change in the instantaneous velocity vector during that time interval: $\Delta \mathbf{v} = \mathbf{v}_2 - \mathbf{v}_1$. Note that $\mathbf{v}_2$ in many cases, such as in Fig. 3–18a, may not be in the same direction as $\mathbf{v}_1$. Hence $\bar{\mathbf{a}}$ may be in a different direction from either $\mathbf{v}_1$ or $\mathbf{v}_2$ (Fig. 3–18b). Furthermore, $\mathbf{v}_2$ and $\mathbf{v}_1$ may have the same magnitude but different directions, and the difference of two such vectors will not be zero. Hence acceleration can result from either a change in the magnitude of the velocity, or from a change in direction of the velocity, or from a change in both.

The **instantaneous acceleration vector** is defined as the limit of the average acceleration vector as the time interval Δt is allowed to approach zero:

$$\mathbf{a} = \lim_{\Delta t \to 0} \frac{\Delta \mathbf{v}}{\Delta t} = \frac{d\mathbf{v}}{dt}, \qquad (3\text{–}11)$$

and is thus the derivative of $\mathbf{v}$ with respect to t. Using components gives us:

$$\mathbf{a} = \frac{d\mathbf{v}}{dt} = \frac{dv_x}{dt}\mathbf{i} + \frac{dv_y}{dt}\mathbf{j} + \frac{dv_z}{dt}\mathbf{k}$$
$$= a_x \mathbf{i} + a_y \mathbf{j} + a_z \mathbf{k}, \qquad (3\text{–}12)$$

where $a_x = dv_x/dt$, etc. The instantaneous acceleration will be nonzero not only when the magnitude of the velocity changes, but also if its direction changes. For example, a person riding in a car traveling at constant speed around a curve, or a child riding on a merry-go-round, will both experience an acceleration because of a change in the direction of the velocity, even though its magnitude may be constant. (More on this later.)

In general, we will use the terms "velocity" and "acceleration" to mean the instantaneous values. If we want to discuss average values, we will use the word "average."

Constant Acceleration

In Chapter 2 we studied the important case of one-dimensional motion for which the acceleration is a constant. We now consider motion in two or three dimensions for which the acceleration vector, $\mathbf{a}$, is constant in magnitude and direction. That is, a_x = constant, a_y = constant, a_z = constant. The average acceleration in this case is equal to the instantaneous acceleration at any moment. The equations we derived in Chapter 2 for one dimension, Eqs. 2–12a, b, and c, apply separately to each perpendicular component of two- or three-dimensional motion. In two dimensions we let $\mathbf{v}_0 = v_{x0}\mathbf{i} + v_{y0}\mathbf{j}$ be the initial velocity, and we apply Eqs. 3–6a, 3–9, and 3–12 for the position vector, $\mathbf{r}$, velocity, $\mathbf{v}$, and acceleration, $\mathbf{a}$. We can then write Eqs. 2–12a, b, and c, for two dimensions as shown in Table 3–1.

TABLE 3–1 Kinematic Equations for Constant Acceleration in 2 Dimensions

x Component (horizontal)		y Component (vertical)
$v_x = v_{x0} + a_x t$	(Eq. 2–12a)	$v_y = v_{y0} + a_y t$
$x = x_0 + v_{x0}t + \frac{1}{2}a_x t^2$	(Eq. 2–12b)	$y = y_0 + v_{y0}t + \frac{1}{2}a_y t^2$
$v_x^2 = v_{x0}^2 + 2a_x(x - x_0)$	(Eq. 2–12c)	$v_y^2 = v_{y0}^2 + 2a_y(y - y_0)$

The first two of the equations in Table 3–1 can be written more formally in vector notation (see Eqs. 3–6a, 3–9, and 3–12):

$$\mathbf{v} = \mathbf{v}_0 + \mathbf{a}t \qquad [\mathbf{a} = \text{constant}] \quad \textbf{(3–13a)}$$

$$\mathbf{r} = \mathbf{r}_0 + \mathbf{v}_0 t + \tfrac{1}{2}\mathbf{a}t^2. \qquad [\mathbf{a} = \text{constant}] \quad \textbf{(3–13b)}$$

Here, $\mathbf{r}$ is the position vector at any time, and $\mathbf{r}_0$ is the position vector at $t = 0$. These equations are the vector equivalent of Eqs. 2–12a and b. In practical situations, we usually use the component form given in Table 3–1.

These equations, and their use, will become clearer as we use them. We next deal with several types of motion in a plane which are common in everyday life: projectile motion, and circular motion.

3–7 | Projectile Motion

In Chapter 2, we studied the motion of objects in one dimension in terms of displacement, velocity, and acceleration, including purely vertical motion of falling bodies undergoing acceleration due to gravity. Now we examine the more general motion of objects moving through the air in two dimensions near the Earth's surface, such as a golf ball, a thrown or batted baseball, kicked footballs, speeding bullets, and athletes doing the long jump or high jump. These are all examples of **projectile motion** (see Fig. 3–19), which we can describe as taking place in two dimensions. Although air resistance is often important, in many cases its effect can be ignored, and we will ignore it in the following analysis. We will not be concerned now with the process by which the object is thrown or projected. We consider only its motion *after* it has been projected and is moving freely through the air under the action of gravity alone. Thus the acceleration of the object is that

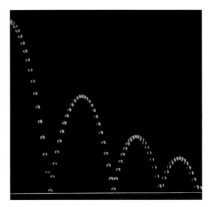

FIGURE 3–19 This strobe photograph of a ball making a series of bounces shows the characteristic "parabolic" path of projectile motion.

due to gravity, which acts downward with magnitude $g = 9.80 \, \text{m/s}^2$, and we assume it is constant.[†]

Galileo first accurately described projectile motion. He showed that it could be understood by analyzing the horizontal and vertical components of the motion separately. For convenience, we assume that the motion begins at time $t = 0$ at the origin of an xy coordinate system (so $x_0 = y_0 = 0$).

Let us look at a (tiny) ball rolling off the end of a table with an initial velocity v_{x0} in the horizontal (x) direction. See Fig. 3–20, where an object falling vertically is also shown for comparison. The velocity vector **v** at each instant points in the direction of the ball's motion at that instant and is always tangent to the path. Following Galileo's ideas, we treat the horizontal and vertical components of the velocity, v_x and v_y, separately, and we can apply the kinematic equations (Eqs. 2–12a through 2–12c) to each.

First we examine the vertical (y) component of the motion. Once the ball leaves the table (at $t = 0$), it experiences a vertically downward acceleration, g, the acceleration due to gravity. Thus v_y is initially zero $(v_{y0} = 0)$ but increases continually in the downward direction (until the ball hits the ground). Let us take y to be positive upwards. Then $a_y = -g$, and from Eq. 2–12a we can write $v_y = -gt$ since we set $y_0 = 0$.

In the horizontal direction, on the other hand, there is no acceleration. So the horizontal component of velocity, v_x, remains constant, equal to its initial value, v_{x0}, and thus has the same magnitude at each point on the path. The two vector components, **v**$_x$ and **v**$_y$, can be added vectorially to obtain the velocity **v** for each point on the path, as shown in Fig. 3–20.

One result of this analysis, which Galileo himself predicted, is that *an object projected horizontally will reach the ground in the same time as an object dropped vertically*. This is because the vertical motions are the same in both cases, as shown in Fig. 3–20. Figure 3–21 is a multiple-exposure photograph of an experiment that confirms this.

[†] This restricts us to objects whose distance traveled and maximum height above the Earth are small compared to the Earth's radius (6400 km).

FIGURE 3–20 Projectile motion. (A vertically falling object is shown at the left for comparison.)

FIGURE 3–21 Multiple-exposure photograph showing positions of two balls at equal time intervals. One ball was dropped from rest at the same time the other was projected horizontally outward. The vertical position of each ball is seen to be the same.

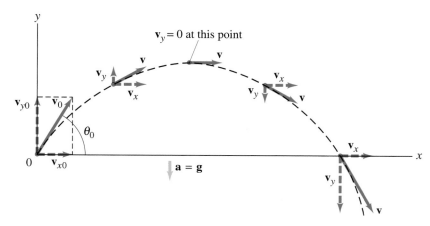

FIGURE 3–22 Path of a projectile fired with initial velocity $\mathbf{v}_0$ at angle θ to the horizontal. Path is shown in black, the velocity vectors are green arrows, and velocity components are dashed.

If an object is projected at an upward angle, as in Fig. 3–22, the analysis is similar, except that now there is an initial vertical component of velocity, v_{y0}. Because of the downward acceleration of gravity, v_y continually decreases until the object reaches the highest point on its path in Fig. 3–22, at which point $v_y = 0$. Then v_y starts to increase in the downward direction, as shown (that is, becoming negative). As before, v_x remains constant.

Object projected upward

Galileo's analysis of almost four centuries ago is exactly equivalent to using Eqs. 2–12 separately for the horizontal (x) and vertical (y) components, as given in Table 3–1 (Section 3–6). The constant acceleration now is that of gravity alone, acting downward. As can be seen in Fig. 3–22 for an object projected at an upward angle θ, the acceleration is in one (constant) direction, whereas the velocity has two components, one of which (v_y) is continuously changing, whereas the other (v_x) stays constant.

We can simplify Eqs. 2–12 (Table 3–1) for use with projectile motion by setting $a_x = 0$. See Table 3–2, which assumes y is positive upward, so $a_y = -g = -9.80 \text{ m/s}^2$. Note also that if θ is chosen as in Fig. 3–22 then the initial velocity has components:

$$v_{x0} = v_0 \cos \theta,$$

$$v_{y0} = v_0 \sin \theta.$$

TABLE 3–2 Kinematic Equations for Projectile Motion
(y positive upward; $a_x = 0$, $a_y = -g = -9.80 \text{ m/s}^2$)

Horizontal Motion $(a_x = 0, v_x = \text{constant})$		Vertical Motion† $(a_y = -g = \text{constant})$
$v_x = v_{x0}$	(Eq. 2–12a)	$v_y = v_{y0} - gt$
$x = x_0 + v_{x0}t$	(Eq. 2–12b)	$y = y_0 + v_{y0}t - \frac{1}{2}gt^2$
	(Eq. 2–12c)	$v_y^2 = v_{y0}^2 - 2gy$

†If y is taken positive downward, the minus $(-)$ signs become $+$ signs.

3-8 | Solving Problems Involving Projectile Motion

We now work through several Examples of projectile motion quantitatively. First, we give a summary of how you can approach these types of problems.

PROBLEM SOLVING | **Projectile Motion**

The approach to solving problems that we discussed in Section 2–6 also applies here. However, solving problems involving projectile motion can require a bit of creativity, and cannot be done just by simply following some rules. Certainly you must avoid just plugging numbers into equations that seem to "work."

As always, **read** carefully and **draw** a careful diagram.

1. Choose an origin and an xy coordinate system.
2. Analyze the horizontal (x) motion and the vertical (y) motion separately. If you are given the initial velocity, you may want to resolve it into its x and y components.

3. List the known and unknown quantities, choosing $a_x = 0$ and $a_y = -g$ or $+g$, where $g = 9.80 \, \text{m/s}^2$, depending on whether you choose y positive up or down. Remember that v_x never changes throughout the trajectory, and that $v_y = 0$ at the highest point of any trajectory that returns downward. The velocity just before hitting the ground is generally not zero.

4. Think for a minute before jumping into the equations. A little planning goes a long way. Apply the relevant equations (Table 3–2), combining equations if necessary. You may need to combine components of a vector to get magnitude and direction (Eqs. 3–3).

FIGURE 3–23 Example 3–4.

EXAMPLE 3–4 **Driving off a cliff.** A movie stunt driver on a motorcycle speeds horizontally off a 50.0-m-high cliff. How fast must the motorcycle leave the cliff-top if it is to land on level ground below, 90.0 m from the base of the cliff (Fig. 3–23) where the cameras are?

SOLUTION We take the y direction to be positive upward, with the top of the cliff as $y_0 = 0$, so the bottom is at $y = -50.0$ m. First, we find how long it takes the motorcycle to reach the ground below. We use Eq. 2–12b for the vertical (y) direction (Table 3–2) with $y_0 = 0$, and $v_{y0} = 0$:

$$y = -\tfrac{1}{2}gt^2$$

We solve for t and set $y = -50.0$ m:

$$t = \sqrt{\frac{2y}{-g}} = \sqrt{\frac{2(-50.0 \, \text{m})}{-9.80 \, \text{m/s}^2}} = 3.19 \, \text{s}.$$

To calculate the initial velocity, v_{x0}, we again use Eq. 2–12b, but this time for the horizontal (x) direction, with $a_x = 0$ and $x_0 = 0$:

$$x = v_{x0}t$$

$$v_{x0} = \frac{x}{t} = \frac{90.0 \, \text{m}}{3.19 \, \text{s}} = 28.2 \, \text{m/s},$$

which is 101 km/h (about 60 mi/h).

EXAMPLE 3–5 **A kicked football.** A football is kicked at an angle $\theta_0 = 37.0°$ with a velocity of 20.0 m/s, as shown in Fig. 3–24. Calculate (*a*) the maximum height, (*b*) the time of travel before the football hits the ground, (*c*) how far away it hits the ground, (*d*) the velocity vector at the maximum height, and (*e*) the acceleration vector at maximum height. Assume the ball leaves the foot at ground level, and ignore air resistance (not very realistic, though).

SOLUTION We take the y direction as positive upward. The components of the initial velocity are (Fig. 3–24):

$$v_{x0} = v_0 \cos 37.0° = (20.0 \, \text{m/s})(0.799) = 16.0 \, \text{m/s}$$

$$v_{y0} = v_0 \sin 37.0° = (20.0 \, \text{m/s})(0.602) = 12.0 \, \text{m/s}.$$

(*a*) At the maximum height, the velocity is horizontal (Fig. 3–24), so $v_y = 0$; and this occurs (see Eq. 2–12a in Table 3–2) at time

$$t = \frac{v_{y0}}{g} = \frac{12.0 \text{ m/s}}{9.80 \text{ m/s}^2} = 1.22 \text{ s}.$$

From Eq. 2–12b, with $y_0 = 0$, we have

$$y = v_{y0}t - \tfrac{1}{2}gt^2$$
$$= (12.0 \text{ m/s})(1.22 \text{ s}) - \tfrac{1}{2}(9.80 \text{ m/s}^2)(1.22 \text{ s})^2 = 7.35 \text{ m}.$$

Alternatively, we could have used Eq. 2–12c, solved for y, and found

$$y = \frac{v_{y0}^2 - v_y^2}{2g} = \frac{(12.0 \text{ m/s})^2 - (0 \text{ m/s})^2}{2(9.80 \text{ m/s}^2)} = 7.35 \text{ m}.$$

(*b*) To find the time it takes for the ball to return to the ground, we use Eq. 2–12b with $y_0 = 0$ and also set $y = 0$ (ground level):

$$y = y_0 + v_{y0}t - \tfrac{1}{2}gt^2$$
$$0 = 0 + (12.0 \text{ m/s})t - \tfrac{1}{2}(9.80 \text{ m/s}^2)t^2$$

which is an equation that can be easily factored:

$$\left[\tfrac{1}{2}(9.80 \text{ m/s}^2)t - 12.0 \text{ m/s}\right]t = 0.$$

There are two solutions, $t = 0$ (which corresponds to the initial point, y_0), and

$$t = \frac{2(12.0 \text{ m/s})}{(9.80 \text{ m/s}^2)} = 2.45 \text{ s},$$

which is the result we sought. Note that the time $t = 2.45$ s is just double the time to reach the highest point, calculated in (*a*). That is, the time to go up equals the time to come down to the same level—ignoring air resistance.

(*c*) The total distance traveled in the x direction is found by applying Eq. 2–12b with $x_0 = 0$, $a_x = 0$, $v_{x0} = 16.0$ m/s:

$$x = v_{x0}t = (16.0 \text{ m/s})(2.45 \text{ s}) = 39.2 \text{ m}.$$

(*d*) At the highest point, there is no vertical component to the velocity. There is only the horizontal component (which remains constant throughout the flight), so $v = v_{x0} = v_0 \cos 37.0° = 16.0$ m/s.

(*e*) The acceleration vector is the same at the highest point as it is throughout the flight, which is 9.80 m/s² downward.

FIGURE 3–24 Example 3–5. (See also Fig. 3–22.)

Note the symmetry

CONCEPTUAL EXAMPLE 3–6 **Where does the apple land?** A child sits upright in a wagon which is moving to the right at constant speed as shown in Fig. 3–25. The child extends her hand and throws an apple straight upward (from her own point of view, Fig. 3–25a), while the wagon continues to travel forward at constant speed. If air resistance is neglected, will the apple land (*a*) behind the wagon, (*b*) in the wagon, or (*c*) in front of the wagon?

RESPONSE The child throws the apple straight up from her own point of view with initial velocity $\mathbf{v}_{0y}$ (Fig. 3–25a). But when viewed by someone on the ground, the apple also has an initial horizontal component of velocity equal to the speed of the wagon, $\mathbf{v}_{0x}$. Thus, to a person on the ground, the apple will follow the path of a projectile as shown in Fig. 3–25b. The apple experiences no horizontal acceleration, so $\mathbf{v}_{0x}$ will stay constant and equal to the speed of the wagon. As the apple follows its arc, the wagon will be directly under the apple at all times because they have the same horizontal velocity. When the apple comes down, it will drop right into the wagon, and into the outstretched hand of the child. The answer is (*b*).

FIGURE 3–25 Conceptual Example 3–6.

(a) Wagon reference frame

(b) Ground reference frame

CONCEPTUAL EXAMPLE 3–7 **The wrong strategy.** A boy on a small hill aims his water-balloon slingshot horizontally, straight at a second boy hanging from a tree branch a distance d away, Fig. 3–26. At the instant the water balloon is released, the second boy lets go and falls from the tree, hoping to avoid being hit. Show that he made the wrong move. (He hadn't studied physics yet.)

RESPONSE Both the water balloon and the boy in the tree start falling at the same instant, and in a time t they each fall the same vertical distance $y = \frac{1}{2}gt^2$ (see Fig. 3–21). In the time it takes the water balloon to travel the horizontal distance d, the balloon will have the same y position as the falling boy. Splat. If the boy had stayed in the tree, he'd have saved himself the humiliation.

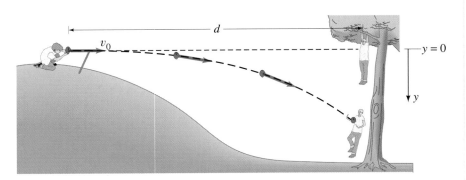

FIGURE 3–26
Example 3–7.

➡ PHYSICS APPLIED

Horizontal range of a projectile

FIGURE 3–27 Example 3–8.
(a) The range R of a projectile;
(b) shows how generally there are two angles θ_0 that will give the same range. Can you show that if one angle is θ_{01}, the other is $\theta_{02} = 90° - \theta_{01}$?

(a)

(b)

Level range formula
$[y \text{ (final)} = y_0]$

EXAMPLE 3–8 **Level horizontal range.** (a) Derive a formula for the horizontal range R of a projectile in terms of its initial velocity v_0 and angle θ_0. The horizontal range is defined as the horizontal distance the projectile travels before returning to its original height, which is typically the ground; that is, $y \text{ (final)} = y_0$. See Fig. 3–27. (b) Suppose one of Napoleon's cannons had a muzzle velocity, v_0, of 60.0 m/s. At what angle should it have been aimed (ignore air resistance) to strike a target 320 m away?

SOLUTION (a) We set $x_0 = 0$ and $y_0 = 0$ at $t = 0$. After the projectile travels a horizontal distance R, it returns to the same level, $y = 0$, the final point. So to find a general expression for R, we set both $y = 0$ and $y_0 = 0$ in Eq. 2–12b for the vertical motion, and obtain

$$v_{y0}t - \tfrac{1}{2}gt^2 = 0.$$

We solve for t, which gives two solutions: $t = 0$ and $t = 2v_{y0}/g$. The first solution corresponds to the initial instant of projection and the second is the time when the projectile returns to $y = 0$. Then the range, R, will be equal to x at the moment t has this value, which we put into Eq. 2–12b for the horizontal motion ($x = v_{x0}t$, with $x_0 = 0$). Thus we have:

$$R = x = v_{x0}t = v_{x0}\left(\frac{2v_{y0}}{g}\right) = \frac{2v_{x0}v_{y0}}{g} = \frac{2v_0^2 \sin\theta_0 \cos\theta_0}{g} \qquad [y = y_0]$$

where we have written $v_{x0} = v_0 \cos\theta_0$ and $v_{y0} = v_0 \sin\theta_0$. This is the result we sought, and it can be rewritten, using the trigonometric identity $2\sin\theta\cos\theta = \sin 2\theta$ (Appendix A), as

$$R = \frac{v_0^2 \sin 2\theta_0}{g}. \qquad [y = y_0]$$

We see that the maximum range, for a given initial velocity v_0, is obtained when the sine takes on its maximum value of 1.0, which occurs for $2\theta_0 = 90°$; so

$$\theta_0 = 45° \text{ for maximum range, and } R_{\text{max}} = v_0^2/g.$$

[When air resistance is important, the range is less for a given v_0, and the maximum range is obtained at an angle smaller than 45°.] Note that the maximum range increases by the square of v_0, so doubling the muzzle velocity of a cannon increases its maximum range by a factor of 4.

(b) From the equation we just derived, Napoleon's cannon should be aimed (assuming, unrealistically, no air resistance) at an angle θ_0 given by

$$\sin 2\theta_0 = \frac{Rg}{v_0^2} = \frac{(320\text{ m})(9.80\text{ m/s}^2)}{(60.0\text{ m/s})^2} = 0.871.$$

We want to solve for an angle θ_0 that is between $0°$ and $90°$, which means $2\theta_0$ in this equation can be as large as $180°$. Thus, $2\theta_0 = 60.6°$ is a solution, but $2\theta_0 = 180° - 60.6° = 119.4°$ is also a solution (see Appendix A). In general we will have two solutions, which in Napoleon's case are given by

$$\theta_0 = 30.3° \text{ or } 59.7°.$$

Either angle gives the same range. Only when $\sin 2\theta_0 = 1$ (so $\theta_0 = 45°$) is there a single solution (that is, both solutions are the same).

EXAMPLE 3–9 A punt. Suppose the football in Example 3–5 was a punt and left the punter's foot at a height of 1.00 m above the ground. How far did the football travel before hitting the ground? Set $x_0 = 0$, $y_0 = 0$.

➡ PHYSICS APPLIED

Sports

SOLUTION We cannot use the range formula from Example 3–8 because it is valid only if y (final) $= y_0$, which is not the case here. We have $y_0 = 0$, and the football hits the ground where $y = -1.00$ m (see Fig. 3–28). We can get x from Eq. 2–12b, $x = v_{x0}t$, since we know that $v_{x0} = 16.0$ m/s. But first we must find t, the time at which the ball hits the ground. With $y = -1.00$ m and $v_{y0} = 12.0$ m/s (see Example 3–5), we use the equation

➡ PROBLEM SOLVING

Do not use any formula unless you are sure its range of validity fits the problem. The range formula does not apply here because $y \neq y_0$

$$y = y_0 + v_{y0}t - \tfrac{1}{2}gt^2,$$

and obtain

$$-1.00\text{ m} = 0 + (12.0\text{ m/s})t - (4.90\text{ m/s}^2)t^2.$$

Rearranging this equation into standard form and using the quadratic formula gives

$$t = \frac{12.0\text{ m/s} \pm \sqrt{(-12.0\text{ m/s})^2 - 4(4.90\text{ m/s}^2)(-1.00\text{ m})}}{2(4.90\text{ m/s}^2)}$$

$$= 2.53\text{ s} \quad \text{or} \quad -0.081\text{ s}.$$

The second solution would correspond to a time previous to the kick, so it doesn't apply here. With $t = 2.53$ s for the time at which the ball touches the ground, the distance the ball traveled is (putting $v_{x0} = 16.0$ m/s, from Example 3–5):

$$x = v_{x0}t = (16.0\text{ m/s})(2.53\text{ s}) = 40.5\text{ m}.$$

Note that our assumption in Example 3–5 that the ball leaves the foot at ground level results in an underestimate of about 1.3 m in the distance traveled.

FIGURE 3–28 Example 3–9: the football leaves the punter's foot at $y = 0$, and reaches the ground where $y = -1.00$ m.

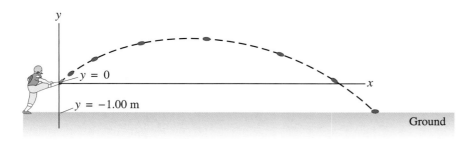

EXAMPLE 3–10 **Rescue plane drops supplies.** A rescue plane wants to drop supplies to isolated mountain climbers on a rocky ridge 200 m below. If the plane is traveling horizontally with a speed of 250 km/h (69 m/s), (*a*) how far in advance of the recipients (horizontal distance) must the goods be dropped (Fig. 3–29a)? (*b*) Suppose, instead, that the plane releases the supplies a horizontal distance of 400 m in advance of the mountain climbers. What vertical velocity should they give the supplies (up or down) so that they arrive precisely at the climbers' position (Fig. 3–29b)? (*c*) With what speed do the supplies land in the latter case?

SOLUTION (*a*) The vertical motion (take +y upwards) doesn't depend on the horizontal motion, so we can find the time to reach the climbers using the vertical distance of 200 m. The supplies are "dropped" so initially they have the velocity of the airplane, $v_{x0} = 69$ m/s, $v_{y0} = 0$. Then, since $y = -\frac{1}{2}gt^2$, we have

$$t = \sqrt{\frac{-2y}{g}} = \sqrt{\frac{-2(-200\text{ m})}{9.80\text{ m/s}^2}} = 6.39\text{ s}.$$

The horizontal motion of the falling supplies is at constant speed of 69 m/s. So

$$x = v_{x0}t = (69\text{ m/s})(6.39\text{ s}) = 440\text{ m}.$$

(*b*) We are given $x = 400$ m, $v_{x0} = 69$ m/s, $y = -200$ m, and we want to find v_{y0} (see Fig. 3–29b). Like most problems, this one can be approached in various ways. Instead of searching for a formula or two, let's try to reason it out in a simple way, based on what we did in part (*a*). If we know t, perhaps we can get v_{y0}. Since the horizontal motion of the supplies is at constant speed (once they are released we don't care what the plane does), we have $x = v_{x0}t$, so

$$t = \frac{x}{v_{x0}} = \frac{400\text{ m}}{69\text{ m/s}} = 5.80\text{ s}.$$

Now let's try to use the vertical motion to get v_{y0}: $y = y_0 + v_{y0}t - \frac{1}{2}gt^2$. Since $y_0 = 0$ and $y = -200$ m, we can solve for v_{y0}:

$$v_{y0} = \frac{y + \frac{1}{2}gt^2}{t} = \frac{-200\text{ m} + \frac{1}{2}(9.80\text{ m/s}^2)(5.80\text{ s})^2}{5.80\text{ s}} = -6.1\text{ m/s}.$$

Thus, in order to arrive at precisely the mountain climbers' position, the supplies must be thrown *downward* from the plane with a speed of 6.1 m/s.

(*c*) We want to know v of the supplies at $t = 5.80$ s. The components are:

$$v_x = v_{x0} = 69\text{ m/s}$$

$$v_y = v_{y0} - gt = -6.1\text{ m/s} - (9.80\text{ m/s}^2)(5.80\text{ s}) = -63\text{ m/s}.$$

So $v = \sqrt{(69\text{ m/s})^2 + (-63\text{ m/s})^2} = 93\text{ m/s}$. (Better not to release the supplies from such an altitude or to use a parachute.)

FIGURE 3–29 Example 3–10.

(a)

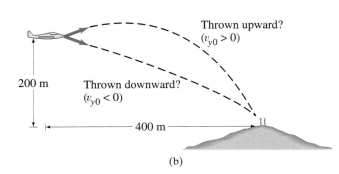

(b)

Projectile Motion Is Parabolic

We now show that the path followed by any projectile is a *parabola*, if we can ignore air resistance and can assume that **g** is constant. To do so, we need to find y as a function of x by eliminating t between the two equations for horizontal and vertical motion (Eq. 2–12b in Table 3–2), and we set $x_0 = y_0 = 0$:

$$x = v_{x0}t$$

$$y = v_{y0}t - \frac{1}{2}gt^2.$$

From the first equation, we have $t = x/v_{x0}$, and we substitute this into the second one to obtain

$$y = \left(\frac{v_{y0}}{v_{x0}}\right)x - \left(\frac{g}{2v_{x0}^2}\right)x^2.$$

We see that y as a function of x has the form

$$y = Ax - Bx^2,$$

Projectile motion equation is a parabola

where A and B are constants for any specific projectile motion. This is the well-known equation for a parabola. See Figs. 3–19 and 3–30.

The idea that projectile motion is parabolic was, in Galileo's day, at the fore-front of physics research. Today we discuss it in Chapter 3 of introductory physics!

FIGURE 3–30 Examples of projectile motion—sparks (small hot glowing pieces of metal), water, and fireworks. All exhibit the parabolic path characteristic of projectile motion, although the effects of air resistance can be seen to significantly alter the path of some trajectories.

3–9 Uniform Circular Motion

An object that moves in a circle at constant speed v is said to undergo **uniform circular motion**. Examples are a ball on the end of a string revolved about one's head, and the nearly uniform circular motion of the Moon around the Earth. The *magnitude* of the velocity remains constant in this case, but the *direction* of the velocity is continually changing (Fig. 3–31). Since acceleration is defined as the rate of change of velocity, a change in direction of velocity means an acceleration is occurring, just as does a change in magnitude. Thus an object undergoing uniform circular motion is accelerating even if the speed remains constant $(v_1 = v_2)$. We now investigate this acceleration quantitatively.

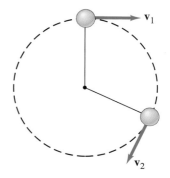

FIGURE 3–31 A particle moving in a circle, showing how the velocity changes direction. Note that at each point, the instantaneous velocity is in a direction tangent to the circular path.

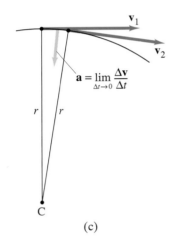

FIGURE 3–32 Determining the change in velocity, $\Delta\mathbf{v}$, for a particle moving in a circle.

Acceleration is defined by

$$\mathbf{a} = \lim_{\Delta t \to 0} \frac{\Delta\mathbf{v}}{\Delta t} = \frac{d\mathbf{v}}{dt},$$

where $\Delta\mathbf{v}$ is the change in velocity during the short time interval Δt. We will eventually consider the situation when Δt approaches zero and thus obtain the instantaneous acceleration. But for purposes of making a clear drawing we consider a nonzero time interval (Fig. 3–32). During the time Δt the particle in Fig. 3–32a moves from point A to point B, covering a small distance Δl *along the arc* which subtends a small angle $\Delta\theta$. The change in the velocity vector is $\mathbf{v}_2 - \mathbf{v}_1 = \Delta\mathbf{v}$, and is shown in Fig. 3–32b.

If we let Δt be very small (approaching zero), then Δl and $\Delta\theta$ are also very small; then $\mathbf{v}_2$ will be almost parallel to $\mathbf{v}_1$, and $\Delta\mathbf{v}$ will be essentially perpendicular to them (Fig. 3–32c). Thus $\Delta\mathbf{v}$ points toward the center of the circle. Since $\mathbf{a}$, by definition, is in the same direction as $\Delta\mathbf{v}$, it too must point toward the center of the circle. Therefore, this acceleration is called **centripetal acceleration** ("center-seeking" acceleration) or **radial acceleration** (since it is directed along the radius, toward the center of the circle), and we denote it by $\mathbf{a}_R$.

We next determine the magnitude of the centripetal (radial) acceleration, a_R. Because CA is perpendicular to $\mathbf{v}_1$, and CB is perpendicular to $\mathbf{v}_2$, it follows that the angle $\Delta\theta$, defined as the angle between CA and CB in Fig. 3–32a, is also the angle between $\mathbf{v}_1$ and $\mathbf{v}_2$. Hence the vectors $\mathbf{v}_2$, $\mathbf{v}_1$, and $\Delta\mathbf{v}$ in Fig. 3–32b form a triangle that is geometrically similar to triangle CAB in Fig. 3–32a. Taking $\Delta\theta$ small (letting Δt be very small) and setting $v = v_1 = v_2$ because the magnitude of the velocity is assumed not to change, we can write

$$\frac{\Delta v}{v} \approx \frac{\Delta l}{r}.$$

This is an exact equality when Δt approaches zero, for then the arc length Δl equals the cord length AB. Since we want to find the instantaneous acceleration, for which Δt approaches zero, we write the above expression as an equality and solve for Δv:

$$\Delta v = \frac{v}{r} \Delta l.$$

To get the centripetal acceleration, a_R, we divide Δv by Δt:

$$a_R = \lim_{\Delta t \to 0} \frac{\Delta v}{\Delta t} = \lim_{\Delta t \to 0} \frac{v}{r} \frac{\Delta l}{\Delta t}.$$

And since

$$\lim_{\Delta t \to 0} \frac{\Delta l}{\Delta t}$$

is the speed, v, of the object, we get

Centripetal acceleration

$$a_R = \frac{v^2}{r}. \tag{3–14}$$

To summarize, *an object moving in a circle of radius r with constant speed v has an acceleration whose direction is toward the center of the circle and whose magnitude is $a_R = v^2/r$.* It is not surprising that this acceleration depends on v and r. For the greater the speed v, the faster the velocity changes direction; and the larger the radius, the less rapidly the velocity changes direction.

The acceleration vector points toward the center of the circle. But the velocity vector always points in the direction of motion, which is tangential to the circle. Thus the velocity and acceleration vectors are perpendicular to each other at every point in the path for uniform circular motion (see Fig. 3–33). This is another example that illustrates the error in thinking that acceleration and velocity are always in the same direction. For an object falling vertically, **a** and **v** are indeed parallel. But in circular motion, **a** and **v** are not parallel—nor are they in projectile motion (Section 3–7), where the acceleration **a** = **g** is always downward but the velocity vector can have various directions (Figs. 3–20 and 3–22).

Acceleration and velocity are not in same direction

Circular motion is often described in terms of the **frequency** f as so many revolutions per second. The **period** T of an object revolving in a circle is the time required for one complete revolution. Period and frequency are related by

$$T = \frac{1}{f}. \tag{3–15}$$

For example, if an object revolves at a frequency of 3 rev/s, then each revolution takes $\frac{1}{3}$ s. For an object revolving in a circle at constant speed v, we can write

$$v = \frac{2\pi r}{T}$$

since in one revolution the object travels one circumference ($= 2\pi r$).

FIGURE 3–33 For uniform circular motion, **a** is always perpendicular to **v**.

EXAMPLE 3–11 **Acceleration of a revolving ball.** A 150-g ball at the end of a string is revolving uniformly in a horizontal circle of radius 0.600 m. The ball makes 2.00 revolutions in a second. What is its centripetal acceleration?

SOLUTION The centripetal acceleration is $a_R = v^2/r$. First, we determine the speed of the ball v. The ball makes two complete revolutions per second, so its period is $T = 0.500$ s. In this time it travels one circumference of the circle, $2\pi r$, so the ball has speed

$$v = \frac{2\pi r}{T} = \frac{2(3.14)(0.600 \text{ m})}{(0.500 \text{ s})} = 7.54 \text{ m/s}.$$

The centripetal acceleration is

$$a_R = \frac{v^2}{r} = \frac{(7.54 \text{ m/s})^2}{(0.600 \text{ m})} = 94.8 \text{ m/s}^2.$$

EXAMPLE 3–12 **Moon's centripetal acceleration.** The Moon's nearly circular orbit about the Earth has a radius of about 384,000 km and a period T of 27.3 days. Determine the acceleration of the Moon toward the Earth.

The Moon's acceleration toward Earth

SOLUTION In orbit around the Earth, the Moon travels a distance $2\pi r$, where $r = 3.84 \times 10^8$ m is the radius of its circular path. The speed of the Moon in its orbit about the Earth is $v = 2\pi r/T$. The period T in seconds is $T = (27.3 \text{ d})(24.0 \text{ h/d})(3600 \text{ s/h}) = 2.36 \times 10^6$ s. Therefore,

$$a_R = \frac{v^2}{r} = \frac{(2\pi r)^2}{T^2 r} = \frac{[2(3.14)(3.84 \times 10^8 \text{ m})]^2}{(2.36 \times 10^6 \text{ s})^2(3.84 \times 10^8 \text{ m})}$$

$$= 0.00272 \text{ m/s}^2 = 2.72 \times 10^{-3} \text{ m/s}^2.$$

We can write this in terms of $g = 9.80$ m/s^2 (the acceleration of gravity at the Earth's surface) as

$$a = 2.72 \times 10^{-3} \text{ m/s}^2 \left(\frac{g}{9.80 \text{ m/s}^2}\right)$$

$$= 2.78 \times 10^{-4} g.$$

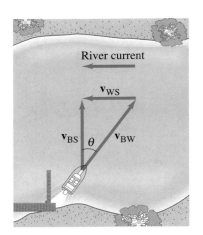

FIGURE 3–34 The boat must head upstream at an angle θ if it is to move directly across the river. Velocity vectors are shown as green arrows:

$\mathbf{v}_{BS}$ = velocity of **B**oat with respect to the **S**hore,

$\mathbf{v}_{BW}$ = velocity of **B**oat with respect to the **W**ater,

$\mathbf{v}_{WS}$ = velocity of the **W**ater with respect to the **S**hore (river current).

➡ PROBLEM SOLVING

Subscripts for adding velocities: first subscript for the object; second for the reference frame

3–10 | Relative Velocity

We now consider how observations made in different frames of reference are related to each other. For example, consider two trains approaching one another, each with a speed of 80 km/h with respect to the Earth. Observers on the Earth beside the train tracks will measure 80 km/hr for the speed of each of the trains. Observers on either one of the trains (a different frame of reference) will measure a speed of 160 km/h for the train approaching them. Similarly, when one car traveling 90 km/h passes a second car traveling in the same direction at 75 km/h, the first car has a speed relative to the second car of 90 km/h − 75 km/h = 15 km/h.

When the velocities are along the same line, simple addition or subtraction is sufficient to obtain the relative velocity. But if they are not along the same line, we must make use of vector addition. We emphasize, as mentioned in Section 2–1, that when specifying a velocity, it is important to specify what the reference frame is.

When determining relative velocity, it is easy to make a mistake by adding or subtracting the wrong velocities. It is useful, therefore, to use a careful labeling process that makes things clear. Each velocity is labeled by *two subscripts: the first refers to the object, the second to the reference frame in which it has this velocity*. For example, suppose a boat is to cross a river to the opposite side, as shown in Fig. 3–34. We let $\mathbf{v}_{BW}$ be the velocity of the **B**oat with respect to the **W**ater. (This is also what the boat's velocity would be relative to the shore if the water were still.) Similarly, $\mathbf{v}_{BS}$ is the velocity of the **B**oat with respect to the **S**hore, and $\mathbf{v}_{WS}$ is the velocity of the **W**ater with respect to the **S**hore (this is the river current). Note that $\mathbf{v}_{BW}$ is what the boat's motor produces (against the water), whereas $\mathbf{v}_{BS}$ is equal to $\mathbf{v}_{BW}$ plus the effect of the current. Therefore, the velocity of the boat relative to the shore is (see vector diagram, Fig. 3–34)

$$\mathbf{v}_{BS} = \mathbf{v}_{BW} + \mathbf{v}_{WS}. \tag{3–16}$$

By writing the subscripts via the convention above, we see that the inner subscripts (the two W's) on the right-hand side of Eq. 3–16 are the same, whereas the outer subscripts on the right of Eq. 3–16 (the B and the S) are the same as the two subscripts for the sum vector on the left, $\mathbf{v}_{BS}$. By following this convention (first subscript for the object, second for the reference frame), one can write down the correct equation relating velocities in different reference frames.† Figure 3–35 gives a derivation of Eq. 3–16, which is valid in general and can be extended to three or more velocities. For example, if a fisherman on a boat walks with a velocity $\mathbf{v}_{FB}$ relative to the boat, his velocity relative to the shore is

† We thus would know by inspection that (for example) the equation $\mathbf{v}_{BW} = \mathbf{v}_{BS} + \mathbf{v}_{WS}$ is wrong.

FIGURE 3–35 Derivation of relative velocity equation (Eq. 3–16), in this case for a person walking along the corridor in a train. We are looking down on the train and two reference frames are shown: xy on the Earth and $x'y'$ fixed on the train. We have

$\mathbf{r}_{PT}$ = position vector of person (P) relative to train (T)

$\mathbf{r}_{PE}$ = position vector of person (P) relative to Earth (E)

$\mathbf{r}_{TE}$ = position vector of train's coordinate system (T) relative to Earth (E)

From the diagram we see that

$$\mathbf{r}_{PE} = \mathbf{r}_{PT} + \mathbf{r}_{TE}.$$

We take the derivative with respect to time to obtain

$$\frac{d}{dt}(\mathbf{r}_{PE}) = \frac{d}{dt}(\mathbf{r}_{PT}) + \frac{d}{dt}(\mathbf{r}_{TE})$$

or, since $d\mathbf{r}/dt = \mathbf{v}$,

$$\mathbf{v}_{PE} = \mathbf{v}_{PT} + \mathbf{v}_{TE}.$$

This is the equivalent of Eq. 3–16 for the present situation (check the subscripts!).

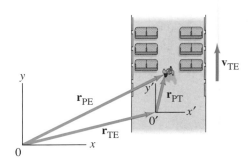

$\mathbf{v}_{FS} = \mathbf{v}_{FB} + \mathbf{v}_{BW} + \mathbf{v}_{WS}$. The equations involving relative velocity will be correct when adjacent inner subscripts are identical and when the outermost ones correspond exactly to the two on the velocity on the left of the equation. But this works only with plus signs (on the right), not minus signs.

It is often useful to remember that for any two objects or reference frames, A and B, the velocity of A relative to B has the same magnitude, but opposite direction, as the velocity of B relative to A:

$$\mathbf{v}_{BA} = -\mathbf{v}_{AB}. \tag{3-17}$$

For example, if a train is traveling 100 km/h relative to the Earth in a certain direction, objects on the Earth (such as trees) appear to an observer on the train to be traveling 100 km/h in the opposite direction.

EXAMPLE 3–13 **Heading upstream.** A boat's speed in still water is $v_{BW} = 1.85$ m/s. If the boat is to travel northward directly across a river whose current has velocity $v_{WS} = 1.20$ m/s toward the west, at what angle must the boat head?

SOLUTION The current will drag the boat westward, so to counteract that motion the boat must head upstream in a northeasterly direction, as shown in Fig. 3–36. Figure 3–36 has been drawn with $\mathbf{v}_{BS}$, the velocity of the **B**oat relative to the **S**hore, pointing directly across the river since this is how the boat is supposed to move. (Note that $\mathbf{v}_{BS} = \mathbf{v}_{BW} + \mathbf{v}_{WS}$.) Thus, $\mathbf{v}_{BW}$ points upstream at an angle θ as shown, where

$$\sin\theta = \frac{v_{WS}}{v_{BW}} = \frac{1.20 \text{ m/s}}{1.85 \text{ m/s}} = 0.6486.$$

Thus $\theta = 40.4°$, so the boat must head upstream at a 40.4° angle.

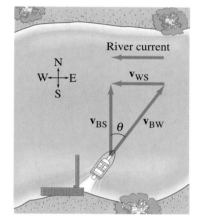

FIGURE 3–36 Example 3–13.

EXAMPLE 3–14 **Heading across the stream.** The same boat $(v_{BW} = 1.85$ m/s$)$ now heads directly across the stream whose current is still 1.20 m/s. (*a*) What is the velocity (magnitude and direction) of the boat relative to the shore? (*b*) If the river is 110 m wide, how long will it take to cross and how far downstream will the boat be then?

SOLUTION (*a*) As shown in Fig. 3–37, the boat is pulled downstream by the current. The boat's velocity with respect to the shore, $\mathbf{v}_{BS}$, is the sum of its velocity with respect to the water, $\mathbf{v}_{BW}$, plus the velocity of the water with respect to the shore, $\mathbf{v}_{WS}$:

$$\mathbf{v}_{BS} = \mathbf{v}_{BW} + \mathbf{v}_{WS},$$

just as before. Since $\mathbf{v}_{BW}$ is directed straight across the stream, it is perpendicular to $\mathbf{v}_{WS}$, and we can get v_{BS} using the theorem of Pythagoras:

$$v_{BS} = \sqrt{v_{BW}^2 + v_{WS}^2} = \sqrt{(1.85 \text{ m/s})^2 + (1.20 \text{ m/s})^2} = 2.21 \text{ m/s}.$$

We can obtain the angle (note how θ is defined in diagram) from:

$$\tan\theta = v_{WS}/v_{BW} = (1.20 \text{ m/s})/(1.85 \text{ m/s}) = 0.6486.$$

A calculator with an INV TAN or TAN^{-1} key gives $\theta = \tan^{-1}(0.6486) = 33.0°$. Note that this angle is *not* equal to the angle calculated in Example 3–13.
(*b*) Given the river's width $D = 110$ m and using the definition of velocity, we solve for $t = D/v_{BW}$, where we use the velocity component in the direction of D; so $t = (110 \text{ m})/(1.85 \text{ m/s}) = 60$ s. The boat will have been carried downstream, in this time, a distance

$$d = v_{WS}t = (1.20 \text{ m/s})(60 \text{ s}) = 72 \text{ m}.$$

FIGURE 3–37 Example 3–14: a boat heading directly across a river whose current moves at 1.20 m/s.

EXAMPLE 3–15 **Car velocities at 90°.** Two automobiles approach a corner at right angles to each other with the same speed of 40.0 km/h (= 11.1m/s), as shown in Fig. 3–38a. What is the relative velocity of one car with respect to the other? That is, determine the velocity of car 1 as seen by car 2.

SOLUTION Figure 3–38a shows the situation in a reference frame fixed to the Earth. But we want to view the situation from a reference frame in which car 2 is at rest, and this is shown in Fig. 3–38b. In this reference frame (the world as seen by the driver of car 2), the Earth moves toward car 2 with velocity $\mathbf{v}_{E2}$ (speed of 40.0 km/h), which is of course equal and opposite to $\mathbf{v}_{2E}$, the velocity of car 2 with respect to the Earth (Eq. 3–17):

$$\mathbf{v}_{2E} = -\mathbf{v}_{E2}.$$

Then the velocity of car 1 as seen by car 2 is

$$\mathbf{v}_{12} = \mathbf{v}_{1E} + \mathbf{v}_{E2}$$

or (since $\mathbf{v}_{E2} = -\mathbf{v}_{2E}$)

$$\mathbf{v}_{12} = \mathbf{v}_{1E} - \mathbf{v}_{2E}.$$

That is, the velocity of car 1 as seen by car 2 is the difference of their velocities, $\mathbf{v}_{1E} - \mathbf{v}_{2E}$, both measured relative to the Earth (see Fig. 3–38c). Since the magnitudes of $\mathbf{v}_{1E}$, $\mathbf{v}_{2E}$, and $\mathbf{v}_{E2}$ are equal (40 km/h = 11.1 m/s) we see (Fig. 3–38b) that $\mathbf{v}_{12}$ points at a 45° angle toward car 2; the speed is

$$v_{12} = \sqrt{(11.1 \text{ m/s})^2 + (11.1 \text{ m/s})^2} = 15.7 \text{ m/s} \, (= 56.5 \text{ km/h}).$$

FIGURE 3–38 Example 3–15.

(a)   (b)   (c)

Summary

A quantity that has both a magnitude and a direction is called a **vector**. A quantity that has only a magnitude is called a **scalar**.

Addition of vectors can be done graphically by placing the tail of each successive arrow (representing each vector) at the tip of the previous one. The sum, or **resultant vector**, is the arrow drawn from the tail of the first to the tip of the last. Two vectors can also be added using the parallelogram method. Vectors can be added more accurately using the analytical method of adding their **components** along chosen axes with the aid of trigonometric functions. A vector of magnitude V making an angle θ with the x axis has components

$$V_x = V \cos\theta \qquad V_y = V \sin\theta.$$

Given the components, we can find the magnitude and direction from

$$V = \sqrt{V_x^2 + V_y^2}, \qquad \tan\theta = \frac{V_y}{V_x}.$$

It is often helpful to express a vector in terms of its components along chosen axes using **unit vectors**, which are vectors of unit length along the chosen coordinate axes; for Cartesian coordinates the unit vectors along the x, y, and z axes are called $\mathbf{i}, \mathbf{j}$, and $\mathbf{k}$.

The general definitions for the instantaneous **velocity**, $\mathbf{v}$, and **acceleration**, $\mathbf{a}$, of a particle (in one, two, or three dimensions) are

$$\mathbf{v} = \frac{d\mathbf{r}}{dt} \quad \text{and} \quad \mathbf{a} = \frac{d\mathbf{v}}{dt}.$$

where **r** is the position vector of the particle. The kinematic equations for motion with constant acceleration can be written for each of the x, y, and z components of the motion and have the same form as for one-dimensional motion (Eqs. 2–12). Or they can be written in the more general vector form:

$$\mathbf{v} = \mathbf{v}_0 + \mathbf{a}t$$

$$\mathbf{r} = \mathbf{r}_0 + \mathbf{v}_0 t + \tfrac{1}{2}\mathbf{a}t^2.$$

Projectile motion of an object moving in the air near the Earth's surface can be analyzed as two separate motions if air resistance can be ignored. The horizontal component of the motion is at constant velocity, whereas the vertical component is at constant acceleration, g, just as for a body falling vertically under the action of gravity.

An object moving in a circle of radius r with constant speed v is said to be in **uniform circular motion** and has a **radial** or **centripetal acceleration** $\mathbf{a}_R$ toward the center of the circle of magnitude:

$$a_R = \frac{v^2}{r}.$$

The frequency f is the number of complete revolutions per second. The period T is the time required for one complete revolution, and is related to the frequency by

$$T = \frac{1}{f}.$$

The velocity of an object relative to one frame of reference can be found by vector addition if its velocity relative to a second frame of reference, and the **relative velocity** of the two reference frames, are known.

Questions

1. One car travels due east at 40 km/h, and a second car travels north at 40 km/h. Are their velocities equal? Explain.

2. Can you conclude that a car is not accelerating if its speedometer indicates a steady 60 km/h?

3. Can you give several examples of an object's motion in which a great distance is traveled but the displacement is zero?

4. Can the displacement vector for a particle moving in two dimensions ever be longer than the length of path traveled by the particle over the same time interval? Can it ever be less? Discuss.

5. During baseball practice, a batter hits a very high fly ball and then runs in a straight line and catches it. Which had the greater displacement, the player or the ball?

6. If $\mathbf{V} = \mathbf{V}_1 + \mathbf{V}_2$, is V necessarily greater than V_1 and/or V_2? Discuss.

7. Two vectors have length $V_1 = 3.5$ km and $V_2 = 4.0$ km. What are the maximum and minimum magnitudes of their vector sum?

8. Can two vectors, of unequal magnitude, add up to give the zero vector? Can *three* unequal vectors? Under what conditions?

9. Can the magnitude of a vector ever (*a*) equal or (*b*) be less than one of its components?

10. Can a particle with constant speed be accelerating? What if it has constant velocity?

11. Can a vector of magnitude zero have a nonzero component?

12. Does the odometer of a car measure a scalar or a vector quantity? What about the speedometer?

13. A child wishes to determine the speed a slingshot imparts to a rock. How can this be done using only a meter stick, a rock, and the slingshot?

14. Is it ever necessary, in projectile motion, to consider the motion in three dimensions if air resistance can be neglected? What if air resistance can't be neglected? Discuss.

15. What physical factors are important for an athlete doing the long jump? What about the high jump?

16. A projectile has the least speed at what point in its path?

17. It was reported in the First World War that a French pilot flying at an altitude of 2 km caught in his bare hands a bullet fired at the plane! Using the fact that a bullet slows down considerably due to air resistance, explain how this incident occurred.

18. A car rounds a curve at a steady 50 km/h. If it rounds the same curve at a steady 70 km/h, will its acceleration be any different? Explain.

19. Will the acceleration of a car be the same if it travels around a sharp curve at 60 km/h as when it travels around a gentle curve at the same speed? Explain.

20. At some amusement parks, to get on a moving "car" the riders first hop on to a moving walkway and then on to the cars themselves. Why is this done?

21. If you are riding on a train that speeds past another train moving in the same direction on an adjacent track, it appears that the other train is moving backwards. Why?

22. If you stand motionless under an umbrella in a rainstorm where the drops fall vertically you remain relatively dry. However, if you start running, the rain begins to hit your legs even if they remain under the umbrella. Why?

23. Two cars with equal speed approach an intersection at right angles to each other. Will they necessarily collide? Show that when the relative velocity of approach is collinear (along the same line) with the relative displacement, we have the nautical maxim "constant bearing means collision."

24. A person sitting in an enclosed train car, moving at constant velocity, throws a ball straight up into the air in her reference frame. (*a*) Where does the ball land? What is your answer if the car (*b*) accelerates, (*c*) decelerates, (*d*) rounds a curve, (*e*) moves with constant velocity but is open to the air?

25. Two rowers, who can row at the same speed in still water, set off across a river at the same time. One heads straight across and is pulled downstream somewhat by the current. The other one heads upstream at an angle so as to arrive at a point opposite the starting point. Which rower reaches the opposite side first?

Problems

Sections 3–1 to 3–5

1. (I) A car is driven 200 km west and then 80 km southwest. What is the displacement of the car from the point of origin (magnitude and direction)? Draw a diagram.

2. (I) A delivery truck travels 18 blocks north, 10 blocks east, and 16 blocks south. What is its final displacement from the origin? Assume the blocks are equal length.

3. (I) Show that the vector labeled "wrong" in Fig. 3–6c is actually the difference of the two vectors. Is it $V_2 - V_1$, or $V_1 - V_2$?

4. (I) If $V_x = 8.80$ units and $V_y = -6.40$ units, determine the magnitude and direction of **V**.

5. (I) Graphically determine the resultant of the following three vector displacements: (1) 14 m, 30° north of east; (2) 18 m, 37° east of north; and (3) 20 m, 30° west of south.

6. (II) Vector V_1 is 6.0 units long and points along the negative x axis. Vector V_2 is 4.5 units long and points at +45° to the positive x axis. (a) What are the x and y components of each vector? (b) Determine the sum $V_1 + V_2$ (magnitude and angle).

7. (II) **V** is a vector 14.3 units in magnitude and points at an angle of 34.8° above the negative x axis. (a) Sketch this vector. (b) Find V_x and V_y. (c) Use V_x and V_y to obtain (again) the magnitude and direction of **V**. [*Note*: Part (c) is a good way to check if you've resolved your vector correctly.]

8. (II) Figure 3–39 shows two vectors, **A** and **B**, whose magnitudes are $A = 6.8$ units and $B = 5.5$ units. Determine **C** if (a) $C = A + B$, (b) $C = A - B$, (c) $C = B - A$. Give the magnitude and direction for each.

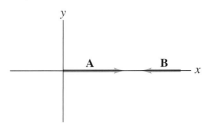

FIGURE 3–39
Problem 8.

9. (II) An airplane is traveling 635 km/h in a direction 41.5° west of north (Fig. 3–40). (a) Find the components of the velocity vector in the northerly and westerly directions. (b) How far north and how far west has the plane traveled after 3.00 h?

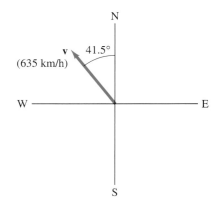

FIGURE 3–40
Problem 9.

10. (II) Let $V_1 = -6.0\mathbf{i} + 8.0\mathbf{j}$ and $V_2 = 4.5\mathbf{i} - 5.0\mathbf{j}$. Determine the magnitude and direction of (a) V_1, (b) V_2, (c) $V_1 + V_2$ and (d) $V_2 - V_1$.

11. (II) (a) Determine the magnitude and direction of the sum of the three vectors $V_1 = 4\mathbf{i} - 8\mathbf{j}$, $V_2 = \mathbf{i} + \mathbf{j}$, and $V_3 = -2\mathbf{i} + 4\mathbf{j}$. (b) Determine $V_1 - V_2 + V_3$.

12. (II) Three vectors are shown in Fig. 3–41. Their magnitudes are given in arbitrary units. Determine the sum of the three vectors. Give the resultant in terms of (a) components, (b) magnitude and angle with x axis.

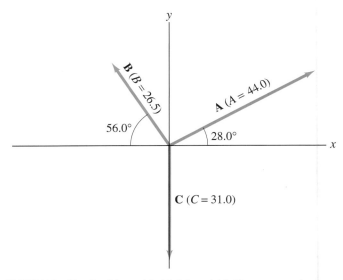

FIGURE 3–41 Problems 12, 13, 14, and 15. Vector magnitudes are given in arbitrary units.

13. (II) (a) Given the vectors **A** and **B** shown in Fig. 3–41, determine $B - A$. (b) Determine $A - B$ without using your answer in (a). Then compare your results and see if they are opposite.

14. (II) For the vectors given in Fig. 3–41, determine (a) $A - B + C$, (b) $A + B - C$, and (c) $C - A - B$.

15. (II) For the vectors shown in Fig. 3–41, determine (a) $B - 2A$, (b) $2A - 3B + 2C$.

16. (II) The summit of a mountain, 2450 m above base camp, is measured on a map to be 4580 m horizontally from the camp in a direction 32.4° west of north. What are the components of the displacement vector from camp to summit? What is its magnitude? Choose the x axis east, y axis north, and z axis up.

17. (III) You are given a vector in the xy plane that has a magnitude of 90.0 units and a y component of -35.0 units. (a) What are the two possibilities for its x component? (b) Assuming the x component is known to be positive, specify the vector which, if you add it to the original one would give a resultant vector that is 80.0 units long and points entirely in the $-x$ direction.

70 CHAPTER 3 Kinematics In Two Dimensions; Vectors

18. (I) The position of a particular particle as a function of time is given by $\mathbf{r} = (7.60t\mathbf{i} + 8.85\mathbf{j} - t^2\mathbf{k})$ m. Determine the particle's velocity and acceleration as a function of time.

19. (I) What was the average velocity of the particle in Problem 18 between $t = 1.00$ s and $t = 3.00$ s? What is the magnitude of the instantaneous velocity at $t = 2.00$ s?

20. (II) What is the shape of the path of the particle of Problem 18?

21. (II) A car is moving with speed 18.0 m/s due south at one moment and 27.5 m/s due east 8.00 s later. Over this time interval, determine (*a*) its average velocity, (*b*) its average acceleration (magnitude and direction for both), and (*c*) its average speed. [*Hint*: Can you determine all these from the information given?]

22. (II) (*a*) A skier is accelerating down a 30.0° hill at 3.80 m/s² (Fig. 3–42). What is the vertical component of her acceleration? (*b*) How long will it take her to reach the bottom of the hill, assuming she starts from rest and accelerates uniformly, if the elevation change is 250 m?

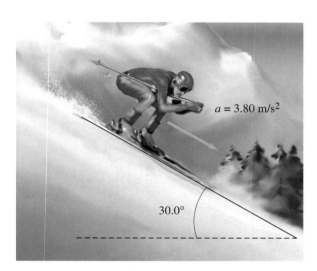

FIGURE 3–42 Problem 22.

23. (II) At $t = 0$, a particle starts from rest and moves in the xy plane with an acceleration $\mathbf{a} = (4.0\mathbf{i} + 3.0\mathbf{j})$ m/s². Determine (*a*) the x and y components of velocity. (*b*) the speed of the particle, and (*c*) the position of the particle, all as a function of time. (*d*) Evaluate all the above at $t = 2.0$ s.

24. (II) A particle starts from the origin at $t = 0$ with an initial velocity of 5.0 m/s along the positive x axis. If the acceleration is $(-3.0\mathbf{i} + 4.5\mathbf{j})$ m/s², determine the velocity and position of the particle at the moment it reaches its maximum x coordinate.

25. (III) The position of a particle is given by $\mathbf{r} = (6.0\cos 3.0t\mathbf{i} + 6.0\sin 3.0t\mathbf{j})$ meters. Determine (*a*) the velocity vector, **v**, and (*b*) the acceleration vector, **a**. (*c*) What is the path of this particle? [*Hint*: Determine $r = |\mathbf{r}|$.] (*d*) What is the relation between r and a (give a formula), and between **r** and **a** (give an angle)? (*e*) Show that $a = v^2/r$.

26. (I) A tiger leaps horizontally from a 6.5-m-high rock with a speed of 4.0 m/s. How far from the base of the rock will she land?

27. (I) A diver running 2.1 m/s dives out horizontally from the edge of a vertical cliff and reaches the water below 3.0 s later. How high was the cliff and how far from its base did the diver hit the water?

28. (I) Determine how much farther a person can jump on the Moon as compared to the Earth if the takeoff speed and angle are the same. The acceleration due to gravity on the Moon is one-sixth what it is on Earth.

29. (I) A fire hose held near the ground shoots water at a speed of 5.5 m/s. At what angle(s) should the nozzle point in order that the water land 3.0 m away (Fig. 3–43)? Why are there two different angles? Sketch the two trajectories.

FIGURE 3–43 Problem 29.

30. (II) A ball is thrown horizontally from the roof of a building 9.0 m tall and lands 8.5 m from the base. What was the ball's initial speed?

31. (II) A football is kicked at ground level with a speed of 18.0 m/s at an angle of 32.0° to the horizontal. How much later does it hit the ground?

32. (II) A ball thrown horizontally at 22.2 m/s from the roof of a building lands 36.0 m from the base of the building. How high is the building?

33. (II) A shotputter throws the shot (mass = 7.3 kg) with an initial speed of 14 m/s at a 40° angle to the horizontal. Calculate the horizontal distance traveled by the shot if it leaves the athlete's hand at a height of 2.2 m above the ground.

34. (II) Show that the time required for a projectile to reach its highest point is equal to the time for it to return to its original height.

35. (II) A projectile is fired with an initial speed of 30.0 m/s. Plot on graph paper its trajectory for initial projection angles of $\theta = 15°, 30°, 45°, 60°, 75°,$ and 90°. Plot at least 10 points for each curve.

36. (II) William Tell must split the apple atop his son's head from a distance of 25.0 m. When he aims directly at the apple, the arrow is horizontal. At what angle must he aim it to hit the apple if the arrow travels at a speed of 22.5 m/s?

37. (II) The pilot of an airplane traveling 160 km/h wants to drop supplies to flood victims isolated on a patch of land 160 m below. The supplies should be dropped how many seconds before the plane is directly overhead?

38. (II) An athlete executing a long jump leaves the ground at a 33.0° angle and travels 7.80 m. (a) What was the takeoff speed? (b) If this speed were increased by just 5.0 percent, how much longer would the jump be?

39. (II) A projectile is fired with an initial speed of 51.2 m/s at an angle of 44.5° above the horizontal on a long flat firing range. Determine (a) the maximum height reached by the projectile, (b) the total time in the air, (c) the total horizontal distance covered (that is, the range), and (d) the velocity of the projectile 1.50 s after firing.

40. (II) A projectile is shot from the edge of a cliff 125 m above ground level with an initial speed of 65.0 m/s at an angle of 37.0° with the horizontal, as shown in Fig. 3–44. (a) Determine the time taken by the projectile to hit point P at ground level. (b) Determine the range X of the projectile as measured from the base of the cliff. At the instant just before the projectile hits point P, find (c) the horizontal and the vertical components of its velocity, (d) the magnitude of the velocity, and (e) the angle made by the velocity vector with the horizontal.

FIGURE 3–44   Problem 40.

41. (II) Revisit Conceptual Example 3–7, and assume that the boy with the slingshot is *below* the boy in the tree (Fig. 3–45) and so aims *upward*, directly at the boy in the tree. Show that again the boy in the tree makes the wrong move by letting go at the moment the water balloon is shot.

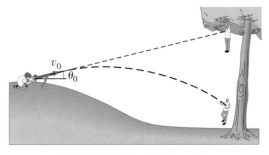

FIGURE 3–45   Problem 41.

42. (II) At what projection angle will the range of a projectile equal its maximum height?

43. (II) Suppose the kick in Example 3–5 is attempted 36.0 m from the goalposts, whose crossbar is 3.00 m above the ground. If the football is directed perfectly between the goalposts, will it pass over the bar and be a field goal? Show why or why not. If not, from what horizontal distance must this kick be made if it is to score?

44. (II) Exactly 3.0 s after a projectile is fired into the air from the ground, it is observed to have a velocity $\mathbf{v} = (7.6\mathbf{i} + 4.8\mathbf{j})$ m/s, where the x axis is horizontal and the y axis is positive upward. Determine (a) the horizontal range of the projectile, (b) its maximum height above the ground, and (c) its speed and angle of motion just before it strikes the ground.

45. (II) A high diver leaves the end of a 5.0-m-high diving board and strikes the water 1.3 s later, 3.0 m beyond the end of the board. Considering the diver as a particle, determine (a) her initial velocity, $\mathbf{v}_0$; (b) the maximum height reached; and (c) the velocity $\mathbf{v}_f$ with which she enters the water.

46. (II) A stunt driver wants to make his car jump over 8 cars parked side by side below a horizontal ramp (Fig. 3–46). (a) With what minimum speed must he drive off the horizontal ramp? The vertical height of the ramp is 1.5 m above the cars and the horizontal distance he must clear is 20 m. (b) What is the new minimum speed if the the ramp is now tilted upward, so that "takeoff angle" is 10° above the horizontal, and nothing else is changed?

FIGURE 3–46   Problem 46.

47. (III) A ball is thrown horizontally from the top of a cliff with initial speed v_0 (at $t = 0$). At any moment, its direction of motion makes an angle θ to the horizontal (Fig. 3–47). Derive a formula for θ as a function of time, t, as the ball follows a projectile's path.

FIGURE 3–47   Problem 47.

48. (III) Derive a formula for the horizontal range R, of a projectile when it lands at a height h above its initial point. (For $h < 0$, it lands a distance $-h$ below the starting point.) Assume it is projected at an angle θ_0 with initial speed v_0.

49. (III) A person stands at the base of a hill that is a straight incline making an angle ϕ with the horizontal (Fig. 3–48). For a given initial speed v_0, at what angle θ (to the horizontal) should objects be thrown so that the distance d they land up the hill is as large as possible?

FIGURE 3–48 Problem 49. Given ϕ and v_0, determine θ to make d maximum.

50. (III) At $t = 0$ a batter hits a baseball with an initial speed of 32 m/s at a 55° angle to the horizontal. An outfielder is 85 m from the batter at $t = 0$, and, as seen from home plate, the line of sight to the outfielder makes a horizontal angle of 22° with the plane in which the ball moves (see Fig. 3–49). What speed and direction must the fielder take in order to catch the ball at the same height from which it was struck? Give angle with respect to the outfielder's line of sight to home plate.

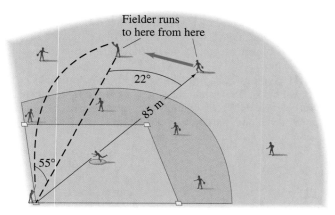

FIGURE 3–49 Problem 50.

Section 3–9

51. (I) A jet plane traveling 1800 km/h (500 m/s) pulls out of a dive by moving in an arc of radius 3.50 km. What is the plane's acceleration in "g's"?

52. (I) What is the centripetal acceleration of a child 3.6 m from the center of a merry-go-round? The child's speed is 0.85 m/s.

53. (I) Calculate the centripetal acceleration of the Earth in its orbit around the Sun. Assume the Earth's orbit is a circle of radius 1.5×10^{11} m.

54. (II) What is the magnitude of the acceleration of a speck of clay on the edge of a potter's wheel turning at 45 rpm (revolutions per minute) if its diameter is 30 cm?

55. (II) Suppose the space shuttle is in orbit 400 km from the Earth's surface, and circles the Earth about once every 90 minutes. Find the centripetal acceleration of the space shuttle in its orbit. Express your answer in terms of g, the gravitational acceleration at the Earth's surface.

56. (II) Because the Earth rotates once per day, the effective acceleration of gravity at the equator is slightly less than it would be if the Earth didn't rotate. Estimate the magnitude of this effect. What fraction of g is this?

57. (II) Use dimensional analysis (Section 1–7) to obtain the form for the centripetal acceleration, $a_R = v^2/r$.

58. (III) The position of a particle moving in the xy plane is given by $\mathbf{r} = \mathbf{i}2.0 \cos 3.0t + \mathbf{j}2.0 \sin 3.0t$, where r is in meters and t is in seconds. (a) Show that this represents circular motion of radius 2.0 m centered at the origin. (b) Determine the velocity and acceleration vectors as functions of time. (c) Determine the speed and magnitude of the acceleration. (d) Show that $a = v^2/r$. (e) Show that the acceleration vector always points toward the center of the circle.

Section 3–10

59. (I) Huck Finn walks at a speed of 1.0 m/s across his raft (that is, he walks perpendicular to the raft's motion relative to the shore). The raft is traveling down the Mississippi River at a speed of 2.5 m/s relative to the river bank (Fig. 3–50). What is the velocity (speed and direction) of Huck relative to the river bank?

FIGURE 3–50 Problem 59.

60. (II) A boat can travel 2.20 m/s in still water. (a) If the boat points its prow directly across a stream whose current is 1.20 m/s, what is the velocity (magnitude and direction) of the boat relative to the shore? (b) What will be the position of the boat, relative to its point of origin, after 3.00 s? (See Fig. 3–37.)

61. (II) Two planes approach each other head-on. Each has a speed of 780 km/h, and they spot each other when they are initially 10.0 km apart. How much time do the pilots have to take evasive action?

62. (II) An airplane is heading due south at a speed of 550 km/h. If a wind begins blowing from the southwest at a speed of 90.0 km/h (average), calculate: (*a*) the velocity (magnitude and direction) of the plane, relative to the ground, and (*b*) how far off course it will be after 12.0 min if the pilot takes no corrective action. [*Hint*: First draw a diagram.]

63. (II) Determine the speed of the boat with respect to the shore in Example 3–13.

64. (II) A passenger on a boat moving at 1.80 m/s on a still lake walks up a flight of stairs at a speed of 0.60 m/s, Fig. 3–51. The stairs are angled at 45° pointing in the direction of motion as shown. What is the velocity of the passenger relative to the water?

FIGURE 3–51 Problem 64.

65. (II) A motorboat whose speed in still water is 3.70 m/s must aim upstream at an angle of 29.5° (with respect to a line perpendicular to the shore) in order to travel directly across the stream. (*a*) What is the speed of the current? (*b*) What is the resultant speed of the boat with respect to the shore? (See Fig. 3–34.)

66. (II) A boat, whose speed in still water is 2.40 m/s, must cross a 280-m-wide river and arrive at a point 120 m upstream from where it starts (Fig. 3–52). To do so, the pilot must head the boat at a 45.0° upstream angle. What is the speed of the river's current?

FIGURE 3–52
Problem 66.

67. (II) A swimmer is capable of swimming 1.00 m/s in still water. (*a*) If she aims her body directly across a 75-m-wide river whose current is 0.80 m/s, how far downstream (from a point opposite her starting point) will she land? (*b*) How long will it take her to reach the other side?

68. (II) At what upstream angle must the swimmer in Problem 67 aim, if she is to arrive at a point directly across the stream?

69. (II) Two cars approach a street corner at right angles to each other. Car 1 travels at 30 km/h and car 2 at 50 km/h. What is the relative velocity of car 1 as seen by car 2? What is the velocity of car 2 relative to car 1?

70. (III) An airplane, whose air speed is 680 km/h, is supposed to fly in a straight path 35.0° N of E. But a steady 120 km/h wind is blowing from the north. In what direction should the plane head?

71. (III) A motorcycle traveling 95.0 km/h approaches a car traveling in the same direction at 75.0 km/h. When the motorcycle is 60.0 m behind the car, the rider accelerates uniformly and passes the car 10.0 s later. What was the acceleration of the motorcycle?

General Problems

72. Two vectors, $\mathbf{V}_1$ and $\mathbf{V}_2$, add to a resultant $\mathbf{V} = \mathbf{V}_1 + \mathbf{V}_2$. Describe $\mathbf{V}_1$ and $\mathbf{V}_2$ if (*a*) $V = V_1 + V_2$, (*b*) $V^2 = V_1^2 + V_2^2$, (*c*) $V_1 + V_2 = V_1 - V_2$.

73. A plumber steps out of his truck, walks 60 m east and 35 m south, and then takes an elevator 12 m into the subbasement of a building where a bad leak is occurring. What is the displacement of the plumber relative to his truck? Give your answer in components, and also in magnitude and angle notation. Assume x is east, y is north, and z is up.

74. On mountainous downhill roads, escape routes are sometimes placed to the side of the road for trucks whose brakes might fail. Assuming a constant upward slope of 30°, calculate the horizontal and vertical components of the acceleration of a truck that slowed from 120 km/h to rest in 12 s. See Fig. 3–53.

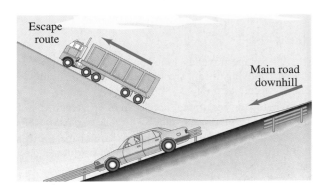

FIGURE 3–53 Problem 74.

75. Romeo is chucking pebbles gently up to Juliet's window, and he wants the pebbles to hit the window with only a horizontal component of velocity. He is standing at the edge of a rose garden 8.0 m below her window and 9.0 m from the base of the wall (Fig. 3–54). How fast are the pebbles going when they hit her window?

FIGURE 3–54 Problem 75.

76. A hunter aims directly at a target (on the same level) 65.0 m away. (*a*) If the bullet leaves the gun at a speed of 145 m/s, by how much will it miss the target? (*b*) At what angle should the gun be aimed so the target will be hit?

77. What is the *y* component of a vector in the *xy* plane whose magnitude is 52.8 and whose *x* component is 46.4? What is the direction of this vector (angle it makes with the *x* axis)?

FIGURE 3–55 Problem 78.

78. Raindrops make an angle θ with the vertical when viewed through a moving train window (Fig. 3–55). If the speed of the train is v_T, what is the speed of the raindrops in the reference frame of the Earth in which they are assumed to fall vertically?

79. A light plane is headed due south with a speed relative to still air of 240 km/h. After 1 hour, the pilot notices that they have covered only 180 km and their direction is not south but southeast. What is the wind velocity?

80. An Olympic long jumper is capable of jumping 8.0 m. Assuming his horizontal speed is 9.2 m/s as he leaves the ground, how long is he in the air and how high does he go? Assume that he lands standing upright—that is, the same way he left the ground.

81. Apollo astronauts took a "nine iron" to the Moon and hit a golf ball about 180 m. Assuming that the swing, launch angle, and so on, were the same as on Earth where the same astronaut could hit it only 30 m, estimate the acceleration due to gravity on the surface of the Moon. (We neglect air resistance in both cases, but on the Moon there is none.)

82. When Babe Ruth hit a homer over the 12-m-high right-field fence 92 m from home plate, roughly what was the minimum speed of the ball when it left the bat? Assume the ball was hit 1.0 m above the ground and its path initially made a 40° angle with the ground.

83. The cliff divers of Acapulco push off horizontally from rock platforms about 35 m above the water, but they must clear rocky outcrops at water level that extend out into the water 5.0 m from the base of the cliff directly under their launch point. See Fig. 3–56. What minimum pushoff speed is necessary to do this? How long are they in the air?

FIGURE 3–56 Problem 83.

84. At serve, a tennis player aims to hit the ball horizontally. What minimum speed is required for the ball to clear the 0.90-m-high net about 15.0 m from the server if the ball is "launched" from a height of 2.50 m? Where will the ball land if it just clears the net (and will it be "good" in the sense that it lands within 7.0 m of the net)? How long will it be in the air? See Fig. 3–57.

FIGURE 3–57 Problem 84.

85. The speed of a boat in still water is v. The boat is to make a round trip in a river whose current travels at speed u. Derive a formula for the time needed to make a round trip of total distance D if the boat makes the round trip (a) by moving upstream and back downstream, and (b) by moving directly across the river and back. We must assume $u < v$; why?

86. A plane whose airspeed is 200 km/h heads due north. But a 100-km/h northeast wind (that is, coming from the northeast) suddenly begins to blow. What is the resulting velocity of the plane with respect to the ground?

87. Spymaster Tim, flying a constant 200 km/h horizontally in a low-flying helicopter, wants to drop secret documents into his contact's open car which is traveling 150 km/h on a level highway 78.0 m below. At what angle (with the horizontal) should the car be in his sights when the packet is released (Fig. 3–58)?

FIGURE 3–58 Problem 87.

88. A basketball leaves a player's hands at a height of 2.10 m above the floor. The basket is 2.60 m above the floor. The player likes to shoot the ball at a 38.0° angle. If the shot is made from a horizontal distance of 11.00 m and must be accurate to ± 0.22 m (horizontally), what is the range of initial speeds allowed to make the basket?

89. What is the speed of a point (a) on the equator of the Earth, (b) at a latitude of 40° N, due to the Earth's daily rotation?

90. A projectile is launched from ground level to the top of a cliff which is 195 m away and 155 m high (see Fig. 3–59). If the projectile lands on top of the cliff 7.6 s after it is fired, find the initial velocity of the projectile (magnitude and direction). Neglect air resistance.

FIGURE 3–59 Problem 90.

91. In hot pursuit, Agent Logan of the FBI must get directly across a 1600-m-wide river in minimum time. The river's current is 0.80 m/s, he can row a boat at 1.50 m/s, and he can run 3.00 m/s. Describe the path he should take (rowing plus running along the shore) for the minimum crossing time, and determine the minimum time.

92. A person going for a morning jog on the deck of a cruise ship is running toward the bow (front) of the ship at 2.0 m/s while the ship is moving ahead at 8.5 m/s. What is the velocity of the jogger relative to the water? Later, the jogger is moving toward the stern (rear) of the ship. What is the jogger's velocity relative to the water now?

93. A projected space station consists of a circular tube that is set rotating about its center (like a tubular bicycle tire, Fig. 3–60). The circle formed by the tube has a diameter of about 1.1 km. What must be the rotation speed (revolutions per day) if an effect equal to gravity at the surface of the Earth (1.0 g) is to be felt?

FIGURE 3–60 Problem 93.

94. A jet pilot takes his aircraft in a vertical loop (Fig. 3–61). If the jet is moving at a speed of 700 km/h at the lowest point on the loop, determine the minimum radius of the circle so that the acceleration at the lowest point does not exceed 6.0 g's.

FIGURE 3–61 Problem 94.

$\mathbf{F}_{\mathrm{PG}}$ m $\mathbf{F}_{\mathrm{PG}}$

This airplane is taking off. It is accelerating, increasing in speed rapidly. To do so, a force must be exerted on it, according to Newton's second law, $\Sigma \mathbf{F} = m\mathbf{a}$. What exerts this force? The two jet engines of this plane exert a strong force on the gases they push out toward the rear of the plane. According to Newton's third law, these ejected gases exert an equal and opposite force on the airplane in the forward direction. The forces exerted on the **p**lane by the **g**ases, labeled $\mathbf{F}_{\mathrm{PG}}$, are what accelerate the plane.

CHAPTER 4

Dynamics: Newton's Laws of Motion

We have discussed how motion is described in terms of velocity and acceleration. Now we deal with the question of *why* objects move as they do: What makes an object at rest begin to move? What causes a body to accelerate or decelerate? What is involved when an object moves in a circle? We can answer in each case that a force is required. In this chapter,[†] we will investigate the connection between force and motion, which is the subject called **dynamics**.

4–1 Force

Intuitively, we experience **force** as any kind of a push or a pull on an object. When you push a stalled car or a grocery cart (Fig. 4–1), you are exerting a force on it. When a motor lifts an elevator, or a hammer hits a nail, or the wind blows the leaves of a tree, a force is being exerted. We say that an object falls because of the *force of gravity*. Forces do not always give rise to motion. For example, you may push very hard on a heavy desk and it may not move.

If an object is at rest, to start it moving requires force—that is, to accelerate it from zero velocity to a non-zero velocity. For an object already moving, if you want to change its velocity—either in direction or in magnitude—again a force is required. In other words, to accelerate an object, a force is required. Indeed, we can define force as an action capable of accelerating an object. In Section 4–4 we will discuss the precise relationship between force and acceleration, which is Newton's second law.

FIGURE 4–1 A force exerted on a grocery cart—in this case exerted by a child.

[†] We treat everyday sort of objects in motion here. The submicroscopic world of atoms and molecules has to be treated separately. Also, velocities that are extremely high, close to the speed of light $(3.0 \times 10^8 \, \text{m/s})$, must be treated using the theory of relativity (Chapter 37).

FIGURE 4–2 A spring scale used to measure a force.

Measuring force

One way to measure the magnitude (or strength) of a force is to make use of a spring scale (Fig. 4–2). Normally, such a spring scale is used to find the weight of an object; by weight we mean the force of gravity acting on the body (Section 4–6). The spring scale, once calibrated, can be used to measure other kinds of forces as well, such as the pulling force shown in Fig. 4–2.

A force has direction as well as magnitude, and is indeed a vector that follows the rules of vector addition discussed in Chapter 3. We can represent any force by an arrow, on a diagram, just as we did with velocity. The direction of the arrow is the direction of the push or pull, and its length is drawn proportional to the magnitude of the force.

4–2 | Newton's First Law of Motion

What is the exact connection between force and motion? Aristotle (384–322 B.C.) believed that a force was required to keep an object moving along a horizontal plane. To Aristotle, the natural state of a body was at rest, and a force was believed necessary to keep a body in motion. Furthermore, Aristotle argued, the greater the force on the body, the greater its speed.

Some 2000 years later, Galileo disagreed, and maintained that it is just as natural for an object to be in horizontal motion with a constant velocity as it is for it to be at rest.

FIGURE 4–3 Photo of an air table. Air issuing from many tiny holes forms a thin air layer between the table and a puck, which, when given an initial shove, will travel at very nearly constant speed in a straight line (until it hits a wall or another puck).

To understand Galileo's idea, consider the following observations involving motion along a horizontal plane. To push an object with a rough surface along a tabletop at constant speed requires a certain amount of force. To push an equally heavy object with a very smooth surface across the table at the same speed will require less force. If a layer of oil or other lubricant is placed between the surface of the object and the table, then almost no force is required to move the object. Notice that in each successive step, less force is required. As the next step, we can imagine a situation in which the object does not rub against the table at all—or there is a perfect lubricant between the object and the table—and theorize that once started, the object would move across the table at constant speed with *no* force applied. A steel ball bearing rolling on a hard horizontal surface approaches this situation. So does a puck on an air table (Fig. 4–3), in which a thin layer of air reduces friction almost to zero.

It was Galileo's genius to imagine such an idealized world—in this case, one where there is no friction—and to see that it could produce a more useful view of the real world. It was this idealization that led him to his remarkable conclusion that if no force is applied to a moving object, it will continue to move with constant speed in a straight line. An object slows down only if a force is exerted on it. Galileo thus interpreted friction as a force akin to ordinary pushes and pulls.

FIGURE 4–4 **F** represents the force applied by the person and **F**ₜᵣ represents the force of friction.

To push an object across a table at constant speed requires a force from your hand only to balance out the force of friction (Fig. 4–4). When the object moves at constant speed, your pushing force is equal in magnitude to the friction force, but these two forces are in opposite directions, so the *net* force on the object (the vector sum of the two forces) is zero. This is consistent with Galileo's viewpoint, for the object moves with constant speed when no net force is exerted on it.

Upon this foundation, Isaac Newton (Fig. 4–5) built his great theory of motion. Newton's analysis of motion is summarized in his famous "three laws of motion." In his great work, the *Principia* (published in 1687), Newton readily acknowledged his debt to Galileo. In fact, **Newton's first law of motion** is very close to Galileo's conclusions. It states that

> **Every body continues in its state of rest or of uniform speed in a straight line unless acted on by a nonzero net force.**

NEWTON'S FIRST LAW OF MOTION

The tendency of a body to maintain its state of rest or of uniform motion in a straight line is called **inertia**. As a result, Newton's first law is often called the **law of inertia**.

Inertia

Inertial Reference Frames

Newton's first law does not hold in every reference frame. For example, if your reference frame is fixed in a car, and the car accelerates, an object such as a cup resting on the dashboard may begin to move toward you (it stayed at rest as long as the car's velocity remained constant). The cup accelerated toward you but neither you nor anything else exerted a force on it in that direction. In such an accelerating reference frame, Newton's first law does not hold. Reference frames in which Newton's first law does hold are called **inertial reference frames**—the law of inertia is valid in them. For most purposes, we can usually assume that reference frames fixed on the Earth are inertial frames. This is not precisely true, due to the Earth's rotation, but usually it is close enough. Any reference frame that moves with constant velocity (say, a car or an airplane) relative to an inertial frame is also an inertial reference frame. Reference frames where the law of inertia does *not* hold, such as the accelerating reference frame discussed above, are called **noninertial** reference frames. How can we be sure a reference frame is inertial or not? By checking to see if Newton's first law holds. Thus Newton's first law serves as the definition of inertial reference frames.

FIGURE 4–5
Isaac Newton (1642–1727).

4–3 | Mass

Newton's second law, which we come to in the next Section, makes use of the concept of mass. Newton used the term *mass* as a synonym for *quantity of matter*. This intuitive notion of the mass of a body is not very precise because the concept "quantity of matter" is not very well defined. More precisely, we can say that **mass** is a *measure of the inertia of a body*. The more mass a body has, the greater the force needed to give it a particular acceleration. It is harder to start it moving from rest, or to stop it when it is moving, or to change its velocity sideways out of a straight-line path. A truck has much more inertia than a baseball moving at the same speed, and it requires a much greater force to change its velocity at the same rate. It therefore has much more mass. How we define mass in terms of inertia is discussed in the next Section.

Mass as inertia

To quantify the concept of mass, we must define a standard. In SI units, the unit of mass is the **kilogram** (kg) as we discussed in Chapter 1, Section 1–4.

The terms *mass* and *weight* are often confused with one another, but it is important to distinguish between them. Mass is a property of a body itself—it is a measure of a body's inertia, or its "quantity of matter". On the other hand, *weight is a force*, the force of gravity acting on a body. To see the difference, suppose we take an object to the Moon. The object will weigh only about one sixth as much as it did on Earth, since the force of gravity is weaker, but its mass will be the same. It will have the same amount of matter and it will have just as much inertia—for in the absence of friction, it will be just as hard to start it moving or to stop it once it is moving. (More on weight in Section 4–6.)

Mass vs. weight

4–4 Newton's Second Law of Motion

FIGURE 4–6 The bobsled accelerates because the team exerts a force.

Newton's first law states that if no net force is acting on a body, it remains at rest, or if moving, it continues moving with constant speed in a straight line. But what happens if a net force is exerted on a body? Newton perceived that the velocity will change (Fig. 4–6). A net force exerted on an object may make its speed increase, or, if the net force is in a direction opposite to the motion, it will reduce the speed. If the net force acts sideways on a moving object, the *direction* of the velocity changes (and the magnitude may as well). Since a change in speed or velocity is an acceleration (Chapter 2, Section 2–4), we can say that *a net force gives rise to acceleration.*

What precisely is the relationship between acceleration and force? Everyday experience can answer this question. Consider the force required to push a cart whose friction is ignorable. If there is friction, consider the *net* force, which is the force you exert minus the force of friction. Now if you push with a gentle but constant force for a certain period of time, you will make the cart accelerate from rest up to some speed, say 3 km/h. If you push with twice the force, you will find that the cart will reach 3 km/h in half the time. That is, the acceleration will be twice as great. If you double the force, the acceleration doubles. If you triple the force, the acceleration is tripled, and so on. Thus, the acceleration of a body is directly proportional to the net applied force.

But the acceleration depends on the mass of the object as well. If you push an empty grocery cart with the same force as you push one that is filled with groceries, you will find that the more massive cart accelerates more slowly. The greater the mass, the less the acceleration for the same net force. The mathematical relation, as Newton argued, is that the acceleration of a body is inversely proportional to its mass. These relationships are found to hold in general and can be summarized as follows:

The acceleration of an object is directly proportional to the net force acting on it and is inversely proportional to its mass. The direction of the acceleration is in the direction of the net force acting on the object.

This is **Newton's second law of motion**. As an equation, it can be written

$$\mathbf{a} = \frac{\Sigma \mathbf{F}}{m},$$

Net force

where $\mathbf{a}$ stands for acceleration, m for the mass, and $\Sigma \mathbf{F}$ for the *net force*. The symbol Σ (Greek "sigma") stands for "sum of"; $\mathbf{F}$ stands for force, so $\Sigma \mathbf{F}$ means the *vector sum of all forces* acting on the body, which we define as the **net force**.

We rearrange this equation to obtain the familiar statement of Newton's second law:

$$\Sigma \mathbf{F} = m\mathbf{a}. \tag{4–1}$$

Newton's second law relates the description of motion to the cause of motion, force. It is one of the most fundamental relationships in physics. From Newton's second law we can make the definition of **force** as *an action capable of accelerating an object.*

Force defined

Every force $\mathbf{F}$ is a vector, with magnitude and direction. Equation 4–1 is a vector equation valid in any inertial reference frame. It can be written in component form in rectangular coordinates as

$$\Sigma F_x = ma_x, \qquad \Sigma F_y = ma_y, \qquad \Sigma F_z = ma_z.$$

If the motion is all along a line (one dimensional), we can leave out the subscripts and simply write $\Sigma F = ma$.

In SI units, with the mass in kilograms, the unit of force is called the **newton** (N). One newton, then, is the force required to impart an acceleration of 1 m/s^2 to a mass of 1 kg. Thus $1 \text{ N} = 1 \text{ kg} \cdot \text{m/s}^2$.

In cgs units, the unit of mass is the gram (g) as mentioned earlier.[†] The unit of force is the *dyne*, which is defined as the net force needed to impart an acceleration of 1 cm/s^2 to a mass of 1 g. Thus $1 \text{ dyne} = 1 \text{ g} \cdot \text{cm/s}^2$. It is easy to show that $1 \text{ dyne} = 10^{-5} \text{ N}$.

In the British system, the unit of force is the *pound* (abbreviated lb), where $1 \text{ lb} = 4.44822 \text{ N} \approx 4.45 \text{ N}$. The unit of mass is the *slug*, which is defined as that mass which will undergo an acceleration of 1 ft/s^2 when a force of 1 lb is applied to it. Thus $1 \text{ lb} = 1 \text{ slug} \cdot \text{ft/s}^2$. Table 4–1 summarizes the units in the different systems.

It is very important that only one set of units be used in a given calculation or problem, with the SI being preferred. If the force is given in, say, newtons, and the mass in grams, then before attempting to solve for the acceleration in SI units, the mass must be changed to kilograms. For example, if the force is given as 2.0 N along the x axis and the mass is 500 g, we change the latter to 0.50 kg, and the acceleration will then automatically come out in m/s^2 when Newton's second law is used:

$$a_x = \frac{\Sigma F_x}{m} = \frac{2.0 \text{ N}}{0.50 \text{ kg}} = \frac{2.0}{0.50} \frac{\text{kg} \cdot \text{m}}{\text{kg} \cdot \text{s}^2} = 4.0 \text{ m/s}^2.$$

Unit of force; the newton

TABLE 4–1
Units for Mass and Force

System	Mass	Force (including weight)
SI	kilogram (kg)	newton (N) ($= \text{kg} \cdot \text{m/s}^2$)
cgs	gram (g)	dyne ($= \text{g} \cdot \text{cm/s}^2$)
British	slug	pound (lb)

Conversion factors: 1 dyne = 10^5 N;
1 lb ≈ 4.45 N.

➡ PROBLEM SOLVING

Use a consistent set of units

EXAMPLE 4–1 ESTIMATE Force to accelerate a fast car. Estimate the net force needed to accelerate (*a*) a 1000-kg car at $\frac{1}{2}g$; (*b*) a 200-g apple at the same rate.

SOLUTION (*a*) The car's acceleration is $a = \frac{1}{2}g = \frac{1}{2}(9.8 \text{ m/s}^2) \approx 5 \text{ m/s}^2$. We use Newton's second law to get the net force needed to achieve this acceleration:

$$\Sigma F = ma \approx (1000 \text{ kg})(5 \text{ m/s}^2) = 5000 \text{ N}.$$

(If you are used to British units, to get an idea of what a 5000 N force is, you can divide by 4.45 N/lb and get a force of about 1000 lb.)
(*b*) For the apple,

$$\Sigma F = ma \approx (0.200 \text{ kg})(5 \text{ m/s}^2) = 1 \text{ N}.$$

EXAMPLE 4–2 Force to stop a car. What constant net force is required to bring a 1500-kg car to rest from a speed of 100 km/h within a distance of 55 m?

SOLUTION We use Newton's second law, $\Sigma F = ma$, but first we must determine the acceleration a, which is constant since the net force is constant. We assume the motion is along the $+x$ axis (Fig. 4–7). We are given the initial velocity $v_0 = 100 \text{ km/h} = 28 \text{ m/s}$, the final velocity $v = 0$, and the distance traveled $x - x_0 = 55 \text{ m}$. From Eq. 2–12c, we have

$$v^2 = v_0^2 + 2a(x - x_0)$$

so

$$a = \frac{v^2 - v_0^2}{2(x - x_0)} = \frac{0 - (28 \text{ m/s})^2}{2(55 \text{ m})} = -7.1 \text{ m/s}^2.$$

The net force required is then

$$\Sigma F = ma = (1500 \text{ kg})(-7.1 \text{ m/s}^2) = -1.1 \times 10^4 \text{ N}.$$

The force must be exerted in the direction *opposite* to the initial velocity, which is what the negative sign tells us.

[†] Be careful not to confuse g for gram with g for the acceleration due to gravity. The latter is always italicized (or bold face as a vector).

$v_0 = 100 \text{ km/h}$ $v = 0$

$x = 0$ $x = 55 \text{ m}$

FIGURE 4–7 Example 4–2.

As mentioned in Section 4–3, we can quantify the concept of mass using its definition as a measure of inertia. How to do this is evident from Eq. 4–1, where we see that the acceleration of a body is inversely proportional to its mass. If the same net force ΣF acts to accelerate each of two masses, m_1 and m_2, then the ratio of their masses can be defined as the inverse ratio of their accelerations:

Mass defined

$$\frac{m_2}{m_1} = \frac{a_1}{a_2}.$$

If one of the masses is known (it could be the standard kilogram) and the two accelerations are precisely measured, then the unknown mass is obtained from this definition. For example, if $m_1 = 1.00\,kg$, and for a particular force $a_1 = 3.00\,m/s^2$ and $a_2 = 2.00\,m/s^2$, then $m_2 = 1.50\,kg$.

Newton's second law, like the first law, is valid only in inertial reference frames. In the non-inertial reference frame of an accelerating car, for example, a cup on the dashboard starts sliding—it accelerates—even though the net force on it is zero; thus $\Sigma \mathbf{F} = m\mathbf{a}$ doesn't work in such an accelerating reference frame.

4–5 | Newton's Third Law of Motion

Newton's second law of motion describes quantitatively how forces affect motion. But where, we may ask, do forces come from? Observations suggest that a force applied to any object is always applied *by another object*. A horse pulls a wagon, a person pushes a grocery cart, a hammer pushes on a nail, a magnet attracts a paper clip. In each of these examples, a force is exerted *on* one body, and that force is exerted *by* another body. For example, the force exerted *on* the nail is exerted *by* the hammer.

A force is exerted **on** *a body and is exerted* **by** *another body*

But Newton realized that things are not so one-sided. True, the hammer exerts a force on the nail (Fig. 4–8). But the nail evidently exerts a force back on the hammer as well, for the hammer's speed is rapidly reduced to zero upon contact. Only a strong force could cause such a rapid deceleration of the hammer. Thus, said Newton, the two bodies must be treated on an equal basis. The hammer exerts a force on the nail, and the nail exerts a force back on the hammer. This is the essence of **Newton's third law of motion**:

NEWTON'S THIRD LAW
OF MOTION

> **Whenever one object exerts a force on a second object, the second exerts an equal and opposite force on the first.**

Action and reaction act on different objects

This law is sometimes paraphrased as "to every action there is an equal and opposite reaction." This is perfectly valid. But to avoid confusion, it is very important to remember that the "action" force and the "reaction" force are acting on *different* objects.

As evidence for the validity of Newton's third law, look at your hand when you push against a grocery cart or against the edge of a desk, Fig. 4–9. Your hand's shape is distorted, clear evidence that a force is being exerted on it. You can *see* the edge of the desk pressing into your hand. You can even *feel* the desk exerting a force on your hand; it hurts! The harder you push against the desk, the harder the desk pushes back on your hand. (Note that you only feel forces exerted *on* you, not forces that you exert on other objects.)

FIGURE 4–8 Multiflash photo of a hammer striking a nail. In accordance with Newton's third law the hammer exerts a force on the nail and the nail exerts a force back on the hammer. The latter force decelerates the hammer and brings it to rest.

FIGURE 4–9 If your hand pushes against the edge of a desk (the force vector is shown in red), the desk pushes back against your hand (this force vector is shown in a different color, violet, to remind us that this force acts on a different object).

Force exerted on hand by desk

Force exerted on desk by hand

Force on skater Force on wall

(a)

(b)

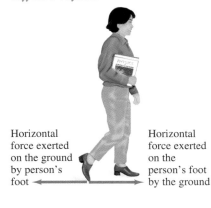

FIGURE 4–10 Two examples of Newton's third law. (a) When an ice-skater pushes against the railing, the railing pushes back and this force causes her to move away. (b) The launch of a rocket. The rocket engine pushes out the gases, and the gases exert an equal and opposite force back on the rocket, accelerating it.

As another demonstration of Newton's third law, consider the ice-skater in Fig. 4–10a. Since there is very little friction between her skates and the ice, she will move freely if a force is exerted on her. She pushes against the railing; and then *she* starts moving backward. Clearly, there had to be a force exerted on her to make her move. The force she exerts on the railing cannot make *her* move, for that force acts on the railing. Something had to exert a force on her to make her start moving, and that force could only have been exerted by the railing. The force with which the railing pushes on her is, by Newton's third law, equal and opposite to the force she exerts on the railing.

When a person throws a package out of a boat (initially at rest), the boat starts moving in the opposite direction. The person exerts a force on the package. The package exerts an equal and opposite force back on the person, and this force propels the person (and the boat) backward slightly. Rocket propulsion also is explained using Newton's third law (Fig. 4–10b). A common misconception is that rockets accelerate because the gases rushing out the back of the engine push against the ground or the atmosphere. Not true. What happens, instead, is that a rocket exerts a strong force on the gases, expelling them; and the gases exert an equal and opposite force *on the rocket*. It is this latter force that propels the rocket forward. Thus, a space vehicle is maneuvered in empty space by firing its rockets in the direction opposite to that in which it needs to accelerate.

Consider how we walk. A person begins walking by pushing with the foot against the ground. The ground then exerts an equal and opposite force back on the person (Fig. 4–11) and it is this force, *on* the person, that moves the person forward. (If you doubt this, try walking on very smooth slippery ice.) In a similar way, a bird flies forward by exerting a force on the air, but it is the air pushing back on the bird's wings that propels the bird forward.

➡ **PHYSICS APPLIED**
How does a rocket accelerate?

FIGURE 4–11 We can walk forward because, when one foot pushes backward against the ground, the ground pushes forward on that foot. The two forces shown *act on different objects.*

Horizontal force exerted on the ground by person's foot

Horizontal force exerted on the person's foot by the ground

Inanimate objects can exert a force

CONCEPTUAL EXAMPLE 4–3 **What exerts the force on a car?** What makes a car go forward?

RESPONSE A common answer is that the engine makes the car move forward. But it is not so simple. The engine makes the wheels go around. But what good is that if they are on slick ice or mud? They just spin. A car moves forward due to the friction force exerted by the ground on the tires, and this force is the reaction to the force exerted on the ground by the tires.

We tend to associate forces with active bodies such as humans, animals, engines, or a moving object like a hammer. It is often difficult to see how an inanimate object at rest, such as a wall or a desk, can exert a force. The explanation lies in the fact that every material, no matter how hard, is elastic, at least to some degree. No one can deny that a stretched rubber band can exert a force on a wad of paper and send it flying across the room. Other materials may not stretch as easily as rubber, but they do stretch when a force is applied to them. And just as a stretched rubber band exerts a force, so does a stretched (or compressed) wall, desk, or car fender.

From the examples discussed above, it is clear that it is quite important to remember *on* what object a given force is exerted and *by* what object that force is exerted. The point is that a force influences the motion of an object only when it is applied *on* that object. A force exerted *by* a body does not influence that body; it only influences the other body *on* which it is exerted. Thus, to avoid confusion, the two prepositions *on* and *by* must always be used—and used with care.

One way to keep clear which force acts on which object is to use double subscripts. For example the force exerted on the **P**erson by the **G**round in Fig. 4–11 can be labeled F_{PG} as shown in Fig. 4–12. And the force exerted on the ground by the person is F_{GP}. Note that we have used different colors for the force vectors when they act on different objects. By Newton's third law

$$\mathbf{F}_{GP} = -\mathbf{F}_{PG}. \tag{4–2}$$

$\mathbf{F}_{GP}$ and $\mathbf{F}_{PG}$ have the same magnitude, and the minus sign reminds us that these two forces are in opposite directions.

Note carefully that the two forces shown in Fig. 4–11 or 4–12 act on different objects—hence we used slightly different colors for the arrows representing the forces. These two forces would never appear together in a sum of forces in Newton's second law, $\Sigma \mathbf{F} = m\mathbf{a}$. Why not? Because **a** is the acceleration of one particular body and $\Sigma \mathbf{F}$ must include only the forces on that body.

CONCEPTUAL EXAMPLE 4–4 | **Third law clarification.** Michelangelo's assistant has been assigned the task of moving a block of marble using a sled (Fig. 4–13). He says to his boss, "When I exert a forward force on the sled, the sled exerts an equal and opposite force backward. So how can I ever start it moving? No matter how hard I pull, the backward reaction force always equals my forward force, so the net force must be zero. I'll never be able to move this load." Is this a case of a little knowledge being dangerous? Explain.

RESPONSE Yes. Although it is true that the action and reaction forces are equal in magnitude, the assistant has forgotten that they are exerted on different objects. The forward ("action") force is exerted by the assistant on the sled (Fig. 4–13), whereas the backward "reaction" force is exerted by the sled on the assistant. To determine if the *assistant* moves or not, we must consider only

NEWTON'S THIRD LAW
OF MOTION

FIGURE 4–12 Newton's third law. Subscripts on forces remind us *on* which body a force acts and *by* which body it is exerted.

FIGURE 4–13 Seventy-year-old Michelangelo has selected a fine block of marble for his next sculpture. Shown here is his assistant pulling it on a sled away from the quarry. Forces on the assistant are shown as red (magenta) arrows. Forces on the sled are purple arrows. Forces acting on the ground are orange arrows. Action–reaction forces that are equal and opposite are labeled by the same subscripts but reversed (such as $\mathbf{F}_{GA}$ and $\mathbf{F}_{AG}$) and are of different colors because they act on different objects. [Only horizontal forces are shown.]

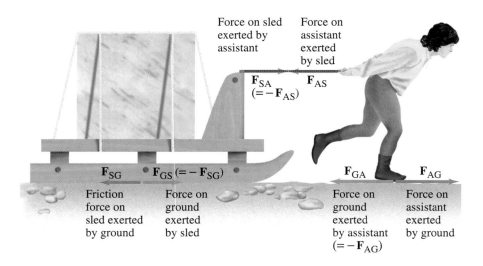

Force on sled exerted by assistant

Force on assistant exerted by sled

$\mathbf{F}_{SA}$ $(= -\mathbf{F}_{AS})$ $\mathbf{F}_{AS}$

$\mathbf{F}_{SG}$ $\mathbf{F}_{GS}$ $(= -\mathbf{F}_{SG})$ $\mathbf{F}_{GA}$ $\mathbf{F}_{AG}$

Friction force on sled exerted by ground

Force on ground exerted by sled

Force on ground exerted by assistant $(= -\mathbf{F}_{AG})$

Force on assistant exerted by ground

the forces *on the assistant* and then apply $\Sigma \mathbf{F} = m\mathbf{a}$, where $\Sigma \mathbf{F}$ is the net force *on the assistant*, $\mathbf{a}$ is the acceleration of the assistant, and m is the assistant's mass. There are two forces on the assistant that affect his forward motion and they are shown as bright red (magenta) arrows in Figs. 4–13 and 4–14: they are (1) the horizontal force $\mathbf{F}_{AG}$ exerted on the assistant by the ground (the harder he pushes backward against the ground, the harder the ground pushes forward on him— Newton's third law), and (2) the force $\mathbf{F}_{AS}$ exerted on the assistant by the sled, pulling backward on him; see Fig. 4–14. When the ground pushes forward on the assistant harder than the sled pulls backward, the assistant accelerates forward (Newton's second law). The sled, on the other hand, accelerates forward when the force on *it* exerted by the assistant is greater than the frictional force acting backward (that is, when $\mathbf{F}_{SA}$ has greater magnitude than $\mathbf{F}_{SG}$ in Fig. 4–13).

Using double subscripts to clarify Newton's third law can become cumbersome, and we won't usually use them in this way. Nevertheless, if there is any confusion in your mind about a given force, go ahead and use them to identify *on* what object and *by* what object the force is exerted.

Force on assistant exerted by sled

$\mathbf{F}_{AS}$

$\mathbf{F}_{AG}$

Force on assistant exerted by ground

FIGURE 4–14 The forces on the assistant of Example 4–4.

| CONCEPTUAL EXAMPLE 4–5 | **Collision using Newton's second and third laws.** A massive truck collides head-on with a small sports car. Which vehicle experiences the greater force of impact? Which experiences the greater acceleration? Which of Newton's laws is useful to obtain the correct answer?

RESPONSE The truck and the car experience the same magnitude impact force (Newton's third law). The car experiences a much greater acceleration in stopping than the truck because its mass is much less (Newton's second law).

4–6 Weight—the Force of Gravity; and the Normal Force

Galileo claimed that objects dropped near the surface of the Earth will all fall with the same acceleration, $\mathbf{g}$, if air resistance can be neglected. The force that gives rise to this acceleration is called the force of gravity. We now apply Newton's second law to the gravitational force; and for the acceleration, $\mathbf{a}$, we use the downward acceleration due to gravity, $\mathbf{g}$. Thus, the force of gravity on an object, $\mathbf{F}_G$, whose magnitude is commonly called its **weight**, can be written as

$$\mathbf{F}_G = m\mathbf{g}. \qquad (4\text{–}3)$$

Weight = force of gravity

The direction of this force is down toward the center of the Earth. If the y axis is upwards, we can write $\mathbf{F}_G = -mg\mathbf{j}$.

In SI units, $g = 9.80 \text{ m/s}^2 = 9.80 \text{ N/kg}$,[†] so the weight of a 1.00-kg mass on Earth is $1.00 \text{ kg} \times 9.80 \text{ m/s}^2 = 9.80 \text{ N}$. We will mainly be concerned with the weight of objects on Earth, but we note that on the Moon, on other planets, or in space, the weight of a given mass will be different. For example, on the Moon the acceleration due to gravity is about one sixth what it is on Earth, and a 1.0 kg mass weighs only 1.7 N. Although we will not have occasion to use British units, we note that for practical purposes on the Earth, a mass of 1 kg weighs about 2.2 lb. (On the Moon, 1 kg weighs only about 0.4 lb.)

The force of gravity acts on an object when it is falling. When an object is at rest on the Earth, the gravitational force on it does not disappear, as we know if we weigh it on a spring scale. The same force, given by Eq. 4–3, continues to act. Why, then, doesn't the object move? From Newton's second law, the net force on an object that remains at rest is zero. There must be another force on the object to balance the gravitational force. For an object resting on a table, the table exerts

[†] Since $1 \text{ N} = 1 \text{ kg} \cdot \text{m/s}^2$ (Section 4–4), $1 \text{ m/s}^2 = 1 \text{ N/kg}$.

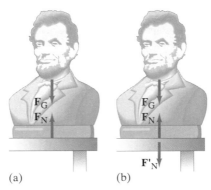

(a)   (b)

FIGURE 4–15 (a) The net force on an object at rest is zero according to Newton's second law. Therefore the downward force of gravity ($\mathbf{F}_G$) on an object must be balanced by an upward force (the normal force $\mathbf{F}_N$) exerted by the table in this case. (b) $\mathbf{F}'_N$ is the force exerted on the table by the statue and is the reaction force to $\mathbf{F}_N$ as per Newton's third law. ($\mathbf{F}'_N$ is shown in a different color to remind us it acts on a different body.) The reaction to $\mathbf{F}_G$ is not shown.

this upward force; see Fig. 4–15a. The table is compressed slightly beneath the object, and due to its elasticity, it pushes up on the object as shown. The force exerted by the table is often called a **contact force**, since it occurs when two objects are in contact. (The force of your hand pushing on a cart is also a contact force.) When a contact force acts *perpendicular* to the common surface of contact, it is usually referred to as the **normal force** ("normal" means perpendicular); hence it is labeled $\mathbf{F}_N$ in the diagram.

The two forces shown in Fig. 4–15a are both acting on the statue, which remains at rest, so the vector sum of these two forces must be zero (Newton's second law: if $\mathbf{a} = 0$, $\Sigma\mathbf{F} = 0$). Hence $\mathbf{F}_G$ and $\mathbf{F}_N$ must be of equal magnitude and in opposite directions. But they are *not* the equal and opposite forces spoken of in Newton's third law. The action and reaction forces of Newton's third law act on *different objects*, whereas the two forces shown in Fig. 4–15a act on the *same* object. For each of the forces shown in Fig. 4–15a, we can ask, "What is the reaction force?" The upward force, $\mathbf{F}_N$, on the statue is exerted by the table. The reaction to this force is a force exerted by the statue on the table. It is shown in Fig. 4–15b, where it is labeled $\mathbf{F}'_N$. This force $\mathbf{F}'_N$, exerted on the table by the statue, is the reaction force to $\mathbf{F}_N$ in accord with Newton's third law. Now, what about the other force on the statue, the force of gravity $\mathbf{F}_G$? Can you guess what the reaction is to this force? [We will see in Chapter 6 that the reaction force is also a gravitational force exerted on the Earth by the statue, and can be considered to act at the Earth's center.]

EXAMPLE 4–6 **Weight, normal force, and a box.** A friend has given you a special gift, a box of mass 10.0 kg with a mystery surprise inside. It's a reward for your fine showing on the physics final. The box, is resting on the smooth (frictionless) horizontal surface of a table, Fig. 4–16a. (*a*) Determine the weight of the box and the normal force acting on it. (*b*) Now your friend pushes down on the box with a force of 40.0 N, as in Fig. 4–16b. Again determine the normal force acting on the box. (*c*) If your friend pulls upward on the box with a force of 40.0 N (Fig. 4–16c), what now is the normal force on the box?

SOLUTION (*a*) The box is resting on the table. The weight of the box is $mg = (10.0\,\text{kg})(9.80\,\text{m/s}^2) = 98.0\,\text{N}$, and this force acts downward. The only other force on the box is the normal force exerted upward on it by the table, as shown in Fig. 4–16a. We chose the upward direction as the positive y direction, and then the net force ΣF_y on the box is $\Sigma F_y = F_N - mg$. Since the box is at rest, the net force on it must be zero ($\Sigma F_y = ma_y$, and $a_y = 0$). Thus

$$\Sigma F_y = F_N - mg = 0,$$

so

$$F_N = mg.$$

The normal force on the box, exerted by the table, is 98.0 N upward, and has magnitude equal to the box's weight.

FIGURE 4–16 Example 4–6. (a) A 10-kg gift box is at rest on a table. (b) A person pushes down on the box with a force of 40.0 N. (c) A person pulls upward on the box with a force of 40.0 N. The forces are all assumed to act along a line; they are shown slightly displaced in order to be distinguishable on the diagram. Only forces acting on the box are shown.

(a)   (b)   (c)

(b) Your friend is pushing down on the box with a force of 40.0 N. So now there are three forces acting on the box, as shown in Fig. 4–16b. The weight of the box is still $mg = 98.0$ N. The net force is $\Sigma F_y = F_N - mg - 40.0$ N, and is equal to zero since the box remains at rest. Thus, since $a = 0$, Newton's second law gives

$$\Sigma F_y = F_N - mg - 40.0 \text{ N} = 0,$$

so the normal force is now

$$F_N = mg + 40.0 \text{ N} = 98.0 \text{ N} + 40.0 \text{ N} = 138.0 \text{ N},$$

which is greater than in (a). The table pushes back with more force.

(c) The box's weight is still 98.0 N and acts downward. The force exerted by your friend and the normal force both act upward (positive direction), as shown in Fig. 4–16c. The box doesn't move since your friend's upward force is less than the weight. The net force, again set to zero, is

$$\Sigma F_y = F_N - mg + 40.0 \text{ N} = 0,$$

so

$$F_N = mg - 40.0 \text{ N} = 98.0 \text{ N} - 40.0 \text{ N} = 58.0 \text{ N}.$$

The table does not push against the full weight of the box because of the upward pull exerted by your friend.

Normal force is not always equal in magnitude to the weight

EXAMPLE 4–7 **Accelerating the box.** What happens when a person pulls upward on the box in Example 4–6 (c) with a force equal to, or greater than, the box's weight, say $F_P = 100.0$ N rather than the 40.0 N shown in Fig. 4–16c?

SOLUTION The net force is now

$$\Sigma F_y = F_N - mg + F_P = F_N - 98.0 \text{ N} + 100.0 \text{ N},$$

and if we set this equal to zero, we would get $F_N = -2.0$ N. This is nonsense, since the negative sign implies F_N points downward, and the table surely cannot pull down on the box (unless there's glue on the table). The least F_N can be is zero, which it will be in this case. What really happens here is clear: the box accelerates upward since the net force is not zero; it is

$$\Sigma F_y = F_P - mg = 100.0 \text{ N} - 98.0 \text{ N} = 2.0 \text{ N}$$

upward. See Fig. 4–17. So the box moves upward with an acceleration of magnitude

$$a_y = \Sigma F_y/m = 2.0 \text{ N}/10.0 \text{ kg} = 0.20 \text{ m/s}^2.$$

FIGURE 4–17 Example 4–7. The box accelerates upwards because $F_P > mg$.

EXAMPLE 4–8 **Unwanted weight loss?** A 65-kg woman descends in an elevator that briefly, when leaving a floor, accelerates at 0.20g. (a) If she stands on a scale during this acceleration, what is her weight and what does the scale read? (b) What does the scale read when the elevator descends at a constant speed of 2.0 m/s?

SOLUTION (a) Figure 4–18 shows all the forces that act on the woman (and only those that act on her). The direction of the acceleration is downward, which we take as positive. From Newton's second law,

$$\Sigma F = ma$$
$$mg - F_N = 0.20mg$$
$$F_N = mg - 0.20mg = 0.80mg.$$

The normal force $\mathbf{F}_N$ is the force the scale exerts on the person, and is equal and opposite to the force she exerts on the scale, $\mathbf{F}'_N = -0.80\ mg$. Her weight (force of gravity on her) is still $mg = (65 \text{ kg})(9.8 \text{ m/s}^2) = 640$ N. But the scale, needing to exert a force of only 0.80mg, will show her mass as $0.80m = 52$ kg.

(b) Now there is no acceleration, $a = 0$, so by Newton's second law, $mg - F_N = 0$ and $F_N = mg$. The scale reads her correct mass of 65 kg.

FIGURE 4–18 Example 4–8.

4–7 Solving Problems with Newton's Laws: Free-Body Diagrams

Newton's second law tells us that the acceleration of an object is proportional to the *net force* acting on the object. The **net force**, as mentioned earlier, is the *vector sum* of all forces acting on the object. Indeed, extensive experiments have shown that forces do add together as vectors precisely according to the rules we developed in Chapter 3. For example, in Fig. 4–19, two forces of equal magnitude (100 N each) are shown acting on an object at right angles to each other. Intuitively, we can see that the object will move at a 45° angle and thus the net force acts at a 45° angle. This is just what the rules of vector addition give. From the theorem of Pythagoras, the magnitude of the resultant force is $F_R = \sqrt{(100\,\text{N})^2 + (100\,\text{N})^2} = 141\,\text{N}$.

FIGURE 4–19 (a) Two forces, $\mathbf{F}_1$ and $\mathbf{F}_2$, act on an object. (b) The sum, or resultant, of $\mathbf{F}_1$ and $\mathbf{F}_2$ is $\mathbf{F}_R$.

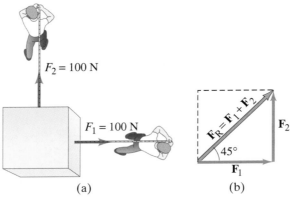

(a)   (b)

FIGURE 4–20 Two force vectors act on a boat (Example 4–9).

(a)

(b)

(c)

EXAMPLE 4–9 Adding force vectors. Calculate the sum of the two forces acting on the boat shown in Fig. 4–20a.

SOLUTION These two forces are shown resolved in Fig. 4–20b. We add the forces using the method of components. The components of $\mathbf{F}_1$ are

$$F_{1x} = F_1 \cos 45.0° = (40.0\,\text{N})(0.707) = 28.3\,\text{N},$$

$$F_{1y} = F_1 \sin 45.0° = (40.0\,\text{N})(0.707) = 28.3\,\text{N}.$$

The components of $\mathbf{F}_2$ are

$$F_{2x} = +F_2 \cos 37.0° = +(30.0\,\text{N})(0.799) = +24.0\,\text{N},$$

$$F_{2y} = -F_2 \sin 37.0° = -(30.0\,\text{N})(0.602) = -18.1\,\text{N}.$$

F_{2y} is negative because it points along the negative y axis. The components of the resultant force are (see Fig. 4–20c)

$$F_{Rx} = F_{1x} + F_{2x} = 28.3\,\text{N} + 24.0\,\text{N} = 52.3\,\text{N},$$

$$F_{Ry} = F_{1y} + F_{2y} = 28.3\,\text{N} - 18.1\,\text{N} = 10.2\,\text{N}.$$

To find the magnitude of the resultant force, we use the Pythagorean theorem:

$$F_R = \sqrt{F_{Rx}^2 + F_{Ry}^2} = \sqrt{(52.3)^2 + (10.2)^2} = 53.3\,\text{N}.$$

The only remaining question is the angle θ that the net force $\mathbf{F}_R$ makes with the x axis. We use:

$$\tan \theta = \frac{F_{Ry}}{F_{Rx}} = \frac{10.2\,\text{N}}{52.3\,\text{N}} = 0.195,$$

and $\tan^{-1}(0.195) = 11.0°$.

When solving problems involving Newton's laws and force, it is very important to draw a diagram showing all the forces acting *on* each object involved. Such a diagram is called a **free-body diagram**, or **force diagram**: draw an arrow to represent each force acting on a given body, being sure to include *every* force acting on that body. Do not show forces that the body exerts on *other* objects. When concerned only about translational motion, all the forces on a given body can be drawn as acting at the center of the object, thus treating the object as a point particle. However, when doing problems involving rotation or statics, the place *where* each force acts is also important, as we shall see.

CONCEPTUAL EXAMPLE 4–10 **The hockey puck.** A hockey puck is sliding at constant velocity across a flat horizontal ice surface that is assumed to be frictionless. Which of the sketches in Fig. 4–21 is the correct free-body diagram for this puck? What would your answer be if the puck slowed down?

RESPONSE Did you choose (a)? If so, can you answer the question: what exerts the horizontal force labeled **F**? If you say that it is the force needed to maintain the motion (as the ancient Greeks said), ask yourself: what exerts this force? Remember that another object must exert any force—and there simply isn't any possibility here. Therefore, (a) is wrong. Besides, the force **F** in Fig. 4–21a would give rise to an acceleration by Newton's second law. It is (b) that is correct, as long as there is no friction. No net force acts on the puck, and the puck slides at constant velocity across the ice.

If someone insists that we come down from the ivory tower of idealized frictionless surfaces, down to the real world where even smooth ice exerts at least a tiny friction force, then (c) is the correct answer. The tiny friction force is in the direction opposite to the motion (it ought to be labeled $\mathbf{F}_{fr}$, not simply **F**), and the puck's velocity decreases, even if very slowly.

FIGURE 4–21 Which is the correct free-body diagram for a hockey puck sliding across frictionless ice (Example 4–10)?

Here now is a brief summary of how to approach solving problems involving Newton's laws:

PROBLEM SOLVING | Newton's Laws; Free-Body Diagrams

1. Draw a sketch of the situation.

2. Consider only one object (at a time), and draw a **free-body diagram** for that body, showing *all* the forces acting *on* that body, including any unknown forces that you have to solve for. Do not show any forces that the body exerts on other bodies. Draw the arrow for each force vector reasonably accurately for direction and magnitude. Label each force, including forces you must solve for, as to its source (gravity, person, friction, and so on). If several bodies are involved, draw a free-body diagram for each body *separately*, showing all the forces acting *on that body* (and *only* forces acting on that body). For each (and every) force, you must be clear about: *on* what object that force acts; and *by* what object that force is exerted. Only forces acting *on* a given body can be included in $\Sigma\mathbf{F} = m\mathbf{a}$ for that body.

3. Newton's second law involves vectors, and it is usually important to resolve vectors into components. Choose an x and a y axis in a way that simplifies the calculation.

4. For each body, Newton's second law can be applied to the x and y components separately. That is, the x component of the net force on that body will be related to the x component of that body's acceleration: $\Sigma F_x = ma_x$, and similarly for the y direction.

5. Solve the equation or equations for the unknown(s).

This problem-solving box should not be considered a prescription. Rather it is a summary of things to do that will start your mind thinking and getting involved in the problem at hand.

$F_P = 40.0\ N$

30.0°

(a)

y

F_P

30.0°

F_N

x

mg

(b)

y

F_N

F_P

x

mg

(c)

FIGURE 4–22 Example 4–11; (b) is the free-body diagram for the box, and (c) is the free-body diagram considering all the forces to act at a point (translational motion only, which is what we have here).

In the Examples that follow, we assume that all surfaces are very smooth so that friction can be ignored. (Friction, and Examples using it, are discussed in the next chapter.

EXAMPLE 4–11 **Pulling the mystery box.** Suppose a friend asks to examine the 10.0-kg box you were given (Example 4–6, Fig. 4–16), hoping to guess what is inside; and you respond, "Sure, pull the box over to you." She then pulls the box by the attached ribbon (or string), as shown in Fig. 4–22a, along the smooth surface of the table. The magnitude of the force exerted by the person is $F_P = 40.0\ N$, and it is exerted at a 30.0° angle as shown. Calculate (a) the acceleration of the box, and (b) the magnitude of the upward force F_N exerted by the table on the box. Assume that friction can be neglected.

SOLUTION Figure 4–22b shows the free-body diagram of the box, which means we show *all* the forces acting on the box and *only* the forces acting on the box. They are: the force of gravity $m\mathbf{g}$; the normal force exerted by the table $\mathbf{F_N}$; and the force exerted by the person $\mathbf{F_P}$. We are interested only in translational motion, so we can show the three forces acting at a point, Fig. 4–22c. With the y axis vertical and the x axis horizontal, the pull of 40.0 N has components

$$F_{Px} = (40.0\ N)(\cos 30.0°) = (40.0\ N)(0.866) = 34.6\ N,$$

$$F_{Py} = (40.0\ N)(\sin 30.0°) = (40.0\ N)(0.500) = 20.0\ N.$$

(a) In the horizontal (x) direction, $\mathbf{F_N}$ and $m\mathbf{g}$ have zero components. Thus the horizontal component of the net force is F_{Px}. From Newton's second law, $\Sigma F_x = ma_x$, we have

$$F_{Px} = ma_x,$$

so

$$a_x = \frac{F_{Px}}{m} = \frac{(34.6\ N)}{(10.0\ kg)} = 3.46\ m/s^2.$$

The acceleration of the box is thus 3.46 m/s² to the right.

(b) In the vertical (y) direction, with upward as positive, again using Newton's second law we have

$$\Sigma F_y = ma_y$$

$$F_N - mg + F_{Py} = ma_y.$$

Now $mg = (10.0\ kg)(9.80\ m/s^2) = 98.0\ N$ and $F_{Py} = 20.0\ N$ as we calculated above. Furthermore, since $F_{Py} < mg$, the box does not move vertically, so $a_y = 0$. Thus

$$F_N - 98.0\ N + 20.0\ N = 0$$

which tells us that the normal force is

$$F_N = 78.0\ N.$$

Notice that F_N is less than mg. The table does not push against the full weight of the box since part of the pull exerted by the person is in the upward direction. Compare this to Example 4–6, part (c).

Tension in a cord

When a flexible cord pulls on an object, the cord is said to be under **tension**, and the force it exerts on the object is the tension $\mathbf{F_T}$. If the cord has negligible mass, the force exerted at one end is transmitted undiminished to each adjacent piece of cord along the entire length to the other end. Why? Because for the cord, $\Sigma \mathbf{F} = m\mathbf{a} = 0$ no matter what $\mathbf{a}$ is if m is zero (negligible); hence the forces pulling on the cord at its two ends must add up to zero (F_T and $-F_T$). Note that flexible ropes and cords can only pull. They can't push because they bend.

FIGURE 4–23 Example 4–12; (a) Two boxes are connected by a cord. A person pulls horizontally on box 1 with force $F_P = 40.0$ N. (b) Free-body diagram for box 1. (c) Free-body diagram for box 2.

EXAMPLE 4–12 **Two boxes connected by a cord.** Two boxes are connected by a lightweight cord and are resting on a smooth table. The boxes have masses of 12.0 kg and 10.0 kg. A horizontal force F_P of 40.0 N is applied by a person to the 10.0-kg box, as shown in Fig. 4–23a. Find (a) the acceleration of each box, and (b) the tension in the cord.

SOLUTION (a) The free-body diagram for each of the boxes is shown in Figs. 4–23b and c. We consider each box by itself so that Newton's second law can be applied to each. The cord is light, so we neglect its mass relative to the mass of the boxes. The force F_P acts on box 1. Box 1 exerts a force F_T on the connecting cord, and the cord exerts an opposite but equal magnitude force F_T back on box 1 (Newton's third law). These forces on box 1 are shown in Fig. 4–23b. Because the cord is considered to be massless, the tension at each end is the same as we saw just above. Hence the cord exerts a force F_T on the second box; Fig. 4–23c shows the forces on box 2. There will be only horizontal motion. We take the positive x axis to the right, and we use subscripts 1 and 2 to refer to the two boxes. Applying $\Sigma F_x = ma_x$ to box 1, we have:

$$\Sigma F_x = F_P - F_T = m_1 a_1. \qquad \text{[box 1]}$$

For box 2, the only horizontal force is F_T, so

$$\Sigma F_x = F_T = m_2 a_2. \qquad \text{[box 2]}$$

The boxes are connected, and if the cord remains taut and doesn't stretch, then the two boxes will have the same acceleration a. Thus $a_1 = a_2 = a$, and we are given $m_1 = 10.0$ kg and $m_2 = 12.0$ kg. We add the two equations above and obtain

$$(m_1 + m_2)a = F_P - F_T + F_T = F_P$$

or

$$a = \frac{F_P}{m_1 + m_2} = \frac{40.0 \text{ N}}{22.0 \text{ kg}} = 1.82 \text{ m/s}^2.$$

This is what we sought. Notice that we would have obtained the same result had we considered a single system, of mass $m_1 + m_2$, acted on by a net horizontal force equal to F_P. (The tension forces F_T would then be considered internal to the system as a whole, and summed together would make zero contribution to the net force on the *whole* system.)

(b) From the equation above for box 2 $\left(F_T = m_2 a_2\right)$, the tension in the cord is

$$F_T = m_2 a = (12.0 \text{ kg})(1.82 \text{ m/s}^2) = 21.8 \text{ N}.$$

Thus, F_T is less than F_P (= 40.0 N), as we expect, since F_T acts to accelerate only m_2.

➡ **PROBLEM SOLVING**

An alternate analysis

FIGURE 4–24 Example 4–13.
(a) Atwood's machine in the form of
an elevator-counterweight system.
(b) and (c) Free-body diagrams for
the two masses.

→ PROBLEM SOLVING

*Check your result by seeing if it
works in situations where the
answer is easily guessed*

EXAMPLE 4–13 **Elevator and counterweight (Atwood's machine).** Two masses suspended over a pulley by a cable, as shown in Fig. 4–24a, is sometimes referred to generically as an *Atwood's machine*. Consider the real-life application of an elevator (m_1) and its counterweight (m_2). To minimize the work done by the motor to raise and lower the elevator safely, m_1 and m_2 are similar in mass. We leave the motor out of the system for this calculation, and assume that the cable's mass is negligible and that the mass of the pulley,† as well as any friction, is small and ignorable. These assumptions assure that the tension F_T in the cord has the same magnitude on both sides of the pulley. Let the mass of the counterweight be $m_2 = 1000$ kg. Assume the mass of the empty elevator is 850 kg, and its mass when carrying four passengers is $m_1 = 1150$ kg. For the latter case $(m_1 = 1150$ kg$)$, calculate (a) the acceleration of the elevator and (b) the tension in the cable.

SOLUTION (a) Figures 4–24b and c show the free-body diagrams for the two masses. It is clear that m_1, being the heavier, will accelerate downward, and m_2 will accelerate upward. The magnitudes of their accelerations will be equal (we assume the cable doesn't stretch). For the counter-weight, $m_2 g = (1000$ kg$)(9.80$ m/s$^2) = 9800$ N, so F_T must be greater than 9800 N (in order that m_2 will accelerate upward). For the elevator, $m_1 g = (1150$ kg$)(9.80$ m/s$^2) = 11{,}300$ N, which must have greater magnitude than F_T so that m_1 accelerates downward. Thus our calculation must give F_T between 9800 N and 11,300 N. To find F_T as well as the acceleration a, we apply $\Sigma F = ma$ to each mass, where we take upward as the positive y direction for both masses. With this choice of axes, $a_2 = a$, and $a_1 = -a$. Thus

$$F_T - m_1 g = m_1 a_1 = -m_1 a$$

$$F_T - m_2 g = m_2 a_2 = +m_2 a.$$

We subtract the first equation from the second to get

$$(m_1 - m_2)g = (m_1 + m_2)a.$$

We solve this for a:

$$a = \frac{m_1 - m_2}{m_1 + m_2} g = \frac{1150 \text{ kg} - 1000 \text{ kg}}{1150 \text{ kg} + 1000 \text{ kg}} g = 0.070g = 0.68 \text{ m/s}^2.$$

The elevator (m_1) accelerates downward (and the counterweight m_2 upward) at $a = 0.070g = 0.68$ m/s^2.
(b) The tension in the cord F_T can be obtained from either of the two $\Sigma F = ma$ equations, setting $a = 0.070g = 0.68$ m/s^2:

$$F_T = m_1 g - m_1 a = m_1(g - a)$$
$$= 1150 \text{ kg} \left(9.80 \text{ m/s}^2 - 0.68 \text{ m/s}^2\right) = 10{,}500 \text{ N},$$

$$F_T = m_2 g + m_2 a = m_2(g + a)$$
$$= 1000 \text{ kg} \left(9.80 \text{ m/s}^2 + 0.68 \text{ m/s}^2\right) = 10{,}500 \text{ N},$$

which are consistent.
    We can check our equation for the acceleration a in this Example by noting that if the masses were equal $(m_1 = m_2)$, then our equation above for a would give $a = 0$, as we should expect. Also, if one of the masses is zero (say, $m_1 = 0$), then the other mass $(m_2 \neq 0)$ would be predicted by our equation to accelerate at $a = g$, again as expected.

---

† We'll see how to deal with a rotating pulley with mass in Chapter 10.

CONCEPTUAL EXAMPLE 4–14 | **The advantage of a pulley.** A mover is trying to lift a piano (slowly) up to a second-story apartment (Fig. 4–25). He is using a rope looped over two pulleys as shown. What force must he exert on the rope to slowly lift the piano's 2000 N weight?

RESPONSE Look at the forces acting on the lower pulley at the piano. The weight of the piano is pulling down. The tension in the rope, looped through this pulley, pulls up *twice*, once on each side of the pulley. Thus, Newton's second law gives

$$2F_T - mg = ma.$$

To move the piano with constant speed ($a = 0$) requires a tension in the cord, and hence a pull on the cord, of $F_T = mg/2$. The mover exerts a force equal to half the piano's weight. We say the pulley has given a **mechanical advantage** of 2, since without the pulley the mover would have to exert twice the force.

FIGURE 4–25 Example 4–14.

EXAMPLE 4–15 **Getting the car out of the mud.** Finding her car stuck in the mud, a bright graduate of a good physics course ties a strong rope to the back bumper of the car, and the other end to a tree, as shown in Fig. 4–26a. She pushes at the midpoint of the rope with her maximum effort, which she estimates to be a force $F_P \approx 300$ N. The car just begins to budge with the rope at an angle θ (see the Figure) which she estimates to be 5°. With what force is the rope pulling on the car? Neglect the mass of the rope.

How to get out of the mud

SOLUTION First, note that the tension in a rope is always along the rope. Any component perpendicular to the rope would cause the rope to bend or buckle (as it does here where $\mathbf{F}_P$ acts)—in other words, a rope can support a tension force only along its length. Let $\mathbf{F}_{T1}$ and $\mathbf{F}_{T2}$ be the forces the rope exerts on the tree and on the car, as shown in Fig. 4–26a. As our "free body," we choose the tiny section of rope where she pushes. The free-body diagram is shown in Fig. 4–26b, which shows $\mathbf{F}_P$ as well as the tensions in the rope (note that we have used Newton's third law). At the moment the car budges, the acceleration is still essentially zero, so $\mathbf{a} = 0$. For the x component of $\Sigma \mathbf{F} = m\mathbf{a} = 0$ on that small section of rope, we have

$$\Sigma F_x = F_{T1} \cos \theta - F_{T2} \cos \theta = 0.$$

Hence $F_{T1} = F_{T2}$, and we can write $F_T = F_{T1} = F_{T2}$. In the y direction, the forces acting are $\mathbf{F}_P$, and the components of $\mathbf{F}_{T1}$ and $\mathbf{F}_{T2}$ that point in the negative y direction (each equal to $F_T \sin \theta$). So for the y component of $\Sigma \mathbf{F} = m\mathbf{a}$, we have

$$\Sigma F_y = F_P - 2F_T \sin \theta = 0.$$

We solve this for F_T and insert $F_P \approx 300$ N, which was given:

$$F_T = \frac{F_P}{2 \sin \theta} \approx \frac{300 \text{ N}}{2 \sin 5°} \approx 1700 \text{ N}.$$

She was able to magnify her effort almost six times using this technique! Notice the symmetry of the problem, which ensures that $F_{T1} = F_{T2}$.

➡ **PROBLEM SOLVING**

Use any symmetry present to simplify a problem

FIGURE 4–26 Example 4–15. Getting a car out of the mud.

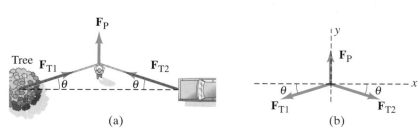

(a)   (b)

EXAMPLE 4–16 **Accelerometer.** A small mass m hangs from a thin string and can swing like a pendulum. You attach it above the window of your car as shown in Fig. 4–27a. When the car is at rest, the string hangs vertically. What angle θ does the string make (*a*) when the car accelerates at a constant $a = 1.20\,\mathrm{m/s^2}$, and (*b*) when the car moves at constant velocity, $v = 90\,\mathrm{km/h}$?

SOLUTION (*a*) Figure 4–27b shows the pendulum at some angle θ and the forces on it: $m\mathbf{g}$ downward and the tension $\mathbf{F}_T$ in the cord. These forces do not add up to zero if $\theta \neq 0$, and since we have an acceleration a, we therefore expect $\theta \neq 0$. Note that θ is the angle relative to the vertical. The acceleration $a = 1.20\,\mathrm{m/s^2}$ is horizontal, so from Newton's second law,

$$ma = F_T\sin\theta$$

for the horizontal component, whereas the vertical component gives

$$0 = F_T\cos\theta - mg.$$

From these two equations, we obtain

$$\tan\theta = \frac{F_T\sin\theta}{F_T\cos\theta} = \frac{ma}{mg} = \frac{a}{g}$$

or

$$\tan\theta = \frac{1.20\,\mathrm{m/s^2}}{9.80\,\mathrm{m/s^2}} = 0.122$$

so

$$\theta = 7.0°.$$

(*b*) The velocity is constant, so $a = 0$ and $\tan\theta = 0$. Hence the pendulum hangs vertically ($\theta = 0°$). This simple device is an **accelerometer**—it can be used to measure acceleration.

FIGURE 4–27 Example 4–16.

(a)   (b)

Now we consider what happens when an object slides down an incline, such as a hill or ramp. Such problems are interesting because gravity is the accelerating force, yet the acceleration is not vertical. Solving problems is usually easier if we choose the xy coordinate system so the x axis points along the incline and the y axis is perpendicular to the incline, as shown in Fig. 4–28a. Note also that the normal force is not vertical, but is perpendicular to the plane, Fig. 4–28b.

EXAMPLE 4–17 **Box slides down an incline.** A box of mass m is placed on a smooth (frictionless) incline that makes an angle θ with the horizontal, as shown in Fig. 4–28a. (a) Determine the normal force on the box. (b) Determine the box's acceleration. (c) Evaluate for a mass $m = 10\,\text{kg}$ and an incline of $\theta = 30°$.

Motion on an incline

SOLUTION The free-body diagram is shown in Fig. 4–28b. The forces on the box are its weight mg vertically downward, which is shown resolved into its components parallel and perpendicular to the incline, and the normal force F_N. The incline acts as a constraint, allowing motion along its surface. The "constraining" force is the normal force. Since we know the motion is along the incline we choose the x axis downward along the incline (the direction of motion), and the y axis perpendicular to the incline upward, as shown. There is no motion in the y direction, so $a_y = 0$. Applying Newton's second law we have

$$F_y = ma_y$$

$$F_N - mg\cos\theta = 0,$$

where F_N and the y component of gravity ($mg\cos\theta$) are all the forces acting on the box in the y direction. Thus the answer to part (a) is that the normal force is given by

$$F_N = mg\cos\theta.$$

Note carefully that unless $\theta = 0°$, F_N has magnitude less than the weight mg. (b) In the x direction the only force acting is the x component of $m\mathbf{g}$, which we see from the diagram is $mg\sin\theta$. The acceleration a is in the x direction so

$F_N < mg\ if\ \theta \neq 0°$

$$F_x = ma_x$$

$$mg\sin\theta = ma,$$

and we see that the acceleration down the plane is

$$a = g\sin\theta.$$

Thus the acceleration along an incline is always less than g, except at $\theta = 90°$, for which $\sin\theta = 1$ and $a = g$. This makes sense, of course, since $\theta = 90°$ is pure vertical fall. For $\theta = 0°$, $a = 0$, which makes sense since $\theta = 0°$ means the plane is horizontal so gravity causes no acceleration. Note too that the acceleration does not depend on the mass m.
(c) For $\theta = 30°$, $\cos\theta = 0.866$ and $\sin\theta = 0.500$, so

$$F_N = 0.866mg = 85\,\text{N},$$

and

$$a = 0.500\,g = 4.9\,\text{m/s}^2.$$

We will discuss more Examples of motion on an incline in the next chapter, where friction will be included.

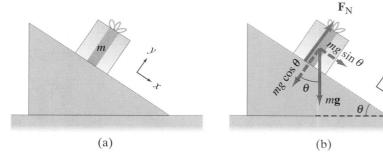

(a)

(b)

FIGURE 4–28 Example 4–17. (a) Box sliding on inclined plane. (b) Free-body diagram of box.

4–8 Problem Solving—A General Approach

A basic part of a physics course is solving problems effectively. The approach discussed here, though emphasizing Newton's laws, can be applied generally for other topics discussed throughout this book.

PROBLEM SOLVING	In General

1. **Read** and reread written problems carefully. A common error is to leave out a word or two when reading, which can completely change the meaning of a problem.

2. **Draw** an accurate picture or diagram of the situation. (This is probably the most overlooked, yet most crucial, part of solving a problem.) Use arrows to represent vectors such as velocity or force, and label the vectors with appropriate symbols. When dealing with forces and applying Newton's laws, make sure to include all forces on a given body, including unknown ones, and make clear what forces act on what body (otherwise you may make an error in determining the *net force* on a particular body). A separate **free-body diagram** needs to be drawn for each body involved, and it must show *all* the forces acting on a given body (and only on that body). Do not show forces that the body exerts on other bodies.

3. Choose a convenient *xy* **coordinate system** (choose one that makes your calculations easier). Vectors are to be resolved into components along these axes. When using Newton's second law, apply $\Sigma \mathbf{F} = m\mathbf{a}$ separately to *x* and *y* components, remembering that *x* direction forces are related to a_x, and similarly for *y*.

4. Note what the unknowns are—that is, what you are trying to determine—and decide what you need in order to find the unknowns. For problems in the present chapter, we use Newton's laws. More generally, it may help to see if there are one or more **relationships** (or **equations**) that relate the unknowns to the knowns. But be sure each relationship is applicable in the given case. It is very important to know the limitations of each formula or relationship—when it is valid and when not. In this book, the more general

equations have been given numbers, but even these can have a limited range of validity (often stated briefly, in brackets, to the right of the equation).

5. Try to solve the problem approximately, to see if it is doable (to check if enough information has been given) and reasonable. Use your intuition, and make **rough calculations**—see "Order of Magnitude Estimating" in Section 1–6. A rough calculation, or a reasonable guess about what the range of final answers might be, is very useful. And a rough calculation can be checked against the final answer to catch errors in calculation, such as in a decimal point or the powers of 10.

6. **Solve** the problem, which may include algebraic manipulation of equations and/or numerical calculations. Recall the mathematical rule that you need as many independent equations as you have unknowns; if you have three unknowns, for example, then you need three independent equations. It is usually best to work out the algebra symbolically before putting in the numbers. Why? Because (*a*) you can then solve a whole class of similar problems with different numerical values; (*b*) you can check your result for cases already understood (say, $\theta = 0°$ or $90°$); (*c*) there may be cancellations or other simplifications; (*d*) there is usually less chance for numerical error; and (*e*) because you may gain better insight into the problem.

7. Be sure to keep track of **units**, for they can serve as a check (they must balance on both sides of any equation).

8. Again consider if your answer is **reasonable**. The use of dimensional analysis, described in Section 1–7, can also serve as a check for many problems.

Summary

Newton's three laws of motion are the basic classical laws describing motion.

Newton's first law (the law of inertia) states that if the net force on an object is zero, an object originally at rest remains at rest, and an object in motion remains in motion in a straight line with constant velocity.

Newton's second law states that the acceleration of a body is directly proportional to the net force acting on it, and inversely proportional to its mass:

$$\Sigma \mathbf{F} = m\mathbf{a}.$$

Newton's second law is one of the most important and fundamental laws in classical physics.

Newton's third law states that whenever one body exerts a force on a second body, the second body always exerts a force on the first body which is equal in magnitude but opposite in direction:

$$\mathbf{F}_{12} = -\mathbf{F}_{21}.$$

The tendency of a body to resist a change in its motion is called **inertia**. **Mass** is a measure of the inertia of a body.

Weight refers to the force of gravity on a body, and is equal to the product of the body's mass m and the acceleration of gravity $\mathbf{g}$:

$$\mathbf{F}_G = m\mathbf{g}.$$

Force, which is a vector, can be considered as a push or pull; or, from Newton's second law, force can be defined as an action capable of giving rise to acceleration. The **net force** on an object is the vector sum of all forces acting on it.

For solving problems involving the forces on one or more bodies, it is essential to draw a **free-body diagram** for each body, showing all the forces acting on only that body. Newton's second law can be applied to the vector components for each body.

Questions

1. Why does a child in a wagon seem to fall backward when you give the wagon a sharp pull?

2. A box rests on the smooth bed of a truck. The truck driver starts the truck and accelerates forward. The box immediately starts to slide toward the rear of the truck bed. Discuss the motion of the box, in terms of Newton's laws, as seen (a) by Andrea standing on the ground beside the truck, and (b) by Dave who is riding on the truck (Fig. 4–29). Assume that the contact surfaces between the box and the truck bed are so smooth that you can ignore friction.

FIGURE 4–29 Question 2.

3. If the acceleration of a body is zero, are no forces acting on it?

4. Why do you push down harder on the pedals of a bicycle when first starting out than when moving at constant speed?

5. Only one force acts on an object. Can the object have zero acceleration? Can it have zero velocity?

6. When a golf ball is dropped to the pavement, it bounces back up. (a) Is a force needed to make it bounce back up? (b) If so, what exerts the force?

7. Which of the following objects weighs about 1 N: (a) an apple, (b) a mosquito, (c) this book, (d) you?

8. Why might your foot hurt if you kick a heavy desk or a wall?

9. When you are running and want to stop quickly, you must decelerate quickly. (a) What is the origin of the force that causes you to stop? (b) Estimate (using your own experience) the maximum rate of deceleration of a person running at top speed to come to rest.

10. A stone hangs by a fine thread from the ceiling, and a section of the same thread dangles from the bottom of the stone (Fig. 4–30). If a person gives a sharp pull on the dangling thread, where is the thread likely to break: below the stone or above it? What if the person gives a slow and steady pull? Explain your answers.

FIGURE 4–30 Question 10.

11. The force of gravity on a 2-kg rock is twice as great as that on a 1-kg rock. Why then doesn't the heavier rock fall faster?

12. Observe the motion of the moving disks of an air hockey game. Explain how Newton's first, second, and third laws apply. The disks float on a layer of air, ejected through tiny holes, so friction is reduced to a very small amount.

13. Compare the effort (or force) needed to lift a 10-kg object when you are on the Moon as compared to lifting it on Earth. Compare the force needed to throw a 2-kg object horizontally with a given speed when on the Moon as compared to on Earth.

14. According to Newton's third law, each team in a tug of war (Fig. 4–31) pulls with equal force on the other team. What, then, determines which team will win?

FIGURE 4–31 A tug of war. Describe the forces on each of the teams and on the rope. Question 14.

15. Whiplash sometimes results from an automobile accident when the victim's car is struck violently from the rear. Explain why the head of the victim seems to be thrown backward in this situation. Is it really?

16. If you walk on a log floating on a lake, why does the log move in the opposite direction?

17. Mary exerts an upward force of 40 N to hold a bag of groceries. Describe the "reaction" force (Newton's third law) by stating (*a*) its magnitude, (*b*) its direction, (*c*) *on* what body it is exerted, and (*d*) *by* what body it is exerted.

18. When you stand still on the ground, how large a force does the ground exert on you? Why doesn't this force make you rise up into the air?

19. A bear sling, Fig. 4–32, is used in some national parks for placing backpackers' food out of the reach of bears. Explain why the force needed to pull the backpack up increases as the backpack gets higher and higher. Is it possible to pull the rope hard enough so that it doesn't sag at all?

FIGURE 4–32 Question 19.

Problems

Sections 4–4 to 4–6

1. (I) Show that a $\frac{1}{4}$-lb cube of butter, or any other "quarter pounder" weighs about 1 N.

2. (I) A net force of 255 N accelerates a bike and rider at 2.20 m/s². What is the mass of the bike and rider?

3. (I) How much force is required to accelerate a 7.0-g object at 10,000 "*g*'s" (say, in a centrifuge)?

4. (I) How much tension must a rope withstand if it is used to accelerate a 1250-kg car horizontally at 1.30 m/s²? Ignore friction.

5. (I) What is the weight of a 58-kg astronaut (*a*) on Earth, (*b*) on the Moon ($g = 1.7$ m/s²), (*c*) on Mars ($g = 3.7$ m/s²), (*d*) in outer space traveling with constant velocity?

6. (II) What average force is required to stop a 1050-kg car in 7.0 s if it is traveling at 90 km/h?

7. (II) What average force is needed to accelerate a 6.25-gram pellet from rest to 155 m/s over a distance of 0.700 m along the barrel of a rifle?

8. (II) A 30.0-kg box rests on a table. (*a*) What is the weight of the box and the normal force acting on it? (*b*) A 20.0-kg box is placed on top of the 30.0-kg box, as shown in Fig. 4–33. Determine the normal force that the table exerts on the 30.0-kg box and the normal force that the 30.0-kg box exerts on the 20.0-kg box.

FIGURE 4–33 Problem 8.

9. (II) Superman must stop a 100-km/h train in 150 m to keep it from hitting a stalled car on the tracks. If the train's mass is 3.6×10^5 kg, how much force must he exert? Compare to the weight of the train.

10. (II) Consider a box resting on a frictionless surface. A constant force acts on the box for a given amount of time, accelerating it to some final speed. Then this process is repeated with another box, which has twice the mass of the first one. Compare the final speed of the heavier box to that of the first box.

11. (II) A fisherman yanks a fish out of the water with an acceleration of 3.5 m/s² using very light fishing line that has a breaking strength of 25 N. The fisherman unfortunately loses the fish as the line snaps. What can you say about the mass of the fish?

12. (II) A 0.140-kg baseball traveling 41.0 m/s strikes the catcher's mitt, which, in bringing the ball to rest, recoils backward 12.0 cm. What was the average force applied by the ball on the glove?

13. (II) Estimate the average force exerted by a shot-putter on a 7.0-kg shot if the shot is moved through a distance of 2.8 m and is released with a speed of 13 m/s.

14. (II) How much tension must a rope withstand if it is used to accelerate a 1200-kg car vertically upward at 0.80 m/s²?

15. (II) A 7.50-kg bucket is lowered by a rope in which there is 63.0 N of tension. What is the acceleration of the bucket? Is it up or down?

16. (II) An elevator (mass 4125 kg) is to be designed so that the maximum acceleration is 0.0600 g. What are the maximum and minimum forces the motor should exert on the supporting cable?

17. (II) A 65-kg petty thief wants to escape from a third-story jail window. Unfortunately, a makeshift rope made of sheets tied together can support a mass of only 57 kg. How might the thief use this "rope" to escape? Give a quantitative answer.

18. (II) A person stands on a bathroom scale in a motionless elevator. When the elevator begins to move, the scale briefly reads only 0.75 of the person's regular weight. Calculate the acceleration of the elevator, and find the direction of acceleration.

19. (II) The cable supporting a 2100-kg elevator has a maximum strength of 21,750 N. What maximum upward acceleration can it give the elevator without breaking?

20. (II) A particular race car can cover a quarter-mile track (402 m) in 6.40 seconds, starting from a standstill. Assuming the acceleration is constant, how many "g's" does the driver experience? If the combined mass of the driver and race car is 280 kg, what horizontal force must the road exert on the tires?

21. (II) A Saturn V rocket has a mass of 2.75×10^6 kg and exerts a force of 33×10^6 N on the gases it expels. Determine (a) the initial vertical acceleration of the rocket, (b) its velocity after 8.0 s, and (c) how long it takes to reach an altitude of 9500 m. Ignore the mass of gas expelled (not realistic) and assume g remains constant.

22. (II) (a) What is the acceleration of two falling sky divers (mass 120.0 kg including parachute) when the upward force of air resistance is equal to one-fourth of their weight? (b) After popping open the parachute, the divers descend leisurely to the ground at constant speed. What now is the force of air resistance on the sky divers and their parachute? See Fig. 4–34.

FIGURE 4–34 Problem 22.

23. (II) An exceptional standing jump would raise a person 0.80 m off the ground. To do this, what force must a 61-kg person exert against the ground? Assume the person crouches a distance of 0.20 m prior to jumping, and thus the upward force has this distance to act over before he leaves the ground.

24. (III) A person jumps from the roof of a house 3.5-m high. When he strikes the ground below, he bends his knees so that his torso decelerates over an approximate distance of 0.70 m. If the mass of his torso (excluding legs) is 43 kg, find (a) his velocity just before his feet strike the ground, and (b) the average force exerted on his torso by his legs during deceleration.

25. (III) The 100-m dash can be run by the best sprinters in 10.0 s. A 62-kg sprinter accelerates uniformly for the first 45 m to reach top speed, which he maintains for the remaining 55 m. (a) What is the average horizontal component of force exerted on his feet by the ground during acceleration? (b) What is the speed of the sprinter over the last 55 m of the race (i.e., his top speed)?

Section 4–7

26. (I) A box weighing 85 N rests on a table. A rope tied to the box runs vertically upward over a pulley and a weight is hung from the other end (Fig. 4–35). Determine the force that the table exerts on the box if the weight hanging on the other side of the pulley weighs (a) 30 N, (b) 60 N, and (c) 90 N.

FIGURE 4–35
Problem 26.

27. (I) A 750-N force acts in a northwesterly direction. A second 750-N force must be exerted in what direction so that the resultant of the two forces points westward?

28. (I) Draw the free-body diagram for a basketball player (a) just before leaving the ground on a jump, and (b) while in the air. (Fig. 4–36).

FIGURE 4–36   Problem 28.

29. (I) Sketch the free-body diagram of a baseball (a) at the moment it is hit by the bat, and again (b) after it has left the bat and is flying toward the outfield.

30. (II) At the instant a race began, a 57-kg sprinter was found to exert a force of 80 N on the starting block at a 22° angle with respect to the ground. (a) What was the horizontal acceleration of the sprinter? (b) If the force was exerted for 0.34 s, with what speed did the sprinter leave the starting block?

31. (II) The two forces F_1 and F_2 shown in Fig. 4–37a and b (looking down) act on a 29.0-kg object on a frictionless tabletop. If $F_1 = 20.2$ N and $F_2 = 26.0$ N, find the net force on the object and its acceleration for each situation, (a) and (b).

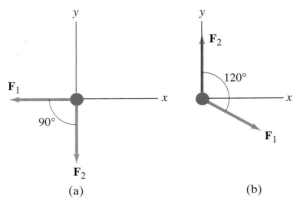

FIGURE 4–37   Problem 31.

32. (II) A person pushes a 13.0-kg lawn mower at constant speed with a force of 78.0 N directed along the handle, which is at an angle of 45.0° to the horizontal (Fig. 4–38). (a) Draw the free-body diagram showing all forces acting on the mower. Calculate (b) the horizontal retarding force on the mower, then (c) the normal force exerted vertically upward on the mower by the ground, and (d) the force the person must exert on the lawn mower to accelerate it from rest to 1.2 m/s in 2.0 seconds (assuming the same retarding force).

FIGURE 4–38   Problem 32.

33. (II) Redo Example 4–13 but (a) set up the equations so that the direction of the acceleration **a** of each object is in the direction of motion of that object. (In Example 4–13, we took **a** as positive upward for both masses.) (b) Solve the equations to obtain the same answers as in Example 4–13.

34. (II) A 7500-kg helicopter accelerates upward at 0.52 m/s² while lifting a 1200-kg car. (a) What is the lift force exerted by the air on the rotors? (b) What is the tension in the cable (ignore its mass) that connects the car to the helicopter?

35. (II) One 3.5-kg paint bucket is hanging by a massless cord from another 3.5-kg paint bucket, also hanging by a massless cord, as shown in Fig. 4–39. (a) If the buckets are at rest, what is the tension in each cord? (b) If the two buckets are pulled upward with an acceleration of 1.60 m/s² by the upper cord, calculate the tension in each cord.

FIGURE 4–39
Problems 35 and 44.

36. (II) A window washer pulls herself upward using the bucket-pulley apparatus shown in Fig. 4–40. (*a*) How hard must she pull downward to raise herself slowly at constant speed? (*b*) If she increases this force by 10 percent, what will her acceleration be? The mass of the person plus the bucket is 58 kg.

FIGURE 4–40 Problem 36.

37. (II) Arlene is to walk across a "high wire" strung horizontally between two buildings 10.0 m apart. The sag in the rope when she is at the mid-point should not exceed 10°, as shown in Fig. 4–41. If her mass is 50.0 kg, what must be the tension in the rope?

FIGURE 4–41 Problem 37.

38. (II) Tom's hang glider supports his weight using the six ropes shown in Fig. 4–42. Each rope is designed to support an equal fraction of Tom's weight. Tom's mass is 70.0 kg. What is the tension in each of the support ropes?

FIGURE 4–42 Problem 38.

39. (II) Two snowcats tow a housing unit to a new location at McMurdo Base, Antarctica, as shown in Fig. 4–43. The sum of the forces F_A and F_B exerted on the unit by the horizontal cables is parallel to the line L, and F_A = 4500 N. Determine F_B and the magnitude of $\mathbf{F}_A + \mathbf{F}_B$.

FIGURE 4–43 Problem 39.

40. (II) The block shown in Fig. 4–44 has mass m = 7.0 kg and lies on a smooth frictionless plane tilted at an angle θ = 22.0° to the horizontal. (*a*) Determine the acceleration of the block as it slides down the plane. (*b*) If the block starts from rest 12.0 m up the plane from its base, what will be the block's speed when it reaches the bottom of the incline?

FIGURE 4–44 Block on inclined plane. Problems 40 and 41.

41. (II) A block is given an initial speed of 4.0 m/s up the 22° plane shown in Fig. 4–44. (*a*) How far up the plane will it go? (*b*) How much time elapses before it returns to its starting point? Ignore friction.

42. (II) A 27-kg chandelier hangs from a ceiling on a vertical 4.0-m-long wire. (*a*) What horizontal force would be necessary to displace its position 0.10 m to one side? (*b*) What will be the tension in the wire?

43. (II) A mass m is at rest at t = 0. Then a constant force F_0 acts on it for a time t_0. Suddenly the force doubles to $2F_0$ and remains constant until t = $2t_0$. Determine the total distance traveled from t = 0 to t = $2t_0$.

44. (II) The cords accelerating the buckets in Problem 35 (Fig. 4–39) are each 1.0 m long and each has a weight of 2.0 N. Determine the tension in each cord at its points of attachment (highest and lowest points).

45. (II) A train locomotive is pulling two cars of the same mass behind it. Show that the tension in the coupling between the locomotive and the first car is twice that between the first car and the second car, for any nonzero acceleration of the train.

FIGURE 4–45 Problem 46.

46. (II) Three blocks on a frictionless horizontal surface are in contact with each other as shown in Fig. 4–45. A force **F** is applied to block 1 (mass m_1). (*a*) Draw a free-body diagram for each block. Determine (*b*) the acceleration of the system (in terms of m_1, m_2, and m_3), (*c*) the net force on each block, and (*d*) the force of contact that each block exerts on its neighbor. (*e*) If $m_1 = m_2 = m_3 = 12.0$ kg and $F = 96.0$ N, give numerical answers to (*b*), (*c*), and (*d*). Do your answers make sense intuitively?

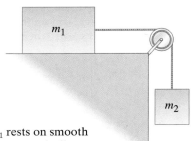

FIGURE 4–46 Mass m_1 rests on smooth horizontal surface, m_2 hangs vertically. Problems 47, 48, 49, and 57.

47. (II) Figure 4–46 shows a block (mass m_1) on a smooth horizontal surface, connected by a thin cord that passes over a pulley to a second block (m_2), which hangs vertically. (*a*) Draw a free-body diagram for each block, showing the force of gravity on each, the force (tension) exerted by the cord, and any normal force. (*b*) Determine formulas for the acceleration of the system and for the tension in the cord. Ignore friction and the masses of the pulley and cord.

48. (II) (*a*) If $m_1 = 13.0$ kg and $m_2 = 6.0$ kg in Fig. 4–46, determine the acceleration of each block. (*b*) If initially m_1 is at rest 1.250 m from the edge of the table, how long does it take to reach the edge of the table if the system is allowed to move freely? (*c*) If $m_2 = 1.0$ kg, how large must m_1 be if the acceleration of the system is to be kept at $\frac{1}{100}g$?

49. (III) Determine a formula for the acceleration of the system shown in Fig. 4–46 (see Problem 47) if the cord has a nonnegligible mass m_C. Specify in terms of l_1 and l_2, the lengths of cord from the respective masses to the pulley. (The total cord length is $l = l_1 + l_2$.)

50. (III) The two masses shown in Fig. 4–47 are each initially 1.80 m above the ground, and the massless frictionless pulley is fixed 4.8 m above the ground. What maximum height does the lighter object reach after the system is released? [*Hint*: First determine the acceleration of the lighter mass and then its velocity at the moment the heavier one hits the ground. This is its "launch" speed. Assume it doesn't hit the pulley.]

FIGURE 4–47 Problem 50.

51. (III) Suppose the cord in Example 4–12 and Fig. 4–23 is a heavy rope of mass 1.0 kg. Calculate the acceleration of each box and the tension at each end of the cord, using the free-body diagrams shown in Fig. 4–48. Assume the cord is fairly rigid so we can neglect any sagging.

FIGURE 4–48 Problem 51. Free-body diagrams for each of the objects of the system shown in Fig. 4–23a. Vertical forces, **F**$_N$ and **F**$_G$, are not shown.

52. (III) Suppose the pulley in Fig. 4–49 is suspended by a cord C. Determine the tension in this cord after the mass is released and before it hits the ground. Ignore the mass of the pulley.

1.2 kg 3.2 kg

FIGURE 4–49
Problem 52.

53. (III) A small block of mass m rests on the sloping side of a triangular block of mass M which itself rests on a horizontal table as shown in Fig. 4–50. Assuming all surfaces are frictionless, determine the force F that must be applied to M so that m remains in a fixed position relative to M (that is, m doesn't move on the incline).

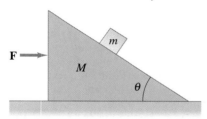

FIGURE 4–50
Problem 53.

54. (III) Determine a formula for the magnitude of the force **F** exerted on the large block (m_3) in Fig. 4–51 so that the mass m_1 does not move relative to m_3. Ignore all friction. Assume m_2 does not make contact with m_3.

FIGURE 4–51   Problem 54.

55. (III) The double Atwood machine shown in Fig. 4–52 has frictionless, massless pulleys and cords. Determine (a) the acceleration of masses m_1, m_2, and m_3, and (b) the tensions F_{T1} and F_{T3} in the cords.

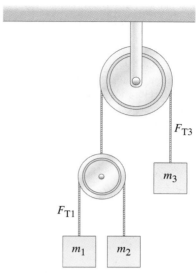

FIGURE 4–52   Problem 55.

*56. (III) A particle of mass m, initially at rest, is accelerated by a force that increases in time as $F = Ct^2$. Determine its velocity v and position x as a function of time.

*57. (III) Determine a formula for the speed v of the masses shown in Fig. 4–46 (see Problem 49) assuming the cord has mass m_C and is uniform. At $t = 0$, $v = 0$, and mass m_2 is at the pulley whereas mass m_1 is a distance l from the pulley. Assume the pulley is very small (ignore its diameter) and ignore friction. [*Hint*: Use the chain rule, $dv/dt = (dv/dy)(dy/dt)$, and integrate.]

*58. (III) A heavy steel cable of length L and mass M passes over a small massless, frictionless pulley. (a) If a length y hangs on one side of the pulley (so $L - y$ hangs on the other side), calculate the acceleration of the cable as a function of y. (b) Assuming the cable starts from rest with length y_0 on one side of pulley, determine the velocity v_f at the moment the whole cable has fallen from the pulley; (c) evaluate v_f for $y_0 = \frac{2}{3}L$. [*Hint*: Use the chain rule, $dv/dt = (dv/dy)(dy/dt)$, and integrate.]

General Problems

59. According to a simplified model of a mammalian heart, at each pulse, approximately 20 g of blood is accelerated from 0.25 m/s to 0.35 m/s during a period of 0.10 s. What is the magnitude of the force exerted by the heart muscle?

60. A person has a reasonable chance of surviving an automobile crash if the deceleration is no more than 30 "g's." Cal-

culate the force on a 70-kg person accelerating at this rate. What distance is traveled if brought to rest at this rate from 90 km/h?

61. A 2.0-kg purse is dropped 55 m from the top of the Leaning Tower of Pisa and reaches the ground with a speed of 29 m/s. What was the average force of air resistance?

62. A fisherman in a boat is using a "10-lb test" fishing line. This means that the line can exert a force of 45 N without breaking (1 lb = 4.45 N). (a) How heavy a fish can the fisherman land if he pulls the fish up vertically at constant speed? (b) If he accelerates the fish upwards at 2.0 m/s², what maximum weight fish can he land?

63. An elevator in a tall building is allowed to reach a maximum speed of 3.5 m/s going down. What must the tension be in the cable to stop this elevator over a distance of 3.0 m if the elevator has a mass of 1300 kg including occupants?

64. A crane's trolley at point P in Fig. 4–53 moves to the right with constant acceleration, and the 800-kg load hangs at a 5.0° angle to the vertical as shown. What is the acceleration of the trolley and load?

FIGURE 4–53   Problem 64.

65. A wet bar of soap (m = 150 g) slides freely down a ramp 3.0 m long inclined at 9.5°. How long does it take to reach the bottom? How would this change if the soap's mass were 300 grams?

66. A block (mass m_1) lying on a frictionless inclined plane is connected to a mass m_2 by a massless cord passing over a pulley, as shown in Fig. 4–54. (a) Determine a formula for the acceleration of the system in terms of m_1, m_2, θ, and g. (b) What conditions apply to masses m_1 and m_2 for the acceleration to be in one direction (say, m_1 down the plane), or in the opposite direction?

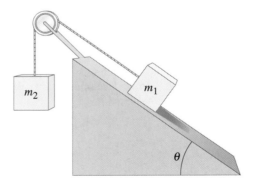

FIGURE 4–54   Problems 66 and 67.

67. (a) In Fig. 4–54, if $m_1 = m_2 = 1.00$ kg and $\theta = 30°$, what will be the acceleration of the system? (b) If $m_1 = 1.00$ kg and $\theta = 30°$, and the system remains at rest, what must the mass m_2 be? (c) Calculate the tension in the cord for (a) and (b).

68. The masses m_1 and m_2 slide on the smooth (frictionless) inclines shown in Fig. 4–55. (a) Determine a formula for the acceleration of the system in terms of m_1, m_2, θ_1, θ_2, and g. (b) If $\theta_1 = 30°$, $\theta_2 = 20°$, and $m_1 = 5.0$ kg, what value of m_2 would keep the system at rest? What would be the tension in the cord in this case?

FIGURE 4–55   Problem 68.

69. If a bicyclist of mass 65 kg (including the bicycle) can coast down a 6.0° hill at a steady speed of 6.0 km/h because of air resistance, how much force must be applied to climb the hill at the same speed (and the same air resistance)?

70. A 75.0-kg person stands on a scale in an elevator. What does the scale read (in N and in kg) when (a) the elevator is at rest, (b) the elevator is climbing at a constant speed of 3.0 m/s, (c) the elevator is falling at 3.0 m/s, (d) the elevator is accelerating upward at 3.0 m/s², (e) the elevator is accelerating downward at 3.0 m/s²?

71. A city planner is working on the redesign of a hilly portion of a city. An important consideration is how steep the roads can be so that even low-powered cars can get up the hills without slowing down. It is given that a particular small car, with a mass of 1100 kg, can accelerate on a level road from rest to 21 m/s (75 km/h) in 14.0 s. Using these data, calculate the maximum steepness of a hill.

72. A bicyclist can coast down a 5.0° hill at a constant speed of 6.0 km/h. If the force of air resistance is proportional to the speed v so that $F_{air} = cv$, calculate (a) the value of the constant c, and (b) the average force that must be applied in order to descend the hill at 20.0 km/h. The mass of the cyclist plus bicycle is 80.0 kg.

73. Francesca, who likes physics experiments, dangles her watch from a thin piece of string while the jetliner she is in takes off from Dulles Airport (Fig. 4–56). She notices that the string makes an angle of 25° with respect to the vertical while the aircraft accelerates for takeoff, which takes about 18 seconds. Estimate the takeoff speed of the aircraft.

FIGURE 4–56   Problem 73.

74. In the design of a supermarket, there are to be several ramps connecting different parts of the store. Customers will have to push grocery carts up the ramps and it is obviously desirable that this not be too difficult. The engineer has done a survey and found that almost no one complains if the force required is no more than 20 N. Ignoring friction, at what maximum angle θ should the ramps be built, assuming a full 20-kg grocery cart?

75. (*a*) What minimum force F is needed to lift the piano (mass M) using the pulley apparatus shown in Fig. 4–57? (*b*) Determine the tension in each section of rope: F_{T1}, F_{T2}, F_{T3}, and F_{T4}.

FIGURE 4–57 Problem 75.

76. A jet aircraft is climbing at an angle of 45° above the horizontal and is also accelerating at 4.5 m/s². What is the total force that the cockpit seat exerts on the 75-kg pilot?

77. In the design process for a child restraint chair, an engineer considers the following set of conditions: A 12-kg child is riding in the chair, which is securely fastened to the seat of an automobile (Fig. 4–58). Assume the automobile is involved in a head-on collision with another vehicle. The initial speed v_0 of the car is 50 km/h and this speed is reduced to zero during the collision time of 0.20 s. Assume a constant car deceleration during the collision and estimate

FIGURE 4–58 Problem 77.

the net horizontal force F that the straps of the restraint chair must exert on the child in order to keep her fixed to the chair. Treat the child as a particle and state any additional assumptions made during your analysis.

78. A helicopter is to lift a 600-kg magnesium frame at a construction site, as shown in Fig. 4–59. As the helicopter accelerates upward at 0.15g, what is the magnitude of the tension F_T in the cord?

FIGURE 4–59 Problem 78.

79. A new high-speed 12-car Italian train has a mass of 660 metric tons (660,000 kg). It can exert a maximum force of 400 kN horizontally against the tracks, whereas at maximum velocity (300 km/h), it exerts a force of about 150 kN. Calculate (*a*) its maximum acceleration, and (*b*) estimate the force of air resistance at top speed.

Newton's laws are fundamental in physics, and this chapter covers some different applications. These photos show two situations of using Newton's laws which involve some new elements in addition to those discussed in the previous chapter. The downhill skier illustrates **friction** on an incline, although at this moment she is not touching the snow, and so is retarded only by air resistance which is a velocity-dependent force (an optional topic in this chapter). The people on the rotating amusement park ride on the right illustrate the dynamics of circular motion.

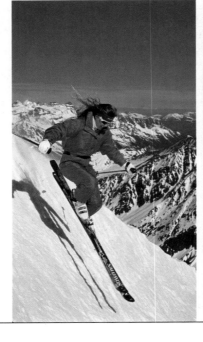

CHAPTER 5

Further Applications of Newton's Laws

This chapter continues our study of Newton's laws and emphasizes their fundamental importance in physics. We cover some important applications of Newton's laws, including friction and the very important subject of the dynamics of circular motion (we studied its kinematics in Chapter 3). Although some material in this chapter may seem to repeat topics covered in Chapter 4, in fact, new elements are involved.

FIGURE 5–1 An object moving to the right on a table or floor. The two surfaces in contact are rough, at least on a microscopic scale.

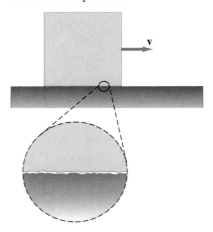

5–1 Applications of Newton's Laws Involving Friction

Until now we have ignored friction, but it must be taken into account in most practical situations. Friction exists between two solid surfaces because even the smoothest looking surface is quite rough on a microscopic scale, Fig. 5–1. When we try to slide an object across another surface, these microscopic bumps impede the motion. In addition, at the atomic level, the atoms on a bump of one surface come so close to the atoms of the other surface that electric forces between the atoms can form chemical bonds, as a tiny weld between the two surfaces. Sliding an object across a surface is often jerky due to the making and breaking of these bonds. Even when a body rolls across a surface, there is still some friction, called *rolling friction*, although it is generally much less than when a body slides across a surface. We will be concerned mainly with sliding friction in this section, and it is usually called **kinetic friction** (*kinetic* is from the Greek for "moving").

When a body is in motion along a rough surface, the force of kinetic friction acts opposite to the direction of the body's velocity. The magnitude of the force of kinetic friction depends on the nature of the two sliding surfaces. For given surfaces, experiment shows that the friction force is approximately proportional to the *normal force* between the two surfaces, which is the force that either object exerts on the other, perpendicular to their common surface of contact (see Fig. 5–2). The force of friction between hard surfaces depends very little on the total surface area of contact; that is, the friction force on this book is roughly the same whether it is being slid on its wide face or on its spine, assuming the surfaces have the same smoothness. We can write the proportionality as an equation by inserting a constant of proportionality, μ_k:

$$F_{fr} = \mu_k F_N.$$

Kinetic friction

This relation is not a fundamental law; it is an experimental relation between the magnitude of the friction force $\mathbf{F}_{fr}$ which acts parallel to the two surfaces, and the magnitude of the normal force $\mathbf{F}_N$ which acts perpendicular to the surfaces. It is *not* a vector equation since the two forces are perpendicular to one another. The term μ_k is called the **coefficient of kinetic friction**, and its value depends on the nature of the two surfaces. Measured values for a variety of surfaces are given in Table 5–1. These are only approximate, however, since μ depends on whether the surfaces are wet or dry, on how much they have been sanded or rubbed, if any burrs remain, and other such factors. But μ_k is roughly independent of the sliding speed.

$\mathbf{F}_{fr} \perp \mathbf{F}_N$

What we have been discussing up to now is *kinetic friction*, when one object slides over another. There is also **static friction**, which refers to a force parallel to the two surfaces that can arise even when they are not sliding. Suppose an object such as a desk is resting on a horizontal floor. If no horizontal force is exerted on the desk, there also is no friction force. But now, suppose you try to push the desk, but it doesn't move. You are exerting a horizontal force, but the desk isn't moving, so there must be another force on the desk keeping it from moving (the net force is zero on an object that doesn't move). This is the force of *static friction* exerted by the floor on the desk. If you push with a greater force without moving the desk, the force of static friction also has increased. If you push hard enough, the desk will eventually start to move, and kinetic friction takes over. At this point, you have exceeded the maximum force of static friction, which is given by $F_{max} = \mu_s F_N$, where μ_s is the **coefficient of static friction** (Table 5–1). Since the force of static friction can vary from zero to this maximum value, we write

FIGURE 5–2 When an object is pulled by an applied force $(\mathbf{F}_A)$ along a surface, the force of friction $\mathbf{F}_{fr}$ opposes the motion. The magnitude of $\mathbf{F}_{fr}$ is proportional to the magnitude of the normal force F_N.

$$F_{fr} \leq \mu_s F_N.$$

Static friction

TABLE 5–1 Coefficients of Friction[†]

Surfaces	Coefficient of Static Friction, μ_s	Coefficient of Kinetic Friction, μ_k
Wood on wood	0.4	0.2
Ice on ice	0.1	0.03
Metal on metal (lubricated)	0.15	0.07
Steel on steel (unlubricated)	0.7	0.6
Rubber on dry concrete	1.0	0.8
Rubber on wet concrete	0.7	0.5
Rubber on other solid surfaces	1–4	1
Teflon® on Teflon in air	0.04	0.04
Teflon on steel in air	0.04	0.04
Lubricated ball bearings	< 0.01	< 0.01
Synovial joints (in human limbs)	0.01	0.01

[†] Values are approximate and are intended only as a guide.

$\mathbf{F}_N$

$\mathbf{F}_A$

$\mathbf{F}_{fr}$

$m\mathbf{g}$

FIGURE 5–2 Repeated for Example 5–1.

FIGURE 5–3 Magnitude of the force of friction as a function of the external force applied to a body initially at rest. As the applied force is increased in magnitude, the force of static friction increases linearly to just match it, until the applied force equals $\mu_s F_N$. If the applied force increases further, the body will begin to move, and the friction force drops to a roughly constant value characteristic of kinetic friction.

You may have noticed that it is often easier to keep a heavy object moving, such as pushing a table, than it is to start it moving in the first place. This is consistent with the fact (see Table 5–1) that μ_s is generally greater than μ_k. (It can never be less. Why?)

EXAMPLE 5–1 **Friction: static and kinetic.** Our 10.0-kg mystery box rests on a horizontal floor. The coefficient of static friction is $\mu_s = 0.40$ and the coefficient of kinetic friction is $\mu_k = 0.30$. Determine the force of friction, F_{fr}, acting on the box if a horizontal external applied force F_A is exerted on it of magnitude: (a) 0, (b) 10 N, (c) 20 N, (d) 38 N, and (e) 40 N.

SOLUTION The free-body diagram of the box is shown in Fig. 5–2. Examine it carefully. In the vertical direction there is no motion, so $\Sigma F_y = ma_y = 0$ yields $F_N - mg = 0$. Hence the normal force for all cases is

$$F_N = mg = (10.0\,\text{kg})(9.80\,\text{m/s}^2) = 98.0\,\text{N}.$$

(a) Since no force is applied in this first case, the box doesn't move, and $F_{fr} = 0$.
(b) The force of static friction will oppose any applied force up to a maximum of

$$\mu_s F_N = (0.40)(98.0\,\text{N}) = 39\,\text{N}.$$

The applied force is $F_A = 10\,\text{N}$. Thus the box will not move; since $\Sigma F_x = F_A - F_{fr} = 0$ then $F_{fr} = 10\,\text{N}$.
(c) An applied force of 20 N is also not sufficient to move the box. Thus $F_{fr} = 20\,\text{N}$ to balance the applied force.
(d) The applied force of 38 N is still not quite large enough to move the box; so the friction force has now increased to 38 N to keep the box at rest.
(e) A force of 40 N will start the box moving since it exceeds the maximum force of static friction, $\mu_s F_N = (0.40)(98.0\,\text{N}) = 39\,\text{N}$. Instead of static friction, we now have kinetic friction, and its magnitude is

$$F_{fr} = \mu_k F_N = (0.30)(98.0\,\text{N}) = 29\,\text{N}.$$

There is now a net (horizontal) force on the box of magnitude $F = 40\,\text{N} - 29\,\text{N} = 11\,\text{N}$, so the box will accelerate at a rate

$$a_x = \Sigma F/m = 11\,\text{N}/10\,\text{kg} = 1.1\,\text{m/s}^2$$

as long as the applied force is 40 N. Figure 5–3 shows a graph that summarizes this Example.

Now we look at some Examples involving kinetic friction in a variety of situations. Note that both the normal force and the friction force are forces exerted by one surface on another; one is perpendicular to the contact surfaces (the normal force), and the other is parallel (the friction force).

CONCEPTUAL EXAMPLE 5–2 **To push or to pull a sled?** Your little sister wants a ride on her sled. If you are on flat ground, will you exert less force if you push her or pull her? See Figs. 5–4a and b. Assume the same angle θ in each case.

RESPONSE Free-body diagrams are shown in Figs. 5–4c and d. If you push her, and $\theta > 0$, there is a vertically downward component to your force. Hence the normal force upward exerted by the ground will be larger than mg (where m is the mass of sister plus sled). If you pull her, your force has a vertically upward component, so the normal force F_N can be less than mg. Because the friction force is proportional to the normal force, it will be less if you pull her. So you exert less force if you pull her.

(a)

(b)

FIGURE 5–4
Example 5–2.

(c)

(d)

CONCEPTUAL EXAMPLE 5–3 | **A box against a wall.** You can hold a box against a rough wall (Fig. 5–5) and prevent it from slipping down by pressing hard horizontally. How does the application of a horizontal force keep an object from moving vertically?

RESPONSE This won't work well if the wall is slippery. You need friction. Even then, if you don't press hard enough, the box will slip. The horizontal force you apply produces a normal force on the box exerted by the wall (why?). The force of gravity, mg, acting downward on the box, can now be balanced by an upward friction force whose magnitude is proportional to the normal force. The harder you push, the greater F_N is and the greater F_{fr} can be. If you don't press hard enough, then $mg > \mu_s F_N$ and the box begins to slide down.

FIGURE 5–5   Example 5–3.

EXAMPLE 5–4 **Pulling against friction.** A 10.0-kg box is pulled along a horizontal surface by a force F_P of 40.0 N which is applied at a 30.0° angle. This is like Example 4–11 in the preceding chapter except now there is friction, and we assume a coefficient of kinetic friction of 0.30. The free-body diagram is shown in Fig. 5–6. Calculate the acceleration.

SOLUTION The force of kinetic friction on the box opposes the direction of motion and is parallel to the surfaces of contact. The calculation for the vertical (y) direction is just the same as before (Example 4–11): there is no motion in the vertical direction because $F_{Py} = F_y \sin 30° = (40.0 \text{ N})(\sin 30.0°) = 20.0 \text{ N}$ is less than the weight of the box $mg = (10.0 \text{ kg})(9.80 \text{ m/s}^2) = 98.0 \text{ N}$. Taking y as positive upward, we have

$$F_N - mg + F_{Py} = ma_y$$

$$F_N - 98.0 \text{ N} + 20.0 \text{ N} = 0$$

and the normal force is $F_N = 78.0 \text{ N}$. Next we apply Newton's second law for the horizontal (x) direction (positive to the right):

$$F_{Px} - F_{fr} = ma_x.$$

FIGURE 5–6   Example 5–4.

The friction force is kinetic as long as $F_{fr} = \mu_k F_N$ is less than F_{Px}, which it is since

$$F_{fr} = \mu_k F_N = (0.30)(78.0\,\text{N}) = 23.4\,\text{N}$$

and

$$F_{Px} = F_P \cos 30.0° = (40.0\,\text{N})(0.866) = 34.6\,\text{N}.$$

[If $\mu_k F_N$ were greater than F_{Px}, what would you conclude?] Hence the box does accelerate:

$$a_x = \frac{F_{Px} - F_{fr}}{m} = \frac{34.6\,\text{N} - 23.4\,\text{N}}{10.0\,\text{kg}} = 1.1\,\text{m/s}^2.$$

In the absence of friction, as we saw in Example 4–11, the acceleration would be much greater. Note that our final answer should have only two significant figures because our least significant input value has two $(\mu_k = 0.30)$

EXAMPLE 5–5 **Two boxes and a pulley.** In Fig. 5–7a, two boxes are connected by a cord running over a pulley. The coefficient of kinetic friction between box I and the table is 0.20. We ignore the mass of the cord and pulley and any friction in the pulley, which means we can assume that a force applied to one end of the cord will have the same magnitude at the other end. We wish to find the acceleration, a, of the system, which will have the same magnitude for both boxes assuming the cord doesn't stretch. As box II moves down, box I moves to the right.

SOLUTION Free-body diagrams are shown for each box in Fig. 5–7b and c. Box I does not move vertically, so the normal force just balances the weight,

$$F_N = m_I g = (5.0\,\text{kg})(9.8\,\text{m/s}^2) = 49\,\text{N}.$$

In the horizontal direction, there are two forces on box I (Fig. 5–7b): F_T, the tension in the cord (whose value we don't know), and the force of friction

$$F_{fr} = \mu_k F_N = (0.20)(49\,\text{N}) = 9.8\,\text{N}.$$

The horizontal acceleration is what we wish to find; we use Newton's second law in the x direction, $\Sigma F_{Ix} = m_I a_x$, which becomes (taking the positive direction to the right and setting $a_{Ix} = a$):

$$\Sigma F_{Ix} = F_T - F_{fr} = m_I a. \qquad \text{[box I]}$$

Next consider box II. The force of gravity $F_G = m_{II} g = 19.6\,\text{N}$ pulls downward; and the cord pulls upward with a force F_T. So we can write Newton's second law for box II (taking the downward direction as positive):

$$\Sigma F_{IIy} = m_{II} g - F_T = m_{II} a. \qquad \text{[box II]}$$

[Note here that if $a \neq 0$, then F_T is not equal to $m_{II} g$.] We have two unknowns, a and F_T, and we also have two equations. We solve the box I equation for F_T:

$$F_T = F_{fr} + m_I a,$$

and substitute this into the box II equation:

$$m_{II} g - F_{fr} - m_I a = m_{II} a.$$

Now we solve for a and put in numerical values:

$$a = \frac{m_{II} g - F_{fr}}{m_I + m_{II}} = \frac{19.6\,\text{N} - 9.8\,\text{N}}{5.0\,\text{kg} + 2.0\,\text{kg}} = 1.4\,\text{m/s}^2,$$

which is the acceleration of box I to the right, and of box II down. If we wish, we can calculate F_T using the first equation:

$$F_T = F_{fr} + m_I a = 9.8\,\text{N} + (5.0\,\text{kg})(1.4\,\text{m/s}^2) = 17\,\text{N}.$$

5.0 kg
I

II
2.0 kg

(a)

$\mathbf{F}_N$

$\mathbf{F}_{fr}$ I $\mathbf{F}_T$

y

x

$m_I \mathbf{g}$

(b)

$\mathbf{F}_T$

II

(c) $m_{II}\mathbf{g}$

FIGURE 5–7 Example 5–5.

Next we take an example of an object moving down an incline, as in Example 4–17 of Chapter 4; but now we include friction.

EXAMPLE 5–6 **The skier.** The skier in Fig. 5–8a has just begun descending the 30° slope. Assuming the coefficient of kinetic friction is 0.10, calculate (a) her acceleration and (b) the speed she will reach after 4.0 s.

SOLUTION First we draw the free-body diagram showing all the forces acting on the skier, Fig. 5–8b: her weight $(\mathbf{F}_G = m\mathbf{g})$ downward, and the two forces exerted on her skis by the snow—the normal force perpendicular to the snow's surface, and the friction force parallel to the surface. These three forces are shown acting at one point in Fig. 5–8b, for convenience. Also for convenience, we choose the x axis parallel to the snow surface with positive direction downhill, and the y axis perpendicular to the surface. With this choice, we have to resolve only one vector into components, the weight, and its components are shown as dashed lines in Fig. 5–8c. They are given by

$$F_{Gx} = mg\sin\theta,$$

$$F_{Gy} = mg\cos\theta,$$

where we have stayed general, using θ rather than 30° for now.
(a) To calculate her acceleration down the hill, a_x, we apply Newton's second law to the x direction:

$$\Sigma F_x = ma_x$$

$$mg\sin\theta - \mu_k F_N = ma_x$$

where the two forces are the component of the gravity force ($+x$ direction) and the friction force ($-x$ direction). We want to find the value of a_x, but we don't yet know F_N in the last equation. Let's see if we can get F_N from the y component of Newton's second law:

$$\Sigma F_y = ma_y$$

$$F_N - mg\cos\theta = ma_y = 0$$

where we set $a_y = 0$ because there is no motion in the y direction (perpendicular to the slope). Thus we can solve for F_N:

$$F_N = mg\cos\theta$$

and we can substitute this into our equation above for ma_x:

$$mg\sin\theta - \mu_k(mg\cos\theta) = ma_x.$$

There is an m in each term which can be canceled out. Thus (setting $\theta = 30°$ and $\mu_k = 0.10$):

$$a_x = g\sin 30° - \mu_k g\cos 30° = 0.50\,g - (0.10)(0.866)g = 0.41\,g.$$

The skier's acceleration is 0.41 times the acceleration of gravity, which in numbers is $a = (0.41)(9.8\,\text{m/s}^2) = 4.0\,\text{m/s}^2$. It is interesting that the mass canceled out here, and so we have the useful conclusion that *the acceleration doesn't depend on the mass*. That such a cancellation sometimes occurs, and thus may give a useful conclusion as well as saving calculation, is a big advantage of working with the algebraic equations and putting in the numbers only at the end.
(b) The speed after 4.0 s is found, since the acceleration is constant, by using Eq. 2–12a:

$$v = v_0 + at = 0 + (4.0\,\text{m/s}^2)(4.0\,\text{s}) = 16\,\text{m/s},$$

where we assumed a start from rest.

➡ **PHYSICS APPLIED**
Sport

(a)

(b)

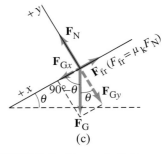

(c)

FIGURE 5–8 Example 5–6: A skier descending a slope.

➡ **PROBLEM SOLVING**
*It is often helpful
to put in numbers
only at the end*

EXAMPLE 5–7 **Measuring** μ_k. Suppose in Example 5–6 that the snow is slushy and the skier moves down the 30° slope at constant speed. What can you say about the coefficient of friction, μ_k?

SOLUTION Now the skier moves down the slope at constant speed, and we want to find μ_k. The free-body diagram and the $\Sigma F = ma$ equations for the x and y components will be the same as above, except that now we are given $a_x = 0$. Thus

$$\Sigma F_y = F_N - mg \cos\theta = ma_y = 0$$

$$\Sigma F_x = mg \sin\theta - \mu_k F_N = ma_x = 0.$$

From the first equation, we have $F_N = mg \cos\theta$; we substitute this into the second equation:

$$mg \sin\theta - \mu_k(mg \cos\theta) = 0.$$

Now we solve for μ_k:

$$\mu_k = \frac{mg \sin\theta}{mg \cos\theta} = \frac{\sin\theta}{\cos\theta} = \tan\theta$$

which for $\theta = 30°$ is

$$\mu_k = \tan\theta = \tan 30° = 0.58.$$

Notice that we could use the equation

$$\mu_k = \tan\theta$$

A method for determining μ_k

to determine μ_k under a variety of conditions. All we need to do is observe at what slope angle the skier descends at constant speed. Here is another reason why it is often useful to plug in numbers only at the end: we obtained a general result useful for other situations as well.

EXAMPLE 5–8 **A plane, a pulley, and two boxes.** A box of mass $m_1 = 10.0$ kg rests on a surface inclined at $\theta = 37°$ to the horizontal. It is connected by a lightweight cord, which passes over a massless and frictionless pulley, to a second box of mass m_2, which hangs freely as shown in Fig. 5–9a. (a) If the coefficient of static friction is $\mu_s = 0.40$, determine what range of values for mass m_2 will keep the system at rest. (b) If the coefficient of kinetic friction is $\mu_k = 0.30$, and $m_2 = 10.0$ kg, determine the acceleration of the system.

SOLUTION (a) Figure 5–9b shows two free-body diagrams for box m_1. The tension force exerted by the cord is labeled $\mathbf{F}_T$. The force of friction can be either up or down the slope, and we show both possibilities in Fig. 5–9b: (i) if $m_2 = 0$ or is sufficiently small, m_1 would tend to slide down the incline, so $\mathbf{F}_{fr}$ would be directed up the incline; (ii) if m_2 is large enough, m_1 will tend to be pulled up the plane, so $\mathbf{F}_{fr}$ would point down the plane. For both cases, Newton's second law for the y direction (perpendicular to the plane) is the same:

$$F_N - m_1 g \cos\theta = m_1 a_y = 0$$

since there is no y motion. So

$$F_N = m_1 g \cos\theta.$$

Now for the x motion. We consider case (i) first for which $\Sigma F = ma$ gives

$$m_1 g \sin\theta - F_T - F_{fr} = m_1 a_x.$$

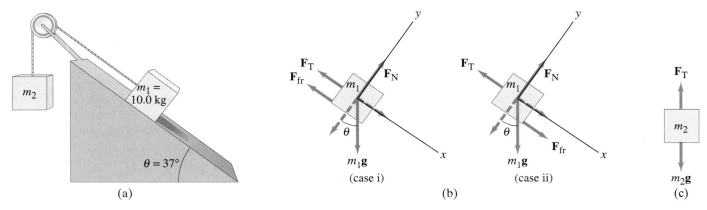

FIGURE 5–9 Example 5–8.

We want $a_x = 0$ and we solve for F_T since F_T is related to m_2 (whose value we are seeking) by $F_T = m_2 g$ (see Fig. 5–9c). Thus

$$m_1 g \sin \theta - F_{fr} = F_T = m_2 g.$$

Since F_{fr} can be at most $\mu_s F_N = \mu_s m_1 g \cos \theta$, the minimum value that m_2 can have to prevent motion $(a_x = 0)$ is, after dividing through by g,

$$m_2 = m_1 \sin \theta - \mu_s m_1 \cos \theta$$
$$= (10.0 \,\text{kg})(\sin 37° - 0.40 \,\cos 37°) = 2.8 \,\text{kg}.$$

Thus if $m_2 < 2.8 \,\text{kg}$, then box 1 will slide down the incline.
Now for case (ii). Newton's second law is

$$m_1 g \sin \theta + F_{fr} - F_T = ma_x = 0.$$

Then the maximum value m_2 can have without causing acceleration is given by

$$F_T = m_2 g = m_1 g \sin \theta + \mu_s m_1 g \cos \theta$$

or

$$m_2 = m_1 \sin \theta + \mu_s m_1 \cos \theta$$
$$= (10.0 \,\text{kg})(\sin 37° + 0.40 \cos 37°) = 9.2 \,\text{kg}.$$

Thus, to prevent motion, we have the condition

$$2.8 \,\text{kg} < m_2 < 9.2 \,\text{kg}.$$

(*b*) If $m_2 = 10.0 \,\text{kg}$ and $\mu_k = 0.30$, then m_2 will fall and m_1 will rise up the plane with an acceleration a given by

$$m_1 a = F_T - m_1 g \sin \theta - \mu_k F_N.$$

Since m_2 also accelerates, $F_T = m_2 g - m_2 a$ (see Fig. 5–9c) and we substitute this into the equation above:

$$m_1 a = m_2 g - m_2 a - m_1 g \sin \theta - \mu_k F_N.$$

We solve for the acceleration a and substitute $F_N = m_1 g \cos \theta$, and then $m_1 = m_2 = 10.0 \,\text{kg}$, to find

$$a = \frac{m_2 g - m_1 g \sin \theta - \mu_k m_1 g \cos \theta}{m_1 + m_2}$$

$$= \frac{(9.8 \,\text{m/s}^2)(10.0 \,\text{kg})(1 - \sin 37° - 0.30 \cos 37°)}{20.0 \,\text{kg}} = 0.079g = 0.78 \,\text{m/s}^2.$$

Friction can be a hindrance. It slows down moving objects and causes heating and binding of moving parts in machinery. Friction can be reduced by using lubricants such as oil. More effective in reducing friction between two surfaces is to maintain a layer of air or other gas between them. Devices using this concept, which is not practical for most situations, include air tracks and air tables (or games) in which the layer of air is maintained by forcing air through many tiny holes. Another technique to maintain the air layer is to suspend objects in air using magnetic fields ("magnetic levitation"). On the other hand, friction can be helpful. Our ability to walk depends on friction between the soles of our shoes (or feet) and the ground. (Does walking involve static friction or kinetic friction?) The movement of a car, and also its stability, depend on friction. When friction is low, such as on ice, safe walking or driving become difficult.

5–2 | Dynamics of Uniform Circular Motion

In Section 3–9 we saw that a particle revolving in a circle of radius r with uniform speed v undergoes a radial (or centripetal) acceleration which at any moment is given by

$$a_R = \frac{v^2}{r}.$$

This acceleration, a_R, is called radial or centripetal acceleration because it is directed toward the center of the circle. For uniform circular motion (v = constant), the magnitude of the acceleration is constant, but its direction is continually changing. Hence the acceleration vector $\mathbf{a}_R$ is a variable acceleration. The direction of $\mathbf{a}_R$ is always perpendicular to the velocity $\mathbf{v}$. [The equation $a_R = v^2/r$ is valid even if v is not constant, and we will treat that case in Section 5–4.]

Force is needed to provide centripetal acceleration

An object moving uniformly in a circle, such as a ball swung horizontally on the end of a string, must have a net force applied to it to keep it moving in that circle, instead of in a straight line. That is, a net force is necessary to give it a centripetal acceleration. The magnitude of the required net force can be calculated using Newton's second law for the radial component, $\Sigma F_R = ma_R$, where $a_R = v^2/r$ is the centripetal acceleration, and ΣF_R is the total (net) force in the radial direction:

$$\Sigma F_R = ma_R = m\frac{v^2}{r}. \qquad \text{[circular motion]} \quad \textbf{(5–1)}$$

Since a_R is directed along the radius toward the center of the circle at any moment, the *net force too must be directed toward the center of the circle* for uniform circular motion (v = constant). A net force is clearly necessary because otherwise, if no net force were exerted on the object, it would not move in a circle but in a straight line, as Newton's first law tells us. To pull an object out of its "natural" straight-line path, a net force to the side is necessary. For uniform circular motion, this sideways net force must act toward the circle's center (see Fig. 5–10). The direction of the force is thus continually changing so that it is always directed toward the center of the circle. This force is sometimes called a centripetal ("aiming toward the center") force. But be aware that "centripetal force" does not indicate some new kind of force. The term merely describes the direction of the force: that the net force is directed toward the circle's center. The force *must be applied by other objects*. For example, when a person swings a ball in a circle on the end of a string, the person pulls on the string and the string exerts the force on the ball.

Careful: Centripetal force is not a new kind of force

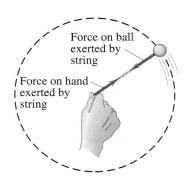

FIGURE 5–10 A force is required to keep an object moving in a circle. If the speed is constant, the force is directed toward the center of the circle.

FIGURE 5–11 Swinging a ball on the end of a string.

There is a common misconception that an object moving in a circle has an outward force acting on it, a so-called centrifugal ("center-fleeing") force. Consider, for example, a person swinging a ball on the end of a string around her head (Fig. 5–11). If you have ever done this yourself, you know that you feel a force pulling outward on your hand. The misconception arises when this pull is interpreted as an outward "centrifugal" force pulling on the ball that is transmitted along the string to your hand. This is not what is happening at all. To keep the ball moving in a circle, you pull inwardly on the string, which in turn exerts the force on the ball. The ball exerts an equal and opposite force on your hand (Newton's third law), and *this* is the force your hand feels (see Fig. 5–11). The force *on the ball* is the one exerted *inwardly* on it by the string.

For even more convincing evidence that a "centrifugal force" does not act on the ball, consider what happens when you let go of the string. If a centrifugal force were acting, the ball would fly outward, as shown in Fig. 5–12a. But it doesn't; the ball flies off tangentially (Fig. 5–12b), in the direction of the velocity it had at the moment it was released, because the inward force no longer acts. Try it and see!

Careful:
Beware a misconception
for a "centrifugal" force

FIGURE 5–12 If centrifugal force existed, the ball would fly off as in (a) when released. In fact, it flies off tangentially as in (b). Similarly, sparks fly in straight lines tangentially from the edge of a rotating grinding wheel (c).

DOESN'T HAPPEN

(a)

HAPPENS

(b)

(c)

FIGURE 5–13 Example 5–9.

EXAMPLE 5–9 **ESTIMATE** **Force on revolving ball (horizontal).** Estimate the force a person must exert on a string attached to a 0.150-kg ball to make the ball revolve in a horizontal circle of radius 0.600 m. The ball makes 2.00 revolutions per second ($T = 0.500$ s), as in our earlier Example 3–11.

SOLUTION First we draw the free-body diagram for the ball, Fig. 5–13, which shows the two forces acting on the ball: the force of gravity, $m\mathbf{g}$; and the tension force $\mathbf{F}_T$ that the string exerts (which occurs because the person exerts that same force on the string). The ball's weight complicates matters and makes it impossible to revolve a ball with the cord horizontal. But if the weight is small enough, we can ignore it. Then $\mathbf{F}_T$ will act nearly horizontally ($\phi \approx 0$ in Fig. 5–13) and provide the force necessary to give the ball its centripetal acceleration. We apply Newton's second law to the radial direction, which now is horizontal, so we call it x:

$$\Sigma F_x = ma_x$$

where $a_x = v^2/r$ and $v = 2\pi r/T = 2\pi(0.600 \text{ m})/(0.500 \text{ s}) = 7.54 \text{ m/s}$. Thus

$$F_{Tx} = m\frac{v^2}{r} = (0.150 \text{ kg})\frac{(7.54 \text{ m/s})^2}{(0.600 \text{ m})} \approx 14 \text{ N},$$

Tension in cord acts to provide the centripetal acceleration

where we have rounded off because our estimate ignores the ball's weight. We keep only two significant figures here in the answer because $mg = (0.150 \text{ kg})(9.80 \text{ m/s}^2) = 1.5 \text{ N}$, being about 1/10 of our result, is small, but not so small as to justify stating a more precise answer since we ignored the effect of $m\mathbf{g}$. [Note: if you want to include the effect of $m\mathbf{g}$ here, resolve $\mathbf{F}_T$ in Fig. 5–13 into components, and set the horizontal component of $\mathbf{F}_T$ equal to mv^2/r and its vertical component equal to mg—something like Example 5–10 below.]

FIGURE 5–14 Example 5–10. Conical pendulum.

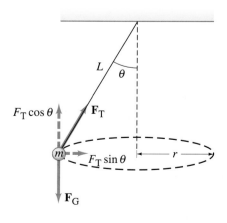

EXAMPLE 5–10 **Conical pendulum.** A small ball of mass m, suspended by a cord of length L, revolves in a circle of radius $r = L \sin \theta$, where θ is the angle the string makes with the vertical (Fig. 5–14). (a) In what direction is the acceleration of the ball, and what causes the acceleration? (b) Calculate the speed and period (time required for one revolution) of the ball in terms of L, θ, and m.

SOLUTION (a) The acceleration points horizontally toward the center of the ball's circular path (not along the cord). The force responsible for the acceleration is the *net* force which here is the vector sum of the forces acting on the mass m: its weight $\mathbf{F}_G$ (of magnitude $F_G = mg$) and the force exerted by the tension in the cord, F_T. The latter has horizontal and vertical components of magnitude $F_T \sin \theta$ and $F_T \cos \theta$, respectively. We apply Newton's second law to the horizontal and vertical directions. In the vertical direction, there is no motion, so the acceleration is zero and the net force in the vertical direction is zero:

$$F_T \cos \theta - mg = 0.$$

In the horizontal direction there is only one force, of magnitude $F_T \sin \theta$, that acts toward the center of the circle and gives rise to the acceleration v^2/r. Newton's second law tells us:

$$F_T \sin \theta = m\frac{v^2}{r}.$$

(b) We solve the two equations above for v by eliminating F_T between them (and using $r = L \sin \theta$):

$$v = \sqrt{\frac{r F_T \sin \theta}{m}} = \sqrt{\frac{r}{m} \left(\frac{mg}{\cos \theta} \right) \sin \theta} = \sqrt{\frac{Lg \sin^2 \theta}{\cos \theta}}.$$

The period T is the time required to make one revolution, a distance of $2\pi r = 2\pi L \sin \theta$. The speed v can thus be written $v = 2\pi L \sin \theta / T$; then

$$T = \frac{2\pi L \sin \theta}{v} = \frac{2\pi L \sin \theta}{\sqrt{\dfrac{Lg \sin^2 \theta}{\cos \theta}}} = 2\pi \sqrt{\frac{L \cos \theta}{g}}.$$

EXAMPLE 5–11 **Revolving ball (vertical circle).** A 0.150-kg ball on the end of a 1.10-m-long cord (negligible mass) is swung in a *vertical* circle. Determine the minimum speed the ball must have at the top of its arc so that it continues moving in a circle. (b) Calculate the tension in the cord at the bottom of the arc if the ball is moving at twice the speed of part (a).

SOLUTION The free-body diagram is shown in Fig. 5–15 for both situations. (a) At the top (point A), two forces can act on the ball: $m\mathbf{g}$, its weight; and $\mathbf{F}_{TA}$, the tension force the cord exerts at point A. Both act downward, and their vector sum acts to give the ball its centripetal acceleration a_R. We apply Newton's second law, for the vertical direction, choosing downward as positive (toward the center):

$$\Sigma F_R = ma_R$$

$$F_{TA} + mg = m\frac{v_A^2}{r}.$$

From this equation we can see that the tension force F_{TA} at A will get larger if v_A (ball's speed at top of circle) is made larger, as expected. But we are asked for the *minimum* speed to keep the ball moving in a circle. The cord will remain taut as long as there is tension in it; but if the tension disappears (because v_A is too small) the cord can go limp, and the ball will fall out of its circular path. Thus, the minimum speed will occur if $F_{TA} = 0$, for which we have $mg = mv_A^2/r$. Then:

$$v_A = \sqrt{gr} = \sqrt{(9.80 \text{ m/s}^2)(1.10 \text{ m})} = 3.28 \text{ m/s}.$$

This is the minimum speed at the top of the circle if the ball is to continue moving in a circular path.

(b) At the bottom of the circle (see Fig. 5–15) the cord exerts its tension force F_{TB} upward whereas the force of gravity, $m\mathbf{g}$, acts downward. So Newton's second law, this time choosing *upward* as positive (toward the center), gives

$$\Sigma F_R = ma_R$$

$$F_{TB} - mg = m\frac{v_B^2}{r}.$$

The speed v_B is given as twice what we found in (a), namely 6.56 m/s. [Note that the speed changes here because gravity acts on the ball all along the path, but Eq. 5–1 still remains valid, $\Sigma F_R = mv^2/r$.] We solve for F_{TB} in the last equation:

$$F_{TB} = m\frac{v_B^2}{r} + mg$$

$$= (0.150 \text{ kg})\frac{(6.56 \text{ m/s})^2}{(1.10 \text{ m})} + (0.150 \text{ kg})(9.80 \text{ m/s}^2) = 7.34 \text{ N}.$$

Note that the cord's tension not only provides the centripetal acceleration, but must be even larger than ma_R to compensate for the downward force of gravity.

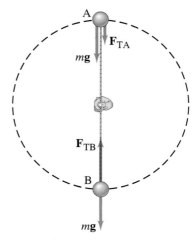

FIGURE 5–15 Example 5–11, with free-body diagrams at the two positions.

Gravity and cord tension together provide centripetal acceleration

Gravity alone provides centripetal acceleration

Cord tension and gravity acting in opposite directions provide centripetal acceleration

The astronauts in the upper left of this photo are working on the space shuttle. As they orbit the Earth—at a pretty high speed—they experience apparent weightlessness. The Moon, in the background, also is orbiting the Earth at high speed. What keeps the Moon and the space shuttle (and its astronauts) from moving off in a straight line away from Earth? It is the force of gravity. According to Newton's law of universal gravitation, all objects attract all other objects with a force proportional to their masses and inversely proportional to the square of the distance between them.

Gravitation and Newton's Synthesis

Sir Isaac Newton was not only the inventor of the three great laws of motion that serve as the foundation for the study of dynamics. He also conceived of another great law to describe one of the basic forces in nature, gravitation, and he applied it to understand the motion of the planets. This new law, published in 1687 in his great book *Philosophiae Naturalis Principia Mathematica* (the *Principia*, for short), is called Newton's law of universal gravitation. It was the capstone of Newton's analysis of the physical world. Indeed, Newtonian mechanics, with its three laws of motion and the law of universal gravitation, was accepted for centuries as a mechanical basis for the way the universe works.

6–1 | Newton's Law of Universal Gravitation

Among his many great accomplishments, Sir Isaac Newton examined the motion of the heavenly bodies—the planets and the Moon. In particular, he wondered about the nature of the force that must act to keep the Moon in its nearly circular orbit around the Earth.

Newton was also thinking about the problem of gravity. Since falling bodies accelerate, Newton had concluded that they must have a force exerted on them, a force we call the force of gravity. Whenever a body has a force exerted *on* it, that force is exerted *by* some other body. But what *exerts* the force of gravity? Every object on the surface of the Earth feels the force of gravity, and no matter where the object is, the force is directed toward the center of the Earth (Fig. 6–1). Newton concluded that it must be the Earth itself that exerts the gravitational force on objects at its surface.

According to an early account, Newton was sitting in his garden and noticed an apple drop from a tree. He is said to have been struck with a sudden inspiration: if the effect of gravity acts at the tops of trees, and even at the tops of mountains, then perhaps it acts all the way to the Moon! Whether this story is true or not, it does seem to capture something of Newton's reasoning and inspiration. With this idea that it is terrestrial gravity that holds the Moon in its orbit, Newton developed his great theory of gravitation. [But there was controversy at the time. Many thinkers had trouble accepting the idea of a force "acting at a distance." Typical forces act through contact—your hand pushes a cart and pulls a wagon, a bat hits a ball, and so on. But gravity acts without contact, said Newton: the Earth exerts a force on a falling apple and on the Moon, even though there is no contact, and the two objects may even be very far apart.]

Newton set about determining the magnitude of the gravitational force that the Earth exerts on the Moon as compared to the gravitational force on objects at the Earth's surface. At the surface of the Earth, the force of gravity accelerates objects at $9.80 \, \text{m/s}^2$. But what is the centripetal acceleration of the Moon? Since the Moon moves with nearly uniform circular motion, the acceleration can be calculated from $a_R = v^2/r$. We already performed this calculation in Example 3–12 and found that $a_R = 0.00272 \, \text{m/s}^2$. In terms of the acceleration of gravity at the Earth's surface, g, this is equivalent to

$$a_R \approx \frac{1}{3600} \, g.$$

That is, the acceleration of the Moon toward the Earth is about $\frac{1}{3600}$ as great as the acceleration of objects at the Earth's surface. The Moon is 384,000 km from the Earth, which is about 60 times the Earth's radius of 6380 km. That is, the Moon is 60 times farther from the Earth's center than are objects at the Earth's surface. But $60 \times 60 = 60^2 = 3600$. Again that number 3600! Newton concluded that the gravitational force exerted by the Earth on any object decreases with the square of its distance, r, from the Earth's center:

$$\text{force of gravity} \propto \frac{1}{r^2}.$$

The Moon, being 60 Earth radii away, feels a gravitational force only $\frac{1}{60^2} = \frac{1}{3600}$ times as strong as it would if it were at the Earth's surface. Any object placed 384,000 km from the Earth would experience the same acceleration due to the Earth's gravity as the Moon experiences: $0.00272 \, \text{m/s}^2$.

Newton realized that the force of gravity on an object depends not only on distance but also on the object's mass. In fact, it is directly proportional to its mass, as we have seen. According to Newton's third law, when the Earth exerts its gravitational force on any body, such as the Moon, that other body exerts an equal and opposite force on the Earth (Fig. 6–2). Because of this symmetry, Newton reasoned, the magnitude of the force of gravity must be proportional to *both* the masses. Thus

$$F \propto \frac{m_E m_B}{r^2}$$

where m_E is the mass of the Earth, m_B the mass of the other body, and r the distance from the Earth's center to the center of the other body.

Newton's apple

FIGURE 6–1 Anywhere on Earth, whether in Alaska, Australia or Peru, the force of gravity acts downward toward the center of the Earth.

The Moon's acceleration toward Earth

FIGURE 6–2 The gravitational force one body exerts on a second body is directed toward the first body, and is equal and opposite to the force exerted by the second body on the first.

Moon

Gravitational force exerted on Moon by Earth

Earth

Gravitational force exerted on Earth by the Moon

Newton went a step further in his analysis of gravity. In his examination of the orbits of the planets, he concluded that the force required to hold the different planets in their orbits around the Sun seems to diminish as the inverse square of their distance from the Sun. This led him to believe that it is also the gravitational force that acts between the Sun and each of the planets to keep them in their orbits. And if gravity acts between these objects, why not between all objects? Thus he proposed his famous **law of universal gravitation**, which we can state as follows:

Every particle in the universe attracts every other particle with a force that is proportional to the product of their masses and inversely proportional to the square of the distance between them. This force acts along the line joining the two particles.

The magnitude of the gravitational force can be written as

$$F = G \frac{m_1 m_2}{r^2}, \tag{6-1}$$

NEWTON'S LAW OF UNIVERSAL GRAVITATION

where m_1 and m_2 are the masses of the two particles, r is the distance between them, and G is a universal constant which must be measured experimentally and has the same numerical value for all objects.

The value of G must be very small, since we are not aware of any force of attraction between ordinary-sized objects, such as between two baseballs. The force between two ordinary objects was first measured, over 100 years after Newton's publication of his law, by Henry Cavendish in 1798. To detect and measure the incredibly small force, he used an apparatus like that shown in Fig. 6–3. Cavendish confirmed Newton's hypothesis that two bodies attract one another, and that Eq. 6–1 accurately describes this force. In addition, because he could measure F, m_1, m_2, and r accurately, he was able to determine the value of the constant G as well. The accepted value today is

$$G = 6.67 \times 10^{-11} \, \text{N} \cdot \text{m}^2/\text{kg}^2.$$

Strictly speaking, Eq. 6–1 gives the magnitude of the gravitational force that one particle exerts on a second particle that is a distance r away. For an extended object (that is, not a point), we must consider how to measure the distance r. You might think that r would be the distance between the centers of the objects. This is often true and often a good approximation even when not quite true; but to do a calculation correctly, each extended body must be considered as a collection of tiny particles, and the total force is the sum of the forces due to all the particles. The sum over all these particles is often best done using integral calculus, which Newton himself invented. Newton showed that for two uniform spheres, Eq. 6–1 gives the correct force where r is the distance between their centers. (The derivation is given in Appendix C.) Also, when extended bodies are small compared to the distance between them (as for the Earth–Sun system), little inaccuracy results from considering them as point particles.

FIGURE 6–3 Schematic diagram of Cavendish's apparatus. Two spheres are attached to a light horizontal rod, which is suspended at its center by a thin fiber. When a third sphere labeled A is brought close to one of the suspended spheres, the gravitational force causes the latter to move, and this twists the fiber slightly. The tiny movement is magnified by the use of a narrow light beam directed at a mirror mounted on the fiber. The beam reflects onto a scale. Previous determination of how large a force will twist the fiber a given amount then allows one to determine the magnitude of the gravitational force between two objects.

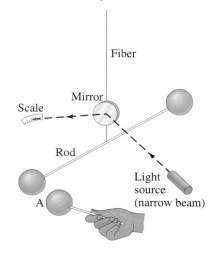

EXAMPLE 6–1 ESTIMATE | **Can you attract another person gravitationally?** A 50-kg person and a 75-kg person are sitting on a bench so that their centers are about 50 cm apart. Estimate the magnitude of the gravitational force each exerts on the other.

SOLUTION We use Eq. 6–1, which gives

$$F = \frac{(6.67 \times 10^{-11} \, \text{N} \cdot \text{m}^2/\text{kg}^2)(50 \, \text{kg})(75 \, \text{kg})}{(0.50 \, \text{m})^2} = 1.0 \times 10^{-6} \, \text{N},$$

which is unnoticeably small unless very delicate instruments are used.

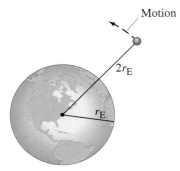

Motion

$2r_E$

r_E

FIGURE 6–4 Example 6–2.

FIGURE 6–5 Orientation of Sun (S), Earth (E), and Moon (M) for Example 6–3 (not to scale).

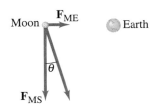

Moon $\mathbf{F}_{ME}$ Earth

θ

$\mathbf{F}_{MS}$

Sun

EXAMPLE 6–2 **Spacecraft at $2r_E$.** What is the force of gravity acting on a 2000-kg spacecraft when it orbits two Earth radii from the Earth's center (that is, a distance $r_E = 6380$ km above the Earth's surface, Fig. 6–4)? The mass of the Earth is $M_E = 5.98 \times 10^{24}$ kg.

SOLUTION We could plug all the numbers into Eq. 6–1, but there is a simpler approach. The spacecraft is twice as far from the Earth's center as when at the surface of the Earth. Therefore, since the force of gravity decreases as the square of the distance $\left(\text{and } \frac{1}{2^2} = \frac{1}{4}\right)$, the force of gravity on it will be only one fourth its weight at the Earth's surface:

$$F_G = \tfrac{1}{4}mg = \tfrac{1}{4}(2000\text{ kg})(9.80\text{ m/s}^2) = 4900\text{ N}.$$

EXAMPLE 6–3 **Force on the Moon.** Find the net force on the Moon $\left(m_M = 7.35 \times 10^{22}\text{ kg}\right)$ due to the gravitational attraction of both the Earth $\left(m_E = 5.98 \times 10^{24}\text{ kg}\right)$ and the Sun $\left(m_S = 1.99 \times 10^{30}\text{ kg}\right)$, assuming they are at right angles to each other (Fig. 6–5).

SOLUTION We must add the two forces vectorially. First we calculate their magnitudes. The Earth is 3.84×10^5 km $= 3.84 \times 10^8$ m from the Moon, so F_{ME} (the force on the Moon due to the Earth) is

$$F_{ME} = \frac{(6.67 \times 10^{-11}\text{ N·m}^2/\text{kg}^2)(7.35 \times 10^{22}\text{ kg})(5.98 \times 10^{24}\text{ kg})}{(3.84 \times 10^8\text{ m})^2}$$
$$= 1.99 \times 10^{20}\text{ N}.$$

The Sun is 1.50×10^8 km from the Earth and the Moon, so F_{MS} (the force on the Moon due to the Sun) is

$$F_{MS} = \frac{(6.67 \times 10^{-11}\text{ N·m}^2/\text{kg}^2)(7.35 \times 10^{22}\text{ kg})(1.99 \times 10^{30}\text{ kg})}{(1.50 \times 10^{11}\text{ m})^2}$$
$$= 4.34 \times 10^{20}\text{ N}.$$

Since the two forces act at right angles in the case we are considering (Fig. 6–5), the total force is

$$F = \sqrt{(1.99)^2 + (4.34)^2} \times 10^{20}\text{ N} = 4.77 \times 10^{20}\text{ N}$$

which acts at an angle $\theta = \tan^{-1}(1.99/4.34) = 24.6°$.

6–2 Vector Form of Newton's Law of Universal Gravitation

We can write Newton's law of universal gravitation in vector form as

$$\mathbf{F}_{12} = -G\frac{m_1 m_2}{r_{21}^2}\hat{\mathbf{r}}_{21}, \tag{6–2}$$

where $\mathbf{F}_{12}$ is the vector force on particle 1 $\left(\text{of mass } m_1\right)$ exerted by particle 2 $\left(\text{of mass } m_2\right)$, which is a distance r_{21} away; $\hat{\mathbf{r}}_{21}$ is a unit vector that points from particle 2 toward particle 1 along the line joining them so that $\hat{\mathbf{r}}_{21} = \mathbf{r}_{21}/r_{21}$, where $\mathbf{r}_{21}$ is the displacement vector as shown in Fig. 6–6. The minus sign in Eq. 6–2 is necessary because the force on particle 1 due to particle 2 points toward m_2, in the direction opposite to $\hat{\mathbf{r}}_{21}$. The displacement vector $\mathbf{r}_{12}$ is a vector of the same magnitude as $\mathbf{r}_{21}$, but it points in the opposite direction so that

$$\mathbf{r}_{12} = -\mathbf{r}_{21}.$$

By Newton's third law, the force $\mathbf{F}_{21}$ acting on m_2 exerted by m_1 must have the

FIGURE 6–6 The displacement vector $\mathbf{r}_{21}$ points from particle of mass m_2 to particle of mass m_1. The unit vector shown, $\hat{\mathbf{r}}_{21}$ is in the same direction as $\mathbf{r}_{21}$, but is defined as having length one.

m_1

$\mathbf{r}_{21}$

$\hat{\mathbf{r}}_{21}$

m_2

same magnitude as $\mathbf{F}_{12}$ but acts in the opposite direction (Fig. 6–7), so that

$$\mathbf{F}_{21} = -\mathbf{F}_{12} = G\frac{m_1 m_2}{r_{21}^2}\hat{\mathbf{r}}_{21}$$

$$= -G\frac{m_2 m_1}{r_{12}^2}\hat{\mathbf{r}}_{12}.$$

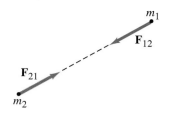

The force of gravity exerted on one particle by a second particle is always directed toward the second particle, as in Fig. 6–6. When many particles interact, the total gravitational force on a given particle is the vector sum of the forces exerted by each of the others. For example, the total force on particle number 1 is

$$\mathbf{F}_1 = \mathbf{F}_{12} + \mathbf{F}_{13} + \mathbf{F}_{14} + \cdots + \mathbf{F}_{1n} = \sum_{i=2}^{n}\mathbf{F}_{1i} \qquad (6\text{–}3)$$

FIGURE 6–7 By Newton's third law, the gravitational force on particle 1 exerted by particle 2, $\mathbf{F}_{12}$, is equal and opposite to that on particle 2 exerted by particle 1, $\mathbf{F}_{21}$; that is $\mathbf{F}_{21} = -\mathbf{F}_{12}$.

where $\mathbf{F}_{1i}$ means the force on particle 1 exerted by particle i, and n is the total number of particles.

This vector notation can be very helpful, especially when sums over many particles are needed. However, in many cases we don't have to be so formal and we can deal with directions by making careful diagrams.

 ## 6–3 Gravity Near the Earth's Surface; Geophysical Applications

When Eq. 6–1 is applied to the gravitational force between the Earth and an object at its surface, m_1 becomes the mass of the Earth, m_E, m_2 becomes the mass of the object, m, and r becomes the distance of the object from the Earth's center,[†] which is the radius of the Earth r_E. This force of gravity due to the Earth is the weight of the object, which we have been writing as mg. Thus,

$$mg = G\frac{mm_E}{r_E^2}.$$

Hence

$$g = G\frac{m_E}{r_E^2}. \qquad (6\text{–}4)$$

g in terms of G

Thus, the acceleration of gravity at the surface of the Earth, g, is determined by m_E and r_E. (Be careful not to confuse G with g; they are very different quantities, but are related by Eq. 6–4.)

Until G was measured, the mass of the Earth was not known. But once G was measured, Eq. 6–4 could be used to calculate the Earth's mass, and Cavendish was the first to do so. Since $g = 9.80\text{ m/s}^2$ and the radius of the Earth is $r_E = 6.38 \times 10^6\text{ m}$, then, from Eq. 6–4, we obtain

$$m_E = \frac{gr_E^2}{G} = \frac{(9.80\text{ m/s}^2)(6.38 \times 10^6\text{ m})^2}{6.67 \times 10^{-11}\text{ N}\cdot\text{m}^2/\text{kg}^2} = 5.98 \times 10^{24}\text{ kg}$$

Mass of the Earth

for the mass of the Earth.

When dealing with the weight of objects at the surface of the Earth, we can continue to use simply mg. If we wish to calculate the force of gravity on an object some distance from the Earth, or the force due to some other heavenly body, such as that exerted by the Moon or a planet, we can calculate the effective value of g from Eq. 6–4, replacing r_E (and m_E) by the appropriate distance (and mass), or we can use Eq. 6–1 directly.

[†] That the distance is measured from the Earth's center does not imply that the force of gravity somehow emanates from that one point. Rather, all parts of the Earth attract gravitationally, but the net effect is a force acting toward the Earth's center. (See Appendix C.)

TABLE 6–1
Acceleration Due to Gravity at Various Locations on Earth

Location	Elevation (m)	g (m/s²)
New York	0	9.803
San Francisco	100	9.800
Denver	1650	9.796
Pikes Peak	4300	9.789
Equator	0	9.780
North Pole (calculated)	0	9.832

➡ **PHYSICS APPLIED**

Geology—mineral and oil exploration

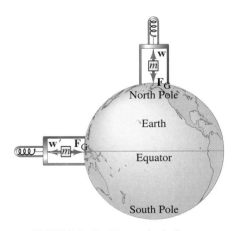

FIGURE 6–8 Example 6–5.

EXAMPLE 6–4 **ESTIMATE** **Gravity on Everest.** Estimate the effective value of g on the top of Mt. Everest, 8848 m (29,028 ft) above the Earth's surface. That is, what is the acceleration due to gravity of objects allowed to fall freely at this altitude?

SOLUTION Let us call the acceleration of gravity at the given point g'. We use Eq. 6–4, with r_E replaced by $r = 6380$ km $+ 8.8$ km $= 6389$ km $= 6.389 \times 10^6$ m:

$$g' = G \frac{m_E}{r^2} = \frac{(6.67 \times 10^{-11}\,\text{N}\cdot\text{m}^2/\text{kg}^2)(5.98 \times 10^{24}\,\text{kg})}{(6.389 \times 10^6\,\text{m})^2} = 9.77\,\text{m/s}^2$$

which is a reduction of about 3 parts in a thousand (0.3%). However, we ignored the mass accumulated under the mountain top.

Note that Eq. 6–4 does not give precise values for g at different locations because the Earth is not a perfect sphere. The Earth not only has mountains and valleys, and bulges at the equator, but its mass is not distributed precisely uniformly. See Table 6–1. The Earth's rotation also has an effect on the value of g, as discussed in Example 6–5 below.

The value of g can vary locally on the Earth's surface because of the presence of irregularities and rocks of different densities. Such variations in g, known as "gravity anomalies," are very small—on the order of 1 part per 10^6 or 10^7 in the value of g. But they can be measured ("gravimeters" today can detect variations in g to 1 part in 10^9). Geophysicists use such measurements as part of their investigations into the structure of the Earth's crust, and in mineral and oil exploration. Mineral deposits, for example, often have a greater density than surrounding material; because of the greater mass in a given volume, g can have a slightly greater value on top of such a deposit than at its flanks. "Salt domes," under which petroleum is often found, have a lower than average density and searches for a slight reduction in the value of g in certain locales have led to the discovery of oil.

EXAMPLE 6–5 **Effect of Earth's rotation on g.** Assuming the Earth is a perfect sphere, determine how the Earth's rotation affects the value of g at the equator compared to its value at the poles.

SOLUTION Figure 6–8 shows a mass m hanging from a spring scale at two places on the Earth. At the North Pole there are two forces acting on the mass m: the force of gravity, $\mathbf{F}_G = mg$, and the force with which the spring pulls up on the mass, $\mathbf{w}$. We call this latter force w because it is what the scale reads as the weight of the object; by Newton's third law it equals the force with which the mass pulls down on the spring. Since the mass is not accelerating, Newton's second law tells us

$$mg - w = 0,$$

so $w = mg$. Thus the weight w that the spring registers equals mg, which is no surprise. Next, at the equator, there *is* an acceleration because the Earth is rotating. The same force of gravity $\mathbf{F}_G = mg$ acts downward (we are letting g represent the acceleration of gravity in the absence of rotation and we ignore the slight bulging of the equator). The spring pulls upward with a force w'; w' is also the force with which the mass pulls on the spring (Newton's third law) and hence is the weight registered on the spring scale. From Newton's second law we now have (see Fig. 6–8)

$$mg - w' = m \frac{v^2}{r_E}$$

where $r_E = 6.38 \times 10^6$ m is the Earth's radius and v is the speed of m due to the Earth's daily rotation [1 day $= (24\,\text{h})(3600\,\text{s/h}) = 8.64 \times 10^4\,\text{s}$] which equals

$$v = 2\pi r_E/1\,\text{day} = (6.28)(6.38 \times 10^6\,\text{m})/(8.64 \times 10^4\,\text{s}) = 4.64 \times 10^2\,\text{m/s}.$$

The effective weight is w', so the effective value of g, which we call g', is $g' = w'/m$. Solving the equation above for w', we have

$$w' = m(g - v^2/r_E),$$

so

$$g' = \frac{w'}{m} = g - \frac{v^2}{r_E}.$$

Hence

$$\Delta g = g - g' = v^2/r_E = (4.64 \times 10^2 \, \text{m/s})^2/(6.38 \times 10^6 \, \text{m}) = 0.0337 \, \text{m/s}^2.$$

In Table 6–1 we see that the difference is actually greater than this: $(9.832 - 9.780) \, \text{m/s}^2 = 0.052 \, \text{m/s}^2$; this discrepancy is due mainly to the Earth's being slightly fatter at the equator (by 21 km) than at the poles. The calculation of the effective value of g at latitudes other than at the poles or equator is a two-dimensional problem because $\mathbf{F}_G$ acts radially toward the Earth's center whereas the centripetal acceleration is directed perpendicular to the axis of rotation, parallel to the equator and that means that a plumb line (the effective direction of g) is not precisely vertical except at the equator and the poles.

6–4 Satellites and "Weightlessness"

Artificial satellites circling the Earth are now commonplace (Fig. 6–9). A satellite is put into orbit by accelerating it to a sufficiently high tangential speed with the use of rockets, as shown in Fig. 6–10. If the speed is too high, the spacecraft will not be confined by the Earth's gravity and will escape, never to return. If the speed is too low, it will return to Earth. Satellites are usually put into circular (or nearly circular) orbits, because they require the least takeoff speed. It is sometimes asked: "What keeps a satellite up?" The answer is: its high speed. If a satellite stopped moving, it would, of course, fall directly to Earth. But at the very high speed a satellite has, it would quickly fly out into space (Fig. 6–11) if it weren't for the gravitational force of the Earth pulling it into orbit. In fact, a satellite *is* falling (accelerating toward Earth), but its high tangential speed keeps it from hitting Earth.

FIGURE 6–9
A satellite circling the Earth.

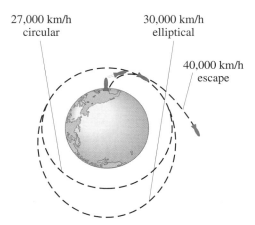

FIGURE 6–10 Artificial satellites launched at different speeds.

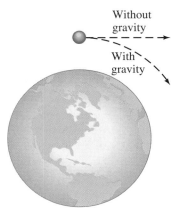

FIGURE 6–11 A moving satellite "falls" out of a straight-line path toward the Earth.

For satellites that move in a circle (at least approximately), the needed acceleration is v^2/r. The force that gives a satellite this acceleration is the force of gravity, and since a satellite may be at a considerable distance from the Earth, we must use Eq. 6–1 for the force acting on it. When we apply Newton's second law, $\Sigma F_R = ma_R$, we find

$$G\frac{mm_E}{r^2} = m\frac{v^2}{r}, \tag{6–5}$$

where m is the mass of the satellite. This equation relates the distance of the satellite from the Earth's center, r, to its speed, v. Note that only one force—gravity—is acting on the satellite, and that r is the sum of the Earth's radius r_E plus the satellite's height h above the Earth: $r = r_E + h$.

➡ PHYSICS APPLIED

*Geosynchronous
satellites*

EXAMPLE 6–6 **Geosynchronous satellite.** A *geosynchronous* satellite is one that stays above the same point on the Earth, which is possible only if it is above a point on the equator. Such satellites are used for such purposes as cable TV transmission, for weather forecasting, and as communication relays. Determine (a) the height above the Earth's surface such a satellite must orbit and (b) such a satellite's speed. (c) Compare to the speed of a satellite orbiting 200 km above Earth's surface.

SOLUTION (a) The only force on the satellite is gravity, so we apply Eq. 6–5 assuming the satellite moves in a circle:

$$G\frac{m_{Sat}\,m_E}{r^2} = m_{Sat}\frac{v^2}{r}.$$

This equation seems to have two unknowns, r and v. But we know that v must be such that the satellite revolves around the Earth with the same period that the Earth rotates on its axis, namely once in 24 hours. Thus the speed of the satellite must be

$$v = \frac{2\pi r}{T}$$

where $T = 1$ day $= (24\,\text{h})(3600\,\text{s/h}) = 86{,}400\,\text{s}$. We put this into the first equation above and obtain (after canceling m_{Sat} on both sides):

$$G\frac{m_E}{r^2} = \frac{(2\pi r)^2}{rT^2}.$$

We solve for r:

$$r^3 = \frac{Gm_E T^2}{4\pi^2} = \frac{(6.67 \times 10^{-11}\,\text{N}\cdot\text{m}^2/\text{kg}^2)(5.98 \times 10^{24}\,\text{kg})(86{,}400\,\text{s})^2}{4\pi^2}$$

$$= 7.54 \times 10^{22}\,\text{m}^3,$$

and, taking the cube root, $r = 4.23 \times 10^7$ m, or 42,300 km from the Earth's center. We subtract the Earth's radius of 6380 km to find that the satellite must orbit about 36,000 km (about 6 r_E) above the Earth's surface.
(b) We solve Eq. 6–5 for v:

$$v = \sqrt{\frac{Gm_E}{r}} = \sqrt{\frac{(6.67 \times 10^{-11}\,\text{N}\cdot\text{m}^2/\text{kg}^2)(5.98 \times 10^{24}\,\text{kg})}{(4.23 \times 10^7\,\text{m})}} = 3070\,\text{m/s}.$$

We get the same result if we use $v = 2\pi r/T$.
(c) The last equation above for v shows $v \propto \sqrt{1/r}$. So for $r = r_E + h = 6380\,\text{km} + 200\,\text{km} = 6580\,\text{km}$, we get

$$v' = v\sqrt{r/r'} = (3070\,\text{m/s})\sqrt{(42{,}300\,\text{km})/(6580\,\text{km})} = 7780\,\text{m/s}.$$

CONCEPTUAL EXAMPLE 6–7 | **Catching a satellite.** You are an astronaut in the space shuttle pursuing a satellite in need of repair. You find yourself in a circular orbit of the same radius as the satellite, but 30 km behind it. How will you catch up with it?

RESPONSE We saw in Example 6–6 (or see Eq. 6–5) that the velocity is proportional to $1/\sqrt{r}$. Thus you need to aim for a smaller orbit and at the same time increase your velocity. Note that you cannot just increase your speed without changing your orbit.

People and other objects in a satellite circling the Earth are said to experience apparent weightlessness. Before tackling the case of a satellite, however, let us first look at the simpler case of a falling elevator. In Fig. 6–12a, we see an elevator at rest with a bag hanging from a spring scale. The scale reading indicates the downward force exerted on it by the bag. This force, exerted *on* the scale, is equal and opposite to the force exerted *by* the scale upward on the bag. We call this force **w**. Since the mass, m, is not accelerating, we apply $\Sigma F = ma$ to the bag and obtain

$$w - mg = 0,$$

"Weightlessness" in a falling elevator

where mg is the weight of the bag. Thus, $w = mg$, and since the scale indicates the force w exerted on it by the bag, it registers a force equal to the weight of the bag as we expect. If, now, the elevator has an acceleration, a, then applying $\Sigma F = ma$ to the bag, we have

$$w - mg = ma.$$

Solving for w, we have

$$w = mg + ma.$$

We have chosen the positive direction up. Thus, if the acceleration a is up, a is positive; and the scale, which measures w, will read more than mg. We call w the *apparent weight* of the bag, which here is greater than its actual weight (mg). If the elevator accelerates downward, a will be negative and w, the apparent weight, will be less than mg. Note that the direction of the velocity v doesn't matter. Only the direction of the acceleration **a** influences the scale reading.

If, for example, the elevator's acceleration is $\frac{1}{2}g$ upward, then we find $w = mg + m(\frac{1}{2}g) = \frac{3}{2}mg$. That is, the scale reads $1\frac{1}{2}$ times the actual weight (Fig. 6–12b). The apparent weight of the bag is $1\frac{1}{2}$ times its real weight. The same is true of the person: her apparent weight (equal to the normal force exerted on her by the elevator floor) is $1\frac{1}{2}$ times her real weight. We can say that she is experiencing $1\frac{1}{2}$ g's, just as astronauts experience so many g's at a rocket's launch.

If, instead, the elevator's acceleration is $-\frac{1}{2}g$ (downward), then $w = mg - \frac{1}{2}mg = \frac{1}{2}mg$. That is, the scale reads one half the actual weight. If the elevator is in *free fall* (for example, if the cables break), then $a = -g$ and $w = mg - mg = 0$. The scale reads zero! See Fig. 6–12c. The bag seems weightless.

$a = 0$

(a)

$a = -\frac{1}{2}\mathbf{g}\,(\text{up})$

(b)

$a = \mathbf{g}\,(\text{down})$

(c)

FIGURE 6–12 (a) An object in an elevator at rest exerts a force on a spring scale equal to its weight. (b) In an elevator accelerating upward at $\frac{1}{2}g$, the object's apparent weight is $1\frac{1}{2}$ times larger. (c) In a freely falling elevator, the object experiences "weightlessness": the scale reads zero.

If the person in the elevator let go of a pencil, say, it would not fall to the floor. True, the pencil would be falling with acceleration g. But so would the floor of the elevator and the person. The pencil would hover right in front of the person. This phenomenon is called *apparent weightlessness* because, in fact, gravity is still acting on the object and its weight is still mg. The objects seem weightless only because the elevator is in free fall.

The "weightlessness" experienced by people in a satellite orbit close to the Earth (Fig. 6–13) is the same apparent weightlessness experienced in a freely falling elevator. It may seem strange, at first, to think of a satellite as freely falling. But a satellite is indeed falling toward the Earth, as was shown in Fig. 6–11. The force of gravity causes it to "fall" out of its natural straight-line path. The acceleration of the satellite must be the acceleration due to gravity at that point, since the only force acting on it is gravity. (We used this to obtain Eq. 6–5.) Thus, although the force of gravity acts on objects within the satellite, the objects experience an apparent weightlessness because they, and the satellite, are accelerating as in free fall.

Figure 6–14 shows some examples of "free fall," or apparent weightlessness, experienced by people on Earth for brief moments.

A completely different situation occurs when a spacecraft is out in space far from the Earth, the Moon, and other attracting bodies. The force of gravity due to the Earth and other heavenly bodies will then be quite small because of the distances involved, and persons in such a spacecraft will experience real weightlessness.

The effects on human beings of weightlessness (whether real or apparent makes no difference) are interesting. In ordinary circumstances, for example, people can become quite tired holding out their arms horizontally. But for a person experiencing weightlessness, no effort is needed. The arms will just "float" there, since there is no sensation of weight. This effect has many applications in athletics (Fig. 6–14). During a jump or a dive, while on a trampoline, and even between strides while running, a person is experiencing apparent weightlessness or free fall, although only for a short time. During these brief periods, limbs can be moved much more easily, since only inertia needs to be overcome. The loss of control because of lack of contact with the ground is compensated for by the increased mobility. Prolonged weightlessness in space, however, can have harmful effects on health. Red blood cells diminish in number, blood collects in the thorax, bones lose calcium and become brittle, and muscles lose their tone. These effects are being carefully studied.

"Weightlessness" in a satellite

FIGURE 6–13 This astronaut is making a repair to the Hubble Space Telescope. He must feel very free because he is experiencing apparent weightlessness.

FIGURE 6–14 Experiencing weightlessness on Earth.

(a)

(b)

(c)

6–5 Kepler's Laws and Newton's Synthesis

More than a half century before Newton proposed his three laws of motion and his law of universal gravitation, the German astronomer Johannes Kepler (1571–1630) had written a number of astronomical works in which we can find a detailed description of the motion of the planets about the Sun. Kepler's work resulted in part from the many years he spent examining data collected by Tycho Brahe (1546–1601) on the positions of the planets in their motion through the heavens. Among Kepler's writings were three empirical findings that we now refer to as **Kepler's laws of planetary motion**. These are summarized as follows, with additional explanation in Figs. 6–15 and 6–16.

Kepler's first law: The path of each planet about the Sun is an ellipse with the Sun at one focus (Fig. 6–15).

Kepler's second law: Each planet moves so that an imaginary line drawn from the Sun to the planet sweeps out equal areas in equal periods of time (Fig. 6–16).

Kepler's third law: The ratio of the squares of the periods of any two planets revolving about the Sun is equal to the ratio of the cubes of their semimajor axes. [The semimajor axis is half the long (major) axis of the orbit, as shown in Fig. 6–15, and represents the planet's average distance from the Sun.[†]] That is, if T_1 and T_2 represent the periods (the time needed for one revolution about the Sun) for any two planets, and s_1 and s_2 represent their semimajor axes, then

$$\left(\frac{T_1}{T_2}\right)^2 = \left(\frac{s_1}{s_2}\right)^3.$$

We can rewrite this as

$$\frac{s_1^3}{T_1^2} = \frac{s_2^3}{T_2^2},$$

meaning that s^3/T^2 should be the same for each planet. Present-day data are given in Table 6–2; see the last column.

[†] The semimajor axis is equal to the planet's average distance from the Sun in the sense that it equals the average of the planet's nearest and farthest distances from the Sun (points Q and R in Fig. 6–15). Most planetary orbits are close to circles, and for a circle the semimajor axis is the radius of the circle.

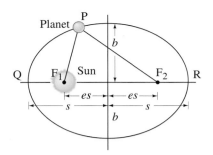

FIGURE 6–15 (a) *Kepler's first law.* An ellipse is a closed curve such that the sum of the distances from any point P on the curve to two fixed points (called the foci, F_1 and F_2) remains constant. That is, the sum of the distances, $F_1 P + F_2 P$, is the same for all points on the curve. A circle is a special case of an ellipse in which the two foci coincide, at the center of the circle. The semimajor axis is s (that is, the long axis is $2s$) and the semiminor axis is b, as shown. The *eccentricity*, e, is defined so that es is the distance from the center to either focus. The Earth and most of the other planets have nearly circular orbits. For Earth $e = 0.017$.

FIGURE 6–16 *Kepler's second law.* The two shaded regions have equal areas. The planet moves from point 1 to point 2 in the same time as it takes to move from point 3 to point 4. Planets move fastest in that part of their orbit where they are closest to the Sun. Exaggerated scale.

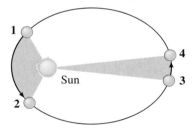

TABLE 6–2	Planetary Data Applied to Kepler's Third Law		
Planet	**Average Distance from Sun, s (10^6 km)**	**Period, T (Earth years)**	**s^3/T^2 (10^{24} km^3/y^2)**
Mercury	57.9	0.241	3.34
Venus	108.2	0.615	3.35
Earth	149.6	1.0	3.35
Mars	227.9	1.88	3.35
Jupiter	778.3	11.86	3.35
Saturn	1427	29.5	3.34
Uranus	2870	84.0	3.35
Neptune	4497	165	3.34
Pluto	5900	248	3.33

Kepler arrived at his laws through careful analysis of experimental data. Fifty years later, Newton was able to show that Kepler's laws could be derived mathematically from the law of universal gravitation and the laws of motion. He also showed that for any reasonable form for the gravitational force law, only one that depends on the inverse square of the distance is fully consistent with Kepler's laws. He thus used Kepler's laws as evidence in favor of his law of universal gravitation, Eq. 6–1.

Derivation of Kepler's third law

We will derive Kepler's second law in Chapter 11 when we study angular momentum. Here we derive Kepler's third law, and we do it for the special case of a circular orbit, in which case the semimajor axis is the radius r of the circle. (Note that most planetary orbits are close to a circle.) First, we write down Newton's second law of motion $\Sigma F = ma$. Then for F we substitute the law of universal gravitation, Eq. 6–1, and for a the centripetal acceleration, v^2/r:

$$\Sigma F = ma$$

$$G \frac{m_1 M_S}{r_1^2} = m_1 \frac{v_1^2}{r_1}.$$

Here m_1 is the mass of a particular planet, r_1 its distance from the Sun, and v_1 its average speed in orbit; M_S is the mass of the Sun, since it is the gravitational attraction of the Sun that keeps each planet in its orbit. The period T_1 of the planet is the time required for one complete orbit, a distance equal to $2\pi r_1$, the circumference of a circle, so

$$v_1 = \frac{2\pi r_1}{T_1}.$$

We substitute this formula for v_1 into the equation above:

$$G \frac{m_1 M_S}{r_1^2} = m_1 \frac{4\pi^2 r_1}{T_1^2}.$$

We rearrange this to get

$$\frac{T_1^2}{r_1^3} = \frac{4\pi^2}{GM_S}. \qquad (6\text{–}6)$$

We derived this for planet 1 (say, Mars). The same derivation would apply for a second planet (say, Saturn):

$$\frac{T_2^2}{r_2^3} = \frac{4\pi^2}{GM_S},$$

where T_2 and r_2 are the period and orbit radius, respectively, for the second planet. Since the right sides of the two previous equations are equal, we have $T_1^2/r_1^3 = T_2^2/r_2^3$ or, rearranging,

Kepler's third law

$$\left(\frac{T_1}{T_2} \right)^2 = \left(\frac{r_1}{r_2} \right)^3, \qquad (6\text{–}7)$$

which is Kepler's third law. Equations 6–6 and 6–7 are valid also for elliptical orbits if we replace r with the semimajor axis s.

The derivations of Eqs. 6–6 and 6–7 (Kepler's third law) are general enough to be applied to other systems. For example, we could determine the mass of the Earth from Eq. 6–6 using the period of the Moon about the Earth and the Moon's distance from the Earth, or the mass of Jupiter from the period and distance of one of its moons (this is indeed how masses are determined; see the Problems). We can also use Eqs. 6–6 and 6–7 to compare objects that orbit other attracting centers, such as the Moon and a weather satellite orbiting Earth. But be careful not to use Eq. 6–7 to compare, say, the Moon's orbit around the Earth to the orbit of Mars around the Sun because they depend on different attracting centers.

In the following examples, we assume the orbits are circles, although it is not quite true in general.

EXAMPLE 6–8 **Where is Mars?** Mars' period (its "year") was first noted by Kepler to be about 687 days (Earth-days), which is $(687\,\text{d}/365\,\text{d}) = 1.88\,\text{yr}$. Determine the distance of Mars from the Sun using the Earth as a reference.

SOLUTION The period of the Earth is $T_E = 1\,\text{yr}$, and the distance of Earth from the Sun is $r_{ES} = 1.50 \times 10^{11}\,\text{m}$. From Kepler's third law (Eq. 6–7):

$$\frac{r_{MS}}{r_{ES}} = \left(\frac{T_M}{T_E}\right)^{\frac{2}{3}} = \left(\frac{1.88\,\text{yr}}{1\,\text{yr}}\right)^{\frac{2}{3}} = 1.52.$$

So Mars is 1.52 times the Earth's distance from the Sun, or $2.28 \times 10^{11}\,\text{m}$.

EXAMPLE 6–9 **The Sun's mass determined.** Determine the mass of the Sun given the Earth's distance from the Sun as $r_{ES} = 1.5 \times 10^{11}\,\text{m}$.

➡ PHYSICS APPLIED

*Determining the
Sun's mass*

SOLUTION We can use Eq. 6–6 and solve for M_S:

$$M_S = \frac{4\pi^2 r_{ES}^3}{G T_E^2} = \frac{4\pi^2 (1.5 \times 10^{11}\,\text{m})^3}{(6.67 \times 10^{-11}\,\text{N}\cdot\text{m}^2/\text{kg}^2)(3.16 \times 10^7\,\text{s})^2} = 2.0 \times 10^{30}\,\text{kg}$$

where we used the fact that

$$T_E = 1\,\text{yr} = \left(365\tfrac{1}{4}\,\text{d}\right)(24\,\text{h/d})(3600\,\text{s/h}) = 3.16 \times 10^7\,\text{s}.$$

EXAMPLE 6–10 ESTIMATE **Geosynchronous satellite, simplified.** A *geosynchronous* satellite of the Earth (as mentioned in Example 6–6) is one that stays above the same point on the equator of the Earth. Estimate the height above the Earth's surface needed for a geosynchronous weather satellite. (This is to be a "lunchtime" calculation done on a napkin, without calculator, as compared to our earlier calculation in Example 6–6.)

SOLUTION To use Kepler's third law we must compare the satellite to some other object that orbits Earth. The simplest choice is the Moon because we know its period and distance. The Moon's period is about $T_M \approx 27\,\text{d}$ and its distance from the Earth about $r_{ME} \approx 380{,}000\,\text{km}$. The period of the weather satellite needs to be $T_{Sat} = 1\,\text{d}$ so that it stays above the same place on the Earth. Hence,

$$r_{Sat} = r_{ME}\left(\frac{T_{Sat}}{T_M}\right)^{\frac{2}{3}} = r_{ME}\left(\frac{1\,\text{d}}{27\,\text{d}}\right)^{\frac{2}{3}} = r_{ME}\left(\frac{1}{3}\right)^2 = \frac{r_{ME}}{9}.$$

(How nice that the Moon's approximate period turns out to be a perfect cube.) A geosynchronous satellite must be $\frac{1}{9}$ the distance to the Moon, which is 42,000 km from the center of the Earth or 36,000 km above the Earth's surface. This is about 6 Earth radii high.

Accurate measurements on the orbits of the planets indicated that they did not precisely follow Kepler's laws. For example, slight deviations from perfectly elliptical orbits were observed. Newton was aware that this was to be expected from the law of universal gravitation ("every body in the universe attracts every other body…") because each planet exerts a gravitational force on the other planets. Since the mass of the Sun is much greater than that of any planet, the force on one planet due to any other planet will be small in comparison to the force on it due to the Sun. (The derivation of perfectly elliptical orbits ignores the forces due to other planets.) But because of this small force, each planetary orbit should depart from a perfect ellipse, especially when a second planet is fairly close to it. Such deviations, or **perturbations**, as they are called, from perfect ellipses are indeed observed. In fact, Newton's recognition of perturbations in the orbit of Saturn was a hint that helped him formulate the law of universal gravitation, that all

➡ PHYSICS APPLIED

*Perturbations and
discovery of planets*

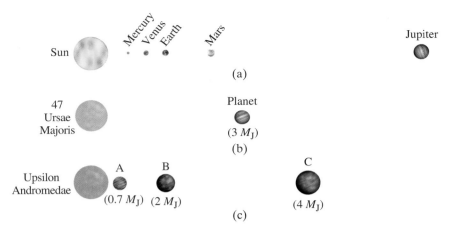

FIGURE 6–17 (a) Our solar system, compared to recently discovered planets orbiting (b) the star 47 Ursae Majoris, and (c) the star Upsilon Andromedae with at least three planets. M_J is the mass of Jupiter. (Sizes not to scale.)

bodies attract gravitationally. Observation of other perturbations later led to the discovery of Neptune and Pluto. Deviations in the orbit of Uranus, for example, could not all be accounted for by perturbations due to the other known planets. Careful calculation in the nineteenth century indicated that these deviations could be accounted for if there were another planet farther out in the solar system. The position of this planet was predicted from the deviations in the orbit of Uranus, and telescopes focused on that region of the sky quickly found it; the new planet was called Neptune. Similar but much smaller perturbations of Neptune's orbit led to the discovery of Pluto in 1930.

More recently, in 1996, planets revolving about distant stars (Fig. 6–17) were inferred from the regular "wobble" of each star due to the gravitational attraction of the revolving planet.

The development by Newton of the law of universal gravitation and the three laws of motion was a major intellectual achievement. For with these laws, Newton was able to describe the motion of objects on Earth and in the heavens. The motions of heavenly bodies and bodies on Earth were seen to follow the same laws (something not previously recognized generally, although Galileo and Descartes had argued in its favor). For this reason, and also because Newton integrated the results of earlier workers into his system, we sometimes speak of Newton's "synthesis."

Newton's work was so encompassing that it constituted a theory of the universe, and influenced philosophy and other fields. The laws formulated by Newton are referred to as **causal laws**. By **causality** we mean the idea that one occurrence can cause another. We have repeatedly observed, for example, that when a rock strikes a window, the window almost immediately breaks. We infer that the rock *caused* the window to break. This idea of "cause and effect" took on more forceful meaning with Newton's laws. For the motion—or rather the acceleration—of any object was seen to be *caused* by the net force acting on it. As a result, the universe came to be pictured by many scientists and philosophers as a big machine whose parts move in a predetermined way—according to natural laws. However, this *deterministic* view of the universe had to be modified by scientists in the twentieth century.

➡ P H Y S I C S   A P P L I E D

Planets around other stars

Newton's synthesis

Causality

6–6 Gravitational Field

Most of the forces we meet in everyday life are contact forces: you push or pull on a lawn mower, a tennis racket exerts a force on a tennis ball when they make contact, or a ball exerts a force on a window when they make contact. But the gravitational force (and even the electromagnetic force, as we shall see later) acts over a

Force acting at a distance

distance: there is a force even when the two objects are not in contact. The Earth, for example, exerts a force on a falling apple. It also exerts a force on the Moon, 384,000 km away. And the Sun exerts a gravitational force on the Earth. The idea of a force *acting at a distance* was a difficult one for early thinkers. Newton himself felt uneasy with this concept when he published his law of universal gravitation.

Another point of view that avoids some of these conceptual difficulties is the concept of the **field**, developed in the nineteenth century by Michael Faraday (1791–1867) to aid understanding of electromagnetism. Only later was it applied to gravity. According to the field concept, a **gravitational field** surrounds every body that has mass, and this field permeates all of space. A second body at a particular location near the first body experiences a force because of the gravitational field that exists there. Because the gravitational field at the location of the second mass is considered to act directly on this mass, we are a little closer to the idea of a contact force. *Field*

To be quantitative, we can define the **gravitational field** as the gravitational force per unit mass at any point in space. If we want to measure the gravitational field at any point, we place a small "test" mass m at that point and measure the force $\mathbf{F}$ exerted on it (making sure only gravitational forces are acting). Then the gravitational field, $\mathbf{g}$, at that point is defined as

$$\mathbf{g} = \frac{\mathbf{F}}{m}.$$ **(6–8)**   *Gravitational field (defined)*

The units of $\mathbf{g}$ are N/kg.

It is clear that the gravitational field which an object experiences has magnitude equal to the acceleration due to gravity at that point. (When we speak of acceleration, however, we use units m/s², which is equivalent to N/kg, since $1\,\text{N} = 1\,\text{kg}\cdot\text{m/s}^2$.)

If the gravitational field is due to a single body of mass M (such as when m is near the Earth's surface), then the gravitational field at a distance r from M has magnitude

$$g = \frac{1}{m}\,G\,\frac{mM}{r^2} = G\,\frac{M}{r^2}.$$

In vector notation we write

$$\mathbf{g} = -\frac{GM}{r^2}\,\hat{\mathbf{r}},$$ *Gravitational field due to a mass M*

where $\hat{\mathbf{r}}$ is a unit vector pointing radially outward from mass M, and the minus sign reminds us that the field points toward mass M (see Eqs. 6–1, 6–2, and 6–4). If several different bodies contribute significantly to the gravitational field, then we write $\mathbf{g}$ as the vector sum of all these contributions. In interplanetary space, for example, $\mathbf{g}$ at any point in space is the vector sum of terms due to the Earth, Sun, Moon, and other bodies that contribute. The gravitational field $\mathbf{g}$ at any point in space does not depend on the value of our test mass, m, placed at that point; $\mathbf{g}$ depends only on the masses (and locations) of the bodies that create the field there.

6–7 | Types of Forces in Nature

We have already discussed that Newton's law of universal gravitation, Eq. 6–1, describes how a particular type of force—gravity—depends on the distance between, and masses of, the objects involved. Newton's second law, $\Sigma\mathbf{F} = m\mathbf{a}$, on the other hand, tells how a body will accelerate due to *any* type of force. But what are the types of forces that occur in nature besides gravity?

In the twentieth century, physicists came to recognize four different fundamental forces in nature: (1) the gravitational force; (2) the electromagnetic force (we shall see later that electric and magnetic forces are intimately related); (3) the strong nuclear force; and (4) the weak nuclear force. In this chapter, we discussed the gravitational force in detail. The nature of the electromagnetic force will be discussed in detail in Chapters 21 to 32. The strong and weak nuclear forces operate at the level of the atomic nucleus, and although they manifest themselves in such phenomena as radioactivity and nuclear energy, they are much less obvious in our daily lives.

Electroweak and GUT

Physicists have been working on theories that would unify these four forces—that is, to consider some or all of these forces as different manifestations of the same basic force. So far, the electromagnetic and weak nuclear forces have been theoretically united to form a successful *electroweak* theory, in which the electromagnetic and weak forces are seen as two different manifestations of a single *electroweak force*. Attempts to further unify the forces, such as in *grand unified theories* (GUT), are hot research topics today.

Contact forces

But where do everyday forces fit into this scheme? Ordinary forces, such as pushes, pulls, and other contact forces like the normal force and friction, are today considered to be due to the electromagnetic force acting at the atomic level. For example, the force your fingers exert on a pencil is the result of electrical repulsion between the outer electrons of the atoms of your finger and those of the pencil.

* 6–8 Gravitational Versus Inertial Mass; the Principle of Equivalence

We have dealt with two aspects of mass. In Chapter 4, we defined mass as a measure of the inertia of a body. Newton's second law relates the force acting on a body to its acceleration and its **inertial mass**, as we call it. We might say that inertial mass represents a resistance to any force. In this chapter we have dealt with mass as a property related to the gravitational force—that is, mass as a quantity that determines the strength of the gravitational force between two bodies. This we call the **gravitational mass**.

Now it is not at all obvious that the inertial mass of a body should be equal to its gravitational mass. (The force of gravity might have depended on a completely different property of a body, just as the electrical force depends on a property called electric charge.) Newton's and Cavendish's experiments indicated that these two types of mass are equal for a body, and modern experiments confirm it to a precision of about 1 part in 10^{12}.

Principle of equivalence

The experimental evidence that gravitational and inertial masses are equal (or at least proportional) is remarkable. This equivalence between gravitational and inertial masses was raised by Albert Einstein (1879–1955) to a principle of nature. Einstein called it simply the **principle of equivalence**, and he used it as a foundation for his general theory of relativity (*c.* 1916). The principle of equivalence, as Einstein gave it, can be stated in another way: there is no experiment observers can perform to distinguish if an acceleration arises because of a gravitational force or because their reference frame is accelerating. If, for example, you were far out in space and an apple fell to the floor of your spacecraft, you might assume a gravitational force was acting on the apple. On the other hand, it would also be possible that the apple fell because your spacecraft accelerated upward (relative to an inertial system). The effects would be indistinguishable, according to the principle of equivalence, because the apple's inertial and gravitational masses—that determine how a body "reacts" to outside influences—are indistinguishable.

(a)

(b)

FIGURE 6–18 (a) Light beam goes straight across an elevator that is not accelerating. (b) The light beam bends (exaggerated) in an elevator accelerating in an upward direction.

Gravitation as Curvature of Space; Black Holes

The principle of equivalence can be used to show that light ought to be deflected due to the gravitational force of a massive body. Let us consider a thought experiment, in an elevator in free space where no gravity acts. If there is a hole in the side of the elevator and a beam of light enters from outside, the beam travels straight across the elevator and makes a spot on the opposite side if the elevator is at rest (Fig. 6–18a). If the elevator is accelerating upward as in Fig. 6–18b, the light beam still travels straight as observed in the original reference frame at rest. In the upwardly accelerating elevator, however, the beam is observed to curve downward. Why? Because during the time the light travels from one side of the elevator to the other, the elevator is moving upward at ever-increasing speed.

Now, according to the equivalence principle, an upwardly accelerating reference frame is equivalent to a downward gravitational field. Hence, we can picture the curved light path in Fig. 6–18b as being the effect of a gravitational field. Thus we expect gravity to exert a force on a beam of light and to bend it out of a straight-line path!

Einstein, in his general theory of relativity, predicted just this, that light would be affected by gravity. It was calculated that light from a distant star would be deflected by 1.75″ of arc (tiny but detectable) as it passed near the Sun, as shown in Fig. 6–19. Such a deflection was measured in 1919 during an eclipse of the Sun. (The eclipse reduced the brightness of the Sun so that the stars in line with its edge at that moment would be visible.)

A light beam travels by the shortest, most direct, path between two points. If it didn't, some other object could travel between the two points in a shorter time and thus have a greater speed than the speed of light, which would contradict the special theory of relativity. If a light beam can follow a curved path (as discussed above), then this curved path must be the shortest distance between the two points. This suggests that *space itself is curved* and that it is the gravitational field that causes the curvature. The curvature is greatest near very massive bodies. To visualize this curvature of space, we might think of space as being like a thin rubber sheet; if a heavy weight is hung from it, it curves as shown in Fig. 6–20. The weight corresponds to a huge mass that causes space (space itself!) to curve.

The extreme curvature of space-time shown in Fig. 6–20, could be produced by a **black hole**. A black hole is a hugely massive star that is so dense and so massive that gravity would be so strong that even light could not escape it. Light would be pulled back in by the force of gravity. Since no light could escape from such a massive star, we could not see it—it would be black. A body might pass by it and be deflected by its gravitational field, but if the body came too close, it would be swallowed up, never to escape. Hence the name for these hypothesized black holes. Experimentally there is good evidence for their existence, although some scientists remain cautious. One possibility is that there may be a giant black hole at the center of our Galaxy and perhaps at the center of other galaxies.

(a)

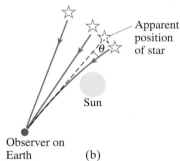

(b)

FIGURE 6–19 (a) Three stars in the sky. (b) If the light from one of these stars passes very near the Sun, whose gravity bends the light beam, the star will appear higher than it actually is.

FIGURE 6–20 Rubber sheet analogy for space (technically space-time) curved by matter.

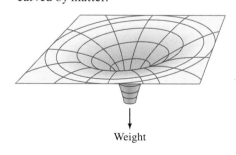

Summary

Newton's **law of universal gravitation** states that every particle in the universe attracts every other particle with a force proportional to the product of their masses and inversely proportional to the square of the distance between them:

$$F = G \frac{m_1 m_2}{r^2}.$$

The direction of this force is along the line joining the two particles. It is this gravitational force that keeps the Moon revolving around the Earth and the planets revolving around the Sun.

The total gravitational force on any body is the vector sum of the forces exerted by all other bodies; frequently the effects of all but one or two bodies can be ignored.

Satellites revolving around the Earth are acted on by gravity, but "stay up" because of their high tangential speed.

Newton's three laws of motion, plus his law of universal gravitation, constituted a wide-ranging theory of the universe. With them, motion of objects on Earth and in the heavens could be accurately described. And they provided a theoretical base for **Kepler's laws** of planetary motion.

According to the **field** concept, a **gravitational field** surrounds every body that has mass, and it permeates all of space. The gravitational field at any point in space is the vector sum of the fields due to all massive bodies and can be defined as

$$\mathbf{g} = \frac{\mathbf{F}}{m}$$

where $\mathbf{F}$ is the force acting on a small "test" mass m placed at that point.

Questions

1. Does an apple exert a gravitational force on the Earth? If so, how large a force? Consider an apple (a) attached to a tree and (b) falling.

2. The Sun's gravitational pull on the Earth is much larger than the Moon's. Yet the Moon's is mainly responsible for the tides. Explain. [*Hint*: Consider the difference in gravitational pull from one side of the Earth to the other.]

3. Will an object weigh more at the equator or at the poles? What two effects are at work? Do they oppose each other?

4. Why is more fuel required for a spacecraft to travel from the Earth to the Moon than it does to return from the Moon to the Earth?

5. The gravitational force on the Moon due to the Earth is only about half the force on the Moon due to the Sun (see Example 6–3). Why isn't the Moon pulled away from the Earth?

6. If you were in a satellite orbiting the Earth, how might you cope with walking, drinking, or putting a pair of scissors on a table?

7. An antenna loosens and becomes detached from a satellite in a circular orbit around the Earth. Describe the antenna's motion subsequently. If it will land on the Earth, describe where; if not, describe how it could be made to land on the Earth.

8. Describe how careful measurements of the variation in g in the vicinity of an ore deposit might be used to estimate the amount of ore present.

9. The Sun is directly below us at midnight, in line with the Earth's center. Are we then heavier at midnight, due to the Sun's gravitational force on us, than we are at noon? Explain.

10. When will your apparent weight be the greatest, as measured by a scale in a moving elevator: when the elevator (a) accelerates downward, (b) when it accelerates upward, (c) when it is in free fall, or (d) when it moves upward at constant speed? In which case would your weight be the least? When would it be the same as when you are on the ground?

11. If the Earth's mass were double what it is, in what ways would the Moon's orbit be different?

12. The source of the Mississippi River is closer to the center of the Earth than is its outlet in Louisiana (since the Earth is fatter at the equator than at the poles). Explain how the Mississippi can flow "uphill."

13. People sometimes ask, "What keeps a satellite up in its orbit around the Earth?" How would you respond?

14. Explain how a runner experiences free fall or apparent weightlessness between steps.

15. The Earth moves faster in its orbit around the Sun in December than in June. Is it closer to the Sun in June or in December? Does this affect the seasons? Explain.

16. The mass of the planet Pluto was not known until it was discovered to have a moon. Explain how this enabled an estimate of Pluto's mass.

17. Does your body directly sense a gravitational field? (Compare to what you would feel in free fall.)

18. Discuss the conceptual differences between $\mathbf{g}$ as acceleration due to gravity and $\mathbf{g}$ as gravitational field.

19. Describe a general procedure to determine the mass of a planet from observations on the orbit of one of its satellites.

20. Which pulls harder gravitationally, the Earth on the Moon, or the Moon on the Earth? Which accelerates more?

Problems

Sections 6–1 to 6–3

1. (I) Calculate the force of gravity on a spacecraft 12,800 km (2 earth radii) above the Earth's surface if its mass is 1400 kg.

2. (I) Calculate the acceleration due to gravity on the Moon. The Moon's radius is about 1.74×10^6 m and its mass is 7.35×10^{22} kg.

3. (I) A hypothetical planet has a radius 2.5 times that of Earth, but has the same mass. What is the acceleration due to gravity near its surface?

4. (I) A hypothetical planet has a mass 3.0 times that of Earth, but the same radius. What is g near its surface?

5. (II) You are explaining to friends why astronauts feel weightless orbiting in the space shuttle, and they respond that they thought gravity was just a lot weaker up there. Convince them and yourself that it isn't so by calculating how much weaker gravity is 300 km above the Earth's surface.

6. (II) Calculate the effective value of g, the acceleration of gravity, at (a) 3200 m, and (b) 3200 km, above the Earth's surface.

7. (II) Four 8.5-kg spheres are located at the corners of a square of side 0.70 m. Calculate the magnitude and direction of the gravitational force on one sphere due to the other three.

8. (II) Every few hundred years most of the planets line up on the same side of the Sun. Calculate the total net force on the Earth due to Venus, Jupiter, and Saturn, assuming all four planets are in a line, Fig. 6–21. The masses are respectively $M_V = 0.815\,M_E$, $M_J = 318\,M_E$, $M_{Sa} = 95.1\,M_E$, and their mean distances from the Sun are 108, 150, 778, and 1430 million km.

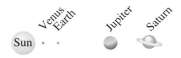

FIGURE 6–21   Problem 8 (not to scale).

9. (II) Four masses are arranged as shown in Fig. 6–22. Determine the x and y components of the gravitational force on the mass at the origin (m). Write the force in vector notation ($\mathbf{i}, \mathbf{j}$).

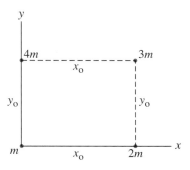

FIGURE 6–22   Problem 9.

10. (II) Suppose a spacecraft travels directly from Earth to Mars (where $g = 3.7$ m/s^2 at the surface) at constant velocity. Graph the weight of (net gravitational force on) a 70-kg passenger as a function of distance between the two planets. Assume any acceleration is negligible.

11. (II) Suppose the mass of the Earth were doubled, but it kept the same density and spherical shape. How would the weight of objects at the Earth's surface change?

12. (II) Equation 6–4 can be applied to other bodies, such as other planets, to obtain useful information, if appropriate variables are used. For example, suppose a spacecraft to Mars has measured g at the surface to be $g_{Mars} = 3.7$ m/s^2, and astronomical observations show the radius of Mars to be $r_{Mars} = 3.4 \times 10^6$ m. Using this information, determine the mass of Mars.

13. (II) At what distance from the Earth will a spacecraft on the way to the Moon experience zero net force due to these two bodies because the Earth and Moon pull with equal and opposite forces?

14. (II) Determine the mass of the Sun using the known value for the period of the Earth and its distance from the Sun. [Note: Compare your answer to that obtained using Kepler's laws, Example 6–9.]

15. (III) (a) Use the binomial expansion

$$(1 \pm x)^n = 1 \pm nx + \frac{n(n-1)}{2}x^2 \pm \cdots$$

to show that the value of g is altered by approximately

$$\Delta g \approx -2g\frac{\Delta r}{r_E}$$

at a height Δr above the Earth's surface, where r_E is the radius of the Earth, as long as $\Delta r \ll r_E$. (b) What is the meaning of the minus sign in this relation? (c) Use this result to compute the effective value of g at 100 km above the Earth's surface. Compare to a direct use of Eq. 6–1.

16. (III) Determine the magnitude and direction of the effective value of $\mathbf{g}$ at a latitude of 45° on the Earth. Assume the Earth is a rotating sphere.

17. (III) A ship steams at speed v along the equator. Show that the apparent weight w of an object as weighed on the ship is given approximately by $w = w_0(1 \pm 4\pi f v/g)$, where f is the frequency of rotation (revolutions/second) of the Earth. Why is there a $\pm$ sign? Let w_0 be the measured weight of the object when the ship is at rest relative to the Earth.

Section 6–4

18. (I) Calculate the velocity of a satellite moving in a stable circular orbit about the Earth at a height of 5200 km.

19. (I) The space shuttle releases a satellite into a circular orbit 600 km above the Earth. How fast must the shuttle be moving (relative to Earth) when the release occurs?

20. (II) A 16.0-kg monkey hangs from a cord suspended from the ceiling of an elevator. The cord can withstand a tension of 200 N and breaks as the elevator accelerates. What was the elevator's minimum acceleration (magnitude and direction)?

21. (II) Calculate the period of a satellite orbiting the Moon, 100 km above the Moon's surface. Ignore effects of the Earth. The radius of the Moon is 1740 km.

22. (II) Determine the time it takes for a satellite to orbit the Earth in a circular "near-Earth" orbit. The definition of "near-Earth" orbit is one which is at a height above the surface of the Earth that is very small compared to the radius of the Earth, and you may take the acceleration due to gravity as essentially the same as that on the surface. Does your result depend on the mass of the satellite?

23. (II) What will a spring scale read for the weight of a 56-kg woman in an elevator that moves (a) with constant speed upward of 3.0 m/s, (b) with constant speed downward of 3.0 m/s, (c) with upward acceleration of 0.33 g, (d) with downward acceleration 0.33 g, and (e) in free fall?

24. (II) A ferris wheel 27.5 m in diameter rotates once every 10.5 s. What is the fractional change in a person's apparent weight (a) at the top and (b) at the bottom, as compared to her weight at rest?

25. (II) What is the apparent weight of a 75-kg astronaut 4100 km from the center of the Moon in a space vehicle (a) moving at constant velocity and (b) accelerating toward the Moon at 2.6 m/s²? State "direction" in each case.

26. (II) How long would a day be if the Earth were rotating so fast that objects at the equator were weightless?

27. (II) Two equal-mass stars maintain a constant distance apart of 8.0×10^{10} m and rotate about a point midway between them at a rate of one revolution every 12.6 yr. (a) Why don't the two stars crash into one another due to the gravitational force between them? (b) What must be the mass of each star?

28. (II) (a) Show that if a satellite orbits very near the surface of a planet with period T, the density (= mass per unit volume) of the planet is $\rho = m/V = 3\pi/GT^2$. (b) Estimate the density of the Earth, given that a satellite near the surface orbits with a period of about 90 minutes.

29. (II) At what horizontal velocity would a satellite have to be launched from the top of Mt. Everest to be placed in a circular orbit around the Earth?

30. (II) You are an astronaut in the space shuttle pursuing a satellite in need of repair. You find yourself in a circular orbit of the same radius as the satellite, but 30 km behind it. (a) How long will it take to reach the satellite if you reduce your orbital radius by 1.0 km? (b) By how much must you reduce your orbital radius to catch up in 8.0 hours?

31. (III) Three bodies of identical mass M form the vertices of an equilateral triangle of side L and rotate in circular orbits about the center of the triangle. They are held in place by their mutual gravitation. What is the speed of each?

32. (III) An inclined plane, fixed to the inside of an elevator, makes a 30° angle with the floor. A mass m slides on the plane without friction. What is its acceleration relative to the plane if the elevator (a) accelerates upward at 0.50 g, (b) accelerates downward at 0.50 g, (c) falls freely, or (d) moves upward at constant speed?

Section 6–5

33. (I) Use Kepler's laws and the period of the Moon (27.4 d) to determine the period of an artificial satellite orbiting very near the Earth's surface.

34. (I) The asteroid Icarus, though only a few hundred meters across, orbits the Sun like the other planets. Its period is about 410 d. What is its mean distance from the Sun?

35. (I) Neptune is an average distance of 4.5×10^9 km from the Sun. Estimate the length of the Neptunian year using the fact that the Earth is 1.50×10^8 km from the Sun on the average.

36. (I) Determine the mass of the Earth from the known period and distance of the Moon.

37. (II) Our Sun rotates about the center of the Galaxy $(M \approx 4 \times 10^{41}$ kg$)$ at a distance of about 3×10^4 light years $(1$ ly $= 3 \times 10^8$ m/s $\times 3.16 \times 10^7$ s/y$)$. What is the period of our orbital motion about the center of the Galaxy?

TABLE 6–3
Principal Moons of Jupiter (Problems 38 and 39)

Moon	Mass (kg)	Period (Earth days)	Mean distance from Jupiter (km)
Io	8.9×10^{22}	1.77	422×10^3
Europa	4.9×10^{22}	3.55	671×10^3
Ganymede	15×10^{22}	7.16	1070×10^3
Callisto	11×10^{22}	16.7	1883×10^3

38. (II) Table 6–3 gives the mean distance, period, and mass for the four largest moons of Jupiter (those discovered by Galileo in 1609). (a) Determine the mass of Jupiter using the data for Io. (b) Determine the mass of Jupiter using data for each of the other three moons. Are the results consistent?

39. (II) Determine the mean distance from Jupiter for each of Jupiter's moons, using the distance of Io and the periods given in Table 6–3. Compare to the values in the Table.

40. (II) The asteroid belt between Mars and Jupiter consists of many fragments, once hypothesized to have been a planet. (a) If the center of mass of the asteroid belt is about 3 times farther from the Sun than the Earth is, how long would it have taken this hypothetical planet to orbit the Sun? (b) Can we use these data to deduce the mass of that planet?

41. (III) A science fiction tale describes an artificial "planet" in the form of a band encircling a sun as shown: Fig. 6–23. The inhabitants live on the inside surface (where it is always noon). Imagine the sun is exactly like our own, that the distance to the band is the same as the Earth–Sun distance (so the climate is temperate), and that the ring rotates quickly enough to produce an apparent gravity of one g as on Earth. What will be the band's period of revolution, this planet's year, in Earth days?

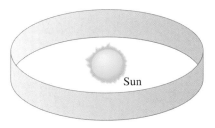

FIGURE 6–23
Problem 41.

42. (III) (a) Use Kepler's second law to show that the ratio of the speeds of a planet at its nearest and farthest points from the Sun is equal to the inverse ratio of the near and far distances: $v_N/v_F = d_F/d_N$. (b) Given that the Earth's distance from the Sun varies from 1.47 to 1.52×10^{11} m, determine the minimum and maximum velocities of the Earth in its orbit around the Sun.

43. (III) The recently discovered comet Hale–Bopp has a period of 3000 years. (a) What is its mean distance from the Sun? (b) At its closest approach, the comet is about 1 A.U. from the Sun (1 A.U. = distance from Earth to the Sun). What is the farthest distance? (c) What is the ratio of the speed at the closest point to the speed at the farthest point?

Section 6–6

44. (II) What is the magnitude and direction of the gravitational field midway between the Earth and Moon? Ignore effects of the Sun.

45. (II) (a) What is the gravitational field at the surface of the Earth due to the Sun? (b) Will this affect your weight significantly?

46. (III) Two identical particles, each of mass m, are located on the x axis at $x = +x_0$ and $x = -x_0$. (a) Determine a formula for the gravitational field due to these two particles for points on the y axis; that is, write $\mathbf{g}$ as a function of y, m, x_0, and so on. (b) At what point (or points) on the y axis is the magnitude of $\mathbf{g}$ a maximum value, and what is its value there? [Hint: Take the derivative $d\mathbf{g}/dy$.]

General Problems

47. How far above the Earth's surface will the acceleration of gravity be half what it is at the surface?

48. At the surface of a certain planet, the gravitational acceleration g has a magnitude of 12.0 m/s². A 3.0-kg brass ball is transported to this planet. What is (a) the mass of the brass ball on the Earth and on the planet, and (b) the weight of the brass ball on the Earth and on the planet?

49. An exotic finish to massive stars is that of a neutron star, which might have as much as five times the mass of our Sun packed into a sphere about 10 km in radius! Estimate the surface gravity on this monster.

50. What is the distance from the Earth's center to a point outside the Earth where the gravitational acceleration due to the Earth is $\frac{1}{10}$ of its value at the Earth's surface?

51. A typical white dwarf star, which once was an average star like our Sun but is now in the last stage of its evolution, is the size of our Moon but has the mass of our Sun. What is the surface gravity on this star?

52. During an *Apollo* lunar landing mission, the command module continued to orbit the Moon at an altitude of about 100 km. How long did it take to go around the Moon once?

53. Estimate what the value of G would need to be if you could actually "feel" yourself gravitationally attracted to someone else near you. Make reasonable assumptions.

54. (a) Calculate the gravitational force of the Sun on the Earth, and the gravitational force of the Moon on the Earth. (b) Why is the gravitational attraction of the Moon rather than the Sun the dominant mechanism for the occurrence of ocean tides?

55. The rings of Saturn are composed of chunks of ice that orbit the planet. The inner radius of the rings is 73,000 km, while the outer radius is 170,000 km. Find the period of an orbiting chunk of ice at the inner radius and the period of a chunk at the outer radius. Compare your numbers with Saturn's mean rotation period of 10 hours and 39 minutes. The mass of Saturn is 5.69×10^{26} kg.

56. The Navstar Global Positioning System (GPS) utilizes a group of 24 satellites orbiting the Earth. Using "triangulation" and signals transmitted by these satellites, the position of a receiver on the Earth can be determined to within an accuracy of a few centimeters. The satellite orbits are distributed evenly around the Earth, with four satellites in each of six orbits, allowing continuous navigational "fixes." The satellites orbit at an altitude of approximately 11,000 nautical miles [1 nautical mile = 1.852 km = 6076 ft]. (a) Determine the speed of each satellite. (b) Determine the period of each satellite.

57. NASA launched the Near Earth Asteroid Rendevous (NEAR), which, after traveling 1.3 billion miles, is meant to orbit the asteroid Eros at a height of about 15 km. Eros is potato-shaped: 40 km × 6 km × 6 km. Assume Eros has a density (mass/volume) of about 2.3×10^3 kg/m³. (a) What will be the period of NEAR as it orbits Eros? (b) Suppose Eros to be a sphere with the same mass and density. What would its radius be? (c) What would g be at the surface of this spherical Eros?

58. Suppose that a binary star system consists of two stars of equal mass. They are observed to be separated by 360 million km and to take 5.0 Earth years to orbit about a point midway between them. What is the mass of each?

59. Halley's comet orbits the Sun roughly once every 76 years. It comes very close to the surface of the Sun on its closest approach (Fig. 6–24). About how far out from the Sun is it at its farthest? Is it still "in" the Solar System? What planet's orbit is nearest when it is out there?

Halley's comet
Sun
FIGURE 6–24
Problem 59.

60. The Sun rotates about the center of the Milky Way Galaxy (Fig. 6–25) at a distance of about 30,000 light years from the center $(1\,\mathrm{ly} = 9.5 \times 10^{15}\,\mathrm{m})$. If it takes about 200 million years to make one rotation, estimate the mass of our Galaxy. Assume that the mass distribution of our galaxy is concentrated mostly in a central uniform sphere. If all the stars had about the mass of our Sun $(2 \times 10^{30}\,\mathrm{kg})$, how many stars would there be in our Galaxy?

FIGURE 6–25 Edge-on view of our galaxy. Problem 60.

61. The planet Jupiter is about 320 times as massive as the Earth. Thus, it has been claimed that a person would be crushed by the force of gravity on Jupiter since people can't survive more than a few g's. Calculate the number of g's a person would experience if she could stand on the equator of Jupiter. Use the following astronomical data for Jupiter: mass $= 1.9 \times 10^{27}\,\mathrm{kg}$, equatorial radius $= 7.1 \times 10^4\,\mathrm{km}$, rotation period 9 hr 55 min. Take the centripetal acceleration into account.

62. Astronomers using the Hubble Space Telescope have recently deduced the presence of an extremely massive core in the distant galaxy M87, so dense that it could well be a black hole (from which no light escapes). They measured the speed of gas clouds orbiting the core to be 780 km/s at a distance of 60 light-years $(5.7 \times 10^{17}\,\mathrm{m})$ from the core. Deduce the mass of the core, and compare it to the mass of our Sun.

63. Derive a formula for the mass of a planet in terms of its radius, r, the acceleration due to gravity at its surface, g_P, and the gravitational constant, G.

64. A plumb bob is deflected from the vertical by an angle θ due to a massive mountain nearby (Fig. 6–26). (a) Find an approximate formula for θ in terms of the mass of the mountain, M_M, the distance to its center, D_M, and the radius and mass of the Earth. (b) Make a rough estimate of the mass of Mt. Everest, assuming it has the shape, say, of an equilateral pyramid (or cone) 4000 m high above its base, and then (c) estimate the angle θ of the pendulum bob if it is 5 km from the center of Mt. Everest.

FIGURE 6–26 Problem 64.

65. Show that the rate of change of your weight is

$$-2G\frac{M_E m}{r^3}v$$

if you are traveling directly away from Earth at constant speed v. Your mass is m, and r is your distance from the center of the Earth at any moment.

66. A geologist searching for oil finds that the gravity at a certain location is 2 parts in 10^7 smaller than average. Assume that the location contains oil and the oil is located 2000 m below the Earth's crust. Estimate the size of the deposit. Take the density (mass per unit volume) of rock to be $3000\,\mathrm{kg/m^3}$ and that of oil to be $1000\,\mathrm{kg/m^3}$.

*67. A particle is released at a height r_E (radius of Earth) above the Earth's surface. Determine its velocity when it hits the Earth. Ignore air resistance. [*Hint*: Use Newton's second law, the law of universal gravitation, the chain rule, and integrate.]

This baseball pitcher is about to accelerate the baseball to a high velocity by exerting a force on it. He will be doing work on the ball as he exerts the force over a displacement of perhaps several meters, from behind his head until he releases the ball with arm outstretched in front of him. The work he does on the ball will be equal to the kinetic energy $\left(\frac{1}{2}mv^2\right)$ he gives to the ball, a result known as the work-energy principle.

CHAPTER 7

Work and Energy

U ntil now we have been studying the translational motion of an object in terms of Newton's three laws of motion. In this analysis, *force* has played a central role as the quantity determining the motion. In this chapter and the two that follow, we discuss an alternative analysis of the motion of an object in terms of the quantities *energy* and *momentum*. The significance of these quantities is that they are *conserved*. That is, in quite general circumstances they remain constant. That conserved quantities exist not only gives us a deeper insight into the nature of the world, but also gives us another way to approach practical problems.

The conservation laws of energy and momentum are especially valuable in dealing with systems of many objects, in which a detailed consideration of the forces involved would be difficult or impossible. These laws are applicable to a wide range of phenomena, including the atomic and subatomic world, where Newton's laws do not apply.

This chapter is devoted to the very important concept of *energy* and the closely related concept of *work*. These two quantities are scalars and so have no direction associated with them, which often makes them easier to work with than vector quantities such as acceleration and force. Energy has great importance for two reasons. First it is conserved. Second, energy is a concept that is useful not only in the study of motion, but in all areas of physics and in other sciences as well. But before discussing energy itself, we first examine the notion of work.

We consider only translational motion for now and, unless otherwise explained, objects are assumed to be rigid and therefore particle-like with no complicating internal motion.

7-1 | Work Done by a Constant Force

The word *work* has a variety of meanings in everyday language. But in physics, work is given a very specific meaning to describe what is accomplished by the action of a force when it acts on an object over some distance. Specifically, the **work** done on an object by a constant force (constant in both magnitude and direction) is defined to be *the product of the magnitude of the displacement times the component of the force parallel to the displacement*. In equation form, we can write

$$W = F_{\parallel} d$$

where $F_{\parallel}$ is the component of the constant force **F** parallel to the displacement **d**. We can also write

Work defined (for constant force)

$$W = Fd \cos \theta, \qquad \text{(7-1)}$$

where F is the magnitude of the constant force, d is the magnitude of the displacement of the object, and θ is the angle between the directions of the force and the displacement. The $\cos \theta$ factor appears in Eq. 7–1 because $F \cos \theta (= F_{\parallel})$ is the component of **F** parallel to **d**. See Fig. 7–1. Work is a scalar quantity—it has only magnitude.

Let's first consider the case in which the motion and the force are in the same direction, so $\theta = 0$ and $\cos \theta = 1$, and then $W = Fd$. For example, if you push a loaded grocery cart a distance of 50 m by exerting a horizontal force of 30 N on the cart, you do 30 N × 50 m = 1500 N · m of work on the cart.

Units for work: the joule

As this example shows, in SI units, work is measured in newton-meters. A special name is given to this unit, the **joule** (J): 1 J = 1 N · m. In the cgs system, the unit of work is called the *erg* and is defined as 1 erg = 1 dyne · cm. In British units, work is measured in foot-pounds. It is easy to show that 1 J = 10^7 erg = 0.7376 ft · lb.

Force without work

A force can be exerted on an object and yet do no work. For example, if you hold a heavy bag of groceries in your hands at rest, you do no work on it. A force is exerted, but the displacement is zero, so the work $W = 0$. You also do no work

FIGURE 7–1 A person pulling a crate along the floor. The work done by the force **F** is $W = Fd \cos \theta$, where **d** is the displacement.

on the bag of groceries if you carry it as you walk horizontally across the floor at constant velocity, as shown in Fig. 7–2. No horizontal force is required to move the package at a constant velocity. However, the person shown in Fig. 7–2 does exert an upward force F_P on the package equal to its weight. But this upward force is perpendicular to the horizontal motion of the package and thus has nothing to do with that motion. Hence, the upward force is doing no work. This conclusion comes from our definition of work, Eq. 7–1: $W = 0$, because $\theta = 90°$ and $\cos 90° = 0$. Thus, when a particular force is perpendicular to the motion, no work is done by that force. (When you start or stop walking, there is a horizontal acceleration and you do briefly exert a horizontal force, and thus do work.)

When dealing with work, as with force, it is necessary to specify whether you are talking about work done *by* a specific object or done *on* a specific object. It is also important to specify whether the work done is due to one particular force (and which one), or the total (net) work done by the *net force* on the object.

EXAMPLE 7–1 Work done on a crate. A 50-kg crate is pulled 40 m along a horizontal floor by a constant force exerted by a person, $F_P = 100$ N, which acts at a 37° angle as shown in Fig. 7–3. The floor is rough and exerts a friction force $F_{fr} = 50$ N. Determine the work done by each force acting on the crate, and the net work done on the crate.

SOLUTION We choose our coordinate system so that **x** can be the vector that represents the 40-m displacement (that is, along the x axis). There are four forces acting on the crate, as shown in Fig. 7–3 which is the free-body diagram: the force exerted by the person $\mathbf{F}_P$; the friction force $\mathbf{F}_{fr}$; the crate's weight $m\mathbf{g}$; and the normal force $\mathbf{F}_N$ exerted upward by the floor. The work done by the gravitational and normal forces is zero, since they are perpendicular to the displacement **x** ($\theta = 90°$ in Eq. 7–1):

$$W_G = mgx \cos 90° = 0$$

$$W_N = F_N x \cos 90° = 0.$$

The work done by $\mathbf{F}_P$ is

$$W_P = F_P x \cos \theta = (100 \text{ N})(40 \text{ m}) \cos 37° = 3200 \text{ J}.$$

The work done by the friction force is

$$W_{fr} = F_{fr} x \cos 180°$$
$$= (50 \text{ N})(40 \text{ m})(-1) = -2000 \text{ J}.$$

The angle between the displacement **x** and $\mathbf{F}_{fr}$ is 180° because they point in opposite directions. Since the force of friction is opposing the motion (and $\cos 180° = -1$), it does *negative* work on the crate.

FIGURE 7–2 Work done on the bag of groceries by the person in this case is zero since $\mathbf{F}_P$ is perpendicular to the displacement **d**.

FIGURE 7–3 Example 7–1: a 50-kg crate being pulled along a floor.

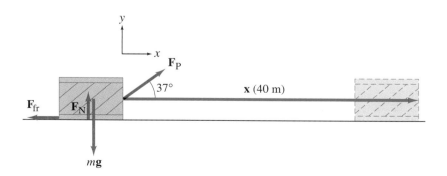

Finally, the net work can be calculated in two equivalent ways.

(1) The net work done on an object is the algebraic sum of the work done by each force, since work is a scalar:

W_{net} *is the work done by all the forces acting on the body*

$$W_{net} = W_G + W_N + W_P + W_{fr}$$
$$= 0 + 0 + 3200\,J - 2000\,J = 1200\,J.$$

(2) The net work can also be calculated by first determining the net force on the object and then taking its component along the displacement: $(F_{net})_x = F_P \cos\theta - F_{fr}$. Then the net work is

$$W_{net} = (F_{net})_x\, x = (F_P \cos\theta - F_{fr})x$$
$$= (100\,N\cos 37° - 50\,N)(40\,m) = 1200\,J.$$

In the vertical (y) direction, there is no displacement and no work done.

Negative work

In Example 7–1 we saw that friction did negative work. In general, the work done by a force is negative whenever the force (or the component of the force, $F_\parallel$) acts in the direction opposite to the direction of motion.

FIGURE 7–4 Example 7–2.

(a)

(b)

(c)

EXAMPLE 7–2 **Work on a carried backpack.** (*a*) Determine the work a hiker must do on a 15.0-kg backpack to carry it up a hill of height $h = 10.0\,m$, as shown in Fig. 7–4a. Determine also (*b*) the work done by gravity on the backpack, and (*c*) the net work done on the backpack. For simplicity, assume the motion is smooth and at constant velocity (i.e., there is negligible acceleration).

SOLUTION (*a*) The forces on the backpack are shown in Fig. 7–4b: the force of gravity, $m\mathbf{g}$, acting downward; and $\mathbf{F}_H$, the force the hiker must exert upward to support the pack. Since we assume there is negligible acceleration, the net horizontal force is negligible. In the vertical (y) direction, we choose up as positive. Newton's second law applied to the backpack gives

$$\Sigma F_y = ma_y$$
$$F_H - mg = 0.$$

Hence,

$$F_H = mg = (15.0\,kg)(9.80\,m/s^2) = 147\,N.$$

To calculate the work done by the hiker on the backpack, Eq. 7–1 can be written

$$W_H = F_H(d\cos\theta),$$

and we note from Fig. 7–4a that $d\cos\theta = h$. So the work done by the hiker can be written:

$$W_H = F_H(d\cos\theta) = F_H h = mgh$$
$$= (147\,N)(10.0\,m) = 1470\,J.$$

Note that the work done depends only on the change in elevation and not on the angle of the hill, θ. The same work would be done to lift the pack vertically the same height h.

(*b*) The work done by gravity is (from Eq. 7–1 and Fig. 7–4c):

$$W_G = (F_G)(d)\cos(180° - \theta).$$

Since $\cos(180° - \theta) = -\cos\theta$, we have

$$W_G = (F_G)(d)(-\cos\theta) = mg(-d\cos\theta)$$
$$= -mgh$$
$$= -(15.0\,kg)(9.80\,m/s^2)(10.0\,m) = -1470\,J.$$

Note that the work done by gravity doesn't depend on the angle of the incline but only on the vertical height h of the hill. This is because gravity does work only in the vertical direction. We will make use of this important result later.

(*c*) The *net* work done on the backpack is $W_{net} = 0$, since the net force on the backpack is zero (it is assumed not to accelerate significantly). We can also determine the net work done by writing

$$W_{net} = W_G + W_H = -1470\,J + 1470\,J = 0$$

which is, as it should be, the same result.

Note in this example that even though the *net* work on the backpack is zero, the hiker nonetheless does do work on the backpack equal to 1470 J.

Work done by gravity depends on the height of the hill and not on the angle of incline

CONCEPTUAL EXAMPLE 7–3 | **Does the Earth do work on the Moon?** The Moon revolves around the Earth in a nearly circular orbit, kept there by the gravitational force exerted by the Earth. Does gravity do (*a*) positive work, (*b*) negative work, or (*c*) no work at all on the Moon?

RESPONSE The gravitational force on the Moon (Fig. 7–5) acts toward the Earth (as a centripetal force), inward along the radius of the Moon's orbit. The Moon's displacement at any moment is along the circle, in the direction of its velocity, perpendicular to the radius and perpendicular to the force of gravity. Hence the angle θ between the force $\mathbf{F}_G$ and the instantaneous displacement of the Moon is 90°, and the work done by gravity is therefore zero ($\cos 90° = 0$). This is why the Moon, as well as artificial satellites, can stay in orbit without expenditure of fuel, since no work needs to be done against the force of gravity.

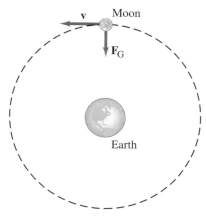

FIGURE 7–5
Conceptual Example 7–3.

PROBLEM SOLVING	Work

1. Choose an xy coordinate system. If the body is in motion, it may be convenient to choose the direction of motion as one of the coordinate directions. [Thus, for an object on an incline, you might choose one coordinate axis to be parallel to the incline.]

2. Draw a free-body diagram showing all the forces acting on the body.

3. Determine any unknown forces using Newton's laws.

4. Find the work done *by* a specific force *on* the body by using $W = Fd \cos \theta$. Note that the work done is negative when a force tends to oppose the displacement.

5. To find the *net* work done on the body, either (a) find the work done by each force and add the results algebraically; or (b) find the net force on the object, F_{net}, and then use it to find the net work done:

$$W_{net} = F_{net}\, d \cos \theta.$$

7–2 | Scalar Product of Two Vectors

Although work is a scalar, it involves the product of two quantities, force and displacement, each of which are vectors. Therefore, we now investigate the multiplication of vectors, which will be useful throughout the book.

Because vectors have direction as well as magnitude, they cannot be multiplied in the same way that scalars are. Instead we must *define* what the operation of vector multiplication means. Among the many possible ways to define how to multiply vectors, there are three ways that we will find useful in physics: (1) multiplication of a vector by a scalar, which was already discussed in Section 3–3; (2) multiplication of one vector by a second vector so as to produce a scalar; (3) multiplication of one vector by a second vector so as to produce another vector. The third type, called the *vector product*, will be discussed later, in Section 11–1.

We now discuss the second type, called the *scalar product*, or *dot product* (because a dot is used to indicate the multiplication). If we have two vectors, **A** and **B**, then their **scalar** (or **dot**) **product** is defined to be

Scalar product
(dot product)

$$\mathbf{A} \cdot \mathbf{B} = AB \cos\theta, \tag{7-2}$$

where A and B are the magnitudes of the vectors and θ is the angle ($< 180°$) between them when their tails touch, Fig. 7–6. Since A, B, and $\cos\theta$ are scalars, then so is the scalar product $\mathbf{A} \cdot \mathbf{B}$ (read "A dot B"). This definition, Eq. 7–2, fits perfectly with our definition of the work done by a constant force, Eq. 7–1. That is, we can write the work done by a constant force as the scalar product of force and displacement:

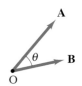

FIGURE 7–6 The scalar product, or dot product, of two vectors **A** and **B**, is $\mathbf{A} \cdot \mathbf{B} = AB \cos\theta$.

$$W = \mathbf{F} \cdot \mathbf{d} = Fd \cos\theta. \tag{7-3}$$

Indeed, the definition of scalar product, Eq. 7–2, is so chosen because many physically important quantities, such as work (and others we will meet later), can be described as the scalar product of two vectors.

An equivalent definition of the scalar product is that it is the product of the magnitude of one vector (say A) and the component of the other vector along the direction of the first ($B \cos\theta$).

Since A, B, and $\cos\theta$ are scalars, it doesn't matter in what order they are multiplied. Hence the scalar product is **commutative**:

Commutative property

$$\mathbf{A} \cdot \mathbf{B} = \mathbf{B} \cdot \mathbf{A}.$$

It is also easy to show that it is **distributive** (see Problem 29 for the proof):

Distributive property

$$\mathbf{A} \cdot (\mathbf{B} + \mathbf{C}) = \mathbf{A} \cdot \mathbf{B} + \mathbf{A} \cdot \mathbf{C}.$$

Let us use these properties and write each vector in terms of its rectangular components using unit vectors (Section 3–5, Eq. 3–5) as

$$\mathbf{A} = A_x\mathbf{i} + A_y\mathbf{j} + A_z\mathbf{k} \quad \text{and} \quad \mathbf{B} = B_x\mathbf{i} + B_y\mathbf{j} + B_z\mathbf{k};$$

then we can see that

Scalar product
(in terms of components)

$$\mathbf{A} \cdot \mathbf{B} = A_x B_x + A_y B_y + A_z B_z, \tag{7-4}$$

because the unit vectors, **i**, **j**, and **k** are perpendicular to each other so that

$$\mathbf{i} \cdot \mathbf{i} = \mathbf{j} \cdot \mathbf{j} = \mathbf{k} \cdot \mathbf{k} = 1$$
$$\mathbf{i} \cdot \mathbf{j} = \mathbf{i} \cdot \mathbf{k} = \mathbf{j} \cdot \mathbf{k} = 0.$$

Equation 7–4 is particularly useful.

If **A** is perpendicular to **B**, then Eq. 7–2 tells us $\mathbf{A} \cdot \mathbf{B} = 0$. But the converse, given that $\mathbf{A} \cdot \mathbf{B} = 0$, can come about in three different ways: $\mathbf{A} = 0$, $\mathbf{B} = 0$, or $\mathbf{A} \perp \mathbf{B}$.

EXAMPLE 7–4 **Using the dot product.** The force shown in Fig. 7–7 has magnitude $F_P = 20\,\text{N}$ and makes an angle of $30°$ to the ground. Calculate the work done by this force using Eq. 7–4 when the wagon is dragged 100 m along the ground.

FIGURE 7–7 Work done by a force $\mathbf{F_P}$ acting at an angle θ to the ground is $W = \mathbf{F_P} \cdot \mathbf{d}$.

SOLUTION We choose the x axis horizontal to the right and the y axis vertically upward. Then

$$\mathbf{F}_P = F_x\mathbf{i} + F_y\mathbf{j} = (F_P\cos 30°)\mathbf{i} + (F_P\sin 30°)\mathbf{j} = (17\,\text{N})\mathbf{i} + (10\,\text{N})\mathbf{j},$$

whereas $\mathbf{d} = (100\,\text{m})\mathbf{i}$. Then, using Eq. 7–4,

$$W = \mathbf{F}_P \cdot \mathbf{d} = (17\,\text{N})(100\,\text{m}) + (10\,\text{N})(0) + (0)(0) = 1700\,\text{J}.$$

Note that by choosing the x axis along $\mathbf{d}$ we simplified the calculation because $\mathbf{d}$ then has only one component.

7–3 | Work Done by a Varying Force

If the force acting on an object is constant, the work done by that force can be calculated using Eq. 7–1. But in many cases, the force varies in magnitude or direction during a process. For example, as a rocket moves away from Earth, work is done to overcome the force of gravity, which varies as the inverse square of the distance from the Earth's center. Other examples are the force exerted by a spring, which increases with the amount of stretch, or the work done by a varying force in pulling a box or cart up an uneven hill.

Figure 7–8 shows the path of an object in the xy plane as it moves from point a to point b. The path has been divided into short intervals each of length Δl_1, $\Delta l_2, \ldots \Delta l_7$. A force $\mathbf{F}$ acts at each point on the path, and is indicated at two points as $\mathbf{F}_1$ and $\mathbf{F}_5$. During each small interval Δl, the force is approximately constant. So, for the first interval, the force does work ΔW of approximately (see Eq. 7–1)

$$\Delta W \approx F_1\cos\theta_1\,\Delta l_1.$$

In the second interval the work done is approximately $F_2\cos\theta_2\,\Delta l_2$, and so on. The total work done in moving the particle the total distance $l = \Delta l_1 + \Delta l_2 + \ldots + \Delta l_7$ is the sum of all these terms:

$$W \approx \sum_{i=1}^{7} F_i\cos\theta_i\,\Delta l_i. \tag{7–5}$$

We can examine this graphically by plotting $F\cos\theta$ versus l as shown in Fig. 7–9a. The distance l has been subdivided into the same seven intervals indicated by the vertical dashed lines. The value of $F\cos\theta$ at the center of each interval is indicated by the horizontal dashed lines. Each of the shaded rectangles has an area $(F_i\cos\theta)(\Delta l_i)$, which is the work done during the interval. Thus the estimate of the work done given by Eq. 7–5 equals the sum of the areas of all the rectangles. If we subdivide the distance into a greater number of intervals so that each Δl_i is smaller, the estimate of the work done given by Eq. 7–5 becomes more accurate (the assumption that F is constant over each interval is more accurate). If we let each Δl_i approach zero (so we approach an infinite number of intervals), then we obtain an exact result for the work done:

$$W = \lim_{\Delta l_i \to 0} \Sigma F_i\cos\theta_i\,\Delta l_i = \int_a^b F\cos\theta\,dl. \tag{7–6}$$

This limit as $\Delta l_i \to 0$ is the *integral* of $(F\cos\theta\,dl)$ from point a to point b. The symbol for the integral, $\int$, is an elongated S to indicate an infinite sum; and Δl has been replaced by dl, meaning an infinitesimal distance.

In this limit as Δl approaches zero, the total area of the rectangles (Fig. 7–9a) approaches the area between the $(F\cos\theta)$ curve and the l axis from a to b as shown shaded in Fig. 7–9b. That is, *the work done by a variable force in moving an object between two points is equal to the area under the $(F\cos\theta)$ versus (l) curve between those two points.*

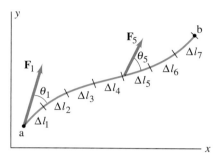

FIGURE 7–8 A particle acted on by a variable force, **F**, moves along the path shown from point a to point b.

FIGURE 7–9 Work done by a force F is (a) approximately equal to the sum of the areas of the rectangles, (b) exactly equal to the area under the curve of $F\cos\theta$ vs. l.

(a)

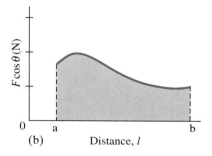

(b)

W = area under
$F\cos\theta$ vs. l curve

In the limit as Δl approaches zero, the infinitesimal distance dl equals[†] the magnitude of the infinitesimal displacement vector dl. The direction of the vector dl is along the tangent to the curve at that point, so θ is the angle between $\mathbf{F}$ and dl at any point. Thus we can rewrite Eq. 7–6, using dot-product notation:

Work defined in general

$$W = \int_a^b F \cos\theta \, dl = \int_a^b \mathbf{F} \cdot dl. \tag{7–7}$$

This is the *most general definition of work*. The integral in Eq. 7–7 is called a *line integral* since it is the integral of $F \cos\theta$ along the line that represents the path of the object. (Equation 7–1 for a constant force is a special case of Eq. 7–7.) In rectangular coordinates, the force can be written

$$\mathbf{F} = F_x \mathbf{i} + F_y \mathbf{j} + F_z \mathbf{k}$$

and the displacement dl is

$$dl = dx \, \mathbf{i} + dy \, \mathbf{j} + dz \, \mathbf{k}.$$

Then the work done can be written

$$W = \int_a^b F_x \, dx + \int_a^b F_y \, dy + \int_a^b F_z \, dz.$$

To actually use Eq. 7–6 or 7–7 to calculate the work, there are several options. (1) If $F \cos\theta$ is known as a function of position, a graph like that of Fig. 7–9b can be made and the area determined graphically. (2) Another possibility is to use numerical integration (numerical summing), perhaps with the aid of a computer or calculator. (3) A third possibility is to use the analytical methods of integral calculus. To do this, we must be able to write $\mathbf{F}$ as a function of position, $F(x, y, z)$, and we must know the path.

For example, let us consider a one-dimensional situation to determine analytically the work done by a coiled spring, such as that shown in Fig. 7–11. For a person to hold a spring either stretched or compressed an amount x from its normal (unstretched) length requires a force F_P that is directly proportional to x. That is,

$$F_P = kx,$$

where k is a constant, called the *spring constant*, and is a measure of the stiffness of the particular spring. The spring itself exerts a force in the opposite direction (Fig. 7–11):

Spring force

$$F_S = -kx. \tag{7–8}$$

This force is sometimes called a "restoring force" because the spring exerts its force in the direction opposite the displacement (hence the minus sign), acting to return it to its normal length. Equation 7–8 is known as the **spring equation** and also as **Hooke's law** (see Chapter 12), and is accurate for springs as long as x is not too great.

[†] The distance along a curve, Δl, is not generally equal to the magnitude of the displacement, $\Delta \mathbf{r}$, as shown in Fig. 7–10 below. But in the limit of infinitesimals they are equal: $dl = dr$, and in this limit the vector $dl = d\mathbf{r}$. Note that we cannot define a vector Δl since we cannot assign it a unique direction when the path curves; we can define a direction for dl—namely, the direction of the tangent to the curve at that point, so $dl = d\mathbf{r}$ and $dl = dr$.

FIGURE 7–10 The displacement vector, $\Delta \mathbf{r}$, has a magnitude that is not necessarily equal to the distance traveled, Δl.

Let us calculate the work a person does to stretch (or compress) a spring from its normal (unstretched) length, $x_a = 0$, to an extra length, $x_b = x$. We assume the stretching is done slowly, so that the acceleration is essentially zero. The force $\mathbf{F_P}$ is exerted parallel to the axis of the spring, along the x axis, so $\mathbf{F_P}$ and $d\mathbf{l}$ are parallel. Hence, since $d\mathbf{l} = dx\,\mathbf{i}$ in this case, the work done by the person is[†]

$$W_P = \int_{x_a=0}^{x_b=x} [F_P(x)\mathbf{i}]\cdot[dx\,\mathbf{i}] = \int_0^x F_P(x)\,dx = \int_0^x kx\,dx = \tfrac{1}{2}kx^2\Big|_0^x = \tfrac{1}{2}kx^2.$$

(As is frequently done, we have used x to represent both the variable of integration, and the particular value of x at the end of the interval $x_a = 0$ to $x_b = x$.) Thus we see that the work needed is proportional to the square of the distance stretched (or compressed), x. This same result can be obtained by computing the area under the F-versus-x graph (with $\cos\theta = 1$ in this case) as shown in Fig. 7–12. Since the area is a triangle of altitude kx and base x, the work a person does to stretch or compress a spring an amount x, equal to the area, is

$$W = \tfrac{1}{2}(x)(kx) = \tfrac{1}{2}kx^2,$$

which is the same result as before.

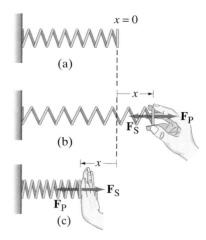

FIGURE 7–11 (a) Spring in normal (unstretched) position. (b) Spring is stretched by a person exerting a force $\mathbf{F_P}$ to the right (positive direction). The spring pulls back with a force $\mathbf{F_S}$ where $F_S = -kx$. (c) Person compresses the spring ($x < 0$), and the spring pushes back with a force $F_S = -kx$ where $F_S > 0$ because $x < 0$.

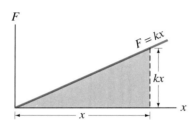

FIGURE 7–12 Work done to stretch a spring a distance x equals the triangular area under the curve $F = kx$. The area of a triangle is $\tfrac{1}{2}\times$ base $\times$ altitude, so $W = \tfrac{1}{2}(x)(kx) = \tfrac{1}{2}kx^2$.

EXAMPLE 7–5 **Work done on a spring.** (*a*) A person pulls on the spring in Fig. 7–11, stretching it 3.0 cm, which requires a maximum force of 75 N. How much work does the person do? (*b*) If, instead, the person compresses the spring 3.0 cm, how much work does the person do?

SOLUTION (*a*) First we need to calculate the spring constant k:

$$k = \frac{F_{max}}{x_{max}} = \frac{75\ \text{N}}{0.030\ \text{m}} = 2.5\times 10^3\ \text{N/m}.$$

Then the work done by the person on the spring is

$$W = \tfrac{1}{2}kx_{max}^2 = \tfrac{1}{2}(2.5\times 10^3\ \text{N/m})(0.030\ \text{m})^2 = 1.1\ \text{J},$$

(*b*) The force that the person exerts is still $F_P = kx$, though now both x and F_P are negative (x is positive to the right). The work done is

$$W_P = \int_{x=0}^{x=-0.030\,\text{m}} F_P(x)\,dx = \int_0^{x=-0.030\,\text{m}} kx\,dx = \tfrac{1}{2}kx^2\Big|_0^{-0.030\,\text{m}}$$

$$= \tfrac{1}{2}(2.5\times 10^3\ \text{N/m})(-0.030\ \text{m})^2 = 1.1\ \text{J},$$

which is the same as for stretching it.

Note that we cannot use $W = Fd$ (Eq. 7–1) for a spring because the force is not constant.

[†] See the Table of Integrals, Appendix B.

EXAMPLE 7–6 **Force as function of x.** A robot arm that controls the position of a video camera (Fig. 7–13) in an automated surveillance system is manipulated by a servo motor that exerts a force on a push-rod. The force is given by

$$F(x) = F_0\left(1 + \frac{1}{6}\frac{x^2}{x_0^2}\right)$$

where $F_0 = 2.0\,\text{N}$, $x_0 = 0.0070\,\text{m}$, and x is the position of the end of the push-rod. If the push-rod moves from $x_1 = 0.010\,\text{m}$ to $x_2 = 0.050\,\text{m}$, how much work did the servo motor do?

SOLUTION The force applied by the motor is not a linear function of x. We can determine the integral $\int F(x)\,dx$, or the area under the $F(x)$ curve (shown in Fig. 7–14). We integrate to find the work done by the motor:

$$W_M = F_0\int_{x_1}^{x_2}\left(1 + \frac{x^2}{6x_0^2}\right)dx = F_0\int_{x_1}^{x_2}dx + F_0\int_{x_1}^{x_2}\frac{x^2\,dx}{6x_0^2}$$

$$= F_0\left(x + \frac{1}{3}\frac{x^3}{6x_0^2}\right)\Big|_{x_1}^{x_2}.$$

We put in the values given and obtain

$$W_M = 2.0\,\text{N}\left[(0.050\,\text{m} - 0.010\,\text{m}) + \frac{(0.050\,\text{m})^3 - (0.010\,\text{m})^3}{(3)(6)(0.0070\,\text{m})^2}\right] = 0.361\,\text{J}.$$

FIGURE 7–13 Robot arm positions a video camera.

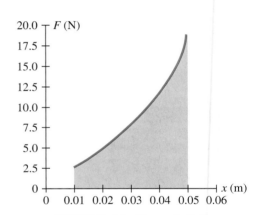

FIGURE 7–14 Example 7–6.

7–4 Kinetic Energy and the Work-Energy Principle

Energy is one of the most important concepts in science. Yet we cannot give a simple general definition of energy in only a few words. Nonetheless, each specific type of energy can be defined fairly simply. In this chapter we define translational kinetic energy; in the next chapter, we take up potential energy. In later chapters we will define other types of energy, such as that related to heat (Chapters 19 and 20). The crucial aspect of all the types of energy is that they can be defined consistently in such a way that the sum of all types, the *total energy*, is the same after any process occurs as it was before: that is, the quantity "energy" can be defined so that it is a conserved quantity. But more on this later.

For the purposes of this chapter, we could define energy in the usual way as "the ability to do work." This simple definition is not very precise, nor is it really valid for all types of energy.[†] However, for mechanical energy which we discuss in

---

[†]Energy associated with heat is often not available to do work, as we will discuss in detail in Chapter 20.

FIGURE 7–15 A constant net force F_{net} accelerates a bus from speed v_1 to speed v_2 over a distance d. The work done is $W = F_{net}d$.

this chapter and the next, it serves to underscore the fundamental connection between work and energy. We now define and discuss one of the basic types of energy, kinetic energy.

A moving object can do work on another object it strikes. A flying cannonball does work on a brick wall it knocks down; a moving hammer does work on a nail it strikes. In either case, a moving object exerts a force on a second object and moves it through a distance. An object in motion has the ability to do work and thus can be said to have energy. The energy of motion is called **kinetic energy**, from the Greek word *kinetikos*, meaning "motion."

To obtain a quantitative definition for kinetic energy, let us consider a rigid object of mass m that is moving in a straight line with an initial speed v_1. To accelerate it uniformly to a speed v_2, a constant net force F_{net} is exerted on it parallel to its motion over a distance d, Fig. 7–15. Then the net work done on the object is $W_{net} = F_{net}d$. We apply Newton's second law, $F_{net} = ma$, and use Eq. 2–12c, which we now write as $v_2^2 = v_1^2 + 2ad$, with v_1 as the initial speed and v_2 the final speed, and we find

$$W_{net} = F_{net}d = mad = m\left(\frac{v_2^2 - v_1^2}{2d}\right)d$$

or

$$W_{net} = \tfrac{1}{2}mv_2^2 - \tfrac{1}{2}mv_1^2. \tag{7–9}$$

We *define* the quantity $\tfrac{1}{2}mv^2$ to be the **translational kinetic energy**, K, of the object:

$$K = \tfrac{1}{2}mv^2. \tag{7–10}$$

Translational kinetic energy defined

(We call this "translational" kinetic energy to distinguish it from rotational kinetic energy, which we will discuss later, in Chapter 10.)

We can rewrite Eq. 7–9 as:

$$W_{net} = K_2 - K_1$$

or

$$W_{net} = \Delta K. \tag{7–11}$$

WORK-ENERGY PRINCIPLE

Equation 7–11 (or Eq. 7–9) is an important result. It can be stated in words:

> **The net work done on an object is equal to the change in its kinetic energy.**

WORK-ENERGY PRINCIPLE

This is known as the **work-energy principle**. Notice, however, that we made use of Newton's second law, $F_{net} = ma$, where F_{net} is the *net* force—the sum of all forces acting on the object. Thus, the work-energy principle is valid only if W is the *net work* done on the object—that is, the work done by all forces acting on the object.

The work-energy principle tells us that if (positive) net work W is done on a body, its kinetic energy increases by an amount W. The principle also holds true for the reverse situation: if negative net work W is done on the body, the body's kinetic energy decreases by an amount W. That is, a net force exerted on a body opposite to the body's direction of motion reduces its speed and its kinetic energy.

FIGURE 7–16 A moving hammer strikes a nail and comes to rest. The hammer exerts a force F on the nail; the nail exerts a force $-F$ on the hammer (Newton's third law). The work done on the nail by the hammer is positive $(W_n = Fd > 0)$. The work done on the hammer is negative $(W_h = -Fd)$.

General derivation of the work-energy principle

An example is a moving hammer (Fig. 7–16) striking a nail. The net force on the hammer ($-F$ in the figure, where **F** is assumed constant for simplicity) acts toward the left, whereas the displacement **d** is toward the right. So the net work done on the hammer, $W_h = (F)(d)(\cos 180°) = -Fd$, is negative and the hammer's kinetic energy decreases (usually to zero).

Figure 7–16 also illustrates how energy can be considered the ability to do work. The hammer, as it slows down, does positive work on the nail: if the nail exerts a force $-F$ on the hammer to slow it down, the hammer exerts a force $+F$ on the nail (Newton's third law) through the distance d. Hence the work done on the nail by the hammer is $W_n = (+F)(+d) = Fd = -W_h$, and W_n is positive. Thus the decrease in kinetic energy of the hammer is equal to the work the hammer can do on another object—which is consistent with energy being the ability to do work.

Note that whereas the translational kinetic energy $(= \frac{1}{2}mv^2)$ is directly proportional to the mass of the object, it is proportional to the *square* of the speed. Thus, if the mass is doubled, the kinetic energy is doubled. But if the speed is doubled, the object has four times as much kinetic energy and is therefore capable of doing four times as much work.

To summarize, the connection between work and kinetic energy (Eq. 7–11) operates both ways. If the net work W done on an object is positive, then the object's kinetic energy increases. If the net work W done on an object is negative, its kinetic energy decreases. If the net work done on the object is zero, its kinetic energy remains constant (which also means its speed is constant).

We derived the work-energy principle, Eq. 7–11, for motion in one dimension with a constant force. It is valid even if the force is variable and the motion is in two or three dimensions, as we now show. Suppose the net force $\mathbf{F}_{net}$ on an object varies in both magnitude and direction, and the path of the object is a curve as in Fig. 7–8. The net force may be considered to be a function of l, the distance along the curve. The net work done is (Eq. 7–6):

$$W_{net} = \int F_{net} \cos\theta \, dl = \int F_{\parallel} \, dl,$$

where $F_{\parallel}$ represents the component of the net force parallel to the curve at any point. By Newton's second law,

$$F_{\parallel} = ma_{\parallel} = m\frac{dv}{dt},$$

where $a_{\parallel}$, the component of a parallel to the curve at any point, is equal to the rate of change of speed, dv/dt. We can think of v as a function of l, and using the chain rule for derivatives, we have

$$\frac{dv}{dt} = \frac{dv}{dl}\frac{dl}{dt} = v\frac{dv}{dl},$$

since dl/dt is the speed v. Thus (letting 1 and 2 refer to the initial and final quantities, respectively):

$$W_{net} = \int_1^2 F_{\parallel} \, dl = \int_1^2 m\frac{dv}{dt} \, dl = \int_1^2 mv\frac{dv}{dl} \, dl = \int_1^2 mv \, dv,$$

which integrates to

$$W_{net} = \frac{1}{2}mv_2^2 - \frac{1}{2}mv_1^2 = \Delta K.$$

This is again the work-energy principle, which we have now derived for motion in three dimensions with a variable net force. Note, incidentally, that the work-energy principle is not a new independent law. Rather, it has been derived from the definitions of work and kinetic energy using Newton's second law.

Notice in this derivation that only the component of $\mathbf{F}_{net}$ parallel to the motion, $F_{\parallel}$, contributes to the work. Indeed, a force (or component of a force) acting perpendicular to the velocity vector does no work. Such a force changes only the direction of the velocity. It does not affect the magnitude of the velocity. One example of this is uniform circular motion in which an object moving with constant speed in a circle has a ("centripetal") force acting on it toward the center of the circle. This force does no work on the object, because (as we saw in Example 7–3) it is always perpendicular to the object's displacement $d\mathbf{l}$.

Centripetal force does no work

Because of the direct connection between work and kinetic energy, Eq. 7–11, energy is measured in the same units as work: joules in SI units, ergs in the cgs, and foot-pounds in the British system. Like work, kinetic energy is a scalar quantity. The kinetic energy of a set of objects is the (scalar) sum of the kinetic energies of the individual objects.

*Energy unit:
the joule*

EXAMPLE 7–7 **A baseball.** A 145-g baseball is thrown with a speed of 25 m/s. (*a*) What is its kinetic energy? (*b*) How much work was done to reach this speed, starting from rest?

SOLUTION (*a*) The kinetic energy is

$$K = \tfrac{1}{2}mv^2 = \tfrac{1}{2}(0.145 \text{ kg})(25 \text{ m/s})^2 = 45 \text{ J}.$$

(*b*) Since the initial kinetic energy was zero, the net work done is just equal to the final kinetic energy, 45 J.

CONCEPTUAL EXAMPLE 7–8 | **Work to stop a car.** An automobile traveling 60 km/h can brake to a stop within a distance of 20 m (Fig. 7–17a). If the car is going twice as fast, 120 km/h, what is its stopping distance (Fig. 7–17b)? The maximum braking force is approximately independent of speed.

RESPONSE We treat the car as if it were a particle or simple rigid body. Since the stopping force F is approximately constant, the work needed to stop the car, Fd, is proportional to the distance traveled. We apply the work-energy principle, noting that $\mathbf{F}$ and $\mathbf{d}$ are in opposite directions and that the final speed of the car is zero:

$$W_{\text{net}} = Fd \cos 180° = -Fd$$
$$= \Delta K = 0 - \tfrac{1}{2}mv^2.$$

Thus, since the force and mass are constant, we can see that the stopping distance, d, increases with the square of the speed:

$$d \propto v^2.$$

*Stopping distance $\propto$
initial speed squared*

If the car's initial speed is doubled, the stopping distance is $(2)^2 = 4$ times as great, or 80 m.

FIGURE 7–17
Example 7–8.

FIGURE 7-18   Example 7-9.

EXAMPLE 7-9   **A compressed spring.**   A horizontal spring has spring constant $k = 360 \, \text{N/m}$. (a) How much work is required to compress it from its uncompressed length ($x = 0$) to $x = 11.0 \, \text{cm}$? (b) If a 1.85-kg block is placed against the spring and the spring is released, what will be the speed of the block when it separates from the spring at $x = 0$? Ignore friction. (c) Repeat part (b) but assume that the block is moving on a table as in Fig. 7-18 and that the coefficient of kinetic friction is $\mu_k = 0.38$.

SOLUTION   (a) We saw in Section 7-3 that the net work, W, needed to stretch or compress a spring by a distance x is $W = \frac{1}{2}kx^2$. Therefore the work required to compress the spring a distance $x = 0.110 \, \text{m}$ is

$$W = \tfrac{1}{2}(360 \, \text{N/m})(0.110 \, \text{m})^2 = 2.18 \, \text{J},$$

where we have converted all units to SI.

(b) In returning to its uncompressed length, the spring does 2.18 J of work on the block (same calculation as in part (a), only in reverse). According to the work-energy principle, the block acquires kinetic energy of 2.18 J. Since $K = \frac{1}{2}mv^2$, the block's speed must be

$$v = \sqrt{\frac{2K}{m}} = \sqrt{\frac{2(2.18 \, \text{J})}{1.85 \, \text{kg}}} = 1.54 \, \text{m/s}.$$

(c) There are two forces on the block: that exerted by the spring and that exerted by friction. The spring does 2.18 J of work on the block. Since the normal force F_N equals the weight mg (there is no vertical motion) the work done by the friction force on the block, $\mu_k F_N = \mu_k mg$, is

$$W_{\text{fr}} = (-\mu_k mg)(x) = -(0.38)(1.85 \, \text{kg})(9.8 \, \text{m/s}^2)(0.110 \, \text{m}) = -0.76 \, \text{J}.$$

This work is negative because the force of friction is in the direction opposite to the displacement x. The net work done on the block is $W_{\text{net}} = 2.18 \, \text{J} - 0.76 \, \text{J} = 1.42 \, \text{J}$. From the work-energy principle, Eq. 7-9 (with $v_2 = v$ and $v_1 = 0$), we have

$$v = \sqrt{\frac{2W_{\text{net}}}{m}} = \sqrt{\frac{2(1.42 \, \text{J})}{1.85 \, \text{kg}}} = 1.24 \, \text{m/s}.$$

FIGURE 7-19   Example 7-10.

$v_1 = 20 \, \text{m/s} \qquad v_2 = 30 \, \text{m/s}$

EXAMPLE 7-10   **Work to accelerate a car.**   How much work is required to accelerate a 1000-kg car from 20 m/s to 30 m/s? See Fig. 7-19.

SOLUTION   We need to be careful, for this is (if looked at in detail) a complicated situation. However, if we treat the car as a particle, then we can write that the net work needed is equal to the increase in its kinetic energy:

$$W_{\text{net}} = \tfrac{1}{2}mv_2^2 - \tfrac{1}{2}mv_1^2$$
$$= \tfrac{1}{2}(1000 \, \text{kg})\big[(30 \, \text{m/s})^2 - (20 \, \text{m/s})^2\big] = 2.5 \times 10^5 \, \text{J}.$$

This conclusion is useful. The difficulty comes if we look deeper. First, we note that the net force that accelerates the car is the friction force that the road exerts on the tires (the reaction to the tires pushing against the road). In reality, this force does no work because the tires are not sliding and have no component of motion parallel to the road or to the friction force. In principle, this Example might be approached by examining the work done by the engine, transmission to the wheels, and so on—sometimes referred to as "internal work"—but it becomes very complicated. What counts here, is that we do get a useful result from our calculation above in which we treated the car as a particle (or simple rigid body) undergoing translational motion.

This last Example, though useful, also shows us that the concept of work is not always helpful. The problem is that we are trying to examine a deformable (non-rigid) object that can have internal motions. To delve deeper into the car example, to deal with the engine, transmission, and connections to the wheels, we need to use the powerful concept of energy. With energy in its various forms, we can talk about the energy stored in the gasoline being used to do work on the piston, and the energy eventually getting transferred to the wheels. In the next chapter we will extend our discussion of energy to other forms of energy, particularly potential energy. Energy is a central topic in physics and it will come up a lot in later chapters as well. An important result discussed in the next chapter is that energy can be transformed from one type to another, but the total energy never increases or decreases. This is the law of conservation of energy, one of the most important laws in physics.

*| 7–5 | Kinetic Energy at Very High Speed

Einstein's theory of special relativity, which we will discuss in Chapter 37, points out some difficulties with classical Newtonian mechanics. First published in 1905, Einstein's great theory has caused us to reevaluate some aspects of mechanics. For now we need only point out that, according to relativity theory, the kinetic energy of a particle of mass m moving with velocity v is given by

$$K = mc^2 \left(\frac{1}{\sqrt{1 - \dfrac{v^2}{c^2}}} - 1 \right) \qquad \text{[Relativistic kinetic energy]} \quad \textbf{(7–12)}$$

where c is the speed of light, $c = 3.00 \times 10^8$ m/s. Einstein's formula, Eq. 7–12, gives a significantly different result from the simpler classical form, $K = \frac{1}{2}mv^2$, only when the object's speed v is extremely high, greater than about $\frac{1}{10}c$. Indeed, Eq. 7–12 reduces to the classical equation when $v \ll c$, which is easy to show using the binomial expansion,

$$(1 + x)^n = 1 + nx + n(n - 1)x^2/2! + \cdots.$$

Thus, setting $n = -\frac{1}{2}$ and $x = -v^2/c^2$, then for $v \ll c$, Eq. 7–12 becomes

$$K = mc^2 \left(\frac{1}{\sqrt{1 - \dfrac{v^2}{c^2}}} - 1 \right) = mc^2 \left(1 + \frac{1}{2} \frac{v^2}{c^2} + \cdots - 1 \right) \approx \frac{1}{2}mv^2$$

where we ignore higher terms in the expansion because they are very small given that $v/c \ll 1$.

Experiments with subatomic particles (Fig. 7–20), such as electrons and protons, confirm Eq. 7–12. An interesting aspect of Eq. 7–12 is that velocities equal to or greater than the speed of light c are not possible. Why? If v were equal to c, then the denominator in the first term, $\sqrt{1 - v^2/c^2}$, would equal zero; so the kinetic energy would be infinite. That is, it would take an infinite amount of work to accelerate an object of non-zero mass to $v = c$. And if v were greater than c, the factor $\sqrt{1 - v^2/c^2}$ would be the square root of a negative number, which is imaginary and, we believe, non-physical. Indeed, experiment confirms that no particle with mass reaches (or exceeds) the speed of light.

FIGURE 7–20 (a) Interior photo of the Stanford linear collider. (b) Photo in Brookhaven bubble chamber that reveals tracks of tiny elementary particles.

(a)

(b)

Summary

Work is done on a body by a force when that force acts on the body as it moves the object through some distance. The **work** W done by a constant force $\mathbf{F}$ on an object whose position changes by a displacement $\mathbf{d}$ is given by

$$W = Fd\cos\theta = \mathbf{F} \cdot \mathbf{d},$$

where θ is the angle between $\mathbf{F}$ and $\mathbf{d}$.

The last expression is called the scalar product of $\mathbf{F}$ and $\mathbf{d}$. In general, the **scalar product** of any two vectors $\mathbf{A}$ and $\mathbf{B}$ is defined as

$$\mathbf{A} \cdot \mathbf{B} = AB\cos\theta$$

where θ is the angle between $\mathbf{A}$ and $\mathbf{B}$. In rectangular coordinates we can also write $\mathbf{A} \cdot \mathbf{B} = A_x B_x + A_y B_y + A_z B_z$.

The work W done by a variable force $\mathbf{F}$ on an object that moves from point a to point b is

$$W = \int_a^b \mathbf{F} \cdot d\mathbf{l} = \int_a^b F\cos\theta \, dl,$$

where $d\mathbf{l}$ represents an infinitesimal displacement along the path of the object and θ is the angle between $d\mathbf{l}$ and $\mathbf{F}$ at each point of the object's path.

The translational **kinetic energy** K of an object of mass m moving with speed v is defined to be

$$K = \tfrac{1}{2}mv^2.$$

The **work-energy principle** states that the net work done on a body by the net resultant force is equal to the change in kinetic energy of the body:

$$W_{net} = \tfrac{1}{2}mv_2^2 - \tfrac{1}{2}mv_1^2.$$

Questions

1. In what ways is the word "work" as used in everyday language the same as defined in physics? In what ways is it different?

2. A woman swimming upstream in a fast river is not moving with respect to the shore. Is she doing any work? If she stops swimming and merely floats, is work done on her?

3. Is the work done by a kinetic friction force always negative? [*Hint*: Consider what happens to the dishes when pulling a tablecloth from under your mom's best china.]

4. Does the scalar product of two vectors depend on the choice of coordinate system?

5. Can a dot product ever be negative? If yes, under what conditions?

6. If $\mathbf{A} \cdot \mathbf{C} = \mathbf{B} \cdot \mathbf{C}$, is it necessarily true that $\mathbf{A} = \mathbf{B}$?

7. Does the dot product of two vectors have direction as well as magnitude?

8. Can the normal force on an object ever do work? Explain.

9. You have two springs that are identical except that spring 1 is stiffer than spring $2(k_1 > k_2)$. On which spring is more work done: (*a*) if they are stretched using the same force; (*b*) if they are stretched the same distance?

10. Can kinetic energy ever be negative? Explain.

11. If the kinetic energy of a particle is doubled, by what factor has its speed increased?

12. If the speed of a particle triples, by what factor does its kinetic energy increase?

13. In Example 7–9, it was stated that the block separates from the compressed spring when the spring reached its equilibrium length ($x = 0$). Explain why separation doesn't take place before (or after) this point.

14. Two bullets are fired at the same time with the same kinetic energy. If one bullet has twice the mass of the other, which has the greater speed and by what factor? Which can do the most work?

15. Does the net work done on a particle depend on the choice of reference frame? How does this affect the work-energy principle?

16. A hand exerts a constant horizontal force on a block that is free to slide on a frictionless surface, as shown below (Fig. 7–21). The block starts from rest at point A, and by the time it has traveled a distance d to point B it is traveling with speed v_B. When the block has traveled another distance d to point C, will its speed be greater than, less than, or equal to $2v_B$? Explain your reasoning.

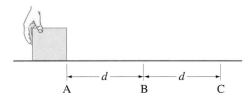

FIGURE 7–21 Question 16.

Problems

Section 7–1

1. (I) How much work is done by the gravitational force when a 250-kg pile driver falls 2.80 m?

2. (I) If the average retarding force on a car is 535 N, how much work is done by this force as the car moves 1.25 km?

3. (I) A 65.0-kg firefighter climbs a flight of stairs 20.0 m high. How much work is required?

4. (I) A 1200-N crate rests on the floor. How much work is required to move it at constant speed (a) 4.0 m along the floor against a friction force of 230 N, and (b) 4.0 m vertically?

5. (I) How much work did the movers do (horizontally) pushing a 160-kg crate 10.3 m across a rough floor without acceleration, if the effective coefficient of friction was 0.50?

6. (I) How high will a 1.85-kg rock go if thrown straight up by someone who does 80.0 J of work on it? Neglect air resistance.

7. (I) Determine the conversion factor between joules and ergs and between joules and foot-pounds.

8. (I) A hammerhead with a mass of 2.0 kg is allowed to fall onto a nail from a height of 0.50 m. What is the maximum amount of work it could do on the nail? Why do people not just "let it fall" but add their own force to the hammer as it falls?

9. (II) Estimate the work you do to mow a lawn 10 m by 20 m. Assume you push with a force of about 15 N.

10. (II) What is the minimum work needed to push a 950-kg car 310 m up along a 9.0° incline? (a) Ignore friction. (b) Assume the effective coefficient of friction retarding the car is 0.25.

11. (II) A lever such as that shown in Fig. 7–22 can be used to lift objects we might not otherwise be able to lift. Show that the ratio of output force, F_O, to input force, F_I, is related to the lengths l_I and l_O from the pivot by $F_O/F_I = l_I/l_O$ (ignoring friction and the mass of the lever), given that the work output equals work input.

(a)

(b)

FIGURE 7–22
A simple lever.
Problem 11.

12. (II) Eight books, each 4.3 cm thick with mass 1.7 kg, lie flat on a table. How much work is required to stack them one on top of another?

13. (II) A 380-kg piano slides 3.5 m down a 27° incline and is kept from accelerating by a man who is pushing back on it *parallel to the incline* (Fig. 7–23). The effective coefficient of kinetic friction is 0.40. Calculate: (a) the force exerted by the man, (b) the work done by the man on the piano, (c) the work done by the friction force, (d) the work done by the force of gravity, and (e) the net work done on the piano.

FIGURE 7–23 Problem 13.

14. (II) A 20,000-kg jet takes off from an aircraft carrier via a catapult (Fig. 7–24a). The jet's engines exert a constant force of 130 kN on the jet; the force exerted on the jet by the catapult is plotted in Fig. 7–24b. Determine: (a) the work done on the jet by its engines during launch of the jet; and (b) the work done on the jet by the catapult during launch of the jet.

(a)

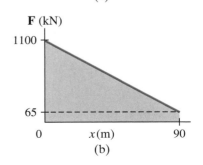

(b)

FIGURE 7–24 Problem 14.

15. (II) A grocery cart with mass of 18 kg is being pushed at constant speed along an aisle by a force $F_P = 14$ N which acts at an angle of 20° to the horizontal. Find the work done by each of the forces $(m\mathbf{g}, \mathbf{F_N}, \mathbf{F_P}, \mathbf{F_{fr}})$ on the cart if the aisle is 15 m long.

16. II (a) Find the force required to give a helicopter of mass M an acceleration of 0.10 g upward. (b) Find the work done by this force as the helicopter moves a distance h upward.

Section 7–2

17. (I) For any vector $\mathbf{V} = V_x\mathbf{i} + V_y\mathbf{j} + V_z\mathbf{k}$ show that

$$V_x = \mathbf{i} \cdot \mathbf{V}, \quad V_y = \mathbf{j} \cdot \mathbf{V}, \quad V_z = \mathbf{k} \cdot \mathbf{V}.$$

18. (I) Calculate the angle between the vectors:

$$\mathbf{A} = 6.8\mathbf{i} + 4.6\mathbf{j} + 6.2\mathbf{k} \quad \text{and} \quad \mathbf{B} = 8.2\mathbf{i} + 2.3\mathbf{j} - 7.0\mathbf{k}.$$

19. (I) Show that $\mathbf{A} \cdot (-\mathbf{B}) = -\mathbf{A} \cdot \mathbf{B}$.

20. (I) Vector $\mathbf{V}_1$ points along the z axis and has magnitude $V_1 = 75$. Vector $\mathbf{V}_2$ lies in the xz plane, has magnitude $V_2 = 50$, and makes a $-48°$ angle with the x axis (points below x axis). What is the scalar product $\mathbf{V}_1 \cdot \mathbf{V}_2$?

21. (II) If $\mathbf{A} = 7.0\mathbf{i} - 8.5\mathbf{j}$, $\mathbf{B} = -8.0\mathbf{i} + 8.1\mathbf{j} + 4.2\mathbf{k}$, and $\mathbf{C} = 6.8\mathbf{i} - 7.2\mathbf{j}$, determine (a) $\mathbf{A} \cdot (\mathbf{B} + \mathbf{C})$; (b) $(\mathbf{A} + \mathbf{C}) \cdot \mathbf{B}$; (c) $(\mathbf{B} + \mathbf{A}) \cdot \mathbf{C}$.

22. (II) Prove that $\mathbf{A} \cdot \mathbf{B} = A_x B_x + A_y B_y + A_z B_z$, starting from Eq. 7–2 and using the distributive law (proved in Problem 29).

23. (II) Given vectors $\mathbf{A} = -4.8\mathbf{i} + 7.8\mathbf{j}$ and $\mathbf{B} = 9.6\mathbf{i} + 6.7\mathbf{j}$, determine the vector $\mathbf{C}$ that lies in the xy plane perpendicular to $\mathbf{B}$ and whose dot product with $\mathbf{A}$ is 20.0.

24. (II) Show that if two non-parallel vectors have the same magnitude, their sum must be perpendicular to their difference.

25. (II) Let $\mathbf{V} = 20.0\mathbf{i} + 12.0\mathbf{j} - 14.0\mathbf{k}$. What angles does this vector make with the x, y, and z axes?

26. (II) Use the scalar product to prove the law of cosines for a triangle:

$$c^2 = a^2 + b^2 - 2ab \cos \theta,$$

where a, b, and c are the lengths of the sides of a triangle and θ is the angle opposite side c.

27. (II) Vectors $\mathbf{A}$ and $\mathbf{B}$ are in the xy plane and their scalar product is 20.0 units. If $\mathbf{A}$ makes a 30° angle with the x axis and has magnitude $A = 12.0$ units, and $\mathbf{B}$ has magnitude $B = 4.0$ units, what can you say about the direction of $\mathbf{B}$?

28. (II) $\mathbf{A}$ and $\mathbf{B}$ are two vectors in the xy plane that make angles α and β with the x axis respectively. Evaluate the scalar product of $\mathbf{A}$ and $\mathbf{B}$ and deduce the following trigonometric identity: $\cos (\alpha - \beta) = \cos \alpha \cos \beta + \sin \alpha \sin \beta$.

29. (III) Show that the scalar product of two vectors is distributive: $\mathbf{A} \cdot (\mathbf{B} + \mathbf{C}) = \mathbf{A} \cdot \mathbf{B} + \mathbf{A} \cdot \mathbf{C}$. [Hint: Use a diagram showing all three vectors in a plane and indicate dot products on the diagram.]

Section 7–3

30. (I) In pedaling a bicycle uphill, a cyclist exerts a downward force of 470 N during each stroke. If the diameter of the circle traced by each pedal is 36 cm, calculate how much work is done in each stroke.

31. (I) A spring has $k = 84$ N/m. Draw a graph like that in Fig. 7–12 and use it to determine the work needed to stretch the spring from $x = 3.0$ cm to $x = 5.5$ cm, where $x = 0$ refers to the spring's unstretched length.

32. (II) If the hill in Example 7–2 (Fig. 7–4) were not an even slope but rather an irregular curve as in Fig. 7–25, show that the same result would be obtained as in Example 7–2: namely, that the work done by gravity depends only on the height of the hill and not on its shape or the path taken.

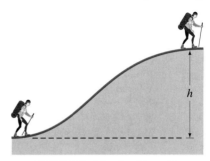

FIGURE 7–25 Problem 32.

33. (II) The net force exerted on a particle acts in the positive x direction. Its magnitude increases linearly from zero at $x = 0$, to 300 N at $x = 3.0$ m. It remains constant at 300 N from $x = 3.0$ m to $x = 7.0$ m and then decreases linearly to zero at $x = 11.0$ m. Determine the work done to move the particle from $x = 0$ to $x = 11.0$ m by graphically determining the area under the F versus x graph.

34. (II) The force on a particle, acting along the x axis, varies as shown in Fig. 7–26. Determine the work done by this force to move the particle along the x axis: (a) from $x = 0.0$ to $x = 10.0$ m; (b) from $x = 0.0$ to $x = 15.0$ m.

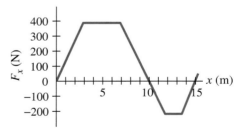

FIGURE 7–26 Problem 34.

35. (II) In Fig. 7–9 assume the distance axis is linear and that $a = 10.0$ m and $b = 30.0$ m. Estimate the work done by this force in moving a 2.50-kg object from a to b.

36. (II) The resistance of a packing material to a sharp object penetrating it is a force proportional to the fourth power of the penetration depth, x: $\mathbf{F} = kx^4\mathbf{i}$. Calculate the work done to force the sharp object a distance d.

37. (II) The force needed to hold a particular spring compressed an amount x from its normal length is given by $F = kx + ax^3 + bx^4$. How much work must be done to compress it by an amount X, starting from $x = 0$?

38. (III) A 3.0-m-long steel chain is stretched out along the top level of a horizontal scaffold at a construction site, in such a way that 2.0 m of the chain remains on the top level and 1.0 m hangs vertically, Fig. 7–27. At this point, the force on the hanging segment is sufficient to pull the entire chain over the edge. Once the chain is moving, the kinetic friction is so small that it can be neglected. How much work is performed on the chain by the force of gravity as the chain falls from the point where 2.0 m remains on the scaffold to the point where the entire chain has left the scaffold? (Assume that the chain has a linear weight density of 20 N/m.)

FIGURE 7–27 Problem 38.

39. (III) A 2500-kg space vehicle, initially at rest, falls vertically from a height of 3000 km above the Earth's surface. (a) Determine how much work is done by the force of gravity in bringing the vehicle to the Earth's surface by first constructing an F versus r graph (using Eq. 7–1), where r is the distance from the Earth's center. Then determine the work graphically to an accuracy of 3 percent. (b) Repeat using integration.

Section 7–4

40. (I) At room temperature, an oxygen molecule, with mass of 5.31×10^{-26} kg, typically has a kinetic energy of about 6.21×10^{-21} J. How fast is it moving?

41. (I) (a) If the kinetic energy of a particle is tripled, by what factor has its speed increased? (b) If the speed of a particle is halved, by what factor does its kinetic energy change?

42. (I) How much work is required to stop an electron $(m = 9.11 \times 10^{-31}$ kg) which is moving with a speed of 1.70×10^6 m/s?

43. (I) How much work must be done to stop a 1300-kg car traveling at 100 km/h?

44. (II) An 85-g arrow is fired from a bow whose string exerts an average force of 105 N on the arrow over a distance of 80 cm. What is the speed of the arrow as it leaves the bow?

45. (II) A baseball $(m = 145$ g) traveling 32 m/s moves a fielder's glove backward 25 cm when the ball is caught. What was the average force exerted by the ball on the glove?

46. (II) If the speed of a car is increased by 50%, by what factor will its minimum braking distance be increased, assuming all else is the same? Ignore the driver's reaction time.

47. (II) At an accident scene on a level road, investigators measure a car's skid mark to be 78 m long. It was a rainy day and the coefficient of friction was estimated to be 0.38. Use these data to determine the speed of the car when the driver slammed on (and locked) the brakes. (Why does the car's mass not matter?)

48. (II) One car has twice the mass of a second car, but only half as much kinetic energy. When both cars increase their speed by 7.0 m/s, they then have the same kinetic energy. What were the original speeds of the two cars?

49. (II) A force of 6.0 N is used to accelerate a mass of 1.0 kg from rest for a distance of 12 m. The force is applied along the direction of travel. The coefficient of kinetic friction is 0.30. What is the work done (a) by the applied force? (b) by friction? (c) What is the kinetic energy at the 12-m mark?

50. (II) A 1200-kg car rolling on a horizontal surface has speed $v = 60$ km/h when it strikes a horizontal coiled spring and is brought to rest in a distance of 2.2 m. What is the spring constant of the spring?

51. (II) At 66.0-kg crate, starting from rest, is pulled across a floor with a constant horizontal force of 225 N. For the first 11.0 m the floor is frictionless, and for the next 10.0 m the coefficient of friction is 0.20. What is the final speed of the crate after being pulled these 21.0 m?

52. (II) A mass m is attached to a spring which is held stretched a distance x by a force F (Fig. 7–28), and then released. The spring compresses, pulling the mass. Assuming there is no friction, determine the speed of the mass m when the spring returns: (a) to its normal length $(x = 0)$; (b) to half its original extension $(x/2)$.

FIGURE 7–28 Problems 52 and 53.

53. (II) Suppose that there is friction in the preceding problem and the mass on the end of the stretched spring in Fig. 7–28, after being released, comes to rest just when it reaches the spring's equilibrium position. Determine the coefficient of friction, μ_k, in terms of F, x, g, and m.

54. (II) A 355-kg load is lifted 33.0 m vertically by a single cable with an acceleration $a = 0.15$ g. Determine (a) the tension in the cable; (b) the net work done on the load; (c) the work done by the cable on the load; (d) the work done by gravity on the load; (e) the final speed of the load assuming it starts from rest.

55. (II) (a) How much work is done by the horizontal force $F_P = 150$ N on the 20-kg block of Fig. 7–29 when the force pushes the block 5.0 m up the 30° frictionless incline? (b) How much work is done by the gravitational force on the block during this displacement? (c) How much work is done by the normal force? (d) What is the speed of the block (assume that it is zero initially) after this displacement?

56. (II) Repeat Problem 55 assuming a coefficient of friction $\mu_k = 0.10$.

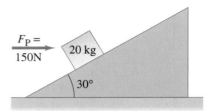

FIGURE 7–29   Problems 55 and 56.

57. (III) An elevator cable breaks when a 755-kg elevator is 22.5 m above the top of a huge spring ($k = 8.00 \times 10^4$ N/m) at the bottom of the shaft. Calculate (a) the work done by gravity on the elevator before it hits the spring; (b) the speed of the elevator just before striking the spring; (c) the amount the spring compresses (note that here work is done by both the spring and gravity).

58. (III) We usually neglect the mass of a spring if it is small compared to the mass attached to the spring. But in some applications, the mass of the spring must be taken into account. Consider a spring of unstretched length L and mass M_S. This mass is uniformly distributed along the length of the spring. A mass m is attached to the end of the spring. One

FIGURE 7–30   Problem 58.

end of the spring is fixed and the mass m is allowed to vibrate horizontally without friction (see Fig. 7–30). Each point on the spring moves with a velocity proportional to the distance from that point to the fixed end. For example, if the mass on the end moves with speed v, the midpoint of the spring moves with speed $v/2$. Show that the kinetic energy of the mass plus spring when the mass is moving with velocity v is

$$K = \tfrac{1}{2}Mv^2$$

where $M = m + \tfrac{1}{3}M_S$ is the "effective mass" of the system.

* Section 7–5

* 59. (II) In two separate experiments, a proton (with mass of 1.67×10^{-27} kg) and an electron (with mass of 9.1×10^{-31} kg) are each accelerated by a device that performs 3.2×10^{-13} J of work on each particle. Using the relativistic formula for kinetic energy, determine the resulting speed of each particle. Compare your calculations with the results that would have been achieved using the classical formula.

* 60. (II) Calculate, to three significant figures, the kinetic energy of a particle of mass m moving with a speed of (a) 3.00×10^4 m/s, (b) 3.00×10^6 m/s (1% of the speed of light), and (c) 3.00×10^7 m/s (= 0.1c). Compare each result to a classical calculation.

General Problems

61. In a certain library the first shelf is 12.0 cm off the ground, and the remaining 4 shelves are each spaced 33.0 cm above the previous one. If the average book has a mass of 1.60 kg with a height of 22.0 cm, and an average shelf holds 25 books (standing vertically), how much work is required to fill this bookshelf from scratch, assuming the books are all laying flat on the floor to start?

62. (a) A 3.0-g locust reaches a speed of 3.0 m/s during its jump. What is its kinetic energy at this speed? (b) If the locust transforms energy with 40 percent efficiency, how much energy is required for the jump?

63. A 6.0-kg block is pushed 7.0 m up a rough 37° inclined plane by a horizontal force of 75 N. If the initial speed of the block is 2.2 m/s up the plane and a constant kinetic friction force of 25 N opposes the motion, calculate (a) the initial kinetic energy of the block; (b) the work done by the 75-N force; (c) the work done by the friction force; (d) the work done by gravity; (e) the work done by the normal force; (f) the final kinetic energy of the block.

64. A mass m slides on a circular horizontal track in a circle of radius R, Fig. 7–31. Its initial speed is v_0, but after one revolution the speed has dropped to $0.75v_0$ because of friction.

Determine (a) the work done by friction during one revolution; (b) the coefficient of friction; (c) the number of revolutions the mass will make before coming to rest.

FIGURE 7–31   Problem 64.

65. Two forces, $\mathbf{F}_1 = (1.50\mathbf{i} - 0.80\mathbf{j} + 0.70\mathbf{k})$ N and $\mathbf{F}_2 = (-0.70\mathbf{i} + 1.20\mathbf{j})$ N, are applied on a moving object of mass 0.20 kg. The displacement vector produced by the two forces is $\mathbf{d} = (8.0\mathbf{i} + 6.0\mathbf{j} + 5.0\mathbf{k})$ m. (a) What is the work done by the two forces? (b) If there is a frictional force given by $\mathbf{F}_{fr} = -0.20\mathbf{F}_1$, what is the net work done? (c) What is the work done by the frictional force in (b)?

66. The barrels of the 16-inch guns (bore diameter = 16 in. = 40 cm) on the World-War II battleship *U.S.S. Massachusetts* were each 15 m long. The shells each had a mass of 1250 kg and were fired with sufficient explosive to provide them with a muzzle velocity of 750 m/s. Use the work-energy principle to determine the explosive force (assumed to be a constant) that was applied to the shell within the barrel of the gun. Express your answer in both newtons and in pounds.

67. A force $\mathbf{F} = (10.0\mathbf{i} + 9.0\mathbf{j} + 12.0\mathbf{k})$kN acts on a small object of mass 100 g. If the displacement of the object is $\mathbf{d} = (5.0\mathbf{i} + 4.0\mathbf{j})$m, find the work done by the force. What is the angle between $\mathbf{F}$ and $\mathbf{d}$?

68. The arrangement of atoms in zinc is an example of "hexagonal close-packed" structure. Three of the nearest neighbors are found at the following (x, y, z) coordinates, given in nanometers $(10^{-9}\,\text{m})$: atom 1 is at $(0, 0, 0)$; atom 2 is at $(0.230, 0.133, 0)$; atom 3 is at $(0.077, 0.133, 0.247)$. Find the angle between two vectors: one that connects atom 1 with atom 2 and another that connects atom 1 with atom 3.

69. A varying force is given by $F = Ae^{-kx}$, where x is the position; A and k are constants that have units of N and m^{-1}, respectively. What is the work done when x goes from 0.10 m to infinity?

70. The force required to compress an imperfect horizontal spring an amount x is given by $F = 150x + 12x^3$, where x is in meters and F in newtons. If the spring is compressed 2.0 m, what speed will it give to a. 3.0-kg ball held against it and then released?

71. A softball having a mass of 0.25 kg is pitched at 110 km/h. By the time it reaches the plate, it may have slowed by 10 percent. Neglecting gravity, estimate the average force of air resistance during a pitch, if the distance between the plate and the pitcher is about 15 m.

72. Today's cars have "5 mi/h (8 km/h) bumpers" that are designed to elastically compress and rebound without any physical damage at speeds below 8 km/h. If the material of the bumpers permanently deforms after a compression of 1.5 cm, but remains like an elastic spring up to that point, what must the effective spring constant of the bumper material be, assuming the car has a mass of 1150 kg and is tested by ramming into a solid wall?

73. What should be the spring constant k of a spring designed to bring a 1300-kg car to rest from a speed of 90 km/h so that the occupants undergo a maximum acceleration of $5.0\,g$?

74. An airplane pilot fell 370 m after jumping without his parachute opening. He landed in a snowbank, creating a crater 1.1 m deep, but survived with only minor injuries. Assuming the pilot's mass was 80 kg and his terminal velocity was 50 m/s, estimate: (a) the work done by the snow in bringing him to rest; (b) the average force exerted on him by the snow to stop him; and (c) the work done on him by air resistance as he fell.

75. Assume a cyclist of weight mg can exert a force on the pedals equal to 0.90 mg on the average. If the pedals rotate in a circle of radius 18 cm, the wheels have a radius of 34 cm, and the front and back sprockets on which the chain runs have 42 and 19 teeth respectively (Fig. 7–32), determine the maximum steepness of hill the cyclist can climb at constant speed. Assume the mass of the bike is 12 kg and that of the rider is 60 kg. Ignore friction. Assume the cyclist's average force is always: (a) downward; (b) tangential to pedal motion.

FIGURE 7–32 Problem 75.

76. A simple pendulum consists of a small object of mass m (the "bob") suspended by a cord of length L (Fig. 7–33) of negligible mass. A force $\mathbf{F}$ is applied in the horizontal direction (so $\mathbf{F} = F\mathbf{i}$), moving the bob very slowly so the acceleration is essentially zero. (Note that the magnitude of $\mathbf{F}$ will need to vary with the angle θ that the cord makes with the vertical at any moment.) (a) Determine the work done by this force, $\mathbf{F}$, to move the pendulum from $\theta = 0$ to $\theta = \theta_0$. (b) Determine the work done by the gravitational force on the bob, $\mathbf{F}_G = m\mathbf{g}$, and the work done by the force $\mathbf{F}_T$ that the cord exerts on the bob.

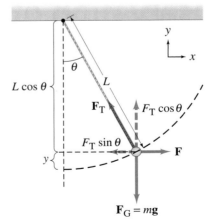

FIGURE 7–33 Problem 76.

A polevaulter running toward the high bar has kinetic energy. When he plants the pole and he puts his weight on it, his kinetic energy gets transformed: first into elastic potential energy of the bent pole and then into gravitational potential energy as his body rises. As he crosses the bar, the pole is straight and has given up all its elastic potential energy to the athlete's gravitational potential energy. Nearly all his kinetic energy has disappeared, also becoming gravitational potential energy of his body at the great height of the bar (world record over 6 m), which is exactly what he wants. In these, and all other energy transformations that continually take place in the world, the total energy is always conserved. Indeed, the conservation of energy is one of the greatest laws of physics, and finds applications in a wide range of other fields.

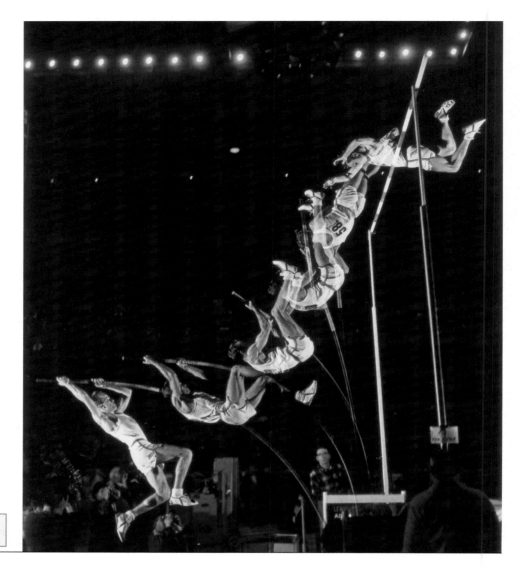

CHAPTER 8

Conservation of Energy

This chapter continues the discussion of the concepts of work and energy begun in the preceding chapter and introduces additional types of energy, in particular potential energy. Now we will see why the concept of energy is so important. The reason, ultimately, is that energy is conserved—the total energy remains constant in any process. That such a quantity can be defined which remains constant, as far as our best experiments can tell, is a remarkable statement about nature. The law of conservation of energy is, in fact, one of the great unifying principles of science.

The law of conservation of energy also gives us another tool, another approach, to solving problems. There are many situations for which an analysis based on Newton's laws would be difficult or impossible—the forces may not be known or accessible to measurement. But often these situations can be dealt with using the law of conservation of energy and in some cases other conservation laws (such as the conservation of momentum).

In this chapter we will mainly treat objects as if they were particles, or rigid objects, that undergo only translational motion, with no internal or rotational motion.

8–1 Conservative and Nonconservative Forces

We will find it important to categorize forces into two types: conservative and non-conservative. By definition, we call any force a **conservative force** if

> *the work done by the force on an object moving from one point to another depends only on the initial and final positions and is independent of the particular path taken.*

We can readily show that the force of gravity is a conservative force. The gravitational force on an object of mass m near the Earth's surface is $\mathbf{F} = m\mathbf{g}$, where $\mathbf{g}$ is a constant. We saw in Chapter 7 (see Example 7–2) that the work done by the gravitational force near the Earth's surface is $W_G = Fd = mgh$, where h is the vertical height through which an object of mass m falls (see Fig. 8–1a). Now suppose instead of moving vertically downward or upward, an object follows some arbitrary path in the xy plane, as shown in Fig. 8–1b. The object starts at a vertical height y_1 and reaches a height y_2, where $y_2 - y_1 = h$. To calculate the work done by gravity, W_G, we use Eq. 7–7:

$$W_G = \int_1^2 \mathbf{F}_G \cdot d\boldsymbol{l} = \int_1^2 mg \cos\theta \, dl.$$

We now let $\phi = 180° - \theta$ be the angle between $d\boldsymbol{l}$ and its vertical component dy, as shown in Fig. 8–1b. Then, since $\cos\theta = -\cos\phi$ and $dy = dl\cos\phi$, we have

$$W_G = -\int_{y_1}^{y_2} mg \, dy$$

$$= -mg(y_2 - y_1). \tag{8–1}$$

Since $(y_2 - y_1)$ is the vertical height h, we see that the work done depends only on the vertical height and does *not* depend on the particular path taken! Hence, by definition, gravity is a conservative force. (Note that in the case shown in Fig. 8–1b, $y_2 > y_1$ and therefore the work done by gravity is negative. If on the other hand $y_2 < y_1$, so that the object is falling, then W_G is positive.)

We can give the definition of a conservative force in another, completely equivalent way: a force is conservative if

> *the net work done by the force on an object moving around any closed path is zero.*

To see why this is equivalent to our earlier definition, consider a small object that moves from point 1 to point 2 via either of two paths labeled A and B in Fig. 8–2a. If we assume a conservative force acts on the object, the work done by this force is the same whether the object takes path A or path B, by our first definition. This work to get from point 1 to point 2 we will call W. Now consider the round trip shown in Fig. 8–2b. The object moves from 1 to 2 via path A and our force does work W. Our object then returns to point 1 via path B. How much work is done during the return? In going from 1 to 2 via path B the work done is W, which by definition equals $\int_1^2 \mathbf{F} \cdot d\boldsymbol{l}$. In doing the reverse, going from 2 to 1, the force $\mathbf{F}$ at each point is the same, but $d\boldsymbol{l}$ is directed in precisely the opposite direction. Consequently $\mathbf{F} \cdot d\boldsymbol{l}$ has the opposite sign at each point so the total work done in making the return trip from 2 to 1 must be $-W$. Hence the total work done in going from 1 to 2 and back to 1 is $W + (-W) = 0$, which proves the equivalence of the two above definitions for a conservative force.

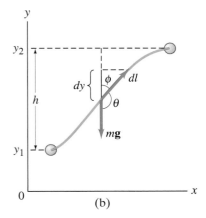

FIGURE 8–1 Object of mass m: (a) falls a height h vertically; (b) is raised along an arbitrary two-dimensional path.

FIGURE 8–2 (a) A tiny object moves between points 1 and 2 via two different paths, A and B. (b) The object makes a round trip, via path A from point 1 to point 2 and via path B back to point 1.

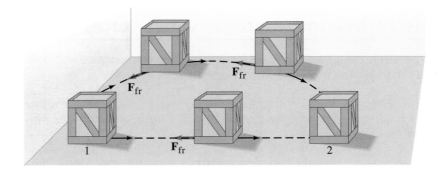

FIGURE 8–3 A crate is pulled across the floor from position 1 to position 2 via two paths, one straight and one curved. The friction force is always in the direction exactly opposed to the direction of motion. Hence for a constant magnitude friction force, $W_{fr} = -F_{fr}\,d$, so if d is greater (as for the curved path), then W is greater.

The second definition of a conservative force illuminates an important aspect of such a force: the *work done by a conservative force is recoverable* in the sense that if positive work is done *by* an object (on something else) on one part of a closed path, an equivalent amount of negative work will be done by our object on its return.

As we saw above, the force of gravity is conservative, and it is easy to show that the elastic force $(F = -kx)$ is also conservative.

But not all forces are conservative. The force of friction, for example, is a **nonconservative** force. The work done in moving a heavy crate across a level floor is equal to the product of the (constant) friction force and the total distance traveled, since the friction force is directed precisely opposite to the direction of motion. Hence the work done to move the object between two points depends on the path length; the work done along a straight line is less than if the path between the two points is curved, as in Fig. 8–3.

Note also in this example of kinetic friction that since the force of friction always opposes the motion, the work done on a body by friction is negative. Thus when an object is moved in a round trip, say from some point 1 to point 2 and back to 1, the total work done by friction is never zero—it is always negative. Thus the work done by a nonconservative force is not recoverable, as it is for a conservative force.

8–2 Potential Energy

In Chapter 7 we discussed the energy associated with a moving object, which is its kinetic energy $K = \frac{1}{2}mv^2$. Now we introduce **potential energy**, which is the energy associated with the position or configuration of an object (or objects). Various types of potential energy can be defined and each type is associated with a particular conservative force.

A wound-up clock spring is an example of potential energy. The clock spring acquired its potential energy because work was done *on* it by the person winding the clock. As the spring unwinds, it exerts a force and does work to move the clock hands around.

Gravitational Potential Energy

Perhaps the most common example of potential energy is *gravitational potential energy*. A heavy brick held high in the air has potential energy because of its position relative to the Earth. It has the ability to do work, for if it is released, it will fall to the ground due to the gravitational force, and can do work on, say, a stake, driving it into the ground. Let us determine quantitatively the gravitational potential energy of an object near the surface of the Earth. In order to lift an object of mass m vertically, an upward force at least equal to its weight, mg, must be exerted on it, say, by a person's hand. In order to lift it without acceleration to a height h, from position y_1 to y_2 in Fig. 8–4 (upward direction chosen positive), a person must

FIGURE 8–4 A person exerts an upward force $F_{ext} = mg$ to lift a brick from y_1 to y_2.

do work equal to the product of the needed external force, $F_{ext} = mg$ upward, and the vertical distance h. That is,

$$W_{ext} = \mathbf{F}_{ext} \cdot \mathbf{d} = mgh \cos 0° = mgh = mg(y_2 - y_1)$$

where both $\mathbf{F}_{ext}$ and $\mathbf{d}$ point upward. Gravity is also acting on the object as it moves from y_1 to y_2, and does work on it equal to

$$W_G = \mathbf{F}_G \cdot \mathbf{d} = mgh \cos 180° = -mgh = -mg(y_2 - y_1).$$

Since $\mathbf{F}_G$ is downward and $\mathbf{d}$ is upward, W_G is negative. If the object follows an arbitrary path, as in Fig. 8–1b, the work done by gravity still depends only on the change in vertical height (see Eq. 8–1):

$$W_G = -mg(y_2 - y_1) = -mgh.$$

Next, if we allow the object to start from rest and fall freely under the action of gravity, it acquires a velocity given by $v^2 = 2gh$ (Eq. 2–12c) after falling a height h. It then has kinetic energy $\frac{1}{2}mv^2 = \frac{1}{2}m(2gh) = mgh$. If it then strikes a stake, the falling object can do work on the stake equal to mgh (work-energy principle, Section 7–4). Thus, to raise an object of mass m to a height h *requires* an amount of work equal to mgh. And once at height h, the object has the *ability* to do an amount of work equal to mgh. We can say that the work done in lifting the object has been stored as gravitational potential energy.

Indeed, we can define the *change in gravitational potential energy U* when an object moves from a height y_1 to a height y_2 as equal to the work done by an external force to accomplish this without acceleration:

$$\Delta U = U_2 - U_1 = W_{ext} = mg(y_2 - y_1).$$

Equivalently, we can define the change in gravitational potential energy as equal to the negative of the work done by gravity itself in the process:

$$\Delta U = U_2 - U_1 = -W_G = mg(y_2 - y_1). \qquad \textbf{(8–2)}$$

Change in gravitational potential energy

Equation 8–2 defines the change in gravitational potential energy between two points. The gravitational potential energy, U, at any point a vertical height y above some reference point (the origin of the coordinate system) can be defined as

$$U = mgy. \qquad \text{[gravity only]} \quad \textbf{(8–3)}$$

Gravitational potential energy

Note that the potential energy is associated with the force of gravity between the Earth and the mass m. Hence U represents the gravitational potential energy, not simply of the mass m alone, but of the mass-Earth system.

We could also define the gravitational potential energy at a point to be

$$U = mgy + C,$$

where C is a constant. This is consistent with Eq. 8–2 (the constants C cancel each other when we subtract U_1 from U_2). We usually choose C to be zero, for convenience, since U depends on the choice of coordinate system anyway (that is, where we choose y to be zero). The gravitational potential energy of a book held high above a table, for example, depends on whether we measure y from the top of the table, from the floor, or from some other reference point. What is physically important in any situation is the *change* in potential energy because that is what is related to the work done. We can thus choose the potential energy to be zero at any point that is convenient, but we must be consistent throughout any given problem. The *change* in potential energy between any two points does not depend on this choice.

Change in potential energy is what is physically meaningful

B

A

10 m

15 m

C

FIGURE 8–5 Example 8–1.

EXAMPLE 8–1 **Potential energy changes for a roller coaster.** A 1000-kg roller-coaster car moves from point A, Fig. 8–5, to point B and then to point C. (*a*) What is its gravitational potential energy at B and C relative to point A? That is, take $y = 0$ at point A. (*b*) What is the change in potential energy when it goes from B to C? (*c*) Repeat parts (*a*) and (*b*), but take the reference point ($y = 0$) to be at point C.

SOLUTION (*a*) We take upward as the positive direction, and measure the heights from point A, which means initially that the potential energy is zero. At point B, where $y_B = 10$ m,

$$U_B = mgy_B = (1000\,\text{kg})(9.8\,\text{m/s}^2)(10\,\text{m}) = 9.8 \times 10^4\,\text{J}.$$

At point C, $y_C = -15$ m, since C is below A. Therefore,

$$U_C = mgy_C = (1000\,\text{kg})(9.8\,\text{m/s}^2)(-15\,\text{m}) = -1.5 \times 10^5\,\text{J}.$$

(*b*) In going from B to C, the potential energy change is

$$U_C - U_B = (-1.5 \times 10^5\,\text{J}) - (9.8 \times 10^4\,\text{J})$$
$$= -2.5 \times 10^5\,\text{J}.$$

The gravitational potential energy decreases by 2.5×10^5 J.
(*c*) In this instance, $y_A = +15$ m at point A, so the potential energy initially (at A) is equal to

$$U_A = (1000\,\text{kg})(9.8\,\text{m/s}^2)(15\,\text{m}) = 1.5 \times 10^5\,\text{J}.$$

At B, $y_B = 25$ m, so the potential energy is

$$U_B = 2.5 \times 10^5\,\text{J}.$$

At C, $y_C = 0$, so $U_C = 0$. The change in potential energy going from B to C is

$$U_C - U_B = 0 - 2.5 \times 10^5\,\text{J} = -2.5 \times 10^5\,\text{J},$$

which is the same as in part (*b*).

Potential Energy in General

We have defined the change in gravitational potential energy (Eq. 8–2) to be equal to the negative of the work done by gravity[†] when the object moves from height y_1 to y_2:

$$\Delta U = -W_G = -\int_1^2 \mathbf{F}_G \cdot d\mathbf{l}.$$

There are other types of potential energy besides gravitational. In general, we define the *change in potential energy associated with a particular conservative force* **F** *as the negative of the work done by that force*:

General definition of potential energy

$$\Delta U = U_2 - U_1 = -\int_1^2 \mathbf{F} \cdot d\mathbf{l} = -W. \qquad (8\text{–}4)$$

However, we cannot use this definition to define a potential energy for all possible forces. It makes sense only for conservative forces such as gravity, for which the

[†] This is equivalent to saying that the change in potential energy, ΔU, equals the work done (not its negative) by a second force (of equal magnitude)—say, exerted by a person—to raise the object against the force of gravity.

integral depends only on the end points and not on the path taken. It does not apply to nonconservative forces like friction, because the integral in Eq. 8–4 would not have a unique value depending on the end points 1 and 2. That is, ΔU would depend on path and we could not say that the U had a particular value at each point in space. Thus the concept of potential energy is meaningless for a nonconservative force.

Elastic Potential Energy

We consider now another type of potential energy, that associated with elastic materials. This includes a great variety of practical applications.

Consider a spring such as the coil spring shown in Fig. 8–6. The spring has potential energy when compressed (or stretched) for when released, it can do work on a ball as shown. Like other elastic materials, a spring is described by Hooke's law (as already discussed in Section 7–3) as long as the displacement x is not too great. Let us take our coordinate system so the end of the uncompressed spring is at $x = 0$ (Fig. 8–6a) and x is positive to the right. To hold the spring compressed (or stretched) a distance x, a person must exert a force $F_P = kx$ (recall Fig. 7–11), and the spring pushes back with a force (Newton's third law),

$$F_S = -kx.$$

The negative sign appears because the force F_S is in the direction opposite to the displacement x (see Fig. 8–6b). From Eq. 8–4, the change in potential energy of the spring between $x_1 = 0$ (its uncompressed position) and $x_2 = x$ is

$$\Delta U = U(x) - U(0) = -\int_1^2 \mathbf{F} \cdot d\mathbf{l} = -\int_0^x (-kx)\, dx = \tfrac{1}{2}kx^2.$$

Here, $U(x)$ means the potential energy at x, and $U(0)$ means U at $x = 0$. It is usually convenient to choose the potential energy at $x = 0$ to be zero: $U(0) = 0$, so the potential energy of a spring compressed or stretched an amount x from equilibrium is

$$U(x) = \tfrac{1}{2}kx^2. \qquad \text{[elastic]} \quad \textbf{(8–5)}$$

FIGURE 8–6 (a) A spring can (b) store energy (elastic potential energy) when compressed, which (c) can be used to do work when released.

Elastic potential energy

Potential Energy Summarized

In each of the preceding examples of potential energy—gravitational or elastic—an object has the capacity or *potential* to do work even though it is not yet actually doing it. That is why we use the term "potential" energy. From these examples we can also see that energy can be stored, for later use, in the form of potential energy. Note that the mathematical form of each type of potential energy depends on the force involved.

Let us summarize here the important aspects of potential energy:

1. A potential energy is always associated with a conservative force, and the difference in potential energy between two points is defined as the negative of the work done by that force, Eq. 8–4.

2. The choice of where $U = 0$ is arbitrary and can be chosen wherever it is most convenient.

3. Since a force is always exerted on one body by another body (the Earth exerts a gravitational force on a falling stone; a compressed spring exerts a force on a ball; and so on), potential energy is not something a body "has" by itself, but rather is associated with the interaction of two (or more) bodies.

In the one-dimensional case, where a conservative force can be written as a function, say, of x, the potential energy can be written

$$U(x) = -\int F(x)\, dx. \qquad (8\text{–}6)$$

This relation tells us how to obtain $U(x)$ when given $F(x)$. If, instead, we are given $U(x)$, we can obtain $F(x)$ by inverting the above equation: that is, we take the derivative of both sides, remembering that integration and differentiation are inverse operations:

$$\frac{d}{dx}\int F(x)\, dx = F(x).$$

Thus

$$F(x) = -\frac{dU(x)}{dx}. \qquad (8\text{–}7)$$

[In three dimensions, we can write the relation between $\mathbf{F}(x, y, z)$ and U as:

$$F_x = -\frac{\partial U}{\partial x}, \quad F_y = -\frac{\partial U}{\partial y}, \quad F_z = -\frac{\partial U}{\partial z},$$

or

$$\mathbf{F}(x, y, z) = -\mathbf{i}\frac{\partial U}{\partial x} - \mathbf{j}\frac{\partial U}{\partial y} - \mathbf{k}\frac{\partial U}{\partial z}.$$

Here, $\partial/\partial x$, and so on, are called partial derivatives; $\partial/\partial x$, for example, means that although U may be a function of x, y, and z, written $U(x, y, z)$, we take the derivative only with respect to x with the other variables held constant.]

EXAMPLE 8–2 **Determine F from U.** Suppose $U(x) = -ax/(b^2 + x^2)$, where a and b are constants. What is F as a function of x?

SOLUTION Since $U(x)$ depends only on x, this is a one-dimensional problem and we don't need to use partial derivatives, so

$$F(x) = -\frac{dU}{dx} = -\frac{d}{dx}\left[-\frac{ax}{b^2 + x^2}\right] = \frac{a}{b^2 + x^2} - \frac{ax}{(b^2 + x^2)^2}\, 2x = \frac{a(b^2 - x^2)}{(b^2 + x^2)^2}.$$

8–3 | Mechanical Energy and Its Conservation

Let us consider a conservative system (meaning only conservative forces do work) in which energy is transformed from kinetic to potential or vice versa. Again, we must consider a system because potential energy does not exist for an isolated body. Our system might be a mass m oscillating on the end of a spring or moving in the Earth's gravitational field.

According to the work-energy principle (Eq. 7–11), the net work W_{net} done on an object is equal to its change in kinetic energy:

$$W_{net} = \Delta K.$$

(If more than one object of our system has work done on it, then W_{net} and ΔK can represent the sum for all of them.) Since we assume a conservative system, we can write the net work done in terms of the total potential energy (see Eqs. 7–7 and 8–4):

$$\Delta U_{\text{total}} = -\int_1^2 \mathbf{F}_{\text{net}} \cdot d\boldsymbol{l} = -W_{\text{net}}. \qquad (8\text{--}8)$$

We combine the previous two equations, letting U be the total potential energy:

$$\Delta K + \Delta U = 0 \qquad \text{[conservative forces only]} \qquad (8\text{--}9a)$$

or

$$\left(K_2 - K_1\right) + \left(U_2 - U_1\right) = 0. \qquad (8\text{--}9b)$$

We now define a quantity E, called the **total mechanical energy** of our system, as the sum of the kinetic energy plus the potential energy of the system at any moment

$$E = K + U,$$

Total mechanical energy defined

Now we can rewrite Eq. 8–9b as

$$K_2 + U_2 = K_1 + U_1 \qquad \text{[conservative forces only]} \qquad (8\text{--}10a)$$

or

$$E_2 = E_1 = \text{constant.} \qquad \text{[conservative forces only]} \qquad (8\text{--}10b)$$

CONSERVATION OF MECHANICAL ENERGY

Equations 8–10 express a useful and profound principle regarding the total mechanical energy—namely, that it is a **conserved quantity**. The total mechanical energy E remains constant as long as no nonconservative forces do work: $(K + U)$ at some initial point 1 is equal to the $(K + U)$ at any later point 2. To say it another way, consider Eq. 8–9 which tells us $\Delta U = -\Delta K$; that is, if the kinetic energy K increases, then the potential energy U must decrease by an equivalent amount to compensate. Thus the total, $K + U$, remains constant. This is called the **principle of conservation of mechanical energy** for conservative forces:

> **If only conservative forces are doing work, the total mechanical energy of a system neither increases nor decreases in any process. It stays constant—it is conserved.**

CONSERVATION OF MECHANICAL ENERGY

We now see the reason for the term "conservative force"—because for such forces, mechanical energy is conserved.

If only one object of a system[†] has significant kinetic energy, then Eqs. 8–10 become

$$E = \tfrac{1}{2}mv^2 + U = \text{constant.} \qquad \text{[conservative forces only]} \qquad (8\text{--}11a)$$

If we let v_1 and U_1 represent the velocity and potential energy at one instant, and v_2 and U_2 represent them at a second instant, then we can rewrite this as

$$\tfrac{1}{2}mv_1^2 + U_1 = \tfrac{1}{2}mv_2^2 + U_2. \qquad \text{[conservative system]} \qquad (8\text{--}11b)$$

From this equation we can see again that it doesn't make any difference where we choose the potential energy to be zero: adding a constant to U (as discussed in Section 8–2) merely adds a constant to both sides of the above equation, and these cancel. A constant also doesn't affect the force obtained from Eq. 8–7, $F = -dU/dx$, since the derivative of a constant is zero. Because we deal only with changes in the potential energy, the absolute value of U doesn't matter.

[†] For an object moving under the influence of Earth's gravity, the kinetic energy of the Earth can usually be ignored, as long as the object's mass is small compared to the Earth's mass. For a mass oscillating at the end of a spring, for example, the mass of the spring, and hence its kinetic energy, can often be ignored.

all potential
energy

$y_1 = h$

half U,
half K h

y

all kinetic $y_2 = 0$
energy

FIGURE 8–7 The rock's potential energy changes to kinetic energy as it falls. Note bar graphs representing potential energy U and kinetic energy K for the three different positions.

8–4 Problem Solving Using Conservation of Mechanical Energy

A simple example of the conservation of mechanical energy is a rock allowed to fall from a height h under gravity (neglecting air resistance), as shown in Fig. 8–7. At the instant it is dropped, a rock, starting at rest, initially has only potential energy. As it falls, its potential energy decreases (because y decreases), but its kinetic energy increases to compensate, so that the sum of the two remains constant. At any point along the path, the total mechanical energy is given by

$$E = K + U = \tfrac{1}{2}mv^2 + mgy$$

where y is the rock's height above the ground at a given instant and v is its speed at that point. If we let the subscript 1 represent the rock at one point along its path (for example, the initial point), and 2 represent it at some other point, then we can write

$$\tfrac{1}{2}mv_1^2 + mgy_1 = \tfrac{1}{2}mv_2^2 + mgy_2. \qquad \text{[gravity only]} \quad \textbf{(8–12)}$$

Just before the rock hits the ground ($y_2 = 0$), the potential energy will be zero: all of the initial potential energy will have been transformed into kinetic energy.

EXAMPLE 8–3 **Falling rock.** If the original height of the stone in Fig. 8–7 is $y_1 = h = 3.0\,\text{m}$, calculate the stone's speed when it has fallen to 1.0 m above the ground.

SOLUTION Since $v_1 = 0$ (the moment of release), $y_1 = 3.0\,\text{m}$, $y_2 = 1.0\,\text{m}$, and $g = 9.8\,\text{m/s}^2$, Eq. 8–12 gives

$$\tfrac{1}{2}mv_1^2 + mgy_1 = \tfrac{1}{2}mv_2^2 + mgy_2$$

$$0 + (m)(9.8\,\text{m/s}^2)(3.0\,\text{m}) = \tfrac{1}{2}mv_2^2 + (m)(9.8\,\text{m/s}^2)(1.0\,\text{m}).$$

The m's cancel out, and solving for v_2^2 (which we see doesn't depend on m), we find

$$v_2^2 = 2\big[(9.8\,\text{m/s}^2)(3.0\,\text{m}) - (9.8\,\text{m/s}^2)(1.0\,\text{m})\big] = 39.2\,\text{m}^2/\text{s}^2,$$

and

$$v_2 = \sqrt{39.2}\,\text{m/s} = 6.3\,\text{m/s}.$$

FIGURE 8–8 A roller-coaster car moving without friction illustrates the conservation of mechanical energy.

y

Equation 8–12 can be applied to any object moving without friction under the action of gravity. For example, Fig. 8–8 shows a roller-coaster car starting from rest at the top of a hill, and coasting without friction to the bottom and up the hill on the other side. True, there is another force besides gravity acting on the car, the normal force exerted by the tracks. But this "constraint" force acts perpendicular to the direction of motion at each point and so does zero work. We ignore rotational motion of the car's wheels and treat the car as a particle undergoing simple translation. Initially, the car has only potential energy. As it coasts down the hill, it loses potential energy and gains in kinetic energy, but the sum of the two remains constant. At the bottom of the hill it has its maximum kinetic energy, and as it climbs up the other side the kinetic energy changes back to potential energy. When the car comes to rest again, all of its energy will be potential energy. Given that the potential energy is proportional to the vertical height, energy conservation tells us that (in the absence of friction) the car comes to rest at a height equal to its original height. If the two hills are the same height, the car will just barely reach the top of the second hill when it stops. If the second hill is lower than the first, not all of the car's kinetic energy will be transformed to potential energy and the car can continue over the top and down the other side. If the second hill is higher, the car will only reach a height on it equal to its original height on the first hill. This is true (in the absence of friction) no matter how steep the hill is, since potential energy depends only on the vertical height.

EXAMPLE 8–4 **Roller-coaster speed using energy conservation.** Assuming the height of the hill in Fig. 8–8 is 40 m, and the roller-coaster car starts from rest at the top, calculate (a) the speed of the roller-coaster car at the bottom of the hill, and (b) at what height it will have half this speed. Take $y = 0$ (and $U = 0$) at the bottom of the hill.

SOLUTION (a) We use Eq. 8–12 with $v_1 = 0$, $y_1 = 40$ m, and $y_2 = 0$. Then

$$\tfrac{1}{2}mv_1^2 + mgy_1 = \tfrac{1}{2}mv_2^2 + mgy_2$$
$$0 + (m)(9.8\,\text{m/s}^2)(40\,\text{m}) = \tfrac{1}{2}mv_2^2 + 0.$$

The m's cancel out and we find $v_2 = \sqrt{2(9.8\,\text{m/s}^2)(40\,\text{m})} = 28$ m/s.

(b) We use the same equation, but now $v_2 = 14$ m/s (half of 28 m/s) and y_2 is unknown:

$$\tfrac{1}{2}mv_1^2 + mgy_1 = \tfrac{1}{2}mv_2^2 + mgy_2$$
$$0 + (m)(9.8\,\text{m/s}^2)(40\,\text{m}) = \tfrac{1}{2}(m)(14\,\text{m/s})^2 + (m)(9.8\,\text{m/s}^2)(y_2).$$

We cancel the m's and solve for y_2 and find $y_2 = 30$ m. That is, the car has a speed of 14 m/s when it is 30 *vertical* meters above the lowest point, both when descending the left-hand hill and when ascending the right-hand hill in Fig. 8–8.

The mathematics of this Example is almost the same as that in Example 8–3. But there is an important difference between them. Example 8–3 could have been solved using force and acceleration. But here, where the motion is not vertical, using $F = ma$ would have been very difficult, whereas energy conservation readily gives us the answer.

CONCEPTUAL EXAMPLE 8–5 **Speeds on two water slides.** Two water slides at a pool are shaped differently but start at the same height h (Fig. 8–9). Two riders, Paul and Kathleen, start from rest at the same time on different slides. (a) Which rider, Paul or Kathleen, is traveling faster at the bottom? (b) Which rider makes it to the bottom first? Ignore friction.

RESPONSE (a) Each rider's initial potential energy mgh gets transformed to kinetic energy, so the speed v at the bottom is obtained from $\tfrac{1}{2}mv^2 = mgh$. The mass cancels in this equation and so the speed will be the same, regardless of the mass of the rider. Since they descend the same vertical height, they will finish with the same speed.

(b) Note that Kathleen is consistently at a lower elevation than Paul for the entire trip. This means she has converted her potential energy to kinetic energy earlier. Consequently, she is traveling faster than Paul for the whole trip, except toward the end where Paul finally gets up to the same speed. Since she was going faster for almost the whole trip, and the distance is roughly the same, Kathleen gets to the bottom first.

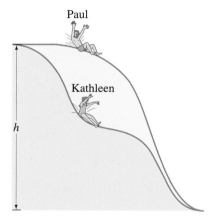

FIGURE 8–9 Example 8–5.

FIGURE 8–10 Transformation of energy during a pole vault.

There are many interesting examples of the conservation of energy in sports, one of which is the pole vault illustrated in Fig. 8–10. We often have to make approximations, but the sequence of events in broad outline for this case is as follows. The kinetic energy of the running athlete is transformed into elastic potential energy of the bending pole and, as the athlete leaves the ground, into gravitational potential energy. When the vaulter reaches the top and the pole has straightened out again, the energy has all been transformed into gravitational potential energy (if we ignore the vaulter's low horizontal speed over the bar). The pole does not supply any energy, but it acts as a device to *store* energy and thus aid in the transformation of kinetic energy into gravitational potential energy, which is the net result. The energy required to pass over the bar depends on how high the center of mass† (CM) of the vaulter must be raised. By bending their bodies, pole vaulters

---

† The center of mass (CM) of a body is that point at which the entire mass of the body can be considered as concentrated, for purposes of describing its translational motion. (This is discussed in Chapter 9.) In Eq. 8–12, y is the position of the CM.

keep their CM so low that it can actually pass slightly beneath the bar (Fig. 8–11), thus enabling them to cross over a higher bar than would otherwise be possible.

➡ PHYSICS APPLIED

Sports

FIGURE 8–11 By bending their bodies, pole vaulters can keep their center of mass so low that it may even pass below the bar. By changing their kinetic energy (of running) into gravitational potential energy (= mgy) in this way, vaulters can cross over a higher bar than if the change in potential energy were accomplished without carefully bending the body.

Conservation of mechanical energy (elastic force only)

EXAMPLE 8–6 **ESTIMATE** **Pole vault.** Estimate the kinetic energy and the speed required for a 70-kg pole vaulter to just pass over a bar 5.0 m high. Assume the vaulter's center of mass is initially 0.90 m off the ground and reaches its maximum height at the level of the bar itself.

SOLUTION We equate the total energy just before the vaulter places the end of the pole onto the ground (and the pole begins to bend and store potential energy) with the vaulter's total energy when passing over the bar (we ignore the small amount of kinetic energy at this point). We choose the initial position of the vaulter's center of mass to be $y_1 = 0$. The vaulter's body must then be raised to a height $y_2 = 5.0 \, \text{m} - 0.9 \, \text{m} = 4.1 \, \text{m}$. Thus, using Eq. 8–12,

$$\tfrac{1}{2}mv_1^2 + 0 = 0 + mgy_2$$

and

$$K_1 = \tfrac{1}{2}mv_1^2 = mgy_2 = (70 \, \text{kg})(9.8 \, \text{m/s}^2)(4.1 \, \text{m}) = 2.8 \times 10^3 \, \text{J}.$$

The speed is

$$v_1 = \sqrt{\frac{2K_1}{m}} = \sqrt{\frac{2(2800 \, \text{J})}{70 \, \text{kg}}} = 8.9 \, \text{m/s}.$$

This is an approximation because we have ignored such things as the vaulter's speed while crossing over the bar, mechanical energy transformed when the pole is planted in the ground, and work done by the vaulter on the pole.

As another example of the conservation of mechanical energy, let us consider a mass m connected to a horizontal spring whose own mass can be neglected and whose stiffness constant is k. The mass m has speed v at any moment and the potential energy of the system is $\tfrac{1}{2}kx^2$, where x is the displacement of the spring from its unstretched length. If neither friction nor any other force is acting, the conservation-of-energy principle tells us that

$$\tfrac{1}{2}mv_1^2 + \tfrac{1}{2}kx_1^2 = \tfrac{1}{2}mv_2^2 + \tfrac{1}{2}kx_2^2 \quad \text{[elastic potential energy only]} \quad \textbf{(8–13)}$$

where the subscripts 1 and 2 refer to the velocity and displacement at two different points.

FIGURE 8–12 Example 8–7. (a) A dart is pushed against a spring, compressing it 6.0 cm. The dart is then released, and in (b) it leaves the spring at high velocity (v_2).

(a)

(b)

EXAMPLE 8–7 **Toy dart gun.** A dart of mass 0.100 kg is pressed against the spring of a toy dart gun as shown in Fig. 8–12. The spring (with spring constant $k = 250 \, \text{N/m}$) is compressed 6.0 cm and released. If the dart detaches from the spring when the latter reaches its normal length ($x = 0$), what speed does the dart acquire?

SOLUTION In the horizontal direction, the only force on the dart (neglecting friction) is the force exerted by the spring. Vertically, gravity is counterbalanced by the normal force exerted on the dart by the gun barrel. (After the dart leaves the barrel it will follow a projectile's path under gravity.) We use Eq. 8–13 with point 1 being at the maximum compression of the spring, so $v_1 = 0$ (dart not yet released) and $x_1 = -0.060 \, \text{m}$. Point 2 we choose to be the instant the dart flies off the end of the spring (Fig. 8–12b), so $x_2 = 0$ and we want to find v_2. Thus Eq. 8–13 can be written

$$0 + \tfrac{1}{2}kx_1^2 = \tfrac{1}{2}mv_2^2 + 0,$$

so

$$v_2 = \sqrt{\frac{kx_1^2}{m}} = \sqrt{\frac{(250 \, \text{N/m})(0.060 \, \text{m})^2}{(0.100 \, \text{kg})}} = 3.0 \, \text{m/s}.$$

EXAMPLE 8–8 **Two kinds of Potential Energy.**   A ball of mass $m = 2.60$ kg, starting from rest, falls a vertical distance $h = 55.0$ cm before striking a vertical coiled spring, which it compresses (see Fig. 8–13) an amount $Y = 15.0$ cm. Determine the spring constant of the spring. Assume the spring has negligible mass. Measure all distances from the point where the ball first touches the uncompressed spring ($y = 0$ at this point).

SOLUTION   Since the motion is vertical, we use y instead of x (y positive upward). We divide this solution into two parts. (See also alternate solution below.) *Part 1*: Let us first consider the energy changes of the ball as it falls from a height $y_1 = h = 0.55$ m, Fig. 8–13a, to $y_2 = 0$, just as it touches the spring, Fig. 8–13b. Our system is the ball acted on by gravity (so far, the spring does nothing) so

$$\tfrac{1}{2}mv_1^2 + mgy_1 = \tfrac{1}{2}mv_2^2 + mgy_2$$
$$0 + mgh = \tfrac{1}{2}mv_2^2 + 0$$

and

$$v_2 = \sqrt{2gh} = \sqrt{2(9.80 \text{ m/s}^2)(0.550 \text{ m})} = 3.28 \text{ m/s}.$$

Part 2: As the ball compresses the spring, Fig. 8–13b to c, there are two conservative forces on the ball—gravity and the spring force. So our energy equation becomes

$$E \text{ (ball touches spring)} = E \text{ (spring compressed)}$$
$$\tfrac{1}{2}mv_2^2 + mgy_2 + \tfrac{1}{2}ky_2^2 = \tfrac{1}{2}mv_3^2 + mgy_3 + \tfrac{1}{2}ky_3^2.$$

We take point 2 to be the instant when the ball just touches the spring, so $y_2 = 0$ and $v_2 = 3.28$ m/s. Point 3 we take to be when the ball comes to rest and the spring is fully compressed, so $v_3 = 0$ and $y_3 = -Y = -0.150$ m (given). Putting these into the above energy equation we get

$$\tfrac{1}{2}mv_2^2 + 0 + 0 = 0 - mgY + \tfrac{1}{2}kY^2.$$

We know m, v_2, and Y, so we can solve for k:

$$k = \frac{2}{Y^2}\left[\tfrac{1}{2}mv_2^2 + mgY\right]$$
$$= \frac{m}{Y^2}\left[v_2^2 + 2gY\right]$$
$$= \frac{2.60 \text{ kg}}{(0.150 \text{ m})^2}\left[(3.28 \text{ m/s})^2 + 2(9.80 \text{ m/s}^2)(0.150 \text{ m})\right] = 1580 \text{ N/m}.$$

Alternate Solution: Instead of dividing the solution into two parts, we can do it all at once. After all, we get to choose what two points are used on the left and right of the energy equation. Let us write the energy equation for points 1 and 3 (Fig. 8–13). Point 1 is the initial point just before the ball starts to fall (Fig. 8–13a), so $v_1 = 0$, $y_1 = h = 0.550$ m; and point 3 is when the spring is fully compressed (Fig. 8–13c), so $v_3 = 0$, $y_3 = -Y = -0.150$ m. The forces on the ball in this process are gravity and (at least part of the time) the spring. So conservation of energy tells us

$$\tfrac{1}{2}mv_1^2 + mgy_1 + \tfrac{1}{2}k(0)^2 = \tfrac{1}{2}mv_3^2 + mgy_3 + \tfrac{1}{2}ky_3^2$$
$$0 + mgh + 0 = 0 - mgY + \tfrac{1}{2}kY^2$$

where we have set $y = 0$ for the spring at point 1 because it is not acting and is not compressed or stretched at this point. We solve for k:

$$k = \frac{2mg(h + Y)}{Y^2} = \frac{2(2.60 \text{ kg})(9.80 \text{ m/s}^2)(0.550 \text{ m} + 0.150 \text{ m})}{(0.150 \text{ m})^2} = 1580 \text{ N/m}$$

just as in our first method of solution.

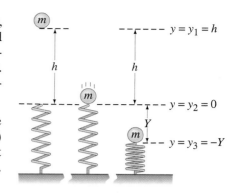

FIGURE 8–13   Example 8–8. (not to scale)

Conservation of energy:
gravity and elastic potential energy

➡ **PROBLEM SOLVING**

Alternate Solution

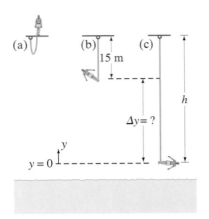

FIGURE 8–14 Example 8–9.
(a) Bungee jumper about to jump.
(b) Bungee cord at its unstretched
length. (c) Maximum stretch of cord.

FIGURE 8–15 Example 8–10: a
simple pendulum; y is measured
positive upwards.

EXAMPLE 8–9 Bungee jump. Dave jumps off a bridge with a bungee cord (a heavy stretchable cord) tied around his ankle, Fig. 8–14. He falls for 15 meters before the bungee cord begins to stretch. Dave's mass is 75 kg and we assume the cord obeys Hooke's law, $F = -kx$, with $k = 50 \, \text{N/m}$. If we neglect air resistance, estimate how far below the bridge Dave will fall before coming to a stop. Ignore the mass of the cord (not realistic, however).

SOLUTION Dave starts out with gravitational potential energy, which during his fall is tranformed into both kinetic energy and elastic potential energy. Assuming no frictional forces act on our system, the total energy at the start must be the same as the total energy at the end. If we define our coordinate system such that $y = 0$ at the lowest point in Dave's jump, and let the stretch of the cord at this point be represented by Δy, then the total fall is (see Fig. 8–14)

$$h = 15 \, \text{m} + \Delta y.$$

Conservation of energy then gives:

$$K_1 + U_1 = K_2 + U_2$$
$$0 + mg(15 \, \text{m} + \Delta y) = 0 + \tfrac{1}{2}k(\Delta y)^2.$$

To solve for Δy, we use the quadratic formula and get two solutions:

$$\Delta y = 40 \, \text{m} \quad \text{and} \quad \Delta y = -11 \, \text{m}.$$

The negative solution is nonphysical, so the distance that Dave drops in his fall is:

$$h = 15 \, \text{m} + 40 \, \text{m} = 55 \, \text{m}.$$

EXAMPLE 8–10 A swinging pendulum. The simple pendulum shown in Fig. 8–15 consists of a small bob of mass m suspended by a massless cord of length l. The bob is released (without a push) at $t = 0$, where the cord makes an angle $\theta = \theta_0$ to the vertical. (a) Describe the motion of the bob in terms of kinetic energy and potential energy. Then determine the speed of the bob: (b) as a function of position θ as it swings back and forth, and (c) at the lowest point of the swing. (d) Find the tension in the cord, $\mathbf{F}_T$. Ignore friction and air resistance.

SOLUTION (a) At the moment of release, the bob is at rest, so $K = 0$. As the bob falls, it loses potential energy and gains kinetic energy. At the lowest point its kinetic energy is a maximum and the potential energy is a minimum. The bob continues its swing until it reaches an equal height and angle (θ_0) on the opposite side, at which point the potential energy is a maximum and $K = 0$. It continues the swinging motion as $U \rightarrow K \rightarrow U$ and so on, but it can never go higher than $\theta = \pm\theta_0$ (conservation of mechanical energy).

(b) The cord is assumed to be massless, so we don't need to be concerned with the energy of the cord but only with the bob's kinetic energy, and the potential energy. The bob has two forces acting on it at any moment: gravity, mg, and the force the cord exerts on it, $\mathbf{F}_T$. The latter (a constraint force) always acts perpendicular to the motion, so it does no work. We need be concerned only with gravity, for which we can write the potential energy. The mechanical energy of the system is

$$E = \tfrac{1}{2}mv^2 + mgy,$$

where y is the vertical height of the bob at any moment. We take $y = 0$ at the lowest point of the bob's swing. Hence at $t = 0$,

$$y = y_0 = l - l\cos\theta_0 = l(1 - \cos\theta_0)$$

as can be seen from the diagram. At the moment of release

$$E = mgy_0,$$

since $v = v_0 = 0$. At any other point along the swing

$$E = \tfrac{1}{2}mv^2 + mgy = mgy_0.$$

We solve this for v:

$$v = \sqrt{2g(y_0 - y)}.$$

In terms of the angle θ of the cord, we can write

$$v = \sqrt{2gl(\cos\theta - \cos\theta_0)},$$

since $y = l - l\cos\theta$ and $y_0 = l - l\cos\theta_0$.
(c) At the lowest point, $y = 0$ so

$$v = \sqrt{2gy_0} \qquad \text{or} \qquad v = \sqrt{2gl(1 - \cos\theta_0)}.$$

(d) The tension in the cord is the force $\mathbf{F}_T$ that the cord exerts on the bob. As we've seen, there is no work done by this force, but we can calculate the force simply by using Newton's second law $\Sigma\mathbf{F} = m\mathbf{a}$ and by noting that at any point the acceleration of the bob in the inward radial direction is v^2/l, since the bob is constrained to move in an arc of a circle. In the radial direction, $\mathbf{F}_T$ acts inward, and a component of gravity equal to $mg\cos\theta$ acts outward. Hence

$$m\frac{v^2}{l} = F_T - mg\cos\theta.$$

We solve for F_T and use the result of part (b) for v^2:

$$F_T = m\left(\frac{v^2}{l} + g\cos\theta\right) = 2mg(\cos\theta - \cos\theta_0) + mg\cos\theta$$

$$= (3\cos\theta - 2\cos\theta_0)mg.$$

8–5 | The Law of Conservation of Energy

We now take into account nonconservative forces such as friction, since they are important in real situations. For example, consider again the roller-coaster car in Fig. 8–8, but this time let us include friction. The car will not in this case reach the same height on the second hill as it had on the first hill because of friction.

In this, and in other natural processes, the mechanical energy (sum of the kinetic and potential energies) does not remain constant but decreases. Because frictional forces reduce the total mechanical energy, they are called **dissipative** *Dissipative forces* **forces**. Historically, the presence of dissipative forces hindered the formulation of a comprehensive conservation of energy law until well into the nineteenth century. It was not until then that heat, which is always produced when there is friction (try rubbing your hands together), was interpreted as a form of energy. Quantitative studies by nineteenth-century scientists (Chapter 19) demonstrated that if heat is considered as energy, now called **thermal energy**, then the total energy is conserved in any process. For example, if the roller-coaster car in Fig. 8–8 is subject to frictional forces, then the initial total energy of the car will be equal to the car's kinetic energy plus the potential energy at any subsequent point along its path plus the amount of thermal energy produced in the process. The thermal energy produced by a constant friction force F_{fr} is equal to the work done by this force. A block sliding freely across a table, for example, comes to rest because of friction. Its initial kinetic energy is all transformed into thermal energy. The block and table are a little warmer as a result of this process. A more obvious example of the transformation of kinetic energy into thermal energy can be observed by vigorously striking a nail several times with a hammer and then gently touching the nail with your finger.

According to the atomic theory, thermal energy represents kinetic energy of rapidly moving molecules. We shall see in Chapter 18 that a rise in temperature corresponds to an increase in the average kinetic energy of the molecules. Because

FIGURE 8–16 The burning of fuel (a chemical reaction) releases energy to boil water in this steam engine. The steam produced expands against a piston to do work in turning the wheels.

LAW OF CONSERVATION OF ENERGY

thermal energy represents the energy of atoms and molecules that make up a body, it is often called *internal energy*. Internal energy,[†] from the atomic point of view, can include not only kinetic energy of molecules but also potential energy (electrical in nature) because of the relative positions of atoms within molecules. On a macroscopic level, internal energy corresponds to nonconservative forces such as friction. But at the atomic level, the energy is kinetic and potential and the corresponding forces are mainly conservative. For example, the energy stored in food or in a fuel such as gasoline can be regarded as potential energy stored by virtue of the relative positions of the atoms within a molecule. For this energy to be used to do work, it must be released, usually through a chemical reaction (Fig. 8–16). This is something like a compressed spring which, when released, can do work.

To establish the more general law of conservation of energy, it required nineteenth-century physicists to recognize electrical, chemical, and other forms of energy in addition to heat and to explore if in fact they could fit into a conservation law. For each type of force, conservative or nonconservative, it has always been found possible to define a type of energy that corresponds to the work done by such a force. And it has been found experimentally that the total energy E always remains constant. That is, the change in the total energy, kinetic plus potential plus all other forms of energy, equals zero:

$$\Delta K + \Delta U + [\text{change in all other forms of energy}] = 0. \qquad \textbf{(8–14)}$$

This is one of the most important principles in physics. It is called the **law of conservation of energy** and can be stated as follows:

The total energy is neither increased nor decreased in any process. Energy can be transformed from one form to another, and transferred from one body to another, but the total amount remains constant.

For conservative mechanical systems, this law can be derived from Newton's laws (Section 8–3) and thus is equivalent to them. But in its full generality, the validity of the law of conservation of energy rests on experimental observation. And even though Newton's laws have been found to fail in the submicroscopic world of the atom, the law of conservation of energy has been found to hold there and in every experimental situation so far tested.

8–6 Energy Conservation with Dissipative Forces: Solving Problems

In Section 8–4 we discussed several Examples of the law of conservation of energy for conservative systems. Now let us consider in detail some examples that involve nonconservative forces.

Suppose, for example, that the roller-coaster car rolling on the hills of Fig. 8–8 is subject to frictional forces. In going from some point 1 to a second point 2, the work done by the friction force $\mathbf{F}_{fr}$ acting on the car is $W_{fr} = \int_1^2 \mathbf{F}_{fr} \cdot d\mathbf{l}$. If $\mathbf{F}_{fr}$ is constant in magnitude, $W_{fr} = -F_{fr}l$, where l is the actual distance along the path traveled by the object from point 1 to point 2. (The minus sign appears because $\mathbf{F}_{fr}$ opposes the motion and is thus in the opposite direction to $d\mathbf{l}$.) From the work-energy principle (Eq. 7–11), the net work W_{net} done on a body is equal to the change in its kinetic energy:

$$\Delta K = W_{net}.$$

The forces that do work on the car in the present case are gravity and friction (the

[†] The term *internal energy* can also be used to refer to kinetic and potential energy of the internal parts of a body, such as vibration, when we are primarily interested in the body's motion as a whole.

normal force exerted by the roadbed or track on the car does no work because it acts perpendicular to the motion). Hence we can write

$$W_{\text{net}} = W_{\text{C}} + W_{\text{NC}}$$

where in general W_{C} represents the work done by conservative forces (gravity for our car) and W_{NC} is the work done by nonconservative forces (friction). We saw in Section 8–2 (Eq. 8–4) that the work done by a conservative force, such as gravity, can be written in terms of potential energy:

$$W_{\text{C}} = \int_1^2 \mathbf{F} \cdot d\boldsymbol{l} = -\Delta U.$$

Hence we can now write

$$\Delta K = -\Delta U + W_{\text{NC}}$$

or

$$\Delta K + \Delta U = W_{\text{NC}}. \qquad\qquad \textbf{(8–15)}$$

Conservation of energy (work-energy principle: general form)

This equation represents the general form of the work-energy principle. It also represents the conservation of energy. For our car, W_{NC} is the work done by the friction force and represents thermal energy. Equation 8–15 tells us that the change in mechanical energy, $\Delta(K + U)$, which is a decrease here since $W_{\text{NC}} < 0$ ($\mathbf{F}_{\text{fr}}$ and $d\boldsymbol{l}$ are in opposite directions), goes into thermal energy. But Eq. 8–15 is valid in general. Based on our derivation from the work-energy principle, W_{NC} on the right side of Eq. 8–15 must be the total work done by all forces that are not included in the potential energy term, ΔU, on the left side.[†] The potential energy term, U, should include all conservative forces acting.

Let us rewrite Eq. 8–15 for our roller-coaster car example of Fig. 8–8, shown here in Fig. 8–17, setting $W_{\text{NC}} = -F_{\text{fr}}l$ as we discussed above:

$$W_{\text{fr}} = -F_{\text{fr}}l = \Delta K + \Delta U = \left(\tfrac{1}{2}mv_2^2 - \tfrac{1}{2}mv_1^2\right) + \left(mgy_2 - mgy_1\right)$$

or

$$\tfrac{1}{2}mv_1^2 + mgy_1 = \tfrac{1}{2}mv_2^2 + mgy_2 + F_{\text{fr}}d. \qquad \begin{bmatrix} \text{gravity and} \\ \text{friction acting} \end{bmatrix} \textbf{(8–16)}$$

Conservation of energy with gravity and friction

We can write this last equation as

initial energy = final energy (including thermal energy).

On the left we have the mechanical energy of the system initially. It equals the mechanical energy at any subsequent point along the path plus the amount of thermal (or internal) energy produced in the process.

EXAMPLE 8–11 **Friction on the roller coaster.** The roller-coaster car in Example 8–4, which starts at a height $y_1 = 40$ m, is found to reach a vertical height of only 25 m on the second hill before coming to a stop (Fig. 8–17). It traveled a total distance of 400 m. Estimate the average friction force (assume constant) on the car, whose mass is 1000 kg.

SOLUTION We use conservation of energy, here in the form of Eq. 8–16, taking point 1 to be the instant when the car started coasting and point 2 to be the instant it stopped. Then $v_1 = 0$, $y_1 = 40$ m, $v_2 = 0$, $y_2 = 25$ m, and $d = 400$ m. Thus

$$0 + (1000\,\text{kg})(9.8\,\text{m/s}^2)(40\,\text{m}) = 0 + (1000\,\text{kg})(9.8\,\text{m/s}^2)(25\,\text{m}) + F_{\text{fr}}(400\,\text{m}).$$

We solve this for F_{fr} to find $F_{\text{fr}} = 370$ N.

FIGURE 8–17 Roller-coaster car coasting, as in Fig. 8–8, but now with friction. Example 8–11.

[†] A conservative force, if desired, could be considered as doing work (and therefore included in W_{NC} on the right in Eq. 8–15) rather than as a change in potential energy, whereas nonconservative forces (such as friction) *must* be included in the work term, W_{NC}. It must be emphasized that all forces acting on a body must be included in one term or the other. But avoid the error of including the same force twice, once in the potential energy term, U, and again in the work term, W.

SECTION 8–6 Energy Conservation with Dissipative Forces **191**

(a)

(b)

FIGURE 8–18 Example 8–12.

EXAMPLE 8–12 **Friction with a spring.** A block of mass m sliding along a rough horizontal surface is traveling at a speed v_0 when it strikes a massless spring head-on (see Fig. 8–18) and compresses the spring a maximum distance X. If the spring has stiffness constant k, determine the coefficient of kinetic friction between block and surface.

SOLUTION At the moment of collision, the block has $K = \frac{1}{2}mv_0^2$ and the spring is presumably uncompressed, so $U = 0$. Initially the mechanical energy of the system is $\frac{1}{2}mv_0^2$. By the time the spring reaches maximum compression, $K = 0$ and $U = \frac{1}{2}kX^2$. In the meantime, the friction force $(= \mu_k F_N = \mu_k mg)$ has done work $W = -\mu_k mgX$ which goes into thermal energy. From conservation of energy we can write

energy (initial) = energy (final)

$$\tfrac{1}{2}mv_0^2 = \tfrac{1}{2}kX^2 + \mu_k mgX.$$

We solve for μ_k and find

$$\mu_k = \frac{v_0^2}{2gX} - \frac{kX}{2mg}.$$

Problem solving is not a process that can be done by simply following a set of rules. The following Problem Solving box, like all others, is thus *not* a prescription, but is a summary to help you get started solving problems related to energy.

PROBLEM SOLVING **Conservation of Energy**

1. Draw a picture.
2. Determine the system for which energy will be conserved: the object or objects and the forces acting. Identify all forces that do work.
3. Ask yourself what quantity you are looking for, and decide what the initial (point 1) and final (point 2) locations are.
4. If the body under investigation changes its height during the problem, then choose a $y = 0$ level for gravitational potential energy. This may be chosen for convenience; the lowest point in the problem is often a good choice.
5. If springs are involved, choose the unstretched spring position to be x (or y) $= 0$.

6. If no friction or other nonconservative forces act, then apply conservation of mechanical energy:

$$K_1 + U_1 = K_2 + U_2.$$

7. Solve for the unknown quantity.
8. If friction or other nonconservative forces are present and significant, then an additional term, W_{NC}, will be needed:

$$K_1 + U_1 = K_2 + U_2 + W_{NC}.$$

To be sure which sign to give W_{NC}, or on which side of the equation to put it, use your intuition: is the total mechanical energy E increased or decreased in the process?

8–7 Gravitational Potential Energy and Escape Velocity

We have been dealing with gravitational potential energy so far in this chapter assuming the force of gravity is constant, $\mathbf{F} = m\mathbf{g}$. This is an accurate assumption for ordinary objects near the surface of the Earth. But to deal with gravity more generally, for points far from the Earth's surface, we must consider that the gravitational force exerted by the Earth on a particle of mass m decreases inversely as the square of the distance r from the Earth's center. The precise relationship is given by Newton's law of universal gravitation (Sections 6–1 and 6–2):

$$\mathbf{F} = -G\frac{mM_E}{r^2}\hat{\mathbf{r}} \qquad\qquad [r > r_E]$$

where M_E is the mass of the Earth and $\hat{\mathbf{r}}$ is a unit vector (at the position of m) directed radially away from the Earth's center. The minus sign indicates that the force on m is directed toward the Earth's center, in the direction opposite to $\hat{\mathbf{r}}$. This equation can also be used to describe the gravitational force on a mass m in the vicinity of other heavenly bodies, such as the Moon, planets, or Sun, in which case M_E must be replaced by that body's mass.

Suppose an object of mass m moves from one position to another along an arbitrary path (Fig. 8–19) so that its distance from the Earth's center changes from r_1 to r_2. The work done by the gravitational force is

$$W = \int_1^2 \mathbf{F} \cdot d\boldsymbol{l} = -GmM_E \int_1^2 \frac{\hat{\mathbf{r}} \cdot d\boldsymbol{l}}{r^2},$$

where $d\boldsymbol{l}$ represents an infinitesimal displacement. Since $\hat{\mathbf{r}} \cdot d\boldsymbol{l} = dr$ is the component of $d\boldsymbol{l}$ along $\hat{\mathbf{r}}$ (see Fig. 8–19), then

$$W = -GmM_E \int_{r_1}^{r_2} \frac{dr}{r^2} = GmM_E\left(\frac{1}{r_2} - \frac{1}{r_1}\right)$$

or

$$W = \frac{GmM_E}{r_2} - \frac{GmM_E}{r_1}.$$

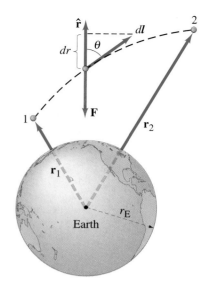

FIGURE 8–19 Arbitrary path of particle of mass m moving from point 1 to point 2.

Work done by gravity

Since the value of the integral depends only on the position of the end points $(r_1$ and $r_2)$ and not on the path taken, the gravitational force is a conservative force. We can therefore use the concept of potential energy for the gravitational force. Since the change in potential energy is always defined (Section 8–2) as the negative of the work done by the force, we have

$$\Delta U = U_2 - U_1 = -\frac{GmM_E}{r_2} + \frac{GmM_E}{r_1}. \qquad (8\text{–}17)$$

From Eq. 8–17 the potential energy at any distance r from the Earth's center can be written:

$$U(r) = -\frac{GmM_E}{r} + C,$$

where C is a constant. It is usual to choose $C = 0$, so that

$$U(r) = -\frac{GmM_E}{r}. \qquad \left[\begin{array}{c} \text{gravity} \\ (r > r_E) \end{array}\right] \quad (8\text{–}18)$$

With this choice for C, $U = 0$ at $r = \infty$. As an object approaches the Earth, its potential energy decreases and is always negative (Fig. 8–20).

Equation 8–17 reduces to Eq. 8–2, $\Delta U = mg(y_2 - y_1)$, for objects near the surface of the Earth (see Problem 40).

The total energy of a particle of mass m, which feels only the force of the Earth's gravity, is conserved since gravity is a conservative force. Therefore we can write

$$\tfrac{1}{2}mv_1^2 - G\frac{mM_E}{r_1} = \tfrac{1}{2}mv_2^2 - G\frac{mM_E}{r_2} = \text{constant}. \qquad \left[\begin{array}{c} \text{gravity} \\ \text{only} \end{array}\right] \quad (8\text{–}19)$$

FIGURE 8–20 Gravitational potential energy plotted as a function of r, the distance from Earth's center. Valid only for points $r > r_E$, the radius of the Earth.

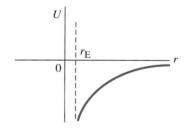

Gravitational potential energy

EXAMPLE 8–13 **Package dropped from high speed rocket.** A box of empty film canisters is dumped from a rocket traveling outward from Earth at a speed of 1800 m/s when 1600 km above the Earth's surface. The package eventually falls to the Earth. Estimate its speed just before impact. Ignore air resistance.

SOLUTION The package initially has a speed relative to Earth equal to the speed of the rocket from which it falls. We use conservation of energy,

$$\tfrac{1}{2}mv_1^2 - G\frac{mM_E}{r_1} = \tfrac{1}{2}mv_2^2 - G\frac{mM_E}{r_2}$$

where $v_1 = 1.80 \times 10^3$ m/s, $r_1 = 1.60 \times 10^6$ m + 6.38×10^6 m = 7.98×10^6 m, and $r_2 = 6.38 \times 10^6$ m (the radius of the Earth). We solve for v_2:

$$v_2 = \sqrt{v_1^2 - 2GM_E\left(\frac{1}{r_1} - \frac{1}{r_2}\right)}$$

$$= \sqrt{\begin{array}{c}(1.80 \times 10^3 \text{ m/s})^2 - 2(6.67 \times 10^{-11}\text{ N} \cdot \text{m}^2/\text{kg}^2)(5.98 \times 10^{24}\text{ kg}) \\ \times \left(\dfrac{1}{7.98 \times 10^6 \text{ m}} - \dfrac{1}{6.38 \times 10^6 \text{ m}}\right)\end{array}}$$

$$= 5320 \text{ m/s}.$$

In reality, the speed will be somewhat less than this because of air resistance. Note, incidentally, that the direction of the velocity never entered into the problem and this is one of the advantages of the energy method. The rocket could have been heading away from the Earth, or toward it, or at some other angle, and the result would be the same.

When a body is projected into the air from the Earth, it will return to Earth unless its speed is very high. But if the speed *is* high enough, it will continue out into space never to return to Earth (barring other forces or collisions). The minimum initial velocity needed to prevent an object from returning to the Earth is called the **escape velocity** from Earth, v_{esc}. To determine v_{esc} from the Earth's surface (ignoring air resistance), we use Eq. 8–19 with $v_1 = v_{esc}$ and $r_1 = r_E = 6.38 \times 10^6$ m, the radius of the Earth. Since we want the minimum speed for escape, we need the object to reach $r_2 = \infty$ with merely zero speed, $v_2 = 0$. Applying Eq. 8–19 we have

Escape velocity

$$\tfrac{1}{2}mv_{esc}^2 - G\frac{mM_E}{r_E} = 0 + 0$$

or

$$v_{esc} = \sqrt{2GM_E/r_E} = 1.12 \times 10^4 \text{ m/s} \qquad (8\text{–}20)$$

or 11.2 km/s. It is important to note that although a mass can escape from the Earth (or solar system) never to return, the force on it due to the Earth's gravitational field is never actually zero for a finite value of r. However, the force does become very small and usually ignorable at great distances.

EXAMPLE 8–14 **Escaping the Earth or the Moon.** (*a*) Compare the escape velocities of a rocket from the Earth and from the Moon. (*b*) Compare the energies required to launch the rockets. For the Moon, $M_M = 7.35 \times 10^{22}$ kg and $r_M = 1.74 \times 10^6$ m, and for the Earth, $M_E = 5.97 \times 10^{24}$ kg and $r_E = 6.38 \times 10^6$ m.

SOLUTION (*a*) Using Eq. 8–20, the ratio of the escape velocities is

$$\frac{v_{esc}(\text{Earth})}{v_{esc}(\text{Moon})} = \sqrt{\frac{M_E}{M_M}\frac{r_M}{r_E}} = 4.7.$$

To escape Earth requires a speed 4.7 times that required to escape the Moon.
(*b*) The fuel that must be burned provides energy proportional to v^2 ($K = \tfrac{1}{2}mv^2$); so to launch a rocket to escape Earth requires $(4.7)^2 = 22$ times as much energy as to escape from the Moon.

8–8 Power

Power is defined as the rate at which work is done. The *average power*, $\overline{P}$, when an amount of work W is done in a time t is

Power defined

$$\overline{P} = \frac{W}{t}. \tag{8–21a}$$

The *instantaneous power*, P, is

$$P = \frac{dW}{dt}. \tag{8–21b}$$

The work done in a process is equal to the energy transformed from one form or from one body to another. For example, as the potential energy stored in the spring of Fig. 8–6b is transformed to kinetic energy of the ball, the spring is doing work on the ball. Similarly, when you throw a ball or push a grocery cart, *whenever work is done, energy is being transformed or transferred from one body to another.* Hence we can also say that power is the *rate at which energy is transformed*:

Energy is transferred whenever work is done

$$P = \frac{dE}{dt}. \tag{8–21c}$$

The power of a horse refers to how much work it can do per unit of time. The power rating of an engine refers to how much chemical or electrical energy can be transformed into mechanical energy per unit of time. In SI units, power is measured in joules per second and this is given a special name, the **watt** (W): $1\text{ W} = 1\text{ J/s}$. We are most familiar with the watt for measuring the rate at which an electric light bulb or heater changes electric energy into light or thermal energy, but it is used for other types of energy transformation as well. In the British system, the unit of power is the foot-pound per second (ft·lb/s). For practical purposes a larger unit is often used, the **horsepower**. One horsepower† (hp) is defined as 550 ft·lb/s, which equals 746 watts.

Units: the watt ($1\text{ W} = 1\text{ J/s}$)

The horsepower ($1\text{ hp} = 746\text{ W}$)

To see the distinction between energy and power, consider the following example. A person is limited in the work he or she can do, not only by the total energy required, but also by the rate this energy is used; that is, by power. For example, a person may be able to walk a long distance or climb many flights of stairs before having to stop because so much energy has been used up. On the other hand, a person who runs very quickly up stairs, may feel exhausted after only a flight or two. He or she is limited in this case by power, the rate at which his or her body can transform chemical energy into mechanical energy.

Energy and power distinguished

EXAMPLE 8–15 **Stair-climbing power.** A 70-kg jogger runs up a long flight of stairs in 4.0 s. The vertical height of the stairs is 4.5 m. (*a*) Estimate the jogger's power output in watts and horsepower. (*b*) How much energy did this require?

SOLUTION (*a*) The work done is against gravity, and equals $W = mgy$. Then the average power output was

$$\overline{P} = \frac{W}{t} = \frac{mgy}{t} = \frac{(70\text{ kg})(9.8\text{ m/s}^2)(4.5\text{ m})}{4.0\text{ s}} = 770\text{ W}.$$

Since there are 746 W in 1 hp, the jogger is doing work at a rate of just over 1 hp. It is worth noting that a human cannot do work at this rate for very long. (*b*) The energy required is $E = \overline{P}t = (770\text{ J/s})(4.0\text{ s}) = 3100\text{ J}$. [Note that the person had to transform more energy than this. The total energy transformed by a person or an engine always includes some thermal energy (recall how hot you get running up stairs).]

---

†The unit was first chosen by James Watt (1736–1819), who needed a way to specify the power of his newly developed steam engines. He found by experiment that a good horse can work all day at an average rate of about 360 ft·lb/s. So as not to be accused of exaggeration in the sale of his steam engines, he multiplied this by roughly $1\frac{1}{2}$ when he defined the hp.

Automobiles do work to overcome the force of friction (and air resistance), to climb hills, and to accelerate. A car is limited by the rate it can do work, which is why automobile engines are rated in horsepower. A car needs power most when it is climbing hills and when accelerating. In the next Example, we will calculate how much power is needed in these situations for a car of reasonable size. Even when a car travels on a level road at constant speed, it needs some power just to do work to overcome the retarding forces of internal friction and air resistance. These forces depend on the conditions and speed of the car, but are typically in the range 400 N to 1000 N.

It is often convenient to write the power in terms of the net force $\mathbf{F}$ applied to an object and its velocity $\mathbf{v}$. Since $P = dW/dt$ and $dW = \mathbf{F} \cdot d\mathbf{l}$ (Eq. 7–7), then

$$P = \frac{dW}{dt} = \mathbf{F} \cdot \frac{d\mathbf{l}}{dt} = \mathbf{F} \cdot \mathbf{v}. \qquad \text{(8–22)}$$

FIGURE 8–21 Example 8–16: calculation of power needed for a car (*a*) to climb a hill, (*b*) to pass another car.

EXAMPLE 8–16 **Power needs of a car.** Calculate the power required of a 1400-kg car under the following circumstances: (*a*) the car climbs a 10° hill (a fairly steep hill) at a steady 80 km/h; and (*b*) the car accelerates along a level road from 90 to 110 km/h in 6.0 s to pass another car. Assume the retarding force on the car is $F_R = 700$ N throughout. See Fig. 8–21. (Be careful not to confuse $\mathbf{F}_R$, which is due to air resistance and friction that retard the motion, with the force $\mathbf{F}$ needed to accelerate the car, which is the frictional force exerted by the road on the tires—the reaction to the motor-driven tires pushing against the road.)

SOLUTION (*a*) To move at a steady speed up the hill, the car must exert a force equal to the sum of the retarding force, 700 N, and the component of gravity parallel to the hill, $mg \sin 10° = (1400 \text{ kg})(9.80 \text{ m/s}^2)(0.174) = 2400$ N. Since $\bar{v} = 80$ km/h = 22 m/s and is parallel to $\mathbf{F}$, then (Eq. 8–22):

$$P = Fv$$
$$= (2400 \text{ N} + 700 \text{ N})(22 \text{ m/s}) = 6.80 \times 10^4 \text{ W} = 91 \text{ hp}.$$

(*b*) The car accelerates from 25.0 m/s to 30.6 m/s (90 to 110 km/h). Thus the car must exert a force that overcomes the 700 N retarding force plus that required to give it the acceleration $\bar{a}_x = (30.6 \text{ m/s} - 25.0 \text{ m/s})/6.0 \text{ s} = 0.93 \text{ m/s}^2$. We apply Newton's second law with x being the direction of motion:

$$ma_x = \Sigma F_x = F - F_R.$$

Then the force required, F, is

$$F = ma_x + F_R$$
$$= (1400 \text{ kg})(0.93 \text{ m/s}^2) + 700 \text{ N} = 2000 \text{ N}.$$

Since $P = \mathbf{F} \cdot \mathbf{v}$, the required power increases with speed and the motor must be able to provide a maximum power output of

$$\bar{P} = (2000 \text{ N})(30.6 \text{ m/s}) = 6.12 \times 10^4 \text{ W} = 82 \text{ hp}.$$

Even taking into account the fact that only 60 to 80 percent of the engine's power output reaches the wheels, it is clear from these calculations that an engine of 100 to 150 hp is more than adequate from a practical point of view.

We mentioned in the Example above that only part of the energy output of a car engine reaches the wheels. Not only is some energy wasted in getting from the engine to the wheels, in the engine itself much of the input energy (from the gasoline) does not end up doing useful work. An important characteristic of all engines is their overall efficiency e, defined as the ratio of the useful power output of the engine, P_{out}, to the power input, P_{in}:

$$e = \frac{P_{out}}{P_{in}}.$$

Efficiency

The efficiency is always less than 1.0 because no engine can create energy, and in fact, cannot even transform energy from one form to another without some going to friction, thermal energy and other nonuseful forms of energy. For example, an automobile engine converts chemical energy released in the burning of gasoline into mechanical energy that moves the pistons and eventually the wheels. But nearly 85% of the input energy is "wasted" as thermal energy that goes out the exhaust pipe, plus friction in the moving parts. Thus car engines are roughly only about 15% efficient. We will discuss efficiency in detail in Chapter 20.

8–9 Potential Energy Diagrams; Stable and Unstable Equilibrium

For an object acted on by only a conservative force, we can learn a great deal about its motion simply by examining a potential energy diagram—the graph of $U(x)$ versus x. An example of a potential energy diagram is shown in Fig. 8–22. The rather complex curve represents some complicated potential energy $U(x)$. The total energy $E = K + U$ is constant and can be represented as a horizontal line on this graph. Four different possible values for E are shown, labeled E_0, E_1, E_2, and E_3. What the actual value of E will be for a given system depends on the initial conditions. (For example, the total energy E of a mass oscillating on the end of a spring depends on the amount the spring is initially compressed or stretched.) Because $E = U + K = $ constant, then $U(x)$ must be less than or equal to E for all situations: $U(x) \leq E$. Thus the minimum value which the total energy can take for the potential energy shown in Fig. 8–22 is that labeled E_0. For this value of E, the mass can only be at rest at $x = x_0$. It has potential energy but no kinetic energy.

FIGURE 8–22 A potential energy diagram.

If the object's total energy E is greater than E_0, say it is E_1 on our plot, the object can have both kinetic and potential energy. Since energy is conserved,

$$K = E - U(x).$$

Since the curve represents $U(x)$ at each x, the kinetic energy at any value of x is represented by the distance between the E line and the curve $U(x)$ at that value of x. In the diagram, the kinetic energy for an object at x_1, when its total energy is E_1, is indicated by the notation K_1.

An object with energy E_1 can oscillate only between the points x_2 and x_3. This is because if $x > x_2$ or $x < x_3$, the potential energy would be greater than E, meaning $K = \frac{1}{2}mv^2 < 0$ and v would be imaginary, and so impossible. At x_2 and x_3 the velocity is zero, since $E = U$ at these points. Hence x_2 and x_3 are called the **turning points** of the motion. If the object is at x_0, say, moving to the right, its kinetic energy (and speed) decreases until it reaches zero at $x = x_2$. The object then reverses direction, proceeding to the left and increasing in speed until it passes x_0 again. It continues to move, decreasing in speed until it reaches $x = x_3$, where again $v = 0$, and the object again reverses direction.

Turning points

If the object has energy $E = E_2$ in Fig. 8–22, there are four turning points. The object can move in only one of the two potential energy "valleys," depending on where it is initially. It cannot get from one valley to the other because of the barrier between them—for example at a point such as x_4, $U > E_2$, which means v would be imaginary.[†] For energy E_3, there is only one turning point since $U(x) < E_3$ for all $x > x_5$. Thus our object, if moving initially to the left, varies in

---

[†]Although this is true according to Newtonian physics, modern quantum mechanics predicts that objects can "tunnel" through such a barrier, and such processes have been observed at the atomic and subatomic level.

speed as it passes the potential valleys but eventually stops and turns around at $x = x_5$. It then proceeds to the right indefinitely, never to return.

How do we know the object reverses direction at the turning points? Because of the force exerted on it. The force F is related to the potential energy U by Eq. 8–7, $F = -dU/dx$. The force F is equal to the negative of the slope of the U-versus-x curve at any point x. At $x = x_2$, for example, the slope is positive so the force is negative, which means it acts to the left (toward decreasing values of x) acting in the direction opposite to the object's motion.

Stable equilibrium

At $x = x_0$ the slope is zero, so $F = 0$. At such a point the particle is said to be in **equilibrium**. This term means simply that the net force on the object is zero. Hence, its acceleration is zero, and so if it is initially at rest, it remains at rest. If the object at rest at $x = x_0$ were moved slightly to the left or right, a nonzero force would act on it in the direction to move it back toward x_0. An object that returns toward its equilibrium point when displaced slightly is said to be at a point of **stable equilibrium**. Any *minimum* in the potential energy curve represents a point of stable equilibrium.

Unstable equilibrium

An object at $x = x_4$ would also be in equilibrium, since $F = -dU/dx = 0$. If the object were displaced a bit to either side of x_4, a force would act to pull the object *away* from the equilibrium point. Points like x_4, where the potential energy curve has a maximum, are points of **unstable equilibrium**. The object will *not* return to equilibrium if displaced slightly, but instead will move farther away.

Neutral equilibrium

When an object is in a region over which U is constant, such as at $x = x_6$ in Fig. 8–22, the force is zero over some distance. The object is in equilibrium and if displaced slightly to one side the force is still zero. The object is said to be in **neutral equilibrium** in this region.

Summary

A **conservative force** is one for which the work done by the force in moving an object from one position to another depends only on the two positions and not on the path taken. The work done by a conservative force is recoverable, which is not true for nonconservative forces, such as friction.

Potential energy is energy associated with the position or configuration of bodies. Examples are: gravitational potential energy

$$U = mgy$$

where the mass m is near the Earth's surface, a height y above some reference point; elastic potential energy

$$U = \tfrac{1}{2}kx^2$$

such as a spring with stiffness constant k stretched or compressed a distance x from equilibrium; and chemical, electrical, and nuclear energy. Potential energy is always associated with a conservative force, and the change in potential energy, ΔU, between two points under the action of a conservative force $\mathbf{F}$ is defined as the negative of the work done by the force:

$$\Delta U = U_2 - U_1 = -\int_1^2 \mathbf{F} \cdot d\boldsymbol{l}.$$

Inversely, we can write, for the one-dimensional case,

$$F = -\frac{dU(x)}{dx}.$$

Only *changes* in potential energy are physically meaningful, so the choice of where $U = 0$ can be chosen for convenience. Potential energy is not a property of a body but is associated with the interaction of two or more bodies.

When only conservative forces act, the total **mechanical energy**, E, defined as the sum of kinetic and potential energies, is conserved:

$$E = K + U = \text{constant}.$$

If nonconservative forces also act, additional types of energy are involved, such as thermal energy. It has been found experimentally that, when all forms of energy are included, the total energy is conserved. This is the **law of conservation of energy**:

$$\Delta K + \Delta U = W_{NC}.$$

The gravitational force as described by Newton's law of universal gravitation is a conservative force. The potential energy of an object of mass m due to the gravitational force exerted on it by the Earth is given by

$$U(r) = -GmM_E/r,$$

where M_E is the mass of the Earth and r is the distance of the object from the Earth's center ($r \geq$ radius of Earth).

Power is defined as the rate at which work is done or the rate at which energy is transformed from one form to another: $P = dW/dt = dE/dt$, or $P = \mathbf{F} \cdot \mathbf{v}$.

Questions

1. List some everyday forces that are not conservative, and explain why they aren't.

2. You lift a heavy book from a table to a high shelf. List the forces on the book during this process, and state whether each is conservative or nonconservative.

3. The net force acting on a particle is conservative and increases the kinetic energy by 300 J. What is the change in (a) the potential energy, and (b) the total energy, of the particle?

4. When a "superball" is dropped, can it rebound to a greater height than its original height?

5. A hill has a height h. A child on a sled (total mass m) slides down starting from rest at the top. Does the velocity at the bottom depend on the angle of the hill if (a) it is icy and there is no friction, and (b) there is friction (deep snow)?

6. Why is it tiring to push hard against a solid wall even though no work is done?

7. Analyze the motion of a simple swinging pendulum in terms of energy, (a) ignoring friction; and (b) taking friction into account. Explain why a grandfather clock has to be wound up.

8. Describe precisely what is "wrong" physically in the famous Escher drawing shown in Fig. 8–23.

FIGURE 8–23   Question 8.

FIGURE 8–24
Question 9.

9. In Fig. 8–24, water balloons are tossed from the roof of a building, all with the same speed but with different launch angles. Which one has the highest speed on impact? Ignore air resistance.

10. Suppose you lift a suitcase from the floor to a table. Does the work you do on the suitcase depend on (a) whether you lift it straight up or along a more complicated path, (b) the time the lifting takes, (c) the height of the table, and/or (d) the weight of the suitcase?

11. A coil spring of mass m rests upright on a table. If you compress the spring by pressing down with your hand and then release it, can the spring actually leave the table? Explain using the law of conservation of energy.

12. What happens to the gravitational potential energy when water at the top of a waterfall falls to the pool below?

13. By approximately how much does your gravitational potential energy change when you jump as high as you can?

14. Experienced hikers prefer to step over a fallen log in their path rather than stepping on top and jumping down on the other side. Explain.

15. Consider two observers who are in different inertial reference frames that move with velocity v with respect to each other. They both observe an object being pulled on a rough horizontal surface. Do they agree as to the value of (a) the object's kinetic energy; (b) the total work done on the object; (c) the amount of energy transformed from mechanical to thermal energy due to friction? Does your answer to (c) contradict (a) and (b)? Explain.

16. (a) Where does the kinetic energy come from when a car accelerates uniformly starting from rest? (b) How is the increase in kinetic energy related to the friction force the road exerts on the tires?

17. The Earth is closest to the Sun in winter (Northern Hemisphere). When is the gravitational potential energy the greatest?

18. Can the total mechanical energy $E = K + U$ ever be negative? Explain.

19. Suppose that you wish to launch a rocket from the surface of the Earth so that it escapes the Earth's gravitational field. You wish to use minimum fuel in doing this. From what point on the surface of the Earth should you make the launch and in what direction? Do the launch location and direction matter? Explain.

20. Recall from Chapter 4, Example 4–14, that you can use a pulley and ropes to decrease the force needed to raise a heavy load (see Fig. 8–25). But for every meter the load is raised, how much rope must be pulled up? Account for this, using energy concepts.

FIGURE 8–25
Question 20.

21. Two identical arrows, one with twice the speed of the other, are fired into a bale of hay. Assuming the hay exerts a constant "frictional" force on the arrows, the faster arrow will penetrate how much farther than the slower arrow? Explain.

22. Why is it easier to climb a mountain via a zigzag trail rather than to climb straight up?

* 23. Give several examples of stable, unstable, and neutral equilibrium.

* 24. In what state of equilibrium is a cube (a) when resting on its face; (b) when on its edge?

* 25. (a) Describe in detail the velocity changes of a particle that has energy E_3 in Fig. 8–22 as it moves from x_6 to x_5 and back to x_6. (b) Where is its kinetic energy the greatest and the least?

*26. Name the type of equilibrium for each position of the balls in Fig. 8–26.

FIGURE 8–26   Question 26.

*27. Figure 8–27 shows a potential energy curve, $U(x)$. (a) At which point does the force have greatest magnitude? (b) For each labeled point, state whether the force acts to the left or to the right, or is zero. (c) Where is there equilibrium and of what type is it?

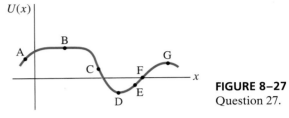

FIGURE 8–27
Question 27.

Problems

Sections 8–1 and 8–2

1. (I) A spring has a spring constant k of 82.0 N/m. How far must this spring be compressed to store 35.0 J of potential energy?

2. (I) A 5.0-kg monkey swings from one branch to another 1.5 m higher. What is the change in gravitational potential energy?

3. (I) By how much does the gravitational potential energy of a 58–kg pole vaulter change if her center of mass rises about 3.8 m during the jump?

4. (I) A 66.5-kg hiker starts at an elevation of 1500 m and climbs to the top of a 2660-m peak. (a) What is the hiker's change in potential energy? (b) What is the minimum work required of the hiker? (c) Can the actual work done be greater than this? Explain.

5. (I) In starting an exercise, a 1.70-m tall person lifts a 2.20-kg book off the ground so it is 2.40 m above the ground. What is the potential energy of the book relative to (a) the ground, and (b) the top of the person's head? (c) How is the work done by the person related to the answers in parts (a) and (b)?

6. (II) If $U = 3x^2 + 2xy + 4y^2 z$, what is the force, $\mathbf{F}$?

7. (II) A particular spring obeys the force law $\mathbf{F} = (-kx + ax^3 + bx^4)\mathbf{i}$. (a) Is this force conservative? Explain why or why not. (b) If it is conservative, determine the form of the potential energy function.

8. (II) Air resistance can be represented by a force proportional to the velocity $\mathbf{v}$ of an object: $\mathbf{F} = -k\mathbf{v}$. Is this force conservative? Explain.

9. (II) (a) A spring of spring constant k is initially compressed a distance x_0 from its unstretched length. What is the change in potential energy if it is then compressed to an amount x from its unstretched length? (b) Suppose the spring is then *stretched* a distance x_0 from the unstretched length. What is the change in potential energy as compared to when it is compressed by an amount x_0?

Sections 8–3 and 8–4

10. (I) Jane, looking for Tarzan, is running at top speed (5.0 m/s) and grabs a vine hanging 4.0 m vertically from a tall tree in the jungle. How high can she swing upward? Does the length of the vine (or rope) affect your answer?

11. (I) A novice skier, starting from rest, slides down a frictionless 32.0° incline whose vertical height is 105 m. How fast is she going when she reaches the bottom?

12. (I) A sled is initially given a shove up a frictionless 25.0° incline. It reaches a maximum vertical height 1.22 m higher than where it started. What was its initial speed?

13. (II) In the high jump, the kinetic energy of an athlete is transformed into gravitational potential energy without the aid of a pole. With what minimum speed must the athlete leave the ground in order to lift his center of mass 2.10 m and cross the bar with a speed of 0.70 m/s?

14. (II) A 75-kg trampoline artist jumps vertically upward from the top of a platform with a speed of 5.0 m/s. (a) How fast is he going as he lands on the trampoline, 2.0 m below (Fig. 8–28)? (b) If the trampoline behaves like a spring of spring constant 5.2×10^4 N/m, how far does he depress it?

2.0 m

FIGURE 8–28
Problem 14.

15. (II) A 60-kg bungee jumper jumps from a bridge. She is tied to a bungee cord that is 12 m long when unstretched, and falls a total of 31 m. (a) Calculate the spring constant k of the bungee cord assuming Hooke's law applies. (b) Calculate the maximum acceleration experienced by the jumper.

16. (II) A roller coaster, shown in Fig. 8–29, is pulled up to point A where it and its screaming occupants are released from rest. Assuming no friction, calculate the speed at points B, C, D.

30 m   12 m   25 m

FIGURE 8–29
Problems 16 and 30.

17. (II) A vertical spring (ignore its mass), whose spring constant is 900 N/m, is attached to a table and is compressed down by 0.150 m. (a) What upward speed can it give to a 0.300-kg ball when released? (b) How high above its original position (spring compressed) will the ball fly?

18. (II) A 0.40-kg ball is thrown with a speed of 10 m/s at an angle of 30°. (a) What is its speed at its highest point, and (b) how high does it go? (Use conservation of energy.)

19. (II) A mass m is attached to the end of a spring (constant k) as shown in Fig. 8–30. The mass is given an initial displacement x_0 from equilibrium, and an initial speed v_0. Ignoring friction and the mass of the spring, use energy methods to find (a) its maximum speed, and (b) its maximum stretch from equilibrium, in terms of the given quantities.

FIGURE 8–30 Problems 19, 33 and 34.

20. (II) A cyclist intends to cycle up a 9.50° hill whose vertical height is 92.0 m. The pedals turn in a circle of diameter 36.0 cm. Assuming the mass of bicycle plus person is 75.0 kg, (a) calculate how much work must be done against gravity. (b) If each complete revolution of the pedals moves the bike 5.10 m along its path, calculate the average force that must be exerted on the pedals tangent to their circular path. Neglect work done by friction and other losses.

21. (II) A pendulum 2.00 m long is released (from rest) at an angle $\theta_0 = 30.0°$ (Fig. 8–15). Determine the speed of the 70.0-g bob (a) at the lowest point ($\theta = 0$); (b) at $\theta = 15.0°$; (c) at $\theta = -15.0°$ (i.e., on the opposite side). (d) Determine the tension in the cord at each of these three points. (e) If the bob is given an initial speed $v_0 = 1.20$ m/s when released at $\theta = 30.0°$, recalculate the speeds for parts (a), (b), and (c).

22. (II) What should be the spring constant k of a spring designed to bring a 1200-kg car to rest from a speed of 100 km/h so that the occupants undergo a maximum acceleration of 5.0 g?

23. (III) An engineer is designing a spring to be placed at the bottom of an elevator shaft. If the elevator cable should happen to break when the elevator is at a height h above the top of the spring, calculate the value that the spring constant k should have so that passengers undergo an acceleration of no more than 5.0 g when brought to rest. Let M be the total mass of the elevator and passengers.

24. (III) A skier of mass m starts from rest at the top of a solid sphere of radius r and slides down its frictionless surface. (a) At what angle θ (Fig. 8–31) will the skier leave the sphere? (b) If friction were present, would the skier fly off at a greater or lesser angle?

FIGURE 8–31
Problem 24.

Sections 8–5 and 8–6

25. (I) Two railroad cars, each of mass 6500 kg and traveling 95 km/h, collide head-on and come to rest. How much thermal energy is produced in this collision?

26. (I) A 16.0-kg child descends a slide 2.50 m high and reaches the bottom with a speed of 2.25 m/s. How much thermal energy due to friction was generated in this process?

27. (II) A ski starts from rest and slides down a 20° incline 100 m long. (a) If the coefficient of friction is 0.090, what is the ski's speed at the base of the incline? (b) If the snow is level at the foot of the incline and has the same coefficient of friction, how far will the ski travel along the level? Use energy methods.

28. (II) A 145-g baseball is dropped from a tree 12.0 m above the ground. (a) With what speed would it hit the ground if air resistance could be ignored? (b) If it actually hits the ground with a speed of 8.00 m/s, what is the average force of air resistance exerted on it?

29. (II) A 90-kg crate, starting from rest, is pulled across a floor with a constant horizontal force of 350 N. For the first 15 m the floor is frictionless, and for the next 15 m the coefficient of friction is 0.25. What is the final speed of the crate?

30. (II) Suppose the roller coaster in Fig. 8–29 passes point A with a speed of 1.70 m/s. If the average force of friction is equal to one fifth of its weight, with what speed will it reach point B? The distance traveled is 45.0 m.

31. (II) A skier traveling 11 m/s reaches the foot of a steady upward 17° incline and glides 12 m up along this slope before coming to rest. What was the average coefficient of friction?

32. (II) Consider the track shown in Figure 8–32. The section AB is one quadrant of a circle of radius 2.0 m and is frictionless. B to C is a horizontal span 3.0 m long with a coefficient of kinetic friction $\mu_k = 0.25$. The section CD under the spring is frictionless. A block of mass 1.0 kg is released from rest at A. After sliding on the track, it is observed to compress the spring by 0.20 m. Determine: (a) the velocity of the block at point B; (b) the work done by friction as the block slides from B to C; (c) the velocity of the block at point C; (d) the stiffness constant k for the spring.

FIGURE 8–32
Problem 32.

33. (II) A 0.620-kg wood block is firmly attached to a very light horizontal spring ($k = 180$ N/m) as shown in Fig. 8–30. It is noted that the block-spring system, when compressed 5.0 cm and released, stretches out 2.3 cm beyond the equilibrium position before stopping and turning back. What is the coefficient of kinetic friction between the block and the table?

34. (II) A 180-g wood block is firmly attached to a very light horizontal spring, Fig. 8–30. The block can slide along a table where the coefficient of friction is 0.30. A force of 22 N compresses the spring 18 cm. If the spring is released from this position, how far beyond its equilibrium position will it stretch on its first swing?

35. (III) A 2.0-kg block slides along a horizontal surface with a coefficient of kinetic friction $\mu_k = 0.30$. The block has a speed $v = 1.3$ m/s when it strikes a massless spring head-on (as in Fig. 8–18). (a) If the spring has force constant $k = 120$ N/m, how far is the spring compressed? (b) What minimum value of the coefficient of static friction, μ_s, will assure that the spring remains compressed at the maximum compressed position? (c) If μ_s is less than this, what is the speed of the block when it detaches from the decompressing spring? [Hint: Detachment occurs when the spring reaches its natural length $(x = 0)$; explain why.]

36. (III) In early test flights for the space shuttle using a "glider" (mass of 1000 kg including pilot), it was noted that after a horizontal launch at 500 km/h at a height of 3500 m, the glider eventually landed at a speed of 200 km/h. (a) What would its landing speed have been in the absence of air resistance? (b) What was the average force of air resistance exerted on it if it came in at a constant glide angle of 10° to the Earth?

Section 8–7

37. (I) For a satellite of mass m_S in a circular orbit of radius r_S, determine (a) its kinetic energy K, (b) its potential energy $U(U = 0$ at infinity), and (c) the ratio K/U.

38. (I) Jill and her friends have built a small rocket that soon after lift-off reaches a speed of 850 m/s. How high above the Earth can it rise? Ignore air friction.

39. (II) Determine the escape velocity from the Sun for an object (a) at the Sun's surface $(r = 7.0 \times 10^5$ km, $M = 2.0 \times 10^{30}$ kg), and (b) at the average distance of the Earth $(1.50 \times 10^8$ km). Compare to the speed of the Earth in its orbit.

40. (II) Show that Eq. 8–17 for gravitational potential energy reduces to Eq. 8–2, $\Delta U = mg(y_2 - y_1)$, for objects near the surface of the Earth.

41. (II) Show that the change in potential energy of an object between the Earth's surface and a height h above is

$$\Delta U = \frac{mgh}{1 + h/r_E},$$

where r_E is the radius of the Earth and h is not necessarily small.

42. (II) (a) Show that the total mechanical energy of a satellite (mass m) orbiting at a distance r from the center of the Earth (mass M_E) is

$$E = -\frac{1}{2}\frac{GmM_E}{r},$$

if $U = 0$ at $r = \infty$. (b) Show that although friction causes the value of E to decrease slowly, kinetic energy must actually increase if the orbit remains a circle.

43. (II) Show that the escape velocity for any satellite in a circular orbit is $\sqrt{2}$ times its velocity.

44. (II) The Earth's distance from the sun varies from 1.471×10^8 km to 1.521×10^8 km during the year. Determine the difference in (a) the potential energy, (b) the Earth's kinetic energy, and (c) the total energy between these extreme points. Take the Sun to be at rest.

45. (II) Take into account the Earth's rotational speed (1 rev/day) and determine the necessary speed, with respect to Earth, for a rocket to escape if fired from the Earth at the equator in a direction (a) eastward; (b) westward; (c) vertically upward.

46. (II) (a) Determine a formula for the maximum height h that a rocket will reach if launched vertically from the Earth's surface with speed $v_0(< v_{esc})$. Express in terms of v_0, r_E, M_E, and G. (b) How high does a rocket go if $v_0 = 8.2$ km/s? Ignore air resistance and the Earth's rotation.

47. (II) (a) Determine the rate at which the escape velocity from the Earth changes with height above the Earth's surface, dv_{esc}/dr. (b) Use the approximation $\Delta v \approx (dv/dr)\Delta r$ to determine the escape velocity for a spacecraft orbiting the Earth at a height of 300 km.

48. (II) A meteor has a speed of 90.0 m/s when 800 km above the Earth. It is falling vertically (ignore air resistance) and strikes a bed of sand in which it is brought to rest in 3.25 m. (a) What is its speed just before striking the sand? (b) How much work does the sand do to stop the meteor (mass = 575 kg)? (c) What is the average force exerted by the sand on the meteor? (d) How much thermal energy is produced?

49. (II) How much work would be required to move a satellite of mass m from a circular orbit of radius $r_1 = 2r_E$ about the Earth to another circular orbit of radius $r_2 = 3r_E$? (r_E is the radius of the Earth.)

50. (II) (a) Suppose we have three masses, m_1, m_2, and m_3, that initially are infinitely far apart from each other. Show that the work needed to bring them to the positions shown in Fig. 8–33 is

$$W = -G\left(\frac{m_1 m_2}{r_{12}} + \frac{m_1 m_3}{r_{13}} + \frac{m_2 m_3}{r_{23}}\right).$$

(b) Can we say that this formula also gives the potential energy of the system, or the potential energy of one or two of the bodies? (c) Is W equal to the binding energy of the system—that is, equal to the energy required to separate the components by an infinite distance? Explain.

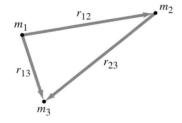

FIGURE 8–33 Problem 50.

51. (II) A NASA satellite has just observed an asteroid that is on a collision course with the Earth. The asteroid has an estimated mass, based on its size, of 5×10^9 kg. It is approaching the Earth on a head-on course with a velocity of 600 m/s relative to the Earth and is now 5.0×10^6 km away. With what speed will it hit the Earth's surface, neglecting friction with the atmosphere?

52. (II) A sphere of radius r_1 has a concentric spherical cavity of radius r_2 (Fig. 8–34). Assume this spherical shell of thickness $r_1 - r_2$ is uniform and has a total mass M. Show that the gravitational potential energy of a mass m at a distance r from the center of the shell $(r > r_1)$ is given by

$$U = -\frac{GmM}{r}.$$

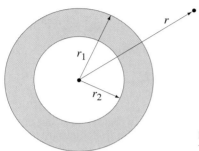

FIGURE 8–34
Problem 52.

53. (III) To escape the solar system, an interstellar spacecraft must overcome the gravitational attraction of both the Earth and Sun. Ignore the effects of other bodies in the solar system. (a) Show that the escape velocity is

$$v = \sqrt{v_E^2 + (v_S - v_0)^2} = 16.7 \text{ km/s},$$

where: v_E is the escape velocity from the Earth (Eq. 8–20); $v_S = \sqrt{2GM_S/r_{SE}}$ is the escape velocity from the gravitational field of the Sun at the orbit of the Earth but far from the Earth's influence (r_{SE} is Sun–Earth distance); and v_0 is the Earth's orbital velocity about the Sun. (b) Show that the energy required is 1.40×10^8 J per kilogram of spacecraft mass. [Hint: Write the energy equation for escape from Earth with v' as the velocity, relative to Earth, but far from Earth; then let $v' + v_0$ be the escape velocity from the Sun.]

Section 8–8

54. (I) How long will it take a 1750-W motor to lift a 285-kg piano to a sixth-story window 16.0 m above?

55. (I) If a car generates 18 hp when traveling at a steady 90 km/h, what must be the average retarding force exerted on the car?

56. (I) (a) Show that a British horsepower (550 ft·lb/s) is equal to 746 W. (b) What is the horsepower rating of a 100-W lightbulb?

57. (I) An 80-kg football player traveling 5.0 m/s is stopped in 1.0 s by a tackler. (a) What is the original kinetic energy of the player? (b) What average power is required to stop him?

58. (II) What minimum horsepower must a motor have to be able to drag a 300-kg box along a level floor at a speed of 1.20 m/s if the coefficient of friction is 0.45?

59. (II) A driver notices that her 1000-kg car slows down from 90 km/h to 70 km/h in about 6.0 s on the level when it is in neutral. Approximately what power (watts and hp) is needed to keep the car traveling at a constant 80 km/h?

60. (II) How much work can a 3.0-hp motor do in 1.0 h?

61. (II) A shot-putter accelerates a 7.3-kg shot from rest to 14 m/s. If this motion takes 1.5 s, what average power was developed?

62. (II) A pump is to lift 18.0 kg of water per minute through a height of 3.50 m. What output rating (watts) should the pump motor have?

63. (II) During a workout, the football players at State U ran up the stadium stairs in 61 s. The stairs are 140 m long and inclined at an angle of 30°. If a typical player has a mass of 105 kg, estimate the average power output on the way up. Ignore friction and air resistance.

64. (II) A 1000-kg car has a maximum power output of 120 hp. How steep a hill can it climb at a constant speed of 70 km/h if the frictional forces add up to 600 N?

65. (II) Squaw Valley ski area in California claims that its lifts can move 47,000 people per hour. If the average lift carries people about 200 m (vertically) higher, estimate the maximum total power needed.

66. (III) A bicyclist coasts down a 7.0° hill at a steady speed of 5.0 m/s. Assuming a total mass of 75 kg (bicycle plus rider), what must be the cyclist's power output to climb the same hill at the same speed?

67. (III) The position of a 280-g object is given (in meters) by $x = 5.0t^3 - 8.0t^2 - 30t$, where t is in seconds. Determine the net rate of work done on this object (a) at $t = 2.0$ s and (b) at $t = 4.0$ s. (c) What is the average net power input during the interval from $t = 0$ s to $t = 2.0$ s, and in the interval from $t = 2.0$ s to 4.0 s?

* Section 8–9

*68. (II) Draw a potential energy diagram and analyze the motion of a mass m resting on a frictionless horizontal table and connected to a horizontal spring with stiffness constant k. The mass is pulled a distance to the right so that the spring is stretched a distance x_0 initially, and then the mass is released from rest.

*69. (II) The spring of Problem 68 has a stiffness constant $k = 160$ N/m. The mass $m = 5.0$ kg is released from rest when the spring is stretched $x_0 = 1.0$ m from equilibrium. Determine (a) the total energy of the system; (b) the kinetic energy when $x = \frac{1}{2}x_0$; (c) the maximum kinetic energy; (d) the maximum speed and at what positions it occurs; (e) the maximum acceleration and where it occurs.

*70. (III) The potential energy of the two atoms in a diatomic (two-atom) molecule can be written

$$U(r) = -\frac{a}{r^6} + \frac{b}{r^{12}},$$

where r is the distance between the two atoms and a and b are positive constants. (a) At what values of r is $U(r)$ a minimum? A maximum? (b) At what values of r is $U(r) = 0$? (c) Plot $U(r)$ as a function of r from $r = 0$ to r at a value large enough for all the features in (a) and (b) to show. (d) Describe the motion of one atom with respect to the second atom when $E < 0$, and for $E > 0$. (e) Let F be the force one atom exerts on the other. For what values of r is $F > 0$, $F < 0$, $F = 0$? (f) Determine F as a function of r.

*71. (III) The binding energy of a two-particle system is defined as the energy required to separate the two particles from their state of lowest energy to $r = \infty$. Determine the binding energy for the molecule discussed in Problem 70.

72. A projectile is fired at an upward angle of 45.0° from the top of a 165-m cliff with a speed of 180 m/s. What will be its speed when it strikes the ground below? (Use conservation of energy.)

73. In a film of Jesse Owens's famous long jump in the 1936 Olympics, it is observed that his center of mass rose 1.1 m from launch point to the top of the arc. What minimum speed did he need at launch if he was also noted to be traveling at 6.5 m/s at the top of the arc?

74. How fast must a cyclist climb a 12° hill to maintain a power output of 0.20 hp? Ignore friction and assume the mass of cyclist plus bicycle is 85 kg.

75. What is the average power output of an elevator that lifts 850 kg a vertical height of 32.0 m in 11.0 s?

76. A 0.20-kg pinecone falls from a branch 18 m above the ground. (a) With what speed would it hit the ground if air resistance could be ignored? (b) If it actually hits the ground with a speed of 10.0 m/s, what was the average force of air resistance exerted on it?

77. A 60-kg skier starts from rest at the top of a ski jump, point A in Fig. 8–35, and travels down the ramp. If friction and air resistance can be neglected, (a) determine her speed v_B when she reaches the horizontal end of the ramp at B. (b) Determine the distance s to where she strikes the ground at C.

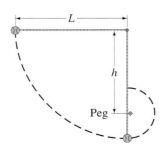

FIGURE 8–35 Problems 77 and 78.

78. Repeat Problem 77, but now assume the skier jumps upward upon reaching point B and acquires a vertical component of velocity (at B) of 3.0 m/s.

79. A ball is attached to a horizontal cord of length L whose other end is fixed, Fig. 8–36. (a) If the ball is released, what will be its speed at the lowest point of its path? (b) A peg is located a distance h directly below the point of attachment of the cord. If $h = 0.80 L$, what will be the speed of the ball when it reaches the top of its circular path about the peg?

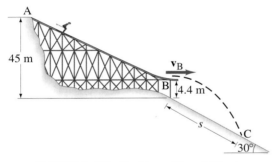

FIGURE 8–36
Problems 79 and 80.

80. Show that only if $h \geq 0.60 L$ can the ball in Fig. 8–36 make a complete circle about the peg.

81. A 65-kg hiker climbs to the top of a 3900-m-high mountain. The climb is made in 5.0 h starting at an elevation of 2200 m. Calculate (a) the work done by the hiker against gravity, (b) the average power output in watts and in horsepower, and (c) assuming the body is 15% efficient, what rate of energy input was required.

82. The small mass m sliding without friction along the looped track shown in Fig. 8–37 is to remain on the track at all times, even at the very top of the loop of radius r. (a) Calculate, in terms of the given quantities, the minimum release height h. Next, if the actual release height is $2h$, calculate (b) the normal force exerted by the track at the bottom of the loop, (c) the normal force exerted by the track at the top of the loop, and (d) the normal force exerted by the track after the block exits the loop onto the flat section.

FIGURE 8–37 Problem 82.

83. Water flows over a dam at the rate of 550 kg/s and falls vertically 80 m before striking the turbine blades. Calculate (a) the speed of the water just before striking the turbine blades (neglect air resistance), and (b) the rate at which mechanical energy is transferred to the turbine blades, assuming 60% efficiency.

84. A bicyclist of mass 75 kg (including the bicycle) can coast down a 4.0° hill at a steady speed of 10 km/h. Pumping hard, the cyclist can descend the hill at a speed of 30 km/h. Using the same power, at what speed can the cyclist climb the same hill? Assume the force of friction is proportional to the square of the speed v; that is, $F_{fr} = bv^2$, where b is a constant.

85. Show that on a roller coaster with a circular vertical loop (Fig. 8–38), the difference in your apparent weight at the top of the loop and the bottom of the loop is 6 g's—that is, six times your weight. Ignore friction. Show also that as long as your speed is above the minimum needed, this answer doesn't depend on the size of the loop or how fast you go through it.

FIGURE 8–38
Problem 85.

86. If you stand on a bathroom scale, the spring inside the scale compresses 0.50 mm, and it tells you your weight is 700 N. Now if you jump on the scale from a height of 1.0 m, what does the scale read at its peak?

87. A 75-kg student runs at 5.0 m/s, grabs a hanging rope, and swings out over a lake (Fig. 8–39). He releases the rope when his velocity is zero. (*a*) What is the angle θ when he releases the rope? (*b*) What is the tension in the rope just before he releases it? (*c*) What is the maximum tension in the rope?

FIGURE 8–39
Problem 87.

88. In the rope climb, a 70-kg athlete climbs a vertical distance of 5.0 m in 9.0 s. What minimum power output was used to accomplish this feat?

89. The nuclear force between two neutrons in a nucleus is described roughly by the Yukawa potential

$$U(r) = -U_0 \frac{r_0}{r} e^{-r/r_0},$$

where r is the distance between the neutrons and U_0 and $r_0 (\approx 10^{-15}\ \text{m})$ are constants. (*a*) Determine the force $F(r)$. (*b*) What is the ratio $F(3r_0)/F(r_0)$? (*c*) Calculate this same ratio for the force between two electrically charged particles where $U(r) = -C/r$, with C a constant. Why is the Yukawa force referred to as a "short-range" force?

90. A 20-kg sled starts up a 30° incline with a speed of 2.4 m/s. The coefficient of kinetic friction is $\mu_k = 0.25$. (*a*) How far up the incline does the sled travel? (*b*) What condition must you put on the coefficient of static friction if the sled is not to get stuck at the point determined in part (*a*)? (*c*) If the sled slides back down, what is its speed when it returns to its starting point?

91. A fire hose for use in urban areas must be able to shoot a stream of water to a maximum height of 30 meters. The water leaves the hose at ground level in a circular stream 3.0 cm in diameter. What minimum power is required to create such a stream of water? Every cubic meter of water has a mass of 1000 kg.

92. Proper design of automobile braking systems must account for heat buildup under heavy braking. Calculate the thermal energy dissipated from brakes in a 1500-kg car that descends a 20° hill. The car begins braking when its speed is 90 km/h and slows to a speed of 30 km/h in a distance of 0.30 km measured along the road.

93. The Lunar Module could make a safe landing if its vertical velocity at impact is 3.0 m/s or less. Suppose that you want to determine the greatest height h at which the pilot could shut off the engine if the velocity of the lander relative to the surface is (*a*) zero; (*b*) 2.0 m/s downward; (*c*) 2.0 m/s upward. Use conservation of energy to determine h in each case. The acceleration due to gravity at the surface of the Moon is $1.62\ \text{m/s}^2$.

94. Some electric power companies use water to store energy. Water is pumped by reversible turbine pumps from a low to a high reservoir. If we wish to store the energy produced in 1.0 hour by a 100-MW (electrical) power plant, how many cubic meters of water will have to be pumped from the lower to the upper pool? Assume the upper pool is 500 m above the lower and we can neglect the small change in depths of each pool. Water has a mass of 1000 kg for every $1.0\ \text{m}^3$.

95. As a city engineer, you need to estimate the power required to pump water from a new well. The well is 400 m deep and the estimated demand is 1,000,000 kg per day. The pump motor is about 80% efficient in converting electrical energy to mechanical energy.

96. Estimate the energy required from fuel to launch a 12,000-kg satellite into orbit 1000 km above the Earth's surface. Consider two cases: (*a*) the satellite is launched into an equatorial orbit from a point on the Earth's equator, and (*b*) it is launched from the north pole into a polar orbit.

97. A satellite is in an elliptic orbit around the Earth (Fig. 8–40). Its velocity at the perigee A is 8650 m/s. (*a*) Use conservation of energy to determine its velocity at B. The radius of the Earth is 6380 km. (*b*) Use conservation of energy to determine the velocity at the apogee C.

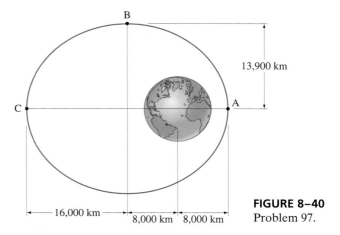

FIGURE 8–40
Problem 97.

*98. A particle moves where its potential energy is given by $U(r) = U_0[(2/r^2) - (1/r)]$. (*a*) Sketch the approximate shape of $U(r)$ versus r. Where does the curve cross the $U(r) = 0$ axis? At what value of r does the minimum value of $U(r)$ occur? (*b*) Suppose that the particle has an energy of $E = -0.050U_0$. Sketch in the approximate "turning points" of the motion of the particle on your diagram. What is the maximum kinetic energy of the particle, and for what value of r does this occur?

Conservation of linear momentum is another of the great conservation laws of physics. Collisions, as between billiard or pool balls, illustrate this vector law very nicely: the total vector momentum $(\Sigma m_i \mathbf{v}_i)$ before the collision equals the total vector momentum just after the collision. In this photo, the moving cue ball strikes the 11 ball at rest. Both balls move after the collision, at angles, but the sum of their vector momenta equals the initial momentum of the incoming cue ball. We will consider both elastic collisions (where kinetic energy is also conserved) and inelastic collisions. We also will examine the concept of center of mass, and how it can make the study of complex motion more readily analyzed and understood.

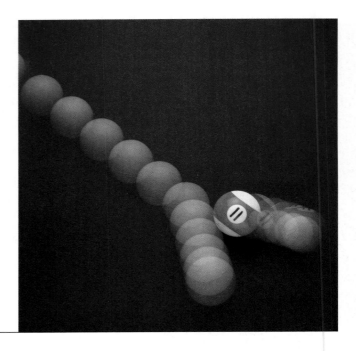

CHAPTER 9

Linear Momentum and Collisions

The law of conservation of energy, which we discussed in the previous chapter, is one of several great conservation laws in physics. Among the other quantities found to be conserved are linear momentum, angular momentum, and electric charge. We will eventually discuss all of these because the conservation laws are among the most important in all of science. In this chapter, we discuss linear momentum and its conservation. We then make use of the laws of conservation of linear momentum and of energy to analyze collisions. Indeed, the law of conservation of momentum is particularly useful when dealing with two or more bodies that interact with each other, as in collisions.

Our focus up to now has been mainly on the motion of a single object. In this chapter we will deal with systems of two or more objects. An important concept for this study is that of center of mass, which we discuss later in the chapter.

9–1 | Momentum and Its Relation to Force

The **linear momentum** (or "momentum" for short) of an object is defined as the product of its mass and its velocity. Momentum (plural is momenta) is usually represented by the symbol **p**. If we let m represent the mass of an object and $\mathbf{v}$ represent its velocity, then its momentum **p** is

Linear momentum

$$\mathbf{p} = m\mathbf{v}. \tag{9–1}$$

Since velocity is a vector, momentum is a vector. The direction of the momentum is the direction of the velocity, and the magnitude of the momentum is $p = mv$. Since $\mathbf{v}$ depends on the reference frame, this frame must be specified. The unit of momentum is simply that of mass $\times$ velocity, which in SI units is kg·m/s. There is no special name for this unit.

Everyday usage of the term *momentum* is in accord with the definition above. For according to Eq. 9–1, a fast-moving car has more momentum than a slow-moving car of the same mass, and a heavy truck has more momentum than a small car moving with the same speed. The more momentum an object has, the harder it is to stop it, and the greater effect it will have if it is brought to rest by impact or collision. A football player is more likely to be stunned if tackled by a heavy opponent running at top speed than by a lighter or slower-moving tackler. A heavy, fast-moving truck can do more damage than a slow-moving motorcycle.

A force is required to change the momentum of an object, whether it is to increase the momentum, to decrease it (such as to bring a moving object to rest), or to change its direction. Newton originally stated his second law in terms of momentum (although he called the product mv the "quantity of motion"). Newton's statement of the **second law of motion**, translated into modern language, is as follows:

> **The rate of change of momentum of an object is equal to the net force applied to it.**

We can write this as an equation,

$$\Sigma \mathbf{F} = \frac{d\mathbf{p}}{dt}, \qquad (9\text{--}2)$$

NEWTON'S SECOND LAW

where $\Sigma \mathbf{F}$ is the net force applied to the object (the vector sum of all forces acting on it). We can readily derive the familiar form of the second law, $\Sigma \mathbf{F} = m\mathbf{a}$, from Eq. 9–2 for the case of constant mass. If $\mathbf{v}_0$ is the initial velocity of an object and $\mathbf{v}$ is its velocity after a time dt has elapsed, then

$$\Sigma \mathbf{F} = \frac{d\mathbf{p}}{dt} = \frac{d(m\mathbf{v})}{dt} = m\frac{d\mathbf{v}}{dt} = m\mathbf{a}, \qquad \text{[constant mass]}$$

because, by definition, $\mathbf{a} = d\mathbf{v}/dt$. Newton's statement, Eq. 9–2, is actually more general than the more familiar one because it includes the situation in which the mass may change. This is important in certain circumstances, such as for rockets which lose mass as they burn fuel (Section 9–10) and in relativity theory (Chapter 37).

EXAMPLE 9–1 **Washing a car: momentum change and force.** Water leaves a hose at a rate of 1.5 kg/s with a speed of 20 m/s and is aimed at the side of a car, which stops it, Fig. 9–1. (That is, we ignore any splashing back.) What is the force exerted by the water on the car?

SOLUTION We take the x direction positive to the right. In each second, water with a momentum of $p_x = mv_x = (1.5 \text{ kg})(20 \text{ m/s}) = 30 \text{ kg} \cdot \text{m/s}$ is brought to rest at the moment it hits the car. The magnitude of the force (assumed constant) that the car must exert to change the momentum of the water by this amount is

$$F = \frac{\Delta p}{\Delta t} = \frac{p_{\text{final}} - p_{\text{initial}}}{\Delta t} = \frac{0 - 30 \text{ kg} \cdot \text{m/s}}{1.0 \text{ s}} = -30 \text{ N}.$$

The minus sign indicates that the force on the water is opposite to the water's original velocity. The car exerts a force of 30 N to the left to stop the water, so by Newton's third law, the water exerts a force of 30 N on the car.

FIGURE 9–1 Example 9–1.

CONCEPTUAL EXAMPLE 9–2 | **Water splashes back.** What if the water splashes back from the car in Example 9–1? Would the force on the car be greater or less?

RESPONSE If the water splashes back toward the hose, the change in momentum will be greater in magnitude, and so the force on the car will be greater in magnitude. Note that $\mathbf{p}_{\text{final}}$ will now point in the negative x direction, as shown in Fig. 9–2 (instead of being zero as in Example 9–1). So the result for F (see displayed equation in Example 9–1), will be minus something of magnitude more than 30 N (i.e., −35 to −40 N, depending on the water's rebound speed). To put it simply, the car exerts not only a force to stop the water, but also an additional force to give it momentum in the opposite direction.

FIGURE 9–2 Example 9–2. Momentum of water before and after splashing back, and $\Delta \mathbf{p}$.

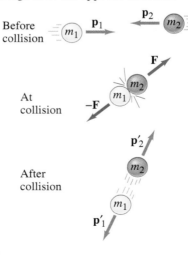

FIGURE 9–3 Momentum is conserved in a collision of two balls.

CONSERVATION OF
LINEAR MOMENTUM
(two bodies colliding)

FIGURE 9–4 Collision of two objects. (a) Their momenta before collision are $\mathbf{p}_1$ and $\mathbf{p}_2$, and after collision are $\mathbf{p}_1'$ and $\mathbf{p}_2'$. At any moment during the collision each exerts a force on the other of equal magnitude but opposite direction.

Before collision m_1 $\mathbf{p}_1$ $\mathbf{p}_2$ m_2

At collision $-\mathbf{F}$ m_1 m_2 $\mathbf{F}$

After collision m_2 $\mathbf{p}_2'$ m_1 $\mathbf{p}_1'$

9–2 Conservation of Momentum

The concept of momentum is particularly important because, under certain circumstances, momentum is a conserved quantity. In the mid-seventeenth century, shortly before Newton's time, it had been observed that the vector sum of the momenta of two colliding objects remains constant. Consider, for example, the head-on collision of two billiard balls, as shown in Fig. 9–3. We assume the net external force on this system of two balls is zero—that is, the only significant forces are those that each ball exerts on the other during the collision. Although the momentum of each of the two balls changes as a result of the collision, the *sum* of their momenta is found to be the same before as after the collision. If $m_1\mathbf{v}_1$ is the momentum of ball number 1 and $m_2\mathbf{v}_2$ the momentum of ball 2, both measured before the collision, then the total momentum of the two balls before the collision is $m_1\mathbf{v}_1 + m_2\mathbf{v}_2$. After the collision, the balls each have a different velocity and momentum, which we will designate by a "prime" on the velocity: $m_1\mathbf{v}_1'$ and $m_2\mathbf{v}_2'$. The total momentum after the collision is $m_1\mathbf{v}_1' + m_2\mathbf{v}_2'$. No matter what the velocities and masses involved are, it is found that the total momentum before the collision is the same as afterwards, whether the collision is head-on or not, as long as no net external force acts:

$$\text{momentum before} = \text{momentum after}$$

or

$$m_1\mathbf{v}_1 + m_2\mathbf{v}_2 = m_1\mathbf{v}_1' + m_2\mathbf{v}_2'$$
$$\mathbf{p}_1 + \mathbf{p}_2 = \mathbf{p}_1' + \mathbf{p}_2'. \tag{9–3}$$

That is, the total vector momentum of the system of two balls is conserved: it stays constant.

Although the law of conservation of momentum was discovered experimentally, it is closely connected to Newton's laws of motion and they can be shown to be equivalent, which we now do.

Let us consider two bodies of mass m_1 and m_2 that have momenta $\mathbf{p}_1$ and $\mathbf{p}_2$ before they collide and $\mathbf{p}_1'$ and $\mathbf{p}_2'$ after they collide, as in Fig. 9–4. During the collision, suppose the force exerted by body 1 on body 2 at any instant is $\mathbf{F}$. Then, by Newton's third law, the force exerted by body 2 on body 1 is $-\mathbf{F}$. During the brief collision time, we assume no other (external) forces are acting (or that $\mathbf{F}$ is much greater than any other external forces acting). We rewrite Eq. 9–2 as $d\mathbf{p} = \mathbf{F}\,dt$ and integrate both sides of this equation over the brief time interval of the collision, from t_i to t_f:

$$\int_{t_i}^{t_f} d\mathbf{p} = \int_{t_i}^{t_f} \mathbf{F}\,dt.$$

We apply this first to body 2, on which the force $\mathbf{F}$ acts (Fig. 9–4),

$$\Delta\mathbf{p}_2 = \mathbf{p}_2' - \mathbf{p}_2 = \int_{t_i}^{t_f} \mathbf{F}\,dt. \tag{9–4a}$$

On body 1, the force is $-\mathbf{F}$, so

$$\Delta\mathbf{p}_1 = \mathbf{p}_1' - \mathbf{p}_1 = -\int_{t_i}^{t_f} \mathbf{F}\,dt. \tag{9–4b}$$

We compare these two equations and see that

$$\Delta\mathbf{p}_1 = -\Delta\mathbf{p}_2$$
$$\mathbf{p}_1' - \mathbf{p}_1 = -(\mathbf{p}_2' - \mathbf{p}_2).$$

Rearranging, we have

$$\mathbf{p}_1 + \mathbf{p}_2 = \mathbf{p}_1' + \mathbf{p}_2',$$

which is Eq. 9–3, the law of conservation of momentum.

We have put this derivation in the context of a collision. As long as no external forces act, Eqs. 9–4 are valid over any time interval, and conservation of momentum is always valid as long as no external forces act. In the real world, external forces do act: friction on billiard balls, gravity acting on a baseball, and so on. So it may seem that conservation of momentum cannot be applied. Or can it? In a collision, the force each body exerts on the other acts only over a very brief time interval, and is very strong. During the brief time, $\mathbf{F}$ in Eqs. 9–4 is very much larger than the other forces acting (gravity, friction). If we measure the momenta just before, and just after, the collision, momentum will be very nearly conserved. We cannot wait for the external forces to produce their effect before measuring $\mathbf{p}'_1$ and $\mathbf{p}'_2$.

For example, when a racket hits a tennis ball or a bat hits a baseball, both before and after the "collision" the ball moves as a projectile under the action of gravity and air resistance. However, when the bat or racket hits the ball, during this brief time of the collision, these external forces are insignificant compared to the collision force the bat exerts on the ball (and vice versa). From Eqs. 9–4 we see then that momentum is conserved (or very nearly so) as long as we measure $\mathbf{p}_1$ and $\mathbf{p}_2$ just before the collision and $\mathbf{p}'_1$ and $\mathbf{p}'_2$ immediately after the collision.

Our derivation of the conservation of momentum via Eqs. 9–4 can be extended to include any number of interacting bodies. Let $\mathbf{P}$ represent the total momentum of a system of n interacting objects:

$$\mathbf{P} = m_1 \mathbf{v}_1 + m_2 \mathbf{v}_2 + \cdots + m_n \mathbf{v}_n = \Sigma \mathbf{p}_i \,.$$

We differentiate with respect to time:

$$\frac{d\mathbf{P}}{dt} = \Sigma \frac{d\mathbf{p}_i}{dt} = \Sigma \mathbf{F}_i \tag{9–5}$$

where $\mathbf{F}_i$ represents the *net* force on the ith object. The forces can be of two types: (1) *external forces* on objects of the system, exerted by objects outside the system, and (2) *internal forces* that objects within the system exert on other objects in the system. By Newton's third law, the internal forces occur in pairs: if one object exerts a force on a second object, the second exerts an equal and opposite force on the first object. Thus, in the sum over all the forces in Eq. 9–5, all the internal forces cancel each other in pairs. Thus we have

$$\frac{d\mathbf{P}}{dt} = \Sigma \mathbf{F}_{\text{ext}} \,, \tag{9–6}$$

NEWTON'S SECOND LAW
(for a system of objects)

where $\Sigma \mathbf{F}_{\text{ext}}$ is the sum of all external forces acting on our system. If the net external force is zero, then $d\mathbf{P}/dt = 0$, so $\Delta \mathbf{P} = 0$ or $\mathbf{P} = $ constant. Thus we have

when the net external force on a system is zero, the total momentum remains constant.

LAW OF

CONSERVATION

OF LINEAR

MOMENTUM

This is the **law of conservation of momentum**. It can also be stated as

the total momentum of an isolated system of bodies remains constant.

By an **isolated system**, we mean one on which no external forces act—the only forces acting are those between objects of the system.

If a net external force acts on a system, then the law of conservation of momentum will not apply. However, if the "system" can be redefined so as to include the other objects exerting these forces, then the conservation of momentum principle can apply. For example, if we take as our system a falling rock, it does not conserve momentum since an external force, the force of gravity exerted by the Earth, is acting on it and its momentum changes. However, if we include the Earth in the system, the total momentum of rock plus Earth is conserved. (This of course means that the Earth comes up to meet the ball. Since the Earth's mass is so great, its upward velocity is very tiny.)

As with energy, the importance of the momentum concept is that momentum is conserved under quite general conditions. Although the law of conservation of momentum follows from Newton's second law, as we have seen, it is in fact more general than Newton's laws. In the tiny world of the atom, Newton's laws fail, but the great conservation laws—those of energy, momentum, angular momentum, and electric charge—have been found to hold in every experimental situation tested. It is for this reason that the conservation laws are considered more basic than Newton's laws.

EXAMPLE 9–3 **Railroad cars collide: momentum conserved.** A 10,000-kg railroad car traveling at a speed of 24.0 m/s strikes an identical car at rest. If the cars lock together as a result of the collision, what is their common speed afterward? See Fig. 9–5.

SOLUTION The initial total momentum is simply $m_1 v_1$ (since $v_2 = 0$), to the right in the $+x$ direction. After the collision, the total momentum will be the same, and it will be shared by both cars. Since the two cars become attached, they will have the same speed, call it v'. Then:

$$(m_1 + m_2)v' = m_1 v_1$$

$$v' = \frac{m_1}{m_1 + m_2} v_1 = \tfrac{1}{2} v_1$$

$$= 12.0 \text{ m/s}^2$$

to the right. Their mutual speed after collision is half the initial speed of car 1.

FIGURE 9–5 Example 9–3.

(a) Before collision

(b) After collision

FIGURE 9–6 (a) A rocket, containing fuel, at rest in some reference frame. (b) In the same reference frame, the rocket fires and gases are expelled at high speed out the rear. The total vector momentum, $\mathbf{p}_{gas} + \mathbf{p}_{rocket}$, remains zero.

The law of conservation of momentum is particularly useful when we are dealing with fairly simple systems such as collisions and certain types of explosions. For example, *rocket propulsion*, which we saw in Chapter 4 can be understood on the basis of action and reaction, can also be explained on the basis of the conservation of momentum. Before a rocket is fired, the total momentum of rocket plus fuel is zero. As the fuel burns, the total momentum remains unchanged: the backward momentum of the expelled gases is just balanced by the forward momentum gained by the rocket itself (see Fig. 9–6). Thus, a rocket can accelerate in empty space. There is no need for the expelled gases to push against the Earth or the air (as is sometimes erroneously thought), as we already discussed in Chapter 4. Similar examples are the recoil of a gun and the throwing of a package from a boat.

EXAMPLE 9–4 **Rifle recoil.** Calculate the recoil velocity of a 5.0-kg rifle that shoots a 0.050-kg bullet at a speed of 120 m/s, Fig. 9–7.

SOLUTION The total momentum of the system is conserved. We let the subscripts B represent the bullet and R the rifle; the final velocities are indicated by primes. Then conservation of momentum in the x direction gives

$$m_B v_B + m_R v_R = m_B v_B' + m_R v_R'$$
$$0 + 0 = (0.050\,\text{kg})(120\,\text{m/s}) + (5.0\,\text{kg})(v_R')$$
$$v_R' = -1.2\,\text{m/s}.$$

Since the rifle has a much larger mass, its (recoil) velocity is much less than that of the bullet. The minus sign indicates that the velocity (and momentum) of the rifle is in the negative x direction, opposite to that of the bullet. Notice that it is the *vector sum* of the momenta that is conserved.

(a) Before shooting

(b) After shooting

FIGURE 9–7 Example 9–4.

EXAMPLE 9–5 **Billiard ball collision in 2-D.** A billiard ball moving with speed $v_1 = 3.0$ m/s in the $+x$ direction (Fig. 9–8) strikes an equal-mass ball initially at rest. The two balls are observed to move off at 45°, ball 1 above the x axis and ball 2 below. That is, $\theta_1' = 45°$ and $\theta_2' = -45°$ in Fig. 9–8. What are the speeds of the two balls after the collision?

SOLUTION We set up an xy coordinate system as shown in Fig. 9–8. From symmetry, we might guess that the two balls have the same speed. But let us not assume that now. Instead, let us apply conservation of momentum. We are given $m_1 = m_2(= m)$ so

$$\mathbf{P}_{\text{initial}} = \mathbf{P}_{\text{final}}$$
$$m\mathbf{v}_1 = m\mathbf{v}_1' + m\mathbf{v}_2'.$$

Vector momentum is conserved, which means that each component is conserved. We can write the x and y components of this vector equation as

$$mv_1 = mv_1' \cos(45°) + mv_2' \cos(-45°)$$

and

$$0 = mv_1' \sin(45°) + mv_2' \sin(-45°).$$

The m's cancel out in both equations. The second equation yields [recall $\sin(-\theta) = -\sin\theta$]:

$$v_2' = -v_1' \frac{\sin(45°)}{\sin(-45°)} = -v_1' \left(\frac{\sin 45°}{-\sin 45°} \right) = v_1',$$

so they do have equal speeds as we guessed at first. The x component equation gives [recall $\cos(-\theta) = \cos\theta$]:

$$v_1 = v_1' \cos(45°) + v_2' \cos(45°) = 2v_1' \cos(45°)$$

so

$$v_1' = v_2' = \frac{v_1}{2\cos(45°)} = \frac{3.0\,\text{m/s}}{2(0.707)} = 2.1\,\text{m/s}.$$

FIGURE 9–8 Example 9–5.

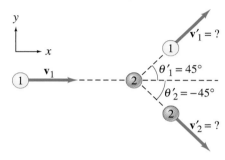

9–3 | Collisions and Impulse

Conservation of momentum is a very useful tool for dealing with collision processes, as we already saw in the Examples of the previous Section. Collisions are a common occurrence in everyday life: a tennis racket or a baseball bat striking a ball, two billiard balls colliding, one railroad car striking another, a hammer hitting a nail. At the subatomic level, scientists learn about the structure of nuclei and their constituents, and about the nature of the forces involved, by careful study of collisions between nuclei and/or elementary particles.

FIGURE 9–9 Tennis racket striking a ball. Note the deformation of both ball and racket due to the large force each exerts on the other.

Impulse

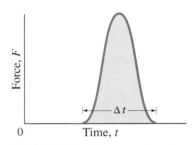

FIGURE 9–10 Force as a function of time during a typical collision.

FIGURE 9–11 The average force $\overline{F}$ over Δt gives the same impulse $(\overline{F} \Delta t)$ as the actual force.

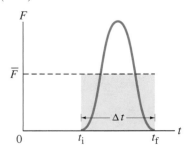

In a collision of two ordinary objects, both objects are deformed, often considerably, because of the large forces involved (Fig. 9–9). When the collision occurs, the force usually jumps from zero at the moment of contact to a very large value within a very short time, and then abruptly returns to zero again. A graph of the magnitude of the force one object exerts on the other during a collision, as a function of time, is something like that shown by the red curve in Fig. 9–10. The time interval Δt is usually very distinct and usually very small.

From Newton's second law, the *net* force on an object is equal to the rate of change of its momentum:

$$\mathbf{F} = \frac{d\mathbf{p}}{dt}.$$

(We have written $\mathbf{F}$ instead of $\Sigma\mathbf{F}$ for the net force, which we assume is entirely due to the brief but large force that acts during the collision.) This equation applies, of course, to *each* of the objects in a collision. During the infinitesimal time interval dt, the momentum changes by

$$d\mathbf{p} = \mathbf{F}\, dt.$$

If we integrate this over the duration of a collision, we have

$$\int_i^f d\mathbf{p} = \mathbf{p}_f - \mathbf{p}_i = \int_{t_i}^{t_f} \mathbf{F}\, dt,$$

where $\mathbf{p}_i$ and $\mathbf{p}_f$ are the momenta of the object just before and just after the collision. The integral of the force over the time interval during which it acts is called the **impulse**, $\mathbf{J}$:

$$\mathbf{J} = \int_{t_i}^{t_f} \mathbf{F}\, dt.$$

Thus the change in momentum of an object, $\Delta\mathbf{p} = \mathbf{p}_f - \mathbf{p}_i$, is equal to the impulse acting on it:

$$\Delta\mathbf{p} = \mathbf{p}_f - \mathbf{p}_i = \int_{t_i}^{t_f} \mathbf{F}\, dt = \mathbf{J}. \tag{9–7}$$

The units for impulse are the same as for momentum, $\text{kg}\cdot\text{m/s}$ (or $\text{N}\cdot\text{s}$) in SI. Since $\mathbf{J} = \int\mathbf{F}\, dt$, we can state that the impulse $\mathbf{J}$ of a force is equal to the area under the F versus t curve, as indicated by the shading in Fig. 9–10.

Equation 9–7 is true only if $\mathbf{F}$ is the *net* force on the object. It is valid for any net force $\mathbf{F}$ where $\mathbf{p}_i$ and $\mathbf{p}_f$ correspond precisely to the times t_i and t_f. But the impulse concept is really most useful for so-called *impulsive forces*—that is, for a force like that shown in Fig. 9–10, which has a very large magnitude over a very short time interval and is essentially zero outside this time interval. For most collision processes, the impulsive force is much larger than any other force acting, and the others can be neglected. Then the impulsive force is essentially the net force, and the change in momentum of an object during a collision is due almost entirely to the impulsive force. For such an impulsive force, the time interval over which we take the integral in Eq. 9–7 is not critical as long as we start before t_i and end after t_f, since $\mathbf{F}$ is essentially zero outside this time interval $\Delta t = t_f - t_i$. (Of course, if the chosen time interval is too large, the effect of the other forces does become significant—such as the flight of a tennis ball which, after the impulsive force administered by the racket, begins to slowly fall under gravity.)

It is sometimes useful to speak of the average force, $\overline{\mathbf{F}}$, during a collision. It is defined as that constant force which, if acting over the same time interval $\Delta t = t_f - t_i$ as the actual force, would produce the same impulse and change in momentum. Thus

$$\overline{\mathbf{F}}\, \Delta t = \int_{t_i}^{t_f} \mathbf{F}\, dt.$$

Figure 9–11 shows the magnitude of the average force, $\overline{F}$, for the impulsive force of Fig. 9–10. The rectangular area $\overline{F}\, \Delta t$ equals the area under the impulsive force curve.

EXAMPLE 9–6 **Bend your knees when landing.** (*a*) Calculate the impulse experienced when a 70-kg person lands on firm ground after jumping from a height of 3.0 m. Then estimate the average force exerted on the person's feet by the ground, if the landing is (*b*) stiff-legged, and (*c*) with bent legs. In the former case, assume the body moves 1.0 cm during impact, and in the second case, when the legs are bent, about 50 cm.

SOLUTION (*a*) We don't know F and can't calculate the impulse $\mathbf{J} = \int \mathbf{F}\, dt$ directly, but we can use the fact that the impulse equals the change in momentum of the object. We need to determine the velocity of the person just before striking the ground, which we can do using conservation of energy (Eq. 8–9), $\Delta K = -\Delta U$:

$$\tfrac{1}{2}mv^2 - 0 = -mg(y - y_0),$$

where we assume he started from rest $(v_0 = 0)$, and $y_0 = 3.0$ m and $y = 0$. Thus, after falling 3.0 m, the person's velocity just before hitting the ground will be

$$v = \sqrt{2g(y_0 - y)} = \sqrt{2(9.8 \text{ m/s}^2)(3.0 \text{ m})} = 7.7 \text{ m/s}.$$

As the person strikes the ground, the momentum is quickly brought to zero, Fig. 9–12. The impulse on the person is

$$J = \bar{F}\,\Delta t = \Delta p = p_{\text{f}} - p_{\text{i}}$$
$$= 0 - (70 \text{ kg})(7.7 \text{ m/s}) = -540 \text{ N} \cdot \text{s}.$$

The negative sign tells us that the force is opposed to the original momentum—that is, the force acts upward.

(*b*) In coming to rest, the body decelerates from 7.7 m/s to zero in a distance $d = 1.0$ cm $= 1.0 \times 10^{-2}$ m. The average speed during this brief period is

$$\bar{v} = (7.7 \text{ m/s} + 0 \text{ m/s})/2 = 3.8 \text{ m/s},$$

so the collision lasts a time

$$\Delta t = \frac{d}{\bar{v}} = \frac{(1.0 \times 10^{-2} \text{ m})}{(3.8 \text{ m/s})} = 2.6 \times 10^{-3} \text{ s}.$$

Since the magnitude of the impulse is $\bar{F}\,\Delta t = 540$ N · s, and $\Delta t = 2.6 \times 10^{-3}$ s, the average net force $\bar{F}$ has magnitude

$$\bar{F} = \frac{J}{\Delta t} = \frac{540 \text{ N} \cdot \text{s}}{2.6 \times 10^{-3} \text{ s}} = 2.1 \times 10^5 \text{ N}.$$

The force $\bar{F}$ is the *net* force upward on the person (we calculated it from Newton's second law). $\bar{F}$ is the sum of the average force upward on the legs exerted by the ground, F_{grd}, which we take as positive, plus the downward force of gravity, $-mg$ (see Fig. 9–13):

$$\bar{F} = F_{\text{grd}} - mg.$$

Since $mg = (70 \text{ kg})(9.8 \text{ m/s}^2) = 690$ N, then

$$F_{\text{grd}} = \bar{F} + mg = 2.1 \times 10^5 \text{ N} + 0.690 \times 10^3 \text{ N} \approx 2.1 \times 10^5 \text{ N}.$$

(*c*) This is like part (*b*), except $d = 0.50$ m, so $\Delta t = (0.50 \text{ m})/(3.8 \text{ m/s}) = 0.13$ s, and

$$\bar{F} = \frac{540 \text{ N} \cdot \text{s}}{0.13 \text{ s}} = 4.2 \times 10^3 \text{ N}.$$

The upward force exerted on the person's feet by the ground is, as in part (*b*):

$$F_{\text{grd}} = \bar{F} + mg = 4.2 \times 10^3 \text{ N} + 0.69 \times 10^3 \text{ N} = 4.9 \times 10^3 \text{ N}.$$

Clearly, the force on the feet and legs is much less when the knees are bent. In fact, the ultimate strength of the leg bone (see Chapter 12, Table 12–2) is not great enough to support the force calculated in part (*b*), so the leg would likely break in such a stiff landing, whereas it probably wouldn't in part (*c*).

FIGURE 9–12 Period during which impulse acts (Example 9–6).

FIGURE 9–13 When the person lands on the ground, the average net force during impact is $\bar{F} = F_{\text{grd}} - mg$, where F_{grd} is the force the ground exerts upward on the person.

9-4 Conservation of Energy and Momentum in Collisions

During most collisions, we usually don't know how the collision force varies over time, and so analysis using Newton's second law becomes difficult or impossible. But we can still determine a lot about the motion after a collision, given the initial motion, by making use of the conservation laws for momentum and energy. We saw in Section 9–2 that in the collision of two objects such as billiard balls, the total momentum is conserved, as long as any other "external" forces can be ignored. If the two objects are very hard and no heat is produced in the collision, then kinetic energy is conserved as well. By this we mean that the sum of the kinetic energies of the two objects is the same after the collision as before. Of course, for the brief moment during which the two objects are in contact, some (or all) of the energy is stored momentarily in the form of elastic potential energy. But if we compare the total kinetic energy before the collision with the total after the collision, they are found to be the same. Such a collision, in which the total kinetic energy is conserved, is called an **elastic collision**. If we use the subscripts 1 and 2 to represent the two objects, we can write the equation for conservation of total kinetic energy as

$$\tfrac{1}{2} m_1 v_1^2 + \tfrac{1}{2} m_2 v_2^2 = \tfrac{1}{2} m_1 v_1'^2 + \tfrac{1}{2} m_2 v_2'^2. \qquad \text{[elastic collision]} \quad \textbf{(9–8)}$$

Here, primed quantities (′) mean after the collision and unprimed mean before the collision, just as in Eq. 9–3 for conservation of momentum.

At the atomic level the collisions of atoms and molecules are often elastic. But in the "macroscopic" world of ordinary objects, an elastic collision is an ideal that is never quite reached, since at least a little thermal energy (and perhaps sound and other forms of energy) is always produced during a collision. Collisions of two air table pucks or two hard elastic balls, such as billiard balls, are very close to perfectly elastic, and we often treat them as such. Even when the kinetic energy is not conserved, the *total* energy is, of course, always conserved.

Collisions in which kinetic energy is not conserved are said to be **inelastic collisions**. The kinetic energy that is lost is changed into other forms of energy, often thermal energy, so that the total energy (as always) is conserved. In this case, we can write that

$$K_1 + K_2 = K_1' + K_2' + \text{thermal and other forms of energy.}$$

See Fig. 9–14.

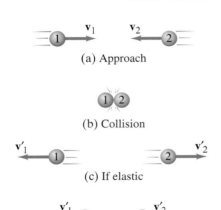

Elastic collision

(a) Approach

(b) Collision

(c) If elastic

(d) If inelastic

FIGURE 9–14 Two equal mass objects (a) approach each other with equal speeds, (b) collide, and then (c) bounce off with equal speeds in the opposite directions if the collision is elastic, or (d) bounce back much less or not at all, if the collision is inelastic.

9-5 Elastic Collisions in One Dimension

We now apply the conservation laws for momentum and kinetic energy to an elastic collision between two small objects that collide head-on, so all the motion is along a line. We deal here only with translational motion, but our analysis works well even for hard billiard balls because at the instant of collision the rolling has little effect. Let us assume that both objects are initially moving with velocities v_1 and v_2 along the x axis, Fig. 9–15a. After the collision, their velocities are v_1' and v_2', Fig. 9–15b. For any $v > 0$, the object is moving to the right (increasing x), whereas for $v < 0$, the object is moving to the left (toward decreasing values of x).

From conservation of momentum, we have

$$m_1 v_1 + m_2 v_2 = m_1 v_1' + m_2 v_2'.$$

Momentum conservation

Because the collision is assumed to be elastic, kinetic energy is also conserved:

$$\tfrac{1}{2} m_1 v_1^2 + \tfrac{1}{2} m_2 v_2^2 = \tfrac{1}{2} m_1 v_1'^2 + \tfrac{1}{2} m_2 v_2'^2.$$

Kinetic energy conservation

We have two equations, so we can solve for two unknowns. If we know the masses and initial velocities, then we can solve these two equations for the velocities after the collision, v_1' and v_2'. We will do this in a moment in some Examples, but first we derive a useful result. To do so we rewrite the momentum equation as

$$m_1(v_1 - v_1') = m_2(v_2' - v_2), \quad \text{(i)}$$

and we rewrite the kinetic energy equation as

$$m_1(v_1^2 - v_1'^2) = m_2(v_2'^2 - v_2^2)$$

or [noting that $(a - b)(a + b) = a^2 - b^2$] we write this as

$$m_1(v_1 - v_1')(v_1 + v_1') = m_2(v_2' - v_2)(v_2' + v_2). \quad \text{(ii)}$$

We divide Eq. (ii) by Eq. (i), and (assuming $v_1 \neq v_1'$ and $v_2 \neq v_2'$)[†] obtain

$$v_1 + v_1' = v_2' + v_2.$$

We can rewrite this equation as

$$v_1 - v_2 = v_2' - v_1'$$
$$= -(v_1' - v_2'). \qquad \text{[head-on elastic collision]} \quad \textbf{(9–9)}$$

This is an interesting result: it tells us that for any elastic head-on collision, the relative speed of the two objects after the collision has the same magnitude as before (but opposite direction), no matter what the masses are.

Now let us look at some special cases of elastic head-on collisions. We assume v_1, v_2, m_1, and m_2 are known, and we wish to solve for v_1' and v_2', the velocities of the two objects after the collision.

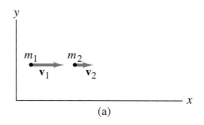

y

m_1 ●→ $\mathbf{v}_1$ m_2 ●→ $\mathbf{v}_2$

x

(a)

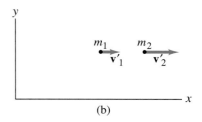

y

m_1 ●→ $\mathbf{v}_1'$ m_2 ●→ $\mathbf{v}_2'$

x

(b)

FIGURE 9–15 Two small objects, of masses m_1 and m_2, (a) before the collision, and (b) after the collision.

EXAMPLE 9–7 **Equal masses.** A billiard ball of mass m moving with speed v collides head-on with a second ball of equal mass. What are the speeds of the two balls after the collision, assuming it is elastic? Assume (a) both balls are moving, (b) ball 2 is initially at rest ($v_2 = 0$).

SOLUTION (a) Since $m_1 = m_2 = m$, then conservation of momentum gives

$$v_1 + v_2 = v_1' + v_2'.$$

We need a second equation, since there are two unknowns. We could use the conservation of kinetic energy equation, or the simpler Eq. 9–9 derived above:

$$v_1 - v_2 = v_2' - v_1'.$$

Now we add these two equations and get

$$v_2' = v_1$$

and then subtract the two equations to get

$$v_1' = v_2.$$

That is, the balls exchange velocities as a result of the collision: ball 2 acquires the velocity that ball 1 had before the collision, and vice versa.

(b) If ball 2 is at rest initially, so that $v_2 = 0$, we have

$$v_2' = v_1 \quad \text{and} \quad v_1' = 0.$$

That is, ball 1 is brought to rest by the collision, whereas ball 2 acquires the original velocity of ball 1. This result is often observed by billiard and pool players, and is valid only if the two balls have equal masses (and no spin is given to the balls). See Fig. 9–16.

FIGURE 9–16 In this multi-flash photo of a head-on collision between two balls of equal mass, the white cue ball is accelerated from rest by the cue stick and then strikes the red ball, initially at rest. The white ball stops in its tracks and the (equal mass) red ball moves off with the same speed as the white ball had before the collision. See Example 9–7.

[†]Note that Eqs. (i) and (ii), which are the conservation laws for momentum and kinetic energy, are both satisfied by the solution $v_1' = v_1$ and $v_2' = v_2$. This is a valid solution, but not very interesting. It corresponds to no collision at all—when the two objects miss each other.

EXAMPLE 9–8 **Unequal masses, target at rest.** A very common practical situation is for a moving object (m_1) to strike a second object $(m_2,$ the "target") at rest $(v_2 = 0)$. Assume the objects have unequal masses and that the collision occurs along a line (head-on). (a) Derive equations for v_2' and v_1' in terms of the initial velocity v_1 of mass m_1 and the masses m_1 and m_2. (b) Determine the final velocities if the moving object is much more massive than the target $(m_1 \gg m_2)$. (c) Determine the final velocities if the moving object is much less massive than the target $(m_1 \ll m_2)$.

SOLUTION (a) We combine the momentum equation, which (with $v_2 = 0$) is

$$m_2 v_2' = m_1(v_1 - v_1'),$$

with Eq. 9–9, rewritten as $v_1 + v_1' = v_2'$, to obtain

$$v_2' = v_1 \left(\frac{2m_1}{m_1 + m_2} \right)$$

$$v_1' = v_1 \left(\frac{m_1 - m_2}{m_1 + m_2} \right).$$

As a check, we let $m_1 = m_2$, and we obtain

$$v_2' = v_1 \quad \text{and} \quad v_1' = 0.$$

This is the same case treated in Example 9–7, and we get the same result: for objects of equal mass, one of which is initially at rest, the velocity of the one moving initially is completely transferred to the object originally at rest.

(b) Let $v_2 = 0$ and $m_1 \gg m_2$. A very heavy moving object strikes a light object at rest, and we have, using the relations for v_2' and v_1' above,

$$v_2' \approx 2v_1$$

$$v_1' \approx v_1.$$

Thus the velocity of the heavy incoming object is practically unchanged, whereas the light object, originally at rest, takes off with twice the velocity of the heavy one. The velocity of a heavy bowling ball, for example, is hardly affected by striking the much lighter bowling pins.

(c) Finally, we let $v_2 = 0$ and $m_1 \ll m_2$. A moving light object strikes a very massive object at rest. In this case, using the equations in part (a)

$$v_2' \approx 0$$

$$v_1' \approx -v_1.$$

The massive object remains essentially at rest and the very light incoming object rebounds with essentially its same speed but in the opposite direction. For example, a tennis ball colliding head-on with a stationary bowling ball will hardly affect the bowling ball, but will rebound with nearly the same speed it had initially, just as if it had struck a hard wall.

It can readily be shown (it is given as Problem 40) for any elastic head-on collision that

$$v_2' = v_1 \left(\frac{2m_1}{m_1 + m_2} \right) + v_2 \left(\frac{m_2 - m_1}{m_1 + m_2} \right)$$

and

$$v_1' = v_1 \left(\frac{m_1 - m_2}{m_1 + m_2} \right) + v_2 \left(\frac{2m_2}{m_1 + m_2} \right).$$

These general equations, however, should not be memorized. They can always be derived quickly from the conservation laws. For many problems, it is simplest just to start from scratch, as we did in the special cases above and as shown in the next Example.

EXAMPLE 9–9 **A nuclear collision.** A proton of mass 1.01 u (unified atomic mass units) traveling with a speed of 3.60×10^4 m/s has an elastic head-on collision with a helium (He) nucleus $(m_{\text{He}} = 4.00\,\text{u})$ initially at rest. What are the velocities of the proton and helium nucleus after the collision? (As mentioned in Chapter 1, 1 u $= 1.66 \times 10^{-27}$ kg, but we won't need this fact.)

⟹ **P H Y S I C S A P P L I E D**

Nuclear collision

SOLUTION Call the initial direction of motion the $+x$ direction. We have $v_2 = v_{\text{He}} = 0$ and $v_1 = v_{\text{p}} = 3.60 \times 10^4$ m/s. We want to find the velocities v'_{p} and v'_{He} after the collision. From conservation of momentum we have

$$m_{\text{p}} v_{\text{p}} + 0 = m_{\text{p}} v'_{\text{p}} + m_{\text{He}} v'_{\text{He}}.$$

Because the collision is elastic, kinetic energy is conserved and we can use Eq. 9–9, which becomes

$$v_{\text{p}} - 0 = v'_{\text{He}} - v'_{\text{p}}.$$

Thus

$$v'_{\text{p}} = v'_{\text{He}} - v_{\text{p}},$$

and substituting this into the momentum equation we get

$$m_{\text{p}} v_{\text{p}} = m_{\text{p}} v'_{\text{He}} - m_{\text{p}} v_{\text{p}} + m_{\text{He}} v'_{\text{He}}.$$

Solving for v'_{He}, we obtain

$$v'_{\text{He}} = \frac{2 m_{\text{p}} v_{\text{p}}}{m_{\text{p}} + m_{\text{He}}} = \frac{2(1.01\,\text{u})(3.60 \times 10^4\,\text{m/s})}{5.01\,\text{u}} = 1.45 \times 10^4\,\text{m/s}.$$

The other unknown is v'_{p}, which we can now obtain from

$$v'_{\text{p}} = v'_{\text{He}} - v_{\text{p}}$$
$$= 1.45 \times 10^4\,\text{m/s} - 3.60 \times 10^4\,\text{m/s} = -2.15 \times 10^4\,\text{m/s}.$$

The minus sign tells us that the proton reverses direction upon collision, and we see that its speed is less than its initial speed (see Fig. 9–17). This makes sense from ordinary experience: the lighter proton would be expected to "bounce back" somewhat from the more massive helium nucleus, but not with its full original velocity as it would from a rigid wall (which would correspond to extremely large, or infinite, mass).

(a)

(b)

FIGURE 9–17 Example 9–9: (a) before collision, (b) after collision.

9–6 | Inelastic Collisions

Collisions in which kinetic energy is not conserved are called inelastic collisions. Some of the initial kinetic energy in such collisions is transformed into other types of energy, such as thermal or potential energy, so the total final kinetic energy is less than the total initial kinetic energy. The inverse can also happen when potential energy (such as chemical or nuclear) is released, in which case the total final kinetic energy can be greater than the initial kinetic energy. Explosives are examples of this type. Typical macroscopic collisions are inelastic, at least to some extent, and often to a large extent. If two objects stick together as a result of a collision, the collision is said to be **completely inelastic**. Two colliding balls of putty

Completely inelastic collision

that stick together or two railroad cars that couple together when they collide are examples of completely inelastic collisions. The kinetic energy in some cases is all transformed to other forms of energy in an inelastic collision, but in other cases only part of it is. In Example 9–3, for instance, we saw that when a traveling railroad car collided with a stationary one, the coupled cars traveled off with some kinetic energy.

Ballistic pendulum

FIGURE 9–18 Ballistic pendulum (Example 9–10).

EXAMPLE 9–10 **Ballistic pendulum.** The *ballistic pendulum* is a device used to measure the speed of a projectile, such as a bullet. The projectile, of mass m, is fired into a large block (of wood or other material) of mass M, which is suspended like a pendulum. (Usually, M is somewhat greater than m.) As a result of the collision, the pendulum-projectile system swings up to a maximum height h, Fig. 9–18. Determine the relationship between the initial horizontal speed of the projectile, v, and the height h.

SOLUTION We analyze this process by dividing it into two parts: (1) the collision itself, and (2) the subsequent motion of the pendulum from the vertical hanging position to height h. In part (1), Fig. 9–18a, we assume the collision time is very short, and so the projectile comes to rest in the block before the block has moved significantly from its position directly below its support. Thus there is no net external force and momentum is conserved:

$$mv = (m + M)v', \qquad \text{(i)}$$

where v' is the speed of the block and embedded projectile just after the collision, before they have moved significantly. Once the pendulum begins to move (part 2, Fig. 9–18b), there will be a net external force (gravity, tending to pull it back to the vertical position). So, for part (2), we cannot use conservation of momentum. But we can use conservation of mechanical energy since the kinetic energy immediately after the collision is changed entirely to gravitational potential energy when the pendulum reaches its maximum height, h. Therefore (letting $y = 0$ for the pendulum in the vertical position):

$$K_1 + U_1 = K_2 + U_2$$

or

$$\tfrac{1}{2}(m + M)v'^2 + 0 = 0 + (m + M)gh, \qquad \text{(ii)}$$

so $v' = \sqrt{2gh}$. We combine equations (i) and (ii) to obtain

$$v = \frac{m + M}{m}v' = \frac{m + M}{m}\sqrt{2gh},$$

which is the final result. To obtain this result, we had to be opportunistic, in that we used whichever conservation laws we could: in (1) we could use only conservation of momentum, since the collision is inelastic and conservation of mechanical energy is not valid[†]; and in (2), conservation of mechanical energy is valid, but not conservation of momentum. In part (1), if there were significant motion of the pendulum during the deceleration of the projectile in the block, then there *would* be an external force during the collision—so conservation of momentum would not be valid, and this would have to be taken into account.

[†] Total energy is conserved, of course.

9–7 | Collisions in Two or Three Dimensions

Conservation of momentum and energy can also be applied to collisions in two or three dimensions, and the vector nature of momentum is especially important. One common type of non-head-on collision is that in which a moving object (called the "projectile") strikes a second object initially at rest (the "target") object. This is the common situation in games such as billiards, and for experiments in atomic and nuclear physics (the projectiles, from radioactive decay or a high-energy accelerator, strike a stationary target nucleus).

Figure 9–19 shows the incoming projectile, m_1 heading along the x axis toward the target, m_2, which is initially at rest. If these are billiard balls, m_1 strikes m_2 and they go off at the angles θ_1' and θ_2', respectively, which are measured relative to m_1's initial direction (the x axis). The objects may begin to deflect even before they touch if electric, magnetic, or nuclear forces act between them.

Let us apply the law of conservation of momentum to a collision like that of Fig. 9–19. We choose the xy plane to be the plane in which the initial and final momenta lie. Because momentum is a vector, and is conserved, its components in the x and y directions remain constant. In the x direction, we have $p_{1x} + p_{2x} = p_{1x}' + p_{2x}'$, or:

$$m_1 v_1 = m_1 v_1' \cos \theta_1' + m_2 v_2' \cos \theta_2'. \qquad \textbf{(9–10a)} \qquad p_x \ conserved$$

Because there is no motion in the y direction initially, the y component of the total momentum is zero:

$$0 = m_1 v_1' \sin \theta_1' + m_2 v_2' \sin \theta_2'. \qquad \textbf{(9–10b)} \qquad p_y \ conserved$$

When we have two independent equations, we can solve for, at most, two unknowns.

If we know that a collision is elastic, we can then apply conservation of kinetic energy and obtain a third equation:

$$K_1 + K_2 = K_1' + K_2'$$

or, for the collision shown in Fig. 9–19,

$$\tfrac{1}{2} m v_1^2 = \tfrac{1}{2} m_1 v_1'^2 + \tfrac{1}{2} m v_2'^2. \qquad \text{[elastic collision]} \quad \textbf{(9–10c)} \qquad Kinetic \ energy \ conserved$$

If the collision is elastic, we have three independent equations and we can solve for three unknowns. If we are given m_1, m_2, v_1 (and v_2, if it is not zero), we cannot, for example, predict the final variables, v_1', v_2', θ_1', and θ_2', because there are four of them. However, if we measure one of these variables, say θ_1', then the other three variables $(v_1', v_2',$ and $\theta_2')$ are uniquely determined, and we can determine them using Eqs. 9–10a, b, c.

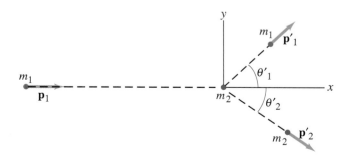

FIGURE 9–19 Object 1, the projectile, collides with object 2, the target. They move off, after the collision, with momenta $\mathbf{p}_1'$ and $\mathbf{p}_2'$ at angles θ_1' and θ_2'.

EXAMPLE 9–11 Proton–proton collision. A proton traveling with speed 8.2×10^5 m/s collides elastically with a stationary proton in a hydrogen target as in Fig. 9–19. One of the protons is observed to be scattered at a 60° angle. At what angle will the second proton be observed, and what will be the velocities of the two protons after the collision?

SOLUTION We saw a two-dimensional collision in Example 9–5, where we needed to use only conservation of momentum. Now we are given less information: we have three unknowns instead of two, so we need the kinetic energy equation as well as the two momentum equations. Since $m_1 = m_2$, Eqs. 9–10a, b, and c become

$$v_1^2 = v_1'^2 + v_2'^2 \qquad \textbf{(i)}$$

$$v_1 = v_1' \cos \theta_1' + v_2' \cos \theta_2' \qquad \textbf{(ii)}$$

$$0 = v_1' \sin \theta_1' + v_2' \sin \theta_2', \qquad \textbf{(iii)}$$

where $v_1 = 8.2 \times 10^5$ m/s and $\theta_1' = 60°$ are given. In the second and third equations, we move the v_1' terms to the left side and square both sides of the equations:

$$v_1^2 - 2v_1 v_1' \cos \theta_1' + v_1'^2 \cos^2 \theta_1' = v_2'^2 \cos^2 \theta_2'$$

$$v_1'^2 \sin^2 \theta_1' = v_2'^2 \sin^2 \theta_2'.$$

We add these two equations and use $\sin^2 \theta + \cos^2 \theta = 1$ to get:

$$v_1^2 - 2v_1 v_1' \cos \theta_1' + v_1'^2 = v_2'^2.$$

Into this equation we substitute $v_2'^2 = v_1^2 - v_1'^2$, from equation (i) above, and get

$$2v_1'^2 = 2v_1 v_1' \cos \theta_1'$$

or

$$v_1' = v_1 \cos \theta_1' = (8.2 \times 10^5 \,\text{m/s})(\cos 60°) = 4.1 \times 10^5 \,\text{m/s}.$$

To obtain v_2', we use equation (i) above (conservation of kinetic energy):

$$v_2' = \sqrt{v_1^2 - v_1'^2} = 7.1 \times 10^5 \,\text{m/s}.$$

Finally, from equation (iii), we have

$$\sin \theta_2' = -\frac{v_1'}{v_2'} \sin \theta_1' = -\left(\frac{4.1 \times 10^5 \,\text{m/s}}{7.1 \times 10^5 \,\text{m/s}}\right)(0.866) = -0.50,$$

so $\theta_2' = -30°$. (The minus sign means particle 2 moves at an angle below the x axis if particle 1 is above the axis, as in Fig. 9–19.) An example of such a collision is shown in the bubble chamber photo of Fig. 9–20. Notice that the two trajectories are at right angles to each other after the collision. This can be shown to be true in general for non-head-on elastic collisions of two particles of equal mass, one of which was at rest initially (see Problem 57).

FIGURE 9–20 Photo of a proton–proton collision in a hydrogen bubble chamber (a device that makes visible the paths of elementary particles). The many lines represent incoming protons which can strike the protons of the hydrogen in the chamber.

PROBLEM SOLVING Momentum Conservation and Collisions

1. Be sure no significant external force acts on your chosen system. That is, the forces that act between the interacting bodies must be the only significant ones if momentum conservation is to be used. [If this is valid for a portion of the problem, you can use momentum conservation for that portion only.]

2. Draw a diagram of the initial situation, just before the interaction (collision, explosion) takes place, and represent the momentum of each object with an arrow and label. Do the same for the final situation, just after the interaction.

3. Choose a coordinate system and "+" and "−" directions. (For a head-on collision, you will need only an

x axis.) It is often convenient to choose the $+x$ axis in the direction of one object's initial velocity.

4. Write momentum conservation equation(s):

   total initial momentum = total final momentum.

   You have one equation for each component (x, y, z); only one equation for a head-on collision.

5. If the collision is elastic, you can also write down a conservation of kinetic energy equation:

   total initial kinetic energy = total final kinetic energy.

   [Alternately, you could use Eq. 9–9: $v_1 - v_2 = v_2' - v_1'$, if the collision is one dimensional (head-on).]

6. Solve algebraically for the unknown(s).

9–8 Center of Mass (CM)

Until now, when we have dealt with an extended body (that is, a body that has size), we have assumed that it could be approximated as a point particle or that it undergoes only translational motion. Real extended bodies, however, can undergo rotational and other types of motion as well. For example, the diver in Fig. 9–21a undergoes only translational motion (all parts of the body follow the same path), whereas the diver in Fig. 9–21b undergoes both translational and rotational motion. We will refer to motion that is not pure translation as *general motion*.

Observations indicate that even if a body rotates, or there are several bodies that move relative to one another, there is one point that moves in the same path that a particle would if subjected to the same net force. This point is called the **center of mass** (abbreviated CM). The general motion of an extended body (or system of bodies) can be considered as *the sum of the translational motion of the* CM, *plus rotational, vibrational, or other types of motion about the* CM.

As an example, consider the motion of the center of mass of the diver in Fig. 9–21; the CM follows a parabolic path even when the diver rotates, as shown in Fig. 9–21b. This is the same parabolic path that a projected particle follows when acted on only by the force of gravity (that is, projectile motion). Other points in the rotating diver's body follow more complicated paths.

Figure 9–22 shows a wrench translating and rotating along a horizontal surface—note that its CM, marked by a red +, moves in a straight line, as shown by the dashed white line.

We show below that the important properties of the CM follow from Newton's laws if the CM is defined in the following way. We can consider any extended body as being made up of many tiny particles. But first we consider a system made up of only two particles (or small objects), of mass m_1 and m_2. We choose a coordinate system so that both particles lie on the x axis at positions x_1 and x_2, Fig. 9–23. The center of mass of this system is defined to be at the position, x_{CM}, given by

$$x_{CM} = \frac{m_1 x_1 + m_2 x_2}{m_1 + m_2} = \frac{m_1 x_1 + m_2 x_2}{M},$$

where $M = m_1 + m_2$ is the total mass of the system. The center of mass lies on the line joining m_1 and m_2. If the two masses are equal $(m_1 = m_2 = m)$, x_{CM} is midway between them, since in this case $x_{CM} = m(x_1 + x_2)/2m = (x_1 + x_2)/2$. If one mass is greater than the other, say, $m_1 > m_2$, then the CM is closer to the larger mass. If all the mass is concentrated at x_2, say, so $m_1 = 0$, then $x_{CM} = (0x_1 + m_2 x_2)/(0 + m_2) = x_2$, as we would expect.

(a)

(b)

FIGURE 9–21 The motion of the diver is pure translation in (a), but is translation plus rotation in (b).

FIGURE 9–22 Translation plus rotation: a wrench moving over a horizontal surface. The CM, marked with a red +, moves in a straight line.

FIGURE 9–23 The center of mass of a two-particle system lies on the line joining the two masses. Here $m_1 > m_2$ so the CM is closer to m_1 than to m_2.

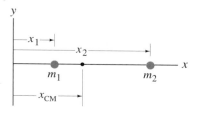

Now let us consider a system consisting of n particles, where n could be very large. This system could be an extended body which we consider as being made up of n tiny particles. If these n particles are all along a straight line (call it the x axis), we define the CM of the system to be located at

$$x_{CM} = \frac{m_1 x_1 + m_2 x_2 + \cdots + m_n x_n}{m_1 + m_2 + \cdots + m_n} = \frac{\sum\limits_{i=1}^{n} m_i x_i}{M},$$

where $m_1, m_2, \ldots m_n$ are the masses of each particle and $x_1, x_2, \ldots x_n$ are their positions. The symbol $\sum\limits_{i=1}^{n}$ is the summation sign meaning to sum over all the particles, where i takes on integer values from 1 to n. (Often we simply write $\sum m_i x_i$, leaving out the $i = 1$ to n.) The total mass of the system is $M = \sum m_i$.

EXAMPLE 9–12 CM of three guys on a raft. Three people of roughly equivalent mass m on a lightweight (air-filled) banana boat sit along the x axis at positions $x_1 = 1.0$ m, $x_2 = 5.0$ m, and $x_3 = 6.0$ m (Fig. 9–24). Find the position of the CM.

SOLUTION We assume each x is the horizontal location of each person's CM. Treating them as particles, the overall CM is:

$$x_{CM} = \frac{mx_1 + mx_2 + mx_3}{m + m + m} = \frac{m(x_1 + x_2 + x_3)}{3m}$$
$$= \frac{(1.0\,\text{m} + 5.0\,\text{m} + 6.0\,\text{m})}{3} = \frac{12.0\,\text{m}}{3} = 4.0\,\text{m}.$$

FIGURE 9–24 Example 9–12.

If the particles are spread out in two or three dimensions, as for a typical extended body, then we define the coordinates of the CM as

Center of mass

$$x_{CM} = \frac{\sum m_i x_i}{M}, \quad y_{CM} = \frac{\sum m_i y_i}{M}, \quad z_{CM} = \frac{\sum m_i z_i}{M}, \quad \textbf{(9–11)}$$

where x_i, y_i, z_i are the coordinates of the particle of mass m_i and again $M = \sum m_i$ is the total mass.

Although from a practical point of view we usually calculate the components of the CM (Eq. 9–11), it is sometimes convenient (for example, for derivations) to write Eq. 9–11 in vector form. If $\mathbf{r}_i = x_i\mathbf{i} + y_i\mathbf{j} + z_i\mathbf{k}$ is the position vector of the ith particle, and $\mathbf{r}_{CM} = x_{CM}\mathbf{i} + y_{CM}\mathbf{j} + z_{CM}\mathbf{k}$ is the position vector of the center of mass, then

$$\mathbf{r}_{CM} = \frac{\sum m_i \mathbf{r}_i}{M}. \quad \textbf{(9–12)}$$

EXAMPLE 9–13 Three particles in 2-D. Three particles, each of mass 2.50 kg, are located at the corners of a right triangle whose sides are 2.00 m and 1.50 m long, as shown in Fig. 9–25. Locate the center of mass.

SOLUTION We choose our coordinate system as shown (to simplify calculations) with m_1 at the origin and m_2 on the x axis. Then m_1 has coordinates $x_1 = y_1 = 0$; m_2 has coordinates $x_2 = 2.0$ m, $y_2 = 0$; and m_3 has coordinates $x_3 = 2.0$ m, $y_3 = 1.5$ m. Then, from Eq. 9–11,

$$x_{CM} = \frac{(2.50\,\text{kg})(0) + (2.50\,\text{kg})(2.00\,\text{m}) + (2.50\,\text{kg})(2.00\,\text{m})}{3(2.50\,\text{kg})} = 1.33\,\text{m}$$

$$y_{CM} = \frac{(2.50\,\text{kg})(0) + (2.50\,\text{kg})(0) + (2.50\,\text{kg})(1.50\,\text{m})}{7.50\,\text{kg}} = 0.50\,\text{m}.$$

The CM and the position vector $\mathbf{r}_{CM}$ are shown in Fig. 9–25.

FIGURE 9–25 Example 9–13.

Note that the numerical values of x_{CM} and y_{CM} depend on the coordinate system chosen, but the physical position of the CM (relative to the three particles) does not depend on the reference frame.

It is often convenient to think of an extended body as made up of a continuous distribution of matter. In other words, we consider the body to be made up of n particles, each of mass Δm_i in a tiny volume around a point x_i, y_i, z_i, and we take the limit of n approaching infinity (Fig. 9–26). Then Δm_i becomes the infinitesimal mass dm at points x, y, z. The summations in Eqs. 9–11 and 9–12 become integrals:

Continuous distribution of mass

$$x_{CM} = \frac{1}{M} \int x\, dm, \quad y_{CM} = \frac{1}{M} \int y\, dm, \quad z_{CM} = \frac{1}{M} \int z\, dm, \qquad (9\text{–}13)$$

where the sum over all the mass elements is $\int dm = M$, the total mass of the object. In vector notation, this becomes

$$\mathbf{r}_{CM} = \frac{1}{M} \int \mathbf{r}\, dm. \qquad (9\text{–}14)$$

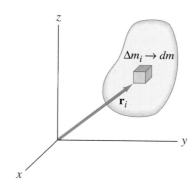

A concept similar to *center of mass* is **center of gravity** (CG). The CG of a body is that point at which the force of gravity can be considered to act. Of course, the force of gravity actually acts on all the different parts or particles of a body, but for purposes of determining the translational motion of a body as a whole, we can assume that the entire weight of the body (which is the sum of the weights of all its parts) acts at the CG. Strictly speaking, there is a conceptual difference between the center of gravity and the center of mass, but for practical purposes, they are generally the same point. [There would be a difference between the two only if a body were large enough so that the acceleration due to gravity was different at different parts of the body.]

FIGURE 9–26 An extended body, here shown in only two dimensions, can be considered to be made up of many tiny particles (n), each having a mass Δm_i. One such particle is shown located at a point $\mathbf{r}_i = x_i\mathbf{i} + y_i\mathbf{j} + z_i\mathbf{k}$. We take the limit of $n \to \infty$ so Δm_i becomes the infinitesimal dm.

EXAMPLE 9–14 CM of a thin rod. (*a*) Show that the CM of a uniform thin rod of length L and mass M is at its center. (*b*) Determine the CM of the rod assuming its linear mass density λ (its mass per unit length) varies linearly from $\lambda = \lambda_0$ at the left end to double that value, $\lambda = 2\lambda_0$, at the right end.

SOLUTION We choose a coordinate system so that the rod lies on the x axis with the left end at $x = 0$, Fig. 9–27. Then $y_{CM} = 0$ and $z_{CM} = 0$.
(*a*) The rod is uniform, so its mass per unit length (linear mass density λ) is constant and we write it as $\lambda = M/L$. We now imagine the rod as divided into infinitesimal elements of length dx, each of which has mass $dm = \lambda\, dx$. We use Eq. 9–13:

FIGURE 9–27 Example 9–14.

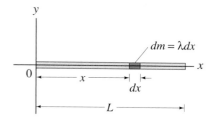

$$x_{CM} = \frac{1}{M} \int_{x=0}^{L} x\, dm = \frac{1}{M} \int_{0}^{L} \lambda x\, dx = \frac{\lambda}{M} \frac{x^2}{2} \Big|_{0}^{L} = \frac{\lambda L^2}{2M} = \frac{L}{2}$$

where we used $\lambda = M/L$. This result, x_{CM} at the center, is what we expected.
(*b*) Now we have $\lambda = \lambda_0$ at $x = 0$ and we are told that λ increases linearly to $\lambda = 2\lambda_0$ at $x = L$. So we write

$$\lambda = \lambda_0(1 + \alpha x)$$

which satisfies $\lambda = \lambda_0$ at $x = 0$ and gives $\lambda = 2\lambda_0$ at $x = L$ if $(1 + \alpha L) = 2$. In other words, $\alpha = 1/L$. Again we use Eq. 9–13:

$$x_{CM} = \frac{1}{M} \int_{x=0}^{L} \lambda x\, dx = \frac{1}{M} \lambda_0 \int_{0}^{L} \left(1 + \frac{x}{L}\right) x\, dx = \frac{\lambda_0}{M} \left(\frac{x^2}{2} + \frac{x^3}{3L}\right)\Big|_{0}^{L} = \frac{5}{6} \frac{\lambda_0}{M} L^2.$$

But what is M in terms of λ_0 and L? We can write

$$M = \int_{x=0}^{L} dm = \int_{0}^{L} \lambda\, dx = \lambda_0 \int_{0}^{L} \left(1 + \frac{x}{L}\right) dx = \lambda_0 \left(x + \frac{x^2}{2L}\right)\Big|_{0}^{L} = \frac{3}{2} \lambda_0 L.$$

Then

$$x_{CM} = \frac{5}{6} \frac{\lambda_0}{M} L^2 = \frac{5}{9} L,$$

which is more than half way along the rod, as we would expect since there is more mass to the right.

FIGURE 9–28 Cylindrical disk with origin of coordinates at geometric center.

FIGURE 9–29 Example 9–15. This L-shaped object has thickness t (not shown on diagram).

For symmetrically shaped bodies of uniform composition, such as spheres, cylinders, and rectangular solids, the CM is located at the geometric center of the body. Consider a uniform circular cylinder, such as a solid circular disk. We expect the CM to be at the center of the circle. To show that it is, we first choose a coordinate system whose origin is at the center of the circle with the z axis perpendicular to the disk (Fig. 9–28). When we take the sum $\Sigma m_i x_i$ in Eq. 9–11, there is as much mass at any $+x_i$ as there is at $-x_i$. So all terms cancel out in pairs and $x_{CM} = 0$. The same is true for y_{CM}. In the vertical (z) direction, the CM must lie halfway between the circular faces: if we choose our origin of coordinates at that point, there is as much mass at any $+z_i$ as at $-z_i$, so $z_{CM} = 0$. For other uniform, symmetrically shaped bodies, we can make similar arguments to show that the CM must lie on a line of symmetry. If a symmetric body is *not* uniform, then these arguments do not hold. For example, the CM of a wheel or disk weighted on one side is not at the geometric center but closer to the weighted side.

To locate the center of mass of a group of extended bodies, we can use Eq. 9–11, where the m_i are the masses of these bodies and x_i, y_i, and z_i are the coordinates of the CM of each of the bodies.

EXAMPLE 9–15 CM of L-shaped flat object. Determine the CM of the uniform thin L-shaped construction brace shown in Fig. 9–29.

SOLUTION Consider the object as two rectangles: rectangle A, which is 2.06 m × 0.20 m, and rectangle B, which is 1.48 m × 0.20 m. With the origin at 0 as shown, the CM of A is at

$$x_A = 1.03 \text{ m}, \quad y_A = 0.10 \text{ m}.$$

The CM of B is at

$$x_B = 1.96 \text{ m}, \quad y_B = -0.74 \text{ m}.$$

The mass of A, whose thickness is t, is

$$M_A = (2.06 \text{ m})(0.20 \text{ m})(t)(\rho) = (0.412 \text{ m}^2)(\rho t),$$

where ρ is the density. The mass of B is

$$M_B = (1.48 \text{ m})(0.20 \text{ m})(\rho t) = (0.296 \text{ m}^2)(\rho t),$$

and the total mass is $M = (0.708 \text{ m}^2)(\rho t)$.

Thus

$$x_{CM} = \frac{M_A x_A + M_B x_B}{M} = \frac{(0.412 \text{ m}^2)(1.03 \text{ m}) + (0.296 \text{ m}^2)(1.96 \text{ m})}{(0.708 \text{ m}^2)} = 1.42 \text{ m},$$

where ρt was canceled out in numerator and denominator. Similarly,

$$y_{CM} = \frac{(0.412 \text{ m}^2)(0.10 \text{ m}) + (0.296 \text{ m}^2)(-0.74 \text{ m})}{(0.708 \text{ m}^2)} = -0.25 \text{ m},$$

which puts the CM approximately at the point so labeled in Fig. 9–29. For an object not perfectly thin, $z_{CM} = t/2$, since the object is assumed to be uniform.

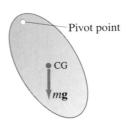

FIGURE 9–30 Determining the CM of a flat uniform body.

FIGURE 9–31 Finding the CG.

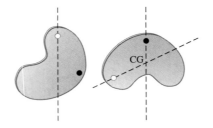

Note in this last Example that the CM can actually lie *outside* the body. Another example is a doughnut whose CM is at the center of the hole.

It is often easier to determine the CM or CG of an extended body experimentally rather than analytically. If a body is suspended from any point, it will swing (Fig. 9–30) unless it is placed so its CG lies on a vertical line directly below the point from which it is suspended. If the object is two-dimensional, or has a plane of symmetry, it need only be hung from two different pivot points and the respective vertical (plumb) lines drawn. Then the CG will be at the intersection of the two lines, as in Fig. 9–31. If the object doesn't have a plane of symmetry, the CG with respect to the third dimension is found by suspending the object from at least three points whose plumb lines do not lie in the same plane.

9–9 Center of Mass and Translational Motion

As mentioned in Section 9–8, a major reason for the importance of the concept of center of mass is that the translational motion of the CM for a system of particles (or extended bodies) is directly related to the net force on the system as a whole. We now show this, by examining the motion of a system of n particles of total mass M, which we assume remains constant. We begin by rewriting Eq. 9–12 as

$$M\mathbf{r}_{CM} = \Sigma m_i \mathbf{r}_i .$$

We differentiate this equation with respect to time:

$$M\frac{d\mathbf{r}_{CM}}{dt} = \Sigma m_i \frac{d\mathbf{r}_i}{dt}$$

or

$$M\mathbf{v}_{CM} = \Sigma m_i \mathbf{v}_i , \qquad\qquad (9\text{–}15)$$

where $\mathbf{v}_i = d\mathbf{r}_i/dt$ is the velocity of the ith particle of mass m_i and $\mathbf{v}_{CM}$ is the velocity of the CM. We take the derivative with respect to time again and obtain

$$M\frac{d\mathbf{v}_{CM}}{dt} = \Sigma m_i \mathbf{a}_i ,$$

where $\mathbf{a}_i = d\mathbf{v}_i/dt$ is the acceleration of the ith particle. Now $d\mathbf{v}_{CM}/dt$ is the acceleration of the CM, $\mathbf{a}_{CM}$. By Newton's second law, $m_i \mathbf{a}_i = \mathbf{F}_i$ where $\mathbf{F}_i$ is the net force on the ith particle. Therefore

$$M\mathbf{a}_{CM} = \mathbf{F}_1 + \mathbf{F}_2 + \cdots + \mathbf{F}_n = \Sigma \mathbf{F}_i . \qquad\qquad (9\text{–}16)$$

That is, the vector sum of all the forces acting on the system is equal to the total mass of the system times the acceleration of its center of mass. Note that our system of n particles could be the n particles that make up one or more extended bodies.

The forces $\mathbf{F}_i$ exerted on the particles of the system can be divided into two types: (1) *external forces* exerted by objects outside the system and (2) *internal forces* that particles within the system exert on one another. By Newton's third law, the internal forces occur in pairs: if one particle exerts a force on a second particle in our system, the second must exert an equal and opposite force on the first. Thus, in the sum over all the forces in Eq. 9–16, these internal forces cancel each other in pairs. We are left, then, with only the external forces on the right side of Eq. 9–16:

$$M\mathbf{a}_{CM} = \Sigma \mathbf{F}_{ext} , \qquad\qquad \text{[constant } M] \quad (9\text{–}17)$$

NEWTON'S SECOND LAW
(for a system)

where $\Sigma \mathbf{F}_{ext}$ is the sum of all the external forces acting on our system, which is the *net force* acting on the system. Thus

> **the sum of all the forces acting on the system is equal to the total mass of the system times the acceleration of its center of mass.**

NEWTON'S SECOND LAW
(for a system)

This is **Newton's second law** for a system of particles. It also applies to an extended body (which can be thought of as a collection of particles), and to a system of bodies. Thus we conclude that

> **the center of mass of a system of particles (or bodies) with total mass M moves like a single particle of mass M acted upon by the same net external force.**

Translational motion of CM

That is, the system moves as if all its mass were concentrated at the CM and all the external forces acted at that point. We can thus treat the *translational motion* of any body or system of bodies as the motion of a particle (see Figs. 9–21 and 9–22).

This theorem clearly simplifies our analysis of the motion of complex systems and extended bodies. Although the motion of various parts of the system may be complicated, we may often be satisfied with knowing the motion of the CM. This theorem also allows us to solve certain types of problems very easily, as illustrated by the following Example.

CONCEPTUAL EXAMPLE 9–16 **A two-stage rocket.** A rocket is shot into the air as shown in Fig. 9–32. At the moment it reaches its highest point, a horizontal distance d from its starting point, a prearranged explosion separates it into two parts of equal mass. Part I is stopped in midair by the explosion and falls vertically to Earth. Where does part II land? Assume $\mathbf{g}$ = constant.

SOLUTION After the rocket is fired, the path of the CM of the system continues to follow the parabolic trajectory of a projectile acted on only by a constant gravitational force. The CM will thus arrive at a point $2d$ from the starting point. Since the masses of I and II are equal, the CM must be midway between them. Therefore, II lands a distance $3d$ from the starting point. (If part I had been given a kick up or down, instead of merely falling, the solution would have been somewhat more complicated.)

FIGURE 9–32 Example 9–16.

We can write Eq. 9–17, $M\mathbf{a}_{CM} = \Sigma\mathbf{F}_{ext}$, in terms of the total momentum $\mathbf{P}$ of a system of particles. $\mathbf{P}$ is defined, as we saw in Section 9–2 as

$$\mathbf{P} = m_1\mathbf{v}_1 + m_2\mathbf{v}_2 + \cdots + m_n\mathbf{v}_n = \Sigma\mathbf{p}_i.$$

From Eq. 9–15 $(M\mathbf{v}_{CM} = \Sigma m_i\mathbf{v}_i)$ we have

Total momentum

$$\mathbf{P} = M\mathbf{v}_{CM}. \qquad (9\text{–}18)$$

Thus, *the total linear momentum of a system of particles is equal to the product of the total mass M and the velocity of the center of mass of the system.* Or, *the linear momentum of an extended body is the product of the body's mass and the velocity of its* CM.

If we differentiate Eq. 9–18 with respect to time, we obtain (assuming the total mass M is constant)

$$\frac{d\mathbf{P}}{dt} = M\frac{d\mathbf{v}_{CM}}{dt} = M\mathbf{a}_{CM}.$$

From Eq. 9–17, we see that

NEWTON'S SECOND LAW
(for a system)

$$\frac{d\mathbf{P}}{dt} = \Sigma\mathbf{F}_{ext}, \qquad (9\text{–}6)$$

where $\Sigma\mathbf{F}_{ext}$ is the net external force on the system. This is just Eq. 9–6 obtained earlier: **Newton's second law for a system of objects.** It is valid for any definite fixed system of particles or objects. If we know $\Sigma\mathbf{F}_{ext}$, we can determine how the total momentum changes.

An interesting example is the discovery since 1996 of nearby stars that seem to "wobble." What could cause such a wobble? Very likely a planet that orbits about the star, each exerting a gravitational force on the other. The planets are too small and too far away to have been observed directly by existing telescopes. But the slight wobble in the motion of the star suggests that both the planet and the star (its sun) orbit about their mutual center of mass, and hence the star appears to have a wobble. Irregularities in the star's motion can be obtained to an accuracy of 3 m/s, and from the data, the size of the planets' orbits can be obtained, as well as their masses (if as large as Jupiter's). See Fig. 6–17 in Chapter 6.

⇒ **PHYSICS APPLIED**

Distant planets discovered

* 9–10 Systems of Variable Mass; Rocket Propulsion

We now treat systems whose mass varies. Such systems could be treated as a type of inelastic collision, but it is simpler to use Eq. 9–6, $d\mathbf{P}/dt = \Sigma\mathbf{F}_{ext}$, where $\mathbf{P}$ is the total momentum of the system and $\Sigma\mathbf{F}_{ext}$ is the net external force exerted on it. Great care must be taken to define the system, and to include all changes in momentum. An important application is to rockets, which propel themselves forward by the ejection of burnt gases: the force exerted by the gases on the rocket accelerates the rocket. The mass M of the rocket decreases during this process, so $dM/dt < 0$. Another application is the dropping of material (gravel, packaged goods) onto a conveyor belt. In this situation, the mass M of the loaded conveyor belt increases, so $dM/dt > 0$.

To treat the general case of a system of variable mass, let us consider the system shown in Fig. 9–33. At some time t, we have a system of mass M and momentum $M\mathbf{v}$; we also have a tiny (infinitesimal) mass dM traveling with velocity $\mathbf{u}$ which is about to enter our system. An infinitesimal time dt later, the mass dM combines with the system. For simplicity we will refer to this as a "collision." So our system has changed in mass from M to $M + dM$ in the time dt. Note that dM can be less than zero, as for a rocket propelled by ejected gases.

In order to apply Eq. 9–6, $d\mathbf{P}/dt = \Sigma\mathbf{F}_{ext}$, we must consider a definite fixed system of particles. That is, in considering the change in momentum, $d\mathbf{P}$, we must consider the momentum of the same particles initially and finally. We will define our *total system* as including M plus dM. Then initially, at time t, the total momentum is $M\mathbf{v} + \mathbf{u}dM$ (Fig. 9–33). At time $t + dt$, after dM has combined with M, the velocity of the whole is now $\mathbf{v} + d\mathbf{v}$ and the total momentum is $(M + dM)(\mathbf{v} + d\mathbf{v})$. So the change in momentum $d\mathbf{P}$ is

$$d\mathbf{P} = (M + dM)(\mathbf{v} + d\mathbf{v}) - (M\mathbf{v} + \mathbf{u}dM)$$

$$= M\,d\mathbf{v} + \mathbf{v}dM + dM\,d\mathbf{v} - \mathbf{u}dM.$$

Then, by Eq. 9–6, we have

$$\Sigma\mathbf{F}_{ext} = \frac{d\mathbf{P}}{dt} = \frac{M\,d\mathbf{v} + \mathbf{v}dM - \mathbf{u}dM}{dt},$$

where we have dropped the term $dM\,d\mathbf{v}/dt$, since in the limit of infinitesimals it is zero ($d\mathbf{v}/dt$ may not approach zero but dM surely does). Thus we get

$$\Sigma\mathbf{F}_{ext} = M\frac{d\mathbf{v}}{dt} - (\mathbf{u} - \mathbf{v})\frac{dM}{dt}. \qquad \textbf{(9–19a)}$$

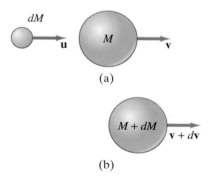

FIGURE 9–33 (a) At time t, a mass dM is about to be added to our system M. (b) At time $t + dt$, the mass dM has been added to our system.

Note that the quantity $(\mathbf{u} - \mathbf{v})$ is the relative velocity, $\mathbf{v}_{rel}$, of dM with respect to M. That is,

$$\mathbf{v}_{rel} = \mathbf{u} - \mathbf{v}$$

is the velocity of the entering mass dM as seen by an observer on M. So we can rearrange Eq. 9–19a:

$$M\frac{d\mathbf{v}}{dt} = \Sigma\mathbf{F}_{ext} + \mathbf{v}_{rel}\frac{dM}{dt}. \qquad \textbf{(9–19b)}$$

We can interpret this equation as follows. $M dv/dt$ is the mass times the acceleration of M. The first term on the right, $\Sigma\mathbf{F}_{ext}$, refers to the external force on the mass M (for a rocket, it would include the force of gravity and air resistance). It does *not* include the force that dM exerts on M as a result of their collision. This is taken care of by the second term on the right, $\mathbf{v}_{rel}(dM/dt)$, which represents the rate at which momentum is being transferred to (or from) the mass M because of the mass that is added to (or leaves) it. It can thus be interpreted as the force exerted on the mass M due to the addition (or ejection) of mass. For a rocket this term is called the *thrust*, since it represents the force exerted on the rocket by the expelled gases.

Thrust on rocket

➡ PHYSICS APPLIED

Moving conveyor belt

FIGURE 9–34 Gravel dropped from hopper onto conveyor belt.

EXAMPLE 9–17 **Conveyor belt.** You are designing a conveyor system for a gravel yard. A hopper drops gravel at a rate of 75.0 kg/s onto a conveyor belt that moves at a constant speed $v = 2.20\ \text{m/s}$ (Fig. 9–34). (*a*) Determine the force needed to keep the conveyor belt moving. (*b*) What power output must the motor have that drives the conveyor belt.

SOLUTION We assume that the hopper is at rest so $u = 0$ and that the hopper has just begun dropping gravel, so $dM/dt = 75.0\ \text{kg/s}$. Since the belt moves at a constant speed ($dv/dt = 0$), we have from Eq. 9–19a:

$$F_{ext} = M\frac{dv}{dt} - (u - v)\frac{dM}{dt}$$

$$= 0 - (0 - v)\frac{dM}{dt}$$

$$= v\frac{dM}{dt} = (2.20\ \text{m/s})(75.0\ \text{kg/s}) = 165\ \text{N}.$$

(*b*) From Eq. 8–22, this force does work at the rate

$$\frac{dW}{dt} = \mathbf{F}_{ext} \cdot \mathbf{v} = v^2\frac{dM}{dt}$$

$$= 363\ \text{W},$$

which is the power output required of the motor.
[But note that this work does not all go into kinetic energy of the gravel, since

$$\frac{dK}{dt} = \frac{d}{dt}\left(\frac{1}{2}Mv^2\right) = \frac{1}{2}\frac{dM}{dt}v^2,$$

which is only half the work done by $\mathbf{F}_{ext}$. The other half of the external work done goes into thermal energy produced by friction between the gravel and the belt (the same friction force that accelerates the gravel).]

228 CHAPTER 9 Linear Momentum and Collisions

EXAMPLE 9–18 **Rocket propulsion.** A fully fueled rocket has a mass of 21,000 kg, of which 15,000 kg is fuel. The burned fuel is spewed out the rear at a rate of 190 kg/s with a speed of 2800 m/s relative to the rocket. If the rocket is fired vertically upward (Fig. 9–35) calculate: (*a*) the thrust of the rocket; (*b*) the net force on the rocket at blastoff, and just before burnout (when all the fuel has been used up); (*c*) the rocket's velocity as a function of time, and (*d*) its final velocity at burnout. Ignore air resistance and assume the acceleration due to gravity is constant at $g = 9.80 \text{ m/s}^2$.

SOLUTION (*a*) The thrust (see discussion after Eq. 9–19b) is:

$$F_{\text{thrust}} = v_{\text{rel}} \frac{dM}{dt} = (-2800 \text{ m/s})(-190 \text{ kg/s}) = 5.3 \times 10^5 \text{ N}.$$

where we have taken upward as positive so v_{rel} is negative because it is downward, and dM/dt is negative because the rocket's mass is diminishing.
(*b*) $F_{\text{ext}} = Mg = (2.1 \times 10^4 \text{ kg})(9.80 \text{ m/s}^2) = 2.1 \times 10^5 \text{ N}$ initially, and $(6.0 \times 10^3 \text{ kg})(9.80 \text{ m/s}^2) = 5.9 \times 10^4 \text{ N}$ at burnout. Hence, the net force on the rocket at blastoff is

$$F_{\text{net}} = 5.3 \times 10^5 \text{ N} - 2.1 \times 10^5 \text{ N} = 3.2 \times 10^5 \text{ N}, \qquad [\text{blastoff}]$$

and just before burnout it is

$$F_{\text{net}} = 5.3 \times 10^5 \text{ N} - 5.9 \times 10^4 \text{ N} = 4.7 \times 10^5 \text{ N}. \qquad [\text{burnout}]$$

After burnout, of course, the net force is that of gravity, $-5.9 \times 10^4 \text{ N}$.
(*c*) From Eq. 9–19b we have

$$dv = \frac{F_{\text{ext}}}{M} dt + v_{\text{rel}} \frac{dM}{M},$$

where $F_{\text{ext}} = -Mg$, and M is the mass of the rocket and is a function of time. Since v_{rel} is constant, we can integrate this easily:

$$\int_{v_0}^{v} dv = -\int_0^t g \, dt + v_{\text{rel}} \int_{M_0}^{M} \frac{dM}{M}$$

or

$$v(t) = v_0 - gt + v_{\text{rel}} \ln \frac{M}{M_0},$$

v as function of t

where $v(t)$ is the rocket's velocity and M its mass at any time t. Note that v_{rel} is negative (-2800 m/s in our case) because it is opposite to the motion, and that $\ln (M/M_0)$ is also negative because $M_0 > M$. Hence, the last term—which represents the thrust—is positive and acts to increase the velocity.
(*d*) The time required to reach burnout is the time needed to use up all the fuel (15,000 kg) at a rate of 190 kg/s; so at burnout,

$$t = \frac{1.50 \times 10^4 \text{ kg}}{190 \text{ kg/s}} = 79.0 \text{ s}.$$

If we take $v_0 = 0$, then using the result of part (*c*):

$$v = -(9.80 \text{ m/s}^2)(79 \text{ s}) + (-2800 \text{ m/s})\left(\ln \frac{6000 \text{ kg}}{21,000 \text{ kg}} \right) = 2730 \text{ m/s}.$$

v at burnout

FIGURE 9–35 Example 9–18; $\mathbf{v}_{\text{rel}} = \mathbf{v}_{\text{gases}} - \mathbf{v}_{\text{rocket}}$. M is the mass of the rocket at any instant and is decreasing until burnout.

Summary

The **linear momentum, p**, of an object is defined as the product of its mass times its velocity,

$$\mathbf{p} = m\mathbf{v}.$$

In terms of momentum, **Newton's second law** can be written as

$$\Sigma\mathbf{F} = \frac{d\mathbf{p}}{dt}.$$

That is, the rate of change of momentum of an object equals the net force exerted on it.

When the net external force on a system of objects is zero, the total momentum remains constant. This is the **law of conservation of momentum**. Stated another way, the total momentum of an isolated system of objects remains constant.

The law of conservation of momentum is very useful in dealing with the class of events known as **collisions**. In a collision two (or more) bodies interact with each other for a very short time and the force between them during this time is very large compared to any other forces acting. The **impulse** of such a force on a body is defined as

$$\mathbf{J} = \int\mathbf{F}\,dt$$

and is equal to the change in momentum of the body as long as $\mathbf{F}$ is the net force on the body:

$$\Delta\mathbf{p} = \mathbf{p}_f - \mathbf{p}_i = \int_{t_i}^{t_f}\mathbf{F}\,dt = \mathbf{J}.$$

Total momentum is conserved in any collision:

$$\mathbf{p}_1 + \mathbf{p}_2 = \mathbf{p}_1' + \mathbf{p}_2'.$$

The total energy is also conserved, but this may not be useful unless the only type of energy transformation involves kinetic energy. In this case, kinetic energy is conserved and the collision is called an **elastic collision**:

$$\tfrac{1}{2}mv_1^2 + \tfrac{1}{2}mv_2^2 = \tfrac{1}{2}m_1v_1'^2 + \tfrac{1}{2}m_2v_2'^2.$$

If kinetic energy is not conserved, the collision is called **inelastic**. If two colliding objects stick together as the result of a collision, the collision is said to be **completely inelastic**.

For a system of particles, or for an extended body that can be considered as having a continuous distribution of matter, the **center of mass** (CM) is defined as

$$x_{CM} = \frac{\Sigma m_i x_i}{M}, \qquad y_{CM} = \frac{\Sigma m_i y_i}{M}, \qquad z_{CM} = \frac{\Sigma m_i z_i}{M}$$

or

$$x_{CM} = \frac{1}{M}\int x\,dm, \quad y_{CM} = \frac{1}{M}\int y\,dm, \quad z_{CM} = \frac{1}{M}\int z\,dm,$$

where M is the total mass of the system.

The center of mass (CM) of a system is important because this point moves like a single particle of mass M acted on by the same net external force, $\Sigma\mathbf{F}_{ext}$. In equation form, this is just Newton's second law for a system of particles (or extended bodies):

$$M\mathbf{a}_{CM} = \Sigma\mathbf{F}_{ext},$$

where M is the total mass of the system, $\mathbf{a}_{CM}$ is the acceleration of the CM of the system, and $\Sigma\mathbf{F}_{ext}$ is the total (net) external force acting on all parts of the system.

For a system of particles of total linear momentum $\mathbf{P} = \Sigma m_i\mathbf{v}_i = M\mathbf{v}_{CM}$, Newton's second law is

$$\frac{d\mathbf{P}}{dt} = \Sigma\mathbf{F}_{ext}.$$

Questions

1. We claim that momentum is conserved. Yet most moving objects eventually slow down and stop. Explain.

2. Two blocks of mass m_1 and m_2 rest on a frictionless table and are connected by a spring. The blocks are pulled apart, stretching the spring, and then released. Describe the subsequent motion of the two blocks.

3. A light body and a heavy body have the same kinetic energy. Which has the greater momentum?

4. When a person jumps from a tree to the ground, what happens to the momentum of the person upon landing?

5. Explain, on the basis of conservation of momentum, how a fish propels itself forward by swishing its tail back and forth.

6. Why, when you release an inflated, untied balloon, does it fly across the room?

7. It is said that in ancient times a rich man with a bag of gold coins was stranded on the surface of a frozen lake and froze to death. Because the ice was frictionless, he could not push himself to shore. What could he have done to save himself had he not been so miserly?

8. If a falling ball were to make a perfectly elastic collision with the floor, would it rebound to its original height? Explain.

9. (a) An empty sled is sliding on frictionless ice when Susan drops vertically from a tree above onto the sled. When she lands, does the sled speed up, slow down, or keep the same speed? (b) Later, Susan falls off the sled. When she drops off, does the sled speed up, slow down, or keep the same speed?

10. Is it easier to hit a home run from a pitched ball than from one tossed in the air by the batter? Explain.

11. The speed of a tennis ball on the return of a serve can be just as fast as the serve, even though the racket isn't swung very fast. How can this be?

12. Is it possible for a body to receive a larger impulse from a small force than from a large force?

13. How could a force give zero impulse over a nonzero time interval even though the force is not zero for at least a part of that time interval?

14. In a collision between two cars, which would you expect to be more damaging to the occupants: if the cars collide and remain together, or if the two cars collide and rebound backward? Explain.

15. A superball is dropped from a height h onto a hard steel plate (fixed to the Earth), from which it rebounds at very nearly its original speed. (a) Is the momentum of the ball conserved during any part of this process? (b) If we consider the ball and the Earth as our system, during what parts of the process is momentum conserved? (c) Answer part (b) for a piece of putty that falls and sticks to the steel plate.

16. It used to be common wisdom to build cars to be as rigid as possible to withstand collisions. Today, though, cars are designed to have "crumple zones" that collapse upon impact. What advantage does this have?

17. A rocket following a parabolic path through the air suddenly explodes into many pieces. What can you say about the motion of this system of pieces?

18. Why is the CM of a 1-m length of pipe at the pipe's midpoint, whereas this is not true for your arm or leg?

19. Show on a diagram how your CM shifts when you move from a lying position to a sitting position.

20. Describe an analytic way of determining the CM of any thin, triangular-shaped, uniform plate.

21. Place yourself facing the edge of an open door. Position your feet astride the door with your nose and abdomen touching the door's edge. Try to rise on your tiptoes. Why can't this be done?

22. If only an external force can change the momentum of the center of mass of an object, how can the internal force of the engine accelerate a car?

23. How can a rocket change direction when it is far out in space and essentially in a vacuum?

24. In observations of nuclear β-decay, the electron and recoil nucleus often do not separate along the same line. Use conservation of momentum in two dimensions to explain why this implies the emission of at least one other particle in the disintegration.

25. Bob and Jim decide to play tug-of-war on a frictionless lake. Jim is considerably stronger than Bob, but Bob weighs 160 lbs while Jim weighs 145 lbs. Who loses by crossing over the midline first?

26. At a carnival game you try to knock over a heavy cylinder by throwing a small ball at it. You have a choice of throwing either a ball that will stick to the cylinder, or a second ball of equal mass and speed that will bounce backward off the cylinder. Which ball is more likely to make the cylinder move?

Problems

Section 9–1

1. (I) Calculate the force exerted on a rocket at takeoff, given that the propelling gases are expelled at a rate of 1200 kg/s with a speed of 50,000 m/s.

2. (I) The momentum of a particle, in SI units, is given by $\mathbf{p} = 4.8t^2\mathbf{i} - 8.0\mathbf{j} - 8.9t\mathbf{k}$. What is the force as a function of time?

3. (II) (a) What is the momentum of a 30.0-g sparrow with a speed of 12 m/s? (b) What will be its momentum 12 s later if a constant 2.0×10^{-2} N force due to air resistance acts on it?

4. (II) A 145-g baseball, moving along the x axis with speed 30.0 m/s, strikes a fence at a 45° angle and rebounds along the y axis with unchanged speed. Give its change in momentum using unit vector notation.

5. (II) The force on a particle of mass m is given by $\mathbf{F} = 26\mathbf{i} - 12t^2\mathbf{j}$ where F is in N and t in s. What will be the change in the particle's momentum between $t = 1.0$ s and $t = 2.0$ s?

6. (II) A 4200-kg rocket is traveling in outer space with a velocity of 120 m/s toward the Sun. It needs to alter its course by 23.0°, which can be done by shooting its rockets briefly in a direction perpendicular to its original motion. If the rocket gases are expelled at a speed of 2200 m/s relative to the rocket, what mass of gas must be expelled?

7. (II) A ball of mass m falls freely a distance h from rest. It hits the floor and rebounds to its starting height. (a) What total time is required for the motion? (Ignore the time of contact with the floor.) (b) What is the speed of the ball as it strikes the floor? (c) What is the change in momentum of the ball as it strikes the floor? (d) Calculate the average force the ball exerts on the floor averaged over the time interval of part (a). Is this result surprising?

8. (III) Air in a 100-km/h wind strikes head-on the face of a building 40 m wide by 60 m high and is brought to rest. If air has a mass of 1.3 kg per cubic meter, determine the average force of the wind on the building.

Section 9–2

9. (I) A 9700-kg boxcar traveling 18 m/s strikes a second car. The two stick together and move off with a speed of 4.0 m/s. What is the mass of the second car?

10. (I) A 130-kg tackler moving at 2.5 m/s meets head-on (and tackles) a 90-kg halfback moving at 5.0 m/s. What will be their mutual speed immediately after the collision?

11. (I) An atomic nucleus at rest decays radioactively into an alpha particle and a smaller nucleus. What will be the speed of this recoiling nucleus if the speed of the alpha particle is 2.5×10^5 m/s? Assume the nucleus has a mass 57 times greater than that of the alpha particle.

12. (I) A 10,500-kg railroad car travels alone on a level frictionless track with a constant speed of 15.0 m/s. An additional 6350 kg load is dropped from a tower onto the car. What then will be its speed?

13. (II) A child in a boat throws a 5.40-kg package out horizontally with a speed of 10.0 m/s, Fig. 9–36. Calculate the velocity of the boat immediately after, assuming it was initially at rest. The mass of the child is 26.0 kg and that of the boat is 55.0 kg.

FIGURE 9–36 Problem 13.

14. (II) A 12-gram bullet traveling 190 m/s penetrates a 2.0-kg block of wood and emerges going 150 m/s. If the block is stationary on a frictionless surface when hit, how fast does it move after the bullet emerges?

15. (II) An explosion breaks an object, originally at rest, into two fragments. One fragment acquires twice the kinetic energy of the other. What is the ratio of their masses? Which is greater?

16. (II) A ball moving with a speed of 17 m/s strikes an identical ball that is initially at rest. After the collision, the incoming ball has been deviated by 45° from its original direction, and the struck ball moves off at 30° from the original direction (Fig. 9–37). What are the speeds of the two balls after the collision?

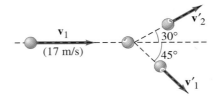

FIGURE 9–37 Problem 16.

17. (II) A rocket of mass m traveling with speed v_0 along the x axis suddenly shoots out one-third its mass parallel to the y axis (as seen by an observer at rest) with speed $2v_0$. Express the velocity of the remainder of the object in $\mathbf{i}, \mathbf{j}, \mathbf{k}$ notation.

18. (II) The decay of a neutron into a proton, an electron, and a neutrino is an example of a three-particle decay process. Use the vector nature of momentum to show that if the neutron is initially at rest, the velocity vectors of the three must be coplanar (that is, all in the same plane). The result is not true for numbers greater than three.

19. (II) A radioactive nucleus at rest decays into a second nucleus, an electron, and a neutrino. The electron and neutrino are emitted at right angles and have momenta of 8.6×10^{-23} kg·m/s and 6.2×10^{-23} kg·m/s, respectively. What is the magnitude and the direction of the momentum of the recoiling nucleus?

20. (II) A 900-kg two-stage rocket is traveling at a speed of 6.50×10^3 m/s away from Earth when a predesigned explosion separates the rocket into two sections of equal mass that then move with a relative speed (relative to each other) of 2.80×10^3 m/s along the original line of motion. (a) What is the speed and direction of each section (relative to Earth) after the explosion? (b) How much energy was supplied by the explosion? [*Hint:* What is the change in kinetic energy as a result of the explosion?]

21. (III) A 200-kg projectile, fired with a speed of 100 m/s at a 60° angle, breaks into three pieces of equal mass at the highest point of its arc. Two of the fragments move with the same equal speed right after the explosion as the entire projectile had just before the explosion; one of these moves vertically downward and the other horizontally. Determine (a) the velocity of the third fragment immediately after the explosion and (b) the energy released in the explosion.

Section 9–3

22. (I) A 0.145-kg baseball pitched at 35.0 m/s is hit on a horizontal line drive straight back at the pitcher at 56.0 m/s. If the contact time between bat and ball is 5.00×10^{-3} s, calculate the force (assumed to be constant) between the ball and bat.

23. (I) The tennis ball may leave the racket of a top player on the serve with a speed of 65.0 m/s. If the ball's mass is 0.0600 kg and is in contact with the racket for 0.0300 s, what is the average force on the ball? Would this force be large enough to lift a 60-kg person?

24. (II) A tennis ball of mass $m = 0.060$ kg and speed $v = 28$ m/s strikes a wall at a 45° angle and rebounds with the same speed at 45° (Fig. 9–38). What is the impulse given the wall?

FIGURE 9–38
Problem 24.

25. (II) Water strikes the turbine blades of a generator so that its rebounding velocity is 0.75 of its original magnitude and reversed in direction. If the flow rate is 60 kg/s and the original water speed is 10 m/s, what is the average force on the blades?

26. (II) A 140-kg astronaut (including space suit) acquires a speed of 2.50 m/s by pushing off with his legs from a 1800-kg space capsule. (a) What is the change in speed of the space capsule? (b) If the push lasts 0.500 s, what is the average force exerted by each on the other? As the reference frame, use the position of the capsule before the push. (c) What is the kinetic energy of each after the push?

27. (II) (a) A molecule of mass m and speed v strikes a wall at right angles and rebounds with the same speed. If the collision time is Δt, what is the average force on the wall during the collision? (b) If molecules, all of this type, strike the wall at intervals a time t apart (on the average) what is the average force on the wall averaged over a long time?

28. (II) Suppose the force acting on a tennis ball (mass 0.060 kg) as a function of time is given by the graph of Fig. 9–39. Use the graph to estimate (a) the total impulse given the ball, and (b) the speed of the ball after being struck, assuming the ball is being served so it is nearly at rest initially.

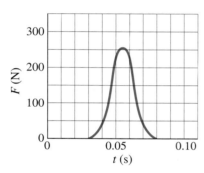

FIGURE 9–39 Problem 28.

29. (II) The force on a bullet is given by the formula $F = \left(780 - 2.6 \times 10^5 t\right)$ N over the time interval $t = 0$ to $t = 3.0 \times 10^{-3}$ s. In this formula t is in seconds and F is in newtons. (a) Plot a graph of F versus t for $t = 0$ to $t = 3.0 \times 10^{-3}$ s. (b) Estimate, using graphs, the impulse given the bullet. (c) Determine the impulse by integration. (d) If the bullet achieves a speed of 300 m/s as a result of this impulse, given to it in the barrel of a gun, what must the bullet's mass be?

30. (III) From what maximum height can a 75-kg person jump without breaking the lower leg bone on either leg? Ignore air resistance and assume the body of the person moves a distance of 0.60 m from the standing to the seated position (that is, in breaking the fall). Assume the breaking strength (force per unit area) of bone is 170×10^6 N/m², and its smallest cross-sectional area is 2.5×10^{-4} m².

31. (III) A scale is adjusted so that when a large, flat pan is placed on it, it reads zero. A water faucet at height $h = 2.5$ m above is turned on and water falls into the pan at a rate $R = 0.12$ kg/s. Determine (a) a formula for the scale reading as a function of time t and (b) the reading for $t = 15$ s. (c) Repeat (a) and (b), but replace the flat pan with a tall, narrow cylindrical container of area $A = 20$ cm² (the level rises in this case).

Sections 9–4 and 9–5

32. (II) A ball of mass 0.540 kg moving east ($+x$ direction) with a speed of 3.90 m/s collides head-on with a 0.320-kg ball at rest. If the collision is perfectly elastic, what will be the speed and direction of each ball after the collision?

33. (II) A 0.450-kg hockey puck, moving east with a speed of 4.20 m/s, has a head-on collision with a 0.900-kg puck initially at rest. Assuming a perfectly elastic collision, what will be the speed and direction of each object after the collision?

34. (II) A 0.060-kg tennis ball, moving with a speed of 7.50 m/s, has a head-on collision with a 0.090-kg ball initially moving away from it at a speed of 3.00 m/s. Assuming a perfectly elastic collision, what is the speed and direction of each ball after the collision?

35. (II) A softball of mass 0.220 kg that is moving with a speed of 6.5 m/s collides head-on and elastically with another ball initially at rest. Afterward it is found that the incoming ball has bounced backward with a speed of 3.8 m/s. Calculate (a) the velocity of the target ball after the collision, and (b) the mass of the target ball.

36. (II) A 0.280-kg croquet ball makes an elastic head-on collision with a second ball initially at rest. The second ball moves off with half the original speed of the first ball. (a) What is the mass of the second ball? (b) What fraction of the original kinetic energy gets transferred to the second ball?

37. (II) An atomic nucleus initially moving at 500 m/s emits an alpha particle in the direction of its velocity, and the new nucleus slows to 450 m/s. If the alpha particle has a mass of 4.0 u and the original nucleus has a mass of 222 u, what speed does the alpha particle have when it is emitted?

38. (II) A ball of mass m makes a head-on-elastic collision with a second ball (at rest) and rebounds with a speed equal to one-fourth its original speed. What is the mass of the second ball?

39. (II) Determine the fraction of kinetic energy lost by a neutron $\left(m_1 = 1.01 \text{ u}\right)$ when it collides head-on and elastically with (a) ^1_1H ($m = 1.01$ u); (b) ^2_1H (heavy hydrogen, $m = 2.01$ u); (c) $^{12}_6\text{C}$ ($m = 12.00$ u); (d) $^{208}_{82}\text{Pb}$ (lead, $m = 208$ u).

40. (II) Show that, in general, for any head-on one-dimensional elastic collision, the speeds after collision are

$$v_2' = v_1\left(\frac{2m_1}{m_1 + m_2}\right) + v_2\left(\frac{m_2 - m_1}{m_1 + m_2}\right)$$

and

$$v_1' = v_1\left(\frac{m_1 - m_2}{m_1 + m_2}\right) + v_2\left(\frac{2m_2}{m_1 + m_2}\right),$$

where v_1 and v_2 are the initial speeds of the two objects of mass m_1 and m_2.

41. (III) A 2.0-kg block slides along a frictionless tabletop at 8.0 m/s toward a second block (at rest) of mass 4.5 kg. A coil spring, which obeys Hooke's law and has spring constant $k = 850$ N/m, is attached to the second block in such a way that it will be compressed when struck by the moving block, Fig. 9–40. (a) What will be the maximum compression of the spring? (b) What will be the final velocities of the blocks after the collision? (c) Is the collision elastic?

FIGURE 9–40 Problem 41.

42. (II) An 18-g rifle bullet traveling 180 m/s buries itself in a 3.6-kg pendulum hanging on a 2.8-m-long string, which makes the pendulum swing upward in an arc. Determine the horizontal component of the pendulum's displacement.

43. (II) (a) Derive a formula for the fraction of kinetic energy lost, $\Delta K/K$, for the ballistic pendulum collision of Example 9–10. (b) Evaluate for $m = 14.0$ g and $M = 380$ g.

44. (II) An explosion breaks an object into two pieces, one of which has 1.5 times the mass of the other. If 17,500 J were released in the explosion, how much kinetic energy did each piece acquire?

45. (II) After a completely inelastic collision between two objects of equal mass, each having initial speed, v, the two move off with speed $v/3$. What was the angle between their initial directions?

46. (II) A 0.95×10^3-kg Toyota collides into the rear end of a 2.2×10^3-kg Cadillac stopped at a red light. The bumpers lock, the brakes are locked, and the two cars skid forward 4.8 m before stopping. The police officer, knowing that the coefficient of kinetic friction between tires and road is 0.40, calculates the speed of the Toyota at impact. What was that speed?

47. (II) A measure of inelasticity in a head-on collision of two bodies is the *coefficient of restitution*, e, defined as

$$e = \frac{v_1' - v_2'}{v_2 - v_1},$$

where $v_1' - v_2'$ is the relative velocity of the two bodies after the collision and $v_2 - v_1$ is their relative velocity before. (a) Show that for a perfectly elastic collision, $e = 1$, and for a completely inelastic collision, $e = 0$. (b) A simple method for measuring the coefficient of restitution for a body colliding with a very hard surface like steel is to drop the body onto a heavy steel plate, as shown in Fig. 9–41. Determine a formula for e in terms of the original height h and the maximum height reached after collision h'.

FIGURE 9–41 Problem 47. Measurement of coefficient of restitution.

48. (II) A wooden block is cut into two pieces, one with three times the mass of the other. A depression is made in both faces of the cut, so that a firecracker can be placed in it with the block reassembled. The reassembled block is set on a rough-surfaced table and the fuse is lit. When the firecracker explodes, the two blocks separate and slide apart. What is the ratio of distances each block travels?

49. (III) A 5.0-kg body moving in the $+x$ direction at 5.5 m/s collides head-on with a 3.0-kg body moving in the $-x$ direction at 4.0 m/s. Find the final velocity of each mass if: (a) the bodies stick together; (b) the collision is elastic; (c) the 5.0-kg body is at rest after the collision; (d) the 3.0-kg body is at rest after the collision; (e) the 5.0-kg body has a velocity of 4.0 m/s in the $-x$ direction after the collision. Are the results in (c), (d), and (e) "reasonable"? Explain.

50. (III) For the ballistic pendulum, let us estimate the accuracy of our approximation that the block (mass M) doesn't move during the collision. Use the symbols of Example 9–10 and Fig. 9–18, and assume the projectile is decelerated uniformly within the block in a distance d. Estimate: (a) the collision time Δt; (b) by how much is momentum not conserved, $\Delta \mathbf{p}$, because there is a net external force acting during this time; and (c) by what fraction the calculated speed of the bullet will be in error if we use the equation given in Example 9–10.

Section 9–7

51. (II) An eagle ($m_1 = 3.3$ kg) moving with speed $v_1 = 7.8$ m/s is on a collision course with a second eagle ($m_2 = 4.6$ kg) moving at $v_2 = 10.2$ m/s in a direction at right angles to the first. After they collide, they hold onto one another. In what direction, and with what speed, are they moving after the collision?

52. (II) A billiard ball of mass $m_A = 0.400$ kg moving with speed $v_A = 1.80$ m/s strikes a second ball, initially at rest, of mass $m_B = 0.500$ kg. As a result of the collision, the first ball is deflected off at an angle of 30.0° with a speed $v_A' = 1.10$ m/s. (a) Taking the x axis to be the original direction of motion of ball A, write down the equations expressing the conservation of momentum for the components in the x and y directions separately. (b) Solve these equations for the speed, v_B', and angle, θ', of ball B. Do not assume the collision is elastic.

53. (II) An atomic nucleus of mass m traveling with speed v collides elastically with a target particle of mass $2m$ (initially at rest) and is scattered at 90°. (a) At what angle does the target particle move after the collision? (b) What are the final speeds of the two particles? (c) What fraction of the initial kinetic energy is transferred to the target particle?

54. (II) Two billiard balls of equal mass move at right angles and meet at the origin of an xy coordinate system. One is moving upward along the y axis at 2.0 m/s, and the other is moving to the right along the x axis with speed 3.7 m/s. After the collision (assumed elastic), the second ball is moving along the positive y axis (Fig. 9–42). What is the final direction of the first ball, and what are their two speeds?

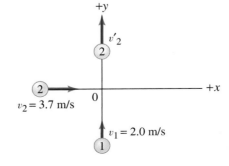

FIGURE 9–42 Problem 54. (Ball 1 after the collision is not shown.)

55. (II) A neutron collides elastically with a helium nucleus (at rest initially) whose mass is four times that of the neutron. The helium nucleus is observed to rebound at an angle $\theta'_2 = 45°$. Determine the angle of the neutron, θ'_1, and the speeds of the two particles, v'_n and v'_{He}, after the collision. The neutron's initial speed is 6.2×10^5 m/s.

56. (III) A neon atom ($m = 20$ u) makes a perfectly elastic collision with another atom. After the impact, the neon atom travels away at a 55.6° angle from its original direction and the unknown atom travels away at a 50.0° angle. What is the mass (in u) of the unknown atom?

57. (III) Prove that in the elastic collision of two objects of identical mass, with one being a target initially at rest, the angle between their final velocity vectors is always 90°.

58. (III) For an elastic collision between a projectile particle of mass m_1 and a target particle (at rest) of mass m_2, show that the scattering angle, θ'_1, of the projectile (a) can take any value, 0 to 180°, for $m_1 < m_2$, but (b) has a maximum angle ϕ given by $\cos^2\phi = 1 - (m_2/m_1)^2$ for $m_1 > m_2$.

Section 9–8

59. (I) The distance between a carbon atom ($m = 12$ u) and an oxygen atom ($m = 16$ u) in the CO molecule is 1.13×10^{-10} m. How far from the carbon atom is the center of mass of the molecule?

60. (I) An empty 1150-kg car has its CM 2.50 m behind the front of the car. How far from the front of the car will the CM be when two people sit in the front seat 2.80 m from the front of the car, and three people sit in the back seat 3.90 m from the front? Assume that each person has a mass of 70.0 kg.

61. (II) Find the center of mass of the ammonia molecule. The chemical formula is NH_3. The hydrogens are at the corners of an equilateral triangle (with sides 0.16 nm) that forms the base of a pyramid, with nitrogen at the apex (0.037 nm vertically above the plane of the triangle).

62. (II) Three cubes, of side l_0, $2l_0$, and $3l_0$, are placed next to one another (in contact) with their centers along a straight line and the $l = 2l_0$ cube in the center (Fig. 9–43). What is the position, along this line, of the CM of this system? Assume the cubes are made of the same uniform material.

FIGURE 9–43   Problem 62.

63. (II) A square uniform raft, 18 m by 18 m, of mass 6200 kg, is used as a ferryboat. If three cars, each of mass 1200 kg, occupy the NE, SE, and SW corners, determine the CM of the loaded ferryboat.

64. (II) A uniform thin machine part is a flat circular plate of radius $2R$ that has a circular hole of radius R cut out of it. The center of the hole is a distance $0.80R$ from the center of the plate, Fig. 9–44. What is the position of the center of mass of the plate? [*Hint:* Think of a solid plate as being the sum of the given plate plus a small one of radius R.]

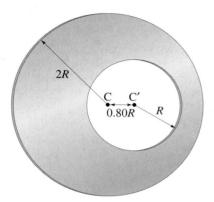

FIGURE 9–44   Problem 64.

65. (II) A uniform thin wire is bent into a semicircle of radius r. Determine the coordinates of its center of mass with respect to an origin of coordinates at the center of the "full" circle.

66. (III) Determine the CM of a thin, uniform, semicircular plate.

67. (III) Determine the CM of a machine part that is a uniform cone of height h and radius R, Fig. 9–45. [*Hint:* Divide the cone into an infinite number of disks of thickness dz, one of which is shown.]

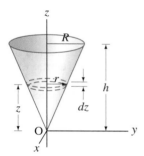

FIGURE 9–45   Problem 67.

68. (III) Determine the CM of a uniform pyramid that has four triangular faces and a square base with equal sides all of length s. [*Hint:* See Problem 67.]

Section 9–9

69. (I) The masses of the Earth and Moon are 5.98×10^{24} kg and 7.35×10^{22} kg, respectively, and their centers are separated by 3.84×10^8 m. (a) Where is the CM of this system located? (b) What can you say about the motion of the Earth-Moon system around the Sun, and of the Earth and Moon separately around the Sun?

70. (II) Two 35-kg masses have velocities (in m/s) of $v_1 = 12i - 16j$ and $v_2 = -20i + 14j$. Determine the velocity of the center of mass of the system.

71. (II) A 50-kg woman and a 70-kg man stand 11.0 m apart on frictionless ice. (a) How far from the man is their CM? (b) If they hold onto the two ends of a rope, and the man pulls on the rope so that he moves 2.8 m, how far from the woman will he now be? (c) How far will the man have moved when he collides with the woman?

72. (II) Suppose that in Example 9–16 (Fig. 9–32) $m_{II} = 3m_I$. (a) Where then would m_{II} land? (b) What if $m_I = 3m_{II}$?

73. (II) A helium balloon and its gondola, of mass M, are in the air and stationary with respect to the ground. A passenger, of mass m, then climbs out and slides down a rope with speed v, measured with respect to the balloon. With what speed and direction (relative to Earth) does the balloon then move? What happens if the passenger stops?

74. (II) Two people, one of mass 75 kg and the other of mass 55 kg, sit in a rowboat of mass 80 kg. With the boat initially at rest, the two people, who have been sitting at opposite ends of the boat, 3.0 m apart from each other, exchange seats. How far and in what direction will the boat move?

75. (III) A 200-kg flatcar 25 m long is moving with a speed of 5.0 m/s along horizontal frictionless rails. A 90-kg man then starts walking from one end of the car to the other in the direction of motion, with speed 2.0 m/s with respect to the car. In the time it takes for him to reach the other end, how far has the flatcar moved?

* Section 9–10

* 76. (II) A 2500-kg rocket is to be accelerated at $3.0\,g$ at take-off from the Earth. If the gases can be ejected at a rate of 30 kg/s, what must be their exhaust speed?

* 77. (II) Suppose the conveyor belt of Example 9–17 is retarded by a friction force of 140 N. Determine the required output power (hp) of the motor as a function of time from the moment gravel first starts falling ($t = 0$) until 3.0 s after the gravel begins to be dumped off the end of the 20-m-long conveyor belt.

* 78. (II) A rocket traveling 1850 m/s away from the Earth at an altitude of 6400 km fires its rockets, which eject gas at a speed of 1200 m/s (relative to the rocket). If the mass of the rocket at this moment is 25,000 kg and an acceleration of 1.7 m/s² is desired, at what rate must the gases be ejected?

* 79. (II) The jet engine of an airplane takes in 100 kg of air per second, which is burned with 4.2 kg of fuel per second. The burned gases leave the plane at a speed of 550 m/s (relative to the plane). If the plane is traveling 270 m/s (600 mi/h), determine (a) the thrust due to ejected fuel; (b) the thrust due to accelerated air passing through the engine; and (c) the power (hp) delivered.

* 80. (III) A sled filled with sand slides without friction down a 30° slope. Sand leaks out a hole in the sled at a rate of 2.0 kg/s. If the sled starts from rest with an initial total mass of 40.0 kg, how long does it take the sled to travel 120 m along the slope?

General Problems

81. A novice pool player is faced with the corner pocket shot shown in Fig. 9–46. The relative size of some of the dimensions (the units aren't important, only their ratios) are also shown. Should the player be worried about this being a "scratch shot," one where the cue ball will also fall into a pocket? Give details.

FIGURE 9–46 Problem 81.

82. During a Chicago storm, winds can whip horizontally at speeds of 100 km/h. If the air strikes a person at the rate of 40 kg/s per square meter and is brought to rest, calculate the force of the wind on a person. Assume the person's area to be 1.50 m high and 0.50 m wide. Compare to the typical maximum force of friction ($\mu \approx 1.0$) between the person and the ground, if the person has a mass of 70 kg.

83. A 0.145-kg pitched baseball moving horizontally at 35.0 m/s strikes a bat and is popped straight up to a height of 55.6 m before turning around. If the contact time is 0.50 ms, calculate the average force on the ball during the contact.

84. In order to convert a tough split in bowling, it is necessary to strike the pin a glancing blow as shown in Fig. 9–47. Assume that the bowling ball, initially traveling at 12.0 m/s, has five times the mass of a pin and that the pin goes off at 80° from the original direction of the ball. Calculate the speed (a) of the pin and (b) of the ball just after collision, and (c) calculate the angle through which the ball was deflected. Assume the collision is elastic and ignore any spin of the ball.

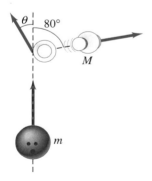

FIGURE 9–47 Problem 84.

85. A 15-g bullet strikes and becomes embedded in a 1.10-kg block of wood placed on a horizontal surface just in front of the gun. If the coefficient of kinetic friction between the block and the surface is 0.25, and the impact drives the block a distance of 9.5 m before it comes to rest, what was the muzzle speed of the bullet?

86. A gun is fired vertically into a 1.40-kg block of wood at rest on a thin horizontal sheet directly above it, Fig. 9–48. If the bullet has a mass of 21.0 g and a speed of 310 m/s, how high will the block rise into the air after the bullet becomes embedded in it?

FIGURE 9–48 Problem 86.

87. A ball of mass m makes a head-on elastic collision with a second ball (at rest) and rebounds with a speed equal to 0.600 of its original speed. What is the mass of the second ball?

88. You have been hired as an expert witness in a court case involving an automobile accident. The accident involved a car of mass 2000 kg (car A) which approached a stationary car of mass 1000 kg (car B). The driver of car A applied his brakes 15 m before he crashed into car B. After the collision, car A slid 15 m while car B slid 30 m. The coefficient of kinetic friction between the locked wheels and the road was measured to be 0.60. Prove to the court that the driver of car A was exceeding the 55-mph speed limit before applying the brakes.

89. For the completely inelastic collision of two railroad cars that we considered in Example 9–3, calculate how much of the initial kinetic energy is transformed to thermal or other forms of energy.

90. A hockey puck of mass $4m$ has been rigged to explode, as part of a practical joke. Initially the puck is at rest on a frictionless ice rink. Then it bursts into three pieces. One chunk, of mass m, slides across the ice at speed v. Another chunk, of mass $2m$, slides across the ice at speed $2v$, in a direction at right angles to the direction of the first piece. From this information, find the velocity of the final chunk.

91. Two blocks of mass m_1 and m_2, resting on a frictionless table, are connected by a stretched spring and then released. (a) Is there a net external force on the system? (b) Determine the ratio of their velocities, v_1/v_2. (c) What is the ratio of their kinetic energies? (d) Describe the motion of the CM of this system. (e) How would the presence of friction alter the above results?

92. A (light) pallet has a load of identical cases of tomato paste (see Fig. 9–49), each of which is a cube of length l. Find the center of gravity in the horizontal plane, so that a crane operator can pick up the load without tipping it.

FIGURE 9–49 Problem 92.

93. A 5800-kg open railroad car coasts along with a constant speed of 8.60 m/s on a level track. Snow begins to fall vertically and fills the car at a rate of 3.50 kg/min. Ignoring friction with the tracks, determine the speed of the car after 60.0 min using momentum conservation.

* 94. Consider the railroad car of Problem 93, which is slowly filling with snow. (a) Determine the speed of the car as a function of time using Eq. 9–19. (b) What is the speed of the car after 60.0 min. Does this agree with the simpler calculation (Problem 93) based on Section 9–2?

95. A meteor whose mass was about 10^8 kg struck the Earth $(m = 6.0 \times 10^{24}$ kg$)$ with a speed of about 15 km/s and came to rest in the Earth. (a) What was the Earth's recoil speed? (b) What fraction of the meteor's kinetic energy was transformed to kinetic energy of the Earth? (c) By how much did the Earth's kinetic energy change as a result of this collision?

96. A block of mass $m = 2.20$ kg slides down a 30.0° incline which is 3.60 m high. At the bottom, it strikes a block of mass $M = 7.00$ kg which is at rest on a horizontal surface, Fig. 9–50. (Assume a smooth transition at the bottom of the incline.) If the collision is elastic, and friction can be ignored, determine (a) the speeds of the two blocks after the collision, and (b) how far back up the incline the smaller mass will go.

FIGURE 9–50 Problems 96 and 97.

97. In Problem 96 (Fig. 9–50), what is the upper limit on the mass m if it is to rebound from M, slide up the incline, stop, slide down the incline, and collide with M again?

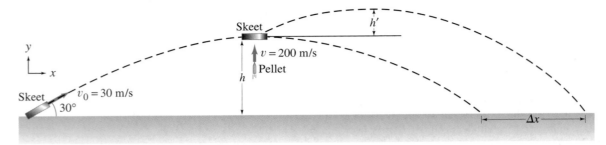

FIGURE 9–51 Problem 98.

98. A 0.25-kg skeet (clay target) is fired at an angle of 30° to the horizon with a speed of 30 m/s (Fig. 9–51). When it reaches the maximum height, it is hit from below by a 15-g pellet traveling vertically upward at a speed of 200 m/s. The pellet is embedded in the skeet. (a) How much higher did the skeet go up? (b) How much extra distance, Δx, does the skeet travel because of the collision?

99. The gravitational slingshot effect in Fig. 9–52, shows the planet Saturn moving in the negative x direction at its orbital speed (with respect to the Sun) of 9.6 km/s. The mass of Saturn is 5.69×10^{26} kg. A spacecraft with mass 825 kg approaches Saturn, moving initially in the $+x$ direction at 10.4 km/s. The gravitational attraction of Saturn (a conservative force) causes the spacecraft to swing around it (orbit shown as dashed line) and head off in the opposite direction. Estimate the final speed of the spacecraft after it is far enough away to be nearly free of Saturn's gravitational pull.

FIGURE 9–52 Problem 99.

100. A mallet consists of a uniform cylindrical head of mass 2.00 kg and a diameter 0.0800 m mounted on a uniform cylindrical handle of mass 0.500 kg and length 0.240 m, as shown in Fig. 9–53. If this mallet is tossed, spinning, into the air, how far above the bottom of the handle is the point that will follow a parabolic trajectory?

24.0 cm

8.00 cm

FIGURE 9–53 Problem 100.

101. A golf ball of mass 0.045 kg is hit off the tee at a speed of 50 m/s. The golf club was in contact with the ball for 5.0×10^{-3} s. Find (a) the impulse imparted to the golf ball, and (b) the average force exerted on the ball by the golf club.

102. The space shuttle launches an 800-kg satellite by ejecting it from the cargo bay. The ejection mechanism is activated and is in contact with the satellite for 4.0 s to give it a velocity of 0.30 m/s in the z-direction relative to the shuttle. The mass of the shuttle is 90,000 kg. (a) Determine the component of velocity v_f of the shuttle in the minus z-direction resulting from the ejection. (b) Find the average force that the shuttle exerts on the satellite during the ejection.

103. Car manufacturers conduct crash tests on new vehicles to ensure the safety of the passengers. In one such test a car of mass 1000 kg traveling at 50 km/h strikes a brick wall. (a) Is this an elastic or inelastic collision? (b) Assume that the front of the vehicle collapsed by 0.70 m. What is the impact time? (c) What is the average impulsive force?

104. A massless spring with spring constant k is placed between a block of mass m and a block of mass $3m$. Initially the blocks are at rest on a frictionless surface and they are held together so that the spring between them is compressed by an amount D from its equilibrium length. The blocks are then released and the spring pushes them off in opposite directions. Find the final speeds of the two blocks.

105. In a physics lab, a small cube slides down a frictionless incline as shown in Fig. 9–54, and elastically strikes a cube that is only one-half its mass. If the incline is 20 cm high and the table is 90 cm off the floor, where does each cube land?

FIGURE 9–54 Problem 105.

When you want to go faster on a bicycle, as Lance Armstrong does in this 1999 Tour de France photo, you exert a strong force on the pedals. The speed of a bike depends on the rotation speed of the wheels and the pedals. The angular velocity of the rotating pedals increases not simply in proportion to the force applied, but rather in proportion to the net torque. Torque is the product of the force times the lever arm (= perpendicular distance from the axis of rotation to the line of action of the force). That is: the angular acceleration of a rotating body is proportional to the net applied torque. When an object is rotating, it has rotational kinetic energy and angular momentum. Angular momentum, and its conservation, play an interesting and often unexpected role in the real world, as we will see.

CHAPTER 10

Rotational Motion About a Fixed Axis

U ntil now, we have been concerned with translational motion. In this chapter and the next, we will deal with rotational motion. The material has been divided so that this Chapter 10 covers all the basic aspects of a rigid body rotating about a fixed axis. Chapter 11, on the other hand, presents a more general approach and some slightly more advanced (optional) topics.

Although we will sometimes deal with systems of many particles, we will mainly be concerned with rigid bodies. By a **rigid body** we mean a body with a definite shape that doesn't change, so that the particles that compose it stay in fixed positions relative to one another. Of course, any real body is capable of vibrating or deforming when a force is exerted on it. But these effects are often small, so the concept of an ideal rigid body is very useful as a good approximation.

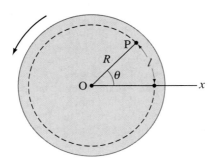

FIGURE 10–1 Looking down on a wheel that is rotating counterclockwise about an axis through the wheel's center at O (perpendicular to the page).

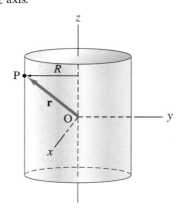

FIGURE 10–2 Showing the distinction between **r** (the position vector) and R (the distance from the rotation axis) for a point P on the edge of a cylinder rotating about the z axis.

The motion of a rigid body can be analyzed as the translational motion of its center of mass plus rotational motion about its center of mass. We have already discussed translational motion in detail, so now we focus our attention on purely rotational motion. By *purely rotational motion* we mean that all points in the body move in circles, such as the point P in the rotating wheel of Fig. 10–1, and that the centers of these circles all lie on a line called the **axis of rotation** (which in Fig. 10–1 is perpendicular to the page and passes through point O). We assume the axis is fixed in an inertial reference frame, but we will not necessarily insist that the axis pass through the center of mass.

10–1 Angular Quantities

For three-dimensional rigid bodies rotating about a fixed axis, we will find it useful to use the symbol R to represent the perpendicular distance of a point or particle from the axis of rotation. We do this to distinguish R from r, which will continue to represent the position of a particle with reference to the origin (a point) of some coordinate system.[†] This distinction is illustrated in Fig. 10–2. For a flat, very thin body, like a wheel, with the origin in the plane of the body (at the center of a wheel, for example), R and r will be the same.

Every point in a body rotating about a fixed axis moves in a circle (shown dashed in Fig. 10–1 for point P) whose center is on the axis and whose radius is R, the distance of that point from the axis of rotation. A straight line drawn from the axis to any point sweeps out the same angle θ in the same time.

To indicate the angular position of the body, or how far it has rotated, we specify the angle θ of some particular line in the body with respect to some reference line, such as the x axis (see Fig. 10–1). A point in the body (such as P in Fig. 10–1) moves through an angle θ when it travels the distance l measured along the circumference of its circular path. Angles are commonly stated in degrees, but the mathematics of circular motion is much simpler if we use the *radian* for angular measure. One **radian** (rad) is defined as the angle subtended by an arc whose length is equal to the radius. For example, in Fig. 10–1, point P is a distance R from the axis of rotation, and it has moved a distance l along the arc of a circle. The arc length l is said to "subtend" the angle θ. If $l = R$, then θ is exactly equal to 1 rad. In general, any angle θ is given by

$$\theta = \frac{l}{R}, \tag{10–1}$$

where R is the radius of the circle and l is the arc length subtended by the angle θ, which is specified in radians. Note that the radian is dimensionless (has no units) since it is the ratio of two lengths.

Radians can be related to degrees in the following way. In a complete circle there are 360°, which of course must correspond to an arc length equal to the circumference of the circle, $l = 2\pi r$. Thus $\theta = l/R = 2\pi R/R = 2\pi$ rad in a complete circle, so

$$360° = 2\pi \text{ rad}.$$

1 rad ≈ 57.3°

One radian is therefore $360°/2\pi \approx 360°/6.28 \approx 57.3°$.

[†] Some books use the symbol ρ instead of R. Since we use ρ to represent density (mass per unit volume), we avoid confusion by using R rather than ρ to represent the perpendicular distance of a point from an axis.

EXAMPLE 10–1 **Birds of prey—in radians.** A particular bird's eye can just distinguish objects that subtend an angle no smaller than about 3×10^{-4} rad. (a) How many degrees is this? (b) How small an object can the bird just distinguish when flying at a height of 100 m (Fig. 10–3a)?

SOLUTION (a) One radian is $360°/2\pi$, so 3×10^{-4} rad is

$$(3 \times 10^{-4} \text{ rad})\left(\frac{360°}{2\pi \text{ rad}}\right) = 0.017°.$$

(b) From Eq. 10–1, $l = R\theta$. For small angles, the arc length and the chord length are approximately the same (Fig. 10–3b). Since $R = 100$ m and $\theta = 3 \times 10^{-4}$ rad, we find

$$l = (100 \text{ m})(3 \times 10^{-4} \text{ rad}) = 3 \times 10^{-2} \text{ m} = 3 \text{ cm}.$$

Had the angle been given in degrees, we would first have had to change it to radians to make this calculation.

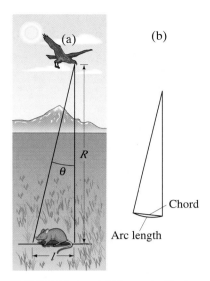

FIGURE 10–3 (a) Example 10–1. (b) For small angles, arc length and the chord length (straight line) are nearly equal. For an angle as large as 15°, the error in making this estimate is only 1 percent. For larger angles the error increases rapidly.

When an object, such as the bicycle wheel in Fig. 10–4, rotates from some initial position, specified by θ_1, to some final position, θ_2, its angular displacement is $\Delta\theta = \theta_2 - \theta_1$. The *angular velocity* (denoted by ω, the Greek lowercase letter omega) is defined in analogy with ordinary linear velocity. Instead of linear displacement, we use the angular displacement. Then the magnitude of the **average angular velocity** is defined as

$$\overline{\omega} = \frac{\Delta\theta}{\Delta t}, \qquad (10\text{–}2a)$$

where $\Delta\theta$ is the angle through which the body has rotated in the time Δt. The magnitude of the **instantaneous angular velocity** is the limit of this ratio as Δt approaches zero:

$$\omega = \lim_{\Delta t \to 0} \frac{\Delta\theta}{\Delta t} = \frac{d\theta}{dt}. \qquad (10\text{–}2b)$$

Angular velocity is generally specified in radians per second (rad/s). Note that *all points in a rigid body rotate with the same angular velocity*, since every position in the body moves through the same angle in the same time interval.

Angular acceleration (denoted by α, the Greek lowercase letter alpha), in analogy to ordinary linear acceleration, is defined as the change in angular velocity divided by the time required to make this change. The **average angular acceleration** is defined as

$$\overline{\alpha} = \frac{\omega_2 - \omega_1}{\Delta t} = \frac{\Delta\omega}{\Delta t}, \qquad (10\text{–}3a)$$

where ω_1 is the angular velocity initially, and ω_2 is the angular velocity after a time Δt. **Instantaneous angular acceleration** is defined as the limit of this ratio as Δt approaches zero:

$$\alpha = \lim_{\Delta t \to 0} \frac{\Delta\omega}{\Delta t} = \frac{d\omega}{dt}. \qquad (10\text{–}3b)$$

Since ω is the same for all points of a rotating body, Eqs. 10–3 tell us that α also will be the same for all points. Thus, ω and α are properties of the rotating body as a whole. With ω measured in radians per second and t in seconds, α will be expressed as radians per second squared (rad/s^2).

FIGURE 10–4 A wheel rotates from (a) initial position θ_1, to (b) final position θ_2. The angular displacement is $\Delta\theta = \theta_2 - \theta_1$.

Angular acceleration (rad/s²)

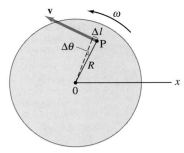

FIGURE 10-5 A particle P on a rotating wheel has a linear velocity **v** at any moment.

Linear and angular velocity related

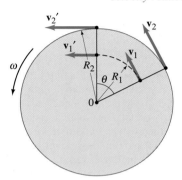

FIGURE 10-6 A wheel rotating uniformly counterclockwise. Two points on the wheel, at distances R_1 and R_2 from the center, have different linear velocities because they travel different distances in the same time interval. Since $R_2 > R_1$, then $v_2 > v_1$ ($v = R\omega$). But the two points have the same angular velocity ω because they travel through the same angle θ in the same time interval.

Centripetal (or radial) acceleration

FIGURE 10-7 On a rotating wheel whose rotation speed is increasing, a point P has both tangential and radial (centripetal) components of acceleration (See also Chapter 5.)

Frequency

Each particle or point of a rotating rigid body has, at any moment, a linear velocity v and a linear acceleration a. We can relate these linear quantities, v and a, of each particle, to the angular quantities, ω and α, of the rotating body as a whole. Consider a particle located a distance R from the axis of rotation, as in Fig. 10-5. If the body rotates with angular velocity ω, any particle will have a linear velocity whose direction is tangent to its circular path. The magnitude of its linear velocity, v, is $v = dl/dt$. From Eq. 10-1, a change in rotation angle $d\theta$ is related to the linear distance traveled by $dl = R\,d\theta$. Hence

$$v = \frac{dl}{dt} = R\frac{d\theta}{dt}$$

or

$$v = R\omega. \tag{10-4}$$

Thus, although ω is the same for every point in a rotating body at any instant, the linear velocity v is greater for points farther from the axis. See Fig. 10-6. Note that Eq. 10-4 is valid both instantaneously and on the average.

If a rotating body speeds up or slows down, so $\alpha \neq 0$, then a point in the object will have not only a (tangential) linear velocity but a tangential acceleration as well. We can use Eq. 10-4 to show that the angular acceleration α is related to the tangential linear acceleration a_{tan}, of a particle in the rotating body by

$$a_{\text{tan}} = \frac{\Delta v}{\Delta t} = R\frac{\Delta \omega}{\Delta t}$$

or

$$a_{\text{tan}} = R\alpha. \tag{10-5}$$

In this equation, R is the radius of the circle in which the particle is moving, and the subscript tan in a_{tan} stands for "tangential" since the acceleration considered here is along the circle (that is, tangent to it).

We saw in Chapter 3 that a particle moving in a circle of radius R with linear speed v has a radial or "centripetal" acceleration given by Eq. 3-14:

$$a_{\text{R}} = \frac{v^2}{R},$$

and this acceleration points toward the center of the circular path (Fig. 10-7). Using Eq. 10-4, we can write this radial acceleration as

$$a_{\text{R}} = \frac{v^2}{R} = \frac{(\omega R)^2}{R} = \omega^2 R. \tag{10-6}$$

Equation 10-6 applies to any particle of a rotating object. Thus the centripetal acceleration is greater the farther you are from the axis of rotation: the children farthest out on a merry-go-round feel the greatest acceleration. The total linear acceleration of a particle in a rotating object is the sum of the two components:

$$\mathbf{a} = \mathbf{a}_{\text{tan}} + \mathbf{a}_{\text{R}}.$$

We can relate the angular velocity ω to the frequency of rotation, f, where by **frequency** we mean the number of complete revolutions (rev) per second. One revolution (of a wheel, say) corresponds to an angle of 2π radians, and thus $1\,\text{rev/s} = 2\pi\,\text{rad/s}$. Hence, in general, the frequency f is related to the angular velocity ω by

$$f = \frac{\omega}{2\pi}$$

or

$$\omega = 2\pi f. \tag{10-7}$$

The unit for frequency, revolutions per second (rev/s), is given the special name, the hertz (Hz). That is

$$1 \text{ Hz} = 1 \text{ rev/s} = 1 \text{ s}^{-1}.$$

The hertz (Hz)

Note that "revolution" is not really a unit, so we can also write $1 \text{ Hz} = 1 \text{ s}^{-1}$.

The time required for one complete revolution is the **period**, T, and it is related to the frequency by

$$T = \frac{1}{f}. \qquad\qquad (10\text{–}8)$$

Period

For example, if a particle rotates at a frequency of three revolutions per second, then each revolution takes $\frac{1}{3}$ s.

EXAMPLE 10–2 **Hard drive.** The platter of the hard disk of a computer rotates at 5400 rpm (revolutions per minute). (*a*) What is the angular velocity of the disk? (*b*) If the reading head of the drive is located 3.0 cm from the rotation axis, what is the speed of the disk below it? (*c*) What is the linear acceleration of this point? (*d*) If a single bit requires 5.0 μm of length along the motion direction, how many bits per second can the writing head write when it is 3.0 cm from the axis? (*e*) If the disk took 3.6 s to spin up to 5400 rpm from rest, what was the average acceleration?

→ **PHYSICS APPLIED**

Hard drive
and bit speed

SOLUTION (*a*) First we find the frequency:

$$f = \frac{(5400 \text{ rev/min})}{(60 \text{ s/min})} = 90 \text{ rev/s} = 90 \text{ Hz}.$$

Then the angular velocity is

$$\omega = 2\pi f = 570 \text{ rad/s}.$$

(*b*) The speed of a point 3.0 cm out from the axis is

$$v = R\omega = (3.0 \times 10^{-2} \text{ m})(570 \text{ rad/s}) = 17 \text{ m/s}.$$

(*c*) The linear acceleration has two components, tangential and radial. Since $\omega = $ constant, then $\alpha = 0$, so $a_{\text{tan}} = r\alpha = 0$. The radial acceleration is

$$a_R = \omega^2 R = (570 \text{ rad/s})^2 (0.030 \text{ m}) = 9700 \text{ m/s}^2$$

toward the axis.

(*d*) Each bit requires 5.0×10^{-6} m, so at a speed of 17 m/s, the number of bits passing the head per second is

$$\frac{17 \text{ m/s}}{5.0 \times 10^{-6} \text{ m}} = 3.4 \times 10^6 \text{ bits per second.}$$

(*e*) The average angular acceleration was

$$\bar{\alpha} = \frac{\Delta\omega}{\Delta t} = \frac{570 \text{ rad/s} - 0}{3.6 \text{ s}} = 160 \text{ rad/s}^2.$$

Thus, on average, during each second the disk increased its angular speed by 160 rad/s or by $(160/2\pi) = 25$ rev/s.

10–2 Kinematic Equations for Uniformly Accelerated Rotational Motion

In Chapter 2, we derived the important equations (2–12) that relate acceleration, velocity, and distance for the situation of uniform linear acceleration. Those equations were derived from the definitions of linear velocity and acceleration, assuming constant acceleration. The definitions of angular velocity and angular acceleration are the same as for their linear counterparts, except that θ has replaced the linear displacement x, ω has replaced v, and α has replaced a. Therefore, the angular equations for **constant angular acceleration** will be analogous to Eqs. 2–12 with x replaced by θ, v by ω, and a by α, and they can be derived in exactly the same way. We summarize them here, opposite their linear equivalents (we've chosen $x_0 = 0$, and $\theta_0 = 0$, for $t = 0$):

	Angular	**Linear**		
Uniformly	$\omega = \omega_0 + \alpha t$	$v = v_0 + at$	[constant α, a]	**(10–9a)**
accelerated	$\theta = \omega_0 t + \frac{1}{2}\alpha t^2$	$x = v_0 t + \frac{1}{2}at^2$	[constant α, a]	**(10–9b)**
rotational	$\omega^2 = \omega_0^2 + 2\alpha\theta$	$v^2 = v_0^2 + 2ax$	[constant α, a]	**(10–9c)**
motion	$\overline{\omega} = \dfrac{\omega + \omega_0}{2}$	$\overline{v} = \dfrac{v + v_0}{2}$	[constant α, a]	**(10–9d)**

Note that ω_0 represents the angular velocity at $t = 0$, whereas θ and ω represent the angular position and velocity, respectively, at time t. Since the angular acceleration is constant, $\alpha = \overline{\alpha}$. These equations are of course also valid for constant angular velocity, for which case $\alpha = 0$ and we have $\omega = \omega_0$, $\theta = \omega_0 t$ and $\overline{\omega} = \omega$.

EXAMPLE 10–3 **Hard drive again.** Through how many revolutions did the hard drive of Example 10–2 turn to reach 5400 rpm during its acceleration period? Assume constant angular acceleration.

SOLUTION We are given that $\omega_0 = 0$, $\omega = 570\ \text{rad/s}$, $\alpha = \overline{\alpha} = 160\ \text{rad/s}^2$, and $t = 3.6\ \text{s}$. We could use either Eq. 10–9b or 10–9c to find θ. The former gives

$$\theta = 0 + \tfrac{1}{2}\left(160\ \text{rad/s}^2\right)(3.6\ \text{s})^2 = 1.04 \times 10^3\ \text{rad}.$$

To find the total number of revolutions, we divide by 2π and obtain

$$\frac{1.04 \times 10^3\ \text{rad}}{2\pi\ \text{rad/rev}} = 165\ \text{rev}.$$

FIGURE 10–8 (a) A wheel rolling to the right. Its center C moves with velocity **v**. (b) The same wheel as seen from a reference frame in which the axle of the wheel C is at rest—that is, we are moving to the right with velocity **v** relative to part (a). Point P, which was at rest in (a), here in (b) moves to the left with velocity −**v** as shown.

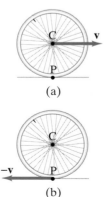

(a)

(b)

10–3 Rolling Motion (without slipping)

The rolling motion of a ball or wheel is familiar in everyday life: a ball rolling across the floor, or the wheels and tires of a car or bicycle rolling along the pavement. Rolling *without slipping* is readily analyzed and depends on static friction between the rolling object and the ground. The friction is static because the rolling object's point of contact with the ground is at rest at each moment. (Kinetic friction comes in if, for example, you brake too hard so the tires skid, or you accelerate so fast that you "burn rubber"—but these are more complicated situations.)

Rolling without slipping involves both rotation and translation. There is then a simple relation between the linear speed v of the axle and the angular velocity ω of the rotating wheel or sphere, namely $v = R\omega$ where R is the radius, as we now show. Figure 10–8a shows a wheel rolling to the right without slipping. At the moment shown, point P on the wheel is in contact with the ground and is momentarily at rest. The velocity of the axle at the wheel's center C is **v**. In Fig. 10–8b we have put ourselves in the reference frame of the wheel—that is, we are moving to the right with velocity **v** relative to the ground. In this reference frame the axle C is at rest, whereas the ground and point P are moving to the left with velocity −**v** as shown. Here we are seeing pure rotation. So we can use Eq. 10–4 to obtain $v = R\omega$ where R is the radius of the wheel. This is the same v as in Fig. 10–8a, so we see that the linear speed v of the axle relative to the ground is related to the angular velocity ω by $v = R\omega$. Note that this is valid only if there is no slipping.

(a) Bike as seen from the ground ($t = 0$).

$v_{ground} = 8.40$ m/s

(b) From rider's reference frame, the ground is moving (to our left) at an initial speed of 8.40 m/s ($t = 0$).

FIGURE 10–9 Example 10–4.

EXAMPLE 10–4 **Bicycle.** A bicycle slows down uniformly from $v_0 = 8.40$ m/s to rest over a distance of 115 m, Fig. 10–9a. Each wheel and tire has an overall diameter of 68.0 cm. Determine (a) the angular velocity of the wheels at the initial instant ($t = 0$), (b) the total number of revolutions each wheel rotates before coming to rest, (c) the angular acceleration of the wheel, and (d) the time it took to come to a stop.

SOLUTION (a) Let us put ourselves in the reference frame of the bike—that is, as if we were riding the bike. Then the ground is going past us, initially, at a speed of 8.40 m/s, Fig. 10–9b. Since the tire is in contact with the ground at any moment, then a point on the rim of the tire (such as that touching the ground) moves at a speed of $v = 8.40$ m/s in this reference frame. Hence the angular velocity of the wheel is

$$\omega_0 = \frac{v_0}{R} = \frac{8.40 \text{ m/s}}{0.340 \text{ m}} = 24.7 \text{ rad/s}.$$

(b) In coming to a stop, 115 m of ground passes beneath the tire. Because the tire is in firm contact with the ground, any point on the edge of the rotating tire travels 115 m total. Each revolution corresponds to a distance of $2\pi R$, so the number of revolutions the wheel makes in coming to a stop is

$$\frac{115 \text{ m}}{2\pi R} = \frac{115 \text{ m}}{(2\pi)(0.340 \text{ m})} = 53.8 \text{ rev}.$$

(c) The angular acceleration of the wheel can be obtained from Eq. 10–9c:

$$\alpha = \frac{\omega - \omega_0^2}{2\theta} = \frac{0 - (24.7 \text{ rad/s})^2}{2(2\pi)(53.8 \text{ rev})} = -0.902 \text{ rad/s}^2,$$

where we have set $\theta = 2\pi$ rad/rev $\times$ 53.8 rev (= 338 rad) because each revolution corresponds to 2π radians. [Alternatively, we could have used Eq. 10–1 to get the total θ: $\theta = l/R = 115$ m/0.340 m = 338 rad.]

(d) Eq.10–9a or b allows us to solve for the time. The first is easier:

$$t = \frac{\omega - \omega_0}{\alpha} = \frac{0 - 24.7 \text{ rad/s}}{-0.902 \text{ rad/s}^2} = 27.4 \text{ s}.$$

* Instantaneous axis

When a wheel rolls without slipping, the point of contact of the wheel with the ground is instantaneously at rest. It is sometimes useful to think of the motion of the wheel as pure rotation about this "instantaneous axis" passing through that point P (Fig. 10–10a). Points close to the ground have a small linear speed, as they are close to this instantaneous axis, whereas points farther away have a greater linear speed. This can be seen in a photograph of a real rolling wheel (Fig. 10–10b): spokes near the top of the wheel appear more blurry because they are moving faster than those near the bottom of the wheel.

FIGURE 10–10 (a) A rolling wheel rotates about the instantaneous axis (perpendicular to the page) passing through the point of contact with the ground, P. The arrows represent the instantaneous velocity of each point. (b) Photograph of a rolling wheel. The spokes are more blurred where the speed is greater.

(a)

(b)

(a)   (b)

FIGURE 10–11 (a) Rotating wheel. (b) Right-hand rule for obtaining direction of **ω**.

Right-hand rule

10–4 | Vector Nature of Angular Quantities

Both **ω** and **α** can be treated as vectors, and we define their directions in the following way. Consider the rotating wheel shown in Fig. 10–11a. The linear velocities of different particles of the wheel point in all different directions. The only unique direction in space associated with the rotation is along the axis of rotation, perpendicular to the actual motion. We therefore choose the axis of rotation to be the direction of the angular velocity vector, **ω**. Actually, there is still an ambiguity since **ω** could point in either direction along the axis of rotation (up or down in Fig. 10–11a). The convention we use, called the **right-hand rule**, is the following: when the fingers of the right hand are curled around the rotation axis and point in the direction of the rotation, then the thumb points in the direction of **ω**. This is shown in Fig. 10–11b. Note that **ω** points in the direction a right-handed screw would move when turned in the direction of rotation. Thus, if the rotation of the wheel in Fig. 10–11a is counterclockwise, the direction of **ω** is upward as shown in Fig. 10–11b. If the wheel rotates clockwise, then **ω** points in the opposite direction, downward.[†] Note that no part of the rotating body moves in the direction of **ω**.

If the axis of rotation is fixed in direction, then **ω** can change only in magnitude. Thus **α** = $d\omega/dt$ must also point along the axis of rotation. If the rotation is counterclockwise as in Fig. 10–11a and the magnitude ω is increasing, then **α** points upward; but if ω is decreasing (the wheel is slowing down), **α** points downward.

FIGURE 10–12 (a) Velocity is a true vector. The reflection of **v** points in the same direction. (b) Angular velocity is a pseudovector since it does not follow this rule. As can be seen, the reflection of the wheel rotates in the opposite direction, so the direction of ω is opposite for the reflection.

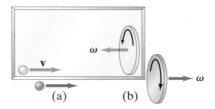

(a)   (b)

[†] Strictly speaking, **ω** and **α** are not quite vectors. The problem is that they do not behave like vectors under reflection. Suppose, as we are looking directly into a mirror, a particle moving with velocity **v** to the right passes in front of and parallel to the mirror. In the reflection of the mirror, **v** still points to the right, Fig. 10–12a. Thus a true vector, like velocity, when pointing parallel to the face of mirror has the same direction in the reflection as in actuality. Now consider a wheel rotating in front of the mirror, so **ω** points to the right. (We will be looking at the edge of the wheel.) As viewed in the mirror, Fig. 10–12b, the wheel will be rotating in the opposite direction. So **ω** will point in the opposite direction (to the left) in the mirror. Because of this difference under reflection between **ω** and true vectors, **ω** is called a *pseudovector* or *axial vector*. The angular acceleration **α** is also a pseudovector, as are all cross products of true vectors (Section 11–1). The difference between true vectors and pseudovectors will not concern us in this book.

If the rotation is clockwise, $\boldsymbol{\alpha}$ will point downward if ω is increasing, and point upward if ω is decreasing. In other words, if ω is increasing, $\boldsymbol{\alpha}$ points in the same direction as $\boldsymbol{\omega}$, whereas if ω is decreasing, $\boldsymbol{\alpha}$ points in the opposite direction.

Since $\boldsymbol{\omega}$ always points along the axis of rotation, if the axis of rotation changes direction, $\boldsymbol{\omega}$ changes direction. In this case $\boldsymbol{\alpha}$ will not point along the axis of rotation. We will see examples of this in Chapter 11, but in this chapter we consider only motion about a fixed axis, so $\boldsymbol{\omega}$ and $\boldsymbol{\alpha}$ are both along the rotation axis.

10–5 | Torque

We have so far discussed rotational kinematics—the description of rotational motion in terms of angle, angular velocity, and angular acceleration. Now we discuss the dynamics, or causes, of rotational motion. Just as we found analogies between linear and rotational motion for the description of motion, so rotational equivalents for dynamics exist as well.

To make an object start rotating about an axis clearly requires a force. But the direction of this force, and where it is applied, are also important. Take, for example, an ordinary situation such as the door in Fig. 10–13. If you apply a force $\mathbf{F}_1$ to the door as shown, you will find that the greater the magnitude, F_1, the more quickly the door opens. But now if you apply the same magnitude force at a point closer to the hinge, say $\mathbf{F}_2$ in Fig. 10–13, you will find that the door will not open so quickly. The effect of the force is less. Indeed, if only this one force acts, it is found that the angular acceleration of the door is proportional not only to the magnitude of the force, but is also directly proportional to *the perpendicular distance from the axis of rotation to the line along which the force acts*. This distance is called the **lever arm**, or **moment arm**, of the force, and is labeled R_1 and R_2 for the two forces in Fig. 10–13. Thus, if R_1 in Fig. 10–13 is three times larger than R_2, then the angular acceleration of the door will be three times as great, assuming that the magnitudes of the forces are the same. To say it another way, if $R_1 = 3R_2$, then F_2 must be three times as large as F_1 to give the same angular acceleration. (Figure 10–14 shows two examples of tools whose long lever arms are very effective.)

The angular acceleration, then, is proportional to the product of the *force times the lever arm*. This product is called the *moment of the force* about the axis, or, more commonly, it is called the **torque**, and is abbreviated τ (Greek lowercase letter tau). Thus, the angular acceleration α of an object is directly proportional to the net applied torque, τ:

$$\alpha \propto \tau,$$

and we see that it is torque that gives rise to angular acceleration. This is the rotational analog of Newton's second law for linear motion, $a \propto F$. (In Section 10–6, we will see what factor is needed to make this proportionality an equation.)

FIGURE 10–13 Applying the same force with different lever arms, R_1 and R_2.

Lever arm

Torque defined

(a)   (b)

FIGURE 10–14 (a) A plumber can exert greater torque using a wrench with a long lever arm. (b) A tire iron too can have a long lever arm.

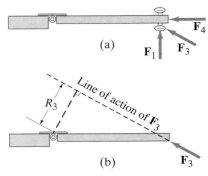

(a)

(b)

FIGURE 10–15 (a) Forces acting at different angles at the doorknob. (b) The lever arm is defined as the perpendicular distance from the axis of rotation (the hinge) to the line of action of the force.

FIGURE 10–16
Torque = $R_\perp F = RF_\perp$.

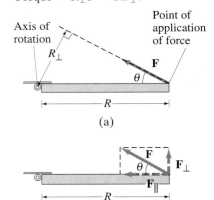

(a)

(b)

Magnitude of a torque

Units for torque: m · N

We defined the lever arm as the *perpendicular* distance of the axis of rotation from the line of action of the force—that is, the distance which is perpendicular both to the axis of rotation and to an imaginary line drawn along the direction of the force. We do this to take into account the effect of forces acting at an angle. It is clear that a force applied at an angle, such as $\mathbf{F}_3$ in Fig. 10–15a, will be less effective than the same magnitude force applied straight on, such as $\mathbf{F}_1$ (Fig. 10–15a). And if you push on the end of the door so that the force is directed at the hinge (the axis of rotation), as indicated by $\mathbf{F}_4$, the door will not rotate at all.

The lever arm for a force such as $\mathbf{F}_3$ is found by drawing a line along the direction of $\mathbf{F}_3$ (this is the "line of action" of $\mathbf{F}_3$), and then drawing another line (perpendicular to the rotation axis) from the axis of rotation perpendicular to this "line of action". The length of this second line is the lever arm for $\mathbf{F}_3$ and is labeled R_3 in Fig. 10–15b.

The torque associated with $\mathbf{F}_3$ is then $R_3 F_3$. This short lever arm and the corresponding smaller torque associated with $\mathbf{F}_3$ is consistent with the observation that $\mathbf{F}_3$ is less effective in accelerating the door than is $\mathbf{F}_1$. When the lever arm is defined in this way, experiment shows that the relation $\alpha \propto \tau$ is valid in general. Notice in Fig. 10–15 that the line of action of the force $\mathbf{F}_4$ passes through the hinge and hence its lever arm is zero. Consequently, zero torque is associated with $\mathbf{F}_4$ and it gives rise to no angular acceleration, in accord with everyday experience.

In general, then, we can write the torque about a given axis as

$$\tau = R_\perp F, \qquad \text{(10–10a)}$$

where $R_\perp$ is the lever arm, and the perpendicular symbol ($\perp$) reminds us that we must use the distance from the axis of rotation that is perpendicular to the line of action of the force (Fig. 10–16a). An alternate but equivalent way of determining the torque associated with a force is to resolve the force into components parallel and perpendicular to the line joining the point of application of the force to the axis, as shown in Fig. 10–16b. Then the torque will be equal to $F_\perp$ times the distance R from the axis to the point of application of the force:

$$\tau = RF_\perp. \qquad \text{(10–10b)}$$

This gives the same result as Eq. 10–10a since $F_\perp = F \sin\theta$ and $R_\perp = R \sin\theta$. So

$$\tau = RF \sin\theta \qquad \text{(10–10c)}$$

in either case. We can use any of Eqs. 10–10 to calculate the torque, which ever is easiest.[†]

Since torque is a distance times a force, it is measured in units of m · N in SI units,[‡] cm · dyne in the cgs system, and ft · lb in the English system.

When more than one torque acts on a body, the acceleration α is found to be proportional to the *net* torque. If all the torques acting on a body tend to rotate it in the same direction, the net torque is the sum of the torques. But if, say, one torque acts to rotate a body in one direction, and a second torque acts to rotate the body in the opposite direction (as in Fig. 10–17), the net torque is the difference of the two torques. We assign a positive sign to torques that act to rotate the body in one direction (say counterclockwise) and a negative sign to torques that act to rotate the body in the opposite direction (clockwise).

[†] We will see in Chapter 11 that Eq. 10–10c gives the magnitude of the torque vector, $\boldsymbol{\tau} = \mathbf{r} \times \mathbf{F}$.

[‡] Note that the units for torque are the same as those for energy. We write the unit for torque here as m · N (in SI) to help distinguish it from energy (N · m) because the two quantities are very different. An obvious difference is that energy is a scalar, whereas torque has a direction and is a vector. The special name *joule* (1 J = 1 N · m) is used only for energy (and for work), *never* for torque.

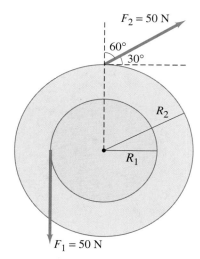

FIGURE 10–17 Example 10–5. The torque due to $\mathbf{F}_1$ tends to accelerate the wheel counterclockwise, whereas the torque due to $\mathbf{F}_2$ tends to accelerate the wheel clockwise.

EXAMPLE 10–5 **Torque on a compound wheel.** Two thin cylindrical wheels, of radii $R_1 = 30\,\text{cm}$ and $R_2 = 50\,\text{cm}$, are attached to each other on an axle that passes through the center of each, as shown in Fig. 10–17. Calculate the net torque on the two-wheel system due to the two forces shown, each of magnitude 50 N.

SOLUTION The force $\mathbf{F}_1$ acts to rotate the system counterclockwise, whereas $\mathbf{F}_2$ acts to rotate it clockwise. So the two forces act in opposition to each other. We must choose one direction of rotation to be positive—say, counterclockwise. Then $\mathbf{F}_1$ exerts a positive torque, $\tau_1 = R_1 F_1$, since the lever arm is R_1. On the other hand, $\mathbf{F}_2$ produces a negative (clockwise) torque and does not act perpendicular to R_2, so we must use its perpendicular component to calculate the torque it produces: $\tau_2 = -R_2 F_{2\perp} = -R_2 F_2 \sin\theta$, where $\theta = 60°$. (Note that θ must be the angle between $\mathbf{F}_2$ and a radial line from the rotation axis.) Hence the net torque is

$$\tau = R_1 F_1 - R_2 F_2 \sin 60°$$
$$= (0.30\,\text{m})(50\,\text{N}) - (0.50\,\text{m})(50\,\text{N})(0.866) = -6.7\,\text{m} \cdot \text{N}.$$

This net torque acts to accelerate the rotation of the wheel in the clockwise direction. Note that the two forces have equal magnitude, yet produce a net torque because their lever arms are different.

10–6 Rotational Dynamics; Torque and Rotational Inertia

We have discussed that the angular acceleration α of a rotating body is proportional to the net torque τ applied to it:

$$\alpha \propto \Sigma\tau$$

where we write $\Sigma\tau$ to remind us that it is the net torque (sum of all torques acting on the body) that is proportional to α. This corresponds to Newton's second law for translational motion, $a \propto \Sigma F$, where torque has taken the place of force, and, correspondingly, the angular acceleration α takes the place of the linear acceleration a. In the linear case, the acceleration is not only proportional to the net force, but it is also inversely proportional to the inertia of the body, which we call its mass, m. Thus we could write $a = \Sigma F/m$. But what plays the role of mass for the rotational case? That is what we now set out to determine. At the same time, we will see that the relation $\alpha \propto \Sigma\tau$ follows directly from Newton's second law, $\Sigma F = ma$.

We first consider a very simple case: a particle of mass m rotating in a circle of radius r at the end of a string or rod whose mass we can ignore (Fig. 10–18), and we assume a single force F acts on it as shown. The torque that gives rise to the angular acceleration is $\tau = RF$. If we make use of Newton's second law for linear quantities, $\Sigma F = ma$, and Eq. 10–5 relating the angular acceleration to the tangential linear acceleration, $a_{\text{tan}} = R\alpha$, we have

$$F = ma$$
$$= mR\alpha.$$

When we multiply both sides by R, the torque $\tau = RF$ is given by

$$\tau = mR^2\alpha. \qquad \text{[single particle]} \quad \textbf{(10–11)}$$

Here we have the direct relation between the angular acceleration and the applied torque τ. The quantity mR^2 represents the *rotational inertia* of the particle and is called its *moment of inertia*.

FIGURE 10–18 A mass m rotating in a circle of radius R about a fixed point.

Now let us consider a rotating rigid body, such as a wheel rotating about an axis through its center, such as an axle. We can think of the wheel as consisting of many particles located at various distances from the axis of rotation. We can apply Eq. 10–11 to each particle of the body; that is, we write $\tau_i = m_i R_i^2 \alpha$ for the ith particle of the body. Then we sum over all the particles. The sum of the various torques is just the total torque, $\Sigma\tau$, so we obtain:

$$\Sigma\tau_i = \left(\Sigma m_i R_i^2\right)\alpha \qquad \text{[axis fixed]} \quad \textbf{(10–12)}$$

where we factored out the α since it is the same for all the particles of the body. The resultant torque, $\Sigma\tau$, represents the sum of all internal torques that each particle exerts on another, plus all external torques applied from the outside: $\Sigma\tau = \Sigma\tau_{\text{ext}} + \Sigma\tau_{\text{int}}$. Now the sum of the internal torques is zero from Newton's third law.[†] Hence $\Sigma\tau$ represents the resultant *external* torque.

The sum $\Sigma m_i R_i^2$ in Eq. 10–12 represents the sum of the masses of each particle in the body multiplied by the square of the distance of that particle from the axis of rotation. If we give each particle a number $(1, 2, 3, \cdots)$, then $\Sigma m_i R_i^2 = m_1 R_1^2 + m_2 R_2^2 + m_3 R_3^2 + \cdots$. This quantity is called the **moment of inertia** (or *rotational inertia*) of the body, I:

$$I = \Sigma m_i R_i^2 = m_1 R_1^2 + m_2 R_2^2 + \cdots. \qquad \textbf{(10–13)}$$

Combining Eqs. 10–12 and 10–13, we can write

$$\Sigma\tau = I\alpha. \qquad \begin{bmatrix} \text{axis fixed in} \\ \text{inertial reference frame} \end{bmatrix} \quad \textbf{(10–14)}$$

This is the rotational equivalent of Newton's second law. It is valid for the rotation of a rigid body about a fixed axis.[‡] It can be shown (see Chapter 11) that Eq. 10–14 is also valid when the body is translating with acceleration, as long as I and α are calculated about the center of mass of the body, and the rotation axis through the CM doesn't change direction. (A ball rolling straight down a ramp is an example.) Then

$$(\Sigma\tau)_{\text{CM}} = I_{\text{CM}}\alpha_{\text{CM}}, \qquad \begin{bmatrix} \text{axis fixed in direction,} \\ \text{but may accelerate} \end{bmatrix} \quad \textbf{(10–15)}$$

where the subscript CM means "calculated about the center of mass".

We see that the moment of inertia, I, which is a measure of the rotational inertia of a body, plays the same role for rotational motion that mass does for translational motion. As can be seen from Eq. 10–13, the rotational inertia of an object depends not only on its mass, but also on how that mass is distributed with respect to the axis. For example, a large-diameter cylinder will have greater rotational inertia than one of equal mass but smaller diameter (and therefore greater length), Fig. 10–19. The former will be harder to start rotating, and harder to stop. When the mass is concentrated farther from the axis of rotation, the rotational inertia is greater. For rotational motion, the mass of a body *cannot* be considered as concentrated at its center of mass.

FIGURE 10–19 A large-diameter wheel has greater rotational inertia than one of smaller diameter but equal mass.

Careful:
Mass can not be considered concentrated at CM for rotational motion

10–7 Solving Problems in Rotational Dynamics

Whenever dealing with torque and angular acceleration (Eq. 10–14), it is important to use a consistent set of units, which in SI is: α in rad/s^2; τ in m·N; and the moment of inertia, I, in kg·m^2.

[†] This depends on the so-called "strong" form of Newton's third law in which not only is the force one particle exerts on a second equal and opposite to the force the second exerts on the first, but these two forces act along the same line.

[‡] That is, the axis is fixed relative to the body and is fixed in an inertial reference frame. This includes an axis moving at uniform velocity in an inertial frame, since the axis can be considered fixed in a second inertial frame that moves with respect to the first.

FIGURE 10–20 Example 10–6: calculating the moment of inertia.

EXAMPLE 10–6 **Two weights on a bar: different axis, different *I*.** Two small "weights" of mass 5.0 kg and 7.0 kg are mounted 4.0 m apart on a light rod (whose mass can be ignored), as shown in Fig. 10–20. Calculate the moment of inertia of the system (*a*) when rotated about an axis halfway between the weights, Fig. 10–20a, and (*b*) when the system rotates about an axis 0.50 m to the left of the 5.0-kg mass (Fig. 10–20b).

SOLUTION (*a*) Both weights are the same distance, 2.0 m, from the axis of rotation. Thus

$$I = \Sigma mR^2 = (5.0\,\text{kg})(2.0\,\text{m})^2 + (7.0\,\text{kg})(2.0\,\text{m})^2$$
$$= 20\,\text{kg}\cdot\text{m}^2 + 28\,\text{kg}\cdot\text{m}^2 = 48\,\text{kg}\cdot\text{m}^2.$$

(*b*) The 5.0-kg mass is now 0.50 m from the axis and the 7.0-kg mass is 4.50 m from the axis. Then

$$I = \Sigma mR^2 = (5.0\,\text{kg})(0.50\,\text{m})^2 + (7.0\,\text{kg})(4.5\,\text{m})^2$$
$$= 1.3\,\text{kg}\cdot\text{m}^2 + 142\,\text{kg}\cdot\text{m}^2 = 143\,\text{kg}\cdot\text{m}^2.$$

The above Example illustrates two important points. First, the moment of inertia of a given system is different for different axes of rotation. Second, we see in part (*b*) that mass close to the axis of rotation contributes little to the total moment of inertia; the 5.0-kg object contributed less than 1 percent to the total in this case.

I depends on axis of rotation and on distribution of mass

For most ordinary bodies, the mass is distributed continuously, and the calculation of the moment of inertia, ΣmR^2, can be difficult. Expressions for the moments of inertia of regularly shaped bodies in terms of their dimensions can, however, be

PROBLEM SOLVING	**Rotational Motion (Axis fixed)**

1. As always, draw a clear and complete diagram.

2. Draw a *free-body diagram* for the body under consideration (or for each body if more than one), showing only (and all) the forces acting on that body and exactly where they act, so you can determine the torque due to each. Gravity acts at the CG of the body.

3. Identify the axis of rotation and calculate the torques about it. Choose positive and negative directions of rotation (counter-clockwise and clockwise), and assign the correct sign to each torque.

4. Apply Newton's second law for rotation, $\Sigma\tau = I\alpha$. If the moment of inertia is not given, and it is not the unknown sought, you need to determine it first. Use consistent units, which in SI are: α in rad/s²; τ in m·N; and I in kg·m².

5. Also apply Newton's second law for translation, $\Sigma\mathbf{F} = m\mathbf{a}$, if needed.

6. Solve the resulting equation(s) for the unknown(s).

7. As always, do a rough estimate to determine if your answer is reasonable: does it make sense?

Object	Location of axis	Moment of inertia
(a) Thin hoop of radius R_0	Through center	MR_0^2
(b) Thin hoop of radius R_0 and width w	Through central diameter	$\frac{1}{2}MR_0^2 + \frac{1}{12}Mw^2$
(c) Solid cylinder of radius R_0	Through center	$\frac{1}{2}MR_0^2$
(d) Hollow cylinder of inner radius R_1 and outer radius R_2	Through center	$\frac{1}{2}M(R_1^2 + R_2^2)$
(e) Uniform sphere of radius r_0	Through center	$\frac{2}{5}Mr_0^2$
(f) Long uniform rod of length l	Through center	$\frac{1}{12}Ml^2$
(g) Long uniform rod of length l	Through end	$\frac{1}{3}Ml^2$
(h) Rectangular thin plate, of length l and width w	Through center	$\frac{1}{12}M(l^2 + w^2)$

FIGURE 10–21
Moments of inertia for various objects of uniform composition.

worked out (using calculus) as discussed in the next Section. Figure 10–21 gives these expressions for a number of solids rotated about the axes specified. The only one for which the result is obvious is that for the thin hoop or ring of radius R_0 rotated about an axis passing through its center perpendicular to the plane of the hoop (Fig. 10–21a). For this object, all the mass is concentrated at the same distance from the axis, R_0. Thus $\Sigma mR^2 = (\Sigma m)R_0^2 = MR_0^2$, where M is the total mass of the hoop. When calculation is difficult, I can be determined experimentally by measuring the angular acceleration α about a fixed axis due to a known net torque, $\Sigma\tau$, and applying Eq. 10–14, $I = \Sigma\tau/\alpha$.

FIGURE 10–22 Example 10–7.

33.0 cm
$(= R_0)$

$\mathbf{F}_T$

EXAMPLE 10–7 **A heavy pulley.** A 15.0-N force (represented by $\mathbf{F}_T$) is applied to a cord wrapped around a pulley of mass $M = 4.00$ kg and radius $R_0 = 33.0$ cm, Fig. 10–22. The pulley is observed to accelerate uniformly from rest to reach an angular speed of 30.0 rad/s in 3.00 s. If there is a frictional torque (at the axle), $\tau_{\text{fr}} = 1.10$ m·N, determine the moment of inertia of the pulley. The pulley is assumed to rotate about its center.

SOLUTION The free-body diagram for the pulley is shown in Fig. 10–22, although the friction force is not shown since we are given only its torque. We can calculate the moment of inertia from Eq. 10–14, $\Sigma\tau = I\alpha$, since from the measurements given we can determine $\Sigma\tau$ and α. The cord leaves the wheel's edge perpendicular to the radius, so the angle between the force $\mathbf{F}_T$ and its lever arm is 90°. The net torque is the applied torque due to $\mathbf{F}_T$ minus the frictional torque; we take positive to be counterclockwise:

$$\Sigma\tau = (0.330\,\text{m})(15.0\,\text{N}) - 1.10\,\text{m}\cdot\text{N} = 3.85\,\text{m}\cdot\text{N}.$$

The angular acceleration is

$$\alpha = \frac{\Delta\omega}{\Delta t} = \frac{30.0\,\text{rad/s} - 0}{3.00\,\text{s}} = 10.0\,\text{rad/s}^2.$$

Hence

$$I = \Sigma\tau/\alpha = (3.85\,\text{m}\cdot\text{N})/(10.0\,\text{rads/s}^2) = 0.385\,\text{kg}\cdot\text{m}^2.$$

EXAMPLE 10–8 **Pulley and bucket.** Consider again the pulley in Fig. 10–22. But this time, suppose that, instead of a constant 15.0-N force being exerted on the cord, we now have a bucket of weight 15.0 N (mass $m = 1.53\,\text{kg}$) hanging from the cord, which we assume not to stretch or slip on the pulley. See Fig. 10–23a. (a) Calculate the angular acceleration α of the pulley and the linear acceleration a of the bucket. (b) Determine the angular velocity ω of the pulley and the linear velocity v of the bucket at $t = 3.00\,\text{s}$ if the pulley (and bucket) start from rest at $t = 0$.

SOLUTION (a) Let F_T be the tension in the cord. Then a force F_T acts at the edge of the pulley, and we use Eq. 10–14 for the rotation of the pulley:

$$I\alpha = \Sigma\tau = F_T R_0 - \tau_{\text{fr}}. \qquad [\text{pulley}]$$

Next we look at the (linear) motion of the bucket of mass m. Figure 10–23b shows a free-body diagram for the bucket. Two forces act on the bucket: the force of gravity mg acts downward, and the tension of the cord F_T pulls upward. So by $\Sigma F = ma$ for the bucket we have (taking downward as positive):

$$mg - F_T = ma. \qquad [\text{bucket}]$$

Note that the tension F_T, which is the force exerted on the edge of the pulley, is *not* equal to the weight of the bucket ($= mg = 15.0\,\text{N}$). There must be a net force (hence $F_T < mg$) if the bucket is accelerating. Indeed, by the last equation above, $F_T = mg - ma$. To obtain α, we note that the tangential acceleration of a point on the edge of the pulley is the same as the acceleration of the bucket if the cord doesn't stretch or slip. Hence we can use Eq. 10–5, $a_{\text{tan}} = a = R_0\alpha$. Substituting $F_T = mg - ma$ into the first equation above, we obtain

$$I\alpha = \Sigma\tau = F_T R_0 - \tau_{\text{fr}} = (mg - mR_0\alpha)R_0 - \tau_{\text{fr}} = mgR_0 - mR_0^2\alpha - \tau_{\text{fr}}.$$

α appears in the middle term on the right, so we bring that term to the left side and solve for α:

$$\alpha = \frac{mgR_0 - \tau_{\text{fr}}}{I + mR_0^2}.$$

Then, since $I = 0.385\,\text{kg}\cdot\text{m}^2$ and $\tau_{\text{fr}} = 1.10\,\text{m}\cdot\text{N}$ (Example 10–7),

$$\alpha = \frac{(15.0\,\text{N})(0.330\,\text{m}) - 1.10\,\text{m}\cdot\text{N}}{0.385\,\text{kg}\cdot\text{m}^2 + (1.53\,\text{kg})(0.330\,\text{m})^2} = 6.98\,\text{rad/s}^2.$$

The angular acceleration is somewhat less in this case than the $10.0\,\text{rad/s}^2$ of Example 10–7. Why? Because $F_T(= mg - ma)$ is somewhat less than the weight of the bucket, mg. The linear acceleration of the bucket is

$$a = R_0\alpha = (0.330\,\text{m})(6.98\,\text{rad/s}^2) = 2.30\,\text{m/s}^2.$$

(b) Since the angular acceleration is constant, after 3.00 s

$$\omega = \omega_0 + \alpha t = 0 + (6.98\,\text{rad/s}^2)(3.00\,\text{s}) = 20.9\,\text{rad/s}.$$

The velocity of the bucket is the same as that of a point on the wheel's edge:

$$v = R_0\omega = (0.330\,\text{m})(20.9\,\text{rad/s}) = 6.91\,\text{m/s}.$$

The same result can also be obtained by using the linear equation $v = v_0 + at = 0 + (2.30\,\text{m/s}^2)(3.00\,\text{s}) = 6.90\,\text{m/s}$. (The difference is due to rounding off.)

FIGURE 10–23 Example 10–8. With free-body diagram for the falling bucket of mass m shown in (b).

FIGURE 10–24 Example 10–9.

EXAMPLE 10–9 **Rotating rod.** A uniform rod of mass M and length l can pivot freely (i.e., we ignore friction) about a hinge or pin attached to the case of a large machine, as in Fig. 10–24. The rod is held horizontally and then released. At the moment of release, determine (a) the angular acceleration of the rod and (b) the linear acceleration of the tip of the rod. Assume the force of gravity acts at the center of mass of the rod, as shown.

SOLUTION (a) The only torque on the rod is that due to gravity, which acts with a force $F = Mg$ downward with a lever arm (at the moment of release) of $l/2$, since the CM is at the center of a uniform rod. (There is also a force on the rod at the pin, but with the pin as axis of rotation, the lever arm of this force is zero.) The moment of inertia of a uniform rod pivoted about its end is (Fig. 10–21g) $I = \frac{1}{3}Ml^2$. Thus, from Eq. 10–14,

$$\alpha = \frac{\tau}{I} = \frac{Mg\dfrac{l}{2}}{\frac{1}{3}Ml^2} = \frac{3}{2}\frac{g}{l}$$

is the initial angular acceleration of the rod. As the rod descends, the force of gravity on it is constant but the torque due to this force is not constant; hence the rod's angular acceleration is not constant.

(b) The linear acceleration of the tip of the rod is found from the relation $a_{\text{tan}} = R\alpha$ (Eq. 10–5) with $R = l$:

$$a_{\text{tan}} = l\alpha = \tfrac{3}{2}g.$$

Thus the tip of the rod falls with an acceleration greater than g! A small object balanced on the tip of the rod would be left behind when the rod is released. In contrast, the CM of the rod, at a distance $l/2$ from the pivot, has acceleration $a_{\text{tan}} = (l/2)\alpha = \tfrac{3}{4}g$.

10–8 Determining Moments of Inertia

By Experiment

The moments of inertia of any body about any axis can be determined experimentally by measuring the net torque $\Sigma\tau$ required to give the body an angular acceleration α. Then, from Eq. 10–14, $I = \Sigma\tau/\alpha$. See Example 10–7.

Using Calculus

For many bodies, or systems of particles, the moment of inertia can be calculated directly, as in Example 10–6. Many bodies can be considered as a continuous distribution of mass. In this case, Eq. 10–13 defining moment of inertia becomes

$$I = \int R^2 \, dm, \tag{10–16}$$

where dm represents the mass of any infinitesimal particle of the body and R is the perpendicular distance of this particle from the axis of rotation. The integral is taken over the whole body. This is easily done only for bodies of simple geometric shape.

EXAMPLE 10–10 **Hollow cylinder.** Show that the moment of inertia of a uniform hollow cylinder of inner radius R_1, outer radius R_2, and mass M, is $I = \frac{1}{2}M(R_1^2 + R_2^2)$, as stated in Fig. 10–21d, if the rotation axis is through the center along the axis of symmetry.

SOLUTION We know that the moment of inertia of a thin ring of radius R is mR^2. So we divide the cylinder into thin concentric cylindrical rings or hoops of thickness dR, one of which is indicated in Fig. 10–25. If the density (mass per unit volume) is ρ, then

$$dm = \rho \, dV,$$

where dV is the volume of the thin ring of radius R, thickness dR, and height h. Since $dV = (2\pi R)(dR)(h)$, we have

$$dm = 2\pi \rho h R \, dR.$$

Then the moment of inertia is obtained by integrating (summing) over all these hoops:

$$I = \int R^2 \, dm$$

$$= \int_{R_1}^{R_2} 2\pi \rho h R^3 \, dR = 2\pi \rho h \left[\frac{R_2^4 - R_1^4}{4} \right],$$

where we are given that the cylinder has uniform density, $\rho = $ constant. (If this were not so, we would have to know ρ as a function of R before the integration could be carried out.) The volume V of this hollow cylinder is $V = (\pi R_2^2 - \pi R_1^2)h$, so its mass M is

$$M = \rho V = \rho \pi (R_2^2 - R_1^2)h.$$

Since $(R_2^4 - R_1^4) = (R_2^2 - R_1^2)(R_2^2 + R_1^2)$, we have

$$I = \frac{\pi \rho h}{2} (R_2^2 - R_1^2)(R_2^2 + R_1^2)$$

$$= \frac{1}{2}M(R_1^2 + R_2^2),$$

as stated in Fig. 10–21d. As a check, note that for a solid cylinder, $R_1 = 0$ and we obtain, with $R_2 = R_0$:

$$I = \frac{1}{2}MR_0^2,$$

which is that given in Fig. 10–21c for a solid cylinder of mass M and radius R_0.

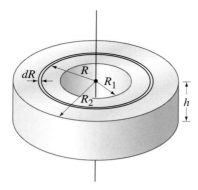

FIGURE 10–25 Determining the moment of inertia of a hollow cylinder (Example 10–10).

The Parallel-axis and Perpendicular-axis Theorems

There are two simple theorems that are helpful in obtaining moments of inertia. The first is called the **parallel-axis theorem**. It states that if I is the moment of inertia of a body of total mass M about any axis, and I_{CM} is the moment of inertia about an axis passing through the center of mass and parallel to the first axis but a distance h away, then

$$I = I_{CM} + Mh^2. \qquad \text{[parallel axis]} \quad \textbf{(10–17)}$$

Thus, for example, if the moment of inertia about an axis through the CM is known, the moment of inertia about any axis parallel to this axis is easily obtained.

EXAMPLE 10–11 **Parallel axis.** Determine the moment of inertia of a solid cylinder of radius R_0, and mass M about an axis tangent to its edge and parallel to its symmetry axis, Fig. 10–26.

SOLUTION We use the parallel axis theorem with $I_{CM} = \frac{1}{2}MR_0$ (Fig. 10–21c). Since $h = R_0$, we have

$$I = I_{CM} + Mh^2 = \frac{3}{2}MR_0^2.$$

Parallel-axis theorem

FIGURE 10–26 Example 10–11.

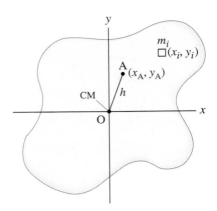

FIGURE 10–27 Derivation of the parallel-axis theorem.

Perpendicular-axis theorem

FIGURE 10–28 The perpendicular axis theorem, and Example 10–12.

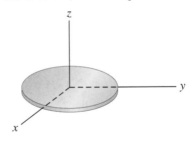

Angular momentum

NEWTON'S SECOND LAW
FOR ROTATION

The proof of the parallel-axis theorem is as follows. We choose our coordinate system so the origin is at the CM, and I_{CM} is the moment of inertia about the z axis. Figure 10–27 shows a cross section of a body of arbitrary shape in the xy plane. We let I represent the moment of inertia of the body about an axis parallel to the z axis that passes through the point A in Fig. 10–27 where the point A has coordinates x_A and y_A. Let x_i, y_i, and m_i represent the coordinates and mass of an arbitrary particle of the body. The square of the distance from this point to A is $[(x_i - x_A)^2 + (y_i - y_A)^2]$. So the moment of inertia, I, about the axis through A is

$$I = \Sigma m_i[(x_i - x_A)^2 + (y_i - y_A)^2]$$
$$= \Sigma m_i(x_i^2 + y_i^2) - 2x_A \Sigma m_i x_i - 2y_A \Sigma m_i y_i + (\Sigma m_i)(x_A^2 + y_A^2).$$

The first term on the right is just $I_{CM} = \Sigma m_i(x_i^2 + y_i^2)$ since the CM is at the origin. The second and third terms are zero since, by definition of the CM, $\Sigma m_i x_i = \Sigma m_i y_i = 0$ because $x_{CM} = y_{CM} = 0$. The last term is Mh^2 since $\Sigma m_i = M$ and $(x_A^2 + y_A^2) = h^2$ where h is the distance of A from the CM. Thus we have proved $I = I_{CM} + Mh^2$, which is Eq. 10–17.

The parallel-axis theorem can be applied to any body. The second theorem, the **perpendicular-axis theorem**, can be applied only to plane figures—that is, to two-dimensional bodies, or bodies of uniform thickness whose thickness can be neglected compared to the other dimensions. This theorem states that the sum of the moments of inertia of a plane body about any two perpendicular axes in the plane of the body is equal to the moment of inertia about an axis through their point of intersection perpendicular to the plane of the object. That is, if the figure is in the xy plane (Fig. 10–28),

$$I_z = I_x + I_y. \qquad \text{[object in } xy \text{ plane]} \quad \textbf{(10–18)}$$

Here I_z, I_x, I_y are moments of inertia about the z, x, and y axes. The proof is simple: since $I_x = \Sigma m_i y_i^2$, $I_y = \Sigma m_i x_i^2$, and $I_z = \Sigma m_i(x_i^2 + y_i^2)$, Eq. 10–18 follows directly.

EXAMPLE 10–12 **Perpendicular axis.** Determine the moment of inertia of a thin circular coin (a cylinder) about an axis through its center in the plane of the coin (Fig. 10–28).

SOLUTION We want to calculate the moment of inertia about the x axis in Fig. 10–28, and we can use the perpendicular-axis theorem. From Fig. 10–21c we have $I_z = \frac{1}{2}MR_0^2$, and, from symmetry, $I_x = I_y$. Hence $2I_x = I_z$, or $I_x = \frac{1}{2}I_z = \frac{1}{4}MR_0^2$.

10–9 Angular Momentum and Its Conservation

Throughout this chapter we have seen that if we use the appropriate angular variables, the kinematic and dynamic equations for rotational motion are analogous to those for ordinary linear motion. In like manner, the linear momentum, $p = mv$, has a rotational analog. It is called **angular momentum**, L, and for a body rotating about a fixed axis with angular velocity ω, it is defined as[†]

$$L = I\omega, \qquad \textbf{(10–19)}$$

where I is the moment of inertia. The SI units for L are $kg \cdot m^2/s$.

We saw in Chapter 9 (Section 9–1) that Newton's second law can be written not only as $\Sigma F = ma$, but also more generally in terms of momentum (Eq. 9–2), $\Sigma F = dp/dt$. In a similar way, the rotational equivalent of Newton's second law, which we saw in Eqs. 10–14 and 10–15 can be written as $\Sigma\tau = I\alpha$, can also be written in terms of angular momentum: since the angular acceleration $\alpha = d\omega/dt$ (Eq. 10–3), then $I\alpha = I(d\omega/dt) = d(I\omega)/dt = dL/dt$, so

$$\Sigma\tau = I\alpha = \frac{dL}{dt}. \qquad \textbf{(10–20)}$$

This simple derivation assumes that the moment of inertia, I, remains constant.

[†]Angular momentum is a vector, and we will see in Chapter 11 that the angular momentum discussed here, $L = I\omega$, is the component along the rotation axis. There may or may not be other components.

However, it is valid even if the moment of inertia changes (see Chapter 11). Equation 10–20 is Newton's second law for rotational motion of a rigid body about a fixed axis, and is also valid for a moving body if its rotation is about an axis passing through its CM (as for Eq. 10–15).

Conservation of Angular Momentum

Angular momentum is an important concept in physics because, under certain conditions, it is a conserved quantity. What are the conditions for which it is conserved? From Eq. 10–20 we see immediately that if the net external torque $\Sigma\tau$ on a body is zero, then dL/dt equals zero. That is, L does not change. This, then, is the **law of conservation of angular momentum** for a rotating body:

> **The total angular momentum of a rotating body remains constant if the net external torque acting on it is zero.**

CONSERVATION OF ANGULAR MOMENTUM

The law of conservation of angular momentum is one of the great laws of physics.

When there is zero net torque acting on a body, and the body is rotating about a fixed axis or about an axis through its CM such that its direction doesn't change, we can write

$$I\omega = I_0\omega_0 = \text{constant}.$$

I_0 and ω_0 are the moment of inertia and angular velocity, respectively, about the axis at some initial time ($t = 0$), and I and ω are their values at some other time. The parts of the body may alter their positions relative to one another, so that I changes. But then ω changes as well and the product $I\omega$ remains constant.

Many interesting phenomena can be understood on the basis of conservation of angular momentum. Consider a skater doing a spin on the tips of her skates, Fig. 10–29. She rotates at a relatively low speed when her arms are outstretched, but when she brings her arms in close to her body, she suddenly spins much faster. By remembering the definition of moment of inertia as $I = \Sigma mR^2$, it is clear that when she pulls her arms in closer to the axis of rotation, R is reduced for the arms so her moment of inertia is reduced. Since the angular momentum $I\omega$ remains constant (we ignore the small torque due to friction), if I decreases, then the angular velocity ω must increase. If the skater reduces her moment of inertia by a factor of 2, she will then rotate with twice the angular velocity.

A similar example is the diver shown in Fig. 10–30. The push as she leaves the board gives her an initial angular momentum about her CM. When she curls herself into the tuck position, she rotates quickly one or more times. She then stretches out again, increasing her moment of inertia, which reduces the angular velocity to a small value, and then she enters the water. The change in moment of inertia from the straight position to the tuck position can be a factor of as much as $3\frac{1}{2}$.

Note that for angular momentum to be conserved, the net torque must be zero, but the net force does not necessarily have to be zero. The net force on the diver in Fig. 10–30, for example, is not zero (gravity is acting), but the net torque on her is zero.

➡ **PHYSICS APPLIED**

Spinning skaters and divers

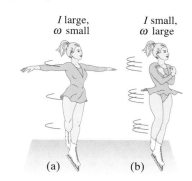

FIGURE 10–29 A skater doing a spin on ice, illustrating conservation of angular momentum: in (a), I is large and ω is small; in (b), I is smaller so ω is larger.

FIGURE 10–30 A diver rotates faster when arms and legs are tucked in than when they are outstretched. Angular momentum is conserved.

(a)

FIGURE 10–31 Example 10–13.

EXAMPLE 10–13 Clutch design. You are designing a clutch assembly for a piece of machinery. The clutch assembly consists of two cylindrical plates, of mass $M_A = 6.0$ kg and $M_B = 9.0$ kg, with equal radii $R_0 = 0.60$ m. They are initially separated (Fig. 10–31). Plate M_A is accelerated from rest to an angular velocity $\omega_1 = 7.2$ rad/s in time $\Delta t = 2.0$ s. Calculate (a) the angular momentum of M_A, and (b) the torque required to have accelerated M_A from rest to ω_1. Next, plate M_B, initially at rest but free to rotate without friction, is allowed to fall vertically (or could be pushed by a spring, as in a car) so it is in firm contact with plate M_A (their contact surfaces are made of high-friction material). Before contact, M_A was rotating at constant ω_1, with no friction or other torque exerted on it. Upon contact, the two plates both rotate at a constant angular velocity ω_2, which is considerably less than ω_1. (c) Why does this happen, and what is ω_2?

SOLUTION (a) The angular momentum of M_A will be

$$L_A = I_A \omega_1 = \tfrac{1}{2} M_A R_0^2 \omega_1 = \tfrac{1}{2}(6.0 \text{ kg})(0.60 \text{ m})^2 (7.2 \text{ rad/s}) = 7.8 \text{ kg} \cdot \text{m}^2/\text{s}.$$

(b) The plate started from rest so the torque, assumed constant, was

$$\tau \approx \frac{\Delta L}{\Delta t} = \frac{7.8 \text{ kg} \cdot \text{m}^2/\text{s} - 0}{2.0 \text{ s}} = 3.9 \text{ m} \cdot \text{N}.$$

(c) Initially, M_A is rotating at constant ω_1 since we assume no torques (such as friction) are acting. When plate B comes in contact, it too is free to rotate, and they rotate together because their surfaces stick together. But why does their rotation speed decrease? You might think in terms of the torque each exerts on the other upon contact. But quantitatively, it's easier to use conservation of angular momentum (which we can, since no external torques are assumed to act). Thus

$$\text{angular momentum before} = \text{angular momentum after}$$

$$I_A \omega_1 = (I_A + I_B)\omega_2.$$

Solving for ω_2 we find

$$\omega_2 = \left(\frac{I_A}{I_A + I_B}\right)\omega_1 = \left(\frac{M_A}{M_A + M_B}\right)\omega_1 = \left(\frac{6.0 \text{ kg}}{15.0 \text{ kg}}\right)(7.2 \text{ rad/s}) = 2.9 \text{ rad/s}.$$

➡ **PHYSICS APPLIED**

Neutron star

EXAMPLE 10–14 ESTIMATE Star collapse. Astronomers often detect stars that are rotating extremely rapidly, known as neutron stars. They are believed to have been formed from the inner core of a larger star that collapsed, due to its own gravitation, to a star of very small radius and very high density. Suppose the core of such a star is the size of our Sun. ($R \approx 7 \times 10^5$ km) before collapse, but of mass 2.0 times as great, and is rotating at a speed of 1.0 revolution every 10 days. If it were to undergo gravitational collapse to a neutron star of radius 10 km, what would its rotation speed be? Assume the star is a uniform sphere at all times.

SOLUTION The star is isolated (no external forces) so we can use conservation of angular momentum for this process:

$$I_i \omega_i = I_f \omega_f$$

where the subscript i and f refer to initial (normal star) and final (neutron star). Then

$$\omega_f = \left(\frac{I_i}{I_f}\right)\omega_i = \left(\frac{\tfrac{2}{5} M_i R_i^2}{\tfrac{2}{5} M_f R_f^2}\right)\omega_i = \frac{R_i^2}{R_f^2}\omega_i,$$

where we assume no mass is lost in the process. The frequency $f = \omega/2\pi$ so

$$f_f = \left(\frac{7 \times 10^5 \text{ km}}{10 \text{ km}}\right)^2 \left(\frac{1 \text{ rev}}{10 \text{ d}(24 \text{ h/d})(3600 \text{ s/h})}\right) \approx 6 \times 10^3 \text{ rev/s}.$$

Vector Nature of Angular Momentum

Although we will not go into the vector nature of angular momentum in detail until the next chapter, we can deal with some simple cases involving conservation of angular momentum here. For a body rotating about a fixed axis, the direction of the angular momentum can be taken as the direction of the angular velocity, $\boldsymbol{\omega}$. That is,

$$\mathbf{L} = I\boldsymbol{\omega}.$$

However, this is strictly true[†] only if the rotation axis is an axis of symmetry of the body or if the body is thin and flat and rotates about an axis perpendicular to the plane of the body (such as a wheel rotating on an axle).

As a simple example, consider a person standing at rest on a circular platform capable of rotating friction-free about an axis through its center (that is, a simplified merry-go-round, Fig. 10–32). If the person now starts to walk along the edge of the platform, the platform starts rotating in the opposite direction. Why? One explanation is that the person's foot exerts a force on it. Another explanation (and this is the most useful analysis here) is that this is an example of the conservation of angular momentum. If the person starts walking counterclockwise, the person's angular momentum will be pointed upward along the axis of rotation (remember how we defined the direction of $\boldsymbol{\omega}$ using the right-hand rule in Section 10–4). The magnitude of the person's angular momentum will be $L = I\omega = (mR^2)(v/R)$, where v is the person's speed (relative to the Earth, not the platform), R is his distance from the rotation axis, m is his mass, and mR^2 is his moment of inertia since all of his mass is roughly the same distance R from the rotation axis. The platform rotates in the opposite direction, so its angular momentum points downward. If the initial total angular momentum was zero (person and platform at rest), it will remain zero after the person starts walking—that is, the upward angular momentum of the person just balances the oppositely directed angular momentum of the platform, Fig. 10–32b, so the total vector angular momentum remains zero. Even though the person exerts a force (and torque) on the platform, the platform exerts an equal and opposite torque on the person. So the net torque on the *system* of person plus platform is zero (ignoring friction) and the total angular momentum remains constant.

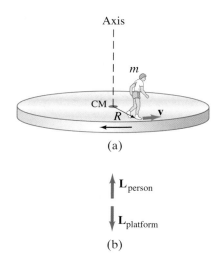

FIGURE 10–32 (a) A person standing on a circular platform, both initially at rest, begins walking along the edge at speed v. The platform, assumed to be mounted on friction-free bearings, begins rotating in the opposite direction, so that the total angular momentum remains zero, as shown in (b).

EXAMPLE 10–15 **Running on a circular platform.** Suppose a 60-kg person stands at the edge of a 6.0-m-diameter circular platform, which is mounted on frictionless bearings and has a moment of inertia of $1800\ \text{kg} \cdot \text{m}^2$. The platform is at rest initially, but when the person begins running at a speed of 4.2 m/s (with respect to the ground) around its edge, the platform begins to rotate in the opposite direction as in Fig. 10–32. Calculate the angular velocity of the platform.

SOLUTION The total angular momentum is zero initially. Since there is no net torque, $\mathbf{L}$ is conserved and will remain zero, as in Fig. 10–32. The person's angular momentum is $L_{\text{per}} = (mR^2)(v/R)$, and we take this as positive. The angular momentum of the platform is $L_{\text{plat}} = -I\omega$. Thus

$$L = L_{\text{per}} + L_{\text{plat}}$$
$$0 = mR^2\left(\frac{v}{R}\right) - I\omega.$$

So

$$\omega = \frac{mRv}{I} = \frac{(60\ \text{kg})(3.0\ \text{m})(4.2\ \text{m/s})}{1800\ \text{kg} \cdot \text{m}^2} = 0.42\ \text{rad/s}.$$

The frequency of rotation is $f = \omega/2\pi = 0.067\ \text{rev/s}$ and the period $T = 1/f = 15\ \text{s}$ per revolution.

[†]For more complicated situations of bodies rotating about a fixed axis, there will be a component of $\mathbf{L}$ along the direction of $\boldsymbol{\omega}$ and its magnitude will be equal to $I\boldsymbol{\omega}$. (There could be other components as well.) If the total angular momentum is conserved, then the component $I\boldsymbol{\omega}$ will also be conserved. So our results here can be applied to any rotation about a fixed axis.

FIGURE 10-33 Example 10-16.

RESPONSE The total angular momentum initially is **L** vertically upward, and that is what the system's angular momentum must be afterward since **L** is conserved. Thus, if the wheel's angular momentum afterward is −**L** downward, then the angular momentum of teacher plus turntable will have to be +2**L** upward. We can safely predict that the teacher will begin spinning around in the same direction the wheel was spinning originally.

10–10 Rotational Kinetic Energy

The quantity $\frac{1}{2}mv^2$ is the kinetic energy of a body undergoing translational motion. A body rotating about an axis is said to have **rotational kinetic energy**. By analogy with translational kinetic energy, we would expect this to be given by the expression $\frac{1}{2}I\omega^2$ where I is the moment of inertia of the body and ω is its angular velocity. We can indeed show that this is true. Consider any rigid rotating object as made up of many tiny particles, each of mass m_i. If we let R_i represent the distance of any one particle from the axis of rotation, then its linear velocity is $v_i = R_i\omega$. The total kinetic energy of the whole body will be the sum of the kinetic energies of all its particles:

$$K = \Sigma\left(\tfrac{1}{2}m_i v_i^2\right) = \Sigma\left(\tfrac{1}{2}m_i R_i^2 \omega^2\right)$$
$$= \tfrac{1}{2}\Sigma\left(m_i R_i^2\right)\omega^2.$$

We have factored out the $\frac{1}{2}$ and the ω^2 since they are the same for every particle of a rigid body.

Since $\Sigma m_i R_i^2 = I$, the moment of inertia, we see that the kinetic energy, K, of an object rotating about a fixed axis is, as expected,

$$K = \tfrac{1}{2}I\omega^2. \qquad \text{[fixed axis]} \quad \textbf{(10–21)}$$

If the axis is not fixed in space, the rotational kinetic energy can take on a more complicated form.

The work done on a body rotating about a fixed axis can be written in terms of angular quantities. Suppose a force **F** is exerted at a point whose distance from the axis of rotation is R, as in Fig. 10–34. The work done by this force is

$$W = \int \mathbf{F} \cdot d\boldsymbol{l} = \int F_\perp R \, d\theta,$$

Rotational kinetic energy

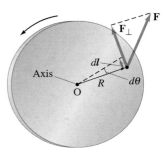

FIGURE 10-34 Calculating the work done by a torque acting on a rigid body rotating about a fixed axis.

Work done by a torque

where $d\boldsymbol{l}$ is an infinitesimal distance perpendicular to R with magnitude $dl = R\,d\theta$, and $F_\perp$ is the component of **F** perpendicular to R and parallel to $d\boldsymbol{l}$ (Fig. 10–34). But $F_\perp R$ is the torque about the axis, so

$$W = \int \tau \, d\theta. \qquad \textbf{(10–22)}$$

The rate of work done, or power P, is

$$P = \frac{dW}{dt} = \tau \frac{d\theta}{dt} = \tau\omega. \qquad \textbf{(10–23)}$$

The work-energy principle holds for rotation of a rigid body about a fixed axis. From Eq. 10–14 we have

$$\tau = I\alpha = I\frac{d\omega}{dt} = I\frac{d\omega}{d\theta}\frac{d\theta}{dt} = I\omega\frac{d\omega}{d\theta},$$

where we used the chain rule and $\omega = d\theta/dt$. Then $\tau\,d\theta = I\omega\,d\omega$ and

$$W = \int_{\theta_1}^{\theta_2} \tau\,d\theta = \int_{\omega_1}^{\omega_2} I\omega\,d\omega = \tfrac{1}{2}I\omega_2^2 - \tfrac{1}{2}I\omega_1^2. \qquad (10\text{–}24)$$

Work-energy principle for rotation

This is the work-energy principle for a body rotating about a fixed axis. It states that the work done in rotating a body through an angle $\theta_2 - \theta_1$ is equal to the change in rotational kinetic energy of the body.

EXAMPLE 10–17 **ESTIMATE** **Flywheel.** Flywheels, which are simply large rotating disks, have been suggested as a means of storing energy for solar-powered generating systems. Estimate the kinetic energy that can be stored in a 20,000-kg (10-ton) flywheel with a diameter of 20 m (a six-story building). Assume it could hold together (without flying apart due to internal stresses) at 100 rpm.

➡ PHYSICS APPLIED

Energy from a flywheel

SOLUTION We are given

$$\omega = 100\,\text{rpm} = \left(100\,\frac{\text{rev}}{\text{min}}\right)\left(\frac{1\,\text{min}}{60\,\text{sec}}\right)\left(\frac{2\pi\,\text{rad}}{\text{rev}}\right) = 10.5\,\text{rad/s}.$$

The kinetic energy stored in the disk $\left(I = \tfrac{1}{2}MR_0^2\right)$ is, from Eq. 10–21,

$$K = \tfrac{1}{2}\left(\tfrac{1}{2}MR_0^2\right)\omega^2 = \tfrac{1}{4}(2.0 \times 10^4\,\text{kg})(10\,\text{m})^2(10.5\,\text{rad/s})^2 = 5.2 \times 10^6\,\text{J}.$$

In terms of kilowatt-hours $\left[1\,\text{kWh} = (1000\,\text{J}\cdot\text{s})(3600\,\text{s/h})(1\,\text{h}) = 3.6 \times 10^6\,\text{J}\right]$, this energy is only about 1.5 kWh, which is not a lot of energy (one 3-kW oven would use it all in $\tfrac{1}{2}$ h). Thus flywheels seem unlikely for this application.

EXAMPLE 10–18 **Rotating rod.** A rod of mass M is pivoted on a frictionless hinge at one end, as shown in Fig. 10–35. The rod is held at rest horizontally and then released. Determine the angular velocity of the rod when it reaches the vertical position, and the speed of the rod's tip at this moment.

SOLUTION We can use the work-energy principle here. The work done is due to gravity, and is equal to the change in gravitational potential energy of the rod. Since the CM of the rod drops a vertical distance $l/2$, the work done by gravity is

$$W = Mg\frac{l}{2}.$$

The initial kinetic energy is zero. Hence, from the work-energy principle,

$$\tfrac{1}{2}I\omega^2 = Mg\frac{l}{2}.$$

Since $I = \tfrac{1}{3}Ml^2$ for a rod pivoted about its end (Fig. 10–21), we can solve for ω:

$$\omega = \sqrt{\frac{3g}{l}}.$$

The tip of the rod will have a linear speed (see Eq. 10–4)

$$v = l\omega = \sqrt{3gl}.$$

By comparison, an object that falls vertically a height l has a speed $v = \sqrt{2gl}$.

FIGURE 10–35 Example 10–18.

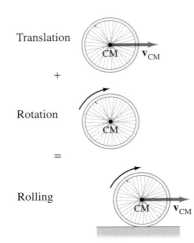

Translation

+

Rotation

=

Rolling

FIGURE 10–36 A wheel rolling without slipping can be considered as translation of the wheel as a whole with velocity $\mathbf{v}_{CM}$ plus rotation about the CM.

➡ **P R O B L E M   S O L V I N G**

*Rotational energy adds
to other forms of energy
to get the total energy
which is conserved*

FIGURE 10–37 A sphere rolling down a hill has both translational and rotational kinetic energy. Example 10–19.

Kinetic Energy $= K_{CM} + K_{rot}$

An object that rotates while its center of mass (CM) undergoes translational motion will have both translational and rotational kinetic energy. Equation 10–21, $K = \frac{1}{2}I\omega^2$, gives the rotational kinetic energy if the rotation axis is fixed. If the object is moving (such as a wheel rolling along the ground, Fig. 10–36), this equation is still valid as long as the rotation axis is fixed in direction. Then the total kinetic energy is

$$K = \tfrac{1}{2}Mv_{CM}^2 + \tfrac{1}{2}I_{CM}\omega^2, \qquad (10\text{–}25)$$

where v_{CM} is the linear velocity of the CM, I_{CM} is the moment of inertia about an axis through the CM, ω is the angular velocity about this axis, and M is the total mass of the body.

Before we prove this important theorem, Eq. 10–25, which is done in the following optional subsection, we first do some Examples to get a good feel for it.

EXAMPLE 10–19 **Sphere rolling down an incline.** What will be the speed of a solid sphere of mass M and radius R_0 when it reaches the bottom of an incline if it starts from rest at a vertical height H and rolls without slipping? See Fig. 10–37. Ignore losses due to dissipative forces, and compare your result to that for an object *sliding* down a frictionless incline.

SOLUTION We use the law of conservation of energy, and we must now include rotational kinetic energy. The total energy at any point a vertical distance y above the base of the incline is

$$\tfrac{1}{2}Mv^2 + \tfrac{1}{2}I_{CM}\omega^2 + Mgy,$$

where v is the speed of the CM. We equate the total energy at the top ($y = H$ and $v = \omega = 0$) to the total energy at the bottom ($y = 0$):

$$0 + 0 + MgH = \tfrac{1}{2}Mv^2 + \tfrac{1}{2}I_{CM}\omega^2 + 0.$$

From Fig. 10–21e, the moment of inertia of a solid sphere about an axis through its CM is $I_{CM} = \frac{2}{5}MR_0^2$. Since the sphere rolls without slipping, the speed, v, of the center of mass with respect to the point of contact (which is momentarily at rest at any instant) is equal to the speed of a point on the edge relative to the center, as we saw in Section 10–3 (Fig. 10–8). We therefore have $\omega = v/R$. Hence

$$\tfrac{1}{2}Mv^2 + \tfrac{1}{2}\left(\tfrac{2}{5}MR_0^2\right)\left(\frac{v^2}{R_0^2}\right) = MgH.$$

Dividing out the M's and R's, we obtain

$$\left(\tfrac{1}{2} + \tfrac{1}{5}\right)v^2 = gH$$

or

$$v = \sqrt{\tfrac{10}{7}gH}.$$

Note first that v is independent of both the mass M and the radius R of the sphere. Also, we can compare this result for the speed of a rolling sphere to that for an object sliding down a plane without rotating and without friction (see Chapter 8, $\frac{1}{2}mv^2 = mgH$), in which case $v = \sqrt{2gH}$, which is greater. An object sliding without friction transforms its initial potential energy entirely into translational kinetic energy (none into rotational kinetic energy), so its speed is greater.

CONCEPTUAL EXAMPLE 10–20 | **Who's fastest?** Several objects roll without slipping down an incline of vertical height H, all starting from rest at the same moment. The objects are a thin hoop (or a plain wedding band), a marble, a solid cylindrical battery (D-cell), an empty soup can, and an unopened soup can. In addition a greased box slides down without friction. In what order do they reach the bottom of the incline?

RESPONSE The sliding box wins every time! As we saw in Example 10–19, the speed of a rolling sphere at the bottom of the incline is less than that for a sliding box (without friction) because the potential energy loss (MgH) is transformed completely into translational kinetic energy for the box, whereas for rolling objects, the initial potential energy is shared between translational and rotational kinetic energy. For each of the rolling objects we can state that the loss in potential energy equals the increase in kinetic energy:

$$MgH = \tfrac{1}{2}Mv^2 + \tfrac{1}{2}I_{CM}\omega^2.$$

First we note that for all our rolling objects, the moment of inertia I_{CM} is a numerical factor times the mass M and the radius R^2 (Fig. 10–21). The mass M is in each term, so the translational speed v doesn't depend on M, nor does it depend on the radius R since $\omega = v/R$ so R^2 cancels out for all the rolling objects, just as in Example 10–19. Thus the speed v at the bottom depends only on that numerical factor in I_{CM} which expresses how the mass is distributed. Consequently, the hoop, with all its mass concentrated at radius R ($I_{CM} = MR^2$), will have the lowest speed and will arrive at the bottom behind the D-cell ($I_{CM} = \tfrac{1}{2}MR^2$) which in turn will be behind the marble ($I_{CM} = \tfrac{2}{5}MR^2$). The empty can, which is mainly a hoop plus a small disk, has its mass concentrated almost all at R; so it will be a bit faster than the pure hoop but slower than the D-cell. See Fig. 10–38. The unopened soup can is more complicated. It cannot be considered a solid cylinder because the soup can move about inside, and that will dissipate some energy; so we expect it to be slower than the D-cell, but that's about all we can safely say. For all the other objects, note that the speed at the bottom does not depend on the object's mass M or radius R, but only on its shape (and the height of the hill H).

FIGURE 10–38 Example 10–20.

If there had been no friction between the sphere (and other rolling objects) and the plane in these Examples, the sphere would have slid rather than rolled. Friction must be present to make a round object roll. We did not need to take friction into account in the energy equation because it is *static* friction and does no work. If we assume the sphere is perfectly rigid, it is in contact with the surface at a point, and the force of friction acts parallel to the plane. But the point of contact of the sphere at each instant does not slide—it moves perpendicular to the plane (first down and then up) as the sphere rolls (Fig. 10–39). Thus, no work is done by the friction force because the force and the motion are perpendicular. The reason the rolling objects in Examples 10–19 and 10–20 move down the slope more slowly than if they were sliding is *not* because friction is doing work. Rather it is because some of the gravitational potential energy is converted to rotational kinetic energy, leaving less for the translational kinetic energy.

Using $\Sigma\tau_{CM} = I_{CM}\alpha_{CM}$

We can examine objects rolling down a plane not only from the point of view of kinetic energy, as we did in Examples 10–19 and 10–20, but also in terms of forces and torques. If we calculate torques about an axis fixed in direction (even though accelerating) which passes through the center of mass of the rolling sphere, then

$$\Sigma\tau_{CM} = I_{CM}\alpha_{CM}$$

is valid, as we discussed in Section 10–6. See Eq. 10–15, whose validity we will show in Chapter 11. Be careful however: Do not assume $\Sigma\tau = I\alpha$ is always valid. You cannot just calculate τ, I, and α about any axis unless the axis is (1) fixed in an inertial reference frame or (2) fixed in direction but passes through the CM of the body.

FIGURE 10–39 A sphere rolling without slipping to the right on a plane surface. The point in contact with the ground at any moment, point P, is momentarily at rest. Point A on the left of P is moving nearly vertically upward at the instant shown, and point B on the right is moving nearly vertically downward. (An instant later, point B will touch the plane and be at rest momentarily.)

Careful:
When is $\Sigma\tau = I\alpha$ valid?

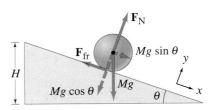

FIGURE 10–40 Example 10–21.

EXAMPLE 10–21 **Analysis of a sphere on an incline using forces.** Analyze the rolling sphere of Example 10–19, Fig. 10–37, in terms of forces and torques. In particular, find the velocity v and the magnitude of the friction force, F_{fr}, Fig. 10–40. [F_{fr} is due to static friction and we cannot assume $F_{fr} = \mu_s F_N$, only $F_{fr} \leq \mu_s F_N$.]

SOLUTION We analyze the motion as translation of the CM plus rotation about the CM. For translation in the x direction we have from $\Sigma F = ma$,

$$Mg \sin \theta - F_{fr} = Ma,$$

and in the y direction

$$F_N - Mg \cos \theta = 0$$

since there is no acceleration perpendicular to the plane. This last equation merely tells us the magnitude of the normal force,

$$F_N = Mg \cos \theta.$$

For the rotational motion about the CM, we use Newton's second law for rotation $\Sigma \tau_{CM} = I_{CM} \alpha_{CM}$ (Eq. 10–15), calculating about an axis passing through the CM but fixed in direction:

$$F_{fr} R_0 = \left(\tfrac{2}{5} M R_0^2\right)\alpha.$$

Since the other forces, $\mathbf{F}_N$ and $M\mathbf{g}$, have lever arms equal to zero, they do not appear here. As we saw in Example 10–19, $\omega = v/R_0$ where v is the speed of the CM. Taking derivatives with respect to time we have $\alpha = a/R_0$ and substituting this into the last equation we find

$$F_{fr} = \tfrac{2}{5} Ma.$$

When we substitute this into the top equation, we get

$$Mg \sin \theta - \tfrac{2}{5} Ma = Ma,$$

or

$$a = \tfrac{5}{7} g \sin \theta.$$

We thus see that the acceleration of the CM of a rolling sphere is less than that for an object sliding without friction ($a = g \sin \theta$). To find the speed v at the bottom we use Eq. 2–12c where the total distance traveled along the plane is $x = H/\sin \theta$, and H is the height of the plane (see Fig. 10–40). Thus

$$v = \sqrt{2ax} = \sqrt{2\left(\frac{5}{7} g \sin \theta\right)\left(\frac{H}{\sin \theta}\right)} = \sqrt{\frac{10}{7} gH}\,.$$

This is the same result obtained in Example 10–19 although less effort was needed there. To get the magnitude of the force of friction, we use the equations obtained above:

$$F_{fr} = \tfrac{2}{5} Ma = \tfrac{2}{5} M\left(\tfrac{5}{7} g \sin \theta\right) = \tfrac{2}{7} Mg \sin \theta.$$

What must μ_s be if the sphere is to roll without slipping?

If the coefficient of static friction is sufficiently small, or θ sufficiently large so that $F_{fr} > \mu_s F_N$ (that is, $\tan \theta > \tfrac{7}{2}\mu_s$), the sphere will not simply roll but will slip as it moves down the plane.

* Derivation of $K = K_{CM} + K_{rot}$

Now we derive a general theorem, that the *total kinetic energy of a moving body will be equal to the translational kinetic energy of its CM plus the kinetic energy of motion relative to the center of mass.* This is a general theorem, and to prove it we proceed as follows. Let $\mathbf{r}_{CM} = x_{CM}\mathbf{i} + y_{CM}\mathbf{j} + z_{CM}\mathbf{k}$ represent the position of the CM

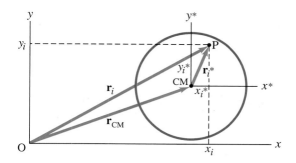

FIGURE 10–41 Point P has coordinates x_i, y_i, z_i, $(\mathbf{r}_i)$ relative to origin O of an inertial frame, and coordinates x_i^*, y_i^*, z_i^*, $(\mathbf{r}_i^*)$ relative to the center of mass (CM).

at any moment in some inertial reference frame. Let $\mathbf{r}_i = x_i\mathbf{i} + y_i\mathbf{j} + z_i\mathbf{k}$ be the position vector of the ith particle of mass m_i in this inertial reference frame, and let $\mathbf{r}_i^* = x_i^*\mathbf{i} + y_i^*\mathbf{j} + z_i^*\mathbf{k}$ be the position vector of this particle with reference to the CM (not necessarily an inertial reference frame). Then (see Fig. 10–41):

$$x_i = x_{\text{CM}} + x_i^*, \qquad y_i = y_{\text{CM}} + y_i^*, \quad \text{and} \quad z_i = z_{\text{CM}} + z_i^*.$$

The velocity of the ith particle in the inertial reference frame is

$$\mathbf{v}_i = \mathbf{v}_{\text{CM}} + \mathbf{v}_i^*,$$

where $\mathbf{v}_{\text{CM}}$ is the velocity of the CM in this reference frame and $\mathbf{v}_i^*$ is the velocity of the ith particle relative to the CM. We can use the vector dot product and write $v^2 = \mathbf{v} \cdot \mathbf{v}$, so that the total kinetic energy K is

$$
\begin{aligned}
K = \tfrac{1}{2}\Sigma m_i v_i^2 &= \tfrac{1}{2}\Sigma m_i(\mathbf{v}_i \cdot \mathbf{v}_i) \\
&= \tfrac{1}{2}\Sigma m_i(\mathbf{v}_{\text{CM}} + \mathbf{v}_i^*) \cdot (\mathbf{v}_{\text{CM}} + \mathbf{v}_i^*) \\
&= \tfrac{1}{2}\Sigma m_i v_{\text{CM}}^2 + \mathbf{v}_{\text{CM}} \cdot (\Sigma m_i \mathbf{v}_i^*) + \tfrac{1}{2}\Sigma m_i v_i^{*2}.
\end{aligned}
$$

Now $\Sigma m_i = M$, the total mass of the body. Also, $\Sigma m_i \mathbf{v}_i^* = 0$, as can be seen by taking the derivative of the definition of the center of mass when the CM is at the origin of coordinates $\left[\mathbf{r}_{\text{CM}} = (1/M)\Sigma m_i \mathbf{r}_i^* = 0\right]$. Thus

$$K = \tfrac{1}{2}M v_{\text{CM}}^2 + \tfrac{1}{2}\Sigma m_i v_i^{*2} \qquad\qquad \textbf{(10–26)}$$ *Total kinetic energy*

and we have proved that the total kinetic energy is the sum of the translational kinetic energy of the center of mass, $\tfrac{1}{2}M v_{\text{CM}}^2$, plus the kinetic energy of motion relative to the CM.

Our proof of Eq. 10–26 is general. We can apply this equation to a rigid body moving in a plane, such as a wheel rolling down a hill. In this case the axis of rotation is fixed in direction (perpendicular to the plane in which the body moves), although it moves along with the center of mass and so is not fixed in position. For each particle, $v_i^* = \omega R_i^*$, where R_i^* is the perpendicular distance of the ith particle from a line passing through the CM and perpendicular to the plane of motion. Then

Rolling wheel

$$\tfrac{1}{2}\Sigma m_i v_i^{*2} = \tfrac{1}{2}\left(\Sigma m_i R_i^{*2}\right)\omega^2 = \tfrac{1}{2}I_{\text{CM}}\omega^2$$

and

$$K = \tfrac{1}{2}M v_{\text{CM}}^2 + \tfrac{1}{2}I_{\text{CM}}\omega^2. \qquad \text{[rigid body, axis fixed in direction]}$$

This is just Eq. 10–25. In this equation, I_{CM} is the moment of inertia of the body about an axis through its center of mass and perpendicular to the plane of motion. Thus the total kinetic energy of a body moving in a plane with both translational and rotational motion, such that the rotation axis doesn't change direction, is the sum of the translational kinetic energy of the CM plus the rotational kinetic energy about the CM, $\tfrac{1}{2}I_{\text{CM}}\omega^2$.

*More Advanced Examples

Here we do three more Examples, all of them fun and interesting. They each make use of $\Sigma\tau = I\alpha$, and again we must remember that this equation is valid only if τ, α and I are calculated about an axis that is either (1) fixed in an inertial reference frame, or (2) passes through the CM of the object and remains fixed in direction.

⇒ PHYSICS APPLIED

Braking distribution of a car

Forces on the tires
for a braking car

EXAMPLE 10–22 **Braking a car.** When the brakes of a car are applied, the force on the front tires is much greater than that on the rear tires. To see why, calculate the magnitude of the friction forces, F_1 and F_2, on the front and rear tires of the car shown in Fig. 10–42 when the car decelerates at a rate $a = 0.50\ g$. The car has mass $M = 1200$ kg, the distance between the front and rear axles is 3.0 m, and its CM (where the force of gravity acts) is midway between the axles 75 cm above the ground.

SOLUTION Figure 10–42 is the free-body diagram showing all the forces on the car. F_1 and F_2 are the frictional forces that decelerate the car. We let F_1 be the sum of the forces on both front tires, and F_2 likewise for the two rear tires. Thus

$$F_1 + F_2 = Ma$$
$$= (1200\ \text{kg})(0.50)(9.8\ \text{m/s}^2) = 5900\ \text{N}. \tag{i}$$

F_{N1} and F_{N2} are the normal forces the road exerts on the tires as shown and, for simplicity, we assume the static friction force acts the same for all the tires, so that F_{N1} and F_{N2} are proportional respectively to F_1 and F_2 $(F_{N1} = F_1/\mu, F_{N2} = F_2/\mu)$.[†] While the car is braking it does not rotate, so the net torque on it is zero. If we calculate the torques about the CM as axis, we see that whereas F_1, F_2, and F_{N2} all act to rotate the car clockwise, only F_{N1} acts to rotate it counterclockwise and so must balance the other three. Hence, F_{N1} must be significantly greater than F_{N2}, and hence F_1 must be significantly greater than F_2. Mathematically, we have for the torques calculated about the CM:

$$(1.5\ \text{m})F_{N1} - (1.5\ \text{m})F_{N2} - (0.75\ \text{m})F_1 - (0.75\ \text{m})F_2 = 0.$$

Since F_{N1} and F_{N2} are proportional to F_1 and F_2, we can write this as

$$(1.5\ \text{m})\frac{(F_1 - F_2)}{\mu} - (0.75\ \text{m})(F_1 + F_2) = 0. \tag{ii}$$

Also, since the car does not accelerate vertically, we have

$$Mg = F_{N1} + F_{N2} = \frac{F_1 + F_2}{\mu}. \tag{iii}$$

Comparing (iii) to (i), we see that $\mu = a/g = 0.50$. We solve (ii) for F_1:

$$F_1 = \frac{1.5\ \text{m}/\mu + 0.75\ \text{m}}{1.5\ \text{m}/\mu - 0.75\ \text{m}} F_2 = \tfrac{5}{3} F_2.$$

Thus F_1 is $1\tfrac{2}{3}$ times greater than F_2. Actual magnitudes, from (i), are $F_1 = 3700$ N and $F_2 = 2200$ N. Because the force on the front tires is generally greater than on the rear tires, cars are often designed with larger brake pads on the front wheels than on the rear. Or, to say it another way, if the brake pads are equal, the front ones wear out a lot faster.

FIGURE 10–42 Forces on a braking car (Example 10–22).

EXAMPLE 10–23 **A falling yo-yo.** String is wrapped around a uniform solid cylinder (something like a yo-yo) of mass M and radius R, and the cylinder starts falling from rest, Fig. 10–43a. As the cylinder falls, find (*a*) its acceleration and (*b*) the tension in the string.

[†] Our proportionality constant μ is not equal to μ_s, the static coefficient of friction $(F_{fr} \leq \mu_s F_N)$, unless the car is just about to skid.

SOLUTION As always we begin with a free–body diagram, Fig. 10–43b, which shows the weight of the cylinder acting at the CM and the tension of the string $\mathbf{F_T}$ acting at the edge of the cylinder. We write Newton's second law for the linear motion (down is positive)

$$Ma = \Sigma F$$
$$= Mg - F_T.$$

Since we do not know the tension in the string, we cannot immediately solve for a. We write Newton's second law for the rotational motion, calculated about the center of mass:

$$\Sigma \tau_{CM} = I_{CM} \alpha_{CM}$$
$$F_T R = \tfrac{1}{2} MR^2 \alpha.$$

Because the cylinder "rolls without slipping" down the string, we have the additional relation that $a = \alpha R$ (Eq. 10–5). Then the torque equation becomes

$$F_T R = \tfrac{1}{2} MR^2 \left(\frac{a}{R} \right) = \tfrac{1}{2} MRa$$

so

$$F_T = \tfrac{1}{2} Ma.$$

Substituting this into the force equation, we obtain

$$Ma = Mg - F_T$$
$$= Mg - \tfrac{1}{2} Ma.$$

Solving for a, we find that $a = \tfrac{2}{3} g$. That is, the linear acceleration is less than what it would be if the cylinder were simply dropped. This makes sense since gravity is not the only vertical force acting; the tension in the string is acting as well. (b) Since $a = \tfrac{2}{3} g$, $F_T = \tfrac{1}{2} Ma = \tfrac{1}{3} Mg$.

(a)

(b)

FIGURE 10–43 Example 10–23.

EXAMPLE 10–24 **What if a rolling ball slips?** A bowling ball of mass M and radius R is thrown along a level surface so that initially ($t = 0$) it slides with a linear speed v_0 but does not rotate. As it slides, it begins to spin, and eventually rolls without slipping. How long does it take to begin rolling without slipping?

Rolling with slipping

SOLUTION The free-body diagram is shown in Fig. 10–44, with the ball moving to the right. Newton's second law for translation gives

$$Ma_x = \Sigma F_x = -\mu_k F_N = -\mu_k Mg.$$

where μ_k is the coefficient of kinetic friction because the ball is sliding. The friction force does two things: it acts to slow down the translational motion of the CM; and it immediately acts to start the ball rotating clockwise. The velocity of the CM is

$$v_{CM} = v_0 + a_x t = v_0 - \mu_k gt.$$

Next we apply Newton's second law for rotation about the CM, $I_{CM} \alpha_{CM} = \Sigma \tau_{CM}$:

$$\tfrac{2}{5} MR^2 \alpha_{CM} = \mu_k MgR.$$

The angular acceleration is thus $\alpha_{CM} = 5\mu_k g / 2R$, which is constant. Then the angular velocity of the ball is (Eq. 10–9a)

$$\omega_{CM} = \omega_0 + \alpha_{CM} t = 0 + \frac{5\mu_k gt}{2R}.$$

The ball starts rolling immediately after it touches the ground, but it rolls and slips at the same time to begin with. It eventually stops slipping, and then rolls without slipping. The condition for rolling without slipping is that

$$v_{CM} = \omega_{CM} R,$$

Eq. 10–4, which is *not* valid if there is slipping. This rolling without slipping begins at a time $t = t_1$ given by (see equations for v_{CM} and ω_{CM} above)

$$v_0 - \mu_k gt_1 = \frac{5\mu_k gt_1}{2R} R \qquad \text{so} \qquad t_1 = \frac{2v_0}{7\mu_k g}.$$

FIGURE 10–44 Example 10–24.

FIGURE 10–45 Sphere rolling to the right.

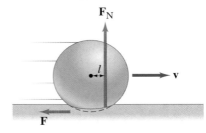

FIGURE 10–46 The normal force, $\mathbf{F}_N$, exerts a torque that slows down the sphere. The deformation of the sphere and the surface it moves on has been exaggerated for detail.

* | 10–12 | **Why Does a Rolling Sphere Slow Down?**

A sphere of mass M and radius R_0 rolling on a horizontal flat surface eventually comes to rest. What force causes it to come to rest? You might think it is friction, but when you examine the problem from a simple, sraightforward point of view, a paradox seems to arise.

Suppose a sphere is rolling to the right as shown in Fig. 10–45, and is slowing down. By Newton's second law, $\Sigma F = Ma$, there must be a force $\mathbf{F}$ (presumably frictional) acting to the left as shown, so that the acceleration $\mathbf{a}$ will also point to the left and v will decrease. Curiously enough, though, if we now look at the torque equation (calculated about the center of mass), $\Sigma\tau_{CM} = I_{CM}\alpha$, we see that the force $\mathbf{F}$ acts to increase the angular acceleration α, and thus to *increase* the velocity of the sphere. Thus the paradox! The force $\mathbf{F}$ acts to decelerate the sphere if we look at the translational motion, but speeds it up if we look at the rotational motion.

The resolution of this apparent paradox is that some other force must be acting. The only other forces acting are gravity, $M\mathbf{g}$, and the normal force $\mathbf{F}_N(= -M\mathbf{g})$. These act vertically and hence do not affect the horizontal translational motion. If we assume the sphere and plane are rigid, so the sphere is in contact at only one point, these forces give rise to no torques about the CM either, since they act through the CM.

The only recourse we have to resolve the paradox is to give up our idealization that the bodies are rigid. In fact, all bodies are deformable to some extent. Our sphere flattens slightly and the level surface also acquires a slight depression where the two are in contact. There is an *area* of contact, not a point. Hence there can be a torque at this area of contact which acts in the opposite direction to the torque associated with $\mathbf{F}$, and thus acts to slow down the rotation of the sphere. This torque is associated with the normal force $\mathbf{F}_N$ that the table exerts on the sphere over the whole area of contact. The net effect is that we can consider $\mathbf{F}_N$ acting vertically a distance l in front of the CM as shown in Fig. 10–46 (where the deformation is greatly exaggerated).

Is it reasonable that the normal force $\mathbf{F}_N$ should effectively act in *front* of the CM as shown in Fig. 10–46? Yes. The sphere is rolling, and the leading edge strikes the surface with a slight impulse. The table therefore pushes upward a bit more strongly on the front part of the sphere than it would if the sphere were at rest. At the back part of the area of contact, the sphere is starting to move upward and so the table pushes upward on it less strongly than when the sphere is at rest. The table pushing up more strongly on the front part of the area of contact gives rise to the necessary torque and justifies the effective acting point of $\mathbf{F}_N$ being in front of the CM.

When other forces are present, the tiny torque τ_N due to $\mathbf{F}_N$ can usually be ignored. For example, when a sphere or cylinder rolls down an incline, the force of gravity has far more influence than τ_N, so the latter can be ignored. For many purposes (but not all), we can assume a hard sphere is in contact with a hard surface at essentially one point.

Summary

When a rigid body rotates about a fixed axis, each point of the body moves in a circular path. Lines drawn perpendicularly from the rotation axis to different points in the body all sweep out the same angle θ in any given time interval.

Angles are conveniently measured in **radians**. One radian is the angle subtended by an arc whose length is equal to the radius, or

$$2\pi \text{ rad} = 360° \quad \text{so} \quad 1 \text{ rad} \approx 57.3°.$$

All parts of a rigid body rotating about a fixed axis have the same **angular velocity** ω and the same **angular**

acceleration α at any instant, where

$$\omega = \frac{d\theta}{dt} \quad \text{and} \quad \alpha = \frac{d\omega}{dt}.$$

The linear velocity and acceleration of any point in a body rotating about a fixed axis are related to the angular quantities by

$$v = R\omega, \quad a_{tan} = R\alpha, \quad a_R = \omega^2 R$$

where R is the perpendicular distance of the point from the rotation axis, and a_{tan} and a_R are the tangential and radial components of the linear acceleration. The frequency f and period T are related to ω by

$$f = \frac{1}{T} = \frac{\omega}{2\pi}.$$

If a rigid body undergoes uniformly accelerated rotational motion ($\alpha =$ constant), equations analogous to those for linear motion are valid:

$$\omega = \omega_0 + \alpha t; \qquad \theta = \omega_0 t + \tfrac{1}{2}\alpha t^2;$$

$$\omega^2 = \omega_0^2 + 2\alpha\theta; \qquad \overline{\omega} = \frac{\omega + \omega_0}{2}.$$

Angular velocity and angular acceleration are vectors. For a rigid body rotating about a fixed axis, both $\boldsymbol{\omega}$ and $\boldsymbol{\alpha}$ point along the rotation axis. The direction of $\boldsymbol{\omega}$ is given by the **right-hand rule**.

The **torque** due to a force $\mathbf{F}$ exerted on a rigid body is equal to

$$\tau = R_\perp F = RF_\perp = RF\sin\theta,$$

where $R_\perp$, called the **lever arm**, is the perpendicular distance from the axis of rotation to the line along which the force acts, and θ is the angle between $\mathbf{F}$ and $\mathbf{R}$.

The rotational equivalent of Newton's second law is

$$\Sigma\tau = I\alpha,$$

where $I = \Sigma m_i R_i^2$ is the **moment of inertia** of the body about the axis of rotation. This relation is valid for a rigid body rotating about an axis fixed in an inertial reference frame, or when τ, I, and α are calculated about the center of mass of a body even if the CM is moving.

The **angular momentum**, L, of a body about a fixed rotation axis is given by

$$L = I\omega.$$

Newton's second law, in terms of angular momentum, becomes

$$\Sigma\tau = \frac{dL}{dt}.$$

If the net torque on the body is zero, $dL/dt = 0$, so $L = $ constant. This is the **law of conservation of angular momentum** for a rotating body.

The **rotational kinetic energy** of a body rotating about a fixed axis with angular velocity ω is

$$K = \tfrac{1}{2}I\omega^2.$$

For a body undergoing both translational and rotational motion, the total kinetic energy is the sum of the translational kinetic energy of the body's CM plus the rotational kinetic energy of the body about its CM:

$$K = \tfrac{1}{2}Mv_{CM}^2 + \tfrac{1}{2}I_{CM}\omega^2$$

as long as the rotation axis is fixed in direction.

Questions

1. A bicycle odometer (which measures distance traveled) is attached near the wheel hub and is designed for 27-in. wheels. What happens if you use it on a bicycle with 24-in. wheels?

2. Suppose a circular platform rotates at constant angular velocity. Does a point on the rim have radial and/or tangential acceleration? If the platform accelerates uniformly, does the point have radial and/or tangential acceleration? What would cause the magnitude of either (linear) acceleration to change?

3. A rotating carousel has one child sitting on a horse near the outer edge and another child seated on a lion halfway out from the center. (a) Which child has the greater linear speed? (b) Which child has the greater angular speed?

4. In what direction is the Earth's angular velocity for its daily rotation on its axis?

5. The angular velocity of a wheel rotating on a horizontal axle points west. In what direction is the linear velocity of a point on the top of the wheel? If the angular acceleration points east, describe the tangential linear acceleration of this point. Is the angular speed increasing or decreasing?

6. If the angular quantities θ, ω and α were specified in terms of degrees rather than radians, how would Eqs. 10–9 for uniformly accelerated rotational motion have to be altered?

7. Can a small force exert a greater torque than a larger force? Explain.

8. If a force $\mathbf{F}$ acts on a body such that its lever arm is zero, does it have any effect on the body's motion?

9. If the net force on a system is zero, is the net torque also zero? If the net torque on a system is zero, is the net force zero?

10. The moment of inertia of this textbook would be the least about what axis?

11. Why is it more difficult to do a situp with your hands behind your head than when they are outstretched in front of you? A diagram may help you to answer this.

12. Expert bicyclists use very lightweight tubular ("sew-up") tires. They claim that reducing the mass of the tires is far more significant than an equal reduction in mass elsewhere on the bicycle. Explain.

13. Why do tightrope walkers carry a long, narrow beam (Fig. 10–47)?

FIGURE 10–47 Question 13.

14. A quarterback leaps into the air to throw a forward pass. As he throws the ball, the upper part of his body rotates. If you look quickly you will notice that his hips and legs rotate in the opposite direction (Fig. 10–48). Explain.

FIGURE 10–48
Quarterback in the air, throwing a pass. Question 14.

15. We claim that momentum and angular momentum are conserved. Yet most moving or rotating bodies eventually slow down and stop. Explain.

16. If there were a great migration of people toward the equator, how would this affect the length of the day?

17. Can the diver of Fig. 10–30 do a somersault without having any initial rotation when she leaves the board?

18. Suppose you are standing on the edge of a large rotating turntable. What happens if you walk toward the center?

19. When a motorcyclist leaves the ground on a jump and leaves the throttle on (so the rear wheel spins), why does the front of the cycle rise up?

20. A stick stands vertically on its end on a frictionless surface. Describe the motion of its CM, and of each end, when it is tipped slightly to one side and falls.

21. Two inclines have the same height but make different angles with the horizontal. The same steel ball is rolled down each incline. On which incline will the speed of the ball at the bottom be greatest? Explain.

22. Two solid spheres simultaneously start rolling (from rest) down an incline. One sphere has twice the radius and twice the mass of the other. Which reaches the bottom of the incline first? Which has the greater speed there? Which has the greater total kinetic energy at the bottom?

23. A sphere and a cylinder have the same radius and the same mass. They start from rest at the top of an incline. Which reaches the bottom first? Which has the greater speed at the bottom? Which has the greater total kinetic energy at the bottom? Which has the greater rotational kinetic energy?

24. A cyclist rides over the top of a hill. Is the bicycle's motion rotational, translational, or a combination of both?

* **25.** The total kinetic energy of a system of particles is equal to the translational kinetic energy of the center of mass of the system plus the kinetic energy of motion relative to the CM (Eq. 10–26). Is it possible or useful to make a similar statement about the total linear momentum?

Problems

Sections 10–1 to 10–3

1. (I) What are the following angles expressed in radians: (*a*) 30°, (*b*) 57°, (*c*) 90°, (*d*) 360°, and (*e*) 420°? Give as numerical values and as fractions of π.

2. (I) The Sun subtends an angle of about 0.5° to us on Earth, 150 million km away. Estimate the radius of the Sun.

3. (I) The Eiffel Tower is 300 m tall. When you are standing at a certain place in Paris, it subtends an angle of 7.5°. How far are you, then, from the Eiffel Tower?

4. (I) A centrifuge rotor is accelerated from rest to 20,000 rpm in 5.0 min. What is its average angular acceleration?

5. (II) Calculate the angular velocity of (*a*) the second hand, (*b*) the minute hand, and (*c*) the hour hand, of a clock. State in rad/s. (*d*) What is the angular acceleration in each case?

6. (II) Calculate the angular velocity of the Earth (*a*) in its orbit around the Sun and (*b*) about its axis.

7. (II) What is the linear speed of a point (*a*) on the Equator, (*b*) on the Arctic Circle (latitude 66.5° N), (*c*) at a latitude of 40.0° N, due to the Earth's rotation?

8. (II) Estimate the angle subtended by the Moon using a ruler and your finger or other object to just blot out the Moon. Describe your measurement and the result obtained and then use it to estimate the diameter of the Moon. The Moon is about 380,000 km from the Earth.

9. (II) (*a*) A 0.35-m diameter grinding wheel rotates at 2500 rpm. Calculate its angular velocity in rad/s. (*b*) What is the linear speed and acceleration of a point on the edge of the grinding wheel?

10. (II) A rotating merry-go-round makes one complete revolution in 4.0 s (Fig. 10–49). (*a*) What is the linear speed of a child seated 1.2 m from the center? (*b*) What is her acceleration?

FIGURE 10–49 Problem 10.

11. (II) A child rolls a ball on a level floor 3.5 m to another child. If the ball makes 15.0 revolutions, what is its diameter?

12. (II) A 60.0-cm-diameter wheel accelerates uniformly from 210 rpm to 380 rpm in 6.5 s. How far will a point on the edge of the wheel have traveled in this time?

13. (II) In traveling to the Moon, astronauts aboard the Apollo spacecraft put it into a slow rotation in order to distribute the Sun's energy evenly. At the start of their trip, they accelerated uniformly during a 10-min time interval from no rotation to one revolution every minute. The spacecraft can be thought of as a cylinder with a diameter of 8.5 m. Determine (a) the angular acceleration, and (b) the radial and tangential components of the linear acceleration of a point on the surface of the spacecraft 5.0 min after it started this acceleration.

14. (II) Using calculus, derive the angular kinematic equations 10–9a and 10–9b for constant angular acceleration. Start with $\alpha = d\omega/dt$.

15. (II) A small rubber wheel is used to drive a large pottery wheel, and they are mounted so that their circular edges touch. If the small wheel has a radius of 2.0 cm and accelerates at the rate of $7.2 \, \text{rad/s}^2$, and it is in contact with the pottery wheel (radius 25.0 cm) without slipping, calculate (a) the angular acceleration of the pottery wheel, and (b) the time it takes the pottery wheel to reach its required speed of 65 rpm.

16. (II) The angle through which a rotating wheel has turned in time t is given by $\theta = 6.0t - 8.0t^2 + 4.5t^4$, where θ is in radians and t in seconds. Determine an expression (a) for the instantaneous angular velocity ω and (b) for the instantaneous angular acceleration α. (c) Evaluate ω and α at $t = 3.0$ s. (d) What is the average angular velocity, and (e) the average angular acceleration between $t = 2.0$ s and $t = 3.0$ s?

17. (II) The angular acceleration of a wheel, as a function of time, is $\alpha = 5.0t^2 - 3.5t$, where α is in rad/s^2 and t in seconds. If the wheel starts from rest ($\theta = \omega = 0$ at $t = 0$), determine a formula for (a) the angular velocity ω and (b) the angular position θ, both as a function of time. (c) Evaluate ω and θ at $t = 2.0$ s.

18. (II) The tires of a car make 85 revolutions as the car reduces its speed uniformly from 90.0 km/h to 60.0 km/h. The tires have a diameter of 0.90 m. (a) What was the angular acceleration? (b) If the car continues to decelerate at this rate, how much more time is required for it to stop?

Section 10–4

19. (II) The axle of a wheel is mounted on supports that rest on a rotating turntable as shown in Fig. 10–50. The wheel has angular velocity $\omega_1 = 50.0$ rad/s about its axle, and the turntable has angular velocity $\omega_2 = 35.0$ rad/s about a vertical axis. (Note arrows showing these motions in the figure.) (a) What are the directions of $\boldsymbol{\omega}_1$ and $\boldsymbol{\omega}_2$? (b) What is the resultant angular velocity of the wheel, as seen by an outside observer, at the instant shown? Give the magnitude and direction. (c) What is the magnitude and direction of the angular acceleration of the wheel at the instant shown? Take the z axis vertically upward and the direction of the axle at the moment shown to be the x axis pointing to the right.

FIGURE 10–50
Problem 19.

Section 10–5

20. (I) A person exerts a force of 38 N on the end of a door 96 cm wide. What is the magnitude of the torque if the force is exerted (a) perpendicular to the door and (b) at a 60.0° angle to the face of the door?

21. (I) The biceps muscle exerts a vertical force on the lower arm as shown in Figs. 10–51a and b. For each case, calculate the torque about the axis of rotation through the elbow joint, assuming the muscle is attached 5.0 cm from the elbow as shown.

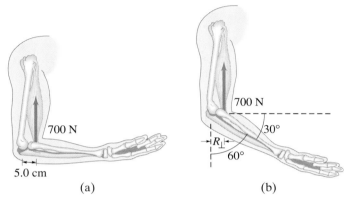

FIGURE 10–51   Problem 21.

22. (II) A wheel of diameter 27.0 cm is constrained to rotate in the xy plane, about the z axis, which passes through its center. A force $\mathbf{F} = (-31.0\mathbf{i} + 38.6\mathbf{j})\,\text{N}$ acts at a point on the edge of the wheel that lies exactly on the x axis at a particular instant. What is the torque about the rotation axis at this instant?

23. (II) Calculate the net torque about the axle of the wheel shown in Fig. 10–52. Assume that a friction torque of $0.30 \, \text{m}\cdot\text{N}$ opposes the motion.

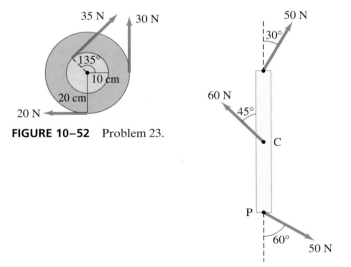

FIGURE 10–52   Problem 23.

FIGURE 10–53   Problem 24.

24. (II) Determine the net torque on the 2.0-m-long beam shown in Fig. 10–53. Calculate about (a) point C, the CM, and (b) point P at one end.

25. (II) The bolts on the cylinder head of certain engines require tightening to a torque of 90 m·N. If a wrench is 26 cm long, what force perpendicular to the wrench must the mechanic exert at its end? If the six-sided bolt is 15 mm in diameter, estimate the force applied near each of the six points by a socket wrench (Fig. 10–54).

FIGURE 10–54 Problem 25.

Sections 10–6 to 10–7

26. (I) Determine the moment of inertia of a 12.0-kg solid sphere of radius 0.80 m when the axis of rotation is through its center.

27. (I) A 1.4-kg grindstone in the shape of a uniform cylinder of radius 0.20 m acquires a rotational rate of 1800 rev/s from rest over a 6.0-s interval at constant angular acceleration. Calculate the torque delivered by the motor.

28. (I) An oxygen molecule consists of two oxygen atoms whose total mass is 5.3×10^{-26} kg and whose moment of inertia about an axis perpendicular to the line joining the two atoms, midway between them, is 1.9×10^{-46} kg·m². Estimate, from these data, the effective distance between the atoms.

29. (II) A 2.4-kg ball on the end of a thin, light rod is rotated in a horizontal circle of radius 1.2 m. Calculate (*a*) the moment of inertia of the ball about the center of the circle, and (*b*) the torque needed to keep the ball rotating at constant angular velocity if air resistance exerts a force of 0.020 N on the ball. Ignore the rod's moment of inertia and air resistance.

30. (II) Calculate the moment of inertia of the array of point objects shown in Fig.10–55 about (*a*) the vertical axis, and (*b*) the horizontal axis. Assume the objects are connected by very light rigid wires. About which axis would it be harder to accelerate this array? In Fig. 10–55, $m = 1.8$ kg and $M = 3.1$ kg. The array is rectangular and it is split through the middle by the horizontal axis.

FIGURE 10–55 Problem 30.

31. (II) Consider a helicopter rotor blade as a long thin rod, as shown in Fig. 10–56. If each of the three helicopter blades is 3.75 m long and has a mass of 160 kg, calculate the moment of inertia of the three rotor blades about the axis of rotation. How much torque must the motor apply to bring the blades up to a speed of 5.0 rev/s in 8.0 s?

FIGURE 10–56
Problem 31.

32. (II) A 0.84-m-diameter solid sphere can be rotated about an axis through its center by a torque of 10.8 m·N which accelerates it uniformly from rest through a total of 180 revolutions in 15.0 s. What is the mass of the sphere?

33. (II) A merry-go-round accelerates from rest to 3.0 rad/s in 24 s. Assuming the merry-go-round is a uniform disk of radius 7.0 m and mass 31,000 kg, calculate the net torque required to accelerate it.

34. (II) Four equal masses M are spaced at equal intervals, *l*, along a horizontal straight rod whose mass can be ignored. The system is to be rotated about a vertical axis passing through the mass at the left end of the rod and perpendicular to it. (*a*) What is the moment of inertia of the system about this axis? (*b*) What minimum force, applied to the farthest mass, will impart an angular acceleration α? (*c*) What is the direction of this force?

35. (II) Suppose the force F_T in the cord hanging from the wheel of Example 10–7, Fig. 10–22, is given by the relation $F_T = 3.00t - 0.20t^2$ (newtons) where *t* is in seconds. If the wheel starts from rest, what is the linear speed of a point on its rim 8.0 s later?

36. (II) A centrifuge rotor rotating at 10,000 rpm is shut off and is eventually brought to rest by a frictional torque of 1.00 m·N. If the mass of the rotor is 4.70 kg and it can be considered a solid cylinder of radius 0.0780 m, through how many revolutions will the rotor turn before coming to rest, and how much time will this take?

37. (II) Two blocks are connected by a light string passing over a pulley of radius 0.25 m and moment of inertia I. The blocks move to the right with an acceleration of 1.00 m/s² on inclines with frictionless surfaces (see Fig. 10–57). (*a*) Draw free-body diagrams for each of the two blocks and the pulley. (*b*) Determine F_{T1} and F_{T2}, the tensions in the two parts of the string. (*c*) Find the net torque acting on the pulley, and determine its moment of inertia, I.

FIGURE 10–57
Problem 37.

38. (II) A string passing over a pulley has a 3.80-kg mass hanging from one end and a 3.40-kg mass hanging from the other end. The pulley is a uniform solid cylinder of radius 3.0 cm and mass 0.80 kg. (a) If the bearings of the pulley were frictionless, what would be the acceleration of the two masses? (b) In fact, it is found that if the heavier mass is given a downward speed of 0.20 m/s, it comes to rest in 6.2 s. What is the average frictional torque acting on the pulley?

39. (II) When discussing moments of inertia, especially for unusual or irregularly shaped objects, it is sometimes convenient to work with the **radius of gyration**, k. This radius is defined so that if all the mass of the object were concentrated at this distance from the axis, the moment of inertia would be the same as that of the original object. Thus, the moment of inertia of any object can be written in terms of its mass M and the radius of gyration as $I = Mk^2$. Determine the radius of gyration for each of the objects (hoop, cylinder, sphere, etc.) shown in Fig. 10–21.

40. (III) A thin rod of length l stands vertically on a table. The rod begins to fall, but its lower end does not slide. (a) Determine the angular velocity of the rod as a function of the angle ϕ it makes with the tabletop. (b) What is the speed of the tip of the rod just before it strikes the table?

41. (III) A hammer thrower accelerates the hammer (mass = 7.30 kg) from rest within four full turns (revolutions) and releases it at a speed of 29.0 m/s. Assuming a uniform rate of increase in angular velocity and a radius of 2.00 m, calculate (a) the angular acceleration, (b) the (linear) tangential acceleration, (c) the centripetal acceleration just before release, (d) the net force exerted on the hammer by the athlete just before release, and (e) the angle of this force with respect to the radius of the circular motion. Neglect the effect of gravity.

Section 10–8

42. (I) Use the parallel-axis theorem to show that the moment of inertia of a thin rod about an axis perpendicular to the rod at one end is $I = \frac{1}{3}Ml^2$, given that if the axis passes through the center, $I = \frac{1}{12}Ml^2$ (Fig. 10–21f and g).

43. (II) Use the perpendicular-axis theorem and Fig. 10–21h to determine a formula for the moment of inertia of a thin, square plate of side s about an axis (a) through its center and along a diagonal of the plate, (b) through the center and parallel to a side.

44. (II) Determine the moment of inertia of a 25-kg door that is 2.5 m high and 1.0 m wide and is hinged along one side. Ignore the thickness of the door.

45. (II) Two uniform solid spheres of mass M and radius R_0 are connected by a thin (massless) rod of length R_0 so that the centers are $3R_0$ apart. (a) Determine the moment of inertia of this system about an axis perpendicular to the rod at its center. (b) What would be the percentage error if the masses of each sphere were assumed to be concentrated at their centers and a very simple calculation made?

46. (II) A ball of mass M and radius R_1 on the end of a thin massless rod is rotated in a horizontal circle of radius R_0 about an axis of rotation AB, as shown in Fig. 10–58. (a) Considering the mass of the ball to be concentrated at

its center of mass, calculate its moment of inertia about AB. (b) Using the parallel-axis theorem and considering the finite radius of the ball, calculate the moment of inertia of the ball about AB. (c) Calculate the percentage error introduced by the point mass approximation for $R_1 = 10$ cm and $R_0 = 1.0$ m.

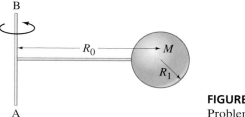

FIGURE 10–58
Problem 46.

47. (II) What is the moment of inertia of a uniform horizontal disk of radius R_0 and mass M about (a) a vertical axis a distance $0.25R_0$, from the center, (b) a horizontal axis through the center, and (c) a horizontal axis tangent to the edge.

48. (II) A thin 7.0-kg wheel of radius 32 cm is weighted to one side by a 1.50 kg weight, small in size, placed 22 cm from the center of the wheel. Calculate (a) the position of the center of mass of the weighted wheel and (b) the moment of inertia about an axis through its center of mass, perpendicular to its face.

49. (III) Derive the formula for the moment of inertia of a uniform sphere of radius r_0 and mass M about an axis through its center (Fig. 10–21e). [*Hint*: Divide the sphere into infinitesimally thin disks (cylinders) of thickness dy. Then use the result of Example 10–10 or Fig. 10–21c, and integrate over these cylindrical disks.]

50. (III) Derive the formula for the moment of inertia of a uniform thin rod of length l about an axis through its center, perpendicular to the rod (see Fig. 10–21f).

51. (III) (a) Derive the formula given in Fig. 10–21h for the moment of inertia of a uniform, flat, rectangular plate of dimensions $l \times w$ about an axis through its center, perpendicular to the plate. (b) What is the moment of inertia about each of the axes through the center that are parallel to the edges of the plate?

Section 10–9

52. (I) (a) What is the angular momentum of a 2.8-kg uniform cylindrical grinding wheel of radius 18 cm when rotating at 1500 rpm? (b) How much torque is required to stop it in 7.0 s?

53. (I) A diver (such as the one in Fig. 10–30) can reduce her moment of inertia by a factor of about 3.5 when changing from the straight position to the tuck position. If she makes two rotations in 1.5 s when in the tuck position, what is her angular speed (rev/s) when in the straight position?

54. (I) A figure skater during her finale can increase her rotation rate from an initial rate of 1.0 rev every 2.0 s to a final rate of 3.0 rev/s. If her initial moment of inertia was 4.6 kg·m², what is her final moment of inertia? How does she physically accomplish this change?

55. (II) A person stands, hands at the side, on a platform that is rotating at a rate of 1.30 rev/s. If the person now raises his arms to a horizontal position, Fig. 10–59, the speed of rotation decreases to 0.80 rev/s. (a) Why does this occur? (b) By what factor has the moment of inertia of the person changed?

FIGURE 10–59 Problem 55.

56. (II) (a) What is the angular momentum of a figure skater spinning (with arms in close to her body) at 3.5 rev/s, assuming her to be a uniform cylinder with a height of 1.5 m, a radius of 15 cm, and a mass of 55 kg. (b) How much torque is required to slow her to a stop in 5.0 s, assuming she does *not* move her arms?

57. (II) Determine the angular momentum of the Earth (a) about its rotation axis (assume the Earth is a uniform sphere), and (b) in its orbit around the Sun (treat the Earth as a particle orbiting the Sun). The Earth has mass = 6.0×10^{24} kg, radius = 6.4×10^6 m, and is 1.5×10^8 km from the Sun.

58. (II) A uniform horizontal rod of mass M and length l rotates with angular velocity ω about a vertical axis through its center. Attached to each end of the rod is a small mass m. Determine the angular momentum of the system about the axis.

59. (II) A 4.8-m-diameter merry-go-round is rotating freely with an angular velocity of 0.80 rad/s. Its total moment of inertia is 1950 kg·m². Four people standing on the ground, each of 65-kg mass, suddenly step onto the edge of the merry-go-round. What is the angular velocity of the merry-go-round now? What if the people were on it initially and then jumped off in a radial direction (relative to the merry-go-round)?

60. (II) A woman of mass m stands at the edge of a solid cylindrical platform of mass M and radius R. At $t = 0$, the platform is rotating with negligible friction at angular velocity ω_0 about a vertical axis through its center, and the person begins walking with speed v (relative to the platform) toward the center of the platform. (a) Determine the angular velocity of the system as a function of time. (b) What will be the angular velocity when the woman reaches the center?

Section 10–10

61. (I) A centrifuge rotor has a moment of inertia of 4.25×10^{-2} kg·m². How much energy is required to bring it from rest to 10,000 rpm?

62. (I) An automobile engine develops a torque of 280 m·N at 4000 rpm. What is the horsepower of the engine?

63. (II) For the star in Example 10–14 undergoing gravitational collapse, by what factor did the original kinetic energy change? Where did this energy come from?

64. (II) A rotating uniform cylindrical platform of mass 280 kg and radius 5.5 m slows down from 3.8 rev/s to rest in 18 s when the driving motor is disconnected. Estimate the power output of the motor (hp) required to maintain a steady speed of 3.8 rev/s.

65. (II) Two masses, $m_1 = 35.0$ kg and $m_2 = 38.0$ kg, are connected by a rope that hangs over a pulley (as in Fig. 10–60). The pulley is a uniform cylinder of radius 0.30 m and mass 4.8 kg. Initially m_1 is on the ground and m_2 rests 2.5 m above the ground. If the system is released, use conservation of energy to determine the speed of m_2 just before it strikes the ground. Assume the pulley bearing is frictionless.

FIGURE 10–60
Problem 65.

66. (II) A uniform thin rod of length l and mass M is suspended freely from one end. It is pulled to the side an angle θ and released. If friction can be ignored, what is its angular velocity, and the speed of its free end, at the lowest point?

67. (II) A 70-kg person stands on a tiny rotating platform with arms outstretched. (a) Estimate the moment of inertia of the person using the following approximations: the body (including head and legs) is a 60-kg cylinder, 12 cm in radius and 1.70 m high; and each arm is a 5.0-kg thin rod, 60 cm long, attached to the cylinder. (b) Using the same approximations, estimate the moment of inertia when the arms are at the person's sides. (c) If one rotation takes 1.5 s when the person's arms are outstretched, what is the time for each rotation with arms at the sides? Ignore the moment of inertia of the lightweight platform. (d) Determine the change in kinetic energy when the arms are lifted from the sides to the horizontal position. (e) From your answer to part (d), would you expect it to be harder or easier to lift your arms when rotating or when at rest?

68. (III) A 4.00-kg mass and a 3.00-kg mass are attached to opposite ends of a thin 50.0-cm-long horizontal rod (Fig. 10–61). The system is rotating at angular speed $\omega = 8.00$ rad/s about a vertical axle at the center of the rod. Determine (a) the kinetic energy K of the system, and (b) the net force on each mass. (c) Repeat parts (a) and (b) assuming that the axle passes through the center of mass of the system.

FIGURE 10–61
Problem 68.

69. (I) Calculate the translational speed of a cylinder when it reaches the foot of an incline 11.8 m high. Assume it starts from rest and rolls without slipping.

70. (I) Estimate the kinetic energy of the Earth with respect to the Sun as the sum of two terms, (a) that due to its daily rotation about its axis and (b) that due to its yearly revolution about the Sun (assume the Earth is a uniform sphere, mass $= 6.0 \times 10^{24}$ kg, radius $= 6.4 \times 10^6$ m, and is 1.5×10^8 km from the Sun).

71. (I) A bowling ball of mass 7.3 kg and radius 9.0 cm rolls without slipping down a lane at 5.3 m/s. Calculate its total kinetic energy.

72. (II) A narrow but solid spool of thread has radius R and mass M. If you pull up on the thread so that the CM of the spool remains suspended in the air at the same place, (a) what force must you exert on the thread? (b) How much work have you done by the time the spool turns with angular velocity ω?

73. (II) A thin, hollow 60.0-g section of pipe of radius 10.0 cm starts rolling (from rest) down a 21.5° incline 5.60 m long. (a) If the pipe rolls without slipping, what will be its speed at the base of the incline? (b) What will be its total kinetic energy at the base of the incline? (c) What minimum value must the coefficient of static friction have if the pipe is not to slip?

74. (II) A solid rubber ball rests on the floor of a railroad car when the car begins moving with acceleration a. Assuming the ball rolls without slipping, what is its acceleration relative to (a) the car and (b) the ground?

75. (II) A ball of radius r_0 rolls on the inside of a track of radius R_0 (see Fig. 10–62). If the ball starts from rest at the vertical edge of the track, what will be its speed when it reaches the lowest point of the track, rolling without slipping?

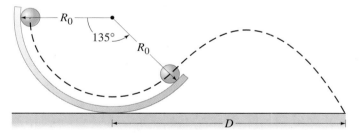

FIGURE 10–62 Problems 75 and 76.

76. (III) A small sphere of radius $r_0 = 1.5$ cm rolls without slipping on the track shown in Fig. 10–62 whose radius is $R_0 = 26.0$ cm. The sphere starts rolling at a height R_0 above the bottom of the track. When it leaves the track after passing through an angle of 135° as shown, (a) what will be its speed, and (b) at what distance D from the base of the track will the sphere hit the ground?

77. (III) The 1100-kg mass of a car includes four tires, each of mass (including wheels) 35 kg and diameter 0.80 m. Assume each tire and wheel combination acts as a solid cylinder. Determine (a) the total kinetic energy of the car when traveling 100 km/h and (b) the fraction of the kinetic energy in the tires and wheels. (c) If the car is initially at rest and is then pulled by a tow truck with a force of 2000 N, what is the acceleration of the car? Ignore frictional losses. (d) What percent error would you make in part (c) if you ignored the rotational inertia of the tires and wheels?

* **78.** (III) Determine the speed of the falling cylinder in Example 10–23 in terms of the distance fallen h. Choose $h = 0$ and $v = 0$ at $t = 0$.

* **79.** (III) In Example 10–24, (a) how far has the ball moved down the lane when it starts rolling without slipping? (b) What are its final linear and rotational speeds?

* **80.** (III) A wheel with rotational inertia $I = \frac{1}{2}MR^2$ about its central axle is set spinning with initial angular speed ω_0 and is then lowered onto the ground so that it touches the ground with no horizontal speed. Initially it slips, but then begins to move forward and eventually rolls without slipping. (a) In what direction does friction act on the slipping wheel? (b) How long does the wheel slip before it begins to roll without slipping? (c) What is the wheel's final translational speed? [Hint: Use $\Sigma\mathbf{F} = m\mathbf{a}$, $\Sigma\tau_{CM} = I_{CM}\alpha_{CM}$, and recall that only when there is rolling without slipping is $v_{CM} = \omega R$.]

* **81.** (III) In Examples 10–19 and 10–21 suppose the sphere does not simply roll, but slips as well. To be concrete, suppose $\theta = 33.0°$ and $\mu = \mu_s = \mu_k = 0.10$ with the incline having a height $H = 2.0$ m and the sphere a radius $R_0 = 11.0$ cm with mass $M = 850$ g. The sphere starts from rest at the top of the incline. Determine: (a) the acceleration of the sphere; (b) the speed of the CM of the sphere when it reaches the foot of the incline; (c) the total kinetic energy of the sphere at the foot; and (d) the loss in mechanical energy. (e) Determine these same quantities for $\mu = 0.30$, for which there is no slipping, and compare. (f) Compare your answers in (a), (b), and (c) to a box of the same mass sliding down the incline. [Hint: Use $\Sigma\mathbf{F} = m\mathbf{a}$, $\Sigma\tau_{CM} = I_{CM}\alpha_{CM}$, and recall that $v_{CM} = \omega R$ is valid only when there is no slipping.]

* **Section 10–12**

* **82.** (I) A rolling ball slows down because the normal force does not pass exactly through the CM of the ball, but passes in front of the CM. Using Fig. 10–46, show that the torque resulting from the normal force $(\tau_N = lF_N$ in Fig. 10–46$)$ is 7/5 of that due to the frictional force, $\tau_{fr} = R_0 F$; that is, show that $\tau_N = \frac{7}{5}\tau_{fr}$.

General Problems

83. Eclipses happen on Earth because of an amazing coincidence. Calculate, using the information inside the front cover, the angular diameter (in radians) of the Sun and the angular diameter of the Moon, as seen on Earth.

84. A hollow cylinder (hoop) is rolling on a horizontal surface at speed $v = 2.10$ m/s when it starts up a 21.5° incline. (a) How far along the incline will it go? (b) How long will it be on the incline before it arrives back at the bottom?

85. A solid wheel of radius R_1 is turned by a circular rubber roller of radius R_2 in contact with it at their outer edges. What is the ratio of their angular velocities, ω_1/ω_2?

86. The Moon orbits the Earth so that the same side always faces the Earth. Determine the ratio of its spin angular momentum (about its own axis) to its orbital angular momentum. (In the latter case, treat the Moon as a particle orbiting the Earth.)

87. A large spool of rope stands on the ground with the end of the rope lying on the top edge of the spool. A person grabs the end of the rope and walks a distance l, holding onto it, Fig. 10–63. The spool rolls behind the person without slipping. What length of rope unwinds from the spool? How far does the spool's CM move?

FIGURE 10–63   Problem 87.   **FIGURE 10–64**   Problem 88.

88. A cyclist accelerates from rest at a rate of 1.00 m/s². How fast will a point on the rim of the tire (diameter = 68 cm) at the top be moving after 3.0 s? See Fig. 10–64.

89. A person stands on a platform, initially at rest, that can rotate freely without friction. The moment of inertia of the person plus the platform is I_p. The person holds a spinning bicycle wheel with axis horizontal. The wheel has moment of inertia I_w and angular velocity ω_w. What will be the angular velocity ω_p of the platform if the person moves the axis of the wheel so that it points (a) vertically upward, (b) at a 60° angle to the vertical, (c) vertically downward? (d) What will ω_p be if the person reaches up and stops the wheel in part (a)?

90. The density (mass per unit length) of a thin rod of length l increases uniformly from λ_0 at one end to $2\lambda_0$ at the other end. Determine the moment of inertia about an axis perpendicular to the rod through its geometric center.

91. (a) For a bicycle (see Fig. 10–65), how is the angular speed of the rear wheel (ω_R) related to that of the pedals and front sprocket (ω_F)? That is, derive a formula for ω_R/ω_F. Let N_F and N_R be the number of teeth on the front and rear sprockets, respectively. The teeth are spaced equally on all sprockets so that the chain meshes properly. (b) Then evaluate the ratio ω_R/ω_F when the front and rear sprockets have 52 and 13 teeth, respectively, and (c) when they have 42 and 28 teeth.

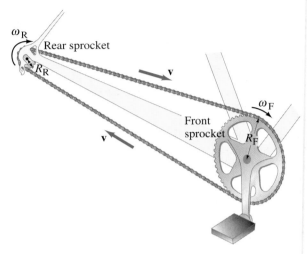

FIGURE 10–65   Problem 91.

92. The forearm in Fig. 10–66 accelerates a 1.00-kg ball at 7.0 m/s² by means of the triceps muscle as shown. Calculate (a) the torque needed and (b) the force that must be exerted by the triceps muscle. Ignore the mass of the arm.

FIGURE 10–66   Problems 92 and 93.

93. Assume that a 1.00-kg ball is thrown solely by the action of the forearm (Fig. 10–66), which rotates about the elbow joint under the action of the triceps muscle. The ball is accelerated from rest to 10.0 m/s in 0.22 s, at which point it is released. Calculate (a) the angular acceleration of the arm and (b) the force required of the triceps muscle. Assume the forearm has a mass of 3.4 kg and rotates like a uniform rod about an axis at its end.

94. A softball player swings a bat, accelerating it from rest to 3.0 rev/s in a time of 0.20 s. Approximate the bat as a 2.2-kg uniform rod of length 0.95 m, and compute the torque the player applies to one end of it.

95. Pilots can be tested for the stresses of flying highspeed jets in a whirling "human centrifuge" which takes 1.0 min to turn through 20 complete revolutions before reaching its final speed. (*a*) What was its angular acceleration, and (*b*) what was its final speed in rpm?

96. A uniform disk turns at 7.0 rev/s around a frictionless spindle. A nonrotating rod, of the same mass as the disk and length equal to the disk's diameter, is dropped onto the freely spinning disk. They then both turn around the spindle with their centers superposed, Fig. 10–67. What is the angular velocity in rev/s of the combination?

FIGURE 10–67
Problem 96.

FIGURE 10–68 Problem 97.

97. In order to get a flat uniform cylindrical satellite spinning at the correct rate, engineers fire four tangential rockets as shown in Fig. 10–68. If the satellite has a mass of 2000 kg and a radius *R* of 3.0 m, what is the required steady force of each rocket if the satellite is to reach 30 rpm in 5.0 min?

98. (*a*) Calculate the translational and rotational speeds of a sphere (radius 20.0 cm and mass 2.20 kg), that rolls without slipping down a 30.0° incline that is 10.0 m long, when it reaches the bottom. Assume it started from rest. (*b*) What is its ratio of translational to rotational kinetic energy at the bottom? Avoid putting in numbers until the end so you can answer: (*c*) do your answers in (*a*) and (*b*) depend on the radius of the sphere or its mass?

99. One possibility for a low-pollution automobile is for it to use energy stored in a heavy rotating flywheel. Suppose such a car has a total mass of 1400 kg, uses a 1.50-m diameter uniform cylindrical flywheel of mass 240 kg, and should be able to travel 300 km without needing a flywheel "spinup." (*a*) Make reasonable assumptions (average frictional retarding force = 500 N, twenty acceleration periods from rest to 90 km/h, equal uphill and downhill—assuming during downhill, energy can be put back into the flywheel), and show that the total energy needed to be stored in the flywheel is about 1.6×10^8 J. (*b*) What is the angular velocity of the flywheel when it has a full "energy charge"? (*c*) About how long would it take a 150-hp motor to give the flywheel a full energy charge before a trip?

100. If the coefficient of static friction between tires and pavement is 0.75, calculate the minimum torque that must be applied to the 66-cm-diameter tire of a 1250-kg automobile in order to "lay rubber" (make the wheels spin, slipping as the car accelerates). Assume each wheel supports an equal share of the weight.

101. The radius of the roll of paper shown in Fig. 10–69 is 7.6 cm and its moment of inertia is $I = 2.9 \times 10^{-3}$ kg·m². A force of 3.2 N is exerted on the end of the roll for 1.3 s, but the paper does not tear so it begins to unroll. A constant friction torque of 0.11 m·N is exerted on the roll which gradually brings it to a stop. Assuming that the paper's thickness is negligible, calculate (*a*) the length of paper that unrolls during the time that the force is applied (1.3 s) and (*b*) the length of paper that unrolls from the time the force ends to the time when the roll has stopped moving.

FIGURE 10–69
F   Problem 101.

102. An *Atwood's machine* consists of two masses, m_1 and m_2, which are connected by a massless inelastic cord that passes over a pulley, Fig. 10–70. If the pulley has radius *R* and moment of inertia *I* about its axle, determine the acceleration of the masses m_1 and m_2, and compare to the situation in which the moment of inertia of the pulley is ignored. [*Hint*: The tensions F_{T1} and F_{T2} are not necessarily equal.]

FIGURE 10–70
Atwood's machine. Problem 102.

103. A person of mass 75 kg stands at the center of a rotating merry-go-round platform of radius 3.0 m and moment of inertia 1000 kg·m². The platform rotates without friction with angular velocity 2.0 rad/s. The person walks radially to the edge of the platform. (*a*) Calculate the angular velocity when the person reaches the edge. (*b*) Compare the rotational kinetic energies of the system of platform plus person before and after the person's walk.

104. A wheel of mass M has radius R. It is standing vertically on the floor, and we want to exert a horizontal force F at its axle so that it will climb a step against which it rests (Fig. 10–71). The step has height h, where $h < R$. What minimum force F is needed?

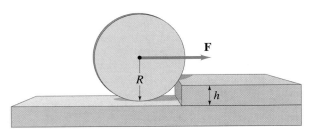

FIGURE 10–71 Problem 104.

105. A cord connected at one end to a block which can slide on an inclined plane has its other end wrapped around a cylinder resting in a depression at the top of the plane as shown in Fig. 10–72. Determine the speed of the block after it has traveled 1.80 m along the plane, starting from rest. Assume (a) there is no friction, (b) the coefficient of friction between all surfaces is $\mu = 0.035$. [Hint: In part (b) first determine the normal force on the cylinder, and make any reasonable assumptions needed.]

FIGURE 10–72 Problem 105.

106. A crucial part of a piece of machinery starts as a flat uniform cylindrical disk of radius R_0 and mass M. It then has a circular hole of radius R_1 drilled into it (Fig. 10–73). The hole's center is a distance h from the center of the disk. Find the moment of inertia of this disk (with off-center hole) when rotated about its center, C. [Hint: Consider a solid disk and "subtract" the hole; use the parallel-axis theorem.]

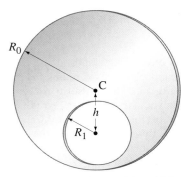

FIGURE 10–73 Problem 106.

107. A marble of mass m and radius r_0 rolls along the looped rough track of Fig. 10–74. What is the minimum value of h if the marble is to reach the highest point of the loop without leaving the track? (a) Assume $r_0 \ll R_0$, (b) do not make this assumption. Ignore frictional losses.

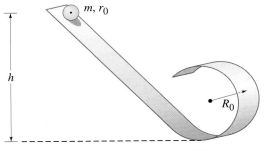

FIGURE 10–74 Problem 107.

108. A solid uniform disk of mass 21.0 kg and radius 85.0 cm is at rest flat on a frictionless surface. Figure 10–75 shows a view from above. A string is wrapped around the rim of the disk and a constant force of 30 N is applied to the string. The string does not slip on the rim. (a) In what direction does the CM move? When the disk has moved a distance of 9.0 m, (b) how fast is it moving? (c) How fast is it spinning (in radians per second)? (d) How much string has unwrapped from around the rim?

FIGURE 10–75 Problem 108, looking down on the disk.

109. (a) A yo-yo is made of two solid cylindrical disks, each of mass 0.050 kg and diameter 0.075 m, joined by a (concentric) thin solid cylindrical hub of mass 0.0050 kg and diameter 0.010 m. Use conservation of energy to calculate the linear speed of the yo-yo when it reaches the end of its 1.0-m-long string, if it is released from rest. (b) What fraction of its kinetic energy is rotational?

The child is seeing the world from a rotating frame of reference. Newton's laws of motion do not hold in such a non-inertial reference frame.
To people on the ground, she and her merry-go-round have angular momentum. The rate of change of angular momentum is proportional to the net applied torque—which, if zero, means the angular momentum is conserved. In this chapter we develop further many of the themes already discussed in Chapter 10.

CHAPTER 11

General Rotation

I n Chapter 10 we dealt with the kinematics and dynamics of the rotation of a rigid body about an axis whose direction is fixed in an inertial reference frame. We analyzed the motion in terms of the rotational equivalent of Newton's laws (where torque plays the role that force does for translational motion) as well as in terms of angular momentum and rotational kinetic energy.

To keep the axis of a rotating body fixed, the body must usually be constrained by external supports (such as bearings at the end of an axle). The motion of bodies that are not constrained to move about a fixed axis, which we introduced in Section 10–11, is more difficult to describe and analyze. Indeed, the complete analysis of the general rotational motion of a body (or system of bodies) in space is very complicated, and we will only look at some aspects of general rotational motion in this chapter. We will prove some general theorems and apply them to some interesting types of motion. In particular, we will deal with the vector nature of torque and angular momentum.

Parts of this chapter may seem like a repeat of material in Chapter 10. But as a careful reading will reveal, this chapter contains a more general and useful treatment of rotational motion, making use of the vector nature of rotational quantities and deriving some of the assertions we made in Chapter 10.

11–1 Vector Cross Product

To deal with the vector nature of angular momentum and torque in general, we will need the concept of the *vector cross product* (often called simply the *vector product* or *cross product*).

In general, the **vector** or **cross product** of two vectors **A** and **B** is defined as another vector $\mathbf{C} = \mathbf{A} \times \mathbf{B}$ *whose magnitude* is

Vector cross product

$$C = |\mathbf{A} \times \mathbf{B}| = AB \sin \theta, \qquad \textbf{(11–1a)}$$

*where θ is the angle ($< 180°$) between **A** and **B**, and whose direction is perpendicular to both **A** and **B** in the sense of the right-hand rule*, Fig. 11–1. The angle θ is measured between **A** and **B** when their tails are together at the same point. According to the right-hand rule, as shown in the figure, you orient your right hand so your fingers point along **A**, and when you bend your fingers they point along **B**. When your hand is correctly oriented in this way, your thumb will point along the direction of $\mathbf{C} = \mathbf{A} \times \mathbf{B}$.

The cross product of two vectors, $\mathbf{A} = A_x\mathbf{i} + A_y\mathbf{j} + A_z\mathbf{k}$, and $\mathbf{B} = B_x\mathbf{i} + B_y\mathbf{j} + B_z\mathbf{k}$, can be written in component form (see Problem 6) as

FIGURE 11–1 The vector $\mathbf{C} = \mathbf{A} \times \mathbf{B}$ is perpendicular to the plane containing **A** and **B**; its direction is given by the right-hand rule.

$$\mathbf{A} \times \mathbf{B} = \begin{vmatrix} \mathbf{i} & \mathbf{j} & \mathbf{k} \\ A_x & A_y & A_z \\ B_x & B_y & B_z \end{vmatrix} \qquad \textbf{(11–1b)}$$

$$= (A_y B_z - A_z B_y)\mathbf{i} + (A_z B_x - A_x B_z)\mathbf{j} + (A_x B_y - A_y B_x)\mathbf{k}. \qquad \textbf{(11–1c)}$$

Equation 11–1b is not actually a determinant, but rather a memory aid—we evaluate Eq. 11–1b using the rules of determinants (and obtain Eq. 11–1c).

Some properties of the cross product are the following:

Properties of the cross product

$$\mathbf{A} \times \mathbf{A} = 0 \qquad \textbf{(11–2a)}$$

$$\mathbf{A} \times \mathbf{B} = -\mathbf{B} \times \mathbf{A} \qquad \textbf{(11–2b)}$$

$$\mathbf{A} \times (\mathbf{B} + \mathbf{C}) = (\mathbf{A} \times \mathbf{B}) + (\mathbf{A} \times \mathbf{C}) \qquad \text{[distributive law]} \quad \textbf{(11–2c)}$$

$$\frac{d}{dt}(\mathbf{A} \times \mathbf{B}) = \frac{d\mathbf{A}}{dt} \times \mathbf{B} + \mathbf{A} \times \frac{d\mathbf{B}}{dt}. \qquad \textbf{(11–2d)}$$

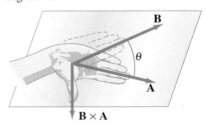

FIGURE 11–2 The vector $\mathbf{B} \times \mathbf{A}$ equals $-\mathbf{A} \times \mathbf{B}$; compare to Fig. 11–1.

Equation 11–2a follows from Eqs. 11–1 (since $\theta = 0$). So does Eq. 11–2b, since the magnitude of $\mathbf{B} \times \mathbf{A}$ is the same as that for $\mathbf{A} \times \mathbf{B}$, but by the right-hand rule the direction is opposite. Thus the order of the two vectors is crucial. If you change the order, you change the result (see Fig. 11–2). That is, the commutative law does *not* hold for the cross product, although it does for the dot product of two vectors and for the product of scalars. The proofs for Eqs. 11–2c and d are given as Problems 4 and 5. Note in Eq. 11–2d that the order of quantities in the two products on the right must not be changed (because of Eq. 11–2b).

11–2 | The Torque Vector

Torque is an example of a quantity that can be expressed as a cross product. To see this, let us take a simple example: the thin wheel shown in Fig. 11–3 which is free to rotate about an axis through its center at point O. A force **F** acts at the edge of the wheel, at a point whose position relative to the center O is given by the position vector **r** as shown. The force **F** tends to rotate the wheel (assumed initially at rest) counterclockwise, so the angular velocity $\boldsymbol{\omega}$ will point out of the page toward the viewer (remember the right-hand rule from Section 10–4). The torque due to **F** will tend to increase $\boldsymbol{\omega}$ so $\boldsymbol{\alpha}$ also points outward along the rotation axis. The relation between angular acceleration and torque that we developed in Chapter 10 for a body rotating about a fixed axis is

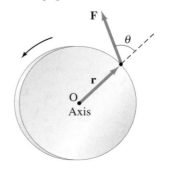

FIGURE 11–3 The torque due to the force **F** starts the wheel rotating counterclockwise so $\boldsymbol{\omega}$ and $\boldsymbol{\alpha}$ point out of the page.

$$\Sigma\tau = I\alpha,$$

(Eq. 10–14) where I is the moment of inertia. This scalar equation is the rotational equivalent of $\Sigma F = ma$, and we would like to make it a vector equation just as $\Sigma\mathbf{F} = m\mathbf{a}$ is a vector equation. To do so in the case of Fig. 11–3 we must have the direction of $\boldsymbol{\tau}$ point outward along the rotation axis, since $\boldsymbol{\alpha}$ ($= d\boldsymbol{\omega}/dt$) does; and the magnitude of the torque must be (see Eqs. 10–10 and Fig. 11–3) $\tau = rF_\perp = rF\sin\theta$. We can achieve this by defining the torque vector to be the cross product of $\mathbf{r}$ and $\mathbf{F}$:

$$\boldsymbol{\tau} = \mathbf{r} \times \mathbf{F}. \qquad (11\text{–}3) \qquad \textit{Vector torque (defined)}$$

From the definition of the cross product above (Eq. 11–1) the magnitude of $\boldsymbol{\tau}$ will be $rF\sin\theta$ and the direction will be along the axis, as required for this special case.

We will see in Sections 11–3 through 11–5 that if we take Eq. 11–3 as the *general definition of torque*, then the vector relation $\Sigma\boldsymbol{\tau} = I\boldsymbol{\alpha}$ will hold in general. Thus we state now that Eq. 11–3 is the general definition of torque. It contains both magnitude and direction information. Note that this definition involves the position vector $\mathbf{r}$ and thus the torque is being calculated about a point. We can choose that point O as we wish.

For a particle of mass m on which a force $\mathbf{F}$ is applied, we define the torque about a point O as

$$\boldsymbol{\tau} = \mathbf{r} \times \mathbf{F}$$

where $\mathbf{r}$ is the position vector of the particle relative to O (Fig. 11–4). If we have a system of particles (which could be the particles making up a rigid body) the total torque $\boldsymbol{\tau}$ on the system will be the sum of the torques on the individual particles:

$$\boldsymbol{\tau} = \Sigma\mathbf{r}_i \times \mathbf{F}_i,$$

where $\mathbf{r}_i$ is the position vector of the ith particle and $\mathbf{F}_i$ is the net force on the ith particle.

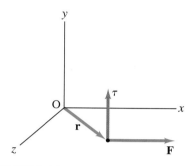

FIGURE 11–4 $\boldsymbol{\tau} = \mathbf{r} \times \mathbf{F}$, where $\mathbf{r}$ is the position vector.

EXAMPLE 11–1 **Torque Vector.** Suppose the vector $\mathbf{r}$ is in the xz plane, as in Fig. 11–4, and is given by $\mathbf{r} = (1.2\,\text{m})\mathbf{i} + (1.2\,\text{m})\mathbf{k}$. Calculate the torque vector $\boldsymbol{\tau}$ if $\mathbf{F} = (150\,\text{N})\mathbf{i}$.

SOLUTION We use the determinant form, Eq. 11–1b:

$$\boldsymbol{\tau} = \mathbf{r} \times \mathbf{F} = \begin{vmatrix} \mathbf{i} & \mathbf{j} & \mathbf{k} \\ 1.2\,\text{m} & 0 & 1.2\,\text{m} \\ 150\,\text{N} & 0 & 0 \end{vmatrix} = 0\mathbf{i} + (180\,\text{m}\cdot\text{N})\mathbf{j} + 0\mathbf{k}.$$

So τ has magnitude $180\,\text{m}\cdot\text{N}$ and points along the positive y axis.

11–3 Angular Momentum of a Particle

The most general way of writing Newton's second law for the translational motion of a particle (or system of particles) is in terms of the linear momentum $\mathbf{p} = m\mathbf{v}$ as given by Eq. 9–2 (or 9–6):

$$\Sigma\mathbf{F} = \frac{d\mathbf{p}}{dt}.$$

The rotational analog of linear momentum is *angular momentum*. Just as the rate of change of $\mathbf{p}$ is related to the net force $\Sigma\mathbf{F}$, so we might expect the rate of change of angular momentum to be related to the net torque. Indeed, we saw this was true in Chapter 10 for the special case of a rigid body rotating about a fixed axis. Now we will see it is true in general. We first treat a single particle.

Suppose a particle of mass m has momentum $\mathbf{p}$ and position vector $\mathbf{r}$ with respect to the origin O in some chosen inertial reference frame. Then the general

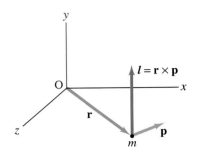

Angular momentum of a particle

FIGURE 11–5 The angular momentum of a particle of mass m is given by $\boldsymbol{l} = \mathbf{r} \times \mathbf{p} = \mathbf{r} \times m\mathbf{v}$.

definition of the angular momentum $\boldsymbol{l}$ of the particle about point O is the vector cross product of $\mathbf{r}$ and $\mathbf{p}$:

$$\boldsymbol{l} = \mathbf{r} \times \mathbf{p}. \qquad \text{[particle]} \quad \textbf{(11–4)}$$

Angular momentum is a vector.[†] Its direction is perpendicular to both $\mathbf{r}$ and $\mathbf{p}$ as given by the right-hand rule (Fig. 11–5). Its magnitude is given by

$$l = rp\sin\theta$$

or

$$l = rp_\perp = r_\perp p$$

where θ is the angle between $\mathbf{r}$ and $\mathbf{p}$ and $p_\perp (= p\sin\theta)$ and $r_\perp (= r\sin\theta)$ are the components of $\mathbf{p}$ and $\mathbf{r}$ perpendicular to $\mathbf{r}$ and $\mathbf{p}$ respectively.

Now let us find the relation between angular momentum and torque for a particle. If we take the derivative of $\boldsymbol{l}$ with respect to time we have

$$\frac{d\boldsymbol{l}}{dt} = \frac{d}{dt}(\mathbf{r} \times \mathbf{p}) = \frac{d\mathbf{r}}{dt} \times \mathbf{p} + \mathbf{r} \times \frac{d\mathbf{p}}{dt}.$$

But

$$\frac{d\mathbf{r}}{dt} \times \mathbf{p} = \mathbf{v} \times m\mathbf{v} = m(\mathbf{v} \times \mathbf{v}) = 0,$$

since $\sin\theta = 0$ for this case. Thus

$$\frac{d\boldsymbol{l}}{dt} = \mathbf{r} \times \frac{d\mathbf{p}}{dt}.$$

If we let $\Sigma\mathbf{F}$ represent the resultant force on the particle, then in an inertial reference frame, $\Sigma\mathbf{F} = d\mathbf{p}/dt$ and

$$\mathbf{r} \times \Sigma\mathbf{F} = \mathbf{r} \times \frac{d\mathbf{p}}{dt} = \frac{d\boldsymbol{l}}{dt}.$$

But $\mathbf{r} \times \Sigma\mathbf{F} = \Sigma\boldsymbol{\tau}$ is the net torque on our particle. Hence

Newton's second law, (particle, rotation)

$$\Sigma\boldsymbol{\tau} = \frac{d\boldsymbol{l}}{dt}. \qquad \text{[particle, inertial frame]} \quad \textbf{(11–5)}$$

The time rate of change of angular momentum of a particle is equal to the net torque applied to it. Equation 11–5 is the rotational equivalent of Newton's second law for a particle, written in its most general form. Equation 11–5 is valid only in an inertial frame since only then is it true that $\Sigma\mathbf{F} = d\mathbf{p}/dt$, which was used in the proof.

FIGURE 11–6 The angular momentum of a particle of mass m rotating in a circle of radius $\mathbf{r}$ with velocity $\mathbf{v}$ is $\boldsymbol{l} = \mathbf{r} \times m\mathbf{v}$ (Example 11–2).

| **EXAMPLE 11–2** | **A particle's angular momentum.** Determine the angular momentum of a particle of mass m moving with speed v in a circle of radius r in a counterclockwise direction.

SOLUTION The value of the angular momentum depends on the choice of the point O. Let us calculate $\boldsymbol{l}$ with respect to the center of the circle, Fig. 11–6. Then $\mathbf{r}$ is perpendicular to $\mathbf{p}$ so $l = |\mathbf{r} \times \mathbf{p}| = rmv$. By the right-hand rule, the direction of $\boldsymbol{l}$ is perpendicular to the plane of the circle, upward and outward toward the viewer. Since $v = \omega r$ and $I = mr^2$ for a single particle rotating about an axis a distance r away, we can write

$$l = mvr = mr^2\omega = I\omega.$$

[†] Actually a pseudovector; see footnote in Section 10–4. Also note that we are using small $\boldsymbol{l}$ for the angular momentum of a particle and capital $\mathbf{L}$ for a collection of particles or an extended body (as in the previous chapter and later in this chapter).

Angular Momentum and Torque for a System of Particles; General Motion

Relation between Angular Momentum and Torque

Consider a system of n particles which have angular momenta $l_1, l_2, \ldots l_n$. The system could be anything from a rigid body to a loose assembly of particles whose positions are not fixed relative to each other. The total angular momentum **L** of the system is defined as the vector sum of the angular momenta of all the particles in the system:

$$\mathbf{L} = \sum_{i=1}^{n} l_i. \tag{11–6}$$

The resultant torque acting on the system is the sum of the net torques acting on all the particles:

$$\boldsymbol{\tau}_{\text{net}} = \sum \boldsymbol{\tau}_i.$$

This sum includes (1) internal torques due to internal forces that particles of the system exert on other particles of the system, and (2) external torques due to forces exerted by objects outside our system. By Newton's third law, the force each particle exerts on another is equal and opposite (and acts along the same line) as the force that the second particle exerts on the first. Hence the sum of all internal torques adds to zero, and

$$\boldsymbol{\tau}_{\text{net}} = \sum_{i} \boldsymbol{\tau}_i = \sum \boldsymbol{\tau}_{\text{ext}}.$$

Now we take the time derivative of Eq. 11–6 and use Eq. 11–5 for each particle to obtain

$$\frac{d\mathbf{L}}{dt} = \sum_i \frac{d l_i}{dt} = \sum \boldsymbol{\tau}_{\text{ext}}$$

or

$$\frac{d\mathbf{L}}{dt} = \sum \boldsymbol{\tau}. \qquad \text{[inertial reference frame]} \tag{11–7a}$$

NEWTON'S SECOND LAW (rotation, system of particles)

(We have dropped the subscript on τ for convenience.) This fundamental result states that the time rate of change of the total angular momentum of a system of particles (or a rigid body) equals the resultant external torque on the system. It is the rotational equivalent of Eq. 9–6, $d\mathbf{P}/dt = \Sigma \mathbf{F}_{\text{ext}}$ for translational motion. Note that **L** and $\Sigma \boldsymbol{\tau}$ must be calculated about the same origin O.

Equation 11–7a is valid when **L** and $\boldsymbol{\tau}$ are calculated with reference to a point fixed in an inertial reference frame. (In the derivation, we used Eq. 11–5 which is valid only in this case.) It is also valid when $\boldsymbol{\tau}$ and **L** are calculated about a point which is moving uniformly in an inertial reference frame since such a point can be considered the origin of a second inertial reference frame. It is *not* valid in general when $\boldsymbol{\tau}$ and **L** are calculated about a point that is *accelerating*, except for one special (and very important) case—when that point is the center of mass of the system:

Accelerating reference frame

$$\frac{d\mathbf{L}_{\text{CM}}}{dt} = \sum \boldsymbol{\tau}_{\text{CM}}. \qquad \text{[even if accelerating]} \tag{11–7b}$$

NEWTON'S SECOND LAW (for CM, even if accelerating)

Equation 11–7b is valid no matter how the CM moves, and $\Sigma \boldsymbol{\tau}_{\text{CM}}$ is the net external torque calculated about the center of mass.

It is because of the validity of Eq. 11–7b that we are justified in describing the general motion of a system of particles, as we did in Chapter 10, as the *translational* motion of the CM plus rotation about the CM. [Eq. 10–25 or 10–26, which tell us the total $K = K_{\text{CM}} + K_{\text{rot}}$, applies to the energy but not to the other aspects of the motion. Equations 11–7b plus 9–6 $(d\mathbf{P}_{\text{CM}}/dt = \Sigma \mathbf{F}_{\text{ext}})$ provide the more general statement of this principle. See also Section 9–9.]

FIGURE 11–7 The position of m_i in the inertial frame is $\mathbf{r}_i$; with regard to the CM (which could be accelerating) it is $\mathbf{r}_i^*$, where $\mathbf{r}_i = \mathbf{r}_i^* + \mathbf{r}_{CM}$ and $\mathbf{r}_{CM}$ is the position of the CM in the inertial frame.

* Derivation of $d\mathbf{L}_{CM}/dt = \Sigma\boldsymbol{\tau}_{CM}$

The proof of Eq. 11–7b is as follows. Let $\mathbf{r}_i$ be the position vector of the ith particle in an inertial reference frame, and $\mathbf{r}_{CM}$ be the position vector of the center of mass of the system in this reference frame. The position of the ith particle with respect to the CM is $\mathbf{r}_i^*$ where (see Fig. 11–7)

$$\mathbf{r}_i = \mathbf{r}_{CM} + \mathbf{r}_i^*.$$

When we take the derivative of this equation, we can write

$$\mathbf{p}_i = m_i \frac{d\mathbf{r}_i}{dt} = m_i \frac{d}{dt}(\mathbf{r}_i^* + \mathbf{r}_{CM}) = \mathbf{p}_i^* + m_i \mathbf{v}_{CM}.$$

The angular momentum with respect to the CM is

$$\mathbf{L}_{CM} = \sum_i (\mathbf{r}_i^* \times \mathbf{p}_i^*) = \sum_i \mathbf{r}_i^* \times (\mathbf{p}_i - m_i \mathbf{v}_{CM}).$$

Then, taking the time derivative, we have

$$\frac{d\mathbf{L}_{CM}}{dt} = \sum_i \left(\frac{d\mathbf{r}_i^*}{dt} \times \mathbf{p}_i^* \right) + \sum_i \left(\mathbf{r}_i^* \times \frac{d\mathbf{p}_i^*}{dt} \right).$$

The first term is $\mathbf{v}_i^* \times m\mathbf{v}_i^* = 0$, since $\mathbf{v}_i^*$ is parallel to itself. Thus

$$\frac{d\mathbf{L}_{CM}}{dt} = \sum_i \mathbf{r}_i^* \times \frac{d}{dt}(\mathbf{p}_i - m_i \mathbf{v}_{CM})$$

$$= \sum_i \mathbf{r}_i^* \times \frac{d\mathbf{p}_i}{dt} - \left(\sum_i m_i \mathbf{r}_i^* \right) \times \frac{d\mathbf{v}_{CM}}{dt}.$$

The second term on the right is zero since, by Eq. 9–12, $\Sigma m_i \mathbf{r}_i^* = M\mathbf{r}_{CM}^*$, and $\mathbf{r}_{CM}^* = 0$ by definition (the position of the CM is at the origin of the CM reference frame). Furthermore, by Newton's second law we have

$$\frac{d\mathbf{p}_i}{dt} = \mathbf{F}_i,$$

where $\mathbf{F}_i$ is the net force on m_i. (Note that $d\mathbf{p}_i^*/dt \neq \mathbf{F}_i$ because the CM may be accelerating and Newton's second law does not hold in a noninertial reference frame.) Consequently

$$\frac{d\mathbf{L}_{CM}}{dt} = \sum_i \mathbf{r}_i^* \times \mathbf{F}_i = \sum_i (\boldsymbol{\tau}_i)_{CM} = \Sigma\boldsymbol{\tau}_{CM},$$

where $\Sigma\boldsymbol{\tau}_{CM}$ is the resultant external torque on the entire system calculated about the CM. (By Newton's third law, the sum over all the $\boldsymbol{\tau}_i$ eliminates the net torque due to internal forces, as we saw earlier.) This last equation is Eq. 11–7b, and this concludes its proof.

Summary

To summarize, the relation

$$\sum \boldsymbol{\tau} = \frac{d\mathbf{L}}{dt}$$

is *not* valid in general. It is true only when $\boldsymbol{\tau}$ and $\mathbf{L}$ are calculated with respect to either (1) the origin of an inertial reference frame or (2) the center of mass of a system of particles (or of a rigid body).

11–5 | Angular Momentum and Torque for a Rigid Body

Let us now consider the rotation of a rigid body about an axis that has a fixed direction in space. We will use the general principles developed earlier in this chapter and will arrive at the conclusions already used in Chapter 10.

Let us calculate the component of angular momentum along the rotation axis of the rotating body. We will call this component L_ω since the angular velocity $\boldsymbol{\omega}$ points along the rotation axis. For each particle of the body,

$$\boldsymbol{l}_i = \mathbf{r}_i \times \mathbf{p}_i.$$

Let θ be the angle between $\boldsymbol{l}_i$ and the rotation axis. (See Fig. 11–8; θ is *not* the angle between $\mathbf{r}_i$ and $\mathbf{p}_i$, which is 90°). Then the component of $\boldsymbol{l}_i$ along the rotation axis is

$$l_{i\omega} = r_i p_i \cos\theta = m_i v_i r_i \cos\theta,$$

where m_i is the mass and v_i the velocity of the ith particle. Now $v_i = R_i \omega$ where ω is the angular velocity of the body and R_i is the perpendicular distance of m_i from the axis of rotation. Furthermore, $R_i = r_i \cos\theta$, as can be seen in Fig. 11–8, so

$$l_{i\omega} = m_i v_i (r_i \cos\theta) = m_i R_i^2 \omega.$$

We sum over all the particles to obtain

$$L_\omega = \sum_i l_{i\omega} = \left(\sum_i m_i R_i^2 \right) \omega.$$

But $\sum m_i R_i^2$ is the moment of inertia I of the body about the axis of rotation. Therefore the component of the total angular momentum along the rotation axis is given by

$$L_\omega = I\omega. \tag{11–8}$$

Note that we would obtain Eq. 11–8 no matter where we choose the point O (for measuring $\mathbf{r}_i$) as long as it is on the axis of rotation. Equation 11–8 is just Eq. 10–19 of Chapter 10, which we have now proved from the general definition of angular momentum.

If the body rotates about a symmetry axis through the CM, then L_ω is the only component of $\mathbf{L}$, as we now show. For each point on one side of the axis there will be a corresponding point on the opposite side. We can see from Fig. 11–8 that each $\boldsymbol{l}_i$ has a component parallel to the axis $(l_{i\omega})$ and a component perpendicular to the axis. The components parallel to the axis add together for each pair of opposite points, but the components perpendicular to the axis for opposite points will have the same magnitude but opposite direction and so will cancel. Hence, for a body rotating about a symmetry axis, the angular momentum vector is parallel to the axis and we can write

$$\mathbf{L} = I\boldsymbol{\omega}, \qquad \text{[rotation axis = symmetry axis, through CM]} \tag{11–9}$$

where $\mathbf{L}$ is measured relative to the CM.

The general relation between angular momentum and torque is Eq. 11–7:

$$\sum \boldsymbol{\tau} = \frac{d\mathbf{L}}{dt}$$

where $\sum \boldsymbol{\tau}$ and $\mathbf{L}$ are calculated either about (1) the origin of an inertial reference frame, or (2) the CM of the system. This is a vector relation, and must therefore be valid for each component. Hence, for a rigid body, the component along the rotation axis is

$$\sum \tau_{\text{axis}} = \frac{dL_\omega}{dt} = \frac{d}{dt}(I\omega) = I\frac{d\omega}{dt} = I\alpha,$$

which is valid for a rigid body rotating about an axis fixed relative to the body and also either (1) fixed in an inertial system or (2) passing through the CM of the body. This is equivalent to Eqs. 10–14 and 10–15, which we now see are special cases of Eq. 11–7, $\sum \boldsymbol{\tau} = d\mathbf{L}/dt$.

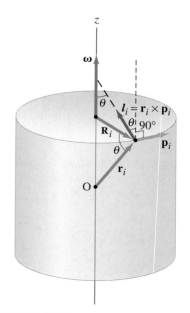

FIGURE 11–8
Calculating $L_z = \Sigma l_{iz}$.

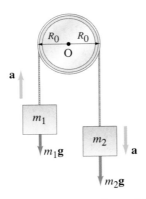

FIGURE 11–9 Atwood's machine, Example 11–3. We also discussed this in Example 4–13.

EXAMPLE 11–3 Atwood's Machine. An *Atwood Machine* consists of two masses, m_1 and m_2, which are connected by a massless inelastic cord that passes over a pulley, Fig. 11–9. If the pulley has radius R_0 and moment of inertia I about its axle, determine the acceleration of the masses m_1 and m_2, and compare to the situation where the moment of inertia of the pulley is ignored.

SOLUTION The angular momentum is calculated about an axis along the axle through the center O of the pulley. The pulley has angular momentum $I\omega$, where $\omega = v/R_0$ and v is the velocity of m_1 and m_2 at any instant. The angular momentum of m_1 is $R_0 m_1 v$ and that of m_2 is $R_0 m_2 v$. The total angular momentum is

$$L = (m_1 + m_2)vR_0 + I\frac{v}{R_0}.$$

The external torque on the system is

$$\tau = m_2 g R_0 - m_1 g R_0$$

(the force on the pulley exerted by the support on its axle gives rise to no torque because the lever arm is zero). We apply Eq. 11–7a:

$$\tau = \frac{dL}{dt}$$

$$(m_2 - m_1)gR_0 = (m_1 + m_2)R_0\frac{dv}{dt} + \frac{I}{R_0}\frac{dv}{dt}.$$

Solving for $a = dv/dt$, we get

$$a = \frac{dv}{dt} = \frac{(m_2 - m_1)g}{(m_1 + m_2) + I/R_0^2}.$$

If we were to ignore I, $a = (m_2 - m_1)g/(m_2 + m_1)$ and we see that the effect of the moment of inertia of the pulley is to slow down the system. This is just what we would expect.

FIGURE 11–10 When you try to tilt a rotating bicycle wheel vertically upward, it swerves to the side.

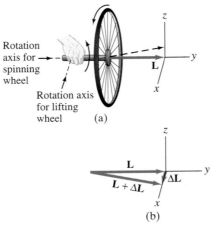

CONCEPTUAL EXAMPLE 11–4 Bicycle wheel. Suppose you are holding a bicycle wheel by a handle connected to its axle as in Fig. 11–10a. The wheel is spinning rapidly so its angular momentum **L** points horizontally as shown. Now you suddenly try to tilt the axle upward as shown by the dashed line in Fig. 11–10a (so the CM moves vertically). You expect the wheel to go up (and it would if it weren't rotating), but it unexpectedly swerves to the right! Explain.

SOLUTION To explain this seemingly odd behavior—you may need to do it to believe it—we only need to use the relation $\tau_{net} = d\mathbf{L}/dt$. In the short time Δt, you exert a net torque (about an axis through your wrist) that points along the x axis perpendicular to **L**. Thus the change in **L** is

$$\Delta\mathbf{L} \approx \tau_{net}\,\Delta t;$$

so $\Delta\mathbf{L}$ must also point (approximately) along the x axis, since τ_{net} does (Fig. 11–10b). Thus the new angular momentum, $\mathbf{L} + \Delta\mathbf{L}$, points to the right, looking along the axis of the wheel, as shown in Fig. 11–10b. Since the angular momentum is directed along the axle of the wheel, we see that the axle, which now is along $\mathbf{L} + \Delta\mathbf{L}$, must move sideways to the right, which is what we observe.

$\mathbf{L} = I\boldsymbol{\omega}$ is not always valid

Although Eq. 11–9, $\mathbf{L} = I\boldsymbol{\omega}$, is often very useful, it is not valid in general if the rotation axis is not along a symmetry axis through the center of mass. Nonetheless, it can be shown that every rigid body, no matter what its shape, has three "principal axes" about which Eq. 11–9 is valid (we won't go into the details here).

As an example of a case where Eq. 11–9 is not valid, consider the nonsymmetrical body shown in Fig. 11–11. It consists of two equal masses, m_1 and m_2, attached to the ends of a rigid (massless) rod which makes an angle ϕ with the axis of rotation. We calculate the angular momentum about the CM at point O. At the moment shown, m_1 is coming toward the viewer, and m_2 is moving away, so $\mathbf{L}_1 = \mathbf{r}_1 \times \mathbf{p}_1$ and $\mathbf{L}_2 = \mathbf{r}_2 \times \mathbf{p}_2$ are as shown. The total angular momentum is $\mathbf{L} = \mathbf{L}_1 + \mathbf{L}_2$, which is clearly *not* along $\boldsymbol{\omega}$.

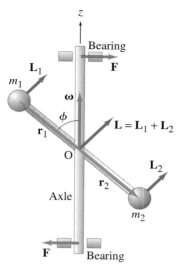

FIGURE 11–11 In this system $\mathbf{L}$ and $\boldsymbol{\omega}$ are not parallel. This is an example of rotational imbalance.

*11–6 Rotational Imbalance

Let us go one step further with the system shown in Fig. 11–11, since it is a fine illustration of $\Sigma\boldsymbol{\tau} = d\mathbf{L}/dt$. If the system rotates with constant angular velocity, ω, the magnitude of $\mathbf{L}$ will not change, but its direction will. As the rod and two masses rotate about the z axis, $\mathbf{L}$ also rotates about the axis. At the moment shown in Fig. 11–11, $\mathbf{L}$ is in the plane of the paper. A time dt later, when the rod has rotated through an angle $d\theta = \omega\,dt$, $\mathbf{L}$ will also have rotated through an angle $d\theta$ (it remains perpendicular to the rod). $\mathbf{L}$ will then have a component pointing into the page. Thus $d\mathbf{L}$ points into the page and so must $d\mathbf{L}/dt$. Since

$$\sum \boldsymbol{\tau} = \frac{d\mathbf{L}}{dt}$$

we see that a net torque, directed into the page at the moment shown, must be applied to the axle on which the rod is mounted. The torque is supplied by bearings (or other constraint) at the end of the axle. The forces $\mathbf{F}$ exerted by the bearings on the axle are shown in Fig. 11–11. The direction of each force $\mathbf{F}$ rotates as the system does, always being in the plane of $\mathbf{L}$ and $\boldsymbol{\omega}$ for this system. If the torque due to these forces were not present, the system would not rotate about the fixed axis as desired.

The axle itself, by Newton's third law, must exert forces of $-\mathbf{F}$ on the bearings. Hence the axle tends to move in the direction of $-\mathbf{F}$ and thus tends to wobble as it rotates. This has many practical applications, such as the vibrations felt in a car whose wheels are not balanced. Consider an automobile wheel that is symmetrical except for an extra mass m_1 on one rim and an equal mass m_2 opposite it on the other rim, as shown in Fig. 11–12. Because of the nonsymmetry of m_1 and m_2, the wheel bearings would have to exert a force perpendicular to the axle at all times simply to keep the wheel rotating, just as in Fig. 11–11. The bearings would wear excessively and the wobble of the wheel would be felt by occupants of the car. When the wheels are balanced, they rotate smoothly without wobble. This is why "dynamic balancing" of automobile wheels and tires is important. The wheel of Fig. 11–12 would balance *statically* just fine. If equal masses m_3 and m_4 are added symmetrically, below m_1 and above m_2, the wheel will be balanced dynamically as well $\left(\mathbf{L}$ will be parallel to $\boldsymbol{\omega}$, and $\boldsymbol{\tau}_{\text{ext}} = 0\right)$.

➡ **PHYSICS APPLIED**
Automobile wheel balancing

FIGURE 11–12
Unbalanced automobile wheel.

FIGURE 11–13 Angular momentum vector looking down along the rotation axis of the system of Fig. 11–11 as it rotates during a time dt.

EXAMPLE 11–5 **Torque on imbalanced system.** Determine the magnitude of the net torque τ_{net} needed to keep the system turning in Fig. 11–11.

SOLUTION Figure 11–13 is a view of the angular momentum vector, looking down the rotation axis (z axis) of the object depicted in Fig. 11–11, as it rotates. $L\cos\phi$ is the component of $\mathbf{L}$ perpendicular to the axle. In a time dt, $\mathbf{L}$ changes by an amount

$$dL = (L\cos\phi)d\theta = L\cos\phi\,\omega\,dt.$$

Hence

$$\tau_{\text{net}} = \frac{dL}{dt} = \omega L\cos\phi.$$

Now $L = L_1 + L_2 = r_1 m_1 v_1 + r_2 m_2 v_2 = r_1 m_1(\omega r_1 \sin\phi) + r_2 m_2(\omega r_2 \sin\phi) = (m_1 r_1^2 + m_2 r_2^2)\omega\sin\phi$. Since $I = (m_1 r_1^2 + m_2 r_2^2)\sin^2\phi$ is the moment of inertia about the axis of rotation, then $L = I\omega/\sin\phi$. So

$$\tau_{\text{net}} = (m_1 r_1^2 + m_2 r_2^2)\omega^2 \sin\phi\cos\phi = I\omega^2/\tan\phi.$$

The situation of Fig. 11–11 illustrates the usefulness of the vector nature of torque and angular momentum. If we had considered only the components of angular momentum and torque along the rotation axis, we could not have calculated the torque due to the bearings (since the forces **F** act at the axle and hence produce no torque along that axis). By using the concept of vector angular momentum we have a far more powerful technique for understanding and for attacking problems.

11–7 Conservation of Angular Momentum

In Chapter 9 we saw that the most general form of Newton's second law for the translational motion of a particle or system of particles is

$$\sum \mathbf{F}_{ext} = \frac{d\mathbf{P}}{dt},$$

where **P** is the (linear) momentum, defined as $m\mathbf{v}$ for a particle, or $M\mathbf{v}_{CM}$ for a system of particles of total mass M whose CM moves with velocity $\mathbf{v}_{CM}$, and $\Sigma\mathbf{F}_{ext}$ is the net external force acting on the particle or system. This relation is valid only in an inertial reference frame.

In this chapter, we have found a similar relation to describe the general rotation of a system of particles (including rigid bodies):

$$\sum \boldsymbol{\tau} = \frac{d\mathbf{L}}{dt},$$

where $\Sigma\boldsymbol{\tau}$ is the net external torque acting on the system, and **L** is the total angular momentum. This relation is valid when $\Sigma\boldsymbol{\tau}$ and **L** are calculated about a point fixed in an inertial reference frame, or about the CM of the system. For a rigid body rotating with angular velocity ω about an axis fixed in direction, the component of angular momentum about this axis is (Eq. 11–8):

$$L_\omega = I\omega,$$

where I is the moment of inertia of the body about this axis.

For translational motion, if the net force on the system is zero, $d\mathbf{P}/dt = 0$, so the total linear momentum of the system remains constant. This is the law of conservation of momentum. For rotational motion, if the net torque on the system is zero, then

$$\frac{d\mathbf{L}}{dt} = 0 \quad \text{and} \quad \mathbf{L} = \text{constant.} \qquad \left[\Sigma\boldsymbol{\tau} = 0\right] \quad \textbf{(11–10)}$$

In words:

CONSERVATION OF
ANGULAR MOMENTUM

The total angular momentum of a system remains constant if the net external torque acting on the system is zero.

This is the **law of conservation of angular momentum**, and ranks with the laws of conservation of energy and linear momentum (and others to be discussed later) as one of the great laws of physics. In Chapter 10 we already saw some examples of this important law applied to the special case of a rigid body rotating about a fixed axis. Here we have it in general form. We use it now in interesting Examples.

EXAMPLE 11–6 **Derivation of Kepler's second law.** *Kepler's second law* states that each planet moves so that a line from the Sun to the planet sweeps out equal areas in equal periods of time (see Section 6–5). Use the law of conservation of angular momentum to show this.

SOLUTION The planet moves in an ellipse as shown in Fig. 11–14. In a time dt, the planet moves a distance $v\,dt$ and sweeps out an area dA equal to the area of a triangle of base r and height $v\,dt\sin\theta$ (shown exaggerated in Fig. 11–14). Hence

$$dA = \tfrac{1}{2}(r)(v\,dt\sin\theta)$$

and

$$\frac{dA}{dt} = \tfrac{1}{2}rv\sin\theta.$$

The magnitude of the angular momentum $\mathbf{L}$ about the Sun is

$$L = |\mathbf{r}\times m\mathbf{v}| = mrv\sin\theta,$$

so

$$\frac{dA}{dt} = \frac{1}{2m}L.$$

But L = constant, since the gravitational force $\mathbf{F}$ is directed toward the Sun so the torque it produces is zero (we ignore the pull of the other planets). Hence dA/dt = constant, which is what we set out to prove.

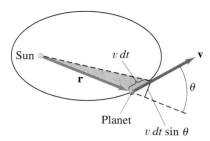

FIGURE 11–14 Kepler's second law of planetary motion (Example 11–6).

EXAMPLE 11–7 **Bullet strikes cylinder edge.** A bullet of mass m moving with velocity v strikes and becomes embedded at the edge of a cylinder of mass M and radius R_0, as shown in Fig. 11–15. The cylinder, initially at rest, begins to rotate about its symmetry axis, which remains fixed in position. Assuming no frictional torque, what is the angular velocity of the cylinder after this collision? Is kinetic energy conserved?

SOLUTION We take as our system the bullet and cylinder, on which there is no net external torque. Thus we can use conservation of angular momentum, and we calculate all angular momenta about the center O of the cylinder. Initially, because the cylinder is at rest, the total angular momentum is solely that of the bullet:

$$L = |\mathbf{r}\times\mathbf{p}| = R_0 mv,$$

since R_0 is the perpendicular distance of $\mathbf{p}$ from O. After the collision, the cylinder $(I_{\text{cyl}} = \tfrac{1}{2}MR_0^2)$ rotates with the bullet $(I = mR_0^2)$ embedded in it at angular velocity ω:

$$L = (I_{\text{cyl}} + mR_0^2)\omega = (\tfrac{1}{2}M + m)R_0^2\omega.$$

Hence, since angular momentum is conserved,

$$\omega = \frac{mv}{(\tfrac{1}{2}M + m)R_0}.$$

Angular momentum is conserved in this collision, but kinetic energy is not:

$$K_f - K_i = \tfrac{1}{2}I_{\text{cyl}}\omega^2 + \tfrac{1}{2}(mR_0^2)\omega^2 - \tfrac{1}{2}mv^2$$

$$= -\frac{mM}{2M + 4m}\,v^2,$$

which is less than zero. Hence $K_f < K_i$. This energy is transformed to thermal energy as a result of the inelastic collision.

FIGURE 11–15 Bullet strikes and becomes embedded in cylinder at its edge (Example 11–7).

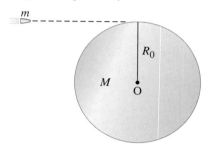

Another example of the conservation of angular momentum is the **gyroscope**, used by mariners and as the basis for guidance systems of aircraft. The rapidly spinning wheel is mounted on a complicated set of bearings, so that even when the mount moves, no net torque acts to change the direction of the angular momentum. Thus the axis of the rotating wheel remains pointed in the same direction in space.

Gyroscope

The motion of a rapidly spinning top, or a gyroscope, is an interesting example of rotational motion and of the use of the vector equation

$$\sum \boldsymbol{\tau} = \frac{d\mathbf{L}}{dt}.$$

Consider a symmetrical top of mass M spinning rapidly about its symmetry axis, as in Fig. 11–16. The top is balanced on its tip at point O in an inertial reference frame. If the axis of the top makes an angle ϕ to the vertical (z axis), when the top is carefully released its axis will move, sweeping out a cone about the vertical as shown by the dashed lines in Fig. 11–16. This type of motion, in which a torque produces a change in the direction of the rotation axis, is called **precession**. The rate at which the rotation axis moves about the vertical (z) axis is called the angular velocity of precession, Ω (capital Greek omega). Let us now try to understand the reasons for this motion, and calculate Ω.

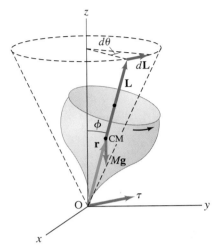

FIGURE 11–16 Spinning top.

If the top were not spinning, it would immediately fall to the ground when released due to the pull of gravity. The apparent mystery of a top is that when it is spinning, it does not immediately fall to the ground but instead precesses—it moves sideways. But this is not really so mysterious if we examine it from the point of view of angular momentum and torque, which we calculate about the point O. When the top is spinning with angular velocity ω about its symmetry axis, it has an angular momentum $\mathbf{L}$ directed along its axis, as shown in Fig. 11–16. (There is also angular momentum due to the precessional motion, so that the total $\mathbf{L}$ is not exactly along the axis of the top; but if $\Omega \ll \omega$, which is usually the case, we can ignore this.) Now to change the angular momentum, a torque is required. If no torque were applied to the top, $\mathbf{L}$ would remain constant in magnitude and direction; the top would neither fall nor precess. But the slightest tip to the side results in a net torque about O, equal to $\boldsymbol{\tau}_{\text{net}} = \mathbf{r} \times M\mathbf{g}$, where $\mathbf{r}$ is the position vector of the top's center of mass with respect to O and M is the mass of the top. The direction of $\boldsymbol{\tau}_{\text{net}}$ is perpendicular to both $\mathbf{r}$ and $M\mathbf{g}$ and by the right-hand rule is, as shown in Fig. 11–16, in the horizontal (xy) plane. The change in $\mathbf{L}$ in a time dt is

$$d\mathbf{L} = \boldsymbol{\tau}_{\text{net}}\, dt,$$

which is perpendicular to $\mathbf{L}$ and horizontal (parallel to $\boldsymbol{\tau}_{\text{net}}$), as shown in Fig. 11–16. Since $d\mathbf{L}$ is perpendicular to $\mathbf{L}$, the magnitude of $\mathbf{L}$ does not change. Only the direction of $\mathbf{L}$ changes. Since $\mathbf{L}$ points along the axis of the top, we see that this axis moves to the right in the figure. That is, the upper end of the top's axis moves in a horizontal direction perpendicular to $\mathbf{L}$. This explains why the top precesses rather than falls. The vector $\mathbf{L}$ and the top's axis move together in a horizontal circle. As they do so, $\boldsymbol{\tau}_{\text{net}}$ and $d\mathbf{L}$ rotate as well so as to be horizontal and perpendicular to $\mathbf{L}$.

To determine Ω, we see from Fig. 11–16 that the angle $d\theta$ (which is in a horizontal plane) is related to dL by

$$dL = L \sin\phi\, d\theta,$$

since $\mathbf{L}$ makes an angle ϕ to the z axis. The angular velocity of precession is $\Omega = d\theta/dt$, which becomes (since $d\theta = dL/L \sin\phi$)

$$\Omega = \frac{1}{L \sin\phi} \frac{dL}{dt} = \frac{\tau}{L \sin\phi}. \qquad \text{[spinning top]} \quad \textbf{(11–11a)}$$

But $\tau_{\text{net}} = |\mathbf{r} \times M\mathbf{g}| = rMg \sin\phi$ [because $\sin(\pi - \phi) = \sin\phi$] so we can also write

$$\Omega = \frac{Mgr}{L}. \qquad \text{[spinning top]} \quad \textbf{(11–11b)}$$

Thus the rate of precession does not depend on the angle ϕ; but it is inversely proportional to the top's angular momentum. The faster the top spins, the greater L is and the slower the top precesses.

Rotating Frames of Reference; Inertial Forces

Inertial and Noninertial Reference Frames

Up to now, we have examined the motion of bodies, including circular and rotational motion, from the outside, as observers fixed on the Earth. Sometimes it is convenient to place ourselves (in theory, if not physically) into a reference frame that is rotating. Let us examine the motion of objects from the point of view, or frame of reference, of persons seated on a rotating platform such as a merry-go-round. It looks to them as if the rest of the world is going around *them*—see chapter opening photo. But let us focus attention on what they observe when they place a tennis ball on the floor of the rotating platform, which we assume is frictionless. If they put the ball down gently, without giving it any push, they will observe that it accelerates from rest and moves outward as shown in Fig. 11–17a. According to Newton's first law, an object initially at rest should stay at rest if no force acts on it. But, according to the observers on the rotating platform, the ball starts moving even though there is no force applied to it. To observers in an inertial reference frame, this is all very clear: the ball has an initial velocity when it is released (because the platform is moving), and it simply continues moving in a straight-line path as shown in Fig. 11–17b, in accordance with Newton's first law.

But what shall we do about the frame of reference of the observers on the rotating platform? Clearly, Newton's first law, the law of inertia, does not hold in this rotating frame of reference. For this reason, such a frame is called a **noninertial reference frame**. An **inertial reference frame** is one in which the law of inertia—Newton's first law—does hold, and so do Newton's second and third laws. In a noninertial reference frame, such as our rotating platform, Newton's second law also does not hold. For instance in the situation described above, there is no net force on the ball; yet, with respect to the rotating platform, the ball accelerates.

Fictitious (Inertial) Forces

Because Newton's laws do not hold when observations are made with respect to a rotating frame of reference, calculation of motion can be complicated. However, we can still make use of Newton's laws in such a reference frame if we make use of a trick. If the ball is kept at rest on the merry-go-round, then to keep it rotating in a circle (as seen from an inertial reference frame) an inward force equal to mv^2/r would be needed, as we saw in Chapter 5 (Eq. 5–1). In the rotating reference frame of the ball, this inward force still acts, and if we want to somehow use Newton's laws, we need to figure out some way to balance it since the ball is at rest in this reference frame. So the trick we use is to write down the equation $\Sigma F = ma$ as if a force equal to mv^2/r (or $m\omega^2 r$) were acting radially outward on the object in addition to any other forces that may be acting. This extra force, which might be designated as "centrifugal force" since it *seems* to act outward, is called a **fictitious force** or **pseudoforce**. It is a pseudoforce ("pseudo" means "false") because there is no object that exerts this force. Furthermore, when viewed from an inertial reference frame, the effect doesn't exist at all. We have made up this pseudoforce so that we can make calculations in a noninertial frame using Newton's second law, $\Sigma F = ma$. Thus the observer in the non-inertial frame of Fig. 11–17a uses Newton's second law for the ball's outward motion by assuming that a pseudoforce equal to mv^2/r acts on it. Such pseudoforces are also called **inertial forces** since they arise only because the reference frame is not an inertial one.

The Earth itself is rotating on its axis. Thus, strictly speaking, Newton's laws are not valid on the Earth. However, the effect of the Earth's rotation is usually so small that it can be ignored, although it does influence the movement of large air masses and ocean currents. Because of the Earth's rotation, the material of the Earth is concentrated slightly more at the equator. The Earth is thus not a perfect sphere but is slightly fatter at the equator than it is at the poles.

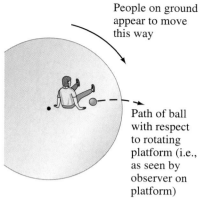

People on ground appear to move this way

Path of ball with respect to rotating platform (i.e., as seen by observer on platform)

(a) Rotating reference frame

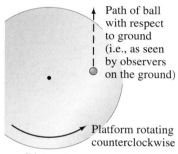

Path of ball with respect to ground (i.e., as seen by observers on the ground)

Platform rotating counterclockwise

(b) Inertial reference frame

FIGURE 11–17 Path of a ball released on a rotating merry-go-round (a) in the reference frame of the merry-go-round, and (b) in an inertial reference frame.

Fictitious force (pseudoforce)

Inertial force

(a) Inertial reference frame

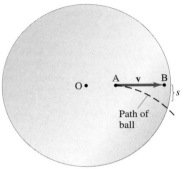

(b) Rotating reference frame

FIGURE 11–18 The origin of the Coriolis effect. Looking down on a rotating platform, (a) as seen from an inertial reference frame, and (b) as seen from the rotating platform as frame of reference.

* **11–10** | **The Coriolis Effect**

In a reference frame that rotates at a constant angular speed ω (relative to an inertial system), there exists another pseudoforce known as the *Coriolis force*. It appears to act on a body in a rotating reference frame only if the body is moving relative to that reference frame, and it acts to deflect the body sideways. It, too, is an effect of the reference frame being noninertial and hence is referred to as an *inertial force*. To see how the Coriolis force arises, consider two people, A and B, at rest on a platform rotating with angular speed ω, as shown in Fig. 11–18a. They are situated at distances r_A and r_B from the axis of rotation (at O). The woman at A throws a ball with a horizontal velocity $\mathbf{v}$ (in her reference frame) radially outward toward the man at B on the outer edge of the platform. In Fig. 11–18a, we view the situation from an inertial reference frame. The ball initially has not only the velocity $\mathbf{v}$ radially outward, but also a tangential velocity $\mathbf{v}_A$ due to the rotation of the platform. Now Eq. 10–4 tells us that $v_A = r_A\omega$, where r_A is the woman's radial distance from the axis of rotation at O. If the man at B had this same velocity v_A, the ball would reach him perfectly. But his speed is $v_B = r_B\omega$, which is greater than v_A because $r_B > r_A$. Thus, when the ball reaches the outer edge of the platform, it passes a point that the man at B has already passed because his speed in that direction is greater than the ball's. So the ball passes behind him.

Figure 11–18b shows the situation as seen from the rotating platform as frame of reference. Both A and B are at rest, and the ball is thrown with velocity $\mathbf{v}$ toward B, but the ball deflects to the right as shown and passes behind B as previously described. This is not a centrifugal-force effect, for the latter acts radially outward. Instead, this effect acts sideways, perpendicular to $\mathbf{v}$, and is called a **Coriolis acceleration**; it is said to be due to the Coriolis force, which is a fictitious, inertial force. Its explanation as seen from an inertial system was given above: it is an effect of being in a rotating system, wherein points that are farther from the rotation axis have higher linear speeds. On the other hand, when viewed from the rotating system, we can describe the motion using Newton's second law, $\Sigma\mathbf{F} = m\mathbf{a}$, if we add a "pseudoforce" term corresponding to this Coriolis effect.

Let us determine the magnitude of the Coriolis acceleration for the simple case described above. (We assume v is large and distances short, so we can ignore gravity.) We do the calculation from the inertial reference frame (Fig. 11–18a). The ball moves radially outward a distance $r_B - r_A$ at speed v in a time t given by

$$r_B - r_A = vt.$$

During this time, the ball moves to the side a distance s_A given by

$$s_A = v_A t.$$

The man at B, in this time t, moves a distance

$$s_B = v_B t.$$

The ball therefore passes behind him a distance s (Fig. 11–18a) given by

$$s = s_B - s_A = (v_B - v_A)t.$$

We saw earlier that $v_A = r_A\omega$ and $v_B = r_B\omega$, so

$$s = (r_B - r_A)\omega t.$$

We substitute $r_B - r_A = vt$ (see above) and get

$$s = \omega v t^2. \tag{11–12}$$

This same s equals the sideways displacement as seen from the noninertial rotating system (Fig. 11–18b).

We see immediately that Eq. 11–12 corresponds to motion at constant acceleration. For as we saw in Chapter 2 (Eq. 2–12b), $y = \frac{1}{2}at^2$ for a constant acceleration (with zero initial velocity in the y direction). Thus, if we write Eq. 11–12 in the form $s = \frac{1}{2}a_{cor}t^2$, we see that the Coriolis acceleration a_{cor} is

$$a_{cor} = 2\omega v. \qquad \qquad \textbf{(11–13)}$$

This relation is valid for any velocity in the plane of rotation—that is, in the plane perpendicular to the axis of rotation[†] (in Fig. 11–18, the axis through point O perpendicular to the page).

Because the Earth rotates, the Coriolis effect has some interesting manifestations on the Earth. It affects the movement of air masses and thus has an influence on weather. In the absence of the Coriolis effect, air would rush directly into a region of low pressure, as shown in Fig. 11–19a. But because of the Coriolis effect, the winds are deflected to the right in the Northern Hemisphere (Fig. 11–19b), since the Earth rotates from west to east. So there tends to be a counterclockwise wind pattern around a low-pressure area. The reverse is true in the Southern Hemisphere. Thus cyclones rotate counterclockwise in the Northern Hemisphere and clockwise in the Southern Hemisphere. The same effect explains the easterly trade winds near the equator: any winds heading south toward the equator will be deflected toward the west (that is, as if coming from the east).

The Coriolis effect also acts on a falling body. A body released from the top of a high tower will not hit the ground directly below the release point, but will be deflected slightly to the east. Equations 11–12 and 11–13 do not apply directly on Earth except at the poles, because they are valid only for velocity in the plane perpendicular to the axis of rotation. A velocity, or component of velocity, parallel to the axis of rotation will not experience a Coriolis acceleration since r (distance from axis) doesn't change. We can alter Eqs. 11–12 and 11–13 for use on Earth by considering only the component of $\mathbf{v}$ perpendicular to the axis of rotation. From Fig. 11–20 we see that this is $v \cos \lambda$ for a vertically falling body, where λ is the latitude of the place on the Earth.

| EXAMPLE 11–8 | **A ball does not fall straight.** A lead ball is dropped vertically from a 110-m-high tower in Florence, Italy (latitude = 44°). How far from the base of the tower is it deflected by the Coriolis force?

SOLUTION The Coriolis acceleration is $a_{cor} = 2\omega v \cos \lambda$, where ω is the angular velocity of rotation of the Earth. Now v is the vertical velocity of free fall, $v = gt$, so

$$a_{cor} = (2\omega g \cos \lambda)t$$

and acts sideways. To obtain the horizontal displacement, which we take to be in the x direction, perpendicular to the fall, we must integrate twice:

$$\frac{dv_x}{dt} = a_{cor} = (2\omega g \cos \lambda)t$$

so

$$\int_0^{v_x} dv_x = \int_0^t (2\omega g \cos \lambda)t \, dt$$

$$v_x = (\omega g \cos \lambda)t^2$$

since $v_{x0} = 0$. Then, since $v_x = dx/dt$, we have

$$x = \int_0^t v_x \, dt = \omega g \cos \lambda \int_0^t t^2 \, dt = \tfrac{1}{3}\omega g \cos \lambda t^3.$$

The value of t is easily obtained, since it is the time for the ball to fall a

(a)

(b)

(c)

(d)

FIGURE 11–19 (a) Winds (moving air masses) would flow directly toward a low-pressure area if the Earth did not rotate. (b) and (c) Because of the Earth's rotation, the winds are deflected to the right in the Northern Hemisphere (as in Fig. 11–18) as if a fictitious (Coriolis) force were acting. (d) The reverse occurs in the Southern Hemisphere.

[†]The Coriolis acceleration can be written in general in terms of the vector cross product as $\mathbf{a}_C = -2\boldsymbol{\omega} \times \mathbf{v}$ where $\boldsymbol{\omega}$ has direction along the rotation axis; its magnitude is $a_C = 2wv_\perp$ where $v_\perp$ is the component of velocity perpendicular to the rotation axis.

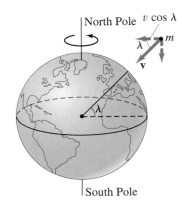

North Pole

$v\cos\lambda$

λ

m

λ

$\mathbf{v}$

South Pole

FIGURE 11–20 Object of mass m falling vertically to Earth at a latitude λ.

distance $h = 110$ m from the top of the tower—namely, $h = \frac{1}{2}gt^2$—so $t = (2h/g)^{\frac{1}{2}}$. Thus the deflection is

$$x = \frac{1}{3}\omega g \cos\lambda \left(\frac{2h}{g}\right)^{\frac{3}{2}} = \frac{1}{3}\omega \cos\lambda \sqrt{\frac{8h^3}{g}}.$$

We set $h = 110$ m and $\lambda = 44°$, and write ω for the rotation of the Earth (1 revolution in 24 h) as

$$\omega = 2\pi f = 2\pi(1\text{ rev})/(24\text{ h})(3600\text{ s/h}) = 7.27 \times 10^{-5}\text{ rad/s}.$$

Then

$$x = \frac{1}{3}(7.27 \times 10^{-5}\text{ rad/s})(\cos 44°)\sqrt{\frac{8(110\text{ m})^3}{9.80\text{ m/s}^2}} = 0.018\text{ m}$$

or about 2 cm. This distance is small and difficult to measure because of wind currents, and so on. Finally, we determine the direction of this deflection. From Fig. 11–20 we see that the ball starts at a point farther from the axis and moves closer to the Earth's axis as it falls, which is the reverse direction of the motion shown in Fig. 11–18. That is, due to the Earth's rotation, the top of the tower is moving slightly faster than the bottom is. Hence the ball deflects to the east (into the page in Fig. 11–20).

Summary

The **vector product** or **cross product** of two vectors $\mathbf{A}$ and $\mathbf{B}$ is another vector $\mathbf{C} = \mathbf{A} \times \mathbf{B}$ whose magnitude is $AB\sin\theta$ and whose direction is perpendicular to both $\mathbf{A}$ and $\mathbf{B}$ in the sense of the right-hand rule.

The **torque** $\boldsymbol{\tau}$ due to a force $\mathbf{F}$ is a vector quantity and is always calculated about some point O (the origin of a coordinate system) as follows:

$$\boldsymbol{\tau} = \mathbf{r} \times \mathbf{F},$$

where $\mathbf{r}$ is the position vector of the point at which the force $\mathbf{F}$ acts.

Angular momentum is also a vector. For a particle having momentum $\mathbf{p} = m\mathbf{v}$, the angular momentum $\boldsymbol{l}$ about some point O is

$$\boldsymbol{l} = \mathbf{r} \times \mathbf{p},$$

where $\mathbf{r}$ is the position vector of the particle relative to the point O at any instant. The net torque $\Sigma\boldsymbol{\tau}$ on a particle is related to its angular momentum by

$$\Sigma\boldsymbol{\tau} = \frac{d\boldsymbol{l}}{dt}.$$

For a system of particles, the total angular momentum $\mathbf{L} = \Sigma\boldsymbol{l}_i$. The total angular momentum of the system is related to the total net torque $\Sigma\boldsymbol{\tau}$ on the system by

$$\Sigma\boldsymbol{\tau} = \frac{d\mathbf{L}}{dt}.$$

This last relation is the vector rotational equivalent of Newton's second law. It is valid when $\mathbf{L}$ and $\Sigma\boldsymbol{\tau}$ are calculated about an origin (1) fixed in an inertial reference system or (2) situated at the CM of the system. For a rigid body rotating about a fixed axis, the component of angular momentum about the rotation axis is given by $L_\omega = I\omega$. If a body rotates about an axis of symmetry, then the vector relation $\mathbf{L} = I\boldsymbol{\omega}$ holds, but this is not true in general.

If the total net torque on a system is zero, then the total vector angular momentum $\mathbf{L}$ remains constant. This is the important **law of conservation of angular momentum**. It applies to the vector $\mathbf{L}$ and therefore also to each of its components.

Questions

1. If all the components of the vectors $\mathbf{V}_1$ and $\mathbf{V}_2$ were reversed, how would this alter $\mathbf{V}_1 \times \mathbf{V}_2$?

2. Name the three different conditions that could make $\mathbf{V}_1 \times \mathbf{V}_2 = 0$.

3. A force $\mathbf{F} = \mathbf{j}F$ is applied to a body at a position $\mathbf{r} = x\mathbf{i} + y\mathbf{j} + z\mathbf{k}$ where the origin is at the CM. Does the torque about the CM depend on x? On y? On z?

4. A particle moves with constant speed along a straight line. How does its angular momentum, calculated about any point not on its path, change in time?

5. If the net force on a system is zero, is the net torque also zero? If the net torque on a system is zero, is the net force zero?

6. Explain how a child "pumps" on a swing to make it go higher.

* 7. For the nonsymmetrical system of Fig. 11–11, is there a point about which the angular momentum ($\mathbf{L}$) could be calculated so that $\mathbf{L}$ would be in the same direction as the angular velocity $\boldsymbol{\omega}$? What if there were only one mass (say, $m_2 = 0$)? If your answer to either question is yes, how do you explain the need for forces at the bearings for both cases and the requirement $\boldsymbol{\tau} = d\mathbf{L}/dt$?

8. An astronaut floats freely in a weightless environment. Describe how the astronaut can move her limbs so as to (a) turn her body upside down and (b) turn her body about face.

9. On the basis of the law of conservation of angular momentum, discuss why a helicopter must have more than one rotor (or propeller). Discuss one or more ways the second propeller can operate in order to keep the body stable.

10. A wheel is rotating freely about a vertical axis with constant angular velocity. Small parts of the wheel come loose and fly off. How does this affect the rotational speed of the wheel? Is angular momentum conserved? Is kinetic energy conserved? Explain.

11. Consider the following vector quantities: displacement, velocity, acceleration, momentum, angular momentum, torque. (a) Which of these are independent of the choice of origin of coordinates? (Consider different points as origin which are at rest with respect to each other.) (b) Which are independent of the velocity of the coordinate system?

12. Describe the torque needed if the person in Fig. 11–10 is to tilt the axle of the rotating wheel directly upward with no swerving to the side.

13. How does a car make a right turn? Where does the torque come from that is needed to change the angular momentum?

14. Suppose you are standing on a turntable that can freely rotate. When you hold a rotating bicycle wheel over your head, with its axis vertical, you are at rest. If you now move the wheel so its axis is horizontal, what happens to you? What happens if you then point the axis of the wheel downward?

*15. The axis of the Earth precesses with a period of about 25,000 years. This is much like the precession of a top. Explain how the Earth's equatorial bulge gives rise to a torque exerted by the Sun and Moon on the Earth; see Fig. 11–21, which is drawn for the winter solstice (December 21). About what axis would you expect the Earth's rotation axis to precess as a result of the torque due to the Sun? Does the torque exist 3 months later? Explain.

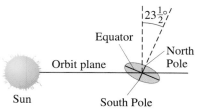

FIGURE 11–21   Question 15.

*16. Why is it that at most locations on the Earth, a plumb bob does not hang precisely in the direction of the Earth's center?

*17. In the battle of the Falkland Islands in 1914, the shots of British gunners initially fell wide of their marks because their calculations were based on naval battles fought in the Northern Hemisphere. The Falklands are in the Southern Hemisphere. Explain the origin of their problem.

Problems

Section 11–1

1. (I) Show that (a) $\mathbf{i} \times \mathbf{i} = \mathbf{j} \times \mathbf{j} = \mathbf{k} \times \mathbf{k} = 0$, (b) $\mathbf{i} \times \mathbf{j} = \mathbf{k}$, $\mathbf{i} \times \mathbf{k} = -\mathbf{j}$, and $\mathbf{j} \times \mathbf{k} = \mathbf{i}$.

2. (I) If vector $\mathbf{A}$ points along the negative x axis and vector $\mathbf{B}$ along the positive z axis, what is the direction of (a) $\mathbf{A} \times \mathbf{B}$ and (b) $\mathbf{B} \times \mathbf{A}$? (c) What is the magnitude of $\mathbf{A} \times \mathbf{B}$ and $\mathbf{B} \times \mathbf{A}$?

3. (II) Consider a particle of a rigid body rotating about a fixed axis. Show that the tangential and radial vector components of the linear acceleration are:
$$\mathbf{a}_{\text{tan}} = \boldsymbol{\alpha} \times \mathbf{r} \quad \text{and} \quad \mathbf{a}_R = \boldsymbol{\omega} \times \mathbf{v}.$$

4. (II) Prove the distributive law for the cross product, Eq. 11–2c.

5. (II) Use the limiting process to obtain Eq. 11–2d. [Hint: Let $\Delta\mathbf{A}$, $\Delta\mathbf{B}$, and Δt approach zero.]

6. (II) (a) Show that the cross product of two vectors, $\mathbf{A} = A_x\mathbf{i} + A_y\mathbf{j} + A_z\mathbf{k}$, and $\mathbf{B} = B_x\mathbf{i} + B_y\mathbf{j} + B_z\mathbf{k}$ is
$$\mathbf{A} \times \mathbf{B} = (A_yB_z - A_zB_y)\mathbf{i} + (A_zB_x - A_xB_z)\mathbf{j} \\ + (A_xB_y - A_yB_x)\mathbf{k}.$$
(b) Then show that the cross product can be written
$$\mathbf{A} \times \mathbf{B} = \begin{vmatrix} \mathbf{i} & \mathbf{j} & \mathbf{k} \\ A_x & A_y & A_z \\ B_x & B_y & B_z \end{vmatrix},$$
where we use the rules for evaluating a determinant. (Note, however, that this is not actually a determinant, but a memory aid.)

7. (II) Use the result of Problem 6 to determine (a) the vector product $\mathbf{A} \times \mathbf{B}$ and (b) the angle between $\mathbf{A}$ and $\mathbf{B}$ if $\mathbf{A} = 7.0\mathbf{i} - 3.5\mathbf{j}$ and $\mathbf{B} = -8.5\mathbf{i} + 7.0\mathbf{j} + 2.0\mathbf{k}$.

8. (III) Show that the velocity $\mathbf{v}$ of any point in a body rotating with angular velocity $\boldsymbol{\omega}$ about a fixed axis can be written
$$\mathbf{v} = \boldsymbol{\omega} \times \mathbf{r}$$
where $\mathbf{r}$ is the position vector of the point relative to an origin O located on the axis of rotation. Can O be anywhere on the rotation axis? Will $\mathbf{v} = \boldsymbol{\omega} \times \mathbf{r}$ if O is located at a point not on the axis of rotation?

9. (III) Let $\mathbf{A}, \mathbf{B}, \mathbf{C}$ be three vectors that do not all lie in the same plane. Show that $\mathbf{A} \cdot (\mathbf{B} \times \mathbf{C}) = \mathbf{B} \cdot (\mathbf{C} \times \mathbf{A}) = \mathbf{C} \cdot (\mathbf{A} \times \mathbf{B})$.

Section 11–2

10. (II) A particle is located at $\mathbf{r} = (4.0\mathbf{i} + 8.0\mathbf{j} + 6.0\mathbf{k})$ m. A force $\mathbf{F} = (16.0\mathbf{j} - 4.0\mathbf{k})$N acts on it. What is the torque, calculated about the origin?

11. (II) The origin of a coordinate system is at the center of a wheel. The wheel can rotate in the xy plane about an axis through the origin. A force $F = 188$ N, which acts at a $+33.0°$ angle to the x axis, is applied to the wheel at the point $x = 22.0$ cm, $y = 33.5$ cm. What are the magnitude and direction of the torque produced by this force about the axis?

12. (II) A thin 43.0-cm-diameter wheel is constrained to rotate about an axis through its center, which we choose to be the z axis (the wheel rotates in the xy plane). A force $\mathbf{F} = 22.8\mathbf{j} - 21.6\mathbf{k}$ (newtons) is exerted at a point on the edge of the wheel which is exactly on the x axis. (a) Determine the torque (magnitude and direction), produced by this force, calculated about the wheel's center. (b) Is the direction of this torque parallel to the direction of $\boldsymbol{\alpha}$? If not, explain how $\boldsymbol{\tau}$ can be proportional to $\boldsymbol{\alpha}$.

13. (II) An engineer estimates that under the most adverse expected weather conditions, the total force on the highway sign in Fig. 11–22 will be $\mathbf{F} = \pm 2.4\mathbf{i} - 3.0\mathbf{j}$ kN, acting at the CM. What torque does this force exert about the base O?

FIGURE 11–22
Problem 13.

Section 11–3

14. (I) What are the x, y, and z components of the angular momentum of a particle located at $\mathbf{r} = x\mathbf{i} + y\mathbf{j} + z\mathbf{k}$ which has momentum $\mathbf{p} = p_x\mathbf{i} + p_y\mathbf{j} + p_z\mathbf{k}$?

15. (I) Show that the kinetic energy K of a particle of mass m, moving in a circular path, is $K = l^2/2I$, where l is its angular momentum and I is its moment of inertia about the center of the circle.

16. (I) Calculate the angular momentum of a particle of mass m moving with constant velocity v for two cases (see Fig. 11–23): (a) about origin O, and (b) about O'.

FIGURE 11–23
Problem 16.

17. (II) Two identical particles have equal but opposite momenta, $\mathbf{p}$ and $-\mathbf{p}$, but they are not traveling along the same line. Show that the total angular momentum of this system does not depend on the choice of origin.

18. (II) Determine the angular momentum of a 60-g particle about the origin of coordinates when the particle is at $x = 7.0$ m, $y = -6.0$ m, and it has velocity $v = (2.0\mathbf{i} - 8.0\mathbf{k})$ m/s.

19. (II) A particle is at the position $(x, y, z) = (1.0, 2.0, 3.0)$ m. It is traveling with a vector velocity $(-5.0, -4.5, -3.1)$ m/s. Its mass is 7.6 kg. What is its vector angular momentum about the origin?

20. (II) Two particles rest a distance d apart on the edge of a table. One of the particles, of mass m, falls off the edge and falls vertically. Using the other particle (still at rest) as the origin, calculate (a) the torque τ on m as a function of time (before hitting the ground below) and (b) the angular momentum l of m as a function of time. (c) Show that $\tau = dl/dt$.

Sections 11–4 to 11–5

21. (II) Four identical particles of mass m are mounted at equal intervals on a thin rod of length l and mass M, with one mass at each end of the rod. If the system is rotated with angular velocity ω about an axis perpendicular to the rod through one of the end masses, determine (a) the kinetic energy and (b) the angular momentum of the system.

22. (II) An Atwood machine (Fig. 11–9) consists of two masses, $m_1 = 7.0$ kg and $m_2 = 8.8$ kg, connected by a cord that passes over a pulley free to rotate about a fixed axis. The pulley is a solid cylinder of radius $R_0 = 0.50$ m and mass 0.80 kg. (a) Determine the acceleration a of each mass. (b) What percentage of error in a would be made if the moment of inertia of the pulley were ignored? Ignore friction in the pulley bearings.

23. (II) Use $\tau = dL/dt$ to determine the acceleration of the bucket in Example 10–8.

24. (II) Two lightweight rods 20 cm in length are mounted perpendicular to an axle and at 180° to each other (Fig. 11–24). At the end of each rod is a 600-g mass. The rods are spaced 40 cm apart along the axle. The axle rotates at 30 rad/s. (a) What is the component of the total angular momentum along the axle? (b) What angle does the vector angular momentum make with the axle? [Hint: Remember that the vector angular momentum must be calculated about the same point for both masses.]

FIGURE 11–24
Problem 24.

FIGURE 11–25
Problems 25 and 26.

25. (II) Figure 11–25 shows two masses connected by a cord passing over a pulley of radius R_0 and moment of inertia I. Mass M_1 slides on a frictionless surface, and M_2 hangs freely. Determine a formula for (a) the angular momentum of the system about the pulley axis, as a function of the speed v of mass M_1 or M_2, and (b) the acceleration of the system.

26. (III) Repeat Problem 25, but this time suppose that the block M_1 in Fig. 11–25 slides on the horizontal surface with coefficient of friction μ_k. [Hint: The cord attaches to M_1 a distance R_0 above the surface. Assume M_1 does not tip.]

27. (III) A uniform thin rod, 7.0 cm long with a mass of 40.0 g, lies on a frictionless horizontal table. It is struck with a horizontal impulse, at right angles to its length, at a point 2.0 cm from one end. If the impulse is 8.5 mN·s, describe the resulting motion of the stick.

28. (III) A thin rod of length L and mass M rotates about a vertical axis through its center with angular velocity ω. The rod makes an angle ϕ with the rotation axis. Determine the magnitude and direction of $\mathbf{L}$.

29. (III) Show that the total angular momentum $\mathbf{L} = \Sigma \mathbf{r}_i \times \mathbf{p}_i$ of a system of particles about the origin of an inertial reference frame can be written as the sum of the angular momentum about the CM, $\mathbf{L}^*$ (spin angular momentum), plus the angular momentum of the CM about the origin (orbital angular momentum): $\mathbf{L} = \mathbf{L}^* + \mathbf{r}_{CM} \times M\mathbf{v}_{CM}$. [Hint: See the derivation of Eq. 10–26.]

* 30. (II) What is the magnitude of the force **F** exerted by each bearing in Fig. 11–11 (Example 11–5)? The bearings are a distance d from point O. Ignore the effects of gravity.

* 31. (II) Suppose in Fig. 11–11 that $m_2 = 0$; that is, only one mass, m_1, is actually present. If the bearings are each a distance d from O, determine the forces F_1 and F_2 at the upper and lower bearings respectively. [*Hint*: Choose an origin—different than O in Fig. 11–11—such that **L** is parallel to **ω**. Ignore effects of gravity.]

* 32. (II) For the system shown in Fig. 11–11, suppose that $m_1 = m_2 = 0.60$ kg, $r_1 = r_2 = 0.30$ m, and the distance between the bearings is 0.20 m. What will be the force that each bearing must exert on the axle if $\phi = 23.0°$ and $\omega = 11.0$ rad/s? [Ignore effects of gravity.]

* 33. (II) What would your answer be to Problem 32 if $m_1 = 0.60$ kg and $m_2 = 0$? [Ignore effects of gravity.]

* 34. (II) A uniform cylindrical wheel 70.0 cm in diameter and of mass 11.8 kg rotates about an axle displaced 1.00 cm from its true center. (*a*) If the bearings are mounted 9.50 cm apart, what is the force exerted on them when the wheel rotates at 11.2 rev/s? (*b*) To balance this wheel, where should a 1.00 kg mass be placed? [Ignore effects of gravity.]

Section 11–7

35. (II) A thin rod of mass M and length l is suspended vertically from a frictionless pivot at its upper end. A mass m of putty traveling horizontally with a speed v strikes the rod at its CM and sticks there. How high does the bottom of the rod swing?

36. (II) A uniform stick 1.0 m long with a total mass of 300 g is pivoted at its center. A 3.0-g bullet is shot through the stick midway between the pivot and one end (Fig. 11–26). The bullet approaches at 250 m/s and leaves at 160 m/s. With what angular speed is the stick spinning after the collision?

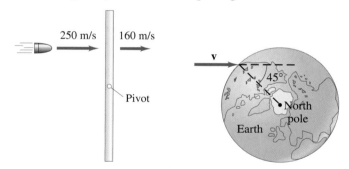

FIGURE 11–26 Problem 36.   **FIGURE 11–27** Problem 37.

37. (II) Suppose a 7.0×10^{10} kg meteor struck the Earth at the equator with a speed $v = 1.0 \times 10^4$ m/s, as shown in Fig. 11–27 and remained stuck. By what factor would this affect the rotational frequency of the Earth (1 rev/day)?

38. (II) A person of mass 55 kg stands at the center of a rotating merry-go-round platform of radius 2.5 m and moment of inertia 670 kg·m². The platform rotates without friction with an angular velocity of 2.0 rad/s. The person walks radially to the edge of the platform. (*a*) Calculate the angular velocity when the person reaches the edge. (*b*) Compare the rotational kinetic energies of the system of platform plus person before and after the person's walk.

39. (III) A 200-kg beam 2.0 m in length slides broadside down the ice with a speed of 18 m/s (Fig. 11–28). A 50-kg man at rest grabs one end as it goes past and hangs on as both he and the beam go spinning down the ice. Assume frictionless motion. (*a*) How fast does the center of mass of the system move after the collision? (*b*) With what angular velocity does the system rotate about its CM?

FIGURE 11–28
Problem 39.

FIGURE 11–29
Problem 40.

40. (III) A thin rod of mass M and length l rests on a frictionless table and is struck at a point $l/4$ from its CM by a clay ball of mass m moving at speed v (Fig. 11–29). The ball sticks to the rod. Determine the translational and rotational motion of the rod after the collision.

* Section 11–8

* 41. (II) A 220-g top spinning at 15 rev/s makes an angle of 30° to the vertical and precesses at a rate of 1 rev per 8.0 s. If its CM is 3.5 cm from its tip along its symmetry axis, what is the moment of inertia of the top?

* 42. (II) A toy gyroscope consists of a 150-g disk with a radius of 5.5 cm mounted at the center of an axle 17 cm long (Fig. 11–30). The gyroscope spins at 70 rev/s. One end of its axle rests on a stand and the other end precesses horizontally about the stand as shown. (*a*) How long does it take the gyroscope to precess once around? (*b*) If all the dimensions of the gyroscope were doubled (radius = 11 cm, axle = 34 cm), how long would it take to precess once?

FIGURE 11–30
A wheel, rotating about a horizontal axle supported at one end, precesses. Problems 42, 43, and 44.

* 43. (II) Suppose the solid wheel of Fig. 11–30 has a mass of 300 g and rotates at 250 rad/s; it has radius 6.0 cm and is mounted at the center of a horizontal thin axle 20 cm long. At what rate does the axle precess?

* 44. (II) If a mass equal to half the mass of the wheel in Problem 43 is placed at the free end of the axle, what will be the precession rate now?

* 45. (II) If a plant is allowed to grow from seed on a rotating platform, it will grow at an angle, pointing inward. Calculate what this angle will be (put yourself in the rotating frame) in terms of $g, r,$ and ω. Why does it grow inward rather than outward?

* 46. (II) In a rotating frame of reference, Newton's first and second laws remain useful if we assume that a pseudoforce equal to $m\omega^2 r$ is acting. What effect does this assumption have on the validity of Newton's third law?

* 47. (III) Let $\mathbf{g}'$ be the effective acceleration of gravity at a point on the rotating Earth, equal to the vector sum of the "true" value $\mathbf{g}$ plus the effect of the rotating reference frame ($m\omega^2 r$ term). See Fig. 11–31. Determine the magnitude and direction of $\mathbf{g}'$ relative to a radial line from the center of the Earth (a) at the North Pole, (b) at a latitude of 45° north, and (c) at the equator. Assume that g (if ω were zero) is a constant 9.80 m/s².

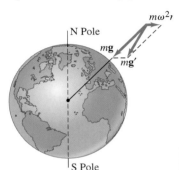

FIGURE 11–31
Problem 47.

* 48. (II) Suppose the man at B in Fig. 11–18 throws the ball toward the woman at A. (a) In what direction is the ball deflected as seen in the noninertial system? (b) Determine a formula for the amount of deflection and for the (Coriolis) acceleration in this case.

* 49. (II) For what directions of velocity would the Coriolis effect on a body moving at the Earth's equator be zero?

* 50. (II) Determine the Coriolis "force" (magnitude and direction) acting on a 1200-kg race car traveling due north at 500 km/h in Utah (latitude 40°). [*Hint*: Use a component of velocity perpendicular to the Earth's axis.]

* 51. (III) A projectile is fired nearly horizontally at high velocity v_0 toward the east. (a) In what direction is it deflected by the Coriolis effect? (b) Determine a formula for the deflection in terms of v_0, ω, λ, and the distance traveled D. (c) Evaluate for $v_0 = 1000$ m/s, $\lambda = 45°$, and $D = 3.0$ km.

* 52. (III) An ant crawls with constant speed along a radial spoke of a wheel rotating at constant angular velocity ω about a vertical axis. Write a vector equation for all the forces (including inertial forces) acting on the ant. Take the x axis along the spoke, y perpendicular to the spoke pointing to the ant's left, and the z axis vertically upward. The wheel rotates counterclockwise as seen from above.

General Problems

53. A particle of mass 0.50 kg is moving with velocity $\mathbf{v} = (8.0\mathbf{i} + 6.0\mathbf{j})$ m/s. (a) Find the angular momentum $\mathbf{L}$ relative to the origin when the particle is at $\mathbf{r} = (2.0\mathbf{j} + 3.0\mathbf{k})$ m. (b) At position $\mathbf{r}$ a force of $\mathbf{F} = 3.0\,\mathrm{N}\mathbf{i}$ is applied to the particle. Find the torque relative to the origin.

54. An asteroid of mass 10^5 kg traveling at a speed of 30 km/s relative to the Earth hits the Earth at the equator. It hits the Earth tangentially and in the direction of the Earth's rotation. Use angular momentum to estimate the fractional change in the angular speed of the Earth as a result of the collision.

55. A boy rolls a tire along a straight level street. The tire has mass 9.0 kg, radius 0.32 m and moment of inertia about its central axis of symmetry of 0.83 kg·m². The boy shoves the tire away from him at a speed of 2.1 m/s and sees that the tire leans 10° to the right (Fig. 11–32). (a) How will the gravitational torque affect the subsequent motion of the tire? (b) Compare the change in angular momentum caused by this torque in 0.20 s to the original magnitude of angular momentum.

56. In the Bohr model of the hydrogen atom, the electron (mass m) is held in a circular orbit about the nucleus (a proton) by the electric force on it, $F = ke^2/r^2$ (where e is the electric charge on the electron and on the proton, and k is a constant), and only certain orbits are allowed: those for which the angular momentum l of the electron about the nucleus is an integer multiple n of $h/2\pi$, where h is a constant called Planck's constant. That is, $l = nh/2\pi$ where n is an integer. Show that the possible radii for electron orbits are given by

$$r = \frac{n^2 h^2}{4\pi^2 kme^2} \qquad n = 1, 2, 3, \cdots.$$

To do this, first show that the radius of an orbit is $r = ke^2/mv^2$ and then apply $l = nh/2\pi$.

57. Water drives a waterwheel (or turbine) of radius $R = 3.0$ m as shown in Fig. 11–33. The water enters at a speed $v_1 = 7.0$ m/s and exits from the waterwheel at a speed $v_2 = 3.0$ m/s. (a) If 150 kg of water passes through per second, what is the rate at which the water delivers angular

FIGURE 11–32
Problem 55.

FIGURE 11–33
Problem 57.

momentum to the waterwheel? (b) What is the torque the water applies to the waterwheel? (c) If the water causes the waterwheel to make one revolution every 5.5 s, how much power is delivered to the wheel?

58. A bicyclist traveling with speed $v = 3.2$ m/s on a flat road is making a turn with a radius $r = 2.8$ m. The forces acting on the cyclist and cycle are the normal force ($\mathbf{F}_N$) and friction force ($\mathbf{F}_{fr}$) exerted by the road on the tires and $m\mathbf{g}$, the total weight of the cyclist and cycle. (a) Explain carefully why the angle θ the bicycle makes with the vertical (Fig. 11–34) must be given by $\tan\theta = F_{fr}/F_N$ if the cyclist is to maintain balance. (b) Calculate θ for the values given. [Hint: Consider the "circular" translational motion of the bicycle and rider.] (c) If the coefficient of static friction between tires and road is $\mu_s = 0.65$, what is the minimum turning radius?

(a)   (b)

FIGURE 11–34 Problem 58.

59. A thin uniform stick of mass M and length l is positioned vertically, with its tip on a frictionless table. It is released and allowed to slip and fall (Fig. 11–35). Determine the speed of its center of mass just before it hits the table.

FIGURE 11–35 Problem 59.

60. A thin string is wrapped around a cylindrical hoop of radius R and mass M. One end of the string is fixed, and the hoop is allowed to fall vertically, starting from rest, as the string unwinds. (a) Determine the angular momentum of the hoop about its CM as a function of time. (b) What is the tension in the string as function of time?

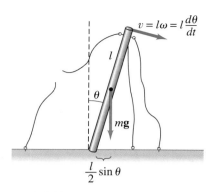

FIGURE 11–36 Problem 61.

61. A radio transmission tower has a mass of 80 kg and is 12 m high. The tower is anchored to the ground by a flexible joint at its base, but it is secured by three cables 120° apart (Fig. 11–36). In an analysis of a potential failure, a mechanical engineer needs to determine the behavior of the tower if one of the cables broke. The tower would fall away from the broken cable, rotating about its base. Determine the speed of the top of the tower as a function of the rotation angle θ. Start your analysis with the rotational dynamics equation of motion $d\mathbf{L}/dt = \boldsymbol{\tau}_{net}$. Approximate the tower as a tall thin rod.

62. A baseball bat has a "sweet spot" where a ball can be hit with almost effortless transmission of energy. A careful analysis of baseball dynamics shows that this special spot is located at the point where an applied force would result in pure rotation of the bat about the handle. Determine the location of the sweet spot of the bat shown in Fig. 11–37. The linear mass density of the bat is given roughly by $(0.56 + 3.5x^2)$ kg/m, where x is in meters measured from the end of the handle. The entire bat is 0.84 m long. The desired rotation point should be 5.0 cm from the end where the bat is held. [Hint: Where is the CM of the bat?]

FIGURE 11–37 Problem 62.

63. Suppose a star the size of our Sun, but of mass 8.0 times as great, were rotating at a speed of 1.0 revolution every 10 days. If it were to undergo gravitational collapse to a neutron star of radius 10 km, losing $\frac{3}{4}$ of its mass in the process, what would its rotation speed be? Assume the star is a uniform sphere at all times. Assume also that the thrown-off mass carries off (a) no angular momentum, (b) its proportional share $\left(\frac{3}{4}\right)$ of the initial angular momentum.

Our whole built environment, from modern bridges to skyscrapers and even 200 year old forts (on the right, beneath the Golden Gate bridge in San Francisco) has required engineers and architects to determine the forces and stresses within these structures. The object is to keep these structures static—that is, not in motion, especially not falling down.

CHAPTER 12

Static Equilibrium;
Elasticity and Fracture

I n this chapter, we will study a special case of motion—when the net force and the net torque on an object, or system of objects, are both zero. In this case the object or system is not accelerating; it is either at rest, or its center of mass is moving at constant velocity. We will be concerned mainly with the first case, when the object or objects are all at rest. Now you may think that the study of objects at rest is not very interesting since the objects will have neither velocity nor acceleration; and the net force and the net torque will be zero. But this does not imply that no forces at all act on the objects. In fact it is virtually impossible to find a body on which no forces act at all. Sometimes the forces may be so great that the object is seriously deformed, or it may even fracture (break)—and avoiding such problems gives this field of **statics** great importance.

This chapter deals with forces within objects at rest

12–1 | Statics—The Study of Forces in Equilibrium

Objects within our experience have at least one force acting on them (gravity), and if they are at rest then there must be other forces acting on them as well so that the net force is zero. An object at rest on a table, for example, has two forces acting on it, the downward force of gravity and the normal force the table exerts upward on it (Fig. 12–1). Since the net force is zero, the upward force exerted by the table must be equal in magnitude to the force of gravity acting downward.

FIGURE 12–1 The book is in equilibrium; the net force on it is zero.

300

Such a body is said to be in **equilibrium** (Latin for "equal forces" or "balance") under the action of these two forces. [Do not confuse the two forces in Fig. 12–1 with the equal and opposite forces of Newton's third law which act on different bodies; here both forces act on the same body.]

The subject of statics is concerned with the calculation of the forces acting on and within structures that are in equilibrium. Determination of these forces, which occupies us in the first part of this chapter, then allows a determination of whether the structures can sustain the forces without significant deformation or fracture, subjects we discuss later in this chapter. These techniques can be applied in a wide range of fields. Architects and engineers must be able to calculate the forces on the structural components of buildings, bridges, machines, vehicles, and other structures, since any material will break or buckle if too much force is applied (Fig. 12–2). In the human body, a knowledge of the forces in muscles and joints is of great value for medicine and physical therapy, and is also valuable for the study of athletic activity.

FIGURE 12–2 Elevated walkway collapse in a Kansas City hotel in 1981. How a simple physics calculation could have prevented the tragic loss of 100 lives is considered in Example 12–14.

12–2 The Conditions for Equilibrium

For an object to be in equilibrium it must not be accelerating, so the *vector sum of all external forces acting on the body must be zero*:

$$\Sigma \mathbf{F} = 0. \qquad \textbf{(12–1a)}$$

This is called the **first condition for equilibrium** and is equivalent to three component equations:

$$\Sigma F_x = 0, \qquad \Sigma F_y = 0, \qquad \Sigma F_z = 0. \qquad \textbf{(12–1b)}$$

First condition for equilibrium: the sum of all forces is zero

EXAMPLE 12–1 **Pullups on a scale.** A 90-kg weakling cannot do even one pull-up. By standing on a scale (Fig. 12–3), he can determine how close he gets. His best effort results in a scale reading of 23 kg. What force is he exerting?

SOLUTION There are three forces acting on our nonathlete, as shown in Fig. 12–3: gravity, $mg = (90 \text{ kg})(9.8 \text{ m/s}^2)$ downward, and two upwards forces which are (1) the force the bar pulls upward on him, F_B (equal and opposite to the force he exerts on the bar), and (2) the force the scale exerts on his feet, F_S. At best, $F_S = (23 \text{ kg})(g)$. The person doesn't move, so the sum of these forces is zero:

$$F_B + F_S - mg = 0.$$

We solve for F_B:

$$F_B = mg - F_S$$
$$= (90 \text{ kg} - 23 \text{ kg})(g) = (67 \text{ kg})(9.8 \text{ m/s}^2) = 660 \text{ N}.$$

That is, he could lift himself if his mass were only 67 kg.

FIGURE 12–3 Example 12–1: (a) A person trying to do a pull-up while standing on a scale. (b) Simple free-body diagram.

(a)   (b)

FIGURE 12-4 Example 12-2.

EXAMPLE 12-2 **Chandelier cord tension.** Calculate the tensions $\mathbf{F}_1$ and $\mathbf{F}_2$ in the two cords that are connected to the cord supporting the 200-kg chandelier in Fig. 12–4.

SOLUTION The three forces, $\mathbf{F}_1$, $\mathbf{F}_2$, and the 200-kg weight of the chandelier, act at the point where the three cords join. We choose this junction point (it could be a knot) as the object for which we write $\Sigma F_x = 0$, $\Sigma F_y = 0$. (We don't bother considering the chandelier itself since only two forces act on it, gravity downward and the equal but opposite force exerted upward by the cord, both of which equal $mg = (200\,\text{kg})(9.8\,\text{m/s}^2) = 1960\,\text{N}$.) There are two unknowns (F_1 and F_2)[†] and we can solve for them using Eqs. 12–1, $\Sigma F_x = 0$, $\Sigma F_y = 0$.

We first resolve $\mathbf{F}_1$ into its horizontal (x) and vertical (y) components. Although we don't know the value of F_1, we can write $F_{1x} = F_1 \cos 60°$ and $F_{1y} = F_1 \sin 60°$. $\mathbf{F}_2$ has only an x component. In the vertical direction, we have only the weight of the chandelier $= (200\,\text{kg})(g)$ acting downward and the vertical component of $\mathbf{F}_1$ upward. Since $\Sigma F_y = 0$, we have

$$\Sigma F_y = F_1 \sin 60° - (200\,\text{kg})(g) = 0$$

so

$$F_1 = \frac{(200\,\text{kg})g}{\sin 60°} = \frac{(200\,\text{kg})g}{0.866} = (231\,\text{kg})g = 2260\,\text{N}.$$

In the horizontal direction,

$$\Sigma F_x = F_2 - F_1 \cos 60° = 0.$$

Thus

$$F_2 = F_1 \cos 60° = (231\,\text{kg})(g)(0.500) = (115\,\text{kg})g = 1130\,\text{N}.$$

The magnitudes of $\mathbf{F}_1$ and $\mathbf{F}_2$ determine the strength of cord or wire that must be used. In this case, the wire must be able to hold more than 230 kg. Note in this Example that we didn't insert the value of g, the acceleration due to gravity, until the end. In this way we found the magnitude of the force in terms of the number of kilograms (which may be a more familiar quantity than newtons) times g.

Although Eqs. 12–1 are a necessary condition for an object to be in equilibrium, they are not a sufficient condition. Figure 12–5 shows an object on which the net force is zero. Although the two forces labeled $\mathbf{F}$ add up to give zero net force on the object, they do give rise to a net torque that will rotate the object. Thus we need a *second condition for equilibrium*: that the vector sum of all external torques acting on the body must add up to zero:

$$\Sigma \boldsymbol{\tau} = 0. \tag{12-2}$$

This will assure that the angular acceleration, α, about any point will be zero. If the body is not rotating initially ($\omega = 0$) it will not start rotating. Equations 12–1 and 12–2 are the only requirements for a body to be in equilibrium.

Equation 12–2 can also be written in component form

$$\Sigma \tau_x = 0, \quad \Sigma \tau_y = 0, \quad \Sigma \tau_z = 0.$$

The torques are calculated about some chosen point O, and τ_x, τ_y, and τ_z are the components along any three chosen axes. However, we will restrict ourselves in most of this chapter to a common situation that is simpler than the general case: the situation when all the external forces are acting in a plane. If we call this plane the xy plane, then we have only two force equations,

$$\Sigma F_x = 0, \quad \Sigma F_y = 0,$$

and one torque equation,

$$\Sigma \tau_z = 0.$$

The torque is calculated about an axis that is perpendicular to the xy plane. The

Second condition for equilibrium: sum of all torques is zero

FIGURE 12-5 Although the net force on it is zero, the ruler will move (rotate). A pair of equal forces acting in opposite directions but at different points on a body (as shown here) is referred to as a *couple*.

[†] The directions of $\mathbf{F}_1$ and $\mathbf{F}_2$ are known, since tension in a rope can only be along the rope—any other direction would cause the rope to bend. Thus, our unknowns are the magnitudes F_1 and F_2.

choice of this axis is arbitrary. Since $\Sigma\tau = I\alpha$ for a rigid body about any fixed axis (Eq. 10–14), $\Sigma\tau = 0$ about any fixed axis if the body is in equilibrium ($\alpha = 0$). Therefore the choice of this axis is arbitrary, so we can choose any axis that makes our calculation easier.

CONCEPTUAL EXAMPLE 12–3 | **A lever.** The bar in Fig. 12–6 is being used as a lever to pry up a large rock. The small rock acts as a fulcrum. The force F_P required at the long end of the bar can be quite a bit smaller than the rock's weight Mg, since it is the *torques* that balance in the rotation about the fulcrum. If, however, the leverage isn't quite good enough, and the rock isn't budged, what are two ways to increase the leverage?

➡ **PHYSICS APPLIED**
The lever

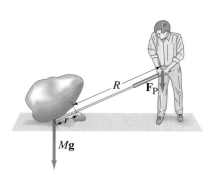

RESPONSE One way is to increase the lever arm of the force F_P by slipping a pipe over the end of the bar and thereby pushing with a longer lever arm. A second, perhaps handier, way is to move the fulcrum closer to the large rock. This may change the long lever arm R only a little, but it changes the short lever arm r by a good fraction and therefore changes the ratio of R/r dramatically. In order to pry the rock, the torque due to F_P must at least balance the torque due to mg, so $mgr = F_P R$ and

$$\frac{r}{R} = \frac{F_P}{Mg}.$$

FIGURE 12–6 Example 12–3.

With r smaller, the weight Mg can be balanced with less force F_P. The ratio R/r is the **mechanical advantage** of the system. A lever is a "simple machine." We discussed another simple machine, the pulley, in Chapter 4, Example 4–14.

12–3 | Solving Statics Problems

This subject of statics is important because it allows us to calculate certain forces on (or within) a structure when some of the forces on it are already known. There is no single technique for attacking such statics problems, but the following procedure, which assumes the force vectors all lie in the xy plane, may be helpful:

PROBLEM SOLVING	Statics

1. Choose one body at a time for consideration, and make a careful *free-body diagram* for it by showing all the forces acting on that body and the points at which these forces act.
2. Choose a convenient coordinate system, and resolve the forces into their components.
3. Using letters to represent unknowns, write down equations for $\Sigma F_x = 0$, $\Sigma F_y = 0$, and $\Sigma\tau = 0$.
4. For the $\Sigma\tau = 0$ equation, choose any axis perpendicular to the xy plane that you like. (For example, you can reduce the number of unknowns in the resulting equation by choosing the axis so that one of the unknown forces passes through the axis; then this force will have

zero lever arm and produce zero torque and so won't appear in the equation.) Pay careful attention to determining the lever arm for each force correctly. Give each torque a + or − sign. If torques that would tend to rotate the object counterclockwise are given a + sign, then torques that would tend to rotate it clockwise are negative.

5. Solve these equations for the unknowns. Three equations allow a maximum of three unknowns to be solved for; they can be forces, distances, even angles. (If an unknown force comes out negative in your solution, it means the direction you originally chose for that force is actually the opposite.)

Another way of stating the torque equilibrium condition is that the sum of all clockwise torques is equal to the sum of all counterclockwise torques. And for the forces, the sum of the upward forces is equal to the sum of the downward forces, and the sum of horizontal forces to the left is equal to the sum of horizontal forces to the right.

Alternate form of
$\Sigma\tau = 0, \Sigma\mathbf{F} = 0$

(a)

Counterweight
M = 9500 kg

←3.4 m→←——7.7 m——→

m = 2800 kg

(b)

$\mathbf{F}_N$

←x→⊢——7.7 m——→
P

$M\mathbf{g}$ $m\mathbf{g}$

(c)

FIGURE 12–7 Example 12–4.

FIGURE 12–8 A 1500-kg beam supports 15,000-kg machine. Example 12–5.

$\mathbf{F}_1$ $\mathbf{F}_2$

P   CG

(1500 kg)$\mathbf{g}$
←——10.0 m——→←5.0 m→←5.0 m→

(15,000 kg)$\mathbf{g}$

One of the forces that acts on bodies is the force of gravity. Our analysis in this chapter is greatly simplified if we use the concept of center of gravity (CG) or center of mass (CM), which are the same point as long as **g** is the same throughout the body (nearly always valid). As we discussed in Section 9–8, we can consider the force of gravity on the body as a whole as acting at its CG. For uniform symmetrically shaped bodies, the CG is at the geometric center. For more complicated bodies, the CG can be determined as discussed in Section 9–8.

EXAMPLE 12–4 **Tower crane.** A tower crane (Fig. 12–7a) must always be carefully balanced so that there is no net torque tending to tip it. A particular crane at a building site is about to lift a 2800-kg air conditioning unit. The crane's dimensions are shown in Fig. 12–7b. (*a*) Where must the crane's 9500-kg counterweight be placed when the load is lifted from the ground? (Note—the counterweight is usually moved automatically via sensors and motors to precisely compensate for the load.) (*b*) Determine the maximum load that can be lifted with this counterweight when it is placed at its full extent. Ignore the mass of the beam.

SOLUTION (*a*) The free-body diagram, showing the forces on the beam, is drawn in Fig. 12–7c. Let us calculate the torques about the pivot point P. Then the upward force $\mathbf{F}_N$ exerted on the beam by the tower (this force keeps the beam up there) has a lever arm of zero and so doesn't appear in the equation. Thus, setting $\Sigma\tau = 0$,

$$\Sigma\tau = (9500\,\text{kg})(g)(x) - (2800\,\text{kg})(g)(7.7\,\text{m}) = 0$$

where x is the position we wish to determine for the counterweight. Solving for x, we find $x = (2800\,\text{kg})(7.7\,\text{m})/(9500\,\text{kg}) = 2.3\,\text{m}$.
(*b*) Figure 12–7b suggests that the counterweight can be at most 3.4 m from the tower. Hence the maximum load m that can be lifted is found using the torque equation

$$\Sigma\tau = (9500\,\text{kg})(g)(3.4\,\text{m}) - mg(7.7\,\text{m}) = 0,$$

which gives $m = (9500\,\text{kg})(3.4\,\text{m})/(7.7\,\text{m}) = 4200\,\text{kg}$.

EXAMPLE 12–5 **Forces on a beam and supports.** A uniform 1500-kg beam, 20.0 m long, supports a 15,000-kg printing press 5.0 m from the right support column (Fig. 12–8). Calculate the force on each of the vertical support columns.

SOLUTION We analyze the forces on the beam, since the force the beam exerts on the columns is equal and opposite to the forces exerted by the columns on the beam. We call the latter $\mathbf{F}_1$ and $\mathbf{F}_2$ in Fig. 12–8. The weight of the beam itself acts at its center of gravity, 10.0 m from either end. Since it doesn't matter which point we choose as the axis for writing the torque equation, we can choose one that is convenient. If we calculate the torques about the point of application of $\mathbf{F}_1$, (labelled P), then $\mathbf{F}_1$ will not enter the equation (its lever arm will be zero) and we will have an equation in only one unknown, F_2. We choose the counterclockwise direction as positive, and $\Sigma\tau_P = 0$ gives

$$\Sigma\tau_P = -(10.0\,\text{m})(1500\,\text{kg})g - (15.0\,\text{m})(15,000\,\text{kg})g + (20.0\,\text{m})F_2 = 0.$$

Solving for F_2 we find $F_2 = (12,000\,\text{kg})g = 118,000\,\text{N}$. To find F_1, we use $\Sigma F_y = 0$:

$$\Sigma F_y = F_1 - (1500\,\text{kg})g - (15,000\,\text{kg})g + F_2 = 0.$$

Putting in $F_2 = (12,000\,\text{kg})g$ we find that $F_1 = (4500\,\text{kg})g = 44,100\,\text{N}$.

Figure 12–9 shows a beam that extends beyond its support like a diving board. Such a beam is called a **cantilever**. The forces acting on the beam in this figure are those due to the supports, $\mathbf{F}_1$ and $\mathbf{F}_2$, and the force of gravity, which acts at the CG, 5.0 m to the right of the right-hand support. If you follow the procedure of

Cantilever

the last Example to calculate F_1 and F_2, assuming they point upward as shown in the figure, you will find that F_1 comes out negative. If the beam has a mass of 1200 kg, then $F_2 = 15,000$ N and $F_1 = -3000$ N (see Problem 13). Whenever an unknown force comes out negative, it merely means that the force actually points in the opposite direction from what you assumed. Thus in Fig. 12–9, $\mathbf{F}_1$ actually points downward. With a little reflection it should become obvious that the left-hand support must indeed pull downward on the beam (by means of bolts, screws, fasteners and/or glue) if the beam is to be in equilibrium; otherwise the sum of the torques about the CM (or about the point where $\mathbf{F}_2$ acts) could not be zero.

Any building, or part of a building, that overhangs its support is said to be cantilevered (Fig. 12–10). An airplane wing is a cantilever. A cantilevered bridge is one whose span between two supports was designed and analyzed as two arms, each fixed to a separate support, that extend out toward each other to meet in the middle.

Our next Example involves a beam attached to a wall by a hinge or "pin," and is supported by a cable or cord (Fig. 12–11). It is important to remember that a flexible cable can support a force only along its length. (If there were a component of force perpendicular to the cable, it would bend because it is flexible.) But for a rigid device, such as the pin in Fig. 12–11, the force can be in any direction and we can know the direction only after solving the problem. The pin is assumed small and smooth, so it can exert no internal torque (about its center) on the beam.

EXAMPLE 12–6 Beam supported by pin and cable.

A uniform beam, 2.20 m long with mass $m = 25.0$ kg, is mounted by a pin on a wall as shown in Fig. 12–11. The beam is held in a horizontal position by a cable that makes an angle $\theta = 30.0°$ as shown. The beam supports a sign of mass $M = 280$ kg suspended from its end. Determine the components of the force $\mathbf{F}_P$ that the pin exerts on the beam, and the tension F_T in the supporting cable.

SOLUTION Figure 12–11 is the free-body diagram for the beam, showing all the forces acting on the beam; it also shows the components of $\mathbf{F}_P$ and $\mathbf{F}_T$. We have three unknowns, F_{Px}, F_{Py}, and F_T (we are given θ), so we will need all three equations, $\Sigma F_x = 0$, $\Sigma F_y = 0$, $\Sigma \tau = 0$. The sum of the forces in the vertical (y) direction is

$$\Sigma F_y = F_{Py} + F_{Ty} - mg - Mg = 0. \qquad \text{(i)}$$

In the horizontal (x) direction, the sum of the forces is

$$\Sigma F_x = F_{Px} - F_{Tx} = 0. \qquad \text{(ii)}$$

For the torque equation, we choose the axis at the point where $\mathbf{F}_T$ and $M\mathbf{g}$ act (so our equation then contains only one unknown, F_{Py}, and we can solve it more quickly); we choose torques that tend to rotate the beam counterclockwise as positive. The weight mg of the (uniform) beam acts at its center, so we have:

$$\Sigma \tau = -(F_{Py})(2.20\text{ m}) + (mg)(1.10\text{ m}) = 0$$

so

$$F_{Py} = \tfrac{1}{2}mg = \tfrac{1}{2}(25.0\text{ kg})(9.80\text{ m/s}) = 123\text{ N}. \qquad \text{(iii)}$$

Next, since the tension $\mathbf{F}_T$ in the cable acts along the cable ($\theta = 30.0°$),

$$F_{Ty} = F_{Tx}\tan\theta = 0.577\, F_{Tx}. \qquad \text{(iv)}$$

From Eqs. (i), (ii), and (iv) we get

$$F_{Ty} = (m + M)g - F_{Py} = (305\text{ kg})(9.80\text{ m/s}^2) - 123\text{ N} = 2870\text{ N}$$

$$F_{Tx} = F_{Ty}/0.577 = 4970\text{ N}$$

$$F_{Px} = F_{Tx} = 4970\text{ N}.$$

The components of $\mathbf{F}_P$ are $F_{Py} = 123$ N and $F_{Px} = 4970$ N. The tension in the cable is $F_T = \sqrt{F_{Tx}^2 + F_{Ty}^2} = 5740$ N.

FIGURE 12 – 9 A cantilever.

FIGURE 12–10 This famous cantilevered house, called "Fallingwater," was designed by the American architect Frank Lloyd Wright.

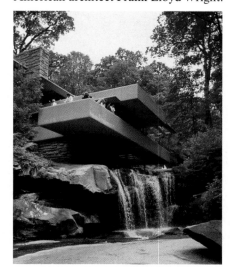

FIGURE 12–11 Beam attached to a wall by a pin. Example 12–6.

EXAMPLE 12–7 **Force exerted by biceps muscle.** How much force must the biceps muscle exert when a 5.0-kg mass is held in the hand (*a*) with the arm horizontal as in Fig. 12–12a, and (*b*) when the arm is at a 30° angle as in Fig. 12–12b? Assume that the mass of forearm and hand together is 2.0 kg and their CG is as shown.

SOLUTION (a) The forces acting on the forearm are shown in Fig. 12–12a and include the upward force F_M exerted by the muscle and a force F_J exerted at the joint by the bone in the upper arm (both assumed to act vertically). We wish to find F_M, which is done most easily by using the torque equation and by choosing our axis through the joint so that F_J does not enter:

$$\Sigma\tau = (0.050\,\text{m})(F_M) - (0.15\,\text{m})(2.0\,\text{kg})(g) - (0.35\,\text{m})(5.0\,\text{kg})(g) = 0.$$

We solve this for F_M and find $F_M = (41\,\text{kg})(g) = 400\,\text{N}$.
(b) The lever arm, as calculated about the joint, is reduced by the factor cos 30° for all three forces. So our torque equation will look like the one just above, except that each term will have a "cos 30°." The latter will cancel out so the same result will be obtained, $F_M = 400\,\text{N}$.

Note that the force required of the muscle (400 N) is quite large compared to the weight of the object lifted (49 N). Indeed, the muscles and joints of the body are generally subjected to quite large forces. The point of insertion of a muscle varies from person to person. A slight increase in the point of insertion of the biceps muscle from 5.0 cm to 5.5 cm can be a considerable advantage for lifting and throwing. Champion athletes are often found to have muscle insertions farther from the joint than the average person.

Leverage with muscles

FIGURE 12–12 Example 12–7.

(a)   (b)

FIGURE 12–13 A ladder leaning against a wall. Example 12–8 (*a*).

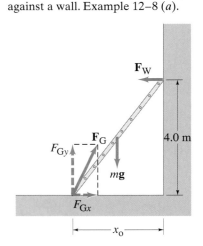

EXAMPLE 12–8 **Ladder.** A 5.0-m-long ladder leans against a wall at a point 4.0 m above the ground as shown in Fig. 12–13. The ladder is uniform and has mass 12.0 kg. Assuming the wall is frictionless (but the ground is not), (*a*) determine the forces exerted on the ladder by the ground and the wall. (*b*) If the coefficient of static friction between the ladder and the ground is $\mu_s = 0.40$, how high up the ladder can a 58-kg painter climb (Fig. 12–14) without the ladder slipping?

SOLUTION (*a*) Figure 12–13 shows the free-body diagram for the ladder, showing all the forces acting on the ladder. The wall, since it is frictionless, can exert a force F_W that must be perpendicular to the wall. The ground, however, can exert both horizontal, F_{Gx}, and vertical, F_{Gy}, force components, the former being frictional and the latter the normal force. Finally, gravity exerts a force $mg = (12.0\,\text{kg})(9.80\,\text{m/s}^2) = 118\,\text{N}$ on the ladder at its midpoint, since the ladder is uniform. The *y* component of the force equation is

$$\Sigma F_y = F_{Gy} - mg = 0,$$

so immediately we have $F_{Gy} = mg = 118\,\text{N}$. The *x* component of the force equation is

$$\Sigma F_x = F_{Gx} - F_W = 0.$$

To determine F_{Gx} and F_W, which are both unknowns, we need another equation, namely a torque equation, which we calculate about the point where the ladder touches the ground. This point is a distance $x_0 = \sqrt{(5.0\,\text{m})^2 - (4.0\,\text{m})^2} = 3.0\,\text{m}$ from the wall. The lever arm for mg is half this, or 1.5 m, and the lever arm for F_W is 4.0 m. Since $\mathbf{F}_G$ acts at the axis, its lever arm is zero and so doesn't enter the equation (we planned it like that), and we get

Thus
$$\Sigma\tau = (4.0\,\text{m})F_W - (1.5\,\text{m})mg = 0.$$

$$F_W = \frac{(1.5\,\text{m})(12.0\,\text{kg})(9.8\,\text{m/s}^2)}{4.0\,\text{m}} = 44\,\text{N}.$$

Then, from the x component of the force equation,

$$F_{Gx} = F_W = 44\,\text{N}.$$

Since the components of $\mathbf{F}_G$ are $F_{Gx} = 44\,\text{N}$ and $F_{Gy} = 118\,\text{N}$, then

$$F_G = \sqrt{(44\,\text{N})^2 + (118\,\text{N})^2} = 126\,\text{N} \approx 130\,\text{N}$$

(rounded off to two significant figures) and it acts at an angle

$$\theta = \tan^{-1}(118\,\text{N}/44\,\text{N}) = 70°$$

to the ground. Note the force $\mathbf{F}_G$ does *not* have to act along the ladder's direction because the ladder is rigid and not flexible like a cord or cable.

(*b*) The free-body diagram of the ladder with the person on it is shown in Fig. 12–14, where x is the horizontal distance from the base of the ladder to a point on the ground directly below the painter. The force equations are

$$\Sigma F_x = 0:\qquad F_{Gx} - F_W = 0 \qquad\qquad\text{(i)}$$

$$\Sigma F_y = 0:\qquad F_{Gy} - (m + M)g = 0, \qquad\qquad\text{(ii)}$$

where $m = 12.0\,\text{kg}$ and $M = 58\,\text{kg}$. The torque equation, calculated as before about the base of the ladder, is

$$(4.0\,\text{m})F_W - (1.5\,\text{m})mg - xMg = 0. \qquad\qquad\text{(iii)}$$

We have four unknowns (F_W, F_{Gx}, F_{Gy}, x) but only three equations; but we get a fourth equation when we use the condition imposed on x, that the painter is at the maximum height without the ladder slipping. This condition limits F_{Gx} to be

$$F_{Gx} = \mu_s F_{Gy} = 0.40 F_{Gy}. \qquad\qquad\text{(iv)}$$

Using Eq. (ii) above, this is

$$F_{Gx} = 0.40(12\,\text{kg} + 58\,\text{kg})g = 28\,g = 270\,\text{N}.$$

Then, from Eq. (i),

$$F_W = F_{Gx} = 28\,g$$

and from Eq. (iii)

$$x = \frac{(4.0\,\text{m})F_W - (1.5\,\text{m})mg}{(58\,\text{kg})g} = 1.62\,\text{m}.$$

Since the base of the ladder is 3.0 m from the wall, the painter can stand at a point that is a fraction $(x/3.0\,\text{m})$ of the distance up the ladder, so her feet can be a height $y = (1.62/3.0)4.0\,\text{m} = 2.2\,\text{m}$ above the ground. If the painter climbs any higher, F_{Gx} will exceed $\mu_s F_{Gy}$ (can you see why?) and the ladder will slip.

➡ **PHYSICS APPLIED**

Safety on a ladder

FIGURE 12–14 Example 12–8 (*b*) with a person on the ladder.

FIGURE 12–15 (a) Stable equilibrium, and (b) unstable equilibrium.

FIGURE 12–16 Equilibrium of a refrigerator resting on a surface.

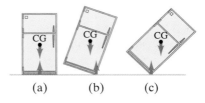

FIGURE 12–17 Humans adjust their posture to achieve stability when carrying loads.

12–4 | Stability and Balance

A body in static equilibrium, if left undisturbed, will undergo no translational or rotational acceleration since the sum of all the forces and the sum of all the torques acting on it are zero. However, if the object is displaced slightly, three different outcomes are possible: (1) the object returns to its original position, in which case it is said to be in **stable equilibrium**; (2) the object moves even farther from its original position, in which case it is said to be in **unstable equilibrium**; or (3) the object remains in its new position, in which case it is said to be in **neutral equilibrium**.

Consider the following examples. A ball suspended freely from a string is in stable equilibrium, for if it is displaced to one side, it will quickly return to its original position (Fig. 12–15a). On the other hand, a pencil standing on its point is in unstable equilibrium. If its CG is directly over its tip (Fig. 12–15b), the net force and net torque on it will be zero. But if it is displaced ever so slightly—say by a slight vibration or tiny air current—there will be a torque on it, and it will continue to fall in the direction of the original displacement. Finally, an example of an object in neutral equilibrium is a sphere resting on a horizontal tabletop. If it is placed slightly to one side, it will remain in its new position.

In most situations, such as in the design of structures and in working with the human body, we are interested in maintaining stable equilibrium or *balance*, as we sometimes say. In general, an object whose CG is below its point of support, such as a ball on a string, will be in stable equilibrium. If the CG is above the base of support, we have a more complicated situation. Consider a standing refrigerator (Fig. 12–16a). If it is tipped slightly, it will return to its original position due to the torque on it as shown in Fig. 12–16b. But if it is tipped too far, Fig. 12–16c, it will fall over. The critical point is reached when the CG is no longer above the base of support. In general, *a body whose CG is above its base of support will be stable if a vertical line projected downward from the CG falls within the base of support.* This is because the normal force upward on the object (which balances out gravity) can be exerted only within the area of contact, so that if the force of gravity acts beyond this area, a net torque will act to topple the object. Stability, then, can be relative. A brick lying on its widest face is more stable than a brick standing on its end, for it will take more of an effort to tip it over. In the extreme case of the pencil in Fig. 12–15b, the base is practically a point and the slightest disturbance will topple it. In general, the larger the base and the lower the CG, the more stable the object.

In this sense, humans are less stable than four-legged mammals, which not only have a larger base of support because of their four legs, but also have a lower center of gravity. Because of their upright position, humans suffer from numerous ailments such as low back pain due to the large forces involved (see Problem 97). When walking and performing other kinds of movement, a person continually shifts the body so that its CG is over the feet, although in the normal adult this requires no conscious thought. Even as simple a movement as bending over requires moving the hips backward so that the CG remains over the feet, and this repositioning is done without thinking about it. To see this, position yourself with your heels and back to a wall and try to touch your toes. You won't be able to do it without falling. Persons carrying heavy loads automatically adjust their posture so that the CG of the total mass is over their feet, Fig. 12–17.

Elasticity and Elastic Moduli; Stress and Strain

In the first part of this chapter we studied how to calculate the forces on objects in equilibrium. In this section we study the effects of these forces, for any object changes shape under the action of applied forces. If the forces are great enough, the object will break or *fracture*—as we will discuss in Section 12–6.

If a force is exerted on an object, such as the vertically suspended metal rod shown in Fig. 12–18, the length of the object changes. If the amount of elongation, ΔL, is small compared to the length of the object, experiment shows that ΔL is proportional to the weight or force exerted on the object. This proportionality can be written as an equation and is sometimes called **Hooke's law**[†] (after Robert Hooke (1635–1703) who first noted it):

$$F = k \, \Delta L. \qquad (12\text{–}3)$$

Here F represents the force (or weight) pulling on the object, ΔL is the increase in length, and k is a proportionality constant. Equation 12–3 is found to be valid for almost any solid material from iron to bone, but it is valid only up to a point. For if the force is too great, the object stretches excessively and eventually breaks. Figure 12–19 shows a typical graph of elongation versus applied force. Up to a point called the **proportional limit**, Eq. 12–3 is a good approximation for many common materials and the curve is a straight line. Beyond this point the graph deviates from a straight line and no simple relationship exists between F and ΔL. Nonetheless, up to a point further along the curve called the **elastic limit**, the object will return to its original length if the applied force is removed. The region from the origin to the elastic limit is called the *elastic region*. If the object is stretched beyond the elastic limit, it enters the *plastic region*: it does not return to the original length upon removal of the external force, but remains permanently deformed (like a bent paper clip). The maximum elongation is reached at the *breaking point*. The maximum force that can be applied without breaking is called the **ultimate strength** of the material (actually force per unit area, as discussed in Section 12–6).

The amount of elongation of an object, such as the rod shown in Fig. 12–18, depends not only on the force applied to it, but also on the material from which it is made and on its dimensions. That is, the constant k in Eq. 12–3 can be written in terms of these factors. If we compare rods made of the same material but of different lengths and cross-sectional areas, it is found that for the same applied force the amount of stretch (again assumed small compared to the total length) is proportional to the original length and inversely proportional to the cross-sectional area. That is, the longer the object, the more it elongates for a given force; and the thicker it is, the less it elongates. These findings can be combined with Eq. 12–3 to yield

$$\Delta L = \frac{1}{E} \frac{F}{A} L_0, \qquad (12\text{–}4)$$

where L_0 is the original length of the object, A is the cross-sectional area, and ΔL is the change in length due to the applied force F. E is a constant of proportionality[‡] known as the **elastic modulus**, or **Young's modulus**, and its value depends only on the material. The value of Young's modulus for various materials is given

FIGURE 12–18 Hooke's law: $\Delta L \propto$ applied force.

FIGURE 12–19 Applied force vs. elongation for a typical metal under tension.

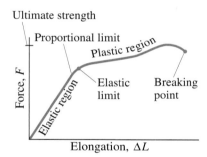

Young's modulus

[†] The term "law" applied to this relation is not really appropriate, since first of all, it is only an approximation, and secondly, it refers only to a limited set of phenomena. Most physicists prefer to reserve the word "law" for those relations that are deeper and more encompassing and precise, such as Newton's laws of motion or the law of conservation of energy.

[‡] That E is in the denominator, so that $1/E$ is the actual proportionality constant, is merely a convention. When we rewrite Eq. 12–4 to get Eq. 12–5, E is in the numerator.

in Table 12–1. Because E is a property only of the material and is independent of the object's size or shape, Eq. 12–4 is far more useful for practical calculation than Eq. 12–3.

TABLE 12–1   Elastic moduli

Material	Young's modulus, E (N/m²)	Shear modulus, G (N/m²)	Bulk modulus, B (N/m²)
Solids			
Iron, cast	100×10^9	40×10^9	90×10^9
Steel	200×10^9	80×10^9	140×10^9
Brass	100×10^9	35×10^9	80×10^9
Aluminum	70×10^9	25×10^9	70×10^9
Concrete	20×10^9		
Brick	14×10^9		
Marble	50×10^9		70×10^9
Granite	45×10^9		45×10^9
Wood (pine) (parallel to grain) (perpendicular to grain)	10×10^9 1×10^9		
Nylon	5×10^9		
Bone (limb)	15×10^9	80×10^9	
Liquids			
Water			2.0×10^9
Alcohol (ethyl)			1.0×10^9
Mercury			2.5×10^9
Gases[†]			
air, H_2, He, CO_2			1.01×10^5

[†] At normal atmospheric pressure, and no temperature variation in the process.

From Eq. 12–4 we see that the change in length of an object is directly proportional to the product of the object's length L_0 and the force per unit area F/A applied to it. It is general practice to define the force per unit area as the **stress**:

Stress (N/m²)

$$\text{stress} = \frac{\text{force}}{\text{area}} = \frac{F}{A},$$

which has units of N/m². Also, the **strain** is defined as the ratio of the change in length to the original length:

Strain

$$\text{strain} = \frac{\text{change in length}}{\text{original length}} = \frac{\Delta L}{L_0},$$

and is dimensionless (no units). Strain is thus the fractional change in length of an object, and is a measure of how much it has been deformed. Stress is applied by outside agents whereas strain is the material's response to the stress. Equation 12–4 can be rewritten as

$$\frac{F}{A} = E \frac{\Delta L}{L_0} \qquad \qquad \textbf{(12–5)}$$

or

Young's modulus (again)

$$E = \frac{F/A}{\Delta L/L_0} = \frac{\text{stress}}{\text{strain}}.$$

Thus we see that the strain is directly proportional to the stress, in the elastic region.

EXAMPLE 12–9 **Tension in a piano wire.** A 1.60-m-long steel piano wire has a diameter of 0.20 cm. How great is the tension in the wire if it stretches 0.30 cm when tightened?

SOLUTION We solve for F in Eq. 12–5 and note that the cross-sectional area $A = \pi r^2 = (3.14)(0.0010\,\text{m})^2 = 3.1 \times 10^{-6}\,\text{m}^2$. Then

$$F = E \frac{\Delta L}{L_0} A$$

$$= (2.0 \times 10^{11}\,\text{N/m}^2)\left(\frac{0.0030\,\text{m}}{1.60\,\text{m}}\right)(3.1 \times 10^{-6}\,\text{m}^2) = 1200\,\text{N},$$

where the value for E was taken from Table 12–1. The strong tension in all the wires in a piano must be supported by a very strong frame, usually of cast iron.

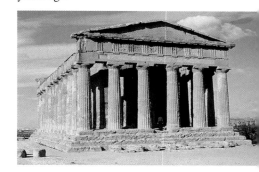

The rod shown in Fig. 12–18 is said to be under **tension** or **tensile stress**. For not only is there a force pulling down on the rod at its lower end, but since the rod is in equilibrium we know that the support at the top is exerting an equal upward force† on the rod at its upper end, Fig. 12–20. In fact, this tensile stress exists throughout the material. Consider, for example, the lower half of a suspended rod as shown in Fig. 12–20b. This lower half is in equilibrium so there must be an upward force on it to balance the downward force at its lower end. What exerts this upward force? It must be the upper part of the rod. Thus we see that external forces applied to an object give rise to internal forces, or stress, within the material itself.

Strain or deformation due to tensile stress is but one type of stress to which materials can be subjected. There are two other common types of stress: compressive and shear. **Compressive stress** is the exact opposite of tensile stress. Instead of being stretched, the material is compressed: the forces act inwardly on the body. Columns that support a weight, such as the columns of a Greek temple (Fig. 12–21) are subjected to compressive stress. Equations 12–4 and 12–5 apply equally well to compression and tension, and the values for E are usually the same.

Figure 12–22 compares tensile and compressive stresses as well as the third type, shear stress. An object under **shear stress** has equal and opposite forces applied *across* its opposite faces. An example is a book or brick, firmly attached to a tabletop, on which a force is exerted parallel to the top surface. The table exerts

FIGURE 12–20 Stress exists *within* the material.

FIGURE 12–21 This Greek temple (in Agrigento, Sicily) was built 2500 years ago.

---

† If we ignore the weight of the bar.

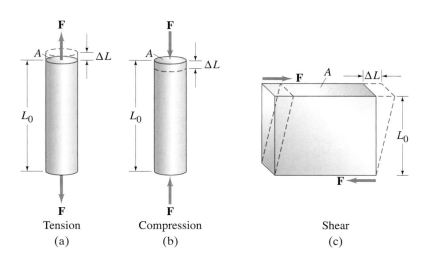

FIGURE 12–22
The three types of stress.

(a)

(b)

FIGURE 12–23 The fatter book (a) shifts more than the thinner book (b) with the same applied shear force.

FIGURE 12–24 Balance of forces and torques for shear stress.

FIGURE 12–25 Fracture as a result of the three types of stress.

an equal and opposite force along the bottom surface. Although the dimensions of the object do not change significantly, the shape of the object does change as shown in Fig. 12–22c. An equation similar to 12–4 can be used for shear strain:

$$\Delta L = \frac{1}{G} \frac{F}{A} L_0 \qquad (12\text{–}6)$$

but ΔL, L_0, and A must be reinterpreted as indicated in Fig. 12–22c. Note that A is the area of the surface *parallel* to the applied force (and not perpendicular as for tension and compression), and ΔL is *perpendicular* to L_0. The constant of proportionality, G, is called the **shear modulus** and is generally one-half to one-third the value of the elastic modulus, E (see Table 12–1). Figure 12–23 illustrates why $\Delta L \propto L_0$: the fatter book shifts more for the same stress.

The rectangular object undergoing shear in Fig. 12–22c would not actually be in equilibrium under the forces shown, for a net torque would exist. If the object is in fact in equilibrium, there must be two more forces acting on it to balance this torque. One force acts vertically upward on the right, and the other acts vertically downward on the left as shown in Fig. 12–24. This is generally true of shear forces. If the object is a brick or book lying on a table, these two additional forces can be exerted by the table and whatever exerts the other horizontal force (such as your hand pushing across the top of a book).

If an object is subjected to a pressure on all sides, its volume will decrease. A common situation is a body submerged in a fluid; for in this case, the fluid exerts a pressure on the object in all directions, as we shall see in Chapter 13. Pressure is defined as force per unit area and thus is the equivalent of "stress." For this situation the change in volume, ΔV, is found to be proportional to the original volume V_0 and to the increase in the pressure ΔP. We thus obtain a relation in the same form as Eq. 12–4 but with a proportionality constant called the **bulk modulus**, B:

$$\frac{\Delta V}{V_0} = -\frac{1}{B} \Delta P$$

or

$$B = -\frac{\Delta P}{\Delta V / V_0}. \qquad (12\text{–}7a)$$

The minus sign is included to indicate that the volume *decreases* with an increase in pressure. The bulk modulus is sometimes given a more general definition in terms of differentials:

$$B = -V \frac{dP}{dV}. \qquad (12\text{–}7b)$$

Values for the bulk modulus are given in Table 12–1. Since liquids and gases do not have a fixed shape, only the bulk modulus applies to them, and not the shear or Young's modulus.

12–6 Fracture

If the stress on a solid object is too great, the object fractures or breaks, Fig. 12–25. Table 12–2 lists the ultimate tensile strengths, compressive strengths, and shear strengths for a variety of materials. These values give the maximum force per unit area that an object can withstand under each of these three types of stress. They are, however, representative values only, and the actual value for a given specimen can differ considerably. It is therefore necessary to maintain a safety factor of from 3 to perhaps 10 or more—that is, the actual stresses on a structure should not exceed one-tenth to one-third the values given in the Table. One sometimes encounters Tables of "allowable stresses" in which appropriate safety factors have already been included.

TABLE 12–2 Ultimate strengths of materials (force/area)

Material	Tensile strength (N/m^2)	Compressive strength (N/m^2)	Shear strength (N/m^2)
Iron, cast	170×10^6	550×10^6	170×10^6
Steel	500×10^6	500×10^6	250×10^6
Brass	250×10^6	250×10^6	200×10^6
Aluminum	200×10^6	200×10^6	200×10^6
Concrete	2×10^6	20×10^6	2×10^6
Brick		35×10^6	
Marble		80×10^6	
Granite		170×10^6	
Wood (pine) (parallel to grain) (perpendicular to grain)	40×10^6	35×10^6 10×10^6	5×10^6
Nylon	500×10^6		
Bone (limb)	130×10^6	170×10^6	

EXAMPLE 12–10 Size and compression of support columns. (*a*) What minimum cross-sectional area should the two columns have to support the beam of Example 12–5 (Fig. 12–8) assuming the columns are made of concrete and a safety factor of 6 is required? We saw in Example 12–5 that the column on the left supports $4.4 \times 10^4\,N$ and that on the right supports $1.2 \times 10^5\,N$. (*b*) How much will the chosen supports compress under the given load?

SOLUTION (*a*) The right-hand column receives the larger force, $F_2 = 1.2 \times 10^5\,N$. It is clearly under compression, and from Table 12–2, we see that the ultimate compressive strength of concrete is $2.0 \times 10^7\,N/m^2$. Using a safety factor of 6, the maximum allowable stress is $\frac{1}{6}(2.0 \times 10^7\,N/m^2) = 3.3 \times 10^6\,N/m^2$, which equals F/A. Since $F = 1.2 \times 10^5\,N$, we can solve for A, and we find:

$$A = \frac{1.2 \times 10^5\,N}{3.3 \times 10^6\,N/m^2} = 3.6 \times 10^{-2}\,m^2, \text{ or } 360\,cm^2.$$

A support $18\,cm \times 20\,cm$ will be adequate.
(*b*) This column will shorten by a fractional amount

$$\frac{\Delta L}{L_0} = \frac{1}{E}\frac{F}{A} = \left(\frac{1}{2.0 \times 10^{10}\,N/m^2}\right)(3.3 \times 10^6\,N/m^2) = 1.7 \times 10^{-4}.$$

Thus, if the support has a length $L_0 = 5.0\,m$, $\Delta L = 0.85 \times 10^{-3}\,m$, or about 1 mm. This calculation was for the right-hand support. If the left-hand support is made of the same cross-sectional area, it will compress less and this should be taken into account.

As can be seen in Table 12–2, concrete (like stone and brick) is reasonably strong under compression but extremely weak under tension. Thus concrete can be used as vertical columns placed under compression but is of little value as a beam since it cannot withstand the tensile forces that arise along the bottom edge when it sags (see Fig. 12–26). *Reinforced concrete*, in which iron rods are embedded in the concrete, is much stronger (Fig. 12–27). But the concrete on the lower edge of a loaded beam still tends to crack because of its weakness under tension. This problem is solved with **prestressed concrete**, which also contains iron rods or a wire mesh, but during the pouring of the concrete, the rods or wire are held under tension. After the concrete dries, the tension on the iron is released, putting the concrete under compression. The amount of compressive stress is carefully predetermined so that when the design loads are applied to the beam, they reduce the compression on the lower edge but never put the concrete into tension.

FIGURE 12–26 A beam sags, at least a little (but is exaggerated here), even under its own weight. The beam thus changes shape so that the upper portion is compressed, and the lower portion is under tension (elongated). Shearing stress also occurs within the beam.

FIGURE 12–27 Steel rods waiting for concrete to be poured around them to form a new highway.

FIGURE 12–28 Example 12–11.

EXAMPLE 12–11 **Shear on a beam.** A uniform pine beam, 3.6 m long and 9.5 cm × 14 cm in cross section, rests on two supports near its ends, as shown in Fig. 12–28. The beam's mass is 25 kg and it is loaded by two vertical roof supports, each 1/3 of the way from each end. What maximum load force F_L can each of the roof supports exert without shearing the pine beam at its supports? Use a safety factor of 5.0.

SOLUTION Each support exerts an upward force F (there is symmetry) that can be at most (see Table 12–2)

$$F = \frac{1}{5} A(5 \times 10^6 \text{ N/m}^2)$$
$$= \frac{1}{5} (0.095 \text{ m})(0.14 \text{ m})(5 \times 10^6 \text{ N/m}^2)$$
$$= 13{,}000 \text{ N}.$$

To determine the maximum load force F_L, we calculate the torque about the left end of the beam (counterclockwise positive):

$$\Sigma\tau = -F_L(1.2 \text{ m}) - (25 \text{ kg})(9.8 \text{ m/s}^2)(1.8 \text{ m}) - F_L(2.4 \text{ m}) + F(3.6 \text{ m}) = 0$$

so

$$F_L = \frac{(13{,}000 \text{ N})(3.6 \text{ m}) - (250 \text{ N})(1.8 \text{ m})}{(1.2 + 2.4)} = 13{,}000 \text{ N}.$$

The total mass of roof the beam can support is $(2)(13{,}000 \text{ N})/(9.8 \text{ m/s}^2) = 2600 \text{ kg}.$

CONCEPTUAL EXAMPLE 12–12 **Maximum stresses.** For the beam in Fig. 12–28, at what points do the shear, compression, and tension stresses reach their maxima?

RESPONSE The shear force is greatest at the ends, given that the beam is uniform and F is slightly greater than F_L (see the last displayed equation in Example 12–11). Compressive stress is greatest at the center on the top side of the beam, and tensile stress is greatest at the center on the bottom side (see Fig. 12–26).

FIGURE 12–29 *Titanic* side wall.

CONCEPTUAL EXAMPLE 12–13 **The *Titanic*.** One hypothesis for why the ship *Titanic* sank when it hit an iceberg on its maiden voyage in 1912 is that the rivet caps were too tall (Fig. 12–29). Explain.

RESPONSE As the ship crashed through an iceberg, the massive iceberg would exert a force on the rivets. With a greater area exposed, more force could be exerted on the rivet heads, causing them to shear off and allowing the side walls to open, letting water in. Note: No rivet heads have ever been found.

➡ PHYSICS APPLIED

A tragic collapse

CONCEPTUAL EXAMPLE 12–14 **A tragic substitution.** Two walkways, one above the other, are suspended from vertical rods attached to the ceiling of a high hotel lobby, as shown in Fig. 12–30a. The original design called for single rods 14 m long, but when such long rods proved to be unwieldy to install, it was decided to replace each long rod with two shorter ones as shown (schematically) in Fig. 12–30b. Determine the net force exerted by the rods on the supporting pin A (assumed the same size) for each design. Assume each vertical rod supports a mass m of each bridge.

(a)

(b)

(c) Force on pin A exerted by vertical rod

(d) Forces on pins at A exerted by vertical rods

FIGURE 12–30 Example 12–14. Parts (c) and (d) show forces on the pins exerted only by the vertical rods.

RESPONSE The single long vertical rod in Fig. 12–30a exerts an upward force of magnitude mg on pin A to support the mass m of the upper bridge, as shown in Fig. 12–30c. Why? Because the pin is in equilibrium, and the only other force on the pin, and which balances this one, is the downward force mg exerted on it by the upper bridge itself. There is thus a shear stress on the pin. The situation when two shorter rods support the bridges (Fig. 12–30b) is shown in Fig. 12–30d in which only the connections at the upper bridge are shown. The lower rod exerts a force of mg downward on the lower of the two pins because it supports the lower bridge. The upper rod exerts a force of $2mg$ on the upper pin (pin A) because the upper rod supports *both* bridges. Thus we see that when the builders substituted two shorter rods for each single long one, the stress in the supporting pin A was *doubled*. What perhaps seemed like a simple substitution did, in fact, lead to a tragic collapse in 1981 (Fig. 12–2) with a loss of over 100 lives. Having a feel for physics, and being able to make simple calculations based on physics, can have a great effect, literally, on people's lives.

*12–7 Trusses and Bridges

A beam used to span a wide space, as for a bridge, is subject to strong stresses of all three types as we saw in Fig. 12–26: compression, tension and shear. A basic engineering device to support large spans is the *truss*, an example of which is shown in Fig. 12–31. Wooden truss bridges were first designed by the great architect Andrea Palladio (1518–1580), who is famous for his design of public buildings and villas. With the introduction of steel in the nineteenth century, much stronger steel trusses came into use, although wood trusses are still used to support the roofs of houses and mountain lodges (Fig. 12–32).

Basically, a **truss** is a framework of rods or struts joined together at their ends by pins or rivets, always arranged as triangles. (Triangles are relatively stable, as compared to a rectangle, which easily becomes a parallelogram under sideways forces and then collapses.) The place where the struts are joined by a pin is called a **joint**.

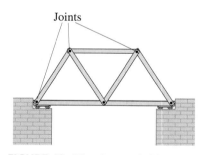

FIGURE 12–31 A truss bridge.

FIGURE 12–32 A roof truss.

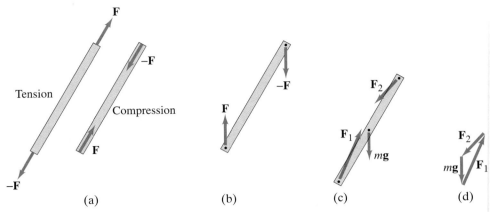

FIGURE 12–33 (a) Each strut (or rod) of a truss is assumed to be under tension or compression. (b) The two equal and opposite forces must be along the same line or a net torque would exist. (c) Real struts have mass, so the forces $\mathbf{F}_1$ and $\mathbf{F}_2$ at the joints do not act precisely along the strut. (d) Vector diagram of part (c).

It is commonly assumed that the struts of a truss are under pure compression or pure tension—that is, the forces act along the length of each strut, Fig. 12–33a. This is an ideal, valid only if a strut has no mass and supports no weight along its length, in which case a strut has only two forces on it, at the ends, as shown in Fig. 12–33a. If the strut is in equilibrium, these two forces must be equal and opposite in direction ($\Sigma\mathbf{F} = 0$). But couldn't they be at an angle, as in Fig. 12–33b? No, because then $\Sigma\tau$ would not be zero. The two forces *must* act along the strut if the strut is in equilibrium. But in a real case of a strut with mass, there are three forces on the strut, as shown in Fig. 12–33c, and $\mathbf{F}_1$ and $\mathbf{F}_2$ do not act along the strut; the vector diagram in Fig. 12–33d shows $\Sigma\mathbf{F} = \mathbf{F}_1 + \mathbf{F}_2 + m\mathbf{g} = 0$. Can you see why $\mathbf{F}_1$ and $\mathbf{F}_2$ both point *above* the strut? (Do $\Sigma\tau$ about each end.)

Consider again the simple beam in Example 12–6, Fig. 12–11. The force $\mathbf{F}_P$ at the pin is *not* along the beam, but acts at an upward angle, albeit only 1.4°. If that beam were massless, we see from Eq. (iii) in Example 12–6 with $m = 0$, that $F_{Py} = 0$, and $\mathbf{F}_P$ would be along the beam.

The assumption that the forces in each strut of a truss are purely along the strut is still very useful whenever the loads act only at the joints and are much greater than the weight of the struts themselves.

➡ PHYSICS APPLIED

A truss bridge

➡ PROBLEM SOLVING

Method of joints

EXAMPLE 12–15 **A truss bridge.** Determine the tension or compression in each of the struts of the truss bridge shown in Fig. 12–34a. The bridge is 64 m long and supports a uniform concrete roadway whose total mass is 1.40×10^6 kg. Use the **method of joints**, which involves (1) drawing a free-body diagram of the truss as a whole, and (2) drawing a free-body diagram for each of the pins (joints), one by one, and setting $\Sigma\mathbf{F} = 0$ for each pin. Ignore the mass of the struts. Assume all triangles are equilateral, and assume that the mass of the roadway acts at the center.

SOLUTION Any bridge has two trusses, one on each side of the roadway. Consider only one truss, Fig. 12–34a, and it will support half the weight of the roadway. That is, our truss supports a total mass $M = 7.0 \times 10^5$ kg. First we draw a free-body diagram for the entire truss as a single unit, which we assume rests on supports at either end that exert upward forces $\mathbf{F}_1$ and $\mathbf{F}_2$, Fig. 12–34b. We assume the mass of the roadway acts entirely at the center, on pin C, as shown. From symmetry [or by doing a torque equation about, say, point A: $(F_2)(l) - Mg(l/2) = 0$] we see that each of the end supports carries half the weight, so

$$F_1 = F_2 = \tfrac{1}{2}Mg.$$

Next we look at pin A and apply $\Sigma \mathbf{F} = 0$ to it. We label the forces on pin A due to each strut with two subscripts: $\mathbf{F}_{AB}$ means the force exerted by the strut AB and $\mathbf{F}_{AC}$ is the force exerted by strut AC. $\mathbf{F}_{AB}$ and $\mathbf{F}_{AC}$ act along their respective struts; but not knowing whether each is compressive or tensile, we could draw four different free-body diagrams, as shown in Fig. 12–34c. Only the one on the left could provide $\Sigma \mathbf{F} = 0$, so we immediately know the directions of $\mathbf{F}_{AB}$ and $\mathbf{F}_{AC}$.[†] These forces act on the pin. The force that pin A exerts on strut AB is opposite in direction to $\mathbf{F}_{AB}$ (Newton's third law), so strut AB is under compression and strut AC is under tension. Now let's calculate the magnitudes of $\mathbf{F}_{AB}$ and $\mathbf{F}_{AC}$. At pin A:

$$\Sigma F_x = F_{AC} - F_{AB} \cos 60° = 0$$
$$\Sigma F_y = F_1 - F_{AB} \sin 60° = 0.$$

Thus

$$F_{AB} = \frac{F_1}{\sin 60°} = \frac{\tfrac{1}{2}Mg}{\tfrac{1}{2}\sqrt{3}} = \frac{1}{\sqrt{3}}Mg,$$

which equals $(7.0 \times 10^5 \,\text{kg})(9.8 \,\text{m/s}^2)/\sqrt{3} = 4.0 \times 10^6$ N; and

$$F_{AC} = F_{AB} \cos 60° = \frac{1}{2\sqrt{3}}Mg.$$

Next we look at pin B, and Fig. 12–34d is the free-body diagram. [Convince yourself that if $\mathbf{F}_{BD}$ or $\mathbf{F}_{BC}$ were in the opposite direction, $\Sigma \mathbf{F}$ could not be zero; note that $\mathbf{F}_{AB}$ here is in the opposite direction from that in Fig. 12–34c because now we are at the opposite end of strut AB.] We see that BC is under tension and BD compression. (Recall that the forces on the struts are opposite to the forces shown which are on the pin.) We set $\Sigma \mathbf{F} = 0$:

$$\Sigma F_x = F_{AB} \cos 60° + F_{BC} \cos 60° - F_{BD} = 0$$
$$\Sigma F_y = F_{AB} \sin 60° - F_{BC} \sin 60° = 0$$

so

$$F_{BC} = F_{AB} = \frac{1}{\sqrt{3}}Mg,$$

and

$$F_{BD} = F_{AB} \cos 60° + F_{BC} \cos 60° = \frac{1}{\sqrt{3}}Mg\left(\tfrac{1}{2}\right) + \frac{1}{\sqrt{3}}Mg\left(\tfrac{1}{2}\right) = \frac{1}{\sqrt{3}}Mg.$$

The solution is complete. By symmetry, $F_{DE} = F_{AB}$, $F_{CE} = F_{AC}$ and $F_{CD} = F_{BC}$. As a check, calculate ΣF_x and ΣF_y for pin C and see if they equal zero. Figure 12–34e shows the free-body diagram.

[†] If we were to choose the direction of a force on a diagram opposite to what it really is, we would get a minus sign, which tells us so.

(a)

(b)

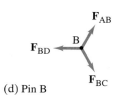

(c) Pin A (different guesses)

(d) Pin B

(e) Pin C

FIGURE 12–34
Example 12–15. (a) A truss bridge. Freebody diagrams:
(b) for the entire truss,
(c) for pin A (different guesses),
(d) for pin B and (e) for pin C.

FIGURE 12–35 (a) Truss with truck of mass m at center of strut AC. (b) Forces on strut AC.

FIGURE 12–36
A suspension bridge (Verrazano Narrows, NY).

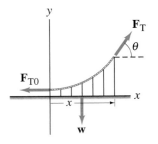

FIGURE 12–37 Example 12–16.

Example 12–15 put the roadway load at the center, C. Now consider a heavy load, such as a heavy truck, supported by strut AC at its middle, as shown in Fig. 12–35a. The strut AC sags under this load, telling us there is shear stress in strut AC. Figure 12–35b shows the forces exerted on strut AC: the weight of the truck $m\mathbf{g}$, and the forces $\mathbf{F}_A$ and $\mathbf{F}_C$ that pins A and C exert on the strut. [Note that $\mathbf{F}_1$ does not appear because it is a force (exerted by external supports) that acts on pin A, not on strut AC.] The forces that pins A and C exert on strut AC will be not only along the strut, but will have vertical components too, perpendicular to the strut, creating shear stress to balance the weight of the truck, $m\mathbf{g}$. The other struts, not bearing weight, remain under pure tension or compression. Problems 64 and 65 deal with this situation, and an early step in their solution is to calculate the forces $\mathbf{F}_A$ and $\mathbf{F}_C$ by using torque equations for the strut.

For very large bridges, truss structures are too heavy. One solution is to build suspension bridges, with the load being carried by relatively light suspension cables under tension, supporting the roadway by means of closely spaced vertical wires, as shown in Fig. 12–36, and in the photo on the first page of this chapter.

EXAMPLE 12–16 **Suspension bridge.** Determine the shape of the cable between the two towers in Fig. 12–36, assuming the weight of the roadway is supported uniformly along its length. Ignore the weight of the cable.

SOLUTION We take $x = 0$, $y = 0$ at the center of the span, as shown in Fig. 12–37. Let $\mathbf{F}_{T0}$ be the tension in the cable at $x = 0$; it acts horizontally as shown. Let F_T be the tension in the cable at some other place where the horizontal coordinate is x, as shown. This section of cable supports a portion of the roadway whose weight w is proportional to the distance x, since the roadway is assumed uniform; that is,

$$w = \lambda x$$

where λ is the weight per unit length. We now set $\Sigma \mathbf{F} = 0$:

$$\Sigma F_x = F_T \cos\theta - F_{T0} = 0$$
$$\Sigma F_y = F_T \sin\theta - w = 0.$$

We divide these two equations,

$$\tan\theta = \frac{w}{F_{T0}} = \frac{\lambda x}{F_{T0}}.$$

The slope of our curve (the cable) at any point is

$$\frac{dy}{dx} = \tan\theta$$

$$= \frac{\lambda}{F_{T0}} x.$$

We integrate this:

$$\int dy = \frac{\lambda}{F_{T0}} \int x\, dx$$

$$y = Ax^2 + B$$

where we set $A = \lambda/F_{T0}$ and B is a constant of integration. This is just the equation of a parabola.

Real bridges have cables that do have mass, so the cables hang only approximately as a parabola, although often it is quite close.

*12–8 Arches and Domes

There are various methods that engineers and architects can use to span interior spaces as well as exterior spaces (bridges). We have already looked briefly at beams, trusses and suspension bridges. In this Section we discuss arches and domes, beginning with a brief history of how builders throughout the centuries have attacked the problem of spanning a space.

The first important architectural invention was the post-and-beam (or post-and-lintel) construction, in which two upright posts support a horizontal beam. Before steel was introduced in the nineteenth century, the length of a beam was quite limited because the strongest building materials were then stone and brick. Hence the width of a span was limited by the size of available stones. Equally important, stone and brick, though strong under compression—are very weak under tension and shear; all three types of stress occur in a beam, as we saw in Fig. 12–26. The minimal space that could be spanned using stone is shown by the closely spaced columns of the great Greek temples (Fig. 12–21).

The introduction of the semicircular **arch** by the Romans (Figs. 12–38 and 12–39), aside from its aesthetic appeal, was a tremendous technological innovation. The advantage of the "true" or semicircular arch is that, if well designed, its wedge-shaped stones experience stress which is mainly compressive even when supporting a large load such as the wall and roof of a cathedral. Because the stones are forced to squeeze against one another, they are mainly under compression (see Fig. 12–40). Note, however, that the arch transfers horizontal as well as vertical forces to the supports. A round arch consisting of many well-shaped stones could span a very wide space. However, considerable buttressing on the sides was needed to support the horizontal components of the forces.

The pointed arch came into use about A.D. 1100 and became the hallmark of the great Gothic cathedrals. It too was an important technical innovation, and was first used to support heavy loads such as the tower of a cathedral, and as the central arch. Apparently the builders realized that, because of the steepness of the pointed arch, the forces due to the weight above could be brought down more nearly vertically, so less horizontal buttressing would be needed. The pointed arch reduced the load on the walls, so there could be more openness and light. The smaller buttressing needed was provided on the outside by graceful flying buttresses (Fig. 12–41).

➡ **P H Y S I C S A P P L I E D**

Architecture: Beams, arches and domes

FIGURE 12–38 Round arches in the Roman Forum. The one in the background is the Arch of Titus.

FIGURE 12–39 An arch is used here to good effect in spanning a chasm on the California coast.

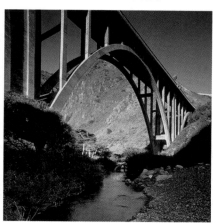

FIGURE 12–40 Stones in a round arch (see Fig. 12–38) are mainly under compression.

FIGURE 12–41 Flying buttresses (on the cathedral of Notre Dame, in Paris).

FIGURE 12–42 (a) Forces in a round arch, compared (b) with those in a pointed arch.

FIGURE 12–43 Interior of the Pantheon in Rome, built almost 2,000 years ago. This view, showing the great dome and its central opening for light, was painted about 1740 by Panini. Photographs do not capture its grandeur as well as this painting does.

FIGURE 12–44 The Louisiana Superdome.

The technical innovation of the pointed arch was achieved not through calculation but through experience and intuition, for it was not until much later that detailed calculations, such as those presented earlier in this chapter, came into use. To make an accurate analysis of a stone arch is quite difficult in practice. But if we make some simplifying assumptions, we can show why the horizontal component of the force at the base is less for a pointed arch than for a round one. Figure 12–42 shows a round arch and a pointed arch, each with an 8.0-m span. The height of the round arch is thus 4.0 m, whereas that of the pointed arch is larger and has been chosen to be 8.0 m. Each arch supports a weight of 12.0×10^4 N $(= 12{,}000 \, \text{kg} \times g)$ which, for simplicity, we have divided into two parts (each 6.0×10^4 N) acting on the two halves of each arch as shown. To be in equilibrium, each of the supports must exert an upward force of 6.0×10^4 N. For rotational equilibrium, each support also exerts a horizontal force, F_H, at the base of the arch, and it is this we want to calculate. We focus only on the right half of each arch. We set equal to zero the total torque calculated about the apex of the arch due to the forces exerted on that half arch. For the round arch, the torque equation is (see Fig. 12–42a)

$$(4.0 \, \text{m})(6.0 \times 10^4 \, \text{N}) - (2.0 \, \text{m})(6.0 \times 10^4 \, \text{N}) - (4.0 \, \text{m})(F_H) = 0.$$

Thus $F_H = 3.0 \times 10^4$ N. For the pointed arch, the torque equation is (see Fig. 12–42b)

$$(4.0 \, \text{m})(6.0 \times 10^4 \, \text{N}) - (2.0 \, \text{m})(6.0 \times 10^4 \, \text{N}) - (8.0 \, \text{m})(F_H) = 0.$$

Solving, we find that $F_H = 1.5 \times 10^4$ N—only half as much! From this calculation we can see that the horizontal buttressing force required for a pointed arch is less because the arch is higher, and there is therefore a longer lever arm for this force. Indeed, the steeper the arch, the less the horizontal component of the force needs to be, and hence the more nearly vertical is the force exerted at the base of the arch.

Whereas an arch spans a two-dimensional space, a **dome**—which is basically an arch rotated about a vertical axis—spans a three-dimensional space. The Romans built the first large domes. Their shape was hemispherical and some still stand, such as that of the Pantheon in Rome (Fig. 12–43).

Only in the twentieth century were larger domes built, the largest being that of the Superdome in New Orleans, completed in 1975 (Fig. 12–44). The 5×10^6-kg dome, over 200 m in diameter, is made of steel trusses and concrete. Its outer thrust (see Example 12–17) is supported by a huge steel tension ring, 200 m in diameter, in the form of a truss, which is attached directly to the dome 50 m above the ground.

EXAMPLE 12–17 **A modern dome.** The 1.2×10^6 kg dome of the Small Sports Palace in Rome (Fig. 12–45a) is supported by 36 buttresses positioned at a 38° angle so they connect smoothly with the dome. Calculate the components of the force, F_H and F_V, that each buttress exerts on the dome so that the force acts purely in compression—that is, at a 38° angle (Fig. 12–45b).

SOLUTION The vertical load on *each* buttress is $\frac{1}{36}$ of the total weight. Thus

$$F_V = \frac{(1.2 \times 10^6 \text{ kg})(9.8 \text{ m/s}^2)}{36} = 3.4 \times 10^5 \text{ N}.$$

The force must act at a 38° angle at the base of the dome in order to be purely compressive. Thus

$$F_H = \frac{F_V}{\tan 38°} = \frac{340,000 \text{ N}}{\tan 38°} = 430,000 \text{ N}.$$

In order that each of the buttresses be able to exert this 430,000-N horizontal force, a prestressed-concrete tension ring surrounds the base of the buttresses beneath the ground (see Problem 68 and Fig. 12–83).

(a)

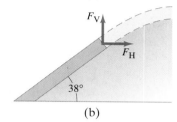

(b)

FIGURE 12–45 (a) The dome of the Small Sports Palace in Rome, built for the 1960 Olympics. (b) Example 12–17.

Summary

A body at rest [or at least not accelerating] is said to be in **equilibrium**. The subject concerned with the determination of the forces within a structure at rest is called **statics**.

The two necessary conditions for a body to be in equilibrium are (1) the vector sum of all the forces on it must be zero, and (2) the sum of all the torques (calculated about any arbitrary axis) must also be zero:

$$\Sigma F_x = 0 \quad \Sigma F_y = 0 \quad \Sigma \tau = 0,$$

when all the forces act in a plane. It is important when doing statics problems to apply the equilibrium conditions to only one body at a time.

A body in static equilibrium is said to be in (*a*) **stable**, (*b*) **unstable**, or (*c*) **neutral equilibrium**, depending on whether a slight displacement leads to (*a*) a return to the original position, (*b*) further movement away from the original position, or (*c*) remaining in the new position. An object in stable equilibrium is also said to be in **balance**.

Hooke's law applies to many elastic solids, and states that the change in length of an object is propor-

tional to the applied force:

$$F = k \, \Delta L.$$

If the force is too great, the object will exceed its **elastic limit**, which means it will no longer return to its original shape when the distorting force is removed. If the force is even greater, the **ultimate strength** of the material can be exceeded and the object fractures.

The force per unit area acting on a body is called the **stress**, and the resulting fractional change in length is called the **strain**.

The stress on a body is present within the body and can be of three types: **compression**, **tension**, and **shear**.

The ratio of stress to strain is called the **elastic modulus** of the material. **Young's modulus** applies for compression and tension, and the **shear modulus** for shear; **bulk modulus** applies to an object whose volume changes as a result of pressure on all sides. All three moduli are constants for a given material when distorted within the elastic region.

Questions

1. Describe several situations where a body is not in equilibrium, even though the net force on it is zero.

2. A bungee jumper momentarily comes to rest at the bottom of the dive before he springs back upward. At that moment, is the bungee jumper in equilibrium? Explain.

3. You can find the center of gravity of a meter stick by resting it horizontally on both your index fingers, and then slowly drawing your fingers together. First the meter stick will slip on one finger, and then on the other, but eventually the fingers meet at the CG. Why does this work?

4. A ladder, leaning against a wall, makes a 60° angle with the ground. When is it more likely to slip: when a person stands near the top or near the bottom? Explain.

5. An earthen retaining wall is shown in Fig. 12–46a. The earth, particularly when wet, can exert a significant force F on the wall. (*a*) What force produces the torque to keep the wall upright? (*b*) Explain why the retaining wall in Fig. 12–46b would be much less likely to overturn.

(a)   (b)

FIGURE 12–46 Question 5.

6. Your doctor's scale uses a slidable weight on an arm to balance your weight, Fig. 12–47. These weights are obviously much lighter than you are. How does this work?

FIGURE 12–47
Question 6.

7. Explain why touching the toes while seated on the floor with outstretched legs produces less stress on the lower spinal column than when touching the toes from a standing position. Use a diagram.

8. Explain under what circumstances a moving body can be in equilibrium.

9. If the net torque on a body is zero when calculated about a point O, will it *necessarily* be zero when calculated about a different point O′?

10. A uniform meter stick supported at the 25-cm mark is in equilibrium when a 1-kg rock is suspended at the 0-cm end, as shown in Fig. 12–48. Is the mass of the meter stick greater than, equal to, or less than the mass of the rock? Explain your reasoning.

FIGURE 12–48
Question 10.

11. Place yourself facing the edge of an open door. Position your feet astride the door with your nose and abdomen touching the door's edge. Try to rise on your tiptoes. Why can't this be done?

12. Which of the configurations of brick, (a) or (b) of Fig. 12–49, is the more likely to be stable? Why?

(a)   (b)

FIGURE 12–49 Question 12. The dots indicate the CG of each brick. The fractions $\frac{1}{4}$ and $\frac{1}{2}$ indicate what portion of each brick is hanging beyond its support.

13. Name the type of equilibrium for each position of the ball in Fig. 12–50.

FIGURE 12–50 Question 13.

14. Why is it not possible to sit upright in a chair and rise to your feet without first leaning forward?

15. Why is it more difficult to do sit-ups when your knees are bent than when your legs are stretched out?

16. Examine how a pair of scissors or shears cuts through a piece of cardboard. Is the name "shears" justified?

Problems

Sections 12–1 to 12–3

1. (I) Three forces are applied to a tree sapling, as shown in Fig. 12–51, to stabilize it. If $\mathbf{F}_1 = 380$ N and $\mathbf{F}_2 = 255$ N, find $\mathbf{F}_3$ in magnitude and direction.

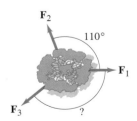

FIGURE 12–51
Problem 1.

2. (I) Two cords support a chandelier in the manner shown in Fig. 12–4 except that the upper wire makes an angle of 45° with the ceiling. If the cords can sustain a force of 1150 N without breaking, what is the maximum chandelier weight that can be supported?

3. (I) Calculate the torque about the front support of a diving board, Fig. 12–52, exerted by a 56-kg person 3.0 m from that support.

4. (I) How far out on the diving board (Fig. 12–52) would a 56-kg diver have to be to exert a torque of 1000 m·N on the board, relative to the left support post?

FIGURE 12–52
Problems 3, 4 and 8.

5. (I) Calculate the mass m needed in order to suspend the leg shown in Fig. 12–53. Assume the leg (with cast) has a mass of 15.0 kg, and its CG is 35.0 cm from the hip joint; the sling is 80.5 cm from the hip joint.

FIGURE 12–53 Problem 5.

6. (I) A 67.0-kg adult sits at one end of a 10.0 m board, on the other end of which sits his 23.0-kg child. Where should the pivot be placed so the board (ignore its mass) is balanced?

7. (II) Repeat Problem 6 taking into account the board's 12.0-kg mass.

8. (II) Calculate the forces F_1 and F_2 that the supports exert on the diving board of Fig. 12–52 when a 56-kg person stands at its tip. (*a*) Ignore the weight of the board. (*b*) Take into account the board's mass of 35 kg. Assume the board's CG is at its center.

9. (II) How close to the edge of the 20.0-kg table shown in Fig. 12–54 can a 66.0-kg person sit without tipping it over?

FIGURE 12–54   Problem 9.

10. (II) A uniform steel beam has a mass of 1100 kg. On it is resting half of an identical beam, as shown in Fig. 12–55. What is the vertical support force at each end?

FIGURE 12–55   Problem 10.

11. (II) Find the tension in the two cords shown in Fig. 12–56. Neglect the mass of the cords, and assume that the angle θ is 30° and the mass m is 200 kg.

FIGURE 12–56
Problem 11.

FIGURE 12–57
Problems 12 and 46.

12. (II) Find the tension in the two wires supporting the traffic light shown in Fig. 12–57.

13. (II) Calculate F_1 and F_2 for the uniform cantilever shown in Fig. 12–9 whose mass is 1200 kg.

14. (II) A 0.60-kg sheet hangs from a massless clothesline as shown in Fig. 12–58. The line on either side of the sheet makes an angle of 3.5° with the horizontal. Calculate the tension in the clothesline on either side of the sheet. Why is the tension so much greater than the weight of the sheet?

FIGURE 12–58   Problem 14.

15. (II) A door, 2.30 m high and 1.30 m wide, has a mass of 13.0 kg. A hinge 0.40 m from the top and another hinge 0.40 m from the bottom each support half the door's weight (Fig. 12–59). Assume that the center of gravity is at the geometrical center of the door, and determine the horizontal and vertical force components exerted by each hinge on the door.

FIGURE 12–59   Problem 15.

16. (II) Figure 12–60 shows a pair of forceps; they are used to firmly hold a thin plastic rod. If each finger squeezes with a force $F_T = F_B = 11.0$ N, what force do the forceps jaws exert on the plastic rod?

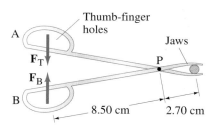

FIGURE 12–60   Problem 16.

17. (II) The mass of the trailer in Fig. 12–61 is 2200 kg. The distances are $a = 2.5$ m and $b = 5.5$ m. The truck is stationary, and the wheels of the trailer can turn freely, meaning the road exerts no horizontal force on them. The hitch at B can be modeled as a pin support. (a) Draw the free-body diagram of the trailer. (b) Determine the total normal force exerted by the road on the rear tires at A and (c) the force exerted on the trailer by the pin support B.

FIGURE 12–61   Problem 17.

18. (II) Approximately what force, F_M, must the extensor muscle in the upper arm exert on the lower arm to hold a 7.3-kg shot put (Fig. 12–62)? Assume the lower arm has a mass of 2.8 kg and its CG is 12 cm from the elbow-joint pivot.

FIGURE 12–62   Problem 18.

19. (II) The Achilles tendon is attached to the rear of the foot as shown in Fig. 12–63. When a person elevates himself just barely off the floor on the "ball of one foot," estimate the tension in the Achilles tendon (pulling upward), and the (downward) force exerted by the lower leg bone on the foot. Assume the person has a mass of 70 kg, and that D is twice as long as d.

FIGURE 12–63   Problem 19.

20. (II) (a) Calculate the force required of the "deltoid" muscle, F_M, to hold up the outstretched arm shown in Fig. 12–64. The total mass of the arm is 3.3 kg. (b) Calculate the magnitude of the force $\mathbf{F}_J$ exerted by the shoulder joint on the upper arm.

FIGURE 12–64   Problems 20 and 21.

21. (II) Suppose the hand in Problem 20 holds a 15-kg mass. What force, F_M, is required of the deltoid muscle, assuming the mass is 52 cm from the shoulder joint?

22. (II) Three children are trying to balance on a seesaw, which consists of a fulcrum rock, acting as a pivot at the center, and a very light board 3.6 m long (Fig. 12–65). Two boys are already on either end. One has a mass of 50 kg, and the other a mass of 35 kg. Where should the third child, whose mass is 25 kg, place herself so as to balance the seesaw?

FIGURE 12–65   Problem 22.

23. (II) A 170-cm-tall person lies on a light (massless) board which is supported by two scales, one under the feet and one beneath the top of the head (Fig. 12–66). The two scales read, respectively, 31.6 and 35.1 kg. Where is the center of gravity of this person?

FIGURE 12–66   Problem 23.

24. (II) A 76.0-kg horizontal beam is supported at each end. A 285-kg piano rests a quarter of the way from one end. What is the vertical force on each of the supports?

25. (II) Calculate F_1 and F_2 for the beam shown in Fig. 12–67. Assume it is uniform and has a mass of 250 kg.

FIGURE 12–67 Problem 25.

FIGURE 12–68
Problem 26.

26. (II) Calculate the tension F_T in the wire that supports the 30.0-kg beam shown in Fig. 12–68, and the force $\mathbf{F_W}$ exerted by the wall on the beam (give magnitude and direction).

27. (II) The two trees in Fig. 12–69 are 7.6 m apart. Calculate the magnitude of the force $\mathbf{F}$ a backpacker must exert to hold a 16-kg backpack so that the rope sags at its midpoint by (a) 1.5 m, (b) 0.15 m.

FIGURE 12–69
Problem 27 and 83.

28. (II) A large 80.0-kg board is propped at a 45° angle against the edge of a barn door that is 2.6 m wide. How great a horizontal force must a person behind the door exert (at the edge) in order to open it? Assume that there is negligible friction between the door and the board but that the board is firmly set against the ground.

29. (II) Repeat Problem 28 assuming the coefficient of friction between the board and the door is 0.45.

30. (II) A shop sign weighing 215 N is supported by a uniform 135-N beam as shown in Fig. 12–70. Find the tension in the guy wire and the horizontal and vertical forces exerted by the pin on the beam.

FIGURE 12–70
Problem 30.

31. (II) A traffic light hangs from a structure as shown in Fig. 12–71. The uniform aluminum pole AB is 7.5 m long and has a mass of 8.0 kg. The mass of the traffic light is 12.0 kg. Determine the tension in the horizontal massless cable CD, and the vertical and horizontal components of the force exerted by the pivot A on the aluminum pole.

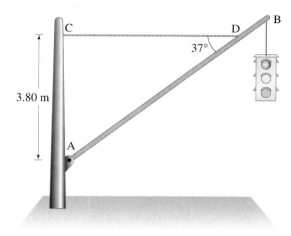

FIGURE 12–71 Problem 31.

32. (II) A uniform ladder of mass m and length l leans at an angle θ against a frictionless wall, Fig. 12–72. If the coefficient of static friction between the ladder and the ground is μ_s, what is the minimum angle at which the ladder will not slip?

FIGURE 12–72
Problem 32 and 82.

33. (II) A meter stick with a mass of 230 g is supported horizontally by two vertical strings, one at the 0-cm mark and the other at the 90-cm mark (Fig. 12–73). (a) What is the tension in the string at 0 cm? (b) What is the tension in the string at 90 cm?

FIGURE 12–73 Problem 33.

34. (II) A uniform rod AB of length 5.0 m and mass $M = 3.0$ kg is hinged at A and held in equilibrium by a light rope, as shown in Fig. 12–74. A load $W = 20$ N hangs from the rod at a distance x so that the tension in the rope is 70 N. (a) Draw a free-body diagram for the rod. (b) Determine the vertical and horizontal forces on the rod exerted by the hinge. (c) Determine x from the appropriate torque equation.

FIGURE 12–74 Problem 34.

35. (III) A cubic crate of side $s = 2.0$ m is top-heavy: its CG is 20 cm above its true center. How steep an incline can the crate rest on without tipping over? What would your answer be if the crate were to slide at constant speed down the plane without tipping over?

36. (III) A refrigerator is approximately a uniform rectangular solid 2.5 m tall, 1.0 m wide, and 1.5 m deep. If it sits upright on a truck with its 1.0-m dimension in the direction of travel, and if the refrigerator cannot slide on the truck, how rapidly can the truck accelerate without tipping the refrigerator over?

37. (III) A 60-kg person stands 2.0 m from the bottom of the stepladder shown in Fig. 12–75. Determine (a) the tension in the horizontal tie rod, which is halfway up the ladder, (b) the normal force the ground exerts on each side of the ladder, and (c) the force (magnitude and direction) that the left side of the ladder exerts on the right side at the hinge on the top. Ignore the mass of the ladder and assume the ground is frictionless. [Hint: Consider free-body diagrams for each section of the ladder.]

FIGURE 12–75
Problem 37.

FIGURE 12–76 Problem 38.

38. (III) A person wants to push a lamp (mass 7.2 kg) across the floor, where the coefficient of friction is 0.20 (Fig. 12–76). Calculate the maximum height above the floor at which the person can push the lamp so that it slides rather than tips.

39. (III) Two guy wires run from the top of a pole 2.6 m tall that supports a volleyball net. The two wires are anchored to the ground 2.0 m apart and each is 2.0 m from the pole (Fig. 12–77). The tension in each wire is 95 N. What is the tension in the net, assumed horizontal and attached at the top of the pole?

FIGURE 12–77 Problem 39.

Section 12–5

40. (I) A nylon tennis string on a racket is under a tension of 250 N. If its diameter is 1.00 mm, by how much is it lengthened from its untensioned length of 30.0 cm?

41. (I) A marble column of cross-sectional area 2.0 m^2 supports a mass of 25,000 kg. (a) What is the stress within the column? (b) What is the strain?

42. (I) By how much is the column in the previous problem shortened if it is 12 m high?

43. (I) A vertical steel girder with a cross-sectional area of 0.15 m^2 has a sign (mass 2000 kg) hanging from its bottom end. (a) What is the stress within the girder? (b) What is the strain on the girder? (c) If the girder is 9.50 m long, how much is it lengthened? (Ignore the mass of the girder itself.)

44. (I) One liter of alcohol (1000 cm^3) in a flexible container is carried to the bottom of the sea, where the pressure is 2.6×10^6 N/m^2. What will be its volume there?

45. (I) A 15-cm-long animal tendon was found to stretch 3.7 mm by a force of 13.4 N. The tendon was approximately round with an average diameter of 8.5 mm. Calculate the elastic modulus of this tendon.

46. (II) If the two wires in Fig. 12–57 (Problem 12) are made of steel wire 1.0 mm in diameter, what is the percentage stretch of each because of the load?

47. (II) How much pressure is needed to compress the volume of an iron block by 0.10 percent? Express answer in N/m^2, and compare it to atmospheric pressure $(1.0 \times 10^5$ N/m$^2)$.

48. (II) At depths of 2000 m in the sea, the pressure is about 200 times atmospheric pressure $(1.0 \times 10^5$ N/m$^2)$. By what percentage does an iron bathysphere's volume change at this depth?

49. (II) Shear forces as shown in Fig. 12–24 are applied to a steel plate that is 0.50 m on a side and 1.1 cm thick. The shear strain produced by these forces is 0.050. Calculate the magnitude of the force.

50. (II) A scallop forces open its shell with an elastic material called abductin, whose elastic modulus is about $2.0 \times 10^6 \, N/m^2$. If this piece of abductin is 3.0 mm thick and has a cross-sectional area of $0.50 \, cm^2$, how much potential energy does it store when compressed 1.0 mm?

51. (III) A pole projects horizontally from the front wall of a shop. A 5.1-kg sign hangs from the pole at a point 2.2 m from the wall (Fig. 12–78). (*a*) What is the torque due to this sign calculated about the point where the pole meets the wall? (*b*) If the pole is not to fall off, there must be another torque exerted to balance it. What exerts this torque? Use a diagram to show how this torque must act. (*c*) Discuss whether compression, tension, and/or shear play a role in part (*b*).

FIGURE 12–78 Problem 51.

Section 12–6

52. (I) The femur bone in the leg has a minimum effective cross section of about $3.0 \, cm^2 \, (= 3.0 \times 10^{-4} \, m^2)$. How much compressive force can it withstand before breaking?

53. (II) What is the maximum tension possible in a 1.00-mm-diameter nylon tennis racket string? If you want tighter strings, what do you do to prevent breakage: go to thinner or thicker strings? Why? What causes strings to break when they are hit by the ball?

54. (II) If a compressive force of $3.6 \times 10^4 \, N$ is exerted on the end of a 20.0-cm-long bone of cross-sectional area $3.6 \, cm^2$, (*a*) will the bone break, and (*b*) if not, by how much does it shorten?

55. (II) (*a*) What is the minimum cross-sectional area required of a vertical steel cable from which is suspended a 320-kg chandelier? Assume a safety factor of 7.0. (*b*) If the cable is 7.5 m long, how much does it elongate?

56. (II) Assume the supports of the cantilever shown in Fig. 12–9 (mass = 2600 kg) are made of wood. Calculate the minimum cross-sectional area required of each, assuming a safety factor of 8.5.

57. (II) An iron bolt is used to connect two iron plates together. The bolt must withstand shear forces up to about 3200 N. Calculate the minimum diameter for the bolt, based on a safety factor of 6.0.

58. (III) A steel cable is to support an elevator whose total (loaded) mass is not to exceed 3100 kg. If the maximum acceleration of the elevator is $1.2 \, m/s^2$, calculate the diameter of cable required. Assume a safety factor of 7.0.

59. (III) Parachutists whose chutes have failed to open have been known to survive if they landed in deep snow. Assume that a 75-kg parachutist hits the ground with an area of impact of $0.30 \, m^2$ at a velocity of $60 \, m/s$, and that the ultimate strength of body tissue is $5 \times 10^5 \, N/m^2$. Assume that the person is brought to rest in 1.0 m of snow. Show that the person can escape injury.

* Section 12–7

***60.** (II) Figure 12–79 shows a simple truss that carries a load at the center (C) of $1.25 \times 10^4 \, N$. (*a*) Calculate the force on each strut at the pins, A, B, C, D, and (*b*) determine which struts (ignore their masses) are under tension and which under compression.

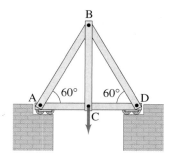

FIGURE 12–79 Problem 60.

***61.** (II) A heavy load *Mg* hangs at point E of the single cantilever truss shown in Fig. 12–80. (*a*) Use a torque equation for the truss as a whole to determine the tension F_T in the support cable, and then determine the force F_A at pin A. (*b*) Determine the force in each member of the truss. Neglect the weight of the trusses, which is small compared to the load.

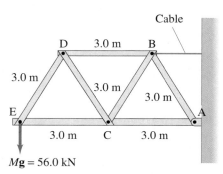

FIGURE 12–80 Problem 61.

***62.** (II) Consider again Example 12–15 but this time assume the roadway is supported uniformly so that $\frac{1}{2}$ its mass *M* acts at the center and $\frac{1}{4}$ *M* at each end support (think of the bridge as two spans, AC and CE, so the center pin supports two spans ends). Calculate the magnitude of the force in each truss member and compare to Example 12–15.

* **63.** (II) (*a*) What minimum cross-sectional area must the trusses have in Example 12–15 if they are of steel (and all the same size for looks), using a safety factor of 6.0? (*b*) If at any time the bridge may carry as many as 50 trucks with an average mass of 1.2×10^4 kg, estimate again the area needed for the truss members.

* **64.** (III) The truss shown in Fig. 12–81 supports a railway bridge. Determine the stress in each strut if a 44 ton $(1 \text{ ton} = 10^3 \text{ kg})$ train locomotive is stopped at the midpoint between the center and one end. Ignore the masses of the rails and truss, and use only $\frac{1}{2}$ the mass of train because there are two trusses. Assume all triangles are equilateral (as in Example 12–15). [*Hint*: See Fig. 12–35.]

FIGURE 12–81 Problem 64.

* **65.** (III) Suppose in Example 12–15, a 28 ton truck $(m = 28 \times 10^3 \text{ kg})$ has its CM located 22 m from the left end of the bridge (point A). Determine the magnitude of the force and type of stress in each strut. [*Hint*: See Fig. 12–35.]

* **66.** (III) For the "Pratt truss" shown in Fig. 12–82, determine the force on each member and whether it is tensile or compressive. Assume the truss is loaded as shown, and give results in terms of *F*. The height is *a* and each of the four lower horizontal spans is *a*.

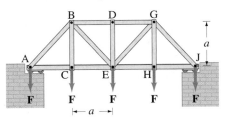

FIGURE 12–82
Problem 66.

* Section 12–8

* **67.** (II) How high must a pointed arch be if it is to span a space 8.0 m wide and exert one third the horizontal force at its base that a round arch would?

* **68.** (II) The subterranean tension ring that exerts the balancing horizontal force on the abutments for the dome in Fig. 12–45 is 36-sided, so each segment makes a 10° angle with the adjacent one (Fig. 12–83). Calculate the tension *F* that must exist in each segment so that the required force of 4.3×10^5 N can be exerted at each corner (Example 12–17).

FIGURE 12–83
Problem 68.

General Problems

69. The mobile in Fig. 12–84 is in equilibrium. The object B has mass of 0.735 kg. Determine the masses of the objects A, C, and D. (Neglect the weights of the crossbars.)

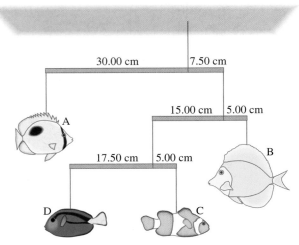

FIGURE 12–84 Problem 69.

70. A tightly stretched "high wire" is 46 m long. It sags 3.4 m when a 60.0-kg tightrope walker stands at its center. What is the tension in the wire? Is it possible to increase the tension in the wire so that there is no sag?

71. What minimum horizontal force *F* is needed to pull a wheel of radius *R* and mass *M* over a step of height *h* as shown in Fig. 12–85 $(R > h)$? (*a*) Assume the force is applied at the top edge as shown. (*b*) Assume the force is applied instead at the wheel's center.

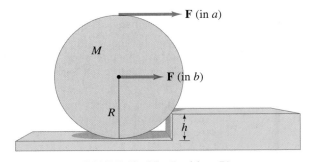

FIGURE 12–85 Problem 71.

72. A 50-story building is being planned. It is to be 200.0 m high with a base 40.0 m by 70.0 m. Its total mass will be about 1.8×10^7 kg and its weight therefore about 1.8×10^8 N. Suppose a 200-km/h wind exerts a force of 950 N/m² over the 70.0-m-wide face (Fig. 12–86). Calculate the torque about the potential pivot point, the rear edge of the building (where $\mathbf{F_E}$ acts in Fig. 12–86), and determine whether the building will topple. Assume the total force of the wind acts at the midpoint of the building's face, and that the building is not anchored in bedrock. [*Hint:* $\mathbf{F_E}$ in Fig. 12–86 represents the force that the Earth exerts on the building in the case where the building is just beginning to tip.]

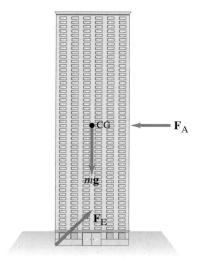

FIGURE 12–86 Force on a building subjected to wind ($\mathbf{F_A}$) and gravity ($m\mathbf{g}$); $\mathbf{F_E}$ is the force on the building due to the Earth in the situation when the building is just about to tip. Problem 72.

73. The center of gravity of a loaded truck depends on how it is packed. If a truck is 4.0 m high and 2.4 m wide, and its CG is 2.2 m above the ground, how steep a slope can the truck be parked on without tipping over sideways (Fig. 12–87)?

FIGURE 12–87 Problem 73.

74. In Example 9–6 in Chapter 9, we calculated the impulse and average force on the leg of a person who jumps 3.0 m down to the ground. If the legs are not bent upon landing, so that the body moves a distance d of only 1.0 cm during collision, determine (*a*) the stress in the tibia bone (area $= 3.0 \times 10^{-4}$ m²), and (*b*) whether or not the bone will break. (*c*) Repeat for a bent-knees landing ($d = 50.0$ cm).

75. The roof of a 9.0 m $\times$ 10.0 m room in a school has a total mass of 12,600 kg. The roof is to be supported by vertical "2 $\times$ 4s" (actually about 4.0 cm $\times$ 9.0 cm) along the 10.0-m sides. How many supports are required on each side and how far apart must they be? Consider only compression and assume a safety factor of 12.

76. In Fig. 12–88, consider the right-hand (northernmost) section of the Golden Gate Bridge, which has a length $d_1 = 343$ m. Assume the CG of this span is halfway between the tower and anchor. Determine F_{T1} and F_{T2} (which act on the northernmost cable) in terms of mg, the weight of the northernmost span, and calculate the tower height h needed for equilibrium. Assume the roadway is supported only by the suspension cables and neglect the mass of the cables. [*Hint:* F_{T3} does not act on this section.]

FIGURE 12–88 Problems 76 and 77.

77. Assume that a single-span suspension bridge such as the Golden Gate Bridge has the configuration indicated in Fig. 12–88. Assume that the roadway is uniform over the length of the bridge and that each segment of the suspension cable provides the sole support for the roadway directly below it. The ends of the cable are anchored to the ground only, not to the roadway. What must the ratio of d_2 to d_1 be so that the suspension cable exerts no net horizontal force on the towers? Neglect the mass of the cables and the fact that the roadway isn't precisely horizontal.

78. A 36-kg round table is supported by three legs placed equal distances apart on the edge. What minimum mass, placed on the table's edge, will cause the table to overturn?

79. The forces acting on a 77,000-kg aircraft flying at constant velocity are shown in Fig. 12–89. The engine thrust, $F_T = 5.0 \times 10^5$ N, acts on a line 1.6 m below the CM. Determine the drag force F_D and the distance above the CM that it acts.

FIGURE 12–89 Problem 79.

80. A uniform flexible steel cable of weight mg is suspended between two equal elevation points as shown in Fig. 12–90, where $\theta = 60°$. Determine the tension in the cable (a) at its lowest point, and (b) at the points of attachment. (c) What is the direction of the tension force in each case?

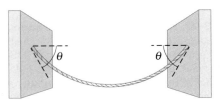

FIGURE 12–90   Problem 80.

81. A 20.0-m-long uniform beam weighing 600 N is supported on walls A and B, as shown in Fig. 12–91. (a) Find the maximum weight a person can be to walk to the extreme end D without tipping the beam. Find the forces that the walls A and B exert on the beam when the person is standing: (b) at D; (c) at a point 2.0 m to the right of B; (d) 2.0 m to the right of A.

FIGURE 12–91   Problem 81.

82. A uniform 7.0-m-long ladder of mass 15.0 kg leans against a smooth wall (so the force exerted by the wall, $\mathbf{F_W}$, is perpendicular to the wall). The ladder makes an angle of 20.0° with the vertical wall (see Fig. 12–72); and the ground is rough. (a) Calculate the components of the force exerted by the ground on the ladder at its base, and (b) determine what the coefficient of static friction at the base of the ladder must be if the ladder is not to slip when a 70.0-kg person stands three-fourths of the way up the ladder.

83. A 23.0 kg backpack is suspended midway between two trees by a light cord as in Fig. 12–69. A bear grabs the backpack and pulls vertically downward with a constant force, so that each section of cord makes an angle of 30° below the horizontal. Initially, without the bear pulling, the angle was 15°; the tension in the cord with the bear pulling is double what it was when he was not. Calculate the force the bear is exerting on the backpack.

84. There is a maximum height of a uniform vertical column made of any material that can support itself without buckling, and it is independent of the cross-sectional area (why?). Calculate this height for (a) steel $(\text{density } 7.8 \times 10^3 \text{ kg/m}^3)$, and (b) granite $(\text{density } 2.7 \times 10^3 \text{ kg/m}^3)$.

85. From what minimum height must a 1.2-kg rectangular brick 15.0 cm × 6.0 cm × 4.0 cm be dropped above a rigid steel floor in order to break the brick? Assume the brick strikes the floor directly on its largest face, and that the compression of the brick is much greater than that of the steel (that is, ignore compression of the steel). State other simplifying assumptions that may be necessary.

86. A cube of side l rests on a rough floor. It is subjected to a steady horizontal pull, F, exerted a distance h above the floor as shown in Fig. 12–92. As F is increased, the block will either begin to slide, or begin to tip over. (a) What must be the coefficient of static faction μ_s so that the block begins to slide rather than tip? (b) What must be the coefficient of static friction so that the block begins to tip? [Hint: Where will the normal force on the block act if it tips?]

FIGURE 12–92   Problem 86.   **FIGURE 12–93**   Problem 87.

87. A 60.0-kg painter is on a scaffold supported from above by ropes (Fig. 12–93). The scaffold has a mass of 25 kg and is uniformly constructed. There is a 4.0-kg pail of paint off to one side, as shown in the figure. Can the painter walk safely to both ends of the scaffold? If not, which end(s) is dangerous and how close to the end can he approach safely?

88. A 95,000-kg train locomotive starts across a 220-m-long bridge at time $t = 0$. The bridge is a uniform beam of mass 23,000 kg and the train travels at 80.0 km/h. What are the magnitudes of the vertical forces, $F_1(t)$ and $F_2(t)$, on the two end supports, written as a function of time during the train's passage?

89. A woman holds a 2.0-m-long uniform 10.0-kg pole as shown in Fig. 12–94. (a) Determine the forces she must exert with each hand (magnitude and direction). To what position should she move her left hand so that neither hand has to exert a force greater than (b) 150 N? (c) 80 N?

FIGURE 12–94
Problem 89.

90. A uniform beam of mass M and length l is mounted on a hinge at a wall as shown in Fig. 12–95. It is held in a horizontal position by a wire making an angle θ as shown. A mass m is placed on the beam a distance x from the wall, and this distance can be varied. Determine, as a function of x, (a) the tension in the wire and (b) the components of the force exerted by the beam on the hinge.

FIGURE 12–95
Problem 90.

91. Two identical, uniform beams are symmetrically set up against each other (Fig. 12–96) on a floor with which they have a coefficient of friction $\mu_s = 0.60$. What is the minimum angle the beams can make with the floor and still not fall?

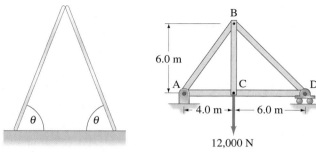

FIGURE 12–96 Problem 91.   **FIGURE 12–97** Problem 92.

*92. Use the method of joints to determine the force in each member of the truss shown in Fig. 12–97. State whether each member is in tension or compression.

93. Four bricks are to be stacked at the edge of a table, each brick overhanging the one below it, so that the top brick extends as far as possible beyond the edge of the table. (*a*) To achieve this, show that successive bricks must extend no more than (starting at the top) 1/2, 1/4, 1/6, and 1/8 of their length beyond the one below (Fig. 12–98a). (*b*) Is the top brick completely beyond the base? (*c*) Determine a general formula for the maximum total distance spanned by *n* bricks if they are to remain stable. (*d*) A builder wants to construct a corbeled arch (Fig. 12–98b) based on the principle of stability discussed in (*a*) and (*c*) above. What minimum number of bricks, each 0.30 m long, is needed if the arch is to span 1.0 m?

(a)   (b)

FIGURE 12–98 Problem 93.

94. One rod of the square frame shown in Fig. 12–99 contains a turnbuckle which, when turned, can put the rod under tension or compression. If the turnbuckle puts rod AB under a compressive force F, determine the forces produced in the other rods. Ignore the mass of the rods and assume the diagonal rods cross each other freely at the center without friction. [*Hint*: Use the symmetry of the situation.]

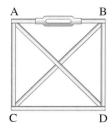

FIGURE 12–99
Problem 94.

95. A 20.0-kg ball is supported from the ceiling by rope A. Rope B pulls downward and to the side on the ball. If the angle of A to the vertical is 20° and if B makes an angle of 50° to the vertical (Fig. 12–100), find the tensions in ropes A and B.

FIGURE 12–100
Problem 95.

FIGURE 12–101 Problem 96.

96. A home mechanic wants to raise the 250-kg engine out of a car. The plan is to stretch a rope vertically from the engine to a branch of a tree 20.0 m above. When the mechanic climbs halfway up the tree and pulls horizontally on the rope at its midpoint, the engine rises out of the car (Fig. 12–101). How much force must the mechanic exert to hold the engine 0.50 m above its normal position?

97. Estimate the magnitude and direction of the force $\mathbf{F}_V$ acting on the fifth lumbar vertebra (exerted by the spine below) in the lower human back. Use the model shown in Fig. 12–102b.

$w_1 = 0.07w$
(head)

$w_2 = 0.12w$
(arms) w = Total weight
                    of person
$w_3 = 0.46w$
(trunk)

(b)

FIGURE 12–102 Problem 97.

Under water, sea creatures and scuba divers experience a buoyant force ($\mathbf{F}_B$) that almost exactly balances their weight $m\mathbf{g}$. The buoyant force is equal to the weight of the volume of fluid displaced (Archimedes' principle) and arises because the pressure increases with depth in the fluid. Sea creatures, and even humans, have a density very close to that of water, so their weight very nearly equals the buoyant force. Actually, humans have a density slightly less than water, so they can float. When fluids flow, interesting effects occur because the pressure in the fluid is lower where the fluid velocity is higher (Bernoulli's principle): airplanes can fly, sailboats can sail against the wind, smoke goes up a chimney, and many others.

CHAPTER 13

Fluids

The three common states, or **phases**, of matter are solid, liquid, and gas. We can distinguish these three phases as follows. A **solid** maintains a fixed shape and a fixed size; even if a large force is applied to a solid, it does not readily change its shape or volume. A **liquid** does not maintain a fixed shape—it takes on the shape of its container—but like a solid it is not readily compressible, and its volume can be changed significantly only by a very large force. A **gas** has neither a fixed shape nor a fixed volume—it will expand to fill its container. For example, when air is pumped into an automobile tire, the air does not all run to the bottom of the tire as a liquid would; it spreads out to fill the whole volume of the tire. Since liquids and gases do not maintain a fixed shape, they both have the ability to flow; they are thus often referred to collectively as **fluids**.

13–1 | Density and Specific Gravity

It is sometimes said that iron is "heavier" than wood. This cannot really be true since a large log clearly weighs more than an iron nail. What we should say is that iron is more *dense* than wood.

The **density**, ρ, of an object (ρ is the lowercase Greek letter "rho") is defined as its mass per unit volume:

Density

$$\rho = \frac{m}{V},$$ (13–1)

where m is the mass of the object and V its volume. Density is a characteristic property of any pure substance. Objects made of a given pure substance, such as pure gold, can have any size or mass, but the density will be the same for each.

(Sometimes we will find Eq. 13–1 useful for writing the mass of an object as $m = \rho V$, and the weight of an object, mg, as $\rho V g$.)

The SI unit for density is kg/m^3. Since $1\,kg/m^3 = 1000\,g/(100\,cm)^3 = 10^{-3}\,g/cm^3$, then a density given in g/cm^3 must be multiplied by 1000 to give the result in kg/m^3. The densities of a variety of substances are given in Table 13–1. The Table specifies temperature and pressure because they affect the density of substances, although the effect is slight for liquids and solids.

EXAMPLE 13–1 **Mass, given volume and density.** What is the mass of a solid iron wrecking ball of radius 18 cm?

SOLUTION The volume of any sphere is $V = \frac{4}{3}\pi r^3$ so we have

$$V = \frac{4}{3}\pi r^3 = \frac{4}{3}(3.14)(0.18\,m)^3 = 0.024\,m^3.$$

From Table 13–1, the density of iron is $\rho = 7800\,kg/m^3$, so we have from Eq. 13–1,

$$m = \rho V = (7800\,kg/m^3)(0.024\,m^3) = 190\,kg.$$

The **specific gravity** of a substance is defined as the ratio of the density of that substance to the density of water at $4.0°C$. Specific gravity (abbreviated SG) is a number, without dimensions or units. Since the density of water is $1.00\,g/cm^3 = 1.00 \times 10^3\,kg/m^3$, the specific gravity of any substance will be equal numerically to its density specified in g/cm^3, or 10^{-3} times its density specified in kg/m^3. For example (see Table 13–1), the specific gravity of lead is 11.3, and that of alcohol is 0.79.

13–2 | Pressure in Fluids

Pressure is defined as force per unit area, where the force F is understood to be acting perpendicular to the surface area A:

$$\text{pressure} = P = \frac{F}{A}. \qquad (13–2)$$

The SI unit of pressure is N/m^2. This unit has the official name **pascal** (Pa), in honor of Blaise Pascal (see Section 13–4); that is, $1\,Pa = 1\,N/m^2$. However, for simplicity, we will often use N/m^2. Other units sometimes used are $dynes/cm^2$, $lb/in.^2$ (sometimes abbreviated "psi"). We will meet several other units shortly, and will discuss conversions between them in Section 13–5.

As an example of calculating pressure, a 60-kg person whose two feet cover an area of $500\,cm^2$ will exert a pressure of

$$F/A = mg/A = (60\,kg)(9.8\,m/s^2)/(0.050\,m^2) = 12 \times 10^3\,N/m^2$$

on the ground. If the person stands on one foot, the force is the same but the area will be half, so the pressure will be twice as much: $24 \times 10^3\,N/m^2$.

The concept of pressure is particularly useful in dealing with fluids. It is an experimental fact that *a fluid exerts a pressure in all directions*. This is well known to swimmers and divers who feel the water pressure on all parts of their bodies. At any point in a fluid at rest, the pressure is the same in all directions. This is illustrated in Fig. 13–1. Consider a tiny cube of the fluid which is so small that we can ignore the force of gravity on it. The pressure on one side of it must equal the pressure on the opposite side. If this weren't true, there would be a net force on the cube and it would start moving. If the fluid is not flowing, then the pressures must be equal.

TABLE 13–1
Densities of Substances†

Substance	Density, $\rho(kg/m^3)$
Solids	
Aluminum	2.70×10^3
Iron and steel	7.8×10^3
Copper	8.9×10^3
Lead	11.3×10^3
Gold	19.3×10^3
Concrete	2.3×10^3
Granite	2.7×10^3
Wood (typical)	$0.3–0.9 \times 10^3$
Glass, common	$2.4–2.8 \times 10^3$
Ice	0.917×10^3
Bone	$1.7–2.0 \times 10^3$
Liquids	
Water (4°C)	1.00×10^3
Sea water	1.025×10^3
Blood, plasma	1.03×10^3
Blood, whole	1.05×10^3
Mercury	13.6×10^3
Alcohol, ethyl	0.79×10^3
Gasoline	0.68×10^3
Gases	
Air	1.29
Helium	0.179
Carbon dioxide	1.98
Water (steam) (100°C)	0.598

†Densities are given at 0°C and 1 atm pressure unless otherwise specified.

FIGURE 13–1 Pressure is the same in every direction in a fluid at a given depth; if it weren't the fluid would start to move.

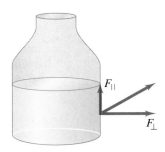

FIGURE 13–2 If there were a component of force parallel to the solid surface, the liquid would move in response to it; for a liquid at rest, $F_\parallel = 0$.

FIGURE 13–3 Calculating the pressure at a depth h in a liquid.

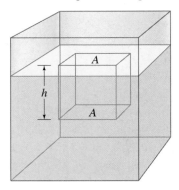

FIGURE 13–4 Forces on a flat, slablike volume of fluid for determining the pressure P at a height y in the fluid.

Another important property of a fluid at rest is that the force due to fluid pressure always acts *perpendicular* to any surface it is in contact with. If there were a component of the force parallel to the surface as shown in Fig. 13–2, then according to Newton's third law, the surface would exert a force back on the fluid that also would have a component parallel to the surface. Such a component would cause the fluid to flow, in contradiction to our assumption that the fluid is at rest. Thus the force due to the pressure is perpendicular to the surface.

Let us now calculate quantitatively how the pressure in a liquid of uniform density varies with depth. Consider a point which is at a depth h below the surface of the liquid (that is, the surface is a height h above this point), as shown in Fig. 13–3. The pressure due to the liquid at this depth h is due to the weight of the column of liquid above it. Thus the force due to the weight of liquid acting on the area A is $F = mg = \rho Ahg$, where Ah is the volume of the column, ρ is the density of the liquid (assumed to be constant), and g is the acceleration of gravity. The pressure P due to the weight of liquid is then

$$P = \frac{F}{A} = \frac{\rho Ahg}{A}$$

$$P = \rho gh. \qquad \text{[liquid]} \quad \textbf{(13–3)}$$

Thus the fluid pressure is directly proportional to the density of the liquid, and to the depth within the liquid. In general, the pressure at equal depths within a uniform liquid is the same.

Equation 13–3 tells us what the pressure is at a depth h in the liquid, due to the liquid itself. But what if there is additional pressure exerted at the surface of the liquid, such as the pressure of the atmosphere or a piston pushing down? And what if the density of the fluid is not constant? Gases are quite compressible and hence their density can vary significantly with depth; and liquids, too, can be compressed, although we can often ignore the variation in density. (One exception is in the depths of the ocean where the great weight of water above significantly compresses the water and increases its density.) To cover these, and other cases, we now treat the general case of determining how the pressure in a fluid varies with depth.

Consider any fluid, and let us determine the pressure at any height y above some reference point (such as the ocean floor or the bottom of a tank or swimming pool), as shown in Fig. 13–4.[†] Within this fluid, at the height y, we consider a tiny, flat, slablike volume of the fluid whose area is A and whose (infinitesimal) thickness is dy, as shown. Let the pressure acting upward on its lower surface (at height y) be P. The pressure acting downward on the top surface of our tiny slab (at height $y + dy$) is designated $P + dP$. The fluid pressure acting on our slab thus exerts a force equal to PA upward on our slab and a force equal to $(P + dP)A$ downward on it. The only other force acting vertically on the slab is the (infinitesimal) force of gravity dF_G, which on our slab of mass dm is

$$dF_G = (dm)g = \rho g\, dV = \rho g A\, dy,$$

where ρ is the density of the fluid at the height y. Since the fluid is assumed to be at rest, our slab is in equilibrium so the net force on it must be zero. Therefore we have

$$PA - (P + dP)A - \rho g A\, dy = 0,$$

which when simplified becomes

$$\frac{dP}{dy} = -\rho g. \qquad \textbf{(13–4)}$$

This relation tells us how the pressure varies with height within the fluid. The minus sign indicates that the pressure decreases with an increase in height; or that the pressure increases with depth (reduced height).

[†] Now we are measuring y positive upwards, the reverse of what we did to get Eq. 13–3 where we measured the depth (i.e. downward as positive).

If the pressure at a height y_1 in the fluid is P_1, and at height y_2 it is P_2, then we can integrate Eq. 13–4 to obtain

$$\int_{P_1}^{P_2} dP = -\int_{y_1}^{y_2} \rho g \, dy$$

$$P_2 - P_1 = -\int_{y_1}^{y_2} \rho g \, dy, \qquad (13\text{–}5)$$

where we assume ρ is a function of height y: $\rho = \rho(y)$. This is a general relation, and we apply it now to two special cases: (1) pressure in liquids of uniform density and (2) pressure variations in the Earth's atmosphere.

For liquids in which any variation in density can be ignored, $\rho = $ constant and Eq. 13–5 is readily integrated:

$$P_2 - P_1 = -\rho g (y_2 - y_1). \qquad (13\text{–}6a)$$

For the everyday situation of a liquid in an open container—such as water in a glass, a swimming pool, a lake, or the ocean—there is a free surface at the top. And it is convenient to measure distances from this top surface. That is, we let h be the *depth* in the liquid where $h = y_2 - y_1$ as shown in Fig. 13–5. If we let y_2 be the position of the top surface, then P_2 represents the atmospheric pressure, P_0, at the top surface. Then, from Eq. 13–6a, the pressure $P(= P_1)$ at a depth h in the fluid is

$$P = P_0 + \rho g h. \qquad [h \text{ is depth in liquid}] \quad (13\text{–}6b)$$

Note that Eq. 13–6b is simply the liquid pressure (Eq. 13–3) plus the pressure P_0 due to the atmosphere above.

FIGURE 13–5 Pressure at a depth $h = (y_2 - y_1)$ in a liquid of density ρ is $P = P_0 + \rho g h$, where P_0 is the external pressure at the liquid's top surface.

EXAMPLE 13–2 **Pressure at a faucet.** The surface of the water in a storage tank is 30 m above a water faucet in the kitchen of a house, Fig. 13–6. Calculate the water pressure at the faucet.

SOLUTION The same atmospheric pressure acts both at the surface of the water in the storage tank, and on the water leaving the faucet. The pressure difference between the inside and outside of the faucet is

$$\Delta P = \rho g h = (1.0 \times 10^3 \, \text{kg/m}^3)(9.8 \, \text{m/s}^2)(30 \, \text{m}) = 2.9 \times 10^5 \, \text{N/m}^2.$$

The height h is sometimes called the **pressure head**. In this Example, the head of the water is 30 m. Note that the very different diameters of the tank and faucet don't affect the result—only pressure does.

➡ **PHYSICS APPLIED**

Water supply

FIGURE 13–6 Example 13–2.

EXAMPLE 13–3 **Force on aquarium window.** Calculate the force due to water pressure exerted on a $1.0 \, \text{m} \times 3.0 \, \text{m}$ aquarium viewing window, Fig. 13–7.

SOLUTION At a depth h, the pressure due to the water is given by Eq. 13–6b. Divide the window up into thin horizontal strips of width $w = 3.0 \, \text{m}$ and thickness dy, as shown in Fig. 13–7. We choose a coordinate system with $y = 0$ at the surface of the water and y is positive downward. (With this choice, the minus sign in Eq. 13–6a becomes plus, or we use Eq. 13–6b with $y = h$.) The force due to water pressure on each strip is $dF = P \, dA = \rho g y w \, dy$. The total force on the window is given by the integral:

$$\int_{y_1=1.0 \, \text{m}}^{y_2=2.0 \, \text{m}} \rho g y w \, dy = \tfrac{1}{2} \rho g w (y_2^2 - y_1^2)$$

$$= \tfrac{1}{2}(1000 \, \text{kg/m}^3)(9.8 \, \text{m/s}^2)(3.0 \, \text{m})[(2.0 \, \text{m})^2 - (1.0 \, \text{m})^2] = 44{,}000 \, \text{N}.$$

FIGURE 13–7 Example 13–3.

Now let us apply Eq. 13–4 or 13–5 to gases. The density of gases is normally quite small, so the difference in pressure at different heights can usually be ignored if $y_2 - y_1$ is not large (which is why, in Example 13–2, we could ignore the difference in air pressure between the faucet and the top of the storage tank). Indeed, for most ordinary containers of gas, we can assume that the pressure is the same throughout. However, if $y_2 - y_1$, is very large, we cannot make this assumption. An interesting example is the Earth's atmosphere, whose pressure at sea level is about $1.013 \times 10^5\,\text{N/m}^2$ and decreases slowly with altitude.

Air pressure variation with altitude

EXAMPLE 13–4 Elevation effect on atmospheric pressure. (*a*) Determine the variation in pressure in the Earth's atmosphere as a function of height y above sea level, assuming g is constant and that the density of the air is proportional to the pressure. (This last assumption is not terribly accurate, in part because temperature and other weather effects are important.) (*b*) At what elevation is the air pressure equal to half the pressure at sea level?

SOLUTION   (*a*) We are assuming that ρ is proportional to P, so we can write

$$\frac{\rho}{\rho_0} = \frac{P}{P_0},$$

where $P_0 = 1.013 \times 10^5\,\text{N/m}^2$ is atmospheric pressure at sea level and $\rho_0 = 1.29\,\text{kg/m}^3$ is the density of air at sea level at $0°\text{C}$ (Table 13–1). From the differential change in pressure with height, Eq. 13–4, we have

$$\frac{dP}{dy} = -\rho g = -P\left(\frac{\rho_0}{P_0}\right)g,$$

so

$$\frac{dP}{P} = -\frac{\rho_0}{P_0}g\,dy.$$

We integrate this from $y = 0$ (Earth's surface) and $P = P_0$, to the height y where the pressure is P:

$$\int_{P_0}^{P}\frac{dP}{P} = -\frac{\rho_0}{P_0}g\int_{0}^{y}dy$$

$$\ln\frac{P}{P_0} = -\frac{\rho_0}{P_0}gy$$

Exponential decrease in pressure with altitude

since $\ln P - \ln P_0 = \ln\left(P/P_0\right)$. Thus

$$P = P_0\,e^{-(\rho_0 g/P_0)y}.$$

So, based on our assumptions, we find that the air pressure in our atmosphere decreases approximately exponentially with height. (Note that the atmosphere does not have a distinct top surface, so there is no natural point from which to measure depth in the atmosphere, as we can do for a liquid.)
(*b*) The constant $\left(\rho_0 g/P_0\right)$ has the value

$$\frac{\rho_0 g}{P_0} = \frac{\left(1.29\,\text{kg/m}^3\right)\left(9.80\,\text{m/s}^2\right)}{\left(1.013 \times 10^5\,\text{N/m}^2\right)}$$

$$= 1.25 \times 10^{-4}\,\text{m}^{-1}.$$

Then, when we set $P = \frac{1}{2}P_0$, we have

$$\tfrac{1}{2} = e^{-(1.25\times 10^{-4}\,\text{m}^{-1})y}$$

or

$$y = (\ln 2.00)/(1.25 \times 10^{-4}\,\text{m}^{-1}) = 5550\,\text{m},$$

where $\ln 2 = 0.693$. Thus, at an elevation of about $5500\,\text{m}$ (about $18{,}000\,\text{ft}$), atmospheric pressure drops to half what it is at sea level. It is not surprising that mountain climbers often use oxygen tanks at very high altitudes.

13–3 | Atmospheric Pressure and Gauge Pressure

The pressure of the Earth's atmosphere varies with altitude, as we have seen. But even at a given altitude, it varies slightly according to the weather. As already mentioned, at sea level the pressure of the atmosphere on the average is $1.013 \times 10^5 \, \text{N/m}^2$ (or $14.7 \, \text{lb/in}^2$). This value is used to define a commonly used unit of pressure, the **atmosphere** (abbreviated atm):

$$1 \, \text{atm} = 1.013 \times 10^5 \, \text{N/m}^2 = 101.3 \, \text{kPa}.$$

One atmosphere (unit)

Another unit of pressure sometimes used (in meteorology and on weather maps) is the **bar**, which is defined as $1 \, \text{bar} = 1.00 \times 10^5 \, \text{N/m}^2$. Thus standard atmospheric pressure is slightly more than 1 bar.

The bar (unit)

The pressure due to the weight of the atmosphere is exerted on all objects immersed in this great sea of air, including our bodies. How does a human body withstand the enormous pressure on its surface? The answer is that living cells maintain an internal pressure that closely equals the external pressure, just as the pressure inside a balloon only slightly exceeds the outside pressure of the atmosphere. An automobile tire, because of its rigidity, can maintain internal pressures much greater than the external pressure.

It is important to note that tire gauges, and most other pressure gauges, register the pressure over and above atmospheric pressure. This is called **gauge pressure**. Thus, to get the absolute pressure, P, one must add the atmospheric pressure, P_A, to the gauge pressure, P_G:

Gauge pressure

$$P = P_A + P_G.$$

Absolute pressure = atmospheric pressure + gauge pressure

If a tire gauge registers 220 kPa, the absolute pressure within the tire is $220 \, \text{kPa} + 101 \, \text{kPa} = 321 \, \text{kPa}$. This is equivalent to about 3.2 atm (2.2 atm gauge pressure).

CONCEPTUAL EXAMPLE 13–5 | **Finger holds water in a straw.** You insert a straw of length L into a tall glass of your favorite beverage. You place your finger over the top of the straw so that no air can get in or out, and then lift the straw from the liquid. You find that the straw retains the liquid such that the distance from the bottom of your finger to the top of the liquid is h. (See Fig. 13–8.) Does the air in the space between your finger and the top of the liquid have a pressure P that is greater than, equal to, or less than, the atmospheric pressure P_A outside the straw?

RESPONSE Consider the forces on the column of liquid. Atmospheric pressure on the outside of the straw pushes upward on the liquid at the bottom of the straw, gravity pulls the liquid downward, and the air pressure inside the top of the straw pushes downward on the liquid. Since the liquid is in equilibrium, the upward force due to atmospheric pressure must balance the two downward forces. The only way this is possible is for the air pressure inside the straw to be rather less than the atmosphere pressure outside the straw.

FIGURE 13–8 Example 13–5.

13–4 | Pascal's Principle

The Earth's atmosphere exerts a pressure on all objects with which it is in contact, including other fluids. External pressure acting on a fluid is transmitted throughout that fluid. For instance, according to Eq. 13–3, the pressure due to the water at a depth of 100 m below the surface of a lake is $P = \rho g h = (1000 \, \text{kg/m}^3)(9.8 \, \text{m/s}^2)(100 \, \text{m}) = 9.8 \times 10^5 \, \text{N/m}^2$, or 9.7 atm. However, the total pressure at this point is due to the pressure of water plus the pressure of the air above it (Eq. 13–6b). Hence the total pressure (if the lake is near sea level) is $9.7 \, \text{atm} + 1.0 \, \text{atm} = 10.7 \, \text{atm}$. This is just one example of a general principle attributed to the French philosopher and scientist Blaise Pascal (1623–1662). **Pascal's principle** states that *pressure applied to a confined fluid increases the pressure throughout by the same amount.*

Pascal's principle

FIGURE 13–9 Application of Pascal's principle: hydraulic lift.

A number of practical devices make use of Pascal's principle. One example is the hydraulic lift, illustrated in Fig. 13–9, in which a small force is used to exert a large force by making the area of the output piston larger than the area of the input piston. To see how this works, we assume the input and output pistons are at the same height (at least approximately). Then the external input force F_{in}, by Pascal's principle, increases the pressure equally throughout so that at the same level (see Fig. 13–9):

$$P_{out} = P_{in}$$

where the input quantities are represented by the subscript "in" and the output by "out." Thus

$$\frac{F_{out}}{A_{out}} = \frac{F_{in}}{A_{in}},$$

or

$$\frac{F_{out}}{F_{in}} = \frac{A_{out}}{A_{in}}.$$

The quantity F_{out}/F_{in} is called the "mechanical advantage" of the hydraulic lift, and is equal to the ratio of the areas. For example, if the area of the output piston is 20 times that of the input cylinder, the force is multiplied by a factor of 20: thus a force of 200 lb could lift a 4000-lb car.

13–5 | Measurement of Pressure; Gauges and the Barometer

Many devices have been invented to measure pressure, some of which are shown in Fig. 13–10. The simplest is the open-tube *manometer* (Fig. 13–10a) which is a U-shaped tube partially filled with a liquid, usually mercury or water. The pressure P being measured is related to the difference in height h of the two levels of the liquid by the relation

$$P = P_0 + \rho g h,$$

where P_0 is atmospheric pressure (acting on the top of the fluid in the left-hand tube), and ρ is the density of the liquid. Note that the quantity $\rho g h$ is the "gauge pressure"—the amount by which P exceeds atmospheric pressure. If the liquid in the left-hand column were lower than that in the right-hand column, this would indicate that P was less than atmospheric pressure (and h would be negative).

FIGURE 13–10 Pressure gauges: (a) open-tube manometer, (b) aneroid gauge, and (c) common tire pressure gauge.

(a) Open-tube manometer

(b) Aneroid gauge (used mainly for air pressure and then called an aneroid barometer)

(c) Tire gauge

Instead of calculating the product $\rho g h$, it is common to simply specify the height h. In fact, pressures are sometimes specified as so many "millimeters of mercury" (mm-Hg), and sometimes as so many "mm of water" (mm-H_2O). The unit mm-Hg is equivalent to a pressure of $133\ N/m^2$, since $\rho g h$ for $1\ mm = 1.0 \times 10^{-3}\ m$ of mercury gives

$$\rho g h = (13.6 \times 10^3\ kg/m^3)(9.80\ m/s^2)(1.00 \times 10^{-3}\ m) = 1.33 \times 10^2\ N/m^2.$$

The unit mm-Hg is also called the **torr** in honor of Evangelista Torricelli (1608–1647), who invented the barometer (see below). Conversion factors among the various units of pressure (an incredible nuisance!) are given in Table 13–2. It is important that only $N/m^2 = Pa$, the proper SI unit, be used in calculations involving other quantities specified in SI units.

The torr (unit)

➥ **PROBLEM SOLVING**

Use SI unit in calculations:
$1\ Pa = 1\ N/m^2$

TABLE 13–2   Conversion Factors Between Different Units of Pressure

In Terms of $1\ Pa = 1\ N/m^2$	Related to 1 atm
$1\ atm = 1.013 \times 10^5\ N/m^2$	$1\ atm = 1.013 \times 10^5\ N/m^2$
$= 1.013 \times 10^5\ Pa = 101.3\ kPa$	
$1\ bar = 1.000 \times 10^5\ N/m^2$	$1\ atm = 1.013\ bar$
$1\ dyne/cm^2 = 0.1\ N/m^2$	$1\ atm = 1.013 \times 10^6\ dyne/cm^2$
$1\ lb/in.^2 = 6.90 \times 10^3\ N/m^2$	$1\ atm = 14.7\ lb/in.^2$
$1\ lb/ft^2 = 47.9\ N/m^2$	$1\ atm = 2.12 \times 10^3\ lb/ft^2$
$1\ cm\text{-}Hg = 1.33 \times 10^3\ N/m^2$	$1\ atm = 76\ cm\text{-}Hg$
$1\ mm\text{-}Hg = 133\ N/m^2$	$1\ atm = 760\ mm\text{-}Hg$
$1\ torr = 133\ N/m^2$	$1\ atm = 760\ torr$
$1\ mm\text{-}H_2O\ (4°C) = 9.81\ N/m^2$	$1\ atm = 1.03 \times 10^4\ mm\text{-}H_2O\ (4°C)$

Another type of pressure gauge is the aneroid gauge (Fig. 13–10b) in which the pointer is linked to the flexible ends of an evacuated thin metal chamber. In an electronic gauge, the pressure may be applied to a thin metal diaphragm whose resulting distortion is translated into an electrical signal by a transducer. How a common tire gauge is constructed is shown in Fig. 13–10c.

Atmospheric pressure is often measured by a modified kind of mercury manometer with one end closed, called a mercury **barometer** (Fig. 13–11). The glass tube is completely filled with mercury and then inverted into the bowl of mercury. If the tube is long enough, the level of the mercury will drop, leaving a vacuum at the top of the tube, since atmospheric pressure can support a column of mercury only about 76 cm high (exactly 76.0 cm at standard atmospheric pressure). That is, a column of mercury 76 cm high exerts the same pressure as the atmosphere:

$$P = \rho g h = (13.6 \times 10^3\ kg/m^3)(9.80\ m/s^2)(0.760\ m) = 1.013 \times 10^5\ N/m^2 = 1.00\ atm.$$

Household barometers are usually of the aneroid type, either mechanical (Fig. 3–10b), or electronic.

A calculation similar to that above will show that atmospheric pressure can maintain a column of water 10.3 m high in a tube whose top is under vacuum (Fig. 13–12). A few centuries ago, it was a source of wonder and frustration that no matter how good a vacuum pump was, it could not lift water more than about 10 m. The only way to pump water out of deep mine shafts, for example, was to use multiple stages for depths greater than 10 m. Galileo studied this problem, and his student Torricelli was the first to explain it. The point is that a pump does not really suck water up a tube—it merely reduces the pressure at the top of the tube. Atmospheric air pressure *pushes* the water up the tube if the top end is at low pressure (under a vacuum), just as it is air pressure that pushes (or maintains) the mercury 76 cm high in a barometer.

FIGURE 13–11   Diagram of a mercury barometer, when the air pressure is 76 cm-Hg.

FIGURE 13–12   A water barometer: water was added at the top, and the spigot at the top then closed; the water level dropped, leaving a vacuum between its upper surface and the spigot. Why? Because air pressure could not support a column of water more than 10 m high.

CONCEPTUAL EXAMPLE 13–6 | **Suction.** You sit in a meeting where a novice NASA engineer proposes suction cup shoes for Space Shuttle astronauts working on the exterior of the spacecraft. Having just studied this chapter, you gently remind him of the fallacy of this plan. What is it?

RESPONSE Suction cups work by pushing out the air underneath the cup. What holds the cup in place is the air pressure outside the cup. (This can be a substantial force on Earth. For example, a 10 cm diameter cup has an area of $7.8 \times 10^{-3} \, \text{m}^2$. The force of the atmosphere on it is $(7.8 \times 10^{-3} \, \text{m}^2)(1.0 \times 10^5 \, \text{N/m}^2) \approx 800 \, \text{N}$, about 180 lbs!). But in outer space, there is no air pressure to hold the suction cup on the spacecraft.

We sometimes mistakenly think of suction as something we actively do. For example, we intuitively think that we pull the soda up through a straw. Instead, all we do is lower the pressure at the top of the straw, and the atmosphere *pushes* the soda up the straw.

13–6 | Buoyancy and Archimedes' Principle

Objects submerged in a fluid appear to weigh less than they do when outside the fluid. For example, a large rock that you would have difficulty lifting off the ground can often be easily lifted from the bottom of a stream. When the rock breaks through the surface of the water, it suddenly seems to be much heavier. Many objects, such as wood, float on the surface of water. These are two examples of *buoyancy*. In each example, the force of gravity is acting downward. But in addition, an upward *buoyant force* is exerted by the liquid.

The buoyant force occurs because the pressure in a fluid increases with depth. Thus the upward pressure on the bottom surface of a submerged object is greater than the downward pressure on its top surface. To see the effect of this, consider a cylinder of height h whose top and bottom ends have an area A and which is completely submerged in a fluid of density ρ_F, as shown in Fig. 13–13. The fluid exerts a pressure $P_1 = \rho_F g h_1$ at the top surface of the cylinder. The force due to this pressure on top of the cylinder is $F_1 = P_1 A = \rho_F g h_1 A$, and it is directed downward. Similarly, the fluid exerts an upward force on the bottom of the cylinder equal to $F_2 = P_2 A = \rho_F g h_2 A$. The net force due to the fluid pressure, which is the **buoyant force**, $\mathbf{F_B}$, acts upward and has the magnitude

$$F_B = F_2 - F_1 = \rho_F g A (h_2 - h_1)$$
$$= \rho_F g A h = \rho_F g V,$$

where $V = Ah$ is the volume of the cylinder. Since ρ_F is the density of the fluid, the product $\rho_F g V = m_F g$ is the weight of fluid which takes up a volume equal to the volume of the cylinder. Thus the buoyant force on the cylinder is equal to the weight of fluid displaced[†] by the cylinder. This result is valid no matter what the shape of the object. Its discovery is credited to Archimedes (287?–212 B.C.), and it is called **Archimedes' principle**: *the buoyant force on a body immersed in a fluid is equal to the weight of the fluid displaced by that object.*

We can derive Archimedes' principle in general by the following simple but elegant argument. The irregularly shaped object D shown in Fig. 13–14a is acted on by the force of gravity (its weight, $m\mathbf{g}$, downward) and the buoyant force, $\mathbf{F_B}$, upward. We wish to determine F_B. To do so, we next consider a body, this time made of our same fluid (D' in Fig. 13–14b) with the same shape and size as the original object, and located at the same depth. You might think of this body of fluid as being separated from the rest of the fluid by an imaginary membrane. The buoyant force F_B on this body of fluid will be exactly the same as that on the

FIGURE 13–13 Determination of the buoyant force.

FIGURE 13–14 Archimedes' principle.

[†] By "fluid displaced," we mean a volume of fluid equal to the volume of the object, or that part of the object submerged if it floats or is only partly submerged (the fluid that used to be where the object is). If the object is placed in a glass or tub initially filled to the brim with water, the water that flows over the top represents the water displaced by the object.

original object since the surrounding fluid, which exerts F_B, is in exactly the same configuration. Now the body of fluid D' is in equilibrium (the fluid as a whole is at rest). Therefore, $F_B = m'g$, where $m'g$ is the weight of the body of fluid. Hence the buoyant force F_B is equal to the weight of the body of fluid whose volume equals the volume of the original submerged object, which is Archimedes' principle.

CONCEPTUAL EXAMPLE 13–7 **Two pails of water.** Consider two identical pails of water filled to the brim. One pail contains only water while the other has a piece of wood floating in it. Which one has the greater weight?

RESPONSE Both pails weigh the same. Recall Archimedes' principle: the wood displaces a volume of water with weight equal to the weight of the wood. Some water will overflow the pail, but the spilled water has weight equal to that of the wood; so the pails have the same weight.

EXAMPLE 13–8 **Recovering a submerged statue.** A 70-kg ancient statue lies at the bottom of the sea. Its volume is $3.0 \times 10^4 \, \text{cm}^3$. How much force is needed to lift it?

SOLUTION The buoyant force on the statue due to the water is equal to the weight of $3.0 \times 10^4 \, \text{cm}^3 = 3.0 \times 10^{-2} \, \text{m}^3$ of water (for seawater $\rho = 1.025 \times 10^3 \, \text{kg/m}^3$):

$$F_B = m_{\text{H}_2\text{O}} \, g = \rho_{\text{H}_2\text{O}} \, gV$$
$$= (1.025 \times 10^3 \, \text{kg/m}^3)(9.8 \, \text{m/s}^2)(3.0 \times 10^{-2} \, \text{m}^3) = 3.0 \times 10^2 \, \text{N}.$$

The weight of the statue is $mg = (70 \, \text{kg})(9.8 \, \text{m/s}^2) = 6.9 \times 10^2 \, \text{N}$. Hence the force needed to lift it is $690 \, \text{N} - 300 \, \text{N} = 390 \, \text{N}$. It is as if the statue had a mass of only $(390 \, \text{N})/(9.8 \, \text{m/s}^2) = 40 \, \text{kg}$.

Archimedes is said to have discovered his principle in his bath while thinking how he might determine whether the king's new crown was pure gold or a fake. Gold has a specific gravity of 19.3, somewhat higher than that of most metals, but a determination of specific gravity or density is not readily done directly because, even if the mass is easily known, the volume of an irregularly shaped object is not easily calculated. However, if the object is weighed in air $(= w)$ and also "weighed" while it is under water $(= w')$, the density can be determined using Archimedes' principle, as the following Example shows. The quantity w' is called the *apparent weight* in water, and is what a scale reads when the object is submerged in water (see Fig. 13–15); w' equals the true weight $(w = mg)$ minus the buoyant force.

EXAMPLE 13–9 **Archimedes: Is the crown gold?** When a crown of mass 14.7 kg is submerged in water, an accurate scale reads only 13.4 kg. Is the crown made of gold?

SOLUTION See analysis in Fig. 13–15. The apparent weight of the submerged object, w' $(= F_T'$ in Fig. 13–15b), equals its actual weight $w(= mg)$ minus the buoyant force F_B as shown:

$$w' = F_T' = w - F_B = \rho_O \, gV - \rho_F \, gV,$$

where V is the volume of the object, ρ_O the object's density, and ρ_F the density of the fluid (water in this case). From this relation, we can see that $F_B = w - w' = \rho_F \, gV$. Then we can write

$$\frac{w}{w - w'} = \frac{\rho_O \, gV}{\rho_F \, gV} = \frac{\rho_O}{\rho_F}.$$

Thus $w/(w - w')$ is equal to the specific gravity of the object if the fluid in which it is submerged is water. For the crown we have

$$\frac{\rho_O}{\rho_{\text{H}_2\text{O}}} = \frac{w}{w - w'} = \frac{(14.7 \, \text{kg})g}{(14.7 \, \text{kg} - 13.4 \, \text{kg})g} = \frac{14.7 \, \text{kg}}{1.3 \, \text{kg}} = 11.3.$$

This corresponds to a density of $11,300 \, \text{kg/m}^3$. The crown seems to be made of lead (see Table 13–1)!

FIGURE 13–15 (a) A scale reads the mass of an object in air—in this case the crown of Example 13–9. All objects are at rest, so the tension F_T in the connecting cord equals the weight w of the object: $F_T = mg$. Note that we show the free-body diagram of the crown, and that F_T is what causes the scale reading (it's equal to the net downward force on the scale). (b) The object submerged has an additional force on it, the buoyant force F_B. The net force is zero, so $F_T' + F_B = mg(= w)$. The scale now reads $m' = 13.4 \, \text{kg}$, where m' is related to the effective weight by $w' = m'g$, where $F_T' = w' = w - F_B$.

FIGURE 13-16 (a) The fully submerged log accelerates upward because $F_B > mg$. It comes to equilibrium (b) when $\Sigma F = 0$, so $F_B = mg = (1200\,\text{kg})g$. Thus 1200 kg, or 1.2 m³, of water is displaced.

(a)

(b)

Floating

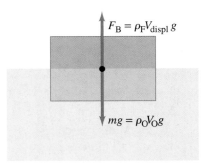

FIGURE 13-17 An object floating in equilibrium: $F_B = mg$.

FIGURE 13-18 A hydrometer. Example 13-10.

Archimedes' principle applies equally well to objects that float, such as wood. In general, *an object floats on a fluid if its density is less than that of the fluid.* This is readily seen from Fig. 13–16a, where a submerged object will experience a net upward force and float to the surface if $F_B > mg$; that is, if $\rho_F Vg > \rho_O Vg$ or $\rho_F > \rho_O$. At equilibrium—that is, when floating—the buoyant force on an object has magnitude equal to the weight of the object. For example, a log whose specific gravity is 0.60 and whose volume is 2.0 m³ has a mass $m = \rho_O V = (0.60 \times 10^3\,\text{kg/m}^3)(2.0\,\text{m}^3) = 1200\,\text{kg}$. If the log is fully submerged, it will displace a mass of water $m_F = \rho_F V = (1000\,\text{kg/m}^3)(2.0\,\text{m}^3) = 2000\,\text{kg}$. Hence the buoyant force on the log will be greater than its weight, and it will float upward to the surface (Fig. 13–16). It will come to equilibrium when it displaces 1200 kg of water, which means that 1.2 m³ of its volume will be submerged. This 1.2 m³ corresponds to 60 percent of the volume of the log (1.2/2.0 = 0.60), so 60 percent of the log is submerged. In general when an object floats, we have $F_B = mg$, which we can write as (see Fig. 13–17)

$$\rho_F V_{\text{displ}}\, g = \rho_O V_O\, g$$

where V_O is the full volume of the object and V_{displ} is the volume of fluid it displaces (= volume submerged). Thus

$$\frac{V_{\text{displ}}}{V_O} = \frac{\rho_O}{\rho_F}.$$

That is, the fraction of the object submerged is given by the ratio of the object's density to that of the fluid.

EXAMPLE 13-10 Hydrometer calibration. A **hydrometer** is a simple instrument used to indicate specific gravity of a liquid by measuring how deeply it sinks in the liquid. A particular hydrometer (Fig. 13–18) consists of a glass tube, weighted at the bottom, which is 25.0 cm long, 2.00 cm² in cross-sectional area, and has a mass of 45.0 g. How far from the end should the 1.000 mark be placed?

SOLUTION The hydrometer has an overall density

$$\rho = \frac{m}{V} = \frac{45.0\,\text{g}}{(2.00\,\text{cm}^2)(25.0\,\text{cm})} = 0.900\,\text{g/cm}^3.$$

Thus, when placed in water, it will come to equilibrium when 0.900 of its volume is submerged. Since it is of uniform cross section, $(0.900)(25.0\,\text{cm}) = 22.5\,\text{cm}$ of its length will be submerged. Since the specific gravity of water is defined to be 1.000, the mark should be placed 22.5 cm from the end.

➡ **PHYSICS APPLIED**

Continental drift—plate tectonics

Archimedes' principle is also useful in geology. According to the theory of plate tectonics and continental drift, the continents can be considered to be floating on a fluid "sea" of slightly deformable rock (mantle rock). Some interesting calculations can be done using very simple models, which we consider in the problems at the end of the chapter.

Weight affected by buoyancy of air

Air is a fluid and it too exerts a buoyant force. Ordinary objects weigh less in air than they do if weighed in a vacuum. Because the density of air is so small, the effect for ordinary solids is slight. There are objects, however, that float in air—helium-filled balloons, for example, because the density of helium is less than the density of air.

EXAMPLE 13–11 **Helium balloon.** What volume V of helium is needed if a balloon is to lift a load of 180 kg (including the weight of the empty balloon)?

SOLUTION The buoyant force on the helium balloon, F_B, which is equal to the weight of displaced air, must be at least equal to the weight of the helium plus the load (Fig. 13–19).

$$F_B = (m_{He} + 180\,kg)g.$$

This equation can be written in terms of density:

$$\rho_{air} Vg = (\rho_{He}V + 180\,kg)g.$$

Solving now for V, we find

$$V = \frac{180\,kg}{\rho_{air} - \rho_{He}} = \frac{180\,kg}{(1.29\,kg/m^3 - 0.18\,kg/m^3)} = 160\,m^3.$$

This is the volume needed near the Earth's surface, where $\rho_{air} = 1.29\,kg/m^3$. To reach a high altitude, a greater volume would be needed since the density of air decreases with altitude.

FIGURE 13–19 Example 13–11.

<table>
<tr><td>13–7</td><td>

Fluids in Motion; Flow Rate and the Equation of Continuity
</td></tr>
</table>

We now turn from the study of fluids at rest to the more complex subject of fluids in motion, which is called **fluid dynamics**, or (especially if the fluid is water) **hydrodynamics**. Many aspects of fluid motion are still being studied today (for example, turbulence as a manifestation of chaos is a "hot" topic today). Nonetheless, with certain simplifying assumptions, a good understanding of this subject can be obtained.

To begin with, we can distinguish two main types of fluid flow. If the flow is smooth, such that neighboring layers of the fluid slide by each other smoothly, the flow is said to be **streamline** or **laminar flow**.[†] In this kind of flow, each particle of the fluid follows a smooth path, called a **streamline**, and these paths do not cross over one another (Fig. 13–20a). Above a certain speed, the flow becomes turbulent. **Turbulent flow** is characterized by erratic, small, whirlpool-like circles called eddy currents or eddies (Fig. 13–20b). Eddies absorb a great deal of energy, and

[†] The word laminar means "in layers."

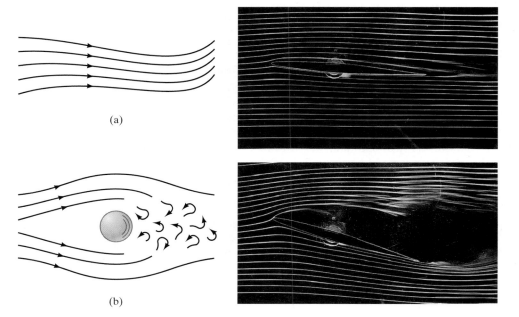

(a)

(b)

FIGURE 13–20 (a) Streamline or laminar flow; (b) turbulent flow.

although a certain amount of internal friction, called **viscosity**, is present even during streamline flow, it is much greater when the flow is turbulent. A few tiny drops of ink or food coloring dropped into a moving liquid can quickly reveal whether the flow is streamline or turbulent.

We will assume in this chapter that the fluid is essentially incompressible (no significant variations in density) and that the flow past any point is steady.

Let us consider the steady laminar flow of a fluid through an enclosed tube or pipe as shown in Fig. 13–21. First we determine how the speed of the fluid changes when the size of the tube changes. The mass **flow rate** is defined as the mass Δm of fluid that passes a given point per unit time Δt: mass flow rate $= \Delta m/\Delta t$. In Fig. 13–21, the volume of fluid passing point 1 (that is, through area A_1) in a time Δt is just $A_1 \Delta l_1$, where Δl_1 is the distance the fluid moves in time Δt. Since the velocity† of fluid passing point 1 is $v_1 = \Delta l_1/\Delta t$, the mass flow rate $\Delta m_1/\Delta t$ through area A_1 is

FIGURE 13–21 Fluid flow through a pipe of varying diameter.

$$\frac{\Delta m_1}{\Delta t} = \frac{\rho_1 \Delta V_1}{\Delta t} = \frac{\rho_1 A_1 \Delta l_1}{\Delta t} = \rho_1 A_1 v_1,$$

where $\Delta V_1 = A_1 \Delta l_1$ is the volume of mass Δm_1, and ρ_1 is the fluid density. Similarly, at point 2 (through area A_2), the flow rate is $\rho_2 A_2 v_2$. Since no fluid flows in or out the sides, the flow rates through A_1 and A_2 must be equal. Thus, since:

$$\frac{\Delta m_1}{\Delta t} = \frac{\Delta m_2}{\Delta t},$$

then

$$\rho_1 A_1 v_1 = \rho_2 A_2 v_2.$$

This is called the **equation of continuity**. If the fluid is incompressible (ρ doesn't change with pressure), which is an excellent approximation for liquids under most circumstances (and sometimes for gases as well), then $\rho_1 = \rho_2$, and the equation of continuity becomes

$$A_1 v_1 = A_2 v_2. \qquad\qquad [\rho = \text{constant}] \quad \textbf{(13–7)}$$

Notice that the product Av represents the *volume rate of flow* (volume of fluid passing a given point per second), since $\Delta V/\Delta t = A\,\Delta l/\Delta t = Av$, which in SI units is m^3/s. Equation 13–7 tells us that where the cross-sectional area is large the velocity is small, and where the area is small the velocity is large. That this is reasonable can be seen by looking at a river. A river flows slowly through a meadow where it is broad, but speeds up to torrential speed when passing through a narrow gorge.

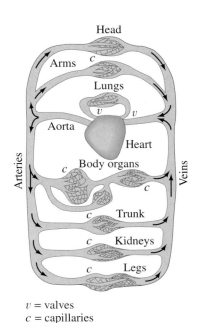

Head
Arms
c
Lungs
v v
Aorta
Heart
Body organs
c
c
c Trunk
c Kidneys
c Legs

Arteries

Veins

v = valves
c = capillaries

FIGURE 13–22
Human circulatory system.

⇒ **PHYSICS APPLIED**

Blood flow

EXAMPLE 13–12 **ESTIMATE** **Blood flow.** In humans, blood flows from the heart into the aorta, from which it passes into the major arteries. These branch into the small arteries (arterioles), which in turn branch into myriads of tiny capillaries, Fig. 13–22. The blood returns to the heart via the veins. The radius of the aorta is about 1.0 cm and the blood passing through it has a speed of about 30 cm/s. A typical capillary has a radius of about 4×10^{-4} cm, and blood flows through it at a speed of about 5×10^{-4} m/s. Estimate how many capillaries there are in the body.

---

† If there were no viscosity, the velocity would be the same across a cross section of the tube. Real fluids have viscosity, and this internal friction causes different layers of the fluid to flow at different speeds. In this case v_1 and v_2 represent the average speeds at each cross section.

SOLUTION Let A_1 be the area of the aorta, and A_2 be the area of *all* the capillaries through which blood flows. Then $A_2 = N\pi r_{cap}^2$ where N is the number of capillaries and $r_{cap} \approx 4 \times 10^{-4}$ cm is the estimated average radius of one capillary. From the equation of continuity (Eq. 13–7), we have

$$v_2 A_2 = v_1 A_1$$

$$v_2 N\pi r_{cap}^2 = v_1 \pi r_{aorta}^2$$

so

$$N = \frac{v_1}{v_2} \frac{r_{aorta}^2}{r_{cap}^2} = \left(\frac{0.30 \text{ m/s}}{5 \times 10^{-4} \text{ m/s}} \right) \left(\frac{1.0 \times 10^{-2} \text{ m}}{4 \times 10^{-6} \text{ m}} \right)^2 \approx 4 \times 10^9,$$

or about 4 billion capillaries.

Another Example that makes use of the equation of continuity and the argument leading up to it is the following.

EXAMPLE 13–13 Heating duct to a room. How large must a heating duct be if air moving 3.0 m/s along it can replenish the air every 15 minutes in a room of 300-m³ volume? Assume the air's density remains constant.

SOLUTION We can apply the equation of continuity (Eq. 13–7) if we consider the room (call it point 2) as a large section of the duct, Fig. 13–23. Reasoning in the same way we did to obtain Eq. 13–7 (changing Δt to t), we see that $A_2 v_2 = A_2 l_2/t = V_2/t$ where V_2 is the volume of the room. Then $A_1 v_1 = A_2 v_2 = V_2/t$ and

$$A_1 = \frac{V_2}{v_1 t} = \frac{300 \text{ m}^3}{(3.0 \text{ m/s})(900 \text{ s})} = 0.11 \text{ m}^2.$$

If the duct is square, then each side has length $l = \sqrt{A} = 0.33$ m or 33 cm. A rectangular duct 20 cm $\times$ 55 cm will also do.

➡ **PHYSICS APPLIED**
Heating duct

FIGURE 13–23 Example 13–13.

13–8 Bernoulli's Equation

Have you ever wondered why smoke goes up a chimney, why a car's convertible top bulges upward at high speeds, or how a sailboat can move against the wind? These are examples of a principle worked out by Daniel Bernoulli (1700–1782) in the early eighteenth century. In essence, **Bernoulli's principle** states that *where the velocity of a fluid is high, the pressure is low, and where the velocity is low, the pressure is high.* For example, if the pressures at points 1 and 2 in Fig. 13–21 are measured, it will be found that the pressure is lower at point 2, where the velocity is greater, than it is at point 1, where the velocity is smaller. At first glance, this might seem strange; you might expect that the greater speed at point 2 would imply a higher pressure. But this cannot be the case. For if the pressure at point 2 were higher than at 1, this higher pressure would slow the fluid down, whereas in fact it has speeded up in going from point 1 to point 2. Thus the pressure at point 2 must be less than at point 1, to be consistent with the fact that the fluid accelerates.

Bernoulli's principle

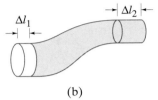

(a)

(b)

FIGURE 13–24 Fluid flow: for derivation of Bernoulli's equation.

Bernoulli developed an equation that expresses this principle quantitatively. To derive Bernoulli's equation, we assume the flow is steady and laminar, the fluid is incompressible, and the viscosity is small enough to be ignored. To be general, we assume the fluid is flowing in a tube of nonuniform cross section that varies in height above some reference level, Fig. 13–24. We will consider the amount of fluid shown in color and calculate the work done to move it from the position shown in (a) to that shown in (b). In this process, fluid at point 1 flows a distance Δl_1 and forces the fluid at point 2 to move a distance Δl_2. The fluid to the left of point 1 exerts a pressure P_1 on our section of fluid and does an amount of work

$$W_1 = F_1 \Delta l_1 = P_1 A_1 \Delta l_1.$$

At point 2, the work done on our section of fluid is

$$W_2 = -P_2 A_2 \Delta l_2;$$

the negative sign is present because the force exerted on the fluid is opposite to the motion (thus the fluid shown in color does work on the fluid to the right of point 2). Work is also done on the fluid by the force of gravity. Since the net effect of the process shown in Fig. 13–24 is to move a mass m of volume $A_1\Delta l_1 (= A_2\Delta l_2$, since the fluid is incompressible) from point 1 to point 2, the work done by gravity is

$$W_3 = -mg(y_2 - y_1),$$

where y_1 and y_2 are heights of the center of the tube above some (arbitrary) reference level. Notice that in the case shown in Fig. 13–24, this term is negative since the motion is uphill against the force of gravity. The net work W done on the fluid is thus:

$$W = W_1 + W_2 + W_3$$

$$W = P_1 A_1 \Delta l_1 - P_2 A_2 \Delta l_2 - mgy_2 + mgy_1.$$

According to the work-energy principle (Section 7–4), the net work done on a system is equal to its change in kinetic energy. Thus

$$\tfrac{1}{2}mv_2^2 - \tfrac{1}{2}mv_1^2 = P_1 A_1 \Delta l_1 - P_2 A_2 \Delta l_2 - mgy_2 + mgy_1.$$

The mass m has volume $A_1\Delta l_1 = A_2\Delta l_2$. Thus we can substitute $m = \rho A_1\Delta l_1 = \rho A_2\Delta l_2$, and also divide through by $A_1 \Delta l_1 = A_2\Delta l_2$, to obtain:

$$\tfrac{1}{2}\rho v_2^2 - \tfrac{1}{2}\rho v_1^2 = P_1 - P_2 - \rho g y_2 + \rho g y_1$$

which we rearrange to get

Bernoulli's equation

$$P_1 + \tfrac{1}{2}\rho v_1^2 + \rho g y_1 = P_2 + \tfrac{1}{2}\rho v_2^2 + \rho g y_2. \qquad \textbf{(13–8)}$$

This is **Bernoulli's equation**. Since points 1 and 2 can be any two points along a tube of flow, Bernoulli's equation can be written:

$$P + \tfrac{1}{2}\rho v^2 + \rho g y = \text{constant}$$

at every point in the fluid, where y is the height of the center of the tube above a fixed reference level. [Note that if there is no flow $(v_1 = v_2 = 0)$, then Eq. 13–8 reduces to the hydrostatic equation, Eq. 13–6a: $P_2 - P_1 = -\rho g(y_2 - y_1)$.]

Bernoulli's equation is an expression of the law of energy conservation, since we derived it from the work-energy principle.

EXAMPLE 13–14 **Flow and pressure in hot-water heating systems.** Water circulates throughout a house in a hot-water heating system. If the water is pumped at a speed of 0.50 m/s through a 4.0-cm-diameter pipe in the basement under a pressure of 3.0 atm, what will be the flow speed and pressure in a 2.6-cm-diameter pipe on the second floor 5.0 m above? Assume the pipes do not divide into branches.

➡ **PHYSICS APPLIED**

Hot-water heating system

SOLUTION We first calculate the flow speed on the second floor, calling it v_2, using the equation of continuity, Eq. 13–7. Noting that the areas are proportional to the radii squared $(A = \pi r^2)$, we call the basement point 1 and obtain

$$v_2 = \frac{v_1 A_1}{A_2} = \frac{v_1 \pi r_1^2}{\pi r_2^2} = (0.50 \text{ m/s}) \frac{(0.020 \text{ m})^2}{(0.013 \text{ m})^2} = 1.2 \text{ m/s}.$$

To find the pressure, we use Bernoulli's equation:

$$
\begin{aligned}
P_2 &= P_1 + \rho g (y_1 - y_2) + \tfrac{1}{2}\rho(v_1^2 - v_2^2) \\
&= (3.0 \times 10^5 \text{ N/m}^2) + (1.0 \times 10^3 \text{ kg/m}^3)(9.8 \text{ m/s}^2)(-5.0 \text{ m}) \\
&\quad + \tfrac{1}{2}(1.0 \times 10^3 \text{ kg/m}^3)\left[(0.50 \text{ m/s})^2 - (1.2 \text{ m/s})^2\right] \\
&= 3.0 \times 10^5 \text{ N/m}^2 - 4.9 \times 10^4 \text{ N/m}^2 - 6.0 \times 10^2 \text{ N/m}^2 \\
&= 2.5 \times 10^5 \text{ N/m}^2,
\end{aligned}
$$

or 2.5 atm. Notice that the velocity term contributes very little in this case.

13–9 | Applications of Bernoulli's Principle: From Torricelli to Sailboats, Airfoils, and TIA

Bernoulli's equation can be applied to many situations. One example is to calculate the velocity, v_1, of a liquid flowing out of a spigot at the bottom of a reservoir, Fig. 13–25. We choose point 2 in Eq. 13–8 to be the top surface of the liquid. Assuming the diameter of the reservoir is large compared to that of the spigot, v_2 will be almost zero. Points 1 (the spigot) and 2 (top surface) are open to the atmosphere so the pressure at both points is equal to atmospheric pressure: $P_1 = P_2$. Then Bernoulli's equation becomes

$$\tfrac{1}{2}\rho v_1^2 + \rho g y_1 = \rho g y_2$$

or

$$v_1 = \sqrt{2g(y_2 - y_1)}. \tag{13–9}$$

FIGURE 13–25 Torricelli's theorem: $v_1 = \sqrt{2g(y_2 - y_1)}$.

This result is called **Torricelli's theorem.** Although it is seen to be a special case of Bernoulli's equation, it was discovered a century before Bernoulli by Evangelista Torricelli, a student of Galileo, hence its name. Equation 13–9 tells us that the liquid leaves the spigot with the same speed that a freely falling object would attain falling the same height. This should not be too surprising since the derivation of Bernoulli's equation relies on the conservation of energy.

Torricelli's theorem

Another special case of Bernoulli's equation arises when a fluid is flowing horizontally with no appreciable change in height; that is, $y_1 = y_2$. Then Eq. 13–8 becomes

$$P_1 + \tfrac{1}{2}\rho v_1^2 = P_2 + \tfrac{1}{2}\rho v_2^2, \tag{13–10}$$

which tells us quantitatively that where the speed is high the pressure is low, and vice versa. It explains many common phenomena, some of which are illustrated in

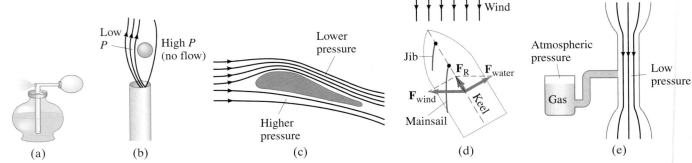

FIGURE 13–26 Examples of Bernoulli's principle: (a) atomizer, (b) Ping-Pong ball in jet of air, (c) airplane wing, (d) sailboat, (e) carburetor barrel.

➥ **PHYSICS APPLIED**

Perfume atomizer

➥ **PHYSICS APPLIED**

Airplanes and dynamic lift

➥ **PHYSICS APPLIED**

Sailing against the wind

FIGURE 13–27 Venturi meter.

➥ **PHYSICS APPLIED**

Smoke up a chimney

Fig. 13–26. The pressure in the air blown at high speed across the top of the vertical tube of a perfume atomizer (Fig. 13–26a) is less than the normal air pressure acting on the surface of the liquid in the bowl. Thus perfume is pushed up the tube because of the reduced pressure at the top. A Ping-Pong ball can be made to float above a blowing jet of air (some vacuum cleaners can blow air), Fig. 13–26b; if the ball begins to leave the jet of air, the higher pressure in the still air outside the jet (Bernoulli's principle) pushes the ball back in.

Airplane wings and other airfoils moving rapidly relative to the air are designed to deflect the air so that, although streamline flow is largely maintained, the streamlines are crowded together above the wing, Fig. 13–26c. Just as the flow lines are crowded together in a pipe constriction where the velocity is high (see Fig. 13–21), so the crowded streamlines above the wing indicate that the air speed is greater above the wing than below it. Hence the air pressure above the wing is less than that below and there is thus a net upward force, called **dynamic lift**. Bernoulli's principle is only one aspect of the lift on a wing. Wings are usually tilted slightly upward so that air striking the bottom surface is deflected downward; the change in momentum of the rebounding air molecules results in an additional upward force on the wing. Turbulence also plays an important role.

A sailboat can move against the wind, Fig. 13–26d and the Bernoulli effect aids in this considerably if the sails are arranged so that the air velocity increases in the narrow constriction between the two sails. The normal atmospheric pressure behind the mainsail is larger than the reduced pressure in front of it (due to the fast moving air in the narrow slot between the sails), and this pushes the boat forward. When going against the wind, the mainsail is set at an angle, as shown in Fig. 13–26d, such that the net force on the sail (wind and Bernoulli) acts nearly perpendicular to the sail ($\mathbf{F}_{wind}$). This would tend to make the boat move sideways if it weren't for the keel that extends vertically downward beneath the water—for the water exerts a force ($\mathbf{F}_{water}$) on the keel nearly perpendicular to the keel. The resultant of these two forces ($\mathbf{F}_R$) is almost directly forward as shown.

A **venturi tube** is essentially a pipe with a narrow constriction (the throat). One example of a venturi tube is the barrel of a carburetor in a car, Fig. 13–26e. The flowing air speeds up as it passes through this constriction, and so the pressure is lower. Because of the reduced pressure, gasoline under atmospheric pressure in the carburetor reservoir is forced into the air stream in the throat and mixes with the air before entering the cylinders.

The venturi tube is also the basis of the *venturi meter*, which is used to measure the flow speed of fluids, Fig. 13–27. Venturi meters can be used to measure the flow velocities of gases and liquids, including blood velocity in arteries.

Why does smoke go up a chimney? It's partly because hot air rises (it's less dense and therefore buoyant). But Bernoulli's principle also plays a role. When wind blows across the top of a chimney, the pressure is less there than inside the house. Hence, air and smoke are pushed up the chimney. Even on an apparently still night there is usually enough ambient air flow at the top of a chimney to assist upward flow of smoke.

In medicine, one of many applications of Bernoulli's principle is to explain a TIA, a transient ischemic attack (meaning a temporary lack of blood supply to the brain), caused by the so-called "subclavian steal syndrome." A person suffering a TIA may experience symptoms such as dizziness, double vision, headache, and weakness of the limbs. A TIA can occur as follows. Blood normally flows up to the brain at the back of the head via the two vertebral arteries—one going up each side of the neck—which meet to form the basilar artery just below the brain, as shown in Fig. 13–28. The vertebral arteries issue from the subclavian arteries, as shown, before the latter pass to the arms. When an arm is exercised vigorously, blood flow increases to meet the needs of the arm's muscles. If the subclavian artery on one side of the body is partially blocked, however, say by arteriosclerosis, the blood velocity will have to be higher on that side to supply the needed blood. (Recall the equation of continuity: smaller area means larger velocity for the same flow rate, Eq. 13–7.) The increased blood velocity past the opening to the vertebral artery results in lower pressure (Bernoulli's principle). Thus, blood rising in the vertebral artery on the "good" side at normal pressure can be *diverted down* into the other vertebral artery because of the low pressure on that side (like the Venturi effect), instead of passing upward into the basilar artery and the brain. Hence the blood supply to the brain can be reduced due to "subclavian steal syndrome": the fast-moving blood in the subclavian artery "steals" the blood away from the brain. The resulting dizziness or weakness usually causes the person to stop the exertions, followed by a return to normal.

The sink where you brush your teeth contains a *trap*—a U-shaped pipe in the drain designed to retain water as a barrier against disagreeable odors (Fig. 13–29a). The air pressure on the two sides of the trap is the same, because of the *main vent*, which usually opens through the roof. However, the water in the trap could be pushed out of the trap, via the Bernoulli effect, if flush water from above passes at high speed—and low pressure—as shown in Fig. 13–29b. Building codes specify that drains of this type must have a separate vent (Fig. 13–29c) that maintains the atmospheric pressure on both sides of the trap, even when flush water passes through the system.

Bernoulli's equation ignores the effects of friction (viscosity) and the compressibility of the fluid. The energy that is transformed to internal (or potential) energy due to compression and to thermal energy by friction can be taken into account by adding terms to the right side of Eq. 13–8. These terms are difficult to calculate theoretically and are normally determined empirically. They do not significantly alter the explanations for the phenomena described above.

➡ PHYSICS APPLIED

Medicine—TIA

FIGURE 13–28 Rear of the head and shoulders showing arteries leading to the brain and to the arms. High blood velocity past the constriction in the left subclavian artery causes low pressure in the left vertebral artery, in which a reverse (downward) blood flow can then result: so-called "subclavian steal syndrome," resulting in a TIA (see text).

➡ PHYSICS APPLIED

Sink trap

FIGURE 13–29 Making sure a trap works.

As already mentioned, real fluids have a certain amount of internal friction which is called **viscosity**. Viscosity exists in both liquids and gases, and is essentially a frictional force between adjacent layers of fluid as the layers move past one another. In liquids, viscosity is due to the cohesive forces between the molecules. In gases, it arises from collisions between the molecules.

Different fluids possess different amounts of viscosity: syrup is more viscous than water; grease is more viscous than engine oil; liquids in general are much more viscous than gases. The viscosity of different fluids can be expressed quantitatively by a *coefficient of viscosity*, η (the Greek lowercase letter eta), which is defined in the following way. A thin layer of fluid is placed between two flat plates. One plate is stationary and the other is made to move, Fig. 13–30. The fluid directly in contact with each plate is held to the surface by the adhesive force between the molecules of the liquid and those of the plate. Thus the upper surface of the fluid moves with the same speed v as the upper plate, whereas the fluid in contact with the stationary plate remains stationary. The stationary layer of fluid retards the flow of the layer just above it, which in turn retards the flow of the next layer, and so on. Thus the velocity varies continuously from 0 to v, as shown. The increase in velocity divided by the distance over which this change is made—equal to v/l—is called the *velocity gradient*. To move the upper plate requires a force, which you can verify by moving a flat plate across a puddle of syrup on a table. For a given fluid, it is found that the force required, F, is proportional to the area of fluid in contact with each plate, A, and to the speed, v, and is inversely proportional to the separation, l, of the plates: $F \propto vA/l$. For different fluids, the more viscous the fluid, the greater is the required force. Hence the proportionality constant for this equation is defined as the *coefficient of viscosity*, η:

FIGURE 13–30
Determination of viscosity.

$$F = \eta A \frac{v}{l}. \qquad \text{(13–11)}$$

Solving for η, we find $\eta = Fl/vA$. The SI unit for η is $\text{N} \cdot \text{s/m}^2 = \text{Pa} \cdot \text{s}$ (pascal·second). In the cgs system, the unit is $\text{dyne} \cdot \text{s/cm}^2$ and this unit is called a *poise* (P). Viscosities are often given in centipoise $(1\ \text{cP} = 10^{-2}\ \text{P})$. Table 13–3 lists the coefficient of viscosity for various fluids. The temperature is also specified, since it has a strong effect; the viscosity of liquids such as motor oil, for example, decreases rapidly as temperature increases.[†]

[†] The Society of Automotive Engineers assigns numbers to represent the viscosity of oils: 30-weight (SAE 30) is more viscous than 10-weight. Multigrade oils, such as 20–50, are designed to maintain viscosity as temperature increases; 20–50 means the oil is 20-wt when cool but is like a 50-wt pure oil when it is hot (engine running temperature).

TABLE 13–3   Coefficient of Viscosity for Various Fluids

Fluid	Temperature (°C)	Coefficient of Viscosity, $\eta(\text{Pa} \cdot \text{s})$[†]
Water	0	1.8×10^{-3}
	20	1.0×10^{-3}
	100	0.3×10^{-3}
Whole blood	37	$\approx 4 \times 10^{-3}$
Blood plasma	37	$\approx 1.5 \times 10^{-3}$
Ethyl alcohol	20	1.2×10^{-3}
Engine oil (SAE 10)	30	200×10^{-3}
Glycerine	20	1500×10^{-3}
Air	20	0.018×10^{-3}
Hydrogen	0	0.009×10^{-3}
Water vapor	100	0.013×10^{-3}

[†] $1\ \text{Pa} \cdot \text{s} = 10\ \text{P} = 1000\ \text{cP}$

*13–11 | Flow in Tubes: Poiseuille's Equation

If a fluid had no viscosity, it could flow through a level tube or pipe without a force being applied. Because of viscosity, a pressure difference between the ends of a tube is necessary for the steady flow of any real fluid, be it water or oil in a pipe, or blood in the circulatory system of a human, even when the tube is level.

The rate of flow of a fluid in a round tube depends on the viscosity of the fluid, the pressure difference, and the dimensions of the tube. The French scientist J. L. Poiseuille (1799–1869), who was interested in the physics of blood circulation (and after whom the "poise" is named), determined how the variables affect the flow rate of an incompressible fluid undergoing laminar flow in a cylindrical tube. His result, known as *Poiseuille's equation*, is:

$$Q = \frac{\pi R^4 (P_1 - P_2)}{8\eta L},$$ (13–12)

Poiseuille's equation for flow rate in a tube

where R is the inside radius of the tube, L is its length, $P_1 - P_2$ is the pressure difference between the ends, η is the coefficient of viscosity, and Q is the volume rate of flow (volume of fluid flowing past a given point per unit time which in SI has units of m^3/s). Equation 13–12 applies only to laminar flow.

Poiseuille's equation tells us that the flow rate Q is directly proportional to the "pressure gradient," $(P_1 - P_2)/L$, and it is inversely proportional to the viscosity of the fluid. This is just what we might expect. It may be surprising, however, that Q also depends on the *fourth* power of the tube's radius. This means that for the same pressure gradient, if the tube radius is halved, the flow rate is decreased by a factor of 16! Thus the rate of flow, or alternately the pressure required to maintain a given flow rate, is greatly affected by only a small change in tube radius. Applying this to blood flow—although only approximately because of the presence of corpuscles and turbulence—we can see how reduction of artery radius by buildup of cholesterol and/or arteriosclerosis requires the heart to work much harder to maintain proper flow rates.

*13–12 | Surface Tension and Capillarity

Up to now in this chapter, we have mainly been studying what happens to fluids as a whole. But the *surface* of a liquid at rest also behaves in an interesting way. A number of common observations suggest that the surface of a liquid acts like a stretched membrane under tension. For example, a drop of water on the end of a dripping faucet, or hanging from a thin branch in the early morning dew (Fig. 13–31), forms into a nearly spherical shape as if it were a tiny balloon filled with water. A steel needle can be made to float on the surface of water even though it is denser than the water. The surface of a liquid acts as if it is under tension, and this tension, acting parallel to the surface, arises from the attractive forces between the molecules. This effect is called *surface tension*. More specifically, a quantity called the *surface tension*, γ (the Greek letter gamma), is defined as the force F per unit length L that acts across any line in a surface, tending to pull the surface closed:

$$\gamma = \frac{F}{L}.$$ (13–13)

To understand this, consider the U-shaped apparatus shown in Fig. 13–32, which encloses a thin film of liquid. Because of surface tension, a force F is required to pull the movable wire and thus increase the surface area of the liquid. The liquid contained by the wire apparatus is a thin film having both a top and a bottom surface. Hence the length of the surface being increased is $2l$, and the surface tension is $\gamma = F/2l$. A delicate apparatus of this type can be used to measure the surface tension of various liquids. The surface tension of water is 0.072 N/m at 20°C. Table 13–4 gives the values for other liquids. Note that temperature has a considerable effect on the surface tension.

FIGURE 13–31 Spherical water droplets, dew on a grass blade.

FIGURE 13–32 U-shaped wire apparatus holding a film of liquid to measure surface tension ($\gamma = F/2l$).

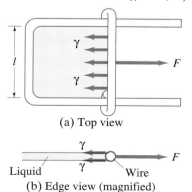

(a) Top view

(b) Edge view (magnified)

TABLE 13–4 Surface Tension of Some Substances

Substance	Surface Tension (N/m)
Mercury (20°C)	0.44
Blood, whole (37°C)	0.058
Blood, plasma (37°C)	0.073
Alcohol, ethyl (20°C)	0.023
Water (0°C)	0.076
(20°C)	0.072
(100°C)	0.059
Benzene (20°C)	0.029
Soap solution (20°C)	≈ 0.025
Oxygen (−193°C)	0.016

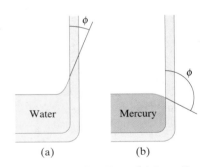

FIGURE 13–33 A water strider.

Because of surface tension, insects (Fig. 13–33) can walk on water; and objects more dense than water, such as a steel needle, can actually float on the surface. Figure 13–34a shows how the surface tension can support the weight w of an object. Actually, the object sinks slightly into the fluid, so w is the "effective weight" of that object—its true weight less the buoyant force. If the object is spherical in shape, the surface tension acts at all points around a horizontal circle of approximately radius r (Fig. 13–34a). Only the vertical component, $\gamma \cos \theta$, acts to balance w. We set the length L equal to the circumference of the circle, $L \approx 2\pi r$, so the net upward force due to surface tension is $F \approx (\gamma \cos \theta)L \approx 2\pi r \gamma \cos \theta$.

EXAMPLE 13–15 **ESTIMATE** **Insect walks on water.** The base of an insect's leg is approximately spherical in shape, with a radius of about 2.0×10^{-5} m. The 0.0030-g mass of the insect is supported equally by the six legs. Estimate the angle θ (see Fig. 13–34) for an insect on the surface of water. Assume the water temperature is 20°C.

SOLUTION Since the insect is in equilibrium, the upward surface tension force is equal to the effective pull of gravity downward on each leg:

$$2\pi r \gamma \cos \theta \approx w,$$

where w is one-sixth the weight of the insect (since it has six legs). Then

$$(6.28)(2.0 \times 10^{-5}\,\text{m})(0.072\,\text{N/m}) \cos \theta \approx \tfrac{1}{6}(3.0 \times 10^{-6}\,\text{kg})(9.8\,\text{m/s}^2)$$

$$\cos \theta \approx \frac{0.49}{0.90} = 0.54.$$

So $\theta \approx 57°$. Notice that if $\cos \theta$ were greater than 1, this would indicate that the surface tension would not be great enough to support the weight.

FIGURE 13–34 Surface tension acting on (a) a sphere, and (b) an insect leg. Example 13–15.

(a)   (b)

Water (a)

Mercury (b)

FIGURE 13–35 Water (a) "wets" the surface of glass, whereas (b) mercury does not "wet" the glass.

FIGURE 13–36 Capillarity.

(a) Glass tube in water

(b) Glass tube in mercury

Soaps and detergents have the effect of lowering the surface tension of water. This is desirable for washing and cleaning since the high surface tension of pure water prevents it from penetrating easily between the fibers of material and into tiny crevices. Substances that reduce the surface tension of a liquid are called *surfactants*.

Surface tension plays a role in another interesting phenomenon, capillarity. It is a common observation that water in a glass container rises up slightly where it touches the glass, Fig. 13–35a. The water is said to "wet" the glass. Mercury, on the other hand, is depressed when it touches the glass, Fig. 13–35b; the mercury does not wet the glass. Whether or not a liquid wets a solid surface is determined by the relative strength of the cohesive forces between the molecules of the liquid compared to the adhesive forces between the molecules of the liquid and those of the container. (*Cohesion* refers to the force between molecules of the same type and *adhesion* to the force between molecules of different types.) Water wets glass because the water molecules are more strongly attracted to the glass molecules than they are to other water molecules. The opposite is true for mercury: the cohesive forces are stronger than the adhesive forces.

In tubes having very small diameters, liquids are observed to rise or fall relative to the level of the surrounding liquid. This phenomenon is called **capillarity**, and such thin tubes are called **capillaries**. Whether the liquid rises or falls (Fig. 13–36) depends on the relative strengths of the adhesive and cohesive forces. Thus water rises in a glass tube whereas mercury falls. The actual amount of rise (or fall) depends on the surface tension—which is what keeps the liquid surface from breaking apart.

Pumps, and the Heart

We conclude this chapter with a brief discussion of pumps of various types, including the heart. Pumps can be classified into categories according to their function. A *vacuum pump* is designed to reduce the pressure (usually of air) in a given vessel. A *force pump*, on the other hand, is a pump that is intended to increase the pressure—for example, to lift a liquid (such as water from a well) or to push a fluid through a pipe. Figure 13–37 illustrates the principle behind a simple reciprocating pump. It could be a vacuum pump, in which case the intake is connected to the vessel to be evacuated. A similar mechanism is used in some force pumps, and in this case the fluid is forced under increased pressure through the outlet. Other kinds of pumps are illustrated in Fig. 13–38. The centrifugal pump, or any force pump, can be used as a *circulating pump*—that is, to circulate a fluid around a closed path, such as the cooling water or lubricating oil in an automobile.

The heart of a human (and of other animals as well) is essentially a circulating pump. The action of a human heart is shown in Fig. 13–39. There are actually two separate paths for blood flow. The longer path takes blood to the parts of the body, via the arteries, bringing oxygen to body tissues and picking up carbon dioxide which it carries back to the heart via veins. This blood is then pumped into the lungs (the second path), where the carbon dioxide is released and oxygen is taken up. The oxygen-laden blood is returned to the heart, where it is again pumped to the tissues of the body.

→ PHYSICS APPLIED

Pumps and the heart

FIGURE 13–37 An example of one kind of pump. As the diagram indicates, the intake valve opens and air (or fluid that is being pumped) fills the empty space when the piston moves to the left. When the piston moves to the right (not shown), the outlet valve opens and fluid is forced out.

FIGURE 13–38 (a) Centrifugal pump: the rotating blades force fluid through the outlet pipe; this kind of pump is used in vacuum cleaners and as a water pump in automobiles. (b) Rotary oil-seal pump, used to obtain vacuums as low as 10^{-4} mm-Hg: gas (usually air) from the vessel to be evacuated diffuses into the space G via the intake pipe I; the rotating off-center cylinder C traps the gas in G and pushes it out the exhaust valve E, in the meantime allowing more gas to diffuse into G for the next cycle. The sliding valve V is kept in contact with C by a spring S, and this prevents the exhaust gas from returning to G. (c) Diffusion pump, used to obtain vacuums as low as 10^{-8} mm-Hg: air molecules from the vessel to be evacuated diffuse into the jet, where the rapidly moving jet of oil sweeps the molecules away. A "forepump" is needed, which is a mechanical pump, such as the rotary type (b), and acts as a first stage in reducing the pressure.

FIGURE 13–39 (a) In the diastole phase, the heart relaxes between beats. Blood moves into the heart; both atria are filled rapidly. (b) When the atria contract, the systole, or pumping, phase begins. The contraction pushes the blood through the mitral and tricuspid valves into the ventricles. (c) The contraction of the ventricles forces the blood through the semilunar valves into the pulmonary artery which leads to the lungs, and to the aorta (the body's largest artery) which leads to the arteries serving all the body. (d) When the heart relaxes, the semilunar valves close; blood fills the atria, beginning the cycle again.

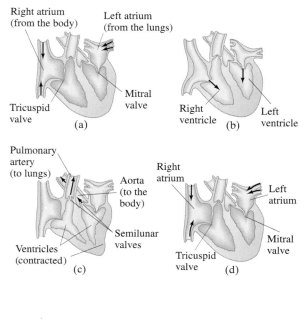

Summary

The three common phases of matter are **solid, liquid**, and **gas**. Liquids and gases are collectively called **fluids**, meaning they have the ability to flow. The **density** of a material is defined as its mass per unit volume. **Specific gravity** is the ratio of the density of the material to the density of water (at 4°C).

Pressure is defined as force per unit area. The pressure at a depth h in a liquid is given by

$$P = \rho g h,$$

where ρ is the density of the liquid and g is the acceleration due to gravity. If the density of a fluid is not uniform, the pressure P varies with height y as

$$\frac{dP}{dy} = -\rho g.$$

Pascal's principle says that an external pressure applied to a confined fluid is transmitted throughout the fluid.

Pressure is measured using a manometer or other type of gauge. A **barometer** is used to measure atmospheric pressure. Standard atmospheric pressure (average at sea level) is $1.013 \times 10^5 \text{ N/m}^2$. **Gauge pressure** is the total pressure less atmospheric pressure.

Archimedes' principle states that an object submerged wholly or partially in a fluid is buoyed up by a force equal to the weight of fluid it displaces.

Fluid flow can be characterized either as **streamline** (sometimes called **laminar**), in which the layers of fluid move smoothly and regularly along paths called streamlines, or as **turbulent**, in which case the flow is not smooth and regular but is characterized by irregularly shaped whirlpools.

Fluid flow rate is the mass or volume of fluid that passes a given point per unit time. The **equation of continuity** states that for an incompressible fluid flowing in an enclosed tube, the product of the velocity of flow and the cross-sectional area of the tube remains constant:

$$Av = \text{constant}.$$

Bernoulli's principle tells us that where the velocity of a fluid is high, the pressure in it is low, and where the velocity is low, the pressure is high. **Bernoulli's equation** for steady laminar flow of an incompressible and nonviscous fluid is

$$P_1 + \tfrac{1}{2}\rho v_1^2 + \rho g y_1 = P_2 + \tfrac{1}{2}\rho v_2^2 + \rho g y_2.$$

for two points along the flow.

Viscosity refers to friction within a fluid that prevents the fluid from flowing freely and is essentially a frictional force between adjacent layers of fluid as they move past one another.

Questions

1. If one material has a higher density than another, does this mean the molecules of the first must be heavier than those of the second? Explain.

2. Airplane travelers often note that their cosmetics bottles and other containers have leaked after a trip. What might cause this?

3. The three containers in Fig. 13–40 are filled with water to the same height and have the same surface area at the base; hence the water pressure, and the total force on the base of each, is the same. Yet the total weight of water is different for each. Explain this "hydrostatic paradox."

FIGURE 13–40 Question 3.

4. Consider what happens when you push both a pin and the blunt end of a pen against your skin with the same force. Decide what determines whether your skin suffers a cut—the net force applied to it or the pressure.

5. It is often said that water seeks its own level. Explain.

6. A small amount of water is boiled in a one-gallon gasoline can. The can is removed from the heat and the lid put on. Shortly thereafter the can collapses. Explain.

7. Explain how the tube in Fig. 13–41, known as a **siphon**, can transfer liquid from one container to a lower one even though the liquid must flow uphill for part of its journey. (Note that the tube must be filled with liquid to start with.)

FIGURE 13–41
A siphon. Question 7.

8. An ice cube floats in a glass of water filled to the brim. What can you say about the density of ice? As the ice melts, will the glass overflow?

9. Will an ice cube float in a glass of alcohol? Why or why not?

10. A barge filled high with sand approaches a low bridge over the river and cannot quite pass under it. Should sand be added to, or removed from, the barge?

11. Will an empty balloon have precisely the same apparent weight on a scale as one that is filled with air? Explain.

12. Does the buoyant force on a diving bell deep beneath the ocean have precisely the same value as when the bell is just beneath the surface? Explain.

13. A small wooden boat floats in a swimming pool, and the level of the water at the edge of the pool is marked. Consider the following situations and determine whether the level of the water will rise, fall, or stay the same. (*a*) The boat is removed from the water. (*b*) The boat in the water holds an iron anchor which is removed from the boat and placed on the shore. (*c*) The iron anchor is removed from the boat and dropped in the pool.

14. Explain why helium weather balloons, which are used to measure atmospheric conditions at high altitude, are normally released while filled to only 10%–20% of their maximum volume.

15. Why do you float more easily in salt water than in fresh?

16. Roofs of houses are sometimes "blown" off (or are they pushed off?) during a tornado or hurricane. Explain, using Bernoulli's principle.

17. If you dangle two pieces of paper vertically, a few inches apart (Fig. 13–42), and blow between them, how do you think the papers will move? Try it and see. Explain.

FIGURE 13–42 Question 17.

18. Why does the canvas top of a convertible bulge out when the car is traveling at high speed?

19. Children are told to avoid standing too close to a rapidly moving train because they might get sucked under it. Is this possible? Explain.

20. Why does a sailboat need a keel? In a small sailboat, the keel (a vertical "board" that extends below the boat into the water) is removed when the boat is anchored. Why?

21. A tall Styrofoam cup is filled with water. Two holes are punched in the cup near the bottom, and water begins rushing out. If the cup is dropped so it falls freely, will the water continue to flow from the holes? Explain.

22. With a little effort, you can blow across a dime on a table and make it land in a cup as shown in Fig. 13–43 without touching either cup or dime. Explain (and try it).

FIGURE 13–43 Question 22.

23. Why do airplanes normally take off into the wind?

24. Why does the stream of water from a faucet become narrower as it falls (Fig. 13–44)?

FIGURE 13–44 Water coming from a faucet. Question 24 and Problem 93.

25. A baseball pitcher puts spin on the ball when throwing a curve. Use Bernoulli's principle to explain in detail why the ball curves. Explain why a spinning ball with a very smooth surface curves in the opposite direction to one with a rough surface (such as a baseball or tennis ball). [This is a challenging question. See "The Physics of Baseball," *Physics Today*, May 1995, p. 29.]

26. Two ships moving in parallel paths close to one another risk colliding. Why?

Problems

Section 13–1

1. (I) The approximate volume of the granite monolith known as El Capitan in Yosemite National Park (Fig. 13–45) is about 10^8 m^3. What is its approximate mass?

FIGURE 13–45
Problem 1.

2. (I) What is the approximate mass of air in a living room 4.8 m × 3.8 m × 2.8 m?

3. (I) If you tried to nonchalantly smuggle gold bricks by filling your backpack, whose dimensions are 60 cm × 25 cm × 15 cm, what would its mass be?

4. (I) Estimate your volume. [*Hint:* Because you can swim on or just under the surface of the water in a swimming pool, you have a pretty good idea of your density.]

5. (II) A bottle has a mass of 35.00 g when empty and 98.44 g when filled with water. When filled with another fluid, the mass is 88.78 g. What is the specific gravity of this other fluid?

6. (II) If 5.0 L of antifreeze solution (specific gravity = 0.80) is added to 4.0 L of water to make a 9.0-L mixture what is the specific gravity of the mixture?

Sections 13–2 to 13–5

7. (I) Estimate the pressure exerted on a floor by (a) a pointed loudspeaker leg (60 kg on four legs) of area = 0.05 cm^2, and compare it (b) to the pressure exerted by a 1500-kg elephant standing on one foot (area = 800 cm^2).

8. (I) (a) Calculate the total force of the atmosphere acting on the top of a table that measures 1.6 m × 2.9 m. (b) What is the total force acting upward on the underside of the table?

9. (II) In a movie, Tarzan is shown evading his captors by hiding underwater for many minutes while breathing through a long thin reed. Assuming the maximum pressure difference lungs can manage and still breathe is −80 mm-Hg, calculate the deepest he could have been.

10. (II) The gauge pressure in each of the four tires of an automobile is 240 kPa. If each tire has a "footprint" of 200 cm^2, estimate the mass of the car.

11. (II) The maximum gauge pressure in a hydraulic lift is 17.0 atm. What is the largest size vehicle (kg) it can lift if the diameter of the output line is 24.5 cm?

12. (II) How high would the level be in an alcohol barometer at normal atmospheric pressure?

13. (II) What is the total force and the absolute pressure on the bottom of a swimming pool 22.0 m by 8.5 m whose uniform depth is 2.0 m? What will be the pressure against the *side* of the pool near the bottom?

14. (II) How high would the atmosphere extend if it were of uniform density, equal to that at sea level throughout?

15. (II) Water and then oil (which don't mix) are poured into a U-shaped tube, open at both ends. They come to equilibrium as shown in Fig. 13–46. What is the density of the oil? [*Hint:* Pressures at points a and b are equal. Why?]

FIGURE 13–46 Problem 15.

16. (II) Determine the water gauge pressure at a house at the bottom of a hill fed by a full tank of water 5.0 m deep and connected to the house by a pipe that is 100 m long at an angle of 60° from the horizontal (Fig. 13–47). Neglect turbulence, and frictional and viscous effects. How high would the water shoot if it came vertically out of a broken pipe in front of the house?

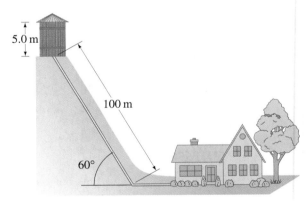

FIGURE 13–47 Problem 16.

17. (II) Estimate the air pressure on the summit of Mt. Everest (8850 m above sea level).

18. (II) Determine the minimum gauge pressure needed in the water pipe leading into a building if water is to come out of a faucet on the twelfth floor, 36.5 m above.

19. (II) An open-tube mercury manometer is used to measure the pressure in an oxygen tank. On a day when the atmospheric pressure is 1040 mbar, what is the absolute pressure (in Pa) in the tank if the height of the mercury in the open tube is (a) 28.0 cm higher, (b) 4.2 cm lower, than the mercury in the tube connected to the tank?

20. (II) A hydraulic press for compacting powdered samples has a large cylinder which is 10.0 cm in diameter, and a small cylinder with a diameter of 2.0 cm (Fig. 13–48). A lever is attached to the small cylinder as shown. The sample, which is placed on the large cylinder, has an area of 4.0 cm². What is the pressure on the sample if 300 N is applied to the lever?

FIGURE 13–48 Problem 20.

21. (II) In working out his principle, Pascal showed dramatically how force can be multiplied with fluid pressure. He placed a long thin tube of 0.30-cm radius vertically into a 20-cm-radius wine barrel, Fig. 13–49. He found that when the barrel was filled with water and the tube filled to a height of 12 m, the barrel burst. Calculate (a) the mass of fluid in the tube, and (b) the net force on the lid of the barrel.

FIGURE 13–49
Problem 21 (Not to scale).

22. (III) A beaker of liquid accelerates from rest, on a horizontal surface, with acceleration a to the right. (a) Show that the surface of the liquid makes an angle $\theta = \tan^{-1} a/g$ with the horizontal. (b) Which edge of the water surface is higher? (c) How does the pressure vary with depth below the surface?

23. (III) Water stands at a height h behind a vertical dam of uniform width b. (a) Use integration to show that the total force of the water on the dam is $F = \frac{1}{2}\rho g h^2 b$. (b) Show that the torque about the base of the dam due to this force can be considered to act with a lever arm equal to h/3. (c) For a freestanding concrete dam of uniform thickness t and height h, what minimum thickness is needed to prevent overturning? Do you need to add in atmospheric pressure for this last part? Explain.

24. (III) Estimate the density of the water 6.0 km deep in the sea. (See Section 12–5 and Table 12–1.) By what fraction does it differ from the density at the surface?

25. (III) A cylindrical bucket of liquid (density ρ) is rotated about its symmetry axis, which is vertical. If the angular velocity is ω, show that the pressure at a distance r from the rotation axis is

$$P = P_0 + \tfrac{1}{2}\rho\omega^2 r^2,$$

where P_0 is the pressure at $r = 0$.

Section 13–6

26. (I) The hydrometer of Example 13–10 sinks to a depth of 22.9 cm when placed in a fermenting vat. What is the density of the brewing liquid?

27. (I) A geologist finds that a moon rock whose mass is 7.85 kg has an apparent mass of 6.18 kg when submerged in water. What is the density of the rock?

28. (I) What fraction of a piece of aluminum will be submerged when it floats in mercury?

29. (II) A spherically shaped balloon has a radius of 7.35 m, and is filled with helium. How large a cargo can it lift, assuming that the skin and structure of the balloon have a mass of 1000 kg? Neglect the buoyant force on the cargo volume itself.

30. (II) A 78-kg person has an apparent mass of 54 kg (because of buoyancy) when standing in water that comes up to the hips. Estimate the mass of each leg. Assume the body has SG = 1.00.

31. (II) What is the likely identity of a metal (see Table 13–1) if a sample has a mass of 63.5 g when measured in air and an apparent mass of 56.4 g when submerged in water?

32. (II) Calculate the true mass (in vacuum) of a piece of aluminum whose apparent mass is 2.0000 kg when weighed in air.

33. (II) An undersea research chamber for aquanauts is spherical with an external diameter of 6.0 m. The mass of the chamber, when occupied, is 75,000 kg. It is anchored to the sea bottom by a cable. What is (a) the buoyant force on the chamber, and (b) the tension in the cable?

34. (II) A scuba diver is diving off the shores of the Cayman Islands. The diver and her gear displace a volume of 65.0 L and have a total mass of 63.0 kg. (a) What is the buoyant force on the diver? (b) Will the diver sink or float?

35. (II) (a) Show that the buoyant force F_B on a partially submerged object such as a ship acts at the center of gravity of the fluid before it is displaced. This point is called the **center of buoyancy**. (b) For a ship to be stable, should its center of buoyancy be above, below, or at the same point as, its center of gravity? Explain (See Fig. 13–50.)

FIGURE 13–50 Problem 35.

36. (II) Archimedes' principle can be used not only to determine the specific gravity of a solid using a known liquid (Example 13–9). The reverse can be done as well. (a) As an example, a 3.40-kg aluminum ball has an apparent mass of 2.10 kg when submerged in a particular liquid: calculate the density of the liquid. (b) Derive a simple formula for determining the density of a liquid using this procedure.

37. (II) A 0.48-kg piece of wood floats in water but is found to sink in alcohol (specific gravity = 0.79) in which it has an apparent mass of 0.047 kg. What is the SG of the wood?

38. (II) The specific gravity of ice is 0.917, whereas that for seawater is 1.025. What fraction of an iceberg is above the surface of the water?

39. (III) A polar bear partially supports herself by pulling part of her body out of the water onto a rectangular slab of ice. The ice sinks down so that only half of what was once exposed now is exposed, and the bear has 70 percent of her volume (and weight) out of the water. Estimate the bear's mass, assuming that the total volume of the ice is $10 \, \text{m}^3$, and the bear's specific gravity is 1.0.

40. (III) A 3.15-kg piece of wood (SG = 0.50) floats on water. What minimum mass of lead, hung from it by a string, will cause it to sink?

41. (III) If an object floats in water, its density can be determined by tying a sinker on it so that both the object and the sinker are submerged. Show that the specific gravity is given by $w/(w_1 - w_2)$, where w is the weight of the object alone in air, w_1 is the apparent weight when a sinker is tied to it and the sinker only is submerged, and w_2 is the apparent weight when both the object and the sinker are submerged.

Sections 13–7 to 13–9

42. (I) Using the data of Example 13–12, calculate the average speed of blood flow in the major arteries of the body which have a total cross-sectional area of about $2.0 \, \text{cm}^2$.

43. (I) A 15-cm-radius air duct is used to replenish the air of a room $9.2 \, \text{m} \times 5.0 \, \text{m} \times 4.5 \, \text{m}$ every 12 min. How fast does the air flow in the duct?

44. (I) Show that Bernoulli's equation reduces to the hydrostatic variation of pressure with depth (Eq. 13–6) when there is no flow $(v_1 = v_2 = 0)$.

45. (I) How fast does water flow from a hole at the bottom of a very wide, 4.6-m-deep storage tank filled with water? Ignore viscosity.

46. (II) A $\frac{5}{8}$-inch (inside) diameter garden hose is used to fill a round swimming pool 6.1 m in diameter. How long will it take to fill the pool to a depth of 1.2 m if water issues from the hose at a speed of 0.33 m/s?

47. (II) What gauge pressure in the water mains is necessary if a firehose is to spray water to a height of 15 m?

48. (II) A 6.0-cm-diameter pipe gradually narrows to 4.0 cm. When water flows through this pipe at a certain rate, the gauge pressure in these two sections is 32 kPa and 24 kPa. What is the volume rate of flow?

49. (II) What is the volume rate of flow of water from a 1.85-cm-diameter faucet if the pressure head is 12.0 m?

50. (II) If wind blows at 25 m/s over your house, what is the net force on the roof if its area is $240 \, \text{m}^2$?

51. (II) What is the lift (in newtons) due to Bernoulli's principle on a wing of area $86 \, \text{m}^2$ if the air passes over the top and bottom surfaces at speeds of 340 m/s and 290 m/s, respectively?

52. (II) Estimate the air pressure inside a category 5 hurricane where the wind speed is 300 km/h. (See Fig. 13–51.)

FIGURE 13–51 Problem 52.

53. (II) Show that the power needed to drive a fluid through a pipe is equal to the volume rate of flow, Q, times the pressure difference, $P_1 - P_2$.

54. (II) Water at a gauge pressure of 3.8 atm at street level flows into an office building at a speed of 0.60 m/s through a pipe 5.0 cm in diameter. The pipes taper down to 2.6 cm in diameter by the top floor, 20 m above (Fig. 13–52). Calculate the flow velocity and the gauge pressure in such a pipe on the top floor. Assume no branch pipes and ignore viscosity.

FIGURE 13–52 Problem 54.

55. (II) In Fig. 13–53, take into account the speed of the top surface of the tank and show that the speed of fluid leaving the opening at the bottom is

$$v_1 = \sqrt{2gh/(1 - A_1^2/A_2^2)},$$

where $h = y_2 - y_1$, and A_1 and A_2 are the areas of the opening and the top surface, respectively. Assume $A_1 \ll A_2$ so that the flow remains nearly steady and laminar.

FIGURE 13–53 Problems 55, 56, 59, and 60.

56. (II) Suppose the top surface of the vessel in Fig. 13–53 is subjected to an external gauge pressure P_2. (a) Derive a formula for the speed, v_1, at which the liquid flows from the opening at the bottom into atmospheric pressure, P_A. Assume the velocity of the liquid surface, v_2, is approximately zero. (b) If $P_2 = 0.85$ atm and $y_2 - y_1 = 2.1$ m, determine v_1 for water.

57. (III) *Thrust of a rocket.* (a) Use Bernoulli's equation and the equation of continuity to show that the emission speed of the propelling gases of a rocket is

$$v = \sqrt{2(P - P_0)/\rho},$$

where ρ is the density of the gas, P is the pressure of the gas inside the rocket, and P_0 is atmospheric pressure just outside the exit orifice. Assume that the gas density stays approximately constant, and that the area of the exit orifice, A_0, is much smaller than the cross-sectional area, A, of the inside of the rocket (take it to be a large cylinder). Assume also that the gas speed is not so high that significant turbulence or non-steady flow sets in. (b) Show that the thrust force on the rocket due to the emitted gases is

$$F = 2A_0(P - P_0).$$

58. (III) (a) Show that the flow velocity measured by a venturi meter is given by the relation

$$v_1 = A_2 \sqrt{2(P_1 - P_2)/\rho(A_1^2 - A_2^2)}.$$

See Fig. 13–27. (b) A venturi tube is measuring the flow of water; it has a main diameter of 3.0 cm tapering down to a throat diameter of 1.0 cm; if the pressure difference is measured to be 18 mm-Hg, what is the velocity of the water?

59. (III) Suppose the opening in the tank of Fig. 13–53 is a height h_1 above the base and the liquid surface is a height h_2 above the base. The tank rests on level ground. (a) At what horizontal distance from the base of the tank will the fluid strike the ground? (b) At what other height, h_1', can a hole be placed so that the emerging liquid will have the same "range"?

60. (III) (a) In Fig. 13–53, show that Bernoulli's principle predicts that the level of the liquid, $h = y_2 - y_1$, drops at a rate

$$\frac{dh}{dt} = -\sqrt{\frac{2ghA_1^2}{A_2^2 - A_1^2}},$$

where A_1 and A_2 are the areas of the opening and the top surface, respectively, assuming $A_1 \ll A_2$, and viscosity is ignored. (b) Determine h as a function of time by integrating. Let $h = h_0$ at $t = 0$. (c) How long would it take to empty a 9.4-cm-tall cylinder filled with 1.0 L of water if the opening is at the bottom and has a 0.50-cm diameter?

* **Section 13–10**

*61. (II) A viscometer consists of two concentric cylinders, 10.20 cm and 10.60 cm in diameter. A particular liquid fills the space between them to a depth of 12.0 cm. The outer cylinder is fixed, and a torque of 0.024 m·N keeps the inner cylinder turning at a steady rotational speed of 62 rev/min. What is the viscosity of the liquid?

* **Section 13–11**

*62. (I) A gardener feels it is taking him too long to water a garden with a $\frac{3}{8}$-in-diameter hose. By what factor will his time be cut if he uses a $\frac{5}{8}$-in-diameter hose? Assume nothing else is changed.

*63. (I) Engine oil (assume SAE 10, Table 13–3) passes through a fine 1.80-mm-diameter tube in a prototype engine. The tube is 5.5 cm long. What pressure difference is needed to maintain a flow rate of 5.6 mL/min?

*64. (II) What must be the pressure difference between the two ends of a 1.9-km section of pipe, 29 cm in diameter, if it is to transport oil $(\rho = 950\ \text{kg/m}^3,\ \eta = 0.20\ \text{Pa·s})$ at a rate of 450 cm^3/s?

*65. (II) What diameter must a 17.5-m-long air duct have if the ventilation and heating system is to replenish the air in a room 9.0 m × 12.0 × 4.0 m every 10 min? Assume the pump can exert a gauge pressure of $0.71 × 10^{-3}$ atm.

*66. (II) Assuming a constant pressure gradient, by what factor does a blood vessel decrease in radius if the blood flow is reduced by 75 percent?

*67. (II) Poiseuille's equation does not hold if the flow velocity is high enough that turbulence sets in. The onset of turbulence occurs when the so-called **Reynolds number**, Re, exceeds approximately 2000. Re is defined as

$$Re = \frac{2\bar{v}r\rho}{\eta},$$

where $\bar{v}$ is the average speed of the fluid, ρ is its density, η is its viscosity, and r is the radius of the tube in which the fluid is flowing. (a) Determine if blood flow through the aorta is laminar or turbulent, when the average speed of blood in the aorta ($r = 1.0$ cm) during the resting part of the heart's cycle is about 30 cm/s. (b) During exercise, the blood-flow speed approximately doubles. Calculate the Reynolds number in this case and determine if the flow is laminar or turbulent.

* 68. (II) Water is to be pumped through a 10.0-cm-diameter pipe over a distance of 300 m. The far end of the pipe is 20 m above the pump and is at normal atmospheric pressure. What gauge pressure must the pump develop for there to be any flow at all?

* 69. (III) A patient is to be given a blood transfusion. The blood is to flow through a tube from a raised bottle to a needle inserted in the vein (Fig. 13–54). The inside diameter of the 4.0-cm-long needle is 0.40 mm and the required flow rate is 4.0 cm³ of blood per minute. How high should the bottle be placed above the needle? Obtain ρ and η from the Tables. Assume the blood pressure is 18 torr above atmospheric pressure.

FIGURE 13–54 Problems 69 and 75.

* Section 13–12

* 70. (I) If the force F needed to move the wire in Fig. 13–32 is 5.1×10^{-3} N, calculate the surface tension γ of the enclosed fluid. Assume $l = 0.070$ m.

* 71. (I) Calculate the force needed to move the wire in Fig. 13–32 if it is immersed in a soapy solution and the wire is 18.2 cm long.

* 72. (II) If the base of an insect's leg has a radius of about 3.0×10^{-5} m and its mass is 0.016 g, would you expect the six-legged insect to remain on top of the water?

* 73. (II) The surface tension of a liquid can be determined by measuring the force F needed to just lift a circular platinum ring of radius r from the surface of the liquid. (a) Find a formula for γ in terms of F and r. (b) At 30°C, if $F = 8.40 \times 10^{-3}$ N and $r = 2.8$ cm, calculate γ for the tested liquid.

* 74. (III) Show that inside a soap bubble, there must be a pressure ΔP in excess of that outside equal to $\Delta P = 4\gamma/r$, where r is the radius of the bubble and γ is the surface tension. [*Hint*: Think of the bubble as two hemispheres in contact with each other; and remember that there are two surfaces to the bubble. Note that this result applies to any kind of membrane, where 2γ is the tension per unit length in that membrane.]

General Problems

75. Intravenous infusions are often made under gravity, as shown in Fig. 13–54. Assuming the fluid has a density of 1.00 g/cm³, at what height h should the bottle be placed so the liquid pressure is (a) 65 mm-Hg, (b) 550 mm-H₂O? (c) If the blood pressure is 18 mm-Hg above atmospheric pressure, how high should the bottle be placed so that the fluid just barely enters the vein?

76. A 2.4-N force is applied to the plunger of a hypodermic needle. If the diameter of the plunger is 1.3 cm and that of the needle 0.20 mm, (a) with what force does the fluid leave the needle? (b) What force on the plunger would be needed to push fluid into a vein where the gauge pressure is 18 mm-Hg? Answer for the instant just before the fluid starts too move.

77. A bicycle pump is used to inflate a tire. The initial tire pressure is 210 kPa (30 psi). At the end of the pumping process, the final pressure is 310 kPa (45 psi). If the diameter of the plunger in the cylinder of the pump is 3.0 cm, what is the range of the force that needs to be applied to the pump handle from beginning to end?

78. Estimate the pressure on the mountains underneath the Antarctic ice pack, which is typically 4 km thick.

79. What is the approximate difference in air pressure between the top and the bottom of the World Trade Center buildings in New York City? They are 410 m tall and are located at sea level. Express as a fraction of atmospheric pressure at sea level.

80. Giraffes are a wonder of cardiovascular engineering. Calculate the difference in pressure (in atmospheres) that the blood vessels in a giraffe's head have to accommodate as it lowers its head from a full upright position to ground level for a drink. The height of an average giraffe is about 6 meters.

81. When you drive up into the mountains, or descend rapidly from the mountains, your ears "pop," which means that the pressure behind the eardrum is being equalized to that outside. If this did not happen, what would be the approximate force on an eardrum of area 0.50 cm² if a change in altitude of 1000 m takes place?

82. One arm of a U-shaped tube (open at both ends) contains water and the other alcohol. If the two fluids meet exactly at the bottom of the U, and the alcohol is at a height of 18.0 cm, at what height will the water be?

83. A simple model (Fig. 13–55) considers a continent as a block (density = 2800 kg/m³) floating in the mantle rock around it (density = 3300 kg/m³). Assuming the continent is 35 km thick (the average thickness of the Earth's crust), estimate the height of the continent above the surrounding rock.

FIGURE 13–55
Problem 83.

84. The contraction of the left ventricle (chamber) of the heart pumps blood to the body. Assuming that the inner surface of the left ventricle has an area of 82 cm² and the maximum pressure in the blood is 120 mm-Hg, estimate the force exerted by the ventricle at maximum pressure.

85. Estimate the total mass of the Earth's atmosphere using the known value of atmospheric pressure at sea level.

86. Suppose a person can reduce the pressure in the lungs to −80 mm-Hg gauge pressure. How high can water then be sucked up a straw?

87. How high should the pressure head be if water is to come from a faucet at a speed of 7.2 m/s? Ignore viscosity.

88. A ship, carrying freshwater to a desert island in the Caribbean, has a horizontal cross-sectional area of 2650 m² at the waterline. When unloaded, the ship rises 8.50 m higher in the sea. How much water was delivered?

89. A raft is made of 10 logs lashed together. Each is 38 cm in diameter and has a length of 6.1 m. How many people can the raft hold before they start getting their feet wet, assuming the average person has a mass of 70 kg? Do *not* neglect the weight of the logs. Assume the specific gravity of wood is 0.60.

90. During each heartbeat, approximately 70 cm³ of blood is pushed from the heart at an average pressure of 105 mm-Hg. Calculate the power output of the heart, in watts, assuming 70 beats per minute.

91. A bucket of water is accelerated upward at 2.4 g. What is the buoyant force on a 3.0-kg granite rock (SG = 2.7) submerged in the water? Will the rock float? Why or why not?

92. The drinking fountain outside your classroom shoots water about 16 cm up in the air from a nozzle of diameter 0.60 cm. The pump at the base of the unit (1.1 m below the nozzle) pushes water into a 1.2-cm-diameter supply pipe that goes up to the nozzle. What gauge pressure does the pump have to provide? Ignore the viscosity; your answer will therefore be an underestimate.

93. The stream of water from a faucet decreases in diameter as it falls (Fig. 13–44). Derive an equation for the diameter of the stream as a function of the distance y below the faucet, given that the water has speed v_0 when it leaves the faucet, whose diameter is D.

94. Four lawn sprinkler heads are fed by a 1.9-cm-diameter pipe. The water comes out of the heads at an angle of 30° to the horizontal and covers a radius of 8.0 m. (a) What is the velocity of the water coming out of the sprinkler head? (Assume zero air resistance.) (b) If the output diameter of each head is 3.0 mm, how many liters of water do the four heads deliver per second? (c) How fast is the water flowing inside the 1.9-cm-diameter pipe?

95. You need to siphon water from a clogged sink. The sink has an area of 0.48 m² and is filled to a height of 4.0 cm. Your siphon tube rises 50 cm above the bottom of the sink and then descends 100 cm to a pail as shown in Fig. 13–56. The siphon tube has a diameter of 2.0 cm. (a) Assuming that the water enters the siphon tube with almost zero velocity, calculate its velocity when it enters the pail. (b) Estimate how long it will take to empty the sink.

FIGURE 13–56
Problems 95 and 96.

96. Consider a siphon which transfers water (20°C) from one vessel to a second (lower) one, as in Fig. 13–56. Determine the rate of flow if the hose has a 1.2-cm-diameter and the difference in water levels of the two containers is 64 cm.

97. An airplane has a mass of 2.0×10^6 kg, and the air flows past the lower surface of the wings at 100 m/s. If the wings have a surface area of 1200 m², how fast must the air flow over the upper surface of the wing if the plane is to stay in the air? Consider only the Bernoulli effect.

98. A hydraulic lift is used to jack a 1000 kg car 10 cm off the floor. The diameter of the output piston is 15 cm and the input force is 250 N. (a) What is the area of the input piston? (b) What is the work done in lifting the car 10 cm? (c) If the travel for each stroke of the input piston is 12 cm, how high does the car move up for each stroke? (d) How many strokes are required to jack the car up 10 cm? (e) Show that energy is conserved.

The pendulum of a clock is an example of oscillatory motion. Many kinds of oscillatory motion are sinusoidal, or nearly so, and are referred to as being simple harmonic motion. Real systems generally have at least some friction, and the motion is damped. When an external sinusoidal force is exerted on a system able to oscillate, resonance occurs if the driving force is at or near the natural frequency of vibration.

CHAPTER 14

Oscillations

Many objects vibrate or oscillate—an object on the end of a spring, a tuning fork, the balance wheel of an old watch, a pendulum, a plastic ruler held firmly over the edge of a table and gently struck, the strings of a guitar or piano. Spiders detect prey by the vibrations of their webs, cars oscillate up and down when they hit a bump, buildings and bridges vibrate when heavy trucks pass or the wind is fierce. Indeed, because most solids are elastic (see Chapter 12), most material objects vibrate (at least briefly) when given an impulse. Electrical oscillations occur in radio and television sets. At the atomic level, atoms vibrate within a molecule, and the atoms of a solid vibrate about their relatively fixed positions. Because it is so common in everyday life and occurs in so many areas of physics, oscillatory (or vibrational) motion is of great importance. Vibrational motion is not really a "new" phenomenon because vibrations of mechanical systems are fully described on the basis of Newtonian mechanics.

14-1 | Oscillations of a Spring

When a **vibration** or an **oscillation** repeats itself, back and forth, over the same path, the motion is **periodic**. The simplest form of periodic motion is represented by an object oscillating on the end of a coil spring. Because many other types of vibrational motion closely resemble this system, we will look at it in detail. We assume that the mass of the spring can be ignored, and that the spring is mounted horizontally (Fig. 14–1a), so that the object of mass m slides without friction on the horizontal surface. Any spring has a natural length at which the net force on the mass m is zero; the position of the mass at this point is called the **equilibrium position**. If the mass is moved either to the left, which compresses the spring, or to the right, which stretches it, the spring exerts a force on the mass that acts in the direction of returning the mass to the equilibrium position; hence it is called a "restoring force." The magnitude of the restoring force F is found to be directly proportional to the displacement x the spring has been stretched or compressed from the equilibrium position (Fig. 14–1b and c):

$$F = -kx. \tag{14-1}$$

Note that the equilibrium position is at $x = 0$. Equation 14–1, which is often referred to as Hooke's law (see Sections 8–2 and 12–5), is accurate as long as the spring is not compressed to the point where the coils come close to touching, or stretched beyond the elastic region (see Fig. 12–19). The minus sign in Eq. 14–1 indicates that the restoring force is always in the direction opposite to the displacement x. For example, if we choose the positive direction to the right in Fig. 14–1, x is positive when the spring is stretched, but the direction of the restoring force is to the left (negative direction). If the spring is compressed, x is negative (to the left) but the force F acts toward the right (Fig. 14–1c).

The proportionality constant k in Eq. 14–1 is called the "spring constant." In order to stretch the spring a distance x, one has to exert an (external) force on the spring at least equal to $F = +kx$. The greater the value of k, the greater the force needed to stretch a spring a given distance.

Note that the force F in Eq. 14–1 is *not* a constant, but varies with position. Therefore the acceleration of the mass m is not constant, so we *cannot* use the equations for constant acceleration developed in Chapter 2.

Let us examine what happens when the spring is initially stretched a distance $x = A$, as shown in Fig. 14–2a, and then released. The spring exerts a force on the mass that pulls it toward the equilibrium position. But because the mass has been accelerated by the force, it passes the equilibrium position with considerable speed. Indeed, as the mass reaches the equilibrium position, the force on it decreases to zero, but its speed at this point is a maximum, Fig. 14–2b. As it moves farther to the left, the force on it acts to slow it down, and it stops momentarily at $x = -A$, Fig. 14–2c. It then begins moving back in the opposite direction, Fig. 14–2d, until it reaches the original starting point, $x = A$, Fig. 14–2e. It then repeats the motion, moving back and forth symmetrically between $x = A$ and $x = -A$.

To discuss vibrational motion, we need to define a few terms. The distance x of the mass from the equilibrium point at any moment is called the **displacement**. The maximum displacement—the greatest distance from the equilibrium point—is called the **amplitude**, A. One **cycle** refers to the complete to-and-fro motion from some initial point back to that same point, say from $x = A$ to $x = -A$ back to $x = A$. The **period**, T, is defined as the time required for one complete cycle. Finally, the **frequency**, f, is the number of complete cycles per second. Frequency is generally specified in hertz (Hz), where 1 Hz = 1 cycle per second (s^{-1}). It is easy to see, from their definitions, that frequency and period are inversely related:

$$f = \frac{1}{T} \quad \text{and} \quad T = \frac{1}{f}; \tag{14-2}$$

for example, if the frequency is 5 cycles per second, then each cycle takes $\frac{1}{5}$ s.

(a)

(b)

(c)

FIGURE 14–1 Mass vibrating at the end of a spring.

FIGURE 14–2 Force on, and velocity of, mass at different positions of its oscillation.

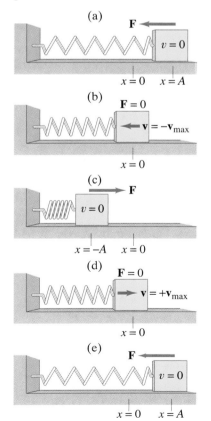

(a)

(b)

(c)

(d)

(e)

x_0
$F = -kx_0$
m
x now
x measured
from here
$m\mathbf{g}$

(a)   (b)

FIGURE 14–3 (a) Free spring, hung vertically. (b) Mass m attached to spring in new equilibrium position, which occurs when $\Sigma F = 0 = kx_0 - mg$.

FIGURE 14–4 Photo of a car's spring. (Also visible is the shock absorber—see Fig. 14–21.)

The oscillation of a spring hung vertically is essentially the same as that of a horizontal spring. Because of the force of gravity, the length of the vertical spring at equilibrium will be longer than when it is horizontal, as shown in Fig. 14–3. The spring is in equilibrium when $\Sigma F = 0 = kx_0 - mg$, so the spring stretches an extra amount $x_0 = mg/k$ to be in equilibrium. If x is measured from this new equilibrium position, Eq. 14–1 can be used directly with the same value of k.

EXAMPLE 14–1 **Car springs.** When a family of four people with a total mass of 200 kg step into their 1200-kg car, the car's springs compress 3.0 cm. (*a*) What is the spring constant of the car's springs (Fig. 14–4), assuming they act as a single spring? (*b*) How far will the car lower if loaded with 300 kg?

SOLUTION (*a*) The added force of $(200\,\text{kg})(9.8\,\text{m/s}^2) = 1960\,\text{N}$ causes the springs to compress $3.0 \times 10^{-2}\,\text{m}$. Therefore, by Eq. 14–1, the spring constant is

$$k = \frac{F}{x} = \frac{1960\,\text{N}}{3.0 \times 10^{-2}\,\text{m}} = 6.5 \times 10^4\,\text{N/m}.$$

(*b*) If the car is loaded with 300 kg,

$$x = \frac{F}{k} = \frac{(300\,\text{kg})(9.8\,\text{m/s}^2)}{(6.5 \times 10^4\,\text{N/m})} = 4.5 \times 10^{-2}\,\text{m},$$

or 4.5 cm. We could have obtained this answer without solving for k: since x is proportional to F, if 200 kg compresses the spring 3.0 cm, then 1.5 times the force will compress the spring 1.5 times as much, or 4.5 cm.

14–2 Simple Harmonic Motion

SHM
SHO

Any vibrating system for which the net restoring force is directly proportional to the negative of the displacement (as in Eq. 14–1, $F = -kx$) is said to exhibit **simple harmonic motion** (SHM). Such a system is often called a **simple harmonic oscillator** (SHO). We saw in Chapter 12 (Section 12–5) that most solid materials stretch or compress according to Eq. 14–1 as long as the displacement is not too great. Because of this, many natural vibrations are simple harmonic or close to it.

Let us now determine the position x as a function of time for a mass attached to the end of a simple spring with spring constant k. To do so, we make use of Newton's second law, $F = ma$. Since the acceleration $a = d^2x/dt^2$, we have

$$ma = \Sigma F$$
$$m\frac{d^2x}{dt^2} = -kx,$$

where m is the mass[†] which is oscillating. We rearrange this to obtain

Equation of motion
(SHM)

$$\frac{d^2x}{dt^2} + \frac{k}{m}x = 0, \qquad (14\text{–}3)$$

which is known as the **equation of motion** for the simple harmonic oscillator. Mathematically it is called a *differential equation*, since it involves derivatives. We want to determine what function of time, $x(t)$, satisfies this equation. We might guess the form of the solution by noting that if a pen were attached to a vibrating mass (Fig. 14–5) and a sheet of paper moved at a steady rate beneath it, the pen would trace the curve shown. The shape of this curve looks a lot like it might be

---

[†] In the case of a mass, m', on the end of a spring, the spring itself also oscillates and at least a part of its mass must be included. It can be shown—see the Problems—that approximately one-third the mass of the spring, m_s, must be included, so $m = m' + \frac{1}{3}m_s$ in our equation. Often m_s is small enough to be ignored.

sinusoidal (such as cosine or sine) as a function of time, and its height is the amplitude A. Let us then guess that the general solution to Eq. 14–3 can be written in a form such as

$$x = A \cos(\omega t + \phi), \qquad \textbf{(14–4)}$$

General solution:
position as functions of time

where we include the constant ϕ in the argument to be general.[†] Let us now put this trial solution into Eq. 14–3 and see if it really works. We need to differentiate the $x = x(t)$ twice:

$$\frac{dx}{dt} = \frac{d}{dt}[A \cos(\omega t + \phi)] = -\omega A \sin(\omega t + \phi)$$

$$\frac{d^2x}{dt^2} = -\omega^2 A \cos(\omega t + \phi).$$

We now put the latter into Eq. 14–3, along with Eq. 14–4 for x:

$$\frac{d^2x}{dt^2} + \frac{k}{m} x = 0$$

$$-\omega^2 A \cos(\omega t + \phi) + \frac{k}{m} A \cos(\omega t + \phi) = 0$$

or

$$\left(\frac{k}{m} - \omega^2\right) A \cos(\omega t + \phi) = 0.$$

Our solution, Eq. 14–4, does indeed satisfy the equation of motion (Eq. 14–3) for any time t. But it does so only if $(k/m - \omega^2) = 0$; hence

$$\omega^2 = \frac{k}{m}. \qquad \textbf{(14–5)}$$

Equation 14–4 is the general solution and it contains two arbitrary constants A and ϕ, which we should expect because the second derivative in Eq. 14–3 implies that two integrations are needed, each yielding a constant. They are "arbitrary" only in a calculus sense, in that they can be anything and still satisfy the differential equation, Eq. 14–3. In real physical situations, however, A and ϕ are determined by the **initial conditions**. Suppose, for example, the mass is started at its maximum displacement and is released from rest. This is, in fact, what is shown in Fig. 14–5, and for this case $x = A \cos \omega t$. Let us confirm it: we are given $v = 0$ at $t = 0$, where

Initial conditions

$$v = \frac{dx}{dt} = \frac{d}{dt}[A \cos(\omega t + \phi)] = -\omega A \sin(\omega t + \phi) = 0. \quad \text{[at } t = 0\text{]}$$

For v to be zero at $t = 0$, then $\sin(\omega t + \phi) = \sin(0 + \phi)$ is zero if $\phi = 0$ (ϕ could also be $\pi, 2\pi$, etc.), and when $\phi = 0$, then

$$x = A \cos \omega t,$$

as we expected. We see immediately that A is the amplitude of the motion, and it is determined initially by how far you pulled the mass m from equilibrium before releasing it.

FIGURE 14–5 Sinusoidal nature of SHM as a function of time. In this case, $x = A \cos(2\pi t/T)$.

[†] Another possible way to write the solution is the combination $x = a \cos \omega t + b \sin \omega t$, where a and b are constants. This is equivalent to Eq. 14–4 as can be seen using the trigonometric identity $\cos(A \pm B) = \cos A \cos B \mp \sin A \sin B$.

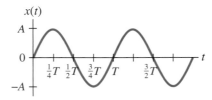

FIGURE 14-6 Special case of SHM where the mass m starts, at $t = 0$, at the equilibrium position $x = 0$ and has initial velocity toward positive values of x ($v > 0$ at $t = 0$).

FIGURE 14-7 A plot of $x = A \cos(\omega t + \phi)$ when $\phi < 0$.

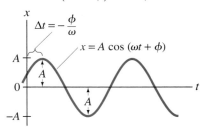

Position as function of time

Consider another interesting case: at $t = 0$, the mass m is at $x = 0$ and is struck, giving it an initial velocity toward increasing values of x. Then at $t = 0$, $x = 0$, so we can write $x = A \cos(\omega t + \phi) = A \cos \phi = 0$, which can happen only if $\phi = \pm \pi/2$ (or $\pm 90°$). Whether $\phi = +\pi/2$ or $-\pi/2$ depends on $v = dx/dt = -\omega A \sin(\omega t + \phi) = -\omega A \sin \phi$, which we are given as positive ($v > 0$ at $t = 0$); hence $\phi = -\pi/2 [\sin(-90°) = -1]$. Thus our solution for this case is

$$x = A \cos\left(\omega t - \frac{\pi}{2} \right)$$
$$= A \sin \omega t,$$

where we used $\cos(\theta - \pi/2) = \sin \theta$. The solution in this case is a pure sine wave, Fig. 14–6, where A is still the amplitude.

Lots of other situations are possible, such as that shown in Fig. 14–7. The constant ϕ is called the **phase angle**, and it tells us how long after (or before) $t = 0$ the peak at $x = A$ is reached. Notice that the value of ϕ does not affect the shape of the $x(t)$ curve, but only affects the displacement at some arbitrary time, $t = 0$. Simple harmonic motion is thus always *sinusoidal*. Indeed, simple harmonic motion is *defined* as motion that is purely sinusoidal.

Since our oscillating mass repeats its motion after a time equal to its period T, it must be at the same position and moving in the same direction at $t = T$ as it was at $t = 0$. And since a sine or cosine function repeats itself after every 2π radians, then from Eq. 14–4, we must have

$$\omega T = 2\pi.$$

Hence

$$\omega = \frac{2\pi}{T} = 2\pi f,$$

where f is the frequency of the motion. We call ω the **angular frequency** (units are rad/s) to distinguish it from the frequency f (units are $\text{s}^{-1} = \text{Hz}$). Thus we can write Eq. 14–4 as

$$x = A \cos\left(\frac{2\pi t}{T} + \phi \right) \tag{14–6a}$$

or

$$x = A \cos(2\pi f t + \phi), \tag{14–6b}$$

where, because of Eq. 14–5

$$f = \frac{1}{2\pi} \sqrt{\frac{k}{m}}, \tag{14–7a}$$

$$T = 2\pi \sqrt{\frac{m}{k}}. \tag{14–7b}$$

f and T are independent of amplitude

Note that the *frequency and period do not depend on the amplitude*. Changing the amplitude of a simple harmonic oscillator does not affect its frequency. Equation 14–7a tells us that the greater the mass, the lower the frequency; and the stiffer the spring, the higher the frequency. This makes sense since a greater mass means more inertia and therefore slower response (or acceleration); and larger k means greater force and therefore quicker response. The frequency f (Eq. 14–7a) at which a SHO oscillates naturally is called its **natural frequency** (to distinguish it from a frequency at which it might be forced to oscillate by an outside force, as discussed in Section 14–8).

The simple harmonic oscillator is important in physics because whenever we have a net restoring force proportional to the displacement ($F = -kx$), which is at least a good approximation for a variety of systems, then the motion is simple harmonic—that is, sinusoidal.

EXAMPLE 14–2 **Car springs again.** What are the period and frequency of the car in Example 14–1 after hitting a bump? Assume the shock absorbers are poor, so the car really oscillates up and down.

SOLUTION From Eq. 14–7b,

$$T = 2\pi \sqrt{\frac{m}{k}} = 6.28 \sqrt{\frac{1400 \text{ kg}}{6.5 \times 10^4 \text{ N/m}}} = 0.92 \text{ s},$$

or slightly less than a second. The frequency $f = 1/T = 1.09$ Hz.

Let us continue our analysis of a simple harmonic oscillator. The velocity and acceleration of the oscillating mass can be obtained by differentiation of Eq. 14–4

$$v = \frac{dx}{dt} = -\omega A \sin(\omega t + \phi) \qquad (14\text{–}8)$$

$$a = \frac{d^2x}{dt^2} = \frac{dv}{dt} = -\omega^2 A \cos(\omega t + \phi). \qquad (14\text{–}9)$$

The velocity and acceleration of a SHO also vary sinusoidally. In Fig. 14–8 we plot the displacement, velocity, and acceleration of a SHO as a function of time for the case when $\phi = 0$. As can be seen, the speed reaches its maximum

$$v_{\text{max}} = \omega A = \sqrt{\frac{k}{m}} A$$

when the oscillating object is passing through its equilibrium point, $x = 0$. And the speed is zero at points of maximum displacement, $x = \pm A$. This is in accord with our discussion of Fig. 14–2. Similarly, the acceleration has its maximum value

$$a_{\text{max}} = \omega^2 A = \frac{k}{m} A$$

which occurs where $x = \pm A$, and a is zero at $x = 0$, as we expect, since $ma = F = -kx$.

For the general case when $\phi \neq 0$, we can relate the constants A and ϕ to the initial values of x, v, and a by setting $t = 0$ in Eqs. 14–4, 14–8, and 14–9:

$$x_0 = x(0) = A \cos \phi$$
$$v_0 = v(0) = -\omega A \sin \phi = -v_{\text{max}} \sin \phi$$
$$a_0 = a(0) = -\omega^2 A \cos \phi = -a_{\text{max}} \cos \phi.$$

FIGURE 14–8 Displacement, x, velocity, dx/dt, and acceleration, d^2x/dt^2, of a simple harmonic oscillator when $\phi = 0$.

(a)

(b)

(c)

EXAMPLE 14–3 **Loudspeaker cone.** The cone of a loudspeaker vibrates in SHM at a frequency of 262 Hz ("middle C"). The amplitude at the center of the cone is $A = 1.5 \times 10^{-4}$ m, and at $t = 0$, $x = A$. (*a*) What is the equation describing the motion of the center of the cone? (*b*) What is its maximum velocity and maximum acceleration? (*c*) What is the position of the cone at $t = 1.00$ ms?

SOLUTION (*a*) The amplitude $A = 1.5 \times 10^{-4}$ m and $\omega = 2\pi f = (6.28 \text{ rad})(262 \text{ s}^{-1}) = 1650$ rad/s. The motion begins ($t = 0$) with the cone at its maximum displacement ($x = A$ at $t = 0$), so we use the cosine function with $\phi = 0$:

$$x = A \cos \omega t = A \cos 2\pi f t = (1.5 \times 10^{-4} \text{ m}) \cos (1650 t).$$

(*b*) From Eq. 14–8,

$$v_{\text{max}} = \omega A = 2\pi A f = 2\pi (1.5 \times 10^{-4} \text{ m})(262 \text{ s}^{-1}) = 0.25 \text{ m/s}.$$

From Eqs. 14–9 and 14–5:

$$a_{\text{max}} = \omega^2 A = (2\pi f)^2 A$$
$$= 4\pi^2 (262 \text{ s}^{-1})^2 (1.5 \times 10^{-4} \text{ m}) = 410 \text{ m/s}^2,$$

which is more than 40 g's.

(*c*) At $t = 1.00 \times 10^{-3}$ s,

$$x = (1.5 \times 10^{-4} \text{ m}) \cos[(1650 \text{ rad/s})(1.00 \times 10^{-3} \text{ s})]$$
$$= (1.5 \times 10^{-4} \text{ m}) \cos(1.65 \text{ rad}) = -1.2 \times 10^{-5} \text{ m}.$$

(a)   (b)   (c)

FIGURE 14–9 Example 14–4.

EXAMPLE 14–4 **Spring calculations.** In a testing device, a spring stretches 0.150 m when a 0.300-kg mass is hung from it (Fig. 14–9). The spring is then stretched an additional 0.100 m from this equilibrium point and released. Determine: (*a*) the values of the spring constant k and the angular frequency ω; (*b*) the amplitude of the oscillation A; (*c*) the maximum velocity, v_{max}; (*d*) the magnitude of the maximum acceleration of the mass; (*e*) the period T and frequency f; (*f*) the displacement x as a function of time; and (*g*) the velocity at $t = 0.150$ s.

SOLUTION (*a*) Since the spring stretches 0.150 m when 0.300 kg is hung from it, we find k from Eq. 14–1 to be

$$k = \frac{F}{x} = \frac{mg}{x} = \frac{(0.300 \text{ kg})(9.80 \text{ m/s}^2)}{0.150 \text{ m}} = 19.6 \text{ N/m}.$$

Also,

$$\omega = \sqrt{\frac{k}{m}} = \sqrt{\frac{19.6 \text{ N/m}}{0.300 \text{ kg}}} = 8.08 \text{ s}^{-1}.$$

(*b*) Since the spring is stretched 0.100 m from equilibrium (Fig. 14–9c) and is given no initial speed, $A = 0.100$ m.
(*c*) From Eq. 14–8, the maximum velocity is

$$v_{max} = \omega A = (8.08 \text{ s}^{-1})(0.100 \text{ m}) = 0.808 \text{ m/s}.$$

(*d*) Since $F = ma$, the maximum acceleration occurs where the force is greatest—that is, when $x = A = 0.100$ m. Thus

$$a_{max} = \frac{kA}{m} = \frac{(19.6 \text{ N/m})(0.100 \text{ m})}{0.300 \text{ kg}} = 6.53 \text{ m/s}^2.$$

(*e*) Equations 14–7a and b give

$$T = 2\pi \sqrt{\frac{m}{k}} = 6.28 \sqrt{\frac{0.300 \text{ kg}}{19.6 \text{ N/m}}} = 0.777 \text{ s}$$

$$f = \frac{1}{T} = 1.29 \text{ Hz}.$$

(*f*) The motion begins at a point of maximum displacement downward. If we take x positive upward, then at $t = 0$, $x = x_0 = -A = -0.100$ m. So we need a sinusoidal curve that has its maximum negative valve at $t = 0$; this is just a negative cosine:

$$x = -A \cos \omega t.$$

To write this in the form of Eq. 14–4 (no minus sign), recall that $\cos \theta = -\cos (\theta - \pi)$; then, putting in numbers,

$$\begin{aligned} x &= -(0.100 \text{ m}) \cos 8.08t \\ &= (0.100 \text{ m}) \cos (8.08t - \pi), \end{aligned}$$

where t is in seconds and x is in meters. Note that the phase angle (Eq. 14–4) is $\phi = \pi$ or $180°$.
(*g*) The velocity at any time t is (see part *c*)

$$v = \frac{dx}{dt} = A\omega \sin \omega t = (0.808 \text{ m/s}) \sin 8.08t.$$

At $t = 0.150$ s, $v = (0.808 \text{ m/s}) \sin(1.21 \text{ rad}) = 0.756 \text{ m/s}$, and is upward (+).

EXAMPLE 14–5 **Spring is started with a push.** Suppose the spring of Example 14–4 is stretched 0.100 m from equilibrium $(x_0 = -0.100\,\text{m})$ but is given an upward shove of $v_0 = 0.400\,\text{m/s}$. Determine (a) the phase angle ϕ, (b) the amplitude A, and (c) the displacement x as a function of time, $x(t)$.

SOLUTION (a) From Eq. 14–8, at $t = 0$, $v_0 = -\omega A \sin\phi$, and from Eq. 14–4, $x_0 = A \cos\phi$. Combining these, we get

$$\tan\phi = \frac{\sin\phi}{\cos\phi} = \frac{(v_0/-\omega A)}{(x_0/A)} = -\frac{v_0}{\omega x_0} = -\frac{0.400\,\text{m/s}}{(8.08\,\text{s}^{-1})(-0.100\,\text{m})} = 0.495.$$

A calculator gives the angle as $26.3°$, but we note from this equation that both the sine and cosine are negative, so our angle is in the third quadrant. Hence

$$\phi = 26.3° + 180° = 206.3° = 3.60\,\text{rad}.$$

(b) Again using Eq. 14–4 at $t = 0$,

$$A = x_0/\cos\phi = (-0.100\,\text{m})/\cos(3.60\,\text{rad}) = 0.112\,\text{m}.$$

(c) $x = A \cos(\omega t + \phi) = 0.112 \cos(8.08t + 3.60)$.

14–3 Energy in the Simple Harmonic Oscillator

When dealing with forces that are not constant, such as here with simple harmonic motion, it is often convenient and useful to use the energy approach.

For a simple harmonic oscillator, such as a mass m oscillating on the end of a massless spring, the restoring force is given by

$$F = -kx.$$

The potential energy function, as we have already seen in Chapter 8, is given by

$$U = -\int F\,dx = \tfrac{1}{2}kx^2,$$

where we set the constant of integration equal to zero so $U = 0$ at $x = 0$ (the equilibrium position).

The total mechanical energy is the sum of the kinetic and potential energies,

$$E = \tfrac{1}{2}mv^2 + \tfrac{1}{2}kx^2,$$

where v is the velocity of the mass m when it is a distance x from the equilibrium position. SHM can occur only if there is no friction, so the total mechanical energy E remains constant. As the mass oscillates back and forth, the energy continuously changes from potential energy to kinetic energy, and back again (Fig. 14–10). At the extreme points, $x = A$ and $x = -A$, all the energy is stored in the spring as potential energy (and is the same whether the spring is compressed or stretched to the full amplitude). At these extreme points, the mass stops momentarily as it changes direction, so $v = 0$ and:

$$E = \tfrac{1}{2}m(0)^2 + \tfrac{1}{2}kA^2 = \tfrac{1}{2}kA^2. \tag{14–10a}$$

Thus, the **total mechanical energy of a simple harmonic oscillator is proportional to the square of the amplitude**. At the equilibrium point, $x = 0$, all the energy is kinetic:

$$E = \tfrac{1}{2}mv^2 + \tfrac{1}{2}k(0)^2 = \tfrac{1}{2}mv_{\text{max}}^2, \tag{14–10b}$$

where v_{max} is the maximum velocity during the motion. At intermediate points the energy is part kinetic and part potential, and because energy is conserved

$$E = \tfrac{1}{2}mv^2 + \tfrac{1}{2}kx^2 = \tfrac{1}{2}kA^2 = \tfrac{1}{2}mv_{\text{max}}^2. \tag{14–10c}$$

We can confirm Eqs. 14–10a and b explicitly by inserting Eqs. 14–4 and 14–8 into this last relation:

$$E = \tfrac{1}{2}m\omega^2 A^2 \sin^2(\omega t + \phi) + \tfrac{1}{2}kA^2 \cos^2(\omega t + \phi).$$

Substituting with $\omega^2 = k/m$, or $kA^2 = m\omega^2 A^2 = mv_{\text{max}}^2$, and noting that $\sin^2(\omega t + \phi) + \cos^2(\omega t + \phi) = 1$, we obtain Eqs. 14–10a and b:

$$E = \tfrac{1}{2}kA^2 = \tfrac{1}{2}mv_{\text{max}}^2.$$

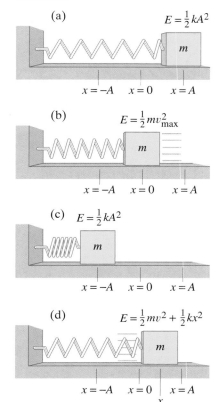

FIGURE 14–10 Energy changes from kinetic energy to potential energy and back again as the spring oscillates.

(a) $E = \tfrac{1}{2}kA^2$

$x = -A \quad x = 0 \quad x = A$

(b) $E = \tfrac{1}{2}mv_{\text{max}}^2$

$x = -A \quad x = 0 \quad x = A$

(c) $E = \tfrac{1}{2}kA^2$

$x = -A \quad x = 0 \quad x = A$

(d) $E = \tfrac{1}{2}mv^2 + \tfrac{1}{2}kx^2$

$x = -A \quad x = 0 \quad x = A$
x

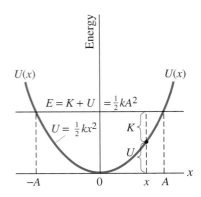

FIGURE 14–11 Graph of potential energy, $U = \frac{1}{2}kx^2$. $K + U = E = $ constant for any point x where $-A \le x \le A$. Values of K and U are indicated for an arbitrary point x.

We can now obtain an equation for the velocity v as a function of x by solving for v^2 in Eq. 14–10c:

$$v = \pm \sqrt{\frac{k}{m}(A^2 - x^2)} \qquad \text{(14–11a)}$$

or, since $v_{max} = A\sqrt{k/m}$,

$$v = \pm v_{max}\sqrt{1 - \frac{x^2}{A^2}}. \qquad \text{(14–11b)}$$

Again we see that v is a maximum at $x = 0$, and is zero at $x = \pm A$.

The potential energy, $U = \frac{1}{2}kx^2$, is plotted in Fig. 14–11. The upper horizontal line represents a particular value of the total energy $E = \frac{1}{2}kA^2$. The distance between the E line and the U curve represents the kinetic energy, K, and the motion[†] is restricted to x values between $-A$ and $+A$. These results are, of course, consistent with our full solution of the previous Section.

Energy conservation is a convenient way to obtain v, for example, if x is given (or vice versa), without having to deal with time t.

EXAMPLE 14–6 **Energy calculations.** For the simple harmonic oscillation of Example 14–4, determine (a) the total energy, (b) the kinetic and potential energies as a function of time, (c) the velocity when the mass is 0.050 m from equilibrium, (d) the kinetic and potential energies at half amplitude ($x = \pm A/2$).

SOLUTION (a) Since $k = 19.6\,\text{N/m}$ and $A = 0.100\,\text{m}$, the total energy E from Eq. 14–10a is

$$E = \tfrac{1}{2}kA^2 = \tfrac{1}{2}(19.6\,\text{N/m})(0.100\,\text{m})^2 = 9.80 \times 10^{-2}\,\text{J}.$$

(b) We have, from parts (f) and (g) of Example 14–4, $x = -(0.100\,\text{m})\cos 8.08t$ and $v = (0.808\,\text{m/s})\sin 8.08t$, so

$$U = \tfrac{1}{2}kx^2 = (9.80 \times 10^{-2}\,\text{J})\cos^2 8.08t$$
$$K = \tfrac{1}{2}mv^2 = (9.80 \times 10^{-2}\,\text{J})\sin^2 8.08t.$$

(c) We use Eq. 14–11b and find

$$v = v_{max}\sqrt{1 - x^2/A^2} = 0.70\,\text{m/s}.$$

(d) At $x = A/2 = 0.050\,\text{m}$, we have

$$U = \tfrac{1}{2}kx^2 = 2.5 \times 10^{-2}\,\text{J}$$
$$K = E - U = 7.3 \times 10^{-2}\,\text{J}.$$

CONCEPTUAL EXAMPLE 14–7 **Doubling the amplitude.** Suppose the spring in Fig. 14–10 is stretched twice as far (to $x = 2A$). What happens to (a) the energy of the system, (b) the maximum velocity, (c) the maximum acceleration?

RESPONSE (a) From Eq. 14–10a, the energy is related to the square of the amplitude, so stretching it twice as far quadruples the energy [You may protest: "I did work stretching the spring from $x = 0$ to $x = A$. Don't I do the same work stretching it from A to $2A$?" No. The force you have to exert for the second leg is more than for the first leg (because $F = -kx$), so the work done is more too.]
(b) From Eq. 14–10b, we can see that since the energy is quadrupled, the maximum velocity must be double what it was before.
(c) Since the force is twice as great when we stretch it twice as far, the acceleration here is also twice as great.

[†] See Section 8–9 for more discussion.

14–4 | Simple Harmonic Motion Related to Uniform Circular Motion

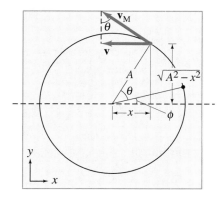

Simple harmonic motion has an interesting simple relationship to a particle rotating in a circle with uniform speed. Consider a mass m rotating in a circle of radius A with speed v_M on top of a table as shown in Fig. 14–12. As viewed from above, the motion is a circle. But a person who looks at the motion from the edge of the table, sees an oscillatory motion back and forth, and this corresponds precisely to SHM as we shall now see. What the person sees, and what we are interested in, is the projection of the circular motion onto the x axis, Fig. 14–12. To see that this motion is analogous to SHM, let us calculate the x component of the velocity v_M which is labeled v in Fig. 14–12. The two right triangles indicated in Fig. 14–12 are similar, so

$$\frac{v}{v_M} = \frac{\sqrt{A^2 - x^2}}{A}$$

or

$$v = v_M \sqrt{1 - \frac{x^2}{A^2}}.$$

This is exactly the equation for the speed of a mass oscillating with SHM, Eq. 14–11b, where $v_M = v_{max}$. Furthermore, we can see from Fig. 14–12 that if the angular displacement at $t = 0$ is ϕ, then after a time t the particle will have rotated through an angle $\theta = \omega t$, and so

$$x = A \cos(\theta + \phi) = A \cos(\omega t + \phi).$$

But what is ω here? The linear velocity v_M of our particle undergoing rotational motion is related to ω by $v_M = \omega A$ where A is the radius of the circle (see Eq. 10–4). To make one revolution requires a time T, so we also have $v_M = 2\pi A/T$ where $2\pi A$ is the circle's circumference. Hence

$$\omega = \frac{v_M}{A} = \frac{2\pi A/T}{A} = 2\pi/T = 2\pi f$$

where T is the time required for one rotation and f is the frequency. This corresponds precisely to the back-and-forth motion of a simple harmonic oscillator. Thus, the projection on the x axis of a particle rotating in a circle has the same motion as a mass undergoing SHM. Indeed, we can say that the projection of circular motion onto a straight line is SHM.

The projection of uniform circular motion onto the y axis is also simple harmonic. Thus uniform circular motion can be thought of as two simple harmonic motions operating at right angles.

(a)

(b)

FIGURE 14–12 Analysis of simple harmonic motion as a side view (b) of circular motion (a).

FIGURE 14–13 Strobe-light photo of an oscillating pendulum.

14–5 | The Simple Pendulum

A **simple pendulum** consists of a small object (the pendulum bob) suspended from the end of a lightweight cord, Fig. 14–13. We assume that the cord doesn't stretch and that its mass can be ignored relative to that of the bob. The motion of a simple pendulum moving back and forth (Fig. 14–13) with negligible friction resembles simple harmonic motion: the pendulum oscillates along the arc of a circle with equal amplitude on either side of its equilibrium point (where it hangs vertically) and as it passes through the equilibrium point it has its maximum speed. But is it really undergoing SHM? That is, is the restoring force proportional to its displacement? Let us find out.

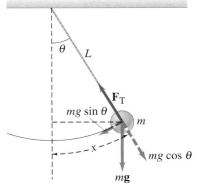

FIGURE 14–14 Simple pendulum.

The displacement of the pendulum along the arc is given by $x = L\theta$, where θ is the angle the cord makes with the vertical and L is the length of the cord, as shown in Fig. 14–14. If the restoring force is proportional to x or to θ, the motion will be simple harmonic. The restoring force is the component of the weight, mg, tangent to the arc:

$$F = -mg \sin \theta,$$

where the minus sign, as in Eq. 14–1, means the force is in the direction opposite to the angular displacement θ. Since F is proportional to the sine of θ and not to θ itself, the motion is *not* SHM. However, if θ is small, then $\sin \theta$ is very nearly equal to θ when the latter is specified in radians. This can be seen by looking at the series expansion[†] of $\sin \theta$ (or by looking at the trigonometry table inside the back cover), or by noting in Fig. 14–14 that the arc length x ($= L\theta$) is nearly the same length as the chord ($= L \sin \theta$) indicated by the straight dashed line, *if θ is small*. For angles less than 15°, the difference between θ (in radians) and $\sin \theta$ is less than 1 percent. Thus, to a very good approximation for small angles,

$$F = -mg \sin \theta \approx -mg\theta.$$

Using $x = L\theta$, we have

$$F \approx -\frac{mg}{L} x.$$

Thus, for small displacements, the motion is essentially simple harmonic, since this equation fits Hooke's law, $F = -kx$ where the effective force constant is $k = mg/L$. Thus we can write

$$\theta = \theta_{\max} \cos(\omega t + \phi)$$

where $\theta_{\max}$ is the maximum angular displacement and $\omega = 2\pi f = 2\pi/T$. To obtain ω we use Eq. 14–5, where for k we substitute mg/L: that is,[‡] $\omega = \sqrt{(mg/L)/m}$, or

$$\omega = \sqrt{\frac{g}{L}}.$$ [θ small] **(14–12a)**

Then the frequency f is

$$f = \frac{\omega}{2\pi} = \frac{1}{2\pi} \sqrt{\frac{g}{L}},$$ [θ small] **(14–12b)**

and the period T is

$$T = \frac{1}{f} = 2\pi \sqrt{\frac{L}{g}}.$$ [θ small] **(14–12c)**

A surprising result is that the period does not depend on the mass of the pendulum bob! You may have noticed this if you pushed a small child and a large one on the same swing. The period *does* depend on the length L.

We saw in Section 14–2 that the period of an object undergoing SHM, including a simple pendulum, does not depend on the amplitude. Galileo is said to have first noted this fact while watching a swinging lamp in the cathedral at Pisa (Fig. 14–15). This discovery led to the pendulum clock, the first really precise timepiece, which became the standard for centuries.

Because a pendulum does not undergo *precisely* SHM, the period does depend slightly on the amplitude, the more so for large amplitudes. The accuracy

FIGURE 14–15 The swinging motion of this lamp, hanging by a very long cord from the ceiling of the cathedral at Pisa, is said to have been observed by Galileo and to have inspired him to the conclusion that the period of a pendulum does not depend on amplitude.

[†] $\sin \theta = \theta - \dfrac{\theta^3}{3!} + \dfrac{\theta^5}{5!} - \dfrac{\theta^7}{7!} + \cdots.$

[‡] Be careful not to think that $\omega = d\theta/dt$ as in rotational motion. Here θ is the angle of the pendulum at any instant (Fig. 14–14), but we use ω now to represent *not* the rate this angle θ changes, but rather as a constant related to the period, $\omega = 2\pi f = \sqrt{g/L}$.

of a pendulum clock would be affected, after many swings, by the decrease in amplitude due to friction; but the mainspring in a pendulum clock (or the falling weight in a grandfather clock) supplies energy to compensate for the friction and to maintain the amplitude constant, so that the timing remains accurate.

EXAMPLE 14–8 **Measuring g.** For calibrating instruments that measure the acceleration of gravity, a geologist uses a simple pendulum whose length is 37.10 cm, and has a frequency of 0.8190 Hz at a particular location on the Earth. What is the acceleration of gravity at this location?

SOLUTION From Eq. 14–12b, we have

$$f = \frac{1}{2\pi}\sqrt{\frac{g}{L}}.$$

Solving for g, we obtain

$$g = (2\pi f)^2 L = (6.283 \times 0.8190\,\text{s}^{-1})^2 (0.3710\,\text{m}) = 9.824\,\text{m/s}^2.$$

*14–6 The Physical Pendulum and the Torsion Pendulum

Physical Pendulum

The term physical pendulum refers to any real extended body which oscillates back and forth, in contrast to the rather idealized simple pendulum where all the mass is assumed concentrated in the tiny pendulum bob. An example of a physical pendulum is a baseball bat suspended from the point O, as shown in Fig. 14–16. The force of gravity acts at the center of gravity (CG) of the body located a distance h from the pivot point O. The physical pendulum is best analyzed using the equations of rotational motion. The torque on a physical pendulum, calculated about point O, is

$$\tau = -mgh\sin\theta.$$

Newton's second law for rotational motion, Eq. 10–14, states that

$$\Sigma\tau = I\alpha = I\frac{d^2\theta}{dt^2},$$

where I is the moment of inertia of the body about the pivot point and $\alpha = d^2\theta/dt^2$ is the angular acceleration. Thus we have

$$I\frac{d^2\theta}{dt^2} = -mgh\sin\theta$$

or

$$\frac{d^2\theta}{dt^2} + \frac{mgh}{I}\sin\theta = 0,$$

where I is calculated about an axis through point O. For small angular amplitude, $\sin\theta \approx \theta$, so we have

$$\frac{d^2\theta}{dt^2} + \left(\frac{mgh}{I}\right)\theta = 0. \qquad \text{[small angular displacement]} \quad \textbf{(14–13)}$$

This is just the equation for SHM, Eq. 14–3, except that θ replaces x and mgh/I replaces k/m. Thus, for small angular displacements, a physical pendulum undergoes SHM, given by

$$\theta = \theta_{\text{max}}\cos(\omega t + \phi),$$

where θ_{max} is the maximum angular displacement and $\omega = 2\pi/T$. The period, T, is (see Eq. 14–7b, replacing m/k with I/mgh):

$$T = 2\pi\sqrt{\frac{I}{mgh}}. \qquad \text{[small angular displacement]} \quad \textbf{(14–14)} \qquad \textit{Period of physical pendulum}$$

FIGURE 14–16 A physical pendulum suspended from point O.

FIGURE 14–17 Example 14–9.

EXAMPLE 14–9 Moment of inertia measurement. An easy way to measure the moment of inertia of an object about any axis is to measure the period of oscillation about that axis. (*a*) Suppose a nonuniform 1.0-kg stick can be balanced at a point 42 cm from one end. If it is pivoted about that end (Fig. 14–17), it oscillates with a period of 3.0 s. What is its moment of inertia about this end? (*b*) What is its moment of inertia about an axis perpendicular to the stick through its center of mass?

SOLUTION (*a*) Given $T = 3.0$ s, and $h = 0.42$ m, we solve Eq. 14–14 for I:

$$I = mghT^2/4\pi^2 = 0.94 \text{ kg} \cdot \text{m}^2.$$

Since $I = \frac{1}{3}ML^2$ for a uniform stick of length L pivoted about one end (Fig. 10–21), do you think our stick is longer or shorter than 84 cm?

(*b*) We use the parallel-axis theorem (Section 10–8). The CM is where the stick balanced, 42 cm from the end, so from Eq. 10–17,

$$I_{CM} = I - Mh^2 = 0.94 \text{ kg} \cdot \text{m}^2 - (1.0 \text{ kg})(0.42 \text{ m})^2 = 0.76 \text{ kg} \cdot \text{m}^2.$$

Since an object does not oscillate about its CM, we can't measure I_{CM} directly, so the parallel-axis theorem provides a convenient method to determine I_{CM}.

Torsion Pendulum

FIGURE 14–18 A torsion pendulum. The disc oscillates in SHM between θ_{max} and $-\theta_{max}$.

Another type of oscillatory motion is a **torsion pendulum**, in which a disc (Fig. 14–18) or a bar (as in Cavendish's apparatus, Fig. 6–3) is suspended from a wire. The twisting (torsion) of the wire serves as the elastic force. The motion here will be SHM since the restoring torque is very closely proportional to the negative of the angular displacement,

$$\tau = -K\theta,$$

where K is a constant that depends of the properties of the system. Then

$$\omega = \sqrt{\frac{K}{I}}.$$

There is no small angle restriction here, as there is for the physical pendulum (where gravity acts), as long as the wire responds linearly in accordance with Hooke's law.

14–7 Damped Harmonic Motion

The amplitude of any real oscillating spring or swinging pendulum slowly decreases in time until the oscillations stop altogether. Figure 14–19 shows a typical graph of the displacement as a function of time. This is called **damped harmonic motion**. The damping[†] is generally due to the resistance of air and to internal friction within the oscillating system. The energy that is dissipated to thermal energy is reflected in a decreased amplitude of oscillation.

[†] To "damp" means to diminish, restrain, or extinguish, as to "dampen one's spirits."

FIGURE 14–19 Damped harmonic motion. The solid red curve represents a cosine times a decreasing exponential (the dashed curves).

Since natural oscillating systems are damped in general, why do we even talk about (undamped) simple harmonic motion? The answer is that SHM is much easier to deal with mathematically. And if the damping is not large, the oscillations can be thought of as simple harmonic motion on which the damping is superposed, as represented by the dashed curves in Fig. 14–19. Although frictional damping does alter the frequency of vibration, the effect is usually small if the damping is small. Let us look at this in more detail.

The damping force depends on the speed of the oscillating object, and opposes the motion. In some simple cases the damping force can be approximated as being directly proportional to the speed:

$$F_{\text{damping}} = -bv,$$

where b is a constant.[†] For a mass oscillating on the end of a spring, the restoring force of the spring is $F = -kx$; so Newton's second law ($ma = \Sigma F$) becomes

$$ma = -kx - bv.$$

We bring all terms to the left side of the equation and substitute $v = dx/dt$ and $a = d^2x/dt^2$ and obtain

$$m\frac{d^2x}{dt^2} + b\frac{dx}{dt} + kx = 0, \qquad \textbf{(14–15)}$$

Equation of motion for damped harmonic motion

which is the equation of motion. To solve this equation, we guess at a solution and then check to see if it works. If the damping constant b is small, x as a function of t is as plotted in Fig. 14–19, which looks like a cosine function times a factor (represented by the dashed lines) that decreases in time. A simple function that does this is the exponential, $e^{-\alpha t}$, and the solution that satisfies Eq. 14–15 is

$$x = Ae^{-\alpha t}\cos\omega' t. \qquad \textbf{(14–16)}$$

General solution: x as function of t

where A, α, and ω' are assumed to be constants, and $x = A$ at $t = 0$. We have called the angular frequency ω' (and <u>not</u> ω) because it is not the same as the ω for SHM without damping $\left(\omega = \sqrt{k/m}\right)$.

If we substitute Eq. 14–16 into Eq. 14–15 (we do this in the optional subsection below) we find that Eq. 14–15 is indeed a solution if α and ω' have the values

$$\alpha = \frac{b}{2m} \qquad \textbf{(14–17)}$$

$$\omega' = \sqrt{\frac{k}{m} - \frac{b^2}{4m^2}}. \qquad \textbf{(14–18)}$$

Thus x as a function of time t for a (lightly) damped harmonic oscillator is

$$x = Ae^{-(b/2m)t}\cos\omega' t. \qquad \textbf{(14–19)}$$

Of course a phase constant, ϕ, can be added to the argument of the cosine in Eq. 14–19. As it stands with $\phi = 0$, it is clear that the constant A in Eq. 14–19 is simply the initial displacement, $x = A$ at $t = 0$. The frequency f is

$$f = \frac{\omega'}{2\pi} = \frac{1}{2\pi}\sqrt{\frac{k}{m} - \frac{b^2}{4m^2}}. \qquad \textbf{(14–20)}$$

The frequency is lower, and the period longer, than for undamped SHM. (In many practical cases of light damping, however, ω' differs only slightly from $\omega = \sqrt{k/m}$.) This makes sense since we expect friction to slow down the motion. Equation 14–20 reduces to Eq. 14–7a, as it should, when there is no friction ($b = 0$). The constant $\alpha = b/2m$ is a measure of how quickly the oscillations decrease toward zero (Fig. 14–19). The time $t_{\text{L}} = 2m/b$ is the time taken for the oscillations to drop to $1/e$ of the original amplitude; t_{L} is called the "mean lifetime" of the oscillations. Note that the larger b is, the more quickly the oscillations die away.

[†]Such velocity-dependent forces were discussed in Section 5–5.

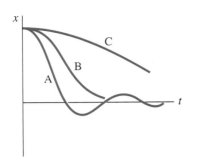

FIGURE 14–20 Underdamped (A), critically damped (B), and overdamped (C) motion.

Attached to car frame

Piston

Viscous fluid

Attached to car axle

FIGURE 14–21 Automobile spring and shock absorber to provide damping so that car won't bounce up and down endlessly.

FIGURE 14–22 These huge dampers placed in a building structure look a lot like huge automobile shock absorbers, and they serve a similar purpose—to reduce the amplitude and the acceleration of movement when the shock of an earthquake hits.

The solution, Eq. 14–19, is not valid if b is so large that

$$b^2 > 4mk$$

since then ω' (Eq. 14–18) would become imaginary. In this case the system does not oscillate at all but returns directly to its equilibrium position, as we now discuss. Three common cases of *heavily damped* systems are shown in Fig. 14–20. Curve C represents the situation when the damping is so large $(b^2 \gg 4mk)$ that it takes a long time to reach equilibrium; the system is **overdamped**. Curve A represents an **underdamped** situation in which the system makes several swings before coming to rest $(b^2 < 4mk)$ and corresponds to a more heavily damped version of Eq. 14–19. Curve B represents **critical damping**: $b^2 = 4mk$; in this case equilibrium is reached in the shortest time. These terms all derive from the use of practical damped systems such as door closing mechanisms and shock absorbers in a car (Fig. 14–21) which are usually designed to give critical damping. But as they wear out, underdamping occurs: a door slams and a car bounces up and down several times whenever it hits a bump.

In many systems, the oscillatory motion is what counts, as in clocks and watches, and damping needs to be minimized. In other systems, oscillations are the problem, such as a car's springs, so a proper amount of damping (i.e., critical) is desired. Well-designed damping is needed for all kinds of applications. Large buildings, especially in California, are now built (or retrofitted) with huge dampers to reduce earthquake damage (Fig. 14–22).

EXAMPLE 14–10 Simple pendulum with damping. A simple pendulum has a length of 1.0 m (Fig. 14–23). It is set swinging with small-amplitude oscillations. After 5.0 minutes, the amplitude is only 50% of what it was initially. (*a*) What is the value of α for the motion? (*b*) By what factor does the frequency, ω', differ from ω, the undamped frequency?

SOLUTION (*a*) The equation of motion for damped harmonic motion is

$$x = Ae^{-\alpha t} \cos \omega' t, \quad \text{where } \alpha = \frac{b}{2m} \quad \text{and} \quad \omega' = \sqrt{\frac{k}{m} - \frac{b^2}{4m^2}},$$

for motion of a mass on the end of a spring. For the simple pendulum without damping, we saw in Section 14–5 that

$$F = -mg\theta.$$

Since $F = ma = mL\dfrac{d^2\theta}{dt^2}$, then

$$L\frac{d^2\theta}{dt^2} + g\theta = 0.$$

Introducing a damping term, $b(d\theta/dt)$, we have

$$L\frac{d^2\theta}{dt^2} + b\frac{d\theta}{dt} + g\theta = 0,$$

which is the same as Eq. 14–15 with θ replacing x, and L and g replacing m and k. Thus

$$\alpha = \frac{b}{2L} \quad \text{and} \quad \omega' = \sqrt{\frac{g}{L} - \frac{b^2}{4L^2}}.$$

At $t = 0$, Eq. 14–16 with θ replacing x is

$$\theta_0 = Ae^{-\alpha \cdot 0} \cos \omega' \cdot 0 = A.$$

Then at $t = 5.0\,\text{min} = 300\,\text{s}$, the amplitude has fallen to $0.50\,A$, so

$$Ae^{-\alpha(300\,\text{s})} = 0.50\,A.$$

We solve this for α and obtain $\alpha = \ln 2.0/(300\,\text{s}) = 2.3 \times 10^{-3}\,\text{s}^{-1}$.
(b) We have $L = 1.0\,\text{m}$, so $b = 2\alpha L = 4.6 \times 10^{-3}\,\text{m/s}$. Thus $(b^2/4L^2)$ is very much less than $g/L(= 9.8\,\text{s}^{-2})$, and the angular frequency of the motion remains almost the same as that of the undamped motion. Specifically,

$$\omega' = \sqrt{\frac{g}{L}\left[1 - \frac{L}{g}\left(\frac{b^2}{4L^2}\right)\right]^{1/2}} \approx \sqrt{\frac{g}{L}}\left[1 - \frac{1}{2}\frac{L}{g}\left(\frac{b^2}{4L^2}\right)\right]$$

where we used the binomial expansion. Then, with $\omega = \sqrt{g/L}$ (Eq. 14–12a),

$$\frac{\omega - \omega'}{\omega} \approx \frac{1}{2}\frac{L}{g}\left(\frac{b^2}{4L^2}\right) = 2.7 \times 10^{-7}.$$

So ω' differs from ω by less than one part in a million.

FIGURE 14–23 Example 14–10.

*Showing $x = Ae^{-\alpha t}\cos\omega' t$ is a Solution

We start with Eq. 14–16, to see if it is a solution to Eq. 14–15. First we take the first and second derivatives

$$\frac{dx}{dt} = -\alpha Ae^{-\alpha t}\cos\omega' t - \omega' Ae^{-\alpha t}\sin\omega' t$$

$$\frac{d^2x}{dt^2} = \alpha^2 Ae^{-\alpha t}\cos\omega' t + \alpha A\omega' e^{-\alpha t}\sin\omega' t + \omega'\alpha Ae^{-\alpha t}\sin\omega' t - \omega'^2 Ae^{-\alpha t}\cos\omega' t.$$

We next substitute these relations back into Eq. 14–15 and reorganize to obtain

$$Ae^{-\alpha t}\left[(m\alpha^2 - m\omega'^2 - b\alpha + k)\cos\omega' t + (2\omega'\alpha m - b\omega')\sin\omega' t\right] = 0. \qquad \textbf{(i)}$$

The left side of this equation must equal zero for all times t, but this can only be so for certain values of α and ω'. To determine α and ω', we choose two values of t that will make their evaluation easy. At $t = 0$, $\sin\omega' t = 0$, so the above relation reduces to $A(m\alpha^2 - m\omega'^2 - b\alpha + k) = 0$, which means[†] that

$$m\alpha^2 - m\omega'^2 - b\alpha + k = 0. \qquad \textbf{(ii)}$$

And at $t = \pi/2\omega'$, $\cos\omega' t = 0$ so Eq. (i) can be valid only if

$$2\alpha m - b = 0. \qquad \textbf{(iii)}$$

From Eq. (iii) we have

$$\alpha = \frac{b}{2m}$$

and from Eq. (ii)

$$\omega' = \sqrt{\alpha^2 - \frac{b\alpha}{m} + \frac{k}{m}} = \sqrt{\frac{k}{m} - \frac{b^2}{4m^2}}.$$

Thus we see that Eq. 14–16 is a solution to the equation of motion for the damped harmonic oscillator as long as α and ω' have these specific values (already given in Eqs. 14–17 and 14–18).

[†]It would also be satisfied by $A = 0$, but this gives the trivial and uninteresting solution $x = 0$ for all t—that is, no oscillation.

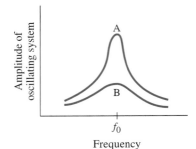

FIGURE 14–24 Resonance for lightly damped (A) and heavily damped (B) systems. (See Fig. 14–27 for a more detailed graph.)

FIGURE 14–25 This goblet breaks as it vibrates in resonance to a trumpet call.

FIGURE 14–26 (a) Large-amplitude oscillations of the Tacoma Narrows Bridge, due to gusty winds, led to its collapse (1940). (b) Collapse of freeway in California, due to the 1989 earthquake.

(a)

(b)

14–8 Forced Vibrations; Resonance

When a vibrating system is set into motion, it vibrates at its natural frequency (Eqs. 14–7a and 14–12b). However, a system may have an external force applied to it that has its own particular frequency and then we have a **forced vibration**.

For example, we might pull the mass on the spring of Fig. 14–1 back and forth at a frequency f. The mass then vibrates at the frequency f of the external force, even if this frequency is different from the **natural frequency** of the spring, which we will now denote by f_0 where (see Eqs. 14–5 and 14–7a)

$$\omega_0 = 2\pi f_0 = \sqrt{\frac{k}{m}}.$$

In a forced vibration, the amplitude of vibration, and hence the energy transferred to the vibrating system, is found to depend on the difference between ω and ω_0 as well as on the amount of damping, reaching a maximum when the frequency of the external force equals the natural frequency of the system—that is, when $f = f_0$. The amplitude is plotted in Fig. 14–24 as a function of the external frequency f. Curve A represents light damping and curve B heavy damping. The amplitude can become large when the driving frequency f is near the natural frequency, $f \approx f_0$, as long as the damping is not too large. When the damping is small, the increase in amplitude near $f = f_0$ is very large (and often dramatic). This effect is known as **resonance**. The natural vibrating frequency f_0 of a system is called its **resonant frequency**.

A simple illustration of resonance is pushing a child on a swing. A swing, like any pendulum, has a natural frequency of oscillation that depends on its length L. If you push on the swing at a random frequency, the swing bounces around and reaches no great amplitude. But if you push with a frequency equal to the natural frequency of the swing, the amplitude increases greatly. The swing clearly illustrates that at resonance, relatively little effort is required to obtain a large amplitude.

The great tenor Enrico Caruso was said to be able to shatter a crystal goblet by singing a note of just the right frequency at full voice. This is an example of resonance, for the sound waves emitted by the voice act as a forced vibration on the glass. At resonance, the resulting vibration of the goblet may be large enough in amplitude that the glass exceeds its elastic limit and breaks (Fig. 14–25).

Since material objects are, in general, elastic, resonance is an important phenomenon in a variety of situations. It is particularly important in structural engineering, although the effects are not always foreseen. For example, it has been reported that a railway bridge collapsed because a nick in one of the wheels of a crossing train set up a resonant vibration in the bridge. Indeed, marching soldiers break step when crossing a bridge to avoid the possibility that their normal rhythmic march might match a resonant frequency of the bridge. The famous collapse of the Tacoma Narrows Bridge (Fig. 14–26a) in 1940 occurred as a result of strong gusting winds driving the span into large-amplitude oscillatory motion. The Oakland freeway collapse in the 1989 California earthquake (Fig. 14–26b) involved resonant oscillation which reached large amplitude on a section that was built on mudfill that readily transmitted that frequency.

We will meet important examples of resonance later. We will also see that vibrating objects often have not one, but many resonant frequencies.

* Equation of Motion and Its Solution

We now look at the equation of motion for a forced vibration, and its solution. Suppose the external force is sinusoidal and can be represented by

$$F_{ext} = F_0 \cos \omega t,$$

where $\omega = 2\pi f$ is the angular frequency applied externally to the oscillator. Then the equation of motion (with damping) is

$$ma = -kx - bv + F_0 \cos \omega t.$$

This can be written as

$$m\frac{d^2x}{dt^2} + b\frac{dx}{dt} + kx = F_0 \cos \omega t. \qquad (14\text{-}21)$$

Equation of motion for forced vibration

The external force, on the right of the equation, is the only term that does not involve x or one of its derivatives. It is left as an exercise (Problem 63) to show that

$$x = A_0 \sin(\omega t + \phi_0) \qquad (14\text{-}22)$$

General solution: x as function of t

is a solution to Eq. 14–21, by direct substitution, where

$$A_0 = \frac{F_0}{m\sqrt{(\omega^2 - \omega_0^2)^2 + b^2\omega^2/m^2}} \qquad (14\text{-}23)$$

Amplitude

and

$$\phi_0 = \tan^{-1}\frac{\omega_0^2 - \omega^2}{\omega(b/m)}. \qquad (14\text{-}24)$$

Phase angle

Actually, the general solution to Eq. 14–21 is Eq. 14–22 plus another term of the form of Eq. 14–19 for the natural damped motion of the oscillator; this second term approaches zero in time, so in many cases we need to be concerned only with Eq. 14–22.

The amplitude of forced harmonic motion, A_0, depends strongly on the difference between the applied and the natural frequency. A plot of A_0 (Eq. 14–23) as a function of the applied frequency, ω, is shown in Fig. 14–27 (a more detailed version of Fig. 14–24) for three specific values of the damping constant b. Curve A $\left(b = \frac{1}{6}m\omega_0\right)$ represents light damping, curve B $\left(b = \frac{1}{2}m\omega_0\right)$ fairly heavy damping, and curve C $\left(b = \sqrt{2}m\omega_0\right)$ overdamped motion. The amplitude can become large when the driving frequency ω is near the natural frequency, $\omega \approx \omega_0$, as long as the damping is not too large. When the damping is small, the increase in amplitude near $\omega = \omega_0$ is very large and, as we saw, is known as *resonance*. The natural vibrating frequency ω_0 of a system is its *resonant frequency*.[†] If $b = 0$, resonance occurs at $\omega = \omega_0$ and the resonant peak (of A_0) becomes infinite; in such a case, energy is being continuously transferred into the system and none is dissipated. For real systems, b is never precisely zero, and the resonant peak is finite. The peak does not occur precisely at $\omega = \omega_0$ (because of the term $b^2\omega^2/m^2$ in the denominator of Eq. 14–23), although it is quite close to ω_0 unless the damping is very large. If the damping is large, there is little or no peak (curve C in Fig. 14–27).

[†] Sometimes the resonant frequency is defined as the actual value of ω at which the amplitude has its maximum value, and this depends somewhat on the damping constant. Except for very heavy damping, this value is quite close to ω_0.

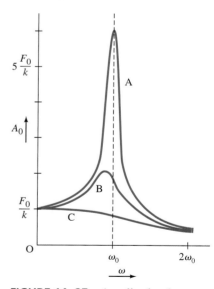

FIGURE 14–27 Amplitude of a forced harmonic oscillator as a function of ω. Curves A, B, and C correspond to light, heavy, and overdamped systems, respectively $\left(Q = m\omega_0/b = 6, 2, 0.71\right)$.

* Q value

The height and narrowness of a resonant peak is often specified by its **quality factor** or **Q value**, defined as

Quality factor or Q-value

$$Q = \frac{m\omega_0}{b}.$$

(14–25)

In Fig. 14–27, curve A has $Q = 6$, curve B has $Q = 2$, and curve C has $Q = 1/\sqrt{2}$. The smaller the damping constant b, the larger the Q value becomes, and the higher the resonance peak. The Q value is also a measure of the width of the peak. To see why, let ω_1 and ω_2 be the frequencies where the square of the amplitude A_0 has half its maximum value (we use the square because the power transferred to the system is proportional A_0^2—see the problems); then $\Delta\omega = \omega_1 - \omega_2$, which is called the *width* of the resonance peak, is related to Q by

$$\frac{\Delta\omega}{\omega_0} = \frac{1}{Q}.$$

(14–26)

(The proof of this relation, which is accurate only for weak damping, is the subject of Problem 67.) The larger the Q value, the narrower will be the resonance peak relative to its height. Thus a large Q value, representing a system of high quality, has a high, narrow resonance peak.

Summary

A vibrating object undergoes **simple harmonic motion** (SHM) if the restoring force is proportional to the displacement,

$$F = -kx.$$

The maximum displacement is called the **amplitude**.

The **period**, T, is the time required for one complete cycle (back and forth), and the **frequency**, f, is the number of cycles per second; they are related by

$$f = \frac{1}{T}.$$

The period of vibration for a mass m on the end of an ideal massless spring is given by

$$T = 2\pi \sqrt{m/k}.$$

SHM is **sinusoidal**, which means that the displacement as a function of time follows a sine or cosine curve. The general solution can be written

$$x = A \cos(\omega t + \phi)$$

where A is the amplitude, ϕ is the **phase angle**, and

$$\omega = 2\pi f = \sqrt{\frac{k}{m}}.$$

The values of A and ϕ depend on the **initial** conditions (x and v at $t = 0$).

During SHM, the total energy $E = \frac{1}{2}mv^2 + \frac{1}{2}kx^2$ is continually changing from potential to kinetic and back again.

A swinging **simple pendulum** of length L approximates SHM if its amplitude is small and friction can be ignored. Its period is then given by (for small amplitudes)

$$T = 2\pi \sqrt{L/g},$$

where g is the acceleration of gravity.

When friction is present (for all real springs and pendulums), the motion is said to be **damped**. The maximum displacement decreases in time, and the mechanical energy is eventually all transformed to thermal energy. If the friction is very large, so no oscillations occur, the system is said to be **overdamped**. If the friction is small enough that oscillations occur, the system is **underdamped**, and the displacement is given by

$$x = Ae^{-\alpha t} \cos \omega' t,$$

where α and ω' are constants. For a **critically damped** system, no oscillations occur and equilibrium is reached in the shortest time.

If an oscillating force is applied to a system capable of vibrating, the amplitude of vibration can be very large if the frequency of the applied force is near the **natural** (or **resonant**) **frequency** of the oscillator; this is called **resonance**.

Questions

1. Give some everyday examples of vibrating objects. Which follow SHM, at least approximately?
2. Contrast the equations for x, v, and a for uniformly accelerated linear motion ($a = $ constant) with those for simple harmonic motion. Discuss their similarities and differences.
3. If a particle undergoes SHM with amplitude A, what is the total distance it travels in one period?
4. Real springs have mass. How will the true period and frequency differ from those given by the equations for a mass oscillating on the end of an idealized massless spring?
5. For a simple harmonic oscillator, when (if ever) are the displacement and velocity vectors in the same direction? When are the displacement and acceleration vectors in the same direction?
6. How could you double the maximum speed of a SHO?
7. A mass m hangs from a spring with stiffness constant k. The spring is cut in half and the same mass hung from it. Will the new arrangement have a higher or a lower stiffness constant than the original spring?
8. Two equal masses are attached to separate identical springs next to one another. One mass is pulled so its spring stretches 20 cm and the other pulled so its spring stretches only 10 cm. The masses are released simultaneously. Which mass reaches the equilibrium point first?
9. A 10-kg fish is attached to the hook of a vertical spring scale, and is then released. Describe the scale reading as a function of time.
10. Is the motion of a piston in an automobile engine simple harmonic? Explain.
11. If a pendulum clock is accurate at sea level, will it gain or lose time when taken to high altitude?

12. A tire swing hangs from a branch nearly to the ground. How could you estimate the height of the branch using only a stopwatch?
13. Does a car bounce on its springs faster when it is empty or when it is fully loaded?
14. What happens to the period of a playground swing if you rise up from sitting to a standing position?
15. Describe the possible motion of a solid object that is suspended so it is free to rotate about its center of gravity. Is it a physical pendulum?
16. A thin uniform rod of mass m is suspended from one end and oscillates with a frequency f. If a small sphere of mass $2m$ is attached to the other end, does the frequency increase or decrease? Explain.
17. Is the acceleration of a simple harmonic oscillator ever zero? If so, when? What about a damped harmonic oscillator?
18. A tuning fork of natural frequency 264 Hz sits on a table at the front of a room. At the back of the room, two tuning forks, one of natural frequency 260 Hz and one of 420 Hz are initially silent, but when the tuning fork at the front of the room is set into vibration, the 260-Hz fork spontaneously begins to vibrate but the 420-Hz fork does not. Explain.
19. Give several everyday examples of resonance.
20. Is a rattle in a car ever a resonance phenomenon? Explain.
21. Over the years, buildings have been able to be built out of lighter and lighter materials. How has this affected the natural vibration frequencies of buildings and the problems of resonance due to passing trucks, airplanes, or by wind and other natural sources of vibration?

Problems

Sections 14–1 and 14–2

1. (I) If a particle undergoes SHM with amplitude 0.15 m, what is the total distance it travels in one period?
2. (I) A fisherman's scale stretches 2.8 cm when a 3.7-kg fish hangs from it. (a) What is the spring constant? (b) What will be the amplitude and frequency of vibration if the fish is pulled down 2.5 cm more and released so that it vibrates up and down?
3. (I) When an 80-kg person climbs into a 1000-kg car, the car's springs compress vertically by 1.40 cm. What will be the frequency of vibration when the car hits a bump? (Ignore damping.)
4. (I) (a) What is the equation describing the motion of a spring that is stretched 8.8 cm from equilibrium and then released, and whose period is 0.75 s? (b) What will be its displacement after 1.8 s?
5. (II) A small fly of mass 0.60 g is caught in a spider's web. The web vibrates predominantly with a frequency of 10 Hz. (a) What is the value of the effective spring constant k for the web? (b) At what frequency would you expect the web to vibrate if an insect of mass 0.40 g were trapped?
6. (II) Determine the phase constant ϕ in Eq. 14–4 if, at $t = 0$, the oscillating mass is at (a) $x = -A$, (b) $x = 0$, (c) $x = A$, (d) $x = \frac{1}{2}A$, (e) $x = -\frac{1}{2}A$, (f) $x = A/\sqrt{2}$.
7. (II) A mass on the end of a spring is stretched a distance x_0 from equilibrium and released. At what distance from equilibrium will it have (a) velocity equal to half its maximum velocity and (b) acceleration equal to half its maximum acceleration?
8. (II) A balsa wood block of mass 50 g floats on a lake, bobbing up and down at a frequency of 2.5 Hz. (a) What is the value of the effective spring constant of the water? (b) A partially filled water bottle of mass 0.25 kg and almost the same size and shape of the balsa block is tossed into the water. At what frequency would you expect the bottle to bob up and down? Assume SHM.
9. (II) At what displacement from equilibrium is the speed of a SHO half the maximum value?

10. (II) A mass m at the end of a spring vibrates with a frequency of 0.88 Hz; when an additional 1.25 kg mass is added to m, the frequency is 0.48 Hz. What is the value of m?

11. (II) A block of mass m is supported by two parallel vertical springs, with spring constant k_1 and k_2 (Fig. 14–28). What will be the frequency of vibration?

FIGURE 14–28 Problem 11.

12. (II) The graph of displacement versus time for a small mass at the end of spring is shown in Fig. 14–29. At $t = 0$, $x = 0.43$ cm. (a) If $m = 14.3$ g, find the spring constant, k. (b) Write the equation for displacement x as a function of time.

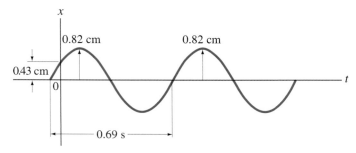

FIGURE 14–29 Problem 12.

13. (II) The position of a SHO as a function of time is given by $x = 3.8 \cos(7\pi t/4 + \pi/6)$ where t is in seconds and x in meters. Find (a) the period and frequency, (b) the position and velocity at $t = 0$, and (c) the velocity and acceleration at $t = 2.0$ s.

14. (II) A tuning fork vibrates at a frequency of 264 Hz and the tip of each prong moves 1.5 mm to either side of center. Calculate (a) the maximum speed and (b) the maximum acceleration of the tip of a prong.

15. (II) A spring vibrates with a frequency of 3.0 Hz when a weight of 0.50 kg is hung from it. What will its frequency be if only 0.35 kg hangs from it?

16. (II) (a) Show that

$$x = a \sin \omega t + b \cos \omega t$$

is a general solution of Eq. 14–3, and (b) determine the constants a and b in terms of the A and ϕ of Eq. 14–4.

17. (II) If a mass m hangs from a vertical spring, as shown in Fig. 14–3, show that $F = -kx$ holds for stretching or compression of the spring, where x is the displacement from the (vertical position) equilibrium point.

18. (II) A mass of 1.62 kg stretches a vertical spring 0.315 m. If the spring is stretched an additional 0.130 m and released, how long does it take to reach the (new) equilibrium position again?

19. (II) A spring of force constant 345 N/m vibrates with an amplitude of 22.0 cm when 0.250 kg hangs from it. (a) What is the equation describing this motion as a function of time? Assume that the mass passes downwards through the equilibrium point at $t = 0$. (b) At what times will the spring have its maximum and minimum extensions? Take y positive upwards.

20. (II) A 450-g object oscillates from a vertically hanging light spring once every 0.55 s. (a) Write down the equation giving its position y (+ upward) as a function of time t, assuming it started by being compressed 10 cm from the equilibrium position (where $y = 0$), and released. (b) How long will it take to get to the equilibrium position for the first time? (c) What will be its maximum speed? (d) What will be its maximum acceleration, and where will it first be attained?

21. (II) A uniform meter stick of mass M is pivoted on a hinge at one end and held horizontal by a spring with spring constant k attached at the other end (Fig. 14–30). If the stick oscillates up and down slightly, what is its frequency? [Hint: Write a torque equation about the hinge.]

FIGURE 14–30
Problem 21.

22. (III) A mass m is at rest on the end of a spring of spring constant k. At $t = 0$ it is given an impulse J by a hammer. Write the formula for the subsequent motion in terms of m, k, J, and t.

23. (III) A mass m is connected to two springs, with spring constants k_1 and k_2, in two different ways as shown in Fig. 14–31a and b. Show that the period for the configuration shown in part (a) is given by

$$T = 2\pi \sqrt{m\left(\frac{1}{k_1} + \frac{1}{k_2}\right)}$$

and for that in part (b) is given by

$$T = 2\pi \sqrt{\frac{m}{k_1 + k_2}}.$$

Ignore friction.

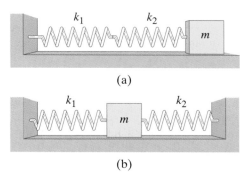

FIGURE 14–31
Problem 23.

24. (III) Two equal masses, m_1 and m_2, are connected by three identical springs of spring constant k as shown in Fig. 14–32. (a) Apply $\Sigma F = ma$ to each mass and obtain two differential equations for the displacements x_1 and x_2. (b) Determine the possible frequencies of vibration by assuming a solution of the form $x_1 = A_1 \cos \omega t$, $x_2 = A_2 \cos \omega t$.

FIGURE 14–32 Problem 24.

25. (III) A spring with spring constant 250 N/m vibrates with an amplitude of 12.0 cm when 0.380 kg hangs from it. (a) What is the equation describing this motion as a function of time? Assume the mass passes through the equilibrium point, toward positive x (upward), at $t = 0.110$ s. (b) At what times will the spring have its maximum and minimum lengths? (c) What is the displacement at $t = 0$? (d) What is the force exerted by the spring at $t = 0$? (e) What is the maximum speed and when is it first reached after $t = 0$?

Section 14–3

26. (I) (a) At what displacement of a SHO is the energy half kinetic and half potential? (b) What fraction of the total energy of a SHO is kinetic and what fraction potential when the displacement is half the amplitude?

27. (I) A 2.00-kg mass vibrates according to the equation $x = 0.650 \cos 8.40t$ where x is in meters and t in seconds. Determine (a) the amplitude, (b) the frequency, (c) the total energy, and (d) the kinetic energy and potential energy when $x = 0.260$ m.

28. (II) A 0.35-kg mass at the end of a spring vibrates 3.0 times per second with an amplitude of 0.15 m. Determine (a) the velocity when it passes the equilibrium point, (b) the velocity when it is 0.10 m from equilibrium, (c) the total energy of the system, and (d) the equation describing the motion of the mass, assuming that at $t = 0$, x was a maximum.

29. (II) It takes a force of 95.0 N to compress the spring of a popgun 0.185 m to "load" a 0.200-kg ball. With what speed will the ball leave the gun?

30. (II) A 0.0125 kg bullet strikes a 0.300-kg block attached to a fixed horizontal spring whose spring constant is 2.25×10^3 N/m and sets it into vibration with an amplitude of 12.4 cm. What was the speed of the bullet if the two objects move together after impact?

31. (II) If one vibration has 10 times the energy of a second one of equal frequency, but the first's spring constant k is twice as large as the second's, how do their amplitudes compare?

32. (II) A mass of 240 g oscillates on a horizontal frictionless surface at a frequency of 3.5 Hz and with amplitude of 4.5 cm. (a) What is the effective spring constant for this motion? (b) How much energy is involved in this motion?

33. (II) A mass sitting on a horizontal, frictionless surface is attached to one end of a spring; the other end is fixed to a wall. 3.0 J of work is required to compress the spring by 0.12 m. If the mass is released from rest with the spring compressed, it experiences a maximum acceleration of 15 m/s². Find the value of (a) the spring constant and (b) the mass.

34. (II) An object with mass 2.1 kg is executing simple harmonic motion, attached to a spring with spring constant $k = 280$ N/m. When the object is 0.020 m from its equilibrium position, it is moving with a speed of 0.55 m/s. (a) Calculate the amplitude of the motion. (b) Calculate the maximum velocity attained by the object.

35. (II) Nikita devised the following method of measuring the muzzle velocity of a rifle (Fig. 14–33). She fires a bullet into a 6.023-kg wooden block resting on a smooth surface, and attached to a spring of spring constant $k = 142.7$ N/m. The bullet, whose mass is 7.870 g, remains embedded in the wooden block. She measures the distance that the block recoils and compresses the spring to be 9.460 cm. What is the speed v of the bullet?

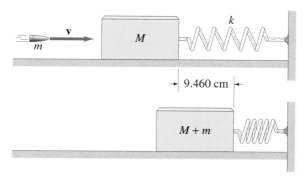

FIGURE 14–33 Problem 35.

36. (II) At $t = 0$, a 650-g mass at rest on the end of a horizontal spring ($k = 184$ N/m) is struck by a hammer which gives it an initial speed of 2.26 m/s. Determine (a) the period and frequency of the motion, (b) the amplitude, (c) the maximum acceleration, (d) the position as a function of time, (e) the total energy, and (f) the kinetic energy when $x = 0.40A$ where A is the amplitude.

37. (II) Obtain the displacement x as a function of time for the simple harmonic oscillator using the conservation of energy, Eq. 14–10. [Hint: Integrate Eq. 14–11a with $v = dx/dt$.]

Section 14–5

38. (I) A pendulum makes 42 vibrations in 50 s. What is its (a) period, and (b) frequency?

39. (I) How long must a simple pendulum be if it is to make exactly one swing per second? That is, one complete vibration takes exactly two seconds.

40. (I) What is the period of a simple pendulum on Mars, where the acceleration of gravity is about 0.37 that on Earth, if the pendulum has a period of 0.80 s on Earth?

41. (II) (a) Determine the length of a simple pendulum whose period is 1.00 s. (b) What would be the period of a 1.00-m-long simple pendulum?

42. (II) What is the period of a simple pendulum 73 cm long (a) on the Earth, and (b) when it is in a freely falling elevator?

43. (II) A simple pendulum is 0.30 m long. At $t = 0$ it is released starting at an angle of 14°. Ignoring friction, what will be the angular position of the pendulum at (a) $t = 0.65$ s, (b) $t = 1.95$ s, and (c) $t = 5.00$ s?

44. (II) Derive a formula for the maximum speed v_0 of a simple pendulum bob in terms of g, the length L, and the maximum angle of swing θ.

45. (II) A simple pendulum vibrates with an amplitude of 10.0°. What fraction of the time does it spend between +5.0 and −5.0°? Assume SHM.

46. (II) The length of a simple pendulum is 0.68 m and it is released at an angle of 12° to the vertical. (a) With what frequency does it vibrate? (b) What is the pendulum bob's speed when it passes through the lowest point of the swing?

* Section 14–6

* 47. (II) A plywood disk of radius 20.0 cm and mass 3.00 kg has a small hole drilled through it, 2.00 cm from its edge (Fig. 14–34). The disk is hung from the wall by means of a metal pin through the hole, and is used as a pendulum. What is the period of this pendulum for small oscillations?

FIGURE 14–34   Problem 47.

* 48. (II) A pendulum consists of a tiny bob of mass M and a uniform cord of mass m and length L. (a) Determine a formula for the period. (b) What would be the fractional error if one used the formula for a simple pendulum, Eq. 14–12a?

* 49. (II) The balance wheel of a watch is a thin ring of radius 0.95 cm and oscillates with a frequency of 3.10 Hz. If a torque of 1.1×10^{-5} m·N causes the wheel to rotate 60°, calculate the mass of the balance wheel.

* 50. (II) (a) Determine the equation of motion (for θ as a function of time) for a torsion pendulum, Fig. 14–18, and show that the motion is simple harmonic. (b) Show that the period T is $T = 2\pi \sqrt{I/K}$. [The balance wheel of a mechanical watch is an example of a torsion pendulum in which the restoring torque is applied by a coil spring.]

* 51. (II) The human leg can be compared to a physical pendulum, with a "natural" swinging period at which walking is easiest. Consider the leg as two rods joined rigidly together at the knee; the axis for the leg is the hip joint. The length of each rod is about the same, 50 cm. The upper rod has a mass of 7.0 kg and the lower rod has a mass of 4.0 kg. (a) Calculate the natural swinging period of the system. (b) Check your answer by standing on a chair and measuring the time for one or more complete back-and-forth swings. The effect of a shorter leg is, of course, a shorter swinging period, enabling a faster, although a shorter, "natural" stride.

* 52. (II) A student wants to use a meter stick as a pendulum. She plans to drill a small hole through the meter stick and suspend it from a smooth pin attached to the wall (Fig. 14–35). Where in the meter stick should she drill the hole to obtain the shortest possible period? How short an oscillation period can she obtain with a meter stick in this way?

FIGURE 14–35   Problem 52.

* 53. (II) A meter stick is hung at its center from a thin wire (Fig. 14–36a). It is twisted and oscillates with a period of 6.0 s. The meter stick is sawed off to a length of 70.0 cm. This piece is again balanced at its center and set in oscillation (Fig. 14–36b). With what period does it oscillate?

FIGURE 14–36   Problem 53.

* 54. (II) An aluminum disk, 12.5 cm in diameter and 500 g in mass, is mounted on a vertical shaft with an air bearing (Fig. 14–37). The disk also floats on an air bearing. One end of a flat coil spring is attached to the disk, the other end to the base of the apparatus. The disk is set into rotational oscillation and the frequency is 0.331 Hz. What is the torsional spring constant K $(\tau = -K\theta)$?

FIGURE 14–37   Problem 54.

Section 14–7

55. (I) A 750-g block oscillates on the end of a spring whose force constant is $k = 56.0 \, \text{N/m}$. The mass moves in a fluid which offers a resistive force $F = -bv$, where $b = 0.162 \, \text{N} \cdot \text{s/m}$. (a) What is the period of the motion? (b) What is the fractional decrease in amplitude per cycle? (c) Write the displacement as a function of time if at $t = 0$, $x = 0$, and at $t = 1.00 \, \text{s}$, $x = 0.120 \, \text{m}$.

56. (II) A physical pendulum consists of an 80-cm-long, 300-g-mass, uniform wooden rod hung from a nail near one end (Fig. 14–38). The motion is damped because of friction in the pivot; the damping force is approximately proportional to $d\theta/dt$. The rod is set in oscillation by displacing it $15°$ from its equilibrium position and releasing it. After 8.0 seconds, the amplitude of the oscillation has been reduced to $5.5°$. If the angular displacement can be written as $\theta = A e^{-\alpha t} \cos \omega' t$, find (a) α, (b) the approximate period of the motion, and (c) how long it takes for the amplitude to be reduced to $\frac{1}{2}$ of its original value.

FIGURE 14–38 Problem 56.

57. (II) A damped harmonic oscillator loses 5.0 percent of its mechanical energy per cycle. (a) By what percentage does its frequency differ from the natural frequency $\omega_0 = \sqrt{k/m}$? (b) After how many periods will the amplitude have decreased to $1/e$ of its original value?

58. (III) (a) Show that the total mechanical energy, $E = \frac{1}{2}mv^2 + \frac{1}{2}kx^2$, as a function of time for a lightly damped harmonic oscillator is

$$E = \frac{1}{2}kA^2 e^{-(b/m)t} = E_0 e^{-(b/m)t},$$

where E_0 is the total mechanical energy at $t = 0$. (Assume $\omega' \gg b/2m$.) (b) Show that the fractional energy lost per period is

$$\frac{\Delta E}{E} = \frac{2\pi b}{m\omega_0} = \frac{2\pi}{Q},$$

where $\omega_0 = \sqrt{k/m}$ and $Q = m\omega_0/b$ is called the *quality factor* or *Q value* of the system. A larger Q value means the system can undergo oscillations for a longer time.

59. (III) A glider on an air track is connected by springs to either end of the track (Fig. 14–39). Both springs have the same spring constant, k, and the glider has mass M. (a) Determine the frequency of the oscillation, assuming no damping, if $k = 100 \, \text{N/m}$ and $M = 200$ grams. (b) It is observed that after 55 oscillations, the amplitude of the oscillation has dropped to one-half of its initial value. Estimate the value of α, using Eq. 14–16. (c) How long does it take the amplitude to decrease to one-quarter of its initial value?

FIGURE 14–39 Problem 59.

Section 14–8

*60. (II) (a) At resonance $(\omega = \omega_0)$, what is the value of the phase angle ϕ_0? (b) What, then, is the displacement at a time when the driving force F_{ext} is a maximum, and at a time when $F_{\text{ext}} = 0$? (c) What is the phase difference (in degrees) between the driving force and the displacement in this case?

*61. (II) Construct an accurate resonance curve, from $\omega = 0$ to $\omega = 2\omega_0$, for $Q = 4.0$.

*62. (II) The amplitude of a driven harmonic oscillator reaches a value of $28.6 \, F_0/m$ at a resonant frequency of $382 \, \text{Hz}$. What is the Q value of this system?

*63. (II) By direct substitution, show that Eq. 14–22, with Eqs. 14–23 and 14–24, is a solution of the equation of motion (Eq. 14–21) for the forced oscillator.

*64. (II) Differentiate Eq. 14–23 to show that the resonant amplitude peaks at

$$\omega = \sqrt{\omega_0^2 - b^2/2m^2}.$$

*65. (III) Consider a simple pendulum (point mass bob) 0.50 m long with a Q of 400. (a) How long does it take for the amplitude (assumed small) to decrease by two-thirds? (b) If the amplitude is 2.0 cm and the bob has mass 0.20 kg, what is the initial energy loss rate of the pendulum in watts? (c) If we are to stimulate resonance with a sinusoidal driving force, how close must the driving frequency be to the natural frequency of the pendulum?

*66. (III) *Power transferred to driven oscillator.* (a) Show that the power input to a forced oscillator due to the external force F_{ext} is

$$P = F_{\text{ext}} v = \frac{F_0^2 \omega \cos \phi_0 \cos^2 \omega t - \frac{1}{2} F_0^2 \omega \sin \phi_0 \sin 2\omega t}{m\sqrt{(\omega^2 - \omega_0^2)^2 + \omega^2 b^2/m^2}}.$$

(b) Show that the average power input, averaged over one (or many) full cycles, is

$$\overline{P} = \frac{\omega F_0^2 \cos \phi_0}{2m\sqrt{(\omega^2 - \omega_0^2)^2 + \omega^2 b^2/m^2}} = \frac{1}{2}\omega A_0 F_0 \cos \phi_0$$

or

$$\overline{P} = \frac{1}{2} F_0 v_{\text{max}} \cos \phi_0,$$

where v_{max} is the maximum value of dx/dt. (c) Plot $\overline{P}$ versus ω from $\omega = 0$ to $\omega = 2\omega_0$ for $Q = 6.0$. Note that although the amplitude is not zero at $\omega = 0$, $\overline{P}$ is zero at $\omega = 0$.

*67. (III) Derive Eq. 14–26.

General Problems

68. A 52-kg person jumps from a window to a fire net 20.0 m below, which stretches the net 1.1 m. Assume that the net behaves like a simple spring, and calculate how much it would stretch if the same person were lying in it. How much would it stretch if the person jumped from 35 m?

69. The length of a simple pendulum is 0.63 m, the pendulum bob has a mass of 365 grams, and it is released at an angle of 15° to the vertical. (a) With what frequency does it vibrate? (b) What is the pendulum bob's speed when it passes through the lowest point of the swing? Assume SHM. (c) What is the total energy stored in this oscillation assuming no losses?

70. A 0.650-kg mass vibrates according to the equation $x = 0.25 \sin (5.50t)$ where x is in meters and t is in seconds. Determine (a) the amplitude, (b) the frequency, (c) the period, (d) the total energy, and (e) the kinetic energy and potential energy when x is 10 cm.

71. A bungee jumper with mass 72.0 kg jumps from a high bridge. After reaching his lowest point, he oscillates up and down, hitting a low point eight more times in 34.7 s. He finally comes to rest 25.0 m below the level of the bridge. Calculate the spring constant and the unstretched length of the bungee cord.

72. (a) A crane has hoisted a 1200-kg car at the junkyard. The steel crane cable is 20.0 m long and has a diameter of 6.4 mm. A breeze starts the car bouncing at the end of the cable. What is the period of the bouncing? [Hint: Refer to Table 12–1]. (b) What amplitude of bouncing will likely cause the cable to snap? (See Table 12–2, and assume Hooke's law holds all the way up to the breaking point.)

73. An energy-absorbing car bumper has a spring constant of 500 kN/m. Find the maximum compression of the bumper if the car, with mass 1500 kg, collides with a wall at a speed of 2.0 m/s (approximately 5 mi/h).

74. A "seconds" pendulum has a period of exactly 2.0 seconds—each one-way swing takes 1.0 s. What is the length of a seconds pendulum in Austin, Texas, where $g = 9.793 \text{ m/s}^2$? If the pendulum is moved to Paris, where $g = 9.809 \text{ m/s}^2$, by how many millimeters must we lengthen the pendulum? What is the length of a "seconds" pendulum on the Moon, where $g = 1.62 \text{ m/s}^2$?

75. A block of jello rests on a cafeteria plate as shown in Fig. 14–40 (which also gives the dimensions of the block). You push it sideways as shown, and then you let go. The jello springs back and begins to vibrate. In analogy to a mass vibrating on a spring, estimate the frequency of this vibration, given that the shear modulus of jello is 520 N/m² and its density is 1300 kg/m³.

FIGURE 14–40 Problem 75.

76. A simple pendulum oscillates with frequency f. What is its frequency if it accelerates at $\frac{1}{2}g$ (a) upward, and (b) downward?

77. A 420-kg wooden raft floats on a lake. When a 75-kg man stands on the raft, it sinks 3.5 cm deeper into the water. When he steps off, the raft vibrates for a while. (a) What is the frequency of vibration? (b) What is the total energy of vibration (ignoring damping)?

78. A 5.0-kg box slides into a spring of spring constant 310 N/m (Fig. 14–41), compressing it 24 cm. (a) What is the incoming speed of the block? (b) How long is the box in contact with the spring before it bounces off in the opposite direction?

FIGURE 14–41 Problem 78.

79. A 1.60-kg table is supported on four springs. A 0.55-kg chunk of modeling clay is held above the table and dropped so that it hits the table with a speed of 1.65 m/s (Fig. 14–42). The clay makes an inelastic collision with the table and the table and clay oscillate up and down. After a long time the table comes to rest 6.0 cm below its original position. (a) What is the effective spring constant of all four springs taken together? (b) With what maximum amplitude does the platform oscillate?

FIGURE 14–42 Problem 79.

80. A diving board oscillates with simple harmonic motion of frequency 5.0 cycles per second. What is the maximum amplitude with which the end of the board can vibrate in order that a pebble placed there (Fig. 14–43) does not lose contact with the board during the oscillation?

FIGURE 14–43 Problem 80.

81. A rectangular block of wood floats in a calm lake. Show that, if friction is ignored, when the block is pushed gently down into the water, it will then vibrate with SHM. Also, determine an equation for the force constant.

82. A mass m is gently placed on the end of a freely hanging spring. The mass then falls 22.0 cm before it stops and begins to rise. What is the frequency of the oscillation?

83. The water in a U-shaped tube is displaced an amount Δx from equilibrium. (The level in one side is $2\Delta x$ above the level in the other side.) If friction is neglected, will the water oscillate harmonically? Determine a formula for the equivalent spring constant k. Does k depend on the density of the liquid, the cross-section of the tube, or the length of the water column?

84. In some diatomic molecules, the force each atom exerts on the other can be approximated by $F = -C/r^2 + D/r^3$, where C and D are positive constants. (a) Graph F versus r from $r = 0$ to $r = 2D/C$. (b) Show that equilibrium occurs at $r = r_0 = D/C$. (c) Let $\Delta r = r - r_0$ be a small displacement from equilibrium, where $\Delta r \ll r_0$. Show that for such small displacements, the motion is approximately simple harmonic, and (d) determine the force constant. (e) What is the period of such motion? [Hint: Assume one atom is kept at rest.]

85. A mass m is connected to two springs with equal spring constants k (Fig. 14–44). In the horizontal position shown, each spring is stretched by an amount Δa. The mass is raised vertically and begins to oscillate up and down. Assuming that the displacement is small, and ignoring gravity, show that the motion is simple harmonic and find the period. [This kind of motion is called transverse oscillation. If we hooked many masses together like this, their transverse oscillations would be a model for a vibrating string in a musical instrument.]

FIGURE 14–44 Problem 85.

86. Carbon dioxide is a linear molecule. The carbon–oxygen bonds in this molecule act very much like springs. Figure 14–45 shows one possible way the oxygen atoms in this molecule can vibrate: the oxygen atoms vibrate symmetrically in and out, while the central carbon atom remains at rest. Hence each oxygen atom acts like a simple harmonic oscillator with a mass equal to the mass of an oxygen atom. It is observed that this oscillation occurs with a frequency of $f = 2.83 \times 10^{13}$ Hz. What is the spring constant of the C–O bond?

FIGURE 14–45 Problem 86, the CO_2 molecule.

87. A thin, straight, uniform rod of length $l = 1.00$ m and mass $m = 160$ g hangs from a pivot at one end. (a) What is its period for small-amplitude oscillations? (b) What is the length of a simple pendulum that will have the same period?

88. Imagine that a 10-cm-diameter circular hole were drilled all the way through the center of the Earth (Fig. 14–46). At one end of the hole, you drop an apple into the hole. Show that, if you assume that the Earth has a constant density, the subsequent motion of the apple is simple harmonic. How long will the apple take to return? Assume that we can ignore all frictional effects.

FIGURE 14–46 Problem 88.

Waves—such as water waves or waves traveling along a cord or slinky—travel away from their source. The waves on a stretched cord shown in these four photographs, however, are "standing waves." They seem not to be traveling; but each can be thought of as the sum of a wave traveling to the right interfering with its reflection traveling back toward the left. We can also think of these standing waves as oscillations of the stretched cord at resonance. Each of these standing waves occurs at a particular frequency. Can you guess how the frequency of each is related to the others?

Wave Motion

When you throw a stone into a lake or pool of water, circular waves form and move outward, Fig. 15–1. Waves will also travel along a cord (or a "slinky") that is stretched out straight on a table if you vibrate one end back and forth as shown in Fig. 15–2. Water waves and waves on a cord are two common examples of wave motion.

Vibrations and wave motion are intimately related subjects. Waves—whether ocean waves, waves on a string, earthquake waves, or sound waves in air—have as their source a vibration. In the case of sound, not only is the source a vibrating object, but so is the detector—the eardrum or the membrane of a microphone.

FIGURE 15–1 A stone thrown into a lake: water waves spread outward from the source.

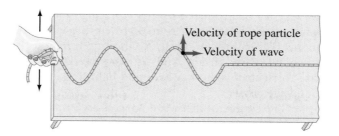

Indeed, the medium through which such "mechanical waves" travel itself vibrates (such as air for sound waves). In this chapter we will mainly be concerned with **mechanical waves** such as waves on water or on a cord. In Chapter 16 we will study sound waves, and in later chapters we will encounter other forms of wave motion, including electromagnetic waves and light.

If you have ever watched ocean waves moving toward shore (before they break), you may have wondered if the waves were carrying water from out at sea into the beach. This is, in fact, not the case.[†] Water waves move with a recognizable velocity. But each particle (or molecule) of the water itself merely oscillates about an equilibrium point. This is clearly demonstrated by observing leaves on a pond as waves move by. The leaves (or a cork) are not carried forward by the waves, but simply oscillate about an equilibrium point because this is the motion of the water itself. Similarly, the wave on the rope or cord of Fig. 15–2 moves to the right, but each piece of the rope only vibrates to and fro. These are general features of mechanical waves: (1) a wave can move over large distances with a particular speed; and (2) each particle of the medium in which the wave travels (the water or the rope) oscillates about an equilibrium point which is simple harmonic if the wave is sinusoidal. Thus, although a wave is not matter, the wave pattern can travel in matter.

Waves carry energy from one place to another. Energy is given to a water wave, for example, by a rock thrown into the water, or by wind far out at sea. The energy is transported by waves to the shore. The oscillating hand in Fig. 15–2 transfers energy to the rope, which is then transported down the rope and can be transferred to an object at the other end. All forms of traveling waves transport energy.

15–1 Characteristics of Wave Motion

Let us look at how a wave is formed and how it comes to "travel." We first look at a single wave bump or **pulse**. A single pulse can be formed on a rope by a quick up and down motion of the hand, Fig. 15–3. The hand pulls up on one end of the rope, and because the end piece is attached to adjacent pieces, these also feel an upward force and they too begin to move upward. As each succeeding piece of rope moves upward, the wave crest moves outward along the rope. Meanwhile, the end piece of rope has been returned to its original position by the hand, and as each succeeding piece of rope reaches its peak position, it too is pulled back down again by the adjacent section of rope. Thus, the source of a traveling wave pulse is a disturbance, and cohesive forces between adjacent pieces of rope cause the pulse to travel outward. Waves in other media are created and propagate outward in a similar fashion.

A **continuous** or **periodic wave**, such as that shown starting in Fig. 15–2, has as its source a disturbance that is continuous and oscillating. That is, the source is a *vibration* or *oscillation*. In Fig. 15–2, a hand oscillates one end of the rope. Water

FIGURE 15–3 Motion of a wave pulse. Arrows indicate velocity of rope particles. Note that the wave velocity is to the right.

(a)

(b)

(c)

(d)

[†]Do not be confused by the "breaking" of ocean waves, which occurs when a wave interacts with the ground in shallow water and hence is no longer a simple wave.

waves may be produced by any vibrating object placed at the surface such as your hand; or the water itself is made to vibrate when wind blows across it or a rock or tennis ball is thrown into it. A vibrating tuning fork or drum membrane gives rise to sound waves in air. And, as we will see later, oscillating electric charges give rise to electromagnetic waves and light. Indeed, almost any vibrating object sends out waves.

The source of any wave, then, is a vibration. And it is the *vibration* that propagates outward and thus constitutes the wave. If the source vibrates sinusoidally in simple harmonic motion (SHM), then the wave itself—if the medium is perfectly elastic—will have a sinusoidal shape both in space and in time. If you were to take a picture of the wave spread throughout space at a given instant of time, the wave would have the shape of a sine or cosine function. On the other hand, if you look at the motion of the medium at one place over a long period of time—for example, if you look between two closely spaced posts of a pier or out of a ship's porthole as water waves pass by—the up and down motion of that small segment of water will be simple harmonic motion—the water moves up and down sinusoidally in time.

Sinusoidal wave:
(a) in space

(b) in time,

Some of the important quantities used to describe a periodic sinusoidal wave are shown in Fig. 15–4. The high points on a wave are called crests, the low points troughs. The **amplitude** is the maximum height of a crest, or depth of a trough, relative to the normal (or equilibrium) level. The total swing from a crest to a trough is twice the amplitude. The distance between two successive crests is called the **wavelength**, λ (the Greek letter lambda). The wavelength is also equal to the distance between *any* two successive identical points on the wave. The **frequency**, f (sometimes called ν, the Greek letter nu) is the number of crests—or complete cycles—that pass a given point per unit time. The **period**, T, is the time required for one complete oscillation or one complete cycle of the wave to pass a given point along the line of travel. As before (Eq. 14–2), $T = 1/f$.

Amplitude

Wavelength, λ

Frequency, f

FIGURE 15–4 Characteristics of a continuous wave.

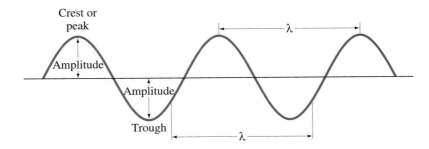

The **wave velocity**, v, is the velocity at which wave crests (or any other part of the waveform) move. The wave velocity is often referred to as the **phase** velocity, and it must be distinguished from the velocity of a particle of the medium itself. For example, for a wave traveling along a cord in Fig. 15–2 the wave velocity is to the right, along the cord, whereas the velocity of a particle of the cord is up or down.

Wave velocity
(don't confuse with
the velocity of a particle)

A wave crest travels a distance of one wavelength, λ, in one period, T. Thus the wave velocity v is equal to λ/T, or (since $1/T = f$):

$v = \lambda f$ (sinusoidal waves)

$$v = \lambda f. \tag{15–1}$$

For example, suppose a wave has a wavelength of 5 m and a frequency of 3 Hz. Since three crests pass a given point per second, and the crests are 5 m apart, the first crest (or any other part of the wave) must travel a distance of 15 m during the 1 s. So its speed is 15 m/s.

15–2 | Wave Types

Transverse and Longitudinal Waves

For a wave that travels down a cord, say from the left to right as in Fig. 15–2, the particles of the cord vibrate up and down in a direction transverse (or perpendicular) to the motion of the wave itself. Such a wave is called a **transverse wave**. A second type of wave is a **longitudinal wave**, in which the vibration of the particles of the medium is along the *same* direction as the motion of the wave. Longitudinal waves are readily formed on a stretched spring or "slinky" by alternately compressing and expanding one end, as shown in Fig. 15–5b (a transverse wave is shown in Fig. 15–5a for comparison). A series of compressions and expansions propagate along the spring. The *compressions* are those areas where the coils are momentarily close together. *Expansions* (sometimes called *rarefactions*) are regions where the coils are momentarily far apart. Compressions and expansions correspond to the crests and troughs of a transverse wave.

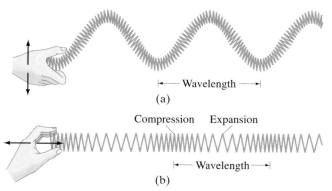

(a)

(b)

FIGURE 15–5 (a) Transverse wave; (b) longitudinal wave.

An important example of a longitudinal wave is a sound wave in air. A vibrating drum head, for example, alternately compresses and rarefies the air just in front of it, producing a longitudinal wave that travels outward in the air, as shown in Fig. 15–6.

As in the case of transverse waves, each particle of the medium in which a longitudinal wave passes oscillates over a very small distance, whereas the wave itself can travel large distances. What are the wavelength, frequency, and wave velocity for a longitudinal wave? The wavelength is the distance between successive compressions (or between successive expansions), and frequency is the number of compressions that pass a given point per second. The wave velocity is the velocity with which each compression appears to move; and it is equal to the product of wavelength and frequency, $v = \lambda f$ (Eq. 15–1).

A longitudinal wave can be represented graphically by plotting the density of air molecules (or coils of a slinky) versus position, as shown in Fig. 15–7b. We will often use such a graphical representation because it is much easier to illustrate what is happening. Note that the graph looks much like a transverse wave.

FIGURE 15–6 Production of a sound wave, which is longitudinal.

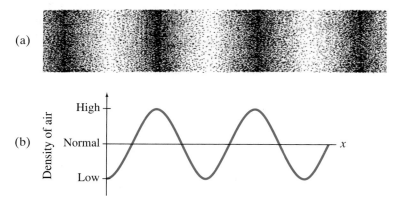

(a)

(b)

FIGURE 15–7 (a) A longitudinal wave with (b) its graphical representation.

Velocity of Transverse Waves

The velocity of a wave depends on the properties of the medium in which it travels. The velocity of a transverse wave on a stretched string or cord, for example, depends on the tension in the cord, F_T, and on the mass per unit length of the cord, μ (the Greek letter mu). For waves of small amplitude, the relationship is

Velocity of wave on a cord

$$v = \sqrt{\frac{F_T}{\mu}}. \qquad \begin{bmatrix} \text{transverse} \\ \text{wave on a cord} \end{bmatrix} \quad \textbf{(15–2)}$$

Before giving a derivation of this formula, it is worth noting that at least qualitatively it makes sense on the basis of Newtonian mechanics. That is, we do expect the tension to be in the numerator and the mass per unit length in the denominator. Why? Because when the tension is greater, we expect the velocity to be greater since each segment of cord is in tighter contact with its neighbor. And, the greater the mass per unit length, the more inertia the cord has and the more slowly the wave would be expected to propagate.

We can make a simple derivation of Eq. 15–2 using a simple model of a cord under a tension F_T as shown in Fig. 15–8a. The cord is pulled upward at a speed v' by the force F_y. As shown in Fig. 15–8b all points of the cord to the left of point A move upward at the speed v', and those to the right are still at rest. The speed of propagation, v, of this wave pulse is the speed of point A, the leading edge of the pulse. Point A moves to the right a distance vt in a time t, whereas the end of the cord moves upward a distance $v't$. By similar triangles we have the approximate relation

$$\frac{F_T}{F_y} = \frac{vt}{v't} = \frac{v}{v'},$$

FIGURE 15–8 Diagram of simple wave pulse on a cord for derivation of Eq. 15–2. The vector shown in (b) as the resultant of $\mathbf{F}_T + \mathbf{F}_y$ has to be directed along the rope because the rope is flexible.

which is accurate for small displacements ($v't \ll vt$) so that F_T does not change appreciably. As we saw in Chapter 9, the impulse given to an object is equal to its change in momentum. During the time t the total upward impulse is $F_y t = (v'/v)F_T t$. The change in momentum of the cord, Δp, is the mass of cord moving upward times its velocity. Since the upward moving segment of cord has mass equal to the mass per unit length μ times its length vt we have

$$F_y t = \Delta p$$

$$\frac{v'}{v} F_T t = (\mu vt)v'.$$

Solving for v we find $v = \sqrt{F_T/\mu}$ which is Eq. 15–2. Although it was derived for a special case, it is valid for any wave shape since other shapes can be considered to be made up of many tiny such lengths. But it is valid only for small displacements (as was our derivation). Experiment is in accord with this result derived from Newtonian mechanics.

EXAMPLE 15–1 **Wave on a wire.** A wave whose wavelength is 0.30 m is traveling down a 300-m-long wire whose total mass is 15 kg. If the wire is under a tension of 1000 N, what are the velocity and frequency of this wave?

SOLUTION From Eq. 15–2, the velocity is

$$v = \sqrt{\frac{1000\,\text{N}}{(15\,\text{kg})/(300\,\text{m})}} = 140\,\text{m/s}.$$

The frequency then is (Eq. 15–1)

$$f = \frac{v}{\lambda} = \frac{140\,\text{m/s}}{0.30\,\text{m}} = 470\,\text{Hz}.$$

Note that a higher tension would increase both v and f, whereas a thicker, denser wire would reduce v and f.

Velocity of Longitudinal Waves

The velocity of a longitudinal wave has a form similar to that for a transverse wave on a cord (Eq. 15–2); that is,

$$v = \sqrt{\frac{\text{elastic force factor}}{\text{inertia factor}}}.$$

In particular, for a longitudinal wave traveling down a long solid rod,

$$v = \sqrt{\frac{E}{\rho}}, \qquad \left[\begin{array}{l}\text{longitudinal} \\ \text{wave in a long rod}\end{array}\right] \quad \textbf{(15–3)}$$

where E is the elastic modulus (Section 12–5) of the material and ρ is its density. For a longitudinal wave traveling in a liquid or gas,

$$v = \sqrt{\frac{B}{\rho}}, \qquad \left[\begin{array}{l}\text{longitudinal wave} \\ \text{in a fluid}\end{array}\right] \quad \textbf{(15–4)}$$

where B is the bulk modulus (Section 12–5) and ρ again is the density.

We now derive Eq. 15–4. Consider a wave pulse traveling in a fluid in a long tube, so that the wave motion is one dimensional. The tube is fitted with a piston at the end and is filled with a fluid which, at $t = 0$, is of uniform density ρ and at uniform pressure P_0, Fig. 15–9a. At this moment the piston is abruptly made to start moving to the right with speed v', compressing the fluid in front of it. In the (short) time t the piston moves a distance $v't$. The compressed fluid itself also moves with speed v', but the leading edge of the compressed region moves to the right at the characteristic speed v of compression waves in that fluid; we assume the wave speed v is much larger than the piston speed v'. The leading edge of the compression in the fluid (which at $t = 0$ was at the piston face) thus moves a distance vt in time t as shown in Fig. 15–9b. Let the pressure in the compression be $P_0 + \Delta P$, which is ΔP higher than in the uncompressed fluid. To move the piston to the right requires an external force $(P_0 + \Delta P)A$ acting to the right, where A is the area of the tube. The *net* force on the compressed region of the fluid is

$$F_{\text{net}} = (P_0 + \Delta P)A - P_0 A = A\,\Delta P$$

since the uncompressed fluid exerts a force $P_0 A$ to the left at the leading edge. Hence the impulse given to the compressed fluid, which equals its change in momentum, is

$$F_{\text{net}}t = \Delta m v'$$
$$A\,\Delta P\,t = (\rho A v t)v',$$

where $(\rho A v t)$ represents the mass of fluid which is given the speed v' (the compressed fluid of area A moves a distance vt, Fig. 15–9, so the volume moved is Avt). Hence we have

$$\Delta P = \rho v v'.$$

From the definition of the bulk modulus, B (Eq. 12–7a):

$$B = -\frac{\Delta P}{\Delta V/V_0} = -\frac{\rho v v'}{\Delta V/V_0},$$

where $\Delta V/V_0$ is the fractional change in volume due to compression. The original volume of the compressed fluid is $V_0 = Avt$ (see Fig. 15–9), and it has been compressed by an amount $\Delta V = -Av't$ (Fig. 15–9b). Thus

$$B = -\frac{\rho v v'}{\Delta V/V_0} = -\rho v v'\left(\frac{Avt}{-Av't}\right) = \rho v^2$$

and so

$$v = \sqrt{\frac{B}{\rho}},$$

which is Eq. 15–4.

The derivation of Eq. 15–3 follows similar lines, but takes into account the expansion of the sides of a rod when the end of the rod is compressed.

FIGURE 15–9 Determining the speed of a one-dimensional longitudinal wave in a fluid contained in a long narrow tube.

EXAMPLE 15–2 **Echolocation.** Echolocation is a form of sensory perception used by animals such as bats, toothed whales and porpoises. The animal emits a pulse of sound (a longitudinal wave) which is reflected from objects; the reflected pulse is detected by the animal. Echolocation waves emitted by whales have frequencies of about 200,000 Hz. (*a*) What is the wavelength of the whale's echolocation wave? (*b*) If an obstacle is 100 m from the whale, how long after the whale emits a wave will the reflected wave return to him?

SOLUTION (*a*) To find the wavelength given the frequency we first need to compute the velocity of longitudinal (sound) waves in sea water, using Eq. 15–4 and Tables 12–1 and 13–1. We find,

$$v = \sqrt{\frac{B}{\rho}} = \sqrt{\frac{2.0 \times 10^9 \, \text{N/m}^2}{1.025 \times 10^3 \, \text{kg/m}^3}} = 1.40 \times 10^3 \, \text{m/s}.$$

Then using Eq. 15–1 we find,

$$\lambda = v/f = (1.40 \times 10^3 \, \text{m/s})/(2.0 \times 10^5 \, \text{Hz}) = 7.0 \, \text{mm}.$$

As we shall see later, waves can be used to "resolve" objects whose sizes are comparable to or larger than the wavelength. Thus, a whale can resolve tiny objects with echolocation.

(*b*) The time required for the round-trip between the whale and the object is

$$t = \frac{\text{distance}}{\text{speed}} = \frac{2(100 \, \text{m})}{1.40 \times 10^3 \, \text{m/s}} = 0.14 \, \text{s}.$$

Other Waves

Both transverse and longitudinal waves are produced when an earthquake occurs. The transverse waves that travel through the body of the Earth are called *S* waves (*S* for shear) and the longitudinal waves are called *P* waves (*P* for pressure). Both longitudinal and transverse waves can travel through a solid since the atoms or molecules can vibrate about their relatively fixed positions in any direction. But in a fluid, only longitudinal waves can propagate, because any transverse motion would experience no restoring force since a fluid is readily deformable. This fact was used by geophysicists to infer that a portion of the Earth's core must be liquid: longitudinal waves are detected diametrically across the Earth, but not transverse waves.

Besides these two types of waves that can pass through the body of the Earth (or other substance), there can also be *surface waves* that travel along the boundary between two materials. A wave on water is actually a surface wave that moves on the boundary between water and air. The motion of each particle of water at the surface is circular or elliptical (Fig. 15–10), so it is a combination of transverse and longitudinal motions. Below the surface, there is also transverse plus longitudinal wave motion, as shown. At the bottom, the motion is only longitudinal. (When a wave approaches shore, the water drags at the bottom and is slowed down, while the crests move ahead at higher speed (Fig. 15–11) and "spill" over the top.)

Surface waves are also set up on the Earth when an earthquake occurs. The waves that travel along the surface are mainly responsible for the damage caused by earthquakes.

Waves which travel along a line in one dimension, such as transverse waves on a stretched string or longitudinal waves in a rod or fluid-filled tube, are *linear* or *one-dimensional waves*. Surface waves, such as the water waves of Fig. 15–1, are *two-dimensional waves*. Finally, waves that move out from a source in all directions, such as sound from a speaker or earthquake waves through the Earth, are *three-dimensional waves*. We will be concerned with all three types, but especially linear waves, since they are simpler and clearer to understand.

FIGURE 15–10 A water wave is an example of a *surface wave*, which is a combination of transverse and longitudinal wave motions.

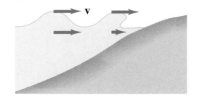

FIGURE 15–11 How a wave breaks. The green arrows represent the local velocity of water molecules.

15–3 | Energy Transported by Waves

Waves transport energy from one place to another. As waves travel through a medium, the energy is transferred as vibrational energy from particle to particle of the medium. For a sinusoidal wave of frequency f, the particles move in SHM as a wave passes, and each particle has an energy $E = \frac{1}{2}kD_M^2$ where D_M is the maximum displacement (amplitude) of its motion, either transversely or longitudinally (see Eq. 14–10a, in which we have replaced A by D_M). Using Eq. 14–7a we can write $k = 4\pi^2 mf^2$, where m is the mass of a particle (or small volume) of the medium. Then in terms of the frequency

$$E = \tfrac{1}{2}kD_M^2 = 2\pi^2 mf^2 D_M^2.$$

For three-dimensional waves traveling in an elastic medium, the mass $m = \rho V$, where ρ is the density of the medium and V is the volume of a small slice of the medium. The volume $V = Al$ where A is the cross-sectional area through which the wave travels (Fig. 15–12), and we can write l as the distance the wave travels in a time t as $l = vt$, where v is the speed of the wave. Thus $m = \rho V = \rho Al = \rho Avt$ and

$$E = 2\pi^2 \rho Avtf^2 D_M^2. \tag{15–5}$$

From this equation we have the important result that the **energy transported by a wave is proportional to the square of the amplitude, and to the square of the frequency**. The average *rate* of energy transferred is the average power P:

$$\overline{P} = \frac{E}{t} = 2\pi^2 \rho Avf^2 D_M^2. \tag{15–6}$$

Finally, the **intensity**, I, of a wave is defined as the average power transferred across unit area perpendicular to the direction of energy flow:

$$I = \frac{\overline{P}}{A} = 2\pi^2 v\rho f^2 D_M^2. \tag{15–7}$$

If a wave flows out from the source in all directions, it is a three-dimensional wave. Examples are sound traveling in the open air, earthquake waves, and light waves. If the medium is isotropic (same in all directions), the wave is said to be a *spherical wave* (Fig. 15–13). As the wave moves outward, the energy it carries is spread over a larger and larger area since the surface area of a sphere of radius r is $4\pi r^2$. Thus the intensity of a wave is

$$I = \frac{\overline{P}}{A} = \frac{\overline{P}}{4\pi r^2}.$$

If the power output $\overline{P}$ is constant, then the intensity decreases as the inverse square of the distance from the source:

$$I \propto \frac{1}{r^2}. \qquad \text{[spherical wave]} \tag{15–8a}$$

If we consider two points at distances r_1 and r_2 from the source, as in Fig. 15–13, then $I_1 = \overline{P}/4\pi r_1^2$ and $I_2 = \overline{P}/4\pi r_2^2$, so

$$\frac{I_2}{I_1} = \frac{r_1^2}{r_2^2}. \tag{15–8b}$$

Thus, for example, when the distance doubles $(r_2/r_1 = 2)$, then the intensity is reduced to $\frac{1}{4}$ its earlier value: $I_2/I_1 = \left(\frac{1}{2}\right)^2 = \frac{1}{4}$.

FIGURE 15–12 Calculating the energy carried by a wave moving with velocity v.

Wave energy $\propto$ (amplitude)^2
Wave energy $\propto$ $(f)^2$

Intensity

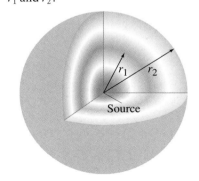

FIGURE 15–13 Wave traveling outward from source has spherical shape. Two different crests (or compressions) are shown, of radius r_1 and r_2.

Intensity $I \propto \dfrac{1}{r^2}$

The amplitude of a wave also decreases with distance. Since the intensity is proportional to the square of the amplitude $\left(I \propto D_{\mathrm{M}}^2,\ \text{Eq. 15-7}\right)$, then the amplitude must decrease as $1/r$, so that $I \propto D_{\mathrm{M}}^2$ will be proportional to $1/r^2$, Eq. 15–8a; that is,

Amplitude $\propto \dfrac{1}{r}$

$$D_{\mathrm{M}} \propto \frac{1}{r}.$$

To see this directly from Eq. 15–6, we consider again two different distances from the source, r_1 and r_2. For constant power output, $A_1 D_{\mathrm{M1}}^2 = A_2 D_{\mathrm{M2}}^2$ where D_{M1} and D_{M2} are the amplitudes of the wave at r_1 and r_2, respectively. Since $A_1 = 4\pi r_1^2$ and $A_2 = 4\pi r_2^2$, we have $\left(D_{\mathrm{M1}}^2 r_1^2\right) = \left(D_{\mathrm{M2}}^2 r_2^2\right)$, or

$$\frac{D_{\mathrm{M2}}}{D_{\mathrm{M1}}} = \frac{r_1}{r_2}.$$

When the wave is twice as far from the source, the amplitude is half as large, as long as we can ignore damping due to friction.

EXAMPLE 15–3 **Earthquake intensity.** If the intensity of an earthquake P wave 100 km from the source is $1.0 \times 10^6\ \mathrm{W/m^2}$, what is the intensity 400 km from the source?

SOLUTION The intensity decreases as the square of the distance from the source. Therefore, at 400 km, the intensity will be $\left(\frac{1}{4}\right)^2 = \frac{1}{16}$ of its value at 100 km, or $6.2 \times 10^4\ \mathrm{W/m^2}$. Alternatively, Eq. 15–8b could be used: $I_2 = I_1 r_1^2 / r_2^2 = \left(1.0 \times 10^6\ \mathrm{W/m^2}\right)(100\ \mathrm{km})^2/(400\ \mathrm{km})^2 = 6.2 \times 10^4\ \mathrm{W/m^2}$.

The situation is different for a one-dimensional wave, such as a transverse wave on a string or a longitudinal wave pulse traveling down a thin uniform metal rod. The area remains constant, so the amplitude D_{M} also remains constant (ignoring friction). Thus the amplitude and the intensity do not decrease with distance.

In practice, frictional damping is generally present, and some of the energy is transformed into thermal energy. Thus the amplitude and intensity of a one-dimensional wave decrease with distance from the source, and for a three-dimensional wave the decrease will be greater than that discussed above, although the effect may often be small.

15–4 Mathematical Representation of a Traveling Wave

Let us now consider a one-dimensional wave traveling along the x axis. It could be, for example, a transverse wave on a cord or a longitudinal wave traveling in a rod or in a fluid-filled tube. Let us assume the wave shape is sinusoidal and has a particular wavelength λ and frequency f. At $t = 0$, suppose the wave shape is given by

$$D(x) = D_{\mathrm{M}} \sin \frac{2\pi}{\lambda} x, \tag{15–9}$$

as shown by the solid curve in Fig. 15–14. $D(x)$ is the **displacement**[†] of the wave (be it a longitudinal or transverse wave) at position x, and D_{M} is the **amplitude**

[†] Some books use $y(x)$ in place of $D(x)$. To avoid confusion, we reserve y (and z) for the coordinate positions of waves in two or three dimensions. Our $D(x)$ can stand for pressure (in longitudinal waves), position displacement (transverse mechanical waves) or—as we will see later—electric or magnetic fields (for electromagnetic waves).

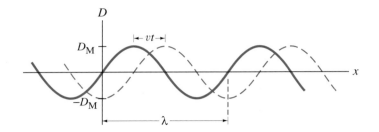

FIGURE 15–14 A traveling wave. In time t, the wave moves a distance vt.

(maximum displacement) of the wave. This relation gives a shape that repeats itself every wavelength, which is what we want so that the displacement is the same at, for example, $x = 0$, $x = \lambda$, $x = 2\lambda$, and so on (since $\sin 4\pi = \sin 2\pi = \sin 0$).

Now suppose the wave is moving to the right with velocity v. Then, after a time t, each part of the wave (indeed, the whole wave "shape") has moved to the right a distance vt; see the dashed curve in Fig. 15–14. Consider any point on the wave at $t = 0$: say, a crest which is at some position x. After a time t, that crest will have traveled a distance vt so its new position is a distance vt greater than its old position. To describe this same point on the wave shape, the argument of the sine function must be the same, so we replace x in Eq. 15–9 by $(x - vt)$:

$$D(x, t) = D_M \sin\left[\frac{2\pi}{\lambda}(x - vt)\right]. \qquad \textbf{(15–10a)}$$

1-D wave, moving in positive x direction

Said another way, if you are riding on a crest, the argument of the sine function, $(2\pi/\lambda)(x - vt)$, remains the same ($= \pi/2, 5\pi/2$, and so on); as t increases, x must increase at the same rate so that $(x - vt)$ remains constant.

Equation 15–10a is the mathematical representation of a sinusoidal wave traveling along the x axis to the right (increasing x). It gives the displacement $D(x, t)$ of the wave at any chosen point x at any time t. The function $D(x, t)$ describes a curve that represents the actual shape of the wave in space at time t. Since $v = \lambda f$ (Eq. 15–1) we can write Eq. 15–10a in other ways that are often convenient:

$$D(x, t) = D_M \sin\left(\frac{2\pi x}{\lambda} - \frac{2\pi t}{T}\right), \qquad \textbf{(15–10b)}$$

1-D wave, moving in positive x direction

where $T = 1/f = \lambda/v$ is the period; and

$$D(x, t) = D_M \sin(kx - \omega t), \qquad \textbf{(15–10c)}$$

1-D wave, moving in positive x direction

where $\omega = 2\pi f = 2\pi/T$ is the angular frequency and

$$k = \frac{2\pi}{\lambda} \qquad \textbf{(15–11)}$$

is called the **wave number**. (Do not confuse the wave number k with the spring constant k; they are very different quantities.) All three forms, Eq. 15–10a, b, and c, are equivalent; Eq. 15–10c is the simplest to write and is perhaps the most common. The quantity $(kx - \omega t)$, and its equivalent in the other two equations, is called the **phase** of the wave. The velocity v of the wave is often called the **phase velocity**, since it describes the velocity of the phase (or shape) of the wave and it can be written in terms of ω and k:

$$v = \lambda f = \left(\frac{2\pi}{k}\right)\left(\frac{\omega}{2\pi}\right) = \frac{\omega}{k}. \qquad \textbf{(15–12)}$$

For a wave traveling along the x axis to the left (decreasing values of x), we start again with Eq. 15–9 and note that a particular point on the wave changes position by $-vt$ in a time t. So x in Eq. 15–9 must be replaced by $(x + vt)$. Thus, for a wave traveling to the left with velocity v,

1-D wave moving in negative x direction

$$D(x, t) = D_M \sin \left[\frac{2\pi}{\lambda} (x + vt) \right] \tag{15–13a}$$

$$= D_M \sin \left(\frac{2\pi x}{\lambda} + \frac{2\pi t}{T} \right) \tag{15–13b}$$

$$= D_M \sin(kx + \omega t). \tag{15–13c}$$

In other words, we simply replace v in Eqs. 15–10 by $-v$.

Let us look for a moment at Eq. 15–13c (or, just as well, at Eq. 15–10c). At $t = 0$ we have

$$D(x, 0) = D_M \sin kx,$$

which is what we started with, a sinusoidal wave shape. If we look at the wave shape in space at a particular later time t_1, then we have

$$D(x, t_1) = D_M \sin(kx + \omega t_1).$$

That is, if we took a picture of the wave at $t = t_1$, we would see a sine wave with a phase constant ωt_1. Thus, for fixed $t = t_1$, the wave has a sinusoidal shape in space. On the other hand, if we consider a fixed point in space, say $x = 0$, we can see how the wave varies in time:

$$D(0, t) = D_M \sin \omega t$$

where we used Eq. 15–13c. This is just the equation for simple harmonic motion (see Section 14–2, Eq. 14–4). For any other fixed value of x, say $x = x_1$, $D = D_M \sin(\omega t + kx_1)$ which differs only by a phase constant kx_1. Thus, at any fixed point in space, the displacement undergoes the oscillations of simple harmonic motion in time. Equations 15–10 and 15–13 combine both these aspects to give us the representation for a **traveling sinusoidal wave** (also called a **harmonic wave**).

The argument of the sine in Eqs. 15–10 and 15–13 can in general contain a phase angle ϕ,

$$D(x, t) = D_M \sin(kx \pm \omega t + \phi),$$

to adjust for the position of the wave at $t = 0$, $x = 0$, just as in Section 14–2 (see Fig. 14–7). If the displacement is zero at $t = 0$, $x = 0$, as in Fig. 14–6, then $\phi = 0$.

Now let us consider a general wave (or wave pulse) of any shape. If frictional losses are small, experiment shows that the wave maintains its shape as it travels. Thus we can make the same arguments as we did right after Eq. 15–9. Suppose our wave has some shape at $t = 0$, given by

$$D(x, 0) = D(x)$$

where $D(x)$ is the displacement of the wave at x and is not necessarily sinusoidal. Then at some later time, if the wave is traveling to the right along the x axis, the wave will have the same shape but all parts will have moved a distance vt where v is the phase velocity of the wave. Hence we must replace x by $x - vt$ to obtain the amplitude at time t:

$$D(x, t) = D(x - vt). \tag{15–14}$$

Similarly, if the wave moves to the left, we must replace x by $x + vt$, so

$$D(x, t) = D(x + vt). \tag{15–15}$$

Thus, any wave traveling along the x axis must have the form of Eq. 15–14 or 15–15.

EXAMPLE 15–4 **A traveling wave.** The left-hand end of a long horizontal stretched cord oscillates transversely in SHM with frequency $f = 250\,\text{Hz}$ and amplitude 2.6 cm. The cord is under a tension of 140 N and has a linear density $\mu = 0.12\,\text{kg/m}$. At $t = 0$, the end of the cord has an upward displacement of 1.6 cm and is falling (Fig. 15–15). Determine (a) the wavelength of waves produced and (b) the equation for the traveling wave.

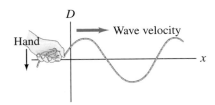

FIGURE 15–15 Example 15–4: The wave at $t = 0$ (the hand is falling). Not to scale.

SOLUTION (a) The wave velocity is

$$v = \sqrt{\frac{F_T}{\mu}} = \sqrt{\frac{140\,\text{N}}{0.12\,\text{kg/m}}} = 34\,\text{m/s}.$$

Then

$$\lambda = \frac{v}{f} = \frac{34\,\text{m/s}}{250\,\text{Hz}} = 0.14\,\text{m} \quad \text{or} \quad 14\,\text{cm}.$$

(b) Let $x = 0$ at the left-hand end of the cord. The phase of the wave at $t = 0$ is not zero in general as was assumed in Eqs. 15–9, 10, and 13. The general form for a wave traveling to the right is then

$$D(x, t) = D_M \sin(kx - \omega t + \phi),$$

where ϕ is the phase angle. In our case, the amplitude $D_M = 2.6\,\text{cm}$; and at $t = 0$, $x = 0$, we are given $D = 1.6\,\text{cm}$. Thus

$$1.6 = 2.6 \sin \phi,$$

so $\phi = 38° = 0.66\,\text{rad}$. We also have $\omega = 2\pi f = 1570\,\text{s}^{-1}$ and $k = 2\pi/\lambda = 45\,\text{m}^{-1}$. Hence

$$D = 0.026 \sin(45x - 1570t + 0.66),$$

where D and x are in meters and t in seconds.

---

* ## 15–5 | The Wave Equation

Many types of waves satisfy an important general equation that is the equivalent of Newton's second law of motion for particles. This "equation of motion for a wave" is called the **wave equation**, and we derive it now for waves traveling on a stretched horizontal string.

We assume the amplitude of the wave is small compared to the wavelength so that each point on the string can be assumed to move only vertically and the tension in the string, F_T, does not vary during a vibration. We apply Newton's second law, $\Sigma F = ma$, to the vertical motion of a tiny section of the string as shown in Fig. 15–16. The amplitude of the wave is small, so the angles θ_1 and θ_2 that the string makes with the horizontal are small. The length of this section is then approximately Δx, and its mass is $\mu\,\Delta x$, where μ is the mass per unit length of the string. The net vertical force on this section of string is $F_T \sin \theta_2 - F_T \sin \theta_1$. So Newton's second law applied to the vertical (y) direction gives

$$\Sigma F_y = ma_y$$

$$F_T \sin \theta_2 - F_T \sin \theta_1 = (\mu\,\Delta x)\frac{\partial^2 D}{\partial t^2}.$$

FIGURE 15–16 Deriving the wave equation from Newton's second law: a segment of string under tension F_T.

We have written the acceleration as $a_y = \partial^2 D/\partial t^2$ since the motion is only vertical, and we use the partial derivative notation because the displacement D is a function of both x and t.

Because the angles θ_1 and θ_2 are assumed small, $\sin\theta \approx \tan\theta$ and $\tan\theta$ is equal to the slope s of the string at each point:

$$\sin\theta \approx \tan\theta = \frac{\partial D}{\partial x} = s.$$

Thus our equation at the bottom of the previous page becomes

$$F_T(s_2 - s_1) = \mu\,\Delta x\,\frac{\partial^2 D}{\partial t^2}$$

or

$$F_T\frac{\Delta s}{\Delta x} = \mu\frac{\partial^2 D}{\partial t^2}, \qquad\qquad \textbf{(a)}$$

where $\Delta s = s_2 - s_1$ is the difference in the slope between the two ends of our tiny section. Now we take the limit of $\Delta x \to 0$, so that

$$F_T\lim_{\Delta x\to 0}\frac{\Delta s}{\Delta x} = F_T\frac{\partial s}{\partial x}$$

$$= F_T\frac{\partial}{\partial x}\left(\frac{\partial D}{\partial x}\right) = F_T\frac{\partial^2 D}{\partial x^2}$$

since the slope $s = \partial D/\partial x$, as we saw above. Substituting this into the equation labelled (a) above gives

$$F_T\frac{\partial^2 D}{\partial x^2} = \mu\frac{\partial^2 D}{\partial t^2}$$

or

$$\frac{\partial^2 D}{\partial t^2} = \frac{F_T}{\mu}\frac{\partial^2 D}{\partial t^2}.$$

We saw earlier in this chapter (Eq. 15–2) that the velocity of waves on a string is given by $v = \sqrt{F_T/\mu}$, so we can write this last equation as

Wave equation (1–D)

$$\frac{\partial^2 D}{\partial x^2} = \frac{1}{v^2}\frac{\partial^2 D}{\partial t^2}. \qquad\qquad \textbf{(15–16)}$$

This is the **one-dimensional wave equation**, and it can describe not only small amplitude waves on a stretched string, but also small amplitude longitudinal waves (such as sound waves) in gases, liquids, and elastic solids, in which case D can refer to the pressure variations. In this case, the wave equation is a direct consequence of Newton's second law applied to a continuous elastic medium. The wave equation also describes electromagnetic waves for which D refers to the electric or magnetic field, as we shall see in Chapter 32. Equation 15–16 applies to waves traveling in one dimension only. For waves spreading out in three dimensions, the wave equation is the same, with the addition of $\partial^2 D/\partial y^2$ and $\partial^2 D/\partial z^2$ to the left side of Eq. 15–16.

The wave equation is a *linear* equation: the displacement D appears singly in each term. There are no terms that contain D^2, or $D(\partial D/\partial x)$, or the like in which D appears more than once. Thus, if $D_1(x, t)$ and $D_2(x, t)$ are two different solutions of a linear equation, such as the wave equation, then the linear combination

Superposition principle

$$D_3(x, t) = aD_1(x, t) + bD_2(x, t),$$

where a and b are constants, is also a solution. This is readily seen by direct substitution into the wave equation. This is the essence of the *superposition principle*, which we discuss in the next Section. Basically it says that if two waves pass through the same region of space at the same time, the actual displacement is the sum of the separate displacements. For waves on a string, or for sound waves, this is valid only for small-amplitude waves. If the amplitude is not small enough, the equations for wave propagation may become nonlinear and the principle of superposition would not hold and other new effects may occur.

EXAMPLE 15–5 **Wave equation solution.** Verify that the sinusoidal wave of Eq. 15–10c, $D(x, t) = D_M \sin(kx - \omega t)$, satisfies the wave equation.

SOLUTION We take the derivative of Eq. 15–10c twice with respect to t:

$$\frac{\partial D}{\partial t} = -\omega D_M \cos(kx - \omega t)$$

$$\frac{\partial^2 D}{\partial t^2} = -\omega^2 D_M \sin(kx - \omega t).$$

With respect to x, the derivatives are

$$\frac{\partial D}{\partial x} = k D_M \cos(kx - \omega t)$$

$$\frac{\partial^2 D}{\partial x^2} = -k^2 D_M \sin(kx - \omega t).$$

If we now divide the second derivatives we get

$$\frac{\partial^2 D / \partial t^2}{\partial^2 D / \partial x^2} = \frac{-\omega^2 D_M \sin(kx - \omega t)}{-k^2 D_M \sin(kx - \omega t)} = \frac{\omega^2}{k^2}.$$

From Eq. 15–12 we have $\omega^2 / k^2 = v^2$, so we see that Eq. 15–10 does satisfy the wave equation (Eq. 15–16).

15–6 | The Principle of Superposition

When two or more waves pass through the same region of space at the same time, it is found that for many waves *the actual displacement is the vector* (or *algebraic*) *sum of the separate displacements*. This is called the **principle of superposition**. It is valid for mechanical waves as long as the displacements are not too large and there is a linear relationship between the displacement and the restoring force of the oscillating medium.[†] If the amplitude of a mechanical wave, for example, is so large that it goes beyond the elastic region of the medium, and Hooke's law is no longer operative, the superposition principle is no longer accurate.[‡] For the most part, we will consider systems for which the superposition principle can be assumed to hold.

One result of the superposition principle is that if two waves pass through the same region of space, they continue to move independently of one another. You may have noticed, for example, that the ripples on the surface of water (two-dimensional waves) that form from two rocks striking the water at different places will pass through each other.

Figure 15–17, shows an example of the superposition principle. In this case there are three waves present, on a stretched string, each of different amplitude and frequency. At any time, such as at the instant shown, the actual amplitude at any position x is the algebraic sum of the amplitude of the three waves at that position. The actual wave is not a simple sinusoidal wave and is called a *composite* (or *complex*) *wave*. (Amplitudes are exaggerated in Fig. 15–17.)

It can be shown that any complex wave can be considered as being composed of many simple sinusoidal waves of different amplitudes, wavelengths, and frequencies. This is known as *Fourier's theorem*. A complex periodic wave of period T can be represented as a sum of pure sinusoidal terms whose frequencies are integral multiples of $f = 1/T$. If the wave is not periodic, the sum becomes an integral (called a *Fourier integral*). Although we will not go into the details here, we see the importance of considering sinusoidal waves (and simple harmonic motion): because any other wave shape can be considered a sum of such pure sinusoidal waves.

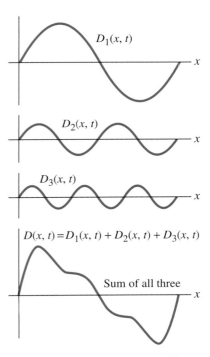

FIGURE 15–17 The superposition principle for one-dimensional waves. Composite wave formed from three sinusoidal waves of different amplitudes and frequencies $(f_0, 2f_0, 3f_0)$ at a certain instant in time. The amplitude of the composite wave at each point in space, at any time, is the algebraic sum of the amplitudes of the component waves. Amplitudes are shown exaggerated; for the superposition principle to hold, they must be small compared to the wavelengths.

[†] For electromagnetic waves in vacuum, Chapter 32, the superposition principle always holds.

[‡] Intermodulation distortion in high-fidelity equipment is an example of the superposition principle not holding when two frequencies do not combine linearly in the electronics.

(a)

(b)

FIGURE 15–18 Example 15–6. Making a square wave.

FIGURE 15–19 Reflection of a wave pulse on a rope when the end of the rope is (a) fixed and (b) free.

EXAMPLE 15–6 **Making a square wave.** At $t = 0$, three waves are given by $D_1 = D_M \cos kx$, $D_2 = \frac{1}{3} D_M \cos 3kx$ and $D_3 = \frac{1}{5} D_M \cos 5kx$ where $D_M = 1.0$ m and $k = 10$ m^{-1}. Plot the sum of the three waves from $x = -0.4$ to $+0.4$ m. (These three waves are the first three Fourier components of a "square wave.")

SOLUTION The first wave, D_1 has amplitude of 1.0 m and wavelength $\lambda = 2\pi/k = 2\pi/10$ m $= 0.628$ m. The second wave, D_2 has amplitude of 0.33 m and wavelength $\lambda = 2\pi/3k = 2\pi/30$ m $= 0.209$ m. The third wave, D_3 has amplitude of 0.20 m and wavelength $\lambda = 2\pi/5k = 2\pi/50$ m $= 0.126$ m. Each wave is plotted in Fig. 15–18a. The sum of the three waves is shown in Fig. 15–18b. The sum begins to resemble a "square wave," shown in blue in Fig. 15–18b.

When the restoring force is not precisely proportional to the displacement for mechanical waves in some continuous medium, the speed of sinusoidal waves depends on the frequency. The variation of speed with frequency is called **dispersion**. The different sinusoidal waves that compose a complex wave will travel with slightly different speeds in such a case. Consequently, a complex wave will change shape as it travels if the medium is "dispersive." A pure sine wave will not change shape under these conditions, however, except by the influence of friction or dissipative forces. If there is no dispersion (or friction), even a complex linear wave doesn't change shape.

15–7 Reflection and Transmission

When a wave strikes an obstacle, or comes to the end of the medium it is traveling in, at least a part of the wave is reflected. You have probably seen water waves reflect off of a rock or the side of a swimming pool. And you may have heard a shout reflected from a distant cliff—which we call an "echo."

A wave pulse traveling down a rope is reflected as shown in Fig. 15–19. You can observe this for yourself (try it with a rope lying on a table) and see that the reflected pulse is inverted as in Fig. 15–19a if the end of the rope is fixed; and returns right side up if the end is free as in Fig. 15–19b. When the end is fixed to a

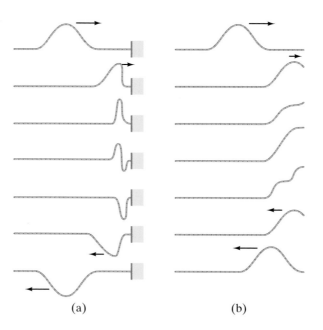

(a)   (b)

support, as in Fig. 15–19a, the pulse reaching that fixed end exerts a force (upward) on the support. The support exerts an equal but opposite force (Newton's third law) downward on the rope. This downward force on the rope is what "generates" the inverted reflected pulse. The inversion of the reflected pulse in part (a) is said to be a phase change of 180°. It is as if the phase shifted by $\frac{1}{2}\lambda$, or 180°, from a crest to a trough. In Fig. 15–19b, the free end is constrained by neither a support nor by additional rope. It therefore tends to overshoot—its displacement being momentarily greater than that of the traveling pulse. The overshooting end exerts an upward pull on the rope, and this is what generates the reflected pulse, which is not inverted (no phase change).

When the wave pulse on the rope in Fig. 15–19a reaches the wall, not all of the energy is reflected. Some of it is absorbed by the wall. Part of the absorbed energy is transformed into thermal energy, and part continues to propagate through the material of the wall. This is more clearly illustrated by considering a pulse that travels down a rope which consists of a light section and a heavy section, as shown in Fig. 15–20. When the wave reaches the boundary between the two sections, part of the pulse is reflected and part is transmitted, as shown. The heavier the second section, the less that is transmitted; and when the second section is a wall or rigid support, very little is transmitted. For a periodic wave, the frequency does not change across the boundary since the boundary point oscillates at that frequency. Thus if the transmitted wave has a lower speed, its wavelength is also less ($\lambda = v/f$).

For a two- or three-dimensional wave, such as a water wave, we are concerned with **wave fronts**, by which we mean all the points along the wave forming the wave crest (what we usually refer to simply as a "wave" when we are at the seashore). A line drawn in the direction of motion, perpendicular to the wave front, is called a **ray**, as shown in Fig. 15–21. Note in Fig. 15–21b that wave fronts far from the source have lost almost all their curvature and are nearly straight, as ocean waves often are; they are then called **plane waves**.

For reflection of a two- or three-dimensional plane wave, as shown in Fig. 15–22, the angle that the incoming or *incident wave* makes with the reflecting surface is equal to the angle made by the reflected wave. This is the **law of reflection: the angle of reflection equals the angle of incidence**. The "angle of incidence" is defined as the angle the incident ray makes with the perpendicular to the reflecting surface (or the wave front makes with a tangent to the surface), and the "angle of reflection" is the corresponding angle for the reflected wave.

Wave fronts

Plane waves

Law of reflection

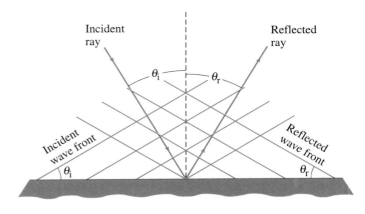

FIGURE 15–20 When a wave pulse traveling to the right (a) reaches a discontinuity, then (b) part is reflected and part is transmitted.

FIGURE 15–21 Rays, signifying the direction of motion, are always perpendicular to the wave fronts (wave crests). (a) Circular or spherical waves near the source. (b) Far from the source, the wave fronts are nearly straight or flat, and are called plane waves.

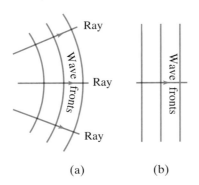

FIGURE 15–22 Law of reflection.

15–8 | Interference

Interference refers to what happens when two waves pass through the same region of space at the same time. It is an example of the superposition principle. Consider for example the two wave pulses on a string traveling toward each other as shown in Fig. 15–23. In part (a) the two pulses have the same amplitude but one is a crest and the other a trough; in part (b) they are both crests. In both cases, the waves meet and pass right by each other. However, when they overlap, the resultant displacement is the *algebraic sum of their separate displacements* (the principle of superposition). In Fig. 15–23a, the two wave amplitudes are opposite one another as they pass by and the result is called **destructive interference**. In Fig. 15–23b, the resultant displacement is greater than that of either pulse and the result is called **constructive interference**.

Destructive interference

Constructive interference

FIGURE 15–23 Two waves pulses pass each other. Where they overlap, interference occurs: (a) destructive; (b) constructive.

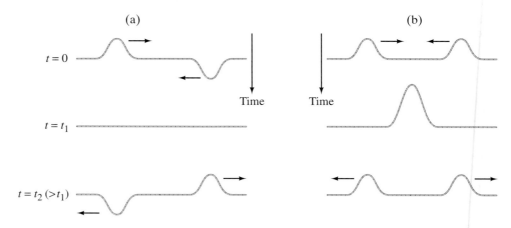

When two rocks are thrown into a pond simultaneously, the two sets of circular waves interfere with one another as shown in Fig. 15–24. In some areas of overlap, crests of one wave meet crests of the other (and troughs meet troughs); this is constructive interference and the water oscillates up and down with greater amplitude than either wave separately. In other areas, destructive interference occurs

FIGURE 15–24 Interference of water waves. Constructive interference occurs where one wave's maximum (a crest) meets the other's maximum. Destructive interference ("flat water") occurs where one wave's maximum (a crest) meets the other's miminum (a trough).

(a)

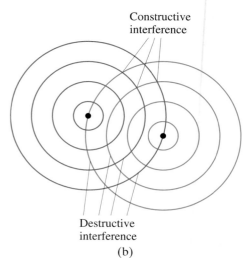

(b)

where the water actually does not move at all—this is where crests of one wave meet troughs of the other, and vice versa. In the first case, of constructive interference, the two waves are **in phase**, whereas in destructive interference the two waves are **out of phase** by one-half wavelength or 180°. Of course the relative phase of the two waves in most areas is intermediate between these two extremes, resulting in partially destructive interference. All three of these situations are shown in Fig. 15–25, where the amplitudes are plotted versus time at a given point in space. We will deal with interference in more detail when we discuss sound and light.

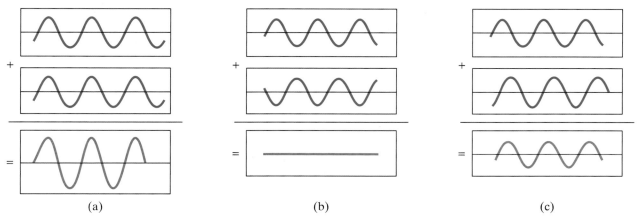

(a)   (b)   (c)

FIGURE 15–25 Two waves interfere: (a) constructively, (b) destructively, (c) partially destructively.

15–9 | Standing Waves; Resonance

If you shake one end of a cord (or slinky) and the other end is kept fixed, a continuous wave will travel down to the fixed end and be reflected back, inverted (Fig. 15–19a). As you continue to vibrate the cord, there will be waves traveling in both directions, and the wave traveling down the cord will interfere with the reflected wave coming back. Usually there will be quite a jumble. But if you vibrate the cord at just the right frequency, the two traveling waves will interfere in such a way that a large-amplitude **standing wave** will be produced, Fig. 15–26. It is called a "standing wave" because it doesn't appear to be traveling. The cord simply appears to have segments that oscillate up and down in a fixed pattern. The points of destructive interference, where the cord remains still at all times, are called **nodes**. Points of constructive interference, where the end oscillates with maximum amplitude, are called **antinodes**. The nodes and antinodes remain in fixed positions for a given frequency.

Standing waves can occur at more than one frequency. The lowest frequency of vibration that produces a standing wave gives rise to the pattern shown in Fig. 15–26a. The standing waves shown in parts (b) and (c) are produced at precisely twice and three times the lowest frequency, respectively, assuming the tension in the cord is the same. The cord can also vibrate with four loops at four times the lowest frequency, and so on.

The frequencies at which standing waves are produced are the **natural frequencies** or **resonant frequencies** of the cord, and the different standing wave patterns shown in Fig. 15–26 are different "resonant modes of vibration." Although a standing wave on a cord is the result of the interference of two waves traveling in opposite directions, it is also a vibrating object at resonance. Standing waves then represent the same phenomenon as the resonance of a vibrating spring or pendulum, which we discussed in Chapter 14. The only difference is that a spring or pendulum has only one resonant frequency, whereas the cord has an infinite number of resonant frequencies, each of which is a whole-number multiple of the lowest resonant frequency.

FIGURE 15–26 Standing waves corresponding to three resonant frequencies.

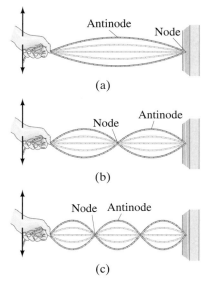

Antinode   Node

(a)

Node   Antinode

(b)

Node   Antinode

(c)

$L = \frac{1}{2}\lambda_1$

Fundamental or first harmonic, f_1

$L = \lambda_2$

First overtone or second harmonic, $f_2 = 2f_1$

$L = \frac{3}{2}\lambda_3$

Second overtone or third harmonic, $f_3 = 3f_1$

(b)

FIGURE 15–27 (a) A string is plucked. (b) Only standing waves corresponding to resonant frequencies persist for long.

Now let us consider a cord stretched between two supports that is plucked like a guitar or violin string, Fig. 15–27a. Waves of a great variety of frequencies will travel in both directions along the string, will be reflected at the ends, and will travel back in the opposite direction. Most of these waves interfere in a random way with each other and quickly die away. However, those waves that correspond to the resonant frequencies of the string will persist. The ends of the string, since they are fixed, will be nodes. There may be other nodes as well. Some of the possible resonant modes of vibration (standing waves) are shown in Fig. 15–27b. Generally, the motion will be a combination of these different resonant modes; but only those frequencies that correspond to a resonant frequency will be present.

To determine the resonant frequencies, we first note that the wavelengths of the standing waves bear a simple relationship to the length L of the string. The lowest frequency, called the **fundamental frequency**, corresponds to one antinode (or loop). And as can be seen in Fig. 15–27b, the whole length corresponds to one-half wavelength. Thus $L = \frac{1}{2}\lambda_1$, where λ_1 stands for the wavelength of the fundamental. The other natural frequencies are called **overtones**; when they are integral multiples of the fundamental (as they are for a simple string), they are also called **harmonics**, with the fundamental being referred to as the **first harmonic**.[†] The next mode after the fundamental has two loops and is called the **second harmonic** (or first overtone); the length of the string L at the second harmonic corresponds to one complete wavelength: $L = \lambda_2$. For the third and fourth harmonics, $L = \frac{3}{2}\lambda_3$, and $L = 2\lambda_4$, respectively, and so on. In general, we can write

$$L = \frac{n\lambda_n}{2}, \qquad \text{where } n = 1, 2, 3, \cdots.$$

The integer n labels the number of the harmonic: $n = 1$ for the fundamental, $n = 2$ for the second harmonic, and so on. We solve for λ_n and find

$$\lambda_n = \frac{2L}{n}, \qquad n = 1, 2, 3, \cdots. \qquad \textbf{(15–17)}$$

In order to find the frequency f of each vibration we use Eq. 15–1, $f = v/\lambda$, and we see that

$$f_n = \frac{v}{\lambda_n} = \frac{nv}{2L} = nf_1,$$

where $f_1 = v/\lambda_1 = v/2L$ is the fundamental frequency. We see that each resonant frequency is an integer multiple of the fundamental frequency.

Because a standing wave is equivalent to two traveling waves moving in opposite directions, the concept of wave velocity still makes sense and is given by Eq. 15–2 in terms of the tension F_T in the string and its mass per unit length μ; that is, $v = \sqrt{F_T/\mu}$ for both traveling waves.

EXAMPLE 15–7 Piano string. A piano string is 1.10 m long and has a mass of 9.00 g. (*a*) How much tension must the string be under if it is to vibrate at a fundamental frequency of 131 Hz? (*b*) What are the frequencies of the first four harmonics?

SOLUTION (*a*) The wavelength of the fundamental is $\lambda = 2L = 2.20$ m (Eq. 15–17). The velocity is then $v = \lambda f = (2.20\text{ m})(131\text{ s}^{-1}) = 288$ m/s. Then, from Eq. 15–2, we have

$$F_T = \mu v^2$$
$$= \left(\frac{9.00 \times 10^{-3}\text{ kg}}{1.10\text{ m}}\right)(288\text{ m/s})^2 = 679\text{ N}.$$

(*b*) The frequencies of the second, third, and fourth harmonics are two, three, and four times the fundamental frequency: 262, 393, and 524 Hz.

[†] The term "harmonic" comes from music, because such integral multiples of frequencies "harmonize."

A standing wave does appear to be standing in place (and a traveling wave appears to move). The term "standing" wave is also meaningful from the point of view of energy. Since the string is at rest at the nodes, no energy flows past these points. Hence the energy is not transmitted down the string but "stands" in place in the string.

Standing waves are produced not only on strings, but on any object that is set into vibration. Even when a rock or a piece of wood is struck with a hammer, standing waves are set up that correspond to the natural resonant frequencies of that object. In general, the resonant frequencies depend on the dimensions of the object, just as for a string they depend on its length. For example a small object does not have as low resonant frequencies as does a large object. All musical instruments depend on standing waves to produce their musical sounds, from stringed instruments to wind instruments (in which a column of air vibrates as a standing wave) to drums and other percussion instruments, as we will discuss in the next chapter.

Mathematical Representation of a Standing Wave

In Section 15–4 we saw how to write an equation for the displacement D of a linear traveling wave as a function of position x and time t. We can do the same for a standing wave on a string. We already discussed that a standing wave can be considered to consist of two traveling waves that move in opposite directions. These can be written (see Eqs. 15–10c and 15–13c)

$$D_1(x, t) = D_M \sin(kx - \omega t)$$

$$D_2(x, t) = D_M \sin(kx + \omega t)$$

since, assuming no damping, the amplitudes are equal as are the frequencies and wavelengths. The sum of these two traveling waves produces a standing wave which can be written mathematically as

$$D = D_1 + D_2 = D_M[\sin(kx - \omega t) + \sin(kx + \omega t)].$$

From the trigonometric identity $\sin\theta_1 + \sin\theta_2 = 2\sin\frac{1}{2}(\theta_1 + \theta_2)\cos\frac{1}{2}(\theta_1 - \theta_2)$, we can rewrite this as

$$D = 2D_M \sin kx \cos \omega t. \qquad \text{(15–18)} \qquad \textit{Standing wave}$$

If we let $x = 0$ at the left-hand end of the string, then the right-hand end is at $x = L$ where L is the length of the string. Since the string is fixed at its two ends (Fig. 15–27), $D(x, t)$ must be zero at $x = 0$ and at $x = L$. Equation 15–18 already satisfies the first condition ($D = 0$ at $x = 0$) and satisfies the second condition if $\sin kL = 0$ which means

$$kL = \pi, 2\pi, 3\pi, \ldots, n\pi, \ldots$$

where n = integer, or, since $k = 2\pi/\lambda$,

$$\lambda = \frac{2L}{n}. \quad (n = \text{integer})$$

This is just Eq. 15–17.

Equation 15–18, with the condition $\lambda = 2L/n$, is the mathematical representation of a standing wave. We see that a particle at any position x vibrates in simple harmonic motion (because of the factor $\cos\omega t$). All particles of the string vibrate with the same frequency $f = \omega/2\pi$, but the amplitude depends on x and equals $2D_M \sin kx$. (Compare this to a traveling wave for which all particles vibrate with the same amplitude.) The amplitude has a maximum, equal to $2D_M$, when $kx = \pi/2, 3\pi/2, 5\pi/2$, and so on—that is, at

$$x = \frac{\lambda}{4}, \frac{3\lambda}{4}, \frac{5\lambda}{4}, \ldots.$$

These are, of course, the positions of the antinodes (see Fig. 15–27).

EXAMPLE 15–8 **Wave forms.** Two waves traveling in opposite directions on a string fixed at $x = 0$ are described by the functions

$$D_1 = (0.20\,\text{m})\sin(2.0x - 4.0t)$$

$$D_2 = (0.20\,\text{m})\sin(2.0x + 4.0t)$$

and they produce a standing wave pattern (x is in m, t is in s). (*a*) Determine the function for the standing wave. (*b*) What is the maximum amplitude at $x = 0.45\,\text{m}$? (*c*) Where is the other end fixed ($x > 0$)? (*d*) What is the maximum amplitude, and where does it occur?

SOLUTION (*a*) The two waves are of the form $D = D_M \sin(kx \pm \omega t)$, so

$$k = 2.0\,\text{m}^{-1} \quad \text{and} \quad \omega = 4.0\,\text{s}^{-1}.$$

These combine to form a standing wave of the form of Eq. 15–18:

$$D = 2D_M \sin kx \cos \omega t = (0.40\,\text{m})\sin(2.0x)\cos(4.0t).$$

(*b*) At $x = 0.45\,\text{m}$,

$$D = (0.40\,\text{m})\sin(0.90)\cos(4.0t) = (0.31\,\text{m})\cos(4.0t).$$

The maximum amplitude at this point is $D = 0.31\,\text{m}$ and occurs when the cosine $= 1$.

(*c*) These waves make a standing wave pattern, so both ends of the string must be nodes. Nodes occur every half wavelength, which for our string is

$$\frac{\lambda}{2} = \frac{1}{2}\frac{2\pi}{k} = \frac{\pi}{2.0}\,\text{m} = 1.57\,\text{m}.$$

If the string includes only one loop, its length is $L = 1.57\,\text{m}$. But since we aren't given more information, it could be twice as long, including two loops, $L = 3.14\,\text{m}$, or any integral number times 1.57 m, and still provide a standing wave pattern for these waves—see Fig. 15–28.

(*d*) The nodes occur at $x = 0$, $x = 1.57\,\text{m}$, and, if the string is longer than $L = 1.57\,\text{m}$, at $x = 3.14\,\text{m}$, 4.71 m, and so on. The maximum amplitude (antinode) is 0.40 m [from part (*b*) above] and occurs midway between the nodes. For $L = 1.57\,\text{m}$, there is only one antinode, at $x = 0.79\,\text{m}$.

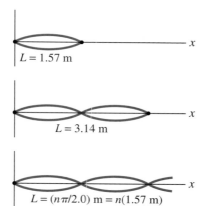

$L = 1.57\,\text{m}$

$L = 3.14\,\text{m}$

$L = (n\pi/2.0)\,\text{m} = n(1.57\,\text{m})$

FIGURE 15–28 Example 15–8: possible lengths for the string.

FIGURE 15–29 Refraction of waves crossing a boundary.

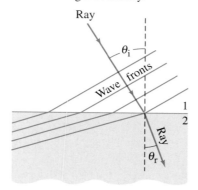

* ## 15–10 Refraction†

When any wave strikes a boundary, some of the energy is reflected and some is transmitted or absorbed. When a two- or three-dimensional wave traveling in one medium crosses a boundary into a medium where its velocity is different, the transmitted wave may move in a different direction than the incident wave, as shown in Fig. 15–29. This phenomenon is known as **refraction**. One example is a water wave; the velocity decreases in shallow water and the waves refract, Fig. 15–30. [When the wave velocity changes gradually, as in Fig. 15–30, without a sharp boundary, the waves change direction (refract) gradually.] In Fig. 15–29, the velocity of the wave in medium 2 is less than in medium 1. In this case, the direction of the wave bends so it travels more nearly perpendicular to the boundary. That is, the *angle of refraction*, θ_r, is less than the *angle of incidence*, θ_i. To see why this is so, and to help us get a quantitative relation between θ_r and θ_i, let us think of each wave front as a row of soldiers. The soldiers are marching from firm

†This Section and the next are covered in more detail in Chapters 33 to 36, on optics.

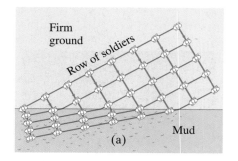

FIGURE 15–30 Water waves refracting as they approach the shore, where their velocity is less. There is no distinct boundary, as in Fig. 15–29, and the wave velocity changes gradually.

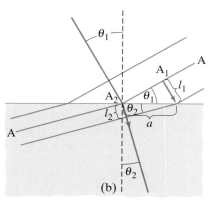

FIGURE 15–31 (a) Soldier analogy, to (b) derive law of refraction for waves.

ground (medium 1) into mud (medium 2) and hence are slowed down. The soldiers that reach the mud first are slowed down first and the row bends as shown in Fig. 15–31a. Let us consider the wave front (or row of soldiers) labeled A in Fig. 15–31b. In the same time t that A_1 moves a distance $l_1 = v_1 t$, we see that A_2 moves a distance $l_2 = v_2 t$. The two triangles shown (one includes θ_1 and l_1, the other θ_2 and l_2) have the side labeled a in common. Thus

$$\sin \theta_1 = \frac{l_1}{a} = \frac{v_1 t}{a}$$

and

$$\sin \theta_2 = \frac{l_2}{a} = \frac{v_2 t}{a}.$$

Dividing these two equations, we find that

$$\frac{\sin \theta_2}{\sin \theta_1} = \frac{v_2}{v_1}. \tag{15–19}$$

Law of refraction

Since θ_1 is the angle of incidence (θ_i), and θ_2 is the angle of refraction (θ_r), Eq. 15–19 gives the quantitative relation between the two. Of course, if the wave were going in the opposite direction, the argument would not be changed. Only θ_1 and θ_2 would change roles: θ_1 would be the angle of refraction and θ_2 the angle of incidence. Clearly then, if the wave travels into a medium where it can move faster, it will bend in the opposite way, $\theta_r > \theta_i$. We see from Eq. 15–19 that if the velocity increases, the angle increases, and vice versa.

Earthquake waves refract within the Earth as they travel through rock of different densities (and therefore the velocity is different) just as water waves do. Light waves refract as well, and when we discuss light we shall find Eq. 15–19 very useful.

EXAMPLE 15–9 Refraction of earthquake wave. An earthquake P wave passes across a boundary in rock where its velocity increases from 6.5 km/s to 8.0 km/s. If it strikes this boundary at 30°, what is the angle of refraction?

Earthquake wave refraction

SOLUTION Since $\sin 30° = 0.50$, Eq. 15–19 yields

$$\sin \theta_2 = \frac{(8.0 \text{ m/s})}{(6.5 \text{ m/s})} (0.50) = 0.62$$

so $\theta_2 = 38°$.

FIGURE 15–32 Wave diffraction. The waves come from the upper left. Note how the waves, as they pass the obstacle, bend around behind it, into the "shadow region."

Waves spread as they travel, and when they encounter an obstacle they bend around it somewhat and pass into the region behind as shown in Fig. 15–32 for water waves. This phenomenon is called **diffraction**.

The amount of diffraction depends on the wavelength of the wave and on the size of the obstacle, as shown in Fig. 15–33. If the wavelength is much larger than the object, as with the grass blades of Fig. 15–33a, the wave bends around them almost as if they were not there. For larger objects, parts (b) and (c), there is more of a "shadow" region behind the obstacle where we might not expect the waves to penetrate—but they do, at least a little. But notice in part (d), where the obstacle is the same as in part (c) but the wavelength is longer, that there is more diffraction into the shadow region. As a rule of thumb, *only if the wavelength is smaller than the size of the object will there be a significant shadow region.* It is worth noting that this rule applies to *reflection* from an obstacle as well. Very little of a wave is reflected unless the wavelength is smaller than the size of the obstacle.

A rough guide to the amount of diffraction is

$$\theta(\text{radians}) \approx \frac{\lambda}{L}$$

where θ is roughly the angular spread of waves after passing through an opening of width L or around an obstacle of width L.

That waves can bend around obstacles, and thus can carry energy to areas behind obstacles, is clearly different from energy carried by material particles. A clear example is the following: if you are standing around a corner on one side of a building, you can't be hit by a baseball thrown from the other side, but you can hear a shout or other sound because the sound waves diffract around the edges.

FIGURE 15–33 Water waves passing objects of various sizes. Note that the larger the wavelength compared to the size of the object, the more diffraction there is into the "shadow region."

(a) Water waves passing blades of grass

(b) Stick in water

(c) Short-wavelength waves passing log

(d) Long-wavelength waves passing log

Summary

Vibrating objects act as sources of **waves** that travel outward from the source. Waves on water and on a string are examples. The wave may be a **pulse** (a single crest) or it may be continuous (many crests and troughs).

The **wavelength** of a continuous wave is the distance between two successive crests (or any two identical points on the wave shape).

The **frequency** is the number of full wavelengths (or crests) that pass a given point per unit time.

The **wave velocity** (how fast a crest moves) is equal to the product of wavelength and frequency,

$$v = \lambda f.$$

The **amplitude** of a wave is the maximum height of a crest, or depth of a trough, relative to the normal (or equilibrium) level.

In a **transverse wave**, the oscillations are perpendicular to the direction in which the wave travels. An example is a wave on a string.

In a **longitudinal wave**, the oscillations are along (parallel to) the line of travel; sound is an example.

The velocity of both longitudinal and transverse waves in matter is proportional to the square root of an elastic force factor divided by an inertia factor (or density).

Waves carry energy from place to place without matter being carried. The energy transported by a wave, the power (energy transported per unit time), and the **intensity** of a wave (energy transported across unit area per unit time) are all proportional to the square of the amplitude of the wave.

For a wave traveling outward in three dimensions from a point source, the intensity (ignoring damping) decreases with the square of the distance from the source,

$$I \propto \frac{1}{r^2}.$$

The amplitude decreases linearly with distance from the source.

A one-dimensional transverse wave traveling to the right along the x axis (x increasing) can be represented by a formula for its amplitude as a function of position and time as

$$D(x, t) = D_M \sin\left[\left(\frac{2\pi}{\lambda}\right)(x - vt)\right] = D_M \sin(kx - \omega t)$$

where

$$k = \frac{2\pi}{\lambda} \quad \text{and} \quad \omega = 2\pi f.$$

If a wave is traveling toward decreasing values of x,

$$D(x, t) = D_M \sin(kx + \omega t).$$

When two or more waves pass through the same region of space at the same time, the displacement at any given point will be the vector sum of the displacements of the separate waves. This is the **principle of superposition**. It is valid for mechanical waves if the amplitudes are small enough that the restoring force of the medium is proportional to displacement.

Waves reflect off objects in their path. When a wave strikes a boundary between two materials in which it can travel, part of the wave is reflected and part is transmitted.

When the **wave front** of a two or three dimensional wave strikes an object, the angle of reflection equals the angle of incidence.

When two waves pass through the same region of space at the same time, they **interfere**. From the superposition principle, the resultant displacement at any point and time is the sum of their separate displacements. This can result in **constructive interference, destructive interference,** or something in between depending on the amplitudes and relative phases of the waves.

Waves traveling on a cord (or other medium) of fixed length interfere with waves that have reflected off the end and are traveling back in the opposite direction. At certain frequencies, **standing waves** can be produced in which the waves seem to be standing still rather than traveling. The cord (or other medium) is vibrating as a whole. This is a resonance phenomenon and the frequencies at which standing waves occur are called **resonant frequencies**. The points of destructive interference (no vibration) are called **nodes**. Points of constructive interference (maximum amplitude of vibration) are called **antinodes**.

The wavelengths of standing waves are given by $\lambda_n = 2L/n$ where n is an integer.

Questions

1. Is the frequency of a simple periodic wave equal to the frequency of its source? Why or why not?

2. Explain the difference between the speed of a transverse wave traveling down a rope and the speed of a tiny piece of the rope.

3. Why do the strings used for the lowest-frequency notes on a piano normally have wire wrapped around them?

4. What kind of waves do you think will travel down a horizontal metal rod if you strike its end (a) vertically from above and (b) horizontally parallel to its length?

5. Since the density of air decreases with an increase in temperature, but the bulk modulus B is nearly independent of temperature, how would you expect the speed of sound waves in air to vary with temperature?

6. Give examples, other than those already mentioned, of one-, two-, and three-dimensional waves.

7. The speed of sound in most solids is somewhat greater than in air, yet the density of solids is much greater $(10^3–10^4$ times$)$. Explain.

8. Give two reasons why circular water waves decrease in amplitude as they travel away from the source.

9. Two linear waves have the same amplitude and otherwise are identical, except one has half the wavelength of the other. Which transmits more energy? By what factor?

10. The intensity of a sound in real-life situations does not decrease precisely with the square of the distance from the source as we might expect from Eq. 15–8. Why not?

11. Will any function of $(x - vt)$—see Eq. 15–14—represent a wave motion? Why or why not? If not, give an example.

12. When a sinusoidal wave crosses the boundary between two sections of rope as in Fig. 15–20, the frequency does not change (although the wavelength and velocity do change). Explain why.

13. If a sinusoidal wave on a two-section string (Fig. 15–20) is inverted upon reflection, does the transmitted wave have a longer or shorter wavelength?

14. Is energy always conserved when two waves interfere? Explain.

15. If a string is vibrating in three segments, are there any places one can touch it with a knife blade without disturbing the motion?

16. When a standing wave exists on a string, the vibrations of incident and reflected waves cancel at the nodes. Does this mean that energy was destroyed? Explain.

17. Why can you make water slosh back and forth in a pan only if you shake the pan at a certain frequency?

18. Can the amplitude of the standing waves in Fig. 15–26 be greater than the amplitude of the vibrations that cause them (up and down motion of the hand)?

19. When a cord is vibrated as in Fig. 15–26 by hand or by a mechanical vibrator, the "nodes" are not quite true nodes (at rest). Explain. [*Hint*: Consider damping and energy flow from hand or vibrator.]

* 20. AM radio signals can usually be heard behind a hill, but FM often cannot. That is, AM signals bend more than FM. Explain. (Radio signals, as we shall see, are carried by electromagnetic waves whose wavelength for AM is typically 200 to 600 m and for FM about 3 m.)

Problems

Sections 15–1 and 15–2

1. (I) A fisherman notices that wave crests pass the bow of his anchored boat every 4.0 s. He measures the distance between two crests to be 9.0 m. How fast are the waves traveling?

2. (I) AM radio signals have frequencies between 550 kHz and 1600 kHz (kilohertz) and travel with a speed of 3.0×10^8 m/s. What are the wavelengths of these signals? On FM the frequencies range from 88 MHz to 108 MHz (megahertz) and travel at the same speed. What are their wavelengths?

3. (I) A sound wave in air has a frequency of 262 Hz and travels with a speed of 330 m/s. How far apart are the wave crests (compressions)?

4. (I) Calculate the speed of longitudinal waves in (a) water and (b) granite.

5. (I) Determine the wavelength of a 5000-Hz sound wave traveling along an iron rod.

6. (II) A rope of mass 0.65 kg is stretched between two supports 30 m apart. If the tension in the rope is 120 N, how long will it take a pulse to travel from one support to the other?

7. (II) A 0.40-kg rope is stretched between two supports, 4.8 m apart. When one support is struck by a hammer, a transverse wave travels down the rope and reaches the other support in 0.85 s. What is the tension in the rope?

8. (II) A sailor strikes the side of his ship just below the surface of the sea. He hears the echo of the wave reflected from the ocean floor directly 3.5 s later. How deep is the ocean at this point?

9. (II) A gondola is connected to the top of a hill by a steel cable of length 600 m and diameter 1.5 cm. As the gondola comes to the end of its run, it bumps into the side and sends a wave pulse along the cable. It is observed that it took 16 s for the pulse to return. (a) What is the speed of the pulse? (b) What is the tension in the cable?

10. (II) The wave on a string shown in Fig. 15–34 is moving to the right with a speed of 1.80 m/s. (a) Draw the shape of the string 1.00 s later and indicate which parts of the string are moving up and which down at that instant. (b) Estimate the vertical speed of point A on the string at the instant shown in the figure.

11. (II) S and P waves from an earthquake travel at different speeds and this difference helps in the determination of the earthquake "epicenter" (where the disturbance took place). (a) Assuming typical speeds of 9.0 km/s and 5.5 km/s for P and S waves, respectively, how far away did the earthquake occur if a particular seismic station detects the arrival of these two types of waves exactly 94 s apart? (b) Is one seismic station sufficient to determine the position of the focus? Explain.

Section 15–3

12. (I) Two earthquake waves have the same frequency as they travel through the same portion of the Earth, but one is carrying twice the energy. What is the ratio of the amplitudes of the two waves?

13. (I) Compare (a) the intensities and (b) the amplitudes of an earthquake P wave as it passes two points 10 km and 20 km from the source.

14. (II) The intensity of a particular earthquake wave is measured to be 2.2×10^6 W/m² at a distance of 100 km from the source. (a) What was the intensity when it passed a point only 4.0 km from the source? (b) What was the total power passing through an area of 5.0 m² at a distance of 4.0 km?

15. (II) Show that if damping is ignored the amplitude D_M of circular water waves decreases as the square root of the distance r from the source: $D_M \propto 1/\sqrt{r}$.

16. (II) (a) Show that the intensity of a wave is equal to the energy density (energy per unit volume) in the wave times the wave speed. (b) What is the energy density 2.0 m from a 100-W light bulb? Light travels 3.0×10^8 m/s.

17. (II) A small steel wire of diameter 1.0 mm is connected to a vibrator and is under a tension of 4.5 N. The frequency of the vibrator is 60.0 Hz and it is observed that the amplitude of the wave on the steel wire is 0.50 cm. (a) What is the power output of the vibrator, assuming that the wave is not reflected back? (b) If the power output stays constant but the frequency is doubled, what is the amplitude of the wave?

FIGURE 15–34 Problem 10.

18. (II) (a) Show that the average rate with which energy is transported along a cord by a mechanical wave of frequency f and amplitude D_M is

$$\overline{P} = 2\pi^2 \mu v f^2 D_M^2,$$

where v is the speed of the wave and μ is the mass per unit length of the cord. (b) If the cord is under a tension $F_T = 100\,\text{N}$ and has mass per unit length $0.10\,\text{kg/m}$, what power is required to transmit 120-Hz transverse waves of amplitude 2.0 cm?

Section 15–4

19. (I) Suppose at $t = 0$, a wave shape is represented by $D = D_M \sin(2\pi x/\lambda + \phi)$; that is, it differs from Eq. 15–9 by a constant phase factor ϕ. What then will be the equation for a wave traveling to the left along the x axis as a function of x and t?

20. (II) A transverse traveling wave on a cord is represented by $D = 0.48 \sin(5.6x + 84t)$ where D and x are in meters and t in seconds. For this wave determine (a) the wavelength, (b) frequency, (c) velocity (magnitude and direction), (d) amplitude, and (e) maximum and minimum speeds of particles of the cord.

21. (II) Consider a point on the string of Example 15–4 that is 1.00 m from the left-hand end. Determine (a) the maximum velocity of this point, and (b) its maximum acceleration. (c) What is its velocity and acceleration at $t = 2.0\,\text{s}$?

22. (II) Show, for a sinusoidal transverse wave traveling on a string, that the slope of the string at any point x is equal to the ratio of the transverse speed of the particle to the speed of the wave at that point.

23. (II) A transverse wave pulse travels to the right along a string with a speed $v = 2.0\,\text{m/s}$. At $t = 0$ the shape of the pulse is given by the function

$$D = 0.45 \cos(3.0x + 1.2),$$

where D and x are in meters. (a) Plot D versus x at $t = 0$. (b) Determine a formula for the wave pulse at any time t assuming there are no frictional losses. (c) Plot $D(x, t)$ versus x at $t = 1.0\,\text{s}$. (d) Repeat parts (b) and (c) assuming the pulse is traveling to the left.

24. (II) A 440-Hz longitudinal wave in air has a speed of 345 m/s. (a) What is the wavelength? (b) How much time is required for the phase to change by 90° at a given point in space? (c) At a particular instant, what is the phase difference (in degrees) between two points 4.4 cm apart?

25. (II) Write down the equation for the wave in Problem 24 if its amplitude is 0.020 cm and at $t = 0$, $x = 0$, and $D = -0.020\,\text{cm}$.

26. (II) A sinusoidal wave traveling on a string in the negative x direction has amplitude 1.00 cm, wavelength 3.00 cm, and frequency 200 Hz. At $t = 0$, the particle of string at $x = 0$ is displaced a distance $D = 0.80\,\text{cm}$ above the origin and is moving upward. (a) Sketch the shape of the wave at $t = 0$ and (b) determine the function of x and t that describes the wave.

* Section 15–5

*27. (II) Determine if the function $D = D_M \sin kx \cos \omega t$ is a solution of the wave equation.

*28. (II) Show by direct substitution that the following functions satisfy the wave equation: (a) $D(x, t) = D_M \ln(x + vt)$; (b) $D(x, t) = (x - vt)^4$.

*29. (II) Show that the wave forms of Eqs. 15–13 and 15–15 satisfy the wave equation, Eq. 15–16.

*30. (II) Let two linear waves be represented by $D_1 = f_1(x, t)$ and $D_2 = f_2(x, t)$. If both these waves satisfy the wave equation (Eq. 15–16), show that any combination $D = C_1 D_1 + C_2 D_2$ does as well, where C_1 and C_2 are constants.

Section 15–7

31. (II) Consider a sine wave traveling down the stretched two-part cord of Fig. 15–20. Determine a formula (a) for the ratio of the speeds of the wave in the two sections, v_2/v_1, and (b) for the ratio of the wavelengths in the two sections. (The frequency is the same in both sections. Why?) (c) Is the wavelength larger in the heavier cord or the lighter?

32. (II) A cord has two sections with linear densities of $0.10\,\text{kg/m}$ and $0.20\,\text{kg/m}$, Fig. 15–35. An incident wave, given by $D = (0.050\,\text{m}) \sin(6.0x - 12.0t)$, where x is in meters and t in seconds, travels along the lighter cord. (a) What is the wavelength on the lighter section of the cord? (b) What is the tension in the cord? (c) What is the wavelength when the wave travels on the heavier section?

$$\mu_1 = 0.10\,\text{kg/m} \qquad \mu_2 = 0.20\,\text{kg/m}$$

$$D = (0.050\,\text{m}) \sin(6.0\,x - 12.0\,t)$$

FIGURE 15–35 Problem 32.

33. (III) A cord stretched to a tension F_T, consists of two sections (as in Fig. 15–21) whose linear densities are μ_1 and μ_2. Take $x = 0$ to be the point (a knot) where they are joined, with μ_1 referring to that section of cord to the left and μ_2 that to the right. A sinusoidal wave, $D = A \sin[k_1(x - v_1 t)]$, starts at the left end of the cord. When it reaches the knot, part of it is reflected and part is transmitted. Let the equation of the reflected wave be $D = A_R \sin[k_1(x + v_1 t)]$ and that for the transmitted wave be $D = A_T \sin[k_2(x - v_2 t)]$. Since the frequency must be the same in both sections, we have $\omega_1 = \omega_2$ or $k_1 v_1 = k_2 v_2$. (a) Using the fact that the rope is continuous—so that a point an infinitesimal distance to the left of the knot has the same displacement at any moment (due to incident plus reflected waves) as a point just to the right of the knot (due to the transmitted wave)—show that $A = A_T + A_R$. (b) Assuming that the slope $(\partial D/\partial x)$ of the string just to the left of the knot is the same as the slope just to the right of the knot, show that the amplitude of the reflected wave is given by

$$A_R = \left(\frac{v_1 - v_2}{v_1 + v_2}\right) A = \left(\frac{k_2 - k_1}{k_2 + k_1}\right) A.$$

(c) What is A_T in terms of A?

34. (I) The two pulses shown in Fig. 15–36 are moving toward each other. (a) Sketch the shape of the string at the moment they directly overlap. (b) Sketch the shape of the string a few moments later. (c) In Fig. 15–23a, at the moment the pulses pass each other, the string is straight. What has happened to the energy at this moment?

FIGURE 15–36 Problem 34.

35. (II) Suppose two linear waves of equal amplitude and frequency have a phase difference ϕ as they travel in the same medium. They can be represented by

$$D_1 = D_M \sin(kx - \omega t)$$

$$D_2 = D_M \sin(kx - \omega t + \phi).$$

(a) Use the trigonometric identity $\sin \theta_1 + \sin \theta_2 = 2 \sin \frac{1}{2}(\theta_1 + \theta_2) \cos \frac{1}{2}(\theta_1 - \theta_2)$ to show that the resultant wave is given by

$$D = \left(2 D_M \cos \frac{\phi}{2} \right) \sin \left(kx - \omega t + \frac{\phi}{2} \right).$$

(b) What is the amplitude of this resultant wave? Is the wave purely sinusoidal, or not? (c) Show that constructive interference occurs if $\phi = 0, 2\pi, 4\pi$, and so on, and destructive interference occurs if $\phi = \pi, 3\pi, 5\pi$, etc. (d) Describe the resultant wave, by equation and in words, if $\phi = \pi/2$.

Section 15–9

36. (I) A violin string vibrates at 294 Hz when unfingered. At what frequency will it vibrate if it is fingered one-fourth of the way down from the end?

37. (I) If a violin string vibrates at 440 Hz as its fundamental frequency, what are the frequencies of the first four harmonics?

38. (I) A particular string resonates in four loops at a frequency of 264 Hz. Give at least three other frequencies at which it will resonate.

39. (I) In an earthquake, it is noted that a footbridge oscillated up and down in a one loop (fundamental standing wave) pattern once every 2.0 s. What other possible resonant periods of motion are there for this bridge? What frequencies do they correspond to?

40. (II) The velocity of waves on a string is 270 m/s. If the frequency of standing waves is 131 Hz, how far apart are the nodes?

41. (II) If two successive harmonics of a vibrating string are 280 Hz and 350 Hz, what is the frequency of the fundamental?

42. (II) A guitar string is 90.0 cm long and has a mass of 3.6 g. From the bridge to the support post ($= L$) is 60.0 cm and the string is under a tension of 520 N. What are the frequencies of the fundamental and first two overtones?

43. (II) Show that the frequency of standing waves on a string of length L and linear density μ, which is stretched to a tension F_T, is given by

$$f = \frac{n}{2L} \sqrt{\frac{F_T}{\mu}}$$

where n is an integer.

44. (II) One end of a horizontal string of linear density 4.8×10^{-4} kg/m is attached to a small amplitude mechanical 60-Hz vibrator. The string passes over a pulley, a distance $L = 1.40$ m away, and weights are hung from this end. What mass must be hung from this end of the string to produce (a) one loop, (b) two loops, and (c) five loops of a standing wave? Assume the string at the vibrator is a node, which is nearly true. Why can the amplitude of the standing wave be much greater than the vibrator amplitude?

45. (II) In Problem 44, the length of the string may be adjusted by moving the pulley. If the hanging mass is fixed at 0.080 kg, how many different standing wave patterns may be achieved by varying L between 10 cm and 1.5 m?

46. (II) The displacement of a standing wave is given by $D = 8.6 \sin(0.60x) \cos(58t)$, where x and D are in centimeters and t is in seconds. (a) What is the distance (cm) between nodes? (b) Give the amplitude, frequency, and speed of each of the component waves. (c) Find the speed of a particle of the string at $x = 3.20$ cm when $t = 2.5$ s.

47. (II) The displacement of a transverse wave traveling on a string is represented by $D = 4.2 \sin(0.71x - 47t + 2.1)$, where D and x are in cm and t in s. (a) Find an equation that represents a wave which, when traveling in the opposite direction, will produce a standing wave when added to this one. (b) What is the equation describing the standing wave?

48. (II) When you slosh the water back and forth in a tub at just the right frequency, the water alternately rises and falls at each end, remaining relatively calm at the center. Suppose the frequency to produce such a standing wave in a 60-cm-wide tub is 0.85 Hz. What is the speed of the water wave?

49. (II) A particular violin string plays at a frequency of 294 Hz. If the tension is increased 10 percent what will the new frequency be?

50. (II) Two traveling waves are described by the functions

$$D_1 = D_M \sin(kx - \omega t)$$

$$D_2 = D_M \sin(kx + \omega t),$$

where $D_M = 0.15$ m, $k = 3.5$ m^{-1}, and $\omega = 1.2$ s^{-1}. (a) Plot these two waves, from $x = 0$ to a point $x(> 0)$ that includes one full wavelength. Choose $t = 1.0$ s. Plot the sum of the two waves and identify the nodes and antinodes in the plot, and compare to the analytic (mathematical) representation.

51. (II) Plot the two waves given in Problem 50, and their sum, as a function of time from $t = 0$ to $t = T$ (one period). Choose (a) $x = 0$ and (b) $x = \lambda/4$. Interpret your results.

52. (II) A standing wave on a 1.80-m-long horizontal string displays three loops when the string vibrates at 120 Hz. The maximum swing of the string (top to bottom) at the center of each loop is 12.0 cm. (a) What is the function describing the standing wave? (b) What are the functions describing the two equal-amplitude waves traveling in opposite directions that make up the standing wave?

*53. (I) An earthquake P wave traveling $8.0\,km/s$ strikes a boundary within the Earth between two kinds of material. If it approaches the boundary at an incident angle of $50°$ and the angle of refraction is $31°$, what is the speed in the second medium?

*54. (I) Water waves approach an underwater "shelf" where the velocity changes from $2.8\,m/s$ to $2.5\,m/s$. If the incident wave crests make a $40°$ angle with the shelf, what will be the angle of refraction?

*55. (II) A longitudinal earthquake wave strikes a boundary between two types of rock at a $25°$ angle. As it crosses the boundary, the specific gravity of the rock changes from 3.7 to 2.8. Assuming that the elastic modulus is the same for both types of rock, determine the angle of refraction.

*56. (II) It is found for any type of wave, say an earthquake wave, that if it reaches a boundary beyond which its speed is increased, there is a maximum incident angle if there is to be a transmitted refracted wave. This maximum incident angle θ_{iM} corresponds to an angle of refraction equal to $90°$. If $\theta_i > \theta_{iM}$, all the wave is reflected at the boundary and none is refracted (because this would correspond to $\sin\theta_r > 1$, where θ_r is the angle of refraction, which is impossible). This phenomenon is referred to as *total internal reflection*. (*a*) Find a formula for θ_{iM} using Eq. 15–19. (*b*) At what angles of incidence will there be only reflection and no transmission for an earthquake P wave traveling $7.5\,km/s$ where it reaches a different kind of rock where its speed is $9.3\,km/s$?

*57. (II) A sound wave is traveling in warm air when it hits a layer of cold dense air. If the sound wave hits the cold air interface at an angle of $25°$, what is the angle of refraction? Assume that the cold air temperature is $-10°C$ and the warm air temperature is $+10°C$. The speed of sound as a function of temperature can be approximated by $v = (331 + 0.60\,T)\,m/s$, where T is in $°C$.

General Problems

58. When you walk with a cup of coffee (diameter 8 cm) at just the right pace of about 1 step per second, the coffee builds up its "sloshing" until eventually, after a few steps, it starts to spill over the top. What is the speed of the waves in the coffee?

59. Two solid rods have the same bulk modulus but one is twice as dense as the other. In which rod will the speed of longitudinal waves be greater, and by what factor?

60. Two waves traveling along a stretched string have the same frequency, but one transports three times the power of the other. What is the ratio of the amplitudes of the two waves?

61. A bug on the surface of a pond is observed to move up and down a total vertical distance of $0.10\,m$, lowest to highest point, as a wave passes. (*a*) What is the amplitude of the wave? (*b*) If the ripples increase to $0.15\,m$, by what factor does the bug's maximum kinetic energy change?

62. A particular guitar string is supposed to vibrate at $200\,Hz$, but it is measured to actually vibrate at $205\,Hz$. By what percentage should the tension in the string be changed to get the frequency to the correct value?

63. An earthquake-produced surface wave can be approximated by a sinusoidal transverse wave. Assuming a frequency of $0.50\,Hz$ (typical of earthquakes, which actually include a mixture of frequencies), what minimum amplitude will cause objects to leave contact with the ground?

64. A uniform cord of length L and mass m is hung vertically from a support. (*a*) Show that the speed of transverse waves in this cord is $\sqrt{gh}$, where h is the height above the lower end. (*b*) How long does it take for a pulse to travel upward from one end to the other?

65. A transverse wave pulse travels to the right along a string with a speed $v = 3.0\,m/s$. At $t = 0$ the shape of the pulse is given by the function

$$D = \frac{4.0}{x^2 - 2.0},$$

where D and x are in meters. (*a*) Plot D versus x at $t = 0$. (*b*) Determine a formula for the wave pulse at any time t assuming there are no frictional losses. (*c*) Plot $D(x, t)$ versus x at $t = 0.50\,s$. (*d*) Repeat parts (*b*) and (*c*) assuming the pulse is traveling to the left.

66. (*a*) Show that if the tension in a stretched string is changed by a small amount ΔF_T, the frequency of the fundamental is changed by an amount $\Delta f = \frac{1}{2}(\Delta F_T/F_T)f$. (*b*) By what percent must the tension in a piano string be increased or decreased to raise the frequency from $438\,Hz$ to $442\,Hz$? (*c*) Does the formula in part (*a*) apply to the overtones as well?

67. Two strings on a musical instrument are tuned to play at $392\,Hz$ (G) and $440\,Hz$ (A). (*a*) What are the first two overtones for each string? (*b*) If the two strings have the same length and are under the same tension, what must be the ratio of their masses (M_G/M_A)? (*c*) If the strings, instead, have the same mass per unit length and are under the same tension, what is the ratio of their lengths (L_G/L_A)? (*d*) If their masses and lengths are the same, what must be the ratio of the tensions in the two strings?

68. A highway overpass was observed to resonate as one full loop $(\frac{1}{2}\lambda)$ when a small earthquake shook the ground vertically at 4.0 Hz. The highway department put a support at the center of the overpass, anchoring it to the ground as shown in Fig. 15–37. What resonant frequency would you now expect for the overpass? It is noted that earthquakes rarely do significant shaking above 5 or 6 Hz. Did the modifications do any good? Explain.

Before modification

Added support

After modification

FIGURE 15–37   Problem 68.

69. A string can have a "free" end if that end is attached to a ring that can slide without friction on a vertical pole (Fig. 15–38). Determine the wavelengths of the resonant vibrations of such a string with one end fixed and the other free.

FIGURE 15–38
Problem 69.

70. A string fixed at two ends is pulled up at the center into the triangular shape shown in Fig. 15–39. Assuming that the tension F_T remains constant, calculate the energy of the vibrations of the string when it is released. [*Hint*: What work does it take to stretch the string up?]

FIGURE 15–39
Problem 70.

71. Figure 15–40 shows the wave shape of a sinusoidal wave traveling to the right at two instants of time. What is the mathematical representation of this wave?

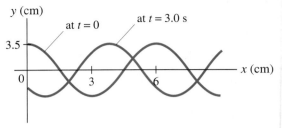

FIGURE 15–40   Problem 71.

72. Estimate the average power of a water wave when it hits the chest of an adult standing in the water at the seashore. Assume that the amplitude of the wave is 0.50 m, the wavelength is 2.5 m, and the period is 5.0 s.

73. Two wave pulses are traveling in opposite directions with the same speed of 5.0 cm/s as shown in Fig. 15–41. At $t = 0$, the leading edges of the two pulses are 15 cm apart. Sketch the wave pulses at $t = 1.0, 2.0$ and 3.0 s.

FIGURE 15–41   Problem 73.

74. For a spherical wave traveling uniformly away from a point source, show that the displacement can be represented by

$$D = \left(\frac{A}{r}\right) \sin(kr - \omega t),$$

where r is the radial distance from the source and A is a constant.

If music be the food of physics, play on.

This orchestra contains stringed instruments whose sound depends on transverse standing waves on strings, and wind instruments whose sound comes from longitudinal standing waves in an air column. Percussive instruments create more complicated standing waves. The human voice utilizes vocal chords (almost like strings) and the throat cavity (vibrating air) to produce its particular timbre.

We also study the decibel intensity scale, the ear's response, sound interference and the Doppler effect.

CHAPTER 16

Sound

Sound is associated with our sense of hearing and, therefore, with the physiology of our ears and the psychology of our brain which interprets the sensations that reach our ears. The term *sound* also refers to the physical sensation that stimulates our ears: namely, longitudinal waves.

We can distinguish three aspects of any sound. First, there must be a *source* for a sound; and as with any wave, the source of a sound wave is a vibrating object. Second, the energy is transferred from the source in the form of longitudinal sound *waves*. And third, the sound is *detected* by an ear or an instrument. Sound is an important factor in the design of buildings, especially theaters and auditoriums, but also in factories and other work places. We start this Chapter by looking at some aspects of sound waves themselves.

16–1 Characteristics of Sound

We already saw in Chapter 15, Fig. 15–6, how a vibrating drumhead produces a sound wave in air. Indeed, we usually think of sound waves traveling in the air, for normally it is the vibrations of the air that force our eardrums to vibrate. But sound waves can also travel in other materials.

Two stones struck together under water can be heard by a swimmer beneath the surface, for the vibrations are carried to the ear by the water. When you put your ear flat against the ground, you can hear an approaching train or truck. In this case the ground does not actually touch your eardrum, but the longitudinal wave transmitted by the ground is called a sound wave just the same, for its vibrations cause the outer ear and the air within it to vibrate. Clearly, sound cannot travel in

TABLE 16–1 Speed of Sound in Various Materials, at 20°C and 1 atm

Material	Speed (m/s)
Air	343
Air (0°C)	331
Helium	1005
Hydrogen	1300
Water	1440
Seawater	1560
Iron and steel	≈ 5000
Glass	≈ 4500
Aluminum	≈ 5100
Hardwood	≈ 4000

➡ **PHYSICS APPLIED**

How far away is the lightning?

Loudness

Pitch

Audible frequency range

➡ **PHYSICS APPLIED**

Autofocusing camera

the absence of matter. For example, a bell ringing inside an evacuated jar cannot be heard, nor does sound travel through the empty reaches of outer space.

The **speed of sound** is different in different materials. In air at $0°C$ and 1 atm, sound travels at a speed of 331 m/s. We saw in Eq. 15–4 $(v = \sqrt{B/\rho})$ that the speed depends on the elastic modulus, B, and the density, ρ, of the material. Thus for helium, whose density is much less than that of air but whose elastic modulus is not greatly different, the speed is about three times as great as in air. In liquids and solids, which are much less compressible and therefore have much greater elastic moduli, the speed is larger still. The speed of sound in various materials is given in Table 16–1. The values depend somewhat on temperature, but this is significant mainly for gases. For example, in air near room temperature, the speed increases approximately 0.60 m/s for each Celsius degree increase in temperature:

$$v \approx (331 + 0.60\,T)\ \text{m/s}, \qquad \text{[Speed of sound in air]}$$

where T is the temperature in °C. Unless stated otherwise, we will assume in this chapter that $T = 20°C$, so that $v = [331 + (0.60)(20)]\ \text{m/s} = 343\ \text{m/s}$.

CONCEPTUAL EXAMPLE 16–1 Distance from a lightning strike. A rule of thumb that tells how close lightning has hit is: "one mile for every five seconds before the thunder is heard." Justify, noting that the speed of light is so high $(3 \times 10^8\ \text{m/s})$ that the time for light to travel is negligible compared to the time for sound.

RESPONSE The speed of sound in air is about 340 m/s, so to travel 1 km = 1000 m takes about 3 seconds. One mile is about 1.6 kilometers, so the time for the thunder to travel a mile is about $(1.6)(3) \approx 5$ seconds.

Two aspects of any sound are immediately evident to a human listener. These are "loudness" and "pitch," and each refers to a sensation in the consciousness of the listener. But to each of these subjective sensations there corresponds a physically measurable quantity. **Loudness** is related to the energy in the sound wave, and we shall discuss it in Section 16–3.

The **pitch** of a sound refers to whether it is high, like the sound of a piccolo or violin, or low, like the sound of a bass drum or string bass. The physical quantity that determines pitch is the frequency, as was first noted by Galileo. The lower the frequency, the lower the pitch, and the higher the frequency, the higher the pitch.[†] The human ear responds to frequencies in the range from about 20 Hz to about 20,000 Hz. (Recall that 1 Hz is 1 cycle per second.) This is called the **audible range**. These limits vary somewhat from one individual to another. One general trend is that as people age, they are less able to hear the high frequencies, so that the high-frequency limit may be 10,000 Hz or less.

Sound waves whose frequencies are outside the audible range may reach the ear, but we are not generally aware of them. Frequencies above 20,000 Hz are called **ultrasonic** (do not confuse with *supersonic*, which is used for an object moving with a speed faster than the speed of sound). Many animals can hear ultrasonic frequencies; dogs, for example, can hear sounds as high as 50,000 Hz, and bats can detect frequencies as high as 100,000 Hz.

EXAMPLE 16–2 Autofocusing with sound waves. Autofocusing cameras emit a pulse of very high frequency (ultrasonic) sound that travels to the object being photographed, and include a sensor that detects the returning reflected sound, as shown in Fig. 16–1. To get an idea of the time sensitivity of the detector, calculate the travel time of the pulse for an object (*a*) 1.0 m away, (*b*) 20 m away.

[†] Although pitch is determined mainly by frequency, it also depends to a slight extent on loudness. For example, a very loud sound may seem slightly lower in pitch than a quiet sound of the same frequency.

FIGURE 16–1 Example 16–2.
Autofocusing camera emits an
ultrasonic pulse. Solid blue lines
represent the moving wave front of
the outgoing wave pulse moving to
the right; dashed lines represent the
wave front of the pulse reflected off
the person's face, returning to the
camera. The time information allows
the camera mechanism to adjust the
lens to focus on the face.

SOLUTION We assume the temperature is about 20°C, so the speed of sound, as calculated above, is 343 m/s. (*a*) The pulse travels 1.0 m to the object and 1.0 m back, for a total of 2.0 m. Since speed = distance/time, we have

$$t = \frac{\text{distance}}{\text{speed}} = \frac{2.0\ \text{m}}{343\ \text{m/s}} = 0.0059\ \text{s} = 5.9\ \text{ms}.$$

(*b*) The total distance now is $2 \times 20\ \text{m} = 40\ \text{m}$, so

$$t = \frac{40\ \text{m}}{343\ \text{m/s}} = 0.12\ \text{s} = 120\ \text{ms}.$$

Sound waves whose frequencies are below the audible range (that is, less than 20 Hz) are called **infrasonic**. Sources of infrasonic waves include earthquakes, thunder, volcanoes, and waves produced by vibrating heavy machinery. This last source can be particularly troublesome to workers, for infrasonic waves—even though inaudible—can cause damage to the human body. These low-frequency waves act in a resonant fashion, causing considerable motion and irritation of internal organs of the body.

16–2 Mathematical Representation of Longitudinal Waves

In Section 15–4, we saw that a one-dimensional sinusoidal wave traveling along the *x* axis can be represented by the relation (Eq. 15–10c)

$$D = D_M \sin(kx - \omega t). \tag{16–1}$$

Here the wave number *k* is related to the wavelength λ by $k = 2\pi/\lambda$, and $\omega = 2\pi f$ where *f* is the frequency; *D* is the displacement at position *x* and time *t*, and D_M is its maximum value, the *amplitude*. For a transverse wave—such as a wave on a string—the displacement *D* is perpendicular to the direction of wave propagation along the *x* axis. But for a longitudinal wave the displacement *D* is *along the direction of wave propagation*. That is, *D* is parallel to *x* and represents the displacement of a tiny volume element of the medium from its equilibrium position.

Longitudinal (sound) waves can also be considered from the point of view of variations in pressure rather than displacement. Indeed, longitudinal waves are often called **pressure waves**. The pressure variation is usually easier to measure than the displacement (see Example 16–4). As can be seen in Fig. 16–2, in a wave "compression" (where molecules are closest together), the pressure is higher than normal, whereas in an expansion (or rarefaction) the pressure is less than normal. Figure 16–3 shows a graphical representation of a sound wave in air in terms of (a) displacement and (b) pressure. Note that the displacement wave is a quarter wavelength, or 90° ($\pi/2$ rad), out of phase with the pressure wave: where the pressure is a maximum or minimum, the molecules are momentarily at rest in equilibrium so the displacement from equilibrium is zero; and where the pressure variation is zero, the displacement is a maximum or minimum.

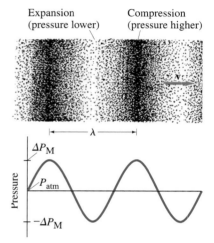

FIGURE 16–2 Longitudinal sound wave traveling to the right, and its graphical representation in terms of pressure.

FIGURE 16–3 Representation of a sound wave in terms of (a) displacement and (b) pressure.

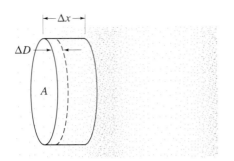

FIGURE 16–4 Longitudinal wave in a fluid moves to the right. A thin layer of fluid, in a thin cylinder of area A and thickness Δx, changes in volume as a result of pressure variation as the wave passes. At the moment shown, the pressure will increase as the wave moves to the right, so the thickness of our layer will decrease, by an amount ΔD.

Pressure Wave Derivation

Let us now derive the mathematical representation of the pressure variation in a traveling longitudinal wave. From the definition of the bulk modulus, B (Eq. 12–7a),

$$\Delta P = -B(\Delta V/V),$$

where $\Delta V/V$ is the fractional change in volume of the medium due to a pressure change ΔP, and ΔP represents the pressure difference from the normal pressure P_0 (no wave present). The negative sign reflects the fact that the volume decreases ($\Delta V < 0$) if the pressure is increased. Consider now a layer of fluid through which the longitudinal wave is passing (Fig. 16–4). If this layer has thickness Δx and area A, then its volume is $V = A \, \Delta x$. As a result of pressure variation in the wave, the volume will change by an amount $\Delta V = A \, \Delta D$ where ΔD is the change in thickness of this layer as it compresses or expands. (Remember that D represents the displacement of the medium.) Thus we have

$$\Delta P = -B \frac{A \, \Delta D}{A \, \Delta x}.$$

To be precise, we take the limit of $\Delta x \to 0$, so we obtain

$$\Delta P = -B \frac{\partial D}{\partial x}, \tag{16–2}$$

where we use the partial derivative notation since D is a function of both x and t. If the displacement D is sinusoidal as given by Eq. 16–1, then we have from Eq. 16–2 that

$$\Delta P = -(BD_M k) \cos(kx - \omega t). \tag{16–3}$$

Thus the pressure varies sinusoidally as well, but is out of phase from the displacement by 90° or a quarter wavelength; see Fig. 16–3. The quantity $BD_M k$ is called the **pressure amplitude**, ΔP_M. It represents the maximum and minimum amounts by which the pressure varies from the normal ambient pressure. We can thus write

$$\Delta P = -\Delta P_M \cos(kx - \omega t), \tag{16–4}$$

where, using $v = \sqrt{B/\rho}$ (Eq. 15–4), and $k = \omega/v = 2\pi f/v$ (Eq. 15–12), then

$$
\begin{aligned}
\Delta P_M &= BD_M k \\
&= \rho v^2 D_M k \\
&= 2\pi \rho v D_M f.
\end{aligned} \tag{16–5}
$$

16–3 Intensity of Sound; Decibels

Intensity

Like pitch, **loudness** is a sensation in the consciousness of a human being. It too is related to a physically measurable quantity, the **intensity** of the wave. Intensity is defined as the energy transported by a wave per unit time across unit area perpendicular to the energy flow, and, as we saw in the previous chapter, is proportional to the square of the wave amplitude. Intensity has units of power per unit area, or watts/meter^2 (W/m^2).

The human ear can detect sounds with an intensity as low as $10^{-12} \, \text{W/m}^2$ and as high as $1 \, \text{W/m}^2$ (and even higher, although above this it is painful). This is an incredibly wide range of intensity, spanning a factor of 10^{12} from lowest to highest. Presumably because of this wide range, what we perceive as loudness is not directly proportional to the intensity. To produce a sound that seems twice as loud requires a sound wave that has about ten times the intensity. This is roughly true at any sound level for frequencies near the middle of the audible range. For example, a sound wave of intensity $10^{-2} \, \text{W/m}^2$ sounds to an average human like it is about twice as loud as one whose intensity is $10^{-3} \, \text{W/m}^2$, and four times as loud as $10^{-4} \, \text{W/m}^2$.

Sound Level

Because of this relationship between the subjective sensation of loudness and the physically measurable quantity "intensity," it is usual to specify sound intensity levels using a logarithmic scale. The unit on this scale is a **bel**, after the inventor, Alexander Graham Bell, or much more commonly, the **decibel** (dB) which is $\frac{1}{10}$ bel (10 dB = 1 bel). The **sound level**, β, of any sound is defined in terms of its intensity, I, as

$$\beta \text{ (in dB)} = 10 \log \frac{I}{I_0}, \qquad \textbf{(16–6)}$$

Sound level (decibels)

where I_0 is the intensity of some reference level and the logarithm is to the base 10. I_0 is usually taken as the minimum intensity audible to an average person, the "threshold of hearing," which is $I_0 = 1.0 \times 10^{-12} \text{ W/m}^2$. Thus, for example, the sound level of a sound whose intensity $I = 1.0 \times 10^{-10} \text{ W/m}^2$ will be

$$\beta = 10 \log \left(\frac{1.0 \times 10^{-10} \text{ W/m}^2}{1.0 \times 10^{-12} \text{ W/m}^2} \right) = 10 \log 100 = 20 \text{ dB},$$

since log 100 is equal to 2.0. Notice that the sound level at the threshold of hearing is 0 dB. That is, $\beta = 10 \log 10^{-12}/10^{-12} = 10 \log 1 = 0$ since log 1 = 0. Notice too that an increase in intensity by a factor of 10 corresponds to a level increase of 10 dB. An increase in intensity by a factor of 100 corresponds to a level increase of 20 dB. Thus a 50-dB sound is 100 times more intense than a 30-dB sound, and so on.

The intensities and sound levels for a number of common sounds are listed in Table 16–2.

TABLE 16–2   Intensity of Various Sounds

Source of the Sound	Sound Level (dB)	Intensity (W/m^2)
Jet plane at 30 m	140	100
Threshold of pain	120	1
Loud rock concert	120	1
Siren at 30 m	100	1×10^{-2}
Auto interior, at 90 km/h	75	3×10^{-5}
Busy street traffic	70	1×10^{-5}
Conversation, at 50 cm	65	3×10^{-6}
Quiet radio	40	1×10^{-8}
Whisper	20	1×10^{-10}
Rustle of leaves	10	1×10^{-11}
Threshold of hearing	0	1×10^{-12}

EXAMPLE 16–3   Loudspeaker response. A high-quality loudspeaker is advertised to reproduce, at full volume, frequencies from 30 Hz to 18,000 Hz with uniform intensity ±3 dB. That is, over this frequency range, the sound level does not vary by more than 3 dB from the average. By what factor does the intensity change for the maximum sound level change of 3 dB?

SOLUTION   Let us call the average intensity I_1 and the average level β_1. Then the maximum intensity, I_2, corresponds to a level $\beta_2 = \beta_1 + 3$ dB. Thus

➡ **PHYSICS APPLIED**

Loudspeaker response (±3 dB)

$$\beta_2 - \beta_1 = 10 \log \frac{I_2}{I_0} - 10 \log \frac{I_1}{I_0}$$

$$3 \text{ dB} = 10 \left(\log \frac{I_2}{I_0} - \log \frac{I_1}{I_0} \right)$$

$$= 10 \log \frac{I_2}{I_1}$$

because $(\log a - \log b) = \log a/b$. Then

$$\log \frac{I_2}{I_1} = 0.30, \qquad \text{or} \qquad \frac{I_2}{I_1} = 10^{0.30} = 2.0,$$

so ±3 dB corresponds to a doubling (or halving) of the intensity.

It is worth noting that a sound level difference of 3 dB (which corresponds to a doubled intensity as we just saw) corresponds to only a very small change in the subjective sensation of apparent loudness. Indeed, the average human can distinguish a difference in level of only about 1 or 2 dB.

The Ear's Response

The ear is not equally sensitive to all frequencies. To hear the same loudness for sounds of different frequencies requires different intensities. Studies averaged over large numbers of people have produced the curves shown in Fig. 16–5. On this graph, each curve represents sounds that seemed to be equally loud. The number labeling each curve represents the **loudness level** (the units are called *phons*), which is numerically equal to the sound level in dB at 1000 Hz. For example, the curve labeled 40 represents sounds that are heard by an average person to have the same loudness as a 1000-Hz sound with a sound level of 40 dB. From this 40-phon curve, we see that a 100-Hz tone must be at a level of about 62 dB to sound as loud as a 1000-Hz tone of only 40 dB. The lowest curve in Fig. 16–5 (labeled 0) represents the sound level, as a function of frequency, for the softest sound that is just audible by a very good ear (the "threshold of hearing"). Note that the ear is most sensitive to sounds of frequency between 2000 and 4000 Hz, which are common in speech and music. Note too that whereas a 1000-Hz sound is audible at a level of 0 dB, a 100-Hz sound must be at least 40 dB to be heard.

Loudness (in "phons")

The top curve in Fig. 16–5, labeled 120, represents the "threshold of feeling or pain." Sounds above this level can actually be felt and cause pain.

Figure 16–5 shows that at lower intensity levels, our ears are less sensitive to the high and low frequencies relative to middle frequencies. The "loudness" control on stereo systems is intended to compensate for this. As the volume is turned down, the loudness control boosts the high and low frequencies relative to the middle frequencies so that the sound will have a more "normal-sounding" frequency balance. Many listeners, however, find the sound more pleasing or natural without the loudness control.

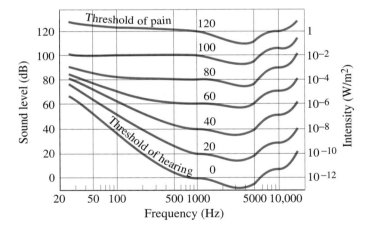

FIGURE 16–5 Sensitivity of the human ear as a function of frequency (see text). Note that the frequency scale is "logarithmic" in order to cover a wide range of frequencies.

Intensity Related to Amplitude

The intensity I is proportional to the square of the amplitude as we saw in Chapter 15. Indeed, using Eq. 15–7, $I = 2\pi^2 v\rho f^2 D_M^2$, we can relate the amplitude quantitatively to the intensity I or level β as the following Example shows.

EXAMPLE 16–4 **How tiny the displacement is.** (*a*) Calculate the maximum displacement of air molecules for a sound at the threshold of hearing, having a frequency of 1000 Hz. (*b*) Determine the maximum pressure variation in such a sound wave.

SOLUTION (a) The "threshold of hearing" at 1000 Hz is (Fig. 16–5) about 0 dB or 1.0×10^{-12} W/m^2. We use Eq. 15–7 of Chapter 15 and solve for D_M:

$$D_M = \frac{1}{\pi f} \sqrt{\frac{I}{2\rho v}}$$

$$= \frac{1}{(3.14)(1.0 \times 10^3 \, \text{s}^{-1})} \sqrt{\frac{1.0 \times 10^{-12} \, \text{W/m}^2}{(2)(1.29 \, \text{kg/m}^3)(343 \, \text{m/s})}} = 1.1 \times 10^{-11} \, \text{m},$$

where we have taken the density of air to be 1.29 kg/m^3 and the speed of sound in air (assumed 20°C) as 343 m/s. We see how incredibly sensitive the human ear is: It can detect displacements of air molecules which are actually less than the diameter of atoms (about 10^{-10} m).

Ear detects displacements smaller than size of atoms

(b) Now we are dealing with sound as a pressure wave (Section 16–2). From Eq. 16–5,

$$\Delta P_M = 2\pi \rho v D_M f = 3.1 \times 10^{-5} \, \text{Pa}$$

or 3.1×10^{-10} atm. Again we see that the human ear is incredibly sensitive.

By combining Eqs. 15–7 and 16–5, we can write the intensity in terms of the pressure amplitude, ΔP_M:

$$I = 2\pi^2 v \rho f^2 D_M^2 = 2\pi^2 v \rho f^2 (\Delta P_M / 2\pi \rho v f)^2$$

$$I = \frac{(\Delta P_M)^2}{2v\rho}. \qquad \qquad \textbf{(16–7)}$$

Intensity related to pressure amplitude

The intensity, when given in terms of pressure amplitude, thus does not depend on frequency.

Normally, the loudness or intensity of a sound decreases as you get farther from the source of the sound. In enclosed rooms, this effect is altered because of absorption or reflection from the walls. However, if a source is in the open so that sound can radiate freely in all directions, the intensity decreases as the inverse square of the distance,

$$I \propto \frac{1}{r^2}$$

as we saw in Section 15–3, Eq. 15–8. Of course, if there is significant reflection from surrounding structures or the ground, the situation will be more complicated.

EXAMPLE 16–5 **Airplane roar.** The sound level of a jet plane at a distance of 30 m is 140 dB. What is the sound level 300 m away? (Ignore reflections from the ground.)

SOLUTION The intensity I at 30 m is found from Eq. 16–6:

$$140 \, \text{dB} = 10 \log \left(\frac{I}{10^{-12} \, \text{W/m}^2} \right).$$

Reversing the log equation to solve for I we have:

$$10^{14} = \frac{I}{10^{-12} \, \text{W/m}^2},$$

so $I = 10^2$ W/m^2. At 300 m, 10 times as far away, the intensity will be $\left(\frac{1}{10}\right)^2 = 1/100$ as much, or 1 W/m^2. Hence, the sound level is

$$\beta = 10 \log \left(\frac{1 \, \text{W/m}^2}{10^{-12} \, \text{W/m}^2} \right) = 120 \, \text{dB}.$$

Even at 300 m, the sound is at the threshold of pain. This is why workers at airports wear ear covers to protect their ears from damage (Fig. 16–6).

➡ **PHYSICS APPLIED**

Jet plane noise

FIGURE 16–6 Airport worker with sound-intensity reducing ear covers (headphones).

TABLE 16–3 Equally
Tempered Chromatic Scale[†]

Note	Frequency (Hz)
C	262
C$^\sharp$ or D$^\flat$	277
D	294
D$^\sharp$ or E$^\flat$	311
E	330
F	349
F$^\sharp$ or G$^\flat$	370
G	392
G$^\sharp$ or A$^\flat$	415
A	440
A$^\sharp$ or B$^\flat$	466
B	494
C$'$	524

[†] Only one octave is included.

➡ **PHYSICS APPLIED**

Musical instrument

FIGURE 16–7 Standing waves on a string—only the lowest three frequencies are shown.

(a)   (b)

FIGURE 16–8 The wavelength of a fingered string (b) is shorter than that of an unfingered string (a). Hence, the frequency of the fingered string is higher. Only one string is shown on this guitar, and only the simplest standing wave, the fundamental, is shown.

16–4 Sources of Sound: Vibrating Strings and Air Columns

The source of any sound is a vibrating object. Almost any object can vibrate and hence be a source of sound. We now discuss some simple sources of sound, particularly musical instruments. In musical instruments, the source is set into vibration by striking, plucking, bowing, or blowing. Standing waves are produced and the source vibrates at its natural resonant frequencies. The vibrating source is in contact with the air (or other medium) and pushes on it to produce sound waves that travel outward. The frequencies of the waves are the same as the source, but the speed and wavelengths can be different. A drum has a stretched membrane that vibrates. Xylophones and marimbas have metal or wood bars that can be set into vibration. Bells, cymbals, and gongs also make use of a vibrating metal. The most widely used instruments make use of vibrating strings, such as the violin, guitar, and piano, or make use of vibrating columns of air, such as the flute, trumpet, and pipe organ. We have already seen that the pitch of a pure sound is determined by the frequency. Typical frequencies for musical notes on the so-called "equally tempered chromatic scale" are given in Table 16–3 for the octave beginning with middle C. Note that one octave corresponds to a doubling of frequency. For example, middle C has frequency of 262 Hz whereas C$'$ (C above middle C) has twice the frequency, 524 Hz.

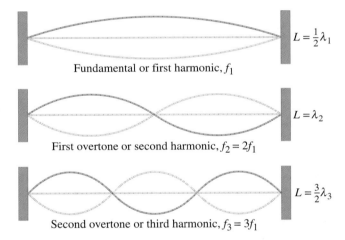

Stringed Instruments

We saw in Chapter 15, Fig. 15–27, how standing waves are established on a string, and we show this again here in Fig. 16–7. This is the basis for all stringed instruments. The pitch is normally determined by the lowest resonant frequency, the **fundamental**, which corresponds to nodes occurring only at the ends. The wavelength of the fundamental on the string is equal to twice the length of the string. Therefore, the fundamental frequency is $f = v/\lambda = v/2L$, where v is the velocity of the wave on the string. When a finger is placed on the string of, say, a guitar or violin, the effective length of the string is shortened. So its fundamental frequency, and pitch, is higher since the wavelength of the fundamental is shorter (Fig. 16–8). The strings on a guitar or violin are all the same length. They sound at a different pitch because the strings have different mass per unit length, μ, which affects the velocity as seen in Eq. 15–2, $v = \sqrt{F_T/\mu}$. Thus the velocity on a heavier string is less and the frequency will be less for the same wavelength. The tension may also be different; adjusting the tension is the means for tuning the instrument. In pianos and harps, the strings are each of different length. For the lower notes the strings are not only longer, but heavier as well, and the reason is illustrated in the following Example.

EXAMPLE 16–6 **Piano strings.** The highest key on a piano corresponds to a frequency about 150 times that of the lowest key. If the string for the highest note is 5.0 cm long, how long would the string for the lowest note have to be if it had the same mass per unit length and was under the same tension?

SOLUTION The velocity would be the same on each string, so the frequency is inversely proportional to the length L of the string ($f = v/\lambda = v/2L$). Thus

$$\frac{L_L}{L_H} = \frac{f_H}{f_L},$$

where the subscripts L and H refer to the lowest and highest notes, respectively. Thus $L_L = L_H(f_H/f_L) = (5.0 \text{ cm})(150) = 750 \text{ cm}$, or 7.5 m. This would be ridiculously long (≈ 25 ft) for a piano. The longer lower strings are made heavier, so that even on grand pianos the strings are no longer than about 3 m.

EXAMPLE 16–7 **Frequencies and wavelengths in the violin.** A 0.32-m-long violin string is tuned to play A above middle C at 440 Hz. (a) What is the wavelength of the fundamental string vibration, and (b) what are the frequency and wavelength of the sound wave produced? (c) Why is there a difference?

SOLUTION (a) From Fig. 16–7 we see that the wavelength of the fundamental is

$$\lambda = 2L = 0.64 \text{ m} = 64 \text{ cm}.$$

This is the wavelength of the standing wave on the string.
(b) The sound wave that travels outward in the air (to reach our ears) has the same frequency, 440 Hz (why?). Its wavelength is

$$\lambda = \frac{v}{f} = \frac{343 \text{ m/s}}{440 \text{ Hz}} = 0.78 \text{ m} = 78 \text{ cm},$$

where v is the speed of sound in air (assumed at 20°C), Section 16–1.
(c) The wavelength of the sound wave is different from that of the standing wave on the string because the speed of sound in air (343 m/s at 20°C) is different from the speed of the wave on the string (= $f\lambda = 440 \text{ Hz} \times 0.64 \text{ m} = 280 \text{ m/s}$), which of course depends on the tension in the string and its mass per unit length.

(a)

(b)

FIGURE 16–9 (a) Sounding box (guitar); (b) sounding board (to which the strings are attached, inside of a piano).

FIGURE 16–10 Wind instruments: clarinet (left) and flute.

Stringed instruments would not be very loud if they relied on their vibrating strings to produce the sound waves since the strings are simply too thin to compress and expand much air. Stringed instruments therefore make use of a kind of mechanical amplifier known as a *sounding board* (piano) or *sounding box* (guitar, violin), which acts to amplify the sound by putting a greater surface area in contact with the air (Fig. 16–9). When the strings are set into vibration, the sounding board or box is set into vibration as well. Since it has much greater area in contact with the air, it can produce a more intense sound wave. On an electric guitar, the sounding box is not so important since the vibrations of the strings are amplified electronically.

Wind Instruments

Instruments such as woodwinds, the brasses, and the pipe organ produce sound from the vibrations of standing waves in a column of air within a tube or pipe (Fig. 16–10). Standing waves can occur in the air of any cavity, but the frequencies present are complicated for any but very simple shapes such as a long, narrow tube. In some instruments, a vibrating reed or the vibrating lip of the player helps to set up vibrations of the air column. In others, a stream of air is directed against one edge of the opening or mouthpiece, leading to turbulence which sets up the vibrations. Because of the disturbance, whatever its source, the air within the tube vibrates with a variety of frequencies; but only frequencies that correspond to standing waves will persist.

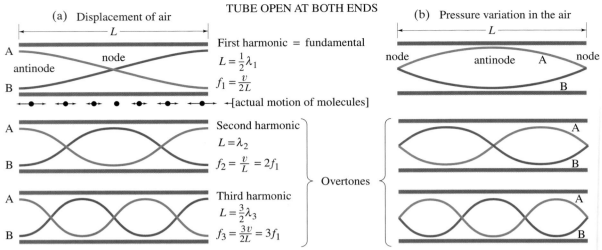

TUBE OPEN AT BOTH ENDS

(a) Displacement of air

First harmonic = fundamental
$$L = \frac{1}{2}\lambda_1$$
$$f_1 = \frac{v}{2L}$$
←[actual motion of molecules]

Second harmonic
$$L = \lambda_2$$
$$f_2 = \frac{v}{L} = 2f_1$$

Third harmonic
$$L = \frac{3}{2}\lambda_3$$
$$f_3 = \frac{3v}{2L} = 3f_1$$

Overtones

(b) Pressure variation in the air

FIGURE 16–11 Modes of vibration (standing waves) for a tube open at both ends ("open tube"). The simplest modes of vibration are shown in (a), on the left, in terms of the motion of the air (displacement), and in (b), on the right, in terms of air pressure. These are graphs, shown placed within the tube, and are labeled A and B where B represents the wave form a half period after the moment when it has the shape labeled A. The actual motion of molecules for one case is shown just below the tube at upper left.

For a string fixed at both ends, Fig. 16–7, the standing waves have nodes (no movement) at the two ends, and one or more antinodes (large amplitude of vibration) in between; a node separates successive antinodes. The lowest-frequency standing wave, the *fundamental*, corresponds to a single antinode. The higher-frequency standing waves are called **overtones** or **harmonics**, as was discussed in Section 15–9. Specifically, the first harmonic is the fundamental, the second harmonic has twice the frequency of the fundamental,[†] and so on.

The situation is similar for a column of air, but we must remember that it is now air itself that is vibrating. We can describe the waves either in terms of the flow of the air—that is, in terms of the *displacement* of air—or in terms of the *pressure* in the air (see Figs. 16–2 and 16–3). In terms of displacement, the air at the closed end of a tube is a displacement node since the air is not free to move there, whereas near the open end of a tube there will be an antinode since the air can move freely. The air within the tube vibrates in the form of longitudinal standing waves. The possible modes of vibration for a tube open at both ends (called an **open tube**), are shown graphically in Fig. 16–11. They are shown for a tube that is open at one end but closed at the other (called a **closed tube**) in Fig. 16–12. [A tube closed at *both* ends, having no connection to the outside air, would be useless as an instrument.] The graphs in part (a) of each Figure (left sides) represent the displacement amplitude of the vibrating air in the tube. Note that these are graphs, and that the air molecules themselves oscillate *horizontally*, parallel to the tube length, as shown by the small arrows in the top diagram of Fig. 16–11a (on the left). The exact position of the antinode near the open end of a tube depends on the diameter of the tube, but if the diameter is small compared to the length, which is the usual case, the antinode occurs very close to the end as shown. We assume this is the case in what follows. (The position of the antinode may also depend slightly on the wavelength and other factors.)

Let us look in detail at the open tube, in Fig. 16–11a, which might be a flute. An open tube has displacement antinodes at both ends since the air is free to move at open ends. Notice that there must be at least one node within an open tube if there is to be a standing wave at all. A single node corresponds to the *fundamental frequency* of the tube. Since the distance between two successive nodes, or between two successive antinodes, is $\frac{1}{2}\lambda$, there is one-half a wavelength within the length of the tube for the simplest case of the fundamental (top diagram in Fig. 16–11a): $L = \frac{1}{2}\lambda$ or $\lambda = 2L$. So the fundamental frequency is $f_1 = v/\lambda = v/2L$, where v is the velocity of sound in air. The standing

†When the resonant frequencies above the fundamental (that is, the overtones) are integral multiples of the fundamental, they are called harmonics. But if the overtones are not integral multiples of the fundamental, as is the case for a vibrating drumhead, for example, they are not harmonics.

426 CHAPTER 16 Sound

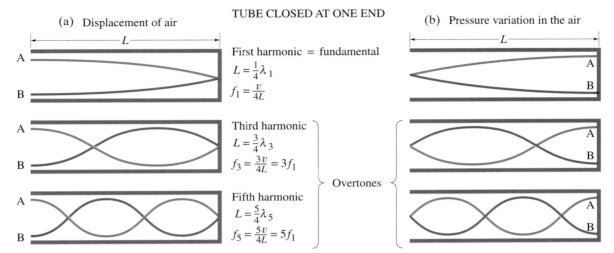

TUBE CLOSED AT ONE END

(a) Displacement of air   (b) Pressure variation in the air

First harmonic = fundamental
$L = \frac{1}{4}\lambda_1$
$f_1 = \frac{v}{4L}$

Third harmonic
$L = \frac{3}{4}\lambda_3$
$f_3 = \frac{3v}{4L} = 3f_1$

Fifth harmonic
$L = \frac{5}{4}\lambda_5$
$f_5 = \frac{5v}{4L} = 5f_1$

Overtones

FIGURE 16–12 Modes of vibration (standing waves) for a tube closed at one end ("closed tube"). See caption for Fig. 16–11.

Open tubes produce all harmonics

wave with two nodes is the *first overtone* or *second harmonic* and has half the wavelength ($L = \lambda$) and twice the frequency. Indeed, the frequency of each overtone is an integral multiple of the fundamental frequency, as shown in Fig. 16–11a. This is just what is found for a string.

For a closed tube, shown in Fig. 16–12a, which could be a clarinet, there is always a displacement node at the closed end (because the air is not free to move) and an antinode at the open end (where the air can move freely). Since the distance between a node and the nearest antinode is $\frac{1}{4}\lambda$, we see that the fundamental in a closed tube corresponds to only one-fourth of a wavelength within the length of the tube: $L = \lambda/4$, and $\lambda = 4L$. The fundamental frequency is thus $f_1 = v/4L$, or half what it is for an open pipe of the same length. There is another difference, for as we can see from Fig. 16–12a, only the odd harmonics are present in a closed pipe: the overtones have frequencies equal to $3, 5, 7, \cdots$ times the fundamental frequency. There is no way for waves with $2, 4, 6, \cdots$ times the fundamental frequency to have a node at one end and an antinode at the other, and thus they cannot exist as standing waves in a closed tube.

Closed tubes produce only odd harmonics

If this description in terms of displacement seems hard to understand, or you want to understand it from another point of view, then consider a description in terms of the *pressure* in the air, shown in part (b) of Figs. 16–11 and 16–12 (right sides). Where the air in a wave is compressed, the pressure is higher, whereas in a wave expansion (or rarefaction), the pressure is less than normal. The open end of a tube is open to the atmosphere. Hence the pressure variation at an open end must be a *node*: the pressure doesn't alternate, but remains at the outside atmospheric pressure. If a tube has a closed end, the pressure at that closed end can readily alternate to be above or below atmospheric pressure. Hence there is a pressure *antinode* at a closed end of a tube. Of course there can be pressure nodes and antinodes within the tube, and some of the possible vibrational modes in terms of pressure for an open tube are shown in Fig. 16–11b, and for a closed tube are shown in Fig. 16–12b.

FIGURE 16–13 The pipe organ used by J. S. Bach in Leipzig, Germany.

Pipe organs (Fig. 16–13) make use of both open and closed pipes. Notes of different pitch are sounded using different pipes with different lengths from a few centimeters to 5 m or more. Other musical instruments act either like a closed tube or an open tube. A flute, for example, is an open tube, for it is open not only where you blow into it, but also at the opposite end as well. The different notes on a flute and many other instruments are obtained by shortening the length of the tube— that is, by uncovering holes along its length. In a trumpet, on the other hand, pushing down on the valves opens additional lengths of tube. In all these instruments, the longer the length of the vibrating air column, the lower the frequency.

EXAMPLE 16–8 **Open and closed organ pipes.** What will be the fundamental frequency and first three overtones for a 26-cm-long organ pipe at 20°C if it is (a) open and (b) closed?

SOLUTION At 20°C, the speed of sound in air is 343 m/s (Section 16–1). (a) For the open pipe, Fig. 16–11, the fundamental frequency is

$$f_1 = \frac{v}{2L} = \frac{343 \text{ m/s}}{2(0.26 \text{ m})} = 660 \text{ Hz}.$$

The overtones, which include all harmonics, are 1320 Hz, 1980 Hz, 2640 Hz, and so on. (b) For a closed pipe, Fig. 16–12, we have

$$f_1 = \frac{v}{4L} = \frac{343 \text{ m/s}}{4(0.26 \text{ m})} = 330 \text{ Hz}.$$

But only the odd harmonics will be present, so the first three overtones will be 990 Hz, 1650 Hz, and 2310 Hz. (The closed pipe plays 330 Hz, which, from Table 16–3, is E above middle C, whereas the open pipe of the same length plays 660 Hz, an octave higher.)

EXAMPLE 16–9 **Flute.** A flute is designed to play middle C (262 Hz) as the fundamental frequency when all the holes are covered. Approximately how long should the distance be from the mouthpiece to the far end of the flute? (Note: This is only approximate since the antinode does not occur precisely at the mouthpiece.) Assume the temperature is 20°C.

SOLUTION The speed of sound in air at 20°C is 343 m/s. Because a flute is open at both ends, we use Fig. 16–11: the fundamental frequency f_1 is related to the length of the vibrating air column by $f = v/2L$. Solving for L, we find

$$L = \frac{v}{2f} = \frac{343 \text{ m/s}}{2(262 \text{ s}^{-1})} = 0.655 \text{ m}.$$

→ PHYSICS APPLIED

Temperature effect on staying in tune

EXAMPLE 16–10 **A cold flute.** If the temperature is only 10°C, what will be the frequency of the note played when all the openings are covered in the flute of Example 16–9?

SOLUTION The length L is still 65.5 cm. But now the velocity of sound is less since it changes by 0.60 m/s per each C°. For a drop of 10 C°, the velocity decreases by 6 m/s to 337 m/s. The frequency will be

$$f = \frac{v}{2L} = \frac{337 \text{ m/s}}{2(0.655 \text{ m})} = 257 \text{ Hz}.$$

We see why players of wind instruments take time to "warm up" their instruments so they will be in tune. The effect of temperature on stringed instruments is much smaller.

FIGURE 16–14 Example 16–11.

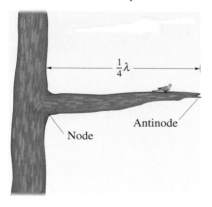

$\frac{1}{4}\lambda$

Antinode

Node

CONCEPTUAL EXAMPLE 16–11 **Wind noise frequencies.** Wind can be noisy—it can "howl" in trees; it can "moan" in chimneys. Why is this so? What is really causing the noise, and about what range of frequencies would you expect to hear?

SOLUTION In each case, jets of air in the wind cause vibrations or oscillations, which produce the sound. The end of a tree limb fixed to the tree trunk is a node, whereas the other end is free to move and therefore is an antinode; the tree limb is thus about $\frac{1}{4}\lambda$ (Fig. 16–14). We estimate $v \approx 4000$ m/s for the speed of sound in wood (Table 16–1). Suppose that a tree limb has length $L \approx 2$ m; then $\lambda = 4L = 8$ m and $f = v/\lambda = (4000 \text{ m/s})/(8 \text{ m}) \approx 500$ Hz.

Wind can excite air oscillations in a chimney, much like in an organ pipe or flute. A chimney is a fairly long tube, perhaps 3 m in length, acting like a tube open at either one end or even both ends. If open at both ends ($\lambda = 2L$), with $v = 340$ m/s, we find $f_1 \approx v/2L \approx 56$ Hz, which is a fairly low note—no wonder chimneys "moan"!

*16–5 Quality of Sound, and Noise

Whenever we hear a sound, particularly a musical sound, we are aware of its loudness, its pitch, and also of a third aspect called "quality." For example, when a piano and then a flute play a note of the same loudness and pitch (say middle C), there is a clear difference in the overall sound. We would never mistake a piano for a flute. This is what is meant by the **quality** of a sound. For musical instruments, the terms *timbre* or *tone color* are also used.

Just as loudness and pitch can be related to physically measurable quantities, so too can quality. The quality of a sound depends on the presence of overtones—their number and their relative amplitudes. Generally, when a note is played on a musical instrument, the fundamental as well as overtones are present simultaneously. We saw in Fig. 15–17 how the superposition of three wave forms, in that case the fundamental and first two overtones (with particular amplitudes), would combine to give a composite *waveform*. Of course, more than two overtones are usually present.

The relative amplitudes of the various overtones are different for different musical instruments, and this is what gives each instrument its characteristic quality or timbre. A graph showing the relative amplitudes of the harmonics produced by an instrument is called a "sound spectrum." Several typical examples for different instruments are shown in Fig. 16–15. Normally, the fundamental has the greatest amplitude and its frequency is what is heard as the pitch, although players can sometimes make the first overtone (for example) sound the loudest.

The manner in which an instrument is played strongly influences the sound quality. Plucking a violin string, for example, makes a very different sound than pulling a bow across it. The sound spectrum at the very start (or end) of a note, as when a hammer strikes a piano string, can be very different from the subsequent sustained tone. This too affects the subjective tone quality of an instrument.

An ordinary sound, like that made by striking two stones together, is a noise that has a certain quality, but a clear pitch is not discernible. A noise such as this is a mixture of many frequencies which bear little relation to one another. If a sound spectrum were made of this noise, it would not show discrete lines like those of Fig. 16–15. Instead it would show a continuous, or nearly continuous, spectrum of frequencies. Such a sound we call "noise" in comparison with the more harmonious sounds which contain frequencies that are simple multiples of the fundamental.

FIGURE 16–15 Sound spectra. The shapes of the spectra change as the instruments play different notes.

16–6 Interference of Sound Waves; Beats

Interference in Space

We saw in Section 15–8 that when two waves simultaneously pass through the same region of space, they interfere with one another. Since this can occur for any kind of wave, we should expect that interference will occur with sound waves, and indeed it does.

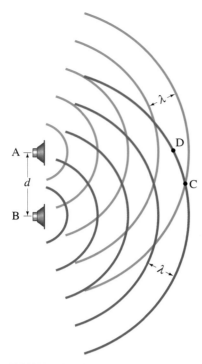

FIGURE 16–16 Sound waves from two loudspeakers interfere.

FIGURE 16–17 Sound waves of a single frequency from loudspeakers A and B (see Fig. 16–16) constructively interfere at C and destructively interfere at D.

(a)

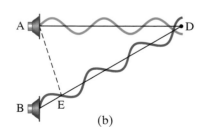

(b)

As a simple example, consider two large loudspeakers, A and B, a distance d apart on the stage of an auditorium as shown in Fig. 16–16. Let us assume the two speakers are emitting sound waves of the same single frequency and that they are in phase: that is, when one speaker is forming a compression, so is the other. (We ignore reflections from walls, floor, etc.) The curved lines in the diagram represent the crests of sound waves from each speaker. Of course, we must remember that for a sound wave, a crest is a compression in the air whereas a trough—which falls between two crests—is a rarefaction. A person or detector at a point such as C, which is the same distance from each speaker, will experience a loud sound because the interference will be constructive. On the other hand, at a point such as D in the diagram, little if any sound will be heard because destructive interference occurs—compressions of one wave meet rarefactions of the other and vice versa (see Fig. 15–24 and related discussion on water waves in Section 15–8).

An analysis of this situation is perhaps clearer if we graphically represent the wave forms as in Fig. 16–17. In Fig. 16–17a it can be seen that at point C, constructive interference occurs since both waves simultaneously have crests or simultaneously have troughs. In Fig. 16–17b we see the situation for point D. The wave from speaker B must travel a greater distance than the wave from A. Thus the wave from B lags behind that from A. In this diagram, point E is chosen so that the distance ED is equal to AD. Thus we see that if the distance BE is equal to precisely one-half the wavelength of the sound, the two waves will be exactly out of phase when they reach D, and destructive interference occurs. This then is the criterion for determining at what points destructive interference occurs: destructive interference occurs at any point whose distance from one speaker is greater than its distance from the other speaker by exactly one-half wavelength. Notice that if this extra distance (BE in Fig. 16–17b) is equal to a whole wavelength (or 2, 3, ··· wavelengths), then the two waves will be in phase and *constructive interference* occurs. If the distance BE equals $\frac{1}{2}, 1\frac{1}{2}, 2\frac{1}{2}, \cdots$ wavelengths, *destructive interference* occurs.

It is important to realize that a person sitting at point D hears nothing at all at this particular frequency, yet sound is coming from both speakers. Indeed, if one of the speakers is turned off, the sound from the other speaker would be clearly heard.

If a loudspeaker emits a whole range of frequencies, only specific ones will destructively interfere completely at a given point.

EXAMPLE 16–12 **Loudspeakers' interference.** Two loudspeakers are 1.00 m apart. A person stands 4.00 m from one speaker. How far must she be from the second speaker in order to detect destructive interference when the speakers emit 1150-Hz sound waves in phase with each other? Assume the temperature is 20°C.

SOLUTION The wavelength of this sound is

$$\lambda = \frac{v}{f} = \frac{343 \text{ m/s}}{1150 \text{ Hz}} = 0.30 \text{ m}.$$

For destructive interference to occur, the person must be one-half wavelength farther from one loudspeaker than from the other, or 0.15 m. Thus the person must be 4.15 m (or 3.85 m) from the second speaker. If the speakers are less than 0.15 m apart, there would be no point that was 0.15 m farther from one speaker than the other, and there would be no point where destructive interference would occur.

Beats—Interference in Time

Beats

We have been discussing interference of sound waves that takes place in space. An interesting and important example of interference that occurs in time is the phenomenon known as **beats**: two sources of sound—say, two tuning forks—are close in frequency but not exactly the same. Sound waves from the two sources interfere with each other and the sound level at a given position alternately rises and falls; the regularly spaced intensity changes are called beats.

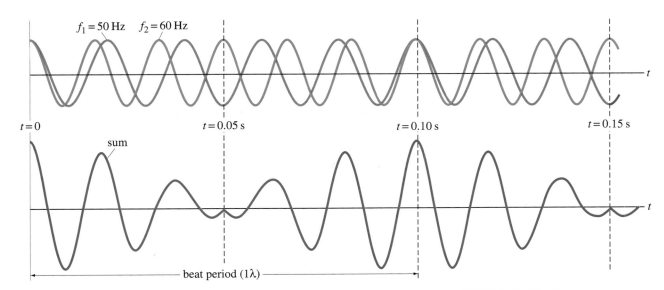

To see how beats arise, consider two equal-amplitude sound waves of frequency $f_1 = 50\,\text{Hz}$ and $f_2 = 60\,\text{Hz}$, respectively. In 1.00 s, the first source makes 50 vibrations whereas the second makes 60. We now examine the waves at one point in space equidistant from the two sources. The waveforms for each wave as a function of time are shown on the top graph of Fig. 16–18; the magenta line represents the 50 Hz wave, the blue line represents the 60 Hz wave. The lower graph in Fig. 16–18 shows the sum of the two waves. At time $t = 0$ the two waves are shown to be in phase and interfere constructively. Because the two waves vibrate at different rates, at time $t = 0.05\,\text{s}$ they are completely out of phase and destructive interference occurs as shown in the figure. At $t = 0.10\,\text{s}$, they are again in phase and the resultant amplitude again is large. Thus the resultant amplitude is large every 0.10 s and in between it drops drastically. This rising and falling of the intensity is what is heard as beats.[†] In this case the beats are 0.10 s apart. That is, the **beat frequency** is ten per second or 10 Hz. This result, that the beat frequency equals the difference in frequency of the two waves can be shown in general as follows.

Beat frequency = difference in the two wave frequencies

Let the two waves, of frequencies f_1 and f_2, be represented at a fixed point in space by

$$D_1 = D_M \sin 2\pi f_1 t$$

and

$$D_2 = D_M \sin 2\pi f_2 t.$$

The resultant displacement, by the principle of superposition, is

$$D = D_1 + D_2 = D_M(\sin 2\pi f_1 t + \sin 2\pi f_2 t).$$

Using the trigonometric identity $\sin A + \sin B = 2 \sin \frac{1}{2}(A + B) \cos \frac{1}{2}(A - B)$, we have

$$D = \left[2D_M \cos 2\pi \left(\frac{f_1 - f_2}{2}\right)t\right] \sin 2\pi \left(\frac{f_1 + f_2}{2}\right)t. \qquad \textbf{(16–8)}$$

We can interpret Eq. 16–8 as follows. The superposition of the two waves results in a wave that vibrates at the average frequency of the two components, $(f_1 + f_2)/2$. This vibration has an amplitude given by the expression in brackets, and this amplitude varies in time, from zero to a maximum of $2D_M$ (the sum of the separate amplitudes), with a frequency of $(f_1 - f_2)/2$. A beat occurs whenever $\cos 2\pi[(f_1 - f_2)/2]t$ equals +1 or −1 (see Fig. 16–18); that is, two beats occur per cycle, so the beat frequency is twice $(f_1 - f_2)/2$ which is just $f_1 - f_2$, the difference in frequency of the component waves.

[†] Beats will be heard even if the amplitudes are not equal, as long as the difference is not great.

The phenomenon of beats can occur with any kind of wave and is a very sensitive method for comparing frequencies. For example, to tune a piano, a piano tuner listens for beats produced between his standard tuning fork and that of a particular string on the piano, and knows it is in tune when the beats disappear. The members of an orchestra can tune up by listening for beats between their instruments and that of a standard tone (usually A above middle C at 440 Hz) produced by a piano or an oboe.

EXAMPLE 16–13 **Beats.** A tuning fork produces a steady 400-Hz tone. When this tuning fork is struck and held near a vibrating guitar string, twenty beats are counted in five seconds. What are the possible frequencies produced by the guitar string?

SOLUTION The beat frequency is

$$f_{\text{beat}} = 20 \text{ vibrations}/5 \text{ s} = 4 \text{ Hz}.$$

This is the difference of the frequencies of the two waves, and because one wave is known to be 400 Hz, the other must be either 404 Hz or 396 Hz.

16–7 Doppler Effect

You may have noticed that the pitch of the siren on a speeding firetruck drops abruptly as it passes you. Or you may have noticed the change in pitch of a blaring horn on a fast-moving car as it passes by. The pitch of the sound from the engine of a race car changes as it passes an observer. When a source of sound is moving toward an observer, the pitch is higher than when the source is at rest; and when the source is traveling away from the observer, the pitch is lower. This phenomenon is known as the **Doppler effect**[†] and occurs for all types of waves. Let us now see why it occurs, and calculate the change in frequency for sound waves.

To be concrete, consider the siren of a firetruck at rest, which is emitting sound of a particular frequency in all directions as shown in Fig. 16–19a. The wave velocity depends only on the medium in which it is traveling, and is independent of the velocity of the source or observer. If our source, the firetruck, is moving, the siren emits sound at the same frequency as it does at rest. But the sound wavefronts it emits forward are closer together than when the firetruck is at rest, as shown in Fig. 16–19b. This is because the firetruck, as it moves, is "chasing after" the previously emitted wavefronts. Thus an observer on the sidewalk will detect more wave crests passing per second, so the frequency heard is higher. The wavefronts emitted behind the truck, on the other hand, are farther apart than when the truck is at rest because the truck is speeding away from them. Hence, fewer wave crests per second pass by an observer behind the truck and the pitch is lower.

[†] After J. C. Doppler (1803–1853).

FIGURE 16–19 (a) Both observers on the sidewalk hear the same frequency from the firetruck at rest. (b) Doppler effect: observer toward whom the firetruck moves hears a higher-frequency sound, and observer behind the firetruck hears a lower frequency.

(a) At rest

(b) Firetruck moving

To calculate the change in frequency, we make use of Fig. 16–20, and we assume the air (or other medium) is at rest in our reference frame. In Fig. 16–20a, the source of the sound, shown as a dot, is at rest; two successive wave crests are shown, the second of which is just in the process of being emitted. The distance between these crests is λ, the wavelength. If the frequency of the source is f, then the time between emissions of wave crests is

Frequency change, moving source

$$T = \frac{1}{f}.$$

In Fig. 16–20b, the source is moving with a velocity v_S. In a time T (as just defined), the first wave crest has moved a distance $d = vT$, where v is the velocity of the sound wave in air (which is, of course, the same whether the source is moving or not). In this same time, the source has moved a distance $d_S = v_S T$. Then the distance between successive wave crests, which is the new wavelength λ', is (since $d = \lambda$)

$$\lambda' = d - d_S$$
$$= \lambda - v_S T$$
$$= \lambda - v_S \frac{\lambda}{v} = \lambda\left(1 - \frac{v_S}{v}\right).$$

The change in wavelength, $\Delta\lambda$, is

$$\Delta\lambda = \lambda' - \lambda = -v_S \frac{\lambda}{v}.$$

So the shift in wavelength is directly proportional to the speed v_S of the source. The new frequency, on the other hand, is given by

$$f' = \frac{v}{\lambda'} = \frac{v}{\lambda\left(1 - \dfrac{v_S}{v}\right)}$$

or, since $v/\lambda = f$,

$$f' = \frac{f}{\left(1 - \dfrac{v_S}{v}\right)}. \qquad \begin{bmatrix}\text{source moving toward}\\\text{stationary observer}\end{bmatrix} \quad \textbf{(16–9a)}$$

Because the denominator is less than 1, $f' > f$. For example, if a source emits a sound of frequency 400 Hz when at rest, then when the source moves toward a fixed observer with a speed of 30 m/s, the observer hears a frequency (at 20°C) of

$$f' = \frac{400\ \text{Hz}}{1 - \dfrac{30\ \text{m/s}}{343\ \text{m/s}}} = 438\ \text{Hz}.$$

For a source moving *away* from the observer at a speed v_S, the new wavelength will be

$$\lambda' = d + d_S,$$

and the change in wavelength will be

$$\Delta\lambda = \lambda' - \lambda = +v_S \frac{\lambda}{v}.$$

The frequency of the wave will be

$$f' = \frac{f}{\left(1 + \dfrac{v_S}{v}\right)}. \qquad \begin{bmatrix}\text{source moving away from}\\\text{stationary observer}\end{bmatrix} \quad \textbf{(16–9b)}$$

In this case, if a source vibrating at 400 Hz is moving away from a fixed observer at 30 m/s, the observer hears a frequency of about 368 Hz.

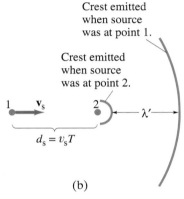

FIGURE 16–20 Determination of frequency change in the Doppler effect (see text). Red dot is the source.

Source ●——— $d = \lambda$ ———

(a)

Crest emitted when source was at point 1.

Crest emitted when source was at point 2.

1   $\mathbf{v}_S$ 2 — λ'—

$d_S = v_S T$

(b)

FIGURE 16–21 Observer moving with speed v_O toward a stationary source "sees" wave crests pass at speed $v' = v + v_O$ where v is the speed of the sound waves in air.

Frequency change, moving observer

The Doppler effect also occurs when the source is at rest and the observer is in motion. If the observer is traveling toward the source, the pitch is higher; and if the observer is traveling away from the source, the pitch is lower. Quantitatively the change in frequency is slightly different than for the case of a moving source. With a fixed source and a moving observer, the distance between wave crests, the wavelength λ, is not changed. But the velocity of the crests with respect to the observer is changed. If the observer is moving toward the source, Fig. 16–21, the speed of the waves relative to the observer is $v' = v + v_O$, where v is the velocity of sound in air (we assume the air is still) and v_O is the velocity of the observer. Hence, the new frequency is

$$f' = \frac{v'}{\lambda} = \frac{v + v_O}{\lambda}$$

or, since $\lambda = v/f$,

$$f' = \left(1 + \frac{v_O}{v}\right)f. \qquad \begin{bmatrix}\text{observer moving toward} \\ \text{stationary source}\end{bmatrix} \quad \textbf{(16–10a)}$$

If the observer is moving away from the source, the relative velocity is $v' = v - v_O$, so

$$f' = \left(1 - \frac{v_O}{v}\right)f. \qquad \begin{bmatrix}\text{observer moving away} \\ \text{from stationary source}\end{bmatrix} \quad \textbf{(16–10b)}$$

EXAMPLE 16–14 **A moving siren.** The siren of a police car at rest emits at a predominant frequency of 1600 Hz. What frequency will you hear if you are at rest and the police car moves at 25.0 m/s (*a*) toward you, (*b*) away from you?

SOLUTION (*a*) We use Eq. 16–9a:

$$f' = \frac{f}{\left(1 - \frac{v_S}{v}\right)} = \frac{1600\,\text{Hz}}{\left(1 - \frac{25.0\,\text{m/s}}{343\,\text{m/s}}\right)} = 1726\,\text{Hz}.$$

(*b*) We use Eq. 16–9b:

$$f' = \frac{f}{\left(1 + \frac{v_S}{v}\right)} = \frac{1600\,\text{Hz}}{\left(1 + \frac{25.0\,\text{m/s}}{343\,\text{m/s}}\right)} = 1491\,\text{Hz}.$$

When a sound wave is reflected from a moving obstacle, the frequency of the reflected wave will, because of the Doppler effect, be different from that of the incident wave. This is illustrated in the following Example.

EXAMPLE 16–15 **Two Doppler shifts.** A 5000-Hz sound wave is emitted by a stationary source toward an object moving 3.50 m/s toward the source (Fig. 16–22). What is the frequency of the wave reflected by the moving object as detected by a detector at rest near the source?

SOLUTION There are actually two Doppler shifts in this situation. First, the emitted wave strikes the moving object which is in effect a moving observer (Fig. 16–22a) that "detects" a sound wave of frequency (Eq. 16–10a):

$$f' = \left(1 + \frac{v_O}{v}\right)f = \left(1 + \frac{3.50 \text{ m/s}}{343 \text{ m/s}}\right)(5000 \text{ Hz}) = 5051 \text{ Hz}.$$

Second, the moving object takes this wave of frequency f' and reemits (or reflects) it, acting effectively as a moving source, so the frequency detected, f'', will be given by Eq. 16–9a:

$$f'' = \frac{f'}{\left(1 - \dfrac{v_S}{v}\right)} = \frac{5051 \text{ Hz}}{\left(1 - \dfrac{3.50 \text{ m/s}}{343 \text{ m/s}}\right)} = 5103 \text{ Hz}.$$

Thus the frequency shifts by 103 Hz.

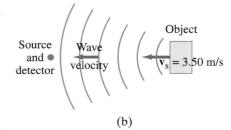

FIGURE 16–22 Example 16–15.

The incident wave and the reflected wave in Example 16–15, when mixed together (say, electronically), interfere with one another and beats are produced. The beat frequency is equal to the difference in the two frequencies, 103 Hz. This Doppler technique is used in a variety of medical applications, usually with ultrasonic waves in the megahertz frequency range. For example, ultrasonic waves reflected from red blood cells can be used to determine the velocity of blood flow. Similarly, the technique can be used to detect the movement of the chest of a young fetus and to monitor its heartbeat.

➡ **PHYSICS APPLIED**
Medical uses

For convenience, we can write Eqs. 16–9 and 16–10 as a single equation that covers all cases of both source and observer in motion:

$$f' = f\left(\frac{v \pm v_O}{v \mp v_S}\right). \tag{16–11}$$

➡ **PHYSICS APPLIED**
Source and observer moving

➡ **PROBLEM SOLVING**
Getting the signs right

The upper signs apply if source and/or observer move toward each other; the lower signs apply if they are moving apart.

Doppler Effect for Light

The Doppler effect occurs for other types of waves as well. Light and other types of electromagnetic waves (such as radar) exhibit the Doppler effect: although the formulas for the frequency shift are not identical to Eqs. 16–9 and 16–10, the effect is similar. One important application is for weather forecasting using radar. The time delay between the emission of radar pulses and their reception after being reflected off raindrops gives the position of precipitation. Measuring the Doppler shift in frequency (as in Example 16–15) tells how fast the storm is moving and in which direction.

➡ **PHYSICS APPLIED**
Doppler effect for EM waves and weather forecasting

Another important application is to astronomy where the velocities of distant galaxies can be determined from the Doppler shift. Light from such galaxies is shifted toward lower frequencies, indicating that the galaxies are moving away from us. This is called the **red shift** since red has the lowest frequency of visible light. The greater the frequency shift, the greater the velocity of recession. It is found that the farther the galaxies are from us, the faster they move away. This observation is the basis for the idea that the universe is expanding, and is one basis for the idea that the universe began with a great explosion, affectionately called the "Big Bang" (see Chapter 45).

➡ **PHYSICS APPLIED**
Red shift in cosmology

*16–8 Shock Waves and the Sonic Boom

An object such as an airplane traveling faster than the speed of sound is said to have a **supersonic speed**. Such a speed is often given as a **Mach**[†] **number**, which is defined as the ratio of the object's speed to that of sound in the medium at that location. For example, a plane traveling 600 m/s high in the atmosphere, where the speed of sound is only 300 m/s, has a speed of Mach 2.

[†] After the Austrian physicist Ernst Mach (1838–1916).

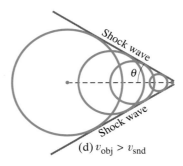

(a) $v_{obj} = 0$ (b) $v_{obj} < v_{snd}$ (c) $v_{obj} = v_{snd}$ (d) $v_{obj} > v_{snd}$

FIGURE 16–23 Sound waves emitted by an object at rest (a) or moving (b, c, and d). If the object's velocity is less than the velocity of sound, the Doppler effect occurs (b); if its velocity is greater than the velocity of sound, a shock wave is produced (d).

Shock wave

FIGURE 16–24 Bow waves produced by a boat.

FIGURE 16–25 (a) The (double) sonic boom has already been heard by person A on the left. It is just being heard by person B in the center. And it will shortly be heard by person C on the right. (b) Special photo of supersonic aircraft showing shock waves produced in the air. (Several closely spaced shock waves are produced by different parts of the aircraft.)

When a source of sound moves at subsonic speeds, the pitch of the sound is altered, as we have seen (the Doppler effect); see also Fig. 16–23a and b. But if a source of sound moves faster than the speed of sound, a more dramatic effect known as a **shock wave** occurs. In this case the source is actually "outrunning" the waves it produces. As shown in Fig. 16–23c, when the source is traveling at the speed of sound, the wave fronts it emits in the forward direction "pile up" directly in front of it. When the object moves at a supersonic speed, the wave fronts pile up on one another along the sides, as shown in Fig. 16–23d. The different wave crests overlap one another and form a single very large crest which is the shock wave. Behind this very large crest there is usually a very large trough. A shock wave is essentially the result of constructive interference of a large number of wave fronts. A shock wave in air is analogous to the bow wave of a boat traveling faster than the speed of the water waves it produces, Fig. 16–24.

When an airplane travels at supersonic speeds, the noise it makes and its disturbance of the air form into a shock wave containing a tremendous amount of sound energy. When the shock wave passes a listener, it is heard as a loud "sonic boom." A sonic boom lasts only a fraction of a second, but the energy it contains is often sufficient to break windows and cause other damage. It can be psychologically unnerving as well. Actually, a sonic boom is made up of two or more booms since major shock waves can form at the front and the rear of the aircraft, as well as at the wings, etc. (Fig. 16–25). Bow waves of a boat are also multiple, as can be seen in Fig. 16–24.

When an aircraft approaches the speed of sound, it encounters a barrier of sound waves in front of it (see Fig. 16–23c). In order to exceed the speed of sound, extra thrust is needed to pass through this "sound barrier." This is called "breaking the sound barrier." Once a supersonic speed is attained, this barrier no longer impedes the motion. It is sometimes erroneously thought that a sonic boom is produced only at the moment an aircraft is breaking through the sound barrier. Actually, a shock wave follows the aircraft at all times it is traveling at supersonic speeds. A series of observers on the ground will each hear a loud "boom" as the shock wave passes, Fig. 16–25. The shock wave consists of a cone whose apex is at the aircraft. The angle of this cone, θ, (see Fig. 16–23 d) is given by

$$\sin \theta = \frac{v_{snd}}{v_{obj}}, \qquad (16\text{–}12)$$

where v_{obj} is the velocity of the object (the aircraft) and v_{snd} is the velocity of sound in the medium (the proof is left as a Problem).

(a)

(b)

* 16–9 Applications; Sonar, Ultrasound, and Ultrasound Imaging

The reflection of sound is used in many applications to determine distance. The **sonar**[†] or pulse-echo technique is used to locate underwater objects. A transmitter sends out a sound pulse through the water, and a detector receives its reflection, or echo, a short time later. This time interval is carefully measured, and from it the distance to the reflecting object can be determined since the speed of sound in water is known. The depth of the sea and the location of reefs, sunken ships, submarines, or schools of fish can be determined in this way. The interior structure of the Earth is studied in a similar way by detecting reflections of waves traveling through the Earth whose source was a deliberate explosion (called "soundings"). An analysis of waves reflected from various structures and boundaries within the Earth reveals characteristic patterns that are also useful in the exploration for oil and minerals.

Sonar generally makes use of **ultrasonic** frequencies: that is, waves whose frequencies are above 20 kHz, beyond the range of human detection. For sonar, the frequencies are typically in the range 20 kHz to 100 kHz. One reason for using ultrasound waves, other than the fact that they are inaudible, is that for shorter wavelengths there is less diffraction, so the beam spreads less and smaller objects can be detected.

The diagnostic use of ultrasound in medicine, in the form of images (sometimes called "sonograms") is an important and interesting application of physical principles. A **pulse-echo technique** is used, much like sonar. A high-frequency sound pulse is directed into the body, and its reflections from boundaries or interfaces between organs and other structures and lesions in the body are then detected. It is even possible to produce "real-time" ultrasound images, as if one were watching a movie of a section of the interior of the body.

The pulse-echo technique for medical imaging works as follows. A brief pulse of ultrasound is emitted by a transducer that transforms an electrical pulse into a sound-wave pulse. Part of the pulse is reflected at each interface surface in the body, and most (usually) continues on. The detection of the reflected pulses by the same transducer can then be displayed on the screen of a display terminal or monitor, as shown in Fig. 16–26a. The time elapsed from when the pulse is emitted to when each reflection (echo) is received is proportional to the distance to the reflecting surface. For example, if the distance from transducer to the vertebra is 25 cm, the pulse travels a round-trip distance of 2 × 25 cm = 0.50 m; the speed of sound in human tissue is about 1540 m/s (close to water), so the time taken is $t = d/v = (0.50 \text{ m})/(1540 \text{ m/s}) = 320 \ \mu\text{s}$.

The *strength* of a reflected pulse depends mainly on the difference in density of the two materials on either side of the interface and can be displayed as a pulse or as a dot (Figs. 16–26b and c). Each echo dot (Fig. 16–26c) can be represented as a point, whose position is given by the time delay and whose brightness depends on the strength of the echo. A two-dimensional image can then be formed out of these dots from a series of scans. The transducer is moved, and at each position it sends out a pulse and receives echoes as shown in Fig. 16–27. Each trace can be plotted, spaced appropriately one below the other, to form an image on a display terminal as shown in Fig. 16–27b. Only 10 lines are shown in Fig. 16–27, so the image is crude. More lines give a more precise image. Photographs of ultrasound images are shown in Fig. 16–28.

[†]Sonar stands for "*so*und *na*vigation *r*anging."

→ **PHYSICS APPLIED**

Sonar and imaging

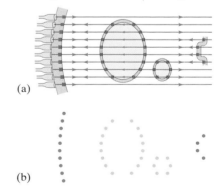

FIGURE 16–26 (a) Ultrasound pulse passes through abdomen, reflecting from surfaces in its path. (b) Reflected pulses plotted as a function of time when received by transducer. The vertical dashed lines point out which reflected pulse goes with which surface. (c) Dot display for the same echoes: brightness of each dot is related to signal strength.

FIGURE 16–27 (a) Ten traces are made across the abdomen by moving the transducer, or by using an array of transducers. (b) The echoes are plotted to produce the image. More closely spaced traces would give a more detailed image.

FIGURE 16–28 (a) Ultrasound image of a human fetus (with head at the left) within the uterus. (b) False-color high-resolution ultrasound image of fetus (different colors represent different intensities of reflected pulses).

Summary

Sound travels as a longitudinal wave in air and other materials. In air, the speed of sound increases with temperature; at 20°C, it is about 343 m/s.

The **pitch** of a sound is determined by the frequency; the higher the frequency, the higher the pitch.

The **audible range** of frequencies for humans is roughly 20 Hz to 20,000 Hz (1 Hz = 1 cycle per second).

The **loudness** or **intensity** of a sound is related to the amplitude of the wave. Because the human ear can detect sound intensities from 10^{-12} W/m² to over 1 W/m², intensity levels are specified on a logarithmic scale. The **sound level** β, specified in decibels, is defined in terms of intensity I as

$$\beta = 10 \log (I/I_0),$$

where the reference intensity I_0 is usually taken to be 10^{-12} W/m².

Musical instruments are simple sources of sound in which *standing waves* are produced.

The strings of a stringed instrument may vibrate as a whole with nodes only at the ends; the frequency at which this occurs is called the **fundamental**. The string can also vibrate at higher frequencies, called **overtones** or **harmonics**, in which there are one or more additional nodes. The frequency of each harmonic is a whole-number multiple of the fundamental.

In wind instruments, standing waves are set up in the column of air within the tube.

The vibrating air in an **open tube** (open at both ends) has displacement antinodes at both ends. The fundamental frequency corresponds to a wavelength equal to twice the tube length: $\lambda_1 = 2L$. The harmonics have frequencies that are 2, 3, 4, $\cdots$ times the fundamental frequency, just as for strings.

For a **closed tube** (closed at one end), the fundamental corresponds to a wavelength four times the length of the tube: $\lambda_1 = 4L$. Only the odd harmonics are present, equal to 1, 3, 5, 7, $\cdots$ times the fundamental frequency.

Sound waves from different sources can interfere with each other. If two sounds are at slightly different frequencies, **beats** can be heard at a frequency equal to the difference in frequency of the two sources.

The **Doppler effect** refers to the change in pitch of a sound due to the motion either of the source or of the listener. If they are approaching each other, the pitch is higher; if they are moving apart, the pitch is lower.

Questions

1. What is the evidence that sound travels as a wave?

2. What is the evidence that sound is a form of energy?

3. Children sometimes play with a homemade "telephone" by attaching a string to the bottoms of two paper cups. When the string is stretched and a child speaks into one of the cups, the sound can be heard at the other cup. Explain clearly how the sound wave travels from one cup to the other. (See Fig. 16–29.)

FIGURE 16–29 Question 3.

4. When a sound wave passes from air into water, do you expect the frequency or wavelength to change?

5. What evidence can you give that the speed of sound in air does not depend significantly on frequency?

6. The voice of a person who has inhaled helium sounds very high-pitched. Why?

7. What is the main reason the speed of sound in hydrogen is greater than the speed of sound in air?

8. The molecules of a gas, such as air, move around randomly at fairly high speeds (Chapter 18). The distance between molecules, on the average, is many times their diameter. When a wave passes through a gas the impulse given to one molecule is given to another only when this distance is traveled and the two collide. Would you therefore expect the speed of sound in a gas to be limited by the average molecular speed?

9. Two tuning forks oscillate with the same amplitude, but one has twice the frequency. Which (if either) produces the more intense sound?

10. How does a rise in air temperature affect the loudness of sound coming from a source of fixed frequency and amplitude? (Assume that atmospheric pressure doesn't change.)

11. What is the reason that catgut strings on some musical instruments are wrapped with fine wire?

12. Explain how a tube might be used as a filter to reduce the amplitude of sounds in various frequency ranges. (An example is a car muffler.)

13. How will the air temperature in a room affect the pitch of organ pipes?

14. Why are the frets on a guitar spaced closer together as you move up the fingerboard toward the bridge?

FIGURE 16–30 Question 14.

15. Standing waves can be said to be due to "interference in space," whereas beats can be said to be due to "interference in time." Explain.

16. In Fig. 16–16, if the frequency of the speakers were lowered, would the points D and C (where destructive and constructive interference occur) move farther apart or closer together?

17. Traditional methods of protecting the hearing of people who work in areas with very high noise levels have consisted mainly of efforts to block or reduce noise levels. With a relatively new technology, headphones are worn that do not block the ambient noise. Instead, a device is used which detects the noise, inverts it electronically, then feeds it to the headphones *in addition to* the ambient noise. How could adding *more* noise actually reduce the sound levels reaching the ears?

18. Suppose a source of sound moves at right angles to the line of sight of a listener at rest in still air. Will there be a Doppler effect? Explain.

19. If a wind is blowing, will this alter the frequency of the sound heard by a person at rest with respect to the source? Is the wavelength or velocity changed?

20. Figure 16–31 shows various positions of a child in motion on a swing. A monitor is blowing a whistle in front of the child on the ground. At which position will the child hear the highest frequency for the sound of the whistle? Explain your reasoning.

FIGURE 16–31 Question 20.

Problems

[Unless stated otherwise, assume $T = 20°C$ and $v_{sound} = 343$ m/s in air.]

Section 16–1

1. (I) A hiker determines the length of a lake by listening for the echo of her shout reflected by a cliff at the far end of the lake. She hears the echo 1.5 s after shouting. Estimate the length of the lake.

2. (I) (*a*) Calculate the wavelengths in air at 20°C for sounds in the maximum range of human hearing, 20 Hz to 20,000 Hz. (*b*) What is the wavelength of a 10-MHz ultrasonic wave?

3. (II) A person sees a heavy stone strike the concrete pavement. A moment later two sounds are heard from the impact: one travels in the air and the other in the concrete, and they are 1.4 s apart. How far away did the impact occur?

4. (II) A fishing boat is drifting just above a school of tuna on a foggy day. Without warning, an engine backfire occurs on another boat 1.0 km away (Fig. 16–32). How much time elapses before the backfire is heard (*a*) by the fish and (*b*) by the fishermen?

FIGURE 16–32 Problem 4.

5. (II) The sound from a very high burst of fireworks takes 4.5 s to arrive at your eardrums. The burst occurred 1500 m above you and traveled vertically through two stratified layers of air, the top one at 0°C and the bottom one at 20°C. How thick is each layer of air?

6. (II) If the camera of Example 16–2 focuses precisely at 20°C, what percent error will there be in distances when used at 0°C?

Section 16–2

7. (I) The pressure amplitude of a sound wave in air $(\rho = 1.29 \text{ kg/m}^3)$ at 0°C is 3.0×10^{-3} Pa. What is the displacement amplitude if the frequency is (*a*) 100 Hz and (*b*) 10 kHz?

8. (I) What must be the pressure amplitude in a sound wave in air (0°C) if the air molecules undergo a maximum displacement equal to the diameter of an oxygen molecule, about 3×10^{-10} m? Assume a sound wave frequency of (*a*) 50 Hz and (*b*) 5.0 kHz.

9. (II) Write an expression that describes the pressure variation as a function of x and t for the waves described in Problem 8.

10. (II) The pressure variation in a sound wave is given by

$$\Delta P = 0.0025 \sin\left(\frac{\pi}{3} x - 1700\pi t\right),$$

where ΔP is in pascals, x in meters, and t in seconds. Determine (*a*) the wavelength, (*b*) the frequency, (*c*) the speed, and (*d*) the displacement amplitude of the wave. Assume the density of the medium to be $\rho = 2.7 \times 10^3$ kg/m^3.

11. (I) (a) What is the sound level of a sound whose intensity is 8.5×10^{-8} W/m^2? (b) What is the intensity of a sound whose sound level is 25 dB?

12. (I) A 6000-Hz tone must have what sound level to seem as loud as a 100-Hz tone that has a 50-dB sound level? (See Fig. 16–5.)

13. (I) What are the lowest and highest frequencies that an average ear can detect when the sound level is 30 dB? (See Fig. 16–5)

14. (I) A stereo cassette player is said to have a signal-to-noise ratio of 63 dB. What is the ratio of intensities of the signal and the background noise?

15. (II) If the amplitude of a sound wave is tripled, (a) by what factor will the intensity increase? (b) By how many dB will the sound level increase?

16. (II) Human beings can typically detect a difference in sound intensity level of 2.0 dB. What is the ratio of the amplitudes of two sounds whose levels differ by this amount?

17. (II) Two sound waves have equal displacement amplitudes, but one has twice the frequency of the other. (a) Which has the greater pressure amplitude and by what factor is it greater? (b) What is the ratio of their intensities?

18. (II) A person standing a certain distance from an airplane with four equally noisy jet engines is experiencing a sound level bordering on pain, 120 dB. What sound level would this person experience if the captain shut down all but one engine?

19. (II) A 40-dB sound wave strikes an eardrum whose area is 5.0×10^{-5} m^2. (a) How much energy is absorbed by the eardrum per second? (b) At this rate, how long would it take your eardrum to receive a total energy of 1.0 J?

20. (II) (a) Estimate the power output of sound from a person speaking in normal conversation. Use Table 16–2. Assume the sound spreads roughly uniformly over a hemisphere in front of the mouth. (b) How many people would produce a total sound output of 100 W of ordinary conversation?

21. (II) If two firecrackers produce a sound level of 90 dB when fired simultaneously at a certain place, what will be the sound level if only one is exploded?

22. (II) What would be the sound level (in dB) of a sound wave in air that corresponds to a displacement amplitude of vibrating air molecules of 1.3 mm at 330 Hz?

23. (II) (a) Calculate the maximum displacement of air molecules when a 210-Hz sound wave passes whose intensity is at the threshold of pain (120 dB). (b) What is the pressure amplitude in this wave?

24. (II) Expensive stereo amplifier A is rated at 250 W per channel, while the more modest amplifier B is rated at 40 W per channel. (a) Estimate the sound level in decibels you would expect at a point 2.5 m from a loudspeaker connected in turn to each amp. (b) Will the expensive amp sound twice as loud as the cheaper one?

25. (II) At a rock concert, a dB meter registered 130 dB when placed 3.4 m in front of a loudspeaker on the stage. (a) What was the power output of the speaker, assuming uniform spherical spreading of the sound and neglecting absorption in the air? (b) How far away would the sound level be a somewhat reasonable 90 dB?

26. (II) A jet plane emits 5.0×10^5 J of sound energy per second. (a) What is the sound level 30 m away? Air absorbs sound at a rate of about 7.0 dB/km; calculate what the sound level will be (b) 1.0 km and (c) 5.0 km away from this jet plane, taking into account air absorption.

27. (II) (a) Show that the sound level, β, can be written in terms of the pressure amplitude, ΔP_M, as

$$\beta(\mathrm{dB}) = 20 \log \frac{\Delta P_M}{\Delta P_{M0}},$$

where ΔP_{M0} is the pressure amplitude at some reference level. (b) The reference pressure amplitude ΔP_{M0} is often taken to be 3.0×10^{-5} N/m^2 corresponding to an intensity of 1.0×10^{-12} W/m^2. What would the sound level be if ΔP_M were 1 atm?

Section 16–4

28. (I) The G string on a violin has a fundamental frequency of 196 Hz. The length of the vibrating portion is 32 cm and has a mass of 0.68 g. Under what tension must the string be placed?

29. (I) (a) What resonant frequency would you expect from blowing across the top of an empty soda bottle that is 15 cm deep? (b) How would that change if it was one-third full of soda?

30. (I) How far from the end of the flute in Example 16–9 should the hole be that must be uncovered to play D above middle C at 294 Hz?

31. (I) If you were to build a pipe organ with open-tube pipes spanning the range of human hearing (20 Hz to 20 kHz), what would be the range of the lengths of pipes required?

32. (I) At a science museum there is a display called a sewer pipe symphony. It consists of many plastic pipes of various lengths, which are open on both ends. (a) If the pipes have lengths of 3.0 m, 2.5 m, 2.0 m, 1.5 m and 1.0 meters, what frequencies will be heard by a visitor's ear placed near the ends of the pipes? (b) Why does this display work better on a noisy day rather than a quiet day?

33. (I) An organ pipe is 78.0 cm long. What are the fundamental and first three audible overtones if the pipe is (a) closed at one end, and (b) open at both ends?

34. (II) An unfingered guitar string is 0.73 m long and is tuned to play E above middle C (330 Hz). (a) How far from the end of this string must the finger be placed to play A above middle C (440 Hz)? (b) What are the frequency and wavelength of the sound wave produced in air at 20°C?

35. (II) (a) Determine the length of an open organ pipe that emits middle C (262 Hz) when the temperature is 21°C. (b) What are the wavelength and frequency of the fundamental standing wave in the tube? (c) What are λ and f in the traveling sound wave produced in the outside air?

36. (II) The human ear canal is approximately 2.5 cm long. It is open to the outside and is closed at the other end by the tympanic membrane. Estimate the frequencies of the standing wave vibrations in the ear canal. What is the relationship of your answer to the information in the graph of Fig. 16–5?

37. (II) An organ is in tune at 20°C. By what percent will the frequency be off at 5.0°C?

38. (II) A particular organ pipe can resonate at 264 Hz, 440 Hz, and 616 Hz, but not at any other intermediate frequencies. (a) Is this an open or closed pipe? (b) What is the fundamental frequency of this pipe?

39. (II) (a) At $T = 15°C$, how long must an open organ pipe be if it is to have a fundamental frequency of 294 Hz? (b) If this pipe were filled with helium, what would its fundamental frequency be?

40. (II) A pipe in air at 20°C is to be designed to produce two successive harmonics at 240 Hz and 280 Hz. How long must the pipe be, and is it open or closed?

41. (II) A uniform narrow tube 1.95 m long is open at both ends. It resonates at two successive harmonics of frequency 275 Hz and 330 Hz. What is the speed of sound in the gas in the tube?

42. (II) How many overtones are present within the audible range for a 2.16 m-long organ pipe at 20°C (a) if it is open, and (b) if it is closed?

* Section 16–5

* **43.** (II) Approximately what are the intensities of the first two overtones of a violin compared to the fundamental? How many decibels softer than the fundamental are the first and second overtones? (See Fig. 16–15.)

Section 16–6

44. (I) A piano tuner hears one beat every 2.0 s when trying to adjust two strings, one of which is sounding 440 Hz. How far off in frequency is the other string?

45. (I) A certain dog whistle operates at 23.5 kHz, while another (brand X) operates at an unknown frequency. If neither whistle can be heard by humans when played separately, but a shrill whine of frequency 5000 Hz occurs when they are played simultaneously, estimate the operating frequency of brand X.

46. (II) Two violin strings are tuned to the same frequency, 294 Hz. The tension in one string is then decreased by 2.0 percent. What will be the beat frequency heard when the two strings are played together?

47. (II) Two piano strings are supposed to be vibrating at 132 Hz, but a piano tuner hears three beats every 2.0 s when they are played together. (a) If one is vibrating at 132 Hz, what must be the frequency of the other (is there only one answer)? (b) By how much (in percent) must the tension be increased or decreased to bring them in tune?

48. (II) How many beats will be heard if two identical flutes each try to play middle C (262 Hz), but one is at 5.0°C and the other at 25.0°C?

49. (II) Two loudspeakers are 1.8 m apart. A person stands 3.0 m from one speaker and 3.5 m from the other. (a) What is the lowest frequency at which destructive interference will occur at this point? (b) Calculate two other frequencies that also result in destructive interference at this point (give the next two highest). Let $T = 20°C$.

50. (II) The two sources of sound in Fig. 16–16 face each other and emit sounds of equal amplitude and equal frequency (250 Hz) but 180° out of phase. For what minimum separation of the two speakers will there be some point at which (a) complete constructive interference occurs and (b) complete destructive interference occurs. (Assume $T = 20°C$)

51. (II) The two sources shown in Fig. 16–16 emit sound waves, in phase, each of wavelength λ and amplitude D_M. Consider a point such as C or D in the diagram, and let r_A and r_B be the distances of this point from source A and source B, respectively. Show that if r_A and r_B are nearly equal $(r_A - r_B \ll r_A)$ then the amplitude varies approximately with position as

$$\left(\frac{2D_M}{r_A}\right) \cos\frac{\pi}{\lambda}(r_A - r_B).$$

52. (II) Two loudspeakers are placed 3.00 m apart, as shown in Fig. 16–33. They emit 440-Hz sounds, in phase. A microphone is placed 3.20 m distant from a point midway between the two speakers, where an intensity maximum is recorded. (a) How far must the microphone be moved to the right to find the first intensity minimum? (b) Suppose the speakers are reconnected so that the 440-Hz sounds they emit are exactly out of phase. At what positions are the intensity maximum and minimum now?

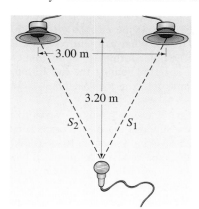

FIGURE 16–33
Problem 52.

53. (II) A guitar string produces 4 beats/s when sounded with a 350-Hz tuning fork and 9 beats/s when sounded with a 355-Hz tuning fork. What is the vibrational frequency of the string? Explain your reasoning.

54. (II) You have three tuning forks, A, B, and C. Fork B has a frequency of 440 Hz; when A and B are sounded together, a beat frequency of 3 Hz is heard. When B and C are sounded together, the beat frequency is 4 Hz. What are the possible frequencies of A and C? What beat frequencies are possible when A and C are sounded together?

55. (II) Show that the two speakers in Fig. 16–16 must be separated by at least a distance d equal to one-half the wavelength λ of sound if there is to be any place where complete destructive interference occurs. The speakers are in phase.

56. (II) A source emits sound of wavelengths 2.64 m and 2.76 m in air. (a) How many beats per second will be heard (assume $T = 20°C$)? (b) How far apart in space are the regions of maximum intensity?

57. (I) The predominant frequency of a certain police car's siren is 1550 Hz when at rest. What frequency do you detect if you move with a speed of 30.0 m/s (a) toward the car, and (b) away from the car?

58. (I) A bat at rest sends out ultrasonic sound waves at 50,000 Hz and receives them returned from an object moving radially away from it at 25.0 m/s. What is the received sound frequency?

59. (II) A bat flies toward a wall at a speed of 5.0 m/s. As it flies, the bat emits an ultrasonic sound wave with frequency 30,000 Hz. What frequency does the bat hear in the reflected wave?

60. (II) In one of the original Doppler experiments, one tuba was played on a moving train car at a frequency of 75 Hz, and a second identical tuba played the same tone while at rest in the railway station. What beat frequency was heard if the train car approached the station at a speed of 10.0 m/s?

61. (II) Two automobiles are equipped with the same single-frequency horn. When one is at rest and the other is moving toward an observer at 15 m/s, a beat frequency of 5.5 Hz is heard. What is the frequency the horns emit? Assume $T = 20°C$.

62. (II) Compare the shift in frequency if a 2000-Hz source is moving toward you at 15 m/s versus if you are moving toward it at 15 m/s. Are the two frequencies exactly the same? Are they close? Repeat the calculation for 150 m/s and then again for 300 m/s. What can you conclude about the asymmetry of the Doppler formulas? Show that at low speeds (relative to the speed of sound), the two formulas—source approaching and detector approaching—yield the same result.

63. (II) A Doppler flow meter uses ultrasound waves to measure blood-flow speeds. Suppose the device emits sound at 3.5 MHz, and the speed of sound in human tissue is taken to be 1540 m/s. What is the expected beat frequency if blood is flowing normally in large leg arteries at 2.0 cm/s directly away from the sound source?

64. (II) The Doppler effect using ultrasonic waves of frequency 2.25×10^6 Hz is used to monitor the heartbeat of a fetus. A (maximum) beat frequency of 500 Hz is observed. Assuming that the speed of sound in tissue is 1.54×10^3 m/s, calculate the maximum velocity of the surface of the beating heart.

65. (II) In Problem 64, the beat frequency is found to appear and then disappear 180 times per minute, which reflects the fact that the heart is beating and its surface changes speed. What is the heartbeat rate?

66. (II) (a) Use the binomial expansion to show that Eqs. 16–9a and 16–10a become essentially the same for small relative velocity between source and observer. (b) What percent error would result if Eq. 16–10a were used instead of Eq. 16–9a for a relative velocity of 22 m/s?

67. (III) A factory whistle emits sound of frequency 570 Hz. On a day when the wind velocity is 12.0 m/s from the north, what frequency will observers hear who are located, at rest, (a) due north, (b) due south, (c) due east, and (d) due west, of the whistle? What frequency is heard by a cyclist heading (e) north or (f) west, toward the whistle at 15.0 m/s? Assume $T = 20°C$.

* 68. (I) (a) How fast is an object moving on land if it is moving at Mach 0.33? (b) A high-flying Concorde passenger jet displays its Mach number on a screen while cruising at 3000 km/h to be 3.2. What is the speed of sound at that altitude?

* 69. (II) Show that the angle θ a sonic boom makes with the path of a supersonic object is given by Eq. 16–12.

* 70. (II) An airplane travels at Mach 2.3 where the speed of sound is 310 m/s. (a) What is the angle the shock wave makes with the direction of the airplane's motion? (b) If the plane is flying at a height of 7100 m, how long after it is directly overhead will a person on the ground hear the shock wave?

* 71. (II) A space probe enters the thin atmosphere of another planet where the speed of sound is only about 35 m/s. (a) What is the probe's Mach number if its initial speed is 15,000 km/h? (b) What is the apex angle of the shock wave it produces?

* 72. (II) A meteorite traveling 8000 m/s strikes the ocean. Determine the shock wave angle it produces (a) in the air just before entering the ocean, and (b) in the water just after entering. Assume $T = 20°C$.

* 73. (II) You look directly overhead and see a plane exactly 1.5 km above the ground flying faster than the speed of sound. By the time you hear the sonic boom, the plane has traveled a horizontal distance of 2.0 km. See Fig. 16–34. Determine (a) the angle of the shock cone, θ, and (b) the speed of the plane (the Mach number). Assume the speed of sound is 330 m/s.

FIGURE 16–34 Problem 73.

* 74. (II) A supersonic jet traveling at Mach 1.8 at an altitude of 10,000 m passes directly over an observer on the ground. Where will the plane be relative to the observer when the latter hears the sonic boom? (See Fig. 16–35.)

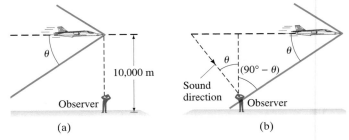

FIGURE 16–35 Problem 74.

General Problems

75. (II) A fish finder uses a sonar device that sends 20,000 Hz sound pulses downwards from the bottom of the boat, and then detects echoes. If the maximum depth for which it is designed to work is 200 meters, what is the minimum time between pulses?

76. Approximately how many octaves are there in the human audible range?

77. A stone is dropped from the top of a cliff. The splash it makes when striking the water below is heard 3.5 s later. How high is the cliff?

78. A single mosquito 5.0 m from a person makes a sound close to the threshold of human hearing (0 dB). What will be the sound level of 1000 such mosquitoes?

79. At the Indianapolis 500, you can estimate the speed of cars just by listening to the difference in pitch of the engine noise between approaching and receding cars. Suppose the sound of a certain car drops by a full octave as it goes by on the straightaway. How fast is it going?

80. A tight guitar string has a frequency of 540 Hz as its third harmonic. What will be its fundamental frequency if it is fingered at a length of only 60 percent of its original length?

81. Each string on a violin is tuned to a frequency $1\frac{1}{2}$ times that of its neighbor. If all the strings are to be placed under the same tension, what must be the mass per unit length of each string relative to that of the lowest string?

82. What is the resultant sound level when an 80-dB sound and an 85-dB sound are heard simultaneously?

83. The sound level 12.0 m from a loudspeaker, placed in the open, is 100 dB. What is the acoustic power output (W) of the speaker, assuming it radiates equally in all directions?

84. The A string of a violin is 32 cm long between fixed points with a fundamental frequency of 440 Hz and a linear density of 6.1×10^{-4} kg/m. (*a*) What are the wave speed and tension in the string? (*b*) What is the length of the tube of a simple wind instrument (say, an organ pipe) closed at one end whose fundamental is also 440 Hz if the speed of sound is 343 m/s in air? (*c*) What is the frequency of the first overtone of each instrument?

85. A stereo amplifier is rated at 150 W output at 1000 Hz. The power output drops by 10 dB at 15 kHz. What is the power output in watts at 15 kHz?

86. A tuning fork is set into vibration above a vertical open tube filled with water (Fig. 16–36). The water level is allowed to drop slowly. As it does so, the air in the tube above the water level is heard to resonate with the tuning fork when the distance from the tube opening to the water level is 0.125 m and again at 0.395 m. What is the frequency of the tuning fork?

0.125 m

0.395 m

FIGURE 16–36
Problem 86.

87. Two loudspeakers face each other at opposite ends of a long corridor. They are connected to the same source which produces a pure tone of 280 Hz. A person walks from one speaker toward the other at a speed of 1.4 m/s. What "beat" frequency does the person hear?

88. Workers around jet aircraft typically wear protective devices over their ears. Assume that the intensity level of a jet airplane engine, at a distance of 30 m, is 140 dB, and that the average human ear has an effective radius of 2.0 cm. What would be the power intercepted by an unprotected ear at a distance of 30 m from a jet airplane engine?

89. The intensity at the "threshold of hearing" for the human ear at a frequency of about 1000 Hz is $I_0 = 1.0 \times 10^{-12}$ W/m^2, for which β, the sound level, is 0 dB. The "threshold of pain" at the same frequency is about 120 dB, or $I = 1.0$ W/m^2, corresponding to an increase of intensity by a factor of 10^{12}. By what factors do the displacement amplitude, D_M, and the pressure amplitude ΔP_M, vary?

90. As Fig. 16–5 shows, the human ear is not equally sensitive to all frequencies; the threshold of hearing is about 60 dB at 35 Hz, 0 dB at 1000 Hz and at 5000 Hz, and 20 dB at 15,000 Hz. Thus, near threshold, the ear is especially insensitive to low frequencies. Estimate the displacement amplitude for each of these four points. At which frequency is the ear most sensitive to displacement?

91. In audio and communications systems, the *gain*, β, in decibels is defined as $\beta = 10 \log\left(P_{\text{out}}/P_{\text{in}}\right)$ where P_{in} is the power input to the system and P_{out} is the power output. A particular stereo amplifier puts out 100 W of power for an input of 1 mW. What is its gain in dB?

92. Two loudspeakers are at opposite ends of a railroad car as it moves past a stationary observer at 10.0 m/s, as shown in Fig. 16–37. If they have identical sound frequencies of 200 Hz, what is the beat frequency heard by the observer when (*a*) he listens from the position A, in front of the car, (*b*) he is between the speakers, at B, and (*c*) he hears the speakers after they have passed him, at C?

$v = 10.0$ m/s

C   B   A

FIGURE 16–37   Problem 92.

93. The frequency of a steam train whistle as it approaches you is 538 Hz. After it passes you, its frequency is measured as 486 Hz. How fast was the train moving (assume constant velocity)?

94. A 75-cm-long guitar string of mass 2.10 g is placed near a tube open at one end, and also 75 cm long. How much tension should be in the string if it is to produce resonance (in its fundamental mode) with the third harmonic in the tube?

95. If the velocity of blood flow in the aorta is normally about 0.32 m/s, what beat frequency would you expect if 5.50-MHz ultrasound waves were directed along the flow and reflected from the red blood cells? Assume that the waves travel with a speed of 1.54×10^3 m/s.

96. A source of sound waves (wavelength λ) is a distance l from a detector. Sound reaches the detector directly, and also by reflecting off an obstacle, as shown in Fig. 16–38. The obstacle is equidistant from source and detector. When the obstacle is a distance d to the right of the line of sight between source and detector, as shown, the two waves arrive in phase. How much farther to the right must the obstacle be moved if the two waves are to be out of phase by $\frac{1}{2}$ wavelength, so destructive interference occurs? (Assume $\lambda \ll l, d$.)

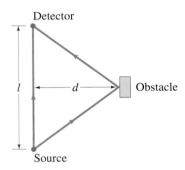

FIGURE 16–38   Problem 96.

97. A person hears a pure tone in the 500–1000 Hz range coming from two sources. The sound is loudest at points equidistant from the two sources. In order to determine exactly what the frequency is, the person moves about and finds that the sound level is minimal at a point 0.31 m farther from one source than the other. What is the frequency of the sound?

98. A bat flies toward a moth at speed 6.5 m/s while the moth is flying toward the bat at speed 5.0 m/s. The bat emits a sound wave of 51.35 kHz. What is the frequency of the wave detected by the bat after it reflects off the moth?

99. A dramatic demonstration, called "singing rods," involves a long, slender aluminum rod held in the hand near the rod's midpoint. The rod is stroked with the other hand. With a little practice, the rod can be made to "sing," or emit a clear, loud, ringing sound. For a 90-cm-long rod: (*a*) what is the fundamental frequency of the sound? (*b*) What is its wavelength in the rod, and (*c*) what is the traveling wavelength in air at 20°C?

100. Room acoustics for stereo listening can be compromised by the presence of standing waves, which can cause acoustic "dead spots" at the locations of the pressure nodes. Consider a living room with dimensions 5.0 m long, 4.0 m wide, and 2.8 m high. Calculate the fundamental frequencies for the standing waves in this room.

101. Assuming that the maximum displacement of the air molecules in a sound wave is about the same as that of the speaker cone that produces the sound (Fig. 16–39), estimate by how much a loudspeaker cone moves for a fairly loud (100 dB) sound of (*a*) 10 kHz, and (*b*) 40 Hz.

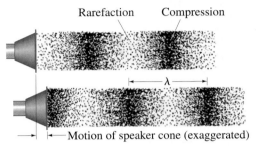

FIGURE 16–39   Problem 101.

102. A Doppler flow meter is used to measure the speed of blood flow. Transmitting and receiving elements are placed on the skin, as shown in Fig. 16–40. Typical sound-wave frequencies of about 5.0 MHz are used, which have a reasonable chance of being reflected from red blood cells. By measuring the frequency of the reflected waves, which are Doppler-shifted because the red blood cells are moving, the speed of the blood flow can be deduced. "Normal" blood flow speed is about 0.1 m/s. Suppose that an artery is partly constricted, so that the speed of the blood flow is increased, and the flow meter measures a Doppler shift of 900 Hz. What is the speed of blood flow in the constricted region? The effective angle between the sound waves (both transmitted and reflected) and the direction of blood flow is 45°. Assume the velocity of sound in tissue is 1540 m/s.

FIGURE 16–40   Problem 102.

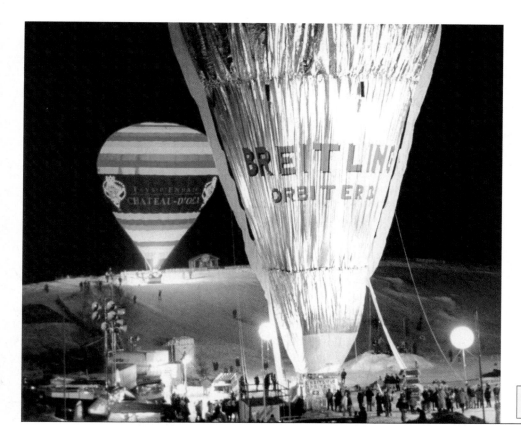

The balloon in the foreground is about to leave on the first successful nonstop trip around the world in a balloon. To the rear, we can see the flame heating the air in another balloon. Heating the air inside a "hot-air" balloon raises the air's temperature and forces some of the air to escape from the opening at the bottom. This is an example of Charles's law: the volume of an enclosed gas at constant pressure increases directly with the Kelvin temperature. The balloon is open at the bottom, so the pressure is atmospheric. The volume increase caused by the heating forces some gas to escape through the opening. The reduced amount of gas inside means its density is lower, so there is a net buoyant force upward on the balloon. In this chapter we study temperature and its effects on matter: thermal expansion, thermal stresses, and the gas laws.

CHAPTER 17

Temperature, Thermal Expansion, and the Ideal Gas Law

The topics of temperature, heat and thermodynamics are the subject of the next four chapters, Chapters 17 through 20. Also included, and closely related, as a means of understanding, is the kinetic theory of gases.

We will often consider a particular **system**, by which we mean a particular object or set of objects; everything else in the universe is called the "environment." We can describe the **state** (or condition) of a particular system—such as a gas in a container—from either a microscopic or macroscopic point of view. A **microscopic** description would involve details of the motion of all the atoms or molecules making up the system, which could be very complicated. A **macroscopic** description is given in terms of quantities that are detectable directly by our senses, such as volume, mass, pressure, and temperature.

Microscopic vs. macroscopic properties

The description of processes in terms of macroscopic quantities is the field of **thermodynamics**. The number of macroscopic variables required to describe the state of a system at any time depends on the type of system. To describe the state of a pure gas in a container, for example, we need only three variables, which could be the volume, the pressure, and the temperature. Quantities such as these that can be used to describe the state of a system are called **state variables**.

The emphasis in this chapter is on the concept of temperature. We will begin, however, with a brief discussion of the theory that matter is made up of atoms and that these atoms are in continuous random motion. This theory is called the *kinetic theory* ("kinetic," you may remember, comes from the Greek word for "moving"), and we will discuss it in more detail in the next chapter.

17–1 | Atomic Theory of Matter

The idea that matter is made of atoms dates back to the ancient Greeks. According to the Greek philosopher Democritus, if a given substance—say, a piece of iron—were cut into smaller and smaller bits, eventually a smallest piece of that substance would be obtained which could not be divided further. This smallest piece was called an **atom**, which comes from the Greek *atomos*, "indivisible."[†] The only real alternative to the atomic theory of matter was the idea that matter is continuous and can be subdivided indefinitely.

Today, we often speak of the relative masses of atoms and molecules—what we call the **atomic mass** or **molecular mass**, respectively.[‡] These are based on assigning the abundant carbon atom, ^{12}C, the value of exactly 12.0000 **unified atomic mass units** (u). In terms of kilograms,

$$1\,u = 1.6605 \times 10^{-27}\,kg.$$

The atomic mass of hydrogen is then 1.0078 u, and the values for other atoms are as listed in the periodic table inside the back cover of this book, and also in Appendix D. The molecular mass of a compound is the sum of the atomic masses of the atoms making up the molecule.

An important piece of evidence for the atomic theory is the so-called **Brownian movement**, named after the biologist Robert Brown, who is credited with its discovery in 1827. While he was observing tiny pollen grains suspended in water under his microscope, Brown noticed that the tiny grains moved about in tortuous paths (Fig. 17–1), even though the water appeared to be perfectly still. The atomic theory easily explains Brownian movement if the further reasonable assumption is made that the atoms of any substance are continually in motion. Then Brown's tiny pollen grains are jostled about by the vigorous barrage of rapidly moving molecules of water.

In 1905, Albert Einstein examined Brownian movement from a theoretical point of view and was able to calculate from the experimental data the approximate size and mass of atoms and molecules. His calculations showed that the diameter of a typical atom is about 10^{-10} m.

At the start of Chapter 13, we distinguished the three common states of matter—solid, liquid, gas—based on macroscopic, or "large-scale," properties. Now let us see how these three phases of matter differ from the atomic, or microscopic, point of view. Clearly, atoms and molecules must exert attractive forces on each other. For how else could a brick or a piece of aluminum stay together in one piece? The attractive forces between molecules are of an electrical nature (more on this in later chapters). If the molecules come too close together, the force between them becomes repulsive (electric repulsion between their outer electrons). Thus molecules maintain a minimum distance from each other. In a solid material, the attractive forces are strong enough that the atoms or molecules are held in more or less fixed positions, often in an array known as a crystal lattice, as shown in Fig. 17–2a. The atoms or molecules in a solid are in motion—they vibrate about their nearly fixed positions. In a liquid, the atoms or molecules are moving more rapidly, or the forces between them are weaker, so that they are sufficiently free to pass over one another, as in Fig. 17–2b. In a gas, the forces are so weak, or the speeds so high, that the molecules do not even stay close together. They move rapidly every which way, Fig. 17–2c, filling any container and occasionally colliding with one another. On the average, the speeds are sufficiently high in a gas that

Atomic and molecular masses

FIGURE 17–1 Path of a tiny particle (pollen grain, for example) suspended in water. The straight lines connect observed positions of the particle at equal time intervals. [The plot has the same general shape independent of the time interval—whether it be every 60 s or 0.1 s, a phenomenon referred to as *fractal* behavior.]

Phases of matter

[†] Today, of course, we don't consider the atom as indivisible, but rather as consisting of a nucleus (containing protons and neutrons) and electrons.

[‡] The terms *atomic weight* and *molecular weight* are popularly used for these quantities, but properly speaking we are comparing masses.

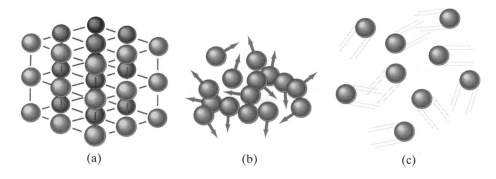

FIGURE 17–2 Atomic arrangements in (a) a crystalline solid, (b) a liquid, and (c) a gas.

(a)   (b)   (c)

when two molecules collide, the force of attraction is not strong enough to keep them close together and they fly off in new directions.

EXAMPLE 17–1 ESTIMATE Distance between atoms. The density of copper is $8.9 \times 10^3 \, \text{kg/m}^3$ and each copper atom has a mass of 63 u, where $1 \, \text{u} = 1.66 \times 10^{-27} \, \text{kg}$. Estimate the average distance between neighboring atoms.

SOLUTION The mass of 1 copper atom is $63 \times 1.66 \times 10^{-27} \, \text{kg} = 1.04 \times 10^{-25} \, \text{kg}$. This means that in a cube of copper 1 m on a side $(\text{volume} = 1 \, \text{m}^3)$, there are

$$\frac{8.9 \times 10^3 \, \text{kg/m}^3}{1.04 \times 10^{-25} \, \text{kg/atom}} = 8.5 \times 10^{28} \, \text{atoms/m}^3.$$

The volume of a cube of side l is $V = l^3$, so on one edge of the 1-m-long cube there are $(8.5 \times 10^{28})^{\frac{1}{3}} \, \text{atoms} = 4.4 \times 10^9$ atoms. Hence the distance between neighboring atoms is

$$\frac{1 \, \text{m}}{4.4 \times 10^9 \, \text{atoms}} = 2.3 \times 10^{-10} \, \text{m between atoms.}$$

17–2 Temperature and Thermometers

In everyday life, **temperature** is a measure of how hot or cold an object is. A hot oven is said to have a high temperature, whereas the ice of a frozen lake is said to have a low temperature.

Many properties of matter change with temperature. For example, most materials expand when heated.[†] An iron beam is longer when hot than when cold. Concrete roads and sidewalks expand and contract slightly according to temperature, which is why compressible spacers or expansion joints (Fig. 17–3) are placed at regular intervals. The electrical resistance of matter changes with temperature (see Chapter 25). So too does the color radiated by objects, at least at high temperatures: you may have noticed that the heating element of an electric stove glows with a red color when hot. At higher temperatures, solids such as iron glow orange or even white. The white light from an ordinary incandescent lightbulb comes from an extremely hot tungsten wire. The surface temperatures of the Sun and other stars can be measured by the predominant color (more precisely, wavelengths) of light they emit.

Instruments designed to measure temperature are called **thermometers**. There are many kinds of thermometers, but their operation always depends on some property of matter that changes with temperature. Most common thermometers

FIGURE 17–3 Expansion joint on a bridge.

Thermometers: measuring temperature

[†] Most materials expand when their temperature is raised, but not all. Water, for example, in the range 0°C to 4°C contracts with an increase in temperature (see Section 17–4).

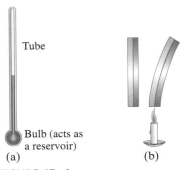

Tube

Bulb (acts as
a reservoir)

(a)   (b)

FIGURE 17–4
(a) Mercury- or alcohol-in-glass
thermometer; (b) bimetallic strip.

Temperature scales

FIGURE 17–5 Photograph of a
thermometer using a coiled
bimetallic strip.

rely on the expansion of a material with an increase in temperature. The first idea for a thermometer, by Galileo, made use of the expansion of a gas.

Common thermometers today consist of a hollow glass tube filled with mercury or with alcohol colored with a red dye. In such liquid-in-glass thermometers, the liquid expands more than the glass when the temperature is increased, so the liquid level rises in the tube (Fig. 17–4a). Although metals also expand with temperature, the change in length of a metal rod, say, is generally too small to measure accurately for ordinary changes in temperature. However, a useful thermometer can be made by bonding together two dissimilar metals whose rates of expansion are different (Fig. 17–4b). When the temperature is increased, the slightly different amounts of expansion cause the bimetallic strip to bend. Often the bimetallic strip is in the form of a coil, one end of which is fixed while the other is attached to a pointer, Fig. 17–5. This kind of thermometer is used as ordinary air thermometers, oven thermometers, automatic-off switch in electric coffeepots, and in room thermostats for determining when the heater or air conditioner should go on or off. Very precise thermometers make use of electrical properties (Chapter 25), such as resistance thermometers, thermocouples, and thermistors, which may have a digital readout.

In order to measure temperature quantitatively, some sort of numerical scale must be defined. The most common scale today is the **Celsius** scale, sometimes called the **centigrade** scale. In the United States, the **Fahrenheit** scale is also common. The most important scale in scientific work is the absolute, or Kelvin, scale, and it will be discussed later in this chapter.

One way to define a temperature scale is to assign arbitrary values to two readily reproducible temperatures. For both the Celsius and Fahrenheit scales these two fixed points are chosen to be the freezing point and the boiling point[†] of water, both taken at atmospheric pressure. On the Celsius scale, the freezing point of water is chosen to be 0°C ("zero degrees Celsius") and the boiling point 100°C. On the Fahrenheit scale, the freezing point is defined as 32°F and the boiling point 212°F. A practical thermometer is calibrated by placing it in carefully prepared environments at each of the two temperatures and marking the position of the mercury or pointer. For a Celsius scale, the distance between the two marks is then divided into one hundred equal intervals separated by small marks representing each degree between 0°C and 100°C (hence the name "centigrade scale" meaning "hundred steps"). For a Fahrenheit scale, the two points are labeled 32°F and 212°F and the distance between them is divided into 180 equal intervals. For temperatures below the freezing point of water and above the boiling point of water the scales can be extended using the same equally spaced intervals. However, ordinary thermometers can be used only over a limited temperature range because of their own limitations—for example, the mercury in a mercury-in-glass thermometer solidifies at some point, below which the thermometer will be useless. It is also rendered useless above temperatures where the fluid vaporizes. For very low or very high temperatures, specialized thermometers are required, some of which we will mention later.

Every temperature on the Celsius scale corresponds to a particular temperature on the Fahrenheit scale, Fig. 17–6. It is easy to convert from one to the other if you remember that 0°C corresponds to 32°F and that a range of 100° on the Celsius scale corresponds to a range of 180° on the Fahrenheit scale. Thus, one Fahrenheit degree (1 F°) corresponds to $100/180 = \frac{5}{9}$ of a Celsius degree (1 C°).

[†] The freezing point of a substance is defined as that temperature at which the solid and liquid phases coexist in equilibrium—that is, without any net liquid changing into the solid or vice versa. Experimentally, this is found to occur at only one definite temperature, for a given pressure. Similarly, the boiling point is defined as that temperature at which the liquid and gas coexist in equilibrium. Since these points vary with pressure, the pressure must be specified (usually it is 1 atm).

That is, $1\,F° = \frac{5}{9}\,C°$. (Notice that when we refer to a specific temperature, we say "degrees Celsius," as in 20°C; but when we refer to a *change* in temperature or a temperature interval, to avoid misunderstanding we will say "Celsius degrees," as in "1 C°.") The conversion between the two temperature scales can be written

$$T(°C) = \tfrac{5}{9}\big[T(°F) - 32\big] \qquad \text{or} \qquad T(°F) = \tfrac{9}{5}T(°C) + 32.$$

Rather than memorizing these relations (it would be easy to confuse them), it is better to simply remember that 0°C = 32°F and 5 C° = 9 F° (or 100°C = 212°F).

EXAMPLE 17–2 **Taking your temperature.** Normal body temperature is 98.6°F. What is this on the Celsius scale?

SOLUTION First we relate the given temperature to the freezing point of water (0°C). That is, 98.6°F is 98.6 − 32.0 = 66.6 F° above the freezing point of water. Since each F° is equal to $\frac{5}{9}$ C°, this corresponds to 66.6 × $\frac{5}{9}$ = 37.0 Celsius degrees above the freezing point. Since the freezing point is 0°C, the temperature is 37.0°C.

Different materials do not expand in quite the same way over a wide temperature range. Consequently, if we calibrate different kinds of thermometers exactly as described above, they will not usually agree precisely. Because of how we calibrated them, they will agree at 0°C and at 100°C. But because of different expansion properties, they may not agree precisely at intermediate temperatures (remember we arbitrarily divided the thermometer scale into 100 equal divisions between 0°C and 100°C). Thus a carefully calibrated mercury-in-glass thermometer might register 52.0°C, whereas a carefully calibrated thermometer of another type might read 52.6°C.

Because of this discrepancy, some standard kind of thermometer must be chosen so that these intermediate temperatures can be precisely defined. The chosen standard for this purpose is the so-called **constant-volume gas thermometer**. As shown in the simplified diagram of Fig. 17–7, this thermometer consists of a bulb filled with a dilute gas connected by a thin tube to a mercury manometer. The volume of the gas is kept constant by raising or lowering the right-hand tube of the manometer so that the mercury in the left tube coincides with the reference mark. An increase in temperature causes a proportional increase in pressure in the bulb or in volume of the gas. The tube must thus be lifted higher to keep the gas volume constant. The height of the mercury in the right-hand column is then a measure of the temperature. This thermometer can be calibrated and gives the same results for all gases in the limit of reducing the gas pressure in the bulb toward zero. The resulting scale is defined as the standard temperature scale (Section 17–10).

FIGURE 17–6 Celsius and Fahrenheit scales compared.

Constant-volume gas thermometer

FIGURE 17–7 Constant-volume gas thermometer.

17–3 | Thermal Equilibrium and the Zeroth Law of Thermodynamics

We are all familiar with the fact that if two objects at different temperatures are placed in thermal contact (meaning thermal energy can pass from one to the other), the two objects will eventually reach the same temperature. They will then be said to be in **thermal equilibrium**. For example, an ice cube placed in a large glass of hot water melts to water, all of which eventually comes to the same temperature. If you put your hand in the water of an icy lake, you can *feel* the temperature of your hand drop. (It's best to pull your hand out before thermal equilibrium is reached!) Two objects are defined to be in *thermal equilibrium* if, when placed in thermal contact, their temperatures don't change.

Thermal equilibrium

Suppose you wanted to determine if two systems, A and B, are in thermal equilibrium, but without putting them in contact. You could do so by making use of a third system, C (which could be considered a thermometer). Suppose that C and A are in thermal equilibrium and that C and B are in thermal equilibrium. Does this imply that A and B are necessarily in thermal equilibrium with each other? Actually, it isn't completely obvious unless you do some experiments, and all experiments indicate that

Zeroth law of thermodynamics

if two systems are in thermal equilibrium with a third system, then they are in thermal equilibrum with each other.

This postulate is called the **zeroth law of thermodynamics**. It has this rather odd name because it was not until after the great first and second laws of thermodynamics (Chapters 19 and 20) were worked out that scientists realized that this apparently obvious postulate needed to be stated first.

Temperature is a property of a system that determines whether the system will be in thermal equilibrium with other systems. When two systems are in thermal equilibrium, their temperatures are, by definition, equal. This is consistent with our everyday notion of temperature, since when a hot body and a cold one are put into contact, they eventually come to the same temperature. Thus the importance of the zeroth law is that it allows a useful definition of temperature.

17–4 | Thermal Expansion

Most substances expand when heated and contract when cooled. However, the amount of expansion or contraction varies, depending on the material.

Linear Expansion

Experiments indicate that the change in length ΔL of almost all solids is, to a very good approximation, directly proportional to the change in temperature ΔT. As might be expected, the change in length is also proportional to the original length of the object, L_0, Fig. 17–8. That is, for the same temperature change, a 4-m-long iron rod will increase in length twice as much as a 2-m-long iron rod. We can write this proportionality as an equation:

$$\Delta L = \alpha L_0 \Delta T, \tag{17–1a}$$

where α, the proportionality constant, is called the *coefficient of linear expansion* for the particular material and has units of $(C°)^{-1}$. This equation can also be written as

$$L = L_0(1 + \alpha \Delta T), \tag{17–1b}$$

where L_0 is the length initially, at temperature T_0, and L is the length after heating or cooling to a temperature T. If the temperature change $\Delta T = T - T_0$ is negative, then $\Delta L = L - L_0$ is also negative, so the length decreases.

The values of α for various materials at 20°C are listed in Table 17–1. It should be noted that α does vary slightly with temperature (which is why thermometers made of different materials do not agree precisely). However, if the temperature range is not too great, the variation can usually be ignored.

FIGURE 17–8 A thin rod of length L_0 at temperature T_0 is heated to a new uniform temperature T and acquires length L, where $L = L_0 + \Delta L$.

Linear expansion

⇒ **PHYSICS APPLIED**

Expansion in structures

EXAMPLE 17–3 **Bridge expansion.** The steel bed of a suspension bridge is 200 m long at 20°C. If the extremes of temperature to which it might be exposed are −30°C to +40°C, how much will it contract and expand?

SOLUTION From Table 17–1, we find that $\alpha = 12 \times 10^{-6}(C°)^{-1}$. The increase in length when it is at 40°C will be

$$\Delta L = (12 \times 10^{-6}/C°)(200\,\text{m})(40°C - 20°C) = 4.8 \times 10^{-2}\,\text{m},$$

or 4.8 cm. When the temperature decreases to −30°C, $\Delta T = -50\,C°$. Then

$$\Delta L = (12 \times 10^{-6}/C°)(200\,\text{m})(-50\,C°) = -12.0 \times 10^{-2}\,\text{m},$$

or a decrease in length of 12 cm.

TABLE 17–1 Coefficients of Expansion, at 20°C

Material	Coefficient of Linear Expansion, α $(C°)^{-1}$	Coefficient of Volume Expansion, β $(C°)^{-1}$
Solids		
Aluminum	25×10^{-6}	75×10^{-6}
Brass	19×10^{-6}	56×10^{-6}
Copper	17×10^{-6}	50×10^{-6}
Iron or steel	12×10^{-6}	35×10^{-6}
Lead	29×10^{-6}	87×10^{-6}
Glass (Pyrex)	3×10^{-6}	9×10^{-6}
Glass (ordinary)	9×10^{-6}	27×10^{-6}
Quartz	0.4×10^{-6}	1×10^{-6}
Concrete and brick	$\approx 12 \times 10^{-6}$	$\approx 36 \times 10^{-6}$
Marble	$1.4–3.5 \times 10^{-6}$	$4–10 \times 10^{-6}$
Liquids		
Gasoline		950×10^{-6}
Mercury		180×10^{-6}
Ethyl alcohol		1100×10^{-6}
Glycerin		500×10^{-6}
Water		210×10^{-6}
Gases		
Air (and most other gases at atmospheric pressure)		3400×10^{-6}

CONCEPTUAL EXAMPLE 17–4 **Do holes expand or contract?** A circular hole is cut from sheet metal, as shown in Fig. 17–9. When the metal is heated in the oven, does the hole get larger or smaller?

RESPONSE You might guess that the metal expands into the hole, making the hole smaller. But instead of cutting a hole, imagine drawing a circle on the sheet metal with a pencil. When the metal expands, the material inside the circle will expand along with the rest of the metal; so the circle expands. Cutting the metal where the circle is makes clear to us that the hole grows as well. The material does *not* expand inward to fill the hole. In a solid object, *all* sections expand outward with increased temperature.

FIGURE 17–9 Conceptual Example 17–4.

EXAMPLE 17–5 **Ring on a rod.** An iron ring is to fit snugly on a cylindrical iron rod. At 20°C, the diameter of the rod is 6.445 cm and the inside diameter of the ring is 6.420 cm. To slip over the rod, the ring must be slightly larger than the rod diameter by about 0.008 cm. To what temperature must the ring be brought if its hole is to be large enough so it will slip over the rod?

SOLUTION The hole in the ring must be increased from a diameter of 6.420 cm to 6.445 cm + 0.008 cm = 6.453 cm. The ring must be heated since the hole diameter will increase linearly with temperature (as in Example 17–4). We solve for ΔT in Eq. 17–1a and find

$$\Delta T = \frac{\Delta L}{\alpha L_0} = \frac{6.453 \text{ cm} - 6.420 \text{ cm}}{(12 \times 10^{-6} \text{ C}°^{-1})(6.420 \text{ cm})} = 430 \text{ C}°.$$

So the ring's temperature must be raised at least to $T = (20°C + 430C°) = 450°C.$

Volume Expansion

The change in *volume* of a material which undergoes a temperature change is given by a relation similar to Eq. 17–1, namely,

Volume expansion

$$\Delta V = \beta V_0 \, \Delta T,$$ (17–2)

where ΔT is the change in temperature, V_0 is the original volume, ΔV is the change in volume, and β is the *coefficient of volume expansion*. The units of β are $(C°)^{-1}$.

Values of β for various materials are given in Table 17–1. Notice that for

$\beta \approx 3\alpha$

solids, β is normally equal to approximately 3α. To see why, consider a rectangular solid of length L_0, width W_0, and height H_0. When its temperature is changed by ΔT, its volume changes from $V_0 = L_0 W_0 H_0$ to

$$V = L_0(1 + \alpha\Delta T)W_0(1 + \alpha\Delta T)H_0(1 + \alpha\Delta T),$$

using Eq. 17–1b and assuming α is the same in all directions. Thus,

$$\Delta V = V - V_0 = V_0(1 + \alpha\Delta T)^3 - V_0 = V_0\big[3\alpha\Delta T + 3(\alpha\Delta T)^2 + (\alpha\Delta T)^3\big].$$

If the amount of expansion is much smaller than the original size of the object, then $\alpha\Delta T \ll 1$ and we can ignore all but the first term and obtain

$$\Delta V \approx (3\alpha)V_0 \, \Delta T.$$

This is Eq. 17–2 with $\beta \approx 3\alpha$. For solids that are not isotropic (having the same properties in all directions), however, the relation $\beta \approx 3\alpha$ is not valid. Note also that linear expansion has no meaning for liquids and gases since they do not have fixed shapes.

Equations 17–1 and 17–2 are accurate only if ΔL (or ΔV) is small compared to L_0 (or V_0). This is of particular concern for liquids and even more so for gases because of the large values of β. Furthermore, β itself varies substantially with temperature for gases. Therefore, a more convenient way of dealing with gases is needed, and this will be discussed starting in Section 17–6.

EXAMPLE 17–6 Gas tank in the sun. The 70-L steel gas tank of a car is filled to the top with gasoline at 20°C. The car is then left to sit in the sun, and the tank reaches a temperature of 40°C (104°F). How much gasoline do you expect to overflow from the tank?

→ PHYSICS APPLIED

Gas tank overflow

SOLUTION The gasoline expands (Eq. 17–2) by

$$\Delta V = \beta V_0 \, \Delta T = \big(950 \times 10^{-6} \, C°^{-1}\big)(70 \, L)(20 \, C°) = 1.3 \, L.$$

The tank also expands. We can think of it as a steel shell that undergoes volume expansion $\big(\beta \approx 3\alpha = 36 \times 10^{-6} \, C°^{-1}\big)$. If the tank were solid, the exterior surface layer (the shell) would expand just the same. Thus the tank increases in volume by

$$\Delta V = \big(36 \times 10^{-6} \, C°^{-1}\big)(70 \, L)(20 \, C°) = 0.050 \, L,$$

so the tank expansion has little effect. If this full tank were left in the sun, over a liter of gas could spill out into the road.

Want to save a few pennies? Fill your gas tank early in the morning when it is cool and the gas is denser—but don't fill the tank quite all the way.

---

* Atomic Theory of Expansion

How can we understand thermal expansion from a microscopic point of view? Let us assume that the atoms in a solid are always in motion, vibrating about their equilibrium positions. We also assume that their average kinetic energy increases with temperature, as we shall discuss in the next chapter. But as temperature increases, does this mean that the average distance between atoms increases?

Experimentally, a solid rod gets longer when you raise its temperature, so we conclude that the average distance between atoms does increase. To understand this, let us look at a simplified potential-energy diagram as shown in Fig. 17–10, which represents the potential energy of two atoms versus their separation r. At large r, we assume the potential energy ≈ 0, and as r decreases, the potential energy decreases, indicating an attractive force as discussed in Section 8–9. For r less than r_0 (the equilibrium position) the potential-energy curve rises, indicating a repulsive force between atoms as they approach each other. The horizontal blue lines in Fig. 17–10 labeled E_2 and E_1 represent the total energy for two different temperatures, T_2 and T_1, where $T_2 > T_1$. The short vertical lines for E_1 and E_2 on the diagram represent the midpoints of the motion at these two temperatures. Because the potential-energy curve is not symmetrical, the average separation of atoms is greater for the higher temperature, as shown. Thus thermal expansion is due to the nonsymmetry of the potential-energy function. If the potential-energy curve were symmetrical, there would be no thermal expansion at all. Indeed, the fact that most substances expand when heated *implies* that the potential-energy curve must be asymmetrical, as in Fig. 17–10.

Anomalous Behavior of Water Below 4°C

Most substances expand more or less uniformly with an increase in temperature, as long as no phase change occurs. Water, however, does not follow the usual pattern. If water at 0°C is heated, it actually *decreases* in volume until it reaches 4°C. Above 4°C water behaves normally and expands in volume as the temperature is increased, Fig. 17–11. Water thus has its greatest density at 4°C. This anomalous behavior of water is of great importance for the survival of aquatic life during cold winters. When the water in a lake or river is above 4°C and begins to cool by contact with cold air, the water at the surface sinks because of its greater density and it is replaced by warmer water from below. This mixing continues until the temperature reaches 4°C. As the surface water cools further, it remains on the surface because it is less dense than the 4°C water below. Water then freezes first at the surface, and the ice remains on the surface since ice (specific gravity = 0.917) is less dense than water. The water at the bottom remains at 4°C until almost the whole body of water is frozen. If water were like most substances, becoming more dense as it cools, the water at the bottom of a lake would be frozen first. Lakes would freeze solid more easily since circulation would bring the warmer water to the surface to be efficiently cooled. The complete freezing of a lake would cause severe damage to its plant and animal life. Because of the unusual behavior of water below 4°C, it is rare for any large body of water to freeze completely, and this is helped by the layer of ice on the surface which acts as an insulator to reduce the flow of heat out of the water into the cold air above. Without this peculiar but wonderful property of water, life on this planet as we know it might not have been possible.

Not only does water expand as it cools from 4°C to 0°C, it expands even more as it freezes to ice. This is why ice cubes float in water and pipes break when water inside them freezes.

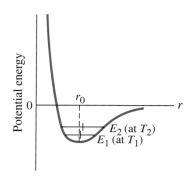

FIGURE 17–10 Typical curve of potential energy versus separation of atoms, r, for atoms in a crystal solid (simplified). Note that the midpoint (short vertical line) of the oscillatory motion of atoms is greater at the higher temperature T_2.

Water is unusual:
it expands when
cooled from 4°C to 0°C

➡ PHYSICS APPLIED

Life under ice

(a) Temperature (°C)

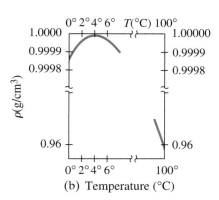

(b) Temperature (°C)

FIGURE 17–11 Behavior of water as a function of temperature near 4°C. (a) Volume, of 1.00000 gram of water, as a function of temperature. (b) Density vs. temperature. [Note the break in each axis.]

*17–5 Thermal Stresses

In many situations, such as in buildings and roads, the ends of a beam or slab of material are rigidly fixed, which greatly limits expansion or contraction. If the temperature should change, large compressive or tensile stresses, called *thermal stresses*, will occur. The magnitude of such stresses can be calculated using the concept of elastic modulus developed in Chapter 12. To calculate the internal stress, we can think of this process as occurring in two steps. The rod expands (or contracts) by an amount ΔL given by Eq. 17–1; then a force is applied to compress (or expand) the material back to its original length. The force F required is given by Eq. 12–4:

$$\Delta L = \frac{1}{E} \frac{F}{A} L_0,$$

where E is Young's modulus for the material. To calculate the internal stress, F/A, we then set ΔL in Eq. 17–1a equal to ΔL in the equation above and find

$$\alpha L_0 \, \Delta T = \frac{1}{E} \frac{F}{A} L_0.$$

Hence, $F/A = \alpha E \, \Delta T$.

EXAMPLE 17–7 **Stress in concrete on a hot day.** A highway is to be made of blocks of concrete 10 m long placed end to end with no space in between them to allow for expansion. If the blocks were placed at a temperature of 10°C, what force of compression would occur if the temperature reached 40°C? The contact area between each block is 0.20 m². Will fracture occur?

SOLUTION We solve for F in the equation above and use the value of E from Table 12–1:

$$F = \alpha \, \Delta T \, E \, A$$
$$= (12 \times 10^{-6}/C°)(30\,C°)(20 \times 10^9\,N/m^2)(0.20\,m^2) = 1.4 \times 10^6\,N.$$

The stress, F/A, is $(1.4 \times 10^6\,N)/(0.20\,m^2) = 7.0 \times 10^6\,N/m^2$. This is not far from the ultimate strength of concrete (Table 12–2) under compression and exceeds it for tension and shear. Hence, assuming the concrete is not perfectly aligned, part of the force will act in shear, and fracture is likely. How much space would you allow between blocks if you expected a temperature range of 0°F to 110°F?

➡ **PHYSICS APPLIED**

Highway buckling

17–6 The Gas Laws and Absolute Temperature

Equation 17–2 is not very useful for describing the expansion of a gas, partly because the expansion can be so great, and partly because gases generally expand to fill whatever container they are in. Indeed, Eq. 17–2 is meaningful only if the pressure is kept constant. The volume of a gas depends very much on the pressure as well as on the temperature. It is therefore valuable to determine a relation between the volume, the pressure, the temperature, and the mass of a gas. Such a relation is called an **equation of state**. (By the word *state*, we mean the physical condition of the system.)

If the state of a system is changed, we will always wait until the pressure and temperature have reached the same values throughout. We thus consider only **equilibrium states** of a system—when the variables that describe it (such as temperature and pressure) are the same throughout the system and are not changing in time. We also note that the results of this Section are accurate only for gases that are not too dense (the pressure is not too high, on the order of an atmosphere or so) and not close to the liquefaction (boiling) point.

For a given quantity of gas it is found experimentally that, to a good approximation, *the volume of a gas is inversely proportional to the pressure applied to it when the temperature is kept constant.* That is,

$$V \propto \frac{1}{P}.$$ [constant T]   *Boyle's law*

where P is the absolute pressure (not "gauge pressure"—see Chapter 13). For example, if the pressure on a gas is doubled, the volume is reduced to half its original volume. This relation is known as **Boyle's law**, after Robert Boyle (1627–1691), who first stated it on the basis of his own experiments. A graph of P vs. V for a fixed temperature is shown in Fig. 17–12. Boyle's law can also be written

$$PV = \text{constant}.$$ [constant T]

That is, at constant temperature, if either the pressure or volume of the gas is allowed to vary, the other variable also changes so that the product PV remains constant.

Temperature also affects the volume of a gas, but a quantitative relationship between V and T was not found until more than a century after Boyle's work. The Frenchman Jacques Charles (1746–1823) found that when the pressure is not too high and is kept constant, the volume of a gas increases with temperature at a nearly constant rate, as in Fig. 17–13a. However, all gases liquefy at low temperatures (for example, oxygen liquefies at $-183°C$) and so the graph cannot be extended below the liquefaction point. Nonetheless, the graph is essentially a straight line and if projected to lower temperatures, as shown by the dashed line, it crosses the axis at about $-273°C$.

Such a graph can be drawn for any gas, and the straight line always projects back to $-273°C$ at zero volume. This seems to imply that if a gas could be cooled to $-273°C$ it would have zero volume, and at lower temperatures a negative volume, which makes no sense, of course. It could be argued that $-273°C$ is the lowest temperature possible, and many other more recent experiments indicate that it is so. This temperature is called the **absolute zero** of temperature. Its value has been determined to be $-273.15°C$.   *Absolute zero*

Absolute zero forms the basis of a temperature scale known as the **absolute** or **Kelvin scale**, and it is used extensively in scientific work. On this scale the temperature is specified as degrees Kelvin or, preferably, simply as kelvins (K) without the degree sign. The intervals are the same as for the Celsius scale, but the zero on this scale (0 K) is chosen as absolute zero itself. Thus the freezing point of water (0°C) is 273.15 K and the boiling point of water is 373.15 K. Indeed, any temperature on the Celsius scale can be changed to kelvins by adding 273.15 to it: *Kelvin scale*

$$T(\text{K}) = T(°C) + 273.15.$$

Now let us look at Fig. 17–13b, where we see that the graph of the volume of a gas versus absolute temperature is a straight line that passes through the origin. Thus,

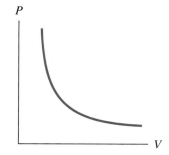

FIGURE 17–12 Pressure vs. volume of a gas at a constant temperature, showing the inverse relationship as given by Boyle's law: as the pressure increases, the volume decreases.

FIGURE 17–13 Volume of a gas as a function of (a) Celsius temperature, and (b) Kelvin temperature, when the pressure is kept constant.

(a) Temperature (°Celsius)

(b) Temperature (kelvins)

to a good approximation, *the volume of a given amount of gas is directly proportional to the absolute temperature when the pressure is kept constant.* This is known as **Charles's law**, and is written

Charles's law

$$V \propto T. \qquad \text{[constant } P\text{]}$$

A third gas law, known as **Gay-Lussac's law**, after Joseph Gay-Lussac (1778–1850), states that *at constant volume*, *the pressure of a gas is directly proportional to the absolute temperature*:

Gay-Lussac's law

$$P \propto T. \qquad \text{[constant } V\text{]}$$

A familiar example is that a closed jar, or aerosol can, thrown into a fire will explode due to the increase in gas pressure inside.

The laws of Boyle, Charles, and Gay-Lussac are not really laws in the sense that we use this term today (precise, deep, wide-ranging validity). They are really only approximations that are accurate for real gases only as long as the pressure and density of the gas are not too high, and the gas is not too close to condensation. The term *law* applied to these three relationships has become traditional, however, so we have stuck with that usage.

17–7 | The Ideal Gas Law

The gas laws of Boyle, Charles, and Gay-Lussac were obtained by means of a technique that is very useful in science: namely, to hold one or more variables constant in order to see clearly the effects of changing only one of the variables. These laws can now be combined into a single more general relation between the pressure, volume, and temperature of a fixed quantity of gas:

$$PV \propto T.$$

This relation indicates how any of the quantities P, V, or T will vary when the other two quantities change. This relation reduces to Boyle's, Charles's, or Gay-Lussac's law when either the temperature, the pressure, or the volume, respectively, is held constant.

Finally, we must incorporate the effect of the amount of gas present. Anyone who has blown up a balloon knows that the more air forced into the balloon, the bigger it gets (Figure 17–14). Indeed, careful experiments show that at constant temperature and pressure, the volume V of an enclosed gas increases in direct proportion to the mass m of gas present. Hence we write

$$PV \propto mT.$$

This proportion can be made into an equation by inserting a constant of proportionality. Experiment shows that this constant has a different value for different gases. However, the constant of proportionality turns out to be the same for all gases if, instead of the mass m, we use the number of moles.

Mole (unit)

One **mole** (abbreviated mol) is defined as the amount of substance that contains as many atoms or molecules as there are in 12.00 grams of carbon 12 (whose atomic mass is exactly 12 u). A simpler but equivalent definition is: 1 mol is that number of grams of a substance numerically equal to the molecular mass (Section 17–1) of the substance. For example, the molecular mass of hydrogen gas (H_2) is 2.0 u (since each molecule contains two atoms of hydrogen and each atom has an atomic mass of 1.0 u). Thus 1 mol of H_2 has a mass of 2.0 g. Similarly, 1 mol of neon gas has a mass of 20 g, and 1 mol of CO_2 has a mass of $[12 + (2 \times 16)] = 44$ g.

FIGURE 17–14 Blowing up a balloon means putting more air (more air molecules) into the balloon, which increases its volume. The pressure is nearly constant (atmospheric) except for the small effect of the balloon's elasticity.

The mole is the official unit in the SI system. In general, the number of moles, n, in a given sample of a pure substance is equal to its mass in grams divided by its molecular mass specified as grams per mole:

$$n \, (\text{mol}) = \frac{\text{mass (grams)}}{\text{molecular mass (g/mol)}}.$$

For example, the number of moles in 132 g of CO_2 is

$$n = \frac{132 \, \text{g}}{44 \, \text{g/mol}} = 3.0 \, \text{mol}.$$

We can now write the proportion discussed above as an equation:

$$PV = nRT, \tag{17–3}$$

IDEAL GAS LAW

where n represents the number of moles and R is the constant of proportionality. R is called the **universal gas constant** because its value is found experimentally to be the same for all gases. The value of R, in several sets of units (only the first is the proper SI unit), is:

$$R = 8.315 \, \text{J/(mol·K)} \qquad \text{[SI units]}$$
$$= 0.0821 \, \text{(L·atm)/(mol·K)}$$
$$= 1.99 \, \text{calories/(mol·K)}.^{\dagger}$$

Equation 17–3 is called the **ideal gas law**, or the **equation of state for an ideal gas**. We use the term "ideal" because real gases do not follow Eq. 17–3 precisely, particularly at high pressure (and density) or when the gas is near the liquefaction point (= boiling point). However, at pressures less than an atmosphere or so, and when T is not close to the liquefaction point of the gas, Eq. 17–3 is quite accurate and useful for real gases.

17–8 | Problem Solving with the Ideal Gas Law

The ideal gas law is an extremely useful tool, and we now consider some Examples. We will often refer to "standard conditions" or "standard temperature and pressure" (STP), which means $T = 273 \, \text{K}$ (0°C) and $P = 1.00 \, \text{atm} = 1.013 \times 10^5 \, \text{N/m}^2 = 101.3 \, \text{kPa}$.

STP = 273 K, 1 atm

EXAMPLE 17–8 **Volume of one mol at STP.** Determine the volume of 1.00 mol of any gas, assuming it behaves like an ideal gas, (*a*) at STP, (*b*) 20°C.

SOLUTION (*a*) We solve for V in Eq. 17–3:

$$V = \frac{nRT}{P} = \frac{(1.00 \, \text{mol})(8.315 \, \text{J/mol·K})(273 \, \text{K})}{(1.013 \times 10^5 \, \text{N/m}^2)}$$
$$= 22.4 \times 10^{-3} \, \text{m}^3.$$

Since 1 liter is 1000 cm³ = $1 \times 10^{-3} \, \text{m}^3$, 1 mol of any gas has a volume of 22.4 L at STP. [How big is 22.4 L? About the size of a cube one foot on a side; more precisely, $\sqrt[3]{22.4 \times 10^{-3} \, \text{m}^3} = 0.28 \, \text{m} = 28 \, \text{cm}$ on a side.]
(*b*) At 20°C, $T = 293 \, \text{K}$ and $V = 24.0 \, \text{L}$.

1 mol of gas at STP has volume = 22.4 L

The value for the volume of 1 mol of an ideal gas at STP (22.4 L) is worth remembering, for it sometimes makes calculation simpler.

---

† Calories are defined in Chapter 19, and sometimes it is useful to use R as given in terms of calories.

➡ PROBLEM SOLVING

Always give T in K, and
P as absolute, not gauge, pressure

➡ PHYSICS APPLIED

What is the mass (and weight)
of the air in a small room?

Always remember, when using the ideal gas law, that temperatures must be given in kelvins (K) and that the pressure P must always be *absolute* pressure, not gauge pressure.

EXAMPLE 17–9 **ESTIMATE** **Mass of air in a room.** Estimate the mass of air in a room whose dimensions are 5 m × 3 m × 2.5 m high, at 20°C.

SOLUTION First we determine the number of moles n, and then we can multiply by the mass of one mole to get the total mass. Example 17–8 told us that 1 mol at 20°C has a volume of 24.0 L. The room's volume is 3 m × 5 m × 2.5 m, so

$$n = \frac{(3\,\text{m})(5\,\text{m})(2.5\,\text{m})}{24.0 \times 10^{-3}\,\text{m}^3} \approx 1600\,\text{mol}.$$

Air is a mixture of about 20% oxygen (O_2) and 80% nitrogen (N_2) whose molecular masses are $2 \times 16\,\text{u} = 32\,\text{u}$ and $2 \times 14\,\text{u} = 28\,\text{u}$ respectively, for an average of about 29 u. Thus, 1 mol of air has a mass of about 29 g = 0.029 kg, so our room has a mass of air

$$m \approx (1600\,\text{mol})(0.029\,\text{kg/mol}) \approx 50\,\text{kg}.$$

That is roughly 100 lbs of air!

Frequently, volume is specified in liters and pressure in atmospheres. Rather than convert these to SI units, we can instead use the value of R given above as $0.0821\,\text{L·atm/mol·K}$.

In many situations it is not necessary to use the value of R at all. For example, many problems involve a change in the pressure, temperature, and volume of a fixed amount of gas. In this case, $PV/T = nR$ = constant, since n and R remain constant. If we now let P_1, V_1, and T_1 represent the appropriate variables initially, and P_2, V_2, T_2 represent the variables after the change is made, then we can write

$$\frac{P_1 V_1}{T_1} = \frac{P_2 V_2}{T_2}.$$

If we know any five of the quantities in this equation, we can solve for the sixth. Or, if one of the three variables is constant $(V_1 = V_2$, or $P_1 = P_2$, or $T_1 = T_2)$ then we can use this equation to solve for one unknown when given the other three quantities.

EXAMPLE 17–10 **Check tires cold.** An automobile tire is filled to a gauge pressure of 200 kPa at 10°C. After driving 100 km, the temperature within the tire rises to 40°C. What is the pressure within the tire now?

SOLUTION Since the volume remains essentially constant, $V_1 = V_2$, and therefore

$$\frac{P_1}{T_1} = \frac{P_2}{T_2}.$$

This is, incidentally, a statement of Gay-Lussac's law. Since the pressure given is the gauge pressure (Section 13–3), we must add atmospheric pressure (= 101 kPa) to get the absolute pressure $P_1 = (200\,\text{kPa} + 101\,\text{kPa}) = 301\,\text{kPa}$. And we convert temperatures to kelvin by adding 273:

$$P_2 = \frac{P_1}{T_1} T_2 = \frac{(3.01 \times 10^5\,\text{Pa})(313\,\text{K})}{(283\,\text{K})} = 333\,\text{kPa}.$$

Subtracting atmospheric pressure, we find the resulting gauge pressure to be 232 kPa, which is a 15 percent increase. This Example shows why car manuals suggest checking the pressure when the tires are cold.

17–9 Ideal Gas Law in Terms of Molecules: Avogadro's Number

The fact that the gas constant, R, has the same value for all gases is a remarkable reflection of simplicity in nature. It was first recognized, although in a slightly different form, by the Italian scientist Amedeo Avogadro (1776–1856). Avogadro stated that *equal volumes of gas at the same pressure and temperature contain equal numbers of molecules.* This is sometimes called **Avogadro's hypothesis**. That this is consistent with R being the same for all gases can be seen as follows. First of all, from Eq. 17–3 we see that for the same number of moles, n, and the same pressure and temperature, the volume will be the same for all gases as long as R is the same. Second, the number of molecules in 1 mole is the same for all gases.[†] Thus Avogadro's hypothesis is equivalent to R being the same for all gases.

Avogadro's hypothesis

The number of molecules in a mole is known as **Avogadro's number**, N_A. Although Avogadro conceived the notion, he was not able to actually determine the value of N_A. Indeed, precise measurements were not done until the twentieth century.

A number of methods have been devised to measure N_A and the accepted value today is

$$N_A = 6.02 \times 10^{23}. \qquad \text{[molecules/mole]}$$

Avogadro's number

Since the total number of molecules, N, in a gas is equal to the number per mole times the number of moles $(N = nN_A)$, the ideal gas law, Eq. 17–3, can be written in terms of the number of molecules present:

$$PV = nRT = \frac{N}{N_A} RT,$$

or

$$PV = NkT, \qquad \text{(17–4)}$$

IDEAL GAS LAW (IN TERMS OF MOLECULES)

where $k = R/N_A$ is called **Boltzmann's constant** and has the value

$$k = \frac{R}{N_A} = 1.38 \times 10^{-23} \, \text{J/K}.$$

EXAMPLE 17–11 **Hydrogen atom mass.** Use Avogadro's number to determine the mass of a hydrogen atom.

SOLUTION One mole of hydrogen (atomic mass = 1.008 u, Section 17–1) has a mass of 1.008×10^{-3} kg and contains 6.02×10^{23} atoms. Thus one atom has a mass

$$m = \frac{1.008 \times 10^{-3} \, \text{kg}}{6.02 \times 10^{23}} = 1.67 \times 10^{-27} \, \text{kg}.$$

Historically, the reverse process was one method used to obtain N_A: that is, from the measured mass of the hydrogen atom.

[†] For example, the molecular mass of H_2 gas is 2.0 atomic mass units (u), whereas that of O_2 gas is 32.0 u. Thus 1 mol of H_2 has a mass of 0.0020 kg and 1 mol of O_2 gas, 0.032 kg. The number of molecules in a mole is equal to the total mass M of a mole divided by the mass m of one molecule; since this ratio (M/m) is the same for all gases by definition of the mole, a mole of any gas must contain the same number of molecules.

SECTION 17–9   Ideal Gas Law in Terms of Molecules: Avogadro's Number   **459**

*17–10 Ideal Gas Temperature Scale—a Standard

It is important to have a very precisely defined temperature scale so that measurements of temperature made at different laboratories around the world can be accurately compared. We now discuss such a scale that has been accepted by the general scientific community.

The standard thermometer for this scale is the constant-volume gas thermometer already discussed in Section 17–2. The scale itself is called the **ideal gas temperature scale**, since it is based on the property of an ideal gas that the pressure is directly proportional to the absolute temperature (Gay-Lussac's law). A real gas, which would have to be used in any real constant-volume gas thermometer, approaches this ideal at low density. In other words, the temperature at any point in space is *defined* as being proportional to the pressure in the (nearly) ideal gas used in the thermometer. To set up a scale we need two fixed points. One fixed point will be $P = 0$ at $T = 0\,K$. The second fixed point is chosen to be the **triple point** of water, which is that point where water in the solid, liquid, and gas states can coexist in equilibrium. This occurs only at a unique temperature and pressure,[†] and can be reproduced at different laboratories with great precision. The pressure at the triple point of water is 4.58 torr and the temperature is 0.01°C. This temperature corresponds to 273.16 K, since absolute zero is about −273.15°C. In fact, the triple point is now *defined* to be exactly 273.16 K.

The absolute or Kelvin temperature T at any point is then defined, using a constant-volume gas thermometer for an ideal gas, as

$$T = (273.16\,\text{K})\left(\frac{P}{P_{\text{tp}}}\right). \qquad \text{[ideal gas; constant volume]} \quad \textbf{(17–5a)}$$

In this relation, P_{tp} is the pressure in the thermometer at the triple point temperature of water, and P is the pressure in the thermometer when it is at the point where T is being determined. Note that if we let $P = P_{\text{tp}}$ in this relation, then $T = 273.16\,K$, as it must.

The definition of temperature, Eq. 17–5a, with a constant-volume gas thermometer filled with a real gas is only approximate because we find that we get different results for the temperature depending on the type of gas that is used in the thermometer. Temperatures determined in this way also vary depending on the amount of gas in the bulb of the thermometer: for example, the boiling point of water at 1.00 atm is found from Eq. 17–5a to be 373.87 K when the gas is O_2 and $P_{\text{tp}} = 1000$ torr. If the amount of O_2 in the bulb is reduced so that at the triple point $P_{\text{tp}} = 500$ torr, the boiling point of water from Eq. 17–5a is then found to be 373.51 K. If H_2 gas is used instead, the corresponding values are 373.07 K and 373.11 K (see Fig. 17–15). But now suppose we use a particular real gas and make a series of measurements in which the amount of gas in the thermometer bulb is reduced to smaller and smaller amounts, so that P_{tp} becomes smaller and smaller. It is found experimentally that an extrapolation of such data to $P_{\text{tp}} = 0$ always gives the *same value* for the temperature of a given system (such as $T = 373.15\,K$ for the boiling point of water at 1.00 atm) as shown in Fig. 17–15. Thus the temperature T at any point in space, determined using a constant-volume gas thermometer containing a real gas, is defined using this limiting process:

$$T = (273.16\,\text{K})\lim_{P_{\text{tp}}\to 0}\left(\frac{P}{P_{\text{tp}}}\right). \qquad \text{[constant volume]} \quad \textbf{(17–5b)}$$

This defines the **ideal gas temperature scale**. One of the great advantages of this

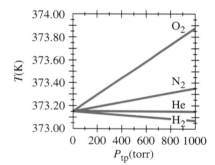

FIGURE 17–15 Temperature readings of a constant-volume gas thermometer for the boiling point of water at 1.00 atm are plotted, for different gases, as a function of the gas pressure in the thermometer at the triple point (P_{tp}). Note that as the amount of gas in the thermometer is reduced, so that $P_{\text{tp}} \to 0$, all gases give the same reading, 373.15 K. For pressure less than 0.10 atm (76 torr), the variation shown is less than 0.07 K.

Ideal gas temperature scale

[†] Liquid water and steam can coexist (the boiling point) at a range of temperatures depending on the pressure. Water boils at a lower temperature when the pressure is less, such as high in the mountains. The triple point represents a more precisely reproducible fixed point than does either the freezing point or boiling point of water at, say, 1 atm. See Section 18–3 for further discussion.

scale is that the value for T does not depend on the kind of gas used. But the scale does depend on the properties of gases in general. Helium has the lowest condensation point of all gases; at very low pressures it liquefies at about 1 K, so temperatures below this cannot be defined on this scale.

Summary

The atomic theory of matter postulates that all matter is made up of tiny entities called **atoms**, which are typically 10^{-10} m in diameter.

Atomic and **molecular masses** are specified on a scale where ordinary carbon (^{12}C) is arbitrarily given the value 12.0000 u (atomic mass units).

The distinction between solids, liquids, and gases can be attributed to the strength of the attractive forces between the atoms or molecules and to their average speed.

Temperature is a measure of how hot or cold a body is. **Thermometers** are used to measure temperature on the **Celsius** (°C), **Fahrenheit** (°F), and **Kelvin** (K) scales. Two standard points on each scale are the freezing point of water (0°C, 32°F, 273.15 K) and the boiling point of water (100°C, 212°F, 373.15 K). A one kelvin change in temperature equals a change of one Celsius degree or $\frac{9}{5}$ Fahrenheit degrees.

The change in length, ΔL, of a solid, when its temperature changes by an amount ΔT, is directly proportional to the temperature change and to its original length L_0. That is,

$$\Delta L = \alpha L_0 \, \Delta T,$$

where α is the *coefficient of linear expansion*.

The change in volume of most solids, liquids, and gases is proportional to the temperature change and to the original volume V_0: $\Delta V = \beta V_0 \, \Delta T$. The *coefficient of volume expansion*, β, is approximately equal to 3α for most solids.

Water is unusual because, unlike most materials whose volume increases with temperature, its volume actually decreases as the temperature increases in the range from 0°C to 4°C.

The **ideal gas law**, or **equation of state for an ideal gas**, relates the pressure P, volume V, and temperature T (in kelvins) of n moles of gas by the equation

$$PV = nRT,$$

where $R = 8.315$ J/mol·K for all gases. Real gases obey the ideal gas law quite accurately if they are not at too high a pressure or near their liquefaction point.

One **mole** of a substance is defined as the number of grams which is numerically equal to the atomic or molecular mass.

Avogadro's number, $N_A = 6.02 \times 10^{23}$, is the number of atoms or molecules in 1 mol of any pure substance.

The ideal gas law can be written in terms of the number of molecules N in the gas as

$$PV = NkT,$$

where $k = R/N_A = 1.38 \times 10^{-23}$ J/K is Boltzmann's constant.

Questions

1. Which has more atoms: 1 kg of iron or 1 kg of aluminum? See the Periodic Table or Appendix D.
2. Name several properties of materials that could be exploited to make a thermometer.
3. Suppose system C is not in equilibrium with system A nor in equilibrium with system B. Does this imply that A and B are not in equilibrium? What can you infer regarding the temperatures of A, B, and C?
4. If system A is in equilibrium with system B, but B is not in equilibrium with system C, what can you say about the temperatures of A, B, and C?
5. A flat bimetallic strip consists of aluminum riveted to a strip of iron. When heated, which metal will be on the outside of the curve?
6. In the relation $\Delta L = \alpha L_0 \Delta T$, should L_0 be the initial length, the final length, or does it matter?
7. Why is it sometimes easier to remove the lid from a tightly closed jar after warming it under hot running water?
8. Long steam pipes often have a section in the shape of a U. Why?
9. Figure 17–16 shows a diagram of a simple *thermostat* used to control a furnace (or other heating or cooling system). The bimetallic strip consists of two strips of different metals bonded together. The electric switch is a glass vessel containing liquid mercury that conducts electricity when it can flow to touch both contact wires. Explain how this device controls the furnace and how it can be set at different temperatures.

FIGURE 17–16 A thermostat (Question 9).

10. The units for the coefficients of expansion α are $(C°)^{-1}$, and there is no mention of a length unit such as meters. Would the expansion coefficient change if we used feet or millimeters instead of meters?

11. Explain why it is advisable to add water to an overheated automobile engine only slowly, and only with the engine running.

12. A glass container may break if one part of it is heated or cooled more rapidly than adjacent parts. Explain.

13. When a cold mercury-in-glass thermometer is first placed in a hot tub of water, the mercury initially descends a bit and then rises. Explain.

14. The principal virtue of Pyrex glass is that its coefficient of linear expansion is much smaller than that for ordinary glass (Table 17–1). Explain why this gives rise to the high "heat resistance" of Pyrex.

15. Will a grandfather clock, accurate at 20°C, run fast or slow on a hot day (30°C)? The clock uses a pendulum supported on a long thin brass rod.

16. Freezing a can of soda will cause its bottom and top to bulge so badly the can will not stand up. What has happened?

17. Why might you expect an alcohol-in-glass thermometer to be more precise than a mercury-in-glass thermometer?

18. Will the buoyant force on an aluminum sphere submerged in water increase or decrease if the temperature is increased from 20°C to 40°C?

19. A flat, uniform cylinder of lead floats in mercury at 0°C. Will the lead float higher or lower when the temperature is raised?

20. Which scale, Fahrenheit, Celsius, or Kelvin, might be considered most "natural" from a scientific point of view? Discuss.

21. If an atom is measured to have a mass of 6.7×10^{-27} kg, what atom do you think it is?

* 22. From a practical point of view, does it really matter what gas is used in a constant-volume gas thermometer? If so, explain. [*Hint*: See Fig. 17–15.]

Problems

Section 17–1

1. (I) How does the number of atoms in a 26.5-gram gold ring compare to the number in a silver ring of the same mass?

2. (I) How many atoms are there in a 3.4-gram copper penny?

Section 17–2

3. (I) (*a*) "Room temperature" is often taken to be 68°F; what is this on the Celsius scale? (*b*) The temperature of the filament in a lightbulb is about 1800°C; what is this on the Fahrenheit scale?

4. (I) (*a*) 15° below zero on the Celsius scale is what Fahrenheit temperature? (*b*) 15° below zero on the Fahrenheit scale is what Celsius temperature?

5. (I) In a foreign country, a thermometer tells you that you have a fever of 40.0°C. What is this in Fahrenheit?

6. (I) In an alcohol-in-glass thermometer, the alcohol column has length 11.82 cm at 0.0°C and length 22.85 cm at 100.0°C. What is the temperature if the column has length (*a*) 16.70 cm, and (*b*) 20.50 cm?

7. (II) At what temperature will the Fahrenheit and Centigrade scales yield the same numerical value?

Section 17–4

8. (I) A concrete highway is built of slabs 12 m long (20°C). How wide should the expansion cracks between the slabs be (at 20°C) to prevent buckling if the range of temperature is −30°C to +50°C?

9. (I) Super Invar, an alloy of iron and nickel, is a strong material with a very low coefficient of thermal expansion $\left[0.2 \times 10^{-6}(C°)^{-1}\right]$. A 2.0-m-long-tabletop of this alloy is used for sensitive laser measurements where extremely high tolerances are required. How much will this alloy table expand along its length if the temperature increases 5.0 C°? Compare to tabletops made of steel and marble.

10. (I) The Eiffel Tower (Fig. 17–17) is built of wrought iron approximately 300 m tall. Estimate how much its height changes between July (average temperature of 25°C) and January (average temperature of 2°C). Ignore the angles of the iron beams and treat the tower as a vertical beam.

FIGURE 17–17 Problem 10: The Eiffel Tower in Paris.

11. (II) To make a secure fit, rivets that are larger than the rivet hole are often used and the rivet is cooled (usually in dry ice) before it is placed in the hole. A steel rivet 1.871 cm in diameter is to be placed in a hole 1.869 cm in diameter. To what temperature must the rivet be cooled if it is to fit in the hole at 20°C?

12. (II) A uniform rectangular plate of length l and width w has coefficient of linear expansion α. Show that, if we neglect very small quantities, the change in area of the plate due to a temperature change ΔT is $\Delta A = 2\alpha l w\, \Delta T$. See Fig. 17–18.

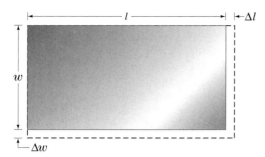

FIGURE 17–18 Rectangular plate is heated. Problem 12.

13. (II) An ordinary glass is filled to the brim with 350.0 mL of water at 100.0°C. If the temperature decreased to 20.0°C, how much water could be added to the glass?

14. (II) It is observed that 55.50 mL of water at 20°C completely fills a container to the brim. When the container and the water are heated to 60°C, 0.35 g of water is lost. (a) What is the coefficient of volume expansion of the container? (b) What is the most likely material of the container? Density of water at 60°C is 0.98324 g/mL.

15. (II) A quartz sphere is 8.75 cm in diameter. What will be its change in volume if it is heated from 30°C to 200°C?

16. (II) A brass plug is to be placed in a ring made of iron. At room temperature, the diameter of the plug is 8.753 cm and that of the inside of the ring is 8.743 cm. They must both be brought to what common temperature in order to fit?

17. (II) If a fluid is contained in a long narrow vessel so it can expand in essentially one direction only, show that the effective coefficient of linear expansion α is approximately equal to the coefficient of volume expansion β.

18. (II) (a) Show that the change in the density ρ of a substance, when the temperature changes by ΔT, is given by $\Delta\rho = -\beta\rho\,\Delta T$. (b) What is the fractional change in density of a lead sphere whose temperature decreases from 25°C to −40°C?

19. (III) The pendulum in a grandfather clock is made of brass and keeps perfect time at 17°C. How much time is gained or lost in a year if the clock is kept at 25°C? (Assume the frequency dependence on length for a simple pendulum applies.)

20. (III) (a) Determine a formula for the change in surface area of a uniform solid sphere of radius r if its coefficient of linear expansion is α (assumed constant) and its temperature is changed by ΔT. (b) What is the increase in area of a solid iron sphere of radius 60.0 cm if its temperature is raised from 20°C to 310°C?

21. (III) A 23.4-kg solid aluminum cylindrical wheel of radius 0.41 m is rotating about its axle in frictionless bearings with angular velocity $\omega = 32.8$ rad/s. If its temperature is now raised from 20.0°C to 75.0°C, what is the fractional change in ω?

* Section 17–5

*22. (I) At what temperature will the ultimate compressive strength of concrete be exceeded for the blocks discussed in Example 17–7?

*23. (I) An aluminum bar has the precisely desired length when at 15°C. How much stress is required to keep it at this length if the temperature increases to 35°C?

*24. (II) (a) A horizontal steel I-beam of cross-sectional area 0.041 m² is rigidly connected to two vertical steel girders. If the beam was installed when the temperature was 30°C, what stress is developed in the beam when the temperature drops to −30°C? (b) Is the ultimate strength of the steel exceeded? (c) What stress is developed if the beam is concrete and has a cross-sectional area of 0.13 m²? Will it fracture?

*25. (III) A barrel of diameter 134.122 cm at 20°C is to be enclosed by an iron band. The circular band has an inside diameter of 134.110 cm at 20°C. It is 7.4 cm wide and 0.65 cm thick. (a) To what temperature must the band be heated so that it will fit over the barrel? (b) What will be the tension in the band when it cools to 20°C?

Section 17–6

26. (I) What are the following temperatures on the Kelvin scale: (a) 86°C, (b) 78°F, (c) −100°C, (d) 5500°C?

27. (I) Absolute zero is what temperature on the Fahrenheit scale?

28. (II) Typical temperatures in the interior of the Earth and Sun are about 4000°C and 15×10^6 °C, respectively. (a) What are these temperatures in kelvins? (b) What percent error is made in each case if a person forgets to change °C to K?

Sections 17–7 and 17–8

29. (I) If 3.00 m³ of a gas initially at STP is placed under a pressure of 3.20 atm, the temperature of the gas rises to 38.0°C. What is the volume?

30. (I) In an internal combustion engine, air at atmospheric pressure and a temperature of about 20°C is compressed in the cylinder by a piston to $\frac{1}{9}$ of its original volume (compression ratio = 9.0). Estimate the temperature of the compressed air, assuming the pressure reaches 40 atm.

31. (II) Calculate the density of oxygen at STP using the ideal gas law.

32. (II) A storage tank contains 21.6 kg of nitrogen (N_2) at an absolute pressure of 3.65 atm. What will the pressure be if the nitrogen is replaced by an equal mass of CO_2?

33. (II) A storage tank at STP contains 18.5 kg of nitrogen (N_2). (a) What is the volume of the tank? (b) What is the pressure if an additional 15.0 kg of nitrogen is added without changing the temperature?

34. (II) If 18.75 mol of helium gas is at 10.0°C and a gauge pressure of 0.350 atm, calculate (a) the volume of the helium gas under these conditions, and (b) the temperature if the gas is compressed to precisely half the volume at a gauge pressure of 1.00 atm.

35. (II) What is the pressure inside a 35.0-L container holding 105.0 kg of argon gas at 20.0°C?

36. (II) A tank contains 30.0 kg of O_2 gas at a gauge pressure of 8.70 atm. If the oxygen is replaced by helium, how many kilograms of the latter will be needed to produce a gauge pressure of 7.00 atm?

37. (II) A hot-air balloon achieves its buoyant lift by heating the air inside the balloon which makes it less dense than the air outside. Suppose the volume of a balloon is 1800 m³ and the required lift is 2700 N (rough estimate of the weight of the equipment and passenger). Calculate the temperature of the air inside the balloon which will produce the required lift. Assume that the outside air temperature is 0° and that air is an ideal gas under these conditions. What factors limit the maximum altitude attainable by this method for a given load? (Neglect variables like wind.)

38. (II) A tire is filled with air at 15°C to a gauge pressure of 220 kPa. If the tire reaches a temperature of 38°C, what fraction of the original air must be removed if the original pressure of 220 kPa is to be maintained?

39. (II) If 61.5 L of oxygen at 18.0°C and an absolute pressure of 2.45 atm are compressed to 48.8 L and at the same time the temperature is raised to 50.0°C, what will the new pressure be?

40. (II) A child's helium-filled balloon escapes at sea level and 20.0°C. When it reaches an altitude of 3000 m, where the temperature is 5.0°C and the pressure only 0.70 atm, how will its volume compare to that at sea level?

41. (III) Compare the value for the density of water vapor at 100°C and 1 atm (Table 13–1) with the value predicted from the ideal gas law. Why would you expect a difference?

42. (III) An air bubble at the bottom of a lake 37.0 m deep has a volume of 1.00 cm³. If the temperature at the bottom is 5.5°C and at the top 21.0°C, what is the volume of the bubble just before it reaches the surface?

Section 17–9

43. (I) Calculate the number of molecules/m³ in an ideal gas at STP.

44. (I) How many moles of water are there in 1.000 L? How many molecules?

45. (I) Estimate the number of molecules you inhale in a 2.0-L breath.

46. (II) Estimate the number of (a) moles, and (b) molecules of water in all the Earth's oceans. Assume water covers 75 percent of the Earth to an average depth of 3 km.

47. (II) A cubic box of volume 5.1×10^{-2} m³ is filled with air at atmospheric pressure at 20°C. The box is closed and heated to 180°C. What is the net force on each side of the box?

48. (III) Estimate how many molecules of air are in each 2.0-L breath you inhale that were also in the last breath Galileo took. [Hint: Assume the atmosphere is about 10 km high and of constant density.]

* Section 17–10

* 49. (I) At the boiling point of sulfur (444.6°C) the pressure in a constant-volume gas thermometer is 187 torr. Estimate (a) the pressure at the triple point of water, (b) the temperature when the pressure in the thermometer is 112 torr.

* 50. (I) In a constant-volume gas thermometer, what is the limiting ratio of the pressure at the boiling point of water at 1 atm to that at the triple point? (Keep five significant figures.)

* 51. (II) Use Fig. 17–15 to determine the inaccuracy of a constant-volume gas thermometer using oxygen if it reads a pressure $P = 268$ torr at the boiling point of water at 1 atm. Express answer (a) in kelvins and (b) as a percentage.

* 52. (II) A constant-volume gas thermometer is being used to determine the temperature of the melting point of a substance. The pressure in the thermometer at this temperature is 218 torr; at the triple point of water, the pressure is 286 torr. Some gas is now released from the thermometer bulb so that the pressure at the triple point of water becomes 163 torr. At the temperature of the melting substance, the pressure is 128 torr. Estimate, as accurately as possible, the melting-point temperature of the substance.

General Problems

53. A precise steel tape measure has been calibrated at 20°C. At 34°C, (a) will it read high or low, and (b) what will be the percentage error?

54. A Pyrex measuring cup was calibrated at normal room temperature. How much error will be made in a recipe calling for 300 mL of cool water, if the water and the cup are hot, at 80°C, instead of at room temperature? Neglect the glass expansion.

55. The pressure in a helium gas cylinder is initially 35 atmospheres. After many balloons have been blown up, the pressure has decreased to 5 atm. What fraction of the original gas remains in the cylinder? Pressures are gauge pressures.

56. Write the ideal gas law in terms of the density of the gas.

57. Estimate the number of air molecules in a room of length 8.0 m, width 6.0 m, and height 4.2 m. Assume the temperature is 20°C. How many moles does that correspond to?

58. The lowest pressure attainable using the best available vacuum techniques is about 10^{-12} N/m². At such a pressure, how many molecules are there per cm³ at 0°C?

59. If a scuba diver fills his lungs to full capacity of 5.5 L when 10 m below the surface, to what volume would his lungs expand if he quickly rose to the surface? Is this advisable?

60. If a rod of original length L_1 has its temperature changed from T_1 to T_2, determine a formula for its new length L_2 in terms of T_1, T_2, and α. Assume (a) α = constant, (b) $\alpha = \alpha(T)$ is some function of temperature, and (c) $\alpha = \alpha_0 + bT$ where α_0 and b are constants.

61. A house has a volume of 770 m³. (a) What is the total mass of air inside the house at 20°C? (b) If the temperature drops to −10°C, what mass of air enters or leaves the house?

62. (a) Use the ideal gas law to show that, for an ideal gas at constant pressure, the coefficient of volume expansion is equal to $\beta = 1/T$ where T is the kelvin temperature. Compare to Table 17–1 for gases at T = 293 K. (b) Show that the bulk modulus (Section 12–5) for an ideal gas held at constant temperature is $B = P$ where P is the pressure.

63. From the known value of atmospheric pressure at the surface of the Earth, estimate the total number of air molecules in the Earth's atmosphere.

64. (a) The tube of a mercury thermometer has an inside diameter of 0.140 mm. The bulb has a volume of 0.315 cm³. How far will the thread of mercury move when the temperature changes from 11.5°C to 33.0°C? Take into account expansion of the Pyrex glass. (b) Determine a formula for the length of the mercury column in terms of relevant variables.

65. What is the average distance between oxygen molecules at STP?

66. An iron cube floats in a bowl of liquid mercury at 0°C. (a) If the temperature is raised to 25°C, will the cube float higher or lower in the mercury? (b) By what percent will the fraction of volume submerged change?

67. If a steel band were to fit snugly around the Earth's equator at 20°C, but then was heated to 35°C, how high above the Earth would the band be (assume equal everywhere)?

68. Estimate the percent change in density of iron when it is still a solid, but deep in the Earth where the temperature is 2000°C and it is under 5000 atm of pressure. Take into account both thermal expansion and changes due to increased outside pressure. Assume both the bulk modulus and the volume coefficient of expansion do not vary with temperature and are the same as at normal room temperature. The bulk modulus for iron is about 90×10^9 N/m².

69. A standard cylinder of oxygen used in a hospital has the following characteristics at room temperature (300 K): gauge pressure = 2000 psi (13,800 kPa), volume = 16 liters (0.016 m³). How long will the cylinder last if the flow rate, measured at atmospheric pressure, is constant at 2.4 liters/min?

70. A helium party balloon, assumed to be a perfect sphere, has a radius of 18.0 cm. At room temperature (20°C), its internal pressure is 1.05 atm. Find the number of moles of helium in the balloon and the mass of helium needed to inflate the balloon to these values.

71. The density of gasoline at 0°C is 0.68×10^3 kg/m³. What is the density on a hot day, when the temperature is 32°C?

72. A brass lid screws tightly onto a glass jar at 20°C. To help open the jar, it can be placed into a bath of hot water. After this treatment, the temperatures of the lid and the jar are both 60°C. The inside diameter of the lid is 8.0 cm. Find the size of the gap (difference in radius) that develops by this procedure.

73. The first length standard, adopted in the 18th century, was a platinum bar with two very fine marks separated by what was defined to be exactly one meter. If this standard bar was to be accurate to within ±1.0 μm, how carefully would the trustees have needed to control the temperature? The coefficient of linear expansion is 9×10^{-6} C°⁻¹.

74. Reinforcing rods in concrete are made of steel, which has nearly the same coefficient of thermal expansion as concrete. What would happen if brass were used instead? Imagine a 2.5-cm diameter brass rod, embedded in a concrete matrix. If the temperature rises by 20 C°, the brass will be under compression and the concrete under tension. Will the concrete stay in one piece? (You will need data from Tables 12–1 and 12–2. Assume the same magnitude of stress in both materials.)

75. A helium balloon has volume V_0 and temperature T_0 at sea level where the pressure is P_0 and the air density is ρ_0. The balloon is allowed to float up in the air to altitude y where the temperature is T_1. (a) Show that the volume occupied by the balloon is then $V = V_0(T_1/T_0)e^{+cy}$ where $c = \rho_0 g/P_0 = 1.25 \times 10^{-4}$ m⁻¹. (b) Show that the buoyant force does not depend on altitude y. Assume that the skin of the balloon maintains the helium pressure at a constant factor of 1.05 times greater than the outside pressure. [Hint: Assume that the pressure change with altitude is $P = P_0 e^{-cy}$, as in Example 13–4, Chapter 13.]

76. A scuba tank when fully charged has a pressure of 200 atmospheres at 20°C. The volume of the tank is 11.3 liters. (a) What would the volume of the air be at 1.00 atmosphere and at the same temperature? (b) Before entering the water, a person consumes 2.0 liters of air in each breath, and breathes 12 times a minute. At this rate, how long would the tank last? (c) At a depth of 20.0 m of sea water and temperature of 10°C, how long would the same tank last assuming the breathing rate does not change?

77. A temperature controller, designed to work in a steam environment, involves a bimetallic strip constructed of brass and steel, connected at their ends by rivets. Each of the metals is 2.0 mm thick. At 20°C, the strip is 10.0 cm long and straight. Find the radius of curvature of the assembly at 100°C. See Fig. 17–19.

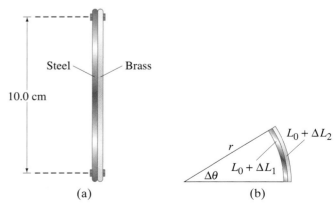

FIGURE 17–19 Problem 77.

Idaho's Salmon River, famous for rafting and fishing, in December. We see the three states of matter in the form of water as a liquid, as a solid (snow and ice), and as a gas (steam or fog). In this chapter we will examine the microscopic theory of matter as atoms or molecules that are always in motion, which we call kinetic theory. We will see that the temperature of a gas is directly related to the average kinetic energy of its molecules. We will consider ideal gases, but we'll also look at real gases and how they change phase including evaporation, vapor pressure, and humidity.

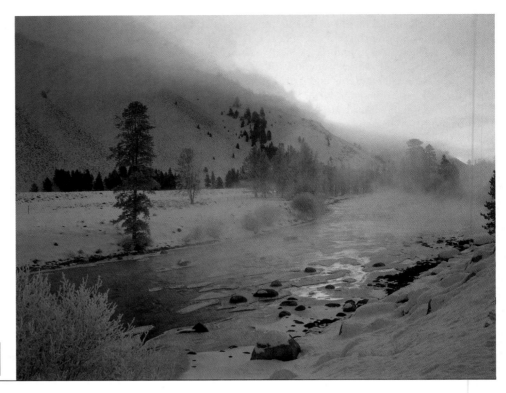

CHAPTER 18

Kinetic Theory of Gases

The analysis of matter in terms of atoms in continuous random motion is called the **kinetic theory**. We now investigate the properties of a gas from the point of view of kinetic theory, which is based on the laws of classical mechanics. But to apply Newton's laws to each one of the vast number of molecules in a gas $(>10^{25}/m^3$ at STP$)$ is far beyond the capability of any present computer. Instead we take a statistical approach and determine averages of certain quantities, and these averages correspond to macroscopic variables. We will, of course, demand that our microscopic description correspond to the macroscopic properties of gases; otherwise our theory would be of little value. Most importantly, we will arrive at an important relation between the average kinetic energy of molecules in a gas and the absolute temperature.

18–1 The Ideal Gas Law and the Molecular Interpretation of Temperature

We make the following assumptions about the molecules in a gas. These assumptions reflect a simple view of a gas, but nonetheless the results they predict correspond well to the essential features of real gases that are at low pressures and far from the liquefaction point. Under these conditions real gases follow the ideal gas law quite closely, and indeed the gas we now describe is referred to as an

ideal gas. The assumptions, which represent the basic postulates of the kinetic theory, are:

1. There are a large number of molecules, N, each of mass m, moving in random directions with a variety of speeds. This assumption is in accord with our observation that a gas fills its container and, in the case of air on the Earth, is kept from escaping only by the force of gravity.

2. The molecules are, on the average, far apart from one another. That is, their average separation is much greater than the diameter of each molecule.

3. The molecules are assumed to obey the laws of classical mechanics, and are assumed to interact with one another only when they collide. Although molecules exert weak attractive forces on each other between collisions, the potential energy associated with these forces is small compared to the kinetic energy, and we ignore it for now.

4. Collisions with another molecule or the wall of the vessel are assumed to be perfectly elastic, like the collisions of perfectly elastic billiard balls (Chapter 9). We assume the collisions are of very short duration compared to the time between collisions. Then we can ignore the potential energy associated with collisions in comparison to the kinetic energy between collisions.

Postulates of kinetic theory

We can see immediately how this kinetic view of a gas can explain Boyle's law (Section 17–6). The pressure exerted on a wall of a container of gas is due to the constant bombardment of molecules. If the volume is reduced by (say) half, the molecules are closer together and twice as many will be striking a given area of the wall per second. Hence we expect the pressure to be twice as great, in agreement with Boyle's law.

Boyle's law explained

Now let us calculate quantitatively the pressure in a gas based on kinetic theory. For purposes of argument, we imagine that the molecules are contained in a rectangular vessel whose ends have area A and whose length is l, as shown in Fig. 18–1a. The pressure exerted by the gas on the walls of its container is, according to our model, due to the collisions of the molecules with the walls. Let us focus our attention on the wall, of area A, at the left end of the container and examine what happens when one molecule strikes this wall, as shown in Fig. 18–1b. This molecule exerts a force on the wall, and the wall exerts an equal and opposite force back on the molecule. The magnitude of this force, according to Newton's second law, is equal to the molecule's rate of change of momentum, $F = dp/dt$. Assuming the collision is elastic, only the x component of the molecule's momentum changes, and it changes from $-mv_x$ (it is moving in the negative x direction) to $+mv_x$. Thus the change in momentum, $\Delta(mv)$, which is the final momentum minus the initial momentum, is

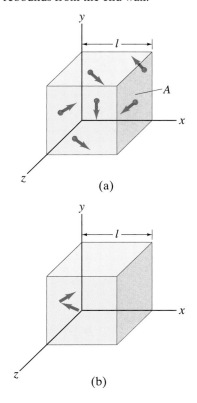

FIGURE 18–1 (a) Molecules of a gas moving about in a cubical container. (b) Arrows indicate the momentum of one molecule as it rebounds from the end wall.

$$\Delta(mv) = mv_x - (-mv_x) = 2mv_x$$

for one collision. This molecule will make many collisions with the wall, each separated by a time Δt, which is the time it takes the molecule to travel across the box and back again, a distance (x component) equal to $2l$. Thus $2l = v_x \Delta t$ or

$$\Delta t = 2l/v_x.$$

The time Δt between collisions is very small, so the number of collisions per second is very large. Thus the average force—averaged over many collisions—will be equal to the force exerted during one collision divided by the time between collisions (Newton's second law):

$$F = \frac{\Delta(mv)}{\Delta t} = \frac{2mv_x}{2l/v_x} = \frac{mv_x^2}{l}. \qquad \text{[due to one molecule]}$$

During its passage back and forth across the container, the molecule may collide with the tops and sides of the container, but this does not alter its x component of momentum and thus does not alter our result. It may also collide with other molecules, which may change its v_x. However, any loss (or gain) of momentum is acquired by the other molecule, and because we will eventually sum over all the molecules, this effect will be included. So our result above is not altered.

Of course, the actual force due to one molecule is intermittent, but because a huge number of molecules are striking the wall per second, the force is, on average, nearly constant. To calculate the force due to *all* the molecules in the box, we have to add the contributions of each. Thus the net force on the wall is

$$F = \frac{m}{l}\left(v_{x1}^2 + v_{x2}^2 + \cdots + v_{xN}^2\right),$$

where v_{x1} means v_x for molecule number 1 (we arbitrarily assign each molecule a number) and the sum extends over the total of N molecules. Now the average value of the square of the x component of velocity is

$$\overline{v_x^2} = \frac{v_{x1}^2 + v_{x2}^2 + \cdots + v_{xN}^2}{N}. \qquad \textbf{(18–1)}$$

Thus we can write the force as

$$F = \frac{m}{l} N \overline{v_x^2}.$$

We know that the square of any vector is equal to the sum of the squares of its components (theorem of Pythagoras). Thus $v^2 = v_x^2 + v_y^2 + v_z^2$ for any velocity v. Taking averages, we obtain

$$\overline{v^2} = \overline{v_x^2} + \overline{v_y^2} + \overline{v_z^2}.$$

Since the velocities of the molecules in our gas are assumed to be random, there is no preference to one direction or another. Hence

$$\overline{v_x^2} = \overline{v_y^2} = \overline{v_z^2}.$$

Combining this relation with the one just above, we get

$$\overline{v^2} = 3\overline{v_x^2}.$$

We substitute this into the equation for net force F:

$$F = \frac{m}{l} N \frac{\overline{v^2}}{3}.$$

The pressure on the wall is then

$$P = \frac{F}{A} = \frac{1}{3} \frac{Nm\overline{v^2}}{Al}$$

Pressure in a gas

or

$$P = \frac{1}{3} \frac{Nm\overline{v^2}}{V}, \qquad \textbf{(18–2)}$$

where $V = lA$ is the volume of the container. This is the result we were seeking, the pressure in a gas expressed in terms of molecular properties.

Equation 18–2 can be rewritten in a clearer form by multiplying both sides by V and slightly rearranging the right side:

$$PV = \tfrac{2}{3} N\left(\tfrac{1}{2} m\overline{v^2}\right). \qquad \textbf{(18–3)}$$

The quantity $\tfrac{1}{2} m\overline{v^2}$ is the average kinetic energy $\left(\overline{K}\right)$ of the molecules in the gas. If we compare Eq. 18–3 with Eq. 17–4, the ideal gas law $PV = NkT$, we see that the two agree if

$$\tfrac{2}{3}\left(\tfrac{1}{2}m\overline{v^2}\right) = kT,$$

or

$$\overline{K} = \tfrac{1}{2}m\overline{v^2} = \tfrac{3}{2}kT. \qquad \text{[Ideal gas]} \quad \textbf{(18–4)}$$

TEMPERATURE RELATED TO AVERAGE KINETIC ENERGY OF MOLECULES

This equation tells us that

> **the average translational kinetic energy of molecules in an ideal gas is directly proportional to the absolute temperature.**

The higher the temperature, according to kinetic theory, the faster the molecules are moving on the average. This relation is one of the triumphs of the kinetic theory.

EXAMPLE 18–1 **Molecular kinetic energy.** What is the average translational kinetic energy of molecules in an ideal gas at 37°C?

SOLUTION We use Eq. 18–4 and change 37°C to 310 K:

$$\overline{K} = \tfrac{3}{2}kT$$
$$= \tfrac{3}{2}(1.38 \times 10^{-23} \text{ J/K})(310 \text{ K}) = 6.42 \times 10^{-21} \text{ J}.$$

Note that a mole of molecules would have a total of translational kinetic energy equal to $(6.42 \times 10^{-21} \text{ J})(6.02 \times 10^{23}) = 3900 \text{ J}$, which equals the kinetic energy of a 1-kg stone traveling faster that 85 m/s.

Equation 18–4 holds not only for gases, but also applies reasonably accurately to liquids and solids. Thus the result of Example 18–1 would apply to molecules within living cells at body temperature (37°C).

We can use Eq. 18–4 to calculate how fast molecules are moving on the average. Notice that the average in Eqs. 18–1 through 18–4 is over the *square* of the velocity. The square root of $\overline{v^2}$ is called the **root-mean-square** velocity, v_{rms} (since we are taking the square *root* of the *mean* of the *square* of the velocity):

Root-mean-square (rms) velocity

$$v_{\text{rms}} = \sqrt{\overline{v^2}} = \sqrt{\frac{3kT}{m}}. \qquad \textbf{(18–5)}$$

rms speed of molecules

The **mean speed**, $\overline{v}$, is the average of the magnitudes of the speeds themselves; $\overline{v}$ is generally not equal to v_{rms}. To see the difference between the mean speed and the rms speed, consider the following Example.

Mean speed

EXAMPLE 18–2 **Mean speed and rms speed.** Eight particles have the following speeds, given in m/s: 1.0, 6.0, 4.0, 2.0, 6.0, 3.0, 2.0, 5.0. Calculate (*a*) the mean speed and (*b*) the rms speed.

SOLUTION (*a*) The mean speed is

$$\overline{v} = \frac{1.0 + 6.0 + 4.0 + 2.0 + 6.0 + 3.0 + 2.0 + 5.0}{8} = 3.6 \text{ m/s}.$$

(*b*) The rms speed is (Eq. 18–1):

$$v_{\text{rms}} = \sqrt{\frac{(1.0)^2 + (6.0)^2 + (4.0)^2 + (2.0)^2 + (6.0)^2 + (3.0)^2 + (2.0)^2 + (5.0)^2}{8}} \text{ m/s}$$
$$= 4.0 \text{ m/s}.$$

We see in this Example that $\overline{v}$ and v_{rms}, are not necessarily equal. In fact, for an ideal gas they differ by about 8 percent. We will see in the next Section how to calculate $\overline{v}$ for an ideal gas. We already have the tool to calculate v_{rms} (Eq. 18–5).

EXAMPLE 18–3 **Speeds of air molecules.** What is the rms speed of air molecules $(O_2$ and $N_2)$ at room temperature (20°C)?

SOLUTION We must apply Eq. 18–5 to oxygen and nitrogen separately since they have different masses. The masses of one molecule of O_2 (molecular mass $= 32$ u) and N_2 (molecular mass $= 28$ u) are $(1\,\text{u} = 1.66 \times 10^{-27}\,\text{kg})$

$$m(O_2) = (32)(1.66 \times 10^{-27}\,\text{kg}) = 5.3 \times 10^{-26}\,\text{kg}$$
$$m(N_2) = (28)(1.66 \times 10^{-27}\,\text{kg}) = 4.7 \times 10^{-26}\,\text{kg}.$$

Thus, for oxygen

$$v_{\text{rms}} = \sqrt{\frac{3kT}{m}} = \sqrt{\frac{(3)(1.38 \times 10^{-23}\,\text{J/K})(293\,\text{K})}{(5.3 \times 10^{-26}\,\text{kg})}}$$
$$= 480\,\text{m/s},$$

and for nitrogen the result is $v_{\text{rms}} = 510\,\text{m/s}$. These speeds are more than 1500 km/h or 1000 mi/h.

Equation 18–4, $\overline{K} = \frac{3}{2}kT$, implies that as the temperature approaches absolute zero the kinetic energy of molecules approaches zero. Modern quantum theory, however, tells us this is not quite so. Instead, as absolute zero is approached, the kinetic energy approaches a very small nonzero minimum value. Even though all real gases become liquid or solid near 0 K, molecular motion does not cease, even at absolute zero.

18–2 Distribution of Molecular Speeds

The Maxwell Distribution

The molecules in a gas are assumed to be in random motion, which means that many molecules have speeds less than the average speed and others have speeds greater than the average. In 1859, James Clerk Maxwell (1831–1879) worked out a formula for the most probable distribution of speeds in a gas containing N molecules. We will not give a derivation here but merely quote his result:

Maxwell distribution

$$f(v) = 4\pi N\left(\frac{m}{2\pi kT}\right)^{\frac{3}{2}} v^2 e^{-\frac{1}{2}\frac{mv^2}{kT}}. \qquad (18–6)$$

$f(v)$ is called the **Maxwell distribution of speeds**, and is plotted in Fig. 18–2. The quantity $f(v)\,dv$ represents the number of molecules that have speed between v and $v + dv$. Notice that $f(v)$ does not give the number of molecules with speed v; $f(v)$ must be multiplied by dv to give the number of molecules (clearly the number of molecules must depend on the "width" or "range" of velocities, dv). In the formula for $f(v)$, m is the mass of a single molecule, T is the absolute temperature, and k is Boltzmann's constant. Since N is the total number of molecules in the gas, when we sum over all the molecules in the gas we must get N; thus we must have

$$\int_0^\infty f(v)\,dv = N.$$

(Problem 15 is an exercise to show that this is true.)

Experiments to determine the distribution of speeds in real gases, starting in the 1920s, confirmed with considerable accuracy the Maxwell distribution (for gases at not too high a pressure) and the direct proportion between average kinetic energy and absolute temperature, Eq. 18–4.

FIGURE 18–2 Distribution of speeds of molecules in an ideal gas. Note that $\overline{v}$ and v_{rms} are not at the peak of the curve (that speed is called the "most probable speed," v_{p}). This is because the curve is skewed to the right: it is not symmetrical.

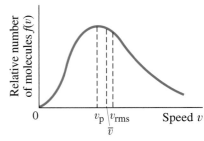

The Maxwell distribution for a given gas depends only on the absolute temperature. Figure 18–3 shows the distributions for two different temperatures. Just as v_{rms} increases with temperature, so the whole distribution curve shifts to the right at higher temperatures.

Figure 18–3 illustrates how kinetic theory can be used to explain why many chemical reactions, including those in biological cells, take place more rapidly as the temperature increases. Most chemical reactions take place in a liquid solution, and the molecules in a liquid have a distribution of speeds close to the Maxwell distribution. Two molecules may chemically react only if their kinetic energy is great enough so that when they collide, they penetrate into each other somewhat. The minimum energy required is called the *activation energy*, E_A, and it has a specific value for each chemical reaction. The molecular speed corresponding to a kinetic energy of E_A for a particular reaction is indicated in Fig. 18–3. The relative number of molecules with energy greater than this value is given by the area under the curve to the right of $v(E_A)$. In Fig. 18–3, the respective areas for two different temperatures are indicated by the two different shadings in the figure. It is clear that the number of molecules that have kinetic energies in excess of E_A increases greatly for only a small increase in temperature. The rate at which a chemical reaction occurs is proportional to the number of molecules with energy greater than E_A, and thus we see why reaction rates increase rapidly with increased temperature.

FIGURE 18–3 Distribution of molecular speeds for two different temperatures.

➡ **PHYSICS APPLIED**

How chemical reactions depend on temperature

*Calculations Using the Maxwell Distribution

Let us see how the Maxwell distribution can be used to obtain some interesting results.

EXAMPLE 18–4 **Determining $\bar{v}$ and v_p.** Determine formulas for (*a*) the average speed, $\bar{v}$, and (*b*) the most probable speed, v_p, of molecules in an ideal gas at temperature T.

SOLUTION (*a*) The average value of any quantity is found by multiplying each possible value of the quantity (say, speed) by the number of molecules that have that value, and then summing all these numbers and dividing by N (the total number). We are given a continuous distribution of speeds (Eq. 18–6), so the sum becomes an integral over the product of v and the number $f(v)\,dv$ that have speed v:

$$\bar{v} = \frac{\int_0^\infty v f(v)\,dv}{N} = 4\pi\left(\frac{m}{2\pi kT}\right)^{\frac{3}{2}} \int_0^\infty v^3 e^{-\frac{1}{2}\frac{mv^2}{kT}}\,dv.$$

We can look up the definite integral in the tables, or integrate by parts, and obtain

$$\bar{v} = 4\pi\left(\frac{m}{2\pi kT}\right)^{\frac{3}{2}}\left(\frac{2k^2 T^2}{m^2}\right) = \sqrt{\frac{8}{\pi}\frac{kT}{m}} \approx 1.60\sqrt{\frac{kT}{m}}.$$

Average speed, $\bar{v}$

(*b*) The *most probable speed* is that speed which occurs more than any others, and thus is that speed where $f(v)$ has its maximum value. Since $df(v)/dv = 0$ at the maximum of the curve, we have

$$\frac{df(v)}{dv} = 4\pi\left(\frac{m}{2\pi kT}\right)^{\frac{3}{2}}\left(2ve^{-\frac{mv^2}{2kT}} - \frac{2mv^3}{2kT}e^{-\frac{mv^2}{2kT}}\right) = 0.$$

Solving for v, we get

$$v_p = \sqrt{\frac{2kT}{m}} \approx 1.41\sqrt{\frac{kT}{m}}.$$

Most probable speed, v_p

(Another solution is $v = 0$, but this corresponds to a minimum, not a maximum.)

In summary,

Most probable speed, v_p

$$v_p = \sqrt{2\frac{kT}{m}} \approx 1.41\sqrt{\frac{kT}{m}} \qquad \textbf{(18–7a)}$$

Mean speed, $\bar{v}$

$$\bar{v} = \sqrt{\frac{8}{\pi}\frac{kT}{m}} \approx 1.60\sqrt{\frac{kT}{m}} \qquad \textbf{(18–7b)}$$

and from Eq. 18–5

rms speed, v_{rms}

$$v_{rms} = \sqrt{3\frac{kT}{m}} \approx 1.73\sqrt{\frac{kT}{m}}.$$

These are all indicated in Fig. 18–2. From Eq. 18–6 and Fig. 18–2, it is clear that the speeds of molecules in a gas vary from zero up to many times the average speed, but as can be seen from the graph, most molecules have speeds that are not far from the average. Less than 1 percent of the molecules exceed four times v_{rms}.

EXAMPLE 18–5 ESTIMATE Using $f(v)$. Suppose a sample of helium gas at temperature $T = 300\,\text{K}$ contains $N = 10^6$ atoms, each of mass $m_{He} = 6.65 \times 10^{-27}\,\text{kg}$. (*a*) What is the most probable speed, v_p, of a helium atom in the sample? (*b*) Estimate how many atoms in the sample have speeds in the range between v_p and $v_p + 40\,\text{m/s}$. (*c*) Now let $v = 10\,v_p$, and calculate the number of molecules with speeds between v and $v + 40\,\text{m/s}$.

SOLUTION (*a*) The most probable speed v_p is (Eq. 18–7a):

$$v_p = \sqrt{2\frac{kT}{m}} = \sqrt{\frac{(2)(1.38 \times 10^{-23}\,\text{J/K})(300\,\text{K})}{(6.65 \times 10^{-27}\,\text{kg})}} = 1.12 \times 10^3\,\text{m/s}.$$

(*b*) The number of molecules having a speed between v and $v + dv$ is equal to $f(v)\,dv$, as we discussed just after Eq. 18–6. For our estimate over a small finite range, $\Delta v = 40\,\text{m/s}$, we replace the differential dv by Δv and write

$$\Delta N = f(v)\,\Delta v.$$

As shown in Fig. 18–4, ΔN is equal to the area under the $f(v)$ curve between v and $v + \Delta v$. We approximate this area by a rectangle $f(v)$ high by Δv wide. A better estimate is obtained using the midpoint of the range between $v = 1120\,\text{m/s}$ and $v + \Delta v = 1160\,\text{m/s}$, namely 1140 m/s, although it doesn't make a great deal of difference here. Thus

$$\Delta N = f(v)\Delta v = 4\pi N\left(\frac{m}{2\pi kT}\right)^{\frac{3}{2}} v^2 e^{-\frac{mv^2}{2kT}}\,\Delta v$$

$$= (4\pi)(10^6)\left(\frac{6.65 \times 10^{-27}\,\text{kg}}{2\pi(1.38 \times 10^{-23}\,\text{J/K})(300\,\text{K})}\right)^{\frac{3}{2}}$$

$$\times (1.14 \times 10^3\,\text{m/s})^2 e^{-\frac{(6.65\times10^{-27}\,\text{kg})(1.14\times10^3\,\text{m/s})^2}{2(1.38\times10^{-23}\,\text{J/K})(300\,\text{K})}}\,(40\,\text{m/s})$$

$$\approx 30{,}000$$

or about 3% of all molecules.
(*c*) If we now replace $v = v_p$ by $v = 10\,v_p = 1.12 \times 10^4\,\text{m/s}$, and repeat the calculation with all else the same, we obtain $\Delta N = f(v)\Delta v = 4 \times 10^{-39}$, which means essentially that *no* molecules have speeds this high.

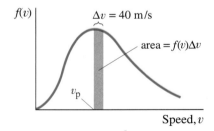

FIGURE 18–4 Calculating the number of molecules ΔN having speeds between v_p and $v_p + 40\,\text{m/s}$. $\Delta N = f(v)\,\Delta v$ is equal to the area shown shaded.

472 CHAPTER 18 Kinetic Theory of Gases

18–3 | Real Gases and Changes of Phase

The ideal gas law

$$PV = NkT$$

is an accurate description of the behavior of a real gas as long as the pressure is not too high and as long as the temperature is far from the liquefaction point. But what happens when these two criteria are not satisfied? First we discuss real gas behavior, and then we will look at how kinetic theory can help us understand this behavior.

Let us look at a graph of pressure plotted against volume for a given amount of gas. On such a "*PV* diagram," Fig. 18–5, each point represents an equilibrium state of the given substance. The various curves (labeled A, B, C, and D) show how the pressure varies as the volume is changed at constant temperature for several different values of the temperature. The dashed curve A′ represents the behavior of a gas as predicted by the ideal gas law; that is, PV = constant. The solid curve A represents the behavior of a real gas at the same temperature. Notice that at high pressure, the volume of a real gas is less than that predicted by the ideal gas law. The curves B and C in Fig. 18–5 represent the gas at successively lower temperatures, and we see that the behavior deviates even more from the curves predicted by the ideal gas law (for example, B′), and the deviation is greater the closer the gas is to liquefying.

To explain this, we note that at higher pressure we expect the molecules to be closer together. And, particularly at lower temperatures, the potential energy associated with the attractive forces between the molecules (which we ignored before) is no longer negligible compared to the now reduced kinetic energy of the molecules. These attractive forces tend to pull the molecules closer together so that at a given pressure, the volume is less than expected from the ideal gas law. At still lower temperatures, these forces cause liquefaction, and the molecules become very close together.

Curve D represents the situation when liquefaction occurs. At low pressure on curve D (on the right in Fig. 18–5), the substance is a gas and occupies a large volume. As the pressure is increased, the volume decreases until point b is reached. Beyond b, the volume decreases with no change in pressure; the substance is gradually changing from the gas to the liquid phase. At point a, all of the substance has changed to liquid. Further increase in pressure reduces the volume only slightly—liquids are nearly incompressible—so the curve is very steep as shown. The area which is colored yellow in Fig. 18–5, within the orange dashed line, represents the region where the gas and liquid phases exist together in equilibrium.

Curve C in Fig. 18–5 represents the behavior of the substance at its **critical temperature**; and the point c (the one point where this curve is horizontal) is called the **critical point**. At temperatures less than the critical temperature (and this is the definition of the term), a gas will change to the liquid phase if sufficient pressure is applied. Above the critical temperature, no amount of pressure can cause a gas to change phase and become a liquid. What happens instead is that the gas becomes denser and denser as the pressure is increased and gradually it acquires properties resembling a liquid, but no liquid surface forms. The critical temperatures for various gases are given in Table 18–1. Scientists tried for many years to liquefy oxygen without success. Only after the discovery of the behavior of substances associated with the critical point was it realized that oxygen can be liquefied only if first cooled below its critical temperature of −118°C.

Often a distinction is made between the terms "gas" and "vapor": a substance below its critical temperature in the gaseous state is called a **vapor**; when above the critical temperature, it is called a **gas**. This is indicated in Fig. 18–5.

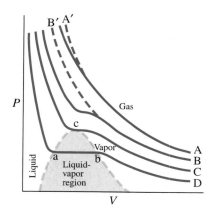

FIGURE 18–5 *PV* diagram for a real substance. Curves A, B, C, and D represent the same gas at different temperatures.

PV diagram

TABLE 18–1
Critical Temperatures and Pressures

Substance	Critical Temperature °C	K	Critical Pressure (atm)
Water	374	647	218
CO_2	31	304	72.8
Oxygen	−118	155	50
Nitrogen	−147	126	33.5
Hydrogen	−239.9	33.3	12.8
Helium	−267.9	5.3	2.3

Vapor vs. gas

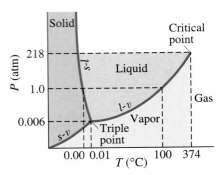

FIGURE 18–6 Phase diagram for water (note that the scales are not linear).

FIGURE 18–7 Phase diagram for carbon dioxide (CO_2).

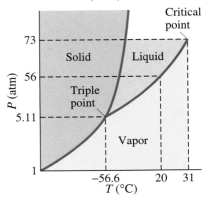

The behavior of a substance can be diagrammed not only on a *PV* diagram but also on a *PT* diagram. A *PT* diagram, often called a **phase diagram**, is particularly convenient for comparing the different phases of a substance. Figure 18–6 is the phase diagram for water. The curve labeled *l-v* represents those points where the liquid and vapor phases are in equilibrium—it is thus a graph of the boiling point versus pressure. Note that the curve correctly shows that at a pressure of 1 atm the boiling point of water is 100°C and that the boiling point is lowered for a decreased pressure. The curve *s-l* represents points where solid and liquid exist in equilibrium, and thus is a graph of the freezing point versus pressure. At 1 atm, the freezing point of water is, of course, 0°C, as shown. Notice also in Fig. 18–6 that at a pressure of 1 atm, the substance is in the liquid phase if the temperature is between 0°C and 100°C, but is in the solid or vapor phase if the temperature is below 0°C or above 100°C. The curve labeled *s-v* is the *sublimation point* versus pressure curve. **Sublimation** refers to the process whereby at low pressures a solid changes directly into the vapor phase without passing through the liquid phase. For water, this occurs at pressures less than 0.0060 atm. Carbon dioxide (CO_2), which in the solid phase is called dry ice, sublimates even at atmospheric pressure.

The intersection of the three curves (in Fig. 18–6) is the **triple point**, and it is only at this point that the three phases can exist together in equilibrium. Because the triple point corresponds to a unique value of temperature and pressure $(273.16\,K, 6.03 \times 10^{-3}\,atm$, for water), it is precisely reproducible and is often used as a point of reference (Section 17–10).

Notice that the *s-l* curve for water slopes upward to the left. This is true only of substances that *expand* upon freezing; for at a higher pressure, a lower temperature is needed to cause the liquid to freeze. More commonly, substances contract upon freezing and the *s-l* curve slopes upward to the right, as shown for CO_2 in Fig. 18–7.

The phase transitions we have been discussing are the common ones. Some substances, however, can exist in several forms in the solid phase. A transition from one of these phases to another occurs at a particular temperature and pressure, just like ordinary phase changes. For example, ice has been observed in at least eight different modifications at very high pressure. Ordinary helium has two distinct liquid phases, called helium I and II. They exist only at temperatures within a few degrees of absolute zero. Helium II exhibits very unusual properties referred to as **superfluidity**. It has essentially zero viscosity and exhibits strange properties such as actually climbing up the sides of an open container.

18–4 Vapor Pressure and Humidity

Evaporation

If a glass of water is left out overnight, the water level will have dropped by morning. We say the water has evaporated, meaning that some of the water has changed to the vapor or gas phase.

This process of **evaporation** can be explained on the basis of kinetic theory. The molecules in a liquid move past one another with a variety of speeds that follow, approximately, the Maxwell distribution. There are strong attractive forces between these molecules, which is what keeps them close together in the liquid phase. A molecule in the upper regions of the liquid may, because of its speed, leave the liquid momentarily. But just as a rock thrown into the air returns to the Earth, so the attractive forces of the other molecules can pull the vagabond molecule back to the liquid surface—that is, if its velocity is not too large. A molecule with a high enough velocity, however, will escape from the liquid entirely, like a rocket escaping the Earth, and become part of the gas phase. Only those molecules that have kinetic energy above a particular value can escape to the gas phase. We have already seen that kinetic theory predicts that the relative number of molecules with kinetic energy above a particular value (such as E_A in Fig. 18–3) increases with temperature. This is in accord with the well-known observation that the evaporation rate is greater at higher temperatures.

Because it is the fastest molecules that escape from the surface, the average speed of those remaining is less. When the average speed is less, the absolute temperature is less. Thus kinetic theory predicts that *evaporation is a cooling process.* You have no doubt noticed this effect when you stepped out of a warm shower and felt cold as the water on your body began to evaporate; and after working up a sweat on a hot day, even a slight breeze makes you feel cool through evaporation.

Vapor Pressure

Air normally contains water vapor (water in the gas phase) and it comes mainly from evaporation. To look at this process in a little more detail, consider a closed container that is partially filled with water (it could just as well be any other liquid) and from which the air has been removed (Fig. 18–8). The fastest moving molecules quickly evaporate into the space above. As they move about, some of these molecules strike the liquid surface and again become part of the liquid phase: this is called **condensation.** The number of molecules in the vapor increases for a time, until a point is reached where the number returning to the liquid equals the number leaving in the same time interval. Equilibrium then exists, and the space is said to be *saturated.* The pressure of the vapor when it is saturated is called the **saturated vapor pressure** (or sometimes simply the vapor pressure).

The saturated vapor pressure does not depend on the volume of the container. If the volume above the liquid were reduced suddenly, the density of molecules in the vapor phase would be increased temporarily. More molecules would then be striking the liquid surface per second. There would be a net flow of molecules back to the liquid phase until equilibrium was again reached, and this would occur at the same value of the saturated vapor pressure, as long as the temperature had not changed.

The saturated vapor pressure of any substance depends on the temperature. At higher temperatures, more molecules have sufficient kinetic energy to break from the liquid surface into the vapor phase. Hence equilibrium will be reached at a higher pressure. The saturated vapor pressure of water at various temperatures is given in Table 18–2. Notice that even solids—for example, ice—have a measurable saturated vapor pressure.

In everyday situations, evaporation from a liquid takes place into the air above it rather than into a vacuum. This does not materially alter the discussion above relating to Fig. 18–8. Equilibrium will still be reached when there are sufficient molecules in the gas phase that the number reentering the liquid equals the number leaving. The concentration of particular molecules (such as water) in the gas phase is not affected by the presence of air, although collisions with air molecules may lengthen the time needed to reach equilibrium. Thus equilibrium occurs at the same value of the saturated vapor pressure as if air weren't there.

Of course, if the container is large or is not closed, all the liquid may evaporate before saturation is reached. And if the container is not sealed—as, for example, a room in your house—it is not likely that the air will become saturated with water vapor (unless it is raining outside).

Boiling

The saturated vapor pressure of a liquid increases with temperature. When the temperature is raised to the point where the saturated vapor pressure at that temperature equals the external pressure, **boiling** occurs (Fig. 18–9). As the boiling point is approached, tiny bubbles tend to form in the liquid, which indicate a change from the liquid to the gas phase. However, if the vapor pressure inside the bubbles is less than the external pressure, the bubbles immediately are crushed. As the temperature is increased, the saturated vapor pressure inside a bubble eventually becomes equal to or exceeds the external air pressure. The bubble will then not collapse but will increase in size and rise to the surface. Boiling has then begun. *A liquid boils when its saturated vapor pressure equals the external pressure.* This occurs for water under 1 atm (760 torr) of pressure at 100°C, as can be seen from Table 18–2.

FIGURE 18–8 Vapor appears above a liquid in a closed container.

TABLE 18–2 Saturated Vapor Pressure of Water

Temp-erature (°C)	Saturated Vapor Pressure	
	torr (= mm-Hg)	Pa (= N/m²)
−50	0.030	4.0
−10	1.95	2.60×10^2
0	4.58	6.11×10^2
5	6.54	8.72×10^2
10	9.21	1.23×10^3
15	12.8	1.71×10^3
20	17.5	2.33×10^3
25	23.8	3.17×10^3
30	31.8	4.24×10^3
40	55.3	7.37×10^3
50	92.5	1.23×10^4
60	149	1.99×10^4
70	234	3.12×10^4
80	355	4.73×10^4
90	526	7.01×10^4
100	760	1.01×10^5
120	1489	1.99×10^5
150	3570	4.76×10^5

At boiling, saturated vapor pressure equals external pressure

The boiling point of a liquid clearly depends on the external pressure. At high elevations, the boiling point of water is somewhat less than at sea level since the air pressure is less. For example, on the summit of Mt. Everest (8850 m) the air pressure is about one-third of what it is at sea level, and from Table 18–2 we can see that water will boil at about 70°C. Cooking food by boiling takes longer at high elevations, since the temperature is less. Pressure cookers, however, reduce cooking time, because they build up a pressure as high as 2 atm, allowing higher boiling temperatures to be attained.

Partial Pressure and Humidity

When we refer to the weather as being dry or humid, we are referring to the water vapor content of the air. In a gas such as air, which is a mixture of several types of gases, the total pressure is the sum of the *partial pressures* of each gas present.[†] By **partial pressure**, we mean the pressure each gas would exert if it alone were present. The partial pressure of water in the air can be as low as zero and can vary up to a maximum equal to the saturated vapor pressure of water at the given temperature. Thus, at 20°C, the partial pressure of water cannot exceed 17.5 torr (see Table 18–2). The **relative humidity** is defined as the ratio of the partial pressure to the saturated vapor pressure at a given temperature. It is usually expressed as a percentage:

FIGURE 18–9 Boiling: bubbles of water vapor float upward from the bottom of the pot (where the temperature is highest).

Relative humidity

$$\text{Relative humidity} = \frac{\text{partial pressure of } H_2O}{\text{saturated vapor pressure of } H_2O} \times 100\%.$$

Thus, when the humidity is close to 100 percent, the air holds nearly all the water vapor it can.

> **EXAMPLE 18–6** **Relative humidity.** On a particular hot day, the temperature is 30°C and the partial pressure of water vapor in the air is 21.0 torr. What is the relative humidity?
>
> **SOLUTION** From Table 18–2, the saturated vapor pressure of water at 30°C is 31.8 torr. Hence the relative humidity is
>
> $$\frac{21.0 \text{ torr}}{31.8 \text{ torr}} \times 100\% = 66\%.$$

Air is saturated with water vapor when the partial pressure of water in the air is equal to the saturated vapor pressure at that temperature. If the partial pressure of water exceeds the saturated vapor pressure, the air is said to be **supersaturated**. This situation can occur when a temperature decrease occurs. For example, suppose the temperature is 30°C and the partial pressure of water is 21 torr, which represents a humidity of 66 percent as we saw above. Suppose now that the temperature falls to, say, 20°C, as might happen at nightfall. From Table 18–2 we see that the saturated vapor pressure of water at 20°C is 17.5 torr. Hence the relative humidity would be greater than 100 percent, and the supersaturated air cannot hold this much water. The excess water condenses and appears as dew—or perhaps instead as fog or rain.

When air containing a given amount of water is cooled, a temperature is reached where the partial pressure of water equals the saturated vapor pressure. This is called the **dew point**. Measurement of the dew point is the most accurate means of determining the relative humidity. One method uses a polished metal surface in contact with air, which is gradually cooled down. The temperature at which moisture begins to appear on the surface is the dew point, and the partial pressure of water can then be obtained from saturated vapor pressure tables. If, for example, on a given day the temperature is 20°C, and the dew point is 5°C, then the partial pressure of water (Table 18–2) in the original air was 6.54 torr, whereas its saturated vapor pressure was 17.5 torr; hence the relative humidity was 6.54/17.5 = 37 percent.

➡ PHYSICS APPLIED
_____

Weather

[†] For example, 78 percent (by volume) of air molecules are nitrogen and 21 percent oxygen, with much smaller amounts of water vapor, argon, and other gases. At an air pressure of 1 atm, oxygen exerts a partial pressure of 0.21 atm and nitrogen 0.78 atm.

Van der Waals Equation of State

In Section 18–3, we discussed how real gases deviate from ideal gas behavior, particularly at high densities or when near condensing to a liquid. We would like to understand these deviations using a microscopic (molecular) point of view. J. D. van der Waals (1837–1923) analyzed this problem and in 1873 arrived at an equation of state which fits real gases more accurately than the ideal gas law. His analysis is based on kinetic theory but takes into account: (1) the finite size of molecules (we previously neglected the actual volume of the molecules themselves, compared to the total volume of the container, and this assumption becomes poorer as the density increases and molecules become closer together); (2) the range of the forces between molecules may be greater than the size of the molecules (we previously assumed that intermolecular forces act only during collision, when the molecules are "in contact"). Let us now look at this analysis and derive the van der Waals equation of state.

Assume the molecules in a gas are spherical with radius r. If we assume these molecules behave like hard spheres, then two molecules collide and bounce off one another if the distance between their centers (Fig. 18–10) gets as small as $2r$. Thus the actual volume in which the molecules can move about is somewhat less than the volume V of the container holding the gas. The amount of "unavailable volume" depends on the number of molecules and on their size. Let b represent the "unavailable volume per mole" of gas. Then in the ideal gas law we replace V by $(V - nb)$, where n is the number of moles, and we obtain

$$P(V - nb) = nRT.$$

If we divide through by n

$$P\left(\frac{V}{n} - b\right) = RT. \tag{18–8}$$

FIGURE 18–10 Molecules, of radius r, colliding.

This relation (sometimes called the **Clausius equation of state**) predicts that for a given temperature T and volume V, the pressure P will be greater than for an ideal gas. This makes sense since the reduced "available" volume means the number of collisions with the walls is increased.

Next we consider the effects of attractive forces between molecules, which are responsible for holding molecules in the liquid and solid states at lower temperatures. These forces are electrical in nature and although they act even when molecules are not touching, we assume their range is small—that is, they act mainly between nearest neighbors. Molecules at the edge of the gas, headed toward a wall of the container, are slowed down by a net force pulling them back into the gas. Thus these molecules will exert less force and less pressure on the wall than if there were no attractive forces. The reduced pressure will be proportional to the density of molecules in the layer of gas at the surface, and also to the density in the next layer, which exerts the inward force.[†] Therefore we expect the pressure to be reduced by a factor proportional to the density squared $(n/V)^2$, here written as moles per volume. If the pressure P is given by Eq. 18–8, then we should reduce this by an amount $a(n/V)^2$ where a is a proportionality constant. Thus we have

$$P = \frac{RT}{(V/n) - b} - \frac{a}{(V/n)^2}$$

or

$$\left(P + \frac{a}{(V/n)^2}\right)\left(\frac{V}{n} - b\right) = RT, \tag{18–9}$$

Van der Waals equation of state

which is the **van der Waals equation of state**.

[†] This is similar to the gravitational force in which the force on mass m_1 due to mass m_2 is proportional to the product of their masses (Newton's law of universal gravitation, Chapter 6).

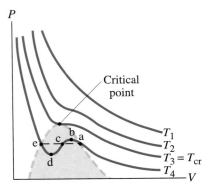

FIGURE 18–11 *PV* diagram for a van der Waals gas, shown for four different temperatures. For T_1, T_2, and T_3 (T_3 is chosen equal to the critical temperature), the curves fit experimental data very well for most gases. The curve labeled T_4, a temperature below the critical point, passes through the liquid–vapor region. The maximum (point b) and minimum (point d) would seem to be artifacts, since we usually see constant pressure, as indicated by the horizontal dashed line. However, for very pure supersaturated vapors or supercooled liquids, the sections ab and ed, respectively, have been observed. (The section bd would be unstable and has not been observed.)

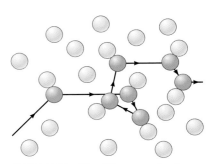

FIGURE 18–12 Zigzag path of a molecule colliding with other molecules.

FIGURE 18–13 Molecule at left moves to the right with speed $\bar{v}$. It collides with any molecule whose center is within the cylinder of radius $2r$.

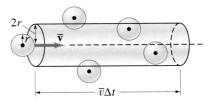

The constants a and b in the van der Waals equation are different for different gases and are determined by fitting to experimental data for each gas. For CO_2 gas, the best fit is obtained for $a = 3.6 \times 10^{-3}$ N·m⁴/mol² and $b = 4.2 \times 10^{-5}$ m³/mol. Figure 18–11 shows a typical *PV* diagram for Eq. 18–9 (a "van der Waals gas") for four different temperatures, with detailed caption, and it should be compared to Fig. 18–5 for real gases.

Neither the van der Waals equation of state nor the many other equations of state that have been proposed are accurate for all gases under all conditions. Yet Eq. 18–9 is a very useful relation. And because it is quite accurate for many situations, its derivation gives us further insight into the nature of gases at the microscopic level. Note that at low densities $a/(V/n)^2 \ll P$ and $b \ll V/n$, so that the van der Waals equation reduces to the equation of state for an ideal gas, $PV = nRT$.

*18–6 Mean Free Path

If gas molecules were truly point particles, they would never collide with one another. Thus, when a person opened a perfume bottle, you would be able to smell it almost instantaneously across the room, since molecules travel hundreds of meters per second. In fact it takes some time before you detect an odor, and according to kinetic theory, this must be due to collisions between molecules of nonzero size.

If we were to follow the path of a particular molecule, we would expect to see it follow a zigzag path as shown in Fig. 18–12. Between each collision the molecule would move in a straight-line path. (Not quite true if we take account of the small intermolecular forces that act between collisions.) An important parameter for a given situation is the **mean free path**, which is defined as the average distance a molecule travels between collisions. We would expect that the greater the gas density, and the larger the molecules, the shorter the mean free path would be. We now determine the nature of this relationship for an ideal gas.

Suppose our gas is made up of molecules which are hard spheres of radius r. A collision will occur whenever the centers of two molecules come within a distance $2r$ of one another. Let us follow a molecule as it traces a straight-line path. In Fig. 18–13, the dashed line represents the path of our particle if it made no collisions. Also shown is a cylinder of radius $2r$. If the center of another molecule lies within this cylinder, a collision will occur. (Of course, when a collision occurs the particle's path would change direction, as would our imagined cylinder, but our result won't be altered by unbending a zigzag cylinder into a straight one for purposes of calculation.) Assume our molecule is an average one, moving at the mean speed $\bar{v}$ in the gas. For the moment, let us assume that the other molecules are not moving, and that the concentration of molecules (number per unit volume) is N/V. Then the number of molecules whose center lies within the cylinder of Fig. 18–13 is N/V times the volume of this cylinder, and this also represents the number of collisions that will occur. In a time Δt, our molecule travels a distance $\bar{v}\Delta t$, so the length of the cylinder is $\bar{v}\Delta t$ and its volume is $\pi(2r)^2\bar{v}\Delta t$. Hence the number of collisions that occur in a time Δt is $(N/V)\pi(2r)^2\bar{v}\Delta t$. We define the **mean free path**, l_M, as the average distance between collisions. This distance is equal to the distance traveled ($\bar{v}\Delta t$) in a time Δt divided by the number of collisions made in time Δt:

$$l_M = \frac{\bar{v}\Delta t}{(N/V)\pi(2r)^2\bar{v}\Delta t} = \frac{1}{4\pi r^2(N/V)}. \qquad \textbf{(18–10a)}$$

Thus we see that l_M is inversely proportional to the cross-sectional area ($= \pi r^2$) of the molecules and to their concentration (number/volume), N/V. However, Eq. 18–10a is not fully correct since we assumed the other molecules are all at rest. In fact, they are moving, and the number of collisions in a time Δt must depend on the *relative* speed of the colliding molecules, rather than on $\bar{v}$. Hence the number of collisions per second is $(N/V)\pi(2r)^2 v_{\text{rel}} \Delta t$ (rather than $(N/V)\pi(2r)^2\bar{v}\,\Delta t$), where v_{rel} is the average relative speed of colliding molecules. A careful calculation shows that for a Maxwellian distribution of speeds $v_{\text{rel}} = \sqrt{2}\bar{v}$. Hence the mean free path is

$$l_M = \frac{1}{4\pi\sqrt{2}r^2(N/V)}. \qquad (18\text{–}10b) \qquad \textit{Mean free path}$$

EXAMPLE 18–7 **ESTIMATE** **Mean free path of air molecules at STP.** Estimate the mean free path of air molecules at STP, standard temperature and pressure (0°C, 1 atm). The diameter of O_2 and N_2 molecules is about 3×10^{-10} m.

SOLUTION We saw in Example 17–8 that 1 mol of an ideal gas occupies a volume of 22.4×10^{-3} m^3 at STP. Hence

$$\frac{N}{V} = \frac{6.02 \times 10^{23} \text{ molecules}}{22.4 \times 10^{-3} \text{ m}^3} = 2.69 \times 10^{25} \text{ molecules/m}^3.$$

Then

$$l_M = \frac{1}{4\pi\sqrt{2}(1.5 \times 10^{-10}\text{ m})^2(2.7 \times 10^{25}\text{ m}^{-3})} \approx 9 \times 10^{-8}\text{ m}$$

which is about 300 times the diameter of a molecule.

At very low densities, such as in an evacuated vessel, the concept of mean free path loses meaning since collisions with walls of the container may occur more frequently than collisions with other molecules. For example, in a cubical box 20 cm on a side containing air at 10^{-7} torr, the mean free path is about 700 m, which means many more collisions are made with the walls than with other molecules. (Note, nonetheless, that the box contains over 10^{12} molecules.) If the concept of mean free path were to include any type of collision, it would be closer to 0.2 m than to the 700 m calculated from Eq. 18–10.

*18–7 Diffusion

If you carefully place a few drops of food coloring in a container of water as in Fig. 18–14, you will find that the color spreads throughout the water. The process may take some time (assuming you don't shake the glass), but eventually the color will become uniform. This mixing, known as **diffusion**, takes place because of the random movement of the molecules. Diffusion occurs in gases too. Common examples include perfume or smoke (or the odor of something cooking on the stove) diffusing in air, although convection (moving air currents) often plays a greater role in spreading odors than does diffusion. Diffusion depends on concentration, by which we mean the number of molecules or moles per unit volume. In general, *the diffusing substance moves from a region where its concentration is high to one where its concentration is low.*

Diffusion occurs from high toward low concentration

FIGURE 18–14 A few drops of food coloring spreads slowly throughout the water, eventually becoming uniform.

(a)   (b)   (c)

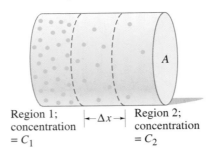

Region 1; $\leftarrow\!\Delta x\!\rightarrow$ Region 2;
concentration concentration
$= C_1$ $= C_2$

FIGURE 18–15 Diffusion occurs from a region of high concentration to one of lower concentration (only one type of molecule is shown).

Diffusion can be readily understood on the basis of kinetic theory and the random motion of molecules. Consider a tube of cross-sectional area A containing molecules in a higher concentration on the left than on the right, Fig. 18–15. We assume the molecules are in random motion. Yet there will be a net flow of molecules to the right. To see why this is true, let us consider the small section of tube of length Δx as shown. Molecules from both regions 1 and 2 cross into this central section as a result of their random motion. The more molecules there are in a region, the more will strike a given area or cross a boundary. Since there is a greater concentration of molecules in region 1 than in region 2, more molecules cross into the central section from region 1 than from region 2. There is, then, a net flow of molecules from left to right, from high concentration toward low concentration. The flow stops only when the concentrations become equal.

You might expect that the greater the difference in concentration, the greater the flow rate. This is indeed the case. In 1855, the physiologist Adolf Fick (1829–1901) determined experimentally that the rate of diffusion (J) is directly proportional to the change in concentration per unit distance $(C_1 - C_2)/\Delta x$ (which is called the **concentration gradient**), and to the cross-sectional area A (see Fig. 18–15):

$$J = DA\frac{C_1 - C_2}{\Delta x},$$

or, in terms of derivatives

Diffusion equation
(Fick's law)

$$J = DA\frac{dC}{dx}. \tag{18–11}$$

D is a constant of proportionality called the **diffusion constant**. Equation 18–11 is known as the **diffusion equation**, or **Fick's law**. If the concentrations are given in mol/m^3, then J is the number of moles passing a given point per second; if the concentrations are given in kg/m^3, then J is the mass movement per second (kg/s).

The diffusion constant D will depend on the properties of the diffusing molecules and on those of the substance in which it is immersed (often water or air), and also on the temperature and the external pressure. The values of D for a variety of substances are given in Table 18–3.

⇒ **PHYSICS APPLIED**

Diffusion time

TABLE 18–3   Diffusion Constants, D (20°C, 1 atm)

Diffusing Molecules	Medium	$D(\text{m}^2/\text{s})$
H$_2$	Air	6.3×10^{-5}
O$_2$	Air	1.8×10^{-5}
O$_2$	Water	100×10^{-11}
Blood hemoglobin	Water	6.9×10^{-11}
Glycine (an amino acid)	Water	95×10^{-11}
DNA (mass 6×10^6 u)	Water	0.13×10^{-11}

EXAMPLE 18–8   ESTIMATE   Diffusion of ammonia in air. To get an idea of the time required for diffusion, estimate how long it might take for ammonia (NH_3) to be detected 10 cm from a bottle after it is opened, assuming only diffusion.

SOLUTION This will be an order-of-magnitude calculation. The rate of diffusion J can be set equal to the number of molecules N diffusing across area A in a time t: $J = N/t$. Then $t = N/J$ and using Eq. 18–11,

$$t = \frac{N}{J} = \frac{N}{AD}\frac{\Delta x}{\Delta C}.$$

The average concentration (midway between bottle and nose) can be approximated by $\overline{C} \approx N/V$, where V is the volume over which the molecules move and is roughly of the order of $V \approx A\,\Delta x$ where Δx is 10 cm. We substitute $N = \overline{C}A\,\Delta x$ into the above equation:

$$t \approx \frac{(\overline{C}A\,\Delta x)\Delta x}{AD\,\Delta C} = \frac{\overline{C}}{\Delta C}\frac{(\Delta x)^2}{D}.$$

The concentration of ammonia is high near the bottle and low near the detecting nose, so $\overline{C} \approx \Delta C/2$ or $(\overline{C}/\Delta C) \approx \frac{1}{2}$. Since NH_3 molecules have a size somewhere between H$_2$ and O$_2$, from Table 18–3, we can estimate $D \approx 4 \times 10^{-5}$ m^2/s. Then

$$t \approx \frac{1}{2}\frac{(0.10\,\text{m})^2}{(4 \times 10^{-5}\,\text{m}^2/\text{s})} \approx 100\text{ s}.$$

or about a minute or two. This seems rather long from experience, suggesting that air currents (convection) are more important than diffusion for transmitting odors.

Summary

According to the **kinetic theory** of gases, which is based on the idea that a gas is made up of molecules that are moving rapidly and at random, the average kinetic energy of the molecules is proportional to the Kelvin temperature T:

$$\overline{K} = \tfrac{1}{2}m\overline{v^2} = \tfrac{3}{2}kT$$

where k is Boltzmann's constant.

At any moment, there exists a wide distribution of molecular speeds within a gas. The **Maxwell distribution of speeds** is derived from simple kinetic theory assumptions, and is in good accord with experiment for gases at not too high a pressure.

The behavior of real gases at high pressure, and/or when near their liquefaction point, deviates from the ideal gas law. These deviations are due to the finite size of molecules and to the attractive forces between molecules.

Below the **critical temperature**, a gas can change to a liquid if sufficient pressure is applied; but if the temperature is higher than the critical temperature, no amount of pressure will cause a liquid surface to form.

The **triple point** of a substance is that unique temperature and pressure at which all three phases—solid, liquid, and gas—can coexist in equilibrium. Because of its precise reproducibility, the triple point of water is often taken as a standard reference point.

Evaporation of a liquid is the result of the fastest moving molecules escaping from the surface. Because the average molecular velocity is less after the fastest molecules escape, the temperature decreases when evaporation takes place.

Saturated vapor pressure refers to the pressure of the vapor above a liquid when the two phases are in equilibrium. The vapor pressure of a substance (such as water) depends strongly on temperature and is equal to atmospheric pressure at the boiling point.

Relative humidity of air at a given place is the ratio of the partial pressure of water vapor in the air to the saturated vapor pressure at that temperature; it is usually expressed as a percentage.

Questions

1. Why doesn't the size of different molecules enter into the gas laws?

2. When a gas is rapidly compressed (say, by pushing down a piston) its temperature increases. When a gas expands against a piston, it cools. Explain these changes in temperature using the kinetic theory, in particular noting what happens to the momentum of molecules when they strike the moving piston.

3. In Section 18–1 we assumed the gas molecules made perfectly elastic collisions with the walls of the container. This assumption is not necessary as long as the walls are at the same temperature as the gas. Why?

4. Explain in words how Charles's law follows from kinetic theory and the relation between average kinetic energy and the absolute temperature.

5. Explain in words how Gay-Lussac's law follows from kinetic theory.

6. As you go higher in the Earth's atmosphere, the ratio of N_2 molecules to O_2 molecules increases. Why?

7. Can you determine the temperature of a vacuum?

8. Is temperature a macroscopic or microscopic variable?

9. Discuss why the Maxwell distribution of speeds (Fig. 18–2) is not a symmetric curve.

10. Explain why the peak of the curve for 310 K in Fig. 18–3 is not as high as for 273 K. (Assume the total number of molecules is the same for both.)

11. Explain, using the Maxwell distribution of speeds, (a) why the Moon has very little atmosphere, (b) why hydrogen, if at one time in the Earth's atmosphere, would probably have escaped.

12. Explain why putting food in the freezer retards spoilage.

13. Would the average kinetic energy of molecules in an ideal gas correspond to $\overline{v}$, v_{rms}, v_p, or to some other value?

14. If the pressure in a gas is doubled while its volume is held constant, by what factor do (a) v_{rms} and (b) $\overline{v}$ change?

15. If a container of gas is at rest, the average velocity of molecules must be zero. Yet the average speed is not zero. Explain.

16. Draw, roughly, the Maxwellian distribution of *velocities*. Discuss what the values would be for mean, rms, and most probable velocities. Would the curve be symmetric?

17. What everyday observation would tell you that not all molecules in a material have the same speed?

18. We saw that the saturated vapor pressure of a liquid (say, water) does not depend on the external pressure. Yet the temperature of boiling does depend on the external pressure. Is there a contradiction? Explain.

19. Alcohol evaporates more quickly than water at room temperature. What can you infer about the molecular properties of one relative to the other?

20. Explain why a hot humid day is far more uncomfortable than a hot dry day at the same temperature.

21. Is it possible to boil water at room temperature (20°C) without heating it? Explain.

22. What exactly does it mean when we say that oxygen boils at −183°C?

23. A length of thin wire is placed over a block of ice (or an ice cube) at 0°C and weights are hung from the ends of the wire. It is found that the wire cuts its way through the ice cube, but leaves a solid block of ice behind it. This process is called *regelation*. Explain how this happens by inferring how the freezing point of water depends on pressure.

24. How do a gas and a vapor differ?

25. (*a*) At suitable temperatures and pressures, can ice be melted by applying pressure? (*b*) At suitable temperatures and pressures, can carbon dioxide be melted by applying pressure?

26. Why does dry ice not last long at room temperature?

27. Under what conditions can liquid CO_2 exist? Be specific. Can it exist as a liquid at normal room temperature?

*28. Name several ways to reduce the mean free path in a gas.

*29. Discuss why sound waves can travel in a gas only if their wavelength is somewhat larger than the mean free path.

Problems

Section 18–1

1. (I) (*a*) What is the average kinetic energy of an oxygen molecule at STP? (*b*) What is the total translational kinetic energy of 2.0 mol of O_2 molecules at 20°C?

2. (I) Calculate the rms speed of helium atoms near the surface of the sun at a temperature of about 6000 K.

3. (I) By what factor will the rms speed of gas molecules increase if the temperature is increased from 0°C to 100°C?

4. (I) A gas is at 20°C. To what temperature must it be raised to double the rms speed of its molecules?

5. (I) Twelve molecules have the following speeds, given in arbitrary units: 6, 2, 4, 6, 0, 4, 1, 8, 5, 3, 7, and 8. Calculate (*a*) the mean speed, and (*b*) the rms speed.

6. (II) The rms speed of molecules in a gas at 20.0°C is to be increased by 1.0%. To what temperature must it be raised?

7. (II) If the pressure in a gas is doubled while its volume is held constant, by what factor does v_{rms} change?

8. (II) Show that the rms speed of molecules in a gas is given by $v_{rms} = \sqrt{3P/\rho}$, where P is the pressure in the gas, and ρ is the gas density.

9. (II) Show that for a mixture of two gases at the same temperature, the ratio of their rms speeds is equal to the inverse ratio of the square roots of their molecular masses.

10. (II) What is the rms speed of nitrogen molecules contained in an 8.5-m^3 volume at 2.1 atm if the total amount of nitrogen is 1300 mol?

11. (II) Calculate (*a*) the rms speed of an oxygen molecule at 0°C and (*b*) determine how many times per second it would move back and forth across a 7.0-m-long room on the average, assuming it made very few collisions with other molecules.

12. (II) What is the average distance between nitrogen molecules at STP?

13. (II) The two isotopes of uranium, ^{235}U and ^{238}U (the superscripts refer to their atomic masses), can be separated by a gas diffusion process by combining them with fluorine to make the gaseous compound UF_6. Calculate the ratio of the rms speeds of these molecules for the two isotopes.

Section 18–2

14. (I) A group of 25 particles have the following speeds: two have speed 10 m/s, seven have 15 m/s, four have 20 m/s, three have 25 m/s, six have 30 m/s, one has 35 m/s, and two have 40 m/s. Determine (*a*) the average speed, (*b*) the rms speed, and (*c*) the most probable speed.

*15. (III) Starting from the Maxwell distribution of speeds, Eq. 18–6, show (*a*) $\int_0^\infty f(v)\, dv = N$, and (*b*)

$$\int_0^\infty v^2 f(v)\, dv / N = 3kT/m.$$

Section 18–3

16. (I) (*a*) At atmospheric pressure, in what phases can CO_2 exist? (*b*) For what range of pressures and temperatures can CO_2 be a liquid? Refer to Fig. 18–7.

17. (I) CO_2 exists in what phase when the pressure is 30 atm and the temperature is 30°C (Fig. 18–7)?

18. (I) Water is in which phase when the pressure is 0.01 atm and the temperature is (*a*) 90°C, (*b*) −20°C?

19. (II) You have a sample of water and are able to control temperature and pressure arbitrarily. (*a*) Using Fig. 18–6, describe the phase changes you would see if you started at a temperature of 100°C, a pressure of 220 atm, and decreased the pressure down to 0.004 atm while keeping the temperature fixed. (*b*) Repeat part (*a*) with the temperature at 0.0°C. Assume that you held the system at the starting conditions long enough for the system to stabilize before making further changes.

Section 18–4

20. (I) What is the dew point (approximately) if the humidity is 50 percent on a day when the temperature is 25°C?

21. (I) What is the air pressure at a place where water boils at 90°C?

22. (I) If the air pressure at a particular place in the mountains is 0.85 atm, estimate the temperature at which water boils.

23. (I) What is the temperature on a day when the partial pressure of water is 530 Pa and the relative humidity is 40 percent?

24. (I) What is the partial pressure of water on a day when the temperature is 25°C and the relative humidity is 35%?

25. (II) What is the approximate pressure inside a pressure cooker if the water is boiling at a temperature of 120°C? Assume no air escaped during the heating process, which started at 20°C.

26. (II) If the humidity in a room of volume 240 m^3 at 25°C is 80 percent, what mass of water can still evaporate from an open pan?

27. (II) An *autoclave* is a device used to sterilize laboratory instruments. It is essentially a high-pressure steam boiler and operates on the same principle as a pressure cooker. However, because hot steam under pressure is more effective in killing microorganisms than moist air at the same temperature and pressure, the air is removed and replaced by steam. Typically, the gauge pressure inside the autoclave is 1.0 atm. What is the temperature of the steam? Assume the steam is in equilibrium with boiling water.

28. (III) Air that is at its dew point of 5°C is drawn into a building where it is heated to 25°C. What will be the relative humidity at this temperature? Assume constant pressure of 1.0 atm. Take into account the expansion of the air.

* Section 18–5

* 29. (II) For oxygen gas, the van der Waals equation of state achieves its best fit for $a = 0.14$ N·m^4/mol^2 and $b = 3.2 \times 10^{-5}$ m^3/mol. Determine the pressure in 1.0 mole of the gas at 0°C if its volume is 0.40 L, calculated using (a) the van der Waals equation, (b) the ideal gas law.

* 30. (II) In the van der Waals equation of state, the constant b represents the amount of "unavailable volume" occupied by the molecules themselves. Thus V is replaced by $(V - nb)$, where n is the number of moles. For oxygen, b is about 3.2×10^{-5} m^3/mol. Estimate the diameter of an oxygen molecule.

* 31. (III) (a) From the van der Waals equation of state, show that the critical temperature and pressure are given by

$$T_{cr} = \frac{8a}{27bR}, \qquad P_{cr} = \frac{a}{27b^2}.$$

[*Hint*: Use the fact that the P versus V curve has an inflection point at the critical point so that the first and second derivatives are zero.] (b) Determine a and b for CO_2 from the measured values of $T_{cr} = 304$ K and $P_{cr} = 72.8$ atm.

* 32. (III) (a) Show that a collision between two spherical molecules each of radius r, is equivalent to a collision between a point particle and a sphere of radius $2r$, and thus the center of one molecule cannot penetrate a volume equal to that of a sphere of radius $2r$ when two molecules collide. (b) Show that the total unavailable volume per mole is $b = 16\pi r^3 N_A/3$, where N_A is Avogadro's number. [*Hint*: In summing the total excluded volume, multiply by $\frac{1}{2}$ to avoid counting each pair of molecules twice.] Note that the unavailable volume is four times the actual molecular volume. (c) Estimate the diameter of a CO_2 molecule, for which $b = 4.2 \times 10^{-5}$ m^3/mol.

* Section 18–6

* 33. (II) At about what pressure would the mean free path of air molecules be (a) 1.0 m and (b) equal to the diameter of air molecules, $\approx 3 \times 10^{-10}$ m?

* 34. (II) (a) The mean free path of CO_2 molecules at STP is measured to be about 5.6×10^{-8} m. Estimate the diameter of a CO_2 molecule. (b) Do the same for He gas for which $l_M \approx 25 \times 10^{-8}$ at STP.

* 35. (II) A cubic box of volume 4.4×10^{-3} m^3 contains 70 marbles, each 1.5 cm in diameter. What is the mean free path for a marble (a) when the box is shaken vigorously and (b) when it is slightly shaken?

* 36. (II) A cubic box 1.20 m on a side is evacuated so the pressure of air inside is 10^{-6} torr. Estimate how many molecular collisions there are per each collision with a wall (0°C).

* 37. (II) A very small amount of hydrogen gas is released into the air. If the air is at 1.0 atm and 25°C, estimate the mean free path for a H_2 molecule. What assumptions did you make?

* 38. (II) Estimate the maximum allowable pressure in a 38-cm-long cathode ray tube if 98 percent of all electrons must hit the screen without first striking an air molecule.

* 39. (II) (a) Show that the number of collisions a molecule makes per second, called the *collision frequency*, f, is given by $f = \bar{v}/l_M$, and thus $f = 4\sqrt{2}\,\pi r^2 \bar{v} N/V$. (b) What is the collision frequency for N_2 molecules in air at $T = 20$°C and $P = 1.0 \times 10^{-2}$ atm?

* 40. (II) We saw in Example 18–7 that the mean free path of air molecules at STP, l_M, is about 9×10^{-8} m. Estimate the collision frequency f, the number of collisions per unit time.

* 41. (II) Suppose that a gas contains two types of molecules in concentrations n_1 and n_2. Their radii are r_1 and r_2. Use arguments similar to those leading to Eq. 18–10a to derive the following relation for the mean free path for type 1 molecules

$$l_{M1} = \frac{1}{4\pi r_1^2 n_1 + \pi(r_1 + r_2)^2 n_2}.$$

* 42. (III) At some instant, suppose we have N_0 identical molecules. Show that the number N of molecules that travel a distance x or more before the next collision is given by $N = N_0 e^{-x/l_M}$, where l_M is the mean free path. This is called the *survival equation*.

* Section 18–7

* 43. (I) Approximately how long would it take for the ammonia of Example 18–8 to be detected 1.5 m from the bottle after it is opened? What does this suggest about the relative importance of diffusion and convection for carrying odors?

* 44. (II) What is the time needed for a glycine molecule (see Table 18–3) to diffuse a distance of 15μm in water at 20°C if its concentration varies over that distance from 1.00 mol/m^3 to 0.40 mol/m^3? Compare this "speed" to its rms (thermal) speed. The molecular mass of glycine is about 75 u.

* 45. (II) Oxygen diffuses from the surface of insects to the interior through tiny tubes called tracheae. An average trachea is about 2 mm long and has cross-sectional area of 2×10^{-9} m^2. Assuming the concentration of oxygen inside is half what is outside in the atmosphere, (a) show that the concentration of oxygen in the air (assume 21 percent is oxygen) at 20°C is about 8.7 mol/m^3, then (b) calculate the diffusion rate J, and (c) estimate the average time for a molecule to diffuse in. Assume the diffusion constant is 1×10^{-5} m^2/s.

General Problems

46. What is the rms speed of nitrogen molecules contained in a 12.8 m^3 volume at 3.42 atm if the total amount of nitrogen is 1800 mol.

47. In outer space the density of matter is about one atom per cm^3, mainly hydrogen atoms, and the temperature is about 2.7 K. Calculate the rms speed of these hydrogen atoms, and the pressure (in atmospheres).

48. Calculate approximately the total translational kinetic energy of all the molecules in an *E. coli* bacterium of mass 2.0×10^{-15} kg at 37°C. Assume 70 percent of the cell, by weight, is water, and the other molecules have an average molecular mass on the order of 10^5 u.

49. (*a*) Calculate the approximate rms speed of an amino acid, whose molecular mass is 89 u, in a living cell at 37°C. (*b*) What would be the rms speed of a protein of molecular mass 50,000 u at 37°C?

50. The escape speed from the Earth is 1.12×10^4 m/s, so that a gas molecule travelling away from Earth near the outer boundary of the Earth's atmosphere would, at this speed, be able to escape from the Earth's gravitational field and be lost to the atmosphere. At what temperature is the average speed of (*a*) oxygen molecules, and (*b*) helium atoms equal to 1.12×10^4 m/s?

51. The second postulate of kinetic theory is that the molecules are, on the average, far apart from one another. That is, their average separation is much greater than the diameter of each molecule. Is this assumption reasonable? To check, calculate the average volume occupied by one molecule of a gas at STP, and compare it to the size of the molecule (diameter of a typical gas molecule is about 0.2 nm). If the molecules were the diameter of ping-pong balls, say 4 cm, how far away, on average, would the next ping-pong ball be?

52. Estimate how many times per second a collision of a molecule of oxygen gas occurs with one of the walls of its container, which is a cube with sides 1.0 m. There are 2.0 mol of oxygen inside at 20°C. [*Hint:* Let $\bar{v}_x \approx \sqrt{\overline{v_x^2}}$.]

53. Consider a container of oxygen gas at a temperature of 20°C that is 0.50 m tall. Compare the gravitational potential energy of a molecule at the top of the container (assuming the potential energy is zero at the bottom) with the average kinetic energy of the molecules. Is it reasonable to neglect the potential energy?

54. In humid climates, people constantly *dehumidify* their cellars in order to prevent rot and mildew. If the cellar in a house (kept at 20°C) has 85 m^2 of floor space and a ceiling height of 2.8 m, what is the mass of water that must be removed from it in order to drop the humidity from 95 percent to a more reasonable 30 percent?

55. The temperature of an ideal gas is increased from 120°C to 290°C while the volume and the number of moles stay constant. By what factor does the pressure change? By what factor does v_{rms} change?

56. A scuba tank has a volume of 2800 cm^3. For very deep dives, the tank is filled with 50 percent (by volume) pure oxygen and 50 percent pure helium. (*a*) How many molecules are there of each type in the tank if it is filled at 20°C to a gauge pressure of 10 atm? (*b*) What is the ratio of the average kinetic energies of the two types of molecule? (*c*) What is the ratio of the rms speeds of the two types of molecule?

57. A space vehicle returning from the Moon enters the atmosphere at a speed of about 40,000 km/h. Molecules (assume nitrogen) striking the nose of the vehicle with this speed correspond to what temperature? (Because of this high temperature, the nose of a space vehicle must be made of special materials; indeed, part of it does vaporize, and this is seen as a bright blaze upon reentry.)

58. At room temperature, it takes approximately 2.45×10^3 J to evaporate 1.00 g of water. Estimate the average speed of evaporating molecules. What multiple of v_{rms} (at 20°C) for water molecules is this? (Assume Eq. 18–4 holds.)

59. Calculate the total water vapor pressure in the air on the following two days: (*a*) a hot summer day, with the temperature 30°C and the relative humidity at 40%; (*b*) a cold winter day, with the temperature 5°C and the relative humidity at 80%.

* 60. At 300 K, an 8.50 mol sample of carbon dioxide occupies a volume of 0.200 m^3. Calculate the gas pressure, first by assuming the ideal gas law, and then by using the van der Waals equation of state. (The values for *a* and *b* are given in the text following Eq. 18–9.) In this range of pressure and volume, the van der Waals equation is very accurate. What percentage error did you make in assuming ideal-gas-law behavior?

* 61. The density of atoms, mostly hydrogen, in interstellar space is about one per cubic centimeter. Estimate the mean free path of the hydrogen atoms, assuming an atomic diameter of 10^{-10} m.

* 62. Using the ideal gas law, find an expression for the mean free path l_M that involves pressure and temperature instead of (N/V). Use this expression to find the mean free path for nitrogen molecules at a pressure of 10 atm and 300 K.

When it is cold, warm clothes act as insulators to reduce heat loss from the body to the outside by conduction and convection.

Heat radiation from a campfire can warm you, and the fire can also transfer energy directly by heat conduction to a cooking pot.

Heat, like work, represents a transfer of energy. Heat is defined as a transfer of energy due to a difference of temperature. Work is a transfer of energy by mechanical means, not due to a temperature difference. The first law of thermodynamics links the two in a general statement of energy conservation: the heat Q added to a system minus the net work W done by the system equals the change in internal energy ΔU of the system: $\Delta U = Q - W$. Internal energy U is the sum total of all the energy of the molecules of the system.

Heat and the First Law of Thermodynamics

W hen a pot of cold water is placed on a hot burner of a stove, the temperature of the water increases. We say that heat flows from the hot burner to the cold water. When two objects at different temperatures are put in contact, heat spontaneously flows from the hotter one to the colder one. The spontaneous flow of heat is in the direction tending to equalize the temperature. If the two objects are kept in contact long enough for their temperatures to become equal, the two bodies are said to be in thermal equilibrium, and there is no further heat flow between them. For example, when the mercury in a fever thermometer is still rising, heat is flowing from the patient's mouth to the thermometer; when the mercury stops, the thermometer is then in equilibrium with the person's mouth, and they are at the same temperature.

Heat and temperature are often confused. They are very different concepts and we will make the clear distinction between them. We begin this Chapter by defining and using the concept of heat. We will also begin our discussion of thermodynamics, which is the name we give to the study of processes in which energy is transferred as heat and as work.

19–1 | Heat as Energy Transfer

We use the term "heat" in everyday life as if we knew what we meant. But the term is often used inconsistently, so it is important for us to define heat clearly, and to clarify the phenomena and concepts related to heat.

We commonly speak of the "flow" of heat—heat flows from a stove burner to a pot of coffee, from the Sun to the Earth, from a person's mouth into a fever thermometer. Heat flows spontaneously from an object at higher temperature to one at lower temperature. Indeed, an eighteenth-century model of heat pictured heat flow as movement of a fluid substance called *caloric*. However, the caloric fluid was never able to be detected. In the nineteenth century, it was found that the various phenomena associated with heat could be described consistently without the need to use the fluid model. The new model viewed heat as being akin to work and energy, as we will discuss in a moment. First we note that a common unit for heat, still in use today, is named after caloric. It is called the **calorie** (cal) and is defined as *the amount of heat necessary to raise the temperature of 1 gram of water by 1 Celsius degree*. [To be precise, the particular temperature range from 14.5°C to 15.5°C is specified because the heat required is very slightly different at different temperatures. The difference is less than 1 percent over the range 0°C to 100°C and we will ignore it for most purposes.] More often used than the calorie is the **kilocalorie** (kcal), which is 1000 calories. Thus *1 kcal is the heat needed to raise 1 kg of water by 1 C°*. Sometimes a kilocalorie is called a **Calorie** (with a capital C), and it is by this unit that the energy value of food is specified (we might also call it the "dietary calorie"). In the British system of units, heat is measured in British thermal units (Btu). One Btu is defined as the heat needed to raise the temperature of 1 lb of water by 1 F°. It can be shown (Problem 4) that 1 Btu = 0.252 kcal = 1055 J.

The idea that heat is related to energy was pursued by a number of scientists in the 1800s, particularly by an English brewer, James Prescott Joule (1818–1889). Joule performed a number of experiments that were crucial for establishing our present-day view that heat, like work, represents a transfer of energy. One of Joule's experiments is shown (simplified) in Fig. 19–1. The falling weight causes the paddle wheel to turn and do work on the water: the friction between the water and the paddle wheel causes the temperature of the water to rise slightly (barely measurable, in fact, by Joule). Of course, the same temperature rise could also be obtained by heating the water on a hot stove. In this and a great many other experiments (some involving electrical energy), Joule determined that a given amount of work done was always equivalent to a particular amount of heat input. Quantitatively, 4.186 joules (J) of work was found to be equivalent to 1 calorie (cal) of heat. This is known as the **mechanical equivalent of heat**:

$$4.186 \text{ J} = 1 \text{ cal}$$
$$4.186 \times 10^3 \text{ J} = 1 \text{ kcal}.$$

As a result of these and other experiments, scientists came to interpret heat not as a substance, and not even as a form of energy. Rather, heat refers to a *transfer of energy*: when heat flows from a hot object to a cooler one, it is energy that is being transferred from the hot to the cold object. Thus, **heat** is *energy that is transferred from one body to another because of a difference in temperature*. In SI units, the unit for heat, as for any form of energy, is the joule. Nonetheless, calories and kcal are still sometimes used. Today the calorie is *defined* in terms of the joule (via the mechanical equivalent of heat, above), rather than in terms of the properties of water, as given previously. The latter is still handy to remember: 1 cal raises 1 g of water by 1 C°, or 1 kcal raises 1 kg of water by 1 C°.

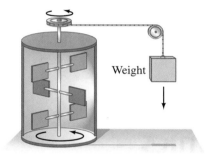

FIGURE 19–1 Joule's experiment on the mechanical equivalent of heat.

calorie (unit)

Kilocalorie (= dietary Calorie)

Btu

Mechanical equivalent of heat

Heat is energy transferred because of a ΔT

➡ PHYSICS APPLIED

Working off Calories

EXAMPLE 19–1 **ESTIMATE** **Working off the extra calories.** A young couple throw caution to the wind one afternoon, eating too much ice cream and cake. They realize that they each overate by about 500 Calories, and to compensate they want to do an equivalent amount of work climbing stairs or a mountain. How much vertical height must each person walk up? Each has a mass of 60 kg.

SOLUTION   500 Calories is 500 kcal, which in joules is

$$(500 \text{ kcal})(4.186 \times 10^3 \text{ J/kcal}) = 2.1 \times 10^6 \text{ J}.$$

The work done to climb a vertical height h is $W = mgh$. We want to solve for h given that $W = 2.1 \times 10^6 \text{ J}$:

$$h = \frac{W}{mg} = \frac{2.1 \times 10^6 \text{ J}}{(60 \text{ kg})(9.80 \text{ m/s}^2)} = 3600 \text{ m}.$$

They need to climb a very high mountain (over 11,000 ft) or many flights of stairs. [The human body does not transform energy with 100 percent efficiency—more like 20 percent—just as no engine does. As we'll see in the next chapter, some energy is always "wasted," so the couple would actually have to climb only about $(0.2)(3600 \text{ m}) \approx 700 \text{ m}.$]

19–2 | Internal Energy

We introduce the concept of internal energy now since it will help clarify ideas about heat. The sum total of all the energy of all the molecules in an object is called its **thermal energy** or **internal energy**. (We will use the two terms interchangeably.) Occasionally, the term "heat content" of a body is used for this purpose, but it is not a good term because it can be confused with heat itself. Heat, as we have seen, is not the energy a body contains, but rather refers to the amount of energy transferred from one body to another at a different temperature.

Internal energy

Distinguishing Temperature, Heat, and Internal Energy

Using the kinetic theory, we can make a clear distinction between temperature, heat, and internal energy. Temperature (in kelvins) is a measure of the *average* kinetic energy of individual molecules. Thermal energy and internal energy refer to the *total* energy of all the molecules in the object. (Thus two equal-mass hot ingots of iron may have the same temperature, but two of them have twice as much thermal energy as one does.) Heat, finally, refers to a *transfer* of energy (such as thermal energy) from one object to another because of a difference in temperature.

Heat vs. internal energy vs. temperature

Notice that the direction of heat flow between two objects depends on their temperatures, not on how much internal energy each has. Thus, if 50 g of water at 30°C is placed in contact (or mixed) with 200 g of water at 25°C, heat flows *from* the water at 30°C *to* the water at 25°C even though the internal energy of the 25°C water is much greater because there is so much more of it.

Direction of heat flow depends on temperature

Internal Energy of an Ideal Gas

Let us calculate the internal energy of n moles of an ideal monatomic (one atom per molecule) gas. The internal energy, U, is the sum of the translational kinetic energies of all the atoms. This sum is just equal to the average kinetic energy per molecule times the total number of molecules, N:

$$U = N(\tfrac{1}{2}m\overline{v^2}).$$

Using Eq. 18–4, $\overline{K} = \tfrac{1}{2}m\overline{v^2} = \tfrac{3}{2}kT$, we can write this as

$$U = \tfrac{3}{2}NkT$$

or

$$U = \tfrac{3}{2}nRT,$$ [monatomic ideal gas]   **(19–1)**

Internal energy of an ideal monatomic gas

where n is the number of moles. Thus, the internal energy of an ideal gas depends only on temperature and the number of moles of gas.

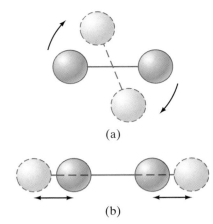

(a)

(b)

FIGURE 19–2 Molecules can have (a) rotational and (b) vibrational energy, as well as translational energy.

Relation between heat flow and temperature change

Specific heat

TABLE 19–1 Specific Heats (at 1 atm constant pressure and 20°C unless otherwise stated)

Substance	Specific Heat, c kcal/kg·C° (= cal/gm·C°)	J/kg·C°
Aluminum	0.22	900
Alcohol (ethyl)	0.58	2400
Copper	0.093	390
Glass	0.20	840
Iron or steel	0.11	450
Lead	0.031	130
Marble	0.21	860
Mercury	0.033	140
Silver	0.056	230
Wood	0.4	1700
Water		
Ice (−5°C)	0.50	2100
Liquid (15°C)	1.00	4186
Steam (110°C)	0.48	2010
Human body (average)	0.83	3470
Protein	0.4	1700

If the gas molecules contain more than one atom, then the rotational and vibrational energy of the molecules (Fig. 19–2) must also be taken into account. The internal energy will be greater at a given temperature than for a monatomic gas, but it will still be a function only of temperature for an ideal gas.

The internal energy of real gases also depends mainly on temperature, but where they deviate from ideal gas behavior, the internal energy also depends somewhat on pressure and volume.

The internal energy of liquids and solids is quite complicated, for it includes electrical potential energy associated with the forces (or "chemical" bonds) between atoms and molecules.

19–3 Specific Heat

If heat is put into an object, its temperature rises. But how much does the temperature rise? That depends. As early as the eighteenth century, experimenters had recognized that the amount of heat Q required to change the temperature of a given material is proportional to the mass m of the material and to the temperature change ΔT. This remarkable simplicity in nature can be expressed in the equation

$$Q = mc\,\Delta T, \tag{19–2}$$

where c is a quantity characteristic of the material called its **specific heat**. Because $c = Q/m\Delta T$, specific heat is specified in units of J/kg·C° (the proper SI unit) or kcal/kg·C°. (The value of c in cal/gm·C° is the same as in kcal/kg·C°.) For water at 15°C and a constant pressure of 1 atm, $c = 1.00$ kcal/kg·C° or 4.19×10^3 J/kg·C°, since, by definition of the cal and the joule, it takes 1 kcal of heat to raise the temperature of 1 kg of water by 1 C°. Table 19–1 gives the values of specific heat for many substances at 20°C. The values of c depend to some extent on temperature (as well as slightly on pressure), but for temperature changes that are not too great, c can often be considered constant.[†]

EXAMPLE 19–2 How heat depends on specific heat. (*a*) How much heat is required to raise the temperature of an empty 20-kg vat made of iron from 10°C to 90°C? (*b*) What if the vat is filled with 20 kg of water?

SOLUTION (*a*) From Table 19–1, the specific heat of iron is 450 J/kg·C°. The change in temperature is $(90°C − 10°C) = 80$ C°. Thus,

$$Q = mc\,\Delta T = (20\,\text{kg})(450\,\text{J/kg·C°})(80\,\text{C°}) = 7.2 \times 10^5\,\text{J} = 720\,\text{kJ}.$$

(*b*) The water alone would require

$$Q = mc\,\Delta T = (20\,\text{kg})(4186\,\text{J/kg·C°})(80\,\text{C°}) = 6.7 \times 10^6\,\text{J} = 6700\,\text{kJ}$$

or almost 10 times what an equal mass of iron requires. The total, for the vat plus the water, is 720 kJ + 6700 kJ = 7400 kJ.

If the iron vat in part (*a*) had been *cooled* from 90°C to 10°C, 720 kJ of heat would have flowed *out* of the iron. In other words, Eq. 19–2 is valid for heat flow either in or out, with a corresponding increase or decrease in temperature. We saw in part (*b*) of Example 19–2 that water requires almost 10 times as much heat as an equal mass of iron to make the same temperature change. Water has one of the highest specific heats of all substances, which makes it an ideal substance for hot-water space-heating systems and other uses that require a minimal drop in temperature for a given amount of heat transfer. It is the water content, too, that causes the apples rather than the crust in hot apple pie to burn our tongues, through heat transfer.

[†] To take into account the dependence of c on T, we can write Eq. 19–2 in differential form: $dQ = mc(T)\,dT$. Then the heat Q required to change the temperature from T_1 to T_2 is

$$Q = \int_{T_1}^{T_2} mc(T)\,dT,$$

where $c(T)$ is a function of temperature.

19–4 Calorimetry—Solving Problems

When different parts of an isolated system are at different temperatures, but are placed in thermal contact, heat will flow from the part at higher temperature to the part at lower temperature. If the system is completely isolated, no energy can flow into or out of it. Again, the *conservation of energy* plays an important role for us; the heat lost by one part of the system is equal to the heat gained by the other part:

$$\text{heat lost} = \text{heat gained.}$$

Energy conservation

Let us take an Example.

EXAMPLE 19–3 **The cup cools the tea.** If $200\,\text{cm}^3$ of tea at $95°C$ is poured into a 150-g glass cup initially at $25°C$ (Fig. 19–3), what will be the final temperature T of the mixture when equilibrium is reached, assuming no heat flows to the surroundings?

SOLUTION Since tea is mainly water, its specific heat is $4186\,\text{J/kg·C°}$ (Table 19–1) and its mass m is its density times its volume ($V = 200\,\text{cm}^3 = 200 \times 10^{-6}\,\text{m}^3$): $m = \rho V = (1.0 \times 10^3\,\text{kg/m}^3)(200 \times 10^{-6}\,\text{m}^3) = 0.20\,\text{kg}$. Applying conservation of energy, we can set

$$\text{heat lost by tea} = \text{heat gained by cup}$$
$$m_{\text{tea}}c_{\text{tea}}(95°C - T) = m_{\text{cup}}c_{\text{cup}}(T - 25°C)$$

where T is the as yet unknown final temperature. Putting in numbers and using Table 19–1, we solve for T, and find

$$(0.20\,\text{kg})(4186\,\text{J/kg·C°})(95°C - T) = (0.15\,\text{kg})(840\,\text{J/kg·C°})(T - 25°C)$$
$$79{,}400\,\text{J} - (836\,\text{J/C°})T = (126\,\text{J/C°})T - 6300\,\text{J}$$
$$T = 89°C.$$

The tea drops in temperature by $6\,\text{C°}$ by coming into equilibrium with the cup.

Alternate Solution: We can set up this Example (and others) by a different approach. We can write that the total heat transferred into or out of the isolated system is zero:

$$\Sigma Q_i = 0.$$

Then each term is written as $Q = mc(T_i - T_f)$ where T_i and T_f are the initial and final temperatures. In the present Example:

$$\Sigma Q = m_{\text{tea}}c_{\text{tea}}(95°C - T) + m_{\text{cup}}c_{\text{cup}}(25°C - T) = 0.$$

Note that the second term is negative because T will be greater than $25°C$. Solving the algebra gives the same result.

The exchange of energy, as exemplified in Example 19–3, is the basis for a technique known as **calorimetry**, which is the quantitative measurement of heat exchange. To make such measurements, a **calorimeter** is used; a simple water calorimeter is shown in Fig. 19–4. It is very important that the calorimeter be well insulated so that only a minimal amount of heat is exchanged with the outside. One important use of the calorimeter is in the determination of specific heats of substances. In the technique known as the "method of mixtures," a sample of the substance is heated to a high temperature, which is accurately measured, and then quickly placed in the cool water of the calorimeter. The heat lost by the sample will be gained by the water and the calorimeter. By measuring the final temperature of the mixture, the specific heat can be calculated, as illustrated in the following Example.

FIGURE 19–3 Example 19–3.

➡ **PROBLEM SOLVING**

Alternate approach

FIGURE 19–4
Simple water calorimeter.

EXAMPLE 19–4 **Unknown specific heat determined by calorimetry.** An engineer wishes to determine the specific heat of a new alloy. A 0.150-kg sample of the alloy is heated to 540°C. It is then quickly placed in 400 g of water at 10.0°C, which is contained in a 200-g aluminum calorimeter cup. (We do not need to know the mass of the insulating jacket since we assume the air space between it and the cup insulates it well, so that its temperature does not change significantly.) The final temperature of the mixture is 30.5°C. Calculate the specific heat of the alloy.

SOLUTION Again we apply the conservation of energy and write that the heat lost equals the heat gained:

$$\left(\begin{array}{c}\text{heat lost}\\\text{by sample}\end{array}\right) = \left(\begin{array}{c}\text{heat gained}\\\text{by water}\end{array}\right) + \left(\begin{array}{c}\text{heat gained by}\\\text{calorimeter cup}\end{array}\right)$$

$$m_s c_s \, \Delta T_s = m_w c_w \, \Delta T_w + m_{cal} c_{cal} \, \Delta T_{cal}$$

where the subscripts s, w, and cal refer to the sample, water, and calorimeter, respectively. When we put in values and use Table 19–1, this equation becomes

$$(0.150 \, \text{kg})(c_s)(540°C - 30.5°C) = (0.40 \, \text{kg})(4186 \, \text{J/kg·C°})(30.5°C - 10.0°C)$$
$$+ (0.20 \, \text{kg})(900 \, \text{J/kg·C°})(30.5°C - 10.0°C)$$
$$76.4 \, c_s = (34,300 + 3700) \, \text{J/kg·C°}$$
$$c_s = 500 \, \text{J/kg·C°}.$$

In making this calculation, we have ignored any heat transferred to the thermometer and the stirrer (which is needed to quicken the heat transfer process and thus reduce heat loss to the outside). It can be taken into account by adding additional terms to the right side of the above equation and will result in a slight correction to the value of c_s (see Problem 14). It should be noted that the quantity $m_{cal} c_{cal}$ is often called the **water equivalent** of the calorimeter—that is, $m_{cal} c_{cal}$ is numerically equal to the mass of water (in kilograms) that would absorb the same amount of heat.

➡ **PHYSICS APPLIED**

Measuring Calorie content

A **bomb calorimeter** is used to measure the heat released when a substance burns. Important applications are the burning of foods to determine their Calorie content, and burning of seeds and other substances to determine their "energy content," or heat of combustion. A carefully weighed sample of the substance, together with an excess amount of oxygen at high pressure, is placed in a sealed container (the "bomb"). The bomb is placed in the water of the calorimeter and a fine wire passing into the bomb is then heated briefly, which causes the mixture to ignite.

19–5 | Latent Heat

When a material changes phase from solid to liquid, or from liquid to gas (see also Section 18–3), a certain amount of energy is involved in this **change of phase**. For example, let us trace what happens when a 1.0-kg block of ice at −40°C has heat added to it at a steady rate until all the ice has changed to water, then the (liquid) water is heated to 100°C and changed to steam above 100°C, all at 1 atm pressure. As shown in the graph of Fig. 19–5, as heat is added to the ice, its temperature rises at a rate of about 2 C°/kcal of heat added (since for ice, $c \approx 0.50 \, \text{kcal/kg·C°}$). However, when 0°C is reached, the temperature stops increasing even though heat is still being added. Now as heat is added, the ice gradually changes to water in the liquid state without any change in temperature. After about 40 kcal have been added at 0°C, half the ice remains and half has changed to water. After about 80 kcal, or 330 kJ, has been added, all the ice has changed to water, still at 0°C. Further addition of heat causes the water's temperature to again increase, now at a rate of 1 C°/kcal. When 100°C is reached, the temperature again remains constant as the heat added changes the liquid water to vapor (steam). About 540 kcal (2260 kJ) is

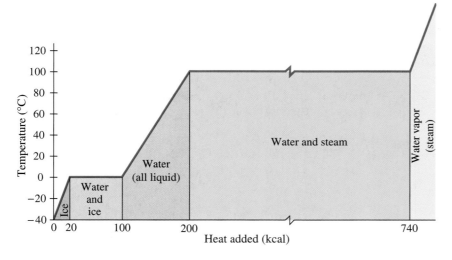

FIGURE 19–5 Temperature as a function of the heat added to bring 1.0 kg of ice at −40°C to steam above 100°C. Note the scale break between 200 and 740 kcal (the page isn't wide enough to fit it).

required to change the 1.0 kg of water completely to steam, after which the curve rises again, indicating that the temperature of the steam now rises as heat is added.

The heat required to change 1.0 kg of a substance from the solid to the liquid state is called the **heat of fusion**; it is denoted by L_F. The heat of fusion of water at 0°C is 79.7 kcal/kg or, in proper SI units, 333 kJ/kg $(= 3.33 \times 10^5$ J/kg$)$. The heat required to change a substance from the liquid to the vapor phase is called the **heat of vaporization**, L_V, and for water at 100°C it is 539 kcal/kg or 2260 kJ/kg. Other substances follow graphs similar to Fig. 19–5, although the melting-point and boiling-point temperatures are different, as are the specific heats and heats of fusion and vaporization. Values for the heats of fusion and vaporization, which are also called the **latent heats**, are given in Table 19–2 for a number of substances.

Heat of fusion

Heat of vaporization

The heats of vaporization and fusion also refer to the amount of heat *released* by a substance when it changes from a gas to a liquid, or from a liquid to a solid. Thus, steam releases 2260 kJ/kg when it changes to water, and water releases 333 kJ/kg when it becomes ice.

Of course, the heat involved in a change of phase depends not only on the latent heat, but also on the total mass of the substance. That is,

$$Q = mL,$$

where L is the latent heat of the particular process and substance, m is the mass of the substance, and Q is the heat required or given off during the phase change. For example, when 5.00 kg of water freezes at 0°C, $(5.00\,\text{kg})(3.33 \times 10^5\,\text{J/kg}) = 1.67 \times 10^6$ J of energy is released.

TABLE 19–2   Latent Heats (at 1 atm)

Substance	Melting Point (°C)	Heat of Fusion		Boiling Point (°C)	Heat of Vaporization	
		kcal/kg[†]	J/kg		kcal/kg[†]	J/kg
Oxygen	−218.8	3.3	0.14×10^5	−183	51	2.1×10^5
Nitrogen	−210.0	6.1	0.26×10^5	−195.8	48	2.00×10^5
Ethyl alcohol	−114	25	1.04×10^5	78	204	8.5×10^5
Ammonia	−77.8	8.0	0.33×10^5	−33.4	33	1.37×10^5
Water	0	79.7	3.33×10^5	100	539	22.6×10^5
Lead	327	5.9	0.25×10^5	1750	208	8.7×10^5
Silver	961	21	0.88×10^5	2193	558	23×10^5
Iron	1808	69.1	2.89×10^5	3023	1520	63.4×10^5
Tungsten	3410	44	1.84×10^5	5900	1150	48×10^5

[†] Numerical values in kcal/kg are the same in cal/g.

Calorimetry sometimes involves a change of state as the following Examples show. Indeed, latent heats are often measured using calorimetry.

EXAMPLE 19–5 **Making ice.** How much energy does a refrigerator have to remove from 1.5 kg of water at 20°C to make ice at −12°C?

SOLUTION Heat must flow out to reduce the water from 20°C to 0°C, to change it to ice, and then to lower the ice from 0°C to −12°C,

$$Q = mc_w(20°C - 0°C) + mL_F + mc_{ice}[0° - (-12°C)]$$
$$= (1.5 \, \text{kg})(4186 \, \text{J/kg}\cdot\text{C}°)(20 \, \text{C}°) + (1.5 \, \text{kg})(3.33 \times 10^5 \, \text{J/kg})$$
$$+ (1.5 \, \text{kg})(2100 \, \text{J/kg}\cdot\text{C}°)(12 \, \text{C}°)$$
$$= 6.6 \times 10^5 \, \text{J} = 660 \, \text{kJ}.$$

EXAMPLE 19–6 **Will all the ice melt?** At a party, a 0.50-kg chunk of ice at −10°C is placed in 3.0 kg of "iced" tea at 20°C. At what temperature and in what phase will the final mixture be? The tea can be considered as water.

➡ PROBLEM SOLVING

First determine (or estimate) the final state

SOLUTION In this situation, before we can write down an equation, we must first check to see if the final state will be all ice, a mixture of ice and water at 0°C, or all water. To bring the 3.0 kg of water at 20°C down to 0°C would require an energy release of

$$m_w c_w(20°C - 0°C) = (3.0 \, \text{kg})(4186 \, \text{J/kg}\cdot\text{C}°)(20 \, \text{C}°) = 250 \, \text{kJ}.$$

On the other hand, to raise the ice from −10°C to 0°C would require

$$m_{ice} c_{ice}[0°C - (-10°C)] = (0.50 \, \text{kg})(2100 \, \text{J/kg}\cdot\text{C}°)(10 \, \text{C}°) = 10.5 \, \text{kJ},$$

and to change the ice to water at 0°C would require

$$m_{ice} L_F = (0.50 \, \text{kg})(333 \, \text{kJ/kg}) = 167 \, \text{kJ},$$

for a total of 10.5 kJ + 167 kJ = 177 kJ. This is not enough energy to bring the 3.0 kg of water at 20°C down to 0°C, so we know that the mixture must end up all water, somewhere between 0°C and 20°C. Now we can determine the final temperature T by applying the conservation of energy and writing

Then determine the final temperature

$$\begin{pmatrix} \text{heat to raise} \\ \text{0.50 kg of ice} \\ \text{from } -10°C \\ \text{to } 0°C \end{pmatrix} + \begin{pmatrix} \text{heat to change} \\ \text{0.50 kg} \\ \text{of ice} \\ \text{to water} \end{pmatrix} + \begin{pmatrix} \text{heat to raise} \\ \text{0.50 kg of water} \\ \text{from } 0°C \\ \text{to } T \end{pmatrix} = \begin{pmatrix} \text{heat lost by} \\ \text{3.0 kg of} \\ \text{water cooling} \\ \text{from } 20°C \text{ to } T \end{pmatrix}$$

so

$$10.5 \, \text{kJ} + 167 \, \text{kJ} + (0.50 \, \text{kg})(4186 \, \text{J/kg}\cdot\text{C}°)(T)$$
$$= (3.0 \, \text{kg})(4186 \, \text{J/kg}\cdot\text{C}°)(20°C - T).$$

Solving for T gives $T = 5.1°C$.

We can make use of kinetic theory to see why energy is needed to melt or vaporize a substance. At the melting point, the latent heat of fusion does not increase the kinetic energy (and the temperature) of the molecules in the solid, but instead is used to overcome the potential energy associated with the forces between the molecules. That is, work must be done against these attractive forces to break the molecules loose from their relatively fixed positions in the solid so they can freely roll over one another in the liquid phase. Similarly, energy is required for molecules held close together in the liquid phase to escape into the gaseous phase. This process is a more violent reorganization of the molecules than is melting (the average distance between the molecules is greatly increased) and hence the heat of vaporization is generally much greater than the heat of fusion for a given substance.

1. Be sure you have enough information to apply energy conservation. Ask yourself: is the system isolated (or very nearly so, enough to get a good estimate)? Do we know or can we calculate all significant sources of heat energy flow?

2. Apply conservation of energy. One approach is to write

$$\text{heat gained} = \text{heat lost}.$$

For each substance in the system, a heat (energy) term will appear on either the left or right of this equation.

3. If no phase changes occur, each term in the energy conservation equation (above) will have the form

$$Q(\text{gain}) = mc(T_f - T_i) \quad \text{or} \quad Q(\text{lost}) = mc(T_i - T_f)$$

where T_i and T_f are the initial and final temperatures of the substance, and m and c are its mass and specific heat.

4. If phase changes do or might occur, there may be terms in the energy conservation equation of the form $Q = mL$, where L is the latent heat. But *before* applying energy conservation, determine (or estimate) in which phase the final state will be, as we did in Example 19–6 by calculating the different contributing values for heat Q.

5. Be sure each term appears on the correct side of the energy equation (heat gained or heat lost) and that each ΔT is positive.

6. Note that when the system reaches thermal equilibrium, the final temperature of each substance will have the *same* value. There is only one T_f.

7. Solve your energy equation for the unknown.

The latent heat to change a liquid to a gas is needed not only at the boiling point. Water can change from the liquid to the gas phase even at room temperature. This process is called **evaporation** (see also Section 18–4). The value of the heat of vaporization increases slightly with a decrease in temperature: at 20°C, for example, it is 2450 kJ/kg (585 kcal/kg) compared to 2260 kJ/kg (539 kcal/kg) at 100°C. When water evaporates, it cools, since the energy required (the latent heat of vaporization) comes from the water itself. So its internal energy, and therefore its temperature, must drop.[†]

19–6 The First Law of Thermodynamics

Up to now in this chapter we have discussed internal energy and heat. But work too is often involved in thermodynamic processes.

In Chapter 8 we saw that work is done when energy is transferred from one body to another by mechanical means. In Section 19–1 we saw that heat is a transfer of energy from one body to a second body at a lower temperature. Thus, heat is much like work. To distinguish them, *heat* is defined as a *transfer of energy due to a difference in temperature*, whereas work is a transfer of energy that is not due to a temperature difference.

Heat distinguished from work

In discussing thermodynamics, we shall often refer to particular systems. A **system** is any object or set of objects that we wish to consider. Everything else in the universe we will refer to as its "environment." There are several categories of systems. A **closed system** is one for which no mass enters or leaves (but energy may be exchanged with the environment). In an **open system**, mass may enter or leave (as well as energy). Many (idealized) systems we study in physics are closed systems. But many systems, including plants and animals, are open systems since they exchange materials (food, oxygen, waste products) with the environment. A closed system is said to be **isolated** if no energy in any form passes across its boundaries; otherwise it is not isolated.

Open and closed systems

[†] According to kinetic theory, evaporation is a cooling process because it is the fastest-moving molecules that escape from the surface. Hence the average speed of the remaining molecules is less, so by Eq. 18–4 the temperature is less.

In Section 19–2, we defined the internal energy of a system as the sum total of all the energy of the molecules of the system. We would expect that the internal energy of a system would be increased if work were done on the system, or if heat were added to it. Similarly the internal energy would be decreased if heat flowed out of the system or if work were done by the system on something else.

Thus, from conservation of energy, it is reasonable to propose an important law: the change in internal energy of a closed system, ΔU, will be equal to the heat added to the system minus the work done by the system; in equation form:

FIRST LAW OF
THERMODYNAMICS

$$\Delta U = Q - W, \tag{19–3}$$

where Q is the net heat *added* to the system and W is the net work done *by* the system. We must be careful and consistent in following the sign conventions for Q and W. Because W in Eq. 19–3 is the work done *by* the system, then if work is done *on* the system, W will be negative and U will increase. [Of course, we could have defined W as the work done *on* the system, in which case there would be a plus sign in Eq. 19–3; but it is conventional to define W and Q as we have done.] Similarly, Q is positive for heat added to the system, so if heat leaves the system, Q is negative. Equation 19–3 is known as the **first law of thermodynamics**. It is one of the great laws of physics, and its validity rests on experiments (such as Joule's) in which no exceptions have been seen. Since Q and W represent energy transferred into or out of the system, the internal energy changes accordingly. Thus, the first law of thermodynamics is a great and broad statement of the *law of conservation of energy*. It is worth noting that the conservation of energy law was not formulated until the nineteenth century, for it depended on the interpretation of heat as a transfer of energy.

Heat added is +
Heat lost is −
Work on system is −
Work by system is +

First law of thermodynamics is conservation of energy

Equation 19–3 applies to a closed system. It also applies to an open system if we take into account the change in internal energy due to the increase or decrease in the amount of matter. For an isolated system, no work is done and no heat enters or leaves the system, so $W = Q = 0$, and hence $\Delta U = 0$.

A given system, in a particular state, can be said to have a certain amount of internal energy, U. This cannot be said of heat or work. A system in a given state does not "have" a certain amount of heat or work. Rather, when work is done on a system (such as compressing a gas), or when heat is added or removed from a system, the state of the system *changes*. Thus, work and heat are involved in *thermodynamic processes* that can change the system from one state to another; they are not characteristic of the state itself, as are pressure P, volume V, temperature T, internal energy U, and the mass m or number of moles n. Because U is a *state variable*, which we saw in Chapter 17 is a variable that depends only on the state of the system and not on how the system arrived in that state, we can write

Internal energy is a property of a system; work and heat are not

$$\Delta U = U_2 - U_1 = Q - W,$$

where U_1 and U_2 represent the internal energy of the system in states 1 and 2, and Q and W are the heat added to the system and work done by the system in going from state 1 to state 2.

It is sometimes useful to write the first law of thermodynamics in differential form:

$$dU = dQ - dW.$$

Here, dU represents an infinitesimal change in internal energy when an infinitesimal amount[†] of heat dQ is added to the system, and the system does an infinitesimal amount of work dW.

[†] The differential form of the first law is often written

$$dU = \bar{d}Q - \bar{d}W,$$

where the bars on the differential sign ($\bar{d}$) are used to remind us that W and Q are not functions of the state variables (such as P, V, T, n). Internal energy, U, is a function of the state variables, and so dU represents the differential (called an *exact differential*) of some function U. The differentials $\bar{d}W$ and $\bar{d}Q$ are not exact differentials (they are not the differential of some mathematical function); they thus only represent infinitesimal amounts. This issue won't really be of concern in this book.

EXAMPLE 19–7 **Using the first law.** (a) An amount of heat equal to 2500 J is added to a system, and 1800 J of work is done on the system. What is the change in internal energy of the system?

SOLUTION We use the first law of thermodynamics, Eq. 19–3. The heat added to the system is $Q = 2500$ J. The work done by the system W is -1800 J, with a minus sign because 1800 J done *on* the system equals -1800 J done *by* the system. Hence

$$\Delta U = 2500 \text{ J} - (-1800 \text{ J}) = 2500 \text{ J} + 1800 \text{ J} = 4300 \text{ J}.$$

You may have intuitively thought that the 2500 J and the 1800 J would need to be added together, since both refer to energy transferred into the system. You would have been right. We did this exercise in detail to emphasize the importance of keeping careful track of signs.

19–7 Applying the First Law of Thermodynamics; Calculating the Work

Let us analyze some simple processes in the light of the first law of thermodynamics.

Isothermal Processes ($\Delta T = 0$)

First we consider an idealized process that is carried out at constant temperature. Such a process is called an **isothermal** process (from the Greek meaning "same temperature"). If the system is an ideal gas, then $PV = nRT$ (Eq. 17–3), so for a fixed amount of gas kept at constant temperature, $PV = $ constant. Thus the process follows a curve like AB on the PV diagram shown in Fig. 19–6, which is a curve for $PV = $ constant. Each point on the curve, such as point A, represents the state of the system at a given moment—that is, its pressure P and volume V. At a lower temperature, another isothermal process would be represented by a curve like A′B′ in Fig. 19–6 (the product $PV = nRT = $ constant is less when T is less). The curves shown in Fig. 19–6 are referred to as *isotherms*.

Let us assume that the gas is enclosed in a container fitted with a movable piston, Fig. 19–7, and that the gas is in contact with a **heat reservoir** (a body whose mass is so large that, ideally, its temperature does not change significantly when heat is exchanged with our system). We also assume that the process of compression (volume decreases) or expansion (volume increases) is done **quasistatically** ("almost statically"), by which we mean extremely slowly, so that all of the gas stays in equilibrium at the same constant temperature. If the gas is initially in a state represented by point A in Fig. 19–6 and an amount of heat Q is added to the system, the system will move to another point, B, on the diagram. If the temperature is to remain constant, the gas must expand and do an amount of work W on the environment (it exerts a force on the piston and moves it through a distance). The temperature and mass are kept constant so, from Eq. 19–1, the internal energy does not change: $\Delta U = \frac{3}{2} nR\Delta T = 0$. Hence, by the first law of thermodynamics, Eq. 19–3, $\Delta U = Q - W = 0$, so $W = Q$: the work done by the gas in an isothermal process equals the heat added to the gas.

Adiabatic Processes ($Q = 0$)

An **adiabatic** process is one in which no heat is allowed to flow into or out of the system: $Q = 0$. This situation can occur if the system is extremely well insulated, or the process happens so quickly that heat—which flows slowly—has no time to flow in or out. The very rapid expansion of gases in an internal combustion engine

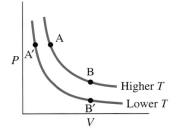

FIGURE 19–6 PV diagram for an ideal gas undergoing isothermal processes at two different temperatures.

Heat reservoir

FIGURE 19–7 An ideal gas in a cylinder fitted with a movable piston.

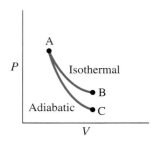

FIGURE 19–8 *PV* diagram for adiabatic (AC) and isothermal (AB) processes on an ideal gas.

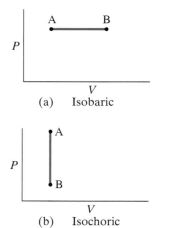

FIGURE 19–9 (a) Isobaric ("same pressure") process. (b) Isochoric ("same volume") process.

FIGURE 19–10 The work done by a gas when its volume increases by $dV = A \, dl$ is $dW = P \, dV$.

is one example of a process that is very nearly adiabatic. A slow adiabatic expansion of an ideal gas follows a curve like that labeled AC in Fig. 19–8. Since $Q = 0$, we have from Eq. 19–3 that $\Delta U = -W$. That is, the internal energy decreases if the gas expands; hence the temperature decreases as well (because $\Delta U = \frac{3}{2} nR\Delta T$). Thus an adiabatic *PV* curve is steeper than an isotherm: in Fig. 19–8 the product $PV(= nRT)$ is less at point C than at point B (curve AB is for an isothermal process, for which $\Delta U = 0$ and $\Delta T = 0$). In an adiabatic compression (going from C to A, for example), work is done *on* the gas, and hence the internal energy increases and the temperature rises. In a diesel engine, the rapid adiabatic compression reduces the volume by a factor of 15 or more; the temperature rise is so great that the air-fuel mixture ignites spontaneously.

| **CONCEPTUAL EXAMPLE 19–8** | **Simple adiabatic process.** Here is an adiabatic process that you can do with just a rubber band. Hold the rubber band loosely with two hands and gauge the temperature with your lips. Stretch the rubber band suddenly and again touch it lightly to your lips. You should notice an increase in temperature. Explain clearly why the temperature increases.

RESPONSE Stretching the rubber band *suddenly* makes the process adiabatic because there is no time for heat to enter or leave the system, so $Q = 0$. You do work on the system, representing an energy input, so by the first law of thermodynamics the internal energy increases, corresponding to an increase in temperature (Eq. 19–1); so the rubber band heats up.

Isobaric and Isochoric Processes

Isothermal and adiabatic processes are just two possible processes that can occur. Two other simple thermodynamic processes are illustrated on the *PV* diagram of Fig. 19–9: (a) an **isobaric** process is one in which the pressure is kept constant, so the process is represented by a horizontal straight line on the *PV* diagram, Fig. 19–9a; (b) an **isochoric** or isovolumetric process is one in which the volume does not change (Fig. 19–9b). In these, and in all other processes, the first law of thermodynamics holds.

Work Done in Volume Changes

We often want to calculate the work done in a process. Suppose we have a gas confined to a cylindrical container fitted with a movable piston (Fig. 19–10). We must always be careful to define exactly what our system is. In this case we choose our system to be the gas; so the container's walls and the piston are parts of the environment. Now let us calculate the work done by the gas when it expands quasistatically, so that P and T are defined for the system at all instants.[†] The gas expands against the piston, whose area is A. The gas exerts a force $F = PA$ on the piston, where P is the pressure in the gas. The work done by the gas to move the piston an infinitesimal displacement $d\mathbf{l}$ is

$$dW = \mathbf{F} \cdot d\mathbf{l} = PA \, dl = P \, dV \qquad (19\text{–}4)$$

since the infinitesimal increase in volume is $dV = A \, dl$. If the gas were *compressed* so that $d\mathbf{l}$ pointed into the gas, the volume would decrease and $dV < 0$. The work done by the gas in this case would then be negative, which is equivalent to saying that positive work was done *on* the gas, not by it. For a finite change in volume from V_A to V_B, the work W done by the gas will be

$$W = \int dW = \int_{V_A}^{V_B} P \, dV. \qquad (19\text{–}5)$$

Equations 19–4 and 19–5 are valid for the work done in any volume change—by a gas, a liquid, or a solid—as long as it is done quasistatically.

[†] If the gas expanded or were compressed quickly, there would be turbulence, and different parts would be at different pressure (and temperature).

In order to integrate Eq. 19–5, we need to know how the pressure varies during the process, and this depends on the type of process. Let us first consider a quasistatic isothermal expansion of an ideal gas. This process is represented by the curve between points A and B on the PV diagram of Fig. 19–11. The work done by the gas in this process, according to Eq. 19–5, is just the area between the PV curve and the V axis, and is shown shaded in Fig. 19–11. We can do the integral in Eq. 19–5 for an ideal gas by using the ideal gas law, $P = nRT/V$. The work done is

$$W = \int_{V_A}^{V_B} P\, dV = nRT \int_{V_A}^{V_B} \frac{dV}{V} = nRT \ln \frac{V_B}{V_A}. \qquad \begin{bmatrix} \text{isothermal process;} \\ \text{ideal gas} \end{bmatrix} \quad \textbf{(19–6)}$$

Let us next consider a different way of taking an ideal gas between the same states A and B. This time, let us lower the pressure in the gas from P_A to P_B, as indicated by the line AD in Fig. 19–12. (In this *isochoric* process, heat must be allowed to flow out of the gas so its temperature drops.) Then let the gas expand from V_A to V_B at constant pressure $(= P_B)$, which is indicated by the line DB in Fig. 19–12. (In this *isobaric* process, heat must be added to the gas to raise its temperature.) No work is done in the isochoric process AD, since $dV = 0$:

$$W = 0. \qquad \text{[isochoric process; ideal gas]}$$

In the isobaric process DB the pressure remains constant, so

$$W = \int_{V_A}^{V_B} P\, dV = P_B(V_B - V_A) = P\,\Delta V. \qquad \begin{bmatrix} \text{isobaric process;} \\ \text{ideal gas} \end{bmatrix} \quad \textbf{(19–7a)}$$

The work done is again represented by the area between the curve (ADB) on the PV diagram, and the V axis as indicated by the shading in Fig. 19–12. Using the ideal gas law, we can also write

$$W = P_B(V_B - V_A) = nRT_B\left(1 - \frac{V_A}{V_B}\right). \qquad \begin{bmatrix} \text{isobaric process;} \\ \text{ideal gas} \end{bmatrix} \quad \textbf{(19–7b)}$$

As can be seen from the shaded areas in Figs. 19–11 and 19–12, or by putting in numbers in Eqs. 19–6 and 19–7 $\left(\text{try it for } V_B = 2V_A\right)$, the work done in these two processes is different. This is a general result. *The work done in taking a system from one state to another depends not only on the initial and final states but also on the type of process (or "path").*

This result reemphasizes the fact that work cannot be considered a property of a system. The same is true of heat. The heat input required to change the gas from state A to state B depends on the process; for the isothermal process of Fig. 19–11, the heat input turns out to be greater than for the process ADB of Fig. 19–12. In general, *the amount of heat added or removed in taking a system from one state to another depends not only on the initial and final states but also on the path or process.*

EXAMPLE 19–9 **First law in isobaric and isochoric processes.** An ideal gas is slowly compressed at a constant pressure of 2.0 atm from 10.0 L to 2.0 L, path B to D in Fig. 19–12. (In this process, some heat flows out and the temperature drops.) Heat is then added to the gas, holding the volume constant, and the pressure and temperature are allowed to rise until the temperature reaches its original value, process D to A in Fig. 19–12. Calculate (*a*) the total work done by the gas in the process BDA, and (*b*) the total heat flow into the gas.

SOLUTION (*a*) Work is done only in the first part, the compression (BD)

$$W = P\,\Delta V = (2.0 \times 10^5 \, \text{N/m}^2)(2.0 \times 10^{-3}\,\text{m}^3 - 10.0 \times 10^{-3}\,\text{m}^3) = -1.6 \times 10^3 \, \text{J}.$$

From D to A no work is done ($\Delta V = 0$); so the total work done by the gas is -1.6×10^3 J, where the minus means that $+1.6 \times 10^3$ J of work is done *on* the gas. (*b*) Since the temperature at the beginning and at the end of the process is the same, there is no change in internal energy of our ideal gas: $\Delta U = 0$. From the first law of thermodynamics we have

$$0 = \Delta U = Q - W,$$

so $Q = W = -1.60 \times 10^3$ J. Since Q is negative, 1600 J of heat flows out of the gas for the whole process, BDA.

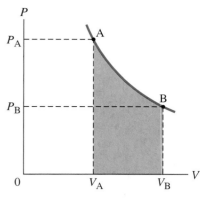

FIGURE 19–11 Work done by an ideal gas in an isothermal process equals the area under the PV curve. Shaded area equals the work done by the gas when it expands from V_A to V_B.

FIGURE 19–12 Process ADB consists of an isochoric (AD) and an isobaric (DB) process.

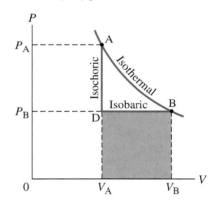

EXAMPLE 19–10 **Work done in an engine.** In an engine, 0.25 moles of gas in the cylinder expands rapidly and adiabatically against the piston. In the process, the temperature drops from 1150 K to 400 K. How much work does the gas do? Assume the gas is ideal.

SOLUTION Instead of calculating the work directly, we can more simply use the first law of thermodynamics, if we can determine ΔU since we know $Q = 0$ because the process is adiabatic. We determine ΔU from Eq. 19–1 for the internal energy of an ideal monatomic gas:

$$\Delta U = U_f - U_i = \tfrac{3}{2}nR(T_f - T_i)$$
$$= \tfrac{3}{2}(0.25 \text{ mol})(8.315 \text{ J/mol·K})(400 \text{ K} - 1150 \text{ K}) = -2300 \text{ J}.$$

Then, from Eq. 19–3, $W = Q - \Delta U = 0 - (-2300 \text{ J}) = 2300 \text{ J}$.

Free Expansion

FIGURE 19–13 Free expansion.

For a free expansion
$Q = 0, W = 0, \Delta U = 0$

One type of adiabatic process is a so-called **free expansion** in which a gas is allowed to expand in volume adiabatically without doing any work. The apparatus to accomplish a free expansion is shown in Fig. 19–13. It consists of two well-insulated compartments (to ensure no heat flow in or out) connected by a valve or stopcock. One compartment is filled with gas, the other is empty. When the valve is opened, the gas expands to fill both containers. No heat flows in or out ($Q = 0$), and no work is done because the gas does not move any other object. Thus $Q = W = 0$ and by the first law of thermodynamics, $\Delta U = 0$. *The internal energy of a gas does not change in a free expansion.* For an ideal gas, $\Delta T = 0$ also, since U depends only on T (Section 19–2). Experimentally, the free expansion has been used to determine if the internal energy of *real gases* depends only on T. The experiments are very difficult to do accurately, but it has been found that the temperature of a real gas drops very slightly in a free expansion. Thus the internal energy of real gases does depend, a little, on pressure or volume as well as on temperature.

Note, incidentally, that a free expansion could not be plotted on a PV diagram, because the process is rapid, not quasistatic. The intermediate states are not equilibrium states, and hence the pressure (and even the volume at some instants) is not clearly defined.

19–8 Molar Specific Heats for Gases, and the Equipartition of Energy

In Section 19–3 we discussed the concept of specific heat and applied it to solids and liquids. Much more than for solids and liquids, the values of the specific heat for gases depends on how the process is carried out. Two important processes are those in which either the volume or the pressure is kept constant. Although for solids and liquids it matters little, Table 19–3 shows that the specific heats of gases at constant volume (c_V) and at constant pressure (c_P) are quite different.

Molar Specific Heats for Gases

The difference in specific heats for gases is nicely explained in terms of the first law of thermodynamics and kinetic theory. Indeed, the values of the specific heats can be calculated using the kinetic theory, and the results are in close agreement with experiment. Our discussion is simplified if we use **molar specific heats**, C_V and C_P, which are defined as the heat required to raise 1 mol of the gas by 1 C° at constant volume and at constant pressure, respectively. That is, in analogy to Eq. 19–2, the heat Q needed to raise the temperature of n moles of gas by ΔT is

Molar specific heats

$$Q = nC_V \Delta T \qquad\qquad\qquad \text{[volume constant]} \qquad \textbf{(19–8a)}$$
$$Q = nC_P \Delta T. \qquad\qquad\qquad \text{[pressure constant]} \qquad \textbf{(19–8b)}$$

TABLE 19–3   Specific heats of gases at 15°C

Gas	Specific heats (kcal/kg·K)		Molar specific heats (cal/mol·K)		$C_P - C_V$ (cal/mol·K)	$\gamma = \dfrac{C_P}{C_V}$
	c_V	c_P	C_V	C_P		
Monatomic						
He	0.75	1.15	2.98	4.97	1.99	1.67
Ne	0.148	0.246	2.98	4.97	1.99	1.67
Diatomic						
N_2	0.177	0.248	4.96	6.95	1.99	1.40
O_2	0.155	0.218	5.03	7.03	2.00	1.40
Triatomic						
CO_2	0.153	0.199	6.80	8.83	2.03	1.30
H_2O (100°C)	0.350	0.482	6.20	8.20	2.00	1.32
Polyatomic						
C_2H_6	0.343	0.412	10.30	12.35	2.05	1.20

It is clear from the definition of molar specific heat (or by comparing Eqs. 19–2 and 19–8) that

$$C_V = Mc_V$$
$$C_P = Mc_P,$$

where M is the molecular mass of the gas ($M = m/n$ in grams/mol). The values for molar specific heats are included in Table 19–3, and we see that the values are nearly the same for different gases that have the same number of atoms per molecule.

Let us now make use of the kinetic theory and see first why the specific heats of gases are higher for constant-pressure processes than for constant-volume processes. Let us imagine that an ideal gas is slowly heated via these two different processes—first at constant volume, and then at constant pressure. In both of these processes, we let the temperature increase by the same amount, ΔT. In the process done at constant volume, no work is done since $\Delta V = 0$. Thus, according to the first law of thermodynamics, the heat added (which we now denote by Q_V) all goes into increasing the internal energy of the gas:

$$Q_V = \Delta U.$$

In the process carried out at constant pressure, work is done, and hence the heat added, Q_P, must not only increase the internal energy but also is used to do the work $W = P\Delta V$. Thus, more heat must be added in this process than in the first process at constant volume. For the process at constant pressure, we have from the first law of thermodynamics

$$Q_P = \Delta U + P\Delta V.$$

Since ΔU is the same in the two processes (ΔT was chosen to be the same), we can combine the two above equations:

$$Q_P - Q_V = P\Delta V.$$

From the ideal gas law, $V = nRT/P$, so for a process at constant pressure we have $\Delta V = nR\Delta T/P$. Putting this into the above equation and using Eqs. 19–8, we find

$$nC_P\,\Delta T - nC_V\,\Delta T = P\left(\frac{nR\Delta T}{P}\right)$$

or, after cancellations,

$$C_P - C_V = R. \tag{19–9}$$

Since the gas constant $R = 8.315\,\text{J/mol·K} = 1.99\,\text{cal/mol·K}$, our prediction is that C_P will be larger than C_V by about 1.99 cal/mol·K. Indeed, this is very close to what is obtained experimentally, as can be seen in the next to last column in Table 19–3.

Let us now calculate the molar specific heat of a monatomic gas using the kinetic theory model of gases. First we consider a process carried out at constant volume. Since no work is done in this process the first law of thermodynamics tells us that if heat Q is added to the gas, the internal energy of the gas changes by

$$\Delta U = Q.$$

For an ideal monatomic gas, the internal energy, U, is the total kinetic energy of all the molecules,

$$U = N(\tfrac{1}{2}m\overline{v^2}) = \tfrac{3}{2}nRT$$

as we saw in Section 19–2. Then, using Eq. 19–8a, we can write $\Delta U = Q$ in the form

$$\Delta U = \tfrac{3}{2}nR\,\Delta T = nC_V\,\Delta T \tag{19–10}$$

or

$$C_V = \tfrac{3}{2}R. \tag{19–11}$$

Since $R = 8.315\,\text{J/mol·K} = 1.99\,\text{cal/mol·K}$, kinetic theory predicts that $C_V = 2.98\,\text{cal/mol·K}$ for an ideal monatomic gas. This is very close to the experimental values for monatomic gases such as helium and neon (Table 19–3). From Eq. 19–9, C_P is predicted to be about $4.97\,\text{cal/mol·K}$, also in agreement with experiment.

Equipartition of Energy

Degrees of freedom

The measured molar specific heats for more complex gases (Table 19–3), such as diatomic (two atoms) and triatomic (three atoms) gases, increase with the increased number of atoms per molecule. We can explain this by assuming that the internal energy includes not only translational kinetic energy but other forms of energy as well. Take, for example, a diatomic gas. As shown in Fig. 19–14 the two atoms can rotate about two different axes (but rotation about a third axis passing through the two atoms would give rise to very little energy since the moment of inertia is so small). The molecules can have rotational as well as translational kinetic energy. It is useful to introduce the idea of **degrees of freedom**, by which we mean the number of independent ways molecules can possess energy. For example, a monatomic gas is said to have three degrees of freedom, since an atom can have velocity along the x axis, the y axis, and the z axis. These are considered to be three independent motions because a change in any one of the components would not affect any of the others. A diatomic molecule has the same three degrees of freedom associated with translational kinetic energy plus two more degrees of freedom associated with rotational kinetic energy, for a total of five degrees of freedom. A quick look at Table 19–3 indicates that the C_V for diatomic gases is about $\tfrac{5}{3}$ times as great as for a monatomic gas—that is, in the same ratio as their degrees of freedom. This led nineteenth-century physicists to an important idea, the **principle of equipartition of energy**. This principle states that energy is shared equally among the active degrees of freedom, and in particular each active degree of freedom of a molecule has on the average an energy equal to $\tfrac{1}{2}kT$. Thus, the average energy for a molecule of a monatomic gas would be $\tfrac{3}{2}kT$ (which we already knew) and of a diatomic gas $\tfrac{5}{2}kT$. Hence the internal energy of a diatomic gas would be $U = N(\tfrac{5}{2}kT) = \tfrac{5}{2}nRT$, where n is the number of moles. Using the same argument we did for monatomic gases, we see that for diatomic gases the molar specific heat at constant volume would be $\tfrac{5}{2}R = 4.97\,\text{cal/mol·K}$, in accordance with measured values. More complex molecules have even more degrees of freedom and thus greater molar specific heats.

Equipartition of energy

The situation was complicated, however, by measurements that showed that for diatomic gases at very low temperatures, C_V has a value of only $\tfrac{3}{2}R$, as if it had only three degrees of freedom. And at very high temperatures, C_V was about $\tfrac{7}{2}R$, as if there were seven degrees of freedom. The explanation is that at low temperatures, nearly all molecules have only translational kinetic energy. That is, no energy goes into rotational energy, so only three degrees of freedom are "active." At very high temperatures, on the other hand, all five degrees of freedom are active plus

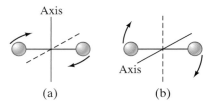

Axis

Axis

(a)   (b)

FIGURE 19–14 A diatomic molecule can rotate about two different axes.

two additional ones. We can interpret the two new degrees of freedom as being associated with the two atoms vibrating as if they were connected by a spring, as shown in Fig. 19–15. One degree of freedom comes from the kinetic energy of the vibrational motion, and the second comes from the potential energy of vibrational motion ($\frac{1}{2}kx^2$). At room temperature, these two degrees of freedom are apparently not active. See Fig. 19–16. Just why fewer degrees of freedom are "active" at lower temperatures was eventually explained by Einstein using the quantum theory. [According to quantum theory, energy does not take on continuous values but is quantized—it can have only certain values, and there is a certain minimum energy. The minimum rotational and vibrational energies are higher than for simple translational kinetic energy, so at lower temperatures and lower translational kinetic energy, there is not enough energy to excite the rotational or vibrational kinetic energy.] Calculations based on kinetic theory and the principle of equipartition of energy (as modified by the quantum theory) give numerical results in accord with experiment.

FIGURE 19–15 A diatomic molecule can vibrate, as if the two atoms were connected by a spring. Of course they are not connected by a spring, but rather exert forces on each other that are electrical in nature, but of a form that resembles a spring force.

FIGURE 19–16 Molar specific heat C_V as a function of temperature for hydrogen molecules (H_2). As the temperature is increased, some of the translational kinetic energy can be transferred in collisions into rotational kinetic energy and, at still higher temperature, into vibrational kinetic energy. [Note: H_2 dissociates into two atoms at about 3200 K, so the last part of the curve is shown dashed.]

Solids

The principle of equipartition of energy can be applied to solids as well. The molar specific heat of any solid, at high temperature, is close to $3R$ (6.0 cal/mol·K), Fig. 19–17. This is called the Dulong and Petit value after the scientists who first measured it in 1819. (Note that Table 19–1 gave the specific heats per kilogram, not per mole.) At high temperatures, each atom apparently has six degrees of freedom, although some are not active at low temperatures. Each atom in a crystalline solid can vibrate about its equilibrium position as if it were connected by springs to each of its neighbors (Fig. 19–18). Thus it can have three degrees of freedom for kinetic energy and three more associated with potential energy of vibration in each of the x, y, and z directions, which is in accord with the measured values.

FIGURE 19–17 Molar heat capacities of solids as a function of temperature.

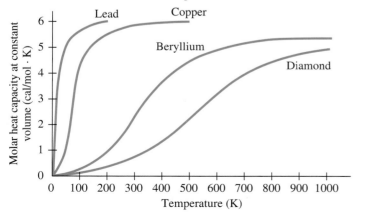

FIGURE 19–18 The atoms in a crystalline solid can vibrate about their equilibrium positions as if they were connected to their neighbors by springs. (The forces between atoms are actually electrical in nature.)

19–9 | Adiabatic Expansion of a Gas

The PV curve for the quasistatic (slow) adiabatic expansion ($Q = 0$) of an ideal gas was shown in Fig. 19–8 (curve AC). It is somewhat steeper than for an isothermal process ($\Delta T = 0$), which indicates that for the same change in volume the change in pressure will be greater. Hence the temperature of the gas must drop during an adiabatic expansion. Conversely, the temperature rises during an adiabatic compression.

We can derive the relation between the pressure P and the volume V of an ideal gas that is allowed to slowly expand adiabatically. We begin with the first law of thermodynamics, written in differential form:

$$dU = dQ - dW = -dW = -P\,dV,$$

since $dQ = 0$ for an adiabatic process. Equation 19–10 gives us a relation between ΔU and C_V, which is valid for any ideal gas process since U is a function only of T for an ideal gas. We write this in differential form:

$$dU = nC_V\,dT.$$

When we combine these last two equations, we obtain

$$nC_V\,dT + P\,dV = 0.$$

We next take the differential of the ideal gas law, $PV = nRT$, allowing $P, V,$ and T to vary:

$$P\,dV + V\,dP = nR\,dT.$$

We solve for dT in this relation and substitute it into the previous relation and get

$$nC_V\left(\frac{P\,dV + V\,dP}{nR}\right) + P\,dV = 0$$

or

$$(C_V + R)P\,dV + C_V V\,dP = 0.$$

We note from Eq. 19–9 that $C_V + R = C_P$, so we have

$$C_P P\,dV + C_V V\,dP = 0,$$

or

$$\frac{C_P}{C_V} P\,dV + V\,dP = 0.$$

We define

$$\gamma = \frac{C_P}{C_V} \tag{19-12}$$

so that our last equation becomes

$$\frac{dP}{P} + \gamma\frac{dV}{V} = 0.$$

This is integrated to become

$$\ln P + \gamma \ln V = \text{constant}.$$

This simplifies (using the rules for addition and multiplication of logarithms) to

$$PV^\gamma = \text{constant.} \qquad \left[\begin{array}{l}\text{quasistatic adiabatic} \\ \text{process; ideal gas}\end{array}\right] \tag{19-13}$$

This is the relation between P and V for a quasistatic adiabatic expansion or contraction. We will find it very useful when we discuss heat engines in the next chapter. Table 19–3 gives values of γ for some real gases. Figure 19–8 compares an adiabatic expansion (Eq. 19–13) in curve AC to an isothermal expansion

($PV = $ constant) in curve AB. It is important to remember that the ideal gas law, $PV = nRT$, continues to hold even for an adiabatic expansion ($PV^\gamma = $ constant); clearly PV is not constant, meaning T is not constant.

EXAMPLE 19–11 **Adiabatic vs. isothermal expansion.** An ideal monatomic gas is allowed to expand slowly until its pressure is reduced to exactly half its original value. By what factor does the volume change if the process is (*a*) adiabatic; (*b*) isothermal?

SOLUTION (*a*) From Eq. 19–13, $P_1 V_1^\gamma = P_2 V_2^\gamma$; hence

$$\frac{V_2}{V_1} = \left(\frac{P_1}{P_2}\right)^{1/\gamma} = (2)^{3/5} = 1.52$$

since $\gamma = C_P/C_V = (5/2)/(3/2) = 5/3$.
(*b*) Equation 19–13 does not apply for an isothermal process. But the ideal gas law, $PV = nRT$, always applies for an ideal gas. Hence, since $T_1 = T_2$, then $P_1 V_1 = P_2 V_2$ and

$$\frac{V_2}{V_1} = \frac{P_1}{P_2} = 2.0.$$

19–10 Heat Transfer: Conduction, Convection, Radiation

Heat is transferred from one place or body to another in three different ways: by *conduction*, *convection*, and *radiation*. We now discuss each of these in turn; but in practical situations, any two or all three may be operating at the same time. We start with conduction.

Three methods of heat transfer

Conduction

When a metal poker is put in a hot fire, or a silver spoon is placed in a hot bowl of soup, the exposed end of the poker or spoon soon becomes hot as well, even though it is not directly in contact with the source of heat. We say that heat has been conducted from the hot end to the cold end.

Heat **conduction** in many materials can be visualized as the result of molecular collisions. As one end of the object is heated, the molecules there move faster and faster. As they collide with their slower-moving neighbors, they transfer some of their energy to these molecules whose speeds thus increase. These in turn transfer some of their energy by collision with molecules still farther along the object. Thus the energy of thermal motion is transferred by molecular collision along the object. In metals, according to modern theory, it is collisions of free electrons within the metal with each other and with the metal lattice atoms that are visualized as being mainly responsible for conduction.

Heat conduction takes place only if there is a difference in temperature. Indeed, it is found experimentally that the rate of heat flow through a substance is proportional to the difference in temperature between its ends. The rate of heat flow also depends on the size and shape of the object. To investigate this quantitatively, let us consider the heat flow through a uniform object, as illustrated in Fig. 19–19. It is found experimentally that the heat flow ΔQ per time interval Δt is given by

$$\frac{\Delta Q}{\Delta t} = kA \frac{T_1 - T_2}{l}, \tag{19-14a}$$

where A is the cross-sectional area of the object, l is the distance between the two ends, which are at temperatures T_1 and T_2, and k is a proportionality constant called the **thermal conductivity**, which is characteristic of the material.

FIGURE 19–19 Heat conduction between areas at temperatures T_1 and T_2. If T_1 is greater than T_2, the heat flows to the right; the rate is given by Eq. 19–14a.

Rate of heat flow by conduction

TABLE 19–4
Thermal Conductivities

Substance	Thermal Conductivity, k	
	kcal/(s·m·C°)	J/(s·m·C°)
Silver	10×10^{-2}	420
Copper	9.2×10^{-2}	380
Aluminum	5.0×10^{-2}	200
Steel	1.1×10^{-2}	40
Ice	5×10^{-4}	2
Glass	2.0×10^{-4}	0.84
Brick	2.0×10^{-4}	0.84
Concrete	2.0×10^{-4}	0.84
Water	1.4×10^{-4}	0.56
Human tissue	0.5×10^{-4}	0.2
Wood	0.3×10^{-4}	0.1
Fiberglass	0.12×10^{-4}	0.048
Cork	0.1×10^{-4}	0.042
Wool	0.1×10^{-4}	0.040
Goose down	0.06×10^{-4}	0.025
Polyurethane	0.06×10^{-4}	0.024
Air	0.055×10^{-4}	0.023

FIGURE 19–20 Example 19–12.

$A = 3.0\,\text{m}^2$

14.0°C   15.0°C

$l = 3.2 \times 10^{-3}\,\text{m}$

Wind can cause much greater heat loss

➡ P H Y S I C S   A P P L I E D

Thermal windows

In some cases (such as when k or A cannot be considered constant) we need to consider the limit of an infinitesimally thin slab of thickness dx. Then Eq. 19–14a becomes

$$\frac{dQ}{dt} = -kA\frac{dT}{dx}, \qquad \text{(19–14b)}$$

where dT/dx is the temperature gradient and the negative sign is included since the heat flow is in the direction opposite to the temperature gradient.[†]

The thermal conductivities, k, for a variety of substances are given in Table 19–4. Substances for which k is large conduct heat rapidly and are said to be good **conductors**. Most metals fall in this category, although there is a wide range even among them as you may observe by holding the ends of a silver spoon and a stainless-steel spoon immersed in the same hot cup of soup. Substances for which k is small, such as fiberglass and down, are poor conductors of heat and are therefore good **insulators**. The relative magnitudes of k can explain simple phenomena such as why a tile floor feels much colder on the feet than a rug-covered floor at the same temperature. Tile is a better conductor of heat than the rug. Heat transferred from your foot to the rug is not conducted away rapidly, so the rug's surface quickly heats up to the temperature of your foot. But the tile conducts the heat away rapidly and thus can take more heat from your foot quickly, so your foot's surface temperature drops.

EXAMPLE 19–12 **Heat loss through windows.** A major source of heat loss from a house is through the windows. Calculate the rate of heat flow through a glass window 2.0 m × 1.5 m in area and 3.2 mm thick, if the temperatures at the inner and outer surfaces are 15.0°C and 14.0°C, respectively (Fig. 19–20).

SOLUTION Since $A = (2.0\,\text{m})(1.5\,\text{m}) = 3.0\,\text{m}^2$, $l = 3.2 \times 10^{-3}\,\text{m}$, and using Table 19–4 to get k, we have from Eq. 19–14a

$$\frac{\Delta Q}{\Delta t} = kA\frac{T_1 - T_2}{l} = \frac{(0.84\,\text{J/s·m·C°})(3.0\,\text{m}^2)(15.0°\text{C} - 14.0°\text{C})}{(3.2 \times 10^{-3}\,\text{m})}$$

$$= 790\,\text{J/s}.$$

This is equivalent to $(790\,\text{J/s})/(4.19 \times 10^3\,\text{J/kcal}) = 0.19\,\text{kcal/s}$, or $(0.19\,\text{kcal/s}) \cdot (3600\,\text{s/h}) = 680\,\text{kcal/h}$.

You might notice in this Example that 15°C is not very warm for the living room of a house. The room itself may indeed be much warmer, and the outside might be colder than 14°C. But the temperatures of 15°C and 14°C were specified as those at the window surfaces, and there is usually a considerable drop in temperature of the air in the vicinity of the window both on the inside and the outside. That is, the layer of air on either side of the window acts as an insulator, and normally the major part of the temperature drop between the inside and outside of the house takes place across the air layer. If there is a heavy wind, the air outside a window will constantly be replaced with cold air; the temperature gradient across the glass will be greater and there will be a much greater rate of heat loss. Increasing the width of the air layer, such as using two panes of glass separated by an air gap, will reduce the heat loss more than simply increasing the glass thickness, since the thermal conductivity of air is much less than for glass.

[†] This is quite similar to the relations describing diffusion (Chapter 18) and the flow of fluids through a tube (Chapter 13). In those cases, the flow of matter was found to be proportional to the concentration gradient or to the pressure gradient. This close similarity is one reason we speak of the "flow" of heat. Yet we must keep in mind that no substance is flowing in this case—it is energy that is being transferred.

The insulating properties of clothing come from the insulating properties of air. Without clothes, our bodies would heat the air in contact with the skin and would soon become reasonably comfortable because air is a very good insulator. But since air moves—there are breezes and drafts, and people themselves move about—the warm air would be replaced by cold air, thus increasing the temperature difference and the heat loss from the body. Clothes keep us warm by holding air so it cannot move readily. It is not the cloth that insulates us, but the air that the cloth traps. Down is a very good insulator because even a small amount of it fluffs up and traps a great amount of air. On this basis, can you see one reason why drapes in front of a window reduce heat loss from a house?

➡ **PHYSICS APPLIED**

*How clothes keep
us warm*

*Clothing
traps air*

For practical purposes, the thermal properties of building materials, particularly when considered as insulation, are commonly specified by R-values (or "thermal resistance"), defined for a given thickness l of material as:

$$R = \frac{l}{k}.$$

The R-value of a given piece of material thus combines the thickness l and the thermal conductivity k in one number. In the United States, R-values are given in British units (although often not stated at all!), as $ft^2 \cdot h \cdot F°/Btu$. Table 19–5 gives R-values for some common building materials: note that R-values increase directly with material thickness. For example, 2 inches of fiberglass has $R = 6\ ft^2 \cdot h \cdot F°/Btu$, half that for 4 inches.

TABLE 19–5 R-values

Material	Thickness	R-value ($ft^2 \cdot h \cdot F°/Btu$)
Glass	$\frac{1}{8}$ inch	1
Brick	$3\frac{1}{2}$ inch	0.6–1
Plywood	$\frac{1}{2}$ inch	0.6
Fiberglass insulation	4 inches	12

Convection

Although liquids and gases are generally not very good conductors of heat, they can transfer heat quite rapidly by convection. **Convection** is the process whereby heat is transferred by the mass movement of molecules from one place to another. Whereas conduction involves molecules (and/or electrons) moving only over small distances and colliding, convection involves the movement of molecules over large distances.

A forced-air furnace, in which air is heated and then blown by a fan into a room, is an example of *forced convection*. *Natural convection* occurs as well, and one familiar example is that hot air rises. For instance, the air above a radiator (or other type of heater) expands as it is heated, and hence its density decreases; because its density is less, it rises, just as a log submerged in water floats upward because its density is less than that of water. Warm or cold ocean currents, such as the balmy Gulf Stream, represent natural convection on a large scale. Wind is another example of convection, and weather in general is a result of convective air currents.

When a pot of water is heated (Fig. 19–21), convection currents are set up as the heated water at the bottom of the pot rises because of its reduced density and is replaced by cooler water from above. This principle is used in many heating systems, such as the hot-water radiator system shown in Fig. 19–22. Water is heated in the furnace and as its temperature increases, it expands and rises, as shown. This causes the water to circulate in the system. Hot water then enters the radiators, heat is transferred by conduction to the air, and the cooled water returns to the furnace. Thus, the water circulates because of convection; pumps are sometimes used to improve circulation. The air throughout the room also becomes heated as a result of convection. The air heated by the radiators rises and is replaced by cooler air, resulting in convective air currents, as shown.

Other types of furnaces also depend on convection. Hot-air furnaces with registers (openings) near the floor often do not have fans but depend on natural convection, which can be appreciable. In other systems, a fan is used. In either case, it is important that cold air can return to the furnace so that convective currents circulate throughout the room if the room is to be uniformly heated.

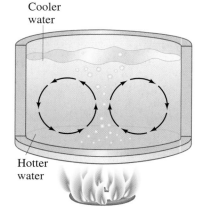

FIGURE 19–21 Convection currents in a pot of water being heated on a stove.

FIGURE 19–22 Convection plays a role in heating a house. The circular green arrows show the convective air currents in the rooms.

Radiation

Convection and conduction require the presence of matter as a medium to carry the heat from the hotter to the colder region. But a third type of heat transfer occurs without any medium at all. All life on Earth depends on the transfer of energy from the Sun, and this energy is transferred to the Earth over empty (or nearly empty) space. This form of energy transfer is heat—since the Sun's temperature is much higher (6000 K) than Earth's—and is referred to as **radiation**. The warmth we receive from a fire is mainly radiant energy (most of the air heated by a fire rises by convection up the chimney and does not reach us).

As we shall see in later chapters, radiation involves electromagnetic waves. Suffice it to say for now that radiation from the Sun consists of visible light plus many other wavelengths that the eye is not sensitive to, including infrared (IR) radiation which is mainly responsible for heating the Earth.

The rate at which an object radiates energy has been found to be proportional to the fourth power of the Kelvin temperature, T. That is, a body at 2000 K as compared to one at 1000 K radiates energy at a rate $2^4 = 16$ times greater. The rate of radiation is also proportional to the area A of the emitting object, so the rate at which energy leaves the object, $\Delta Q/\Delta t$, is

Radiation $\propto T^4$

$$\frac{\Delta Q}{\Delta t} = e\sigma A T^4. \tag{19-15}$$

This is called the **Stefan-Boltzmann equation**, and σ is a universal constant called the **Stefan-Boltzmann constant** which has the value

Stefan-Boltzmann constant

$$\sigma = 5.67 \times 10^{-8} \, \text{W/m}^2 \cdot \text{K}^4.$$

The factor e, called the **emissivity**, is a number between 0 and 1 that is characteristic of the material. Very black surfaces, such as charcoal, have emissivity close to 1, whereas bright shiny surfaces have e close to zero and thus emit correspondingly less radiation. The value of e depends somewhat on the temperature of the body.

Not only do light shiny surfaces emit less radiation, but they absorb little of the radiation that falls upon them (most is reflected). Black and very dark objects, on the other hand, absorb nearly all the radiation that falls on them—which is why light-colored clothing is usually preferable to dark clothing on a hot day. Thus, **a good absorber is also a good emitter.**

Good absorber is good emitter

Any object not only emits energy by radiation, but also absorbs energy radiated by other bodies. If an object of emissivity e and area A is at a temperature T_1, it radiates energy at a rate $e\sigma A T_1^4$. If the object is surrounded by an environment at temperature T_2 and high emissivity (≈ 1), the rate the surroundings radiate energy is proportional to T_2^4, and the rate that energy is absorbed by the object is proportional to T_2^4. The *net* rate of radiant heat flow from the object is given by the equation

Net flow rate of heat radiation

$$\frac{\Delta Q}{\Delta t} = e\sigma A\big(T_1^4 - T_2^4\big), \tag{19-16}$$

where A is the surface area of the object, T_1 its temperature and e its emissivity (at temperature T_1), and T_2 is the temperature of the surroundings. Notice in this equation that the rate of heat absorption by an object was taken to be $e\sigma A T_2^4$. That is, the proportionality constant is the same for both emission and absorption. This must be true to correspond with the experimental fact that equilibrium between object and surroundings is reached when they come to the same temperature. That is, $\Delta Q/\Delta t$ must equal zero when $T_1 = T_2$, so the coefficients of emission and absorption terms must be the same. This confirms the idea that a good emitter is a good absorber.

Because both the object and its surroundings radiate energy, there is a net transfer of energy from one to the other unless everything is at the same temperature. From Eq. 19–16 it is clear that if $T_1 > T_2$, the net flow of heat is from the body to the surroundings, so the body cools. But if $T_1 < T_2$, the net heat flow is from the surroundings into the body, and its temperature rises. If different parts of the surroundings are at different temperatures, Eq. 19–16 becomes more complicated.

EXAMPLE 19–13 ESTIMATE   Two teapots.   A ceramic teapot ($e = 0.70$) and a shiny one ($e = 0.10$) each hold 0.75 L of tea at 95°C. (*a*) Estimate the rate of heat loss from each, and (*b*) estimate the temperature drop after 30 min for each. Consider only radiation and assume the surroundings are at 20°C.

SOLUTION   (*a*) A teapot that holds 0.75 L can be approximated by a cube 10 cm on a side, with five sides exposed, so its surface area would be about $5 \times 10^{-2} \text{ m}^2$. The rate of heat loss would be about

$$\frac{\Delta Q}{\Delta t} = e\sigma A(T_1^4 - T_2^4)$$
$$= e(5.67 \times 10^{-8} \text{ W/m}^2\cdot\text{K}^4)(5 \times 10^{-2} \text{ m}^2)[(368 \text{ K})^4 - (293 \text{ K})^4]$$
$$= e(30) \text{ W},$$

or about 20 W for the ceramic pot ($e = 0.70$) and 3 W for the shiny one ($e = 0.10$). (*b*) To estimate the temperature drop, we use the concept of specific heat and ignore the contribution of the pots compared to that of the 0.75 L of water. Then, using Eq. 19–2,

$$\frac{\Delta T}{\Delta t} = \frac{\Delta Q/\Delta t}{mc} = \frac{e(30) \text{ J/s}}{(0.75 \text{ kg})(4.19 \times 10^3 \text{ J/kg}\cdot\text{C}°)} = e(0.010) \text{ C}°/\text{s},$$

which for 30 min (1800 s) represents about 12 C° for the ceramic pot and about 2 C° for the shiny one. The shiny one clearly has an edge, at least as far as radiation is concerned. However, convection and conduction could play a greater role than radiation.

Heating of an object by radiation from the Sun cannot be calculated using Eq. 19–16 since this equation assumes a uniform temperature, T_2, of the environment surrounding the object, whereas the Sun is essentially a point source. Hence the Sun must be treated as a separate source of energy. Heating by the Sun is calculated using the fact that about 1350 J of energy strikes the atmosphere of the Earth from the Sun per second per square meter of area at right angles to the Sun's rays. This number, 1350 W/m², is called the **solar constant**. The atmosphere may absorb as much as 70 percent of this energy before it reaches the ground, depending on the cloud cover. On a clear day, about 1000 W/m² reaches the Earth's surface. An object of emissivity e with area A facing the Sun absorbs heat at a rate, in watts, of about

$$\frac{\Delta Q}{\Delta t} = (1000 \text{ W/m}^2)eA \cos\theta,$$

where θ is the angle between the Sun's rays and a line perpendicular to the area A (Fig. 19–23). That is, $A \cos\theta$ is the "effective" area, at right angles to the Sun's rays. The explanations for the seasons and the polar ice caps (Fig. 19–24), and why the Sun heats the Earth more at midday than at sunrise or sunset, are also related to this $\cos\theta$ factor.

Notice that if a person wears light-colored clothing, e is much smaller, so the energy absorbed is less.

An interesting application of thermal radiation to diagnostic medicine is **thermography**. A special instrument, the thermograph, scans the body, measuring the intensity of radiation from many points and forming a picture that resembles an X-ray (Fig. 19–25). Areas where metabolic activity is high, such as in tumors, can often be detected on a thermogram as a result of their higher temperature and consequent increased radiation.

FIGURE 19–23   Radiant energy striking a body at an angle θ.

Radiation from the Sun

FIGURE 19–24   June sun makes an angle of about 23° with the equator. Thus (a) θ in the southern United States is near 0° (direct summer sun), whereas (b) in the southern hemisphere, θ is 50° or 60°, and less heat can be absorbed—hence it is winter. (c) At the poles, there is never strong direct sun; $\cos\theta$ varies from about $\frac{1}{2}$ in summer to 0 in winter; with so little heating, ice can form.

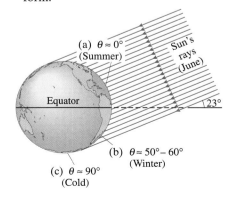

FIGURE 19–25 Thermograms of a healthy person's arms and hands (a) before smoking, (b) after smoking a cigarette, showing temperature decrease due to impaired blood circulation associated with smoking. The thermograms have been color-coded according to temperature; the scale on the right goes from blue (cold) to white (hot).

(a)

(b)

➡ **PHYSICS APPLIED**

Astronomy—size of a star

EXAMPLE 19–14 **ESTIMATE** **Star radius.** The giant star Betelgeuse emits radiant energy at a rate 10^4 times greater than our Sun, whereas its surface temperature is only half (2900 K) that of our Sun. Estimate the radius of Betelgeuse, assuming $e = 1$. The Sun's radius is $r_S = 7 \times 10^8$ m.

SOLUTION We assume both Betelgeuse and the Sun are spherical, with surface area $4\pi r^2$. We solve Eq. 19–15 for A:

$$4\pi r^2 = A = \frac{(\Delta Q/\Delta t)}{e\sigma T^4}.$$

Then

$$\frac{r_B^2}{r_S^2} = \frac{(\Delta Q/\Delta t)_B}{(\Delta Q/\Delta t)_S} \cdot \frac{T_S^4}{T_B^4} = (10^4)(2^4) = 16 \times 10^4.$$

Hence $r_B = \sqrt{16 \times 10^4}\, r_S = (400)(7 \times 10^8 \text{ m}) \approx 3 \times 10^{11}$ m. If Betelgeuse were our Sun, it would envelop us $\left(\text{Earth is } 1.5 \times 10^{11} \text{ m from the Sun}\right)$.

Summary

Thermal energy, or **internal energy**, U, refers to the total energy of all the molecules in a body. For an ideal monatomic gas

$$U = \tfrac{3}{2}NkT = \tfrac{3}{2}nRT.$$

Heat refers to the transfer of energy from one body to another because of a difference of temperature. Heat is thus measured in energy units, such as joules.

Heat and thermal energy are also sometimes specified in calories or kilocalories, where

$$1 \text{ cal} = 4.186 \text{ J}$$

is the amount of heat needed to raise the temperature of 1 g of water by 1 C°.

The **specific heat**, c, of a substance is defined as the energy (or heat) required to change the temperature of unit mass of substance by 1 degree; as an equation,

$$Q = mc\,\Delta T,$$

where Q is the heat absorbed or given off, ΔT is the temperature increase or decrease, and m is the mass of the substance.

When heat flows within an isolated system, conservation of energy tells us that the heat gained by one part of the system is equal to the heat lost by the other part of the system; this is the basis of **calorimetry**, which is the quantitative measurement of heat exchange.

Exchange of energy occurs, without a change in temperature, whenever a substance changes phase. The **heat of fusion** is the heat required to melt 1 kg of a solid into the liquid phase; it is also equal to the heat given off when the substance changes from liquid to solid. The **heat of vaporization** is the energy required to change 1 kg of a substance from the liquid to the vapor phase; it is also the energy given off when the substance changes from vapor to liquid.

The **first law of thermodynamics** states that the change in internal energy, ΔU, of a system is equal to the heat added to the system, Q, minus the work, W, done by the system:

$$\Delta U = Q - W.$$

This important law is a broad restatement of the conservation of energy and is found to hold for all processes.

Two simple thermodynamic processes are **isothermal**, which is a process carried out at constant temperature, and **adiabatic**, a process in which no heat is exchanged. The work done by (or on) a gas to change its volume by dV is $dW = P\,dV$, where P is the pressure.

Work and heat are not functions of the state of a system (as are P, V, T, n, and U) but depend on the type of process that takes a system from one state to another.

The **molar specific heat** of an ideal gas at constant volume, C_V, and at constant pressure, C_P, are related by $C_P - C_V = R$, where R is the gas constant. For a monatomic ideal gas, $C_V = \frac{3}{2}R$.

For ideal gases made up of diatomic or more complex molecules, C_V is equal to $\frac{1}{2}R$ times the number of **degrees of freedom** of the molecule. Unless the temperature is very high, some of the degrees of freedom may not be active and so do not contribute. According to the **principle of equipartition of energy**, energy is shared equally among the active degrees of freedom in an amount $\frac{1}{2}kT$ per molecule on average.

When an ideal gas expands (or contracts) adiabatically $(Q = 0)$, the relation $PV^\gamma = $ constant holds, where $\gamma = C_P/C_V$.

Heat is transferred from one place (or body) to another in three different ways. In **conduction**, energy is transferred from higher-kinetic energy molecules or electrons to lower-kinetic energy neighbors when they collide.

Convection is the transfer of energy by the mass movement of molecules over considerable distances.

Radiation, which does not require the presence of matter, is energy transfer by electromagnetic waves, such as from the Sun. All bodies radiate energy in an amount that is proportional to the fourth power of their Kelvin temperature (T^4) and to their surface area. The energy radiated (or absorbed) also depends on the nature of the surface (dark surfaces absorb and radiate more than bright shiny ones), which is characterized by the emissivity, e.

Questions

1. Is heat really involved in the Joule experiment, Fig. 19–1?
2. What happens to the work done when a jar of orange juice is vigorously shaken?
3. When a hot object warms a cooler object, does temperature flow between them? Are the temperature changes of the two objects equal?
4. (a) If two objects of different temperature are placed in contact, will heat naturally flow from the object with higher internal energy to the object with lower internal energy? (b) Is it possible for heat to flow even if the internal energies of the two objects are the same? Explain.
5. In warm regions where tropical plants grow but the temperature may drop below freezing a few times in the winter, the destruction of sensitive plants due to freezing can be reduced by watering them in the evening. Explain.
6. The specific heat of water is quite large. Explain why this fact makes water particularly good for heating systems (that is, hot-water radiators).
7. Why does water in a canteen stay cooler if the cloth jacket surrounding the canteen is kept moist?
8. Explain why burns caused by steam on the skin are often more severe than burns caused by water at 100°C.
9. Explain, using the concepts of latent heat and internal energy, why water cools (its temperature drops) when it evaporates.
10. Will potatoes cook faster if the water is boiling faster?
11. The temperature very high in the Earth's atmosphere can be 700°C. Yet an animal there would freeze to death rather than roast. Explain.
12. What happens to the internal energy of water vapor in the air that condenses on the outside of a cold glass of water? Is work done or heat exchanged? Explain.
13. Use the conservation of energy to explain why the temperature of a gas increases when it is compressed—say, by pushing down on a cylinder—whereas the temperature decreases when the gas expands.
14. In an isothermal process, 3700 J of work is done by an ideal gas. Is this enough information to tell how much heat has been added to the system? If so, how much?
15. One liter of air is cooled at constant pressure until its volume is halved. Then it is allowed to expand isothermally back to its original volume. Draw the process on a PV diagram.
16. Is it possible for the temperature of a system to remain constant even though heat flows into or out of it? If so, give examples.
17. Discuss how the first law of thermodynamics can apply to metabolism in humans. In particular, note that a person does work W, but very little heat Q is added to the body (rather, it tends to flow out). Why then doesn't the internal energy drop drastically in time?
18. Explain in words why C_P is greater than C_V.
19. Explain why the temperature of a gas increases when it is adiabatically compressed.
20. An ideal monatomic gas is allowed to expand slowly to twice its volume (1) isothermally; (2) adiabatically; (3) isobarically. Plot each on a PV diagram. In which process is ΔU the greatest, and in which is ΔU the least? In which is W the greatest and the least? In which is Q the greatest and the least?
21. Ceiling fans are sometimes reversible, so that they drive the air down in one season and pull it up in another season. Which way should you set the fan for summer? For winter?
22. Down sleeping bags and parkas are often specified as so many inches or centimeters of *loft*, the actual thickness of the garment when it is fluffed up. Explain.
23. Microprocessor chips nowadays have a "heat sink" glued on top that looks like a series of fins. Why is it shaped like that?
24. Sea breezes are often encountered on sunny days at the shore of a large body of water. Explain in light of the fact that the temperature of the land rises more rapidly than that of the nearby water.
25. The Earth cools off at night much more quickly when the weather is clear than when cloudy. Why?
26. Explain why air-temperature readings are always taken with the thermometer in the shade.
27. A premature baby in an incubator can be dangerously cooled even when the air temperature in the incubator is warm. Explain.

28. Heat loss occurs through windows by the following processes: (1) ventilation around edges; (2) through the frame, particularly if it is metal; (3) through the glass panes and (4) radiation. (*a*) For the first three, what is (are) the mechanism(s): conduction, convection, or radiation? (*b*) Heavy curtains reduce which of these heat losses? Explain in detail.

29. A piece of wood lying in the Sun absorbs more heat than a piece of shiny metal. Yet the wood feels less hot than the metal when you pick it up. Explain.

30. Explain why cities situated on the ocean tend to have less extreme temperatures than inland cities at the same latitude.

31. Early in the day, after the sun has reached the slope of a mountain, there tends to be a gentle upward movement of air. Later, after a slope goes into shadow, there is a gentle downdraft. Explain.

32. In the Northern Hemisphere the amount of heat required to heat a room where the windows face north is much higher than that required where the windows face south. Explain.

Problems

Section 19–1

1. (I) How much heat (joules) is required to raise the temperature of 30.0 kg of water from 15°C to 95°C?

2. (I) To what temperature will 7700 J of heat raise 3.0 kg of water initially at 10.0°C?

3. (II) When a 3.0-g bullet, traveling with a speed of 400 m/s, passes through a tree, its speed is reduced to 200 m/s. How much heat Q is produced and shared by the bullet and the tree?

4. (II) A British thermal unit (Btu) is a unit of heat in the British system of units. One Btu is defined as the heat needed to raise 1 pound of water by 1 F°. Show that
$$1\,\text{Btu} = 0.252\,\text{kcal} = 1055\,\text{J}.$$

5. (II) A water heater can generate 7200 kcal/h. How much water can it heat from 15°C to 50°C per hour?

6. (II) A small immersion heater is rated at 350 W. Estimate how long it will take to heat a cup of soup (assume this is 250 mL of water) from 20°C to 60°C.

7. (II) How many kilocalories of heat are generated when the brakes are used to bring a 1000-kg car to rest from a speed of 95 km/h?

Sections 19–3 and 19–4

8. (I) What is the specific heat of a metal substance if 135 kJ of heat is needed to raise 5.1 kg of the metal from 20°C to 30°C?

9. (I) An automobile cooling system holds 16 L of water. How much heat does it absorb if its temperature rises from 20°C to 90°C?

10. (II) A 35-g glass thermometer reads 21.6°C before it is placed in 135 mL of water. When the water and thermometer come to equilibrium, the thermometer reads 39.2°C. What was the original temperature of the water?

11. (II) The 1.20-kg head of a hammer has a speed of 6.5 m/s just before it strikes a nail (Fig. 19–26) and is brought to rest. Estimate the temperature rise of a 14-g iron nail generated by ten such hammer blows done in quick succession. Assume the nail absorbs all the energy.

12. (II) What will be the equilibrium temperature when a 245-g block of copper at 300°C is placed in a 150-g aluminum calorimeter cup containing 820 g of water at 12.0°C?

13. (II) A hot iron horseshoe (mass = 0.40 kg) which has just been forged, is dropped into 1.35 L of water in a 0.30-kg iron pot initially at 20°C. If the final equilibrium temperature is 25°C, estimate the initial temperature of the hot horseshoe.

14. (II) When 215 g of a substance is heated to 330°C and then plunged into a 100-g aluminum calorimeter cup containing 150 g of water at 12.5°C, the final temperature, as registered by a 17-g glass thermometer, is 35.0°C. What is the specific heat of the substance?

15. (II) How long does it take a 750-W coffeepot to bring to a boil 0.75 L of water initially at 8.0°C? Assume that the part of the pot which is heated with the water is made of 360 g of aluminum, and that no water boils away.

16. (II) Estimate the Calorie content of 100 g of Mullaney's fudge brownie from the following measurements. A 10-g sample of the brownie is allowed to dry before putting it in a bomb calorimeter. The aluminum bomb has a mass of 0.615 kg and is placed in 2.00 kg of water contained in an aluminum calorimeter cup of mass 0.524 kg. The initial temperature of the mixture is 15.0°C and its temperature after ignition is 36.0°C.

17. (II) (*a*) Show that if the specific heat varies as a function of temperature, $c(T)$, the heat needed to raise the temperature of a substance from T_1 to T_2 is given by
$$Q = \int_{T_1}^{T_2} mc(T)\,dT.$$
(*b*) Suppose $c(T) = c_0(1 + aT)$ for some substance, where $a = 2.0 \times 10^{-3}\,\text{C}^{\circ-1}$ and T is the Celsius temperature. Determine the heat required to raise the temperature from T_1 to T_2. (*c*) What is the mean value of c over the range T_1 to T_2 for part (*b*), expressed in terms of c_0, the specific heat at 0°C?

FIGURE 19–26
Problem 11.

Section 19–5

18. (I) How much heat is needed to melt 15.50 kg of silver that is initially at 20°C?

19. (I) During exercise, a person may give off 180 kcal of heat in 30 min by evaporation of water from the skin. How much water has been lost?

20. (I) If 2.80×10^5 J of energy is supplied to a flask of liquid oxygen at −183°C, how much oxygen will evaporate?

21. (II) What will be the final result when equal amounts of ice at 0°C and steam at 100°C are mixed together?

22. (II) A 40-g ice cube at its melting point is dropped into an insulated container of liquid nitrogen. How much nitrogen evaporates if it is at its boiling point of 77 K and has a latent heat of vaporization of 200 kJ/kg? Assume for simplicity that the specific heat of ice is a constant and is equal to its value near its melting point.

23. (II) A cube of ice is taken from the freezer at −8.5°C and placed in a 75-g aluminum calorimeter filled with 300 g of water at room temperature of 20°C. The final situation is observed to be all water at 17°C. What was the mass of the ice cube?

24. (II) An iron boiler of mass 230 kg contains 760 kg of water at 20°C. A heater supplies energy at the rate of 52,000 kJ/h. How long does it take for the water (*a*) to reach the boiling point, and (*b*) to all have changed to steam?

25. (II) On a hot day's race, a bicyclist consumes 8.0 L of water over the span of four hours. Making the approximation that all of the cyclist's energy goes into evaporating this water as sweat, how much energy in kcal did the rider use during the ride? (Since the efficiency of the rider is only about 20 percent, most of the energy consumed does go to heat, so our approximation is not far off.)

26. (II) What mass of steam at 100°C must be added to 1.00 kg of ice at 0°C to yield liquid water at 20°C?

27. (II) The specific heat of mercury is 138 J/kg·C°. Determine the latent heat of fusion of mercury using the following calorimeter data: 1.00 kg of solid Hg at its melting point of −39.0°C is placed in a 0.620-kg aluminum calorimeter with 0.430 kg of water at 12.80°C; the resulting equilibrium temperature is 5.06°C.

28. (II) A 54.0-kg ice-skater moving at 4.8 m/s glides to a stop. Assuming the ice is at 0°C and that 50 percent of the heat generated by friction is absorbed by the ice, how much ice melts?

29. (II) At a crime scene, the forensic investigator notes that the 8.2-g lead bullet that was stopped in a door-frame apparently melted completely on impact. Assuming the bullet was fired at room temperature (20°C), what does the investigator calculate the *minimum* muzzle velocity of the gun was?

Sections 19–6 and 19–7

30. (I) In Example 19–9, if the heat lost from the gas in the process BD is 2.78×10^3 J, what is the change in internal energy of the gas?

31. (I) An ideal gas expands isothermally, performing 5.00×10^3 J of work in the process. Calculate (*a*) the change in internal energy of the gas, and (*b*) the heat absorbed during this expansion.

32. (I) 1.0 L of air initially at 6.0 atm of (absolute) pressure is allowed to expand isothermally until the pressure is 1.0 atm. It is then compressed at constant pressure to its initial volume and lastly is brought back to its original pressure by heating at constant volume. Draw the process on a *PV* diagram, including numbers and labels for the axes.

33. (I) Sketch a *PV* diagram of the following process: 2.0 L of ideal gas at atmospheric pressure are cooled at constant pressure to a volume of 1.0 L, and then expanded isothermally back to 2.0 L, whereupon the pressure is increased at constant volume until the original pressure is reached.

34. (I) A gas is enclosed in a cylinder fitted with a light frictionless piston and maintained at atmospheric pressure. When 1400 kcal of heat is added to the gas, the volume is observed to increase slowly from 12.0 m³ to 18.2 m³. Calculate (*a*) the work done by the gas and (*b*) the change in internal energy of the gas.

35. (II) An ideal gas has its pressure cut in half slowly, while being kept in a rigid wall container. In the process, 1300 kJ of heat left the gas. (*a*) How much work was done during this process? (*b*) What was the change in internal energy of the gas during this process?

36. (II) In an engine, an ideal gas is compressed adiabatically to half its volume. In doing so, 2350 J of work is done on the gas. (*a*) How much heat flows into or out of the gas? (*b*) What is the change in internal energy of the gas? (*c*) Does its temperature rise or fall?

37. (II) An ideal gas expands at a constant pressure of 5.0 atm from 400 mL to 710 mL. Heat then flows out of the gas, at constant volume, and the pressure and temperature are allowed to drop until the temperature reaches its original value. Calculate (*a*) the total work done by the gas in the process, and (*b*) the total heat flow into the gas.

38. (II) Consider the following two-step process. Heat is allowed to flow out of an ideal gas at constant volume so that its pressure drops from 2.2 atm to 1.5 atm. Then the gas expands at constant pressure, from a volume of 6.8 L to 10.0 L, where the temperature reaches its original value. See Fig. 19–27. Calculate (*a*) the total work done by the gas in the process, (*b*) the change in internal energy of the gas in the process, and (*c*) the total heat flow into or out of the gas.

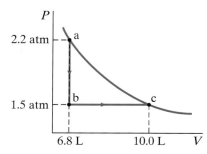

FIGURE 19–27 Problem 38.

39. (II) Suppose 2.00 mol of an ideal gas of volume $V_1 = 3.50$ m³ at $T_1 = 300$ K is allowed to expand isothermally to $V_2 = 7.00$ m³ at $T_2 = 300$ K. Determine (*a*) the work done by the gas, (*b*) the heat added to the gas, and (*c*) the change in internal energy of the gas.

40. (II) Determine (*a*) the work done and (*b*) the change in internal energy of 1.00 kg of water when it is all boiled to steam at 100°C. Assume a constant pressure of 1.00 atm.

41. (II) How much work is done to slowly compress, isothermally, 2.50 L of nitrogen at 0°C and 1 atm to 1.50 L at 0°C?

42. (II) The *PV* diagram in Fig. 19–28 shows two possible states of a system containing 1.45 moles of a monatomic ideal gas. $(P_1 = P_2 = 450 \text{ N/m}^2, \quad V_1 = 2.00 \text{ m}^3, \quad V_2 = 8.00 \text{ m}^3)$. (*a*) Draw the process which depicts an isobaric expansion from state 1 to state 2 and label this process (A). (*b*) Find the work done by the gas and the change in internal energy of the gas in process (A). (*c*) Draw the process which depicts an isothermal expansion from state 1 to the volume V_2 followed by an isochoric increase in temperature to state 2 and label this two-step process (B). (*d*) Find the change in internal energy of the gas for the two-step process (B).

FIGURE 19–28 Problem 42.

43. (II) Show that Eqs. 19–4 and 19–5 are valid for any shape volume that changes. To do so, draw an arbitrary closed curve to represent the boundary of the volume; then draw a slightly larger curve to represent an increase in volume; choose a small section of the original boundary, of area ΔA, and show that $dW = P \,\Delta A \, dl = P \, dV$ for this section; then integrate over the whole boundary, and finally integrate over a finite volume.

44. (II) When a gas is taken from a to c along the curved path in Fig. 19–29, the work done by the gas is $W = -35$ J and the heat added to the gas is $Q = -63$ J. Along path abc, the work done is $W = -48$ J. (*a*) What is Q for path abc? (*b*) If $P_c = \frac{1}{2}P_b$, what is W for path cda? (*c*) What is Q for path cda? (*d*) What is $U_a - U_c$? (*e*) If $U_d - U_c = 5$ J, what is Q for path da?

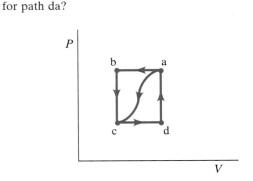

FIGURE 19–29 Problems 44, 45, and 46.

45. (III) In the process of taking a gas from state a to state c along the curved path shown in Fig. 19–29, 80 J of heat leave the system and 55 J of work are done *on* the system. (*a*) Determine the change in internal energy, $U_a - U_c$. (*b*) When the gas is taken along the path cda, the work done by the gas is $W = 38$ J. How much heat Q is added to the gas in the process cda? (*c*) If $P_a = 2.5P_d$, how much work is done by the gas in the process abc? (*d*) What is Q for path abc? (*e*) If $U_a - U_b = 10$ J, what is Q for the process bc? Here is a summary of what is given:

$$Q_{a\to c} = -80 \text{ J} \qquad U_a - U_b = 10 \text{ J}$$
$$W_{a\to c} = -55 \text{ J} \qquad P_a = 2.5P_d$$
$$W_{cda} = 38 \text{ J}$$

46. (III) Suppose a gas is taken clockwise around the rectangular cycle shown in Fig. 19–29, starting at b, then to a, to d, to c, and returning to b. Using the values given in Problem 45. (*a*) describe each leg of the process, and then calculate (*b*) the net work done during the cycle, (*c*) the net heat flow during the cycle, and (*d*) the total internal energy change during the cycle. (*e*) What percentage of the *intake* heat was turned into usable work: i.e., how efficient is this "rectangular" cycle (give as a percentage)?

*47. (III) Determine the work done by 1.00 mol of a van der Waals gas (Section 18–5) when it expands from volume V_1 to V_2 isothermally.

Section 19–8

48. (I) What is the internal energy of 4.50 mol of an ideal diatomic gas at 600 K, assuming all degrees of freedom are active?

49. (I) If a heater supplies 1.8×10^6 J/h to a room 6.5 m $\times$ 4.6 m $\times$ 3.0 m containing air at 20°C and 1.0 atm, by how much will the temperature rise in one hour, assuming no heat losses to the outside?

50. (I) Show that if the molecules of a gas have n degrees of freedom, then theory predicts $C_V = (n/2)R$ and $C_P = [(n + 2)/2]R$.

51. (I) Estimate the molar specific heat and specific heat at both constant pressure and constant volume for hydrogen gas (H_2) at room temperature.

52. (II) Show that the work done by n moles of an ideal gas when it expands adiabatically is $W = nC_V(T_1 - T_2)$, where T_1 and T_2 are the initial and final temperatures, and C_V is the molar specific heat at constant volume.

53. (II) A certain gas has specific heat $c_V = 0.0356$ kcal/kg·C°, which changes little over a wide temperature range. What is the atomic mass of this gas? What gas is it?

54. (II) The specific heat at constant volume of a particular gas is 0.182 kcal/kg·K at room temperature, and its molecular mass is 34. (*a*) What is its specific heat at constant pressure? (*b*) What do you think is the molecular structure of this gas?

55. (II) An audience of 2500 fills a concert hall of volume 30,000 m³. If there were no ventilation, by how much would the temperature rise over a period of 2.0 h due to the metabolism of the people (70 W/person)?

56. (II) A sample of 770 mol of nitrogen gas is maintained at a constant pressure of 1.00 atm in a flexible container. The gas is heated from 40°C to 180°C. Calculate (a) the heat added to the gas, (b) the work done by the gas, and (c) the change in internal energy.

57. (II) One mole of N_2 gas at 0°C is heated to 100°C at constant pressure (1.00 atm). Determine (a) the change in internal energy, (b) the work the gas does, and (c) the heat added to it.

58. (III) A 1.00-mol sample of an ideal diatomic gas at a pressure of 1.00 atm and temperature of 490 K undergoes a process in which its pressure increases linearly with temperature. The final temperature and pressure are 720 K and 1.60 atm. Determine (a) the change in internal energy, (b) the work done by the gas, and (c) the heat added to the gas. (Assume 5 active degrees of freedom.)

Section 19–9

59. (I) A 1.00-mol sample of an ideal diatomic gas, originally at 1.00 atm and 20°C, expands adiabatically to twice its volume. What are the final pressure and temperature for the gas? (Assume no molecular vibration.)

60. (II) Show, using Eqs. 19–4 and 19–13, that the work done by a gas that slowly expands adiabatically from pressure P_1 and volume V_1, to P_2 and V_2, is given by $W = (P_1 V_1 - P_2 V_2)/(\gamma - 1)$.

61. (II) 1.0 mol of an ideal monatomic gas at 300 K and 3.0 atm expands adiabatically to a final pressure of 1.0 atm. How much work does the gas do in the expansion?

62. (II) An ideal gas at 400 K is expanded adiabatically to 4.2 times its original volume. Determine its resulting temperature if the gas is (a) monatomic; (b) diatomic (no vibrations); (c) diatomic (molecules do vibrate).

63. (II) A 4.65-mol sample of an ideal diatomic gas expands adiabatically from a volume of $0.1210\ m^3$ to $0.750\ m^3$. Initially the pressure was 1.00 atm. Determine: (a) the initial and final temperatures; (b) the change in internal energy; (c) the heat lost by the gas; (d) the work done on the gas. (Assume no molecular vibration.)

64. (II) An ideal monatomic gas, consisting of 2.6 mol of volume $0.084\ m^3$, expands adiabatically. The initial and final temperatures are 25°C and −68°C. What is the final volume of the gas?

65. (III) A 1.00-mol sample of an ideal monatomic gas, originally at a pressure of 1.00 atm, undergoes a three-step process: (1) it is expanded adiabatically from $T_1 = 550$ K to $T_2 = 389$ K; (2) it is compressed at constant pressure until its temperature reaches T_3; (3) it then returns to its original pressure and temperature by a constant-volume process. (a) Plot these processes on a PV diagram. (b) Determine T_3. (c) Calculate the change in internal energy, the work done by the gas, and the heat added to the gas for each process, and (d) for the complete cycle.

Section 19–10

66. (I) Calculate the rate of heat flow by conduction in Example 19–12 assuming there are strong gusty winds and that the external temperature is −5°C.

67. (I) (a) How much power is radiated by a tungsten sphere (emissivity $e = 0.35$) of radius 18.0 cm at a temperature of 25°C? (b) If the sphere is enclosed in a room whose walls are kept at −5°C, what is the net flow rate of energy out of the sphere?

68. (I) Over what distance must there be heat flow by conduction from the blood capillaries beneath the skin to the surface if the temperature difference is 0.50 C°? Assume 200 W must be tranferred through the whole body's surface area of $1.5\ m^2$.

69. (I) What is the rate of energy absorption from the Sun by a person lying flat on the beach on a clear day if the Sun makes a 30° angle with the vertical? Assume that $e = 0.70$, the area of the body exposed to the Sun is $0.80\ m^2$, and that $1000\ W/m^2$ reaches the Earth's surface.

70. (II) Two rooms, each a cube 4.0 m on a side, share a 15-cm thick brick wall. Because of a number of 100-W light-bulbs in one room, the air is at 30°C, while in the other room it is at 10°C. How many of the 100-W light-bulbs are needed to maintain the temperature difference across the wall?

71. (II) How long does it take the Sun to melt a block of ice at 0°C with an area of $1.0\ m^2$ and thickness 1.6 cm? Assume that the Sun's rays make an angle of 30° with the normal to the area, and that the emissivity of ice is 0.050.

72. (II) A copper rod and an aluminum rod of the same length and cross-sectional area are attached end to end (Fig. 19–30). The copper end is placed in a furnace which is maintained at a constant temperature of 250°C. The aluminum end is placed in an ice bath held at constant temperature of 0.0°C. Calculate the temperature at the point where the two rods are joined.

FIGURE 19–30 Problem 72.

73. (II) (a) Estimate, using the solar constant, the rate at which the whole Earth receives energy from the Sun. (b) Assume the Earth radiates an equal amount back into space (that is, the Earth is in equilibrium). Then, assuming the Earth is a perfect emitter ($e = 1.0$), estimate its average surface temperature.

74. (II) Suppose the insulating qualities of the wall of a house come mainly from a 4.0-inch layer of brick and an *R*-19 layer of insulation, as shown in Fig. 19–31. What is the total rate of heat loss through such a wall, if its total area is 240 ft² and the temperature difference across it is 10 F°?

FIGURE 19–31 Two layers of insulation of a wall of a building. Problem 74.

FIGURE 19–32 Problem 75.

75. (II) A double-glazed window is one with two panes of glass separated by an air space, Fig. 19–32. (*a*) Show that the rate of heat flow by conduction is given by

$$\frac{\Delta Q}{\Delta t} = \frac{A(T_2 - T_1)}{l_1/k_1 + l_2/k_2 + l_3/k_3},$$

where $k_1, k_2,$ and k_3 are the thermal conductivities for glass, air, and glass, respectively. (*b*) Generalize this expression for any number of materials placed next to one another.

76. (III) Approximately how long should it take 8.50 kg of ice at 0°C to melt when it is placed in a carefully sealed Styrofoam icebox of dimensions 25 cm × 35 cm × 50 cm whose walls are 1.5 cm thick? Assume that the conductivity of Styrofoam is double that of air and that the outside temperature is 30°C.

77. (III) A house thermostat is normally set to 22°C, but at night it is turned down to 12°C for 7.0 h. Estimate how much more heat would be needed (state as a percentage of daily usage) if the thermostat were not turned down at night. Assume that the outside temperature averages 0°C for the 7.0 h at night and 8°C for the remainder of the day, and that the heat loss from the house is proportional to the difference in temperature inside and out. To obtain an estimate from the data, you will have to make other simplifying assumptions; state what these are.

78. (III) A cylindrical pipe has inner radius R_1 and outer radius R_2. The interior of the pipe carries hot water at temperature T_1. The temperature outside is T_2 ($< T_1$). (*a*) Show that the rate of heat loss for a length *L* of pipe is

$$\frac{dQ}{dt} = \frac{2\pi k(T_1 - T_2)L}{\ln(R_2/R_1)},$$

where *k* is the thermal conductivity of the pipe. (*b*) Suppose the pipe is steel with $R_1 = 3.0$ cm, $R_2 = 4.0$ cm, and $T_2 = 30$°C. If the pipe holds still water at $T_1 = 71$°C, what will be the initial rate of change of its temperature? (*c*) Suppose water at 71°C enters the pipe and moves at a speed of 8.0 cm/s. What will be its temperature drop per centimeter of travel?

General Problems

79. To get an idea of how much thermal energy is contained in the world's oceans, estimate the heat liberated when a cube, 1 km on a side, of ocean water is cooled by 1 K. (Approximate the ocean water by pure water for this estimate.)

80. A 15-g lead bullet is tested by firing it into a fixed block of wood with a mass of 0.92 kg. If the block and imbedded bullet together absorb all the heat energy generated and, after thermal equilibrium has been reached, the system has a temperature rise measured as 0.020 C°, estimate the entering speed of the bullet.

81. The specific heat of mercury is 0.033 kcal/kg·C°. When 1.00 kg of solid mercury at its melting point of −39°C is placed in a 0.50-kg aluminum calorimeter filled with 1.20 kg of water at 20.0°C, the final temperature of the mixture is found to be 16.5°C. What is the heat of fusion of mercury in kcal/kg?

82. A scuba diver releases a 3.00-cm-diameter (spherical) bubble of air from a depth of 14.0 m. Assume the temperature is constant at 298 K, and that the air behaves as a perfect gas. (*a*) How large is the bubble when it reaches the surface? (*b*) Sketch a *PV* diagram for the process. (*c*) Apply the first law of thermodynamics to the bubble, and find the work done by the air in rising to the surface, the change in its internal energy, and the heat added or removed from the air in the bubble as it rises. Take the density of water to be 1000 kg/m³.

83. A reciprocating compressor is a device that compresses air by a back-and-forth straight-line motion, like a piston in a cylinder. Consider a reciprocating compressor running at 400 rpm. During a compression stroke, 1.00 mol of air is compressed. The initial temperature of the air is 390 K, the engine of the compressor is supplying 10 kW of power to compress the air, and heat is being removed at the rate of 1.5 kW. Calculate the temperature change per compression stroke.

84. Suppose 2.0 mol of neon (an ideal monatomic gas) at STP are compressed slowly and isothermally to $\frac{1}{5}$ the original volume. The gas is then allowed to expand quickly and adiabatically back to its original volume. Find the highest and lowest temperatures and pressures attained by the gas, and show on a *PV* diagram where these values occur.

85. At very low temperatures, the molar specific heat of many substances varies as the cube of the absolute temperature:

$$C = k\frac{T^3}{T_0^3},$$

which is sometimes called Debye's law. For rock salt, $T_0 = 281$ K and $k = 1940$ J/mol·K. Determine the heat needed to raise 3.5 mol of salt from 22.0 K to 55.0 K.

86. (*a*) Find the total power radiated into space by the Sun, assuming it to be a perfect emitter at $T = 5500$ K. The Sun's radius is 7.0×10^8 m. (*b*) From this, determine the power per unit area arriving at the Earth, 1.5×10^{11} m away (Fig. 19–33).

$r = 1.5 \times 10^{11}$ m

Sun ——————————————— Earth **FIGURE 19–33**
Problem 86.

87. During light activity, a 70-kg person may generate 200 kcal/h. Assuming that 20 percent of this goes into useful work and the other 80 percent is converted to heat, calculate the temperature rise of the body after 1.00 h if none of this heat were transferred to the environment.

88. The *heat capacity*, *C*, of an object is defined as the amount of heat needed to raise its temperature by $1 \text{ C}°$. Thus, to raise the temperature by ΔT requires heat Q given by

$$Q = C \, \Delta T.$$

(*a*) Write the heat capacity *C* in terms of the specific heat, *c*, of the material. (*b*) What is the heat capacity of 1.0 kg of water? (*c*) Of 25 kg of water?

89. A very long 2.00-cm-diameter lead rod absorbs 410 kJ of heat. By how much does its length change? What would happen if the rod were only 2.0 cm long?

90. A mountain climber wears down clothing 3.5 cm thick with total surface area 1.9 m². The temperature at the surface of the clothing is $-20°$C and at the skin is $34°$C. Determine the rate of heat flow by conduction through the clothing (*a*) assuming it is dry and that the thermal conductivity, *k*, is that of down, and (*b*) assuming the clothing is wet, so that *k* is that of water and the jacket has matted down to 0.50 cm thickness.

91. A marathon runner has an average metabolism rate for the race of about 1000 kcal/h. If the runner has a mass of 65.0 kg, how much water would the runner lose to evaporation from the skin for a race that lasts 2.5 h?

92. Estimate the rate at which heat can be conducted from the interior of the body to the surface. Assume that the thickness of tissue is 4.0 cm, that the skin is at $34°$C and the interior at $37°$C, and that the surface area is 1.5 m². Compare this to the measured value of about 230 W that must be dissipated by a person working lightly. This clearly shows the necessity of convective cooling by the blood.

93. *Newton's law of cooling* states that for small temperature differences, if a body at a temperature T_1 is in surroundings at a temperature T_2, the body cools at a rate given by

$$\frac{\Delta Q}{\Delta t} = K(T_1 - T_2)$$

where *K* is a constant. It includes the effects of conduction, convection, and radiation. That this linear relationship should hold is obvious if only conduction is considered. Show that it is also approximately true for radiation by showing that Eq. 19–16 reduces to

$$\Delta Q/\Delta t = 4\sigma e A T_2^3 (T_1 - T_2)$$
$$= \text{constant} \times (T_1 - T_2)$$

if $(T_1 - T_2)$ is small.

94. A house has well-insulated walls 15.5 cm thick (assume conductivity of air) and area 410 m², a roof of wood 6.5 cm thick and area 280 m², and plain glass windows 0.65 cm thick and total area 33 m². (*a*) Assuming that the heat loss is only by conduction, calculate the rate at which heat must be supplied to this house to maintain its temperature at $23°$C if the outside temperature is $-10°$C. (*b*) If the house is initially at $10°$C, estimate how much heat must be supplied to raise the temperature to $23°$C within 30 min. Assume that only the air needs to be heated and that its volume is 750 m³. Take the specific heat of air to be 0.24 kcal/kg·C°.

95. A leaf of area 40 cm² and mass 4.5×10^{-4} kg directly faces the Sun on a clear day. The leaf has an emissivity of 0.85 and a specific heat of 0.80 kcal/kg·K. (*a*) Estimate the rate of rise of the leaf's temperature. (*b*) Calculate the temperature the leaf would reach if it lost all its heat by radiation (the surroundings are at $20°$C). (*c*) In what other ways can the heat be dissipated by the leaf?

96. An iron meteorite melts when it enters the Earth's atmosphere. If its initial temperature, outside of the atmosphere, was $-125°$C, calculate the minimum velocity the meteorite must have had before it entered Earth's atmosphere.

97. Write a formula for the density of a gas when it is allowed to expand (*a*) as a function of temperature when the pressure is kept constant and (*b*) as a function of pressure when temperature is kept constant.

98. If 2.0 moles of an ideal monatomic gas expand adiabatically, performing 7500 J of work in the process, what is the change in temperature of the gas?

99. When 5.30×10^4 J of heat are added to a gas enclosed in a cylinder fitted with a light frictionless piston maintained at atmospheric pressure, the volume is observed to increase from 2.2 m³ to 4.1 m³. Calculate (*a*) the work done by the gas, and (*b*) the change in internal energy of the gas. (*c*) Graph this process on a *PV* diagram.

100. A diesel engine accomplishes ignition without a spark plug by an adiabatic compression of air to a temperature above the ignition temperature of the diesel fuel, which is injected into the cylinder at the peak of the compression. Suppose air is taken into the cylinder at 300 K and volume V_1 and is compressed adiabatically to $560°$C ($\approx 1000°$F) and volume V_2. Assuming that the air behaves as an ideal gas whose ratio of C_P to C_V is 1.4, calculate the compression ratio V_1/V_2 of the engine.

101. An ice sheet forms on a lake. The air above the sheet is at $-15°$C, whereas the water is at $0°$C. Assume that the heat of fusion of the water freezing on the lower surface is conducted through the sheet to the air above. How much time will it take to form a sheet of ice 25 cm thick?

102. The temperature within the Earth's crust increases about 1.0 C° for each 30 m of depth. The thermal conductivity of the crust is 0.80 W/C°·m. (*a*) Determine the heat transferred from the interior to the surface for the entire Earth in one day. (*b*) Compare this heat to the amount of energy incident on the Earth in one day due to radiation from the Sun.

103. A 100-W lightbulb generates 95 W of heat, which is dissipated through a glass bulb that has a radius of 3.0 cm and is 1.0 mm thick. What is the difference in temperature between the inner and outer surfaces of the glass?

Two uses for a heat engine: a modern coal-burning power plant, and an old steam locomotive. Both produce steam which does work—on turbines to generate electricity, and on a piston that moves linkage to turn locomotive wheels. The efficiency of any engine—no matter how carefully engineered—is limited by nature as described in the second law of thermodynamics. This great law is best stated in terms of a quantity called entropy, which is unlike any other. Entropy is *not* conserved, but instead is constrained always to increase in any real process. Entropy is a measure of disorder. The second law of thermodynamics tells us that as time moves forward, the disorder in the universe increases.

We also discuss many practical matters such as heat engines, heat pumps, and refrigeration.

CHAPTER 20

Second Law of Thermodynamics

In this final chapter on heat and thermodynamics, we discuss the famous second law of thermodynamics, and the quantity "entropy" that arose from this fundamental law and is its quintessential expression. We also discuss heat engines—the engines that transform heat into work in power plants, trains, and motor vehicles—because they first showed us that a new law was needed. Finally, we briefly discuss the third law of thermodynamics.

20–1 The Second Law of Thermodynamics—Introduction

The first law of thermodynamics states that energy is conserved. There are, however, many processes we can imagine that conserve energy but are not observed to occur in nature. For example, when a hot object is placed in contact with a cold object, heat flows from the hotter one to the colder one, never spontaneously the reverse. If heat were to leave the colder object and pass to the hotter one, energy could still be conserved. Yet it doesn't happen spontaneously.[†] As a second example, consider what happens when you drop a rock and it hits the ground. The initial

---

[†] By spontaneously, we mean by itself without input of work of some sort. (A refrigerator does move heat from a cold environment to a warmer one, but only by doing work.)

potential energy of the rock changes to kinetic energy as the rock falls, and when the rock hits the ground this energy in turn is transformed into internal energy (thermal energy) of the rock and the ground in the vicinity of the impact; the molecules move faster and the temperature rises slightly. But have you seen the reverse happen—a rock at rest on the ground suddenly rise up in the air because the thermal energy of molecules is transformed into kinetic energy of the rock as a whole? Energy could be conserved in this process, yet we never see it happen.

The first law of thermodynamics, conservation of energy, would not be violated if either of these processes occurred in reverse. To explain this lack of reversibility, scientists in the latter half of the nineteenth century came to formulate a new principle known as the **second law of thermodynamics**. This law is a statement about which processes occur in nature and which do not. It can be stated in a variety of ways, all of which are equivalent. One statement, due to R. J. E. Clausius (1822–1888), is that

heat flows naturally from a hot object to a cold object; heat will not flow spontaneously from a cold object to a hot object.

Since this statement applies to one particular process, it is not obvious how it applies to other processes. A more general statement is needed that will include other possible processes in a more obvious way.

The development of a general statement of the second law was based partly on the study of heat engines. A **heat engine** is any device that changes thermal energy into mechanical work, such as steam engines and automobile engines. We now examine heat engines, both from a practical point of view and to show their importance in developing the second law of thermodynamics.

20–2 | Heat Engines

It is easy to produce thermal energy by doing work—for example, by simply rubbing your hands together briskly, or indeed by any frictional process. But to get work from thermal energy is more difficult, and the invention of a practical device to do this came only about 1700 with the development of the steam engine.

The basic idea behind any heat engine is that mechanical energy can be obtained from thermal energy only when heat is allowed to flow from a high temperature to a low temperature. In the process, some of the heat can then be transformed to mechanical work, as diagrammed schematically in Fig. 20–1. That is, a heat input $|Q_H|$ at a high temperature T_H is partly transformed into work $|W|$ and partly exhausted as heat $|Q_L|$ at a lower temperature T_L. By conservation of energy, $|Q_H| = |W| + |Q_L|$. The high and low temperatures, T_H and T_L, are called the **operating temperatures** of the engine. We will be interested only in engines that run in a repeating *cycle* (that is, the system returns repeatedly to its starting point) and thus can run continuously.

We use absolute value signs around Q_L, Q_H and W because we are interested only in the magnitudes, and thus avoid worrying about the sign conventions established in Section 19–6. These quantities are thus positive, and arrows on diagrams (as in Fig. 20–1) tell us the direction of energy transfers.

The temperatures T_H and T_L may or may not be precisely constant, depending on the cycle.

FIGURE 20–1 Schematic diagram of energy transfers for a heat engine.

High temperature

Steam

Water

Boiler

Intake valve (open during expansion)

Piston

Exhaust valve (closed during expansion)

Pump

Water

Condenser

Low temperature

(a) Reciprocating type

High-pressure steam, from boiler

Low-pressure steam, exhausted to condenser

(b) Turbine (boiler and condenser not shown)

FIGURE 20–2 Steam engines.

Steam Engine and Internal Combustion Engine

➥ PHYSICS APPLIED

Engines

The operation of two practical engines, the steam engine and the internal combustion engine (used in most automobiles), is illustrated in Figs. 20–2 and 20–3. Steam engines are of two main types, each making use of steam heated by combustion of coal, oil, or gas (or nuclear energy). In the so-called reciprocating type, Fig. 20–2a, the heated steam passes through the open intake valve and expands against a piston, forcing it to move. When the piston reaches the end of its stroke and begins to return to its original position, the exhaust valve opens and the piston forces the gases out. In a steam turbine, Fig. 20–2b, everything is essentially the same, except that the reciprocating piston is replaced by a rotating turbine that resembles a paddlewheel with many sets of blades. Most of our electricity today is generated using

FIGURE 20–3 Four-cycle internal combustion engine: (a) the gasoline–air mixture flows into the cylinder as the piston moves down; (b) the piston moves upward and compresses the gas; (c) firing of the spark plug ignites the gasoline–air mixture, raising it to a high temperature; (d) the gases, now at high temperature and pressure, expand against the piston in this, the power stroke; (e) the burned gases are pushed out to the exhaust pipe; the intake valve then opens, and the whole cycle repeats.

Intake valve (open)   Exhaust valve (closed)

Gas-air mixture from carburetor

Connecting rod

Crankshaft

(a) Intake

Both valves closed

Cylinder

Piston

(b) Compression

Both valves closed

Spark plug (firing)

(c) Ignition

Both valves closed

Rings

(d) Expansion (power stroke)

Exhaust valve (open)

To exhaust pipe

(e) Exhaust

steam turbines.[†] The material that is heated and cooled, steam in this case, is called the **working substance**. In a steam engine, the high temperature is obtained by burning coal, oil, or other fuel to heat the steam. In an internal combustion engine, the high temperature is achieved by burning the gasoline–air mixture in the cylinder itself (ignited by the spark plug), Fig. 20–3.

Why a ΔT is needed to drive a heat engine

To see why a *temperature difference* is required to run an engine, let us examine the steam engine. In the reciprocating engine, for example (Fig. 20–2a), suppose there were no condenser or pump, and that the steam was at the same temperature throughout the system. This would mean that the pressure of the gas being exhausted would be the same as that on intake. Thus, although work would be done by the gas *on* the piston when it expanded, an equal amount of work would have to be done *by* the piston to force the steam out the exhaust; hence, no net work would be done. In a real engine, the exhausted gas is cooled to a lower temperature and condensed so that the exhaust pressure is less than the intake pressure. Thus, although the piston must do work on the gas to expel it on the exhaust stroke, it is less than the work done by the gas on the piston during the intake. So a net amount of work can be obtained—but only if there is a difference of temperature. Similarly, in the gas turbine if the gas were not cooled, the pressure on each side of the blades would be the same; by cooling the gas on the exhaust side, the pressure on the front side of the blade is greater and hence the turbine turns.

The **efficiency**, e, of any heat engine can be defined as the ratio of the work it does, $|W|$, to the heat input at the high temperature, $|Q_H|$ (Fig. 20–1):

$$e = \frac{|W|}{|Q_H|}.$$

This is a sensible definition since $|W|$ is the output (what you get from the engine), whereas $|Q_H|$ is what you put in and pay for in fuel that burns. Because energy is conserved, the heat input $|Q_H|$ must equal the work done plus the heat that flows out at the low temperature $(|Q_L|)$:

$$|Q_H| = |W| + |Q_L|.$$

Thus $|W| = |Q_H| - |Q_L|$, and the efficiency of an engine is

$$e = \frac{|W|}{|Q_H|} = \frac{|Q_H| - |Q_L|}{|Q_H|} = 1 - \frac{|Q_L|}{|Q_H|}. \qquad \textbf{(20–1)}$$

Efficiency of any heat engine

EXAMPLE 20–1 **Car efficiency.** An automobile engine has an efficiency of 20 percent and produces an average of 23,000 J of mechanical work per second during operation. How much heat is discharged from this engine per second?

SOLUTION The output heat is $|Q_L|$. We are given $e = 0.20$, so from Eq. 20–1 we find that

$$\frac{|Q_L|}{|Q_H|} = 1 - e = 0.80.$$

We also know that $e = |W|/|Q_H|$, by definition, so in one second

$$|Q_H| = \frac{|W|}{e} = \frac{23,000 \text{ J}}{0.20} = 1.15 \times 10^5 \text{ J}.$$

Thus

$$|Q_L| = 0.80\,|Q_H| = (0.80)(1.15 \times 10^5 \text{ J}) = 9.2 \times 10^4 \text{ J}.$$

The engine discharges 9.2×10^4 J/s = 92,000 watts.

[†] Even nuclear power plants utilize steam turbines; the nuclear fuel—uranium—merely serves as fuel to heat the steam.

It is clear from Eq. 20–1 that the efficiency of an engine will be greater if $|Q_L|$ can be made small. However, from experience with a wide variety of systems, it has not been found possible to reduce $|Q_L|$ to zero. If $|Q_L|$ could be reduced to zero we would have a 100 percent efficient engine, as diagrammed in Fig. 20–4. That such a perfect engine (running continuously in a cycle) is not possible is another way of expressing the second law of thermodynamics. This can be stated formally as follows:

No device is possible whose sole effect is to transform a given amount of heat completely into work.

This is known as the **Kelvin-Planck statement of the second law of thermodynamics**. Said another way, *there can be no perfect (100 percent efficient) heat engine* such as that diagrammed in Fig. 20–4.

If the second law were not true, so that a perfect engine could be built, some rather remarkable things could happen. For example, if the engine of a ship did not need a low-temperature reservoir to exhaust heat into, the ship could sail across the ocean using the vast resources of the internal energy of the ocean water. Indeed, we would have no fuel problems at all!

20–3 Reversible and Irreversible Processes; Carnot Engine

In the early nineteenth century, the French scientist N. L. Sadi Carnot (1796–1832) studied in detail the process of transforming heat into mechanical energy. His aim had been to determine how to increase the efficiency of heat engines, but his studies soon led him to investigate the foundations of thermodynamics itself.

As an aid in this pursuit, Carnot, in 1824, invented (on paper) an idealized type of engine which we now call the "Carnot engine." The importance of the Carnot engine is not really as a practical engine, but rather as an aid to the understanding of heat engines in general, and also because Carnot and his engine contributed to the establishment and understanding of the second law of thermodynamics.

Reversible and Irreversible Processes

The Carnot engine involves *reversible processes*, so before we discuss it we must discuss what is meant by reversible and irreversible processes. A **reversible process** is one that is carried out infinitely slowly, so that the process can be considered as a series of equilibrium states, and the whole process could be done in reverse with no change in magnitude of the work done or heat exchanged. For example, a gas contained in a cylinder fitted with a tight, movable, but frictionless piston could be compressed isothermally in a reversible way if done infinitely slowly. Not all very slow (quasistatic) processes are reversible, however. If there is friction present, for example (as between the movable piston and cylinder just mentioned), the work done in one direction (going from some state A to state B) will not be the negative of the work done in the reverse direction (state B to state A). Such a process would not be considered reversible. Of course a perfectly reversible process is not possible in reality since it would require an infinite time; reversible processes can be approached arbitrarily closely, however, and they are very important theoretically.

All real processes are **irreversible**: they are not done infinitely slowly. There could be turbulence in the gas, friction would be present, and so on; any process could not be done precisely in reverse since the heat lost to friction would not reverse itself, the turbulence would be different, and so on. For any given volume there would not be a well-defined pressure P and temperature T since the system would not always be in an equilibrium state. Thus a real, irreversible, process cannot be plotted on a PV diagram (except insofar as it may approach an ideal

SECOND LAW OF THERMODYNAMICS
(Kelvin-Planck statement)

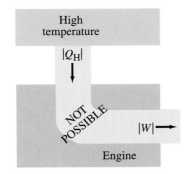

FIGURE 20–4 Schematic diagram of a hypothetical perfect heat engine in which all the heat input would be used to do work.

Reversible process

Irreversible process

reversible process). But a reversible process (since it is a quasistatic series of equilibrium states) always can be plotted on a *PV* diagram; and a reversible process, when done in reverse, retraces the same path on a *PV* diagram. Although all real processes are irreversible, reversible processes are conceptually important, just as the concept of an ideal gas is.

Carnot's Engine

Now let us look at Carnot's idealized engine. The Carnot engine makes use of a **reversible cycle**, by which we mean a series of reversible processes that take a given substance (the *working substance*) from an initial equilibrium state through many other equilibrium states and returns it again to the same initial state. In particular, the Carnot engine utilizes the **Carnot cycle**, which is illustrated in Fig. 20–5, with the working substance assumed to be an ideal gas. Let us take point a as the initial state. The gas is first expanded isothermally, and reversibly, path ab, at temperature T_H. To do so, we can imagine the gas to be in contact with a heat reservoir at a constant temperature T_H which delivers heat $|Q_H|$ to our working substance. Next the gas is expanded adiabatically and reversibly, path bc; no heat is exchanged and the temperature of the gas is reduced to T_L. The third step is a reversible isothermal compression, path cd, in contact with a heat reservoir at a constant low temperature, T_L, during which heat $|Q_L|$ flows out of the working substance. Finally, the gas is compressed adiabatically, path da, back to its original state. Thus a Carnot cycle consists of two isothermal and two adiabatic processes.

Carnot cycle

It is easy to see that the net work done in one cycle by a Carnot engine (or any other type of engine using a reversible cycle) is equal to the area enclosed by the curve representing the cycle on the *PV* diagram, the curve abcd in Fig. 20–5. (See Section 19–7.)

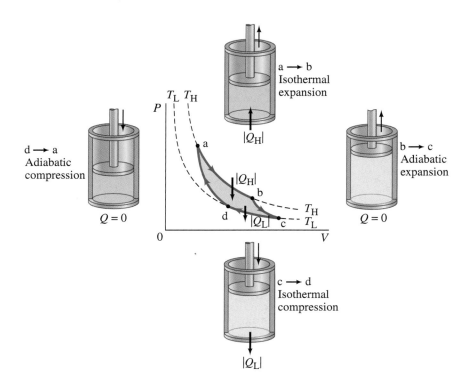

FIGURE 20–5 The Carnot cycle. Heat engines work in a cycle, and the cycle for the Carnot engine begins at point a on this *PV* diagram. (1) The gas is first expanded isothermally, with the addition of heat $|Q_H|$, along the path ab at temperature T_H. (2) Next the gas expands adiabatically from b to c—no heat is exchanged, but the temperature drops to T_L. (3) The gas is then compressed at constant temperature T_L, path c to d, and heat $|Q_L|$ flows out. (4) Finally, the gas is compressed adiabatically, path da, back to its original state. No Carnot engine actually exists, but as a theoretical engine it played an important role in the development of thermodynamics.

Carnot Efficiency and the Second Law of Thermodynamics

The efficiency of a Carnot engine, like any heat engine, is given by Eq. 20–1:

$$e = 1 - \frac{|Q_L|}{|Q_H|}.$$

For a Carnot engine, however, we can show that the efficiency depends only on the temperatures of the heat reservoirs, T_H and T_L. In the first isothermal process, ab in Fig. 20–5, the work done by the gas is (see Eq. 19–6)

$$W_{ab} = nRT_H \ln \frac{V_b}{V_a},$$

where n is the number of moles of the ideal gas used as working substance. Since the internal energy of an ideal gas does not change when the temperature remains constant, the first law of thermodynamics tells us that the heat added to the gas equals the work done by the gas:

$$|Q_H| = nRT_H \ln \frac{V_b}{V_a}.$$

Similarly, the heat lost by the gas in the isothermal process cd is

$$|Q_L| = nRT_L \ln \frac{V_c}{V_d}.$$

Since paths bc and da are adiabatic, we have (see Eq. 19–13)

$$P_b V_b^\gamma = P_c V_c^\gamma \quad \text{and} \quad P_d V_d^\gamma = P_a V_a^\gamma.$$

Also, from the ideal gas law,

$$\frac{P_b V_b}{T_H} = \frac{P_c V_c}{T_L} \quad \text{and} \quad \frac{P_d V_d}{T_L} = \frac{P_a V_a}{T_H}.$$

When we divide these last equations, term by term, into the corresponding set of equations on the line above, we obtain

$$T_H V_b^{\gamma-1} = T_L V_c^{\gamma-1} \quad \text{and} \quad T_L V_d^{\gamma-1} = T_H V_a^{\gamma-1}.$$

Next we divide the equation on the left by the one on the right and get

$$\left(\frac{V_b}{V_a}\right)^{\gamma-1} = \left(\frac{V_c}{V_d}\right)^{\gamma-1}.$$

Hence

$$\frac{V_b}{V_a} = \frac{V_c}{V_d}$$

or

$$\ln \frac{V_b}{V_a} = \ln \frac{V_c}{V_d}.$$

Using this result in our equations for $|Q_H|$ and $|Q_L|$ above we have

$$\frac{|Q_L|}{|Q_H|} = \frac{T_L}{T_H}. \qquad \text{[Carnot cycle]} \quad \textbf{(20–2)}$$

Hence the efficiency of a reversible Carnot engine is

Carnot (ideal) efficiency

$$e_{\text{ideal}} = 1 - \frac{|Q_L|}{|Q_H|} = 1 - \frac{T_L}{T_H}. \qquad \left[\begin{array}{l}\text{Carnot efficiency;}\\ \text{Kelvin temperatures}\end{array}\right] \quad \textbf{(20–3)}$$

The temperatures T_L and T_H are the absolute or Kelvin temperatures as measured on the ideal gas temperature scale. Thus the efficiency of a Carnot engine depends only on the temperatures T_L and T_H.

We could imagine other possible reversible cycles that could be used for an ideal reversible engine. According to a theorem stated by Carnot:

> **All reversible engines operating between the same two constant temperatures T_H and T_L have the same efficiency. Any irreversible engine operating between the same two fixed temperatures will have an efficiency less than this.**

Carnot's theorem

This is known as **Carnot's theorem**.[†] It tells us that Eq. 20–3, $e = 1 - (T_L/T_H)$, applies to any ideal reversible engine with fixed input and exhaust temperatures, T_H and T_L, and that this equation represents a maximum possible efficiency for a real (i.e., irreversible) engine.

In practice, the efficiency of real engines is always less than the Carnot efficiency. Well-designed engines reach perhaps 60 to 80 percent of Carnot efficiency.

EXAMPLE 20–2 Steam engine efficiency. A steam engine operates between 500°C and 270°C. What is the maximum possible efficiency of this engine?

SOLUTION We must first change the temperature to kelvins. Thus, $T_H = 773$ K and $T_L = 543$ K. Then from Eq. 20–3,

$$e_{ideal} = 1 - \frac{543}{773} = 0.30.$$

Thus, the maximum (or Carnot) efficiency is 30 percent. Realistically, an engine might attain 0.70 of this value or 21 percent. Note in this Example that the exhaust temperature is still rather high, 270°C. Steam engines are often arranged in series so that the exhaust of one engine is used as intake by a second or third engine.

EXAMPLE 20–3 A phony claim? An engine manufacturer makes the following claims: The heat input per second of the engine is 9.0 kJ at 475 K. The heat output per second is 4.0 kJ at 325 K. Do you believe these claims?

SOLUTION The efficiency of the engine is (Eq. 20–1)

$$e = \frac{|Q_H| - |Q_L|}{|Q_H|} = \frac{9.0 \text{ kJ} - 4.0 \text{ kJ}}{9.0 \text{ kJ}} = 0.56.$$

However, the maximum possible efficiency is given by the Carnot efficiency, Eq. 20–3:

$$e_{ideal} = \frac{T_H - T_L}{T_H} = \frac{475 \text{ K} - 325 \text{ K}}{475 \text{ K}} = 0.32.$$

So the manufacturer's claims violate the second law of thermodynamics and cannot be believed.

It is clear from Eq. 20–3 that a 100 percent efficient engine is not possible. Only if the exhaust temperature, T_L, were at absolute zero would 100 percent efficiency be obtainable. But reaching absolute zero is a practical (as well as theoretical) impossibility.[‡] Thus we can state, as we already did in Section 20–2, that **no device is possible whose sole effect is to transform a given amount of heat completely into work.** As we saw in Section 20–2, this is known as the *Kelvin-Planck statement of the second law of thermodynamics.* It tells us that there can be no perfect (100 percent efficient) heat engine such as the one diagrammed in Fig. 20–4.

SECOND LAW OF THERMODYNAMICS (Kelvin-Planck statement)

[†] Carnot's theorem can be shown to follow directly from either the Clausius or Kelvin-Planck statements of the second law of thermodynamics.

[‡] This result is known as the *third law of thermodynamics*, as discussed in Section 20–10.

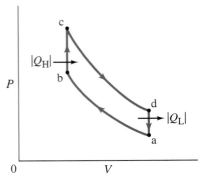

P

0   V

FIGURE 20–6 The Otto cycle.

* The Otto Cycle

The operation of an automobile internal combustion engine (see Fig. 20–3) can be approximated by a reversible cycle known as the *Otto cycle*, whose *PV* diagram is shown in Fig. 20–6. Unlike the Carnot cycle, the input and exhaust temperatures of the Otto cycle are *not* constant. Paths ab and cd are adiabatic, and paths bc and da are at constant volume. The gas (gasoline-air mixture) enters the cylinder at point a and is compressed adiabatically (compression stroke) to point b. At b ignition occurs (spark plug) and the burning of the gas adds heat $|Q_H|$ to the system at constant volume (approximately in a real engine). The temperature and pressure rise, and then in the power stroke, cd, the gas expands adiabatically. In the exhaust stroke, da, heat $|Q_L|$ is ejected to the environment (in a real engine, the gas leaves the engine and is replaced by a new mixture of air and fuel).

EXAMPLE 20–4 The Otto cycle. (*a*) For an ideal gas as working substance, show that the efficiency of an Otto cycle engine is

$$e = 1 - \left(\frac{V_a}{V_b}\right)^{1-\gamma}$$

where γ is the ratio of specific heats (Sections 19–8 and 19–9) and V_a/V_b is the *compression ratio*. (*b*) Calculate the efficiency for a compression ratio $V_a/V_b = 8.0$ assuming a diatomic gas like O_2 and N_2.

SOLUTION The heat exchanges take place at constant volume in the ideal Otto cycle, so from Eq. 19–8a:

$$|Q_H| = nC_V(T_c - T_b) \quad \text{and} \quad |Q_L| = nC_V(T_d - T_a).$$

Then from Eq. 20–1,

$$e = 1 - \frac{|Q_L|}{|Q_H|} = 1 - \left[\frac{T_d - T_a}{T_c - T_b}\right].$$

To get this in terms of the compression ratio, V_a/V_b, we use the result from Section 19–9, Eq. 19–13, $PV^\gamma = $ constant during the adiabatic processes ab and cd. Thus

$$P_a V_a^\gamma = P_b V_b^\gamma \quad \text{and} \quad P_c V_c^\gamma = P_d V_d^\gamma.$$

We use the ideal gas law, $P = nRT/V$, and substitute P into these two equations

$$T_a V_a^{\gamma-1} = T_b V_b^{\gamma-1} \quad \text{and} \quad T_c V_c^{\gamma-1} = T_d V_d^{\gamma-1}.$$

Then the efficiency (see above) is

$$e = 1 - \left[\frac{T_d - T_a}{T_c - T_b}\right] = 1 - \left[\frac{T_c(V_c/V_d)^{\gamma-1} - T_b(V_b/V_a)^{\gamma-1}}{T_c - T_b}\right].$$

But processes bc and da are at constant volume, so $V_c = V_b$ and $V_d = V_a$. Hence $V_c/V_d = V_b/V_a$ and

$$e = 1 - \left[\frac{(V_b/V_a)^{\gamma-1}(T_c - T_b)}{T_c - T_b}\right] = 1 - \left(\frac{V_b}{V_a}\right)^{\gamma-1} = 1 - \left(\frac{V_a}{V_b}\right)^{1-\gamma}.$$

(*b*) For diatomic molecules (Section 19–8), $\gamma = C_P/C_V = 1.4$ so

$$e = 1 - (8.0)^{1-\gamma} = 1 - (8.0)^{-0.4} = 0.56.$$

Real engines do not reach this high efficiency because they do not follow perfectly the Otto cycle, plus there is friction, turbulence, heat loss and incomplete combustion of the gases.

(a)   (b)   (c)

FIGURE 20–7 (a) An array of mirrors focuses sunlight on a boiler to produce steam at a solar energy installation. (b) A fossil-fuel steam plant. (c) Large cooling towers at an electric generating plant.

*Thermal Pollution

Much of the energy we utilize in everyday life—from motor vehicles to most of the electricity produced by power plants—makes use of a heat engine. Electricity produced by falling water at dams, by windmills, or by solar cells (Fig. 20–7a) does not involve a heat engine. But over 90 percent of the electric energy produced in the U.S. is done at fossil-fuel steam plants (Fig. 20–7b), and they make use of a heat engine (essentially steam engines). In electric power plants, the steam drives the turbines and generators whose output is electric energy. Even nuclear power plants use nuclear fuel to run a steam engine. The heat output $|Q_L|$ from every heat engine, from power plants to cars, is referred to as **thermal pollution** because this heat $|Q_L|$ must be absorbed by the environment—such as by water from rivers or lakes or via large cooling towers (Fig. 20–7c). This heat raises the temperature of the cooling water, altering the natural ecology of aquatic life (largely because warmer water holds less oxygen); or, in the case of air cooling towers, the output heat raises the temperature of the atmosphere which affects the weather. Air pollution—by which we mean the chemicals released in the burning of fossil fuels in cars, power plants, and industrial furnaces—gives rise to smog and other problems $\left(CO_2$ buildup in the atmosphere, causing the greenhouse effect and global warming$\right)$ that can be controlled to some extent and hopefully more so in the coming years. But thermal pollution is unavoidable. Engineers can try to design and build engines that are more efficient, but they cannot surpass the Carnot efficiency and must live with T_L being at best the ambient temperature of water or air. The second law of thermodynamics tells us the limit imposed by nature. What we can do, in the light of the second law of thermodynamics, is use less energy and conserve our fuel resources.

20–4 | Refrigerators, Air Conditioners, and Heat Pumps

The operating principle of refrigerators, air conditioners, and heat pumps is just the reverse of a heat engine. Each operates to transfer heat out of a cool environment into a warm environment. As diagrammed in Fig. 20–8, by doing work $|W|$, heat is taken from a low-temperature region, T_L (inside a refrigerator, say), and a greater amount of heat is exhausted at a high temperature, T_H (the room). You can often feel this heated air blowing out beneath a refrigerator. The

FIGURE 20–8 Schematic diagram of energy transfers for a refrigerator or air conditioner.

FIGURE 20–9 (a) Typical refrigerator system. The compressor motor forces a gas at high pressure through a heat exchanger (condenser) on the outside walls of the refrigerator where $|Q_H|$ is given off and the gas cools to become liquid. The liquid passes from a high-pressure region, via a valve, to low-pressure tubes on the inside walls of the refrigerator; the liquid evaporates at this lower pressure and thus absorbs heat ($|Q_L|$) from the inside of the refrigerator. The fluid returns to the compressor where the cycle begins again. (b) Schematic diagram, like Fig. 20–8.

work $|W|$ is usually done by a compressor motor which compresses a fluid, as illustrated in Fig. 20–9.

A perfect refrigerator—one in which no work is required to take heat from the low-temperature region to the high-temperature region—is not possible. This is **the Clausius statement of the second law of thermodynamics**, already mentioned in Section 20–1, which can be stated formally as

> **No device is possible whose sole effect is to transfer heat from one system at one temperature into a second system at a higher temperature.**

To make heat flow from a low-temperature body (or system) to one at a higher temperature, work must be done. Thus, *there can be no perfect refrigerator.*

Coefficient of performance (CP)

The **coefficient of performance (CP)** of a refrigerator is defined as the heat $|Q_L|$ removed from the low-temperature area (inside a refrigerator) divided by the work $|W|$ done to remove the heat (Fig. 20–8 or 20–9b):

$$CP = \frac{|Q_L|}{|W|}. \qquad \left[\begin{array}{c}\text{Refrigerator and} \\ \text{air conditioner}\end{array}\right] \quad \textbf{(20–4a)}$$

This makes sense since the more heat, $|Q_L|$, that can be removed from the inside of the refrigerator for a given amount of work, the better (more efficient) the refrigerator. Energy is conserved, so from the first law of thermodynamics we can write (see Fig. 20–8 or 20–9b) $|Q_L| + |W| = |Q_H|$, or $|W| = |Q_H| - |Q_L|$. Then Eq. 20–4a becomes

CP for

$$CP = \frac{|Q_L|}{|W|} = \frac{|Q_L|}{|Q_H| - |Q_L|}. \qquad \textbf{(20–4b)}$$

refrigerator

For an ideal refrigerator (not a perfect one, which is impossible), the best one could do would be

and

$$CP_{ideal} = \frac{T_L}{T_H - T_L}, \qquad \textbf{(20–4c)}$$

air conditioner

as for an ideal (Carnot) engine (Eq. 20–3).

➡ **PHYSICS APPLIED**

Air conditioner

An air conditioner works very much like a refrigerator, although the actual construction details are different because an air conditioner takes heat $|Q_L|$ from

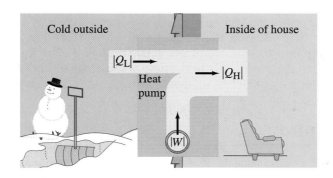

FIGURE 20–10 A heat pump "pumps" heat from the cold outside to the warm inside of a house.

inside a room or building at a low temperature, and deposits heat $|Q_H|$ outside to the environment at a higher temperature. Equations 20–4 also describe the coefficient of performance for an air conditioner.

Heat naturally flows from high temperature to low temperature. Refrigerators and air conditioners do work to accomplish the opposite: to make heat flow from cold to hot. We might say they "pump" heat from cold areas to hotter areas, against the natural tendency of heat to flow from hot to cold, just as water can be pumped uphill, against the natural tendency to flow downhill. The term **heat pump** is usually reserved for a device that can heat a house in winter by taking heat $|Q_L|$ from the outside at low temperature and delivering heat $|Q_H|$ to the warmer inside of the house, by doing work $|W|$; see Fig. 20–10. As in a refrigerator, there is an indoor and an outdoor heat exchanger (coils of the refrigerator) and a compressor. The operating principle is like that for a refrigerator or air conditioner; but the objective of a heat pump is to heat (deliver $|Q_H|$), rather than to cool (remove $|Q_L|$). Thus, the coefficient of performance of a heat pump is defined differently than for an air conditioner because it is the heat $|Q_H|$ delivered to the inside of the house that is important now:

$$\text{CP} = \frac{|Q_H|}{|W|}. \qquad \text{[heat pump]} \quad \textbf{(20–5)}$$

CP for heat pump

Most heat pumps can, however, be "turned around" and used as air conditioners in the summer.

EXAMPLE 20–5 **Heat pump.** A heat pump has a coefficient of performance of 3.0 and is rated to do work at 1500 W. (*a*) How much heat can it add to a room per second? (*b*) If the heat pump were turned around to act as an air conditioner in the summer, what would you expect its coefficient of performance to be, assuming all else stays the same?

SOLUTION (*a*) We use Eq. 20–5 for the heat pump and, since our device does 1500 J of work per second, it can pour heat into the room at a rate of

$$|Q_H| = \text{CP} \times |W| = 3.0 \times 1500\,\text{J} = 4500\,\text{J}$$

per second, or at a rate of 4500 W.
(*b*) If our device is turned around in summer, it can take heat $|Q_L|$ from inside the house and do 1500 J of work per second, and then dumps $|Q_H| = 4500$ J per second to the hot outside. Energy is conserved, so $|Q_L| + |W| = |Q_H|$ (see Fig. 20–10, but reverse inside and outside of house), so

$$|Q_L| = |Q_H| - |W| = 4500\,\text{J} - 1500\,\text{J} = 3000\,\text{J}.$$

The coefficient of performance as an air conditioner would thus be (Eq. 20–4a)

$$\text{CP} = \frac{|Q_L|}{|W|} = \frac{3000\,\text{J}}{1500\,\text{J}} = 2.0.$$

A good heat pump can be a money saver and an energy saver. Compare, for example, our heat pump in this Example to, say, a 1500-watt electric heater. We plug the latter into the wall, it draws 1500 watts of electricity and delivers 1500 watts of heat to the room. Our heat pump when plugged into the wall also draws 1500 W of electricity (which is what we pay for), but it delivers 4500 W of heat!

20–5 Entropy

We have seen several aspects of the second law of thermodynamics; but we have not yet arrived at a general statement of it. Both the Clausius and Kelvin-Planck statements deal with rather specific situations. Yet, as mentioned at the beginning of this chapter, there are a great many processes that simply are not observed in nature, even though they would not violate the first law of thermodynamics if they did occur. To cover all these processes, a more general statement of the second law of thermodynamics is needed. This general statement can be made in terms of a quantity, introduced by Clausius in the 1860s, called *entropy*, which we now discuss.

In our study of the Carnot cycle we found (Eq. 20–2) that $|Q_L|/|Q_H| = T_L/T_H$. We rewrite this as

$$\frac{|Q_H|}{T_H} = \frac{|Q_L|}{T_L}.$$

In this relation, both $|Q_H|$ and $|Q_L|$ are positive since they are absolute values. Let us now remove the absolute value signs and recall our original convention as used in the first law (see Section 19–6), that Q is positive when it represents a heat flow into the system (as Q_H) and negative for a heat flow out of the system (as Q_L). Then this relation becomes

$$\frac{Q_H}{T_H} + \frac{Q_L}{T_L} = 0. \qquad \text{[Carnot cycle]} \quad \textbf{(20–6)}$$

Now consider *any* reversible cycle, as represented by the smooth (oval-shaped) curve in Fig. 20–11. Any reversible cycle can be approximated as a series of Carnot cycles. Figure 20–11 shows only six—the isotherms (dashed lines) are connected by adiabatic paths for each—and the approximation becomes better and better if we increase the number of Carnot cycles. Equation 20–6 is valid for each of these cycles, so we can write

$$\Sigma \frac{Q}{T} = 0 \qquad \text{[Carnot cycles]} \quad \textbf{(20–7)}$$

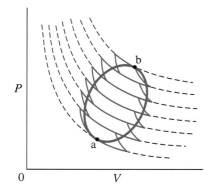

FIGURE 20–11 Any reversible cycle can be approximated as a series of Carnot cycles. (The dashed lines represent isotherms.)

for the sum of all these cycles. But note that the heat output Q_L of one cycle is approximately equal to the negative of the heat input, Q_H, of the cycle below it (actual equality in the limit of an infinite number of infinitely thin Carnot cycles). Hence the heat flows on the inner paths of all these Carnot cycles cancel out, so the net heat transferred, and the work done, is the same for the series of Carnot cycles as for the original cycle. Hence, in the limit of infinitely many Carnot cycles, Eq. 20–7 applies to any reversible cycle. In this case Eq. 20–7 becomes

$$\oint \frac{dQ}{T} = 0, \qquad \text{[reversible cycle]} \quad \textbf{(20–8)}$$

where dQ represents an infinitesimal heat flow.[†] The symbol $\oint$ means take the integral around a closed path; the integral can be started at any point on the path such

[†] dQ is often written $đQ$: see footnote at end of Section 19–6.

as at a or b in Fig. 20–11, and proceed in either direction. Let us divide the cycle of Fig. 20–11 into two parts as indicated in Fig. 20–12. Then we rewrite Eq. 20–8 as

$$\int_{a}^{b} \frac{dQ}{T} + \int_{b}^{a} \frac{dQ}{T} = 0.$$

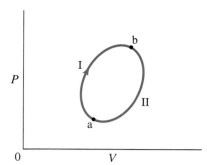

The first term is the integral from point a to point b along path I in Fig. 20–12, and the second term is the integral from b back to a along path II. (Path I plus path II is the whole cycle.) If one path is taken in reverse, say path II, dQ at each point becomes $-dQ$, since the path is reversible. We can therefore write

$$\int_{a}^{b} \frac{dQ}{T} = \int_{a}^{b} \frac{dQ}{T}. \qquad \text{[reversible paths]} \quad (20-9)$$

Since our cycle is arbitrary, Eq. 20–9 tells us that the integral of dQ/T between any two equilibrium states, a and b, does not depend on the path of the process. We therefore define a new quantity, called the **entropy**, S, by the relation

$$dS = \frac{dQ}{T}. \qquad (20-10)$$

FIGURE 20–12 The integral, $\oint dS$, of the entropy for a reversible cycle is zero. Hence the difference in entropy between states a and b, $S_b - S_a = \int_{a}^{b} dS$, is the same for path I as for path II.

Entropy defined

From Eq. 20–8, we have that

$$\oint dS = 0. \qquad \text{[reversible cycle]} \quad (20-11)$$

The quantity ΔS, where

$$\Delta S = S_b - S_a = \int_{a}^{b} dS = \int_{a}^{b} \frac{dQ}{T}, \qquad \text{[reversible processes]} \quad (20-12)$$

is seen from Eq. 20–9 to be *independent of the path between the two points a and b.* This is an important result. It tells us that the difference in entropy, $S_b - S_a$, between two equilibrium states of a system does not depend on how you get from one state to the other. Thus *entropy is a state variable*—its value depends only on the state of the system, and not on the process or the past history of how it got there.[†] This is in clear distinction to Q and W which are *not* state variables; their values do depend on the processes undertaken.

Entropy is a state variable

20–6 Entropy and the Second Law of Thermodynamics

We have defined a new quantity, S, the entropy, which can be used to describe the state of the system, along with P, T, V, U, and n. But what does this rather abstract quantity have to do with the second law of thermodynamics? To answer this, let us take some examples in which we calculate the entropy changes during particular processes. But note first that Eq. 20–12 can be applied only to reversible processes. How then do we calculate $\Delta S = S_b - S_a$ for a real process that is irreversible? What we can do is this: we figure out some other *reversible* process that takes the system between the same two states, and calculate ΔS for this reversible process. This will equal ΔS for the irreversible process since ΔS depends only on the initial and final states of the system.

How to determine ΔS
(using a reversible process)

[†]Equation 20–12 says nothing about the absolute value of S; it only gives the change in S. This is much like potential energy (Chapter 8). However, one form of the so-called *third law of thermodynamics* (see also Section 20–10) states that as $T \rightarrow 0$, $S \rightarrow 0$.

EXAMPLE 20–6 **Entropy change in melting.** A 1.00-kg piece of ice at 0°C melts very slowly to water at 0°C. Assume the ice is in contact with a heat reservoir whose temperature is only infinitesimally greater than 0°C. Determine the entropy change of (a) the ice cube and (b) the heat reservoir.

SOLUTION (a) The process is carried out at a constant temperature $T = 273$ K and is done reversibly, so we can use Eq. 20–12:

$$\Delta S_{\text{ice}} = \int \frac{dQ}{T} = \frac{1}{T} \int dQ = \frac{Q}{T}.$$

Since the heat needed to melt the ice is $Q = mL$, where the heat of fusion $L = 79.7$ kcal/kg $= 3.33 \times 10^5$ J/kg, we have

$$\Delta S_{\text{ice}} = \frac{mL}{T} = \frac{(1.00 \text{ kg})(79.7 \text{ kcal/kg})}{273 \text{ K}} = 0.292 \text{ kcal/K},$$

or 1220 J/K.
(b) Heat $Q = mL$ is *removed* from the heat reservoir, so (since $T = 273$ K and is constant)

$$\Delta S_{\text{res}} = -\frac{Q}{T} = -0.292 \text{ kcal/K}.$$

Note that the *total* entropy change, $\Delta S_{\text{ice}} + \Delta S_{\text{res}}$, is zero.

EXAMPLE 20–7 **ESTIMATE** **Entropy change when mixing water.** A sample of 50.0 kg of water at 20.0°C is mixed with 50.0 kg of water at 24.0°C. Estimate the change in entropy without using calculus.

SOLUTION The final temperature of the mixture will be 22.0°C, since we started with equal amounts of water. A quantity of heat,

$$Q = mc \, \Delta T = (50.0 \text{ kg})(1.00 \text{ kcal/kg} \cdot \text{K})(2.0 \text{ C}°) = 100 \text{ kcal},$$

flows out of the hot water as it cools down from 24°C to 22°C, and this heat flows into the cold water as it warms from 20°C to 22°C. The total change in entropy, ΔS, will be the sum of the changes in entropy of the hot water, ΔS_H, and that of the cold water, ΔS_C:

$$\Delta S = \Delta S_H + \Delta S_C.$$

We estimate entropy changes by writing $\Delta S = Q/T_{\text{av}}$, where T_{av} is an "average" temperature for each process, which ought to give a reasonable estimate since the temperature change is small. For the hot water we use an average temperature of 23°C (296 K), and for the cold water an average temperature of 21°C (294 K). Thus

$$\Delta S_H \approx -\frac{100 \text{ kcal}}{296 \text{ K}} = -0.338 \text{ kcal/K}$$

and

$$\Delta S_C \approx \frac{100 \text{ kcal}}{294 \text{ K}} = 0.340 \text{ kcal/K}.$$

Note that the entropy of the hot water (S_H) decreases since heat flows out of the hot water. But the entropy of the cold water (S_C) increases by a greater amount. The total change in entropy is

$$\Delta S = \Delta S_H + \Delta S_C \approx -0.338 \text{ kcal/K} + 0.340 \text{ kcal/K}$$
$$\approx +0.002 \text{ kcal/K}.$$

We see that although the entropy of one part of the system decreased, the entropy of the other part increased by a greater amount so that the net change in entropy of the whole system is positive.

We can now easily show in general that for an isolated system of two bodies, the flow of heat from the higher-temperature (T_H) body to the lower-temperature (T_L) body always results in an increase in the total entropy. The two bodies eventually come to some intermediate temperature, T_M. The heat lost by the hotter body $(Q_H = -Q$ where Q is positive$)$ is equal to the heat gained by the colder one $(Q_L = Q)$, so the total change in entropy is

$$\Delta S = \Delta S_H + \Delta S_L = -\frac{Q}{T_{HM}} + \frac{Q}{T_{LM}},$$

where T_{HM} is some intermediate temperature between T_H and T_M for the hot body as it cools from T_H to T_M, and T_{LM} is the counterpart for the cold body; since the temperature of the hot body is, at all times during the process, greater than that of the cold body, then $T_{HM} > T_{LM}$. Hence

$$\Delta S = Q\left(\frac{1}{T_{LM}} - \frac{1}{T_{HM}}\right) > 0.$$

One body decreases in entropy, while the other gains in entropy, but the *total* change is positive.

EXAMPLE 20–8 **Entropy changes in a free expansion.** Consider the *adiabatic free expansion* of n moles of an ideal gas from volume V_1 to volume V_2, where $V_2 > V_1$, as was discussed in Section 19–7, Fig. 19–13. Calculate the change in entropy (a) of the gas and (b) of the surrounding environment. (c) Evaluate ΔS for 1.00 mole, with $V_2 = 2.00\, V_1$.

SOLUTION As we saw in Section 19–7, the gas is initially in a closed container of volume V_1, and, with the opening of a valve, adiabatically it expands into a previously empty container; the total volume of the two containers is V_2. The whole apparatus is thermally insulated from the surroundings, so no heat flows into the gas, $Q = 0$. The gas does no work, $W = 0$, so there is no change in internal energy, $\Delta U = 0$, and the temperature of the initial and final states is the same, $T_2 = T_1 = T$.
(a) The process takes place very quickly, and so is irreversible. So we cannot apply Eq. 20–12 to this process. Instead we must think of a reversible process that will take the gas from volume V_1 to V_2 at the same temperature, and use Eq. 20–12 on this reversible process to get ΔS. A reversible isothermal process will do the trick; in such a process, the internal energy does not change, so (from the first law)

$$dQ = dW = P\, dV.$$
Then
$$\Delta S_{gas} = \int \frac{dQ}{T} = \frac{1}{T}\int_{V_1}^{V_2} P\, dV;$$

from the ideal gas law $P = nRT/V$, so

$$\Delta S_{gas} = \frac{nRT}{T}\int_{V_1}^{V_2} \frac{dV}{V} = nR \ln \frac{V_2}{V_1}.$$

Since $V_2 > V_1$, $\Delta S_{gas} > 0$.
(b) Since no heat is transferred to the surrounding environment, there is no change of the state of the environment due to this process. Hence $\Delta S_{env} = 0$. Note that the total change in entropy, $\Delta S_{gas} + \Delta S_{env}$ is greater than zero.
(c) Since $n = 1.00$ and $V_2 = 2.00\, V_1$, then $\Delta S_{gas} = R \ln 2.00 = 5.76\, \text{J/K}$.

EXAMPLE 20–9 **Heat conduction.** A red-hot 2.0-kg piece of iron at temperature $T_1 = 880$ K is thrown into a huge lake whose temperature is $T_2 = 280$ K. Assume the lake is so large that its temperature rise is insignificant. Determine the change in entropy (a) of the iron and (b) of the surrounding environment (the lake).

SOLUTION (a) The process is irreversible, but the same entropy change will occur for a reversible process, and we use the concept of specific heat, Eq. 19–2. We assume the specific heat of the iron is constant at $c = 0.11$ kcal/kg·K. Then $dQ = mc\, dT$ and in a quasistatic reversible process

$$\Delta S_{\text{iron}} = \int \frac{dQ}{T} = mc \int_{T_1}^{T_2} \frac{dT}{T} = mc \ln \frac{T_2}{T_1} = -mc \ln \frac{T_1}{T_2}.$$

Putting in numbers, we find

$$\Delta S_{\text{iron}} = -(2.0\,\text{kg})(0.11\,\text{kcal/kg·K}) \ln \frac{880\,\text{K}}{280\,\text{K}} = -0.25\,\text{kcal/K}$$

or $\Delta S = -1100$ J/K.
(b) The initial and final temperatures of the lake are the same, $T = 280$ K. The lake receives from the iron an amount of heat

$$Q = mc(T_2 - T_1) = (2.0\,\text{kg})(0.11\,\text{kcal/kg·K})(880\,\text{K} - 280\,\text{K}) = 130\,\text{kcal}.$$

Strictly speaking, this is an irreversible process (the lake heats up locally before equilibrium is reached), but is equivalent to a reversible isothermal transfer of heat $Q = 130$ kcal at $T = 280$ K. Hence

$$\Delta S_{\text{env}} = \frac{130\,\text{kcal}}{280\,\text{K}} = 0.46\,\text{kcal/K}$$

or 1900 J/K. Thus, although the entropy of the iron actually decreases, the *total* change in entropy of iron plus environment is positive: 0.46 kcal/K − 0.25 kcal/K = +0.21 kcal/K, or 800 J/K.

In each of these examples, the entropy of our system plus that of the environment (or surroundings) either stayed constant or increased. For any *reversible* process, such as that in Example 20–6, the total entropy change is zero. This can be seen in general as follows: any reversible process can be considered as a series of quasistatic isothermal transfers of heat ΔQ between a system and the environment, which differ in temperature only by an infinitesimal amount. Hence the change in entropy of either the system or environment is $\Delta Q/T$ and that of the other is $-\Delta Q/T$, so the total is

$$\Delta S = \Delta S_{\text{syst}} + \Delta S_{\text{env}} = 0. \qquad \text{[any reversible process]}$$

In Examples 20–7, 20–8, and 20–9, we found that the total entropy of system plus environment increases. Indeed, it has been found that for all real (irreversible) processes, the total entropy increases. No exceptions have been found. We can thus make the *general statement* of the **second law of thermodynamics** as follows:

The entropy of an isolated system never decreases. It either stays constant (reversible processes) or increases (irreversible processes).

Since all real processes are irreversible, we can equally well state the second law as:

The total entropy of any system plus that of its environment increases as a result of any natural process:

$$\Delta S = \Delta S_{\text{syst}} + \Delta S_{\text{env}} > 0. \tag{20–13}$$

Although the entropy of one part of the universe may decrease in any process (see the Examples above), the entropy of some other part of the universe always increases by a greater amount, so the total entropy always increases.

Now that we finally have a quantitative general statement of the second law of thermodynamics, we can see that it is an unusual law. It differs considerably from other laws of physics, which are typically equalities (such as $F = ma$) or conservation laws (such as for energy and momentum). The second law introduces a new quantity, the entropy, S, but does not tell us it is conserved. Quite the opposite. Entropy always increases in time.

Entropy increases in all real processes

The second law of thermodynamics summarizes, very succinctly, which processes are observed in nature, and which are not. Or, said another way, it tells us about the *direction* processes go. For the reverse of any of the processes in the last few Examples, the entropy would decrease; and we never observe them. For example, we never observe heat flowing spontaneously from a cold body to a hot body, the reverse of Example 20–9. Nor do we ever observe a gas spontaneously compressing itself into a smaller volume, the reverse of Example 20–8 (gases always expand to fill their containers). Nor do we see thermal energy transform into kinetic energy of a rock so the rock rises spontaneously from the ground. Any of these processes would be consistent with the first law of thermodynamics (conservation of energy). It is the second law of thermodynamics they are not consistent with, and this is why we need the second law. If you were to see a movie run backward, you would probably realize it immediately because you would see odd occurrences—such as rocks rising spontaneously from the ground, or air rushing in from the atmosphere to fill an empty balloon (the reverse of free expansion). When watching a movie or video, we are tipped off to a reversal of time by observing whether entropy is increasing or decreasing. Hence entropy has been called **time's arrow**, for it tells us in which direction time is going.

Time's arrow

Is the general statement of the second law of thermodynamics—"the principle of entropy increase"—consistent with the Clausius and Kelvin-Planck statements? Yes, it is. This is easy to see, for if a process occurred in which heat flowed spontaneously out of a low-temperature (T_L) reservoir (so its entropy decreased), and all of it flowed into a high-temperature (T_H) reservoir (which increased in entropy), in violation of the Clausius statement, then the total change in entropy $\Delta S = Q/T_H - Q/T_L$ would be less than zero, since $T_L < T_H$. Thus the principle of entropy increase implies the Clausius statement. Can you show the equivalence of the entropy principle to the Kelvin-Planck statement?

Equivalence of second law statements

20–7 | Order to Disorder

The concept of entropy, as we have discussed it so far, may seem rather abstract. But we can relate it to the more ordinary concepts of *order* and *disorder*. In fact, the entropy of a system can be considered a *measure of the disorder of the system*. Then the second law of thermodynamics can be stated simply as:

Natural processes tend to move toward a state of greater disorder.

SECOND LAW OF THERMODYNAMICS (general statement)

Exactly what we mean by disorder may not always be clear, so we now consider a few examples. Some of these will show us how this very general statement of the second law actually applies beyond what we usually consider as thermodynamics.

Let us look at some very simple processes. First, a jar containing separate layers of salt and pepper is more orderly than when the salt and pepper are all mixed up. Shaking a jar containing separate layers results in a mixture, and no amount of shaking brings the layers back again. The natural process is from a state of relative order (layers) to one of relative disorder (a mixture), not the reverse. That is, disorder increases. Second, a solid coffee cup is a more "orderly"

(a)

(b)

(c)

FIGURE 20–13 Have you ever observed this process, a broken cup spontaneously reassembling and rising up to a table?

object than the pieces of a broken cup. Cups break when they fall, but they do not spontaneously mend themselves (Fig. 20–13). Again, the normal course of events is an increase of disorder.

Now let us consider some processes for which we have actually calculated the entropy change, and see that an increase in entropy results in an increase in disorder (or vice versa). When ice melts to water at 0°C, the entropy of the water increases, as we saw in Example 20–6. Intuitively, we can think of solid water, ice, as being more ordered than the less orderly fluid state which can flow all over the place. This change from order to disorder can be seen more clearly from the molecular point of view: the orderly arrangement of water molecules in an ice crystal has changed to the disorderly and somewhat random motion of the molecules in the fluid state.

When a hot object is put in contact with a cold object, heat flows from the high temperature to the low until the two objects reach the same intermediate temperature. At the beginning of the process we can distinguish two classes of molecules—those with a high average kinetic energy and those with a low average kinetic energy. After the process, all the molecules are in one class with the same average kinetic energy, and we no longer have the more orderly arrangement of molecules in two classes. Order has gone to disorder. Furthermore, note that the separate hot and cold objects could serve as the hot- and cold-temperature regions of a heat engine, and thus could be used to obtain useful work. But once the two objects are put in contact and reach the same temperature, no work can be obtained. Disorder has increased, since a system that has the ability to perform work must surely be considered to have a higher order than a system no longer able to do work.

An interesting example of the increase in entropy relates to the theory of biological evolution and to the growth of organisms. Clearly, a human being is a highly ordered organism. The process of evolution from the early macromolecules and simple forms of life to *Homo sapiens* represents increasing order. So, too, the development of an individual from a single cell to a grown person is a process of increasing order. Do these processes violate the second law of thermodynamics? No, they do not. In the processes of evolution and growth, and even during the life of an individual, waste products are eliminated. These small molecules that remain as a result of metabolism are simple molecules without much order in comparison to the macromolecules of life such as DNA and proteins. Thus they represent relatively higher disorder or entropy. Indeed, the total entropy of the molecules cast aside by organisms during the processes of evolution and growth is greater than the decrease in entropy associated with the order of the growing individual or evolving species.

20–8 Energy Availability; Heat Death

In the process of heat conduction from a hot body to a cold one, we have seen that entropy increases and that order goes to disorder. The separate hot and cold objects could serve as the high- and low-temperature regions for a heat engine and thus could be used to obtain useful work. But after the two objects are put in contact with each other and reach the same uniform temperature, no work can be obtained from them. With regard to being able to do useful work, order has gone to disorder in this process.

The same can be said about a falling rock that comes to rest upon striking the ground. Just before hitting the ground, all the kinetic energy of the rock could have been used to do useful work. But once the rock's mechanical kinetic energy becomes thermal energy, this is no longer possible.

Both these examples illustrate another important aspect of the second law of thermodynamics—*in any natural process, some energy becomes unavailable to do useful work*. In any process, no energy is ever lost (it is always conserved). Rather, it becomes less useful—it can do less useful work. As time goes on, **energy is degraded**, in a sense; it goes from more orderly forms (such as mechanical) eventually to the least orderly form, internal or thermal energy. Entropy is a factor here because the amount of energy that becomes unavailable to do work is proportional to the change in entropy during any process.[†]

Degradation of energy

A natural outcome of this is the prediction that as time goes on, the universe will approach a state of maximum disorder. Matter will become a uniform mixture, heat will have flowed from high-temperature regions to low-temperature regions until the whole universe is at one temperature. No work can then be done. All the energy of the universe will have become degraded to thermal energy. All change will cease. This, the so-called **heat death** of the universe, has been much discussed by philosophers. This final state seems an inevitable consequence of the second law of thermodynamics, although it lies very far in the future. Yet it is based on the assumption that the universe is finite, which cosmologists are not fully sure of. The answers are not all in yet; and that makes science interesting.

"Heat death"

* | **20–9** | **Statistical Interpretation of Entropy and the Second Law**

The ideas of entropy and disorder are made clearer with the use of a statistical or probabilistic analysis of the molecular state of a system. This statistical approach, which was first applied toward the end of the nineteenth century by Ludwig Boltzmann (1844–1906), makes a distinction between the "macrostate" and the "microstate" of a system. The **microstate** of a system would be specified when the position and velocity of every particle (or molecule) is given. The **macrostate** of a system is specified by giving the macroscopic properties of the system—the temperature, pressure, number of moles, and so on. In reality, we can know only the macrostate of a system. There are generally far too many molecules in a system to be able to know the velocity and position of every one at a given moment. Nonetheless, it is important to recognize that a great many different microstates can correspond to the *same* macrostate.

Microstates and macrostates

Let us take a simple example. Suppose you repeatedly shake four coins in your hand and drop them on the table. Specifying the number of heads and the number of tails that appear on a given throw is the macrostate of this system. Specifying each coin as being a head or a tail is the microstate of the system. In the following table we see the number of microstates that correspond to each macrostate:

Macrostate	Possible microstates (H = heads, T = tails)	Number of microstates
4 heads	H H H H	1
3 heads, 1 tail	H H H T, H H T H, H T H H, T H H H	4
2 heads, 2 tails	H H T T, H T H T, T H H T, H T T H, T H T H, T T H H	6
1 head, 3 tails	T T T H, T T H T, T H T T, H T T T	4
4 tails	T T T T	1

[†]It can be shown that the amount of energy that becomes unavailable to do useful work is equal to $T_L \Delta S$, where T_L is the lowest available temperature and ΔS is the total increase in entropy during the process.

A basic principle behind the statistical approach is that *each microstate is equally probable*. Thus the number of microstates that give the same macrostate corresponds to the relative probability of that macrostate occurring. The macrostate of two heads and two tails is the most probable one in our case of tossing four coins. Out of the total of 16 possible microstates, six correspond to two heads and two tails, so the probability of throwing two heads and two tails is 6 out of 16, or 38 percent. The probability of throwing one head and three tails is 4 out of 16, or 25 percent. The probability of four heads is only 1 in 16, or 6 percent. Of course if you threw the coins 16 times, you might not find that two heads and two tails appears exactly 6 times, or four tails exactly once. These are only probabilities or averages. But if you made 1600 throws, very nearly 38 percent of them would be two heads and two tails. The greater the number of tries, the closer are the percentages to the calculated probabilities.

If we consider tossing more coins, say 100, the relative probability of throwing all heads (or all tails) is greatly reduced. There is only one microstate corresponding to all heads. For 99 heads and 1 tail, there are 100 microstates since each of the coins could be the one tail. The relative probabilities for other macrostates are given in Table 20–1. There are a total of about 10^{30} microstates possible.[†] Thus the relative probability of finding all heads is 1 in 10^{30}, an incredibly unlikely event! The probability of obtaining 50 heads and 50 tails (see Table 20–1) is $1.0 \times 10^{29}/10^{30} = 0.10$ or 10 percent. The probability of obtaining between 45 and 55 heads is 90 percent.

Thus we see that as the number of coins increases, the probability of obtaining an orderly arrangement (all heads or all tails) becomes extremely unlikely. The least orderly arrangement (half heads, half tails) is the most probable; and the probability of being within a certain percentage (say 5 percent) of the most probable arrangement greatly increases as the number of coins increases. These same ideas can be applied to the molecules of a system. For example the most probable state of a gas (say the air in a room) is one in which the molecules take up the whole space and move about randomly; this corresponds to the Maxwellian distribution, Fig. 20–14a (see Section 18–2). On the other hand, the very orderly arrangement of all the molecules located in one corner of the room and all moving with the same velocity (Fig. 20–14b) is extremely unlikely.

From these examples, it is clear that probability is directly related to disorder and hence to entropy. That is, the most probable state is the one with greatest entropy or greatest disorder and randomness. Boltzmann showed that, consistent with Clausius's definition $(dS = dQ/T)$ that the entropy of a system in a given (macro) state can be written

$$S = k \ln \mathcal{W}, \tag{20–14}$$

where k is Boltzmann's constant $\left(k = R/N_A = 1.38 \times 10^{-23}\,\text{J/K}\right)$ and $\mathcal{W}$ is the number of microstates corresponding to the given macrostate. That is, $\mathcal{W}$ is

TABLE 20–1 Probabilities of various macrostates for 100 coin tosses

Macrostate		Number of microstates
Heads	Tails	$\mathcal{W}$
100	0	1
99	1	1.0×10^2
90	10	1.7×10^{13}
80	20	5.4×10^{20}
60	40	1.4×10^{28}
55	45	6.1×10^{28}
50	50	1.0×10^{29}
45	55	6.1×10^{28}
40	60	1.4×10^{28}
20	80	5.4×10^{20}
10	90	1.7×10^{13}
1	99	1.0×10^2
0	100	1

Entropy in terms of probability

[†] Each coin has two possibilities, heads or tails. Then the possible number of microstates is $2 \times 2 \times 2 \times \ldots = 2^{100} = 1.27 \times 10^{30}$ (using a calculator or logarithms).

FIGURE 20–14 (a) Most probable distribution of molecular speeds in a gas (Maxwellian, or random); (b) orderly, but highly unlikely, distribution of speeds in which all molecules have nearly the same speed.

(a)

(b)

proportional to the probability of occurrence of that state. $\mathcal{W}$ is called the **thermodynamic probability**, or, sometimes, the disorder parameter.

> **EXAMPLE 20–10 Free expansion—statistical determination of entropy.**
> Use Eq. 20–14 to determine the change in entropy for the adiabatic *free expansion* of a gas, a calculation we did macroscopically in Example 20–8. Assume $\mathcal{W}$, the number of microstates for each macrostate, is the number of possible positions.
>
> **SOLUTION** Assume the number of moles is $n = 1$ and then the number of molecules is $N = nN_A = N_A$, and assume the volume doubles, just as in Example 20–8. Because the volume doubles, the number of possible positions for each molecule doubles. Since there are N_A molecules, the total number of microstates increases by a factor of 2^{N_A} when the volume doubles. That is
>
> $$\frac{\mathcal{W}_2}{\mathcal{W}_1} = 2^{N_A}.$$
>
> The change in entropy is, from Eq. 20–14,
>
> $$\Delta S = S_2 - S_1 = k(\ln \mathcal{W}_2 - \ln \mathcal{W}_1) = k \ln \frac{\mathcal{W}_2}{\mathcal{W}_1} = k \ln 2^{N_A} = kN_A \ln 2 = R \ln 2$$
>
> which is the same result we got in Example 20–8.

In terms of probability, the second law of thermodynamics—which tells us that entropy increases in any process—reduces to the statement that those processes occur which are most probable. The second law thus becomes a trivial statement. However there is an additional element now. The second law in terms of probability does not *forbid* a decrease in entropy. Rather, it says the probability is extremely low. It is not impossible that salt and pepper should separate spontaneously into layers, or that a broken tea cup should mend itself. It is even possible that a lake should freeze over on a hot summer day (that is, heat flow out of the cold lake into the warmer surroundings). But the probability for such events occurring is extremely small. In our coin examples above, we saw that increasing the number of coins from 4 to 100 reduced drastically the probability of large deviations from the average, or most probable, arrangement. In ordinary systems we are dealing with incredibly large numbers of molecules: in 1 mol alone there are 6×10^{23} molecules. Hence the probability of deviation far from the average is incredibly tiny. For example, it has been calculated that the probability that a stone resting on the ground should transform 1 cal of thermal energy into mechanical energy and rise up into the air is much less likely than the probability that a group of monkeys typing randomly would by chance produce the complete works of Shakespeare.

*20–10 Thermodynamic Temperature Scale; Absolute Zero, and the Third Law of Thermodynamics

In Section 20–3 we discussed the Carnot engine and other (ideal) reversible engines. We saw that the efficiency of any reversible engine operating between two heat reservoirs depends only on the temperatures of these two reservoirs; and the efficiency does not depend on the working substance—it could be helium, water, or something else, and the efficiency would be the same. The efficiency is given by

$$e = 1 - \frac{|Q_L|}{|Q_H|},$$

where $|Q_H|$ is the heat absorbed from the high-temperature reservoir and $|Q_L|$ is the heat exhausted to the low-temperature reservoir. Since $|Q_L/Q_H|$ is the same for any reversible engine operating between the same two temperatures, Kelvin suggested using this fact to define an absolute temperature scale. That is, the ratio of the temperatures of the two reservoirs, T_H and T_L, is defined as the ratio of the heats, $|Q_H|$ and $|Q_L|$, exchanged with them by a Carnot or other reversible engine:

$$\frac{T_H}{T_L} = \frac{|Q_H|}{|Q_L|}. \tag{20–15}$$

This is the basis for the **Kelvin** or **thermodynamic temperature scale**.

In Section 20–3, we saw that this same relation, $T_H/T_L = |Q_H/Q_L|$, holds for a Carnot engine (Eq. 20–2) when the temperatures are based on the ideal gas temperature scale (Section 17–10) which we have been using up to now. Indeed, to complete the definition of the thermodynamic scale, we assign the value $T_{tp} = 273.16\,\text{K}$ to the triple point of water so that

$$T = (273.16\,\text{K})\left(\frac{|Q|}{|Q_{tp}|}\right),$$

where $|Q|$ and $|Q_{tp}|$ are the heats exchanged by a Carnot engine with reservoirs at temperatures T and T_{tp}. This corresponds precisely to the definition of the ideal gas scale. Thus, the thermodynamic scale is identical to the ideal gas scale over the range of validity of the latter (below about 1 K, no substance is a gas, so the ideal gas scale cannot be used). The thermodynamic scale is now considered the standard scale since it can be used over the entire range of possible temperatures, and it is also independent of the substance used. From a practical point of view, it is especially useful at very low temperatures.

Very low temperatures are difficult to obtain experimentally. In fact, it is found experimentally that the closer the temperature is to absolute zero, the more difficult it is to reduce the temperature further. And it is generally accepted that it is not possible to actually reach absolute zero in any finite number of processes. This last statement is one way to state[†] the **third law of thermodynamics**. Since the maximum efficiency that any heat engine can have is the Carnot efficiency

Third law of thermodynamics

$$e = 1 - \frac{T_L}{T_H},$$

and since T_L can never be zero, we see that a 100-percent efficient heat engine is not possible.

[†] See also the statement in the footnote on page 529.

Summary

A **heat engine** is a device for changing thermal energy, by means of heat flow, into useful work.

The **efficiency** of a heat engine is defined as the ratio of the work W done by the engine to the heat input $|Q_H|$. Because of conservation of energy, the work output equals $|Q_H| - |Q_L|$, where $|Q_L|$ is the heat exhausted to the environment; hence the efficiency

$$e = \frac{|W|}{|Q_H|} = 1 - \frac{|Q_L|}{|Q_H|}.$$

Carnot's (idealized) engine consists of two isothermal and two adiabatic processes in a reversible cycle. For a **Carnot engine**, or any reversible engine operating between two temperatures, T_H and T_L, the efficiency is

$$e_{\text{ideal}} = 1 - \frac{T_L}{T_H}.$$

Irreversible (real) engines always have an efficiency less than this.

All heat engines give rise to thermal pollution because they exhaust heat to the environment.

The operation of **refrigerators** and **air conditioners** is the reverse of that of a heat engine: work is done to extract heat from a cool region and exhaust it to a region at a higher temperature. A **heat pump** does work to bring heat from a cold exterior into a warmer interior.

The **second law of thermodynamics** can be stated in several equivalent ways:

(a) heat flows spontaneously from a hot object to a cold one, but not the reverse;

(b) there can be no 100 percent efficient heat engine—that is, one that can change a given amount of heat completely into work;

(c) natural processes tend to move toward a state of greater disorder or greater **entropy**.

Statement (c) is the most general statement of the second law of thermodynamics, and can be restated as: the total entropy, S, of any system plus that of its environment increases as a result of any natural process:

$$\Delta S > 0.$$

Entropy, which is a state variable, is a quantitative measure of the disorder of a system. The change in entropy of a system during a reversible process is given by $\Delta S = \int dQ/T$.

The second law of thermodynamics tells us in which direction processes tend to proceed; hence entropy is called "time's arrow."

As time goes on, energy is degraded to less useful forms—that is, it is less available to do useful work.

Questions

1. Is it possible to cool down a room on a hot summer day by leaving the refrigerator door open? Explain.

2. Can mechanical energy ever be transformed completely into heat or internal energy? Can the reverse happen? In each case, if your answer is no, explain why not; if yes, give examples.

3. Would a definition of heat engine efficiency as $e = |W|/|Q_L|$ be a useful one? Explain.

4. What plays the role of high-temperature and low-temperature reservoirs in (a) an internal combustion engine, (b) a steam engine? Are they, strictly speaking, heat reservoirs?

5. Which will give the greater improvement in the efficiency of a Carnot engine, a 10°C increase in the high-temperature reservoir, or a 10°C decrease in the low-temperature reservoir?

6. The oceans contain a tremendous amount of thermal energy. Why, in general, is it not possible to put this energy to useful work?

7. Discuss the factors that keep real engines from reaching Carnot efficiency.

8. The expansion valve in a refrigeration system, Fig. 20–9, is crucial for cooling the fluid. Explain how the cooling occurs.

9. Describe a process in nature that is nearly reversible.

10. (a) Describe how heat could be added to a system reversibly. (b) Could you use a stove burner to add heat to a system reversibly? Explain.

11. Powdered milk is very slowly (quasistatically) added to water while being stirred. Is this a reversible process?

12. Two identical systems are taken from state a to state b by two different *irreversible* processes. Will the change in entropy for the system be the same for each process? For the environment? Answer carefully and completely.

13. It can be said that the *total change in entropy during a process is a measure of the irreversibility of the process.* Discuss why this is valid, starting with the fact that $\Delta S = 0$ for a reversible process.

14. Use arguments, other than the principle of entropy increase, to show that for an adiabatic process, $\Delta S = 0$ if it is done reversibly and $\Delta S > 0$ if done irreversibly.

15. A gas is allowed to expand (a) adiabatically and (b) isothermally. In each process, does the entropy increase, decrease, or stay the same?

16. Entropy is often called "time's arrow" because it tells us in which direction natural processes occur. If a movie film were run backward, name some processes that you might see that would tell you that time was "running backward."

17. Give three examples, other than those mentioned in this chapter, of naturally occurring processes in which order goes to disorder. Discuss the observability of the reverse process.

18. Which do you think has the greater entropy, 1 kg of solid iron or 1 kg of liquid iron? Why?

19. What happens if you remove the lid of a bottle containing chlorine gas? Does the reverse process ever happen? Why or why not?

20. Think up several processes (other than those already mentioned in the text) that would obey the first law of thermodynamics, but, if they actually occurred, would violate the second law.

21. Describe how a free expansion, for which the entropy increases, can be considered as a process in which order goes to disorder. [*Hint*: consider a stack of papers dropped into a large box versus dropped all over the floor.]

22. Suppose you collect a lot of papers strewn all over the floor and put them in a neat stack; does this violate the second law of thermodynamics? Explain.

23. The first law of thermodynamics is sometimes whimsically stated as, "You can't get something for nothing," and the second law as, "You can't even break even." Explain how these statements could be equivalent to the formal statements.

24. Give three examples of naturally occurring processes that illustrate the degradation of usable energy into internal energy.

25. Living organisms convert relatively simple food molecules into a complex structure as they grow. Is this a violation of the second law of thermodynamics? Explain.

Problems

Section 20–2

1. (I) A heat engine exhausts 8500 J of heat while performing 2700 J of useful work. What is the efficiency of this engine?

2. (I) A heat engine does 8200 J of work in each cycle while absorbing 18.0 kcal of heat from a high-temperature reservoir. What is the efficiency of this engine?

3. (I) A typical power plant puts out 500 MW of electric power. Estimate the heat discharged per second, assuming that the plant has an efficiency of 38 percent.

4. (II) A four-cylinder gasoline engine has an efficiency of 0.25 and delivers 180 J of work per cycle per cylinder. The engine fires at 25 cycles per second. (*a*) Determine the work done per second. (*b*) What is the total heat input per second from the gasoline? (*c*) If the energy content of gasoline is 130 MJ per gallon, how long does one gallon last?

5. (II) The burning of gasoline in a car releases about 3.0×10^4 kcal/gal. If a car averages 38 km/gal when driving 90 km/h, which requires 20 hp, what is the efficiency of the engine under those conditions?

6. (II) Figure 20–15 is a *PV* diagram for a reversible heat engine in which 1.0 mol of argon, a nearly ideal monatomic gas, is initially at STP (point a). Points b and c are on an isotherm at $T = 423$ K. Process ab is at constant volume, process ac at constant pressure. (*a*) Is the path of the cycle carried out clockwise or counterclockwise? (*b*) What is the efficiency of this engine?

7. (II) A 38-percent-efficient power plant puts out 810 MW (megawatts) of power (electrical). Cooling towers are used to take away the exhaust heat. If the air temperature is allowed to rise 7.5 C°, what volume of air (km^3) is heated per day? Will the local climate be heated significantly? If the heated air were to form a layer 200 m thick, how large an area would it cover for 24 h of operation? (The heat capacity of air is about 7.0 cal/mol·C° at constant pressure.)

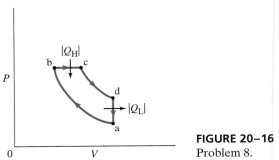

FIGURE 20–16
Problem 8.

8. (III) The operation of a *Diesel engine* can be idealized by the cycle shown in Fig. 20–16. Air is drawn into the cylinder during the intake stroke (not part of the idealized cycle). The air is compressed adiabatically, path ab. At point b diesel oil is injected into the cylinder which immediately burns since the temperature is very high. Combustion is slow, and during the first part of the power stroke, the gas expands at (nearly) constant pressure, path bc. After burning, the rest of the power stroke is adiabatic, path cd. Path da corresponds to the exhaust stroke. (*a*) Show that, for a quasistatic reversible engine undergoing this cycle using an ideal gas, the ideal efficiency is

$$e = 1 - \frac{(V_a/V_c)^{-\gamma} - (V_a/V_b)^{-\gamma}}{\gamma[(V_a/V_c)^{-1} - (V_a/V_b)^{-1}]},$$

where V_a/V_b is the "compression ratio", V_a/V_c is the "expansion ratio", and γ is defined by Eq. 19–12. (*b*) If $V_a/V_b = 15$ and $V_a/V_c = 5.0$, calculate the efficiency assuming the gas is diatomic $(\text{like } N_2 \text{ and } O_2)$ and ideal.

Section 20–3

9. (I) What is the maximum efficiency of a heat engine whose operating temperatures are 530°C and 305°C?

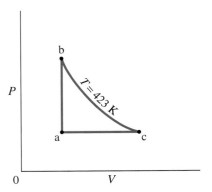

FIGURE 20–15 Problem 6.

10. (I) The exhaust temperature of a heat engine is 220°C. What must be the high temperature if the Carnot efficiency is to be 36 percent?

11. (II) (a) Show that the work done by a Carnot engine is equal to the area enclosed by the Carnot cycle on a PV diagram, Fig. 20–5. (See Section 19–7.) (b) Generalize this to any reversible cycle.

12. (II) A heat engine exhausts its heat at 360°C and has a Carnot efficiency of 35 percent. What exhaust temperature would enable it to achieve a Carnot efficiency of 50 percent?

13. (II) A nuclear power plant operates at 75 percent of its maximum theoretical (Carnot) efficiency between temperatures of 660°C and 360°C. If the plant produces electric energy at the rate of 1.1 GW, how much exhaust heat is discharged per hour?

14. (II) An engine that operates at half its theoretical (Carnot) efficiency operates between 525°C and 280°C while producing work at the rate of 850 kW. How much heat is wasted per hour?

15. (II) Assume that a hiker needs 4000 kcal of energy to supply a day's worth of metabolism. Estimate the maximum height the person can climb in one day, using only this amount of energy. As a rough prediction, treat the person as an isolated heat engine, operating between the internal temperature of 37°C (98.6°F) and the ambient air temperature of 20°C.

16. (II) A Carnot engine performs work at the rate of 570 kW while using 1350 kcal of heat per second. If the temperature of the heat source is 580°C, at what temperature is the waste heat exhausted?

17. (II) A heat engine utilizes a heat source at 580°C and has a Carnot efficiency of 29 percent. To increase the efficiency to 35 percent, what must be the temperature of the heat source?

18. (II) At a steam power plant, steam engines work in pairs, the heat output of the first one being the approximate heat input of the second. The operating temperatures of the first are 680°C and 430°C, and of the second 415°C and 280°C. If the heat of combustion of coal is 2.8×10^7 J/kg, at what rate must coal be burned if the plant is to put out 900 MW of power? Assume the efficiency of the engines is 65 percent of the ideal (Carnot) efficiency.

19. (II) Water is used to cool the power plant in Problem 18. If the water temperature is allowed to increase by no more than 5.5 C°, estimate how much water must pass through the plant per hour.

20. (II) Show that if two different adiabatic paths intersected at a single point on a PV diagram, they could be connected by an isotherm to form a cycle, and that an engine run on this cycle would violate the second law of thermodynamics. What do you conclude about the crossing of adiabatic lines?

21. (III) A Carnot cycle, shown in Fig. 20–5, has the following conditions: $V_a = 6.0$ L, $V_b = 15.0$ L, $T_H = 470°C$, and $T_L = 290°C$. The gas used in the cycle is 0.50 mole of a diatomic gas, $\gamma = 1.4$. Calculate (a) the pressures at a and b; (b) the volumes at c and d. (c) What is the work done along

process ab? (d) What is the heat lost along process cd? (e) Calculate the net work done for the whole cycle. (f) What is the efficiency of the cycle, using the definition $e = |W|/|Q_H|$? Show that this is the same as given by Eq. 20–3.

22. (III) One mole of monatomic gas undergoes a Carnot cycle with $T_H = 350°C$ and $T_L = 210°C$. The initial pressure is 10 atm. During the isothermal expansion, the volume doubles. (a) Find the values of the pressure and volume at the points a, b, c, and d (see Fig. 20–5). (b) Determine Q, W, and ΔU for each segment of the cycle. (c) Calculate the efficiency of the cycle.

Section 20–4

23. (I) The low temperature of a freezer cooling coil is −15°C and the discharge temperature is 30°C. What is the maximum theoretical coefficient of performance?

24. (I) If an ideal refrigerator keeps its contents at −15°C when the house temperature is 22°C, what is its coefficient of performance?

25. (I) A restaurant refrigerator has a coefficient of performance of 5.0. If the temperature in the kitchen outside the refrigerator is 29°C, what is the lowest temperature that could be obtained inside the refrigerator if it were ideal?

26. (II) A heat pump is used to keep a house warm at 22°C. How much work is required of the pump to deliver 2800 J of heat into the house if the outdoor temperature is (a) 0°C, (b) −15°C? Assume ideal (Carnot) behavior.

27. (II) An ideal engine has an efficiency of 35 percent. If it were run backward as a heat pump, what would be its coefficient of performance?

28. (II) The engine of Example 20–2 is run in reverse. How long would it take to freeze a tray of a dozen 40-g compartments of liquid water at room temperature (20°C) into a dozen ice cubes at the freezing point, assuming that it takes 450 W of input electric power to run it? Assume ideal (Carnot) behavior.

29. (II) A "Carnot" refrigerator (reverse of a Carnot engine) absorbs heat from the freezer compartment at a temperature of −17°C and exhausts it into the room at 25°C. (a) How much work must be done by the refrigerator to change 0.50 kg of water at 25°C into ice at −17°C? (b) If the compressor output is 200 W, what minimum time is needed to take 0.50 kg of 25°C water and freeze it at 0°C?

30. (II) (a) Given that the coefficient of performance of a refrigerator is defined (Eq. 20–4a) as

$$CP = \frac{|Q_L|}{|W|},$$

show that for an ideal (Carnot) refrigerator,

$$CP_{ideal} = \frac{T_L}{T_H - T_L}.$$

(b) Write the CP in terms of the efficiency e of the reversible heat engine obtained by running the refrigerator backward. (c) What is the coefficient of performance for an ideal refrigerator that maintains a freezer compartment at −16°C when the condenser's temperature is 22°C?

31. (II) What volume of water at 0°C can a freezer make into ice cubes in one hour, if the coefficient of performance of the cooling unit is 7.0 and the power input is 1.0 kilowatt?

32. (II) A central heat pump operating as an air conditioner draws 36,000 Btu per hour from a building and operates between the temperatures of 24°C and 38°C. (a) If its coefficient of performance is 24% of that of a Carnot air conditioner, what is the effective coefficient of performance? (b) What is the power required of the compressor motor? (c) What is the power in terms of hp?

Sections 20–5 and 20–6

33. (I) A 10.0-kg box having an initial speed of 3.0 m/s slides along a rough table and comes to rest. Estimate the total change in entropy of the universe. Assume all objects are at room temperature (293 K).

34. (I) What is the change in entropy of 1.00 m³ of water at 0°C when it is frozen to ice at 0°C?

35. (II) If the water in Problem 34 were frozen by being in contact with a great deal of ice at −10°C, what would be the total change in entropy of the process?

36. (II) If 1.00 kg of water at 100°C is changed by a reversible process to steam at 100°C, determine the change in entropy of (a) the water, (b) the surroundings, and (c) the universe as a whole. (d) How would your answers differ if the process were irreversible?

37. (II) An aluminum rod conducts 7.50 cal/s from a heat source maintained at 240°C to a large body of water at 27°C. Calculate the rate at which entropy increases in this process.

38. (II) A 3.8-kg piece of aluminum at 30°C is placed in 1.0 kg of water in a Styrofoam container at room temperature (20°C). Estimate the net change in entropy of the system.

39. (II) What is the total change in entropy when 2.5 kg of water at 0°C is frozen to ice at 0°C by being in contact with 450 kg of ice at −15°C?

40. (II) When 2.0 kg of water at 20°C is mixed with 3.0 kg of water at 80°C in a well-insulated container, what is the change in entropy of the system?

41. (II) Show that the principle of entropy increase is equivalent to the Kelvin-Planck statement of the second law of thermodynamics.

42. (II) The temperature of 2.0 mol of an ideal diatomic gas goes from 25°C to 45°C at a constant volume. What is the change in entropy?

43. (II) (a) Calculate the change in entropy of 1.00 kg of water when it is heated from 0°C to 100°C. (b) Does the entropy of the surroundings change? If so, by how much?

44. (II) A 150-g insulated aluminum cup at 20°C is filled with 240 g of water at 100°C. (a) What is the final temperature of the mixture? (b) What is the total change in entropy as a result of the mixing process?

45. (II) Two samples of an ideal gas are initially at the same temperature and pressure; they are each compressed reversibly from a volume V to volume V/2, one isothermally, the other adiabatically. (a) In which sample is the final pressure greater? (b) Determine the change in entropy of the gas for each process. (c) What is the entropy change of the environment for each process?

46. (II) One mole of nitrogen (N₂) gas and one mole of argon (Ar) gas are in separate, equal-sized, insulated containers at the same temperature. The containers are then connected and the gases (assumed ideal) allowed to mix. (a) What is the change in entropy of the system and (b) of the environment? (c) Repeat part (a) but assume one container is twice as large as the other.

47. (II) (a) Why would you expect the total entropy change in a Carnot cycle to be zero? (b) Do a calculation to show that it is zero.

48. (III) The specific heat per mole of potassium at low temperatures is given by $C_V = aT + bT^3$, where $a = 2.08$ mJ/mol·K² and $b = 2.57$ mJ/mol·K⁴. Calculate the entropy change of 0.25 mol of potassium when its temperature is lowered from 3.0 K to 1.0 K.

Section 20–8

49. (III) A general theorem states that the amount of energy that becomes unavailable to do useful work in any process is equal to $T_L \Delta S$, where T_L is the lowest temperature available and ΔS is the total change in entropy during the process. Show that this is valid in the specific cases of (a) a falling rock that comes to rest when it hits the ground; (b) the free adiabatic expansion of an ideal gas; and (c) the conduction of heat, Q, from a high-temperature (T_H) reservoir to a low-temperature (T_L) reservoir. [Hint: In part (c) compare to a Carnot engine.]

50. (III) Determine the work available in a 5.0 kg block of copper at 420 K if the surroundings are at 290 K. Use results of Problem 49.

* Section 20–9

* 51. (II) Suppose you repeatedly shake six coins in your hand and drop them on the table. Construct a table showing the number of microstates that correspond to each macrostate. What is the probability of obtaining (a) three heads and three tails and (b) six heads?

* 52. (II) Calculate the relative probabilities, when you throw two dice, of obtaining (a) a 7, (b) an 11, (c) a 5.

* 53. (II) Rank the following five-card hands in order of increasing probability: (a) four aces and a king; (b) six of hearts, eight of diamonds, queen of clubs, three of hearts, jack of spades; (c) two jacks, two queens, and an ace; and (d) any hand having no two equal-value cards. Explain your ranking using microstates and macrostates.

* 54. (II) (a) Suppose you have four coins, all with tails up. You now rearrange them so two heads and two tails are up. What was the change in entropy of the coins? (b) Suppose your system is the 100 coins of Table 20–1; what is the change in entropy of the coins if they are mixed randomly initially, 50 heads and 50 tails, and you arrange them so all 100 are heads? (c) Compare these entropy changes to ordinary thermodynamic entropy changes, such as Examples 20–6, 7, 8 and 9.

General Problems

55. It is not necessary that a heat engine's hot environment be hotter than ambient temperature. Liquid nitrogen (90 K) is about as cheap as bottled water. What would be the efficiency of an engine that made use of heat transferred from air at room temperature (293 K) to the liquid nitrogen "fuel"?

56. It has been suggested that a heat engine could be developed that made use of the temperature difference between water at the surface of the ocean and that several hundred meters deep. In the tropics, the temperatures may be 27°C and 4°C, respectively. What is the maximum efficiency such an engine could have? Why might such an engine be feasible in spite of the low efficiency? Can you imagine any adverse environmental effects that might occur?

57. Two 1100-kg cars are traveling 95 km/h in opposite directions when they collide and are brought to rest. Estimate the change in entropy of the universe as a result of this collision. Assume $T = 20°C$.

58. A 120-g insulated aluminum cup at 15°C is filled with 210 g of water at 50°C. After a few minutes, equilibrium is reached. Determine (a) the final temperature, and (b) the total change in entropy.

59. (a) What is the coefficient of performance of an ideal heat pump that extracts heat from 6°C air outside and deposits heat inside your house at 24°C? (b) If this heat pump operates on 1000 W of electrical energy, what is the maximum heat it can deliver into your house each hour?

60. An inventor claims to have designed and built an engine that produces 1.50 MW (megawatts) of usable work while taking in 3.00 MW of thermal energy at 425 K, and rejecting 1.50 MW of thermal energy at 215 K. Is there anything fishy about his claim? Explain.

61. Suppose a power plant delivers energy at 900 MW using steam turbines. The steam goes into the turbines superheated at 600 K and deposits its unused heat in river water at 285 K. Assume that the turbine operates as an ideal Carnot engine. (a) If the river flow rate is 37 m³/s, calculate the average temperature increase of the river water downstream from the power plant. (b) What is the entropy increase per kilogram of the downstream river water in J/kg·K?

62. A 100-hp car engine operates at about 15 percent efficiency. Assume the engine's water temperature of 85°C is its cold-temperature (exhaust) reservoir and 500°C is its thermal "intake" temperature (the temperature of the exploding gas/air mixture). (a) Calculate its efficiency relative to its maximum possible (Carnot) efficiency. (b) Estimate how much power (in watts) goes into moving the car, and how much heat, in joules and in kcal, is exhausted to the air in 1.0 h.

63. A falling rock has kinetic energy K just before striking the ground and coming to rest. What is the total change in entropy of the rock plus environment as a result of this collision?

64. An aluminum can, with negligible heat capacity, is filled with 500 g of water at 0°C and then is brought into thermal contact with a similar can filled with 500 g of water at 50°C. Find the change in entropy of the system if no heat is allowed to exchange with the surroundings.

65. (II) Thermodynamic processes can be represented not only on PV and PT diagrams; another useful one is a TS (temperature-entropy) diagram. (a) Draw a TS diagram for a Carnot cycle. (b) What does the area within the curve represent?

66. (II) A real heat engine working between heat reservoirs at 400 K and 850 K produces 600 J of work per cycle for a heat input of 1600 J. (a) Compare the efficiency of this real engine to that of a Carnot engine. (b) Calculate the total entropy change of the universe for each cycle of the real engine. (c) Calculate the total entropy change of the universe for a Carnot engine operating between the same two temperatures. (d) Show that the difference in work done by these two engines per cycle is $T_L \Delta S$, where T_L is the temperature of the low-temperature reservoir (400 K) and ΔS is the entropy increase per cycle of the real engine. (See also Problem 49 and Section 20–8.)

67. The *Stirling cycle*, shown in Fig. 20–17, is useful to describe external combustion engines as well as solar-power systems. Find the efficiency of the cycle in terms of the parameters shown, assuming a monatomic gas as the working substance. The processes ab and cd are isothermal whereas bc and da are at constant volume. How does it compare to the Carnot efficiency?

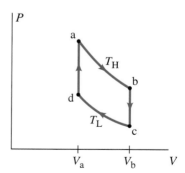

FIGURE 20–17 Problem 67.

(Problems continue on the next page.)

68. One mole of an ideal monatomic gas at STP first undergoes an isothermal expansion so that the volume at b is 2.5 times the volume at a (Fig. 20–18). Next, heat is extracted at a constant volume so that the pressure drops. The gas is then compressed adiabatically back to the original state. (a) Calculate the pressures at b and c. (b) Determine the temperature at c. (c) Determine the work done, heat input or extracted, and the change in entropy for each process. (d) What is the efficiency of this cycle?

FIGURE 20–18 Problem 68.

69. As a mechanical engineer, you must design a cooling unit for a new freezer. The enclosure has an inner surface area of 6.0 m² and is bounded by walls that are 10 cm thick, with a thermal conductivity of 0.050 W/m·K. The inside must be kept at −10 C° in a room that is at 20°C. The motor for the cooling unit must run no more than 15% of the time. What is the minimum power requirement of the cooling motor?

70. A gas turbine operates under the *Brayton cycle*, which is depicted in the *PV* diagram of Fig. 20–19. In process ab the air-fuel mixture undergoes an adiabatic compression. This is followed, in process bc, with an isobaric (constant pressure) heating, by combustion. Process cd is an adiabatic expansion with expulsion of the products to the atmosphere. The return step, da, takes place at constant pressure. If the working gas behaves like an ideal gas, show that the efficiency of the Brayton Cycle is

$$ e = 1 - \left(\frac{P_b}{P_a} \right)^{\frac{1-\gamma}{\gamma}}. $$

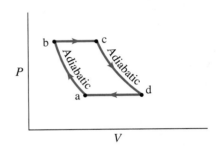

FIGURE 20–19 Problem 70.

*71. Estimate the probability that a bridge player will be dealt (a) all four aces (among 13 cards), and (b) all 13 cards of one suit.

This comb has been passed through hair, or rubbed by a cloth or paper towel, which gave it a static electric charge. The electrical charge on the comb induces a polarization (separation of charge) in all those scraps of paper, and thus attracts them.

Our introduction to electricity in this chapter covers conductors and insulators, and Coulomb's law which relates the force between two point charges as a function of their distance apart. We introduce the concept of electric field, which is easy to understand and powerful to use.

CHAPTER 21

Electric Charge and Electric Field

The word "electricity" may evoke an image of complex modern technology: computers, lights, motors, electric power. But the electric force would seem to play an even deeper role in our lives. According to atomic theory, electric forces between atoms and molecules hold them together to form liquids and solids, and electric forces are also involved in the metabolic processes that occur within our bodies. Many of the forces we have dealt with so far, such as elastic forces, the normal force, and other contact forces (pushes and pulls) are now considered to result from electric forces acting at the atomic level. Gravity, on the other hand, is a separate force.[†]

The earliest studies on electricity date back to the ancients, but it has been only in the past two centuries that electricity was studied in detail. We will discuss the development of ideas about electricity, including practical devices, as well as the relation to magnetism, in the next twelve chapters.

[†] As we discussed in Section 6–7, physicists in this century came to recognize four different fundamental forces in nature: (1) gravitational force, (2) electromagnetic force (we will see later that electric and magnetic forces are intimately related), (3) strong nuclear force, and (4) weak nuclear force. The last two forces operate at the level of the nucleus of an atom. A recent theory has combined the electromagnetic and weak nuclear forces so they are now considered to have a common origin known as the electroweak force.

(a)　　　　　　　　　　(b)

FIGURE 21–1 Rub a plastic ruler (a) and bring it close (b) to some tiny pieces of paper.

FIGURE 21–2 Unlike charges attract, whereas like charges repel one another.

(a) Two charged plastic rulers repel

(b) Two charged glass rods repel

(c) Charged glass rod attracts charged plastic ruler

Likes repel;
unlikes attract

21–1 | Static Electricity; Electric Charge and Its Conservation

The word *electricity* comes from the Greek word *elektron*, which means "amber." Amber is petrified tree resin, and the ancients knew that if you rub an amber rod with a piece of cloth, the amber attracts small pieces of leaves or dust. A piece of hard rubber, a glass rod, or a plastic ruler rubbed with a cloth will also display this "amber effect," or **static electricity** as we call it today. You can readily pick up small pieces of paper with a plastic comb or ruler that you've just vigorously rubbed with even a paper towel. See the photo on the previous page and Fig. 21–1. You have probably experienced static electricity when combing your hair or when taking a synthetic blouse or shirt from a clothes dryer. And you may have felt a shock when you touched a metal doorknob after sliding across a car seat or walking across a nylon carpet. In each case, an object becomes "charged" due to a rubbing process and is said to possess a net **electric charge**.

Is all electric charge the same, or is it possible that there is more than one type? In fact, there are *two* types of electric charge, as the following simple experiments show. A plastic ruler is suspended by a thread and rubbed vigorously with a cloth to charge it. When a second ruler, which has also been charged in the same way, is brought close to the first, it is found that the one ruler *repels* the other. This is shown in Fig. 21–2a. Similarly, if a rubbed glass rod is brought close to a second charged glass rod, again a repulsive force is seen to act, Fig. 21–2b. However, if the charged glass rod is brought close to the charged plastic ruler, it is found that they *attract* each other, Fig. 21–2c. The charge on the glass must therefore be different from that on the plastic. Indeed, it is found experimentally that all charged objects fall into one of two categories. Either they are attracted to the plastic and repelled by the glass, just as glass is; or they are repelled by the plastic and attracted to the glass, just as the plastic ruler is. Thus there seem to be two, and only two, types of electric charge. Each type of charge repels the same type but attracts the opposite type. That is: **unlike charges attract; like charges repel**.

The two types of electric charge were referred to as *positive* and *negative* by the American statesman, philosopher, and scientist Benjamin Franklin (1706–1790). The choice of which name went with which type of charge was of course arbitrary. Franklin's choice set the charge on the rubbed glass rod to be positive charge, so the charge on a rubbed plastic ruler (or amber) is called negative charge. We still follow this convention today.

Franklin argued that whenever a certain amount of charge is produced on one body in a process, an equal amount of the opposite type of charge is produced on another body. The positive and negative are to be treated *algebraically*, so that during any process, the net change in the amount of charge produced is zero. For example, when a plastic ruler is rubbed with a paper towel, the plastic acquires a negative charge and the towel an equal amount of positive charge. The charges are separated, but the sum of the two is zero. This is an example of a law that is now well established: the **law of conservation of electric charge**, which states that

the net amount of electric charge produced in any process is zero.

LAW OF CONSERVATION OF ELECTRIC CHARGE

If one object or one region of space acquires a positive charge, then an equal amount of negative charge will be found in neighboring areas or objects. No violations have ever been found, and this conservation law is as firmly established as those for energy and momentum.

21–2 | Electric Charge in the Atom

Only within the past century has it become clear that an understanding of electricity begins inside the atom itself. In later chapters we will discuss atomic structure and the ideas that led to our present view of the atom in more detail. But it will help our understanding of electricity if we discuss it briefly now.

A simplified model of an atom shows it as having a tiny but heavy, positively charged nucleus surrounded by one or more negatively charged electrons (Fig. 21–3). The nucleus contains protons, which are positively charged, and neutrons, which have no net electric charge. All protons and all electrons have exactly the same magnitude of electric charge, but their signs are opposite. Hence, neutral atoms, having no net charge, contain equal numbers of protons and electrons. Sometimes, as we shall see, an atom may lose one or more of its electrons, or may gain extra electrons. In this case the atom will have a net positive or negative charge, and is called an **ion**.

In solid materials the nuclei tend to remain close to fixed positions, whereas some of the electrons move quite freely. The charging of a solid object by rubbing can be explained by the transfer of electrons from one material to the other. When a plastic ruler becomes negatively charged by rubbing with a paper towel, the transfer of electrons from the towel to the plastic leaves the towel with a positive charge equal in magnitude to the negative charge acquired by the plastic. In liquids and gases, nuclei or ions can move as well as electrons.

Normally when objects are charged by rubbing, they hold their charge only for a limited time and eventually return to the neutral state. Where does the charge go? In some cases it is neutralized by charged ions in the air (formed, for example, by collisions with charged particles known as cosmic rays that reach the Earth from space). Often more importantly, the charge can "leak off" onto water molecules in the air. This is because water molecules are **polar**—that is, even though they are neutral, their charge is not distributed uniformly, Fig. 21–4. Thus the extra electrons on, say, a charged plastic ruler can "leak off" into the air because they are attracted to the positive end of water molecules. A positively charged object, on the other hand, can be neutralized by transfer of loosely held electrons from water molecules in the air. On dry days, static electricity is much more noticeable since the air contains fewer water molecules to allow leakage. On humid or rainy days, it is difficult to make any object hold a net charge for long.

FIGURE 21–3 Simple model of the atom.

FIGURE 21–4 Diagram of a water molecule. Because it has opposite charges on different ends, it is called a "polar" molecule.

21–3 | Insulators and Conductors

Suppose we have two metal spheres, one highly charged and the other electrically neutral (Fig. 21–5a). If we now place a metal object, such as a nail, so that it touches both the spheres (Fig. 21–5b), it is found that the previously uncharged sphere quickly becomes charged. If, instead, we connect the two spheres together by a wooden rod or a piece of rubber (Fig. 21–5c), the uncharged ball does not become noticeably charged. Materials like the iron nail are said to be **conductors** of electricity, whereas wood and rubber are **nonconductors** or **insulators**.

FIGURE 21–5 (a) A charged metal sphere and a neutral metal sphere. (b) The two spheres connected by a metal nail, which conducts charge from one sphere to the other. (c) The two spheres connected by an insulator (wood); almost no charge is conducted.

Charged   Neutral

(a)

Metal

(b) Conductor

Wood

(c) Insulator

Metals are good conductors

Metals are generally good conductors whereas most other materials are insulators (although even insulators conduct electricity very slightly). It is interesting that nearly all natural materials fall into one or the other of these two quite distinct categories. There are some materials, however (notably silicon, germanium, and carbon), that fall into an intermediate (but distinct) category known as **semiconductors**.

From the atomic point of view, the electrons in an insulating material are bound very tightly to the nuclei. In a good conductor, on the other hand, some of the electrons are bound very loosely and can move about freely within the material (although they cannot *leave* the object easily) and are often referred to as *free electrons* or *conduction electrons*. When a positively charged object is brought close to or touches a conductor, the free electrons in the conductor are attracted by this positive charge and move quickly toward it. On the other hand, the free electrons move swiftly away from a negative charge that is brought close. In a semiconductor, there are very few free electrons, and in an insulator, almost none.

(a) Neutral metal rod

(b) Metal rod acquires charge by contact

FIGURE 21–6 (a) Neutral metal rod acquires a charge (b) when placed in contact with a charged metal object.

FIGURE 21–7 Charging by induction.

(a) Neutral metal rod

(b) Metal rod still neutral, but with a separation of charge

FIGURE 21–8 Inducing a charge on an object connected to ground.

(a)

(b)

(c)

21–4 Induced Charge; the Electroscope

Suppose a positively charged metal object is brought close to an uncharged metal object. If the two touch, the free electrons in the neutral one are attracted to the positively charged object and some will pass over to it, Fig. 21–6. Since the second object is now missing some of its negative electrons, it will have a net positive charge. This process is called "charging by conduction," or "by contact," and the two objects end up with the same sign of charge.

Now suppose a positively charged object is brought close to a neutral metal rod, but does not touch it. Although the electrons of the metal rod do not leave the rod, they still move within the metal toward the charged object, which leaves a positive charge at the opposite end, Fig. 21–7. A charge is said to have been *induced* at the two ends of the metal rod. Of course no net charge has been created in the rod; charges have merely been *separated*. The net charge on the metal rod is still zero. However, if the metal were broken into two pieces, we could have two charged objects, one charged positively and one charged negatively.

Another way to induce a net charge on a metal object is to connect it with a conducting wire to the ground (or a conducting pipe leading into the ground) as shown in Fig. 21–8a ($\perp$ means "ground"). The object is then said to be "grounded" or "earthed." Now the Earth, since it is so large and can conduct, can easily accept or give up electrons; hence it acts like a reservoir for charge. If a charged object—say negative this time—is brought up close to the metal, free electrons in the metal are repelled and many of them move down the wire into the Earth, Fig. 21–8b. This leaves the metal positively charged. If the wire is now cut, the metal will have a positive induced charge on it (Fig. 21–8c). If the wire were cut after the negative object is moved away, the electrons would all have moved back into the metal and it would be neutral.

An **electroscope** is a device that can be used for detecting charge. As shown in Fig. 21–9, inside of a case are two movable metal leaves, often made of gold. (Sometimes only one leaf is movable.) The leaves are connected by a conductor to a metal knob on the outside of the case, but are insulated from the case itself. If a positively charged object is brought close to the knob, a separation of charge is induced, as electrons are attracted up into the knob, leaving the leaves positively charged, Fig. 21–10a. The two leaves repel each other as shown. If, instead, the

FIGURE 21–9 Electroscope.

(a)   (b)

FIGURE 21–10 Electroscope charged (a) by induction, (b) by conduction.

knob is charged by conduction, the whole apparatus acquires a net charge as shown in Fig. 21–10b. In either case, the greater the amount of charge, the greater the separation of the leaves.

Note, however that you cannot tell the sign of the charge in this way, since a negative charge will cause the leaves to separate just as much as an equal magnitude positive charge—in either case the two leaves repel each other. An electroscope can, however, be used to determine the sign of the charge if it is first charged by conduction, say negatively, as in Fig. 21–11a. Now if a negative object is brought close, as in Fig. 21–11b, more electrons are induced to move down into the leaves and they separate further. On the other hand, if a positive charge is brought close, the electrons are induced to flow upward, leaving the leaves less negative and their separation is reduced, Fig. 21–11c.

The electroscope was much used in the early studies of electricity. The same principle, aided by some electronics, is used in much more sensitive modern **electrometers**.

(a)   (b)   (c)

FIGURE 21–11 A previously charged electroscope can be used to determine the sign of a given charge.

21–5 Coulomb's Law

We have seen that an electric charge exerts a force on other electric charges. What factors affect the magnitude of this force? To answer this, the French physicist Charles Coulomb (1736–1806) investigated electric forces in the 1780s using a torsion balance (Fig. 21–12) much like that used by Cavendish for his studies of the gravitational force (Chapter 6).

Although precise instruments for the measurement of electric charge were not available in Coulomb's time, he was able to prepare small spheres with different magnitudes of charge in which the *ratio* of the charges was known. He reasoned that if a charged conducting sphere is placed in contact with an identical uncharged sphere, the charge on the first would be shared equally by the two of them because of symmetry. He thus had a way to produce charges equal to $\frac{1}{2}, \frac{1}{4}$, and so on, of the original charge. Although he had some difficulty with induced charges, Coulomb was able to argue that the force one tiny charged object exerted on a second tiny charged object is directly proportional to the charge on each of them. That is, if the charge on either one of the objects was doubled, the force was doubled; and if the charge on both of the objects was doubled, the force increased to four times the original value. This was the case when the distance between the two charges remained the same. If the distance between them was allowed to increase, he found that the force decreased with the *square of the distance* between

FIGURE 21–12 Principle of Coulomb's apparatus. It is similar to Cavendish's, which was used for the gravitational force. When a charged sphere is placed close to the one on the suspended bar, the bar rotates slightly. The suspending fiber resists the twisting motion and the angle of twist is proportional to the force applied. By the use of this apparatus, Coulomb investigated how the electric force varies as a function of the magnitude of the charges and of the distance between them.

fiber

rod

Q_1 Q_2

FIGURE 21–13 Coulomb's law, Eq. 21–1, gives the force between two point charges, Q_1 and Q_2, a distance r apart.

COULOMB'S LAW

them. That is, if the distance was doubled, the force fell to one-fourth of its original value. Thus, Coulomb concluded, the force one tiny charged object exerts on a second one is proportional to the product of the magnitude of the charge on one, Q_1, times the magnitude of the charge on the other, Q_2, and inversely proportional to the square of the distance r between them (Fig. 21–13). As an equation, we can write **Coulomb's law** as

$$F = k\frac{Q_1 Q_2}{r^2}, \qquad (21-1)$$

where k is a proportionality constant.

Equation 21–1 gives the *magnitude* of the electric force that either object exerts on the other. The *direction* of the electric force *is always along the line joining the two objects*. If the two charges have the same sign, the force on either object is directed away from the other. If the two charges have opposite signs, the force on one is directed toward the other, Fig. 21–14. Notice that the force one charge exerts on the second is equal but opposite to that exerted by the second on the first, in accord with Newton's third law.

The validity of Coulomb's law today rests on precision measurements that are much more sophisticated than Coulomb's original experiment. The exponent, 2, in Coulomb's law has been shown to be accurate to 1 part in 10^{16} [that is, $2 \pm (1 \times 10^{-16})$].

Since we are dealing here with a new quantity (electric charge), we could choose its unit so that the proportionality constant k in Eq. 21–1 would be one. Indeed, such a system of units was once common.[†] However, the most widely used unit now is the **coulomb** (C), which is the SI unit. The precise definition of the coulomb today is in terms of electric current and magnetic field, and will be discussed later (Section 28–3). In SI units, k has the value

$$k = 8.988 \times 10^9 \, \text{N}\cdot\text{m}^2/\text{C}^2 \approx 9.0 \times 10^9 \, \text{N}\cdot\text{m}^2/\text{C}^2.$$

Thus, 1 C is that amount of charge which, if placed on each of two point objects that are 1.0 m apart, will result in each object exerting a force of $(9.0 \times 10^9 \, \text{N}\cdot\text{m}^2/\text{C}^2)(1.0\,\text{C})(1.0\,\text{C})/(1.0\,\text{m})^2 = 9.0 \times 10^9 \, \text{N}$ on the other. This would be an enormous force, equal to the weight of almost a million tons. We don't normally encounter charges as large as a coulomb.

Charges produced by rubbing ordinary objects (such as a comb or plastic ruler) are typically around a microcoulomb ($1 \, \mu\text{C} = 10^{-6} \, \text{C}$) or less. Objects that carry a positive charge have a deficit of electrons, whereas negatively charged objects have an excess of electrons. The magnitude of the charge on one electron has been determined to be about $1.602 \times 10^{-19} \, \text{C}$, and its sign is negative. This is the smallest charge found in nature,[‡] and because of its fundamental nature, it is given the symbol e and is often referred to as the *elementary charge*:

$$e = 1.602 \times 10^{-19} \, \text{C}.$$

Note that e is defined as a positive number, so the charge on the electron is $-e$. (The charge on a proton, on the other hand, is $+e$). Since an object cannot gain or lose a fraction of an electron, the net charge on any object must be an integral multiple of this charge. Electric charge is thus said to be **quantized** (existing only in discrete amounts: $1e, 2e, 3e$, etc.). Because e is so small, however, we normally don't notice this discreteness in macroscopic charges ($1 \, \mu\text{C}$ requires about 10^{13} electrons), which thus seem continuous.

Unit for charge: the coulomb

FIGURE 21–14 Direction of the force depends on whether the charges have the same sign, (a) and (b), or opposite signs (c).

F_{12} = force on 1 due to 2 F_{21} = force on 2 due to 1

$\mathbf{F}_{12}$ ◄— ⊕   1   ⊕ —► $\mathbf{F}_{21}$ 2

(a)

$\mathbf{F}_{12}$ ◄— ⊖   1   ⊖ —► $\mathbf{F}_{21}$ 2

(b)

$\mathbf{F}_{12}$ —► ⊕   1 $\mathbf{F}_{21}$ ◄— ⊖   2

(c)

Charge on electron (the elementary charge)

Electric charge is quantized

[†] This is a cgs system of units, and the unit of electric charge is called the *electrostatic unit* (esu) or the statcoulomb. One esu is defined as that charge, on each of two point objects 1 cm apart, that gives rise to a force of 1 dyne.

[‡] According to the standard model of elementary particle physics, subnuclear particles called quarks have a smaller charge than that on the electron, equal to $\frac{1}{3}e$ or $\frac{2}{3}e$. Quarks have not been detected directly as isolated objects, and theory indicates that free quarks may not be detectable.

Note the similarity of Coulomb's law to the law of universal gravitation, Eq. 6–1. Both are inverse square laws $\left(F \propto 1/r^2\right)$. Both also have a proportionality to a product of a property of each body—mass for gravity, electric charge for electricity. A major difference between the two laws is that gravity is always an attractive force, whereas the electric force can be either attractive or repulsive.

The constant k in Eq. 21–1 is often written in terms of another constant, ϵ_0, called the **permittivity of free space**. It is related to k by $k = 1/4\pi\epsilon_0$. Coulomb's law can then be written

$$F = \frac{1}{4\pi\epsilon_0} \frac{Q_1 Q_2}{r^2},$$ (21–2)

where

$$\epsilon_0 = \frac{1}{4\pi k} = 8.85 \times 10^{-12}\,\mathrm{C^2/N \cdot m^2}.$$

COULOMB'S LAW
(in terms of ϵ_0)

Equation 21–2 looks more complicated than Eq. 21–1, but other fundamental equations we haven't seen yet are simpler in terms of ϵ_0 rather than k. It doesn't matter which form we use, of course, since Eqs. 21–1 and 21–2 are equivalent.

It should be recognized that Eqs. 21–1 and 21–2 apply to objects whose size is much smaller than the distance between them. Ideally, it is precise for **point charges** (spatial size negligible compared to other distances). For finite-sized objects, it is not always clear what value to use for r, particularly since the charge may not be distributed uniformly on the objects. If the two objects are spheres and the charge is known to be distributed uniformly on each, then r is the distance between their centers.

Coulomb's law describes the force between two charges when they are at rest. Additional forces come into play when charges are in motion, and these will be discussed in later chapters. In this chapter we discuss only charges at rest, the study of which is called **electrostatics**.

When calculating with Coulomb's law, we can usually ignore the signs of the charges and determine direction based on whether the force is attractive or repulsive.

EXAMPLE 21–1 **Electric force on electron by proton.** Determine the magnitude of the electric force on the electron of a hydrogen atom exerted by the single proton $\left(Q_2 = +e\right)$ that is its nucleus. Assume the electron "orbits" the proton at its average distance of $r = 0.53 \times 10^{-10}\,\mathrm{m}$, Fig. 21–15.

SOLUTION We use Coulomb's law, $F = k\, Q_1 Q_2/r^2$ (Eq. 21–1), with $r = 0.53 \times 10^{-10}\,\mathrm{m}$, and $Q_1 = Q_2 = 1.6 \times 10^{-19}\,\mathrm{C}$ (ignoring the signs of the charges):

$$F = \frac{\left(9.0 \times 10^9\,\mathrm{N \cdot m^2/C^2}\right)\left(1.6 \times 10^{-19}\,\mathrm{C}\right)\left(1.6 \times 10^{-19}\,\mathrm{C}\right)}{\left(0.53 \times 10^{-10}\,\mathrm{m}\right)^2}$$

$$= 8.2 \times 10^{-8}\,\mathrm{N}.$$

The direction of the force on the electron is toward the proton, since the charges have opposite signs and the force is attractive.

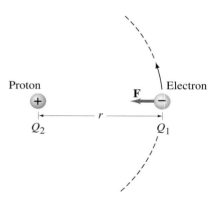

FIGURE 21–15 Example 21–1.

CONCEPTUAL EXAMPLE 21–2 **Which charge exerts the greater force?** Two positive point charges, $Q_1 = 50\,\mu\mathrm{C}$ and $Q_2 = 1\,\mu\mathrm{C}$, are separated by a distance l, Fig. 21–16. Which is larger in magnitude, the force that Q_1 exerts on Q_2, or the force that Q_2 exerts on Q_1?

FIGURE 21–16 Example 21–2.

$Q_1 = 50\,\mu\mathrm{C}$ ●———— l ————● $Q_2 = 1\,\mu\mathrm{C}$

RESPONSE From Coulomb's law, the force on Q_1 exerted by Q_2 is:

$$F_{12} = k\frac{Q_1 Q_2}{l^2}.$$

The force on Q_2 exerted by Q_1 is the same except that Q_1 and Q_2 are reversed. The equation is symmetric with respect to the two charges, so $F_{21} = F_{12}$. Newton's third law also tells us that these two forces must have equal magnitude.

It is very important to keep in mind that Eq. 21–1 (or 21–2) gives the force on a charge due to only *one* other charge. If several (or many) charges are present, the *net force on any one of them will be the vector sum of the forces due to each of the others*. This **principle of superposition** is based on experiment, and tells us that electric force vectors add like any other vector. For continuous distributions of charge, the sum becomes an integral.

Superposition principle: electric forces add as vectors

When dealing with several charges, it is helpful to use double subscripts on each of the forces involved. The first subscript refers to the particle *on* which the force acts; the second refers to the particle that exerts the force. For example, if we have three charges, $\mathbf{F}_{31}$ means the force exerted *on* particle 3 *by* particle 1.

As in all problem solving, it is very important to draw a diagram, in particular a free-body diagram (Chapter 4) for each body, showing all the forces acting on that body. In applying Coulomb's law, we can deal with charge magnitudes only (leaving out minus signs) to get the magnitude of each force. Then determine the direction of the force physically (along the line joining the two particles: like charges repel, unlike charges attract), and show the force on the diagram. Finally, add all the forces on one object together as vectors.

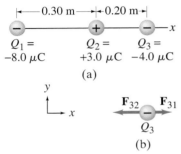

FIGURE 21–17 Diagram for Example 21–3.

EXAMPLE 21–3 **Three charges in a line.** Three charged particles are arranged in a line, as shown in Fig. 21–17a. Calculate the net electrostatic force on particle 3 (the $-4.0\ \mu\text{C}$ on the right) due to the other two charges.

SOLUTION The net force on particle 3 will be the vector sum of the force $\mathbf{F}_{31}$ exerted by particle 1 and the force $\mathbf{F}_{32}$ exerted by particle 2: $\mathbf{F} = \mathbf{F}_{31} + \mathbf{F}_{32}$. The magnitudes of these two forces are

$$F_{31} = \frac{(9.0 \times 10^9\ \text{N}\cdot\text{m}^2/\text{C}^2)(4.0 \times 10^{-6}\ \text{C})(8.0 \times 10^{-6}\ \text{C})}{(0.50\ \text{m})^2} = 1.2\ \text{N}$$

$$F_{32} = \frac{(9.0 \times 10^9\ \text{N}\cdot\text{m}^2/\text{C}^2)(4.0 \times 10^{-6}\ \text{C})(3.0 \times 10^{-6}\ \text{C})}{(0.20\ \text{m})^2} = 2.7\ \text{N}.$$

Since we were calculating the magnitudes of the forces, we omitted the signs of the charges; but we must be aware of them to get the direction of each force. Let the line joining the particles be the x axis, and we take it positive to the right. Then, because $\mathbf{F}_{31}$ is repulsive and $\mathbf{F}_{32}$ is attractive, the directions of the forces are as shown in Fig. 21–17b: F_{31} points in the positive x direction and F_{32} points in the negative x direction. The net force on particle 3 is then

$$F = -F_{32} + F_{31} = -2.7\ \text{N} + 1.2\ \text{N} = -1.5\ \text{N}.$$

The magnitude of the net force is 1.5 N, and it points to the left.

Notice in this Example that the charge in the middle (Q_2) in no way blocks the effect of the other charge (Q_1).

EXAMPLE 21–4 **Electric force using vector components.** Calculate the net electrostatic force on charge Q_3 shown in Fig. 21–18a due to the charges Q_1 and Q_2.

SOLUTION The forces $\mathbf{F}_{31}$ and $\mathbf{F}_{32}$ have the directions shown in the diagram since Q_1 exerts an attractive force and Q_2 a repulsive force. The magnitudes of $\mathbf{F}_{31}$ and $\mathbf{F}_{32}$ are (ignoring signs since we know the directions)

$$F_{31} = \frac{(9.0 \times 10^9\ \text{N}\cdot\text{m}^2/\text{C}^2)(6.5 \times 10^{-5}\ \text{C})(8.6 \times 10^{-5}\ \text{C})}{(0.60\ \text{m})^2} = 140\ \text{N},$$

$$F_{32} = \frac{(9.0 \times 10^9\ \text{N}\cdot\text{m}^2/\text{C}^2)(6.5 \times 10^{-5}\ \text{C})(5.0 \times 10^{-5}\ \text{C})}{(0.30\ \text{m})^2} = 330\ \text{N}.$$

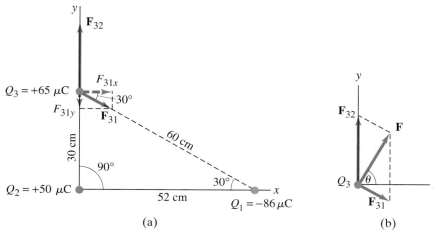

(a)   (b)

FIGURE 21–18 Determining the forces for Example 21–4.

We resolve $\mathbf{F}_{31}$ into its components along the x and y axes, as shown in Fig. 21–18a:

$$F_{31x} = F_{31}\cos 30° = 120\,\text{N},$$

$$F_{31y} = -F_{31}\sin 30° = -70\,\text{N}.$$

The force $\mathbf{F}_{32}$ has only a y component. So the net force $\mathbf{F}$ on Q_3 has components

$$F_x = F_{31x} = 120\,\text{N}$$

$$F_y = F_{32} + F_{31y} = 330\,\text{N} - 70\,\text{N} = 260\,\text{N}.$$

Thus the magnitude of the net force is

$$F = \sqrt{F_x^2 + F_y^2} = \sqrt{(120\,\text{N})^2 + (260\,\text{N})^2} = 290\,\text{N};$$

and it acts at an angle θ (see Fig. 21–18b) given by $\tan\theta = F_y/F_x = 260\,\text{N}/120\,\text{N} = 2.2$, so $\theta = 65°$.

* Vector Form of Coulomb's Law

Coulomb's law can be written in vector form (as we did for Newton's law of universal gravitation in Chapter 6, Section 6–2) as

$$\mathbf{F}_{12} = k\frac{Q_1 Q_2}{r_{21}^2}\,\hat{\mathbf{r}}_{21},$$

where $\mathbf{F}_{12}$ is the vector force on charge Q_1 due to Q_2 and $\hat{\mathbf{r}}_{21}$ is the unit vector pointing from Q_2 toward Q_1. That is $\mathbf{r}_{21}$ points from the "source" charge (Q_2) toward the charge on which we want to know the force (Q_1). See Fig. 21–19. The charges Q_1 and Q_2 can be either positive or negative, and this will affect the direction of the electric force. If Q_1 and Q_2 have the same sign, the product $Q_1 Q_2 > 0$ and the force on Q_1 points away from Q_2—that is, it is repulsive. If Q_1 and Q_2 have opposite signs, $Q_1 Q_2 < 0$ and $\mathbf{F}_{12}$ points toward Q_2—that is, it is attractive.

FIGURE 21–19 Determining the force on Q_1 due to Q_2, showing the direction of the unit vector $\hat{\mathbf{r}}_{21}$.

FIGURE 21–20 An electric field surrounds every charge. P is an arbitrary point.

FIGURE 21–21 Force exerted by charge $+Q$ on a small test charge, q, placed at points a, b, and c.

Both the gravitational force and the electrical force act over a distance: there is a force even when the two objects are not touching. The idea of a force *acting at a distance* was a difficult one for early thinkers. Newton himself felt uneasy with this idea when he published his law of universal gravitation. A helpful way to look at the situation uses the idea of the **field**, developed by the British scientist Michael Faraday (1791–1867). In the electrical case, according to Faraday, an *electric field* extends outward from every charge and permeates all of space (Fig. 21–20). When a second charge is placed near the first charge, it feels a force because of the electric field that is there (say, at point P in Fig. 21–20). The electric field at the location of the second charge is considered to interact directly with this charge to produce the force.

We can in principle investigate the electric field surrounding a charge or group of charges by measuring the force on a small positive **test charge**. By a test charge we mean a charge so small that the force it exerts does not significantly alter the distribution of the charges that create the field being measured. The force on a tiny positive test charge q placed at various locations in the vicinity of a single positive charge Q would be as shown in Fig. 21–21. The force at b is less than at a because the distance is greater (Coulomb's law); and the force at c is smaller still. In each case, the force is directed radially outward from Q. The electric field is defined in terms of the force on such a positive test charge. In particular, the **electric field**, **E**, at any point in space is defined as the force **F** exerted on a tiny positive test charge at that point divided by the magnitude of the test charge q:

Definition of electric field

$$\mathbf{E} = \frac{\mathbf{F}}{q}. \qquad (21\text{–}3)$$

Ideally, **E** is defined as the limit of $\mathbf{F}/q$ as q is taken smaller and smaller, approaching zero. From this definition (Eq. 21–3), we see that the electric field at any point in space is a vector whose direction is the direction of the force on a positive test charge at that point, and whose magnitude is the *force per unit charge*. Thus **E** is measured in units of newtons per coulomb (N/C).

The reason for defining **E** as $\mathbf{F}/q$ (with $q \to 0$) is so that **E** does not depend on the magnitude of the test charge q. This means that **E** describes only the effect of the charges creating the electric field at that point.

The electric field at any point in space can be measured, based on the definition, Eq. 21–3. For simple situations involving one or several point charges, we can calculate what **E** will be. For example, the electric field at a distance r from a single point charge Q would have magnitude

$$E = \frac{F}{q} = \frac{kqQ/r^2}{q}$$

$$= k\frac{Q}{r^2}; \qquad \text{[single point charge]} \quad (21\text{–}4\mathbf{a})$$

or, in terms of ϵ_0 as in Eq. 21–2 $(k = 1/4\pi\epsilon_0)$:

Electric field due to one point charge

$$E = \frac{1}{4\pi\epsilon_0}\frac{Q}{r^2}. \qquad \text{[single point charge]} \quad (21\text{–}4\mathbf{b})$$

Notice that E is independent of q—that is, it depends only on the charge Q which produces the field, and not on the value of the test charge q. Equations 21–4 could be referred to as the electric field form of Coulomb's law.

EXAMPLE 21-5 **Electrostatic copier.** An electrostatic copier works by selectively arranging positive charges (in a pattern to be copied) on the surface of a nonconducting drum, then gently sprinkling negatively charged dry toner (ink) particles onto the drum. The toner particles temporarily stick to the pattern on the drum and are later transferred to paper and "melted" to produce the copy. Suppose each toner particle has a mass of 9.0×10^{-16} kg and carries an average of 20 extra electrons to provide an electric charge. Assuming that the electric force on a toner particle must exceed twice its weight in order to ensure sufficient attraction, compute the required electric field strength near the surface of the drum. See Fig. 21–22.

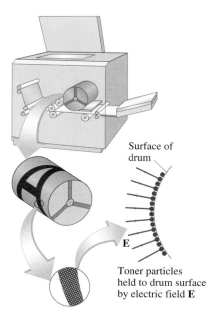

SOLUTION The minimum value of electric field satisfies the relation

$$qE = 2mg$$

where $q = 20e$. Hence

$$E = \frac{2mg}{q} = \frac{2(9.0 \times 10^{-16}\,\text{kg})(9.8\,\text{m/s}^2)}{20(1.6 \times 10^{-19}\,\text{C})}$$

$$= 5.5 \times 10^3\,\text{N/C}.$$

FIGURE 21–22 Example 21–5.

EXAMPLE 21-6 **Electric field of a single point charge.** Calculate the magnitude and direction of the electric field at a point P which is 30 cm to the right of a point charge $Q = -3.0 \times 10^{-6}$ C.

SOLUTION The magnitude of the electric field due to a single point charge is given by Eq. 21–4:

$$E = k\frac{Q}{r^2} = \frac{(9.0 \times 10^9\,\text{N·m}^2/\text{C}^2)(3.0 \times 10^{-6}\,\text{C})}{(0.30\,\text{m})^2} = 3.0 \times 10^5\,\text{N/C}.$$

The direction of the electric field is *toward* the charge Q as shown in Fig. 21–23a since we defined the direction as that of the force on a positive test charge. If Q had been positive, the electric field would have pointed away, as in Fig. 21–23b.

FIGURE 21–23 Example 21–6. Electric field at point P (a) due to a negative charge Q, and (b) due to a positive charge Q.

|←——30 cm——→|

$Q = -3.0 \times 10^{-6}$ C   ←———•P $E = 3.0 \times 10^5$ N/C

(a)

$Q = +3.0 \times 10^{-6}$ C   P•——→ $E = 3.0 \times 10^5$ N/C

(b)

This Example illustrates a general result: The electric field due to a positive charge points away from the charge, whereas **E** due to a negative charge points toward that charge.

If the field is due to more than one charge, the individual fields (call them $\mathbf{E}_1$, $\mathbf{E}_2$, etc.) due to each charge are added vectorially to get the total field at any point:

$$\mathbf{E} = \mathbf{E}_1 + \mathbf{E}_2 + \cdots. \tag{21–5}$$

The validity of this **superposition principle** for electric fields is fully confirmed by experiment.

If we are given the electric field **E** at a given point in space, then we can calculate the force **F** on any charge q (even if not small) placed at that point by writing (see Eq. 21–3):

$$\mathbf{F} = q\mathbf{E}.$$

If q is positive, **F** and **E** will point in the same direction. If q is negative, **F** and **E** point in opposite directions. See Fig. 21–24.

FIGURE 21–24 (a) Electric field at a given point in space. (b) Force on a positive charge. (c) Force on a negative charge.

$Q_1 = -25\,\mu C$ P $Q_2 = +50\,\mu C$

$r_1 = 2.0\,\text{cm}$ $r_2 = 8.0\,\text{cm}$

(a)

Q_1 E_1 Q_2

E_2 (b)

FIGURE 21–25 Example 21–7. In (b), we don't know the relative lengths of $\mathbf{E}_1$ and $\mathbf{E}_2$ until we do the calculation.

EXAMPLE 21–7 **E in between two point charges.** Two point charges are separated by a distance of 10.0 cm. One has a charge of $-25\,\mu C$ and the other $+50\,\mu C$. (a) What is the direction and magnitude of the electric field at a point P in between them, that is 2.0 cm from the negative charge (Fig. 21–25a)? (b) If an electron is placed at rest at P, what will its acceleration (direction and magnitude) be initially?

SOLUTION (a) The field will be a combination of two fields both pointing to the left: the field due to the negative charge Q_1 points toward Q_1, and the field due to the positive charge Q_2 points away from Q_2, again to the left, Fig. 21–25b. Thus, we can add the magnitudes of the two fields together algebraically:

$$E = k\frac{Q_1}{r_1^2} + k\frac{Q_2}{r_2^2} = k\left(\frac{Q_1}{r_1^2} + \frac{Q_2}{r_2^2}\right) = k\frac{Q_1}{r_1^2}\left[1 + \frac{(Q_2/Q_1)}{(r_2^2/r_1^2)}\right]$$

where in the last step we factored out (Q_1/r_1^2). We substitute in $r_1 = 2.0\,\text{cm} = 2.0 \times 10^{-2}\,\text{m}$ and $r_2 = 8.0 \times 10^{-2}\,\text{m}$:

$$E = (9.0 \times 10^9\,\text{N}\cdot\text{m}^2/\text{C}^2)\frac{(25 \times 10^{-6}\,\text{C})}{(2.0 \times 10^{-2}\,\text{m})^2}\left[1 + \frac{(50/25)}{(8.0/2.0)^2}\right]$$

$$= 5.6 \times 10^8 \left[1 + \tfrac{1}{8}\right]\text{N/C} = 6.3 \times 10^8\,\text{N/C}.$$

Notice how factoring out Q_1/r_1^2 on the first line allowed us to see the relative strengths of the two contributing fields—namely that Q_2's field is only $\tfrac{1}{8}$ of Q_1's (or $\tfrac{1}{9}$ of the total).

(b) The electron will feel a force to the *right* since it is negatively charged and the acceleration will therefore be to the right. From the definition of electric field, Eq. 21–3, the force on any charge q (even if not small) placed in an electric field E is given by $F = qE$. Hence the magnitude of the acceleration is

$$a = \frac{F}{m} = \frac{qE}{m} = \frac{(1.60 \times 10^{-19}\,\text{C})(6.3 \times 10^8\,\text{N/C})}{9.1 \times 10^{-31}\,\text{kg}} = 1.1 \times 10^{20}\,\text{m/s}^2.$$

EXAMPLE 21–8 **E above two point charges.** Calculate the total electric field (a) at point A and (b) at point B in Fig. 21–26 due to both charges, Q_1 and Q_2.

SOLUTION (a) The calculation is much like that of Example 21–4, but now we are dealing with electric fields. The electric field at A is the vector sum of the fields $\mathbf{E}_{A1}$ due to Q_1, and $\mathbf{E}_{A2}$ due to Q_2; for each point charge $E = kQ/r^2$, so

$$E_{A1} = \frac{(9.0 \times 10^9\,\text{N}\cdot\text{m}^2/\text{C}^2)(50 \times 10^{-6}\,\text{C})}{(0.60\,\text{m})^2} = 1.25 \times 10^6\,\text{N/C},$$

$$E_{A2} = \frac{(9.0 \times 10^9\,\text{N}\cdot\text{m}^2/\text{C}^2)(50 \times 10^{-6}\,\text{C})}{(0.30\,\text{m})^2} = 5.0 \times 10^6\,\text{N/C}.$$

The directions are as shown, so the total electric field at A, $\mathbf{E}_A$, has components

$$E_{Ax} = E_{A1}\cos 30° = 1.1 \times 10^6\,\text{N/C},$$
$$E_{Ay} = E_{A2} - E_{A1}\sin 30° = 4.4 \times 10^6\,\text{N/C}.$$

→ **PROBLEM SOLVING**

Ignore signs of charges and determine direction physically, showing directions on diagram

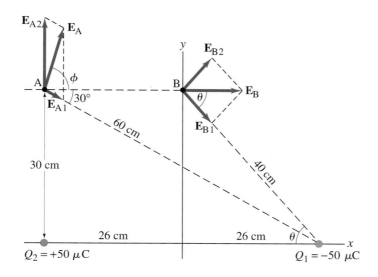

FIGURE 21–26 Calculation of the electric field at points A and B for Example 21–8.

Thus the magnitude of $\mathbf{E}_A$ is

$$E_A = \sqrt{(1.1)^2 + (4.4)^2} \times 10^6 \,\text{N/C} = 4.5 \times 10^6 \,\text{N/C},$$

and its direction is ϕ given by $\tan\phi = E_{Ay}/E_{Ax} = 4.4/1.1 = 4.0$, so $\phi = 76°$.
(b) Because B is equidistant (40 cm by the Pythagorean theorem) from the two equal charges, the magnitudes of E_{B1} and E_{B2} are the same; that is,

➡ **PROBLEM SOLVING**

Use symmetry to save work, when possible

$$E_{B1} = E_{B2} = \frac{kQ}{r^2} = \frac{(9.0 \times 10^9 \,\text{N·m}^2/\text{C}^2)(50 \times 10^{-6} \,\text{C})}{(0.40 \,\text{m})^2}$$

$$= 2.8 \times 10^6 \,\text{N/C}.$$

Also, because of the symmetry, the y components are equal and opposite. Hence the total field E_B is horizontal and equals $E_{B1}\cos\theta + E_{B2}\cos\theta = 2E_{B1}\cos\theta$; from the diagram, $\cos\theta = 26\,\text{cm}/40\,\text{cm} = 0.65$. Then

$$E_B = 2E_{B1}\cos\theta = 2(2.8 \times 10^6 \,\text{N/C})(0.65) = 3.6 \times 10^6 \,\text{N/C},$$

and the direction of $\mathbf{E}_B$ is along the $+x$ direction.

PROBLEM SOLVING | Electrostatics

Solving electrostatics problems follows, to a large extent, the general problem-solving procedure discussed in Section 4–8. In particular,

1. Draw a careful diagram—namely, a free-body diagram for each object, showing all the forces acting on that object, or the electric field at a point due to all sources.

2. Apply Coulomb's law to get the magnitude of the force that each contributing charge exerts on a charged object, or the electric field at a point. Deal only with magnitudes of charges (leaving out minus signs), and obtain the magnitude of each force or electric field. Then determine the direction of each force or electric field physically (like charges repel each other, unlike charges attract). Show and label each vector force or field on your diagram. Then add vectorially all the forces on an object, or the contributing fields at a point, to get the resultant.

3. Use symmetry (say, in the geometry) whenever possible.

In many cases we can treat[†] charge as being distributed continuously. We can divide up a charge distribution into infinitesimal charges dQ, each of which will act as a tiny point charge. The contribution to the electric field at a distance r from each dQ is

$$dE = \frac{1}{4\pi\epsilon_0}\frac{dQ}{r^2}. \qquad (21\text{–}6a)$$

Then the electric field, **E**, at any point is obtained by summing over all the infinitesimal contributions, which is the integral

$$\mathbf{E} = \int d\mathbf{E}. \qquad (21\text{–}6b)$$

Note that $d\mathbf{E}$ is a vector (Eq. 21–6a gives its magnitude). [In situations where Eq. 21–6b is difficult to evaluate, other techniques (discussed in the next two chapters) can often be used instead to determine **E**. Numerical integration can also be used in many cases.]

EXAMPLE 21–9 **A ring of charge.** A thin, ring-shaped object of radius a holds a total charge Q distributed uniformly around it. Determine the electric field at a point P on its axis, a distance x from the center. See Fig. 21–27. Let λ be the charge per unit length (C/m).

SOLUTION The electric field, $d\mathbf{E}$, due to a particular segment of the ring of length dl has magnitude

$$dE = \frac{1}{4\pi\epsilon_0}\frac{dQ}{r^2}.$$

The whole ring has length (circumference) of $2\pi a$, so the charge on a length dl is

$$dQ = Q\left(\frac{dl}{2\pi a}\right) = \lambda\, dl$$

where $\lambda = Q/2\pi a$ is the charge per unit length. Now we write dE as

$$dE = \frac{1}{4\pi\epsilon_0}\frac{\lambda\, dl}{r^2}.$$

The vector $d\mathbf{E}$ has components dE_x along the x axis and $dE_\perp$ perpendicular to the x axis (Fig. 21–27). We are going to sum (integrate) around the entire ring. We note that an equal-length segment diametrically opposite the dl shown will produce a $d\mathbf{E}$ whose component perpendicular to the x axis will just cancel the $dE_\perp$ shown. This is true for all segments of the ring, so by symmetry **E** will be

➡ **PROBLEM SOLVING**

Use symmetry when possible

[†] Because we believe there is a minimum charge (e), the treatment here is only for convenience; it is nonetheless useful and accurate since e is usually very much smaller than macroscopic charges.

FIGURE 21–27 Example 21–9.

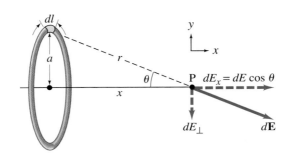

directed along the x axis and we need only sum the x components, dE_x. The total field is then

$$E = E_x = \int dE_x = \int dE \cos\theta = \frac{1}{4\pi\epsilon_0}\lambda\int \frac{dl}{r^2}\cos\theta.$$

Since $\cos\theta = x/r$, where $r = (x^2 + a^2)^{\frac{1}{2}}$, we have

$$E = \frac{\lambda}{(4\pi\epsilon_0)}\frac{x}{(x^2 + a^2)^{\frac{3}{2}}}\int_0^{2\pi a} dl = \frac{1}{4\pi\epsilon_0}\frac{\lambda x(2\pi a)}{(x^2 + a^2)^{\frac{3}{2}}}$$

$$= \frac{1}{4\pi\epsilon_0}\frac{Qx}{(x^2 + a^2)^{\frac{3}{2}}}.$$

At great distances, $x \gg a$, this reduces to $E = Q/4\pi\epsilon_0 x^2$. We would expect this result since at great distances the ring would appear to be a point charge ($1/r^2$ dependence).

Note in this Example three important problem solving techniques or "tricks" that can be used elsewhere: (1) The use of symmetry to reduce the complexity of the problem; (2) expressing the charge dQ in terms of a charge density (here linear, $\lambda = Q/2\pi a$); and (3) the checking of the answer at the limit of large r, which serves as an indication (but not proof) of the correctness of the answer—if the result did not check at large r, the result no doubt would be wrong entirely.

➥ **PROBLEM SOLVING**

Check result by noting that at a great distance the ring looks like a point charge

➥ **PROBLEM SOLVING**

Hints

EXAMPLE 21–10 **Long line of charge.** Determine the magnitude of the electric field at any point P a distance x from a very long line (a wire, say) of uniformly distributed charge, Fig. 21–28. Assume x is much smaller than the length of the wire, and let λ be the charge per unit length (C/m).

SOLUTION We set up a coordinate system so the wire is on the y axis with origin 0 as shown. A segment of wire dy has charge $dQ = \lambda\, dy$. The field $d\mathbf{E}$ at P due to such a length of wire at y has magnitude

$$dE = \frac{1}{4\pi\epsilon_0}\frac{dQ}{r^2} = \frac{1}{4\pi\epsilon_0}\frac{\lambda\, dy}{(x^2 + y^2)},$$

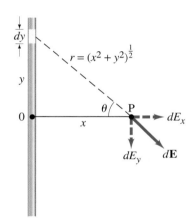

FIGURE 21–28 Example 21–10.

where $r = (x^2 + y^2)^{\frac{1}{2}}$ as shown in Fig. 21–28. The vector $d\mathbf{E}$ has components dE_x and dE_y as shown where $dE_x = dE\cos\theta$ and $dE_y = dE\sin\theta$. If the wire is extremely long in both directions (so distant contributions have little effect compared to nearby ones), or if 0 is at the midpoint of the wire (even if the wire is short), the y component of $\mathbf{E}$ will be zero since there will be equal contributions to $E_y = \int dE_y$ from above and below point 0, so

$$E_y = \int dE\sin\theta = 0.$$

Then we have

$$E = E_x = \int dE\cos\theta = \frac{\lambda}{4\pi\epsilon_0}\int \frac{\cos\theta\, dy}{x^2 + y^2}.$$

The integration here is over y, along the wire, with x treated as constant. We must now write θ as a function of y, or y as a function of θ. We do the latter: since $y = x\tan\theta$, $dy = x\, d\theta/\cos^2\theta$ and $(x^2 + y^2) = x^2/\cos^2\theta$. Then

$$E = \frac{\lambda}{4\pi\epsilon_0}\frac{1}{x}\int_{-\pi/2}^{\pi/2}\cos\theta\, d\theta = \frac{\lambda}{4\pi\epsilon_0 x}(\sin\theta)\Big|_{-\pi/2}^{\pi/2} = \frac{1}{2\pi\epsilon_0}\frac{\lambda}{x},$$

where we have assumed the wire is extremely long in both directions ($y \to \pm\infty$) which corresponds to the limits $\theta = \pm\pi/2$. Thus the field of a long straight line of charge decreases inversely as the first power of the distance from the wire. This result, obtained for an infinite wire, is a good approximation for a wire of finite length as long as x is small compared to the distance of P from the ends of the wire.

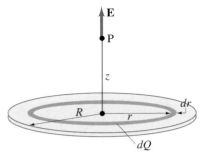

FIGURE 21–29 Example 21–11: a uniformly charged flat disk of radius R.

EXAMPLE 21–11 **Uniformly charged disk.** Charge is distributed uniformly over a thin circular disk of radius R. The charge per unit area (C/m^2) is σ. Calculate the electric field at a point P on the axis of the disk, a distance z above its center, Fig. 21–29.

SOLUTION We can think of the disk as a set of concentric rings. We can then apply the result of Example 21–9 to each of these rings, and then sum over all the rings. For the ring of radius r shown in Fig. 21–29, the electric field has magnitude

$$dE = \frac{1}{4\pi\epsilon_0} \frac{z\, dQ}{(z^2 + r^2)^{\frac{3}{2}}}$$

where we have used the result of Example 21–9 and have written dE (instead of E) for this thin ring of total charge dQ. The ring has area $(dr)(2\pi r)$ and we insert the charge per unit area $\sigma = dQ/(2\pi r\, dr)$:

$$dE = \frac{1}{4\pi\epsilon_0} \frac{z\sigma 2\pi r\, dr}{(z^2 + r^2)^{\frac{3}{2}}} = \frac{z\sigma r\, dr}{2\epsilon_0(z^2 + r^2)^{\frac{3}{2}}}.$$

Now we sum over all the rings, starting at $r = 0$ out to the largest with $r = R$:

$$E = \frac{z\sigma}{2\epsilon_0} \int_0^R \frac{r\, dr}{(z^2 + r^2)^{\frac{3}{2}}} = \frac{z\sigma}{2\epsilon_0} \left[-\frac{1}{(z^2 + r^2)^{\frac{1}{2}}} \right]_0^R$$

$$= \frac{\sigma}{2\epsilon_0} \left[1 - \frac{z}{(z^2 + R^2)^{\frac{1}{2}}} \right].$$

This gives the magnitude of **E** at any point z along the axis of the disk. The direction of each $d\mathbf{E}$ due to each ring is along the z axis (as in Example 21–9), and therefore the direction of **E** is along z. If Q (and σ) are positive, **E** points away from the disk; if Q (and σ) are negative, **E** points toward the disk.

If the radius of the disk in Example 21–11 is much greater than the distance of our point P from the disk (i.e., $z \ll R$) then we can obtain a very useful result: the second term in the solution above becomes very small, so

Electric field near a thin large surface

$$E = \frac{\sigma}{2\epsilon_0}. \qquad \text{[infinite plane]} \quad \textbf{(21–7)}$$

This result is valid for any point above (or below) an infinite plane of any shape holding a uniform charge density σ. It is also valid for points close to a finite plane, as long as the point is close to the plane compared to the distance to the plane's edge. Thus the field near a large uniformly charged plane is uniform, and directed outward if the plane is positively charged.

It is interesting to compare here the distance dependence of the electric field due to a point charge $(E \sim 1/r^2)$, due to a very long uniform line of charge $(E \sim 1/r)$, and due to a very large uniform plane of charge (E does not depend on r).

EXAMPLE 21–12 **Two parallel plates.** Determine the electric field between two large parallel plates or sheets, which are very thin and are separated by a distance d which is small compared to their height and width. One plate carries a uniform surface charge density σ and the other carries a uniform surface charge density $-\sigma$, as shown in Fig. 21–30.

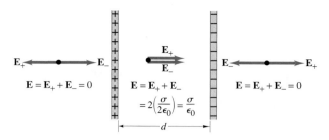

$E = E_+ + E_- = 0$

$E = E_+ + E_-$
$= 2\left(\dfrac{\sigma}{2\epsilon_0}\right) = \dfrac{\sigma}{\epsilon_0}$

$E = E_+ + E_- = 0$

FIGURE 21–30 Example 21–12.

SOLUTION From Eq. 21–7, each plate sets up an electric field of magnitude $E = \pm\sigma/2\epsilon_0$. The field due to the positive plate points away from that plate whereas the field due to the negative plate points toward that plate. Hence, in the region between the plates, the fields add together as shown:

$$E = E_+ + E_- = \frac{\sigma}{2\epsilon_0} + \frac{\sigma}{2\epsilon_0} = \frac{\sigma}{\epsilon_0}.$$

The field is uniform, since the plates are very large compared to their separation, so this result is valid for any point, whether near one or the other of the plates, or midway between them as long as the point is far from the ends. Outside the plates, the fields cancel,

$$E = E_+ + E_- = \frac{\sigma}{2\epsilon_0} - \frac{\sigma}{2\epsilon_0} = 0,$$

as shown in the diagram. These results are valid ideally for infinitely large plates; they are a good approximation for finite plates if the separation is much less than the dimensions of the plate and for points not too close to the edge. These useful and extraordinary results illustrate the principle of superposition and its great power.

Electric field between two large oppositely charged parallel plates

21–8 Field Lines

Since the electric field is a vector, it is sometimes referred to as a *vector field*. We could indicate the electric field with arrows at various points in a given situation, such as at a, b, and c in Fig. 21–31. The directions of $\mathbf{E}_a$, $\mathbf{E}_b$, and $\mathbf{E}_c$ are the same as that of the forces shown earlier in Fig. 21–21, but the lengths (magnitudes) are different since we divide by q. However, the relative lengths of $\mathbf{E}_a$, $\mathbf{E}_b$, and $\mathbf{E}_c$ are the same as for the forces since we divide by the same q each time. To indicate the electric field in such a way at *many* points, however, would result in many arrows, which might appear complicated or confusing. To avoid this, we use another technique, that of field lines.

In order to visualize the electric field, we draw a series of lines to indicate the direction of the electric field at various points in space. These **electric field lines** (sometimes called *lines of force*) are drawn so that they indicate the direction of the force due to the given field on a positive test charge. The lines of force due to a single isolated positive charge are shown in Fig. 21–32a and for a single isolated negative charge in Fig. 21–32b. In part (a) the lines point radially outward from the charge, and in part (b) they point radially inward toward the charge because that is the direction the force would be on a positive test charge in each case (as in Fig. 21–23). Only a few representative lines are shown. One could just as well draw lines in between those shown since the electric field exists there as well. We can always draw the lines so that the *number of lines starting on a positive charge, or ending on a negative charge, is proportional to the magnitude of the charge.* Notice that near the charge, where the electric field is greatest, the lines are closer together. This is a general property of electric field lines: *the closer the lines are together, the stronger the electric field in that region.* In fact the lines can always be drawn so that the number of lines crossing unit area perpendicular to $\mathbf{E}$ is proportional to the magnitude of the electric field.

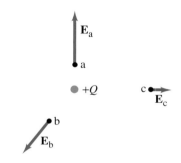

FIGURE 21–31 Electric field vector shown at three points, due to a single point charge Q. (Compare to Fig. 21–21.)

FIGURE 21–32 Electric field lines (a) near a single positive point charge, (b) near a single negative point charge.

(a)　　　　(b)

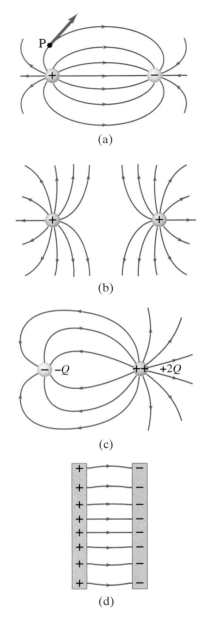

(a)

(b)

$-Q$ $+2Q$

(c)

(d)

FIGURE 21–33 Electric field lines for four arrangements of charges.

FIGURE 21–34 The Earth's gravitational field.

Figure 21–33a shows the electric field lines due to two charges of opposite sign. The electric field lines are curved in this case and they are directed from the positive charge to the negative charge. The direction of the field at any point is directed tangentially as shown by the arrow at point P. To satisfy yourself that this is the correct pattern for the electric field lines, you can make a few calculations such as those done in Example 21–8 for just this case (see Fig. 21–26). Figures 21–33b and c show the electric field lines (b) for two equal positive charges, and (c) for unequal charges, $+2Q$ and $-Q$; note that twice as many lines leave $+2Q$ as there are lines entering $-Q$ (number of lines is proportional to magnitude of Q). Finally, in (d), we see the field between two oppositely charged parallel plates. Notice that the electric field lines between the two plates start out perpendicular to the surface of the metal plates (we'll see why this is always true in the next Section) and go directly from one plate to the other, as we expect because a positive test charge placed between the plates would feel a strong repulsion from the positive plate and a strong attraction to the negative plate. The field lines between the plates are parallel and equally spaced in the central region far from the edges (as in Example 21–12), but fringe outward near the edges. Thus, in the central region, the electric field has the same magnitude at all points and we can write (see Example 21–12)

$$E = \text{constant} = \frac{\sigma}{\epsilon_0}. \qquad \text{[between two closely spaced parallel plates]} \qquad (21\text{–}8)$$

The fringing of the field near the edges can often be ignored, particularly if the separation of the plates is small compared to their size.

We summarize the properties of field lines as follows:

1. The field lines indicate the direction of the electric field; the field points in the direction tangent to the field line at any point.
2. The lines are drawn so that the magnitude of the electric field, E, is proportional to the number of lines crossing unit area perpendicular to the lines. The closer the lines, the stronger the field.
3. Electric field lines start on positive charges and end on negative charges; and the number starting or ending is proportional to the magnitude of the charge.

Also note that field lines never cross. Why not? Because it would not make sense for the electric field to have two values at the same point.

The field concept can also be applied to the gravitational force as mentioned in Chapter 6. Thus we can say that a **gravitational field** exists for every object that has mass. One object attracts another by means of the gravitational field. The Earth, for example, can be said to possess a gravitational field (Fig. 21–34) which is responsible for the gravitational force on objects. The *gravitational field* is defined as the *force per unit mass*. The magnitude of the Earth's gravitational field at any point is then (GM_E/r^2), where M_E is the mass of the Earth, r is the distance of the point from the Earth's center, and G is the gravitational constant (Chapter 6). At the Earth's surface, r is simply the radius of the Earth and the gravitational field is simply equal to g, the acceleration due to gravity (since $F/m = mg/m = g$). Beyond the Earth, the gravitational field can be calculated at any point as a sum of terms due to Earth, Sun, Moon, and other bodies that contribute significantly.

21–9 Electric Fields and Conductors

We now discuss some properties of conductors. First, *the electric field inside a conductor is zero in the static situation*—that is, when the charges are at rest. If there were an electric field within a conductor, there would be a force on its free electrons. The electrons would move until they reached positions where the electric field, and therefore the electric force on them, did become zero.

This reasoning has some interesting consequences. For one, *any net charge on a conductor distributes itself on the surface*. For a negatively charged conductor, you can imagine that the negative charges repel one another and race to the surface to get as far from one another as possible. Another consequence is the following. Suppose that a positive charge Q is surrounded by an isolated uncharged metal conductor whose shape is a spherical shell, Fig. 21–35. Because there can be no field within the metal, the lines leaving the positive charge must end on negative charges on the inner surface of the metal. Thus an equal amount of negative charge, $-Q$, is induced on the inner surface of the spherical shell. Then, since the shell is neutral, a positive charge of the same magnitude, $+Q$, must exist on the outer surface of the shell. Thus, although no field exists in the metal itself, an electric field exists outside of it, as shown in Fig. 21–35, as if the metal were not even there.

A related property of static electric fields and conductors is that *the electric field is always perpendicular to the surface outside of a conductor*. If there were a component of **E** parallel to the surface (Fig. 21–36), electrons at the surface would move along the surface in response to this force until they reached positions where no net force was exerted on them parallel to the surface—that is, until the electric field was perpendicular to the surface.

These properties pertain only to conductors. Inside a nonconductor, which does not have free electrons, an electric field can exist as we will see later, in Chapter 24. And the electric field outside a nonconductor does not necessarily make an angle of 90° to the surface.

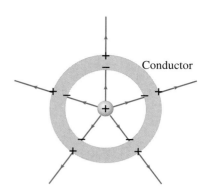

FIGURE 21–35 A charge placed inside a spherical shell. Charges are induced on the conductor surfaces. The electric field exists even beyond the shell but not within the conductor itself.

FIGURE 21–36 If the electric field **E** at the surface of a conductor had a component parallel to the surface, $\mathbf{E}_\parallel$, the latter would accelerate electrons into motion. In the static case (no charges are in motion), $\mathbf{E}_\parallel$ must be zero, and so the electric field must be perpendicular to the conductor's surface: $\mathbf{E} = \mathbf{E}_\perp$.

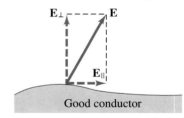

| CONCEPTUAL EXAMPLE 21–13 | **Shielding, and safety in a storm.** A hollow metal box is placed between two parallel charged plates as shown in Fig. 21–37a. What is the field like inside the box?

RESPONSE If our metal box were solid, and not hollow, the electrons in the box, even if it were neutral overall, would redistribute themselves along the surface so that the field lines would not penetrate the conducting metal of the box. For a hollow box, the external field is not changed since the electrons in the metal can move just as freely as before to the surface. Hence we conclude that the field inside the hollow metal box is zero. So the field lines are something like those shown in Fig. 21–37b. A conducting box used in this way is an effective device for shielding delicate instruments and electronic circuits from unwanted external electric fields. We also can see that a relatively safe place to be during a lightning storm is inside a car, surrounded by metal. See also Fig. 21–38, where we see that a person inside a porous "cage" is protected from a strong electric discharge.

FIGURE 21–37 Example 21–13.

FIGURE 21–38 A strong electric field exists in the vicinity of this "Faraday cage," so strong that electrons are pulled off atoms of the air, and charge flows to (or from) the metal cage. Yet the person inside the cage is not affected.

21–10 Motion of a Charged Particle in an Electric Field

If an object having an electric charge q is at a point in space where the electric field is **E**, the force on the object is given by

$$\mathbf{F} = q\mathbf{E}$$

(see Eq. 21–3). In the past few Sections we have seen how to determine **E** for some particular situations. Now let us suppose we know **E** and we want to find the force on a charged object and its subsequent motion.

FIGURE 21–39 Example 21–14.

EXAMPLE 21–14 **Electron accelerated by electric field.** An electron (mass $m = 9.1 \times 10^{-31}$ kg) is accelerated in the uniform field **E** ($E = 2.0 \times 10^4$ N/C) between two parallel charged plates. The separation of the plates is 1.5 cm. The electron is accelerated from rest near the negative plate and passes through a tiny hole in the positive plate, Fig. 21–39. (a) With what speed does it leave the hole? (b) Show that the gravitational force can be ignored. Assume the hole is so small that it does not affect the uniform field between the plates.

SOLUTION (a) The magnitude of the force on the electron is

$$F = qE$$

and is directed to the right. The magnitude of the electron's acceleration is

$$a = \frac{F}{m} = \frac{qE}{m}.$$

Between the plates **E** is uniform so the electron undergoes uniformly accelerated motion with acceleration

$$a = \frac{(1.6 \times 10^{-19}\,\text{C})(2.0 \times 10^4\,\text{N/C})}{(9.1 \times 10^{-31}\,\text{kg})} = 3.5 \times 10^{15}\,\text{m/s}^2.$$

It travels a distance $x = 1.5 \times 10^{-2}$ m before reaching the hole, and since its initial speed was zero, we can use the kinematic equation, $v^2 = v_0^2 + 2ax$ (Eq. 2–12c), with $v_0 = 0$:

$$v = \sqrt{2ax} = \sqrt{2(3.5 \times 10^{15}\,\text{m/s}^2)(1.5 \times 10^{-2}\,\text{m})} = 1.0 \times 10^7\,\text{m/s}.$$

There is no electric field outside the plates, so after passing through the hole, the electron moves with this speed, which is now constant.
(b) The magnitude of the electric force on the electron is

$$qE = (1.6 \times 10^{-19}\,\text{C})(2.0 \times 10^4\,\text{N/C}) = 3.2 \times 10^{-15}\,\text{N}.$$

The gravitational force is

$$mg = (9.1 \times 10^{-31}\,\text{kg})(9.8\,\text{m/s}^2) = 8.9 \times 10^{-30}\,\text{N},$$

which is 10^{14} times smaller! Note that the electric field due to the electron does not enter the problem (since a particle cannot exert a force on itself).

EXAMPLE 21–15 **Electron moving perpendicular to E.** Suppose an electron (say from the Example 21–14) traveling with speed $v_0 = 1.0 \times 10^7$ m/s enters a uniform electric field **E** at right angles to $\mathbf{v}_0$ as shown in Fig. 21–40. Describe its motion by giving the equation of its path while in the electric field. Ignore gravity.

SOLUTION When the electron enters the electric field (at $x = y = 0$) it has velocity $\mathbf{v}_0 = v_0\mathbf{i}$ in the x direction. The electric field **E**, pointing vertically upward, imparts a uniform vertical acceleration to the electron of

$$a_y = \frac{F}{m} = \frac{qE}{m} = -\frac{eE}{m},$$

where we set $q = -e$ for the electron. Its vertical position is given by

$$y = \frac{1}{2}a_y t^2 = -\frac{eE}{2m}t^2$$

since the motion is at constant acceleration. The horizontal position is given by

$$x = v_0 t$$

since $a_x = 0$. We eliminate t between these two equations and obtain

$$y = -\frac{eE}{2mv_0^2}x^2,$$

which is the equation of a parabola (just as in projectile motion, Section 3–7).

FIGURE 21–40 Example 21–15.

21–11 | Electric Dipoles

The combination of two equal charges of opposite sign, $+Q$ and $-Q$, separated by a distance l, is referred to as an **electric dipole**. The quantity Ql is called the **dipole moment** and is represented[†] by the symbol p. The dipole moment can be considered to be a vector **p**, of magnitude Ql, that points from the negative to the positive charge as shown in Fig. 21–41. Many molecules, such as the diatomic molecule CO, have a dipole moment (C has a small positive charge and O a small negative charge of equal magnitude), and are referred to as **polar molecules**. Even though the molecule as a whole is neutral, there is a separation of charge that results from an uneven sharing of electrons by the two atoms.[‡] (Symmetric diatomic molecules, like O_2, have no dipole moment.) The water molecule, with its uneven sharing of electrons (O is negative, the two H are positive), also has a dipole moment—see Figs. 21–4 and 21–42.

Dipole moment

FIGURE 21–41 A dipole consists of equal but opposite charges, $+Q$ and $-Q$, separated by a distance l. The dipole moment $\mathbf{p} = Ql$ and points from the negative to the positive charge.

[†] Be careful not to confuse this p for dipole moment with p for momentum.

[‡] The value of the separated charges may be a fraction of e (say $\pm 0.2e$ or $\pm 0.4e$) but note that such charges don't violate what we said about e being the smallest charge. These charges less than e cannot be isolated and merely represent how much time electrons spend around one atom or the other.

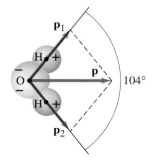

FIGURE 21–42 In the water molecule (H_2O), the electrons spend more time around the oxygen atom than around the two hydrogen atoms. The net dipole moment **p** can be considered as the vector sum of two dipole moments $\mathbf{p}_1$ and $\mathbf{p}_2$ that point from the O to each H as shown: $\mathbf{p} = \mathbf{p}_1 + \mathbf{p}_2$.

FIGURE 21–43 An electric dipole in a uniform electric field.

Dipole in an External Field

First let us consider a dipole, of dipole moment $p = Ql$, that is placed in a uniform electric field **E**, as shown in Fig. 21–43. If the field is uniform, the force $Q\mathbf{E}$ on the positive charge and the force $-Q\mathbf{E}$ on the negative charge result in no net force on the dipole. There will, however, be a *torque* on the dipole which has magnitude (calculated about the center, 0, of the dipole)

$$\tau = QE\frac{l}{2}\sin\theta + QE\frac{l}{2}\sin\theta = pE\sin\theta. \tag{21–9a}$$

This can be written in vector notation as

$$\boldsymbol{\tau} = \mathbf{p} \times \mathbf{E}. \tag{21–9b}$$

The effect of the torque is to try to turn the dipole so **p** is parallel to **E**. The work, W, done on the dipole by the electric field to change the angle θ from θ_1 to θ_2, is given by (see Eq. 10–22)

$$W = \int_{\theta_1}^{\theta_2} \tau\, d\theta = pE\int_{\theta_1}^{\theta_2} \sin\theta\, d\theta = -pE\cos\theta\Big|_{\theta_1}^{\theta_2} = pE(\cos\theta_1 - \cos\theta_2).$$

Work done by the field decreases the potential energy, U, of the dipole in this field. If we choose $U = 0$ when **p** is perpendicular to **E** ($\theta = 90°$), then

Potential energy of dipole in electric field

$$U = -W = -pE\cos\theta = -\mathbf{p}\cdot\mathbf{E}. \tag{21–10}$$

If the electric field is *not* uniform, the force on the $+Q$ of the dipole may not have the same magnitude as the force on the $-Q$, so there may be a net force, as well as a torque.

Electric Field Produced by a Dipole

We have just seen how an external electric field affects an electric dipole. Now let us suppose that there is no external field, and we want to determine the electric field produced *by* the dipole. For brevity, we restrict ourselves to points that are on the perpendicular bisector of the dipole, such as point P in Fig. 21–44 which is a distance r above the midpoint of the dipole. Note that r in Fig. 21–44 is not the distance from either charge to P; the latter distance is $\left(r^2 + l^2/4\right)^{\frac{1}{2}}$ and this is what must be used in Eq. 21–4. The total field at P is

$$\mathbf{E} = \mathbf{E}_+ + \mathbf{E}_-,$$

FIGURE 21–44 Electric field due to an electric dipole.

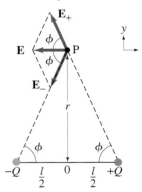

where $\mathbf{E}_+$ and $\mathbf{E}_-$ are the fields due to the $+$ and $-$ charges respectively. The magnitudes E_+ and E_- are equal:

$$E_+ = E_- = \frac{1}{4\pi\epsilon_0}\frac{Q}{r^2 + l^2/4}.$$

Their y components cancel at point P, so the magnitude of the total field **E** is

$$E = 2E_+\cos\phi = \frac{1}{2\pi\epsilon_0}\left(\frac{Q}{r^2 + l^2/4}\right)\frac{l}{2\left(r^2 + l^2/4\right)^{\frac{1}{2}}}$$

or

$$E = \frac{1}{4\pi\epsilon_0}\frac{p}{\left(r^2 + l^2/4\right)^{\frac{3}{2}}}. \qquad \begin{bmatrix}\text{on perpendicular bisector}\\ \text{of dipole}\end{bmatrix} \tag{21–11}$$

Far from the dipole, $r \gg l$, this reduces to

$$E = \frac{1}{4\pi\epsilon_0}\frac{p}{r^3}. \qquad \begin{bmatrix}\text{on perpendicular bisector}\\ \text{of dipole; } r \gg l\end{bmatrix} \tag{21–12}$$

So the field decreases more rapidly for a dipole than for a single point charge $(1/r^3$ versus $1/r^2)$, which we expect since at large distances the two opposite charges appear so close together as to neutralize each other. This $1/r^3$ dependence also applies for points not on the perpendicular bisector (see the Problems).

EXAMPLE 21–16 Dipole in a field. The dipole moment of a water molecule is 6.1×10^{-30} C·m. A water molecule is placed in a uniform electric field with magnitude 2.0×10^5 N/C. (a) What is the magnitude of the maximum torque that the field can exert on the molecule? (b) What is the potential energy when the torque is at its maximum? (c) In what position will the potential energy take on its greatest value? Why is this different than the position where the torque is maximized?

SOLUTION (a) From Eq. 21–9 we see that τ is maximized when θ is 90°. Then $\tau = pE = (6.1 \times 10^{-30}\,\text{C·m})(2.0 \times 10^5\,\text{N/C}) = 1.2 \times 10^{-24}$ N·m.
(b) The potential energy is given by Eq. 21–10. For $\theta = 90°$ the potential energy is zero. Note that the potential energy is negative for smaller values of θ, so U is not a minimum for $\theta = 90°$.
(c) The potential energy will be a maximum when $\cos \theta = -1$, so $\theta = 180°$, meaning $\mathbf{E}$ and $\mathbf{p}$ are antiparallel. The potential energy is maximized when the dipole is oriented so that it has to rotate through the largest angle, 180°, to reach the equilibrium position at $\theta = 0°$. The torque on the other hand is maximized when the electric forces are perpendicular to $\mathbf{p}$.

Summary

There are two kinds of **electric charge**, positive and negative. These designations are to be taken algebraically—that is, any charge is plus or minus so many coulombs (C), in SI units.

Electric charge is **conserved**: if a certain amount of one type of charge is produced in a process, an equal amount of the opposite type is also produced; thus the *net* charge produced is zero.

We can better understand electricity by appealing to atomic theory, which theorizes an atom as consisting of a positively charged nucleus surrounded by negatively charged electrons. Each electron has a charge $-e = -1.6 \times 10^{-19}$ C.

Conductors are those materials in which many electrons are relatively free to move, whereas electric **insulators** are those in which very few electrons are free to move.

An object is negatively charged when it has an excess of electrons, and positively charged when it has less than its normal amount of electrons. The charge on any object is thus a whole number times $+e$ or $-e$. That is, charge is **quantized**.

An object can become charged by rubbing (in which electrons are transferred from one material to another), by conduction (which is transfer of charge from one charged object to another by touching), or by induction (the separation of charge within an object because of the close approach of another charged object but without touching).

Electric charges exert a force on each other. If two charges are of opposite types, one positive and one negative, they each exert an attractive force on the other. If the two charges are the same type, each repels the other.

The magnitude of the force one point charge exerts on another is proportional to the product of their charges, and inversely proportional to the square of the distance between them:

$$F = k\frac{Q_1 Q_2}{r^2};$$

this is **Coulomb's law**. In SI units, k is often written as $1/4\pi\epsilon_0$.

We think of an **electric field** as existing in space around any charge or group of charges. The force on another charged object is then said to be due to the electric field present at its location.

The *electric field*, $\mathbf{E}$, at any point in space due to one or more charges, is defined as the force per unit charge that would act on a test charge q placed at that point:

$$\mathbf{E} = \frac{\mathbf{F}}{q}.$$

Electric fields are represented by **electric field lines** that start on positive charges and end on negative charges. Their direction indicates the direction the force would be on a tiny positive test charge placed at a point. The lines can be drawn so that the number per unit area is proportional to the magnitude of E.

The static electric field (that is, no charges moving) inside a conductor is zero, and the electric field lines just outside a charged conductor are perpendicular to its surface.

An **electric dipole** is a combination of two equal but opposite charges, $+Q$ and $-Q$, separated by a distance l. The **dipole moment** is $p = Ql$. A dipole placed in a uniform electric field feels no net force but does feel a net torque (unless $\mathbf{p}$ is parallel to $\mathbf{E}$). The electric field produced by a dipole decreases as the third power of the distance r from the dipole $(E \propto 1/r^3)$ for r large compared to l.

Questions

1. If you charge a pocket comb by rubbing it with a silk scarf, how can you determine if the comb is positively or negatively charged?

2. Why does a shirt or blouse taken from a clothes dryer sometimes cling to your body?

3. Explain why fog or rain droplets tend to form around ions or electrons in the air.

4. A positively charged rod is brought close to a neutral piece of paper, which it attracts. Draw a diagram showing the separation of charge and explain why attraction occurs.

5. Why does a plastic ruler that has been rubbed with a cloth have the ability to pick up small pieces of paper? Why is this difficult to do on a humid day?

6. Contrast the *net charge* on a conductor to the "free charges" in the conductor.

7. Figures 21–7 and 21–8 show how a charged rod placed near an uncharged metal object can attract (or repel) electrons. There are a great many electrons in the metal, yet only some of them move as shown. Why not all of them?

8. When an electroscope is charged, the two leaves repel each other and remain at an angle. What balances the electric force of repulsion so that the leaves don't separate further?

9. The form of Coulomb's law is very similar to that for Newton's law of universal gravitation. What are the differences between these two laws? Compare also gravitational mass and electric charge.

10. We are not normally aware of the gravitational or electric force between two ordinary objects. What is the reason in each case? Give an example where we are aware of each one and why.

11. Is the electric force a conservative force? Why or why not? (See Chapter 8.)

12. What experimental observations mentioned in the text rule out the possibility that the numerator in Coulomb's law contains the sum $(Q_1 + Q_2)$ rather than the product $Q_1 Q_2$?

13. When a charged ruler attracts small pieces of paper, sometimes a piece jumps quickly away after touching the ruler. Explain.

14. Explain why we use *small* test charges when measuring electric fields.

15. When determining an electric field, must we use a *positive* test charge, or would a negative one do as well? Explain.

16. Draw the electric field lines surrounding two negative electric charges a distance l apart.

17. Assume that the two opposite charges in Fig. 21–33a are 12.0 cm apart. Consider the magnitude of the electric field 2.5 cm from the positive charge. On which side of this charge—top, bottom, left, or right—is the electric field the strongest? The weakest?

18. Consider the electric field at the three points indicated by the letters A, B, and C in Fig. 21–45. First draw an arrow at each point indicating the direction of the net force that a positive test charge would experience if placed at that point, then list the letters in order of *decreasing* field strength (strongest first).

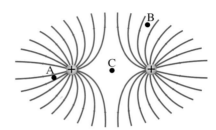

FIGURE 21–45 Question 18.

19. Why can electric field lines never cross?

20. Show, using the three rules for field lines given in Section 21–8, that the electric field lines starting or ending on a single point charge must be symmetrically spaced around the charge.

21. Given two point charges, Q and $2Q$, a distance l apart, is there a point along the straight line that passes through them where $E = 0$ when their signs are (*a*) opposite, (*b*) the same? If yes, state roughly where this point will be.

22. Suppose the ring of Fig. 21–27 has a uniformly distributed negative charge Q. What is the magnitude and direction of **E** at point P?

23. Consider a small positive test charge located on an electric field line at some point, such as point P in Fig. 21–33a. Is the direction of the velocity and/or acceleration of the test charge along this line? Discuss.

24. We wish to determine the electric field at a point near a positively charged metal sphere (a good conductor). We do so by bringing a small test charge, q_0, to this point and measure the force F_0 on it. Will F_0/q_0 be greater than, less than, or equal to, the electric field **E** as it was at that point before the test charge was present?

25. In what ways does the electron motion in Example 21–15 resemble projectile motion (Section 3–7)? In which ways not?

26. Describe the motion of the dipole shown in Fig. 21–43 if it is released from rest at the position shown.

27. Explain why there can be a net force on an electric dipole placed in a nonuniform electric field.

Problems

Section 21-5

1. (I) Calculate the magnitude of the force between two 2.50-C point charges 3.0 m apart.

2. (I) How many electrons make up a charge of $-30.0 \mu C$?

3. (I) What is the magnitude of the electric force of attraction between an iron nucleus ($q = +26e$) and its innermost electron if the distance between them is 1.5×10^{-12} m?

4. (I) What is the repulsive electrical force between two protons in a nucleus that are 5.0×10^{-15} m apart from each other?

5. (I) What is the magnitude of the force a $+25\text{-}\mu C$ charge exerts on a $+3.0\text{-mC}$ charge 35 cm away? ($1 \mu C = 10^{-6}$ C, 1 mC $= 10^{-3}$ C.)

6. (II) Two charged smoke particles exert a force of 4.2×10^{-2} N on each other. What will be the force if they are moved so they are only one eighth as far apart?

7. (II) Two charged balls are 15.0 cm apart. They are moved, and the force on each of them is found to have been tripled. How far apart are they now?

8. (II) A person scuffing her feet on a wool rug on a dry day accumulates a net charge of $-40 \mu C$. How many excess electrons does this person get, and by how much does her mass increase?

9. (II) What is the total charge of all the electrons in 1.0 kg of H_2O?

10. (II) Particles of charge $+70$, $+48$, and $-80 \mu C$ are placed in a line (Fig. 21-46). The center one is 0.35 m from each of the others. Calculate the net force on each charge due to the other two.

$70 \mu C \quad 48 \mu C \quad -80 \mu C$
$\overline{\;\;0.35 \text{ m}\;\;0.35 \text{ m}\;\;}$

FIGURE 21-46 Problem 10.

11. (II) Three positive particles of charges $11.0 \mu C$ are located at the corners of an equilateral triangle of side 15.0 cm (Fig. 21-47). Calculate the magnitude and direction of the net force on each particle.

$11.0 \mu C$

15.0 cm   15.0 cm

$11.0 \mu C \quad 15.0 \text{ cm} \quad 11.0 \mu C$

FIGURE 21-47 Problem 11.

12. (II) A charge of 6.00 mC is placed at each corner of a square 0.100 m on a side. Determine the magnitude and direction of the force on each charge.

13. (II) Repeat Problem 12 for the case when two of the positive charges, on opposite corners, are replaced by negative charges of the same magnitude (Fig. 21-48).

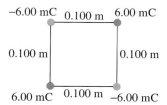

-6.00 mC   0.100 m   6.00 mC

0.100 m   0.100 m

6.00 mC   0.100 m -6.00 mC

FIGURE 21-48 Problem 13.

14. (II) At each corner of a square of side l there are point charges of magnitude $Q, 2Q, 3Q,$ and $4Q$ (Fig. 21-49). Determine the force on each charge due to the other three charges.

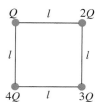

$Q \quad l \quad 2Q$

$l \qquad l$

$4Q \quad l \quad 3Q$

FIGURE 21-49 Problem 14.

15. (II) Repeat Problem 14 assuming the $3Q$ charge is replaced by a $-3Q$ charge (i.e., its sign is changed).

16. (II) Compare the electric force holding the electron in orbit ($r = 0.53 \times 10^{-10}$ m) around the proton nucleus of the hydrogen atom, with the gravitational force between the same electron and proton. What is the ratio of these two forces?

17. (II) Two positive point charges are a fixed distance apart. The sum of their charges is Q_T. What charge must each have in order to (a) maximize the electric force between them, and (b) minimize it?

18. (II) Two point charges have a total charge of $560 \mu C$. When placed 1.10 m apart, the force each exerts on the other is 22.8 N and is repulsive. What is the charge on each?

19. (II) Two charges, $-Q_0$ and $-3Q_0$, are a distance l apart. These two charges are free to move but do not because there is a third charge nearby. What must be the charge and placement of the third charge for the first two to be in equilibrium?

20. (III) A $+7.7 \mu C$ and a $-3.5 \mu C$ charge are placed 18.5 cm apart. Where can a third charge be placed so that it experiences no net force?

21. (III) Two small nonconducting spheres have a total charge of $90.0 \mu C$. When placed 1.16 m apart, the force each exerts on the other is 12.0 N and is repulsive. What is the charge on each? What if the force were attractive?

22. (III) Two small charged spheres hang from cords of equal length l as shown in Fig. 21–50 and make small angles θ_1 and θ_2 with the vertical. (a) If $Q_1 = Q$, $Q_2 = 2Q$, and $m_1 = m_2 = m$, determine the ratio θ_1/θ_2. (b) If $Q_1 = Q$, $Q_2 = 2Q$, $m_1 = m$, and $m_2 = 2m$, determine the ratio θ_1/θ_2. (c) Estimate the distance between the spheres for each case.

FIGURE 21–50
Problem 22.

FIGURE 21–51
Problem 23.

23. (III) On each corner of a cube of side l there is a point charge Q. What is the force on the charge at the origin 0 due to the others? Give answer in vector notation for the charge at the origin in Fig. 21–51.

Sections 21–6 to 21–8

24. (I) What is the magnitude of the acceleration experienced by an electron in an electric field of 600 N/C? How does the direction of the acceleration depend on the direction of the field at that point?

25. (I) What is the magnitude and direction of the electric force on an electron in a uniform electric field of strength 1360 N/C that points due east?

26. (I) A proton is released in a uniform electric field, and it experiences an electric force of 2.75×10^{-14} N toward the south. What are the magnitude and direction of the electric field?

27. (I) What is the magnitude and direction of the electric field 20.0 cm directly above an isolated 33.0×10^{-6}-C charge?

28. (I) What is the magnitude and direction of the electric field at a point midway between a -8.0-μC and a $+7.0$-μC charge 8.0 cm apart? Assume no other charges are nearby.

29. (I) The electric force on a $+4.20$-μC charge is $\mathbf{F} = 5.85 \times 10^{-4}$ N$\mathbf{j}$. What is the electric field at the position of the charge?

30. (I) What is the electric field at a point when the force on a 1.25-μC charge placed at that point is $\mathbf{F} = (3.0\mathbf{i} - 5.0\mathbf{j}) \times 10^{-3}$ N?

31. (II) Electric field lines can always be drawn so that their number per unit area perpendicular to the lines is proportional to the electric field magnitude E. However, if Coulomb's law were not valid—that is, if the field due to a single point did not fall off precisely as $1/r^2$ (so the exponent on r was not precisely 2)—this property of field lines would not be true. Show why. [Hint: Take the example of a single point charge.]

32. (II) Draw, approximately, the electric field lines about two point charges, $+Q$ and $-3Q$, which are a distance l apart.

33. (II) Draw, approximately, the electric field lines emanating from a uniformly charged straight wire whose length l is not great. The spacing between lines near the wire should be somewhat less than l. [Hint: Also consider points very far from the wire.]

34. (II) What is the electric field strength at a point in space where a proton $(m = 1.67 \times 10^{-27}$ kg) experiences an acceleration of 1 million "g's"?

35. (II) An electron is released from rest in a uniform electric field and accelerates to the north at a rate of 145 m/s². What is the magnitude and direction of the electric field?

36. (II) The electric field midway between two equal but opposite point charges is 845 N/C, and the distance between the charges is 16.0 cm. What is the magnitude of the charge on each?

37. (II) Use Coulomb's law to determine the magnitude and direction of the electric field at points A and B in Fig. 21–52 due to the two positive charges $(Q = 7.0 \, \mu$C) shown. Is your result consistent with Fig. 21–33b?

FIGURE 21–52
Problem 37.

38. (II) Calculate the electric field at the center of a square 52.5 cm on a side if one corner is occupied by a $+45.0$-μC charge and the other three are occupied by -27.0-μC charges.

39. (II) Calculate the electric field at one corner of a square 1.00 m on a side if the other three corners are occupied by 3.25×10^{-6}-C charges.

40. (II) (a) Determine the electric field $\mathbf{E}$ at the origin 0 in Fig. 21–53 due to the two charges at A and B. (b) Repeat, but let the charge at B be reversed in sign.

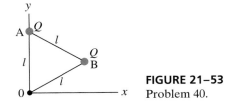

FIGURE 21–53
Problem 40.

41. (II) You are given two unknown point charges, Q_1 and Q_2. At a point on the line joining them, one-third of the way from Q_1 to Q_2, the electric field is zero (Fig. 21–54). What is the ratio Q_1/Q_2?

FIGURE 21–54
Problem 41.

42. (II) Two parallel circular rings of radius R have their centers on the x axis separated by a distance l as shown in Fig. 21–55. If each ring carries a uniformly distributed charge Q, find the electric field, $\mathbf{E}(x)$, at points along the x axis.

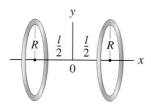

FIGURE 21–55
Problem 42.

43. (II) (a) Two equal charges Q are positioned at points $(x = l, y = 0)$ and $(x = -l, y = 0)$. Determine the electric field as a function of y for points along the y axis. (b) Show that the field is a maximum at $y = \pm l/\sqrt{2}$.

44. (II) At what position, $x = x_M$, is the magnitude of the electric field along the axis of the ring of Example 21–9 a maximum?

45. (II) Suppose the charge Q on the ring of Fig. 21–27 was all distributed uniformly on only the upper half of the ring, and no charge was on the lower half. Determine the electric field $\mathbf{E}$ at P. (Take y vertically upward.)

46. (II) Estimate the electric field at a point 2.8 cm perpendicular to the midpoint of a 2.0-m-long thin wire carrying a total charge of 4.75 μC.

47. (II) The uniformly charged straight wire in Fig. 21–28 has the length L, where point 0 is at the midpoint. Show that the field at point P, a perpendicular distance x from 0, is given by

$$E = \frac{\lambda}{2\pi\epsilon_0} \frac{L}{x(L^2 + 4x^2)^{\frac{1}{2}}},$$

where λ is the charge per unit length.

48. (III) A thin rod of length l carries a total charge Q distributed uniformly along its length. See Fig. 21–56. Determine the electric field along the axis of the rod starting at one end—that is, find $E(x)$ for $x \geq 0$ in Fig. 21–56.

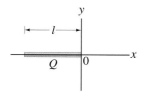

FIGURE 21–56
Problem 48.

49. (III) A thin rod bent into the shape of an arc of a circle of radius R carries a uniform charge per unit length λ. The arc subtends a total angle $2\theta_0$, symmetric about the x axis, as shown in Fig. 21–57. Determine the electric field $\mathbf{E}$ at the origin 0.

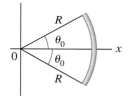

FIGURE 21–57
Problems 49 and 50.

50. (III) (a) Suppose the circular arc shown in Fig. 21–57, carries a charge per unit length λ that varies with θ as $\lambda = \lambda_0 \cos \theta$, where θ is measured from the x axis. Determine the electric field $\mathbf{E}$ at the origin 0. (b) Repeat, assuming $\lambda = \lambda_0 \sin \theta$.

51. (III) Suppose the uniformly charged wire of Fig. 21–28 starts at point 0 and rises vertically to a length L. (a) Determine the components of the electric field E_x and E_y at point P, a distance x from 0. (That is, calculate $\mathbf{E}$ near one end of a long wire, in the plane perpendicular to the wire.) (b) If the wire extends from $y = 0$ to $y = \infty$, so that $L = \infty$, show that $\mathbf{E}$ makes a 45° angle to the horizontal for any x.

52. (III) Suppose in Example 21–10 that $x = 0.250$ m, $Q = 3.15 \mu$C, and that the uniformly charged wire is only 6.0 m long and extends along the y axis from $y = -4.0$ m to $y = +2.0$ m. (a) Calculate E_x and E_y at point P. (b) Determine what the error would be if you simply used the result of Example 21–10, $E = \lambda/2\pi\epsilon_0 x$. Express this error as $(E_x - E)/E$ and E_y/E.

53. (III) *Uniform plane of charge.* Charge is distributed uniformly over a large square plane of side L, as shown in Fig. 21–58. The charge per unit area (C/m^2) is σ. Calculate the electric field at a point P a distance z above the center of the plane, in the limit $L \to \infty$. [*Hint:* Divide the plane into long narrow strips of width dy, and use the result of Example 21–10 to sum the fields due to each strip to get the total field.]

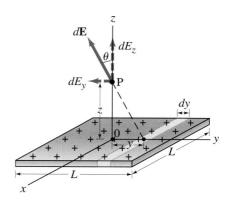

FIGURE 21–58 Problem 53.

Section 21–10

54. (II) An electron with speed $v_0 = 21.5 \times 10^6$ m/s is traveling parallel to an electric field $(\mathbf{v_0}\|\mathbf{E})$ of magnitude $E = 11.4 \times 10^3$ N/C. (a) How far will it travel before it stops? (b) How much time will elapse before it returns to its starting point?

55. (II) An electron has an initial velocity $v_0 = 8.0 \times 10^4$ m/s $\mathbf{i}$. It enters a region where $\mathbf{E} = (2.0\mathbf{i} + 8.0\mathbf{j}) \times 10^4$ N/C. (a) Determine the vector acceleration of the electron as a function of time. (b) At what angle θ is it moving (relative to its initial direction) at $t = 1.0$ ns?

56. (II) A water droplet of radius 0.020 mm remains stationary, in the air. If the electric field of the Earth is 150 N/C downward, how many excess electron charges must the water droplet have?

57. (II) At what angle will the electrons in Example 21–15 leave the uniform electric field at the end of the parallel plates (point P in Fig. 21–40)? Assume the plates are 6.0 cm long and $E = 5.0 \times 10^3$ N/C. Ignore fringing of the field.

58. (II) Suppose electrons enter the uniform electric field midway between two plates, moving at an upward 45° angle as shown in Fig. 21–59. What maximum speed can the electrons have if they are to avoid striking the upper plate? Ignore fringing of the field.

FIGURE 21–59 Problem 58.

59. (III) A positive charge q is placed at the center of a circular ring of radius R. The ring carries a uniformly distributed negative charge of total magnitude $-Q$. (a) If the charge q is displaced from the center a small distance x as shown in Fig. 21–60, show that it will undergo simple harmonic motion when released. (b) If its mass is m, what is its period?

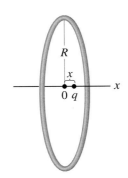

FIGURE 21–60 Problem 59.

Section 21–11

60. (II) A dipole consists of charges $+e$ and $-e$ separated by 0.68 nm. It is in an electric field $E = 2.7 \times 10^4$ N/C. (a) What is the value of the dipole moment? (b) What is the torque on the dipole when it is perpendicular to the field? (c) What is the torque on the dipole when it is at an angle of 45° to the field? (d) What is the work required to rotate the dipole from being oriented parallel to the field to being antiparallel to the field?

61. (II) The HCl molecule has a dipole moment of about 3.4×10^{-30} C·m. The two atoms are separated by about 1.0×10^{-10} m. (a) What is the net charge on each atom? (b) Is this equal to an integral multiple of e? If not, explain. (c) What maximum torque would this dipole experience in a 2.5×10^4 N/C electric field? (d) How much energy would be needed to rotate one molecule 45° from its equilibrium position of lowest potential energy?

62. (II) Suppose both charges in Fig. 21–44 were positive. (a) Show that the field on the perpendicular bisector, for $r \gg l$, is given by $(1/4\pi\epsilon_0)(2Q/r^2)$. (b) Explain why the field decreases as $1/r^2$ here whereas for a dipole it decreases as $1/r^3$.

63. (II) An electric dipole, of dipole moment p and moment of inertia I, is placed in a uniform electric field **E**. (a) If displaced by an angle θ as shown in Fig. 21–43 and released, under what conditions will it oscillate in simple harmonic motion? (b) What will be its frequency?

64. (III) Suppose a dipole **p** is placed in a nonuniform electric field **E** = E**i** that points along the x axis. If E depends only on x, show that the net force on the dipole is

$$\mathbf{F} = \left(\mathbf{p} \cdot \frac{d\mathbf{E}}{dx} \right) \mathbf{i},$$

where $d\mathbf{E}/dx$ is the gradient of the field in the x direction.

65. (III) (a) Show that at points along the axis of a dipole (along the same line that contains $+Q$ and $-Q$), the electric field has magnitude

$$E = \frac{1}{4\pi\epsilon_0} \frac{2p}{r^3}$$

for $r \gg l$ (Fig. 21–44), where r is the distance from the point to the center of the dipole. (b) In what direction does **E** point?

General Problems

66. How close must two electrons be if the electric force between them is equal to the weight of either at the Earth's surface?

67. Imagine that space invaders could deposit extra electrons in equal amounts on the Earth and on your car, which has a mass of 1050 kg. Note that the rubber tires would provide some insulation. How much charge Q would need to be placed on your car (same amount on the Earth) in order to levitate it (overcome gravity)? [Hint: Assume that the Earth's charge is spread uniformly so it acts as if it were located at the Earth's center, and then the separation distance is the radius of the Earth.]

68. A 3.0-g copper penny has a positive charge of 5.5 μC. What fraction of its electrons has it lost?

69. Suppose that electrical attraction, rather than gravity, were responsible for holding the Moon in orbit around the Earth. If equal and opposite charges Q were placed on the Earth and the Moon, what should be the value of Q to maintain the present orbit? Use these data: mass of Earth = 5.97×10^{24} kg, mass of Moon = 7.35×10^{22} kg, radius of orbit = 3.84×10^8 m. Treat the Earth and Moon as point particles.

70. One type of *electric quadrupole* consists of two dipoles placed end to end with their negative charges (say) overlapping; that is, in the center is $-2Q$ flanked (on a line) by a $+Q$ to either side (Fig. 21–61). Determine the electric field **E** at points along the perpendicular bisector and show that E decreases as $1/r^4$.

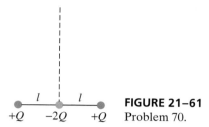

FIGURE 21–61
Problem 70.

71. Three charged particles are placed at the corners of an equilateral triangle of side 1.20 m (Fig. 21–62). The charges are $+4.0\,\mu\text{C}$, $-8.0\,\mu\text{C}$, and $-6.0\,\mu\text{C}$. Calculate the magnitude and direction of the net force on each due to the other two.

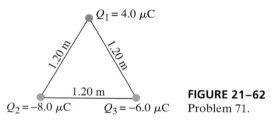

FIGURE 21–62
Problem 71.

72. A proton $(m = 1.67 \times 10^{-27}\,\text{kg})$ is suspended at rest in a uniform electric field **E**. Take into account gravity and determine **E**.

73. Calculate the magnitude of the electric field at the center of a square with sides 35 cm long if the corners, taken in rotation, have charges of $1.0\,\mu\text{C}$, $2.0\,\mu\text{C}$, $3.0\,\mu\text{C}$, and $4.0\,\mu\text{C}$ (all positive).

74. In a simple model of the hydrogen atom, the electron revolves in a circular orbit around the proton with a speed of $1.1 \times 10^6\,\text{m/s}$. What is the radius of the electron's orbit?

75. Two charges, $-Q_0$ and $-4Q_0$, are a distance l apart. These two charges are free to move but do not because there is a third charge nearby. What must be the magnitude of the third charge and its placement in order for the first two to be in equilibrium?

76. A point charge $(m = 1.0\,\text{g})$ at the end of an insulating string of length 55 cm is observed to be in equilibrium in a uniform horizontal electric field of 10,000 N/C, when the pendulum's position is as shown in Fig. 21–63, with the charge 12 cm above the lowest (vertical) position. If the field points to the right in Fig. 21–63, determine the magnitude and sign of the point charge.

FIGURE 21–63
Problem 76.

77. A positive point charge $Q_1 = 1.85 \times 10^{-5}\,\text{C}$ is fixed at the origin of coordinates, and a negative charge $Q_2 = -7.65 \times 10^{-6}\,\text{C}$ is fixed to the x axis at $x = +2.00\,\text{m}$. Find the location of the place(s) along the x axis where the electric field due to these two charges is zero.

78. Estimate the net force between the CO group and the HN group shown in Fig. 21–64. The C and O have effective charges $\pm 0.40e$ and the H and N have effective charges $\pm 0.20e$ where $e = 1.6 \times 10^{-19}\,\text{C}$. [*Hint*: Do not include the "internal" forces between C and O, or between H and N.]

FIGURE 21–64
Problem 78.

79. An electron moving with a speed of $2.0 \times 10^6\,\text{m/s}$ to the right enters a uniform electric field region where the field is known to be parallel to its direction of motion. If the electron is to be brought to rest in the space of 5.4 cm, (*a*) in what direction is the electric field, and (*b*) what is the magnitude of the field?

80. The two strands of the helix-shaped DNA molecule (the genetic material in living cells) are held together by electrostatic forces as shown in Fig. 21–65. Assume that the net average charge indicated on H and N atoms is effectively $0.2e$, and on the indicated C and O atoms is $0.4e$, that atoms on each molecule are separated by $1.0 \times 10^{-10}\,\text{m}$, and that all relevant angles are 120°. Estimate the net force between: (*a*) a thymine and an adenine; and (*b*) a cytosine and a guanine. (*c*) Estimate the total force for a DNA molecule containing 10^5 pairs of such molecules.

(a)   (b)

FIGURE 21–65 Problem 80. (a) Section of a DNA double helix. (b) "Close-up" view of the helix, showing how A and T attract each other and how G and C attract each other through electrostatic forces, to hold the double helix together. The red dots are used to indicate the electrostatic attraction (often called a "weak bond" or "hydrogen bond"). Note that there are two weak bonds between A and T, and three between C and G. The distance unit is the angstrom $(1\,\text{Å} = 10^{-10}\,\text{m})$.

81. Suppose electrons enter the electric field midway between two plates at an angle θ_0 to the horizontal, as shown in Fig. 21–66. The path is symmetrical, so they leave at the same angle θ_0 and just barely miss the top plate. What is θ_0? Ignore fringing of the field.

FIGURE 21–66   Problem 81.

82. Two point charges, $Q_1 = -6.7\,\mu C$ and $Q_2 = 1.3\,\mu C$, are located between two oppositely charged parallel plates, as shown in Fig. 21–67. The two point charges are separated by a distance of $x = 0.34\,m$. Assume that the electric field produced by the charged plates is uniform and equal to $E = 73{,}000\,N/C$. Calculate the net electrostatic force on Q_1 and give its direction.

FIGURE 21–67   Problem 82.

83. A small lead ball is encased in insulating plastic and suspended vertically from an ideal spring $(k = 126\,N/m)$ above a lab table, Fig. 21–68. The total mass of the coated ball is 0.800 kg, and its center lies 15.0 cm above the tabletop when in equilibrium. The ball is pulled down 5.00 cm below equilibrium, an electric charge $Q = -3.00 \times 10^{-6}\,C$ is deposited on the ball, and the system is released. Using what you know about harmonic oscillation, write an expression for the electric field strength as a function of time that would be measured at the point on the tabletop (P) directly below the ball.

FIGURE 21–68   Problem 83.

84. A large electroscope is made with "leaves" that are 75-cm-long wires with 22-g balls at the ends. When charged, nearly all the charge resides on the balls. If the wires each make a 30° angle with the vertical (Fig. 21–69), what total charge Q must have been applied to the electroscope?

FIGURE 21–69   Problem 84.

85. Three very large square planes of charge are arranged as shown (on edge) in Fig. 21–70. From left to right, the planes have charge densities per unit area of $-0.50\,\mu C/m^2$, $+0.10\,\mu C/m^2$, and $-0.35\,\mu C/m^2$. Find the total electric field (direction and magnitude) at the points A, B, C, and D. Assume the planes are much larger than the distance AD.

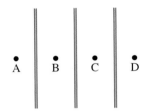

FIGURE 21–70   Problem 85.

86. What is the total charge of all the electrons in a 15-kg bar of aluminum? What is the net charge of the bar? (Aluminum has 13 electrons per atom and an atomic mass of 27 u.)

87. Given the two charges shown in Fig. 21–71, at what position(s) x is the electric field zero? Is the field zero at any other points, not on the x axis?

FIGURE 21–71   Problem 87.

88. An electron moves in a circle of radius r around a very long uniformly charged wire in a vacuum chamber, as shown in Fig. 21–72. The charge density on the wire is $\lambda = 0.14\,\mu C/m$. (a) What is the electric field at the electron (magnitude and direction)? (b) What is the speed of the electron?

FIGURE 21–72   Problem 88.

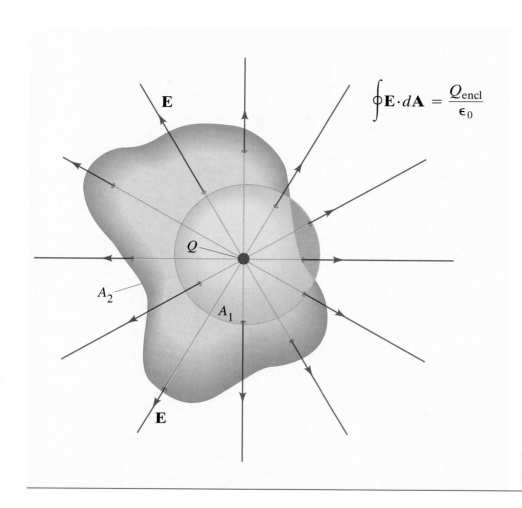

$$\oint \mathbf{E} \cdot d\mathbf{A} = \frac{Q_{\text{encl}}}{\epsilon_0}$$

Gauss's law is an elegant relation between electric charge and electric field. It is more general than Coulomb's law. Gauss's law involves an integral of the electric field $\mathbf{E}$ at each point on a closed surface. The surface is only imaginary, and we choose the shape and placement of the surface so that we can figure out the integral. In this drawing, two different surfaces are shown, both enclosing a point charge Q. Gauss's law states that the product $\mathbf{E} \cdot d\mathbf{A}$, where $d\mathbf{A}$ is an infinitesimal area of the surface, integrated over the entire surface, equals the charge enclosed by the surface Q_{encl} divided by ϵ_0. Both surfaces here enclose the same charge Q. Hence $\oint \mathbf{E} \cdot d\mathbf{A}$ will give the same result for both.

CHAPTER 22

Gauss's Law

Gauss's law, which we develop and discuss in this chapter, is a statement of the relation between electric charge and electric field. It is a more general and elegant form of Coulomb's law.

We can, in principle, determine the electric field due to any given distribution of electric charge using Coulomb's law. The total electric field at any point will be the vector sum (or integral) of contributions from all charges present (see Eqs. 21–5 and 21–6). Except for some simple cases, the sum or integral can be quite complicated to evaluate. For situations in which an analytic solution (such as we carried out in the Examples of Sections 21–6 and 21–7) is not possible, a computer can be used.

In some cases, however, the electric field due to a given charge distribution can be calculated more easily or more elegantly using Gauss's law, as we shall see in this chapter. But the major importance of Gauss's law is that it gives us additional insight into the nature of electrostatic fields, and a more general relationship between charge and field.

Before discussing Gauss's law itself, we first discuss the concept of *flux*.

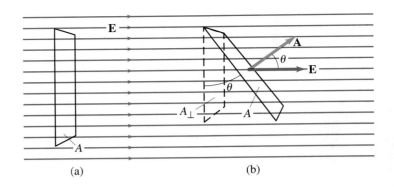

FIGURE 22–1 A uniform field **E** (indicated by the parallel field lines) passing through a surface of area A: (a) perpendicular to **E**; (b) not perpendicular to **E**. The dashed surface of area $A_\perp$ in (b) is the projection of **A** perpendicular to the field **E**.

(a)   (b)

22–1 | Electric Flux

Imagine a surface of area A through which a uniform electric field **E** passes, Fig. 22–1. The surface could be a rectangle (as shown), a circle, or any other shape. If the electric field direction is perpendicular to the surface, as in Fig. 22–1a, the **electric flux**, Φ_E, through this surface is defined as the product

$$\Phi_E = EA.$$

If the area A is not perpendicular to **E**, but rather makes an angle θ as shown in Fig. 22–1b, fewer field lines will pass through the area. In this case we define the electric flux through the surface as

$$\Phi_E = EA_\perp = EA\cos\theta, \qquad \text{[E uniform]} \quad \textbf{(22–1a)}$$

where $A_\perp$ is the projection of the area A on a surface perpendicular to **E** as shown. The area A of a surface can be represented by a vector **A** whose magnitude is A and whose direction is perpendicular to the surface, as shown in Fig. 22–1b. The angle θ is the angle between **E** and **A**, so the electric flux can also be written

$$\Phi_E = \mathbf{E} \cdot \mathbf{A}. \qquad \text{[E uniform]} \quad \textbf{(22–1b)}$$

Because of how we defined it, the electric flux has a simple intuitive interpretation in terms of field lines. We saw in Section 21–8 that field lines can always be drawn so that the number (N) passing through unit area perpendicular to the field ($A_\perp$) is proportional to the magnitude of the field (E): that is, $E \propto N/A_\perp$. Hence,

Electric flux is proportional to the number of field lines passing through the area

$$N \propto EA_\perp = \Phi_E,$$

so the flux through an area is proportional to the number of lines passing through that area.

EXAMPLE 22–1 **Electric flux.** (a) Calculate the electric flux through the rectangle shown in Fig. 22–1a. The rectangle is 10 cm by 20 cm and the electric field is uniform at 200 N/C. (b) What is the flux in Fig. 22–1b if the angle θ is 30°?

SOLUTION (a) The electric flux is

$$\begin{aligned}
\Phi_E &= EA\cos\theta \\
&= (200\,\text{N/C})(0.10\,\text{m} \times 0.20\,\text{m})\cos 0° = 4.0\,\text{N}\cdot\text{m}^2/\text{C}.
\end{aligned}$$

(b) In this case the flux is

$$\Phi_E = (200\,\text{N/C})(0.10\,\text{m} \times 0.20\,\text{m})\cos 30° = 3.5\,\text{N}\cdot\text{m}^2/\text{C}.$$

Now let us consider the more general case, when the electric field **E** is not uniform and the surface is not flat, Fig. 22–2. We divide up the chosen surface into n small elements of surface whose areas are $\Delta A_1, \Delta A_2, \ldots \Delta A_n$. We choose the division so that each ΔA_i is small enough that (1) it can be considered flat, and (2) the electric field varies so little over this small area that it can be considered uniform over this tiny area. Then the electric flux through the entire surface is approximately

$$\Phi_E \approx \sum_{i=1}^{n} \mathbf{E}_i \cdot \Delta \mathbf{A}_i,$$

where $\mathbf{E}_i$ is the field passing through ΔA_i. In the limit as we let $\Delta A_i \to 0$, the sum becomes an integral over the entire surface and the relation becomes mathematically exact:

$$\Phi_E = \int \mathbf{E} \cdot d\mathbf{A}. \tag{22–2}$$

FIGURE 22–2 Electric flux through a curved surface. One small area of the surface, ΔA_i, is indicated.

Electric flux defined

In many cases (in particular, for Gauss's law) we deal with the flux through a *closed* surface—that is, a surface that completely encloses a volume (like a sphere or the surface of a football), Fig. 22–3. In this case, the net flux through the surface is given by

$$\Phi_E = \oint \mathbf{E} \cdot d\mathbf{A}, \tag{22–3}$$

Electric flux through a closed surface

where the integral sign is written $\oint$ to indicate that the integral is over the value of **E** on an enclosing surface.

Up to now we have not been concerned with the fact that there is an ambiguity in the direction of the vector **A** that represents a surface. For example, in Fig. 22–1, the vector **A** could point upward and to the right (as shown) or downward to the left and still be perpendicular to the surface. For a closed surface, we define (arbitrarily) the direction of **A**, or of $d\mathbf{A}$, to point *outward* from the enclosed volume, Fig. 22–4. For a line leaving the enclosed volume (on the right in Fig. 22–4), the angle θ between **E** and $d\mathbf{A}$ must be less than $\pi/2$ (= 90°), so $\cos \theta > 0$. For a line entering the volume (on the left in Fig. 22–4) $\theta > \pi/2$, so $\cos \theta < 0$. Hence, *flux entering the enclosed volume is negative* $\left(\int E \cos \theta \, dA < 0 \right)$, whereas *flux leaving the volume is positive*. Consequently, Eq. 22–3 gives the net flux *out* of the volume. If Φ_E is negative, there is a net flux *into* the volume.

Flux entering is negative
Flux leaving is positive

In Figs. 22–3 and 22–4, each field line that enters the volume also leaves the volume. Hence $\Phi_E = \oint \mathbf{E} \cdot d\mathbf{A} = 0$. There is no net flux into or out of this surface. The flux, $\oint \mathbf{E} \cdot d\mathbf{A}$, will be nonzero only if one or more lines start or end within the surface.

FIGURE 22–3 Electric flux through a closed surface.

FIGURE 22–4 The direction of an element of area $d\mathbf{A}$ is taken to point outward from an enclosed surface.

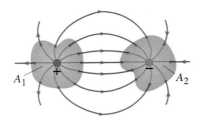

FIGURE 22–5 An electric dipole. Flux through surface A_1 is positive. Flux through A_2 is negative.

FIGURE 22–6 Net flux through surface A is negative.

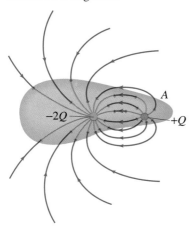

FIGURE 22–7 A single point charge Q at the center of an imaginary sphere of radius r (our "gaussian surface"—that is, the closed surface we choose to use for applying Gauss's law in this case).

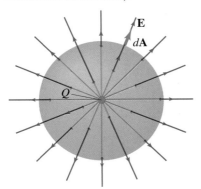

Since electric field lines start and stop only on electric charges, the flux will be nonzero only if the surface encloses a net charge. For example, the surface labeled A_1 in Fig. 22–5 encloses a positive charge and there is a net outward flux through this surface $(\Phi_E > 0)$. The surface A_2 encloses an equal magnitude negative charge and there is a net inward flux $(\Phi_E < 0)$. For the configuration shown in Fig. 22–6, the flux through the surface shown is negative (count the lines). The value of Φ_E depends on the charge enclosed by the surface, and this is what Gauss's law is all about.

[The concept of flux applies equally well to fluid flow, and thus makes an interesting analogy. (Indeed, the word "flux" comes from the Latin word for "flow.") The electric field **E** at each point corresponds to the fluid flow velocity **v**, so the electric field lines correspond to streamlines of a fluid flow. The flux, Φ, through a surface for a fluid is the volume rate of flow and is given by $\Phi = \int \mathbf{v} \cdot d\mathbf{A}$. In Figs. 22–1, 22–2, and 22–3, the lines can correspond to the steady streamline flow of a fluid with no sources (such as a faucet) or sinks (such as a leak or drain). In this case, the net flux through a closed surface, as in Fig. 22–3, is zero: what flows in also flows out. In Figs. 22–5 and 22–6, there are sources (corresponding to a positive charge) where flow lines start, and also sinks (corresponding to negative charge) where flow lines end. Although this comparison between electric flux and fluid flux is interesting, and perhaps offers some insight, do not get them confused—an electric flux is not a flow of any substance. Flux can be defined for any vector field, and we will later use it also for the magnetic field.]

22–2 Gauss's Law

The precise relation between the electric flux through a closed surface and the net charge Q_{encl} enclosed within that surface is given by **Gauss's law**:

$$\oint \mathbf{E} \cdot d\mathbf{A} = \frac{Q_{encl}}{\epsilon_0}, \qquad (22\text{–}4)$$

where ϵ_0 is the same constant (permittivity of free space) that appears in Coulomb's law. The integral on the left is over the value of **E** on a closed surface, which we choose for our convenience in any given situation. The charge Q_{encl} is the net charge *enclosed* by that surface. It doesn't matter where or how the charge is distributed within the surface. Any charge outside this surface must not be included. A charge outside the chosen surface may affect the position of the electric field lines, but will not affect the net number of lines entering or leaving the surface. For example, Q_{encl} for the gaussian surface A_1 in Fig. 22–5 would be the positive charge enclosed by A_1; the negative charge does contribute to the electric field at A_1 but it is *not* enclosed by surface A_1 and so is not included in Q_{encl}.

Before we discuss the validity of Gauss's law, we note that the integral is often rather difficult to carry out in practice. We rarely need to do it except for some fairly simple situations that will be discussed shortly (Section 22–3).

Now let us see how Gauss's law is related to Coulomb's law.[†] First, we show that Coulomb's law follows from Gauss's law. In Fig. 22–7 we have a single charge Q.

[†] Note that Gauss's law would look more complicated in terms of the constant $k = 1/4\pi\epsilon_0$ that we originally used in Coulomb's law (Eq. 21–1 or 21–4a):

Coulomb's law	Gauss's law
$E = k\dfrac{Q}{r^2}$	$\oint \mathbf{E} \cdot d\mathbf{A} = 4\pi k Q$
$E = \dfrac{1}{4\pi\epsilon_0}\dfrac{Q}{r^2}$	$\oint \mathbf{E} \cdot d\mathbf{A} = \dfrac{Q}{\epsilon_0}.$

Gauss's law has a simpler form using ϵ_0; Coulomb's law is simpler using k. The normal convention is to use ϵ_0 rather than k because Gauss's law is considered more general and therefore it is preferable to have it in simpler form.

For our "gaussian surface," we choose an imaginary sphere of radius r centered on the charge. Because Gauss's law is supposed to be valid for any surface, we have chosen one that will make our calculation easy. Because of the symmetry of this (imaginary) sphere about the charge at its center, we know that $\mathbf{E}$ must have the same magnitude at any point on the surface, and that $\mathbf{E}$ points radially outward (or inward) parallel to $d\mathbf{A}$, an element of the surface area. Hence we write the integral in Gauss's law as

$$\oint \mathbf{E} \cdot d\mathbf{A} = \oint E \, dA = E \oint dA = E(4\pi r^2)$$

since the surface area of a sphere of radius r is $4\pi r^2$, and the magnitude of $\mathbf{E}$ is the same at all points on this spherical surface. Then Gauss's law becomes, with $Q_{encl} = Q$,

$$\frac{Q}{\epsilon_0} = \oint \mathbf{E} \cdot d\mathbf{A} = E(4\pi r^2).$$

Solving for E we obtain

$$E = \frac{Q}{4\pi\epsilon_0 r^2},$$

which is the electric field form of Coulomb's law, Eq. 21–4b.

Now let us do the reverse, and derive Gauss's law from Coulomb's law for static electric charges. First we consider a single point charge Q surrounded by an imaginary spherical surface as in Fig. 22–7. Coulomb's law tells us that the electric field at the spherical surface is $E = (1/4\pi\epsilon_0)(Q/r^2)$. Reversing the argument we just used, we have

$$\oint \mathbf{E} \cdot d\mathbf{A} = \oint \frac{1}{4\pi\epsilon_0} \frac{Q}{r^2} \, dA = \frac{Q}{4\pi\epsilon_0 r^2}(4\pi r^2) = \frac{Q}{\epsilon_0}.$$

This is Gauss's law, with $Q_{encl} = Q$, and we derived it for the special case of a spherical surface enclosing a point charge at its center. But what about some other surface, such as the irregular surface labeled A_2 in Fig. 22–8? The same number of field lines (due to our charge Q) pass through surface A_2, as pass through the spherical surface, A_1. Therefore, because the flux through a surface is proportional to the number of lines through it as we saw in Section 22–1, the flux through A_2 is the same as through A_1:

$$\oint_{A_2} \mathbf{E} \cdot d\mathbf{A} = \oint_{A_1} \mathbf{E} \cdot d\mathbf{A} = \frac{Q}{\epsilon_0}.$$

Hence we can expect that

$$\oint \mathbf{E} \cdot d\mathbf{A} = \frac{Q}{\epsilon_0}$$

would be valid for *any* surface surrounding a single point charge Q.

Finally, let us look at the case of more than one charge. For each charge, Q_i, enclosed by the chosen surface,

$$\oint \mathbf{E}_i \cdot d\mathbf{A} = \frac{Q_i}{\epsilon_0},$$

where $\mathbf{E}_i$ refers to the electric field produced by Q_i alone. By the superposition principle for electric fields (Eq. 21–5), the total field $\mathbf{E}$ is equal to the sum of the fields due to each separate charge, $\mathbf{E} = \Sigma\mathbf{E}_i$. Hence

$$\oint \mathbf{E} \cdot d\mathbf{A} = \oint (\Sigma \mathbf{E}_i) \cdot d\mathbf{A} = \Sigma \frac{Q_i}{\epsilon_0} = \frac{Q_{encl}}{\epsilon_0},$$

where $Q_{encl} = \Sigma Q_i$ is the total net charge enclosed within the surface. Thus we see, based on this simple argument, that Gauss's law follows from Coulomb's law for any distribution of electric charge enclosed within a closed surface of any shape.

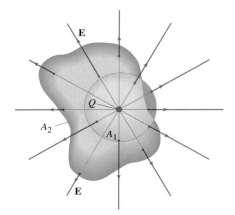

FIGURE 22–8 A single point charge surrounded by a spherical surface A_1, and an irregular surface, A_2.

FIGURE 22–9 Electric flux through a closed surface. (Same as Fig. 22–3.) No electric charge is enclosed by this surface ($Q_{encl} = 0$).

The derivation of Gauss's law from Coulomb's law is valid for electric fields produced by static electric charges. We will see later that electric fields can also be produced by changing magnetic fields. Coulomb's law cannot be used to describe such electric fields. But Gauss's law *is* found to hold also for electric fields generated in this way. Hence *Gauss's law is a more general law than Coulomb's law.* It holds for any electric field whatsoever.

Even for the case of static electric fields we are treating in this chapter, it is important to recognize that **E** on the left side of Gauss's law is not necessarily due only to the charge Q_{encl} that appears on the right. For example, in Fig. 22–9 there is an electric field **E** at all points on the imaginary gaussian surface, but it is not due to the charge enclosed by the surface (which is $Q_{encl} = 0$ in this case). The electric field **E** which appears on the left side of Gauss's law is the *total* electric field at each point, on the gaussian surface chosen, not just that due to the charge Q_{encl}, which appears on the right side. Gauss's law has been found to be valid for the total field at any surface. It tells us that any *difference* between the input and output flux of the electric field over any surface is due to charge within that surface.

CONCEPTUAL EXAMPLE 22–2 | **Flux from Gauss's law.** Consider the two gaussian surfaces, A_1 and A_2, shown in Fig. 22–10. The only charge present is the charge Q at the center of surface A_1. What is the net flux through each surface, A_1 and A_2?

RESPONSE The surface A_1 encloses the charge $+Q$. By Gauss's law, the net flux through A_1 is then Q/ϵ_0. For surface A_2, the charge $+Q$ is outside the surface. Surface A_2 encloses zero net charge, so the net electric flux through A_2 is zero, by Gauss's law. Note that all field lines that enter the volume enclosed by surface A_2 also leave it.

FIGURE 22–10 Example 22–2. Two gaussian surfaces.

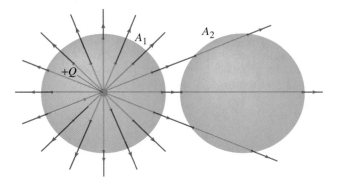

22–3 | Applications of Gauss's Law

Gauss's law is a very compact and elegant way to write the relation between electric charge and electric field. It also offers a simple way to determine the electric field when the charge distribution is simple and/or possesses a high degree of symmetry. In order to apply Gauss's law however, we must choose the "gaussian" surface very carefully (for the integral on the left side of Gauss's law) so we can determine **E**. We normally try to think of a surface that has just the symmetry needed so that E will be constant on all or on parts of its surface. Sometimes we choose a surface so the flux through part of the surface is zero.

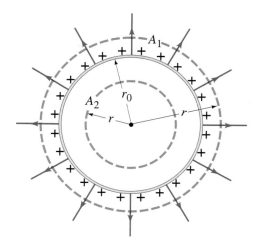

FIGURE 22–11 Cross sectional drawing of a thin spherical shell of radius r_0, carrying a net charge Q uniformly distributed. A_1 and A_2 represent two Gaussian surfaces we use to determine **E**.

EXAMPLE 22–3 **Spherical conductor.** A thin spherical shell of radius r_0 possesses a total net charge Q that is uniformly distributed on it, Fig. 22–11. Determine the electric field at points (*a*) outside the shell, and (*b*) inside the shell. (*c*) What if the conductor were a solid sphere?

SOLUTION (*a*) Because the charge is distributed symmetrically, the electric field must also be symmetric. Thus the field must be directed radially outward (inward if $Q < 0$) and must depend only on r, not on angle (spherical coordinates). First we want to find **E** outside the spherical shell, so we choose our imaginary gaussian surface to be a sphere of radius r $(r > r_0)$ concentric with the shell, and shown in Fig. 22–11 as a dashed circle A_1 outside the shell. The electric field **E** then has the same magnitude at all points on the surface, and because **E** is perpendicular to this surface, the cosine of the angle between **E** and $d\mathbf{A}$ is always 1. Gauss's law then gives $\left(\text{with } Q_{\text{encl}} = Q\right)$

$$\oint \mathbf{E} \cdot d\mathbf{A} = E(4\pi r^2) = \frac{Q}{\epsilon_0}$$

or

$$E = \frac{1}{4\pi\epsilon_0} \frac{Q}{r^2}. \qquad\qquad [r > r_0]$$

Thus the field outside a uniform spherical shell of charge is the same as if all the charge were concentrated at the center as a point charge.

Field outside spherical shell is same as for point charge at center

(*b*) Inside the shell, the field must also be symmetric. So E must again have the same value at all points on a spherical gaussian surface $(A_2$ in Fig. 22–11) concentric with the shell. Thus E can be factored out of the integral and, with $Q_{\text{encl}} = 0$, we have

$$\oint \mathbf{E} \cdot d\mathbf{A} = E(4\pi r^2) = 0.$$

Hence

$$E = 0 \qquad\qquad [r < r_0]$$

E = 0 inside uniformly charged spherical shell

inside a uniform spherical shell of charge.

(*c*) These same results also apply to a uniformly charged solid spherical conductor, since all the charge would lie in a thin layer at the surface.

Above results apply also for solid conducting sphere

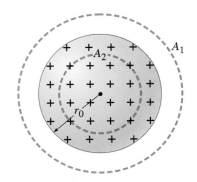

FIGURE 22–12 A solid sphere of uniform charge density.

Uniform sphere produces same field as a point charge at center

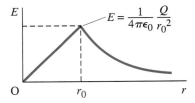

FIGURE 22–13 Magnitude of the electric field as a function of the distance r from the center of a uniformly charged solid sphere.

Electric field inside a uniformly charged nonconducting sphere

EXAMPLE 22–4 **Solid sphere of charge.** An electric charge Q is distributed uniformly throughout a nonconducting sphere of radius r_0, Fig. 22–12. Determine the electric field (a) outside the sphere $(r > r_0)$ and (b) inside the sphere $(r < r_0)$.

SOLUTION Since the charge is distributed symmetrically in the sphere, the electric field at all points must again be symmetric. **E** depends only on r and is directed radially outward (or inward if $Q < 0$).

(a) For our gaussian surface we choose a sphere of radius r $(r > r_0)$, labeled A_1 in Fig. 22–12. Since E depends only on r, Gauss's law gives, with $Q_{encl} = Q$,

$$\oint \mathbf{E} \cdot d\mathbf{A} = E(4\pi r^2) = \frac{Q}{\epsilon_0}$$

or

$$E = \frac{1}{4\pi\epsilon_0} \frac{Q}{r^2}.$$

Again, the field outside a spherically symmetric distribution of charge is the same as that for a point charge of the same magnitude located at the center of the sphere.

(b) Inside the sphere, we choose for our gaussian surface a concentric sphere of radius r $(r < r_0)$, labeled A_2 in Fig. 22–12. From symmetry, the magnitude of **E** is the same at all points on A_2, and **E** is perpendicular to the surface, so

$$\oint \mathbf{E} \cdot d\mathbf{A} = E(4\pi r^2).$$

We must equate this to Q_{encl}/ϵ_0 where Q_{encl} is the charge enclosed by A_2. Q_{encl} is not the total charge Q but only a portion of it. We define the **charge density**, ρ_E, as the charge per unit volume $(\rho_E = dQ/dV)$, and here we are given that $\rho_E =$ constant. So the charge enclosed by the gaussian surface A_2, a sphere of radius r, is

$$Q_{encl} = \left(\frac{\frac{4}{3}\pi r^3 \rho_E}{\frac{4}{3}\pi r_0^3 \rho_E} \right) Q = \frac{r^3}{r_0^3} Q.$$

Hence, from Gauss's law,

$$E(4\pi r^2) = \frac{Q_{encl}}{\epsilon_0} = \frac{r^3}{r_0^3} \frac{Q}{\epsilon_0}$$

or

$$E = \frac{1}{4\pi\epsilon_0} \frac{Q}{r_0^3} r. \qquad [r < r_0]$$

Thus the field increases linearly with r, until $r = r_0$. It then decreases as $1/r^2$, as plotted in Fig. 22–13.

The results above would have been difficult to obtain from Coulomb's law by integrating over the sphere. Using Gauss's law and the symmetry of the situation, this result is obtained rather easily, and shows the great power of Gauss's law. However, its use in this way is limited mainly to cases where the charge distribution has a high degree of symmetry. In such cases, we *choose* a simple surface on which $E =$ constant, so the integration is simple. Gauss's law holds, of course, for any surface. The next two Examples are symmetric cases that we did treat before, using Coulomb's law, but we get the result more easily using Gauss's law.

EXAMPLE 22–5 Long uniform line of charge. A very long straight wire possesses a uniform positive charge per unit length, λ. Calculate the electric field at points near (but outside) the wire, far from the ends.

SOLUTION Because of the symmetry, we expect the field to be directed radially outward and to depend only on the perpendicular distance, r, from the wire. Because of the cylindrical symmetry, the field will be the same at all points on a gaussian surface that is a cylinder with the wire along its axis, Fig. 22–14. **E** is perpendicular to this surface at all points. For Gauss's law, we need a closed surface, so we include the flat ends of the cylinder. Since **E** is parallel to the ends, there is no flux through the ends (the cosine of the angle between **E** and $d\mathbf{A}$ on the ends is $\cos 90° = 0$). So Gauss's law tells us

$$\oint \mathbf{E} \cdot d\mathbf{A} = E(2\pi r l) = \frac{Q_{encl}}{\epsilon_0} = \frac{\lambda l}{\epsilon_0},$$

where l is the length of our chosen gaussian surface ($l \ll$ length of wire), and $2\pi r$ is its circumference. Hence

$$E = \frac{1}{2\pi\epsilon_0} \frac{\lambda}{r}.$$

This is the same result as we got in Example 21–10 using Coulomb's law (we used x there instead of r), but here it took much less effort. Again we see the great power of Gauss's law.[†]

FIGURE 22–14 Calculation of **E** due to a very long line of charge.

EXAMPLE 22–6 Infinite plane of charge. Charge is distributed uniformly, with a surface charge density σ (σ = charge per unit area = dQ/dA), over a very large but very thin nonconducting flat plane surface. Determine the electric field at points near the plane.

SOLUTION We choose as our gaussian surface a small, closed cylinder whose axis is perpendicular to the plane and which extends through the plane as shown in Fig. 22–15. Because of the symmetry, we expect **E** to be directed perpendicular to the plane on both sides as shown, and to be uniform over the end caps of the cylinder, each of whose area is A. Since no flux passes through the curved sides of the cylinder, all the flux is through the two end caps. So Gauss's law gives

$$\oint \mathbf{E} \cdot d\mathbf{A} = 2EA = \frac{Q_{encl}}{\epsilon_0} = \frac{\sigma A}{\epsilon_0},$$

where $Q_{encl} = \sigma A$ is the charge enclosed by our gaussian cylinder. The electric field is then

$$E = \frac{\sigma}{2\epsilon_0}.$$

This is the same result we obtained much more laboriously in Chapter 21, Eq. 21–7. The field is uniform for points far from the ends of the plane, and close to its surface.

FIGURE 22–15 Calculation of the electric field outside a large uniformly charged nonconducting plane surface.

Electric field near a thin uniformly charged plane

[†] But note that the method of Example 21–10 allows calculation of E also for a short line of charge by using the appropriate limits for the integral, whereas Gauss's law is not readily adapted due to lack of symmetry.

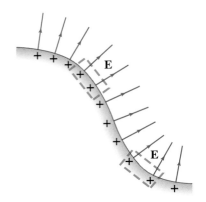

FIGURE 22–16 Electric field near surface of a conductor.

Electric field at surface of a conductor

When is $E = \sigma/\epsilon_0$ and when is $E = \sigma/2\epsilon_0$

FIGURE 22–17 Thin flat charged conductor with surface charge density σ at each surface, but for the conductor as a whole, the charge density is $\sigma' = 2\sigma$.

EXAMPLE 22–7 **Electric field near any conducting surface.** Show that the electric field just outside the surface of any good conductor of arbitrary shape is given by

$$E = \frac{\sigma}{\epsilon_0},$$

where σ is the surface charge density on the conductor's surface at that point.

SOLUTION We choose as our gaussian surface a small cylindrical box, as we did in the previous example. We choose the cylinder to be very small in height, so that one of its circular ends is just above the conductor (Fig. 22–16). The other end is just below the conductor's surface, and the sides are perpendicular to it. The electric field is zero inside a conductor and is perpendicular to the surface just outside it (Section 21–9), so electric flux passes only through the outside end of our cylindrical box. We choose the area A (of the flat cylinder end) small enough so that E is essentially uniform over it. Then Gauss's law gives

$$\oint \mathbf{E} \cdot d\mathbf{A} = EA = \frac{Q_{\text{encl}}}{\epsilon_0} = \frac{\sigma A}{\epsilon_0},$$

so that

$$E = \frac{\sigma}{\epsilon_0}. \qquad \text{[at surface of conductor]} \quad \textbf{(22–5)}$$

This is a useful result which applies for a conductor of any shape.

Why is it that the field outside a large plane nonconductor is $E = \sigma/2\epsilon_0$ (Example 22–6) whereas outside a conductor it is $E = \sigma/\epsilon_0$ (Example 22–7)? The reason for the factor of 2 comes not so much from conductor versus nonconductor as from what we mean by the surface charge density σ. For a conductor, the charge lies at the surface and all the electric field lines leave on one side of the surface. For a thin plane nonconductor, the lines leave both sides, Fig. 22–15. If we had a large, thin, flat conducting plane, the charge would accumulate on both surfaces (Fig. 22–17) and the field would emanate from both sides. If we called σ' the surface charge for the plane as a *whole*, each face of the plane would have surface charge $\sigma = \sigma'/2$, and so the result of Example 22–7 would give $E = (\sigma'/2)/\epsilon_0 = \sigma'/2\epsilon_0$, the same as for a nonconducting plane. Normally, however, we use σ to apply to *each* face of a conducting plane and then we would get $E = \sigma/\epsilon_0$. Thus the factor of 2 between Examples 22–6 and 22–7 comes from two different ways of defining σ.

We saw in Section 21–9 that in the static situation, the electric field inside any conductor must be zero even if it has a net charge. (Otherwise, the free charges in the conductor would move—until the net force on each, and hence **E**, were zero.) We also mentioned there that any net electric charge on a conductor must all reside on its outer surface. This is readily shown using Gauss's law. Consider any charged conductor of any shape, such as that shown in Fig. 22–18, which carries a net charge Q. We choose the gaussian surface, shown dashed in the diagram, so that it lies just below the surface of the conductor. Our gaussian surface can be arbitrarily close to the surface, but still *inside* the conductor. The electric field is zero at all points on this gaussian surface since it is inside the conductor. Hence, from Gauss's law, Eq. 22–4, the net charge within the surface must be zero. Hence, there can be no net charge within the conductor. Any net charge must lie on the surface of the conductor.

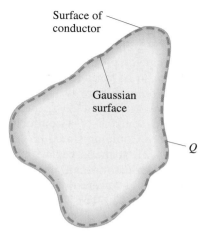

Surface of conductor

Gaussian surface

Q

Q

FIGURE 22–18 An insulated charged conductor of arbitrary shape, showing gaussian surface (dashed) just below the surface of the conductor.

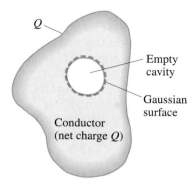

Q

Empty cavity

Gaussian surface

Conductor (net charge Q)

FIGURE 22–19 An empty cavity inside a charged conductor carries zero net charge.

If there is an empty cavity inside a conductor, can charge accumulate on that (inner) surface too? As shown in Fig. 22–19, if we imagine a gaussian surface (shown dashed) just inside the conductor above the cavity, we know that **E** must be zero everywhere on this surface since it is inside the conductor. Hence, by Gauss's law, *there can be no net charge at the surface of the cavity.*

But what if the cavity is not empty and there is a charge inside it?

CONCEPTUAL EXAMPLE 22–8 **Conductor with charge inside a cavity.** Suppose a conductor carries a net charge $+Q$ and contains a cavity, inside of which resides a point charge $+q$. What can you say about the charges on the inner and outer surfaces of the conductor?

RESPONSE As shown in Fig. 22–20, a gaussian surface just inside the conductor surrounding the cavity must contain zero net charge ($E = 0$ in a conductor). Thus a net charge of $-q$ must exist on the cavity surface. The conductor itself carries a net charge $+Q$, so its outer surface must carry a charge equal to $Q + q$.

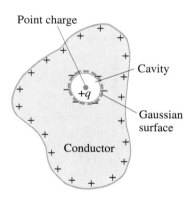

Point charge

Cavity

+q

Gaussian surface

Conductor

FIGURE 22–20 Example 22–8.

PROBLEM SOLVING	**Gauss's Law for Symmetric Charge Distributions**

1. First identify the symmetry of the charge distribution: spherical, cylindrical, planar. This identification should suggest a gaussian surface for which **E** will be constant and/or zero on all or on parts of the surface: a sphere for spherical symmetry, a cylinder for cylindrical symmetry and a small cylinder or "pillbox" for planar symmetry.

2. Draw the appropriate gaussian surface making sure it passes through the point where you want to know the electric field.

3. Use the symmetry of the charge distribution to determine the direction of **E** on the gaussian surface.

4. Evaluate the flux, $\oint \mathbf{E} \cdot d\mathbf{A}$. With the appropriate gaussian surface, the dot product $\mathbf{E} \cdot d\mathbf{A}$ should be zero or equal to $\pm E \, dA$, with the magnitude of E being constant.

5. Calculate the charge *enclosed* by the gaussian surface. Remember it's the enclosed charge that matters. Ignore all the charge outside the gaussian surface.

6. Equate the flux to the enclosed charge and solve for E.

(a)

Insulator

(b)

(c)

FIGURE 22–21 (a) A charged conductor (metal ball) is lowered into an insulated metal can (a good conductor) carrying zero net charge. (b) The charged ball is touched to the can and all of its charge quickly flows to the outer surface of the can. (c) When the ball is then removed, it is found to carry zero net charge.

* | 22–4 | Experimental Basis of Gauss's and Coulomb's Law

Gauss's law predicts that any net charge on a conductor must lie only on its surface. But is this true in real life? Let us see how it can be verified experimentally. And in confirming this prediction of Gauss's law, Coulomb's law is also confirmed since the latter follows from Gauss's law, as we saw in Section 22–2. Indeed, the earliest observation that charge resides only on the outside of a conductor was recorded by Benjamin Franklin some 30 years before Coulomb stated his law.

A simple experiment is illustrated in Fig. 22–21. A metal can with a small opening at the top rests on an insulator. The can, a conductor, is initially uncharged (Fig. 22–21a). A charged metal ball (also a conductor) is lowered by an insulating thread into the can, and is allowed to touch the can (Fig. 22–21b). The ball and can now form a single conductor. Gauss's law, as discussed above, predicts that all the charge will flow to the outer surface of the can. (The flow of charge in such situations does not occur instantaneously, but the time involved is usually negligible.) These predictions are confirmed in experiments by (1) connecting an electroscope to the can, which will show that the can is charged, and (2) connecting an electroscope to the ball after it has been withdrawn from the can (Fig. 22–21c), which will show that the ball carries zero charge.

The precision with which Coulomb's and Gauss's laws hold can be stated quantitatively by writing Coulomb's law as

$$F = k \frac{Q_1 Q_2}{r^{2+\delta}}.$$

For a perfect inverse-square law, $\delta = 0$. The most recent and precise experiments (1971) give $\delta = (2.7 \pm 3.1) \times 10^{-16}$. Thus Coulomb's and Gauss's laws are found to be valid to an extremely high precision!

Summary

The **electric flux** passing through a flat area A for a uniform electric field $\mathbf{E}$ is

$$\Phi_E = \mathbf{E} \cdot \mathbf{A}.$$

If the field is not uniform, the flux is determined from the integral

$$\Phi_E = \int \mathbf{E} \cdot d\mathbf{A}.$$

The direction of the vector $\mathbf{A}$ or $d\mathbf{A}$ is chosen to be perpendicular to the surface whose area is A or dA, and points outward from an enclosed surface. The flux through a surface is proportional to the number of field lines passing through it.

Gauss's law states that the net flux passing out of any closed surface is equal to the net charge Q_{encl} enclosed by the surface divided by ϵ_0:

$$\oint \mathbf{E} \cdot d\mathbf{A} = \frac{Q_{\text{encl}}}{\epsilon_0}.$$

Gauss's law can in principle be used to determine the electric field due to a given charge distribution, but its usefulness is mainly limited to a small number of cases, usually where the charge distribution displays much symmetry. The real importance of Gauss's law is that it is a more general and elegant statement (than Coulomb's law) for the relation between electric charge and electric field. It is one of the basic equations of electromagnetism.

Questions

1. If the electric flux through a closed surface is zero, is the electric field necessarily zero at all points on the surface? What about the converse: If $\mathbf{E} = 0$ at all points on the surface is the flux through the surface zero?

2. Is the electric field $\mathbf{E}$ in Gauss's law, $\oint \mathbf{E} \cdot d\mathbf{A} = Q_{\text{encl}}/\epsilon_0$, that due only to the charge Q_{encl}?

3. A point charge is surrounded by a spherical gaussian surface of radius r. If the sphere is replaced by a cube of side r, will Φ_E be larger, smaller, or the same?

4. What can you say about the flux through a closed surface that encloses an electric dipole?

5. The electric field $\mathbf{E}$ is zero at all points on a closed surface; is there necessarily no net charge within the surface? If a surface encloses zero net charge, is the electric field necessarily zero at all points on the surface?

6. Define gravitational flux in analogy to electric flux. Are there "sources" and "sinks" for the gravitational field as there are for the electric field? Discuss.

7. Would Gauss's law be helpful in determining the electric field due to an electric dipole?

8. A spherical basketball (a nonconductor) is given a charge Q distributed uniformly over its surface. What can you say about the electric field inside the ball? A person now steps on the ball, collapsing it, and forcing most of the air out without altering the charge. What can you say about the field inside now?

9. In Example 22–5, it may seem that the electric field calculated is due only to the charge on the wire that is enclosed by the cylinder chosen as our gaussian surface. In fact, the entire charge along the whole length of the wire contributes to the field. Explain how the charge outside the cylindrical gaussian surface of Fig. 22–14 contributes to E at the gaussian surface. [*Hint*: Compare to what the field would be due to a short wire.]

10. Suppose the line of charge in Example 22–5 extended only a short way beyond the ends of the cylinder shown in Fig. 22–14. How would the result of Example 22–5 be altered?

11. A point charge Q is surrounded by a spherical surface of radius r_0, whose center is at Q. Later, the charge is moved to the right a distance $\frac{1}{2}r_0$, but the sphere remains where it was, Fig. 22–22. How is the electric flux Φ_E through the sphere changed? Is the electric field at the surface of the sphere changed? For each "yes" answer, describe the change.

FIGURE 22–22
Question 11.

FIGURE 22–23
Question 12.

12. A conductor carries a net positive charge Q. There is a hollow cavity within the conductor, at whose center is a negative point charge $-q$ (Fig. 22–23). What is the charge on (*a*) the outer surface of the conductor and (*b*) the inner surface of the conductor?

13. A point charge q is placed at the center of the cavity of a thin metal shell which is neutral. Will a charge Q placed outside the shell feel an electric force? Explain.

14. In Fig. 22–24, two objects, O_1 and O_2, have charges $+1.0\ \mu C$ and $-2.0\ \mu C$ respectively, and a third object, O_3, is electrically neutral. (*a*) What is the electric flux through the surface A_1 that encloses all the three objects? (*b*) What is the electric flux through the surface A_2 that encloses the third object only?

FIGURE 22–24
Question 14.

Problems

Section 22–1

1. (I) A flat circle of radius 15 cm is placed in a uniform electric field of magnitude $5.8 \times 10^2\ \text{N/C}$. What is the electric flux through the circle when its face is (*a*) perpendicular to the field lines, (*b*) at 45° to the field lines, and (*c*) parallel to the field lines.

2. (I) The Earth possesses an electric field of (average) magnitude 150 N/C near its surface. The field points radially inward. Calculate the net electric flux outward through a spherical surface surrounding, and just beyond, the Earth's surface.

3. (II) A cube of side l is placed in a uniform field $E = 6.50 \times 10^3\ \text{N/C}$ with edges parallel to the field lines. What is the net flux through the cube? What is the flux through each of its six faces?

4. (II) A uniform field $\mathbf{E}$ is parallel to the axis of a hollow hemisphere of radius R, Fig. 22–25. (*a*) What is the electric flux through the hemispherical surface? (*b*) What is the result if $\mathbf{E}$ is instead perpendicular to the axis?

FIGURE 22–25
Problem 4.

5. (I) The total electric flux from a cubical box 28.0 cm on a side is $1.45 \times 10^3\ \text{N} \cdot \text{m}^2/\text{C}$. What charge is enclosed by the box?

6. (I) Figure 22–26 shows five closed surfaces that surround various charges as indicated. Determine the electric flux through each surface, S_1, S_2, S_3, S_4, and S_5. The surfaces are flat "pill-box" surfaces that extend only slightly above and below the plane of the paper.

FIGURE 22–26 Problem 6.

FIGURE 22–27 Problem 7.

7. (II) In a certain region of space, the electric field is constant in direction (say horizontally, in the x direction), but its magnitude decreases from $E = 560\ \text{N/C}$ at $x = 0$ to $E = 410\ \text{N/C}$ at $x = 30\ \text{m}$. Determine the charge within a cubical box of side $l = 30\ \text{m}$, where the box is oriented so that four of its sides are parallel to the field lines (Fig. 22–27).

8. (II) A point charge Q is placed at the center of a cube of side l. What is the flux through one face of the cube?

Section 22–3

9. (I) The field just outside a 3.50-cm-radius metal ball is $2.75 \times 10^2\ \text{N/C}$ and points toward the ball. What charge resides on the ball?

10. (I) Starting from the result of Example 22–3, show that the electric field just outside a uniformly charged spherical conductor is $E = \sigma/\epsilon_0$, consistent with Example 22–7.

11. (I) A long thin wire, hundreds of meters long, carries a uniformly distributed charge of $-2.8\ \mu\text{C}$ per meter of length. What are the magnitude and direction of the electric field at points (a) 5.0 m and (b) 2.0 m from the wire?

12. (II) A solid metal sphere of radius 3.00 m carries a total charge of $-3.50\ \mu\text{C}$. What is the magnitude of the electric field at a distance from the sphere's center of (a) 0.15 m, (b) 2.90 m, (c) 3.10 m, and (d) 6.00 m? How would the answers differ if the sphere were (e) a thin shell, or (f) a solid nonconductor uniformly charged throughout?

13. (II) A 15.0-cm-diameter nonconducting sphere carries a total charge of $12.0\ \mu\text{C}$ distributed uniformly throughout its volume. Graph the electric field E as a function of the distance r from the center of the sphere from $r = 0$ to $r = 30\ \text{cm}$.

14. (II) A flat square sheet of thin aluminum foil, 25 cm on a side, carries a uniformly distributed 35 nC charge. What, approximately, is the electric field (a) 1.0 cm above the sheet and (b) 20 m above the sheet?

15. (II) A spherical cavity of radius 4.50 cm is at the center of a metal sphere of radius 18.0 cm. A point charge $Q = 5.50\ \mu\text{C}$ rests at the very center of the cavity, whereas the metal conductor carries no net charge. Determine the electric field at a point (a) 3.0 cm from the center of the cavity and (b) 6.0 cm from the center of the cavity.

16. (II) A point charge Q rests at the center of an uncharged thin spherical conducting shell. What is the electric field E as a function of r (a) for r less than the radius of the shell, (b) inside the shell, and (c) beyond the shell? (d) Does the shell affect the field due to Q alone? Does the charge Q affect the shell?

17. (II) A solid metal cube has a spherical cavity at its center as shown in Fig. 22–28. At the center of the cavity there is a point charge $Q = +8.00\ \mu\text{C}$. The metal cube carries a net charge $q = -7.00\ \mu\text{C}$ (not including Q). Determine (a) the total charge on the surface of the spherical cavity and (b) the total charge on the outer surface of the cube.

FIGURE 22–28 Problem 17.

FIGURE 22–29
Problems 18, 19, and 20.

18. (II) Two large, flat metal plates are separated by a distance that is very small compared to their height and width. The conductors are given equal but opposite uniform surface charge densities $\pm\sigma$. Ignore edge effects and use Gauss's law to show that for points far from the edges, (a) the electric field between the plates is $E = \sigma/\epsilon_0$ and (b) that outside the plates on either side the field is zero. (c) How would your results be altered if the two plates were nonconductors? (See Fig. 22–29.)

19. (II) Suppose the two conducting plates in Problem 18 have the *same* sign of charge. What then will be the electric field (a) between them and (b) outside them on either side? (c) What if the planes are nonconducting?

20. (II) The electric field between two square metal plates is $100\ \text{N/C}$. The plates are 1.0 m on a side and are separated by 3.0 cm, as in Fig. 22–29. What is the charge on each plate? Neglect edge effects.

21. (II) Two thin concentric spherical shells of radii r_1 and r_2 $(r_1 < r_2)$ contain uniform surface charge densities σ_1 and σ_2, respectively (see Fig. 22–30). Determine the electric field for (a) $r < r_1$, (b) $r_1 < r < r_2$, and (c) $r > r_2$. (d) Under what conditions will $E = 0$ for $r > r_2$? (e) Under what conditions will $E = 0$ for $r_1 < r < r_2$?

FIGURE 22–30 Two spherical shells (Problem 21).

FIGURE 22–31
Problems 22, 23, and 34.

22. (II) Suppose the nonconducting sphere of Example 22–4 has a spherical cavity of radius r_1 centered at the sphere's center (Fig. 22–31). Assuming the charge Q is distributed uniformly in the "shell" (between $r = r_1$ and $r = r_0$), determine the electric field as a function of r for (a) $r < r_1$, (b) $r_1 < r < r_0$, and (c) $r > r_0$.

23. (II) Suppose in Fig. 22–31, Problem 22, there is also a charge q at the center of the cavity. Determine the electric field for (a) $0 < r < r_1$, (b) $r_1 < r < r_0$, and (c) $r > r_0$.

24. (II) Suppose the thick spherical shell of Problem 22 is a conductor. It carries a total net charge Q and at its center there is a point charge q. What total charge is found on (a) the inner surface of the shell and (b) the outer surface of the shell? Determine the electric field for (c) $0 < r < r_1$, (d) $r_1 < r < r_0$, and (e) $r > r_0$.

25. (II) Suppose that at the center of the cavity inside the shell (charge Q) of Fig. 22–11 (and Example 22–3), there is a point charge q $(\neq Q)$. Determine the electric field for (a) $r < r_0$ and (b) $r > r_0$. What are your answers if (c) $q = Q$ and (d) $q = -Q$?

26. (II) A spherical rubber balloon carries a total charge Q uniformly distributed on its surface. At $t = 0$ the nonconducting balloon has radius r_0 and the balloon is then slowly blown up so that r increases linearly to $2r_0$ in a time T. Determine the electric field as a function of time (a) just outside the balloon surface and (b) at $r = 4r_0$.

27. (II) A long cylindrical shell of radius R_0 and length L $(R_0 \ll L)$ possesses a uniform surface charge density (charge per unit area) σ (Fig. 22–32). Determine the electric field at points (a) outside the cylinder $(r > R_0)$ and (b) inside the cylinder $(r < R_0)$; assume the points are far from the ends and not too far from the shell $(r \ll L)$. (c) Compare to the result for a long line of charge, Example 22–5.

FIGURE 22–32 Problem 27.

FIGURE 22–33
Problem 28.

28. (II) A very long solid nonconducting cylinder of radius R_0 and length L $(R_0 \ll L)$ possesses a uniform volume charge density ρ_E (C/m^3), Fig. 22–33. Determine the electric field at points (a) outside the cylinder $(r > R_0)$ and (b) inside the cylinder $(r < R_0)$. Do only for points far from the ends and for which $r \ll L$.

29. (II) A thin cylindrical shell of radius R_1 is surrounded by a second concentric cylindrical shell of radius R_2 (Fig. 22–34). The inner shell has a total charge $+Q$ and the outer shell $-Q$. Assuming the length L of the shells is much greater than R_1 or R_2, determine the electric field as a function of r (the perpendicular distance from the common axis of the cylinders) for (a) $r < R_1$, (b) $R_1 < r < R_2$, and (c) $r > R_2$. (d) What is the kinetic energy of an electron if it moves between (and concentric with) the shells in a circular orbit of radius $(R_1 + R_2)/2$?

FIGURE 22–34 Problems 29, 30, 31, and 32.

30. (II) In Problem 29 (a) under what conditions will $E = 0$ for $r > R_2$? (b) Under what conditions will $E = 0$ for $R_1 < r < R_2$?

31. (II) A thin cylindrical shell of radius $R_1 = 5.0\,\text{cm}$ is surrounded by a second cylindrical shell of radius $R_2 = 9.0\,\text{cm}$, as in Fig. 22–34. Both cylinders are 5.0 m long and the inner one carries a total charge $Q_1 = -3.8\,\mu\text{C}$ and the outer one $Q_2 = +3.2\,\mu\text{C}$. For points far from the ends of the cylinders, determine the electric field at a radial distance r from the central axis of (a) 3.0 cm, (b) 6.0 cm, and (c) 12.0 cm.

32. (II) (a) If an electron $(m = 9.1 \times 10^{-31}\,\text{kg})$ escaped from the surface of the inner cylinder in Problem 31 (Fig. 22–34) with negligible speed, what would be its speed when it reached the outer cylinder? (b) If a proton $(m = 1.67 \times 10^{-27}\,\text{kg})$ revolves in a circular orbit of radius $r = 6.0\,\text{cm}$ about the axis (i.e., between the cylinders), what must be its speed?

33. (II) A very long solid nonconducting cylinder of radius R_1 is uniformly charged with a charge density ρ_E. It is surrounded by a concentric cylindrical tube of inner radius R_2 and outer radius R_3 as shown in Fig. 22–35, and it too carries a uniform charge density ρ_E. Determine the electric field as a function of the distance r from the center of the cylinders for (a) $0 < r < R_1$, (b) $R_1 < r < R_2$, (c) $R_2 < r < R_3$, and (d) $r > R_3$. (e) If $\rho_E = 15\,\mu\text{C/m}^3$ and $R_1 = \frac{1}{2}R_2 = \frac{1}{3}R_3 = 5.0\,\text{cm}$, plot E as a function of r from $r = 0$ to $r = 20.0\,\text{cm}$. Assume the cylinders are very long compared to R_3.

FIGURE 22–35
Problem 33.

34. (III) Suppose the density of charge between r_1 and r_0 of the hollow sphere of Problem 22 (Fig. 22–31) varies as $\rho_E = \rho_0 r_1/r$. Determine the electric field as a function of r for (a) $r < r_1$, (b) $r_1 < r < r_0$, and (c) $r > r_0$. (d) Plot E versus r from $r = 0$ to $r = 2r_0$.

35. (III) A point charge Q is on the axis of a cylinder at its center. The diameter of the cylinder is equal to its length L (Fig. 22–36). What is the total flux through the curved sides of the cylinder? [*Hint:* First calculate the flux through the ends.]

FIGURE 22–36 Problem 35. FIGURE 22–37 Problem 36.

36. (III) A flat slab of nonconducting material (Fig. 22–37) carries a uniform charge per unit volume, ρ_E. The slab has thickness d which is small compared to the height and breadth of the slab. Determine the electric field as a function of x (a) inside the slab and (b) outside the slab (at distances much less than the slab's height or breadth). Take the origin at the center of the slab.

General Problems

37. Write Gauss's law for the gravitational field **g** (see Section 6–6).

38. The Earth is surrounded by an electric field, pointing inward at every point, of magnitude $E \approx 150\,\text{N/C}$ near the surface. (*a*) What is the net charge on the Earth? (*b*) How many excess electrons per square meter on the Earth's surface does this correspond to?

39. A cube of side l has one corner at the origin of coordinates, and extends along the positive x, y, and z axes. Suppose the electric field in this region is given by $E = (a + by)\mathbf{j}$. Determine the charge inside the cube.

40. A solid nonconducting sphere of radius r_0 has a total charge Q which is distributed according to $\rho_E = br$, where ρ_E is the charge per unit volume, or charge density (C/m^3), and b is a constant. Determine (*a*) b in terms of Q, (*b*) the electric field at points inside the sphere, and (*c*) the electric field at points outside the sphere.

41. A point charge of 3.50 nC is located at the origin and a second charge of -5.00 nC is located on the x axis at $x = 1.50$ m. Calculate the electric flux through a sphere centered at the origin with radius 1.00 m. Repeat the calculation for a sphere of radius 2.00 m.

42. A point charge produces an electric flux of $+500\,\text{N}\cdot\text{m}^2/\text{C}$ through a gaussian sphere of radius 15.0 cm centered on the charge. (*a*) What is the flux through a gaussian sphere with a radius 35.0 cm? (*b*) What is the magnitude and sign of the charge?

43. A point charge Q is placed a distance $r_0/2$ above the surface of an imaginary spherical surface of radius r_0 (Fig. 22–38). (*a*) What is the electric flux through the sphere? (*b*) What range of values does E have at the surface of the sphere? (*c*) Is **E** perpendicular to the sphere at all points? (*d*) Is Gauss's law useful for obtaining E at the surface of the sphere?

FIGURE 22–38 Problem 43. **FIGURE 22–39** Problem 44.

44. Three large but thin charged sheets are parallel to each other as shown in Fig. 22–39. Sheet I has a total surface charge density of $9.0\,\text{nC/m}^2$, sheet II a charge of $-2.0\,\text{nC/m}^2$, and sheet III a charge of $5.0\,\text{nC/m}^2$. What is the force per unit area on each sheet, in N/m^2?

45. Neutral hydrogen can be modeled as a positive point charge $+e$ surrounded by a distribution of negative charge with volume density given by $\rho_E(r) = -Ae^{-2r/a_0}$ where $a_0 = 0.53 \times 10^{-10}$ m is called the *Bohr radius*, and A is a constant such that the total amount of negative charge is $-e$. (*a*) What is the net charge inside a sphere of radius a_0? (*b*) What is the strength of the electric field at a distance a_0 from the nucleus?

46. A very large thin plane has uniform surface charge density σ. Touching it on the right (Fig. 22–40) is a long and wide slab of thickness d with uniform volume charge density ρ_E. Determine the electric field (*a*) to the left of the plane, (*b*) to the right of the slab, and (*c*) everywhere inside the slab.

 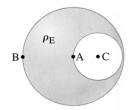

FIGURE 22–40 Problem 46. **FIGURE 22–41** Problem 47.

47. A sphere of radius r_0 carries a volume charge density ρ_E (Fig. 22–41). A spherical cavity of radius $r_0/2$ is then scooped out and left empty, as shown. (*a*) What is the magnitude and direction of the electric field at point A? (*b*) What is the direction and magnitude of the electric field at point B?

48. Dry air will break down and generate a spark if the electric field exceeds about $3 \times 10^6\,\text{N/C}$. How much charge could be packed onto a green pea (diameter 0.75 cm) before the pea spontaneously discharges?

49. Three very large sheets are separated by equal distances of 20.0 cm (Fig. 22–42). The first and third sheets are very thin and nonconducting and have surface charge densities of $+5.00\,\mu\text{C/m}^2$ and $-5.00\,\mu\text{C/m}^2$ respectively. The middle sheet is conducting but has no net charge. (*a*) What is the electric field inside the middle sheet? What is the electric field (*b*) between the left and middle sheets, and (*c*) between the middle and right sheets? (*d*) What is the charge density on the surface of the middle sheet facing the left sheet, and (*e*) on the surface facing the right sheet?

 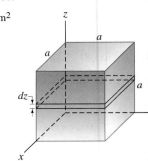

FIGURE 22–42 Problem 49. **FIGURE 22–43** Problem 50.

50. Careful measurement of the electric field in the ground can help provide useful information about charge. In a particular region, a technician has determined that the electric field in a cubical volume, 1.00 m on a side, is

$$E = E_0\left(1 + \frac{z}{a}\right)\hat{\mathbf{i}} + E_0\left(\frac{z}{a}\right)\hat{\mathbf{j}}$$

where $E_0 = 1.00\,\text{N/C}$ and $a = 1.00$ m. The cube has its sides parallel to the coordinate axes, Fig. 22–43. Determine the net charge within the cube.

Lightning: The potential difference (voltage) between clouds and the Earth can become so high that electrons are pulled off atoms of the air by the large electric field. The air becomes a conductor as the ionized atoms and freed electrons flow rapidly, colliding with more atoms, and causing more ionization. The massive flow of charge reduces the potential difference and the "discharge" quickly ceases. The light represents energy released when the ions and electrons recombine to form atoms.

CHAPTER 23

Electric Potential

We saw in Chapters 7 and 8 that the concept of energy was extremely valuable in dealing with mechanical problems. For one thing, energy is a conserved quantity and is thus an important tool for understanding nature. Furthermore, we saw that many problems could be solved using the energy concept even though a detailed knowledge of the forces involved was not possible, or when a calculation involving Newton's laws would have been too difficult.

The energy point of view can be used in electricity, and it is especially useful. It not only extends the law of conservation of energy, but it gives us another way to view electrical phenomena; and it is a tool in solving problems more easily, in many cases, than by using forces and electric fields.

23–1 | Electric Potential and Potential Difference

To apply conservation of energy, we need to define electric potential energy as for other types of potential energy (Chapter 8). As we saw, potential energy can be defined only for a conservative force. Recall that the work done by a conservative force in moving an object between any two positions is independent of the path

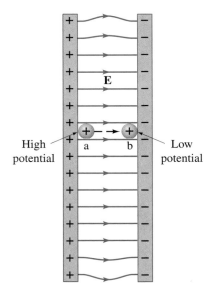

FIGURE 23–1 Work is done by the electric field in moving the positive charge from position a to position b.

High potential

Low potential

Potential is potential energy per unit charge

taken. It is easy to see that the electrostatic force between any two charges $(F = kQ_1Q_2/r^2)$ is conservative: the dependence on position is $1/r^2$ just as for the gravitational force, which we saw in Section 8–7 is conservative. Hence the electrostatic force given by Coulomb's law is conservative and we can define potential energy U for it.

We define the change in electric potential energy, $U_b - U_a$, when a charge q moves from some point a to a second point b, as the negative of the work done by the electric force to move the charge from a to b. For example, consider the electric field between two equally but oppositely charged parallel plates whose separation is small compared to their width and height, so the field **E** will be uniform over most of the region, Fig. 23–1. Now consider a tiny positive point charge q placed at point a very near the positive plate as shown. This charge q is so small it doesn't affect **E**. If this charge q at point a is released, the electric force will do work on the charge and accelerate it toward the negative plate. In the process, the charged particle will have its kinetic energy K increased. By conservation of energy, the potential energy will decrease by an equal amount, equal to the negative of the work done by the electric force. In accord with the conservation of energy, electric potential energy is transformed into kinetic energy, and the total energy is conserved. Note that the positive charge q has its greatest potential energy at point a, near the positive plate,[†] so $(U_b - U_a) < 0$. The reverse is true for a negative charge: its potential energy is greatest near the negative plate.

We defined the electric field (Chapter 21) as the force per unit charge. Similarly, it is useful to define the **electric potential** (or simply the **potential** when "electric" is understood) as the *potential energy per unit charge*. Electric potential is given the symbol V. If a positive test charge q has electric potential energy U_a at some point a (relative to some zero potential energy), the electric potential V_a at this point is

$$V_a = \frac{U_a}{q}.$$

Potential difference

As we discussed in Chapter 8, only differences in potential energy are physically meaningful. Hence only the **difference in potential**, or the **potential difference**, between two points a and b (such as between a and b in Fig. 23–1) is measurable. When the electric force does positive work on a charge, the kinetic energy increases and the potential energy decreases. The difference in potential energy, $U_b - U_a$, is equal to the negative of the work, W_{ba}, done by the electric force to move the charge from point a to point b, so the potential difference V_{ba} is

$$V_{ba} = V_b - V_a = \frac{U_b - U_a}{q} = -\frac{W_{ba}}{q}. \qquad \textbf{(23–1)}$$

Note that electric potential, like electric field, does not depend on our test charge q. V depends on the other charges that create the field, not on q; q acquires potential energy by being in the potential V due to the other charges.

We can see from our definition that the positive plate in Fig. 23–1 is at a higher potential than the negative plate. Thus a positively charged object moves naturally from a high potential to a low potential. A negative charge does the reverse.

The volt (1 V = 1 J/C)

The unit of electric potential, and of potential difference, is joules/coulomb and is given a special name, the **volt**, in honor of Alessandro Volta (1745–1827; he is best known for having invented the electric battery). The volt is abbreviated V, so $1\,\text{V} = 1\,\text{J/C}$. Potential difference, since it is measured in volts, is often referred to as **voltage**.

Voltage = potential difference

If we wish to speak of the potential, V_a, at some point a, we must be aware that V_a depends on where the potential is chosen to be zero. The zero point for electric potential in a given situation, just as for potential energy, can be chosen arbitrarily

V = 0 chosen arbitrarily

[†] At this point it has its greatest ability to do work (on some other object or system).

since only differences in potential energy can be measured. Often the ground, or a conductor connected directly to the ground (the Earth), is taken as zero potential, and other potentials are given with respect to ground. (Thus, a point where the voltage is 50 V is one where the difference of potential between it and ground is 50 V.) In other cases, as we shall see, we may choose the potential to be zero at an infinite distance ($r = \infty$).

CONCEPTUAL EXAMPLE 23–1 **A negative charge.** Suppose a negative charge, such as an electron, is placed at point b in Fig. 23–1. If the electron is free to move, will its electric potential energy increase or decrease? How will the electric potential change?

RESPONSE An electron placed at point b will move toward the positive plate. (An electron placed at point a would not move.) As the electron moves to the left, its potential energy *decreases* as its kinetic energy gets larger. But note that the electron moves from a point b at low potential to a point a at higher potential: $\Delta V = V_a - V_b > 0$. (The potentials V_a and V_b are due to the charges on the plates, not due to the electron.)

Since the electric potential difference is defined as the potential energy difference per unit charge, then the change in potential energy of a charge q when moved between two points a and b is

$$U_b - U_a = q(V_b - V_a) = qV_{ba}.$$ **(23–2)**

Electric potential and potential energy

That is, if an object with charge q moves through a potential difference V_{ba}, its potential energy changes by an amount qV_{ba}. For example, if the potential difference between the two plates in Fig. 23–1 is 6 V, then a $+1$ C charge moved (say by an external force) from b to a will gain $(1\,\text{C})(6\,\text{V}) = 6\,\text{J}$ of electric potential energy. (And it will lose 6 J of electric potential energy if it moves from a to b.) Similarly, a 2-C charge will gain 12 J, and so on. Thus, electric potential difference is a measure of how much energy an electric charge can acquire in a given situation. And, since energy is the ability to do work, the electric potential difference is also a measure of how much work a given charge can do. The exact amount depends both on the potential difference and on the charge.

To better understand electric potential, let's make a comparison to the gravitational case when a rock falls from the top of a cliff. The greater the height, h, of a cliff, the more potential energy ($= mgh$) the rock has at the top of the cliff, relative to the bottom, and the more kinetic energy it will have when it reaches the bottom. The actual amount of kinetic energy it will acquire, and the amount of work it can do, depends both on the height of the cliff and the mass m of the rock. A large rock and a small rock can be at the same height h (Fig. 23–2a) and thus have the same "gravitational potential," but the larger rock has the greater potential energy. The electrical case is similar (Fig. 23–2b): the potential energy change, or the work that can be done, depends both on the potential difference (corresponding to the height of the cliff) and on the charge (corresponding to mass), Eq. 23–2. [But note a significant difference: electric charge comes in two types, $+$ and $-$, whereas gravitational mass is always $+$.]

Potential likened to height of a cliff

(a)

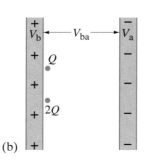

(b)

FIGURE 23–2 (a) Two rocks are at the same height. The larger rock has more potential energy. (b) Two charges have the same electric potential. The $2Q$ charge has more potential energy.

TABLE 23–1
Some Typical Voltages

Source	Voltage (approx.)
Thundercloud to ground	10^8 V
High-voltage power line	10^6 V
Power supply for TV tube	10^4 V
Automobile ignition	10^4 V
Household outlet	10^2 V
Automobile battery	12 V
Flashlight battery	1.5 V
Resting potential across nerve membrane	10^{-1} V
Potential changes on skin (EKG and EEG)	10^{-4} V

FIGURE 23–3 Electron accelerated in TV picture tube. Example 23–2.

Practical sources of electrical energy such as batteries and electric generators are meant to maintain a potential difference. The actual amount of energy used or transformed depends on how much charge flows. For example, consider an automobile headlight connected to a 12.0-V battery. The amount of energy transformed (into light and thermal energy) is proportional to how much charge flows, which in turn depends on how long the light is on. If over a given period of time 5.0 C of charge flows through the light, the total energy transformed is $(5.0 \, \text{C})(12.0 \, \text{V}) = 60 \, \text{J}$. If the headlight is left on twice as long, 10.0 C of charge will flow and the energy transformed is $(10.0 \, \text{C})(12.0 \, \text{V}) = 120 \, \text{J}$.

Table 23–1 presents some typical voltages.

EXAMPLE 23–2 Electron in TV tube. Suppose an electron in the picture tube of a television set is accelerated from rest through a potential difference $V_{ba} = +5000 \, \text{V}$ (Fig. 23–3). (a) What is the change in potential energy of the electron? (b) What is the speed of the electron $(m = 9.1 \times 10^{-31} \, \text{kg})$ as a result of this acceleration? (c) Repeat for a proton $(m = 1.67 \times 10^{-27} \, \text{kg})$ that accelerates through a potential difference of $V_{ba} = -5000 \, \text{V}$.

SOLUTION (a) The charge on an electron is $e = -1.6 \times 10^{-19} \, \text{C}$. Therefore its change in potential energy (Eq. 23–2) is equal to

$$\Delta U = qV_{ba} = (-1.6 \times 10^{-19} \, \text{C})(+5000 \, \text{V})$$
$$= -8.0 \times 10^{-16} \, \text{J}.$$

The minus sign indicates that the potential energy decreases. (The potential difference, V_{ba}, has a positive sign since the final potential is higher than the initial potential; that is, negative electrons are attracted from a negative electrode to a positive one.)

(b) The potential energy lost by the electron becomes kinetic energy $(= K)$. From conservation of energy (Eq. 8–9), $\Delta K + \Delta U = 0$, so

$$\Delta K = -\Delta U$$
$$\tfrac{1}{2}mv^2 - 0 = -qV_{ba},$$

where the initial kinetic energy is zero since we assume the electron started from rest. We solve for v and put in the mass of the electron $m = 9.1 \times 10^{-31} \, \text{kg}$:

$$v = \sqrt{-\frac{2qV_{ba}}{m}}$$
$$= \sqrt{-\frac{2(-1.6 \times 10^{-19} \, \text{C})(5000 \, \text{V})}{9.1 \times 10^{-31} \, \text{kg}}}$$
$$= 4.2 \times 10^7 \, \text{m/s}.$$

[Note: For such a high speed, which is $\tfrac{1}{7}$ the speed of light, we really should use the theory of relativity, Chapter 37, to get a more precise result.]

(c) The proton has the same magnitude of charge as the electron, though of opposite sign. Hence for the same magnitude of V_{ba} we expect the same change in U, but a lesser speed since the proton's mass is greater. Thus:

$$\Delta U = qV_{ba} = (+1.6 \times 10^{-19} \, \text{C})(-5000 \, \text{V}) = -8.0 \times 10^{-16} \, \text{J},$$

and

$$v = \sqrt{-\frac{2qV_{ba}}{m}} = \sqrt{-\frac{2(1.6 \times 10^{-19} \, \text{C})(-5000 \, \text{V})}{(1.67 \times 10^{-27} \, \text{kg})}}$$
$$= 9.8 \times 10^5 \, \text{m/s}.$$

Note that the energy doesn't depend on the mass, only on the charge and voltage. The speed *does* depend on m.

23-2 Relation Between Electric Potential and Electric Field

The effects of any charge distribution can be described either in terms of electric field or in terms of electric potential. Electric potential is often easier to use because it is a scalar, as compared to electric field which is a vector. There is a crucial connection between the electric potential produced by a given arrangement of charges and the electric field due to those charges, which we now examine.

We start by recalling the relation between a conservative force $\mathbf{F}$ and the potential energy U associated with that force. As discussed in Section 8–2, the difference in potential energy between any two points in space, a and b, is given by Eq. 8–4:

$$U_b - U_a = -\int_a^b \mathbf{F} \cdot d\mathbf{l},$$

where $d\mathbf{l}$ is an infinitesimal increment of displacement, and the integral is taken along any path in space from point a to point b. For the electrical case, we are more interested in the potential difference, given by Eq. 23–1, $V_{ba} = V_b - V_a = (U_b - U_a)/q$, rather than in the potential energy itself. Also, the electric field $\mathbf{E}$ at any point in space is defined as the force per unit charge (Eq. 21–3): $\mathbf{E} = \mathbf{F}/q$. Putting these two relations in the above equation gives us

$$V_{ba} = V_b - V_a = -\int_a^b \mathbf{E} \cdot d\mathbf{l}. \qquad \textbf{(23–3)}$$

V related to $\mathbf{E}$

This is the general relation between electric field and potential difference. See Fig. 23–4. If we are given the electric field due to some arrangement of electric charge, we can use Eq. 23–3 to determine V_{ba}.

A simple special case is when the field is uniform. In Fig. 23–1, for example, a path parallel to the electric field lines from point a at the positive plate to point b at the negative plate gives (since $\mathbf{E}$ and $d\mathbf{l}$ are in the same direction at each point),

$$V_b - V_a = -\int_a^b \mathbf{E} \cdot d\mathbf{l} = -E \int_a^b dl = -Ed$$

or

$$V_{ba} = -Ed \qquad \text{[only if } E \text{ is uniform]} \quad \textbf{(23–4)}$$

where d is the distance, parallel to the field lines, between points a and b. Be careful not to use Eq. 23–4 unless you are sure the electric field is uniform.

From either Eq. 23–3 or 23–4 we can see that the units for electric field intensity can be written as volts per meter (V/m) as well as newtons per coulomb (N/C). These are equivalent in general, since $1\,\text{N/C} = 1\,\text{N·m/C·m} = 1\,\text{J/C·m} = 1\,\text{V/m}$.

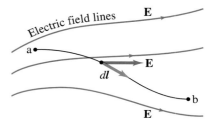

FIGURE 23–4 Integrating $\mathbf{E} \cdot d\mathbf{l}$ from point a to point b in a nonuniform electric field $\mathbf{E}$.

EXAMPLE 23–3 **Special case: Uniform electric field obtained from voltage.** Two parallel plates are charged to a voltage of 50 V. If the separation between the plates is 5.0 cm, calculate the electric field between them, ignoring any fringing.

SOLUTION We have from Eq. 23–4, dealing only with magnitudes for convenience,

$$E = \frac{V_{ba}}{d} = \frac{50\,\text{V}}{0.050\,\text{m}} = 1000\,\text{V/m}.$$

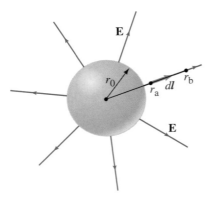

FIGURE 23–5 Example 23–4: integrating $\mathbf{E} \cdot d\mathbf{l}$ for the field outside a spherical conductor.

FIGURE 23–6 (a) E versus r, and (b) V versus r, for a uniformly charged solid conducting sphere of radius r_0 (the charge distributes itself on the surface); r is the distance from the center of the sphere.

(a)

(b)

EXAMPLE 23–4 **Charged conducting sphere.** Determine the potential at a distance r from the center of a uniformly charged conducting sphere of radius r_0 for (a) $r > r_0$, (b) $r = r_0$, (c) $r < r_0$. The total charge on the sphere is Q.

SOLUTION (a) The charge Q is distributed over the surface of the sphere since it is a conductor. We saw in Example 22–3 that the electric field outside a conducting sphere is

$$E = \frac{1}{4\pi\epsilon_0} \frac{Q}{r^2} \qquad [r > r_0]$$

and points radially outward (inward if $Q < 0$). Since we know $\mathbf{E}$, we can use Eq. 23–3 and integrate along a radial line with $d\mathbf{l}$ parallel to $\mathbf{E}$ (Fig. 23–5) between two points which are distances r_a and r_b from the sphere's center:

$$V_b - V_a = -\int_{r_a}^{r_b} \mathbf{E} \cdot d\mathbf{l} = -\frac{Q}{4\pi\epsilon_0} \int_{r_a}^{r_b} \frac{dr}{r^2} = \frac{Q}{4\pi\epsilon_0} \left(\frac{1}{r_b} - \frac{1}{r_a} \right).$$

If we let $V = 0$ for $r = \infty$ (say $V_b = 0$ at $r_b = \infty$), then at any other point r (for $r > r_0$) we have

$$V = \frac{1}{4\pi\epsilon_0} \frac{Q}{r}. \qquad [r > r_0]$$

We will see in the next Section that this same equation applies for the potential a distance r from a single point charge. Thus the electric potential outside a spherical conductor with a uniformly distributed charge is the same as if all the charge were at its center.

(b) As r approaches r_0, we see that

$$V = \frac{1}{4\pi\epsilon_0} \frac{Q}{r_0} \qquad [r = r_0]$$

at the surface of the conductor.

(c) For points within the conductor, $E = 0$. Thus the integral, $\int \mathbf{E} \cdot d\mathbf{l}$, between $r = r_0$ and any point within the conductor gives zero change in V. Hence V is constant within the conductor:

$$V = \frac{1}{4\pi\epsilon_0} \frac{Q}{r_0}. \qquad [r \leq r_0]$$

The whole conductor, not just its surface, is at this same potential. Plots of both E and V as a function of r are shown in Fig. 23–6 for a conducting sphere.

EXAMPLE 23–5 **Breakdown voltage.** In many kinds of equipment, very high voltages are used. A problem with high voltage is that the air can become ionized due to the high electric fields: free electrons in the air (produced by cosmic rays, for example) can be accelerated by such high fields to speeds sufficient to ionize O_2 and N_2 molecules by collision, knocking out one or more of their electrons. The air then becomes conducting and the high voltage cannot be maintained as charge flows. The breakdown of air occurs for electric fields of about 3×10^6 V/m. (a) Show that the breakdown voltage for a spherical conductor in air is proportional to the radius of the sphere, and (b) estimate the breakdown voltage in air for a sphere of diameter 1.0 cm.

SOLUTION (a) The electric potential at the surface of a spherical conductor of radius r_0 (Example 23–4), and the electric field just outside its surface, are

$$V = \frac{1}{4\pi\epsilon_0} \frac{Q}{r_0} \quad \text{and} \quad E = \frac{1}{4\pi\epsilon_0} \frac{Q}{r_0^2}.$$

Combining these we obtain

$$V = r_0 E. \qquad \text{[at surface of spherical conductor]}$$

(b) For $r_0 = 5 \times 10^{-3}$ m, the breakdown voltage in air is

$$V = (5 \times 10^{-3} \text{ m})(3 \times 10^6 \text{ V/m}) \approx 15,000 \text{ V}.$$

Example 23–5 makes clear why large rounded terminals without sharp edges are used for high-voltage equipment. It also explains why breakdown, or sparks, occur at rough edges or points (regions with small radius of curvature) on a conductor, and why conductors are usually made very smooth.

23–3 | Electric Potential Due to Point Charges

The electric potential at a distance r from a single point charge Q can be derived directly from Eq. 23–3, $V_b - V_a = -\int \mathbf{E} \cdot d\mathbf{l}$. The electric field due to a single point charge has magnitude (Eq. 21–4)

$$E = \frac{1}{4\pi\epsilon_0}\frac{Q}{r^2} \quad \text{or} \quad E = k\frac{Q}{r^2}$$

(where $k = 1/4\pi\epsilon_0 = 8.99 \times 10^9\,\text{N}\cdot\text{m}^2/\text{C}^2$), and is directed radially outward from the charge (inward if $Q < 0$). We take the integral in Eq. 23–3 along a (straight) field line (Fig. 23–7) from point a, a distance r_a from Q, to point b, a distance r_b from Q. Then $d\mathbf{l}$ will be parallel to $\mathbf{E}$ and $dl = dr$. Thus

$$V_b - V_a = -\int_{r_a}^{r_b} \mathbf{E} \cdot d\mathbf{l} = -\frac{Q}{4\pi\epsilon_0}\int_{r_a}^{r_b}\frac{1}{r^2}\,dr = \frac{1}{4\pi\epsilon_0}\left(\frac{Q}{r_b} - \frac{Q}{r_a}\right).$$

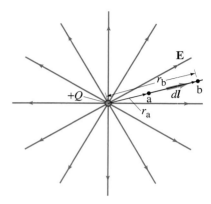

FIGURE 23–7 We integrate Eq. 23–3 along the straight line (shown in black) from point a to point b. The line ab is parallel to a field line.

As mentioned earlier, only differences in potential have physical meaning. We are free, therefore, to choose the value of the potential at some one point to be whatever we please. It is common to choose the potential to be zero at infinity (let $V_b = 0$ at $r_b = \infty$). Then the electric potential V at a distance r from a single point charge is

$$V = \frac{1}{4\pi\epsilon_0}\frac{Q}{r}. \qquad \left[\begin{array}{c}\text{single point charge;}\\ V = 0 \text{ at } r = \infty\end{array}\right] \quad \textbf{(23–5)}$$

We can think of V here as representing the absolute potential, where $V = 0$ at $r = \infty$, or we can think of V as the potential difference between r and infinity. Notice that the potential V decreases with the first power of the distance, whereas the electric field (Eq. 21–4) decreases as the *square* of the distance. The potential near a positive charge is large, and it decreases toward zero at very large distances (Fig. 23–8). For a negative charge, the potential is negative and increases toward zero at large distances (Fig. 23–9).

In Example 23–4 we found that the potential due to a uniformly charged sphere is given by the same relation, Eq. 23–5, for points outside the sphere. Thus we see that the potential outside a uniformly charged sphere is the same as if all the charge were concentrated at its center.

FIGURE 23–8 Potential V as a function of distance r from a single point charge Q when the charge is positive.

FIGURE 23–9 Potential V as a function of distance r from a single point charge Q when the charge is negative.

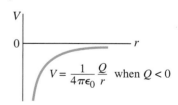

EXAMPLE 23–6 **Work to force two + charges close together.** What minimum work is required by an external force to bring a charge $q = 3.00\,\mu\text{C}$ from a great distance away (take $r = \infty$) to a point 0.500 m from a charge $Q = 20.0\,\mu\text{C}$?

SOLUTION The work done by the electric field is equal to the negative of the change in potential energy:

$$W = -qV_{ba} = -\frac{q}{4\pi\epsilon_0}\left(\frac{Q}{r_b} - \frac{Q}{r_a}\right),$$

where $r_b = 0.500$ m and $r_a = \infty$. The second term is zero ($1/\infty = 0$), so

$$W = -(3.00 \times 10^{-6}\,\text{C})\frac{(8.99 \times 10^9\,\text{N}\cdot\text{m}^2/\text{C}^2)(2.00 \times 10^{-5}\,\text{C})}{(0.500\,\text{m})} = -1.08\,\text{J}.$$

The electric field does negative work in this case. In order to bring the charge to this point, an *external* force would have to do work $W = +1.08$ J, assuming no acceleration of the charges.

To determine the electric field surrounding a collection of two or more point charges requires adding up the electric fields due to each charge. Since the electric field is a vector, this can often be a chore. To find the electric potential due to a collection of point charges is far easier, since the electric potential is a scalar, and hence you only need to add numbers together without concern for direction. This is a major advantage in using electric potential. We do have to include the signs of charges, however.

Potentials add as scalars
(Fields add as vectors)

EXAMPLE 23–7 **Potential above two charges.** Calculate the electric potential at points A and B in Fig. 23–10 due to the two charges shown. (This is the same situation as Example 21–8, Fig. 21–26, where we calculated the electric field at these points.) Assume $V = 0$ at $r = \infty$.

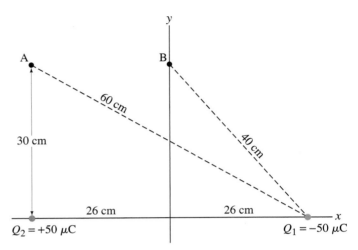

FIGURE 23–10 Example 23–7. (See also Example 21–8, Fig. 21–26).

SOLUTION The potential at point A is the sum of the potentials due to the + and − charges, and we use Eq. 23–5 for each:

$$V_A = V_{A2} + V_{A1}$$
$$= \frac{(9.0 \times 10^9 \, \text{N} \cdot \text{m}^2/\text{C}^2)(5.0 \times 10^{-5} \, \text{C})}{0.30 \, \text{m}}$$
$$+ \frac{(9.0 \times 10^9 \, \text{N} \cdot \text{m}^2/\text{C}^2)(-5.0 \times 10^{-5} \, \text{C})}{0.60 \, \text{m}}$$
$$= 1.50 \times 10^6 \, \text{V} - 0.75 \times 10^6 \, \text{V}$$
$$= 7.5 \times 10^5 \, \text{V}.$$

At point B:

$$V_B = V_{B2} + V_{B1}$$
$$= \frac{(9.0 \times 10^9 \, \text{N} \cdot \text{m}^2/\text{C}^2)(5.0 \times 10^{-5} \, \text{C})}{0.40 \, \text{m}}$$
$$+ \frac{(9.0 \times 10^9 \, \text{N} \cdot \text{m}^2/\text{C}^2)(-5.0 \times 10^{-5} \, \text{C})}{0.40 \, \text{m}}$$
$$= 0 \, \text{V}.$$

It should be clear that the potential will be zero everywhere on the plane equidistant between the two charges. Thus this plane is an equipotential surface with $V = 0$.

A summation like these can be easily performed for any number of point charges.

23–4 | Potential Due to Any Charge Distribution

If we know the electric field in a region of space due to any distribution of electric charge, we can determine the difference in potential between two points in the region using Eq. 23–3, $V_{ba} = -\int_a^b \mathbf{E} \cdot d\mathbf{l}$. In many cases we don't know $\mathbf{E}$ as a function of position, and it may be difficult to calculate. We can calculate the potential V due to a given charge distribution in another way, often easier, using the potential due to a single point charge, Eq. 23–5:

$$V = \frac{1}{4\pi\epsilon_0} \frac{Q}{r},$$

where $V = 0$ at $r = \infty$. Then we can sum over all the charges. If we have n individual point charges, the potential at some point a (relative to $V = 0$ at $r = \infty$) is

$$V_a = \sum_{i=1}^{n} V_i = \frac{1}{4\pi\epsilon_0} \sum_{i=1}^{n} \frac{Q_i}{r_{ia}}, \qquad \textbf{(23–6a)}$$

Potential due to many point charges

where r_{ia} is the distance from the ith charge (Q_i) to the point a. (We already used this approach in Example 23–7.) If the charge distribution can be considered continuous, then

$$V = \frac{1}{4\pi\epsilon_0} \int \frac{dq}{r}, \qquad \textbf{(23–6b)}$$

Potential due to a continuous charge distribution

where r is the distance from a tiny element of charge, dq, to the point where V is being determined.

EXAMPLE 23–8 **Potential due to a ring of charge.** A thin circular ring of radius R carries a uniformly distributed charge Q. Determine the electric potential at a point P on the axis of the ring a distance x from its center, Fig. 23–11.

SOLUTION Each point on the ring is equidistant from point P, and this distance is $(x^2 + R^2)^{\frac{1}{2}}$. So the potential at P is:

$$V = \frac{1}{4\pi\epsilon_0} \int \frac{dq}{r} = \frac{1}{4\pi\epsilon_0} \frac{1}{(x^2 + R^2)^{\frac{1}{2}}} \int dq = \frac{1}{4\pi\epsilon_0} \frac{Q}{(x^2 + R^2)^{\frac{1}{2}}}.$$

Note that for points very far away from the ring, $x \gg R$, this result reduces to $(1/4\pi\epsilon_0)(Q/x)$, the potential of a point charge, as we should expect.

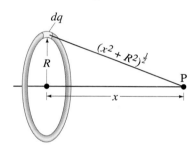

FIGURE 23–11 Calculating the potential at point P, a distance x from the center of a uniform ring of charge (Example 23–8).

EXAMPLE 23–9 **Potential due to a charged disk.** A thin flat disk, of radius R, carries a uniformly distributed charge Q, Fig. 23–12. Determine the potential at a point P on the axis of the disk, a distance x from its center.

FIGURE 23–12 Calculating the electric potential at point P on the axis of a uniformly charged thin disk. Example 23–9.

SOLUTION Divide the disk into thin rings of radius r and thickness dr. The charge Q is distributed uniformly, so the charge contained in each ring is proportional to its area. The disk has area πR^2 and each thin ring has area $dA = (2\pi r)(dr)$. Hence

$$\frac{dq}{Q} = \frac{2\pi r \, dr}{\pi R^2}$$

so

$$dq = Q \frac{(2\pi r)(dr)}{\pi R^2} = \frac{2Qr \, dr}{R^2}.$$

Then the potential at P, using Eq. 23–6b in which r is replaced by $(x^2 + r^2)^{\frac{1}{2}}$, is

$$V = \frac{1}{4\pi\epsilon_0} \int \frac{dq}{(x^2 + r^2)^{\frac{1}{2}}} = \frac{2Q}{4\pi\epsilon_0 R^2} \int_0^R \frac{r \, dr}{(x^2 + r^2)^{\frac{1}{2}}} = \frac{Q}{2\pi\epsilon_0 R^2} (x^2 + r^2)^{\frac{1}{2}} \Big|_{r=0}^{r=R}$$

$$= \frac{Q}{2\pi\epsilon_0 R^2} \left[(x^2 + R^2)^{\frac{1}{2}} - x \right].$$

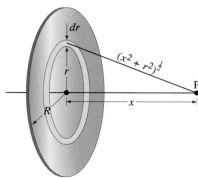

The electric potential can be represented graphically by drawing **equipotential lines** or, in three dimensions, **equipotential surfaces**. An equipotential surface is one on which all points are at the same potential. That is, the potential difference between any two points on the surface is zero, and no work is required to move a charge from one point to the other. An *equipotential surface must be perpendicular to the electric field* at any point. If this were not so—that is, if there were a component of **E** parallel to the surface—it would require work to move the charge along the surface against this component of **E**; and this would contradict the idea that it is an equipotential surface. This can also be seen from Eq. 23–3, $\Delta V = -\int \mathbf{E} \cdot d\mathbf{l}$. On a surface where V is constant, $\Delta V = 0$, so we must have either $\mathbf{E} = 0$, $d\mathbf{l} = 0$, or $\cos\theta = 0$ where θ is the angle between **E** and $d\mathbf{l}$. Thus in a region where **E** is not zero, the path $d\mathbf{l}$ along an equipotential must have $\cos\theta = 0$, meaning $\theta = 90°$ and **E** is perpendicular to the equipotential.

The fact that the electric field lines and equipotential surfaces are mutually perpendicular helps us locate the equipotentials when the electric field lines are known. In a normal two-dimensional drawing, we show equipotential *lines*, which are the intersections of equipotential surfaces with the plane of the drawing. In Fig. 23–13, a few of the equipotential lines are drawn (dashed green lines) for the electric field (red lines) between two parallel plates at a potential difference of 20 V. The negative plate is arbitrarily chosen to be zero volts and the potential of each equipotential line is indicated. Note that **E** points toward lower values of V. The equipotential lines for the case of two equal but oppositely charged particles are shown in Fig. 23–14 as green dashed lines. Equipotential lines and surfaces, unlike field lines, are always continuous and never end, and so continue beyond the borders of Figs. 23–13 and 23–14.

We saw in Section 21–9 that there can be no electric field within a conductor in the static case, for otherwise the free electrons would feel a force and would move. Indeed, the entire volume of *a conductor must be entirely at the same potential in the static case*, and the surface of a conductor is then an equipotential surface. (If it weren't, the free electrons at the surface would move, since whenever there is a potential difference between two points, free charges will move.) This is fully consistent with our result, discussed earlier, that the electric field at the surface of a conductor must be perpendicular to the surface.

FIGURE 23–13 Equipotential lines (the green dashed lines) between two oppositely charged parallel plates. Note that they are perpendicular to the electric field lines (solid red lines).

Conductors are equipotential surfaces

FIGURE 23–14 Equipotential lines (green, dashed) are always perpendicular to the electric field lines (solid red) shown here for two equal but oppositely charged particles.

FIGURE 23–15 **FIGURE 23–15** A topographic map (here, a portion of the Sierra Nevada in California) shows continuous contour lines, each of which is at a fixed height above sea level. Here they are at 80 ft (25 m) intervals. If you walk along one contour line, you neither climb nor descend. If you cross lines, and especially (maximally), if you climb perpendicular to the lines, you will be changing your gravitational potential (rapidly, if the lines are close together).

A useful analogy for equipotential lines is a topographic map: the contour lines are essentially gravitational equipotential lines (Fig. 23–15).

23–6 | Electric Dipoles

Two equal point charges Q, of opposite sign, separated by a distance l, are called an **electric dipole**, as we already saw in Section 21–11. Also, the two charges we saw in Figs. 23–10 and 23–14 constitute an electric dipole, and the latter shows the electric field lines and equipotential surfaces for a dipole. Because electric dipoles occur often in physics, as well as in other fields, it is useful to examine them more closely.

Let us calculate the electric potential at an arbitrary point P due to a dipole, as shown in Fig. 23–16. As usual, we take $V = 0$ at $r = \infty$. Since V is the sum of the potentials due to each of the two charges, we have

$$V = \frac{1}{4\pi\epsilon_0} \frac{Q}{r} + \frac{1}{4\pi\epsilon_0} \frac{(-Q)}{r + \Delta r} = \frac{1}{4\pi\epsilon_0} Q\left(\frac{1}{r} - \frac{1}{r + \Delta r}\right) = \frac{Q}{4\pi\epsilon_0} \frac{\Delta r}{r(r + \Delta r)},$$

where r is the distance from P to the positive charge and $r + \Delta r$ is the distance to the negative charge. This equation becomes simpler if we consider points P whose distance from the dipole is much larger than the separation of the two charges—that is, for $r \gg l$. From the diagram we can see that in this case, $\Delta r \approx l \cos\theta$. Since $r \gg \Delta r$, we can neglect Δr in the denominator as compared to r. Therefore, we obtain

$$V = \frac{1}{4\pi\epsilon_0} \frac{Ql\cos\theta}{r^2} = \frac{1}{4\pi\epsilon_0} \frac{p\cos\theta}{r^2} \qquad \text{[dipole; } r \gg l\text{]} \quad (23\text{–}7)$$

where $p = Ql$ is called the **dipole moment**. When θ is between $0°$ and $90°$, V is positive. If θ is between $90°$ and $180°$, V is negative (since $\cos\theta$ is then negative). This makes sense since in the first case P is closer to the positive charge and in the second case it is closer to the negative charge. At $\theta = 90°$, the potential is zero ($\cos 90° = 0$), in agreement with the result of Example 23–7 (point B). From Eq. 23–7, we see that the potential decreases as the *square* of the distance from the dipole, whereas for a single point charge the potential decreases with the first power of the distance (Eq. 23–5). It is not surprising that the potential should fall off faster for a dipole; for when you are far from a dipole, the two equal but opposite charges appear so close together as to tend to neutralize each other.

Table 23–2 gives the dipole moments for several molecules. The $+$ and $-$ signs indicate on which atoms these charges lie. The last two entries are a part of many organic molecules and play an important role in molecular biology.

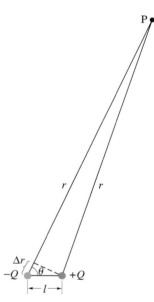

FIGURE 23–16 Electric dipole. Calculation of potential V at point P.

TABLE 23–2 Dipole Moments of Selected Molecules

Molecule	Dipole Moment (C·m)
$H_2^{(+)}O^{(-)}$	6.1×10^{-30}
$H^{(+)}Cl^{(-)}$	3.4×10^{-30}
$N^{(-)}H_3^{(+)}$	5.0×10^{-30}
$>N^{(-)}{-}H^{(+)\ddagger}$	$\approx 3.0 \times 10^{-30}$
$>C^{(+)}{=}O^{(-)\ddagger}$	$\approx 8.0 \times 10^{-30}$

‡These groups often appear on larger molecules; hence the value for the dipole moment will vary somewhat, depending on the rest of the molecule.

FIGURE 23–17 Electron "cloud" around C and O in the C=O group. The C=O group has a dipole moment because two electrons originally on the carbon atom spend some of their time in the vicinity of the oxygen atom.

EXAMPLE 23–10 **The C=O group dipole.** The distance between the carbon (+) and oxygen (−) atoms in the group C=O which occurs in many organic molecules is about 1.2×10^{-10} m and the dipole moment of this group is about 8.0×10^{-30} C·m. Calculate (a) the effective charge Q on the C (carbon) and O (oxygen) atoms, and (b) the potential 9.0×10^{-10} m from the dipole along its axis, with the oxygen being the nearer atom (that is, to the left in Fig. 23–16, so $\theta = 180°$). (c) What would the potential be at this point if only the oxygen (O) were charged?

SOLUTION (a) The dipole moment $p = Ql$. Therefore

$$Q = \frac{p}{l} = \frac{8.0 \times 10^{-30} \text{ C·m}}{1.2 \times 10^{-10} \text{ m}} = 6.7 \times 10^{-20} \text{ C}.$$

Although this charge is less than e, the smallest known charge, it is not a charge that can be isolated, but is the effective charge that results from unequal sharing of the electrons, Fig. 23–17.
(b) Since $\theta = 180°$, we have, using Eq. 23–7:

$$V = \frac{1}{4\pi\epsilon_0} \frac{p \cos\theta}{r^2}$$
$$= \frac{(9.0 \times 10^9 \text{ N·m}^2/\text{C}^2)(8.0 \times 10^{-30} \text{ C·m})(-1.00)}{(9.0 \times 10^{-10} \text{ m})^2} = -0.089 \text{ V}.$$

(c) If we assume that the oxygen has charge $Q = -6.7 \times 10^{-20}$ C, as in part (a) above, and that the carbon is not charged, we use the formula for a single charge:

$$V = \frac{1}{4\pi\epsilon_0} \frac{Q}{r} = \frac{(9.0 \times 10^9 \text{ N·m}^2/\text{C}^2)(-6.7 \times 10^{-20} \text{ C})}{9.0 \times 10^{-10} \text{ m}} = -0.67 \text{ V}.$$

Of course, we expect the potential of a single charge to have greater magnitude than that of a dipole of equal charge at the same distance.

23–7 | *E* Determined from *V*

We can use Eq. 23–3, $V_b - V_a = -\int_a^b \mathbf{E} \cdot d\mathbf{l}$, to determine the difference in potential between two points if the electric field is known in the region between those two points. By inverting Eq. 23–3, we can write the electric field in terms of the potential. Then the electric field can be determined from a knowledge of V. Let us see how to do this.

We write Equation 23–3 in differential form as

$$dV = -\mathbf{E} \cdot d\mathbf{l} = -E_l dl,$$

where dV is the infinitesimal difference in potential between two points a distance dl apart, and E_l is the component of the electric field in the direction of the infinitesimal displacement $d\mathbf{l}$. We can then write

$$E_l = -\frac{dV}{dl}. \tag{23–8}$$

Thus *the component of the electric field in any direction is equal to the negative of the rate of change of the electric potential with distance in that direction.* The quantity dV/dl is called the gradient of V in a particular direction. If the direction is not specified, the term *gradient* refers to that direction in which V changes most rapidly; this would be the direction of **E** at that point, so we can write

$$E = -\frac{dV}{dl}. \qquad [\text{if } d\mathbf{l} \parallel \mathbf{E}]$$

If **E** is written as a function of x, y, and z, and we let l refer to the x, y, and z axes, then Eq. 23–8 becomes

$$E_x = -\frac{\partial V}{\partial x}, \qquad E_y = -\frac{\partial V}{\partial y}, \qquad E_z = -\frac{\partial V}{\partial z}. \qquad \textbf{(23–9)} \qquad \textit{E related to V}$$

Here, $\partial V/\partial x$ is the "partial derivative" of V with respect to x, with y and z held constant.[†]

EXAMPLE 23–11 **E for ring and disk.** Determine the electric field at point P on the axis of (a) a circular ring of charge (Fig. 23–11) and (b) a uniformly charged disk (Fig. 23–12).

SOLUTION (a) From Example 23–8,

$$V = \frac{1}{4\pi\epsilon_0} \frac{Q}{(x^2 + R^2)^{\frac{1}{2}}}.$$

Then

$$E_x = -\frac{\partial V}{\partial x} = -\frac{1}{4\pi\epsilon_0} \frac{Qx}{(x^2 + R^2)^{\frac{3}{2}}}$$

$$E_y = E_z = 0.$$

This is the same result we obtained in Example 21–9, but here we didn't have to break the vector electric field into components and then integrate.

(b) From Example 23–9,

$$V = \frac{Q}{2\pi\epsilon_0 R^2} \left[(x^2 + R^2)^{\frac{1}{2}} - x \right],$$

so

$$E_x = -\frac{\partial V}{\partial x} = \frac{Q}{2\pi\epsilon_0 R^2} \left[1 - \frac{x}{(x^2 + R^2)^{\frac{1}{2}}} \right]$$

$$E_y = E_z = 0.$$

For points very close to the disk, $x \ll R$, this can be approximated by

$$E_x \approx \frac{Q}{2\pi\epsilon_0 R^2} = \frac{\sigma}{2\epsilon_0}$$

where $\sigma = Q/\pi R^2$ is the surface charge density. We also obtained these results in Chapter 21, Example 21–11 and Eq. 21–7.

If we compare this last Example with Examples 21–9 and 21–11, we see that here, as for many charge distributions, it is easier to calculate V first, and then **E** from Eq. 23–9, than to calculate **E** due to each charge from Coulomb's law. This is because V due to many charges is a scalar sum, whereas **E** is a vector sum.

23–8 | Electrostatic Potential Energy; the Electron Volt

Suppose a point charge q is moved between two points in space, a and b, where the electric potential due to other charges is V_a and V_b, respectively. The change in electrostatic potential energy of q in the field of these other charges is, according to Eq. 23–2,

$$\Delta U = U_b - U_a = q(V_b - V_a) = q V_{ba}.$$

[†]Equation 23–9 can be written as a vector equation,

$$\mathbf{E} = -\text{grad } V = -\boldsymbol{\nabla} V = -\left(\mathbf{i}\frac{\partial}{\partial x} + \mathbf{j}\frac{\partial}{\partial y} + \mathbf{k}\frac{\partial}{\partial z} \right) V$$

where the symbol $\boldsymbol{\nabla}$ is called the *del* or *gradient operator*: $\boldsymbol{\nabla} = \mathbf{i}\frac{\partial}{\partial x} + \mathbf{j}\frac{\partial}{\partial y} + \mathbf{k}\frac{\partial}{\partial z}.$

Now suppose we have a system of several point charges. What is the electrostatic potential energy of the system? It is most convenient to choose the electric potential energy to be zero when the charges are very far (ideally infinitely far) apart. A single point charge, Q_1, in isolation, has no potential energy, because if there are no other charges around, no electric force can be exerted on it. If a second point charge Q_2 is brought close to Q_1, the potential due to Q_1 at the position of this second charge is

$$V = \frac{1}{4\pi\epsilon_0}\frac{Q_1}{r_{12}},$$

where r_{12} is the distance between the two. The potential energy of the two charges, relative to $V = 0$ at $r = \infty$, is

Potential energy,
2 point charges

$$U = Q_2 V = \frac{1}{4\pi\epsilon_0}\frac{Q_1 Q_2}{r_{12}}. \qquad \textbf{(23–10)}$$

This represents the work that needs to be done by an external force to bring Q_2 from infinity ($V = 0$) to a distance r_{12} from Q_1. It is also the negative of the work needed to separate them to infinity.

If the system consists of three charges, the total potential energy will be the work needed to bring all three together. Equation 23–10 represents the work needed to bring Q_2 close to Q_1; to bring a third charge Q_3 so that it is a distance r_{13} from Q_1 and r_{23} from Q_2 requires work equal to

$$\frac{1}{4\pi\epsilon_0}\frac{Q_1 Q_3}{r_{13}} + \frac{1}{4\pi\epsilon_0}\frac{Q_2 Q_3}{r_{23}}.$$

So the potential energy of a system of three point charges is

Potential energy,
3 point charges

$$U = \frac{1}{4\pi\epsilon_0}\left(\frac{Q_1 Q_2}{r_{12}} + \frac{Q_1 Q_3}{r_{13}} + \frac{Q_2 Q_3}{r_{23}}\right). \qquad [V = 0 \text{ at } r = \infty]$$

For a system of four charges, the potential energy would contain six such terms, and so on. (Caution must be used when making such sums to avoid double counting of the different pairs.)

The Electron Volt Unit

The joule is a very large unit for dealing with energies of electrons, atoms, or molecules (see Example 23–2), and for this purpose, the unit **electron volt** (eV) is used. One electron volt is defined as the energy acquired by a particle carrying a charge equal to that on an electron ($q = e$) when it moves through a potential difference of 1 V. Since $e = 1.6 \times 10^{-19}$ C, and since the change in potential energy equals qV, 1 eV is equal to $(1.6 \times 10^{-19}$ C$)(1.0$ V$) = 1.6 \times 10^{-19}$ J:

Electron volt
(energy unit)

$$1\,\text{eV} = 1.6 \times 10^{-19}\,\text{J}.$$

An electron that accelerates through a potential difference of 1000 V will lose 1000 eV of potential energy and will thus gain 1000 eV or 1 keV (kilo-electron volt) of kinetic energy. On the other hand, if a particle has a charge equal to twice the charge on the electron ($= 2e = 3.2 \times 10^{-19}$ C), when it moves through a potential difference of 1000 V its energy will change by 2000 eV.

Although the electron volt is handy for *stating* the energies of molecules and elementary particles, it is not a proper SI unit. For calculations it should be converted to joules using the conversion factor given above. In Example 23–2, for example, the electron acquired a kinetic energy of 8.0×10^{-16} J. We normally would quote this energy as 5000 eV ($= 8.0 \times 10^{-16}$ J/1.6×10^{-19} J/eV). But in determining its speed in SI units we had to use the kinetic energy in J.

EXAMPLE 23–12 **Disassembling a hydrogen atom.** Calculate the work needed to "disassemble" a hydrogen atom. Assume that the proton and electron are initially separated by a distance equal to the "average" radius of the hydrogen atom in its ground state, 0.529×10^{-10} m, and that they end up an infinite distance apart from each other.

From Eq. 23–10 we have initially

$$U = \frac{1}{4\pi\epsilon_0}\frac{Q_1 Q_2}{r} = -\frac{1}{4\pi\epsilon_0}\frac{e^2}{r} = \frac{-(8.99 \times 10^9\,\text{N}\cdot\text{m}^2/\text{C}^2)(1.60 \times 10^{-19}\,\text{C})^2}{(0.529 \times 10^{-10}\,\text{m})}$$

$$= -27.2(1.60 \times 10^{-19})\,\text{J} = -27.2\,\text{eV}.$$

This represents the potential energy. The total energy must include also the kinetic energy of the electron moving in an orbit of radius $r = 0.529 \times 10^{-10}$ m. From $F = ma$ for centripetal acceleration, we have $(1/4\pi\epsilon_0)(e^2/r^2) = mv^2/r$. Then

$$K = \tfrac{1}{2}mv^2 = \tfrac{1}{2}\left(\frac{1}{4\pi\epsilon_0}\right)\frac{e^2}{r}$$

which equals $-\tfrac{1}{2}U$ (as calculated above), so $K = +13.6\,\text{eV}$. The total energy initially is $E = K + U = 13.6\,\text{eV} - 27.2\,\text{eV} = -13.6\,\text{eV}$. To separate a stable hydrogen atom into a proton and an electron at rest very far apart ($U = 0$ at $r = \infty$, $K = 0$ because $v = 0$) requires $+13.6\,\text{eV}$. This is, in fact, the measured ionization energy for hydrogen.

* 23–9 ## Cathode Ray Tube: TV and Computer Monitors, Oscilloscope

An important device that makes use of voltage, and that allows us to "visualize" voltages in the sense of displaying graphically how a voltage changes in time, is the **cathode ray tube (CRT)**. A CRT used in this way is an *oscilloscope*—but an even more common use of a CRT is as the picture tube of television sets and computer monitors.

The operation of a CRT depends first of all on the phenomenon of **thermionic emission**, discovered by Thomas Edison (1847–1931) in the course of experiments on developing the electric light bulb. To understand how thermionic emission occurs, consider two small plates (electrodes) inside an evacuated "bulb" or "tube" as shown in Fig. 23–18, to which is applied a potential difference (by a battery, say). The negative electrode is called the **cathode**, the positive one the **anode**. If the negative cathode is heated (usually by an electric current, as in a lightbulb) so that it becomes hot and glowing, it is found that negative charge leaves the cathode and flows to the positive anode. These negative charges are now called electrons, but originally they were called **cathode rays** since they seemed to come from the cathode.

We can understand how electrons might be "boiled off" a hot metal plate if we treat electrons like molecules in a gas. This makes sense if electrons are relatively free to move about inside a metal, which is consistent with metals being good conductors. However, electrons don't readily escape from the metal. If an electron were to escape outside the metal surface, a net positive charge would remain behind, and this would attract the electron back. To escape, an electron needs a certain minimum kinetic energy, just as molecules in a liquid must have a minimum kinetic energy to "evaporate" into the gaseous state. We saw in Chapter 18 that the average kinetic energy ($\overline{K}$) of molecules in a gas is proportional to the absolute temperature T. We can apply this idea, but only very roughly, to free electrons in a metal as if they made up an "electron gas." Of course, some electrons have more kinetic energy than average and others less. At room temperature, very few electrons would have sufficient energy to escape. At high temperature, $\overline{K}$ is larger and many electrons escape—just as molecules evaporate from liquids, which occurs more readily at high temperatures. Thus, significant thermionic emission occurs only at elevated temperatures.

FIGURE 23–18 If the cathode inside the evacuated glass tube is heated to glowing, negatively charged "cathode rays" (electrons) are "boiled off" and flow across to the anode (+) to which they are attracted.

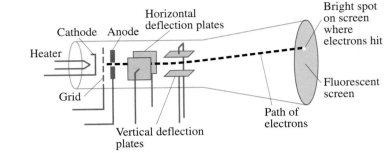

FIGURE 23–19 A cathode-ray tube. Magnetic deflection coils are often used in place of the electric deflection plates. The relative positions of the elements have been exaggerated for clarity.

FIGURE 23–20 Electron beam sweeps across a television screen in a succession of horizontal lines.

FIGURE 23–21 An electrocardiogram (ECG) trace displayed on a CRT.

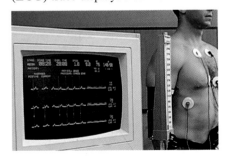

The **cathode-ray tube** (CRT) derives its name from the fact that inside an evacuated glass tube, a beam of cathode rays (electrons) is directed to various parts of a screen to produce a "picture." A simple CRT is diagrammed in Fig. 23–19. Electrons emitted by the heated cathode are accelerated by a high voltage (5,000–50,000 V) applied to the anode. The electrons pass out of this "electron gun" through a small hole in the anode. The inside of the tube face is coated with a fluorescent material that glows when struck by electrons. A tiny bright spot is thus visible where the electron beam strikes the screen. Two horizontal and two vertical plates deflect the beam of electrons when a voltage is applied to them. The electrons are deflected toward whichever plate is positive. By varying the voltage on the deflection plates, the bright spot can be placed at any point on the screen. Today many CRTs use magnetic deflection coils (Chapter 27) instead of electric plates.

In the picture tube or monitor for a computer or television set, the electron beam is made to sweep over the screen in the manner shown in Fig. 23–20. The beam is swept horizontally by the horizontal deflection plates or coils. When the horizontal deflecting field is maximum in one direction, the beam is at one edge of the screen. As the field decreases to zero, the beam moves to the center; and as the field increases to a maximum in the opposite direction, the beam approaches the opposite edge. When the beam reaches this edge, the voltage or current abruptly changes to return the beam to the opposite side of the screen. Simultaneously, the beam is deflected downward slightly by the vertical deflection plates (or coils), and then another horizontal sweep is made. For standard television in the United States, 525 lines constitutes a complete sweep over the entire screen. (High-definition TV provides more than double this number of lines, giving greater picture sharpness.) The complete picture of 525 lines is swept out in $\frac{1}{30}$ s. Actually, a single vertical sweep takes $\frac{1}{60}$ s and involves every other line. The lines in between are then swept out over the next $\frac{1}{60}$ s (called interlacing). We see a picture because the image is retained by the fluorescent screen and by our eyes for about $\frac{1}{20}$ s. The picture we see consists of the varied brightness of the spots on the screen. The brightness at any point is controlled by the grid (a "porous" electrode, such as a wire grid, that allows passage of electrons) which can limit the flow of electrons by means of the voltage applied to it: the more negative this voltage, the more electrons are repelled and the fewer pass through. The voltage on the grid is determined by the video signal (a voltage) sent out by the TV station and received by the TV set. Accompanying this signal are signals that synchronize the grid voltage to the horizontal and vertical sweeps.

An **oscilloscope** is a device for amplifying, measuring, and visually observing an electrical signal (a "signal" is usually a time-varying voltage), especially rapidly changing signals. The signal is displayed on the screen of a CRT. In normal operation, the electron beam is swept horizontally at a uniform rate in time by using a time varying potential difference applied to the horizontal deflection plates. The signal to be displayed is applied, after amplification, to the vertical deflection plates. The visible "trace" on the screen, which could be an ECG (Fig. 23–21), a voltage in an electronic device being repaired, or a signal from an experiment, is thus a plot of the signal voltage (vertically) versus time (horizontally).

Summary

Electric potential is defined as electric potential energy per unit charge. That is, the **electric potential difference** between any two points in space is defined as the difference in potential energy of a test charge q placed at those two points, divided by the charge q:

$$V_{ba} = \frac{U_b - U_a}{q}.$$

Potential difference is measured in volts ($1\ \text{V} = 1\ \text{J/C}$) and is sometimes referred to as **voltage**.

The change in potential energy of a charge q when it moves through a potential difference V_{ba} is

$$\Delta U = q V_{ba}.$$

The potential difference V_{ba} between two points, a and b, is given by the relation

$$V_{ba} = -\int_a^b \mathbf{E} \cdot d\boldsymbol{l}.$$

Thus V_{ba} can be found in any region where $\mathbf{E}$ is known. If the electric field is uniform, the integral is easy: $V_{ba} = -Ed$, where d is the distance (parallel to the field lines) between the two points.

When V is known, the components of $\mathbf{E}$ can be found from the inverse of the above relation, namely

$$E_x = -\frac{\partial V}{\partial x}, \quad E_y = -\frac{\partial V}{\partial y}, \quad E_z = -\frac{\partial V}{\partial z}.$$

An **equipotential line** or **surface** is all at the same potential, and is perpendicular to the electric field at all points.

The electric potential due to a single point charge Q, relative to zero potential at infinity, is given by

$$V = \frac{1}{4\pi\epsilon_0} \frac{Q}{r}.$$

The potential due to any charge distribution can be obtained by summing (or integrating) over the potentials for all the charges.

Questions

1. If two points are at the same potential, does this mean that no work is done in moving a test charge from one point to the other? Does this imply that no force must be exerted?

2. If a negative charge is initially at rest in an electric field, will it move toward a region of higher potential or lower potential? What about a positive charge? How does the potential energy of the charge change in each of these two instances?

3. State clearly the difference (*a*) between electric potential and electric field, (*b*) between electric potential and electric potential energy.

4. An electron is accelerated by a potential difference of, say, 0.10 V. How much greater would its final speed be if it were accelerated with four times as much voltage?

5. Can a particle ever move from a region of low electric potential to one of high potential and yet have its electric potential energy decrease? Explain.

6. If $V = 0$ at a point in space, must $\mathbf{E} = 0$? If $\mathbf{E} = 0$ at some point, must $V = 0$ at that point? Explain. Give examples for each.

7. When dealing with practical devices, we often take the ground (the Earth) to be 0 V. (*a*) If instead we said the ground was -10 V, how would this affect V and E at other points? (*b*) Does the fact that the Earth carries a net charge affect the choice of V at its surface?

8. Can two equipotential lines cross? Explain.

9. Draw in a few equipotential lines in Fig. 21–33b and c.

10. What can you say about the electric field in a region of space that has the same potential throughout?

11. A satellite orbits the Earth along a gravitational equipotential line. What shape must the orbit be?

12. Suppose the charged ring of Example 23–8 was not uniformly charged, so that the density of charge was twice as great near the top as near the bottom. Assuming the total charge Q is unchanged, would this affect the potential at point P on the axis (Fig. 23–11)? Would it affect the value of $\mathbf{E}$ at that point? Is there a discrepancy here? Explain.

13. Consider a metal conductor in the shape of a football. If it carries a total charge Q, where would you expect the charge density σ to be greatest, at the ends or along the flatter sides? Explain. (*Hint:* Near the surface of a conductor, $E = \sigma/\epsilon_0$.)

14. A conducting sphere carries a charge Q and a second identical conducting sphere is neutral. The two are initially insulated, but then they are placed in contact. (*a*) What can you say about the potential of each when they are in contact? (*b*) Will charge flow from one to the other? If so, how much? (*c*) If the spheres do not have the same radius, how are your answers to parts (*a*) and (*b*) altered?

15. At a particular point, the electric field points due north. In what direction(s) will the rate of change of potential be (*a*) greatest, (*b*) least, and (*c*) zero?

16. If you know V at a point in space, can you calculate $\mathbf{E}$ at that point? If you know $\mathbf{E}$ at a point can you calculate V at that point? If not, what else must be known in each case?

17. Equipotential lines are spaced 1.00 V apart. Does the distance between the lines in different regions of space tell you anything about the relative strengths of $\mathbf{E}$ in those regions? If so, what?

18. If the electric field $\mathbf{E}$ is uniform in a region, what can you infer about the electric potential V? If V is uniform in a region of space, what can you infer about $\mathbf{E}$?

19. Is the electric potential energy of two unlike charges positive or negative? What about two like charges? What is the significance of the sign of the potential energy in each case?

Problems

1. (I) How much work is needed to move a -7.0-μC charge from ground to a point whose potential is $+6.00$ V higher?

2. (I) How much work is needed to move a proton from a point with a potential of $+100$ V to a point where it is -50 V?

3. (I) How much kinetic energy will an electron gain (in joules) if it falls through a potential difference of 21,000 V in a TV picture tube?

4. (I) An electron acquires 16.4×10^{-16} J of kinetic energy when it is accelerated by an electric field from plate A to plate B. What is the potential difference between the plates, and which plate is at the higher potential?

5. (II) The work done by an external force to move a -8.10-μC charge from point a to point b is 8.00×10^{-4} J. If the charge was started from rest and had 2.10×10^{-4} J of kinetic energy when it reached point b, what must be the potential difference between a and b?

Section 23–2

6. (I) The electric field between two parallel plates connected to a 45-V battery is 1500 V/m. How far apart are the plates?

7. (I) An electric field of 640 V/m is desired between two parallel plates 11.0 mm apart. How large a voltage should be applied?

8. (I) How strong is the electric field between two parallel plates 5.0 mm apart if the potential difference between them is 110 V?

9. (I) What is the maximum amount of charge that a spherical conductor of radius 5.0 cm can hold in air?

10. (I) What minimum radius must a large conducting sphere of an electrostatic generating machine have if it is to carry 30,000 V without discharge into the air? How much charge will it carry?

11. (II) A uniform electric field $\mathbf{E} = -300$ N/C$\mathbf{i}$ points in the negative x direction as shown in Fig. 23–22. The x and y coordinates of points A, B, and C are given on the diagram (in meters). Determine the differences in potential (a) V_{BA}, (b) V_{CB}, and (c) V_{CA}.

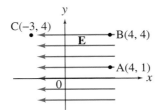

FIGURE 23–22
Problem 11.

12. (II) The electric potential of a very large flat metal plate is V_0. It carries a uniform distribution of charge of surface density σ(C/m^2). Determine V at a distance x from the plate. Consider the point x to be far from the edges and assume x is much smaller than the plate dimensions.

13. (II) The Earth produces an inwardly directed electric field of magnitude 150 V/m near its surface. (a) What is the potential of the Earth's surface relative to $V = 0$ at $r = \infty$? (b) If the potential of the Earth is chosen to be zero, what is the potential at infinity? (Ignore the fact that positive charge in the ionosphere approximately cancels the Earth's net charge; how would this affect your answer?)

14. (II) A 32-cm-diameter conducting sphere is charged to 500 V relative to $V = 0$ at $r = \infty$. (a) What is the surface charge density σ? (b) At what distance will the potential due to the sphere be only 10 V?

15. (II) An insulated spherical conductor of radius r_1 carries a charge Q. A second conducting sphere of radius r_2 and initially uncharged is then connected to the first by a long conducting wire. (a) After the connection, what can you say about the electric potential of each sphere? (b) How much charge is transferred to the second sphere? Assume the connected spheres are far apart compared to their radii. (Why make this assumption?)

16. (II) Determine the difference in potential between two points that are distances R_a and R_b from a very long ($\gg R_a$ or R_b) straight wire carrying a uniform charge per unit length λ.

17. (II) A nonconducting sphere of radius r_0 carries a total charge Q distributed uniformly throughout its volume. Determine the electric potential as a function of the distance r from the center of the sphere for (a) $r > r_0$ and (b) $r < r_0$. Take $V = 0$ at $r = \infty$. (c) Plot V versus r and E versus r.

18. (III) Repeat Problem 17 assuming the charge density ρ_E increases as the square of the distance from the center of the sphere, and $\rho_E = 0$ at the center.

19. (III) A very long conducting cylinder (length L) of radius R_0 $(R_0 \ll L)$ carries a uniform surface charge density σ(C/m^2). The cylinder is at an electric potential V_0. What is the potential, at points far from the end, at a distance r from the center of the cylinder? Determine for (a) $r > R_0$ and (b) $r < R_0$. (c) Is $V = 0$ at $r = \infty$ (assume $L = \infty$)? Explain.

20. (III) A hollow spherical conductor, carrying a net charge $+Q$, has inner radius r_1 and outer radius $r_2 = 2r_1$ (Fig. 23–23). At the center of the sphere is a point charge $+Q/2$. (a) Write the electric field strength E in all three regions as a function of r. Then determine the potential as a function of r, the distance from the center, for (b) $r > r_2$, (c) $r_1 < r < r_2$, and (d) $0 < r < r_1$. (e) Plot both V and E as a function of r from $r = 0$ to $r = 2r_2$.

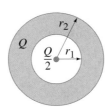

FIGURE 23–23
Problem 20.

Section 23–3

21. (I) (a) What is the electric potential 0.50×10^{-10} m from a proton (charge $+e$)? Let $V = 0$ at $r = \infty$. (b) What is the potential energy of an electron at this point?

22. (I) A charge Q creates an electric potential of $+125$ V at a distance of 15 cm. What is Q?

23. (II) A $+25$-μC charge is placed 6.0 cm from an identical $+25$-μC charge. How much work would be required by an external force to move a $+0.10$-μC test charge from a point midway between them to a point 1.0 cm closer to either of the charges?

24. (II) A 3.0-μC and a -2.0-μC charge are placed 4.0 cm apart. At what points along the line joining them is (a) the electric field zero and (b) the potential zero? Let $V = 0$ at $r = \infty$.

25. (II) How much voltage must be used to accelerate a proton (radius 1.2×10^{-15} m) so that it has sufficient energy to just penetrate a silicon nucleus? A silicon nucleus has a charge of $+14e$ and its radius is about 3.6×10^{-15} m. Assume the potential is that for point charges.

26. (II) Consider point a which is 70 cm north of a -3.8-μC point charge, and point b which is 80 cm west of the charge (Fig. 23–24). Determine (a) $V_{ba} = V_b - V_a$, and (b) $\mathbf{E}_b - \mathbf{E}_a$ (magnitude and direction).

FIGURE 23–24
Problem 26.

27. (II) An electron starts from rest 72.5 cm from a fixed point charge with $Q = -0.125\,\mu$C. How fast will the electron be moving when it is very far away?

28. (II) Two identical $+7.5$-μC point charges are initially spaced 5.5 cm from each other. If they are released at the same instant from rest, how fast will they be moving when they are very far away from each other? Assume they have identical masses of 1.0 mg.

29. (II) Two equal but opposite charges are separated by a distance d, as shown in Fig. 23–25. Determine a formula for $V_{BA} = V_B - V_A$ for points B and A on the line between the charges situated as shown.

FIGURE 23–25   Problem 29.

Section 23–4

30. (II) Three point charges are arranged at the corners of a square of side L as shown in Fig. 23–26. What is the potential at the fourth corner (point A), taking $V = 0$ at a great distance.

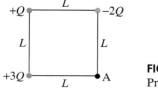

FIGURE 23–26
Problem 30.

31. (II) A flat ring of inner radius R_1 and outer radius R_2, Fig. 23–27, carries a uniform surface charge density σ. Determine the electric potential at points along the axis (the x axis).

FIGURE 23–27
Problem 31.

32. (II) A thin rod of length $2L$ is centered on the x axis as shown in Fig. 23–28. The rod carries a uniformly distributed charge Q. Determine the potential V as a function of y for points along the y axis. Let $V = 0$ at infinity.

FIGURE 23–28   Problems 32, 33, 34, and 48.

33. (III) Determine the potential $V(x)$ for points along the x axis outside the rod of Fig. 23–28 (Problem 32).

34. (III) The charge on the rod of Fig. 23–28 has a nonuniform linear charge distribution, $\lambda = ax$. Determine the potential V for (a) points along the y axis and (b) points along the x axis outside the rod.

35. (III) Suppose the flat circular disk of Fig. 23–12 (Example 23–9) has a nonuniform surface charge density $\sigma = ar^2$, where r is measured from the center of the disk. Find the potential $V(x)$ at points along the x axis, relative to $V = 0$ at $x = \infty$.

Section 23–5

36. (I) Draw a conductor in the shape of a football. This conductor carries a net negative charge, $-Q$. Draw in a dozen or so electric field lines and equipotential lines.

37. (II) Equipotential surfaces are to be drawn 100 V apart near a very large uniformly charged plate carrying a charge density $\sigma = 0.55\,\mu$C/m^2. How far apart (in space) are the equipotential surfaces?

38. (II) A metal sphere of radius $r_0 = 0.30$ m carries a charge $Q = 0.50\,\mu$C. Equipotential surfaces are to be drawn for 100-V intervals outside the sphere. Determine the radius r of (a) the first, (b) the tenth, and (c) the 100th equipotential from the surface.

Section 23–6

39. (I) An electron and a proton are 0.53×10^{-10} m apart. (a) What is their dipole moment if they are at rest? (b) What is the average dipole moment if the electron revolves about the proton in a circular orbit?

40. (II) Calculate the electric potential due to a dipole whose dipole moment is 4.8×10^{-30} C·m at a point 1.1×10^{-9} m away if this point is far from the dipole, and: (a) along the axis of the dipole nearer the positive charge; (b) 45° above the axis but nearer the positive charge; (c) 45° above the axis but nearer the negative charge.

41. (II) (a) In Example 23–10 part b, calculate the electric potential without using the dipole approximation, Eq. 23–7; that is, don't assume $r \gg l$. (b) What is the percent error in this case when the dipole approximation is used?

42. (III) Show that if an electric dipole is placed in a uniform electric field, then a torque is exerted on it equal to $pE \sin \phi$, where ϕ is the angle between the dipole moment vector and the direction of the electric field as shown in Fig. 23–29. What is the net force on the dipole? How are your answers affected if the field is nonuniform? Note that the dipole moment vector **p** is defined so that its magnitude is Ql and its direction is pointing from the negative end to the positive end as shown.

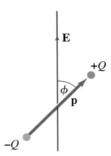

FIGURE 23–29
Problem 42.

43. (III) The dipole moment, considered as a vector, points from the negative to the positive charge. The water molecule, Fig. 23–30, has a dipole moment **p** which can be considered as the vector sum of the two dipole moments $\mathbf{p}_1$ and $\mathbf{p}_2$ as shown. The distance between each H and the O is about 0.96×10^{-10} m; the lines joining the center of the O atom with each H atom make an angle of 104° as shown, and the net dipole moment has been measured to be $p = 6.1 \times 10^{-30}$ C·m. (a) Determine the effective charge q on each H atom. (b) Determine the electric potential, far from the molecule, due to each dipole, $\mathbf{p}_1$ and $\mathbf{p}_2$, and show that

$$ V = \frac{1}{4\pi\epsilon_0} \frac{p \cos\theta}{r^2}, $$

where p is the magnitude of the net dipole moment, $\mathbf{p} = \mathbf{p}_1 + \mathbf{p}_2$, and V is the total potential due to both $\mathbf{p}_1$ and $\mathbf{p}_2$. Take $V = 0$ at $r = \infty$.

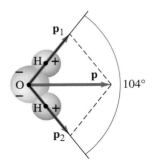

FIGURE 23–30
Problem 43.

Section 23–7

44. (I) What is the potential gradient just outside the surface of a uranium nucleus ($Q = +92e$) whose diameter is about 15×10^{-15} m?

45. (I) Show that the electric field of a single point charge (Eq. 21–4) follows from Eq. 23–5, $V = (1/4\pi\epsilon_0)(Q/r)$.

46. (II) The electric potential in a region of space varies as $V = ay/(b^2 + y^2)$. Determine **E**.

47. (II) In a certain region of space, the electric potential is given by $V = y^2 + 2xy - 4xyz$. Determine the electric field vector, **E**, in this region.

48. (III) Use the results of Problems 32 and 33 to determine the electric field due to the uniformly charged rod of Fig. 23–28 for points (a) along the y axis and (b) along the x axis.

Section 23–8

49. (I) Determine the mutual electrostatic potential energy (in electron volts) of two protons in a uranium (^{235}U) nucleus (a) if they are at the surface, on opposite sides of the nucleus, and (b) if one is at the center and the other is at the surface. The diameter of a ^{235}U nucleus is about 15×10^{-15} m. Ignore the other protons for this calculation.

50. (I) How much work must be done to bring three electrons from a great distance apart to within 1.0×10^{-10} m from one another?

51. (I) What potential difference is needed to give a helium nucleus ($Q = 3.2 \times 10^{-19}$ C) 48 keV of kinetic energy?

52. (I) What is the speed of (a) a 3.5 keV (kinetic energy) electron and (b) a 3.5 keV proton?

53. (II) Write the total electrostatic potential energy, U, for (a) four point charges and (b) five point charges. Draw a diagram defining all quantities.

54. (II) An alpha particle (which is a helium nucleus, $Q = +2e$, $m = 6.64 \times 10^{-27}$ kg) is emitted in a radioactive decay with kinetic energy 5.53 MeV. What is its speed?

55. (II) An electron starting from rest acquires 2.0 keV of kinetic energy in moving from point A to point B. (a) How much kinetic energy would a proton acquire, starting from rest at B and moving to point A? (b) Determine the ratio of their speeds at the end of their respective trajectories.

56. (II) Four equal point charges, Q, are fixed at the corners of a square of side b. (a) What is their total electrostatic potential energy? (b) How much potential energy will a fifth charge, Q, have at the center of the square (relative to $V = 0$ at $r = \infty$)? (c) If constrained to remain in that plane, is the fifth charge in stable or unstable equilibrium? If unstable, what maximum kinetic energy could it acquire? (d) Repeat part (c) for a negative ($-Q$) charge.

57. (II) Repeat Problem 56, parts a and b, assuming that two of the charges, on diagonally opposite corners, are replaced by $-Q$ charges.

58. (II) Determine the total electrostatic potential energy of a conducting sphere of radius r_0 that carries a total charge Q distributed uniformly on its surface.

59. (III) Determine the total electrostatic potential energy of a nonconducting sphere of radius r_0 carrying a total charge Q distributed uniformly throughout its volume.

*60. (I) Use the ideal gas as a model to estimate the rms speed of a free electron in a metal at 300 K, and at 2500 K (the typical temperature of the cathode in a CRT).

*61. (III) In a given CRT, electrons are accelerated horizontally by 15 kV. They then pass through a uniform electric field E for a distance of 2.8 cm which deflects them upward so they reach the top of the screen 22 cm away, 11 cm above the center. Estimate the value of E.

General Problems

62. Sketch the electric field and equipotential lines for two charges of the same sign and magnitude separated by a distance d.

63. By rubbing a nonconducting material, a charge of 10^{-8} C can readily be produced. If this is done to a sphere of radius 10 cm, estimate the potential produced at the surface. Let $V = 0$ at $r = \infty$.

64. If the electrons in a single raindrop, 3 mm in diameter, could be removed from the Earth (without removing the atomic nuclei), by how much would the potential of the Earth increase?

65. A lightning flash transfers 4.0 C of charge and 4.2 MJ of energy to the Earth. (a) Between what potential difference did it travel? (b) How much water could this boil, starting from room temperature?

66. At each corner of a cube of side l there is a point charge Q, Fig. 23–31. (a) What is the potential at the center of the cube ($V = 0$ at $r = \infty$)? (b) What is the potential at each corner due to the other seven charges? (c) What is the total potential energy of this system?

FIGURE 23–31 Problem 66.

67. How much voltage must be used to accelerate a proton so that it has just sufficient energy to touch the surface of an iron nucleus? An iron nucleus has a charge 26 times that of the proton ($= e$) and its radius is about 4.0×10^{-15} m, whereas the proton has a radius of about 1.2×10^{-15} m. Assume the nucleus is spherical and uniformly charged.

*68. Electrons are accelerated by 14 kV in a CRT. The screen is 30 cm wide and is 34 cm from the 2.6-cm-long deflection plates. Over what range must the horizontally deflecting electric field vary to sweep the beam fully across the screen?

69. Suppose a uniform layer of electrons was held near the surface of the Earth by the Earth's gravity. What is the maximum number of electrons that could be held in this manner? Ignore all other electric charges and fields except those of the electrons themselves.

70. Four point charges are located at the corners of a square that is 8.0 cm on a side. The charges, going in rotation around the square, are Q, $2Q$, $-3Q$, and $2Q$, where $Q = 4.8 \mu C$ (Fig. 23–32). What is the total electric potential energy stored in the system, relative to $U = 0$ at infinite separation?

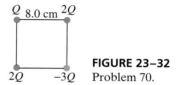

FIGURE 23–32 Problem 70.

71. In a television picture tube, electrons are accelerated by thousands of volts through a vacuum. If a television set were laid on its back, would electrons be able to move upward against the force of gravity? What potential difference, acting over a distance of 3.0 cm, would be needed to balance the downward force of gravity so that an electron would remain stationary? Assume that the electric field is uniform.

72. A proton starting from rest acquires 5.2 keV of kinetic energy in moving from point P to point Q. (a) How much kinetic energy would an electron acquire, starting from rest at Q and moving to point P? (b) Determine the ratio of their speeds at the end of their respective trajectories.

73. In a photocell, ultraviolet (UV) light provides enough energy to some electrons in barium metal to eject them from the surface at high speed. See Fig. 23–33. To measure the maximum energy of the electrons, another plate above the barium surface is kept at a negative enough potential that the emitted electrons are slowed down and stopped, and return to the barium surface. If the plate voltage is -3.02 V (compared to the barium) when the fastest electrons are stopped, what was the speed of these electrons when they were emitted?

FIGURE 23–33 Problem 73.

74. Near the surface of the Earth there is an electric field of about 150 V/m which points downward. Two identical balls with mass $m = 0.540$ kg are dropped from a height of 2.00 m, but one of the balls is positively charged with $q_1 = 550 \mu C$, and the second is negatively charged with $q_2 = -550 \mu C$. Use conservation of energy to determine the difference in the speed of the two balls when they hit the ground. (Neglect air resistance.)

75. Three charges are at the corners of an equilateral triangle (side L) as shown in Fig. 23–34. Determine the potential at the midpoint of each of the sides.

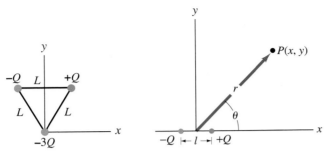

FIGURE 23–34 Problem 75. **FIGURE 23–35** Problem 76.

76. Determine the components of the electric field, E_x and E_y, at any point P in the xy plane due to a dipole, Fig. 23–35, starting with Eq. 23–7. Assume $r = (x^2 + y^2)^{\frac{1}{2}} \gg l$.

77. (*a*) What is the electric potential a distance of 2.5×10^{-15} m away from a proton? Let $V = 0$ at $r = \infty$. (*b*) What is the electric potential energy of a system that consists of two protons 2.5×10^{-15} m apart—as might occur inside a typical nucleus?

78. A thin flat nonconducting disk, with radius R and charge Q, has a hole with a radius $R/2$ in its center. Find the electric potential $V(x)$ at points along the symmetry (x) axis of the disk (a line perpendicular to the disk, passing through its center). Let $V = 0$ at $x = \infty$.

79. A Geiger counter is used to detect charged particles emitted by radioactive nuclei. It consists of a thin, positively charged central wire of radius R_a surrounded by a concentric conducting cylinder of radius R_b with an equal negative charge (Fig. 23–36). The charge per unit length on the inner wire is λ (units C/m). The cylindrical assembly is filled with low-pressure inert gas. Charged particles ionize some of these gas atoms; the resulting free electrons are attracted toward the central wire. If the radial electric field is strong enough, the freed electrons gain enough energy to ionize other atoms, causing an "avalanche" of electrons to strike the central wire, generating an electric "signal." Find the expression for the electric field between the wire and the cylinder, and show that the potential difference between R_a and R_b is

$$V_a - V_b = \Delta V = \frac{\lambda}{2\pi\epsilon_0 \ln(R_b/R_a)}.$$

FIGURE 23–36 Problem 79.

80. A Van de Graaff generator (Fig. 23–37) can develop a very large potential difference, even millions of volts. Electrons are pulled off the belt by the high voltage pointed electrode at A, leaving the belt positively charged. The belt carries the positive charge up inside the spherical shell where it passes to the pointed conductor at B, and races to the outer surface of the conducting sphere. As more charge is brought up, the sphere reaches extremely high voltage. Consider a Van de Graaff generator with a sphere of radius 0.15 m. (*a*) What is the electric potential on the surface of the sphere when electrical breakdown occurs? (Assume that $V = 0$ at $r = \infty$.) (*b*) What is the charge on the sphere for the potential found in part (*a*)?

FIGURE 23–37
Problem 80.

81. Show that if two dipoles with dipole moments p_1 and p_2 are in line with one another (Fig. 23–38), the potential energy of one in the presence of the other (their "interaction energy") is given by

$$U = -\frac{1}{2\pi\epsilon_0} \frac{p_1 p_2}{r^3}$$

where r is the distance between the two dipoles. Assume that r is much greater than the length of either dipole.

FIGURE 23–38
Problem 81.

82. Show that the electrostatic potential energy of two dipoles in a plane as shown in Fig. 23–39 is given by

$$U = \frac{1}{4\pi\epsilon_0} \frac{p_1 p_2}{r^3} \left[\cos(\theta_1 - \theta_2) - 3\cos\theta_1 \cos\theta_2 \right].$$

Assume r is much larger than the length of each dipole. The vector dipole moments, $\mathbf{p}_1$ and $\mathbf{p}_2$, point from the negative charge toward the positive charge of the dipole.

FIGURE 23–39
Problem 82.

83. A nonconducting sphere of radius r_2 contains a concentric spherical cavity of radius r_1. The material between r_1 and r_2 carries a uniform charge density ρ_E (C/m^3). Determine the electric potential V, relative to $V = 0$ at $r = \infty$, as a function of the distance r from the center for (*a*) $r > r_2$, (*b*) $r_1 < r < r_2$, and (*c*) $r < r_1$. Is V continuous at r_1 and r_2?

Capacitors come in a wide range of sizes and shapes, only a few of which are shown here. A capacitor is basically two conductors that do not touch, and which therefore can store charge of opposite sign on its two conductors. Capacitors are used in a wide variety of circuits, as we shall see in this and later chapters.

Capacitance, Dielectrics, Electric Energy Storage

This chapter will complete our study of electrostatics. It deals first of all with an important device, the capacitor, which is used in nearly all electronic circuits. We will also discuss electric energy storage and the effects of an insulator, or dielectric, on electric fields and potential differences.

24–1 Capacitors

A **capacitor**, sometimes called a *condenser*, is a device that can store electric charge, and usually consists of two conducting objects (usually plates or sheets) placed near each other but not touching. Capacitors are widely used in electronic circuits: they store charge for later use, as in a camera flash, and as energy backup in computers if the power fails; capacitors block surges of charge and energy to protect circuits; they form part of the tuner of a radio; very tiny capacitors serve as memory for the "ones" and "zeroes" of the binary code in the random access memory (RAM) of computers; and capacitors serve many other applications, some of which we will discuss.

➡ **PHYSICS APPLIED**

Uses of capacitors

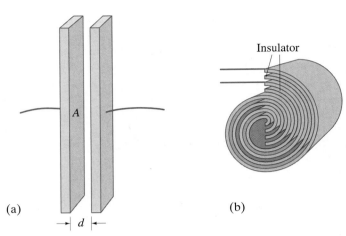

(a)   (b)

FIGURE 24–1 Capacitors: Diagrams of (a) parallel plate, (b) cylindrically shaped (rolled up parallel plate).

A simple capacitor consists of a pair of parallel plates of area A separated by a small distance d (Fig. 24–1a). Often the two plates are rolled into the form of a cylinder with paper or other insulator separating the plates, Fig. 24–1b. On the previous page is a photo of some actual capacitors used for various applications. In a diagram, a capacitor is represented by the symbol

[capacitor symbol]

Another symbol for a capacitor you may encounter is ⊣⊢. A battery, which is a source of voltage, is indicated by the symbol:

[battery symbol]

with unequal arms.

If a voltage is applied to a capacitor, say by connecting the capacitor to a battery as in Fig. 24–2, it quickly becomes charged. One plate acquires a negative charge, the other an equal amount of positive charge. Each battery terminal, connecting wire, and plate of the capacitor are conductors and at the same potential; hence the full battery voltage appears across the capacitor. For a given capacitor, it is found that the amount of charge Q acquired by each plate is proportional to the magnitude of the potential difference V_{ba} between them:

$$Q = CV_{ba}. \qquad (24\text{–}1)$$

The constant of proportionality, C, in the above relation is called the **capacitance** of the capacitor. The unit of capacitance is coulombs per volt and this unit is called a **farad** (F). Most capacitors have capacitance in the range of 1 pF (picofarad = 10^{-12} F) to 1 μF (microfarad = 10^{-6} F). The relation, Eq. 24–1, was first suggested by Volta in the late eighteenth century. The capacitance C does not in general depend on Q or V. Its value depends only on the size, shape, and relative position of the two conductors, and also on the material that separates them.

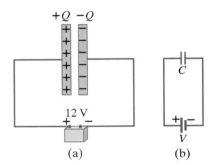

(a)   (b)

FIGURE 24–2 (a) Parallel-plate capacitor connected to a battery. (b) Same circuit shown using symbols.

Capacitance

Unit is farad (1 F = 1 C/V)

24–2 | Determination of Capacitance

The capacitance of a given capacitor can be determined experimentally directly from Eq. 24–1, by measuring the charge Q on either conductor for a given potential difference V_{ba}.

For capacitors whose geometry is simple, we can determine C analytically, and in this Section we assume the conductors are separated by a vacuum or air. To illustrate this, we now determine C for a parallel-plate capacitor, Fig. 24–3. Each plate has area A and the two plates are separated by a distance d. We assume d is small compared to the dimensions of each plate so that the electric field $\mathbf{E}$ is uniform between them and we can ignore fringing (lines of $\mathbf{E}$ not straight) at the edges. We saw earlier (Example 21–12) that the electric field between two closely spaced parallel plates has magnitude $E = \sigma/\epsilon_0$ and its direction is perpendicular to the plates. Since σ is the charge per unit area, $\sigma = Q/A$, then the field between the plates is

$$E = \frac{Q}{\epsilon_0 A}.$$

The relation between electric field and electric potential, as given by Eq. 23–4, is

$$V_{ba} = -\int_a^b \mathbf{E} \cdot d\mathbf{l}.$$

FIGURE 24–3 Parallel-plate capacitor, each of whose plates has area A. Fringing of the field is ignored.

We can take the line integral along a path antiparallel to the field lines, from one plate to the other; then $\theta = 180°$ and $\cos 180° = -1$, so

$$V_{ba} = V_b - V_a = -\int_a^b E\, dl \cos 180° = +\int_a^b E\, dl = \frac{Q}{\epsilon_0 A} \int_a^b dl = \frac{Qd}{\epsilon_0 A}.$$

This relates Q to V_{ba}, and from it we can get the capacitance C in terms of the geometry of the plates:

$$C = \frac{Q}{V_{ba}} = \epsilon_0 \frac{A}{d}. \qquad \text{[parallel-plate capacitor]} \quad \textbf{(24–2)}$$

Parallel plate capacitance formula

This relation makes sense intuitively: a larger area A means that for a given number of charges (electrons), there will be less repulsion between them (they're farther apart), so we expect that more charge can be held on each plate. And a greater separation d means the charge on each plate exerts less attractive force on the other plate, so less charge is drawn from the battery, and the capacitance is less.

Note also from Eq. 24–2 that the value of C does not depend on Q or V_{ba}, so Q is predicted to be proportional to V_{ba} as is found experimentally.

EXAMPLE 24–1 **Capacitor calculations.** (*a*) Calculate the capacitance of a capacitor whose plates are 20 cm $\times$ 3.0 cm and are separated by a 1.0-mm air gap. (*b*) What is the charge on each plate if the capacitor is connected to a 12-V battery? (*c*) What is the electric field between the plates? (*d*) Estimate the area of the plates needed to achieve a capacitance of 1 F, given the same air gap d.

SOLUTION (*a*) The area $A = (20 \times 10^{-2}\,\text{m})(3.0 \times 10^{-2}\,\text{m}) = 6.0 \times 10^{-3}\,\text{m}^2$. The capacitance C is then

$$C = \epsilon_0 \frac{A}{d} = (8.85 \times 10^{-12}\,\text{C}^2/\text{N}\cdot\text{m}^2)\frac{6.0 \times 10^{-3}\,\text{m}^2}{1.0 \times 10^{-3}\,\text{m}} = 53\,\text{pF}.$$

(*b*) The charge on each plate (where we write simply V in place of V_{ba}) is:

$$Q = CV = (53 \times 10^{-12}\,\text{F})(12\,\text{V}) = 6.4 \times 10^{-10}\,\text{C}.$$

(*c*) From Eq. 23–4 for a uniform electric field, the magnitude of E is

$$E = \frac{V}{d} = \frac{12\,\text{V}}{1.0 \times 10^{-3}\,\text{m}} = 1.2 \times 10^4\,\text{V/m}.$$

(*d*) We solve for A in Eq. 24–2 and find that to make a 1-F capacitor with a 1.0-mm air gap requires plates with an area

$$A = \frac{Cd}{\epsilon_0} \approx \frac{(1\,\text{F})(1.0 \times 10^{-3}\,\text{m})}{(9 \times 10^{-12}\,\text{C}^2/\text{N}\cdot\text{m}^2)} \approx 10^8\,\text{m}^2.$$

This is the area of a square $10^4\,\text{m} = 10\,\text{km}$ on a side: the size of a city like San Francisco or Boston!

Ten or fifteen years ago, a capacitance greater than $1 \, \mu F$ was unusual. Today there are capacitors available that are 1 or 2 F, yet they are physically small, a few cm on a side. Such capacitors are used as power backups for low voltage applications, such as computer memory and VCR's where the time and date can be maintained through tiny charge flow. [Capacitors are superior to rechargable batteries for this purpose because they can be recharged more than 10^5 times with no degradation.] How are these high capacitance capacitors made? One type uses activated carbon which has very high porosity, so that the surface area is very large; one tenth of a gram of activated carbon can have a surface area of $100 \, m^2$. Furthermore, the equal and opposite charges exist in an electric "double layer" which is a layer of charge at the interface between the carbon particles and the sulfuric acid surrounding them. The positive charge resides at the edge of the carbon and the negative charge at the edge of the acid with a spacing between them of about $10^{-9} \, m$. Thus, the capacitance of $0.1 \, g$ of activated carbon, whose internal area can be $10^2 \, m^2$, is $C = \epsilon_0 A / d = (8.85 \times 10^{-12} \, C^2/N \cdot m^2)(10^2 \, m^2)/(10^{-9} \, m) \approx 1 \, F$.

One type of computer keyboard operates by capacitance. As shown in Fig. 24–4, each key is connected to the upper plate of a capacitor. The upper plate moves down when the key is pressed, reducing the spacing between the capacitor plates, and increasing the capacitance (Eq. 24–2: smaller d, larger C). The *change* in capacitance becomes an electric signal that is detected by an electronic circuit.

The proportionality, $C \propto A/d$ in Eq. 24–2, is valid also for a parallel-plate capacitor that is rolled up into a spiral cylinder, as in Fig. 24–1b. However, the constant factor, ϵ_0, must be replaced if an insulator such as paper separates the plates, as is usual, and this is discussed in Section 24–5. For a true cylindrical capacitor—consisting of two long coaxial cylinders—the result is somewhat different as the next Example shows.

Key —
Movable plate —
Insulator (flexible) —
Fixed plate —

Computer keys

} Capacitor

FIGURE 24–4 Key on a computer keyboard. Pressing the key reduces the capacitor spacing thus increasing the capacitance which can be detected electronically.

FIGURE 24–5 (a) Cylindrical capacitor consists of two coaxial cylindrical conductors. (b) The electric field lines are shown in cross-sectional view.

R_a

L

R_b

(a)

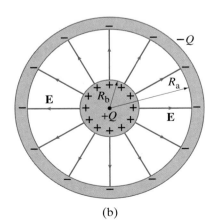

$-Q$

R_a

E

$+R_b$

$+Q$

E

(b)

EXAMPLE 24–2 **Cylindrical capacitor.** A cylindrical capacitor consists of a cylinder (or wire) of radius R_b surrounded by a coaxial cylindrical shell of inner radius R_a, Fig. 24–5a. Both cylinders have length L which we assume is much greater than the separation of the cylinders, $R_a - R_b$, so we can neglect end effects. The capacitor is charged (say, by connecting it to a battery) so that one cylinder has a charge $+Q$ (say, the inner one) and the other one a charge $-Q$. Determine a formula for the capacitance.

SOLUTION To obtain $C = Q/V_{ba}$, we need to determine the potential difference between the cylinders, V_{ba}, in terms of Q. We can use our earlier result (Example 21–10 or 22–5) that the electric field outside a long wire is directed radially outward and has magnitude $E = (1/2\pi\epsilon_0)(\lambda/r)$, where r is the distance from the axis and λ is the charge per unit length, Q/L; then $E = (1/2\pi\epsilon_0)(Q/Lr)$ for points between the cylinders.

To obtain V_{ba} in terms of Q, we use this result in Eq. 23–3, $V_{ba} = -\int_a^b \mathbf{E} \cdot d\mathbf{l}$, and write the line integral from the outer cylinder to the inner one (so $V_{ba} > 0$) along a radial line[†]:

$$V_{ba} = -\int_a^b \mathbf{E} \cdot d\mathbf{l} = -\frac{Q}{2\pi\epsilon_0 L} \int_{R_a}^{R_b} \frac{dr}{r} = -\frac{Q}{2\pi\epsilon_0 L} \ln \frac{R_b}{R_a} = \frac{Q}{2\pi\epsilon_0 L} \ln \frac{R_a}{R_b}.$$

Q and V_{ba} are proportional, and the capacitance C is

$$C = \frac{Q}{V_{ba}} = \frac{2\pi\epsilon_0 L}{\ln(R_a/R_b)}. \qquad \text{[cylindrical capacitor]}$$

Does the dependence on L, R_a, and R_b make sense intuitively? (See discussion immediately after Eq. 24–2.)

[†] Note that **E** points outward in Fig. 24–5b, but $d\mathbf{l}$ points inward for our chosen direction of integration; the angle between **E** and $d\mathbf{l}$ is 180° and cos 180° = −1. Also, $dl = -dr$ because dr increases outward. These two minus signs cancel.

EXAMPLE 24–3 **Spherical capacitor.** A spherical capacitor consists of two thin concentric spherical conducting shells, of radius r_a and r_b as shown in Fig. 24–6. The inner shell carries a uniformly distributed charge Q on its surface, and the outer shell an equal but opposite charge $-Q$. Determine the capacitance of the two shells.

SOLUTION In Example 22–3 we used Gauss's law to show that the electric field outside a uniformly charged conducting sphere is $E = Q/4\pi\epsilon_0 r^2$ as if all the charge were concentrated at the center. Now we use Eq. 23–3, $V_{ba} = -\int_a^b \mathbf{E} \cdot d\mathbf{l}$, and integrate along a radial line to obtain the potential difference between the two conducting shells:

$$V_{ba} = -\int_a^b \mathbf{E} \cdot d\mathbf{l} = -\frac{Q}{4\pi\epsilon_0} \int_{r_a}^{r_b} \frac{1}{r^2} \, dr$$

$$= \frac{Q}{4\pi\epsilon_0}\left(\frac{1}{r_b} - \frac{1}{r_a}\right) = \frac{Q}{4\pi\epsilon_0}\left(\frac{r_a - r_b}{r_a r_b}\right).$$

Finally,

$$C = \frac{Q}{V_{ba}} = 4\pi\epsilon_0\left(\frac{r_a r_b}{r_a - r_b}\right).$$

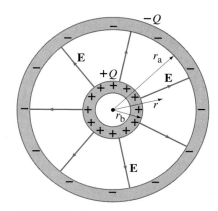

FIGURE 24–6 Cross section of a spherical capacitor. The thin inner shell has radius r_b and the thin outer shell has radius r_a.

A single isolated conductor can also be said to have a capacitance, C. In this case, C can still be defined as the ratio of the charge to absolute potential V on the conductor (relative to $V = 0$ at $r = \infty$), so that the relation

$$Q = CV$$

remains valid. For example, the potential of a single conducting sphere of radius r_b can be obtained from our results in Example 24–3 by letting r_a become infinitely large. As $r_a \to \infty$, then

$$V = \frac{Q}{4\pi\epsilon_0}\left(\frac{1}{r_b} - \frac{1}{r_a}\right) = \frac{1}{4\pi\epsilon_0}\frac{Q}{r_b};$$

so its capacitance is

$$C = \frac{Q}{V} = 4\pi\epsilon_0 r_b.$$

But note that a single conductor alone is not considered a capacitor. In practical cases, a single conductor may be near other conductors or the Earth (which can be thought of as the other "plate" of a capacitor), and these will affect the value of the capacitance.

24–3 Capacitors in Series and Parallel

Capacitors are found in many electric circuits. By electric circuit we mean a closed path of conductors, usually wires connecting capacitors and/or other devices, in which charge can flow and which includes a source of voltage such as a battery. The battery voltage is usually given the symbol V, which means that V represents a potential *difference*. Capacitors can be connected together in various ways. Two common ways are in *series*, or in *parallel*, and we now discuss both.

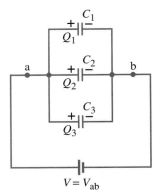

FIGURE 24–7 Capacitors in parallel: $C = C_1 + C_2 + C_3$.

A circuit containing three capacitors connected in **parallel** is shown in Fig. 24–7. They are in "parallel" because when a battery of voltage V is connected to points a and b, this voltage $V = V_{ab}$ exists across each of the capacitors. That is, since the left-hand plates of all the capacitors are connected by conductors, they all reach the same potential V_a when connected to the battery; and the right-hand plates each reach potential V_b. Each capacitor plate acquires a charge given by $Q_1 = C_1 V$, $Q_2 = C_2 V$, and $Q_3 = C_3 V$. The total charge Q that must leave the battery is then

$$Q = Q_1 + Q_2 + Q_3 = C_1 V + C_2 V + C_3 V.$$

Let us try to find a single equivalent capacitor that will hold the same charge Q at the same voltage $V = V_{ab}$. It will have a capacitance C_{eq} given by

$$Q = C_{eq} V.$$

Combining the two previous equations, we have

$$C_{eq} V = C_1 V + C_2 V + C_3 V = (C_1 + C_2 + C_3)V$$

or

Capacitors in parallel

$$C_{eq} = C_1 + C_2 + C_3. \qquad \text{[parallel]} \quad \textbf{(24–3)}$$

The net effect of connecting capacitors in parallel is thus to *increase* the capacitance. This makes sense because we are essentially increasing the area of the plates where charge can accumulate (see, for example, Eq. 24–2).

Capacitors can also be connected in **series**. That is, end to end, as shown in Fig. 24–8. A charge $+Q$ flows from the battery to one plate of C_1, and $-Q$ flows to one plate of C_3. The regions A and B between the capacitors were originally neutral; so the net charge there must still be zero. The $+Q$ on the left plate of C_1 attracts a charge of $-Q$ on the opposite plate. Because region A must have a zero net charge, there is thus $+Q$ on the left plate of C_2. The same considerations apply to the other capacitors, so we see the charge on each capacitor is the same value Q. A single capacitor that could replace these three in series without affecting the circuit (that is, Q and V the same) would have a capacitance C_{eq} where

FIGURE 24–8 Capacitors in series:
$$\frac{1}{C} = \frac{1}{C_1} + \frac{1}{C_2} + \frac{1}{C_3}$$

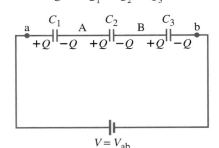

$$Q = C_{eq} V.$$

Now the total voltage V across the three capacitors in series must equal the sum of the voltages across each capacitor:

$$V = V_1 + V_2 + V_3.$$

We also have $Q = C_1 V_1$, $Q = C_2 V_2$, and $Q = C_3 V_3$, so we substitute for V, V_1, V_2, and V_3 into the last equation and get

$$\frac{Q}{C_{eq}} = \frac{Q}{C_1} + \frac{Q}{C_2} + \frac{Q}{C_3} = Q\left(\frac{1}{C_1} + \frac{1}{C_2} + \frac{1}{C_3}\right)$$

or

Capacitors in series

$$\frac{1}{C_{eq}} = \frac{1}{C_1} + \frac{1}{C_2} + \frac{1}{C_3}. \qquad \text{[series]} \quad \textbf{(24–4)}$$

Note that the equivalent capacitance C_{eq} is smaller than the smallest contributing capacitance.

Other connections of capacitors can be analyzed similarly using charge conservation, and often simply in terms of series and parallel connections.

EXAMPLE 24-4 **Equivalent capacitance.** Determine the capacitance of a single capacitor that will have the same effect as the combination shown in Fig. 24–9. Take $C_1 = C_2 = C_3 = C$.

SOLUTION C_2 and C_3 are connected in parallel, so they are equivalent to a single capacitor having capacitance

$$C_{23} = C_2 + C_3 = 2C.$$

C_{23} is in series with C_1, so the equivalent capacitance, C_{eq}, is given by

$$\frac{1}{C_{eq}} = \frac{1}{C_1} + \frac{1}{C_{23}} = \frac{1}{C} + \frac{1}{2C} = \frac{3}{2C}.$$

Hence $C_{eq} = \frac{2}{3}C$, which is the equivalent capacitance of the entire combination.

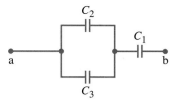

FIGURE 24-9 Example 24-4.

➡ **PROBLEM SOLVING**

Remember to take the reciprocal

EXAMPLE 24-5 **Capacitor combination.** (*a*) Determine the equivalent capacitance of the combination shown in Fig. 24–10a (that is, the capacitance between points a and b). Take $C_1 = 6.0\,\mu\text{F}$, $C_2 = 4.0\,\mu\text{F}$, and $C_3 = 8.0\,\mu\text{F}$. (*b*) If the capacitors are charged by a 12-V battery placed between a and b, determine the charge on each capacitor and the potential difference across each.

SOLUTION (*a*) C_2 and C_3 are connected in parallel, so they are equivalent to a single capacitor of capacitance (Eq. 24–3)

$$C_{23} = C_2 + C_3 = 4.0\,\mu\text{F} + 8.0\,\mu\text{F} = 12.0\,\mu\text{F}.$$

C_{23} is in series with C_1 as shown in Fig. 24–10b. So the equivalent capacitance C of the entire combination is given by (Eq. 24–4)

$$\frac{1}{C_{eq}} = \frac{1}{C_1} + \frac{1}{C_{23}} = \frac{1}{6.0\,\mu\text{F}} + \frac{1}{12.0\,\mu\text{F}} = \frac{3}{12.0\,\mu\text{F}}.$$

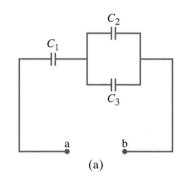

Hence $C = 12.0\,\mu\text{F}/3 = 4.0\,\mu\text{F}$.
(*b*) The total charge that flows from the battery is

$$Q = CV = (4.0 \times 10^{-6}\,\text{F})(12\,\text{V}) = 4.8 \times 10^{-5}\,\text{C}.$$

Both C_1 and C_{23} carry this charge Q. The voltage across C_1 is then

$$V_1 = \frac{Q}{C_1} = \frac{4.8 \times 10^{-5}\,\text{C}}{6.0 \times 10^{-6}\,\text{F}} = 8.0\,\text{V}.$$

The voltage across the combination C_{23} is

$$V_{23} = \frac{Q}{C_{23}} = \frac{4.8 \times 10^{-5}\,\text{C}}{12.0 \times 10^{-6}\,\text{F}} = 4.0\,\text{V}.$$

FIGURE 24-10 Example 24-5.

Since the actual capacitors C_2 and C_3 are in parallel, this represents the voltage across each of them:

$$V_2 = V_3 = 4.0\,\text{V}.$$

The charges on C_2 and C_3 are

$$Q_2 = C_2 V_2 = (4.0 \times 10^{-6}\,\text{F})(4.0\,\text{V}) = 1.6 \times 10^{-5}\,\text{C}$$
$$Q_3 = C_3 V_3 = (8.0 \times 10^{-6}\,\text{F})(4.0\,\text{V}) = 3.2 \times 10^{-5}\,\text{C}.$$

To summarize:

$$V_1 = 8.0\,\text{V} \qquad Q_1 = 48\,\mu\text{C}$$
$$V_2 = 4.0\,\text{V} \qquad Q_2 = 16\,\mu\text{C}$$
$$V_3 = 4.0\,\text{V} \qquad Q_3 = 32\,\mu\text{C}.$$

Note that $Q_2 + Q_3 = Q_1 = Q$, as it should.

(a) Initial configuration.

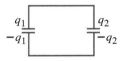

(b) At the instant of reconnection only.

(c) A short time later.

FIGURE 24–11 Example 24–6.

EXAMPLE 24–6 **Capacitors reconnected.** Two capacitors, $C_1 = 2.2\,\mu\text{F}$ and $C_2 = 1.2\,\mu\text{F}$ are connected in parallel to a 24-volt source as shown in Fig. 24–11a. After they are charged they are disconnected from the source and from each other and reconnected directly to each other, with plates of opposite sign connected together (see Fig. 24–11b). Find the charge on each capacitor and the potential across each after equilibrium is established.

SOLUTION First we need to calculate how much charge has been placed on each capacitor after the power source has charged them fully. Using Eq. 24–1 for each capacitor we find:

$$Q_1 = C_1 V = (2.2\,\mu\text{F})(24\,\text{V}) = 52.8\,\mu\text{C},$$
$$Q_2 = C_2 V = (1.2\,\mu\text{F})(24\,\text{V}) = 28.8\,\mu\text{C}.$$

Now we examine Fig. 24–11b. The capacitors are connected in parallel, and the potential difference across each must quickly reach the same value. Thus, the charge cannot remain as shown in Fig. 24–11b, but the charge must rearrange itself so that the upper plates at least have the same sign of charge, with the lower plates having the opposite charge as shown in Fig. 24–11c. Equation 24–1 applies for each:

$$q_1 = C_1 V' \qquad \text{and} \qquad q_2 = C_2 V'$$

where V' is the voltage across each capacitor after the charges have rearranged themselves. We don't know q_1, q_2, or V', so we need a third equation. This is provided by charge conservation. The charges have rearranged themselves between Figs. 24–11b and c. The total charge on the upper plates in those two figures must be the same, so we have

$$q_1 + q_2 = Q_1 - Q_2 = 24.0\,\mu\text{C}.$$

Combining the last three equations we find:

$$V' = (q_1 + q_2)/(C_1 + C_2) = 24.0\,\mu\text{C}/3.4\,\mu\text{F} = 7.06\,\text{V} \approx 7.1\,\text{V}$$
$$q_1 = C_1 V' = (2.2\,\mu\text{F})(7.06\,\text{V}) = 15.5\,\mu\text{F} \approx 16\,\mu\text{F}$$
$$q_2 = C_2 V' = (1.2\,\mu\text{F})(7.06\,\text{V}) = 8.5\,\mu\text{F}$$

where we have kept only two significant figures in our final answers.

24–4 Electric Energy Storage

A charged capacitor stores electrical energy. The energy stored in a capacitor will be equal to the work done to charge it. The net effect of charging a capacitor is to remove charge from one plate and add it to the other plate. This is what a battery does when it is connected to a capacitor. A capacitor does not become charged instantly. It takes time (Section 26–4). Initially, when the capacitor is uncharged, it requires no work to move the first bit of charge over. When some charge is on each plate, it requires work to add more charge of the same sign because of the electric repulsion. The more charge already on a plate, the more work is required to add additional charge. The work needed to add a small amount of charge dq, when a potential difference V is across the plates, is $dW = V\,dq$. Since $V = q/C$ at any moment (Eq. 24–1), where C is the capacitance, the work needed to store a total charge Q is

$$W = \int_0^Q V\,dq = \frac{1}{C}\int_0^Q q\,dq = \frac{1}{2}\frac{Q^2}{C}.$$

Thus we can say that the energy "stored" in a capacitor is

$$U = \frac{1}{2}\frac{Q^2}{C}$$

when the capacitor C carries charges $+Q$ and $-Q$ on its two conductors. Since $Q = CV$, where V is the potential difference across the capacitor, we can also write

Energy stored in a capacitor

$$U = \frac{1}{2}\frac{Q^2}{C} = \frac{1}{2}CV^2 = \frac{1}{2}QV. \tag{24–5}$$

EXAMPLE 24–7 **Energy stored in a capacitor.** A camera flash unit stores energy in a 150 μF capacitor at 200 V. How much electric energy can be stored?

SOLUTION From Eq. 24–5, we have

$$U = \text{energy} = \tfrac{1}{2}CV^2 = \tfrac{1}{2}(150 \times 10^{-6}\,\text{F})(200\,\text{V})^2 = 3.0\,\text{J}.$$

Notice how the units work out: $\text{FV}^2 = \left(\dfrac{\text{C}}{\text{V}}\right)(\text{V}^2) = \text{CV} = \text{C}\left(\dfrac{\text{J}}{\text{C}}\right) = \text{J}.$

If this energy could be released in $\tfrac{1}{1000}$ of a second $(10^{-3}\,\text{s})$ the power output would be equivalent to 3000 W.

It is useful to think of the energy stored in a capacitor as being stored in the electric field between the plates. As an example let us calculate the energy stored in a parallel-plate capacitor in terms of the electric field.

We have seen (Eq. 23–4) that the electric field **E** between two close parallel plates is (approximately) uniform and its magnitude is related to the potential difference by $V = Ed$ where d is the separation. Also, Eq. 24–2 tells us $C = \epsilon_0 A/d$ for a parallel-plate capacitor. Thus

$$U = \tfrac{1}{2}CV^2 = \tfrac{1}{2}\left(\frac{\epsilon_0 A}{d}\right)(E^2 d^2)$$

$$= \tfrac{1}{2}\epsilon_0 E^2\,Ad.$$

The quantity Ad is the volume between the plates in which the electric field E exists. If we divide both sides by the volume, we obtain an expression for the energy per unit volume or **energy density**, u:

$$u = \text{energy density} = \tfrac{1}{2}\epsilon_0 E^2. \tag{24–6}$$

Energy stored per unit volume in electric field

The *electric energy stored per unit volume in any region of space is proportional to the square of the electric field* in that region. We derived Eq. 24–6 for the special case of a parallel-plate capacitor. But it can be shown to be true for any region of space where there is an electric field.

24–5 | Dielectrics

In most capacitors there is an insulating sheet of material (such as paper or plastic) called a **dielectric** between the plates. This serves several purposes. First of all, dielectrics break down (allowing electric charge to flow) less readily than air, so higher voltages can be applied without charge passing across the gap. Furthermore, a dielectric allows the plates to be placed closer together without touching, thus allowing an increased capacitance because d is less in Eq. 24–2. Finally, it is found experimentally that if the dielectric fills the space between the two conductors, it increases the capacitance by a factor K which is known as the **dielectric constant**. Thus

$$C = KC_0, \tag{24–7}$$

Dielectric constant (defined)

where C_0 is the capacitance when the space between the two conductors of the capacitor is a vacuum and C is the capacitance when the space is filled with a material whose dielectric constant is K.

TABLE 24–1
Dielectric constants (at 20° C)

Material	Dielectric constant K	Dielectric strength (V/m)
Vacuum	1.0000	
Air (1 atm)	1.0006	3×10^6
Paraffin	2.2	10×10^6
Polystyrene	2.6	24×10^6
Rubber, neoprene	6.7	12×10^6
Vinyl (plastic)	2–4	50×10^6
Paper	3.7	15×10^6
Quartz	4.3	8×10^6
Oil	4	12×10^6
Glass, Pyrex	5	14×10^6
Porcelain	6–8	5×10^6
Mica	7	150×10^6
Water (liquid)	80	
Strontium titanate	300	8×10^6

Energy density in electric field inside a dielectric

The values of the dielectric constant for various materials are given in Table 24–1. Note that for air (at 1 atm pressure) $K = 1.0006$, which differs so little from 1.0000 that the capacitance for an air-filled capacitor hardly differs from that for a vacuum. Also shown in Table 24–1 is the **dielectric strength**, the maximum electric field before breakdown (charge flow) occurs.

For a parallel-plate capacitor (see Eq. 24–2),

$$C = K\epsilon_0 \frac{A}{d} \qquad \text{[parallel-plate capacitor]} \quad \textbf{(24–8)}$$

when the space between the plates is completely filled with a dielectric whose dielectric constant is K. (The situation when the dielectric only partially fills the space will be discussed shortly in Example 24–9.) The quantity $K\epsilon_0$ appears so often in formulas that we define a new quantity

$$\epsilon = K\epsilon_0 \qquad \textbf{(24–9)}$$

called the **permittivity** of a material. Then the capacitance of a parallel-plate capacitor becomes

$$C = \epsilon \frac{A}{d}.$$

Note that ϵ_0 represents the permittivity of free space (a vacuum).

The energy density stored in an electric field E (Section 24–4) in a dielectric is given by (see Eq. 24–6)

$$u = \tfrac{1}{2}K\epsilon_0 E^2 = \tfrac{1}{2}\epsilon E^2.$$

Two simple experiments illustrate the effect of a dielectric. In the first, Fig. 24–12a, a battery of voltage V_0 is kept connected to a capacitor as a dielectric is inserted between the plates. If the charge on the plates without dielectric is Q_0, then when the dielectric is inserted, it is found experimentally (first by Faraday) that the charge Q on the plates is increased by a factor K,

$$Q = KQ_0. \qquad \text{[voltage constant]}$$

The capacitance has increased to $C = Q/V_0 = KQ_0/V_0 = KC_0$, which is Eq. 24–7. In a second experiment, Fig. 24–12b, a battery V_0 is connected to a capacitor C_0 which then holds a charge $Q_0 = C_0V_0$. The battery is then disconnected, leaving the capacitor isolated with charge Q_0 and still at voltage V_0. Next a dielectric is inserted between the plates of the capacitor. The charge remains Q_0 (there is nowhere for the charge to go) but the voltage is found experimentally to drop by a factor K:

$$V = \frac{V_0}{K}. \qquad \text{[charge constant]}$$

Note that the capacitance changes to $C = Q_0/V = Q_0/(V_0/K) = KQ_0/V_0 = KC_0$, so this experiment too confirms Eq. 24–7.

FIGURE 24–12 Two experiments with a capacitor. Dielectric inserted with (a) voltage held constant, (b) charge held constant.

(a) Voltage constant

(b) Charge constant

The electric field within a dielectric is also altered. When no dielectric is present, the electric field between the plates of a parallel-plate capacitor is given by Eq. 23–4:

$$E_0 = \frac{V_0}{d},$$

where V_0 is the potential difference between the plates and d is their separation. If the capacitor is isolated so that the charge remains fixed on the plates when a dielectric is inserted, filling the space between the plates, the potential difference drops to $V = V_0/K$. So the electric field in the dielectric is now

$$E = E_D = \frac{V}{d} = \frac{V_0}{Kd}$$

or

$$E_D = \frac{E_0}{K}. \qquad \text{[in a dielectric]} \quad \textbf{(24–10)}$$

The electric field within a dielectric is thus also reduced by a factor equal to the dielectric constant. Although the field is reduced in a dielectric (or insulator), it is not reduced all the way to zero as in a conductor.

EXAMPLE 24–8 **Dielectric removal.** A parallel-plate capacitor, filled with a dielectric with $K = 3.4$, is connected to a 100-V battery (Fig. 24–13a). After the capacitor is fully charged, the battery is disconnected. The plates have area $A = 4.0 \, m^2$, and are separated by $d = 4.0 \, mm$. (a) Find the capacitance, the charge on the capacitor, the electric field strength, and the energy stored in the capacitor. (b) The dielectric is carefully removed, without changing the plate separation nor does any charge leave the capacitor (Fig. 24–13b). Find the new values of capacitance, electric field strength, voltage between the plates, and the energy stored on the capacitor.

SOLUTION (a) First we find the capacitance, with dielectric:

$$C = \frac{K\epsilon_0 A}{d} = \frac{3.4(8.85 \times 10^{-12} \, C^2/N \cdot m^2)(4.0 \, m^2)}{4.0 \times 10^{-3} \, m} = 3.0 \times 10^{-8} \, F.$$

The charge Q on the plates is

$$Q = CV = (3.0 \times 10^{-8} \, F)(100 \, V) = 3.0 \times 10^{-6} \, C.$$

The electric field between the plates is

$$E = \frac{V}{d} = \frac{100 \, V}{4.0 \times 10^{-3} \, m} = 25 \, kV/m.$$

Finally, the total energy stored in the capacitor is

$$U = \tfrac{1}{2}CV^2 = \tfrac{1}{2}(3.0 \times 10^{-8} \, F)(100 \, V)^2 = 1.5 \times 10^{-4} \, J.$$

(b) The capacitance without dielectric is

$$C_0 = \frac{C}{K} = \frac{(3.0 \times 10^{-8} \, F)}{(3.4)} = 8.8 \times 10^{-9} \, F.$$

The charge Q doesn't change, so $V = Q/C$ increases by a factor $K = 3.4$ to 340 V. The electric field is

$$E = \frac{V}{d} = \frac{340 \, V}{4.0 \times 10^{-3} \, m} = 85 \, kV/m.$$

The energy stored is

$$U = \tfrac{1}{2}CV^2 = \tfrac{1}{2}(8.8 \times 10^{-9} \, F)(340 \, V)^2 = 5.1 \times 10^{-4} \, \textbf{J}.$$

Where did all this extra energy come from? The energy increased because work had to be done to remove the dielectric. The work required was $W = 5.1 \times 10^{-4} \, J - 1.5 \times 10^{-4} \, J = 3.6 \times 10^{-4} \, J$. (We will see in the next Section that work is required because of the force of attraction between induced charge on the dielectric and the charges on the plates—see Fig. 24–14c.)

FIGURE 24–13 Example 24–8.

Molecular Description of Dielectrics

Molecular description
of dielectrics

Let us now examine, from the molecular point of view, why the capacitance of a capacitor should increase when a dielectric is inserted between the plates. Consider a capacitor whose plates are separated by an air gap. This capacitor has a charge $+Q$ on one plate and $-Q$ on the other (Fig. 24–14a). The capacitor is isolated (not connected to a battery) so charge cannot flow to or from the plates. The potential difference between the plates, V_0, is given by Eq. 24–1: $Q = C_0 V_0$; the subscripts $(_0)$ refer to the situation when only air is between the plates. Now we insert a dielectric between the plates (Fig. 24–14b). The molecules of the dielectric may be *polar*. That is, although the molecules are neutral, they may have a permanent dipole moment (as water does). Because of the electric field between the plates, the molecules will tend to become oriented as shown in Fig. 24–14b; they won't be perfectly aligned because of thermal motion (Chapter 18), but they will usually be at least partially aligned (the stronger the electric field the more alignment).

Even if the molecules are not polar, the electric field between the plates will induce some separation of charge in the molecules (induced dipole moment). Although the electrons do not leave the molecules, they will move slightly within the molecules toward the positive plate. So the situation is still as illustrated in Fig. 24–14b.

The net effect in either case is as if there were a net negative charge on the outer edge of the dielectric facing the positive plate, and a net positive charge on the opposite side, as shown in Fig. 24–14c.

We can visualize that some of the electric field lines do not pass through the dielectric but instead end on charges induced on the surface of the dielectric as shown. Hence the electric field within the dielectric is less than in air. Now imagine a positive test charge within the dielectric. Because the electric field is less, the force a test charge feels is reduced by some factor K (equal, as we shall see, to the dielectric constant). Because the force on our test charge is reduced by a factor K, the work needed to move it from one plate to the other is reduced by a factor K. (We assume

FIGURE 24–14 Molecular view of the effects of a dielectric.

(a)

(b)

(c)

(d)

that the dielectric fills all the space between the plates, even though Fig. 24–14 leaves a space so we can show the field there.) The voltage, which is the work done per unit charge, must therefore also have decreased by the factor K. That is, the voltage between the plates is now

$$V = \frac{V_0}{K}.$$

Now the charge Q on the plates has not changed, because they are isolated. So we have

$$Q = CV,$$

where C is the capacitance when the dielectric is present. When we combine this with the relation, $V = V_0/K$, we obtain

$$C = \frac{Q}{V} = \frac{Q}{V_0/K} = \frac{QK}{V_0} = KC_0,$$

since $C_0 = Q/V_0$. Thus we see why the capacitance is increased by a factor K.

As shown in Fig. 24–14d, the electric field within the dielectric E_D can be considered as the vector sum of the electric field $\mathbf{E}_0$ due to the "free" charges on the conducting plates, and the field $\mathbf{E}_{ind}$ due to the induced charge on the surfaces of the dielectric. Since these two fields are in opposite directions, the net field within the dielectric, $E_0 - E_{ind}$, is less than E_0. The precise relationship is given by Eq. 24–10:

$$E_D = E_0 - E_{ind} = \frac{E_0}{K},$$

so

$$E_{ind} = E_0\left(1 - \frac{1}{K}\right).$$

The electric field between two parallel plates is related to the surface charge density, σ (see Section 22–3 and Example 22–7), by $E = \sigma/\epsilon_0$. Thus

$$E_0 = \frac{\sigma}{\epsilon_0}$$

where $\sigma = Q/A$ is the surface charge density on the conductor; Q is the net charge on the conductor and is often called the **free charge** (since charge is free to move in a conductor). Similarly, we define an equivalent induced surface charge density σ_{ind} on the dielectric; then

$$E_{ind} = \frac{\sigma_{ind}}{\epsilon_0}$$

where E_{ind} is the electric field due to the induced charge $Q_{ind} = \sigma_{ind}A$ on the surface of the dielectric, Fig. 24–14d. Q_{ind} is often called the **bound charge**, since it is on an insulator and is not free to move. Since $E_{ind} = E_0(1 - 1/K)$ as shown above, we now have

$$\sigma_{ind} = \sigma\left(1 - \frac{1}{K}\right) \tag{24–11a}$$

and

$$Q_{ind} = Q\left(1 - \frac{1}{K}\right). \tag{24–11b}$$

Since K is always greater than 1, we see that the charge induced on the dielectric is always less than the free charge on each of the capacitor plates.

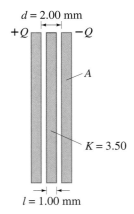

$d = 2.00$ mm

$+Q$ ⟶ $-Q$

A

$K = 3.50$

$l = 1.00$ mm

FIGURE 24–15 Example 24–9.

EXAMPLE 24–9 **Dielectric partially fills capacitor.** A parallel-plate capacitor has plates of area $A = 250 \text{ cm}^2$ and separation $d = 2.00$ mm. The capacitor is charged to a potential difference $V_0 = 150$ V. Then the battery is disconnected (the charge Q on the plates then won't change), and a dielectric sheet ($K = 3.50$) of the same area A but thickness $l = 1.00$ mm is placed between the plates as shown in Fig. 24–15. Determine (a) the initial capacitance of the air-filled capacitor, (b) the charge on each plate before the dielectric is inserted, (c) the charge induced on each face of the dielectric after it is inserted, (d) the electric field in the space between each plate and the dielectric, (e) the electric field in the dielectric, (f) the potential difference between the plates after the dielectric is added, and (g) the capacitance after the dielectric is in place.

SOLUTION (a) Before the dielectric is in place, the capacitance is

$$C_0 = \epsilon_0 \frac{A}{d} = (8.85 \times 10^{-12} \text{ C}^2/\text{N} \cdot \text{m}^2)\left(\frac{2.50 \times 10^{-2} \text{ m}^2}{2.00 \times 10^{-3} \text{ m}}\right) = 111 \text{ pF}.$$

(b) The charge on each plate is

$$Q = C_0 V_0 = (1.11 \times 10^{-10} \text{ F})(150 \text{ V}) = 1.66 \times 10^{-8} \text{ C}.$$

(c) From Eq. 24–11b,

$$Q_{\text{ind}} = Q\left(1 - \frac{1}{K}\right) = (1.66 \times 10^{-8} \text{ C})\left(1 - \frac{1}{3.50}\right) = 1.19 \times 10^{-8} \text{ C}.$$

(d) The electric field in the gaps between the plates and the dielectric (see Fig. 24–14c) is the same as in the absence of the dielectric since the charge on the plates has not been altered. Gauss's law, as applied in Example 22–7, could be used here, which gives $E_0 = \sigma/\epsilon_0$. Or we can note that, in the absence of the dielectric, $E_0 = V_0/d = Q/C_0 d$ (since $V_0 = Q/C_0$) $= Q/\epsilon_0 A$ (since $C_0 = \epsilon_0 A/d$) which is the same result. Thus

$$E_0 = \frac{Q}{\epsilon_0 A} = \frac{1.66 \times 10^{-8} \text{ C}}{(8.85 \times 10^{-12} \text{ C}^2/\text{N} \cdot \text{m}^2)(2.50 \times 10^{-2} \text{ m}^2)} = 7.50 \times 10^4 \text{ V/m}.$$

(e) In the dielectric the electric field is (Eq. 24–10)

$$E_{\text{D}} = \frac{E_0}{K} = \frac{7.50 \times 10^4 \text{ V/m}}{3.50} = 2.14 \times 10^4 \text{ V/m}.$$

(f) To obtain the potential difference in the presence of the dielectric we use Eq. 23–3, and integrate along a straight line parallel to the field lines:

$$V = -\int \mathbf{E} \cdot d\mathbf{l} = E_0(d - l) + E_{\text{D}}l,$$

which can be simplified to

$$V = E_0\left(d - l + \frac{l}{K}\right)$$

$$= (7.50 \times 10^4 \text{ V/m})\left(1.00 \times 10^{-3} \text{ m} + \frac{1.00 \times 10^{-3} \text{ m}}{3.50}\right)$$

$$= 96.4 \text{ V}.$$

(g) In the presence of the dielectric, the capacitance is

$$C = \frac{Q}{V} = \frac{1.66 \times 10^{-8} \text{ C}}{96.4 \text{ V}} = 172 \text{ pF}.$$

Note that if the dielectric filled the space between the plates, the answers to (f) and (g) would be 42.9 V and 387 pF, respectively.

Summary

A **capacitor** is a device used to store charge and consists of two separated conductors. The two conductors generally carry equal and opposite charges, Q, and the ratio of this charge to the potential difference V between the conductors is called the **capacitance**, C; so

$$Q = CV.$$

The capacitance of a parallel-plate capacitor is proportional to the area A of each plate and inversely proportional to their separation d:

$$C = \epsilon_0 \frac{A}{d}.$$

The space between the conductors contains a nonconducting material such as air, paper, or plastic. The latter materials are referred to as **dielectrics**, and the capacitance is proportional to a property of dielectrics called the **dielectric constant**, K (nearly equal to 1 for air). For a parallel-plate capacitor

$$C = K\epsilon_0 \frac{A}{d} = \epsilon \frac{A}{d}$$

where $\epsilon = K\epsilon_0$ is called the **permittivity** of the dielectric material.

When capacitors are connected in **parallel**, the equivalent capacitance is the sum of the individual capacitances:

$$C_{eq} = C_1 + C_2 + \cdots.$$

When capacitors are connected in **series**, the reciprocal of the equivalent capacitance equals the sum of the reciprocals of the individual capacitances:

$$\frac{1}{C_{eq}} = \frac{1}{C_1} + \frac{1}{C_2} + \cdots.$$

A charged capacitor stores an amount of electric energy given by

$$U = \tfrac{1}{2}QV = \tfrac{1}{2}CV^2 = \tfrac{1}{2}\frac{Q^2}{C}.$$

This energy can be thought of as stored in the electric field between the plates. In any electric field $\mathbf{E}$ in free space the **energy density** u (energy per unit volume) is

$$u = \tfrac{1}{2}\epsilon_0 E^2.$$

If a dielectric is present, the energy density is

$$u = \tfrac{1}{2}K\epsilon_0 E^2 = \tfrac{1}{2}\epsilon E^2.$$

Questions

1. Suppose two nearby conductors carry the same negative charge. Can there be a potential difference between them? If so, can the definition of capacitance, $C = Q/V$, be used here?

2. Suppose the separation of plates d in a parallel-plate capacitor is not very small compared to the dimensions of the plates. Would you expect Eq. 24–2 to give an overestimate or underestimate of the true capacitance? Explain.

3. Suppose one of the plates of a parallel-plate capacitor was moved so that the area of overlap was reduced by half, but they are still parallel. How would this affect the capacitance?

4. Explain how the relation for the capacitance of a cylindrical capacitor, Example 24–2, makes sense intuitively. Use arguments such as those just after Eq. 24–2.

5. Describe a simple method of measuring ϵ_0 using a capacitor.

6. When a battery is connected to a capacitor, why do the two plates acquire charges of the same magnitude? Will this be true if the two conductors are different sizes or shapes?

7. A large copper sheet of thickness l is placed between the parallel plates of a capacitor, but does not touch the plates. How will this affect the capacitance?

8. Suppose three identical capacitors are connected to a battery. Will they store more energy if connected in series or in parallel?

9. The parallel plates of an isolated capacitor carry opposite charges, Q. If the separation of the plates is increased, is a force required? Is the potential difference changed? What happens to the work done in the pulling process?

10. How does the energy in a capacitor change if (a) the potential difference is doubled, (b) the charge on each plate is doubled, and (c) the separation of the plates is doubled, as the capacitor remains connected to a battery?

11. For dielectrics consisting of polar molecules, how would you expect the dielectric constant to change with temperature?

12. An isolated charged capacitor has horizontal plates. If a thin dielectric is inserted a short way between the plates, Fig. 24–16, how will it move when it is then released?

Dielectric

FIGURE 24–16
Question 12.

13. Suppose a battery remains connected to the capacitor in Question 12. What then will happen when the dielectric is released?

14. A dielectric is pulled from between the plates of a capacitor which remains connected to a battery. What changes occur to the capacitance, charge on the plates, potential difference, energy stored, and electric field?

15. How does the energy stored in a capacitor change when a dielectric is inserted if (a) the capacitor is isolated so Q doesn't change; (b) the capacitor remains connected to a battery so V doesn't change?

16. We have seen that the capacitance C depends on the size, shape, and position of the two conductors, as well as on the dielectric constant K. What then did we mean when we said that C is a constant in Eq. 24–1?

17. What value might we assign to the dielectric constant for a good conductor? Explain.

18. *Dissolving Power of Water.* The very high dielectric constant of water, $K = 80$ (Table 24–1), has a profound effect on materials in that it allows many of them to be dissolved in water. For example, ordinary table salt, NaCl (sodium chloride), whose crystal structure (Fig. 24–17a) is held together by the attractive forces between the ions Na^+ and Cl^-, is easily dissolved when placed in water. Explain why we would expect that the electric field produced by each ion would be reduced by a factor equal to the dielectric constant; that is, discuss the extension of Eq. 24–10 to the field of a point charge in a dielectric, and thus explain (using this simple model) how salt is dissolved (see Fig. 24–17b).

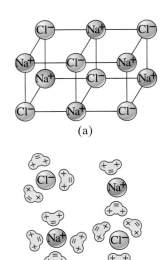

(a)

(b)

FIGURE 24–17 (a) Sodium chloride crystal; (b) sodium chloride dissolving in water. Question 18.

Problems

Section 24–1

1. (I) The two plates of a capacitor hold $+2500\,\mu C$ and $-2500\,\mu C$ of charge, respectively, when the potential difference is 950 V. What is the capacitance?

2. (I) A 12,000-pF capacitor holds $28.0 \times 10^{-8}\,C$ of charge. What is the voltage across the capacitor?

3. (I) The potential difference between two parallel wires in air is 12.0 V. They carry equal and opposite charge of magnitude 75 pC. What is the capacitance of the two wires?

4. (I) How much charge flows from a 12-V battery when it is connected to a 15.6-μF capacitor?

5. (I) The charge on a capacitor increases by $16\,\mu C$ when the voltage across it increases from 28 V to 48 V. What is the capacitance of the capacitor?

6. (II) A capacitor C_1 carries a charge Q_0. It is then connected directly to a second, initially uncharged, capacitor C_2. What charge will each carry now? What will be the potential difference across each?

7. (II) It takes 25 J of energy to move a 0.20-mC charge from one plate of a 16-μF capacitor to the other. How much charge is on each plate?

8. (II) A 2.40-μF capacitor is charged to 880 V and a 4.00-μF capacitor is charged to 560 V. (a) These capacitors are then disconnected from their batteries, and the positive plates are now connected to each other and the negative plates are connected to each other. What will be the potential difference across each capacitor and the charge on each? (b) What is the voltage and charge for each capacitor if plates of opposite sign are connected?

Section 24–2

9. (I) A 0.40-μF capacitor is desired. What area must the plates have if they are to be separated by a 4.0-mm air gap?

10. (I) What is the capacitance per unit length (F/m) of a coaxial cable whose inner conductor has a 1.0-mm diameter and the outer cylindrical sheath has a 5.0-mm diameter? Assume the space between is filled with air.

11. (I) Determine the capacitance of the Earth, assuming it to be a spherical conductor.

12. (II) Use Gauss's law to show that $\mathbf{E} = 0$ inside the inner conductor of a cylindrical capacitor (see Fig. 24–5 and Example 24–2) as well as outside the outer cylinder.

13. (II) Dry air will break down if the electric field exceeds about 3.0×10^6 V/m. What amount of charge can be placed on a capacitor if the area of each plate is 8.5 cm²?

14. (II) An electric field of 2.80×10^5 V/m is desired between two parallel plates each of area 21.0 cm² and separated by 0.250 cm of air. What charge must be on each plate?

15. (II) In the limit of a very small separation of the two cylinders of a cylindrical capacitor $(R_a - R_b \ll R_a$ in Fig. 24–5) show that the relation derived in Example 24–2 reduces to that of a parallel-plate capacitor (Eq. 24–2).

16. (II) Suppose a capacitor carries a charge of $\pm 4.2\,\mu C$, and an electric field of 2.0 kV/mm is desired between the plates which are separated by 4.0 mm of air. What must each plate's area be?

17. (II) How strong is the electric field between the plates of a 0.80-μF air-gap capacitor if they are 2.0 mm apart and each has a charge of $72\,\mu C$?

18. (II) Show that for a spherical capacitor (Example 24–3), if the spacing between shells is very small $(r_a - r_b \ll r_a)$, the formula reduces to that for a parallel-plate capacitor.

19. (II) A large metal sheet of thickness l is placed between, and parallel to, the plates of the parallel-plate capacitor of Fig. 24–3. It does not touch the plates, and extends beyond their edges. (a) What is now the net capacitance in terms of A, d, and l? (b) If $l = \frac{2}{3}d$, by what factor does the capacitance change when the sheet is inserted?

Section 24–3

20. (I) Six 1.8-μF capacitors are connected in parallel. What is the equivalent capacitance? What is their equivalent capacitance if connected in series?

21. (I) The capacitance of a portion of a circuit is to be reduced from 3600 pF to 1600 pF. What capacitance can be added to the circuit to produce this effect without removing existing circuit elements? Must any existing connections be broken to accomplish this?

22. (II) Suppose three parallel-plate capacitors, whose plates have areas A_1, A_2, and A_3 and separations d_1, d_2, and d_3, are connected in parallel. Show, using only Eq. 24–2, that Eq. 24–3 is valid.

23. (II) (a) Determine the equivalent capacitance of the circuit shown in Fig. 24–18. (b) If $C_1 = C_2 = 2C_3 = 14.0\,\mu$F, how much charge is stored on each capacitor when $V = 25.0$ V?

FIGURE 24–18 Problems 23, 24, and 43.

24. (II) In Fig. 24–18, suppose $C_1 = C_2 = C_3 = 16.0\,\mu$F. If the charge on C_2 is $Q_2 = 24.0\,\mu$C, determine the charge on each of the other capacitors, the voltage across each capacitor, and the voltage V_{ab} across the entire combination.

25. (II) A 3.00-μF and a 4.00-μF capacitor are connected in series and this combination is connected in parallel with a 2.00-μF capacitor (see Fig. 24–19). (a) What is the net capacitance? (b) If 26.0 V is applied across the whole network, calculate the voltage across each capacitor.

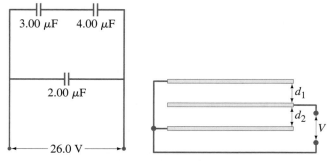

FIGURE 24–19 Problem 25.   **FIGURE 24–20** Problem 26.

26. (II) Three conducting plates, each of area A, are connected as shown in Fig. 24–20. (a) Are the two capacitors thus formed connected in series or in parallel? (b) Determine C as a function of d_1, d_2, and A. Assume $d_1 + d_2$ is much less than the dimensions of the plates. (c) The middle plate can be moved (changing the values of d_1 and d_2), so as to vary the capacitance. What are the minimum and maximum values of the net capacitance?

27. (II) Consider three capacitors, of capacitance 3000 pF, 5000 pF, and 0.010 μF. What are the maximum and minimum capacitances that you can form from these? How do you make the connection in each case?

28. (II) A 0.20-μF and a 0.30-μF capacitor are connected in series to a 9.0-V battery. Calculate (a) the potential difference across each capacitor and (b) the charge on each. (c) Repeat parts (a) and (b) assuming the two capacitors are in parallel.

29. (II) In Fig. 24–21, suppose $C_1 = C_2 = C_3 = C_4 = C$. (a) Determine the equivalent capacitance between points a and b. (b) Determine the charge on each capacitor and the potential difference across each if $V_{ba} = V$.

30. (II) Suppose in Fig. 24–21 that $C_1 = C_2 = C_3 = 16.0\,\mu$F and $C_4 = 36.0\,\mu$F. If the charge on C_2 is $Q_2 = 12.4\,\mu$C, determine the charge on each of the other capacitors, the voltage across each capacitor, and the voltage V_{ab} across the entire combination.

FIGURE 24–21 Problems 29, 30, and 44.   **FIGURE 24–22** Problem 31.

31. (II) The switch S in Fig. 24–22 is connected downward so that capacitor C_2 becomes fully charged by the battery of voltage V_0. If the switch is then connected upward, determine the charge on each capacitor after the switching.

32. (II) (a) Determine the equivalent capacitance between points a and b for the combination of capacitors shown in Fig. 24–23. (b) Determine the charge on each capacitor and the voltage across each if $V_{ab} = V$.

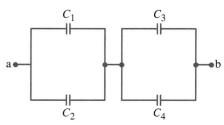

FIGURE 24–23 Problems 32 and 33.

33. (II) Suppose in Problem 32, Fig. 24–23, that $C_1 = C_3 = 8.0\,\mu$F, $C_2 = C_4 = 16\,\mu$F, and $Q_3 = 30\,\mu$C. Determine (a) the charge on each of the other capacitors, (b) the voltage across each capacitor, and (c) the voltage V_{ba} across the combination.

34. (II) Two capacitors connected in parallel produce an equivalent capacitance of 35.0 μF but when connected in series the equivalent capacitance is only 4.0 μF. What is the individual capacitance of each capacitor?

35. (II) In the *capacitance bridge* shown in Fig. 24–24, a voltage V_0 is applied and the variable capacitor C_1 is adjusted until there is zero voltage between points a and b as measured on the voltmeter ($\bullet\!-\!\text{V}\!-\!\bullet$). Determine the unknown capacitance C_x if $C_1 = 8.9\,\mu\text{F}$ and the fixed capacitors have $C_2 = 18.0\,\mu\text{F}$ and $C_3 = 6.0\,\mu\text{F}$.

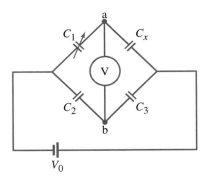

FIGURE 24–24
Problem 35.

36. (II) Two capacitors, $C_1 = 3200\,\text{pF}$ and $C_2 = 2200\,\text{pF}$, are connected in series to a 12.0-V battery. The capacitors are later disconnected from the battery and connected directly to each other, positive plate to positive plate, and negative plate to negative plate. What then will be the charge on each capacitor?

37. (III) Suppose one plate of a parallel-plate capacitor were tilted so it made a small angle θ with the other plate, as shown in Fig. 24–25. Determine a formula for C in terms of A, d, and θ, where A is the area of each plate and θ is small. Assume the plates are square. [*Hint*: Imagine the capacitor as many infinitesimal capacitors in parallel.]

FIGURE 24–25
Problem 37.

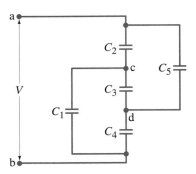

FIGURE 24–26 Problem 38.

38. (III) A voltage V is applied to the capacitor network shown in Fig. 24–26. (*a*) What is the equivalent capacitance? [*Hint*: Assume a potential difference V_{ab} exists across the network as shown; write potential differences for various pathways through the network from a to b in terms of the charges on the capacitors and the capacitances.] (*b*) Determine the equivalent capacitance if $C_2 = C_4 = 8.0\,\mu\text{F}$ and $C_1 = C_3 = C_5 = 6.0\,\mu\text{F}$.

Section 24–4

39. (I) 1200 V is applied to a 2800-pF capacitor. How much electric energy is stored?

40. (I) There is an electric field near the Earth's surface whose intensity is about 150 V/m. How much energy is stored per cubic meter in this field?

41. (I) How much energy is stored by the electric field between two square plates, 8.0 cm on a side, separated by a 1.5 mm air gap? The charges on the plates are equal and opposite and of magnitude $420\,\mu\text{C}$.

42. (II) A parallel-plate capacitor has fixed charges $+Q$ and $-Q$. The separation of the plates is then doubled. (*a*) By what factor does the energy stored in the electric field change? (*b*) How much work must be done if the separation of the plates is doubled from d to $2d$? The area of each plate is A.

43. (II) In Fig. 24–18, let $V = 10.0\,\text{V}$ and $C_1 = C_2 = C_3 = 2200\,\text{pF}$. How much energy is stored in the capacitor network?

44. (II) What is the total energy stored in the network of capacitors in Fig. 24–21, Problem 29?

45. (II) How much energy must a 12-V battery expend to fully charge a $0.15\text{-}\mu\text{F}$ and a $0.20\text{-}\mu\text{F}$ capacitor when they are placed (*a*) in parallel, (*b*) in series? (*c*) How much charge flowed from the battery in each case?

46. (II) (*a*) Suppose the outer radius R_a of a cylindrical capacitor was doubled, but the charge was kept constant. By what factor would the stored energy change? Where would the energy come from? (*b*) Repeat, assuming the voltage remains constant.

47. (II) A $3.0\text{-}\mu\text{F}$ capacitor is charged by a 12-V battery. It is disconnected from the battery and then connected to an uncharged $5.0\text{-}\mu\text{F}$ capacitor. Determine the total stored energy (*a*) before the two capacitors are connected and (*b*) after they are connected. (*c*) What is the change in energy? (*d*) Is energy conserved? Explain why.

48. (II) How much work would be required to remove the metal sheet from between the plates of the capacitor in Problem 19, assuming: (*a*) the battery remains connected so the voltage remains constant; (*b*) the battery is disconnected so the charge remains constant?

49. (II) (*a*) Show that each plate of a parallel-plate capacitor exerts a force

$$F = \frac{1}{2}\frac{Q^2}{\epsilon_0 A}$$

on the other, by calculating dW/dx where dW is the work needed to increase the separation by dx. (*b*) Why does using $F = QE$ give the wrong answer?

50. (II) Show that the electrostatic energy stored in the electric field outside an isolated spherical conductor of radius R carrying a net charge Q is

$$U = \frac{1}{8\pi\epsilon_0} \frac{Q^2}{R}.$$

Do this in three ways: (*a*) Use Eq. 24–6 for the energy density in an electric field [*Hint*: Consider spherical shells of thickness dr]; (*b*) use Eq. 24–5 together with the capacitance of an isolated sphere (Section 24–2); (*c*) by calculating the work needed to bring all the charge Q up from infinity in infinitesimal bits dq.

Section 24–5

51. (I) What is the capacitance of a pair of circular plates with a radius of 5.0 cm separated by 3.2 mm of mica?

52. (I) What is the capacitance of two square parallel plates 5.5 cm on a side that are separated by 1.8 mm of paraffin?

53. (II) A 3500-pF air-gap capacitor is connected to a 22-V battery. If a piece of mica is placed between the plates, how much charge will flow from the battery?

54. (II) Suppose the capacitor in Example 24–8 remains connected to the battery as the dielectric is removed. What will be the work required to remove the dielectric in this case?

55. (II) How much energy would be stored in the capacitor of Problem 41 if a mica dielectric is placed between the plates? Assume the mica is 1.5 mm thick (and therefore fills the space between the plates).

FIGURE 24–27 Problem 56.

56. (II) Two different dielectrics each fill half the space between the plates of a parallel-plate capacitor as shown in Fig. 24–27. Determine a formula for the capacitance in terms of K_1, K_2, the area A of the plates, and the separation d. [*Hint*: Can you consider this capacitor as two capacitors in series or in parallel?]

57. (II) Two different dielectrics fill the space between the plates of a parallel-plate capacitor as shown in Fig. 24–28. Determine a formula for the capacitance in terms of K_1, K_2, the area A, of the plates, and the separation $d_1 = d_2 = d/2$. [*Hint*: Can you consider this capacitor as two capacitors in series or in parallel?]

FIGURE 24–28 Problems 57 and 58.

58. (II) Repeat Problem 57 (Fig. 24–28) but assume the separation $d_1 \neq d_2$.

59. (II) Two identical capacitors are connected in parallel and each acquires a charge Q_0 when connected to a source of voltage V_0. The voltage source is disconnected and then a dielectric ($K = 4.0$) is inserted to fill the space between the plates of one of the capacitors. Determine (*a*) the charge now on each capacitor, and (*b*) the voltage now across each capacitor.

60. (III) A slab of width d and dielectric constant K is inserted a distance x between the square parallel plates (of side l) of a capacitor as shown in Fig. 24–29. Determine, as a function of x, (*a*) the capacitance, (*b*) the energy stored if the potential difference is V_0, and (*c*) the magnitude and direction of the force exerted on the slab (assume V_0 is constant).

FIGURE 24–29 Problem 60.

* Section 24–6

*61. (II) Repeat Example 24–9 assuming the battery remains connected when the dielectric is inserted. Also, what is the free charge on the plates after the dielectric is added (let this be part (*h*) of the problem)?

*62. (II) Show that the capacitor in Example 24–9 with dielectric inserted can be considered as equivalent to three capacitors in series, and using this assumption show that the same value for the capacitance is obtained as was obtained in part (*g*) of the Example.

*63. (II) In Example 24–9 what percent of the stored energy is stored in the electric field in the dielectric?

*64. (II) Using Example 24–9 as a model, derive a formula for the capacitance of a parallel-plate capacitor whose plates have area A, separation d, with a dielectric of thickness $l(l < d)$ with dielectric constant K placed between the plates.

*65. (III) The capacitor shown in Fig. 24–30 is connected to a 90.0-V battery. Calculate (and sketch) the electric field everywhere between the capacitor plates. Find both the free charge on the capacitor plate and the induced charge on the faces of the glass dielectric plate.

FIGURE 24–30 Problem 65.

▢ General Problems

66. An electric circuit was accidentally constructed using a 5.0-μF capacitor instead of the required 16-μF value. What can a technician add to correct this circuit?

67. A *cardiac defibrillator* is used to shock a heart that is beating erratically. A capacitor in this device is charged to 6000 V and stores 200 J of energy. What is its capacitance?

68. A homemade capacitor is assembled by placing two 9-inch pie pans 10 cm apart and connecting them to the opposite terminals of a 9-V battery. Estimate (*a*) the capacitance, (*b*) the charge on each plate, (*c*) the electric field halfway between the plates, (*d*) the work done by the battery to charge the plates. (*e*) Which of the above values change if a dielectric is inserted?

69. How does the energy stored in a capacitor change if (*a*) the potential difference is doubled, (*b*) the charge on each plate is doubled, and (*c*) the separation of the plates is doubled, as the capacitor remains connected to a battery?

70. A huge 7.0-F capacitor has enough stored energy to heat 2.5 kg of water from 20°C to 95°C. What is the potential difference across the plates?

71. An uncharged capacitor is connected to a 24.0-V battery until it is fully charged, after which it is disconnected from the battery. A slab of paraffin is then inserted between the plates. What will now be the voltage between the plates?

72. It takes 18.5 J of energy to move a 13.0-mC charge from one plate of a 12.0-μF capacitor to the other. How much charge is on each plate?

73. A coaxial cable, Fig. 24–31, consists of an inner cylindrical conducting wire of radius R_b surrounded by a dielectric insulator. Surrounding the dielectric insulator is an outer conducting sheath of radius R_a, which is usually "grounded." (*a*) Determine an expression for the capacitance per unit length of a cable whose insulator has dielectric constant K. (*b*) For a given cable, $R_b = 3.5\,\text{mm}$ and $R_a = 9.0\,\text{mm}$. The dielectric constant of the dielectric insulator is $K = 2.6$. Suppose that there is a potential of 1.0 kV between the inner conducting wire and the outer conducting sheath. Find the capacitance per meter of the cable.

Outer conducting sheath, radius R_a
Dielectric
Outer covering
Radius R_b inner conductor

FIGURE 24–31 Problem 73.

74. The electric field between the plates of a paper-separated ($K = 3.75$) capacitor is $9.21 \times 10^4\,\text{V/m}$. The plates are 1.95 mm apart and the charge on each plate is 0.475 μC. Determine the capacitance of this capacitor and the area of each plate.

75. A parallel-plate capacitor is isolated with a charge $\pm Q$ on each plate. If the separation of the plates is halved and a dielectric (constant K) is inserted in place of air, by what factor does the energy storage change? To what do you attribute the change in stored potential energy? How does the new value of the electric field between the plates compare with the original value?

76. A 3.5-μF capacitor is charged by a 12.4-V battery and then is disconnected from the battery. When this capacitor (C_1) is then connected to a second (initially uncharged) capacitor, C_2, the voltage on the first drops to 5.2 V. What is the value of C_2?

77. The power supply for a pulsed nitrogen laser has a 0.060-μF capacitor with a maximum voltage rating of 25 kilovolts. (*a*) Estimate how much energy could be stored in this capacitor. (*b*) If 10 percent of this stored electrical energy is converted to light energy in a pulse that is 10 microseconds long, what is the power of the laser pulse?

78. The variable capacitance of an old radio tuner consists of four plates connected together placed alternately between four other plates, also connected together (Fig. 24–32). Each plate is separated from its neighbor by 1.0 mm of air. One set of plates can move so that the area of overlap of each plate varies from 2.0 cm² to 9.0 cm². (*a*) Are these seven capacitors connected in series or in parallel? (*b*) Determine the range of capacitance values.

FIGURE 24–32
Problems 78 and 79.

79. A high-voltage supply can be constructed from a variable capacitor with interleaving plates which can be rotated as in Fig. 24–32. A version of this type of capacitor with more plates has a capacitance which can be varied from 10 pF to 1 pF. (*a*) Initially, this capacitor is charged by a 10,000-volt power supply when the capacitance is 10 pF. It is then disconnected from the power supply and the capacitance reduced to 1.0 pF by rotating the plates. What is the voltage across the capacitor now? (*b*) What is a major disadvantage of this as a high-voltage power supply?

80. A 150-pF capacitor is connected in series with an unknown capacitor, and as a series combination they are connected to a 25.0-V battery. If the 150-pF capacitor stores 125 pC of charge on its plates, what is the unknown capacitance?

81. A circuit contains a single 330-pF capacitor hooked across a battery. It is desired to store three times as much energy by adding a single capacitor to this one. How would you hook it up and what would its value be?

82. What are the values of effective capacitance which can be obtained by connecting four identical capacitors, each having a capacitance C?

83. In the circuit shown in Fig. 24–33, $C_1 = 1.0\,\mu\text{F}$, $C_2 = 2.0\,\mu\text{F}$, $C_3 = 3.0\,\mu\text{F}$, and a voltage $V_{ab} = 24\,\text{V}$ is applied across points a and b. After C_1 is fully charged the switch is thrown to the right. What is the final charge and potential difference on each capacitor?

FIGURE 24–33 Problem 83.

84. The long cylindrical capacitor shown in Fig. 24–34 consists of 4 concentric cylinders, with respective radii R_a, R_b, R_c, and R_d. The cylinders b and c are joined by metal strips, as indicated. Determine the capacitance per unit length of this arrangement.

FIGURE 24–34 Problem 84.

85. A parallel plate capacitor has plate area A, plate separation x, and has a charge Q stored on its plates (Fig. 24–35). Find the amount of work required to double the plate separation to $2x$, assuming the charge remains constant at Q. Show that your answer is consistent with the change in energy stored by the capacitor.

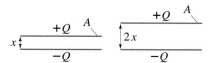

FIGURE 24–35 Problem 85.

86. Consider the use of capacitors as memory cells. A charged capacitor would represent a one and an uncharged capacitor a zero. Suppose these capacitors were fabricated on a silicon chip and had a capacitance of 30 femto-farads each $(1\,\text{fF} = 10^{-15}\,\text{F}.)$ The dielectric filling the space between the parallel plates has dielectric constant $K = 1.00 \times 10^4$ and a dielectric strength of $5.0 \times 10^7\,\text{V/m}$. (a) If the operating voltage is 3.0 volts, how many electrons would be stored on one of these capacitors when charged? (b) If no safety factor is allowed, how thin a dielectric layer could we use for operation at 3.0 volts? (c) Using the layer thickness from your answer to part (b), what would be the area of the capacitor plates?

87. A parallel plate capacitor with plate area $A = 2.5\,\text{m}^2$ and plate separation $d = 3.0\,\text{mm}$ is connected to a 45 V battery (Fig. 24–36). (a) Determine the charge on the capacitor, the electric field, the capacitance, and the energy stored in the capacitor. (b) With the capacitor still connected to the battery, a slab of plastic with dielectric strength $K = 3.6$ is placed between the plates of the capacitor, so that the gap is completely filled with the dielectric. What are the new values of charge, electric field, capacitance, and the energy U stored in the capacitor?

FIGURE 24–36 Problem 87.

88. A smooth conducting sphere of radius r_0 carries a charge Q. Half of the energy stored in its electric field is contained in a volume of what radius?

89. Paper has a dielectric constant of $K = 3.7$ and a dielectric strength of $15 \times 10^6\,\text{V/m}$. Suppose that a typical sheet of paper has a thickness of 0.030 mm. You make a "home made" capacitor by placing a sheet of 8.5×11 inch paper between two aluminum foil sheets (Fig. 24–37). The thickness of the aluminum foil is 0.040 mm. (a) What is the capacitance C_0 of your device? (b) About how much charge could you store on your capacitor before it would break down? (c) Show in a sketch how you could overlay sheets of paper and aluminum for a parallel combination. If you made 100 such capacitors, and connected the ends of the sheets in parallel, so that you have a single large capacitor of capacitance $100\,C_0$, how thick would your new large capacitor be? (d) What is the maximum voltage you can apply to this $100\,C_0$ capacitor without breakdown?

FIGURE 24–37 Problem 89.

90. In lightning storms, the potential difference between the Earth and the bottom of the thunderclouds can be as high as 35,000,000 V. The bottoms of the thunderclouds are typically 1500 m above the Earth, and can have an area of 110 km². For the purpose of this problem, model the Earth-cloud system as a huge capacitor and calculate (a) the capacitance of the Earth-cloud system, (b) the charge stored in the "capacitor," and (c) the energy stored in the "capacitor."

The glow of the thin wire filament of a light bulb is caused by the electric current passing through it. Electric energy is transformed to thermal energy (via collisions between moving electrons and atoms of the wire), which causes the wire's temperature to become so high that it glows. Electric current and electric power in electric circuits are of basic importance in everyday life. We examine both dc and ac in this chapter, and include the microscopic analysis of electric current, as well as a look at electric hazards.

Electric Currents and Resistance

n the previous four chapters we have been studying static electricity: electric charges at rest. In this chapter we begin our study of charges in motion, and we call a flow of charge an electric current.

In everyday life we are familiar with electric currents flowing in wires and other conductors. Indeed, most practical electrical devices depend on electric current: current flowing through a light bulb, current in the heating element of a stove or electric heater, and of course currents in electronic devices. Electric currents can exist in conductors such as wires and also in other devices, such as the CRT of a television or computer monitor whose charged electrons flow through space (Section 23–9).

In electrostatic situations, we saw (Sections 21–9 and 22–3) that the electric field must be zero inside a conductor (if it weren't, the charges would move). But when charges are moving in a conductor, there can be an electric field in the conductor. Indeed, an electric field is needed to get charges into motion, and to keep them in motion in any normal conductor.

We first look at electric current from a macroscopic point of view: that is, current as measured in a laboratory. Later in the chapter we look at currents from a microscopic (theoretical) point of view as a flow of electrons in a wire.

We can control the flow of charge using electric fields and electric potential (voltage), concepts we have just been discussing. In order to have a current in a wire, a potential difference is needed, which can be provided by a battery.

Until the year 1800, the technical development of electricity consisted mainly of producing a static charge by friction. It all changed in 1800 when Alessandro Volta (1745–1827; Fig. 25–1) invented the electric battery, and with it produced the first steady flow of electric charge—that is, a steady electric current.

FIGURE 25–1 Alessandro Volta. In this portrait, Volta exhibits his battery to Napoleon in 1801.

The events that led to the discovery of the battery are interesting. For not only was this an important discovery, but it also gave rise to a famous scientific debate.

In the 1780s, Luigi Galvani (1737–1798), professor at the University of Bologna, carried out a series of experiments on the contraction of a frog's leg muscle through electricity produced by static electricity. Galvani found that contraction of the muscle could also be produced when dissimilar metals were inserted into the frog. Galvani believed that the source of the electric charge was in the frog muscle or nerve itself, and that the metal merely transmitted the charge to the proper points. When he published his work in 1791, he termed this charge "animal electricity." Many wondered, including Galvani himself, if he had discovered the long-sought "life-force."

Volta, at the University of Pavia 200 km away, was skeptical of Galvani's results, and came to believe that the source of the electricity was not in the animal itself, but rather in the *contact between the dissimilar metals*. Volta realized that a moist conductor, such as a frog muscle or moisture at the contact point of two dissimilar metals, was necessary in the circuit if it was to be effective. He also saw that the contracting frog muscle was a sensitive instrument for detecting electric "tension" or "electromotive force" (his words for what we now call potential), in fact more sensitive than the best available electroscopes that he and others had developed.[†]

Volta's research found that certain combinations of metals produced a greater effect than others, and, using his measurements, he listed them in order of effectiveness. (This "electrochemical series" is still used by chemists today.) He also found that carbon could be used in place of one of the metals.

Volta then conceived his greatest contribution to science. Between a disc of zinc and one of silver, he placed a piece of cloth or paper soaked in salt solution or dilute acid and piled a "battery" of such couplings, one on top of another, as shown in Fig. 25–2. This "pile" or "battery" produced a much increased potential difference. Indeed, when strips of metal connected to the two ends of the pile were brought close, a spark was produced. Volta had designed and built the first electric battery.

A battery produces electricity by transforming chemical energy into electrical energy. Today a great variety of electric cells and batteries are available, from flashlight batteries to the storage battery of a car. The simplest batteries contain two plates or rods made of dissimilar metals (one can be carbon) called **electrodes**. The electrodes are immersed in a solution, such as a dilute acid, called the **electrolyte**. Such a device is properly called an **electric cell**, and several cells connected together is a **battery**, although today even a single cell is called a **battery**. The chemical reactions involved in most electric cells are quite complicated. Here we describe how one very simple cell works, emphasizing the physical aspects.

The cell shown in Fig. 25–3 uses dilute sulfuric acid as the electrolyte. One of the electrodes is made of carbon, the other of zinc. That part of each electrode outside the solution is called the **terminal**, and connections to wires and circuits are made here. The acid reacts with the zinc electrode and tends to dissolve it. Each zinc atom leaves two electrons behind and enters the solution as a positive ion. The zinc electrode thus acquires a negative charge. As the electrolyte becomes positively charged, electrons are pulled off the carbon electrode. Thus the carbon electrode becomes positively charged. Because there is an opposite charge on the two electrodes, there is a potential difference between the two terminals. In a cell

FIGURE 25–2 A voltaic battery: Taken from Volta's original publication.

Cells and batteries

FIGURE 25–3 Simple electric cell.

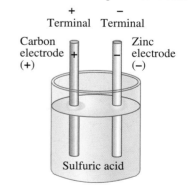

[†] Volta's most sensitive electroscope measured about 40 V per degree (angle of leaf separation). Nonetheless, he was able to estimate the potential differences produced by dissimilar metals in contact: for a silver-zinc contact he got about 0.7 V, remarkably close to today's value of 0.78 V.

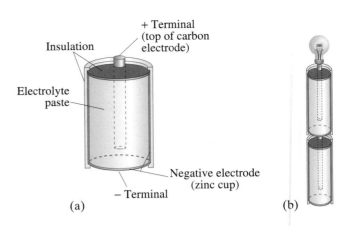

FIGURE 25–4 (a) Diagram of an ordinary dry cell (like a D-cell or AA). The cylindrical zinc cup is covered on the sides; its flat bottom is the negative terminal. (b) Two dry cells (AA type) connected in series. Note that the positive terminal of one cell pushes against the negative terminal of the other.

whose terminals are not connected, only a small amount of the zinc is dissolved, for as the zinc electrode becomes increasingly negative, any new positive zinc ions produced are attracted back to the electrode. Thus, a particular potential difference or voltage is maintained between the two terminals. If charge is allowed to flow between the terminals, say, through a wire (or a lightbulb), then more zinc can be dissolved. After a time, one or the other electrode is used up and the cell becomes "dead."

The voltage that exists between the terminals of a battery depends on what the electrodes are made of and their relative ability to be dissolved or give up electrons.

When two or more cells are connected so that the positive terminal of one is connected to the negative terminal of the next, they are said to be connected in *series* and their voltages add up. Thus, the voltage between the ends of two 1.5-V flashlight batteries connected in series is 3.0 V, whereas the six 2-V cells of an automobile storage battery give 12 V. Figure 25–4 shows (a) a diagram of a common "dry cell" or "flashlight battery" used in portable radios, Walkmans, flashlights, etc., and (b) shows two of them in series.

25–2 | Electric Current

Electric circuit

Battery

When a continuous conducting path is connected between the terminals of a battery, we have an electric **circuit**, Fig. 25–5a. On any diagram of a circuit, as in Fig. 25–5b, we represent a battery by the symbol

$$\dashv\!\vdash_{\;-}^{\;+}\;.$$ [battery symbol]

The device powered by the battery could be a lightbulb (which is just a fine wire inside an evacuated glass bulb), a heater, a radio, or whatever. When such a circuit is formed, charge can flow through the wires of the circuit, from one terminal of the battery to the other. Any flow of charge such as this is called an **electric current**. Electric current can flow whenever there is a potential difference between the ends of a conductor—or, more simply, if there are opposite charges at the two ends of the conductor, or even in empty space.

FIGURE 25–5 (a) A simple electric circuit. (b) Schematic drawing of the same circuit.

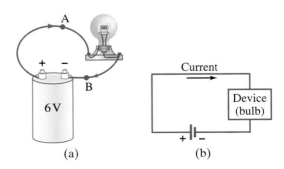

(a) (b)

More precisely, the electric current in a wire is defined as the net amount of charge that passes through the wire's full cross section at any point per unit time. Thus, the average current $\bar{I}$ is defined as

$$\bar{I} = \frac{\Delta Q}{\Delta t},$$
(25–1a)

Electric current

where ΔQ is the amount of charge that passes through the conductor at any location during the time interval Δt. The instantaneous current is defined by the differential limit

$$I = \frac{dQ}{dt}.$$
(25–1b)

Electric current is measured in coulombs per second; this is given a special name, the **ampere** (abbreviated amp or A), after the French physicist André Ampère (1775–1836). Thus, $1 \text{ A} = 1 \text{ C/s}$. Smaller units of current are often used, such as the milliampere $(1 \text{ mA} = 10^{-3} \text{ A})$ and microampere $(1 \mu\text{A} = 10^{-6} \text{ A})$.

The ampere $(1 \text{ A} = 1 \text{ C/s})$

In any single circuit, with only a single path for current to follow such as in Fig. 25–5, a steady current at any instant is the same at one point (say point A) as at any other point (such as B). This follows from the conservation of electric charge (charge doesn't disappear).

> **EXAMPLE 25–1** **Current is flow of charge.** A steady current of 2.5 A flows in a wire for 4.0 min. (*a*) How much charge passed by any point in the circuit? (*b*) How many electrons would this be?
>
> **SOLUTION** (*a*) Since the current was 2.5 A, or 2.5 C/s, then in 4.0 minutes (= 240 seconds) the total charge that flowed was, from Eq. 25–1,
>
> $$\Delta Q = I \Delta t$$
> $$= (2.5 \text{ C/s})(240 \text{ s}) = 600 \text{ C}.$$
>
> (*b*) The charge on one electron is 1.60×10^{-19} C, so 600 C would consist of
>
> $$\frac{600 \text{ C}}{1.6 \times 10^{-19} \text{ C/electron}} = 3.8 \times 10^{21} \text{ electrons}.$$

We saw in Chapter 21 that conductors contain many free electrons. Thus, if a continuous conducting wire is connected to the terminals of a battery, negatively charged electrons flow in the wire. When the wire is first connected, the potential difference between the terminals of the battery sets up an electric field inside the wire and parallel to it. Thus free electrons at one end of the wire are attracted into the positive terminal, and at the same time, electrons leave the negative terminal of the battery and enter the wire at the other end. There is a continuous flow of electrons through the wire that begins as soon as the wire is connected to *both* terminals. However, when the conventions of positive and negative charge were devised two centuries ago, it was assumed that positive charge flowed in a wire. For nearly all purposes, positive charge flowing in one direction is exactly equivalent to negative charge flowing in the opposite direction,[†] as shown in Fig. 25–6. Today, we still use the historical convention of positive current when discussing the direction of a current. So when we speak of the current in a circuit, we mean the direction positive charge would flow. This is sometimes referred to as **conventional current**. When we want to speak of the direction of electron flow, we will specifically state it is the electron current. In liquids and gases, both positive and negative charges (ions) can move.

FIGURE 25–6 Conventional current from + to − is equivalent to a negative (electron) current flowing from − to +.

Conventional current

[†] An exception is discussed in Section 27–8.

25-3 | Ohm's Law: Resistance and Resistors

To produce an electric current in a circuit, a difference in potential is required. One way of producing a potential difference is by a battery. It was Georg Simon Ohm (1787–1854) who established experimentally that the current in a metal wire is proportional to the potential difference V applied to its two ends:

$$I \propto V.$$

If, for example, we connect a wire to a 6-V battery, the current flow will be twice what it would be if the wire were connected to a 3-V battery. For simplicity, we are using V, rather than V_{ba}, to represent potential difference or voltage. It is also found that reversing the sign of the voltage does not affect the magnitude of the current.

Water analogy   It is helpful to compare an electric current to the flow of water in a river or a pipe acted on by gravity. If the pipe (or river) is nearly level, the flow rate is small. But if one end is somewhat higher than the other, the flow rate—or current—is greater. The greater the difference in height, the swifter the current. We saw in Chapter 23 that electric potential is analogous, in the gravitational case, to the height of a cliff. This applies in the present case to the height through which the fluid flows. Just as an increase in height can cause a greater flow of water, so a greater electric potential difference, or voltage, causes a greater electric current flow.

Exactly how much current flows in a wire depends not only on the voltage, but also on the resistance the wire offers to the flow of electrons. The walls of a pipe, or the banks of a river and rocks in the middle, offer resistance to the flow of current. Similarly, electrons are slowed down because of interactions with the atoms of the wire. The higher this resistance, the less the current for a given voltage V. We then define *resistance* so that the current is inversely proportional to the resistance: that is,

$$R = \frac{V}{I} \qquad \textbf{(25–2a)}$$

where R is the **resistance** of a wire or other device, V is the potential difference across the device, and I is the current that flows through it. Eq. 25–2a is often written as

$$V = IR. \qquad \textbf{(25–2b)}$$

As mentioned above, Ohm found experimentally that in metal conductors R is a constant independent of V, a result known as **Ohm's law**. Also, Eq. 25–2b, $V = IR$, is itself sometimes called Ohm's law, but only when referring to materials or devices for which R is a constant independent of V. But R is not a constant for many substances, nor for devices such as diodes, vacuum tubes, transistors, and so on. Thus "Ohm's law" is not a fundamental law, but rather a description of a certain class of materials (metal conductors). Materials or devices that do not follow Ohm's law ($R = $ constant) are said to be *nonohmic*. See Fig. 25–7.

The unit for resistance is called the **ohm** and is abbreviated Ω (Greek capital omega). Because $R = V/I$, we see that $1.0\,\Omega$ is equivalent to $1.0\,\text{V/A}$.

FIGURE 25–7 Graphs of current vs. voltage for (a) a metal conductor which obeys Ohm's law, and (b) for a nonohmic device, in this case a semiconductor diode.

The ohm $(1\,\Omega = 1\,\text{V/A})$

FIGURE 25–8 Example 25–2.

![Circuit diagram with resistor R between points A and B, current I flowing to the right]

| CONCEPTUAL EXAMPLE 25–2 | **Current and potential.** Current I enters a resistor R as shown in Fig. 25–8. (*a*) Is the potential higher at point A or at point B? (*b*) Is the current greater at point A or at point B?

RESPONSE (*a*) Positive charge always flows from + to −, from high potential to low potential. Think again of the gravitational analogy: a mass will fall down from high gravitational potential to low. So for positive current I, point A is at a higher potential than point B. (*b*) Conservation of charge requires that whatever current flows into the resistor at point A, emerges at point B. Current does not get "used up" by a resistor, just as an object that falls through a gravitational potential difference does not gain or lose mass. So the current is the same at A and B.

EXAMPLE 25–3 **Flashlight bulb resistance.** A small flashlight bulb (Fig. 25–9) draws 300 mA from its 1.5-V battery. (a) What is the resistance of the bulb? (b) If the voltage dropped to 1.2 V, how would the current change?

SOLUTION (a) From Eq. 25–2,

$$R = \frac{V}{I} = \frac{1.5\,\text{V}}{0.30\,\text{A}} = 5.0\,\Omega.$$

(b) If the resistance didn't change, the current would be

$$I = \frac{V}{R} = \frac{1.2\,\text{V}}{5.0\,\Omega} = 0.24\,\text{A},$$

or a drop of 60 mA. Actually, resistance does depend on temperature (Section 25–4), so this is only a rough approximation.

FIGURE 25–9 Flashlight (Example 25–3). Note how circuit is completed along the side strip.

All electric devices, from heaters to lightbulbs to stereo amplifiers, offer resistance to the flow of current. The filaments of lightbulbs and electric heaters are special types of wires whose resistance results in their becoming very hot. Generally, the connecting wires have very low resistance in comparison to the resistance of the wire filaments or coils. In many circuits, particularly in electronic devices, **resistors** are used to control the amount of current. Resistors have resistances from less than an ohm to millions of ohms (see Figs. 25–10 and 25–11). The main types are "wire-wound" resistors which consist of a coil of fine wire; "composition" resistors which are usually made of the semiconductor carbon; and thin metal films.

When we draw a diagram of a circuit, we indicate a resistance with the symbol

-\/\/\/- .   [resistor symbol]

Wires whose resistance is negligible, however, are shown simply as straight lines.

FIGURE 25–10 Photo of resistors (mostly).

Resistor Color Code

Color	Number	Multiplier	Tolerance (%)
Black	0	1	
Brown	1	10^1	
Red	2	10^2	
Orange	3	10^3	
Yellow	4	10^4	
Green	5	10^5	
Blue	6	10^6	
Violet	7	10^7	
Gray	8	10^8	
White	9	10^9	
Gold		10^{-1}	5%
Silver		10^{-2}	10%
No color			20%

First digit
Second digit
Multiplier
Tolerance

FIGURE 25–11 The resistance value of a given resistor is written on the exterior, or may be given as a color code, as shown above and in the table: the first two colors represent the first two digits in the value of the resistance, the third color represents the power of ten that it must be multiplied by, and the fourth is the manufactured tolerance. For example, a resistor whose four colors are red, green, orange, and silver has a resistance of 25,000 Ω (25 kΩ), give or take 10 percent.

TABLE 25–1 Resistivity and Temperature Coefficients (at 20°C)

Material	Resistivity, ρ ($\Omega \cdot m$)	Temperature Coefficient, α (C°)$^{-1}$
Conductors		
Silver	1.59×10^{-8}	0.0061
Copper	1.68×10^{-8}	0.0068
Gold	2.44×10^{-8}	0.0034
Aluminum	2.65×10^{-8}	0.00429
Tungsten	$5.6 \ \times 10^{-8}$	0.0045
Iron	9.71×10^{-8}	0.00651
Platinum	$10.6 \ \times 10^{-8}$	0.003927
Mercury	$98 \ \ \times 10^{-8}$	0.0009
Nichrome (alloy of Ni, Fe, Cr)	$100 \ \ \times 10^{-8}$	0.0004
Semiconductors[†]		
Carbon (graphite)	$(3-60) \times 10^{-5}$	−0.0005
Germanium	$(1-500) \times 10^{-3}$	−0.05
Silicon	0.1–60	−0.07
Insulators		
Glass	$10^9 - 10^{12}$	
Hard rubber	$10^{13} - 10^{15}$	

[†]Values depend strongly on presence of even slight amounts of impurities.

25–4 Resistivity

It is found experimentally that the resistance R of a metal wire is directly proportional to its length l and inversely proportional to its cross-sectional area A. That is,

$$R = \rho \frac{l}{A}, \tag{25–3}$$

Resistivity

where ρ, the constant of proportionality, is called the **resistivity** and depends on the material used. Typical values of ρ, whose units are $\Omega \cdot m$ (see Eq. 25–3), are given for various materials in the middle column of Table 25–1. The values depend somewhat on purity, heat treatment, temperature, and other factors. Notice that silver has the lowest resistivity and is thus the best conductor (although it is expensive). Copper is close and much less expensive, so it is clear why most wires are made of copper. Aluminum, although it has a higher resistivity, is much less dense than copper; it is thus preferable to copper in some situations, such as transmission lines, because its resistance for the same weight is less than that for copper.

The reciprocal of the resistivity, called the **conductivity** σ, is

Conductivity

$$\sigma = \frac{1}{\rho}, \tag{25–4}$$

and has units of $(\Omega \cdot m)^{-1}$.

EXAMPLE 25–4 Speaker wires. Suppose you want to connect your stereo to remote speakers (Fig. 25–12). (*a*) If each wire must be 20 m long, what diameter copper wire should you use to keep the resistance less than $0.10 \, \Omega$ per wire? (*b*) If the current to each speaker is 4.0 A, what is the voltage drop across each wire?

SOLUTION   (*a*) We solve Eq. 25–3 for the area A and use Table 25–1:

$$A = \rho\frac{l}{R} = \frac{(1.68 \times 10^{-8}\,\Omega\cdot\text{m})(20\,\text{m})}{(0.10\,\Omega)} = 3.4 \times 10^{-6}\,\text{m}^2.$$

The cross-sectional area A of a circular wire is related to its diameter d by $A = \pi d^2/4$. The diameter must then be at least

$$d = \sqrt{\frac{4A}{\pi}} = 2.1 \times 10^{-3}\,\text{m} = 2.1\,\text{mm}.$$

(*b*) From $V = IR$ we have

$$V = IR = (4.0\,\text{A})(0.10\,\Omega) = 0.40\,\text{V}.$$

FIGURE 25–12   Example 25–4.

CONCEPTUAL EXAMPLE 25–5 | **Stretching changes resistance.** A wire of resistance R is stretched uniformly until it is twice its original length. What happens to its resistance?

RESPONSE   If the length l doubles then the cross-sectional area A halves, so that the volume ($V = Al$) of the wire remains the same. From Eq. 25–3 we see that the resistance would increase by a factor of four $\left(2/\frac{1}{2} = 4\right)$.

Temperature Dependence of Resistivity

The resistivity of a material depends somewhat on temperature. In general, the resistance of metals increases with temperature. This is not surprising, for at higher temperatures, the atoms are moving more rapidly and are arranged in a less orderly fashion. So they might be expected to interfere more with the flow of electrons. If the temperature change is not too great, the resistivity of metals usually increases nearly linearly with temperature. That is,

$$\rho_T = \rho_0\left[1 + \alpha(T - T_0)\right] \qquad \textbf{(25–5)}$$

Effect of temperature

where ρ_0 is the resistivity at some reference temperature T_0 (such as 0°C or 20°C), ρ_T is the resistivity at a temperature T and α is the *temperature coefficient of resistivity*. Values for α are given in Table 25–1. Note that the temperature coefficient for semiconductors can be negative. Why? It seems that at higher temperatures, some of the electrons that are not normally free in a semiconductor become free and can contribute to the current. Thus, the resistance of a semiconductor can decrease with an increase in temperature, although this is not always the case.

EXAMPLE 25–6 | **Resistance thermometer.** The variation in electrical resistance with temperature can be used to make precise temperature measurements. Platinum is commonly used since it is relatively free from corrosive effects and has a high melting point. Suppose at 20°C the resistance of a platinum resistance thermometer is 164.2 Ω. When placed in a particular solution, the resistance is 187.4 Ω. What is the temperature of this solution?

➡ **PHYSICS APPLIED**

Resistance thermometer

SOLUTION   Since the resistance R is directly proportional to the resistivity ρ, we can combine Eq. 25–3 with Eq. 25–5:

$$R = R_0\left[1 + \alpha(T - T_0)\right].$$

Here $R_0 = \rho_0 L/A$ is the resistance of the wire at $T_0 = 20$°C. We solve this equation for T and find (see Table 25–1 for α)

$$T = T_0 + \frac{R - R_0}{\alpha R_0} = 20\text{°C} + \frac{187.4\,\Omega - 164.2\,\Omega}{(3.927 \times 10^{-3}(\text{C°})^{-1})(164.2\,\Omega)} = 56.0\text{°C}.$$

FIGURE 25–13   A thermistor shown next to a millimeter ruler for scale.

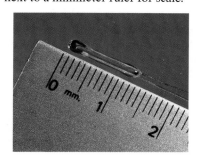

More convenient for some applications is a *thermistor* (Fig. 25–13), which consists of a metal oxide or semiconductor whose resistance also varies in a repeatable way with temperature. They can be made quite small and respond very quickly to temperature changes. Resistance thermometers have another advantage in that they can be used at very high or low temperatures where gas or liquid thermometers would be useless.

The value of α in Eq. 25–5 itself can depend on temperature, so it is important to check the temperature range of validity of any value (say, in a handbook of physical data). If the temperature range is wide, Eq. 25–5 is not adequate and terms proportional to the square and cube of the temperature are needed, but they are generally very small except when $T - T_0$ is large.

25–5 Electric Power

Electric energy is useful to us because it can be easily transformed into other forms of energy. Motors, whose operation we will examine in Chapter 27, transform electric energy into mechanical work.

In other devices such as electric heaters, stoves, toasters, and hair dryers, electric energy is transformed into thermal energy in a wire resistance known as a "heating element." And in an ordinary lightbulb, the tiny wire filament (Fig. 25–14) becomes so hot it glows; only a few percent of the energy is transformed into visible light, and the rest, over 90 percent, into thermal energy. Lightbulb filaments and heating elements in household appliances have resistances typically of a few ohms to a few hundred ohms.

Electric energy is transformed into thermal energy or light in such devices, and there are many collisions between the moving electrons and the atoms of the wire. In each collision, part of the electron's kinetic energy is transferred to the atom with which it collides. As a result, the kinetic energy of the wire's atoms increases and hence the temperature of the wire element increases. The increased thermal energy can be transferred as heat by conduction and convection to the air in a heater or to food in a pan, by radiation to bread in a toaster, or radiated as light.

To find the power transformed by an electric device, recall that the energy transformed when an infinitesimal charge dq moves through a potential difference V is $dU = dq\,V$ (Eq. 23–2). If dt is the time required for an amount of charge dq to move through the potential difference V, the power P, which is the rate energy is transformed, is

$$P = \frac{dU}{dt} = \frac{dq}{dt}V.$$

The charge that flows per second, dq/dt, is the electric current I. Thus we have

Electric power (general)

$$\boxed{P = IV.} \tag{25–6}$$

This general relation gives us the power transformed by any device, where I is the current passing through it and V is the potential difference across it. It also gives the power delivered by a source such as a battery. The SI unit of electric power is the same as for any kind of power, the **watt** (1 W = 1 J/s).

The rate of energy transformation in a resistance R can be written, using $V = IR$, in two other ways:

Electric power (in resistance R)

$$P = IV \tag{25–7a}$$
$$= I(IR) = I^2 R \tag{25–7b}$$
$$= \left(\frac{V}{R}\right)V = \frac{V^2}{R}. \tag{25–7c}$$

Equations 25–7b and c apply only to resistors, whereas Eq. 25–7a, $P = IV$, applies to any device.

Filament

Insulator

FIGURE 25–14 Incandescent lightbulb.

EXAMPLE 25–7 **Headlights.** Calculate the resistance of a 40-W automobile headlight designed for 12 V (Fig 25–15).

SOLUTION Since we are given $P = 40$ W and $V = 12$ V, we can use Eq. 25–7c and solve for R:

$$R = \frac{V^2}{P} = \frac{(12\,\text{V})^2}{(40\,\text{W})} = 3.6\,\Omega.$$

40-W Headlight

FIGURE 25–15 Example 25–7.

This is the resistance when the bulb is burning brightly at 40 W. When the bulb is cold, the resistance is much lower, as we saw in Eq. 25–5. Since the current is high when the resistance is low, lightbulbs burn out most often when first turned on.

It is energy, not power, that you pay for on your electric bill. Since power is the *rate* energy is transformed, the total energy used by any device is simply its power consumption multiplied by the time it is on. If the power is in watts and the time is in seconds, the energy will be in joules since $1\,\text{W} = 1\,\text{J/s}$. Electric companies usually specify the energy with a much larger unit, the **kilowatt-hour** (kWh). One kWh $= (1000\,\text{W})(3600\,\text{s}) = 3.60 \times 10^6$ J.

You pay for energy

Kilowatt-hour (unit of energy)

EXAMPLE 25–8 **Electric heater.** An electric heater draws a steady 15.0 A on a 120-V line. How much power does it require and how much does it cost per month (30 days) if it operates 3.0 h per day and the electric company charges 10.5 cents per kWh?

SOLUTION The power is

$$P = IV = (15.0\,\text{A})(120\,\text{V}) = 1800\,\text{W}$$

or 1.80 kW. To operate it for $(3.0\,\text{h/d})(30\,\text{d}) = 90$ h would cost $(1.80\,\text{kW})(90\,\text{h})(\$0.105) = \$17$.

EXAMPLE 25–9 **ESTIMATE** **Lightning bolt.** Lightning is a spectacular example of electric current in a natural phenomenon (Fig. 25–16). There is much variability to lightning bolts, but a typical event can transfer 10^9 J of energy across a potential difference of perhaps 5×10^7 V during a time interval of about 0.2 s. Use this information to estimate the total amount of charge transferred, the current, and the average power over the 0.2 s.

SOLUTION From Eq. 23–2, energy = QV, so

$$Q \approx \frac{10^9\,\text{J}}{5 \times 10^7\,\text{V}} = 20\,\text{C}.$$

The current over the 0.2 s is about

$$I = \frac{Q}{t} \approx \frac{20\,\text{C}}{0.2\,\text{sec}} = 100\,\text{A}.$$

Lightning

FIGURE 25–16 Example 25–9: a lightning bolt. See caption for photo at start of Chapter 23.

Since most lightning bolts consist of several stages, it is possible that individual parts could carry currents much higher than this. The average power delivered is

$$\overline{P} = \frac{\text{energy}}{\text{time}} = \frac{10^9\,\text{J}}{0.2\,\text{sec}} = 5 \times 10^9\,\text{W} = 5\,\text{GW}.$$

We can also use Eq. 25–7a:

$$P = IV = (100\,\text{A})(5 \times 10^7\,\text{V}) = 5\,\text{GW}.$$

(a) Fuses   (b) Circuit breaker (closed)   (c) Circuit breaker (open)

FIGURE 25–17 (a) Fuses. When the current exceeds a certain value, the metallic ribbon melts and the circuit opens. Then the fuse must be replaced. (b) A circuit breaker. Electric current passes to a circuit via a bimetallic strip. When the current is great enough (i.e., too great to be safe), the bimetallic strip is heated and bends so far to the left that the notch in the spring-loaded metal strip drops down over the end of the bimetallic strip; (c) the circuit then opens at the contact points (one is attached to the metal strip) and the outside switch is also flipped. As soon as the bimetallic strip cools down, it can be reset using the outside switch.

25–6 Power in Household Circuits

FIGURE 25–18 Connection of household appliances.

The electric wires that carry electricity to lights and other electric appliances have some resistance, although usually it is quite small. Nonetheless, if the current is large enough, the wires will heat up and produce thermal energy at a rate equal to I^2R, where R is the wire's resistance. One possible hazard is that the current-carrying wires in the wall of a building may become so hot as to start a fire. Thicker wires, of course, have less resistance (see Eq. 25–3) and thus can carry more current without becoming too hot. When a wire carries more current than is safe, it is said to be "overloaded." To prevent overloading, fuses or circuit breakers are installed in circuits. They are basically switches (Fig. 25–17) that open the circuit when the current exceeds some particular value. A 20-A fuse or circuit breaker, for example, opens when the current passing through it exceeds 20 A. If a circuit repeatedly burns out a fuse or opens a circuit breaker, there are two possibilities: there may be too many devices drawing current in that circuit; or there is a fault somewhere, such as a "short." A short, or "short circuit," means that two wires have crossed (perhaps because the insulation has worn down) so the path of the current is shortened. The resistance of the circuit is then very small, so the current will be very large. Short circuits, of course, should be remedied immediately.

Household circuits are designed with the various devices connected so that each receives the standard voltage (usually 120 V in the United States) from the electric company (Fig. 25–18). Circuits with the devices arranged as in Fig. 25–18 are called *parallel circuits*, as we will discuss more fully in the next chapter. When a fuse blows or circuit breaker opens, the total current being drawn on that circuit should be checked.

EXAMPLE 25–10 Will a fuse blow? Determine the total current drawn by all the devices in the circuit of Fig. 25–18.

SOLUTION The circuit in Fig. 25–18 draws the following currents: the lightbulb draws $I = P/V = 100\,\text{W}/120\,\text{V} = 0.8\,\text{A}$; the heater draws $1800\,\text{W}/120\,\text{V} = 15.0\,\text{A}$; the stereo draws a maximum of $350\,\text{W}/120\,\text{V} = 2.9\,\text{A}$; and the hair dryer draws $1200\,\text{W}/120\,\text{V} = 10.0\,\text{A}$. The total current drawn, if all devices are used at the same time, is

$$0.8\,\text{A} + 15.0\,\text{A} + 2.9\,\text{A} + 10.0\,\text{A} = 28.7\,\text{A}.$$

If the circuit in Fig. 25–18 is designed for a 20-A fuse, the fuse should blow, and we hope it will, to prevent overloaded wires from getting hot enough to start a fire. Something will have to be turned off to get this circuit below 20 A. (Houses and apartments usually have several circuits, each with its own fuse or circuit breaker; try moving one of the devices to another circuit.) If the circuit is designed for a 30-A fuse, it shouldn't blow, so if it does, a short may be the problem. (The most likely place is in the cord of one of the devices.) Proper fuse size is selected according to the wire used to supply the current; a properly rated fuse should *never* be replaced by a higher-rated one. A fuse blowing or a circuit breaker opening is acting like a switch, making an "open circuit." By an open circuit, we mean that there is no longer a complete conducting path, so no current can flow; it is as if $R = \infty$.

25–7 | Alternating Current

When a battery is connected to a circuit, the current flows steadily in one direction. This is called a **direct current**, or **dc**. Electric generators at electric power plants, however, produce **alternating current**, or **ac**. (Sometimes capital letters are used, DC and AC.) An alternating current reverses direction many times per second and is commonly sinusoidal, as shown in Fig. 25–19. The electrons in a wire first move in one direction and then in the other. The current supplied to homes and businesses by electric companies is ac throughout virtually the entire world. We will discuss and analyze ac circuits in detail in Chapter 31. But because ac circuits are so common in real life, we will discuss some of their simpler aspects here.

DC and AC

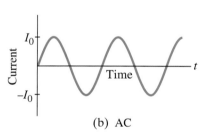

FIGURE 25–19 (a) Direct current. (b) Alternating current.

The voltage produced by an ac electric generator is sinusoidal, as we shall see later. The current it produces is thus sinusoidal (Fig. 25–19b). We can write the voltage as a function of time as

$$V = V_0 \sin 2\pi f t = V_0 \sin \omega t.$$

The potential V oscillates between $+V_0$ and $-V_0$. V_0 is referred to as the **peak voltage**. The frequency f is the number of complete oscillations made per second, and $\omega = 2\pi f$. In most areas of the United States and Canada, f is 60 Hz (the unit "hertz," as we saw in Chapter 14, means cycles per second). In many countries, 50 Hz is used.

From Eq. 25–2, $V = IR$, if a voltage V exists across a resistance R, then the current I is

$$I = \frac{V}{R} = \frac{V_0}{R} \sin \omega t = I_0 \sin \omega t. \qquad (25\text{–}8)$$

The quantity $I_0 = V_0/R$ is the **peak current**. The current is considered positive when the electrons flow in one direction and negative when they flow in the opposite direction. It is clear from Fig. 25–19b that an alternating current is as often positive as it is negative. Thus, the average current is zero. This does not mean, however, that no power is needed or that no heat is produced in a resistor. Electrons do move back and forth, and do produce heat. Indeed, the power delivered to a resistance R at any instant is

$$P = I^2 R = I_0^2 R \sin^2 \omega t.$$

Because the current is squared, we see that the power is always positive, Fig. 25–20. The quantity $\sin^2 \omega t$ varies between 0 and 1; and it is not too difficult to show[†] that its average value is $\frac{1}{2}$ as can be seen graphically in the figure. Thus, the *average power* developed, $\overline{P}$, is

$$\overline{P} = \frac{1}{2} I_0^2 R.$$

FIGURE 25–20 Power delivered to a resistor in an ac circuit.

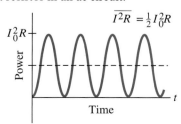

[†] A graph of $\cos^2 \omega t$ versus t is identical to that for $\sin^2 \omega t$ in Fig. 25–20 except that the points are shifted (by $\frac{1}{4}$ cycle) on the time axis. Hence the average value of $\sin^2$ and $\cos^2$, averaged over one or more full cycles, will be the same: $\overline{\sin^2 \omega t} = \overline{\cos^2 \omega t}$. From the trigonometric identity $\sin^2 \theta + \cos^2 \theta = 1$, we can write

$$\overline{(\sin^2 \omega t)} + \overline{(\cos^2 \omega t)} = 2\overline{(\sin^2 \omega t)} = 1.$$

Hence the average value of $\sin^2 \omega t$ is $\frac{1}{2}$.

Since power can also be written $P = V^2/R = (V_0^2/R)\sin^2 \omega t$, we also have that the average power is

$$\overline{P} = \tfrac{1}{2}\frac{V_0^2}{R}.$$

The average or mean value of the *square* of the current or voltage is thus what is important for calculating average power: $\overline{I^2} = \tfrac{1}{2}I_0^2$ and $\overline{V^2} = \tfrac{1}{2}V_0^2$. The square root of each of these is the **rms** (root-mean-square) value of the current or voltage:

rms current
$$I_{rms} = \sqrt{\overline{I^2}} = \frac{I_0}{\sqrt{2}} = 0.707\,I_0, \qquad \textbf{(25–9a)}$$

rms voltage
$$V_{rms} = \sqrt{\overline{V^2}} = \frac{V_0}{\sqrt{2}} = 0.707\,V_0. \qquad \textbf{(25–9b)}$$

The rms values of V and I are sometimes called the "effective values." They are useful because they can be substituted directly into the power formulas, Eqs. 25–7, to get the average power:

$$\overline{P} = \tfrac{1}{2}I_0^2 R = I_{rms}^2 R \qquad \textbf{(25–10a)}$$

$$\overline{P} = \tfrac{1}{2}\frac{V_0^2}{R} = \frac{V_{rms}^2}{R}. \qquad \textbf{(25–10b)}$$

Thus, a direct current whose values of I and V equal the rms values of I and V for an alternating current will produce the same power. Hence it is usually the rms value of current that is specified or measured. For example, in the United States and Canada, standard line voltage[†] is 120 V ac. The 120 V is V_{rms}; the peak voltage V_0 is

$$V_0 = \sqrt{2}\,V_{rms} = 170\text{ V}.$$

In most of Europe the rms voltage is 240 V, so the peak voltage is 340 V.

EXAMPLE 25–11 **Hair dryer.** (*a*) Calculate the resistance and the peak current in a 1000-W hair dryer (Fig. 25–21) connected to a 120-V line. (*b*) What happens if it is connected to a 240-V line in Britain?

SOLUTION (*a*) We can apply Eq. 25–7a using rms values. Then the rms current is

$$I_{rms} = \frac{\overline{P}}{V_{rms}} = \frac{1000\text{ W}}{120\text{ V}} = 8.33\text{ A}.$$

Thus $I_0 = \sqrt{2}\,I_{rms} = 11.8$ A. Then the resistance is

$$R = \frac{V_{rms}}{I_{rms}} = \frac{120\text{ V}}{8.33\text{ A}} = 14.4\ \Omega.$$

The resistance could equally well be calculated using peak values: $R = V_0/I_0 = 170\text{ V}/11.8\text{ A} = 14.4\ \Omega$.
(*b*) When connected to a 240-V line, more current would flow and the resistance would change with the increased temperature (Section 25–4). But let us make an estimate based on the same 14.4 Ω resistance. The average power delivered would be

$$\overline{P} = \frac{V_{rms}^2}{R} = \frac{(240\text{ V})^2}{(14.4\ \Omega)} = 4000\text{ W}.$$

This is four times the dryer's power rating and would undoubtedly melt the heating element or the wire coils of the motor. Be sure your hair dryer (or electric shaver) has a 120/240 V switch before traveling too far from home, or carry a transformer (Section 29–6) for travelers.

Motor
Fan
Heating coils
Switch
Cord

FIGURE 25–21 A hair dryer. Most of the current goes through the heating coils, a pure resistance; a small part goes to the motor to turn the fan. Example 25–11.

[†]The line voltage can vary, depending on the total load; the frequency of 60 Hz, however, remains extremely steady.

25–8 Microscopic View of Electric Current: Current Density and Drift Velocity

Up to now in this chapter we have dealt mainly with a macroscopic, everyday-world, view of electric current. We did mention, however, that according to atomic theory, the electric current in metal wires is carried by negatively charged electrons, and that in liquid solutions current can also be carried by positive and/or negative ions. Let us now look at this microscopic picture in more detail.

When a potential difference is applied to the two ends of a wire of uniform cross-section, the direction of the electric field **E** is parallel to the walls of the wire (Fig. 25–22). The existence of **E** within the conducting wire does not contradict our earlier result that **E** = 0 inside a conductor in the electrostatic case. We are no longer dealing with the static case. Charges are free to move in a conductor, and hence can move under the action of the electric field. If all the charges are at rest, then **E** must be zero (electrostatics).

We now define a new microscopic quantity, the **current density**, **j**. It is defined as the *electric current per unit cross-sectional area* at any point in space. If the current density **j** in a wire of cross-sectional area A is uniform over the cross section, then j is related to the electric current by

$$j = \frac{I}{A} \quad \text{or} \quad I = jA. \tag{25–11}$$

If the current density is not uniform, then the general relation is

$$I = \int \mathbf{j} \cdot d\mathbf{A}, \tag{25–12}$$

where $d\mathbf{A}$ is an element of surface and I is the current through the surface over which the integration is taken. The direction of the current density at any point is the direction that a positive charge would move when placed at that point—that is, the direction of **j** at any point is generally the same as the direction of **E**, Fig. 25–22. (Inertial effects can usually be ignored.) The current density exists for any *point* in space. The current I, on the other hand, refers to a conductor as a whole, and hence is a macroscopic quantity.

The direction of **j** is chosen to represent the direction of flow of positive charge. In a conductor, it is negatively charged electrons that move, so they move in the direction of −**j**, or −**E** (to the left in Fig. 25–22). We can imagine the free electrons as moving about randomly at high speeds, bouncing off the atoms of the wire (somewhat like the molecules of a gas—Chapter 18). When an electric field exists in the wire, Fig. 25–23, the electrons feel a force and initially begin to accelerate. But they soon reach a more or less steady average speed (due to collisions with atoms in the wire), known as their **drift velocity**, v_d. The drift velocity is normally very much smaller than the electrons' average random speed.

We can relate v_d to the macroscopic current I in the wire. In a time Δt, the electrons will travel a distance $l = v_d \Delta t$ on average. Suppose the wire has cross-sectional area A. Then in time Δt, electrons in a volume $V = Al = Av_d \Delta t$ will pass through the cross section A of wire, as shown in Fig. 25–24. If there are n free electrons (each of charge $-e$) per unit volume ($n = N/V$), then the total charge ΔQ that passes through the area A in a time Δt is

$$\Delta Q = (\text{no. of charges}, N) \times (\text{charge per particle})$$
$$= (nV)(-e) = -(nAv_d \, \Delta t)(e).$$

The current I in the wire is thus

$$I = \frac{\Delta Q}{\Delta t} = -neAv_d. \tag{25–13}$$

FIGURE 25–22 Electric field **E** in a uniform wire of cross-sectional area A carrying a current I. The current density $j = I/A$.

Current

density

defined

FIGURE 25–23 Electric field **E** in a wire gives electrons in random motion a drift velocity v_d.

Drift velocity

FIGURE 25–24 Electrons in the volume Al will all pass through the cross-section indicated in a time Δt, where $l = v_d \Delta t$.

Current (microscopic variables)

The current density, $j = I/A$, is

Current density

in terms of

drift velocity

$$j = -ne v_d. \qquad (25\text{–}14)$$

In vector form, this is written

$$\mathbf{j} = -ne\mathbf{v}_d, \qquad (25\text{–}15)$$

where the minus sign indicates that the direction of (positive) current flow is opposite to the drift velocity of electrons.

We can generalize Eq. 25–15 to any type of charge flow, such as flow of ions in an electrolyte. If there are several types of ions (which can include free electrons), each of density n_i (number per unit volume), charge q_i ($q_i = -e$ for electrons) and drift velocity $\mathbf{v}_{di}$, then the net current density at any point is

$$\mathbf{j} = \sum_i n_i q_i \mathbf{v}_{di}. \qquad (25\text{–}16)$$

The total current I passing through an area A perpendicular to a uniform $\mathbf{j}$ is then

$$I = \sum_i n_i q_i v_{di} A.$$

EXAMPLE 25–12 **Electron speeds in a wire.** A copper wire 3.2 mm in diameter, carries a 5.0-A current. Determine (*a*) the current density in the wire, and (*b*) the drift velocity of the free electrons. (*c*) Estimate the rms speed of electrons assuming they behave like an ideal gas at 20°C. Assume that one electron per Cu atom is free to move (the others remain bound to the atom).

SOLUTION (*a*) The cross-sectional area of the wire is

$$A = \pi r^2 = (3.14)(1.60 \times 10^{-3}\,\text{m})^2 = 8.0 \times 10^{-6}\,\text{m}^2.$$

The current density is then

$$j = \frac{I}{A} = \frac{5.0\,\text{A}}{8.0 \times 10^{-6}\,\text{m}^2} = 6.2 \times 10^5\,\text{A/m}^2.$$

(*b*) Since we assume there is one free electron per atom, the density of free electrons, n, is the same as the density of Cu atoms. The atomic mass of Cu is 63.5 u (see Periodic Table inside the back cover), so 63.5 g of Cu contains one mole or 6.02×10^{23} free electrons. The mass density of copper (Table 13–1) is $\rho_D = 8.9 \times 10^3\,\text{kg/m}^3$, where $\rho_D = m/V$. (We use ρ_D to distinguish it here from ρ for resistivity.) So the number of free electrons per unit volume is

$$n = \frac{N}{V} = \frac{N}{m/\rho_D} = \frac{N(1\,\text{mole})}{m(1\,\text{mole})}\rho_D$$

$$n = \left(\frac{6.02 \times 10^{23}\,\text{electrons}}{63.5 \times 10^{-3}\,\text{kg}} \right)(8.9 \times 10^3\,\text{kg/m}^3) = 8.4 \times 10^{28}\,\text{m}^{-3}.$$

Then, by Eq. 25–14, the drift velocity is

$$v_d = \frac{j}{ne} = \frac{6.2 \times 10^5\,\text{A/m}^2}{(8.4 \times 10^{28}\,\text{m}^{-3})(1.6 \times 10^{-19}\,\text{C})} = 4.6 \times 10^{-5}\,\text{m/s},$$

which is only about 0.05 mm/s.

(*c*) If we model the free electrons as an ideal gas (a rather rough approximation), we use Eq. 18–5 to estimate the random rms speed of an electron as it darts around:

$$v_{rms} = \sqrt{\frac{3kT}{m}} = \sqrt{\frac{3(1.38 \times 10^{-23}\,\text{J/K})(293\,\text{K})}{9.11 \times 10^{-31}\,\text{kg}}} = 1.2 \times 10^5\,\text{m/s}.$$

Thus we see that the drift velocity (average speed in the direction of the current) is very much less than the rms thermal speed of the electrons, by a factor of about 10^9. [Note: The result in (*c*) is an underestimate. Quantum theory calculations, and experiments, give the rms speed in copper to be about $1.6 \times 10^6\,\text{m/s}$.]

The drift velocity of electrons in a wire is clearly very slow, only about 0.05 mm/s for the example above, which means it takes an electron 20×10^3 s, or $5\frac{1}{2}$ h, to travel only 1 m. This is not, of course, how fast "electricity travels": when you flip a light switch, the light—even if many meters away—goes on nearly instantaneously, for electric fields travel essentially at the speed of light $(3 \times 10^8 \, \text{m/s})$. We can think of electrons in a wire as being like a pipe full of water: when a little water enters one end of the pipe, almost immediately some water issues forth at the other end.

Electricity's "speed"

Equation 25–2b, $V = IR$, can be written in terms of microscopic quantities as follows. We write the resistance R in terms of the resistivity ρ:

$$R = \rho \frac{l}{A};$$

and we write V and I as

$$I = jA$$

and

$$V = El.$$

The last relation follows from Eq. 23–3, where we assume the electric field is uniform within the wire and l is the length of the wire (or a portion of the wire) between whose ends the potential difference is V. Thus, from $V = IR$, we have

$$V = IR$$

$$El = (jA)\left(\rho \frac{l}{A}\right) = j\rho l$$

so

$$j = \frac{1}{\rho} E = \sigma E, \qquad \qquad \text{(25–17)}$$

where $\sigma = 1/\rho$ is the *conductivity* (Eq. 25–4). For a metal conductor, ρ and σ do not depend on V (and hence not on E). Therefore the current density $\mathbf{j}$ is proportional to the electrical field $\mathbf{E}$ in the conductor. This is the "microscopic" statement of Ohm's law. Equation 25–17, which can be written in vector form as

$$\mathbf{j} = \sigma \mathbf{E} = \frac{1}{\rho} \mathbf{E},$$

is sometimes taken as the definition of conductivity σ and resistivity ρ.

EXAMPLE 25–13 **Electric field inside a wire.** What is the electric field inside the wire of Example 25–12?

SOLUTION Table 25–1 gives $\rho = 1.68 \times 10^{-8} \, \Omega \cdot \text{m}$ for copper. Since $j = 6.2 \times 10^5 \, \text{A/m}^2$,

$$E = \rho j = (1.68 \times 10^{-8} \, \Omega \cdot \text{m})(6.2 \times 10^5 \, \text{A/m}^2)$$

$$= 1.0 \times 10^{-2} \, \text{V/m}.$$

For comparison, the electric field between the plates of a capacitor is often much larger; in Example 24–1, for example, E is on the order of $10^4 \, \text{V/m}$. Thus we see that only a modest electric field is needed for current flow in practical cases.

FIGURE 25–25 A superconducting material has zero resistivity when its temperature is below T_C, its "transition temperature." At T_C, the resistivity jumps to a "normal" non-zero value and increases with temperature as most materials do (Eq. 25–5).

High-temperature superconductors

FIGURE 25–26 An experimental train in Japan, supported by the magnetic field produced by current in coils beneath the tracks (in the red containers).

*25–9 Superconductivity

At very low temperatures, near absolute zero, the resistivity (Section 25–4) of certain metals and certain compounds or alloys becomes zero as measured by the highest-precision techniques. Materials in such a state are said to be **superconducting**. This phenomenon was first observed by H. K. Onnes (1853–1926) in 1911 when he cooled mercury below 4.2 K (−269°C). He found that at this temperature, the resistance of mercury suddenly dropped to zero. In general, superconductors become superconducting only below a certain *transition temperature* T_C, which is usually within a few degrees of absolute zero. Current in a ring-shaped superconducting material has been observed to flow for years in the absence of a potential difference, with no measurable decrease. Measurements show that the resistivity ρ of superconductors is less than 4×10^{-25} $\Omega \cdot m$, which is over 10^{16} times smaller than that for copper, and is considered to be zero in practice. See Fig. 25–25.

Much research has been done on superconductivity in recent years to try to understand why it occurs, and to find materials that superconduct at more reasonable temperatures to reduce the cost and inconvenience of refrigeration at the required very low temperature. Before 1986 the highest temperature at which a material was found to superconduct was 23 K, and this required liquid helium to keep the material cold. In 1987, a compound of yttrium, barium, copper, and oxygen was developed that can be superconducting at 90 K. Since this is above the temperature of liquid nitrogen, 77 K, boiling liquid nitrogen is sufficiently cold to keep the material superconducting. This was an important breakthrough since liquid nitrogen is much more easily and cheaply obtained than is the liquid helium needed for previous superconductors. Since then, superconductivity at temperatures in the vicinity of 160 K have been reported, though in fragile compounds.

Considerable research is being done to develop high-T_C superconductors as wires that can carry currents strong enough to be practical. Most applications today use a bismuth-strontium-calcium-copper oxide, known (for short) as BSCCO. A major use today of superconductors is for carrying the current in electromagnets (we shall see in Chapter 27 that electric currents produce magnetic fields). In large non-superconducting magnets, a great amount of energy is needed just to maintain the current, and this energy is wasted as heat.

A major problem is how to make a useable, bendable wire out of the BSCCO, which is very brittle. One solution is to embed tiny filaments of the high-T_C superconductor in a metal alloy matrix. The first major commercial use of high-T_C superconductors embeds the filaments in silver; the wires are formed into a cable to carry very high currents for the city of Detroit's electric distribution grid. The superconducting wire is wrapped around a tube carrying liquid nitrogen to keep the BSCCO below T_C. The wire is not resistanceless, because of the silver connections, but the resistance is much less than that of a conventional copper cable. Indeed, the 100 kg of this 130-m-long superconducting cable can carry as much current as 8000 kg of the copper cable it replaces.

Electric motors, generators, and transformers using superconductors are also being worked on, and will be much smaller and lighter than conventional ones. Prototype motors under development are half the size and weight of non-superconducting motors.

Superconductors could make electric cars more practical, make computers much faster, and have great potential in devices to store energy for use at peak demand. Superconductors are being studied for use in high-speed ground transportation: the magnetic fields produced by superconducting magnets would be used to "levitate" vehicles over tracks so there is essentially no friction (Fig. 25–26). The levitation arises from the repulsive force between the magnet (say, on the train) and the eddy currents produced in the track below (or, vice versa).

25–10 | Electric Hazards; Leakage Currents

An electric shock can damage the body or may even be fatal. The severity of a shock depends on the magnitude of the current, how long it acts, and through what part of the body it passes. A current passing through vital organs such as the heart or brain is especially serious. Electric current heats tissue and can cause burns. A current also stimulates nerves and muscles, and we feel a "shock."

➡ **P H Y S I C S   A P P L I E D**

Electric shock

Most people can "feel" a current of about 1 mA. Currents of a few mA cause pain but rarely much damage in a healthy person. However, currents above 10 mA cause severe contraction of the muscles, and a person may not be able to release the source of the current (say, a faulty appliance or wire). Death from paralysis of the respiratory system can occur. Artificial respiration, however, can sometimes revive a victim. If a current above about 70 mA passes across the torso so that a portion passes through the heart for a second or more, the heart muscles will begin to contract irregularly and blood will not be properly pumped. This condition is called "ventricular fibrillation." If it lasts for long, death results. Strangely enough, if the current is much larger, on the order of 1 A, the damage may be less and death by heart failure may be less likely[†] under some conditions.

The seriousness of a shock depends on the effective resistance of the body. Living tissue has quite low resistance since the fluid of cells contains ions that can conduct quite well. However, the outer layers of skin, when dry, offer much resistance. The effective resistance between two points on opposite sides of the body when the skin is dry is in the range of 10^4 to $10^6 \, \Omega$. However, when the skin is wet, the resistance may be $10^3 \, \Omega$ or less. A person in good contact with the ground who touches a 120-V dc line with wet hands can suffer a current

$$I = \frac{120 \, \text{V}}{1000 \, \Omega} = 120 \, \text{mA}.$$

As we saw above, this could be lethal.

Figure 25–27 shows how the circuit is completed when a person touches an electric wire. One side of a 120-V source is connected to ground by a wire connected to a buried conductor, such as a water pipe. Thus the current passes from the high-voltage wire, through the person, to the ground; it passes through the ground back to the other terminal of the source to complete the circuit. If the person in Fig. 25–27 stands on a good insulator—thick-soled shoes or a dry wood floor—there will be much more resistance in the circuit and consequently much less current will flow. However, if the person stands with bare feet on the ground, or is sitting in a bathtub, there is considerable danger because the resistance is much less. In a bathtub, not only are you wet, but the water is in contact with the drain pipe that leads to the ground. That is why it is strongly recommended not to touch anything electrical in such a situation.

FIGURE 25–27 A person receives an electric shock when the circuit is completed.

A principal danger comes from touching a bare wire whose insulation has worn off, or from a bare wire inside an appliance when you're tinkering with it. (Always unplug an electrical device before investigating its insides!) Sometimes a wire inside a device breaks or loses its insulation and comes in contact with the case. If the case is metal, it will conduct electricity. A person could then suffer a

➡ **P H Y S I C S   A P P L I E D**

Grounding and shocks

[†] Apparently, larger currents bring the entire heart to a standstill. Upon release of the current, the heart returns to its normal rhythm. This may not happen when fibrillation occurs since it is often hard to stop once it starts. Fibrillation may also occur as a result of a heart attack or during heart surgery. A device known as a *defibrillator* can apply a brief high current to the heart; this causes complete heart stoppage and is often followed by resumption of normal beating.

(a) (b) (c)

FIGURE 25–28 (a) An electric appliance operating normally with a two-prong plug. (b) Short to the case with ungrounded case: shock. (c) Short to the case with the case grounded with third prong.

FIGURE 25–29 Human body modeled electrically as resistance and capacitor in parallel when a voltage is applied.

Leakage current

severe shock merely by touching the case, as shown in Fig. 25–28b. To prevent an accident, metal cases are supposed to be connected directly to ground, so they cannot become "hot." Then if a "hot" wire touches the grounded case, a short circuit to ground immediately occurs internally, as shown in Fig. 25–28c; most of the current passes through the low resistance ground wire rather than through the person. Furthermore, the high current immediately opens the fuse or circuit breaker in the circuit. Grounding a metal case is best done by a separate ground wire connected to the third (round) prong of a 3-prong plug. It can also be done by connecting the case to the larger prong of a so-called "polarized" 2-prong plug. Of course not only the device, but also the outlets must be wired correctly to ground.

The human body acts as if it had capacitance in parallel with its resistance (Fig. 25–29). A dc current can pass through the resistance, but not the capacitance. An ac current, like the changing currents discussed in Section 25–7, can exist also in the capacitive branch. Because of the additional path allowing current flow, the ac current for a given V_{rms} will be greater than for the same dc voltage. Thus an ac voltage is more dangerous than an equal dc voltage.

Another danger is *leakage current*, by which we mean a current along an unintended path. Leakage currents are often capacitively coupled. For example, a wire in a lamp forms a capacitor with the metal case; charges moving in one conductor attract or repel charge in the other, so there is a current. Typical electrical codes limit leakage currents to 1 mA for any device. A 1-mA leakage current is usually harmless. It can be very dangerous, however, to a hospital patient with implanted electrodes connected to ground through the apparatus. This is because the current can pass directly through the heart as compared to the usual situation where the current enters at the hands and spreads out through the body. Although 70 mA may be needed to cause heart fibrillation when entering through the hands (very little of it actually passes through the heart), as little as 0.02 mA has been known to cause fibrillation when passing directly to the heart. Thus, a "wired" patient is in considerable danger from leakage current even from as simple an act as touching a lamp.

Summary

An electric battery serves as a source of potential difference by transforming chemical energy into electric energy. A simple battery consists of two electrodes made of different metals immersed in a solution or paste known as an electrolyte.

Electric current, I, refers to the rate of flow of electric charge and is measured in **amperes** (A): 1 A equals a flow of 1 C/s past a given point.

The direction of **conventional current** flow is that of positive charge. In a wire, it is actually negatively charged electrons that move, so they flow in a direction opposite to the direction of the conventional current. Positive conventional current always flows from a high potential to a low potential.

The **resistance** R of a device is defined by the relation

$$V = IR,$$

where I is the current in the device when a potential difference V is applied across it. For materials such as metals, R is a constant independent of V (thus $I \propto V$), a result known as Ohm's law.

The unit of resistance is the **ohm** (Ω), where $1\,\Omega = 1\,\text{V/A}$. See Table 25–2.

TABLE 25–2	Summary of Units
Current	$1\,\text{A} = 1\,\text{C/s}$
Potential difference	$1\,\text{V} = 1\,\text{J/C}$
Power	$1\,\text{W} = 1\,\text{J/s}$
Resistance	$1\,\Omega = 1\,\text{V/A}$

The resistance R of a wire is inversely proportional to its cross-sectional area A, and directly proportional to its length l and to a property of the material called its resistivity: $R = \rho l / A$. The **resistivity**, ρ, increases with temperature for metals, but for semiconductors it may decrease.

The rate at which energy is transformed in a resistance R from electric to other forms of energy (such as heat and light) is equal to the product of current and voltage. That is, the **power** transformed, measured in watts, is given by

$$P = IV$$

and for resistors can be written as

$$P = I^2 R = \frac{V^2}{R}.$$

The total electric energy transformed in any device equals the product of power and the time during which the device is operated. In SI units, energy is given in joules ($1\,\text{J} = 1\,\text{W} \cdot \text{s}$), but electric companies use a larger unit, the **kilowatt-hour** $\left(1\,\text{kWh} = 3.6 \times 10^6\,\text{J}\right)$.

Electric current can be **direct current (dc)**, in which the current is steady in one direction; or it can be **alternating current (ac)**, in which the current reverses direction at a particular frequency f, typically 60 Hz. Alternating currents are often sinusoidal in time, $I = I_0 \sin \omega t$, where $\omega = 2\pi f$, and are produced by an alternating voltage.

The rms values of sinusoidally alternating currents and voltages are given by

$$I_{\text{rms}} = \frac{I_0}{\sqrt{2}} \quad \text{and} \quad V_{\text{rms}} = \frac{V_0}{\sqrt{2}},$$

respectively, where I_0 and V_0 are the peak values. The power relationship, $P = IV = I^2 R = V^2/R$, is valid for the average power in alternating currents when the rms values of V and I are used.

Current density $\mathbf{j}$ is the current per cross-sectional area. From a microscopic point of view, the current density is related to the number of charge carriers per unit volume, n, their charge, q, and their drift velocity, $\mathbf{v}_{\text{d}}$, by $\mathbf{j} = nq\mathbf{v}_{\text{d}}$. The electric field within a wire is related to $\mathbf{j}$ by $\mathbf{j} = \sigma \mathbf{E}$ where $\sigma = 1/\rho$ is the **conductivity**.

At very low temperatures certain materials become **superconducting**, which means their electrical resistance becomes zero.

Electric shocks are caused by current passing through the body. To avoid shocks, the body must not become part of a circuit by allowing different parts of the body to touch objects at different potentials.

Questions

1. Car batteries can be rated in ampere-hours ($\text{A} \cdot \text{h}$). What aspect of the battery is being rated?

2. When an electric cell is connected to a circuit, electrons flow away from the negative terminal in the circuit. But within the cell, electrons flow *to* the negative terminal. Explain.

3. Develop an analogy between blood circulation and an electrical circuit. Discuss what plays the role of the heart for the electric case, and so on.

4. In a car, one terminal of the battery is said to be connected to "ground." Since it is not really connected to the ground, what is meant by this expression?

5. When you turn on a water faucet, the water usually flows immediately. You don't have to wait for water to flow from the faucet valve to the spout. Why not? Is the same thing true when you connect a wire to the terminals of a battery?

6. Can a copper wire and an aluminum wire of the same length have the same resistance? Explain.

7. If a rectangular solid made of carbon has sides of length a, $2a$, $3a$, how would you connect the wires from a battery so as to obtain (a) the least resistance, (b) the greatest resistance?

8. The equation $P = V^2/R$ indicates that the power dissipated in a resistor decreases if the resistance is increased, whereas the equation $P = I^2R$ implies the opposite. Is there a contradiction here? Explain.

9. What happens when a lightbulb burns out?

10. Explain why lightbulbs almost always burn out just as they are turned on and not after they have been on for some time.

11. Which draws more current, a 100-W lightbulb or a 75-W bulb? Which has the higher resistance?

12. Electric power is transferred over large distances at very high voltages. Explain how the high voltage reduces power losses in the transmission lines.

13. Why is it dangerous to replace a 15-A fuse that blows repeatedly with a 25-A fuse?

14. When electric lights are operated on low-frequency ac (say, 10 Hz) they flicker noticeably. Why?

15. Driven by ac power, the same electrons pass back and forth through your reading lamp over and over again. Explain why the light stays lit instead of going out after the first pass of electrons.

16. The heating element in a toaster is made of Nichrome wire. Immediately after the toaster is turned on, is the current (I_{rms}) in the wire increasing, decreasing, or staying constant? Explain.

17. Is current used up in a resistor?

18. A voltage V is connected across a wire of length l and radius r. How is the electron drift velocity affected if (a) l is doubled, (b) r is doubled, (c) V is doubled.

19. Compare the drift velocities and electric currents in two wires that are geometrically identical and the density of atoms is similar, but the number of free electrons per atom in the material of one wire is twice that in the other.

20. Why is it more dangerous to turn on an electric appliance when you are standing outside in bare feet than when you are inside wearing shoes with thick soles?

Problems

Sections 25–2 and 25–3

1. (I) A current of 1.50 A flows in a wire. How many electrons are flowing past any point in the wire per second? The charge on one electron is 1.60×10^{-19} C.

2. (I) A service station charges a battery using a current of 5.7 A for 7.0 h. How much charge passes through the battery?

3. (I) What is the current in amperes if 1000 Na^+ ions were to flow across a cell membrane in 7.5 μs? The charge on the sodium is the same as on an electron, but positive.

4. (I) What is the resistance of a toaster if 110 V produces a current of 4.2 A?

5. (I) What voltage will produce 0.25 A of current through a 3000-Ω resistor?

6. (II) An electric device draws 5.50 A at 110 V. (a) If the voltage drops by 10 percent, what will be the current, assuming nothing else changes? (b) If the resistance of the device were reduced by 10 percent, what current would be drawn at 110 V?

7. (II) A 9.0-V battery is connected to a bulb whose resistance is 1.6 Ω. How many electrons leave the battery per minute?

8. (II) If a 12-V battery pushes a current of 0.50 A through a resistor, what is its resistance, and how many joules of energy does the battery lose in a minute?

9. (II) A hair dryer draws 7.5 A when plugged into a 120-V line. (a) What is its resistance? (b) How much charge passes through it in 15 min? (Assume direct current.)

10. (II) A bird stands on a dc electric transmission line carrying 2500 A (Fig. 25–30). The line has 2.5×10^{-5} Ω resistance per meter and the bird's feet are 4.0 cm apart. What potential difference does the bird feel?

FIGURE 25–30 Problem 10.

Section 25–4

11. (I) What is the diameter of a 1.00-m length of tungsten wire whose resistance is 0.22 Ω?

12. (I) What is the resistance of a 3.5-m length of copper wire 1.5 mm in diameter?

13. (II) Compare the resistance of 10.0 m of aluminum wire 2.0 mm in diameter with 20.0 m of copper filament wire 2.5 mm in diameter.

14. (II) Can a 2.5-mm-diameter copper wire have the same resistance as a tungsten wire of the same length? Give numerical details.

15. (II) A certain copper wire has a resistance of 10.0 Ω. At what point along the length must the wire be cut so that the resistance of one piece is 5.0 times the resistance of the other? What is the resistance of each piece.

16. (II) How much would you have to raise the temperature of a copper wire (originally at 20°C) to increase its resistance by 20 percent?

17. (II) A length of aluminum wire is connected to a precision 10.00-V power supply, and a current of 0.4212 A is precisely measured at 20.0°C. The wire is placed in a new environment of unknown temperature where the measured current is 0.3618 A. What is the unknown temperature?

18. (II) Estimate at what temperature copper will have the same resistivity as tungsten does at 20°C.

19. (II) A 100-W lightbulb has a resistance of about 12 Ω when off (20°C) and 140 Ω when on (hot). Estimate the temperature of the filament when "on" assuming an average temperature coefficient of resistivity $\alpha = 0.0060\ (\text{C}°)^{-1}$.

20. (II) A rectangular solid made of carbon has sides lying along the x, y, and z axes, whose lengths are 1.0 cm, 2.0 cm, and 4.0 cm, respectively (Fig. 25–31). Determine the resistance for current that flows through the solid in (a) the x direction, (b) the y direction, and (c) the z direction. Assume the resistivity is $\rho = 3.0 \times 10^{-5}\ \Omega \cdot \text{m}$.

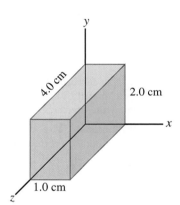

FIGURE 25–31
Problem 20.

21. (II) A length of wire is cut in half and the two lengths are wrapped together side by side to make a thicker wire. How does the resistance of this new combination compare to the resistance of the original wire?

22. (II) For some applications, it is important that the value of a resistance not change with temperature. For example, suppose you made a 4.70-kΩ resistor from a carbon resistor and a Nichrome wire-wound resistor connected together so the total resistance is the sum of their separate resistances. What value should each of these resistors have (at 0°C) so that the combination is temperature independent?

23. (II) (a) Show that if a straight wire of cross-sectional area A lies along the x axis, the rate at which charge flows is given by

$$\frac{dq}{dt} = -\sigma A \frac{dV}{dx},$$

where dV/dx is the potential gradient, and σ is the conductivity. (b) Make an analogy to heat conduction (Chapter 19). Would you expect σ and k (thermal conductivity) to be related?

24. (III) A hollow cylindrical resistor with inner radius r_1 and outer radius r_2, and length l, is made of a material whose resistivity is ρ (Fig. 25–32). (a) Show that the resistance is given by

$$R = \frac{\rho}{2\pi l} \ln \frac{r_2}{r_1}$$

for current that flows radially outward. [*Hint*: Divide the resistor into concentric cylindrical shells and integrate.] (b) Evaluate the resistance R for such a resistor made of carbon whose inner and outer radii are 1.0 mm and 1.8 mm and whose length is 3.0 cm. (c) What is the resistance in part (b) for current flowing *parallel* to the axis?

FIGURE 25–32 Problem 24.

25. (III) Determine a formula for the total resistance of a spherical shell made of material whose conductivity is σ and whose inner and outer radii are r_1 and r_2. Assume the current flows radially outward.

26. (III) The filament of a light bulb has a resistance of 12 Ω at 20°C and 140 Ω when hot (as in Problem 19). (a) Calculate the temperature of the filament when it is hot, and take into account the change in length and area of the filament due to thermal expansion (assume tungsten for which the thermal expansion coefficient is $\approx 5 \times 10^{-6}\ \text{C}°^{-1}$). (b) In this temperature range, what is the percentage change in resistance due to thermal expansion, and what is the percentage change in resistance due solely to the change in ρ? Use Eq. 25–5.

Sections 25–5 and 25–6

27. (I) What is the maximum power consumption of a 9.0-V portable cassette player that draws a maximum of 350 mA of current?

28. (I) The element of an electric oven is designed to produce 3.1 kW of heat when connected to a 240-V source. What must be the resistance of the element?

29. (I) What is the maximum voltage that can be applied to a 5.4-kΩ resistor rated at $\frac{1}{4}$ watt?

30. (I) A hair dryer has two settings: 600 W and 1200 W. (a) At which setting do you expect the resistance to be higher? After making a guess, determine the resistance at (b) the lower setting, (c) the higher setting.

31. (I) (a) What is the resistance and current through a 60-W lightbulb if it is connected to its proper source voltage of 120 V? (b) Repeat for a 150-W bulb.

32. (I) You buy a 60-W lightbulb in Europe, where electricity is delivered to homes at 240 V. If you use the lightbulb in the United States at 120 V (assume its resistance doesn't change), how bright will it be relative to 60-W 120-V bulbs? Estimate how much power it will consume.

33. (II) How many kWh of energy does a 550-W toaster use in the morning if it is in operation for a total of 10 min? At a cost of 12 cents/kWh, how much would this add to your monthly electric energy bill if you made toast five mornings per week?

34. (II) At $0.110 per kWh, what does it cost to leave a 60-W porch light on day and night for a year?

35. (II) What is the total amount of energy stored in a 12-V, 90-A·h car battery when it is fully charged?

36. (II) A transistor to be used in a circuit is rated for a maximum current of 28 mA if operated at 9.0 V. (a) What is the maximum acceptable power to the transistor, and (b) what would be the limiting current if the voltage applied were actually only 7.0 V?

37. (II) How many 100-W lightbulbs, connected to 120 V as in Fig. 25–18, can be used without blowing a 2.5-A fuse?

38. (II) What is the efficiency of a 0.50-hp electric motor that draws 4.6 A from a 120-V line?

39. (II) A power station delivers 520 kW of power to a factory through wires of total resistance of 3.0 Ω. How much less power is wasted if the electricity is delivered at 50,000 V rather than 12,000 V?

40. (II) A 2800-W oven is hooked to a 240-V source. (a) What is the resistance of the oven? (b) How long will it take to boil 100 mL of water assuming 80 percent efficiency? (c) How much will this cost at 10 cents/kWh?

41. (III) The current in an electromagnet connected to a 240-V line is 14.5 A. At what rate must cooling water pass over the coils if the water temperature is to rise by no more than 6.50 C°?

42. (III) A small immersion heater can be used in a car to heat a cup of water for coffee. If the heater can heat 150 mL of water from 5°C to 95°C in 6.0 min, approximately how much current does it draw from the 12-V battery, and what is its resistance? Assume the manufacturer's claim of 60 percent efficiency.

Section 25–7

43. (I) Calculate the peak current in a 1.8-kΩ resistor connected to a 120-V rms ac source.

44. (I) An ac voltage, whose peak value is 180 V, is across a 330-Ω resistor. What is the value of the peak and rms currents in the resistor?

45. (I) What is the resistance of the circuits in your house as seen by the power company, when (a) everything electrical is turned off, and (b) there is a lone 75-W lightbulb burning on the porch?

46. (II) The peak value of an alternating current passing through a 1500-W device is 6.0 A. What is the rms voltage across it?

47. (II) Calculate the peak voltage across, and peak current through, an 1800-W arc welder connected to a 450-V ac line.

48. (II) What is the maximum instantaneous power dissipated by, and maximum current passing through, a 3.0-hp pump connected to a 240-V ac power source?

49. (II) A heater coil connected to a 240-V ac line has a resistance of 38 Ω. (a) What is the average power used? (b) What are the maximum and minimum values of the instantaneous power?

50. (II) Suppose a current is given by the equation $I = 1.80 \sin 210t$, where I is in amperes, and t in seconds. (a) What is the frequency? (b) What is the rms value of the current? (c) If this is the current through a 42.0-Ω resistor, what is the equation that describes the voltage as a function of time?

Section 25–8

51. (II) A 0.55-mm-diameter copper wire carries a tiny current of 2.5 μA. Estimate (a) the electron drift velocity in the wire, (b) the current density, and (c) the electric field.

52. (II) A 5.00-m length of 2.0-mm-diameter wire carries a 750-mA current when 22.0 mV is applied to its ends. If the drift velocity has been measured (by the Hall effect— Section 27–8) to be 1.7×10^{-5} m/s, determine (a) the resistance R of the wire, (b) the resistivity ρ, (c) the current density j, (d) the electric field inside the wire, and (e) the number n of free electrons per unit volume.

53. (III) At a point high in the Earth's atmosphere, He^{2+} ions in a concentration of $2.8 \times 10^{12}/m^3$ are moving due north at a speed of 2.0×10^6 m/s. Also, an $8.0 \times 10^{11}/m^3$ concentration of O_2^- ions is moving due south at a speed of 7.2×10^6 m/s. Determine the magnitude and direction of the current density **j** at this point.

General Problems

54. How many coulombs are there in 1.00 ampere-hour?

55. A person accidentally leaves a car with the lights on. If each of the two front lights uses 40 W and each of the two rear lights 6 W, for a total of 92 W, how long will a fresh 12-V battery last if it is rated at 90 A·h? Assume the full 12 V appears across each bulb.

56. What is the average current drawn by a 1.5-hp 120-V motor?

57. The *conductance* G of an object is defined as the reciprocal of the resistance R: that is, $G = 1/R$. The unit of conductance is the mho $(= \text{ohm}^{-1})$, which is also called the siemens (S). What is the conductance (in siemens) of an object that draws 800 mA of current at 12.0 V?

58. 10.0 m of wire consists of 5.0 m of copper followed by 5.0 m of aluminum of equal diameter (both 1.0 mm). A potential difference of 25 V is placed across the composite wire. (a) What is the total resistance of the wire? (b) What is the current flow through the wire? (c) What are the voltages across the aluminum part and across the copper part?

59. An ordinary flashlight uses 2 D-cell 1.5 V batteries connected in series as in Fig. 25–4b. The bulb draws 350 mA when turned on. (a) Calculate the resistance of the bulb and the power dissipated. (b) By what factor would the power increase if 4 D-cells in series were used with the same bulb? (Neglect heating effects of the filament.) Why shouldn't you try this?

60. The heating element of a 110-V, 900-W heater is 5.4 m long. If it is made of iron, what must its diameter be?

61. (a) A particular household uses a 1.8-kW heater 3.0 h/day ("on" time), four 100-W lightbulbs 6.0 h/day, a 3.0-kW electric stove element for a total of 1.4 h/day, and miscellaneous power amounting to 2.0 kWh/day. If electricity costs $0.105 per kWh, what will be their monthly bill (30 days)? (b) How much coal (which produces 7000 kcal/kg) must be burned by a 35-percent-efficient power plant to provide the yearly needs of this household?

62. A small city requires about 10 MW of power. Suppose that instead of using high-voltage lines to supply the power, the power is delivered at 120 V. Assuming a two-wire line of 0.50-cm-diameter copper wire, estimate the cost of the energy lost to heat per hour per meter. Assume the cost of electricity is about 10 cents per kWh.

63. A 1200-W hair dryer is designed for 117 V. (a) What will be the percentage change in power output if the voltage drops to 105 V? Assume no change in resistance. (b) How would the actual change in resistivity with temperature affect your answer?

64. The wiring in a house must be thick enough so it doesn't become so hot as to start a fire. What diameter must a copper wire be if it is to carry a maximum current of 30 A and produce no more than 1.6 W of heat per meter of length?

65. An air conditioner draws 12 A at 120 V ac. The connecting cord is copper wire which has a diameter of 1.628 mm. (a) How much power does the air conditioner draw? (b) If the total length of wire is 15 m, how much power is dissipated in the wiring? (c) If no. 12 wire, with a diameter of 2.053 mm were used instead, how much power would be dissipated? (d) Assuming that the air conditioner is run 12 hours per day, how much money per day would be saved by using no. 12 wire? Assume that the cost of electricity is 10 cents per kWh.

66. In a "brownout" situation, the voltage supplied by the electric company falls. Assuming the percent drop is small, show that the power output of a given appliance falls by approximately twice that percent, assuming the resistance does not change. How much of a voltage drop does it take for a 60-W lightbulb to begin acting like a 50-W bulb?

67. A microwave oven running at 60 percent efficiency delivers 900 W of energy per second to the interior. Determine (a) the power drawn from the source, and (b) the rms current drawn. Assume a source voltage of 120 Vrms.

68. A 1.00-Ω wire is drawn out to 3.00 times its original length. What is its resistance now?

69. 220 V is applied to two different conductors made of the same material. One conductor is twice as long and twice as thick as the second. What is the ratio of the power transformed in the first relative to the second?

70. An electric heater is used to heat a room of volume 68 m³. Air is brought into the room at 5°C and is changed completely twice per hour. Heat loss through the walls amounts to approximately 850 kcal/h. If the air is to be maintained at 20°C, what minimum wattage must the heater have? (The specific heat of air is about 0.17 kcal/kg·C°.)

71. A 6.50-Ω resistor is made from a coil of copper wire whose total mass is 18.0 g. What is the diameter of the wire and how long is it?

72. The new EV-1 electric car makes use of storage batteries as its source of energy. Its mass is 1300 kg and it is powered by 26 batteries, each 12 V, 52 A·h. Assume that the car is driven on the level at an average speed of 40 km/h, and the average retarding force is 240 N. Assume 100 percent efficiency and neglect energy used for acceleration. Note that no energy is consumed when the vehicle is stopped since the engine doesn't need to idle. (a) Determine the horsepower required. (b) After approximately how many kilometers must the batteries be recharged?

73. A 100-W, 120-V lightbulb has a resistance of 12 Ω when cold (20°C) and 140 Ω when on (hot). Calculate its power consumption at (a) the instant it is turned on, and (b) after a few moments when it is hot.

74. A capacitor is often used in electronics to keep energy flowing even if there is a momentary loss of power from the electric company. What capacitance would be required for a television, operating at an internal dc voltage of 120 V at 150 W, to provide sufficient energy during a 0.10 s lapse in power?

75. The Tevatron accelerator at Fermilab (Illinois) is designed to carry an 11 mA beam of protons traveling at very nearly the speed of light $(3.0 \times 10^8 \text{ m/s})$ around a ring 6300 m in circumference. How many protons are stored in the beam?

76. How far does an average electron move along the wires of a 500-W toaster during an alternating current cycle? The power cord has copper wires of diameter 1.8 mm and is plugged into a standard 60 Hz 120 V ac outlet. [Hint: The maximum current in the cycle is related to the maximum drift velocity. The maximum velocity in an oscillation is related to the maximum displacement; see Chapter 14.]

77. For the wire in Fig. 25–33, whose diameter varies uniformly from a to b as shown, suppose a current $I = 2.0$ A enters at a. If $a = 3.0$ mm and $b = 4.0$ mm, what is the current density (assume uniform) at each end?

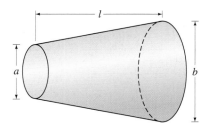

FIGURE 25–33 Problems 77 and 78.

78. The cross section of a portion of wire increases uniformly as shown in Fig. 25–34 so it has the shape of a truncated cone. The diameter at one end is a and at the other it is b, and the total length along the axis is l. If the material has resistivity ρ, determine the resistance R between the two ends in terms of a, b, l, and ρ. Assume that the current flows uniformly through each section, and that the taper is small, i.e., $(b - a) \ll l$.

This "Walkman" contains circuits within it that are dc, at least in part. (The audio signal is ac.) The circuit diagram below shows a possible amplifier circuit (actually, two identical circuits are used, one for each stereo channel). Although the large triangle is an amplifier circuit board containing transistors (not discussed in this chapter), the other circuit elements are ones we have met, resistors and capacitors, and we discuss them in circuits in this chapter. We also discuss voltmeters and ammeters, and how they are built and used.

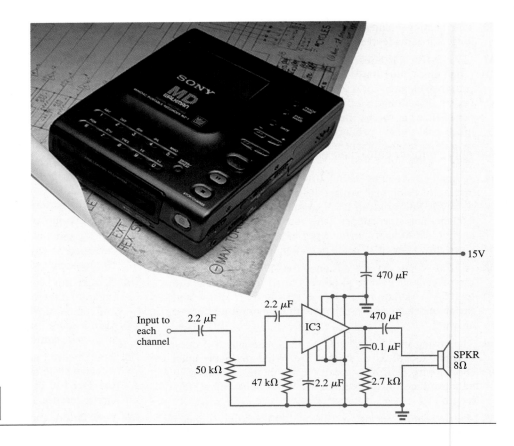

CHAPTER 26

DC Circuits

TABLE 26-1
Symbols for Circuit Elements

Symbol	Device
─┤├─	Battery
─┤├─	Capacitor
─\/\/\─	Resistor
────	Wire with negligible resistance
⏚ or ↓	Ground

Electric circuits are basic parts of all electronic gear from radio and TV sets to computers and automobiles. Scientific measurements, from physics to biology and medicine, make use of electric circuits. In Chapter 25, we discussed the basic principles of electric current. Now we will apply these principles to analyze dc circuits and to understand the operation of a number of useful instruments.[†]

When we draw a diagram for a circuit, we represent batteries, capacitors, and resistors by the symbols shown in Table 26–1. Wires whose resistance is negligible compared to other resistance in the circuit are drawn simply as straight lines. Some circuit diagrams show a ground symbol ($\perp$ or $\downarrow$) which may mean a real connection to the ground, perhaps via a metal pipe, or it may simply mean a common connection, such as the frame of a car.

For the most part in this chapter, except in Section 26–4, we will be interested in circuits operating in their steady state—that is, we won't be looking at a circuit at the moment a change is made in it, such as when a battery or resistor is connected or disconnected, but rather a short time later when the currents have reached their steady values.

[†] Ac circuits that contain only a voltage source and resistors can be analyzed like the dc circuits in this chapter. However, ac circuits that contain capacitors and other circuit elements are more complicated, and we discuss them in Chapter 31.

26–1 EMF and Terminal Voltage

To have current in an electric circuit, we need a device such as a battery or an electric generator that transforms one type of energy (chemical, mechanical, light, and so on) into electric energy. Such a device is called a *source* of *electromotive force* or of *emf*. (The term "electromotive force" is a misnomer since it does not refer to a "force" that is measured in newtons. Hence, to avoid confusion, we prefer to use the abbreviation, emf.) The potential difference between the terminals of such a source, when no current flows to an external circuit, is called the **emf** of the source. The symbol $\mathscr{E}$ is usually used for emf (don't confuse it with E for electric field).

emf

You may have noticed in your own experience that when a current is drawn from a battery, the voltage across its terminals drops below its rated emf. For example, if you start a car with the headlights on, you may notice the headlights dim. This happens because the starter draws a large current, and the battery voltage drops as a result. The voltage drop occurs because the chemical reactions in a battery cannot supply charge fast enough to maintain the full emf. For one thing, charge must flow (within the electrolyte) between the electrodes of the battery, and there is always some hindrance to completely free flow. Thus, a battery itself has some resistance, which is called its **internal resistance**; it is usually designated r. A real battery is then modeled as if it were a perfect emf $\mathscr{E}$ in series with a resistor r, as shown in Fig. 26–1. Since this resistance r is inside the battery, we can never separate it from the battery. The two points a and b in the diagram represent the two terminals of the battery. What we measure is the **terminal voltage** $V_{ab} = V_a - V_b$. When no current is drawn from the battery, the terminal voltage equals the emf, which is determined by the chemical reactions in the battery: $V_{ab} = \mathscr{E}$. However, when a current I flows naturally from the battery there is an internal drop in voltage equal to Ir. Thus the terminal voltage (the actual voltage delivered) is[†]

Why battery voltage isn't constant

FIGURE 26–1 Diagram for an electric cell or battery.

Terminal voltage

$$V_{ab} = \mathscr{E} - Ir.$$

For example, if a 12-V battery has an internal resistance of 0.1 Ω, then when 10 A flows from the battery, the terminal voltage is $12\,\text{V} - (10\,\text{A})(0.1\,\Omega) = 11\,\text{V}$. The internal resistance of a battery is usually small. For example, an ordinary flashlight battery when fresh may have an internal resistance of perhaps 0.05 Ω. (However, as it ages and the electrolyte dries out, the internal resistance increases to many ohms.) Car batteries have even lower internal resistance.

EXAMPLE 26–1 **Battery with internal resistance.** A 65.0-Ω resistor is connected to the terminals of a battery whose emf is 12.0 V and whose internal resistance is 0.5 Ω, Fig. 26–2. Calculate (*a*) the current in the circuit, (*b*) the terminal voltage of the battery, V_{ab}, and (*c*) the power dissipated in the resistor R and in the battery's internal resistance r.

SOLUTION (*a*) From the equation above relating emf $\mathscr{E}$ to terminal voltage, we have

$$V_{ab} = \mathscr{E} - Ir,$$

where $V_{ab} = IR$ (Eq. 25–2). Hence $IR = \mathscr{E} - Ir$ or $\mathscr{E} = I(R + r)$, and so

$$I = \frac{\mathscr{E}}{R + r} = \frac{12.0\,\text{V}}{65.5\,\Omega} = 0.183\,\text{A}.$$

(*b*) The terminal voltage is

$$V_{ab} = \mathscr{E} - Ir = 12.0\,\text{V} - (0.183\,\text{A})(0.5\,\Omega) = 11.9\,\text{V}.$$

(*c*) The power dissipated is

$$P_R = I^2 R = (0.183\,\text{A})^2(65.0\,\Omega) = 2.18\,\text{W}$$
$$P_r = I^2 r = (0.183\,\text{A})^2(0.5\,\Omega) = 0.02\,\text{W}.$$

FIGURE 26–2 Example 26–1.

[†]When a battery is being charged, a current is forced to pass through it and we have to write $V_{ab} = \mathscr{E} + Ir$. See Example 26–10.

In much of what follows, unless stated otherwise, we assume that the battery's internal resistance is negligible, and that the battery voltage given is its terminal voltage, which we will usually write simply as V rather than V_{ab}.

26–2 Resistors in Series and in Parallel

When two or more resistors are connected end to end as shown in Fig. 26–3, they are said to be connected in **series**. The resistors could be simple resistors as were pictured in Fig. 25–10, or they could be lightbulbs, heating elements, or other resistive devices. Any charge that passes through R_1 in Fig. 26–3a will also pass through R_2 and then R_3. Hence the same current I passes through each resistor. (If it did not, this would imply that charge was accumulating at some point in the circuit, which does not happen in the steady state.) We let V represent the voltage across all three resistors. We assume all other resistance in the circuit can be ignored, and so V equals the terminal voltage of the battery. We let V_1, V_2, and V_3 be the potential differences across each of the resistors, R_1, R_2, and R_3, respectively, as shown in Fig. 26–3a. From $V = IR$, we can write $V_1 = IR_1$, $V_2 = IR_2$, and $V_3 = IR_3$. Because the resistors are connected end to end, energy conservation tells us that the total voltage V is equal to the sum of the voltages across each resistor:

Series circuit: voltages add; current the same in each R

$$V = V_1 + V_2 + V_3 = IR_1 + IR_2 + IR_3. \qquad \text{[series]} \quad (26\text{–}1)$$

To see in more detail why this is true, note that an electric charge q passing through R loses potential energy by qV_1. In passing through R_2 and R_3, the potential energy U decreases by qV_2 and qV_3, for a total $\Delta U = qV_1 + qV_2 + qV_3$; this sum must equal the energy given to q by the battery, qV, so that energy is conserved. Hence $qV = q(V_1 + V_2 + V_3)$, and so $V = V_1 + V_2 + V_3$, which is Eq. 26–1.

Now let us determine the equivalent single resistance R_{eq} that would draw the same current as our combination; see Fig. 26–3c. Such a single resistance R_{eq} would be related to V by

$$V = IR_{eq}.$$

We equate this expression with Eq. 26–1, $V = I(R_1 + R_2 + R_3)$, and find

Resistances in series

$$R_{eq} = R_1 + R_2 + R_3. \qquad \text{[series]} \quad (26\text{–}2)$$

This is, in fact, what we expect. When we put several resistances in series, the total resistance is the sum of the separate resistances. This applies to any number of resistances in series. Note that when you add more resistance to the circuit, the current will decrease. For example, if a 12-V battery is connected to a 4-Ω resistor, the current will be 3 A. But if the 12-V battery is connected to three 4-Ω resistors in series, the total resistance is 12 Ω and the current will be only 1 A.

FIGURE 26–3 (a) Resistances connected in series: $R_{eq} = R_1 + R_2 + R_3$. (b) Resistances could be lightbulbs, or any other type of resistance. (c) Equivalent single resistance R_{eq} that draws the same current.

Another simple way to connect resistors is in **parallel**, so that the current from the source splits into separate branches, as shown in Fig. 26–4. The wiring in houses and buildings is arranged so all electric devices are in parallel, as we already saw in Chapter 25, Fig. 25–18. With parallel wiring, if you disconnect one device (say R_1 in Fig. 26–4), the current to the others is not interrupted. But in a series circuit, if one device (say R_1 in Fig. 26–3) is disconnected, the current *is* stopped to all the others.

In a parallel circuit, Fig. 26–4a, the total current I that leaves the battery breaks into three branches. We let I_1, I_2, and I_3 be the currents through each of the resistors, R_1, R_2, and R_3, respectively. Because electric charge is conserved, the current flowing into a junction (where different wires or conductors meet) must equal the current flowing out of the junction. Thus, in Fig. 26–4a,

$$I = I_1 + I_2 + I_3.$$ [parallel]

Parallel circuit:
currents add;
voltage the same across each R

When resistors are connected in parallel, each experiences the same voltage. (Indeed, any two points in a circuit connected by a wire of negligible resistance are at the same potential.) Hence the full voltage of the battery is applied to each resistor in Fig. 26–4a, so

$$I_1 = \frac{V}{R_1}, \qquad I_2 = \frac{V}{R_2}, \qquad \text{and} \qquad I_3 = \frac{V}{R_3}.$$

Let us now determine what single resistor R_{eq} (Fig. 26–4c) will draw the same current I as these three resistances in parallel. This equivalent resistance R_{eq} must satisfy

$$I = \frac{V}{R_{\text{eq}}}.$$

We now combine the equations above:

$$I = I_1 + I_2 + I_3,$$
$$\frac{V}{R_{\text{eq}}} = \frac{V}{R_1} + \frac{V}{R_2} + \frac{V}{R_3}.$$

When we divide out the V from each term, we have

$$\frac{1}{R_{\text{eq}}} = \frac{1}{R_1} + \frac{1}{R_2} + \frac{1}{R_3}.$$ [parallel] **(26–3)**

Resistances in parallel

For example, suppose you connect two 4-Ω loudspeakers to a single set of output terminals of your stereo amplifier or receiver. (Ignore the other channel for a moment—our two speakers are both connected to the left channel, say.) The equivalent resistance will then be found from

$$\frac{1}{R_{\text{eq}}} = \frac{1}{4\,\Omega} + \frac{1}{4\,\Omega} = \frac{2}{4\,\Omega} = \frac{1}{2\,\Omega}$$

and so $R_{\text{eq}} = 2\,\Omega$. Thus the net resistance is *less* than that of each single resistance. This may at first seem surprising. But remember that when you put resistors in parallel, you are giving the current additional paths to follow. Hence the net resistance will be less.

FIGURE 26–4 (a) Resistances connected in parallel: $1/R_{\text{eq}} = 1/R_1 + 1/R_2 + 1/R_3$, which could be (b) lightbulbs; (c) shows the equivalent circuit with R_{eq} obtained from Eq. 26–3.

(a)   (b)   (c)

FIGURE 26–5 Water pipes in parallel—analogy to electric currents in parallel.

FIGURE 26–6 Example 26–2.

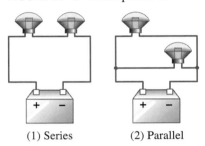

(1) Series   (2) Parallel

FIGURE 26–7 Example 26–3.

(a)

(b)

An analogy may help here. Consider two pipes taking in water near the top of a dam and releasing it below as shown in Fig. 26–5. The gravitational potential difference, proportional to the height h, is the same for both pipes, just as in the electrical case of parallel resistors. If both pipes are open, rather than only one, twice as much water current will flow through. That is, with two equal pipes open, the net resistance to the flow of water will be reduced, by half. Note that if both pipes are closed, the dam offers infinite resistance to the flow of water. This corresponds in the electrical case to an open circuit—when no current flows—so the electrical resistance is infinite.

Notice that the forms of the equations for resistors, Eqs. 26–2 and 26–3, are just the reverse of their counterparts for capacitors, Chapter 24, Eqs. 24–3 and 24–4. That is, the formula for resistors in series has the same form as the formula for capacitors in parallel, and vice versa.

CONCEPTUAL EXAMPLE 26–2 **Series or Parallel?** (a) The lightbulbs in Fig. 26–6 are identical and have identical resistance R. Which configuration produces more light? (b) Which way do you think the headlights of a car are wired?

RESPONSE (a) The parallel combination has lower resistance ($= R/2$) than the series combination ($= 2R$). There will be more total current in configuration (2). The total power transformed, which is proportional to the light produced, is $P = IV$, so the greater current in (2) means more light.
(b) In parallel (2), because if one bulb goes out, the other bulb can stay lit. If they were in series (1), when one bulb burned out (the filament broke), the circuit would be open and no current would flow, so even the good bulb would not light.

EXAMPLE 26–3 **Series and parallel resistors.** Two 100-Ω resistors are connected (a) in series, and (b) in parallel, to a 24.0-V battery. See Fig. 26–7. What is the current through each resistor and what is the equivalent resistance of each circuit?

SOLUTION (a) All the current that flows out of the battery passes first through R_1 and then R_2. So the current I is the same in both resistors; and the potential difference across the battery, V, equals the total change in potential across the two resistors:

$$V = V_1 + V_2 = IR_1 + IR_2.$$

Hence

$$I = \frac{V}{R_1 + R_2} = \frac{24.0 \text{ V}}{100 \text{ }\Omega + 100 \text{ }\Omega} = 0.120 \text{ A}.$$

The equivalent resistance, using Eq. 26–2, is $R_{eq} = R_1 + R_2 = 200 \text{ }\Omega$. We could also get R_{eq} by thinking from the point of view of the battery: the total resistance R_{eq} must equal the battery voltage divided by the current it puts out: $R_{eq} = V/I = 24.0 \text{ V}/0.120 \text{ A} = 200 \text{ }\Omega$.
[Note that the voltage across R_1 is $V_1 = IR_1 = (0.120 \text{ A})(100 \text{ }\Omega) = 12.0 \text{ V}$, and that across R_2 is $V_2 = IR_2 = 12.0 \text{ V}$, each being half of the battery voltage. A simple circuit like Fig. 26–7a is thus often called a simple "voltage divider."]
(b) Any given charge (or electron) can flow through only one or the other of the two resistors. Just as a river may break into two streams when going around an island, here too the total current I from the battery (Fig. 26–7b) equals the sum of the separate currents through the two resistors:

$$I = I_1 + I_2.$$

The potential difference across each resistor is the battery voltage $V = 24.0 \text{ V}$. Hence

$$I = I_1 + I_2 = \frac{V}{R_1} + \frac{V}{R_2} = \frac{24.0 \text{ V}}{100 \text{ }\Omega} + \frac{24.0 \text{ V}}{100 \text{ }\Omega} = 0.24 \text{ A} + 0.24 \text{ A} = 0.48 \text{ A}.$$

The equivalent resistance is

$$R_{eq} = \frac{V}{I} = \frac{24.0\,\text{V}}{0.48\,\text{A}} = 50\,\Omega.$$

We could also have obtained this result from Eq. 26–3:

$$\frac{1}{R_{eq}} = \frac{1}{100\,\Omega} + \frac{1}{100\,\Omega} = \frac{2}{100\,\Omega} = \frac{1}{50\,\Omega},$$

and so $R_{eq} = 50\,\Omega$.

EXAMPLE 26–4 **Circuit with series and parallel.** How much current flows from the battery shown in Fig. 26–8a?

SOLUTION The current I that flows out of the battery all passes through the 400-Ω resistor, but then it splits into I_1 and I_2 passing through the 500-Ω and 700-Ω resistors. The latter two are in parallel. We look for simplicity, something that we already know how to treat. So let's start by finding the equivalent resistance, R_P, of the parallel resistors, 500-Ω and 700-Ω:

$$\frac{1}{R_P} = \frac{1}{500\,\Omega} + \frac{1}{700\,\Omega} = 0.0020\,\Omega^{-1} + 0.0014\,\Omega^{-1} = 0.0034\,\Omega^{-1}.$$

This is $1/R_P$, so we take the reciprocal to find R_P. (It is a common mistake to forget to take this reciprocal. Notice that the units of reciprocal ohms, Ω^{-1}, are a reminder.) Thus

$$R_P = \frac{1}{0.0034\,\Omega^{-1}} = 290\,\Omega.$$

This 290 Ω is the equivalent resistance of the two parallel resistors, and is in series with the 400-Ω resistor as shown in the equivalent circuit of Fig. 26–8b. To find the total equivalent resistance R_{eq}, we add the 400-Ω and 290-Ω resistances together, since they are in series, and find

$$R_{eq} = 400\,\Omega + 290\,\Omega = 690\,\Omega.$$

The total current flowing from the battery is then

$$I = \frac{V}{R_{eq}} = \frac{12.0\,\text{V}}{690\,\Omega} = 0.017\,\text{A} = 17\,\text{mA}.$$

FIGURE 26–8 (a) Circuit for Examples 26–4 and 26–5. (b) Equivalent circuit, showing the equivalent resistance of 290 Ω for the two parallel resistors in (a).

EXAMPLE 26–5 **Current in one branch.** What is the current flowing through the 500-Ω resistor in Fig. 26–8a?

SOLUTION We need to find the voltage V_{bc} across the 500-Ω resistor, and then apply $V = IR$ to get the current. First we find the voltage across the 400-Ω resistor, V_{ab}, since we know that 17 mA passes through it; V_{ab} can be found using $V = IR$:

$$V_{ab} = (0.017\,\text{A})(400\,\Omega) = 6.8\,\text{V}.$$

Since the total voltage across the network of resistors is $V_{ac} = 12.0\,\text{V}$, then V_{bc} must be $12.0\,\text{V} - 6.8\,\text{V} = 5.2\,\text{V}$. Then Ohm's law tells us that the current I_1 through the 500-Ω resistor is

$$I_1 = \frac{5.2\,\text{V}}{500\,\Omega} = 1.0 \times 10^{-2}\,\text{A} = 10\,\text{mA}.$$

This is the answer we wanted. We can also calculate the current I_2 through the 700-Ω resistor since the voltage across it is also 5.2 V:

$$I_2 = \frac{5.2\,\text{V}}{700\,\Omega} = 7\,\text{mA}.$$

Notice that when I_1 combines with I_2 to form the total current I (at point c in Fig. 26–8a), their sum is $10\,\text{mA} + 7\,\text{mA} = 17\,\text{mA}$. This is, of course, the total current as calculated in Example 26–4.

FIGURE 26–9 Example 26–6, three identical lightbulbs.

FIGURE 26–10 Example 26–7.

FIGURE 26–11 Currents can be calculated using Kirchhoff's rules.

CONCEPTUAL EXAMPLE 26–6 | **Bulb brightness in a circuit.** The circuit shown in Fig. 26–9 has three identical lightbulbs, each of resistance R. How will the brightness of bulbs A and B compare with that of bulb C?

RESPONSE The current that passes through bulb C splits into two equal parts when it reaches the junction leading to bulbs A and B, because the resistance of bulbs A and B are equal. Thus, bulbs A and B receive half the current, and will be less bright than bulb C.

EXAMPLE 26–7 | **Resistor "ladder."** (a) Estimate the equivalent resistance of the "ladder" of equal 100-Ω resistors shown in Fig. 26–10a. In other words, what resistance would an ohmmeter read if connected between points A and B? (b) What is the current through each of the three resistors on the left if a 50.0 V battery is connected between points A and B?

SOLUTION It may seem that none of these resistors is in series or parallel. But there is a place to start: the three resistors on the far right are definitely in series with one another. Each has resistance $R (= 100\,\Omega)$, so those last three have a net resistance of $3R (= 300\,\Omega)$. Next we can see that this combination is in parallel with the next resistor to the left, as shown in the dashed box in Fig. 26–10b. The equivalent resistance of the resistors in the dashed box (b), call it R_{eq1}, is given by

$$\frac{1}{R_{eq1}} = \frac{1}{R} + \frac{1}{3R} = \frac{4}{3R}, \quad \text{so} \quad R_{eq1} = \frac{3R}{4}$$

which numerically is $300\,\Omega/4 = 75\,\Omega$. Next, this equivalent resistance of $3R/4$ is in series with the next two (Fig. 26–10c). The resistors in the dashed box (c) are in series, and are equivalent to $2R + 3R/4 = 11R/4$ (which equals $1100\,\Omega/4 = 275\,\Omega$). Now this $11R/4$ is in parallel with the next step on the ladder (Fig. 26–10d). The resistance in the dashed box (d) is equivalent to

$$\frac{1}{R_{eq2}} = \frac{1}{R} + \frac{4}{11R} = \frac{15}{11R}, \quad \text{so} \quad R_{eq2} = \frac{11R}{15}.$$

This in turn is in series with two more, yielding the final equivalent resistance of

$$R_{eq} = \frac{11R}{15} + R + R = \frac{41}{15}R.$$

We are given that $R = 100\,\Omega$, so $R_{eq} = 273\,\Omega$. Notice again the value of using algebra all along: we can have an answer that enables us to calculate the overall resistance regardless of the value of the individual resistor.

(b) If a 50.0 V battery is placed between points A and B, then the current issuing from the battery, and passing through the two resistors on the left, is $I = 50.0\,\text{V}/273\,\Omega = 0.183\,\text{A}$. The current through the first cross resistance is less than this since this R is in parallel with a net resistance of $2\frac{3}{4}R = \frac{11}{4}R$ (Fig. 26–10d). Let I_1 be the current through this R and I_2 the current in the $\frac{11}{4}R$, with $I_1 + I_2 = I$. The potential difference across R is the same as that across the $\frac{11}{4}R$, so $I_1 R = I_2(\frac{11}{4}R)$, and so $I_2 = \frac{4}{11}I_1$. Then

$$I = I_1 + I_2 = \left(1 + \tfrac{4}{11}\right)I_1 = \tfrac{15}{11}I_1.$$

Hence $I_1 = \frac{11}{15}I = \frac{11}{15}(0.183\,\text{A}) = 0.134\,\text{A}$.

26–3 Kirchhoff's Rules

In the last few Examples we have been able to find the currents flowing in circuits by combining resistances in series and parallel. This technique can be used for many circuits. However, some circuits are too complicated for that analysis. For example, we cannot find the currents flowing in each part of the circuit shown in Fig. 26–11 simply by combining resistances as we did before.

To deal with such complicated circuits, we use Kirchhoff's rules, devised by G. R. Kirchhoff (1824–1887) in the mid-nineteenth century. There are two of them, and they are simply convenient applications of the laws of conservation of charge and energy. **Kirchhoff's first** or **junction rule** is based on the conservation of charge, and we already used it in deriving the rule for parallel resistors. It states that

> **at any junction point, the sum of all currents entering the junction must equal the sum of all currents leaving the junction.**

Junction rule
(conservation of charge)

That is, what goes in must come out. For example, at the junction point a in Fig. 26–11, I_3 is entering whereas I_1 and I_2 are leaving. Thus Kirchhoff's junction rule states that $I_3 = I_1 + I_2$. We already saw an instance of this at the end of Example 26–5.

      Kirchhoff's second or **loop rule** is based on the conservation of energy. It states

> **the sum of the changes in potential around any closed path of a circuit must be zero.**

Loop rule
(conservation of energy)

To see why this should hold, consider a rough analogy with the potential energy of a roller coaster on its track. When it starts from the station, it has a particular potential energy. As it climbs the first hill, its potential energy increases and reaches a peak at the top. As it descends the other side, its potential energy decreases and reaches a local minimum at the bottom of the hill. As the roller coaster continues on its path, its potential energy goes through more changes. But when it arrives back at the starting point, it has exactly as much potential energy as it had when it started at this point. Another way of saying this is that there was as much uphill as there was downhill.

      Similar reasoning can be applied to an electric circuit. We will do the circuit of Fig. 26–11 shortly but first we consider the simpler circuit in Fig. 26–12. We have chosen it to be the same as the equivalent circuit of Fig. 26–8b already discussed. The current in this circuit is $I = (12.0 \text{ V})/(690 \, \Omega) = 0.017 \text{ A}$, as we calculated in Example 26–4. The positive side of the battery, point e in Fig. 26–12a, is at a high potential compared to point d at the negative side of the battery. That is, point e is like the top of a hill for a roller coaster. We can now follow the current around the circuit starting at any point we choose. Let us start at point e and follow a positive test charge completely around this circuit. As we go, we will note all changes in potential. When the test charge returns to point e, the potential there will be the same as when we started, so the total change in potential will be zero. It is useful to plot the changes in voltage around the circuit, and we do this in Fig. 26–12b; point d is arbitrarily taken as zero. As our positive test charge goes from point e to point a, there is no change in potential since there is no source of potential nor any resistance. However, as the charge passes through the 400-Ω resistor to get to point b, there is a decrease in potential of $V = IR = (0.017 \text{ A})(400 \, \Omega) = 6.8 \text{ V}$. In effect, the positive test charge is flowing "downhill" since it is heading toward the negative terminal of the battery. This is indicated in the graph of Fig. 26–12b. The decrease in potential between the two ends of a resistor ($= IR$) is called a **voltage drop**. Because this is a *decrease* in potential, we use a *negative* sign when applying Kirchhoff's loop rule; that is,

$$V_{ba} = V_b - V_a = -6.8 \text{ V}.$$

As the charge proceeds from b to c there is another voltage drop of $(0.017 \text{ A}) \times (290 \, \Omega) = 5.2 \text{ V}$, and since this is a decrease in potential, we write

$$V_{cb} = -5.2 \text{ V}.$$

There is no change in potential as our test charge moves from c to d. But when it moves from d, which is the negative or low potential side of the battery, to point e which is the positive terminal, the potential *increases* by 12.0 V. That is,

$$V_{ed} = +12.0 \text{ V}.$$

The sum of all the changes in potential in going around the circuit of Fig. 26–12 is

$$-6.8 \text{ V} - 5.2 \text{ V} + 12.0 \text{ V} = 0.$$

And this is exactly what Kirchhoff's loop rule said it would be.

(a)

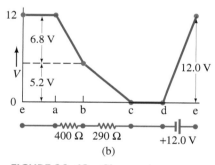

(b)

FIGURE 26–12 Changes in potential around the circuit in (a) are plotted in (b).

➡ **PROBLEM SOLVING**

Be consistent with signs

FIGURE 26–13 Currents can be calculated using Kirchhoff's rules. See Example 26–8.

When using Kirchhoff's rules, we will designate the current in each separate branch of the given circuit by a different subscript, such as I_1, I_2, and I_3 in Fig. 26–13 (this is the same circuit as in Fig. 26–11). You do not have to know in advance in which direction these currents actually are moving. You make a guess and calculate the potentials around the circuit for that direction. If the current actually flows in the opposite direction, your answer will have a negative sign.

EXAMPLE 26–8 Using Kirchhoff's rules. Calculate the currents I_1, I_2, and I_3 in each of the branches of the circuit in Fig. 26–13.

SOLUTION Since (positive) current tends to move away from the positive terminal of a battery, we assume I_2 and I_3 to have the directions shown in Fig. 26–13. The direction of I_1 is not obvious in advance, so we arbitrarily chose the direction indicated. We have three unknowns and therefore we need three equations. We first apply Kirchhoff's junction rule to the currents at point a, where I_3 enters and I_2 and I_1 leave:

$$I_3 = I_1 + I_2. \quad (a)$$

This same equation holds at point d, so we get no new information there. We now apply Kirchhoff's loop rule to two different closed loops. First we apply it to the loop ahdcba. We start (and end) at point a. From a to h we have a voltage drop $V_{ha} = -(I_1)(30\,\Omega)$. From h to d there is no change, but from d to c the potential increases by 45 V: that is, $V_{cd} = +45$ V. From c to a the voltage drops through the two resistances by an amount $V_{ac} = -(I_3)(40\,\Omega + 1\,\Omega)$. Thus we have $V_{ha} + V_{cd} + V_{ac} = 0$, or

$$-30I_1 + 45 - (40 + 1)I_3 = 0 \quad (b)$$

where we have omitted the units. For our second loop, we take the complete circuit ahdefga. (We could have just as well taken abcdefg instead.) Again we start at point a and have $V_{ha} = -(I_1)(30\,\Omega)$, and $V_{dh} = 0$. But when we take our positive test charge from d to e, it actually is going uphill, against the flow of current—or at least against the *assumed* direction of the current, which is what counts in this calculation. Thus $V_{ed} = I_2(20\,\Omega)$ has a *positive* sign. Similarly, $V_{fe} = I_2(1\,\Omega)$. From f to g there is a decrease in potential of 80 V since we go from the high potential terminal of the battery to the low. Thus $V_{gf} = -80$ V. Finally, $V_{ag} = 0$, and the sum of the potential charges around this loop is then

$$-30I_1 + (20 + 1)I_2 - 80 = 0. \quad (c)$$

The physics is now done. The rest is algebra. We have three equations—labeled (a), (b), and (c)—in three unknowns. From Eq. (c) we have

$$I_2 = \frac{80 + 30I_1}{21} = 3.8 + 1.4I_1. \quad (d)$$

From Eq. (b) we have

$$I_3 = \frac{45 - 30I_1}{41} = 1.1 - 0.73I_1. \quad (e)$$

We substitute these into Eq. (a) and solve for I_1:

$$I_1 = I_3 - I_2 = 1.1 - 0.73I_1 - 3.8 - 1.4I_1$$
$$3.1I_1 = -2.7$$
$$I_1 = -0.87 \text{ A}.$$

The negative sign indicates that the direction of I_1 is actually opposite to that initially assumed and shown in Fig. 26–13. Note that the answer automatically comes out in amperes because all values were in volts and ohms. From Eq. (d) we have

$$I_2 = 3.8 + 1.4I_1 = 2.6 \text{ A},$$

and from Eq. (e)

$$I_3 = 1.1 - 0.73I_1 = 1.7 \text{ A}.$$

This completes the solution.

1. Label + and − for each battery. The long side of a battery symbol is +.

2. Label the current in each branch of the circuit with a symbol and an arrow (as in Fig. 26–13): The direction of the arrow can be chosen arbitrarily. If the current is actually in the opposite direction, it will come out with a minus sign in the solution.

3. Apply Kirchhoff's junction rule at one or more junctions, and the loop rule for one or more loops. You will need as many independent equations as there are unknowns. You may write down more equations than this, but you will find that some of the equations will be redundant (that is, not be independent in the sense of providing new information). You may use $V = IR$ for each resistor, which sometimes will reduce the number of unknowns.

4. In applying the loop rule, follow each loop in one direction only. Pay careful attention to subscripts, and to signs:

   (a) For a resistor, the sign of the potential difference is negative if your chosen loop direction is the same as the chosen current direction through that resistor; the sign is positive if you are moving opposite to the chosen current direction.

   (b) For a battery, the sign of the potential difference is positive if your loop direction moves from the negative terminal toward the positive; the sign is negative if you are moving from the positive terminal toward the negative terminal.

5. Solve the equations algebraically for the unknowns. Be careful in manipulating equations not to err with signs. At the end, check your answers by plugging them into the original equations, or even by using any additional equations not used previously (either loop or junction rule equations).

EXAMPLE 26–9 Wheatstone bridge. A wheatstone bridge is a type of "bridge circuit" used to make measurements of resistance. The unknown resistance to be measured, R_x, is placed in the circuit with accurately known resistances R_1, R_2, and R_3. One of these, R_3, is a variable resistor which is adjusted so that when the switch is closed momentarily, the ammeter Ⓐ shows zero current flow. (We will see later in this Chapter how an ammeter works.) (a) Determine R_x in terms of R_1, R_2, and R_3. (b) If a Wheatstone bridge is "balanced" when $R_1 = 630\,\Omega$, $R_2 = 972\,\Omega$, and $R_3 = 42.6\,\Omega$, what is the value of the unknown resistance?

SOLUTION (a) We are told that R_3 has been adjusted until no current flows through the ammeter. Hence points B and D in Fig. 26–14 are at the same potential, so $V_{AB} = V_{AD}$ or

$$I_3 R_3 = I_1 R_1 .$$

I_1 is the current that passes through R_1 and also through R_2 when the bridge is balanced; I_3 is the current through R_3 and R_x. When the bridge is balanced the voltage across R_x equals that across R_2, so

$$I_3 R_x = I_1 R_2 .$$

We divide these two equations and find

$$R_x = \frac{R_2}{R_1} R_3 .$$

In practice, the ammeter is very sensitive, and so when R_3 is being adjusted the switch is closed only momentarily to check if the current is zero or not.

(b) $$R_x = \frac{R_2}{R_1} R_3 = \left(\frac{972\,\Omega}{630\,\Omega} \right)(42.6\,\Omega) = 65.7\,\Omega .$$

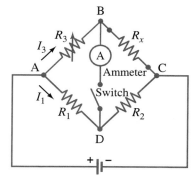

FIGURE 26–14 Example 26–9, Wheatstone bridge.

EMFs in Series and in Parallel; Charging a Battery

When two or more sources of emf, such as batteries, are arranged in series, as in Fig. 26–15a, the total voltage is the algebraic sum of their respective voltages. On the other hand, when a 20-V and a 12-V battery are connected oppositely, as shown in Fig. 26–15b, the net voltage V_{ca} is 8 V. That is, a positive test charge moved from a to b gains in potential by 20 V, but when it passes from b to c it drops by 12 V. So the net change is 20 V − 12 V = 8 V. You might think that connecting batteries in reverse like this would be wasteful. And for most purposes that would be true. But such a reverse arrangement is precisely how a battery charger works. In Fig. 26–15b, the 20-V source is charging up the 12-V battery. Because of its greater voltage, the 20-V source is forcing charge back into the 12-V battery: electrons are being forced into its negative terminal and removed from its positive terminal. An automobile alternator keeps the car battery charged in the same way. A voltmeter placed across the terminals of a (12-V) car battery with the engine running fairly fast can tell you whether or not the alternator is charging the battery. If it is, the voltmeter reads 13 or 14 V. If the battery is not being charged, the voltage will be 12 V, or less if the battery is discharging. Car batteries can be recharged, but other batteries may not be rechargeable, since the chemical reactions in many cannot be reversed. In such cases, the arrangement of Fig. 26–15b would simply waste energy.

FIGURE 26–15 Batteries in series, (a) and (b), and in parallel, (c).

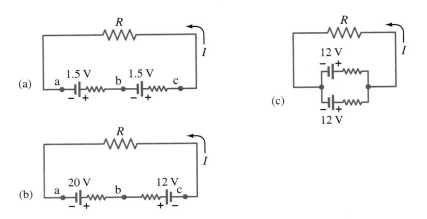

Sources of emf can also be arranged in parallel, Fig. 26–15c, which is useful normally only if the emfs are the same. A parallel arrangement is not used to increase voltage, but rather to provide more energy when large currents are needed. Each of the cells in parallel has to produce only a fraction of the total current, so the loss due to internal resistance is less than for a single cell; and the batteries will go dead less quickly.

FIGURE 26–16 Example 26–10, a jump start.

EXAMPLE 26–10 Jump starting a car. A good car battery is being used to jump start a car with a weak battery. The good battery has an emf of 12.5 V and internal resistance 0.020 Ω. Suppose the weak battery has an emf of 10.1 V and internal resistance 0.10 Ω. Each copper jumper cable is 3.0 m long and 0.50 cm in diameter, and can be attached as shown in Fig. 26–16. Assume the starter motor can be represented as a resistor $R_s = 0.15\,\Omega$. Determine the current through the starter motor (a) if only the weak battery is connected to it, and (b) if the good battery is also connected, as shown in Fig. 26–16.

SOLUTION (a) The circuit is simple: an emf of 10.1 V connected to two resistances in series, 0.10 Ω + 0.15 Ω = 0.25 Ω. Hence the current is $I = V/R = (10.1\ \text{V})/(0.25\ \Omega) = 40\ \text{A}$.

(b) We need to find the resistance of the jumper cables that connect the good battery. From Eq. 25–3, each has resistance $R_J = \rho L/A = (1.68 \times 10^{-8}\,\Omega\cdot\text{m})(3.0\,\text{m})/(\pi)(0.25 \times 10^{-2}\,\text{m})^2 = 0.0026\,\Omega$. Kirchhoff's loop rule for the full outside loop gives

$$12.5\,\text{V} - I_1(2R_J + r) - I_3 R_S = 0$$
$$12.5\,\text{V} - I_1(0.025\,\Omega) - I_3(0.15\,\Omega) = 0. \qquad (a)$$

The loop rule for the lower loop, including the weak battery and the starter, gives

$$10.1\,\text{V} - I_3(0.15\,\Omega) - I_2(0.10\,\Omega) = 0. \qquad (b)$$

The junction rule at point B gives

$$I_1 + I_2 = I_3. \qquad (c)$$

We have three equations in three unknowns. We combine equations (a) and (c), eliminating I_1, to obtain

$$12.5\,\text{V} - (I_3 - I_2)(0.025\,\Omega) - I_3(0.15\,\Omega) = 0$$
$$12.5\,\text{V} - I_3(0.175\,\Omega) + I_2(0.025\,\Omega) = 0.$$

Combining this last with (b) gives $I_3 = 71\,\text{A}$. Quite a bit better than in (a). The other currents are $I_2 = -6.2\,\text{A}$ and $I_1 = 77\,\text{A}$.

The circuit shown in Fig. 26–16, without the starter motor, is how a battery can be charged. The stronger battery pushes charge back into the weaker battery. Note in this Example that $I_2 = -6.2\,\text{V}$ and so is in the opposite direction from that assumed in Fig. 26–16. The terminal voltage of the weak 10.1 V battery is thus $V_{\text{BA}} = 10.1\,\text{V} + (6.2\,\text{A})(0.10\,\Omega) = 10.7\,\text{V}$

EXAMPLE 26–11 **Jumper cables reversed.** What would happen if the jumper cables of Example 26–10 were mistakenly connected in reverse, the positive terminal of each battery connected to the negative terminal of the other battery? Why could this be dangerous?

SOLUTION The circuit is shown in Fig. 26–17. Even before the starter motor is engaged (the switch S in Fig. 26–17 is open) there is trouble if the batteries are connected in this way. By Kirchhoff's loop rule in the single (upper) loop carrying current I, we have

$$12.5\,\text{V} - I(2R_J + 0.10\,\Omega + 0.02\,\Omega) + 10.1\,\text{V} = 0$$

where each $R_J = 0.0026\,\Omega$. We solve for I:

$$I = \frac{22.6\,\text{V}}{0.125\,\Omega} = 180\,\text{A}.$$

The extremely high current through the batteries could cause them to become very hot and explode. For example, the power dissipated in the weak battery would be $P = I^2 r = (180\,\text{A})^2(0.10\,\Omega) = 3200\,\text{W}!$

FIGURE 26–17 Example 26–11.

Circuits Containing Resistor and Capacitor (RC Circuits)

Our study of circuits in this chapter has, until now, dealt with steady currents that don't change in time. Now we examine circuits that contain both resistance and capacitance. Such a circuit is called an **RC circuit**. RC circuits are common in everyday life: they are used to control the speed of a car's windshield wiper, and the timing of the change of a traffic light from red to green. They are used in camera flashes and in heart pacemakers.

RC circuit

Charging the capacitor

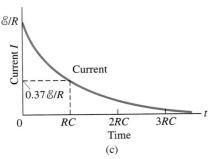

FIGURE 26–18 For the *RC* circuit shown in (a), the charge on the capacitor increases with time as shown in (b), and the current through the resistor decreases with time as shown in (c).

Charge on charging capacitor

Voltage across charging capacitor

Time constant = RC

Let us now examine the simple *RC* circuit shown in Fig. 26–18a. When the switch S is closed, current immediately begins to flow through the circuit. Electrons will flow out from the negative terminal of the battery, through the resistor *R*, and accumulate on the upper plate of the capacitor. And electrons will flow into the positive terminal of the battery, leaving a positive charge on the other plate of the capacitor. As charge accumulates on the capacitor, the potential difference across it increases; and the current is reduced until eventually the voltage across the capacitor equals the emf of the battery, $\mathscr{E}$. There is then no potential difference across the resistor, and no further current flows. The charge *Q* on the capacitor thus increases gradually as shown in Fig. 26–18b and reaches a maximum value equal to $C\mathscr{E}$ (Eq. 24–1, $Q_{max} = CV_{ba} = C\mathscr{E}$). The mathematical form of this curve—that is, *Q* as a function of time—can be derived using conservation of energy (or Kirchhoff's loop rule). The emf $\mathscr{E}$ of the battery will equal the sum of the voltage drops across the resistor (*IR*) and the capacitor (*Q/C*):

$$\mathscr{E} = IR + \frac{Q}{C}. \tag{26-4}$$

The resistance *R* includes all resistance in the circuit, including the internal resistance of the battery; *I* is the current in the circuit at any instant, and *Q* is the charge on the capacitor at that same instant. Although $\mathscr{E}$, *R*, and *C* are constants, both *Q* and *I* are functions of time. The rate at which charge flows through the resistor ($I = dQ/dt$) is equal to the rate at which charge accumulates on the capacitor. Thus we can write

$$\mathscr{E} = R\frac{dQ}{dt} + \frac{1}{C}Q.$$

This equation can be solved by rearranging it:

$$\frac{dQ}{C\mathscr{E} - Q} = \frac{dt}{RC}.$$

We now integrate from $t = 0$, when $Q = 0$, to time *t* when a charge *Q* is on the capacitor:

$$\int_0^Q \frac{dQ}{C\mathscr{E} - Q} = \frac{1}{RC}\int_0^t dt$$

$$-\ln(C\mathscr{E} - Q) - (-\ln C\mathscr{E}) = \frac{t}{RC}$$

or

$$\ln(C\mathscr{E} - Q) - \ln(C\mathscr{E}) = -\frac{t}{RC}$$

so

$$\ln\left(1 - \frac{Q}{C\mathscr{E}}\right) = -\frac{t}{RC}.$$

We take the exponential of both sides

$$1 - \frac{Q}{C\mathscr{E}} = e^{-t/RC}$$

or

$$Q = C\mathscr{E}(1 - e^{-t/RC}). \tag{26-5a}$$

The potential difference across the capacitor is $V_C = Q/C$, so

$$V_C = \mathscr{E}(1 - e^{-t/RC}). \tag{26-5b}$$

From Eqs. 26–5 we see that the charge *Q* on the capacitor, and the voltage V_C across it, increase from zero at $t = 0$ to maximum values $Q_{max} = C\mathscr{E}$ and $V_C = \mathscr{E}$ after a very long time. The quantity *RC* that appears in the exponent is called the **time constant** τ of the circuit:

$$\tau = RC.$$

(The units of *RC* are $\Omega \cdot \text{F} = (\text{V/A})(\text{C/V}) = \text{C}/(\text{C/s}) = \text{s.}$) It represents the time required for the capacitor to reach $(1 - e^{-1}) = 0.63$ or 63 percent of its full charge. Thus the product RC is a measure of how quickly the capacitor gets charged. In a circuit, for example, where $R = 200\,k\Omega$ and $C = 3.0\,\mu F$, the time constant is $(2.0 \times 10^5\,\Omega)(3.0 \times 10^{-6}\,F) = 0.60\,s$. If the resistance is much lower, the time constant is much smaller. This makes sense, since a lower resistance will retard the flow of charge less. All circuits contain some resistance (if only in the connecting wires), so a capacitor never can be charged instantaneously when connected to a battery.

From Eqs. 26–5, it appears that Q and V_C never quite reach their maximum values within a finite time. However, they reach 86 percent of maximum in $2RC$, 95 percent in $3RC$, 98 percent in $4RC$, and so on. Q and V_C approach their maximum values asymptotically. For example, if $R = 20\,k\Omega$ and $C = 0.30\,\mu F$, the time constant is $(2.0 \times 10^4\,\Omega)(3.0 \times 10^{-7}\,F) = 6.0 \times 10^{-3}\,s$. So the capacitor is more than 98 percent charged in less than $\frac{1}{40}$ of a second.

The current I through the circuit of Fig. 26–18a at any time t can be obtained by differentiating Eq. 26–5a:

$$I = \frac{dQ}{dt} = \frac{\mathscr{E}}{R} e^{-t/RC}. \qquad (26\text{–}6)$$

Current in resistor

Thus, at $t = 0$, the current is $I = \mathscr{E}/R$, as expected for a circuit containing only a resistor (there is not yet a potential difference across the capacitor). The current then drops exponentially in time with a time constant equal to RC. This is shown in Fig. 26–18c. The time constant RC represents the time required for the current to drop to $1/e \approx 0.37$ of its initial value.

EXAMPLE 26–12 **RC circuit, with emf.** The capacitance in the circuit of Fig. 26–18a is $C = 0.30\,\mu F$, the total resistance is $20\,k\Omega$, and the battery emf is 12 V. Determine (a) the time constant, (b) the maximum charge the capacitor could acquire, (c) the time it takes for the charge to reach 99 percent of this value, (d) the current I when the charge Q is half its maximum value, (e) the maximum current, and (f) the charge Q when, the current I is 0.20 its maximum value.

SOLUTION (a) The time constant is $RC = (2.0 \times 10^4\,\Omega)(3.0 \times 10^{-7}\,F) = 6.0 \times 10^{-3}\,s$.
(b) The maximum charge would be $Q = C\mathscr{E} = (3.0 \times 10^{-7}\,F)(12\,V) = 3.6\,\mu C$.
(c) In Eq. 26–5a, we set $Q = 0.99C\mathscr{E}$:

$$0.99C\mathscr{E} = C\mathscr{E}(1 - e^{-t/RC}),$$

or

$$e^{-t/RC} = 1 - 0.99 = 0.01.$$

Then

$$\frac{t}{RC} = -\ln(0.01) = 4.6$$

so

$$t = 4.6RC = 28 \times 10^{-3}\,s$$

or 28 ms (less than $\frac{1}{30}\,s$).
(d) From part (b) the maximum charge is $3.6\,\mu C$. When the charge is half this value, $1.8\,\mu C$, the current I in the circuit can be found using the original differential equation, or Eq. 26–4:

$$I = \frac{1}{R}\left(\mathscr{E} - \frac{Q}{C}\right) = \frac{1}{2.0 \times 10^4\,\Omega}\left(12\,V - \frac{1.8 \times 10^{-6}\,C}{0.30 \times 10^{-6}\,F}\right) = 300\,\mu A.$$

(e) The current is a maximum when there is no charge on the capacitor ($Q = 0$):

$$I_{max} = \frac{\mathscr{E}}{R} = \frac{12\,V}{2.0 \times 10^4\,\Omega} = 600\,\mu A.$$

(f) Again using Eq. 26–4, with $I = 0.20 I_{max} = 120\,\mu A$, we have

$$Q = C(\mathscr{E} - IR) = (3.0 \times 10^{-7}\,F)\left[12\,V - (1.2 \times 10^{-4}\,A)(2.0 \times 10^4\,\Omega)\right] = 2.9\,\mu C.$$

FIGURE 26-19 For the *RC* circuit

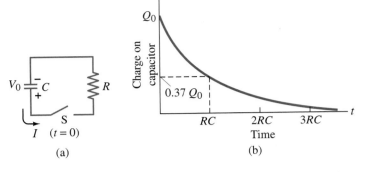

shown in (a), the charge *Q* on the
capacitor decreases with time, as
shown in (b), after the switch S is
closed at $t = 0$. The voltage across
the capacitor follows the same curve
since $V \propto Q$.

The circuit just discussed involved the *charging* of a capacitor by a battery through a resistance. Now let us look at another situation: when a capacitor is already charged (say to a voltage V_0), and it is allowed to *discharge* through a resistance *R* as shown in Fig. 26–19a. (In this case there is no battery.) When the switch S is closed, charge begins to flow through resistor *R* from one side of the capacitor toward the other side, until the capacitor is fully discharged. The voltage across the resistor at any instant equals that across the capacitor:

Capacitor discharges

$$IR = \frac{Q}{C}.$$

The rate at which charge leaves the capacitor equals the negative of the current in the resistor, $I = -dQ/dt$, because the capacitor is discharging (*Q* is decreasing). So we write the above equation as

$$-\frac{dQ}{dt}R = \frac{Q}{C}.$$

We rearrange this to

$$\frac{dQ}{Q} = -\frac{dt}{RC}$$

and integrate it from $t = 0$ when the charge on the capacitor is Q_0, to some time t when the charge is Q:

$$\ln\frac{Q}{Q_0} = -\frac{t}{RC}$$

or

Charge on discharging capacitor

$$Q = Q_0 e^{-t/RC}. \quad (26\text{–}7)$$

Thus the charge on the capacitor decreases exponentially in time with a time constant *RC*. This is shown in Fig. 26–19b. The current is

Current in resistor

$$I = -\frac{dQ}{dt} = \frac{Q_0}{RC}e^{-t/RC} = I_0 e^{-t/RC}, \quad (26\text{–}8)$$

and it too is seen to decrease exponentially in time with the same time constant *RC*. Both the charge on the capacitor and the voltage across it ($V_C = Q/C$), as well as the current in the resistor, decrease to 37 percent of their original value in one time constant $t = \tau = RC$.

EXAMPLE 26–13 Discharging *RC* circuit. In the *RC* circuit shown in Fig. 26–20, the battery has fully charged the capacitor, so $Q_0 = C\mathscr{E}$. Then at $t = 0$ the switch is thrown from position a to b. The battery emf is 20.0 V, and the capacitance $C = 1.02\ \mu$F. The current *I* is observed to decrease to 0.50 of its initial value in 40 μs. (*a*) What is the value of *R*? (*b*) What is the value of *Q*, the charge on the capacitor, at $t = 0$? (*c*) What is *Q* at $t = 60\ \mu$s?

FIGURE 26–20 Example 26–13.

SOLUTION (*a*) At $t = 0$, the battery is removed from the circuit and the capacitor begins discharging through the resistor, as in Fig. 26–19. At any time t later (Eq. 26–7) we have

$$Q = Q_0 e^{-t/RC} = C\mathscr{E}e^{-t/RC},$$

and

$$I = -\frac{dQ}{dt} = \frac{\mathscr{E}}{R}e^{-t/RC} = I_0 e^{-t/RC}.$$

At $t = 40\,\mu s$, $I = 0.50 I_0$. Hence

$$0.50 I_0 = I_0 e^{-t/RC}$$

or, taking natural logs on both sides ($\ln 0.50 = -0.693$):

$$0.693 = \frac{t}{RC}$$

and

$$R = \frac{t}{(0.693)C} = \frac{(40 \times 10^{-6}\,\text{s})}{(0.693)(1.02 \times 10^{-6}\,\text{F})} = 57\,\Omega.$$

(*b*) At $t = 0$,

$$Q = Q_0 = C\mathscr{E} = (1.02 \times 10^{-6}\,\text{F})(20.0\,\text{V}) = 20.4\,\mu C.$$

(*c*) At $t = 60\,\mu s$,

$$Q = Q_0 e^{-\frac{60 \times 10^{-6}\,\text{s}}{(57\,\Omega)(1.02 \times 10^{-6}\,\text{F})}} = 7.3\,\mu C.$$

Applications of *RC* Circuits

The charging and discharging in an *RC* circuit can be used to produce voltage pulses at a regular frequency. The charge on the capacitor increases to a particular voltage, and then discharges. A simple way of initiating the discharge is by the use of a gas-filled tube that breaks down when the voltage across it reaches a certain value V_0. After the discharge is finished, the tube no longer conducts current and the recharging process repeats itself, starting at V_0'. Figure 26–21 shows a possible circuit, and the "sawtooth" voltage it produces.

An automobile turn signal indicator can be an application of a sawtooth oscillator circuit. Here the emf is supplied by the car battery ($\mathscr{E} = 12$ V); the neon bulb, which flashes on at a rate of perhaps 2 cycles per second, is the turn signal indicator. The main component of the "flasher unit" is a moderately large capacitor.

The intermittent windshield wipers of a car can also use an *RC* circuit. The *RC* time constant, which can be changed using a multi-positioned switch for different values of R with fixed C, determines the rate at which the wipers come on.

Another interesting use of an *RC* circuit is the electronic heart pacemaker, which can make a stopped heart start beating again by applying an electric stimulus through electrodes attached to the chest. The stimulus can be repeated at the normal heartbeat rate if necessary. The heart itself contains *pacemaker* cells, which send out tiny electric pulses at a rate of 60 to 80 per minute. These signals induce the start of each heartbeat. In some forms of heart disease, the natural pacemaker fails to function properly, and the heart loses its beat. People suffering from this ailment now commonly make use of *electronic pacemakers* which produce a regular voltage pulse that starts and controls the frequency of the heartbeat. The electrodes are implanted in or near the heart and the circuit usually contains a capacitor and a resistor. The charge on the capacitor increases to a certain point and then discharges. Then it starts charging again. The pulsing rate depends on the values of R and C.

➡ **PHYSICS APPLIED**

Sawtooth voltage; turn signals and wipers; pacemakers

FIGURE 26–21 (a) An *RC* circuit, coupled with a gas-filled tube as a switch, can produce a repeating "sawtooth" voltage, as shown in (b).

(a)

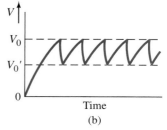

(b)

* 26–5 DC Ammeters and Voltmeters

FIGURE 26–22 A multimeter used as a voltmeter.

Ammeter uses shunt resistor

An **ammeter** is used to measure current, and a **voltmeter** measures potential difference or voltage. The crucial part of an analog ammeter or voltmeter, in which the reading is by a pointer on a scale (Fig. 26–22), is a *galvanometer*. The galvanometer works on the principle of the force between a magnetic field and a current-carrying coil of wire, and will be discussed in Chapter 27. For now, we merely need to know that the deflection of the needle of a galvanometer is proportional to the current flowing through it. The *full-scale current sensitivity*, I_m, of a galvanometer is the current needed to make the needle deflect full scale.

A galvanometer can be used directly to measure small dc currents. For example, a galvanometer whose sensitivity I_m is $50\,\mu\text{A}$ can measure currents from about $1\,\mu\text{A}$ (currents smaller than this would be hard to read on the scale) up to $50\,\mu\text{A}$. To measure larger currents, a resistor is placed in parallel with the galvanometer. Thus an ammeter, represented by the symbol •—Ⓐ—•, consists of a galvanometer (•—Ⓖ—•) in parallel with a resistor called the **shunt resistor**, ("shunt" is a synonym for "in parallel"), as shown in Fig. 26–23. The shunt resistance is R_{sh}, and the resistance of the galvanometer coil, through which current passes, is r. The value of R_{sh} is chosen according to what full-scale deflection is desired and is normally very small, giving an ammeter a very small internal resistance.

FIGURE 26–23 An ammeter is a galvanometer in parallel with a (shunt) resistor with low resistance, R_{sh}.

EXAMPLE 26–14 **Ammeter design.** Design an ammeter to read $1.0\,\text{A}$ at full scale using a galvanometer with a full-scale sensitivity of $50\,\mu\text{A}$ and a resistance $r = 30\,\Omega$. Check if the scale is linear.

SOLUTION When the total current I entering the ammeter is $1.0\,\text{A}$, we want the current I_G through the galvanometer to be precisely $50\,\mu\text{A}$ (to give full-scale deflection). See Fig. 26–23. Thus, when $1.0\,\text{A}$ flows into the meter, we want $0.999950\,\text{A}\,(= I_R)$ to pass through the shunt resistor R_{sh}. Since the potential difference across the shunt is the same as across the galvanometer,

$$I_R R_{sh} = I_G r,$$

then

$$R_{sh} = \frac{I_G r}{I_R} = \frac{(5.0 \times 10^{-5}\,\text{A})(30\,\Omega)}{(0.999950\,\text{A})} = 1.5 \times 10^{-3}\,\Omega,$$

or $0.0015\,\Omega$. The shunt resistor must thus have a very low resistance so that most of the current passes through it.

If the current I into the meter is $0.50\,\text{A}$, this will produce a current to the galvanometer equal to $I_G = I_R R_{sh}/r = (0.50\,\text{A})(1.5 \times 10^{-3}\,\Omega)/30\,\Omega = 25\,\mu\text{A}$, which gives a deflection half of full scale, as required.

Voltmeter uses series resistor

A voltmeter (•—Ⓥ—•) also consists of a galvanometer and a resistor. But the resistor R_{ser} is connected in series, Fig. 26–24, and it is usually large, giving a voltmeter a high internal resistance.

FIGURE 26–24 A voltmeter is a galvanometer in series with a resistor with high resistance, R_{ser}.

EXAMPLE 26–15 **Voltmeter design.** Using the same galvanometer with internal resistance $r = 30\,\Omega$ and full-scale current sensitivity of $50\,\mu\text{A}$, design a voltmeter that reads from 0 to 15 V. Is the scale linear?

SOLUTION When a potential difference of 15 V exists across the terminals of our voltmeter, we want $50\,\mu\text{A}$ to be passing through it so as to give a full-scale deflection. From Ohm's law, $V = IR$, we have (see Fig. 26–24)

$$15\,\text{V} = (50\,\mu\text{A})(r + R_{\text{ser}}),$$

so

$$R_{\text{ser}} = \frac{15\,\text{V}}{5.0 \times 10^{-5}\,\text{A}} - r = 300\,\text{k}\Omega - 30\,\Omega = 300\,\text{k}\Omega.$$

Notice that $r = 30\,\Omega$ is so small compared to the value of R_{ser} that it doesn't influence the calculation significantly.

The scale will again be linear: if the voltage to be measured is 6.0 V, the current passing through the voltmeter will be $(6.0\,\text{V})/(3.0 \times 10^5\,\Omega) = 2.0 \times 10^{-5}\,\text{A}$, or $20\,\mu\text{A}$. This will produce two-fifths of full-scale deflection, as required $(6.0\,\text{V}/15.0\,\text{V} = 2/5)$.

The meters discussed above are for direct current. A dc meter can be modified to measure ac with the addition of diodes which allow current to flow in one direction only. An ac meter can be calibrated to read rms or peak values.

*Use of Voltmeters and Ammeters

Suppose you wish to determine the current I in the circuit shown in Fig. 26–25a, and the voltage V across the resistor R_1. How exactly are ammeters and voltmeters connected to the circuit being measured?

Because an ammeter is used to measure the current flowing in the circuit, it must be inserted directly into the circuit, in series with the other elements, as shown in Fig. 26–25b. The smaller its internal resistance, the less it will affect the circuit.

A voltmeter, on the other hand, is connected in parallel with the circuit element across which the voltage is to be measured. It is used to measure the potential difference between two points and its two wire leads (connecting wires) are connected to the two points, as shown in Fig. 26–25c where the voltage across R_1 is being measured. The larger its internal resistance ($R_{\text{ser}} + r$ in Fig. 26–24) the less it affects the circuit being measured.

Voltmeters and ammeters can have several series or shunt resistors to offer a choice of range. **Multimeters** can measure voltage, current, and resistance. Sometimes they are called VOMs (Volt-Ohm-Meter). Meters with digital readout are called digital voltmeters (DVM) or digital multimeters (DMM), Fig. 26–26.

To measure resistance, the meter must contain a battery of known voltage connected in series to a resistor (R_{ser}) and to an ammeter. This makes an **ohmmeter**, Fig. 26–27. The resistor whose resistance is to be measured completes the circuit. The deflection is inversely proportional to the resistance. The calibration of the scale depends on the value of the series resistor. Since an ohmmeter sends a current through the device whose resistance is to be measured, it should not be used on very delicate devices that could be damaged by the current.

The **sensitivity** of a meter is generally specified on the face. It may be given as so many ohms per volt, which indicates how many ohms of resistance there are in the meter per volt of full-scale reading. For example, if the sensitivity is $30,000\,\Omega/\text{V}$, this means that on the 10-V scale the meter has a resistance of $300,000\,\Omega$. The full-scale current sensitivity, I_m, discussed earlier, is just the reciprocal of the sensitivity in Ω/V.

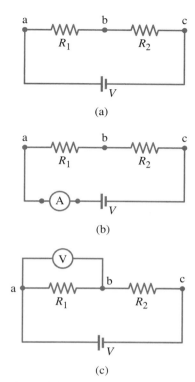

(a)

(b)

(c)

FIGURE 26–25 Measuring current and voltage.

FIGURE 26–26 A digital multimeter being used to measure resistance.

FIGURE 26–27 An ohmmeter.

* Effects of Meter Resistance

It is important to know the sensitivity of a meter, for in many cases the resistance of the meter can seriously affect your results. Take the following Example.

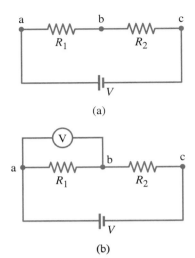

(a)

(b)

FIGURE 26–28   Example 26–16.

EXAMPLE 26–16   Voltage reading versus true voltage.   Suppose you are testing an electronic circuit which has two resistors, R_1 and R_2, each 15 kΩ, connected in series as shown in Fig. 26–28a. The battery maintains 8.0 V across them and has negligible internal resistance. A voltmeter whose sensitivity is 10,000 Ω/V is put on the 5.0-V scale. What voltage does the meter read when connected across R_1, and what error is caused by the finite resistance of the meter?

SOLUTION   On the 5.0-V scale, the voltmeter has an internal resistance of $(5.0\,\text{V})(10,000\,\Omega/\text{V}) = 50,000\,\Omega$. When connected across R_1, as in Fig. 26–28b, we have this 50 kΩ in parallel with $R_1 = 15\,\text{k}\Omega$. The net resistance R_{eq} of these two is given by

$$\frac{1}{R_{eq}} = \frac{1}{50\,\text{k}\Omega} + \frac{1}{15\,\text{k}\Omega} = \frac{13}{150\,\text{k}\Omega};$$

so $R_{eq} = 11.5\,\text{k}\Omega$. This $R_{eq} = 11.5\,\text{k}\Omega$ is in series with $R_2 = 15\,\text{k}\Omega$, so the total resistance of the circuit is now 26.5 kΩ. Hence the current from the battery is

$$I = \frac{8.0\,\text{V}}{26.5\,\text{k}\Omega} = 0.30\,\text{mA}.$$

Then the voltage drop across R_1, which is the same as that across the voltmeter, is $(3.0 \times 10^{-4}\,\text{A})(11.5 \times 10^3\,\Omega) = 3.5\,\text{V}$. [The voltage drop across R_2 is $(3.0 \times 10^{-4}\,\text{A})(15 \times 10^3\,\Omega) = 4.5\,\text{V}$, for a total of 8.0 V.] If we assume the meter is precise, it will read 3.5 V. In the normal circuit, without the meter, $R_1 = R_2$ so the voltage across R_1 is half that of the battery, or 4.0 V. Thus the voltmeter, because of its internal resistance, gives a low reading. In this case it is off by 0.5 V, or more than 10 percent.

Example 26–16 illustrates how seriously a meter can affect a circuit and give a misleading reading. If the resistance of a voltmeter is much higher than the resistance of the circuit, however, it will have little effect and its readings can be trusted, at least to the manufactured precision of the meter, which for ordinary analog meters is typically 3 to 4 percent of full-scale deflection. An ammeter also can interfere with a circuit, but the effect is minimal if its resistance is much less than that of the circuit as a whole. For both voltmeters and ammeters, the more sensitive the galvanometer the less effect it will have. A 50,000-Ω/V meter is far better than a 1,000-Ω/V meter.

Electronic voltmeters using transistors, including digital meters, have very high input resistance (usually specified in ohms) in the range 10^6 to $10^8\,\Omega$, and even higher. Hence they have very little effect on most circuits and their readings are reliable for most circuits. The precision of digital meters is typically one part in 10^4 (= 0.01 percent) or better. State-of-the-art instruments reach a precision of one part in 10^6.

* 26–6   Transducers and the Thermocouple

A *transducer* is a device that converts one type of energy into another. A high fidelity loudspeaker is one kind of transducer—it transforms electric energy into sound energy (see Chapter 27). So is a microphone, which changes sound into an electrical signal.

FIGURE 26–29 Wire strain gauge.

Transducers are often used for measuring particular quantities. In this case they can be thought of as changing one kind of signal into another kind, usually electrical. The *resistance thermometer* and the *thermistor* mentioned in the previous chapter are examples; basically they translate a change in temperature into a change in an electrical property, resistance.

A *strain gauge*, Fig. 26–29, uses the elasticity of a wire that will stretch an amount proportional to the stress (or force) applied to it. When stretched, its resistance must increase since it is longer and its cross-sectional area is reduced (see Eq. 25–3). The fine wire of a strain gauge is usually bonded to a flexible backing material. When fixed tightly to a structure, the resistance of the strain gauge changes in direct proportion to any change in stress on the structure. Since the wire must not be strained beyond its elastic limit, the change in length is generally quite small. Hence the change in resistance is quite small (less than one part per thousand) and a very sensitive Wheatstone bridge is used to measure it. Strain gauges must be carefully calibrated and are used in many applications, such as by architects and engineers on models of proposed structures to determine the stress at critical points. A wire strain gauge can be attached to a membrane or diaphragm. A change in pressure against the membrane causes strain in the wire. Thus a strain gauge can be used to measure pressure, and is then called a *pressure transducer*.

Another kind of pressure transducer makes use of the *piezoelectric effect*. This effect occurs in certain crystals, such as quartz, that become polarized (Section 24–5) when a mechanical force is applied and produce an emf proportional to the force.

The *thermocouple* is a device that produces an electrical signal when subjected to differing temperatures. Unlike the resistance thermometer, in which it is the resistance that changes with temperature, a thermocouple produces an emf and is based on the "thermoelectric effect." When two dissimilar metals, say iron and copper, are joined at the ends, as shown in Fig. 26–30a, it is found that an emf is produced if the two junctions are at different temperatures.[†] The magnitude of this emf depends on the temperature difference. In operation, one junction of a thermocouple is kept at a known temperature. This "reference temperature" is often 0°C. The other junction, called the "test junction," is placed where the desired temperature is to be measured. The emf is measured by some precise means, Fig. 26–30b, whose terminals need to be made of the same metal, and kept at the same temperature, so that no additional emfs are produced. Often the thermocouple junctions are connected to lead-in wires, and these additional connections must also be kept at the same temperature.

A *microphone* is a transducer that changes a sound wave into an electrical signal. One type of microphone transducer is the capacitor (or condenser) microphone, Fig. 26–31. The changing air pressure in a sound wave causes one plate of the capacitor C to move back and forth. We saw in Chapter 24 that the capacitance is inversely proportional to the separation of the plates. Thus a sound wave causes the capacitance to change. This in turn causes the charge Q on the plates to change ($Q = CV$) so that an electric current is generated at the same frequencies as the incoming sound wave.

[†]Part of the theoretical explanation is that electrons in one metal occupy lower energy states than in the other, so some of them flow across the junction. This leaves one metal slightly more positive than the other, and so a *contact potential* exists between them. If the two junctions are at the same temperature, the same contact potential exists at each; they balance each other and no current flows. But if one of the junctions is at a higher temperature, the energy states are altered and the contact potentials will be different. In this case there will be a net emf and a current will flow.

➡ **PHYSICS APPLIED**

Strain gauge, piezoelectric, thermocouple, microphones

FIGURE 26–30 Thermocouple.

FIGURE 26–31 Diagram of a capacitor microphone.

Summary

A device that transforms one type of energy into electrical energy is called a **source** of **emf**. A battery behaves like a source of emf in series with an **internal resistance**. The emf is the potential difference determined by the chemical reactions in the battery and equals the terminal voltage when no current is drawn. When a current is drawn, the voltage at the battery's terminals is less than its emf by an amount equal to the Ir drop across the internal resistance.

When resistances are connected in **series** (end to end), the equivalent resistance is the sum of the individual resistances:

$$R_{eq} = R_1 + R_2 + \cdots.$$

When resistors are connected in **parallel**, the reciprocal of the total resistance equals the sum of the reciprocals of the individual resistances:

$$\frac{1}{R_{eq}} = \frac{1}{R_1} + \frac{1}{R_2} + \cdots.$$

In a parallel connection, the net resistance is less than any of the individual resistances.

Kirchhoff's rules are helpful in determining the currents and voltages in circuits. Kirchhoff's **junction rule** is based on conservation of electric charge and states that the sum of all currents entering any junction equals the sum of all currents leaving that junction. The second, or **loop rule**, is based on conservation of energy and states that the algebraic sum of the voltage changes around any closed path of the circuit must be zero.

When an **RC circuit** containing a resistor R in series with a capacitance C is connected to a dc source of emf, the voltage across the capacitor rises gradually in time characterized by an exponential of the form $\left(1 - e^{-t/RC}\right)$ where the **time constant**, $\tau = RC$, is the time it takes for the voltage to reach 63 percent of its maximum value. The current through the resistor decreases as $e^{-t/RC}$.

A capacitor discharging through a resistor is characterized by the same time constant: in a time $\tau = RC$, the voltage across the capacitor drops to 37 percent of its initial value. The charge on the capacitor, and voltage across it, decreases as $e^{-t/RC}$, as does the current.

Questions

1. Explain why birds can sit on power lines safely, while leaning a metal ladder up against one to fetch a stuck kite is extremely dangerous.
2. Discuss the advantages and disadvantages of Christmas tree lights connected in parallel versus those connected in series.
3. If all you have is a 120-V line, would it be possible to light several 6-V lamps without burning them out? How?
4. Two lightbulbs of resistance R_1 and R_2 $(>R_1)$ are connected in series. Which is brighter? What if they are connected in parallel?
5. Describe carefully the difference between emf and potential difference.
6. Household outlets are often double outlets. Are these connected in series or parallel? How do you know?
7. With two identical lightbulbs and two identical batteries, how would you arrange the bulbs and batteries in a circuit in order to get the maximum possible total power out. (Assume that the batteries have negligible internal resistance.)
8. Explain why Kirchhoff's junction rule is based on conservation of electric charge.
9. Explain why Kirchhoff's loop rule is a result of the conservation of energy.
10. How does the overall resistance of your room's electric circuit change when instead of having a single 60-W lightbulb on, you turn on an additional 100-W bulb?
11. Given the circuit shown in Fig. 26–32, use the words "increases," "decreases," or "stays the same" to complete the following statements:
    (a) If R_7 increases, the potential difference between A and E (assume no resistance in Ⓐ and ℰ) _____.

FIGURE 26–32
Question 11.

(b) If R_7 increases, the potential difference between A and E (assume Ⓐ and ℰ have resistance) _____.
(c) If R_7 increases, the voltage drop across R_4 _____.
(d) If R_2 decreases, the current through R_1 _____.
(e) If R_2 decreases, the current through R_6 _____.
(f) If R_2 decreases, the current through R_3 _____.
(g) If R_5 increases, the voltage drop across R_2 _____.
(h) If R_5 increases, the voltage drop across R_4 _____.
(i) If R_2, R_5, and R_7 increase, ℰ _____.

12. Why are batteries connected in series? Why in parallel? Does it matter if the batteries are nearly identical or not in either case?

13. Can the terminal voltage of a battery ever exceed its emf? Explain.

14. The 18-V source in Fig. 26–33 is "charging" the 12-V battery. Explain how it does this.

FIGURE 26–33 Questions 14 and 18; and Problem 24.

15. Explain in detail how you could measure the internal resistance of a battery.

16. Compare and discuss the formulas for the equivalent values for resistors and for capacitors when connected in series and in parallel.

17. Suppose that three identical capacitors are connected to a battery. Will they store more energy if connected in series or in parallel?

18. When applying Kirchhoff's loop rule (such as in Fig. 26–33), does the sign (or direction) of a battery's emf depend on the direction of current through the battery?

19. In an *RC* circuit, current flows from the battery until the capacitor is completely charged. Is the total energy supplied by the battery equal to the total energy stored by the capacitor? If not, where does the extra energy go?

20. Design a circuit in which two different switches of the type shown in Fig. 26–34 can be used to operate the same lightbulb from opposite sides of a room.

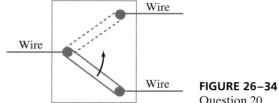

FIGURE 26–34
Question 20.

* **21.** What is the main difference between a voltmeter and an ammeter?

* **22.** What would happen if you mistakenly used an ammeter where you needed to use a voltmeter?

* **23.** Explain why an ideal ammeter would have zero resistance and an ideal voltmeter infinite resistance.

Problems

Section 26–1

1. (I) Calculate the terminal voltage for a battery with an internal resistance of 0.900 Ω and an emf of 8.50 V when the battery is connected in series with (*a*) a 68.0-Ω resistor, and (*b*) a 680-Ω resistor.

2. (I) Four 2.0-V cells are connected in series to a 12-Ω lightbulb. If the resulting current flow is 0.62 A, what is the internal resistance of each cell, assuming they are identical and neglecting the wires?

3. (II) A 1.5-V dry cell can be tested by connecting it to a low-resistance ammeter. It should be able to supply at least 25 A. What is the internal resistance of the cell in this case?

4. (II) What is the internal resistance of a 12.0-V car battery whose terminal voltage drops to 9.8 V when the starter draws 60 A? What is the resistance of the starter?

Section 26–2

In the following Problems neglect the internal resistance of a battery unless the Problem refers to it.

5. (I) Four 90-Ω lightbulbs are connected in series. What is the total resistance of the circuit? What is their resistance if they are connected in parallel?

6. (I) Three 40-Ω lightbulbs and three 80-Ω lightbulbs are connected in series. (*a*) What is the total resistance of the circuit? (*b*) What is their resistance if all six are wired in parallel?

7. (I) Given only one 25-Ω and one 70-Ω resistor, list all possible values of resistance that can be obtained.

8. (I) Suppose that you have a 500-Ω, a 900-Ω, and a 1.40-kΩ resistor. What is (*a*) the maximum, and (*b*) the mininium resistance you can obtain by combining these?

9. (II) Suppose that you have a 6.0-V battery and you wish to apply a voltage of only 4.0 V. Given an unlimited supply of 1.0-Ω resistors, how could you connect them so as to make a "voltage divider" that produced a 4.0-V output for a 6.0-V input?

10. (II) Three 1.20-kΩ resistors can be connected together in four different ways, making combinations of series and/or parallel circuits. What are these four ways and what is the net resistance in each case?

11. (II) What is the net resistance of the circuit connected to the battery in Fig. 26–35? Each resistance has *R* = 2.8 kΩ.

FIGURE 26–35
Problems 11 and 18.

12. (II) Eight lights are connected in series across a 110-V line. (a) What is the voltage across each bulb? (b) If the current is 0.60 A, what is the resistance of each bulb and the power dissipated in each?

13. (II) Eight lights are connected in parallel to a 110-V source by two long leads of total resistance 1.5 Ω. If 340 mA flows through each bulb, what is the resistance of each, and what fraction of the total power is wasted in the leads?

14. (II) Eight 7.0-W Christmas tree lights are connected in series to each other and to a 110-V source. What is the resistance of each bulb?

15. (II) A close inspection of an electric circuit reveals that a 480-Ω resistor was inadvertently soldered in the place where a 320-Ω resistor is needed. How can this be fixed without removing anything from the existing circuit?

16. (II) Two resistors when connected in series to a 110-V line use one fourth the power that is used when they are connected in parallel. If one resistor is 1.6 kΩ, what is the resistance of the other?

17. (II) A 75-W, 110-V bulb is connected in parallel with a 40-W, 110-V bulb. What is the net resistance?

18. (II) Calculate the current through each resistor in Fig. 26–35 if each resistance $R = 2.20$ kΩ. What is the potential difference between points A and B?

19. (II) Consider the network of resistors shown in Fig. 26–36. Answer qualitatively: (a) What happens to the voltage across each resistor when the switch S is closed? (b) What happens to the current through each when the switch is closed? (c) What happens to the power output of the battery when the switch is closed? (d) Let $R_1 = R_2 = R_3 = R_4 = 100$ Ω and $V = 45.0$ V. Determine the current through each resistor before and after closing the switch. Are your qualitative predictions confirmed?

FIGURE 26–36
Problem 19.

20. (II) Three equal resistors (R) are connected to a battery, as shown in Fig. 26–37. Qualitatively, what happens to (a) the voltage drop across each of these resistors, (b) the current flow through each, and (c) the terminal voltage of the battery, when the switch S is opened, after having been closed for a long time? (d) If the emf of the battery is 12.0 V, what is its terminal voltage when the switch is closed if the internal resistance is 0.50 Ω and $R = 5.50$ Ω? (e) What is the terminal voltage when the switch S is open?

FIGURE 26–37
Problem 20.

21. (II) A battery with an emf of 12.0 V shows a terminal voltage of 11.8 V when operating in a circuit with two lightbulbs, each rated at 3.0 W (at 12.0 V) which are connected in parallel with it. What is the battery's internal resistance?

22. (III) A 3.8-kΩ and a 2.1-kΩ resistor are connected in parallel; this combination is connected in series with a 1.8-kΩ resistor. If each resistor is rated at $\frac{1}{2}$ W, what is the maximum voltage that can be applied across the whole network?

Section 26–3

23. (I) Calculate the current in the circuit of Fig. 26–38 and show that the sum of all the voltage changes around the circuit is zero.

FIGURE 26–38
Problem 23.

24. (II) Determine the terminal voltage of each battery in Fig. 26–33.

25. (II) Determine the magnitudes and directions of the currents through R_1 and R_2 in Fig. 26–39.

26. (II) Repeat Problem 25, now assuming that each battery has an internal resistance $r = 1.2$ Ω.

FIGURE 26–39
Problems 25 and 26.

FIGURE 26–40
Problems 27 and 28.

27. (II) Determine the magnitudes and directions of the currents through each resistor shown in Fig. 26–40. The batteries have emfs of $\mathscr{E}_1 = 9.0$ V and $\mathscr{E}_2 = 12.0$ V, and the resistors have values of $R_1 = 15$ Ω, $R_2 = 20$ Ω, and $R_3 = 40$ Ω.

28. (II) Repeat Problem 27 assuming each battery has internal resistance $r = 1.0$ Ω.

29. (II) Determine the current through each of the resistors in Fig. 26–41.

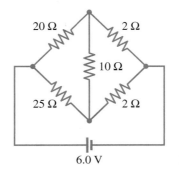

FIGURE 26–41
Problems 29 and 30.

30. (II) If the 20-Ω resistor in Fig. 26–41 is shorted out (resistance = 0), what then would be the current through the 10-Ω resistor?

31. (II) The current through the 4.0-kΩ resistor in Fig. 26–42 is 3.50 mA. What is the terminal voltage V_{ba} of the "unknown" battery? (There are two answers. Why?)

FIGURE 26–42 Problem 31.

32. (II) Suppose the 10-Ω resistor in Fig. 26–43 were replaced by an unknown resistance R. If the current through this unknown resistance is measured to be $I_2 = 0.90\,\text{A}$ to the right, what is the value of R?

33. (II) Suppose the 6.0-V battery in Fig. 26–43 is replaced by an unknown emf $\mathscr{E}$. If the current through the 10-Ω resistor is $I_2 = 0.30\,\text{A}$ to the left, what is $\mathscr{E}$? Assume $r = 1.0\,\Omega$.

FIGURE 26–43
Problems 32, 33, 34, and 35.

34. (III) Determine the currents I_1, I_2, and I_3 in Fig. 26–43. Assume the internal resistance of each battery is $r = 1.0\,\Omega$. What is the terminal voltage of the 6.0-V battery?

35. (III) What would the current I_1 be in Fig. 26–43 ($r = 1.0\,\Omega$) if the 18-Ω resistor were shorted out?

36. (III) Determine the net resistance of the network shown in Fig. 26–44 (a) between points a and c, and (b) between points a and b. Assume $R' = R$. [*Hint*: Use symmetry.]

FIGURE 26–44
Problems 36 and 39.

37. (III) A voltage V is applied to n resistors connected in parallel. If the resistors are instead all connected in series with the applied voltage, show that the power transformed is decreased by a factor n^2.

38. (III) For the circuit shown in Fig. 26–45, determine (a) the current through the 14-V battery and (b) the potential difference between points a and b.

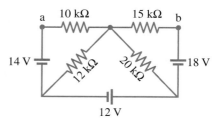

FIGURE 26–45 Problem 38.

39. (III) Determine the net resistance in Fig. 26–44 (a) between points a and c, and (b) between points a and b. Assume $R' \neq R$. [*Hint*: Apply an emf and determine currents; use symmetry at junctions.]

40. (III) Twelve resistors, each of resistance R, are connected as the edges of a cube as shown in Fig. 26–46. Determine the equivalent resistance (a) between points a and b, the ends of a side; (b) between points a and c, the ends of a face diagonal; (c) between points a and d, the ends of the volume diagonal. [*Hint*: Apply an emf and determine currents; use symmetry at junctions.]

FIGURE 26–46
Problem 40.

Section 26–4

41. (II) In Fig. 26–18a, the total resistance is 15 kΩ, and the battery's emf is 24.0 V. If the time constant is measured to be 55 μs, calculate (a) the total capacitance of the circuit and (b) the time it takes for the voltage across the resistor to reach 16.0 V

42. (II) The RC circuit of Fig. 26–19a has $R = 6.7\,\text{k}\Omega$ and $C = 6.0\,\mu\text{F}$. The capacitor is at voltage V_0 at $t = 0$, when the switch is closed. How long does it take the capacitor to discharge to 1.0 percent of its initial voltage?

43. (II) How long does it take for the energy stored in a capacitor in a series RC circuit (Fig. 26–18a) to reach half its maximum value? Express answer in terms of the time constant $\tau = RC$.

44. (II) Two 6.0-μF capacitors, two 2.2-kΩ resistors, and a 12.0-V source are connected in series. Starting from the uncharged state, how long does it take for the current to drop from its initial value to 1.50 mA?

45. (III) Determine the time constant for charging the capacitor in the circuit of Fig. 26–47. [*Hint*: Use Kirchhoff's rules.] (*b*) What is the maximum charge on the capacitor?

FIGURE 26–47
Problem 45.

46. (III) Two resistors and two uncharged capacitors are arranged as shown in Fig. 26–48. Then a potential difference of 24 V is applied across the combination as shown. (*a*) What is the potential at point a with *S* open? (Let *V* = 0 at the negative terminal of the source.) (*b*) What is the potential at point b with the switch open? (*c*) When the switch is closed, what is the final potential of point b? (*d*) How much charge flows through the switch S after it is closed?

47. (III) Suppose the switch S in Fig. 26–48 is closed. What is the time constant (or time constants) for charging the capacitors after the 24 V is applied.

FIGURE 26–48
Problems 46 and 47.

* Section 26–5

* **48.** (I) What is the resistance of a voltmeter on the 250-V scale if the meter sensitivity is 50,000 Ω/V?

* **49.** (I) An ammeter has a sensitivity of 20,000 Ω/V. What current passing through the galvanometer produces full-scale deflection?

* **50.** (II) A galvanometer has an internal resistance of 30 Ω and deflects full scale for a 50-μA current. Describe how to use this galvanometer to make (*a*) an ammeter to read currents up to 30 A, and (*b*) a voltmeter to give a full-scale deflection of 1000 V.

* **51.** (II) A galvanometer has a sensitivity of 35 kΩ/V and internal resistance 20.0 Ω. How could you make this into (*a*) an ammeter that reads 2.0 A full scale, or (*b*) a voltmeter reading 1.00 V full scale?

* **52.** (II) A milliammeter reads 20 mA full scale. It consists of a 0.20-Ω resistor in parallel with a 30-Ω galvanometer. How can you change this ammeter to a voltmeter giving a full-scale reading of 10 V without taking the ammeter apart? What will be the sensitivity (Ω/V) of your voltmeter?

* **53.** (II) A 45-V battery of negligible internal resistance is connected to a 37-kΩ and a 28-kΩ resistor in series. What reading will a voltmeter, of internal resistance 100 kΩ, give when used to measure the voltage across each resistor? What is the percent inaccuracy due to meter resistance for each case?

* **54.** (II) An ammeter whose internal resistance is 60 Ω reads 4.25 mA when connected in a circuit containing a battery and two resistors in series whose values are 700 Ω and 400 Ω. What is the actual current when the ammeter is absent?

* **55.** (II) A battery with $\mathscr{E} = 12.0$ V and internal resistance $r = 1.0 \Omega$ is connected to two 9.0-kΩ resistors in series. An ammeter of internal resistance 0.50 Ω measures the current and at the same time a voltmeter with internal resistance 11.5 kΩ measures the voltage across one of the 9.0-kΩ resistors in the circuit. What do the ammeter and voltmeter read?

* **56.** (II) A 12.0-V battery (assume the internal resistance = 0) is connected to two resistors in series. A voltmeter whose internal resistance is 15.0 kΩ measures 5.5 V and 4.0 V, respectively, when connected across each of the resistors. What is the resistance of each resistor?

* **57.** (II) Two 8.4-kΩ resistors are placed in series and connected to a battery. A voltmeter of sensitivity 1000 Ω/V is on the 3.0-V scale and reads 2.0 V when placed across either of the resistors. What is the emf of the battery? (Ignore its internal resistance.)

* **58.** (III) The voltage across a 120-kΩ resistor in a circuit containing additional resistance (R_2) in series with a battery (V) is measured, by a 20,000-Ω/V meter on the 100-V scale, to be 25 V. On the 30-V scale, the reading is 23 V. What is the actual voltage in the absence of the voltmeter? What is R_2?

* **59.** (III) (*a*) A voltmeter and an ammeter can be connected as shown in Fig. 26–49a to measure a resistance *R*. The value of *R* will not quite be V/I where *V* is the voltmeter reading and *I* is the ammeter reading since some of the current actually goes through the voltmeter. Show that the actual value of *R* is given by

$$\frac{1}{R} = \frac{I}{V} - \frac{1}{R_v}$$

where R_v is the voltmeter resistance. Note that $R \approx V/I$ if $R_v \gg R$. (*b*) A voltmeter and an ammeter can also be connected as shown in Fig. 26–49b to measure a resistance *R*. Show in this case that

$$R = \frac{V}{I} - R_A$$

where *V* and *I* are the voltmeter and ammeter readings and R_A is the resistance of the ammeter. Note that $R \approx V/I$ if $R_A \ll R$.

(a)

(b)

FIGURE 26–49
Problem 59.

* 60. (I) A copper-constantan thermocouple produces emfs of about 40 μV/C°. If the reference temperature is 25°C, what must be the temperature of the test junction (assume a temperature increase) if the emf produced is 1.72 mV?

* 61. (II) For iron-copper junctions near room temperature, the emf produced by a thermocouple is about 14 μV/C°. If emfs can be detected as low as 0.50 μV, to what accuracy can the temperature be read?

* 62. (II) The *strain factor* K of a strain gauge is defined as the fractional change in resistance ($\Delta R/R$) divided by the fractional change in length:

$$K = \frac{\Delta R/R}{\Delta L/L}$$

and is relatively constant at a given temperature. In other words, the change in resistance ΔR is proportional to the change in length of the wire ΔL. (a) Show that this linear relationship is reasonable for small $\Delta L/L$. (b) A strain gauge with a strain factor of 1.8 is attached to a small muscle about 4.5 mm wide. The gauge is connected as the unknown arm in a Wheatstone bridge (Example 26–9). When the muscle is relaxed, the Wheatstone bridge is balanced when $R_2/R_1 = 1.4800$ and $R_3 = 40.700\ \Omega$. When the muscle contracts, balance occurs for $R_3 = 40.736\ \Omega$. How much has the muscle widened?

General Problems

63. Suppose that you wish to apply a 0.25-V potential difference between two points on the body. The resistance is about 2000 Ω, and you only have a 6.0-V battery. How can you connect up one or more resistors so that you can produce the desired voltage?

64. A three-way lightbulb can produce 50 W, 100 W, or 150 W, at 120 V. Such a bulb contains two filaments that can be connected to the 120 V individually or in parallel. Describe how the connections to the two filaments are made to give each of the three wattages. What must be the resistance of each filament?

65. Suppose you want to run some apparatus that is 115 m from an electric outlet. Each of the wires connecting your apparatus to the 120-V source has a resistance per unit length of 0.0065 Ω/m. If your apparatus draws 3.0 A, what will be the voltage drop across the connecting wires and what voltage will be applied to your apparatus?

66. Electricity can be a hazard in hospitals, particularly to patients who are connected to electrodes, such as an ECG. For example, suppose that the motor of a motorized bed shorts out to the bed frame, and the bed frame's connection to a ground has broken (or was not there in the first place). If a nurse touches the bed and the patient at the same time, she becomes a conductor and a complete circuit can be made through the patient to ground through the ECG apparatus. This is shown schematically in Fig. 26–50. Calculate the current through the patient.

67. A heart pacemaker is designed to operate at 72 beats/min using a 7.5-μF capacitor in a simple RC circuit. What value of resistance should be used if the pacemaker is to fire (capacitor discharge) when the voltage reaches 45 percent of maximum?

68. The internal resistance of a 1.35-V mercury cell is 0.030 Ω, whereas that of a 1.5-V dry cell is 0.35 Ω. Explain why three mercury cells can more effectively power a 2-W hearing aid that requires 4.0 V than can three dry cells.

69. Suppose that a person's body resistance is 1100 Ω. (a) What current passes through the body when the person accidentally is connected to 110 V? (b) If there is an alternative path to ground whose resistance is 40 Ω, what current passes through the person? (c) If the voltage source can produce at most 1.5 A, how much current passes through the person in case (b)?

70. An unknown length of platinum wire 0.920 mm in diameter is placed as the unknown resistance in a Wheatstone bridge (Example 26–9 and Fig. 26–51). Arms 1 and 2 have resistance of 38.0 Ω and 46.0 Ω, respectively. Balance is achieved when R_3 is 3.48 Ω. How long is the platinum wire?

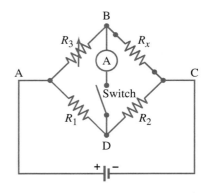

FIGURE 26–51 Wheatstone bridge. Problem 70.

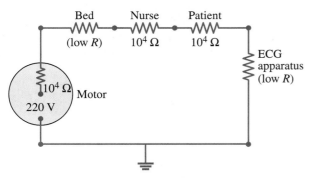

FIGURE 26–50 Problem 66.

71. Suppose two batteries, with unequal emfs of 2.0 V and 3.0 V, are connected as shown in Fig. 26–52. If each internal resistance is $r = 0.10\,\Omega$, and $R = 4.0\,\Omega$, what is the voltage across the resistor R?

FIGURE 26–52
Problem 71.

72. Electrocardiographs are often connected as shown in Fig. 26–53. The leads are said to be capacitively coupled. A time constant of 3.0 s is typical and allows rapid changes in potential to be accurately recorded. If $C = 3.0\,\mu F$, what value must R have?

FIGURE 26–53 Problem 72.

73. A battery produces 40.8 V when 7.40 A are drawn from it and 44.5 V when 2.20 A are drawn. What is the emf and internal resistance of the battery?

74. How many $\frac{1}{2}$-W resistors, each of the same resistance, must be used to produce an equivalent 1.2-kΩ, 5-W resistor? What is the resistance of each, and how must they be connected?

75. Some light dimmer switches use a variable resistor as shown in Fig. 26–54. The slide moves from position $x = 0$ to $x = 1$, and the resistance up to slide position x is proportional to x (the total resistance is $R_{pot} = 100\,\Omega$). What is the power expended in the lightbulb if (a) $x = 1.00$, (b) $x = 0.50$, (c) $x = 0.25$?

FIGURE 26–54 Problem 75.

76. A **potentiometer** is a device to precisely measure potential differences or emf, using a "null" technique. In the simple potentiometer circuit shown in Fig. 26–55, R' represents the total resistance of the resistor from A to B (which could be a long uniform "slide" wire), whereas R represents the resistance of only the part from A to the movable contact at C. When the unknown emf to be measured, $\mathcal{E}_x$, is placed into the circuit as shown, the movable contact C is moved until the galvanometer G gives a null reading (i.e., zero) when the switch S is closed. The resistance between A and C for this situation we call R_x. Next, a standard emf, $\mathcal{E}_s$, which is known precisely, is inserted into the circuit in place of $\mathcal{E}_x$ and again the contact C is moved until zero current flows through the galvanometer when the switch S is closed. The resistance between A and C now is called R_s. (a) Show that the unknown emf is given by

$$\mathcal{E}_x = \left(\frac{R_x}{R_s}\right)\mathcal{E}_s$$

where R_x, R_s, and $\mathcal{E}_s$ are all precisely known. The working battery is assumed to be fresh and give a constant voltage. (b) A slide-wire potentiometer is balanced against a 1.0182-V standard cell when the slide wire is set at 25.4 cm out of a total length of 100.0 cm. For an unknown source, the setting is 45.8 cm. What is the emf of the unknown? (c) The galvanometer of a potentiometer has an internal resistance of 30 Ω and can detect a current as small as 0.015 mA. What is the minimum uncertainty possible in measuring an unknown voltage? (d) Explain the advantage of using this "null" method of measuring emf.

FIGURE 26–55
Potentiometer circuit.
Problem 76.

77. Electronic devices often use an RC circuit to protect against power outages as shown in Fig. 26–56. (a) If the device is supposed to keep the supply voltage at least 70 percent of nominal for as long as 0.20 s, how big a resistance is needed? The capacitor is $14\,\mu F$. (b) Between which two terminals should the device be connected, a and b, b and c, or a and c?

Power supply   Power outage protector

FIGURE 26–56
Problem 77.

78. For the circuit shown in Fig. 26–19a, show that the decrease in energy stored in the capacitor from $t = 0$ until one time constant has elapsed equals the energy dissipated as heat in the resistor.

79. A solar cell, 3.0 cm square, has an output of 350 mA at 0.80 V when exposed to full sunlight. A solar panel that delivers 1.0 A of current at an emf of 100 V to an external load is needed. How many cells will you need? How big a panel will you need, and how should you connect the cells to one another? How can you optimize the output of your solar panel?

80. Determine the current in each resistor of the circuit shown in Fig. 26–57.

FIGURE 26–57 Problem 80.

81. In the circuit shown in Fig. 26–58, switch S is closed at time $t = 0$. (*a*) What is the current I_0 leaving the battery at $t = 0$, immediately after the switch is closed? (*b*) What is the current I a "long time" later? (*c*) What charge has accumulated on the capacitor after this long time? (*d*) If, finally, switch S is opened again, how long will it take after the switch is opened for the capacitor to lose 80 percent of its charge?

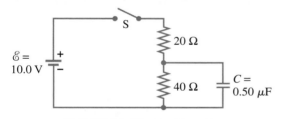

FIGURE 26–58 Problem 81.

82. A power supply has a fixed output voltage of 12.0 V, but you need $V_T = 3.0$ V for an experiment. (*a*) Using the voltage divider shown in Fig. 26–59, what should R_2 be if R_1 is 10.0 Ω? (*b*) What will the terminal voltage V_T be if you connect a load to the 3.0-volt terminal, which has a resistance of 7.0 Ω?

FIGURE 26–59 Problem 82.

83. In the circuit shown in Fig. 26–60, switch S is closed at time $t = 0$. (*a*) After the capacitor is fully charged, what is the voltage across it? How much charge is on it? (*b*) Switch S is now opened. How long does it now take for the capacitor to discharge until it has only 5.0 percent of its initial charge?

FIGURE 26–60 Problem 83.

84. Figure 26–61 shows the circuit for a simple *sawtooth oscillator*. At time $t = 0$, its switch S is closed. The neon bulb has initially infinite resistance until the voltage across it reaches 90.0 V, and then it begins to conduct with very little resistance (essentially zero). It stops conducting (its resistance becomes essentially infinite) when the voltage drops down to 70.0 V. (*a*) At what time t_1 does the neon bulb reach 90.0 V and start conducting? (*b*) At what time t_2 does the bulb reach 90.0 V for a second time and again become conducting? (*c*) Sketch the sawtooth waveform between $t = 0$ and $t = 0.50$ seconds.

FIGURE 26–61 Problem 84.

Magnets produce magnetic fields, but so do electric currents. An electric current flowing in this straight wire produces a magnetic field which causes the tiny pieces of iron (iron "filings") to align in the field. We will see in this chapter how magnetic field is defined, and that the magnetic field direction is along the iron filings. The magnetic field lines due to the electric field in this long wire form circles around the wire. We will also discuss how magnetic fields exert forces on electric currents and on charged particles, as well as useful applications of the interaction between magnetic fields and electric currents and moving electric charges.

CHAPTER 27

Magnetism

The history of magnetism begins thousands of years ago, with the ancient civilizations in Asia Minor. It was in a region of Asia Minor known as Magnesia that rocks were found that could attract each other. These rocks were called "magnets" after their place of discovery.

Not until the nineteenth century was it seen that magnetism and electricity are closely related. A crucial discovery was that electric currents produce magnetic effects (we will say "magnetic fields") like magnets do. All kinds of practical devices depend on magnetism, as we shall see in this and later chapters: from compasses to motors, loudspeakers, computer memory, and electric generators.

27–1 | Magnets and Magnetic Fields

Poles of a magnet

We have all observed a magnet attract paper clips, nails, and other objects made of iron. Any magnet, whether it is in the shape of a bar or a horseshoe, has two ends or faces, called **poles**, which is where the magnetic effect is strongest. If a magnet is suspended from a fine thread, it is found that one pole of the magnet will always point toward the north. It is not known for sure when this fact was discovered, but it is known that the Chinese were making use of it as an aid to navigation by the

27–7 Discovery and Properties of the Electron

The electron plays a basic role in our understanding of electricity and magnetism today. But its existence was not suggested until the 1890s. We discuss it here because magnetic fields were crucial for measuring its properties.

Toward the end of the nineteenth century, studies were being done on the discharge of electricity through rarefied gases. One apparatus, diagrammed in Fig. 27–28, was a glass tube fitted with electrodes and evacuated so only a small amount of gas remained inside. When a very high voltage was applied to the electrodes, a dark space seemed to extend outward from the cathode (negative electrode) toward the opposite end of the tube; and that far end of the tube would glow. If one or more screens containing a small hole were inserted as shown, the glow was restricted to a tiny spot on the end of the tube. It seemed as though something being emitted by the cathode traveled to the opposite end of the tube. These "somethings" were named **cathode rays**.

There was much discussion at the time about what these rays might be. Some scientists thought they might resemble light. But the observation that the bright spot at the end of the tube could be deflected to one side by an electric or magnetic field suggested that cathode rays could be charged particles; and the direction of the deflection was consistent with a negative charge. Furthermore, if the tube contained certain types of rarefied gas, the path of the cathode rays was made visible by a slight glow.

Estimates of the charge e of the (assumed) cathode-ray particles, as well as of their charge-to-mass ratio e/m, had been made by 1897. But in that year, J. J. Thomson (1856–1940) was able to measure e/m directly, using the apparatus shown in Fig. 27–29. Cathode rays are accelerated by a high voltage and then pass between a pair of parallel plates built into the tube. The voltage applied to the plates produces an electric field, and a pair of coils produces a magnetic field. When only the electric field is present, say with the upper plate positive, the cathode rays are deflected upward as in path a in Fig. 27–29. If only a magnetic field exists, say inward, the rays are deflected downward along path c. These observations are just what is expected for a negatively charged particle. The force on the rays due to the magnetic field is $F = evB$, where e is the charge and v is the velocity of the cathode rays. In the absence of an electric field, the rays are bent into a curved path, so we have, from $F = ma$,

$$evB = m\frac{v^2}{r},$$

and thus

$$\frac{e}{m} = \frac{v}{Br}.$$

The radius of curvature r can be measured and so can B. The velocity v can be found by applying an electric field in addition to the magnetic field. The electric field E is adjusted so that the cathode rays are undeflected and follow path b in Fig. 27–29. This

FIGURE 27–28 Discharge tube. In some models, one of the screens is the anode (positive plate).

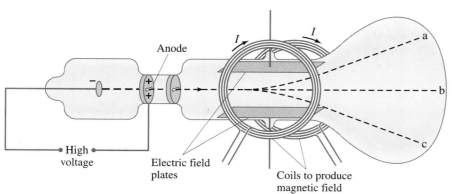

FIGURE 27–29 Cathode rays deflected by electric and magnetic fields.

is just like the velocity selector of Example 27–7 where the force due to the electric field, $F = eE$, is balanced by the force due to the magnetic field, $F = evB$. Thus $eE = evB$ and $v = E/B$. Combining this with the above equation we have

e/m measured

$$\frac{e}{m} = \frac{E}{B^2 r}.$$ **(27–13)**

The quantities on the right side can all be measured so that although e and m could not be determined separately, the ratio e/m could be determined. The accepted value today is $e/m = 1.76 \times 10^{11}$ C/kg. Cathode rays soon came to be called **electrons**.

"Discovery" of the electron

It is worth noting that the "discovery" of the electron, like many others in science, is not quite so obvious as discovering gold or oil. Should the discovery of the electron be credited to the person who first saw a glow in the tube? Or to the person who first called them cathode rays? Perhaps neither one, for they had no conception of the electron as we know it today. In fact, the credit for the discovery is generally given to Thomson, but not because he was the first to see the glow in the tube. Rather it is because he believed that this phenomenon was due to tiny negatively charged particles and made careful measurements on them. Furthermore he argued that these particles were constituents of atoms, and not ions or atoms themselves as many thought, and he developed an electron theory of matter. His view is close to what we accept today, and this is why Thomson is credited with the "discovery." Note, however, that neither he nor anyone else ever actually saw an electron itself. We discuss this briefly, for it illustrates the fact that discovery in science is not always a clear-cut matter. In fact some philosophers of science think the word "discovery" is often not appropriate, such as in this case.

Thomson believed that an electron was not an atom, but rather a constituent, or part, of an atom. Convincing evidence for this came soon with the determination of the charge and the mass of the cathode rays. Thomson's student J. S. Townsend made the first direct (but rough) measurements of e in 1897. But it was the more refined **oil-drop experiment** of Robert A. Millikan (1868–1953) that yielded a precise value for the charge on the electron and showed that charge comes in discrete amounts. In this experiment, tiny droplets of mineral oil carrying an electric charge were allowed to fall under gravity between two parallel plates, Fig. 27–30. The electric field E between the plates was adjusted until the drop was suspended in midair. The downward pull of gravity, mg, was then just balanced by the upward force due to the electric field. Thus $qE = mg$, so the charge $q = mg/E$. The mass of the droplet was determined by measuring its terminal velocity in the absence of the electric field. Sometimes the drop was charged negatively, and sometimes positively, suggesting that the drop had acquired or lost electrons (by friction, leaving the atomizer). Millikan's painstaking observations and analysis presented convincing evidence that any charge was an integral multiple of a smallest charge, e, that was ascribed to the electron, and that the value of e was 1.6×10^{-19} C. (Today's value of e, as mentioned in Chapter 21, is $e = 1.602 \times 10^{-19}$ C.) This value of e, combined with the measurement of e/m, gives the mass of the electron to be $(1.6 \times 10^{-19}\ \text{C})/(1.76 \times 10^{11}\ \text{C/kg}) = 9.1 \times 10^{-31}$ kg. This mass is less than a thousandth the mass of the smallest atom, and thus confirmed the idea that the electron is only a part of an atom. The accepted value today for the mass of the electron is $m_e = 9.11 \times 10^{-31}$ kg.

Millikan oil-drop experiment

Atomizer

Telescope

FIGURE 27–30 Millikan's oil-drop experiment.

CRT, Revisited

The cathode ray tube (CRT), which is the picture tube of TV sets, oscilloscopes, and computer monitors, was discussed in Chapter 23. There, in Fig. 23–19, we saw a design using electric deflection plates to maneuver the electron beam. Many CRTs, however, make use of the magnetic field produced by coils to maneuver the electron beam. They operate much like the coils shown in Fig. 27–29. Both types—electrostatic deflection and magnetic deflection—are in use today.

The Hall Effect

When a current-carrying conductor is held firmly in a magnetic field, the field exerts a sideways force on the charges moving in the conductor. For example, if electrons move to the right in the rectangular conductor shown in Fig. 27–31a, the inward magnetic field will exert a downward force on the electrons $\mathbf{F}_B = -e\mathbf{v}_d \times \mathbf{B}$, where $\mathbf{v}_d$ is the drift velocity of the electrons (Section 25–8). So the electrons will tend to move nearer face D than face C. There will thus be a potential difference between faces C and D of the conductor. This potential difference builds up until the electric field $\mathbf{E}_H$ it produces exerts a force, $e\mathbf{E}_H$, on the moving charges that is equal and opposite to the magnetic force. This effect is called the **Hall effect** after E. H. Hall, who discovered it in 1879. The difference of potential produced is called the **Hall emf**.

The electric field due to the separation of charge is called the *Hall field*, $\mathbf{E}_H$, and points downward in Fig. 27–31a, as shown. In equilibrium, the force due to this electric field is balanced by the magnetic force $ev_d B$, so

$$eE_H = ev_d B.$$

Hence $E_H = v_d B$. The Hall emf is then (assuming the conductor is long and thin so E_H is uniform)

$$\mathcal{E}_H = E_H l = v_d B l, \qquad (27\text{–}14)$$

where l is the width of the conductor.

A current of negative charges moving to the right is equivalent to positive charges moving to the left, at least for most purposes. But the Hall effect can distinguish these two. As can be seen in Fig. 27–31b, positive particles moving to the left are deflected downward, so that the bottom surface is positive relative to the top surface. This is the reverse of part (a). Indeed, the direction of the emf in the Hall effect first revealed that it is negative particles that move in metal conductors.

The magnitude of the Hall emf is proportional to the strength of the magnetic field. The Hall effect can thus be used to measure magnetic field strengths. First the conductor, called a *Hall probe*, is calibrated with known magnetic fields. Then, for the same current, its emf output will be a measure of B. Hall probes can be made very small and are convenient and accurate to use.

The Hall effect can also be used to measure the drift velocity of charge carriers when the external magnetic field B is known. Such a measurement also allows us to determine the density of charge carriers in the material.

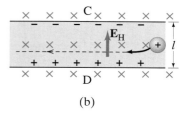

FIGURE 27–31 The Hall effect. (a) Negative charges moving to the right as the current. (b) Positive charges moving to the left as the current.

EXAMPLE 27–10 **Drift velocity using the Hall effect.** A long copper strip 1.8 cm wide and 1.0 mm thick is placed in a 1.2-T magnetic field as in Fig. 27–31a. When a steady current of 15 A passes through it, the Hall emf is measured to be 1.02 μV. Determine the drift velocity of the electrons and the density of free (conducting) electrons (number per unit volume) in the copper.

SOLUTION The drift velocity (Eq. 27–14) is

$$v_d = \frac{\mathcal{E}_H}{Bl} = \frac{1.02 \times 10^{-6}\,\text{V}}{(1.2\,\text{T})(1.8 \times 10^{-2}\,\text{m})} = 4.7 \times 10^{-5}\,\text{m/s}.$$

The density of charge carriers n is obtained from Eq. 25–13, $I = nev_d A$, where A is the cross-sectional area through which the current I flows. Then

$$n = \frac{I}{ev_d A} = \frac{15\,\text{A}}{(1.6 \times 10^{-19}\,\text{C})(4.7 \times 10^{-5}\,\text{m/s})(1.8 \times 10^{-2}\,\text{m})(1.0 \times 10^{-3}\,\text{m})}$$

$$= 11 \times 10^{28}\,\text{m}^{-3}.$$

This value for the density of free electrons in copper, $n = 11 \times 10^{28}$ per m^3, is the experimentally measured value. It represents *more* than one free electron per atom, which as we saw in Example 25–12 is $8.4 \times 10^{28}\,\text{m}^{-3}$.

* 27-9 | Mass Spectrometer

PHYSICS APPLIED

The mass spectrometer

FIGURE 27–32 Bainbridge mass spectrometer. The magnetic fields B and B' point out of the paper (indicated by the dots).

Various methods were developed in the early part of this century to measure the masses of atoms. One of the most accurate was the **mass spectrometer** of Fig. 27–32. Ions are produced by heating, or by an electric current, in the source S. Often the particles are accelerated and then pass through slit S_1 into a velocity selector of crossed electric and magnetic fields (as in Example 27–7). Only those ions whose speed is $v = E/B$ will pass through undeflected and emerge through slit S_2. In the second region, after S_2, there is only a magnetic field B' so the ions follow a circular path. The radius of their path can be measured because the ions darken the photographic plate where they strike. Since $qvB' = mv^2/r$ and $v = E/B$, then we have

$$m = \frac{qB'r}{v} = \frac{qBB'r}{E}.$$

By measuring the quantities on the right, m can be determined. Note that for ions of the same charge, the mass of each is proportional to the radius of its path.

The masses of many atoms were measured in this way. When a pure substance was used, it was sometimes found that two or more closely spaced marks would appear on the film. For example, neon produced two marks whose radii corresponded to atoms of mass 20 and 22 atomic mass units (u). Impurities were ruled out and it was concluded that there must be two types of neon with different masses. These different forms were called **isotopes**. It was soon found that most elements are mixtures of isotopes, and that the difference in mass is due to different numbers of neutrons. Mass spectrometers can be used to separate not only different elements and isotopes, but different molecules as well.

EXAMPLE 27–11 **Mass spectrometry.** Carbon atoms of atomic mass 12.0 u are found to be mixed with another, unknown, element. In a mass spectrometer, the carbon traverses a path of radius 22.4 cm and the unknown's path has a 26.2 cm radius. What is the unknown element? Assume they have the same charge.

SOLUTION Since mass is proportional to the radius, we have

$$\frac{m_x}{m_C} = \frac{26.2 \text{ cm}}{22.4 \text{ cm}} = 1.17.$$

Thus $m_x = 1.17 \times 12.0 \text{ u} = 14.0 \text{ u}$. The other element is probably nitrogen (see the Periodic Table, inside the back cover). However, it could also be an isotope of carbon or oxygen. Further physical or chemical analysis would be needed.

Summary

A magnet has two **poles**, north and south. The north pole is that end which points toward geographic north when the magnet is freely suspended. Unlike poles of two magnets attract each other, whereas like poles repel.

We can imagine that a **magnetic field** surrounds every magnet. The SI unit for magnetic field is the **tesla** (T).

Electric currents produce magnetic fields. For example, the lines of magnetic field due to a current in a straight wire form circles around the wire and the field exerts a force on magnets placed near it.

A magnetic field exerts a force on an electric current. The force on an infinitesimal length of wire $d\mathbf{l}$ carrying a current I in a magnetic field $\mathbf{B}$ is

$$d\mathbf{F} = I \, d\mathbf{l} \times \mathbf{B}.$$

If the field $\mathbf{B}$ is uniform over a straight length l of wire, then the force is

$$\mathbf{F} = I\mathbf{l} \times \mathbf{B}$$

which has magnitude

$$F = IlB \sin \theta$$

where θ is the angle between magnetic field $\mathbf{B}$ and the wire. The direction of the force is perpendicular to the wire and to the magnetic field, and is given by the right-hand rule. This relation serves as the definition of magnetic field $\mathbf{B}$.

Similarly, a magnetic field **B** exerts a force on a charge q moving with velocity **v** given by

$$\mathbf{F} = q\mathbf{v} \times \mathbf{B}.$$

The magnitude of the force is

$$F = qvB \sin\theta,$$

where θ is the angle between **v** and **B**. The path of a charged particle moving perpendicular to a uniform magnetic field is a circle.

If both electric and magnetic fields are present, the force is

$$\mathbf{F} = q\mathbf{E} + q\mathbf{v} \times \mathbf{B}.$$

The torque on a current loop in a magnetic field **B** is

$$\boldsymbol{\tau} = \boldsymbol{\mu} \times \mathbf{B},$$

where $\boldsymbol{\mu}$ is the **magnetic dipole moment** of the loop:

$$\boldsymbol{\mu} = NI\mathbf{A}.$$

Here N is the number of coils carrying current I in the loop and **A** is a vector perpendicular to the plane of the loop (use right-hand rule) and has magnitude equal to the area of the loop.

The measurement of the charge-to-mass ratio (e/m) of the electron was done using magnetic and electric fields. The charge e on the electron was first measured in the Millikan oil-drop experiment and then its mass was obtained from the measured value of the e/m ratio.

Questions

1. A compass needle is not always balanced parallel to the Earth's surface, but one end may dip downward. Explain.

2. Draw the magnetic field lines around a straight section of wire carrying a current horizontally to the left.

3. In what direction are the magnetic field lines surrounding a straight wire carrying a current that is moving directly toward you?

4. A horseshoe magnet is held vertically with the north pole on the left and south pole on the right. A wire passing between the poles, equidistant from them, carries a current directly away from you. In what direction is the force on the wire?

5. In the relation, $\mathbf{F} = I\boldsymbol{l} \times \mathbf{B}$, which pairs of the vectors ($\mathbf{F}, \boldsymbol{l}, \mathbf{B}$) are always at 90°? Which can be at other angles?

6. The magnetic field due to current in wires in your home can affect a compass. Discuss the effect in terms of currents, including if they are ac or dc.

7. If a negatively charged particle enters a region of uniform magnetic field which is perpendicular to the particle's velocity, will the kinetic energy of the particle increase, decrease, or stay the same. Explain your answer. (Neglect gravity.)

8. In Fig. 27–33, charged particles move in the vicinity of a current-carrying wire. For each charged particle the arrow indicates the direction of motion of the particle and the + or − indicates the sign of the charge. For each of the particles, indicate the direction of the magnetic force due to the magnetic field produced by the wire.

10. Note that the pattern of magnetic field lines surrounding a bar magnet is similar to that of the electric field around an electric dipole. From this fact, predict how the magnetic field will change with distance (a) when near one pole of a very long bar magnet, and (b) when far from a magnet as a whole.

11. Explain why a strong magnet held near a television screen causes the picture to become distorted. Also, explain why the picture sometimes goes completely black where the field is the strongest. [But don't risk damage to your TV by trying this.]

12. Describe the trajectory of a negatively charged particle in the velocity filter of Fig. 27–20 if its speed exceeds E/B. What is its trajectory if $v < E/B$? Would it make any difference if the particle were positively charged?

13. Can you set a resting electron into motion with a magnetic field? With an electric field?

14. A charged particle is moving in a circle under the influence of a uniform magnetic field. If an electric field that points in the same direction as the magnetic field is turned on, describe the path the charged particle will take.

15. The force on a particle in a magnetic field is the idea behind *electromagnetic pumping*. It is used to pump metallic fluids (such as sodium) and more recently to pump blood in artificial heart machines. The basic design is shown in Fig. 27–35. An electric field is applied perpendicular to a blood vessel and to a magnetic field. Explain how ions are caused to move. Do positive and negative ions feel a force in the same direction?

FIGURE 27–33 Question 8.   **FIGURE 27–34** Question 9.

9. A positively charged particle in a nonuniform magnetic field follows the trajectory shown in Fig. 27–34. Indicate the direction of the magnetic field everywhere in space, assuming the path is always in the plane of the page, and indicate the relative magnitudes of the field in each region.

FIGURE 27–35 Electromagnetic pumping in a blood vessel. Question 15.

16. A beam of electrons is directed toward a horizontal wire carrying a current from left to right (Fig. 27–36). In what direction is the beam deflected?

$\longrightarrow I$

↑ Electron direction

FIGURE 27–36
Question 16.

17. What kind of field or fields surround a moving electric charge?

18. Could we have defined the direction of the magnetic field **B** to be in the direction of the force on a moving charged particle? Explain.

19. A charged particle moves in a straight line through a particular region of space. Could there be a nonzero magnetic field in this region? If so, give two possible situations.

20. If a moving charged particle is deflected sideways in some region of space, can we conclude, for certain, that $\mathbf{B} \neq 0$ in that region? Explain.

21. In a particular region of space there is a uniform magnetic field **B**. Outside this region, $B = 0$. Can you inject an electron from outside into the field perpendicularly so it will move in a closed circular path in the field? What if the electron is injected near the center?

22. How could you tell whether moving electrons in a certain region of space are being deflected by an electric field or by a magnetic field (or by both)?

23. How can you make a compass without using iron or other ferromagnetic material?

24. Describe how you could determine the dipole moment of a bar magnet or compass needle.

25. In what positions (if any) will a current loop placed in a uniform magnetic field be in (a) stable equilibrium, and (b) unstable equilibrium?

*   **26.** A rectangular piece of semiconductor is inserted in a magnetic field and a battery is connected to its ends as shown in Fig. 27–37. When a sensitive voltmeter is connected between points a and b, it is found that point a is at a higher potential than b. What is the sign of the charge carriers in this semiconductor material?

FIGURE 27–37 Question 26.

*   **27.** Two ions have the same mass, but one is singly ionized and the other is doubly ionized. How will their positions on the film of the mass spectrometer of Fig. 27–32 differ?

Problems

Section 27–3

1. (I) (a) What is the force per meter of length on a straight wire carrying a 7.40-A current when perpendicular to a 0.90-T uniform magnetic field? (b) What if the angle between the wire and field is 45.0°?

2. (I) Calculate the magnetic force on a 240-m length of wire stretched between two towers carrying a 150-A current. The Earth's magnetic field of 5.0×10^{-5} T makes an angle of 60° with the wire.

3. (I) How much current is flowing in a wire 4.20 m long if the maximum force on it is 0.900 N when placed in a uniform 0.0800-T field?

4. (I) A 1.5-m length of wire carrying 4.5 A of current is oriented horizontally. At that point on the Earth's surface the dip angle of the Earth's magnetic field makes an angle of 40° to the wire. Estimate the magnetic force on the wire due to the Earth's magnetic field of 5.5×10^{-5} T at this point.

5. (I) The force on a wire carrying 8.75 A is a maximum of 1.18 N when placed between the pole faces of a magnet. If the pole faces are 55.5 cm in diameter, what is the approximate strength of the magnetic field?

6. (II) The magnetic force per meter on a wire is measured to be only 45 percent of its maximum possible value. Sketch the relationship of the wire and the field if the force were a maximum, and sketch the relationship as it actually is, calculating the angle between the wire and the magnetic field.

7. (II) The force on a wire is a maximum of 5.30 N when placed between the pole faces of a magnet. The current flows horizontally to the right and the magnetic field is vertical. The wire is observed to "jump" toward the observer when the current is turned on. (a) What type of magnetic pole is the top pole face? (b) If the pole faces have a diameter of 10.0 cm, estimate the current in the wire if the field is 0.15 T. (c) If the wire is tipped so that it now makes an angle of 10° with the horizontal, what force will it now feel?

8. (II) Suppose the straight wires connected to the conductor bent into a semicircle in Fig. 27–15 were bent outwardly, still in the plane of the page, so they were horizontal at the base of the semicircle. If a length L of each remained in the field **B**, what would be the total force on the conductor as a whole?

9. (II) A straight 2.0-mm-diameter copper wire can just "float" horizontally in air because of the force of the Earth's magnetic field **B**, which is horizontal, perpendicular to the wire, and of magnitude 5.0×10^{-5} T. What current does the wire carry?

10. (II) A long wire stretches along the x axis and carries a 3.0-A current to the right $(+x)$. The wire passes through a uniform magnetic field $\mathbf{B} = (0.20\mathbf{i} - 0.30\mathbf{j} + 0.25\mathbf{k})$T. Determine the components of the force on the wire per cm of length.

11. (III) A curved wire, connecting two points a and b, lies in a plane perpendicular to a uniform magnetic field **B** and carries a current I. Show that the resultant magnetic force on the wire, no matter what its shape, is the same as that on a straight wire connecting the two points carrying the same current I. See Fig. 27–38.

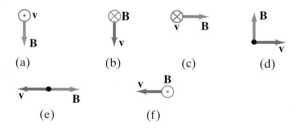

FIGURE 27–38 Problem 11.   FIGURE 27–39 Problem 12.

12. (III) A circular loop of wire, of radius r, carries current I. It is placed in a magnetic field whose straight lines seem to diverge from a point a distance d below the ring on its axis. (That is, the field makes an angle θ with the loop at all points, Fig. 27–39, where $\tan \theta = r/d$.) Determine the force on the loop.

Section 27–4

13. (I) Determine the magnitude and direction of the force on an electron traveling 7.75×10^5 m/s horizontally to the east in a vertically upward magnetic field of strength 0.85 T.

14. (I) Find the direction of the force on a negative charge for each diagram shown in Fig. 27–40, where **v** is the velocity of the charge and **B** is the direction of the magnetic field. ($\otimes$ means the vector points inward. $\odot$ means it points outward, toward the viewer.)

FIGURE 27–40 Problem 14.

15. (I) Determine the direction of **B** for each case in Fig. 27–41, where **F** represents the force on a positively charged particle moving with velocity **v**.

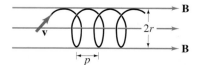

FIGURE 27–41
Problem 15.

16. (I) An electron is projected vertically upward with a speed of 1.80×10^6 m/s into a uniform magnetic field of 0.250 T that is directed horizontally away from the observer. Describe the electron's path in this field.

17. (I) A particle of charge q moves in a circular path of radius r in a uniform magnetic field **B**. Show that its momentum is $p = qBr$.

18. (II) What is the velocity of a beam of electrons that go undeflected when passing through crossed electric and magnetic fields of magnitude 8.8×10^3 V/m and 3.5×10^{-3} T, respectively? What is the radius of the electron orbit if the electric field is turned off?

19. (II) For a particle of mass m and charge q moving in a circular path in a magnetic field B, (a) show that its kinetic energy is proportional to r^2, the square of the radius of curvature of its path, and (b) show that its angular momentum is $L = qBr^2$, about the center of the circle.

20. (II) An electron moves with velocity $\mathbf{v} = (4.0\mathbf{i} - 6.0\mathbf{j}) \times 10^4$ m/s in a magnetic field $\mathbf{B} = (-0.80\mathbf{i} + 0.60\mathbf{j})$T. Determine the magnitude and direction of the force on the electron.

21. (II) A 5.0-MeV (kinetic energy) proton enters a 0.20-T field, in a plane perpendicular to the field. What is the radius of its path?

22. (II) An electron experiences the greatest force as it travels 2.9×10^6 m/s in a magnetic field when it is moving northward. The force is upward and of magnitude 7.2×10^{-13} N. What is the magnitude and direction of the magnetic field?

23. (II) A doubly charged helium atom whose mass is 6.6×10^{-27} kg is accelerated by a voltage of 2100 V. (a) What will be its radius of curvature if it moves in a plane perpendicular to a uniform 0.340-T field? (b) What is its period of revolution?

24. (II) A 3.40-g bullet moves with a speed of 160 m/s perpendicular to the Earth's magnetic field of 5.00×10^{-5} T. If the bullet possesses a net charge of 13.5×10^{-9} C, by what distance will it be deflected from its path due to the Earth's magnetic field after it has traveled 1.00 km?

25. (II) Suppose the Earth's magnetic field at the equator has magnitude 0.40×10^{-4} T and a northerly direction at all points. How fast must a singly ionized uranium ion ($m = 238$ u, $q = e$) move so as to circle the Earth 5.0 km above the equator? Can you ignore gravity?

26. (II) A proton (mass m_p), a deuteron ($m = 2m_p$, $Q = e$), and an alpha particle ($m = 4m_p$, $Q = 2e$) are accelerated by the same potential difference V and then enter a uniform magnetic field **B** where they move in circular paths perpendicular to **B**. Determine the radius of the paths for the deuteron and alpha particle in terms of that for the proton.

27. (II) A proton moves through a region of space where there is a magnetic field, $\mathbf{B} = (0.45\mathbf{i} + 0.20\mathbf{j})$T and an electric field $\mathbf{E} = (3.0\mathbf{i} - 4.2\mathbf{j}) \times 10^3$ V/m. At a given instant, the proton's velocity is $\mathbf{v} = (6.0\mathbf{i} + 3.0\mathbf{j} - 5.0\mathbf{k}) \times 10^3$ m/s. Determine the components of the total force on the proton.

28. (II) An electron experiences a force $\mathbf{F} = (3.8\mathbf{i} - 2.7\mathbf{j}) \times 10^{-13}$ N when passing through a magnetic field $\mathbf{B} = (0.35\,\text{T})\mathbf{k}$. Determine the electron's velocity.

29. (II) An electron enters a uniform magnetic field $B = 0.23$ T at a 45° angle to **B**. Determine the radius r and pitch p (distance between loops) of the electron's helical path assuming its speed is 3.0×10^6 m/s. See Fig. 27–42.

FIGURE 27–42
Problem 29.

30. (II) The path of protons emerging from an accelerator must be bent by 90° by a "bending magnet" so as not to strike a barrier in their path a distance l from their exit hole in the accelerator. Show that the field **B** in the bending magnet, which we assume is uniform and can extend over an area $l \times l$, must have magnitude of at least $B \geq (2mK/e^2l^2)^{\frac{1}{2}}$, where m is the mass of a proton and K is its kinetic energy.

31. (II) A proton moving with speed $v = 2.0 \times 10^5$ m/s in a field-free region abruptly enters an essentially uniform magnetic field $B = 0.850$ T as shown in Fig. 27–43 (**B** $\perp$ **v**). If the proton enters the magnetic field region at a 45° angle as shown, (a) at what angle does it leave and (b) at what distance x does it exit from the field?

FIGURE 27–43
Problem 31.

Section 27–5

32. (I) A 13.0-cm-diameter circular loop of wire is placed with its face parallel to the uniform magnetic field between the pole pieces of a large magnet. When 7.10 A flows in the coil, the torque on it is 0.185 m·N. What is the magnetic field strength?

33. (I) How much work is required to rotate the current loop (Fig. 27–21) in a uniform magnetic field **B** from (a) $\theta = 0°$ ($\boldsymbol{\mu} \| \mathbf{B}$) to $\theta = 180°$, (b) $\theta = 90°$ to $\theta = -90°$?

34. (II) Show that the magnetic dipole moment $\boldsymbol{\mu}$ of an electron orbiting the proton nucleus of a hydrogen atom is related to the orbital momentum L of the electron by

$$\mu = \frac{e}{2m} L.$$

35. (II) A circular coil 17.0 cm in diameter and containing twelve loops lies flat on the ground. The Earth's magnetic field at this location has magnitude 5.50×10^{-5} T and points into the Earth at an angle of 66.0° below a line pointing due north. If a 7.10-A clockwise current passes through the coil, (a) determine the torque on the coil, and (b) which edge of the coil rises up, north, east, south, or west.

36. (II) A 20-loop circular coil 20 cm in diameter lies in the xy plane. The current in each loop of the coil is 7.6 A clockwise, and an external magnetic field $\mathbf{B} = (0.80\mathbf{i} + 0.60\mathbf{j} - 0.65\mathbf{k})$ T passes through the coil. Determine: (a) the magnetic moment of the coil, $\boldsymbol{\mu}$; (b) the torque on the coil due to the external magnetic field; (c) the potential energy U of the coil in the field (take the same zero for U as we did in our discussion of Fig. 27–21).

37. (III) Suppose a nonconducting rod of length l carries a uniformly distributed charge Q. It is rotated with angular velocity ω about an axis perpendicular to the rod at one end. Show that the magnetic dipole moment of this rod is $\frac{1}{6}Q\omega l^2$. [*Hint*: Consider the motion of each infinitesimal length of the rod.]

* 38. (I) A galvanometer needle deflects full scale for a 63.0-μA current. What current will give full-scale deflection if the magnetic field weakens to 0.860 of its original value?

* 39. (I) If the restoring spring of a galvanometer weakens by 20 percent over the years, what current will give full-scale deflection if it originally required 36 μA?

* 40. (I) If the current to a motor drops by 18 percent, by what factor does the output torque change?

Section 27–7

41. (I) What is the value of q/m for a particle that moves in a circle of radius 8.0 mm in a 0.46-T magnetic field if a crossed 200-V/m electric field will make the path straight?

42. (II) An oil drop whose mass is determined to be 3.3×10^{-15} kg is held at rest between two large plates separated by 1.0 cm as in Fig. 27–30. If the potential difference between the plates is 340 V, how many excess electrons does this drop have?

* 43. (II) A rectangular sample of a metal is 3.0 cm wide and 500 μm thick. When it carries a 42-A current and is placed in a 0.80-T magnetic field it produces a 6.5-μV Hall emf. Determine: (a) the Hall field in the conductor; (b) the drift speed of the conduction electrons; (c) the density of free electrons in the metal.

* 44. (II) In a probe that uses the Hall effect to measure magnetic fields, a 12.0-A current passes through a 1.50-cm-wide 1.30-mm-thick strip of sodium metal. If the Hall emf is 2.42 μV, what is the magnitude of the magnetic field (take it perpendicular to the flat face of the strip)? Assume one free electron per atom of Na, and take its specific gravity to be 0.971.

* 45. (II) The Hall effect can be used to measure blood flow rate because the blood contains ions that constitute an electric current. (a) Does the sign of the ions influence the emf? (b) Determine the flow velocity in an artery 3.3 mm in diameter if the measured emf is 0.10 mV and B is 0.070 T. (In actual practice, an alternating magnetic field is used.)

* 46. (I) In a mass spectrometer, germanium atoms have radii of curvature equal to 21.0, 21.6, 21.9, 22.2, and 22.8 cm. The largest radius corresponds to an atomic mass of 76 u. What are the atomic masses of the other isotopes?

* 47. (II) Suppose the electric field between the electric plates in the mass spectrometer of Fig. 27–32 is 2.48×10^4 V/m and the magnetic fields $B = B' = 0.58$ T. The source contains carbon isotopes of mass numbers 12, 13, and 14 from a long-dead piece of a tree. (To estimate atomic masses, multiply by 1.66×10^{-27} kg.) How far apart are the lines formed by the singly charged ions of each type on the photographic film? What if the ions were doubly charged?

*48. (II) A mass spectrometer is being used to monitor air pollutants. It is difficult, however, to separate molecules with nearly equal mass such as CO (28.0106 u) and N_2 (28.0134 u). How large a radius of curvature must a spectrometer have if these two molecules are to be separated on the film by 0.50 mm?

*49. (II) One form of mass spectrometer accelerates ions by a voltage V before they enter a magnetic field B. The ions are assumed to start from rest. Show that the mass of an ion is $m = qB^2R^2/2V$, where R is the radius of the ions' path in the magnetic field and q is their charge.

General Problems

50. Protons move in a circle of radius 5.10 cm in a 0.725-T magnetic field. What value of electric field could make their paths straight? In what direction must the electric field point?

51. Protons with momentum 4.8×10^{-16} kg·m/s are magnetically steered clockwise in a circular path 2.0 km in diameter at Fermi National Accelerator Laboratory in Illinois. What is the magnitude and direction of the field in the magnets surrounding the beam pipe?

52. A proton and an electron have the same kinetic energy upon entering a region of constant magnetic field. What is the ratio of the radii of their circular paths?

53. Near the equator, the Earth's magnetic field points almost horizontally to the north and has magnitude $B = 0.50 \times 10^{-4}$ T. What should be the magnitude and direction for the velocity of an electron if its weight is to be exactly balanced by the magnetic force?

54. Calculate the force on an airplane which has acquired a net charge of 1550 μC and moves with a speed of 120 m/s perpendicular to the Earth's magnetic field of 5.0×10^{-5} T.

55. The power cable for an electric trolley carries a horizontal dc current of 330 A toward the east. The Earth's magnetic field has a strength 5.0×10^{-5} T and makes an angle of dip of 22° at this location. Calculate the magnitude and direction of the magnetic force on a 10-m length of this cable.

56. Two stiff parallel wires a distance l apart in a horizontal plane act as rails to support a light metal rod of mass m (perpendicular to each rail), Fig. 27–44. A magnetic field **B**, directed vertically upward (outward in diagram), acts throughout. At $t = 0$, wires connected to the rails are connected to a constant current source and a current I begins to flow through the system. Determine the speed of the rod, which starts from rest at $t = 0$, as a function of time (a) assuming no friction between the rod and the rails, and (b) if the coefficient of friction is μ_k. (c) In which direction does the rod move, east or west, if the current through it heads north?

57. Suppose the rod in Fig. 27–44 (Problem 56) has mass $m = 0.40$ kg and length 22 cm and the current through it is $I = 40$ A. If the coefficient of static friction is $\mu_s = 0.50$, determine the minimum magnetic field **B** (not necessarily vertical) that will just cause the rod to slide. Give the magnitude of **B** and its direction relative to the vertical.

58. Estimate the approximate maximum deflection of the electron beam near the center of a TV screen due to the Earth's 5.0×10^{-5} T field. Assume the screen is 20 cm from the electron gun where the electrons are accelerated (a) by 2.0 kV, or (b) by 30 kV. Note that in color TV sets, the beam must be directed accurately to within less than 1 mm in order to strike the correct phosphor. Because the Earth's field is significant here, mu-metal shields are used to reduce the Earth's field in the CRT. (See Section 23–9.)

59. The **cyclotron** (Fig. 27–45) is a device used to accelerate elementary particles such as protons to high speeds. Particles starting at point A with some initial velocity travel in circular orbits in the magnetic field B. The particles are accelerated to higher speeds each time they pass in the gap between the metal "dees," where there is an electric field E. (There is no electric field within the cavity of the metal dees.) The electric field changes direction each half-cycle, owing to an ac voltage $V = V_0 \sin 2\pi ft$, so that the particles are increased in speed at each passage through the gap. (a) Show that the frequency f of the voltage must be $f = Bq/2\pi m$, where q is the charge on the particles and m their mass. (b) Show that the kinetic energy of the particles increases by $2qV_0$ each revolution, assuming that the gap is small. (c) If the radius of the cyclotron is 2.0 m and the magnetic field strength is 0.50 T, what will be the maximum kinetic energy of accelerated protons in MeV?

FIGURE 27–44 Looking down on a rod sliding on rails. Problems 56 and 57.

FIGURE 27–45 A cyclotron. Problem 59.

60. The rectangular loop of wire shown in Fig. 27–21 has mass m and carries current I. Show that if the loop is oriented at an angle $\theta \ll 1$ (in radians), then when it is released it will execute simple harmonic motion about $\theta = 0$. Calculate the period of the motion.

61. Magnetic fields are very useful in particle accelerators for "beam steering"; that is, the magnetic fields can be used to change the beam's direction without altering its speed (Fig. 27–46). Discuss how this could work with a beam of protons. What happens to protons that are not moving with the speed that the magnetic field is designed for? If the field extends over a region 5.0 cm wide and has a magnitude of 0.33 T, by approximately what angle will a beam of protons traveling at 0.75×10^7 m/s be bent?

Magnet

B

Evacuated tubes, inside of which the protons move with velocity indicated by the green arrows.

FIGURE 27–46   Problem 61.

62. A square loop of aluminum wire is 20.0 cm on a side. It is to carry 25.0 A and rotate in a uniform 1.65-T magnetic field as shown in Fig. 27–47. (a) Determine the minimum diameter of the wire so that it will not fracture from tension or shear. Assume a safety factor of 10. (See Table 12–2.) (b) What is the resistance of a single loop of this wire?

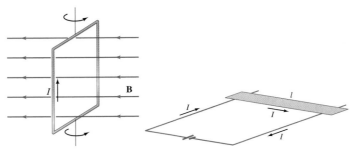

B

FIGURE 27–47
Problem 62.

FIGURE 27–48   Problem 63.

63. A sort of "projectile launcher" is shown in Fig. 27–48. A large current moves in a closed loop composed of fixed rails, a power supply, and a very light, almost frictionless bar touching the rails. A magnetic field is perpendicular to the plane of the circuit. If the bar has a length of 20 cm, a mass of 1.5 g, and is placed in a field of 1.7 T, what constant current flow is needed in order for it to accelerate to 30 m/s in a distance of 1.0 m? In what direction must the field point?

64. (a) What value of magnetic field would make a beam of electrons, traveling to the right at a speed of 4.8×10^6 m/s, go undeflected through a region where there is a uniform electric field of 10,000 V/m pointing vertically up? (b) What is the direction of the magnetic field if it is known to be perpendicular to the electric field? (c) What is the frequency of the circular orbit of the electrons if the electric field is turned off?

65. In a certain cathode ray tube, electrons are accelerated horizontally by 25 kV. They then pass through a uniform magnetic field B for a distance of 3.5 cm, which deflects them upward so they reach the top of the screen 22 cm away, 11 cm above the center. Estimate the value of B.

66. **Zeeman effect**. In the Bohr model of the hydrogen atom, the electron is held in its circular orbit of radius r about its proton nucleus by electrostatic attraction. If the atoms are placed in a weak magnetic field **B**, the rotation frequency of electrons rotating in a plane perpendicular to **B** is changed by an amount

$$\Delta f = \pm \frac{eB}{4\pi m}$$

where e and m are the charge and mass of an electron. (a) Derive this result, assuming the force due to **B** is much less than that due to electrostatic attraction of the nucleus. (b) What does the $\pm$ sign indicate?

67. A proton follows a spiral path through a gas in a magnetic field of 0.010 T, perpendicular to the plane of the spiral, as shown in Fig. 27–49. In two successive loops, P and Q, the radii are 10.0 mm and 8.5 mm, respectively. Calculate the change in the kinetic energy of the proton as it travels from P to Q.

FIGURE 27–49
Problem 67.

FIGURE 27–50   Problem 68.

68. The net force on a current loop in a uniform magnetic field is zero, since contributions to the net force from opposite sides of the loop cancel. However, if the field varies in magnitude from one side of the loop to the other, then a net force can be generated on the loop. Consider a square loop with sides whose length is a, located with one side at $x = b$ in the xy plane (Fig. 27–50). A magnetic field is directed along z, with a magnitude that varies with x according to

$$B = B_0\left(1 - \frac{x}{d}\right).$$

If the current in the loop circulates counter-clockwise (that is, the magnetic dipole moment of the loop is along the z axis), find an expression for the net force on the loop.

A long coil of wire with many closely spaced loops is called a solenoid. When a solenoid carries an electric current, a nearly uniform magnetic field is produced within the loops as suggested by the alignment of the iron filings in this photo. The magnitude of the field inside a solenoid is readily found using Ampère's law, one of the great general laws of electromagnetism, relating magnetic fields and electric currents. We examine these connections in detail in this chapter, as well as other means for producing magnetic fields.

CHAPTER 28

Sources of Magnetic Field

In the previous chapter, we discussed primarily the effects (forces and torques) that a magnetic field has on electric currents and on moving electric charges. We did see, however, that magnetic fields are produced not only by magnets but also by electric currents (Oersted's great discovery). It is this aspect of magnetism, the production of magnetic fields, that we discuss in this chapter. We will now see how magnetic field strengths are determined for some simple situations, and discuss some general relations between magnetic fields and their sources. We begin with the simplest case, the magnetic field created by a long straight wire carrying a steady electric current. We then look at how such a field, created by one wire, exerts a force on a second current-carrying wire. Interestingly enough, this interaction is used for the precise definitions of both the units of electric current and electric charge, the ampere and the coulomb.

Then we develop an elegant general approach to finding the connection between current and magnetic field known as Ampère's law, one of the fundamental equations of physics. We also examine a second technique for determining the magnetic field due to a current, known as the Biot-Savart law; though harder to visualize intuitively, the Biot-Savart law does allow us to solve problems more readily in many cases than Ampère's law.

Finally, we end this chapter with a discussion of what we understand about iron and other magnetic materials and how they produce magnetic fields.

FIGURE 28–1 Same as Fig. 27–9a, magnetic field lines around a long straight wire carrying an electric current I.

Magnetic field due to current in a long straight wire

FIGURE 28–2 Example 28–1.

➡ P H Y S I C S   A P P L I E D

A compass, near a current, may not point North

FIGURE 28–3 (a) Two parallel conductors carrying currents I_1 and I_2. (b) Magnetic field produced by I_1. (Field produced by I_2 is not shown.)

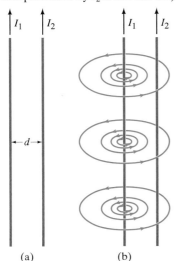

(a)   (b)

28–1 Magnetic Field Due to a Straight Wire

We saw in Section 27–2, Fig. 27–9, that the magnetic field due to the electric current in a long straight wire is such that the field lines are circles with the wire at the center (Fig. 28–1). You might expect that the field strength at a given point would be greater if the current flowing in the wire were greater; and that the field would be less at points farther from the wire. This is indeed the case. Careful experiments show that the magnetic field B at a point near a long straight wire is directly proportional to the current I in the wire and inversely proportional to the distance r from the wire:

$$B \propto \frac{I}{r}.$$

This relation is valid as long as r, the perpendicular distance to the wire, is much less than the distance to the ends of the wire (i.e., the wire is long).

The proportionality constant is written[†] as $\mu_0/2\pi$; thus,

$$B = \frac{\mu_0}{2\pi}\frac{I}{r}. \qquad \text{[outside a long straight wire]} \quad \textbf{(28–1)}$$

The value of the constant μ_0, which is called the **permeability of free space**, is $\mu_0 = 4\pi \times 10^{-7}\,\text{T·m/A}$ (see Section 28–3).

EXAMPLE 28–1 Calculation of B near a wire. A vertical electric wire in the wall of a building carries a dc current of 25 A upward. What is the magnetic field at a point 10 cm due north of this wire (Fig. 28–2)?

SOLUTION According to Eq. 28–1:

$$B = \frac{\mu_0 I}{2\pi r} = \frac{(4\pi \times 10^{-7}\,\text{T·m/A})(25\,\text{A})}{(2\pi)(0.10\,\text{m})} = 5.0 \times 10^{-5}\,\text{T},$$

or 0.50 G. By the right-hand rule (Fig. 27–9b), the field points to the west (into the page in Fig. 28–2) at this point. Since this field has about the same magnitude as Earth's, a compass placed at this location would not point north, but in a northwesterly direction.

28–2 Force Between Two Parallel Wires

We have seen that a wire carrying a current produces a magnetic field (magnitude given by Eq. 28–1 for a long straight wire), and furthermore that such a wire feels a force when placed in a magnetic field (Section 27–3, Eq. 27–1). Thus, we expect that two current-carrying wires would exert a force on each other.

Consider two long parallel conductors separated by a distance d, as in Fig. 28–3a. They carry currents I_1 and I_2, respectively. Each current produces a magnetic field that is "felt" by the other so that each must exert a force on the other, as Ampère first pointed out. For example, the magnetic field B_1 produced by I_1 is given by Eq. 28–1. At the location of the second conductor, the magnitude of the magnetic field produced by I_1 is

$$B_1 = \frac{\mu_0}{2\pi}\frac{I_1}{d}.$$

See Fig. 28–3b where the field due *only* to I_1 is shown. According to Eq. 27–2, the

[†]The constant is chosen in this complicated way so that Ampère's law (Section 28–4), which is considered more fundamental, will have a simple and elegant form.

force F per unit length l on the conductor carrying current I_2, when the field and current are perpendicular, is

$$\frac{F}{l} = I_2 B_1.$$

Note that the force on I_2 is due only to the field produced by I_1. Of course I_2 also produces a field, but it does not exert a force on itself. We substitute in the above formula for B_1 and find that the force per unit length on I_2 is

$$\frac{F}{l} = \frac{\mu_0}{2\pi}\frac{I_1 I_2}{d}. \qquad \textbf{(28–2)}$$

If we use the right-hand rule of Fig. 27–9b, we see that the lines of B_1 are as shown in Fig. 28–3b. Then using the right-hand rule of Fig. 27–12c, we see that the force exerted on I_2 will be to the left in Fig. 28–3b. That is, I_1 exerts an attractive force on I_2 (Fig. 28–4a). This is true as long as the currents are in the same direction. If I_2 is in the opposite direction, the right-hand rule indicates that the force is in the opposite direction. That is, I_1 exerts a repulsive force on I_2 (Fig. 28–4b).

Reasoning similar to that above shows that the magnetic field produced by I_2 exerts an equal but opposite force on I_1. We expect this to be true also, of course, from Newton's third law. Thus, as shown in Fig. 28–4, parallel currents in the same directions attract each other, whereas parallel currents in opposite directions repel.

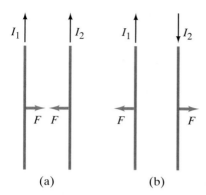

FIGURE 28–4 (a) Parallel currents in the same direction exert an attractive force on each other. (b) Antiparallel currents (in opposite directions) exert a repulsive force on each other.

EXAMPLE 28–2 **Force between two current carrying wires.** The two wires of a 2.0-m-long appliance cord are 3.0 mm apart and carry a current of 8.0 A dc. Calculate the force between these wires.

SOLUTION Equation 28–2 gives us

$$F = \frac{(2.0 \times 10^{-7}\,\text{T·m/A})(8.0\,\text{A})^2(2.0\,\text{m})}{(3.0 \times 10^{-3}\,\text{m})} = 8.5 \times 10^{-3}\,\text{N},$$

where we have written $\mu_0/2\pi = 2.0 \times 10^{-7}\,\text{T·m/A}$. Since the currents are in opposite directions, the force would tend to spread them apart.

EXAMPLE 28–3 **Suspending a current with a current.** A horizontal wire carries a current $I_1 = 80$ A dc. A second parallel wire 20 cm below it (Fig. 28–5) must carry how much current I_2 so that it doesn't fall due to gravity? The lower wire has a mass of 0.12 g per meter of length.

SOLUTION The force of gravity on wire 2 is downward, and per each meter of length has magnitude

$$\frac{F}{l} = \frac{mg}{l} = \frac{(0.12 \times 10^{-3}\,\text{kg})(9.8\,\text{m/s}^2)}{1.0\,\text{m}} = 1.18 \times 10^{-3}\,\text{N/m}.$$

The magnetic force on wire 2 must be upward $\bigl($hence I_2 must have the same direction as $I_1\bigr)$ and, with $d = 0.20$ m and $I_1 = 80$ A, has magnitude

$$\frac{F}{l} = \frac{\mu_0}{2\pi}\frac{I_1 I_2}{d}.$$

We solve for I_2 and find

$$I_2 = \frac{2\pi d}{\mu_0 I_1}\left(\frac{F}{l}\right) = \frac{2\pi(0.20\,\text{m})}{(4\pi \times 10^{-7}\,\text{T·m/A})(80\,\text{A})}(1.18 \times 10^{-3}\,\text{N/m}) = 15\,\text{A}.$$

FIGURE 28–5 Example 28–3.

28-3 Operational Definitions of the Ampere and the Coulomb

You may have wondered how the constant μ_0 in Eq. 28–1 could be exactly $4\pi \times 10^{-7}$ T·m/A. Here is how it happened. With an older definition of the ampere, μ_0 was measured experimentally to be very close to this value. Today, however, μ_0 is *defined* to be exactly $4\pi \times 10^{-7}$ T·m/A. This, of course, could not be done if the ampere were defined independently. The ampere, the unit of current, is now defined in terms of the magnetic field B it produces using the defined value of μ_0.

In particular, we use the force between two parallel current-carrying wires, Eq. 28–2, to define the ampere precisely. If $I_1 = I_2 = 1$ A exactly, and the two wires are exactly 1 m apart, then

$$\frac{F}{l} = \frac{\mu_0}{2\pi} \frac{I_1 I_2}{d} = \frac{(4\pi \times 10^{-7}\,\text{T·m/A})}{(2\pi)} \frac{(1\,\text{A})(1\,\text{A})}{(1\,\text{m})} = 2 \times 10^{-7}\,\text{N/m}.$$

Definitions

of ampere

and coloumb

Thus, *one* **ampere** *is defined as that current flowing in each of two long parallel conductors 1 m apart, which results in a force of exactly* 2×10^{-7} N/m *of length of each conductor.*

This is the precise definition of the ampere. The **coulomb** is then defined as being *exactly* one ampere-second: 1 C = 1 A·s. The value of k or ϵ_0 in Coulomb's law (Section 21–5) is obtained from experiment.

This may seem a rather roundabout way of defining quantities. The reason behind it is the desire for **operational definitions** of quantities—that is, definitions of quantities that can actually be measured given a definite set of operations to carry out. For example, the unit of charge, the coulomb, could be defined in terms of the force between two equal charges after defining a value for ϵ_0 or k in Eqs. 21–1 or 21–2. However, to carry out an actual experiment to measure the force between two charges is very difficult. For one thing, any desired amount of charge is not easily obtained precisely; and charge tends to leak from objects into the air. The amount of current in a wire, on the other hand, can be varied accurately and continuously (by putting a variable resistor in a circuit). Thus the force between two current-carrying conductors is far easier to measure precisely. And this is why the ampere is defined first and then the coulomb in terms of the ampere. At the National Institute of Standards and Technology in Maryland, precise measurement of current is made using circular coils of wire rather than straight lengths because it is more convenient and accurate.

Electric and magnetic field strengths are also defined operationally: the electric field in terms of the measurable force on a charge, via Eq. 21–3; and the magnetic field in terms of the force per unit length on a current-carrying wire, via Eq. 27–2.

28-4 Ampère's Law

In Section 28–1 we saw that Eq. 28–1 gives the relation between the current in a long straight wire and the magnetic field it produces. This equation is valid only for a long straight wire. The following question arises: is there a general relation between a current in a wire of any shape and the magnetic field around it? The answer is yes: the French scientist André Marie Ampère (1775–1836), proposed such a relation shortly after Oersted's discovery. Consider an arbitrary closed path around a current as shown in Fig. 28–6, and imagine this path as being made up of short segments each of length Δl. First, we take the product of the length of each segment times the component of **B** parallel to that segment (call this component $B_\parallel$). If we now sum all these terms, according to Ampère, the result will be equal to μ_0 times the net current I_{encl} that passes through the surface enclosed by the path:

$$\sum B_\parallel \Delta l = \mu_0 I_{\text{encl}}.$$

The lengths Δl are chosen so that $B_\parallel$ is essentially constant along each length. The

FIGURE 28–6 Arbitrary path enclosing a current, for Ampère's law. The path is broken down into segments of equal length Δl.

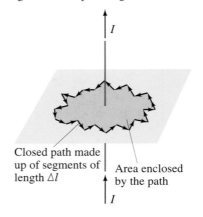

Closed path made up of segments of length Δl

Area enclosed by the path

sum must be made over a *closed path*; and I_{encl} is the net current passing through the surface bounded by this closed path. In the limit $\Delta l \to 0$, this relation becomes

$$\oint \mathbf{B} \cdot d\mathbf{l} = \mu_0 I_{encl}, \qquad \text{(28–3)}$$

where $d\mathbf{l}$ is an infinitesimal length vector and the vector dot product assures that the parallel component of **B** is taken. Equation 28–3 is known as **Ampère's law**. The integrand in Eq. 28–3 is taken around a closed path, and I_{encl} is the current passing through the area enclosed by the chosen path.

To understand Ampère's law better, let us apply it to the simple case of a long straight wire carrying a current I which we've already examined, and which served as an inspiration for Ampère himself. Suppose we want to find the magnitude of **B** at some point A which is a distance r from the wire (Fig. 28–7). We know the magnetic field lines are circles with the wire at their center. So to apply Eq. 28–3 we choose as our path of integration a circle of radius r. The choice of path is ours, so we choose one that will be convenient: at any point on this circular path, **B** will be tangent to the circle. Furthermore, since all points on the path are the same distance from the wire, by symmetry we expect B to have the same magnitude at each point. Thus for any short segment of the circle (Fig. 28–7), **B** will be parallel to that segment, and hence

$$\mu_0 I = \oint \mathbf{B} \cdot d\mathbf{l} = \oint B \, dl = B \oint dl = B(2\pi r),$$

where $\oint dl = 2\pi r$, the circumference of the circle, and $I_{encl} = I$. We solve for B and obtain

$$B = \frac{\mu_0 I}{2\pi r}.$$

This is just Eq. 28–1 for the field near a long straight wire as discussed earlier.

Ampère's law thus works for this simple case. A great many experiments indicate that Ampère's law is valid in general. However, as with Gauss's law for the electric field, its practical value as a means to calculate the magnetic field is limited mainly to simple or symmetric situations. Its importance is that it relates the magnetic field to the current in a direct and mathematically elegant way. Ampère's law is thus considered one of the basic laws of electricity and magnetism. It is valid for any situation where the currents and fields are steady and not changing in time, and no magnetic materials are present.

We now can see why the constant in Eq. 28–1 is written $\mu_0 / 2\pi$. This is done so that only μ_0 appears in Eq. 28–3, rather than, say, $2\pi k$ if we had used k in Eq. 28–1. In this way, the more fundamental equation, Ampère's law, has the simpler form.

It should be noted, that the **B** in Ampère's law is not necessarily due only to the current I_{encl}. Ampère's law, like Gauss's law for the electric field, is valid in general. **B** is the field at each point in space along the chosen path due to all sources—including the current I enclosed by the path, but also due to any other sources. For example, the field surrounding two parallel current-carrying wires is the vector sum of the fields produced by each, and the field lines are shown in Fig. 28–8. If the path chosen for the integral (Eq. 28–3) is a circle centered on one of the wires with radius less than the distance between the wires (the dashed line in Fig. 28–8), only the current (I_1) in the encircled wire is included on the right side of Eq. 28–3. **B** on the left side of the equation must be the total **B** at each point due to both wires. Note also that $\oint \mathbf{B} \cdot d\mathbf{l}$ for the path shown in Fig. 28–8 is the same whether the second wire is present or not (in both cases, it equals, $\mu_0 I_1$). How can this be? It can be so because although the fields due to the two wires tend to cancel one another at points between them, such as point N in the diagram ($\mathbf{B} = 0$ at a point midway between the wires if $I_1 = I_2$), at a point such as M in the Figure, the fields add together to produce a larger field. In the *sum*, $\oint \mathbf{B} \cdot d\mathbf{l}$, these effects just balance so that $\oint \mathbf{B} \cdot d\mathbf{l} = \mu_0 I_1$, whether the second wire is there or not. The integral $\oint \mathbf{B} \cdot d\mathbf{l}$ will be the same in each case, even though **B** will not be the same at every point for each of the two cases.

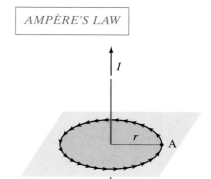

FIGURE 28–7 Circular path of radius r.

B for straight wire using Ampère's law

FIGURE 28–8 Magnetic field lines around two long parallel wires whose equal currents, I_1 and I_2, are coming out of the paper toward the viewer.

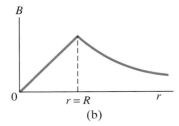

(a)

(b)

FIGURE 28–9 Magnetic field inside and outside a cylindrical conductor (Example 28–4).

EXAMPLE 28–4 **Field inside and outside a wire.** A long straight cylindrical wire conductor of radius R carries a current I of uniform current density in the conductor. Determine the magnetic field at (a) points outside the conductor $(r > R)$, and (b) points inside the conductor $(r < R)$. See Fig. 28–9. Assume that r, the radial distance from the axis, is much less than the length of the wire. (c) If $R = 2.0\,\text{mm}$ and $I = 60\,\text{A}$, what is B at $r = 1.0\,\text{mm}$, $r = 2.0\,\text{mm}$, and $r = 3.0\,\text{mm}$?

SOLUTION (a) Because the wire is long, straight, and cylindrical, we expect from the symmetry of the situation that the magnetic field must be the same at all points that are the same distance from the center of the conductor—there is no reason why any such point should have preference over the others at the same distance from the wire (they are physically equivalent), and so B must have the same value at all points the same distance from the center. We also expect **B** to be tangent to circles around the wire (Fig. 28–1), so let us choose a circular path of integration outside the wire $(r > R)$, but concentric with it, as we did in Fig. 28–7. Then $I_{\text{encl}} = I$, so

$$\oint \mathbf{B} \cdot d\mathbf{l} = B(2\pi r) = \mu_0 I_{\text{encl}}$$

or

$$B = \frac{\mu_0 I}{2\pi r}. \qquad\qquad [r > R]$$

which is the same result as for a thin wire.

(b) Inside the wire $(r < R)$, we again choose a circular path concentric with the cylinder; we expect **B** to be tangential to this path, and again, because of the symmetry, it will have the same magnitude at all points on the circle. The current enclosed in this case is less than I by a factor of the ratio of the areas:

$$I_{\text{encl}} = I \frac{\pi r^2}{\pi R^2}.$$

So Ampère's law gives

$$\oint \mathbf{B} \cdot d\mathbf{l} = \mu_0 I_{\text{encl}}$$

$$B(2\pi r) = \mu_0 I \left(\frac{\pi r^2}{\pi R^2}\right)$$

so

$$B = \frac{\mu_0 I r}{2\pi R^2}. \qquad\qquad [r < R]$$

The field is zero at the center of the conductor and increases linearly with r until $r = R$; beyond $r = R$, B decreases as $1/r$. This is shown in Fig. 28–9b. Note that these results are valid only for points close to the center of the conductor as compared to its length. For a current to flow, there must be connecting wires (to a battery, say), and the field due to these conducting wires, if not very far away, will destroy the assumed symmetry.

(c) At $r = 2.0\,\text{mm}$, the surface of the wire, $r = R$, so

$$B = \frac{\mu_0 I}{2\pi R} = \frac{(4\pi \times 10^{-7}\,\text{T·m/A})(60\,\text{A})}{(2\pi)(2.0 \times 10^{-3}\,\text{m})} = 6.0 \times 10^{-3}\,\text{T}.$$

We saw in (b) that inside the wire B is linear in r. So at $r = 1.0\,\text{mm}$, B will be half what it is at $r = 2.0\,\text{mm}$ or $3.0 \times 10^{-3}\,\text{T}$. Outside the wire, B falls off as $1/r$, so at $r = 3.0\,\text{mm}$ it will be two-thirds as great as at $r = 2.0\,\text{mm}$, or $B = 4.0 \times 10^{-3}\,\text{T}$. To check, we use our result in (a), $B = \mu_0 I/2\pi r$, which gives the same result.

CONCEPTUAL EXAMPLE 28–5 | **Coaxial cable.** A *coaxial cable* is a single wire surrounded by a cylindrical metallic braid, as shown in Fig. 28–10. The two conductors are separated by an insulator. The central wire carries current to the other end of the cable, and the outer braid carries the return current and is usually considered ground. Describe the magnetic field (*a*) in the space between the conductors, and (*b*) outside the cable.

RESPONSE (*a*) In the space between the conductors, we can apply Ampère's law for a circular path around the center wire, just as we did for the case shown in Fig. 28–7, and the magnitude is as given by Eq. 28–1. The current in the outer conductor has no bearing on this result. (Ampère's law uses only the current enclosed *inside* the path; as long as the currents outside the path don't affect the symmetry of the field, they do not contribute to the field along the path at all). (*b*) Outside the cable, we can draw a similar circular path, for we expect the field to have the same circular symmetry. Now, however, there are two currents enclosed by the path, and they add up to zero. The field outside the cable is zero.

The nice feature of coaxial cables is that they are self-shielding: no stray magnetic fields escape outside the cable. The outer cylindrical conductor also shields external electric fields from coming in (see also Example 21–13). This makes them ideal for carrying signals near sensitive equipment. Audiophiles use coaxial cables between stereo equipment components and even to the loudspeakers.

➥ **PHYSICS APPLIED**
Coaxial cable (shielding)

FIGURE 28–10 Coaxial cable. Conceptual Example 28–5.

EXAMPLE 28–6 | **A nice use for Ampère's law.** Use Ampère's law to show that in any region of space where there are no currents the magnetic field cannot be both unidirectional and nonuniform as shown in Fig. 28–11a.

SOLUTION The wider spacing of lines near the top of Fig. 28–11a indicates the field has a smaller magnitude at the top than it does lower down. We now apply Ampère's law to the rectangular path abcd shown dashed in the diagram. Since no current is enclosed by this path,

$$\oint \mathbf{B} \cdot d\mathbf{l} = 0.$$

The integral along sections ab and cd is zero, since $\mathbf{B} \perp d\mathbf{l}$. Thus

$$\oint \mathbf{B} \cdot d\mathbf{l} = B_{bc}l - B_{da}l = (B_{bc} - B_{da})l,$$

which is not zero since the field B_{bc} along the path bc is less than the field B_{da} along path da. Hence we have a contradiction: $\oint \mathbf{B} \cdot d\mathbf{l}$ cannot be both zero (since $I = 0$) and nonzero. Thus we have shown that a nonuniform unidirectional field is not consistent with Ampère's law. A nonuniform field whose direction also changes, as in Fig. 28–11b, is consistent with Ampère's law (convince yourself this is so), and possible. The fringing of a permanent magnet's field (Fig. 27–7) has this shape.

FIGURE 28–11 Example 28–6.

(a)

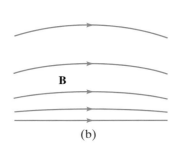

(b)

1. Ampère's Law, like Gauss' law, is always a valid statement. But as a calculation tool it is limited primarily to systems with a high degree of symmetry. The first step in applying Ampère's Law is to identify any useful symmetry.

2. Choose an integration path that reflects the symmetry (see the Examples to get a feel for this). In particular, search for paths where B has constant magnitude along the entire path or along segments of the path. Make sure your integration path passes through the point where you wish to evaluate the magnetic field.

3. Use symmetry to determine the direction of **B** along the integration path. With a smart choice of path, **B** will be either parallel or perpendicular to the path.

4. Evaluate the right-hand side of Ampère's Law by determining the enclosed current. Be careful with signs. Let the fingers of your right hand curl along the direction of **B** so that your thumb shows the direction of positive current. If the problem involves a solid conductor and your integration path does not enclose the full current, calculate the enclosed current using the current density (current per unit area) multiplied by the enclosed area (as in Example 28–4).

28–5 Magnetic Field of a Solenoid and a Toroid

A long coil of wire consisting of many loops is called a **solenoid**. Each loop produces a magnetic field as was shown in Fig. 27–10. In Fig. 28–12a, we see the field due to a solenoid when the coils are far apart. Near each wire, the field lines are very nearly circles as for a straight wire (that is, at distances that are small compared to the curvature of the wire). Between any two wires, the fields due to each loop tend to cancel. Toward the center of the solenoid, the fields add up to give a field that can be fairly large and fairly uniform. For a long solenoid, with closely packed coils, the field is nearly uniform and parallel to the solenoid axes within the entire cross section, as shown in Fig. 28–12b. The field outside the solenoid is very small compared to the field inside, except near the ends. Note that the same number of field lines that are concentrated inside the solenoid, spread out into the vast open space outside.

(a)

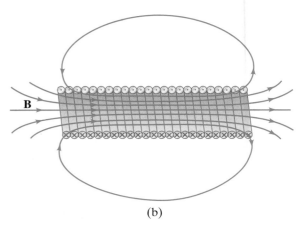
(b)

FIGURE 28–12 Magnetic field due to a solenoid: (a) loosely spaced turns, (b) closely spaced turns.

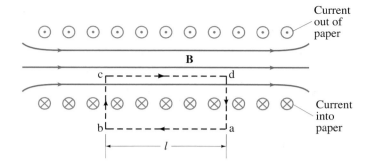

We now use Ampère's law to determine the magnetic field inside a very long (ideally, infinitely long) closely packed solenoid. We choose the path abcd shown in Fig. 28–13, far from either end, for applying Ampère's law. We will consider this path as made up of four segments, the sides of the rectangle: ab, bc, cd, da. Then the left side of Eq. 28–3, Ampère's law, becomes

$$\oint \mathbf{B} \cdot d\mathbf{l} = \int_a^b \mathbf{B} \cdot d\mathbf{l} + \int_b^c \mathbf{B} \cdot d\mathbf{l} + \int_c^d \mathbf{B} \cdot d\mathbf{l} + \int_d^a \mathbf{B} \cdot d\mathbf{l}.$$

The field outside the solenoid is so small as to be negligible compared to the field inside. Thus the first term in this sum will be zero. Furthermore, **B** is perpendicular to the segments bc and da inside the solenoid, and is nearly zero between and outside the coils, so these terms too are zero. Therefore we have reduced the integral to the segment cd where **B** is the nearly uniform field inside the solenoid, and is parallel to $d\mathbf{l}$, so

$$\oint \mathbf{B} \cdot d\mathbf{l} = \int_c^d \mathbf{B} \cdot d\mathbf{l} = Bl,$$

where l is the length cd. Now we determine the current enclosed by this loop for the right side of Ampère's law, Eq. 28–3. If a current I flows in the wire of the solenoid, the total current enclosed by our path abcd is NI where N is the number of loops our path encircles (five in Fig. 28–13). Thus Ampère's law gives us

$$Bl = \mu_0 NI.$$

If we let $n = N/l$ be the *number of loops per unit length*, then

$$B = \mu_0 nI. \qquad \text{[solenoid]} \qquad \textbf{(28–4)}$$

Magnetic field inside a solenoid

This is the magnitude of the magnetic field within a solenoid. Note that B depends only on the number of loops per unit length, n, and the current I. The field does not depend on position within the solenoid, so B is uniform. This is strictly true only for an infinite solenoid, but it is a good approximation for real ones for points not close to the ends.

EXAMPLE 28–7 **Field inside a solenoid.** A thin 10-cm-long solenoid used for fast electromechanical switching has a total of 400 turns of wire and carries a current of 2.0 A. Calculate the field inside near the center.

SOLUTION The number of turns per unit length is $n = 400/0.10\,\text{m} = 4.0 \times 10^3\,\text{m}^{-1}$. Thus

$$B = \mu_0 nI = (12.57 \times 10^{-7}\,\text{T·m/A})(4.0 \times 10^3\,\text{m}^{-1})(2.0\,\text{A})$$
$$= 1.0 \times 10^{-2}\,\text{T}.$$

A close look at Fig. 28–12 shows that the field outside of a solenoid is much like that of a bar magnet (Fig. 27–3). Indeed, a solenoid acts like a magnet, with one end acting as a north pole and the other as south pole, depending on the direction of the current in the loops. Since magnetic field lines leave the north pole of a magnet, the north poles of the solenoids in Fig. 28–12 are on the right.

Solenoids have many practical applications, and we discuss some of them later in the chapter, in Section 28–8.

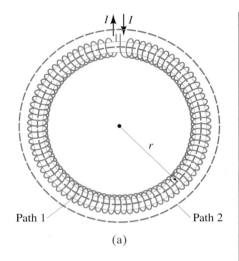

Path 1   Path 2

(a)

(b)

FIGURE 28–14 (a) A toroid. (b) A section of the toroid showing direction of the current for three loops: ⊙ means current toward viewer, ⊗ means current away from viewer.

EXAMPLE 28–8 **Toroid.** Use Ampère's law to determine the magnetic field (a) inside and (b) outside a toroid, which is like a solenoid bent into the shape of a circle as shown in Fig. 28–14a.

SOLUTION The magnetic field lines inside the toroid will be circles concentric with the toroid. (If you think of the toroid as a solenoid bent into a circle, the field lines bend along with the solenoid.) We choose as our path of integration one of these field lines of radius r inside the toroid as shown by the dashed line labeled "path 1" in Fig. 28–14a. We make this choice to use the symmetry of the situation, so B must be the same at all points along the path (although it is not necessarily the same across the whole cross-section of the toroid); so Ampère's law

$$\oint \mathbf{B} \cdot d\mathbf{l} = \mu_0 I_{\text{encl}}$$

becomes

$$B(2\pi r) = \mu_0 N I,$$

where N is the total number of coils and I is the current in each of the coils. Thus

$$B = \frac{\mu_0 N I}{2\pi r}.$$

The magnetic field B is not uniform within the toroid: it is largest along the inner edge (where r is smallest) and smallest at the outer edge. However, if the toroid is large, but thin (so that the difference between the inner and outer radii is small compared to the average radius) the field will be essentially uniform within the toroid. In this case, the formula for B reduces to that for a straight solenoid $B = \mu_0 n I$ where $n = N/(2\pi r)$ is the number of coils per unit length. (b) Outside the toroid, we choose as our path of integration a circle concentric with the toroid, "path 2" in Fig. 28–14a. This path encloses N loops carrying current I in one direction and N loops carrying the same current in the opposite direction. (Figure 28–14b shows the directions of the current for the parts of the loop on the inside and outside of the toroid.) Thus the net current enclosed by path 2 is zero. For a very tightly packed toroid, all points on path 2 are equidistant from the toroid and equivalent, so we expect B to be the same at all points along the path. Hence, Ampère's law gives

$$\oint \mathbf{B} \cdot d\mathbf{l} = \mu_0 I_{\text{encl}}$$

$$B(2\pi r) = 0$$

or

$$B = 0.$$

The same is true for a path taken at a radius smaller than that of the toroid. So there is no field exterior to a very tightly wound toroid. It is all inside the loops.

28–6 Biot-Savart Law

The usefulness of Ampère's law for determining the magnetic field **B** due to particular electric currents is restricted to situations where the symmetry of the given currents allows us to evaluate $\oint \mathbf{B} \cdot d\mathbf{l}$ readily. This does not, of course, invalidate Ampère's law nor does it reduce its fundamental importance. Recall the electric case, where Gauss's law is considered fundamental but is limited in its use for actually calculating **E**. We must often determine the electric field **E** by another method summing over contributions due to infinitesimal charge elements dq via Coulomb's law: $dE = (1/4\pi\epsilon_0)(dq/r^2)$. A magnetic equivalent to this infinitesimal form of Coulomb's law would be helpful for currents that do not have great symmetry. Such a law was developed by Jean Baptiste Biot (1774–1862) and Felix Savart (1791–1841) shortly after Oersted's discovery in 1820 that a current produces a magnetic field.

According to Biot and Savart, a current I flowing in any path can be considered as many tiny (infinitesimal) current elements, such as in the wire of Fig. 28–15. If $d\mathbf{l}$ represents any infinitesimal length along which the current is flowing, then the magnetic field, $d\mathbf{B}$, at any point P in space, due to this element of current, is given by

$$d\mathbf{B} = \frac{\mu_0 I}{4\pi} \frac{d\mathbf{l} \times \hat{\mathbf{r}}}{r^2},$$ (28–5) *Biot-Savart law*

where **r** is the displacement vector from the element $d\mathbf{l}$ to the point P, and $\hat{\mathbf{r}} = \mathbf{r}/r$ is the unit vector in the direction of **r** (see Fig. 28–15). Equation 28–5 is known as the **Biot-Savart law**. The magnitude of $d\mathbf{B}$ is

$$dB = \frac{\mu_0 I \, dl \sin \theta}{4\pi r^2},$$ (28–6)

where θ is the angle between $d\mathbf{l}$ and **r** (Fig. 28–15). The total magnetic field at point P is then found by summing (integrating) over all current elements:

$$\mathbf{B} = \int d\mathbf{B}.$$

Note that this is a *vector* sum. The Biot-Savart law is the magnetic equivalent of Coulomb's law in its infinitesimal form. It is even an inverse square law, like Coulomb's law.

An important difference between the Biot-Savart law and Ampère's law (Eq. 28–3) is that in Ampère's law $\left[\oint \mathbf{B} \cdot d\mathbf{l} = \mu_0 I_{\text{encl}}\right]$, **B** is not necessarily due only to the current enclosed by the path of integration. But in the Biot-Savart law the field $d\mathbf{B}$ in Eq. 28–5 is due only, and entirely, to the current element $I \, d\mathbf{l}$. To find the total **B** at any point in space, it is necessary to include *all* currents.

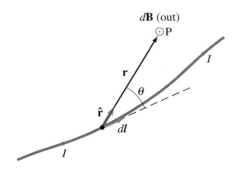

FIGURE 28–15 Biot-Savart law: the field at P due to current element $I \, d\mathbf{l}$ is $d\mathbf{B} = (\mu_0 I/4\pi)(d\mathbf{l} \times \hat{\mathbf{r}}/r^2)$.

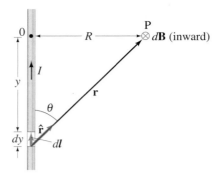

FIGURE 28–16 Determining **B** due to a long straight wire using the Biot-Savart law.

EXAMPLE 28–9 **B due to current I in straight wire.** For the field near a long straight wire carrying a current I, show that the Biot-Savart law gives the same result as Eq. 28–1, $B = \mu_0 I / 2\pi r$.

SOLUTION We calculate the magnetic field in Fig. 28–16 at point P, which is a perpendicular distance R from an infinitely long wire. The current is moving upwards, and both dl and $\hat{\mathbf{r}}$, which appear in the cross product of Eq. 28–5, are in the plane of the page. Hence the direction of the field $d\mathbf{B}$ due to each element of current must be directed into the plane of the page as shown (right-hand rule for the cross product $dl \times \hat{\mathbf{r}}$). Thus all the $d\mathbf{B}$ have the same direction at point P, and add up to give **B** the same direction consistent with our previous results (Figs. 28–1, 28–7, and 28–9). The magnitude of **B** will be

$$B = \frac{\mu_0 I}{4\pi} \int_{y=-\infty}^{+\infty} \frac{dy \sin\theta}{r^2},$$

where $dy = dl$ and $r^2 = R^2 + y^2$. Note that we are integrating over y (the length of the wire) so R is considered constant. Both y and θ are variables, but they are not independent. In fact, $y = -R/\tan\theta$. Note that we measure y as positive upward from point 0, so for the current element we are considering $y < 0$. Then

$$dy = +R \csc^2\theta \, d\theta = \frac{R \, d\theta}{\sin^2\theta} = \frac{R \, d\theta}{(R/r)^2} = \frac{r^2 \, d\theta}{R}.$$

So our integral becomes

$$B = \frac{\mu_0 I}{4\pi} \frac{1}{R} \int_{\theta=0}^{\pi} \sin\theta \, d\theta = -\frac{\mu_0 I}{4\pi R} \cos\theta \Big|_0^{\pi} = \frac{\mu_0 I}{2\pi R}.$$

This is just Eq. 28–1 for the field near a long wire, where R has been used instead of r.

FIGURE 28–17 Determining **B** due to a current loop.

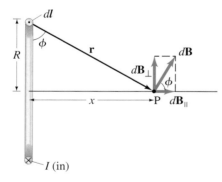

EXAMPLE 28–10 **Current loop.** Determine **B** for points on the axis of a circular loop of wire of radius R carrying a current I, Fig. 28–17.

SOLUTION For an element of current at the top of the loop, the magnetic field $d\mathbf{B}$ at point P on the axis has the direction shown, and magnitude (Eq. 28–5)

$$dB = \frac{\mu_0 I \, dl}{4\pi r^2}$$

since dl is perpendicular to **r** so $|dl \times \hat{\mathbf{r}}| = dl$. We can break $d\mathbf{B}$ down into components $dB_\parallel$ and $dB_\perp$, which are parallel and perpendicular to the axis as shown. When we sum over all the elements of the loop symmetry tells us that the perpendicular components will cancel on opposite sides, so $B_\perp = 0$. Hence, the total **B** will point along the axis, and will have magnitude

$$B = B_\parallel = \int dB \cos\phi = \int dB \frac{R}{r} = \int dB \frac{R}{(R^2 + x^2)^{\frac{1}{2}}},$$

where x is the distance of P from the center of the ring, and $r^2 = R^2 + x^2$. Now we put in dB from the equation above and integrate around the current loop, noting that all segments dl of current are the same distance, $(R^2 + x^2)^{\frac{1}{2}}$, from point P:

$$B = \frac{\mu_0 I}{4\pi} \frac{R}{(R^2 + x^2)^{\frac{3}{2}}} \int dl = \frac{\mu_0 I R^2}{2(R^2 + x^2)^{\frac{3}{2}}}$$

since $\int dl = 2\pi R$, the circumference of the loop. At the very center of the loop (where $x = 0$) the field has its maximum value

$$B = \frac{\mu_0 I}{2R}. \qquad \text{[at center of loop]}$$

Recall from Section 27–5 that a current loop, such as that just discussed (Fig. 28–17) is considered a *magnetic dipole*. We saw there that a magnetic dipole has a dipole moment

$$\mu = NIA,$$

where A is the area of the loop and N is the number of coils in the loop, each carrying current I. We also saw in Chapter 27 that a magnetic dipole placed in an external magnetic field experiences a torque and possesses potential energy, just like an electric dipole. In Example 28–10, we have looked at another aspect of a magnetic dipole: the magnetic field *produced by* a magnetic dipole has magnitude, along the dipole axis, of

$$B = \frac{\mu_0 I R^2}{2(R^2 + x^2)^{\frac{3}{2}}}.$$

We can write this in terms of the magnetic dipole moment $\mu = IA = I\pi R^2$ (for a single loop $N = 1$):

$$B = \frac{\mu_0}{2\pi} \frac{\mu}{(R^2 + x^2)^{\frac{3}{2}}}. \qquad \text{[magnetic dipole]} \quad \textbf{(28–7a)}$$

(Be careful to distinguish μ for dipole moment from μ_0, the magnetic permeability constant.) For distances far from the loop, $x \gg R$, this becomes

$$B \approx \frac{\mu_0}{2\pi} \frac{\mu}{x^3}. \qquad \left[\begin{matrix}\text{on axis,}\\ \text{magnetic dipole, } x \gg R\end{matrix}\right] \quad \textbf{(28–7b)}$$

The magnetic field on the axis of a magnetic dipole decreases with the cube of the distance, just as for an electric dipole. B decreases as the cube of the distance also for points not on the axis, although the multiplying factor is not the same. The magnetic field due to a current loop can be determined at various points using the Biot-Savart law and the results are in accord with experiment. The field lines around a current loop are shown in Fig. 28–18.

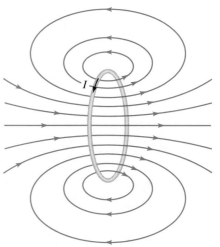

FIGURE 28–18 Magnetic field due to a circular loop of wire. (Same as Fig. 27–10.)

EXAMPLE 28–11 B due to a wire segment. One quarter of a circular loop of wire carries a current I as shown in Fig. 28–19. The current I enters and leaves on straight segments of wire, as shown; the straight wires are along the radial direction from the center C of the circular portion. Find the magnetic field at point C.

SOLUTION The current in the straight sections produces no magnetic field at point C because $d\mathbf{l}$ and $\hat{\mathbf{r}}$ in the Biot-Savart law (Eq. 28–5) are parallel and therefore $d\mathbf{l} \times \hat{\mathbf{r}} = 0$. (Said another way, the magnitude of $d\mathbf{l} \times \mathbf{r}$ is $(dl)(r)\sin\theta$ where $\theta = 0$.) Each piece $d\mathbf{l}$ of the curved section of wire produces a field $d\mathbf{B}$ that points into the paper at C (right-hand rule). The magnitude of each $d\mathbf{B}$ is (Eq. 28–6)

$$dB = \frac{\mu_0 I \, dl}{4\pi R^2}$$

where $r = R$ is the radius of the curved section, and $\sin\theta$ in Eq. 28–6 is $\sin 90° = 1$. Because $r = R$ for all pieces $d\mathbf{l}$, then

$$B = \int dB = \frac{\mu_0 I}{4\pi R^2} \int dl = \frac{\mu_0 I}{4\pi R^2}\left(\frac{1}{4} 2\pi R\right) = \frac{\mu_0 I}{8R}.$$

FIGURE 28–19 Example 28–11.

Magnetic fields can be produced (*a*) by magnetic materials (magnets) and (*b*) by electric currents. We have studied the latter extensively in this chapter. Now we take a brief look at magnetic materials, which are so present in everyday life: ordinary magnets, iron cores in motors and electromagnets, magnetic recording tape and computer data storage disks, even the magnetic stripe on credit cards. We saw in Section 27–1 that iron (and a few other materials) can be made into strong magnets. These materials are said to be **ferromagnetic**. We now look more deeply into the sources of ferromagnetism.

A bar magnet, with its two opposite poles near either end, resembles an electric dipole (equal-magnitude positive and negative charges separated by a distance). Indeed, a bar magnet is sometimes referred to as a "magnetic dipole." There are opposite "poles" separated by a distance. And the magnetic field lines of a bar magnet form a pattern much like that for the electric field of an electric dipole: compare Fig. 21–33a with Fig. 27–3b.

Microscopic examination reveals that a magnet is actually made up of tiny regions known as **domains**, which are at most about 1 mm in length or width. Each domain behaves like a tiny magnet with a north and a south pole. In an unmagnetized piece of iron, these domains are arranged randomly, as shown in Fig, 28–20a. The magnetic effects of the domains cancel each other out, so this piece of iron is not a magnet. In a magnet, the domains are preferentially aligned in one direction as shown in Fig. 28–20b (downward in this case). A magnet can be made from an unmagnetized piece of iron by placing it in a strong magnetic field. (You can make a needle magnetic, for example, by stroking it with one pole of a strong magnet.) Careful observations show in this case that the magnetization of domains may actually rotate slightly so as to be more nearly parallel to the external field. Or, more commonly, the borders of domains move so that those domains whose magnetic orientation is parallel to the external field grow in size at the expense of other domains, as can be seen by comparing Figs. 28–20a and b. This explains how a magnet can pick up unmagnetized pieces of iron like paper clips or bobby pins. The magnet's field causes a slight alignment of the domains in the unmagnetized object so that the object becomes a temporary magnet with its north pole facing the south pole of the permanent magnet, and vice versa; thus, attraction results. In the same way, elongated iron filings will arrange themselves in a magnetic field just as a compass needle does, and will reveal the shape of the magnetic field, Fig. 28–21.

An iron magnet can remain magnetized for a long time, and thus it is referred to as a "permanent magnet." However, if you drop a magnet on the floor or strike it with a hammer, you may jar the domains into randomness. The magnet can thus lose some or all of its magnetism. Heating a magnet too can cause a loss of magnetism, for raising the temperature increases the random thermal motion of the atoms which tends to randomize the domains. Above a certain temperature known as the **Curie temperature** (1043 K for iron), a magnet cannot be made at all.[†]

The striking similarity between the fields produced by a bar magnet (Fig. 27–3b) and by a loop of electric current (Fig. 28–18) suggests that the magnetic field produced by a current may have something to do with ferromagnetism, an idea proposed by Ampère in the nineteenth century. According to modern atomic theory, the atoms that make up any material can be roughly visualized as containing electrons that orbit around a central nucleus. Since the electrons are charged, they constitute an electric current and therefore produce a magnetic

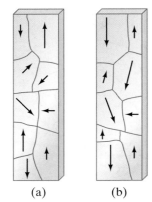

FIGURE 28–20 (a) An unmagnetized piece of iron is made up of domains that are randomly arranged. Each domain is like a tiny magnet; the arrows represent the magnetization direction, with the arrowhead being the N pole. (b) In a magnet, the domains are preferentially aligned in one direction, and may be altered in size by the magnetization process.

FIGURE 28–21 Iron filings line up along magnetic field lines.

[†] Iron, nickel, cobalt, gadolinium, and certain alloys are ferromagnetic at room temperature; several other elements and alloys have low Curie temperature and thus are ferromagnetic only at low temperatures.

field. But if there is no external field, the electron orbits in different atoms are arranged randomly, so the magnetic effects due to the many orbits in all the atoms in a material cancel out. However, electrons by themselves produce an additional intrinsic magnetic field—they have an intrinsic magnetic moment referred to as their "spin" magnetic moment.[†] It is the magnetic field due to electron spin that is now believed to produce ferromagnetism. In most materials, the magnetic fields due to electron spin cancel out because they are oriented at random. But in iron and other ferromagnetic materials, a complicated cooperative mechanism, known as "exchange coupling," operates. The result is that the spin of the electrons contributing to the ferromagnetism in a domain point in the same direction. Thus the tiny magnetic fields due to each of these electrons add up to give the magnetic field of a domain. And when the domains are aligned, as we have seen, a strong magnet results.

* | 28–8 | Electromagnets and Solenoids

A long coil of wire consisting of many loops of wire, as discussed in Section 28–5, is called a solenoid. The magnetic field within a solenoid can be fairly large since it will be the sum of the fields due to the current in each loop (see Fig. 28–22). The solenoid acts like a magnet; one end can be considered the north pole and the other the south pole, depending on the direction of the current in the loops (use the right-hand rule). Since the magnetic field lines leave the north pole of a magnet, the north pole of the solenoid in Fig. 28–22 is on the right.

If a piece of iron is placed inside a solenoid, the magnetic field is increased greatly because the domains of the iron are aligned by the magnetic field produced by the current. The resulting magnetic field is the sum of that due to the current and that due to the iron, and can be hundreds or thousands of times that due to the current alone (see Section 28–9). This arrangement is called an **electromagnet**. The iron used in electromagnets acquires and loses its magnetism quite readily when the current is turned on or off, and so is referred to as "soft iron." (It is "soft" only in a magnetic sense.) Iron that holds its magnetism even when there is no externally applied field is called "hard iron." Hard iron is used in permanent magnets. Soft iron is usually used in electromagnets so that the field can be turned on and off readily. Whether iron is hard or soft depends on heat treatment and other factors.

Electromagnets find use in many practical applications, from use in motors and generators to producing large magnetic fields for research. For some applications an iron core is not present—the magnetic field comes only from the current in the wire coils. When the current flows continuously in a normal electromagnet, a great deal of waste heat (I^2R power) can be produced. Cooling coils, which are tubes carrying water, must be used to absorb the heat in bigger installations. For some applications, superconducting magnets are used. The current-carrying wires are made of superconducting material (Section 25–9) kept below the transition temperature. Very high fields can be produced in the absence of an iron core, higher than what could be reached with an iron core (due to saturation of the iron—see the next Section). No electric power is needed to maintain large current in the superconducting coils, which means large savings of electricity, nor must huge amounts of heat be dissipated. Of course, energy is needed to keep the superconducting coils at the necessary low temperature.

Another useful device consists of a solenoid into which a rod of iron is partially inserted. This combination is also referred to as a solenoid. One simple use is

FIGURE 28–22 Magnetic field of a solenoid. The north pole of this solenoid, thought of as a magnet, is on the right, and the south pole is on the left.

➡ **PHYSICS APPLIED**

Electromagnets and solenoids

[†]The name "spin" comes from an early suggestion that this intrinsic magnetic moment arises from the electron "spinning" on its axis (as well as "orbiting" the nucleus) to produce the extra field. However this view of a spinning electron is oversimplified and not valid.

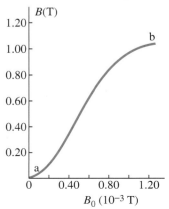

FIGURE 28–23 Solenoid used as a doorbell.

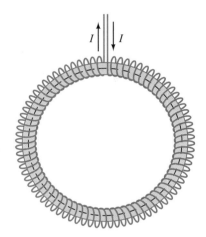

FIGURE 28–24 Iron-core toroid.

FIGURE 28–25 Total magnetic field B in an iron-core toroid as a function of the external field B_0 (B_0 is caused by the current I in the coil).

as a doorbell (Fig. 28–23). When the circuit is closed by pushing the button, the coil effectively becomes a magnet and exerts a force on the iron rod. The rod is pulled into the coil and strikes the bell. A larger solenoid is used in the starters of cars; when you engage the starter, you are closing a circuit that not only turns the starter motor, but activates a solenoid that first moves the starter into direct contact with the gears on the engine's flywheel. Solenoids are used as switches in many other devices, such as tape recorders. They have the advantage of moving mechanical parts quickly and accurately.

* 28–9 **Magnetic Fields in Magnetic Materials; Hysteresis**

The field of a long solenoid is directly proportional to the current. Indeed, Eq. 28–4 tells us that the field B_0 inside a solenoid is given by

$$B_0 = \mu_0 nI.$$

This is valid if there is only air inside the coil. If we put a piece of iron or other ferromagnetic material inside the solenoid, the field will be greatly increased, often by hundreds or thousands of times. This occurs because the domains in the iron become preferentially aligned by the external field. The resulting magnetic field is the sum of that due to the current and that due to the iron. It is sometimes convenient to write the total field in this case as a sum of two terms:

$$\mathbf{B} = \mathbf{B}_0 + \mathbf{B}_M. \tag{28–8}$$

Here, $\mathbf{B}_0$ refers to the field due only to the current in the wire (the "external field"). It is the field that would be present in the absence of a ferromagnetic material. Then $\mathbf{B}_M$ represents the additional field due to the ferromagnetic material itself; often $\mathbf{B}_M \gg \mathbf{B}_0$.

The total field inside a solenoid in such a case can also be written by replacing the constant μ_0 in Eq. 28–4 by another constant, μ, characteristic of the material inside the coil:

$$B = \mu nI; \tag{28–9}$$

μ is called the **magnetic permeability** of the material (do not confuse it with μ for magnetic moment). For ferromagnetic materials, μ is much greater than μ_0. For all other materials, its value is very close to μ_0 (Section 28–10). The value of μ, however, is not constant for ferromagnetic materials; it depends on the value of the external field B_0, as the following experiment shows.

Measurements on magnetic materials are generally done using a toroid, which is essentially a long solenoid bent into the shape of a circle (Fig. 28–24), so that practically all the lines of **B** remain within the toroid. Suppose the toroid has an iron core that is initially unmagnetized and there is no current in the windings of the toroid. Then the current I is slowly increased, and B_0 increases linearly with I. The total field B also increases, but follows the curved line shown in the graph of Fig. 28–25. (Note the different scales: $B \gg B_0$.) Initially, point a, the domains (Section 28–7) are randomly oriented. As B_0 increases, the domains become more and more aligned until at point b, nearly all are aligned. The iron is said to be approaching **saturation**. Point b is typically 70 percent of full saturation. (If B_0 is increased further, the curve continues to rise very slowly, and reaches 98 percent saturation only when B_0 reaches a value about a thousandfold above that at point b; the last few domains are very difficult to align.) Next, suppose the external field B_0 is reduced by decreasing the current in the coils. As the current is reduced to zero, shown as point c in Fig. 28–26, the domains do not become completely random. Some permanent magnetism remains. If the current is then reversed in direction, enough domains can be turned around so $B = 0$ (point d). As the reverse current is increased further, the iron approaches saturation in the opposite direction (point e). Finally, if the current is again reduced to zero and then increased in the original direction, the total field follows the path efgb, again approaching saturation at point b.

Notice that the field did not pass through the origin (point a) in this cycle. The fact that the curves do not retrace themselves on the same path is called **hysteresis**. The curve bcdefgb is called a **hysteresis loop**. In such a cycle, much energy is transformed to thermal energy (friction) due to realigning of the domains. It can be shown that the energy dissipated in this way is proportional to the area of the hysteresis loop.

At points c and f, the iron core is magnetized even though there is no current in the coils. These points correspond to a permanent magnet. For a permanent magnet, it is desired that ac and af be as large as possible. Materials for which this is true are said to have high **retentivity**.

Materials with a broad hysteresis curve as in Fig. 28–26 are said to be magnetically "hard." On the other hand, a hysteresis curve such as that in Fig. 28–27 occurs for "soft" iron. This is preferred for *electromagnets* and transformers (Section 29–6) since the field can be more readily switched off, and the field can be reversed with less loss of energy.

A ferromagnetic material can be demagnetized—that is, made unmagnetized. This can be done by reversing the magnetizing current repeatedly while decreasing its magnitude. This results in the curve of Fig. 28–28. The heads of a tape recorder are demagnetized in this way. The alternating magnetic field acting at the heads due to a demagnetizer is strong when the demagnetizer is placed near the heads and decreases as it is moved slowly away. Video and audio tapes themselves can be erased and ruined by a magnetic field, as can computer disks.

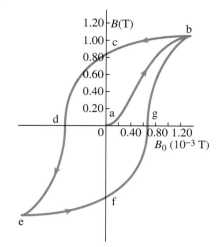

FIGURE 28–26 Hysteresis curve.

FIGURE 28–27 Hysteresis curve for soft iron.

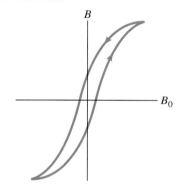

FIGURE 28–28 Successive hysteresis loops during demagnetization.

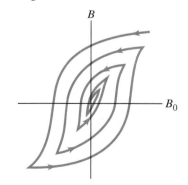

*28–10 Paramagnetism and Diamagnetism

All materials are magnetic to some extent. Nonferromagnetic materials fall into two principal classes: *paramagnetic*, in which the magnetic permeability μ is slightly greater than μ_0; and *diamagnetic*, in which, μ is slightly less than μ_0. The ratio of μ to μ_0 for any material is called the **relative permeability** K_m:

$$K_m = \frac{\mu}{\mu_0}.$$

Another useful parameter is the **magnetic susceptibility** χ_m defined as

$$\chi_m = K_m - 1.$$

Paramagnetic substances have $K_m > 1$ and $\chi_m > 0$, whereas diamagnetic substances have $K_m < 1$ and $\chi_m < 0$. See Table 28–1.

TABLE 28–1 Paramagnetism and Diamagnetism: Magnetic Susceptibilities

Paramagnetic substance	χ_m	Diamagnetic substance	χ_m
Aluminum	2.3×10^{-5}	Copper	-9.8×10^{-6}
Calcium	1.9×10^{-5}	Diamond	-2.2×10^{-5}
Magnesium	1.2×10^{-5}	Gold	-3.6×10^{-5}
Oxygen (STP)	2.1×10^{-6}	Lead	-1.7×10^{-5}
Platinum	2.9×10^{-4}	Nitrogen (STP)	-5.0×10^{-9}
Tungsten	6.8×10^{-5}	Silicon	-4.2×10^{-6}

The difference between paramagnetic and diamagnetic materials can be understood theoretically at the molecular level on the basis of whether or not the molecules have a permanent magnetic dipole moment. **Paramagnetism** occurs in materials whose molecules (or ions) have a permanent magnetic dipole moment.[†] In the absence of an external field, the molecules are randomly oriented and no magnetic effects are observed. However, when an external magnetic field is applied, say, by putting the material in a solenoid, the applied field exerts a torque on the magnetic dipoles (Section 27–5), tending to align them parallel to the field. The total magnetic field (external plus that due to aligned magnetic dipoles) will be slightly greater than B_0. The thermal motion of the molecules reduces the alignment, however. A useful quantity is the **magnetization vector**, **M**, defined as the magnetic dipole moment per unit volume,

Paramagnetism

$$\mathbf{M} = \frac{\boldsymbol{\mu}}{V},$$

where $\boldsymbol{\mu}$ is the magnetic dipole moment of the sample and V its volume. It is found experimentally that M is directly proportional to the external magnetic field (tending to align the dipoles) and inversely proportional to the Kelvin temperature T (tending to randomize dipole directions). This is called *Curie's law*, after Pierre Curie (1859–1906), who first noted it:

$$M = C\frac{B}{T},$$

where C is a constant. If the ratio B/T is very large (B very large or T very small) Curie's law is no longer accurate; as B is increased (or T decreased), the magnetization approaches some maximum value, M_{max}. This makes sense, of course, since M_{max} corresponds to complete alignment of all the permanent magnetic dipoles. However, even for very large magnetic fields, ≈ 2.0 T, deviations from Curie's law are normally noted only at very low temperatures, on the order of a few kelvins.

Ferromagnetic materials, as mentioned in Section 28–7, are no longer ferromagnetic above a characteristic temperature called the Curie temperature (1043 K for iron). Above this Curie temperature, they generally are paramagnetic.

Diamagnetism

Diamagnetic materials (for which μ_m is slightly less than μ_0) are made up of molecules that have no permanent magnetic dipole moment. When an external magnetic field is applied, magnetic dipoles are induced, but the induced magnetic dipole moment is in the direction opposite to that of the field. Hence the total field will be slightly less than the external field. The effect of the external field—in the crude model of electrons orbiting nuclei—is to increase the "orbital" speed of electrons revolving in one direction, and to decrease the speed of electrons revolving in the other direction; the net result is a net dipole moment opposing the external field. Diamagnetism is present in all materials, but is weaker even than paramagnetism and so is overwhelmed by paramagnetic and ferromagnetic effects in materials that display these other forms of magnetism.

[†] Other types of paramagnetism also occur whose origin is different from that described here, such as in metals where free electrons can contribute.

Summary

Ampère's law states that the line integral of the magnetic field **B** around any closed loop is equal to μ_0 times the total net current I_{encl} enclosed by the loop:

$$\oint \mathbf{B} \cdot d\mathbf{l} = \mu_0 I_{encl}.$$

The magnetic field B at a distance r from a long straight wire is directly proportional to the current I in the wire and inversely proportional to r. The magnetic field lines are circles centered at the wire.

The magnetic field inside a long tightly wound solenoid is $B = \mu_0 n I$ where n is the number of coils per unit length and I is the current in each coil.

The force that one long current-carrying wire exerts on a second parallel current-carrying wire a distance l away serves as the definition of the ampere unit, and ultimately of the coulomb as well.

The **Biot-Savart law** is useful for determining the magnetic field due to a known arrangement of currents. It states that

$$d\mathbf{B} = \frac{\mu_0 I}{4\pi} \frac{d\mathbf{l} \times \hat{\mathbf{r}}}{r^2},$$

where $d\mathbf{B}$ is the contribution to the total field at some point P due to a current I along an infinitesimal length $d\mathbf{l}$ of its path, and $\hat{\mathbf{r}}$ is the unit vector along the direction of the displacement vector $\mathbf{r}$ from $d\mathbf{l}$ to P. The total field **B** will be the integral over all $d\mathbf{B}$.

Iron and a few other materials can be made into strong permanent magnets. They are said to be **ferromagnetic**. Ferromagnetic materials are made up of tiny **domains**—each a tiny magnet—which are preferentially aligned in a permanent magnet, but randomly aligned in a nonmagnetized sample.

When a ferromagnetic material is placed in a magnetic field B_0 due to a current, say inside a solenoid or toroid, the material becomes magnetized. When the current is turned off, however, the material remains magnetized, and when the current is increased in the opposite direction (and then again reversed), a graph of the total field B versus B_0 is a **hysteresis loop**, and the fact that the curves do not retrace themselves is called **hysteresis**.

Questions

1. The magnetic field due to current in wires in your home can affect a compass. Discuss the problem in terms of currents, depending on whether they are ac or dc, and their distance away.

2. Compare and contrast the magnetic field due to a long straight current and the electric field due to a long straight line of electric charge at rest (Section 21–7).

3. Two long wires carrying equal currents I are at right angles to each other, but don't quite touch. Describe the magnetic force one exerts on the other.

4. A horizontal wire carries a large current. A second wire carrying a current in the same direction is suspended below. Can the current in the upper wire hold the lower wire in suspension against gravity? Under what conditions will it be in equilibrium?

5. A horizontal current-carrying wire, free to move in Earth's gravitational field, is suspended directly above a second, parallel, current-carrying wire. (a) In what direction is the current in the lower wire? (b) Can the upper wire be held in stable equilibrium due to the magnetic force of the lower wire? Explain.

6. (a) Write Ampère's law for a path that surrounds both conductors in Fig. 28–8. (b) Repeat, assuming the lower current I_2, is in the opposite direction $(I_2 = -I_1)$.

7. Suppose the cylindrical conductor of Fig. 28–9a has a concentric cylindrical hollow cavity inside it (so it looks like a pipe). What can you say about **B** in the cavity?

8. Explain why a field such as that shown in Fig. 28–11b is consistent with Ampère's law. Could the lines curve upward instead of downward?

9. What would be the effect on B inside a long solenoid if (a) the diameter of all the loops was doubled, or (b) the spacing between loops was doubled, or (c) the solenoid's length was doubled along with a doubling in the total number of loops.

10. Use the Biot-Savart law to convince yourself that the field of the current loop in Fig. 28–18 is correct as shown for points off the axis.

11. Do you think **B** will be the same for all points in the plane of the current loop of Fig. 28–18?

12. Why does twisting the lead-in wires to electrical devices reduce the magnetic effects of the leads?

13. Compare the Biot-Savart law with Coulomb's law: What are the similarities and differences?

14. A type of magnetic switch similar to a solenoid is a **relay**. A relay is an electromagnet (the iron rod inside the coil doesn't move) which, when activated, attracts a piece of soft iron on a pivot. Design a relay (a) to make a doorbell and (b) to close an electrical switch. A relay is used in the latter case when you need to switch on a circuit carrying a very large current but you do not want that large current flowing through the main switch. For example the starter switch of a car is connected to a relay so that the large current needed for the starter doesn't pass to the dashboard switch.

15. How might you measure the magnetic dipole moment of the Earth?

16. How might you define or determine the magnetic pole strength (the magnetic equivalent of a single electric charge) for (*a*) a bar magnet, (*b*) a current loop?

17. A heavy magnet attracts, from rest, a heavy block of iron. Before striking the magnet the block has acquired considerable kinetic energy. (*a*) What is the source of this kinetic energy? (*b*) When the block strikes the magnet, some of the latter's domains are jarred into randomness; describe the energy transformations.

18. Will a magnet attract any metallic object, or only those made of iron? (Try it and see.) Why is this so?

19. An unmagnetized nail will not attract an unmagnetized paper clip. However, if one end of the nail is in contact with a magnet, the other end *will* attract a paper clip. Explain.

20. How do you suppose the first magnets found in Magnesia were formed?

21. Why will either pole of a magnet attract an unmagnetized piece of iron?

22. Suppose you have three iron rods, two of which are magnetized but the third is not. How would you determine which two are the magnets without using any additional objects?

23. Two iron bars attract each other no matter which ends are placed close together. Are both magnets? Explain.

* 24. Describe the magnetization curve for (*a*) paramagnetic substance and (*b*) a diamagnetic substance, and compare to that for a ferromagnetic substance (Fig. 28–26).

* 25. Can all materials be considered (*a*) diamagnetic, (*b*) paramagnetic, (*c*) ferromagnetic?

Problems

Sections 28–1 and 28–2

1. (I) Jumper cables used to start a stalled vehicle often carry a 65-A current. How strong is the magnetic field 7.5 cm away? Compare to the Earth's magnetic field.

2. (I) If an electric wire is allowed to produce a magnetic field no larger than that of the Earth $(0.55 \times 10^{-4}\,\text{T})$ at a distance of 25 cm, what is the maximum current the wire can carry?

3. (I) What is the magnitude and direction of the force between two parallel wires 45 m long and 6.0 cm apart, each carrying 35 A in the same direction?

4. (I) A vertical straight wire carrying an upward 22-A current exerts an attractive force per unit length of $8.8 \times 10^{-4}\,\text{N/m}$ on a second parallel wire 7.0 cm away. What current (magnitude and direction) flows in the second wire?

5. (I) In Fig. 28–29, a long straight wire carries current I out of the page toward the viewer. Indicate, with appropriate arrows, the direction of **B** at each of the points C, D, and E in the plane of the page.

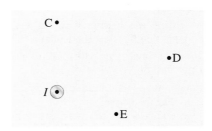

FIGURE 28–29 Problem 5.

6. (II) An experiment on the Earth's magnetic field is being carried out 1.00 m from an electric cable. What is the maximum allowable current in the cable if the experiment is to be accurate to ±1.0 percent?

7. (II) Two long thin parallel wires 15.0 cm apart carry 25-A currents in the same direction. Determine the magnetic field strength at a point 12.0 cm from one wire and 5.0 cm from the other (Fig. 28–30).

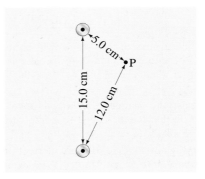

FIGURE 28–30
Problem 7.

8. (II) A horizontal compass is placed 20 cm due south from a straight vertical wire carrying a 40-A current downward. In what direction does the compass needle point at this location? Assume the horizontal component of the Earth's field at this point is $0.45 \times 10^{-4}\,\text{T}$ and the magnetic declination is 0°.

9. (II) A long horizontal wire carries 22.0 A of current due north. What is the net magnetic field 20.0 cm due west of the wire if the Earth's field there points downward, 40° below the horizontal, and has magnitude $5.0 \times 10^{-5}\,\text{T}$?

10. (II) A straight stream of protons passes a given point in space at a rate of 1.5×10^{9} protons/s. What magnetic field do they produce 2.0 m from the beam?

11. (II) Determine the magnetic field midway between two long straight wires 2.0 cm apart in terms of the current I in one when the other carries 15 A. Assume these currents are (*a*) in the same direction, and (*b*) in opposite directions.

12. (II) A long pair of wires serves to conduct 25.0 A of dc current to and from an instrument. If the wires are of negligible diameter but are 2.8 mm apart, what is the magnetic field 10.0 cm from their midpoint, in their plane (Fig. 28–31)? Compare to the magnetic field of the Earth.

FIGURE 28–31 Problem 12.

13. (II) A compass needle points 20° E of N outdoors. However, when it is placed 12.0 cm to the east of a vertical wire inside a building, it points 55° E of N. What is the magnitude and direction of the current in the wire? The Earth's field there is 0.50×10^{-4} T and is horizontal.

14. (II) A rectangular loop of wire is placed next to a straight wire, as shown in Fig. 28–32. There is a current of 2.5 A in both wires. What is the magnitude and direction of the net force on the loop?

FIGURE 28–32 Problem 14.

15. (II) Let two long parallel wires, a distance d apart, carry equal currents I in the same direction. One wire is at $x = 0$, the other at $x = d$, Fig. 28–33. Determine **B** between the wires as a function of x.

FIGURE 28–33 Problems 15 and 16.

16. (II) Repeat Problem 15 if the wire at $x = 0$ carries twice the current $(2I)$ as the other wire, and in the opposite direction.

17. (II) Two long wires are oriented so that they are perpendicular to each other, and at their closest, they are 20.0 cm apart (Fig. 28–34). What is the magnitude of the magnetic field at a point midway between them if the top one carries a current of 20.0 A and the bottom one carries 5.0 A?

FIGURE 28–34
Problem 17.

18. (II) Two long parallel wires 7.00 cm apart carry 16.5-A currents in the same direction. Determine the magnetic field strength at a point P 12.0 cm from one wire and 13.0 cm from the other. See Fig. 28–35.

FIGURE 28–35
Problem 18.

19. (III) A very long flat conducting strip of width L and negligible thickness lies in a horizontal plane and carries a uniform current I across its cross section. (a) Show that at points a distance y directly above its center, the field is given by

$$B = \frac{u_0 I}{\pi L} \tan^{-1} \frac{L}{2y},$$

assuming the strip is infinitely long. [*Hint*: Divide the strip into many thin "wires," and sum (integrate) over these.] (b) What value does B approach for $y \gg L$? Does this make sense? Explain.

20. (III) An electron is moving in a plane that also contains a long straight current-carrying wire. The electron is heading at a 45° angle toward the wire, with a speed of 3.4×10^6 m/s, when it is 50 cm away. The electron reaches only as close as 1.0 cm, before being repelled away, always moving in the same plane. What is the current in the wire?

Sections 28–4 and 28–5

21. (I) A 40.0-cm long solenoid 1.35 cm in diameter is to produce a field of 0.385 mT at its center. How much current should the solenoid carry if it has 1000 turns of wire?

22. (I) A 32-cm-long solenoid, 1.8 cm in diameter, is to produce a 0.30-T magnetic field at its center. If the maximum current is 5.7 A, how many turns must the solenoid have?

23. (I) A 2.5-mm-diameter copper wire carries a 40-A current. Determine the magnetic field: (a) at the surface of the wire; (b) inside the wire, 0.50 mm below the surface; (c) outside the wire 2.5 mm from the surface.

24. (II) A toroid (Fig. 28–14) has a 50.0-cm inner diameter and a 54.0-cm outer diameter. It carries a 25.0 A current in its 500 coils. Determine the range of values for B inside the toroid.

25. (II) A 20.0 m long copper wire, 2.00 mm in diameter including insulation, is tightly wrapped in a single layer with adjacent coils touching, to form a solenoid of diameter 2.50 cm. What is (a) the length of the solenoid and (b) the field at the center when the current in the wire is 20.0 A?

26. (II) (a) Use Eq. 28–1, and the vector nature of **B**, to show that the magnetic field lines around two long parallel wires carrying equal currents $I_1 = I_2$ are as shown in Fig. 28–8. (b) Draw the equipotential lines around two stationary positive electric charges. (c) Are these two diagrams similar? Identical? Why or why not?

27. (II) A coaxial cable consists of a solid inner conductor of radius R_1, surrounded by a concentric cylindrical tube or inner radius R_2 and outer radius R_3 (Fig. 28–36). The conductors carry equal and opposite currents I_0 distributed uniformly across their cross-sections. Determine the magnetic field at a distance R from the axis for: (a) $R < R_1$; (b) $R_1 < R < R_2$; (c) $R_2 < R < R_3$; (d) $R > R_3$.

FIGURE 28–36 Problems 27 and 28.

28. (III) Suppose the current in the coaxial cable of Problem 27, Fig. 28–36, is not uniformly distributed, but instead the current density j varies linearly with distance from the center: $j_1 = C_1 R$ for the inner conductor and $j_2 = C_2 R$ for the outer conductor. Each conductor still carries the same total current I_0, in opposite directions. Determine the magnetic field in terms of I_0 in the same four regions of space as in Problem 27.

Section 28–6

29. (I) The Earth's magnetic field is essentially that of a magnetic dipole. If the field near the North Pole is about 1.0×10^{-4} T, what will it be (approximately) 13,000 km above the surface at the North Pole?

30. (II) A wire, in a plane, has the shape shown in Fig. 28–37, two arcs of a circle connected by radial lengths of wire. Determine **B** at point C in terms of R_1, R_2, θ, and the current I.

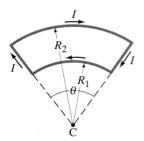

FIGURE 28–37 Problem 30.

31. (II) A circular conducting ring of radius R is connected to two exterior straight wires ending at two ends of a diameter (Fig. 28–38). The current I splits into unequal portions while passing through the ring as shown. What is **B** at the center of the ring?

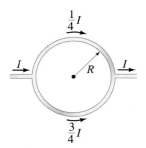

FIGURE 28–38 Problem 31.

32. (II) A small loop of wire of radius 1.8 cm is placed at the center of a 25.0-cm wire loop. The planes of the loops are perpendicular to each other, and a 7.0-A current flows in each. Estimate the torque the large loop exerts on the smaller one. What simplifying assumption did you make?

33. (II) A wire is formed into the shape of two half circles connected by equal-length straight sections as shown in Fig. 28–39. A current I flows in the circuit clockwise as shown. Determine (a) the magnitude and direction of the magnetic field at the center, C, and (b) the magnetic dipole moment of the circuit.

FIGURE 28–39 Problem 33.

34. (II) Use the Biot-Savart law to show that the magnetic field **B**, due to a single point charge q moving with velocity **v**, at a point P whose position vector relative to the charge is **r** (Fig. 28–40) is given by

$$\mathbf{B} = \frac{\mu_0}{4\pi} \frac{q\mathbf{v} \times \mathbf{r}}{r^3}.$$

(Assume v is much less than the speed of light.)

FIGURE 28–40   Problem 34.

35. (II) A nonconducting circular disk, of radius R, carries a uniformly distributed electric charge Q. The plate is set spinning with angular velocity ω about an axis perpendicular to the plate through its center (Fig. 28–41). Determine (a) its magnetic dipole moment and (b) the magnetic field at points on its axis a distance x from its center; (c) does Eq. 28–7b apply in this case for $x \gg R$?

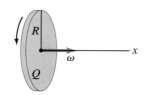

FIGURE 28–41   Problem 35.

36. (II) Consider a straight section of wire of length l, as in Fig. 28–42, which carries a current I. (a) Show that the magnetic field at a point P a distance R from the wire along its perpendicular bisector is

$$B = \frac{\mu_0 I}{2\pi R} \frac{l}{(l^2 + 4R^2)^{\frac{1}{2}}}.$$

(b) Show that this is consistent with Example 28–9 for an infinite wire.

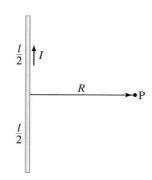

FIGURE 28–42   Problem 36.

37. (II) A segment of wire of length l carries a current I as shown in Fig. 28–43. (a) Show that for points along the positive x axis (the axis of the wire), such as point Q, the magnetic field **B** is zero. (b) Determine a formula for the field at points along the y axis, such as point P.

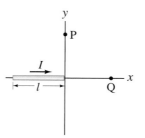

FIGURE 28–43
Problem 37.

38. (II) Use the result of Problem 37 to find the magnetic field at point P in Fig. 28–44 due to the current in the square loop.

FIGURE 28–44
Problem 38.

39. (II) A wire is bent into the shape of a regular polygon with n sides whose vertices are a distance R from the center. (See Fig. 28–45, which shows the special case of $n = 6$.) If the wire carries a current I_0, (a) determine the magnetic field at the center; (b) if n is allowed to become very large ($n \rightarrow \infty$), show that the formula in part (a) reduces to that for a circular loop (Example 28–10).

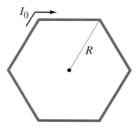

FIGURE 28–45
Problem 39.

40. (III) Start with the result of Example 28–10 for the magnetic field along the axis of a single loop to obtain the field inside a very long solenoid (Eq. 28–4).

41. (III) A single rectangular loop of wire, with sides a and b, carries a current I. An xy coordinate system has its origin at the lower left corner of the rectangle with the x axis parallel to side b (Fig. 28–46). Determine the magnetic field B at all points (x, y) within the loop.

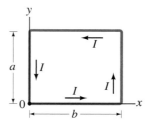

FIGURE 28–46
Problem 41.

42. (III) A square loop of wire, of side l, carries a current I. (a) Determine the magnetic field B at points on a line perpendicular to the plane of the square which passes through the center of the square (Fig. 28–47). Express B as a function of x, the distance along the line from the center of the square. (b) For $x \gg l$, does the square appear to be a magnetic dipole? If so, what is its dipole moment?

FIGURE 28–47
Problem 42.

* Section 28–7

* 43. (II) An iron atom has a magnetic dipole moment of about $1.8 \times 10^{-23}\,\text{A}\cdot\text{m}^2$. (a) Determine the dipole moment of an iron bar 12 cm long, 1.2 cm wide, and 1.2 cm thick, if it is 100 percent saturated. (b) What torque would be exerted on this bar when placed in a 1.2-T field acting at right angles to the bar?

* Section 28–9

* 44. (I) The following are some values of B and B_0 for a piece of annealed iron as it is being magnetized:

$B_0(10^{-4}\,\text{T})$	0	0.13	0.25	0.50	0.63	0.78	1.0	1.3
$B(\text{T})$	0	0.0042	0.010	0.028	0.043	0.095	0.45	0.67
$B_0(10^{-4}\,\text{T})$		1.9	2.5	6.3	13.0	130	1,300	10,000
$B(\text{T})$		1.01	1.18	1.44	1.58	1.72	2.26	3.15

Determine the magnetic permeability μ for each value and plot a graph of μ versus B_0.

* 45. (I) A large thin toroid has 400 loops of wire per meter, and a 20-A current flows through the wire. If the relative permeability of the iron is 3000, what is the total field B inside the toroid?

* 46. (II) An iron-core solenoid is 38 cm long, 1.8 cm in diameter, and has 600 turns of wire. A magnetic field of 2.2 T is produced when 48 A flows in the wire. What is the permeability μ at this high field strength?

General Problems

47. Three long parallel wires are 38.0 cm from one another. (Looking along them, they are at three corners of an equilateral triangle.) The current in each wire is 8.00 A, but that in wire M is opposite to that in wires N and P (Fig. 28–48). Determine the magnetic force per unit length on each wire due to the other two.

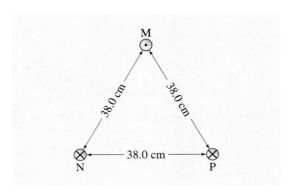

FIGURE 28–48 Problems 47 and 48.

48. In Fig. 28–48 the top wire is 2.00-mm-diameter copper wire and is suspended in air due to the two magnetic forces from the bottom two wires. The current is 20.0 A in each of the two bottom wires. Calculate the required current flow in the suspended wire.

49. An electron enters a large solenoid at a 7.0° angle to the axis. If the field is a uniform $3.3 \times 10^{-2}\,\text{T}$, determine the radius and pitch (distance between loops) of the electron's helical path if its speed is $1.3 \times 10^7\,\text{m/s}$.

50. A rectangular loop of wire carries a 2.0-A current and lies in a plane which also contains a very long straight wire carrying a 10.0-A current as shown in Fig. 28–49. Determine (a) the net force and (b) the net torque on the loop due to the straight wire.

FIGURE 28–49 Problem 50.

51. A very large flat conducting sheet of thickness t carries a uniform current density $\mathbf{j}$ throughout (Fig. 28–50). Determine the magnetic field (magnitude and direction) at a distance y above the plane. (Assume the plane is infinitely long and wide.)

FIGURE 28–50
Problem 51.

52. A long horizontal wire carries a current of 48 A. A second wire, made of 2.5-mm-diameter copper wire and parallel to the first but 15 cm below it, is held in suspension magnetically (Fig. 28–51). (a) What is the magnitude and direction of the current in the lower wire? (b) Is the lower wire in stable equilibrium? (c) Repeat parts (a) and (b) if the second wire is suspended 15 cm *above* the first due to the first's magnetic field.

$I = 48$ A

15 cm

$I = ?$

FIGURE 28–51
Problem 52.

53. You have 1.0 kg of copper and want to make a practical solenoid that produces the greatest possible magnetic field. Consider variables such as solenoid diameter, length, and so on, to determine if you should make your copper wire long and thin, short and fat, or something else.

54. For two long parallel wires separated by a distance L, carrying currents I_1 and I_2 as in Fig. 28–8, show that for the circular path of radius r ($r < L$) centered on I_1, that

$$\oint \mathbf{B} \cdot d\mathbf{l} = \mu_0 I_1$$

in accord with Ampère's law. (But do not use Ampère's law.)

55. Near the Earth's poles the magnetic field is about 1 G $(1 \times 10^{-4}\,\text{T})$. Imagine a simple model in which the Earth's field is produced by a single current loop around the equator. What current would this loop carry?

56. A square loop of wire, of side L, carries a current I. Show that the magnetic field at the center of the square is

$$B = \frac{2\sqrt{2}\,\mu_0 I}{\pi L}.$$

[*Hint*: Determine $\mathbf{B}$ for each segment of length L.]

57. In Problem 56, if you reshaped the square wire into a circle, would B increase or decrease at the center? Explain.

58. Two large coils of wire, each with N turns carrying a current I and separated by a distance equal to the radius R of the coils, are called *Helmholtz coils* (see Fig. 28–52). (a) Determine B at points x along the line joining their centers. Let $x = 0$ at the center of one coil, $x = R$ at the center of the other. (b) Plot B versus x from $x = 0$ to $x = R$. (c) Determine B on the axis midway between the coils $(x = R/2)$ if $R = 20.0$ cm, $I = 35$ A, and each coil contains $N = 350$ turns. (d) Show that the field midway between the coils is particularly uniform by showing that when the separation of the coils is R, then $dB/dx = 0$ and $d^2B/dx^2 = 0$ at the midpoint.

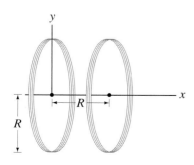

FIGURE 28–52
Problem 58.

59. A 175-g model airplane charged to 18.0 mC and traveling at 2.8 m/s passes within 8.6 cm of a wire, nearly parallel to its path, carrying a 30-A current. What acceleration (in g's) does this interaction give the airplane?

60. Suppose that an electromagnet uses a coil 2.0 m in diameter made from square copper wire 2.0 mm on a side; the power supply produces 50 V at a maximum power output of 1.0 kW. (a) How many turns are needed to run the power supply at maximum power? (b) What is the magnetic field strength at the center of the coil? (c) If you use a greater number of turns and this same power supply, will a greater magnetic field strength result? Explain.

61. Four long straight parallel wires located at the corners of a square of side l carry equal currents I_0 perpendicular to the page as shown in Fig. 28–53. Determine the magnitude and direction of $\mathbf{B}$ at the center of the square.

FIGURE 28–53   Problem 61.

62. Determine the magnetic field at the point P due to the long wire with a square bend shown in Fig. 28–54. The point P is halfway between the two corners. [*Hint*: You can use the results of Problems 36 and 37.]

FIGURE 28–54   Problem 62.

63. You want to get an idea of the magnitude of magnetic fields produced by overhead power lines. You estimate that the two wires run about 30 m above the ground about 3 m apart. A call to the local power company provides the information that the lines operate at 10 kilovolts and provide a maximum of 40 MW to the local area. Estimate the maximum magnetic field you might experience walking under these power lines, and compare to the Earth's field. [Note that if the current is ac, the magnetic field will be changing too.]

One of the great laws of physics is Faraday's law of induction, which says that a changing magnetic field produces an induced emf. This photo shows a bar magnet moving inside a coil, and the galvanometer registers an induced current. This phenomenon of electromagnetic induction is the basis for many practical devices, from generators to alternators to transformers, tape recording and reading computer memory.

29

Electromagnetic Induction and Faraday's Law

I n Chapter 27, we discussed two ways in which electricity and magnetism are related: (1) an electric current produces a magnetic field; and (2) a magnetic field exerts a force on an electric current or moving electric charge. These discoveries were made in 1820–1821. Scientists then began to wonder: if electric currents produce a magnetic field, is it possible that a magnetic field can produce an electric current? Ten years later the American Joseph Henry (1797–1878) and the Englishman Michael Faraday (1791–1867) independently found that it was possible. Henry actually made the discovery first. But Faraday published his results earlier and investigated the subject in more detail. We now discuss this phenomenon and some of its world-changing applications such as the electric generator.

29–1 Induced EMF

In his attempt to produce an electric current from a magnetic field, Faraday used an apparatus like that shown in Fig. 29–1. A coil of wire, X, was connected to a battery. The current that flowed through X produced a magnetic field that was intensified by the iron core. Faraday hoped that by using a strong enough battery, a steady current in X would produce a great enough magnetic field to produce a current in a second coil Y. This second circuit, Y, contained a galvanometer to detect any current but contained no battery. He met no success with steady currents.

*Constant **B** induces no emf*

FIGURE 29–1 Faraday's experiment to induce an emf.

But the long-sought effect was finally observed when Faraday saw the galvanometer in circuit Y deflect strongly at the moment he closed the switch in circuit X. And the galvanometer deflected strongly in the opposite direction when he opened the switch. A *steady* current in X had produced *no* current in Y. Only when the current in X was starting or stopping was a current produced in Y.

Faraday concluded that although a steady magnetic field produces no current, a *changing* magnetic field can produce an electric current! Such a current is called an **induced current**. When the magnetic field through coil Y changes, a current flows as if there were a source of emf in the circuit. We therefore say that

an induced emf is produced by a changing magnetic field.

Changing B induces an emf

Faraday did further experiments on **electromagnetic induction**, as this phenomenon is called. For example, Fig. 29–2 shows that if a magnet is moved quickly into a coil of wire, a current is induced in the wire. If the magnet is quickly removed, a current is induced in the opposite direction. Furthermore, if the magnet is held steady and the coil of wire is moved toward or away from the magnet, or the coil is rotated, again an emf is induced and a current flows. Motion or change is required to induce an emf. It doesn't matter whether the magnet or the coil moves. It is the relative motion that counts.

FIGURE 29–2 (a) A current is induced when a magnet is moved toward a coil. (b) The induced current is opposite when the magnet is moved away from the coil. Note that the galvanometer zero is at the center of the scale and the needle deflects left or right, depending on the direction of the current. In (c) no current is induced if the magnet does not move relative to the coil. It is the relative motion that counts: the magnet can be held steady and the coil moved, which also induces an emf.

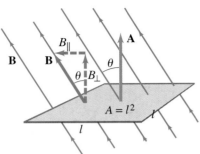

FIGURE 29–3 Determining the flux through a flat loop of wire. This loop is square, of side l and area $A = l^2$.

FIGURE 29–4 Magnetic flux Φ_B is proportional to the number of lines of **B** that pass through the loop.

$\theta = 90°$ $\theta = 45°$ $\theta = 0°$
(a)   (b)   (c)

FARADAY'S LAW OF INDUCTION

Lenz's law

Faraday investigated quantitatively what factors influence the magnitude of the emf induced. He found first of all that the more rapidly the magnetic field changes, the greater the induced emf. But the emf is not simply proportional to the rate of change of the magnetic field, **B**. Rather the emf is proportional to the rate of change of the **magnetic flux**, Φ_B, passing through the circuit or loop of area A. Magnetic flux for a uniform magnetic field is defined as

$$\Phi_B = B_\perp A = BA \cos\theta = \mathbf{B} \cdot \mathbf{A}. \qquad [\mathbf{B}\text{ uniform}] \quad (29\text{–}1a)$$

Here $B_\perp$ is the component of the magnetic field **B** perpendicular to the face of the loop, and θ is the angle between **B** and the vector **A** (representing the area) whose direction is perpendicular to the face of the loop. These quantities are shown in Fig. 29–3 for a square loop of side l whose area is $A = l^2$. If the area is of some other shape, or **B** is not uniform, the magnetic flux can be written[†]

$$\Phi_B = \int \mathbf{B} \cdot d\mathbf{A}. \qquad (29\text{–}1b)$$

As we saw earlier, the lines of **B** (like lines of **E**) can be drawn such that the number of lines per unit area is proportional to the field strength. Then the flux Φ_B can be thought of as being proportional to the *total number of lines passing through the loop*. This is illustrated in Fig. 29–4, where the loop is viewed from the side (on edge). For $\theta = 90°$, no lines pass through the loop and $\Phi_B = 0$, whereas Φ_B is a maximum when $\theta = 0°$. The unit of magnetic flux is the tesla-meter²; this is called a **weber**: $1\,\text{Wb} = 1\,\text{T}\cdot\text{m}^2$.

With this definition of the flux, we can now write down the results of Faraday's investigations—namely, that the emf induced in a circuit is equal to the rate of change of magnetic flux through the circuit:

$$\mathscr{E} = -\frac{d\Phi_B}{dt}. \qquad (29\text{–}2a)$$

This fundamental result is known as **Faraday's law of induction**, and is one of the basic laws of electromagnetism.

If the circuit contains N closely wrapped loops, the emfs induced in each add together, so

$$\mathscr{E} = -N\frac{d\Phi_B}{dt}. \qquad (29\text{–}2b)$$

The minus sign in Eq. 29–2 is placed there to remind us in which direction the induced emf acts. Experiments show that:

An induced emf gives rise to a current whose magnetic field opposes the original change in flux.

This is known as **Lenz's law**. Said another way, valid even if no current can flow (as when a circuit is not complete), is:

An induced emf is always in a direction that opposes the original change in flux that caused it.

Let us apply Lenz's law to the case of relative motion between a magnet and a

---

[†]The integral is taken over an open surface—that is, one bounded by a closed curve such as a circle or square. In the present discussion, the area is that enclosed by the loop under discussion. The area is not an enclosed surface as we used in Gauss's law, Chapter 22.

coil, Fig. 29–2. The changing flux through the coil induces an emf, which produces a current in the coil. And this induced current produces its own magnetic field. In Fig. 29–2a the distance between the coil and the magnet decreases. The magnetic field (and number of field lines), and therefore the flux, through the coil increases. The magnetic field of the magnet points upward. To oppose this upward increase, the field inside the coil produced by the induced current points *downward*. Thus, Lenz's law tells us that the current moves as shown (use the right-hand rule). In Fig. 29–2b, the flux *decreases* (because the magnet is moved away), so the induced current produces an *upward* magnetic field inside the coil that is "trying" to maintain the status quo. Thus the current is as shown.

Let us consider what would happen if Lenz's law were not true, but were just the reverse. The induced current in this imaginary situation would produce a flux in the same direction as the original change. This greater change in flux would produce an even larger current followed by a still greater change in flux, and so on. The current would continue to grow indefinitely, producing power $(= I^2R)$ even after the original stimulus ended. This would violate the conservation of energy. Such "perpetual motion" devices do not exist. Thus, Lenz's law as stated above (and not its opposite) is consistent with the law of conservation of energy.

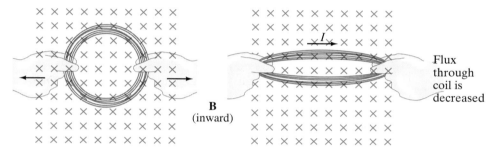

FIGURE 29–5 A current can be induced by changing the area of the coil. In both this case and that of Fig. 29–6, the flux through the coil is reduced. Here the brief induced current acts in the direction shown so as to try to maintain the original flux ($\Phi = BA$) by producing its own magnetic field into the page. That is, as the area A decreases, the current acts to increase B in the original (inward) direction.

It is important to note that an emf is induced whenever there is a change in flux. Since magnetic flux $\Phi_B = \int \mathbf{B} \cdot d\mathbf{A} = \int B \cos\theta \, dA$, we see that an emf can be induced in three ways: (1) by a changing magnetic field B; (2) by changing the area A of the loop in the field; or (3) by changing the loop's orientation θ with respect to the field. Figures 29–1 and 29–2 illustrated case 1. Examples of cases 2 and 3 are illustrated in Figs. 29–5 and 29–6, respectively.

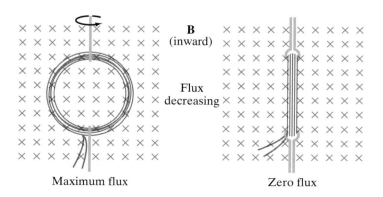

Maximum flux   Zero flux

FIGURE 29–6 A current can be induced by rotating a coil in a magnetic field.

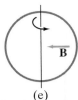

(a)	(b)	(c)	(d)	(e)
N magnetic pole moving toward loop into the page	N magnetic pole moving toward the loop in the plane of the page	Pulling the loop to the right out of a magnetic field that points out of the page	Shrinking a loop in a magnetic field pointing into the page	Rotating the loop about the vertical diameter by pulling the left side toward the reader and pushing the right side away from the reader in a magnetic field that points from right to left in the plane of the page

FIGURE 29–7 Example 29–1.

CONCEPTUAL EXAMPLE 29–1 **Practice with Lenz's law.** In which direction is the current induced in the loop for each situation in Fig. 29–7?

RESPONSE (*a*) Magnetic field lines point out from the N pole of a magnet, so as the magnet moves toward the loop, the field points into the page and is getting stronger. The current will be induced in the counterclockwise direction to produce a field **B** *out* of the page so that its own flux counteracts the externally imposed change.
(*b*) The field is in the plane of the page, so the flux through the loop is zero throughout the process; hence there is no change in magnetic flux with time, and there will be no induced emf or current in the loop.
(*c*) Initially, the magnetic flux pointing out of the page passes through the loop. If you pull the loop out of the field, the induced current will be in a direction to make up the deficiency: the current flow will be counterclockwise to produce an outward (toward the reader) magnetic field.
(*d*) The flux is into the page and the coil area shrinks so the flux will decrease; hence the induced current will be clockwise to try to produce its own flux into the page to make up for the flux decrease.
(*e*) Initially there is no flux through the loop (why?). When you start to rotate the loop, the flux begins passing through the loop increasing to the left. To counteract this, the loop will have current induced in a counterclockwise direction so as to produce its own flux to the right.

FIGURE 29–8 Example 29–2. The square coil in a magnetic field $B = 0.600\,\text{T}$ is pulled abruptly to the right to a region where $B = 0$.

EXAMPLE 29–2 **Pulling a coil from a magnetic field.** A square coil of wire with side 5.00 cm contains 100 loops and is positioned perpendicular to a uniform 0.600-T magnetic field, as shown in Fig. 29–8. It is quickly and uniformly pulled from the field (moving perpendicular to **B**) to a region where B drops abruptly to zero. At $t = 0$, the right edge of the coil is at the edge of the field. It takes 0.100 s for the whole coil to reach the field-free region. Find (*a*) the rate of change in flux through the coil, (*b*) the emf and current induced, and (*c*) how much energy is dissipated in the coil if its resistance is $100\,\Omega$. (*d*) What was the average force required?

SOLUTION (*a*) First we find how the magnetic flux changes during the time interval $\Delta t = 0.100\,\text{s}$. The area of the coil is $A = (5.00 \times 10^{-2}\,\text{m})^2 = 2.50 \times 10^{-3}\,\text{m}^2$. The flux is initially $\Phi_B = BA = (0.600\,\text{T})(2.50 \times 10^{-3}\,\text{m}^2) = 1.50 \times 10^{-3}\,\text{Wb}$. After 0.100 s, the flux is zero. The rate of change in flux is constant (because the coil is square), equal to

$$\frac{\Delta\Phi_B}{\Delta t} = \frac{0 - 1.50 \times 10^{-3}\,\text{Wb}}{0.100\,\text{s}} = -1.50 \times 10^{-2}\,\text{Wb/s}.$$

(*b*) The emf induced (Eq. 29–2) during this period is

$$\mathscr{E} = -N\frac{d\Phi_B}{dt} = -(100)(-1.50 \times 10^{-2}\,\text{Wb/s}) = 1.50\,\text{V}.$$

The current is

$$I = \frac{\mathscr{E}}{R} = \frac{1.50\,\text{V}}{100\,\Omega} = 15.0\,\text{mA},$$

and, by Lenz's law, must be clockwise to oppose the decreasing flux into the page. (*c*) The total energy dissipated is

$$E = Pt = I^2Rt = (1.50 \times 10^{-2}\,\text{A})^2(100\,\Omega)(0.100\,\text{s}) = 2.25 \times 10^{-3}\,\text{J}.$$

(*d*) We can calculate the force directly using $\mathbf{F} = I\mathbf{l} \times \mathbf{B}$, Eq. 27–3, for constant $\mathbf{B}$. The force the magnetic field exerts on the top and bottom sections of the square loop of Fig. 29–8 are in opposite directions and cancel each other. The magnetic force $\mathbf{F_M}$ exerted on the left vertical section of the square loop acts to the left as shown because the current is up (clockwise). The right side of the loop is in the region where $\mathbf{B} = 0$. Hence the needed external force, to the right, has magnitude

$$F_{\text{ext}} = NIlB = (100)(0.0150\,\text{A})(0.0500\,\text{m})(0.600\,\text{T}) = 0.0450\,\text{N},$$

since there are $N = 100$ current-carrying loops.
We can also calculate the average force using the result of part (*c*): From conservation of energy, the energy dissipated E is equal to the work W needed to pull the coil out of the field. Because $W = \bar{F}d$ where $d = 5.00\,\text{cm}$, then

$$\bar{F} = \frac{W}{d} = \frac{2.25 \times 10^{-3}\,\text{J}}{5.00 \times 10^{-2}\,\text{m}} = 0.0450\,\text{N},$$

which is the same answer, as we expect.

29–3 | EMF Induced in a Moving Conductor

Another way to induce an emf is shown in Fig. 29–9, and this situation helps illuminate the nature of the induced emf. Assume that a uniform magnetic field $\mathbf{B}$ is perpendicular to the area bounded by the U-shaped conductor and the movable rod resting on it. If the rod is made to move at a speed v, it travels a distance $dx = v\,dt$ in a time dt. Therefore, the area of the loop increases by an amount $dA = l\,dx = lv\,dt$ in a time dt. By Faraday's law, there is an induced emf $\mathscr{E}$ whose magnitude is given by

$$\mathscr{E} = \frac{d\Phi_B}{dt} = \frac{B\,dA}{dt} = \frac{Blv\,dt}{dt} = Blv. \tag{29–3}$$

This equation is valid as long as B, l, and v are mutually perpendicular. (If they are not, we use only the components of each that are mutually perpendicular.) An emf induced on a conductor moving in a magnetic field is sometimes called a **motional emf**.

We can also obtain Eq. 29–3 without using Faraday's law. We saw in Chapter 27 that a charged particle moving perpendicular to a magnetic field B with speed v experiences a force $\mathbf{F} = q\mathbf{v} \times \mathbf{B}$. When the rod of Fig. 29–9a moves to the right with speed v, the electrons in the rod move with this same speed. Therefore, since $\mathbf{v} \perp \mathbf{B}$, each electron feels a force $F = qvB$, which acts upward as shown in Fig. 29–9b. If the rod were not in contact with the U-shaped conductor, free electrons would collect at the upper end of the rod, leaving the lower end positive. There must thus be an induced emf. If the rod does slide on the U-shaped conductor, the electrons will flow into it. There will then be a clockwise (conventional) current flowing in the loop. To calculate the emf, we determine the work W needed to move a charge q from one end of the rod to the other against this potential difference: $W = \text{force} \times \text{distance} = (qvB)(l)$. The emf equals the work done per unit charge, so $\mathscr{E} = W/q = qvBl/q = Blv$, just as above.[†]

[†]This argument, which is basically the same as for the Hall effect (Section 27–8), explains this one way of inducing an emf. It does not explain the general case of electromagnetic induction.

FIGURE 29–9 (a) A conducting rod is moved to the right on a U-shaped conductor in a uniform magnetic field $\mathbf{B}$ that points out of the page. (b) Rod only, showing force on one electron.

This emf produces an electric field E in the rod which moves the electrons along (as in Section 25–8). Assuming a uniform E in the rod, then $E = \mathscr{E}/l = Bv$.

FIGURE 29–10 Example 29 – 3.

EXAMPLE 29–3 **Does a moving airplane develop a large emf?** An airplane travels 1000 km/h in a region where the Earth's magnetic field is 5.0×10^{-5} T and is nearly vertical (Fig. 29–10). What is the potential difference induced between the wing tips that are 70 m apart?

SOLUTION Since $v = 1000$ km/h $= 280$ m/s, and $\mathbf{v} \perp \mathbf{B}$, we have
$$\mathscr{E} = Blv = (5.0 \times 10^{-5}\,\text{T})(70\,\text{m})(280\,\text{m/s}) = 1.0\,\text{V}.$$
Not much to worry about.

EXAMPLE 29–4 **Force on the rod.** To make the rod of Fig. 29–9a move to the right at speed v, you need to apply a force. (a) Explain and determine the magnitude of the required force. (b) What external power is needed to move the rod?

SOLUTION (a) When the rod moves to the right, a current flows downward in the rod, as already discussed. We can see this also from Lenz's law: the upward magnetic flux through the loop is increasing, so the induced current must oppose the increase. Thus the current is clockwise so as to produce a magnetic field into the page (right-hand rule). The force on the moving rod is $\mathbf{F} = I\mathbf{l} \times \mathbf{B}$ for a constant $\mathbf{B}$ (Eq. 27–3). The right-hand rule tells us this force is to the left, and is thus a "drag force" opposing our effort to move the rod to the right.
The magnitude of the external force, to the right, needs to be $F = IlB$, where the current $I = \mathscr{E}/R = Blv/R$. The resistance R is that of the whole circuit, the rod and the U-shaped conductor. The force F required to move the rod is thus
$$F = IlB = \frac{B^2 l^2}{R} v.$$
If B, l, and R are constant, then a constant speed v is produced by a constant force. Constant R implies that all the resistance is in the rod and none in the U-shaped conductor.
(b) The external power needed to move the rod for constant R is
$$P = Fv = \frac{B^2 l^2 v^2}{R}.$$
The power dissipated in the resistance is $P = I^2 R$. With $I = \mathscr{E}/R = Blv/R$,
$$P = I^2 R = \frac{B^2 l^2 v^2}{R}$$
so the power input equals that dissipated in the resistance at any moment.

29–4 Electric Generators

FIGURE 29–11 An ac generator.

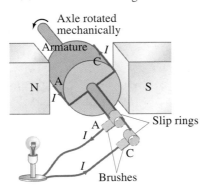

Probably the most important practical result of Faraday's great discovery was the development of the **electric generator** or **dynamo**. A generator transforms mechanical energy into electric energy. This is just the opposite of what a motor does. Indeed, a generator is basically the inverse of a motor. A simplified diagram of an **ac generator** is shown in Fig. 29–11. A generator consists of many coils of wire (only one is shown) wound on an armature that can rotate in a magnetic field. The axle is turned by some mechanical means (falling water, car motor belt), and an emf is induced in the rotating coil. An electric current is thus the *output* of a generator. In Fig. 29–11, the equation $\mathbf{F} = q\mathbf{v} \times \mathbf{B}$ tells us that, with the armature rotating counterclockwise, the (conventional) current in the wire labeled A on the armature is outward; therefore it is outward at brush A, as shown. (Each brush presses against a continuous slip ring.) After one-half revolution, wire A will be where wire C is now in the drawing, and the current then at brush A will be inward. Thus the current produced is alternating. Let us look at this in more detail.

Let us assume the loop is being made to rotate in a uniform magnetic field **B** with constant angular velocity ω. From Faraday's law (Eq. 29–2a), the induced emf is

$$\mathscr{E} = -\frac{d\Phi_B}{dt} = -\frac{d}{dt}\int \mathbf{B} \cdot d\mathbf{A} = -\frac{d}{dt}[BA\cos\theta]$$

where A is the area of the loop and θ is the angle between **B** and **A**. Since $\omega = d\theta/dt$, then $\theta = \theta_0 + \omega t$. We arbitrarily take $\theta_0 = 0$, so

$$\mathscr{E} = -BA\frac{d}{dt}(\cos\omega t) = BA\omega\sin\omega t.$$

If the rotating coil contains N loops,

$$\mathscr{E} = NBA\omega\sin\omega t$$
$$= \mathscr{E}_0\sin\omega t. \qquad \textbf{(29–4)}$$

Thus the output emf is sinusoidal (Fig. 29–12) with amplitude $\mathscr{E}_0 = NBA\omega$. Such a rotating coil in a magnetic field is the basic operating principle of an ac generator.

Over 99 percent of the electricity used in the United States is produced from generators (Fig. 29–13). The frequency $f = \omega/2\pi$ is 60 Hz for general use in the United States and Canada, although 50 Hz is used in many countries. In electric power generating plants, the armature is mounted on a heavy axle connected to a turbine, which is the modern equivalent of a waterwheel. Water pressure at a dam can turn the turbine at a hydroelectric plant. Most of the power generated at present in the United States, however, is done at steam plants, where the burning of fossil fuels (coal, oil, natural gas) boils water to produce high-pressure steam that turns the turbines. Likewise, at nuclear power plants, the nuclear energy released is used to produce steam to turn turbines. Thus, a heat engine (Chapter 20) connected to a generator is the principal means of generating electric power.

The frequency of 60 Hz is maintained very precisely by power companies, and in doing Problems, we will assume it is at least as precise as other numbers given.

EXAMPLE 29–5 **An ac generator.** The armature of a 60-Hz ac generator rotates in a 0.15-T magnetic field. If the area of the coil is 2.0×10^{-2} m², how many loops must the coil contain if the peak output is to be $\mathscr{E}_0 = 170$ V?

SOLUTION From Eq. 29–4 we see that the maximum emf is $\mathscr{E}_0 = NBA\omega$. Since $\omega = 2\pi f = (6.28)(60\,\text{s}^{-1}) = 377\,\text{s}^{-1}$, we have

$$N = \frac{\mathscr{E}_0}{BA\omega} = \frac{170\,\text{V}}{(0.15\,\text{T})(2.0\times10^{-2}\,\text{m}^2)(377\,\text{s}^{-1})} = 150 \text{ turns.}$$

A **dc generator** is much like an ac generator, except the slip rings are replaced by split-ring commutators, Fig. 29–14a, just as in a dc motor. The output of such a generator is as shown and can be smoothed out by placing a capacitor in parallel with the output (Section 26–4). More common is the use of many armature windings, as in Fig. 29–14b, which produces a smoother output.

FIGURE 29–12 An ac generator produces an alternating current. The output emf $\mathscr{E} = \mathscr{E}_0\sin\omega t$, where $\mathscr{E}_0 = NAB\omega$ (Eq. 29–4).

➡ **PHYSICS APPLIED**

Power plants

FIGURE 29–13 Water-driven generators at the base of Boulder Dam, Nevada.

➡ **PHYSICS APPLIED**

DC generator

FIGURE 29–14 (a) A dc generator with one set of commutators, and (b) a dc generator with many sets of commutators and windings.

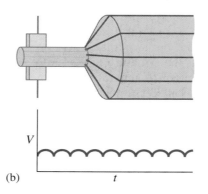

In the past, automobiles used dc generators. More common now, however, are ac generators or **alternators**, which avoid the problems of wear and electrical arcing (sparks) across the split-ring commutators of dc generators. Alternators differ from the generators discussed above in the following way. In an alternator, current from the battery produces a magnetic field in an electromagnet, called the *rotor*, which is made to rotate by a belt from the engine. Surrounding the rotating rotor are a set of stationary coils called the *stator*, Fig. 29–15. The magnetic field of the rotor passes through the stator coils and, since the rotor is rotating, the field through the fixed stator coils is changing. Hence an alternating current is induced in the stator coils, which is the output. This ac output is changed to dc for charging the battery by the use of semiconductor diodes, which allow current flow in one direction only.

Output emf of alternator

FIGURE 29–15 (a) Simplified schematic diagram of an alternator. The input electromagnet current to the rotor is connected through continuous slip rings. Sometimes the rotor electromagnet is replaced by a permanent magnet. (b) Actual shape of an alternator. The rotor is made to turn by a belt from the engine. The current in the wire coil of the rotor produces a magnetic field inside it on its axis that points horizontally from left to right, thus making north and south poles of the plates attached at either end. These end plates are made with triangular fingers that are bent over the coil—hence there are alternating N and S poles quite close to one another, with magnetic field lines between them as shown by the blue lines. As the rotor turns, these field lines pass through the fixed stator coils (shown on the right for clarity, but in operation the rotor rotates within the stator), inducing a current in them, which is the output.

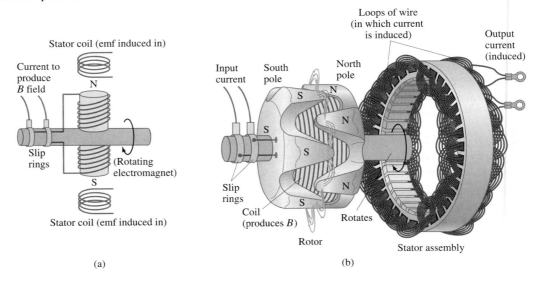

(a)   (b)

---

* | 29–5 | Counter EMF and Torque; Eddy Currents

* Counter EMF

Back emf

A motor turns and produces mechanical energy when a current is made to flow in it. From our description in Section 27–6 of a simple dc motor, you might expect that the armature would accelerate indefinitely due to the torque on it. However, as the armature of the motor turns, the magnetic flux through the coil changes and an emf is generated. This induced emf acts to oppose the motion (Lenz's law) and is called the **back emf** or **counter emf**. The greater the speed of the motor, the greater the counter emf. A motor normally turns and does work on something, but if there were no load, the motor's speed would increase until the counter emf equaled the input voltage. When there is a mechanical load, the speed of the motor may be limited also by the load. The counter emf will then be less than the external applied voltage. The greater the mechanical load, the slower the motor rotates and the lower is the counter emf ($\mathscr{E} \propto \omega$, Eq. 29–4).

EXAMPLE 29–6 **Counter emf in a motor.** The armature windings of a dc motor have a resistance of 5.0 Ω. The motor is connected to a 120-V line, and when the motor reaches full speed against its normal load, the counter emf is 108 V. Calculate (*a*) the current into the motor when it is just starting up, and (*b*) the current when it reaches full speed.

SOLUTION (*a*) Initially, the motor is not turning (or turning very slowly), so there is no induced counter emf. Hence, from Ohm's law, the current is

$$I = \frac{V}{R} = \frac{120\,\text{V}}{5.0\,\Omega} = 24\,\text{A}.$$

(*b*) At full speed, the counter emf is a source of emf that opposes the exterior source. We represent this counter emf as a battery in the equivalent circuit shown in Fig. 29–16. In this case, Ohm's law (or Kirchhoff's rule) gives

$$120\,\text{V} - 108\,\text{V} = I\,(5.0\,\Omega).$$

Therefore

$$I = \frac{12\,\text{V}}{5.0\,\Omega} = 2.4\,\text{A}.$$

This result shows that the current can be very high when a motor first starts up. This is why the lights in your house may dim when the motor of the refrigerator (or other large motor) starts up. The large initial current causes the voltage at the outlets to drop, since the house wiring has resistance and there is some voltage drop across it when large currents are drawn.

FIGURE 29–16 Circuit of a motor showing induced counter emf.

Effect of back emf on current

CONCEPTUAL EXAMPLE 29–7 **Motor overload.** When using an appliance such as a blender, electric drill, or sewing machine, if the appliance is overloaded or jammed so that the motor slows appreciably or stops while the power is still connected, the device can burn out and be ruined. Explain why this happens.

RESPONSE The motors are designed to run at a certain speed for a given applied voltage and the designer must take the expected counter emf into account. If the rotation speed is reduced, the counter emf will not be as high as expected ($\mathcal{E} \propto \omega$, Eq. 29–4), and the current will increase, and may become large enough that the windings of the motor heat up to the point of ruining the motor.

➥ **PHYSICS APPLIED**
Burning out a motor

* Counter Torque

In a generator, the situation is the reverse of that for a motor. As we saw, the mechanical turning of the armature induces an emf in the loops, which is the output. If the generator is not connected to an external circuit, the emf exists at the terminals but no current flows. In this case, it takes little effort to turn the armature. But if the generator *is* connected to a device that draws current, then a current flows in the coils of the armature. Because this current-carrying coil is in a magnetic field, there will be a torque exerted on it (as in a motor), and this torque opposes the motion (use the right-hand rule for the force on a wire, in Fig. 29–11). This is called a **counter torque**. The greater the electrical load—that is, the more current that is drawn—the greater will be the counter torque. Hence the external applied torque will have to be greater to keep the generator turning. This of course makes sense from the conservation-of-energy principle. More mechanical-energy input is needed to produce more electrical-energy output.

Counter torque

(a)

B (inward)

(b)

FIGURE 29–17 Production of eddy currents in a rotating wheel.

FIGURE 29–18 Repairing a step-down transformer on a utility pole.

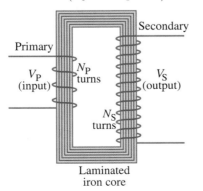

FIGURE 29–19 Step-up transformer ($N_P = 4$, $N_S = 12$).

* Eddy Currents

Induced currents are not always confined to well-defined paths such as in wires. Consider, for example, the rotating metal wheel in Fig. 29–17a. A magnetic field is applied to a limited area as shown and points into the paper. The section of wheel in the magnetic field has an emf induced in it because the conductor is moving, carrying electrons with it. The flow of (conventional) current is upward in the region of the magnetic field (Fig. 29–17b), and the current follows a downward return path outside that region. Why? According to Lenz's law, the induced currents oppose the change that causes them. Consider the part of the wheel labeled c in Fig. 29–17b, where the magnetic field is zero but is just about to enter a region where **B** points into the page. To oppose this change, the induced current is counterclockwise to produce a field pointing out of the page (right-hand rule). Similarly region d is about to move to e, where **B** is zero; hence the current is clockwise to produce an inward field opposed to this change. These currents are referred to as **eddy currents** and can be present in any conductor that is moving across a magnetic field or through which the magnetic flux is changing.

In Fig. 29–17, the magnetic field exerts a force **F** on the induced currents it has created, and that force opposes the rotational motion. Eddy currents can be used in this way as a smooth braking device on, say, a rapid-transit car. In order to stop the car, an electromagnet can be turned on that applies its field either to the wheels or to the moving steel rail below. Eddy currents can also be used to dampen (reduce) the oscillation of a vibrating system. Eddy currents, however, can be a problem. For example, eddy currents induced in the armature of a motor or generator produce heat ($P = I\mathscr{E}$) and waste energy. To reduce the eddy currents, the armatures are *laminated*; that is, they are made of very thin sheets of iron that are well insulated from one another. (See Fig. 29–19 in the next Section.) Thus the total path length of the eddy currents is confined to each slab, which increases the total resistance; hence the current is less and there is less wasted energy.

29–6 | Transformers and Transmission of Power

A transformer is a device for increasing or decreasing an ac voltage. Transformers are found everywhere: in TV sets to give the high voltage needed for the picture tube, in converters for plugging in a portable "Walkman," on utility poles (Fig. 29–18) to reduce the high voltage from the electric company to that usable in houses (110 V or 220 V), and in many other applications. A **transformer** consists of two coils of wire known as the **primary** and **secondary** coils. The two coils can be interwoven (with insulated wire); or they can be linked by a soft iron core which is laminated to prevent eddy-current losses (Section 29–5), as shown in Fig. 29–19. Transformers are designed so that (nearly) all the magnetic flux produced by the current in the primary also passes through the secondary coil, and we assume this is true in what follows. We also assume that energy losses in the resistance of the coils and hysteresis in the iron can be ignored—a good approximation for real transformers, which are often better than 99 percent efficient.

When an ac voltage is applied to the primary, the changing magnetic field it produces will induce an ac voltage of the same frequency in the secondary. However, the voltage will be different according to the number of loops in each coil. From Faraday's law, the voltage or emf induced in the secondary is

$$V_S = N_S \frac{d\Phi_B}{dt},$$

where N_S is the number of turns in the secondary coil, and $d\Phi_B/dt$ is the rate at which the magnetic flux changes.

The input primary voltage, V_P, is also related to the rate at which the flux changes

$$V_P = N_P \frac{d\Phi_B}{dt},$$

where N_P is the number of turns in the primary coil. This follows because the changing flux produces a counter emf, $N_P d\Phi_B/dt$, in the primary that exactly balances the applied voltage V_P if the resistance of the primary can be ignored (Kirchhoff's rules). We divide these two equations, assuming little or no flux is lost, to find

$$\frac{V_S}{V_P} = \frac{N_S}{N_P}. \qquad \qquad \textbf{(29-5)} \qquad \textit{Transformer equation}$$

This *transformer equation* tells how the secondary (output) voltage is related to the primary (input) voltage; V_S and V_P in Eq. 29–5 can be the rms values for both, or peak values for both. (DC voltages don't work in a transformer because there would be no changing magnetic flux.)

If N_S is greater than N_P, we have a **step-up transformer**. The secondary voltage is greater than the primary voltage. For example, if the secondary has twice as many turns as the primary, then the secondary voltage will be twice that of the primary. If N_S is less than N_P, we have a **step-down transformer**.

Although ac voltage can be increased (or decreased) with a transformer, we don't get something for nothing. Energy conservation tells us that the power output can be no greater than the power input. A well-designed transformer can be greater than 99 percent efficient, so little energy is lost to heat. The power output thus essentially equals the power input. Since power $P = VI$ (Eq. 25–6), we have

$$V_P I_P = V_S I_S,$$

or

$$\frac{I_S}{I_P} = \frac{N_P}{N_S}. \qquad \qquad \textbf{(29-6)} \qquad \textit{Transformer equation II}$$

EXAMPLE 29–8 **Portable radio transformer.** A transformer for home use of a portable radio reduces 120-V ac to 9.0-V ac. (Such a device also contains diodes to change the 9.0-V ac to dc.) The secondary contains 30 turns and the radio draws 400 mA. Calculate: (*a*) the number of turns in the primary; (*b*) the current in the primary; and (*c*) the power transformed.

SOLUTION (*a*) This is a step-down transformer, and from Eq. 29–5 we have

$$N_P = N_S \frac{V_P}{V_S} = \frac{(30)(120\,\text{V})}{(9.0\,\text{V})} = 400 \text{ turns.}$$

(*b*) From Eq. 29–6,

$$I_P = I_S \frac{N_S}{N_P} = (0.40\,\text{A})\left(\frac{30}{400}\right) = 0.030\,\text{A.}$$

(*c*) The power transformed is

$$P = I_S V_S = (9.0\,\text{V})(0.40\,\text{A}) = 3.6\,\text{W,}$$

which is, assuming 100 percent efficiency, the same as the power in the primary, $P = (120\,\text{V})(0.030\,\text{A}) = 3.6\,\text{W.}$

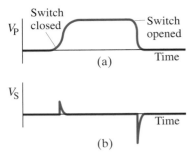

(a)

(b)

FIGURE 29–20 A dc voltage turned on and off as shown in (a) produces voltage pulses in the secondary (b). Voltage scales in (a) and (b) are not necessarily the same.

A transformer operates only on ac. A dc current in the primary does not produce a changing flux and therefore induces no emf in the secondary. However, if a dc voltage is applied to the primary through a switch, at the instant the switch is opened or closed there will be an induced current in the secondary. For example, if the dc is turned on and off as shown in Fig. 29–20a, the voltage induced in the secondary is as shown in Fig. 29–20b. Notice that the secondary voltage drops to zero when the dc voltage is steady. This is basically how, in the ignition system of an automobile, the high voltage is created to produce the spark across the gap of a spark plug that ignites the gas-air mixture. The transformer is referred to simply as the "ignition coil," and transforms the 12 V of the battery (when switched off) into a spike of as much as 25 kV.

FIGURE 29–21
The transmission of electric power from power plants to homes makes use of transformers at various stages.

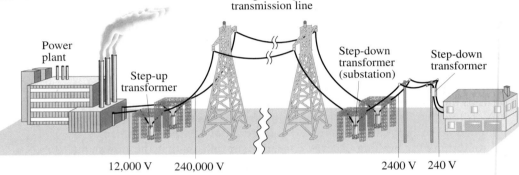

Transformers play an important role in the transmission of electricity. Power plants are often situated some distance from metropolitan areas, so electricity must often be transmitted over long distances (Fig. 29–21). There is always some power loss in the transmission lines, and this loss can be minimized if the power is transmitted at high voltage, using transformers, as the following Example shows.

EXAMPLE 29–9 Transmission lines. An average of 120 kW of electric power is sent to a small town from a power plant 10 km away. The transmission lines have a total resistance of 0.40 Ω. Calculate the power loss if the power is transmitted at (*a*) 240 V and (*b*) 24,000 V.

SOLUTION We cannot use $P = V^2/R$ because if R is the resistance of the transmission lines, we don't know the voltage drop along them; the given voltages are applied across the lines plus the load (the town). But for each case we can determine the current I in the lines, and then find the power loss from $P = I^2R$. (*a*) If 120 kW is sent at 240 V, the total current will be

$$I = \frac{P}{V} = \frac{1.2 \times 10^5 \text{ W}}{2.4 \times 10^2 \text{ V}} = 500 \text{ A}.$$

The power loss in the lines, P_L, is then

$$P_L = I^2R = (500 \text{ A})^2(0.40 \ \Omega) = 100 \text{ kW}.$$

Thus, over 80 percent of all the power would be wasted as heat in the power lines!

(b) If 120 kW is sent at 24,000 V, the total current will be

$$I = \frac{P}{V} = \frac{1.2 \times 10^5 \, \text{W}}{2.4 \times 10^4 \, \text{V}} = 5.0 \, \text{A}.$$

The power loss in the lines is then

$$P_L = I^2 R = (5.0 \, \text{A})^2 (0.40 \, \Omega) = 10 \, \text{W},$$

which is less than $\frac{1}{100}$ of 1 percent. We see that the greater the voltage, the less the current and thus the less power is wasted in the transmission lines. It is for this reason that power is usually transmitted at very high voltages, as high as 700 kV.

The great advantage of ac, and a major reason it is in nearly universal use, is that the voltage can easily be stepped up or down by a transformer. The output voltage of an electric generating plant is stepped up prior to transmission. Upon arrival in a city, it is stepped down in stages at electric substations prior to distribution. The voltage in lines along city streets is typically 2400 V and is stepped down to 240 V or 120 V for home use by transformers (Figs. 29–18 and 29–21).

29–7 | A Changing Magnetic Flux Produces an Electric Field

We have seen in earlier chapters (especially Chapter 25, Section 25–8) that when an electric current flows in a wire, there is an electric field in the wire that does the work of moving the electrons in the wire. In this chapter we have seen that a changing magnetic flux induces a current in the wire, which implies that there is an electric field in the wire induced by the changing magnetic flux. Thus we come to the important conclusion that

a changing magnetic flux produces an electric field.

This result applies not only to wires and other conductors, but is actually a general result that applies to any region in space. Indeed, an electric field will be produced at any point in space where there is a changing magnetic field.

Faraday's Law—General Form

We can put these ideas into mathematical form by generalizing our relation between an electric field and the potential difference between two points a and b: $V_{ab} = \int_a^b \mathbf{E} \cdot d\mathbf{l}$ (Eq. 23–3) where $d\mathbf{l}$ is an element of displacement along the path of integration. The emf $\mathscr{E}$ induced in a circuit is equal to the work done per unit charge by the electric field, which equals the integral of $\mathbf{E} \cdot d\mathbf{l}$ around the closed path:

$$\mathscr{E} = \oint \mathbf{E} \cdot d\mathbf{l}. \tag{29–7}$$

We combine this with Eq. 29–2a, to obtain a more elegant and general form of Faraday's law

$$\oint \mathbf{E} \cdot d\mathbf{l} = -\frac{d\Phi_B}{dt} \tag{29–8}$$

FARADAY'S LAW (GENERAL FORM)

which relates the changing magnetic flux to the electric field it produces. The integral on the left is taken around a path enclosing the area through which the magnetic flux Φ_B is changing. This more elegant statement of Faraday's law (Eq. 29–8) is valid not only in conductors, but in any region of space. To illustrate this, let us take an Example.

(a)

(b)

(c)

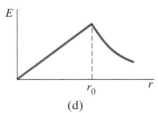

(d)

FIGURE 29–22 Example 29–10.
(a) Side view of nearly constant **B**.
(b) Top view, for determining the electric field **E** at point P. (c) Lines of **E** produced by increasing **B** (pointing outward). (d) Graph of E vs. r.

EXAMPLE 29–10 **E produced by changing B.** A magnetic field **B** between the pole faces of an electromagnet is nearly uniform at any instant over a circular area of radius r_0 as shown in Figs. 29–22a and b. The current in the windings of the electromagnet is increasing in time so that **B** changes in time at a constant rate $d\mathbf{B}/dt$ at each point. Beyond the circular region $(r > r_0)$, we assume **B** = 0 at all times. Determine the electric field **E** at any point P a distance r from the center of the circular area.

SOLUTION The changing magnetic flux through a circle of radius r, shown dashed in Fig. 29–22b, will produce an emf around this circle. Because all points on the dashed circle are equivalent physically, the electric field too will show this symmetry and will be in the plane perpendicular to **B**. Thus we can expect **E** to be perpendicular to **B** and to be tangent to the circle of radius r; the direction of **E** will be as shown in Fig. 29–22b and c, since by Lenz's law the induced **E** needs to be capable of producing a current that generates a magnetic field opposing the original change in **B**. By symmetry, we also expect **E** to have the same magnitude at all points on the circle of radius r. We therefore take this circle as our path of integration in Eq. 29–8 (ignoring the minus sign so we can concentrate on magnitude since we got the direction of **E** from Lenz's law) and obtain

$$E(2\pi r) = (\pi r^2)\frac{dB}{dt}, \qquad [r < r_0]$$

since $\Phi_B = BA = B(\pi r^2)$ at any instant. We solve for E and obtain

$$E = \frac{r}{2}\frac{dB}{dt}. \qquad [r < r_0]$$

This expression is valid up to the edge of the circle $(r \le r_0)$, beyond which **B** = 0. If we now consider a point where $r > r_0$, the flux through a circle of radius r is $\Phi_B = \pi r_0^2 B$. Then Eq. 29–8 gives

$$E(2\pi r) = \pi r_0^2 \frac{dB}{dt} \qquad [r > r_0]$$

or

$$E = \frac{r_0^2}{2r}\frac{dB}{dt}. \qquad [r > r_0]$$

Thus the magnitude of the electric field increases linearly from zero at the center of the magnet to $E = (dB/dt)(r_0/2)$ at the edge, and then decreases inversely with distance in the region beyond the edge of the magnetic field. The electric field lines are circles as shown in Fig. 29–22c. A graph of E vs. r is shown in Fig. 29–22d.

* Forces Due to Changing B are Nonconservative

Example 29–10 illustrates an important difference between electric fields produced by changing magnetic fields and electric fields produced by electric charges at rest (electrostatic fields). Electric field lines produced in the electrostatic case (Chapters 21 to 24) start and stop on electric charges. But the electric field lines produced by a changing magnetic field are continuous; they form closed loops. This distinction goes even further and is an important one. In the electrostatic case, the potential difference between two points is given by

$$V_{ab} = V_a - V_b = \int_a^b \mathbf{E} \cdot d\mathbf{l}.$$

If the integral is around a closed loop, so points a and b are the same, then $V_{ab} = 0$. Hence the integral of $\mathbf{E} \cdot d\mathbf{l}$ around a closed path is zero:

$$\oint \mathbf{E} \cdot d\mathbf{l} = 0. \qquad \text{[electrostatic field]}$$

This followed from the fact that the electrostatic force (Coulomb's law) is a conservative force, and so a potential energy function could be defined. Indeed, the relation above, $\oint \mathbf{E} \cdot d\mathbf{l} = 0$, tells us that the work done per unit charge around any closed path is zero (or the work done between any two points is independent of path—see Chapter 8), which is a property only of a conservative force. But in the nonelectrostatic case, when the electric field is produced by a changing magnetic field, the integral around a closed path is *not* zero, but is given by Eq. 29–8:

$$\oint \mathbf{E} \cdot d\mathbf{l} = -\frac{d\Phi_B}{dt}.$$

We thus come to the conclusion that the forces due to changing magnetic fields are *nonconservative*. We are not able therefore to define a potential energy, or potential function, at a given point in space for the nonelectrostatic case. Although static electric fields are *conservative fields*, the electric field produced by a changing magnetic field is a **nonconservative field**.

* 29–8 Applications of Induction: Sound Systems, Computer Memory, the Seismograph

There are various types of *microphones*, and many operate on the principle of induction. In one form, a microphone is just the inverse of a loudspeaker (Section 27–6). A small coil connected to a membrane is suspended close to a small permanent magnet, as shown in Fig. 29–23. The coil moves in the magnetic field when sound waves strike the membrane. The frequency of the induced emf will be just that of the impinging sound waves, and this emf is the "signal" that can be amplified and sent to loudspeakers, or sent to a tape recorder to be recorded on tape. In a "ribbon" microphone, a thin metal ribbon is suspended between the poles of a permanent magnet. The ribbon vibrates in response to sound waves, and the emf induced in the ribbon is proportional to its velocity.

Tape recording and tape playback is done by *tape heads* inside a tape recorder (or cassette deck). Recording tape for use in audio and video tape recorders contains a thin layer of magnetic oxide on a thin plastic tape. During recording, the audio or video signal voltage is sent to the recording head, which acts as a tiny electromagnet (Fig. 29–24) that magnetizes the tiny section of tape passing over the narrow gap in the head at each instant. In playback, the changing magnetism of the moving tape at the gap causes corresponding changes in the magnetic field within the soft-iron head, which in turn induces an emf in the coil (Faraday's law). This induced emf is the output signal that can be amplified and sent to a loudspeaker (or, in the case of a video signal, to the picture tube). In audio and video recorders, the signals may be *analog*—they vary continuously in amplitude over time. The variation in degree of magnetization of the tape at any point reflects the variation in amplitude of the audio or video signal.

Digital information, such as used on computer disks (floppy disks or hard disks) or on magnetic computer tape and some types of digital tape recorders, is read and written using heads that are basically the same as just described (Fig. 29–24). The essential difference is in the signals, which are not analog, but are digital, and in particular binary, meaning that only two values are possible for each of the extremely high number of predetermined spaces on the tape or disk. The two possible values are usually referred to as 1 and 0. The signal voltage does not vary continuously but rather takes on only two values, say +5 volts and 0 volts, corresponding to the 1 or 0. Thus, information is carried as a series of "bits," each of which can have only one of two values, 1 or 0.

In another field, geophysics, an important device based on electromagnetic induction is one type of *seismograph* or *geophone*. A seismograph is placed in direct contact with the Earth and converts the motion of the Earth—whether due to an earthquake or to an explosion (such as for mineral prospecting or for detecting a bomb test)—into an electrical signal. A seismograph contains a magnet and a

➡ **PHYSICS APPLIED**

Microphones

Membrane   Small coil of wire

To tape recorder or amplifier

FIGURE 29–23 Diagram of a microphone that works by induction.

FIGURE 29–24 Recording and/or playback head for tape or disk. In recording (or "writing"), the electric input signal to the head, which acts as an electromagnet, magnetizes the passing tape or disk. In playback (or "reading"), the changing magnetic field of the passing tape or disk induces a changing magnetic field in the head, which in turn induces in the coil an emf that is the output signal.

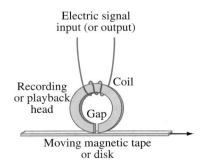

Electric signal input (or output)

Recording or playback head   Coil

Gap

Moving magnetic tape or disk

coil of wire, one of which is fixed rigidly to the case which moves as the Earth does where it is planted. The other element is inertial and is suspended from the case by a spring. In the type shown in Fig. 29–25, the coil moves with the Earth, and the relative motion of the magnet and coil produces an induced emf in the coil, which is the output of the device. In many geophones, the coil is inertial and the magnet moves with the Earth.

FIGURE 29–25 (a) A seismograph or geophone. (b) A seismograph reading (Northridge, California earthquake, January 17, 1994).

(a)

(b)

Summary

The **magnetic flux** passing through a loop is equal to the product of the area of the loop times the perpendicular component of the (uniform) magnetic field: $\Phi_B = B_\perp A = BA\cos\theta$. If **B** is not uniform, then

$$\Phi_B = \int \mathbf{B} \cdot d\mathbf{A}.$$

If the magnetic flux through a coil of wire changes in time, an emf is induced in the coil. The magnitude of the induced emf equals the time rate of change of the magnetic flux through the loop times the number N of loops in the coil:

$$\mathscr{E} = -N\frac{d\Phi_B}{dt}.$$

This is **Faraday's law of induction**.

The induced emf, and the current it produces, are in a direction that opposes the change in flux which caused them (**Lenz's law**).

We can also see from Faraday's law that a straight wire of length l moving with speed v perpendicular to a magnetic field of strength B has an emf induced between its ends equal to:

$$\mathscr{E} = Blv.$$

Faraday's law also tells us that *a changing magnetic field produces an electric field*. The mathematical relation is

$$\oint \mathbf{E} \cdot d\mathbf{l} = -\frac{d\Phi_B}{dt}$$

and is the general form of Faraday's law. The integral on the left is taken around the loop through which the magnetic flux Φ_B is changing.

An electric **generator** changes mechanical energy into electrical energy. Its operation is based on Faraday's law: a coil of wire is made to rotate uniformly by mechanical means in a magnetic field, and the changing flux through the coil induces a sinusoidal current, which is the output of the generator.

A motor, which operates in the reverse of a generator, acts like a generator in that a **counter emf** is induced in its rotating coil; since this counter emf opposes the input voltage, it can act to limit the current in a motor coil.

Similarly, a generator acts somewhat like a motor in that a **counter torque** acts on its rotating coil.

A **transformer**, which is a device to change the magnitude of an ac voltage, consists of a primary and a secondary coil. The changing flux due to an ac voltage in the primary induces an ac voltage in the secondary. In a 100 percent efficient transformer, the ratio of output to input voltages (V_S/V_P) equals the ratio of the number of turns N_S in the secondary to the number N_P in the primary:

$$\frac{V_S}{V_P} = \frac{N_S}{N_P}.$$

The ratio of secondary to primary current is in the inverse ratio of turns:

$$\frac{I_S}{I_P} = \frac{N_P}{N_S}.$$

Questions

1. What would be the advantage, in Faraday's experiments (Fig. 29–1), of using coils with many turns?

2. Suppose you are holding a circular loop of wire and suddenly thrust a magnet, south pole first, toward the center of the circle. Is a current induced in the wire? Is a current induced when the magnet is held steady within the loop? Is a current induced when you withdraw the magnet? In each case, if your answer is yes, specify the direction.

3. Suppose you are looking along a line through the centers of two circular (but separate) wire loops, one behind the other. A battery is suddenly connected to the front loop, establishing a clockwise current. (*a*) Will a current be induced in the second loop? (*b*) If so, when does this current start? (*c*) When does it stop? (*d*) In what direction is this current? (*e*) Is there a force between the two loops? (*f*) If so, in what direction?

4. The battery discussed in Question 3 is disconnected. Will a current be induced in the second loop? If so, when does it start and stop? In what direction is this current?

5. Two loops of wire are moving in the vicinity of a very long straight wire carrying a steady current as shown in Fig. 29–26. Find the direction of the induced current in each loop.

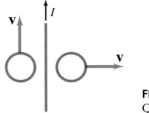

FIGURE 29–26
Question 5.

6. Is there a force between the two loops discussed in Question 5? If so, in what direction?

7. In what direction will the current flow in Fig. 29–9 if the rod moves to the left, which decreases the area of the loop to the left?

8. Some modern stove burners are based on induction. That is, an ac current passes around a coil that is the "burner," a burner that never gets hot. Explain why it will heat a metal pan but not a glass container.

9. A region where no magnetic field is desired is surrounded by a sheet of low-resistivity metal. (*a*) Will this sheet shield the interior from a rapidly changing magnetic field outside? Explain. (*b*) Will it act as a shield to a static magnetic field? (*c*) What if the sheet is superconducting (resistivity = 0)?

10. Show, using Lenz's law, that the emf induced in the moving rod in Fig. 29–9 is positive at the bottom and negative at the top so that the current flows clockwise in the circuit loop on the left.

11. What is the advantage of placing the two insulated electric wires carrying ac close together or even twisted about each other?

12. Explain why, exactly, the lights may dim briefly when a refrigerator motor starts up. When an electric heater is turned on, the lights may stay dimmed as long as it is on. Explain the difference.

* **13.** Use Fig. 29–11 and the right-hand rules to show why the counter torque in a generator *opposes* the motion.

* **14.** Will an eddy-current brake (Fig. 29–17) work on a copper or aluminum wheel, or must it be ferromagnetic?

* **15.** It has been proposed that eddy currents be used to help sort solid waste for recycling. The waste is first ground into tiny pieces and iron removed with a dc magnet. The waste then is allowed to slide down an incline over permanent magnets. How will this aid in the separation of nonferrous metals (Al, Cu, Pb, brass) from nonmetallic materials?

* **16.** The pivoted metal bar with slots in Fig. 29–27 falls much more quickly through a magnetic field than does a solid bar. Explain in detail.

FIGURE 29–27   Question 16.

* **17.** If an aluminum sheet is held between the poles of a large bar magnet, it requires some force to pull it out of the magnetic field even though the sheet is not ferromagnetic and does not touch the pole faces. Explain.

* **18.** A metal bar, pivoted at its upper end, oscillates freely in the absence of a magnetic field; but in a magnetic field, its oscillations are quickly damped out. Explain. (This *magnetic damping* is used in a number of practical devices.)

19. An enclosed transformer has four wire leads coming from it. How could you determine the ratio of turns on the two coils without taking the transformer apart? How would you know which wires paired with which?

20. A transformer designed for a 120-V ac input will often "burn out" if connected to a 120-V dc source. Explain. [*Hint*: The resistance of the primary coil is usually very low.]

* **21.** Since a magnetic microphone is basically like a loudspeaker, could a loudspeaker (Section 27–6) actually serve as a microphone? That is, could you speak into a loudspeaker and obtain an output signal that could be amplified? Explain. Discuss, in light of your response, how a microphone and loudspeaker differ in construction.

Problems

Sections 29–1 and 29–2

1. (I) The magnetic flux through a coil of wire containing two loops changes uniformly from $-80\,\text{Wb}$ to $+58\,\text{Wb}$ in 0.72 s. What is the emf induced in the coil?

2. (I) A 26-cm-diameter circular loop of wire lies in a plane perpendicular to a 0.90-T magnetic field. It is removed from the field in 0.15 s. What is the average induced emf?

3. (I) The rectangular loop shown in Fig. 29–28 is pushed into the magnetic field which points inward as shown. In what direction is the induced current? Explain.

FIGURE 29–28   Problem 3.

FIGURE 29–29   Problem 4.

4. (I) The north pole of the magnet in Fig. 29–29 is being inserted into the coil. In which direction is the induced current flowing through the resistor R? Explain.

5. (I) A 7.2-cm-diameter loop of wire is initially oriented perpendicular to a 1.3-T magnetic field. It is rotated so that its plane is parallel to the field direction in 0.20 s. What is the average induced emf in the loop?

6. (I) A 9.2-cm-diameter wire coil is initially oriented so that its plane is perpendicular to a magnetic field of 0.63 T pointing up. During the course of 0.15 s, the field is changed to one of 0.25 T pointing down. What is the average induced emf in the coil?

7. (II) What is the direction of the induced current in the circular loop due to the current shown in each part of Fig. 29–30? Explain.

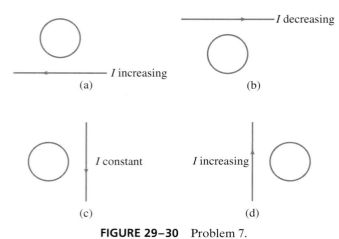

(a)   (b)

(c)   (d)

FIGURE 29–30 Problem 7.

8. (II) (a) If the resistance of the resistor in Fig. 29–31 is slowly increased, what is the direction of the current induced in the small circular loop inside the larger loop? (b) What would it be if the small loop were placed outside and to the left of the larger one? Explain.

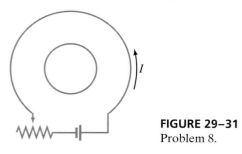

FIGURE 29–31
Problem 8.

9. (II) If the solenoid in Fig. 29–32 is being pulled away from the loop shown, in what direction is the induced current in the loop? Explain.

FIGURE 29–32
Problem 9.

10. (II) The magnetic field perpendicular to a circular loop of wire 20 cm in diameter is changed from +0.52 T to −0.45 T in 180 ms, where + means the field points away from an observer and − toward the observer. (a) Calculate the induced emf. (b) In what direction does the induced current flow?

11. (II) An elastic circular loop in the plane of the paper lies in a 0.75 T magnetic field pointing into the paper. If the loop's diameter changes from 20.0 cm to 6.0 cm in 0.50 s, (a) what is the direction of the induced current, (b) what is the magnitude of the average induced emf, and (c) if the loop's resistance is 2.5 Ω, what is the average induced current during the 0.50 s?

12. (II) A square loop of wire 15 cm on a side is rotated uniformly about an axis through its center and parallel to two sides. It rotates 360° in a magnetic field **B** perpendicular to the axis in 45 ms. If the rms induced emf is 70 mV, what is the average value of B?

13. (II) A 20-cm-diameter circular loop of wire has a resistance of 150 Ω. It is initially in a 0.40-T magnetic field, with its plane perpendicular to **B**, but is removed from the field in 100 ms. Calculate the electric energy dissipated in this process.

14. (II) A single rectangular loop of wire with the dimensions shown in Fig. 29–33 is situated so that part is inside a region of uniform magnetic field of 0.450 T and part is outside the field. The total resistance of the loop is 0.230 Ω. Calculate the force required to pull the loop from the field (to the right) at a constant velocity of 3.40 m/s. Neglect gravity.

0.350 m

0.750 m

FIGURE 29–33 Problem 14.

15. (II) The magnetic field perpendicular to a single 15.6-cm-diameter circular loop of copper wire decreases uniformly from 0.550 T to zero. If the wire is 2.05 mm in diameter, how much charge moves through the coil during this operation?

16. (II) The magnetic flux through each loop of a 60-loop coil is given by $(8.8t - 0.51t^3) \times 10^{-2}$ T·m², where the time t is in seconds. (a) Determine the emf $\mathcal{E}$ as a function of time. (b) What is $\mathcal{E}$ at $t = 1.0$ s and $t = 5.0$ s?

17. (II) A 35.0-cm-diameter coil consists of 20 turns of circular copper wire 2.0 mm in diameter. A uniform magnetic field, perpendicular to the plane of the coil, changes at a rate of 3.20×10^{-3} T/s. Determine (a) the current in the loop, and (b) the rate at which thermal energy is produced.

18. (II) A single circular loop of wire is placed inside a long solenoid with its plane perpendicular to the axis of the solenoid. The area of the loop is A_1 and that of the solenoid, which has n turns per unit length, is A_2. A current $I = I_0 \cos \omega t$ flows in the solenoid turns. What is the induced emf in the small loop?

19. (II) The area of an elastic circular loop decreases at a constant rate, $dA/dt = -3.50 \times 10^{-2}$ m²/s. The loop is in a magnetic field $B = 0.48$ T whose direction is perpendicular to the plane of the loop. At $t = 0$, the loop has area $A = 0.285$ m². Determine the induced emf at $t = 0$, and at $t = 2.00$ s.

20. (II) Suppose the radius of the elastic loop in Problem 19 increases at a constant rate, $dr/dt = 7.00\,\text{cm/s}$. Determine the emf induced in the loop at $t = 0$ and at $t = 1.00\,\text{s}$.

21. (III) Determine the magnetic flux through a square loop of side a (Fig. 29–34) if one side is parallel to, and a distance a from, a straight wire that carries a current I.

FIGURE 29–34
Problem 21.

Section 29–3

22. (I) The moving rod in Fig. 29–9 is 19.0 cm long and moves with a speed of 25.0 cm/s. If the magnetic field is 0.750 T, calculate the emf developed.

23. (II) In Fig. 29–9 the rod moves with a speed of 1.8 m/s, is 24.0 cm long, and has 2.2 Ω resistance. The magnetic field is 0.35 T and the resistance of the U-shaped conductor is 26.0 Ω at a given instant. Calculate: (a) the emf induced; (b) the current flowing in the U-shaped conductor, and (c) the external force needed to keep the rod's velocity constant at that instant.

24. (II) If the U-shaped conductor in Fig. 29–9 has resistivity ρ, whereas that of the moving rod is negligible, derive a formula for the current I as a function of time. Assume the rod has length l, starts at the bottom of the U at $t = 0$, and moves with uniform speed v in the magnetic field B. The cross-sectional area of the rod and all parts of the U is A.

25. (II) A conducting rod rests on two long frictionless parallel rails in a magnetic field $\mathbf{B}$ ($\perp$ to the rails and rod) as in Fig. 29–35. (a) If the rails are horizontal and the rod is given an initial push, will the rod travel at constant speed even though a magnetic field is present? (b) Suppose at $t = 0$, when the rod has speed $v = v_0$, the two rails are connected electrically by a wire from point a to point b. Assuming the rod has resistance R and the rails have negligible resistance, determine the speed of the rod as a function of time. Discuss your answer.

FIGURE 29–35
Problems 25 and 26.

26. (III) Suppose a conducting rod (mass m, resistance R) rests on two frictionless and resistanceless parallel rails a distance l apart in a uniform magnetic field $\mathbf{B}$ ($\perp$ to the rails and the rod) as in Fig. 29–35. At $t = 0$, the rod is at rest and a source of emf is connected to the points a and b. Determine the speed of the rod as a function of time (a) if the source puts out a constant current I, (b) the source puts out a constant emf $\mathcal{E}_0$. (c) Does the rod reach a terminal speed in either case? If so, what is it?

27. (III) A short section of wire, of length a, is moving with velocity $\mathbf{v}$, parallel to a very long wire carrying a current I as shown in Fig. 29–36. The near end of the wire section is a distance b from the long wire. Assuming the vertical wire is very long compared to $a + b$, determine the emf between the ends of the short section. Assume $\mathbf{v}$ is (a) in the same direction as I, (b) in the opposite direction to I.

FIGURE 29–36
Problem 27.

Section 29–4

28. (I) The generator of a car idling at 950-rpm produces 12.4 V. What will the output be at a rotation speed of 2500 rpm assuming nothing else changes?

29. (I) Show that the rms output of an ac generator is $V_{\text{rms}} = NAB\omega/\sqrt{2}$.

30. (II) A simple generator has a 420-loop square coil 21.0 cm on a side. How fast must it turn in a 0.350-T field to produce a 120-V peak output?

31. (II) A 350-loop circular armature coil with a diameter of 10.0 cm rotates at 60 rev/s in a uniform magnetic field of strength 0.45 T. What is the rms voltage output of the generator? What would you do to the rotation frequency in order to double the rms voltage output?

* Section 29–5

* 32. (I) A motor has an armature resistance of 3.75 Ω. If it draws 9.20 A when running at full speed and connected to a 120-V line, how large is the counter emf?

* 33. (I) The counter emf in a motor is 72 V when operating at 1800 rpm. What would be the counter emf at 2500 rpm if the magnetic field is unchanged?

* 34. (II) The counter emf in a motor is 100 V when the motor is operating at 1000 rpm. How would you change the motor's magnetic field if you wanted to reduce the counter emf to 75 V when the motor was running at 2500 rpm?

* 35. (II) What will be the current in the motor of Example 29–6 if the load causes it to run at half speed?

* 36. (II) A dc generator is rated at 10 kW, 200 V, and 50 A when it rotates at 1000 rpm. The resistance of the armature windings is 0.40 Ω. (a) Calculate the "no-load" voltage at 1000 rpm (when there is no circuit hooked up to the generator). (b) Calculate the full-load voltage (i.e. at 50 A) when the generator is run at 800 rpm. Assume that the magnitude of the magnetic field remains constant.

37. (I) A transformer is designed to change 120 V into 8500 V, and there are 500 turns in the primary. How many turns are in the secondary, assuming 100 percent efficiency?

38. (I) A transformer has 720 turns in the primary and 120 in the secondary. What kind of transformer is this and, assuming 100 percent efficiency, by what factor does it change the voltage? By what factor does it change the current?

39. (I) A step-up transformer increases 22 V to 120 V. What is the current in the secondary as compared to the primary? Assume 100 percent efficiency.

40. (I) Neon signs require 12 kV for their operation. To operate from a 220-V line, what must be the ratio of secondary to primary turns of the transformer? What would the voltage output be if the transformer were connected backward?

41. (II) A model-train transformer plugs into 120 V ac, and draws 0.65 A while supplying 15 A to the train. (a) What voltage is present across the tracks? (b) Is the transformer step-up or step-down?

42. (II) The output voltage of a 100-W transformer is 12 V and the input current is 26 A. (a) Is this a step-up or a step-down transformer? (b) By what factor is the voltage multiplied?

43. (II) A transformer has 330 primary turns and 1510 secondary turns. The input voltage is 120 V and the output current is 15.0 A. What is the output voltage and input current assuming 100 percent efficiency?

44. (II) If 30 MW of power at 45 kV (rms) arrives at a town from a generator via 3.0-Ω transmission lines, calculate (a) the emf at the generator end of the lines, and (b) the fraction of the power generated that is lost in the lines.

45. (III) If 65 kW is to be transmitted over two 0.100-Ω lines, estimate how much power is saved if the voltage is stepped up from 120 V to 1200 V and then down again, rather than simply transmitting at 120 V. Assume the transformers are each 99 percent efficient.

46. (III) Design a dc transmission line that can transmit 300 MW of electricity 200 km with only a 2 percent loss. The wires are to be made of aluminum and the voltage is 600 kV.

47. (I) Determine the electric field in the moving rod of Problem 22.

48. (II) In Fig. 29–22c, draw two identical small circles, one placed near the center of the magnetic field region and the other near its edge (but not beyond the edge). Show that the emf around each circle is the same, even though E is greater in the region of the outer circle.

49. (II) The *betatron*, a device used to accelerate electrons to high energy, consists of a circular vacuum tube placed in a magnetic field (Fig. 29–37), into which electrons are injected. The electromagnet produces a field that (1) keeps the electrons in their circular orbit inside the tube, and (2) increases the speed of the electrons when B changes. (a) Explain how the electrons are accelerated. (See Fig. 29–37.) (b) In what direction are the electrons moving in Fig. 29–37 (give directions as if looking down from above)? (c) Should B increase or decrease to accelerate the electrons? (d) The magnetic field is actually 60 Hz ac; show that the electrons can be accelerated only during $\frac{1}{4}$ of a cycle ($\frac{1}{240}$ s). (During this time they make hundreds of thousands of revolutions and acquire very high energy.)

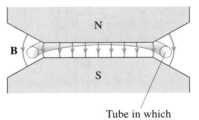

Tube in which electrons orbit

FIGURE 29–37 Problems 49 and 50.

50. (III) Show that the electrons in a betatron, Problem 49 and Fig. 29–37, are accelerated at constant radius if the magnetic field B_0 at the position of the electron orbit in the tube is equal to half the average value of the magnetic field (B_{av}) over the area of the circular orbit at each moment: $B_0 = \frac{1}{2}B_{av}$. (This is the reason the pole faces have a rather odd shape, as indicated in Fig. 29–37.)

51. (III) Find a formula for the net electric field in the moving rod of Problem 26 as a function of time for each case, (a) and (b).

General Problems

52. Suppose you are looking at two current loops in the plane of the page as shown in Fig. 29–38. When the switch is thrown in the left-hand coil, (a) what is the direction of the induced current in the other loop? (b) What is the situation after a "long" time? (c) What is the direction of the induced current in the second loop if the second loop is quickly pulled horizontally to the right?

FIGURE 29–38 Problem 52.

53. A simple generator is used to generate a peak output voltage of 24.0 V. The square armature consists of windings that are 7.0 cm on a side and rotates in a field of 0.420 T at a rate of 60 rev/s. How many loops of wire should be wound on the square armature?

54. A square loop 24.0 cm on a side has a resistance of 6.50 Ω. It is initially in a 0.755-T magnetic field, with its plane perpendicular to **B**, but is removed from the field in 40.0 ms. Calculate the electric energy dissipated in this process.

55. Two conducting rails of negligible resistance 30 cm apart rest on a 5.0° ramp. They are joined at the bottom by a 0.60 Ω resistor, and at the top a copper bar of mass 0.040 kg is laid across the rails. The whole apparatus is immersed in a vertical 0.55 T field. What is the terminal (steady) velocity of the bar as it slides frictionlessly down the rails?

56. A pair of power transmission lines each have a 0.80-Ω resistance and carry 700 A over 9.0 km. If the rms input voltage is 42 kV, calculate (a) the voltage at the other end, (b) the power input, (c) power loss in the lines, and (d) the power output.

57. Power is generated at 24 kV at a generating plant which is located 100 km from a town that requires 50 MW of power at 12 kV. The transmission lines from the plant to the town have a total resistance of 0.10 Ω/km. What should the output voltage of the transformer at the generating plant be for an overall transmission efficiency of 98.5 percent, assuming a perfect transformer?

58. A coil with 150 turns, a radius of 5.2 cm and a resistance of 11.0 Ω surrounds a solenoid with 200 turns/cm and a radius of 4.5 cm. The current in the solenoid changes at a constant rate from 0 A to 2.0 A in 0.10 s. Calculate the magnitude and direction of the induced current in the coil.

59. A ring with a radius of 3.0 cm and a resistance of 0.025 Ω is rotated about an axis through its diameter by 90° in a magnetic field of 0.15 T perpendicular to that axis. What is the largest number of electrons that would flow past a fixed point in the ring as this process is accomplished?

60. Calculate the peak output voltage of a simple generator whose square armature windings are 6.60 cm on a side if the armature contains 125 loops and rotates in a field of 0.200 T at a rate of 120 rev/s.

61. A small electric car overcomes a 250-N friction force when traveling 30 km/h. The electric motor is powered by ten 12-V batteries connected in series and is coupled directly to the wheels whose diameters are 50 cm. The 300 armature coils are rectangular, 10 cm by 15 cm, and rotate in a 0.60-T magnetic field. (a) How much current does the motor draw to produce the required torque? (b) What is the back emf? (c) How much power is dissipated in the coils? (d) What percent of the input power is used to drive the car?

62. A **search coil** for measuring B (also called a *flip coil*) is a small coil with N turns, each of cross-sectional area A. It is connected to a so-called **ballistic galvanometer**, which is a device to measure the total charge Q that passes through it in a short time. The flip coil is placed in the magnetic field to be measured with its face perpendicular to the field. It is then quickly rotated 180°. Show that the total charge Q that flows in the induced current during this short "flip" time is proportional to the magnetic field B. In particular, show that B is given by

$$B = \frac{QR}{2NA}$$

where R is the total resistance of the circuit, including that of the coil and that of the ballistic galvanometer which measures the charge Q.

63. The primary windings of a transformer, which has an 80 percent efficiency, are connected to 110 V ac. The secondary windings are connected across a 2.4 Ω, 75 W lightbulb. (a) Calculate the current through the primary windings of the transformer. (b) Calculate the ratio of the number of primary windings of the transformer to the number of secondary windings of the transformer.

64. What is the energy dissipated as a function of time in a circular loop of ten turns of wire having a radius of 10.0 cm and a resistance of 2.0 Ω if the plane of the loop is perpendicular to a magnetic field given by

$$B(t) = B_0 e^{-t/\tau}$$

with $B_0 = 0.50$ T and $\tau = 0.10$ s?

65. A thin metal rod of length L rotates with angular velocity ω about an axis through one end. The rotation axis is perpendicular to the rod and is parallel to a uniform magnetic field **B**. Determine the emf developed between the ends of the rod.

66. High-intensity desk lamps are rated at 40 W but require only 12 V. They contain a transformer that converts 120 V household voltage. (a) Is the transformer step-up or step-down? (b) What is the current in the secondary when the lamp is on? (c) What is the current in the primary? (d) What is the resistance of the bulb when on?

67. Show that the power loss in transmission lines, P_L, is given by $P_L = (P_T)^2 R_L/V^2$, where P_T is the power transmitted to the user, V is the delivered voltage, and R_L is the resistance of the power lines.

* 68. The magnetic field of a "shunt-wound" dc motor is produced by field coils placed in parallel with the armature coils. Suppose that the field coils have a resistance of 36.0 Ω and the armature coils 3.00 Ω. The back emf at full speed is 105 V when the motor is connected to 115 V dc. (a) Draw the equivalent circuit for the situations when the motor is just starting and when it is running full speed. (b) What is the total current drawn by the motor at start up? (c) What is the total current drawn when the motor runs at full speed?

69. Apply Faraday's law, in the form of Eq. 29–8, to show that the static electric field between the plates of a parallel-plate capacitor cannot drop abruptly to zero at the edges, but must, in fact, fringe. Use the path shown dashed in Fig. 29–39.

FIGURE 29–39
Problem 69.

70. A circular metal disk of radius R rotates with angular velocity ω about an axis through its center perpendicular to its face. The disk rotates in a uniform magnetic field B whose direction is parallel to the rotation axis. Determine the emf induced between the center and the edges.

71. What is the magnitude and direction of the electric field at each point in the rotating disk of Problem 70?

* 72. Suppose the "magnetic brake" of Fig. 29–17 acts on a circular metal disk of radius R and thickness d, whose electrical resistivity is ρ. The magnetic field **B**, perpendicular to the disk, acts over a small area A whose center is a distance l from the center of the wheel. At a moment when the disk is rotating with angular speed ω about an axis through its center, work out an approximate formula for the torque acting to slow it down.

General Problems **755**

A spark plug in a car receives a high voltage, which produces a high enough electric field in the air across its gap to pull electrons off the atoms in the air-gasoline mixture and form a spark. The high voltage is produced, from the basic 12 V of the car battery, by an induction coil which is basically a transformer or mutual inductance. Any coil of wire has a self-inductance, and a changing current in it causes an emf to be induced. Such inductors are useful in many circuits.

CHAPTER 30

Inductance; and Electromagnetic Oscillations

We discussed in the last chapter how a changing magnetic flux through a circuit induces an emf in that circuit. Before that we saw that an electric current produces a magnetic field. Combining these two ideas, we expect that a changing current in one circuit ought to induce an emf and a current in a second nearby circuit and even induce an emf in itself. We already saw an example in the previous chapter (transformers), but now we will treat this effect in a more general way in terms of what we will call mutual inductance and self-inductance. The concept of inductance also gives us a springboard to treat energy storage in a magnetic field. This chapter concludes with an analysis of circuits that contain inductance as well as resistance and/or capacitance.

30–1 Mutual Inductance

If two coils of wire are placed near each other, as in Fig. 30–1, a changing current in one will induce an emf in the other. According to Faraday's law, the emf $\mathscr{E}_2$ induced in coil 2 is proportional to the rate of change of flux passing through it. This flux is due to the current I_1 in coil 1, and it is often convenient to express the emf in coil 2 in terms of the current in coil 1.

We let Φ_{21} be the magnetic flux in each loop of coil 2 created by the current in coil 1. If coil 2 contains N_2 closely packed loops, then $N_2\Phi_{21}$ is the total flux passing through coil 2. If the two coils are fixed in space, $N_2\Phi_{21}$ is proportional to the current I_1 in coil 1; the proportionality constant is called the **mutual inductance**, M_{21}, defined by

$$M_{21} = \frac{N_2\Phi_{21}}{I_1}.$$ (30–1)

The emf $\mathscr{E}_2$ induced in coil 2 due to a changing current in coil 1 is, by Faraday's law,

$$\mathscr{E}_2 = -N_2\frac{d\Phi_{21}}{dt}.$$

We combine this with Eq. 30–1 rewritten as $\Phi_{21} = M_{21}I_1/N_2$ (and take its

Coil 1   Coil 2

I_1

$\mathscr{E}_2$
(induced)

FIGURE 30–1 A changing current in one coil will induce a current in the second coil.

Mutual inductance

derivative) and obtain

$$\mathcal{E}_2 = -M_{21}\frac{dI_1}{dt}.$$ (30-2)

emf due to mutual inductance

This relates the change in current in coil 1 to the emf it induces in coil 2. The mutual inductance of coil 2 with respect to coil 1, M_{21}, is a "constant" in that it does not depend on I_1; M_{21} depends on "geometric" factors such as the size, shape, number of turns, and relative positions of the two coils, and also on whether iron (or other ferromagnetic material) is present. For example, the farther apart the two coils are in Fig. 30–1, the fewer lines of flux can pass through coil 2, so M_{21} will be less. For some arrangements, the mutual inductance can be calculated (see Example 30–1). More often it is determined experimentally.

Suppose, now, we consider the reverse situation: when a changing current in coil 2 induces an emf in coil 1. In this case,

$$\mathcal{E}_1 = -M_{12}\frac{dI_2}{dt}$$

where M_{12} is the mutual inductance of coil 1 with respect to coil 2. It is possible to show, although we will not prove it here, that $M_{12} = M_{21}$. Hence, for a given arrangement we do not need the subscripts and we can let

$$M = M_{12} = M_{21},$$

so that

$$\mathcal{E}_1 = -M\frac{dI_2}{dt}$$ (30-3a)

and

$$\mathcal{E}_2 = -M\frac{dI_1}{dt}.$$ (30-3b)

The SI unit for mutual inductance is the henry (H), where $1\,\text{H} = 1\,\text{V}\cdot\text{s/A} = 1\,\Omega\cdot\text{s}$.

EXAMPLE 30–1 **Solenoid and coil.** A long thin solenoid of length l and cross-sectional area A contains N_1 closely packed turns of wire. Wrapped around it is an insulated coil of N_2 turns, Fig. 30–2. Assume all the flux from coil 1 (the solenoid) passes through coil 2, and calculate the mutual inductance.

SOLUTION We first determine the flux produced by the solenoid, all of which passes uniformly through coil N_2: The magnetic field inside the solenoid was derived in Chapter 28, Eq. 28–4:

$$B = \mu_0 \frac{N_1}{l} I_1,$$

where I_1 is the current in the solenoid. The solenoid is close packed, so we assume all the flux in the solenoid links the coil. Then the flux Φ_{21} through coil 2 is

$$\Phi_{21} = BA = \mu_0 \frac{N_1}{l} I_1 A.$$

Hence the mutual inductance is

$$M = \frac{N_2 \Phi_{21}}{I_1} = \frac{\mu_0 N_1 N_2 A}{l}.$$

Note that we calculated M_{21}; if we had tried to calculate M_{12}, it would have been difficult. Given $M_{12} = M_{21} = M$, we can do the simpler calculation to obtain M. Note also that M depends only on geometric factors, and not on the currents.

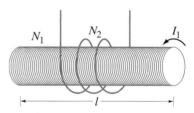

FIGURE 30–2 Example 30–1.

A transformer is an example of mutual inductance in which the coupling is maximized so nearly all flux lines pass through both coils. However, mutual inductance has other uses as well. For example some pacemakers, which are used to regulate the heartbeat so that blood flow is maintained in heart patients, are powered externally. Power in an external coil is transmitted via mutual inductance to a second coil in the pacemaker at the heart. This has the advantage over battery-powered pacemakers in that surgery is not needed to replace a battery when it wears out.

Mutual inductance can sometimes be a problem, however. Any changing current in a circuit can induce an emf in another part of the same circuit or in a different circuit even though the conductors are not in the shape of a coil. The mutual inductance M is usually small unless coils with many turns and/or iron cores are involved. However, in situations where small signals are present, problems due to mutual inductance often arise. Shielded cable, in which an inner conductor is surrounded by a cylindrical grounded conductor, is often used to reduce the problem.

30-2 | Self-Inductance

The concept of inductance applies also to a single isolated coil of N turns. When a changing current passes through a coil (or solenoid), a changing magnetic flux is produced inside the coil, and this in turn induces an emf in that same coil. This induced emf opposes the change in flux (Lenz's law). For example, if the current through the coil is increasing, the increasing magnetic flux induces an emf that opposes the original current and tends to retard its increase. If the current is decreasing in the coil, the decreasing flux induces an emf in the same direction as the current, thus tending to maintain the original current.

The magnetic flux Φ_B passing through the N turns of a coil is proportional to the current I in the coil, so we define the **self-inductance** L (in analogy to mutual inductance, Eq. 30–1) as

$$L = \frac{N\Phi_B}{I}. \tag{30-4}$$

Then the emf $\mathscr{E}$ induced in a coil of self-inductance L is, from Faraday's law,

$$\mathscr{E} = -N\frac{d\Phi_B}{dt} = -L\frac{dI}{dt}. \tag{30-5}$$

Like mutual inductance, self-inductance is measured in henrys. The magnitude of L depends on the geometry and on the presence of a ferromagnetic material. Self-inductance can be defined, as above, for any circuit or part of a circuit.

Circuits always contain some inductance, but often it is quite small unless the circuit contains a coil of many turns. A coil that has significant self-inductance L is called an **inductor**. It is shown on circuit diagrams by the symbol

$$\overset{}{-\text{⟋⟋⟋⟋⟍}-} .$$ [inductor symbol]

It can serve a useful purpose in certain circuits. Often, inductance is to be avoided in a circuit. Precision resistors are normally wire wound and thus would have inductance as well as resistance. The inductance can be minimized by winding the insulated wire back on itself in the opposite sense so that the current going in opposite directions produces little net magnetic flux; this is called a "noninductive winding."

If an inductor has negligible resistance, it is the inductance (or induced emf) that controls a changing current. If a source of changing or alternating voltage is applied to the coil, this applied voltage will just be balanced by the induced emf of the coil given by Eq. 30–5. Thus we can see from Eq. 30–5 that, for a given $\mathcal{E}$, if the inductance L is large, the change in the current—and therefore the current itself if it is ac—will be small. The greater the inductance, the less the ac current. An inductance thus acts something like a resistance to impede the flow of alternating current. We use the term *reactance* or *impedance* for this quality of an inductor. We will discuss reactance and impedance more fully in Chapter 31, and we shall see that it depends not only on L, but also on the frequency. Here we mention one example of its importance. The resistance of the primary in a transformer is usually quite small, perhaps less than $1\,\Omega$. If resistance alone limited the current in a transformer, tremendous currents would flow when a high voltage was applied. Indeed, a dc voltage applied to a transformer can burn it out. It is the induced emf (or reactance) of the coil that limits the current to a reasonable value.

Common inductors have inductances in the range from about $1\,\mu\mathrm{H}$ to about $1\,\mathrm{H}$ (where $1\,\mathrm{H} = 1\,\text{henry} = 1\,\Omega\cdot\mathrm{s}$).

| CONCEPTUAL EXAMPLE 30–2 | **Direction of emf in inductor.** (a) Suppose current passes through the coil in Fig. 30–3 from left to right as shown, and is increasing with time. In which direction is the induced emf? (b) If the current passes through the coil in the same direction, but is decreasing in time, what then is the direction of the induced emf?

(a)

(b)

FIGURE 30–3 Example 30–2. The $+$ and $-$ signs refer to the induced emf, due to the changing current, as if points A and B were the terminals of a battery.

RESPONSE (a) From Lenz's Law we know that the induced emf must oppose the change in magnetic flux. If the current is increasing, so is the magnetic flux. The induced emf acts to oppose the increasing flux, which means it acts like a source of emf that opposes the outside source of emf driving the current. So the induced emf in the coil acts to oppose I in Fig. 30–3a. In other words the inductor might be thought of as a battery with a positive terminal at point A on the left and negative at point B on the right.
(b) If the current is decreasing, then by Lenz's Law the induced emf acts to bolster the flux—like a source of emf reinforcing the external emf. The induced emf acts to increase I in Fig. 30–3b, so in this situation you can think of the induced emf as a battery with its positive terminal on the right, at point B.

EXAMPLE 30–3 **Solenoid inductance.** (a) Determine a formula for the self-inductance L of a tightly wrapped solenoid (a long coil) containing N turns of wire in its length l and whose cross-sectional area is A. (b) Calculate the value of L if $N = 100$, $l = 5.0\,\mathrm{cm}$, $A = 0.30\,\mathrm{cm}^2$ and the solenoid is air filled. (c) Calculate L if the solenoid has an iron core with $\mu = 4000\,\mu_0$.

SOLUTION (a) To determine the inductance L, it is usually simplest to start with Eq. 30–4, so we need to first determine the flux. According to Eq. 28–4, the magnetic field inside a solenoid (ignoring end effects) is constant: $B = \mu_0 nI$ where $n = N/l$. The flux is $\Phi_B = BA = \mu_0 NIA/l$, so

$$L = \frac{N\Phi_B}{I} = \frac{\mu_0 N^2 A}{l}.$$

(b) Since $\mu_0 = 4\pi \times 10^{-7}\,\mathrm{T}\cdot\mathrm{m/A}$

$$L = \frac{(4\pi \times 10^{-7}\,\mathrm{T}\cdot\mathrm{m/A})(100)^2(3.0 \times 10^{-5}\,\mathrm{m}^2)}{(5.0 \times 10^{-2}\,\mathrm{m})} = 7.5\,\mu\mathrm{H}.$$

(c) Here we replace μ_0 by $\mu = 4000\,\mu_0$ so L will be 4000 times larger: $L = 0.030\,\mathrm{H} = 30\,\mathrm{mH}$.

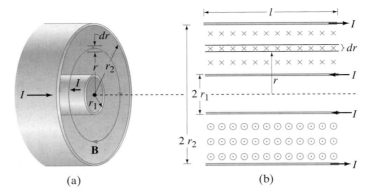

FIGURE 30–4 Example 30–4. Coaxial cable: (a) end view, (b) side view (cross-section).

(a) (b)

EXAMPLE 30–4 **Coaxial cable inductance.** Determine the inductance per unit length of a coaxial cable whose inner conductor has a radius r_1 and the outer conductor has a radius r_2, Fig. 30–4. Assume the conductors are thin, so that the magnetic field within them can be ignored. The conductors carry equal currents I in opposite directions.

SOLUTION Again we need to find the magnetic flux, $\Phi_B = \int \mathbf{B} \cdot d\mathbf{A}$, this time between the conductors. The lines of $\mathbf{B}$ are circles surrounding the inner conductor (only one is shown in Fig. 30–4a). From Ampère's law, $\oint \mathbf{B} \cdot d\mathbf{l} = \mu_0 I$, the magnitude of the field at a distance r from the center, when the inner conductor carries a current I, is (see also Example 28–4):

$$B = \frac{\mu_0 I}{2\pi r}.$$

The magnetic flux through a rectangle of width dr and length l (along the cable, Fig. 30–4b), a distance r from the center, is

$$d\Phi_B = B(l\,dr) = \frac{\mu_0 I}{2\pi r} l\,dr.$$

The total flux in a length l of cable is

$$\Phi_B = \int d\Phi_B = \frac{\mu_0 Il}{2\pi} \int_{r_1}^{r_2} \frac{dr}{r} = \frac{\mu_0 Il}{2\pi} \ln \frac{r_2}{r_1}.$$

Since the current I all flows in one direction in the inner conductor, and the same current I all flows in the opposite direction in the outer conductor, we have only one turn, so $N = 1$. Hence the self-inductance for a length l is

$$L = \frac{\Phi_B}{I} = \frac{\mu_0 l}{2\pi} \ln \frac{r_2}{r_1}.$$

The inductance per unit length is

$$\frac{L}{l} = \frac{\mu_0}{2\pi} \ln \frac{r_2}{r_1}.$$

Note that L depends only on geometric factors and not on the current I.

30–3 Energy Stored in a Magnetic Field

When an inductor of inductance L is carrying a current I which is changing at a rate dI/dt, energy is being supplied to the inductor at a rate

$$P = I\mathscr{E} = LI\frac{dI}{dt}$$

where P stands for power and we used[†] Eq. 30–5. Let us calculate the work needed to increase the current in an inductor from zero to some value I. Using this

[†] No minus sign here because we are supplying power to oppose the emf of the inductor.

last equation the work dW done in a time dt is

$$dW = P\,dt = LI\,dI.$$

Then the total work done to increase the current from zero to I is

$$W = \int dW = \int_0^I LI\,dI = \tfrac{1}{2}LI^2.$$

This work done is equal to the energy U stored in the inductor when it is carrying a current I (and we take $U = 0$ when $I = 0$):

$$U = \tfrac{1}{2}LI^2. \qquad\qquad (30\text{--}6)$$

Energy stored in inductor

This can be compared to the energy stored in a capacitor, C, when the potential difference across it is V (see Section 24–4):

$$U = \tfrac{1}{2}CV^2.$$

Just as the energy stored in a capacitor can be considered to reside in the electric field between its plates, so the energy in an inductor can be considered to be stored in its magnetic field. To write the energy in terms of the magnetic field, let us use the result of Example 30–3, that the inductance of an ideal solenoid (end effects ignored) is $L = \mu_0 N^2 A/l$. Now the magnetic field B in a solenoid is related to the current I by $B = \mu_0 NI/l$. Thus

$$U = \tfrac{1}{2}LI^2 = \frac{1}{2}\left(\frac{\mu_0 N^2 A}{l}\right)\left(\frac{Bl}{\mu_0 N}\right)^2$$

$$= \frac{1}{2}\frac{B^2}{\mu_0}Al.$$

We can think of this energy as residing in the volume enclosed by the windings, which is Al. Then the energy per unit volume or *energy density* is

$$u = \text{energy density} = \frac{1}{2}\frac{B^2}{\mu_0}. \qquad\qquad (30\text{--}7)$$

Energy density in magnetic field

This formula, which was derived for the special case of a solenoid, can be shown to be valid for any region of space where a magnetic field exists. If a ferromagnetic material is present, μ_0 is replaced by μ. This equation is analogous to that for an electric field, $\tfrac{1}{2}\epsilon_0 E^2$, Eq. 24–6.

EXAMPLE 30–5 **Energy stored in a coaxial cable.** (*a*) How much energy is being stored per unit length in a coaxial cable whose conductors have radii r_1 and r_2 (Example 30–4, Fig. 30–4) and which carry a current I? (*b*) Where is the energy density highest?

SOLUTION (*a*) The inductance per unit length, as we saw in Example 30–4, is

$$\frac{L}{l} = \frac{\mu_0}{2\pi}\ln\frac{r_2}{r_1}.$$

We want to find the energy stored per unit length, which is

$$\frac{U}{l} = \frac{\tfrac{1}{2}LI^2}{l} = \frac{\mu_0 I^2}{4\pi}\ln\frac{r_2}{r_1}.$$

(*b*) Since $B = \mu_0 I/2\pi r$, the field is greatest near the surface of the inner conductor, so the energy density, $u = B^2/2\mu_0$, will be greatest there.

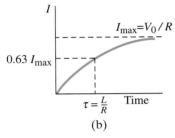

FIGURE 30–5 (a) *LR* circuit;
(b) growth of current when
connected to battery.

Current in LR circuit
when battery connected at $t = 0$

Inductive time constant, $\tau = L/R$

FIGURE 30–6 The switch is flipped
quickly so the battery is removed but
we still have a circuit. The current at
this moment (call it $t = 0$) is I_0.

30–4 | *LR* Circuits

Any inductor will have some resistance. We represent an inductor by drawing its inductance L and its resistance R separately, as in Fig. 30–5a. The resistance R could also include a separate resistor connected in series. Now we ask, what happens when a battery or other source of dc voltage V_0 is connected in series to such an *LR* circuit? At the instant the switch connecting the battery is closed, the current starts to flow. The current increases starting from zero. It is opposed by the induced emf in the inductor which means point B in Fig. 30–5 is positive relative to point C. However, as soon as current starts to flow, there is also a voltage ($= IR$) across the resistance. Hence the voltage applied across the inductance is reduced and the current increases less rapidly. The current thus rises gradually as shown in Fig. 30–5b, and approaches the steady value $I_{max} = V_0/R$ when all the voltage drop is across the resistance.

We can show this analytically by applying Kirchhoff's loop rule to the circuit of Fig. 30–5a. The emf's in the circuit are the battery voltage V_0, and the emf $\mathscr{E} = -L(dI/dt)$ in the inductor opposing the increasing current. Hence the sum of the potential changes around the loop is

$$V_0 - L\frac{dI}{dt} - IR = 0,$$

where I is the current in the circuit at any instant. We rearrange this to obtain

$$L\frac{dI}{dt} + RI = V_0. \qquad (30\text{–}8)$$

This is a linear differential equation and can be integrated in the same way we did in Section 26–4 for an *RC* circuit. We rewrite Eq. 30–8 and then integrate:

$$\int_{I=0}^{I} \frac{dI}{V_0 - IR} = \int_0^t \frac{dt}{L}.$$

Then

$$-\frac{1}{R}\ln\left(\frac{V_0 - IR}{V_0}\right) = \frac{t}{L}$$

or,

$$I = \frac{V_0}{R}\left(1 - e^{-t/\tau}\right) \qquad (30\text{–}9)$$

where

$$\tau = \frac{L}{R} \qquad (30\text{–}10)$$

is the **time constant** of the *LR* circuit. The symbol τ represents the time required for the current I to reach $(1 - 1/e) = 0.63$ or 63 percent of its maximum value (V_0/R). Equation 30–9 is plotted in Fig. 30–5b. (Compare to the *RC* circuit, Section 26–4.)

Now let us flip the switch in Fig. 30–5a so that the battery is taken out of the circuit, and points A and C are connected together as shown in Fig. 30–6 at the moment when the switching occurs (call it $t = 0$) and the current is I_0. Then the differential equation (Eq. 30–8) becomes (since $V_0 = 0$):

$$L\frac{dI}{dt} + RI = 0.$$

We rearrange this equation and integrate:

$$\int_{I_0}^{I} \frac{dI}{I} = -\int_0^t \frac{R}{L} dt$$

where $I = I_0$ at $t = 0$, and $I = I$ at time t.

We integrate this last equation to obtain

$$\ln \frac{I}{I_0} = -\frac{R}{L} t$$

or

$$I = I_0 e^{-t/\tau} \qquad\qquad (30\text{–}11)$$

where again the time constant is $\tau = L/R$. The current thus decays exponentially to zero as shown in Fig. 30–7.

This analysis shows that there is always some "reaction time" when an electromagnet, for example, is turned on or off. We also see that an LR circuit has properties similar to an RC circuit (Section 26–4). Unlike the capacitor case, however, the time constant here is *inversely* proportional to R.

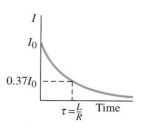

FIGURE 30–7 Decay of the current in Fig. 30–6 in time after the battery is removed from the circuit.

EXAMPLE 30–6 **An *LR* circuit.** At $t = 0$, a 12.0-V battery is connected in series with a 30-Ω resistor and a 220-mH inductor, as shown in Fig. 30–8. (*a*) What is the current at $t = 0$? (*b*) What is the time constant? (*c*) What is the maximum current? (*d*) How long will it take the current to reach half its maximum possible value? (*e*) At this instant, at what rate is energy being delivered by the battery, and (*f*) at what rate is energy being stored in the inductor's magnetic field?

SOLUTION (*a*) The current cannot instantaneously jump from zero to some other value when the switch is closed because the inductor opposes the change $(\mathscr{E}_L = -L(dI/dt))$. Hence just after the switch is closed I is still zero at $t = 0$ and then begins to increase.
(*b*) The time constant is, from Eq. 30–10, $\tau = L/R = (0.22\,\text{H})/(30\,\Omega) = 7.3\,\text{ms}$.
(*c*) The current reaches its maximum steady value after a long time, when $dI/dt = 0$ so $I_{max} = V_0/R = 12.0\,\text{V}/30\,\Omega = 0.40\,\text{A}$.
(*d*) We set $I = \frac{1}{2}I_{max} = V_0/2R$ in Eq. 30–9, which gives us

$$1 - e^{-t/\tau} = \tfrac{1}{2}$$

or

$$e^{-t/\tau} = 1 - \tfrac{1}{2} = \tfrac{1}{2}.$$

We solve for t:

$$t = \tau \ln 2 = (7.3 \times 10^{-3}\,\text{s})(0.69) = 5.0\,\text{ms}.$$

(*e*) At this instant, $I = I_{max}/2 = 200\,\text{mA}$, so the power being delivered by the battery is

$$P = IV = (0.20\,\text{A})(12\,\text{V}) = 2.4\,\text{W}.$$

(*f*) From Eq. 30–6, the energy stored in an inductor L at any instant is

$$U = \tfrac{1}{2}LI^2$$

where I is the current in the inductor at that instant. The *rate* at which the energy changes is

$$\frac{dU}{dt} = LI\frac{dI}{dt}.$$

We can differentiate Eq. 30–9 to obtain dI/dt, or use the differential equation, Eq. 30–8, directly:

$$\frac{dU}{dt} = I\left(L\frac{dI}{dt}\right) = I(V_0 - RI)$$

$$= (0.20\,\text{A})\big[12\,\text{V} - (30\,\Omega)(0.20\,\text{A})\big] = 1.2\,\text{W}.$$

Since only part of the battery's power is feeding the inductor at this instant, where is the rest going?

FIGURE 30–8 Example 30–7.

$L = 220\,\text{mH}$

S

$R = 30\Omega$

$V_0 = 12.0\,\text{V}$

30–5 LC Circuits and Electromagnetic Oscillations

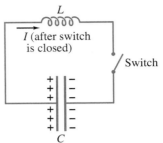

FIGURE 30–9 An LC circuit.

In any electric circuit, there can be three basic components: resistance, capacitance, and inductance, in addition to a source of emf. (There can also be more complex components, such as diodes or transistors.) We have previously discussed both RC and LR circuits. Now we look at an LC circuit, one that contains only a capacitance C and an inductance, L, Fig. 30–9. This is an idealized circuit in which we assume there is no resistance in the inductor; in the next Section we include resistance. Let us suppose the capacitor in Fig. 30–9 is initially charged so that one plate has charge Q_0 and the other plate has charge $-Q_0$. Suppose that at $t = 0$, the switch is closed. The capacitor immediately begins to discharge. As it does so, the current I through the inductor increases. We now apply Kirchhoff's loop rule (sum of potential changes around a loop is zero):

$$-L\frac{dI}{dt} + \frac{Q}{C} = 0.$$

Because charge leaves the positive plate on the capacitor to produce the current I as shown in Fig. 30–9, the charge Q on the (positive) plate of the capacitor is decreasing, so $I = -dQ/dt$. We can then rewrite the above equation as

$$\frac{d^2Q}{dt^2} + \frac{Q}{LC} = 0. \tag{30–12}$$

This is a familiar differential equation. It has the same form as the equation for simple harmonic motion (Chapter 14, Eq. 14–3). The solution of Eq. 30–12 is

$$Q = Q_0\cos(\omega t + \phi) \tag{30–13}$$

where Q_0 and ϕ are constants that depend on the initial conditions. We insert Eq. 30–13 into Eq. 30–12, noting that $d^2Q/dt^2 = -\omega^2 Q_0\cos(\omega t + \phi)$; thus

$$-\omega^2 Q_0\cos(\omega t + \phi) + \frac{1}{LC}Q_0\cos(\omega t + \phi) = 0$$

or

$$\left(-\omega^2 + \frac{1}{LC}\right)\cos(\omega t + \phi) = 0.$$

This relation can be true for all times t only if $\left(-\omega^2 + 1/LC\right) = 0$, which tells us that

LC angular frequency

$$\omega = 2\pi f = \sqrt{\frac{1}{LC}}. \tag{30–14}$$

FIGURE 30–10 Charge Q and current I in an LC circuit. The period $T = \dfrac{1}{f} = \dfrac{2\pi}{\omega} = 2\pi\sqrt{LC}$.

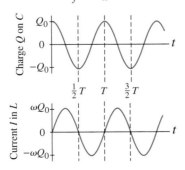

Equation 30–13 shows that the charge on the capacitor in an LC circuit oscillates sinusoidally. The current in the inductor is

$$I = -\frac{dQ}{dt} = \omega Q_0\sin(\omega t + \phi)$$

$$= I_0\sin(\omega t + \phi); \tag{30–15}$$

so the current too is sinusoidal. The maximum value of I is $I_0 = \omega Q_0 = Q_0/\sqrt{LC}$. Equations 30–14 and 30–15 for Q and I when $\phi = 0$, are plotted in Fig. 30–10.

Now let us look at LC oscillations from the point of view of energy. The energy stored in the electric field of the capacitor at any time t is (see Eq. 24–5):

$$U_E = \frac{1}{2}\frac{Q^2}{C} = \frac{Q_0^2}{2C}\cos^2(\omega t + \phi).$$

The energy stored in the magnetic field of the inductor at the same instant is (Eq. 30–6)

$$U_B = \frac{1}{2}LI^2 = \frac{L\omega^2 Q_0^2}{2}\sin^2(\omega t + \phi) = \frac{Q_0^2}{2C}\sin^2(\omega t + \phi)$$

where we used Eq. 30–14. If we let $\phi = 0$, then at times $t = 0$, $t = \frac{1}{2}T$, $t = T$, and so on (where T is the period $= 1/f = 2\pi/\omega$), we have $U_E = Q_0^2/2C$ and $U_B = 0$. That is, all the energy is stored in the electric field of the capacitor. But at $t = \frac{1}{4}T, \frac{3}{4}T$, and so on, $U_E = 0$ and $U_B = Q_0^2/2C$, and so all the energy is stored in the magnetic field of the inductor. At any time t, the total energy is

$$U = U_E + U_B = \frac{1}{2}\frac{Q^2}{C} + \frac{1}{2}LI^2$$

$$= \frac{Q_0^2}{2C}\left[\cos^2(\omega t + \phi) + \sin^2(\omega t + \phi)\right] = \frac{Q_0^2}{2C}. \qquad \textbf{(30–16)} \qquad \textit{Energy}$$

Hence the total energy is constant, and energy is conserved.

What we have in this LC circuit is an **LC oscillator** or **electromagnetic oscillation**. The charge Q oscillates back and forth, from one plate of the capacitor to the other, and repeats this continuously. Likewise, the current oscillates back and forth as well. They are also energy oscillations: when Q is a maximum, the energy is all stored in the electric field of the capacitor; but when Q reaches zero, the current I is a maximum and all the energy is stored in the magnetic field of the inductor. Thus the energy oscillates between being stored in the electric field of the capacitor and in the magnetic field of the inductor. See Fig. 30–11.

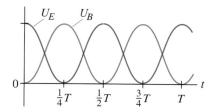

FIGURE 30–11 Energy U_E (red line) and U_B (blue line) stored in the capacitor and the inductor as a function of time. Note how the energy oscillates between electric and magnetic.

EXAMPLE 30–7 **LC circuit.** A 1200-pF capacitor is fully charged by a 500-V dc power supply. It is disconnected from the power supply and is connected, at $t = 0$, to a 75-mH inductor. Determine: (*a*) the initial charge on the capacitor; (*b*) the maximum current; (*c*) the frequency f and period T of oscillation; and (*d*) the total energy oscillating in the system.

SOLUTION (*a*) The 500-V power supply, before being disconnected, charged the capacitor to a charge of

$$Q_0 = CV = (1.2 \times 10^{-9}\,\text{F})(500\,\text{V}) = 6.0 \times 10^{-7}\,\text{C}.$$

(*b*) The maximum current, I_{max}, is (see Eq. 30–15)

$$I_{max} = \omega Q_0 = \frac{Q_0}{\sqrt{LC}} = \frac{(6.0 \times 10^{-7}\,\text{C})}{\sqrt{(0.075\,\text{H})(1.2 \times 10^{-9}\,\text{F})}} = 63\,\text{mA}.$$

(*c*) Equation 30–14 gives us the frequency:

$$f = \frac{\omega}{2\pi} = \frac{1}{(2\pi\sqrt{LC})} = 17\,\text{kHz},$$

and the period T is

$$T = \frac{1}{f} = 6.0 \times 10^{-5}\,\text{s}.$$

(*d*) Finally the total energy (Eq. 30–16) is

$$U = \frac{Q_0^2}{2C} = \frac{(6.0 \times 10^{-7}\,\text{C})^2}{2(1.2 \times 10^{-9}\,\text{F})} = 1.5 \times 10^{-4}\,\text{W}.$$

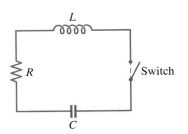

FIGURE 30–12 An *LRC* circuit.

FIGURE 30–13 Charge Q on the capacitor in an *LRC* circuit as a function of time: curve A is for underdamped oscillation $(R^2 < 4L/C)$, curve B is for critically damped, and curve C is for overdamped.

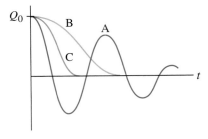

Angular frequency (damped LC)

30–6 | *LC* Oscillations with Resistance (*LRC* Circuit)

The *LC* circuit discussed in the previous Section is an idealization. There is always some resistance R in any circuit, and so we now discuss such a simple *LRC* circuit, Fig. 30–12.

Suppose again that the capacitor is initially given a charge Q_0 and the battery or other source is then removed from the circuit. The switch is closed at $t = 0$. Since there is now a resistance in the circuit, we expect some of the energy to be converted to thermal energy, and so we don't expect undamped oscillations as in a pure *LC* circuit. Indeed, if we use Kirchhoff's loop rule around this circuit, we obtain

$$-L\frac{dI}{dt} - IR + \frac{Q}{C} = 0,$$

which is the same equation we had in Section 30–5 with the addition of the voltage drop IR across the resistor. Since $I = -dQ/dt$, as we saw in Section 30–5, this equation becomes

$$L\frac{d^2Q}{dt^2} + R\frac{dQ}{dt} + \frac{1}{C}Q = 0. \tag{30–17}$$

This second-order differential equation in the variable Q has precisely the same form as that for the damped harmonic oscillator, Eq. 14–15:

$$m\frac{d^2x}{dt^2} + b\frac{dx}{dt} + kx = 0.$$

Hence we can analyze our *LRC* circuit in the same way as for damped harmonic motion, Section 14–7. Our system may undergo damped oscillations, curve A in Fig. 30–13 (underdamped system), or it may be critically damped (curve B), or overdamped (curve C), depending on the relative values of R, L, and C. Using the results of Section 14–7, with m replaced by L, b by R, and k by C^{-1}, we find that the system will be underdamped when

$$R^2 < \frac{4L}{C},$$

and overdamped for $R^2 > 4L/C$. Critical damping (curve B in Fig. 30–13) occurs when $R^2 = 4L/C$. If R is smaller than $\sqrt{4L/C}$, the angular frequency, ω', will be

$$\omega' = \sqrt{\frac{1}{LC} - \frac{R^2}{4L^2}} \tag{30–18}$$

(compare to Eq. 14–20). And the charge Q as a function of time will be

$$Q = Q_0 e^{-\frac{R}{2L}t}\cos(\omega't + \phi) \tag{30–19}$$

where ϕ is a phase constant (compare to Eq. 14–19).

Oscillators are an important element in many electronic devices: radios and television sets use them for tuning, tape recorders use them (the "bias frequency") when recording, and so on. Because some resistance is always present, electrical oscillators generally need a periodic input of power to compensate for the energy converted to thermal energy in the resistance.

EXAMPLE 30–8 **Damped oscillations.** At $t = 0$, a 40-mH inductor is placed in series with a resistance $R = 3.0\,\Omega$ and a charged capacitor $C = 4.8\,\mu\text{F}$. (*a*) Show that this circuit will oscillate. (*b*) Determine the frequency. (*c*) What is the time required for the charge amplitude to drop to half its starting value? (*d*) What is the current amplitude? (*e*) What value of R will make the circuit nonoscillating?

SOLUTION (a) To be underdamped, we must have $R^2 < 4L/C$. Since $R^2 = 9.0\,\Omega^2$ and $4L/C = 4(0.040\,\text{H})/(4.8 \times 10^{-6}\,\text{F}) = 3.3 \times 10^4\,\Omega^2$, this relation is satisfied, so the circuit will oscillate.

(b) We use Eq. 30–18:

$$f' = \frac{\omega'}{2\pi} = \frac{1}{2\pi}\sqrt{\frac{1}{LC} - \frac{R^2}{4L^2}} = 2.3 \times 10^3\,\text{Hz}.$$

(c) From Eq. 30–19, the amplitude will be half when

$$e^{-\frac{R}{2L}t} = \tfrac{1}{2}$$

or

$$t = \frac{2L}{R}\ln 2 = 18\,\text{ms}.$$

(d) We differentiate Eq. 30–19 to obtain $I = -dQ/dt$ (see comment just before Eq. 30–12), noting that $\phi = 0$ since $Q = Q_0$ at $t = 0$:

$$I = -\frac{dQ}{dt} = Q_0 e^{-\frac{R}{2L}t}\left(\frac{R}{2L}\cos\omega't + \frac{1}{\sqrt{LC}}\sin\omega't\right)$$

where we approximated $\omega' \approx \sqrt{1/LC}$. Because R is much less than $\sqrt{4L/C}$ (see part *a* above), we can ignore the $\cos\omega't$ term, so

$$I \approx \frac{Q_0}{\sqrt{LC}}\,e^{-\frac{R}{2L}t}\sin\omega't.$$

Hence the initial current amplitude is $Q_0/\sqrt{LC}$. We are not given Q_0, but if it were, say, $1.0\,\mu\text{C}$, then $I_0 = (1.0 \times 10^{-6}\,\text{C})/\sqrt{(40 \times 10^{-3}\,\text{H})(4.8 \times 10^{-6}\,\text{F})} = 2.3\,\text{mA}$.

(e) To make the circuit critically damped or overdamped, we must use the criterion $R^2 \geq 4L/C = 3.3 \times 10^4\,\Omega^2$. Hence we must have $R \geq 180\,\Omega$.

☐ Summary

A changing current in a coil of wire will induce an emf in a second coil placed nearby. The **mutual inductance**, M, is defined as the proportionality constant between the induced emf $\mathscr{E}_2$ in the second coil and the time rate of change of current in the first:

$$\mathscr{E}_2 = -M\,dI_1/dt.$$

We can also write M as

$$M = \frac{N\Phi_B}{I}$$

where Φ_B is the magnetic flux through one coil (or circuit) with N loops produced by the current I in a second coil or circuit.

Within a single coil, a changing current induces an opposing emf, $\mathscr{E}$, so a coil has a **self-inductance** L defined by

$$\mathscr{E} = -L\,dI/dt.$$

This induced emf acts as an *impedance* to the flow of an alternating current. We can also write L as

$$L = N\frac{\Phi_B}{I}$$

where Φ_B is the flux through the inductance when a current I flows in its N loops.

When the current in an inductance L is I, the energy stored in the inductance is given by

$$U = \tfrac{1}{2}LI^2.$$

This energy can be thought of as being stored in the mag-netic field of the inductor. The energy density u in any magnetic field B is given by

$$u = \tfrac{1}{2}\frac{B^2}{\mu},$$

where μ is the magnetic permeability in that region.

When an inductance L and resistor R are connected in series to a source of emf, V_0, the current rises according to an exponential of the form

$$I = \frac{V_0}{R}\left(1 - e^{-t/\tau}\right),$$

where

$$\tau = L/R$$

is the time constant. The current eventually levels out at $I = V_0/R$. If the battery is suddenly switched out of the **LR circuit**, and the circuit remains complete, the current drops exponentially, $I = I_0 e^{-t/\tau}$, with the same time constant τ.

The current in a pure **LC circuit** (or charge on the capacitor) would oscillate sinusoidally. The energy too would oscillate back and forth between electric and magnetic, from the capacitor to the inductor, and back again. The current in a series **LRC** circuit (or charge on the capacitor), in which the capacitor at some instant is charged, can undergo damped oscillations or undergo critically damped or overdamped oscillations.

Questions

1. In situations where a small signal must travel over a distance, a "shielded cable" is used in which the signal wire is surrounded by an insulator and then enclosed by a cylindrical conductor. Why is a "shield" necessary?

2. What is the advantage of placing the two electric wires carrying ac close together?

3. The primary of a transformer on a telephone pole has a resistance of $0.10\ \Omega$ and the input voltage is 2400 V ac. Can you estimate the current that will flow? Will it be 24,000 A? Explain.

4. A transformer designed for a 120-V ac input will often "burn out" if connected to a 120-V dc source. Explain. (*Hint*: The resistance of the primary coil is usually very low.)

5. How would you arrange two flat circular coils so that their mutual inductance was (*a*) greatest, (*b*) least (without separating them by a great distance)?

6. If the two coils in Fig. 30–1 were connected electrically, would there still be mutual inductance?

7. Suppose the second coil of N_2 turns in Fig. 30–2 were moved so it was near the end of the solenoid. How would this affect the mutual inductance?

8. Would two circuits with mutual inductance also have self-inductance? Explain.

9. Is the self-inductance per unit length of a solenoid greater near its center or near its ends?

10. Is the energy density greatest near the ends of a solenoid or near its center?

11. If you are given a fixed length of wire, how would you shape it to obtain the greatest self-inductance? The least?

12. In a battery, when the current is in the same direction as the emf, the energy of the battery decreases, whereas if the current is in the opposite direction, the energy of the battery increases (as in charging a battery). Is this also true for an inductor?

13. Two solenoids have the same length and circular cross-sectional area. Both consist of tightly packed turns of wire, but one uses thicker wire than the other. Which has the greater inductance? Which has the greater inductive time constant?

14. Does the emf of the battery in Fig. 30–5a affect the time needed for the LR circuit to reach (*a*) a given fraction of its maximum possible current, (*b*) a given value of current?

15. A circuit with large inductive time constant carries a steady current. If a switch is opened, there can be a very large (and sometimes dangerous) spark or "arcing over." Explain.

16. At the instant the battery is connected into the LR circuit of Fig. 30–5a, the emf in the inductor has its maximum value even though the current is zero. Explain.

17. Explain physically why we might expect the time constant of an LR circuit to be proportional to L and inversely proportional to R, Eq. 30–10.

18. Explain how a solenoid whose ends are connected together can by itself oscillate like an LC circuit. Where does the capacitance come from?

19. What keeps an LC circuit oscillating even after the capacitor has discharged completely?

20. Is the current in the inductor always the same as that in the resistor of the LRC circuit of Fig. 30–12?

Problems

Section 30–1

1. (I) Suppose the second coil of Example 30–1 (Fig. 30–2) has twice the diameter of the solenoid, but is still concentric with it. What then will be the mutual inductance? Assume the solenoid is very long.

2. (II) A 2.44-m-long coil containing 300 loops is wound on an iron core (average $\mu = 2000\mu_0$) along with a second coil of 100 loops. The loops of each coil have a radius of 2.00-cm. If the current in the first coil drops uniformly from 12.0 A to zero in 98.0 ms, determine: (*a*) the mutual inductance M; (*b*) the emf induced in the second coil.

3. (II) Determine the mutual inductance per unit length between two long solenoids, one inside the other, whose radii are r_1 and $r_2 (r_2 < r_1)$ and whose turns per unit length are n_1 and n_2.

4. (III) A small thin coil with N_2 loops, each of area A_2, is placed inside a long solenoid, near its center. The solenoid has N_1 loops in its length l and has area A_1. Determine the mutual inductance as a function of θ, the angle between the plane of the small coil and the axis of the solenoid.

5. (III) A long straight wire and a small rectangular wire loop lie in the same plane, Fig. 30–14. Determine the mutual inductance in terms of l_1, l_2, and w. Assume the wire is very long compared to l_1, l_2 and w, and that the rest of its circuit is very far away compared to l_1, l_2 and w.

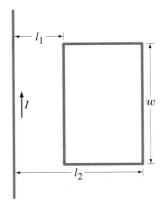

FIGURE 30–14 Problem 5.

6. (I) If the current in a 180-mH coil changes steadily from 20.0 mA to 38.0 mA in 340 ms, what is the induced emf?

7. (I) Estimate the inductance L of a 0.45-m-long air-filled coil 3.7 cm in diameter containing 20,000 loops.

8. (I) What is the inductance of a coil if it produces an emf of 8.50 V when the current in it changes uniformly from -22.0 mA to $+23.0$ mA in 21.0 ms?

9. (I) What is the inductance of a coaxial cable 22.0 m long if its inner and outer conductors have diameters of 2.0 mm and 3.5 mm?

10. (I) A 35-V emf is induced in a 150-mH coil by a current that rises uniformly from zero to I_0 in 3.0 ms. What is the value of I_0?

11. (II) If the outer conductor of a coaxial cable has radius 3.0 mm, what should be the radius of the inner conductor so that the inductance per unit length does not exceed 40 nH per meter?

12. (II) (a) Show that if two circuits, such as the coils in Fig. 30–1, carry currents I_1 and I_2, the magnetic flux through each is $\Phi_1 = L_1 I_1 + M I_2$ and $\Phi_2 = L_2 I_2 + M I_1$. (b) Determine a formula for the emf induced in each coil in terms of the rate of change of current in the two coils.

13. (II) The wire of a tightly wound solenoid is unwound and used to make another tightly wound solenoid of 3.0 times the diameter. By what factor does the inductance change?

14. (II) A coil has 2.70-Ω resistance and 0.418-H inductance. If the current is 5.00 A and is increasing at a rate of 4.50 A/s, what is the potential difference across the coil at this moment?

15. (II) Ignoring any mutual inductance, what is the equivalent inductance of two inductors connected (a) in series, (b) in parallel?

16. (II) A toroid has a rectangular cross-section as shown in Fig. 30–15. Show that the self-inductance is

$$L = \frac{\mu_0 N^2 h}{2\pi} \ln \frac{r_2}{r_1}$$

where N is the total number of turns and r_1, r_2, and h are the dimensions shown in Fig. 30–15. [Hint: Use Ampère's law to get B as a function of r inside the toroid, and integrate.]

FIGURE 30–15 Problems 16 and 22. A toroid of rectangular cross-section, with N turns carrying a current I.

17. (I) The magnetic field inside an air-filled solenoid 32.0 cm long and 2.10 cm in diameter is 0.600 T. Approximately how much energy is stored in this field?

18. (I) How much energy is stored in a 400-mH inductor at an instant when the current is 9.0 A?

19. (I) Typical large values for electric and magnetic fields attained in laboratories are about 1.0×10^4 V/m and 2.0 T. (a) Determine the energy density for each field and compare. (b) What magnitude electric field would be needed to produce the same energy density as the 2.0-T magnetic field?

20. (II) What is the energy density at the center of a circular loop of wire carrying a 30-A current if the radius of the loop is 28.0 cm?

21. (II) Calculate the magnetic and electric energy densities at the surface of a 3.0-mm-diameter copper wire carrying a 25-A current.

22. (II) For the toroid of Fig. 30–15, determine the energy density in the magnetic field as a function of $r(r_1 < r < r_2)$ and integrate this over the volume to obtain the total energy stored in the toroid, which carries a current I in each of its N loops.

23. (II) Determine the total energy stored per unit length in the magnetic field between the coaxial cylinders of a coaxial cable by using Eq. 30–7 for the energy density and integrating over the volume. Compare your answer to that obtained in Example 30–5.

24. (II) After how many time constants does the current in Fig. 30–5 reach within (a) 10 percent, (b) 1.0 percent, and (c) 0.10 percent of its maximum value?

25. (II) How many time constants does it take for the potential difference across the resistor in an LR circuit like that in Fig. 30–6 to drop to 1.0 percent of its original value?

26. (II) In Example 30–6, where is the power delivered by the battery going besides into the inductor? Do a calculation to show that energy is conserved.

27. (II) Determine dI/dt at $t = 0$ (when the battery is connected) for the circuit of Fig. 30–5a and show that if I continued to increase at this rate, it would reach its maximum value in one time constant.

28. (II) It takes 2.56 ms for the current in an LR circuit to increase from zero to half its maximum value. Determine (a) the time constant of the circuit, (b) the resistance of the circuit if $L = 310$ H.

29. (II) (a) Determine the energy stored in the inductor L as a function of time for the LR circuit of Fig. 30–5a. (b) After how many time constants does the stored energy reach 99 percent of its maximum value?

30. (II) In the circuit of Fig. 30–16, determine the current in each resistor (I_1, I_2, I_3) at the moment (a) the switch is closed, (b) a long time after the switch is closed. After the switch has been closed a long time, and reopened, what is each current (c) just after it is opened, (d) after a long time?

FIGURE 30–16
Problem 30.

Section 30–5

31. (I) The variable capacitor in the tuner of an AM radio has a capacitance of 1800 pF when the radio is tuned to a station at 550 kHz. (a) What must be the capacitance for a station at 1600 kHz? (b) What is the inductance (assumed constant)?

32. (I) (a) If the initial conditions of an LC circuit were $I = I_0$ and $Q = 0$ at $t = 0$, write Q as a function of time. (b) Practically, how could you set up these initial conditions?

33. (I) Use the definitions of the farad and henry to show that $1/\sqrt{LC}$ has units of s^{-1}.

34. (II) A 760-pF capacitor is charged to 135 V and then quickly connected to a 175-mH inductor. Determine (a) the frequency of oscillation, (b) the peak value of the current, and (c) the maximum energy stored in the magnetic field of the inductor.

35. (II) At $t = 0$, $Q = Q_0$ and $I = 0$ in an LC circuit. At the first moment when the energy is shared equally by the inductor and the capacitor, (a) what is the charge on the capacitor? (b) How much time has elapsed (in terms of the period T)?

Section 30–6

36. (II) In an oscillating LRC circuit, how much time does it take for the energy stored in the fields of the capacitor and inductor to fall to half its initial value? (See Fig. 30–12; assume $R \ll \sqrt{4L/C}$.)

37. (II) A damped LC circuit loses 5.5 percent of its electromagnetic energy per cycle to thermal energy. If $L = 65$ mH and $C = 1.00\ \mu$F, what is the value of R?

38. (II) (a) Obtain a differential equation for the current I in the LRC circuit of Fig. 30–12 by differentiating Eq. 30–17. (b) Obtain a general solution for I, assuming the circuit is connnected at $t = 0$ with charge Q_0 on the capacitor. Assume $R \ll \sqrt{4L/C}$. (c) Compare your result to part (d) of Example 30–8, and note any differences. (d) What complications would arise if R were not much less than $\sqrt{4L/C}$, say $R \approx \sqrt{L/C}$?

39. (II) Starting with Eq. 30–19, with $\phi = 0$, show that the current I in a lightly damped LRC circuit is given by

$$I \approx \frac{Q_0}{\sqrt{LC}}\, e^{-\frac{R}{2L}t} \sin(\omega' t + \delta)$$

where

$$\delta = \tan^{-1}\frac{R}{2L\omega'}.$$

40. (II) Show that the phase constant ϕ in Eq. 30–19 is given by

$$\phi = -\cot^{-1}\left(\frac{4L}{R^2 C}\right)^{\frac{1}{2}}$$

if $I = 0$ at $t = 0$ in the circuit of Fig. 30–12. Assume that $R^2 \ll 4L/C$.

41. (III) How much resistance must be added to a pure LC circuit $(L = 300\ \text{mH},\ C = 1800\ \text{pF})$ to change the oscillator's frequency by 0.10 percent? Will it be increased or decreased?

42. (III) Suppose a battery of fixed voltage V_0 is connected in the LRC circuit of Fig. 30–12. At $t = 0$, the switch is closed. Assuming this is a very overdamped circuit $(R^2 \gg 4L/C)$, (a) determine the current as a function of time. (b) Plot I vs. t. (c) Compare your result to a pure RC circuit, for which $L = 0$. Real RC circuits always have some inductance, so the result of this Problem is more realistic than a pure RC circuit.

General Problems

43. An air-filled cylindrical inductor has 3000 turns, and it is 2.5 cm in diameter and 28.2 cm long. Ignoring end effects, what is its inductance? How many turns would you need to generate the same inductance if the core were iron-filled instead? Assume the magnetic permeability of iron is about 1000 times that of free space.

44. At $t = 0$, the current through a 60.0 mH inductor is 50.0 mA and is increasing at the rate of 100 mA/s. What is the initial energy stored in the inductor, and how long does it take for the energy to increase by a factor of 10 from the initial value?

45. A 3000-pF capacitor is charged to 120 V and then quickly connected to an inductor. The frequency of oscillation is observed to be 20 kHz. Determine (a) the inductance, (b) the peak value of the current, and (c) the maximum energy stored in the magnetic field of the inductor.

46. Estimate the mutual inductance M between two small loops, of radius r_1 and r_2, which are separated by a distance l that is large compared to r_1 and r_2, Fig. 30–17. Give M as a function of θ, the angle between the planes of the two coils. Assume the line joining their centers is perpendicular to the plane of coil 1.

FIGURE 30–17   Problem 46.

47. Approximately how many turns does an air-filled coil have if it is 2.2 cm in diameter and 17.0 cm long and its inductance is 25 mH? How many turns are needed if it has an iron core and $\mu = 10^3 \mu_0$?

48. (a) Show that the self-inductance L of a toroid (Fig. 30–18) of radius r_0 containing N loops each of diameter d is

$$ L \approx \frac{\mu_0 N^2 d^2}{8 r_0} $$

if $r_0 \gg d$. Assume the field is uniform inside the toroid; is this actually true? Is this result consistent with L for a solenoid? Should it be? (b) Calculate the inductance L of a large toroid if the diameter of the coils is 2.0 cm and the diameter of the whole ring is 50 cm. Assume the field inside the toroid is uniform. There are a total of 550 loops of wire.

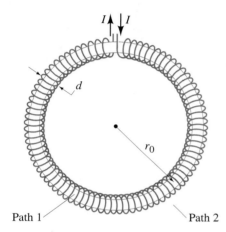

FIGURE 30–18 A toroid, Problem 48.

49. A pair of straight parallel thin wires, such as a lamp cord, each of radius r, are a distance l apart and carry current to a circuit some distance away. Ignoring the field within each wire, show that the inductance per unit length is $(\mu_0/\pi) \ln[(l - r)/r]$.

50. The potential difference across a given coil is 2.55 V at an instant when the current is 360 mA and is changing at a rate of 340 mA/s. At a later instant, the potential difference is 1.82 V while the current is 420 mA and is decreasing at a rate of 180 mA/s. Determine the inductance and resistance of the coil.

51. (a) Show that if two thin coils (inductance L_1 and L_2) are connected in series and placed close to each other, the net self-inductance is

$$ L = L_1 + L_2 \pm 2M $$

where M is their mutual inductance. Explain the ± sign. (b) How can M be made zero (or nearly so)? (c) Determine the net inductance if the two coils are connected in parallel, assuming the mutual inductance is ignorable. If M cannot be ignored, how would it affect your answer?

52. Assuming the Earth's magnetic field averages about 0.50×10^{-4} T near the surface of the Earth, estimate the total energy stored in this field in the first 10 km above the Earth's surface.

53. Show that the fraction of electromagnetic energy lost (to thermal energy) per cycle in a lightly damped $(R^2 \ll 4L/C)$ LRC circuit is approximately

$$ \frac{\Delta U}{U} = \frac{2\pi R}{L\omega} = \frac{2\pi}{Q}. $$

The quantity Q, defined as $Q = L\omega/R$, is called the *Q-value*, or *quality factor*, of the circuit and is a measure of the damping present. A high Q-value means smaller damping and less energy input required to maintain oscillations.

54. Two tightly wound solenoids have the same length and circular cross-sectional area. But solenoid 1 uses wire that is half as thick as solenoid 2. (a) What is the ratio of their inductances? (b) What is the ratio of their inductive time constants (assuming no other resistance in the circuits)?

55. (a) For an underdamped LRC circuit, determine a formula for the energy $U = U_E + U_B$ stored in the electric and magnetic fields as a function of time. Give in terms of the initial charge Q_0 on the capacitor. (b) Show how dU/dt is related to the rate energy is transformed in the resistor, $I^2 R$.

56. An electronic device needs to be protected against sudden surges in current. In particular, after the power is turned on the current should rise to no more than 7.5 mA in the first 100 μs. The device has resistance 150 Ω and is designed to operate at 50 mA. How would you protect this device?

57. The circuit shown in Fig. 30–19a can integrate (in the calculus sense) the input voltage V_{in}, if the time constant L/R is large compared with the time during which V_{in} varies. Explain how this integrator works and sketch its output for the square wave signal input shown in Fig. 30–19b. [*Hint:* Write Kirchhoff's loop rule for the circuit. Multiply each term in this differential equation (in I) by a factor $e^{Rt/L}$ to make it easier to integrate.]

(a)

(b)

FIGURE 30–19 Problem 57.

A stereo system is a very complex ac circuit. The power source for home stereo is ac. The audio signal is ac at sound frequencies. The circuits inside a stereo contain resistors, capacitors, and inductors, plus other more complicated elements like diodes and transistors. A loudspeaker also contains an ac circuit, called a crossover (we'll look at this too in this chapter) to divide the signal between woofer and tweeters.

CHAPTER 31

AC Circuits

In earlier chapters we discussed circuits that contain combinations of resistor, capacitor, and inductor (or all three), but only when they are connected to a dc source of emf or to no source as in the discharge of a capacitor in an *RC* circuit or oscillation of an *LC* or *LRC* circuit. See Sections 26–4, 30–4, 30–5, and 30–6. Now we discuss these circuit elements when connected to a source of alternating voltage that produces an alternating current (ac). Such **ac circuits** are important first of all because the output of most generators (Section 29–4) is sinusoidal and most electricity generated and transmitted over wires is ac. Secondly, any voltage that varies in time, no matter how complex it is, can be written as a sum of sine and cosine terms of different frequencies in a Fourier series. Thus, the response of resistors, capacitors, and inductors to a sinusoidal source of emf is of basic importance.

31–1 | Introduction: AC Circuits

We briefly discussed alternating currents in Section 25–7, and saw there that for sinusoidally varying currents and voltages, the rms and peak values are related by

$$V_{\text{rms}} = \frac{V_0}{\sqrt{2}}, \qquad I_{\text{rms}} = \frac{I_0}{\sqrt{2}}.$$

We now examine how a resistor, a capacitor, and an inductor behave when connected to a source of alternating voltage, whose symbol is

$$\bullet\!-\!\sim\!\!-\!\bullet \quad , \qquad\qquad \text{[symbol for ac source]}$$

which produces a sinusoidal voltage of frequency f.

We first look at the R, C, and L separately, and then later all of them together. In each case we assume the voltage gives rise to a current

$$I = I_0 \sin 2\pi f t = I_0 \sin \omega t \tag{31–1}$$

where $\omega = 2\pi f$.

31–2 | AC Circuit Containing only Resistance R

When an ac source is connected to a resistor as in Fig. 31–1a, the current increases and decreases with the alternating voltage, and Kirchhoff's loop rule tells us $V - IR = 0$. Hence

$$V = I_0 R \sin \omega t = V_0 \sin \omega t$$

where $V_0 = I_0 R$ is the peak voltage. Figure 31–1b shows the voltage (red curve) and the current (blue curve). Because the current is zero when the voltage is zero and the current reaches a peak when the voltage does, we say that the current and voltage are **in phase**. Energy is transformed into heat (Section 25–7), at an average rate

$$\overline{P} = \overline{IV} = I_{rms}^2 R = V_{rms}^2 / R.$$

31–3 | AC Circuit Containing only Inductance L

In Fig. 31–2a an inductor of inductance L, represented by the symbol , is connected to the ac source. We ignore any resistance it might have (it is usually small). The voltage applied to the inductor will be equal to the emf generated in the inductor by the changing current as given by Eq. 30–5. This is because the sum of the electric potential changes around any closed circuit must add up to zero, as Kirchhoff's rule tells us. Thus

$$V - L \frac{dI}{dt} = 0$$

or

$$V = L \frac{dI}{dt} = \omega L I_0 \cos \omega t. \tag{31–2}$$

Using the identity $\cos \theta = \sin(\theta + 90°)$, we can write

$$V = \omega L I_0 \sin(\omega t + 90°) = V_0 \sin(\omega t + 90°) \tag{31–3a}$$

where

$$V_0 = I_0 \omega L \tag{31–3b}$$

is the peak voltage. The current I and voltage V across the inductor as a function of time are graphed in Fig. 31–2b. It is clear from this graph, as well as from Eqs. 31–3, that the current and voltage are out of phase by a quarter cycle, which is equivalent to $\pi/2$ radians or 90°. We see from the graph that

in an inductor, the current lags the voltage by 90°.

That is, the current in an inductor reaches its peaks a quarter cycle after the voltage does. Alternatively, we can say that the voltage leads the current by 90°.

Because the current and voltage are out of phase by 90°, no energy is transformed to other forms of energy in an inductor on the average; in particular, no energy is dissipated as thermal energy. This can be seen as follows from Fig. 31–2b. From point c to d, the voltage is increasing from zero to its maximum. The current, however, is in the opposite direction to the voltage and is approaching zero. The average power over this interval, VI, is negative. From d to e, however, both V and I are positive so VI is positive; this contribution just balances the negative contribution of the previous quarter cycle. Similar considerations apply to the rest of the cycle. Thus, the average power transformed over one or many cycles is zero. We can see that energy from the source passes into the magnetic field of the inductor, where it is stored temporarily. Then the field decreases and the energy is transferred back to the source. None is dissipated in this process. Compare this to a resistor where the current is always in the same direction as the voltage and energy is transferred out of the source and never back into it. (The product VI is never negative.) This energy is not stored in the resistor, but is transformed to thermal energy.

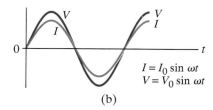

FIGURE 31–1 Resistor connected to ac source. Current is in phase with the voltage across a resistor.

FIGURE 31–2 Inductor connected to an ac source. Current (blue curve) lags voltage (red curve) by a quarter cycle or 90°.

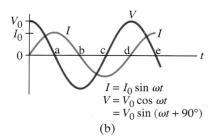

Inductor: current lags voltage

SECTION 31–3   **773**

Just as a resistor impedes the flow of charge, so too an inductor impedes the flow of charge in an alternating current due to the back emf produced. For a resistor R, the peak current and peak voltage are related by $V_0 = I_0 R$. We can write a similar relation for an inductor:

$$V_0 = I_0 X_L \qquad \begin{bmatrix} \text{maximum values} \\ \text{not an any instant} \end{bmatrix} \qquad \textbf{(31–4a)}$$

where, from Eq. 31–3b

Reactance (impedance) of inductor

$$X_L = \omega L. \qquad \textbf{(31–4b)}$$

The term X_L is called the *inductive reactance,* or *impedance,* of the inductor, and it is easy to show it has units of ohms. Normally, we use the term "reactance" to refer solely to the inductive properties. We then reserve the term "impedance" to include the total "impeding" qualities of the coil—its inductance as well as any resistance it may have (more on this in Section 31–5). In the absence of any resistance (or capacitance), the impedance is the same as the reactance.

Eq. 31–4a NOT valid at any instant

The quantities V_0 and I_0 in Eq. 31–4a refer to peak values. (It is also valid for rms values, $V_{\text{rms}} = I_{\text{rms}} X_L$.) Note, however, that although this equation relates the peak values, the peak current and voltage are not reached at the same time; so Eq. 31–4a is *not valid at a particular instant,* as is the case for a resistor ($V = IR$).

Note from Eq. 31–4b that if $\omega = 2\pi f = 0$ (so the current is dc), there is no back emf and no impedance to the flow of charge.

EXAMPLE 31–1 **Reactance of a coil.** A coil has a resistance $R = 1.00\,\Omega$ and an inductance of 0.300 H. Determine the current in the coil if (a) 120-V dc is applied to it; (b) 120-V ac (rms) at 60.0 Hz is applied.

SOLUTION (a) There is no inductive reactance ($X_L = 0$ since $f = 0$), so we can write for the resistance:

$$I = \frac{V}{R} = \frac{120\,\text{V}}{1.00\,\Omega} = 120\,\text{A}.$$

(b) The inductive reactance in this case is:

$$X_L = 2\pi f L = (6.28)(60.0\,\text{s}^{-1})(0.300\,\text{H}) = 113\,\Omega.$$

In comparison to this, the 1.00 Ω resistance can be ignored. Thus

$$I_{\text{rms}} \approx \frac{V_{\text{rms}}}{X_L} = \frac{120\,\text{V}}{113\,\Omega} = 1.06\,\text{A}.$$

[It might be tempting to say that the total impedance is $113\,\Omega + 1\,\Omega = 114\,\Omega$. This might imply that about 1 percent of the voltage drop is across the resistor, or about 1 V; and that across the inductance is 119 V. Although the 1 Vrms across the resistor is accurate, the other statements are *not* true because of the alteration in phase in an inductor. This will be discussed in Section 31–5.]

31–4 AC Circuit Containing only Capacitance C

When a capacitor is connected to a battery, the capacitor plates quickly acquire equal and opposite charges; but no steady current flows in the circuit. A capacitor prevents the flow of a dc current. However, if a capacitor is connected to an alternating source of voltage, as in Fig. 31–3a, an alternating current will flow continuously. This can happen because when the ac voltage is first turned on, charge begins to flow so that one plate acquires a negative charge and the other a positive charge. But when the voltage reverses itself, the charges flow in the opposite direction. Thus, for an alternating applied voltage, an ac current is present in the circuit continuously.

Let us look at this in more detail. By Kirchhoff's loop rule, the applied source voltage must equal the voltage V across the capacitor at any moment:

$$V = \frac{Q}{C}$$

where C is the capacitance and Q is the charge on the capacitor plates. Now the current I at any instant (given as $I = I_0 \sin \omega t$) is

$$I = \frac{dQ}{dt} = I_0 \sin \omega t.$$

Hence the charge Q on the plates at any instant is given by

$$Q = \int_0^t dQ = \int_0^t I_0 \sin \omega t \, dt = -\frac{I_0}{\omega} \cos \omega t.$$

Then the voltage across the capacitor is

$$V = \frac{Q}{C} = -I_0 \left(\frac{1}{\omega C} \right) \cos \omega t.$$

Using the trigonometric identity $\cos \theta = -\sin(\theta - 90°)$ we can rewrite this as

$$V = I_0 \left(\frac{1}{\omega C} \right) \sin(\omega t - 90°) = V_0 \sin(\omega t - 90°) \qquad \textbf{(31–5a)}$$

where

$$V_0 = I_0 \left(\frac{1}{\omega C} \right) \qquad \textbf{(31–5b)}$$

is the peak voltage. The current $I(= I_0 \sin \omega t)$ and voltage V (Eq. 31–5a) across the capacitor are graphed in Fig. 31–3b. It is clear from this graph, as well as a comparison of Eq. 31–5a with Eq. 31–1, that the current and voltage are out of phase by a quarter cycle or 90° ($\pi/2$ radians):

The current leads the voltage across a capacitor by 90°.

Capacitor: current leads voltage

Alternatively we can say that the voltage lags the current by 90°. This is the opposite of what happens for an inductor, where the current lags the voltage by 90°.

Because the current and voltage are out of phase by 90°, the average power dissipated is zero, just as for an inductor. Energy from the source is fed to the capacitor, where it is stored in the electric field between its plates. As the field decreases, the energy returns to the source.

Thus, in an ac circuit, *only a resistance will dissipate energy* to thermal energy.

Only R (not C or L) dissipates electric energy

A relationship between the applied voltage and the current in a capacitor can be written just as for an inductance:

$$V_0 = I_0 X_C \qquad \begin{bmatrix} \text{maximum values} \\ \text{not at any instant} \end{bmatrix} \quad \textbf{(31–6a)}$$

where X_C is the **capacitive reactance** (or *impedance*) of the capacitor, and has units of ohms. X_C is given by (see Eq. 31–5b)

$$X_C = \frac{1}{\omega C}. \qquad \textbf{(31–6b)}$$

Reactance (impedance) of capacitor

Equation 31–6a relates the peak values of V and I, or the rms values $(V_{\text{rms}} = I_{\text{rms}} X_C)$. But it is not valid at a particular instant because I and V are not in phase.

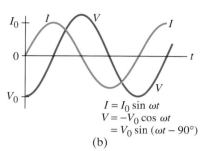

$I = I_0 \sin \omega t$
$V = -V_0 \cos \omega t$
$= V_0 \sin(\omega t - 90°)$

(b)

FIGURE 31–3 Capacitor connected to an ac source. Current leads voltage by a quarter cycle, or 90°.

Note from Eq. 31–6b that for dc conditions, $\omega = 2\pi f = 0$ and X_C becomes infinite. This is as it should be, since a pure capacitor does not pass dc current. Also, note that the reactance of an inductor increases with frequency, but that of a capacitor decreases with frequency.

EXAMPLE 31–2 **Capacitor reactance.** What are the peak and rms currents in the circuit of Fig. 31–3a if $C = 1.0\,\mu\text{F}$ and $V_{\text{rms}} = 120\,\text{V}$? Calculate for (a) $f = 60\,\text{Hz}$, and then for (b) $f = 6.0 \times 10^5\,\text{Hz}$.

SOLUTION (a) $V_0 = \sqrt{2}\,V_{\text{rms}} = 170\,\text{V}$. Then

$$X_C = \frac{1}{2\pi f C} = \frac{1}{(6.28)(60\,\text{s}^{-1})(1.0 \times 10^{-6}\,\text{F})} = 2.7\,\text{k}\Omega.$$

Thus

$$I_0 = \frac{V_0}{X_C} = \frac{170\,\text{V}}{2.7 \times 10^3\,\Omega} = 63\,\text{mA},$$

$$I_{\text{rms}} = \frac{V_{\text{rms}}}{X_C} = \frac{120\,\text{V}}{2.7 \times 10^3\,\Omega} = 44\,\text{mA}.$$

(b) For $f = 6.0 \times 10^5\,\text{Hz}$, X_C will be $0.27\,\Omega$, $I_0 = 630\,\text{A}$, and $I_{\text{rms}} = 440\,\text{A}$. The dependence on f is dramatic.

Capacitors are used for a variety of purposes. some of which have already been described. Two other applications are illustrated in Fig. 31–4. In Fig. 31–4a, circuit A is said to be capacitively coupled to circuit B. The purpose of the capacitor is to prevent a dc voltage from passing from A to B but allowing an ac signal to pass relatively unimpeded. If C is sufficiently large, the ac signal will not be significantly attenuated, whereas dc is filtered out. The capacitor in Fig. 31–4b also passes ac but not dc. In this case, a dc voltage can be maintained between circuits A and B. If the capacitance C is large enough, the capacitor offers little impedance to an ac signal leaving A. Such a signal then passes to ground instead of into B. Thus the capacitor in Fig. 31–4b acts like a *filter* when a constant dc voltage is required; any sharp variation in voltage will pass to ground instead of into circuit B. Capacitors used in these two ways are very common in circuits.

Loudspeakers that have separate speakers for low- and high-frequency sounds can use a simple "cross-over" that uses inductors and capacitors to filter frequencies reaching each speaker. Low-frequency sounds go to a large-diameter speaker, the *woofer*, and high-frequency sounds go to a small-diameter *tweeter*. (Recall standing waves from Chapter 16—small vibrating objects have high resonant frequencies whereas large vibrating objects have lower frequencies.) The simple diagram of Fig. 31–5 shows how the output signal from the amplifier is divided and goes to both tweeter and woofer, in parallel. A capacitor in the tweeter circuit impedes low-frequency signals (much like the capacitor in Fig. 31–4a). An inductor in the woofer circuit impedes high-frequency signals ($X_L = 2\pi f L$) so mainly low-frequency sounds are emitted by the woofer.

⇒ **PHYSICS APPLIED**

Capacitors as filters

FIGURE 31–4 Two common uses for a capacitor.

FIGURE 31–5
Loudspeaker cross-over.

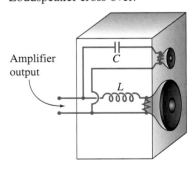

31–5 *LRC* Series AC Circuit

We now examine a circuit containing all three elements in series, a resistor R, an inductor L, and a capacitor C, Fig 31–6. If a given circuit contains only two of these elements, we can still use the results of this Section by setting $R = 0$, $L = 0$, or $C = \infty$ (infinity) as needed. We let V_R, V_L, and V_C represent the voltage across each element at a *given instant* in time; and V_{R0}, V_{L0}, and V_{C0} represent the *maximum* (peak) values of these voltages. The voltage across each of the elements will follow the phase relations we discussed in the previous Sections. That is, V_R will be

FIGURE 31–6 An *LRC* circuit.

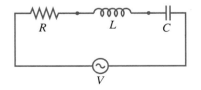

in phase with the current; V_L will lead the current by 90°; and V_C will lag behind the current by 90°. And at any instant the voltage V supplied by the source will be, by Kirchhoff's loop rule,

$$V = V_R + V_L + V_C. \qquad (31\text{-}7)$$

However, because the various voltages are not in phase, they do not reach their peak values at the same time, and so the peak voltage of the source, V_0, will *not* equal $V_{R0} + V_{L0} + V_{C0}$. Likewise the rms voltages (which are what ac voltmeters usually read) will not simply add up to give the rms voltage of the source.

Let us now examine the circuit in detail. What we would like to find in particular is the impedance of the circuit as a whole, the peak current I_0 that flows, and the phase difference between the source voltage and the current.

First we note that the current at any instant must be the same at all points in the circuit. Thus, *the currents in each element are in phase with each other, although the voltages are not.* We choose our origin in time ($t = 0$) so that the current I at any time t is

$$I = I_0 \sin \omega t$$

just as we did in the previous Sections of this chapter.

We will analyze the LRC circuit using[†] a so-called **phasor diagram**: Arrows (acting somewhat like vectors) are drawn in an xy coordinate system to represent the amplitude of each voltage. The *length of each arrow represents the magnitude of the peak voltage across each element:*

$$V_{R0} = I_0 R, \qquad V_{L0} = I_0 X_L, \qquad V_{C0} = I_0 X_C.$$

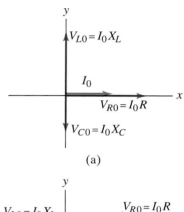

(a)

The angle of each arrow represents the phase of each voltage, relative to the current, and the arrows rotate at angular frequency ω to take into account the time dependence of the voltages and current. In particular, *the projection of each arrow on the y axis will represent the voltage across each element at a given time.* Let us see how this works.

First we draw the phasor diagram for time $t = 0$. The current at $t = 0$ is $I = I_0 \sin \omega t = 0$, and we draw the arrow representing I_0 along the positive x axis in our phasor diagram, Fig. 31–7a. (Note that the projection of I_0 on the y axis is zero, which corresponds to $I = 0$ at $t = 0$.) The voltage across a resistor is always in phase with the current, so the arrow representing V_{R0} is drawn parallel to I_0 along the x axis as shown. Since the voltage across the inductor, V_L, leads the current by 90°, V_{L0} leads V_{R0} by 90°, and is drawn perpendicular to it as shown. V_C lags the current by 90°, so V_{C0} lags V_{R0} by 90°, and hence V_{C0} is drawn perpendicular to V_{R0}, but downward (Fig. 31–7a). Now if we let this diagram rotate as a whole at angular frequency ω, then we obtain the diagram shown in Fig. 31–7b: after a time t, each arrow has rotated through an angle ωt. Then, as mentioned above, the projections of each arrow on the y axis represent the voltages across each element at the instant t. See Fig. 31–7c. For example, the projection of V_{R0} on the y axis is $V_{R0} \sin \omega t$ ($= I_0 R \sin \omega t = IR$ since $I = I_0 \sin \omega t$). The projections of V_{L0} and V_{C0} on the y axis are $V_L = V_{L0} \cos \omega t = V_L \sin(\omega t + 90°)$, and $V_C = -V_{C0} \cos \omega t = V_{C0} \sin(\omega t - 90°)$. These results are consistent with our earlier results as shown in Figs. 31–1, 31–2, and 31–3. See also Eqs. 31–3a and 31–5a. Maintaining the 90° angle between each vector ensures the correct phase relations. Although these facts show the validity of a phasor diagram, what we are really interested in is how to add the voltages.

(b)

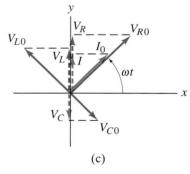

(c)

FIGURE 31–7 Phasor diagram for a series LRC circuit.

[†]We could instead do our analysis by rewriting Eq. 31–7 as a differential equation (setting $V_C = Q/C$, $V_R = IR = (dQ/dt)R$, and $V_L = L\,dI/dt$) and trying to solve the differential equation. The differential equation we would get would look like Eq. 14–21 in Section 14–8 (on forced vibrations), and would be solved in the same way. Phasor diagrams are easier, and at the same time give us some physical insight into the situation.

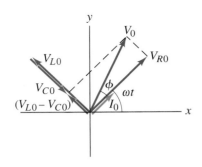

FIGURE 31–8 Phasor diagram for a series *LRC* circuit showing the sum vector, V_0.

The sum of the projections of the three vectors on the y axis is equal to the projection of their sum. But the sum of the projections represents the instantaneous voltage across the whole circuit, which is the source voltage V. We can then let the vector sum of these vectors be the vector that represents the peak source voltage V_0 on a phasor diagram. This is shown in Fig. 31–8, where it is seen that V_0 makes an angle ϕ with V_{R0} and I_0. As time passes, V_0 rotates with the other vectors, so the instantaneous voltage V (projection of V_0 on y axis) is (see Fig. 31–8):

$$V = V_0 \sin(\omega t + \phi).$$

Thus we see that the voltage from the source is out of phase[†] with the current by an angle ϕ.

From this analysis we can now draw some useful conclusions. First we determine the total **impedance** Z of the circuit, which is defined by the relation

$$V_{\text{rms}} = I_{\text{rms}} Z, \qquad \text{or} \qquad V_0 = I_0 Z. \tag{31–8}$$

From Fig. 31–8, we see, using the Pythagorean theorem (V_0 is the hypoteneuse), that

$$
\begin{aligned}
V_0 &= \sqrt{V_{R0}^2 + (V_{L0} - V_{C0})^2} \\
&= \sqrt{I_0^2 R^2 + (I_0 X_L - I_0 X_C)^2} \\
&= I_0 \sqrt{R^2 + (X_L - X_C)^2}.
\end{aligned}
$$

Thus, from Eq. 31–8, and then Eqs. 31–4b and 31–6b,

Total impedance of LRC ac circuit

$$Z = \sqrt{R^2 + (X_L - X_C)^2} \tag{31–9a}$$

$$= \sqrt{R^2 + \left(\omega L - \frac{1}{\omega C}\right)^2}. \tag{31–9b}$$

This gives the total impedance Z of the circuit. Also from Fig. 31–8 we can find the phase angle ϕ:

Phase angle of LRC ac circuit

or

$$\tan\phi = \frac{V_{L0} - V_{C0}}{V_{R0}} = \frac{I_0(X_L - X_C)}{I_0 R} = \frac{X_L - X_C}{R} \tag{31–10a}$$

$$\cos\phi = \frac{V_{R0}}{V_0} = \frac{I_0 R}{I_0 Z} = \frac{R}{Z}. \tag{31–10b}$$

Note that Fig. 31–8 was drawn for the case $X_L > X_C$, and the current lags the source voltage by ϕ. If the reverse is true, $X_L < X_C$, then ϕ in Eq. 31–10 is less than zero, and the current leads the source voltage.

Finally, we can determine the power dissipated in the circuit. We saw earlier that power is dissipated only by a resistance; none is dissipated by inductance or capacitance. Therefore, the average power $\overline{P} = I_{\text{rms}}^2 R$. But from Eq. 31–10b, $R = Z \cos\phi$. Therefore

Power factor

$$
\begin{aligned}
\overline{P} &= I_{\text{rms}}^2 Z \cos\phi \\
&= I_{\text{rms}} V_{\text{rms}} \cos\phi.
\end{aligned}
\tag{31–11}
$$

The factor $\cos\phi$ is referred to as the **power factor** of the circuit. For a pure resistor, $\cos\phi = 1$ and $\overline{P} = I_{\text{rms}} V_{\text{rms}}$. For a capacitor or inductor alone, $\phi = -90°$ or $+90°$, respectively, so $\cos\phi = 0$ and no power is dissipated.

[†] As a check, note that if $R = X_C = 0$, then $\phi = 90°$, and V_0 would lead the current by $90°$, as it must for an inductor alone. Similarly, if $R = L = 0$, $\phi = -90°$ and V_0 would lag the current by $90°$, as it must for a capacitor alone.

The test of this analysis is, of course, in experiment; and experiment is in full agreement with these results.[†]

EXAMPLE 31-3 **LRC Circuit.** Suppose that $R = 25.0\,\Omega$, $L = 30.0\,\text{mH}$, and $C = 12.0\,\mu\text{F}$ in Fig. 31–6, and that they are connected to a 90.0-V ac (rms) 500-Hz source. Calculate (a) the current in the circuit, (b) the voltmeter readings (rms) across each element, (c) the phase angle ϕ, and (d) the power dissipated in the circuit.

SOLUTION (a) First, we find the individual impedances at $f = 500\,\text{Hz} = 500\,\text{s}^{-1}$:

$$X_L = 2\pi f L = 94.2\,\Omega,$$

$$X_C = \frac{1}{2\pi f C} = 26.5\,\Omega.$$

Then

$$Z = \sqrt{R^2 + (X_L - X_C)^2}$$
$$= \sqrt{(25.0\,\Omega)^2 + (94.2\,\Omega - 26.5\,\Omega)^2} = 72.2\,\Omega.$$

From Eq. 31–8,

$$I_{\text{rms}} = \frac{V_{\text{rms}}}{Z} = \frac{90.0\,\text{V}}{72.2\,\Omega} = 1.25\,\text{A}.$$

(b) The rms voltage across each element is

$$(V_R)_{\text{rms}} = I_{\text{rms}} R = (1.25\,\text{A})(25.0\,\Omega) = 31.2\,\text{V}$$

$$(V_L)_{\text{rms}} = I_{\text{rms}} X_L = (1.25\,\text{A})(94.2\,\Omega) = 118\,\text{V}$$

$$(V_C)_{\text{rms}} = I_{\text{rms}} X_C = (1.25\,\text{A})(26.5\,\Omega) = 33.1\,\text{V}.$$

Notice that these do *not* add up to give the source voltage, 90.0 V (rms). Indeed, the rms voltage across the inductance *exceeds* the source voltage. This can happen because the different voltages are out of phase with each other, and at any instant one voltage can be negative, to compensate for a large positive voltage of another. The rms voltages, however, are always positive by definition. Although the rms voltages do not have to add up to the source voltage, the instantaneous voltages at any time do add up, of course, to the source voltage at that instant.

(c) The phase angle ϕ is given by Eq. 31–10b,

$$\cos\phi = \frac{R}{Z} = \frac{25.0\,\Omega}{72.2\,\Omega} = 0.346,$$

so $\phi = 69.7°$. Note that ϕ is positive because $X_L > X_C$ in this case, so $V_{L0} > V_{C0}$ in Fig. 31–8.

(d) $\overline{P} = I_{\text{rms}} V_{\text{rms}} \cos\phi = (1.25\,\text{A})(90.0\,\text{V})(25.0\,\Omega/72.2\,\Omega) = 39.0\,\text{W}.$

[†] At the beginning of this Section, we chose the phase of the current so that $I = I_0 \sin\omega t$. This choice is, of course, arbitrary. (What is physically important is the phase difference, ϕ, between current and voltage.) If we had chosen, instead,

$$V = V_0 \sin\omega t,$$

then the current I would be

$$I = I_0 \sin(\omega t - \phi)$$

where ϕ and I_0 have the same values as given by Eqs. 31–8, 31–9, and 31–10.

31-6 Resonance in AC Circuits

The rms current in an LRC series circuit is given by (see Eqs. 31–8 and 31–9b):

$$I_{rms} = \frac{V_{rms}}{Z} = \frac{V_{rms}}{\sqrt{R^2 + \left(\omega L - \dfrac{1}{\omega C}\right)^2}}. \qquad (31\text{-}12)$$

Because the impedance of inductors and capacitors depends on the frequency f ($= \omega/2\pi$) of the source, the current in an LRC circuit will depend on frequency. From Eq. 31–12 we can see that the current will be maximum at a frequency such that

$$\left(\omega L - \frac{1}{\omega C}\right) = 0.$$

We solve this for ω and call the solution ω_0:

$$\omega_0 = \sqrt{\frac{1}{LC}}. \qquad (31\text{-}13)$$

When $\omega = \omega_0$, the circuit is in **resonance**, and $f_0 = \omega_0/2\pi$ is the **resonant frequency** of the circuit. At this frequency, $X_C = X_L$, so the impedance is purely resistive and $\cos\phi = 1$. A graph of I_{rms} versus ω is shown in Fig. 31–9 for particular values of R, L, and C. For smaller R compared to X_L and X_C, the resonance peak will be higher and sharper. When R is very small, the circuit approaches the pure LC circuit we discussed in Section 30–5.

This electrical resonance is analogous to mechanical resonance, which we discussed in Chapter 14. The energy transferred to the system by the source is a maximum at resonance whether it is electrical resonance, the oscillation of a spring, or pushing a child on a swing (Section 14–8). That this is true in the electrical case can be seen from Eq. 31–11. At resonance, $\cos\phi = 1$, and I_{rms} is a maximum. For constant voltage V_{rms}, the power then is a maximum at resonance. A graph of power versus frequency peaks like that for the current, Fig. 31–9.

Electric resonance is used in many circuits. Radio and TV sets, for example, use resonant circuits for tuning in a station. Many frequencies reach the circuit, but a significant current flows only for those at or near the resonant frequency. Either L or C is variable so that different stations can be tuned in.

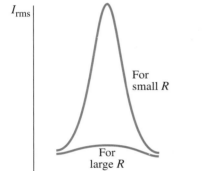

Resonance

FIGURE 31–9 Current in LRC circuit as a function of frequency, showing resonance peak at $\omega = \omega_0 = \sqrt{1/LC}$.

EXAMPLE 31–4 **Radio station oscillator.** A radio station is authorized to broadcast on a frequency of 1040 kHz. If you are designing a receiver circuit to pick up this station, and already have a coil with an inductance of 4.0 mH, what capacitance will you need?

SOLUTION The tuned circuit must have a resonant frequency of 1040 kHz. Using an inductor in series with a capacitor will accomplish the job. Since the resonance will be at

$$f_0 = \frac{1}{2\pi}\sqrt{\frac{1}{LC}}$$

you will need to find a capacitor with a value of

$$C = \frac{1}{L(2\pi f_0)^2} = \frac{1}{(4.0 \times 10^{-3}\,\text{H})(2\pi \times 1.04 \times 10^6\,\text{s}^{-1})^2} = 5.85 \times 10^{-12}\,\text{F} = 5.85\,\text{pF}$$

Impedance Matching

It is common to connect one electric circuit to a second circuit. For example, a TV antenna is connected to a TV set; an FM tuner is connected to an amplifier; the output of an amplifier is connected to a speaker; electrodes for an ECG or EEG (electrocardiogram and electroencephalogram—electrical traces of heart and brain signals) are connected to an amplifier or a recorder. In many cases it is important that the maximum power be transferred from one to the other, with a minimum of loss. This can be achieved when the output impedance of the one device matches the input impedance of the second.

To show why this is true, we consider simple circuits that contain only resistance. In Fig. 31–10 the source in circuit 1 could represent a power supply, the output of an amplifier, or the signal from an antenna, a laboratory probe, or a set of electrodes. R_1 represents the resistance of this device and includes the internal resistance of the source. R_1 is called the output impedance (or resistance) of circuit 1. The output of circuit 1 is across the terminals a and b which are connected to the input of circuit 2 which may be very complicated. We let R_2 be the equivalent "input resistance" of circuit 2.

The power delivered to circuit 2 is $P = I^2 R_2$ where $I = V/(R_1 + R_2)$. Thus

$$P = I^2 R_2 = \frac{V^2 R_2}{(R_1 + R_2)^2}.$$

We divide the top and bottom of the right side by R_1^2 and find

$$P = \frac{V^2}{R_1} \frac{\left(\dfrac{R_2}{R_1}\right)}{\left(1 + \dfrac{R_2}{R_1}\right)^2}.$$

The question is, if the resistance of the source is R_1, what value should R_2 have so that the maximum power is transferred to circuit 2? To determine this, we take the derivative of P with respect to R_2 and set it equal to zero:

$$0 = \frac{dP}{dR_2} = \frac{V^2}{R_1^2} \frac{\left(1 - R_2/R_1\right)}{\left(1 + R_2/R_1\right)^3}.$$

This expression can be zero only if $\left(1 - R_2/R_1\right) = 0$, or

$$R_2 = R_1.$$

Thus, the maximum power is transmitted when the *output impedance* of one device *equals the input impedance* of the second. This is called **impedance matching**.

In an ac circuit that contains capacitors and inductors, the different phases are important and the analysis is more complicated. However, the same result holds: to maximize power transfer it is important to match impedances $\left(Z_2 = Z_1\right)$.

In addition, one must be aware that it is possible to seriously distort a signal. For example, when a second circuit is connected, it may put the first circuit into resonance, or take it out of resonance for a certain frequency.

Without proper consideration of the impedances involved, one can make measurements that are completely meaningless. These considerations are normally examined by engineers when designing an integrated set of apparatus. It has happened that researchers have connected several components to one another without regard for impedance matching, and made a "new discovery" that later was found (embarrassingly) to be due to impedance mismatch rather than the natural phenomenon they had thought.

In some cases, a transformer is used to alter an impedance, so it can be matched to that of a second circuit. If Z_s is the secondary impedance and Z_p the

FIGURE 31–10 Output of the circuit on the left is input to the circuit on the right.

Impedance matching

primary impedance, then $V_s = I_s Z_s$ and $V_p = I_p Z_p$ (I and V are either peak or rms values of current and voltage). Hence

$$\frac{Z_p}{Z_s} = \frac{V_p I_s}{V_s I_p} = \left(\frac{N_p}{N_s}\right)^2$$

where we have used Eqs. 29–5 and 29–6 for a transformer. Thus the impedance can be changed with a transformer.

Some instruments, such as oscilloscopes, require only a signal voltage but very little power. Maximum power transfer is then not important and such instruments can have a high input impedance, which has the advantage that the instrument draws very little current and disturbs the original circuit as little as possible.

* 31–8 | Three-Phase AC

Transmission lines typically consist of four wires, rather than two as you might have guessed. One of these wires is the ground; the remaining three are used to transmit three-phase ac power which is a superposition of three ac voltages 120° out of phase with each other:

$$V_1 = V_0 \sin \omega t$$
$$V_2 = V_0 \sin (\omega t + 2\pi/3)$$
$$V_3 = V_0 \sin (\omega t + 4\pi/3).$$

(See Fig. 31–11.) Why is three-phase power used? We saw in Fig. 25–20 that single-phase ac (i.e., the voltage V_1 by itself) delivers power to the load in pulses. A much smoother flow of power can be delivered if we use three-phase power. Suppose that each of the three voltages making up the three-phase source is hooked up to a resistor R. Then the power delivered is:

$$P = \frac{1}{R}(V_1^2 + V_2^2 + V_3^2).$$

You can show that this power is a constant equal to $3V_0^2/2R$, which is three times the rms power delivered by a single-phase source. This smooth flow of power makes electrical equipment run smoothly. Although houses use single-phase ac power, most industrial-grade machinery is wired for three-phase power.

FIGURE 31–11 The three voltages, out of phase by 120° ($=\frac{2}{3}\pi$ radians), in a three-phase power line.

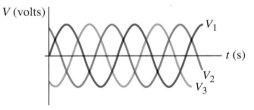

EXAMPLE 31–5 **Three-phase circuit.** In a three-phase circuit there is 266 V rms between line 1 and ground. What is the rms voltage between lines 2 and 3?

SOLUTION We are given $V_{rms} = V_0/\sqrt{2} = 266$ V. Hence $V_0 = 376$ V. Now $V_3 - V_2 = V_0[\sin(\omega t + 4\pi/3) - \sin(\omega t + 2\pi/3)] = 2V_0 \sin\frac{1}{2}\left(\frac{2\pi}{3}\right)\cos\frac{1}{2}(2\omega t)$ where we used the identity: $\sin A - \sin B = 2\sin\frac{1}{2}(A - B)\cos\frac{1}{2}(A + B)$. The rms voltage is

$$(V_3 - V_2)_{rms} = \frac{1}{\sqrt{2}}2V_0 \sin\frac{\pi}{3} = \sqrt{2}(376\,\text{V})(0.866) = 460\,\text{Vrms}.$$

Summary

Capacitance and inductance offer *impedance* to the flow of alternating current just as resistance does. This impedance is referred to as **reactance**, X. For capacitance and inductance the reactance is defined, as for resistors, as the proportionality constant between voltage and current (either the rms or peak values). Across a capacitor,

$$V_0 = I_0 X_C,$$

and across an inductor,

$$V_0 = I_0 X_L.$$

The reactance of a capacitor decreases with frequency:

$$X_C = 1/\omega C,$$

where $\omega = 2\pi f$ and f is the frequency.

The reactance of an inductor increases with frequency:

$$X_L = \omega L.$$

Whereas the current through a resistor is always in phase with the voltage across it, this is not true for inductors and capacitors: in an inductor, the current lags the voltage by 90°, and in a capacitor the current leads the voltage by 90°.

In an **LRC series circuit**, the total impedance Z is defined by the equivalent of $V = IR$ for resistance, namely $V_0 = I_0 Z$ or $V_{rms} = I_{rms} Z$; Z is related to R, C, and L by

$$Z = \sqrt{R^2 + (X_L - X_C)^2}.$$

The current in the circuit lags (or leads) the source voltage by an angle ϕ given by $\cos\phi = R/Z$. Only the resistor in an *LRC* circuit dissipates energy, and at a rate

$$\overline{P} = I_{rms}^2 Z \cos\phi$$

where the factor $\cos\phi$ is referred to as the **power factor**.

An *LRC* series circuit **resonates** at a frequency given by

$$\omega_0 = \frac{1}{\sqrt{LC}} \quad \text{or} \quad f_0 = \frac{\omega_0}{2\pi} = \frac{1}{2\pi\sqrt{LC}}.$$

The rms current in the circuit is largest when the applied voltage has a frequency equal to f_0. The lower the resistance R, the higher and sharper the resonance peak.

Questions

1. Under what conditions is the impedance in an *LRC* circuit a minimum?

2. Why can we assume that the current in an *LRC* circuit will have the same frequency as the applied emf?

3. In an *LRC* circuit, if $X_L > X_C$, the circuit is said to be predominantly "inductive." And if $X_C > X_L$, the circuit is said to be predominantly "capacitive." Discuss the reasons for these terms. In particular, do they say anything about the relative values of L and C at a given frequency?

4. Do the results of Section 31–5 approach the proper expected results when ω approaches zero? What are the expected results?

5. When an ac generator is connected to an *LRC* circuit, where does the energy come from ultimately? Where does it go? How do the values of L, C, and R affect the energy supplied by the generator?

6. Discuss the validity of both of Kirchhoff's rules (Section 26–3) when applied to ac circuits that contain several loops.

7. Is it possible for the instantaneous power output of an ac generator connected to an *LRC* circuit ever to be negative? Explain.

8. Can you tell whether the current in an *LRC* circuit leads or lags the applied voltage from a knowledge of the power factor, $\cos\phi$?

9. If $\cos\phi$ were less than zero, Eq. 31–11 tells us that $\overline{P} < 0$. Can this happen? Can $\cos\phi$ be made negative? Explain.

10. Does the power factor, $\cos\phi$, depend on frequency? Does the power dissipated in an *LRC* circuit depend on frequency?

11. What is the significance of the sign of $\phi(+$ or $-)$? Is this a convention, or is it a fixed rule?

12. Describe briefly how the frequency of the source emf affects the impedance of (a) a pure resistance, (b) a pure capacitance, (c) a pure inductance, (d) an *LRC* circuit near resonance (R small), (e) an *LRC* circuit far from resonance (R small).

13. Discuss the response of an *LRC* circuit as $R \rightarrow 0$ when the frequency is (a) at resonance, (b) near resonance, (c) far from resonance. Is there energy dissipation in each case? Discuss the transformations of energy that occur in each case.

14. Can you tell whether or not a circuit is in resonance if you are given the value of the power factor, $\cos\phi$?

15. An *LRC* resonance circuit is often called an *oscillator* circuit. What is it that oscillates?

16. Compare the oscillations of an *LRC* circuit to the vibration of a mass m on a spring. What do L and C correspond to in the mechanical system?

Sections 31–1 to 31–4

1. (I) What is the reactance of a 7.2 μF capacitor at a frequency of (a) 60 Hz, (b) 1.0 MHz?

2. (I) At what frequency will a 22.0-mH inductor have a reactance of 660 Ω?

3. (I) At what frequency will a 2.40-μF capacitor have a reactance of 6.70 kΩ?

4. (I) Plot a graph of the impedance of a 5.8-μF capacitor as a function of frequency from 10 Hz to 1000 Hz.

5. (I) Plot a graph of the impedance of a 5.0-mH inductor as a function of frequency from 100 Hz to 10,000 Hz.

6. (I) Calculate the impedance of, and rms current in, a 36.0-mH radio coil connected to 750-V (rms) 33.3-kHz ac line. Ignore resistance.

7. (II) What is the inductance L of the primary of a transformer whose input is 110 V at 60 Hz and the current drawn is 2.2 A? Assume no current in the secondary.

8. (II) (a) What is the impedance of a well-insulated 0.036-μF capacitor connected to a 22-kV (rms), 600-Hz line? (b) What will be the peak value of the current and its frequency?

9. (II) Instead of starting our analysis of ac circuits with Eq. 31–1, suppose we assumed that the external voltage was given as $V = V_0 \sin \omega t$. Show that when this voltage is (a) connected only to a capacitor C, the current would be $I = \omega C V_0 \cos \omega t = \omega C V_0 \sin (\omega t + 90°)$, and (b) connected only to an inductor L, the current would be $I = -(V_0/\omega L) \cos \omega t = (V_0/\omega L) \sin (\omega t - 90°)$.

10. (II) A current $I = 1.80 \cos 377 t$ (I in amps, t in seconds and the "angle" is in radians) flows in a series LR circuit in which $L = 3.85$ mH and $R = 260 \Omega$. What is the average power dissipation?

11. (II) A capacitor is placed in parallel with some device, B, as in Fig. 31–4b, to filter out stray high-frequency signals, but to allow ordinary 60-Hz ac to pass through with little loss. Suppose that circuit B in Fig. 31–4b is a resistance $R = 400 \Omega$ connected to ground, and that $C = 0.35 \mu$F. What percent of the incoming current will pass through C rather than R if (a) it is 60 Hz; (b) 60,000 Hz?

Section 31–5

12. (I) A 1.20-kΩ resistor and a 6.8-μF capacitor are connected in series to an ac source. Calculate the impedance of the circuit if the source frequency is (a) 60 Hz; (b) 60,000 Hz.

13. (I) A 9.0-kΩ resistor is in series with a 26.0-mH inductor and an ac source. Calculate the impedance of the circuit if the source frequency is (a) 50 Hz; (b) 30,000 Hz.

14. (I) For a 120-V, 60-Hz voltage, a current of 70 mA passing through the body for 1.0 s could be lethal. What must be the impedance of the body for this to occur?

15. (II) (a) What is the rms current in a series RC circuit if $R = 6.0$ kΩ, $C = 0.80 \mu$F, and the rms applied voltage is 120 V at 60 Hz? (b) What is the phase angle between voltage and current? (c) What is the power dissipated by the circuit? (d) What are the voltmeter readings across R and C?

16. (II) (a) What is the rms current in a series LR circuit when a 60.0-Hz, 120-V rms ac voltage is applied, where $R = 765 \Omega$ and $L = 250$ mH? (b) What is the phase angle between voltage and current? (c) How much power is dissipated? (d) What are the rms voltage readings across R and L?

17. (II) A 35-mH inductor with 2.0-Ω resistance is connected in series to a 20-μF capacitor and a 60-Hz, 45-V source. Calculate (a) the rms current, (b) the phase angle, and (c) the power dissipated in this circuit.

18. (II) A 40-mH coil whose resistance is 0.80 Ω is connected to a capacitor C and a 360-Hz source voltage. If the current and voltage are to be in phase, what value must C have?

19. (II) What is the resistance of a coil if its impedance is 335 Ω and its reactance is 45.5 Ω?

20. (II) In the LRC circuit of Fig. 31–6, suppose $I = I_0 \sin \omega t$ and $V = V_0 \sin (\omega t + \phi)$. Determine the instantaneous power dissipated in the circuit from $P = IV$ using these equations and show that on the average, $\overline{P} = \frac{1}{2} V_0 I_0 \cos \phi$, which confirms Eq. 31–11.

21. (II) If $V = V_0 \sin \omega t$, what is the average value of V over (a) a whole cycle, (b) each half cycle? How do these compare to V_{rms}?

22. (II) A circuit consists of a 150-Ω resistor in series with a 40.0-mH inductor and a 60.0-V ac generator. The power dissipated by the circuit is 15.5 W. What is the frequency of the generator?

23. (II) What is the total impedance, phase angle, and rms current in an LRC circuit connected to a 10.0-kHz, 800-V (rms) source if $L = 32.0$ mH, $R = 8.70$ kΩ, and $C = 5000$ pF?

Section 31–6

24. (I) A 3200-pF capacitor is connected in series to a 26.0-μH coil of resistance 2.00 Ω. What is the resonant frequency of this circuit?

25. (I) What is the resonant frequency of the LRC circuit of Example 31–3? At what rate is energy taken from the generator, on the average, at this frequency?

26. (II) An LRC circuit has $L = 4.15$ mH and $R = 220 \Omega$. (a) What value must C have to produce resonance at 33.0 kHz? (b) What will be the maximum current at resonance if the peak external voltage is 136 V?

27. (II) What will be the peak current in Problem 26 if the capacitor is chosen so that the resonant frequency is twice the applied frequency of 33.0 kHz?

28. (II) A resonant circuit using a 220-μF capacitor is to resonate at 18.0 kHz. The air-core inductor is to be a solenoid with closely packed coils made from 12.0 m of insulated wire 1.1 mm in diameter. How many loops will the inductor contain?

29. (II) (a) Show that oscillation of charge Q on the capacitor of an LRC circuit has amplitude

$$Q_0 = \frac{V_0}{\sqrt{(\omega R)^2 + \left(\omega^2 L - \dfrac{1}{C}\right)^2}}.$$

(b) At what angular frequency, ω', will Q_0 be a maximum? (c) Compare to a damped harmonic oscillator, and discuss. (See also Question 16 in this chapter.)

30. (II) Show that the width of a sharp resonance peak, defined as the difference in (angular) frequency between the two frequencies where $I = \frac{1}{2} I_0$, is given by $\Delta\omega \approx \sqrt{3}\, R/L$.

31. (II) (a) Determine a formula for the average power $\overline{P}$ dissipated in an LRC circuit in terms of L, R, C, ω, and V_0. (b) At what frequency is the power a maximum? (c) Find an approximate formula for the width of the resonance peak in average power, $\Delta\omega$, which is the difference in the two (angular) frequencies where $\overline{P}$ has half its maximum value. Assume a sharp peak.

* Section 31–7

*32. (I) The output of an ECG amplifier has an impedance of $35\,k\Omega$. It is to be connected to an 8.0-Ω loudspeaker through a transformer. What should be the turns ratio of the transformer?

*33. (I) An audio amplifier has output connections for $4\,\Omega$, $8\,\Omega$, and $16\,\Omega$. If two 8-Ω speakers are to be connected in parallel, to which output terminals should they be connected?

General Problems

34. Suppose circuit B in Fig. 31–4a consists of a resistance $R = 800\,\Omega$, and a capacitance $C = 1.2\,\mu\text{F}$. Will this capacitor act to eliminate 60-Hz ac but pass a high-frequency signal of frequency $60{,}000\,\text{Hz}$? To check this, determine the voltage drop across R for a 130-mV signal of frequency (a) 60 Hz; (b) 60,000 Hz.

35. A 230-mH coil, whose resistance is $18.5\,\Omega$, is connected to a capacitor C and a 3360-Hz source voltage. If the current and voltage are to be in phase, what value must C have?

36. A circuit contains two elements, but it is not known if they are L, R, or C. The current in this circuit when connected to a 120-V 60-Hz source is 5.6 A and lags the voltage by 50°. What are the two elements and what are their values?

37. An inductance coil operates at 240 V and 60 Hz. It draws 22.8 A. What is the coil's inductance?

38. (a) What is the impedance of a well-insulated 0.038-μF capacitor connected to a 4.0-kV (rms) 700-Hz line? (b) What will be the peak value of the current?

39. A 3.5-kΩ resistor in series with a 620-mH inductor is driven by an ac power supply. At what frequency is the impedance double that of the impedance at 60 Hz?

40. (a) What is the rms current in an RC circuit if $R = 8.80\,k\Omega$, $C = 1.80\,\mu\text{F}$, and the rms applied voltage is 120 V at 60.0 Hz? (b) What is the phase angle between voltage and current? (c) What is the power dissipated by the circuit? (d) What are the voltmeter readings across R and C?

41. An inductance coil draws 2.5 A dc when connected to a 36-V battery. When connected to a 60-Hz 120-V (rms) source, the current drawn is 3.8 A (rms). Determine the inductance and resistance of the coil.

42. The **Q factor** of a resonance circuit can be defined as the ratio of the voltage across the capacitor (or inductor) to the voltage across the resistor, at resonance. The larger the Q factor, the sharper the resonance curve will be and the sharper the tuning. (a) Show that the Q factor is given by the equation $Q = (1/R)\sqrt{L/C}$. (b) At a resonant frequency $f_0 = 1.0\,\text{MHz}$, what must be the value of L and R to produce a Q factor of 1000? Assume that $C = 0.010\,\mu\text{F}$. (c) What is the Q factor of the circuit of Example 31–3? [Note: see also Problem 53 in Chapter 30.]

43. In a series LRC circuit, the inductance is 20 mH, the capacitance is 50 nF, and the resistance is $200\,\Omega$. At what frequencies is the power factor equal to 0.17?

44. In our analysis of a series LRC circuit, Fig. 31–6, suppose we chose $V = V_0 \sin\omega t$. (a) Construct a phasor diagram, like that of Fig. 31–8, for this case. (b) Write a formula for the current I, defining all terms.

45. A voltage $V = 0.95\sin 754t$ is applied to an LRC circuit (I is in amperes, t is in seconds and the "angle" is in radians) which has $L = 22.0\,\text{mH}$, $R = 23.2\,k\Omega$, and $C = 0.30\,\mu\text{F}$. (a) What is the impedance and phase angle? (b) How much power is dissipated in the circuit? (c) What is the rms current and voltage across each element?

46. *Filter circuit.* Figure 31–12 shows a simple filter circuit designed to pass dc voltages with minimal attenuation and to remove, as much as possible, any ac components (such as 60-Hz line voltage that could cause hum in a stereo receiver, for example). Assume $V_{\text{in}} = V_1 + V_2$ where V_1 is dc and $V_2 = V_{20}\sin\omega t$, and that any resistance is very small. (a) Determine the current through the capacitor: give amplitude and phase (assume $R = 0$ and $X_L > X_C$). (b) Show that the ac component of the output voltage, $V_{2\,\text{out}}$, equals $(Q/C) - V_1$, where Q is the charge on the capacitor at any instant, and determine the amplitude and phase of $V_{2\,\text{out}}$. (c) Show that the attenuation of the ac voltage is greatest when $X_C \ll X_L$, and calculate the ratio of the output to input ac voltage in this case. (d) Compare the dc output to input voltage.

FIGURE 31–12 Problems 46 and 47.

47. Show that if the inductor L in the filter circuit of Fig. 31–12 (Problem 46) is replaced by a large resistor R, there will still be significant attenuation of the ac voltage and little attenuation of the dc voltage if the input dc voltage is high and the current (and power) are low.

48. A resistor R, capacitor C, and inductor L are connected in parallel across an ac generator as shown in Fig. 31–13. The source emf is $V = V_0 \sin \omega t$. Determine the current as a function of time (including amplitude and phase) (*a*) in the resistor, (*b*) in the inductor, (*c*) in the capacitor. (*d*) What is the total current leaving the source? (Give amplitude I_0 and phase.) (*e*) Determine the impedance Z defined as $Z = V_0/I_0$. (*f*) What is the power factor?

FIGURE 31–13   Problem 48.

49. Suppose a series LRC circuit has two resistors, R_1 and R_2, two capacitors, C_1 and C_2, and two inductors, L_1 and L_2, all in series. Calculate the total impedance of the circuit.

50. Determine the inductance L of the primary of a transformer whose input is 220 V at 60 Hz when the current drawn is 5.8 A. Assume no current in the secondary.

51. You have a small electromagnet that consumes 300 W from a residential circuit operating at 120 V at 60 Hz. Using your ac multimeter, you determine that the unit draws 4.0 A rms. What are the values of the inductance and the internal resistance?

52. An inductor L in series with a resistor R, driven by a sinusoidal voltage source, responds as described by the following differential equation:

$$V_0 \sin \omega t = L\frac{dI}{dt} + RI.$$

Show that a current of the form $I = I_0 \sin(\omega t - \phi)$ flows through the circuit by direct substitution into the differential equation. Determine the amplitude of the current (I_0) and the phase difference ϕ between the current and the voltage source.

53. For the circuit shown in Fig. 31–14, $V = V_0 \sin \omega t$. Calculate the current in each element of the circuit, as well as the total impedance.

FIGURE 31–14
Problem 53.

54. Show that if the condition $R_1 R_2 = L/C$ is satisfied in the circuit shown in Fig. 31–15, then the potential difference between points a and b is zero for all frequencies.

FIGURE 31–15
Problem 54.

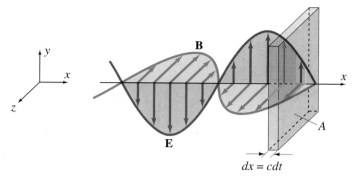

FIGURE 32–14 Electromagnetic wave carrying energy through area A.

$dx = c dt$

Now let us determine the energy the wave transports per unit time per unit area. This is given by a vector $\mathbf{S}$, which is called the **Poynting vector**.[†] The units of $\mathbf{S}$ are W/m^2. The direction of $\mathbf{S}$ is the direction in which the energy is transported, which is the direction in which the wave is moving. Let us imagine the wave is passing through an area A perpendicular to the x axis as shown in Fig. 32–14. In a short time dt, the wave moves to the right a distance $dx = c dt$ where c is the wave speed. The energy that passes through A in the time dt is the energy that occupies the volume $dV = A dx = Ac dt$. The energy density u is $u = \epsilon_0 E^2$ where E is the electric field in this volume at the given instant. So the total energy dU contained in this volume dV is the energy density u times the volume: $dU = u dV = (\epsilon_0 E^2)(Ac dt)$. Therefore the energy crossing the area A per time dt is

$$S = \frac{1}{A}\frac{dU}{dt} = \epsilon_0 c E^2. \tag{32–17}$$

Since $E = cB$ and $c = 1/\sqrt{\epsilon_0 \mu_0}$, this can also be written:

$$S = \epsilon_0 c E^2 = \frac{cB^2}{\mu_0} = \frac{EB}{\mu_0}.$$

The direction of $\mathbf{S}$ is along $\mathbf{v}$, perpendicular to $\mathbf{E}$ and $\mathbf{B}$, so the Poynting vector $\mathbf{S}$ can be written

$$\mathbf{S} = \frac{1}{\mu_0}(\mathbf{E} \times \mathbf{B}). \tag{32–18}$$

Poynting vector

Equation 32–17 or 32–18 gives the energy transported per unit area per unit time at any *instant*. We often want to know the *average* over an extended period of time since the frequencies are usually so high we don't detect the rapid time variation. If E and B are sinusoidal, then $\overline{E^2} = E_0^2/2$, just as for electric currents and voltages (Section 25–7), where E_0 is the *maximum* value of E. Thus we can write for the magnitude of the Poynting vector, on the average,

$$\overline{S} = \frac{1}{2}\epsilon_0 c E_0^2 = \frac{1}{2}\frac{c}{\mu_0}B_0^2 = \frac{E_0 B_0}{2\mu_0}, \tag{32–19a}$$

where B_0 is the maximum value of B. This time averaged value of $\mathbf{S}$ is the **intensity**, defined as the average power transferred across unit area (Section 15–3). We can also write

Intensity

$$\overline{S} = \frac{E_{rms} B_{rms}}{\mu_0} \tag{32–19b}$$

where E_{rms} and B_{rms} are the rms values $\left(E_{rms} = \sqrt{\overline{E^2}}, B_{rms} = \sqrt{\overline{B_{rms}^2}}\right)$.

[†] After J. H. Poynting (1852–1914).

EXAMPLE 32–4 **E and B from the Sun.** Radiation from the Sun reaches the Earth (above the atmosphere) at a rate of about 1350 W/m². Assume that this is a single EM wave and calculate the maximum values of E and B.

SOLUTION Since $\overline{S} = 1350 \, \text{W/m}^2 = \epsilon_0 c E_0^2 / 2$, then

$$E_0 = \sqrt{\frac{2\overline{S}}{\epsilon_0 c}}$$

$$= \sqrt{\frac{2(1350 \, \text{W/m}^2)}{(8.85 \times 10^{-12} \, \text{C}^2/\text{N·m}^2)(3.0 \times 10^8 \, \text{m/s})}}$$

$$= 1.01 \times 10^3 \, \text{V/m}.$$

From Eq. 32–11, $B = E/c$, so

$$B_0 = \frac{E_0}{c} = \frac{1.01 \times 10^3 \, \text{V/m}}{3.00 \times 10^8 \, \text{m/s}} = 3.37 \times 10^{-6} \, \text{T}.$$

This Example illustrates that B has a small numerical value compared to E. This is because of the different units for E and B and the way these units are defined. But, as we saw earlier, B contributes the same energy to the wave as E does.

*32–8 Radiation Pressure

If electromagnetic waves carry energy, as we have just seen, then we might expect them to also carry linear momentum. When an electromagnetic wave encounters the surface of a material object and is absorbed or reflected, there will be a force exerted on the surface as a result of the momentum transfer $(F = dp/dt)$ just as when a moving object strikes a surface. The force per unit area exerted by the waves is called **radiation pressure**, and its existence was predicted by Maxwell. He showed that if a beam of EM radiation (light, for example) is completely absorbed by an object, then the momentum transferred is,

$$\Delta p = \frac{\Delta U}{c}$$

where ΔU is the energy absorbed by the object in a time Δt and c is the speed of light. If instead, the radiation is fully reflected (suppose the object is a mirror) then the momentum transferred is twice as great, just as when a ball bounces elastically off a surface (Chapter 7):

$$\Delta p = \frac{2\Delta U}{c}.$$

If a surface absorbs some of the energy, and reflects some of it, then $\Delta p = k \, \Delta U/c$, where k is a constant between 1 and 2.

Using Newton's second law we can calculate the force and the pressure exerted by radiation on the object. The force F is given by

$$F = \frac{dp}{dt}.$$

The average rate that energy is delivered to the object is related to the Poynting vector by

$$\frac{dU}{dt} = \overline{S}A,$$

where A is the cross-sectional area of the object which intercepts the radiation.

The radiation pressure P (assuming full absorption) is given by

$$P = \frac{F}{A} = \frac{1}{A}\frac{dp}{dt} = \frac{1}{Ac}\frac{dU}{dt} = \frac{\overline{S}}{c}.$$

Radiation pressure (absorbed)

If the light is fully reflected, the pressure is twice as great:

$$P = \frac{2\overline{S}}{c}.$$

Radiation pressure (reflected)

EXAMPLE 32–5 **ESTIMATE** **Solar pressure.** Radiation from the sun that reaches the Earth's surface transports energy at a rate of about 1000 W/m^2. Estimate the pressure and force exerted by the Sun on your outstretched hand.

SOLUTION The radiation is partially reflected and partially absorbed, so let us estimate simply $P = \overline{S}/c$, which gives

$$P \approx \frac{\overline{S}}{c} = \frac{1000 \text{ W/m}^2}{3 \times 10^8 \text{ m/s}} \approx 3 \times 10^{-6} \text{ N/m}^2.$$

An estimate of the area of your outstretched hand might be about 10 cm by 20 cm, so $A = 0.02 \text{ m}^2$. Then the force is

$$F = PA \approx (3 \times 10^{-6} \text{ N/m}^2)(0.02 \text{ m}^2) \approx 6 \times 10^{-8} \text{ N}.$$

These numbers are tiny. The force of gravity on your hand, for comparison, is maybe a half pound, or with $m = 0.2 \text{ kg}$, $mg \approx (0.2 \text{ kg})(9.8 \text{ m/s}^2) \approx 2 \text{ N}$. We see that the radiation pressure on your hand is imperceptible.

Although you can not directly feel the effects of radiation pressure, the phenomenon is quite dramatic when applied to atoms irradiated by a laser beam. An atom has a mass on the order of 10^{-27} kg, and a laser can deliver energy at a rate of 1000 W/m^2. This is the same intensity used in Example 32–5, above, but here a radiation pressure of 10^{-6} N/m^2 would be very significant on a molecule whose mass might be 10^{-23} to 10^{-26} kg. It is possible, in fact, to move atoms and molecules around by steering them with a laser beam, in a device called "optical tweezers." Optical tweezers have some remarkable applications. They are of great interest to biologists, especially since optical tweezers can manipulate live organisms without damaging them. Optical tweezers have been used to measure the elastic properties of DNA by pulling each end of the molecule with a tweezers.

➡ **PHYSICS APPLIED**

Optical tweezers

---

*32–9 Radio and Television

Electromagnetic waves offer the possibility of transmitting information over long distances. Among the first to realize this and put it into practice was Guglielmo Marconi (1874–1937) who, in the 1890s, invented and developed the wireless telegraph. With it, messages could be sent hundreds of kilometers at the speed of light without the use of wires. The first signals were merely long and short pulses that could be translated into words by a code, such as the "dots" and "dashes" of the Morse code. The next decade saw the development of vacuum tubes. Out of this early work radio and television were born. We now discuss briefly (1) how radio and TV signals are transmitted, and (2) how they are received at home.

➡ **PHYSICS APPLIED**

"Wireless" transmission

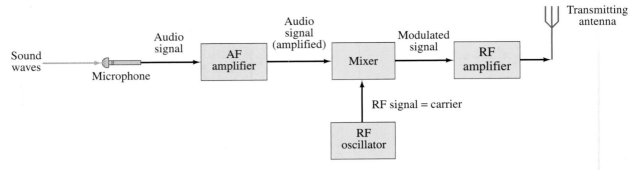

FIGURE 32–15 Block diagram of a radio transmitter.

Transmission of radio waves

The process by which a radio station transmits information (words and music) is outlined in Fig. 32–15. The audio (sound) information is changed into an electrical signal of the same frequencies by, say, a microphone or tape recorder head. This electrical signal is called an audiofrequency (AF) signal, since the frequencies are in the audio range (20 to 20,000 Hz). The signal is amplified electronically and is then mixed with a radio-frequency (RF) signal called its **carrier frequency**. AM radio stations have carrier frequencies from about 530 kHz to 1600 kHz. For example, "710 on your dial" means a station whose carrier frequency is 710 kHz. FM radio stations have much higher carrier frequencies, between 88 MHz and 108 MHz. The carrier frequencies for TV stations in the United States lie between 54 MHz and 88 MHz for channels 2 through 6, and between 174 MHz and 216 MHz for channels 7 through 13; UHF (ultra-high-frequency) stations have even higher carrier frequencies, between 470 MHz and 890 MHz.

Carrier frequency

PHYSICS APPLIED

AM and FM

The mixing of the audio and carrier frequencies is done in two ways. In **amplitude modulation** (AM), the amplitude of the higher frequency carrier wave is made to vary in proportion to the amplitude of the audio signal, as shown in Fig. 32–16. It is called "amplitude modulation" because the amplitude of the carrier is altered ("modulate" means to change or alter). In **frequency modulation** (FM), the *frequency* of the carrier wave is made to change in proportion to the audio signal's amplitude, as shown in Fig. 32–17.

The mixed signal is amplified further and is then sent to the antenna, where the complex mixture of frequencies is sent out in the form of EM waves.

A television transmitter works in a similar way, using frequency modulation, except that both audio and video signals are mixed with carrier frequencies.

FIGURE 32–16 In amplitude modulation (AM), the amplitude of the carrier signal is made to vary in proportion to the audio signal's amplitude.

FIGURE 32–17 In frequency modulation (FM), the frequency of the carrier signal is made to change in proportion to the audio signal's amplitude. This method is used by FM radio and television.

Receiving
antenna

RF tuner and amplifier → RF signal → Detector (demodulation) → Audio signal → AF amplifier → Loudspeaker

FIGURE 32–18 Block diagram of a simple radio receiver.

Now let us look at the other end of the process, the reception of radio and TV programs at home. A simple radio receiver is diagrammed in Fig. 32–18. The EM waves sent out by all stations are received by the antenna. One kind of antenna consists of one or more conducting rods; the electric field in the EM waves exerts a force on the electrons in the conductor, causing them to move back and forth at the frequencies of the waves (Fig. 32–19a). A second type of antenna consists of a tubular coil of wire, often found in AM radios, or the simple loop of a UHF television antenna. These antennas detect the magnetic field of the wave, for the changing B field induces an emf in the coil (Fig. 32–19b).

Radio and TV receivers

Antennas

FIGURE 32–19 Antennas. (a) Electric field of EM wave produces a current in an antenna consisting of straight wire or rods. (b) Changing magnetic field induces an emf and current in a loop antenna.

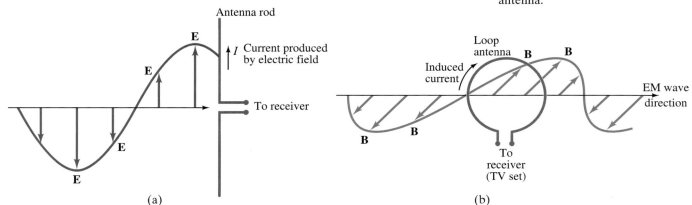

(a)   (b)

The signal the antenna detects and sends to the receiver is very small and contains frequencies from many different stations. The receiver selects out a particular RF frequency (actually a narrow range of frequencies) corresponding to a particular station using a resonant LC circuit (Sections 30–5 and 31–6) with a variable capacitor or inductor. A simple example is shown in Fig. 32–20. A particular station is "tuned-in" by adjusting L and C so that the resonant frequency of the circuit equals that of the station's carrier frequency. The signal, containing both audio and carrier frequencies, next goes to the *detector* (Fig. 32–18) where "demodulation" takes place—that is, the RF carrier frequency is separated from the audio signal. The audio signal is amplified and sent to a loudspeaker or headphones.

Modern receivers have more stages than those shown. Various means are used to increase the sensitivity and selectivity (ability to detect weak signals and distinguish them from other stations), and to minimize distortion of the original signal.[†]

A television receiver does similar things to both the audio and the video signals. The audio signal goes finally to the loudspeaker, and the video signal to the picture tube, a *cathode ray tube* (CRT) whose operation was discussed in Sections 23–9 and 27–7.

FIGURE 32–20 Simple tuning stage of a radio.

[†]For *FM stereo broadcasting*, two signals are carried by the carrier wave. One of these contains frequencies up to about 15 kHz, which includes most audio frequencies. The other signal includes the same range of frequencies, but 19 kHz is added to it. A stereo receiver subtracts this 19,000-Hz signal and distributes the two signals to the left and right channels. The first signal actually consists of the sum of left and right channels ($L + R$), so mono radios detect all the sound. The second signal is the difference between left and right ($L - R$). Hence the receiver must add and subtract the two signals to get pure left and right signal for each channel.

EXAMPLE 32–6 **Tuning a station.** An FM radio station transmits at 100 MHz. Calculate (a) the transmitting wavelength, and (b) the value of the capacitance in the tuning circuit if $L = 0.40\,\mu H$.

SOLUTION (a) The carrier frequency is $f = 100\,MHz = 1.0 \times 10^8\,s^{-1}$, so

$$\lambda = \frac{c}{f} = \frac{(3.0 \times 10^8\,m/s)}{(1.0 \times 10^8\,s^{-1})} = 3.0\,m.$$

The wavelengths of other FM signals (88 MHz to 108 MHz) are close to this. FM antennas are typically 1.5 m long, or about a half wavelength. This length is chosen so that the antenna reacts in a resonant fashion and thus is more sensitive to FM frequencies.

(b) According to Eq. 30–14 or 31–13, the resonant frequency is $f_0 = 1/(2\pi\sqrt{LC})$. Therefore,

$$C = \frac{1}{4\pi^2 f_0^2 L} = \frac{1}{4(3.14)^2(1.0 \times 10^8\,s^{-1})^2(4.0 \times 10^{-7}\,H)}$$

$$= 6.3\,pF.$$

Of course, the capacitor or inductor is variable, so other stations can be selected too.

The various regions of the radio-wave spectrum are assigned by governmental agencies for various purposes. Besides those mentioned above, there are "bands" assigned for use by ships, airplanes, police, military, amateurs, satellites and space, and radar. Cellular telephones, for example, occupy a band from 824 MHz to 894 MHz.

Summary

James Clerk Maxwell synthesized an elegant theory in which all electric and magnetic phenomena could be described using four equations, now called **Maxwell's equations**. They are based on earlier ideas, but Maxwell added one more—that a changing electric field produces a magnetic field. Maxwell's equations are

$$\oint \mathbf{E} \cdot d\mathbf{A} = \frac{Q}{\epsilon_0} \qquad \oint \mathbf{E} \cdot d\mathbf{l} = -\frac{d\Phi_B}{dt}$$

$$\oint \mathbf{B} \cdot d\mathbf{A} = 0 \qquad \oint \mathbf{B} \cdot d\mathbf{l} = \mu_0 I + \mu_0 \epsilon_0 \frac{d\Phi_E}{dt}.$$

The two on the left are Gauss's laws for electricity and for magnetism; the two on the right are Faraday's law and Ampère's law (as extended by Maxwell).

Maxwell's theory predicted that transverse **electromagnetic (EM) waves** would be produced by accelerating electric charges, and that these waves would propagate through space at the speed of light c, given by

$$c = \frac{1}{\sqrt{\epsilon_0 \mu_0}}.$$

The oscillating electric and magnetic fields in an EM wave are perpendicular to each other and to the direction of propagation.

After EM waves were experimentally detected in the late 1800s, the idea that light is an EM wave (although of much higher frequency than those detected directly) became generally accepted. The **electromagnetic spectrum** includes EM waves of a wide variety of wavelengths, from microwaves and radio waves to visible light to X-rays and gamma rays, all of which travel through space at a speed $c = 3.00 \times 10^8\,m/s$.

The energy carried by EM waves can be described by the **Poynting vector**

$$\mathbf{S} = \frac{1}{\mu_0} \mathbf{E} \times \mathbf{B}$$

which gives the rate energy is carried across unit area per unit time when the electric and magnetic fields in an EM wave in free space are **E** and **B**.

Questions

1. Suppose you are looking along the same direction as an electric field **E** that is increasing. Will the induced magnetic field be clockwise or counterclockwise? What if **E** points toward you and is decreasing?

2. What is the direction of the displacement current in Fig. 32–3? (Note: The capacitor is discharging.)

3. Why is it that the magnetic field of a displacement current in a capacitor is so much harder to detect than the magnetic field of a conduction current?

4. Are there any good reasons for calling the term $\mu_0 \epsilon_0 \, d\Phi_E/dt$ in Eq. 32–1 a "current"? Explain.

5. The electric field in an EM wave traveling north oscillates in an east–west plane. Describe the direction of the magnetic field vector in this wave.

6. Is sound an electromagnetic wave? If not, what kind of wave is it?

7. Can EM waves travel through a perfect vacuum? Can sound waves?

8. When you flip a light switch, does the overhead light go on immediately? Explain.

9. Are the wavelengths of radio and television signals longer or shorter than those detectable by the human eye?

10. What does the result of Example 32–2 tell you about the phase of a 60-Hz ac current that starts at a power plant as compared to its phase at a house 200 km away?

11. When you connect two loudspeakers to the output of a stereo amplifier, should you be sure the lead wires are equal in length so that there will not be a time lag between speakers? Explain.

12. In the electromagnetic spectrum, what type of EM wave would have a wavelength of 10^3 km? 1 km? 1 m? 1 cm? 1 mm? 1 μm?

* 13. A lost person may signal by flashing a flashlight on and off using Morse code. This is actually a modulated EM wave. Is it AM or FM? What is the frequency of the carrier, approximately?

* 14. Can two radio or TV stations broadcast on the same carrier frequency? Explain.

* 15. If a radio transmitter has a vertical antenna, should a receiver's antenna (rod type) be vertical or horizontal to obtain best reception?

* 16. The carrier frequencies of FM broadcasts are much higher than for AM broadcasts. On the basis of what you learned about diffraction in Chapter 15, explain why AM signals can be detected more readily than FM signals behind low hills or buildings.

17. Discuss how cordless telephones make use of EM waves. What about cellular telephones?

Problems

Section 32–1

1. (I) Determine the rate at which the electric field changes between the round plates of a capacitor, 6.0 cm in diameter, if the plates are spaced 1.3 mm apart and the voltage across them is changing at a rate of 120 V/s.

2. (I) Calculate the displacement current I_D between the square plates, 3.8 cm on a side, of a capacitor if the electric field is changing at a rate of 2.0×10^6 V/m·s.

3. (II) At a given instant, a 1.8-A current flows in the wires connected to a parallel-plate capacitor. What is the rate at which the electric field is changing between the plates if the square plates are 1.60 cm on a side?

4. (II) A 1200-nF capacitor with circular parallel plates 2.0 cm in diameter is accumulating charge at the rate of 35.0 mC/s at some instant in time. What will be the induced magnetic field strength 10.0 cm radially outward from the center of the plates? What will be the value of the field strength after the capacitor is fully charged?

5. (II) Show that the displacement current through a parallel-plate capacitor can be written $I_D = C \, dV/dt$, where V is the voltage across the capacitor at any instant.

6. (II) Suppose an air-gap capacitor has circular plates of radius $R = 2.5$ cm and separation $d = 2.0$ mm. A 96.0-Hz emf, $\mathscr{E} = \mathscr{E}_0 \cos \omega t$, is applied to the capacitor. The maximum displacement current is 35 μA. Determine (a) the maximum conduction current I, (b) the value of $\mathscr{E}_0$, (c) the maximum value of $d\Phi_E/dt$ between the plates. Neglect fringing.

Section 32–3

7. (II) If magnetic monopoles existed, which of Maxwell's equations would be altered, and what would be their new form? Let Q_m be the strength of a magnetic monopole, analogous to electric charge Q.

Section 32–5

8. (I) If the magnetic field in a traveling EM wave has a peak value of 17.5 nT, what is the peak value of the electric field strength?

9. (I) If the electric field in an EM wave has a peak value of 0.43×10^{-4} V/m, what is the peak value of the magnetic field strength?

10. (I) In an EM wave traveling west, the B field oscillates vertically and has a frequency of 80.0 kHz and an rms strength of 6.75×10^{-9} T. What is the frequency and rms strength of the electric field and what is its direction?

11. (II) The electric field of a plane EM wave is given by $E_x = E_0 \cos(kz + \omega t)$, $E_y = E_z = 0$. Determine (a) the magnitude and direction of **B**, and (b) the direction of propagation.

Section 32–6

12. (I) What is the frequency of a microwave whose wavelength is 1.80 cm?

13. (I) (a) What is the wavelength of a 27.75×10^9 Hz radar signal? (b) What is the frequency of an X-ray with wavelength 0.10 nm?

14. (I) An EM wave has a wavelength of 850 nm. What is its frequency, and how would we classify it?

15. (I) An EM wave has frequency 9.56×10^{14} Hz. What is its wavelength, and how would we classify it?

16. (I) A light-year is a measure of distance (not time). How many meters does light travel in a year?

17. (II) How long would it take a message sent as radio waves from Earth to reach (a) Mars, (b) a spacecraft near Saturn?

18. (II) Pulsed lasers used for science and medicine produce very short bursts of electromagnetic energy. If the laser light wavelength is 1062 nm (this corresponds to a Neodymium-YAG laser), and the pulse lasts for 30 picoseconds, how many wavelengths are found within the laser pulse? How short would the pulse need to be to fit only one wavelength?

Section 32–7

19. (I) The **E** field in an EM wave has a peak of 36.5 mV/m. What is the average rate at which this wave carries energy across unit area per unit time?

20. (II) The magnetic field in a traveling EM wave has an rms strength of 32.5 nT. How long does it take to deliver 335 J of energy to 1.00 cm^2 of a wall that it hits perpendicularly?

21. (II) How much energy is transported across a 1.00 cm^2 area per hour by an EM wave whose E field has an rms strength of 28.6 mV/m?

22. (II) A spherically spreading EM wave comes from a 1000-W source. At a distance of 10.0 m, what is the intensity, and what is the rms value of the electric field?

23. (II) What is the energy contained in a 1.00-m^3 volume near the Earth's surface due to radiant energy from the Sun? See Example 32–4.

24. (II) A 12.8-mW laser puts out a narrow beam 2.00 mm in diameter. What are the average (rms) values of E and B in the beam?

25. (II) Estimate the average power output of the Sun, given that about 1350 W/m^2 reaches the upper atmosphere of the Earth.

26. (II) A high-energy pulsed laser emits a 1.0-ns-long pulse of average power 2.5×10^{11} W. The beam is 2.2×10^{-3} m in radius. Determine (a) the energy delivered in each pulse, and (b) the rms value of the electric field.

27. (III) (a) Show that the Poynting vector **S** points radially inward toward the center of a circular parallel-plate capacitor when it is being charged as in Example 32–1. (b) Integrate **S** over the cylindrical boundary of the capacitor gap to show that the rate at which energy enters the capacitor is equal to the rate at which electrostatic energy is being stored in the electric field of the capacitor (Section 24–4). Ignore fringing of **E**.

* Section 32–8

* 28. (II) Estimate the radiation pressure due to a 100 W bulb at a distance of 8.0 cm from the center of the bulb. Estimate the force exerted on your fingertip if you place it at this point.

* 29. (III) When the Sun became hot and luminous long ago, it is believed that it ejected dust particles and individual atoms out of the solar system using radiation pressure. Calculate how small the dust particles had to be in order to be ejected by comparing the radiation force with the gravitational force on the particles. Assume the particles are spherical, have a density of $2.0 \times 10^3 \text{ kg/m}^3$, and totally absorb the Sun's radiation. The Sun has an average power output of 3.8×10^{26} W.

* 30. (III) We discussed in the text two different equations for the momentum transfer due to radiation, one valid for total absorption, the other valid for total reflection. The latter equation is identical to the former with the exception of a factor of two. Give a simple mechanical argument for this factor of two by treating light as massless particles (photons) each with momentum p. In the case of absorption, light particles collide completely inelastically with a massive object, whereas in the case of reflection they collide elastically. Analyze these collisions and explain the factor of two.

* Section 32–9

* 31. (I) The variable capacitor in the tuner of an AM radio has a capacitance of 2400 pF when the radio is tuned to a station at 550 kHz. What must the capacitance be for a station at the other end of the dial, 1550 kHz?

* 32. (I) The oscillator of a 96.1-MHz FM station has an inductance of 1.8 μH. What value must the capacitance be?

* 33. (II) A certain FM radio tuning circuit has a fixed capacitor $C = 840$ pF. Tuning is done by a variable inductance. What range of values must the inductance have to tune stations from 88 MHz to 108 MHz?

* 34. (II) An amateur radio operator wishes to build a receiver that can tune a range from 14.0 MHz to 15.0 MHz. A variable capacitor has a minimum capacitance of 92 pF. (a) What is the required value of the inductance? (b) What is the maximum capacitance used on the variable capacitor?

General Problems

35. (a) What is the wavelength of an AM station at 680 kHz? (b) An FM station broadcasts at 100.7 MHz. What is the wavelength of this wave?

36. Compare 940 on the AM dial to 94 on the FM dial. Which has the longer wavelength, and by what factor is it larger?

37. Television broadcast frequencies range between 54.0 MHz for Channel 2 to 806 MHz for Channel 69 (and beyond). What are the wavelengths for these two channels?

38. If the Sun were to disappear or somehow radically change its output, how long would it take for us on Earth to learn about it?

39. (a) How long did it take for a message sent from Earth to reach the first astronauts on the Moon? (b) How long will it take for a message from Earth to reach the first astronauts who arrive on Mars; assume Mars is at its closest approach to Earth $(78 \times 10^6 \text{ km})$?

40. A radio voice signal from the Apollo crew on the Moon is beamed to a listening crowd from a radio speaker. If you are standing 50 m from the loudspeaker, what is the total time lag between when you hear the sound and when the sound left the Moon?

41. A point source emits light energy uniformly in all directions at an average rate P_0 with a single frequency f. Show that the peak electric field in the wave is given by

$$E_0 = \sqrt{\frac{\mu_0 c P_0}{2 \pi r^2}}.$$

42. What are E_0 and B_0 2.00 m from a 100-W light source? Assume the bulb emits radiation of a single frequency uniformly in all directions.

43. Estimate the rms electric field in the sunlight that hits Mars, knowing that the Earth receives about 1350 W/m² and that Mars is 1.52 times farther away from the Sun (on average) than is the Earth.

44. At a given instant in time, a traveling EM wave is noted to have its maximum magnetic field pointing west and its maximum electric field pointing south. In which direction is the wave traveling? If the rate of energy flow is 500 W/m², what are the maximum values for the two fields?

45. Who will hear the voice of a singer first: a person in the balcony 50 m away from the stage (see Fig. 32–21), or a person 3000 km away at home whose ear is next to the radio? Roughly how much sooner? Assume the microphone is a few centimeters from the singer and the temperature is 20°C.

FIGURE 32–21   Problem 45.

46. A 1.80-m-long FM receiving antenna is oriented parallel to the electric field of an EM wave. How large must the **E** field strength be to produce a 1.0-mV rms voltage between the ends of the antenna? What is the rate of energy transport per unit area?

47. Suppose a 50-kW radio station emits EM waves uniformly in all directions. (a) How much energy per second crosses a 1.0-m² area 100 m from the transmitting antenna? (b) What is the rms magnitude of the **E** field at this point, assuming the station is operating at full power? (c) What is the voltage induced in a 1.0-m-long vertical car antenna at this distance?

48. Repeat Problem 47 for a distance of 100 km from the station.

49. Referring to Problem 47, what is the maximum power level of the radio station so as to avoid electrical breakdown of air at a distance of 1.0 m from the antenna? Assume the antenna is a point source. Air breaks down in an electric field of about 3×10^6 V/m. [Hint: See Problem 41.]

50. How large an emf (rms) will be generated in an antenna that consists of a 380-loop circular coil of wire 2.2 cm in diameter if the EM wave has a frequency of 810 kHz and is transporting energy at an average rate of 1.0×10^{-4} W/m² at the antenna? [Hint: You can use Eq. 29–4 for a generator, since it could be applied to an observer moving with the coil so that the magnetic field is oscillating with the frequency $f = \omega/2\pi$.]

* 51. The variable capacitance of a radio tuner consists of six plates connected together placed alternately between six other plates, also connected together (Fig. 32–22). Each plate is separated from its neighbor by 1.1 mm of air. One set of plates can move so that the area of overlap varies from 1.0 cm² to 9.0 cm². (a) Are these capacitors connected in series or in parallel? (b) Determine the range of capacitance values. (c) What value of inductor is needed if the radio is to tune AM stations from 550 kHz to 1600 kHz?

FIGURE 32–22   Problem 51.

52. A cylindrical conductor of radius r and conductivity σ carries a steady current I distributed uniformly across its cross-section. (a) Determine **E** inside the conductor. (b) Determine **B** just outside the conductor. (c) Determine the Poynting vector **S** at the surface of the conductor and show that it is normal to the surface and points inward. (d) Integrate over **S** to show that the rate at which electromagnetic energy enters the sides of the conductor is equal to the rate at which energy is dissipated, I^2R. Thus we can think of the energy as entering the conductor in the form of electromagnetic fields from the sides rather than through the ends.

Reflection from still water, as from a glass mirror, can be analyzed using the ray model of light.

Is this picture right side up? How can you tell? What are the clues? See Example 33–2.

In this first chapter on light and optics, we will use the ray model of light to understand the formation of images by mirrors, both plane and curved (spherical), and we will begin examining the refraction of light in transparent media, which will lead us to a study of lenses in the next chapter.

Light: Reflection and Refraction

The sense of sight is extremely important to us, for it provides us with a large part of our information about the world. How do we see? What is the something called *light* that enters our eyes and causes the sensation of sight? How does light behave so that we can see everything that we do? We saw in Chapter 32 that light can be considered a form of electromagnetic radiation. We now examine the subject of light in detail in the next several chapters.

We see an object in one of two ways: (1) the object may be a source of light, such as a lightbulb, a flame, or a star, in which case we see the light emitted directly from the source; or (2), more commonly, we see an object by light reflected from it. In the latter case, the light may have originated from the sun, artificial lights, or a campfire. An understanding of how bodies *emit* light was not achieved until the 1920s, and this will be discussed in Chapter 38. How light is *reflected* from objects was understood much earlier, and will be discussed in Section 33–3.

33–1 The Ray Model of Light

A great deal of evidence suggests that *light travels in straight lines* under a wide variety of circumstances. For example, a point source of light like the sun casts distinct shadows, and the beam of a flashlight appears to be a straight line. In fact, we infer the positions of objects in our environment by assuming that light moves from the object to our eyes in straight-line paths. Our orientation to the physical world is based on this assumption.

This reasonable assumption has led to the **ray model** of light. This model assumes that light travels in straight-line paths called light **rays**. Actually, a ray is an idealization; it is meant to represent an extremely narrow beam of light. When we see an object, according to the ray model, light reaches our eyes from each point on the object. Although light rays leave each point in many different directions, normally only a small bundle of these rays can enter an observer's eye, as shown in Fig. 33–1. If the person's head moves to one side, a different bundle of rays will enter the eye from each point.

We saw in Chapter 32 that light can be considered as an electromagnetic wave. Although the ray model of light does not deal with this aspect of light (we discuss the wave nature of light in Chapters 35 and 36), the ray model has been very successful in describing many aspects of light such as reflection, refraction, and the formation of images by mirrors and lenses. Because these explanations involve straight-line rays at various angles, this subject is referred to as **geometric optics**. In ignoring the wave properties of light we must be careful that when the light rays pass by objects or through apertures, these must be large compared to the wavelength of the light (so the wave phenomena of interference and diffraction, as discussed in Chapter 15, can be ignored), and we ignore what happens to the light at the edges of objects until we get to Chapters 35 and 36.

Light rays

FIGURE 33–1 Light rays come from each single point on an object. A small bundle of rays leaving one point is shown entering a person's eye.

33–2 The Speed of Light and Index of Refraction

Galileo attempted to measure the speed of light by trying to measure the time required for light to travel a known distance between two hilltops. He stationed an assistant on one hilltop and himself on another and ordered the assistant to lift the cover from a lamp the instant he saw a flash from Galileo's lamp. Galileo measured the time between the flash of his lamp and when he received the light from his assistant's lamp. The time was so short that Galileo concluded it merely represented human reaction time, and that the speed of light must be extremely high.

The first successful determination that the speed of light is finite was made by the Danish astronomer Ole Roemer (1644–1710). Roemer had noted that the carefully measured period of Io, one of Jupiter's moons with an average period of 42.5 h, varied slightly, depending on the relative motion of Earth and Jupiter. When Earth was moving away from Jupiter, the period of the moon was slightly longer, and when Earth was moving toward Jupiter, the period was slightly shorter. He attributed this variation to the extra time needed for light to travel the increasing distance to Earth when Earth is receding, or to the shorter travel time for the decreasing distance when the two planets are approaching one another. Roemer concluded that the speed of light—though great—is finite.

Since then a number of techniques have been used to measure the speed of light. Among the most important were those carried out by the American Albert

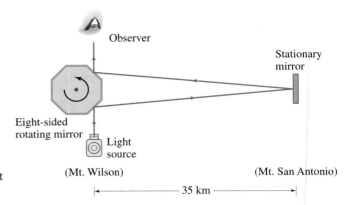

FIGURE 33–2 Michelson's speed-of-light apparatus (not to scale).

Michelson (1852–1931). Michelson used the rotating mirror apparatus diagrammed in Fig. 33–2 for a series of high-precision experiments carried out from 1880 to the 1920s. Light from a source was directed at one face of a rotating eight-sided mirror. The reflected light traveled to a stationary mirror a large distance away and back again as shown. If the rotating mirror was turning at just the right rate the returning beam of light would reflect from one face of the mirror into a small telescope through which the observer looked. If the speed of rotation was only slightly different, the beam would be deflected to one side and would not be seen by the observer. From the required speed of the rotating mirror and the known distance to the stationary mirror, the speed of light could be calculated. In the 1920s, Michelson set up the rotating mirror on the top of Mt. Wilson in southern California and the stationary mirror on Mt. San Antonio, 35 km away. He later measured the speed of light in vacuum using a long evacuated tube.

The accepted value today for the speed of light, c, in vacuum is[‡]

$$c = 2.99792458 \times 10^8 \, \text{m/s}.$$

We usually round this off to

$$3.00 \times 10^8 \, \text{m/s}$$

when extremely precise results are not required. In air, the speed is only slightly less. In other transparent materials such as glass and water, the speed is always less than that in vacuum. For example, in water it travels at about $\frac{3}{4}c$. The ratio of the speed of light in vacuum to the speed v in a given material is called the **index of refraction** n of that material:

$$n = \frac{c}{v}. \tag{33-1}$$

Index of refraction

The index of refraction is never less than 1, and its value for various materials is given in Table 33–1. For example, since $n = 2.42$ for diamond, the speed of light in diamond is

$$v = \frac{c}{n} = \frac{(3.00 \times 10^8 \, \text{m/s})}{2.42} = 1.24 \times 10^8 \, \text{m/s}.$$

As we shall see later, n varies somewhat with the wavelength of the light—except in vacuum—so a particular wavelength is specified; in Table 33–1 it is that of yellow light with wavelength $\lambda = 589 \, \text{nm}$.

TABLE 33–1
Indices of refraction[†]

Material	$n = \dfrac{c}{v}$
Vacuum	1.0000
Air (at STP)	1.0003
Water	1.33
Ethyl alcohol	1.36
Glass	
Fused quartz	1.46
Crown glass	1.52
Light flint	1.58
Lucite or Plexiglas	1.51
Sodium chloride	1.53
Diamond	2.42

[†] $\lambda = 589 \, \text{nm}$

33–3 Reflection; Image Formation by a Plane Mirror

When light strikes the surface of an object, some of the light is reflected. The rest is either absorbed by the object (and transformed to thermal energy) or, if the object is transparent like glass or water, part of it is transmitted through. For a very shiny object such as a silvered mirror, over 95 percent of the light may be reflected.

[‡]When the meter was redefined in 1983, the speed of light was given this fixed value, and the meter then defined in terms of it (see Section 1–4).

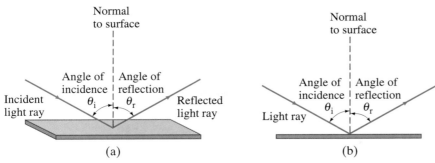

FIGURE 33–3 Law of reflection: (a) Shows an incident ray being reflected at the top of a flat surface; (b) shows a side or "end-on" view, which we will usually use because of its clarity.

When a narrow beam of light strikes a flat surface (Fig. 33–3) we define the **angle of incidence**, θ_i, to be the angle an incident ray makes with the normal to the surface, and the **angle of reflection**, θ_r, to be the angle the reflected ray makes with the normal. It is found that the *incident and reflected rays lie in the same plane with the normal to the surface*, and that

the angle of incidence equals the angle of reflection.

This is the **law of reflection** and is indicated in Fig. 33–3. It was known to the ancient Greeks, and you can confirm it yourself by shining a narrow flashlight beam at a mirror in a darkened room.

When light is incident upon a rough surface, even microscopically rough such as this page, it is reflected in many directions, Fig. 33–4. This is called **diffuse reflection**. The law of reflection still holds, however, at each small section of the surface. Because of diffuse reflection in all directions, an ordinary object can be seen from many different angles. When you move your head to the side, different reflected rays reach your eye from each point on the object (such as this page), Fig. 33–5a. Let us compare diffuse reflection to reflection from a mirror, which is known as *specular* reflection. ("Speculum" is Latin for mirror.) When a narrow beam of light shines on a mirror, the light will not reach your eye unless your eye is positioned at just the right place where the law of reflection is satisfied, as shown in Fig. 33–5b. This is what gives rise to the special image-forming properties of mirrors.

When you look straight in a mirror, you see what appears to be yourself as well as various objects around and behind you, Fig. 33–6. Your face and the other objects look as if they are in front of you, beyond the mirror; but, of course, they are not. What you see in the mirror is an **image** of the objects.

Angle of incidence and Angle of reflection (angle with the normal)

Law of reflection

FIGURE 33–4 Diffuse reflection from a rough surface.

Mirrors (specular reflection)

FIGURE 33–5 A beam of light from a flashlight shines on (a) white paper, and (b) a mirror. In part (a), you can see the white light reflected at various points because of diffuse reflection. But in part (b), you see the reflected light only when your eye is placed correctly ($\theta_r = \theta_i$); this is known as specular reflection. (Galileo, using similar arguments, showed that the Moon must have a rough surface rather than a highly polished surface like a mirror, as some people thought.)

FIGURE 33–6 When you look in a mirror, you see an image of yourself and objects around you. Note that you don't see yourself as others see you, because left and right appear reversed in the image.

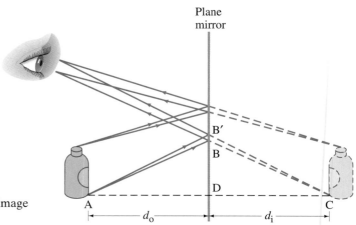

FIGURE 33–7 Formation of a virtual image by a plane mirror.

Ray model of how images are formed

Each point of image is where rays intersect

Image distance = object distance

Real and virtual images

Figure 33–7 shows how an image is formed by a plane mirror (that is, flat), according to the ray model. We are viewing the mirror, on edge, in the diagram of Fig. 33–7, and the rays are shown reflecting from the front surface. (Good mirrors are generally made by putting a highly reflective metallic coating on one surface of a very flat piece of glass.) Rays from two different points on an object are shown in Fig. 33–7: rays leaving from a point on the top of the bottle, and from a point on the bottom. Rays leave each point on the object going in many directions, but only those that enclose the bundle of rays that reach the eye from the two points are shown. The diverging rays that enter the eye *appear* to come from behind the mirror as shown by the dashed lines. That is, our eyes and brain interpret any rays that enter an eye as having traveled a straight-line path. The point from which each bundle of rays seems to come is one point on the image. For each point on the object, there is a corresponding image point. Let us concentrate on the two rays that leave point A on the object and strike the mirror at points B and B′. The angles ADB and CDB are right angles; and angles ABD and CBD are equal because of the law of reflection. Therefore, the two triangles ABD and CBD are congruent, and the length AD = CD. Thus the image appears as far behind the mirror as the object is in front: the **image distance**, d_i (distance from mirror to image, Fig. 33–7), equals the **object distance**, d_o. From the geometry, we also see that the height of the image is the same as that of the object.

The light rays do not actually pass through the image location itself. It merely *seems* as though the light is coming from the image because our brains interpret any light entering our eyes as having come in a *straight-line path from in front of us*. Because the rays do not actually pass through the image, the image would not appear on paper or film placed at the location of the image. Therefore, it is called a **virtual image**. This is to distinguish it from a **real image** in which the light does pass through the image and which therefore could appear on paper or film placed at the image position. We will see that curved mirrors and lenses can form real images. A movie projector lens, for example, produces a real image that is visible on the screen.

CONCEPTUAL EXAMPLE 33–1 | **How tall must a full-length mirror be?**
A woman 1.60 m tall stands in front of a vertical plane mirror. What is the minimum height of the mirror, and how high must its lower edge be above the floor, if she is to see her whole body? Assume her eyes are 10 cm below the top of her head.

RESPONSE The situation is diagrammed in Fig. 33–8. First consider the ray from her foot, AB, which upon reflection becomes BE and enters the eye E. The light from point A (her foot) enters the eye after reflecting at B; so the mirror needs to extend no lower than B. Because the angle of reflection equals the angle of incidence, the height BD is half of the height AE. Because AE = 1.60 m − 0.10 m = 1.50 m, then BD = 0.75 m. Similarly, if the woman is to see the top of her head, the top edge of the mirror only needs to reach point F,

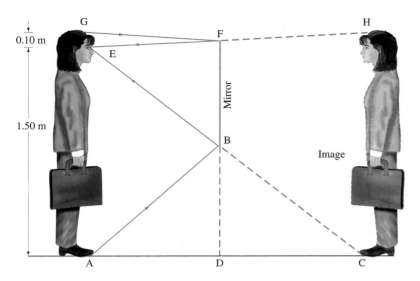

FIGURE 33–8 Seeing oneself in a mirror. Example 33–1.

which is 5 cm below the top of her head (half of GE = 10 cm). Thus, DF = 1.55 m, and the mirror need have a vertical height of only (1.55 m − 0.75 m) = 0.80 m. And the mirror's bottom edge must be 0.75 m above the floor. In general, a mirror need be only half as tall as a person for that person to see all of himself or herself. Does this result depend on the person's distance from the mirror?

CONCEPTUAL EXAMPLE 33–2 | **Is the photo upside down?** Close examination of the photograph on the first page of this chapter reveals that in the top portion, the image of the Sun is seen clearly, whereas in the lower portion, the image of the Sun is partially blocked by the tree branches. Why isn't the reflection in the water an exact replica of the real scene? Illustrate your answer by drawing a sketch of this situation, showing the Sun, the camera, the branch, and two rays going from the Sun to the camera (one direct and one reflected). Does your sketch tell you if the photograph is right side up?

RESPONSE We need to draw two diagrams, one assuming the photo on p. 810 is right side up, and another assuming it is upside down. Figure 33–9 is drawn assuming the photo is upside down. In this case, the Sun blocked by the tree would be the direct view, and the full view of the Sun the reflection. Thus, as diagrammed in Fig. 33–9, the ray which reflects off the water and into the camera travels at an angle below the branch, whereas the ray that travels directly to the camera passes through the branches. This works. Try to draw a diagram assuming the photo is right side up (thus assuming that the image of the Sun in the reflection is higher above the horizon than it is as viewed directly). It won't work. The photo on p. 810 is upside down.

Also, what about the people in the photo? Try to draw a diagram showing why they don't appear in the reflection. [*Hint*: Assume they are not sitting on the edge of poolside, but back from the edge a bit.] Then try to draw a diagram of the reverse (i.e., assume the photo is right side up so the people are visible only in the reflection). In general, reflected images are not perfect replicas when different planes (distances) are involved.

FIGURE 33–9 Example 33–2.

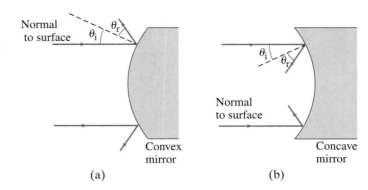

FIGURE 33–10 Mirrors with convex and concave spherical surfaces. Note that $\theta_r = \theta_i$ for each ray.

(a)   (b)

FIGURE 33–11 (a) A concave cosmetic mirror gives a magnified image. (b) A convex mirror in a store demagnifies and so includes a wide field of view.

(a)

(b)

33–4 Formation of Images by Spherical Mirrors

Reflecting surfaces do not have to be flat. The most common *curved* mirrors are *spherical*, which means they form a section of a sphere. A spherical mirror is called **convex** if the reflection takes place on the outer surface of the spherical shape so that the center of the mirror surface bulges out toward the viewer (Fig. 33–10a). A mirror is called **concave** if the reflecting surface is on the inner surface of the sphere so that the center of the mirror sinks away from the viewer (like a "cave"), Fig. 33–10b. Concave mirrors are used as shaving or cosmetic mirrors (Fig. 33–11a), and convex mirrors are sometimes used on cars and trucks (rearview mirrors) and in shops (to watch for thieves), because they take in a wide field of view (Fig. 33–11b).

These rays are the only ones shown that will strike the mirror, and they are essentially parallel if the object is far away.

FIGURE 33–12 If the object's distance is large compared to the size of the mirror (or lens), the rays are nearly parallel. They are parallel for an object at infinity (∞).

To see how spherical mirrors form images, we first consider an object that is very far from a concave mirror. For a distant object, as shown in Fig. 33–12, the rays from each point on the object that reach the mirror will be nearly parallel. *For an object infinitely far away* (the Sun and stars approach this), *the rays would be precisely parallel.* Now consider such parallel rays falling on a concave mirror as in Fig. 33–13. The law of reflection holds for each of these rays at the point each strikes the mirror. As can be seen, they are not all brought to a single point. In

FIGURE 33–13 Parallel rays striking a concave spherical mirror do not focus at precisely a single point.

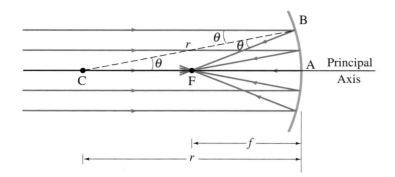

FIGURE 33–14 Rays parallel to the principal axis of a spherical mirror come to a focus at F, called the focal point, as long as the mirror is small in width as compared to its radius of curvature, r, so that the rays are "paraxial"—that is, make only small angles with the axis.

order to form a sharp image, the rays must come to a point. Thus a spherical mirror will not make as sharp an image as a plane mirror will. However, as we show below, if the mirror is small compared to its radius of curvature, so that the reflected rays make only a *small angle* upon reflection, then the rays will cross each other at very nearly a single point, or **focus**, as shown in Fig. 33–14. In the case shown, the rays are parallel to the **principal axis**, which is defined as the straight line perpendicular to the curved surface at its center (line CA in the diagram). The point F, where rays parallel to the principal axis come to a focus, is called the **focal point** of the mirror. The distance between F and the center of the mirror, length FA, is called the **focal length**, f, of the mirror. Another way of defining the focal point is to say that it is the *image point for an object infinitely far away* along the principal axis. The image of the Sun, for example, would be at F.

Small angle approximation

Focal point
Focal length

Now we will show, for a mirror whose reflecting surface is small compared to the radius of curvature, that the rays very nearly meet at a common point, F, and we will also calculate the focal length f. In this approximation, we consider only rays that make a small angle with the principal axis; such rays are called **paraxial rays**, and their angles are exaggerated in Fig. 33–14 only so the labeling will be clear. First we consider a ray that strikes the mirror at B in Fig. 33–14. The point C is the center of curvature of the mirror (the center of the sphere of which the mirror is a part). So the dashed line CB is equal to r, the radius of curvature, and CB is normal to the mirror's surface at B. The incoming ray that hits the mirror at B makes an angle θ with this normal, and hence the reflected ray, BF, also makes an angle θ with the normal. Note also from the geometry that angle BCF is also θ as shown. The triangle CBF is isosceles because two of its angles are equal. Thus length CF = BF. We assume the mirror has a width or diameter that is small compared to its radius of curvature, so the angles are small, and the length FB is nearly equal to length FA. In this approximation, FA = FC. But FA = f, the focal length, and CA = 2 FA = r. Thus the focal length is half the radius of curvature:

$$f = \frac{r}{2}. \qquad (33–2)$$

*Focal length
of mirror*

This argument assumed only that the angle θ was small, so the same result applies for all other paraxial rays. Thus all paraxial rays pass through the same point F in the paraxial approximation.

Since it is only approximately true that the rays come to a perfect focus at F, the more curved the mirror, the worse the approximation (Fig. 33–13) and the more blurred the image. This "defect" of spherical mirrors is called **sperical aberration**; we will discuss it more with regard to lenses in Chapter 34. A *parabolic* reflector, on the other hand, will reflect the rays to a perfect focus. However, because parabolic shapes are much harder to make and thus much more expensive, spherical mirrors are used for most purposes. (Many astronomical telescopes use parabolic reflectors.) We consider here only spherical mirrors and we will assume that they are small compared to their radius of curvature so that the image is sharp and Eq. 33–2 holds.

Parabolic mirror

We saw that for an object at infinity the image is located at the focal point of a concave spherical mirror, where $f = r/2$. But where does the image lie for an object not at infinity? First consider the object shown as an arrow in Fig. 33–15, which is placed at point O between F and C. Let us determine where the image will be for a given point O′ on the object. To do this we can draw several rays and make sure these reflect from the mirror such that the angle of reflection equals the angle of incidence. Determining the image position is simplified if we deal with three particularly simple rays. These are the rays labeled 1, 2, and 3 in Fig. 33–15 and we draw them leaving object point O′ as follows:

➡ RAY DIAGRAM

*Finding the image position
for a curved mirror*

Ray 1 is drawn parallel to the axis; therefore after reflection it must pass along a line through F (as we saw in Fig. 33–14, and drawn here in Fig. 33–15a).

Ray 2 leaves O′ and is made to pass through F; therefore it must reflect so it is parallel to the axis (Fig. 33–15b).

Ray 3 is chosen to be perpendicular to the mirror, and so is drawn so that it passes through C, the center of curvature; it is along a radius of the spherical surface, and because it is perpendicular to the mirror it will be reflected back on itself (Fig. 33–15c).

The point at which these rays cross is the image point I′. All other rays from the same object point will pass through this image point. To find the image point for any object point, only these three particular rays need to be used. Actually, only two of these rays are needed, but the third serves as a check.

We have shown the image point in Fig. 33–15 only for a single point on the object. Other points on the object are imaged nearby, so a complete image of the object is formed, as shown by the dashed arrow in Fig. 33–15c. Because the light actually passes through the image itself, this is a real image that will appear on a piece of paper or film placed there. This can be compared to the virtual image formed by a plane mirror (the light does not actually pass through that image, Fig. 33–7).

Real image

The image in Fig. 33–15 can be seen by the eye when the eye is placed to the left of the image so that some of the rays diverging from each point on the image (as point I′) can enter the eye as shown in Fig. 33–15c. (See also Figs. 33–1 and 33–7.)

FIGURE 33–15 Rays leave point O′ on the object (an arrow). Shown are the three most useful rays for determining where the image I′ is formed. [Note that our mirror's height is not small compared to f, so our diagram will not give the precise position of the image.]

(a) Ray 1 goes out from O′ parallel to the axis and reflects through F.

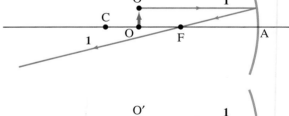

(b) Ray 2 goes through F and then reflects back parallel to the axis.

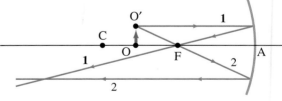

(c) Ray 3 heads out perpendicular to mirror and then reflects back on itself and goes through C (center of curvature)

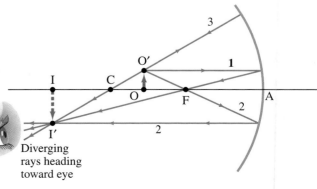

Diverging rays heading toward eye

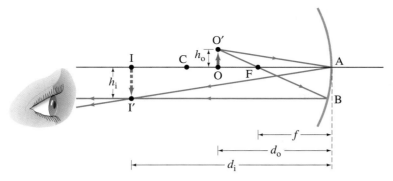

FIGURE 33-16 Diagram for deriving the mirror equation. For the derivation, we assume the mirror size is small compared to its radius of curvature.

Image points can be determined, roughly, by drawing the three rays as described above, but high accuracy is hard to obtain. For one thing it is difficult to draw, maintaining small angles for the "paraxial" rays as we assumed. However, it is possible to derive an equation that gives the image distance if the object distance and radius of curvature of the mirror are known. To do this, we refer to Fig. 33–16. The distance of the object from the center of the mirror, called the **object distance**, is labeled d_o. The **image distance** is labeled d_i. The height of the object OO′ is called h_o and the height of the image, I′I, is h_i. Two rays are shown, O′FBI′ (same as ray 2 in the previous figure) and O′AI′. The ray O′AI′ obeys the law of reflection, of course, so the two right triangles O′AO and I′AI are similar. Therefore, we have

$$\frac{h_o}{h_i} = \frac{d_o}{d_i}.$$

For the other ray, O′FBI′, the triangles O′FO and AFB are also similar since the length AB = h_i (in our approximation of a mirror that is small compared to its radius) and FA = f, the focal length of the mirror. Therefore,

$$\frac{h_o}{h_i} = \frac{OF}{FA} = \frac{d_o - f}{f}.$$

The left sides of the two preceding expressions are the same, so we can equate the right sides:

$$\frac{d_o}{d_i} = \frac{d_o - f}{f}.$$

We now divide both sides by d_o and rearrange to obtain

$$\frac{1}{d_o} + \frac{1}{d_i} = \frac{1}{f}. \tag{33–3}$$

Mirror equation

This is the equation we were seeking. It is called the **mirror equation** and relates the object and image distances to the focal length f (where $f = r/2$).

The **lateral magnification**, m, of a mirror is defined as the height of the image divided by the height of the object. From our first set of similar triangles above, we can write:

$$m = \frac{h_i}{h_o} = -\frac{d_i}{d_o}. \tag{33–4}$$

Lateral magnification for curved mirror

The minus sign in Eq. 33–4 is inserted as a convention. Indeed, we must be careful about the signs of all quantities in Eqs. 33–3 and 33–4. Sign conventions are chosen so as to give the correct locations and orientations of images, as predicted by ray diagrams. The sign conventions we use are: the image height h_i is positive if the image is upright, and negative if inverted, relative to the object (assuming h_o is taken as positive); d_i and d_o are positive if image and object are on the reflecting side of the mirror (as in Fig. 33–16), but if either image or object is behind the mirror, the corresponding distance is negative (an example can be seen in Fig. 33–17, Example 33–5). Thus the magnification (Eq. 33–4) is positive for an upright image and negative for an inverted image. We will summarize sign conventions more fully after discussing convex mirrors later in this Section.

Sign conventions

EXAMPLE 33–3 **Image in a concave mirror.** A 1.50-cm-high diamond ring is placed 20.0 cm from a concave mirror whose radius of curvature is 30.0 cm. Determine (a) the position of the image, and (b) its size.

SOLUTION The focal length $f = r/2 = 15.0$ cm. The ray diagram is basically like that shown in Fig. 33–15 and Fig. 33–16, since the object is between F and C. Referring to either Fig. 33–15 or Fig. 33–16, we have CA = r = 30.0 cm, FA = f = 15.0 cm, and OA = d_o = 20.0 cm. (a) From Eq. 33–3

$$\frac{1}{d_i} = \frac{1}{f} - \frac{1}{d_o} = \frac{1}{15.0 \text{ cm}} - \frac{1}{20.0 \text{ cm}} = 0.0167 \text{ cm}^{-1}.$$

➡ **PROBLEM SOLVING**

Remember to take the reciprocal

So $d_i = 1/0.0167 \text{ cm}^{-1} = 60.0$ cm. The image is 60.0 cm from the mirror on the same side as the object.
(b) From Eq. 33–4, the lateral magnification is

$$m = -\frac{60.0 \text{ cm}}{20.0 \text{ cm}} = -3.00.$$

Therefore the image height is

$$h_i = mh_o = (-3.00)(1.5 \text{ cm}) = -4.5 \text{ cm}.$$

The minus sign reminds us that the image is inverted, as in Figs. 33–15 and 33–16.

CONCEPTUAL EXAMPLE 33–4 **Reversible rays.** If the object in Example 33–3 is placed instead where the image is (see Fig. 33–16), where will the new image be?

RESPONSE The mirror equation is symmetric in d_o and d_i. Thus the new image will be where the old object was. Indeed, in Fig. 33–16 we need only reverse the direction of the rays to get our new situation.

EXAMPLE 33–5 **Object closer to concave mirror.** A 1.00-cm-high object is placed 10.0 cm from a concave mirror whose radius of curvature is 30.0 cm. (a) Draw a ray diagram to locate (approximately) the position of the image. (b) Determine the position of the image and the magnification analytically.

SOLUTION (a) Since $f = r/2 = 15.0$ cm, the object is between the mirror and the focal point. We draw the three rays as described earlier (Fig. 33–15) and this is shown in Fig. 33–17. Ray 1 leaves the tip of our object heading toward the mirror parallel to the axis, and reflects through F. Ray 2 cannot head toward F because it would not strike the mirror; so ray 2 must head as if it started at F (dashed line) and heads to the mirror and then is reflected parallel to the principal axis. Ray 3 is perpendicular to the mirror, as before. The rays reflected from the mirror diverge and so never meet at a point. They appear, however, to be coming from a point behind the mirror. This point is the image which is thus behind the mirror and *virtual*. (Why?)
(b) We use Eq. 33–3 to find d_i when d_o = 10.0 cm:

$$\frac{1}{d_i} = \frac{1}{15.0 \text{ cm}} - \frac{1}{10.0 \text{ cm}} = \frac{2 - 3}{30.0 \text{ cm}} = -\frac{1}{30.0 \text{ cm}}.$$

Therefore, $d_i = -30.0$ cm. The minus sign means the image is behind the mirror. The lateral magnification is $m = -d_i/d_o = -(-30.0 \text{ cm})/(10.0 \text{ cm}) = +3.00$. So the image is 3.00 times larger than the object; the plus sign indicates that the image is upright (which is consistent with the ray diagram, Fig. 33–17).
[Note that the image distance cannot be obtained accurately by measuring on Fig. 33–17, because our diagram violates the paraxial ray assumption (we had to, to make all rays clearly visible).]

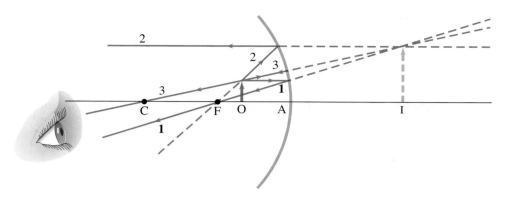

FIGURE 33–17 Object placed within the focal point F. The image is *behind* the mirror and is *virtual*, Example 33–5. [Note that the vertical scale (height of object = 1.0 cm) is different from the horizontal (OA = 10.0 cm) for ease of drawing, and this will affect the precision of the drawing.]

It is useful to compare Figs. 33–15 and 33–17. We can see that if the object is within the focal point, as in Fig. 33–17, the image is virtual, upright, and magnified. This is how a shaving or cosmetic mirror is used—you must place your head within the focal point if you are to see yourself right side up (Fig. 33–11a). If the object is *beyond* the focal point, as in Fig. 33–15, the image is real and inverted (upside down—and hard to use!). Whether the magnification is greater or less than 1.0 in the latter case depends on the position of the object relative to the center of curvature, point C.

The mirror equation also holds for a plane mirror: the focal length is $f = r/2 = \infty$, and Eq. 33–3 gives $d_i = -d_o$.

The analysis used for concave mirrors can be applied to **convex** mirrors. Even the mirror equation (Eq. 33–3) holds for a convex mirror, although the quantities involved must be carefully defined. Figure 33–18a shows parallel rays falling on a convex mirror. Again spherical aberration will be present, but we assume the mirror's size is small compared to its radius of curvature. The reflected rays diverge, but seem to come from point F behind the mirror. This is the **focal point**, and its distance from the center of the mirror is the **focal length**, f. It is easy to show that again $f = r/2$. We see that an object at infinity produces a virtual image in a convex mirror. Indeed, no matter where the object is placed on the reflecting side of a convex mirror, the image will be virtual and erect, as indicated in Fig. 33–18b. To find the image we draw rays 1 and 3 according to the rules used before on the concave mirror, as shown in Fig. 33–18b. Note that although rays 1 and 3 don't actually pass through points F and C, the line along which they are drawn do (shown dashed).

➥ P H Y S I C S   A P P L I E D

Magnifying mirror

Convex mirrors—analysis

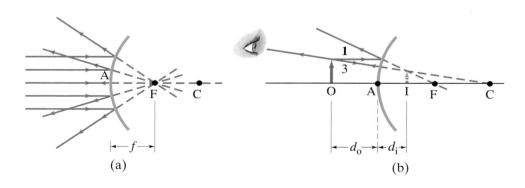

(a)

(b)

FIGURE 33–18 Convex mirror: (a) the focal point is at F, behind the mirror; (b) the image I of the object at O is virtual, upright, and smaller than the object. [Not to scale for Example 33–6.]

The mirror equation, Eq. 33–3, holds for convex mirrors but the focal length f must be considered negative, as must the radius of curvature. The proof is left as a Problem. It is also left as a Problem to show that Eq. 33–4 for the magnification is also valid.

PHYSICS APPLIED

Convex rearview mirror

EXAMPLE 33–6 **Convex rearview mirror.** A convex rearview car mirror has a radius of curvature of 16 m. Determine the location of the image and its magnification for an object 10.0 m from the mirror.

SOLUTION The ray diagram will be like Fig. 33–18, but the large object distance $(d_o = 10.0\,\text{m})$ makes a precise drawing difficult. We have a convex mirror, so r is negative by convention. Specifically, $r = -16\,\text{m}$, so $f = -8.0\,\text{m}$. The mirror equation gives

$$\frac{1}{d_i} = \frac{1}{f} - \frac{1}{d_o} = \frac{1}{-8.0\,\text{m}} - \frac{1}{10.0\,\text{m}} = -\frac{18}{80\,\text{m}}.$$

So $d_i = -80\,\text{m}/18 = -4.4\,\text{m}$, or 4.4 m behind the mirror. The lateral magnification is $m = -d_i/d_o = -(-4.4\,\text{m})/(10.0\,\text{m}) = 0.44$.

So the upright image is a little less than half the size of the object. Rearview mirrors sometimes come with warnings that objects look farther away (appear smaller) than they really are; judging distances in a spherical mirror requires skill.

PROBLEM SOLVING **Spherical Mirrors**

1. Always draw a ray diagram even though you are going to make an analytic calculation—the diagram serves as a check, even if not precise. Draw at least two, and preferably three, of the easy-to-draw rays as described in Fig. 33–15. Generally draw the rays starting from a point on the object to the left of the mirror and moving to the right.

2. Use Eqs. 33–3 and 33–4; it is crucially important to follow the sign conventions.

3. **Sign Conventions**

(a) When the object, image, or focal point is on the reflecting side of the mirror (on the left in all our drawings), the corresponding distance is considered positive. If any of these points is behind the mirror (on the right) the corresponding distance is negative.[†]

(b) The image height h_i is positive if the image is upright, and negative if inverted, relative to the object (that is, if h_o is taken as positive).

[†] Object distances are positive for material objects, but can be negative in systems with more than one mirror or lens—see Section 34–3.

33–5 | Refraction: Snell's Law

Angle of refraction

When light passes from one medium into another with a different index of refraction (Section 33–2), part of the incident light is reflected at the boundary. The remainder passes into the new medium. If a ray of light is incident at an angle to the surface (other than perpendicular), the ray is bent as it enters the new medium. This bending is called **refraction**. Figure 33–19a shows a ray passing from air into water. The angle θ_1 is the angle the incident ray makes with the perpendicular to the surface and is called the **angle of incidence**. The angle θ_2 is the **angle of refraction**, the angle the refracted ray makes with the perpendicular. Notice that the ray bends toward the normal when entering the water. This is always the case

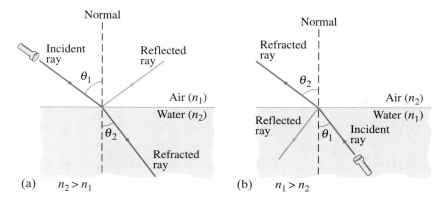

FIGURE 33–19 Refraction.
(a) Light refracted when passing from air (n_1) into water (n_2): $n_2 > n_1$.
(b) Light refracted when passing from water (n_1) into air (n_2): $n_1 > n_2$.

when the ray enters a medium where the speed of light is *less*. If light travels from one medium into a second where its speed is *greater*, the ray bends away from the normal; this is shown in Fig. 33–19b for a ray traveling from water to air.

Refraction is responsible for a number of common optical illusions. For example, a person standing in waist-deep water appears to have shortened legs. As shown in Fig. 33–20, the rays leaving the person's foot are bent at the surface. The observer's brain assumes the rays to have traveled a straight-line path, and so the feet appear to be higher than they really are. Similarly, when you put a pencil in water, it appears to be bent (Fig. 33–21).

→ PHYSICS APPLIED

Optical illusions

The angle of refraction depends on the speed of light in the two media and on the incident angle. An analytical relation between θ_1 and θ_2 was arrived at experimentally about 1621 by Willebrord Snell (1591–1626). It is known as **Snell's law** and is written:

$$n_1 \sin \theta_1 = n_2 \sin \theta_2. \qquad (33-5)$$

*Snell's law
(law of refraction)*

θ_1 is the angle of incidence and θ_2 is the angle of refraction; n_1 and n_2 are the respective indices of refraction in the materials. See Fig. 33–19. The incident and refracted rays lie in the same plane, which also includes the perpendicular to the surface. Snell's law is the basic **law of refraction**. Snell's law can be derived from the wave theory of light (Chapter 35), and in fact we did derive it in Section 15–10 where Eq. 15–19 is just a combination of Eqs. 33–5 and 33–1.

It is clear from Snell's law that if $n_2 > n_1$, then $\theta_2 < \theta_1$. That is, if light enters a medium where n is greater (and its speed less), then the ray is bent toward the normal. And if $n_2 < n_1$, then $\theta_2 > \theta_1$, so the ray bends away from the normal. This is what we saw in Fig. 33–19.

FIGURE 33–20 Ray diagram showing why a person's legs look shorter when standing in waist-deep water: the path of light traveling from the bather's foot to the observer's eye bends at the water's surface, and our brain interprets the light as having traveled in a straight line, from higher up (dashed line).

FIGURE 33–21 A pencil in water looks bent even when it isn't.

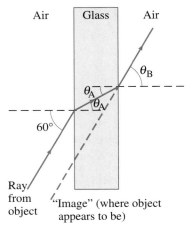

Air   Glass   Air

θ_B

θ_A

θ_A

60°

Ray from object

"Image" (where object appears to be)

FIGURE 33–22 Light passing through a piece of glass (Example 33–7).

FIGURE 33–23 Example 33–8.

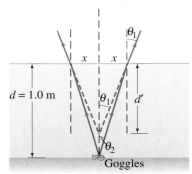

θ_1

x x

$d = 1.0$ m

θ_1 d'

θ_2

Goggles

EXAMPLE 33–7 **Refraction through flat glass.** Light strikes a flat piece of uniformly thick glass at an incident angle of 60°, as shown in Fig. 33–22. If the index of refraction of the glass is 1.50, (a) what is the angle of refraction θ_A in the glass; (b) what is the angle θ_B at which the ray emerges from the glass?

SOLUTION (a) We assume the incident ray is in air, so $n_1 = 1.00$ and $n_2 = 1.50$. Then, from Eq. 33–5 we have

$$\sin \theta_A = \frac{1.00}{1.50} \sin 60° = 0.577,$$

so $\theta_A = 35.2°$.
(b) Since the faces of the glass are parallel, the incident angle at the second surface is just θ_A, so $\sin \theta_A = 0.577$. This time $n_1 = 1.50$ and $n_2 = 1.00$. Thus, $\theta_B\ (= \theta_2)$ is

$$\sin \theta_B = \frac{1.50}{1.00} \sin \theta_A = 0.866,$$

and $\theta_B = 60.0°$. The direction of the beam is thus unchanged by passing through a flat piece of glass of uniform thickness. It should be clear that this works for any angle of incidence. The ray is displaced slightly to one side, however. You can observe this by looking through a piece of glass (near its edge) at some object and then moving your head to the side so that you see the object directly.

EXAMPLE 33–8 **Apparent depth of a pool.** A swimmer has dropped her goggles in the shallow end of a pool, marked as 1.0 m deep. But the goggles don't look that deep. Why? How deep do the goggles appear to be when you look straight down into the water?

SOLUTION The ray diagram of Fig. 33–23 shows why the water seems less deep than it actually is. Rays traveling upward from the goggles on the bottom are refracted *away* from the normal as they exit the water. The rays appear to be diverging from a point higher in the water. To calculate the apparent depth d', given a real depth $d = 1.0$ m, we use Snell's law with $n_1 = 1$ for air and $n_2 = 1.33$ for water: $\sin \theta_1 = n_2 \sin \theta_2$. We are considering only small angles, so $\sin \theta \approx \tan \theta \approx \theta$, with θ in radians. So Snell's law becomes

$$\theta_1 \approx n_2 \theta_2.$$

From Fig. 33–23, we see that

$$\theta_1 \approx \tan \theta_1 = \frac{x}{d'} \quad \text{and} \quad \theta_2 \approx \tan \theta_2 = \frac{x}{d}.$$

Putting these into Snell's law, $\theta_1 \approx n_2 \theta_2$, we get

$$\frac{x}{d'} \approx n_2 \frac{x}{d} \quad \text{or} \quad d' \approx \frac{d}{n_2} = \frac{1.0\ \text{m}}{1.33} = 0.75\ \text{m}.$$

The pool seems only three-fourths as deep as it actually is.

33–6 Visible Spectrum and Dispersion

An obvious property of visible light is its color. Color is related to the wavelengths or frequencies of the light. (How this was discovered will be discussed in Chapter 35.) Visible light—that to which our eyes are sensitive—falls in the wavelength range of about 400 nm to 750 nm.[†] This is known as the **visible spectrum**, and within it lie

---

[†] Sometimes the angstrom (Å) unit is used when referring to light: $1\ \text{Å} = 1 \times 10^{-10}\ \text{m}$. Then visible light falls in the wavelength range of 4000 Å to 7500 Å.

the different colors from violet to red, as shown in Fig. 33–24. Light with wavelength shorter than 400 nm is called **ultraviolet** (UV), and light with wavelength greater than 750 nm is called **infrared** (IR).[‡] Although human eyes are not sensitive to UV or IR, some types of photographic film do respond to them.

A prism separates white light into a rainbow of colors, as shown in Fig. 33–25. This happens because the index of refraction of a material depends on the wavelength, as shown for several materials in Fig. 33–26. White light is a mixture of all visible wavelengths, and when incident on a prism, as in Fig. 33–27, the different wavelengths are bent to varying degrees. Because the index of refraction is greater for the shorter wavelengths, violet light is bent the most and red the least as indicated. This spreading of white light into the full spectrum is called **dispersion**.

FIGURE 33–25 White light passing through a prism is broken down into its constituent colors.

FIGURE 33–26 Index of refraction as a function of wavelength for various transparent solids.

FIGURE 33–27 White light dispersed by a prism into the visible spectrum.

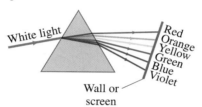

Rainbows are a spectacular example of dispersion—by drops of water. You can see rainbows when you look at falling water droplets with the Sun at your back. Figure 33–28 shows how red and violet rays are bent by spherical water droplets and are reflected off the back surface. Red is bent the least and so reaches the observer's eyes from droplets higher in the sky, as shown in the diagram. Thus the top of the rainbow is red.

➡ **PHYSICS APPLIED**

Rainbows

[‡]The complete electromagnetic spectrum is illustrated in Fig. 32–12.

(a)

(b)

FIGURE 33–28 (a) Ray diagram explaining how a rainbow (b) is formed.

33–7 | Total Internal Reflection; Fiber Optics

When light passes from one material into a second material where the index of refraction is less (say, from water into air), the light bends away from the normal, as for rays I and J in Fig. 33–29. At a particular incident angle, the angle of refraction will be 90°, and the refracted ray would skim the surface (ray K) in this case. The incident angle at which this occurs is called the **critical angle**, θ_C. From Snell's law, θ_C is given by

Critical angle

$$\sin \theta_C = \frac{n_2}{n_1} \sin 90° = \frac{n_2}{n_1}.$$ (33–6)

For any incident angle less than θ_C there will be a refracted ray, although part of the light will also be reflected at the boundary. However, for incident angles greater than θ_C, Snell's law would tell us that $\sin \theta_2$ is greater than 1.00. Yet the sine of an angle can never be greater than 1.00. In this case there is no refracted ray at all, and *all of the light is reflected*, as for ray L in Fig. 33–29. This effect is called **total internal reflection**. But note that total internal reflection can occur only when light strikes a boundary where the medium beyond has a lower index of refraction.

Total internal reflection

FIGURE 33–29 Since $n_2 < n_1$, light rays are totally internally reflected if $\theta > \theta_C$ as for ray L. If $\theta < \theta_C$, as for rays I and J, only a part of the light is reflected, and the rest is refracted.

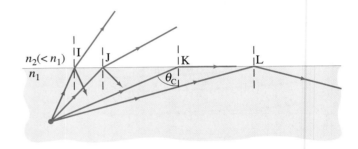

CONCEPTUAL EXAMPLE 33–9 | **View up from under water.** Describe what a person would see who looked up at the world from beneath the perfectly smooth surface of a lake or swimming pool.

RESPONSE For an air–water interface, the critical angle is given by

$$\sin \theta_C = \frac{1.00}{1.33} = 0.750.$$

Therefore, $\theta_C = 49°$. Thus the person would see the outside world compressed into a circle whose edge makes a 49° angle with the vertical. Beyond this angle, the person would see reflections from the sides and bottom of the pool or lake (Fig. 33–30).

FIGURE 33–30 (a) Light rays, and (b) view looking upward from beneath the water (the surface of the water must be very smooth).

(a)

(b)

Diamonds

Diamonds achieve their brilliance from a combination of dispersion (previous Section) and total internal reflection. Because diamonds have a very high index of refraction of about 2.4, the critical angle for total internal reflection is

FIGURE 33–31 Total internal reflection of light by prisms in binoculars.

FIGURE 33–32 Light reflected totally at the interior surface of a glass or transparent plastic fiber.

FIGURE 33–33 Total internal reflection within the tiny fibers of this light pipe makes it possible to transmit light in complex paths with minimal loss.

FIGURE 33–34 (a) How a fiber-optic image is made. (b) Example of a fiber-optic device inserted through the nose, and the image seen.

(a)

(b)

only 25°. Incident light therefore strikes many of the internal surfaces before it strikes one at less than 25° and emerges. After many such reflections, the light has traveled far enough that the colors have become sufficiently separated to be seen individually and brilliantly by the eye after leaving the crystal.

Many optical instruments, such as binoculars, use total internal reflection within a prism to reflect light. The advantage is that very nearly 100 percent of the light is reflected, whereas even the best mirrors reflect somewhat less than 100 percent. Thus the image is brighter. For glass with $n = 1.50$, $\theta_C = 41.8°$. Therefore, 45° prisms will reflect all the light internally, if oriented as shown in the binoculars of Fig. 33–31.

Total internal reflection is the principle behind **fiber optics**. Glass and plastic fibers as thin as a few micrometers in diameter can now be made. A bundle of such tiny fibers is called a **light pipe** or cable, and light can be transmitted along it with almost no loss because of total internal reflection. Figure 33–32 shows how light traveling down a thin fiber makes only glancing collisions with the walls so that total internal reflection occurs. Even if the light pipe is bent into a complicated shape, the critical angle still won't be exceeded, so light is transmitted practically undiminished to the other end (see Fig. 33–33). Very small losses do occur, mainly by reflection at the ends and absorption within the fiber.

Important applications of fiber-optic cables are in telecommunications and medicine. They are being used to transmit telephone calls, video signals, and computer data. The signal is a modulated light beam (a light beam whose intensity can be varied) and is transmitted at a much higher rate and with less loss and less interference than an electrical signal in a copper wire. Fibers have been developed that can support over one hundred separate wavelengths, each modulated to carry up to 10 gigabits $(10^{10}$ bits$)$ of information per second. That amounts to a terabit $(10^{12}$ bits$)$ per second for the full one hundred wavelengths. The sophisticated use of fiber optics to transmit a clear picture, is particularly useful in medicine, Fig. 33–34. For example, a patient's lungs can be examined by inserting a light pipe known as a bronchoscope through the mouth and down the bronchial tube. Light is sent down an outer set of fibers to illuminate the lungs. The reflected light returns up a central core set of fibers. Light directly in front of each fiber travels up that fiber. At the opposite end, a viewer sees a series of bright and dark spots, much like a TV screen—that is, a picture of what lies at the opposite end. Lenses are used at each end: at the object end to bring the rays in parallel, and at the viewing end as a telescope. The image may be viewed directly or on a TV monitor or film. The fibers must be optically insulated from one another, usually by a thin coating of material whose refractive index is less than that of the fiber. The fibers must be arranged precisely parallel to one another if the picture is to be clear. The more fibers there are, and the smaller they are, the more detailed the picture. Such instruments, including bronchoscopes, colonoscopes (for viewing the colon) and endoscopes (stomach or other organs) are extremely useful for examining hard-to-reach places.

FIGURE 33–35 Rays from a point O on an object will be focused at a single image point I by a spherical boundary between two transparent materials $(n_2 > n_1)$, as long as the rays make small angles with the axis.

* 33–8 Refraction at a Spherical Surface

We now examine the refraction of rays at the spherical surface of a transparent material. Such a surface could be one face of a lens or the cornea of the eye. To be general, let us consider an object which is located in a medium whose index of refraction is n_1, and rays from each point on the object can enter a medium whose index of refraction is n_2. The radius of curvature of the spherical boundary is R, and its center of curvature is at point C, Fig. 33–35. We now show that all rays leaving a point O on the object will be focused at a single point I, the image point, if we consider only paraxial rays, rays that make a small angle with the axis. To do so, we consider a single ray that leaves point O as shown in Fig. 33–36. From Snell's law, Eq. 33–5, we have

$$n_1 \sin \theta_1 = n_2 \sin \theta_2.$$

We are assuming that angles θ_1, θ_2, α, β, and γ are small, so $\sin \theta \approx \theta$ (in radians), and Snell's law becomes, approximately,

$$n_1 \theta_1 = n_2 \theta_2.$$

Also, $\beta + \phi = 180°$ and $\theta_2 + \gamma + \phi = 180°$, so

$$\beta = \gamma + \theta_2.$$

Similarly for triangle OPC,

$$\theta_1 = \alpha + \beta.$$

These three relations can be combined to yield

$$n_1 \alpha + n_2 \gamma = (n_2 - n_1)\beta.$$

Since we are considering only the case of small angles, we can write, approximately,

$$\alpha = \frac{h}{d_o}, \qquad \beta = \frac{h}{R}, \qquad \gamma = \frac{h}{d_i},$$

where d_o and d_i are the object and image distances and h is the height as shown in Fig. 33–36. We substitute these into the previous equation, divide through by h, and obtain

$$\frac{n_1}{d_o} + \frac{n_2}{d_i} = \frac{n_2 - n_1}{R}. \tag{33–7}$$

For a given object distance d_o, this equation tells us d_i, the image distance, does not depend on the angle of a ray. Hence all paraxial rays meet at the same point I. This

FIGURE 33–36 Diagram for proving that all paraxial rays from O focus at the same point I $(n_2 > n_1)$.

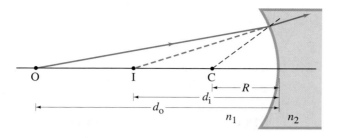

FIGURE 33–37 Rays from O refracted by a concave surface form a virtual image $(n_2 > n_1)$.

is true, of course, only for rays that make small angles with the axis and with each other. This is equivalent to assuming that the width of the refracting spherical surface is small compared to its radius of curvature. If this assumption is not true, the rays will not converge to a point; there will be spherical aberration, just as for a mirror (see Fig. 33–13), and the image will be blurry. (Spherical aberration will be discussed further in Section 34–10.)

We derived Eq. 33–7 using Fig. 33–36 for which the spherical surface is convex (as viewed by the incoming ray). It is also valid for a concave surface—as can be seen using Fig. 33–37—if we make the following conventions:

1. If the surface is convex (so the center of curvature C is on the side of the surface opposite to that from which the light comes), R is positive; if the surface is concave (C on the same side from which the light comes) R is negative.
2. The image distance, d_i, follows the same convention: positive if on the opposite side from where the light comes, negative if on the same side.
3. The object distance is positive if on the same side from which the light comes (this is the normal case, although when several surfaces bend the light it may not be so), otherwise it is negative.

For the case shown in Fig. 33–37 with a concave surface, both R and d_i are negative when used in Eq. 33–7. Note, in this case, that the image is virtual.

EXAMPLE 33–10 Apparent depth II. A person looks vertically down into a 1.0-m-deep pool. How deep does the pool appear to be?

SOLUTION Example 33–8 solved this problem using Snell's law. Here we use Eq. 33–7. A ray diagram is shown in Fig. 33–38. Point O represents a point on the pool's bottom. The rays diverge and appear to come from point I, the image. We have $d_o = 1.0$ m and, for a flat surface $R = \infty$. Then Eq. 33–7 becomes

$$\frac{1.33}{1.0 \text{ m}} + \frac{1.00}{d_i} = \frac{(1.00 - 1.33)}{\infty} = 0$$

Hence $d_i = -(1.0 \text{ m})/(1.33) = -0.75$ m. So the pool appears to be only three-fourths as deep as it actually is, the same result we got in Example 33–8. The minus sign tells us the image point I is on the same side of the surface as O, and the image is virtual. At angles other than vertical, this conclusion must be modified.

FIGURE 33–38 Example 33–10.

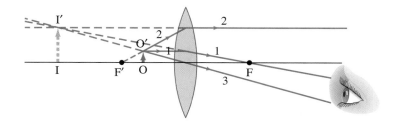

FIGURE 34–12 An object placed within the focal point of a converging lens produces a virtual image. Example 34–2.

SOLUTION (a) Given $f = 15\,\text{cm}$ and $d_o = 10\,\text{cm}$, then

$$\frac{1}{d_i} = \frac{1}{15\,\text{cm}} - \frac{1}{10\,\text{cm}} = -\frac{1}{30\,\text{cm}},$$

and $d_i = -30\,\text{cm}$. (Remember to take the reciprocal!) Because d_i is negative, the image must be virtual and on the same side of the lens as the object. The magnification

$$m = -\frac{d_i}{d_o} = -\frac{-30\,\text{cm}}{10\,\text{cm}} = 3.0.$$

➥ **PROBLEM SOLVING**

Don't forget to take the reciprocal

The image is three times as large as the object and is upright. This lens is being used as a simple magnifying glass, which we discuss in more detail in Section 34–7. (b) The ray diagram is shown in Fig. 34–12, and confirms the result in part (a). For point O′ on the top of the object, ray 1 is easy to draw. But ray 2 may take some thought: if we draw it heading toward F′, it is going the wrong way—so we have to draw it as if coming from F′ (and so dashed), striking the lens, and then going out parallel to the principal axis. We project it backward with a dashed line, as we must do also for ray 1, in order to find where they meet. Ray 3 is easy to draw, through the lens center, and it meets the other two at the image point, I′.

From this last Example and Fig. 34–12, we readily see that, whenever an object is placed between a converging lens and its focal point, the image is virtual.

EXAMPLE 34–3 **Diverging lens.** Where must a small insect be placed if a 25-cm-focal-length diverging lens is to form a virtual image 20 cm in front of the lens?

SOLUTION The ray diagram is basically that of Fig. 34–10 because our lens here is diverging and our image is in front of the lens within the focal distance. (It would be a valuable exercise to draw the ray diagram to scale, precisely, now.) Since $f = -25\,\text{cm}$ and $d_i = -20\,\text{cm}$, then Eq. 34–2 gives

$$\frac{1}{d_o} = \frac{1}{f} - \frac{1}{d_i} = -\frac{1}{25\,\text{cm}} + \frac{1}{20\,\text{cm}} = \frac{-4 + 5}{100\,\text{cm}} = \frac{1}{100\,\text{cm}}.$$

So the object must be 100 cm in front of the lens.

34–3 | Combinations of Lenses

We now consider Examples illustrating how to deal with lenses used in combination. In general, when light passes through more than one lens, we find the image formed by the first lens as if it were alone. This image becomes the *object* for the second lens, and we find the image then formed by this second lens, which is the final image if there are only two lenses. The total magnification will be the product of the separate magnifications of each lens as we shall see.

Multiple lenses: image formed by first lens is object for second lens

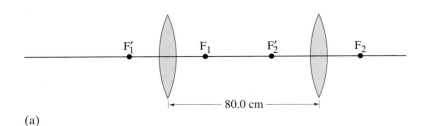

(a)

(b)

FIGURE 34–13   Example 34–4.

EXAMPLE 34–4   A two-lens system. Two converging lenses, with focal lengths $f_1 = 20.0$ cm and $f_2 = 25.0$ cm, are placed 80.0 cm apart, as shown in Fig. 34–13a. An object is placed 60.0 cm in front of the first lens as shown in Fig. 34–13b. Determine (a) the position, and (b) the magnification, of the final image formed by the combination of the two lenses.

SOLUTION (a) The object is a distance $d_{o1} = +60.0$ cm from the first lens, and this lens forms an image whose position can be calculated using the lens equation:

$$\frac{1}{d_{i1}} = \frac{1}{f_1} - \frac{1}{d_{o1}} = \frac{1}{20.0 \text{ cm}} - \frac{1}{60.0 \text{ cm}} = \frac{3 - 1}{60.0 \text{ cm}} = \frac{1}{30.0 \text{ cm}}.$$

So the first image I_1 is at $d_{i1} = 30.0$ cm behind the first lens. This image becomes the object for the second lens. It is a distance $d_{o2} = 80.0 \text{ cm} - 30.0 \text{ cm} = 50.0$ cm in front of lens 2, as shown in Fig. 34–13b. The image formed by the second lens, again using the lens equation, is at a distance d_{i2} from the second lens:

$$\frac{1}{d_{i2}} = \frac{1}{f_2} - \frac{1}{d_{o2}} = \frac{1}{25.0 \text{ cm}} - \frac{1}{50.0 \text{ cm}} = \frac{4 - 2}{100.0 \text{ cm}} = \frac{2}{100.0 \text{ cm}}.$$

Hence $d_{i2} = 50.0$ cm behind lens 2. This is the final image—see Fig. 34–13b.
(b) The first lens has a magnification (Eq. 34–3)

$$m_1 = -\frac{d_{i1}}{d_{o1}} = -\frac{30.0 \text{ cm}}{60.0 \text{ cm}} = -0.500.$$

Thus, the first image is inverted and is half as high as the object: again Eq. 34–3,

$$h_{i1} = m_1 h_{o1} = -0.500 \, h_{o1}.$$

The second lens takes this image as object and changes its height by a factor

$$m_2 = -\frac{d_{i2}}{d_{o2}} = -\frac{50.0 \text{ cm}}{50.0 \text{ cm}} = -1.000.$$

The final image height is (remember h_{o2} is the same as h_{i1}):

$$h_{i2} = m_2 h_{o2} = m_2 h_{i1} = m_2 m_1 h_{o1}.$$

We see from this equation that the total magnification is the product of m_1 and m_2, which here equals $(-1.000)(-0.500) = +0.500$, or $\frac{1}{2}$ the original height and upright.

Careful!
*Note that object distance for second lens is **not** equal to the image distance for first lens*

Total magnification is
$m = m_1 m_2$

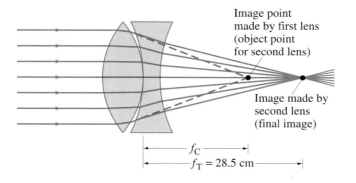

FIGURE 34–14 Determining the focal length of a diverging lens. Example 34–5.

EXAMPLE 34–5 **Measuring _f_ for a diverging lens.** To measure the focal length of a diverging lens, a converging lens is placed in contact with it, as shown in Fig. 34–14. The Sun's rays are focused by this combination at a point 28.5 cm, behind the lenses as shown. If the converging lens has a focal length f_C of 16.0 cm, what is the focal length f of the diverging lens? Assume both lenses are thin and the space between them is negligible.

SOLUTION Rays from the Sun are focused 28.5 cm behind the combination, so the focal length of the total combination is $f_T = 28.5$ cm. If the diverging lens were absent, the converging lens would form the image at its focal point—that is, at a distance $f_C = 16.0$ cm behind it (dashed lines in Fig. 34–14). When the diverging lens is placed next to the converging lens, we treat the image formed by the first lens as the *object* for the second lens. Since this object lies to the right of the diverging lens, this is a situation where d_o is negative (see the sign conventions, page 841). Thus, for the diverging lens, the object is virtual and $d_o = -16.0$ cm. The diverging lens forms the image of this virtual object at a distance $d_i = 28.5$ cm away (this was given). Thus,

$$\frac{1}{f_D} = \frac{1}{d_o} + \frac{1}{d_i} = \frac{1}{-16.0 \text{ cm}} + \frac{1}{28.5 \text{ cm}} = -0.0274 \text{ cm}^{-1}.$$

We take the reciprocal to find $f_D = -1/(0.0274 \text{ cm}^{-1}) = -36.5$ cm. Note that the converging lens must be "stronger" than the diverging lens—that is, it must have a focal length whose magnitude is less than that of the diverging lens—if this technique is to work.

34–4 Lensmaker's Equation

In this Section, we will prove that parallel rays are brought to a focus at a *single* point for a thin lens. At the same time, we will also derive an equation that relates the focal length of a lens to the radii of curvature of its two surfaces, which is known as the lensmaker's equation.

In Fig. 34–15, a ray parallel to the axis of a lens is refracted at the front surface of the lens at point A_1 and is refracted at the back surface at point A_2. This ray then passes through point F, which we call the focal point for this ray. Point A_1 is a height h_1 above the axis, and point A_2 is height h_2 above the axis. C_1 and C_2 are the centers of curvature of the two lens surfaces; so the length $C_1A_1 = R_1$, the radius of curvature of the front surface, and $C_2A_2 = R_2$ is the radius of the second surface. The thickness of the lens has been grossly exaggerated so the various angles would be clear. But we will assume that the lens is actually very thin and that angles between the rays and the axis are small. In this approximation, $h_1 \approx h_2$, and the sines and tangents of all the angles will be equal to the angles themselves in radians. For example, $\sin\theta_1 \approx \tan\theta_1 \approx \theta_1$ (radians).

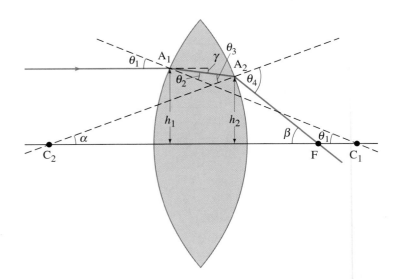

FIGURE 34–15 Diagram of a ray passing through a lens for derivation of the lensmaker's equation.

To this approximation, then, Snell's law tells us that

$$\theta_1 = n\theta_2$$

$$\theta_4 = n\theta_3$$

where n is the index of refraction of the glass, and we assume that the lens is surrounded by air ($n = 1$). Notice also in Fig. 34–15 that

$$\theta_1 \approx \sin\theta_1 = \frac{h_1}{R_1}$$

$$\alpha \approx \frac{h_2}{R_2}$$

$$\beta \approx \frac{h_2}{f}.$$

The last follows because the distance from F to the lens (assumed very thin) is f. From the diagram, the angle γ is

$$\gamma = \theta_1 - \theta_2.$$

A careful examination of Fig. 34–15 shows also that

$$\alpha = \theta_3 - \gamma.$$

This can be seen by drawing a horizontal line to the left from point A_2, which divides the angle θ_3 into two parts. The upper part equals γ and the lower part equals α. (The opposite angles between an oblique line and two parallel lines are equal.) Thus, $\theta_3 = \gamma + \alpha$. Finally, by drawing a horizontal line to the right from point A_2, we divide θ_4 into two parts. The upper part is α and the lower is β. Thus

$$\theta_4 = \alpha + \beta.$$

We now combine all these equations:

$$\alpha = \theta_3 - \gamma = \frac{\theta_4}{n} - (\theta_1 - \theta_2) = \frac{\alpha}{n} + \frac{\beta}{n} - \theta_1 + \theta_2,$$

or

$$\frac{h_2}{R_2} = \frac{h_2}{nR_2} + \frac{h_2}{nf} - \frac{h_1}{R_1} + \frac{h_1}{nR_1}.$$

Because the lens is thin, $h_1 \approx h_2$ and all h's can be canceled from all the numerators.

We then multiply through by n and rearrange to find that

$$\frac{1}{f} = (n - 1)\left(\frac{1}{R_1} + \frac{1}{R_2}\right). \qquad\qquad (34\text{--}4) \qquad \textit{Lensmaker's equation}$$

This is called the **lensmaker's equation**. It relates the focal length of a lens to the radii of curvature of its two surfaces and its index of refraction. Notice that f does not depend on h_1 or h_2. Thus the position of the point F does not depend on where the ray strikes the lens. Hence, all rays parallel to the axis of a thin lens will pass through the same point F, which we wished to prove.

In our derivation, both surfaces are convex and R_1 and R_2 are considered positive.[†] Equation 34–4 also works for lenses with one or both surfaces concave; but for a concave surface, the radius must be considered *negative*.

Notice in Eq. 34–4 that the equation is symmetrical in R_1 and R_2. Thus, if a lens is turned around so that light impinges on the other surface, the focal length is the same even if the two lens surfaces are different.

EXAMPLE 34–6 **Calculating f for a converging lens.** A convex meniscus lens (Figs. 34–1a and 34–16) is made from glass with $n = 1.50$. The radius of curvature of the convex surface is 22.4 cm and that of the concave surface is 46.2 cm. (*a*) What is the focal length? (*b*) Where will it focus an object 2.00 m away?

SOLUTION (*a*) $R_1 = 22.4$ cm and $R_2 = -46.2$ cm; the latter is negative because it refers to the concave surface. Then

$$\frac{1}{f} = (1.50 - 1.00)\left(\frac{1}{22.4\text{ cm}} - \frac{1}{46.2\text{ cm}}\right)$$
$$= 0.0115\text{ cm}^{-1}.$$

So

$$f = \frac{1}{0.0115\text{ cm}^{-1}} = 87\text{ cm}$$

and the lens is converging. Notice that if we turn the lens around so that $R_1 = -46.2$ cm and $R_2 = +22.4$ cm, we get the same result.
(*b*) From the lens equation, with $f = 0.87$ m and $d_o = 2.00$ m, we have

$$\frac{1}{d_i} = \frac{1}{f} - \frac{1}{d_o} = \frac{1}{0.87\text{ m}} - \frac{1}{2.00\text{ m}}$$
$$= 0.65\text{ m}^{-1},$$

so $d_i = 1/0.65\text{ m}^{-1} = 1.53$ m.

FIGURE 34–16 Example 34–6.

EXAMPLE 34–7 **Calculating f for a diverging lens.** A Lucite planoconcave lens (see Fig. 34–1b) has one flat surface and the other has $R = -18.4$ cm. What is the focal length?

SOLUTION From Table 33–1, n for Lucite is 1.51. A plane surface has infinite radius of curvature; if we call this R_1, then $1/R_1 = 0$. Therefore,

$$\frac{1}{f} = (1.51 - 1.00)\left(-\frac{1}{18.4\text{ cm}}\right).$$

So $f = (-18.4\text{ cm})/0.51 = -36$ cm, and the lens is diverging.

[†]Some books use a different convention—for example, R_1 and R_2 are considered positive if their centers of curvature are to the right of the lens, in which case a minus sign appears in their equivalent of Eq. 34–4.

In the remainder of this chapter, we discuss briefly some of the more common optical instruments, including the camera, our eyes and corrective lenses, the magnifying glass, telescopes, and microscopes. We start with the camera.

The basic elements of a **camera** are a lens, a light-tight box, a shutter to let light pass through the lens only briefly, and a sensitized plate or piece of film (Fig. 34–17). When the shutter is opened, light from external objects in the field of view is focused by the lens as an image on the film. The film contains light-sensitive chemicals that undergo change when light strikes them. In the development process, chemical reactions cause the changed areas to turn opaque so that the image is recorded on the film.[†] You can see the image yourself if you can remove the back of the camera and then view through a piece of tissue or wax paper (on which the image can form) placed at the position of the film with the shutter open.

There are three main adjustments on good-quality cameras: shutter speed, *f*-stop, and focusing (see Fig. 34–18), and we now discuss them.

Shutter speed. This refers to how long the shutter is open and the film exposed. It may vary from a second or more ("time exposures") to $\frac{1}{1000}$ s or less. To avoid blurring from camera movement, speeds faster than $\frac{1}{100}$ s are normally used. If the object is moving, faster shutter speeds are needed to "stop" the action. A shutter can be "behind the lens," as in Fig. 34–17, or a "focal plane" shutter which is a moveable curtain just in front of the film.

***f*-stop.** The amount of light reaching the film must be carefully controlled to avoid **underexposure** (too little light for any but the brightest objects to show up) or **overexposure** (too much light, so that all bright objects look the same, with a consequent lack of contrast and a "washed-out" appearance). To control the exposure, a "stop" or iris diaphragm, whose opening is of variable diameter, is placed behind the lens (Fig. 34–17). The size of the opening is varied to compensate for bright or dark days, the sensitivity of the film[‡] used, and for different shutter speeds. The size of the opening is specified by the ***f*-stop**, defined as

$$f\text{-stop} = \frac{f}{D},$$

where f is the focal length of the lens and D is the diameter of the opening. For example, when a 50-mm-focal-length lens has an opening $D = 25$ mm, we say it is set at $f/2$. When the lens is set at $f/8$, the opening is only $6\frac{1}{4}$ mm $(50/6\frac{1}{4} = 8)$. The faster the shutter speed, or the darker the day, the greater the opening must be to get a proper exposure. This corresponds to a smaller *f*-stop number. The smaller the *f*-stop number, the more light passes through the lens to the film. The smallest *f*-number of a lens (largest opening) is referred to as the *speed* of the lens. It is common to find $f/2.0$ lenses today, and some even faster. The advantage of a fast lens is that it allows pictures to be taken under poor lighting conditions. Good quality lenses consist of several elements to reduce the defects present in simple thin lenses (Section 34–10). Standard *f*-stop markings on good lenses are 1.0, 1.4, 2.0, 2.8, 4.0, 5.6, 8, 11, 16, 22, and 32. Each of these stops corresponds to a diameter reduction by a factor of about $\sqrt{2} = 1.4$. Because the amount of light reaching the film is proportional to the *area* of the opening, and therefore proportional to the diameter squared, each standard *f*-stop corresponds to a factor of 2 in light intensity reaching the film.

The camera

FIGURE 34–17 A simple camera.

FIGURE 34–18 On this camera, the *f*-stops and the focusing ring are on the camera lens. Shutter speeds are selected on the small wheel on top of the camera body.

[†] This is called a *negative*, because the black areas correspond to bright objects and vice versa. The same process occurs during printing to produce a black-and-white "positive" picture from the negative. Color film makes use of three dyes corresponding to the primary colors.

[‡] Different films have different sensitivities to light, referred to as the "film speed," and specified as an "ASA" number; a "faster" film is more sensitive and needs less light to produce a good image.

(a)   (b)

FIGURE 34–19 Photos with camera focused (a) on nearby object with distant object blurry, and (b) on more distant object with nearby object blurry.

Focusing. Focusing is the operation of placing the lens at the correct position relative to the film for the sharpest image. The image distance is a minimum for objects at infinity (the symbol ∞ is used for infinity) and is equal to the focal length. For closer objects, the image distance is greater than the focal length, as can be seen from the lens equation, $1/f = 1/d_o + 1/d_i$. To focus on nearby objects, the lens must therefore be moved away from the film, and this is usually done by turning a ring on the lens.

Focusing

  If the lens is focused on a nearby object, a sharp image of it will be formed, but distant objects may be blurry (Fig. 34–19). The rays from a point on the distant object will be out of focus—they will form a circle on the film as shown (exaggerated) in Fig. 34–20. The distant object will thus produce an image consisting of overlapping circles and will be blurred. These circles are called **circles of confusion**. To include near and distant objects in the same photo, you can try setting the lens focus at an intermediate position. For a given distance setting, there is a range of distances over which the circles of confusion will be small enough that the images will be reasonably sharp. This is called the **depth of field**. For a particular choice of circle of confusion diameter as upper limit (typically taken to be 0.03 mm for 35-mm cameras), the depth of field varies with the lens opening. If the lens opening is smaller, only rays through the central part of the lens are accepted, and these form smaller circles of confusion for a given object distance. Hence, at smaller lens openings, a greater range of object distances will fit within the circle of confusion criterion, so that the depth of field is greater.

Depth of field

Rays from nearby object

Rays from distant object

"Circle of confusion" for distant object (greatly exaggerated)

FIGURE 34–20 When the lens is positioned to focus on a nearby object, points on a distant object produce circles and are therefore blurred. (The effect is shown greatly exaggerated.)

Camera lenses are categorized into normal, telephoto, and wide angle, according to focal length and film size. A **normal lens** is one that covers the film with a field of view that corresponds approximately to that of normal vision. A normal lens for 35-mm film has a focal length in the vicinity of 50 mm.[†] A **telephoto lens**, as its name implies, acts like a telescope to magnify images. They have longer focal lengths than a normal lens. As we saw in Section 34–2 (Eq. 34–3), the height of the image for a given object distance is proportional to the image distance, and the image distance will be greater for a lens with longer focal length. For distant objects, the image height is very nearly proportional to the focal length (can you prove this?). Thus a 200-mm telephoto lens for use with a 35-mm camera gives a 4× magnification over the normal 50-mm lens. A **wide-angle lens** has a shorter focal length than normal: a wider field of view is included and objects appear smaller. A **zoom lens** is one whose focal length can be changed so that you seem to zoom up to, or away from, the subject as you change the focal length.

Telephoto and wide-angle lenses

CONCEPTUAL EXAMPLE 34–8 **Shutter speed.** To improve the depth of field, you stop down your camera lens by two f-stops (say, from $f/4$ to $f/8$). What should you do to the shutter speed to maintain the same exposure?

RESPONSE The amount of light admitted by the lens is proportional to the area of the lens opening. Reducing the lens opening by one f-stop reduces the diameter by a factor of $\sqrt{2}$, and the area by a factor of 2. Stopping down by two f-stops reduces the area of the lens opening by a factor of 4. To maintain the same exposure, the shutter must be open 4 times as long. So if the shutter speed had been $\frac{1}{250}$ s, you have to increase it to $\frac{1}{60}$ s.

34–6 | The Human Eye; Corrective Lenses

Ciliary muscles
Retina
Iris
Fovea
Pupil
Cornea
Lens
Optic nerve
Ciliary muscles

FIGURE 34–21 Diagram of a human eye.

Focusing the eye

The human eye resembles a camera in its basic structure (Fig. 34–21), but is far more sophisticated. The eye is an enclosed volume into which light passes through a lens. A diaphragm, called the **iris** (the colored part of your eye), adjusts automatically to control the amount of light entering the eye. The hole in the iris through which light passes (the **pupil**) is black because no light is reflected from it (it's a hole), and very little light is reflected back out from the interior of the eye. The **retina**, which plays the role of the film in a camera, is on the curved rear surface. It consists of a complex array of nerves and receptors known as *rods* and *cones* which act to change light energy into electrical signals that travel along the nerves. The reconstruction of the image from all these tiny receptors is done mainly in the brain, although some analysis is apparently done in the complex interconnected nerve network at the retina itself. At the center of the retina is a small area called the **fovea**, about 0.25 mm in diameter, where the cones are very closely packed and the sharpest image and best color discrimination are found.

Unlike a camera, the eye contains no shutter. The equivalent operation is carried out by the nervous system, which analyzes the signals to form images at the rate of about 30 per second. This can be compared to motion picture or television cameras, which operate by taking a series of still pictures at a rate of 24 (movies) or 30 (U.S. television) per second. Their rapid projection on the screen gives the appearance of motion.

The lens of the eye does little of the bending of the light rays. Most of the refraction is done at the front surface of the **cornea** (index of refraction = 1.376), which also acts as a protective covering. The lens acts as a fine adjustment for

[†]Note that a "35-mm camera" uses film that is 35 mm wide; that 35 mm is not to be confused with a focal length.

focusing at different distances. This is accomplished by the ciliary muscles (Fig. 34–21), which change the curvature of the lens so that its focal length is changed. To focus on a distant object, the muscles are relaxed and the lens is thin, Fig. 34–22a, and parallel rays focus at the focal point (on the retina). To focus on a nearby object, the muscles contract, causing the center of the lens to be thicker, Fig. 34–22b, thus shortening the focal length so that images of nearby objects can be focused on the retina, behind the focal point. This focusing adjustment is called **accommodation**.

The closest distance at which the eye can focus clearly is called the **near point** of the eye. For young adults it is typically 25 cm, although younger children can often focus on objects as close as 10 cm. As people grow older, the ability to accommodate is reduced and the near point increases. A given person's **far point** is the farthest distance at which an object can be seen clearly. For some purposes it is useful to speak of a **normal eye** (a sort of average over the population), which is defined as one having a near point of 25 cm and a far point of infinity. To check your own near point, place this book close to your eye and slowly move it away until the type is sharp.

The "normal" eye is more of an ideal than a commonplace. A large part of the population have eyes that do not accommodate within the normal range of 25 cm to infinity, or have some other defect. Two common defects are nearsightedness and farsightedness. Both can be corrected to a large extent with lenses—either eyeglasses or contact lenses.

Nearsightedness, or *myopia*, refers to an eye that can focus only on nearby objects. The far point is not infinity but some shorter distance, so distant objects are not seen clearly. It is usually caused by an eyeball that is too long, although sometimes it is the curvature of the cornea that is too great. In either case, images of distant objects are focused in front of the retina. A diverging lens, because it causes parallel rays to diverge, allows the rays to be focused at the retina (Fig. 34–23a) and thus corrects this defect.

Farsightedness, or *hyperopia*, refers to an eye that cannot focus on nearby objects. Although distant objects are usually seen clearly, the near point is somewhat greater than the "normal" 25 cm, which makes reading difficult. This defect is caused by an eyeball that is too short or (less often) by a cornea that is not sufficiently curved. It is corrected by a converging lens, Fig. 34–23b. Similar to hyperopia is *presbyopia*, which refers to the lessening ability of the eye to accommodate as one ages, and the near point moves out. Converging lenses also compensate for this.

Astigmatism is usually caused by an out-of-round cornea or lens so that point objects are focused as short lines, which blurs the image. Astigmatism is corrected with the use of a cylindrical lens which, for eyes that are nearsighted or farsighted, is superimposed on the spherical surface, so that the radius of curvature of the correcting lens is different in different planes.

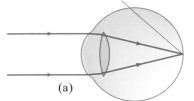

Focal point of lens and cornea

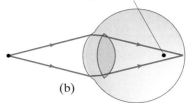

Focal point of lens and cornea

FIGURE 34–22 Accommodation by a normal eye: (a) lens relaxed, focused at infinity; (b) lens thickened, focused on a nearby object.

➡ **P H Y S I C S   A P P L I E D**

Corrective lenses

Nearsightedness

Farsightedness

Astigmatism

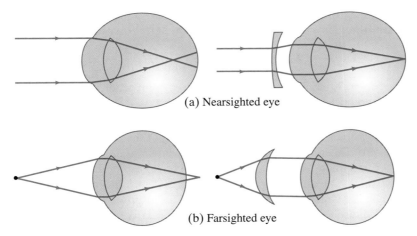

(a) Nearsighted eye

(b) Farsighted eye

FIGURE 34–23 Correcting eye defects with lenses: (a) a nearsighted eye, which cannot focus clearly on distant objects, can be corrected by use of a diverging lens; (b) a farsighted eye, which cannot focus clearly on nearby objects, can be corrected by use of a converging lens.

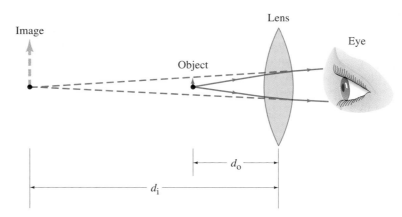

FIGURE 34–24 Lens of reading glasses (Example 34–9).

EXAMPLE 34–9 **Farsighted eye.** A particular farsighted person has a near point of 100 cm. Reading glasses must have what lens power so that this person can read a newspaper at a distance of 25 cm? Assume the lens is very close to the eye.

SOLUTION When the object is placed 25 cm from the lens, we want the image to be 100 cm away on the *same* side of the lens, and so it will be virtual, Fig. 34–24. Thus, $d_o = 25$ cm, $d_i = -100$ cm, and the lens equation gives

$$\frac{1}{f} = \frac{1}{25 \text{ cm}} + \frac{1}{-100 \text{ cm}} = \frac{4 - 1}{100 \text{ cm}} = \frac{1}{33 \text{ cm}}.$$

So $f = 33$ cm $= 0.33$ m. The power P of the lens is $P = 1/f = +3.0$ D. The plus sign indicates that it is a converging lens.

EXAMPLE 34–10 **Nearsighted eye.** A nearsighted eye has near and far points of 12 cm and 17 cm, respectively. (*a*) What lens power is needed for this person to see distant objects clearly, and (*b*) what then will be the near point? Assume that the lens is 2.0 cm from the eye (typical for eyeglasses).

SOLUTION (*a*) First we determine the power of the lens needed to focus objects at infinity, when the eye is relaxed. For a distant object $(d_o = \infty)$, as shown in Fig. 34–25, the lens must put the image 17 cm from the eye (its far point), which is 15 cm in front of the lens; hence $d_i = -15$ cm. We use the lens equation to solve for the focal length of the needed lens:

$$\frac{1}{f} = -\frac{1}{15 \text{ cm}} + \frac{1}{\infty} = -\frac{1}{15 \text{ cm}}.$$

FIGURE 34–25 Example 34–10a.

FIGURE 34–26 Example 34–10b.

So $f = -15$ cm $= -0.15$ m or $P = 1/f = -6.7$ D. The minus sign indicates that it must be a diverging lens.

(*b*) To determine the near point when wearing the glasses, we note that a sharp image will be 12 cm from the eye (its near point, see Fig. 34–26), which is 10 cm from the lens; so $d_i = -0.10$ m and the lens equation gives

$$\frac{1}{d_o} = \frac{1}{f} - \frac{1}{d_i} = -\frac{1}{0.15 \text{ m}} + \frac{1}{0.10 \text{ m}} = \frac{1}{0.30 \text{ m}}.$$

So $d_o = 30$ cm, which means the near point when the person is wearing glasses is 30 cm in front of the lens.

Contact lenses

Contact lenses could be used to correct the eye in Example 34–10. Since contacts are placed directly on the cornea, we would not subtract out the 2.0 cm for the image distances. That is, for distant objects $d_i = -17$ cm, so $P = 1/f = -5.9$ D (diopters). The new near point would be 41 cm. Thus we see that a contact lens and an eyeglass lens will require slightly different powers, or focal lengths, for the same eye because of their different placements relative to the eye.

34–7 Magnifying Glass

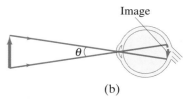

Much of the remainder of this chapter will deal with optical devices that are used to produce magnified images of objects. We first discuss the **simple magnifier**, or **magnifying glass**, which is simply a converging lens (see chapter opening photo).

How large an object appears, and how much detail we can see on it, depends on the size of the image it makes on the retina. This, in turn, depends on the angle subtended by the object at the eye. For example, a penny held 30 cm from the eye looks twice as tall as one held 60 cm away because the angle it subtends is twice as great (Fig. 34–27). When we want to examine detail on an object, we bring it up close to our eyes so that it subtends a greater angle. However, our eyes can accommodate only up to a point (the near point), and we will assume a standard distance of 25 cm as the near point in what follows.

A magnifying glass allows us to place the object closer to our eye so that it subtends a greater angle. As shown in Fig. 34–28a, the object is placed at the focal point or just within it. Then the converging lens produces a virtual image, which must be at least 25 cm from the eye if the eye is to focus on it. If the eye is relaxed, the image will be at infinity, and in this case the object is exactly at the focal point. (You make this slight adjustment yourself when you "focus" on the object by moving the magnifying glass.)

FIGURE 34–27 When the same object is viewed at a shorter distance, the image on the retina is greater, so the object appears larger and more detail can be seen. The angle θ that the object subtends in (a) is greater than in (b).

FIGURE 34–28 Leaf viewed (a) through a magnifying glass, and (b) with the unaided eye, with the eye focused at its near point.

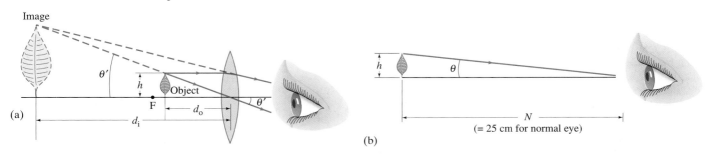

A comparison of part (a) of Fig. 34–28 with part (b), in which the same object is viewed at the near point with the unaided eye, reveals that the angle the object subtends at the eye is much larger when the magnifier is used. The **angular magnification** or **magnifying power**, M, of the lens is defined as the ratio of the angle subtended by an object when using the lens, to the angle subtended using the unaided eye, with the object at the near point N of the eye ($N = 25$ cm for the normal eye):

$$M = \frac{\theta'}{\theta}, \tag{34–5}$$

where θ and θ' are shown in Fig. 34–28. We can write M in terms of the focal length by noting that $\theta = h/N$ (Fig. 34–28b) and $\theta' = h/d_o$ (Fig. 34–28a), where h is the height of the object and we assume the angles are small so θ and θ' equal their sines and tangents. If the eye is relaxed (for least eye strain), the image will be at infinity and the object will be precisely at the focal point; see Fig. 34–29. Then $d_o = f$ and $\theta' = h/f$. Thus

$$M = \frac{\theta'}{\theta} = \frac{h/f}{h/N} = \frac{N}{f}. \qquad \left[\begin{array}{l} \text{eye focused at } \infty; \\ N = 25 \text{ cm for normal eye} \end{array} \right] \tag{34–6a}$$

We see that the shorter the focal length of the lens, the greater the magnification.

FIGURE 34–29 With the eye relaxed, the object is placed at the focal point, and the image is at infinity. Compare to Fig. 34–28a where the image is at the eye's near point.

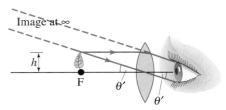

Magnification of a simple magnifier

[However, simple single-lens magnifiers are limited to about 2 or 3× because of distortion due to spherical aberration (Section 34–10).] The magnification of a given lens can be increased a bit by moving the lens and adjusting your eye so it focuses on the image at the eye's near point. In this case, $d_i = -N$ (see Fig. 34–28a) if your eye is very near the magnifier. Then the object distance d_o is given by

$$\frac{1}{d_o} = \frac{1}{f} - \frac{1}{d_i} = \frac{1}{f} + \frac{1}{N}.$$

We see from this equation that $d_o < f$, as shown in Fig. 34–28a, since $N/(f + N)$ must be less than 1. With $\theta' = h/d_o$ the magnification is

$$M = \frac{\theta'}{\theta} = \frac{h/d_o}{h/N} = \frac{N}{d_o} = N\left(\frac{1}{f} + \frac{1}{N}\right)$$

or

$$M = \frac{N}{f} + 1. \qquad \left[\begin{array}{l}\text{eye focused at near point, } N; \\ N = 25 \text{ cm for normal eye}\end{array}\right] \quad \textbf{(34–6b)}$$

We see that the magnification is slightly greater when the eye is focused at its near point, rather than relaxed.

EXAMPLE 34–11 ESTIMATE | **A jeweler's loupe.** An 8-cm-focal-length converging lens is used as a "jeweler's loupe," which is a magnifying glass. Estimate (a) the magnification when the eye is relaxed, and (b) the magnification if the eye is focused at its near point $N = 25$ cm.

SOLUTION (a) With the relaxed eye focused at infinity, $M = N/f = 25 \text{ cm}/8 \text{ cm} \approx 3\times$. (b) The magnification when the eye is focused at its near point ($N = 25$ cm), and the lens is near the eye, is:

$$M = 1 + \frac{N}{f} = 1 + \frac{25}{8} \approx 4\times.$$

* ### 34–8 | Telescopes

A telescope is used to magnify objects that are very far away. In most cases, the object can be considered to be at infinity.

Galileo, although he did not invent it,[†] developed the telescope into a usable and important instrument. He was the first to examine the heavens with the telescope (Fig. 34–30), and he made world-shaking discoveries (the moons of Jupiter, the phases of Venus, sunspots, the structure of the Moon's surface, that the Milky Way is made up of a huge number of individual stars, among others).

Several types of **astronomical telescope** exist. The common **refracting** type, sometimes called **Keplerian**, contains two converging lenses located at opposite ends of a long tube, as diagrammed in Fig. 34–31. The lens closest to the object is called the **objective lens** and forms a real image I_1 of the distant object in the plane of its focal point F_o (or near it if the object is not at infinity). Although this image, I_1, is smaller than the original object, it subtends a greater angle and is very close to the second lens, called the **eyepiece**, which acts as a magnifier. That is, the eye-

FIGURE 34–30 (a) Objective lens (mounted now in an ivory frame) from the telescope with which Galileo made his world-shaking discoveries, including the moons of Jupiter. (b) Later telescopes made by Galileo.

(a)

(b)

[†]Galileo built his first telescope in 1609 after having heard of such an instrument existing in Holland. The first telescopes magnified only 3 to 4 times, but Galileo soon made a 30-power instrument. The first Dutch telescope seems to date from about 1604, but there is a reference suggesting it may have been copied from an Italian telescope built as early as 1590. Kepler gave a ray description (1611) of the Keplerian telescope, which is named for him because he first described it, although he did not build it.

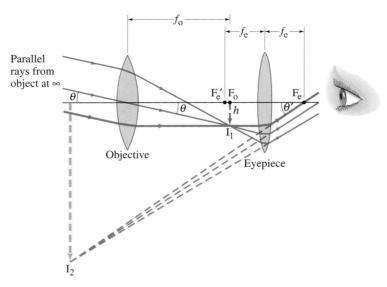

piece magnifies the image produced by the objective to produce a second, greatly magnified image, I_2, which is virtual, and inverted. If the viewing eye is relaxed, the eyepiece is adjusted so the image I_2 is at infinity. Then the real image I_1 is at the focal point F'_e of the eyepiece, and the distance between the lenses is $f_o + f_e$ for an object at infinity.

To find the total magnification of this telescope, we note that the angle an object subtends as viewed by the unaided eye is just the angle θ subtended at the telescope objective. From Fig. 34–31 we can see that $\theta \approx h/f_o$, where h is the height of the image I_1 and we assume θ is small so that $\tan\theta \approx \theta$. Note, too, that the thickest of the three rays drawn in Fig. 34–31 is parallel to the axis before it strikes the eyepiece and therefore passes through the eyepiece focal point F_e. Thus, $\theta' \approx h/f_e$ and the total magnifying power (angular magnification) of this telescope is

$$M = \frac{\theta'}{\theta} = -\frac{f_o}{f_e}, \tag{34–7}$$

Telescope magnification

where we have inserted a minus sign to indicate that the image is inverted. To achieve a large magnification, the objective lens should have a long focal length and the eyepiece a short focal length.

EXAMPLE 34–12 Telescope magnification. The largest refracting telescope in the world is located at the Yerkes Observatory in Wisconsin, Fig. 34–32. It is referred to as a "40-inch" telescope, meaning that the diameter of the objective is 40 inches, or 102 cm. The objective has a focal length of 19 m, and the eyepiece has a focal length of 10 cm. (*a*) Calculate the total magnifying power of this telescope. (*b*) Estimate the length of the telescope.

SOLUTION (*a*) From Eq. 34–7 we find,

$$M = -\frac{f_o}{f_e} = -\frac{19\,\text{m}}{0.10\,\text{m}} = -190\times.$$

(*b*) For a relaxed eye, the image I_1 is at the focal point of both the eyepiece and the objective lenses. The distance between the two lenses is thus $f_o + f_e \approx 19\,\text{m}$, which is essentially the length of the telescope.

FIGURE 34–32 This large refracting telescope was built in 1897 and is housed at Yerkes Observatory in Wisconsin. The objective lens is 102 cm (40 inches) in diameter, and the telescope tube is about 19 m long.

For an astronomical telescope to produce bright images of distant stars, the objective lens must be large to allow in as much light as possible. Indeed, the diameter of the objective (and hence its "light-gathering power") is an important parameter for an astronomical telescope, which is why the largest ones are specified by giving the objective diameter (such as the 200-inch Hale telescope on Palomar Mountain). The construction and grinding of large lenses is very difficult. Therefore,

(a) (b) (c) (d)

FIGURE 34–33 A concave mirror can be used as the objective of an astronomical telescope. Either a lens (a) or a mirror (b) can be used as the eyepiece. Arrangement (a) is called the Newtonian focus and (b) the Cassegrainian focus. Other arrangements are also possible. (c) The 200-inch (mirror diameter) Hale telescope on Palomar Mountain in California. (d) The 10-meter Keck telescope on Mauna Kea, Hawaii. The Keck combines thirty-six 1.8 meter six-sided mirrors into the equivalent of a very large (10-m diameter) single reflector.

Reflecting telescopes

the largest telescopes are **reflecting telescopes** that use a curved mirror as the objective, Fig. 34–33, since a mirror has only one surface to be ground and can be supported along its entire surface[†] (a large lens, supported at its edges, would sag under its own weight). Normally, the eyepiece lens or mirror (see Fig. 34–33) is removed so that the real image formed by the objective mirror can be recorded directly on film.

A **terrestrial telescope** (for use in viewing objects on Earth), unlike its astronomical counterpart, must provide an upright image. Two designs are shown in Fig. 34–34. The **Galilean** type shown in part (a), which Galileo used for his great astronomical discoveries, has a diverging lens as eyepiece which intercepts the converging rays from the objective lens before they reach a focus, and acts to form a virtual upright image. This design is often used in opera glasses. The tube is reasonably short, but the field of view is small. The second type, shown in Fig. 34–34b, is often called a **spyglass** and makes use of a third lens ("field lens") that acts to make the image upright as shown. A spyglass must be quite long. The most practical design today is the **prism binocular** which was shown in Fig. 33–31. The objective and eyepiece are converging lenses. The prisms reflect the rays by total internal reflection and shorten the physical size of the device, and they also act to produce an upright image. One prism reinverts the image in the vertical plane, the other in the horizontal plane.

FIGURE 34–34 Terrestrial telescopes that produce an upright image: (a) Galilean; (b) spyglass, or field-lens, type.

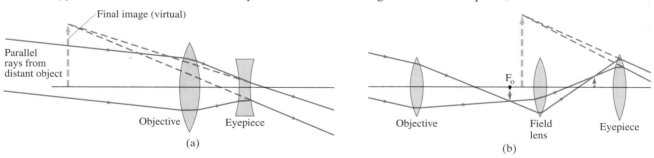

(a) (b)

* 34–9 Compound Microscope

Microscopes

The compound **microscope**, like the telescope, has both objective and eyepiece (or ocular) lenses, Fig. 34–35. The design is different from that for a telescope because a microscope is used to view objects that are very close, so the object distance is

[†]Another advantage of mirrors is that they exhibit no chromatic aberration because the light doesn't pass through them; and they can be ground in a parabolic shape to correct for spherical aberration (Section 34–10). The reflecting telescope was first proposed by Newton.

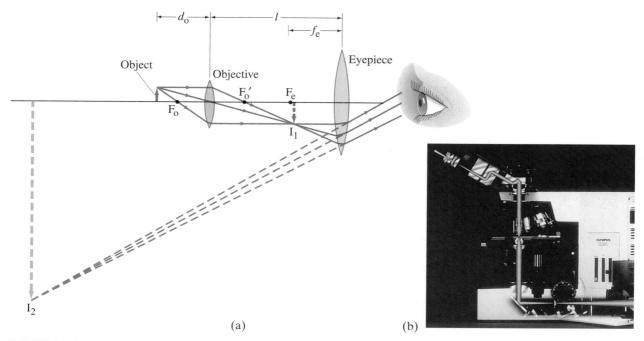

(a)   (b)

FIGURE 34–35 Compound microscope: (a) ray diagram, (b) photograph (illumination comes from the lower right, then up through the slide holding the object).

very small. The object is placed just beyond the objective's focal point as shown in Fig. 34–35a. The image I_1 formed by the objective lens is real, quite far from the lens, and much enlarged. This image is magnified by the eyepiece into a very large virtual image, I_2, which is seen by the eye and is inverted.

The overall magnification of a microscope is the product of the magnifications produced by the two lenses. The image I_1 formed by the objective is a factor m_o greater than the object itself. From Fig. 34–35a and Eq. 34–3 for the lateral magnification of a simple lens, we have

$$m_o = \frac{h_i}{h_o} = \frac{d_i}{d_o} = \frac{l - f_e}{d_o}, \tag{34–8}$$

where d_o and d_i are the object and image distances for the objective lens, l is the distance between the lenses (equal to the length of the barrel), and we ignored the minus sign in Eq. 34–3 which only tells us that the image is inverted. We set $d_i = l - f_e$, which is true only if the eye is relaxed, so that the image I_1 is at the eyepiece focal point F_e. The eyepiece acts like a simple magnifier. If we assume that the eye is relaxed, the eyepiece angular magnification M_e is (from Eq. 34–6a)

$$M_e = \frac{N}{f_e}, \tag{34–9}$$

where the near point $N = 25\,\text{cm}$ for the normal eye. Since the eyepiece enlarges the image formed by the objective, the overall angular magnification M is the product of the lateral magnification of the objective lens, m_o, times the angular magnification, M_e, of the eyepiece lens (Eqs. 34–8 and 34–9):

$$M = M_e m_o = \left(\frac{N}{f_e}\right)\left(\frac{l - f_e}{d_o}\right) \tag{34–10a}$$

$$\approx \frac{Nl}{f_e f_o}. \tag{34–10b}$$

Magnification of microscope

The approximation, Eq. 34–10b, is accurate when f_e and f_o are small compared to l, so $l - f_e \approx l$ and $d_o \approx f_o$ (Fig. 34–35a). This is a good approximation for large magnifications, since these are obtained when f_o and f_e are very small (they are in the denominator of Eq. 34–10b). In order to make lenses of very short focal length, which can be done best for the objective, compound lenses involving several elements must be used to avoid serious aberrations, as discussed in the next Section.

EXAMPLE 34–13 **Microscope.** A compound microscope consists of a $10\times$ eyepiece and a $50\times$ objective 17.0 cm apart. Determine (a) the overall magnification, (b) the focal length of each lens, and (c) the position of the object when the final image is in focus with the eye relaxed. Assume a normal eye, so $N = 25$ cm.

SOLUTION (a) The overall magnification is $(10\times)(50\times) = 500\times$.
(b) The eyepiece focal length is (Eq. 34–9) $f_e = N/M_e = 25\,\text{cm}/10 = 2.5$ cm. For the objective lens, it is easier to next find d_o (part (c)) before we find f_o because we can use Eq. 34–8. Solving for d_o, we find

$$d_o = \frac{l - f_e}{m_o} = \frac{(17.0\,\text{cm} - 2.5\,\text{cm})}{50} = 0.29\,\text{cm}.$$

Then, from the lens equation with $d_i = l - f_e = 14.5$ cm (see Fig. 34–35a),

$$\frac{1}{f_o} = \frac{1}{d_o} + \frac{1}{d_i} = \frac{1}{0.29\,\text{cm}} + \frac{1}{14.5\,\text{cm}} = 3.52;$$

so $f_o = 0.28$ cm.
(c) We just calculated $d_o = 0.29$ cm, which is very close to f_o.

---

* ## 34–10 Aberrations of Lenses and Mirrors

Earlier in this chapter, we developed a theory of image formation by a thin lens. We found, for example, that all rays from each point on an object are brought to a single point as the image point. This, and other results, were based on approximations such as that all rays make small angles with one another and we can use $\sin\theta \approx \theta$. Because of these approximations, we expect deviations from the simple theory and these are referred to as **lens aberrations**. There are several types of aberration; we will briefly discuss each of them separately but all may be present at one time.

Spherical aberration

Consider an object at any point (even at infinity) on the axis of a lens. Rays from this point that pass through the outer regions of the lens are brought to a focus at a different point from those that pass through the center of the lens. This is called **spherical aberration**, and is shown exaggerated in Fig. 34–36. Consequently, the image seen on a piece of film (for example) will not be a point but a tiny circular patch of light. If the film is placed at the point C, as indicated, the circle will have its smallest diameter, which is referred to as the **circle of least confusion**. Spherical aberration is present whenever spherical surfaces are used. It can be corrected by using nonspherical lens surfaces, but to grind such lenses is difficult and expensive. It can be minimized with spherical surfaces by choosing the curvatures so that equal amounts of bending occur at each lens surface; a lens can be designed like this for only one particular object distance. Spherical aberration is usually corrected (by which we mean reduced greatly) by the use of several lenses in combination, and by using only the central part of lenses.

FIGURE 34–36 Spherical aberration (exaggerated). Circle of least confusion is at C.

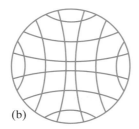

(a)　　　(b)

FIGURE 34–37 Distortion. Lenses may image a square grid of perpendicular lines to produce (a) barrel distortion or (b) pincushion distortion. These distortions can be seen in the photographs of Fig. 34–1d.

For object points off the lens axis, additional aberrations occur. Rays passing through the different parts of the lens cause spreading of the image that is noncircular. We won't go into the details but merely point out that there are two effects: **coma** (because the image of a point is comet-shaped rather than a circle) and **off-axis astigmatism**.[†] Furthermore, the image points for objects off the axis but at the same distance from the lens do not fall on a flat plane but on a curved surface—that is, the focal plane is not flat. (We expect this because the points on a flat plane, such as the film in a camera, are not equidistant from the lens.) This aberration is known as **curvature of field** and is obviously a problem in cameras and other devices where the film is placed in a flat plane. In the eye, however, the retina is curved, which compensates for this effect. Another aberration, known as **distortion**, is a result of variation of magnification at different distances from the lens axis. Thus a straight line object some distance from the axis may form a curved image. A square grid of lines may be distorted to produce "barrel distortion," or "pincushion distortion," Fig. 34–37. The latter is common in extreme wide-angle lenses.

All the above aberrations occur for monochromatic light and hence are referred to as *monochromatic aberrations*. Normal light is not monochromatic, and there will also be **chromatic aberration**. This aberration arises because of dispersion—the variation of index of refraction of transparent materials with wavelength (Section 33–6). For example, blue light is bent more than red light by glass. So if white light is incident on a lens, the different colors are focused at different points, Fig. 34–38, and there will be colored fringes in the image. Chromatic aberration can be eliminated for any two colors (and reduced greatly for all others) by the use of two lenses made of different materials with different indices of refraction and dispersion. Normally one lens is converging and the other diverging, and they are often cemented together (Fig. 34–39). Such a lens combination is called an **achromatic doublet** (or "color-corrected" lens).

It is not possible to fully correct all aberrations. Combining two or more lenses together can reduce them. High-quality lenses used in cameras, microscopes, and other devices are **compound lenses** consisting of many simple lenses (referred to as **elements**). A typical high-quality camera lens may contain six to eight (or more) elements.

For simplicity we will normally indicate lenses in diagrams as if they were simple lenses. But it must be remembered that good-quality lenses are compound.

The human eye is also subject to aberrations, but they are minimal. Spherical aberration, for example, is minimized because (1) the cornea is less curved at the edges than at the center, and (2) the lens is less dense at the edges than at the center. Both effects cause rays at the outer edges to be bent less strongly, and thus help to reduce spherical aberration. Chromatic aberration is partially compensated for because the lens absorbs the shorter wavelengths appreciably and the retina is less sensitive to the blue and violet wavelengths. This is just the region of the spectrum where dispersion—and thus chromatic aberration—is greatest (Fig. 33–26).

Spherical mirrors (Section 33–4) also suffer aberrations including spherical aberration (see Fig. 33–13). Mirrors can be ground in a parabolic shape to correct for spherical aberration, but they are much harder to make and therefore very expensive. Spherical mirrors do not, however, exhibit chromatic aberration because the light does not pass through them.

Off-axis aberrations

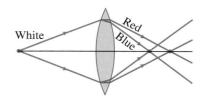

FIGURE 34–38 Chromatic aberration. Different colors are focused at different points.

FIGURE 34–39 Achromatic doublet.

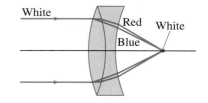

[†] Although the effect is the same as for astigmatism in the eye (Section 34–6), the cause is different. Off-axis astigmatism is no problem in the eye because objects are clearly seen only at the fovea, on the lens axis.

Summary

A lens uses refraction to produce a real or virtual image. Parallel rays of light are focused to a point, called the **focal point**, by a converging lens. The distance of the focal point from the lens is called the **focal length** f of the lens. After parallel rays pass through a diverging lens, they appear to diverge from a point, its focal point; and the corresponding focal length is considered negative.

The **power** P of a lens, which is $P = 1/f$, is given in diopters, which are units of inverse meters (m^{-1}).

For a given object, the position and size of the image formed by a lens can be found by ray tracing. Algebraically, the relation between image and object distances, d_i and d_o, and the focal length f, is given by the **lens equation**:

$$\frac{1}{d_o} + \frac{1}{d_i} = \frac{1}{f}.$$

The ratio of image height to object height, which equals the magnification m, is

$$m = \frac{h_i}{h_o} = -\frac{d_i}{d_o}.$$

When using the various equations of geometric optics, it is important to remember the **sign conventions** for all quantities involved: carefully review them when doing problems.

A **camera** lens forms an image on film by allowing light in through a shutter. The lens is focused by moving it relative to the film, and its f-stop (or lens opening) must be adjusted for the brightness of the scene and the chosen shutter speed. The f-stop is defined as the ratio of the focal length to the diameter of the lens opening.

The human **eye** also adjusts for the available light—by opening and closing the iris. It focuses not by moving the lens, but by adjusting the shape of the lens to vary its focal length. The image is formed on the retina, which contains an array of receptors known as rods and cones.

Diverging eyeglass or contact lenses are used to correct the defect of a nearsighted eye, which cannot focus well on distant objects. Converging lenses are used to correct for defects in which the eye cannot focus on close objects.

A **simple magnifier** is a converging lens that forms a virtual image of an object placed at (or within) the focal point. The **angular magnification**, when viewed by a relaxed normal eye, is

$$M = \frac{N}{f},$$

where f is the focal length of the lens and N is the near point of the eye (25 cm for a "normal" eye).

An **astronomical telescope** consists of an **objective** lens or mirror and an **eyepiece** that magnifies the real image formed by the objective. The **magnification** is equal to the ratio of the objective and eyepiece focal lengths, and the image is inverted:

$$M = -\frac{f_o}{f_e}.$$

A compound **microscope** also uses objective and eyepiece lenses, and the final image is inverted. The total magnification is the product of the magnifications of the two lenses and is approximately

$$M \approx \left(\frac{N}{f_e}\right)\left(\frac{l}{f_o}\right),$$

where l is the distance between the lenses, N is the near point of the eye, and f_o and f_e are the focal lengths of objective and eyepiece, respectively.

Microscopes, telescopes, and other optical instruments are limited in the formation of sharp images by **lens aberrations**. These include **spherical aberration**, in which rays passing through the edge of a lens are not focused at the same point as those that pass near the center; and **chromatic aberration**, in which different colors are focused at different points. Compound lenses, consisting of several elements, can largely correct for aberrations.

Questions

1. Where must the film be placed if a camera lens is to make a sharp image of an object far away?

2. A photographer moves closer to his subject and then refocuses. Does the camera lens move farther away from or nearer to the film?

3. Can a diverging lens form a real image under any circumstances? Explain.

4. Show that a real image formed by a thin lens is always inverted, whereas a virtual image is always upright if the object is real.

5. Light rays are said to be "reversible." Is this consistent with the lens equation?

6. Can real images be projected on a screen? Can virtual images? Can either be photographed? Discuss carefully.

7. A thin converging lens is moved closer to a nearby object. Does the real image formed change (a) in position, (b) in size? If yes, describe how.

8. Compare the mirror equation with the lens equation. Discuss similarities and differences, especially the sign conventions for the quantities involved.

9. A lens is made of a material with an index of refraction $n = 1.30$. In air, it is a converging lens. Will it still be a converging lens if placed in water? Explain, using a ray diagram.

10. A dog stands facing a converging lens with its tail in the air. If the nose and the tail are each focused on a screen in turn, which will have the greater magnification?

11. A cat stands facing a converging lens with its tail in the air. Under what circumstances (if any) would the image of the nose be virtual and the image of the tail be real? Where would the image of the rest of the cat be?

12. Why, in Example 34–5, must the converging lens have a shorter focal length than the diverging lens if the latter's focal length is to be determined by combining them?

13. An unsymmetrical lens (say, planoconvex) forms an image of a nearby object. Does the image point change if the lens is turned around?

14. The thicker a double convex lens is in the center as compared to its edges, the shorter its focal length for a given lens diameter. Explain.

15. Does the focal length of a lens depend on the fluid in which it is immersed? What about the focal length of a spherical mirror? Explain.

16. An underwater lens consists of a carefully shaped thin-walled plastic container filled with air. What shape should it have in order to be (a) converging (b) diverging? Use ray diagrams to support your answer.

* 17. Why is the depth of field greater, and the image sharper, when a camera lens is "stopped down" to a larger *f*-number? Ignore diffraction.

* 18. Explain why swimmers with good eyes see distant objects as blurry when they are underwater. Use a diagram and also show why goggles correct this problem.

* 19. Will a nearsighted person who wears corrective lenses be able to see clearly underwater when wearing glasses? Use a diagram to show why or why not.

* 20. You can tell whether a person is nearsighted or farsighted by looking at the width of the face through their glasses. If the person's face appears narrower through the glasses, is the person farsighted or nearsighted?

* 21. The human eye is much like a camera—yet, when a camera shutter is left open and the camera moved, the image will be blurred; but when you move your head with your eyes open, you still see clearly. Explain.

* 22. In attempting to discern distant details, people will sometimes squint. Why does this help?

* 23. Is the image formed on the retina of the human eye upright or inverted? Discuss the implications of this for our perception of objects.

* 24. Reading glasses use converging lenses. A simple magnifier is also a converging lens. Are reading glasses therefore magnifiers? Discuss the similarities and differences between converging lenses as used for these two different purposes.

* 25. Spherical aberration in a thin lens is minimized if rays are bent equally by the two surfaces. If a planoconvex lens is used to form a real image of an object at infinity, which surface should face the object? Use ray diagrams to show why.

Problems

Sections 34–1 and 34–2

1. (I) A sharp image is located 88.0 mm behind a 65.0-mm-focal-length converging lens. Find the object distance (a) using a ray diagram, (b) by calculation.

2. (I) A certain lens focuses an object 2.85 m away as an image 48.3 cm on the other side of the lens. What type of lens is it and what is its focal length? Is the image real or virtual?

3. (I) (a) What is the power of a 27.5-cm-focal-length lens? (b) What is the focal length of a −6.25-diopter lens? Are these lenses converging or diverging?

4. (II) A stamp collector uses a converging lens with focal length 24 cm to view a stamp 18 cm in front of the lens. (a) Where is the image located? (b) What is the magnification?

5. (II) How large is the image of the Sun on the film used in a camera with (a) a 28-mm-focal-length lens, (b) a 50-mm-focal-length lens, and (c) a 200-mm-focal-length lens? The Sun's diameter is 1.4×10^6 km and it is 1.5×10^8 km away.

6. (II) A 35-mm slide (picture size is actually 24 by 36 mm) is to be projected on a screen 1.80 by 2.70 m placed 8.00 m from the projector. What focal-length lens should be used if the image is to cover the screen?

7. (II) An 80-mm-focal-length lens is used to focus an image on the film of a camera. The maximum distance allowed between the lens and the film plane is 120 mm. (a) How far ahead of the film should the lens be if the object to be photographed is 10.0 m away? (b) 3.0 m away? (c) 1.0 m away? (d) What is the closest object this lens could photograph sharply?

8. (II) A −6.0-diopter lens is held 12.5 cm from an ant 1.0 mm high. What is the position, type, and height of the image?

9. (II) It is desired to magnify reading material by a factor of 2.5× when a book is placed 8.0 cm behind a lens. (a) Draw a ray diagram and describe the type of image this would be. (b) What type of lens is needed for this? (c) What is the power of the lens in diopters?

10. (II) (a) How far from a 50.0-mm-focal-length lens must an object be placed if its image is to be magnified 2.00× and be real? (b) What if the image is to be virtual and magnified 2.00×?

11. (II) (a) An object 37.5 cm in front of a certain lens is imaged 8.20 cm in front of that lens (on the same side as the object). What type of lens is this and what is its focal length? Is the image real or virtual? (b) If the image were located, instead, 46.0 cm in front of the lens, what type of lens would it be and what focal length would it have?

12. (II) (a) A 2.20-cm-high insect is 1.10 m from a 135-mm-focal-length lens. Where is the image, how high is it, and what type is it? (b) What if $f = -135$ mm?

13. (II) A bright object and a viewing screen are separated by a distance of 76.0 cm. At what location(s) between the object and the screen should a lens of focal length 16.0 cm be placed in order to produce a crisp image on the screen? [Hint: First draw a diagram.]

14. (II) How far apart are an object and an image formed by a 75-cm-focal-length converging lens if the image is 2.75× larger than the object and is real?

15. (II) Show analytically that the image formed by a converging lens is real and inverted if the object is beyond the focal point $(d_o > f)$, and is virtual and upright if the object is within the focal point $(d_o < f)$. Describe the image if the object is itself an image formed by another lens, so its position is beyond the lens (on the side of the lens opposite from the incoming light), for which $-d_o > f$, and for which $0 < -d_o < f$.

16. (III) (a) Show that the lens equation can be written in the Newtonian form:

$$xx' = f^2,$$

where x is the distance of the object from the focal point on the front side of the lens, and x' is the distance of the image to the focal point on the other side of the lens. Calculate the location of an image if the object is placed 45.0 cm in front of a convex lens with a focal length of 32.0 cm using (b) the standard form of the thin lens formula, and (c) the Newtonian form, derived above.

Section 34–3

17. (II) Two 27.0-cm-focal-length converging lenses are placed 16.5 cm apart. An object is placed 35.0 cm in front of one. Where will the final image formed by the second lens be located? What is the total magnification?

18. (II) A diverging lens with $f = -31.5$ cm is placed 14.0 cm behind a converging lens with $f = 20.0$ cm. Where will an object at infinity be focused?

19. (II) The two converging lenses of Example 34–4 are placed so they are now only 20.0 cm apart. The object is still 60.0 cm in front of the first lens as in Fig. 34–13. In this case, determine (a) the position of the final image, and (b) the overall magnification. (c) Sketch the ray diagram for this system.

20. (II) A 31.0-cm-focal-length converging lens is 24.0 cm behind a diverging lens. Parallel light strikes the diverging lens. After passing through the converging lens, the light is again parallel. What is the focal length of the diverging lens? [Hint: First draw a ray diagram.]

21. (II) A diverging lens is placed next to a converging lens of focal length f_C, as in Fig. 34–14. If f_T represents the focal length of the combination, show that the focal length of the diverging lens, f_D, is given by

$$\frac{1}{f_D} = \frac{1}{f_T} - \frac{1}{f_C}.$$

Section 34–4

22. (I) A double concave lens has surface radii of 33.4 cm and 23.8 cm. What is the focal length if $n = 1.58$?

23. (I) Both surfaces of a double convex lens have radii of 31.0 cm. If the focal length is 28.9 cm, what is the index of refraction of the lens material?

24. (I) Show that if the lens of Example 34–7 is reversed so the light enters the curved face, the focal length is unchanged.

25. (I) A planoconvex lens (Fig. 34–1a) is to have a focal length of 17.5 cm. If made from fused quartz, what must be the radius of curvature of the convex surface?

26. (I) A glass ($n = 1.50$) planoconcave lens has a focal length of -21.5 cm. What is the radius of the concave surface?

27. (II) A prescription for a corrective lens calls for +2.50 diopters. The lensmaker grinds the lens from a "blank" with $n = 1.56$ and a preformed convex front surface of radius of curvature of 20.0 cm. What should be the radius of curvature of the other surface?

28. (II) An object is placed 90.0 cm from a glass lens ($n = 1.56$) with one concave surface of radius 22.0 cm and one convex surface of radius 18.5 cm. Where is the final image? What is the magnification?

29. (III) A glass lens ($n = 1.50$) in air has a power of +4.5 diopters. What would its power be if it were submerged in water?

* Section 34–5

* 30. (I) A 45-mm-focal-length lens has f-stops ranging from $f/1.4$ to $f/22$. What is the corresponding range of lens diaphragm diameters?

* 31. (I) A television camera lens has a 14-cm focal length and a lens diameter of 6.0 cm. What is its f-number?

* 32. (I) A properly exposed photograph is taken at $f/16$ and $\frac{1}{60}$ s. What lens opening would be required if the shutter speed were $\frac{1}{1000}$ s?

* 33. (II) A "pinhole" camera uses a tiny pinhole instead of a lens. Show, using ray diagrams, how reasonably sharp images can be formed using such a pinhole camera. In particular, consider two point objects 2.0 cm apart that are 1.0 m from a 1.0-mm-diameter pinhole. Show that on a piece of film 7.0 cm behind the pinhole the two objects produce two separate circles that do not overlap.

* 34. (II) Suppose that a correct exposure is $\frac{1}{250}$ s at $f/11$. Under the same conditions, what exposure time would be needed for a pinhole camera if the pinhole diameter is 1.0 mm and the film is 7.0 cm from the hole?

* 35. (II) A nature photographer wishes to photograph a 32-m tall tree from a distance of 55 m. What focal-length lens should be used if the image is to fill the 24-mm height of the film?

36. (I) A human eyeball is about 2.0 cm long and the pupil has a maximum diameter of about 5.0 mm. What is the "speed" of this lens?

37. (II) A person struggles to read by holding a book at arm's length, a distance of 50 cm away. What power of reading glasses should be prescribed for him, assuming they will be placed 2.0 cm from the eye and he wants to read at the "normal" near point of 25 cm?

38. (II) Reading glasses of what power are needed for a person whose near point is 120 cm, so that he can read a computer screen at 50 cm? Assume a lens–eye distance of 1.8 cm.

39. (II) Show that if the nearsighted person in Example 34–10 wore contact lenses corrected for the far point ($= \infty$), that the near point would be 41 cm. (Would glasses be better in this case?)

40. (II) A person's left eye is corrected by a -4.0-diopter lens, 2.0 cm from the eye. (a) Is this person near- or farsighted? (b) What is this person's far point without glasses?

41. (II) A person's right eye can see objects clearly only if they are between 25 cm and 75 cm away. (a) What power of contact lens is required so that objects far away are sharp? (b) What will be the near point with the lens in place?

42. (II) About how much longer is the nearsighted eye in Example 34–10 than the 2.0 cm of a normal eye?

43. (II) One lens of a nearsighted person's eyeglasses has a focal length of -25.0 cm and the lens is 1.8 cm from the eye. If the person switches to contact lenses that are placed directly on the eye, what should be the focal length of the corresponding contact lens?

44. (II) What is the focal length of the eye lens system when viewing an object (a) at infinity, and (b) 30 cm from the eye? Assume that the lens–retina distance is 2.0 cm.

45. (II) A nearsighted person has near and far points of 10.0 and 20.0 cm respectively. If she puts on a pair of glasses with power $P = -4.0$ D, what are her new near and far points? Neglect the distance between her eye and the lens.

Section 34–7

46. (I) What is the magnification of a lens used with a relaxed eye if its focal length is 11 cm?

47. (I) What is the focal length of a magnifying glass of 3.0× magnification for a relaxed normal eye?

48. (I) A magnifier is rated at 2.5× for a normal eye focusing on an image at the near point. (a) What is its focal length? (b) What is its focal length if the 2.5× refers to a relaxed eye?

49. (II) Sherlock Holmes is using an 8.50-cm-focal-length lens as his magnifying glass. To obtain maximum magnification, where must the object be placed (assume a normal eye), and what will be the magnification?

50. (II) A 3.70-mm-wide bolt is viewed with a 9.00-cm-focal-length lens. A normal eye views the image at its near point. Calculate (a) the angular magnification, (b) the width of the image, and (c) the object distance from the lens.

51. (II) A small insect is placed 5.35 cm from a +6.00-cm-focal-length lens. Calculate (a) the position of the image, and (b) the angular magnification.

52. (II) A magnifying glass with a focal length of 9.5 cm is used to read print placed at a distance of 8.5 cm. Calculate: (a) the position of the image; (b) the linear magnification; and (c) the angular magnification.

53. (II) A magnifying glass is rated at 3.0× for a normal eye that is relaxed. What would be the magnification for a relaxed eye whose near point is (a) 60 cm, and (b) 18 cm? Explain the differences.

* Section 34–8

* 54. (I) What is the magnification of an astronomical telescope whose objective lens has a focal length of 72 cm, and whose eyepiece has a focal length of 2.8 cm? What is the overall length of the telescope when adjusted for a relaxed eye?

* 55. (I) The overall magnification of an astronomical telescope is desired to be 25×. If an objective of 80 cm focal length is used, what must be the focal length of the eyepiece? What is the overall length of the telescope when adjusted for use by the relaxed eye?

* 56. (I) An 8.0× binocular has 3.0-cm-focal-length eyepieces. What is the focal length of the objective lenses?

* 57. (II) An astronomical telescope has an objective with focal length 95 cm and a +35 D eyepiece. What is the total magnification?

* 58. (II) An astronomical telescope has its two lenses spaced 76.0 cm apart. If the objective lens has a focal length of 74.5 cm, what is the magnification of this telescope? Assume a relaxed eye.

* 59. (II) A Galilean telescope adjusted for a relaxed eye is 33.0 cm long. If the objective lens has a focal length of 36.0 cm, what is the magnification?

* 60. (II) What is the magnifying power of an astronomical telescope using a reflecting mirror whose radius of curvature is 6.2 m and an eyepiece whose focal length is 2.8 cm?

* 61. (II) The Moon's image appears to be magnified 120× by a reflecting astronomical telescope with an eyepiece having a focal length of 3.3 cm. What are the focal length and radius of curvature of the main mirror?

* 62. (II) A 130× astronomical telescope is adjusted for a relaxed eye when the two lenses are 1.25 m apart. What is the focal length of each lens?

* 63. (III) A 7.0× pair of binoculars has an objective focal length of 26 cm. If the binoculars are focused on an object 4.0 m away (from the objective), what is the magnification? (The 7.0× refers to objects at infinity; Eq. 34–7 holds only for objects at infinity and not for nearby ones.)

* Section 34–9

* 64. (I) A microscope uses an eyepiece with a focal length of 1.6 cm. Using a normal eye with a final image at infinity, the tube length is 17.5 cm and the focal length of the objective lens is 0.65 cm. What is the magnification of the microscope?

* 65. (I) A 650× microscope uses a 0.40-cm-focal-length objective lens. If the tube length is 17.5 cm, what is the focal length of the eyepiece? Assume a normal eye and that the final image is at infinity.

* 66. (II) A microscope has a 12.0× eyepiece and a 56.0× objective 20.0 cm apart. Calculate (a) the total magnification, (b) the focal length of each lens, and (c) where the object must be for a normal relaxed eye to see it in focus.

* 67. (II) A microscope has a 1.8-cm-focal-length eyepiece and 0.80-cm objective. Assuming a relaxed normal eye calculate (a) the position of the object if the distance between the lenses is 16.0 cm, and (b) the total magnification.

* 68. (II) Repeat Problem 69 assuming that the final image is located 25 cm from the eyepiece (near point of a normal eye).

* 69. (II) The eyepiece of a compound microscope has a focal length of 2.70 cm and the objective has $f = 0.740$ cm. If an object is placed 0.790 cm from the objective lens, calculate (a) the distance between the lenses when the microscope is adjusted for a relaxed eye, and (b) the total magnification.

* Section 34–10

* 70. (II) An achromatic lens is made of two very thin lenses placed in contact that have focal lengths of $f_1 = -28$ cm and $f_2 = +23$ cm. (a) Is the combination converging or diverging? (b) What is the net focal length?

* 71. (III) Let's examine spherical aberration in a particular situation. A planoconvex lens of index of refraction 1.50 and a radius of curvature $R = 12.0$ cm is shown in Fig. 34–40. Consider an incoming ray parallel to the principle axis and a height h above it as shown. Determine the distance d, from the flat face of the lens, to where this ray crosses the principle axis if (a) $h = 1.0$ cm, and (b) $h = 6.0$ cm. (c) How far apart are these "focal points"? (d) How large is the "circle of confusion" produced by the $h = 6.0$ cm ray at the "focal point" for $h = 1.0$ cm?

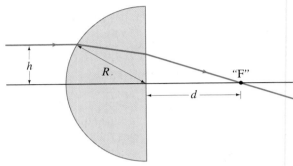

FIGURE 34–40 Problem 71.

General Problems

72. If a 135-mm telephoto lens is designed to cover object distances from 1.40 m to ∞, over what distance must the lens move relative to the plane of the film?

73. A 200-mm-focal-length lens can be adjusted so that it is 200.0 mm to 206.0 mm from the film. For what range of object distances can it be adjusted?

74. Show that for objects very far away (assume infinity), the magnification of a camera lens is proportional to its focal length.

75. For a camera equiped with a 50-mm-focal-length lens, what is the object distance if the image height equals the object height? How far is the object from the image on the film?

76. What is the magnifying power of a +8.0-D lens used as a magnifier? Assume a relaxed normal eye.

77. A lucite planoconvex lens (Fig. 34–1a) has one flat surface and one with $R = 18.4$ cm. It is used to view an object, located 66.0 cm away from the lens, which is a mixture of red and yellow. The index of refraction of the lucite is 1.5106 for red light and 1.5226 for yellow light. What are the locations of the red and yellow images formed by the lens?

78. A movie star catches a reporter shooting pictures of her at home. She claims the reporter was trespassing and to prove her point, she gives as evidence the film she seized. Her 1.75-m height is 8.25 mm high on the film, and the focal length of the camera lens was 200 mm. How far away from the subject was the reporter standing?

79. In a slide or movie projector, the film acts as the object whose image is projected on a screen (Fig. 34–41). If a 100-mm-focal-length lens is to project an image on a screen 7.50 m away, how far from the lens should the slide be? If the slide is 36 mm wide, how wide will the picture be on the screen?

Slide

Lens

Screen

FIGURE 34–41 Slide projector, Problem 79.

80. A converging lens with focal length of 12.0 cm is placed in contact with a diverging lens with a focal length of 20.0 cm. What is the focal length of the combination, and is the combination converging or diverging?

81. Show analytically that a diverging lens can never form a real image of a real object. Can you describe a situation in which a diverging lens can form a real image?

82. A lighted candle is placed 30 cm in front of a converging lens of focal length $f_1 = 15$ cm, which in turn is 50 cm in front of another converging lens of focal length $f_2 = 10$ cm (see Fig. 34–42). (a) Draw a ray diagram and estimate the location and the size of the final image. (b) Calculate the position and size of the final image.

$f_1 = 15$ cm $f_2 = 10$ cm

30 cm   50 cm

FIGURE 34–42 Problem 82.

83. A bright object is placed on one side of a converging lens of focal length f, and a white screen for viewing the image is on the opposite side. The distance $d_T = d_i + d_o$ between the object and the screen is kept fixed, but the lens can be moved. Show that (a) if $d_T > 4f$, there will be two positions where the lens can be placed and a sharp image will be produced on the screen, and (b) if $d_T < 4f$, there will be no lens position where a sharp image is formed. (c) Determine the distance between the two lens positions in part (a), and the ratio of the image sizes.

84. (a) Show that if two thin lenses of focal lengths f_1 and f_2 are placed in contact with each other, the focal length of the combination is given by $f_T = f_1 f_2 / (f_1 + f_2)$. (b) Show that the power P of the combination of two lenses is the sum of their separate powers, $P = P_1 + P_2$.

85. A lens whose index of refraction is n is submerged in a material whose index of refraction is n' ($n' \neq 1$). Derive the equivalent of Eqs. 34–2, 34–3 and 34–4 for this lens.

86. Sam purchases +2.5-diopter eyeglasses which correct his faulty vision to put his near point at 25 cm. (Assume he wears the lenses 2.0 cm from his eyes.) (a) Is Sam near-sighted or farsighted? (b) Calculate the focal length of Sam's glasses. (c) Calculate Sam's near point without glasses. (d) Pam, who has normal eyes with near point at 25 cm, puts on Sam's glasses. Calculate Pam's near point with Sam's glasses on.

87. A 50-year-old man uses +2.5-diopter lenses to be able to read a newspaper 25 cm away. Ten years later, he finds that he must hold the paper 35 cm away to see clearly with the same lenses. What power lenses does he need now? (Distances are measured from the lens.)

88. A woman can see clearly with her right eye only when objects are between 40 cm and 150 cm away. Prescription bifocals should have what powers so that she can see distant objects clearly (upper part) and be able to read a book 25 cm away (lower part)? Assume that the glasses will be 2.0 cm from the eye.

89. A child has a near point of 15 cm. What is the maximum magnification the child can obtain using an 8.0-cm-focal-length magnifier? Compare to that for a normal eye.

*90. As early morning passed toward midday, and the sunlight got more intense, a photographer who was taking repeated shots of the same subject noted that, if she kept her shutter speed constant, she had to change the f-number from $f/5.6$ to $f/22$. By how much had the sunlight intensity increased during that time?

*91. A person tries to make a telescope using the lenses from reading glasses. They have powers of +2.0 D and +5.0 D, respectively. (a) What maximum magnification telescope is possible? (b) Which lens should be used as the eyepiece?

*92. Exposure times must be increased for pictures taken at very short distances, because of the increased distance of the lens from the film for a focused image. (a) Show that when the object is so close to the camera that the image height equals the object height, the exposure time must be four times longer than when the object is a long distance away (say, ∞), given the same illumination and f-stop. (b) Show that if d_o is at least four or five times the focal length f of the lens, the exposure time is increased negligibly relative to the same object being a great distance away.

*93. The objective lens and the eyepiece of a telescope are spaced 85 cm apart. If the eyepiece is +25 diopters, what is the total magnification of the telescope?

The wave nature of light nicely explains how light reflected from the front and back surfaces of this very thin film of soapy water interferes constructively to produce the bright colors. Which color we see depends on the thickness of the film at that point—the film is very thin at the top and gets thicker going down toward the bottom (because of gravity). Can you use the clues (colors of light) to determine the thickness of the film at any point? You should be able to after reading this chapter.

The Wave Nature of Light; Interference

That light carries energy is obvious to anyone who has focused the Sun's rays with a magnifying glass on a piece of paper and burned a hole in it. But how does light travel, and in what form is this energy carried? In our discussion of waves in Chapter 15, we noted that energy can be carried from place to place in basically two ways: by particles or by waves. In the first case, material bodies or particles can carry energy, such as an avalanche or rushing water. In the second case, water waves and sound waves, for example, can carry energy over long distances even though mass itself does not travel these distances. Then, what can we say about the nature of light: does light travel as a stream of particles away from its source, or does it travel in the form of waves that spread outward from the source?

Historically, this question has turned out to be a difficult one. For one thing, light does not reveal itself in any obvious way as being made up of tiny particles nor do we see tiny light waves passing by as we do water waves. The evidence seemed to favor first one side and then the other until about 1830, when most physicists had accepted the wave theory. By the end of the nineteenth century, light was considered to be an *electromagnetic wave* (Chapter 32). In the early twentieth century, light was shown to have a particle nature as well, as we shall discuss in Chapter 38. Nonetheless, the wave theory of light remains valid and has proved very successful. We now investigate the evidence for the wave theory and how it has explained a wide range of phenomena.

35–1 | Huygens' Principle and Diffraction

The Dutch scientist Christiaan Huygens (1629–1695), a contemporary of Newton, proposed a wave theory of light that had much merit. Still useful today is a technique he developed for predicting the future position of a wave front when an earlier position is known. This is known as **Huygens' principle** and can be stated as follows: *Every point on a wave front can be considered as a source of tiny wavelets that spread out in the forward direction at the speed of the wave itself. The new wave front is the envelope of all the wavelets—that is, the tangent to all of them.*

As a simple example of the use of Huygens' principle, consider the wave front AB in Fig. 35–1, which is traveling away from a source S. We assume the medium is *isotropic*—that is, the speed v of the waves is the same in all directions. To find the wave front a short time t after it is at AB, tiny circles are drawn with radius $r = vt$. The centers of these tiny circles are on the original wave front AB and the circles represent Huygens' (imaginary) wavelets. The tangent to all these wavelets, the line CD, is the new position of the wave front.

Huygens' principle is particularly useful when waves impinge on an obstacle and the wave fronts are partially interrupted. Huygens' principle predicts that waves bend in behind an obstacle, as shown in Fig. 35–2. This is just what water waves do, as we saw in Chapter 15 (Figs. 15–32 and 15–33). The bending of waves behind obstacles into the "shadow region" is known as **diffraction**. Since diffraction occurs for waves, but not for particles, it can serve as one means for distinguishing the nature of light.

Does light actually exhibit diffraction? In the mid-seventeenth century, a Jesuit priest, Francesco Grimaldi (1618–1663), had observed that when sunlight entered a darkened room through a tiny hole in a screen, the spot on the opposite wall was larger than would be expected from geometric rays. He also observed that the border of the image was not clear but was surrounded by colored fringes. Grimaldi attributed this to the diffraction of light.

Note that the ray model (Chapter 33) cannot account for diffraction, and it is important to be aware of the limitations of the ray model. Geometric optics using rays is so successful in its limited sphere because normal openings and obstacles are much larger than the wavelength of the light, and so relatively little diffraction or bending occurs.

Huygens' principle

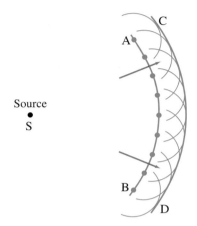

FIGURE 35–1 Huygens' principle used to determine the wave front CD when the wave front AB is given.

(a)   (b)   (c)

FIGURE 35–2 Huygens' principle is consistent with diffraction (a) around the edge of an obstacle, (b) through a large hole, (c) through a small hole whose size is on the order of the wavelength of the wave.

35–2 | Huygens' Principle and the Law of Refraction

The laws of reflection and refraction were well known in Newton's time. The law of reflection could not distinguish between the two theories: waves versus particles, that we discussed on the previous page. For when waves reflect from an obstacle, the angle of incidence equals the angle of reflection (Fig. 15–22). The same is true of particles—think of a tennis ball without spin striking a flat surface.

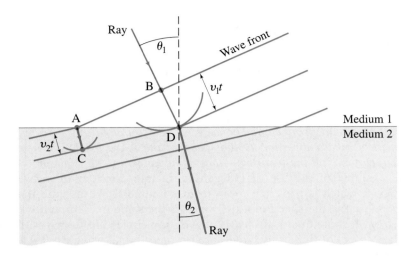

FIGURE 35–3 Refraction explained, using Huygens' principle.

The law of refraction is another matter. Consider light entering a medium where it is bent toward the normal, as when it travels from air into water. As shown in Fig. 35–3, this effect can be constructed using Huygens' principle if we assume the speed of light is less in the second medium $(v_2 < v_1)$. In time t, the point B on wave front AB goes a distance $v_1 t$ to reach point D. Point A, on the other hand, travels a distance $v_2 t$ to reach point C, and $v_2 t < v_1 t$. Huygens' principle is applied to points A and B to obtain the curved wavelets shown at C and D. The wave front is tangent to these two wavelets, so the new wave front is the line CD. Hence the rays, which are perpendicular to the wave fronts, bend toward the normal if $v_2 < v_1$, as drawn.[†] Newton favored a particle theory of light which predicted the opposite result, that the speed of light would be greater in the second medium $(v_2 > v_1)$. Thus the wave theory predicts that the speed of light in water, for example, is less than in air; and Newton's particle theory predicts the reverse. An experiment to actually measure the speed of light in water was performed in 1850 by the French physicist Jean Foucault, and it confirmed the wave-theory prediction. By then, however, the wave theory was already fully accepted, as we shall see in the next Section.

It is easy to show that Snell's law of refraction follows directly from Huygens' principle, given that the speed of light v in any medium is related to the speed in a vacuum, c, and the index of refraction, n, by Eq. 33–1, $v = c/n$. From the Huygens' construction of Fig. 35–3, angle ADC is equal to θ_2 and angle BAD is equal to θ_1. Then for the two triangles that have the common side AD, we have

$$\sin \theta_1 = \frac{v_1 t}{\text{AD}}, \qquad \sin \theta_2 = \frac{v_2 t}{\text{AD}}.$$

We divide these two equations and obtain:

$$\frac{\sin \theta_1}{\sin \theta_2} = \frac{v_1}{v_2}.$$

Then, since $v_1 = c/n_1$ and $v_2 = c/n_2$,

$$n_1 \sin \theta_1 = n_2 \sin \theta_2,$$

which is Snell's law of refraction, Eq. 33–5. (The law of reflection can be derived from Huygens' principle in a similar way, and this is given as Problem 1 at the end of the chapter.)

[†] This is basically the same as the discussion around Fig. 15–31.

When a light wave travels from one medium to another, its frequency does not change, but its wavelength does. This can be seen from Fig. 35–3, where we assume each of the blue lines representing a wave front corresponds to a crest (peak) of the wave. Then

Wavelength depends on n

$$\frac{\lambda_2}{\lambda_1} = \frac{v_2 t}{v_1 t} = \frac{v_2}{v_1} = \frac{n_1}{n_2},$$

where, in the last step, we used Eq. 33–1, $v = c/n$. If medium 1 is a vacuum (or air), so $n_1 = 1$, $v_1 = c$, and we call λ_1 simply λ, then the wavelength in another medium of index of refraction $n (= n_2)$ will be

$$\lambda_n = \frac{\lambda}{n}. \tag{35–1}$$

This result is consistent with the frequency f being unchanged since $c = f\lambda$. Combining this with $v = f\lambda_n$ in a medium where $v = c/n$ gives $\lambda_n = v/f = c/nf = f\lambda/nf = \lambda/n$, which checks.

Wave fronts can be used to explain how mirages are produced by refraction of light. Let us explain why, for example, on a hot day motorists sometimes see a mirage of water on the highway ahead of them, with distant vehicles seemingly reflected in it (Fig. 35–4a). On a hot day, there can be a layer of very hot air next to the roadway (made hot by the sun beating on the road). Hot air is less dense than cooler air, so the index of refraction is slightly lower in the hot air. In Fig. 35–4b, we see a diagram of light coming from one point on a distant car (on the right) heading left toward the observer. Wave fronts and two rays are shown. Ray A heads directly at the observer and follows a straight-line path, and represents the normal view of the car. Ray B is a ray initially directed slightly downward. But it does not hit the ground. Instead, ray B is bent slightly as it moves through layers of air of different index of refraction. The wave fronts, shown in blue in Fig. 35–4b, move slightly faster in the layers of air nearer the ground (see Fig. 35–3, and also the soldier analogy in Fig. 15–31). Thus ray B is bent as shown, and seems to the observer to be coming from below (dashed line) as if reflected off the road. Hence the mirage.

→ PHYSICS APPLIED

Highway mirages

FIGURE 35–4 (a) A highway mirage. (b) Drawing (greatly exaggerated) showing wave fronts and rays to explain highway mirages. Note how the sections of the wavefronts near the ground move faster and so are further apart.

(a)

(b)

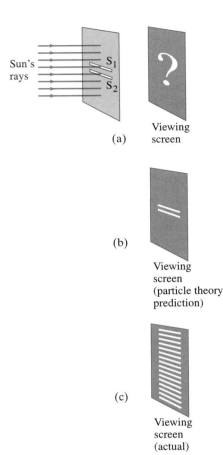

Sun's rays

(a) Viewing screen

(b) Viewing screen (particle theory prediction)

(c) Viewing screen (actual)

FIGURE 35–5 (a) Young's double-slit experiment. (b) If light consists of particles, we would expect to see two bright lines on the screen behind the slits. (c) Young observed many lines.

35–3 Interference—Young's Double-Slit Experiment

In 1801, the Englishman Thomas Young (1773–1829) obtained convincing evidence for the wave nature of light and was even able to measure the wavelengths for visible light. Figure 35–5a shows a schematic diagram of Young's famous double-slit experiment. Light from a single source (Young used the Sun) falls on a screen containing two closely spaced slits S_1 and S_2. If light consists of tiny particles, we might expect to see two bright lines on a screen placed behind the slits as in (b). But Young observed instead a series of bright lines as in (c). Young was able to explain this result as a **wave-interference** phenomenon. To see this, imagine plane waves of light of a single wavelength—called **monochromatic**, meaning "one color"—falling on the two slits as shown in Fig. 35–6. Because of diffraction, the waves leaving the two small slits spread out as shown. This is equivalent to the interference pattern produced when two rocks are thrown into a lake (Fig. 15–24), or when sound from two loudspeakers interferes (Fig. 16–16).

FIGURE 35–6 If light is a wave, light passing through one of two slits should interfere with light passing through the other slit.

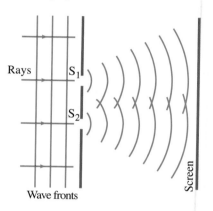

To see how an interference pattern is produced on the screen, we make use of Fig. 35–7. Waves of wavelength λ are shown entering the slits S_1 and S_2, which are a distance d apart. The waves spread out in all directions after passing through the slits, but they are shown only for three different angles θ. In Fig. 35–7a, the waves reaching the center of the screen are shown ($\theta = 0$). The waves from the two slits travel the same distance, so they are in phase: a crest of one wave arrives at the same time as a crest of the other wave. Hence the amplitudes of the two waves add to

FIGURE 35–7 How the wave theory explains the pattern of lines seen in the double-slit experiment. (a) At the center of the screen the waves from each slit have traveled the same distance and are in phase. (b) At this angle θ, the lower wave travels an extra distance of one whole wavelength, and the waves are in phase; note from the shaded triangle that the extra distance equals $d \sin \theta$. (c) For this angle θ, the lower wave travels an extra distance equal to one-half wavelength, so the two waves arrive at the screen fully out of phase. (d) A more detailed diagram showing the geometry for parts (b) and (c).

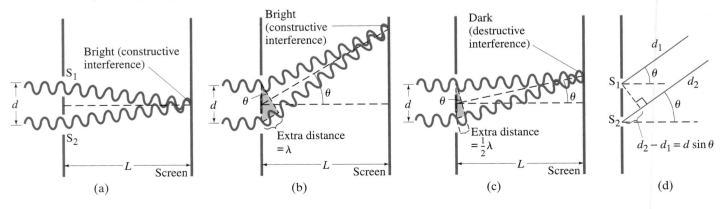

form a larger amplitude as shown in Fig. 35–8a. This is **constructive interference**, and there is a bright spot at the center of the screen. Constructive interference also occurs when the paths of the two rays differ by one wavelength (or any whole number of wavelengths), as shown in Fig. 35–7b. But if one ray travels an extra distance of one-half wavelength (or $\frac{3}{2}\lambda$, $\frac{5}{2}\lambda$, and so on), the two waves are exactly out of phase when they reach the screen: the crests of one wave arrive at the same time as the troughs of the other wave, and so they add to produce zero amplitude (Fig. 35–8b). This is **destructive interference**, and the screen is dark, Fig. 35–7c. Thus, there will be a series of bright and dark lines (or **fringes**) on the viewing screen.

To determine exactly where the bright lines fall, first note that Fig. 35–7 is somewhat exaggerated; in real situations, the distance d between the slits is very small compared to the distance L to the screen. The rays from each slit for each case will therefore be essentially parallel and θ is the angle they make with the horizontal, as shown in Fig. 35–7d. From the right triangle shown in Fig. 35–7, we can see that the extra distance traveled by the lower ray is $d \sin\theta$. Constructive interference will occur, and a bright fringe will appear on the screen, when the *path difference*, $d \sin\theta$, equals a whole number of wavelengths:

$$d \sin\theta = m\lambda, \qquad m = 0, 1, 2, \cdots. \qquad \left[\begin{array}{c}\text{constructive}\\\text{interference}\end{array}\right] \quad \textbf{(35–2a)}$$

The value of m is called the **order** of the interference fringe. The first order ($m = 1$), for example, is the first fringe on each side of the central fringe (at $\theta = 0$). Destructive interference occurs when the path difference, $d \sin\theta$, is $\frac{1}{2}$, $\frac{3}{2}$, and so on, wavelengths:

$$d \sin\theta = \left(m + \tfrac{1}{2}\right)\lambda, \qquad m = 0, 1, 2, \cdots. \qquad \left[\begin{array}{c}\text{destructive}\\\text{interference}\end{array}\right] \quad \textbf{(35–2b)}$$

The intensity of the bright fringes is greatest for the central fringe ($m = 0$) and decreases for higher orders, as shown in Fig. 35–9.

(a)

(b)

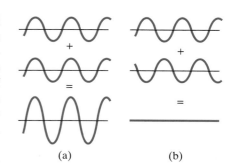

FIGURE 35–8
(a) Constructive interference.
(b) Destructive interference. (See also Section 15–8.)

FIGURE 35–9 (a) Interference fringes produced by a double-slit experiment and detected by photographic film placed on the viewing screen. The arrow marks the central fringe. (b) Intensity of light in the interference pattern. Also shown are values of m for Eq. 35–2a (constructive interference) and Eq. 35–2b (destructive interference).

CONCEPTUAL EXAMPLE 35–1 | **Interference pattern lines.** (*a*) Will there be an infinite number of points on the viewing screen where constructive and destructive interference occur, or only a finite number of points? (*b*) Are neighboring points of constructive interference uniformly spaced, or is the spacing between neighboring points of constructive interference not uniform?

RESPONSE (*a*) When you look at Eqs. 35–2a and b you might be tempted to say, given the statement $m = 0, 1, 2, \cdots$ beside the equations, that there are an infinite number of points of constructive and destructive interference. However, recall that $\sin\theta$ cannot exceed 1. Thus, there is an upper limit to the values of m that can be used in these equations. For Eq. 35–2a, the maximum value of m is the integer closest in value but smaller than d/λ. So there are a finite number of points of constructive and destructive interference spread out over an even infinitely tall screen. (*b*) The spacing between neighboring points of constructive or destructive interference is not uniform: The spacing gets larger as θ gets larger, and you can verify this statement mathematically. For small values of θ the spacing is nearly uniform as you will see in Example 35–2.

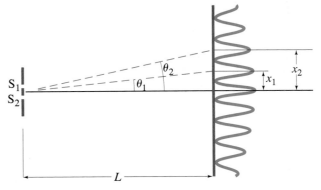

FIGURE 35–10 For small angles, the interference fringes occur at distance $x = \theta L$ above the center fringe ($m = 0$); θ_1 and x_1 are for the first order fringe ($m = 1$), θ_2 and x_2 are for $m = 2$.

EXAMPLE 35–2 **Line spacing for double-slit interference.** A screen containing two slits 0.100 mm apart is 1.20 m from the viewing screen. Light of wavelength $\lambda = 500$ nm falls on the slits from a distant source. Approximately how far apart will the bright interference fringes be on the screen?

SOLUTION Given $d = 0.100$ mm $= 1.00 \times 10^{-4}$ m, $\lambda = 500 \times 10^{-9}$ m, and $L = 1.20$ m, the first-order fringe ($m = 1$) occurs at an angle θ given by

$$\sin \theta_1 = \frac{m\lambda}{d} = \frac{(1)(500 \times 10^{-9}\,\text{m})}{1.00 \times 10^{-4}\,\text{m}} = 5.00 \times 10^{-3}.$$

This is a very small angle, so we can take $\sin \theta = \theta$, with θ in radians. The first-order fringe will occur a distance x_1 above the center of the screen (see Fig. 35–10) given by $x_1/L = \tan \theta_1 = \theta_1$, so

$$x_1 = L\theta_1 = (1.20\,\text{m})(5.00 \times 10^{-3}) = 6.00\,\text{mm}.$$

The second-order fringe ($m = 2$) will occur at

$$x_2 = L\theta_2 = L\frac{2\lambda}{d} = 12.0\,\text{mm}$$

above the center, and so on. Thus the lower order fringes are 6.00 mm apart.

CONCEPTUAL EXAMPLE 35–3 **Changing the wavelength.** (*a*) What happens to the interference pattern shown in Fig. 35–10, Example 35–2, if the incident light (500 nm) is replaced by light of wavelength 700 nm? (*b*) What happens instead if the slits are moved farther apart?

RESPONSE (*a*) When λ increases in Eq. 35–2a but d stays the same, then the angle θ for maxima increases and the interference pattern spreads out. (*b*) Increasing the slit spacing d reduces θ for each order, so the lines are closer together.

Wavelength (or frequency) determines color

From Eqs. 35–2 we can see that, except for the zeroth-order fringe at the center, the position of the fringes depends on wavelength. Consequently, when white light falls on the two slits, as Young found in his experiments, the central fringe is white, but the first- (and higher-) order fringes contain a spectrum of colors like a rainbow; θ was found to be smallest for violet light and largest for red. By measuring the position of these fringes, Young was the first to determine the wavelengths of visible light (using Eqs. 35–2). In doing so, he showed that what distinguishes different colors physically is their wavelength, an idea put forward earlier by Grimaldi in 1665.

EXAMPLE 35–4 Wavelengths from double-slit interference. White light passes through two slits 0.50 mm apart and an interference pattern is observed on a screen 2.5 m away. The first-order fringe resembles a rainbow with violet and red light at either end. The violet light falls about 2.0 mm and the red 3.5 mm from the center of the central white fringe (Fig. 35–11). Estimate the wavelengths of the violet light and the red light.

White

|←2.0 mm→|

|←——— 3.5 mm ———→|

FIGURE 35–11 Example 35–4.

SOLUTION We use Eq. 35–2a with $m = 1$ and $\sin\theta = \theta$. Then for violet light, $x = 2.0$ mm, so (see also Fig. 35–10)

$$\lambda = \frac{d\theta}{m} = \frac{d}{m}\frac{x}{L} = \left(\frac{5.0 \times 10^{-4}\,\text{m}}{1}\right)\left(\frac{2.0 \times 10^{-3}\,\text{m}}{2.5\,\text{m}}\right) = 4.0 \times 10^{-7}\,\text{m},$$

or 400 nm. For red light, $x = 3.5$ mm, so

$$\lambda = \frac{d}{m}\frac{x}{L} = \left(\frac{5.0 \times 10^{-4}\,\text{m}}{1}\right)\left(\frac{3.5 \times 10^{-3}\,\text{m}}{2.5\,\text{m}}\right) = 7.0 \times 10^{-7}\,\text{m} = 700\,\text{nm}.$$

35–4 Coherence

The two slits in Fig. 35–7 act as if they were two sources of radiation. They are called **coherent sources** because the waves leaving them bear the same phase relationship to each other at all times. This happens because the waves come from a single source to the left of the two slits in Fig. 35–7. An interference pattern is observed only when the sources are coherent. If two tiny lightbulbs replaced the two slits, an interference pattern would not be seen. The light emitted by one lightbulb would have a random phase with respect to the second bulb, and the screen would be more or less uniformly illuminated. Two such sources, whose output waves have phases that bear no fixed relationship to each other over time, are called **incoherent sources**.

Coherent and incoherent sources

The subject of coherence is rather complicated, and we discuss it only briefly. Two light beams do not have to be in phase to be coherent and produce an interference pattern. For example, suppose a piece of glass were placed in front of the lower slit in Fig. 35–7 and suppose the glass is just thick enough to slow down the light so that it enters the lower slit a half wavelength behind the light entering the upper slit. The two beams would be a constant 180° out of phase, but there would still be an interference pattern on the screen. (Can you guess what it would look like? Hint: The central point would be dark instead of bright.) Two beams can be coherent whether they are in phase or out of phase; the important thing is that they have a *constant* phase relation to each other over time.

Coherent sources of water or sound waves are easier to obtain than are coherent sources of light—two loudspeakers receiving the same pure frequency signal from an amplifier will be coherent sources. And two antennas connected to the same LC oscillator can be coherent sources of low-frequency electromagnetic waves. But LC oscillators at the high frequencies of visible light $(10^{15}\,\text{Hz})$ don't exist since L and C can't be made small enough. For sources of visible light we have to rely on the oscillations (or acceleration) of electric charge within atoms. In an incandescent light bulb, for example, the atoms in the filament are excited by heating, and give off "wave trains" of light, each of which lasts only about 10^{-8} s. The light we see is the sum of a great many such wave trains that bear a random phase relation to each other. Two light bulbs are thus not coherent, and an interference pattern would not be seen. It wasn't until the 1950s that a really coherent source of light was developed, the *laser*. Because of its coherence, laser light on a double slit produces a very "clean" interference pattern.

Coherence is a relative concept. Perfect coherence of a beam would correspond to light that is perfectly sinusoidal (of one frequency) for all times, whereas complete incoherence occurs when there are waves whose phase relations are completely random over time. A measure of the "relative coherence" can be defined in terms of the sharpness of a two-slit interference pattern.

35-5 | Intensity in the Double-Slit Interference Pattern

We saw in Section 35–3 that the interference pattern produced by the coherent light from two slits, S_1 and S_2 (Figs. 35–7 and 35–9), produces a series of bright and dark fringes. If the two monochromatic waves of wavelength λ are in phase at the slits, the maxima (brightest points) occur at angles θ given by

$$d \sin \theta = m\lambda,$$

and the minima (darkest points) when

$$d \sin \theta = \left(m + \tfrac{1}{2}\right)\lambda,$$

where m is an integer ($m = 0, 1, 2, \ldots$).

We now determine the intensity of the light at all points in the pattern (that is, for all θ). For simplicity, let us assume that if either slit were covered, the light passing through the other would diffract sufficiently to illuminate a large portion of the screen uniformly. The intensity I of the light at any point is proportional to the square of its wave amplitude (Section 15–3). Treating light as an electromagnetic wave, I is proportional to the square of the electric field E or magnetic field B (Section 32–7). Since E and B are proportional to each other, it doesn't matter which we use, but it is conventional to use E and write the intensity as $I \propto E^2$. The electric field $\mathbf{E}$ at any point P (see Fig. 35–12) will be the sum of the electric field vectors of the waves coming from each of the two slits, $\mathbf{E}_1$ and $\mathbf{E}_2$. Since $\mathbf{E}_1$ and $\mathbf{E}_2$ are essentially parallel (on a screen far away compared to the slit separation), the magnitude of the electric field at angle θ (that is, at point P) will be

$$E_\theta = E_1 + E_2.$$

Both E_1 and E_2 vary sinusoidally with frequency $f = c/\lambda$, but they differ in phase, depending on their different travel distances from the slits. The electric field at P can then be written for the light from each of the two slits, using $\omega = 2\pi f$, as

$$E_1 = E_{10} \sin \omega t$$
$$E_2 = E_{20} \sin (\omega t + \delta)$$

(35–3)

where E_{10} and E_{20} are their respective amplitudes and δ is the phase difference. The value of δ depends on the angle θ, so let us now determine δ as a function of θ.

At the center of the screen (point 0), $\delta = 0$. If the difference in path length from P to S_1 and S_2 is $d \sin \theta = \lambda/2$, the two waves are exactly out of phase so $\delta = \pi$ (or 180°). If $d \sin \theta = \lambda$, the two waves differ in phase by $\delta = 2\pi$. In general, then, δ is related to θ by

$$\frac{\delta}{2\pi} = \frac{d \sin \theta}{\lambda}$$

or

$$\delta = \frac{2\pi}{\lambda} d \sin \theta.$$

(35–4)

To determine $E_\theta = E_1 + E_2$, we add the two scalars E_1 and E_2 which are sine functions differing by the phase δ. One way to determine the sum of E_1 and E_2 is to use a **phasor diagram**. (We used this technique before, in Chapter 31.) As shown in

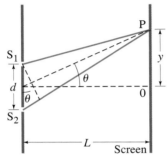

FIGURE 35–12 Determining the intensity in a double-slit interference pattern. Not to scale: in fact $L \gg d$, and the two rays become essentially parallel.

Fig. 35–13, we draw an arrow of length E_{10} to represent the amplitude of E_1 (Eq. 35–3); and the arrow of length E_{20}, which we draw to make a fixed angle δ with E_{10}, represents the amplitude of E_2. When the diagram rotates at angular frequency ω about the origin, the projections of E_{10} and E_{20} on the vertical axis represent E_1 and E_2 as a function of time (see Eq. 35–3). We let $E_{\theta 0}$ be the "vector" sum[†] of E_{10} and E_{20}; it is the amplitude of the sum $E_\theta = E_1 + E_2$, and the projection of $E_{\theta 0}$ on the vertical axis is just E_θ. If the two slits provide equal illumination, so that $E_{10} = E_{20} = E_0$, then from Fig. 35–13 the angle $\phi = \delta/2$, and we can write

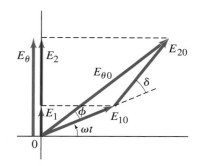

FIGURE 35–13 Phasor diagram for double-slit interference pattern.

$$E_\theta = E_{\theta 0} \sin\left(\omega t + \frac{\delta}{2}\right). \qquad \textbf{(35–5a)}$$

From Fig. 35–13 we can also see that

$$E_{\theta 0} = 2E_0 \cos\phi = 2E_0 \cos\frac{\delta}{2}. \qquad \textbf{(35–5b)}$$

Combining Eqs. 35–5a and b, we obtain

$$E_\theta = 2E_0 \cos\frac{\delta}{2}\sin\left(\omega t + \frac{\delta}{2}\right), \qquad \textbf{(35–5c)}$$

where δ is given by Eq. 35–4. We can also derive Eq. 35–5c, when $E_{10} = E_{20}$, using the trigonometric identity

$$\sin A + \sin B = 2\sin\left(\frac{A+B}{2}\right)\cos\left(\frac{A-B}{2}\right).$$

That is, we add the two Eqs. 35–3 and obtain Eq. 35–5c:

$$E_\theta = E_1 + E_2 = 2E_0 \sin\left(\omega t + \frac{\delta}{2}\right)\cos\frac{\delta}{2}.$$

We are not really interested in E_θ as a function of time, since for visible light the frequency $\left(10^{14} \text{ to } 10^{15}\,\text{Hz}\right)$ is much too high to be noticeable. We are interested in the average intensity, which is proportional to the amplitude squared, $E_{\theta 0}^2$. We now drop the word "average," and we let $I_\theta\left(I_\theta \propto E_\theta^2\right)$ be the intensity at any point P at an angle θ to the horizontal. We let I_0 be the intensity at point 0, the center of the screen, where $\theta = \delta = 0$, so $I_0 \propto \left(E_{10} + E_{20}\right)^2 = \left(2E_0\right)^2$. Then the ratio I_θ/I_0 is equal to the ratio of the squares of the electric-field amplitudes at these two points, so

$$\frac{I_\theta}{I_0} = \frac{E_{\theta 0}^2}{\left(2E_0\right)^2} = \cos^2\frac{\delta}{2}$$

where we used Eq. 35–5b. Thus the intensity I_θ at any point is related to the maximum intensity at the center of the screen by

$$\begin{aligned} I_\theta &= I_0 \cos^2\frac{\delta}{2} \\ &= I_0 \cos^2\left(\frac{\pi d \sin\theta}{\lambda}\right) \end{aligned} \qquad \textbf{(35–6)}$$

where δ was given by Eq. 35–4. This is the relation we sought. From Eq. 35–6 we see that maxima occur where $\cos\delta/2 = \pm 1$, which corresponds to $\delta = 0, 2\pi, 4\pi, \cdots$; from Eq. 35–4, δ has these values when

$$d\sin\theta = m\lambda, \qquad m = 0, 1, 2, \cdots.$$

Minima occur where $\delta = \pi, 3\pi, 5\pi, \ldots$, which corresponds to

$$d\sin\theta = \left(m + \tfrac{1}{2}\right)\lambda, \qquad m = 0, 1, 2, \cdots.$$

These are the same results we obtained in Section 35–3. But now we know not only the position of maxima and minima, but from Eq. 35–6 we can determine the intensity at all points.

[†] We are not adding the actual electric field vectors; instead we are adding the "phasors" in order to get the amplitude, taking into account the phase difference of the two waves.

In the usual situation where the distance L to the screen from the slits is large compared to the slit separation d ($L \gg d$), if we consider only points P whose distance y from the center (point 0) is small compared to L ($y \ll L$)—see Fig. 35–12—then

$$\sin \theta = \frac{y}{L}.$$

From this it follows (see Eq. 35–4) that

$$\delta = \frac{2\pi}{\lambda} \frac{d}{L} y.$$

Equation 35–6 then becomes

$$I_\theta = I_0 \left[\cos \left(\frac{\pi d}{\lambda L} y \right) \right]^2. \qquad\qquad [y \ll L, d \ll L] \quad \textbf{(35–7)}$$

The intensity I_θ as a function of the phase difference δ is plotted in Fig. 35–14. In the approximation of Eq. 35–7, the horizontal axis could as well be y, the position on the screen.

The intensity pattern expressed in Eqs. 35–6 and 35–7, and plotted in Fig. 35–14, shows a series of maxima of equal height, and is based on the assumption that each slit (alone) would illuminate the screen uniformly. This is never quite true, as we shall see when we discuss diffraction in the next chapter. We will see that the center maximum is strongest and each succeeding maximum to each side is less strong.

FIGURE 35–14 Intensity I as a function of phase difference δ and position on screen y (assuming $y \ll L$).

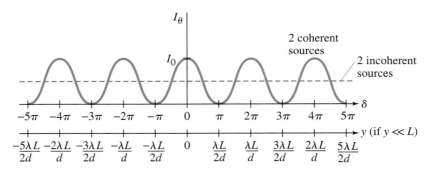

FIGURE 35–15 Example 35–5. The two dots represent the antennas.

EXAMPLE 35–5 **Radar antenna intensity.** Two radio antennas are located close to each other as shown in Fig. 35–15, separated by a distance d. The antennas radiate in phase with each other, emitting waves of intensity I_0 at wavelength λ. (a) Calculate the net intensity as a function of θ for points very far from the antennas. (b) For $d = \lambda$, determine I and find in which directions I is a maximum and a minimum. (c) Repeat part b when $d = \lambda/2$.

SOLUTION (a) This setup is similar to Young's double-slit experiment. Points of constructive and destructive interference are still given by Eqs. 35–2a and b, and the net intensity as a function of position is given by Eq. 35–6 or 35–7. (b) We let $d = \lambda$ in Eq. 35–6, and find for the intensity,

$$I = I_0 \cos^2 (\pi \sin \theta).$$

I is a maximum, equal to I_0, when $\sin \theta = 0$, 1 or -1, meaning $\theta = 0$, $90°$, $180°$ and $270°$. I is zero when $\sin \theta = \frac{1}{2}$ and $-\frac{1}{2}$, for which $\theta = 30°$, $150°$, $210°$ and $330°$. (c) For $d = \lambda/2$ I is maximized for $\theta = 0$ and $180°$ and minimized for $90°$ and $270°$.

Antenna arrays are important when a directional signal needs to be trasmitted such as in radar. We saw in Example 35–5 *b* and *c* that the intensity is maximized along particular directions and falls off slowly to zero at other angles. A narrow beam would be useful to concentrate radar energy on a target, and can be achieved using an array of more than two antennas. If the phases of the antennas are varied relative to each other, then the radiation pattern will shift direction, without having to physically rearrange the antennas. Such an antenna array is called an *electronically steered phase array*.

* Intensity for Incoherent Sources

Let us consider, briefly, the intensity pattern if the two slits of Fig. 35–12 were replaced by two *incoherent* sources of equal strength $(E_{10} = E_{20} = E_0)$. There would be no interference pattern, but rather a uniform illumination of the screen. The phase difference δ between the two waves varies randomly at any point P. Thus we must use the time average over $\cos^2 \delta/2$ in Eq. 35–6, which is $\frac{1}{2}$. The intensity, I_{inc}, of the two incoherent sources, will be

$$I_{inc} = \tfrac{1}{2} I_{coh}$$

where I_{coh} is the intensity due to two coherent sources at a maximum $(I_{coh}$ is the I_0 of Eq. 35–6). This result, which is shown as the dashed line in Fig. 35–14, is confirmed by experiment and can be obtained in another way: when the sources are coherent, we add their wave amplitudes and then square to get the intensity

$$I_{coh} \propto (E_{10} + E_{20})^2.$$

But if the sources are incoherent, each wave produces an intensity unrelated to the other, and the two intensities add up:

$$I_{inc} \propto E_{10}^2 + E_{20}^2.$$

That is, we square the amplitudes and *then* add them. If $E_{10} = E_{20} = E_0$, then the two relations above become

$$I_{coh} \propto (2E_0)^2 = 4E_0^2$$

$$I_{inc} \propto E_0^2 + E_0^2 = 2E_0^2$$

so

$$I_{inc} = \tfrac{1}{2} I_{coh}$$

which is our result above.

35–6 | Interference in Thin Films

Interference of light gives rise to many everyday phenomena such as the bright colors reflected from soap bubbles and from thin oil or gasoline films on water, Fig. 35–16. In these and other cases, the colors are a result of constructive interference between light reflected from the two surfaces of the thin film.

FIGURE 35–16 Thin film interference patterns seen in (a) soap bubbles, (b) thin films of soapy water, and (c) a thin layer of gasoline on water.

(a)

(b)

(c)

FIGURE 35–17 Light reflected from the upper and lower surfaces of a thin film of oil lying on water. Analysis assumes the light strikes the surface perpendicularly, but is shown here at a slight angle for clarity.

Newton's rings

To see how this happens, consider a smooth surface of water on top of which is a thin uniform layer of another substance, say an oil whose index of refraction is less than that of water (we'll see why we assume this in a moment); see Fig. 35–17. Assume for the moment that the incident light is of a single wavelength. Part of the incident light is reflected at A on the top surface, and part of the light transmitted is reflected at B on the lower surface. The part reflected at the lower surface must travel the extra distance ABC. If this *path difference* ABC is equal to one or a whole number of wavelengths in the film (λ_n), the two waves will reach the eye in phase and interfere constructively. Hence the region AC on the surface film will appear bright. But if ABC equals $\frac{1}{2}\lambda_n, \frac{3}{2}\lambda_n$, and so on, the two waves will be exactly out of phase, destructive interference occurs, and the area AC on the film will be dark. The wavelength λ_n is *the wavelength in the film:* $\lambda_n = \lambda/n$ where n is the index of refraction in the film and λ is the wavelength in vacuum.

When white light falls on such a film, the path difference ABC will equal λ_n (or $m\lambda_n$, with $m =$ an integer) for only one wavelength at a given viewing angle. The color corresponding to λ (λ in air) will be seen as very bright. For light viewed at a slightly different angle, the path difference ABC will be longer or shorter and a different color will undergo constructive interference. Thus, for an extended (non-point) source emitting white light, a series of bright colors will be seen next to one another. Variations in thickness of the film will also alter the path difference ABC and therefore affect the color of light that is most strongly reflected.

When a curved glass surface is placed in contact with a flat glass surface, Fig. 35–18, a series of concentric rings is seen when illuminated from above by monochromatic light. These are called **Newton's rings**[†] and they are due to interference between rays reflected by the top and bottom surfaces of the very thin *air gap* between the two pieces of glass. Because this gap (which is equivalent to a thin film) increases in width from the central contact point out to the edges, the extra path length for the lower ray (equal to BCD) varies; where it equals $0, \frac{1}{2}\lambda, \lambda, \frac{3}{2}\lambda, 2\lambda$, and so on, it corresponds to constructive and destructive interference; and this gives rise to the series of bright and dark lines seen in Fig. 35–18b.

The point of contact of the two glass surfaces (A in Fig. 35–18a) is dark in Fig. 35–18b. Since the path difference is zero here, we expect the rays reflected from each surface to be in phase and so this central point ought to be bright. But it is dark, which tells us the two rays must be completely out of phase. This can happen only because one of the waves undergoes a change in phase of 180° upon reflection, corresponding to $\frac{1}{2}$ cycle. Indeed, this and other experiments reveal that *a beam of light reflected by a material whose index of refraction is greater than that of the material in which it is traveling, changes phase by 180° or $\frac{1}{2}$ cycle.*

[†]Although Newton gave an elaborate description of them, they had been first observed and described by his contemporary, Robert Hooke.

FIGURE 35–18 Newton's rings.

(a)

(b)

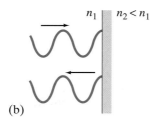

(a)   (b)

FIGURE 35–19 (a) Reflected ray changes phase by 180° or $\frac{1}{2}$ cycle if $n_2 > n_1$, but (b) does not if $n_2 < n_1$.

See Fig. 35–19. If the refractive index is less than that of the material in which the light is traveling, no phase change occurs.[†] [This can be derived from Maxwell's equations. It corresponds to the reflection of a wave traveling along a rope when it reaches the end; as we saw in Fig. 15–19, if the end is tied down, the wave changes phase and the pulse flips over, but if the end is free, no phase change occurs.] Thus the ray reflected by the curved surface above the air gap in Fig. 35–18a undergoes no change in phase. The ray reflected at the lower surface, where the beam in air strikes the glass, undergoes a 180° $\left(\frac{1}{2}\lambda\right)$ phase change. Thus the two rays reflected at the point of contact A of the two glass surfaces (where the air gap approaches zero thickness) will be 180° $\left(\frac{1}{2}\lambda\right)$ out of phase, and a dark spot occurs. Other dark bands will occur when the path difference BCD in Fig. 35–18a is equal to an integral number of wavelengths. Bright bands will occur when the path difference is $\frac{1}{2}\lambda$, $\frac{3}{2}\lambda$, and so on, because the phase change at one surface effectively adds another $\frac{1}{2}\lambda$.

FIGURE 35–20 (a) Light rays reflected from the upper and lower surfaces of a thin wedge of air interfere to produce bright and dark bands. (b) Pattern observed when glass plates are optically flat; (c) pattern when plates are not so flat. See Example 35–6.

| **EXAMPLE 35–6** | **Thin film of air, wedge-shaped.** A very fine wire 7.35×10^{-3} mm in diameter is placed between two flat glass plates as in Fig. 35–20a. Light whose wavelength in air is 600 nm falls (and is viewed) perpendicular to the plates, and a series of bright and dark bands is seen, Fig. 35–20b. How many light and dark bands will there be in this case? Will the area next to the wire be bright or dark?

SOLUTION The thin film is the wedge of air between the two glass plates. Because of the phase change at the lower surface, there will be a dark band when the path difference is 0, λ, 2λ, 3λ, and so on. Since the light rays are perpendicular to the plates, the extra path length equals $2t$, where t is the thickness of the air gap at any point; so dark bands occur where

$$2t = m\lambda, \qquad m = 0, 1, 2, \cdots.$$

Bright bands occur when $2t = \left(m + \frac{1}{2}\right)\lambda$, where m is an integer. At the position of the wire, $t = 7.35 \times 10^{-6}$ m. At this point there will be $2t/\lambda = (2)(7.35 \times 10^{-6}\,\text{m})/(6.00 \times 10^{-7}\,\text{m}) = 24.5$ wavelengths. This is a "half integer," so the area next to the wire will be bright. There will be a total of 25 dark lines along the plates, corresponding to path lengths of $0\lambda, 1\lambda, 2\lambda, 3\lambda, \cdots, 24\lambda$, including the one at the point of contact A ($m = 0$). Between them, there will be 24 bright lines plus the one at the end, or 25. The bright and dark bands will be straight only if the glass plates are extremely flat. If they are not, the pattern is uneven, as in Fig. 35–20c. Thus we see a very precise way of testing a glass surface for flatness. Curved lens surfaces can be tested for precision similarly by placing the lens on a flat glass surface and observing Newton's rings (Fig. 35–18b) for perfect circularity.

(a)

(b)

(c)

[†] Note that in Fig. 35–17, the light reflecting at both interfaces, air–oil and oil–water, underwent a phase change of $\frac{1}{2}\lambda$, since we assumed $n_{\text{water}} > n_{\text{oil}} > n_{\text{air}}$. Since the phase changes were equal, they didn't affect our analysis.

If the wedge between the two glass plates of Example 35–6 is filled with some transparent substance other than air—say, water—the pattern shifts because the wavelength of the light changes. In a material where the index of refraction is n, the wavelength is $\lambda_n = \lambda/n$ where λ is the wavelength in vacuum (see Eq. 35–1). For instance, if the thin wedge of Example 35–6 were filled with water, $\lambda_n = 600\,\text{nm}/1.33 = 450\,\text{nm}$; instead of 25 dark lines, there would be 33.

When white light (rather than monochromatic light) is incident on the thin wedge of Figs. 35–18a or 35–20a, a colorful series of fringes is seen. This is because constructive interference occurs in the reflected light at different locations along the wedge for different wavelengths. Such a difference in thickness is part of the reason bright colors appear when light is reflected from a soap bubble or a thin layer of oil or gasoline on a puddle or lake (Fig. 35–16). Which wavelengths appear brightest also depends on the viewing angle, as we saw earlier.

→ **PROBLEM SOLVING**

A formula is not enough: you must also check for phase changes at surfaces

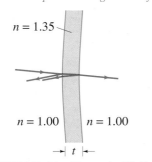

$n = 1.35$

$n = 1.00$ $n = 1.00$

t

FIGURE 35–21 Example 35–7.

→ **PHYSICS APPLIED**

Lens coatings

FIGURE 35–22 A coated lens. Note color of light reflected from the front lens surface.

EXAMPLE 35–7 **Thickness of soap bubble skin.** A soap bubble appears green ($\lambda = 540\,\text{nm}$) at the point on its front surface nearest the viewer. What is its minimum thickness? Assume $n = 1.35$.

SOLUTION The light is reflected perpendicularly from the point on a spherical surface nearest the viewer, Fig. 35–21. Therefore the path difference is $2t$, where t is the thickness of the soap film. Light reflected from the first (outer) surface undergoes a $\frac{1}{2}\lambda$ phase change (index of refraction of soap is greater than that of air), whereas that at the second (inner) surface does not. Therefore, green light is bright when the minimum path difference equals $\frac{1}{2}\lambda_n$. Thus, $2t = \lambda/2n$, so

$$t = \frac{\lambda}{4n} = \frac{(540\,\text{nm})}{(4)(1.35)} = 100\,\text{nm}.$$

This is the minimum thickness. The front surface would also appear green if $2t = 3\lambda/2n$, and, in general, if $2t = (2m + 1)\lambda/2n$ where m is an integer. Note that green is seen in air, so $\lambda = 540\,\text{nm}$ (not λ/n).

An important application of thin-film interference is in the coating of glass to make it "nonreflecting," particularly for lenses. A glass surface reflects about 4 percent of the light incident upon it. Good-quality cameras, microscopes, and other optical devices may contain six to ten thin lenses. Reflection from all these surfaces can reduce the light level considerably, and multiple reflections produce a background haze that reduces the quality of the image. By reducing reflection, transmission is increased. A very thin coating on the lens surfaces can reduce reflections considerably: the thickness of the film is chosen so that light (at least for one wavelength) reflecting from the front and rear surfaces of the film destructively interferes. The amount of reflection at a boundary depends on the difference in index of refraction between the two materials. Ideally, the coating material should have an index of refraction which is the geometric mean of those for air and glass, so that the amount of reflection at each surface is about equal. Then destructive interference can occur nearly completely for one particular wavelength depending on the thickness of the coating. Nearby wavelengths will at least partially destructively interfere, but it is clear that a single coating cannot eliminate reflections for all wavelengths. Nonetheless, a single coating can reduce total reflection from 4 percent to 1 percent of the incident light. Often the coating is designed to eliminate the center of the reflected spectrum (around 550 nm). The extremes of the spectrum—red and violet—will not be reduced as much. Since a mixture of red and violet produces purple, the light seen reflected from such coated lenses is purple (Fig. 35–22). Lenses containing two or three separate coatings can more effectively reduce a wider range of reflecting wavelengths.

EXAMPLE 35–8 **Nonreflective coating.** What is the thickness of an optical coating of MgF_2, whose index of refraction is $n = 1.38$, which is designed to eliminate reflected light at wavelengths centered at 550 nm when incident normally on glass for which $n = 1.50$?

SOLUTION Figure 35–23 shows an incoming ray and two rays reflected from the front and rear surfaces of the coating on the lens. The rays are drawn not quite perpendicular to the lens so we can see each of them. To eliminate reflection, we want the reflected rays 1 and 2 to be $\frac{1}{2}$ wavelength out of phase with each other so they destructively interfere. Rays 1 and 2 *both* undergo a change of phase by $\frac{1}{2}\lambda$ when they reflect, respectively, from the front and rear surfaces of the coating. Therefore, we want the extra distance traveled by ray 2 ($= 2t$) to be a half integral number of wavelengths. That is, $2t = \left(m + \frac{1}{2}\right)\lambda_n$, where m is an integer and λ_n is the wavelength inside the MgF_2 coating. The minimum thickness ($m = 0$) is usually chosen because destructive interference will then occur over the widest angle. Then

$$t = \frac{\lambda_n}{4} = \frac{\lambda}{4n} = \frac{(550 \text{ nm})}{(4)(1.38)} = 99.6 \text{ nm}.$$

FIGURE 35–23 Incident ray of light is partially reflected at the front surface of a lens coating (ray 1) and again partially reflected at the rear surface of the coating (ray 2), with most of the energy passing as the transmitted ray into the glass.

*35–7 | Michelson Interferometer

A useful instrument involving wave interference is the **Michelson interferometer** (Fig. 35–24),[†] invented by the American Albert A. Michelson (Section 33–2). Monochromatic light from a single point on an extended source is shown striking a half-silvered mirror M_S. This **beam splitter** mirror M_S has a thin layer of silver that reflects only half the light that hits it, so that half of the beam passes through to a fixed mirror M_2, where it is reflected back. The other half is reflected by M_S up to a mirror M_1 that is movable (by a fine-thread screw), where it is also reflected back. Upon its return, part of beam 1 passes through M_S and reaches the eye; and part of beam 2, on its return, is reflected by M_S into the eye. If the two path lengths are identical, the two coherent beams entering the eye constructively interfere and brightness will be seen. If the movable mirror is moved a distance $\lambda/4$, one beam will travel an extra distance equal to $\lambda/2$ (because it travels back and forth over the distance $\lambda/4$). In this case, the two beams will destructively interfere and darkness will be seen. As M_1 is moved farther, brightness will recur (when the path difference is λ), then darkness, and so on.

Very precise length measurements can be made with an interferometer. The motion of mirror M_1 by only $\frac{1}{4}\lambda$ produces a clear difference between brightness and darkness. For $\lambda = 400 \text{ nm}$, this means a precision of 100 nm or 10^{-4} mm! If mirror M_1 is tilted very slightly, the bright or dark spots are seen instead as a series of bright and dark lines or "fringes." By counting the number of fringes, or fractions thereof, extremely precise length measurements can be made.

Michelson saw that the interferometer could be used to determine the length of the standard meter in terms of the wavelength of a particular light. In 1960, that standard was chosen to be a particular orange line in the spectrum of krypton-86 (krypton atoms with atomic mass 86). Careful repeated measurements of the old standard meter (the distance between two marks on a platinum–iridium bar kept in Paris) were made to establish 1 meter as being 1,650,763.73 wavelengths of this light, which was *defined* to be the meter. In 1983, the meter was redefined in terms of the speed of light (Section 1–4).

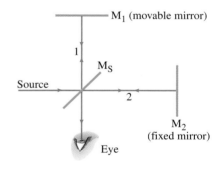

FIGURE 35–24 Michelson interferometer.

[†]There are other types of interferometer, but Michelson's is the best known.

*35–8 Luminous Intensity

Although the *intensity* of light, as for any electromagnetic wave, is measured by the Poynting vector in W/m^2, and the total power output of a source can be measured in watts (the *radiant flux*), these are not adequate for measuring the visual sensation we call brightness. The reason is that we are really interested here only in the visible spectrum, whereas the two quantities just mentioned would take into account all wavelengths present. It is also important to take into account the eye's sensitivity to different wavelengths—the eye is most sensitive in the central, 550-nm (yellow), portion of the spectrum; so a yellow source would appear brighter than a red or blue source of the same power output.

These factors are taken into account in the quantity **luminous flux**, F_l, whose unit is the **lumen** (lm). One lumen is defined experimentally as the brightness of $\frac{1}{60}$ cm^2 of platinum surface at its melting temperature (1770°C). It is equivalent to $\frac{1}{683}$ watts of 555-nm light.

Since the luminous flux from a source may not be uniform over all directions, we define the **luminous intensity** I_l as the luminous flux per unit solid angle (steradian). Its unit is the **candela** (cd) where $1\text{ cd} = 1\text{ lm/sr}$.

Candela (unit)

The **illuminance**, E_l, is the luminous flux incident on a surface per unit area of the surface: $E_l = F_l/A$. Its unit is the lumen per square meter (lm/m^2) and is a measure of the illumination falling on a surface.[†]

We will not go into this subject in any more detail. We have introduced it here for completeness since the *luminous intensity*, as measured by the candela in SI units, is one of the seven basic quantities in the SI. (See Section 1–4 and Table 1–5.) The other six (and their units) we have already met: length (m), time (s), mass (kg), electric current (A), temperature (K), and amount of substance (mol).

EXAMPLE 35–9 **Light bulb illuminance.** The brightness of a particular type of 100-W light bulb is rated at 1700 lm. Determine (*a*) the luminous intensity and (*b*) the illuminance at a distance of 2.0 m. Assume the light output is uniform in all directions.

SOLUTION (*a*) A full sphere corresponds to 4π sr. Hence, $I_l = 1700\text{ lm}/4\pi\text{ sr} = 135\text{ cd}$. It does not depend on distance. (*b*) At $d = 2.0\text{ m}$ from the source, the luminous flux per unit area is

$$E_l = \frac{F_l}{4\pi d^2} = \frac{1700\text{ lm}}{(4\pi)(2.0\text{ m})^2} = 34\text{ lm/m}^2.$$

The illuminance decreases as the square of the distance.

PROBLEM	Interference
SOLVING	

1. Interference effects depend on the simultaneous arrival of two or more waves at the same point in space.
2. Constructive interference occurs when waves arrive in phase with each other: a crest of one wave arrives at the same time as a crest of the other wave. The amplitudes of the waves then add to form a larger amplitude. Constructive interference also occurs when the phase difference is exactly one full wavelength or any integer multiple of a full wavelength: $1\lambda, 2\lambda, 3\lambda, \cdots$.
3. Destructive interference occurs when a crest of one wave arrives at the same time as a trough of the other wave. The amplitudes add, but they are of opposite sign, so the total amplitude is reduced to zero if the two amplitudes are equal. Destructive interference occurs whenever the phase difference is a half-integral number of wavelengths. Thus, the total amplitude will be zero if two identical waves arrive one-half wavelength out of phase, or $\left(m + \frac{1}{2}\right)\lambda$ out of phase where m is an integer.
4. For thin-film interference, don't forget an extra one-half wavelength phase shift that occurs when light reflects from an optically more dense medium (going from a medium of lesser toward greater index of refraction).

    [†]The British unit is the foot-candle, or lumen per square foot.

Summary

The wave theory of light is strongly supported by the observations that light exhibits **interference** and **diffraction**. Wave theory also explains the refraction of light and the fact that light travels more slowly in transparent solids and liquids than it does in air. The wavelength of light in a medium with index of refraction n is

$$\lambda_n = \frac{\lambda}{n},$$

where λ is the wavelength in vacuum. The frequency is not changed.

Young's double-slit experiment clearly demonstrated the interference of light. The observed bright spots of the interference pattern were explained as constructive interference between the beams coming through the two slits, where the beams differ in path length by an integral number of wavelengths. The dark areas in between are due to destructive interference when the path lengths differ by $\frac{1}{2}\lambda, \frac{3}{2}\lambda$, and so on. The angles θ at which **constructive interference** occurs are given by

$$\sin\theta = m\frac{\lambda}{d},$$

where λ is the wavelength of the light, d the separation of the slits, and m an integer $(0, 1, 2, \cdots)$. **Destructive interference** occurs at angles θ given by

$$\sin\theta = \left(m + \tfrac{1}{2}\right)\frac{\lambda}{d}$$

where m is an integer $(0, 1, 2, \cdots)$.

The light intensity I_θ at any point in a double-slit interference pattern can be calculated using a phasor diagram, which predicts that

$$I_\theta = I_0 \cos^2 \frac{\delta}{2}$$

where I_0 is the intensity at $\theta = 0$ and the phase angle δ is

$$\delta = \frac{2\pi d}{\lambda}\sin\theta.$$

Two sources of light are perfectly **coherent** if the waves leaving them are sinusoidal, of the same single frequency, and maintain the same phase relationship at all times. If the light waves from the two sources have a random phase with respect to each other over time (as for two incandescent light bulbs) the two sources are **incoherent**.

Light reflected from the front and rear surfaces of a thin film of transparent material can interfere constructively or destructively, depending on the optical path difference. A phase change of 180° or $\frac{1}{2}\lambda$ occurs when the light reflects at a surface where the index of refraction increases. Such **thin-film interference** has many practical applications, such as lens coatings and using Newton's rings to check the uniformity of glass surfaces.

Questions

1. Does Huygens' principle apply to sound waves? To water waves?

2. What is the evidence that light carries energy?

3. Why is light sometimes described as rays and sometimes as waves?

4. We can hear sounds around corners but we cannot see around corners, yet both sound and light are waves. Explain the difference.

5. Can the wavelength of light be determined from reflection or refraction measurements?

6. Monochromatic red light is incident on a double slit and the interference pattern is viewed on a screen some distance away. Explain how the fringe pattern would change if the red light source is replaced by a blue light source.

7. Suppose white light falls on the two slits of Fig. 35–7, but one slit is covered by a red filter (700 nm) and the other by a blue (450 nm) filter. Describe the pattern on the screen.

8. Compare a double-slit experiment for sound waves to that for light waves. Discuss the similarities and differences.

9. Two rays of light from the same source destructively interfere if their path lengths differ by how much?

10. If Young's double-slit experiment were submerged in water, how would the fringe pattern be changed?

11. Why doesn't the light from the two headlights of a distant car produce an interference pattern?

12. Why are interference fringes noticeable only for a *thin* film like a soap bubble and not for a thick piece of glass, say?

13. Why are Newton's rings (Fig. 35–18) closer together farther from the center?

14. Some coated lenses appear greenish yellow when seen by reflected light. What wavelengths do you suppose the coating is designed to eliminate completely?

15. A drop of oil on a pond appears bright at its edges where its thickness is much less than the wavelengths of visible light. What can you say about the index of refraction of the oil?

*16. Describe how a Michelson interferometer could be used to measure the index of refraction of air.

Problems

1. (II) Derive the law of reflection—namely, that the angle of incidence equals the angle of reflection from a flat surface—using Huygens' principle for waves.

Section 35–3

2. (I) Monochromatic light falling on two slits 0.016 mm apart produces the fifth-order fringe at a 9.8° angle. What is the wavelength of the light used?

3. (I) The third-order fringe of 610 nm light is observed at an angle of 18° when the light falls on two narrow slits. How far apart are the slits?

4. (II) Monochromatic light falls on two very narrow slits 0.048 mm apart. Successive fringes on a screen 5.00 m away are 6.5 cm apart near the center of the pattern. What is the wavelength and frequency of the light?

5. (II) A parallel beam of light from a He-Ne laser, with a wavelength 656 nm, falls on two very narrow slits 0.060 mm apart. How far apart are the fringes in the center of the pattern if the screen is 3.6 m away?

6. (II) Light of wavelength 680 nm falls on two slits and produces an interference pattern in which the fourth-order fringe is 38 mm from the central fringe on a screen 2.0 m away. What is the separation of the two slits?

7. (II) If 720-nm and 660-nm light passes through two slits 0.58 mm apart, how far apart are the second-order fringes for these two wavelengths on a screen 1.0 m away?

8. (II) Suppose a thin piece of glass were placed in front of the lower slit in Fig. 35–7 so that the two waves enter the slits 180° out of phase (Fig. 35–25). Describe in detail the interference pattern on the screen.

FIGURE 35–25
Problem 8.

9. (II) In a double-slit experiment it is found that blue light of wavelength 460 nm gives a second-order maximum at a certain location on the screen. What wavelength of visible light would have a minimum at the same location?

10. (II) Light of wavelength 480 nm in air falls on two slits 6.00×10^{-2} mm apart. The slits are immersed in water, as is a viewing screen 40.0 cm away. How far apart are the fringes on the screen?

11. (III) A very thin sheet of plastic ($n = 1.60$) covers one slit of a double-slit apparatus illuminated by 640-nm light. The center point on the screen, instead of being a maximum, is dark. What is the (minimum) thickness of the plastic?

Section 35–5

12. (I) If one slit in Fig. 35–12 is covered, by what factor does the intensity at the center of the screen change?

13. (II) Show that the angular full width at half maximum of the central peak in a double-slit interference pattern is given by $\Delta\theta = \lambda/2d$ if $\lambda \ll d$.

14. (II) Suppose that one slit of a double-slit apparatus is wider than the other so that the intensity of light passing through it is twice as great. Determine the intensity I as a function of position (θ) on the screen for coherent light.

15. (III) (a) Consider three equally spaced and equal-intensity coherent sources of light (such as adding a third slit to the two slits of Fig. 35–12). Use the phasor method to obtain the intensity as a function of the phase difference δ (Eq. 35–4). (b) Determine the positions of maxima and minima.

16. (III) Apply the phasor method to four parallel slits separated by equal distances d. Assume coherent light and determine the intensity as a function of position on the screen and find the positions of maxima and minima.

Section 35–6

17. (I) If a soap bubble is 120 nm thick, what color will appear at the center when illuminated normally by white light? Assume that $n = 1.34$.

18. (I) How far apart are the dark fringes in Example 35–6 if the glass plates are each 26.5 cm long?

19. (II) What is the minimum thickness (>0) of a soap film ($n = 1.34$) that would appear black if illuminated with 480-nm light? Assume there is air on both sides of the soap film.

20. (II) A lens appears greenish yellow ($\lambda = 570$ nm is strongest) when white light reflects from it. What minimum thickness of coating ($n = 1.28$) do you think is used on such a (glass) lens, and why?

21. (II) A total of 28 bright and 28 dark Newton's rings (not counting the dark spot at the center) are observed when 650-nm light falls normally on a planoconvex lens resting on a flat glass surface (Fig. 35–18). How much thicker is the center than the edges?

22. (II) A fine metal foil separates one end of two pieces of optically flat glass, as in Fig. 35–20. When light of wavelength 670 nm is incident normally, 25 dark lines are observed (with one at each end). How thick is the foil?

23. (II) How thick (minimum) should the air layer be between two flat glass surfaces if the glass is to appear bright when 480-nm light is incident normally? What if the glass is to appear dark?

24. (II) A thin film of alcohol ($n = 1.36$) lies on a flat glass plate ($n = 1.51$). When monochromatic light, whose wavelength can be changed, is incident normally, the reflected light is a minimum for $\lambda = 512$ nm and a maximum for $\lambda = 640$ nm. What is the thickness of the film?

25. (II) When a Newton's ring apparatus (Fig. 35–18) is immersed in a liquid, the diameter of the eighth dark ring decreases from 2.92 cm to 2.60 cm. What is the refractive index of the liquid?

26. (II) A planoconvex lucite lens 3.4 cm in diameter is placed on a flat piece of glass as in Fig. 35–18. When 580-nm light is incident normally, 48 bright rings are observed, the last one right at the edge. What is the radius of curvature of the lens surface, and the focal length of the lens?

27. (II) Show that the radius r of the mth dark Newton's ring, as viewed from directly above (Fig. 35–18), is given by $r = \sqrt{m\lambda R}$ where R is the radius of curvature of the curved glass surface and λ is the wavelength of light used. Assume that the thickness of the air gap is much less than R at all points and that $r \ll R$.

28. (III) Use the result of Problem 27 to show that the distance between adjacent dark Newton's rings is

$$\Delta r \approx \sqrt{\frac{\lambda R}{4\,m}}$$

for the mth ring, assuming $m \gg 1$.

29. (III) A single optical coating reduces reflection to zero for $\lambda = 550\ \text{nm}$. By what factor is the intensity reduced by the coating for $\lambda = 450\ \text{nm}$ and $\lambda = 650\ \text{nm}$ as compared to no coating? Assume normal incidence.

* Section 35–7

* 30. (II) What is the wavelength of the light entering an interferometer if 344 bright fringes are counted when the movable mirror moves 0.125 mm?

* 31. (II) How far must the mirror M_1 in a Michelson interferometer be moved if 750 fringes of 589-nm light are to pass by a reference line?

FIGURE 35–26   Problem 32.

* 32. (III) One of the beams of an interferometer (Fig. 35–26) passes through a small glass container containing a cavity 1.30 cm deep. When a gas is allowed to slowly fill the container, a total of 186 dark fringes are counted to move past a reference line. The light used has a wavelength of 610 nm. Calculate the index of refraction of the gas at its final density, assuming that the interferometer is in vacuum.

* 33. (III) The yellow sodium D lines have wavelengths of 589.0 and 589.6 nm. When they are used to illuminate a Michelson interferometer, it is noted that the interference fringes disappear and reappear periodically as the mirror M_1 is moved. Why does this happen? How far must the mirror move between one disappearance and the next?

* Section 35–8

* 34. (I) The illuminance of direct sunlight on Earth is about $10^5\ \text{lm/m}^2$. Estimate the luminous flux and luminous intensity of the Sun.

* 35. (II) The *luminous efficiency* of a light bulb is the ratio of luminous flux to electric power input. (*a*) What is the luminous efficiency of a 100-W, 1700-lm bulb? (*b*) How many 40-W, 60-lm/W fluorescent lamps would be needed to provide an illuminance of 250 lm/m^2 on a factory floor of area 25 m × 30 m? Assume the lights are 10 m above the floor and that half their flux reaches the floor.

General Problems

36. Light of wavelength λ strikes a screen containing two slits a distance d apart at an angle θ_i to the normal. Determine the angle θ_m at which the mth-order maximum occurs.

37. Television and radio waves can reflect from nearby mountains or from airplanes, and the reflections can interfere with the direct signal from the station. (*a*) Determine what kind of interference will occur when 75-MHz television signals arrive at a receiver directly from a distant station, and are reflected from an airplane 118 m directly above the receiver. (Assume $\frac{1}{2}\lambda$ change in phase of the signal upon reflection.) (*b*) What kind of interference will occur if the plane is 22 m closer to the receiver?

38. A radio station operating at 102.1 MHz broadcasts from two identical antennae at the same elevation but separated by a 7.0-m horizontal distance d, Fig. 35–27. A maximum signal is found along the midline, perpendicular to d at its midpoint and extending horizontally in both directions. If the midline is taken as 0°, at what other angle(s) θ is a maximum signal detected? A minimum signal? Assume all measurements are made much farther than 7 m from the antenna towers.

FIGURE 35–27
Problem 38.

39. Light of wavelength 690 nm passes through two narrow slits 0.60 mm apart. The screen is 1.50 m away. A second source of unknown wavelength produces its second-order fringe 1.13 mm closer to the central maximum than the 690-nm light. What is the wavelength of the unknown light?

40. What is the index of refraction of a clear material if a minimum of 150 nm thickness of it, when laid on glass, is needed to reduce reflection to nearly zero when light of 600 nm is incident normally upon it? Do you have a choice for an answer?

41. Monochromatic light of variable wavelength is incident normally on a thin sheet of plastic film in air. The reflected light is a minimum only for $\lambda = 510$ nm and $\lambda = 680$ nm in the visible spectrum. What is the thickness of the film ($n = 1.58$)?

42. In a water tank experiment, water waves are generated with their crests 3.5 cm apart and parallel. They pass through two openings 6.0 cm apart in a long wooden board. If the end of the tank is 2.0 m beyond the boards, where would you stand, relative to the "straight-through" direction, so that you received little or no wave action?

43. Compare the minimum thickness needed for an antireflective coating ($n = 1.38$) applied to a glass lens in order to eliminate (a) blue (450 nm), or (b) red (700 nm) reflections for light at normal incidence.

44. Suppose you viewed the light *transmitted* through a thin film on a flat piece of glass. Draw a diagram, similar to Fig. 35–17 or 35–23, and describe the conditions required for maxima and minima. Consider all possible values of index of refraction. Discuss the relative size of the minima compared to the maxima and to zero.

45. Stealth aircraft are designed to not reflect radar, whose wavelength is typically 2 cm, by using an antireflecting coating. Ignoring any change in wavelength in the coating, estimate its thickness.

46. Very highly reflective mirrors for a particular wavelength can be made by alternating many layers of *transparent* materials of indices of refraction n_1 and $n_2(1 < n_1 < n_2)$. What should the minimum thicknesses d_1 and d_2 of Fig. 35–28 be, in terms of the incident wavelength λ, to maximize reflection?

FIGURE 35–28   Problem 46.

47. When a charged particle, such as a proton, travels through a transparent medium at a speed v_p faster than the speed of light in that medium ($v = c/n$), it emits electromagnetic radiation (light) called **Čerenkov radiation**. It is the electromagnetic equivalent of a shock wave (see Section 16–8), and is confined to a particular angle that depends on v_p and v. Determine this angle for a proton traveling 2.21×10^8 m/s in a plastic whose index of refraction is 1.52.

48. What is the minimum (non-zero) thickness for the air layer between two flat glass surfaces if the glass is to appear dark when 640-nm light is incident normally? What if the glass is to appear bright?

49. *Lloyd's mirror* provides one way of obtaining a double-slit interference pattern from a single source so the light is coherent; as shown in Fig. 35–29, the light that reflects from the plane mirror appears to come from the virtual image of the slit. Describe in detail the interference pattern on the screen.

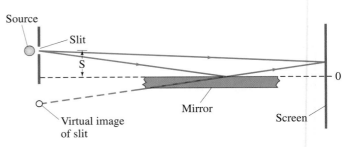

FIGURE 35–29   Problem 49.

50. Consider the antenna array of Example 35–5, Fig. 35–15. Let $d = \lambda/2$, and suppose that the two antennas are now 180° out of phase with each other. Find the directions for constructive and destructive interference, and compare with the case when the sources are in phase. (These results illustrate the basis for directional antennas.)

* 51. Suppose the mirrors in a Michelson interferometer are perfectly aligned and the path lengths to mirrors M_1 and M_2 are identical. With these initial conditions, an observer sees a bright maximum at the center of the viewing area. Now one of the mirrors is moved a distance x. Determine a formula for the intensity at the center of the viewing area as a function of x, the distance the movable mirror is moved from the initial conditions.

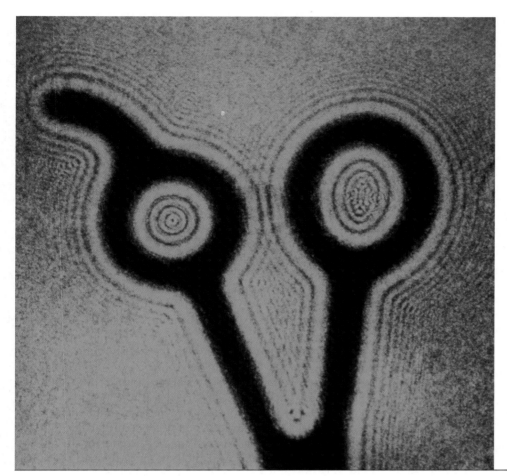

Parallel coherent light from a laser, which acts as nearly a point source, illuminates these shears. Instead of a clean shadow, there is a dramatic diffraction pattern, which is a strong confirmation of the wave theory of light. Diffraction patterns are washed out when typical extended sources of light are used, and hence are not seen, although a careful examination of shadows will reveal fuzziness. We will examine diffraction by a single slit, and how it affects the double-split pattern, as well as diffraction gratings and x-rays. We will see how diffraction affects the resolution of optical instruments, and that the ultimate resolution can never be greater than the wavelength of the radiation used. Finally we study the polarization of light.

Diffraction and Polarization

Young's double-slit experiment put the wave theory of light on a firm footing. But full acceptance came only with studies on diffraction more than a decade later.

We have already discussed diffraction briefly with regard to water waves (Section 15–11) as well as for light (Section 35–1) and we have seen that it refers to the spreading or bending of waves around edges. Now we look at diffraction more closely, including its important practical effects of limiting the amount of detail, or *resolution*, that can be obtained with any optical instrument including telescopes, cameras, and the eye.

A part of the history of the wave theory of light belongs to Augustin Fresnel (1788–1827) who in 1819 presented to the French Academy a wave theory of light that predicted and explained interference and diffraction effects. Almost immediately Siméon Poisson (1781–1840) pointed out a counter-intuitive inference: according to Fresnel's wave theory, if light from a point source were to fall on a solid disk, part of the incident light would be diffracted around the edges and would constructively interfere at the center of the shadow (Fig. 36–1). That prediction seemed very unlikely. But when the experiment was actually carried out by François Arago, the bright spot was seen at the very center of the shadow (Fig. 36–2a). This was strong evidence for the wave theory.

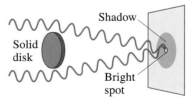

FIGURE 36–1 If light is a wave, a bright spot will appear at the center of the shadow of a solid disk illuminated by a point source of monochromatic light.

Figure 36–2a is a photograph of the shadow cast by a coin using a (nearly) point source of light, a laser in this case. The bright spot is clearly present at the center. Note that there also are bright and dark fringes beyond the shadow. These resemble the interference fringes of a double slit. Indeed, they are due to interference of waves diffracted around the disk, and the whole is referred to as a **diffraction pattern**. A diffraction pattern exists around any sharp-edged object illuminated by a point source, as shown in Fig. 36–2b and c. We are not always aware of diffraction because most sources of light in everyday life are not points, so light from different parts of the source washes out the pattern.

36–1 | Diffraction by a Single Slit

To see how a diffraction pattern arises, we will analyze the important case of monochromatic light passing through a narrow slit. We will assume that parallel rays (plane waves) of light fall on the slit of width a, and pass through to a viewing screen very far away. If the viewing screen is not far away, lenses can be used to make the rays parallel.[†] As we know from studying water waves and from Huygens' principle, the waves passing through the slit spread out in all directions. We will now examine how the waves passing through different parts of the slit interfere with each other.

Parallel rays of monochromatic light pass through the narrow slit as shown in Fig. 36–3a. The light falls on a screen which is assumed to be very far away, so the rays heading for any point are essentially parallel. First we consider rays that pass straight through as in Fig. 36–3a. They are all in phase, so there will be a central bright spot on the screen. In Fig. 36–3b, we consider rays moving at an angle θ such that the ray from the top of the slit travels exactly one wavelength farther than the ray from the bottom edge. The ray passing through the very center of the slit will travel one-half wavelength farther than the ray at the bottom of the slit. These two rays will be exactly out of phase with one another and so will destructively interfere. Similarly, a ray slightly above the bottom one will cancel a ray that is the same distance above the central one. Indeed, each ray passing through the lower half of the slit will cancel with a corresponding ray passing through the upper half. Thus, all the

[†] Such a diffraction pattern, involving parallel rays, is called *Fraunhofer diffraction*. If the screen is close and no lenses are used, it is called *Fresnel diffraction*. The analysis in the latter case is rather involved, so we consider only the limiting case of Fraunhofer diffraction.

FIGURE 36–2 Diffraction pattern of (a) a circular disk (a coin), (b) a razor blade, (c) a single slit, each illuminated by a (nearly) point source of monochromatic light.

(a)   (b)   (c)

(a) $\theta = 0$
Bright

(b) $\sin\theta = \dfrac{\lambda}{a}$
Dark

(c) $\sin\theta = \dfrac{3\lambda}{2a}$
Bright

(d) $\sin\theta = \dfrac{2\lambda}{a}$
Dark

FIGURE 36–3 Analysis of diffraction pattern formed by light passing through a narrow slit.

rays destructively interfere in pairs, and so no light will reach the viewing screen at this angle. The angle θ at which this takes place can be seen from the diagram to occur when $\lambda = a\sin\theta$, so

$$\sin\theta = \frac{\lambda}{a}.$$ [first minimum] (36–1)

Diffraction equation
(angular width of central spot)

The light intensity is a maximum at $\theta = 0°$ and decreases to a minimum (intensity = zero) at the angle θ given by Eq. 36–1.

Now consider a larger angle θ such that the top ray travels $\frac{3}{2}\lambda$ farther than the bottom ray, as in Fig. 36–3c. In this case, the rays from the bottom third of the slit will cancel in pairs with those in the middle third because they will be $\lambda/2$ out of phase. However, light from the top third of the slit will still reach the screen, so there will be a bright spot centered near $\sin\theta \approx 3\lambda/2a$, but it will not be nearly as bright as the central spot at $\theta = 0°$. For an even larger angle θ such that the top ray travels 2λ farther than the bottom ray, Fig. 36–3d, rays from the bottom quarter of the slit will cancel with those in the quarter just above it because the path lengths differ by $\lambda/2$. And the rays through the quarter of the slit just above center will cancel with those through the top quarter. At this angle there will again be a minimum of zero intensity in the diffraction pattern. A plot of the intensity as a function of angle is shown in Fig. 36–4. This corresponds well with the photo of Fig. 36–2c. Notice that minima (zero intensity) occur at

$$a\sin\theta = m\lambda, \qquad m = 1, 2, 3, \cdots,$$ [minima] (36–2)

but *not* at $m = 0$ where there is the strongest maximum. Between the minima, smaller intensity maxima occur at approximately (but not exactly) $m \approx \frac{3}{2}\lambda, \frac{5}{2}\lambda, \cdots$. [Note that the *minima* for a diffraction pattern, Eq. 36–2, satisfy a criterion very similar to that for the *maxima* (bright spots) for double-slit interference, Eq. 35–2a.]

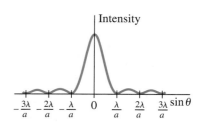

FIGURE 36–4 Intensity in the diffraction pattern of a single slit as a function of $\sin\theta$. Note that the central maximum is not only much higher than the maxima to each side, but it is also twice as wide ($2\lambda/a$ wide) as any of the others (only λ/a wide each).

Single slit diffraction minima

FIGURE 36–5 Example 36–1.

EXAMPLE 36–1 **Single-slit diffraction maximum.** Light of wavelength 750 nm passes through a slit 1.0×10^{-3} mm wide. How wide is the central maximum (*a*) in degrees, and (*b*) in centimeters, on a screen 20 cm away?

SOLUTION (*a*) The first minimum occurs at

$$\sin\theta = \frac{\lambda}{a} = \frac{7.5 \times 10^{-7}\,\text{m}}{1.0 \times 10^{-6}\,\text{m}} = 0.75.$$

So $\theta = 49°$. This is the angle between the center and the first minimum, Fig. 36–5. The angle subtended by the whole central maximum, between the minima above and below the center, is twice this, or 98°.
(*b*) The width of the central maximum is $2x$, where $\tan\theta = x/20$ cm. So $2x = 2(20\,\text{cm})(\tan 49°) = 46$ cm. A large width of the screen will be illuminated, but it will not normally be very bright since the amount of light that passes through such a small slit will be small and it is spread over a large area.

FIGURE 36-6 Example 36-2.

CONCEPTUAL EXAMPLE 36-2 **Diffraction spreads.** Light shines through a rectangular hole that is narrower in the vertical direction than the horizontal, Fig. 36-6. (a) Would you expect the diffraction pattern to be more spread out in the vertical direction or in the horizontal direction? (b) Should a rectangular loudspeaker horn at a stadium be high and narrow, or wide and flat?

RESPONSE (a) From Eq. 36-2 we can see that if we make the slit (width a) narrower, the pattern spreads out more. This is consistent with our earlier study of waves in Chapter 15. Hence the diffraction through the rectangular hole will be wider vertically, since the aperture in that direction is smaller.
(b) For the loudspeaker, the sound pattern desired is one spread out horizontally, so the horn should be tall and narrow (rotate Fig. 36-6 by 90°).

36-2 | Intensity in Single-Slit Diffraction Pattern

We have determined the positions of the minima in the diffraction pattern produced by light passing through a single slit, Eq. 36-2. We now discuss a method for predicting the amplitude and intensity at any point in the pattern using the phasor technique already discussed in Section 35-5.

Let us consider the slit divided into N very thin strips of width Δy as indicated in Fig. 36-7. Each strip sends light in all directions toward a screen on the right. Again we take the rays heading for any particular point on the distant screen to be parallel, all making an angle θ with the horizontal as shown. We choose the strip width Δy to be much smaller than the wavelength λ of the monochromatic light falling on the slit, so all the light from a given strip is in phase. The strips are of equal size, and if the whole slit is uniformly illuminated, we can take the electric field wave amplitudes ΔE_0 from each thin strip to be equal as long as θ is not too large. However, the separate amplitudes from the different strips will differ in phase. The phase difference in the light coming from adjacent strips will be (see Section 35-5, Eq. 35-4)

$$\Delta\beta = \frac{2\pi}{\lambda}\Delta y \sin\theta \tag{36-3}$$

since the difference in path length is $\Delta y \sin\theta$.

FIGURE 36-7 Slit of width a divided into N strips of width Δy.

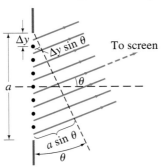

The total amplitude on the screen at any angle θ will be the sum of the separate wave amplitudes due to each strip. These wavelets have the same amplitude ΔE_0 but differ in phase. To obtain the total amplitude, we can use a phasor diagram as we did in Section 35-5 (Fig. 35-13). The phasor diagrams for four different angles θ are shown in Fig. 36-8. At the center of the screen, $\theta = 0$, the waves from each strip are all in phase ($\Delta\beta = 0$, Eq. 36-3), so the arrows representing each ΔE_0 line up as shown in Fig. 36-8a. The total amplitude of the light arriving at the center of the screen is then $E_0 = N\Delta E_0$. At a small angle θ, for a point on the distant screen not far from the center, Fig. 36-8b shows how the wavelets of amplitude ΔE_0 add up to give E_θ, the total amplitude on the screen at this angle θ. Note that each wavelet differs in phase from the adjacent one by $\Delta\beta$. The phase difference between the wavelets from the top and bottom edges of the slit is

$$\beta = N\,\Delta\beta = \frac{2\pi}{\lambda}N\,\Delta y\sin\theta = \frac{2\pi}{\lambda}a\sin\theta \tag{36-4}$$

where $a = N\,\Delta y$ is the total width of the slit. Although the "arc" in Fig. 36-8b has length $N\,\Delta E_0$, and so equals E_0 (total amplitude at $\theta = 0$), the amplitude of the total wave E_θ at angle θ is the *vector* sum of each wavelet amplitude and so is equal to the length of the chord as shown. The chord is shorter than the arc, so $E_\theta < E_0$.

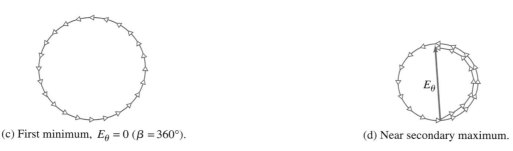

(c) First minimum, $E_\theta = 0$ $(\beta = 360°)$.   (d) Near secondary maximum.

FIGURE 36–8 Phasor diagram for single-slit diffraction, giving the total amplitude E_θ at four different angles θ.

For greater θ, we eventually come to the case, illustrated in Fig. 36–8c, where the chain of arrows closes on itself. In this case the vector sum is zero, so $E_\theta = 0$ for this angle θ. This corresponds to the first minimum. Since $\beta = N\,\Delta\beta$ is 360° or 2π in this case, we have from Eq. 36–3,

$$2\pi = N\,\Delta\beta = N\!\left(\frac{2\pi}{\lambda}\,\Delta y \sin\theta\right)$$

or, since the slit width $a = N\,\Delta y$,

$$\sin\theta = \frac{\lambda}{a}.$$

Thus the first minimum $\left(E_\theta = 0\right)$ occurs where $\sin\theta = \lambda/a$, which is the same result we obtained in the previous Section, Eq. 36–1.

For even greater values of θ, the chain of arrows spirals beyond 360°. Figure 36–8d shows the case near the secondary maximum next to the first minimum. Here $\beta = N\,\Delta\beta \approx 360° + 180° = 540°$ or 3π. (Note that although β may be 540°, θ can still be a very small angle, depending on the values of a and λ.) When $\beta = 4\pi$, we have a double circle and again a minimum, where $\sin\theta = 2\lambda/a$, corresponding to $m = 2$ in Eq. 36–2. When greater angles θ are considered, new maxima and minima occur. But since the total length of the coil remains constant, equal to $N\,\Delta E_0\left(= E_0\right)$, each succeeding maximum is smaller and smaller as the coil winds in on itself.

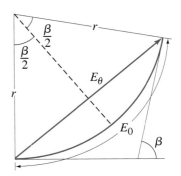

FIGURE 36–9 Determining amplitude E_θ as a function of θ for single-slit diffraction.

To obtain a quantitative expression for the amplitude (and intensity) for any point on the screen (that is, for any angle θ), we now consider the limit $N \to \infty$ so Δy become the infinitesimal width dy. In this case, the diagrams of Fig. 36–8 become smooth curves, one of which is shown in Fig. 36–9. For any angle θ, the wave amplitude on the screen is E_θ, equal to the chord in Fig. 36–9. The length of the arc is E_0, as before. If r is the radius of curvature of the arc, then

$$\frac{E_\theta}{2} = r \sin \frac{\beta}{2}.$$

Using radian measure for $\beta/2$, we also have

$$\frac{E_0}{2} = r \frac{\beta}{2}.$$

We combine these to obtain

$$E_\theta = E_0 \frac{\sin \beta/2}{\beta/2}. \tag{36–5}$$

The angle β is the phase difference between the waves from the top and bottom edges of the slit. The path difference for these two rays is $a \sin \theta$ (see Fig. 36–7 as well as Eq. 36–4), so

$$\beta = \frac{2\pi}{\lambda} a \sin \theta. \tag{36–6}$$

Intensity is proportional to the square of the wave amplitude, so the intensity I_θ at any angle θ is, from Eq. 36–5,

$$I_\theta = I_0 \left(\frac{\sin \beta/2}{\beta/2} \right)^2 \tag{36–7}$$

where $I_0 (\propto E_0^2)$ is the intensity at $\theta = 0$ (the central maximum). We can combine Eqs. 36–7 and 36–6 (although it is often simpler to leave them as separate equations) to obtain

Intensity in single-slit diffraction pattern

$$I_\theta = I_0 \left[\frac{\sin \left(\dfrac{\pi a \sin \theta}{\lambda} \right)}{\left(\dfrac{\pi a \sin \theta}{\lambda} \right)} \right]^2. \tag{36–8}$$

According to Eq. 36–8, minima $(I_\theta = 0)$ occur where $\sin(\pi a \sin \theta/\lambda) = 0$, which means $\pi a \sin \theta / \lambda$ must be $\pi, 2\pi, 3\pi$, and so on, or

$$a \sin \theta = m\lambda, \qquad m = 1, 2, 3, \cdots \qquad \text{[minima]}$$

which is what we have obtained previously, Eq. 36–2. Notice that m cannot be zero: when $\beta/2 = \pi a \sin \theta / \lambda = 0$, the denominator as well as the numerator in Eqs. 36–7 or 36–8 vanishes. We can evaluate the intensity in this case by taking the limit as $\theta \to 0$ (or $\beta \to 0$); for very small angles, $\sin \beta/2 \approx \beta/2$, so $(\sin \beta/2)/(\beta/2) \to 1$ and $I_\theta = I_0$, the maximum at the center of the pattern.

The intensity I_θ as a function of θ, as given by Eq. 36–8, corresponds to the diagram of Fig. 36–4.

EXAMPLE 36–3 **ESTIMATE** | **Intensity at secondary maxima.** Estimate the intensities of the first two secondary maxima to either side of the central maximum.

SOLUTION The secondary maxima occur close to halfway between the minima, at about

$$\frac{\beta}{2} = \frac{\pi a \sin \theta}{\lambda} \approx \left(m + \tfrac{1}{2}\right)\pi. \qquad m = 1, 2, 3, \cdots.$$

The actual maxima are not quite at these points—their positions can be determined by differentiating Eq. 36–7 (see Problem 10)—but we are only seeking an estimate. Using these values for β in Eq. 36–7 or 36–8, with $\sin\left(m + \tfrac{1}{2}\right)\pi = 1$, gives

$$I_\theta = \frac{I_0}{\left(m + \tfrac{1}{2}\right)^2 \pi^2}. \qquad m = 1, 2, 3, \cdots.$$

For $m = 1$ and 2, we get

$$I_\theta = \frac{I_0}{22.2} = 0.045\, I_0 \qquad\qquad [m = 1]$$

$$I_\theta = \frac{I_0}{61.7} = 0.016\, I_0. \qquad\qquad [m = 2]$$

The first maximum to the side of the central peak has only 1/22, or 4.5 percent, the intensity of the central intensity, and succeeding ones are smaller still, just as we can see in Fig. 36–4 and the photo of Fig. 36–2c.

Diffraction by a circular opening produces a similar pattern (though circular rather than rectangular) and is of great practical importance, since lenses are essentially circular apertures through which light passes. We will discuss this in Section 36–4 and see how diffraction limits the resolution (or sharpness) of images.

*| 36–3 | Diffraction in the Double-Slit Experiment

When we analyzed Young's double-slit experiment in Section 35–5, we assumed that the central portion of the screen was uniformly illuminated. This is equivalent to assuming the slits are infinitesimally narrow, so that the central diffraction peak is spread out over the whole screen. This can never be the case for real slits; diffraction reduces the intensity of the bright interference fringes to the side of center so they are not all of the same height as they were shown in Fig. 35–14. (They were shown more correctly in Fig. 35–9b.)

To calculate the intensity in a double-slit interference pattern, including diffraction, let us assume the slits have equal widths a and their centers are separated by a distance d. Since the distance to the screen is large compared to the slit separation d, the wave amplitude due to each slit is essentially the same at each point on the screen. Then the total wave amplitude at any angle θ will no longer be

$$E_{\theta 0} = 2E_0 \cos \frac{\delta}{2},$$

as was given by Eq. 35–5b. Rather, it must be modified, because of diffraction, by Eq. 36–5, so that

$$E_{\theta 0} = 2E_0 \left(\frac{\sin \beta/2}{\beta/2} \right) \cos \frac{\delta}{2}.$$

(Note that up to now in this chapter we have dealt only with the amplitude of E, and so we merely wrote E_θ rather than $E_{\theta 0}$ as was done in Chapter 35.) Thus the intensity will be given by

$$I_\theta = I_0 \left(\frac{\sin \beta/2}{\beta/2} \right)^2 \left(\cos \frac{\delta}{2} \right)^2 \tag{36–9}$$

where, from Eqs. 36–6 and 35–4,

$$\frac{\beta}{2} = \frac{\pi}{\lambda} a \sin \theta \quad \text{and} \quad \frac{\delta}{2} = \frac{\pi}{\lambda} d \sin \theta.$$

The first term in parentheses in Eq. 36–9 is sometimes called the "diffraction factor" and the second one the "interference factor." These two factors are plotted in Fig. 36–10a and b for the case when $d = 6a$, and $a = 10\lambda$. (Figure 36–10b is essentially the same as Fig. 35–14.) Figure 36–10c shows the product of these two curves, times I_0, which is the actual intensity as a function of θ (or as a function of position on the screen for θ not too large) as given by Eq. 36–9. As indicated by the dashed lines in Fig. 36–10c, the diffraction factor acts as a sort of envelope that limits the interference peaks.

EXAMPLE 36–4 **Diffraction plus interference.** Show why the central diffraction peak in Fig. 36–10c contains 11 interference fringes.

SOLUTION The first minimum in the diffraction pattern occurs where

$$\sin \theta = \frac{\lambda}{a}.$$

Since $d = 6a$,

$$d \sin \theta = 6a \left(\frac{\lambda}{a} \right) = 6\lambda.$$

From Eq. 35–2a, interference peaks (maxima) occur for $d \sin \theta = m\lambda$ where m can be 0, 1, $\cdots$ or any integer. Thus the diffraction minimum ($d \sin \theta = 6\lambda$) coincides with $m = 6$ in the interference pattern, so the $m = 6$ peak won't appear. Hence the central diffraction peak encloses the central interference peak ($m = 0$) and five peaks ($m = 1$ to 5) on each side for a total of 11. Since the sixth order doesn't appear, it is said to be a "missing order."

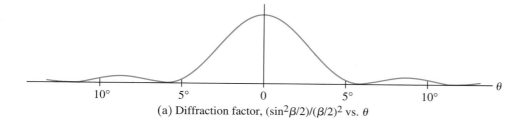

(a) Diffraction factor, $(\sin^2 \beta/2)/(\beta/2)^2$ vs. θ

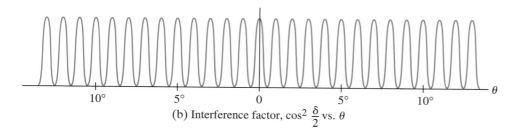

(b) Interference factor, $\cos^2 \dfrac{\delta}{2}$ vs. θ

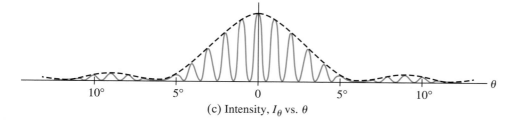

(c) Intensity, I_θ vs. θ

FIGURE 36–10 (a) Diffraction factor, (b) interference factor, and (c) the resultant intensity I_θ, plotted as a function of θ for $d = 6a = 60\lambda$.

Notice from this last Example that the number of interference fringes in the central diffraction peak depends only on the ratio d/a. It does not depend on wavelength λ. The actual spacing (in angle, or in position on the screen) does depend on λ. For the case illustrated, $a = 10\lambda$, and so the first diffraction minimum occurs at $\sin \theta = \lambda/a = 0.10$ or about $6°$.

The decrease in intensity of the interference fringes away from the center, as graphed in Fig. 36–10, is shown in Fig. 36–11.

The patterns due to interference and diffraction arise from the same phenomenon—the superposition of coherent waves of different phase. The distinction between them is thus not so much physical as for convenience of description, as in this Section where we analyzed the two-slit pattern in terms of interference and diffraction separately. In general, we use the word "diffraction" when referring to an analysis by superposition of many infinitesimal and usually contiguous sources, such as when we subdivide a source into infinitesimal parts. We use the term "interference" when we superpose the wave from a finite (and usually small) number of coherent sources.

FIGURE 36–11 Photographs of a double-slit interference pattern showing effects of diffraction. In both cases $d = 0.50$ mm, whereas $a = 0.040$ mm in (a) and 0.080 mm in (b).

(a)

(b)

FIGURE 36–12 Photographs of images (greatly magnified) formed by a lens, showing the diffraction pattern of an image for: (a) a single point object; (b) two point objects whose images are barely resolved.

The ability of a lens to produce distinct images of two point objects very close together is called the **resolution** of the lens. The closer the two images can be and still be seen as distinct (rather than overlapping blobs), the higher the resolution. The resolution of a camera lens, for example, is often specified as so many lines per millimeter,[†] and can be determined by photographing a standard set of parallel lines on fine-grain film. The minimum spacing of lines distinguishable on film using the lens gives the resolution.

Two principal factors limit the resolution of a lens. The first is lens aberrations. As we saw, because of spherical and other aberrations, a point object is not a point on the image but a tiny blob. Careful design of compound lenses can reduce aberrations significantly, but they cannot be eliminated entirely. The second factor that limits resolution is *diffraction*, which cannot be corrected for because it is a natural result of the wave nature of light. We discuss it now.

In Section 36–1, we saw that because light travels as a wave, light from a point source passing through a slit is spread out into a diffraction pattern (Figs. 36–2 and 36–4). A lens, because it has edges, acts like a slit. When a lens forms the image of a point object, the image is actually a tiny diffraction pattern. Thus *an image would be blurred even if aberrations were absent.*

In the analysis that follows, we assume that the lens is free of aberrations, so that we can focus our attention on diffraction effects and how much they limit the resolution of a lens. In Fig. 36–4 we saw that the diffraction pattern produced by light passing through a rectangular slit has a central maximum in which most of the light falls. This central peak falls to a minimum on either side of its center at an angle $\theta \approx \sin\theta = \lambda/a$ (this is Eq. 36–1) where a is the width of the slit, λ is the wavelength of light used, and we assume θ is small. There are also low-intensity fringes beyond. For a lens, or any circular hole, the image of a point object will consist of a *circular* central peak (called the *diffraction spot* or *Airy disk*) surrounded by faint circular fringes, as shown in Fig. 36–12a. The central maximum has an angular half width given by

$$\theta = \frac{1.22\lambda}{D}$$

where D is the diameter of the circular opening.

This formula differs from that for a slit (Eq. 36–1) by the factor 1.22. This factor appears because the width of a circular hole is not uniform (like a rectangular slit) but varies from its diameter D to zero. A careful analysis shows that the "average" width is $D/1.22$. Hence we get the equation above rather than Eq. 36–1.

[†]This may be specified at the center of the field of view as well as at the edges, where it is usually less because of off-axis aberrations.

FIGURE 36–13 Intensity of light across the diffraction pattern of a circular hole.

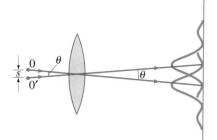

FIGURE 36–14 The *Rayleigh criterion.* Two images are just resolvable when the center of the diffraction peak of one is directly over the first minimum in the diffraction pattern of the other. The two point objects 0 and 0′ subtend an angle θ at the lens; only one ray is drawn for each object, to indicate the center of the diffraction pattern of its image.

The intensity of light in the diffraction pattern of light from a point source passing through a circular opening is shown in Fig. 36–13. The image for a non-point source is a superposition of such patterns. For most purposes we need consider only the central spot, since the concentric rings are so much dimmer.

If two point objects are very close, the diffraction patterns of their images will overlap as shown in Fig. 36–12b. As the objects are moved closer, a separation is reached where you can't tell if there are two overlapping images or a single image. The separation at which this happens may be judged differently by different observers. However, a generally accepted criterion is one proposed by Lord Rayleigh (1842–1919). This **Rayleigh criterion** states that *two images are just resolvable when the center of the diffraction disk of one image is directly over the first minimum in the diffraction pattern of the other.* This is shown in Fig. 36–14. Since the first minimum is at an angle $\theta = 1.22\lambda/D$ from the central maximum, Fig. 36–14 shows that two objects can be considered just resolvable if they are separated by this angle θ:

$$\theta = \frac{1.22\lambda}{D}.$$ (36–10)

Rayleigh criterion (resolution limit)

This is the limit on resolution set by the wave nature of light due to diffraction.

EXAMPLE 36–5 Hubble space telescope. The Hubble Space Telescope (HST) is a reflecting telescope that was placed in orbit above the Earth's atmosphere, so its resolution would not be limited by turbulence in the atmosphere (Fig. 36–15). Its objective diameter is 2.4 m. For visible light, say $\lambda = 550\,\text{nm}$, estimate the improvement in resolution the Hubble offers over Earth-bound telescopes, which are limited in resolution by movement of the Earth's atmosphere to about half an arc second (each degree is divided into 60 minutes each containing 60 seconds, so $1° = 3600$ arc seconds).

SOLUTION Earth-bound telescopes are limited to an angular resolution of

$$\theta = \tfrac{1}{2}\left(\frac{1}{3600}\right)^{\!\circ}\left(\frac{2\pi\,\text{rad}}{360°}\right) = 2.4 \times 10^{-6}\,\text{rad}.$$

The Hubble, on the other hand, is limited by diffraction (Eq. 36–10) which for $\lambda = 550\,\text{nm}$ is

$$\theta = \frac{1.22(550 \times 10^{-9}\,\text{m})}{2.4\,\text{m}} = 2.8 \times 10^{-7}\,\text{rad},$$

which is almost ten times better resolution.

FIGURE 36–15 Hubble Space Telescope, with Earth in the background. The flat orange panels are solar cells that collect energy from the Sun.

36-5 Resolution of Telescopes and Microscopes; the λ Limit

You might think that a microscope or telescope could be designed to produce any desired magnification, depending on the choice of focal lengths and quality of the lenses. But this is not possible, because of diffraction. An increase in magnification above a certain point merely results in magnification of the diffraction patterns. This could be highly misleading since we might think we are seeing details of an object when we are really seeing details of the diffraction pattern. To examine this problem, we apply the Rayleigh criterion: two objects (or two nearby points on one object) are just resolvable if they are separated by an angle θ (Fig. 36-14) given by Eq. 36-10:

Resolution

$$\theta = \frac{1.22\lambda}{D}.$$

This is valid for either a microscope or a telescope, where D is the diameter of the objective lens. For a telescope, the resolution is specified by stating θ as given by this equation.[†]

For a microscope, it is more convenient to specify the actual distance, s, between two points that are just barely resolvable, Fig. 36-14. Since objects are normally placed near the focal point of the microscope objective, $\theta = s/f$, or $s = f\theta$. If we combine this with Eq. 36-10, we obtain for the **resolving power (RP)**:

Resolving power (microscope)

$$\text{RP} = s = f\theta = \frac{1.22\lambda f}{D}. \qquad (36\text{-}11)$$

This distance s is called the resolving power of the lens because it is the minimum separation of two object points that can just be resolved, assuming the highest quality lens since this limit is imposed by the wave nature of light.

FIGURE 36-16 The 300-meter radiotelescope in Arecibo, Puerto Rico, uses radio waves (Fig. 32-12) instead of visible light.

EXAMPLE 36-6 **Telescope resolution (radio wave vs. visible light).** What is the theoretical minimum angular separation of two stars that can just be resolved by: (a) the 200-inch telescope on Palomar Mountain (Fig. 34-33c); and (b) the Arecibo radiotelescope (Fig. 36-16), whose diameter is 300 m and whose radius of curvature is also 300 m. Assume $\lambda = 550$ nm for the visible-light telescope in part (a), and $\lambda = 4$ cm (the shortest wavelength at which the radiotelescope has been operated) in part (b).

SOLUTION (a) Since $D = 200$ inch $= 5.1$ m, we have from Eq. 36-10 that

$$\theta = \frac{1.22\lambda}{D} = \frac{(1.22)(5.50 \times 10^{-7}\,\text{m})}{(5.1\,\text{m})} = 1.3 \times 10^{-7}\,\text{rad},$$

or 0.75×10^{-5} deg. (Note that this is equivalent to resolving two points less than 1 cm apart from a distance of 100 km!) This is the limit set by diffraction. The resolution is not this good because of aberrations and, more importantly, turbulence in the atmosphere. In fact, large-diameter objectives are not justified by increased resolution, but by their greater light-gathering ability—they allow more light in, so fainter objects can be seen.

(b) Radiotelescopes are not hindered by atmospheric turbulence, and for radio waves with $\lambda = 0.04$ m the resolution is

$$\theta = \frac{(1.22)(0.04\,\text{m})}{(300\,\text{m})} = 1.6 \times 10^{-4}\,\text{rad}.$$

[†]Telescopes with large-diameter objectives are usually limited not by diffraction but by other effects such as turbulence in the atmosphere. The resolution of a high-quality microscope, on the other hand, normally *is* limited by diffraction because microscope objectives are complex compound lenses containing many elements of small diameter (since f is small).

Diffraction sets an ultimate limit on the detail that can be seen on any object. In Eq. 36–11 we note that the focal length of a lens cannot be made less than (approximately) the radius of the lens, and even that is very difficult—see the lensmaker's equation (Eq. 34–4). In this best case, Eq. 36–11 gives, with $f \approx D/2$,

$$\mathrm{RP} \approx \frac{\lambda}{2}. \tag{36–12}$$

Thus we can say, to within a factor of 2 or so, that

it is not possible to resolve detail of objects smaller than the wavelength of the radiation being used.

Resolution limited to λ

This is an important and useful rule of thumb.

Compound lenses are now designed so well that the actual limit on resolution is often set by diffraction—that is, by the wavelength of the light used. To obtain greater detail, one must use radiation of shorter wavelength. The use of UV radiation can increase the resolution by a factor of perhaps 2. Far more important, however, was the discovery in the early twentieth century that electrons have wave properties (Chapter 38) and that their wavelengths can be very small. The wave nature of electrons is utilized in the electron microscope (Section 38–7), which can magnify 100 to 1000 times more than a visible-light microscope because of the much shorter wavelengths. X-rays, too, have very short wavelengths and are often used to study objects in great detail (Section 36–10).

*36–6 Resolution of the Human Eye and Useful Magnification

The resolution of the human eye is limited by several factors, all of roughly the same order of magnitude. The resolution is best at the fovea, where the cone spacing is smallest, about $3\,\mu\mathrm{m}$ ($= 3000\,\mathrm{nm}$). The diameter of the pupil varies from about 0.1 cm to about 0.8 cm. So for $\lambda = 550\,\mathrm{nm}$ (where the eye's sensitivity is greatest), the diffraction limit is about $\theta \approx 1.22\lambda/D \approx 8 \times 10^{-5}\,\mathrm{rad}$ to $6 \times 10^{-4}\,\mathrm{rad}$. Since the eye is about 2 cm long, this corresponds to a resolving power of $s \approx (8 \times 10^{-5}\,\mathrm{rad})(2 \times 10^{-2}\,\mathrm{m}) \approx 2\,\mu\mathrm{m}$ at best, to about $15\,\mu\mathrm{m}$ at worst (pupil small). Spherical and chromatic aberration also limit the resolution to about $10\,\mu\mathrm{m}$. The net result is that the eye can resolve objects whose angular separation is about $5 \times 10^{-4}\,\mathrm{rad}$ at best. This corresponds to objects separated by 1 cm at a distance of about 20 m.

The typical near point of a human eye is about 25 cm. At this distance, the eye can just resolve objects that are $(25\,\mathrm{cm})(5 \times 10^{-4}\,\mathrm{rad}) \approx 10^{-4}\,\mathrm{m} = \frac{1}{10}\,\mathrm{mm}$ apart. Since the best light microscopes can resolve objects no smaller than about 200 nm at best (Eq. 36–12 for violet light, $\lambda = 400\,\mathrm{nm}$), the useful magnification $[= (\text{resolution by naked eye})/(\text{resolution by microscope})]$ is limited to about

$$\frac{10^{-4}\,\mathrm{m}}{200 \times 10^{-9}\,\mathrm{m}} = 500\times.$$

In practice, magnifications of about $1000\times$ are often used to minimize eye-strain. Any greater magnification would simply make visible the diffraction pattern produced by the microscope objective.

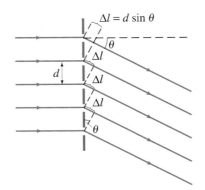

FIGURE 36–17 Diffraction grating.

A large number of equally spaced parallel slits is called a **diffraction grating**, although the term "interference grating" might be as appropriate. Gratings can be made by precision machining of very fine parallel lines on a glass plate. The untouched spaces between the lines serve as the slits. Photographic transparencies of an original grating serve as inexpensive gratings. Gratings containing 10,000 lines per centimeter are common today, and are very useful for precise measurements of wavelengths. A diffraction grating containing slits is called a **transmission grating.** **Reflection gratings** are also used, which are made by ruling fine lines on a metallic or glass surface from which light is reflected and analyzed. The analysis is basically the same as for a transmission grating, which we now discuss.

The analysis of a diffraction grating is much like that of Young's double-slit experiment. We assume parallel rays of light are incident on the grating as shown in Fig. 36–17. We also assume that the slits are narrow enough so that diffraction by each of them spreads light over a very wide angle on a distant screen behind the grating, and interference can occur with light from all the other slits. Light rays that pass through each slit without deviation ($\theta = 0°$) interfere constructively to produce a bright line at the center of the screen. Constructive interference also occurs at an angle θ such that rays from adjacent slits travel an extra distance of $\Delta l = m\lambda$, where m is an integer. Thus, if d is the distance between slits, then we see from Fig. 36–17 that $\Delta l = d \sin \theta$, and

Diffraction grating maxima (m = order)

$$\sin \theta = \frac{m\lambda}{d}, \qquad m = 0, 1, 2, \cdots \qquad \text{[principal maxima]} \quad \textbf{(36–13)}$$

is the criterion to have a brightness maximum. This is the same equation as for the double-slit situation, and again m is called the order of the pattern.

There is an important difference between a double-slit and a multiple-slit pattern, however. The bright maxima are much *sharper* and *narrower* for a grating. Why this happens can be seen as follows. Suppose that the angle θ is increased just slightly beyond that required for a maximum. In the case of only two slits, the two waves will be only slightly out of phase, so nearly full constructive interference occurs. This means the maxima are wide (see Fig. 35–9). For a grating, the waves from two adjacent slits will also not be significantly out of phase. But waves from one slit and those from a second one a few hundred slits away may be exactly out of phase; all or nearly all the light will cancel in pairs in this way. For example, suppose the angle θ is different from its first-order maximum so that the extra path length for a pair of adjacent slits is not exactly λ but rather 1.0010λ. The wave through one slit and another one 500 slits below will be out of phase by 1.5000λ, or exactly $1\frac{1}{2}$ wavelengths, so the two will cancel. A pair of slits, one below each of these, will also cancel. That is, the light from slit 1 cancels with that from slit 501; light from slit 2 cancels with that from slit 502, and so on. Thus even for a tiny angle[†] corresponding to an extra path length of $\frac{1}{1000}\lambda$, there is much destructive interference, and so the maxima are very narrow. The more lines there are in a grating, the sharper will be the peaks (see Fig. 36–18). Because a grating produces much sharper (and brighter) lines than two slits alone, it is a far more precise device for measuring wavelengths.

Why more slits yield sharper peaks

FIGURE 36–18 Intensity as a function of viewing angle θ (or position on the screen) for (a) two slits, (b) six slits. For a diffraction grating, the number of slits is very large ($\sim 10^4$) and the peaks are narrower still.

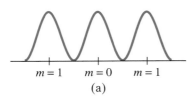

$m = 1 \qquad m = 0 \qquad m = 1$
(a)

$m = 1 \qquad m = 0 \qquad m = 1$
(b)

Suppose the light striking a diffraction grating is not monochromatic, but rather consists of two or more distinct wavelengths. Then for all orders other than $m = 0$, each wavelength will produce a maximum at a different angle (Fig. 36–19a), just as for a double slit. If white light strikes a grating, the central ($m = 0$) maximum will be a sharp white peak. But for all other orders, there will be a distinct

[†]Depending on the total number of slits, there may or may not be complete cancellation for such an angle, so there will be very tiny peaks between the main maxima (see Fig. 36–18b), but they are usually much too small to be seen.

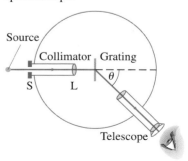

FIGURE 36–19 Spectrum produced by a grating: (a) two wavelengths, 400 nm and 700 nm; (b) white light. The second order will normally be dimmer than the first order. (Higher orders are not shown.) If the grating spacing is small enough, the second and higher orders will be missing.

spectrum of colors spread out over a certain angular width, Fig. 36–19b. Because a diffraction grating spreads out light into its component wavelengths, the resulting pattern is called a **spectrum**.

EXAMPLE 36–7 **Diffraction grating: lines.** Calculate the first- and second-order angles for light of wavelength 400 nm and 700 nm if the grating contains 10,000 lines/cm.

SOLUTION The grating contains 10^4 lines/cm $= 10^6$ lines/m, which means the separation between slits is $d = (1/10^6)$ m $= 1.0 \times 10^{-6}$ m. In first order $(m = 1)$, the angles are

$$\sin\theta_{400} = \frac{m\lambda}{d} = \frac{(1)(4.0 \times 10^{-7}\,\text{m})}{1.0 \times 10^{-6}\,\text{m}} = 0.400$$

$$\sin\theta_{700} = 0.700$$

so $\theta_{400} = 23.6°$ but $\theta_{700} = 44.4°$. In second order,

$$\sin\theta_{400} = \frac{(2)(4.0 \times 10^{-7}\,\text{m})}{1.0 \times 10^{-6}\,\text{m}} = 0.800$$

$$\sin\theta_{700} = 1.40$$

so $\theta_{400} = 53.1°$, but the second order does not exist for $\lambda = 700$ nm because $\sin\theta$ cannot exceed 1. No higher orders will appear.

The diffraction grating is the essential component of a spectroscope, a device for precise measurement of wavelengths, and we discuss it next.

*36–8 The Spectrometer and Spectroscopy

FIGURE 36–20 Spectrometer or spectroscope.

A **spectrometer** or **spectroscope**, Fig. 36–20, is a device[†] to measure wavelengths accurately using a diffraction grating, or a prism, to separate different wavelengths of light. Light from a source passes through a narrow slit S in the collimator. The slit is at the focal point of the lens L, so parallel light falls on the grating. The movable telescope can bring the rays to a focus. Nothing will be seen in the viewing telescope unless it is positioned at an angle θ that corresponds to a diffraction peak (first order is usually used) of a wavelength emitted by the source. The angle θ can be measured

---

[†]If the spectrum of a source is recorded (say, on film) rather than viewed by the eye, the device is called a **spectrometer** or spectrograph, as compared to a **spectroscope**, which is for viewing only; but these terms are often used interchangeably. Devices that can also measure the intensity of light of a given wavelength are called **spectrophotometers**.

Atomic hydrogen

Mercury

Sodium

Solar absorption spectrum

FIGURE 36–21 Line spectra for the gases indicated, and spectrum from the Sun showing absorption lines.

to very high accuracy, so the wavelength of a line can be determined to high accuracy using Eq. 36–13:

$$\lambda = \frac{d}{m} \sin \theta,$$

where m is an integer representing the order, and d is the distance between grating lines. The line you see in a spectrometer corresponding to each wavelength is actually an image of the slit S. The narrower the slit, the narrower—but dimmer—the line is, and the more precisely we can measure its angular position. If the light contains a continuous range of wavelengths, then a continuous spectrum is seen in the spectroscope.

Line spectra

An important use of a spectrometer is for the identification of atoms or molecules. When a gas is heated or a large electric current is passed through it, the gas emits a characteristic **line spectrum**. That is, only certain discrete wavelengths of light are emitted, and these are different for different elements and compounds. Figure 36–21 shows the line spectra for a number of elements in the gas state. Line spectra occur only for gases at high temperatures and low pressure. The light from heated solids, such as a lightbulb filament, and even from a dense gaseous object such as the Sun, produces a **continuous spectrum** including a wide range of wavelengths.

Figure 36–21 also shows the Sun's "continuous spectrum," which contains a number of *dark* lines (only the most prominent are shown), called **absorption lines**. Atoms and molecules can absorb light at the same wavelengths at which they emit light. The Sun's absorption lines are due to absorption by atoms and molecules in the cooler outer atmosphere of the Sun, as well as by atoms and molecules in the Earth's atmosphere. A careful analysis of all these thousands of lines reveals that at least two-thirds of all elements are present in the Sun's atmosphere. The presence of elements in the atmosphere of other planets, in interstellar space, and in stars is also determined by spectroscopy.

Peak Widths and Resolving Power for a Diffraction Grating

We now look at the pattern of maxima produced by a multiple-slit grating using phasor diagrams. We can determine a formula for the width of each peak, and we will see why there are tiny maxima between the principal maxima, as indicated in Fig. 36–18b. First of all, it should be noted that the two-slit and six-slit patterns shown in Fig. 36–18 were drawn assuming very narrow slits so that diffraction does not limit the height of the peaks. For real diffraction gratings, this is not normally the case: the slit width a is often not much smaller than the slit separation d, and diffraction thus limits the intensity of the peaks so the central peak ($m = 0$) is brighter than the side peaks. We won't worry about this effect on intensity except to note that if a diffraction minimum coincides with a particular order of the interference pattern, that order will not appear. (For example, if $d = 2a$, all the even orders, $m = 2, 4, \cdots$, will be missing. Can you see why? Hint: See Example 36–4.)

Figures 36–22 and 36–23 show phasor diagrams for a two-slit and a six-slit grating, respectively. Each short arrow represents the amplitude of a wave from a single slit, and their vector sum (as phasors) represents the total amplitude for a given viewing angle θ. Part (a) of each Figure shows the phasor diagram at $\theta = 0°$, at the center of the pattern, which is the central maximum ($m = 0$). Part (b) of each Figure shows the condition for the adjacent minimum: where the arrows first close on themselves (add to zero) so the amplitude E_θ is zero. For two slits, this occurs when the two separate amplitudes are 180° out of phase. For six slits, it occurs when each amplitude makes a 60° angle with its neighbor. For two slits, the minimum occurs when the phase between slits is $2\pi/2$ (in radians); for six slits it occurs when the phase δ is $2\pi/6$; and in the general case of N slits, the minimum occurs for a phase difference between adjacent slits of

$$\delta = \frac{2\pi}{N}. \tag{36–14}$$

What does this correspond to in θ? First note that δ is related to θ by

$$\frac{\delta}{2\pi} = \frac{d \sin \theta}{\lambda} \quad \text{or} \quad \delta = \frac{2\pi}{\lambda} d \sin \theta \tag{36–15}$$

just as in Eq. 35–4. Let us call $\Delta\theta_0$ the angular position of the minimum next to the peak at $\theta = 0$. This corresponds to an extra path length between adjacent slits (see Fig. 36–17) of $\Delta l = d \sin \Delta\theta_0$, so

$$\frac{\delta}{2\pi} = \frac{\Delta l}{\lambda} = \frac{d \sin \Delta\theta_0}{\lambda}.$$

We insert Eq. 36–14 for δ and find

$$\sin \Delta\theta_0 = \frac{\lambda}{Nd}. \tag{36–16a}$$

Since $\Delta\theta_0$ is usually small (N is usually very large for a grating), $\sin \Delta\theta_0 \approx \Delta\theta_0$, so in the small angle limit we can write

$$\Delta\theta_0 = \frac{\lambda}{Nd}. \tag{36–16b} \qquad \textit{Peak half width (m = 0)}$$

It is clear from either of the last two relations that the larger N is, the narrower will be the central peak. (For $N = 2$, $\sin \Delta\theta_0 = \lambda/2d$, which is what we obtained earlier for the double slit, Eq. 35–2b with $m = 0$.)

FIGURE 36–22 Phasor diagram for two slits (a) at the central maximum, (b) at the nearest minimum.

FIGURE 36–23 Phasor diagram for six slits (a) at the central maximum, (b) at the nearest minimum.

FIGURE 36–24 Phasor diagram for the secondary peak.

Either of Eqs. 36–16 shows why the peaks become narrower for larger N. The origin of the small secondary maxima between the principal peaks (see Fig. 36–18b) can be deduced from the diagram of Fig. 36–24. This is just a continuation of Fig. 36–23b (where $\delta = 60°$); but now the phase δ has been increased to almost 90°, where E_θ is a relative maximum. Note that E_θ is much less than E_0 (Fig. 36–23a), so the intensity in this secondary maximum is much smaller than in a principal peak. As δ (and θ) is increased further, E_θ again decreases to zero (a "double circle"), then reaches another tiny maximum, and so on. Eventually the diagram unfolds again and when $\delta = 360°$, all the amplitudes again lie in a straight line (as in Fig. 35–23a) corresponding to the next principal maximum ($m = 1$ in Eq. 36–13).

Equation 36–16b gives the half width of the central ($m = 0$) peak. To determine the half width of higher order peaks, $\Delta\theta_m$ for order m, we differentiate Eq. 36–15 so as to relate the change $\Delta\delta$ in δ, to the change $\Delta\theta$ in the angle θ:

$$\Delta\delta \approx \frac{d\delta}{d\theta}\,\Delta\theta = \frac{2\pi d}{\lambda}\cos\theta\,\Delta\theta.$$

If $\Delta\theta_m$ represents the half width of a peak of order m ($m = 1, 2, \cdots$)—that is, the angle between the peak maximum and the minimum to either side—then $\Delta\delta = 2\pi/N$ as given by Eq. 36–14. We insert this into the above relation and find

Peak half width (order m)

$$\Delta\theta_m = \frac{\lambda}{Nd\cos\theta_m}, \tag{36–17}$$

where θ_m is the angular position of the mth peak as given by Eq. 36–13. This derivation is valid, of course, only for small $\Delta\delta\,(= 2\pi/N)$ which is indeed the case for real gratings since N is on the order of 10^4 or more.

An important property of any diffraction grating used in a spectroscope is its ability to resolve two very closely spaced wavelengths. The **resolving power** R of a grating is defined as

Resolving power (grating)

$$R = \frac{\lambda}{\Delta\lambda}. \tag{36–18}$$

With a little work, using Eq. 36–17, we can show that $\Delta\lambda = \lambda N/m$ where N is the total number of grating lines and m is the order. Then we have

$$R = Nm. \tag{36–19}$$

The larger the value of R, the closer two wavelengths can be and still be resolvable. If R is given, the minimum separation $\Delta\lambda$ between two wavelengths near λ, is (Eq. 36–18)

$$\Delta\lambda = \frac{\lambda}{R}.$$

EXAMPLE 36–8 **Resolving two close lines.** Yellow sodium light, which consists of two wavelengths, $\lambda_1 = 589.00\,\text{nm}$ and $\lambda_2 = 589.59\,\text{nm}$, falls on a 7500-line/cm diffraction grating. Determine (a) the maximum order m that will be present for sodium light, (b) the width of grating necessary to resolve the two sodium lines, (c) the angular width of each sodium line.

SOLUTION (a) Since $d = 1\,\text{cm}/7500 = 1.33 \times 10^{-6}\,\text{m}$, then the maximum value of m for $\lambda = 589\,\text{nm}$ can be found from Eq. 35–13 with $\sin\theta \le 1$:

$$m = \frac{d}{\lambda}\sin\theta \le \frac{1.33 \times 10^{-6}\,\text{m}}{5.89 \times 10^{-7}\,\text{m}} = 2.25,$$

so $m = 2$ is the maximum order present.

(b) The resolving power needed is

$$R = \frac{\lambda}{\Delta\lambda} = \frac{589 \text{ nm}}{0.59 \text{ nm}} = 1000.$$

From Eq. 36–19, the total number N of lines needed is $N = R/m = 1000/2 = 500$, so the grating need only be $500/7500 \text{ cm}^{-1} = 0.0667 \text{ cm}$ wide. A typical grating is a few centimeters wide, and so will easily resolve the two lines.

(c) We need to use Eq. 36–17 for $m = 2$, but what is $\cos\theta_2$? From Eq. 36–13,

$$\sin\theta_2 = m\frac{\lambda}{d} = \frac{2(589 \times 10^{-9} \text{ m})}{1.33 \times 10^{-6} \text{ m}} = 0.886.$$

Then $\cos\theta_2 = (1 - \sin^2\theta_2)^{\frac{1}{2}} = 0.464$. For a grating with the minimum 500 lines, the angular width (Eq. 36–17) for $m = 2$ is

$$\Delta\theta_2 = \frac{\lambda}{Nd\cos\theta_2} = \frac{589 \times 10^{-9} \text{ m}}{(500)(1.33 \times 10^{-6} \text{ m})(0.464)} = 0.0019 \text{ rad},$$

or 0.11°.

*36–10 X-Rays and X-Ray Diffraction

In 1895, W. C. Roentgen (1845–1923) discovered that when electrons were accelerated by a high voltage in a vacuum tube and allowed to strike a glass (or metal) surface inside the tube, fluorescent minerals some distance away would glow, and photographic film would become exposed. Roentgen attributed these effects to a new type of radiation (different from cathode rays). They were given the name **X-rays** after the algebraic symbol x, meaning an unknown quantity. He soon found that X-rays penetrated through some materials better than through others, and within a few weeks he presented the first X-ray photograph (of his wife's hand). The production of X-rays today is usually done in a tube (Fig. 36–25) similar to Roentgen's, using voltages of typically 30 kV to 150 kV.

Investigations into the nature of X-rays indicated they were not charged particles (such as electrons) since they could not be deflected by electric or magnetic fields. It was suggested that they might be a form of invisible light. However, they showed no diffraction or interference effects using ordinary gratings. Of course, if their wavelengths were much smaller than the typical grating spacing of $10^{-6} \text{ m} (= 10^3 \text{ nm})$, no effects would be expected. Around 1912, it was suggested by Max von Laue (1879–1960) that if the atoms in a crystal were arranged in a regular array (see Fig. 17–2a), such a crystal might serve as a diffraction grating for very short wavelengths on the order of the spacing between atoms, estimated to be about $10^{-10} \text{ m} (= 10^{-1} \text{ nm})$. Experiments soon showed that X-rays scattered from a crystal did indeed show the peaks and valleys of a diffraction pattern (Fig. 36–26). Thus it was shown, in a single blow, that X-rays have a wave nature and that atoms are arranged in a regular way in crystals. Today, X-rays are recognized as electromagnetic radiation with wavelengths in the range of about 10^{-2} nm to 10 nm, the range readily produced in an X-ray tube.

X-rays

FIGURE 36–25 X-ray tube. Electrons emitted by a heated filament in a vacuum tube are accelerated by a high voltage. When they strike the surface of the anode, the "target," X-rays are emitted.

FIGURE 36–26 This X-ray diffraction pattern is one of the first observed by Max von Laue in 1912 when he aimed a beam of X-rays at a zinc sulfide crystal. The diffraction pattern was detected directly on a photographic plate.

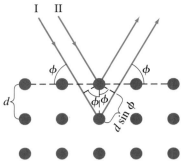

FIGURE 36–27 X-ray diffraction by a crystal.

We saw in Section 36–5 that light of shorter wavelength provides greater resolution when we are examining an object microscopically. Since X-rays have much shorter wavelengths than visible light, in principle they should offer much greater resolution. However, there seems to be no effective material to use as lenses for the very short wavelengths of X-rays. Instead, the clever but complicated technique of **X-ray diffraction** (or **crystallography**) has proved very effective for examining the microscopic world of atoms and molecules. In a simple crystal such as NaCl, the atoms are arranged in an orderly cubical fashion, Fig. 36–27, with atoms spaced a distance d apart. Suppose that a beam of X-rays is incident on the crystal at an angle ϕ to the surface, and that the two rays shown are reflected from two subsequent planes of atoms as shown. The two rays will constructively interfere if the extra distance ray I travels is a whole number of wavelengths farther than what ray II travels. This extra distance is $2d \sin \phi$. Therefore, constructive interference will occur when

$$m\lambda = 2d \sin \phi, \qquad m = 1, 2, 3, \cdots, \qquad (36\text{–}20)$$

Bragg equation

where m can be any integer. (Notice that ϕ is *not* the angle with respect to the normal to the surface.) This is called the **Bragg equation** after W. L. Bragg (1890–1971), who derived it and who, together with his father W. H. Bragg (1862–1942), developed the theory and technique of X-ray diffraction by crystals in 1912–1913. Thus, if the X-ray wavelength is known and the angle ϕ at which constructive interference occurs is measured, d can be obtained. This is the basis for X-ray crystallography.

Actual X-ray diffraction patterns are quite complicated. First of all, a crystal is a three-dimensional object, and X-rays can be diffracted from different planes at different angles within the crystal, as shown in Fig. 36–28. Although the analysis is complex, a great deal can be learned about any substance that can be put in crystalline form. If the substance is not a single crystal but a mixture of many tiny crystals—as in a metal or a powder—then instead of a series of spots, as in Fig. 36–26, a series of circles is obtained, Fig. 36–29, each corresponding to diffraction of a certain order m from a particular set of parallel planes.

X-ray diffraction has been very useful in determining the structure of biologically important molecules. For example, it was with the help of X-ray diffraction that, in 1953, J. D. Watson and F. H. C. Crick worked out the double-helix structure of DNA.

FIGURE 36–28 There are many possible planes existing within a crystal from which X-rays can be diffracted.

FIGURE 36–29 (a) Diffraction of X-rays from a polycrystalline substance produces a set of circular rings as in (b), which is for polycrystalline sodium acetoacetate.

(a)

(b)

36–11 Polarization

An important and useful property of light is that it can be *polarized*. To see what this means, let us examine waves traveling on a rope. A rope can be set into oscillation in a vertical plane as in Fig. 36–30a, or in a horizontal plane as in Fig. 36–30b. In either case, the wave is said to be **linearly polarized** or **plane-polarized**—that is, the oscillations are in a plane.

If we now place an obstacle containing a vertical slit in the path of the wave, Fig. 36–31, a vertically polarized wave passes through, but a horizontally polarized wave will not. If a horizontal slit were used, the vertically polarized wave would be stopped. If both types of slit were used, both types of wave would be stopped. Note that polarization can exist *only* for *transverse waves*, and not for longitudinal waves such as sound. The latter oscillate only along the direction of motion, and neither orientation of slit would stop them.

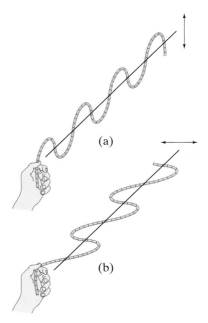

FIGURE 36–30 (above) Transverse waves on a rope polarized (a) in a vertical plane and (b) in a horizontal plane.

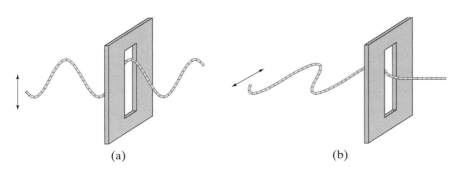

(a)   (b)

FIGURE 36–31 Vertically polarized wave passes through a vertical slit, but a horizontally polarized wave will not. [Note: This diagram would apply to the magnetic field **B** in an EM wave—see footnote below.]

Maxwell's theory of light as electromagnetic (EM) waves predicted that light can be polarized since an EM wave is a transverse wave. The direction of polarization in a plane-polarized EM wave is taken as the direction of the electric field vector **E**.

Light is not necessarily polarized. It can be **unpolarized**, which means that the source has oscillations in many planes at once, as shown in Fig. 36–32. An ordinary incandescent lightbulb emits unpolarized light, as does the Sun.

Polaroids

Plane-polarized light can be obtained from unpolarized light using certain crystals such as tourmaline. Or, more commonly today, we can use a **Polaroid sheet**. (Polaroid materials were invented in 1929 by Edwin Land.) A Polaroid sheet consists of complicated long molecules arranged parallel to one another. Such a Polaroid acts like a series of parallel slits to allow one orientation of polarization to pass through nearly undiminished (this direction is called the *axis* of the Polaroid), whereas a perpendicular polarization is absorbed almost completely.[†]

FIGURE 36–32 Oscillation of the electric field vectors in unpolarized light. The light is traveling into or out of the page.

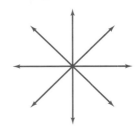

[†] How this occurs can be explained at the molecular level. An electric field **E** that oscillates parallel to the long molecules can set electrons into motion along the molecules, thus doing work on them and transferring energy. Hence, if **E** is parallel to the molecules, it gets absorbed. An electric field **E** perpendicular to the long molecules does not have this possibility of doing work and transferring its energy, and so passes through freely. When we speak of the *axis* of a Polaroid, we mean the direction for which **E** is passed, so a Polaroid axis is perpendicular to the long molecules. (If we want to think of there being slits between the parallel molecules in the sense of Fig. 36–31, then Fig. 36–31 would apply for the **B** field in the EM wave, not the **E** field.)

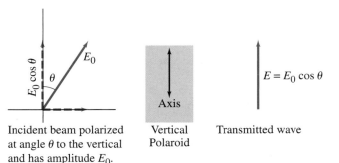

Incident beam polarized at angle θ to the vertical and has amplitude E_0.

Vertical Polaroid

Transmitted wave

FIGURE 36–33 Vertical Polaroid transmits only the vertical component of a wave (electric field) incident upon it.

If a beam of plane-polarized light strikes a Polaroid whose axis is at an angle θ to the incident polarization direction, the beam will emerge plane-polarized parallel to the Polaroid axis and its amplitude will be reduced by $\cos\theta$, Fig. 36–33. Thus, a Polaroid passes only that component of polarization (the electric field vector, **E**) that is parallel to its axis. Because the intensity of a light beam is proportional to the square of the amplitude (Sections 15–3 and 32–7), we see that the intensity of a plane-polarized beam transmitted by a polarizer is

Intensity of plane polarized wave reduced by polarizer

$$I = I_0 \cos^2\theta, \tag{36–21}$$

where θ is the angle between the polarizer axis and the plane of polarization of the incoming wave, and I_0 is the incoming intensity.[†]

A Polaroid can be used as a **polarizer** to produce plane-polarized light from unpolarized light, since only the component of light parallel to the axis is transmitted. A Polaroid can also be used as an **analyzer** to determine (1) if light is polarized and (2) what is the plane of polarization. A Polaroid acting as an analyzer will pass the same amount of light independent of the orientation of its axis if the light is unpolarized; try rotating one lens of a pair of Polaroid sunglasses while looking through it at a lightbulb. If the light is polarized, however, when you rotate the Polaroid the transmitted light will be a maximum when the plane of polarization is parallel to the Polaroid's axis, and a minimum when perpendicular to it. If you do this while looking at the sky, preferably at right angles to the Sun's direction, you will see that skylight is polarized. (Direct sunlight is unpolarized, but don't look directly at the Sun, even through a polarizer, for damage to the eye may occur.) If the light trasmitted by an analyzer Polaroid falls to zero at one orientation, then the light is 100 percent plane-polarized. If it merely reaches a minimum, the light is *partially polarized.*

Unpolarized light consists of light with random directions of polarization. Each of these polarization directions can be resolved into components along two mutually perpendicular directions. On average, an unpolarized beam can be thought of as two plane-polarized beams of equal magnitude perpendicular to one another. When unpolarized light passes through a polarizer, one of the components is eliminated. So the intensity of the light passing through is reduced by half since half the light is eliminated, $I = \frac{1}{2}I_0$ (Fig. 36–34).

When two Polaroids are *crossed*—that is, their axes are perpendicular to one another—unpolarized light can be entirely stopped. As shown in Fig. 36–35, unpolarized light is made plane-polarized by the first Polaroid (the polarizer). The second Polaroid, the analyzer, then eliminates this component since its axis is perpendicular to the first. You can try this with Polaroid sunglasses (Fig. 36–36). Note that Polaroid sunglasses eliminate 50 percent of unpolarized light because of their polarizing property; they absorb even more because they are colored.

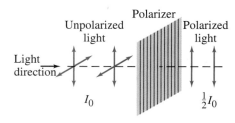

FIGURE 36–34 Unpolarized light has both vertical and horizontal components. After passing through a polarizer, one of these components is eliminated. The intensity of the light is reduced to half.

[†]Equation 36–21 is often referred to as **Malus' law**, after Etienne Malus, a contemporary of Fresnel.

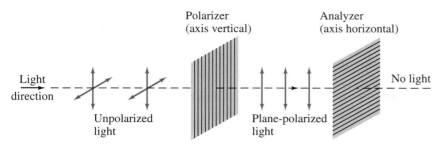

FIGURE 36–35 Crossed Polaroids completely eliminate light.

FIGURE 36–36 Crossed Polaroids. When the two polarized sunglass lenses overlap, with axes perpendicular, almost no light passes through.

EXAMPLE 36–9 **Two Polaroids at 60°.** Unpolarized light passes through two Polaroids; the axis of one is vertical and that of the other is at 60° to the vertical. What is the orientation and intensity of the transmitted light?

SOLUTION The first Polaroid eliminates half the light so the intensity is reduced by half: $I_1 = \frac{1}{2}I_0$. The light reaching the second polarizer is vertically polarized and so is reduced in intensity (Eq. 36–21) to

$$I_2 = I_1(\cos 60°)^2 = \frac{1}{4}I_1.$$

Thus, $I_2 = \frac{1}{8}I_0$. The transmitted light has an intensity one-eighth that of the original and is plane-polarized at a 60° angle to the vertical.

CONCEPTUAL EXAMPLE 36–10 **Three Polaroids.** We saw in Fig. 36–35 that when unpolarized light falls on two crossed Polaroids (axes at 90°), no light passes through. What happens if a third Polaroid, with axis at 45° to each of the other two, is placed between them?

RESPONSE We start just as in Example 36–9. The first Polaroid changes the unpolarized light to plane-polarized and reduces the intensity from I_0 to $I = \frac{1}{2}I_0$. The second polarizer further reduces the intensity by $(\cos 45°)^2$, Eq. 36–21:

$$I_2 = I_1(\cos 45°)^2 = \frac{1}{2}I_1 = \frac{1}{4}I_0.$$

The light leaving the second polarizer is plane polarized at 45° (Fig. 36–37) relative to the third polarizer, so the latter reduces the intensity to

$$I_3 = I_2(\cos 45°)^2 = \frac{1}{2}I_2$$

or $I_3 = \frac{1}{8}I_0$. Thus $\frac{1}{8}$ of the original intensity gets transmitted. But if we don't put the 45° Polaroid in at all, zero intensity results (Fig. 36–35). The 45° Polaroid must be inserted *between* the other two if transmission is to occur. Placing it before or after the other two results in zero intensity.

FIGURE 36–37 Example 36–10.

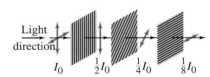

An important use of polarizers is in a **liquid crystal display** (LCD). LCDs are used in many displays, from digital watches to small TV sets. Each pixel (picture element) consists of a small liquid crystal contained between two glass plates whose outer surfaces have a thin film of polarizer. Liquid crystals are materials that, at room temperature, exist in a phase neither fully solid nor fully liquid. Useful ones are composed of rod-like molecules that interact only weakly with each other and tend to align with each other. The overall direction of alignment can be controlled using electric fields. Each pixel can be bright or dark depending on that alignment relative to the polarizers on the ends. Color displays can be obtained by using color filters.

➡ **PHYSICS APPLIED**
Liquid crystals (LCD)

FIGURE 36–38 Light reflected from a nonmetallic surface, such as the smooth surface of water in a lake, is partially polarized parallel to the surface.

FIGURE 36–39 Photographs of a river, (a) allowing all light into the camera lens, and (b) using a polarizer which is adjusted to absorb most of the (polarized) light reflected from the water's surface, allowing the dimmer light from the bottom of the river, and any fish swimming there, to be seen more readily.

FIGURE 36–40 At θ_p the reflected light is plane-polarized parallel to the surface, and $\theta_p + \theta_r = 90°$, where θ_r is the refraction angle. (The large dots represent vibrations perpendicular to the page.)

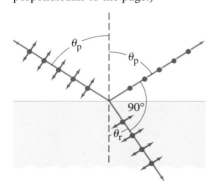

Polarization by Reflection

Another means of producing polarized light from unpolarized light is by reflection. When light strikes a nonmetallic surface at any angle other than perpendicular, the reflected beam is polarized preferentially in the plane parallel to the surface, Fig. 36–38. In other words, the component with polarization in the plane perpendicular to the surface is preferentially transmitted or absorbed. You can check this by rotating Polaroid sunglasses while looking through them at a flat surface of a lake or road. Since most outdoor surfaces are horizontal, Polaroid sunglasses are made with their axes vertical to eliminate the more strongly reflected horizontal component, and thus reduce glare. Fishermen wear Polaroids to eliminate reflected glare from the surface of a lake or stream and thus see beneath the water more clearly (Fig. 36–39).

(a)   (b)

The amount of polarization in the reflected beam depends on the angle, varying from no polarization at normal incidence to 100 percent polarization at an angle known as the **polarizing angle**, θ_p.[†] This angle is related to the index of refraction of the two materials on either side of the boundary by the equation

$$\tan\theta_p = \frac{n_2}{n_1}, \qquad (36\text{--}22a)$$

where n_1 is the index of refraction of the material in which the beam is traveling, and n_2 is that of the medium beyond the reflecting boundary. If the beam is traveling in air, $n_1 = 1$, and Eq. 36–22a becomes

$$\tan\theta_p = n. \qquad (36\text{--}22b)$$

The polarizing angle θ_p is also called **Brewster's angle**, and Eqs. 36–22 called *Brewster's law*, after the Scottish physicist David Brewster (1781–1868), who worked it out experimentally in 1812. Equations 36–22 can be derived from the electromagnetic wave theory of light. It is interesting that at Brewster's angle, the reflected and transmitted rays make a 90° angle to each other; that is, $\theta_p + \theta_r = 90°$, Fig. 36–40. This can be seen by substituting Eq. 36–22a, $n_2 = n_1\tan\theta_p = n_1\sin\theta_p/\cos\theta_p$, into Snell's law, $n_1\sin\theta_p = n_2\sin\theta_r$, and get $\cos\theta_p = \sin\theta_r$ which can only hold if $\theta_p = 90° - \theta_r$.

EXAMPLE 36–11 Polarizing angle. (a) At what incident angle is sunlight reflected from a lake plane-polarized? (b) What is the refraction angle?

SOLUTION (a) We use Eq. 36–22b with $n = 1.33$, so $\tan\theta_p = 1.33$ and $\theta_p = 53.1°$.
(b) $\theta_r = 90.0° - \theta_p = 36.9°$.

[†]Only a fraction of the incident light is reflected at the surface of the transparent medium. Although this reflected light is 100 percent polarized (if $\theta = \theta_p$), the remainder of the light, which is transmitted into the new medium, is only partially polarized.

Sunsets are red, the sky is blue, and skylight is polarized (at least partially). These phenomena can be explained on the basis of the *scattering* of light by the molecules of the atmosphere. In Fig. 36–41 we see unpolarized light from the Sun impinging on a molecule of the Earth's atmosphere. The electric field of the EM wave sets the electric charges within the molecule into motion, and the molecule absorbs some of the incident radiation. But it quickly reemits this light since the charges are oscillating. As discussed in Section 32–4, oscillating electric charges produce EM waves. The electric field of these waves is in a plane that includes the line of oscillation. The intensity is strongest along a line perpendicular to the oscillation, and drops to zero along the line of oscillation (Section 32–4). In Fig. 36–41 the motion of the charges is resolved into two components. An observer at right angles to the direction of the sunlight, as shown, will see plane-polarized light because no light is emitted along the line of the other component of the oscillation. (Another way to understand this is to note that when viewing along the line of oscillation, one doesn't see the oscillation, and hence sees no waves made by it. Furthermore, the EM wave produces no oscillation along its own direction of motion.) At other viewing angles, both components will be present; one will be stronger, however, so the light appears partially polarized. Thus, the process of scattering explains the polarization of skylight.

Scattering of light by the Earth's atmosphere depends on λ. For particles much smaller than the wavelength of light (such as molecules of air), the particles will be less of an obstruction to long wavelengths than to short ones. The scattering decreases, in fact, as $1/\lambda^4$. Blue and violet light are thus scattered much more than red and orange, which is why the sky looks blue. At sunset, the Sun's rays pass through a maximum length of atmosphere. Much of the blue has been taken out by scattering. The light that reaches the surface of the Earth, and reflects off clouds and haze, is thus lacking in blue which is why sunsets appear reddish.

The dependence of scattering on $1/\lambda^4$ is valid only if the scattering objects are much smaller than the wavelength of the light. This is valid for oxygen and nitrogen molecules whose diameters are about 0.2 nm. Clouds, however, contain water droplets or crystals that are much larger than λ. They scatter all frequencies of light nearly uniformly. Hence clouds appear white (or gray, if shadowed).

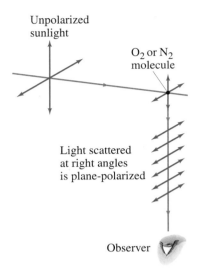

FIGURE 36–41 Unpolarized sunlight scattered by molecules of the air. An observer at right angles sees plane-polarized light, since the component of oscillation along the line of sight emits no light along that line.

➡ **PHYSICS APPLIED**

Why the sky is blue
Why sunsets are red

Summary

Diffraction refers to the fact that light, like other waves, bends around objects it passes, and spreads out after passing through narrow slits. This bending gives rise to a **diffraction pattern** due to interference between rays of light that travel different distances.

Light passing through a very narrow slit of width a will produce a pattern with a bright central maximum of half-width θ given by

$$\sin \theta = \frac{\lambda}{a},$$

flanked by fainter lines to either side.

The minima in the diffraction pattern occur at

$$a \sin \theta = m\lambda$$

where $m = 1, 2, 3, \cdots$, but not $m = 0$ (for which the pattern has its strongest maximum).

The **intensity** at any point in the single-slit diffraction pattern can be calculated using **phasor** diagrams. The same technique can be used to determine the intensity of the pattern produced by two slits.

The pattern for two-slit interference can be described as a series of maxima due to interference of light from the two slits, modified by an "envelope" due to diffraction at each slit.

The wave nature of light limits the sharpness or **resolution** of images. Because of diffraction, it is not possible to *discern details smaller than the wavelength* of the radiation being used. The useful magnification of a light microscope is limited by diffraction to about $1000\times$.

A **diffraction grating** consists of many parallel slits or lines, each separated from its neighbors by a distance d. The peaks of constructive interference occur at angles θ given by

$$\sin\theta = \frac{m\lambda}{d},$$

where $m = 0, 1, 2, \cdots$. The peaks are much brighter and sharper for a diffraction grating than for the simple two-slit apparatus. Peak width is inversely proportional to the total number of lines in the grating.

A diffraction grating (or a prism) is used in a **spectroscope** to separate different colors or to observe **line spectra**, since for a given order m, θ depends on λ. Precise determination of wavelength can be done with a spectroscope by careful measurement of θ.

In **unpolarized light**, the electric field vectors oscillate transversely at all angles. If the electric vector oscillates only in one plane the light is said to be **plane-polarized**. Light can also be partially polarized.

When an unpolarized light beam passes through a Polaroid sheet, the emerging beam is plane-polarized. When a light beam is polarized and passes through a Polaroid, the intensity varies as the Polaroid is rotated. Thus a Polaroid can act as a polarizer or as an analyzer.

The intensity of a plane-polarized beam incident on a Polaroid is reduced by the factor

$$\cos^2\theta$$

where θ is the angle between the axis of the Polaroid and the initial plane of polarization.

Light can also be partially or fully **polarized by reflection**. If light traveling in air is reflected from a medium of index of refraction n, the reflected beam will be *completely* plane-polarized if the incident angle θ_p is given by $\tan\theta_p = n$. The fact that light can be polarized shows that it must be a transverse wave.

Questions

1. Radio waves and light are both electromagnetic waves. Why can we hear a radio behind a hill when we cannot see the transmitting antenna?

2. Hold one hand close to your eye and focus on a distant light source through a narrow slit between two fingers. (Adjust your fingers to obtain the best pattern.) Describe the pattern that you see.

3. A rectangular slit is twice as high as it is wide. Will the light spread more in the horizontal or in the vertical plane? Describe the pattern.

4. Explain why diffraction patterns are more difficult to observe with an extended light source than for a point source. Compare also a monochromatic source to white light.

5. For diffraction by a single slit, what is the effect of increasing (a) the slit width, (b) the wavelength?

6. Describe the single-slit diffraction pattern produced when white light falls on a slit having a width of (a) 50 nm, (b) 50,000 nm.

7. What happens to the diffraction pattern of a single slit if the whole apparatus is immersed in (a) water, (b) a vacuum, instead of in air.

8. Describe clearly the difference between the angles θ and β, as in Eq. 36–6.

9. In the single-slit diffraction pattern, why does the first off-center maximum not occur at exactly $\sin\theta = \frac{3}{2}\lambda/a$?

* 10. Figure 36–10 shows a two-slit interference pattern for the case when d is larger than a. Can the reverse case occur, when d is less than a?

* 11. When both diffraction and interference are taken into account in the double-slit experiment, discuss the effect of increasing (a) the wavelength, (b) the slit separation, (c) the slit width.

12. Discuss the similarities, and differences, of double-slit interference and single-slit diffraction.

13. Does diffraction limit the resolution of images formed by (a) spherical mirrors, (b) plane mirrors?

14. Do diffraction effects occur for virtual as well as real images?

15. What are the advantages (give at least two) for the use of large reflecting mirrors in astronomical telescopes?

16. Atoms have diameters of about 10^{-8} cm. Can visible light be used to "see" an atom? Why or why not?

17. Which color of visible light would give the best resolution in a microscope?

18. If monochromatic light were used in a microscope, would the color affect the resolution? Explain.

19. Could a diffraction grating just as well be called an interference grating? Discuss.

20. For light consisting of wavelengths between 400 nm and 700 nm, incident normally on a diffraction grating, for what orders (if any) would there be overlap in the observed spectrum? Does your answer depend on the slit width?

21. What is the difference in the interference patterns formed (i) by two slits 10^{-4} cm apart, (ii) by a diffraction grating containing 10^4 lines/cm?

22. White light strikes (a) a diffraction grating and (b) a prism. A rainbow appears on a wall just below the direction of the horizontal incident beam in each case. What is the color of the top of the rainbow in each case?

23. Explain why there are tiny peaks between the main peaks produced by a diffraction grating illuminated with monochromatic light. Why are the peaks so tiny?

24. What does polarization tell us about the nature of light?

25. How can you tell if a pair of sunglasses is polarizing or not?

* 26. What would be the color of the sky if the Earth had no atmosphere?

* 27. If the Earth's atmosphere were 50 times denser than it is, would sunlight still be white, or would it be some other color?

Problems

Section 36–1

1. (I) If 680-nm light falls on a slit 0.0345 mm wide, what is the angular width of the central diffraction peak?

2. (I) Monochromatic light falls on a slit that is 3.00×10^{-3} mm wide. If the angle between the first dark fringes on either side of the central maximum is 37.0° (dark fringe to dark fringe), what is the wavelength of the light used?

3. (II) Light of wavelength 550 nm falls on a slit that is 3.50×10^{-3} mm wide. Estimate how far from the central maximum the first diffraction maximum fringe is if the screen is 10.0 m away?

4. (II) Monochromatic light of wavelength 689 nm falls on a slit. If the angle between the first bright fringes on either side of the central maximum is 38°, estimate the slit width.

5. (II) If a slit diffracts 550-nm light so that the diffraction maximum is 8.0 cm wide on a screen 2.50 m away, what will be the width of the diffraction maximum for light with a wavelength of 400 nm?

6. (II) (a) For a given wavelength λ, what is the maximum slit width for which there will be no diffraction minima? (b) What is the maximum slit width so that no visible light exhibits a diffraction minimum?

7. (II) How wide is the central diffraction peak on a screen 3.50 m behind a 0.0655-mm-wide slit illuminated by 400-nm light?

8. (II) If parallel light falls on a single slit of width a at a 30° angle to the normal, describe the diffraction pattern.

Section 36–2

9. (II) If you double the width of a single slit, the intensity of the light passing through the slit is doubled. (a) Show, however, that the intensity at the center of the screen increases by a factor of 4. (b) Explain why this does not violate conservation of energy.

10. (III) (a) Explain why the secondary maxima in the single-slit diffraction pattern do not occur precisely at $\beta/2 = (m + \frac{1}{2})\pi$ where $m = 1, 2, 3, \cdots$. (b) By differentiating Eq. 36–7, show that the secondary maxima occur when $\beta/2$ satisfies the relation $\tan(\beta/2) = \beta/2$. (c) Carefully and precisely plot the curves $y = \beta/2$ and $y = \tan \beta/2$. From their intersections, determine the values of β for the first and second secondary maxima. What is the percent difference from $\beta/2 = (m + \frac{1}{2})\pi$?

11. (III) Determine, approximately, the angular width at half maximum (where $I = \frac{1}{2}I_0$) of the central diffraction peak for a single slit. [Hint: Use graphical methods, or trial and error; the Problem cannot be solved analytically.] To be concrete, assume $\lambda = 550$ nm and $a = 2.60 \times 10^{-3}$ mm.

* Section 36–3

* 12. (II) Design a double-slit apparatus so that the central diffraction peak contains precisely fifteen fringes.

* 13. (II) If a double-slit pattern contains exactly seven fringes in the central diffraction peak, what can you say about the slit width and separation?

* 14. (II) Suppose $d = a$ in a double-slit apparatus, so that the two slits merge into one slit of width $2a$. Show that Eq. 36–9 reduces to the correct equation for single-slit diffraction.

* 15. (II) Two 0.010-mm-wide slits are 0.030 mm apart (center to center). Determine (a) the spacing between interference fringes for 550 nm light on a screen 1.0 m away and (b) the distance between the two diffraction minima on either side of the central maximum of the envelope.

* 16. (II) How many fringes are contained in the central diffraction peak for a double-slit pattern if (a) $d = 2.00a$, (b) $d = 12.0a$, (c) $d = 4.50a$, (d) $d = 7.20a$.

* 17. (III) Draw phasor diagrams (as in Fig. 36–8) for the double-slit experiment, including both interference and diffraction. Make a diagram for each of several crucial points in the pattern (see Fig. 36–10c) such as at the center, first minimum, next maximum, and at $\sin \theta = \lambda/a$.

* 18. (III) (a) Derive an expression for the intensity in the interference pattern for three equally spaced slits. Express in terms of $\delta = 2\pi d \sin \theta/\lambda$ where d is the distance between adjacent slits and assume the slit width $a \approx \lambda$. (b) Show that there is only one secondary maximum between principal peaks.

Sections 36–4 and 36–5

19. (I) What is the angular resolution limit set by diffraction for the 100-inch (mirror diameter) Mt. Wilson telescope ($\lambda = 500$ nm)?

20. (II) Two stars 10 light-years away are barely resolved by a 90 cm (mirror diameter) telescope. How far apart are the stars? Assume $\lambda = 550$ nm and that the resolution is limited by diffraction.

21. (II) The normal lens on a 35-mm camera has a focal length of 50 mm. Its aperture diameter varies from a maximum of 25 mm ($f/2$) to a minimum of 3.0 mm ($f/16$). Determine the resolution limit set by diffraction for $f/2$ and $f/16$. Specify as the number of lines per millimeter resolved on the film. Take $\lambda = 500$ nm.

22. (II) Suppose that you wish to construct a telescope that can resolve features 7.0 km across on the moon, 384,000 km away. You have a 2.0-m-focal-length objective lens whose diameter is 11.0 cm. What focal-length eyepiece is needed if your eye can resolve objects 0.10 mm apart at a distance of 25 cm? What is the resolution limit set by the size of the objective lens (that is, by diffraction)? Use $\lambda = 500$ nm.

Section 36–7

23. (I) At what angle will 440-nm light produce a third-order maximum when falling on a grating whose slits are 1.35×10^{-3} cm apart?

24. (I) How many lines per centimeter does a grating have if the third order occurs at a 13.0° angle for 650-nm light?

25. (I) A grating has 6600 lines/cm. How many spectral orders can be seen when it is illuminated by white light?

26. (I) A 3500-line/cm grating produces a third-order fringe at a 22.0° angle. What wavelength of light is being used?

27. (I) A source produces first-order lines when incident normally on a 10,000-line/cm diffraction grating at angles $29.8°$, $37.7°$, $39.6°$, and $48.9°$. What are the wavelengths?

28. (II) White light containing wavelengths from 400 nm to 750 nm falls on a grating with 7800 lines/cm. How wide is the first-order spectrum on a screen 2.80 m away?

29. (II) Show that the second- and third-order spectra of white light produced by a diffraction grating always overlap. What wavelengths overlap?

30. (II) Two first-order spectrum lines are measured by a 9550 line/cm spectroscope at angles, on each side of center, of $+26°38'$, $+41°02'$ and $-26°18'$, $-40°27'$. What are the wavelengths?

31. (II) Suppose the angles measured in Problem 30 were produced when the spectrometer (but not the source) was submerged in water. What then would be the wavelengths (in air)?

32. (II) The first-order line of 589-nm light falling on a diffraction grating is observed at a $15.5°$ angle. How far apart are the slits? At what angle will the third order be observed?

33. (II) Monochromatic light falls on a transmission diffraction grating at an angle ϕ to the normal. Show that Eq. 36–13 for diffraction maxima must be replaced by

$$d(\sin\phi \pm \sin\theta) = m\lambda. \qquad m = 0, 1, 2, \cdots.$$

Explain the $\pm$ sign.

* Section 36–9

* 34. (II) Missing orders occur for a diffraction grating when a diffraction minimum coincides with an interference maximum. Let a be the width of each slit and d the separation of slits and show (a) that if $d = 2a$, all even orders ($m = 2, 4, 6, \cdots$) are missing. (b) Show that there will be missing orders whenever

$$\frac{d}{a} = \frac{m_1}{m_2}$$

where m_1 and m_2 are integers. (c) Discuss the case $d = a$, the limit in which the space between slits becomes negligible.

* 35. (II) Let 580-nm light be incident normally on a diffraction grating for which $d = 3.00a = 1200$ nm. (a) How many orders (principal maxima) are present? (b) If the grating is 1.80 cm wide, what is the full angular width of each principal maximum?

* 36. (II) A 6500-line/cm diffraction grating is 3.61 cm wide. If light with wavelengths near 624 nm falls on the grating, how close can two wavelengths be if they are to be resolved in any order? What order gives the best resolution?

* 37. (II) A diffraction grating has 16,000 rulings in its 2.4 cm width. Determine (a) its resolving power in first and second orders, and (b) the minimum wavelength resolution ($\Delta\lambda$) it can yield for $\lambda \approx 410$ nm.

* 38. (II) For a fixed wavelength and diffraction angle, show that the resolving power of a diffraction grating depends on the total width, Nd, where N is the total number of slits each of width d.

* 39. (II) Determine a formula for the minimum difference in frequency, Δf, that a diffraction grating can resolve when two frequencies, $f_1 \approx f_2 = f$, are incident on it.

* Section 36–10

* 40. (II) X-rays of wavelength 0.138 nm fall on a crystal whose atoms, lying in planes, are spaced 0.265 nm apart. At what angle must the X-rays be directed if the first diffraction maximum is to be observed?

* 41. (II) First-order Bragg diffraction is observed at $26.2°$ from a crystal with spacing between atoms of 0.24 nm. (a) At what angle will second order be observed? (b) What is the wavelength of the X-rays?

* 42. (II) If X-ray diffraction peaks corresponding to the first three orders ($m = 1, 2,$ and 3) are measured, can both the X-ray wavelength λ and lattice spacing d be determined? Prove your answer.

Section 36–11

43. (I) Two polarizers are oriented at $75°$ to one another. Unpolarized light falls on them. What fraction of the light intensity is transmitted?

44. (I) Two Polaroids are aligned so that the light passing through them is a maximum. At what angle should one of them be placed so the intensity is subsequently reduced by half?

45. (I) What is Brewster's angle for an air–glass ($n = 1.56$) surface?

46. (I) What is Brewster's angle for a diamond submerged in water if the light is hitting the diamond while traveling in the water?

47. (II) At what angle should the axes of two Polaroids be placed so as to reduce the intensity of the incident unpolarized light to (a) $\frac{1}{3}$, (b) $\frac{1}{10}$?

48. (II) The critical angle for total internal reflection at a boundary between two materials is $52°$. What is Brewster's angle at this boundary?

49. (II) What would Brewster's angle be for reflections off the surface of water for light coming from beneath the surface? Compare to the angle for total internal reflection, and to Brewster's angle from above the surface.

50. (II) Two polarizers are oriented at $34.0°$ to one another. Light polarized at a $17.0°$ angle to each polarizer passes through both. What reduction in intensity takes place?

51. (II) Unpolarized light passes through five successive Polaroid sheets each of whose axis makes a $45°$ angle with the previous one. What is the intensity of the transmitted beam?

52. (II) Describe how to rotate the plane of polarization of a plane-polarized beam of light by $90°$ and produce only a 10 percent loss in intensity, using polarizers.

53. (III) The percent polarization P of a partially polarized beam of light is defined as

$$P = \frac{I_{\max} - I_{\min}}{I_{\max} + I_{\min}} \times 100$$

where $I_{\max}$ and $I_{\min}$ are the maximum and minimum intensities that are obtained when the light passes through a polarizer that is slowly rotated. Such light can be considered as the sum of two unequal plane-polarized beams of intensities $I_{\max}$ and $I_{\min}$ perpendicular to each other. Show that the light transmitted by a polarizer, whose axis makes an angle ϕ to the direction in which $I_{\max}$ is obtained, has intensity

$$\frac{1 + p\cos 2\phi}{1 + p} I_{\max}$$

where $p = P/100$ is the "fractional polarization."

General Problems

54. When violet light of wavelength 415 nm falls on a single slit, it creates a central diffraction peak that is 9.20 cm wide on a screen that is 2.55 m away. How wide is the slit?

55. A teacher stands well back from an outside doorway 0.88 m wide, and blows a whistle of frequency 750 Hz. Ignoring reflections, estimate at what angle(s) it is *not* possible to hear the whistle clearly on the playground outside the doorway.

56. The wings of a certain beetle have a series of parallel lines across them. When normally incident 460-nm light is reflected from the wing, the wing appears bright when viewed at an angle of 50°. How far apart are the lines?

57. How many lines per centimeter must a grating have if there is to be no second-order spectrum for any visible wavelength?

58. Light is incident on a diffraction grating with 7500 lines per centimeter and the pattern is viewed on a screen located 2.5 m from the grating. The incident light beam consists of two wavelengths, $\lambda_1 = 4.4 \times 10^{-7}$ m and $\lambda_2 = 6.3 \times 10^{-7}$ m. Calculate the linear distance between the first-order bright fringes of these two wavelengths on the screen.

59. If parallel light falls on a single slit of width a at a 20° angle to the normal, describe the diffraction pattern.

60. When yellow sodium light, $\lambda = 589$ nm, falls on a diffraction grating, its first-order peak on a screen 60.0 cm away falls 3.32 cm from the central peak. Another source produces a line 3.71 cm from the central peak. What is its wavelength? How many lines/cm are on the grating?

61. What is the highest spectral order that can be seen if a grating with 6000 lines per cm is illuminated with 633-nm laser light? Assume normal incidence.

62. Two and only two full spectral orders can be seen on either side of the central maximum when white light is sent through a diffraction grating. What is the maximum number of lines per cm for the grating?

63. Two of the lines of the atomic hydrogen spectrum have wavelengths of 656 nm and 410 nm. If these fall at normal incidence on a grating with 7600 lines per cm, what will be the angular separation of the two wavelengths in the first-order spectrum?

64. Light falling normally on a 9850 line/cm grating is revealed to contain three lines in the first-order spectrum at angles of 31.2°, 36.4°, and 47.5°. What wavelengths are these?

65. (a) How far away can a human eye distinguish two car headlights 2.0 m apart? Consider only diffraction effects and assume an eye diameter of 5.0 mm and a wavelength of 500 nm. (b) What is the minimum angular separation an eye could resolve when viewing two stars, considering only diffraction effects? In reality, it is about 1' of arc. Why is it not equal to your answer in (b)?

66. At what angle above the horizon is the Sun when light reflecting off a smooth lake is polarized most strongly?

67. Unpolarized light falls on two polarizer sheets whose axes are at right angles. (a) What fraction of the incident light intensity is transmitted? (b) What fraction is transmitted if a third polarizer is placed between the first two so that its axis makes a 60° angle with the axis of the first polarizer? (c) What if the third polarizer is in front of the other two?

68. Four polarizers are placed in succession with their axes vertical, at 30° to the vertical, at 60° to the vertical, and at 90° to the vertical. (a) Calculate what fraction of the incident unpolarized light is transmitted by the four polarizers. (b) Can the transmitted light be *decreased* by removing one of the polarizers? If so, which one? (c) Can the transmitted light intensity be extinguished by removing polarizers? If so, which one(s)?

69. At what angle should the axes of two Polaroids be placed so as to reduce the intensity of the incident unpolarized light by an additional factor (after the first Polaroid cuts it in half) of (a) 25 percent, (b) 10 percent, (c) 1 percent?

70. Two polarizers are oriented at 40° to each other and plane-polarized light is incident on them. If only 15 percent of the light gets through both of them, what was the initial polarization direction of the incident light?

71. Show that if two equally intense sources of light produce light that is plane-polarized, but with their planes of polarization perpendicular to each other, then they cannot produce an interference pattern even if they are in phase at all moments.

72. Spy planes fly at extremely high altitudes (25 km) to avoid interception. Their cameras are reportedly able to discern features as small as 5 cm. What must be the minimum aperture of the camera lens to afford this resolution? (Use $\lambda = 550$ nm.)

* 73. X-rays of wavelength 0.0973 nm are directed at an unknown crystal. The second diffraction maximum is recorded at 23.4°. What is the spacing between crystal planes?

* 74. X-rays of wavelength 0.10 nm fall on a microcrystalline powder sample as in Fig. 36–29. The sample is located 10 cm from the photographic film. The crystal structure of the sample has an atomic spacing of 0.25 nm. Calculate the radii of the diffraction rings corresponding to first- and second-order scattering.

Albert Einstein (1879–1955), one of the great minds of the twentieth century, was the creator of the special and general theories of relativity.

In this chapter we examine the special theory of relativity, which includes a number of non-classical results, including length contraction and time dilation for moving reference frames, and new formulas for momentum and energy.

Special Theory of Relativity

FIGURE 37–1 Albert Einstein and his second wife.

Classical vs. modern physics

P hysics at the end of the nineteenth century looked back on a period of great progress. The theories developed over the preceding three centuries had been very successful in explaining a wide range of natural phenomena. Newtonian mechanics beautifully explained the motion of objects on Earth and in the heavens. Furthermore, it formed the basis for successful treatments of fluids, wave motion, and sound. Kinetic theory explained the behavior of gases and other materials. Maxwell's theory of electromagnetism not only brought together and explained electric and magnetic phenomena, but it predicted the existence of electromagnetic (EM) waves that would behave in every way just like light—so light came to be thought of as an electromagnetic wave. Indeed, it seemed that the natural world, as seen through the eyes of physicists, was very well explained. A few puzzles remained, but it was felt that these would soon be explained using already known principles.

But it did not turn out so simply. Instead, these few puzzles were to be solved only by the introduction, in the early part of the twentieth century, of two revolutionary new theories that changed our whole conception of nature: the *theory of relativity* and *quantum theory*.

Physics as it was known at the end of the nineteenth century (what we've covered up to now in this book) is referred to as **classical physics**. The new physics that grew out of the great revolution at the turn of the twentieth century is now called **modern physics**. In this chapter, we present the special theory of relativity, which was first proposed by Albert Einstein (1879–1955; Fig. 37–1) in 1905. In the following chapter, we introduce the equally momentous quantum theory.

37–1 | Galilean–Newtonian Relativity

Einstein's special theory of relativity deals with how we observe events, particularly how objects and events are observed from different frames of reference. This subject had, of course, already been explored by Galileo and Newton.

The special theory of relativity deals with events that are observed and measured from so-called **inertial reference frames** which (Sections 4–2 and 11–9) are reference frames in which Newton's first law is valid: if an object experiences no net force the object either remains at rest or continues in motion with constant speed in a straight line. It is easiest to analyze events when they are observed and measured from inertial frames, and the Earth, though not quite an inertial frame (it rotates), is close enough that for most purposes we can consider it an inertial frame. Rotating or otherwise accelerating frames of reference are noninertial frames, and won't concern us in this chapter (they are dealt with in Einstein's general theory of relativity).

A reference frame that moves with constant velocity with respect to an inertial frame is itself also an inertial frame, since Newton's laws hold in it as well. When we say that we observe or make measurements from a certain reference frame, it means that we are at rest in that reference frame.

Both Galileo and Newton were aware of what we now call the **relativity principle** applied to mechanics: that *the basic laws of physics are the same in all inertial reference frames.* You may have recognized its validity in everyday life. For example, objects move in the same way in a smoothly moving (constant-velocity) train or airplane as they do on Earth. (This assumes no vibrations or rocking—for they would make the reference frame noninertial.) When you walk, drink a cup of soup, play Ping-Pong, or drop a pencil on the floor while traveling in a train, airplane, or ship moving at constant velocity, the bodies move just as they do when you are at rest on Earth. Suppose you are in a car traveling rapidly along at constant velocity. If you release a coin from above your head inside the car, how will it fall? It falls straight downward with respect to the car, and hits the floor directly below the point of release, Fig. 37–2a. (If you drop the coin out the car's window, this won't happen because the moving air drags the coin backward relative to the car.) This is just how objects fall on the Earth—straight down—and thus our experiment in the moving car is in accord with the relativity principle.

Note in this example, however, that to an observer on the Earth, the coin follows a curved path, Fig. 37–2b. The actual path followed by the coin is different as viewed from different frames of reference. This does not violate the relativity principle because this principle states that the *laws* of physics are the same in all inertial frames. The same law of gravity, and the same laws of motion, apply in both reference frames. And the acceleration of the coin is the same in both reference frames. The difference in Figs. 37–2a and b is that in the Earth's frame of reference, the coin has an initial velocity (equal to that of the car). The laws of physics therefore predict it will follow a parabolic path like any projectile. In the car's ref-

Relativity principle: the laws of physics are the same in all inertial reference frames

FIGURE 37–2 A coin is dropped by a person in a moving car. (a) In the reference frame of the car, the coin falls straight down. (b) In a reference frame fixed on the Earth, the coin follows a curved (parabolic) path. The upper views show the moment of the coin's release, and the lower views are a short time later.

(a)
Reference frame = car

(b)
Reference frame = Earth

erence frame, there is no initial velocity, and the laws of physics predict that the coin will fall straight down. The laws are the same in both reference frames, although the specific paths are different.

Galilean–Newtonian relativity involves certain unprovable assumptions that make sense from everyday experience. It is assumed that the lengths of objects are the same in one reference frame as in another, and that time passes at the same rate in different reference frames. In classical mechanics, then, space and time are considered to be **absolute**: their measurement doesn't change from one reference frame to another. The mass of an object, as well as all forces, are assumed to be unchanged by a change in inertial reference frame.

The position of an object is, of course, different when specified in different reference frames, and so is velocity. For example, a person may walk inside a bus toward the front with a speed of 5 km/h. But if the bus moves 40 km/h with respect to the Earth, the person is then moving with a speed of 45 km/h with respect to the Earth. The acceleration of a body, however, is the same in any inertial reference frame according to classical mechanics. This is because the change in velocity, and the time interval, will be the same. For example, the person in the bus may accelerate from 0 to 5 km/h in 1.0 seconds, so $a = 5$ km/h/s in the reference frame of the bus. With respect to the Earth, the acceleration is $(45 \text{ km/h} - 40 \text{ km/h})/(1.0 \text{ s}) = 5 \text{ km/h/s}$, which is the same.

Since neither F, m, nor a changes from one inertial frame to another, then Newton's second law, $F = ma$, does not change. Thus Newton's second law satisfies the relativity principle. It is easily shown that the other laws of mechanics also satisfy the relativity principle.

That the laws of mechanics are the same in all inertial reference frames implies that no one inertial frame is special in any sense. We express this important conclusion by saying that **all inertial reference frames are equivalent** for the description of mechanical phenomena. No one inertial reference frame is any better than another. A reference frame fixed to a car or an aircraft traveling at constant velocity is as good as one fixed on the Earth. When you travel smoothly at constant velocity in a car or airplane, it is just as valid to say you are at rest and the Earth is moving as it is to say the reverse. There is no experiment you can do to tell which frame is "really" at rest and which is moving. Thus, there is no way to single out one particular reference frame as being at absolute rest.

All inertial reference frames are equally valid

A complication arose, however, in the last half of the nineteenth century. When Maxwell presented his comprehensive and very successful theory of electromagnetism (Chapter 32), he showed that light can be considered an electromagnetic wave. Maxwell's equations predicted that the velocity of light c would be 3.00×10^8 m/s; and this is just what is measured, within experimental error. The question then arose: in what reference frame does light have precisely the value predicted by Maxwell's theory? For it was assumed that light would have a different speed in different frames of reference. For example, if observers were traveling on a rocket ship at a speed of 1.0×10^8 m/s away from a source of light, we might expect them to measure the speed of the light reaching them to be 3.0×10^8 m/s $- 1.0 \times 10^8$ m/s $= 2.0 \times 10^8$ m/s. But Maxwell's equations have no provision for relative velocity. They predicted the speed of light to be $c = 3.0 \times 10^8$ m/s. This seemed to imply there must be some special reference frame where c would have this value.

We discussed in Chapters 15 and 16 that waves travel on water and along ropes or strings, and sound waves travel in air and other materials. Nineteenth-century physicists viewed the material world in terms of the laws of mechanics, so it was natural for them to assume that light too must travel in some *medium*. They called this transparent medium the **ether** and assumed it permeated all space.[†] It was therefore assumed that the velocity of light given by Maxwell's equations must be with respect to the ether.

The "ether"

However, it appeared that Maxwell's equations did *not* satisfy the relativity principle. They were not the same in all inertial reference frames. They were simplest in the frame where $c = 3.00 \times 10^8$ m/s; that is, in a reference frame at rest in the ether. In any other reference frame, extra terms would have to be added to take into account the relative velocity. Thus, although most of the laws of physics obeyed the relativity principle, the laws of electricity and magnetism apparently did not. Instead, they seemed to single out one reference frame that was better than any other—a reference frame that could be considered absolutely at rest.

Scientists soon set out to determine the speed of the Earth relative to this absolute frame, whatever it might be. A number of clever experiments were designed. The most direct were performed by A. A. Michelson and E. W. Morley in the 1880s. The details of their experiment are discussed in the next Section. Briefly, what they did was measure the difference in the speed of light in different directions. They expected to find a difference depending on the orientation of their apparatus with respect to the ether. For just as a boat has different speeds relative to the land when it moves upstream, downstream, or across the stream, so too light would be expected to have different speeds depending on the velocity of the ether past the Earth.

Strange as it may seem, they detected no difference at all. This was a great puzzle. A number of explanations were put forth over a period of years, but they led to contradictions or were otherwise not generally accepted.

Then in 1905, Albert Einstein proposed a radical new theory that reconciled these many problems in a simple way. But at the same time, as we shall soon see, it completely changed our ideas of space and time.

* | 37–2 | The Michelson–Morley Experiment

The Michelson–Morley experiment was designed to measure the speed of the *ether*—the medium in which light was assumed to travel—with respect to the Earth. The experimenters thus hoped to find an absolute reference frame, one that could be considered to be at rest.

One of the possibilities nineteenth-century scientists considered was that the ether is fixed relative to the Sun, for even Newton had taken the Sun as the center of the universe. If this were the case (there was no guarantee, of course), the Earth's speed of about 3×10^4 m/s in its orbit around the Sun would produce a change of 1 part in 10^4 in the speed of light $(3.0 \times 10^8 \text{ m/s})$. Direct measurement of the speed of light to this accuracy was not possible. But A. A. Michelson, later with the help of E. W. Morley, was able to use his interferometer (Section 35–7) to measure the difference in the speed of light in different directions to this accuracy.

[†]The medium for light waves could not be air, since light travels from the Sun to Earth through nearly empty space. Therefore, another medium was postulated, the ether. The ether was not only transparent, but, because of difficulty in detecting it, was assumed to have zero density.

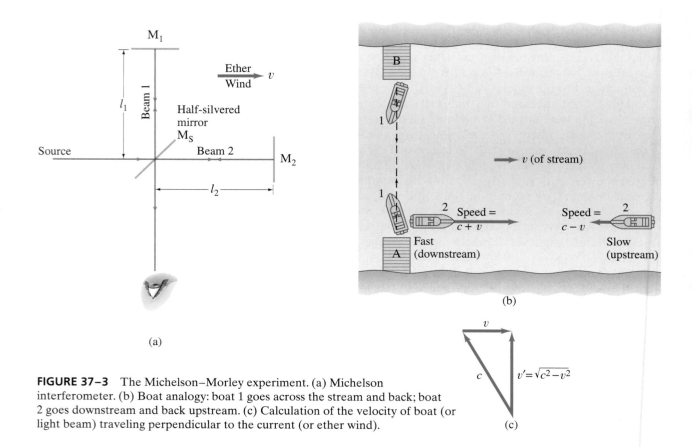

FIGURE 37–3 The Michelson–Morley experiment. (a) Michelson interferometer. (b) Boat analogy: boat 1 goes across the stream and back; boat 2 goes downstream and back upstream. (c) Calculation of the velocity of boat (or light beam) traveling perpendicular to the current (or ether wind).

This famous experiment is based on the principle shown in Fig. 37–3. Part (a) is a diagram of the Michelson interferometer, and it is assumed that the "ether wind" is moving with speed v to the right. (Alternatively, the Earth is assumed to move to the left with respect to the ether at speed v.) The light from the source is split into two beams by the half-silvered mirror M_S. One beam travels to mirror M_1 and the other to mirror M_2. The beams are reflected by M_1 and M_2 and are joined again after passing through M_S. The now superposed beams interfere with each other and the resultant is viewed by the observer's eye as an interference pattern (discussed in Section 35–7).

Whether constructive or destructive interference occurs at the center of the interference pattern depends on the relative phases of the two beams after they have traveled their separate paths. To examine this let us consider an analogy of a boat traveling up and down, and across, a river whose current moves with speed v, as shown in Fig. 37–3b. In still water, the boat can travel with speed c (not the speed of light in this case).

First we consider beam 2 in Fig. 37–3a, which travels parallel to the "ether wind." In its journey from M_S to M_2, we expect the light to travel with speed $c + v$, just as a boat traveling downstream (see Fig. 37–3b) adds the speed of the river to its own speed. Since the beam travels a distance l_2, the time it takes to go from M_S to M_2 is $t = l_2/(c + v)$. To make the return trip from M_2 to M_S, the light must move against the ether wind (like the boat going upstream), so its relative speed is expected to be $c - v$. The time for the return trip is $l_2/(c - v)$. The total time required for beam 2 to travel from M_S to M_2 and back to M_S is

$$ t_2 \;=\; \frac{l_2}{c + v} + \frac{l_2}{c - v} \;=\; \frac{2l_2}{c(1 - v^2/c^2)}. $$

Now let us consider beam 1, which travels crosswise to the ether wind. Here the boat analogy (part b) is especially helpful. The boat is to go from wharf A to wharf B directly across the stream. If it heads directly across, the stream's current will drag it downstream. To reach wharf B, the boat must head at an angle upstream. The precise angle depends on the magnitudes of c and v, but is of no interest to us in itself. Part (c) of Fig. 37–3 shows how to calculate the velocity v' of the boat relative to Earth as it crosses the stream. Since c, v, and v' form a right triangle, we have that $v' = \sqrt{c^2 - v^2}$. The boat has the same speed when it returns. If we now apply these principles to light beam 1 in Fig. 37–3a, we see that the beam travels with a speed $\sqrt{c^2 - v^2}$ in going from M_S to M_1 and back again. The total distance traveled is $2l_1$, so the time required for beam 1 to make the round trip is $2l_1/\sqrt{c^2 - v^2}$, or

$$t_1 = \frac{2l_1}{c\sqrt{1 - v^2/c^2}}.$$

Notice that the denominator in this equation for t_1 involves a square root, whereas that for t_2 does not.

If $l_1 = l_2 = l$, we see that beam 2 will lag behind beam 1 by an amount

$$\Delta t = t_2 - t_1 = \frac{2l}{c}\left(\frac{1}{1 - v^2/c^2} - \frac{1}{\sqrt{1 - v^2/c^2}}\right).$$

If $v = 0$, then $\Delta t = 0$, and the two beams will return in phase since they were initially in phase. But if $v \neq 0$, then $\Delta t \neq 0$, and the two beams will return out of phase. If this change of phase from the condition $v = 0$ to that for $v = v$ could be measured, then v could be determined. But the Earth cannot be stopped. Furthermore, we should not be too quick to assume that lengths are not affected by motion and therefore to assume $l_1 = l_2$.

Michelson and Morley realized that they could detect the difference in phase (assuming that $v \neq 0$) if they rotated their apparatus by 90°, for then the interference pattern between the two beams should change. In the rotated position, beam 1 would now move parallel to the ether and beam 2 perpendicular to it. Thus the roles could be reversed, and in the rotated position the times (designated by primes) would be

$$t_1' = \frac{2l_1}{c(1 - v^2/c^2)} \quad \text{and} \quad t_2' = \frac{2l_2}{c\sqrt{1 - v^2/c^2}}.$$

The time lag between the two beams in the nonrotated position (unprimed) would be

$$\Delta t = t_2 - t_1 = \frac{2l_2}{c(1 - v^2/c^2)} - \frac{2l_1}{c\sqrt{1 - v^2/c^2}}.$$

In the rotated position, the time difference would be

$$\Delta t' = t_2' - t_1' = \frac{2l_2}{c\sqrt{1 - v^2/c^2}} - \frac{2l_1}{c(1 - v^2/c^2)}.$$

When the rotation is made, the fringes of the interference pattern (Section 35–7) will shift an amount determined by the difference:

$$\Delta t - \Delta t' = \frac{2}{c}(l_1 + l_2)\left(\frac{1}{1 - v^2/c^2} - \frac{1}{\sqrt{1 - v^2/c^2}}\right).$$

This expression can be considerably simplified if we assume that $v/c \ll 1$. For in this case we can use the binomial expansion (Appendix A), so

$$\frac{1}{1 - v^2/c^2} \approx 1 + \frac{v^2}{c^2} \quad \text{and} \quad \frac{1}{\sqrt{1 - v^2/c^2}} \approx 1 + \frac{1}{2}\frac{v^2}{c^2}.$$

Then

$$\Delta t - \Delta t' \approx \frac{2}{c}(l_1 + l_2)\left(1 + \frac{v^2}{c^2} - 1 - \frac{1}{2}\frac{v^2}{c^2}\right)$$

$$\approx (l_1 + l_2)\frac{v^2}{c^3}.$$

Now we take $v = 3.0 \times 10^4$ m/s, the speed of the Earth in its orbit around the Sun. In Michelson and Morley's experiments, the arms l_1 and l_2 were about 11 m long. The time difference would then be about

$$\frac{(22\ \text{m})(3.0 \times 10^4\ \text{m/s})^2}{(3.0 \times 10^8\ \text{m/s})^3} \approx 7.0 \times 10^{-16}\ \text{s}.$$

For visible light of wavelength $\lambda = 5.5 \times 10^{-7}$ m, say, the frequency would be $f = c/\lambda = (3.0 \times 10^8\ \text{m/s})/(5.5 \times 10^{-7}\ \text{m}) = 5.5 \times 10^{14}$ Hz, which means that wave crests pass by a point every $1/(5.5 \times 10^{14}\ \text{Hz}) = 1.8 \times 10^{-15}$ s. Thus, with a time difference of 7.0×10^{-16} s, Michelson and Morley should have noted a movement in the interference pattern of $(7.0 \times 10^{-16}\ \text{s})/(1.8 \times 10^{-15}\ \text{s}) = 0.4$ fringe. They could easily have detected this, since their apparatus was capable of observing a fringe shift as small as 0.01 fringe.

The null result   But they found *no significant fringe shift whatever*! They set their apparatus at various orientations. They made observations day and night so that they would be at various orientations with respect to the Sun (due to the Earth's rotation). They tried at different seasons of the year (the Earth at different locations due to its orbit around the Sun). Never did they observe a significant fringe shift.

This **null result** was one of the great puzzles of physics at the end of the nineteenth century. To explain it was a difficult challenge. One possibility to explain the null result was put forth independently by G. F. Fitzgerald and H. A. Lorentz (in the 1890s) in which they proposed that any length (including the arm of an interferometer) contracts by a factor $\sqrt{1 - v^2/c^2}$ in the direction of motion through the ether. According to Lorentz, this could be due to the ether affecting the forces between the molecules of a substance, which were assumed to be electrical in nature. This theory was eventually replaced by the far more comprehensive theory proposed by Albert Einstein in 1905—the special theory of relativity.

37–3 | Postulates of the Special Theory of Relativity

The problems that existed at the turn of the century with regard to electromagnetic theory and Newtonian mechanics were beautifully resolved by Einstein's introduction of the theory of relativity in 1905. Einstein, however, was apparently not influenced directly by the null result of the Michelson–Morley experiment. What motivated Einstein were certain questions regarding electromagnetic theory and light waves. For example, he asked himself: "What would I see if I rode a light beam?" The answer was that instead of a traveling electromagnetic wave, he would see alternating electric and magnetic fields at rest whose magnitude changed in space, but did not change in time. Such fields, he realized, had never been detected and indeed were not consistent with Maxwell's electromagnetic theory. He argued, therefore, that it was unreasonable to think that the speed of light relative to any observer could be reduced to zero, or in fact reduced at all. This idea became the second postulate of his theory of relativity.

Einstein concluded that the inconsistencies he found in electromagnetic theory were due to the assumption that an absolute space exists. In his famous 1905 paper, he proposed doing away completely with the idea of the ether and the accompanying assumption of an absolute reference frame at rest. This proposal was embodied in two postulates. The first postulate was an extension of the Newtonian relativity principle to include not only the laws of mechanics but also those of the rest of physics, including electricity and magnetism:

> *First postulate (the relativity principle)*: **The laws of physics have the same form in all inertial reference frames.**

The second postulate is consistent with the first:

> *Second postulate (constancy of the speed of light)*: **Light propagates through empty space with a definite speed c independent of the speed of the source or observer.**

The two postulates of special relativity

These two postulates form the foundation of Einstein's **special theory of relativity**. It is called "special" to distinguish it from his later "general theory of relativity," which deals with noninertial (accelerating) reference frames. The special theory, which is what we discuss here, deals only with inertial frames.

The second postulate may seem hard to accept, for it violates commonsense notions. First of all, we have to think of light traveling through empty space. Giving up the ether is not too hard, however, for after all, it had never been detected. But the second postulate also tells us that the speed of light in vacuum is always the same, 3.00×10^8 m/s, no matter what the speed of the observer or the source. Thus, a person traveling toward or away from a source of light will measure the same speed for that light as someone at rest with respect to the source. This conflicts with our everyday notions, for we would expect to have to add in the velocity of the observer. Part of the problem is that in our everyday experience, we do not measure velocities anywhere near as large as the speed of light. Thus we can't expect our everyday experience to be helpful when dealing with such a high velocity. On the other hand, the Michelson–Morley experiment is fully consistent with the second postulate.[†]

Einstein's proposal has a certain beauty. For by doing away with the idea of an absolute reference frame, it was possible to reconcile classical mechanics with Maxwell's electromagnetic theory. The speed of light predicted by Maxwell's equations *is* the speed of light in vacuum in *any* reference frame.

Einstein's theory required giving up commonsense notions of space and time, and in the following Sections we will examine some strange but interesting consequences of special relativity. Our arguments for the most part will be simple ones. We will use a technique that Einstein himself did: we will imagine very simple experimental situations in which little mathematics is needed. In this way, we can see many of the consequences of relativity theory without getting involved in detailed calculations. Einstein called these "thought" experiments. Starting in Section 37–8 we will look more fully at the mathematics of relativity.

[†]The Michelson–Morley experiment can also be considered as evidence for the first postulate, for it was intended to measure the motion of the Earth relative to an absolute reference frame. Its failure to do so implies the absence of any such preferred frame.

37–4 | Simultaneity

One of the important consequences of the theory of relativity is that we can no longer regard time as an absolute quantity. No one doubts that time flows onward and never turns back. But, as we shall see in this Section and the next, the time interval between two events, and even whether two events are simultaneous, depends on the observer's reference frame.

Two events are said to occur simultaneously if they occur at exactly the same time. But how do we know if two events occur precisely at the same time? If they occur at the same point in space—such as two apples falling on your head at the same time—it is easy. But if the two events occur at widely separated places, it is more difficult to know whether the events are simultaneous since we have to take into account the time it takes for the light from them to reach us. Because light travels at finite speed, a person who sees two events must calculate back to find out when they actually occurred. For example, if two events are *observed* to occur at the same time, but one actually took place farther from the observer than the other, then the more distant one must have occurred earlier, and the two events were not simultaneous.

FIGURE 37–4 A moment after lightning strikes points A and B, the pulses of light are traveling toward the observer O, but O "sees" the lightning only when the light reaches O.

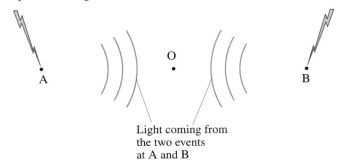

Light coming from the two events at A and B

We will now make use of a simple thought experiment. We assume an observer, called O, is located exactly halfway between points A and B where two events occur, Fig. 37–4. The two events may be lightning that strikes the points A and B, as shown, or any other type of events. For brief events like lightning, only short pulses of light will travel outward from A and B and reach O. O "sees" the events when the pulses of light reach point O. If the two pulses reach O at the same time, then the two events had to be simultaneous. This is because the two light pulses travel at the same speed (postulate 2), and since the distance OA equals OB, the time for the light to travel from A to O and B to O must be the same. Observer O can then definitely state that the two events occurred simultaneously. On the other hand, if O sees the light from one event before that from the other, then it is certain the former event occurred first.

The question we really want to examine is this: if two events are simultaneous to an observer in one reference frame, are they also simultaneous to another observer moving with respect to the first? Let us call the observers O_1 and O_2 and assume they are fixed in reference frames 1 and 2 that move with speed v relative to one another. These two reference frames can be thought of as trains (Fig. 37–5). O_2 says that O_1 is moving to the right with speed v, as in Fig. 37–5a; and O_1 says O_2 is moving to the left with speed v, as in Fig. 37–5b. Both viewpoints are legitimate according to the relativity principle. (There is, of course, no third point of view which will tell us which one is "really" moving.)

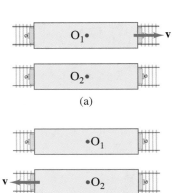

FIGURE 37–5 Observers O_1 and O_2, on two different trains (two different reference frames), are moving with relative velocity v. O_2 says that O_1 is moving to the right (a); O_1 says that O_2 is moving to the left (b). Both viewpoints are legitimate—it all depends on your reference frame.

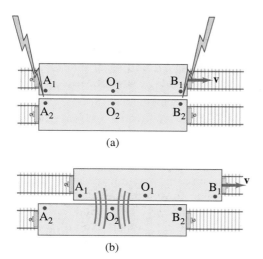

(a)

(b)

FIGURE 37–6 Thought experiment on simultaneity. To observer O_2, the reference frame of O_1 is moving to the right. In (a), one lightning bolt strikes the two reference frames at A_1 and A_2, and a second lightning bolt strikes at B_1 and B_2. (b) A moment later, the light from the two events reaches O_2 at the same time, so according to observer O_2, the two bolts of lightning strike simultaneously. But in O_1's reference frame, the light from B_1 has already reached O_1, whereas the light from A_1 has not yet reached O_1. So in O_1's reference frame, the event at B_1 must have preceded the event at A_1. Time is not absolute.

Now suppose two events occur that are observed and measured by both observers. Let us assume again that the two events are the striking of lightning and that the lightning marks both trains where it struck: at A_1 and B_1 on O_1's train, and at A_2 and B_2 on O_2's train. For simplicity, we assume that O_1 happens to be exactly halfway between A_1 and B_1, and that O_2 is halfway between A_2 and B_2. We now put ourselves in one reference frame or the other, from which we make our observations and measurements. Let us put ourselves in O_2's reference frame, so we observe O_1 moving to the right with speed v. Let us also assume that the two events occur *simultaneously* in O_2's frame, and just at the instant when O_1 and O_2 are opposite each other, Fig. 37–6a. A short time later, Fig. 37–6b, the light from A_2 and B_2 reaches O_2 at the same time (we assumed this). Since O_2 knows (or measures) the distances O_2A_2 and O_2B_2 as equal, O_2 knows the two events are simultaneous in the O_2 reference frame.

But what does observer O_1 observe and measure? From our (O_2) reference frame, we can predict what O_1 will observe. We see that O_1 moves to the right during the time the light is traveling to O_1 from A_1 and B_1. As shown in Fig. 37–6b, we can see from our O_2 reference frame that the light from B_1 has already passed O_1, whereas the light from A_1 has not yet reached O_1. Therefore, it is clear that O_1 will observe the light coming from B_1 before he observes the light coming from A_1. Now O_1's frame is as good as O_2's. Light travels at the same speed c for O_1 as for O_2 (the second postulate)[†]; and in the O_1 reference frame, this speed c is of course the same for light traveling from A_1 to O_1 as it is for light traveling from B_1 to O_1. Furthermore the distance O_1A_1 equals O_1B_1. Hence, since O_1 observes the light from B_1 before he observes the light from A_1 (we established this above, looking from the O_2 reference frame, Fig. 37–6b), then observer O_1 can only conclude that the event at B_1 occurred before the event at A_1. The two events are not simultaneous for O_1, even though they are for O_2.

We thus find that two events which are simultaneous to one observer are not necessarily simultaneous to a second observer.

It may be tempting to ask: "Which observer is right, O_1 or O_2?" The answer, according to relativity, is that they are *both* right. There is no "best" reference frame we can choose to determine which observer is right. Both frames are equally good. We can only conclude that *simultaneity is not an absolute concept*, but is relative. We are not aware of it in everyday life, however, because the effect is noticeable only when the relative speed of the two reference frames is very large (near c), or the distances involved are very large.

Simultaneity is relative

[†]Note that O_1 does not see himself catching up with one light beam and running away from the other (that is O_2's viewpoint of what happens for O_1). O_1 sees both light beams traveling at the same speed, c.

Because of the principle of relativity, the argument we gave for the thought experiment of Fig. 37–6 can be done from O_1's reference frame as well. In this case, O_1 will be at rest and will see event B_1 occur before A_1. But O_1 will recognize (by drawing a diagram equivalent to Fig. 37–6—try it and see!) that O_2, who is moving with speed v to the left, will see the two events as simultaneous.

37–5 | Time Dilation and the Twin Paradox

The fact that two events simultaneous to one observer may not be simultaneous to a second observer suggests that time itself is not absolute. Could it be that time passes differently in one reference frame than in another? This is, indeed, just what Einstein's theory of relativity predicts, as the following thought experiment shows.

Figure 37–7 shows a spaceship traveling past Earth at high speed. The point of view of an observer on the spaceship is shown in part (a), and that of an observer on Earth in part (b). Both observers have accurate clocks. The person on the spaceship (a) flashes a light and measures the time it takes the light to travel across the spaceship and return after reflecting from a mirror. The light travels a distance $2D$ at speed c, so the time required, which we call Δt_0, is

$$\Delta t_0 = \frac{2D}{c}.$$

This is the time interval as measured by the observer on the spaceship.

The observer on Earth, Fig. 37–7b, observes the same process. But to this observer, the spaceship is moving. So the light travels the diagonal path shown going across the spaceship, reflecting off the mirror, and returning to the sender. Although the light travels at the same speed to this observer (the second postulate), it travels a greater distance. Hence the time required, as measured by

FIGURE 37–7 Time dilation can be shown by a thought experiment: the time it takes for light to travel over and back on a spaceship is longer for the observer on Earth (b) than for the observer on the spaceship (a).

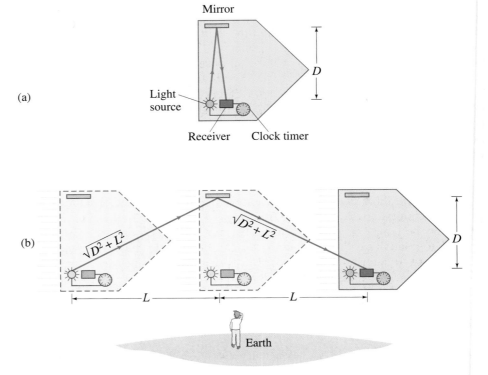

the observer on Earth, will be *greater* than that measured by the observer on the spaceship. The time interval, Δt, observed by the observer on Earth can be calculated as follows. In the time Δt, the spaceship travels a distance $2L = v \Delta t$ where v is the speed of the spaceship (Fig. 37–7b). Thus, the light travels a total distance on its diagonal path of $2\sqrt{D^2 + L^2}$, and therefore

$$c = \frac{2\sqrt{D^2 + L^2}}{\Delta t} = \frac{2\sqrt{D^2 + \dfrac{v^2(\Delta t)^2}{4}}}{\Delta t}.$$

We square both sides, and then solve for Δt, to find

$$c^2 = \frac{4D^2}{(\Delta t)^2} + v^2,$$

$$\Delta t = \frac{2D}{c\sqrt{1 - v^2/c^2}}.$$

We combine this with the formula above for $\Delta t_0 \left(\Delta t_0 = 2D/c\right)$ and find:

$$\Delta t = \frac{\Delta t_0}{\sqrt{1 - v^2/c^2}}.$$
(37–1)   *Time-dilation formula*

Since $\sqrt{1 - v^2/c^2}$ is always less than 1, we see that $\Delta t > \Delta t_0$. That is, the time interval between the two events (the sending of the light, and its reception on the spaceship) is *greater* for the observer on Earth than for the observer on the spaceship. This is a general result of the theory of relativity, and is known as **time dilation**. Stated simply, the time-dilation effect says that

clocks moving relative to an observer are measured by that observer to run more slowly (as compared to clocks at rest).

Time dilation: moving clocks run slow

However, we should not think that the clocks are somehow at fault. Time is actually measured to pass more slowly in any moving reference frame as compared to your own. This remarkable result is an inevitable outcome of the two postulates of the theory of relativity.

The concept of time dilation may be hard to accept, for it violates our commonsense understanding. We can see from Eq. 37–1 that the time dilation effect is negligible unless v is reasonably close to c. If v is much less than c, then the term v^2/c^2 is much smaller than the 1 in the denominator of Eq. 37–1, and then $\Delta t \approx \Delta t_0$ (see Example 37–2). The speeds we experience in everyday life are much smaller than c, so it is little wonder we don't ordinarily notice time dilation. Experiments have tested the time-dilation effect, and have confirmed Einstein's predictions. In 1971, for example, extremely precise atomic clocks were flown around the world in jet planes. The speed of the planes $\left(10^3 \text{ km/h}\right)$ was much less than c, so the clocks had to be accurate to nanoseconds $\left(10^{-9} \text{ s}\right)$ in order to detect any time dilation. They were this accurate, and they confirmed Eq. 37–1 to within experimental error. Time dilation had been confirmed decades earlier, however, by observation on "elementary particles" which have very small masses (typically 10^{-30} to 10^{-27} kg) and so require little energy to be accelerated to speeds close to the speed of light, c. Many of these elementary particles are not stable and decay after a time into lighter particles. One example is the muon, whose mean lifetime is $2.2 \mu s$ when at rest. Careful experiments showed that when a muon is traveling at high speeds, its lifetime is measured to be longer than when it is at rest, just as predicted by the time-dilation formula.

Why we don't usually notice time dilation

EXAMPLE 37–1 **Lifetime of a moving muon.** (a) What will be the mean lifetime of a muon as measured in the laboratory if it is traveling at $v = 0.60c = 1.8 \times 10^8$ m/s with respect to the laboratory? Its mean life at rest is 2.2×10^{-6} s. (b) How far does a muon travel in the laboratory, on average, before decaying?

SOLUTION (a) If an observer were to move along with the muon (the muon would be at rest to this observer), the muon would have a mean life of 2.2×10^{-6} s. To an observer in the lab, the muon lives longer because of time dilation. From Eq. 37–1 with $v = 0.60c$, we have

$$\Delta t = \frac{\Delta t_0}{\sqrt{1 - \dfrac{v^2}{c^2}}} = \frac{2.2 \times 10^{-6}\,\text{s}}{\sqrt{1 - \dfrac{0.36\,c^2}{c^2}}} = \frac{2.2 \times 10^{-6}\,\text{s}}{\sqrt{0.64}} = 2.8 \times 10^{-6}\,\text{s}.$$

(b) At a speed of 1.8×10^8 m/s, classical physics would tell us that with a mean life of $2.2\,\mu$s, an average muon would travel $d = vt = (1.8 \times 10^8\,\text{m/s})(2.2 \times 10^{-6}\,\text{s}) = 400$ m. But relativity predicts an average distance of $(1.8 \times 10^8\,\text{m/s})(2.8 \times 10^{-6}\,\text{s}) = 500$ m, and it is this longer distance that is measured experimentally.

We need to make a comment about the use of Eq. 37–1 and the meaning of Δt and Δt_0. The equation is true only when Δt_0 represents the time interval between the two events in a reference frame where the two events occur at *the same point in space* (as in Fig. 37–7a where the two events are the light flash being sent and being received). This time interval, Δt_0, is called the **proper time**. Then Δt in Eq. 37–1 represents the time interval between the two events as measured in a reference frame moving with speed v with respect to the first. In Example 37–1 above, Δt_0 (and not Δt) was set equal to 2.2×10^{-6} s because it is only in the rest frame of the muon that the two events ("birth" and "decay") occur at the same point in space.

Proper time

EXAMPLE 37–2 **Time dilation at 100 km/h.** Let's check time dilation for everyday speeds. A car traveling 100 km/h covers a certain distance in 10.00 s according to the driver's watch. What does an observer on Earth measure for the time interval?

SOLUTION The car's speed relative to Earth is 100 km/h = $(1.00 \times 10^5\,\text{m})/(3600\,\text{s}) = 27.8$ m/s. We set $\Delta t_0 = 10.00$ s in the time-dilation formula (the driver is at rest in the reference frame of the car), and then Δt is

$$\Delta t = \frac{\Delta t_0}{\sqrt{1 - \dfrac{v^2}{c^2}}} = \frac{10.00\,\text{s}}{\sqrt{1 - \left(\dfrac{27.8\,\text{m/s}}{3.00 \times 10^8\,\text{m/s}}\right)^2}} = \frac{10.00\,\text{s}}{\sqrt{1 - 8.59 \times 10^{-15}}}.$$

If you put these numbers into a calculator, you will obtain $\Delta t = 10.00$ s, since the denominator differs from 1 by such a tiny amount. Indeed, the time measured by an observer on Earth would be no different from that measured by the driver, even with the best of today's instruments. A computer that could calculate to a large number of decimal places could reveal a difference between Δt and Δt_0. But we can estimate the difference quite easily using the binomial expansion (Appendix A),

$\Rightarrow$ **PROBLEM SOLVING**

Use of the binomial expansion

$$(1 \pm x)^n \approx 1 \pm nx. \qquad\qquad [\text{for } x \ll 1]$$

In our time-dilation formula, we have the factor $\left(1 - v^2/c^2\right)^{-\frac{1}{2}}$. Thus

$$\Delta t = \Delta t_0 \left(1 - \frac{v^2}{c^2}\right)^{-\frac{1}{2}} \approx \Delta t_0 \left(1 + \frac{1}{2}\frac{v^2}{c^2}\right)$$

$$\approx 10.00\,\text{s}\left[1 + \frac{1}{2}\left(\frac{27.8\,\text{m/s}}{3.00 \times 10^8\,\text{m/s}}\right)^2\right] \approx 10.00\,\text{s} + 4 \times 10^{-15}\,\text{s}.$$

So the difference between Δt and Δt_0 is predicted to be $4 \times 10^{-15}\,\text{s}$, an extremely small amount.

Time dilation has aroused interesting speculation about space travel. According to classical (Newtonian) physics, to reach a star 100 light-years away would not be possible for ordinary mortals (1 light-year is the distance light can travel in 1 year $= 3.0 \times 10^8\,\text{m/s} \times 3.15 \times 10^7\,\text{s} = 9.5 \times 10^{15}\,\text{m}$). Even if a spaceship could travel at close to the speed of light, it would take over 100 years to reach such a star. But time dilation tells us that the time involved would be less for an astronaut. In a spaceship traveling at $v = 0.999c$, the time for such a trip would be only about $\Delta t_0 = \Delta t\sqrt{1 - v^2/c^2} = (100\,\text{yr})\sqrt{1 - (0.999)^2} = 4.5\,\text{yr}$. Thus time dilation allows such a trip, but the enormous practical problems of achieving such speeds will not be overcome in the near future.

Notice, in this example, that whereas 100 years would pass on Earth, only 4.5 years would pass for the astronaut on the trip. Is it just the clocks that would slow down for the astronaut? The answer is no. All processes, including aging and other life processes, run more slowly for the astronaut according to the Earth observer. But to the astronaut, time would pass in a normal way. The astronaut would experience 4.5 years of normal sleeping, eating, reading, and so on. And people on Earth would experience 100 years of ordinary activity.

Not long after Einstein proposed the special theory of relativity, an apparent paradox was pointed out. According to this **twin paradox**, suppose one of a pair of 20-year-old twins takes off in a spaceship traveling at very high speed to a distant star and back again, while the other twin remains on Earth. According to the Earth twin, the traveling twin will age less. Whereas 20 years might pass for the Earth twin, perhaps only 1 year (depending on the spacecraft's speed) would pass for the traveler. Thus, when the traveler returns, the earthbound twin could expect to be 40 years old whereas the traveling twin would be only 21.

Twin paradox

This is the viewpoint of the twin on the Earth. But what about the traveling twin? If all inertial reference frames are equally good, won't the traveling twin make all the claims the Earth twin does, only in reverse? Can't the astronaut twin claim that since the Earth is moving away at high speed, time passes more slowly on Earth and the twin on Earth will age less? This is the opposite of what the Earth twin predicts. They cannot both be right, for after all the spacecraft returns to Earth and a direct comparison of ages and clocks can be made.

There is, however, not a paradox at all. The consequences of the special theory of relativity—in this case, time dilation—can be applied only by observers in inertial reference frames. The Earth is such a frame (or nearly so), whereas the spacecraft is not. The spacecraft accelerates at the start and end of its trip and, more importantly, when it turns around at the far point of its journey. During these acceleration periods, the twin on the spacecraft is not maintaining a steady inertial reference frame, so the spacecraft twin's predictions based on special relativity are not valid. The twin on Earth is in an inertial frame and can make valid predictions. Thus, there is no paradox. The traveling twin's point of view expressed above is not correct. The predictions of the Earth twin *are* valid according to the theory of relativity, and the prediction that the traveling twin returns having aged less is the proper one. Einstein's general theory of relativity, which deals with accelerating reference frames, confirms this result.

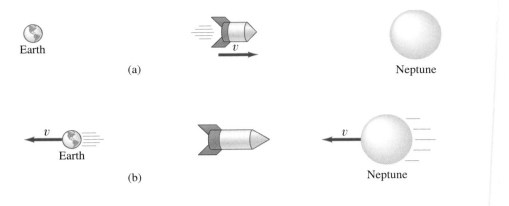

FIGURE 37–8 (a) A spaceship traveling at very high speed from Earth to Neptune, as seen from Earth's frame of reference. (b) As viewed by an observer on the spaceship, Earth and Neptune are moving at the very high velocity v: Earth leaves the spaceship, and a time Δt_0 later planet Neptune arrives at the spaceship. [Note that from the spaceship, part (b), each planet does not look shortened because at high speeds observers see the trailing edge (as in Fig. 37–10), and the net effect is to leave its appearance as a circle.]

37–6 | Length Contraction

Not only time intervals are different in different reference frames. Space intervals—lengths and distances—are different as well, according to the special theory of relativity, and we illustrate this with a thought experiment.

Observers on Earth watch a spacecraft traveling at speed v from Earth to, say, Neptune, Fig. 37–8a. The distance between the planets, as measured by the Earth observers, is L_0. The time required for the trip, measured from Earth, is $\Delta t = L_0/v$. In Fig. 37–8b we see the point of view of observers on the spacecraft. In this frame of reference, the spaceship is at rest; Earth and Neptune move with speed v. (We assume v is much greater than the relative speed of Neptune and Earth, so the latter can be ignored.) The time between the departure of Earth and arrival of Neptune (as observed from the spacecraft) is the "proper time" (since the two events occur at the same point in space—i.e., on the spacecraft). Therefore the time interval is less for the spacecraft observers than for the Earth observers, because of time dilation. From Eq. 37–1, the time for the trip as viewed by the spacecraft is $\Delta t_0 = \Delta t\sqrt{1 - v^2/c^2}$. Since the spacecraft observers measure the same speed but less time between these two events, they must also measure the distance as less. If we let L be the distance between the planets as viewed by the spacecraft observers, then $L = v\,\Delta t_0$. We have already seen that $\Delta t_0 = \Delta t\sqrt{1 - v^2/c^2}$ and $\Delta t = L_0/v$, so we have $L = v\,\Delta t_0 = v\,\Delta t\sqrt{1 - v^2/c^2} = L_0\sqrt{1 - v^2/c^2}$. That is,

Length-contraction formula

$$L = L_0\sqrt{1 - \frac{v^2}{c^2}}. \tag{37–2}$$

This is a general result of the special theory of relativity and applies to lengths of objects as well as to distance. The result can be stated most simply in words as:

the length of an object is measured to be shorter when it is moving relative to the observer than when it is at rest.

Length contraction: moving objects are shorter (in the direction of motion)

This is called **length contraction**. The length L_0 in Eq. 37–2 is called the **proper length**. It is the length of the object—or distance between two points whose *positions are measured at the same time*—as determined by observers at rest with respect to it. Equation 37–2 gives the length L that will be measured by observers when the object travels past them at speed v. It is important to note, however, that length contraction occurs *only along the direction of motion*. For example, the moving spaceship in Fig. 37–8a is shortened in length, but its height is the same as when it is at rest.

Length contraction, like time dilation, is not noticeable in everyday life because the factor $\sqrt{1 - v^2/c^2}$ in Eq. 37–2 differs from 1.00 significantly only when v is very large.

(a)

(b)

FIGURE 37-9 Example 37-3.

The appearance of moving objects

EXAMPLE 37-3 **Painting's contraction.** A rectangular painting measures 1.00 m tall and 1.50 m wide. It is hung on the side wall of a spaceship which is moving past the Earth at a speed of 0.90 c. See Fig. 37-9a. (*a*) What are the dimensions of the picture according to the captain of the spaceship? (*b*) What are the dimensions as seen by an observer on the Earth?

SOLUTION (*a*) The painting (as well as everything else in the spaceship) looks perfectly normal to everyone on the spaceship, so the captain sees a 1.00 m by 1.50 m painting.
(*b*) Only the dimension in the direction of motion is shortened, so the height is unchanged at 1.00 m, Fig. 37-9b. The length, however, is contracted to

$$L = L_0 \sqrt{1 - \frac{v^2}{c^2}}$$
$$= (1.50 \text{ m}) \sqrt{1 - (0.90)^2} = 0.65 \text{ m}.$$

So the picture has dimensions 1.00 m × 0.65 m.

The Appearance of Objects Moving at Relativistic Speeds

Equation 37-2 tells us what the length of an object will be *measured* to be when traveling at speed *v*. The *appearance* of the object is another matter. Suppose, for example, you are traveling to the left past a small building at speed $v = 0.85 c$. This is equivalent to the building moving past you to the right at speed *v*. The building will look narrower (and the same height), but you will also be able to see the side of the building even if you are directly in front of it. This is shown in Fig. 37-10b—part (a) shows the building at rest. The fact that you see the side is not really a relativistic effect, but is due to the finite speed of light. To see how this occurs, we look at Fig. 37-10c which is a top view of the building, looking down. At the instant shown, the observer O is directly in front of the building. Light from points A and B reach O at the same time. If the building were at rest, light from point C could never reach O. But the building is moving at very high speed and does "get out of the way" so that light from C can reach O. Indeed, at the instant shown, light from point C when it was at an earlier location (C′ on the diagram) can reach O because the building has moved. In order to reach the observer at the same time as light from A and B, light from C had to leave at an earlier time since it must travel a greater distance. Thus it is light from C′ that reaches the observer at the same time as light from A and B. This, then, is how an observer might see both the front and side of an object at the same time even when directly in front of it.[†] It can be shown, by the same reasoning, that spherical objects will actually still have a circular outline even at high speeds. That is why the planets in Fig. 37-8b are drawn round rather than contracted.

[†] It would be an error to think that the building in Fig. 37-10b would look rotated. This is not correct since in that case side A would look shorter than side B. In fact, if the observer is directly in front, these sides appear equal in height. Thus the building looks contracted in its front face, but we also see the side, as described above. Also, though not shown in Fig. 37-10b, the walls of the building would appear curved, because of differing distances from the observer's eye of the various points from top to bottom along a vertical wall.

(a)

(b)

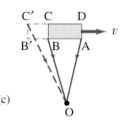
(c)

FIGURE 37-10 Building seen (a) at rest, and (b) moving at high speed. (c) Diagram explains why the side of the building is seen (see the text).

(a)

(b)

FIGURE 37–11 According to an accurate clock on a fast-moving train, a person (a) begins dinner at 7:00 and (b) finishes at 7:15. At the beginning of the meal, observers on Earth set their watches to correspond with the clock on the train. These observers measure the eating time as 20 minutes.

37–7 Four-Dimensional Space–Time

Let us imagine a person is on a train moving at a very high speed, say $0.65c$, Fig. 37–11. This person begins a meal at 7:00 and finishes at 7:15, according to a clock on the train. The two events, beginning and ending the meal, take place at the same point on the train. So the proper time between these two events is 15 min. To observers on Earth, the meal will take longer—20 min according to Eq. 37–1. Let us assume that the meal was served on a 20-cm-diameter plate. To observers on the Earth, the plate is only 15 cm wide (length contraction). Thus, to observers on the Earth, the meal looks smaller but lasts longer.

In a sense these two effects, time dilation and length contraction, balance each other. When viewed from the Earth, what the meal seems to lose in size it gains in length of time it lasts. Space, or length, is exchanged for time.

Considerations like this led to the idea of **four-dimensional space–time**: space takes up three dimensions and time is a fourth dimension. Space and time are intimately connected. Just as when we squeeze a balloon we make one dimension larger and another smaller, so when we examine objects and events from different reference frames, a certain amount of space is exchanged for time, or vice versa.

Although the idea of four dimensions may seem strange, it refers to the idea that any object or event is specified by four quantities—three to describe where in space, and one to describe when in time. The really unusual aspect of four-dimensional space–time is that space and time can intermix: a little of one can be exchanged for a little of the other when the reference frame is changed.

It is difficult for most of us to understand the idea of four-dimensional space–time. Somehow we feel, just as physicists did before the advent of relativity, that space and time are completely separate entities. Yet we have found in our thought experiments that they are not completely separate. Our difficulty in accepting this is reminiscent of the situation in the seventeenth century at the time of Galileo and Newton. Before Galileo, the vertical direction, that in which objects fall, was considered to be distinctly different from the two horizontal dimensions. Galileo showed that the vertical dimension differs only in that it happens to be the direction in which gravity acts. Otherwise, all three dimensions are equivalent, a viewpoint we all accept today. Now we are asked to accept one more dimension, time, which we had previously thought of as being somehow different. This is not to say that there is no distinction between space and time. What relativity has shown is that space and time determinations are not independent of one another.

37–8 Galilean and Lorentz Transformations

We now examine in detail the mathematics of relating quantities in one inertial reference frame to the equivalent quantities in another. In particular, we will see how positions and velocities *transform* (that is, change) from one frame to the other.

We begin with the classical or Galilean viewpoint. Consider two reference frames S and S′ which are each characterized by a set of coordinate axes, Fig. 37–12.

FIGURE 37–12 Inertial reference frame S′ moves to the right at speed v with respect to frame S.

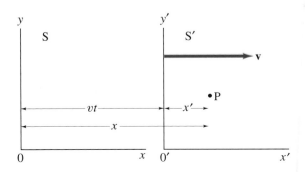

The axes x and y (z is not shown) refer to S and x' and y' to S'. The x' and x axes overlap one another, and we assume that frame S' moves to the right (in the x direction) at speed v with respect to S. And for simplicity let us assume the origins 0 and 0' of the two reference frames are superimposed at time $t = 0$.

Now consider an event that occurs at some point P (Fig. 37–12) represented by the coordinates x', y', z' in reference frame S' at the time t'. What will be the coordinates of P in S? Since S and S' overlap precisely initially, after a time t', S' will have moved a distance vt'. Therefore, at time t', $x = x' + vt'$. The y and z coordinates, on the other hand, are not altered by motion along the x axis; thus $y = y'$ and $z = z'$. Finally, since time is assumed to be absolute in Galilean–Newtonian physics, clocks in the two frames will agree with each other; so $t = t'$. We summarize these in the following **Galilean transformation equations**:

$$x = x' + vt'$$
$$y = y'$$
$$z = z'$$
$$t = t'.$$

[Galilean] **(37–3)** *Galilean transformation*

These equations give the coordinate of an event in the S frame when those in the S' frame are known. If those in the S frame are known, then the S' coordinates are obtained from

$$x' = x - vt, \qquad y' = y, \qquad z' = z, \qquad t' = t. \qquad \text{[Galilean]}$$

These four equations are the "inverse" transformation and are very easily obtained from Eqs. 37–3. Notice that the effect is merely to exchange primed and unprimed quantities and replace v by $-v$. This makes sense because from the S' frame, S moves to the left (negative x direction) with speed v.

Now suppose the point P in Fig. 37–12 represents a particle that is moving. Let the components of its velocity vector in S' be u'_x, u'_y, u'_z. (We use u to distinguish it from the relative velocity of the two frames, v.) Now $u'_x = dx'/dt'$, $u'_y = dy'/dt'$ and $u'_z = dz'/dt'$. The velocity of P as seen from S will have components u_x, u_y, and u_z. We can show how these are related to the velocity components in S' by differentiating Eqs. 37–3. For u_x we get

$$u_x = \frac{dx}{dt} = \frac{d(x' + vt')}{dt'} = u'_x + v$$

since v is assumed constant. For the other components, $u'_y = u_y$ and $u'_z = u_z$, so we have

$$u_x = u'_x + v$$
$$u_y = u'_y$$
$$u_z = u'_z.$$

[Galilean] **(37–4)** *Galilean velocity transformation*

These are known as the **Galilean velocity transformation equations**. We see that the y and z components of velocity are unchanged, but the x components differ by v: $u_x = u'_x + v$. This is just what we have used before when dealing with relative velocity.

The Galilean transformations, Eqs. 37–3 and 37–4, are valid only when the velocities involved are much less than c. We can see, for example, that the first of Eqs. 37–4 will not work for the speed of light; for light traveling in S' with speed $u'_x = c$ will have speed $c + v$ in S, whereas the theory of relativity insists it must be c in S. Clearly, then, a new set of transformation equations is needed to deal with relativistic velocities.

We will derive the required equations in a simple way, again looking at Fig. 37–12.

We assume that the transformation is linear and of the form

$$x = \gamma(x' + vt'), \qquad y = y', \qquad z = z'. \tag{i}$$

That is, we modify the first of Eqs. 37–3 by multiplying by a constant γ which is yet to be determined ($\gamma = 1$ non-relativistically). But we assume the y and z equations are unchanged since there is no length contraction in these directions. We will not assume a form for t, but will derive it. The inverse equations must have the same form with v replaced by $-v$. (The principle of relativity demands it, since S′ moving to the right with respect to S is equivalent to S moving to the left with respect to S′.) Therefore

$$x' = \gamma(x - vt). \tag{ii}$$

Now if a light pulse leaves the common origin of S and S′ at time $t = t' = 0$, after a time t it will have traveled a distance $x = ct$ or $x' = ct'$ along the x axis. Therefore, from Eqs. (i) and (ii) above,

$$ct = \gamma(ct' + vt') = \gamma(c + v)t', \tag{iii}$$
$$ct' = \gamma(ct - vt) = \gamma(c - v)t. \tag{iv}$$

We substitute t' from Eq. (iv) into Eq. (iii) and find $ct = \gamma(c + v)\gamma(c - v)(t/c) = \gamma^2(c^2 - v^2)t/c$. We cancel out the t on each side and solve for γ to find

$$\gamma = \frac{1}{\sqrt{1 - v^2/c^2}}.$$

Now that we have found γ, we need only find the relation between t and t'. To do so, we combine $x' = \gamma(x - vt)$ with $x = \gamma(x' + vt')$:

$$x' = \gamma(x - vt) = \gamma(\gamma[x' + vt'] - vt).$$

We solve for t and find $t = \gamma(t' + vx'/c^2)$. In summary,

$$
\begin{aligned}
x &= \gamma(x' + vt') &&= \frac{1}{\sqrt{1 - v^2/c^2}}(x' + vt') \\
y &= y' \\
z &= z' && \tag{37–5} \\
t &= \gamma\left(t' + \frac{vx'}{c^2}\right) &&= \frac{1}{\sqrt{1 - v^2/c^2}}\left(t' + \frac{vx'}{c^2}\right).
\end{aligned}
$$

These are called the **Lorentz transformation equations**. They were first proposed, in a slightly different form, by Lorentz in 1904 to explain the null result of the Michelson–Morley experiment and to make Maxwell's equations take the same form in all inertial reference frames. A year later Einstein derived them independently based on his theory of relativity. Notice that not only is the x equation modified as compared to the Galilean transformation, but so is the t equation; indeed, we see directly in this last equation how the space and time coordinates mix.

EXAMPLE 37–4 **Deriving length contraction.** Derive the length contraction formula, Eq. 37–2, from the Lorentz transformation equations.

SOLUTION Let an object of length L_0 be at rest on the x axis in S. The coordinates of its two end points are x_1 and x_2, so that $x_2 - x_1 = L_0$. At any instant in S′, the end points will be at x'_1 and x'_2 as given by the Lorentz transformation equations. The length measured in S′ is $L = x'_2 - x'_1$. An observer in S′ measures this length by measuring x'_2 and x'_1 at the same time (in the S′ reference frame), so $t'_2 = t'_1$. Then, from the first of Eqs. 37–5,

$$L_0 = x_2 - x_1 = \frac{1}{\sqrt{1 - v^2/c^2}}(x'_2 + vt'_2 - x'_1 - vt'_1).$$

Since $t'_2 = t'_1$, we have

$$L_0 = \frac{1}{\sqrt{1 - v^2/c^2}}(x'_2 - x'_1) = \frac{L}{\sqrt{1 - v^2/c^2}},$$

or

$$L = L_0\sqrt{1 - v^2/c^2},$$

which is Eq. 37–2.

EXAMPLE 37–5 **Deriving time dilation.** Derive the time-dilation formula, Eq. 37–1, using the Lorentz transformation equations.

SOLUTION The time Δt_0 between two events that occur at the same place $(x_2' = x_1')$ in S' is measured to be $\Delta t_0 = t_2' - t_1'$. Since $x_2' = x_1'$, then from the last of Eqs. 37–5, the time Δt between the events as measured in S is

$$\Delta t = t_2 - t_1 = \frac{1}{\sqrt{1 - v^2/c^2}}\left(t_2' + \frac{vx_2'}{c^2} - t_1' - \frac{vx_1'}{c^2}\right)$$

$$= \frac{1}{\sqrt{1 - v^2/c^2}}(t_2' - t_1') = \frac{\Delta t_0}{\sqrt{1 - v^2/c^2}},$$

which is Eq. 37–1. Notice that we chose S' to be the frame in which the two events occur at the same place, so that $x_1' = x_2'$ and the terms containing x_1' and x_2' cancel out.

The relativistically correct velocity equations are readily obtained by differentiating Eqs. 37–5 with respect to time. For example (using $\gamma = 1/\sqrt{1 - v^2/c^2}$ and the chain rule for derivatives):

$$u_x = \frac{dx}{dt} = \frac{d}{dt}\left[\gamma(x' + vt')\right]$$

$$= \frac{d}{dt'}\left[\gamma(x' + vt')\right]\frac{dt'}{dt} = \gamma\left[\frac{dx'}{dt'} + v\right]\frac{dt'}{dt}.$$

But $dx'/dt' = u_x'$ and $dt'/dt = 1/(dt/dt') = 1/\left[\gamma(1 + vu_x'/c^2)\right]$ where we have differentiated the last of Eqs. 37–5 with respect to time. Therefore

$$u_x = \frac{\left[\gamma(u_x' + v)\right]}{\left[\gamma(1 + vu_x'/c^2)\right]} = \frac{u_x' + v}{1 + vu_x'/c^2}.$$

The others are obtained in the same way and we collect them here:

$$u_x = \frac{u_x' + v}{1 + vu_x'/c^2} \tag{37–6a}$$

Relativistic

$$u_y = \frac{u_y'\sqrt{1 - v^2/c^2}}{1 + vu_x'/c^2} \tag{37–6b}$$

velocity

$$u_z = \frac{u_z'\sqrt{1 - v^2/c^2}}{1 + vu_x'/c^2}. \tag{37–6c}$$

transformation

Note that even though the relative velocity **v** is in the x direction, the transformation of all the components of a particle's velocity are affected by v and the x component of the particle's velocity; this was not true for the Galilean transformation, Eqs. 37–4.

FIGURE 37–13 Rocket 2 is fired from rocket 1 with speed $u' = 0.60c$. What is the speed of rocket 2 with respect to the Earth?

EXAMPLE 37–6 **Adding velocities.** Calculate the speed of rocket 2 in Fig. 37–13 with respect to Earth.

SOLUTION Rocket 2 moves with speed $u' = 0.60c$ with respect to rocket 1. Rocket 1 has speed $v = 0.60c$ with respect to Earth. The velocities are along the same straight line which we take to be the x (and x') axis. We need use only the first of Eqs. 37–6. Then the speed of rocket 2 with respect to Earth is

$$u = \frac{u' + v}{1 + \dfrac{vu'}{c^2}} = \frac{0.60c + 0.60c}{1 + \dfrac{(0.60c)(0.60c)}{c^2}} = \frac{1.20c}{1.36} = 0.88c.$$

(The Galilean transformation would have given $u = 1.20c$.)

Notice that Eqs. 37–6 reduce to the classical (Galilean) forms for velocities small compared to the speed of light, since $1 + vu'/c^2 \approx 1$ for v and $u' \ll c$. At the other extreme, let rocket 1 in Fig. 37–13 send out a beam of light, so that $u' = c$. Then Eq. 37–6a tells us the speed of light relative to Earth is

$$u = \frac{0.60\,c + c}{1 + \dfrac{(0.60\,c)(c)}{c^2}} = c,$$

which is consistent with the second postulate of relativity.

37–9 | Relativistic Momentum and Mass

So far in this chapter, we have seen that two basic mechanical quantities, length and time intervals, need modification because they are relative—their value depends on the reference frame from which they are measured. We might expect that other physical quantities might need some modification according to the theory of relativity, such as momentum, energy, and mass.

Let us first examine momentum, which classically is defined as mass times velocity, $\mathbf{p} = m\mathbf{v}$. Classically, momentum is a conserved quantity. It would be a great benefit if the law of conservation of momentum were still valid in the relativistic domain, and we will try to insist that it does as we investigate any possible modification to the definition of momentum. A good guess as to how momentum might be altered would be the use of the factor $\gamma = 1/\sqrt{1 - v^2/c^2}$, as in Eqs. 37–1, 2, and 37–5. But let us be a little more general and let momentum be defined by $p = fmv$ where f is some function of v, $f(v)$. Now we consider a hypothetical collision between two objects—a thought experiment—and see what form $f(v)$ must take if momentum is to be conserved.

Our thought experiment involves the elastic collision of two *identical* balls. If the two balls travel at the same speed, v, we can safely say that they will have the same magnitude of momentum.

The collision in this thought experiment takes place as follows. We consider two inertial reference frames, S and S′, moving along the x axis with a speed v with respect to each other. In reference frame S, a ball (call it ball A) is thrown with speed u along the y axis. In reference frame S′, a second ball (B) is thrown with speed u along the negative y' axis. The two balls are thrown at just the right time so that they collide. We assume that they rebound elastically and, from symmetry, that each moves with the same speed u back along the y axis in its thrower's reference frame. Figure 37–14a shows the collision as seen by the person in reference frame S; and Fig. 37–14b shows the collision as seen from reference frame S′. In reference frame S ball A has velocity $+u$ along the y axis before the collision and $-u$ along the y axis after the collision. In frame S′, ball A has, both before and after the collision, an x component of velocity equal to v, and a y component (see Eq. 37–6b with $u'_x = 0$) of magnitude

$$u\sqrt{1 - v^2/c^2}.$$

The same holds true for ball B, except in reverse. The velocity components are indicated in Fig. 37–14. Let us further assume that $u \ll v$, so that the speed of ball A is essentially v as seen in reference frame S′, and thus A's momentum can be written $f(v)mv$. Similarly, the momentum of B in reference frame S is $f(v)mv$. We now apply the law of conservation of momentum, which we hope remains valid in relativity, even if momentum has to be redefined. That is, we assume that the total

(a)

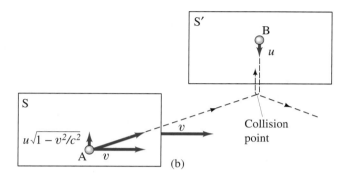

(b)

FIGURE 37–14 Deriving the momentum formula. Collision as seen (a) from reference frame S, (b) from reference frame S'.

momentum before the collision is equal to the total after the collision. We apply this to the y component of momentum in reference frame S (Fig. 37–14a):

$$f(u)mu - f(v)mu\sqrt{1 - v^2/c^2} = -f(u)mu + f(v)mu\sqrt{1 - v^2/c^2},$$

where we use $f(u)$ for ball A because its speed in S is only u.

We solve this for $f(v)$ and obtain

$$f(v) = \frac{f(u)}{\sqrt{1 - v^2/c^2}}.$$

This relation is valid for any u and v, (as long as $u \ll v$), and to simplify it so we can solve for f let us consider what happens if we let u become very small so that it approaches zero (this corresponds to a glancing collision with one of the balls essentially at rest and the other moving with speed v). Then the momentum terms $f(u)mu$ are in the nonrelativistic realm and take on the classical form, simply mu, meaning that $f(u) = 1$. So the previous equation becomes

$$f(v) = \frac{1}{\sqrt{1 - v^2/c^2}}.$$

We see that $f(v)$ comes out to be the factor we used before and called γ, and here has been shown to be valid for ball A. Using Fig. 37–14b we can derive the same relation for ball B. Thus we can conclude that we need to define the relativistic momentum of a particle as

Relativistic momentum

$$\mathbf{p} = \frac{m\mathbf{v}}{\sqrt{1 - v^2/c^2}} = \gamma m\mathbf{v}. \tag{37–7}$$

With this definition, as we saw in our thought experiment, the law of conservation of momentum will remain valid even in the relativistic realm. This relativistic momentum formula (Eq. 37–7) has been tested countless times on tiny elementary particles and been found valid.

This relativistic definition of momentum, Eq. 37–7, is sometimes interpreted as an increase in the mass of an object. That is, we can keep the form of our classical definition of momentum as

$$\mathbf{p} = m_{rel}\mathbf{v},$$

but only if we interpret m_{rel} to be the **relativistic mass**, which increases with speed according to

$$m_{rel} = \frac{m}{\sqrt{1 - v^2/c^2}}. \tag{37–8}$$

Rest mass In this "mass-increase" formula, m is referred to as the **rest mass** of the object—the mass it has as measured in a reference frame in which it is at rest; and m_{rel} is the object's relativistic mass in a reference frame in which it moves at speed v. With this interpretation, *the mass of an object appears to increase as its speed increases*. But we must be careful in the use of relativistic mass. We cannot just plug it into formulas like $F = ma$ or $K = \frac{1}{2}mv^2$ (as we can here for momentum, obtaining Eq. 37–7). For example, if we substitute Eq. 37–8 into $F = ma$, we obtain a formula that does not agree with experiment. If however, we write Newton's second law in its more general form, $\mathbf{F} = d\mathbf{p}/dt$, we do get a correct result. That is, Newton's second law, stated in its most general form is

$$\mathbf{F} = \frac{d\mathbf{p}}{dt} = \frac{d}{dt}(\gamma m\mathbf{v}) = \frac{d}{dt}\left(\frac{m\mathbf{v}}{\sqrt{1 - v^2/c^2}}\right) \tag{37–9}$$

and is valid relativistically.[†]

Whenever we talk about the mass of an object, we will always mean its rest mass (a fixed value). On the rare occasion when we want to refer to an object's relativistic mass, we will say so explicitly.

37–10 The Ultimate Speed

A basic result of the special theory of relativity is that the speed of an object cannot equal or exceed the speed of light. That the speed of light is a natural speed limit in the universe can be seen from any of Eqs. 37–1, 37–2, 37–7, or the addition of velocities formula. It is perhaps easiest to see from Eq. 37–7. As an object is accelerated to greater and greater speeds, its momentum becomes larger and larger. Indeed, if v were to equal c, the denominator in this equation would be zero, and the momentum would be infinite. To accelerate an object up to $v = c$ would thus require infinite energy, and so is not possible.

37–11 Energy and Mass; $E = mc^2$

When a steady net force is applied to an object, the object increases in speed. Since the force is acting through a distance, work is done on the object and its energy increases. As the speed of the object approaches c, the speed cannot increase indefinitely since it cannot exceed c. On the other hand, the relativistic mass of the object increases with increasing speed. That is, the work done on an object not only increases its speed but also contributes to increasing its inertia. Normally, the work done on an object increases its energy. This new twist from the theory of relativity leads to the idea that mass is a form of energy, a crucial part of Einstein's theory of relativity.

[†] Clearly, Newton's second law written as $\mathbf{F} = m\mathbf{a}$ is not valid, even if we use the relativistic mass. There is an extra term:

$$\mathbf{F} = \frac{d\mathbf{p}}{dt} = \frac{d(\gamma m\mathbf{v})}{dt} = \frac{d(m_{rel}\mathbf{v})}{dt} = m_{rel}\mathbf{a} + \frac{dm_{rel}}{dt}\mathbf{v}.$$

To find the mathematical relationship between mass and energy, we assume that the work-energy theorem is still valid in relativity, and we take the motion to be along the x axis. The work done to increase a particle's speed from zero to v is

$$W = \int_i^f F \, dx = \int_i^f \frac{dp}{dt} \, dx = \int_i^f \frac{dp}{dt} \, v \, dt = \int_i^f v \, dp$$

where i and f refer to the initial $(v = 0)$ and final $(v = v)$ states. Since $d(pv) = p \, dv + v \, dp$ we can write

$$v \, dp = d(pv) - p \, dv$$

so

$$W = \int_i^f d(pv) - \int_i^f p \, dv.$$

Since integration is the exact inverse of differentiation, the first term on the right becomes

$$\int_i^f d(pv) = pv \Big|_i^f = (\gamma m v) v = \frac{m v^2}{\sqrt{1 - v^2/c^2}}.$$

The second term in our equation for W above is easily integrated since

$$\frac{d}{dv} \left(\sqrt{1 - v^2/c^2} \right) = -(v/c^2)/\sqrt{1 - v^2/c^2},$$

and so becomes

$$-\int_i^f p \, dv = -\int_0^v \frac{mv}{\sqrt{1 - v^2/c^2}} \, dv = mc^2 \sqrt{1 - v^2/c^2} \Big|_0^v$$

$$= mc^2 \sqrt{1 - v^2/c^2} - mc^2.$$

Finally, we have for W:

$$W = \frac{m v^2}{\sqrt{1 - v^2/c^2}} + mc^2 \sqrt{1 - v^2/c^2} - mc^2.$$

We multiply the second term by $\sqrt{1 - v^2/c^2}/\sqrt{1 - v^2/c^2} = 1$, and obtain

$$W = \frac{mc^2}{\sqrt{1 - v^2/c^2}} - mc^2.$$

By the work-energy theorem, the work done must equal the final kinetic energy K since the particle started from rest. Therefore

$$K = \frac{mc^2}{\sqrt{1 - v^2/c^2}} - mc^2 \qquad \textbf{(37–10a)}$$

$$= \gamma mc^2 - mc^2 = (\gamma - 1)mc^2. \qquad \textbf{(37–10b)}$$

Relativistic kinetic energy

Clearly, the kinetic energy K is not $\frac{1}{2}mv^2$ at high speeds. $\left[\frac{1}{2}mv^2 \text{ does not work for } m \text{ as rest mass, nor for relativistic mass.} \right]$

Equations 37–10 require some interpretation. First of all, what does the second term in Eq. 37–10a mean, the mc^2? Consistent with the idea that mass is a form of energy, Einstein called mc^2 the **rest energy** of the object. We can rearrange Eq. 37–10a to get

Rest energy, mc^2

$$E = K + mc^2, \qquad \textbf{(37–11a)}$$

where

$$E = \gamma mc^2 = \frac{mc^2}{\sqrt{1 - v^2/c^2}} \qquad \textbf{(37–11b)}$$

Total energy (defined)

is called the *total energy E* of the particle (assuming no potential energy), and equals the rest energy plus the kinetic energy.

For a particle at rest in a given reference frame, K is zero in Eq. 37–11a, so its total energy is its rest energy:

$$E = mc^2. \tag{37–12}$$

Here we have Einstein's famous formula, $E = mc^2$. This formula mathematically relates the concepts of energy and mass. But if this idea is to have any meaning from a practical point of view, then mass ought to be convertible to energy and vice versa. That is, if mass is just one form of energy, then it should be convertible to

*Mass and energy
interchangeable*

other forms of energy just as other types of energy are interconvertible. Einstein suggested that this might be possible, and indeed changes of mass to other forms of energy, and vice versa, have been experimentally confirmed countless times. The interconversion of mass and energy is most easily detected in nuclear and elementary particle physics. For example, the neutral pion (π^0) of rest mass 2.4×10^{-28} kg is observed to decay into pure electromagnetic radiation (photons). The π^0 completely disappears in the process. The amount of electromagnetic energy produced is found to be exactly equal to that predicted by Einstein's formula, $E = mc^2$. The reverse process is also commonly observed in the laboratory: electromagnetic radiation under certain conditions can be converted into material particles such as electrons (see Section 38–4 on pair production). On a larger scale, the energy produced in nuclear power plants is a result of the loss in rest mass of the uranium fuel as it undergoes the process called fission. Even the radiant energy we receive from the Sun is an example of $E = mc^2$; the Sun's mass is continually decreasing as it radiates electromagnetic energy outward.

The relation $E = mc^2$ is now believed to apply to all processes, although the changes are often too small to measure. That is, when the energy of a system changes by an amount ΔE, the mass of the system changes by an amount Δm given by

$$\Delta E = (\Delta m)(c^2).$$

In a chemical reaction where heat is gained or lost, the masses of the reactants and the products will be different. Even when water is heated on a stove, the mass of the water increases very slightly.[†]

EXAMPLE 37–7 **Pion's kinetic energy.** A π^0 meson $(m = 2.4 \times 10^{-28}$ kg) travels at a speed $v = 0.80\,c = 2.4 \times 10^8$ m/s. What is its kinetic energy? Compare to a classical calculation.

SOLUTION We substitute values into Eq. 37–10b

$$K = (\gamma - 1)mc^2$$

where

$$\gamma = \frac{1}{\sqrt{1 - v^2/c^2}} = \frac{1}{\sqrt{1 - (0.80)^2}} = 1.67.$$

Then

$$K = (1.67 - 1)(2.4 \times 10^{-28}\,\text{kg})(3.0 \times 10^8\,\text{m/s})^2$$
$$= 1.4 \times 10^{-11}\,\text{J}.$$

Notice that the units of mc^2 are kg·m²/s², which is the joule. A classical calculation would give $K = \frac{1}{2}mv^2 = \frac{1}{2}(2.4 \times 10^{-28}\,\text{kg})(2.4 \times 10^8\,\text{m/s})^2 = 6.9 \times 10^{-12}$ J, about half as much, but this is not a correct result.

EXAMPLE 37–8 **Energy from nuclear decay.** The energy required or released in nuclear reactions and decays comes from a change in mass between the initial and final particles. In one type of radioactive decay, an atom of uranium $(m = 232.03714\,\text{u})$ decays to an atom of thorium $(m = 228.02873\,\text{u})$ plus an atom of helium $(m = 4.00260\,\text{u})$ where the masses (always rest masses) given are in atomic mass units $(1\,\text{u} = 1.6605 \times 10^{-27}\,\text{kg})$. Calculate the energy released in this decay.

[†] This last example is also easy to understand from the point of view of kinetic theory (Chapter 18): as heat is added, the temperature and therefore the average speed of the molecules increases, and Eq. 37–8 tells us that the relativistic mass of the molecules also increases.

SOLUTION The initial mass is 232.03714 u, and after the decay it is 228.02873 u + 4.00260 u = 232.03133 u, so there is a decrease in mass of 0.00581 u. This mass decrease, which equals $(0.00581\text{ u})(1.66 \times 10^{-27}\text{ kg}) = 9.64 \times 10^{-30}\text{ kg}$, is changed into kinetic energy. Thus

Energy released in nuclear process

$$E = \Delta mc^2 = (9.64 \times 10^{-30}\text{ kg})(3.0 \times 10^8\text{ m/s})^2 = 8.68 \times 10^{-13}\text{ J}.$$

Since $1\text{ MeV} = 1.60 \times 10^{-13}\text{ J}$, the energy released is 5.4 MeV.

EXAMPLE 37–9 **Mass change in a chemical reaction.** When two moles of hydrogen and one mole of oxygen react to form two moles of water, the energy released is 484 kJ. How much does the mass decrease in this reaction?

SOLUTION Using Eq. 37–12 we have for the change in mass Δm:

$$\Delta m = \frac{\Delta E}{c^2} = \frac{(-484 \times 10^3\text{ J})}{(3.00 \times 10^8\text{ m/s})^2} = -5.38 \times 10^{-12}\text{ kg}.$$

The initial mass of the system is $0.002\text{ kg} + 0.016\text{ kg} = 0.018\text{ kg}$. Thus the change in mass is relatively very tiny and can normally be neglected. [Conservation of mass is usually a reasonable principle to apply to chemical reactions.]

Equation 37–10a for the kinetic energy is

$$K = mc^2 \left(\frac{1}{\sqrt{1 - v^2/c^2}} - 1 \right).$$

At low speeds, $v \ll c$, we can expand the square root in the denominator using the binomial expansion, $(1 \pm x)^n = 1 \pm nx + n(n - 1)x^2/2! + \cdots$. With $n = -\frac{1}{2}$, we get

$$K \approx mc^2 \left(1 + \frac{1}{2}\frac{v^2}{c^2} + \cdots - 1 \right)$$
$$\approx \tfrac{1}{2}mv^2,$$

where the dots in the first expression represent very small terms in the expansion which we have neglected since we assumed that $v \ll c$. Thus at low speeds, the relativistic form for kinetic energy reduces to the classical form, $K = \frac{1}{2}mv^2$. This is, of course, what we would like. It makes relativity a more viable theory in that it can predict accurate results at low speed as well as at high. Indeed, the other equations of special relativity also reduce to their classical equivalents at ordinary speeds: length contraction, time dilation, and modifications to momentum as well as kinetic energy, all disappear for $v \ll c$ since $\sqrt{1 - v^2/c^2} \approx 1$.

A useful relation between the total energy E of a particle and its momentum p can also be derived. The momentum of a particle of mass m and speed v is given by Eq. 37–7

$$p = \gamma mv = \frac{mv}{\sqrt{1 - v^2/c^2}}.$$

Relativistic momentum

The total energy is

$$E = K + mc^2$$

or

$$E = \gamma mc^2 = \frac{mc^2}{\sqrt{1 - v^2/c^2}}.$$

We square this equation (and we insert "$v^2 - v^2$" which is zero, but will help us):

$$E^2 = \frac{m^2c^2(v^2 - v^2 + c^2)}{1 - v^2/c^2}$$
$$= p^2c^2 + \frac{m^2c^4(1 - v^2/c^2)}{1 - v^2/c^2}$$

or

$$E^2 = p^2c^2 + m^2c^4. \tag{37–13}$$

Energy related to momentum

Thus, the total energy can be written in terms of the momentum p, or in terms of the kinetic energy (Eq. 37–11), where we have assumed there is no potential energy.

We can rewrite Eq. 37–13 as $E^2 - p^2c^2 = m^2c^4$. Since the rest mass m of a given particle is the same in any reference frame, we see that the quantity $E^2 - p^2c^2$ must also be the same in any reference frame. Thus, at any given moment the total energy E and momentum p of a particle will be different in different reference frames, but the quantity $E^2 - p^2c^2$ will have the same value in all inertial reference frames. We say that the quantity $E^2 - p^2c^2$ is **invariant** under a Lorentz transformation.

$E^2 - p^2c^2$
is invariant

In Section 16–7 we discussed how the frequency and wavelength of sound are altered if the source of the sound and the observer are moving toward or away from each other. When a source is moving toward us, the frequency is higher than when the source is at rest. If the source moves away from us, the frequency is lower. We obtained four different equations for the Doppler shift (Eqs. 16–9a and b, Eqs. 16–10a and b), depending on the direction of the relative motion and whether the source or the observer is moving. The Doppler effect occurs also for light; but the shifted frequency or wavelength is given by slightly different equations, and there are only two of them, because for light—according to special relativity—we can make no distinction between motion of the source and motion of the observer. (Recall that sound travels in a medium such as air, whereas light does not—there is no evidence for an ether.)

To derive the Doppler shift for light, let us consider a light source and an observer that move toward each other, and let their relative velocity be v as measured in the reference frame of either the source or the observer. Figure 37–15a shows a source at rest emitting light waves of frequency f_0 and wavelength $\lambda_0 = c/f_0$. Two wavecrests are shown, a distance λ_0 apart, the second crest just having been emitted. In Fig. 37–15b, the source is shown moving at speed v toward a stationary observer who will see the wavelength λ being somewhat less than λ_0. (This is much like Fig. 16–20 for sound.) Let Δt represent the time between crests as detected by the observer, whose reference frame is shown in Fig. 37–15b. From Fig. 37–15b we see that

$$\lambda = c\,\Delta t - v\,\Delta t$$

where $c\,\Delta t$ is the distance crest 1 has moved in the time Δt after it was emitted, and $v\,\Delta t$ is the distance the source has moved in time Δt. So far our derivation has not differed from that for sound (Section 16–7). Now we invoke the theory of relativity. The time between emission of wavecrests has undergone time dilation:

$$\Delta t = \Delta t_0 / \sqrt{1 - v^2/c^2}$$

where Δt_0 is the time between emissions of wavecrests in the reference frame where the source is at rest (the "proper" time). In the source's reference frame (Fig. 37–15a), we have

$$\Delta t_0 = \frac{1}{f_0} = \frac{\lambda_0}{c}$$

(Eqs. 3–15 and 32–14). Thus

$$\lambda = (c - v)\,\Delta t = (c - v)\frac{\Delta t_0}{\sqrt{1 - v^2/c^2}} = \frac{(c - v)}{\sqrt{c^2 - v^2}}\lambda_0$$

or

$$\lambda = \lambda_0 \sqrt{\frac{c - v}{c + v}}. \qquad \left[\begin{array}{c}\text{source and observer}\\ \text{moving toward}\\ \text{each other}\end{array}\right] \quad \textbf{(37–14a)}$$

The frequency f is $\left(\text{recall } \lambda_0 = c/f_0\right)$

$$f = \frac{c}{\lambda} = f_0 \sqrt{\frac{c + v}{c - v}}. \qquad \left[\begin{array}{c}\text{source and observer}\\ \text{moving toward}\\ \text{each other}\end{array}\right] \quad \textbf{(37–14b)}$$

Doppler

shift

for

light

Source

Observer

λ_0

S

(a)

Crest emitted when
source was at point 2

Crest emitted
when source
was at point 1

1   2
S v S λ

$v\Delta t$

$c\Delta t$

(b)

FIGURE 37–15 Doppler shift for light. (a) Reference frame of light source. (b) Reference frame of observer, toward whom source is moving.

Here f_0 is the frequency of the light as seen in the source's reference frame, and f is the frequency as measured by an observer moving toward the source or toward whom the source is moving. Equations 37–14 depend only on the relative velocity v. For relative motion *away* from each other we set $v < 0$ in Eqs. 37–14, and obtain

$$\lambda = \lambda_0 \sqrt{\frac{c + v}{c - v}}$$

$$f = f_0 \sqrt{\frac{c - v}{c + v}}.$$

$$\begin{bmatrix} \text{source and observer} \\ \text{moving away from} \\ \text{each other} \end{bmatrix}$$

(37–15a)

(37–15b)

From Eqs. 37–14 and 37–15 we see that light from a source moving toward us will have a higher frequency and shorter wavelength, whereas if a light source moves away from us, we will see a lower frequency and a longer wavelength. In the latter case, visible light will have its wavelength lengthened toward the red end of the visible spectrum (Fig. 33–24), an effect called a **redshift**. As we will see in the next chapter, all atoms have their own distinctive signature in terms of the frequencies of the light they emit. In 1929 the American astronomer Edwin Hubble (1889–1953) found that radiation from atoms in many galaxies is redshifted. That is, the frequencies of light emitted are lower than those emitted by stationary atoms on Earth, suggesting that the galaxies are receding from us. This is the origin of the idea that the universe is expanding.

37–13 | The Impact of Special Relativity

A great many experiments have been performed to test the predictions of the special theory of relativity. Within experimental error, no contradictions have been found. Scientists have therefore accepted relativity as an accurate description of nature.

At speeds much less than the speed of light, the relativistic formulas reduce to the old classical ones, as we have discussed. We would, of course, hope—or rather, insist—that this be true since Newtonian mechanics works so well for objects moving with speeds $v \ll c$. This insistence that a more general theory (such as relativity) give the same results as a more restricted theory (such as classical mechanics which works for $v \ll c$) is called the **correspondence principle**. The two theories must correspond where their realms of validity overlap. Relativity thus does not contradict classical mechanics. Rather, it is a more general theory, of which classical mechanics is now considered to be a limiting case.

Correspondence principle

The importance of relativity is not simply that it gives more accurate results, especially at very high speeds. Much more than that, it has changed the way we view the world. The concepts of space and time are now seen to be relative, and intertwined with one another, whereas before they were considered absolute and separate. Even our concepts of matter and energy have changed: either can be converted to the other. The impact of relativity extends far beyond physics. It has influenced the other sciences, and even the world of art and literature; it has, indeed, entered the general culture.

From a practical point of view, we do not have much opportunity in our daily lives to use the mathematics of relativity. For example, the factor $\sqrt{1 - v^2/c^2}$, which appears in many relativistic formulas, has a value of 0.995 when $v = 0.10\,c$. Thus, for speeds even as high as $0.10\,c = 3.0 \times 10^7$ m/s, the factor $\sqrt{1 - v^2/c^2}$ in relativistic formulas gives a numerical correction of less than 1 percent. For speeds less than $0.10\,c$, or unless mass and energy are interchanged, we don't usually need to use the more complicated relativistic formulas, and can use the simpler classical formulas.

Summary

An **inertial reference frame** is one in which Newton's law of inertia holds. Inertial reference frames can move at constant velocity relative to one another; accelerating reference frames are noninertial.

The **special theory of relativity** is based on two principles: the **relativity principle**, which states that the laws of physics are the same in all inertial reference frames, and the principle of the **constancy of the speed of light**, which states that the speed of light in empty space has the same value in all inertial reference frames.

One consequence of relativity theory is that two events that are simultaneous in one reference frame may not be simultaneous in another. Other effects are **time dilation**: moving clocks are measured to run slow; and **length contraction**: the length of a moving object is measured to be shorter (in its direction of motion) than when it is at rest. Quantitatively,

$$L = L_0 \sqrt{1 - v^2/c^2}$$

$$\Delta t = \frac{\Delta t_0}{\sqrt{1 - v^2/c^2}}$$

where L and Δt are the length and time interval of objects (or events) observed as they move by at the speed v; L_0 and Δt_0 are the **proper length** and **proper time**—that is, the same quantities as measured in the rest frame of the objects or events.

The **Lorentz transformations** relate the positions and times of events in one inertial reference frame to their positions and times in a second inertial reference frame.

$$x = \gamma(x' + vt')$$
$$y = y'$$
$$z = z'$$
$$t = \gamma\left(t' + \frac{vx'}{c^2}\right)$$

where $\gamma = 1/\sqrt{1 - v^2/c^2}$.

Velocity addition also must be done in a special way. All these relativistic effects are significant only at high speeds, close to the speed of light, which itself is the ultimate speed in the universe.

The theory of relativity has changed our notions of space and time, and of momentum, energy, and mass. Space and time are seen to be intimately connected, with time being the fourth dimension in addition to the three dimensions of space.

The **momentum** of an object is given by

$$\mathbf{p} = \gamma m\mathbf{v} = \frac{m\mathbf{v}}{\sqrt{1 - v^2/c^2}}.$$

This formula can be interpreted as a *mass increase*, where the relativistic mass is $m_{rel} = \gamma m$ and m is the **rest mass** of the object ($v = 0$).

Mass and energy are interconvertible. The equation

$$E = mc^2$$

tells how much energy E is needed to create a mass m, or vice versa. Said another way, $E = mc^2$ is the amount of energy an object has because of its mass m. The law of conservation of energy must include mass as a form of energy.

The kinetic energy K of an object moving at speed v is given by

$$K = (\gamma - 1)mc^2 = \frac{mc^2}{\sqrt{1 - v^2/c^2}} - mc^2$$

where m is the rest mass of the object. The total energy E is

$$E = K + mc^2 = \gamma mc^2.$$

The momentum p of an object is related to its total energy E (assuming no potential energy) by

$$E^2 = p^2c^2 + m^2c^4.$$

Questions

1. You are in a windowless car in an exceptionally smooth train. Is there any physical experiment you can do in the train car to determine whether you are moving?

2. You might have had the experience of being at a red light when, out of the corner of your eye, you see the car beside you creep forward. Instinctively you stomp on the brake pedal, thinking that you are rolling backward. What does this say about absolute and relative motion?

3. A worker stands on top of a moving railroad car, and throws a heavy ball straight up (from his point of view). Ignoring air resistance, will the ball land on the car or behind it?

4. Does the Earth really go around the Sun? Or is it also valid to say that the Sun goes around the Earth? Discuss in view of the first principle of relativity (that there is no best reference frame).

5. If you were on a spaceship traveling at $0.5c$ away from a star, at what speed would the starlight pass you?

6. Will two events that occur at the same place and same time for one observer be simultaneous to a second observer moving with respect to the first?

7. Analyze the thought experiment of Section 37–4 from O_1's point of view. (Make a diagram analogous to Fig. 37–6.)

8. The time-dilation effect is sometimes expressed as "moving clocks run slowly." Actually, this effect has nothing to do with motion affecting the functioning of clocks. What then does it deal with?

9. Does time dilation mean that time actually passes more slowly in moving reference frames or that it only *seems* to pass more slowly?

10. A young-looking woman astronaut has just arrived home from a long trip. She rushes up to an old gray-haired man and in the ensuing conversation refers to him as her son. How might this be possible?

11. If you were traveling away from Earth at speed $0.5c$, would you notice a change in your heartbeat? Would your mass, height, or waistline change? What would observers on Earth using telescopes say about you?

12. Do time dilation and length contraction occur at ordinary speeds, say 90 km/h?

13. Suppose the speed of light were infinite. What would happen to the relativistic predictions of length contraction and time dilation?

14. Explain how the length-contraction and time-dilation formulas might be used to indicate that c is the limiting speed in the universe.

15. Does the equation $E = mc^2$ conflict with the conservation of energy principle? Explain.

16. If mass is a form of energy, does this mean that a spring has more mass when compressed than when relaxed?

17. It is not correct to say that "matter can neither be created nor destroyed." What must we say instead?

18. Is our intuitive notion that velocities simply add, as we did in Section 3–10, completely wrong?

Problems

Sections 37–4 to 37–6

1. (I) Lengths and time intervals depend on the factor
$$\sqrt{1 - v^2/c^2}$$
according to the theory of relativity (Eqs. 37–1 and 37–2). Evaluate this factor for speeds of: (*a*) $v = 20{,}000$ m/s (typical speed of a satellite); (*b*) $v = 0.0100c$; (*c*) $v = 0.100c$; (*d*) $v = 0.900c$; (*e*) $v = 0.990c$; (*f*) $v = 0.999c$.

2. (I) A spaceship passes you at a speed of $0.750c$. You measure its length to be 28.2 m. How long would it be when at rest?

3. (I) A certain type of elementary particle travels at a speed of 2.70×10^8 m/s. At this speed, the average lifetime is measured to be 4.76×10^{-6} s. What is the particle's lifetime at rest?

4. (I) If you were to travel to a star 100 light-years from Earth at a speed of 2.50×10^8 m/s, what would you measure this distance to be?

5. (II) What is the speed of a pion if its average lifetime is measured to be 4.10×10^{-8} s? At rest, its mean lifetime is 2.60×10^{-8} s.

6. (II) Suppose you decide to travel to a star 75 light-years away. How fast would you have to travel so the distance would be only 25 light-years?

7. (II) At what speed do the relativistic formulas for length and time intervals differ from classical values by 1.00 percent? (This is a reasonable way to estimate when to do relativistic calculations rather than classical.)

8. (II) Suppose a news report stated that starship *Enterprise* had just returned from a 5-year voyage while traveling at $0.84c$. (*a*) If the report meant 5.0 years of *Earth time*, how much time elapsed on the ship? (*b*) If the report meant 5.0 years of *ship time*, how much time passed on Earth?

9. (II) A certain star is 95.0 light-years away. How long would it take a spacecraft traveling $0.960c$ to reach that star from Earth, as measured by observers: (*a*) on Earth, (*b*) on the spacecraft? (*c*) What is the distance traveled according to observers on the spacecraft? (*d*) What will the spacecraft occupants compute their speed to be from the results of (*b*) and (*c*)?

10. (II) A friend of yours travels by you in her fast sports car at a speed of $0.660c$. It is measured in your frame to be 4.80 m long and 1.25 m high. (*a*) What will be its length and height at rest? (*b*) How many seconds would you say elapsed on your friend's watch when 20.0 s passed on yours? (*c*) How fast did you appear to be traveling according to your friend? (*d*) How many seconds would she say elapsed on your watch when she saw 20.0 s pass on hers?

11. (II) How fast must an average pion be moving to travel 15 m before it decays? The average lifetime, at rest, is 2.6×10^{-8} s.

Section 37–8

12. (I) Suppose in Fig. 37–12 that the origins of S and S′ overlap at $t = t' = 0$ and that S′ moves at speed $v = 30$ m/s with respect to S. In S′, a person is resting at a point whose coordinates are $x' = 25$ m, $y' = 20$ m, and $z' = 0$. Calculate this person's coordinates in S (x, y, z) at (*a*) $t = 2.5$ s, (*b*) $t = 10.0$ s. Use the Galilean transformation.

13. (I) Repeat Problem 12 using the Lorentz transformation and a relative speed $v = 1.80 \times 10^8$ m/s, but choose the time to be (*a*) 2.5 μs and (*b*) 10.0 μs.

14. (I) A person on a rocket traveling at $0.50c$ (with respect to the Earth) observes a meteor come from behind and pass her at a speed she measures as $0.50c$. How fast is the meteor moving with respect to the Earth?

15. (II) Two spaceships leave Earth in opposite directions, each with a speed of $0.50c$ with respect to Earth. (a) What is the velocity of spaceship 1 relative to spaceship 2? (b) What is the velocity of spaceship 2 relative to spaceship 1?

16. (II) A spaceship leaves Earth traveling at $0.71c$. A second spaceship leaves the first at a speed of $0.87c$ with respect to the first. Calculate the speed of the second ship with respect to Earth if it is fired (a) in the same direction the first spaceship is already moving, (b) directly backward toward Earth.

17. (II) In Problem 12, suppose that the person moves with a velocity whose components are $u_x' = u_y' = 25.0$ m/s, What will be her velocity with respect to S? (Give magnitude and direction.)

18. (II) In Problem 13, suppose that the person moves with a velocity (with a rocket) whose components are $u_x' = u_y' = 2.0 \times 10^8$ m/s. What will be her velocity (magnitude and direction) with respect to S?

19. (II) A spaceship traveling at $0.66c$ away from Earth fires a module with a speed of $0.82c$ at right angles to its own direction of travel (as seen by the spaceship). What is the speed of the module, and its direction of travel (relative to the spaceship's direction), as seen by an observer on Earth?

20. (II) If a particle moves in the xy plane of system S (Fig. 37–12) with speed u in a direction that makes an angle θ with the x axis, show that it makes an angle θ' in S' given by $\tan\theta' = (\sin\theta)\sqrt{1 - v^2/c^2}/(\cos\theta - v/u)$.

21. (II) A stick of length L_0, at rest in reference frame S, makes an angle θ with the x axis. In reference frame S', which moves to the right with velocity $\mathbf{v} = v\mathbf{i}$ with respect to S, determine (a) the length L of the stick, and (b) the angle θ' it makes with the x' axis.

22. (III) In the old West, a marshal riding on a train traveling 50 m/s sees a duel between two men standing on the Earth 50 m apart parallel to the train. The marshal's instruments indicate that in his reference frame the two men fired simultaneously. (a) Which of the two men, the first one the train passes (A) or the second one (B) should be arrested for firing the first shot? That is, in the gunfighter's frame of reference, who fired first? (b) How much earlier did he fire? (c) Who was struck first?

23. (III) A farm boy studying physics believes that he can fit a 13.0-m long pole into a 10.0-m long barn if he runs fast enough, carrying the pole. Can he do it? Explain in detail. How does this fit with the idea that when he is running the barn looks even shorter than 10.0 m?

Section 37–9

24. (I) What is the momentum of a proton traveling at $v = 0.85c$?

25. (I) At what speed will an object's relativistic mass be twice its rest mass?

26. (II) A particle of rest mass m travels at a speed $v = 0.20c$. At what speed will its momentum be doubled?

27. (II) (a) A particle travels at $v = 0.10c$. By what percentage will a calculation of its momentum be wrong if you use the classical formula? (b) Repeat for $v = 0.50c$.

28. (II) What is the percent change in momentum of a proton that accelerates (a) from $0.45c$ to $0.90c$, (b) from $0.90c$ to $0.98c$?

Section 37–11

29. (I) A certain chemical reaction requires 4.82×10^4 J of energy input for it to go. What is the increase in mass of the products over the reactants?

30. (I) When a uranium nucleus at rest breaks apart in the process known as fission in a nuclear reactor, the resulting fragments have a total kinetic energy of about 200 MeV. How much mass was lost in the process?

31. (I) Calculate the rest energy of an electron in joules and in MeV (1 MeV $= 1.60 \times 10^{-13}$ J).

32. (I) Calculate the rest mass of a proton in MeV/c^2.

33. (I) The total annual energy consumption in the United States is about 8×10^{19} J. How much mass would have to be converted to energy to fuel this need?

34. (II) How much energy can be obtained from conversion of 1.0 gram of mass? How much mass could this energy raise to a height of 100 m?

35. (II) Show that when the kinetic energy of a particle equals its rest energy, the speed of the particle is about $0.866c$.

36. (II) At what speed will an object's kinetic energy be 25 percent of its rest energy?

37. (II) (a) How much work is required to accelerate a proton from rest up to a speed of $0.997c$? (b) What would be the momentum of this proton?

38. (II) Calculate the kinetic energy and momentum of a proton traveling 2.60×10^8 m/s.

39. (II) What is the momentum of a 750-MeV proton (that is, its kinetic energy is 750 MeV)?

40. (II) What is the speed of a proton accelerated by a potential difference of 95 MV?

41. (II) What is the speed of an electron whose kinetic energy is 1.00 MeV?

42. (II) What is the speed of an electron when it hits a television screen after being accelerated by the 25,000 V of the picture tube?

43. (II) Two identical particles of rest mass m approach each other at equal and opposite speeds, v. The collision is completely inelastic and results in a single particle at rest. What is the rest mass of the new particle? How much energy was lost in the collision? How much kinetic energy is lost in this collision?

44. (II) Calculate the speed of a proton ($m = 1.67 \times 10^{-27}$ kg) whose kinetic energy is exactly half its total energy.

45. (II) What is the speed and momentum of an electron ($m = 9.11 \times 10^{-31}$ kg) whose kinetic energy equals its rest energy?

46. (II) Suppose a spacecraft of mass 27,000 kg is accelerated to $0.21c$. (a) How much kinetic energy would it have? (b) If you used the classical formula for kinetic energy, by what percentage would you be in error?

47. (II) Calculate the kinetic energy and momentum of a proton ($m = 1.67 \times 10^{-27}$ kg) traveling 8.4×10^7 m/s. By what percentages would your calculations have been in error if you had used classical formulas?

48. (II) The americium nucleus, $^{241}_{95}$Am, decays to a neptunium nucleus, $^{237}_{93}$Np, by emitting an alpha particle of mass 4.00260 u and kinetic energy 5.5 MeV. Estimate the mass of the neptunium nucleus, ignoring its recoil, given that the americium mass is 241.05682 u.

49. (II) An electron $(m = 9.11 \times 10^{-31}\,\text{kg})$ is accelerated from rest to speed v by a conservative force. In this process, its potential energy decreases by $5.60 \times 10^{-14}\,\text{J}$. Determine the electron's speed, v.

50. (II) Make a graph of the kinetic energy versus momentum for (a) a particle of nonzero rest mass, and (b) a particle with zero rest mass.

51. (II) What magnetic field intensity is needed to keep 900-GeV protons revolving in a circle of radius 1.0 km (at, say, the Fermilab synchrotron)? Use the relativistic mass. The proton's rest mass is $0.938\,\text{GeV}/c^2$. $(1\,\text{GeV} = 10^9\,\text{eV}.)$

52. (II) A negative muon traveling at 33 percent the speed of light collides head on with a positive muon traveling at 50 percent the speed of light. The two muons (each of mass $105.7\,\text{MeV}/c^2$) annihilate, and produce how much electromagnetic energy?

53. (II) Show that the energy of a particle of charge e revolving in a circle of radius r in a magnetic field B is given by $E(\text{in eV}) = Brc$ in the relativistic limit $(v \approx c)$.

54. (II) Show that the kinetic energy K of a particle of rest mass m is related to its momentum p by the equation

$$p = \sqrt{K^2 + 2Kmc^2}\big/c.$$

55. (III) (a) In reference frame S, a particle has momentum $\mathbf{p} = p_x\mathbf{i}$ along the positive x axis. Show that in frame S′, which moves with speed v as in Fig. 37–12, the momentum has components

$$p'_x = \frac{p_x - vE/c^2}{\sqrt{1 - v^2/c^2}}$$

$$p'_y = p_y$$

$$p'_z = p_z$$

$$E' = \frac{E - p_x v}{\sqrt{1 - v^2/c^2}}.$$

(These transformation equations hold, actually, for any direction of $\mathbf{p}$.) (b) Show that p_x, p_y, p_z, E/c transform according to the Lorentz transformation in the same way as x, y, z, ct.

* Section 37–12

* 56. (II) A certain galaxy has a Doppler shift given by $f_0 - f = 0.797 f_0$. How fast is it moving away from us?

* 57. (II) A quasar emits familiar hydrogen lines whose wavelengths are 3.0 times longer than what we measure in the laboratory. (a) What is the speed of this quasar? (b) What result would you obtain if you used the "classical" Doppler shift discussed in Chapter 16?

* 58. (II) A spaceship moving toward Earth at $0.80\,c$ transmits radio signals at 95.0 MHz. At what frequency should Earth receivers be tuned?

* 59. (II) Starting from Eq. 37–15a, show that the Doppler shift in wavelength is

$$\frac{\Delta\lambda}{\lambda} = \frac{v}{c}$$

if $v \ll c$.

General Problems

60. As a rule of thumb, anything traveling faster than about $0.1\,c$ is called *relativistic*—i.e., for which the correction using special relativity is a significant effect. Is the electron in a hydrogen atom (radius $0.5 \times 10^{-10}\,\text{m}$) relativistic? (Treat the electron as though it were in a circular orbit around the proton.)

61. An atomic clock is taken to the North Pole, while another stays at the Equator. How far will they be out of synchronization after a year has elapsed?

62. The nearest star to Earth is Proxima Centauri, 4.3 light-years away. (a) At what constant velocity must a spacecraft travel from Earth if it is to reach the star in 4.0 years, as measured by travelers on the spacecraft? (b) How long does the trip take according to Earth observers?

63. Derive a formula showing how the apparent density of an object changes with speed v relative to an observer. [*Hint*: Use relativistic mass.]

64. An airplane travels 1500 km/h around the world, returning to the same place, in a circle of radius essentially equal to that of the Earth. Estimate the difference in time to make the trip as seen by Earth and airplane observers. [*Hint*: Use the binomial expansion, Appendix A.]

65. (a) What is the speed of an electron whose kinetic energy is 10,000 times its rest energy? Such speeds are reached in the Stanford Linear Accelerator, SLAC. (b) If the electrons travel in the lab through a tube 3.0 km long (as at SLAC), how long is this tube in the electrons' reference frame?

66. How many grams of matter would have to be totally destroyed to run a 100-W lightbulb for 1 year?

67. What minimum amount of electromagnetic energy is needed to produce an electron and a positron together? A positron is a particle with the same mass as an electron, but has the opposite charge. (Note that electric charge is conserved in this process. See Section 38–4.)

68. A 1.68-kg mass oscillates on the end of a spring whose spring constant is $k = 48.7 \, \text{N/m}$. If this system is in a spaceship moving past Earth at $0.900 \, c$, what is its period of oscillation according to (a) observers on the ship, and (b) observers on Earth?

69. An electron ($m = 9.11 \times 10^{-31} \, \text{kg}$) enters a uniform magnetic field $B = 1.8 \, \text{T}$, and moves perpendicular to the field lines with a speed $v = 0.92 \, c$. What is the radius of curvature of its path?

70. An observer on Earth sees an alien vessel approach at a speed of $0.60 \, c$. The *Enterprise* comes to the rescue (Fig. 37–16), overtaking the aliens while moving directly toward Earth at a speed of $0.90 \, c$ relative to Earth. What is the relative speed of one vessel as seen by the other?

FIGURE 37–16 Problem 70.

71. A free neutron can decay into a proton, an electron, and a neutrino. Assume the neutrino's rest mass is zero, and the other masses can be found in the table inside the front cover. Determine the total kinetic energy shared among the three particles when a neutron decays at rest.

72. The Sun radiates energy at a rate of about $4 \times 10^{26} \, \text{W}$. (a) At what rate is the Sun's mass decreasing? (b) How long does it take for the Sun to lose a mass equal to that of Earth? (c) Estimate how long the Sun could last if it radiated constantly at this rate.

73. An unknown particle is measured to have a negative charge and a speed of $2.24 \times 10^8 \, \text{m/s}$. Its momentum is determined to be $3.07 \times 10^{-22} \, \text{kg·m/s}$. Identify the particle by finding its mass.

74. How much energy would be required to break a helium nucleus into its constituents, two protons and two neutrons? The rest masses of a proton (including an electron), a neutron, and helium are, respectively, 1.00783 u, 1.00867 u, and 4.00260 u. (This is called the *total binding energy* of the ^4_2He nucleus.)

75. What is the percentage increase in the (relativistic) mass of a car traveling 110 km/h as compared to at rest?

76. Two protons, each having a speed of $0.935 \, c$ in the laboratory, are moving toward each other. Determine (a) the momentum of each proton in the laboratory, (b) the total momentum of the two protons in the laboratory, and (c) the momentum of one proton as seen by the other proton.

77. Show analytically that a particle with momentum p and energy E has a speed given by

$$v = \frac{pc^2}{E} = \frac{pc}{\sqrt{m^2c^2 + p^2}}.$$

78. A slab of glass moves to the right with speed v. A flash of light is emitted at point A (Fig. 37–17) and passes through the glass arriving at point B a distance L away. The glass has thickness d in the reference frame where it is at rest, and the speed of light in the glass is c/n. How long does it take the light to go from point A to point B according to an observer at rest with respect to points A and B? Check your answer for the cases $v = 0$, $n = 1$, and for $v = c$.

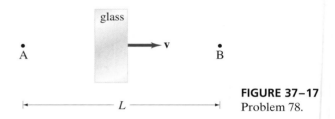

FIGURE 37–17
Problem 78.

79. Show that the space-time "distance" $(c\Delta t)^2 - (\Delta x)^2$ is invariant, meaning that all observers in all inertial reference frames calculate the same number for this quantity for any pair of events.

80. The fictional starship *Enterprise* obtains its power by combining matter and antimatter, achieving complete conversion of mass into energy. If the mass of the *Enterprise* is approximately $5 \times 10^9 \, \text{kg}$, how much mass must be converted into kinetic energy to accelerate it from rest to one-tenth the speed of light?

81. A spacecraft (reference frame S′) moves past Earth (reference frame S) at velocity $\mathbf{v}$, which points along the x and x' axes. The spacecraft emits light along its y' axis as shown in Fig. 37–18. (a) What angle does this light make with the x axis in the Earth's reference frame? (b) Show that the light moves with speed c also in the Earth's reference frame (i.e., given c in frame S′). (c) Compare these relativistic results to what you would have obtained classically (Galilean transformations).

FIGURE 37–18 Problem 81.

Hydrogen

Barium

Calcium

Solar spectrum with absorption levels

Atoms are tiny, and not observable using visible light. Trying to figure out the structure of atoms is a fascinating task. A major source of information about atoms comes from a study of the light emitted by the atoms of a pure material. Shown here are the line spectra (visible part) emitted by thin gases of hydrogen (top), barium, and calcium. At the bottom is the spectrum of the sun (visible light) in which are seen dark lines, indicating absorption at those wavelengths. What information do these spectra give us about atoms? We will see in this chapter that an entirely new theory is needed, the quantum theory.

CHAPTER 38

Early Quantum Theory and Models of the Atom

The second aspect of the revolution that shook the world of physics in the early part of the twentieth century (the first part was Einstein's theory of relativity) was the quantum theory. Unlike the special theory of relativity, the revolution of quantum theory required almost three decades to unfold, and many scientists contributed to its development. It began in 1900 with Planck's quantum hypothesis, and culminated in the mid-1920s with the theory of quantum mechanics of Schrödinger and Heisenberg which has been so effective in explaining the structure of matter.

38–1 | Planck's Quantum Hypothesis

One of the observations that was unexplained at the end of the nineteenth century was the spectrum of light emitted by hot objects. We saw in Chapter 19 that all objects emit radiation whose total intensity is proportional to the fourth power of the Kelvin temperature (T^4). At normal temperatures, we are not aware of this electromagnetic radiation because of its low intensity. At higher temperatures, there is sufficient infrared radiation that we can feel heat if we are close to the object. At still higher temperatures (on the order of 1000 K), objects actually glow, such as a red-hot electric stove burner or the element in a toaster. At temperatures above 2000 K, objects glow with a yellow or whitish color, such as white-hot iron and the filament of a lightbulb. As the temperature increases, the electromagnetic radiation emitted by bodies is most intense at higher and higher frequencies.

FIGURE 38–1 Spectrum of frequencies emitted by a blackbody at two different temperatures.

Blackbody radiation

The spectrum of light emitted by a hot dense object is shown in Fig. 38–1 for an idealized **blackbody**. A blackbody is a body that would absorb all the radiation falling on it (and so would appear black under reflection when illuminated from outside). The radiation such a blackbody would emit when hot and luminous, called **blackbody radiation** (though not necessarily black in color), is the easiest to deal with, and the radiation approximates that of many real objects. As can be seen, the spectrum contains a continuous range of frequencies. Such a continuous spectrum is emitted by any heated solid or liquid, and even by dense gases. The 6000-K curve in Fig. 38–1, corresponding to the temperature of the surface of the Sun, peaks in the visible part of the spectrum. For lower temperatures, the total radiation drops considerably and the peak occurs at longer wavelengths. Hence the blue end of the visible spectrum (and the UV) is relatively weaker. (This is why objects glow with a red color at around 1000 K.) It is found experimentally that the wavelength at the peak of the spectrum, λ_P, is related to the Kelvin temperature T by

$$\lambda_P T = 2.90 \times 10^{-3}\,\text{m} \cdot \text{K}. \tag{38-1}$$

This is known as **Wien's law**.

EXAMPLE 38–1 **The Sun's surface temperature.** Estimate the temperature of the surface of our Sun, given that the Sun emits light whose peak intensity occurs in the visible spectrum at around 500 nm.

SOLUTION Wien's law gives

$$T = \frac{2.90 \times 10^{-3}\,\text{m} \cdot \text{K}}{\lambda_P} = \frac{2.90 \times 10^{-3}\,\text{m} \cdot \text{K}}{500 \times 10^{-9}\,\text{m}} \approx 6000\,\text{K}.$$

EXAMPLE 38–2 **Star color.** Suppose a star has a surface temperature of 32,500 K. What color would this star appear?

SOLUTION From Wien's law we have,

$$\lambda_P = \frac{2.90 \times 10^{-3}\,\text{m} \cdot \text{K}}{T} = \frac{2.90 \times 10^{-3}\,\text{m} \cdot \text{K}}{3.25 \times 10^{4}\,\text{K}} = 89.2\,\text{nm}.$$

The peak is in the UV range of the spectrum. In the visible region, the curve will be descending (see Fig. 38–1), so the shortest visible wavelengths will be strongest. Hence the star will appear bluish (or blue-white).

A major problem facing scientists in the 1890s was to explain blackbody radiation. Maxwell's electromagnetic theory had predicted that oscillating electric charges produce electromagnetic waves, and the radiation emitted by a hot object could be due to the oscillations of electric charges in the molecules of the material. Although this would explain where the radiation came from, it did not correctly predict the observed spectrum of emitted light. Two important theoretical curves based on classical ideas were those proposed by W. Wien (in 1896) and by Lord Rayleigh (in 1900). The latter was modified later by J. Jeans and since then has been known as the Rayleigh–Jeans theory. As experimental data came in, it became clear that neither Wien's nor the Rayleigh–Jeans formulations were in accord with experiment. Wien's was accurate at short wavelengths but deviated from experiment at longer wavelengths, whereas the reverse was true for the Rayleigh–Jeans theory (see Fig. 38–2).

The break came in late 1900 when Max Planck (1858–1947) proposed an empirical formula that nicely fit the data:

$$I(\lambda, T) = \frac{2\pi hc^2\lambda^{-5}}{e^{hc/\lambda kT} - 1}.$$

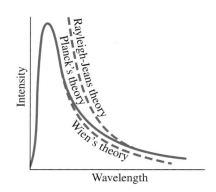

FIGURE 38–2 Comparison of the Wien and the Rayleigh–Jeans theories to that of Planck, which closely follows experiment.

$I(\lambda, T)$ is the radiation intensity as a function of wavelength λ at the temperature T; k is Boltzman's constant, c is the speed of light, and h is a new constant, now called **Planck's constant**. The value of h was estimated by Planck by fitting his formula for the blackbody radiation curve to experiment. The value accepted today is

$$h = 6.626 \times 10^{-34}\,\text{J}\cdot\text{s}.$$

Planck then sought a theoretical basis for the formula and within two months found that he could obtain the formula by making a new and radical (though not so recognized at the time) assumption: that the energy distributed among the oscillations of atoms within molecules is not continuous but instead consists of a finite number of very small discrete amounts, each related to the frequency of oscillation by

$$E_{\text{min}} = hf.$$

Planck's assumption suggests that the energy of any molecular vibration could be only some whole number multiple of hf:

$$E = nhf, \qquad n = 1, 2, 3, \cdots. \tag{38–2}$$

This idea is often called **Planck's quantum hypothesis** ("quantum" means "discrete amount" as opposed to "continuous"), although little attention was brought to this point at the time. In fact, it appears that Planck considered it more as a mathematical device to get the "right answer" rather than as a discovery comparable to those of Newton. Planck himself continued to seek a classical explanation for the introduction of h. The recognition that this was an important and radical innovation did not come until later, after about 1905 when others, particularly Einstein, entered the field.

The quantum hypothesis, Eq. 38–2, states that the energy of an oscillator can be $E = hf$, or $2hf$, or $3hf$, and so on, but there cannot be vibrations whose energy lies between these values. That is, energy would not be a continuous quantity as had been believed for centuries; rather it is **quantized**—it exists only in discrete amounts. The smallest amount of energy possible (hf) is called the **quantum of energy**. Recall from Chapter 14 that the energy of an oscillation is proportional to the amplitude squared. Thus another way of expressing the quantum hypothesis is that not just any amplitude of vibration is possible. The possible values for the amplitude are related to the frequency f.

A simple analogy may help. A stringed instrument such as a violin or guitar can be played over a continuous range of frequencies by moving your finger along the string. A flute or piano, on the other hand, is "quantized" in the sense that only certain frequencies (notes) can be played. Or compare a ramp, on which a box can be placed at any height, to a flight of stairs on which the box can have only certain discrete amounts of potential energy, as shown in Fig. 38–3.

Planck's quantum hypothesis

FIGURE 38–3 Ramp versus stair analogy. (a) On a ramp, a box can have continuous values of energy. (b) But on stairs, the box can have only discrete (quantized) values of energy.

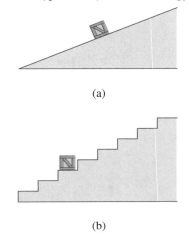

(a)

(b)

38–2 Photon Theory of Light and the Photoelectric Effect

In 1905, the same year that he introduced the special theory of relativity, Einstein made a bold extension of the quantum idea by proposing a new theory of light. Planck's work had suggested that the vibrational energy of molecules in a radiating object is quantized with energy $E = nhf$, where n is an integer. Einstein argued that therefore, when light is emitted by a molecular oscillator, the molecule's vibrational energy of nhf must decrease by an amount hf (or by $2hf$, etc.) to another integer times hf, namely $(n - 1)hf$. Then to conserve energy, the light ought to be emitted in packets or quanta, each with an energy

$$E = hf. \tag{38–3}$$

Again h is Planck's constant. Since all light ultimately comes from a radiating source, this suggests that perhaps *light is transmitted as tiny particles*, or **photons**, as they are now called, rather than as waves. This, too, was a radical departure from classical ideas. Einstein proposed a test of the quantum theory of light: quantitative measurements on the photoelectric effect.

The **photoelectric effect** is the phenomenon that when light shines on a metal surface, electrons are emitted from the surface. (The photoelectric effect occurs in other materials, but is most easily observed with metals.) It can be observed using the apparatus shown in Fig. 38–4. A metal plate P and a smaller electrode C are placed inside an evacuated glass tube, called a **photocell**. The two electrodes are connected to an ammeter and a source of emf, as shown. When the photocell is in the dark, the ammeter reads zero. But when light of sufficiently high frequency illuminates the plate, the ammeter indicates a current flowing in the circuit. To explain completion of the circuit, we can imagine electrons, ejected by the impinging radiation, flowing across the tube from the plate to the "collector" C as shown in the diagram.

That electrons should be emitted when light shines on a metal is consistent with the electromagnetic (EM) wave theory of light: the electric field of an EM wave could exert a force on electrons in the metal and eject some of them. Einstein pointed out, however, that the wave theory and the photon theory of light give very different predictions on the details of the photoelectric effect. For example, one thing that can be measured with the apparatus of Fig. 38–4 is the maximum kinetic energy (K_{max}) of the emitted electrons. This can be done by using a variable voltage source and reversing the terminals so that electrode C is negative and P is positive. The electrons emitted from P will be repelled by the negative electrode, but if this reverse voltage is small enough, the fastest electrons will still reach C and there will be a current in the circuit. If the reversed voltage is increased, a point is reached where the current reaches zero—no electrons have sufficient kinetic energy to reach C. This is called the *stopping potential*, or *stopping voltage*, V_0, and from its measurement, K_{max} can be determined using conservation of energy:

$$K_{max} = eV_0.$$

Now let us examine the details of the photoelectric effect from the point of view of the wave theory versus Einstein's particle theory. First the wave theory, assuming monochromatic light. The two important properties of a light wave are its intensity and its frequency (or wavelength). When these two quantities are varied, the wave theory makes the following predictions:

1. If the light intensity is increased, the number of electrons ejected and their maximum kinetic energy should be increased because the higher intensity means a greater electric field amplitude, and the greater electric field should eject electrons with higher speed.

2. The frequency of the light should not affect the kinetic energy of the ejected electrons. Only the intensity should affect K_{max}.

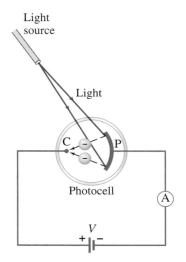

Photoelectric effect

FIGURE 38–4 The photoelectric effect.

Wave

theory

predictions

The photon theory makes completely different predictions. First we note that in a monochromatic beam, all photons have the same energy ($= hf$). Increasing the intensity of the light beam means increasing the number of photons in the beam, but does not affect the energy of each photon as long as the frequency is not changed. According to Einstein's theory, an electron is ejected from the metal by a collision with a single photon. In the process, all the photon energy is transferred to the electron and the photon ceases to exist. Since electrons are held in the metal by attractive forces, some minimum energy W_0 (called the **work function**, which is on the order of a few electron volts for most metals) is required just to get an electron out through the surface. If the frequency f of the incoming light is so low that hf is less than W_0, then the photons will not have enough energy to eject any electrons at all. If $hf > W_0$, then electrons will be ejected and energy will be conserved in the process. That is, the input energy (of the photon), hf, will equal the outgoing kinetic energy K of the electron plus the energy required to get it out of the metal, W:

$$hf = K + W. \tag{38-4a}$$

The least tightly held electrons will be emitted with the most kinetic energy (K_{max}), in which case W in this equation becomes the work function W_0, and K becomes K_{max}:

$$hf = K_{max} + W_0. \tag{38-4b}$$

Many electrons will require more energy than the bare minimum (W_0) to get out of the metal, and thus the kinetic energy of such electrons will be less than the maximum.

From these considerations, the photon theory makes the following predictions:

1. An increase in intensity of the light beam means more photons are incident, so more electrons will be ejected; but since the energy of each photon is not changed, the maximum kinetic energy of electrons is not changed by an increase in intensity.

2. If the frequency of the light is increased, the maximum kinetic energy of the electrons increases linearly, according to Eq. 38-4b. That is,

$$K_{max} = hf - W_0.$$

This relationship is plotted in Fig. 38-5.

3. If the frequency f is less than the "cutoff" frequency f_0, where $hf_0 = W_0$, no electrons will be ejected at all, no matter how great the intensity.

Photon

theory

predictions

These predictions of the photon theory are clearly very different from the predictions of the wave theory. In 1913–1914, careful experiments were carried out by R. A. Millikan. The results were fully in agreement with Einstein's photon theory.

One other aspect of the photoelectric effect also confirmed the photon theory. If extremely low light intensity is used, the wave theory predicts a time delay before electron emission so that an electron can absorb enough energy to exceed the work function. The photon theory predicts no such delay—it only takes one photon (if its frequency is high enough) to eject an electron—and experiments showed no delay. This too confirmed Einstein's photon theory.

FIGURE 38–5 Photoelectric effect: the maximum kinetic energy of ejected electrons increases linearly with the frequency of incident light. No electrons are emitted if $f < f_0$.

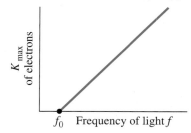

EXAMPLE 38–3 **Photon energy.** Calculate the energy of a photon of blue light, $\lambda = 450\ \text{nm}$.

SOLUTION Since $f = c/\lambda$, we have

$$E = hf = \frac{hc}{\lambda} = \frac{(6.63 \times 10^{-34}\ \text{J}\cdot\text{s})(3.0 \times 10^8\ \text{m/s})}{(4.5 \times 10^{-7}\ \text{m})} = 4.4 \times 10^{-19}\ \text{J},$$

or $(4.4 \times 10^{-19}\ \text{J})/(1.6 \times 10^{-19}\ \text{J/eV}) = 2.7\ \text{eV}$.

EXAMPLE 38–4 **ESTIMATE** **Photons from a lightbulb.** Estimate how many visible light photons a 100-W lightbulb emits per second.

SOLUTION Let's assume an average wavelength in the middle of the visible spectrum, $\lambda \approx 500\,\text{nm}$. The energy emitted in one second ($= 100\,\text{J}$) is $E = nhf$ when n is the number of photons emitted per second and $f = c/\lambda$. Hence

$$n = \frac{E}{hf} = \frac{E\lambda}{hc} = \frac{(100\,\text{J})(500 \times 10^{-9}\,\text{m})}{(6.63 \times 10^{-34}\,\text{J}\cdot\text{s})(3.0 \times 10^8\,\text{m/s})} = 2.5 \times 10^{20}.$$

This is an overestimate since much of the 100 J of electric energy input is transformed into heat rather than light. If the efficiency is between 1 percent and 10 percent, then the number of photons emitted is on the order of 10^{19}, still an enormous number.

EXAMPLE 38–5 **Photoelectron speed and energy.** What is the maximum kinetic energy and speed of an electron ejected from a sodium surface whose work function is $W_0 = 2.28\,\text{eV}$ when illuminated by light of wavelength: (a) 410 nm; (b) 550 nm?

SOLUTION (a) For $\lambda = 410\,\text{nm}$,

$$hf = \frac{hc}{\lambda} = 4.85 \times 10^{-19}\,\text{J} \quad \text{or} \quad 3.03\,\text{eV}.$$

From Eq. 38–4b, $K_{\text{max}} = 3.03\,\text{eV} - 2.28\,\text{eV} = 0.75\,\text{eV}$, or $1.2 \times 10^{-19}\,\text{J}$. Since $K = \frac{1}{2}mv^2$ where $m = 9.1 \times 10^{-31}\,\text{kg}$,

$$v = \sqrt{\frac{2K}{m}} = 5.1 \times 10^5\,\text{m/s}.$$

Notice that we used the nonrelativistic equation for kinetic energy. If v had turned out to be more than about $0.1c$, our calculation would have been inaccurate by more than a percent or so, and we would probably prefer to redo it using the relativistic form (Eq. 37–10).
(b) For $\lambda = 550\,\text{nm}$, $hf = 3.61 \times 10^{-19}\,\text{J} = 2.26\,\text{eV}$. Since this photon energy is less than the work function, no electrons are ejected.

Applications of the Photoelectric Effect

The photoelectric effect, besides playing an important historical role in confirming the photon theory of light, also has many practical applications. Burglar alarms and automatic door openers often make use of the photocell circuit of Fig. 38–4. When a person interrupts the beam of light, the sudden drop in current in the circuit activates a switch—often a solenoid—which operates a bell or opens the door. UV or IR light is sometimes used in burglar alarms because of its invisibility. Many smoke detectors use the photoelectric effect to detect tiny amounts of smoke that interrupt the flow of light and so alter the electric current. Photographic light meters use this circuit as well. Photocells are used in many other devices, such as absorption spectrophotometers, to measure light intensity. One type of film sound track is a variably shaded narrow section at the side of the film. Light passing through the film is thus "modulated," and the output electrical signal of the photocell detector follows the frequencies on the sound track. See Fig. 38–6. For many applications today, the vacuum-tube photocell of Fig. 38–4 has been replaced by a semiconductor device known as a **photodiode**. In these semiconductors, the absorption of a photon liberates a bound electron, which changes the conductivity of the material, so the current through a photodiode is altered.

FIGURE 38–6 Optical sound track on movie film. In the projector, light from a small source (different from that for the picture) passes through the sound track on the moving film. The light and dark areas on the sound track vary the intensity of the transmitted light which reaches the photocell, whose current output is then a replica of the original sound. This output is amplified and sent to the loudspeakers. High-quality projectors can show movies containing several parallel sound tracks to go to different speakers around the theater.

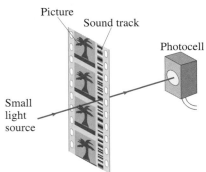

EXAMPLE 38–6 **Photosynthesis.** In *photosynthesis*, the process by which pigments such as chlorophyll in plants capture the energy of sunlight to change CO_2 to useful carbohydrate, about nine photons are needed to trasform one molecule of CO_2 to carbohydrate and O_2. Assuming light of wavelength $\lambda = 670\ nm$ (chlorophyll absorbs most strongly in the range 650 nm to 700 nm), how efficient is the photosynthetic process? The reverse chemical reaction releases an energy of 4.9 eV/molecule of CO_2.

➡ **PHYSICS APPLIED**

Photosynthesis

SOLUTION The energy of nine photons, each of energy $hf = hc/\lambda$ is $(9)(6.6 \times 10^{-34}\ J \cdot s)(3.0 \times 10^8\ m/s)/(6.7 \times 10^{-7}\ m) = 2.7 \times 10^{-18}\ J$ or 17 eV. Thus the process is $(4.9\ eV/17\ eV) = 29$ percent efficient.

38–3 | Photons and the Compton Effect

A number of other experiments were carried out in the early twentieth century which also supported the photon theory. One of these was the **Compton effect** (1923) named after its discoverer, A. H. Compton (1892–1962). Compton scattered short wavelength light (actually X-rays) from various materials. He found that the scattered light had a slightly longer wavelength than did the incident light, and therefore a slightly lower frequency indicating a loss of energy. This finding, he showed, could be explained on the basis of the photon theory as incident photons colliding with electrons of the material, Fig. 38–7. He applied the laws of conservation of energy and momentum to such collisions and found that the predicted energies of scattered photons were in accord with experimental results.

Let us look at the Compton effect in detail and seek to predict the shifted wavelength using the photon theory of light and Fig. 38–7. First we note that the photon is truly a relativistic particle—it travels at the speed of light. Thus we must use relativistic formulas for dealing with its mass, energy, and momentum. The momentum of any particle of rest mass m is given by $p = \gamma m v = mv/\sqrt{1 - v^2/c^2}$. Since $v = c$ for a photon, the denominator is zero. So the rest mass of a photon must also be zero, or its momentum would be infinite. Of course a photon is never at rest. The momentum of a photon, from Eq. 37–13 with $m = 0$, is given by $E^2 = p^2c^2$, or

$$p = \frac{E}{c}.$$

Since $E = hf$, the momentum of a photon is related to its wavelength by

$$p = \frac{hf}{c} = \frac{h}{\lambda}. \tag{38–5}$$

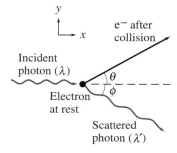

FIGURE 38–7 The Compton effect. A single photon of wavelength λ strikes an electron in some material, knocking it out of its atom. The scattered photon has less energy (since some is given to the electron) and hence has a longer wavelength λ'. Experiments found scattered X-rays of just the wavelengths predicted by conservation of energy and momentum using the photon model.

If the incoming photon in Fig. 38–7 has wavelength λ, then its total energy and momentum are

$$E = hf = \frac{hc}{\lambda} \quad \text{and} \quad p = \frac{h}{\lambda}.$$

After the collision of Fig. 38–7, the photon scattered at the angle ϕ has a wavelength which we call λ'. Its energy and momentum are

$$E' = \frac{hc}{\lambda'} \quad \text{and} \quad p' = \frac{h}{\lambda'}.$$

The electron, assumed at rest before the collision but free to move when struck is scattered at an angle θ as shown in Fig. 38–7.

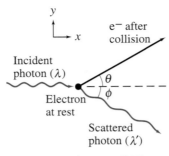

FIGURE 38-7 (repeated) The Compton effect.

FIGURE 38-8 Plots of intensity of radiation scattered from a target such as graphite (carbon), for three different angles. The values for λ' match Eq. 38-6. For (a) $\phi = 0°$, $\lambda' = \lambda_0$. In (b) and (c) a peak is found not only at λ' due to photons scattered from free electrons (or very nearly free) but also a peak at almost precisely λ_0. The latter is due to scattering from electrons very tightly bound to their atoms so the mass in Eq. 38-6 becomes very large (mass of the atom) and $\Delta\lambda$ becomes very small.

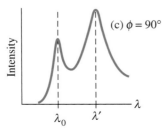

The electron's kinetic energy is (see Eq. 37–10):

$$K_e = \left(\frac{1}{\sqrt{1 - v^2/c^2}} - 1 \right) m_e c^2$$

where m_e is the rest mass of the electron and v is its speed. The electron's momentum is

$$p_e = \frac{1}{\sqrt{1 - v^2/c^2}} m_e v.$$

We apply conservation of energy to the collision (see Fig. 38–7):

incoming photon $\rightarrow$ scattered photon + electron

$$\frac{hc}{\lambda} = \frac{hc}{\lambda'} + \left(\frac{1}{\sqrt{1 - v^2/c^2}} - 1 \right) m_e c^2.$$

And we apply conservation of momentum to the x and y components of momentum:

$$\frac{h}{\lambda} = \frac{h}{\lambda'} \cos\phi + \frac{m_e v \cos\theta}{\sqrt{1 - v^2/c^2}}$$

$$0 = \frac{h}{\lambda'} \sin\phi - \frac{m_e v \sin\theta}{\sqrt{1 - v^2/c^2}}.$$

We can combine these three equations to eliminate v and θ (the algebra is left as an exercise—see Problem 25) and we obtain, as Compton did, an equation for the wavelength of the scattered photon in terms of its scattering angle ϕ:

$$\lambda' = \lambda + \frac{h}{m_e c} (1 - \cos\phi).$$

For $\phi = 0$, the wavelength is unchanged (there is no collision for this case of the photon passing straight through). At any other angle, λ' is longer than λ. The difference in wavelength,

$$\Delta\lambda = \lambda' - \lambda = \frac{h}{m_e c} (1 - \cos\phi), \tag{38-6}$$

is called the **Compton shift.** The quantity $h/m_e c$, which has the dimensions of length, is called the **Compton wavelength** λ_C of a free electron,

$$\lambda_C = \frac{h}{m_e c} = 2.43 \times 10^{-3} \text{ nm} = 2.43 \text{ pm.} \qquad \text{[electron]}$$

Equation 38–6 predicts that λ' depends on the angle ϕ at which they are detected. Compton's measurements of 1923 were consistent with this formula, confirming the value of λ_C and the dependence of λ' on ϕ. See Fig. 38–8. The wave theory of light predicts no wavelength shift: an incoming electromagnetic wave of frequency f should set electrons into oscillation at the same frequency f, and such oscillating electrons should reemit EM waves of this same frequency f (Chapter 32), and would not change with the angle ϕ. Hence the Compton effect adds to the firm experimental foundation for the photon theory of light.

EXAMPLE 38–7 **X-ray scattering.** X-rays of wavelength 0.140 nm are scattered from a block of carbon. What will be the wavelengths of X-rays scattered at (a) 0°, (b) 90°, (c) 180°?

SOLUTION (a) For $\phi = 0°$, $\cos\phi = 1$, and Eq. 38–6 gives $\lambda' = \lambda = 0.140$ nm. This makes sense since for $\phi = 0°$, there really isn't any collision as the photon goes straight through without interacting.
(b) For $\phi = 90°$, $\cos\phi = 0$, so

$$\lambda' = \lambda + \frac{h}{m_e c} = 0.140\,\text{nm} + \frac{6.63 \times 10^{-34}\,\text{J·s}}{(9.11 \times 10^{-31}\,\text{kg})(3.00 \times 10^8\,\text{m/s})}$$
$$= 0.140\,\text{nm} + 2.4 \times 10^{-12}\,\text{m} = 0.142\,\text{nm};$$

that is, the wavelength is longer by one Compton wavelength (= 0.0024 nm for an electron).
(c) For $\phi = 180°$, which means the photon is scattered backward, returning in the direction from which it came (a direct "head-on" collision), $\cos\phi = -1$, so

$$\lambda' = \lambda + 2\frac{h}{m_e c} = 0.140\,\text{nm} + 2(0.0024\,\text{nm}) = 0.145\,\text{nm}.$$

The Compton effect has been used to diagnose bone disease such as osteoporosis. Gamma rays, which are photons of even shorter wavelength than X-rays, coming from a radioactive source are scattered off bone material. The total intensity of the scattered radiation is proportional to the density of electrons, which is in turn proportional to the bone density. Changes in the density of bone material can indicate the onset of osteoporosis.

➡ **PHYSICS APPLIED**
Measuring bone density

38–4 | Photon Interactions; Pair Production

When a photon passes through matter, it interacts with the atoms and electrons. There are four important types of interactions that a photon can undergo:

1. The photon can be scattered from an electron (or a nucleus) and in the process lose some energy; this is the *Compton effect* (Fig. 38–7). But notice that the photon is not slowed down. It still travels with speed c, but its frequency will be lower.

2. The *photoelectric effect*: a photon may knock an electron out of an atom and in the process itself disappear.

3. The photon may knock an atomic electron to a higher energy state in the atom if its energy is not sufficient to knock the electron out altogether. In this process the photon also disappears, and all its energy is given to the atom. Such an atom is then said to be in an *excited state*, and we shall discuss this more later.

4. *Pair production*: A photon can actually create matter, such as the production of an electron and a positron, Fig. 38–9. (A positron has the same mass as an electron, but the opposite charge, $+e$.)

This last process is called **pair production**, and the photon disappears in the process of creating the electron–positron pair. This is an example of rest mass being created from pure energy, and it occurs in accord with Einstein's equation $E = mc^2$. Notice that a photon cannot create an electron alone since electric charge would not then be conserved. The inverse of pair production also occurs: if an electron collides with a positron, the two **annihilate** each other and their energy, including their mass, appears as electromagnetic energy of photons. Because of this process positrons usually do not last long in nature.

Photon

interactions

FIGURE 38–9 Pair production: a photon disappears and produces an electron and a positron.

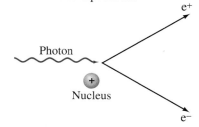

Pair production

EXAMPLE 38–8 **Pair production.** What is (a) the minimum energy of a photon that can produce an electron–positron pair? (b) What is this photon's wavelength?

SOLUTION (a) Because $E = mc^2$, and the mass created is equal to two electron masses, the photon must have energy

$$E = 2(9.11 \times 10^{-31}\,\text{kg})(3.0 \times 10^8\,\text{m/s})^2 = 1.64 \times 10^{-13}\,\text{J} = 1.02\,\text{MeV}.$$

A photon with less energy cannot undergo pair production.
(b) Since $E = hf = hc/\lambda$, the wavelength of a 1.02-MeV photon is

$$\lambda = \frac{hc}{E} = \frac{(6.6 \times 10^{-34}\,\text{J·s})(3.0 \times 10^8\,\text{m/s})}{(1.64 \times 10^{-13}\,\text{J})} = 1.2 \times 10^{-12}\,\text{m},$$

which is 0.0012 nm. Such photons are in the gamma-ray (or very short X-ray) region of the electromagnetic spectrum.

Pair production cannot occur in empty space, for energy and momentum could not simultaneously be conserved. In Example 38–8, for instance, energy is conserved, but only enough energy was provided to create the electron–positron pair at rest and thus with no momentum to carry away the initial momentum of the photon. Indeed, it can be shown that at any energy, an additional massive object, such as an atomic nucleus, must take part in the interaction to carry off some of the momentum.

38–5 Wave–Particle Duality; the Principle of Complementarity

The photoelectric effect, the Compton effect, and other experiments have placed the particle theory of light on a firm experimental basis. But what about the classic experiments of Young and others (Chapters 35 and 36) on interference and diffraction which showed that the wave theory of light also rests on a firm experimental basis?

We seem to be in a dilemma. Some experiments indicate that light behaves like a wave; others indicate that it behaves like a stream of particles. These two theories seem to be incompatible, but both have been shown to have validity. Physicists have finally come to the conclusion that this duality of light must be accepted as a fact of life. It is referred to as the **wave–particle duality**. Apparently, light is a more complex phenomenon than just a simple wave or a simple beam of particles.

To clarify the situation, the great Danish physicist Niels Bohr (1885–1962, Fig. 38–10) proposed his famous **principle of complementarity**. It states that to understand any given experiment, we must use either the wave or the photon theory, but not both. Yet we must be aware of both the wave and particle aspects of light if we are to have a full understanding of light. Therefore these two aspects of light complement one another.

It is not possible to "visualize" this duality. We cannot picture a combination of wave and particle. Instead, we must recognize that the two aspects of light are different "faces" that light shows to experimenters.

Part of the difficulty stems from how we think. Visual pictures (or models) in our minds are based on what we see in the everyday world. We apply the concepts of waves and particles to light because in the macroscopic world we see that energy is transferred from place to place by these two methods. We cannot see directly whether light is a wave or particle—so we do indirect experiments. To explain the experiments, we apply the models of waves or of particles to the nature of light. But these are abstractions of the human mind. When we try to conceive of what light really "is," we insist on a visual picture. Yet there is no reason why light should conform to these models (or visual images) taken from the macroscopic

FIGURE 38–10 Niels Bohr, walking with Enrico Fermi along the Appian Way outside Rome.

Wave–particle duality

Principle of complementarity

world. The "true" nature of light—if that means anything—is not possible to visualize. The best we can do is recognize that our knowledge is limited to the indirect experiments, and that in terms of everyday language and images, light reveals both wave and particle properties.

It is worth noting that Einstein's equation $E = hf$ itself links the particle and wave properties of a light beam. In this equation, E refers to the energy of a particle; and on the other side of the equation, we have the frequency f of the corresponding wave.

38–6 | Wave Nature of Matter

In 1923, Louis de Broglie (1892–1987) extended the idea of the wave–particle duality. He much appreciated the symmetry in nature, and argued that if light sometimes behaves like a wave and sometimes like a particle, then perhaps those things in nature thought to be particles—such as electrons and other material objects—might also have wave properties. De Broglie proposed that the wavelength of a material particle would be related to its momentum in the same way as for a photon, Eq. 38–5, $p = h/\lambda$. That is, for a particle having linear momentum p, the wavelength λ is given by

$$\lambda = \frac{h}{p}.$$

(38–7)   *de Broglie wavelength*

This is sometimes called the **de Broglie wavelength** of a particle.

EXAMPLE 38–9 **Wavelength of a ball.** Calculate the de Broglie wavelength of a 0.20-kg ball moving with a speed of 15 m/s. Discuss.

SOLUTION $\lambda = \dfrac{h}{p} = \dfrac{h}{mv} = \dfrac{(6.6 \times 10^{-34}\,\text{J·s})}{(0.20\,\text{kg})(15\,\text{m/s})} = 2.2 \times 10^{-34}\,\text{m}.$

This is an incredibly small wavelength. Even if the speed were extremely small, say 10^{-4} m/s, the wavelength would be about 10^{-29} m. Indeed, the wavelength of any ordinary object is much too small to be measured and detected. The problem is that the properties of waves, such as interference and diffraction, are significant only when the size of objects or slits is not much larger than the wavelength. And there are no known objects or slits to diffract waves only 10^{-30} m long, so the wave properties of ordinary objects go undetected.

But tiny elementary particles, such as electrons, are another matter. Since the mass m appears in the denominator in Eq. 38–7, a very small mass should have a much larger wavelength.

EXAMPLE 38–10 **Wavelength of an electron.** Determine the wavelength of an electron that has been accelerated through a potential difference of 100 V.

SOLUTION We assume that the speed of the electron will be much less than c, so we use nonrelativistic mechanics. (If this assumption were to come out wrong, we would have to recalculate using relativistic formulas—see Sections 37–9 and 37–11.) The gain in kinetic energy will equal the loss in potential energy, so $\frac{1}{2}mv^2 = eV$ and

$$v = \sqrt{\frac{2\,eV}{m}} = \sqrt{\frac{(2)(1.6 \times 10^{-19}\,\text{C})(100\,\text{V})}{(9.1 \times 10^{-31}\,\text{kg})}} = 5.9 \times 10^{6}\,\text{m/s}.$$

Then

$$\lambda = \frac{h}{mv} = \frac{(6.6 \times 10^{-34}\,\text{J·s})}{(9.1 \times 10^{-31}\,\text{kg})(5.9 \times 10^{6}\,\text{m/s})} = 1.2 \times 10^{-10}\,\text{m},$$

or 0.12 nm.

FIGURE 38–11 Diffraction pattern of electrons scattered from aluminum foil, as recorded on film.

From this Example, we see that electrons can have wavelengths on the order of 10^{-10} m. Although small, this wavelength can be detected: the spacing of atoms in a crystal is on the order of 10^{-10} m and the orderly array of atoms in a crystal could be used as a type of diffraction grating, as was done earlier for X-rays (see Section 36–10). C. J. Davisson and L. H. Germer performed the crucial experiment; they scattered electrons from the surface of a metal crystal and, in early 1927, observed that the electrons were scattered into a pattern of regular peaks. When they interpreted these peaks as a diffraction pattern, the wavelength of the diffracted electron wave was found to be just that predicted by de Broglie, Eq. 38–7. In the same year, G. P. Thomson (son of J. J. Thomson, who is credited with the discovery of the particle nature of electrons as we saw in Section 27–7), using a different experimental arrangement, also detected diffraction of electrons. (See Fig. 38–11. You can compare it to X-ray diffraction, Section 36–10.) Later experiments showed that protons, neutrons, and other particles also have wave properties.

Thus the wave–particle duality applies to material objects as well as to light. The principle of complementarity applies to matter as well. That is, we must be aware of both the particle and wave aspects in order to have an understanding of matter, including electrons. But again we must recognize that a visual picture of a "wave–particle" is not possible.

What is an electron?

We might ask ourselves: "What is an electron?" The early experiments of J. J. Thomson (Section 27–7) indicated a glow in a tube, and that glow moved when a magnetic field was applied. The results of these and other experiments were best interpreted as being caused by tiny negatively charged particles which we now call electrons. No one, however, has actually seen an electron directly. The drawings we sometimes make of electrons as tiny spheres with a negative charge on them are merely convenient pictures (now recognized to be inaccurate). Again we must rely on experimental results, some of which are best interpreted using the particle model and others using the wave model. These models are mere pictures that we use to extrapolate from the macroscopic world to the tiny microscopic world of the atom. And there is no reason to expect that these models somehow reflect the reality of an electron. We thus use a wave or a particle model (whichever works best in a situation) so that we can talk about what is happening. But we should not be led to believe that an electron *is* a wave or a particle. Instead we could say that an electron is the set of its properties that we can measure. Bertrand Russell said it well when he wrote that an electron is "a logical construction."

↪ **PHYSICS APPLIED**

Electron diffraction

EXAMPLE 38–11 **Electron diffraction.** The wave nature of electrons is manifested in experiments where an electron beam interacts with the atoms on the surface of a solid. By studying the angular distribution of the diffracted electrons, one can indirectly measure the geometrical arrangement of atoms. Assume that the electrons strike perpendicular to the surface of a solid (see Fig. 38–12), and that their energy is low, $K = 100$ eV, so that they interact only with the surface layer of atoms. If the smallest angle at which a diffraction maximum occurs is at 24°, what is the separation d between the atoms on the surface?

FIGURE 38–12 Exmple 38–11.

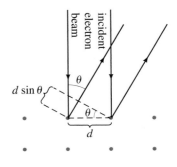

SOLUTION Treating the electrons as waves, we need to determine the condition where the difference in path traveled by the wave diffracted from adjacent atoms is an integer multiple of the de Broglie wavelength, so that constructive interference occurs. The path length difference is $d \sin \theta$; so for the smallest value of θ we must have

$$d \sin \theta = \lambda.$$

However, λ is related to the (non-relativistic) kinetic energy K by

$$K = \frac{p^2}{2m_e} = \frac{h^2}{2m_e \lambda^2}.$$

Thus

$$\lambda = \frac{h}{\sqrt{2m_e K}} = \frac{\left(6.63 \times 10^{-34}\,\text{J·s}\right)}{\sqrt{2\left(9.11 \times 10^{-31}\,\text{kg}\right)\left(100\,\text{eV}\right)\left(1.6 \times 10^{-19}\,\text{J/eV}\right)}} = 0.123\,\text{nm}.$$

The surface inter-atomic spacing is

$$d = \frac{\lambda}{\sin \theta} = \frac{0.123\,\text{nm}}{\sin 24°} = 0.30\,\text{nm}.$$

* 38–7 Electron Microscopes

The idea that electrons have wave properties led to the development of the **electron microscope**, which can produce images of much greater magnification than a light microscope. Figures 38–13 and 38–14 are diagrams of two types, the **transmission electron microscope**, which produces a two-dimensional image, and the **scanning electron microscope**, which produces images with a three-dimensional quality. In each design, the objective and eyepiece lenses are actually magnetic fields that exert forces on the electrons to bring them to a focus. The fields are produced by carefully designed current-carrying coils of wire.

As discussed in Section 36–5, the best resolution of details on an object is on the order of the wavelength of the radiation used to view it. Electrons accelerated by voltages on the order of $10^5\,\text{V}$ have wavelengths on the order of 0.004 nm. The maximum resolution obtainable would be on this order, but in practice, aberrations in the magnetic lenses limit the resolution in transmission electron microscopes to at best about 0.1 to 0.5 nm. This is still 10^3 times finer than that attainable with a light microscope, and corresponds to a useful magnification of about a million. Such magnifications are difficult to attain, and more common magnifications are 10^4 to 10^5. The maximum resolution attainable with a scanning electron microscope is somewhat less, about 5 to 10 nm at best.

➡ **PHYSICS APPLIED**

Electron microscope

FIGURE 38–13 Transmission electron microscope. The magnetic-field coils are designed to be "magnetic lenses," which bend the electron paths and bring them to a focus, as shown.

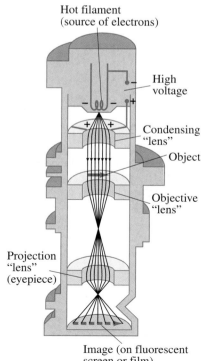

FIGURE 38–14 Scanning electron microscope. Scanning coils move an electron beam back and forth across the specimen. Secondary electrons produced when the beam strikes the specimen are collected and modulate the intensity of the beam in the CRT to produce a picture.

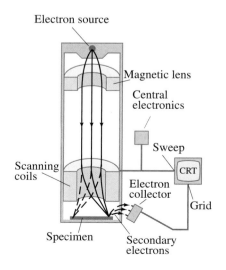

Early Models of the Atom

The idea that matter is made up of atoms was accepted by most scientists by 1900. With the discovery of the electron in the 1890s, scientists began to think of the atom itself as having a structure and electrons as part of that structure. We now introduce our modern approach to the atom, and of the quantum theory with which it is intertwined.[†]

A typical model of the atom in the 1890s visualized the atom as a homogeneous sphere of positive charge inside of which there were tiny negatively charged electrons, a little like plums in a pudding, Fig. 38–15. J. J. Thomson, soon after his discovery of the electron in 1897, argued that the electrons in this model should be moving.

Around 1911, Ernest Rutherford (1871–1937) and his colleagues performed experiments whose results contradicted Thomson's model of the atom. In these experiments a beam of positively charged "alpha (α) particles" was directed at a thin sheet of metal foil such as gold, Fig. 38–16a. (These newly discovered α particles were emitted by certain radioactive materials and were soon shown to be doubly ionized helium atoms—that is, having a charge of $+2e$.) It was expected from Thomson's model that the alpha particles would not be deflected significantly since electrons are so much lighter than alpha particles, and the alpha particles should not have encountered any massive concentration of positive charge in that model to strongly repel them. The experimental results completely contradicted these predictions. It was found that most of the alpha particles passed through the foil unaffected, as if the foil were mostly empty space. And of those deflected, a few were deflected at very large angles—some even nearly back in the direction from which they had come. This could happen, Rutherford reasoned, only if the positively charged alpha particles were being repelled by a massive positive charge concentrated in a very small region of space (see Fig. 38–16b). He theorized that the atom must consist of a tiny but massive positively charged nucleus, containing over 99.9 percent of the mass of the atom, surrounded by electrons some distance away. The electrons would be moving in orbits about the nucleus—much as the planets move around the Sun—because if they were at rest, they would fall into the nucleus

[†] Some readers may say: "Tell us the facts as we know them today, and don't bother us with the historical background and its outmoded theories." Such an approach would ignore the creative aspect of science and thus give a false impression of how science develops. Moreover, it is not really possible to understand today's view of the atom without insight into the concepts that led to it

FIGURE 38–15 Plum-pudding model of the atom.

$\approx 10^{-10}$ m

Positively charged material

Rutherford's planetary model

FIGURE 38–16 (a) Experimental setup for Rutherford's experiment: α particles emitted by radon strike a metallic foil and some rebound backward; (b) backward rebound of α particles explained as the repulsion from a heavy positively charged nucleus.

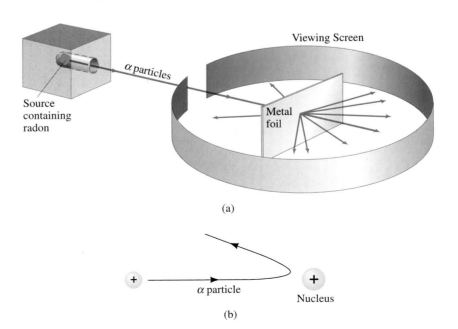

Viewing Screen

α particles

Source containing radon

Metal foil

(a)

$+$ α particle $+$

Nucleus

(b)

due to electrical attraction, Fig. 38–17. Rutherford's experiments suggested that the nucleus must have a radius of about 10^{-15} to 10^{-14} m. From kinetic theory, and especially Einstein's analysis of Brownian movement (see Section 17–1), the radius of atoms was estimated to be about 10^{-10} m. Thus the electrons would seem to be at a distance from the nucleus of about 10,000 to 100,000 times the radius of the nucleus itself (if the nucleus were the size of a baseball, the atom would have the diameter of a big city several kilometers across). So an atom would be mostly empty space.

Rutherford's "planetary" model of the atom (also called the "nuclear model of the atom") was a major step toward how we view the atom today. It was not, however, a complete model and presented some major problems, as we shall see.

FIGURE 38–17 Rutherford's model of the atom, in which electrons orbit a tiny positive nucleus (not to scale). The atom is visualized as mostly empty space.

38–9 | Atomic Spectra: Key to the Structure of the Atom

Earlier in this chapter we saw that heated solids (as well as liquids and dense gases) emit light with a continuous spectrum of wavelengths. This radiation is assumed to be due to oscillations of atoms and molecules, which are largely governed by the interaction of each atom or molecule with its neighbors.

FIGURE 38–18 Gas-discharge tube: (a) is a diagram; (b) photo of actual discharge tube for hydrogen.

Rarefied gases can also be excited to emit light. This is done by intense heating, or more commonly by applying a high voltage to a "discharge tube" containing the gas at low pressure, Fig. 38–18. The radiation from excited gases had been observed early in the nineteenth century, and it was found that the spectrum was not continuous, but *discrete*. Since excited gases emit light of only certain wavelengths, when this light is analyzed through the slit of a spectroscope or spectrometer, a **line spectrum** is seen rather than a continuous spectrum. The line spectra emitted by a number of elements in the visible region are shown here in Fig. 38–19,

FIGURE 38–19 Emission spectra of the gases (a) atomic hydrogen, (b) helium, and (c) the *solar absorption* spectrum.

(a)

(b)

(c)

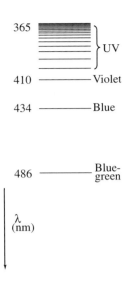

365 — UV

410 — Violet

434 — Blue

486 — Blue-green

λ (nm)

656 — Red

FIGURE 38–20 Balmer series of lines for hydrogen.

and also at the start of this chapter and in Chapter 35. The **emission spectrum** is characteristic of the material and can serve as a type of "fingerprint" for identification of the gas.

We also saw (Chapter 35) that if a continuous spectrum passes through a gas, dark lines are observed in the emerging spectrum, at wavelengths corresponding to lines normally emitted by the gas. This is called an **absorption spectrum** (Fig. 38–19c), and it became clear that gases can absorb light at the same frequencies at which they emit. Using film sensitive to ultraviolet and to infrared light, it was found that gases emit and absorb discrete frequencies in these regions as well as in the visible.

For our purposes here, the importance of the line spectra is that they are emitted (or absorbed) by gases at low pressure and low density. In such low density gases, the atoms are far apart on the average and hence the light emitted or absorbed is assumed to be by *individual atoms* rather than through interactions between atoms, as in a solid, liquid, or dense gas. Thus the line spectra serve as a key to the structure of the atom: any theory of atomic structure must be able to explain why atoms emit light only of discrete wavelengths, and it should be able to predict what these wavelengths are.

Hydrogen is the simplest atom—it has only one electron orbiting its nucleus. It also has the simplest spectrum. The spectrum of most atoms shows little apparent regularity. But the spacing between lines in the hydrogen spectrum decreases in a regular way, Fig. 38–20. Indeed, in 1885, J. J. Balmer (1825–1898) showed that the four visible lines in the hydrogen spectrum (with measured wavelengths 656 nm, 486 nm, 434 nm, and 410 nm) fit the following formula

$$\frac{1}{\lambda} = R\left(\frac{1}{2^2} - \frac{1}{n^2}\right), \qquad n = 3, 4, \cdots, \tag{38–8}$$

where n takes on the values 3, 4, 5, 6 for the four lines, and R, called the **Rydberg constant**, has the value $R = 1.0974 \times 10^7 \, \text{m}^{-1}$. Later it was found that this **Balmer series** of lines extended into the UV region, ending at $\lambda = 365 \, \text{nm}$, as shown in Fig. 38–20. Balmer's formula, Eq. 38–8, also worked for these lines with higher integer values of n. The lines near 365 nm became too close together to distinguish, but the limit of the series at 365 nm corresponds to $n = \infty$ (so $1/n^2 = 0$ in Eq. 38–8).

Later experiments on hydrogen showed that there were similar series of lines in the UV and IR regions, and each series had a pattern just like the Balmer series, but at different wavelengths, Fig. 38–21. Each of these series was found to fit a formula with the same form as Eq. 38–8 but with the $1/2^2$ replaced by $1/1^2$, $1/3^2$, $1/4^2$, and so on. For example, the so-called **Lyman series** contains lines with wavelengths from 91 nm to 122 nm (in the UV region) and fits the formula

$$\frac{1}{\lambda} = R\left(\frac{1}{1^2} - \frac{1}{n^2}\right), \qquad n = 2, 3, \cdots.$$

And the wavelengths of the **Paschen series** (in the IR region) fit

$$\frac{1}{\lambda} = R\left(\frac{1}{3^2} - \frac{1}{n^2}\right), \qquad n = 4, 5, \cdots.$$

FIGURE 38–21 Line spectrum of atomic hydrogen. Each series fits the formula $\frac{1}{\lambda} = R\left(\frac{1}{n'^2} - \frac{1}{n^2}\right)$ where $n' = 1$ for the Lyman series, $n' = 2$ for the Balmer series, $n' = 3$ for the Paschen series, and so on; n can take on all integer values from $n = n' + 1$ up to infinity. The only lines in the visible region of the electromagnetic spectrum are part of the Balmer series.

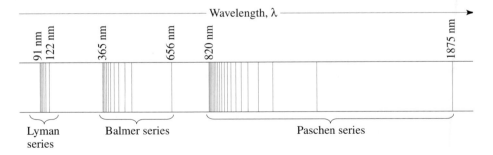

Wavelength, λ

91 nm 122 nm   365 nm   656 nm   820 nm   1875 nm

Lyman series   Balmer series   Paschen series

The Rutherford model, as it stood, was unable to explain why atoms emit line spectra. It had other difficulties as well. According to the Rutherford model, electrons orbit the nucleus, and since their paths are curved the electrons are accelerating. Hence they should give off light like any other accelerating electric charge (Chapter 32). Then, since energy is conserved, the electron's own energy must decrease to compensate. Hence electrons would be expected to spiral into the nucleus. As they spiraled inward, their frequency would increase in a short time and so too would the frequency of the light emitted. Thus the two main difficulties of the Rutherford model are these: (1) it predicts that light of a continuous range of frequencies will be emitted, whereas experiment shows line spectra; (2) it predicts that atoms are unstable—electrons should quickly spiral into the nucleus—but we know that atoms in general are stable, since the matter around us is stable.

Clearly Rutherford's model was not sufficient. Some sort of modification was needed, and it was Niels Bohr who provided it by adding an essential idea—the quantum hypothesis.

38–10 The Bohr Model

Bohr had studied in Rutherford's laboratory for several months in 1912 and was convinced that Rutherford's planetary model of the atom had validity. But in order to make it work, he felt that the newly developing quantum theory would somehow have to be incorporated in it. The work of Planck and Einstein had shown that in heated solids, the energy of oscillating electric charges must change discontinuously—from one discrete energy state to another, with the emission of a quantum of light. Perhaps, Bohr argued, the electrons in an atom also cannot lose energy continuously, but must do so in quantum "jumps." In working out his theory during the next year, Bohr postulated that electrons move about the nucleus in circular orbits, but that only certain orbits are allowed. He further postulated that an electron in each orbit would have a definite energy and would move in the orbit *without radiating energy* (even though this violated classical ideas since accelerating electric charges are supposed to emit EM waves; see Chapter 32). He thus called the possible orbits **stationary states**. Light is emitted, he hypothesized, only when an electron jumps from one stationary state to another of lower energy. When such a jump occurs, a single photon of light is emitted whose energy, since energy is conserved, is given by

Stationary state

$$hf = E_u - E_l, \qquad (38-9)$$

where E_u refers to the energy of the upper state and E_l the energy of the lower state. See Fig. 38–22.

Bohr set out to determine what energies these orbits would have, since the spectrum of light emitted could then be predicted from Eq. 38–9. In the Balmer formula he had the key he was looking for. Bohr quickly found that his theory would be in accord with the Balmer formula if he assumed that the electron's angular momentum L is quantized and equal to an integer n times $h/2\pi$. As we saw in Chapter 10, angular momentum is given by $L = I\omega$, where I is the moment of inertia and ω is the angular velocity. For a single particle of mass m moving in a circle of radius r with speed v, $I = mr^2$ and $\omega = v/r$; hence, $L = I\omega = (mr^2)(v/r) = mvr$. Bohr's **quantum condition** is

$$L = mvr_n = n\frac{h}{2\pi}, \qquad n = 1, 2, 3, \cdots, \qquad (38-10)$$

where n is an integer and r_n is the radius of the nth possible orbit. The allowed orbits are numbered 1, 2, 3, ..., according to the value of n, which is called the **quantum number** of the orbit.

FIGURE 38–22 An atom emits a photon (energy = hf) when its energy changes from E_u to a lower energy E_l.

Angular momentum quantized

Quantum number, n

Equation 38–10 did not have a firm theoretical foundation. Bohr had searched for some "quantum condition," and such tries as $E = hf$ (where E represents the energy of the electron in an orbit) did not give results in accord with experiment. Bohr's reason for using Eq. 38–10 was simply that it worked; and we now look at how. In particular, let us determine what the Bohr theory predicts for the measurable wavelengths of emitted light.

An electron in a circular orbit of radius r_n (Fig. 38–23) would have a centripetal acceleration v^2/r_n produced by the electrical force of attraction between the negative electron and the positive nucleus. This force is given by Coulomb's law,

$$F = \frac{1}{4\pi\epsilon_0}\frac{(Ze)(e)}{r^2}.$$

The charge on the nucleus is $+Ze$, where Z is the number of positive charges[†] (i.e., protons). For the hydrogen atom, $Z = +1$.

In Newton's second law, $F = ma$, we substitute $a = v^2/r_n$ and Coulomb's law for F, and obtain

$$F = ma$$

$$\frac{1}{4\pi\epsilon_0}\frac{Ze^2}{r_n^2} = \frac{mv^2}{r_n}.$$

We solve this for r_n, and then substitute for v from Eq. 38–10 (which says $v = nh/2\pi m r_n$):

$$r_n = \frac{Ze^2}{4\pi\epsilon_0 mv^2} = \frac{Ze^2 4\pi^2 m r_n^2}{4\pi\epsilon_0 n^2 h^2}.$$

We solve for r_n (it appears on both sides, so we cancel one of them) and find

$$r_n = \frac{n^2 h^2 \epsilon_0}{\pi m Z e^2} = \frac{n^2}{Z}r_1 \qquad \text{(38–11)}$$

where

$$r_1 = \frac{h^2 \epsilon_0}{\pi m e^2}. \qquad \text{(38–12a)}$$

Equation 38–11 gives the radii of all possible orbits. The smallest orbit is for $n = 1$, and for hydrogen ($Z = 1$) has the value

$$r_1 = \frac{(6.626 \times 10^{-34}\,\text{J}\cdot\text{s})^2(8.85 \times 10^{-12}\,\text{C}^2/\text{N}\cdot\text{m}^2)}{(3.14)(9.11 \times 10^{-31}\,\text{kg})(1.602 \times 10^{-19}\,\text{C})^2}$$

or

$$r_1 = 0.529 \times 10^{-10}\,\text{m}. \qquad \text{(38–12b)}$$

The radius of the smallest orbit in hydrogen, r_1, is sometimes called the **Bohr radius**. From Eq. 38–11, we see that the radii of the larger orbits increase as n^2, so

$$r_2 = 4r_1 = 2.12 \times 10^{-10}\,\text{m},$$

$$r_3 = 9r_1 = 4.76 \times 10^{-10}\,\text{m},$$

$$\vdots$$

$$r_n = n^2 r_1.$$

The first four orbits are shown in Fig. 38–24. Notice that, according to Bohr's model, an electron can exist only in the orbits given by Eq. 38–11. There are no allowable orbits in between.

FIGURE 38–23 Electric force (Coulomb's law) keeps the negative electron in orbit around the positively charged nucleus.

Bohr radius

FIGURE 38–24 Possible orbits in the Bohr model of hydrogen; $r_1 = 0.529 \times 10^{-10}$ m.

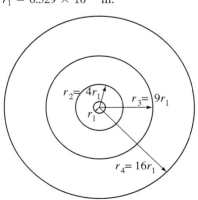

[†] We include Z in our derivation so that we can treat other single-electron ("hydrogenlike") atoms such as the ions He$^+$ ($Z = 2$) and Li^{2+} ($Z = 3$). Helium in the neutral state has two electrons; if one electron is missing, the remaining He$^+$ ion consists of one electron revolving around a nucleus of charge $+2e$. Similarly, doubly ionized lithium, Li^{2+}, also has a single electron, and in this case $Z = 3$.

In each of its possible orbits, the electron would have a definite energy, as the following calculation shows. The total energy equals the sum of the kinetic and potential energies. The potential energy of the electron is given by $U = qV = -eV$, where V is the potential due to a point charge $+Ze$ as given by Eq. 23–5: $V = (1/4\pi\epsilon_0)(Q/r) = (1/4\pi\epsilon_0)(Ze/r)$. So

$$U = -eV = -\frac{1}{4\pi\epsilon_0}\frac{Ze^2}{r}.$$

The total energy E_n for an electron in the nth orbit of radius r_n is the sum of the kinetic and potential energies:

$$E_n = \tfrac{1}{2}mv^2 - \frac{1}{4\pi\epsilon_0}\frac{Ze^2}{r_n}.$$

When we substitute v from Eq. 38–10 and r_n from Eq. 38–11 into this equation, we obtain

$$E_n = -\left(\frac{Z^2e^4m}{8\epsilon_0^2h^2}\right)\left(\frac{1}{n^2}\right) \qquad n = 1, 2, 3, \cdots \qquad \textbf{(38–13)} \qquad \textit{Energy levels}$$

or

$$E_n = \frac{Z^2}{n^2}E_1 \qquad n = 1, 2, 3, \cdots$$

where E_1 is the lowest energy level ($n = 1$) for hydrogen ($Z = 1$). The value of E_1 is

$$E_1 = -\frac{me^4}{8\epsilon_0^2h^2} = -2.17 \times 10^{-18}\,\text{J} = -13.6\,\text{eV}, \qquad \textit{Ground state of hydrogen}$$

where we have converted joules to electron volts, as is customary in atomic physics. Since n^2 appears in the denominator of Eq. 38–13, the energies of the larger orbits in hydrogen ($Z = 1$) are given by

$$E_n = \frac{-13.6\,\text{eV}}{n^2}.$$

For example,

$$E_2 = \frac{-13.6\,\text{eV}}{4} = -3.40\,\text{eV}, \qquad \textit{Excited states (first two)}$$

$$E_3 = \frac{-13.6\,\text{eV}}{9} = -1.51\,\text{eV}.$$

We see that not only are the orbit radii quantized, but from Eq. 38–13 so is the energy. The quantum number n that labels the orbit radii also labels the energy levels. The lowest **energy level** or **energy state** has energy E_1, and is called the **ground state**. The higher states, E_2, E_3, and so on, are called **excited states**.

Notice that although the energy for the larger orbits has a smaller numerical value, all the energies are less than zero. Thus, $-3.4\,\text{eV}$ is a higher energy than $-13.6\,\text{eV}$. Hence the orbit closest to the nucleus (r_1) has the lowest energy. The reason the energies have negative values has to do with the way we defined the zero for potential energy (U). For two point charges, $U = kq_1q_2/r$ corresponds to zero potential energy when the two charges are infinitely far apart. Thus, an electron that can just barely be free from the atom by reaching $r = \infty$ (or, at least, far from the nucleus) with zero kinetic energy will have $E = 0$, corresponding to $n = \infty$ in Eq. 38–13. If an electron is free and has kinetic energy, then $E > 0$. To remove an electron that is part of an atom requires an energy input (otherwise atoms would not be stable). Since $E \geq 0$ for a free electron, then an electron bound to an atom must have $E < 0$. That is, energy must be added to bring its energy up, from a negative value, to at least zero in order to free it.

The minimum energy required to remove an electron from the ground state of an atom is called the **binding energy** or **ionization energy**. The ionization energy for hydrogen has been measured to be 13.6 eV, and this corresponds precisely to removing an electron from the lowest state, $E_1 = -13.6$ eV, up to $E = 0$ where it can be free.

It is useful to show the various possible energy values as horizontal lines on an energy-level diagram.[†] This is shown for hydrogen in Fig. 38–25. The electron in a hydrogen atom can be in any one of these levels according to Bohr theory. But it could never be in between, say at -9.0 eV. At room temperature, nearly all H atoms will be in the ground state ($n = 1$). At higher temperatures, or during an electric discharge when there are many collisions between free electrons and atoms, many atoms can be in excited states ($n > 1$). Once in an excited state, an atom's electron can jump down to a lower state, and give off a photon in the process. This is, according to the Bohr model, the origin of the emission spectra of excited gases. The vertical arrows in Fig. 38–25 represent the transitions or jumps that correspond to the various observed spectral lines. For example, an electron jumping from the level $n = 3$ to $n = 2$ would give rise to the 656-nm line in the Balmer series, and the jump from $n = 4$ to $n = 2$ would give rise to the 486-nm line (see Fig. 38–20). We can predict wavelengths of the spectral lines emitted by combining Eq. 38–9 with Eq. 38–13. Since $hf = hc/\lambda$, we have from Eq. 38–9

$$\frac{1}{\lambda} = \frac{hf}{hc} = \frac{1}{hc}(E_n - E_{n'}),$$

and then using Eq. 38–13,

$$\frac{1}{\lambda} = \frac{Z^2 e^4 m}{8\epsilon_0^2 h^3 c}\left(\frac{1}{(n')^2} - \frac{1}{(n)^2}\right), \tag{38–14}$$

where n refers to the upper state and n' to the lower state. This theoretical formula

[†]Note that above $E = 0$, an electron is free and can have any energy (E is not quantized). Thus there is a continuum of energy states above $E = 0$, as indicated in the energy-level diagram of Fig. 38–25.

Binding energy
Ionization energy

Line spectra
emission
explained

FIGURE 38–25 Energy-level diagram for the hydrogen atom, showing the origin of spectral lines for the Lyman, Balmer, and Paschen series.

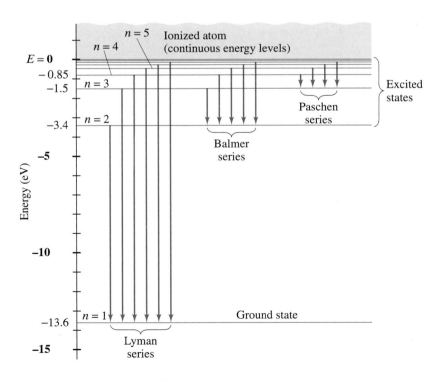

has the same form as the experimental Balmer formula, Eq. 38–8, with $n' = 2$. Thus we see that the Balmer series of lines corresponds to transitions or "jumps" that bring the electron down to the second energy level. Similarly, $n' = 1$ corresponds to the Lyman series and $n' = 3$ to the Paschen series (see Fig. 38–25). When the constant $(Z^2 e^4 m / 8\epsilon_0^2 h^3 c)$ in Eq. 38–14 is evaluated with $Z = 1$, it is found to have the measured value of the Rydberg constant, $R = 1.0974 \times 10^7 \, \text{m}^{-1}$ in Eq. 38–8, in accord with experiment (see Problem 47).

The great success of Bohr's theory is that it explains why atoms emit line spectra and accurately predicts, for hydrogen, the wavelengths of emitted light. The Bohr theory also explains absorption spectra: photons of just the right wavelength can knock an electron from one energy level to a higher one. To conserve energy, only photons that have just the right energy (and frequency) will be absorbed. This explains why a continuous spectrum of light entering a gas will emerge with dark (absorption) lines at frequencies that correspond to emission lines.

Absorption lines explained

The Bohr theory also ensures the stability of atoms. It establishes stability by fiat: the ground state is the lowest state for an electron and there is no lower energy level to which it can go and emit more energy. Finally, as we saw above, the Bohr theory accurately predicts the ionization energy of 13.6 eV for hydrogen. However, the Bohr theory was not so successful for other atoms, as we shall see.

EXAMPLE 38–12 Wavelength of a Lyman line. Use Fig. 38–25 to determine the wavelength of the first Lyman line, the transition from $n = 2$ to $n = 1$. In what region of the electromagnetic spectrum does this lie?

SOLUTION In this case, $hf = E_2 - E_1 = 13.6 \, \text{eV} - 3.4 \, \text{eV} = 10.2 \, \text{eV} = 1.63 \times 10^{-18} \, \text{J}$. Since $\lambda = c/f$, we have

$$\lambda = \frac{c}{f} = \frac{hc}{E_2 - E_1} = \frac{(6.63 \times 10^{-34} \, \text{J} \cdot \text{s})(3.00 \times 10^8 \, \text{m/s})}{1.63 \times 10^{-18} \, \text{J}} = 1.22 \times 10^{-7} \, \text{m},$$

or 122 nm, which is in the UV region. See Fig. 38–21.

EXAMPLE 38–13 Wavelength of a Balmer line. Determine the wavelength of light emitted when a hydrogen atom makes a transition from the $n = 6$ to the $n = 2$ energy level according to the Bohr model.

SOLUTION We can use Eq. 38–14 or its equivalent, Eq. 38–8, with $R = 1.097 \times 10^7 \, \text{m}^{-1}$. Thus

$$\frac{1}{\lambda} = (1.097 \times 10^7 \, \text{m}^{-1})\left(\frac{1}{4} - \frac{1}{36}\right) = 2.44 \times 10^6 \, \text{m}^{-1}.$$

So $\lambda = 1/(2.44 \times 10^6 \, \text{m}^{-1}) = 4.10 \times 10^{-7} \, \text{m}$ or 410 nm. This is the fourth line in the Balmer series, Fig. 38–20, and is violet in color.

EXAMPLE 38–14 Absorption wavelength. Use Fig. 38–25 to determine the maximum wavelength that hydrogen in its ground state can absorb. What would be the next smaller wavelength that would work?

SOLUTION Maximum λ corresponds to minimal energy, and thus the jump from the ground state to the first excited state (Fig. 38–25) for which the energy is $13.6 \, \text{eV} - 3.4 \, \text{eV} = 10.2 \, \text{eV}$; the required wavelength, as we saw in Example 38–12, is 122 nm. The next possibility is to jump from the ground state to the second excited state, which requires $13.6 \, \text{eV} - 1.5 \, \text{eV} = 12.1 \, \text{eV}$ and corresponds to a wavelength

$$\lambda = \frac{c}{f} = \frac{hc}{E_3 - E_1}$$

$$= \frac{(6.63 \times 10^{-34} \, \text{J} \cdot \text{s})(3.00 \times 10^8 \, \text{m/s})}{(12.1 \, \text{eV})(1.60 \times 10^{-19} \, \text{J/eV})} = 103 \, \text{nm}.$$

EXAMPLE 38–15 **Ionization energy.** Use the Bohr model to determine the ionization energy of the He$^+$ ion, which has a single electron. Also calculate the minimum wavelength a photon must have to cause ionization.

SOLUTION We want to determine the minimum energy required to lift the electron from its ground state and to barely reach the free state at $E = 0$. The ground state energy of He$^+$ is given by Eq. 38–13 with $n = 1$ and $Z = 2$. Since all the symbols in Eq. 38–13 are the same as for the calculation for hydrogen, except that Z is 2 instead of 1, we see that E_1 will be $Z^2 = 2^2 = 4$ times the E_1 for hydrogen. That is,

$$E_1 = 4(-13.6 \text{ eV}) = -54.4 \text{ eV}.$$

Thus, to ionize the He$^+$ ion should require 54.4 eV, and this value agrees with experiment. The minimum wavelength photon that can cause ionization will have energy $hf = 54.4 \text{ eV}$ and wavelength $\lambda = c/f = hc/hf = (6.63 \times 10^{-34} \text{ J·s})(3.00 \times 10^8 \text{ m/s})/(54.4 \text{ eV})(1.60 \times 10^{-19} \text{ J/eV}) = 22.8 \text{ nm}$. If the atom absorbed a photon of greater energy (wavelength shorter than 22.8 nm), the atom could still be ionized and the freed electron would have kinetic energy of its own.

In this last Example, we saw that E_1 for the He$^+$ ion is four times more negative than that for hydrogen. Indeed the energy-level diagram for He$^+$ looks just like that for hydrogen, Fig. 38–25, except that the numerical values for each energy level are four times larger. It is important to note, however, that we are talking here about the He$^+$ *ion*. Normal (neutral) helium has two electrons and its energy level diagram is entirely different.

CONCEPTUAL EXAMPLE 38–16 **Hydrogen at 20°C.** Estimate the average kinetic energy of hydrogen atoms (or molecules) at room temperature, and use the result to explain why nearly all H atoms are in the ground state at room temperature, and hence emit no light.

RESPONSE According to kinetic theory (Chapter 18), the average kinetic energy of atoms or molecules in a gas is given by Eq. 18–4,

$$\overline{K} = \tfrac{3}{2}kT,$$

where $k = 1.38 \times 10^{-23} \text{ J/K}$ is Boltzmann's constant, and T is the kelvin (absolute) temperature. Room temperature is about $T = 300 \text{ K}$, so

$$\overline{K} = \tfrac{3}{2}(1.38 \times 10^{-23} \text{ J/K})(300 \text{ K}) = 6.2 \times 10^{-21} \text{ J},$$

or, in electron volts:

$$\overline{K} = \frac{6.2 \times 10^{-21} \text{ J}}{1.6 \times 10^{-19} \text{ J/eV}} = 0.04 \text{ eV}.$$

The average kinetic energy is thus very small compared to the energy between the ground state and the next higher energy state ($13.6 \text{ eV} - 3.4 \text{ eV} = 10.2 \text{ eV}$). Any atoms in excited states emit light and eventually fall to the ground state. Once in the ground state, collisions with other atoms can transfer energy of only 0.04 eV on the average. A small fraction of atoms can have much more energy (see Section 18–2 on the distribution of molecular speeds), but even kinetic energy that is 10 times the average is not nearly enough to excite atoms above the ground state. Thus, at room temperature, nearly all atoms are in the ground state. Atoms can be excited to upper states by very high temperatures, or by passing a current of high energy electrons through the gas, as in a discharge tube (Fig. 38–18).

We should note that Bohr made some radical assumptions that were at variance with classical ideas. He assumed that electrons in fixed orbits do not radiate light even though they are accelerating (moving in a circle), and he assumed that angular momentum is quantized. Furthermore, he was not able to say how an electron moved when it made a transition from one energy level to another. On the other hand, there is no real reason to expect that in the tiny world of the atom electrons would behave as ordinary-sized objects do. Nonetheless, he felt that where quantum theory overlaps with the macroscopic world, it should predict classical results. This is the **correspondence principle**, already mentioned in regard to relativity (Section 37–13). This principle does work for Bohr's theory of the hydrogen atom. The orbit sizes and energies are quite different for $n = 1$ and $n = 2$, say. But orbits with $n = 100,000,000$ and $100,000,001$ would be very close in size and energy (see Fig. 38–25). Indeed, jumps between such large orbits, which would approach macroscopic sizes, would be imperceptible. Such orbits would thus appear to be continuously spaced, which is what we expect in the everyday world.

Correspondence principle

Finally, be careful not to believe that the well-defined orbits of the Bohr model actually exist. The Bohr model is only a model, not reality. The idea of electron orbits was rejected a few years later, and today electrons are better thought of as forming "clouds."

38–11 de Broglie's Hypothesis Applied to Atoms

Bohr's theory was largely of an *ad hoc* nature. Assumptions were made so that theory would agree with experiment. But Bohr could give no reason why the orbits were quantized, nor why there should be a stable ground state. Finally, ten years later, a reason was proposed by de Broglie.

We saw in Section 38–6 that in 1923, Louis de Broglie proposed that material particles, such as electrons, have a wave nature; and that this hypothesis was confirmed by experiment several years later.

One of de Broglie's original arguments in favor of the wave nature of electrons was that it provided an explanation for Bohr's theory of the hydrogen atom. According to de Broglie, a particle of mass m moving with a nonrelativistic speed v would have a wavelength (Eq. 38–7) of

$$\lambda = \frac{h}{mv}.$$

Each electron orbit in an atom, he proposed, is actually a standing wave. As we saw in Chapter 15, when a violin or guitar string is plucked, a vast number of wavelengths are excited. But only certain ones—those that have nodes at the ends—are sustained. These are the *resonant* modes of the string. Waves with other wavelengths interfere with themselves upon reflection and their amplitudes quickly drop to zero. With electrons moving in circles, according to Bohr's theory, de Broglie argued that the electron wave was a *circular* standing wave that closes on itself, Fig. 38–26. If the wavelength of a wave does not close on itself, as in Fig. 38–27, destructive interference takes place as the wave travels around the loop, and the wave quickly dies out. Thus, the only waves that persist are those for which the circumference of the circular orbit contains a whole number of wavelengths, Fig. 38–28. The circumference of a Bohr orbit of radius r_n is $2\pi r_n$, so we have

$$2\pi r_n = n\lambda, \qquad n = 1, 2, 3, \cdots.$$

When we substitute $\lambda = h/mv$, we get $2\pi r_n = nh/mv$, or

$$mvr_n = \frac{nh}{2\pi}.$$

This is just the *quantum condition* proposed by Bohr on an *ad hoc* basis, Eq. 38–10.

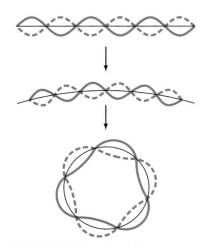

FIGURE 38–26 An ordinary standing wave compared to a circular standing wave.

FIGURE 38–27 When a wave does not close (and hence interferes destructively with itself), it rapidly dies out.

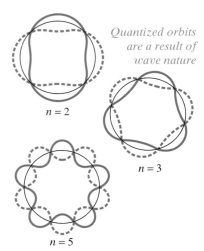

Quantized orbits are a result of wave nature

$n = 2$

$n = 3$

$n = 5$

FIGURE 38–28 Standing circular waves for two, three, and five wavelengths on the circumference; n, the number of wavelengths, is also the quantum number.

And it is from this equation that the discrete orbits and energy levels were derived. Thus we have an explanation for the quantized orbits and energy states in the Bohr model: they are due to the wave nature of the electron, and that only resonant "standing" waves can persist. This implies that the *wave–particle duality* is at the root of atomic structure.

It should be noted in viewing the circular electron waves of Fig. 38–28 that the electron is not to be thought of as following the oscillating wave pattern. In the Bohr model of hydrogen, the electron, considered as a particle, moves in a circle. The circular wave, on the other hand, represents the *amplitude* of the electron "matter wave," and in Fig. 38–28 the wave amplitude is shown superimposed on the circular path of the particle orbit for convenience.

Bohr's theory worked well for hydrogen and for one-electron ions. But it did not prove as successful for multielectron atoms. Bohr theory could not predict line spectra even for the next simplest atom, helium. It could not explain why some emission lines are brighter than others, nor why some lines are actually split into two or more closely spaced lines ("fine structure"). A new theory was needed and was indeed developed in the 1920s. This new and radical theory is called *quantum mechanics*, and finally solved the problem of atomic structure; but it gives us a very different view of the atom: the idea of electrons in well-defined orbits was replaced with the idea of electron "clouds." And this new theory of quantum mechanics has given us a wholly different view of the basic mechanisms underlying physical processes.

Summary

Quantum theory has its origins in **Planck's quantum hypothesis** that molecular oscillations are quantized: their energy E can only be integer (n) multiples of hf, where h is Planck's constant and f is the natural frequency of oscillation: $E = nhf$. This hypothesis explained the spectrum of radiation emitted by (black) bodies at high temperature.

Einstein proposed that for some experiments, light could be pictured as being emitted and absorbed as quanta (particles), which we now call **photons**, each with energy

$$E = hf.$$

He proposed the photoelectric effect as a test for the photon theory of light. In the **photoelectric effect**, the photon theory says that each incident photon can strike an electron in a material and eject it if it has sufficient energy. The maximum energy of ejected electrons is then linearly related to the frequency of the incident light. The photon theory is also supported by the **Compton effect** and the observation of electron–positron **pair production**.

The wave–particle duality refers to the idea that light and matter (such as electrons) have both wave and particle properties. The wavelength of a material object is given by

$$\lambda = \frac{h}{mv},$$

where mv is the momentum of the object. The **principle of complementarity** states that we must be aware of both the particle and wave properties of light and of matter for a complete understanding of them.

Early models of the atom include the plum-pudding model and Rutherford's planetary (or nuclear) model. Rutherford's model, which was created to explain the backscattering of

alpha particles from thin metal foils, assumes that an atom consists of a tiny but massive positively charged nucleus surrounded (at a relatively great distance) by electrons.

To explain the line spectra emitted by atoms, as well as the stability of atoms, **Bohr theory** postulated (1) that electrons bound in an atom can only occupy orbits for which the angular momentum is quantized, which results in discrete values for the radius and energy; (2) that an electron in such a **stationary state** emits no radiation; (3) that, if an electron jumps to a lower state, it emits a photon whose energy equals the difference in energy between the two states; (4) that the angular momentum L of atomic electrons is quantized by the rule

$$L = \frac{nh}{2\pi},$$

where n is an integer called a **quantum number**. The $n = 1$ state in hydrogen is the **ground state**, which has an energy $E_1 = -13.6\,\text{eV}$; higher values of n correspond to **excited states** and their energies are

$$E_n = -\frac{13.6\,\text{eV}}{n^2}.$$

Atoms are excited to these higher states by collisions with other atoms or electrons or by absorption of a photon of just the right frequency.

De Broglie's hypothesis that electrons (and other matter) have a wavelength $\lambda = h/mv$ gave an explanation for Bohr's quantized orbits by bringing in the wave–particle duality: the orbits correspond to circular standing waves in which the circumference of the orbit equals a whole number of wavelengths.

Questions

1. What can be said about the relative temperature of whitish-yellow, reddish, and bluish stars?

2. If energy is radiated by all objects, why can we not see them in the dark? (See also Section 19–10.)

3. Does a lightbulb at a temperature of 2500 K produce as white a light as the Sun at 6000 K? Explain.

4. An ideal black body can be approximated by a small hole in an otherwise enclosed cavity. Explain. [*Hint*: The pupil of your eye is an approximate case.]

5. Darkrooms for developing black-and-white film are sometimes lit by a red bulb. Why red? Would such a bulb work in a darkroom for developing color photographs?

6. If the threshold wavelength in the photoelectric effect increases when the emitting metal is changed to a different metal, what can you say about the work functions of the two metals?

7. Explain why the existence of a cutoff frequency in the photoelectric effect more strongly favors a particle theory rather than a wave theory of light.

8. UV light causes sunburn, whereas visible light does not. Explain.

9. If an X-ray photon is scattered by an electron, does its wavelength change? If so, does it increase or decrease?

10. In both the photoelectric effect and in the Compton effect we have a photon colliding with an electron causing the electron to fly off. What then, is the difference between the two processes?

11. Show that the units of Planck's constant are those for angular momentum.

12. Consider a point source of light. How would the intensity of light vary with distance from the source according to (*a*) wave theory, (*b*) particle (photon) theory? Would this help to distinguish the two theories?

13. Explain how the photoelectric circuit of Fig. 38–4 could be used in (*a*) a burglar alarm, (*b*) a smoke detector, (*c*) a photographic light meter, and (*d*) a spectrophotometer (see Section 36–8).

14. Why do we say that light has wave properties? Why do we say that light has particle properties?

15. Why do we say that electrons have wave properties? Why do we say that electrons have particle properties?

16. What is the difference between a photon and an electron? Be specific: make a list.

17. If an electron and a proton travel at the same speed, which has the shorter wavelength?

18. In Rutherford's planetary model of the atom, what keeps the electrons from flying off into space?

19. Which of the following can emit a line spectrum: (*a*) gases, (*b*) liquids, (*c*) solids? Which can emit a continuous spectrum?

20. Why doesn't the O_2 gas in the air around us give off light?

21. How can you tell if there is oxygen near the surface of the Sun?

22. When a wide spectrum of light passes through hydrogen gas at room temperature, absorption lines are observed that correspond only to the Lyman series. Why don't we observe the other series?

23. Explain how the closely spaced energy levels for hydrogen near the top of Fig. 38–25 correspond to the closely spaced spectral lines at the top of Fig. 38–20.

24. Discuss the differences between Rutherford's and Bohr's theory of the atom.

25. Is it possible for the de Broglie wavelength of a "particle" to be greater than the dimensions of the particle? To be smaller? Is there any direct connection?

26. Approximately how large would h have to be so that quantum effects would be noticeable in the macroscopic world of everyday life?

27. Does it make sense that the potential energy of the electron in a hydrogen atom is negative and greater in magnitude than the kinetic energy, as given by the Bohr theory? What is the significance of this?

28. In a helium atom, which contains two electrons, do you think that on the average the electrons are closer to the nucleus or farther away than in a hydrogen atom? Why?

29. How can the spectrum of hydrogen contain so many lines when hydrogen contains only one electron?

30. The Lyman series is brighter than the Balmer series, because this series of transitions ends up in the most common state for hydrogen, the ground state. Why then was the Balmer series discovered first?

31. Use conservation of momentum to explain why photons emitted by hydrogen atoms have slightly less energy than that predicted by Eq. 38–9

Problems

Section 38–1

1. (I) How hot is a metal being welded if it radiates most strongly at 440 nm?

2. (I) (*a*) What is the temperature if the peak of a blackbody spectrum is at 25.0 nm? (*b*) What is the wavelength at the peak of a blackbody spectrum if the body is at a temperature of 2800 K?

3. (I) An HCl molecule vibrates with a natural frequency of 8.1×10^{13} Hz. What is the difference in energy (in joules and electron volts) between possible values of the oscillation energy?

4. (I) Estimate the peak wavelength for radiation from (*a*) ice at $0°$ C, (*b*) a floodlamp at 3300 K, (*c*) helium at 4 K, assuming blackbody emission. In what region of the EM spectrum is each?

5. (II) The steps of a flight of stairs are 20.0 cm high (vertically). If a 58.0-kg person stands with both feet on the same step, what is the gravitational potential energy of this person, relative to the ground, on (*a*) the first step, (*b*) the second step, (*c*) the third step, (*d*) the nth step? (*e*) What is the change in energy as the person descends from step 6 to step 2?

6. (II) Estimate the peak wavelength of light issuing from the pupil of the human eye (which approximates a blackbody) assuming normal body temperature.

7. (III) Planck's radiation law is given by:

$$I(\lambda, T) = \frac{2\pi hc^2 \lambda^{-5}}{e^{hc/\lambda kT} - 1}$$

where $I(\lambda, T)$ is the rate energy is radiated per unit surface area per unit wavelength interval at wavelength λ and Kelvin temperature T. (a) Show that Wien's displacement law follows from this relationship. (b) Determine the value of h from the experimental value of $\lambda_P T$ given in the text. (c) Derive the Stefan-Boltzmann law (the T^4 dependence of the rate at which energy is radiated—Eq. 19–15), by integrating Planck's formula over all wavelengths; that is, show that

$$\int I(\lambda, T) \, d\lambda \propto T^4.$$

Section 38–2

8. (I) What is the energy range in eV, of photons in the visible spectrum, of wavelength 400 nm to 750 nm?

9. (I) What is the energy of photons (in eV) emitted by an 88.5-MHz FM radio station?

10. (I) A typical gamma ray emitted from a nucleus during radioactive decay may have an energy of 300 keV. What is its wavelength? Would we expect significant diffraction of this type of light when it passes through an everyday opening, like a door?

11. (I) About 0.1 eV is required to break a "hydrogen bond" in a protein molecule. What are the minimum frequency and maximum wavelength of a photon that can accomplish this?

12. (I) What minimum frequency of light is needed to eject electrons from a metal whose work function is 4.3×10^{-19} J?

13. (II) What is the longest wavelength of light that will emit electrons from a metal whose work function is 3.10 eV?

14. (II) The work functions for sodium, cesium, copper and iron are 2.3, 2.1, 4.7 and 4.5 eV respectively. Which of these metals will not emit electrons when visible light shines on it?

15. (II) In a photoelectric-effect experiment it is observed that no current flows unless the wavelength is less than 570 nm. (a) What is the work function of this material? (b) What is the stopping voltage required if light of wavelength 400 nm is used?

16. (II) What is the maximum kinetic energy of electrons ejected from barium $(W_0 = 2.48 \text{ eV})$ when illuminated by white light, $\lambda = 400$ to 750 nm?

17. (II) When UV light of wavelength 255 nm falls on a metal surface, the maximum kinetic energy of emitted electrons is 1.40 eV. What is the work function of the metal?

18. (II) The threshold wavelength for emission of electrons from a given surface is 350 nm. What will be the maximum kinetic energy of ejected electrons when the wavelength is changed to (a) 280 nm, (b) 380 nm?

19. (II) A certain type of film is sensitive only to light whose wavelength is less than 660 nm. What is the energy (eV and kcal/mol) needed for the chemical reaction to occur which causes the film to change?

20. (II) When 230-nm light falls on a metal, the current through a photoelectric circuit (Fig. 38–4) is brought to zero at a reverse voltage of 1.64 V. What is the work function of the metal?

Section 38–3

21. (II) The quantity h/mc, which has the dimensions of length, is called the *Compton wavelength*. Determine the Compton wavelength for (a) an electron, (b) a proton. (c) Show that if a photon has wavelength equal to the Compton wavelength of a particle, the photon's energy is equal to the rest energy of the particle.

22. (II) X-rays of wavelength $\lambda = 0.120 \text{ nm}$ are scattered from a carbon block. What is the Compton wavelength shift for photons detected at angles (relative to the incident beam) of (a) 45°, (b) 90°, (c) 180°?

23. (II) For each of the scattering angles in the previous Problem, determine (a) the fractional energy loss of the photon, and (b) the energy given to the scattered electron.

24. (II) (a) Show that the fractional energy loss of scattered photons in the Compton effect can be written as $\Delta\lambda/\lambda$. (b) Is this an exact expression? If not, specify what quantity limits the accuracy and what its value can be so this formula is accurate to 0.1 percent.

25. (III) In the Compton effect (see Fig. 38–7), use the relativistic equations for conservation of energy and of linear momentum to show that the Compton shift in wavelength is given by Eq. 38–6.

26. (III) In the Compton effect, a 0.100-nm photon strikes a free electron in a head-on collision and knocks it into the forward direction. The rebounding photon recoils directly backward. Use conservation of (relativistic) energy and momentum to determine (a) the kinetic energy of the electron, and (b) the wavelength of the recoiling photon. (Note: Use Eq. 38–5, but not Eq. 38–6.)

Section 38–4

27. (I) How much total kinetic energy will an electron–positron pair have if produced by a 2.84-MeV photon?

28. (II) What is the longest wavelength photon that could produce a proton–antiproton pair? (Each has a mass of $1.67 \times 10^{-27} \text{ kg}$.)

29. (II) What is the minimum photon energy needed to produce a $\mu^+ - \mu^-$ pair? The mass of each μ (muon) is 207 times the mass of the electron. What is the wavelength of such a photon?

30. (II) A gamma-ray photon produces an electron–positron pair, each with a kinetic energy of 345 keV. What was the energy and wavelength of the photon?

Section 38–6

31. (I) Calculate the wavelength of a 0.21-kg ball traveling at 0.10 m/s.

32. (I) What is the wavelength of a neutron $(m = 1.67 \times 10^{-27} \text{ kg})$ traveling at 5.5×10^4 m/s?

33. (I) Through how many volts of potential difference must an electron be accelerated to achieve a wavelength of 0.28 nm?

34. (II) Calculate the de Broglie wavelength of an electron in your TV picture tube if it is accelerated by 30,000 V. Is it relativistic? How does its wavelength compare to the size of the "neck" of the tube, typically 5 cm? Do we have to worry about diffraction problems blurring our picture on the screen?

35. (II) What is the wavelength of an electron of energy (a) 10 eV, (b) 100 eV, (c) 1.0 keV?

36. (II) Show that if an electron and a proton have the same nonrelativistic kinetic energy, the proton has the shorter wavelength.

37. (II) Calculate the ratio of the kinetic energy of an electron to that of a proton if their wavelengths are equal. Assume that the speeds are non-relativistic.

38. (II) What is the wavelength of an O_2 molecule in the air at room temperature? [Hint: See Chapter 18.]

39. (III) A Cadillac with a mass of 2000 kg approaches a freeway underpass that is 10 m across. At what speed must the car be moving, in order for it to have a wavelength such that it might somehow "diffract" after passing through this single "slit"? How do these conditions compare to normal freeway speeds of 30 m/s?

*Section 38–7

* 40. (II) What voltage is needed to produce electron wavelengths of 0.10 nm? (Assume that the electrons are nonrelativistic.)

* 41. (II) Electrons are accelerated by 2250 V in an electron microscope. What is the maximum possible resolution?

Sections 38–9 and 38–10

42. (I) For the three hydrogen transitions indicated below, with n being the initial state and n' being the final state, is the transition an absorption or an emission? Which is higher, the initial state energy or the final state energy of the atom? Finally, which of these transitions involves the largest energy photon? (a) $n = 1$, $n' = 3$ (b) $n = 6$, $n' = 2$ (c) $n = 4$, $n' = 5$.

43. (I) How much energy is needed to ionize a hydrogen atom in the $n = 2$ state?

44. (I) The third longest wavelength in the Paschen series in hydrogen (Fig. 38–25) corresponds to what transition?

45. (I) Calculate the ionization energy of doubly ionized lithium, Li^{2+}, which has $Z = 3$.

46. (I) (a) Determine the wavelength of the second Balmer line ($n = 4$ to $n = 2$ transition) using Fig. 38–25. Determine likewise (b) the wavelength of the second Lyman line and (c) the wavelength of the third Balmer line.

47. (I) Evaluate the Rydberg constant R using Bohr theory (compare Eqs. 38–8 and 38–14) and show that its value is $R = 1.0974 \times 10^7 \, m^{-1}$.

48. (II) What is the longest wavelength light capable of ionizing a hydrogen atom in the ground state?

49. (II) What wavelength photon would be required to ionize a hydrogen atom in the ground state and give the ejected electron a kinetic energy of 10.0 eV?

50. (II) In the Sun, an ionized helium (He^+) atom makes a transition from the $n = 6$ state to the $n = 2$ state, emitting a photon. Can that photon be absorbed by hydrogen atoms present in the Sun? If so, between what energy states will the hydrogen atom jump?

51. (II) Construct the energy-level diagram for the He^+ ion (see Fig. 38–25).

52. (II) Construct the energy-level diagram for doubly ionized lithium, Li^{2+}.

53. (II) What is the potential energy and the kinetic energy of an electron in the ground state of the hydrogen atom?

54. (II) An excited hydrogen atom could, in principle, have a radius of 1.00 mm. What would be the value of n for a Bohr orbit of this size? What would its energy be?

55. (II) Is the use of nonrelativistic formulas justified in the Bohr atom? To check, calculate the electron's velocity, v, in terms of c for the ground state of hydrogen, and then calculate $\sqrt{1 - v^2/c^2}$.

56. (II) Suppose an electron was bound to a proton, as in the hydrogen atom, by the gravitational force rather than by the electric force. What would be the radius, and energy, of the first Bohr orbit?

57. (II) Show that the magnitude of the potential energy of an electron in any Bohr orbit of a hydrogen atom is twice the magnitude of its kinetic energy in that orbit.

58. (III) Correspondence principle: Show that for large values of n, the difference in radius Δr between two adjacent orbits (with quantum numbers n and $n - 1$) is given by

$$\Delta r = r_n - r_{n-1} \approx \frac{2r_n}{n},$$

so $\Delta r/r_n \to 0$ as $n \to \infty$, in accordance with the correspondence principle. [Note that we can check the correspondence principle by either considering large values of $n (n \to \infty)$ or by letting $h \to 0$. Are these equivalent?]

59. (III) (a) For very large values of n, show that when an electron jumps from the level n to the level $n - 1$, the frequency of the light emitted is equal to

$$f = \frac{v}{2\pi r_n}.$$

(b) Show that this is also just the frequency predicted by classical theory for an electron revolving in a circular orbit of radius r_n with speed v. (c) Explain why this is consistent with the correspondence principle.

General Problems

60. The Big Bang theory states that the beginning of the Universe was accompanied by a huge burst of photons. Those photons are still present today and make up the so called Cosmic Microwave Background Radiation. The Universe radiates like a blackbody with a temperature of about 2.7 K. Calculate the peak wavelength of this radiation.

61. At low temperatures, nearly all the atoms in hydrogen gas will be in the ground state. What minimum frequency photon is needed if the photoelectric effect is to be observed?

62. A beam of 85-eV electrons is scattered from a crystal, as in X-ray diffraction, and a first-order peak is observed at $\theta = 38°$. What is the spacing between planes in the diffracting crystal? (See Section 36–10).

63. Show that the energy E (in electron volts) of a photon whose wavelength is λ (meters) is given by
$$E = 1.24 \times 10^{-6}/\lambda.$$

64. A microwave oven produces electromagnetic radiation at $\lambda = 12.2$ cm and produces a power of 760 W. Calculate the number of microwave photons produced by the microwave oven each second.

65. Sunlight reaching the Earth's surface has an intensity of about 1000 W/m². Estimate how many photons per square meter per second this represents. Take the average wavelength to be 550 nm.

66. A beam of red laser light ($\lambda = 633$ nm) hits a black wall and is fully absorbed. If this exerts a total force $F = 5.5$ nN on the wall, how many photons per second are hitting the wall?

67. If a 100-W light bulb emits 3.0 percent of the input energy as visible light (average wavelength 550 nm) uniformly in all directions, estimate how many photons per second of visible light will strike the pupil (4.0 mm diameter) of the eye of an observer 1.0 km away.

68. An electron and a positron collide head on, annihilate, and create two 0.90 MeV photons traveling in opposite directions. What were the initial kinetic energies of electron and positron?

69. In Compton's original experiment he saw X-rays scattered where the wavelength shifted from 0.0711 nm to 0.0735 nm. Through what angle did these X-rays scatter?

70. By what potential difference must (a) a proton ($m = 1.67 \times 10^{-27}$ kg), and (b) an electron ($m = 9.11 \times 10^{-31}$ kg), be accelerated to have a wavelength $\lambda = 5.0 \times 10^{-12}$ m?

71. In some of Rutherford's experiments (Fig. 38–16) the α particles (mass $= 6.64 \times 10^{-27}$ kg) had a kinetic energy of 4.8 MeV. How close could they get to a gold nucleus (charge $= +79e$)? Ignore the recoil motion of the nucleus.

72. By what fraction does the mass of an H atom decrease when it makes an $n = 3$ to $n = 1$ transition?

73. For what maximum kinetic energy is a collision between an electron and a hydrogen atom in its ground state definitely elastic?

74. Using Bohr theory, derive an equation for the angular velocity ω, and frequency f, of an electron in a hydrogen atom. Determine (a) for the ground state and (b) for the first excited state ($n = 2$).

75. Calculate the ratio of the gravitational to electric force for the electron in a hydrogen atom. Can the gravitational force be safely ignored?

76. A child's swing has a natural frequency of 0.75 Hz. (a) What is the separation between possible energy values (in joules)? (b) If the swing reaches a vertical height of 45 cm above its lowest point and has a mass of 20 kg (including the child), what is the value of the quantum number n? (c) What is the fractional change in energy between levels whose quantum numbers are n (as just calculated) and $n + 1$? Would quantization be measurable in this case?

77. Electrons accelerated by a potential difference of 12.3 V pass through a gas of hydrogen atoms at room temperature. What wavelengths of light will be emitted?

78. Atoms can be formed in which a muon (mass = 207 times the mass of an electron) replaces one of the electrons in an atom. Calculate, using Bohr theory, the energy of the photon emitted when a muon makes a transition from $n = 2$ to $n = 1$ in a muonic $^{208}_{82}$Pb atom (lead whose nucleus has a mass 208 times the proton mass and charge $+ 82e$).

79. In a particular photoelectric experiment a stopping potential of 2.10 volts is measured when ultraviolet light having a wavelength of 290 nm is incident on the metal. Using the same setup, what will the new stopping potential be if blue light with a wavelength of 440 nm is used, instead of the ultraviolet light.

80. In an X-ray tube (see Fig. 36–25 and discussion in Section 36–10), the high voltage between filament and target is V. After being accelerated through this voltage, an electron strikes the target where it is decelerated (by positively charged nuclei) and in the process one or more X-ray photons are emitted. (a) Show that the photon of shortest wavelength will have
$$\lambda_0 = \frac{hc}{eV}.$$
(b) What is the shortest wavelength of X-ray emitted when accelerated electrons strike the face of a 30-kV television picture tube?

81. Show that the wavelength of a particle of mass m with kinetic energy K is given by the relativistic formula $\lambda = hc/\sqrt{K^2 + 2mc^2 K}$.

82. What is the kinetic energy and wavelength of a "thermal" neutron (one that is in equilibrium at room temperature–see Chapter 18)?

83. What is the theoretical limit of resolution for an electron microscope whose electrons are accelerated through 60 kV? (Relativistic formulas should be used.)

84. The intensity of the Sun's light in the vicinity of the Earth is about 1000 W/m². Imagine a spacecraft with a mirrored square sail of dimension 1.0 km. Estimate how much thrust (in newtons) this craft will experience due to collisions with the Sun's photons. [Hint: Assume the photons bounce off the sail with no change in magnitude of their momentum.]

85. The human eye can respond to as little as 10^{-18} J of light energy. If this comes with a wavelength at the peak of visual sensitivity, 550 nm, how many photons lead to an observable flash?

86. Light with a wavelength of 300 nm strikes a metal whose work function is 2.2 eV. What is the shortest de Broglie wavelength for the electrons that are produced as photoelectrons?

Participants in the 1927 Solvay Conference. Albert Einstein, seated at front center, had difficulty accepting that nature could behave according to the rules of quantum mechanics, which we study in this chapter starting with the wave function and the uncertainty principle. We examine the Schrödinger equation and its solutions for some simple cases: free particles, the square well, and tunnelling through a barrier.

A. Piccard   E. Henriot   P. Ehrenfest   Ed. Herzen   Th. De Donder   E. Schrödinger   E. Verschaffelt   W. Pauli   W. Heisenberg   R. H. Fowler   L. Brillouin
P. Debye   M. Knudsen   W. L. Bragg   H. A. Kramers   P. A. M. Dirac   A. H. Compton   L. de Broglie   M. Born   N. Bohr
I. Langmuir   M. Planck   Mme Curie   H. A. Lorentz   A. Einstein   P. Langevin   Ch. E. Guye   C. T. R. Wilson   O. W. Richardson

CHAPTER 39

Quantum Mechanics

Bohr's model of the atom gave us a first (though rough) picture of what an atom is like. It proposed an explanation for why there should be emission and absorption of light by atoms at discrete wavelengths, as well as for the stability of atoms. The wavelengths of the line spectra and the ionization energy for hydrogen (and one-electron ions) that are predicted are in excellent agreement with experiment. But the Bohr theory had important limitations. It was not able to predict the line spectra for more complex atoms—not even for the neutral helium atom, which has only two electrons. Nor could it explain why emission lines, when viewed with great precision, consist of two or more very closely spaced lines (referred to as *fine structure*). The Bohr theory also did not explain why some spectral lines were brighter than others. And it could not explain the bonding of atoms in molecules or in solids and liquids.

From a theoretical point of view, too, the Bohr theory was not satisfactory. For it was a strange mixture of classical and quantum ideas. Moreover, the wave–particle duality was still not really resolved.

We mention these limitations of the Bohr theory not to disparage it—for it was a landmark in the history of science. Rather, we mention them to show why, in the early 1920s, it became increasingly evident that a new, more comprehensive theory was needed. It was not long in coming. Less than two years after de Broglie

FIGURE 39–1 Erwin Schrödinger with Lise Meitner (see Chapter 43).

FIGURE 39–2 Werner Heisenberg (center) on Lake Como with Wolfgang Pauli (right) and Enrico Fermi (left).

gave us his matter–wave hypothesis, Erwin Schrödinger (1887–1961; Fig. 39–1) and Werner Heisenberg (1901–1976; Fig. 39–2) independently developed a new comprehensive theory. Their separate approaches were quite different but were soon shown to be fully compatible. In this chapter we look at this new theory, quantum mechanics. In the following two chapters we will discuss the results of quantum mechanics applied to the hydrogen atom, to other atoms, and to molecules and solids.

39–1 | Quantum Mechanics—A New Theory

The new theory, called **quantum mechanics**, unifies the wave–particle duality into a single consistent theory. As a theory, quantum mechanics has been extremely successful. It has successfully dealt with the spectra emitted by complex atoms, even the fine details. It explains the relative brightness of spectral lines and how atoms form molecules. It is also a much more general theory that covers all quantum phenomena from blackbody radiation to atoms and molecules. It has explained a wide range of natural phenomena and from its predictions many new practical devices have become possible. Indeed, it has been so successful that it is accepted today by nearly all physicists as the fundamental theory underlying physical processes.

Quantum mechanics deals mainly with the microscopic world of atoms and light. But in our everyday macroscopic world, we do perceive light and we accept that ordinary objects are made up of atoms. This new theory must therefore also account for the verified results of classical physics. That is, when it is applied to macroscopic phenomena, quantum mechanics must be able to produce the old classical laws. This, the **correspondence principle** (already mentioned in Section 37–13), is met fully by quantum mechanics. This doesn't mean we throw away classical theories such as Newton's laws. In the everyday world, the latter are far easier to apply and they give an accurate description. But when we deal with high speeds, close to the speed of light, we must use the theory of relativity; and when we deal with the tiny world of the atom, we use quantum mechanics.

Correspondence principle

Although we won't go into the detailed mathematics of quantum mechanics, we will discuss the main ideas and how they involve the wave and particle properties of matter to explain atomic structure and other applications.

The important properties of any wave are its wavelength, frequency, and amplitude. For an electromagnetic wave, the wavelength determines whether the light is in the visible spectrum or not, and if so, what color it is. We also have seen that the wavelength (or frequency) is a measure of the energy of the corresponding photon ($E = hf$). The amplitude or displacement of an electromagnetic wave at any point is the strength of the electric (or magnetic) field at that point, and is related to the intensity of the wave (the brightness of the light).

For material particles such as electrons, quantum mechanics relates the wavelength to momentum according to de Broglie's formula, $\lambda = h/p$. But what corresponds to the amplitude or displacement of a matter wave? In quantum mechanics, this role is played by the **wave function**, which is given the symbol Ψ (the Greek capital letter psi, pronounced "sigh"). Thus Ψ represents the displacement as a function of time and position, of a new kind of field which we might call a "matter" field or a matter wave. *Wave function*

To calculate the wave function Ψ in a given situation (say, for an electron in an atom) is one of the basic tasks of quantum mechanics. Indeed, the development of an equation to do so was Schrödinger's great contribution. The *Schrödinger wave equation*, as it is called, is considered to be the basic equation for the description of nonrelativistic material particles. We will discuss the Schrödinger equation and its solutions for some simple cases later in this chapter. Now let us ask: What is the meaning of the wave function Ψ for a particle? One way to interpret Ψ is simply as the displacement at any point in space and time of a "matter wave," so that it plays the role that E (the electric field) does for an electromagnetic wave. Another interpretation is possible, however, based on the wave–particle duality. To understand this, we make an analogy with light.

We saw in Chapter 15 that the intensity I of any wave is proportional to the square of the displacement amplitude. This holds true for light waves as well, and, as we saw in Chapter 32, the energy density in a light wave is proportional to E^2, where E is the electric field strength. From the *particle* point of view, the energy density of a light beam is proportional to the number of photons per unit volume, n. The more photons (of a given wavelength) per unit volume there are, the greater the energy density. Hence

$$n \propto E^2.$$

That is, the density of photons is proportional to the square of the electric field strength.

If the light beam is very weak, only a few photons will be involved. Indeed, it is possible to "build up" a photograph on film using very weak light so that the effect of individual photons can be seen. If we are dealing with only one photon, the relationship above ($n \propto E^2$) can be interpreted in a slightly different way. At any point the square of the electric field strength, E^2, is a measure of the *probability* that a photon will be found within a unit volume at that location. Said another way, if dV represents an infinitesimal volume element around that point, then $E^2 \, dV$ is proportional to the probability of finding a photon in the volume dV. At points where E^2 is large, there is a high probability of finding the photon; where E^2 is small, the probability of finding the photon is low.

We can interpret matter waves in the same way. The wave function Ψ may vary in magnitude from point to point in space and time. If Ψ describes a collection of many electrons, then $|\Psi|^2 \, dV$ will be proportional to the number of electrons expected to be found in the volume dV around any given point. When

Probability $\propto |\Psi|^2$

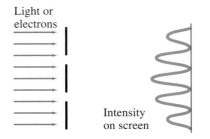

FIGURE 39–3 Parallel beam, of light or electrons, falls on two slits whose sizes are comparable to the wavelength. An interference pattern is observed.

FIGURE 39–4 Young's double-slit experiment done with electrons—note that the pattern is not evident with only a few electrons (top photo), but with more and more electrons (second and third photos), the familiar double-slit interference pattern (Chapter 35) is seen.

dealing with small numbers of electrons we can't make very exact predictions, so $|\Psi|^2$ takes on the character of a probability. If Ψ, which depends on time and position,[†] represents a single electron (say in an atom), then $|\Psi|^2$ is interpreted as follows: $|\Psi|^2 \, dV$ *at a certain point in space and time represents the probability of finding the electron within the volume dV about the given position at that time.* Thus, $|\Psi|^2$ is often referred to as the **probability density** or **probability distribution**, since it is the probability of finding the particle per unit volume.

To understand this better, we take as a thought experiment the familiar double-slit experiment, and consider it both for light and for electrons.

Consider two slits whose size and separation are on the order of the wavelength of whatever we direct at them, either light or electrons, Fig. 39–3. We know very well what would happen in this case for light, since this is just Young's double-slit experiment (Section 35–3): an interference pattern would be seen on the screen behind. If light were replaced by electrons with wavelength comparable to the slit size, they too would produce an interference pattern (recall Fig. 38–11). In the case of light, the pattern would be visible to the eye or could be recorded on film. For electrons, a fluorescent screen could be used (it glows where an electron strikes).

Now, if we reduced the flow of electrons (or photons) so that they passed one at a time through the slits, we would see a flash each time one struck the screen. At first, the flashes would seem random. Indeed, there is no way to predict just where any one electron would hit the screen. If we let the experiment run for a long time, however, and kept track of where each electron hit the screen, we would soon see a pattern emerging—namely the interference pattern predicted by the wave theory; see Fig. 39–4. Thus, although we could not predict where a given electron would strike the screen, we could predict probabilities. (The same can be said for photons.) The probability, as mentioned before, is proportional to $|\Psi|^2$. Where $|\Psi|^2$ is zero, we would get a minimum in the interference pattern. And where $|\Psi|^2$ is a maximum, we would get a peak in the interference pattern.

Since the interference pattern would occur even when electrons (or photons) passed through the slits one at a time, it is clear that the interference pattern would not arise from the interaction of one electron with another. It is as if the electron passed through both slits at the same time, interfering with itself. This is possible because, remember, an electron is not precisely a particle. It is as much a wave as it is a particle, and a wave could certainly travel through both slits at once. But what would happen if we covered one of the slits so we knew that the electrons passed through the other one, and a little later we covered the second slit so that electrons had to have passed through the first? The result would be that no interference pattern would be seen. We would see, instead, two bright areas (or diffraction patterns) on the screen behind the slits. This confirms our idea that if both slits are open, the screen shows an interference pattern as if each electron passed through both slits, like a wave. Yet each electron would make a tiny spot on the screen as if it were a particle.

The main point of this discussion is this: If we treat electrons (and other matter) as if they were waves, then Ψ represents the wave amplitude. If we treat them as particles, then we must treat them on a *probabilistic* basis. The square of the wave function, $|\Psi|^2$, gives the probability per unit volume of finding a given electron at a given point. We cannot predict—or even follow—the path of a single electron precisely through space and time.

[†]The wave function Ψ is generally a complex quantity (that is, it involves $i = \sqrt{-1}$) and hence is not directly observable. On the other hand, $|\Psi|^2$, the absolute value of Ψ squared, is always a real quantity and it is to $|\Psi|^2$ that we can give a physical interpretation.

The Heisenberg Uncertainty Principle

Whenever a measurement is made, some uncertainty or error is always involved. For example, you cannot make an absolutely exact measurement of the length of a table. Even with a measuring stick that has markings 1 mm apart, there will be an inaccuracy of about $\frac{1}{2}$ mm or so. More precise instruments will produce more precise measurements. But there is always some uncertainty involved in a measurement no matter how good the measuring device. We expect that by using more precise instruments, the uncertainty in a measurement can be made indefinitely small.

But according to quantum mechanics, there is actually a limit to the accuracy of certain measurements. This limit is not a restriction on how well instruments can be made; rather, it is inherent in nature. It is the result of two factors: the wave–particle duality, and the unavoidable interaction between the thing observed and the observing instrument. Let us look at this in more detail.

Measurement uncertainty inherent in nature

To make a measurement on an object without somehow disturbing it, at least a little, is not possible. Consider trying to locate a Ping-pong ball in a completely dark room. You grope about trying to find its position; and just when you touch it with your finger, it bounces away. Whenever we measure the position of an object, whether it's a Ping-pong ball or an electron, we always touch it with something else that gives us the information about its position. To locate a lost Ping-pong ball in a dark room, you could probe about with your hand or a stick; or you could shine a light and detect the light reflecting off the ball. When you search with your hand or a stick, you find the ball's position when you touch it. But when you touch the ball you unavoidably bump it, and give it some momentum. Thus you won't know its *future* position. The same would be true, but to a much lesser extent, if you observe the Ping-pong ball using light. In order to "see" the ball, at least one photon must scatter from it, and the reflected photon must enter your eye or some other detector. When a photon strikes an ordinary-sized object, it does not appreciably alter the motion or position of the object. But when a photon strikes a very tiny object like an electron, it can transfer momentum to the object and thus greatly change the object's motion and position in an unpredictable way. The mere act of measuring the position of an object at one time makes our knowledge of its future position imprecise.

Now let us see where the wave–particle duality comes in. Imagine a thought experiment in which we are trying to measure the position of an object, say an electron, with photons, Fig. 39–5. (The arguments would be similar if we were using, instead, an electron microscope.) As we saw in Chapter 36, objects can be seen to an accuracy at best of about the wavelength of the radiation used. If we want an accurate position measurement, we must use a short wavelength. But a short wavelength corresponds to high frequency and large momentum ($p = h/\lambda$); and the more momentum the photons have, the more momentum they can give the object when they strike it. If photons of longer wavelength, and correspondingly smaller momentum are used, the object's motion when struck by the photons will not be affected as much. But the longer wavelength means lower resolution, so the object's position will be less accurately known. Thus the act of observing produces a significant uncertainty in either the *position* or the *momentum* of the electron. This is the essence of the *uncertainty principle* first enunciated by Heisenberg in 1927.

FIGURE 39–5 Thought experiment for observing an electron with a powerful light microscope. At least one photon must scatter from the electron (transferring some momentum to it) and enter the microscope.

Quantitatively, we can make an approximate calculation of the magnitude of this effect. If we use light of wavelength λ, the position can be measured at best to an accuracy of about λ. That is, the uncertainty in the position measurement, Δx, is approximately

$$\Delta x \approx \lambda.$$

Suppose that the object can be detected by a single photon. The photon has a momentum $p = h/\lambda$. When the photon strikes our object, it will give some or all of this momentum to the object, Fig. 39–5. Therefore, the final momentum of our object will be uncertain in the amount

$$\Delta p \approx \frac{h}{\lambda}$$

since we can't tell beforehand how much momentum will be transferred. The product of these uncertainties is

$$(\Delta x)(\Delta p) \approx h.$$

Of course, the uncertainties could be worse than this, depending on the apparatus and the number of photons needed for detection. In Heisenberg's more careful calculation, he found that at the very best

$$(\Delta x)(\Delta p_x) \gtrsim \frac{h}{2\pi},$$

(39–1)

where Δp_x is the uncertainty of the momentum in the x direction.[†] This is a mathematical statement of the **Heisenberg uncertainty principle**, or, as it is sometimes called, the **indeterminancy principle**. It tells us that we cannot measure both the position and momentum of an object precisely at the same time. The more accurately we try to measure the position, so that Δx is small, the greater will be the uncertainty in momentum, Δp_x. If we try to measure the momentum very precisely, then the uncertainty in the position becomes large. The uncertainty principle does not forbid individual exact measurements, however. For example, in principle we could measure the position of an object exactly. But then its momentum would be completely unknown. Thus, although we might know the position of the object exactly at one instant, we could have no idea at all where it would be a moment later.

Another useful form of the uncertainty principle relates energy and time, and we examine this as follows. The object to be detected has an uncertainty in position $\Delta x \approx \lambda$. Now the photon used to detect it travels with speed c, and it takes a time $\Delta t \approx \Delta x/c \approx \lambda/c$ to pass through the distance of uncertainty. Hence, the measured time when our object is at a given position is uncertain by about

$$\Delta t \approx \frac{\lambda}{c}.$$

Since the photon can transfer some or all of its energy ($= hf = hc/\lambda$) to our object, the uncertainty in energy of our object as a result is

$$\Delta E \approx \frac{hc}{\lambda}.$$

The product of these two uncertainties is

$$(\Delta E)(\Delta t) \approx h.$$

Heisenberg's more careful calculation gives

$$(\Delta E)(\Delta t) \gtrsim \frac{h}{2\pi}.$$

(39–2)

This form of the uncertainty principle tells us that the energy of an object can be uncertain, or may even be nonconserved, by an amount ΔE for a time $\Delta t \approx h/(2\pi \, \Delta E)$.

[†]Note, however, that quantum mechanics does allow simultaneous precise measurements of p_x and y: that is $(\Delta y)(\Delta p_x) \gtrsim 0$.

The quantity $(h/2\pi)$ appears so often in quantum mechanics that for convenience it is given the symbol $\hbar$ ("h-bar"). That is

$$\hbar = \frac{h}{2\pi} = \frac{6.626 \times 10^{-34}\,\text{J}\cdot\text{s}}{2\pi} = 1.055 \times 10^{-34}\,\text{J}\cdot\text{s}.$$

By using this notation, Eqs. 39–1 and 39–2 for the uncertainty principle can be written

$$(\Delta x)(\Delta p_x) \gtrsim \hbar \qquad \text{and} \qquad (\Delta E)(\Delta t) \gtrsim \hbar.$$

We have been discussing the position and velocity of an electron as if it were a particle. But it isn't a particle. Indeed, we have the uncertainty principle because an electron—and matter in general—has wave as well as particle properties. What the uncertainty principle really tells us is that if we insist on thinking of the electron as a particle, then there are certain limitations on this simplified view— namely, that the position and velocity cannot both be known precisely at the same time; and that the energy can be uncertain (or nonconserved) in the amount ΔE for a time $\Delta t \approx \hbar/\Delta E$.

Because Planck's constant, h, is so small, the uncertainties expressed in the uncertainty principle are usually negligible on the macroscopic level. But at the level of the atom, the uncertainties are significant. Because we consider ordinary objects to be made up of atoms containing nuclei and electrons, the uncertainty principle is relevant to our understanding of all of nature. The uncertainty principle expresses, perhaps most clearly, the probabilistic nature of quantum mechanics. It thus is often used as a basis for philosophic discussion.

EXAMPLE 39–1 **Position uncertainty of electron.** An electron moves in a straight line with a constant speed $v = 1.10 \times 10^6$ m/s which has been measured to a precision of 0.10 percent. What is the maximum precision with which its position could be simultaneously measured?

SOLUTION The momentum of the electron is $p = mv = (9.11 \times 10^{-31}\,\text{kg})\cdot(1.10 \times 10^6\,\text{m/s}) = 1.00 \times 10^{-24}\,\text{kg}\cdot\text{m/s}$. The uncertainty in the momentum is 0.10 percent of this, or $\Delta p = 1.0 \times 10^{-27}\,\text{kg}\cdot\text{m/s}$. From the uncertainty principle, the best simultaneous position measurement will have an uncertainty of

$$\Delta x = \frac{\hbar}{\Delta p} = \frac{1.06 \times 10^{-34}\,\text{J}\cdot\text{s}}{1.0 \times 10^{-27}\,\text{kg}\cdot\text{m/s}} = 1.1 \times 10^{-7}\,\text{m},$$

or 110 nm. This is about 1000 times the diameter of an atom.

EXAMPLE 39–2 **Position uncertainty of a baseball.** What is the uncertainty in position, imposed by the uncertainty principle, on a 150-g baseball thrown at (93 ± 2) mph $= (42 \pm 1)$ m/s?

SOLUTION The uncertainty in the momentum is

$$\Delta p = m\,\Delta v = (0.150\,\text{kg})(1\,\text{m/s}) = 0.15\,\text{kg}\cdot\text{m/s}.$$

Hence the uncertainty in a position measurement could be as small as

$$\Delta x = \frac{\hbar}{\Delta p} = \frac{1.06 \times 10^{-34}\,\text{J}\cdot\text{s}}{0.15\,\text{kg}\cdot\text{m/s}} = 7 \times 10^{-34}\,\text{m},$$

which is a distance incredibly smaller than any we could imagine observing or measuring. Indeed, the uncertainty principle sets no relevant limit on measurement for macroscopic objects.

EXAMPLE 39–3 **ESTIMATE** **J/ψ lifetime calculated.** The J/ψ meson, discovered in 1974, was measured to have an average mass of 3100 MeV/c^2 (note the use of energy units since $E = mc^2$) and an intrinsic width of 63 keV/c^2. By this we mean that the masses of different J/ψ mesons were actually measured to be slightly different from one another. This mass "width" is related to the very short lifetime of the J/ψ before it decays into other particles. Estimate its lifetime using the uncertainty principle.

SOLUTION The uncertainty of 63 keV/c^2 in the J/ψ's mass is an uncertainty in its rest energy, which in joules is

$$\Delta E = (63 \times 10^3 \, \text{eV})(1.60 \times 10^{-19} \, \text{J/eV}) = 1.01 \times 10^{-14} \, \text{J}.$$

Then we expect its lifetime $\tau \ (= \Delta t$ here) to be

$$\tau \approx \frac{\hbar}{\Delta E} = \frac{1.06 \times 10^{-34} \, \text{J} \cdot \text{s}}{1.01 \times 10^{-14} \, \text{J}} \approx 1.0 \times 10^{-20} \, \text{s}.$$

Lifetimes this short are difficult to measure directly, and the assignment of very short lifetimes depends on this use of the uncertainty principle. (See Chapter 44.)

The uncertainty principle applies also for angular variables:

$$(\Delta L_z)(\Delta \phi) \gtrsim \hbar$$

where L is the component of angular momentum along a given axis (z) and ϕ is the angular position in a plane perpendicular to that axis.

39–4 Philosophic Implications; Probability versus Determinism

The classical Newtonian view of the world is a deterministic one (see Section 6–5). One of its basic ideas is that once the position and velocity of an object are known at a particular time, its future position can be predicted if the forces on it are known. For example, if a stone is thrown a number of times with the same initial velocity and angle, and the forces on it remain the same, the path of the projectile will always be the same. If the forces are known (gravity and air resistance, if any), the stone's path can be precisely predicted. This mechanistic view implies that the future unfolding of the universe, assumed to be made up of particulate bodies, is completely determined.

This classical deterministic view of the physical world has been radically altered by quantum mechanics. As we saw in the analysis of the double-slit experiment (Section 39–2), electrons all prepared in the same way will not all end up in the same place. According to quantum mechanics, certain probabilities exist that an electron will arrive at different points. This is very different from the classical view, in which the path of a particle is precisely predictable from the initial position and velocity and the forces exerted on it. According to quantum mechanics, the position and velocity of an object cannot even be known accurately at the same time. This is expressed in the uncertainty principle, and arises because basic entities, such as electrons, are not considered simply as particles: they have wave properties as well. Quantum mechanics allows us to calculate only the probability[†] that, say, an electron (when thought of as a particle) will be observed at various places. Quantum mechanics says there is some inherent unpredictability in nature.

[†]Note that these probabilities can be calculated precisely, just like exact predictions of probabilities at dice or playing cards, but unlike predictions of probabilities at sporting events or for natural or man-made disasters, which are only estimates.

Since matter is considered to be made up of atoms, even ordinary-sized objects are expected to be governed by probability, rather than by strict determinism. For example, quantum mechanics predicts a finite (but negligibly small) probability that when you throw a stone, its path will suddenly curve upward instead of following the downward-curved parabola of normal projectile motion. Quantum mechanics predicts with extremely high probability that ordinary objects will behave just as the classical laws of physics predict. But these predictions are considered probabilities, not certainties. The reason that macroscopic objects behave in accordance with classical laws with such high probability is due to the large number of molecules involved: when large numbers of objects are present in a statistical situation, deviations from the average (or most probable) approach zero. It is the average configuration of vast numbers of molecules that follows the so-called fixed laws of classical physics with such high probability, and gives rise to an apparent "determinism." Deviations from classical laws are observed when small numbers of molecules are dealt with. We can say, then, that although there are no precise deterministic laws in quantum mechanics, there are statistical laws based on probability.

It is important to note that there is a difference between the probability imposed by quantum mechanics and that used in the nineteenth century to understand thermodynamics and the behavior of gases in terms of molecules (Chapters 18 and 20). In thermodynamics, probability is used because there are far too many particles to keep track of. But the molecules are still assumed to move and interact in a deterministic way following Newton's laws. Probability in quantum mechanics is quite different; it is seen as *inherent* in nature, and not as a limitation on our abilities to calculate or to measure.

Although a few physicists have not given up the deterministic view of nature and have refused to accept quantum mechanics as a complete theory—one was Einstein—nonetheless, the vast majority of physicists do accept quantum mechanics and the probabilistic view of nature. This view, which as presented here is the generally accepted one, is called the **Copenhagen interpretation** of quantum mechanics in honor of Niels Bohr's home, since it was largely developed there through discussions between Bohr and other prominent physicists.

Copenhagen interpretation

Because electrons are not simply particles, they cannot be thought of as following particular paths in space and time. This suggests that a description of matter in space and time may not be completely correct. This deep and far-reaching conclusion has been a lively topic of discussion among philosophers. Perhaps the most important and influential philosopher of quantum mechanics was Bohr. He argued that a space–time description of actual atoms and electrons is not possible. Yet a description of experiments on atoms or electrons must be given in terms of space and time and other concepts familiar to ordinary experience, such as waves and particles. We must not let our *descriptions* of experiments lead us into believing that atoms or electrons themselves actually move in space and time as particles.

39–5 The Schrödinger Equation in One Dimension—Time-Independent Form

In order to describe physical systems quantitatively using quantum mechanics, we must have a means of determining the wave function Ψ mathematically. As mentioned in Section 39–2, the basic equation (in the nonrelativistic realm) for determining Ψ is the *Schrödinger equation*. We cannot, however, derive the Schrödinger equation from some higher principles, just as Newton's second law, for example, cannot be derived. The relation $\mathbf{F} = m\mathbf{a}$ was *invented* by Newton to describe how the motion of a body is related to the net applied force. As we saw early in this book, Newton's second law works exceptionally well. In the realm of classical physics it is the starting point for analytically solving a wide range of problems, and the solutions it yields are fully consistent with experiment. The validity of any fundamental equation resides in its agreement with experiment. The Schrödinger equation forms part of a new theory, and it too had to be *invented*—and then checked against experiment, a test that it passed splendidly.

The Schrödinger equation can be written in two forms: the time-dependent version and the time-independent version. We will mainly be interested in steady-state situations—that is, when there is no time dependence—and so we mainly deal with the time-independent version. (We briefly discuss the time-dependent version in the next, optional, section.) The time-independent version involves a wave function with only spatial dependence which we represent by lowercase psi, $\psi(x)$.

In classical mechanics, we solved problems using two approaches: via Newton's laws with the concept of force, and by using the energy concept with the conservation laws. The Schrödinger equation is based on the energy approach. Even though the Schrödinger equation cannot be derived, we can suggest what form it might take by using conservation of energy and considering a very simple case: that of a free particle on which no forces act, so that its potential energy U is constant. We assume that our particle moves along the x axis, and since no force acts on it, its momentum remains constant and its wavelength ($\lambda = h/p$) is fixed. To describe a wave for a free particle such as an electron, we expect that its wave function will satisfy a differential equation that is akin to (but not identical to) the classical wave equation. Let us see what we can infer about this equation. Consider a simple traveling wave of a single wavelength λ whose wave displacement, as we saw in Chapter 15 for mechanical waves and in Chapter 32 for electromagnetic waves, is given by $A \sin(kx - \omega t)$, or more generally as a superposition of sine and cosine: $A \sin(kx - \omega t) + B \cos(kx - \omega t)$. We are only interested in the spatial dependence, so we consider the wave at a specific moment, say $t = 0$. Thus we write as the wave function for our free particle

$$\psi(x) = A \sin kx + B \cos kx, \tag{39-3a}$$

where A and B are constants[†] and $k = 2\pi/\lambda$ (Eq. 15–11). For a particle of mass m and velocity v, the de Broglie wavelength is $\lambda = h/p$, where $p = mv$ is the particle's momentum. Hence

$$k = \frac{2\pi}{\lambda} = \frac{2\pi p}{h} = \frac{p}{\hbar}, \tag{39-3b}$$

One requirement for our wave equation, then, is that it have the wave function $\psi(x)$ as given by Eq. 39–3 as a solution for a free particle. A second requirement is that it be consistent with the conservation of energy, which we can express as

$$\frac{p^2}{2m} + U = E,$$

where E is the total energy, U is the potential energy, and (since we are considering the nonrelativistic realm) the kinetic energy K of our particle of mass m is $K = \frac{1}{2}mv^2 = p^2/2m$. Since $p = \hbar k$ (Eq. 39–3b), we can write the conservation of energy condition as

$$\frac{\hbar^2 k^2}{2m} + U = E. \tag{39-4}$$

Thus we are seeking a differential equation that satisfies conservation of energy

[†] In quantum mechanics, constants can be complex (i.e., with a real and/or imaginary part).

(Eq. 39–4) when $\psi(x)$ is its solution. Now, note that if we take two derivatives of our expression for $\psi(x)$, Eq. 39–3a, we get a factor $-k^2$ multiplied by $\psi(x)$:

$$\frac{d\psi(x)}{dx} = \frac{d}{dx}(A \sin kx + B \cos kx) = k(A \cos kx - B \sin kx)$$

$$\frac{d^2\psi(x)}{dx^2} = k\frac{d}{dx}(A \cos kx - B \sin kx) = -k^2(A \sin kx + B \cos kx) = -k^2\psi(x).$$

Can this last term be related to the k^2 term in Eq. 39–4? Indeed, if we multiply this last relation by $-\hbar^2/2m$, we obtain

$$-\frac{\hbar^2}{2m}\frac{d^2\psi(x)}{dx^2} = \frac{\hbar^2 k^2}{2m}\psi(x).$$

The right side is just the first term on the left in Eq. 39–4 multiplied by $\psi(x)$. If we multiply Eq. 39–4 through by $\psi(x)$, and make this substitution, we obtain

$$-\frac{\hbar^2}{2m}\frac{d^2\psi(x)}{dx^2} + U(x)\psi(x) = E\psi(x). \qquad \textbf{(39–5)}$$

SCHRÖDINGER EQUATION
(time-independent form)

This is, in fact, the one-dimensional **time-independent Schrödinger equation**, where for generality we have written $U = U(x)$. It is the basis for solving problems in non-relativistic quantum mechanics. For a particle moving in three dimensions there would be additional derivatives with respect to y and z (see Chapter 40).

But note that we have by no means derived the Schrödinger equation. Although we have made a good argument in its favor, other arguments could also be made which might or might not lead to the same equation. The Schrödinger equation as written (Eq. 39–5) is useful and valid only because it has given results in accord with experiment for a wide range of situations.

There are some requirements we impose on any wave function that is a solution of the Schrödinger equation in order that it be physically meaningful. First, we insist that it be a continuous function; after all, if $|\psi|^2$ represents the probability of finding a particle at a certain point, we expect that probability to be continuous from point to point and not to take discontinuous jumps. Second, we want the wave function to be *normalized*. By this we mean that for a single particle, the probability of finding the particle at one point or another (i.e., the probabilities summed over all space) must be exactly 1 (or 100 percent). For a single particle, $|\psi|^2$ represents the probability of finding the particle in unit volume. Then

$$|\psi|^2 \, dV$$

is the probability of finding the particle within a volume dV, where ψ is the value of the wave function in this infinitesimal volume dV. For the one-dimensional case, $dV = dx$. Then the sum of the probabilities over all space—that is, the probability of finding the particle at one point or another—becomes

$$\int_{\text{all space}} |\psi|^2 \, dV = \int |\psi|^2 \, dx = 1. \qquad \textbf{(39–6)}$$

Normalization

This is called the **normalization condition**, and the integral is taken over whatever region of space in which the particle has a chance of being found, which is often all of space, from $x = -\infty$ to $x = \infty$.

The more general form of the Schrödinger equation, including time dependence, for a particle of mass m moving in one dimension, is

Schrödinger equation (time-dependent form)

$$-\frac{\hbar^2}{2m}\frac{\partial^2 \Psi(x,t)}{\partial x^2} + U(x)\Psi(x,t) = i\hbar\frac{\partial \Psi(x,t)}{\partial t}.$$ **(39–7)**

This is the **time-dependent Schrödinger equation**; here $U(x)$ is the potential energy of the particle as a function of position, and i is the imaginary number $i = \sqrt{-1}$. For a particle moving in three dimensions, there would be additional derivatives with respect to y and z, just as for the classical wave equation discussed in Section 15–5. Indeed, it is worth noting the similarity between the Schrödinger wave equation for zero potential energy ($U = 0$) and the classical wave equation: $\partial^2 D/\partial t^2 = v^2 \partial^2 D/\partial x^2$, where D is the wave displacement (equivalent of the wave function). In both equations there is the second derivative with respect to x; but in the Schrödinger equation there is only the first derivative with respect to time, whereas the classical wave equation has the second derivative for time.

As we pointed out in the preceding Section, we cannot derive the time-dependent Schrödinger equation. But we can show how the time-independent Schrödinger equation (Eq. 39–5) is obtained from it. For many problems in quantum mechanics, it is possible to write the wave function as a product of separate functions of space and time:

$$\Psi(x,t) = \psi(x)f(t).$$

Substituting this into the time-dependent Schrödinger equation (Eq. 39–7), we get:

$$-\frac{\hbar^2}{2m}f(t)\frac{d^2\psi(x)}{dx^2} + U(x)\psi(x)f(t) = i\hbar\psi(x)\frac{df(t)}{dt}.$$

We divide both sides of this equation by $\psi(x)f(t)$ and obtain an equation that involves only x on one side and only t on the other:

Separation of variables

$$-\frac{\hbar^2}{2m}\frac{1}{\psi(x)}\frac{d^2\psi(x)}{dx^2} + U(x) = i\hbar\frac{1}{f(t)}\frac{df(t)}{dt}.$$

This *separation of variables* is very convenient. Since the left side is a function only of x, and the right side is a function only of t, the equality can be valid for all values of x and all values of t only if each side is equal to a constant (the same constant, of course), which we call C:

$$-\frac{\hbar^2}{2m}\frac{1}{\psi(x)}\frac{d^2\psi(x)}{dx^2} + U(x) = C$$ **(39–8a)**

$$i\hbar\frac{1}{f(t)}\frac{df(t)}{dt} = C.$$ **(39–8b)**

We multiply the first of these (Eq. 39–8a) through by $\psi(x)$ and obtain

$$-\frac{\hbar^2}{2m}\frac{d^2\psi(x)}{dx^2} + U(x)\psi(x) = C\psi(x).$$ **(39–8c)**

This we recognize immediately as the time-independent Schrödinger equation, Eq. 39–5, where the constant C equals the total energy E. Thus we have obtained the time-independent form of Schrödinger's equation from the time-dependent form.

Equation 39–8b is easy to solve. Putting $C = E$, we rewrite Eq. 39–8b as

$$\frac{df(t)}{dt} = -i \frac{E}{\hbar} f(t)$$

(note that since $i^2 = -1$, $i = -1/i$), and then as

$$\frac{df(t)}{f(t)} = -i \frac{E}{\hbar} dt.$$

We integrate both sides to obtain

$$\ln f(t) = -i \frac{E}{\hbar} t$$

or

$$f(t) = e^{-i\left(\frac{E}{\hbar}\right)t}.$$

Thus the total wave function is

$$\Psi(x, t) = \psi(x) e^{-i\left(\frac{E}{\hbar}\right)t}, \tag{39–9}$$

where $\psi(x)$ satisfies Eq. 39–5. It is, in fact, the solution of the time-independent Schrödinger equation (Eq. 39–5) that is the major task of nonrelativistic quantum mechanics. Nonetheless, we should note that in general the wave function $\Psi(x, t)$ is a complex function since it involves $i = \sqrt{-1}$. It has both a real and an imaginary part.[†] Since $\Psi(x, t)$ is not purely real, it cannot itself be physically measurable. Rather it is only $|\Psi|^2$, which *is* real, that can be measured physically.

Note also that

$$\left| f(t) \right| = \left| e^{-i\left(\frac{E}{\hbar}\right)t} \right| = 1,$$

so $|f(t)|^2 = 1$. Hence the probability density in space does not depend on time:

$$\left| \Psi(x, t) \right|^2 = \left| \psi(x) \right|^2.$$

We thus will be interested only in the time-independent Schrödinger equation, Eq. 39–5, which we will now examine for a number of simple situations.

39–7 Free Particles; Plane Waves and Wave Packets

A **free particle** is one that is not subject to any force, and we can therefore take its potential energy to be zero. (Although we dealt with the free particle in Section 39–5 in arguing for Schrödinger's equation, here we treat it directly using Schrödinger's equation as the basis.) Schrödinger's equation (Eq. 39–5) with $U(x) = 0$ becomes

$$-\frac{\hbar^2}{2m} \frac{d^2\psi(x)}{dx^2} = E\psi(x),$$

which can be written

$$\frac{d^2\psi}{dx^2} + \frac{2mE}{\hbar^2} \psi = 0.$$

This is a familiar equation that we encountered in Chapter 14 (Eq. 14–3) in connection with the simple harmonic oscillator. The solution to this equation, but with

[†] Recall that $e^{-i\theta} = \cos\theta - i\sin\theta$.

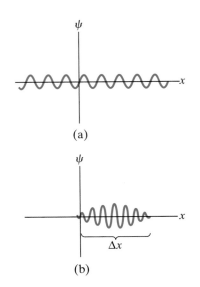

FIGURE 39–6 (a) A plane wave describing a free particle. (b) A wave packet of "width" Δx.

(a)

(b)

Wave packet

appropriate variable changes[†] for our case here, is

$$\psi = A \sin kx + B \cos kx, \qquad (39\text{–}10)$$

where

$$k = \sqrt{\frac{2mE}{\hbar^2}}. \qquad (39\text{–}11a)$$

Since $U = 0$, the total energy E of the particle is $E = \frac{1}{2}mv^2 = p^2/2m$ (where p is the momentum); thus

$$k = \frac{p}{\hbar} = \frac{h}{\lambda\hbar} = \frac{2\pi}{\lambda}. \qquad (39\text{–}11b)$$

So a free particle of momentum p and energy E can be represented by a plane wave that varies sinusoidally. If we are not interested in the phase, we can choose $B = 0$ in Eq. 39–10, and we show this sine wave in Fig. 39–6a.

Note in Fig. 39–6a that the sine wave will extend indefinitely[‡] in the $+x$ and $-x$ directions. Thus, since $|\psi|^2$ represents the probability of finding the particle, the particle could be anywhere between $x = -\infty$ and $x = \infty$. This is fully consistent with the uncertainty principle (Section 39–3): the momentum of the particle was given and hence is known precisely ($p = \hbar k$), so the particle's position must be totally unpredictable. Mathematically, if $\Delta p = 0$, $\Delta x \gtrsim \hbar/\Delta p = \infty$.

To describe a particle whose position is well localized—that is, it is known to be within a small region of space—we can use the concept of a **wave packet**. Figure 39–6b shows an example of a wave packet whose width is about Δx as shown, meaning that the particle is most likely to be found within this region of space. A well-localized particle moving through space can thus be represented by a moving wave packet.

A wave packet can be represented mathematically as the sum of many plane waves (sine waves) of slightly different wavelengths. That this will work can be seen by looking carefully at Fig. 16–18. There we combined only two nearby frequencies (to explain why there are "beats") and found that the sum of two sine waves looked like a series of wave packets. If we add additional waves with other nearby frequencies, we can eliminate all but one of the packets and arrive at Fig. 39–6b. Thus a wave packet consists of waves of a *range* of wavelengths; hence it does not have a definite momentum $p(= h/\lambda)$, but rather, a range of momenta. This is, of course, consistent with the uncertainty principle: we have made Δx small, so the momentum cannot be precise; that is, Δp cannot be zero. Instead, our particle can be said to have a range of momenta, Δp, or to have an uncertainty in its momentum, Δp. It is not hard to show, even for this simple situation (see Problem 15), that $\Delta p \approx h/\Delta x$, in accordance with the uncertainty principle.

39–8 Particle in an Infinitely Deep Square Well Potential (a Rigid Box)

The Schrödinger equation can be solved analytically only for a few possible forms of the potential energy U. We consider two simple cases here which at first may not seem realistic, but have simple solutions that can be used as approximations to understand a variety of phenomena.

[†] In Eq. 14–3, t becomes x and ω becomes $k = \sqrt{2mE/\hbar^2}$. (Don't confuse this k with the spring constant k of Chapter 14.)

[‡] Such an infinite wave makes problems for normalization since $\int_{-\infty}^{\infty} |\psi|^2\, dx = A^2 \int_{-\infty}^{\infty} \sin^2 kx\, dx$ is infinite for any nonzero value for A. For practical purposes we can usually normalize the waves ($A \neq 0$) by assuming that the particle is in a large but finite region of space. The region can be chosen large enough so that momentum is still rather precisely fixed.

In our first case, we assume that a particle of mass m is confined to a one-dimensional box of width L whose walls are perfectly rigid. (This can serve as a very crude approximation for an electron in a metal, for example.) The particle is trapped in this box and collisions with the walls are perfectly elastic. The potential energy for this situation, which is commonly known as an **infinitely deep square well potential** or **rigid box**, is shown in Fig. 39–7. We can write the potential energy $U(x)$ as

$$U(x) = 0 \qquad 0 < x < L$$
$$U(x) = \infty \qquad x \leq 0 \quad \text{and} \quad x \geq L.$$

For the region $0 < x < L$, where $U(x) = 0$, we already know the solution of the Schrödinger equation from our discussion in Section 39–7: it is just Eq. 39–10,

$$\psi(x) = A \sin kx + B \cos kx,$$

where (from Eq. 39–11a)

$$k = \sqrt{\frac{2mE}{\hbar^2}}.$$

Outside the well $U(x) = \infty$, so $\psi(x)$ must be zero. (If it weren't, the product $U\psi$ in the Schrödinger equation wouldn't be finite; besides, if $U = \infty$, we can't expect a particle of finite total energy to be in such a region.) So we are concerned only with the wave function within the well, and we must determine the constants A and B as well as any restrictions on the value of k (and hence on the energy E).

We have insisted that the wave function must be continuous. Hence, if $\psi = 0$ outside the well, it must be zero at $x = 0$ and at $x = L$:

$$\psi(0) = 0 \quad \text{and} \quad \psi(L) = 0.$$

These are the **boundary conditions** for this problem. At $x = 0$, $\sin kx = 0$ but $\cos kx = 1$, so

$$0 = \psi(0) = A \sin 0 + B \cos 0 = 0 + B.$$

Thus B must be zero. Our solution is reduced to

$$\psi(x) = A \sin kx.$$

Now we apply the other boundary condition, $\psi = 0$ at $x = L$:

$$0 = \psi(L) = A \sin kL.$$

We don't want $A = 0$ or we won't have a particle at all ($|\psi|^2 = 0$ everywhere). Therefore, we set

$$\sin kL = 0,$$

which, since the sine is zero for angles of $0, \pi, 2\pi, 3\pi \cdots$ radians, can happen only if $kL = 0, \pi, 2\pi, 3\pi, \cdots$. In other words,

$$kL = n\pi \qquad n = 1, 2, 3, \cdots, \tag{39–12}$$

where n is an integer. We eliminate the case $n = 0$ since this, too, produces $x = 0$ everywhere. Thus k, and hence E, cannot have just any value; rather, k is limited to values

$$k = \frac{n\pi}{L}.$$

Putting this expression in Eq. 39–11a (and substituting $h/2\pi$ for $\hbar$), we find that E can have only the values

$$E = n^2 \frac{h^2}{8mL^2}. \tag{39–13}$$

FIGURE 39–7 A plot of potential energy U vs x for an infinitely deep square well potential.

Boundary conditions

Energy levels, quantized (∞ potential well)

FIGURE 39–8 Possible energy levels for a particle in a box with perfectly rigid walls (infinite square well potential).

Zero-point energy

FIGURE 39–9 Wave functions corresponding to the quantum number n being 1, 2, 3 and 10 for a particle confined to a rigid box.

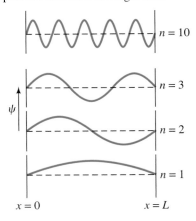

FIGURE 39–10 (below) The probability distribution for a particle in a rigid box for the states with $n = 1, 2, 3,$ and 10.

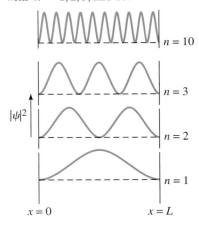

A particle trapped in a rigid box thus can only have certain *quantized energies*. The lowest energy (the ground state) has $n = 1$ and is given by

$$E_1 = \frac{h^2}{8mL^2}.$$

The next highest energy ($n = 2$) is

$$E_2 = 4E_1,$$

and for higher energies (see Fig. 39–8),

$$E_3 = 9E_1$$
$$\vdots$$
$$E_n = n^2 E_1.$$

The integer n is called the **quantum number** of the state. That the lowest energy, E_1, is not zero means that the particle in the box can never be at rest. This is contrary to classical ideas, according to which a particle can have $E = 0$. E_1 is called the **zero-point energy**. One outcome of this result is that even at a temperature of absolute zero ($0\,\text{K}$), quantum mechanics predicts that particles in a box would not be at rest but would have a zero-point energy.

We can also note that the energy E_1 and momentum $p_1 = \hbar k = \hbar\pi/L$ in the ground state are related inversely to the width of the box. The smaller the width L, the larger the momentum (and energy). This can be considered a direct result of the uncertainty principle (see Problem 21). The wave function $\psi = A \sin kx$ for each of the quantum states is (since $k = n\pi/L$)

$$\psi_n = A \sin\left(\frac{n\pi}{L}x\right). \tag{39–14}$$

We can determine the constant A by imposing the normalization condition (Eq. 39–6):

$$1 = \int_{-\infty}^{\infty} \psi^2\, dx = \int_0^L A^2 \sin^2\left(\frac{n\pi}{L}x\right) dx, \tag{39–15}$$

where the integral needs to be done only over the range $0 < x < L$ because outside these limits $\psi = 0$. The integral (see Example 39–5) is equal to $A^2 L/2$, so we have

$$A = \sqrt{\frac{2}{L}} \quad \text{and} \quad \psi_n = \sqrt{\frac{2}{L}} \sin\left(\frac{n\pi}{L}x\right).$$

The amplitude A is the same for all the quantum numbers. Figure 39–9 shows the wave functions (Eq. 39–14) for $n = 1, 2, 3,$ and 10. They look just like standing waves on a string—see Fig. 15–27. This is not surprising since the wave function solutions, Eq. 39–14, are the same as for the standing waves on a string, and the condition $kL = n\pi$ is the same in the two cases.

Figure 39–10 shows the probability distribution, $|\psi|^2$, for the same states ($n = 1, 2, 3, 10$) for which ψ is shown in Fig. 39–9. We see immediately that the particle is more likely to be found in some places than in others. For example, in the ground state ($n = 1$), the electron is much more likely to be found near the center of the box than near the walls. This is clearly at variance with classical ideas, which predict a uniform probability density—the particle would be as likely to be found at one point in the box as at any other. The quantum-mechanical probability densities for higher states are even more complicated, with areas of low probability not only near the walls but also at regular intervals in between.

EXAMPLE 39-4 **Electron in an infinite potential well.** (a) Calculate the three lowest energy levels for an electron trapped in an infinitely deep square well potential of width $L = 1.00 \times 10^{-10}$ m (about the diameter of a hydrogen atom in its ground state). (b) If a photon were emitted when the electron jumps from the $n = 2$ state to the $n = 1$ state, what would its wavelength be?

SOLUTION (a) The ground state ($n = 1$) has energy

$$E_1 = \frac{h^2}{8mL^2} = \frac{(6.63 \times 10^{-34} \text{ J·s})^2}{8(9.11 \times 10^{-31} \text{ kg})(1.00 \times 10^{-10} \text{ m})^2} = 6.03 \times 10^{-18} \text{ J}.$$

In electron volts this is

$$E_1 = \frac{6.03 \times 10^{-18} \text{ J}}{1.60 \times 10^{-19} \text{ J/eV}} = 37.7 \text{ eV}.$$

Then

$$E_2 = (2)^2 E_1 = 151 \text{ eV}$$
$$E_3 = (3)^2 E_1 = 339 \text{ eV}.$$

(b) The energy difference is $E_2 - E_1 = 151 \text{ eV} - 38 \text{ eV} = 113 \text{ eV}$ or 1.81×10^{-17} J, and this would equal the energy of the emitted photon (energy conservation). Its wavelength would be

$$\lambda = \frac{c}{f} = \frac{hc}{E} = \frac{(6.63 \times 10^{-34} \text{ J·s})(3.00 \times 10^8 \text{ m/s})}{1.81 \times 10^{-17} \text{ J}} = 1.10 \times 10^{-8} \text{ m}$$

or 11.0 nm, which is in the ultraviolet region of the spectrum.

EXAMPLE 39-5 **Calculating a normalization constant.** Show that the normalization constant A for all wave functions describing a particle in an infinite potential well of width L has a value of $A = \sqrt{2/L}$.

SOLUTION The wave functions are

$$\psi = A \sin \frac{n\pi x}{L}.$$

To normalize ψ, we must have (Eq. 39-15)

$$1 = \int_0^L |\psi|^2 \, dx = \int_0^L A^2 \sin^2 \frac{n\pi x}{L} \, dx.$$

We need integrate only from 0 to L since $\psi = 0$ for all other values of x. To evaluate this integral we let $\theta = n\pi x/L$ and use the trigonometric identity $\sin^2 \theta = \frac{1}{2}(1 - \cos 2\theta)$. Then, with $dx = L \, d\theta/n\pi$, we have

$$1 = A^2 \int_0^{n\pi} \sin^2 \theta \left(\frac{L}{n\pi} \right) d\theta = \frac{A^2 L}{2n\pi} \int_0^{n\pi} (1 - \cos 2\theta) \, d\theta$$

$$= \frac{A^2 L}{2n\pi} \left(\theta - \frac{1}{2} \sin 2\theta \right) \Big|_0^{n\pi}$$

$$= \frac{A^2 L}{2}.$$

Thus $A^2 = 2/L$ and

$$A = \sqrt{\frac{2}{L}}.$$

EXAMPLE 39–6 **ESTIMATE** **Confined bacterium.** A tiny bacterium with a mass of about 10^{-14} kg is confined between two rigid walls 0.1 mm apart. (a) Estimate its minimum speed. (b) If, instead, its speed is about 1 mm in 100 s, estimate the quantum number of its state.

SOLUTION (a) The minimum speed occurs in the ground state, $n = 1$; since $E = \frac{1}{2}mv^2$, we have

$$v = \sqrt{\frac{2E}{m}} = \sqrt{\frac{h^2}{4m^2L^2}} = \frac{h}{2mL} = \frac{6.6 \times 10^{-34}\,\text{J}\cdot\text{s}}{2(10^{-14}\,\text{kg})(10^{-4}\,\text{m})} \approx 3 \times 10^{-16}\,\text{m/s}.$$

This is a speed so small that we could not measure it and the object would seem at rest, consistent with classical physics.

(b) Given $v = 10^{-3}\,\text{m}/100\,\text{s} = 10^{-5}\,\text{m/s}$, the kinetic energy of the bacterium is

$$E = \frac{1}{2}mv^2 = \frac{1}{2}(10^{-14}\,\text{kg})(10^{-5}\,\text{m/s})^2 = 0.5 \times 10^{-24}\,\text{J}.$$

From Eq. 39–13, the quantum number of this state is

$$n = \sqrt{E\left(\frac{8mL^2}{h^2}\right)} = \sqrt{\frac{(0.5 \times 10^{-24}\,\text{J})(8)(10^{-14}\,\text{kg})(10^{-4}\,\text{m})^2}{(6.6 \times 10^{-34}\,\text{J}\cdot\text{s})^2}}$$

$$\approx \sqrt{1 \times 10^{21}} \approx 3 \times 10^{10}.$$

This number is so large that we could never distinguish between adjacent energy states (between $n = 3 \times 10^{10}$ and $3 \times 10^{10} + 1$). The energy states would appear to form a continuum. Thus, even though the energies involved here are small ($\ll$ 1 eV), we are still dealing with a macroscopic object (though visible only under a microscope) and the quantum result is not distinguishable from a classical one. This is in accordance with the correspondence principle.

* 39–9 Finite Potential Well

FIGURE 39–11 Potential energy U vs x for a finite one-dimensional square well.

Let us now look at a particle in a box whose walls are not perfectly rigid. That is, the potential energy outside the box or well is not infinite, but rises to some level U_0, as shown in Fig. 39–11. This is called a **finite potential well**. It can serve as an approximation for, say, a neutron in a nucleus. There are some significant new features that arise for the finite well as compared to the infinite well. We divide the well into three regions as shown in Fig. 39–11. In region II, inside the well, the Schrödinger equation is the same as before ($U = 0$), although the boundary conditions will be different. So we write the solution for region II as

$$\psi_{\text{II}} = A \sin kx + B \cos kx \qquad (0 < x < L)$$

but we don't immediately set $B = 0$ or assume that k is given by Eq. 39–12.

In regions I and III, the Schrödinger equation, now with $U(x) = U_0$, is

$$-\frac{\hbar^2}{2m}\frac{d^2\psi}{dx^2} + U_0\psi = E\psi.$$

We rewrite this as

$$\frac{d^2\psi}{dx^2} - \left[\frac{2m(U_0 - E)}{\hbar^2}\right]\psi = 0.$$

Let us assume that E is less than U_0, so the particle is "trapped" in the well (at least classically). There might be only one such **bound state**, or several, or even none, as we shall discuss later. We define the constant G by

$$G^2 = \frac{2m(U_0 - E)}{\hbar^2} \qquad\qquad \textbf{(39–16)}$$

and rewrite the Schrödinger equation as

$$\frac{d^2\psi}{dx^2} - G^2\psi = 0.$$

This equation has the general solution

$$\psi_{I,\,III} = Ce^{Gx} + De^{-Gx},$$

which can be confirmed by direct substitution, since

$$\frac{d^2}{dx^2}\left(e^{\pm Gx}\right) = G^2 e^{\pm Gx}.$$

In region I, x is always negative, so D must be zero (otherwise, $\psi \to \infty$ as $x \to -\infty$, giving an unacceptable result). Similarly in region III, where x is always positive, C must be zero. Hence

$$\psi_I = Ce^{Gx} \qquad (x < 0)$$
$$\psi_{III} = De^{-Gx} \qquad (x > L).$$

In regions I and III, the wave function decreases exponentially with distance from the well. The mathematical forms of the wave function inside and outside the well are different, but we insist that the wave function be continuous even at the two walls. We also insist that the slope of ψ, which is its first derivative, be continuous at the walls. Hence we have the boundary conditions:

$$\psi_I = \psi_{II} \quad \text{and} \quad \frac{d\psi_I}{dx} = \frac{d\psi_{II}}{dx} \qquad \text{at } x = 0$$

$$\psi_{II} = \psi_{III} \quad \text{and} \quad \frac{d\psi_{II}}{dx} = \frac{d\psi_{III}}{dx} \qquad \text{at } x = L.$$

At the left-hand wall ($x = 0$) these boundary conditions become

$$Ce^0 = A\sin 0 + B\cos 0 \qquad \text{or} \qquad C = B$$

and

$$GCe^0 = kA\cos 0 - kB\sin 0 \qquad \text{or} \qquad GC = kA.$$

These are two of the relations that link the constants A, B, C, D and the energy E. We get two more relations from the boundary conditions at $x = L$, and a fifth relation from normalizing the wave functions over all space, $\int_{-\infty}^{\infty}|\psi|^2\,dx = 1$. These five relations allow us to solve for the five unknowns, including the energy E. We will not go through the detailed mathematics, but we will discuss some of the results.

Figure 39–12a shows the wave function ψ for the three lowest possible states, and Fig. 39–12b shows the probability distributions $|\psi|^2$. We see that the wave functions are smooth at the walls of the well. Within the well ψ has the form of a sinusoidal wave; for the ground state, there is less than a half wavelength. Compare this to the infinite well (Fig. 39–9), where the ground-state wave function is exactly a half wavelength: $\lambda = 2L$. For our finite well, $\lambda > 2L$. Thus for a finite well the momentum of a particle ($p = h/\lambda$), and hence its ground-state energy, will be less than for an infinite well of the same width L.

Outside the finite well we see that the wave function drops off exponentially on either side of the walls. That ψ is not zero beyond the walls means that the particle can sometimes be found outside the well. This completely contradicts classical ideas. Outside the well, the potential energy of the particle is greater than its total energy: $U_0 > E$. This violates conservation of energy. But we clearly see in Fig. 39–12b that the particle can spend some time outside the well, where $U_0 > E$ (although the penetration into this classically forbidden region is generally not far since $|\psi|^2$ decreases exponentially with distance from either wall). The penetration

FIGURE 39–12 The wave functions (a), and probability distributions (b), for the three lowest possible states of a particle in a finite potential well. Each of the ψ and $|\psi|^2$ curves has been superposed on its energy level (dashed lines) for convenience.

(a)

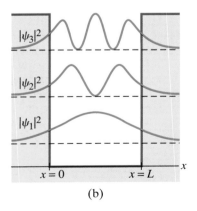

(b)

of a particle into a classically forbidden region is a very important result of quantum mechanics. But how can it be? How can we accept this nonconservation of energy? We can look to the uncertainty principle, in the form

$$\Delta E \, \Delta t \gtrsim \hbar.$$

It tells us that the energy is uncertain, and thus can even be nonconserved, by an amount ΔE for very short times $\Delta t \sim \hbar / \Delta E$.

Now let us consider the situation when the total energy E of the particle is greater than U_0. In this case the particle is a free particle and everywhere its wave function is sinusoidal, Fig. 39–13. Its wavelength is different outside the well than inside, as shown. Since $K = \frac{1}{2}mv^2 = p^2/2m$, the wavelength in region II is

$$\lambda = \frac{h}{p} = \frac{h}{\sqrt{2mE}} \qquad 0 < x < L,$$

whereas in regions I and III, where $p^2/2m = K = E - U_0$, the wavelength is

$$\lambda = \frac{h}{p} = \frac{h}{\sqrt{2m(E - U_0)}} \qquad x < 0 \quad \text{and} \quad x > L.$$

For $E > U_0$, any energy E is possible. But for $E < U_0$, as we saw above, the energy is quantized and only certain states are possible.

FIGURE 39–13 Particle of energy E traveling over a potential well whose depth U_0 is less than E (measured in the well).

39–10 Tunneling through a Barrier

We saw in Section 39–9 that according to quantum mechanics, a particle such as an electron can penetrate a barrier into a region forbidden by classical mechanics. There are a number of important applications of this phenomenon, particularly as applied to penetration of a thin barrier.

We consider a particle of mass m and energy E traveling to the right along the x axis in free space where the potential energy $U = 0$ so the energy is all kinetic energy ($E = K$). The particle encounters a narrow potential barrier whose height U_0 (in energy units) is greater than E, and whose thickness is L (distance units); see Fig. 39–14a. Since $E < U_0$, we would expect from classical physics that the particle could not penetrate the barrier but would simply be reflected and would return in the opposite direction. Indeed, this is what happens for macroscopic objects. But quantum mechanics predicts a nonzero probability for finding the particle on the other side of the barrier. We can see how this can happen in part (b) of Fig. 39–14, which shows the wave function. The approaching particle has a sinusoidal wave function. Within the barrier the solution to the Schrödinger equation is a decaying exponential just as for the finite well of Section 39–9. However, before the exponential dies away to zero, the barrier ends (at $x = L$),

FIGURE 39–14 (a) A potential barrier of height U_0 and thickness L. (b) The wave function for a particle of energy $E (< U_0)$ that approaches from the left. The curve for ψ is superposed, for convenience, on the energy level line (dashed).

(a)

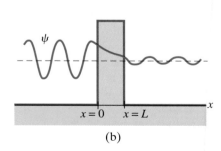

(b)

and for $x > L$ there is again a sinusoidal wave function, since $U = 0$ and $E = K > 0$. But it is a sine wave of greatly reduced amplitude. Nonetheless, because $|\psi|^2$ is nonzero beyond the barrier, we see that there is a nonzero probability that the particle penetrates the barrier. This process is called **tunneling** through the barrier, or **barrier penetration**. Although we cannot observe the particle within the barrier (it would violate conservation of energy), we can detect it after it has penetrated the barrier.

Quantitatively, we can describe the tunneling probability with a *transmission coefficient*, T, and a *reflection coefficient*, R. Suppose, for example, that $T = 0.03$ and $R = 0.97$; then if 100 particles struck the barrier, on the average 3 would tunnel through and 97 would be reflected. Note that $T + R = 1$, since an incident particle must either reflect or tunnel through. The transmission coefficient can be determined by writing the wave function for each of the three regions, just as we did for the finite well, and then applying the boundary conditions that ψ and $d\psi/dx$ must be continuous at the edges of the barrier ($x = 0$ and $x = L$). The calculation shows (see Problem 36) that if T is small ($\ll 1$), then

$$T \approx e^{-2GL}, \qquad \qquad \text{(39–17a)}$$

where

$$G = \sqrt{\frac{2m(U_0 - E)}{\hbar^2}}. \qquad \qquad \text{(39–17b)}$$

(This is the same G as in Section 39–9, Eq. 39–16.) We note that increasing the height of the barrier, U_0, or increasing its thickness, L, will drastically reduce T. Indeed, for macroscopic situations, T is extremely small, in accord with classical physics, which predicts no tunneling (again the correspondence principle).

EXAMPLE 39–7 **Barrier penetration.** A 50-eV electron approaches a square barrier 70 eV high and (*a*) 1.0 nm thick, (*b*) 0.10 nm thick. What is the probability that the electron will tunnel through?

SOLUTION (*a*) First we write the energy in SI units.

$$U_0 - E = (70 \text{ eV} - 50 \text{ eV})(1.6 \times 10^{-19} \text{ J/eV}) = 3.2 \times 10^{-18} \text{ J}.$$

Then, using Eq. 39–17, we have

$$2GL = 2\sqrt{\frac{2(9.11 \times 10^{-31} \text{ kg})(3.2 \times 10^{-18} \text{ J})}{(1.06 \times 10^{-34} \text{ J·s})^2}}(1.0 \times 10^{-9} \text{ m}) = 46$$

and

$$T = e^{-2GL} = e^{-46} = 1.1 \times 10^{-20},$$

which is extremely small.
(*b*) For $L = 0.10$ nm, $2GL = 4.6$ and

$$T = e^{-4.6} = 0.010.$$

Thus the electron has a 1 percent chance of penetrating a 0.1-nm-thick barrier, but only 1 chance in 10^{20} to penetrate a 1-nm barrier. By reducing the barrier thickness by a factor of 10, the probability of tunneling through was increased 10^{18} times! Clearly the transmission coefficient is extremely sensitive to the values of L, $U_0 - E$, and m.

Tunneling is a result of the wave properties of material particles, and also occurs for classical waves. For example, we saw in Section 33–7 that, when light traveling in glass strikes a glass–air boundary at an angle greater than the critical angle, the light is 100 percent totally reflected. We studied this phenomenon of

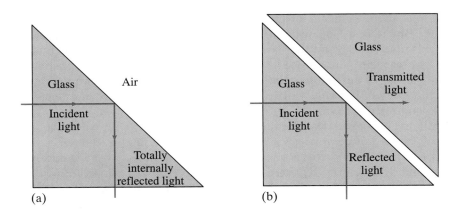

FIGURE 39–15 (a) Light traveling in glass strikes the interface with air at an angle greater than the critical angle, and is totally internally reflected. (b) A small amount of light tunnels through a narrow air gap between two pieces of glass.

FIGURE 39–16 Potential energy seen by an alpha particle (charge q) in presence of nucleus (charge Q), showing the wave function for tunneling out.

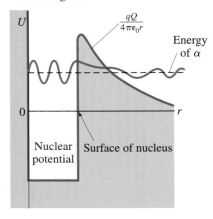

FIGURE 39–17 Probe tip of a scanning tunneling microscope moves up and down to maintain a constant tunneling current, producing an image of the surface.

total internal reflection from the point of view of ray optics, and we show it here in Fig. 39–15a. The wave theory, however, predicts that waves actually penetrate the air for a few wavelengths—almost as if they "needed" to pass the interface to find out there is air beyond and hence need to be totally reflected. Indeed, if a second piece of glass is brought near the first as shown in Fig. 39–15b, a transmitted wave that has tunneled through the air gap can be experimentally observed. You can actually observe this for yourself by looking down into a glass of water at an angle such that light entering your eye has been totally internally reflected from the (outer) glass surface (it will look silvery). If you press a moistened fingertip against the glass, you can see the whorls of the ridges on your fingerprints, because at the ridges you have interfered with the total internal reflection at the outer surface of the glass. So you see light that has penetrated the gap and reflected off the ridges on your finger.

Applications of Tunneling

Tunneling thus occurs even for classical waves. What is new in quantum mechanics is that material particles have wave properties and hence can tunnel. Tunneling has provided the basis for a number of useful devices, as well as helped to explain a number of important phenomena, some of which we mention briefly now.

Some atomic nuclei undergo **radioactive decay** by the emission of an alpha (α) particle, which consists of two protons and two neutrons. Inside a radioactive nucleus, we can imagine that the protons and neutrons are moving about, and sometimes two of each come together and form this stable entity, the alpha particle. We will study alpha decay in more detail in Chapter 42, but for now we note that the potential energy diagram for the alpha particle inside this type of nucleus looks something like Fig. 39–16. The square well represents the attractive nuclear force that holds the nucleus together. To this is added the $1/r$ Coulomb potential energy of repulsion between the positive alpha particle and the remaining positively charged nucleus. The barrier that results is called the **Coulomb barrier**. The wave function for the tunneling particle shown must have energy greater than zero (or the barrier would be infinitely wide and tunneling could not occur), but less than the height of the barrier. If the alpha particle had energy higher than the barrier, it would always be free and the original nucleus wouldn't exist. Thus the barrier keeps the nucleus together, but occasionally a nucleus of this type can decay by the tunneling of an alpha particle. The probability of an α-particle escaping, and hence the "lifetime" of a nucleus, depends on the height and width of the barrier, and can take on a very wide range of values for only a limited change in barrier width as we saw in Ex. 39–7. Lifetimes of α-decaying radioactive nuclei range from less than 1 μs to 10^{10} yr.

A so-called **tunnel diode** is an electronic device made of two types of semiconductor carrying opposite-sign charge carriers, separated by a very thin neutral region. Current can tunnel through this thin barrier and can be controlled by the voltage applied to it, which affects the height of the barrier.

The **scanning tunneling electron microscope** (STM), developed in the 1980s, makes use of tunneling through a vacuum. A tiny probe, whose tip may be only one (or a few) atoms wide, is moved across the specimen to be examined in a series of linear passes, like those made by the electron beam in a TV tube or CRT. The tip, as it scans, remains very close to the surface of the specimen, about 1 nm above it, Fig. 39–17. A small voltage applied between the probe and the surface causes electrons to tunnel through the vacuum between them. This tunneling current is very sensitive to the gap width (see Example 39–7), so that a feedback mechanism can be used to raise and lower the probe to maintain a constant electron tunneling current. The probe's vertical motion, following the surface of the specimen, is then plotted as a function of position, producing a three-dimensional image of the surface. (See, for example, Fig. 39–18.) Surface features as fine as the size of an atom can be resolved: a resolution better than 0.1 nm laterally and 10^{-2} to 10^{-3} nm vertically. This kind of resolution was not available previously and has given a great impetus to the study of the surface structure of materials. The "topographic" image of a surface actually represents the distribution of electron charge (electron wave probability distributions).

The new **atomic force microscope** (AFM) is in many ways similar to an STM, but can be used on a wider range of sample materials. Instead of detecting an electric current, the AFM measures the force between a cantilevered tip and the sample, a force which depends strongly on the tip–sample separation at each point. The tip is moved as for the STM.

FIGURE 39–18 Image of cellular DNA, magnified about 2 million times, taken with a scanning tunneling microscope. Three turns of the DNA double helix can be seen in this false-color image.

☐ Summary

In 1925, Schrödinger and Heisenberg separately worked out a new theory, **quantum mechanics**, which is now considered to be the basic theory at the atomic level. It is a statistical theory rather than a deterministic one.

An important aspect of quantum mechanics is the Heisenberg **uncertainty principle**. It results from the wave-particle duality and the unavoidable interaction between the observed object and the observer.

One form of the uncertainty principle states that the position x and momentum p_x of an object cannot both be measured precisely at the same time. The products of the uncertainties, $(\Delta x)(\Delta p_x)$, can be no less than $\hbar(= h/2\pi)$:

$$(\Delta p_x)(\Delta x) \gtrsim \hbar.$$

Another form states that the energy can be uncertain, or nonconserved, by an amount ΔE for a time Δt where

$$(\Delta E)(\Delta t) \gtrsim \hbar.$$

A particle such as an electron is represented by a **wave function** ψ. The square of the wave function, $|\psi|^2$, at any point in space represents the **probability** of finding the particle at that point. The wave function must be **normalized**, meaning that $\int |\psi|^2 \, dV$ over all space must equal 1, since the particle must be found at one place or another.

In nonrelativistic wave mechanics, ψ satisfies the **Schrödinger equation**:

$$-\frac{\hbar^2}{2m}\frac{d^2\psi}{dx^2} + U\psi = E\psi,$$

here in its one-dimensional time-independent form, where U is the potential energy as a function of position and E is the total energy of the particle.

A **free particle** subject to no forces has a sinusoidal wave function $\psi = A \sin kx + B \cos kx$ with $k = p/\hbar$ and p is the particle's momentum. Such a wave of fixed momentum is spread out indefinitely in space as a plane wave.

A **wave packet**, localized in space, is a superposition of sinusoidal waves with a range of momenta.

For a particle confined to an **infinitely deep square well potential**, or **rigid box**, the Schrödinger equation gives the wave functions

$$\psi = A \sin kL,$$

where $A = \sqrt{2/L}$, as solutions inside the well. The energy is quantized,

$$E = \frac{\hbar^2 k^2}{2m} = \frac{n^2 h^2}{8mL^2},$$

where n is an integer.

In a **finite potential well**, the wave function extends into the classically forbidden region where the total energy is less than the potential energy. That this is possible is consistent with the uncertainty principle. The solutions to the Schrödinger equation in these areas are decaying exponentials.

Because quantum-mechanical particles can penetrate such classically forbidden areas, they can **tunnel** through thin barriers even though the potential energy in the barrier is greater than the total energy of the particle.

Questions

1. Compare a matter wave ψ to (a) a wave on a string, (b) an EM wave. Discuss similarities and differences.
2. Explain why Bohr's theory of the atom is not compatible with quantum mechanics, particularly the uncertainty principle.
3. Explain why it is that the more massive an object is, the easier it becomes to predict its future position.
4. In view of the uncertainty principle, why does a baseball seem to have a well-defined position and speed, whereas an electron does not?
5. Would it ever be possible to balance a very sharp needle precisely on its point? Explain.
6. When you check the pressure in a tire, doesn't some air inevitably escape? Is it possible to avoid this escape of air altogether? What is the relation to the uncertainty principle?
7. It has been said that the ground-state energy in the hydrogen atom can be precisely known but that the excited states have some uncertainty in their values (an "energy width"). Is this consistent with the uncertainty principle in its energy form? Explain.
8. If Planck's constant were much larger than it is, how would this affect our everyday life?
9. In what ways is Newtonian mechanics contradicted by quantum mechanics?
10. Discuss the connection between the zero-point energy for a particle in a box and the uncertainty principle.
11. The wave function for a particle in a box is zero at points within the box (except for $n = 1$). Does this mean that the probability of finding the particle at these points is zero? Does it mean that the particle cannot pass by these points? Explain.
12. What does the probability density look like for a particle in an infinite potential well for large values of n, say $n = 100$ or $n = 1000$? As n becomes very large, do your predictions approach classical predictions in accord with the correspondence principle?
13. For a particle in an infinite potential well the separation between energy states increases as n increases (see Eq. 39–13). But doesn't the correspondence principle require closer spacing between states as n increases so as to approach a classical (nonquantized) situation? Explain.
14. A particle is trapped in an infinite potential well. Describe what happens to the particle's ground-state energy and wave function as the potential walls become finite and get lower and lower until they finally reach zero ($U = 0$ everywhere).
15. A hydrogen atom and a helium atom, each with 4 eV of kinetic energy, approach a thin barrier 6 MeV high. Which has the greater probability of tunneling through?

Problems

Section 39–2

1. (II) The neutrons in a parallel beam, each having kinetic energy $\frac{1}{40}$ eV, are directed through two slits 1.0 mm apart. How far apart will the interference peaks be on a screen 1.0 m away?
2. (II) Bullets of mass 2.0 g are fired in parallel paths with speeds of 120 m/s through a hole 3.0 mm wide. How far from the hole must you be to detect a 1.0-cm spread in the beam?

Section 39–3

3. (I) A proton is traveling with a speed of $(4.825 \pm 0.012) \times 10^5$ m/s. With what maximum accuracy can its position be ascertained?
4. (I) An electron remains in an excited state of an atom for typically 10^{-8} s. What is the minimum uncertainty in the energy of the state (in eV)?
5. (I) If an electron's position can be measured to an accuracy of 1.6×10^{-8} m, how accurately can its velocity be known?
6. (II) A 12-g bullet leaves a rifle at a speed of 150 m/s. (a) What is the wavelength of this bullet? (b) If the position of the bullet is known to an accuracy of 0.55 cm (radius of the barrel), what is the minimum uncertainty in its vertical momentum? (c) If the accuracy of the bullet were determined only by the uncertainty principle (an unreasonable assumption), by how much might the bullet miss a pinpoint target 300 m away?
7. (II) An electron and a 150-g baseball are each traveling 75 m/s measured to an accuracy of 0.065 percent. Calculate and compare the uncertainty in position of each.

8. (II) Use the uncertainty principle to show that if an electron were present in the nucleus ($r \approx 10^{-15}$ m), its kinetic energy (use relativity) would be hundreds of MeV. (Since such electron energies are not observed, we conclude that electrons are not present in the nucleus.) [*Hint*: a particle can have energy as large as its uncertainty.]
9. (II) An electron in the $n = 2$ state of hydrogen remains there on the average about 10^{-8} s before jumping to the $n = 1$ state. (a) Estimate the uncertainty in the energy of the $n = 2$ state. (b) What fraction of the transition energy is this? (c) What is the wavelength, and width (in nm), of this line in the spectrum of hydrogen?
10. (II) How accurately can the position of a 2.50-keV electron be measured assuming its energy is known to 1.00 percent?
11. (III) In a double-slit experiment on electrons (or photons), suppose that we use indicators to determine which slit each electron went through (Section 39–2). These indicators must tell us the y coordinate to within $d/2$, where d is the distance between slits. Use the uncertainty principle to show that the interference pattern will be destroyed [*Note*: First show that the angle θ between maxima and minima of the interference pattern is given by $\frac{1}{2}\lambda/d$, Fig. 39–19.]

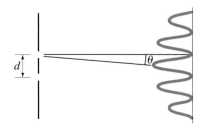

FIGURE 39–19
Problem 11.

*12. (III) (a) Show that $\Psi(x, t) = Ae^{i(kx-\omega t)}$ is a solution to the time-dependent Schrödinger equation for a free particle $[U(x) = U_0 = \text{constant}]$ but that $\Psi(x, t) = A\cos(kx - \omega t)$ and $\Psi(x, t) = A\sin(kx - \omega t)$ are not. (b) Show that the valid solution of part (a) satisfies conservation of energy if the de Broglie relations hold, $\lambda = h/p$, $\omega = E/\hbar$. That is, show that direct substitution in Eq. 39–7 gives

$$\hbar\omega = \frac{\hbar^2 k^2}{2m} + U_0.$$

Section 39–7

13. (I) Write the wave function for (a) a free electron and (b) a free proton, each having a constant velocity $v = 4.0 \times 10^5$ m/s.

14. (I) A free electron has a wave function $\psi(x) = A\sin(1.0 \times 10^{10}x)$, where x is given in meters. Determine the electron's (a) wavelength, (b) momentum, (c) speed, and (d) kinetic energy.

15. (II) Show that the uncertainty principle holds for a "wave packet" that is formed by two waves of similar wavelength λ_1 and λ_2. To do so, follow the argument leading up to Eq. 16–8, but use as the two waves $\psi_1 = A\sin k_1 x$ and $\psi_2 = A\sin k_2 x$. Then show that the width of each "wave packet" is $\Delta x = 2\pi/(k_1 - k_2) = 2\pi/\Delta k$ (from $t = 0.05$ s to $t = 0.15$ s in Fig. 16–18). Finally, show that $\Delta x\,\Delta p = h$ for this simple situation.

Section 39–8

16. (II) Show that for a particle in a perfectly rigid box, the wavelength of the wave function for any state is the de Broglie wavelength.

17. (II) What is the minimum speed of an electron trapped in a 0.10-nm-wide infinitely deep square well?

18. (II) An $n = 3$ to $n = 1$ transition for an electron trapped in a rigid box produces a 240-nm photon. What is the width of the box?

19. (II) An electron trapped in an infinitely deep square well has a ground-state energy $E = 8.0$ eV. (a) What is the longest wave-length photon this system can emit, and (b) what is the width of the well?

20. (II) The longest-wavelength line in the spectrum emitted by an electron trapped in an infinitely deep square well is 690 nm. What is the width of the well?

21. (II) For a particle in a box with rigid walls, determine whether our results for the ground state are consistent with the uncertainty principle by calculating the product $\Delta p\,\Delta x$. Take $\Delta x \approx L$, since the particle is known to be at least within the box. For Δp, note that although p is known $(= \hbar k)$, the direction of $\mathbf{p}$ is not known, so the x component could vary from $-p$ to $+p$; hence take $\Delta p \approx 2p$.

22. (II) Write a formula for the positions of (a) the maxima and (b) the minima in $|\psi|^2$ for a particle in the nth state in an infinite square well.

23. (II) Determine the lowest four energy levels and wave functions for an electron trapped in an infinitely deep potential well of width 2.0 nm.

24. (II) An electron is trapped in a rigid box 0.50 nm wide. (a) Determine the energies and wave functions for the four lowest states. (b) Determine the wavelengths of photons emitted for all possible transitions between these states.

25. (II) Consider an atomic nucleus to be a rigid box of width 10^{-14} m. What would be the ground-state energy for (a) an electron, (b) a neutron, and (c) a proton?

26. (II) Suppose that the walls of an infinite square well are at $x = -L/2$ and $x = L/2$. What will be the wave functions for the four lowest energy levels? Sketch the wave functions and probability densities.

27. (II) An electron is trapped in an infinitely deep potential well of width L. Determine the probability of finding the electron within $\frac{1}{4}L$ of either wall if it is (a) in the ground state, (b) in the $n = 4$ state. [Hint: Evaluate $\int_0^{L/4}|\psi|^2\,dx + \int_{3L/4}^{L}|\psi|^2\,dx$.] (c) What is the classical prediction?

28. (II) An electron is trapped in a 1.00-nm-wide rigid box. Determine the probability of finding the electron within 0.10 nm of the center of the box (on either side of center) for (a) $n = 1$, (b) $n = 5$, and (c) $n = 20$. (d) Compare to the classical prediction.

*29. (II) Sketch the wave functions and the probability distributions for the $n = 4$ and $n = 5$ states for a particle trapped in a finite square well.

*30. (II) An electron with 100 eV of kinetic energy in free space passes over a finite potential well 50 eV deep that stretches from $x = 0$ to $x = 0.50$ nm. What is the electron's wavelength (a) in free space, (b) when over the well? (c) Draw a diagram showing the potential energy and total energy as a function of x, and on the diagram sketch a possible wave function.

*31. (II) An electron is trapped in a 0.10-nm-wide finite square well of height $U_0 = 1.0$ keV. Estimate at what distance outside the walls of the well the ground state wave function drops to 1 percent of its value at the walls.

*32. (II) Suppose that a particle of mass m is trapped in a finite potential well that has a rigid wall at $x = 0$ ($U = \infty$ for $x < 0$) and a finite wall of height $U = U_0$ at $x = L$, Fig. 39–20. (a) Sketch the wave functions for the lowest three states. (b) What is the form of the wave function in the ground state in the three regions $x < 0$, $0 < x < L$, $x > L$?

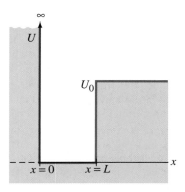

FIGURE 39–20 Problem 32.

33. (II) An electron approaches a potential barrier 10 eV high and 0.50 nm wide. If the electron has a 1.0 percent probability of tunneling through the barrier, what must its energy be?

34. (II) A 1.0-mA current of 1.0-MeV protons strikes a 2.0-MeV-high potential barrier 2.0×10^{-13} m thick. Estimate the transmitted current.

35. (II) For part (b) of Example 39–7, what effect will there be on the transmission coefficient if (a) the barrier height is raised 1 percent, (b) the barrier thickness is increased by 1 percent?

36. (III) Show that the transmission coefficient is given roughly by Eqs. 39–17 for a high or thick barrier, by calculating $|\psi(x = L)|^2/|\psi(0)|^2$. [*Hint*: Assume that ψ is a decaying exponential inside the barrier.]

37. (III) A uranium-238 nucleus ($Z = 92$) lasts about 5×10^9 years before it decays by emission of an alpha particle ($Z = 2$, $M = 4M_{\text{proton}}$). (a) Assuming that the α particle is a point, and the nucleus is roughly 8 fm in radius, estimate the height of the Coulomb barrier (the peak in Fig. 39–16). (b) The alpha particle, when free, has kinetic energy ≈ 4 MeV. Estimate the width of the barrier. (c) Assuming that the square well has $U = 0$ inside (where $U = 0$ far from the nucleus), calculate the speed of the alpha particle and how often it hits the barrier, and from this (and Eq. 39–17) estimate its lifetime. [*Hint*: Replace the $1/r$ coulomb barrier with an "averaged" rectangular barrier (as in Fig. 39–14) of width equal to $\frac{1}{3}$ that calculated in (b).]

General Problems

38. If an electron's position can be measured to an accuracy of 2.0×10^{-8} m, how accurately can its velocity be known?

39. Estimate the lowest possible energy of a neutron contained in a typical nucleus of radius 1.0×10^{-15} m. [*Hint*: A particle can have an energy at least as large as its uncertainty.]

40. The Z^0 boson, discovered in 1985, is the mediator of the weak nuclear force, and it typically decays very quickly. Its average rest energy is 91.19 GeV, but its short lifetime shows up as an intrinsic width of 2.5 GeV. What is the lifetime of this particle?

41. What is the uncertainty in the mass of a muon ($m = 105.7$ MeV/c^2), specified in eV/c^2, given its average lifetime of 2.20 μs?

42. A free neutron ($m = 1.67 \times 10^{-27}$ kg) has a mean life of 900 s. What is the uncertainty in its mass (in kg)?

43. Use the uncertainty principle to estimate the position uncertainty for the electron in the ground state of the hydrogen atom. [*Hint*: Determine the momentum using the Bohr model of Section 38–10 and assume the momentum can be anywhere between this value and zero.] How does this compare to the Bohr radius?

44. A neutron is trapped in an infinitely deep potential well 2.0 fm in width. Determine (a) the four lowest possible energy states and (b) their wave functions. (c) What is the wavelength and energy of a photon emitted when the neutron makes a transition between the two lowest states? In what region of the EM spectrum does this photon lie? [*Note*: This is a rough model of an atomic nucleus.]

45. Estimate the kinetic energy and speed of an alpha particle ($q = 2e$, $m = 4$ proton masses) trapped in a nucleus 10^{-14} m wide. Assume an infinitely deep square well potential.

46. *Simple Harmonic Oscillator.* Suppose that an electron is trapped not in a square well, but one whose potential energy is that of a simple harmonic oscillator: $U(x) = \frac{1}{2}Cx^2$. That is, if the electron is displaced from $x = 0$, a restoring force $F = -Cx$ acts on it, where C is constant. (a) Sketch this potential energy. (b) Show that $x = Ae^{-Bx^2}$ is a solution to the Schrödinger equation and that the energy of this state is $E = \frac{1}{2}\hbar\omega$, where $\omega = \sqrt{2\hbar B/m}$, and A and B are constants. [*Note*: This is the ground state and this energy $\frac{1}{2}\hbar\omega$ is the zero-point energy for a harmonic oscillator. The energies of higher states are $E_n = (n + \frac{1}{2})\hbar\omega$, where n is an integer.]

47. A 10-gram pencil, 20 cm long, is balanced on its point. Classically, this is a configuration of (unstable) equilibrium, so the pencil could remain there forever if it were perfectly placed. A quantum mechanical analysis shows that the pencil must fall. (a) Why is this the case? (b) Estimate (within a factor of 2) how long it will take the pencil to hit the table if it is initially positioned as well as possible? [*Hint*: Use the uncertainty principle in its angular form to obtain an expression for the initial angle $\phi_0 \approx \Delta\phi$.]

48. The probability density for finding a particle at a particular location is $|\psi|^2$. The average value $\bar{x}$, the location of a particle averaged over many measurements, is given by

$$\bar{x} = \int_0^L x|\psi|^2 \, dx$$

for a particle in a rigid box. Calculate the average value of x for any value of n.

49. By how much does the tunneling current through the tip of an STM change if the tip rises 0.010 nm from some initial height above a sodium surface with a work function $W_0 = 2.28$ eV? [*Hint*: Let the work function (see Section 38–2) equal the energy needed to raise the electron to the top of the barrier.]

50. A small ball of mass 1.0×10^{-6} kg is dropped on a table from a height of 2.0 m. After each bounce the ball rises to 80 percent of its height before the bounce because of its inelastic collision with the table. Estimate how many bounces occur before the uncertainty principle plays a role in the problem. [*Hint*: Determine when the uncertainty in the ball's speed is comparable to its speed of impact on the table.]

A neon tube is a thin glass tube, moldable into various shapes, filled with neon (or other) gas that glows with a particular color when a current at high voltage passes through it. Gas atoms, excited to upper energy levels, jump down to lower energy levels and emit light (photons) whose wavelengths (color) are characteristic of the type of gas.

In this chapter we study what quantum mechanics tells us about atoms, their wave functions and energy levels, including the effect of the exclusion principle. We also discuss interesting applications such as lasers and holography.

CHAPTER 40

Quantum Mechanics of Atoms

At the beginning of Chapter 39 we discussed the limitations of the Bohr theory of atomic structure and why a new theory was needed. Although the Bohr theory had great success in predicting the wavelengths of light emitted and absorbed by the hydrogen atom, it could not do so for more complex atoms. Nor did it explain *fine structure*, the splitting of emission lines into two or more closely spaced lines. And, as a theory, it was an uneasy mixture of classical and quantum ideas.

Quantum mechanics came to the rescue in 1925 and 1926, and in this chapter we examine the quantum-mechanical theory of atomic structure, which is far more complete than the old Bohr theory.

40–1 Quantum-Mechanical View of Atoms

Although the Bohr model has been discarded as an accurate description of nature, nonetheless, quantum mechanics reaffirms certain aspects of the older theory, such as that electrons in an atom exist only in discrete states of definite energy, and that a photon of light is emitted (or absorbed) when an electron makes a transition from one state to another. But quantum mechanics is a much deeper theory, and has provided us with a very different view of the atom. According to quantum mechanics, electrons do not exist in well-defined circular orbits as in the Bohr theory. Rather, the electron (because of its wave nature) can be thought of as spread out in space as if it were a "cloud." The size and shape of the electron cloud can be calculated for a given state of an atom. For the ground state in the hydrogen atom,

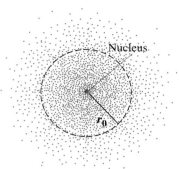

FIGURE 40–1 Electron cloud or "probability distribution" for the ground state of the hydrogen atom, as seen from afar. The circle represents the Bohr radius.

Probability distribution

the solution of the Schrödinger equation, as we will discuss in more detail in Section 40–3, gives

$$\psi(r) = \frac{1}{\sqrt{\pi r_0^3}} e^{-\frac{r}{r_0}}.$$

Here $\psi(r)$ is the wave function as a function of position, and it depends only on the radial distance r from the center, and not on angle θ or ϕ. (The constant r_0 has a value that happens to come out equal to the first Bohr radius.) Thus the electron cloud, whose density is $|\psi|^2$, for the ground state of hydrogen is spherically symmetric as shown in Fig. 40–1. The extent of the electron cloud at its higher densities roughly indicates the "size" of an atom, but just as a cloud may not have a distinct border, atoms do not have a precise boundary or a well-defined size. Not all electron clouds have a spherical shape, as we shall see later in this chapter. But note that $\psi(r)$, while becoming extremely small for large r (see the equation above), does not equal zero in any finite region. So quantum mechanics suggests that an atom is not mostly empty space. (Indeed, since $\psi \to 0$ only for $r \to \infty$, we might question the idea that there is any truly empty space in the universe.)

The electron cloud can be interpreted from either the particle or the wave viewpoint. Remember that by a particle we mean something that is localized in space—it has a definite position at any given instant. By contrast, a wave is spread out in space. The electron cloud, spread out in space as in Fig. 40–1, is a result of the wave nature of electrons. Electron clouds can also be interpreted as **probability distributions** for a particle. As we saw in Section 39–3, we cannot predict the path an electron will follow. After one measurement of its position we cannot predict exactly where it will be at a later time. We can only calculate the probability that it will be found at different points. If you were to make 500 different measurements of the position of an electron, considering it as a particle, the majority of the results would show the electron at points where the probability is high (dark area in Fig. 40–1). Only occasionally would the electron be found where the probability is low.

40–2 Hydrogen Atom: Schrödinger Equation and Quantum Numbers

The hydrogen atom is the simplest of all atoms, consisting of a single electron of charge $-e$ moving around a central nucleus (a single proton) of charge $+e$. It is with hydrogen that a study of atomic structure must begin.

The Schrödinger equation (see Eq. 39–5) includes a term containing the potential energy. For the hydrogen (H) atom, the potential energy is due to the Coulomb force between electron and proton:

$$U = -\frac{1}{4\pi\epsilon_0} \frac{e^2}{r}$$

where r is the radial distance from the proton (situated at $r = 0$) to the electron. See Fig. 40–2. The (time-independent) Schrödinger equation, which must now be written in three dimensions, is then

$$-\frac{\hbar^2}{2m}\left(\frac{\partial^2\psi}{\partial x^2} + \frac{\partial^2\psi}{\partial y^2} + \frac{\partial^2\psi}{\partial z^2}\right) - \frac{1}{4\pi\epsilon_0}\frac{e^2}{r}\psi = E\psi, \qquad \textbf{(40–1)}$$

where $\partial^2\psi/\partial x^2$, $\partial^2\psi/\partial y^2$, and $\partial^2\psi/\partial z^2$ are partial derivatives with respect to x, y, and z. To solve the Schrödinger equation for the H atom, it is usual to write it in terms of spherical coordinates (r, θ, ϕ). We will not, however, actually go through the process of solving it. Instead, we look at the properties of the solutions, and (in the next Section) at the wave functions themselves.

FIGURE 40–2 Potential energy $U(r)$ for the hydrogen atom. The radial distance r of the electron from the nucleus is given in terms of the Bohr radius r_0.

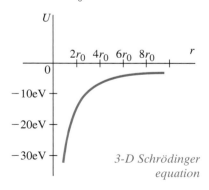

3-D Schrödinger equation

Recall from Chapter 39 that the solutions of the Schrödinger equation in one dimension for the infinite square well were characterized by a single quantum number, which we called n, which arises from applying the boundary conditions. In the three-dimensional problem of the H atom, the solutions of the Schrödinger equation are characterized by three quantum numbers corresponding to boundary conditions applied in the three dimensions. However, four different quantum numbers are actually needed to specify each state in the H atom, the fourth coming from a relativistic treatment. We now discuss each of these quantum numbers. Much of our analysis here will also apply to more complex atoms, which we discuss starting in Section 40–4.

Quantum mechanics predicts the same energy levels (Fig. 38–25) for the H atom as does the Bohr theory. That is,

$$E_n = -\frac{13.6\,\text{eV}}{n^2} \qquad n = 1, 2, 3, \ldots, \tag{40–2}$$

where n is an integer called the **principal quantum number**. This is the same as the quantum number n that appeared in Bohr theory. It can have any integer value from 1 to ∞. The total energy of a state in the H atom depends on n, Eq. 40–2.

The **orbital quantum number**, l, is related to the magnitude of the orbital angular momentum of the electron; l can take on integer values from 0 to $(n - 1)$. For the ground state, $n = 1$, l can only be zero.[†] But for $n = 3$, say, l can be 0, 1, or 2. The actual magnitude of the orbital angular momentum L is related to the quantum number l by

$$L = \sqrt{l(l + 1)}\,\hbar. \tag{40–3}$$

The value of l has almost no effect on the total energy in the hydrogen atom; only n does to any appreciable extent.[‡] But in atoms with two or more electrons, the energy does depend on l as well as n.

The **magnetic quantum number**, m_l, is related to the direction of the electron's orbital angular momentum, and it can take on integer values ranging from $-l$ to $+l$. For example, if $l = 2$, then m_l can be $-2, -1, 0, +1$, or $+2$. Since angular momentum is a vector, it is not surprising that both its magnitude and its direction would be quantized. For $l = 2$, the five different directions allowed can be represented by the diagram of Fig. 40–3. This limitation on the direction of **L** is often called **space quantization**. In quantum mechanics, the direction of the angular momentum is usually specified by giving its component along the z axis (this choice is arbitrary). Then L_z is related to m_l by the equation

$$L_z = m_l\hbar. \tag{40–4}$$

The values of L_x and L_y are not definite, however (see Problem 65). The name for m_l derives not from theory (which relates it to L_z) but from experiment. It was found that when a gas discharge tube (Fig. 38–18) was placed in a magnetic field, the spectral lines were split into several very closely spaced lines. This splitting, known as the **Zeeman effect**, implies that the energy levels must be split (Fig. 40–4), and thus that the energy of a state depends not only on n but also on m_l when a magnetic field is applied—hence the name "magnetic quantum number." (Why the energy should depend on the direction of **L** can be seen from a semiclassical view of a moving electron as an electric current which interacts with the magnetic field—see Chapter 27, Section 27–5, and Section 40–7.)

Finally, there is the **spin quantum number**, m_s, which for an electron can have only two values, $m_s = +\frac{1}{2}$ and $m_s = -\frac{1}{2}$. The existence of this quantum number did not come out of Schrödinger's original theory, as did n, l, and m_l. Instead, a subsequent modification by P. A. M. Dirac (1902–1984) explained its presence as a relativistic effect. The first hint that m_s was needed, however, came from experiment. A careful study of the spectral lines of hydrogen showed that each actually consisted of two (or more) very closely spaced lines even in the absence of an external

FIGURE 40–3 Quantization of angular momentum direction for $l = 2$.

FIGURE 40–4 When a magnetic field is applied, an $n = 3$, $l = 2$ energy level is split into five separate levels, corresponding to the five values of m_l (2, 1, 0, −1, −2). An $n = 2$, $l = 1$ level is split into three levels ($m_l = 1, 0, -1$). Transitions can occur between levels (not all transitions are shown), with photons of several slightly different frequencies being given off (the Zeeman effect).

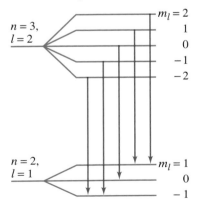

[†] Contrast this with the Bohr theory, which assigned $l = 1$ to the ground state (Eq. 38–10).

[‡] See discussion of *fine structure* starting at the bottom of this page.

magnetic field. It was at first hypothesized that this tiny splitting of energy levels, called **fine structure**, was due to angular momentum associated with a spinning of the electron. That is, the electron might spin on its axis as well as orbit the nucleus, just as the Earth spins on its axis as it orbits the Sun. The interaction between the tiny current of the spinning electron could then interact with the magnetic field due to the orbiting charge and cause the small observed splitting of energy levels. (So the energy depends slightly on m_l and m_s.) Today we consider this picture of a spinning electron as not legitimate. We cannot even view an electron as a localized object, much less a spinning one. What is important is that the electron can have two different states due to some intrinsic property that behaves as an angular momentum, and we still call this property "spin." The electron is said to have a spin quantum number $s = \frac{1}{2}$, which produces a spin angular momentum S given by

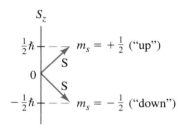

Spin angular momentum

$$S = \sqrt{s(s+1)}\,\hbar = \frac{\sqrt{3}}{2}\,\hbar.$$

(Compare Eq. 40–3.) This spin can have two different directions, $m_s = +\frac{1}{2}$ or $m_s = -\frac{1}{2}$, which are often said to be "spin up" and "spin down" (see Fig. 40–5). A state with spin down $(m_s = -\frac{1}{2})$ has slightly lower energy than one with spin up. (Note that we include m_s, but not s, in our list of quantum numbers since s is the same for all electrons.)

The possible values of the four quantum numbers for an electron in the hydrogen atom are summarized in Table 40–1.

FIGURE 40–5 The spin angular momentum S can take on only two directions, $m_s = +\frac{1}{2}$ or $-\frac{1}{2}$, called "spin up" and "spin down."

TABLE 40–1   Quantum Numbers for an Electron

Name	Symbol	Possible Values
Principal	n	$1, 2, 3, \ldots, \infty$.
Orbital	l	For a given n: l can be $0, 1, 2, \ldots, n - 1$.
Magnetic	m_l	For given n and l: m_l can be $l, l - 1, \ldots, 0, \ldots, -l$.
Spin	m_s	For each set of n, l, and m_l: m_s can be $+\frac{1}{2}$ or $-\frac{1}{2}$.

CONCEPTUAL EXAMPLE 40–1 | **Possible states for $n = 3$.** How many different states are possible for an electron whose principal quantum number is $n = 3$?

SOLUTION For $n = 3$, l can have the values $l = 2, 1, 0$. For $l = 2$, m_l can be $2, 1, 0, -1, -2$, which is five different possibilities. For each of these, m_s can be either up or down ($+\frac{1}{2}$ or $-\frac{1}{2}$), so for $l = 2$ there are $2 \times 5 = 10$ states. For $l = 1$, m_l can be $1, 0, -1$, and since m_s can be $+\frac{1}{2}$ or $-\frac{1}{2}$ for each of these, we have 6 more possible states. Finally, for $l = 0$, m_l can only be 0, and there are only 2 states corresponding to $m_s = +\frac{1}{2}$ and $-\frac{1}{2}$. The total number of states is $10 + 6 + 2 = 18$, as detailed in the following table:

n	l	m_l	m_s	n	l	m_l	m_s
3	2	2	$\frac{1}{2}$	3	2	-2	$-\frac{1}{2}$
3	2	2	$-\frac{1}{2}$	3	1	1	$\frac{1}{2}$
3	2	1	$\frac{1}{2}$	3	1	1	$-\frac{1}{2}$
3	2	1	$-\frac{1}{2}$	3	1	0	$\frac{1}{2}$
3	2	0	$\frac{1}{2}$	3	1	0	$-\frac{1}{2}$
3	2	0	$-\frac{1}{2}$	3	1	-1	$\frac{1}{2}$
3	2	-1	$\frac{1}{2}$	3	1	-1	$-\frac{1}{2}$
3	2	-1	$-\frac{1}{2}$	3	0	0	$\frac{1}{2}$
3	2	-2	$\frac{1}{2}$	3	0	0	$-\frac{1}{2}$

EXAMPLE 40–2 *E* and *L* for *n* = 3. Determine (*a*) the energy and (*b*) the orbital angular momentum for each of the states in Example 40–1.

SOLUTION (*a*) The energy of a state depends only on *n*, except for the very small corrections mentioned above, which we will ignore. Since *n* = 3 for all these states, they all have the same energy,

$$E_3 = -\frac{13.6 \text{ eV}}{(3)^2} = -1.51 \text{ eV}.$$

(*b*) For *l* = 0,

$$L = \sqrt{l(l+1)}\hbar = \sqrt{0(0+1)}\hbar = 0.$$

For *l* = 1,

$$L = \sqrt{1(1+1)}\hbar = \sqrt{2}\hbar$$
$$= 1.49 \times 10^{-34} \text{ J} \cdot \text{s}.$$

Atomic angular momenta are generally given as a multiple of $\hbar$ ($\sqrt{2}\hbar$ in this case), rather than in SI units. But note that for *l* = 1, *L* is on the order of 10^{-34} J·s. This means that macroscopic angular momenta will have such extremely high quantum numbers that the quantization of angular momentum will not be detectable: *L* will appear continuous, in accordance with the correspondence principle. Finally, for *l* = 2,

$$L = \sqrt{2(2+1)}\hbar = \sqrt{6}\hbar.$$

Another prediction of quantum mechanics is that when a photon is emitted or absorbed, transitions can occur only between states with values of *l* that differ by one unit:

$$\Delta l = \pm 1.$$

Selection rule

According to this **selection rule**, an electron in an *l* = 2 state can jump only to a state with *l* = 1 or *l* = 3. It cannot jump to a state with *l* = 2 or *l* = 0. A transition such as *l* = 2 to *l* = 0 is called a **forbidden transition**. Actually, such a transition is not absolutely forbidden and can occur, but only with very low probability compared to **allowed transitions**—those that satisfy the selection rule $\Delta l = \pm 1$. Since the orbital angular momentum of an H atom must change by one unit when it emits a photon, conservation of angular momentum tells us that the photon must carry off angular momentum. Indeed, experimental evidence of many sorts shows that the photon can be assigned a spin quantum number of 1.

"forbidden" transition

40–3 | Hydrogen Atom Wave Functions

The solution of the Schrödinger equation for the ground state of hydrogen—the state with lowest energy *E*—has an energy $E_1 = -13.6$ eV, as we have seen. The wave function for the ground state depends only on *r* and so is spherically symmetric. As already mentioned in Section 40–1, its form is

$$\psi_{100} = \frac{1}{\sqrt{\pi r_0^3}} e^{-\frac{r}{r_0}}$$ **(40–5a)**

Ground state in hydrogen

where $r_0 = h^2 \epsilon_0 / \pi m e^2 = 0.0529$ nm is the Bohr radius (Section 38–10). The subscript 100 on ψ represents the quantum numbers *n*, *l*, m_l:

$$\psi_{nlm_l}.$$

For the ground state, *n* = 1, *l* = 0, m_l = 0, and there is only one wave function that serves for both $m_s = +\frac{1}{2}$ and $m_s = -\frac{1}{2}$ (the value of m_s does not affect the

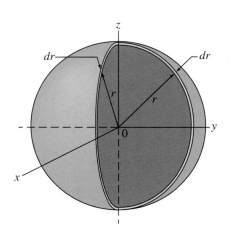

FIGURE 40–6 A spherical shell of thickness dr, inner radius r, and outer radius $r + dr$. Its volume is $dV = 4\pi r^2 \, dr$.

Hydrogen ground state radial probability distribution

FIGURE 40–7 The radial probability distribution P_r for the ground state of hydrogen, $n = 1$, $l = 0$. The peak occurs at $r = r_0$, the Bohr radius.

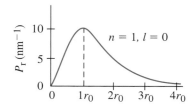

spatial dependence of the wave function for any state, since spin is an *internal* or *intrinsic* property of the electron). The probability density for the ground state is

$$|\psi_{100}|^2 = \frac{1}{\pi r_0^3} e^{-\frac{2r}{r_0}} \qquad (40\text{–}5b)$$

which falls off exponentially with r. Note that ψ_{100}, as well as all other wave functions we discuss, has been normalized:

$$\int_{\text{all space}} |\psi_{100}|^2 \, dV = 1.$$

The quantity $|\psi|^2 \, dV$ gives the probability of finding the electron in a volume dV about a given point. It is often more useful to specify the **radial probability distribution**, P_r, which is defined so that $P_r \, dr$ is the probability of finding the electron at a radial distance between r and $r + dr$ from the nucleus. That is, $P_r \, dr$ specifies the probability of finding the electron within a thin shell of thickness dr of inner radius r and outer radius $r + dr$, regardless of direction (see Fig. 40–6). The volume of this shell is the product of its surface area, $4\pi r^2$, and its thickness, dr:

$$dV = 4\pi r^2 \, dr.$$

Hence

$$|\psi|^2 \, dV = |\psi|^2 4\pi r^2 \, dr$$

and the radial probability distribution is

$$P_r = 4\pi r^2 |\psi|^2. \qquad (40\text{–}6)$$

For the ground state of hydrogen, P_r becomes

$$P_r = 4 \frac{r^2}{r_0^3} e^{-\frac{2r}{r_0}} \qquad (40\text{–}7)$$

and is plotted in Fig. 40–7. The peak of the curve is the "most probable" value of r and occurs for $r = r_0$, the Bohr radius, which we now prove.

EXAMPLE 40–3 **Most probable electron radius in hydrogen.** Determine the most probable distance r from the nucleus at which to find the electron in the ground state of hydrogen.

SOLUTION The peak of the curve in Fig. 40–7 corresponds to the most probable value of r. At this point the curve has zero slope, so we take the derivative of Eq. 40–7 and set it equal to zero:

$$\frac{d}{dr}\left(4 \frac{r^2}{r_0^3} e^{-\frac{2r}{r_0}}\right) = 0$$

$$\left(8 \frac{r}{r_0^3} - \frac{8r^2}{r_0^4}\right) e^{-\frac{2r}{r_0}} = 0.$$

Since $e^{-\frac{2r}{r_0}}$ goes to zero only at $r = \infty$, it is the term in parentheses that must be zero:

$$8 \frac{r}{r_0^3} - 8 \frac{r^2}{r_0^4} = 0.$$

Therefore,

$$\frac{r}{r_0^3} = \frac{r^2}{r_0^4}$$

or

$$r = r_0.$$

The most probable radial distance of the electron from the nucleus according to quantum mechanics is at the Bohr radius, an interesting coincidence.

EXAMPLE 40–4 **Calculating probability.** Determine the probability of finding the electron in the ground state of hydrogen within two Bohr radii of the nucleus.

SOLUTION We need to integrate P_r from $r = 0$ out to $r = 2r_0$: that is, we want to find

$$P_r = \int_{r=0}^{2r_0} |\psi|^2 \, dV = \int 4 \frac{r^2}{r_0^3} e^{-\frac{2r}{r_0}} \, dr.$$

We first make the substitution

$$x = 2\frac{r}{r_0}$$

and then integrate by parts $\left(\int u \, dv = uv - \int v \, du \right)$ letting $u = x^2$ and $dv = e^{-x} \, dx$ (and note that $dx = 2 \, dr/r_0$):

$$P_r = \frac{1}{2} \int_{x=0}^{4} x^2 e^{-x} \, dx = \frac{1}{2} \left[-x^2 e^{-x} + \int 2x e^{-x} \, dx \right]\Bigg|_0^4.$$

The second term we also integrate by parts with $u = 2x$ and $dv = e^{-x} \, dx$:

$$P_r = \frac{1}{2} \left[-x^2 e^{-x} - 2x e^{-x} + 2 \int e^{-x} \, dx \right]\Bigg|_0^4$$

$$= \left(-\tfrac{1}{2} x^2 - x - 1 \right) e^{-x} \Bigg|_0^4.$$

We evaluate this at $x = 0$ and at $x = 2(2r_0)/r_0 = 4$:

$$P_r = (-8 - 4 - 1) e^{-4} + e^0 = 0.76$$

or 76 percent. Thus the electron would be found 76 percent of the time within 2 Bohr radii of the nucleus and 24 percent of the time farther away. Note that this result depends on our wave function being properly normalized, which it is, as is readily shown by letting $r \to \infty$ and integrating over all space: $\int_0^\infty |\psi|^2 \, dV = 1$.

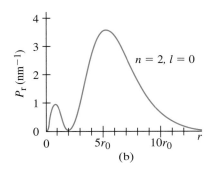

$|\psi_{200}|^2$

(a)

The first excited state in hydrogen has $n = 2$. For $l = 0$, the solution of the Schrödinger equation (Eq. 40–1) is a wave function that is again spherically symmetric:

$$\psi_{200} = \frac{1}{\sqrt{32\pi r_0^3}} \left(2 - \frac{r}{r_0} \right) e^{-\frac{r}{2r_0}}. \qquad \textbf{(40–8)}$$

Figure 40–8a shows the probability distribution $|\psi_{200}|^2$ and Fig. 40–8b shows[†] a plot of the radial probability distribution

$$P_r = \frac{1}{8} \frac{r^2}{r_0^3} \left(2 - \frac{r}{r_0} \right)^2 e^{-\frac{r}{r_0}}.$$

There are two peaks in this curve; the second, at $r \approx 5r_0$, is higher and corresponds to the most probable value for r in the $n = 2$, $l = 0$ state. We see that the

FIGURE 40–8 (a) The probability distribution $|\psi_{200}|^2$ and (b) the radial probability distribution P_r for the $n = 2$, $l = 0$ state in hydrogen.

[†] Just as for a particle in a deep square well potential (see Figs. 39–9 and 39–10), the higher the energy, the more nodes there are in ψ and $|\psi|^2$ also for the H atom.

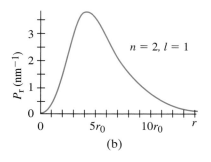

FIGURE 40–9 (a) The probability distribution for the three states with $n = 2$, $l = 1$. (b) Radial probability distribution for the sum of the three states with $n = 2$, $l = 1$, and $m_l = +1$, 0, or -1.

electron tends to be somewhat farther from the nucleus in the $n = 2$, $l = 0$ state than in the $n = 1$, $l = 0$ state.

For the state with $n = 2$, $l = 1$, there are three possible wave functions, corresponding to $m_l = +1$, 0, or −1:

$$\psi_{210} = \frac{z}{\sqrt{32\pi r_0^5}} e^{-\frac{r}{2r_0}}$$

$$\psi_{211} = \frac{x + iy}{\sqrt{64\pi r_0^5}} e^{-\frac{r}{2r_0}} \tag{40–9}$$

$$\psi_{21-1} = \frac{x - iy}{\sqrt{64\pi r_0^5}} e^{-\frac{r}{2r_0}}.$$

where i is the imaginary number $i = \sqrt{-1}$. These wave functions are *not* spherically symmetric. The probability distributions, $|\psi|^2$, are shown in Fig. 40–9a, where we can see their directional orientation.

You may wonder how such non-spherically symmetric wave functions arise when the potential energy in the Schrödinger equation has spherical symmetry. Indeed, how could an electron select one of these states? In the absence of any external influence, such as a magnetic field in a particular direction, all three of these states are equally likely, and they all have the same energy. Thus an electron can be considered to spend one-third of its time in each of these states. The net effect, then, is the sum of these three wave functions squared:

$$|\psi_{210}|^2 + |\psi_{211}|^2 + |\psi_{21-1}|^2,$$

which is spherically symmetric, since $x^2 + y^2 + z^2 = r^2$. The radial probability distribution for this sum is shown in Fig. 40–9b.

Although the spatial distributions of the electron can be calculated for the various states, it is difficult to measure them experimentally. Indeed, most of the experimental information about atoms has come from a careful examination of the emission spectra under various conditions.

40–4 Complex Atoms; the Exclusion Principle

We have discussed the hydrogen atom in detail because it is the simplest to deal with. Now we briefly discuss more complex atoms, those that contain more than one electron, and whose energy levels can be determined experimentally from an analysis of the emission spectra. The energy levels are *not* the same as in the H atom, since the electrons interact with each other as well as with the nucleus. For atoms with more than one electron, the energy levels depend on both n and l.

Atomic number

The number of electrons in a neutral atom is called its **atomic number**, Z; Z is also the number of positive charges (protons) in the nucleus, and determines what kind of atom it is. That is, Z determines most of the properties that distinguish one atom from another.

Although modifications of the Bohr theory had been attempted in order to deal with complex atoms, the development of quantum mechanics in the years after 1925 proved far more successful. The mathematics becomes very difficult, however, since in multi-electron atoms, each electron is not only attracted to the nucleus but is repelled by the other electrons.

The simplest approach has been to treat each electron in an atom as occupying a particular state characterized by the quantum numbers n, l, m_l, and m_s. But to understand the possible arrangements of electrons in an atom, a new principle was needed. It was introduced by Wolfgang Pauli (1900–1958; Fig. 39–2) and is called the **Pauli exclusion principle**. It states:

No two electrons in an atom can occupy the same quantum state.

Pauli exclusion principle

Thus, no two electrons in an atom can have exactly the same set of the quantum numbers n, l, m_l, and m_s. The Pauli exclusion principle[†] forms the basis not only for understanding complex atoms, but also for understanding molecules and bonding, and other phenomena as well.

    Let us now look at the structure of some of the simpler atoms when they are in the ground state. After hydrogen, the next simplest atom is *helium*, with two electrons. Both electrons can have $n = 1$, since one can have spin up $\left(m_s = +\frac{1}{2}\right)$ and the other spin down $\left(m_s = -\frac{1}{2}\right)$, thus satisfying the exclusion principle. Of course, since $n = 1$, l and m_l must be zero (Table 40–1). Thus the two electrons have the quantum numbers indicated in the table in the margin.

    Lithium has three electrons, two of which can have $n = 1$. But the third cannot have $n = 1$ without violating the exclusion principle. Hence the third electron must have $n = 2$. Since it happens that the $n = 2$, $l = 0$ level has a lower energy than $n = 2$, $l = 1$, the electrons in the ground state have the quantum numbers indicated in the table in the margin. Of course, the quantum numbers of the third electron could also be, say, $\left(3, 1, -1, \frac{1}{2}\right)$. But the atom in this case would be in an excited state since it would have greater energy. It would not be long before it jumped to the ground state with the emission of a photon. At room temperature, unless extra energy is supplied (as in a discharge tube), the vast majority of atoms are in the ground state because the average thermal energy $\left(K = \frac{3}{2}kT \approx 0.04\,\text{eV} \text{ at } T = 300\,\text{K}\right)$ is much less than the energy needed to excite atoms (1 to 10 eV).

    We can continue in this way to describe the quantum numbers of each electron in the ground state of larger and larger atoms. That for sodium, with its eleven electrons, is shown in the table in the margin.

    Figure 40–10 shows a simple energy level diagram where occupied states are shown as up or down arrows $\left(m_s = +\frac{1}{2} \text{ or } -\frac{1}{2}\right)$, and possible empty states are shown as small circles.

    The ground-state configuration for all atoms is given in the **periodic table**, which is displayed inside the back cover of this book, and discussed in the next Section.

[†] The exclusion principle applies to identical particles whose spin quantum number is a half-integer $\left(\frac{1}{2}, \frac{3}{2}, \text{and so on}\right)$, including electrons, protons, and neutrons; such particles are called **fermions** (after E. Fermi who derived a statistical theory describing them—see Section 41–6.). The exclusion principle does not apply to particles with integer spin quantum number (0, 1, 2, and so on), such as the photon and π meson, all of which are referred to as **bosons** (after S. N. Bose, who derived a statistical theory for them).

Helium, $Z = 2$

n	l	m_l	m_s
1	0	0	$-\frac{1}{2}$
1	0	0	$\frac{1}{2}$

Lithium, $Z = 3$

n	l	m_l	m_s
1	0	0	$-\frac{1}{2}$
1	0	0	$\frac{1}{2}$
2	0	0	$-\frac{1}{2}$

Sodium, $Z = 11$

n	l	m_l	m_s
1	0	0	$-\frac{1}{2}$
1	0	0	$\frac{1}{2}$
2	0	0	$-\frac{1}{2}$
2	0	0	$\frac{1}{2}$
2	1	1	$-\frac{1}{2}$
2	1	1	$\frac{1}{2}$
2	1	0	$-\frac{1}{2}$
2	1	0	$\frac{1}{2}$
2	1	-1	$-\frac{1}{2}$
2	1	-1	$\frac{1}{2}$
3	0	0	$-\frac{1}{2}$

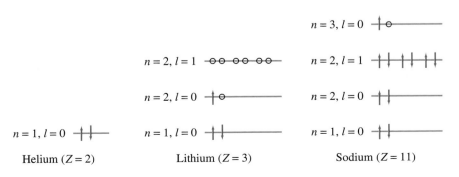

FIGURE 40–10 Energy level diagram showing occupied states (arrows) and unoccupied states ($\circ$) for He, Li, and Na. Note that we have shown the $n = 2$, $l = 1$ level of Li even though it is empty.

TABLE 40–2 Values of l

Value of l	Letter Symbol	Maximum Number of Electrons in Subshell
0	s	2
1	p	6
2	d	10
3	f	14
4	g	18
5	h	22
⋮	⋮	⋮

TABLE 40–3
Electron Configuration
of Some Elements

Z (No. of Electrons)	Element[†]	Ground State Configuration (outer electrons)
1	H	$1s^1$
2	He	$1s^2$
3	Li	$2s^1$
4	Be	$2s^2$
5	B	$2s^2 2p^1$
6	C	$2s^2 2p^2$
7	N	$2s^2 2p^3$
8	O	$2s^2 2p^4$
9	F	$2s^2 2p^5$
10	Ne	$2s^2 2p^6$
11	Na	$3s^1$
12	Mg	$3s^2$
13	Al	$3s^2 3p^1$
14	Si	$3s^2 3p^2$
15	P	$3s^2 3p^3$
16	S	$3s^2 3p^4$
17	Cl	$3s^2 3p^5$
18	Ar	$3s^2 3p^6$
19	K	$4s^1$
20	Ca	$4s^2$
21	Sc	$3d^1 4s^2$
22	Ti	$3d^2 4s^2$
23	V	$3d^3 4s^2$
24	Cr	$3d^5 4s^1$
25	Mn	$3d^5 4s^2$
26	Fe	$3d^6 4s^2$

[†]Names of elements can be found in Appendix D.

40–5 The Periodic Table of Elements

More than a century ago, Dmitri Mendeleev (1834–1907) arranged the then known elements into what we now call the **periodic table** of the elements. The atoms were arranged according to increasing mass, but also so that elements with similar chemical properties would fall in the same column. Today's version is shown inside the back cover. Each square contains the atomic number Z, the symbol for the element, and the atomic mass (in atomic mass units). Finally, in the lower left corner the configuration of the ground state of the atom is given. This requires some explanation. Electrons with the same value of n are referred to as being in the same **shell**. Electrons with $n = 1$ are in one shell (the K shell), those with $n = 2$ are in a second shell (the L shell), those with $n = 3$ are in the third (M) shell, and so on. Electrons with the same values of n and l are referred to as being in the same **subshell**. Letters are often used to specify the value of l as shown in Table 40–2. That is, $l = 0$ is the s subshell; $l = 1$ is the p subshell; $l = 2$ is the d subshell; beginning with $l = 3$, the letters follow the alphabet, f, g, h, i, and so on. (The first letters s, p, d, and f were originally abbreviations of "sharp," "principal," "diffuse," and "fundamental," experimental terms referring to the spectra.)

The Pauli exclusion principle limits the number of electrons possible in each shell and subshell. For any value of l, there are $2l + 1$ different m_l values (m_l can be any integer from 1 to l, from -1 to $-l$, or zero), and two different m_s values. There can be, therefore, at most $2(2l + 1)$ electrons in any l subshell. For example, for $l = 2$, five m_l values are possible $(2, 1, 0, -1, -2)$, and for each of these, m_s can be $+\frac{1}{2}$ or $-\frac{1}{2}$ for a total of $2(5) = 10$ states. Table 40–2 lists the maximum number of electrons that can occupy each subshell.

Since the energy levels depend almost entirely on the values of n and l, it is customary to specify the electron configuration simply by giving the n value and the appropriate letter for l, with the number of electrons in each subshell given as a superscript. The ground-state configuration of sodium, for example, is written as $1s^2 2s^2 2p^6 3s^1$. This is simplified in the periodic table by specifying the configuration only of the outermost electrons and any other nonfilled subshells (see Table 40–3 here, and the periodic table inside the back cover).

| **CONCEPTUAL EXAMPLE 40–5** | **Electron configurations.** Which of the following electron configurations are possible, and which forbidden? (a) $1s^2 2s^2 2p^6 3s^3$; (b) $1s^2 2s^2 2p^6 3s^2 3p^5 4s^2$; (c) $1s^2 2s^2 2p^6 2d^1$.

RESPONSE (a) This is not allowed, because there are too many electrons in the s subshell in the M ($n = 3$) shell. The s subshell has $m_l = 0$, with two slots only: for "spin up" and "spin down" electrons. (b) This is allowed, but it is an excited state. One of the electrons in the 3p subshell has jumped up to the 4s subshell. Since there are 19 electrons, the element is potassium. (c) This is not allowed, because there is no d ($l = 2$) subshell in the $n = 2$ shell (Table 40–1). The outermost electron will have to be (at least) in the $n = 3$ shell.

The grouping of atoms in the periodic table is according to increasing atomic number, Z. There is also a strong regularity according to chemical properties. And although this is treated in chemistry textbooks, we discuss it here briefly because it is a result of quantum mechanics. See the periodic table on the inside back cover.

All the noble gases (in the last column of the periodic table) have completely filled shells or subshells. That is, their outermost subshell is completely full, and the electron distribution is spherically symmetric. With such full spherical symmetry, other electrons are not attracted nor are electrons readily lost (ionization energy is high). This is why the noble gases are nonreactive (more on this when we discuss molecules and bonding in Chapter 41). Column seven contains the **halogens**, which lack one electron from a filled shell. Because of the shapes of the orbits (see Section 41–1), an additional electron can be accepted from another atom, and hence these elements are quite reactive. They have a valence of −1, meaning that when an extra electron is acquired, the resulting ion has a net charge of −1e. At the left of the periodic table, column I contains the **alkali metals**, all of which have a single outer s electron. This electron spends most of its time outside the inner closed shells and subshells which shield it from most of the nuclear charge. Indeed, it is relatively far from the nucleus and is attracted to it by a net charge of only about +1e, because of the shielding effect of the other electrons. Hence this outer electron is easily removed and can spend much of its time around another atom, forming a molecule. This is why the alkali metals are highly reactive and have a valence of +1. The other columns of the periodic table can be treated similarly.

The presence of the transition elements in the center of the table, as well as the lanthanides (rare earths) and actinides below, is a result of incomplete inner shells. For the lowest Z elements, the subshells are filled in a simple order: first 1s, then 2s, followed by 2p, 3s, and 3p. You might expect that 3d ($n = 3, l = 2$) would be filled next, but it isn't. Instead, the 4s level actually has a slightly lower energy than the 3d (due to electrons interacting with each other), so it fills first (K and Ca). Only then does the 3d shell start to fill up, beginning with Sc. (The 4s and 3d levels are close, so some elements have only one 4s electron, such as Cr.) Most of the chemical properties of these **transition elements** are governed by the relatively loosely held 4s electrons and hence they usually have valences of +1 or +2. See also Table 40–3. A similar effect is responsible for the rare earths, which are shown at the bottom of the periodic table for convenience. All have very similar chemical properties, which are determined by their two outer 6s or 7s electrons, whereas the different numbers of electrons in the unfilled inner shells have little effect.

40–6 X-Ray Spectra and Atomic Number

The line spectra of atoms in the visible, UV, and IR regions of the EM spectrum are mainly due to transitions between states of the outer electrons. Much of the charge of the nucleus is shielded from these electrons by the negative charge on the inner electrons. But the innermost electrons in the $n = 1$ shell "see" the full charge of the nucleus. Since the energy of a level is proportional to Z^2 (see Eq. 38–13), for an atom with $Z = 50$, we would expect wavelengths about $50^2 = 2500$ times shorter than those found in the Lyman series of hydrogen (around 100 nm), or 10^{-2} to 10^{-1} nm. Such short wavelengths lie in the X-ray region of the spectrum.

X-rays are produced when electrons accelerated by a high voltage strike the metal target inside the X-ray tube (Section 36–10). If we look at the spectrum of wavelengths emitted by an X-ray tube, we see that the spectrum consists of two parts: a continuous spectrum with a cutoff at some λ_0 which depends only on the voltage across the tube, and a series of peaks superimposed. A typical example is shown in Fig. 40–11. The smooth curve and the cutoff wavelength λ_0 move to the left as the voltage across the tube increases. The peaks (labeled K_α and K_β in Fig. 40–11), however, remain at the same wavelength when the voltage is changed, although they are located at different wavelengths when different target materials are used. This observation suggests that the peaks are characteristic of the material used. Indeed, we can explain them by imagining that the electrons accelerated by the high voltage of the tube can reach sufficient energies that when they collide with the atoms of the target, they can knock out one of the very tightly held inner electrons. Then we explain these **characteristic X-rays** (the peaks in Fig. 40–11) as photons emitted when an electron in an upper state drops down to fill the vacated lower state. The K lines result from transitions *into* the K shell ($n = 1$). The K_α line is a transition that originates from the $n = 2$ (L) shell, the K_β line from the $n = 3$ (M) shell. An L line is due to a transition into the L shell, and so on.

FIGURE 40–11 Spectrum of X-rays emitted from a molybdenum target in an X-ray tube operated at 50 kV.

Characteristic X-rays

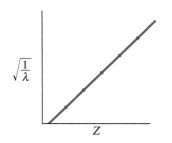

FIGURE 40–12 Plot of $\sqrt{1/\lambda}$ vs. Z for K_α X-ray lines.

Measurement of the characteristic X-ray spectra has allowed a determination of the inner energy levels of atoms. It has also allowed the determination of Z values for many atoms, since (as we have seen) the wavelength of the shortest X-rays emitted will be inversely proportional to Z^2. Actually, for an electron jumping from, say, the $n = 2$ to the $n = 1$ level, the wavelength is inversely proportional to $(Z - 1)^2$ because the nucleus is shielded by the one electron that still remains in the 1s level. In 1914, H. G. J. Moseley (1887–1915) found that a plot of $\sqrt{1/\lambda}$ versus Z produced a straight line, Fig. 40–12. The Z values of a number of elements were determined by fitting them to such a **Moseley plot**. The work of Moseley put the concept of atomic number on a firm experimental basis.

EXAMPLE 40–6 **X-ray wavelength.** Estimate the wavelength for an $n = 2$ to $n = 1$ transition in molybdenum ($Z = 42$). What is the energy of such a photon?

SOLUTION We use the Bohr formula, Eq. 38–14, with Z^2 replaced by $(Z - 1)^2 = (41)^2$. Or, more simply, we can use the result of Example 38–12 for the $n = 2$ to $n = 1$ transition in hydrogen ($Z = 1$). Since

$$\lambda \propto \frac{1}{(Z - 1)^2},$$

we will have

$$\lambda = \frac{(1.22 \times 10^{-7}\,\text{m})}{(41)^2} = 0.073\,\text{nm}.$$

This is close to the measured value (Fig. 40–11) of 0.071 nm. Each of these photons would have energy (in eV) of:

$$E = hf = \frac{hc}{\lambda} = \frac{(6.63 \times 10^{-34}\,\text{J}\cdot\text{s})(3.00 \times 10^8\,\text{m/s})}{(7.3 \times 10^{-11}\,\text{m})(1.60 \times 10^{-19}\,\text{J/eV})} = 17\,\text{keV}.$$

EXAMPLE 40–7 **Determining atomic number.** High-energy photons are used to bombard an unknown material. The strongest peak is found for X-rays emitted with an energy of 66 keV. Guess what the material is.

SOLUTION The strongest X-rays are generally for the K_α line (see Fig. 40–11) which occurs when photons knock out K shell electrons (the innermost orbit) and their place is taken by electrons from the L shell. We use the Bohr model, and assume the electrons "see" a nuclear charge of $Z - 1$ (screened by one electron). The hydrogen transition $n = 2$ to $n = 1$ would yield about 10.2 eV (see Fig. 38–25 or Example 38–12). Then since energy E is proportional to Z^2 (Eq. 38–13), or rather $(Z - 1)^2$ as we've just discussed, we can write

$$\frac{(Z - 1)^2}{1^2} = \frac{66 \times 10^3\,\text{eV}}{10.2\,\text{eV}} = 6.5 \times 10^3,$$

so $Z - 1 = \sqrt{6500} = 81$, and $Z = 82$, which makes it lead.

Now we briefly analyze the continuous part of an X-ray spectrum (Fig. 40–11) based on the photon theory of light. When electrons strike the target, they collide with atoms of the material and give up most of their energy as heat (about 99 percent, so X-ray tubes must be cooled). However, electrons can also give up energy by emitting a photon of light. An electron can be decelerated by interaction with atoms of the target (Fig. 40–13). But an accelerating charge can emit radiation (Chapter 32), and in this case it is called **bremsstrahlung** (German for "braking

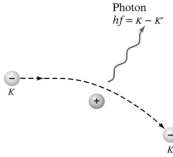

FIGURE 40–13 Bremsstrahlung photon produced by an electron decelerated by interaction with a target atom.

radiation"). Because energy is conserved, the energy of the emitted photon, hf, must equal the loss of kinetic energy K of the electron, $\Delta K = K - K'$, so

$$hf = \Delta K.$$

An electron may lose all or a part of its energy in such a collision. The continuous X-ray spectrum (Fig. 40–11) is explained as being due to such bremsstrahlung collisions in which varying amounts of energy are lost by the electrons. The shortest-wavelength X-ray (the highest frequency) must be due to an electron that gives up *all* its kinetic energy to produce one photon in a single collision. Since the initial kinetic energy of an electron is equal to the energy given it by the accelerating voltage, V, then $K = eV$. In a single collision in which the electron is brought to rest, we have

$$hf_0 = eV$$

or

$$\lambda_0 = \frac{hc}{eV}, \qquad\qquad\qquad (40\text{–}10)$$

where $\lambda_0 = c/f_0$ is the cutoff wavelength, Fig. 40–11. This prediction for λ_0 corresponds precisely with that observed experimentally. This result is further evidence that X-rays are a form of electromagnetic radiation (light)[†] and that the photon theory of light is valid.

EXAMPLE 40–8 **Cutoff wavelength.** What is the shortest-wavelength X-ray photon emitted in an X-ray tube subjected to 50 kV?

SOLUTION From Eq. 40–10,

$$\lambda_0 = \frac{(6.6 \times 10^{-34}\,\text{J·s})(3.0 \times 10^8\,\text{m/s})}{(1.6 \times 10^{-19}\,\text{C})(5.0 \times 10^4\,\text{V})} = 2.5 \times 10^{-11}\,\text{m},$$

or 0.025 nm. This agrees well with experiment, Fig. 40–11.

---

* **40–7** **Magnetic Dipole Moments; Total Angular Momentum**

An electron orbiting the nucleus of an atom can be considered as a current loop, classically, and thus might be expected to have a **magnetic dipole moment** as discussed in Chapter 27. Indeed, in Example 27–9 we did a classical calculation of the magnetic dipole moment of the electron in the ground state of hydrogen based, essentially, on the Bohr model, and found it to give

$$\mu = IA = \tfrac{1}{2}evr.$$

Here v is the orbital velocity of the electron, and for a particle moving in a circle of radius r, its angular momentum is

$$L = mvr.$$

So we can write

$$\mu = \frac{1}{2}\frac{e}{m}L.$$

The direction of the angular momentum $\mathbf{L}$ is perpendicular to the plane of the current loop. So is the direction of the magnetic dipole moment vector $\boldsymbol{\mu}$, although in

---

[†] If X-rays were not photons but rather neutral particles with rest mass m_0, Eq. 40–10 would not hold.

the opposite direction since the electron's charge is negative. Hence we can write the vector equation

Magnetic dipole moment, H atom

$$\boldsymbol{\mu} = -\frac{1}{2}\frac{e}{m}\mathbf{L}. \qquad (40\text{--}11)$$

This rough semiclassical derivation was based on the Bohr theory. The same result (Eq. 40–11), is obtained using quantum mechanics. Since $\mathbf{L}$ is quantized in quantum mechanics, the magnetic dipole moment, too, must be quantized.

As we saw in Section 27–5, a magnetic dipole moment in a magnetic field $\mathbf{B}$ experiences a torque, and the potential energy U of such a system depends on $\mathbf{B}$ and the orientation of $\boldsymbol{\mu}$ relative to $\mathbf{B}$ (Eq. 27–12):

$$U = -\boldsymbol{\mu} \cdot \mathbf{B}.$$

If the magnetic field $\mathbf{B}$ is in the z direction, then $U = -\mu_z B_z$ and from Eq. 40–4 $(L_z = m_l\hbar)$ and Eq. 40–11, we have

$$\mu_z = -\frac{e\hbar}{2m}m_l.$$

(Be careful here not to confuse the electron mass m with the magnetic quantum number, m_l.) It is useful to define the quantity

Bohr magneton

$$\mu_B = \frac{e\hbar}{2m} \qquad (40\text{--}12)$$

which is called the **Bohr magneton** and has the value $\mu_B = 9.27 \times 10^{-24}\,\text{J/T}$ (joule/tesla). Then we can write

$$\mu_z = -\mu_B m_l, \qquad (40\text{--}13)$$

where m_l has integer values from 0 to $\pm l$ (see Table 40–1). An atom placed in a magnetic field would have its energy split into levels that differ by $\Delta U = \mu_B B$; this is the *Zeeman effect*, and was shown in Fig. 40–4.

Zeeman effect

* Stern-Gerlach Experiment

The first evidence of this *space quantization* (Section 40–2) came in 1922 in a famous experiment known as the **Stern-Gerlach experiment**. Silver atoms (and later others) were heated in an oven from which they escaped as shown in Fig. 40–14. The atoms were made to pass through a collimator, which eliminated all but a narrow beam. The beam then passed into a *nonhomogeneous* magnetic field. The field was deliberately made nonhomogeneous so that it would exert a force on atomic magnetic moments: remember that the potential energy (in this case $-\boldsymbol{\mu} \cdot \mathbf{B}$) must change in space if there is to be a force ($F_x = -dU/dx$, etc., Section 8–2). If $\mathbf{B}$ has a gradient along the z axis, as in Fig. 40–14, then the force is along z:

$$F_z = -\frac{dU}{dz} = \mu_z\frac{dB_z}{dz}.$$

FIGURE 40–14 The Stern-Gerlach experiment, which is done inside a vacuum chamber.

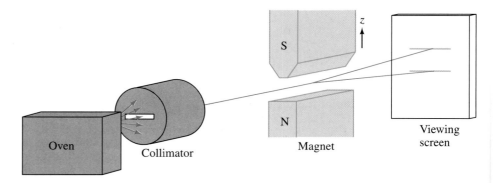

Oven   Collimator   Magnet   Viewing screen

Thus the silver atoms would be deflected up or down depending on the value of μ for each atom. Classically, we would expect to see a continuous distribution on the viewing screen, since we would expect the atoms to have randomly oriented magnetic moments. But Stern and Gerlach saw instead two distinct lines for silver (and for other atoms sometimes more than two lines). These observations were the first evidence for space quantization, though not fully explained until a few years later. If the lines were due to orbital angular momentum, there should have been an odd number of them, corresponding to the possible values of m_l (since $\mu_z = -\mu_B m_l$). For $l = 0$, there is only one possibility, $m_l = 0$. For $l = 1$, m_l can be 1, 0, or -1, and we would expect three lines, and so on. Why there are only two lines was eventually explained by the concept of electron spin. With a spin of $\frac{1}{2}$, the electron spin can have only two orientations in space, as we saw in Fig. 40–5. Hence a magnetic dipole moment associated with spin would have only two positions. Thus, the two states for silver seen in the Stern-Gerlach experiment must be due to the spin of its one valence electron. Silver atoms must thus have zero orbital angular momentum but a total spin of $\frac{1}{2}$ due to this one valence electron (of its 47 electrons, the spins of the first 46 cancel). Two lines were found also for the H atom in its ground state, again due to the spin $\frac{1}{2}$ of its electron since the orbital angular momentum is zero.

The Stern-Gerlach deflection is proportional to the magnetic dipole moment, μ_z, and for a spin $\frac{1}{2}$ particle we expect $\mu_z = -\mu_B m_s = -(\frac{1}{2})(e\hbar/2m)$ as for the case of orbital angular momentum, Eq. 40–13. Instead, μ_z for spin was found to be about twice as large:

$$\mu_z = -g\mu_B m_s, \qquad \qquad (40\text{–}14)$$

Magnetic dipole moment for spin

where g, called the **g-factor** or **gyromagnetic ratio**, has been measured to be slightly larger than 2: $g = 2.0023 \cdots$ for a free electron. This unexpected factor of (about) 2 clearly indicates that spin cannot be viewed as a classical angular momentum. It is a purely quantum-mechanical effect. Equation 40–14 is the same as Eq. 40–13 for orbital angular momentum with m_s replacing m_l. But for the orbital case, $g = 1$.

* Total Angular Momentum J

An atom can have both orbital and spin angular momenta. For example, in the 2p state of hydrogen $l = 1$ and $s = \frac{1}{2}$. In the 4d state, $l = 2$ and $s = \frac{1}{2}$. The **total angular momentum** is the vector sum of the orbital angular momentum **L** and the spin **S**:

$$\mathbf{J} = \mathbf{L} + \mathbf{S}.$$

Total angular momentum

According to quantum mechanics, the magnitude of the total angular momentum **J** is quantized:

$$J = \sqrt{j(j + 1)}\,\hbar. \qquad \qquad (40\text{–}15)$$

For the single electron in the H atom, quantum mechanics gives the result that j can be

$$j = l + s = l + \tfrac{1}{2}$$

or

$$j = l - s = l - \tfrac{1}{2}$$

but never less than zero, just as for l and s. For the 1s state, $l = 0$ and $j = \frac{1}{2}$ is the only possibility. For p states, say the 2p state, $l = 1$ and j can be either $\frac{3}{2}$ or $\frac{1}{2}$. The z component for j is quantized in the usual way:

$$m_j = j, j - 1, \cdots, -j.$$

For a 2p state with $j = \frac{1}{2}$, m_j can be $\frac{1}{2}$ or $-\frac{1}{2}$; for $j = \frac{3}{2}$, m_j can be $\frac{3}{2}, \frac{1}{2}, -\frac{1}{2}, -\frac{3}{2}$, for a total of four states. Note that the state of a single electron can be specified by giving n, l, m_l, m_s, or by giving n, j, l, m_j (only one of these descriptions at a time).

The interaction of magnetic fields with atoms, as in the Zeeman effect and the Stern-Gerlach experiment, involves the *total* angular momentum. Thus the Stern-Gerlach experiment on H atoms in the ground state shows two lines (for $m_j = +\frac{1}{2}$ and $-\frac{1}{2}$), but for the first excited state it shows four lines corresponding to the four possible m_j values $(\frac{3}{2}, \frac{1}{2}, -\frac{1}{2}, -\frac{3}{2})$.

* Spectroscopic Notation

We can specify the state of an atom, including the total angular momentum quantum number j, using the following **spectroscopic notation**. For a single electron state we can write

$$nL_j,$$

where the value of L is specified using the same letters as in Table 40–2, but in upper case:

$$L = 0 \quad 1 \quad 2 \quad 3 \quad 4 \cdots$$

$$\text{letter} = S \quad P \quad D \quad F \quad G \cdots .$$

So the $2P_{3/2}$ state has $n = 2$, $l = 1$, $j = \frac{3}{2}$, whereas $1S_{1/2}$ specifies the ground state in hydrogen.

* Fine Structure; Spin-Orbit Interaction

It is also a magnetic effect that produces the *fine structure* splitting mentioned in Section 40–2, which occurs in the absence of any external field. Instead, it is due to a magnetic field produced by the atom itself. We can see how it occurs by putting ourselves in the reference frame of the electron, in which case we see the nucleus revolving about us as a moving charge or electric current that produces a magnetic field, B_n. The electron has an intrinsic magnetic dipole moment μ_s (Eq. 40–14) and hence its energy will be altered by an amount (Eq. 27–12)

$$\Delta U = -\boldsymbol{\mu}_s \cdot \mathbf{B}_n .$$

Since μ_s takes on quantized values according to the values of m_s, the energy of a single electron state will split into two closely spaced energy levels (for $m_s = \frac{1}{2}$ and $-\frac{1}{2}$). This tiny splitting of energy levels produces a tiny splitting in spectral lines. For example, in the H atom, the $2P \rightarrow 1S$ transition is split into two lines corresponding to $2P_{1/2} \rightarrow 1S_{1/2}$ and $2P_{3/2} \rightarrow 1S_{1/2}$. The difference in energy between these two is only about $5 \times 10^{-5}\,\text{eV}$, which is very small compared to the $2P \rightarrow 1S$ transition energy of $13.6\,\text{eV} - 3.4\,\text{eV} = 10.2\,\text{eV}$.

The magnetic field $\mathbf{B}_n$ produced by the orbital motion is proportional to the orbital angular momentum $\mathbf{L}$, and since $\boldsymbol{\mu}_s$ is proportional to the spin $\mathbf{S}$, then $\Delta U = -\boldsymbol{\mu}_s \cdot \mathbf{B}_n$ can be written

Spin-orbit interaction

$$\Delta U \propto \mathbf{L} \cdot \mathbf{S}.$$

This interaction, which produces the fine structure, is thus called the **spin-orbit interaction**. Its magnitude is related to a dimensionless constant known as the **fine structure constant**,

Fine-structure constant
$$\alpha = \frac{e^2}{2\epsilon_0 hc} \approx \frac{1}{137},$$

which also appears elsewhere in atomic physics.

* | 40–8 | Fluorescence and Phosphorescence

When an atom is excited from one energy state to a higher one by the absorption of a photon, it may return to the lower level in a series of two (or more) jumps if there is an energy level in between (Fig. 40–15). The photons emitted will consequently have lower energy and frequency than the absorbed photon. When the absorbed photon is in the UV and the emitted photons are in the visible region of the spectrum, this phenomenon is called **fluorescence** (Fig. 40–16).

The wavelength for which fluorescence will occur depends on the energy levels of the particular atoms. Because the frequencies are different for different substances, and because many substances fluoresce readily, fluorescence is a powerful tool for identification of compounds. It is also used for assaying—determining how much of a substance is present—and for following substances along a natural pathway as in plants and animals. For detection of a given compound, the stimulating light must be monochromatic, and solvents or other materials present must not fluoresce in the same region of the spectrum. Sometimes the observation of fluorescent light being emitted is sufficient to detect a compound. In other cases, spectrometers are used to measure the wavelengths and intensities of the emitted light.

Fluorescent lightbulbs work in a two-step process. The applied voltage accelerates electrons that strike atoms of the gas in the tube and cause them to be excited. When the excited atoms jump down to their normal levels, they emit UV photons which strike a fluorescent coating on the inside of the tube. The light we see is a result of this material fluorescing in response to the UV light striking it.

Materials such as those used for luminous watch dials are said to be **phosphorescent**. When an atom is raised to a normal excited state, it drops back down within about 10^{-8} s. In phosphorescent substances, atoms can be excited by photon absorption to energy levels, said to be **metastable**, which are states that last much longer—even a few seconds or longer. In a collection of such atoms, many of the atoms will descend to the lower state fairly soon, but many will remain in the excited state for over an hour. Hence light will be emitted even after long periods. When you put your watch dial close to a bright lamp, it excites many atoms to metastable states, and you can see the glow a long time after.

* | 40–9 | Lasers

A laser is a device that can produce a very narrow intense beam of monochromatic coherent light. (By *coherent*, we mean that across any cross section of the beam, all parts have the same phase.) The emitted beam is a nearly perfect plane wave. An ordinary light source, on the other hand, emits light in all directions (so the intensity decreases rapidly with distance), and the emitted light is incoherent (the different parts of the beam are not in phase with each other). The excited atoms that emit the light in an ordinary lightbulb act independently, so each photon emitted can be considered as a short wave train that lasts about 10^{-8} s. These wave trains bear no phase relation to one another. Just the opposite is true of lasers.

The action of a laser is based on quantum theory. We have seen that a photon can be absorbed by an atom if (and only if) its energy hf corresponds to the energy difference between an occupied energy level of the atom and an available excited state, Fig. 40–17a. This is, in a sense, a resonant situation. If the atom is already in the excited state, it may of course jump spontaneously (i.e., no apparent stimulus) to the lower state with the emission of a photon. However, if a photon with this same energy strikes the excited atom, it can stimulate the atom to make the transition sooner to the lower state, Fig. 40–17b. This phenomenon is called **stimulated emission**, and it can be seen that not only do we still have the original photon, but also a second one of the same frequency as a result of the atom's transition. And these two photons are

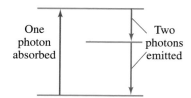

FIGURE 40–15 Fluorescence.

FIGURE 40–16 When UV light illuminates these various "fluorescent" rocks, they fluoresce in the visible region of the spectrum.

FIGURE 40–17 (a) Absorption of a photon. (b) Stimulated emission.

Stimulated emission

exactly *in phase*, and they are moving in the same direction. This is how coherent light is produced in a laser. Hence the name "laser," which is an acronym for light **a**mplification by **s**timulated **e**mission of **r**adiation.

The natural population of atoms in thermal equilibrium at any temperature T (in K) is given by the **Boltzmann distribution** (or **Boltzmann factor**):

$$N_n = Ce^{-\frac{E_n}{kT}}, \qquad \text{(40–16a)}$$

where N_n is the number of atoms in the state with energy E_n. For two states n and n', the ratio of the number of atoms in the two states is

$$\frac{N_n}{N_{n'}} = e^{-\left(\frac{E_n - E_{n'}}{kT}\right)} \qquad \text{(40–16b)}$$

Thus most atoms are in the ground state unless the temperature is very high. In the two-level system of Fig. 40–17, most atoms are normally in the lower state, so the majority of incident photons will be absorbed. In order to obtain the coherent light from stimulated emission, two conditions must be satisfied. First, atoms must be excited to the higher state, so that an **inverted population** is produced, one in which more atoms are in the upper state than in the lower one (Fig. 40–18). Then *emission* of photons will dominate over absorption. Hence the system will not be in thermal equilibrium. And second, the higher state must be a **metastable state**—a state in which the electrons remain longer than usual[†] so that the transition to the lower state occurs by stimulated emission rather than spontaneously. How these conditions are achieved for different lasers will be discussed shortly. For now, we assume that the atoms have been excited to an upper state. Figure 40–19 is a schematic diagram of a laser: the "lasing" material is placed in a long narrow tube at the ends of which are two mirrors, one of which is partially transparent (perhaps 1 or 2 percent). Some of the excited atoms drop down fairly soon after being excited. One of these is the atom shown on the far left in Fig. 40–19. If the emitted photon strikes another atom in the excited state, it stimulates this atom to emit a photon of the *same* frequency, moving in the *same* direction, and *in phase* with it. These two photons then move on to strike other atoms causing more stimulated emission. As the process continues, the number of photons multiplies. When the photons strike the end mirrors, most are reflected back, and as they move in the opposite direction, they continue to stimulate more atoms to emit photons. As the photons move back and forth between the mirrors, a small percentage passes through the partially transparent mirror at one end. These photons make up the narrow coherent external laser beam.

Inside the tube, some spontaneously emitted photons will be emitted at an angle to the axis, and these will merely go out the side of the tube and not affect the narrowness of the main beam. In a well-designed laser, the spreading of the beam is limited only by diffraction, so the angular spread is $\approx \lambda/a$ (see Eq. 36–1) where a is the diameter of the end mirror. The diffraction spreading can be incred-

FIGURE 40–18 Two energy levels for a collection of atoms. Each dot represents the energy state of one atom. (a) A normal situation; (b) an inverted population.

[†] An excited atom may land in such a state and can jump to a lower state only by a so-called forbidden transition (discussed in Section 40–2), which is why its lifetime is longer than normal.

FIGURE 40–19 Schematic laser diagram, showing excited atoms stimulated to emit light.

ibly small. The light energy, instead of spreading out in space as it does for an ordinary light source, is directed in a pencil-thin beam.

The excitation of the atoms in a laser can be done in several ways to produce the necessary inverted population. In a ruby laser, the lasing material is a ruby rod consisting of Al_2O_3 with a small percentage of aluminum (Al) atoms replaced by chromium (Cr) atoms. The Cr atoms are the ones involved in lasing. The atoms are excited by strong flashes of light of wavelength 550 nm, which corresponds to a photon energy of 2.2 eV. As shown in Fig. 40–20, the atoms are excited from state E_0 to state E_2. This process is called **optical pumping**. The atoms quickly decay either back to E_0 or to the intermediate state E_1, which is metastable with a lifetime of about 3×10^{-3} s (compared to 10^{-8} s for ordinary levels). With strong pumping action, more atoms can be forced into the E_1 state than are in the E_0 state. Thus we have the inverted population needed for lasing. As soon as a few atoms in the E_1 state jump down to E_0, they emit photons that produce stimulated emission of the other atoms and the lasing action begins. A ruby laser thus emits a beam whose photons have energy 1.8 eV and a wavelength of 694.3 nm (or "ruby-red" light).

In a helium–neon (He–Ne) laser, the lasing material is a gas, a mixture of about 15 percent He and 85 percent Ne. The atoms are excited by applying a high voltage to the tube so that an electric discharge takes place within the gas. In the process, some of the He atoms are raised to the metastable state E_1 shown in Fig. 40–21, which corresponds to a jump of 20.61 eV, almost exactly equal to an excited state in neon, 20.66 eV. The He atoms do not quickly return to the ground state by spontaneous emission, but instead often give their excess energy to a Ne atom when they collide—see Fig. 40–21. In such a collision, the He drops to the ground state and the Ne atom is excited to the state E_3' (the prime refers to neon states). The slight difference in energy (0.05 eV) is supplied by the kinetic energy of the moving molecules ($\frac{3}{2}kT \approx 0.04$ eV at $T = 300$ K). In this manner, the E_3' state in Ne—which is metastable—becomes more populated than the E_2' level. This inverted population between E_3' and E_2' is what is needed for lasing.

Other types of laser include: chemical lasers, in which the energy input comes from the chemical reaction of highly reactive gases; dye lasers, whose frequency is tunable; CO_2 gas lasers, capable of high power output in the infrared; rare-earth solid-state lasers such as the high-power Nd:Yag laser; and the *pn* junction laser in which the transitions occur between the bottom of the conduction band and the upper part of the valence band (Section 41–7).

The excitation of the atoms in a laser can be done continuously or in pulses. In a **pulsed laser**, the atoms are excited by periodic inputs of energy. The multiplication of photons continues until all the atoms have been stimulated to jump down

Ruby laser

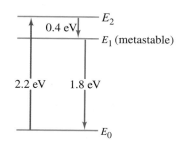

FIGURE 40–20 Energy levels of chromium in a ruby crystal. Photons of energy 2.2 eV "pump" atoms from E_0 to E_2, followed by decay to the metastable state E_1. Lasing action occurs by stimulated emission of photons in transition from E_1 to E_0.

He–Ne laser

Other lasers

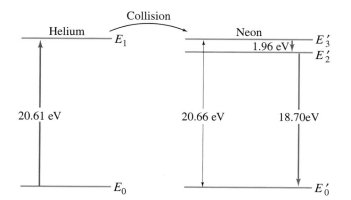

FIGURE 40–21 Energy levels for He and Ne. He is excited in the electric discharge to the E_1 state. This energy is transferred to the E_3' level of the Ne by collision. E_3' is metastable and decays to E_2' by stimulated emission, producing the output beam of the laser.

to the lower state, and the process is repeated with each input pulse. In a **continuous laser**, the energy input is continuous so that as atoms are stimulated to jump down to the lower level, they are soon excited back up to the upper level so that the output is a continuous laser beam. Any laser, of course, is not a source of energy. Energy must be put in, and the laser converts a part of this input energy into an intense narrow beam output.

The unique feature of light from a laser is, as mentioned before, that it is a coherent narrow beam of a single frequency (or several distinct frequencies). Because of this feature, the laser has found many applications. Lasers are a useful surgical tool. The narrow intense beam can be used to destroy tissue in a localized area, or to break up gallstones and kidney stones. Because of the heat produced, a laser beam can be used to "weld" broken tissue, such as a detached retina (Fig. 40–22). For some types of internal surgery, the laser beam can be carried by an optical fiber (Section 33–7) to the surgical point. An example is the removal of plaque clogging human arteries. Tiny organelles within a living cell have been destroyed using lasers by researchers studying how the absence of that organelle affects the behavior of the cell. Laser beams have been used to destroy cancerous and precancerous cells; at the same time, the heat seals off capillaries and lymph vessels, thus "cauterizing" the wound in the process to prevent spread of the disease. The intense heat produced in a small area by a laser beam is also used for welding and machining metals and for drilling tiny holes in hard materials. The beam of a laser is narrow in itself (typically, a few mm). But because the beam is coherent, monochromatic, and essentially parallel and narrow, lenses can be used to focus the light into incredibly small areas without the usual aberration problems. The limiting factor thus becomes diffraction, and the energy crossing unit area per unit time can be very large. The precise straightness of a laser beam is also useful to surveyors for lining up equipment precisely, especially in inaccessible places.

In everyday life, lasers are used as bar-code readers (at store checkout stands) and in compact disc (CD) players. The laser beam reflects off the stripes and spaces of a bar code, and off the tiny pits of a CD as shown in Fig. 40–23. The recorded information on a CD is a series of pits and spaces representing 0s and 1s (or "off" and "on") of a digitized code that is decoded electronically before being sent to the audio or video system. A bar-code reader is similar.

FIGURE 40–22 Laser being used in eye surgery.

FIGURE 40–23 Reading a CD. The fine beam of a laser, focused even more finely with lenses, is directed at the undersurface of a rotating compact disc. The beam is reflected back from the areas between pits but reflects much less from pits. The reflected light is detected as shown, reflected by a half-reflecting mirror MS. The strong and weak reflections correspond to the 0s and 1s of the binary code representing the audio or video signal.

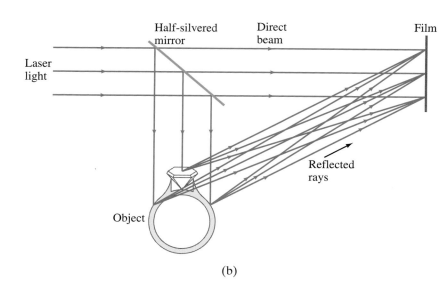

(a) (b)

FIGURE 40–24   (a) Photo of a hologram of coins. (b) Making a hologram. Light reflected from various points on the object interferes (at the film) with light from the direct beam.

* | 40–10 | **Holography**

One of the most interesting applications of laser light is the production of three-dimensional images called **holograms** (see Fig. 40–24). In an ordinary photograph, the film simply records the intensity of light reaching it at each point. When the photograph or transparency is viewed, light reflecting from it or passing through it gives us a two-dimensional picture. In holography, three-dimensional images are formed by interference, without lenses. When a laser hologram is made on film, a broadened laser beam is split into two parts by a half-silvered mirror, Fig. 40–24b. One part goes directly to the film; the rest passes to the object to be photographed, from which it is reflected to the film. Light from every point on the object reaches each point on the film, and the interference of the two beams allows the film to record both the intensity and relative phase of the light at each point. It is crucial that the light be coherent—that is, in phase at all points—which is why a laser is used. After the film is developed, it is placed again in a laser beam and a three-dimensional image of the object is created. You can walk around such an image and see it from different sides as if it were the original object. Yet, if you try to touch it with your hand, there will be nothing material there.

The details of how the image is formed are quite complicated. But we can get the basic idea by considering one single point on the object. In Fig. 40–25a the rays OA and OB have reflected from one point on our object. The rays CA and DB come directly from the source and interfere with OA and OB at points A and B on the film. A set of interference fringes is produced as shown in Fig. 40–25b. The spacing between the fringes changes from top to bottom as shown. Why this

➡ **PHYSICS APPLIED**

Holography

FIGURE 40–25   Light from point O on the object interferes with light of the direct beam (rays CA and DB).

(a) (b)

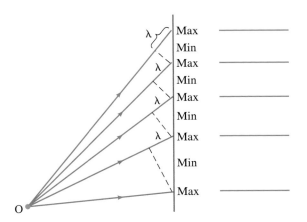

FIGURE 40–26 Each of the rays shown leaving point O is one wavelength shorter than the one above it. If the top ray is in phase with the direct beam (not shown), which has the same phase at all points on the screen, all the rays shown produce constructive interference. From this diagram it can be seen that the fringe spacing increases toward the bottom.

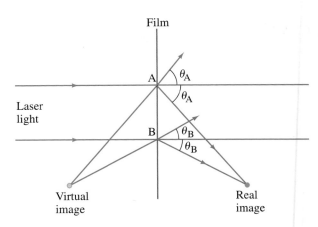

FIGURE 40–27 Reconstructing the image of one point on the object. Laser beam strikes film that is like a diffraction grating of variable spacing. Rays corresponding to the first diffraction maxima are shown emerging. The angle $\theta_A > \theta_B$ because the spacing at B is greater than at A ($\sin\theta = \lambda/d$). Hence real and virtual images of the point are reproduced as shown.

happens is explained in Fig. 40–26. Thus the hologram of a single point object would have the pattern shown in Fig. 40–25b. The film in this case looks like a diffraction grating with variable spacing. Hence, when coherent laser light is passed back through the developed film to reconstruct the image, the diffracted rays in the first order maxima occur at slightly different angles because the spacing changes. (Remember Eq. 36–13, $\sin\theta = \lambda/d$: where the spacing d is greater, the angle θ is less.) Hence, the rays diffracted upward (in first order) seem to diverge from a single point, Fig. 40–27. This is a virtual image of the original object, which can be seen with the eye. Rays diffracted in first order *downward* converge to make a real image, which can be seen and also photographed. (Note that the straight-through undiffracted rays are of no interest.) Of course real objects consist of many points, so a hologram will be a complex interference pattern which, when laser light is incident on it, will reproduce an image of the object. Each image point will be at the correct (three-dimensional) position with respect to other points, so the image accurately represents the original object. And it can be viewed from different angles as if viewing the original object. Holograms can be made in which a viewer can walk entirely around the image (360°) and see all sides of it.

White-light holograms   So-called **volume** or **white-light holograms** do not require a laser to see the image, but can be viewed with ordinary white light (preferably a nearby point source, such as the Sun or a clear bulb with a small bright filament). Such holograms must be made, however, with a laser. They are made not on thin film, but on a *thick* emulsion. The interference pattern, instead of being two-dimensional as for an ordinary hologram, is actually three-dimensional (hence the name "volume hologram"). The interference pattern in the film emulsion can be thought of as an array of bands or ribbons (consisting of the silver grains from the development process) where constructive interference occurred. This array, and the reconstruction of the image, can be compared to Bragg scattering of X-rays from the atoms in a crystal (see Section 36–10). White light can reconstruct the image because the Bragg condition (Eq. 36–20) selects out the appropriate single wavelength. If the hologram is originally produced by lasers emitting the three additive primary colors (red, green, and blue), the three-dimensional image can be seen in full color when viewed with white light.

Summary

In the quantum mechanical view of the atom, the electrons do not have well-defined orbits, but instead exist as a "cloud." Electron clouds can be interpreted as an electron wave spread out in space, or as a **probability distribution** for electrons considered as particles.

For the simplest atom, hydrogen, the Schrödinger equation contains the potential energy $U = -(1/4\pi\epsilon_0)(e^2/r)$. The solutions give the same values of energy as the old Bohr theory.

According to quantum mechanics, the state of an electron in an atom is specified by four **quantum numbers**: n, l, m_l, and m_s:

the principal quantum number, n, can take on any integer value $(1, 2, 3, \cdots)$ and corresponds to the quantum number of the old Bohr theory;

l can take on values from 0 up to $n - 1$;

m_l can take on integer values from $-l$ to $+l$;

m_s can be $+\frac{1}{2}$ or $-\frac{1}{2}$.

The energy levels in the hydrogen atom depend on n, whereas in other atoms they depend on n and l.

When an external magnetic field is applied, the spectral lines are split (the **Zeeman effect**), indicating that the energy depends also on m_l in this case.

Even in the absence of a magnetic field, precise measurements of spectral lines show a tiny splitting of the lines called **fine structure**, whose explanation is that the energy depends very slightly on m_l and m_s.

Transitions between states that obey the selection rule $\Delta l = \pm 1$ are far more probable than other so-called "forbidden" transitions.

The ground-state wave function in hydrogen has spherical symmetry, as do other $l = 0$ states. States with $l > 0$ have some directionality in space.

The probability density, $|\psi|^2$, and the *radial probability density*, $P_r = 4\pi r^2|\psi|^2$, are both useful to illustrate the spatial extent of the electron cloud.

The arrangement of electrons in multi-electron atoms is governed by the **Pauli exclusion principle**, which states that no two electrons can occupy the same quantum state—that is, they cannot have the same set of quantum numbers n, l, m_l, and m_s.

As a result, electrons in multi-electron atoms are grouped into **shells** (according to the value of n) and **subshells** (according to l).

Electron configurations are specified using the numerical values of n, and using letters for l: s, p, d, f, etc., for $l = 0, 1, 2, 3$, and so on, plus a superscript for the number of electrons in that subshell. Thus, the ground state of hydrogen is $1s^1$, whereas that for oxygen is $1s^2 2s^2 2p^4$. The **periodic table** arranges the elements in horizontal rows according to increasing atomic number (number of electrons in the neutral atom). The shell structure gives rise to a periodicity in the properties of the elements, so that each vertical column can contain elements with similar chemical properties.

X-rays, which are a form of electromagnetic radiation of very short wavelength, are produced when high-speed electrons strike a target. The spectrum of X-rays so produced consists of two parts, a continuous spectrum produced when the electrons are decelerated by atoms of the target, and peaks representing photons emitted by atoms of the target after being excited by collision with the high-speed electrons. Measurement of these peaks allows determination of inner energy levels of atoms and determination of Z.

Questions

1. Discuss the differences between Bohr's view of the atom and the quantum-mechanical view.

2. The probability density $|\psi|^2$ is a maximum at the center of the H atom ($r = 0$) for the ground state, whereas the radial probability density $P_r = 4\pi r^2|\psi|^2$ is zero at this point. Explain why.

3. Why do three quantum numbers come out of the Schrödinger theory (rather than, say, two or four)?

4. In Fig. 40–4, why do the upper and lower levels have different energy splittings in a magnetic field?

5. Which model of the hydrogen atom, the Bohr model or the quantum-mechanical model, predicts that the electron spends more time near the nucleus?

6. The size of atoms varies by only a factor of three or so from largest to smallest, yet the number of electrons varies from one to over 100. Why?

7. Excited hydrogen and excited helium atoms both radiate light as they jump down to the $n = 1$, $l = 0$, $m_l = 0$ state. Yet the two elements have very different emission spectra. Why?

8. The 589-nm yellow line in sodium is actually two very closely spaced lines. This splitting is due to an "internal" Zeeman effect. Can you explain this? [*Hint*: Put yourself in the reference frame of the electron.]

9. Which of the following electron configurations are forbidden? (*a*) $1s^2 2s^2 2p^4 3s^2 4p^2$; (*b*) $1s^2 2s^2 2p^8 3s^1$, (*c*) $1s^2 2s^2 2p^6 3s^2 3p^5 4s^2 4d^5 4f^1$.

10. Give the complete electron configuration for a uranium atom (careful scrutiny across the periodic table on the inside back cover will provide useful hints).

11. In what column of the periodic table would you expect to find the atom with each of the following configurations? (a) $1s^2 2s^2 2p^6 3s^2$; (b) $1s^2 2s^2 2p^6 3s^2 3p^6$; (c) $1s^2 2s^2 2p^6 3s^2 3p^6 4s^1$; (d) $1s^2 2s^2 2p^5$.

12. On what factors does the periodicity of the periodic table depend? Consider the exclusion principle, quantization of angular momentum, spin, and any others you can think of.

13. How would the periodic table look if there were no electron spin but otherwise quantum mechanics were valid? Consider the first 20 elements or so.

14. The ionization energy for neon ($Z = 10$) is 21.6 eV and that for sodium ($Z = 11$) is 5.1 eV. Explain the large difference.

15. Why do chlorine and iodine exhibit similar properties?

16. Explain why potassium and sodium exhibit similar properties.

17. Why are the chemical properties of the rare earths so similar?

18. Why do we not expect perfect agreement between measured values of X-ray line wavelengths and those calculated using Bohr theory, as in Example 40–6?

19. Why does the Bohr theory, which does not work at all well for normal transitions involving the outer electrons for He and more complex atoms, nevertheless predict reasonably well the atomic X-ray spectra for transitions deep inside the atom?

20. Why does the cutoff wavelength in Fig. 40–11 imply a photon nature for light?

21. How would you figure out which lines in an X-ray spectrum correspond to K_α, K_β, L, etc., transitions?

22. Why do the characteristic X-ray spectra vary in a systematic way with Z, whereas the visible spectra (Fig. 36–21) do not?

23. Why do we expect electron transitions deep within an atom to produce shorter wavelengths than transitions by outer electrons?

* **24.** Why is the direction of the magnetic dipole moment of an electron opposite to that of its orbital angular momentum?

* **25.** Why is a nonhomogeneous field used in the Stern-Gerlach experiment?

* **26.** Compare spontaneous emission to stimulated emission.

* **27.** Does the intensity of light from a laser fall off as the inverse square of the distance?

* **28.** How does laser light differ from ordinary light? How is it the same?

* **29.** Explain how a 0.0005-W laser beam, photographed at a distance, can seem much stronger than a 1000-W street lamp.

Problems

Section 40–2

1. (I) For $n = 6$, what values can l have?

2. (I) For $n = 5$, $l = 3$, what are the possible values of m_l and m_s?

3. (I) How many different states are possible for an electron whose principal quantum number is $n = 4$? Write down the quantum numbers for each state.

4. (I) If a hydrogen atom has $m_l = -3$, what are the possible values of n, l, m_s?

5. (I) A hydrogen atom is known to have $l = 4$. What are the possible values for n, m_l, and m_s?

6. (I) Calculate the magnitude of the angular momentum of an electron in the $n = 4$, $l = 2$ state of hydrogen.

7. (II) A hydrogen atom is in the 6g state. Determine (a) the principal quantum number, (b) the energy of the state, (c) the orbital angular momentum and its quantum number l, and (d) the possible values for the magnetic quantum number.

8. (II) (a) Show that the number of different states possible for a given value of l is equal to $2(2l + 1)$. (b) What is this number for $l = 0, 1, 2, 3, 4, 5$, and 6?

9. (II) Show that the number of different electron states possible for a given value of n is $2n^2$. (See Problem 8.)

10. (II) An excited H atom is in a 6d state. (a) Name all the states (n, l) to which the atom is "allowed" to jump with the emission of a photon. (b) How many different wavelengths are there (ignoring fine structure)?

Section 40–3

11. (I) Show that the ground-state wave function, Eq. 40–5, is normalized. [Hint: See Example 40–4.]

12. (II) For the ground state of hydrogen, what is the value of (a) ψ, (b) $|\psi|^2$, and (c) P_r at $r = r_0$?

13. (II) For the $n = 2$, $l = 0$ state of hydrogen, what is the value of (a) ψ, (b) $|\psi|^2$, and (c) P_r at $r = 5r_0$?

14. (II) Show that ψ_{200} as given by Eq. 40–8 is normalized.

15. (II) By what factor is it more likely to find the electron in the ground state of hydrogen at the Bohr radius (r_0) than at twice the Bohr radius $(2r_0)$?

16. (II) For the ground state of hydrogen, what is the probability of finding the electron within a spherical shell of inner radius $0.99r_0$ and outer radius $1.01r_0$?

17. (II) For the $n = 2$, $l = 0$ state of hydrogen, what is the probability of finding the electron within a spherical shell of inner radius $4.00r_0$ and outer radius $5.00r_0$?

18. (II) (a) Show that the probability of finding the electron in the ground state of hydrogen at less than one Bohr radius from the nucleus is 32 percent. (b) What is the probability of finding a 1s electron between $r = r_0$ and $r = 2r_0$?

19. (II) Determine the radius r of a sphere centered on the nucleus within which the probability of finding the electron for the ground state of hydrogen is (a) 50 percent, (b) 90 percent, (c) 99 percent.

20. (II) (a) Estimate the probability of finding an electron, in the ground state of hydrogen, within the nucleus assuming it to be a sphere of radius $r = 1.1$ fm. (b) What would be the probability if the electron were replaced with a muon, which is very similar to an electron (Chapter 44) except that its mass is 207 times greater?

21. (II) Determine the average radial probability distribution P_r for the $n = 2$, $l = 1$ state in hydrogen by calculating

$$P_r = 4\pi r^2 [\tfrac{1}{3}|\psi_{210}|^2 + \tfrac{1}{3}|\psi_{211}|^2 + \tfrac{1}{3}|\psi_{21-1}|^2].$$

22. (II) Use the result of Problem 21 to show that the most probable distance r from the nucleus for an electron in the 2p state of hydrogen is $r = 4r_0$, which is just the second Bohr radius (Eq. 38–11, Fig. 38–24).

23. (II) Show that the mean value of r for an electron in the ground state of hydrogen is $\bar{r} = \tfrac{3}{2}r_0$, by calculating

$$\bar{r} = \int_{\text{all space}} r|\psi_{100}|^2 \, dV = \int_0^\infty r|\psi_{100}|^2 4\pi r^2 \, dr.$$

24. (III) Show that the probability of finding the electron within 1 Bohr radius of the nucleus in the hydrogen atom is (a) 3.4 percent for the $n = 2$, $l = 0$ state, and (b) 0.37 percent for the $n = 2$, $l = 1$ state.

25. (III) Show that ψ_{100} (Eq. 40–5a) satisfies the Schrödinger equation (Eq. 40–1) with the Coulomb potential, for energy $E = -me^4/8\epsilon_0^2 h^2$.

26. (III) For the $n = 2$, $l = 0$ state in hydrogen, what is the probability of finding the electron within the smaller peak in the radial probability distribution, Fig. 40–8b?

27. (III) The wave function for the $n = 3$, $l = 0$ state in hydrogen is

$$\psi_{300} = \frac{1}{\sqrt{27\pi r_0^3}}\left(1 - \frac{2r}{3r_0} + \frac{2r^2}{27r_0^2}\right)e^{-\frac{r}{3r_0}}.$$

(a) Determine the radial probability distribution P_r for this state, and (b) draw the curve for it on a graph. (c) Determine the most probable distance from the nucleus for an electron in this state.

Sections 40–4 and 40–5

28. (I) List the quantum numbers for each electron in the ground state of nitrogen ($Z = 7$).

29. (I) List the quantum numbers for each electron in the ground state of (a) carbon ($Z = 6$), (b) magnesium ($Z = 12$).

30. (I) How many electrons can be in the $n = 6$, $l = 3$ subshell?

31. (II) What is the full electron configuration for (a) selenium (Se), (b) gold (Au), (c) uranium (U)? [Hint: See the periodic table inside the back cover.]

32. (II) For each of the following atomic transitions, state whether the transition is allowed or forbidden, and why: (a) 4p → 3p; (b) 2p → 1s; (c) 3d → 2d; (d) 4d → 3s; (e) 4s → 2p.

33. (II) Using the Bohr formula for the radius of an electron orbit, estimate the average distance from the nucleus for an electron in the innermost ($n = 1$) orbit in uranium ($Z = 92$). Approximately how much energy would be required to remove this innermost electron?

34. (II) Estimate the binding energy of the third electron in lithium using Bohr theory. [Hint: This electron has $n = 2$ and "sees" a net charge of approximately $+1e$.] The measured value is 5.36 eV.

35. (II) Show that the total angular momentum is zero for a filled subshell.

36. (II) Let us apply the exclusion principle to an infinitely high square well (Section 39–8). Let there be five electrons confined to this rigid box whose width is L. Find the lowest energy state of this system, by placing the electrons in the lowest available levels, consistent with the Pauli exclusion principle.

Section 40–6

37. (I) What are the shortest-wavelength X-rays emitted by electrons striking the face of a 30-kV TV picture tube? What are the longest wavelengths?

38. (I) If the shortest-wavelength bremsstrahlung X-rays emitted from an X-ray tube have $\lambda = 0.029$ nm, what is the voltage across the tube?

39. (I) Show that the cutoff wavelength λ_0 is given by

$$\lambda_0 = \frac{1240 \text{ nm}}{V},$$

where V is the X-ray tube voltage in volts.

40. (II) Use the result of Example 40–6 to estimate the X-ray wavelength emitted when a cobalt atom ($Z = 27$) jumps from $n = 2$ to $n = 1$.

41. (II) Estimate the wavelength for an $n = 2$ to $n = 1$ transition in iron ($Z = 26$).

42. (II) Use Bohr theory to estimate the wavelength for an $n = 3$ to $n = 1$ transition in molybdenum. The measured value is 0.063 nm. Why do we not expect perfect agreement?

43. (II) A mixture of iron and an unknown material are bombarded with electrons. The wavelength of the K_α lines are 194 pm for iron and 229 pm for the unknown material. What is the unknown material?

44. (II) Use conservation of energy and momentum to show that a moving electron cannot give off an X-ray photon unless there is a third body present, such as an atom or nucleus.

* Section 40–7

* 45. (I) Verify that the Bohr magneton has the value $\mu_B = 9.27 \times 10^{-24}$ J/T (see Eq. 40–12).

* 46. (II) Suppose that the splitting of energy levels shown in Fig. 40–4 was produced by a 2.0-T magnetic field. (a) What is the separation in energy between adjacent m_l levels for the same l? (b) How many different wavelengths will there be for 3d to 2p transitions, if m_l can change only by ±1 or 0? (c) What is the wavelength for each of these transitions?

* 47. (II) In a Stern-Gerlach experiment, Ag atoms issued from the oven with an average speed of 700 m/s and passed through a magnetic field gradient $dB/dz = 1.5 \times 10^3$ T/m for a distance of 4.0 cm. (a) What is the separation of the two beams as they emerge from the magnet? (b) What would the separation be if the g-factor were 1 for electron spin?

*48. (II) (a) Write down the quantum numbers for each electron in the aluminum atom. (b) Which subshells are filled? (c) The last electron is in the 3p state; what are the possible values of the total angular momentum quantum number, j, for this electron? (d) Explain why the angular momentum of this last electron also represents the total angular momentum for the entire atom (ignoring any angular momentum of the nucleus). (e) How could you use a Stern-Gerlach experiment to determine which value of j the atom has?

*49. (II) What are the possible values of j for an electron in (a) the 3p, (b) the 4f, and (c) the 4d state of hydrogen? What is J in each case?

*50. (II) For an electron in a 5g state, what are all the possible values of j, m_j, J, and J_z?

*51. (II) The difference between the $2P_{3/2}$ and $2P_{1/2}$ energy levels in hydrogen is about 5×10^{-5} eV, due to the spin-orbit interaction. (a) Taking the electron's (orbital) magnetic moment to be 1 Bohr magneton, estimate the internal magnetic field due to the electron's orbital motion. (b) Estimate the internal magnetic field using a simple model of the nucleus revolving in a circle about the electron.

* Section 40–9

*52. (II) A laser used to weld detached retinas puts out 25-ms-long pulses of 640-nm light which average 0.65 W output during a pulse. How much energy can be deposited per pulse and how many photons does each pulse contain?

*53. (II) Estimate the angular spread of a laser beam due to diffraction if the beam emerges through a 3.0-mm-diameter mirror. Assume that $\lambda = 694$ nm. What would be the diameter of this beam if it struck a satellite 300 km above the Earth?

*54. (II) Suppose that the energy level system in Fig. 40–20 is not being pumped and is in thermal equilibrium. Determine the fraction of atoms in levels E_2 and E_1 relative to E_0 at $T = 300$ K.

*55. (II) To what temperature would the system in Fig. 40–20 have to be raised (see Problem 54) so that in thermal equilibrium the level E_2 would have half as many atoms as E_0? (Note that pumping mechanisms do not maintain thermal equilibrium.)

*56. (II) Show that a population inversion for two levels (as in a pumped laser) corresponds to a negative Kelvin temperature in the Boltzmann distribution. Explain why such a situation does not contradict the idea that negative Kelvin temperatures cannot be reached in the normal sense of temperature.

General Problems

57. The ionization (binding) energy of the outermost electron in boron is 8.26 eV. (a) Use the Bohr model to estimate the "effective charge," Z_{eff}, seen by this electron. (b) Estimate the average orbital radius.

58. Show that there can be 18 electrons in a "g" subshell.

59. What is the full electron configuration in the ground state for elements with Z equal to (a) 27, (b) 36, (c) 38? [Hint: See the periodic table inside the back cover.]

60. What are the largest and smallest possible values for the angular momentum L of an electron in the $n = 5$ shell?

61. Estimate (a) the quantum number l for the orbital angular momentum of the Earth about the Sun, and (b) the number of possible orientations for the plane of Earth's orbit.

62. Use the Bohr theory (especially Eq. 38–14) to show that the Moseley plot (Fig. 40–12) can be written

$$\sqrt{\frac{1}{\lambda}} = a(Z - b),$$

where $b \approx 1$, and evaluate a.

63. Determine the most probable distance from the nucleus of an electron in the $n = 2$, $l = 0$ state of hydrogen.

64. In the so-called *vector model* of the atom, space quantization of angular momentum (Fig. 40–3) is illustrated as shown in Fig. 40–28. The angular momentum vector of magnitude $L = \sqrt{l(l + 1)}\,\hbar$ is thought of as precessing around the z axis (like a spinning top or gyroscope) in such a way that the z component of angular momentum, $L_z = m_l \hbar$, also stays constant. Calculate the possible values for the angle θ between $\mathbf{L}$ and the z axis (a) for $l = 1$, (b) $l = 2$, and (c) $l = 3$. (d) Determine the minimum value of θ for $l = 100$ and $l = 10^6$. Is this consistent with the correspondence principle?

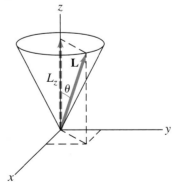

FIGURE 40–28 The vector model for orbital angular momentum. The orbital angular momentum vector $\mathbf{L}$ is imagined to precess about the z axis; L and L_z remain constant, but L_x and L_y continually change. Problems 64 and 65.

65. The vector model (Problem 64) gives some insight into the uncertainty principle for angular momentum, which is

$$\Delta L_z \, \Delta \phi \geq \hbar$$

for the z component. Here ϕ is the angular position measured in the plane perpendicular to the z axis. Once m_l for an atom is known, L_z is known precisely, so $\Delta L_z = 0$. (*a*) What does this tell us about ϕ? (*b*) What can you say about L_x and L_y, which are *not* quantized (only L and L_z are)? (*c*) Show that although L_x and L_y are not quantized, nonetheless $(L_x^2 + L_y^2)^{1/2} = [l(l+1) - m_l^2]^{1/2} \hbar$ is.

* 66. Show that the diffractive spread of a laser beam, $\approx \lambda/a$ as described in Section 40–9, is precisely what you might expect from the uncertainty principle. [*Hint*: Since the beam's width is constrained by the dimension of the aperture a, the component of the light's momentum perpendicular to the laser axis is uncertain.]

* 67. Silver atoms (spin $= \frac{1}{2}$) are placed in a 1.0-T magnetic field which splits the ground state into two close levels. (*a*) What is the difference in energy between these two levels, and (*b*) what wavelength photon could cause a transition from the lower level to the upper one? (*c*) How would your answer differ if the atoms were hydrogen?

* 68. Estimate the angular spread of a laser beam due to diffraction if the beam emerges through a 4.0-mm diameter mirror. Assume that $\lambda = 694$ nm. What would be the diameter of this beam if it struck (*a*) a satellite 1000 km above the Earth, or (*b*) the Moon?

* 69. *Populations in the H atom.* Use the Boltzmann factor (Eq. 40–16) to estimate the fraction of H atoms in the $n = 2$ and $n = 3$ levels (relative to the ground state) for thermal equilibrium at (*a*) $T = 300$ K and (*b*) $T = 6000$ K. [Note: Since there are eight states with $n = 2$ and only two with $n = 1$, multiply your result for $n = 2$ by $\frac{8}{2} = 4$; do similarly for $n = 3$.] (*c*) Given 1.0 g of hydrogen, estimate the number of atoms in each state at $T = 6000$ K. (*d*) Estimate the number of $n = 3$ to $n = 1$ and $n = 2$ to $n = 1$ photons that will be emitted per second at $T = 6000$ K. Assume that the lifetime of each excited state is 10^{-8} s.

70. A very simple model of a "one-dimensional" metal consists of N electrons confined to a rigid box of width L. We neglect the Coulomb interaction between the electrons (this is called the "independent electron model" and does remarkably well in predicting many properties of real metals). The Fermi energy of a metal is the energy of the most energetic electron when the metal is in its ground state ($T = 0$ K). (*a*) Calculate the Fermi energy for this one-dimensional metal, taking into account the Pauli exclusion principle. You can assume for simplicity that N is even. (*b*) What is the smallest amount of energy that this metal can absorb? Consider the limit of your answer for large N. (The relatively small value you should obtain is indicative of how metals can readily conduct).

71. If the principal quantum number n were limited to the range from 1 to 6, how many elements would we find in nature?

72. It is possible for atoms to be excited into states with very high values of the principal quantum number. Electrons in these so-called *Rydberg states* have very small ionization energies and huge orbital radii. This makes them particularly sensitive to external perturbation, as would be the case if the atom were in an electric field. Consider the $n = 50$ state of the hydrogen atom. Determine the binding energy, the radius of the orbit, and the effective cross sectional area of this Rydberg state.

* 73. In the analytical technique known as *electron spin resonance* (ESR), materials with atoms having one or more unpaired electrons are placed in a variable magnetic field plus a constant radio frequency (RF) electric field. The magnetic field splits the energy of the unpaired electrons according to the two possible spin orientations. If the splitting is the same as the photon energy which is available from the RF radiation, a resonant absorption takes place. This can be detected, and is a measure of the *g*-factor of the electrons in those atoms. The *g*-factor for molecules and solids can be different than for electrons in atoms, due to the chemical interactions in the material. Therefore, ESR provides a fingerprint of the system. In a particular ESR experiment, the RF radiation has a fixed wavelength of 2.0 cm. As the magnetic field is varied, a resonance is observed at 0.476 T. What is the *g*-factor for the unpaired electrons in the sample?

This Pentium chip is one of the leading processors used in computers today. It contains over 3 million transistors, not to mention diodes and other semiconductor electronic elements. Before discussing semiconductors and their applications later in this chapter, we begin with a study of how quantum theory describes how atoms bond together to form molecules, and how molecules behave. We then examine how atoms and molecules come together to form solids, with emphasis on metals and semi-conductors, and their use in electricity and electronics.

CHAPTER 41

Molecules and Solids

Since its development in the 1920s, quantum mechanics has had a profound influence on our lives, both intellectually and technologically. Even the way we view the world has changed, as we have seen in the last couple of chapters. In the present chapter, we will discuss how quantum mechanics has given us an understanding of the structure of molecules and matter in bulk, as well as a number of important applications including semiconductor devices. Our discussions will necessarily be qualitative for the most part.

41–1 | Bonding in Molecules

One of the great successes of quantum mechanics was to give scientists, at last, an understanding of the nature of chemical bonds. Since it is based in physics, and because this understanding is so important in many fields, we discuss it here.

By a molecule, we mean a group of two or more atoms that are strongly held together so as to function as a single unit. When atoms make such an attachment, we say that a chemical **bond** has been formed. There are two main types of strong chemical bond: covalent and ionic. Many bonds are actually intermediate between these two types.

Covalent Bonds

Covalent bond

To understand how **covalent bonds** are formed, we take the simplest case, the bond that holds two hydrogen atoms together to form the hydrogen molecule, H_2. The mechanism is basically the same for other covalent bonds. As two H atoms approach each other, the electron clouds begin to overlap, and the electrons from each atom can "orbit" both nuclei. (This is sometimes called "sharing" electrons.) If both electrons are in the ground state ($n = 1$) of their respective atoms, there are two possibilities: their spins can be parallel (both up or both down), in which case the total spin is $S = \frac{1}{2} + \frac{1}{2} = 1$; or their spins can be opposite ($m_s = +\frac{1}{2}$ for one, $m_s = -\frac{1}{2}$ for the other), so that the total spin $S = 0$. We shall now see that a bond is formed only for the $S = 0$ state, when the spins are opposite. First we consider the $S = 1$ state, for which the spins are the same. The two electrons cannot both be in the lowest energy state and be attached to the same atom, for then they would have identical quantum numbers in violation of the exclusion principle. The exclusion principle tells us that since no two electrons can occupy the same quantum state, if two electrons have the same quantum numbers, they must be different in some other way—namely, by being in different places in space (for example, attached to different atoms). When the two hydrogen atoms approach, the electrons will stay away from each other as shown by the probability distribution of Fig. 41–1. The positively charged nuclei then repel each other, and no bond is formed.

For the $S = 0$ state, on the other hand, the spins are opposite and the two electrons are consequently in different quantum states. Hence they can come close together spatially. In this case, the probability distribution looks like Fig. 41–2. As can be seen, the electrons spend much of their time between the two positively charged nuclei, and the latter are attracted to the negatively charged electron cloud between them. This attraction, which holds the two atoms together to form a molecule, constitutes a *covalent bond*.

The probability distributions of Figs. 41–1 and 41–2 can perhaps be better understood on the basis of waves. What the exclusion principle requires is that when the spins are the same, there is destructive interference of the electron wave functions in the region between the two atoms. But when the spins are opposite, constructive interference occurs in the region between the two atoms, resulting in a large amount of negative charge there. Thus a covalent bond can be said to be the result of constructive interference of the electron wave functions in the space between the two atoms, and of the electrostatic attraction of the two positive nuclei for the negative charge concentration between them.

Why a bond is formed can also be understood from the energy point of view. When the two H atoms approach close to one another, if the spins of their electrons are opposite, the electrons can occupy the same space, as discussed above. This means that each electron can now move about in the space of two atoms instead of in the volume of only one. Because each electron now occupies more space, it is less well localized. From the uncertainty principle, with Δx increased, the momentum, and hence the energy, can be less. Another way of understanding this is from the point of view of de Broglie waves (Section 38–11). Because each electron has a larger "orbit," its wavelength λ can be longer, so its momentum $p = h/\lambda$ (Eq. 38–7) can be less. With less momentum, each electron has less energy when the two atoms combine than when they are separate. That is, the molecule has less energy than the two separate atoms, and so is more stable. An energy input is required to break the H_2 molecule into two separate H atoms, so the H_2 molecule is a stable entity. This is what we mean by a *bond*. The energy required to break a bond is called the **bond energy**, the **binding energy**, or the **dissociation energy**. For the hydrogen molecule, H_2, the bond energy is 4.5 eV.

Ionic Bonds

An **ionic bond** is, in a sense, a special case of the covalent bond. Instead of the electrons being shared equally, they are shared unequally. For example, in sodium chloride (NaCl), the outer electron of the sodium spends nearly all its time around the chlorine (Fig. 41–3). The chlorine atom acquires a net negative charge as a result of the extra electron, whereas the sodium atom is left with a net positive charge. The electrostatic attraction between these two charged atoms holds them together. The resulting bond is called an *ionic bond* because it is created by the attraction between the two ions (Na^+ and Cl^-). But to understand the ionic bond,

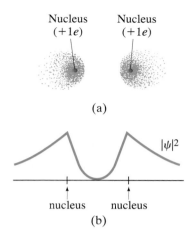

FIGURE 41–1 Electron probability distribution (electron cloud) for two H atoms when the spins are the same ($S = 1$): (a) electron cloud; (b) projection of $|\psi|^2$ along the line through the centers of the two atoms.

FIGURE 41–2 Electron probability distribution for two H atoms when the spins are opposite ($S = 0$): (a) electron cloud; (b) projection of $|\psi|^2$ along the line through the centers of the atoms. In this case a bond is formed because the positive nuclei are attracted to the concentration of negative charge between them. This is a hydrogen molecule, H_2.

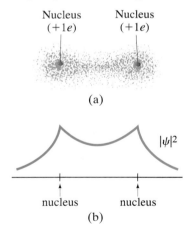

FIGURE 41–3 (below) Probability distribution $|\psi|^2$ for the last electron of Na in NaCl.

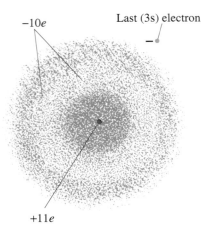

Last (3s) electron

−10e

+11e

FIGURE 41–4 In a neutral sodium atom, the 10 inner electrons shield the nucleus, so the single outer electron is attracted by a net charge of +1e.

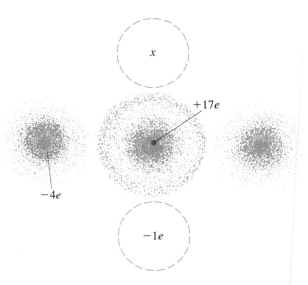

x

+17e

−4e

−1e

FIGURE 41–5 Neutral chlorine atom. The +17e of the nucleus is shielded by the 12 electrons in the inner shells and subshells. Four of the five 3p electrons are shown in doughnut-shaped clouds, and the fifth is in the (dashed-line) cloud concentrated about the vertical axis. An extra electron at x will be attracted by a net charge that can be as much as +5e.

we must understand why the extra electron from the sodium spends so much of its time around the chlorine. After all, the chlorine is neutral; why should it attract another electron?

The answer lies in the probability distributions of the two neutral atoms. Sodium contains 11 electrons, 10 of which together form spherically symmetric closed shells (Fig. 41–4). The last electron spends most of its time beyond these closed shells. Because the closed shells have a total charge of −10e and the nucleus has charge +11e, the outermost electron in sodium "feels" a net attraction due to +1e. It is not held very strongly. On the other hand, 12 of chlorine's 17 electrons form closed shells, or subshells (corresponding to $1s^2 2s^2 2p^6 3s^2$). These 12 form a spherically symmetric shield around the nucleus. The other five electrons are in 3p states whose probability distributions are not spherically symmetric and have a form similar to those for the 2p states in hydrogen shown in Fig. 40–9a. Four of these 3p electrons can have "doughnut-shaped" distributions symmetric about the z axis, as shown in Fig. 41–5. The fifth can have a "barbell-shaped" distribution (as for $m_l = 0$ in Fig. 40–9a), which in Fig. 41–5 is shown only as a dashed outline because it is half empty. That is, the exclusion principle allows one more electron to be in this state (it will have spin opposite to that of the electron already there). If an extra electron—say from a Na atom—happens to be in the vicinity, it can be in this state, say at point x in Fig. 41–5. It could experience an attraction due to as much as +5e because the +17e of the chlorine nucleus is partly shielded at this point by the 12 inner electrons. Thus, the outer electron of a Na atom will be more strongly attracted by the +5e of the chlorine atom that it "sees" at certain points rather than by the +1e of its own atom. This, combined with the strong attraction between the two ions when the extra electron stays with the Cl⁻, produces the charge distribution of Fig. 41–3, and hence the ionic bond.

Partial Ionic Character of Covalent Bonds

A pure covalent bond in which the electrons are shared equally occurs mainly in symmetrical molecules such as H_2, O_2, and Cl_2. When the atoms involved are different from each other, it is usual to find that the shared electrons are more likely to be in the vicinity of one atom than the other. The extreme case is an ionic bond; in intermediate cases the covalent bond is said to have a *partial ionic character*. The molecules themselves are **polar**—that is, they have an electric dipole moment because one part (or parts) of the molecule has a net positive charge and other parts a net negative charge. An example is the water molecule, H_2O (Fig. 41–6). The shared electrons are more likely to be found around the oxygen atom than around the two hydrogens. The reason is similar to that discussed above in con-

FIGURE 41–6
The water molecule is polar.

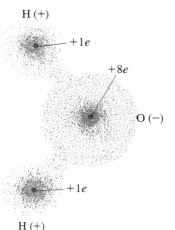

H (+)

+1e

+8e

O (−)

+1e

H (+)

nection with ionic bonds. Oxygen has eight electrons $(1s^2 2s^2 2p^4)$, of which four form a spherically symmetric core and the other four could have, for example, a doughnut-shaped distribution. The barbell-shaped distribution on the z axis (like that shown dashed in Fig. 41–5) could be empty, so electrons from hydrogen atoms can be attracted by a net charge of $+4e$. They are also attracted by the H nuclei, so they partly orbit the H atoms as well as the O atom. The net effect is that there is a net positive charge on each H atom (less than $+1e$), because the electrons spend only part of their time there. And, there is a net negative charge on the O atom.

41–2 Potential-Energy Diagrams for Molecules

It is useful to analyze the interaction between two objects—say, between two atoms or molecules—with the use of a potential-energy diagram, a plot of the potential energy versus the separation distance.

For the simple case of two point charges, q_1 and q_2, the potential energy U is given by (see Chapter 23):

$$U = \frac{1}{4\pi\epsilon_0} \frac{q_1 q_2}{r},$$

where r is the distance between the charges, and the constant $(1/4\pi\epsilon_0)$ is equal to $9.0 \times 10^9\,\text{N·m}^2/\text{C}^2$. If the two charges have the same sign, the potential energy is positive for all values of r, and a graph of U versus r in this case is shown in Fig. 41–7a. The force is repulsive (the charges have the same sign) and the curve rises as r decreases; this makes sense since work is done to bring the charges together, thereby increasing their potential energy. If, on the other hand, the two charges are of opposite sign, the potential energy is negative because the product $q_1 q_2$ is negative. The force is attractive in this case and the graph of $U(\propto -1/r)$ versus r looks like Fig. 41–7b. The potential energy becomes more *negative* as r decreases.

Now let us look at the potential energy diagram for the formation of a covalent bond, such as for the hydrogen molecule. The potential energy of one H atom in the presence of the other is plotted in Fig. 41–8. The potential energy U decreases as the atoms approach, because the electrons concentrate between the two nuclei (Fig. 41–2), so attraction occurs. However, at very short distances, the electrons would be "squeezed out"—there is no room for them between the two nuclei. Without the electrons between them, each nucleus would feel a repulsive force due to the other, so the curve rises as r decreases further. There is an optimum separation of the atoms, r_0 in Fig. 41–8, at which the energy is lowest. This is the point of greatest stability for the hydrogen molecule, and r_0 is the average separation of atoms in the H_2 molecule. The depth of this "well" is the *binding energy*,[†] as shown. This is how much energy must be put into the system to separate the two atoms to infinity, where the potential energy $U = 0$. For the H_2 molecule, the binding energy is about 4.5 eV, and $r_0 = 0.074\,\text{nm}$. In molecules made

[†]The binding energy corresponds not quite to the bottom of the potential-energy curve, but to the lowest energy state, slightly above it, as shown in Fig. 41–8. Compare to a square well, Fig. 39–8.

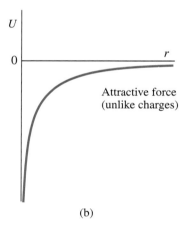

FIGURE 41–7 Potential energy U as a function of the separation for two point charges of (a) like sign and (b) opposite sign.

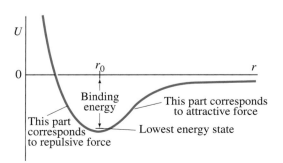

FIGURE 41–8 Potential-energy diagram for H_2 molecule; r is the separation of the two H atoms. The binding energy (the energy difference between $U = 0$ and the lowest energy state near the bottom of the well) is 4.5 eV, and $r_0 = 0.074\,\text{nm}$.

of larger atoms, say, oxygen or nitrogen, repulsion also occurs at short distances, because the closed inner electron shells begin to overlap and the exclusion principle forbids their coming too close. The repulsive part of the curve rises even more steeply than $1/r$. A reasonable approximation to the potential energy, at least in the vicinity of r_0, is

$$U = -\frac{A}{r^m} + \frac{B}{r^n},\qquad\qquad\qquad \textbf{(41–1)}$$

where A and B are constants associated with the attractive and repulsive parts of the potential energy and m and n are small integers. For ionic and some covalent bonds, the attractive term can often be written with $m = 1$ (Coulomb potential).

For many bonds, the potential-energy curve has the shape shown in Fig. 41–9. There is still an optimum distance r_0 at which the molecule is stable. But when the atoms approach from a large distance, the force is initially repulsive rather than attractive. The atoms thus do not interact spontaneously. Instead, some additional energy must be injected into the system to get it over the "hump" (or barrier) in the potential-energy diagram. This required energy is called the **activation energy**.

The curve of Fig. 41–9 is much more common than that of Fig. 41–8. The activation energy often reflects a need to break other bonds, before the one under discussion can be made. For example, to make water from O_2 and H_2, the H_2 and O_2 molecules must first be broken into H and O atoms by an input of energy; this is what the activation energy represents. Then the H and O atoms can combine to form H_2O with the release of a great deal more energy than was put in initially. The initial activation energy can be provided by applying an electric spark to a mixture of H_2 and O_2, breaking a few of these molecules into H and O atoms. The resulting explosive release of energy when these atoms combine to form H_2O quickly provides the activation energy needed for further reactions, so additional H_2 and O_2 molecules are broken up and recombined to form H_2O.

The potential-energy diagrams for ionic bonds can have similar shapes. In NaCl, for example, the Na^+ and Cl^- ions attract each other at distances a bit larger than some r_0, but at shorter distances the overlapping of inner electron shells gives rise to repulsion. The two atoms thus are most stable at some intermediate separation, r_0, and there often is an activation energy.

FIGURE 41–9 Potential energy diagram for a bond requiring an activation energy.

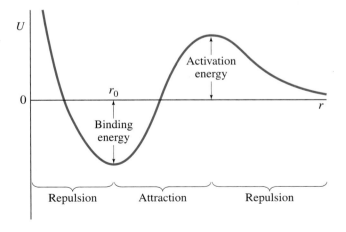

EXAMPLE 41–1 **ESTIMATE** **Sodium chloride bond.** A potential-energy diagram for the NaCl ionic bond is shown in Fig. 41–10, where we have set $U = 0$ for free Na and Cl neutral atoms (which are represented on the right in Fig. 41–10). Measurements show that 5.14 eV are required to remove an electron from a neutral Na atom to produce the Na^+ ion; and 3.61 eV of energy is released when an electron is "grabbed" by a Cl atom to form the Cl^- ion. Thus, forming Na^+ and Cl^- ions from neutral Na and Cl atoms requires 5.14 eV − 3.61 eV = 1.53 eV of energy, a form of activation energy. This is shown as the "bump" in Fig. 41–10. But note that the potential-energy diagram from here out to the right is not really a function of distance—it is drawn dashed to remind us that it only represents the energy difference between the ions and the neutral atoms (for which we have chosen $U = 0$). (a) Calculate the separation distance, r_1, at which the potential of the Na^+ and Cl^- ions drops to zero (measured value is $r_1 = 0.94$ nm). (b) Estimate the binding energy of the NaCl bond, which occurs at a separation $r_0 = 0.24$ nm. Ignore the repulsion of the overlapping electron shells that occurs at this distance (and causes the rise of the potential-energy curve for $r < r_0$, Fig. 41–10).

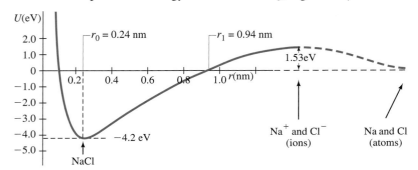

FIGURE 41–10 Potential-energy diagram for the NaCl bond. Beyond about $r = 1.2$ nm, the diagram is schematic only, and represents the energy difference between ions and neutral atoms. $U = 0$ is chosen for the two separated atoms Na and Cl (not for the ions). [For the two ions, Na^+ and Cl^-, the zero of potential energy at $r = \infty$, corresponds to $U \approx 1.53$ eV on this diagram.]

SOLUTION (a) The potential energy of two point charges is given by Coulombs' law:

$$U' = \frac{1}{4\pi\epsilon_0} \frac{q_1 q_2}{r},$$

where we distinguish U' from the U in Fig. 41–10. This formula works for our two ions if we set $U' = 0$ at $r = \infty$, which in the plot of Fig. 41–10 corresponds to $U = +1.53$ eV (Fig. 41–10 is drawn for $U = 0$ for the free *atoms*). The point r_1 in Fig. 41–10 corresponds to $U' = -1.53$ eV relative to the two free ions. Since $q_1 = +e$ and $q_2 = -e$, we have

$$r_1 = \frac{1}{4\pi\epsilon_0} \frac{q_1 q_2}{U'} = \frac{(9.0 \times 10^9 \,\mathrm{N \cdot m^2/C^2})(-1.60 \times 10^{-19}\,\mathrm{C})(+1.60 \times 10^{-19}\,\mathrm{C})}{(-1.53\,\mathrm{eV})(1.60 \times 10^{-19}\,\mathrm{J/eV})}$$
$$= 0.94 \,\mathrm{nm},$$

which is just the measured value.

(b) At $r_0 = 0.24$ nm, the potential energy of the two ions (relative to $r = \infty$ for the two ions) is

$$U' = \frac{1}{4\pi\epsilon_0} \frac{q_1 q_2}{r}$$
$$= \frac{(9.0 \times 10^9 \,\mathrm{N \cdot m^2/C^2})(-1.60 \times 10^{-19}\,\mathrm{C})(+1.60 \times 10^{-19}\,\mathrm{C})}{(0.24 \times 10^{-9}\,\mathrm{m})(1.60 \times 10^{-19}\,\mathrm{J/eV})} = -6.0 \,\mathrm{eV}.$$

Thus, we estimate that 6.0 eV of energy is given up when Na^+ and Cl^- ions form a NaCl bond. Put another way, it takes 6.0 eV to break the NaCl bond and form the Na^+ and Cl^- ions. To get the binding energy—the energy to separate the NaCl into Na and Cl atoms—we need to subtract out the 1.53 eV (the "bump" in Fig. 41–10) needed to ionize them:

$$\text{binding energy} = 6.0 \,\mathrm{eV} - 1.53 \,\mathrm{eV} = 4.5 \,\mathrm{eV}.$$

The measured value (shown on Fig. 41–10) is 4.2 eV. The difference can be attributed to the energy associated with the repulsion of the electron shells at this distance.

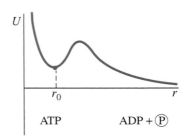

FIGURE 41–11 Potential energy diagram for the formation of ATP from ADP and phosphate (Ⓟ).

FIGURE 41–12 The C^+—O^- and H^+—N^- dipoles attract each other. (These dipoles may be part of larger molecules. See Fig. 41–13.)

Sometimes the potential energy of a bond looks like that of Fig. 41–11. In this case, the energy of the bonded molecule, at a separation r_0, is greater than when there is no bond ($r = \infty$). That is, an energy *input* is required to make the bond (hence the binding energy is negative), and there is energy release when the bond is broken. Such a bond is stable only because there is the barrier of the activation energy. This type of bond is important in living cells, for it is in such bonds that energy can be stored efficiently in certain molecules, particularly ATP (adenosine triphosphate). The bond that connects the last phosphate group (designated Ⓟ in Fig. 41–11) to the rest of the molecule (ADP, meaning adenosine diphosphate, since it contains only two phosphates) is of the form shown in Fig. 41–11. Energy is actually stored in this bond. When the bond is broken $\left(\text{ATP} \rightarrow \text{ADP} + Ⓟ\right)$, energy is released and this energy can be used to make other chemical reactions "go."

41–3 Weak (van der Waals) Bonds

Once a bond between two atoms or ions is made, energy must normally be supplied to break the bond and separate the atoms. As mentioned in Section 41–1, this energy is called the *bond energy* or *binding energy*. The binding energy for covalent and ionic bonds is typically 2 to 5 eV. These bonds, which hold atoms together to form molecules, are often called **strong bonds** to distinguish them from so-called "weak bonds." The term **weak bond**, as we use it here, refers to an attachment between molecules due to simple electrostatic attraction—such as *between* polar molecules (not *within* a polar molecule, which is a strong bond). The strength of the attachment is much less than for the strong bonds. Binding energies are typically in the range 0.04 to 0.3 eV—hence their name "weak bonds."

Weak bonds are generally the result of attraction between dipoles. For example, Fig. 41–12 shows two molecules, which have permanent dipole moments, attracting one another. Besides such **dipole–dipole bonds**, there can also be **dipole–induced dipole bonds**, in which a polar molecule with a permanent dipole moment can induce a dipole moment in an otherwise electrically balanced (nonpolar) molecule, just as a single charge can induce a separation of charge in a nearby object (see Fig. 21–7). There can even be an attraction between two nonpolar molecules. Even though a molecule may not have a permanent dipole moment on the average, we can think of its electrons as moving about so that at any instant there may be a separation of charge. Such transient dipoles can induce a dipole moment in a nearby molecule, creating a weak attraction. All these weak bonds are referred to as **van der Waals bonds**, and the forces involved **van der Waals forces**. The potential energy has the general shape shown in Fig. 41–8, with the attractive van der Waals potential energy varying as $1/r^6$.

When one of the atoms in a dipole–dipole bond is hydrogen, as in Fig. 41–12, it is called a **hydrogen bond**. A hydrogen bond is generally quite strong for a so-called "weak bond." This is because the hydrogen atom is the smallest atom and thus can be approached more closely. A hydrogen bond also has a partial "covalent" character: that is, electrons between the two dipoles may be shared to a small extent, thus making a stronger, more lasting, bond.

Weak bonds are important in liquids and solids when strong bonds are absent (see Section 41–5). They are also very important for understanding the activities of cells, such as the double-helix shape of DNA (Fig. 41–13), and DNA replication. The average kinetic energy of molecules in a cell is around $\frac{3}{2}kT \approx 0.04$ eV, about the magnitude of weak bonds. This means that a weak bond can readily be broken just by a molecular collision. Hence weak bonds are not very permanent—they are, instead, brief attachments. But because of this, they play an important role in the cell. On the other hand, strong bonds—those that hold molecules together—are almost never broken simply by molecular collision.

(a)

(b)

FIGURE 41–13 (a) Section of a DNA double helix. The red dots represent hydrogen bonds between the two chains. (b) "Close-up" view of helix: cytosine (C) and guanine (G) molecules on separate chains of a DNA double helix are held together by the hydrogen bonds (red dots) involving an H^+ on one molecule attracted to an N^- or C^+—O^- on the other molecule.

Thus they are relatively permanent. They can be broken by chemical action (the making of even stronger bonds), and this usually happens in the cell with the aid of an enzyme, which is a protein molecule.

EXAMPLE 41–2 **Nucleotide energy.** Calculate the interaction energy between the C=O dipole of thymine and the H—N dipole of adenine, assuming that the two dipoles are lined up as shown in Fig. 41–12. Dipole moment measurements (see Table 23–2) give $q_H = -q_N = 0.19e = 3.0 \times 10^{-20}$ C, and $q_C = -q_O = 0.42e = 6.7 \times 10^{-20}$ C.

SOLUTION The interaction energy will be equal to the potential energy of one dipole in the presence of the other, since this will be equal to the work needed to pull the two dipoles infinitely far apart. U will consist of four terms:

$$U = U_{CH} + U_{CN} + U_{OH} + U_{ON},$$

where U_{CH} means the potential energy of C in the presence of H, and similarly for the other terms. We do not have terms corresponding to C and O, or N and H, because the two dipoles are assumed to be stable entities (the bonds C=O and H—N remain intact). The potential energy for two point charges is $U_{12} = (1/4\pi\epsilon_0)(q_1 q_2/r)$, where r is the distance between them. So

$$U = \frac{1}{4\pi\epsilon_0}\left(\frac{q_C q_H}{r_{CH}} + \frac{q_C q_N}{r_{CN}} + \frac{q_O q_H}{r_{OH}} + \frac{q_O q_N}{r_{ON}}\right).$$

Using the distances shown in Fig. 41–12, we get:

$$U = (9.0 \times 10^9 \, \text{N} \cdot \text{m}^2/\text{C})\left(\frac{(6.7)(3.0)}{0.31} + \frac{(6.7)(-3.0)}{0.41} + \frac{(-6.7)(3.0)}{0.19}\right.$$
$$\left. + \frac{(-6.7)(-3.0)}{0.29}\right)\frac{(10^{-20} \, \text{C})^2}{(10^{-9} \, \text{m})}$$
$$= -1.82 \times 10^{-20} \, \text{J} = -0.11 \, \text{eV}.$$

The potential energy is negative, meaning 0.11 eV of work (or energy input) is required to separate the molecules. That is, the binding energy of this "weak" or hydrogen bond is 0.11 eV. This is only an estimate, of course, since other charges in the vicinity would have an influence too.

41–4 Molecular Spectra

When atoms combine to form molecules, the wave functions (and probability distributions) of the outer electrons overlap, and this interaction alters the energy levels. Nonetheless, molecules can undergo transitions between electron energy levels just as atoms do. For example, the H_2 molecule can absorb a photon of just the

FIGURE 41–14 The individual energy levels of an isolated atom become bands of closely spaced levels in molecules, as well as in solids and liquids.

3p

2s

Isolated atom   Atom in a molecule

right frequency to excite one of its ground state electrons to an excited state. The excited electron can then return to the ground state, emitting a photon. The energy of photons emitted by molecules is of the same order of magnitude as for atoms, typically 1 to 10 eV.

Additional energy levels become possible for molecules (but not for atoms) because the molecule as a whole can rotate, and the atoms of the molecule can vibrate relative to each other. The energy levels for both rotational and vibrational levels are quantized, and are generally spaced much more closely $(10^{-3}$ to $10^{-1}\,\text{eV})$ than the electronic levels. Each atomic energy level thus becomes a set of closely spaced levels corresponding to the vibrational and rotational motions, Fig. 41–14. Transitions from one level to another appear as many very closely spaced lines. In fact, the lines are not always distinguishable, and these spectra are called **band spectra**. Each type of molecule has its own characteristic spectrum, which can be used for identification and for determination of structure.

Rotational Energy Levels in Molecules

Let us now look in more detail at rotational and vibrational states in molecules. We begin with rotation, considering here only diatomic molecules, although the analysis can be extended to polyatomic molecules. When a diatomic molecule rotates about an axis through the center of mass perpendicular to the line joining the two atoms as shown in Fig. 41–15, its kinetic energy of rotation (see Section 10–10) is

$$E_{\text{rot}} = \frac{1}{2}I\omega^2 = \frac{(I\omega)^2}{2I},$$

where $(I\omega)$ is the angular momentum (Section 10–9). Quantum mechanics predicts quantization of angular momentum just as in atoms (see Eq. 40–3):

$$I\omega = \sqrt{L(L+1)}\,\hbar, \qquad L = 0, 1, 2, \cdots,$$

where L is an integer called the **rotational angular momentum quantum number**. Thus the rotational energy is quantized:

$$E_{\text{rot}} = \frac{(I\omega)^2}{2I} = L(L+1)\frac{\hbar^2}{2I}, \qquad L = 0, 1, 2, \cdots. \qquad \textbf{(41–2)}$$

Transitions between rotational energy levels are subject to the *selection rule* (see the end of Section 40–2):

Selection rule

$$\Delta L = \pm 1.$$

The energy of a photon emitted or absorbed for a transition between rotational states with angular momentum quantum number L and $L-1$ will be

$$\Delta E_{\text{rot}} = E_L - E_{L-1} = \frac{\hbar^2}{2I}L(L+1) - \frac{\hbar^2}{2I}(L-1)(L)$$

$$= \frac{\hbar^2}{I}L. \qquad \begin{bmatrix} L \text{ is for the} \\ \text{upper state} \end{bmatrix} \quad \textbf{(41–3)}$$

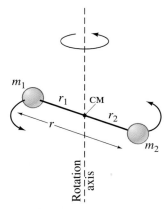

FIGURE 41–15 Diatomic molecule rotating about a vertical axis. The two atoms, of mass m_1 and m_2, are separated by a distance $r = r_1 + r_2$ where r_1 and r_2 are their respective distances from the center of mass.

We see that the transition energy increases directly with L. Figure 41–16 shows some of the allowed rotational energy levels and transitions. Measured absorption lines fall in the microwave or far-infrared regions of the spectrum, and their frequencies are generally 2, 3, 4, $\cdots$ times higher than the lowest one, as predicted by Eq. 41–3.

The moment of inertia of the molecule in Fig. 41–15 rotating about its center of mass (Section 10–6) is

$$I = m_1 r_1^2 + m_2 r_2^2,$$

where r_1 and r_2 are the distances of each atom from their common center of mass. We show in Example 41–3 that I can be written

$$I = \frac{m_1 m_2}{m_1 + m_2} r^2 = \mu r^2, \tag{41–4}$$

where $r = r_1 + r_2$ is the distance between the two atoms of the molecule and $\mu = m_1 m_2 / (m_1 + m_2)$ is called the **reduced mass**. If $m_1 = m_2$, then $\mu = \frac{1}{2} m_1 = \frac{1}{2} m_2$.

FIGURE 41–16 Rotational energy levels and allowed transitions (emission and absorption) for a diatomic molecule. Upward-pointing arrows represent absorption of a photon, and downward arrows represent emission.

EXAMPLE 41–3 **Reduced mass.** Show that the moment of inertia of a diatomic molecule rotating about its center of mass can be written

$$I = \mu r^2,$$

where

$$\mu = \frac{m_1 m_2}{m_1 + m_2}$$

is the reduced mass, Eq. 41–4, and r is the distance between the atoms.

SOLUTION The moment of inertia of a single particle of mass m a distance r from the rotation axis is $I = mr^2$. Hence for our diatomic molecule (Fig. 41–15)

$$I = m_1 r_1^2 + m_2 r_2^2.$$

Now $r = r_1 + r_2$ and $m_1 r_1 = m_2 r_2$ because the axis of rotation passes through the center of mass. Hence

$$r_1 = r - r_2 = r - \frac{m_1}{m_2} r_1,$$

so

$$r_1 = \frac{r}{1 + \dfrac{m_1}{m_2}} = \frac{m_2 r}{m_1 + m_2}.$$

Similarly,

$$r_2 = \frac{m_1 r}{m_1 + m_2}.$$

Then

$$I = m_1 \left(\frac{m_2 r}{m_1 + m_2} \right)^2 + m_2 \left(\frac{m_1 r}{m_1 + m_2} \right)^2 = \frac{m_1 m_2 (m_1 + m_2) r^2}{(m_1 + m_2)^2}$$

$$= \frac{m_1 m_2}{m_1 + m_2} r^2 = \mu r^2,$$

where

$$\mu = \frac{m_1 m_2}{m_1 + m_2},$$

which is what we wished to show.

EXAMPLE 41–4 **Rotational transition.** A rotational transition $L = 1$ to $L = 0$ for the molecule CO has a measured absorption wavelength $\lambda = 2.60 \, \text{mm}$ (microwave region). Use this to calculate (a) the moment of inertia of the CO molecule, and (b) the CO bond length, r. (c) Calculate the wavelengths of the next three rotational transitions, and the energies of the photon emitted for each of these four transitions.

SOLUTION (a) From Eq. 41–3, we can write

$$\frac{\hbar^2}{I} L = \Delta E = hf = \frac{hc}{\lambda_1}.$$

With $L = 1$ (the upper state) in this case, we solve for I:

$$I = \frac{\hbar^2 L}{hc} \lambda_1 = \frac{h\lambda_1}{4\pi^2 c} = \frac{(6.63 \times 10^{-34} \, \text{J·s})(2.60 \times 10^{-3} \, \text{m})}{4\pi^2 (3.00 \times 10^8 \, \text{m/s})}$$

$$= 1.46 \times 10^{-46} \, \text{kg·m}^2.$$

(b) The masses of C and O are 12.0 and 16.0 u, respectively, where $1 \, \text{u} = 1.66 \times 10^{-27} \, \text{kg}$. Thus the reduced mass is

$$\mu = \frac{m_1 m_2}{m_1 + m_2} = \frac{(12.0)(16.0)}{28.0} (1.66 \times 10^{-27} \, \text{kg}) = 1.14 \times 10^{-26} \, \text{kg}$$

or 6.86 u. Then, from Eq. 41–4, the bond length is

$$r = \sqrt{\frac{I}{\mu}} = \sqrt{\frac{1.46 \times 10^{-46} \, \text{kg·m}^2}{1.14 \times 10^{-26} \, \text{kg}}} = 1.13 \times 10^{-10} \, \text{m} = 0.113 \, \text{nm}.$$

(c) From Eq. 41–3, $\Delta E \propto L$. Hence $\lambda = c/f = hc/\Delta E$ is proportional to $1/L$. Thus, for $L = 2$ to $L = 1$ transitions, $\lambda_2 = \frac{1}{2}\lambda_1 = 1.30 \, \text{mm}$. For $L = 3$ to $L = 2$, $\lambda_3 = \frac{1}{3}\lambda_1 = 0.87 \, \text{mm}$. And for $L = 4$ to $L = 3$, $\lambda_4 = 0.65 \, \text{mm}$. All are close to measured values. The energies of the photons, $hf = hc/\lambda$, are respectively $4.8 \times 10^{-4} \, \text{eV}$, $9.5 \times 10^{-4} \, \text{eV}$, $1.4 \times 10^{-3} \, \text{eV}$, and $1.9 \times 10^{-3} \, \text{eV}$.

Vibrational Energy Levels in Molecules

The potential energy of the two atoms in a typical diatomic molecule has the shape shown in Fig. 41–8 or 41–9, and Fig. 41–17 again shows the potential energy for the H_2 molecule. We note that the potential energy, at least in the vicinity of the equilibrium separation r_0, closely resembles the potential energy of a harmonic oscillator, $U = \frac{1}{2}kx^2$, which is shown superposed in dashed lines. Thus, for small displacements from r_0, each atom experiences a restoring force approximately proportional to the displacement, and the molecule vibrates as a simple harmonic oscillator (SHO). The classical frequency of vibration is

Classical vibration frequency

$$f = \frac{1}{2\pi} \sqrt{\frac{k}{\mu}}, \tag{41–5}$$

FIGURE 41–17 Potential energy for the H_2 molecule and for a simple harmonic oscillator ($U_{SHO} = \frac{1}{2}kx^2$, with $|x| = |r - r_0|$). The 0.50 eV energy height marked is for use in Example 41–5 to estimate k. [Note that $U_{SHO} = 0$ is not the same as $U = 0$ for the molecule.]

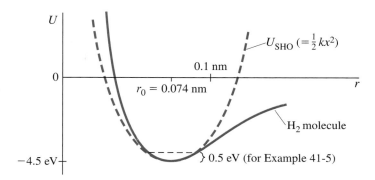

where k is the "stiffness constant" (as for a spring, Chapter 14) and instead of the mass m we must again use the reduced mass $\mu = m_1 m_2/(m_1 + m_2)$. (This is shown in Problem 59.) The Schrödinger equation for the SHO potential energy yields solutions for energy that are quantized according to

$$E_{\text{vib}} = \left(v + \tfrac{1}{2}\right)hf \qquad v = 0, 1, 2, \cdots, \qquad \textbf{(41–6)}$$

where f is given by Eq. 41–5 and v is an integer called the **vibrational quantum number**. The lowest energy state ($v = 0$) is not zero (as for rotation) but has $E = \tfrac{1}{2}hf$. This is called the **zero-point energy**.[†] Higher states have energy $\tfrac{3}{2}hf, \tfrac{5}{2}hf$, and so on, as shown in Fig. 41–18. Transitions are subject to the selection rule

$$\Delta v = \pm 1,$$

so allowed transitions occur only between adjacent states and all give off photons of energy

$$\Delta E_{\text{vib}} = hf. \qquad \textbf{(41–7)}$$

This is very close to experimental values for small v, but for higher energies, the potential-energy curve (Fig. 41–17) begins to deviate from a perfect SHO curve, which affects the wavelengths and frequencies of the transitions. Typical transition energies are on the order of $10^{-1}\,\text{eV}$, about 10 to 100 times larger than for rotational transitions, with wavelengths in the infrared region of the spectrum $(\approx 10^{-5}\,\text{m})$.[‡]

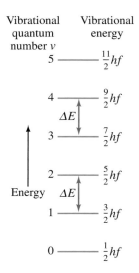

FIGURE 41–18 Allowed vibrational energies of a diatomic molecule, where f is the fundamental frequency of vibration, given by Eq. 41–5. Transitions are allowed only between adjacent levels ($\Delta v = \pm 1$), which are equally spaced.

EXAMPLE 41–5 ESTIMATE Wavelength for H$_2$. (*a*) Use the curve of Fig. 41–17 to estimate the value of the stiffness constant k for the H$_2$ molecule, and then (*b*) estimate the fundamental wavelength for vibrational transitions.

SOLUTION (*a*) We choose arbitrarily an energy height of 0.50 eV which is indicated in Fig. 41–17. By measuring directly on the graph, we find that this energy corresponds to a vibration on either side of $r_0 = 0.074\,\text{nm}$ of about $x = \pm 0.017\,\text{nm}$. For SHO, $U_{\text{SHO}} = \tfrac{1}{2}kx^2$ and $U_{\text{SHO}} = 0$ at $x = 0$ $(r = r_0)$; then

$$k = \frac{2U_{\text{SHO}}}{x^2} \approx \frac{2(0.50\,\text{eV})(1.6 \times 10^{-19}\,\text{J/eV})}{(1.7 \times 10^{-11}\,\text{m})^2} \approx 550\,\text{N/m}.$$

This value of k would also be reasonable for a macroscopic spring. (*b*) The reduced mass is $\mu = m_1 m_2/(m_1 + m_2) = m_1/2 = \tfrac{1}{2}(1.0\,\text{u})(1.66 \times 10^{-27}\,\text{kg}) = 0.83 \times 10^{-27}\,\text{kg}$. Hence, using Eq. 41–5,

$$\lambda = \frac{c}{f} = 2\pi c \sqrt{\frac{\mu}{k}} = 2\pi(3.0 \times 10^8\,\text{m/s})\sqrt{\frac{0.83 \times 10^{-27}\,\text{kg}}{550\,\text{N/m}}}$$

$$= 2300\,\text{nm},$$

which is in the infrared region of the spectrum.

Experimentally, we do the inverse process: The wavelengths of vibrational transitions for a given molecule are measured, and from this the stiffness constant k can be calculated. The values of k calculated in this way are a measure of the strength of the molecular bond.

[†] Recall this phenomenon for a square well, Fig. 39–8.

[‡] Forbidden transitions with $\Delta v = 2$ are emitted somewhat more weakly, but their observation can be important in some cases, such as in astronomy.

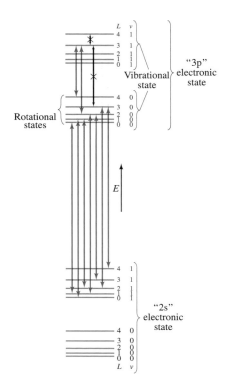

FIGURE 41–19 Combined electronic, vibrational, and rotational energy levels. Transitions marked with an × are not allowed by the selection rules.

EXAMPLE 41–6 **Vibrational energy levels in hydrogen.** Given that the hydrogen molecule emits infrared radiation of wavelength around 2300 nm, (*a*) what is the separation in energy between different vibrational levels, and (*b*) what is the lowest vibrational energy state?

SOLUTION (*a*)

$$\Delta E_{\text{vib}} = hf = \frac{hc}{\lambda} = \frac{(6.63 \times 10^{-34}\,\text{J}\cdot\text{s})(3.00 \times 10^{8}\,\text{m/s})}{(2300 \times 10^{-9}\,\text{m})(1.60 \times 10^{-19}\,\text{J/eV})} = 0.54\,\text{eV}.$$

(*b*) The lowest vibrational energy has $\nu = 0$ in Eq. 41–6: $E = \frac{1}{2}hf = 0.27\,\text{eV}$.

Rotational plus Vibrational Levels

When energy is imparted to a molecule, both the rotational and vibrational modes can be excited. Because rotational energies are an order of magnitude or so smaller than vibrational energies, which in turn are smaller than the electronic energy levels, we can represent the grouping of levels as shown in Fig. 41–19. Transitions between energy levels, with emission of a photon, are subject to the *selection rules*:

$$\Delta \nu = \pm 1$$

and

$$\Delta L = \pm 1.$$

Some allowed and forbidden (marked ×) transitions are indicated in Fig. 41–19. Not all transitions and levels are shown, and the separation between vibrational levels, and (even more) between rotational levels, has been exaggerated. But we can clearly see the origin of the very closely spaced lines that give rise to the band spectra, as mentioned with reference to Fig. 41–14 earlier in this Section.

The spectra are quite complicated, so we consider briefly only transitions within the same electronic level, such as those at the top of Fig. 41–19. A transition from a state with quantum numbers ν and L, to one with quantum numbers $\nu + 1$ and $L \pm 1$ (see the selection rules above), will absorb[†] a photon of energy:

$$
\begin{aligned}
\Delta E &= \Delta E_{\text{vib}} + \Delta E_{\text{rot}} \\[4pt]
&= hf + (L + 1)\frac{\hbar^2}{I} \quad \begin{bmatrix} L \to L + 1 \\ (\Delta L = +1) \end{bmatrix}, \quad L = 0, 1, 2, \cdots \\[4pt]
&= hf - L\frac{\hbar^2}{I} \quad\quad \begin{bmatrix} L \to L - 1 \\ (\Delta L = -1) \end{bmatrix}, \quad L = 1, 2, 3, \cdots,
\end{aligned}
\tag{41–8}
$$

where we have used Eqs. 41–3 and 41–7. Note that for $L \to L - 1$ transitions, L cannot be zero since there is then no state with $L = -1$. Equations 41–8 predict an absorption spectrum like that shown schematically in Fig. 41–20, with transi-

[†] This is for absorption. For emission of a photon, the transition would be $\nu \to \nu - 1$, $L \to L \pm 1$.

FIGURE 41–20 Expected spectrum for transitions between combined rotational and vibrational states.

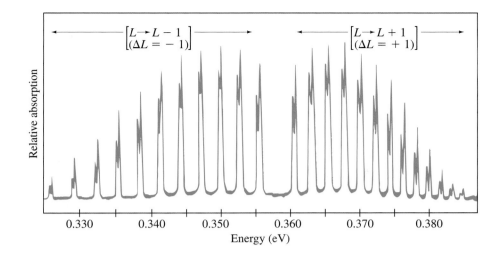

FIGURE 41–21 Absorption spectrum for HCl molecules. Lines on the left correspond to transitions where $L \rightarrow L - 1$; those on the right are for $L \rightarrow L + 1$. Each line has a double peak because chlorine has two isotopes of different mass and different moment of inertia.

tions $L \rightarrow L - 1$ on the left and $L \rightarrow L + 1$ on the right. Figure 41–21 shows the molecular absorption spectrum of HCl, which follows this pattern very well. (Each line in that spectrum is split into two because Cl consists of two isotopes of different mass; hence there are two kinds of HCl molecule with different moments of inertia I.)

EXAMPLE 41–7 **ESTIMATE** **The HCl molecule.** Estimate the moment of inertia of the HCl molecule using the absorption spectrum shown in Fig. 41–21. For the purposes of a rough estimate you can ignore the difference between the two isotopes.

SOLUTION The locations of the peaks in Fig. 41–21 should correspond to Eqs. 41–8. We don't know what value of L each peak corresponds to in Fig. 41–21, but we can estimate the energy difference between peaks to be about $\Delta E' = 0.0025\,\text{eV}$. Then from Eqs. 41–8, the energy difference between two peaks is given by

$$\Delta E' = \Delta E_{L+1} - \Delta E_L$$
$$= \frac{\hbar^2}{I}.$$

Then

$$I = \frac{\hbar^2}{\Delta E'} = \frac{\left(6.626 \times 10^{-34}\,\text{J·s}/2\pi\right)^2}{(0.0025\,\text{eV})\left(1.6 \times 10^{-19}\,\text{J/eV}\right)} = 2.8 \times 10^{-47}\,\text{kg·m}^2.$$

If we use the results of Ex. 41–3, we can get a better idea of what this number means. We write $I = \mu r^2$ and calculate μ:

$$\mu = \frac{m_1 m_2}{m_1 + m_2} = \frac{(1.0\,\text{u})(35\,\text{u})}{36\,\text{u}}\left(1.66 \times 10^{-27}\,\text{kg/u}\right) = 1.6 \times 10^{-27}\,\text{kg}.$$

Then the bond length is given by (Eq. 41–4)

$$r = \left(\frac{I}{\mu}\right)^{\frac{1}{2}} = \left(\frac{2.8 \times 10^{-47}\,\text{kg·m}^2}{1.6 \times 10^{-27}\,\text{kg}}\right)^{\frac{1}{2}} = 1.3 \times 10^{-10}\,\text{m},$$

which is the expected order of magnitude for a bond length.

(a)

(b)

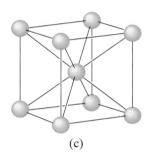

(c)

FIGURE 41–22 Arrangement of atoms in (a) a simple cubic crystal, (b) face-centered cubic crystal (note the atom at the center of each face) and (c) body-centered cubic crystal. Each shows the relationship of the bonds. Each of these "cells" is repeated in three dimensions to the edges of the macroscopic crystal.

FIGURE 41–23 Diagram of an NaCl crystal, showing the "packing" of atoms.

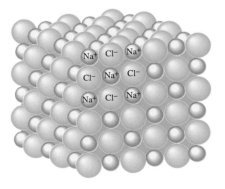

41–5 Bonding in Solids

Quantum mechanics has been a great tool for understanding the structure of solids. This active field of research today is called **solid-state physics**, or **condensed-matter physics** so as to include liquids as well. The rest of this chapter is devoted to this subject, and we begin with a brief look at the structure of solids and the bonds that hold them together.

Although some solid materials are *amorphous* in structure, in that the atoms and molecules show no long-range order, we will be interested here in the large class of *crystalline* substances whose atoms, ions, or molecules are generally believed to form an orderly array known as a *lattice*. Figure 41–22 shows three of the possible arrangements of atoms in a crystal: simple cubic, face-centered cubic, and body-centered cubic. The NaCl crystal is face-centered cubic (see Fig. 41–23), with one Na^+ ion or one Cl^- ion at each lattice point (i.e., considering Na and Cl separately).

The molecules of a solid are held together in a number of ways. The most common are by *covalent* bonding (such as between the carbon atoms of the diamond crystal) or *ionic* bonding (as in a NaCl crystal). Often the bonds are partially covalent and partially ionic. Our discussion of these bonds earlier in this chapter for molecules applies equally well here to solids.

Let us look for a moment at the NaCl crystal of Fig. 41–23. Each Na^+ ion feels an attractive Coulomb potential due to each of the six "nearest neighbor" Cl^- ions surrounding it. Note that one Na^+ does not "belong" exclusively to one Cl^-, so we must not think of ionic solids as consisting of individual molecules. Each Na^+ also feels a repulsive Coulomb potential due to other Na^+ ions, although this is weaker since the other Na^+ ions are farther away. Thus we expect a net attractive potential

$$U = -\alpha \frac{1}{4\pi\epsilon_0} \frac{e^2}{r}.$$

The factor α is called the *Madelung constant*. If each Na^+ were surrounded by only the six Cl^- ions, α would be 6, but the influence of all the other ions reduces it to a value $\alpha = 1.75$ for the NaCl crystal. The potential must also include a term representing the repulsive force when the wave functions of the inner shells and subshells overlap, and this has the form $U = B/r^m$, where m is a small integer. Thus

$$U = -\frac{\alpha}{4\pi\epsilon_0} \frac{e^2}{r} + \frac{B}{r^m}, \tag{41–9}$$

which has the same form as Eq. 41–1 for molecules (Section 41–2). It can be shown (Problem 19) that, at the equilibrium distance r_0,

$$U = U_0 = -\frac{\alpha}{4\pi\epsilon_0} \frac{e^2}{r_0} \left(1 - \frac{1}{m}\right).$$

This U_0 is known as the *ionic cohesive energy*; it is a sort of "binding energy"—the energy (per ion) needed to take the solid apart into separated ions.

A different type of bond occurs in metals. Metal atoms have relatively loosely held outer electrons. Present-day **metallic bond** theories propose that in a metallic solid, these outer electrons roam rather freely among all the metal atoms which, without their outer electrons, act like positive ions. The electrostatic attraction between the metal ions and this negative electron "gas" is what is believed, at least in part, to hold the solid together. The binding energy of metal bonds are typically 1 to 3 eV, somewhat weaker than ionic or covalent bonds (5 to 10 eV in solids). The "free electrons," according to this theory, are responsible for the high electrical and thermal conductivity of metals (see Sections 41–6 and 41–7). This theory also nicely accounts for the shininess of smooth metal surfaces: the electrons are free and can vibrate at any frequency, so when light of a range of frequencies falls on a metal, the electrons can vibrate in response and re-emit light of those same frequencies. Hence the reflected light will consist largely of the same frequencies as the incident light. Compare this to nonmetallic materials that have a distinct color—the atomic electrons exist only in certain energy states, and when white light falls on them, the atoms absorb at certain frequencies, and reflect other frequencies which make up the color we see.

The atoms or molecules of some materials, such as the noble gases, can form only **weak bonds** with each other. As we saw in Section 41–3, weak bonds have very low binding energies and would not be expected to hold atoms together as a liquid or solid at room temperature. The noble gases condense only at very low temperatures, where the atomic kinetic energy is small and the weak attraction can then hold the atoms together.

41–6 Free-Electron Theory of Metals

Let us look more closely at the free-electron theory of metals mentioned in the preceding Section. Let us imagine the electrons trapped within the metal as being in a potential well: inside the metal, the potential energy is zero, but at the edges of the metal there are high potential walls. Since very few electrons leave the metal at room temperature, we can imagine the walls as being infinitely high (as in Section 39–8). At higher temperatures, electrons do leave the metal (we know that thermionic emission occurs, Section 23–9), so we must recognize that the well is of finite depth. In this simple model, the electrons are trapped within the metal, but are free to move about inside the well whose size is macroscopic—the size of the piece of metal. The energy will be quantized, but the spacing between energy levels will be very tiny (see Eq. 39–13) since L is very large. Indeed, for a cube 1 cm on a side, the number of states with energy between, say, 5.0 and 5.5 eV, is on the order of 10^{22} (see Example 41–8).

To deal with such vast numbers of states, which are so closely spaced as to seem continuous, we need to use statistical methods. We define a quantity known as the **density of states**, $g(E)$, whose meaning is similar to the Maxwell distribution, Eq. 18–6 (see Section 18–2). That is, the quantity $g(E)\,dE$ represents the number of states per unit volume that have energy between E and $E + dE$. A careful calculation (see Problem 35), which must treat the potential well as three dimensional, shows that

$$g(E) = \frac{8\sqrt{2}\,\pi m^{\frac{3}{2}}}{h^3}\,E^{\frac{1}{2}}. \qquad \textbf{(41–10)}$$

This is plotted in Fig. 41–24.

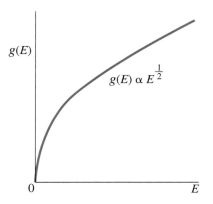

FIGURE 41–24 Density of states $g(E)$ as a function of energy E (Eq. 41–10).

Density of states

EXAMPLE 41–8 **Electron states in copper.** Estimate the number of states in the range 5.0 to 5.5 eV available to electrons in a 1.0-cm cube of copper metal.

SOLUTION Since $g(E)$ is the number of states per unit volume per unit energy interval, the number N of states is approximately (it is approximate because ΔE is not small)

$$N \approx g(E)V\,\Delta E,$$

where the volume $V = 1.0\,\text{cm}^3 = 1.0 \times 10^{-6}\,\text{m}^3$ and $\Delta E = 0.50\,\text{eV}$. We evaluate $g(E)$ at 5.25 eV, and find (Eq. 41–10):

$$N \approx g(E)V\,\Delta E = \frac{8\sqrt{2}\pi\left(9.1 \times 10^{-31}\,\text{kg}\right)^{\frac{3}{2}}}{\left(6.63 \times 10^{-34}\,\text{J}\cdot\text{s}\right)^3}\sqrt{(5.25\,\text{eV})(1.6 \times 10^{-19}\,\text{J/eV})}$$

$$\times\;(1.0 \times 10^{-6}\,\text{m}^3)(0.50\,\text{eV})(1.6 \times 10^{-19}\,\text{J/eV})$$

$$\approx 8 \times 10^{21}$$

in 1.0 cm³. Note that the type of metal did not enter the calculation.

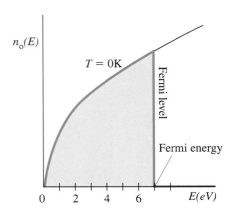

$n_0(E)$

$T = 0\,\text{K}$

Fermi level

Fermi energy

$E(eV)$

0   2   4   6

FIGURE 41–25 At $T = 0\,\text{K}$, all states up to energy E_F, called the Fermi energy, are filled.

Fermi level

Fermi energy

FIGURE 41–26 The Fermi–Dirac probability function for two temperatures, $T = 0\,\text{K}$ (black line) and $T = 1200\,\text{K}$ (blue curve). For $f(E) = 1$, a state with energy E is certainly occupied. For $f(E) = 0.5$, which occurs at $E = E_F$, the state with E_F has a 50% chance of being occupied.

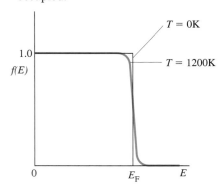

$T = 0\,\text{K}$

$T = 1200\,\text{K}$

1.0

$f(E)$

0   E_F E

Fermi factor

Equation 41–10 gives us the density of states. Now we must ask: How are the states available to an electron gas actually populated? Let us first consider the situation at absolute zero, $T = 0\,\text{K}$. For a classical ideal gas, all the particles would be in the lowest state, with zero kinetic energy $(= \frac{3}{2}kT = 0)$. But the situation is vastly different for an electron gas because electrons obey the exclusion principle. Electrons do not obey classical statistics but rather a quantum statistics called **Fermi–Dirac statistics**[†] that takes into account the exclusion principle. All particles that have spin $\frac{1}{2}$ (or other half-integral spin: $\frac{3}{2}, \frac{5}{2}$, etc.), such as electrons, protons, and neutrons, obey Fermi–Dirac statistics and are referred to as **fermions**.[‡] The electron gas in a metal is often called a **Fermi gas**. According to the exclusion principle, no two electrons in the metal can have the same set of quantum numbers. Therefore, in each of the states of our potential well, there can be at most two electrons: one with spin up $(m_s = +\frac{1}{2})$ and one with spin down $(m_s = -\frac{1}{2})$. (This factor of 2 has already been included in Eq. 41–10.) Thus, at $T = 0\,\text{K}$, the possible energy levels will be filled, two electrons each, up to a maximum level called the **Fermi level**. This is shown in Fig. 41–25, where the vertical axis is labeled $n_0(E)$ for "density of occupied states." The energy of the state at the Fermi level is called the **Fermi energy**, E_F. To determine E_F, we integrate Eq. 41–10 from $E = 0$ to $E = E_F$ (all states up to E_F are filled at $T = 0\,\text{K}$):

$$\frac{N}{V} = \int_0^{E_F} g(E)\,dE, \tag{41–11}$$

where N/V is the number of conduction electrons per unit volume in the metal. Then, solving for E_F, the result (see Example 41–9) is

$$E_F = \frac{h^2}{8m}\left(\frac{3}{\pi}\frac{N}{V}\right)^{\frac{2}{3}}. \tag{41–12}$$

The average energy in this distribution (see Problem 31) is

$$\overline{E} = \tfrac{3}{5}E_F. \tag{41–13}$$

For copper, $E_F = 7.0\,\text{eV}$ (see Example 41–9) and $\overline{E} = 4.2\,\text{eV}$. This is very much greater than the energy of thermal motion at room temperature $(\frac{3}{2}kT \approx 0.04\,\text{eV})$. Clearly, all motion does not stop at absolute zero.

Thus, at $T = 0$, all states with energy below E_F are occupied, and all states above E_F are empty. What happens for $T > 0$? We expect that some (at least) of the electrons will increase in energy due to thermal motion. Classically, the distribution of occupied states would be given by the Boltzmann factor, $e^{-E/kT}$ (see Eqs. 40–16). But for our electron gas, a quantum-mechanical system obeying the exclusion principle, the probability of a given state of energy E being occupied is given by the **Fermi–Dirac probability function** (or **Fermi factor**):

$$f(E) = \frac{1}{e^{(E-E_F)/kT} + 1}, \tag{41–14}$$

where E_F is the Fermi energy. This function is plotted in Fig. 41–26 for two temperatures, $T = 0\,\text{K}$ and $T = 1200\,\text{K}$ (just below the melting point of copper). At $T = 0$ (or as T approaches zero) the factor $e^{(E-E_F)/kT}$ in Eq. 41–14 is zero if $E < E_F$ and is ∞ if $E > E_F$. Thus

$$f(E) = \begin{cases} 1 & E < E_F \\ 0 & E > E_F \end{cases} \quad \text{at} \quad T = 0.$$

This is what is plotted in black in Fig. 41–26 for $T = 0$ and is consistent with Fig. 41–25: all states up to the Fermi level are occupied [probability $f(E) = 1$]

[†] Developed independently by Enrico Fermi (Figs. 39–2, 39–10, and 42–7) in early 1926 and by P.A.M. Dirac a few months later.

[‡] Particles with integer spin (0, 1, 2, etc.), such as the photon, obey *Bose–Einstein* statistics and are called *bosons*.

and all states above are unoccupied. For $T = 1200\,\text{K}$, the Fermi factor changes only a little, as shown in Fig. 41–26 as the blue curve. Note that at any temperature T, when $E = E_F$, then Eq. 41–14 gives $f(E) = 0.50$, meaning the state at $E = E_F$ has a 50% chance of being occupied. To see how $f(E)$ affects the actual distribution of electrons in energy states, we must weight the density of possible states, $g(E)$, by the probability that those states will be occupied, $f(E)$. The product of these two functions then gives the **density of occupied states**,

$$n_o(E) = g(E)f(E) = \frac{8\sqrt{2}\,\pi m^{\frac{3}{2}}}{h^3}\frac{E^{\frac{1}{2}}}{e^{(E-E_F)/kT} + 1}. \qquad \textbf{(41–15)}$$

Density of occupied states

Then $n_o(E)\,dE$ represents the number of electrons per unit volume with energy between E and $E + dE$ in thermal equilibrium at temperature T. This is plotted in Fig. 41–27 for $T = 1200\,\text{K}$, a temperature at which a metal is so hot it would glow. We see immediately that the distribution differs very little from that at $T = 0$. We see also that the changes that do occur are concentrated about the Fermi level. A few electrons from slightly below the Fermi level move to energy states slightly above it. The average energy of the electrons increases only very slightly when the temperature is increased from $T = 0\,\text{K}$ to $T = 1200\,\text{K}$. This is very different from the behavior of an ideal gas, for which kinetic energy increases directly with T. Nonetheless, this behavior is readily understood as follows. Energy of thermal motion at $T = 1200\,\text{K}$ is about $\frac{3}{2}kT \approx 0.1\,\text{eV}$. The Fermi level, on the other hand, is on the order of several eV: for copper it is $E_F \approx 7.0\,\text{eV}$. An electron at $T = 1200\,\text{K}$ may have $7\,\text{eV}$ of energy, but it can acquire at most only a few times $0.1\,\text{eV}$ of energy by a (thermal) collision with the lattice. Only electrons very near the Fermi level would find vacant states close enough to make such a transition. Essentially none of the electrons could increase in energy by, say, $3\,\text{eV}$, so electrons farther down in the electron gas are unaffected. Only electrons near the top of the energy distribution can be thermally excited to higher states. And their new energy is on the average only slightly higher than their old energy.

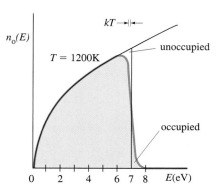

FIGURE 41–27 The density of occupied states for the electron gas in copper. The width kT represents thermal energy at $T = 1200\,\text{K}$.

EXAMPLE 41–9 **The Fermi level.** For the metal copper, determine (a) the Fermi energy, (b) the average energy of electrons, and (c) the speed of electrons at the Fermi level (this is called the *Fermi speed*).

SOLUTION (a) We combine Eqs. 41–10 and 41–11:

$$\frac{N}{V} = \frac{8\sqrt{2}\,\pi m^{\frac{3}{2}}}{h^3}\int_0^{E_F} E^{\frac{1}{2}}\,dE = \frac{8\sqrt{2}\,\pi m^{\frac{3}{2}}}{h^3}\frac{2}{3}E_F^{\frac{3}{2}}.$$

Solving for E_F, we obtain

$$E_F = \frac{h^2}{8m}\left(\frac{3}{\pi}\frac{N}{V}\right)^{\frac{2}{3}}$$

[note: $(\sqrt{2})^{\frac{2}{3}} = 2(\sqrt{2}/2^{\frac{3}{2}})^{\frac{2}{3}} = 2(\frac{1}{2})^{\frac{2}{3}}$], and this is Eq. 41–12. We calculated N/V, the number of conduction electrons per unit volume in copper, in Example 25–12 to be $N/V = 8.4 \times 10^{28}\,\text{m}^{-3}$. Hence for copper

$$E_F = \frac{(6.63 \times 10^{-34}\,\text{J}\cdot\text{s})^2}{8(9.1 \times 10^{-31}\,\text{kg})}\left[\frac{3(8.4 \times 10^{28}\,\text{m}^{-3})}{\pi}\right]^{\frac{2}{3}}\frac{1}{1.6 \times 10^{-19}\,\text{J/eV}} = 7.0\,\text{eV}.$$

(b) From Eq. 41–13, $\overline{E} = \frac{3}{5}E_F = 4.2\,\text{eV}$.
(c) In our model, we took $U = 0$ inside the metal, so $E = $ kinetic energy $= \frac{1}{2}mv^2$. Therefore, at the Fermi level, the Fermi speed is

$$v_F = \sqrt{\frac{2E_F}{m}} = \sqrt{\frac{2(7.0\,\text{eV})(1.6 \times 10^{-19}\,\text{J/eV})}{9.1 \times 10^{-31}\,\text{kg}}} = 1.6 \times 10^6\,\text{m/s},$$

a very high speed. The temperature of a classical gas would have to be extremely high to produce an average particle speed this large.

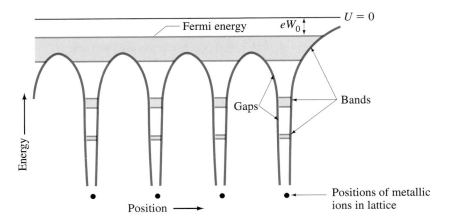

FIGURE 41–28 Potential energy for an electron in a metal crystal, with deep potential wells in the vicinity of each ion in the crystal lattice.

The simple model of the electron gas presented here provides good explanations for the electrical and thermal properties of conductors. But it does not explain why some materials are good conductors and others are good insulators. To provide an explanation, our model of a metal gas as a uniform potential well needs to be refined to include the effect of the lattice. Figure 41–28 shows a "periodic" potential that takes into account the attraction of electrons for each atomic ion in the lattice. Here we have taken $U = 0$ for an electron free of the metal; so within the metal, electron energies are less than zero (just as for molecules, or for the H atom, for example, in which the ground state has $E = -13.6 \, \text{eV}$). The quantity eW_0 represents the minimum energy to remove an electron from the metal, where W_0 is the *work function* (see Section 38–2). The crucial outcome of putting a periodic potential (more easily approximated with narrow square wells) into the Schrödinger equation is that the allowed energy states are divided into *bands*, with energy gaps in between. Only electrons in the highest band, close to the Fermi level, are able to move about freely within the metal crystal. In the next Section we will see physically why there are bands and how they explain the properties of conductors, insulators, and semiconductors.

41–7 Band Theory of Solids

We saw in Section 41–1 that when two hydrogen atoms approach each other, the wave functions overlap, and the two 1s states (one for each atom) divide into two states of different energy. (As we saw, only one of these states, $S = 0$, has low enough energy to give a bound H_2 molecule.) Figure 41–29a shows this situation for 1s and 2s states for two atoms: as the atoms get closer, the 1s and 2s states split

FIGURE 41–29 The splitting of 1s and 2s atomic energy levels as (a) two atoms approach each other (the atomic separation decreases, moving toward the left); (b) the same for six atoms, and (c) for many atoms when they come together to form a solid.

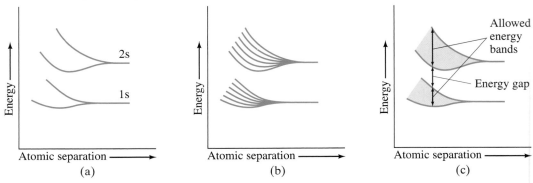

into two levels. If six atoms come together, as in Fig. 41–29b, each of the states splits into six levels. If a large number of atoms come together to form a solid, then each of the original atomic levels becomes a **band** as shown in Fig. 41–29c. The energy levels are so close together in each band that they seem essentially continuous. This is why the spectrum of heated solids (Section 38–1) appears continuous.

The crucial aspect of a good **conductor** is that the highest energy band containing electrons is only partially filled. Consider sodium, for example, whose energy bands are shown in Fig. 41–30. The 1s, 2s, and 2p bands are full (just as in a Na atom) and don't concern us. The 3s band, however, is only half full. To see why, recall that the exclusion principle stipulates that in an atom, only two electrons can be in the 3s state, one with spin up and one with spin down. These two states have slightly different energy. For a solid consisting of N atoms, the 3s band will contain $2N$ possible energy states. Now a sodium atom has a single 3s electron, so in a sample of sodium metal containing N atoms, there are N electrons in the 3s band, and N unoccupied states. When a potential difference is applied across the metal, electrons can respond by accelerating and increasing their energy, since there are plenty of unoccupied states of slightly higher energy available. Hence, a current flows readily and sodium is a good conductor. The characteristic of all good conductors is that the highest energy band is only partially filled, or two bands overlap so that unoccupied states are available. An example of the latter is magnesium, which has two 3s electrons, so its 3s band is filled. But the unfilled 3p band overlaps the 3s band in energy, so there are lots of available states for the electrons to move into. Thus magnesium, too, is a good conductor.

In a material that is a good **insulator**, on the other hand, the highest band containing electrons, called the **valence band**, is completely filled. The next higher energy band, called the **conduction band**, is separated from the valence band by a "forbidden" **energy gap**, (or **band gap**), E_g, of typically 5 to 10 eV. So at room temperature (300 K), where thermal energies (that is, average kinetic energy—see Chapter 18) are on the order of $\frac{3}{2}kT \approx 0.04$ eV, almost no electrons can acquire the 5 eV needed to reach the conduction band. When a potential difference is applied across the material, no available states are accessible to the electrons, and no current flows. Hence, the material is a good insulator.

Figure 41–31 compares the relevant energy bands (a) for conductors, (b) for insulators, and also (c) for the important class of materials known as **semiconductors**. The bands for a pure (or **intrinsic**) semiconductor, such as silicon or germanium, are like those for an insulator, except that the unfilled conduction band is separated from the filled valence band by a much smaller energy gap, E_g, typically on the order of 1 eV. At room temperature, a few electrons can acquire enough thermal energy to reach the conduction band, and so a very small current may flow when a voltage is applied. At higher temperatures, more electrons have enough energy to jump the gap. This effect can often more than offset the effects of more frequent collisions due to increased disorder at higher temperature, so that the resistivity of semiconductors can *decrease* with increasing temperature (see Table 25–1). But this is not the whole story of semiconductor conduction. When a potential difference is applied to a semiconductor, the few electrons in the conduction band move toward the positive electrode. Electrons in the valence band try to do the same thing, and a few can because there are a small number of unoccupied

3s

2p

2s

1s

FIGURE 41–30 Energy bands for sodium.

Insulators

Valence and conduction bands

Energy gap

Semiconductors (pure)

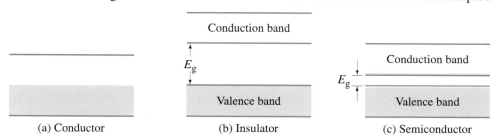

(a) Conductor

Conduction band

E_g

Valence band

(b) Insulator

Conduction band

E_g

Valence band

(c) Semiconductor

FIGURE 41–31 Energy bands for (a) a conductor, (b) an insulator, and (c) a semiconductor. Shading represents occupied states. Pale shading in part (c) represents electrons that can pass from the valence band to the conduction band due to thermal agitation at room temperature (exaggerated).

states which were left empty by the electrons reaching the conduction band. Such unfilled electron states are called **holes**. Each electron in the valence band that fills a hole in this way as it moves toward the positive electrode leaves behind a hole, so that the holes migrate toward the negative electrode. As the electrons tend to accumulate at one side of the material, the holes tend to accumulate on the opposite side. We will look at this phenomenon in more detail in the next Section.

Holes (in semiconductor)

EXAMPLE 41–10 **Calculating the energy gap.** It is found that the conductivity of a certain semiconductor increases when light of wavelength 345 nm or shorter strikes it, suggesting that electrons are being promoted from the valence band to the conduction band. What is the energy gap, E_g, for this semiconductor?

SOLUTION The longest wavelength, or lowest energy, photon to cause an increase in conductivity has $\lambda = 345$ nm, and it can transfer to an electron an energy

$$E_g = hf = \frac{hc}{\lambda} = \frac{(6.63 \times 10^{-34}\,\text{J}\cdot\text{s})(3.00 \times 10^8\,\text{m/s})}{(345 \times 10^{-9}\,\text{m})(1.60 \times 10^{-19}\,\text{J/eV})} = 3.6\,\text{eV}.$$

EXAMPLE 41–11 **ESTIMATE** **Free electrons in semiconductors and insulators.** Use the Fermi–Dirac probability function, Eq. 41–14, to estimate the order of magnitude of the numbers of free electrons in the conduction band of a solid containing 10^{21} atoms, assuming the solid is at room temperature ($T = 300$ K) and is (a) a semiconductor with $E_g \approx 1.1$ eV, (b) an insulator with $E_g \approx 5$ eV. Compare to a conductor.

SOLUTION At $T = 0$, all states above the Fermi energy E_F are empty, and all those below are full. So for semiconductors and insulators we can take E_F to be about midway between the valence and conduction bands, Fig. 41–32, and it does not change significantly as we go to room temperature. (a) For the semiconductor, the gap $E_g \approx 1.1$ eV, so $E - E_F \approx 0.55$ eV for the lowest states in the conduction band. Since at room temperature we have $kT \approx 0.026$ eV, then $(E - E_F)/kT \approx 0.55\,\text{eV}/0.026\,\text{eV} \approx 21$ and

$$f(E) = \frac{1}{e^{(E-E_F)/kT} + 1} \approx \frac{1}{e^{21}} \approx 10^{-9}.$$

Thus about 1 atom in 10^9 can contribute an electron to the conductivity. (b) For the insulator with $E - E_F \approx 5.0\,\text{eV} - \frac{1}{2}(5.0\,\text{eV}) = 2.5$ eV, we get

$$f(E) \approx \frac{1}{e^{2.5/0.026} + 1} \approx \frac{1}{e^{96}} \approx 10^{-42}.$$

Thus in an ordinary sample containing 10^{21} atoms, there would be no free electrons in an insulator $(10^{21} \times 10^{-42} = 10^{-21})$, about 10^{12} $(10^{21} \times 10^{-9})$ free electrons in a semiconductor, and about 10^{21} free electrons in a good conductor.

CONCEPTUAL EXAMPLE 41–12 **Which is transparent?** The energy gap for silicon is 1.14 eV at room temperature while that of zinc sulfide (ZnS) is 3.6 eV. Which one of these is opaque and which is transparent to visible light?

RESPONSE Visible light photons span energies from roughly 1.8 eV to 3.2 eV ($E = hf = hc/\lambda$ where $\lambda = 400$ nm to 700 nm and 1 eV $= 1.6 \times 10^{-19}$ J). Light is absorbed by the electrons in a material. Silicon's gap is small enough to absorb these photons, thus bumping electrons well up into the conduction band, and so silicon is opaque. On the other hand, zinc sulfide's energy gap is too wide to absorb visible photons, and so the light can pass through the material; it can be transparent.

Conduction band

E_g - - - - - - - - - - - - - - E_F

Valence band

FIGURE 41–32 The Fermi energy is midway between the valence band and the conduction band.

41–8 | Semiconductors and Doping

The most commonly used semiconductors in modern electronics are silicon (Si) and germanium (Ge). An atom of silicon or germanium has four outer electrons that act to hold the atoms in the regular lattice structure of the crystal, shown schematically in Fig. 41–33a. Germanium and silicon acquire useful properties for use in electronics only when a tiny amount of impurity is introduced into the crystal structure (perhaps 1 part in 10^6 or 10^7). This is called **doping** the semiconductor. Two kinds of doped semiconductor can be made, depending on the type of impurity used. If the impurity is an element whose atoms have five outer electrons, such as arsenic, we have the situation shown in Fig. 41–33b, with the arsenic atoms holding positions in the crystal lattice where normally silicon atoms would be. Only four of arsenic's electrons fit into the bonding structure. The fifth does not fit in and can move relatively freely, somewhat like the electrons in a conductor. Because of this small number of extra electrons, a doped semiconductor becomes slightly conducting. The density of conduction electrons in an intrinsic (pure) semiconductor is about 1 per 10^9 atoms, as we saw in Example 41–11. With an impurity concentration of 1 in 10^6 or 10^7 when doped, the conductivity will be much higher and it can be controlled with great precision. An arsenic-doped silicon crystal is called an **n-type semiconductor** because *n*egative charges (electrons) carry the electric current.

Doped semiconductors

n-type

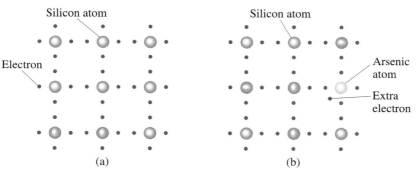

(a)   (b)

p-type

In a **p-type semiconductor**, a small percentage of semiconductor atoms are replaced by atoms with three outer electrons—such as gallium. As shown in Fig. 41–34a, there is a "hole" in the lattice structure near a gallium atom since it has only three outer electrons. Electrons from nearby silicon atoms can jump into this hole and fill it. But this leaves a hole where that electron had previously been, Fig. 41–34b. The vast majority of atoms are silicon, so holes are almost always next to a silicon atom. Since silicon atoms require four outer electrons to be neutral, this means that there is a net positive charge at the hole. Whenever an electron moves to fill a hole, the positive hole is then at the previous position of that electron. Another electron can then fill this hole, and the hole thus moves to a new location; and so on. This type of semiconductor is called *p-type* because it is the *p*ositive holes that seem to carry the electric current. Note, however, that both *p*-type and *n*-type semiconductors have *no net charge* on them.

Holes are positive

FIGURE 41–33 Two-dimensional representation of a silicon crystal. (a) Four (outer) electrons surround each silicon atom. (b) Silicon crystal doped with a few arsenic atoms: the extra electron doesn't fit into the crystal lattice and so is free to move about. This is an *n*-type semiconductor.

FIGURE 41–34 A *p*-type semiconductor, gallium-doped silicon. (a) Gallium has only three outer electrons, so there is an empty spot, or *hole*, in the structure. (b) Electrons from silicon atoms can jump into the hole and fill it. As a result, the hole moves to a new location (to the right in this figure), to where the electron used to be.

(a)   (b)

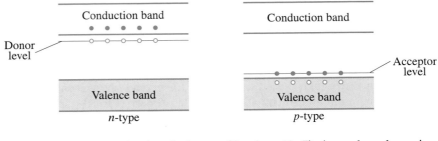

FIGURE 41–35 Impurity energy levels in doped semiconductors.

According to the band theory (Section 41–7), in a doped semiconductor, the impurity provides additional energy states between the bands, as shown in Fig. 41–35. In an *n*-type semiconductor, the impurity energy level lies just below the conduction band. Electrons in this energy level need only about 0.05 eV (in Si; even less in Ge) of energy to reach the conduction band; this is on the order of the thermal energy, $\frac{3}{2}kT$ (= 0.04 eV at 300 K), so transitions occur readily at room temperature. This energy level can thus supply electrons to the conduction band, so it is called a **donor** level. In *p*-type semiconductors, the impurity energy level is just above the valence band (Fig. 41–35). It is called an **acceptor** level because electrons from the valence band can easily jump into it. Positive holes are left behind in the valence band, and as other electrons move into these holes, the holes move about as discussed earlier.

(a)

(b)

FIGURE 41–36 Schematic diagram showing how a semiconductor diode operates. Current flows when the voltage is connected in forward bias, as in (a), but not when connected in reverse bias, as in (b).

FIGURE 41–37 Current through a diode as a function of applied voltage.

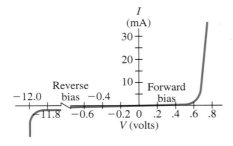

* 41–9 Semiconductor Diodes

Semiconductor diodes and transistors are essential components of modern electronic devices. The miniaturization achieved today allows many thousands of diodes, transistors, resistors, and so on, to be placed on a single *chip* only a millimeter on a side. We now discuss, briefly and qualitatively, the operation of diodes and transistors.

When an *n*-type semiconductor is joined to a *p*-type, a ***p-n* junction diode** is formed. Separately, the two semiconductors are electrically neutral. When joined, a few electrons near the junction diffuse from the *n*-type into the *p*-type semiconductor, where they fill a few of the holes. The *n*-type is left with a positive charge, and the *p*-type acquires a net negative charge. Thus a potential difference is established, with the *n* side positive relative to the *p* side, and this prevents further diffusion of electrons.

If a battery is connected to a diode with the positive terminal to the *p* side and the negative terminal to the *n* side as in Fig. 41–36a, the externally applied voltage opposes the internal potential difference and the diode is said to be **forward biased**. If the voltage is great enough (about 0.3 V for Ge, 0.6 V for Si at room temperature), a current will flow. The positive holes in the *p*-type semiconductor are repelled by the positive terminal of the battery and the electrons in the *n*-type are repelled by the negative terminal of the battery. The holes and electrons meet at the junction, and the electrons cross over and fill the holes. A current is flowing. Meanwhile, the positive terminal of the battery is continually pulling electrons off the *p* end, forming new holes, and electrons are being supplied by the negative terminal at the *n* end. Consequently, a large current flows through the diode.

When the diode is **reverse biased**, as in Fig. 41–36b, the holes in the *p* end are attracted to the battery's negative terminal and the electrons in the *n* end are attracted to the positive terminal. The current carriers do not meet near the junction and, ideally, no current flows.

A graph of current versus voltage for a typical diode is shown in Fig. 41–37. As can be seen, a real diode does allow a small amount of reverse current to flow due to thermal motion $\left(\frac{3}{2}kT\right)$ of electron pairs. For most practical purposes, this is negligible (at room temperature, a few μA in Ge, a few pA in Si; the reverse current increases rapidly with temperature, however, and may render a diode ineffective above 200°C).

| **EXAMPLE 41–13** | **A diode.** The diode whose current–voltage characteristics are shown in Fig. 41–37 is connected in series with a 4.0-V battery and a resistor. If a current of 10 mA is to pass through the diode, what resistance must the resistor have?

SOLUTION In Fig. 41–37, we see that the voltage drop across the diode is about 0.7 V when the current is 10 mA. Therefore, the voltage drop across the resistor is 4.0 V − 0.7 V = 3.3 V, so $R = V/I = (3.3\,\text{V})/(1.0 \times 10^{-2}\,\text{A}) = 330\,\Omega$.

If the voltage across a diode connected in reverse bias is increased greatly, a point is reached where breakdown occurs. The electric field across the junction becomes so large that ionization of atoms results. The electrons thus pulled off their atoms contribute to a larger and larger current as breakdown continues. The voltage remains constant over a wide range of currents. This is shown on the far left in Fig. 41–37. This property of diodes can be used to accurately regulate a voltage supply. A diode designed for this purpose is called a **zener diode**. When placed across the output of an unregulated power supply, a zener diode can maintain the voltage at its own breakdown voltage as long as the supply voltage is always above this point. Zener diodes can be obtained corresponding to voltages of a few volts to hundreds of volts.

Since a *p-n* junction diode allows current to flow only in one direction (as long as the voltage is not too high), it can serve as a **rectifier**—to change ac into dc. A simple rectifier circuit is shown in Fig. 41–38a where the arrow inside the symbol for a diode indicates the direction in which a diode conducts conventional (+) current. The ac source applies a voltage across the diode alternately positive and negative. Only during half of each cycle will a current pass through the diode; so only then is there a current through the resistor R. Hence, a graph of the voltage V_{ab} across R as a function of time looks like Fig. 41–38b. This **half-wave rectification** is not exactly dc, but it is unidirectional. More useful is a **full-wave rectifier** circuit, which uses two diodes (or sometimes four) as shown in Fig. 41–39a. At any given instant, either one diode or the other will conduct current to the right. Therefore, the output across the load resistor R will be as shown in Fig. 41–39b. Actually this is the voltage if the capacitor C were not in the circuit. The capacitor tends to store charge, and thus helps to smooth out the current as shown in Fig. 41–39c.

Rectifier circuits are important because most line voltage is ac, and most electronic devices require a dc voltage for their operation. Hence, diodes are found in nearly all electronic devices including radio and TV sets, calculators, and computers.

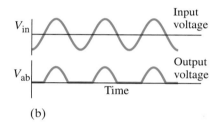

FIGURE 41–38 (a) A simple (half-wave) rectifier circuit using a semiconductor diode. (b) AC source input voltage, and output voltage across R, as functions of time.

FIGURE 41–39 (a) Full-wave-rectifier circuit (including a transformer so the magnitude of the voltage can be changed). (b) Output voltage in the absence of capacitor C. (c) Output voltage with the capacitor in the circuit.

(a)

(b) Without capacitor

(c) With capacitor

Another useful device is a **light-emitting diode** (LED). When a *p-n* junction is forward biased, a current begins to flow. Electrons cross from the *n* region into the *p* region and combine with holes, and a photon can be emitted with an energy approximately equal to the band gap, E_g (see Figs. 41–31c and 41–35). Often the energy, and hence the wavelength, is in the red region of the visible spectrum, producing the familiar LED displays on VCRs, CD players, car instrument panels, clocks, and so on. Infrared (i.e., nonvisible) LEDs are used in remote controls for TV, VCRs, and stereos.

Photodiodes and **solar cells** are *p-n* junctions used in the reverse way. Photons are absorbed, creating electron–hole pairs if the photon energy is greater than the band gap energy, E_g. The created electrons and holes produce a current that, when connected to an external circuit, becomes a source of emf and power.

A diode is called a **nonlinear device** because the current is not proportional to the voltage; that is, a graph of current versus voltage (Fig. 41–37) is not a straight line as it is for a resistor (which ideally *is* linear). Transistors are also *nonlinear* devices.

* | 41–10 | Transistors and Integrated Circuits

A simple **junction transistor** consists of a crystal of one type of doped semiconductor sandwiched between two crystals of the opposite type. Both *pnp* and *npn* transistors are made, and they are shown schematically in Fig. 41–40a. The three semiconductors are given the names *collector*, *base*, and *emitter*. The symbols for *npn* and *pnp* transistors are shown in Fig. 41–40b. The arrow is always placed on the emitter and indicates the direction of (conventional) current flow in normal operation.

The operation of a transistor can be analyzed qualitatively—very briefly—as follows. Consider an *npn* transistor connected as shown in Fig. 41–41. A voltage V_{CE} is maintained between the collector and emitter by the battery $\mathscr{E}_C$. The voltage applied to the base is called the *base bias voltage*, V_{BE}. If V_{BE} is positive, conduction electrons in the emitter are attracted into the base. Since the base region is very thin (perhaps 1 μm), most of these electrons flow right across into the collector, which is maintained at a positive voltage. A large current, I_C, flows between collector and emitter and a much smaller current, I_B, through the base. A small variation in the base voltage due to an input signal causes a large change in the collector current and therefore a large change in the voltage drop across the output resistor R_C. Hence a transistor can *amplify* a small signal into a larger one.

FIGURE 41–40 (a) Schematic diagram of *npn* and *pnp* transistors. (b) Symbols for *npn* and *pnp* transistors.

FIGURE 41–41 An *npn* transistor used as an amplifier.

Normally a small ac signal is to be amplified, and when added to the base bias voltage causes the voltage and current at the collector to vary at the same rate but magnified. Thus, what is important for amplification is the *change* in collector current for a given input *change* in base current. The **current gain** is defined as the ratio

$$\beta_I = \frac{\text{output (collector) ac current}}{\text{input (base) ac current}} = \frac{i_C}{i_B} \qquad \textit{Current gain}$$

where i_B and i_C are the ac base current (input signal) and ac collector current (output). β_I is typically on the order of 10 to 100. Similarly, the **voltage gain** is

$$\beta_V = \frac{\text{output (collector) ac voltage}}{\text{input (base) ac voltage}}. \qquad \textit{Voltage gain}$$

Transistors are the basic elements in modern electronic **amplifiers** of all sorts.

A *pnp* transistor operates like an *npn*, except that holes move instead of electrons. The collector voltage is negative, and so is the base voltage in normal operation.

In Fig. 41–41 two batteries were shown: $\mathcal{E}_C$ supplied the collector voltage and $\mathcal{E}_B$ the base bias voltage. In practice, only one source is often used, and the base bias voltage can be obtained using a resistance voltage divider as in Fig. 41–42. Transistors can be connected in many other ways as well, and many new and innovative uses have been found for them.

Transistors were a great advance in miniaturization of electronic circuits. Although individual transistors are very small compared to the once used vacuum tubes, they are huge compared to **integrated circuits** or **chips** (see photo at start of this chapter). Tiny amounts of impurities can be placed at particular locations within a single silicon crystal. These can be arranged to form diodes, transistors, and resistors (undoped semiconductors). Capacitors and inductors can also be formed, although they are often connected separately. A tiny chip, only 1 mm on a side, may contain thousands of transistors and other circuit elements. Integrated circuits are now the heart of computers, television, calculators, cameras, and the electronic instruments that control aircraft, space vehicles, and automobiles. The "miniaturization" produced by integrated circuits not only allows extremely complicated circuits to be placed in a small space, but also has allowed a great increase in the speed of operation of, say, computers, because the distances the electronic signals travel are so tiny.

FIGURE 41–42 Typical transistor circuit involving an *npn* transistor. $\mathcal{E}_C = +12\,\text{V}$ and $\mathcal{E}_B$ is determined by the resistors.

Summary

Quantum mechanics explains the bonding together of atoms to form **molecules**. In a **covalent bond**, the electron clouds of two or more atoms overlap because of constructive interference between the electron waves. The positive nuclei are attracted to this concentration of negative charge between them, forming the bond. An **ionic bond** is an extreme case of a covalent bond in which one or more electrons from one atom spend much more time around the other atom than around their own. The atoms then act as oppositely charged ions that attract each other, forming the bond.

These **strong bonds** hold molecules together, and also hold atoms and molecules together in solids. Also important are **weak bonds** (or **van der Waals bonds**), which are generally dipole attractions between molecules.

When atoms combine to form molecules, the energy levels of the outer electrons are altered because they now interact with each other. Additional energy levels also become possible because the atoms can vibrate with respect to each other, and the molecule as a whole can rotate. The energy levels for both vibrational and rotational motion are quantized, and are very close together (typically, $10^{-1}\,\text{eV}$ to $10^{-3}\,\text{eV}$ apart). Each atomic energy level thus becomes a set of closely spaced levels corresponding to the vibrational and rotational motions. Transitions from one level to another appear as many very closely spaced lines. The resulting spectra are called **band spectra**.

The quantized rotational energy levels are given by

$$E_{\text{rot}} = L(L + 1)\frac{\hbar^2}{2I}, \qquad L = 0, 1, 2, \cdots,$$

where I is the moment of inertia of the molecule.

The energy levels for vibrational motion are given by

$$E_{\text{vib}} = \left(\nu + \tfrac{1}{2}\right)hf, \qquad \nu = 0, 1, 2, \cdots,$$

where f is the classical natural frequency of vibration for the molecule. Transitions between energy levels are subject to the selection rules $\Delta L = \pm 1$ and $\Delta \nu = \pm 1$.

Some **solids** are bound together by covalent and ionic bonds, just as molecules are. In metals, the electrostatic force between free electrons and positive ions helps form the **metallic bond**.

In the free-electron theory of metals, electrons occupy the possible energy states according to the exclusion principle. At $T = 0\,\text{K}$, all possible states are filled up to a maximum energy level called the *Fermi energy*, E_F, the magnitude of which is typically a few eV. All states above E_F are vacant at $T = 0\,\text{K}$. At normal temperatures $(300\,\text{K})$ the distribution of occupied states is only slightly altered and is given by the **Fermi–Dirac probability function**

$$f(E) = \frac{1}{e^{(E - E_F)/kT} + 1}.$$

In a crystalline solid, the possible energy states for electrons are arranged in **bands**. Within each band the levels are very close together, but between the bands there may be forbidden **energy gaps**. Good conductors are characterized by the highest occupied band (the **conduction band**) being only partially full, so there are many accessible states available to electrons to move about and accelerate when a voltage is applied. In a good insulator, the highest occupied energy band (the **valence band**) is completely full, and there is a large energy gap (5 to 10 eV) to the next highest band, the *conduction band*. At room temperature, molecular kinetic energy (thermal energy) available due to collisions is only about 0.04 eV, so almost no electrons can jump from the valence to the conduction band. In a **semiconductor**, the gap between valence and conduction bands is much smaller, on the order of 1 eV, so a few electrons can make the transition from the essentially full valence band to the nearly empty conduction band.

In a **doped** semiconductor, a small percentage of impurity atoms with five or three valence electrons replace a few of the normal silicon atoms with their four valence electrons. A five-electron impurity produces an **n-type** semiconductor with negative electrons as carriers of current. A three-electron impurity produces a **p-type** semiconductor in which positive **holes** carry the current. The energy level of impurity atoms lies slightly below the conduction band in an *n*-type semiconductor, and acts as a **donor** from which electrons readily pass into the conduction band. The energy level of impurity atoms in a *p*-type semiconductor lies slightly above the valence band and acts as an **acceptor** level, since electrons from the valence band easily reach it, leaving holes behind to act as charge carriers.

A semiconductor **diode** consists of a **p-n junction** and allows current to flow in one direction only; it can be used as a **rectifier** to change ac to dc. Common **transistors** consist of three semiconductor sections, either as **pnp** or **npn**. Transistors can amplify electrical signals and find many other uses. An integrated circuit consists of a tiny semiconductor crystal or "chip" on which many transistors, diodes, resistors, and other circuit elements have been constructed using careful placement of impurities.

Questions

1. What type of bond would you expect for (*a*) the N_2 molecule, (*b*) the HCl molecule, (*c*) Fe atoms in a solid?
2. Describe how the molecule $CaCl_2$ could be formed.
3. Does the H_2 molecule have a permanent dipole moment? Does O_2? Does H_2O? Explain.
4. Although the molecule H_3 is not stable, the ion H_3^+ is. Explain, using the Pauli exclusion principle.
5. The energy of a molecule can be divided into four categories. What are they?
6. If conduction electrons are free to roam about in a metal, why don't they leave the metal entirely?
7. A silicon semiconductor is doped with phosphorus. Will these atoms be donors or acceptors? What type of semiconductor will this be?
8. Explain why the resistivity of metals increases with temperature whereas the resistivity of semiconductors may decrease with increasing temperature.
9. Discuss the differences between an ideal gas and a Fermi electron gas.
10. Which aspects of Fig. 41–27 are peculiar to copper, and which are valid in general for other metals?

* **11.** Can a diode be used to amplify a signal?

* **12.** Figure 41–43 shows a "bridge-type" full-wave rectifier. Explain how the current is rectified and how current flows during each half cycle.

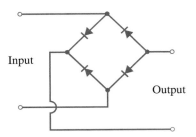

FIGURE 41–43 Question 12.

* **13.** Compare the resistance of a *p-n* junction diode connected in forward bias to its resistance when connected in reverse bias.

* **14.** Explain how a transistor could be used as a switch.

* **15.** If $\mathscr{E}_C$ were reversed in Fig. 41–41, how would the amplification be altered?

* **16.** Describe how a *pnp* transistor can operate as an amplifier.

* **17.** Do diodes and transistors obey Ohm's law?

* **18.** In a transistor, the base–emitter junction and the base–collector junction are essentially diodes. Are these junctions reverse-biased or forward-biased in the application shown in Fig. 41–41?

* **19.** What purpose does the capacitor in Fig. 41–42 serve?

Problems

Sections 41–1 to 41–3

1. (I) Estimate the binding energy of a KCl molecule by calculating the electrostatic potential energy when the K^+ and Cl^- ions are at their stable separation of 0.28 nm. Assume each has a charge of magnitude 1.0e.

2. (I) Binding energies are often measured experimentally in kcal per mole, and then the binding energy in eV per molecule is calculated from that result. What is the conversion factor in going from kcal per mole to eV per molecule? What is the binding energy of KCl (= 4.43 eV) in kcal per mole?

3. (II) The measured binding energy of KCl is 4.43 eV. From the result of Problem 1, estimate the contribution to the binding energy of the repelling electron clouds at the equilibrium distance $r_0 = 0.28$ nm.

4. (II) Estimate the binding energy of the H_2 molecule, assuming the two H nuclei are 0.074 nm apart and the two electrons spend 33 percent of their time midway between them.

5. (III) Apply reasoning similar to that in the text for the $S = 0$ and $S = 1$ states in the formation of the H_2 molecule to show why the molecule He_2 is *not* formed. Show also why the He_2^+ molecular ion *is* formed (with a binding energy of 3.1 eV at $r_0 = 0.11$ nm).

Section 41–4

6. (I) Show that the quantity $\hbar^2/I$ has units of energy.

7. (I) What is the reduced mass of the molecules (a) NaCl; (b) N_2; (c) HCl?

8. (II) The so-called "characteristic rotational energy," $\hbar^2/2I$, for N_2 is 2.48×10^{-4} eV. Calculate the N_2 bond length.

9. (II) The fundamental vibration frequency for the CO molecule is 6.42×10^{13} Hz. Determine (a) the reduced mass, and (b) the effective value of the constant k. Compare to k for the H_2 molecule.

10. (II) Explain why there is no transition for $E = hf$ in Fig. 41–20 (and Fig. 41–21).

11. (II) (a) Calculate the characteristic rotational energy, $\hbar^2/2I$, for the O_2 molecule whose bond length is 0.121 nm. (b) What are the energy and wavelength of photons emitted in a $L = 2$ to $L = 1$ transition?

12. (II) The equilibrium separation of H atoms in the H_2 molecule is 0.074 nm (Fig. 41–8). Calculate the energies and wavelengths of photons for the rotational transitions (a) $L = 1$ to $L = 0$, (b) $L = 2$ to $L = 1$, and (c) $L = 3$ to $L = 2$.

13. (II) Calculate the bond length for the NaCl molecule given that three successive wavelengths for rotational transitions are 23.1 mm, 11.6 mm, and 7.71 mm.

14. (II) Derive Eqs. 41–8.

Section 41–5

15. (I) Estimate the ionic cohesive energy for NaCl taking $\alpha = 1.75$, $m = 8$, and $r_0 = 0.28$ nm.

16. (II) The spacing between "nearest neighbor" Na and Cl ions in a NaCl crystal is 0.24 nm. What is the spacing between two nearest neighbor Na ions?

17. (II) Common salt, NaCl, has a density of 2.165 g/cm^3. The molecular weight of NaCl is 58.44. Estimate the distance between nearest neighbor Na and Cl ions. [*Hint: Each* ion can be considered to have one "cube" or "cell" of side s (our unknown) extending out from it.]

18. (II) Repeat the previous Problem for KCl whose density is 1.99 g/cm^3.

19. (III) (a) Starting from Eq. 41–9, show that the ionic cohesive energy is given by $U_0 = -(\alpha e^2/4\pi\epsilon_0 r_0)(1 - 1/m)$. Determine U_0 for (b) NaI ($r_0 = 0.33$ nm) and (c) MgO ($r_0 = 0.21$ nm). Assume $m = 10$. (d) If you used $m = 8$ instead, how far off would your answers be?

20. (III) For a long one-dimensional chain of alternating positive and negative ions, show that the Madelung constant would be $\alpha = 2 \ln 2$ [*Hint:* Use a series expansion for $\ln(1 + x)$.]

Section 41–6

21. (I) Estimate the number of states available to electrons in a 1.0-cm^3 cube of copper between 6.90 and 7.00 eV.

22. (II) What, roughly, is the ratio of the density of molecules in an ideal gas at 300 K and 1 atm (say O_2) to the density of free electrons (assume one per atom) in a metal (copper) also at 300 K?

23. (II) Calculate the energy which has 90 percent occupancy probability for copper at (a) $T = 300$ K; (b) $T = 1200$ K.

24. (II) Calculate the energy which has 10 percent occupancy probability for copper at (a) $T = 300$ K; (b) $T = 1200$ K.

25. (II) What is the occupancy probability for a conduction electron in copper at $T = 300$ K for an energy $E = 1.010 E_F$?

26. (II) Calculate the number of possible electron states in a 1.00-cm^3 cube of silver between $0.99 E_F$ and $E_F (= 5.48$ eV).

27. (II) Calculate the Fermi energy and Fermi speed for sodium, which has a density of 0.97×10^3 kg/m^3 and has one conduction electron per atom.

28. (II) The atoms in zinc metal ($\rho = 7.1 \times 10^3$ kg/m^3) have two free electrons. Calculate (a) the density of conduction electrons, (b) their Fermi energy, and (c) their Fermi speed.

29. (II) Given that the Fermi energy of aluminum is 11.63 eV, (a) calculate the density of free electrons using Eq. 41–12, and (b) estimate the valence of aluminum using this model and the known density $(2.70 \times 10^3 \, \text{kg/m}^3)$ and atomic weight (27.0) of aluminum.

30. (II) The neutrons in a neutron star (Chapter 45) can be treated as a Fermi gas with neutrons in place of the electrons in our model of an electron gas. Determine the Fermi energy for a neutron star of radius 10 km and mass twice that of our Sun. Assume that the star is made entirely of neutrons and is of uniform density.

31. (II) Show that the average energy of conduction electrons in a metal at $T = 0 \, \text{K}$ is $\bar{E} = \frac{3}{5}E_F$ (Eq. 41–13) by calculating

$$\bar{E} = \frac{\int E \, n_o(E) \, dE}{\int n_o(E) \, dE}.$$

32. (II) Show that the probability for the state at the Fermi energy being occupied is exactly $\frac{1}{2}$, independent of temperature.

33. (II) (a) For copper at room temperature $(T = 300 \, \text{K})$, calculate the Fermi factor, Eq. 41–14, for an electron with energy 0.10 eV above the Fermi energy. This represents the probability that this state is occupied. Is this reasonable? (b) What is the probability that a state 0.10 eV below the Fermi energy is occupied? (c) What is the probability that the state in part (b) is unoccupied?

34. (II) For a one-dimensional potential well, start with Eq. 39–13 and show that the number of states per unit energy interval for an electron gas is given by

$$g_L(E) = \sqrt{\frac{8mL^2}{h^2 E}}.$$

Remember that there can be two electrons (spin up and spin down) for each value of n. [Hint: Write the quantum number n in terms of E. Then $g_L(E) = 2 \, dn/dE$ where dn is the number of energy levels between E and $E + dE$.]

35. (III) Proceed as follows to derive the density of states, $g(E)$, the number of states per unit volume per unit energy interval, Eq. 41–10. Let the metal be a cube of side L. Extend the discussion of Section 39–8 for an infinite well to three dimensions, giving energy levels

$$E = \frac{h^2}{8mL^2}(n_1^2 + n_2^2 + n_3^2).$$

(Explain the meaning of n_1, n_2, n_3.) Each set of values for the quantum numbers n_1, n_2, n_3 corresponds to one state. Imagine a space where n_1, n_2, n_3 are the axes, and each state is represented by a point on a cubic lattice in this space, each separated by 1 unit along an axis. Consider the octant $n_1 > 0, n_2 > 0, n_3 > 0$. Show that the number of states N within a radius $R = (n_1^2 + n_2^2 + n_3^2)^{\frac{1}{2}}$ is $2(\frac{1}{8})(\frac{4}{3}\pi R^3)$. Then, to get Eq. 41–10, set $g(E) = (1/V)(dN/dE)$, where $V = L^3$ is the volume of the metal.

Section 41–7

36. (I) A semiconductor, bombarded with light of slowly increased frequency, begins to conduct when the wavelength of light is 640 nm; estimate the size of the energy gap E_g.

37. (I) Explain on the basis of energy bands why the sodium chloride crystal is a good insulator. [Hint: Consider the shells of Na^+ and Cl^- ions.]

38. (II) We saw that there are $2N$ possible electron states in the 3s band of Na, where N is the total number of atoms. How many possible electron states are there in the (a) 2s band, (b) 2p band, and (c) 3p band? (d) State a general formula for the total number of possible states in any given electron band.

39. (II) Calculate the longest-wavelength photon that can cause an electron in silicon $(E_g = 1.1 \, \text{eV})$ to jump from the valence band to the conduction band.

40. (II) The energy gap E_g in germanium is 0.72 eV. When used as a photon detector, roughly how many electrons can be made to jump from the valence to the conduction band by the passage of a 710-keV photon that loses all its energy in this fashion?

Section 41–8

41. (II) Suppose that a silicon semiconductor is doped with phosphorus so that one silicon atom in 10^6 is replaced by a phosphorus atom. Assuming that the "extra" electron in every phosphorus atom is donated to the conduction band, by what factor is the density of conduction electrons increased? The density of silicon is $2330 \, \text{kg/m}^3$, and the density of conduction electrons in pure silicon is about $10^{16} \, \text{m}^{-3}$ at room temperature.

* Section 41–9

* 42. (I) At what wavelength will an LED radiate if made from a material with an energy gap $E_g = 1.4 \, \text{eV}$?

* 43. (I) If an LED emits light of wavelength $\lambda = 650 \, \text{nm}$, what is the energy gap (in eV) between valence and conduction bands?

* 44. (II) A silicon diode, whose current–voltage characteristics are given in Fig. 41–37, is connected in series with a battery and a 760-Ω resistor. What battery voltage is needed to produce a 12-mA current?

* 45. (II) Suppose that the diode of Fig. 41–37 is connected in series to a 100-Ω resistor and a 2.0-V battery. What current flows in the circuit? [Hint: Draw a line on Fig. 41–37 representing the current in the resistor as a function of the voltage across the diode; the intersection of this line with the characteristic curve will give the answer.]

* 46. (II) Sketch the resistance as a function of current, for $V > 0$, for the diode shown in Fig. 41–37.

* 47. (II) An ac voltage of 120 V rms is to be rectified. Estimate very roughly the average current in the output resistor R (28 kΩ) for (a) a half-wave rectifier (Fig. 41–38), and (b) a full-wave rectifier (Fig. 41–39) without capacitor.

* 48. (III) A silicon diode passes significant current only if the forward-bias voltage exceeds about 0.6 V. Make a rough estimate of the average current in the output resistor R of (a) a half-wave rectifier (Fig. 41–38), and (b) a full-wave rectifier (Fig. 41–39) without a capacitor. Assume that $R = 150 \, \Omega$ in each case and that the ac voltage is 9.0 V rms in each case.

*49. (III) A 120-V rms 60-Hz voltage is to be rectified with a full-wave rectifier as in Fig. 41–39, where $R = 18\,k\Omega$, and $C = 25\,\mu F$. (a) Make a rough estimate of the average current. (b) What happens if $C = 0.10\,\mu F$? [*Hint:* See Section 26–4.]

* Section 41–10

*50. (II) Suppose that the current gain of the transistor in Fig. 41–41 is $\beta = i_C/i_B = 80$. If $R_C = 3.3\,k\Omega$, calculate the ac output voltage for a time-varying input current of $2.0\,\mu A$.

*51. (II) If the current gain of the transistor amplifier in Fig. 41–41 is $\beta = i_C/i_B = 100$, what value must R_C have if a 1.0-μA ac base current is to produce an ac output voltage of 0.40 V?

*52. (II) A transistor, whose current gain $\beta = i_C/i_B = 70$, is connected as in Fig. 41–41 with $R_B = 3.2\,k\Omega$ and $R_C = 6.8\,k\Omega$. Calculate (a) the voltage gain, and (b) the power amplification.

*53. (II) An amplifier has a voltage gain of 80 and a 15-kΩ load (output) resistance. What is the peak output current through the load resistor if the input voltage is an ac signal with a peak of 0.080 V?

General Problems

54. Estimate the binding energy of the H_2 molecule by calculating the difference in kinetic energy of the electrons between when they are in separate atoms and when they are in the molecule, using the uncertainty principle. Take Δx for the electrons in the separated atoms to be the radius of the first Bohr orbit, 0.053 nm, and for the molecule take Δx to be the separation of the nuclei, 0.074 nm.

55. The average translational kinetic energy of an atom or molecule is about $\bar{K} = \frac{3}{2}kT$ (see Chapter 18), where $k = 1.38 \times 10^{-23}\,J/K$ is Boltzmann's constant. At what temperature T will $\bar{K}$ be on the order of the bond energy (and hence the bond able to be broken by thermal motion) for (a) a covalent bond (say H_2) of binding energy 4.5 eV, and (b) a "weak" hydrogen bond of binding energy 0.15 eV?

56. In the ionic salt KF, the separation distance between ions is about 0.27 nm. (a) Estimate the electrostatic potential energy between the ions assuming them to be point charges (magnitude 1e). (b) It is known that F releases 4.07 eV of energy when it "grabs" an electron, and 4.34 eV is required to ionize K. Find the binding energy of KF relative to free K and F atoms, neglecting the energy of repulsion.

57. The fundamental vibration frequency for the HCl molecule is $8.66 \times 10^{13}\,Hz$. Determine (a) the reduced mass, and (b) the effective value of the constant k. Compare to k for the H_2 molecule.

58. Imagine the two atoms of a diatomic molecule as if they were connected by a spring, Fig. 41–44. Show that the classical frequency of vibration is given by Eq. 41–5. [*Hint:* Let x_1 and x_2 be the displacements of each mass from initial equilibrium positions; then $m_1 d^2x_1/dt^2 = -kx$, and $m_2 d^2x_2/dt^2 = -kx$, where $x = x_1 + x_2$. Find another relationship between x_1 and x_2, assuming that the center of mass of the system stays at rest, and then show that $\mu d^2x/dt^2 = -kx$.]

59. Explain, using the Boltzmann factor (Eq. 40–16), why the heights of the peaks in Fig. 41–21 are different from one another. Explain also why the lines are not equally spaced. [*Hint:* Does the moment of inertia necessarily remain constant?]

60. Do we need to consider quantum effects for everyday rotating objects? Estimate the differences between rotational energy levels for a spinning baton compared to the energy of the baton. Assume the baton consists of a 30-cm-long bar with a mass of 200 g and two end masses (each points), each of mass 300 g, and that it rotates at 1.6 rev/s.

61. When solid argon melts at −189°C, its latent heat of fusion goes directly into breaking the bonds between the atoms. Solid argon is a weakly bound cubic lattice, with each atom connected to six neighbors, each bond having a binding energy of 3.9×10^{-3} eV. What is the latent heat of fusion for argon, in J/kg? [*Hint:* Show that in a simple cubic lattice (Fig. 41–45), there are *three* times as many bonds as there are atoms, when the number of atoms is large.]

FIGURE 41–44 Problem 58.

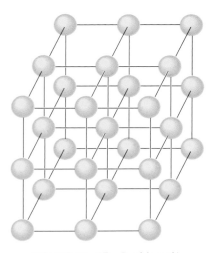

FIGURE 41–45 Problem 61.

62. A strip of silicon 1.5 cm wide and 1.0 mm thick is immersed in a magnetic field of strength 1.6 T perpendicular to the strip (Fig. 41–46). When a current of 0.20 mA is run through the strip, there is a resulting Hall effect voltage of 18 mV across the strip (Section 27–8). How many electrons per silicon atom are in the conduction band? The density of silicon is 2330 kg/m^3.

FIGURE 41–46 Problem 62.

63. Most of the Sun's radiation has wavelengths shorter than 1000 nm. For a solar cell to absorb all this, what energy gap ought the material have?

64. The energy gap between valence and conduction bands in germanium is 0.72 eV. What range of wavelengths can a photon have to excite an electron from the top of the valence band into the conduction band?

65. A TV remote control emits IR light. If the detector on the TV set is *not* to react to visible light, could it make use of silicon as a "window" with its energy gap $E_g = 1.14$ eV? What is the shortest-wavelength light that can strike silicon without causing electrons to jump from the valence band to the conduction band?

66. The *Fermi temperature* T_F is defined as that temperature at which the thermal energy kT (without the $\frac{3}{2}$) is equal to the Fermi energy: $kT_F = E_F$. (a) Determine the Fermi temperature for copper. (b) Show that for $T \gg T_F$, the Fermi factor (Eq. 41–14) approaches the Boltzmann factor. (Note: This last result is not very useful for understanding conductors. Why?)

67. For an arsenic donor atom in a doped silicon semiconductor, assume that the "extra" electron moves in a Bohr orbit about the arsenic ion. For this electron in the ground state, take into account the dielectric constant $K = 12$ of the Si lattice (which represents the weakening of the Coulomb force due to all the other atoms or ions in the lattice), and estimate (a) the binding energy, and (b) the orbit radius for this extra electron. [*Hint*: Substitute $\epsilon = K\epsilon_0$ in Coulomb's law.]

* 68. A full-wave rectifier (Fig. 41–39) uses two diodes to rectify a 75-V rms 60 Hz ac voltage. If $R = 8.8$ kΩ and $C = 30\,\mu$F, what will be the approximate percent variation in the output voltage? The variation in output voltage (Fig. 41–39c) is called *ripple voltage*. [*Hint*: See Section 26–4 and assume the discharge of the capacitor is approximately linear.]

* 69. A zener diode voltage regulator is shown in Fig. 41–47. Suppose that $R = 1.80$ kΩ, and that the diode breakdown voltage is 130 V: the diode is rated at a maximum current of 110 mA. (a) If $R_{load} = 16.0$ kΩ, over what range of supply voltages will the circuit maintain the voltage at 130 V? (b) If the supply voltage is 200 V, over what range of load resistance will the voltage be regulated?

FIGURE 41–47 Problem 69.

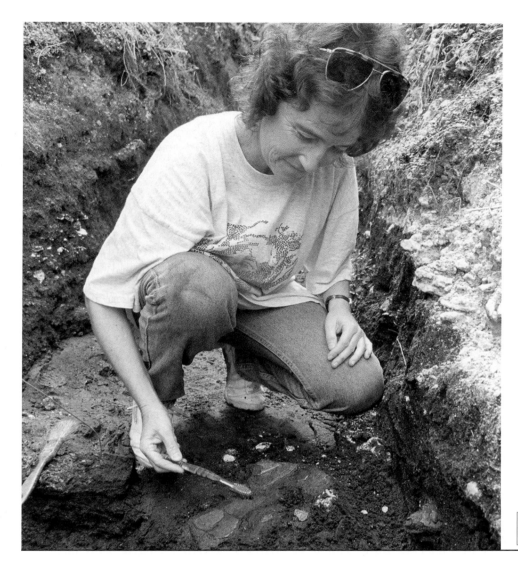

This archeologist has unearthed the remains of a sea-turtle within an ancient man-made stone circle. Carbon dating of the remains can tell her when humans inhabited the site.

In this chapter we begin our discussion of nuclear physics including the properties of nuclei, the various forms of radioactivity, and how radioactive decay can be used in a variety of fields to determine the age of old objects, from bones and trees to rocks and other mineral substances, and obtain information on the history of the Earth.

CHAPTER 42

Nuclear Physics and Radioactivity

In the early part of the twentieth century, Rutherford's experiments led to the idea that at the center of an atom there is a tiny but massive nucleus. At the same time that the quantum theory was being developed and scientists were attempting to understand the structure of the atom and its electrons, investigations into the nucleus itself had also begun. In this chapter and the next, we take a brief look at *nuclear physics*.

42–1 | Structure and Properties of the Nucleus

An important question to physicists in the early part of this century was whether the nucleus had a structure, and what that structure might be. It turns out that the nucleus is a complicated entity and is not fully understood even today. However, by the early 1930s, a model of the nucleus had been developed that is still useful.

According to this model, a nucleus is considered as an aggregate of two types of particles: protons and neutrons. (Of course, we must remember that these "particles" also have wave properties, but for ease of visualization and language, we often refer to them simply as "particles.") A **proton** is the nucleus of the simplest atom, hydrogen. It has a positive charge $(= +e = +1.60 \times 10^{-19}\,\text{C})$ and a mass

Proton

$$m_p = 1.67262 \times 10^{-27}\,\text{kg}.$$

The **neutron**, whose existence was ascertained only in 1932 by the Englishman James Chadwick (1891–1974), is electrically neutral $(q = 0)$, as its name implies. Its mass, which is almost identical to that of the proton, is

Neutron

$$m_n = 1.67493 \times 10^{-27}\,\text{kg}.$$

Nucleons

These two constituents of a nucleus, neutrons and protons, are referred to collectively as **nucleons**.

Although the hydrogen nucleus consists of a single proton alone, the nuclei of all other elements consist of both neutrons and protons. The different types of nuclei are often referred to as **nuclides**. The number of protons in a nucleus (or

Z and A

nuclide) is called the **atomic number** and is designated by the symbol Z. The total number of nucleons, neutrons plus protons, is designated by the symbol A and is called the **atomic mass number**. This name is used since the mass of a nucleus is very closely A times the mass of one nucleon. A nuclide with 7 protons and 8 neutrons thus has $Z = 7$ and $A = 15$. The **neutron number** N is $N = A - Z$.

To specify a given nuclide, we need give only A and Z. A special symbol is commonly used which takes the form

$$^{A}_{Z}\text{X},$$

where X is the chemical symbol for the element (see Appendix D, and the periodic table inside the back cover), A is the atomic mass number, and Z is the atomic number. For example, $^{15}_{7}\text{N}$ means a nitrogen nucleus containing 7 protons and 8 neutrons for a total of 15 nucleons. In a neutral atom, the number of electrons orbiting the nucleus is equal to the atomic number Z (since the charge on an electron has the same magnitude but opposite sign to that of a proton). The main properties of an atom, and how it interacts with other atoms, are largely determined by the number of electrons. Hence Z determines what kind of atom it is: carbon, oxygen, gold, or whatever. It is redundant to specify both the symbol of a nucleus and its atomic number Z as described above. If the nucleus is nitrogen, for example, we know immediately that $Z = 7$. The subscript Z is thus sometimes dropped and $^{15}_{7}\text{N}$ is then written simply ^{15}N; in words we say "nitrogen fifteen."

For a particular type of atom (say, carbon), nuclei are found to contain different numbers of neutrons, although they all have the same number of protons. For example, carbon nuclei always have 6 protons, but they may have 5, 6, 7, 8, 9, or 10 neutrons. Nuclei that contain the same number of protons but different num-

Isotopes

bers of neutrons are called **isotopes**. Thus, $^{11}_{6}\text{C}$, $^{12}_{6}\text{C}$, $^{13}_{6}\text{C}$, $^{14}_{6}\text{C}$, $^{15}_{6}\text{C}$, and $^{16}_{6}\text{C}$ are all isotopes of carbon. Of course, the isotopes of a given element are not all equally common. For example, 98.9 percent of naturally occurring carbon (on Earth) is the isotope $^{12}_{6}\text{C}$, and about 1.1 percent is $^{13}_{6}\text{C}$. These percentages are referred to as the **natural abundances**.[†] Many isotopes that do not occur naturally can be produced in the laboratory by means of nuclear reactions (more on this later). Indeed, all elements beyond uranium $(Z > 92)$ do not occur naturally and are only produced artificially, as are many nuclides with $Z \leq 92$.

The approximate size of nuclei was determined originally by Rutherford from the scattering of charged particles. Of course, we cannot speak about a definite size for nuclei because of the wave–particle duality: their spatial extent must remain somewhat fuzzy. Nonetheless a rough "size" can be measured by scattering high-speed electrons off nuclei. It is found that nuclei have a roughly spherical

---

[†] The mass value for each element as given in the periodic table (inside back cover) is an average weighted according to the natural abundances of its isotopes.

shape with a radius that increases with A according to the approximate formula

$$r \approx (1.2 \times 10^{-15} \, \text{m})(A^{\frac{1}{3}}).$$ **(42–1)** *Nuclear radii*

Since the volume of a sphere is $V = \frac{4}{3}\pi r^3$, we see that the volume of a nucleus is proportional to the number of nucleons, $V \propto A$. This is what we would expect if nucleons were like impenetrable billiard balls: if you double the number of balls, you double the total volume. Hence, all nuclei have nearly the same density, and it is enormous (see Problem 5).

EXAMPLE 42–1 **ESTIMATE** **Nuclear sizes.** Estimate the diameter of the following nuclei: (a) $^1_1$H, (b) $^{40}_{20}$Ca, (c) $^{208}_{82}$Pb, (d) $^{235}_{92}$U.

SOLUTION (a) For hydrogen, $A = 1$, Eq. 42–1 gives

$$d = \text{diameter} = 2r \approx 2.4 \times 10^{-15} \, \text{m}$$

since $A^{\frac{1}{3}} = 1^{\frac{1}{3}} = 1$.

(b) For calcium $d = 2r \approx (2.4 \times 10^{-15} \, \text{m})(40)^{\frac{1}{3}} = 8.2 \times 10^{-15} \, \text{m}$.

(c) For lead $d \approx (2.4 \times 10^{-15} \, \text{m})(208)^{\frac{1}{3}} = 14 \times 10^{-15} \, \text{m}$.

(d) For uranium $d \approx (2.4 \times 10^{-15} \, \text{m})(235)^{\frac{1}{3}} = 15 \times 10^{-15} \, \text{m}$.

The range of nuclear diameters is only from 2.4 fm to 15 fm.

The masses of nuclei can be determined, in one method, by measuring the radius of curvature of fast-moving nuclei in a magnetic field using a mass spectrometer, as discussed in Section 27–9. Indeed, as mentioned there, the existence of different isotopes of the same element was discovered using this device. Nuclear masses can be specified in **unified atomic mass units** (u). On this scale, a neutral $^{12}_{6}$C atom is given the precise value 12.000000 u. A neutron then has a measured mass of 1.008665 u, a proton 1.007276 u, and a neutral hydrogen atom, $^1_1$H (proton plus electron) 1.007825 u. The masses of many nuclides are given in Appendix D. It should be noted that the masses in this table, as is customary, are for the *neutral atom*, and not for a bare nucleus.

➡ **PROBLEM SOLVING**

Masses are for neutral atom

Masses are often specified using the electron-volt energy unit. This can be done because mass and energy are related, and the precise relationship is given by Einstein's equation $E = mc^2$ (Chapter 37). Since the mass of a proton is 1.67262×10^{-27} kg, or 1.007276 u, then

$$1.0000 \, \text{u} = \left(\frac{1.0000 \, \text{u}}{1.007276 \, \text{u}} \right)(1.67262 \times 10^{-27} \, \text{kg})$$
$$= 1.66054 \times 10^{-27} \, \text{kg};$$

this is equivalent to an energy (see table inside front cover)

$$E = mc^2 = \frac{(1.66054 \times 10^{-27} \, \text{kg})(2.9979 \times 10^8 \, \text{m/s})^2}{(1.6022 \times 10^{-19} \, \text{J/eV})}$$
$$= 931.5 \, \text{MeV}.$$

Thus

$$1 \, \text{u} = 1.6605 \times 10^{-27} \, \text{kg} = 931.5 \, \text{MeV}/c^2.$$ *Atomic mass unit*

The rest masses of some of the basic particles are given in Table 42–1.

TABLE 42–1 Rest Masses in Kilograms, Unified Atomic Mass Units, and MeV/c^2

Object	Mass		
	kg	**u**	**MeV/c^2**
Electron	9.1094×10^{-31}	0.00054858	0.51100
Proton	1.67262×10^{-27}	1.007276	938.27
$^1_1$H atom	1.67353×10^{-27}	1.007825	938.78
Neutron	1.67493×10^{-27}	1.008665	939.57

Just as an electron has intrinsic spin and angular momentum quantum numbers, so too do nuclei and their constituents, the proton and neutron. Both the proton and the neutron are spin-$\frac{1}{2}$ particles. A nucleus, made up of protons and neutrons, has a **nuclear spin** quantum number, I, that can be either integer or half integer, depending on whether it is made up of an even or an odd number of nucleons. The *nuclear angular momentum* of a nucleus is given, as might be expected (see Section 40–2), by $\sqrt{I(I+1)}\,\hbar$.

Nuclear magnetic moments are measured in terms of the **nuclear magneton**

$$\mu_N = \frac{e\hbar}{2m_p},\tag{42–2}$$

which is defined by analogy with the Bohr magneton for electrons ($\mu_B = e\hbar/2m_e$, Section 40–7). Since μ_N contains the proton mass, m_p, instead of the electron mass, it is about 2000 times smaller. The electron magnetic moment is about 1 Bohr magneton, so we might expect the proton to have a magnetic moment μ_p of about $1\,\mu_N$. Instead, it is

$$\mu_p = 2.7928\mu_N.$$

There is no satisfactory explanation for this large factor. Also surprising is the fact that the neutron has a magnetic moment:

$$\mu_n = -1.9135\mu_N.$$

This suggests that, although the neutron carries no net charge, there may be some sort of electric current within the neutron. The minus sign for μ_n indicates that its magnetic moment is opposite to its spin.

Important applications based on nuclear spin are nuclear magnetic resonance (NMR) and magnetic resonance imaging (MRI). They are discussed in the next chapter (Section 43–10).

42–2 Binding Energy and Nuclear Forces

The total mass of a stable nucleus is always less than the sum of the masses of its separate protons and neutrons, as the following Example shows.

EXAMPLE 42–2 $^4_2$**He mass compared to its constituents.** Compare the mass of a $^4_2$He nucleus to that of its constituent nucleons.

SOLUTION The mass of a neutral $^4_2$He atom, from Appendix D, is 4.002603 u. The mass of two neutrons and two protons (including the two electrons) is

$$2m_n = 2.017330\,\text{u}$$
$$2m(^1_1\text{H}) = \underline{2.015650\,\text{u}}$$
$$4.032980\,\text{u}.$$

We almost always deal with masses of neutral atoms—that is, nuclei with Z electrons—since this is how masses are measured. We must therefore be sure to balance out the electrons when we compare masses, which is why we used the mass of $^1_1$H in this Example rather than that of the proton alone.

PROBLEM SOLVING

Keep track of electron masses

Binding energy

Thus the mass of $^4_2$He is measured to be $4.032980\,\text{u} - 4.002603\,\text{u} = 0.030377\,\text{u}$ less than the masses of its constituents. Where has this lost mass gone?

It has, in fact, gone into energy of another kind (such as radiation, or kinetic energy, for example). The mass (or energy) difference in the case of $^4_2$He, given in energy units, is $(0.030377\,\text{u})(931.5\,\text{MeV/u}) = 28.30\,\text{MeV}$. This difference is referred to as the **total binding energy** of the nucleus. The total binding energy represents the amount of energy that must be put into a nucleus in order to break it apart into its constituent protons and neutrons. If the mass of, say, a $^4_2$He nucleus were exactly equal to the mass of two neutrons plus two protons, the nucleus could fall apart without any input of energy. To be stable, the mass of a nucleus *must* be less than that of its constituent nucleons, so that energy input *is* needed to break it apart. Note that the binding energy is not something a nucleus has—it is energy it "lacks" relative to the total mass of its separate constituents.

Nuclear binding energy can be compared to the binding energy of electrons in an atom. We saw in Chapter 38 that the binding energy of the one electron in the hydrogen atom, for example, is 13.6 eV. The mass of a $^1_1$H atom is less than that of a single proton plus a single electron by 13.6 eV. Compared to the total mass of the atom (939 MeV), this is incredibly small (1 part in 10^8), and for practical purposes the mass difference can be ignored. The binding energies of nuclei are on the order of 10^6 times greater than the binding energies of electrons in atoms.

The **average binding energy per nucleon** is defined as the total binding energy of a nucleus divided by A, the total number of nucleons. For $^4_2$He, it is 28.3 MeV/4 = 7.1 MeV. Figure 42–1 shows the average binding energy per nucleon as a function of A for stable nuclei. The curve rises as A increases and reaches a plateau at about 8.7 MeV per nucleon above about $A \approx 40$. Beyond about $A \approx 80$, the curve decreases slowly, indicating that larger nuclei are held together a little less tightly than those in the middle of the periodic table. (We will see later that these characteristics allow the release of nuclear energy in the processes of fission and fusion.)

Binding energy per nucleon

FIGURE 42–1 Average binding energy per nucleon as a function of mass number A for stable nuclei.

EXAMPLE 42–3 **Binding energy for iron.** Calculate the total binding energy and the average binding energy per nucleon for $^{56}_{26}$Fe, the most common stable isotope of iron.

SOLUTION $^{56}_{26}$Fe has 26 protons and 30 neutrons whose separate masses are

$$(26)(1.007825 \, \text{u}) = 26.2035 \, \text{u (includes electrons)}$$
$$(30)(1.008665 \, \text{u}) = \underline{30.2600 \, \text{u}}$$
$$\text{Total} = 56.4635 \, \text{u}.$$
$$\text{Subtract mass of } ^{56}_{26}\text{Fe:} \quad \underline{-55.9349 \, \text{u}} \, (\text{Appendix D})$$
$$\Delta m = 0.5286 \, \text{u}.$$

The total binding energy is thus

$$(0.5286 \, \text{u})(931.5 \, \text{MeV/u}) = 492.4 \, \text{MeV}$$

and the average binding energy per nucleon is

$$\frac{492.4 \, \text{MeV}}{56 \, \text{nucleons}} = 8.8 \, \text{MeV}.$$

EXAMPLE 42–4 **Binding energy of last neutron.** What is the binding energy of the last neutron in $^{13}_{6}C$?

SOLUTION We compare the mass of $^{13}_{6}C$ to that of the atom with one less neutron, $^{12}_{6}C$, plus a free neutron (Appendix D):

$$
\begin{aligned}
\text{Mass } ^{12}_{6}C &= 12.000000 \text{ u} \\
\text{Mass } ^{1}_{0}n &= \underline{1.008665 \text{ u}} \\
\text{Total} &= 13.008665 \text{ u.} \\
\text{Subtract mass of } ^{13}_{6}C: & \underline{-13.003355 \text{ u}} \\
\Delta m &= 0.005310 \text{ u}
\end{aligned}
$$

which in energy is $(931.5 \text{ MeV/u})(0.005310 \text{ u}) = 4.95 \text{ MeV}$. That is, it would require 4.95 MeV input of energy to remove one neutron from $^{13}_{6}C$.

We can analyze nuclei not only from the point of view of energy, but also from the point of view of the forces that hold them together. We would not expect a collection of protons and neutrons to come together spontaneously, since protons are all positively charged and thus exert repulsive forces on each other. Indeed, the question arises as to how a nucleus stays together at all in view of the fact that the electric force between protons would tend to break it apart. Since stable nuclei *do* stay together, it is clear that another force must be acting. Because this new force is stronger than the electric force (which, in turn, is much stronger than gravity at the nuclear level) it is called the **strong nuclear force**. The strong nuclear force is an attractive force that acts between all nucleons—protons and neutrons alike. Thus protons attract each other via the nuclear force at the same time they repel each other via the electric force. Neutrons, since they are electrically neutral, only attract other neutrons or protons via the nuclear force.

The nuclear force turns out to be far more complicated than the gravitational and electromagnetic forces. A precise mathematical description is not yet possible. Nonetheless, a great deal of work has been done to try to understand the nuclear force. One important aspect of the strong nuclear force is that it is a **short-range** force: it acts only over a very short distance. It is very strong between two nucleons if they are less than about 10^{-15} m apart, but it is essentially zero if they are separated by a distance greater than this. Compare this to electric and gravitational forces, which can act over great distances and are therefore called **long-range** forces. The strong nuclear force has some strange quirks. For example, if a nuclide contains too many or too few neutrons relative to the number of protons, the binding of the nucleons is reduced; nuclides that are too unbalanced in this regard are unstable. As shown in Fig. 42–2, stable nuclei tend to have the same number of protons as neutrons ($N = Z$) up to about $A \approx 30$ or 40. Beyond this, stable nuclei contain more neutrons than protons. This makes sense since, as Z increases, the electrical repulsion increases, so a greater number of neutrons—which exert only the attractive nuclear force—are required to maintain stability. For very large Z, no number of neutrons can overcome the greatly increased electric repulsion. Indeed, there are no completely stable nuclides above $Z = 82$.

What we mean by a stable nucleus is one that stays together indefinitely. What then is an unstable nucleus? It is one that comes apart; and this results in radioactive decay. Before we discuss the important subject of radioactivity (the next Section), we note that there is a second type of nuclear force that is much weaker than the strong nuclear force. It is called the **weak nuclear force**, and we are aware of its existence only because it shows itself in certain types of radioactive decay. These two nuclear forces, the strong and the weak, together with the gravitational and electromagnetic forces, comprise the four known types of force in nature (more on this in Chapter 44).

Strong nuclear force

Long- and short-range forces

FIGURE 42–2 Number of neutrons versus number of protons for stable nuclides, which are represented by dots. The straight line represents $N = Z$.

Weak force

42–3 | Radioactivity

Nuclear physics had its beginnings in 1896. In that year, Henri Becquerel (1852–1908) made an important discovery: in his studies of phosphorescence, he found that a certain mineral (which happened to contain uranium) would darken a photographic plate even when the plate was wrapped to exclude light. It was clear that the mineral emitted some new kind of radiation that, unlike X-rays, occurred without any external stimulus. This new phenomenon eventually came to be called **radioactivity**.

Discovery of radioactivity

Soon after Becquerel's discovery, Marie Curie (1867–1934) and her husband, Pierre Curie (1859–1906), isolated two previously unknown elements that were very highly radioactive (Fig. 42–3). These were named polonium and radium. Other radioactive elements were soon discovered as well. The radioactivity was found in every case to be unaffected by the strongest physical and chemical treatments, including strong heating or cooling and the action of strong chemical reagents. It was clear that the source of radioactivity must be deep within the atom, that it must emanate from the nucleus. And it became apparent that radioactivity is the result of the *disintegration* or *decay* of an unstable nucleus. Certain isotopes are not stable under the action of the nuclear force, and they decay with the emission of some type of radiation or "rays."

FIGURE 42–3 Marie and Pierre Curie in their laboratory (about 1906) where radium was discovered.

Many unstable isotopes occur in nature, and such radioactivity is called "natural radioactivity." Other unstable isotopes can be produced in the laboratory by nuclear reactions (Section 43–1); these are said to be produced "artificially" and to have "artificial radioactivity."

Rutherford and others began studying the nature of the rays emitted in radioactivity about 1898. They found that the rays could be classified into three distinct types according to their penetrating power. One type of radiation could barely penetrate a piece of paper. The second type could pass through as much as 3 mm of aluminum. The third was extremely penetrating: it could pass through several centimeters of lead and still be detected on the other side. They named these three types of radiation alpha (α), beta (β), and gamma (γ), respectively, after the first three letters of the Greek alphabet.

FIGURE 42–4 Alpha and beta rays are bent in opposite directions by a magnetic field, whereas gamma rays are not bent at all.

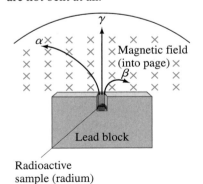

Each type of ray was found to have a different charge and hence is bent differently in a magnetic field, Fig. 42–4; α rays are positively charged, β rays are negatively charged, and γ rays are neutral. It was soon found that all three types of radiation consisted of familiar kinds of particles. Gamma rays are very high-energy photons whose energy is even higher than that of X-rays. Beta rays are electrons, identical to those that orbit the nucleus (but they are created within the nucleus itself). Alpha rays (or α particles) are simply the nuclei of helium atoms, ^4_2He; that is, an α ray consists of two protons and two neutrons bound together.

We now discuss each of these three types of radioactivity, or decay, in more detail.

42–4 Alpha Decay

When a nucleus emits an α particle ($^4_2$He), it is clear that the remaining nucleus will be different from the original: for it has lost two protons and two neutrons. Radium 226 ($^{226}_{88}$Ra), for example, is an α emitter. It decays to a nucleus with $Z = 88 - 2 = 86$ and $A = 226 - 4 = 222$. The nucleus with $Z = 86$ is radon (Rn) —see Appendix D or the periodic table. Thus the radium decays to radon with the emission of an α particle. This is written

α decay

$$^{226}_{88}\text{Ra} \rightarrow {}^{222}_{86}\text{Rn} + {}^4_2\text{He}.$$

See Fig. 42–5.

FIGURE 42–5 Radioactive decay of radium to radon with emission of an alpha particle.

Daughter nucleus
Parent nucleus
Transmutation

It is clear that when α decay occurs, a new element is formed. The **daughter** nucleus ($^{222}_{86}$Rn in this case) is different from the **parent** nucleus ($^{226}_{88}$Ra in this case). This changing of one element into another is called **transmutation**.

Alpha decay can be written

$$^A_Z N \rightarrow {}^{A-4}_{Z-2} N' + {}^4_2\text{He} \qquad\qquad [\alpha \text{ decay}]$$

where N is the parent, N' the daughter, and Z and A are the atomic number and atomic mass number, respectively, of the parent.

Alpha decay occurs because the strong nuclear force is unable to hold very large nuclei together. Because the nuclear force is a short-range force, it acts only between neighboring nucleons. But the electric force can act all the way across a large nucleus. For very large nuclei, the large Z means the repulsive electric force becomes very large (Coulomb's law); and it acts between all protons. The strong nuclear force, since it acts only between neighboring nucleons, is overpowered and is unable to hold the nucleus together.

Why the strong nuclear force cannot hold a nucleus together

We can express the instability in terms of energy (or mass): the mass of the parent nucleus is greater than the mass of the daughter nucleus plus the mass of the α particle. The mass difference appears as kinetic energy, which is carried away by the α particle and the recoiling daughter nucleus. The total energy released is called the **disintegration energy**, Q, or the **Q-value** of the decay. From conservation of energy,

$$M_\text{P} c^2 = M_\text{D} c^2 + m_\alpha c^2 + Q,$$

so

Q-value

$$Q = M_\text{P} c^2 - (M_\text{D} + m_\alpha) c^2 \qquad\qquad \textbf{(42–3)}$$

where M_P, M_D, and m_α are the masses of the parent, daughter, and α particle, respectively. If the parent had *less* mass than the daughter plus the α particle (so $Q < 0$), the decay could not occur, for the conservation of energy law would be violated.

EXAMPLE 42–5 **Uranium decay energy release.** Calculate the disintegration energy when $^{232}_{92}\text{U}$ (mass = 232.037146 u) decays to $^{228}_{90}\text{Th}$ (228.028731 u) with the emission of an α particle. (As always, masses are for neutral atoms.)

SOLUTION Since the mass of the ^4_2He is 4.002603 u (Appendix D), the total mass in the final state is

$$228.028731 \text{ u} + 4.002603 \text{ u} = 232.031334 \text{ u}.$$

The mass lost when the $^{232}_{92}\text{U}$ decays is

$$232.037146 \text{ u} - 232.031334 \text{ u} = 0.005812 \text{ u}.$$

Since $1 \text{ u} = 931.5 \text{ MeV}$, the energy Q released is

$$Q = (0.005812 \text{ u})(931.5 \text{ MeV/u}) \approx 5.4 \text{ MeV},$$

and this energy appears as kinetic energy of the α particle and the daughter nucleus. (Using conservation of momentum, it can be shown that the α particle emitted by a $^{232}_{92}\text{U}$ nucleus at rest has a kinetic energy of about 5.3 MeV. Thus, the daughter nucleus—which recoils in the opposite direction from the emitted α particle—has about 0.1 MeV of kinetic energy. See Problem 68.)

If the mass of the daughter nucleus plus the mass of the α particle is less than the mass of the parent nucleus (so the parent is energetically allowed to decay), why are there any parent nuclei at all? That is, why haven't radioactive nuclei all decayed long ago, right after they were formed in the early stages of the universe? We can understand decay using a model of a nucleus inside of which there is an alpha particle (at least some of the time) bouncing around. The potential energy "seen" by the α particle would have a shape something like that shown in Fig. 42–6. The potential energy well (approximately square) between $r = 0$ and $r = R_0$ represents the short-range attractive nuclear force. Beyond the nuclear radius, R_0, the Coulomb repulsion dominates (since the nuclear force drops to zero) and we see the characteristic $1/r$ dependence of the Coulomb potential. The α particle, trapped within the nucleus, can be thought of as bouncing back and forth between the potential walls. Since the potential energy just beyond $r = R_0$ is greater than the energy of the α particle (dashed line), the α particle could not escape the nucleus if it were governed by classical physics. But according to quantum mechanics, there is a certain probability that the α particle can tunnel through the Coulomb barrier, from point A to point B in Fig. 42–6, as we discussed in Section 39–10. The height and width of the barrier affect the rate at which the nuclei decay (Section 42–8). Because of this barrier, the lifetimes of α-unstable nuclei can be quite long, from a fraction of a microsecond to over 10^{10} years. Note in Fig. 42–6 that the Q-value represents the total kinetic energy when the α particle is far from the nucleus.

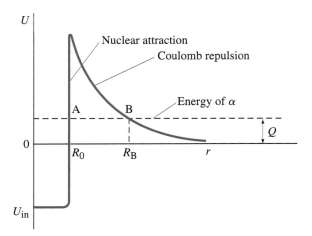

FIGURE 42–6 Potential energy for alpha particle and (daughter) nucleus, showing the Coulomb barrier through which the α particle must tunnel to escape. The Q-value of the reaction is also shown.

Why, you may wonder, do nuclei emit this combination of four nucleons called an α particle? Why not just four separate nucleons, or even one? The answer is that the α particle is very strongly bound, so that its mass is significantly less than that of four separate nucleons. As we saw in Example 42–2, two protons and two neutrons separately have a total mass of about 4.032980 u. The total mass of a $^{228}_{90}$Th nucleus plus four separate nucleons is 232.061711 u, which is greater than the mass of the parent nucleus. Such a decay could not occur because it would violate the conservation of energy. Similarly, it is almost always true that the emission of a single nucleon is energetically not possible. Or, to put it another way, for any nuclide for which it could be possible (mass of parent > mass of daughter + mass of nucleon), the decay happens so fast after formation of the parent that we don't see the parent in nature.

One widespread application of nuclear physics is present in nearly every home in the form of an ordinary **smoke detector**. The most common type of detector contains about 0.2 mg of the radioactive americium isotope, $^{241}_{95}$Am, in the form of AmO_2. The radiation ionizes the nitrogen and oxygen molecules in the air space between two oppositely charged plates. The resulting conductivity allows a small steady current. If smoke enters, the radiation is absorbed by the smoke particles rather than by the air molecules, thus reducing the current. The current drop is detected by the device's electronics and sets off the alarm. The radiation dose that escapes from an intact americium smoke detector is much less than the natural radioactive background, and so can be considered harmless. There is no question that smoke detectors save lives and reduce property damage.

42–5 | Beta Decay

Transmutation of elements also occurs when a nucleus decays by β decay—that is, with the emission of an electron or $β^-$ particle. The nucleus $^{14}_{6}$C, for example, emits an electron when it decays:

$$^{14}_{6}\text{C} \rightarrow {}^{14}_{7}\text{N} + e^- + \text{a neutrino},$$

where e^- is the symbol for the electron. (The symbol $_{-1}^{0}e$ is sometimes used for the electron whose charge corresponds to $Z = -1$ and, since it is not a nucleon and has very small mass, has $A = 0$.) The particle known as the neutrino, with rest mass $m = 0$ and charge $q = 0$, was not initially detected and was only later hypothesized to exist, as we shall discuss later in this Section. No nucleons are lost when an electron is emitted, and the total number of nucleons, A, is the same in the daughter nucleus as in the parent. But because an electron has been emitted from the nucleus itself, the charge on the daughter nucleus is $+1e$ greater than that on the parent. The parent nucleus had $Z = +6$, so from charge conservation the nucleus remaining behind must have a charge of $+7e$. So the daughter nucleus has $Z = 7$, which is nitrogen.

It must be carefully noted that the electron emitted in β decay is *not* an orbital electron. Instead, the electron is created *within the nucleus itself*. What happens is that one of the neutrons changes to a proton and in the process (to conserve charge) throws off an electron. Indeed, free neutrons actually do decay in this fashion:

$$n \rightarrow p + e^- + \text{a neutrino}.$$

Because of their origin in the nucleus, the electrons emitted in β decay are often referred to as "β particles," rather than as electrons, to remind us of their origin. They are, nonetheless, indistinguishable from orbital electrons.

EXAMPLE 42–6 **Energy release in $^{14}_{6}$C decay.** How much energy is released when $^{14}_{6}$C decays to $^{14}_{7}$N by β emission? Use Appendix D.

SOLUTION The masses given in Appendix D are those of the neutral atom, and we have to keep track of the electrons involved. Assume the parent nucleus has six orbiting electrons so it is neutral, and its mass is 14.003242 u. The daughter, which in this decay is $^{14}_{7}$N, is not neutral since it has the same six electrons circling it but the nucleus has a charge of $+7e$. However, the mass of this daughter with its six electrons, plus the mass of the emitted electron (which makes a total of seven electrons), is just the mass of a neutral nitrogen atom. That is, the total mass in the final state is

(mass of $^{14}_{7}$N nucleus + 6 electrons) + (mass of 1 electron),

and this is equal to

mass of neutral $^{14}_{7}$N (includes 7 electrons),

which, from Appendix D is a mass of 14.003074 u. (Note that the neutrino doesn't contribute to either the mass or charge balance since it has $m = 0$ and $q = 0$.) Hence the mass after decay is 14.003074 u, whereas before decay, it was 14.003242 u. So the mass difference is 0.000168 u, which corresponds to 0.156 MeV or 156 keV.

According to this Example, we would expect the emitted electron to have a kinetic energy of 156 keV. (The daughter nucleus, because its mass is very much larger than that of the electron, recoils with very low velocity and hence gets very little of the kinetic energy.) Indeed, very careful measurements indicate that a few emitted β particles do have kinetic energy close to this calculated value. But the vast majority of emitted electrons have somewhat less energy. In fact, the energy of the emitted electron can be anywhere from zero up to the maximum value as calculated above. This range of electron kinetic energy was found for any β decay. It was as if the law of conservation of energy was being violated, and indeed Bohr actually considered this possibility. Careful experiments indicated that linear momentum and angular momentum also did not seem to be conserved. Physicists were troubled at the prospect of having to give up these laws, which had worked so well in all previous situations. In 1930, Wolfgang Pauli proposed an alternate solution: perhaps a new particle that was very difficult to detect was emitted during β decay in addition to the electron. This hypothesized particle could be carrying off the energy, momentum, and angular momentum required to maintain the conservation laws. This new particle was named the **neutrino**—meaning "little neutral one"—by the great Italian physicist Enrico Fermi (1901–1954; Fig. 42–7), who in 1934 worked out a detailed theory of β decay. (It was Fermi who, in this theory, postulated the existence of the fourth force in nature, which we call the weak nuclear force.) The electron neutrino has zero charge, spin of $\frac{1}{2}\hbar$, and seems to have zero rest mass, although we cannot yet rule out the possibility that it might have a very tiny rest mass. If its rest mass is zero, it is much like a photon in that it is neutral and travels at the speed of light. But the neutrino is far more difficult to detect. In 1956, complex experiments produced further evidence for the existence of the neutrino; but by then, most physicists had already accepted its existence.

The symbol for the neutrino is the Greek letter nu (ν). The correct way of writing the decay of $^{14}_{6}$C is then

$$^{14}_{6}\text{C} \rightarrow {}^{14}_{7}\text{N} + e^- + \bar{\nu}.$$

The bar ($^-$) over the neutrino symbol is to indicate that it is an "antineutrino." (Why this is called an antineutrino rather than simply a neutrino need not concern us now; it is discussed in Chapter 44.)

FIGURE 42–7 Enrico Fermi. Fermi contributed significantly to both theoretical and experimental physics, a feat almost unique in this century.

β^- decay

Many isotopes decay by electron emission. They are always isotopes that have too many neutrons compared to the number of protons. That is, they are isotopes that lie above the stable isotopes plotted in Fig. 42–2. But what about unstable isotopes that have too few neutrons compared to their number of protons—those that fall below the stable isotopes of Fig. 42–2? These, it turns out, decay by emitting a **positron** instead of an electron. A positron (sometimes called an e^+ or β^+ particle) has the same mass as the electron, but it has a positive charge of $+1e$. Because it is so like an electron, except for its charge, the positron is called the **antiparticle**[†] to the electron. An example of a β^+ decay is that of $^{19}_{10}\text{Ne}$:

Positron (β^+) decay

$$^{19}_{10}\text{Ne} \rightarrow {}^{19}_{9}\text{F} + e^+ + \nu,$$

where e^+ (or ^0_1e) stands for a positron. Note that the ν emitted here is a neutrino, whereas that emitted in β^- decay is called an antineutrino. Thus an antielectron (= positron) is emitted with a neutrino, whereas an antineutrino is emitted with an electron; this is discussed in Chapter 44.

We can write β^- and β^+ decay, in general, as follows:

$$^A_Z N \rightarrow {}^{A}_{Z+1}N' + e^- + \bar{\nu} \qquad [\beta^- \text{ decay}]$$

$$^A_Z N \rightarrow {}^{A}_{Z-1}N' + e^+ + \nu, \qquad [\beta^+ \text{ decay}]$$

where N is the parent nucleus and N' is the daughter.

Besides β^- and β^+ emission, there is a third related process. This is **electron capture** (abbreviated EC in Appendix D) and occurs when a nucleus absorbs one of its orbiting electrons. An example is ^7_4Be, which as a result becomes ^7_3Li. The process is written

Electron capture

$$^7_4\text{Be} + e^- \rightarrow {}^{7}_{3}\text{Li} + \nu,$$

or, in general,

$$^A_Z N + e^- \rightarrow {}^{A}_{Z-1}N' + \nu. \qquad [\text{electron capture}]$$

K-capture

Usually it is an electron in the innermost (K) shell that is captured, in which case it is called "K-capture." The electron disappears in the process and a proton in the nucleus becomes a neutron; a neutrino is emitted as a result. This process is inferred experimentally by detection of emitted X-rays (due to electrons jumping down to fill the empty state) of just the proper energy.

In β decay, it is the weak nuclear force that plays the crucial role. The neutrino is unique in that it interacts with matter only via the weak force, which is why it is so hard to detect.

42–6 Gamma Decay

Gamma rays are photons having very high energy. They have their origin in the decay of a nucleus, much like emission of photons by excited atoms. Like an atom, a nucleus itself can be in an excited state. When it jumps down to a lower energy state, or to the ground state, it emits a photon which we call a γ ray. The possible energy levels of a nucleus are much farther apart than those of an atom: on the order of keV or MeV, as compared to a few eV for electrons in an atom. Hence, the emitted photons have energies that can range from a few keV to several MeV. For a given decay, the γ ray always has the same energy. Since a γ ray carries no charge, there is no change in the element as a result of a γ decay.

[†]Discussed in Chapter 44. Briefly, an antiparticle has the same mass as its corresponding particle, but opposite charge.

How does a nucleus get into an excited state? It may occur because of a violent collision with another particle. More commonly, the nucleus remaining after a previous radioactive decay may be in an excited state. A typical example is shown in the energy-level diagram of Fig. 42–8. $^{12}_5$B can decay by β decay directly to the ground state of $^{12}_6$C; or it can go by β decay to an excited state of $^{12}_6$C, which then decays by emission of a 4.4 MeV γ ray to the ground state.

We can write γ decay as

$$^A_Z N^* \rightarrow {}^A_Z N + \gamma, \qquad \text{[γ decay]}$$

where the asterisk means "excited state" of that nucleus.

In some cases, a nucleus may remain in an excited state for some time before it emits a γ ray. The nucleus is then said to be in a **metastable state** and is called an **isomer**.

An excited nucleus can sometimes return to the ground state by another process known as **internal conversion** with no γ ray emitted. In this process, the excited nucleus interacts with one of the orbital electrons and ejects this electron from the atom with the same kinetic energy (minus the binding energy of the electron) that an emitted γ ray would have had.

What, you may wonder, is the difference between a γ ray and an X-ray? They both are electromagnetic radiation (photons) and, though γ rays usually have higher energy than X-rays, their range of energies overlap to some extent. The difference is not intrinsic. We use the term X-ray if the photon is produced by an electron–atom interaction, and γ ray if the photon is produced in a nuclear process.

FIGURE 42–8 Energy-level diagram showing how $^{12}_5$B can decay to the ground state of $^{12}_6$C by β decay (total energy released = 13.4 MeV), or can instead β decay to an excited state of $^{12}_6$C (indicated by *), which subsequently decays to its ground state by emitting a 4.4-MeV γ ray.

Internal conversion

42–7 Conservation of Nucleon Number and Other Conservation Laws

In all three types of radioactive decay, the classical conservation laws hold. Energy, linear momentum, angular momentum, and electric charge are all conserved. These quantities are the same before the decay as after. But a new conservation law is also revealed, the **law of conservation of nucleon number**. According to this law, the total number of nucleons (A) remains constant in any process, although one type can change into the other type (protons into neutrons or vice versa). This law holds in all three types of decay. Table 42–2 gives a summary of α, β, and γ decay.

42–8 Half-Life and Rate of Decay

A macroscopic sample of any radioactive isotope consists of a vast number of radioactive nuclei. These nuclei do not all decay at one time. Rather, they decay one by one over a period of time. This is a random process: we can not predict exactly when a given nucleus will decay. But we can determine, on a probabilistic basis, approximately how many nuclei in a sample will decay over a given time period, by assuming that each nucleus has the same probability of decaying in each second that it exists.

The number of decays ΔN that occur in a very short time interval Δt is then proportional to Δt and to the total number N of radioactive nuclei present:

$$\Delta N = -\lambda N \, \Delta t. \qquad \textbf{(42–4)}$$

In this equation, λ is a constant of proportionality called the **decay constant**, which is different for different isotopes. The greater λ is, the greater the rate of decay and the more "radioactive" that isotope is said to be for a given number of nuclei. The number of decays that occur in the short time interval Δt is designated ΔN

TABLE 42–2 The Three Types of Radioactive Decay

α decay:
$$^A_Z N \rightarrow {}^{A-4}_{Z-2} N' + {}^4_2 He$$

β decay:
$$^A_Z N \rightarrow {}^A_{Z+1} N' + e^- + \bar{\nu}$$
$$^A_Z N \rightarrow {}^A_{Z-1} N' + e^+ + \nu$$
$$^A_Z N + e^- \rightarrow {}^A_{Z-1} N' + \nu \quad [EC]^\dagger$$

γ decay:
$$^A_Z N^* \rightarrow {}^A_Z N + \gamma$$

*Indicates the excited state of a nucleus.
†Electron capture.

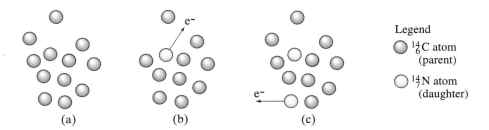

FIGURE 42-9 Radioactive nuclei decay one by one. Hence, the number of parent nuclei in a sample is continually decreasing. When a $^{14}_6C$ nucleus emits the electron, it becomes a $^{14}_7N$ nucleus.

(a)   (b)   (c)

Legend
$^{14}_6C$ atom (parent)
$^{14}_7N$ atom (daughter)

because each decay that occurs corresponds to a decrease by one in the number N of nuclei present. That is, radioactive decay is a "one-shot" process, Fig. 42–9. Once a particular parent nucleus decays into its daughter, it cannot do it again. The minus sign in Eq. 42–4 is needed to indicate that N is decreasing.

If we take the limit $\Delta t \to 0$ in Eq. 42–4, ΔN will be small compared to N, and we can write the equation in infinitesimal form as

$$dN = -\lambda N\, dt. \tag{42-5}$$

We can determine N as a function of t by rearranging this equation to

$$\frac{dN}{N} = -\lambda\, dt$$

and then integrating from $t = 0$ to $t = t$:

$$\int_{N_0}^{N} \frac{dN}{N} = -\int_{0}^{t} \lambda\, dt,$$

where N_0 is the number of parent nuclei present at $t = 0$ and N is the number remaining at time t. The integration gives

$$\ln \frac{N}{N_0} = -\lambda t$$

or

Radioactive decay law

$$N = N_0 e^{-\lambda t}. \tag{42-6}$$

Equation 42–6 is called the **radioactive decay law**. It tells us that the number of radioactive nuclei in a given sample decreases exponentially in time. This is shown in Fig. 42–10a, for the case of $^{14}_6C$ whose decay constant is $\lambda = 3.8 \times 10^{-12}\ \text{s}^{-1}$.

FIGURE 42-10 (a) The number N of parent nuclei in a given sample of $^{14}_6C$ decreases exponentially. (b) The number of decays per second also decreases exponentially. The half-life of $^{14}_6C$ is about 5730 yr, which means that the number of parent nuclei, N, and the rate of decay, dN/dt, decrease by half every 5730 yr.

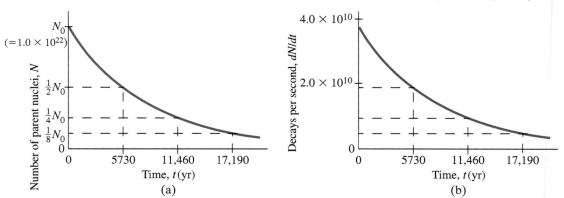

The rate of decay, or number of decays per second, in a pure sample is

$$\frac{dN}{dt}$$

which is called the **activity** of a given sample. From Eqs. 42–5 and 42–6,

$$\frac{dN}{dt} = -\lambda N = -\lambda N_0 e^{-\lambda t}. \qquad \text{(42–7a)}$$

At $t = 0$, the activity is

$$\left(\frac{dN}{dt}\right)_0 = -\lambda N_0. \qquad \text{(42–7b)}$$

Hence

$$\frac{dN}{dt} = \left(\frac{dN}{dt}\right)_0 e^{-\lambda t}, \qquad \text{(42–7c)} \qquad \textit{Activity}$$

so the activity also decreases exponentially in time at the same rate as for N (Fig. 42–10b).

The rate of decay of any isotope is often specified by giving its half-life rather than the decay constant λ. The **half-life** of an isotope is defined as the time it takes for half the original amount of isotope in a given sample to decay. For example, the half-life of $^{14}_6 C$ is about 5730 yr. If at some time a piece of petrified wood contains, say, 1.00×10^{22} $^{14}_6 C$ nuclei, then 5730 yr later it will contain only 0.50×10^{22} nuclei. After another 5730 yr it will contain 0.25×10^{22} nuclei, and so on. This is characteristic of the exponential function, and is shown in Fig. 42–10a. Since the rate of decay dN/dt is proportional to N, it too decreases by a factor of 2 every half-life, Fig. 42–10b.

The half-lives of known radioactive isotopes vary from as short as 10^{-22} s to about 10^{28} s (about 10^{21} yr). The half-lives of many isotopes[†] are given in Appendix D. It should be clear that the half-life (which we designate $T_{\frac{1}{2}}$) bears an inverse relationship to the decay constant. The longer the half-life of an isotope, the more slowly it decays, and hence λ is smaller. The precise relation is obtained from Eq. 42–6 by setting $N = N_0/2$ at $t = T_{\frac{1}{2}}$:

$$\frac{N_0}{2} = N_0 e^{-\lambda T_{\frac{1}{2}}} \qquad \text{or} \qquad e^{\lambda T_{\frac{1}{2}}} = 2.$$

We take natural logs of both sides ("ln" and "e" are inverse operations, meaning $\ln(e^x) = x$) and find

$$\ln(e^{\lambda T_{\frac{1}{2}}}) = \ln 2, \qquad \text{so} \qquad \lambda T_{\frac{1}{2}} = \ln 2$$

and

$$T_{\frac{1}{2}} = \frac{\ln 2}{\lambda} = \frac{0.693}{\lambda}. \qquad \text{(42–8)} \qquad \textit{Half-life}$$

We can then write Eq. 42–6 as

$$N = N_0 e^{-0.693 t / T_{\frac{1}{2}}}.$$

Now let us use these concepts and equations in worked-out Examples.

[†] You may find the *mean life* of an isotope quoted. The mean life τ is defined as $\tau = 1/\lambda$ (see Problem 76), so that Eq. 42–6 can be written $N = N_0 e^{-t/\tau}$ just as for RC and LR circuits (Chapters 26 and 30) where τ is called the time constant. Since

$$\tau = \frac{1}{\lambda} = \frac{T_{\frac{1}{2}}}{0.693}$$

the mean life and half-life differ significantly in numerical value, so confusing them can cause serious error.

EXAMPLE 42–7 Sample activity. The isotope $^{14}_{6}C$ has a half-life of 5730 yr. If at some time a sample contains 1.00×10^{22} carbon-14 nuclei, what is the activity of the sample?

SOLUTION First we calculate the decay constant λ from Eq. 42–8, and obtain

$$\lambda = \frac{0.693}{T_{\frac{1}{2}}} = \frac{0.693}{(5730\,\text{yr})(3.156 \times 10^7\,\text{s/yr})} = 3.83 \times 10^{-12}\,\text{s}^{-1},$$

since the number of seconds in a year is $(60)(60)(24)(365\frac{1}{4}) = 3.156 \times 10^7\,\text{s}$. From Eq. 42–5, the magnitude of the activity or rate of decay is

$$\frac{dN}{dt} = \lambda N = (3.83 \times 10^{-12}\,\text{s}^{-1})(1.00 \times 10^{22})$$

$$= 3.83 \times 10^{10}\,\text{decays/s}.$$

(The unit "decays/s" is often written simply as s^{-1} since "decays" is not a unit but refers only to the number.) Note that the graph of Fig. 42–10b starts at this value, corresponding to the original value of $N = 1.0 \times 10^{22}$ nuclei in Fig. 42–10a.

EXAMPLE 42–8 A sample of radioactive $^{13}_{7}N$. A laboratory has $1.49\,\mu\text{g}$ of pure $^{13}_{7}N$, which has a half-life of 10.0 min (600 s). (a) How many nuclei are present initially? (b) What is the activity initially? (c) What is the activity after 1.00 h? (d) After approximately how long will the activity drop to less than one per second?

SOLUTION (a) Since the atomic mass is 13.0, then 13.0 g will contain 6.02×10^{23} nuclei (Avogadro's number). Since we have only $1.49 \times 10^{-6}\,\text{g}$, the number of nuclei, N_0, that we have initially is given by the ratio

$$\frac{N_0}{1.49 \times 10^{-6}\,\text{g}} = \frac{6.02 \times 10^{23}}{13.0\,\text{g}},$$

so $N_0 = 6.90 \times 10^{16}$ nuclei.
(b) From Eq. 42–8, $\lambda = (0.693)/(600\,\text{s}) = 1.16 \times 10^{-3}\,\text{s}^{-1}$. Then, at $t = 0$ (Eq. 42–7b)

$$\left(\frac{dN}{dt}\right)_0 = \lambda N_0 = (1.16 \times 10^{-3}\,\text{s}^{-1})(6.90 \times 10^{16}) = 8.00 \times 10^{13}\,\text{s}^{-1}.$$

(c) After 1.00 h = 3600 s, the activity will be (Eq. 42–7c)

$$\frac{dN}{dt} = \left(\frac{dN}{dt}\right)_0 e^{-\lambda t}$$

$$= (8.00 \times 10^{13}\,\text{s}^{-1})e^{-(1.16 \times 10^{-3}\,\text{s}^{-1})(3600\,\text{s})} = 1.23 \times 10^{12}\,\text{s}^{-1}.$$

This result can be obtained in another way: since 1.00 h represents six half-lives (6×10.0 min), the activity will decrease to $(\frac{1}{2})(\frac{1}{2})(\frac{1}{2})(\frac{1}{2})(\frac{1}{2})(\frac{1}{2}) = (\frac{1}{2})^6 = \frac{1}{64}$ of its original value, or $(8.00 \times 10^{13}\,\text{s}^{-1})/64 = 1.25 \times 10^{12}\,\text{s}^{-1}$. (The slight discrepancy between the two values arises because we kept only three significant figures.)
(d) We want to determine the time t when $dN/dt = 1.00\,\text{s}^{-1}$. From Eq. 42–7c, we have

$$e^{-\lambda t} = \frac{(dN/dt)}{(dN/dt)_0} = \frac{1.00\,\text{s}^{-1}}{8.00 \times 10^{13}\,\text{s}^{-1}} = 1.25 \times 10^{-14}.$$

We take the natural log (ln) of both sides and divide by λ to find

$$t = -\frac{\ln(1.25 \times 10^{-14})}{\lambda} = \frac{32.0}{1.16 \times 10^{-3}\,\text{s}^{-1}} = 2.76 \times 10^4\,\text{s} = 7.67\,\text{h}.$$

It is often the case that one radioactive isotope decays to another isotope that is also radioactive. Sometimes this daughter decays to yet a third isotope which also is radioactive. Such successive decays are said to form a **decay series**. An important example is illustrated in Fig. 42–11. As can be seen, $^{238}_{92}\text{U}$ decays by α emission to $^{234}_{90}\text{Th}$, which in turn decays by β decay to $^{234}_{91}\text{Pa}$. The series continues as shown, with several possible branches near the bottom. For example, $^{218}_{84}\text{Po}$ can decay either by α decay to $^{214}_{82}\text{Pb}$ or by β decay to $^{218}_{85}\text{At}$. The series ends at the stable lead isotope $^{206}_{82}\text{Pb}$. Other radioactive series also exist.

Because of such decay series, certain radioactive elements are found in nature that otherwise would not be. For when the solar system acquired its present form about 5 billion years ago, it is believed that nearly all nuclides were formed (by the fusion process, Sections 43–4 and 45–2). Many isotopes with short half-lives decayed quickly and no longer exist in nature today. But long-lived isotopes, such as $^{238}_{92}\text{U}$ with a half-life of 4.5×10^9 yr, still do exist in nature today. Indeed, about half of the original $^{238}_{92}\text{U}$ still remains (assuming that the origin of the solar system was about 5×10^9 yr ago). We might expect, however, that radium $\left(^{226}_{88}\text{Ra}\right)$, with a half-life of 1600 yr, would long since have disappeared from the Earth. Indeed, the original $^{226}_{88}\text{Ra}$ nuclei must by now have all decayed. However, because $^{238}_{92}\text{U}$ decays (in several steps) to $^{226}_{88}\text{Ra}$, the supply of $^{226}_{88}\text{Ra}$ is continually replenished, which is why it is still found on Earth today. The same can be said for many other radioactive nuclides.

FIGURE 42–11 Decay series beginning with $^{238}_{92}\text{U}$. Nuclei in the series are specified by a dot representing A and Z values. Half-lives are given in seconds (s), minutes (min), hours (h), days (d), or years (yr). Note that a horizontal arrow represents β decay (A does not change), whereas a diagonal line represents α decay (A changes by 4, Z changes by 2).

CONCEPTUAL EXAMPLE 42–9 | **Decay chain.** The decay chain starting with ^{234}U in Fig. 42–11 has nuclides with half-lives of 250,000 yr, 75,000 yr, 1600 yr, and a little under 4 days, respectively. Each decay in the chain has an alpha particle of a characteristic energy, and so we can monitor the radioactive decay rate of each nuclide. Given a sample that was pure ^{234}U a million years ago, which alpha decay would you expect to have the highest activity rate in the sample?

RESPONSE The first instinct is to say that the process with the shortest half-life would show the highest activity. Surprisingly, however, the activity rates in this sample are all the same! The reason is that in each case the decay of the parent acts as a bottleneck to the decay of the daughter. Compared to the 1600-yr half-life of ^{226}Ra, for example, its daughter ^{222}Rn decays almost immediately, but it cannot decay until it is made. (This is like an automobile assembly line: If worker A takes 20 minutes to do a task and then worker B takes only 1 minute to do the next task, worker B still does only one car every 20 minutes).

| 42–10 | Radioactive Dating |

Radioactive decay has many interesting applications. One is the technique of *radioactive dating* by which the age of ancient materials can be determined.

The age of any object made from once-living matter, such as wood, can be determined using the natural radioactivity of $^{14}_{6}$C. All living plants absorb carbon dioxide (CO_2) from the air and use it to synthesize organic molecules. The vast majority of these carbon atoms are $^{12}_{6}$C, but a small fraction, about 1.3×10^{-12}, is the radioactive isotope $^{14}_{6}$C. The ratio of $^{14}_{6}$C to $^{12}_{6}$C in the atmosphere has remained roughly constant over many thousands of years, in spite of the fact that $^{14}_{6}$C decays with a half-life of about 5730 yr. This is because neutrons in the cosmic radiation that impinges on the Earth from outer-space collide with atoms of the atmosphere. In particular, collisions with nitrogen nuclei produce the following nuclear transformation: $n + {}^{14}_{7}N \rightarrow {}^{14}_{6}C + p$. That is, a neutron strikes and is absorbed by a $^{14}_{7}$N nucleus, and a proton is knocked out in the process. The remaining nucleus is $^{14}_{6}$C. This continual production of $^{14}_{6}$C in the atmosphere roughly balances the loss of $^{14}_{6}$C by radioactive decay. As long as a plant or tree is alive, it continually uses the carbon from carbon dioxide in the air to build new tissue and to replace old. Animals eat plants, so they too are continually receiving a fresh supply of carbon for their tissues. Organisms cannot distinguish[†] $^{14}_{6}$C from $^{12}_{6}$C, and since the ratio of $^{14}_{6}$C to $^{12}_{6}$C in the atmosphere remains nearly constant, the ratio of the two isotopes within the living organism remains nearly constant as well. But when an organism dies, carbon dioxide is no longer absorbed and utilized. Because the $^{14}_{6}$C decays radioactively, the ratio of $^{14}_{6}$C to $^{12}_{6}$C in a dead organism decreases in time. Since the half-life of $^{14}_{6}$C is about 5730 yr, the $^{14}_{6}$C/$^{12}_{6}$C ratio decreases by half every 5730 yr. If, for example, the $^{14}_{6}$C/$^{12}_{6}$C ratio of an ancient wooden tool is half of what it is in living trees, then the object must have been made from a tree that was felled about 5700 years ago. Actually, corrections must be made for the fact that the $^{14}_{6}$C/$^{12}_{6}$C ratio in the atmosphere has not remained precisely constant over time. The determination of what this ratio has been over the centuries has required using techniques such as comparing the expected ratio to the actual ratio for objects whose age is known, such as very old trees whose annual rings can be counted.

[†] Organisms operate almost exclusively via chemical reactions—which involve only the outer orbital electrons of the atom; extra neutrons in the nucleus have essentially no effect.

EXAMPLE 42–10 **An ancient animal.** An animal bone fragment found in an archeological site has a carbon mass of 200 g. It registers an activity of 16 decays/s. What is the age of the bone?

➡ PHYSICS APPLIED
Archeological dating

SOLUTION When the animal was alive, the ratio of $^{14}_6C$ to $^{12}_6C$ in the 200-g piece of bone was 1.3×10^{-12}. The number of $^{14}_6C$ nuclei at that time was

$$N_0 = \left(\frac{6.02 \times 10^{23} \text{ atoms}}{12 \text{ g}} \right)(200 \text{ g})(1.3 \times 10^{-12}) = 1.3 \times 10^{13}.$$

From Eq. 42–7b, considering only the magnitude,

$$\left(\frac{dN}{dt} \right)_0 = \lambda N_0,$$

where $\lambda = 3.83 \times 10^{-12} \text{ s}^{-1}$ (Example 42–7). So the original activity was

$$\left(\frac{dN}{dt} \right)_0 = (3.83 \times 10^{-12} \text{ s}^{-1})(1.3 \times 10^{13}) = 50 \text{ s}^{-1}.$$

From Eq. 42–7c,

$$\frac{dN}{dt} = \left(\frac{dN}{dt} \right)_0 e^{-\lambda t},$$

and we rewrite this as

$$e^{\lambda t} = \left[\frac{(dN/dt)_0}{(dN/dt)} \right].$$

Now we take the natural log (ln) of both sides to get

$$t = \frac{1}{\lambda} \ln \left[\frac{(dN/dt)_0}{(dN/dt)} \right] = \frac{1}{3.83 \times 10^{-12} \text{ s}^{-1}} \ln \left[\frac{50 \text{ s}^{-1}}{16 \text{ s}^{-1}} \right]$$
$$= 2.98 \times 10^{11} \text{ s} = 9400 \text{ yr},$$

which is the time elapsed since the death of the animal.

Carbon dating is useful only for determining the age of objects less than about 60,000 years old. The amount of $^{14}_6C$ remaining in older objects is usually too small to measure accurately, although new techniques are allowing detection of even smaller amounts of $^{14}_6C$, pushing the time frame further back. On the other hand, radioactive isotopes with longer half-lives can be used in certain circumstances to obtain the age of older objects. For example, the decay of $^{238}_{92}U$, because of its long half-life of 4.5×10^9 years, is useful in determining the ages of rocks on a geologic time scale. When molten material solidified into rock as the temperature dropped, different compounds solidified according to the melting points, and thus different compounds separated to some extent. Uranium present in a material became fixed in position and the daughter nuclei that result from the decay of uranium were also fixed in that position. Thus, by measuring the amount of $^{238}_{92}U$ remaining in the material relative to the amount of daughter nuclei, the time when the rock solidified can be determined.

➡ PHYSICS APPLIED
Geological dating

Radioactive dating methods using $^{238}_{92}U$ and other isotopes have shown the age of the oldest Earth rocks to be about 4×10^9 yr. The age of rocks in which the oldest fossilized organisms are embedded indicates that life appeared at least 3 billion years ago. The earliest fossilized remains of mammals are found in rocks 200 million years old, and the first humanlike creatures seem to have appeared about 2 million years ago. Radioactive dating has been indispensable for the reconstruction of Earth's history.

➡ PHYSICS APPLIED
Oldest Earth rocks and earliest life

Individual particles such as electrons, protons, α particles, neutrons, and γ rays are not detected directly by our senses. Consequently, a variety of instruments have been developed to detect them.

One of the most common is the **Geiger counter**. As shown in Fig. 42–12, it consists of a cylindrical metal tube filled with a certain type of gas. A long wire runs down the center and is kept at a high positive voltage $(\approx 10^3 \text{ V})$ with respect to the outer cylinder. The voltage is just slightly less than that required to ionize the gas atoms. When a charged particle enters through the thin "window" at one end of the tube, it ionizes a few atoms of the gas. The freed electrons are attracted toward the positive wire and as they are accelerated they strike and ionize additional atoms. An "avalanche" of electrons is quickly produced, and when it reaches the wire anode, it produces a voltage pulse. The pulse, after being amplified, can be sent to an electronic counter, which counts how many particles have been detected. Or the pulses can be sent to a loudspeaker and each detection of a particle is heard as a "click."

A **scintillation counter** makes use of a solid, liquid, or gas known as a **scintillator** or **phosphor**. The atoms of a scintillator are easily excited when struck by an incoming particle and emit visible light when they return to their ground states. Typical scintillators are crystals of NaI and certain plastics. One face of a solid scintillator is cemented to a photomultiplier tube, and the whole is wrapped with opaque material to keep it light tight or is placed within a light-tight container. The **photomultiplier** (PM) **tube** converts the energy of the scintillator-emitted photon(s) into an electric signal. A PM tube is a vacuum tube containing several electrodes (typically 8 to 14), called *dynodes*, which are maintained at successively higher voltages as shown in Fig. 42–13. At its top surface is a photoelectric surface, called the *photocathode*, whose work function (Section 38–2) is low enough that an electron is easily released when struck by a photon from the scintillator. Such an electron is accelerated toward the first dynode. When it strikes the first dynode, the electron has acquired sufficient kinetic energy so that it can eject two to five more electrons. These, in turn, are accelerated to the second dynode, and a multiplication process begins. The number of electrons striking the last dynode may be 10^6 or more. Thus the passage of a particle through the scintillator results in an electric signal at the output of the PM tube that can be sent to an electronic counter just as for a Geiger tube. Because a scintillator crystal is a solid and therefore much more dense than the gas of a Geiger counter, it is a more efficient detector—especially for γ rays, which interact less with matter than do β rays.

In tracer work (Section 43–8), **liquid scintillators** are often used. Radioactive samples taken at different times or from different parts of an organism are placed directly in small bottles containing the liquid scintillator. This is particularly convenient for detection of β rays from $_{1}^{3}\text{H}$ and $_{6}^{14}\text{C}$, which have very low energies and have difficulty passing through the outer covering of a crystal scintillator or Geiger tube. A PM tube is still used to produce the electric signal.

A **semiconductor detector** consists of a reverse-biased *p-n* junction diode (Sections 41–8 and 9). A particle passing through the junction can excite electrons into the conduction band, leaving holes in the valence band. The freed charges produce a short electrical pulse that can be counted just as for Geiger and scintillation counters.

In Chapter 44 we will discuss additional detectors, those that allow visualization of particle tracks and are used in connection with nuclear reactions and elementary particles.

FIGURE 42–12 Diagram of a Geiger counter.

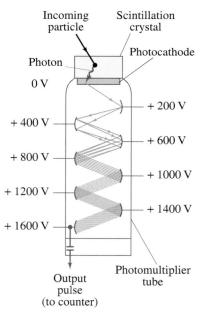

FIGURE 42–13 Scintillation counter with a photomultiplier tube.

Summary

Nuclear physics is the study of atomic nuclei. Nuclei contain **protons** and **neutrons**, which are collectively known as **nucleons**. The total number of nucleons, A, is the **atomic mass number**. The number of protons, Z, is the **atomic number**. The number of neutrons equals $A - Z$. **Isotopes** are nuclei with the same Z, but with different numbers of neutrons. For an element X, an isotope of given Z and A is represented by

$$_Z^A X.$$

The nuclear radius is proportional to $A^{\frac{1}{3}}$, indicating that all nuclei have about the same density. Nuclear masses are specified in **unified atomic mass units** (u), where the mass of $_6^{12}C$ (including its 6 electrons) is defined as exactly 12.000000 u, or in terms of their energy equivalent (because $E = mc^2$), where

$$1\,u = 931.5\,\text{MeV}/c^2 = 1.66 \times 10^{-27}\,\text{kg}.$$

The mass of a stable nucleus is less than the sum of the masses of its constituent nucleons. The difference in mass (times c^2) is the **total binding energy**. It represents the energy needed to break the nucleus into its constituent nucleons. The **binding energy per nucleon** averages about 8 MeV per nucleon, and is lowest for low mass and high mass nuclei.

Unstable nuclei undergo **radioactive decay**; they change into other nuclei with the emission of an α, β, or γ particle. An α particle is a $_2^4$He nucleus; a β particle is an electron or positron; and a γ ray is a high-energy photon. In β decay, a **neutrino** is also emitted. The transformation of the parent into the daughter nucleus is called **transmutation** of the elements. Radioactive decay occurs spontaneously only when the rest mass of the products is less than the mass of the parent nucleus. The loss in mass appears as kinetic energy of the products.

Nuclei are held together by the **strong nuclear force**. The **weak nuclear force** makes itself apparent in β decay. These two forces, plus the gravitational and electromagnetic forces, are the four known types of force. Electric charge, linear and angular momentum, mass–energy, and **nucleon number** are **conserved** in all decays.

Radioactive decay is a statistical process. For a given type of radioactive nucleus, the number of nuclei that decay (ΔN) in a time Δt is proportional to the number N of parent nuclei present:

$$\Delta N = -\lambda N\,\Delta t.$$

The proportionality constant, λ, is called the **decay constant** and is characteristic of the given nucleus. The number N of nuclei remaining after a time t decreases exponentially

$$N \propto e^{-\lambda t}$$

as does the **activity**, dN/dt. The **half-life**, $T_{\frac{1}{2}}$, is the time required for half the nuclei of a radioactive sample to decay. It is related to the decay constant by $T_{\frac{1}{2}} = 0.693/\lambda$.

Questions

1. What do different isotopes of a given element have in common? How are they different?

2. What are the elements represented by the X in the following: (a) $_{92}^{232}X$; (b) $_7^{18}X$; (c) $_1^1X$; (d) $_{38}^{82}X$; (e) $_{97}^{247}X$?

3. How many protons and how many neutrons do each of the isotopes in Question 2 have?

4. Why are the atomic masses of many elements (see the periodic table) not close to whole numbers?

5. How do we know there is such a thing as the strong nuclear force?

6. What are the similarities and the differences between the strong nuclear force and the electric force?

7. What is the experimental evidence in favor of radioactivity being a nuclear process?

8. The isotope $_{29}^{64}Cu$ is unusual in that it can decay by γ, β^-, and β^+ emission. What is the resulting nuclide for each case?

9. A $_{92}^{238}U$ nucleus decays to a nucleus containing how many neutrons?

10. Describe, in as many ways as possible, the difference between α, β, and γ rays.

11. What element is formed by the radioactive decay of (a) $_{11}^{24}Na$ (β^-); (b) $_{11}^{22}Na$ (β^+); (c) $_{84}^{210}Po$ (α)?

12. What element is formed by the decay of (a) $_{15}^{32}P$ (β^-); (b) $_{16}^{35}S$ (β^-); (c) $_{83}^{211}Bi$ (α)?

13. Fill in the missing particle or nucleus:

(a) $_{20}^{45}Ca \rightarrow ? + e^- + \bar{\nu}$ (b) $_{29}^{58}Cu \rightarrow ? + \gamma$

(c) $_{24}^{46}Cr \rightarrow _{23}^{46}V + ?$ (d) $_{94}^{234}Pu \rightarrow ? + \alpha$

(e) $_{93}^{239}Np \rightarrow _{94}^{239}Pu + ?$

14. Explain α decay by tunneling using the uncertainty principle.

15. Immediately after a $_{92}^{238}U$ nucleus decays to $_{90}^{234}Th + _2^4He$, the daughter thorium nucleus still has 92 electrons circling it. Since thorium normally holds only 90 electrons, what do you suppose happens to the two extra ones?

16. When a nucleus undergoes either β^- or β^+ decay, what happens to the energy levels of the atomic electrons? What is likely to happen to these electrons following the decay?

17. The alpha particles from a given alpha-emitting nuclide are monoenergetic; that is, they all have the same kinetic energy. But the beta particles from a beta-emitting nuclide have a spectrum of energies. Explain the difference between these two cases.

18. Do isotopes that undergo electron capture generally lie above or below the line of stability in Fig. 42–2?

19. Can hydrogen or deuterium emit an α particle?

20. Why are many artificially produced radioactive isotopes rare in nature?

21. An isotope has a half-life of one month. After two months, will a given sample of this isotope have completely decayed? If not, how much remains?

22. Describe how the potential energy curve for an α particle in an α-emitting nucleus differs from that for a stable nucleus.

23. Explain the absence of β^+ emitters in the radioactive decay series of Fig. 42–11.

24. Can $^{14}_{6}C$ dating be used to measure the age of stone walls and tablets of ancient civilizations?

25. What assumptions are made in carbon-dating? What do you think could affect these assumptions?

☐ Problems

Section 42–1

1. (I) What is the rest mass of an α particle in MeV/c^2?

2. (I) A pi meson has a mass of $139\ \text{MeV}/c^2$. What is this in atomic mass units?

3. (I) What is the approximate radius of an alpha particle (^4_2He)?

4. (I) By what percentage is the radius of the isotope $^{14}_{6}C$ greater than that of its sister $^{12}_{6}C$?

5. (II) (a) Determine the density of nuclear matter in kg/m^3, and show that it is essentially the same for all nuclei. (b) What would be the radius of the Earth if it had its actual mass but had the density of nuclei? (c) What would be the radius of a $^{238}_{92}U$ nucleus if it had the density of the Earth?

6. (II) (a) What is the approximate radius of a $^{64}_{29}Cu$ nucleus? (b) Approximately what is the value of A for a nucleus whose radius is $3.9 \times 10^{-15}\ \text{m}$?

7. (II) How much energy must an α particle have to just "touch" the surface of a $^{238}_{92}U$ nucleus?

8. (II) If an alpha particle were released from rest near the surface of a $^{243}_{95}Am$ nucleus, what would its kinetic energy be when far away?

9. (II) What stable nucleus has approximately half the radius of a uranium nucleus? [*Hint*: Find A and use Appendix D to get Z.]

Section 42–2

10. (I) Estimate the total binding energy for $^{40}_{20}Ca$, using Fig. 42–1.

11. (I) Use Fig. 42–1 to estimate the total binding energy of (a) $^{238}_{92}U$, and (b) $^{84}_{36}Kr$.

12. (II) Use Appendix D to calculate the binding energy of ^2_1H (deuterium).

13. (II) Calculate the binding energy per nucleon for a $^{14}_{7}N$ nucleus.

14. (II) Determine the binding energy of the last neutron in a $^{40}_{19}K$ nucleus.

15. (II) Calculate the total binding energy, and the binding energy per nucleon, for (a) ^6_3Li, (b) $^{208}_{82}Pb$. Use Appendix D.

16. (II) Calculate the binding energy of (a) the last proton, and (b) the last neutron, in a $^{12}_{6}C$ nucleus. Use Appendix D.

17. (II) Compare the binding energy of a neutron in $^{23}_{11}Na$ to that in $^{24}_{11}Na$.

18. (II) How much energy is required to remove (a) a proton, (b) a neutron, from $^{16}_{8}O$? Explain the difference in your answers.

19. (II) (a) Show that the nucleus ^8_4Be (mass = 8.005305 u) is unstable to decay into two α particles. (b) Is $^{12}_{6}C$ stable against decay into three α particles? Show why or why not.

Sections 42–3 to 42–7

20. (I) How much energy is released when tritium, ^3_1H, decays by β^- emission?

21. (I) What is the maximum kinetic energy of an electron emitted in the β decay of a free neutron?

22. (I) Show that the decay $^{11}_{6}C \rightarrow {}^{10}_{5}B + p$ is not possible because energy would not be conserved.

23. (II) $^{22}_{11}Na$ is radioactive. Is it a β^- or β^+ emitter? Write down the decay reaction, and estimate the maximum kinetic energy of the emitted β.

24. (II) Give the result of a calculation that shows whether or not the following decays are possible:
   (a) $^{236}_{92}U \rightarrow {}^{235}_{92}U + n$; (b) $^{16}_{8}O \rightarrow {}^{15}_{8}O + n$;
   (c) $^{23}_{11}Na \rightarrow {}^{22}_{11}Na + n$.

25. (II) A $^{232}_{92}U$ nucleus emits an α particle with kinetic energy = 5.32 MeV. What is the daughter nucleus and what is the approximate atomic mass (in u) of the daughter atom? Ignore recoil of the daughter nucleus.

26. (II) When $^{23}_{10}Ne$ (mass = 22.9945 u) decays to $^{23}_{11}Na$ (mass = 22.9898 u), what is the maximum kinetic energy of the emitted electron? What is its minimum energy? What is the energy of the neutrino in each case?

27. (II) The nuclide $^{32}_{15}P$ decays by emitting an electron whose maximum kinetic energy can be 1.71 MeV. (a) What is the daughter nucleus? (b) What is its atomic mass (in u)?

28. (II) The isotope $^{218}_{84}Po$ can decay by either α or β^- emission. What is the energy release in each case? The mass of $^{218}_{84}Po$ is 218.008965 u.

29. (II) How much energy is released in electron capture by beryllium: $^7_4Be + {}^{\ 0}_{-1}e \rightarrow {}^7_3Li + \nu$?

30. (II) Decay series, such as that shown in Fig. 42–11, can be classified into four families, depending on whether the mass numbers have the form $4n$, $4n + 1$, $4n + 2$, or $4n + 3$, where n is an integer. Justify this statement and show that for a nuclide in any family, all its daughters will be in the same family.

31. (III) What is the energy of the α particle emitted in the decay $^{210}_{84}Po \rightarrow {}^{206}_{82}Pb + \alpha$? Take into account the recoil of the daughter nucleus.

32. (III) The α particle emitted when $^{238}_{92}U$ decays has 4.20 MeV of kinetic energy. Calculate the recoil kinetic energy of the daughter nucleus and the Q-value of the decay.

33. (III) (a) Show that when a nucleus decays by β^+ decay, the total energy released is equal to

$$(M_P - M_D - 2m_e)c^2,$$

where M_P and M_D are the masses of the parent and daughter atoms (neutral), and m_e is the mass of an electron or positron. (b) Determine the maximum kinetic energy of β^+ particles released when $^{11}_6C$ decays to $^{11}_5B$. What is the maximum energy the neutrino can have? What is its minimum energy?

34. (III) (a) Calculate the kinetic energy of the α particle emitted when $^{236}_{92}U$ decays. (b) Use Eq. 42–1 to estimate the radius of an α particle and $^{232}_{90}Th$ nucleus. Use this to estimate (c) the maximum height of the Coulomb barrier, and (d) its width AB in Fig. 42–6.

Sections 42–8 to 42–10

35. (I) A radioactive material produces 1280 decays per minute at one time, and 6 h later produces 320 decays per minute. What is its half-life?

36. (I) (a) What is the decay constant of $^{238}_{92}U$ whose half-life is 4.5×10^9 yr? (b) The decay constant of a given nucleus is 8.2×10^{-5} s^{-1}. What is its half-life?

37. (I) What is the activity of a sample of $^{14}_6C$ that contains 3.1×10^{20} nuclei?

38. (I) What fraction of a sample of $^{68}_{32}Ge$, whose half-life is about 9 months, will remain after 3.0 yr?

39. (I) How many nuclei of $^{238}_{92}U$ remain in a rock if the activity registers 875 decays per second?

40. (II) What fraction of a sample is left after (a) exactly 4 half-lives, (b) exactly $4\frac{1}{2}$ half-lives?

41. (II) In a series of decays, the nuclide $^{235}_{92}U$ becomes $^{207}_{82}Pb$. How many α and β^- particles are emitted in this series?

42. (II) The iodine isotope $^{131}_{53}I$ is used in hospitals for diagnosis of thyroid function. If 632 μg are ingested by a patient, determine the activity (a) immediately, (b) 1.0 h later when the thyroid is being tested, and (c) 6 months later. Use Appendix D.

43. (II) $^{124}_{55}Cs$ has a half-life of 30.8 s. (a) If we have 9.8 μg initially, how many Cs nuclei are present? (b) How many are present 2.0 min later? (c) What is the activity at this time? (d) After how much time will the activity drop to less than about 1 per second?

44. (II) Calculate the activity of a pure 7.7-μg sample of $^{32}_{15}P$ ($T_{\frac{1}{2}} = 1.23 \times 10^6$ s).

45. (II) The activity of a sample of $^{35}_{16}S$ ($T_{\frac{1}{2}} = 7.56 \times 10^6$ s) is 2.65×10^5 decays per second. What is the mass of the sample?

46. (II) A sample of $^{233}_{92}U$ ($T_{\frac{1}{2}} = 1.59 \times 10^5$ yr) contains 7.50×10^{19} nuclei. (a) What is the decay constant? (b) Approximately how many disintegrations will occur per minute?

47. (II) The activity of a sample drops by a factor of 10 in 8.6 minutes. What is its half-life?

48. (II) A 185-g sample of pure carbon contains 1.3 parts in 10^{12} (atoms) of $^{14}_6C$. How many disintegrations occur per second?

49. (II) A sample of $^{40}_{19}K$ is decaying at a rate of 8.70×10^2 decays/s. What is the mass of the sample?

50. (II) The rubidium isotope $^{87}_{37}Rb$, a β emitter with a half-life of 4.75×10^{10} yr, is used to determine the age of rocks and fossils. Rocks containing fossils of early animals contain a ratio of $^{87}_{38}Sr$ to $^{87}_{37}Rb$ of 0.0160. Assuming that there was no $^{87}_{38}Sr$ present when the rocks were formed, calculate the age of these fossils.

51. (II) Use Fig. 42–11 and calculate the relative decay rates for α decay of $^{218}_{84}Po$ and $^{214}_{84}Po$.

52. (II) ^7_4Be decays with a half-life of about 53 d. It is produced in the upper atmosphere, and filters down onto the Earth's surface. If a plant leaf is detected to have 250 decays/s of ^7_4Be, how long do we have to wait for the decay rate to drop to 10/s? Estimate the initial mass of ^7_4Be on the leaf.

53. (II) Which radioactive isotope of lead is being produced in a reaction where the measured activity of a sample drops to 1.050 percent of its original activity in 4.00 h?

54. (II) An ancient club is found that contains 190 g of carbon and has an activity of 5.0 decays per second. Determine its age assuming that in living trees the ratio of $^{14}C/^{12}C$ atoms is about 1.3×10^{-12}.

55. (II) At $t = 0$, a pure sample of radioactive nuclei contains N_0 nuclei whose decay constant is λ. Determine a formula for the number of daughter nuclei, N_D, as a function of time. Assume the daughter is stable and that $N_D = 0$ at $t = 0$.

56. (III) At $t = 0$, a pure sample of a radioactive nuclide (the parent) contains N_{P0} nuclei whose half-life is T_P. The daughter nuclide is also radioactive, with half-life T_D. (a) Determine the number of daughter nuclei, N_D, as a function of time, assuming that $N_D = 0$ at $t = 0$. (b) Plot N_D versus t for the cases $T_P = T_D$, $T_P = 3T_D$, $T_P = \frac{1}{3}T_D$. [Hint: Eq. 42–5 must be modified.]

General Problems

57. (a) What is the fraction of the hydrogen atom's mass that is in the nucleus? (b) What is the fraction of the hydrogen atom's volume that is occupied by the nucleus? (c) What is the density of nuclear matter? Compare this with water.

58. Using the uncertainty principle and the radius of a nucleus, estimate the minimum possible kinetic energy of a nucleon in, say, iron. Ignore relativistic corrections. [Hint: A particle can have an energy at least as large as its uncertainty.]

59. How much recoil energy does a $^{40}_{19}K$ nucleus get when it emits a 1.46 MeV gamma ray?

60. An old wooden tool is found to contain only 9.0 percent of $^{14}_6C$ that a sample of fresh wood would. How old is the tool?

61. A neutron star consists of neutrons at approximately nuclear density. Estimate, for a 10-km-diameter neutron star, (a) its mass number, (b) its mass (kg), and (c) the acceleration of gravity at its surface.

62. The $^3_1$H isotope of hydrogen, which is called *tritium* (because it contains three nucleons), has a half-life of 12.33 yr. It can be used to measure the age of objects up to about 100 yr. It is produced in the upper atmosphere by cosmic rays and brought to Earth by rain. As an application, determine approximately the age of a bottle of wine whose $^3_1$H radiation is about $\frac{1}{10}$ that present in new wine.

63. Some elementary particle theories (Section 44–11) suggest that the proton may be unstable, with a half-life $\geq 10^{32}$ yr. How long would you expect to wait for one proton in your body to decay (consider that your body is all water)?

64. When water is placed near an intense neutron source, the neutrons can be slowed down by collisions with the water molecules and eventually captured by a hydrogen nucleus to form the stable isotope called deuterium, $^2_1$H, giving off a gamma ray. What is the energy of the gamma ray?

65. How long must you wait (in half-lives) for a radioactive sample to drop to 1.00 percent of its original activity?

66. If the potassium isotope $^{40}_{19}$K gives 60 decays/s in a liter of milk, estimate how much $^{40}_{19}$K and regular $^{39}_{19}$K are in a liter of milk. Use Appendix D.

67. Show, using the decays given in Section 42–5, that the neutrino must have spin $\frac{1}{2}$.

68. In α decay of, say, a $^{226}_{88}$Ra nucleus, show that the nucleus carries away a fraction $1/(1 + A_D/4)$ of the total energy available, where A_D is the mass number of the daughter nucleus. [*Hint*: Use conservation of momentum as well as conservation of energy.] Approximately what percentage of the energy available is thus carried off by the α particle in the case cited?

69. Strontium-90 is produced as a nuclear fission product of uranium in both reactors and atomic bombs. Look at its location in the periodic table to see what other elements it might be similar to chemically, and tell why you think it might be dangerous to ingest. It has too many neutrons, and it decays with a half-life of about 29 yr. How long will we have to wait for the amount of $^{90}_{38}$Sr on the Earth's surface to reach 1 percent of its current level, assuming no new material is scattered about? Write down the decay reaction, including the daughter nucleus. Is the daughter radioactive? If so, write down its decay. Finish the decay scheme until you reach a stable nucleus.

70. The nuclide $^{191}_{76}$Os decays with β^- energy of 0.14 MeV accompanied by γ rays of energy 0.042 MeV and 0.129 MeV. (*a*) What is the daughter nucleus? (*b*) Draw an energy-level diagram showing the ground states of the parent and daughter and excited states of the daughter. To which of the daughter states does β^- decay of $^{191}_{76}$Os occur?

71. Use the uncertainty principle to argue why electrons are unlikely to be found in the nucleus. Use relativity. [*Hint*: A particle can have an energy at least as large as its uncertainty.]

72. Estimate the total binding energy for copper and then estimate the energy, in joules, needed to break a 3.0-g copper penny into its constituent nucleons.

73. Instead of giving atomic masses for nuclides as in Appendix D, some tables give the *mass excess*, Δ, defined as $\Delta = M - A$, where A is the atomic number and M is the mass in u. Determine the mass excess, in u and in MeV/c^2, for: (*a*) $^4_2$He; (*b*) $^{12}_6$C; (*c*) $^{107}_{47}$Ag; (*d*) $^{235}_{92}$U. (*e*) From a glance at Appendix D, can you make a generalization about the sign of Δ as a function of Z or A?

74. (*a*) A 100-gram sample of natural carbon contains the usual fraction of $^{14}_6$C. How long will it take on average before there is only one $^{14}_6$C nucleus left? (*b*) How does the answer in (*a*) change if the sample is 200 grams? What does this tell you about the limits of carbon dating?

75. If the mass of the proton were just a little closer to the mass of the neutron, the following reaction would be possible even at low collision energies:

$$e^- + p \rightarrow n + \nu.$$

Why would this situation be catastrophic? By what percentage would the proton's mass have to be increased to make this reaction possible?

76. (*a*) Show that the *mean life*, or *average lifetime* of a radioactive nuclide, defined as

$$\tau = \frac{\int_0^\infty N(t)t\,dt}{\int_0^\infty N(t)\,dt}$$

is $\tau = 1/\lambda$. (*b*) What fraction of the original number of nuclei remains after one mean life?

77. Two of the naturally occurring radioactive decay sequences start with $^{232}_{90}$Th, and $^{235}_{92}$U. The first five decays of these two sequences are:

$$\alpha, \beta, \beta, \alpha, \alpha$$

and

$$\alpha, \beta, \alpha, \beta, \alpha.$$

With this information determine the resulting intermediate daughter nuclei in each case.

78. What is the ratio of the kinetic energies for an alpha particle and a beta particle if both make tracks with the same radius of curvature in a magnetic field, oriented perpendicular to the paths of the particles.

79. In a rough theoretical model of α decay, the α particle moves back and forth within the parent nucleus at speed v_{in}, making a collision with the Coulomb barrier (Fig. 42–6) every $2R_0/v_{in}$ seconds, where R_0 is the radius of the nucleus. (*a*) Show that the decay constant λ in this model can be written roughly (using Eq. 39–17) as $\lambda = (v_{in}/2R_0)e^{-2GL}$, where L is the "average" barrier thickness (because the barrier is not square), and $G = \sqrt{2m(U_0 - E)}/\hbar^2$. (*b*) Use this result, plus estimates of the barrier height and width (as in Problem 34) to estimate the halflife of $^{236}_{92}$U. For the barrier thickness (Fig. 42–6) try first $L = \frac{1}{2}(R_B - R_0)$, and then (*c*) set $L = \frac{1}{3}(R_B - R_0)$ as an average thickness since the barrier is not square. Choose the potential U_{in} inside the well to be zero. (*d*) What barrier width would provide, in this simple model, the measured half-life of 2×10^7 yr?

Interior of the Tokamak Fusion Test Reactor at Princeton, shown with a scientist inside (in yellow). A plasma of electrons and light nuclei can be heated to high temperatures that rival the Sun. Confining the plasma by magnetic fields has proved difficult, and intense research is needed if the fusion of light nuclei is to fulfill its great promise as a source of abundant and relatively clean power. In this chapter we also discuss fission of large nuclei which is used in present-day nuclear power plants. We also examine nuclear reactions, dosimetry, and uses of radiation including therapy and imaging.

CHAPTER 43

Nuclear Energy; Effects and Uses of Radiation

We continue our study of nuclear physics in this chapter. We begin with a discussion of nuclear reactions, after which we examine the important energy-releasing processes of fission and fusion. The remainder of the chapter deals with the effects of nuclear radiation when it passes through matter, including us, and how radiation is used medically for therapy and diagnosis, including valuable imaging techniques.

43–1 Nuclear Reactions and the Transmutation of Elements

When a nucleus undergoes α or β decay, the daughter nucleus is that of a different element from the parent. The transformation of one element into another, called *transmutation*, also occurs by means of nuclear reactions. A **nuclear reaction** is said to occur when a given nucleus is struck by another nucleus, or by a simpler particle such as a γ ray or neutron, so that an interaction takes place. Ernest Rutherford was the first to report seeing a nuclear reaction. In 1919 he observed that some of the α particles passing through nitrogen gas were absorbed and protons emitted. He concluded that nitrogen nuclei had been transformed into oxygen nuclei via the reaction

$$^4_2\text{He} + {}^{14}_7\text{N} \rightarrow {}^{17}_8\text{O} + {}^1_1\text{H},$$

where ^4_2He is an α particle, and ^1_1H is a proton.

1085

Since then, a great many nuclear reactions have been observed. Indeed, many of the radioactive isotopes used in the laboratory are made by means of nuclear reactions. Nuclear reactions can be made to occur in the laboratory, but they also occur regularly in nature. In Chapter 42 we saw an example of this: $^{14}_6C$ is continually being made in the atmosphere via the reaction $n + {}^{14}_7N \rightarrow {}^{14}_6C + p$.

Nuclear reactions are sometimes written in a shortened form: for example, the reaction

$$n + {}^{14}_7N \rightarrow {}^{14}_6C + p$$

is written

$$^{14}_7N\,(n, p)\,^{14}_6C.$$

The symbols outside the parentheses on the left and right represent the initial and final nuclei, respectively. The symbols inside the parentheses represent the bombarding particle (first) and the emitted small particle (second).

In any nuclear reaction, both electric charge and nucleon number are conserved. These conservation laws are valuable in "balancing" nuclear reactions, as the following Example shows.

Deuterium

CONCEPTUAL EXAMPLE 43–1 | **Deuterium reaction.** A neutron is observed to strike an $^{16}_8O$ nucleus and a deuteron is given off. (A **deuteron** is the nucleus of **deuterium**, the isotope of hydrogen containing one proton and one neutron, ^2_1H; it is sometimes given the symbol d or D.) What is the nucleus that results?

RESPONSE We have the reaction $n + {}^{16}_8O \rightarrow ? + {}^2_1H$. The total number of nucleons initially is $1 + 16 = 17$, and the total charge is $0 + 8 = 8$. The same totals apply to the right side of the reaction. Hence the product nucleus must have $Z = 7$ and $A = 15$. From the periodic table, we find that it is nitrogen that has $Z = 7$, so the nucleus produced is $^{15}_7N$. The reaction can be written $^{16}_8O\,(n, d)\,^{15}_7N$, where d represents deuterium, ^2_1H.

Energy (as well as momentum) is conserved in nuclear reactions, and we can use this to determine whether a given reaction can occur or not. For example, if the total mass of the products is less than the total mass of the initial particles, this decrease in mass (recall $E = mc^2$) appears as kinetic energy (K) of the outgoing particles. But if the total mass of the products is greater than the total mass of the initial reactants, the reaction requires an energy input. The reaction will then not occur unless the bombarding particle has sufficient kinetic energy. Consider a nuclear reaction of the general form

$$a + X \rightarrow Y + b,$$

where a is a projectile particle (or small nucleus) that strikes nucleus X, producing nucleus Y and particle b (typically, p, n, α, γ). We define the **reaction energy**, or **Q-value**, in terms of the masses involved, as

Q-value
$$Q = (M_a + M_X - M_b - M_Y)c^2. \tag{43–1}$$

Since energy is conserved, Q is equal to the change in kinetic energy (final minus initial):

$$Q = K_b + K_Y - K_a - K_X. \tag{43–2}$$

Normally, $K_X = 0$ since X is the target nucleus at rest (or nearly so) struck by an incoming particle a. For $Q > 0$, the reaction is said to be *exothermic* or *exoergic*. Energy is released in the reaction, so the total kinetic energy is greater after the reaction than before. If Q is negative $(Q < 0)$, the reaction is said to be *endothermic* or *endoergic*. In this case the final total kinetic energy is less than the initial kinetic energy, and an energy input is required to make the reaction happen. The energy input comes from the kinetic energy of the initial colliding particles (a and X).

EXAMPLE 43–2 **A slow-neutron reaction.** The nuclear reaction

$$n + {}^{10}_{5}B \rightarrow {}^{7}_{3}Li + {}^{4}_{2}He$$

is observed to occur even when very slow-moving neutrons $(M_n = 1.0087\,u)$ strike a boron atom at rest. For a particular reaction in which $K_n \approx 0$, the helium $(M_{He} = 4.0026\,u)$ is observed to have a speed of $9.30 \times 10^6\,m/s$. Determine (*a*) the kinetic energy of the lithium $(M_{Li} = 7.0160\,u)$, and (*b*) the Q-value of the reaction.

SOLUTION (*a*) Since the neutron and boron are both essentially at rest, the total momentum before the reaction is zero, and afterward is also zero. Therefore,

$$M_{Li} v_{Li} = M_{He} v_{He}.$$

We solve this for v_{Li} and substitute it into the equation for kinetic energy. We can use classical kinetic energy with little error, rather than relativistic formulas, because $v_{He} = 9.30 \times 10^6\,m/s$ is not close to the speed of light c, and v_{Li} will be even less since $M_{Li} > M_{He}$. Thus we can write:

$$K_{Li} = \frac{1}{2} M_{Li} v_{Li}^2 = \frac{1}{2} M_{Li} \left(\frac{M_{He} v_{He}}{M_{Li}} \right)^2 = \frac{M_{He}^2 v_{He}^2}{2 M_{Li}}.$$

We put in numbers, changing the mass in u to kg and recalling that $1.60 \times 10^{-13}\,J = 1\,MeV$:

$$K_{Li} = \frac{(4.0026\,u)^2 (1.66 \times 10^{-27}\,kg/u)^2 (9.30 \times 10^6\,m/s)^2}{2(7.0160\,u)(1.66 \times 10^{-27}\,kg/u)}$$

$$= 1.64 \times 10^{-13}\,J = 1.02\,MeV.$$

(*b*) We are given the data $K_a = K_X = 0$ in Eq. 43–2, so $Q = K_{Li} + K_{He}$, where

$$K_{He} = \tfrac{1}{2} M_{He} v_{He}^2 = \tfrac{1}{2} (4.0026\,u)(1.66 \times 10^{-27}\,kg/u)(9.30 \times 10^6\,m/s)^2$$

$$= 2.87 \times 10^{-13}\,J = 1.80\,MeV.$$

Hence, $Q = 1.02\,MeV + 1.80\,MeV = 2.82\,MeV.$

EXAMPLE 43–3 **Will the reaction "go"?** Can the reaction ${}^{13}_{6}C\,(p, n)\,{}^{13}_{7}N$ occur when ${}^{13}_{6}C$ is bombarded by 2.0-MeV protons?

SOLUTION We use Eq. 43–1, looking up the masses of the nuclei in Appendix D. The total masses before and after the reaction are:

Before	After
$M({}^{13}_{6}C) = 13.003355$	$M({}^{13}_{7}N) = 13.005739$
$M({}^{1}_{1}H) = 1.007825$	$M(n) = 1.008665$
14.011180	14.014404

(We must use the mass of the ${}^{1}_{1}H$ atom rather than that of the bare proton because the masses of ${}^{13}_{6}C$ and ${}^{13}_{7}N$ include the electrons, and we must include an equal number of electrons on each side of the equation since none are created or destroyed.) The products have an excess mass of $(14.014404 - 14.011180)u = 0.003224\,u \times 931.5\,MeV/u = 3.00\,MeV$. Thus $Q = -3.00\,MeV$, and the reaction is endoergic. This reaction requires energy, and the 2.0 MeV protons do not have enough to make it go.

The proton in this last Example would have to have somewhat more than 3.00 MeV of kinetic energy to make this reaction go; 3.00 MeV would be enough to conserve

(a) Neutron captured by $^{238}_{92}U$.

(b) $^{239}_{92}U$ decays by β decay to neptunium-239.

(c) $^{239}_{93}Np$ itself decays by β decay to produce plutonium-239.

FIGURE 43–1 Neptunium and plutonium are produced in this series of reactions, after bombardment of $^{238}_{92}U$ by neutrons.

FIGURE 43–2 Projectile particles fall on a target of area A and thickness t made up of n nuclei per unit volume.

energy, but a proton of this energy would produce the $^{13}_{7}N$ and n with no kinetic energy and hence no momentum. Since an incident 3.0-MeV proton has momentum, conservation of momentum would be violated. A more complicated calculation shows that to conserve both energy and momentum, the minimum proton energy, called the **threshold energy**, is 3.23 MeV in this case (see Problem 14).

The artificial transmutation of elements took a great leap forward in the 1930s when Enrico Fermi realized that neutrons would be the most effective projectiles for causing nuclear reactions and in particular for producing new elements. Because neutrons have no net electric charge, they are not repelled by positively charged nuclei as are protons or alpha particles. Hence the probability of a neutron reaching the nucleus and causing a reaction is much greater than for charged projectiles,[†] particularly at low energies. Between 1934 and 1936, Fermi and his co-workers in Rome produced many previously unknown isotopes by bombarding different elements with neutrons. Fermi realized that if the heaviest known element, uranium, were bombarded with neutrons, it might be possible to produce new elements whose atomic numbers were greater than that of uranium. After several years of hard work, it was suspected that two new elements had been produced, neptunium ($Z = 93$) and plutonium ($Z = 94$). The full confirmation that such "transuranic" elements could be produced came several years later at the University of California, Berkeley. The reactions are shown in Fig. 43–1.

It was soon shown that what Fermi actually had observed when he bombarded uranium was an even stranger process—one that was destined to play an extraordinary role in the world at large. We discuss it in Section 43–3.

43–2 | Cross Section

Some reactions have a higher probability of occurring than others. The reaction probability is specified by a quantity called the *cross section*. Although the size of a nucleus, like that of an atom, is not a clearly defined quantity since the edges are not distinct like those of a tennis ball or baseball, we can nonetheless define a **cross section** for nuclei undergoing collisions by using an analogy. Suppose that projectile particles strike a stationary target of total area A and thickness t, as shown in Fig. 43–2. Suppose the target is made up of identical objects (such as marbles or nuclei), each of which has a cross-sectional area σ. We assume that the objects are fairly far apart, so we don't have to worry about overlapping; this is often a reasonable assumption because nuclei have diameters on the order of 10^{-14} m but are at least 10^{-10} m (atomic size) apart even in solids. If there are n such objects per unit volume, the total area of all these tiny targets is

$$A' = nAt\sigma$$

since $nAt = (n)(\text{volume})$ is the total number of targets and σ is the area of each. If $A' \ll A$, most of the incident projectile particles will pass through the target without colliding. If R_0 is the rate at which the projectile particles strike the target (number/second), the rate at which collisions occur, R, is

$$R = R_0 \frac{A'}{A} = R_0 \frac{nAt\sigma}{A}$$

so

$$R = R_0 nt\sigma.$$

Projectiles

[†] That is, positively charged particles. Electrons rarely cause nuclear reactions because they do not interact via the strong nuclear force.

Thus, by measuring the collision rate, R, we can determine σ:

$$\sigma = \frac{R}{R_0 nt}. \qquad\qquad \textbf{(43–3)}$$

Cross section

If nuclei were simple billiard balls, and R the number of particles that are deflected per second, σ would represent the cross-sectional area of each ball. But nuclei are complicated objects that cannot be considered to have distinct boundaries. Furthermore, collisions can be either elastic or inelastic, and reactions can occur in which the nature of the particles can change. By measuring R for each possible process, we can determine an **effective cross section**, σ, for each process. None of these cross sections is necessarily related to a geometric cross-sectional area. Rather, σ is an "effective" target area. It is a *measure of the probability of a collision or of a particular reaction occurring* per target nucleus, independent of the dimensions of the entire target. The concept of cross section is useful because σ depends only on the properties of the interacting particles, whereas R depends on the thickness and area of the physical (macroscopic) target, on the number of particles in the incident beam, and so on.

We can define the elastic cross section σ_{el}, using Eq. 43–3, where R for a given experimental setup is the rate of elastic collisions (or **elastic scattering**), by which we mean collisions for which the final particles are the same as the initial particles ($a = b, X = Y$) and $Q = 0$. Similarly, the inelastic cross section, σ_{inel}, is related to the rate of inelastic collisions, or **inelastic scattering**, which involves the same final and initial particles but $Q \neq 0$, usually because excited states are involved. Then for each possible reaction there is a particular cross section. For protons (p) incident on $^{13}_{6}C$, for example, we could have various reactions, such as $p + {}^{13}_{6}C \rightarrow {}^{13}_{7}N + n$ or $p + {}^{13}_{6}C \rightarrow {}^{10}_{5}B + {}^{4}_{2}He$, and so on. The sum of all the separate reaction cross sections is called the **total reaction cross section**, σ_R. The **total cross section**, σ_T, is

Elastic scattering

Inelastic scattering

$$\sigma_T = \sigma_{el} + \sigma_{inel} + \sigma_R$$

Total cross section

and is a measure of all possible interactions or collisions. Said another way, σ_T is a measure of how many particles interact in some way and hence are eliminated from the incident beam. It is also possible to define "differential cross sections," which represent the probability of the deflected (or emitted) particles leaving at particular angles.

It is said that when one of the first nuclear cross sections was measured, a physicist, surprised that it was as large as it was ($\approx 10^{-28}\,m^2$), remarked, "it's as big as a barn." Ever since then nuclear cross sections have been measured in "barns," where 1 barn (bn) $= 10^{-28}\,m^2$.

The barn

The value of σ for a given reaction depends on, among other things, the incident kinetic energy. Typical nuclear cross sections are on the order of barns, but they can vary from millibarns to 1000 bn or more. Figure 43–3 shows the cross section for neutron capture in cadmium ($n + {}^{114}Cd \rightarrow {}^{115}Cd + \gamma$) as a function of neutron kinetic energy. Neutron cross sections for most materials are greater at low energies, as in Fig. 43–3. To produce nuclear reactions at a high rate it is therefore desirable that the bombarding neutrons have low energy. Neutrons that have been slowed down and have reached equilibrium with matter at room temperature ($\frac{3}{2}kT \approx 0.04\,eV$ at $T = 300\,K$) are called **thermal neutrons**.

Thermal neutrons

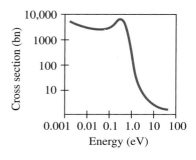

FIGURE 43–3 The neutron cross section for cadmium, which is extraordinarily large for $K \lesssim 1\,eV$. Note that both scales are logarithmic.

EXAMPLE 43–4 Using cross section. The reaction

$$p + {}^{56}_{26}Fe \rightarrow {}^{56}_{27}Co + n$$

has a cross section of 0.65 bn for a particular incident proton energy. Suppose the iron target has an area of 1.5 cm², and is 2.0 μm thick. The density of iron is 7.8×10^3 kg/m³. If the protons are incident at a rate of 2.0×10^{13} particles/s, calculate the rate at which neutrons are produced.

SOLUTION We use Eq. 43–3 in the form:

$$R = R_0 n t \sigma.$$

Here $R_0 = 2.0 \times 10^{13}$ particles/s, $t = 2.0 \times 10^{-6}$ m, and $\sigma = 0.65$ bn. Recalling from Chapter 17 that one mole (mass = 56 g for iron) contains 6.02×10^{23} molecules, then the number of iron atoms per unit volume is

$$n = \frac{(7.8 \times 10^3 \text{ kg/m}^3)(6.02 \times 10^{23} \text{ atoms/mole})}{(56 \times 10^{-3} \text{ kg/mole})} = 8.4 \times 10^{28} \text{ atoms/m}^3.$$

Then we find that the rate at which neutrons are produced is

$$R = (2.0 \times 10^{13} \text{ particles/s})(8.4 \times 10^{28} \text{ atoms/m}^3)(2.0 \times 10^{-6} \text{ m})(0.65 \times 10^{-28} \text{ m}^2)$$
$$= 2.2 \times 10^8 \text{ particles/s}.$$

FIGURE 43–4 Fission of a ${}^{235}_{92}U$ nucleus after capture of a neutron, according to the liquid-drop model.

${}^{235}_{92}U$

(a)

(b) ${}^{236}U$ (excited)

Liquid-drop model

(c)

Fission fragments

N_1 N_2

(d)

43–3 Nuclear Fission; Nuclear Reactors

In 1938, the German scientists Otto Hahn and Fritz Strassmann made an amazing discovery. Following up on Fermi's work, they found that uranium bombarded by neutrons sometimes produced smaller nuclei that were roughly half the size of the original uranium nucleus. Lise Meitner and Otto Frisch, two refugees from Nazi Germany working in Scandinavia, quickly realized what had happened: the uranium nucleus, after absorbing a neutron, actually had split into two roughly equal pieces. This was startling, for until then the known nuclear reactions involved knocking out only a tiny fragment (for example, n, p, or α) from a nucleus.

This new phenomenon was named **nuclear fission** because of its resemblance to biological fission (cell division). It occurs much more readily for ${}^{235}_{92}U$ than for the more common ${}^{238}_{92}U$. The process can be visualized by imaging the uranium nucleus to be like a liquid drop. According to this **liquid-drop model**, the neutron absorbed by the ${}^{235}_{92}U$ nucleus gives the nucleus extra internal energy (like heating a drop of water). This intermediate state, or **compound nucleus**, is ${}^{236}_{92}U$ (because of the absorbed neutron). The extra energy of this nucleus—it is in an excited state— appears as increased motion of the individual nucleons, which causes the nucleus to take on abnormal elongated shapes, Figure 43–4b. When the nucleus elongates (in this model) into the shape shown in Fig. 43–4c, the attraction of the two ends via the short-range nuclear force is greatly weakened by the increased separation distance, and the electric repulsive force becomes dominant. So the nucleus splits in two. The two resulting nuclei, N_1 and N_2, are called **fission fragments**, and in the process a number of neutrons (typically two or three) are also given off. The reaction can be written

$$n + {}^{235}_{92}U \rightarrow {}^{236}_{92}U \rightarrow N_1 + N_2 + \text{neutrons}. \tag{43–4}$$

The compound nucleus, ${}^{236}_{92}U$, exists for less than 10^{-12} s, so the process occurs very quickly. The two fission fragments more often split the original uranium mass as about 40 percent–60 percent rather than precisely half and half. A typical fission reaction is

$$n + {}^{235}_{92}U \rightarrow {}^{141}_{56}Ba + {}^{92}_{36}Kr + 3n, \qquad \textbf{(43–5)}$$

although many others also occur. Figure 43–5 shows the distribution of fission fragments according to mass. Note that only rarely $(\sim 1$ in $10^4)$ does a fission result in equal mass fragments (small arrow in Fig. 43–5).

A tremendous amount of energy is released in a fission reaction because the mass of ${}^{235}_{92}U$ is considerably greater than the total mass of the fission fragments plus neutrons. This can be seen from the binding-energy-per-nucleon curve of Fig. 42–1; the binding energy per nucleon for uranium is about 7.6 MeV/nucleon, but for fission fragments that have intermediate mass (in the center portion of the graph, $A \approx 100$), the average binding energy per nucleon is about 8.5 MeV/nucleon. Since the fission fragments are more tightly bound, they have less mass. The difference in mass, or energy, between the original uranium nucleus and the fission fragments is about $8.5 - 7.6 = 0.9$ MeV per nucleon. Since there are 236 nucleons involved in each fission, the total energy released per fission is

$$(0.9 \text{ MeV/nucleon})(236 \text{ nucleons}) \approx 200 \text{ MeV}.$$

This is an enormous amount of energy for one single nuclear event. At a practical level, the energy from one fission is, of course, tiny. But if many such fissions could occur in a short time, an enormous amount of energy at the macroscopic level would be available. A number of physicists, including Fermi, recognized that the neutrons released in each fission (Eqs. 43–4 and 5) could be used to create a **chain**   *Chain reaction* **reaction**. That is, one neutron initially causes one fission of a uranium nucleus; the two or three neutrons released can go on to cause additional fissions, so the process multiplies as shown schematically in Fig. 43–6. If a **self-sustaining chain reaction** was actually possible in practice, the enormous energy available in fission could be released on a larger scale. Fermi and his co-workers (at the University of

FIGURE 43–5 Mass distribution of fission fragments from ${}^{235}_{92}U$ + n. The small arrow indicates equal mass fragments $(\frac{1}{2} \times (236 - 2) = 117)$. Note that the vertical scale is logarithmic.

FIGURE 43–6 Chain reaction.

FIGURE 43–7 Color painting of the first nuclear reactor, built by Fermi under the grandstand of Stagg Field at the University of Chicago. (There are no photographs of the original reactor because of military secrecy.) Natural uranium was used with graphite as moderator. On December 2, 1942, Fermi slowly withdrew the cadmium control rods and the reactor went critical. This first self-sustaining chain reaction was announced to Washington, by telephone, by Arthur Compton who witnessed the event and reported: "The Italian navigator has just landed in the new world."

Moderator

FIGURE 43–8 If the amount of uranium exceeds the critical mass, as in (b), a sustained chain reaction is possible. If the mass is less than critical, as in (a), most neutrons escape before additional fissions occur, and the chain reaction is not sustained.

(a)

(b)

Critical mass

Chicago) showed it was possible by constructing the first **nuclear reactor** in 1942 (Fig. 43–7).

Several problems have to be overcome to make any nuclear reactor function. First, the probability that a $^{235}_{92}U$ nucleus will absorb a neutron is large only for slow neutrons, but the neutrons emitted during a fission, and which are needed to sustain a chain reaction, are moving very fast. A substance known as a **moderator** must be used to slow down the neutrons. The most effective moderator will consist of atoms whose mass is as close as possible to that of the neutrons. (To see why this is true, recall from Chapter 9, especially Examples 9–7 and 9–8, that a billiard ball striking an equal mass ball at rest can itself be stopped in one collision; but a billiard ball striking a heavy object bounces off with nearly the same speed it had.) The best moderator would thus contain ^1_1H atoms. Unfortunately, ^1_1H tends to absorb neutrons. But the isotope of hydrogen called *deuterium*, ^2_1H, does not absorb many neutrons and is thus an almost ideal moderator. Either ^1_1H or ^2_1H can be used in the form of water. In the latter case, it is **heavy water**, in which the hydrogen atoms have been replaced by deuterium. Another common moderator is *graphite*, which consists of $^{12}_6C$ atoms.

A second problem is that the neutrons produced in one fission may be absorbed and produce other nuclear reactions with other nuclei in the reactor, rather than produce further fissions. In a "light-water" reactor, the ^1_1H nuclei absorb neutrons, as does $^{238}_{92}U$ to form $^{239}_{92}U$ in the reaction $n + ^{238}_{92}U \rightarrow ^{239}_{92}U + \gamma$. Naturally occurring uranium[†] contains 99.3 percent $^{238}_{92}U$ and only 0.7 percent fissionable $^{235}_{92}U$. To increase the probability of fission of $^{235}_{92}U$ nuclei, natural uranium can be **enriched** to increase the percentage of $^{235}_{92}U$ using processes such as diffusion or centrifugation. Enrichment is not usually necessary for reactors using heavy water as moderator since heavy water doesn't absorb neutrons.

The third problem is that some neutrons will escape through the surface of the reactor core before they cause further fissions (Fig. 43–8). Thus the mass of fuel must be sufficiently large for a self-sustaining chain reaction to take place. The minimum mass of uranium needed is called the **critical mass**. The value of the critical mass depends on the moderator, the fuel ($^{239}_{94}Pu$ may be used instead of $^{235}_{92}U$), and how much the fuel is enriched, if at all. Typical values are on the order of a few kilograms (that is, not grams nor thousands of kilograms).

To have a self-sustaining chain reaction, it is clear that on the average at least one neutron produced in each fission must go on to produce another fission. The

[†] $^{238}_{92}U$ will fission, but only with fast neutrons ($^{238}_{92}U$ is more stable than $^{235}_{92}U$). The probability of absorbing a fast neutron and producing a fission is too low to produce a self-sustaining chain reaction.

average number of neutrons per each fission that do go on to produce further fissions is called the **multiplication factor**, f. For a self-sustaining chain reaction, we must have $f \geq 1$. If $f < 1$, the reactor is "subcritical." If $f > 1$, it is "supercritical." Reactors are equipped with movable **control rods** (usually of cadmium or boron), whose function is to absorb neutrons and maintain the reactor at just barely "critical," $f = 1$. The release of neutrons and subsequent fissions caused by them occurs so quickly that manipulation of the control rods to maintain $f = 1$ would not be possible if it weren't for the small percentage (~1 percent) of so-called **delayed neutrons**. They come from the decay of neutron-rich fission fragments (or their daughters) having lifetimes on the order of seconds—sufficient to allow enough reaction time to operate the control rods and maintain $f = 1$.

Critical reaction

Delayed neutrons

Types of nuclear reactor

Nuclear reactors have been built for use in research and to produce electric power. Fission produces many neutrons and a "research reactor" is basically an intense source of neutrons. These neutrons can be used as projectiles in nuclear reactions to produce nuclides not found in nature, including isotopes used as tracers and for therapy. A "power reactor" is used to produce electric power. The energy released in the fission process appears as heat, which is used to boil water and produce steam to drive a turbine connected to an electric generator (Fig. 43–9). The **core** of a nuclear reactor consists of the fuel and a moderator (water in most U.S. commercial reactors). The fuel is usually uranium enriched so that it contains 2 to 4 percent $^{235}_{92}\text{U}$. Water at high pressure or other liquid (such as liquid sodium) is allowed to flow through the core. The thermal energy it absorbs is used to produce steam in the heat exchanger, so the fissionable fuel acts as the heat input for a heat engine (Chapter 20).

Many problems are associated with nuclear power plants. Besides the usual thermal pollution associated with any heat engine (page 525), there is the problem of disposal of the radioactive fission fragments produced in the reactor, plus radioactive nuclides produced by neutrons interacting with the structural parts of the reactor. Fission fragments, like their uranium or plutonium parents, have about 50 percent more neutrons than protons. Nuclei with atomic number in the typical range for fission fragments ($Z \approx 30$ to 60) are stable only if they have more nearly equal numbers of protons and neutrons (see Fig. 42–2). Hence the highly neutron-rich fission fragments are very unstable and decay radioactively. The accidental release of highly radioactive fission fragments into the atmosphere poses a serious threat to human health (Section 43–5), as does possible leakage of the radioactive

FIGURE 43–9 A nuclear reactor. The heat generated by the fission process in the fuel rods is carried off by hot water or liquid sodium and is used to boil water to steam in the heat exchanger. The steam drives a turbine to generate electricity and is then cooled in the condenser.

FIGURE 43–10 Devastation around Chernobyl in Russia, after the nuclear power plant disaster in 1986.

wastes when they are disposed of. The accidents at Three Mile Island (1979) and at Chernobyl (1986) have illustrated some of the dangers and have shown that nuclear plants must be constructed, maintained, and operated with great care and precision (Fig. 43–10). Finally, the lifetime of nuclear power plants is limited to 30-some years, due to buildup of radioactivity and the fact that the structural materials themselves are weakened by the intense conditions inside. Decommissioning of a power plant could take a number of forms, but the cost of any method of decommissioning a large plant will be very great.

So-called **breeder reactors** were proposed as a solution to the problem of limited supplies of fissionable uranium. A breeder reactor is one in which some of the neutrons produced in the fission of $^{235}_{92}U$ are absorbed by $^{238}_{92}U$, and $^{239}_{94}Pu$ is produced via the set of reactions shown in Fig. 43–1. $^{239}_{94}Pu$ is fissionable with slow neutrons, so after separation it can be used as a fuel in a nuclear reactor. Thus a breeder reactor "breeds" new fuel[†] $\left(^{239}_{94}Pu\right)$ from otherwise useless $^{238}_{92}U$. Since natural uranium is 99.3 percent $^{238}_{92}U$, this means that the supply of fissionable fuel could be increased by more than a factor of 100. But breeder reactors not only have the same problems as other reactors, but in addition present other serious problems. Not only is plutonium considered by many to be a serious health hazard in itself (radioactive with a half-life of 24,000 years), but plutonium produced in a reactor can readily be used in a bomb. Thus the use of a breeder reactor, even more than a conventional uranium reactor, presents the danger of nuclear proliferation, and the possibility of theft of fuel by terrorists who could produce a bomb.

It is clear that nuclear power presents many risks. Other large-scale energy-conversion methods, such as conventional coal-burning steam plants, also present health and environmental hazards, some of which were discussed in Chapter 20, including air pollution, oil spills, and the release of CO_2 gas which may be trapping heat as in a greenhouse and raising the Earth's temperature. The solution to the world's needs for energy is not only technological, but economic and political as well. A major factor surely is to "conserve"—to not waste energy and use as little as possible.

EXAMPLE 43–5 **ESTIMATE** **Uranium fuel amount.** Estimate the minimum amount of $^{235}_{92}U$ that needs to undergo fission in order to run a 1000-MW power reactor per year of continuous operation. Assume an efficiency (Chapter 20) of about 33 percent.

SOLUTION For 1000 MW output, the total power generation is 3000 MW, of which 2000 MW is dumped as "waste" heat. Thus the total energy release in 1 yr $(3 \times 10^7\,\text{s})$ from fission is about

$$(3 \times 10^9\,\text{J/s})(3 \times 10^7\,\text{s}) \approx 10^{17}\,\text{J}.$$

If each fission releases 200 MeV of energy, the number of fissions required is

$$\frac{(10^{17}\,\text{J})}{(2 \times 10^8\,\text{eV/fission})(1.6 \times 10^{-19}\,\text{J/eV})} \approx 3 \times 10^{27}\ \text{fissions}.$$

The mass of a single uranium atom is about $(235\,\text{u})(1.66 \times 10^{-27}\,\text{kg/u}) \approx 4 \times 10^{-25}\,\text{kg}$, so the total mass needed is $(4 \times 10^{-25}\,\text{kg})(3 \times 10^{27}\ \text{fissions}) \approx 1000\,\text{kg}$, or about a ton. Since $^{235}_{92}U$ is only a fraction of normal uranium, and even when enriched it is never more than 10 percent of the total, the yearly requirement for uranium is on the order of tens of tons. This is orders of magnitude less than coal, both in mass and volume (see Problem 18 in Chapter 20).

[†] A breeder reactor does *not* produce more fuel than it uses.

The first use of fission, however, was not to produce electric power. Instead, it was first used as a fission bomb (the "atomic bomb"). In early 1940, with Europe already at war, Hitler banned the sale of uranium from the Czech mines he had recently taken over. Research into the fission process suddenly was enshrouded in secrecy. Physicists in the United States were alarmed. A group of them approached Einstein—a man whose name was a household word—to send a letter to President Roosevelt about the possibilities of using nuclear fission for a bomb far more powerful than any previously known, and inform him that Germany might already have begun development of such a bomb. Roosevelt responded by authorizing the program known as the Manhattan Project, to see if a bomb could be built. Work began in earnest after Fermi's demonstration in 1942 that a sustained chain reaction was possible. A new secret laboratory was developed on an isolated mesa in New Mexico known as Los Alamos. Under the direction of J. Robert Oppenheimer (1904–1967; Fig. 43–11), it became the home of famous scientists from all over Europe and the United States.

To build a bomb that was subcritical during transport but that could be made supercritical (to produce a chain reaction) at just the right moment, two pieces of uranium were used, each less than the critical mass but together greater than the critical mass. The two masses would be kept separate until the moment of detonation arrived. Then a kind of gun would force the two pieces together very quickly, a chain reaction of explosive proportions would occur, and a tremendous amount of energy would be released very suddenly. The first fission bomb was tested in the New Mexico desert in July 1945. It was successful. In early August, a fission bomb using uranium was dropped on Hiroshima and a second, using plutonium, was dropped on Nagasaki (Fig. 43–12). World War II ended shortly thereafter.

Besides its great destructive power, a fission bomb produces many highly radioactive fission fragments, as does a nuclear reactor. When a fission bomb explodes, these radioactive isotopes are released into the atmosphere and are known as *radioactive fallout*.

Testing of nuclear bombs in the atmosphere after World War II was a cause of concern, for the movement of air masses spread the fallout all over the globe. Radioactive fallout eventually settles to the Earth, particularly in rainfall, and is absorbed by plants and grasses and enters the food chain. This is a far more serious problem than the same radioactivity on the exterior of our bodies, since α and β particles are largely absorbed by clothing and the outer (dead) layer of skin. But once inside our bodies via food, the isotopes are in direct contact with living cells. One particularly dangerous radioactive isotope is $^{90}_{38}\text{Sr}$, which is chemically much like calcium and becomes concentrated in bone, where it causes bone cancer and destruction of bone marrow. The 1963 treaty signed by over 100 nations that bans nuclear weapons testing in the atmosphere was motivated because of the hazards of fallout.

Atom bomb

FIGURE 43–11 J. Robert Oppenheimer, on the left, with General Leslie Groves, who was the administrative head of Los Alamos during the war. The photograph was taken at the Trinity site in the New Mexico desert, where the first atomic bomb was exploded.

FIGURE 43–12 Photo taken a month after the bomb was dropped on Nagasaki. The shacks were constructed afterwards from debris in the ruins.

43–4 | Fusion

The mass of every stable nucleus is less than the sum of the masses of its constituent protons and neutrons. For example, the mass of the helium isotope ^4_2He is less than the mass of two protons plus the mass of two neutrons, as we saw in Example 42–2. Thus, if two protons and two neutrons were to come together to form a helium nucleus there would be a loss of mass. This mass loss is manifested in the release of a large amount of energy.

FIGURE 43–13 Average binding energy per nucleon as a function of mass number A for stable nuclei. Same as Fig. 42–1.

The process of building up nuclei by bringing together individual protons and neutrons, or building larger nuclei by combining small nuclei, is called **nuclear fusion**. A glance at Fig. 43–13 (same as Fig. 42–1) shows why small nuclei can combine to form larger ones with the release of energy: it is because the binding energy per nucleon is smaller for light nuclei than it is for those of increasing mass (up to about $A \approx 60$). It is believed that many of the elements in the universe were originally formed through the process of fusion (see Chapter 45), and that today, fusion is continually taking place within the stars, including our Sun, producing the prodigious amounts of radiant energy they emit.

EXAMPLE 43–6 **Fusion energy release.** One of the simplest fusion reactions involves the production of deuterium, ^2_1H, from a neutron and a proton: $^1_1\text{H} + \text{n} \rightarrow {}^2_1\text{H} + \gamma$. How much energy is released in this reaction?

SOLUTION From Appendix D, the initial rest mass is $1.007825\,\text{u} + 1.008665\,\text{u} = 2.016490\,\text{u}$, and after the reaction the mass is that of the ^2_1H, namely $2.014102\,\text{u}$. The mass difference is $0.002388\,\text{u}$, so the energy released is $(0.002388\,\text{u})(931.5\,\text{MeV/u}) = 2.22\,\text{MeV}$, and it is carried off by the ^2_1H nucleus and the γ ray.

The energy output of our Sun is believed to be due principally to the following sequence of fusion reactions:

Fusion reactions

$$^1_1\text{H} + {}^1_1\text{H} \rightarrow {}^2_1\text{H} + \text{e}^+ + \nu \qquad\qquad (0.42\,\text{MeV}) \quad \textbf{(43–6a)}$$

in the Sun

$$^1_1\text{H} + {}^2_1\text{H} \rightarrow {}^3_2\text{He} + \gamma \qquad\qquad (5.49\,\text{MeV}) \quad \textbf{(43–6b)}$$

(proton–proton cycle)

$$^3_2\text{He} + {}^3_2\text{He} \rightarrow {}^4_2\text{He} + {}^1_1\text{H} + {}^1_1\text{H} \qquad\qquad (12.86\,\text{MeV}) \quad \textbf{(43–6c)}$$

where the energy released (Q-value) for each reaction is given in parentheses. The net effect of this sequence, which is called the **proton–proton cycle**, is for four protons to combine to form one ^4_2He nucleus plus two positrons, two neutrinos, and two gamma rays:

$$4\,{}^1_1\text{H} \rightarrow {}^4_2\text{He} + 2\text{e}^+ + 2\nu + 2\gamma. \qquad\qquad \textbf{(43–7)}$$

Note that it takes two of each of the first two reactions (Eqs. 43–6a and b) to produce the two ^3_2He for the third reaction, so the total energy release for the net reaction, Eq. 43–7, is $(2 \times 0.42\,\text{MeV} + 2 \times 5.49\,\text{MeV} + 12.86\,\text{MeV}) = 24.7\,\text{MeV}$.

However, each of the two e^+ (formed in reaction Eq. 43–6a) quickly annihilates with an electron to produce $2m_e c^2 = 1.02\ \text{MeV}$, so the total energy released is $(24.7\ \text{MeV} + 2 \times 1.02\ \text{MeV}) = 26.7\ \text{MeV}$. The first reaction, the formation of deuterium from two protons (Eq. 43–6a), has a very low probability, and the infrequency of that reaction serves to limit the rate at which the Sun produces energy.

In stars hotter than the Sun, it is more likely that the energy output comes principally from the **carbon cycle**, which comprises the following sequence of reactions:

$$\begin{aligned}
{}^{12}_{6}\text{C} + {}^{1}_{1}\text{H} &\rightarrow {}^{13}_{7}\text{N} + \gamma \\
{}^{13}_{7}\text{N} &\rightarrow {}^{13}_{6}\text{C} + e^+ + \nu \\
{}^{13}_{6}\text{C} + {}^{1}_{1}\text{H} &\rightarrow {}^{14}_{7}\text{N} + \gamma \\
{}^{14}_{7}\text{N} + {}^{1}_{1}\text{H} &\rightarrow {}^{15}_{8}\text{O} + \gamma \\
{}^{15}_{8}\text{O} &\rightarrow {}^{15}_{7}\text{N} + e^+ + \nu \\
{}^{15}_{7}\text{N} + {}^{1}_{1}\text{H} &\rightarrow {}^{12}_{6}\text{C} + {}^{4}_{2}\text{He}.
\end{aligned}$$

Carbon

cycle

(some stars)

(See Problem 38.) The theory of the proton–proton cycle and of the carbon cycle as the source of energy for the Sun and stars was first worked out by Hans Bethe (1906–   ) in 1939.

CONCEPTUAL EXAMPLE 43–7 | **Stellar fusion.** What is the heaviest element likely to be produced in fusion processes in stars?

RESPONSE Fusion is possible as long as the final product has more binding energy (less mass) than the reactants, for then there is net release of energy. Since the binding energy curve in Fig. 43–13 (or 42–1) peaks near $A \approx 56$ to 58, which corresponds to iron or nickel, it would not be energetically favorable to produce elements heavier than that. Nevertheless, in the center of massive stars or in supernova explosions, there is enough energy available to drive endothermic reactions that produce heavier elements, as well.

The possibility of utilizing the energy released in fusion to make a power reactor is very attractive. The fusion reactions most likely to succeed in a reactor involve the isotopes of hydrogen, ${}^{2}_{1}\text{H}$ (deuterium) and ${}^{3}_{1}\text{H}$ (tritium), and are as follows, with the energy released given in parentheses:

Fusion reactor

$$\begin{aligned}
{}^{2}_{1}\text{H} + {}^{2}_{1}\text{H} &\rightarrow {}^{3}_{1}\text{H} + {}^{1}_{1}\text{H} & (4.03\ \text{MeV}) & \quad \textbf{(43–8a)} \\
{}^{2}_{1}\text{H} + {}^{2}_{1}\text{H} &\rightarrow {}^{3}_{2}\text{He} + \text{n} & (3.27\ \text{MeV}) & \quad \textbf{(43–8b)} \\
{}^{2}_{1}\text{H} + {}^{3}_{1}\text{H} &\rightarrow {}^{4}_{2}\text{He} + \text{n}. & (17.59\ \text{MeV}) & \quad \textbf{(43–8c)}
\end{aligned}$$

Fusion reactions

for possible

reactor

Comparing these energy yields with that for the fission of ${}^{235}_{92}\text{U}$, we can see that the energy released in fusion reactions can be greater for a given mass of fuel than in fission. Furthermore, as fuel, a fusion reactor could use deuterium, which is very plentiful in the water of the oceans (the natural abundance of ${}^{2}_{1}\text{H}$ is 0.015 percent, or about 1 g of deuterium per 60 L of water). The simple proton–proton reaction of Eq. 43–6a, which could use a much more plentiful source of fuel, ${}^{1}_{1}\text{H}$, has such a small probability of occurring that it cannot be considered a possibility on Earth.

Although a successful fusion reactor has not yet been achieved, considerable progress has been made in overcoming the inherent difficulties. The problems are associated with the fact that all nuclei have a positive charge and thus repel each other. However, if they can be brought close enough together so that the short-range attractive nuclear force can come into play, the latter can pull the nuclei together and fusion will occur. Thus, in order for the nuclei to get close enough together, they must have rather large kinetic energy to overcome the electric repulsion. High kinetic energies are easily attainable with particle accelerators such as the cyclotron, but the number of particles involved is too small. To produce realistic amounts of energy, we must deal with matter in bulk, for which high kinetic energy means higher temperatures. Indeed, very high temperatures are required for fusion to occur, and fusion devices are often referred to as **thermonuclear devices**. The Sun and other stars are very hot, many millions of degrees, so the nuclei are moving fast enough for fusion to take place, and the energy released keeps the temperature high so that further fusion reactions can occur. The Sun and the stars represent huge self-sustaining thermonuclear reactors that stay together because of their great gravitational mass; but on Earth, containment of the plasma at the high temperatures and densities required has proven difficult.

It was realized after World War II that the temperature produced within a fission (or "atomic") bomb was close to 10^8 K. This suggested that a fission bomb could be used to ignite a fusion bomb (popularly known as a thermonuclear or hydrogen bomb) to release the vast energy of fusion. The uncontrollable release of fusion energy in an H-bomb was relatively easy to obtain. But to realize usable energy from fusion at a slow and controlled rate turned out to be difficult.

EXAMPLE 43–8 **ESTIMATE** **Temperature needed for d–t fusion.** Estimate the temperature required for deuterium–tritium fusion (d–t) to occur.

SOLUTION We assume the nuclei approach head-on, each with kinetic energy K, and that the nuclear force comes into play when the distance between their centers equals the sum of their radii as given by Eq. 42–1, namely $r_d \approx 1.5$ fm and $r_t \approx 1.7$ fm. The electrostatic potential energy (Chapter 23) of the two particles at this distance must equal the total kinetic energy of the two particles when far apart:

$$2K \approx \frac{1}{4\pi\epsilon_0} \frac{e^2}{(r_d + r_t)}$$

$$\approx \left(9.0 \times 10^9 \, \frac{\text{N} \cdot \text{m}^2}{\text{C}^2}\right) \frac{(1.6 \times 10^{-19} \, \text{C})^2}{(3.2 \times 10^{-15} \, \text{m})(1.6 \times 10^{-19} \, \text{J/eV})}$$

$$\approx 0.45 \, \text{MeV}.$$

Thus, $K \approx 0.22$ MeV, and if we ask that the average kinetic energy be this high, then from Eq. 18–4, $\frac{3}{2}kT = \bar{K}$, we have

$$T = \frac{2K}{3k} = \frac{2(0.22 \, \text{MeV})(1.6 \times 10^{-13} \, \text{J/MeV})}{3(1.38 \times 10^{-23} \, \text{J/K})} \approx 2 \times 10^9 \, \text{K}.$$

More careful calculations show that the temperature required for fusion is actually about an order of magnitude less than this rough estimate, partly because it is not necessary that the *average* kinetic energy be 0.22 MeV—a small percentage with this much energy (particles in the high-energy tail of the Maxwell distribution, Fig. 18–3) would be sufficient—and because of tunneling through the Coulomb barrier. Reasonable estimates for a usable fusion reactor are in the range $T \gtrsim 2$ to 4×10^8 K.

It is not only a high temperature that is required for a fusion reactor. There must also be a high density of nuclei to ensure a sufficiently high collision rate. A real difficulty with controlled fusion is to contain nuclei long enough and at a high enough density for sufficient reactions to occur that a usable amount of energy is obtained. At the temperatures needed for fusion, the atoms are ionized, and the *Plasma* resulting collection of nuclei and electrons is referred to as a **plasma**. Ordinary materials vaporize at a few thousand degrees at best, and hence cannot be used to contain a high-temperature plasma. Two major containment techniques are being investigated at present: *magnetic confinement* and *inertial confinement*.

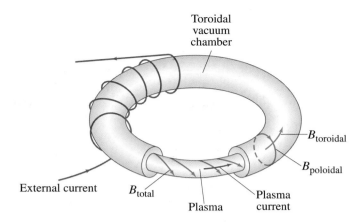

FIGURE 43–14 "Magnetic bottle" used to confine a plasma.

FIGURE 43–15 Tokamak configuration showing total **B** field due to external current plus current in the plasma.

In **magnetic confinement**, magnetic fields are used to try to contain the hot plasma. One possibility is the "magnetic bottle" shown in Fig. 43–14. The paths of the charged particles in the plasma are bent by the magnetic field, and where the lines are close together, the force on the particles is such that they are reflected back toward the center by this "magnetic mirror." Unfortunately, magnetic bottles develop "leaks" and the charged particles slip out before sufficient fusion takes place. A more complex device called the **tokamak**, first developed in the USSR, shows considerable promise. A tokamak (Fig. 43–15) is toroid-shaped and involves complicated magnetic fields: first, current-carrying conductors produce a magnetic field directed along the axis of the toroid ("toroidal" field); an additional field is produced by currents within the plasma itself ("poloidal" field). The combination of these fields produces a helical field as shown in Fig. 43–15, confining the plasma, at least briefly, so it doesn't touch the vacuum chamber's metal walls.

In 1957, J. D. Lawson showed that the product of ion density n and confinement time τ must satisfy, at a minimum, approximately

$$n\tau \gtrsim 3 \times 10^{20}\,\text{s/m}^3.$$

Magnetic confinement

Lawson criterion

This **Lawson criterion** must be reached to produce **ignition**, by which we mean a self-sustaining thermonuclear "burn" that continues after all external heating is turned off. To reach **break-even**, the point at which the energy output due to fusion is equal to the energy input to heat the plasma, requires an $n\tau$ about an order of magnitude less. The break-even point was very closely approached at the Tokamak Fusion Test Reactor (TFTR) at Princeton, and the very high temperature needed for ignition $(4 \times 10^8\,\text{K})$ was exceeded—although not both of these at the same time. It is believed that a working fusion reactor might be a reality in the first decade of this century.

Ignition

Break-even

The second method for containing the fuel for fusion is **Inertial confinement**: a small pellet of deuterium and tritium is struck simultaneously from several directions by very intense laser beams. The intense influx of energy in such a laser fusion device heats and ionizes the pellet into a plasma. The outer layers evaporate, but the collisions they make with ions in the core of the pellet drive the latter inward. This implosion raises the density to about 10^3 times normal and further heats the core to temperatures at which fusion occurs. The confinement time is very short, on the order of 10^{-11} to 10^{-9} s, during which time the ions do not move appreciably because of their own inertia. Very soon thereafter fusion takes place and the pellet explodes.

Inertial confinement

FIGURE 43–16 (a) Target chamber (5 m in diameter) of NOVA laser at Lawrence Livermore Laboratory, into which 10 laser beams converge on a target. (b) A 1-mm-diameter DT (deuterium–tritium) target, on its support, at the center of the target chamber.

(a)

(b)

The NOVA laser used for fusion research at the Lawrence Livermore Laboratory (Fig. 43–16) can deposit 10^5 J into the pellet over a 10^{-9} s pulse. This is a power input of 10^{14} W, more than the total electric-generator capacity of the United States! Of course, it is sustained only over an extremely short interval of time. A future reactor might implode 100 such pellets per second, thus requiring an input of 10^5 J $\times$ $100 \, \text{s}^{-1} = 10^7$ W on the average. Inertial confinement depends on high particle density, n, over very short time intervals, τ. Recently, the Lawson criterion has been achieved, but at a temperature less than that needed for ignition.

Besides the problems of confinement, the building of a practical fusion reactor will require the development of materials used in its construction that can withstand high temperatures and high levels of radiation.

43–5 Passage of Radiation Through Matter; Radiation Damage

When we speak of *radiation*, we include α, β, γ, and X-rays, as well as protons, neutrons, and other particles such as pions (see Chapter 44). Because charged particles can ionize the atoms or molecules of any material they pass through, they are referred to as **ionizing radiation**. And because radiation produces ionization, it can cause considerable damage to materials, particularly to biological tissue.

Charged particles, such as α and β rays and protons, cause ionization because of the electric force. That is, when they pass through a material, they can attract or repel electrons strongly enough to remove them from the atoms of the material. Since the α and β rays emitted by radioactive substances have energies on the order of a MeV (10^4 to 10^7 eV), whereas ionization of atoms and molecules requires on the order of 10 eV, it is clear that a single α or β particle can cause thousands of ionizations.

Neutral particles also give rise to ionization when they pass through materials. For example, X-ray and γ-ray photons can ionize atoms by knocking out electrons by means of the photoelectric and Compton effects (Chapter 38). Furthermore, if a γ ray has sufficient energy (greater than 1.02 MeV), it can undergo pair production: an electron and a positron are produced (Section 38–4). The charged particles produced in all three of these processes can themselves go on to produce further ionization. Neutrons, on the other hand, interact with matter mainly by collisions with nuclei, with which they interact strongly. Often the nucleus is broken apart by such a collision, altering the molecule of which it was a part. And the fragments produced can in turn cause ionization.

Radiation passing through matter can do considerable damage. Metals and other structural materials become brittle and their strength can be weakened if the radiation is very intense, as in nuclear reactor power plants and for space vehicles that must pass through areas of intense cosmic radiation.

The radiation damage produced in biological organisms is due primarily to ionization produced in cells. Several related processes can occur. Ions or radicals are produced that are highly reactive and take part in chemical reactions that interfere with the normal operation of the cell. All forms of radiation can ionize atoms by knocking out electrons. If these are bonding electrons, the molecule may break apart, or its structure may be altered so that it does not perform its normal function or performs a harmful function. If many cells die, the organism may not be able to recover. On the other hand, a cell may survive but be defective. It may go on dividing and produce many more defective cells, to the detriment of the whole organism. Thus radiation can cause cancer—the rapid production of defective cells.

Biological damage

Radiation damage to biological organisms is often separated into categories. *Somatic damage* refers to any part of the body except the reproductive organs, and affects that particular organism, causing cancer and, at high doses, radiation sickness (characterized by nausea, fatigue, loss of body hair, and other symptoms) or even death. *Genetic damage* refers to damage to reproductive cells, causing mutations, the majority of which are harmful, which are transmitted to future generations. The possible damage done by the medical use of X-rays and other radiation must be balanced against the medical benefits and prolongation of life as a result of their use.

43–6 Measurement of Radiation—Dosimetry

Although the passage of ionizing radiation through the human body can cause considerable damage, radiation can also be used to treat certain diseases, particularly cancer, often by using very narrow beams directed at the cancerous tumor to destroy it (Section 43–7). It is therefore important to be able to quantify the amount, or **dose**, of radiation. This is the subject of **dosimetry**.

The strength of a source can be specified at a given time by stating the **source activity**, or how many disintegrations occur per second. The traditional unit is the **curie** (Ci), defined as

Source activity

$$1 \text{ Ci} = 3.70 \times 10^{10} \text{ disintegrations per second.}$$

(This figure comes from the original definition as the activity of exactly 1 gram of radium.) Although the curie is still in common use, the proper SI unit for source activity is the **becquerel** (Bq), defined as

$$1 \text{ Bq} = 1 \text{ disintegration/s.}$$

Commercial suppliers of **radionuclides** (radioactive nuclides) specify the activity at a given time. Since the activity decreases over time, particularly for short-lived isotopes, it is important to take this into account.

The source activity (dN/dt) is related to the half-life, $T_{\frac{1}{2}}$, by (see Section 42–8):

$$\left| \frac{dN}{dt} \right| = \lambda N = \frac{0.693}{T_{\frac{1}{2}}} N.$$

EXAMPLE 43–9 **Radioactivity taken up by cells.** In a certain experiment, $0.016 \, \mu\text{Ci}$ of $^{32}_{15}\text{P}$ is injected into a medium containing a culture of bacteria. After 1 h, the cells are washed and a detector that is 70 percent efficient (counts 70 percent of emitted β particles) records 720 counts per minute from all the cells. What percentage of the original $^{32}_{15}\text{P}$ was taken up by the cells?

SOLUTION The halflife of $^{32}_{15}\text{P}$ being about 14 days, the loss of activity over the 1 hour can be ignored. The total number of disintegrations per second originally was $(0.016 \times 10^{-6})(3.7 \times 10^{10}) = 590$. The counter could be expected to count 70 percent of this, or 410 per second. Since it counted $720/60 = 12$ per second, then $12/410 = 0.029$ or 2.9 percent was incorporated into the cells.

Another type of measurement is the exposure or **absorbed dose**—that is, the effect the radiation has on the absorbing material. The earliest unit of dosage was the **roentgen** (R), which was defined in terms of the amount of ionization produced by the radiation (1.6×10^{12} ion pairs per gram of dry air at standard conditions). Today, 1 R is defined as the amount of X- or γ radiation that deposits 0.878×10^{-2} J of energy per kilogram of air. The roentgen was largely superseded by another unit of absorbed dose applicable to any type of radiation, the **rad**: *1 rad is that amount of radiation which deposits energy at a rate of 1.00×10^{-2} J/kg in any absorbing material.* (This is quite close to the roentgen for X- and γ rays.) The proper SI unit for absorbed dose is the **gray** (Gy):

The rad (unit)

The gray (unit)

$$1 \, \text{Gy} = 1 \, \text{J/kg} = 100 \, \text{rad}, \tag{43–9}$$

TABLE 43–1
Quality Factor (QF) of Different Kinds of Radiation

Type	QF
X and γ rays	≈ 1
β (electrons)	≈ 1
Fast protons	1
Slow neutrons	≈ 3
Fast neutrons	Up to 10
α particles and heavy ions	Up to 20

and is now coming into use. The absorbed dose depends not only on the strength of a given radiation beam (number of particles per second) and the energy per particle, but also on the type of material that is absorbing the radiation. Since bone, for example, is denser than flesh and absorbs more of the radiation normally used, the same beam passing through a human body deposits a greater dose (in rads or grays) in bone than in flesh.

The gray and the rad are physical units of dose—the energy deposited per unit mass of material. They are, however, not the most meaningful units for measuring the biological damage produced by radiation. This is because equal doses of different types of radiation cause differing amounts of damage. For example, 1 rad of α radiation does 10 to 20 times the amount of damage as 1 rad of β or γ rays. This difference arises largely because α rays (and other heavy particles such as protons and neutrons) move much more slowly than β and γ rays of equal energy due to their greater mass. Hence, ionizing collisions occur closer together, so more irreparable damage can be done. The **relative biological effectiveness** (RBE) or **quality factor** (QF) of a given type of radiation is defined as the number of rads of X or γ radiation that produces the same biological damage as 1 rad of the given radiation. Table 43–1 gives the QF for several types of radiation. The numbers are approximate since they depend somewhat on the energy of the particles and on the type of damage that is used as the criterion.

RBE or QF

Effective dose

The **effective dose** can be given as the product of the dose in rads and the QF, and this unit is known as the **rem** (which stands for *rad equivalent man*):

The rem (unit)

$$\text{effective dose (in rem)} = \text{dose (in rad)} \times \text{QF.} \tag{43–10a}$$

This unit is being replaced by the SI unit for "effective dose," the **sievert** (Sv):

The sievert (unit)

$$\text{effective dose (Sv)} = \text{dose (Gy)} \times \text{QF.} \tag{43–10b}$$

By this definition, 1 rem of any type of radiation does approximately the same amount of biological damage. For example, 50 rem of fast neutrons does the same damage as 50 rem of γ rays. But note that 50 rem of fast neutrons is only 5 rads, whereas 50 rem of γ rays is 50 rads.

Natural radioactive background

We are constantly exposed to low-level radiation from natural sources: cosmic rays, natural radioactivity in rocks and soil, and naturally occurring radioactive isotopes in our food, such as $^{40}_{19}\text{K}$. Radon, $^{222}_{86}\text{Rn}$, is of considerable concern today. It is the product of radium decay and is an intermediate in the decay series from uranium (see Fig. 42–11). Most intermediates remain in the rocks where formed, but radon is a gas that can escape from rock (and from building material like concrete) to enter the atmosphere we breathe. Although radon is inert chemically (it is a noble gas), it is not inert physically—it decays by alpha emission, and its products, also radioactive, are *not* chemically inert and can attach to the interior of the lung.

The natural radioactive background averages about 0.36 rem (3.6 mSv) per year per person, although there are large variations. From medical X-rays, the average person receives about 40 mrem (0.4 mSv) per year. The U.S. government specifies the recommended upper limit of allowed radiation for an individual in the general populace at about 0.5 rem (5 mSv) per year, exclusive of natural sources. It is not known if low doses of radiation increase the chances of cancer or genetic defects, so the attitude today is to play safe and keep the radiation dose low.

The upper limit for people who work around radiation—in hospitals, power plants, research—has been set somewhat higher, on the order of 5 rem/yr whole-body dose (presumably because such people know what they are getting into). People who work around radiation generally carry some type of dosimeter, typically a **radiation film badge**, which is a piece of film wrapped in light-tight material. The passage of ionizing radiation through the film changes it so that the film is darkened upon development, and so indicates the received dose.

Large doses of radiation can cause a large number of unpleasant symptoms such as nausea, fatigue, and loss of body hair. Such effects are sometimes referred to as **radiation sickness**. Large doses can also be fatal, although the time span of the dose is important. A short dose of 1000 rem (10 Sv) is nearly always fatal. A 400-rem (4-Sv) dose in a short period of time is fatal in 50 percent of the cases. However, the body possesses remarkable repair processes, so that a 400-rem dose spread over several weeks is not usually fatal. It will, nonetheless, cause considerable damage to the body.

Radiation sickness

EXAMPLE 43–10 **Whole-body dose.** What whole-body dose is received by a 70-kg laboratory worker exposed to a 40 mCi $^{60}_{27}$Co source, assuming the person's body has cross-sectional area 1.5 m^2 and is, on average, 4.0 m from the source for 4.0 h per day? $^{60}_{27}$Co emits γ rays of energy 1.33 MeV and 1.17 MeV in quick succession. Approximately 50 percent of the γ rays interact in the body and deposit all their energy. (The rest pass through.)

SOLUTION The total γ-ray energy per decay is $(1.33 + 1.17)$MeV $= 2.50$ MeV, so the total energy emitted by the source is

$$(0.040 \, \text{Ci})(3.7 \times 10^{10} \, \text{decays/Ci·s})(2.50 \, \text{MeV}) = 3.7 \times 10^{9} \, \text{MeV/s}.$$

The proportion of this intercepted by the body is its 1.5 m^2 area divided by the area of a sphere of radius 4.0 m (Fig. 43–17)

$$\frac{1.5 \, \text{m}^2}{4\pi r^2} = \frac{1.5 \, \text{m}^2}{4\pi (4.0 \, \text{m})^2} = 7.5 \times 10^{-3}.$$

So the rate energy is deposited in the body (remembering that only $\frac{1}{2}$ of the γ rays interact in the body) is

$$E = \left(\tfrac{1}{2}\right)(7.5 \times 10^{-3})(3.7 \times 10^{9} \, \text{MeV/s})(1.6 \times 10^{-13} \, \text{J/MeV})$$
$$= 2.2 \times 10^{-6} \, \text{J/s}.$$

Since 1 Gy = 1 J/kg, the whole-body dose rate for this 70-kg person is $(2.2 \times 10^{-6} \, \text{J/s})/(70 \, \text{kg}) = 3.1 \times 10^{-8}$ Gy/s. In the space of 4.0 h, this amounts to a dose of $(4.0 \, \text{h})(3600 \, \text{s/h})(3.1 \times 10^{-8} \, \text{Gy/s}) = 4.5 \times 10^{-4}$ Gy. Since QF ≈ 1 for gammas, the effective dose (Eq. 43–10) is 450 μSv or (see Eq. 43–9):

$$(100 \, \text{rad/Gy})(4.5 \times 10^{-4} \, \text{Gy})(1) = 45 \, \text{mrem}.$$

This 45-mrem effective dose is nearly 10 percent of the normal allowed dose for a whole year (500 mrem/yr), or 1 percent of the yearly allowance for radiation workers. This worker should not receive such a dose every day and should seek ways to reduce it (shield the source, vary the work, work farther away, etc.).

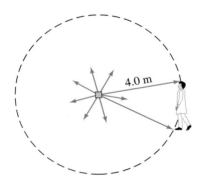

FIGURE 43–17 Radiation spreads out in all directions. A person 4.0 m away intercepts only a fraction: her cross-sectional area divided by the area of a sphere of radius 4.0 m.

* 43–7 Radiation Therapy

The applications of radioactivity and radiation to human beings and other organisms is a vast field that has filled many books. In the medical field there are two basic aspects: (1) *radiation therapy*—the treatment of disease (mainly cancer)—which we discuss in this Section; and (2) the *diagnosis* of disease, which we discuss in the following Sections of this chapter.

Radiation can cause cancer. It can also be used to treat it. Rapidly growing cancer cells are especially susceptible to destruction by radiation. Nonetheless, large doses are needed to kill the cancer cells, and some of the surrounding normal cells are inevitably killed as well. It is for this reason that cancer patients receiving radiation therapy often suffer side effects characteristic of radiation sickness. To minimize the destruction of normal cells, a narrow beam of γ or X-rays is often used when the cancerous tumor is well localized. The beam is directed at the tumor, and the source (or body) is rotated so that the beam passes through various parts of the body to keep the dose at any one place as low as possible—except at the tumor and its immediate surroundings, where the beam passes at all times (Fig. 43–18). The radiation may be from a radioactive source such as $^{60}_{27}\text{Co}$, or it may be from an X-ray machine that produces photons in the range $200\,\text{keV}$ to $5\,\text{MeV}$. Protons, neutrons, electrons, and pions, which are produced in particle accelerators (Section 44–2), are also being used in cancer therapy.

In some cases, a tiny radioactive source may be inserted directly inside a tumor, which will eventually kill the majority of the cells. A similar technique is used to treat cancer of the thyroid with the radioactive isotope $^{131}_{53}\text{I}$. The thyroid gland tends to concentrate any iodine present in the bloodstream; so when $^{131}_{53}\text{I}$ is injected into the blood, it becomes concentrated in the thyroid, particularly in any area where abnormal growth is taking place. The intense radioactivity emitted can then destroy the defective cells.

➡ **PHYSICS APPLIED**

Radiation therapy

FIGURE 43–18 Radiation source rotates so that the beam always passes through the diseased tissue, but minimizing the dose in the rest of the body.

FIGURE 43–19 Collimated gamma-ray detector for scanning (moving) over patient. The collimator (to "collimate" means to "make straight") is necessary to select γ rays that come in a straight line from the patient. Without the collimator, γ rays from all parts of the body could strike the scintillator, producing a very poor image.

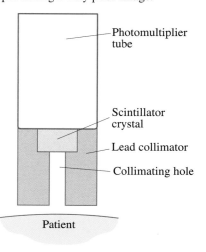

Photomultiplier tube

Scintillator crystal

Lead collimator

Collimating hole

Patient

* 43–8 Tracers

Radioactive isotopes are commonly used in biological, chemical, and medical research as **tracers**. A given compound is artificially synthesized using a radioactive isotope such as $^{14}_{6}\text{C}$ or $^{3}_{1}\text{H}$. Such "tagged" molecules can then be traced as they move through an organism or as they undergo chemical reaction. The presence of these tagged molecules (or parts of them, if they undergo chemical change) can be detected by a Geiger or scintillation counter (see Section 42–11).

For medical diagnosis, the radionuclide commonly used today is $^{99\text{m}}_{43}\text{Tc}$, a long-lived excited state of technetium-99 (the "m" in the symbol stands for "metastable" state). It is formed when $^{99}_{42}\text{Mo}$ decays. The great usefulness of $^{99\text{m}}_{43}\text{Tc}$ derives from its convenient half-life of 6 h (short, but not too short) and the fact that it can combine with a large variety of compounds. The compound to be labeled with the radionuclide is so chosen because it concentrates in the organ or region of the anatomy to be studied. Detectors outside the body then record, or image, the distribution of the radioactively labeled compound. The detection can be done by a single detector (Fig. 43–19) which is moved across the body, measuring the intensity of radioactivity at a large number of points. The image represents the relative intensity of radioactivity at each point. The relative radioactivity is a diagnostic tool. For example, high or low radioactivity may represent overactivity or underactivity of an organ or part of an organ, or in another case may represent a lesion or tumor. More complex *gamma cameras* make use of many detectors which simultaneously record the radioactivity at many points. The measured intensities can be displayed on a CRT (TV monitor or oscilloscope screen), and allow "dynamic" studies (that is, images that change in time) to be performed.

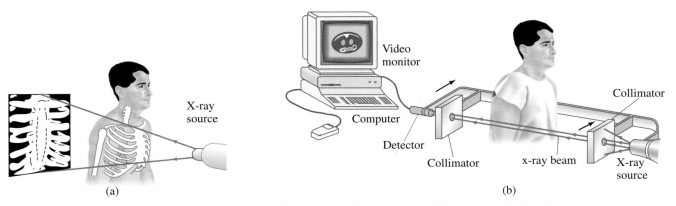

FIGURE 43–20 (a) Conventional X-ray imaging, which is essentially shadowing. (b) Tomographic imaging: the X-ray source and detector move together across the body, the transmitted intensity being measured at a large number of points. Then the source–detector assembly is rotated slightly (say, 1°) and another scan is made. This process is repeated for perhaps 180°. The computer reconstructs the image of the slice and it is presented on a TV monitor (cathode-ray tube).

*43–9 Imaging by Tomography: CAT Scans and Emission Tomography

Normal X-Ray Image

For a conventional medical (or dental) X-ray photograph, the X-rays emerging from the tube (Fig. 36–25, Section 36–10) pass through the body and are detected on photographic film or a fluorescent screen, Fig. 43–20a. The rays travel in very nearly straight lines through the body with minimal deviation since at X-ray wavelengths there is little diffraction or refraction. There is absorption (and scattering), however; and the difference in absorption by different structures in the body is what gives rise to the image produced by the transmitted rays. The less the absorption, the greater the transmission and the darker the film. The image is, in a sense, a "shadow" of what the rays have passed through.

Computerized Axial Tomography (CAT Scan)

In conventional X-ray images, the entire thickness of the body is projected onto the film; structures overlap and in many cases are difficult to distinguish. In the 1970s, a revolutionary new technique called **computerized tomography** (CT) using X-rays was developed, which produces an image of a *slice* through the body. (The word **tomography** comes from the Greek: *tomos* = slice, *graph* = picture.) Structures and lesions previously impossible to visualize can now be seen with remarkable clarity. The principle behind CT is shown in Fig. 43–20b: a thin collimated beam of X-rays passes through the body to a detector that measures the transmitted intensity. Measurements are made at a large number of points as the source and detector are moved past the body together. The apparatus is then rotated slightly about the body axis and again scanned; this is repeated at (perhaps) 1° intervals for 180°. The intensity of the transmitted beam for the many points of each scan, and for each angle, are sent to a computer that reconstructs the image of the slice. Note that the imaged slice is perpendicular to the long axis of the body. For this reason, CT is sometimes called **computerized axial tomography** (CAT), although the abbreviation CAT, as in CAT scan, can also be read as **computer-assisted tomography**.

The use of a single detector as in Fig. 43–20b would require some time to make the many scans needed to form a complete image. Much faster scanners use

→ **P H Y S I C S   A P P L I E D**

CT images

CAT scans

FIGURE 43–21 (a) Fan-beam scanner. Rays transmitted through the entire body are measured simultaneously at each angle. The source and detector rotate to take measurements at different angles. In another type of fan-beam scanner, there are detectors around the entire 360° of the circle which remain fixed as the source moves. (b) In another type, a beam of electrons from the source is directed by magnetic fields at tungsten targets surrounding the patient.

(a)

(b)

FIGURE 43–22 Two CT images, with different resolutions, each showing a cross section of a brain. Photo (a) is of low resolution; photo (b), of higher resolution, shows a brain tumor (dark area on the right).

(a)

(b)

➡ **PHYSICS APPLIED**

Medical imaging

SPET
PET

a fan beam, Fig. 43–21a, in which beams passing through the entire cross section of the body are detected simultaneously by many detectors. The source and detectors are then rotated about the patient, and an image requires only a few seconds. Even faster, and therefore useful for heart scans, are fixed source machines wherein an electron beam is directed (by magnetic fields) to tungsten targets surrounding the patient, creating the X-rays. See Fig. 43–21b.

But how is the image formed? We can think of the slice to be imaged as being divided into many tiny picture elements (or **pixels**), which could be squares. For CT, the width of each pixel is chosen according to the width of the detectors and/or the width of the X-ray beams, and this determines the resolution of the image, which might be 1 mm. An X-ray detector measures the intensity of the transmitted beam. Subtracting this value from the intensity of the beam at the source, yields the total absorption (called a "projection") along that beam line. Complicated mathematical techniques, such as Fourier transforms, are used to analyze all the absorption projections for the huge number of beam scans measured, obtaining the absorption at each pixel and assigning each a "grayness value" according to how much radiation was absorbed. The image is made up of tiny spots (pixels) of varying shades of gray. Often the amount of absorption is color-coded. The colors in the resulting ("false-color") image have nothing to do, however, with the actual color of the object.

Figure 43–22 illustrates what actual CT images look like. It is generally agreed that CT scanning has revolutionized some areas of medicine by providing much less invasive, and/or more accurate, diagnosis.

Computerized tomography can also be applied to ultrasound imaging (Section 16–9) and to emissions from radioisotopes and nuclear magnetic resonance, which we discuss next.

Emission Tomography

It is possible to image the emissions of a radioactive tracer (see Section 43–8) in a single plane or slice through a body using computed tomography techniques. A basic gamma detector (Fig. 43–19) can be moved around the patient to measure the radioactive intensity from the tracer at many points and angles; the data are processed in much the same way as for X-ray CT scans. This technique is referred to as **single photon emission tomography** (SPET).[†]

Another important technique is **positron emission tomography** (PET), which makes use of positron emitters such as $^{11}_{6}C$, $^{13}_{7}N$, $^{15}_{8}O$, and $^{18}_{9}F$. These isotopes are incorporated into molecules that, when inhaled or injected, accumulate in the organ or region of the body to be studied. When such a nuclide undergoes β^+ decay, the emitted positron travels at most a few millimeters before it collides with a normal electron. In this collision, the positron and electron are annihilated, producing two γ rays $(e^+ + e^- \rightarrow 2\gamma)$, each having energy of 511 keV. The two γ rays fly off in oppo-

[†] Also known as SPECT, "single photon emission computed tomography."

site directions ($180° \pm 0.25°$) since they must have almost exactly equal and opposite momenta to conserve momentum (the momenta of the e^+ and e^- are essentially zero compared to the momenta of the γ rays). Because the photons travel along the same line in opposite directions, their detection in coincidence by rings of detectors surrounding the patient (Fig. 43–23) readily establishes the line along which the emission took place. If the difference in time of arrival of the two photons could be determined accurately, the actual position of the emitting nuclide along that line could be calculated. Present-day electronics can measure times to at best ± 300 ps, so at the γ ray's speed $(c = 3 \times 10^8 \, \text{m/s})$, the actual position could be determined to an accuracy on the order of about $vt \approx (3 \times 10^8 \, \text{m/s})(300 \times 10^{-12} \, \text{s}) \approx 10$ cm, which is not very useful. Although there may be future potential for time-of-flight measurements to determine position, today computed tomography techniques are used instead, similar to those for X-ray CT, which can reconstruct PET images with a resolution on the order of 3–5 mm. One big advantage of PET is that no collimators are needed (as for detection of a single photon—see Fig. 43–19). Thus, fewer photons are "wasted" and lower doses can be administered to the patient with PET.

Both PET and SPET systems can give images related to biochemistry, metabolism, and function. This is to be compared to X-ray CT scans, whose images reflect shape and structure—that is, the anatomy of the imaged region.

FIGURE 43–23 Positron emission tomography (PET) system showing a ring of detectors to detect the two annihilation γ rays $(e^+ + e^- \rightarrow 2\gamma)$ emitted at 180° to each other.

* 43–10 **Nuclear Magnetic Resonance (NMR) and Magnetic Resonance Imaging (MRI)**

Nuclear magnetic resonance (NMR) is a phenomenon which soon after its discovery in 1946 became a powerful research tool in a variety of fields from physics to chemistry and biochemistry. It is also an important medical imaging technique. We first briefly discuss the phenomenon, and then look at its applications.

Nuclear Magnetic Resonance (NMR)

We saw in Chapter 40 that the energy levels in atoms are split when they are placed in a magnetic field B (the Zeeman effect) according to the angular momentum or spin of the state. The splitting is proportional to B and to the magnetic moment, μ. Nuclei too have magnetic moments (Section 42–1), and we examine only the simplest, the hydrogen (^1_1H) nucleus which consists of a single proton. Its spin angular momentum (and its magnetic moment), like that of the electron, can take on only two values when placed in a magnetic field: spin up (parallel to the field) and spin down (antiparallel to the field) as suggested in Fig. 43–24. When a magnetic field is present, an energy state splits into two levels as shown in Fig. 43–25 with the spin-up state (parallel to field) having the lower energy. The spin-down state acquires an additional energy μB_T and the spin-up state $-\mu B_\text{T}$ (Eq. 27–12 and Section 40–7), where B_T is the total magnetic field at the nucleus. The difference in energy between the two states (Fig. 43–25) is thus

$$\Delta E = 2\mu_\text{p} B_\text{T},$$

where μ_p is the magnetic moment of the proton.

In a standard nuclear magnetic resonance (NMR) setup, the sample to be examined is placed in a static magnetic field. A radiofrequency (RF) pulse of electromagnetic radiation (that is, photons) is applied to the sample. If the frequency, f, of this pulse corresponds precisely to the energy difference between the two energy levels (Fig. 43–25), so that

$$hf = \Delta E = 2\mu_\text{p} B_\text{T} \qquad \textbf{(43–11)}$$

then the photons of the RF beam will be absorbed, exciting many of the nuclei from the lower state to the upper state. This is a resonance phenomenon since there is significant absorption only if f is very near $f = 2\mu_\text{p} B_\text{T}/h$. Hence the name "nuclear magnetic resonance." For free ^1_1H nuclei, the frequency is 42.58 MHz for a field $B_\text{T} = 1.0$ T (Example 43–11). If the H atoms are bound in a

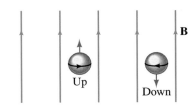

FIGURE 43–24 Schematic picture of a proton represented in a magnetic field **B** (pointing upward) with its two possible states of spin, up and down.

FIGURE 43–25 Energy E_0 in the absence of a magnetic field splits into two levels in the presence of a magnetic field.

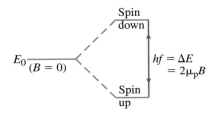

molecule, the total magnetic field B_T at the H nuclei will be the sum of the external applied field (B_{ext}) plus the local magnetic field (B_{loc}) due to electrons and nuclei of neighboring atoms. Since f is proportional to B_T, the value of f for a given external field will be slightly different for bound H atoms than for free atoms:

$$hf = 2\mu_p(B_{ext} + B_{loc}).$$

This change in frequency, which can be measured, is called the "chemical shift." A great deal has been learned about the structure of molecules and bonds using this NMR technique.

EXAMPLE 43–11 **NMR for free protons.** Calculate the resonant frequency for free protons in a 1.000-T magnetic field.

SOLUTION We saw in Section 42–1 that the magnetic moment of the proton is

$$\mu_p = 2.7928\mu_N = 2.7928\left(\frac{e\hbar}{2m_p}\right).$$

Then

$$f = \frac{\Delta E}{h} = \frac{2\mu_p B}{h} = (2.7928)\left(\frac{eB}{2\pi m_p}\right) = 2.7928\left[\frac{(1.6022 \times 10^{-19}\,\text{C})(1.000\,\text{T})}{2\pi(1.6726 \times 10^{-27}\,\text{kg})}\right]$$

$$= 42.58\,\text{MHz}.$$

Magnetic Resonance Imaging (MRI)

For producing medically useful NMR images—now commonly called MRI, or **magnetic resonance imaging**—the element most used is hydrogen since it is the commonest element in the human body and gives the strongest NMR signals. The experimental apparatus is shown in Fig. 43–26. The large coils set up the static magnetic field, and the RF coils produce the RF pulse of electromagnetic waves (photons) that cause the nuclei to jump from the lower state to the upper one (Fig. 43–25). These same coils (or another coil) can detect the absorption of energy or the emitted radiation (also of frequency $f = \Delta E/h$) when the nuclei jump back down to the lower state.

The formation of a two-dimensional or three-dimensional image can be done using techniques similar to those for computed tomography (Section 43–9). The simplest thing to measure for creating an image is the intensity of absorbed and/or reemitted radiation from many different points of the body, and this would be a measure of the density of H atoms at each point. But how do we determine from what part of the body a given photon comes? One technique is to give the static

FIGURE 43–26 Typical MRI imaging setup: (a) diagram; (b) photograph.

Magnets

RF coils

(a)

(b)

magnetic field a gradient; that is, instead of applying a uniform magnetic field, B_T, the field is made to vary with position across the width of the sample (or patient). Since the frequency absorbed by the H nuclei is proportional to B_T (Eq. 43–11), only one plane within the body will have the proper value of B_T to absorb photons of a particular frequency f. By varying f, absorption by different planes can be measured. Alternately, if the field gradient is applied *after* the RF pulse, the frequency of the emitted photons will be a measure of where they were emitted. See Fig. 43–27. If a magnetic field gradient in one direction is applied during excitation (absorption of photons) and photons of a single frequency are transmitted, only H nuclei in one thin slice will be excited. By applying a gradient in a different direction, perpendicular to the first, during reemission, the frequency f of the reemitted radiation will represent depth in that slice. Other ways of varying the magnetic field throughout the volume of the body can be used in order to correlate NMR frequency with position.

A reconstructed image based on the density of H atoms (that is, the intensity of absorbed or emitted radiation) is not very interesting. More useful are images based on the rate at which the nuclei decay back to the ground state, and such images can produce resolution of 1 mm or better. This NMR technique (sometimes called **spin-echo**) is producing images of great diagnostic value, both in the delineation of structure (anatomy) and in the study of metabolic processes. An NMR image is shown in Fig. 43–28.

NMR imaging is considered to be noninvasive. We can calculate the energy of the photons involved: as determined in Example 43–11, in a 1.0 T magnetic field, $f = 42.58\,\text{MHz}$ for ^1_1H. This corresponds to an energy of $hf = (6.6 \times 10^{-34}\,\text{J·s})(43 \times 10^6\,\text{Hz}) \approx 3 \times 10^{-26}\,\text{J}$ or about $10^{-7}\,\text{eV}$. Since molecular bonds are on the order of 1 eV, it is clear that the RF photons can cause little cellular disruption. This should be compared to X- or γ rays whose energies are 10^4 to $10^6\,\text{eV}$ and thus can cause significant damage. The static magnetic fields, though often large (~ 0.1 to 1 T), are believed to be harmless (except for people wearing heart pacemakers), although not a great deal of research has been done in this area.

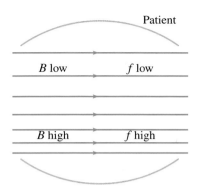

FIGURE 43–27 A static field that is stronger at the bottom than at the top. The frequency of absorbed or emitted radiation is proportional to B in NMR.

FIGURE 43–28 False-color NMR image (MRI) of a vertical section through the head showing structures in the normal brain.

Summary

A **nuclear reaction** occurs when two nuclei collide and two or more other nuclei (or particles) are produced. In this process, as in radioactivity, **transmutation** (change) of elements occurs.

The **reaction energy** or **Q-value** of a reaction a + X → Y + b is

$$Q = (M_a + M_X - M_Y - M_b)c^2$$
$$= K_b + K_Y - K_a - K_X.$$

The effective **cross section** for a reaction is a measure of the reaction probability per target nucleus.

In **fission** a heavy nucleus such as uranium splits into two intermediate-sized nuclei after being struck by a neutron. $^{235}_{92}\text{U}$ is fissionable by slow neutrons, whereas some fissionable nuclei require fast neutrons. Much energy is released in fission because the binding energy per nucleon is lower for heavy nuclei than it is for intermediate-sized nuclei, so the mass of a heavy nucleus is greater than the total mass of its fission products. The fission process releases neutrons, so that a **chain reaction** is possible. The **critical mass** is the minimum mass of fuel needed to sustain a chain reaction. In a **nuclear reactor** or nuclear bomb, a moderator is needed to slow down the released neutrons.

The **fusion** process, in which small nuclei combine to form larger ones, also releases energy. The energy from our Sun is believed to originate in the fusion reactions known as the **proton–proton cycle** in which four protons fuse to form a ^4_2He nucleus producing over 25 MeV of energy. A working fusion reactor for power generation has not yet proved possible because of the difficulty in containing the fuel (e.g., deuterium) long enough at the high temperature required. Nonetheless, great progress has been made in confining the collection of charged ions known as a **plasma**. The two main methods are **magnetic confinement**, using a magnetic field in a device such as the toroidal-shaped **tokamak**, and **inertial confinement**, in which intense laser beams compress a fuel pellet of deuterium and tritium.

Radiation can cause damage to materials, including biological tissue. Quantifying amounts of radiation is the subject of **dosimetry**. The **curie** (Ci) and the **becquerel** (Bq) are units that measure the **source activity** or rate of decay of a sample: $1\,\text{Ci} = 3.70 \times 10^{10}$ disintegrations per second, whereas $1\,\text{Bq} = 1$ disintegration/s. The **absorbed dose**, often specified in **rads**, measures the amount of energy deposited per unit mass of absorbing material: 1 rad is the amount of radiation that deposits energy at the rate of $10^{-2}\,\text{J/kg}$ of material. The SI unit of absorbed dose is the **gray**: $1\,\text{Gy} = 1\,\text{J/kg} = 100\,\text{rad}$. The **effective dose** is often specified by the **rem** = rad × QF, where QF is the "quality factor" of a given type of radiation; 1 rem of any type of radiation does approximately the same amount of biological damage. The average dose received per person per year in the United States is about 0.36 rem. The SI unit for effective dose is the **sievert**: $1\,\text{Sv} = 10^2\,\text{rem}$.

Questions

1. Fill in the missing particles or nuclei: (a) $^{137}_{56}\text{Ba}(n, \gamma)?$; (b) $^{137}_{56}\text{Ba}(n, ?)^{137}_{55}\text{Cs}$; (c) $^{2}_{1}\text{H}(d, ?)^{4}_{2}\text{He}$; (d) $^{197}_{79}\text{Au}(\alpha, d)?$ where d stands for deuterium.

2. The isotope $^{32}_{15}\text{P}$ is produced by an (n, p) reaction. What must be the target nucleus?

3. When $^{22}_{11}\text{Na}$ is bombarded by deuterons ($^{2}_{1}\text{H}$), an α particle is emitted. What is the resulting nuclide?

4. Why are neutrons such good projectiles for producing nuclear reactions?

5. A proton strikes a $^{20}_{10}\text{Ne}$ nucleus, and an α particle is observed to emerge. What is the residual nucleus? Write down the reaction equation.

6. Are fission fragments β^+ or β^- emitters?

7. If $^{235}_{92}\text{U}$ released only 1.5 neutrons per fission on the average, would a chain reaction be possible? If so, what would be different?

8. $^{235}_{92}\text{U}$ releases an average of 2.5 neutrons per fission compared to 2.9 for $^{239}_{94}\text{Pu}$. Pure samples of which of these two nuclei do you think would have the smaller critical mass?

9. The energy from nuclear fission appears in the form of thermal energy—but the thermal energy of what?

10. Why can't uranium be enriched by chemical means?

11. How can a neutron, with practically no kinetic energy, excite a nucleus to the extent shown in Fig. 43–4?

12. Why would a porous block of uranium be more likely to explode if kept under water rather than in air?

13. A reactor that uses highly enriched uranium can use ordinary water (instead of heavy water) as a moderator and still have a self-sustaining chain reaction. Explain.

14. Why must the fission process release neutrons if it is to be useful?

15. Discuss the relative merits and disadvantages, including pollution and safety, of power generation by fossil fuels, nuclear fission, and nuclear fusion.

16. What is the reason for the "secondary system" in a nuclear reactor, Fig. 43–9? That is, why is the water heated by the fuel in a nuclear reactor not used directly to drive the turbines?

17. Discuss how the course of history might have been changed if, during World War II, scientists had refused to work on developing a nuclear bomb. Do you think it would have been possible to delay the building of a bomb indefinitely?

18. Research in molecular biology is moving toward the ability to perform genetic manipulations on human beings. The moral implications of future discoveries along these lines has led to a warning that this may be the molecular biologists' "Hiroshima." Discuss.

19. Does $E = mc^2$ apply in (a) fission, (b) fusion, (c) nuclear reactions?

20. Light energy emitted by the Sun and stars comes from the fusion process. What conditions in the interior of stars makes this possible?

21. How do stars, and our Sun, maintain confinement of the plasma for fusion?

22. What is the basic difference between fission and fusion?

23. People who work around metals that emit alpha particles are trained that there is little danger from proximity or even touching the material, but that they must take extreme precautions against ingesting it. Hence, there are strong rules against eating and drinking while working, and against machining the metal. Why?

24. Why is the recommended maximum radiation dose higher for women beyond the child-bearing age than for younger women?

25. A higher temperature is required for deuterium–deuterium ignition than for deuterium–tritium. Explain.

26. Radiation is sometimes used to sterilize medical supplies and even food. Explain how it works.

* 27. How might radioactive tracers be used to find a leak in a pipe?

Problems

Section 43–1

1. (I) Natural aluminum is all $^{27}_{13}\text{Al}$. If it absorbs a neutron, what does it become? Does it decay by β^+ or β^-? What will be the product nucleus?

2. (I) Determine whether the reaction $^{2}_{1}\text{H}(d, n)^{3}_{2}\text{He}$ requires a threshold energy. (d stands for deuterium, $^{2}_{1}\text{H}$.)

3. (I) Is the reaction $^{238}_{92}\text{U}(n, \gamma)^{239}_{92}\text{U}$ possible with slow neutrons? Explain.

4. (II) Does the reaction $^7_3Li(p, \alpha)^4_2He$ require energy or does it release energy? How much energy?

5. (II) Calculate the energy released (or energy input required) for the reaction $^9_4Be(\alpha, n)^{12}_6C$.

6. (II) (a) Can the reaction $^{24}_{12}Mg(n, d)^{23}_{11}Na$ occur if the bombarding particles have 10.00 MeV of kinetic energy? (b) If so, how much energy is released?

7. (II) (a) Can the reaction $^7_3Li(p, \alpha)^4_2He$ occur if the incident proton has kinetic energy = 2500 keV? (b) If so, what is the total kinetic energy of the products?

8. (II) In the reaction $^{14}_7N(\alpha, p)^{17}_8O$, the incident α particles have 7.68 MeV of kinetic energy. (a) Can this reaction occur? (b) If so, what is the total kinetic energy of the products? The mass of $^{17}_8O$ is 16.999131 u.

9. (II) Calculate the Q-value for the "capture" reaction $^{16}_8O(\alpha, \gamma)^{20}_{10}Ne$.

10. (II) Calculate the total kinetic energy of the products of the reaction $^{13}_6C(d, n)^{14}_7N$ if the incoming deuteron (d) has $K = 36.3$ MeV.

11. (II) An example of a "stripping" nuclear reaction is $^6_3Li(d, p)X$. (a) What is X, the resulting nucleus? (b) Why is it called a "stripping" reaction? (c) What is the Q-value of this reaction? Is the reaction endothermic or exothermic?

12. (II) An example of a "pick-up" nuclear reaction is $^{12}_6C(^3_2He, \alpha)X$. (a) Why is it called a "pickup" reaction? (b) What is the resulting nucleus? (c) What is the Q-value of this reaction? Is the reaction endothermic or exothermic?

13. (II) (a) Complete the following nuclear reaction, $?(p, \gamma)^{32}_{16}S$. (b) What is the Q-value?

14. (III) Use conservation of energy and momentum to show that a bombarding proton must have an energy of 3.23 MeV in order to make the reaction $^{13}_6C(p, n)^{13}_7N$ occur. (See Example 43–3.)

Section 43–2

15. (I) What is the effective cross section for the collision of two hard spheres of radius R_1 and R_2.

16. (I) The cross section for the reaction $n + ^{10}_5B \rightarrow ^7_3Li + ^4_2He$ is about 40 bn for an incident neutron of low energy (kinetic energy ≈ 0). The boron is contained in a gas with $n = 1.7 \times 10^{21}$ nuclei/m^3 and the target has thickness $t = 12.0$ cm. What fraction of incident neutrons will be scattered?

17. (II) When the target is thick, the rate at which projectile particles collide with nuclei in the rear of the target is less than in the front of the target, since some scattering (i.e., collisions) takes place in the front layers. Let R_0 be the rate at which incident particles strike the front of the target, and R_x be the rate at a distance x into the target ($R_x = R_0$ at $x = 0$). Then show that the rate at which particles are scattered (and therefore lost from the incident beam) in a thickness dx is $-dR_x = R_x n\sigma dx$, where the minus sign means that R_x is decreasing and n is the number of nuclei per unit volume. Then show that $R_x = R_0 e^{-n\sigma x}$, where σ is the total cross section. If the thickness of the target is t, what does $R_x = R_0 e^{-n\sigma t}$ represent?

18. (II) A 1.0-cm-thick lead target reduces a beam of gamma rays to 30 percent of its original intensity. What thickness of lead will allow only one γ in 10^6 to penetrate (see Problem 17)?

19. (II) Use Fig. 43–3 to estimate what thickness of Cd $(\rho = 8650$ kg/m$^3)$ will cause a 1 percent reaction rate $(R/R_0 = 0.01)$ for (a) 0.1-eV neutrons (b) 10-eV neutrons.

Section 43–3

20. (I) Calculate the energy released in the fission reaction $n + ^{235}_{92}U \rightarrow ^{88}_{38}Sr + ^{136}_{54}Xe + 12n$. Use Appendix D, and assume the initial kinetic energy of the neutron is very small.

21. (I) What is the energy released in the fission reaction of Eq. 43–5? (The masses of $^{141}_{56}Ba$ and $^{92}_{36}Kr$ are 140.91440 u and 91.92630 u, respectively.)

22. (I) How many fissions take place per second in a 200-MW reactor? Assume 200 MeV is released per fission.

23. (II) Suppose that the average electric power consumption, day and night, in a typical house is 300 W. What initial mass of $^{235}_{92}U$ would have to undergo fission to supply the electrical needs of such a house for a year? (Assume 200 MeV is released per fission, as well as 100% efficiency.)

24. (II) What initial mass of $^{235}_{92}U$ is required to operate a 500-MW reactor for 1 yr? Assume 40 percent efficiency.

25. (II) If a 1.0-MeV neutron emitted in a fission reaction loses one-half of its kinetic energy in each collision with moderator nuclei, how many collisions must it make to reach thermal energy $(\frac{3}{2}kT = 0.040$ eV)?

26. (II) Suppose that the neutron multiplication factor is 1.0004. If the average time between successive fissions in a chain of reactions is 1.0 ms, by what factor will the reaction rate increase in 1.0 s?

27. (II) Estimate the ratio of the height of the Coulomb barrier for α decay to that for fission of $^{236}_{92}U$. (Both are described by a potential energy diagram of the shape shown in Fig. 42–6.)

28. (II) Assuming a fission of $^{236}_{92}U$ into two roughly equal fragments, estimate the electric potential energy just as the fragments separate from each other. Assume that the fragments are spherical (see Eq. 42–1) and compare your calculation to the nuclear fission energy released, about 200 MeV.

Section 43–4

29. (I) What is the average kinetic energy of protons at the center of a star where the temperature is 10^7 K?

30. (II) Show that the energy released in the fusion reaction $^2_1H + ^3_1H \rightarrow ^4_2He + n$ is 17.59 MeV.

31. (II) Show that the energy released when two deuterium nuclei fuse to form ^3_2He with the release of a neutron is 3.27 MeV.

32. (II) Verify the Q-value stated for each of the reactions of Eqs. 43–6. [Hint: Be careful with electrons.]

33. (II) Calculate the energy release per gram of fuel for the reactions of Eqs. 43–8a, b, and c. Compare to the energy release per gram of uranium in fission.

34. (II) If a typical house requires 300 W of electric power on average, what minimum amount of deuterium fuel would have to be used in a year to supply these electrical needs? Assume the reaction of Eq. 43–8b.

35. (II) Show that the energies carried off by the $^4_2$He nucleus and the neutron for the reaction of Eq. 43–8c are about 3.5 MeV and 14 MeV, respectively. Are these fixed values, independent of the plasma temperature?

36. (II) Suppose a fusion reactor ran on "d–d" reactions, Eqs. 43–8a and b. Estimate how much water, for fuel, would be needed per hour to run a 1000-MW reactor, assuming 30 percent efficiency.

37. (II) How much energy (J) is contained in 1.00 kg of water if its natural deuterium is used in the fusion reaction of Eq. 43–8a? Compare to the energy obtained from the burning of 1.0 kg of gasoline, about 5×10^7 J.

38. (III) The energy output of massive stars is believed to be due to the *carbon cycle* (see text). (a) Show that no carbon is consumed in this cycle and that the net effect is the same as for the proton–proton cycle. (b) What is the total energy release? (c) Determine the energy output for each reaction and decay. (d) Why does the carbon cycle require a higher temperature $(\approx 2 \times 10^7$ K$)$ than the proton–proton cycle $(\approx 1.5 \times 10^7$ K$)$?

39. (III) (a) Compare the energy needed for the first reaction of the carbon cycle to that for a deuteron–tritium reaction (Example 43–8). (b) If a deuteron–tritium reaction requires $T \approx 3 \times 10^8$ K, estimate the temperature needed for the first carbon-cycle reaction.

40. (III) The deuterium–tritium pellet in a laser fusion device contains equal numbers of $^2_1$H and $^3_1$H atoms raised to a density of 200×10^3 kg/m^3 by the laser pulses. Estimate (a) the density of particles in this compressed state, and (b) the length of time τ the pellet must be confined to meet the Lawson criterion for ignition.

Section 43–6

41. (I) A dose of 4.0 Sv of γ rays in a short period would be lethal to about half the people subjected to it. How many grays is this?

42. (I) Fifty rads of α-particle radiation is equivalent to how many rads of X-rays in terms of biological damage?

43. (I) How many rads of slow neutrons will do as much biological damage as 50 rads of fast neutrons?

44. (I) How much energy is deposited in the body of a 70-kg adult exposed to a 50-rad dose?

45. (II) A 0.025-μCi sample of $^{32}_{15}$P is injected into an animal for tracer studies. If a Geiger counter intercepts 20 percent of the emitted β particles, what will be the counting rate, assumed 90 percent efficient.

46. (II) A 1.0-mCi source of $^{32}_{15}$P (in NaHPO$_4$), a β^- emitter, is implanted in an organ where it is to administer 36 Gy. The half-life of $^{32}_{15}$P is 14.3 days and 1 mCi delivers about 10 mGy/min. Approximately how long should the source remain implanted?

47. (II) About 35 eV is required to produce one ion pair in air. Show that this is consistent with the two definitions of the roentgen given in the text.

48. (II) $^{57}_{27}$Co emits 122-keV γ rays. If a 70-kg person swallowed 1.85 μCi of $^{57}_{27}$Co, what would be the dose rate (Gy/day) averaged over the whole body? Assume that 50 percent of the γ-ray energy is deposited in the body.

49. (II) What is the mass of a 1.00-μCi $^{14}_{6}$C source?

50. (II) Huge amounts of radioactive $^{131}_{53}$I were released in the accident at Chernobyl in 1986. Chemically, iodine goes to the human thyroid. (Doctors can use it for diagnosis and treatment of thyroid problems.) In a normal thyroid, $^{131}_{53}$I absorption can cause damage to the thyroid. (a) Write down the decay scheme for $^{131}_{53}$I. (b) Its half-life is 8.0 d; how long would it take for ingested $^{131}_{53}$I to become 10 percent of the initial value? (c) Absorbing 1 mCi of $^{131}_{53}$I can be harmful; what mass of iodine is this?

51. (II) Assume a liter of milk typically has an activity of 2000 pCi due to $^{40}_{19}$K. If a person drinks two glasses (0.5 L) per day, estimate the total dose (in Sv and in rem) received in a year. As a crude model, assume the milk stays in the stomach 12 hr and is then released. Assume also that very roughly 10 percent of the 1.5 MeV released per decay is absorbed by the body. Compare your result to the normal allowed dose per year. Make your estimate for (a) a 50-kg adult, and (b) a 5-kg baby.

52. (II) Radon gas, $^{222}_{86}$Rn, is considered a serious health hazard (see discussion in text). It decays by α-emission. (a) What is the daughter nucleus? (b) Is the daughter nucleus stable or radioactive? If the latter, how does it decay, and what is its lifetime? (c) Is the daughter nucleus also a noble gas, or is it chemically reacting? (d) Suppose 1.0 ng of $^{222}_{86}$Rn seeps into a basement. What will be its activity? If the basement is then sealed, what will be the activity 1 month later?

* Section 43–9

* 53. (II) (a) Suppose for a conventional X-ray image that the X-ray beam consisted of parallel rays. What would be the magnification of the image? (b) Suppose, instead, that the X-rays came from a point source (as in Fig. 43–20a) that is 15 cm in front of a human body 25 cm thick, and the film is pressed against the person's back. Determine and discuss the range of magnifications that result.

* Section 43–10

* 54. (I) Calculate the wavelength of photons needed to produce NMR transitions in free protons in a 1.000-T field. In what region of the spectrum does it lie?

General Problems

55. J. Chadwick discovered the neutron by bombarding $^9_4$Be with the popular projectile of the day, alpha particles. (a) If one of the reaction products was the then unknown neutron, what was the other product? (b) What is the Q of this reaction?

56. Fusion temperatures are often given in keV. Determine the conversion factor from kelvins to keV using, as is common in this field, $\bar{K} = kT$ without the factor $\frac{3}{2}$.

57. One means of enriching uranium is by diffusion of the gas UF_6. Calculate the ratio of the speeds of molecules of this gas containing $^{235}_{92}$U and $^{238}_{92}$U, on which this process depends.

58. (a) What mass of $^{235}_{92}$U was actually fissioned in the first atomic bomb, whose energy was the equivalent of about 20 kilotons of TNT (1 kiloton of TNT releases 5×10^{12} J)? (b) What was the actual mass transformed to energy?

59. In a certain town the average yearly background radiation consists of 21 mrad of X-rays and γ rays plus 3.0 mrad of particles having a QF of 10. How many rem will a person receive per year on the average?

60. Deuterium makes up 0.015 percent of natural hydrogen. Make a rough estimate of the total deuterium in the Earth's oceans and estimate the total energy released if all of it were used in fusion reactors.

61. A shielded γ-ray source yields a dose rate of 0.052 rad/h at a distance of 1.0 m for an average-sized person. If workers are allowed a maximum dose rate of 5.0 rem/yr, how close to the source may they operate, assuming a 40-h work week? Assume that the intensity of radiation falls off as the square of the distance. (It actually falls off more rapidly than $1/r^2$ because of absorption in the air, so the answer above will give a better-than-permissible value.)

62. Radon gas, $^{222}_{86}$Rn, is formed by α decay. (a) Write the decay equation. (b) Ignoring the kinetic energy of the daughter nucleus (it's so massive), estimate the kinetic energy of the α particle produced. (c) Estimate the momentum of the alpha and of the daughter nucleus. (d) Estimate the kinetic energy of the daughter, and show that your approximation in (b) was valid.

63. The reaction $^{18}_8$O(p, n)$^{18}_9$F requires an input of energy equal to 2.453 MeV. What is the mass of $^{18}_9$F?

64. Consider a system of nuclear power plants that produce 3400 MW. (a) What total mass of $^{235}_{92}$U fuel would be required to operate these plants for 1 yr, assuming that 200 MeV is released per fission? (b) Typically 6 percent of the $^{235}_{92}$U nuclei that fission produce $^{90}_{38}$Sr, a β^- emitter with a half-life of 29 yr. What is the total radioactivity of the $^{90}_{38}$Sr, in curies, produced in 1 yr? (Neglect the fact that some of it decays during the 1-yr period.)

65. In the net reaction, Eq. 43–7, for the proton–proton cycle in the Sun, the neutrinos escape from the Sun with energy of about 0.5 MeV. The remaining energy, 26.2 MeV, is available within the Sun. Use this value to calculate the "heat of combustion" per kilogram of hydrogen fuel and compare it to the heat of combustion of coal, about 3×10^7 J/kg.

66. Energy reaches the Earth from the Sun at a rate of about 1400 W/m^2. Calculate (a) the total energy output of the Sun, and (b) the number of protons consumed per second in the reaction of Eq. 43–7, assuming that this is the source of all the Sun's energy. (c) Assuming that the Sun's mass of 2.0×10^{30} kg was originally all protons and that all could be involved in nuclear reactions in the Sun's core, how long would you expect the Sun to "glow" at its present rate?

67. Estimate how many solar neutrinos pass through a 100 m^2 ceiling in a room at latitude 40°, in a year. [Hint: See Problems 65 and 66.]

68. Show, using the laws of conservation of energy and momentum, that for a nuclear reaction requiring energy, the minimum kinetic energy of the bombarding particle (the threshold energy) is equal to $[-Qm_{pr}/(m_{pr} - m_b)]$, where $-Q$ is the energy required (difference in total mass between products and reactants), m_b is the rest mass of the bombarding particle, and m_{pr} the total rest mass of the products. Assume the target nucleus is at rest before an interaction takes place, and that all particles are nonrelativistic.

69. The early scattering experiments performed around 1910 in Ernest Rutherford's laboratory in England produced the first evidence that an atom consists of a heavy nucleus surrounded by electrons. In one of these experiments, α particles struck a gold-foil target 4.0×10^{-5} cm thick in which there were 5.9×10^{28} gold atoms per cubic meter. Although most α particles either passed straight through the foil or were scattered at small angles, approximately 1.6×10^{-3} percent were scattered at angles greater than 90°—that is, in the backward direction. (a) Calculate the cross section, in barns, for backward scattering. (b) Rutherford concluded that such backward scattering could occur only if an atom consisted of a very tiny, massive, and positively charged nucleus with electrons orbiting some distance away. Assuming that backward scattering occurs for nearly direct collisions (i.e., $\sigma \approx$ area of nucleus), estimate the diameter of a gold nucleus.

70. Some stars, in a later stage of evolution, may begin to fuse two $^{12}_6$C nuclei into one $^{24}_{12}$Mg nucleus. (a) How much energy would be released in such a reaction? (b) What kinetic energy must two carbon nuclei each have when far apart, if they can then approach each other to within 6.0 fm, center-to-center? (c) Approximately what temperature would this require?

71. An average adult body contains about 0.10 μCi of $^{40}_{19}$K, which comes from food. (a) How many decays occur per second? (b) The potassium decays produce beta particles with energies of around 1.4 MeV. Calculate the dose per year in sieverts for a 50-kg adult. Is this a significant fraction of the 3.6 mSv/year background rate?

72. When the nuclear reactor accident occurred at Chernobyl in 1986, 2.0×10^7 Ci were released into the atmosphere. Assuming that this radiation was distributed uniformly over the surface of the Earth, what was the activity per square meter? (The actual activity was not uniform; even within Europe wet areas received more radioactivity from rainfall).

This computer-generated reconstruction of a proton–antiproton collision at Fermilab (Fig. 44–4) occurred at a combined energy of nearly 2 TeV. It is one of the events that provided evidence for the top quark, announced in 1995. The particle tracks are curved, due to a magnetic field, and the radius of curvature is a measure of each particle's momentum (Chapter 27). The top quark (t) has too brief a lifetime ($\approx 10^{-23}$ s) to be detected itself, so we look for its possible decay products. Analysis indicates the following interaction and subsequent decays:

The tracks in the photo include jets (groups of particles moving in roughly the same direction), and a muon (μ^-) whose track is the pink one enclosed by a yellow rectangle to make it stand out. After reading this chapter, try to give the name for each symbol above and comment on whether all conservation laws hold.

CHAPTER 44

Elementary Particles

In the final two chapters of this book we discuss two of the most exciting areas of contemporary physics: (1) elementary particles, and (2) cosmology and astrophysics. These are subjects at the forefront of knowledge—one treats the smallest objects in the universe, the other the largest (and oldest) aspects of the universe. The reader who wants an understanding of the great beauties of present-day science—and/or wants to be a good citizen—will want to read these chapters, even if there is not time to cover them in a physics course.

In this penultimate chapter we discuss *elementary particle* physics, which represents the human endeavor to understand the basic building blocks of all matter. In the years after World War II, it was found that if the incoming particle in a nuclear reaction has sufficient energy, new types of particles can be produced. The earliest experiments used **cosmic rays**—particles that impinge on the Earth from space. But in order to produce high-energy particles in the laboratory, various types of particle accelerators have been constructed. Most commonly they accelerate protons or electrons, although heavy ions can also be accelerated, depending on the design. These high-energy accelerators have been used to probe the nucleus more deeply, to produce and study new particles, and to give us information about the basic forces and constituents of nature. Because the projectile particles are at high-energy, this field is sometimes called **high-energy physics**.

44–1 | High-Energy Particles

Particles accelerated to high energy are projectiles which can probe the interior of nuclei and nucleons that they strike. An important factor is that faster-moving projectiles can reveal more detail. The wavelength of projectile particles is given by de Broglie's wavelength formula (Eq. 38–7),

$$\lambda = \frac{h}{p}, \tag{44–1}$$

de Broglie
wavelength

showing that the greater the momentum p, of the bombarding particle, the shorter its wavelength. As discussed in Chapter 36 on diffraction, resolution of details in images is limited by the wavelength: the shorter the wavelength, the finer the detail that can be obtained. This is one reason why particle accelerators of higher and higher energy have been built in recent years.

EXAMPLE 44–1 **High resolution with electrons.** What is the wavelength, and hence the expected resolution, for a beam of 1.3-GeV electrons?

SOLUTION The 1.3 GeV = 1300 MeV refers to the kinetic energy, and is about 2500 times the rest energy (mass) of the electron $(mc^2 = 0.51 \text{ MeV})$. We are clearly dealing with relativistic speeds here, and it is easily shown (see Eqs. 37–10 and 37–13) that the speed of the electron is nearly $c = 3.0 \times 10^8 \text{ m/s}$. Therefore $K = (\gamma - 1) mc^2 \approx \gamma(mc^2) = 1300 \text{ MeV}$, and

$$\lambda = \frac{h}{p} = \frac{h}{\gamma m v} \approx \frac{h}{\gamma m c} = \frac{hc}{\gamma mc^2} \approx \frac{hc}{K}$$

$$= \frac{(6.6 \times 10^{-34} \text{ J} \cdot \text{s})(3.0 \times 10^8 \text{ m/s})}{(1.3 \times 10^9 \text{ eV})(1.6 \times 10^{-19} \text{ J/eV})} = 0.96 \times 10^{-15} \text{ m},$$

or 0.96 fm. The maximum possible resolution of this beam of electrons is far greater than for a light beam in a light microscope ($\lambda \approx 500 \text{ nm}$). Indeed, this resolution of about 1 fm is on the order of the size of nuclei (see Eq. 42–1).

Another major reason for building high-energy accelerators is that new particles of greater mass can be produced at higher energies, as we will discuss shortly. Now we look briefly at several types of particle accelerator.

44–2 | Particle Accelerators and Detectors

Van de Graaff Accelerator

The key component of a Van de Graaff accelerator (Fig. 44–1) is a Van de Graaff generator, invented in 1931. A large, hollow, spherical conductor is charged to a high potential by a nonconducting moving belt in the following way: A high voltage of typically 50,000 V is applied to a pointed conductor A, which "sprays" positive charge onto the moving belt. (Actually, electrons are pulled off the belt onto the electrode A.) The belt carries the positive charge into the interior of the sphere, where it is "wiped off" the belt at B and races to the outer surface of the spherical conductor. (Remember that charge collects on the outer surface of any conductor, since the charges repel each other and try to get as far from each other as possible.) Very high potential differences can be obtained in this way, often giving the accelerated particles as much as 30 MeV of kinetic energy when generators are used in tandem. Connected to the Van de Graaff generator is an evacuated tube which serves as the particle accelerator. A source of H or He ions (p or α) is located inside the tube, and the large positive voltage repels them so they are accelerated toward the grounded target at the far end of the tube (Fig. 44–1).

FIGURE 44–1 Van de Graaff generator and accelerator.

Cyclotron frequency

FIGURE 44–3 Diagram of a cyclotron. The magnetic field, applied by a large electromagnet, points into the page. A is the ion source. The field lines shown are for the electric field in the gap.

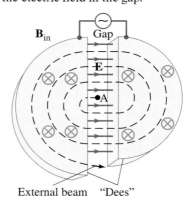

Cyclotron

The cyclotron, developed in 1930 by E. O. Lawrence (1901–1958; Fig. 44–2) at the University of California, Berkeley, uses a magnetic field to maintain charged ions—usually protons—in nearly circular paths (Chapter 27). The protons move within two D-shaped cavities, as shown in Fig. 44–3. Each time they pass into the gap between the "dees," a voltage accelerates them via the electric force, increasing their speed and increasing the radius of curvature of their path. After many revolutions, the protons acquire high kinetic energy and reach the outer edge of the cyclotron. They then either strike a target placed inside the cyclotron or leave the cyclotron with the help of a carefully placed "bending magnet" and are directed to an external target. Acceleration only occurs when the protons are in the gap *between* the dees, and the voltage must be alternating. When the protons are moving to the right across the gap in Fig. 44–3, the right dee must be electrically negative and the left one positive. A half-cycle later, the protons are moving to the left, so the left dee must be negative in order to accelerate them. The frequency, f, of the applied voltage must be equal to that of the circulating protons, and this is just the **cyclotron frequency** determined in Section 27–4, Eq. 27–6:

$$f = \frac{1}{T} = \frac{qB}{2\pi m}, \tag{44–2}$$

where q and m are the charge and mass of the particles moving in the magnetic field B.

EXAMPLE 44–2 Cyclotron. A small cyclotron of maximum radius $R = 0.25\,\text{m}$ accelerates protons in a 1.7-T magnetic field. Calculate (*a*) what frequency is needed for the applied alternating voltage, and (*b*) the kinetic energy of protons when they leave the cyclotron.

SOLUTION (*a*) From Eq. 44–2,

$$f = \frac{qB}{2\pi m}$$
$$= \frac{(1.6 \times 10^{-19}\,\text{C})(1.7\,\text{T})}{(6.28)(1.67 \times 10^{-27}\,\text{kg})} = 2.6 \times 10^{7}\,\text{Hz} = 26\,\text{MHz},$$

which is in the radio wave region of the EM spectrum (Fig. 32–12).
(*b*) The protons leave the cyclotron at $r = R = 0.25\,\text{m}$. From $F = ma$, we have (Eq. 27–5) $qvB = mv^2/r$, so $v = qBr/m$, and the kinetic energy is

$$K = \frac{1}{2}mv^2 = \frac{1}{2}m\frac{q^2B^2R^2}{m^2} = \frac{q^2B^2R^2}{2m}$$
$$= \frac{(1.6 \times 10^{-19}\,\text{C})^2(1.7\,\text{T})^2(0.25\,\text{m})^2}{(2)(1.67 \times 10^{-27}\,\text{kg})} = 1.4 \times 10^{-12}\,\text{J} = 8.7\,\text{MeV}.$$

Note that the magnitude of the voltage applied to the dees does not affect the final energy. But the higher this voltage, the fewer revolutions are required to bring the protons to full energy.

An important aspect of the cyclotron is that the frequency of the applied voltage, as given by Eq. 44–2, does not depend on the radius r. That is, the frequency does not have to be changed as the ions start from the source and are accelerated to paths of larger and larger radii. But this is only true at nonrelativistic energies. At higher speeds, the momentum is (Eq. 37–7) $p = mv/\sqrt{1 - v^2/c^2} = \gamma mv$, so we must replace m in Eq. 44–2 with γm. One way to build a machine for higher speeds is to use a magnetic field that increases with radius so as to keep $B/\gamma m$ constant as the particles move to larger radius and momentum, thus allowing the frequency to be kept constant. Another approach is to keep B uniform while decreasing the frequency in time as a packet of ions increase in speed and mass, and reach larger orbits. Such a machine is called a **synchrocyclotron**.

Synchrocyclotron

Synchrotron

Synchrotron

Another way to deal with the relativistic increase in effective mass (γm) with speed is to increase the magnetic field B in time as the particles speed up. Such devices are called **synchrotrons**, and today they can be enormous. The *Tevatron* accelerator at Fermilab (the Fermi National Accelerator Laboratory) at Batavia, Illinois, has a radius of 1.0 km, and that at CERN (European Center for Nuclear Research) in Geneva, Switzerland, is 4.3 km in radius. The Tevatron uses superconducting magnets to accelerate protons to about $1000 \text{ GeV} = 1 \text{ TeV}$ (hence its name; $1 \text{ TeV} = 10^{12} \text{ eV}$). These large synchrotrons use a narrow ring of magnets (see Fig. 44–4) with each magnet placed at the same radius from the center of the circle. The magnets are interrupted by gaps where high voltage accelerates the particles. Thus, once the particles are injected, they must move in a circle of constant radius. This is accomplished by giving them considerable energy initially in a much smaller accelerator, and then slowly increasing the magnetic field as they speed up in the large synchrotron.

One problem of any accelerator is that accelerating electric charges radiate electromagnetic energy (see Chapter 32). Since ions or electrons are accelerated in an accelerator, we can expect considerable energy to be lost by radiation. The effect increases with speed and is especially important in circular machines where centripetal acceleration is present, particularly in synchrotrons, and hence is called **synchrotron radiation**. Synchrotron radiation can actually be useful, however. Intense beams of photons are sometimes needed, and they are usually obtained from an electron synchrotron.

FIGURE 44–4 (a) Aerial view of Fermilab at Batavia, Illinois; the accelerator is a circular ring 1.0 km in radius. (b) The interior of the tunnel of the main accelerator at Fermilab. The upper (rectangular-shaped) ring of magnets is for the older 500-GeV accelerator. Below it is the ring of superconducting magnets for the 1-TeV Tevatron.

(a)

(b)

Linear Accelerators

Linac

A Van de Graaff accelerator is essentially a linear accelerator since the ions move in a linear path. But the name *linear accelerator* is usually reserved for a more complex arrangement in which particles are accelerated many, many times along a straight-line path. Figure 44–5a is a diagram of a simple "linac." The ions pass through a series of tubular conductors. The voltage applied to the tubes must be alternating so that when positive ions (say) reach a gap, the tube in front of them is negative and the one they just left is positive. This assures that they are accelerated at each gap. As the ions increase in speed, they cover more distance in the same amount of time. Consequently, the tubes must be longer the farther they are from the source. Linear accelerators are of particular importance for accelerating electrons. Because of their small mass, electrons reach high speeds very quickly; an electron linac such as the one shown in Fig. 44–5a would have tubes nearly equal in length, since the electrons would be traveling close to $c = 3.0 \times 10^8$ m/s for almost the entire distance. The amount of energy radiated away by electrons in a linear machine is much less than for a circular machine (see mention of synchrotron radiation above). The largest electron linear accelerator is that at Stanford (Stanford Linear Accelerator Center, or SLAC), Fig. 44–5b. It is about 3 km (2 miles) long and can accelerate electrons to 50 GeV. Many hospitals have 10-MeV electron linacs that produce photons to irradiate tumors.

FIGURE 44–5 (a) Diagram of a simple linear accelerator. (b) Photo of the Stanford Linear Accelerator (SLAC) in California.

(a)

(b)

Colliding Beams

Colliders

Originally, high-energy physics experiments were carried out by allowing a beam of particles from an accelerator to strike a stationary target. A newer, very powerful technique allows us to obtain the maximum possible collision energy from a given accelerator: two beams of particles are accelerated to very high energy and are steered so that they collide head-on. One way to accomplish such **colliding beams** with a single accelerator is through the use of **storage rings**, in which the particles of one beam can continue to circulate while the second beam is being accelerated. The two beams are then steered so that they intersect and collide head-on. For example, in the experiments that provided strong evidence for the top quark (see chapter opening photo and Section 44–9), the Fermilab Tevatron accelerated protons and antiprotons each to 900 GeV, so that the combined energy of head-on collisions was 1.8 TeV. The largest collider accelerator today is the Large Electron-Positron (LEP) Collider at CERN, which has a circumference of 26.7 km (Fig. 44–6). It produces oppositely revolving beams of e^+ and e^- each of energy 95 GeV, for a total interaction energy of 190 GeV. The most powerful accelerator-collider will soon be the Large Hadron Collider (LHC) at CERN, scheduled to be completed about 2005. The two colliding beams will each carry 7 TeV protons for a total interaction energy of 14 TeV.

FIGURE 44–6 The large circle represents the position of the tunnel now containing the 8.5 km diameter LEP collider, about 100 m below the ground, at CERN (near Geneva) on the French-Swiss border. The LHC will use the same tunnel. The smaller circle shows the position of the Super Proton Synchrotron that will be used for accelerating protons in the LHC.

EXAMPLE 44–3 **Protons at relativistic speeds.** Determine the energy required to accelerate a proton in a high-energy accelerator (a) from rest to $v = 0.900c$, and (b) from $v = 0.900c$ to $v = 0.999c$. (c) What is the kinetic energy achieved by each proton?

SOLUTION We use the work-energy principle, which is still valid relativistically as we saw in Section 37–11: $W = \Delta K$. The kinetic energy of a proton is given by Eq. 37–10

$$K = (\gamma - 1)mc^2,$$

where the relativistic factor γ is

$$\gamma = \frac{1}{\sqrt{1 - v^2/c^2}}.$$

The work-energy theorem becomes

$$W = \Delta K = (\Delta \gamma)mc^2$$

since m (the rest mass) and c are constant.
(a) For $v = 0$, $\gamma = 1$; and for $v = 0.900c$

$$\gamma = \frac{1}{\sqrt{1 - (0.900)^2}} = 2.29.$$

Hence, with rest mass of the proton $mc^2 = 938\ \text{MeV}$, the work needed is

$$W = \Delta K = (\Delta \gamma)mc^2 = (2.29 - 1.00)(938\ \text{MeV}) = 1.21\ \text{GeV}.$$

(b) For $v = 0.999c$,

$$\gamma = \frac{1}{\sqrt{1 - (0.999)^2}} = 22.4.$$

So the work needed to accelerate a proton from $0.900c$ to $0.999c$ is

$$W = \Delta K = (\Delta \gamma)mc^2 = (22.4 - 2.29)(938\ \text{MeV}) = 18.9\ \text{GeV}.$$

(c) The kinetic energy reached by the proton in (a) is just equal to the work done on it, $K = 1.21\ \text{GeV}$. The final kinetic energy of the proton in (b), moving at $v = 0.999c$, is

$$K = (\gamma - 1)mc^2 = (21.4)(938\ \text{MeV}) = 20.1\ \text{GeV},$$

which makes sense since, starting from rest, we did work $W = 1.21\ \text{GeV} + 18.9\ \text{GeV} = 20.1\ \text{GeV}$ on it.

FIGURE 44-7 (a) In a cloud or bubble chamber, droplets or bubbles are formed around ions produced by the passage of a charged particle. (b) Bubble-chamber photo of particle tracks, here revealing evidence for the Ω^- particle. Note the curvature of the tracks in the magnetic field.

Path of particle

(a)

(b)

Particle Detectors

Detectors of particles such as scintillators, photomultipliers, and semiconductor detectors, which we discussed in Section 42–11, are also used in the study of nuclear reactions and in elementary particle physics. More sophisticated devices have been developed that allow the track of a charged particle to be *seen*. The simplest is the **photographic emulsion** (which, being small and simple and therefore portable, is now used mainly for cosmic-ray studies from balloons). A particle passing through a layer of photographic emulsion ionizes the atoms along its path. This results in a chemical change at these points, and when the emulsion is developed, the particle's path is revealed.

In a **cloud chamber**, a gas is cooled to a temperature slightly below its usual condensation point. (It is said to be "supercooled.") The gas molecules begin to condense on any ionized molecules present. Thus the ions produced when a charged particle passes through serve as centers on which tiny droplets form (Fig. 44–7a). Light scatters more from these droplets than from the gas background, so a photo of the cloud chamber at the right moment shows the track of the particle. An important instrument in the early days of nuclear physics, it is little used today.

The **bubble chamber**, invented in 1952 by D. A. Glaser (1926–), makes use of a superheated liquid. The liquid is kept close to its normal boiling point and the bubbles characteristic of boiling form around ions produced by the passage of a charged particle (Fig. 44–7b). A photograph of the interior of the chamber thus reveals the paths of particles that recently passed through. Because the bubble chamber uses a liquid—often liquid hydrogen—the density of atoms is much greater than in a cloud chamber. Hence it is a much more efficient device for observing the tracks of charged particles and their interactions with the nuclei of the liquid. Usually, a magnetic field is applied across the chamber and the momentum of the moving particles can be determined from the radius of curvature of their paths.

Much more used today is the **wire drift chamber**, which consists of a set of closely spaced fine wires immersed in a gas. Many wires are grounded, and others in between are kept at very high voltage. A charged particle passing through produces ions in the gas. The freed electrons drift toward the nearest high voltage wires, creating an "avalanche," and producing an electric pulse or signal at that wire. The positions of the particles can be determined both by the position of the wire and by the time it takes the pulses to reach detectors at the ends of the wires. The paths of the particles are reconstructed electronically with computers which can then "draw" a picture of the tracks, as shown in the photo at the start of the chapter: the white dots are the wires, and the colored lines are the particle tracks. The device that produced that photo is shown in Fig. 44–8. (Another example is shown in Fig. 44–12).

FIGURE 44-8 Photo of the large and complicated CDF detector at Fermilab which, utilizing wire drift chambers, detected the particles and produced the image of the tracks shown at the start of this chapter.

44–3 | Beginnings of Elementary Particle Physics— Particle Exchange

By the mid-1930s, it was recognized that all atoms can be considered to be made up of neutrons, protons, and electrons. The basic constituents of the universe were no longer considered to be atoms but rather the proton, neutron, and electron. Besides these three *elementary particles*, several others were also known: the positron (a positive electron), the neutrino, and the γ particle (or photon), for a total of six elementary particles.

In the decades that followed, hundreds of other subnuclear particles were discovered. The properties and interactions of these particles, and which ones should be considered as fundamental or "elementary," became the substance of research in **elementary particle physics**.

Today, the standard model for elementary particles views even more fundamental entities, *quarks* and *leptons*, as the basic constituents of matter. In order to understand the standard model, we need to begin with the ideas leading up to its formulation.[†]

Elementary particle physics might be said to have begun in 1935 when the Japanese physicist Hideki Yukawa (1907–1981) predicted the existence of a new particle that would in some way mediate the strong nuclear force. To understand Yukawa's idea, we first consider the electromagnetic force. When we first discussed electricity, we saw that the electric force acts over a distance, without contact. To better perceive how a force can act over a distance, we saw that Faraday introduced the idea of a **field**. The force that one charged particle exerts on a second can be said to be due to the electric field set up by the first. Similarly, the magnetic field can be said to carry the magnetic force. Later (Chapter 32), we saw that electromagnetic (EM) fields can travel through space as waves. Finally, in Chapter 38, we saw that electromagnetic radiation (light) can be considered as either a wave or as a collection of particles called *photons*. Because of this wave–particle duality, it is possible to imagine that the electromagnetic force between charged particles is due to

(1) the EM field set up by one charged particle and felt by the other, or

(2) an exchange of photons or γ particles between them.

It is (2) that we want to concentrate on here, and a crude analogy for how an exchange of particles could give rise to a force is suggested in Fig. 44–9. In part (a), two children start throwing heavy pillows at each other; each catch results in the child being pushed backward by the impulse. This is the equivalent of a repulsive force. On the other hand, if the two children exchange pillows by grabbing them out of the other person's hand, they will be pulled toward each other, as when an attractive force acts.

For the electromagnetic force, it is photons that are exchanged between two charged particles that give rise to the force between them. A simple diagram describing this photon exchange is shown in Fig. 44–10. Such a diagram, called a **Feynman diagram** (after its inventor, the American physicist, Richard Feynman (1918–1988)), is based on the theory of **quantum electrodynamics** (QED).

Figure 44–10 represents the simplest case, in which a single photon is exchanged. One of the charged particles emits the photon and recoils somewhat as a result; and the second particle absorbs the photon. In any collision or *interaction*, energy and momentum are transferred from one charged particle to the other, carried by the photon. Because the photon is absorbed by the second particle very shortly after it is emitted by the first, it is not observable, and is referred to as a *virtual* photon, in contrast to one that is free and can be detected by instruments. The photon is said to *mediate*, or *carry*, the electromagnetic force.

(a) Repulsive force (children throwing pillows)

(b) Attractive force (children grabbing pillows from each other's hands)

FIGURE 44–9 Forces equivalent to particle exchange. (a) Repulsive force (children throwing pillows at each other). (b) Attractive force (children grabbing pillows from each other's hands).

FIGURE 44–10 Feynman diagram showing a photon acting as the carrier of the electromagnetic force between two electrons. This is sort of an *x* vs. *t* graph, with *t* increasing upward. Starting at the bottom, two electrons approach each other (the distance between them decreases in time). As they get close, momentum and energy get transferred from one to the other, carried by a photon (or, perhaps, by more than one), and the two electrons bounce apart.

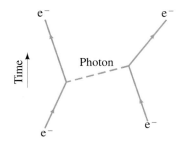

Particles that mediate or "carry" forces

[†] Just telling you how it is today would not be a scientific discussion but one of dogma—see footnote on p. 962.

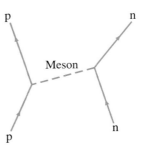

p   n

Meson

p   n

FIGURE 44–11 Meson exchange when a proton and neutron interact via the strong nuclear force.

By analogy with photon exchange that mediates the electromagnetic force, Yukawa argued in this early theory that there ought to be a particle that mediates the strong nuclear force—the force that holds nucleons together in the nucleus.

Yukawa called this predicted particle a **meson** (meaning "of medium mass"), and Fig. 44–11 is a Feynman diagram of meson exchange. We can make a rough

Mass estimate of exchange particle

estimate of the mass of the meson as follows. Suppose the proton on the left in Fig. 44–11 is at rest. For it to emit a meson would require energy (to make the meson's mass) that would have to come from nowhere, which would violate conservation of energy. But the uncertainty principle allows nonconservation of energy by an amount ΔE if it occurs only for a time Δt given by $(\Delta E)(\Delta t) \approx h/2\pi$. We set ΔE equal to the energy needed to create the mass m of the meson: $\Delta E = mc^2$. Conservation of energy is violated only as long as the meson exists, which is the time Δt required for the meson to pass from one nucleon to the other. If we assume the meson travels at relativistic speed, close to the speed of light c, then Δt need be at most about $\Delta t = d/c$, where d is the maximum distance that can separate the interacting nucleons. Thus we can write

$$\Delta E \, \Delta t \approx \frac{h}{2\pi}$$

$$mc^2\left(\frac{d}{c}\right) \approx \frac{h}{2\pi}$$

or

Mass of exchange particle

$$mc^2 \approx \frac{hc}{2\pi d}. \tag{44–3}$$

The range of the strong nuclear force (the maximum distance away it can be felt), is small—not much more than the size of a nucleon or small nucleus (see Eq. 42–1)—so let us take $d \approx 1.5 \times 10^{-15}$ m. Then from Eq. 44–3,

$$mc^2 \approx \frac{hc}{2\pi d} = \frac{(6.6 \times 10^{-34}\,\text{J·s})(3.0 \times 10^8\,\text{m/s})}{(6.28)(1.5 \times 10^{-15}\,\text{m})}$$

$$\approx 2.1 \times 10^{-11}\,\text{J} = 130\,\text{MeV}.$$

The mass of the predicted meson is roughly 130 MeV/c^2 or about 250 times the electron mass of 0.51 MeV/c^2. [Note, incidentally, that since the electromagnetic force has infinite range ($d = \infty$), Eq. 44–3 tells us that the exchanged particle for the electromagnetic force, the photon, will have zero rest mass.]

The particle predicted by Yukawa was discovered in cosmic rays by C. F. Powell and G. Occhialini in 1947. It is called the "π" or pi meson, or simply

Pion

the **pion**. It comes in three charge states: +, −, or 0. The π^+ and π^- have mass of 139.6 MeV/c^2 and the π^0 a mass of 135.0 MeV/c^2. All three interact strongly with matter. Reactions observed in the laboratory, using a particle accelerator, include

$$\text{p} + \text{p} \rightarrow \text{p} + \text{p} + \pi^0,$$

$$\text{p} + \text{p} \rightarrow \text{p} + \text{n} + \pi^+. \tag{44–4}$$

The incident proton from the accelerator must have sufficient energy to produce the additional mass of the free pion.

(a)

(b)

FIGURE 44–12 (a) Computer reconstruction of a Z-particle decay into an electron and a positron $(Z^0 \rightarrow e^+ + e^-)$ whose tracks are shown in white, which took place in the UA1 detector at CERN. (b) Photo of the UA1 detector at CERN as it was being built.

Yukawa's theory of pion exchange as carrier of the strong force is now out of date, and has been replaced by *quantum chromodynamics* in which the basic entities are *quarks*, and between them are exchanged *gluons* as the basic carriers of the strong force, as we shall discuss shortly. But the basic idea of the earlier theory, that forces can be understood as the exchange of particles, remains valid, as does Eq. 44–3 for estimating the mass of exchange particles.

There are four known types of force—or interaction—in nature. What about the other two: the weak nuclear force, and gravity? Theorists believe that these are also mediated by particles. The particles presumed to transmit the weak force are referred to as the W^+, W^-, and Z^0, and were only detected in 1983 (see Fig. 44–12). The quantum (or carrier) of the gravitational force is called the **graviton**, and if it exists it has not yet been observed. A comparison of the four forces is given in Table 44–1, where they are listed according to their (approximate) relative strengths. Notice that although gravity may be the most obvious force in daily life (because of the huge mass of the Earth), on a nuclear scale, it is much the weakest of the four forces and its effect at the nuclear or atomic level can nearly always be ignored.

Graviton

TABLE 44–1 The Four Forces in Nature

Type	Relative Strength (approx., for 2 protons in nucleus)	Field Particle
Strong nuclear	1	Gluons[†] (mesons)
Electromagnetic	10^{-2}	Photon
Weak nuclear	10^{-6}	$W^{\pm}$ and Z^0
Gravitational	10^{-38}	Graviton (?)

[†]Until the 1970s, thought to be mesons, but now gluons (see Section 44–10).

FIGURE 44–13 Liquid-hydrogen bubble-chamber photograph of an antiproton ($\overline{p}$) colliding with a proton, producing a Xi–anti-Xi pair ($\overline{p} + p \rightarrow \Xi^- + \overline{\Xi}^+$) that subsequently decay into other particles. The drawing indicates the assignment of particles to each track, which is based on how or if that particle decays, and on mass values estimated from measurement of momentum (curvature of track in magnetic field) and energy (heaviness of track, for example). Neutral particle paths are shown by dashed lines since neutral particles produce no bubbles and hence no tracks.

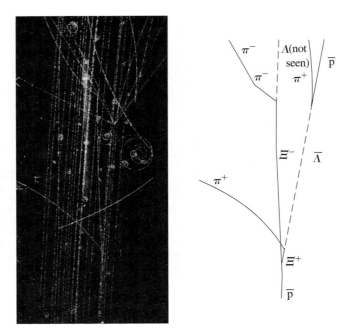

44–4 Particles and Antiparticles

The positron, as we have seen, is basically a positive electron. That is, many of its properties are the same as for the electron, such as mass, but it has the opposite charge. The positron is said to be the **antiparticle** to the electron. After the positron was discovered in 1932, it was predicted that other particles also ought to have antiparticles. In 1955 the antiparticle to the proton was found, the **antiproton** ($\overline{p}$), which carries a negative charge; see Fig. 44–13. (The bar over the p is used to indicate antiparticle.) Soon after, the antineutron ($\overline{n}$) was found. All other particles also have antiparticles. But a few particles, like the photon and the π^0, do not have distinct antiparticles—we say that they are their own antiparticles.

Antiparticles are produced in nuclear reactions when there is sufficient energy available, and they do not live very long in the presence of matter. For example, a positron is stable when by itself, but if it encounters an electron, the two annihilate each other. The energy of their vanished mass, plus any kinetic energy they possessed, is converted to the energy of γ rays or of other particles. Annihilation also occurs for all other particle–antiparticle pairs.

44–5 Particle Interactions and Conservation Laws

One of the important uses of high-energy accelerators is to study the interactions of elementary particles with each other. As a means of ordering this subnuclear world, the conservation laws are indispensable. The laws of conservation of energy, of momentum, of angular momentum, and of electric charge are found to hold precisely in all particle interactions.

A study of particle interactions has revealed a number of new conservation laws which (just like the old ones) are ordering principles: they help to explain why some reactions occur and others do not. For example, the following reaction has never been found to occur:

$$p + n \nrightarrow p + p + \overline{p}$$

even though charge, energy, and so on, are conserved ($\overline{p}$ means an antiproton and $\nrightarrow$ means the reaction does not occur). To understand why such a reaction does not occur, physicists hypothesized a new conservation law, the conservation of **baryon number**. (Baryon number is a generalization of nucleon number, which we saw earlier is conserved in nuclear reactions and decays.) An important addition to this law is the proposal that whereas all nucleons have baryon number $B = +1$, all antinucleons (antiprotons, antineutrons) have $B = -1$. The reaction above does not conserve baryon number since on the left side we have $B = (+1) + (+1) = +2$, and on the right $B = (+1) + (+1) + (-1) = +1$. On the other hand, the following reaction does conserve B and *does* occur if the incoming proton has sufficient energy:

$$p + p \rightarrow p + p + \bar{p} + p,$$
$$B = +1 + 1 = +1 + 1 - 1 + 1.$$

Baryon number

As indicated, $B = +2$ on both sides of this equation. From these and other reactions, the conservation of baryon number has been established as a basic law of physics.

Also useful are the conservation laws for the three **lepton numbers**, associated with weak interactions including decays. In ordinary β decay, an electron or positron is emitted along with a neutrino or antineutrino. In a similar type of decay, a particle known as a **muon** can be emitted instead of an electron. A muon seems to be much like an electron, except its mass is 207 times as large ($106\ \text{MeV}/c^2$). The neutrino (ν_e) that accompanies an emitted electron is found to be different from the neutrino (ν_μ) that accompanies an emitted muon. Each of these neutrinos has an antiparticle: $\bar{\nu}_e$ and $\bar{\nu}_\mu$. In ordinary β decay we have, for example,

Lepton numbers

$$n \rightarrow p + e^- + \bar{\nu}_e$$

but never $n \rightarrow p + e^- + \bar{\nu}_\mu$ nor $n \rightarrow p + e^- + \bar{\nu}_e + \nu_e$. To explain why these do not occur, the concept of electron lepton number, L_e, was invented. If the electron (e^-) and the electron neutrino (ν_e) are assigned $L_e = +1$, and e^+ and $\bar{\nu}_e$ are assigned $L_e = -1$, whereas all other particles have $L_e = 0$, then all observed decays conserve L_e. For example, in $n \rightarrow p + e^- + \bar{\nu}_e$, $L_e = 0$ initially, and $L_e = 0 + (+1) + (-1) = 0$ after the decay. Decays that do not conserve L_e but would obey the other conservation laws are not observed to occur. Hence it is believed that L_e is conserved in all[†] interactions.

In a decay involving muons, such as

$$\pi^+ \rightarrow \mu^+ + \nu_\mu,$$

a second quantum number, muon lepton number (L_μ), is conserved. The μ^- and ν_μ are assigned $L_\mu = +1$, and μ^+ and $\bar{\nu}_\mu$ have $L_\mu = -1$, whereas other particles have $L_\mu = 0$. It is believed that L_μ is also conserved in all interactions or decays. Similar assignments can be made for a third lepton number, L_τ, associated with the more recently discovered τ lepton and its neutrino, ν_τ.

Keep in mind that antiparticles have not only opposite electric charge from their particles, but also opposite B, L_e, L_μ, and L_τ.

Antiparticles have opposite Q, B, L

CONCEPTUAL EXAMPLE 44–4 | **Lepton number in muon decay.** Which of the following decay schemes is possible for muon decay: (a) $\mu^- \rightarrow e^- + \bar{\nu}_e$; (b) $\mu^- \rightarrow e^- + \bar{\nu}_e + \nu_\mu$; (c) $\mu^- \rightarrow e^- + \nu_e$? All of these particles have $L_\tau = 0$.

RESPONSE A μ^- has $L_\mu = +1$ and $L_e = 0$. This is the initial state, and the final state (after decay) must also have $L_\mu = +1$, $L_e = 0$. In (a), the final state has $L_\mu = 0 + 0 = 0$, and $L_e = +1 - 1 = 0$; L_μ would not be conserved and indeed this decay is not observed to occur. The final state of (b) has $L_\mu = 0 + 0 + 1 = +1$ and $L_e = +1 - 1 + 0 = 0$, so both L_μ and L_e are conserved. This is in fact the most common decay mode of the μ^-. Finally, (c) does not occur because L_e ($= +2$ in the final state) is not conserved, nor is L_μ.

[†] Recent evidence hints that under certain circumstances, lepton number may not always be conserved.

EXAMPLE 44–5 **Energy and momentum are conserved.** In addition to the "number" conservation laws which help explain the decay schemes of particles, we can also apply the laws of conservation of energy and momentum. The decay of a Σ^+ particle with a rest mass of 1189 MeV/c^2 (see Table 44–2 in the next Section) commonly yields a proton (rest mass of 938 MeV/c^2) and a neutral pion, π^0 (rest mass of 135 MeV/c^2). What are the kinetic energies of the decay products, assuming the Σ^+ parent particle was at rest?

SOLUTION The energy released through the decay can be calculated from the change in the rest energies, as we did for nuclear processes (Eq. 42–3 or 43–1):

$$Q = [m_{\Sigma^+} - (m_p + m_{\pi^0})]c^2 = [1189 - (938 + 135)]\,\text{MeV} = 116\,\text{MeV}.$$

This energy Q becomes the kinetic energy of the resulting decay particles, p and π^0:

$$Q = K_p + K_{\pi^0}$$

with each particle's kinetic energy related to its momentum by (Eqs. 37–11 and 13):

$$K_p = E_p - m_p c^2 = \sqrt{(p_p c)^2 + (m_p c^2)^2} - m_p c^2,$$

and similarly for the pion. From momentum conservation, the proton and pion have the same magnitude of momentum since the original particle was at rest: $p_p = p_\pi = p$. Then

$$Q = 116\,\text{MeV} = \left[\sqrt{(pc)^2 + (938\,\text{MeV})^2} - 938\,\text{MeV}\right]$$
$$+ \left[\sqrt{(pc)^2 + (135\,\text{MeV})^2} - 135\,\text{MeV}\right].$$

We solved this for pc, which gives $pc = 189$ MeV. Substituting into the expression for the kinetic energy, first for the proton, then for the pion we obtain $K_p = 19$ MeV and $K_{\pi^0} = 97$ MeV.

44–6 Particle Classification

In the decades following the discovery of the π meson in the late 1940s, a great many other subnuclear particles were discovered, now numbering in the hundreds. It is useful to arrange the particles in categories according to their properties. One way of doing this is according to their interactions, since not all particles interact by means of all four of the forces known in nature (though all interact via gravity). Table 44–2 lists some of the more common particles classified in this way along with many of their properties. The particles listed are those that are stable, and many that are unstable. At the top of the table are the **gauge bosons** (so-named[†] after the theory that describes them, "gauge theory"), which include the *photon*, and the W and Z particles, that mediate the electromagnetic and weak interactions, respectively.

Gauge bosons

Leptons

Next in Table 44–2 are the **leptons**, which are particles that do not interact via the strong force but do interact via the weak nuclear force (as well as the much weaker gravitational force); those that carry electric charge also interact via the electromagnetic force. The leptons include the electron, the muon, and the tau (or τ, discovered in 1976 and more than 3000 times heavier than the electron), and three types of neutrino: the electron neutrino (ν_e), the muon neutrino (ν_μ), and the tau neutrino (ν_τ). They each have antiparticles, as indicated in Table 44–2.

Hadrons

The third category of particle in Table 44–2 is the **hadron**. Hadrons are those particles that interact via the strong nuclear force. Hence they are said to be **strongly interacting particles**. They also interact via the other forces, but the strong

[†]Bosons are particles that are not governed by the Pauli exclusion principle (see Sections 40–4, 41–6, and 44–10).

TABLE 44–2 Particles (stable under strong decay)[†]

Category	Forces involved	Particle name	Symbol	Anti-particle	Spin	Rest Mass (MeV/c^2)	B	L_e	L_μ	L_τ	S	Lifetime (s)	Principal Decay Modes
Gauge bosons	em	Photon	γ	Self	1	0	0	0	0	0	0	Stable	
	w, em	W	W^+	W^-	1	80.41×10^3	0	0	0	0	0	3×10^{-25}	$e\nu_e, \mu\nu_\mu, \tau\nu_\tau$, hadrons
	w	Z	Z^0	Self	1	91.19×10^3	0	0	0	0	0	3×10^{-25}	$e^+e^-, \mu^+\mu^-, \tau^+\tau^-$, hadrons
Leptons	w, em[‡]	Electron	e^-	e^+	$\frac{1}{2}$	0.511	0	+1	0	0	0	Stable	
		Neutrino (e)	ν_e	$\bar{\nu}_e$	$\frac{1}{2}$	$0(<4 \times 10^{-6})$[‡]	0	+1	0	0	0	Stable	
		Muon	μ^-	μ^+	$\frac{1}{2}$	105.7	0	0	+1	0	0	2.20×10^{-6}	$e^- \bar{\nu}_e \nu_\mu$
		Neutrino (μ)	ν_μ	$\bar{\nu}_\mu$	$\frac{1}{2}$	$0(<0.17)$[‡]	0	0	+1	0	0	Stable	
		Tau	τ^-	τ^+	$\frac{1}{2}$	1777	0	0	0	+1	0	2.91×10^{-13}	$\mu^- \bar{\nu}_\mu \nu_\tau, e^- \bar{\nu}_e \nu_\tau$, hadrons $+ \nu_\tau$
		Neutrino (τ)	ν_τ	$\bar{\nu}_\tau$	$\frac{1}{2}$	$0(<18)$[‡]	0	0	0	+1	0	Stable	
Hadrons (selected)													
Mesons	s, em, w	Pion	π^+	π^-	0	139.6	0	0	0	0	0	2.60×10^{-8}	$\mu^+ \nu_\mu$
			π^0	Self	0	135.0	0	0	0	0	0	0.84×10^{-16}	2γ
		Kaon	K^+	K^-	0	493.7	0	0	0	0	+1	1.24×10^{-8}	$\mu^+ \nu_\mu, \pi^+ \pi^0$
			K_S^0	$\bar{K}_S^0$	0	497.7	0	0	0	0	+1	0.89×10^{-10}	$\pi^+ \pi^-, 2\pi^0$
			K_L^0	$\bar{K}_L^0$	0	497.7	0	0	0	0	+1	5.17×10^{-8}	$\pi^\pm e^\mp \overset{(-)}{\nu_e}, \pi^\pm \mu^\mp \overset{(-)}{\nu_\mu}, 3\pi$
		Eta	η^0	Self	0	547.3	0	0	0	0	0	5×10^{-19}	$2\gamma, 3\pi^0, \pi^+ \pi^- \pi^0$
		and others											
Baryons	s, em, w	Proton	p	$\bar{p}$	$\frac{1}{2}$	938.3	+1	0	0	0	0	Stable	
		Neutron	n	$\bar{n}$	$\frac{1}{2}$	939.6	+1	0	0	0	0	887	$pe^- \bar{\nu}_e$
		Lambda	Λ^0	$\bar{\Lambda}^0$	$\frac{1}{2}$	1115.7	+1	0	0	0	−1	2.63×10^{-10}	$p\pi^-, n\pi^0$
		Sigma	Σ^+	$\bar{\Sigma}^-$	$\frac{1}{2}$	1189.4	+1	0	0	0	−1	0.80×10^{-10}	$p\pi^0, n\pi^+$
			Σ^0	$\bar{\Sigma}^0$	$\frac{1}{2}$	1192.6	+1	0	0	0	−1	7.4×10^{-20}	$\Lambda^0 \gamma$
			Σ^-	$\bar{\Sigma}^+$	$\frac{1}{2}$	1197.4	+1	0	0	0	−1	1.48×10^{-10}	$n\pi^-$
		Xi	Ξ^0	$\bar{\Xi}^0$	$\frac{1}{2}$	1314.9	+1	0	0	0	−2	2.90×10^{-10}	$\Lambda^0 \pi^0$
			Ξ^-	$\bar{\Xi}^+$	$\frac{1}{2}$	1321.3	+1	0	0	0	−2	1.64×10^{-10}	$\Lambda^0 \pi^-$
		Omega	Ω^-	Ω^+	$\frac{1}{2}$	1672.5	+1	0	0	0	−3	0.82×10^{-10}	$\Xi^0 \pi^-, \Lambda^0 K^-, \Xi^- \pi^0$
		and others											

[†] See also Table 44–4 for particles with charm and bottomness.
[‡] Neutrinos partake only in the weak interaction. Experimental upper limits on neutrino masses are given in parentheses. Recent evidence (2000) suggests that at least one of the neutrinos may have mass, although probably quite small ($<1\,\text{eV}$).

force predominates at short distances. The hadrons include nucleons, pions, and a large number of other particles. They are divided into two subgroups: **baryons**, which are those particles that have baryon number +1 (or −1 in the case of their antiparticles); and **mesons**, which have baryon number $= 0$.

Baryons

Mesons

Notice that the baryons Λ, Σ, Ξ, and Ω all decay to lighter-mass baryons, and eventually to a proton or neutron. All these processes conserve baryon number. Since there is no lighter particle than the proton with $B = +1$, if baryon number is strictly conserved, the proton itself cannot decay and is stable. (But see Section 44–11.)

44–7 Particle Stability and Resonances

Many of the particles listed in Table 44–2 are unstable. The lifetime of an unstable particle depends on which force is most active in causing the decay. When we say the strong nuclear force is stronger than the electromagnetic, we mean that two particles will interact more strongly and more quickly if this force is acting. When a stronger force influences a decay, that decay occurs more quickly. Decays caused by the weak force typically have lifetimes of $10^{-13}\,\text{s}$ or longer. Particles that decay via the electromagnetic force have much shorter lifetimes, typically about 10^{-16} to $10^{-19}\,\text{s}$. (Exceptions to this scheme are the W and Z particles, which decay via the

Lifetime depends on which force is acting

FIGURE 44–14 Number of π^+ particles scattered by a proton target as a function of the incident π^+ kinetic energy. The resonance shape represents the formation of a short-lived particle, the Δ, which has a charge in this case of $+2e(\Delta^{++})$.

weak interaction but have very short lifetimes due to their special nature as exchange particles.) The unstable particles listed in Table 44–2 decay either via the weak or the electromagnetic interaction. Decays that involve a γ (photon) are electromagnetic (such as $\pi^0 \to 2\gamma$). The other decays shown take place via the weak interaction, often accompanied by a neutrino which interacts only via the weak interaction: examples are $\pi^+ \to \mu^+\nu_\mu$ and $\Sigma^- \to n\pi^-$.

A great many particles have been found that decay via the strong interaction, and these are not listed in Table 44–2. Such particles decay into other strongly interacting particles (say, n, p, π, but not involving γ, e, ν, and so on) and their lifetimes are very short, typically about 10^{-23} s. In fact their lifetimes are so short that they do not travel far enough to be detected before decaying. Their decay products can be detected, however, and it is from them that the existence of such short-lived particles is inferred. To see how this is done, we consider the first such particle discovered (by Fermi). Fermi used a beam of π^+ directed through a hydrogen target (protons) with varying amounts of energy. A graph of the number of interactions (π^+ scattered) versus the pion's kinetic energy is shown in Fig. 44–14. The large peak around 200 MeV was much higher than expected and certainly much higher than the number of interactions at neighboring energies. This led Fermi to conclude that the π^+ and proton combined momentarily to form a short-lived particle before coming apart again, or at least that they resonated together for a short time. Indeed, the large peak in Fig. 44–14 resembles a resonance curve (see Figs. 14–24, 14–27, and 31–9), and this new "particle"—now called the Δ—is referred to as a **resonance**. Hundreds of other resonances have been found in a similar way. Many resonances are regarded as excited states of other particles such as of the nucleon (proton or neutron).

The width of a resonance—in Fig. 44–14 the width of the Δ peak is on the order of 100 MeV—is an interesting application of the uncertainty principle. If a particle lives only 10^{-23} s, then its mass (i.e., its rest energy) will be uncertain by an amount $\Delta E \approx h/2\pi\Delta t \approx (6.6 \times 10^{-34}\,\text{J}\cdot\text{s})/(6)(10^{-23}\,\text{s}) \approx 10^{-11}\text{J} \approx 100\,\text{MeV}$, which is what is observed. Actually, the lifetimes of $\approx 10^{-23}$ s for such resonances are inferred by the reverse process: from the measured width being ≈ 100 MeV.

Very short-lived particles are inferred from their decay products

Resonance

44–8 | Strange Particles

In the early 1950s, certain of the newly found particles, namely, the K, Λ, and Σ, were found to behave rather strangely in two ways. First, they were always produced in pairs. For example, the reaction

$$\pi^- + p \rightarrow K^0 + \Lambda^0$$

occurred with high probability, but the reaction $\pi^- + p \rightarrow K^0 + n$ was never observed to occur. This seemed strange because the unobserved reaction would not have violated any known conservation law, and plenty of energy was available. The second feature of these **strange particles** (as they came to be called) was that, although they were clearly produced via the strong interaction (that is, at a high rate), they did not decay at a rate characteristic of the strong interaction even though they decayed into strongly interacting particles (for example, K $\rightarrow 2\pi$, $\Sigma^+ \rightarrow p + \pi^0$). Instead of lifetimes of 10^{-23} s as expected for strongly interacting particles, strange particles have lifetimes of 10^{-10} to 10^{-8} s, which are characteristic of the weak interaction.

To explain these observations, a new quantum number, **strangeness**, and a new conservation law, conservation of strangeness, were introduced. By assigning the strangeness numbers (S) indicated in Table 44–2, the production of strange particles in pairs was readily explained. Antiparticles were assigned opposite strangeness from their particles: one of each pair was assigned $S = +1$ and the other $S = -1$ (see Table 44–2). For example, in the reaction $\pi^- + p \rightarrow K^0 + \Lambda^0$, the initial state has strangeness $S = 0 + 0 = 0$, and the final state has $S = +1 - 1 = 0$, so strangeness is conserved. But for $\pi^- + p \rightarrow K^0 + n$, the initial state has $S = 0$ and the final state has $S = +1 + 0 = +1$, so strangeness would not be conserved; and this reaction is not observed.

Strangeness and its conservation

To explain the decay of strange particles, it is assumed that strangeness is conserved in the strong interaction but is *not* conserved in the weak interaction. Thus, although strange particles were forbidden by strangeness conservation to decay to nonstrange particles of lower mass via the strong interaction, they could undergo such decay by means of the weak interaction. This would occur much more slowly, of course, which accounts for their longer lifetimes of 10^{-10} to 10^{-8} s.

Strangeness conserved in strong interactions but not in weak

The conservation of strangeness was the first example of a "partially conserved" quantity. In this case, the quantity strangeness is conserved by strong interactions but not by weak.

CONCEPTUAL EXAMPLE 44–6 | **Guess the missing particle.** Using the conservation laws for particle interactions, determine the other particle as a result of the reaction

$$\pi^- + p \rightarrow K^0 + ?,$$

in addition to $K^0 + \Lambda^0$.

SOLUTION We write equations for the conserved numbers in this reaction, with B, L_e, S, and Q as unknowns whose determination will reveal what the possible particle might be:

Baryon number: $0 + 1 = 0 + B$
Lepton number: $0 + 0 = 0 + L_e$
Charge: $-1 + 1 = 0 + Q$
Strangeness: $0 + 0 = 1 + S.$

The unknown product particle would have to have these characteristics:

$$B = +1 \qquad L_e = 0 \qquad Q = 0 \qquad S = -1.$$

In addition to Λ^0, a neutral sigma particle, Σ^0, is also consistent with these numbers.

SECTION 44–8   **1129**

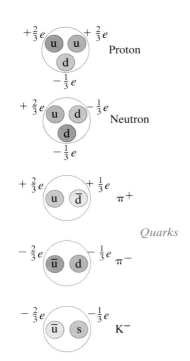

FIGURE 44–15 Quark compositions for several particles.

44-9 Quarks

We saw in our discussion of Table 44–2 that all particles, except the gauge bosons, fall into either of two categories: leptons and hadrons. The principal difference between these two groups is that the hadrons interact via the strong interaction, whereas the leptons do not. Another important difference that physicists had to deal with in the 1960s was that there were only four known leptons (e^-, μ^-, ν_e, ν_μ; the τ and ν_τ were not yet discovered), but there were well over a hundred hadrons.

The leptons are considered to be truly elementary particles since they do not seem to break down into smaller entities, do not show any internal structure, and have no measurable size. (Attempts to determine the size of leptons have put an upper limit of about 10^{-18} m.)

The hadrons, on the other hand, are more complex. Experiments indicate they do have an internal structure. And the fact that there are so many of them suggests that they can't all be elementary. To deal with this problem, M. Gell-Mann and G. Zweig in 1963 independently proposed that none of the hadrons so far observed, not even the proton and neutron, was elementary. Instead, they proposed that the hadrons are made up of combinations of three, more fundamental, pointlike entities called **quarks**.[†] Today, the quark theory is well-accepted, and quarks are considered the truly elementary particles, like leptons. The three quarks were labeled u, d, s, and given the names *up*, *down*, and *strange*. They were assumed to have fractional charge ($\frac{1}{3}$ or $\frac{2}{3}$ the charge on the electron—that is, less than the previously thought smallest charge). Other properties of quarks and anti-quarks are indicated in Table 44–3. All hadrons known at the time could be constructed in theory from these three types of quark. Mesons would consist of a quark–antiquark pair. For example, a π^+ meson is considered a $u\bar{d}$ pair (note that

[†] Gell-Mann chose the word from a phrase in James Joyce's *Finnegans Wake*.

TABLE 44–3 Properties of Quarks and Antiquarks

Quarks								
Name	Symbol	Spin	Charge	Baryon Number	Strangeness	Charm	Bottomness	Topness
Up	u	$\frac{1}{2}$	$+\frac{2}{3}e$	$\frac{1}{3}$	0	0	0	0
Down	d	$\frac{1}{2}$	$-\frac{1}{3}e$	$\frac{1}{3}$	0	0	0	0
Strange	s	$\frac{1}{2}$	$-\frac{1}{3}e$	$\frac{1}{3}$	-1	0	0	0
Charmed	c	$\frac{1}{2}$	$+\frac{2}{3}e$	$\frac{1}{3}$	0	$+1$	0	0
Bottom	b	$\frac{1}{2}$	$-\frac{1}{3}e$	$\frac{1}{3}$	0	0	-1	0
Top	t	$\frac{1}{2}$	$+\frac{2}{3}e$	$\frac{1}{3}$	0	0	0	$+1$

Antiquarks								
Name	Symbol	Spin	Charge	Baryon Number	Strangeness	Charm	Bottomness	Topness
Up	$\bar{u}$	$\frac{1}{2}$	$-\frac{2}{3}e$	$-\frac{1}{3}$	0	0	0	0
Down	$\bar{d}$	$\frac{1}{2}$	$+\frac{1}{3}e$	$-\frac{1}{3}$	0	0	0	0
Strange	$\bar{s}$	$\frac{1}{2}$	$+\frac{1}{3}e$	$-\frac{1}{3}$	$+1$	0	0	0
Charmed	$\bar{c}$	$\frac{1}{2}$	$-\frac{2}{3}e$	$-\frac{1}{3}$	0	-1	0	0
Bottom	$\bar{b}$	$\frac{1}{2}$	$+\frac{1}{3}e$	$-\frac{1}{3}$	0	0	$+1$	0
Top	$\bar{t}$	$\frac{1}{2}$	$-\frac{2}{3}e$	$-\frac{1}{3}$	0	0	0	-1

TABLE 44–4 Partial List of Hadrons Associated with Charm and Bottomness ($L_e = L_\mu = L_\tau = 0$)

Category	Particle	Anti-particle	Spin	Rest Mass (MeV/c^2)	Baryon Number	Strange-ness	Charm	Bottom-ness	Lifetime (s)	Principal Decay Modes
Mesons	D^+	D^-	0	1869.4	0	0	+1	0	10.6×10^{-13}	K + others, e + others
	D^0	$\overline{D}^0$	0	1864.6	0	0	+1	0	4.2×10^{-13}	K + others, μ or e + others
	D_s^+	D_s^-	0	1969	0	+1	+1	0	4.7×10^{-13}	K + others
	J/ψ (3097)	Self	1	3096.9	0	0	0	0	0.8×10^{-20}	Hadrons, e^+e^-, $\mu^+\mu^-$
	Υ (9460)	Self	1	9460.4	0	0	0	0	1.3×10^{-20}	Hadrons, $\mu^+\mu^-$, e^+e^-, $\tau^+\tau^-$
	B^-	B^+	0	5279	0	0	0	−1	1.5×10^{-12}	D^0 + others
	B^0	$\overline{B}^0$	0	5279	0	0	0	−1	1.5×10^{-12}	D^0 + others
Baryons	Λ_c^+	Λ_c^-	$\frac{1}{2}$	2285	+1	0	+1	0	2.0×10^{-13}	Hadrons (e.g., Λ + others)
	Σ_c^{++}	Σ_c^{--}	$\frac{1}{2}$	2453	+1	0	+1	0	?	$\Lambda_c^+ \pi^+$
	Σ_c^+	Σ_c^-	$\frac{1}{2}$	2454	+1	0	+1	0	?	$\Lambda_c^+ \pi^0$
	Σ_c^0	$\overline{\Sigma}_c^0$	$\frac{1}{2}$	2452	+1	0	+1	0	?	$\Lambda_c^+ \pi^-$
	Λ_b^0	$\overline{\Lambda}_b^0$	$\frac{1}{2}$	5640	+1	0	0	−1	1.1×10^{-12}	$J/\psi \Lambda^0$, $pD^0\pi^-$, $\Lambda_c^+\pi^+\pi^-\pi^-$

for the $u\overline{d}$ pair, $Q = \frac{2}{3}e + \frac{1}{3}e = +1e$, $B = \frac{1}{3} - \frac{1}{3} = 0$, $S = 0 + 0 = 0$, as they must for a π^+); and a $K^+ = u\overline{s}$, with $Q = +1$, $B = 0$, $S = +1$. Baryons, on the other hand, would consist of three quarks. For example, a neutron is $n = ddu$, whereas an antiproton is $\overline{p} = \overline{u}\overline{u}\overline{d}$. See Fig. 44–15.

Soon after the quark theory was proposed, physicists began looking for these fractionally charged particles, but direct detection has not been successful. Indeed, current models suggest that quarks may be so tightly bound together that they may not ever exist singly in the free state.

In 1964 several physicists proposed that there ought to be a fourth quark. Their argument was based on the expectation that there exists a deep symmetry in nature, including a connection between quarks and leptons. If there are four leptons (as was thought in the 1960s), then symmetry in nature would suggest there should also be four quarks. The fourth quark was said to be *charmed*. Its charge would be $+\frac{2}{3}e$ and it would have another property to distinguish it from the other three quarks. This new property, or quantum number, was called **charm** (see Table 44–3). Charm was assumed to be like strangeness: it would be conserved in strong and electromagnetic interactions, but would not be conserved by the weak. The new charmed quark would have charm $C = +1$ and its antiquark $C = -1$. The first charmed particle, the J/ψ meson, was discovered in 1974.

Charm

Charmed quark

Also in the 1970s strong evidence appeared for the tau (τ) lepton, with a mass of 1777 MeV/c^2. This lepton, like the electron and muon, presumably has a neutrino associated with it. Thus, the family of leptons is at present believed to have six members. This would upset the balance between leptons and quarks, the presumed basic building blocks of matter, unless two additional quarks also exist. Indeed, theoretical physicists postulated the existence of a fifth and sixth quark, named **top** and **bottom**. The names apply also to the new properties (quantum numbers) that distinguish the new quarks from the old quarks (see Table 44–3), and which (like strangeness) are conserved in strong, but not weak, interactions. New mesons involving b quarks were soon detected (Table 44–4). Convincing evidence for the top quark came in 1995 (see photo at start of this chapter) after years of searching. Its extremely high mass of about 175 GeV/c^2 contributed to its elusiveness because of the huge energy required to produce it.

Two more leptons

t and b quarks

TABLE 44–5 The Elementary Particles[†] as Seen Today

	First generation	Second generation	Third generation
Quarks	u, d	s, c	b, t
Leptons	e, ν_e	μ, ν_μ	τ, ν_τ
Gauge bosons	γ(photon)	$W^\pm$, Z^0	gluons

[†]Note that the quarks and leptons are arranged into three generations each, and the gauge particles are arranged in groups for the forces they mediate.

The elementary particles

Today, the truly elementary particles are considered to be the six quarks, the six leptons, and the gauge bosons that carry the fundamental forces. See Table 44–5, where the quarks and leptons are arranged in three groups (generations).

CONCEPTUAL EXAMPLE 44–7 | Quark combinations. Find the baryon number, charge, and strangeness for the following quark combinations and identify the hadron particle that is made up of these quark combinations: (*a*) udd, (*b*) u$\bar{u}$, (*c*) uss, (*d*) sdd, and (*e*) b$\bar{u}$.

RESPONSE We use Table 44–3 to get the properties of the quarks, then Table 44–2 or 44–4 to find the particle that has these properties. (*a*) udd has

$$Q = +\tfrac{2}{3}e - \tfrac{1}{3}e - \tfrac{1}{3}e = 0,$$
$$B = \tfrac{1}{3} + \tfrac{1}{3} + \tfrac{1}{3} = 1,$$
$$S = 0 + 0 + 0 = 0,$$

as well as $C = 0$, bottomness = 0, topness = 0. The only baryon ($B = +1$) that has $Q = 0$, $S = 0$, etc., is the neutron (Table 44–2).
(*b*) u$\bar{u}$ has $Q = \tfrac{2}{3}e - \tfrac{2}{3}e = 0$, $B = 0$, and all other quantum numbers = 0. Sounds like a π^0.
(*c*) uss has $Q = 0$, $B = +1$, $S = -2$, others = 0. This is a Ξ^0.
(*d*) sdd has $Q = -1$, $B = +1$, $S = -1$, so must be a Σ^-.
(*e*) b$\bar{u}$ has $Q = -1$, $B = 0$, $S = 0$, $C = 0$, bottomness = -1, topness = 0. This must be a B^- meson (Table 44–4).

44–10 The "Standard Model": Quantum Chromodynamics (QCD) and the Electroweak Theory

Not long after the quark theory was proposed, it was suggested that quarks have another property (or quality) called **color**. The distinction between the six quarks (u, d, s, c, b, t) was referred to as **flavor**. According to theory, each of the flavors of quark can have three colors, usually designated red, green, and blue. (These are the three primary colors which, when added together in equal amounts, as on a TV screen, produce white.) Note that the names "color" and "flavor" have nothing to do with our senses, but are purely whimsical—as are other names, such as charm, in this new field. (We did, however, "color" the quarks in Fig. 44–15.) The antiquarks are colored antired, antigreen, and antiblue. Baryons are made up of three quarks, one of each color. Mesons consist of a quark–antiquark pair of a particular color and its anticolor. Thus both baryons and mesons are colorless.

Originally, the idea of quark color was proposed to preserve the Pauli exclusion principle (Section 40–4). Not all particles obey the exclusion principle. Those that do, such as electrons, protons, and neutrons, are called **fermions**. Those that don't are called **bosons**. These two categories are distinguished also in their spin (Section 40–2): bosons have integer spin (0, 1, 2, etc.) whereas fermions have half-integer spin, usually $\tfrac{1}{2}$, as for electrons and nucleons, but can also be $\tfrac{3}{2}, \tfrac{5}{2}$, etc. Matter is made up mainly of fermions, but the carriers of the forces (γ, W, Z, and gluons, as we'll see) are all bosons. Quarks are fermions (they have spin $\tfrac{1}{2}$) and

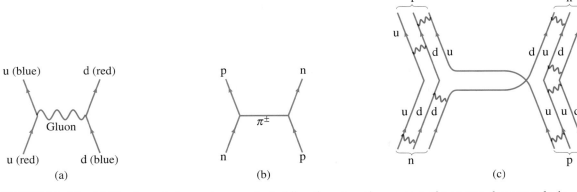

FIGURE 44–16 (a) The force between two quarks holding them together as part of a proton, for example, is carried by a gluon, which in this case involves a change in color. (b) Strong interaction n + p → n + p with the exchange of a charged π meson (+ or −, depending on whether it is considered moving to the left or to the right). (c) Quark representation of the same interaction n + p → n + p. The wavy lines between quarks represent gluon exchanges holding the hadrons together.

therefore should obey the exclusion principle. Yet for three particular baryons (uuu, ddd, and sss), all three quarks would have the same quantum numbers, and at least two of them have their spin in the same direction (since there are only two choices, spin up $[m_s = +\frac{1}{2}]$ or spin down $[m_s = -\frac{1}{2}]$). This would seem to violate the exclusion principle; but if quarks have an additional quantum number (color), which could be different for each quark, it would serve to distinguish them and the exclusion principle would hold. Although quark color, and the resulting threefold increase in the number of quarks, was thus originally an *ad hoc* idea, it also served to bring the theory into better agreement with experiment, such as predicting the correct lifetime of the π^0 meson. The idea of color soon became, in addition, a central feature of the theory as determining the force binding quarks together in a hadron. Each quark is assumed to carry a *color charge*, analogous to electric charge, and the strong force between quarks is often referred to as the **color force**. This new theory of the strong force is called **quantum chromodynamics** (*chroma* = color in Greek), or **QCD**, to indicate that the force acts between color charges (and not between, say, electric charges). The strong force between two hadrons[†] is considered to be a force between the quarks that make them up, as suggested in Fig. 44–16. The particles that transmit the color force (analogous to photons for the EM force) are called **gluons** (a play on "glue"). They are included in Table 44–5. There are eight gluons, according to the theory, all massless, and six of them have color charge.[‡] Thus gluons have replaced mesons (Table 44–1) as the particles carrying the strong (color) force.

QCD

Gluons

The color force has the interesting property that its strength increases with increasing distance (as does the force exerted by a coiled spring, Chapter 14); as two quarks approach each other very closely (equivalently, have high energy), the force between them becomes very small. This aspect is referred to as **asymptotic freedom**.

The weak force, as we have seen, is thought to be mediated by the W^+, W^-, and Z^0 particles. It acts between the "weak charges" that each particle has. Each elementary particle can thus have electric charge, weak charge, color charge, and gravitational mass, although one or more of these could be zero. For example, all leptons have color charge of zero, so they do not interact via the strong force.

[†] The strong force between hadrons appears feeble, however, in comparison to the force directly between quarks within hadrons.

[‡] Compare to the EM interaction, where the photon has no electric charge. Because gluons have color charge, they could attract each other and form composite particles (photons cannot). Such "glueballs" are being searched for.

To summarize, the standard model says that the truly elementary particles (Table 44–5) are the leptons, the quarks, and the gauge bosons (photon, W and Z, and the gluons). Some theories suggest there may be other bosons as well. The photon and the leptons are observed in experiments, and finally so too were the W^+, W^-, and Z^0. But so far only combinations of quarks (baryons and mesons) have been observed, and it seems likely that free quarks and gluons are unobservable.

One important aspect of new theoretical work is the attempt to find a unified basis for the different forces in nature. This was a long-held hope of Einstein, which he was never able to fulfill. A so-called **gauge theory** that unifies the weak and electromagnetic interactions was put forward in the 1960s by S. Weinberg, S. Glashow, and A. Salam. In this **electroweak theory**, the weak and electromagnetic forces are seen as two different manifestations of a single, more fundamental, *electroweak* interaction. The electroweak theory has had many successes, including the prediction of the $W^\pm$ particles as carriers of the weak force, with masses of $81 \pm 2\ GeV/c^2$ in excellent agreement with the measured values of $80.41 \pm 0.10\ GeV/c^2$ (and similar accuracy for the Z^0). The electroweak theory plus QCD for the strong interaction are often referred to today as the **Standard Model**.

Theoreticians have wondered why the W and Z have large masses rather than being massless like the photon. Electroweak theory suggests an explanation by means of a new **Higgs field** and its particle, the **Higgs boson**, which interact with the W and Z to "slow them down." In being forced to go slower than the speed of light, they must acquire mass.

Electroweak theory

Standard Model

44–11 | Grand Unified Theories

With the success of the unified electroweak theory, attempts have recently been made to incorporate it and QCD for the strong (color) force into a so-called **grand unified theory** (GUT). One type of such a grand unified theory of the electromagnetic, weak, and strong forces has been worked out in which there is only one class of particle—leptons and quarks belong to the same family and are able to change freely from one type to the other—and the three forces are different aspects of a single underlying force. The unity is predicted to occur, however, only on a scale of less than about $10^{-30}\ m$. If two elementary particles (leptons or quarks) approach each other to within this **unification scale**, the apparently fundamental distinction between them would not exist at this level, and a quark could readily change to a lepton, or vice versa. Baryon and lepton numbers would not be conserved. The weak, electromagnetic, and strong (color) force would blend to a force of a single strength.

GUT

Unification of forces

How could a lepton become a quark, or vice versa? The theory predicts the existence of particles, called X bosons, that can be exchanged between a quark and a lepton allowing one to change into the other, somewhat like the charged pion exchanged between the p and n in Fig. 44–16b allows the proton on the right to become a neutron. The mass of the proposed X boson, consistent with the uncertainty principle as applied earlier in this chapter (see Eq. 44–3 with $d \approx 10^{-30}\ m$), would be about $10^{14}\ GeV/c^2$, or 10^{14} times the proton mass. With such an incredibly large mass, there is little hope of seeing them in the laboratory. It is also this huge mass that would keep baryon and lepton numbers conserved in observed reactions, since the likelihood of producing such a massive particle, even as a virtual exchange particle, is extremely small at even the highest laboratory energies.

What happens between the unification distance of $10^{-30}\ m$ and more normal (larger) distances is referred to as **symmetry breaking**. As an analogy, consider an atom in a crystal. Deep within the atom, there is much symmetry—in the innermost regions the electron cloud is spherically symmetric (Chapter 40). Farther out, this symmetry breaks down—the electron clouds are distributed preferentially along the lines (bonds) joining the atoms in the crystal. In a similar way, at $10^{-30}\ m$ the force between elementary particles is theorized to be a single force—it is symmetrical

Symmetry breaking

and does not single out one type of "charge" over another. But at larger distances, that symmetry is broken and we see three distinct forces. (In the "standard model" of electroweak interactions, Section 44–10, the symmetry breaking between the electromagnetic and the weak interactions occurs at about 10^{-18} m.)

CONCEPTUAL EXAMPLE 44–8 **Symmetry.** The table in Fig. 44–17 has four identical place settings. Four people sit down to eat. Describe the symmetry of this table and what happens to it when someone starts the meal.

RESPONSE The table has several kinds of symmetry. It is symmetric to rotations of 90°: that is, the table will look the same if everyone moved one chair to the left or to the right. It is also north–south symmetric and east–west symmetric, so that swaps across the table don't affect the way the table looks. It also doesn't matter whether any person picks up the fork to the left of the plate or the fork to the right. But once that first person picks up either fork, the choice is set for all the rest at the table as well. The symmetry has been *broken*. The underlying symmetry is still there—the water glasses could still be chosen either way—but some choice must get made and at that moment the symmetry of the diners is broken.

FIGURE 44–17 Symmetry around a table. Example 44–8.

Proton decay?

Since unification occurs at such tiny distances and huge energies, the theory is difficult to test experimentally. But it is not completely impossible. One possibly testable prediction is the basis for the idea that the proton might decay (via, for example, $p \rightarrow \pi^0 + e^+$) and violate conservation of baryon number. This could happen if two quarks approached to within 10^{-31} m of each other. But it is very unlikely at normal temperature and energy, so the decay of a proton can only be an unlikely process. In the simplest form of GUT, the theoretical estimate of the proton lifetime for the decay mode $p \rightarrow \pi^0 + e^+$ is $\approx 10^{31}$ yr, and this has just come within the realm of testability. Proton decays have still not been seen and experiments put the lower limit on the proton lifetime for the above mode to be about 2.5×10^{32} yr, an order of magnitude greater than this prediction. This may seem a disappointment, but on the other hand, it presents a challenge. Indeed more complex GUTs are not affected by this result.

EXAMPLE 44–9 **ESTIMATE** **Proton decay.** An experiment uses 3300 tons of water waiting to see a proton decay of the type $p \rightarrow \pi^0 + e^+$. If the experiment is run for four years without detecting a decay, estimate the lower limit on the proton half-life.

SOLUTION As with radioactive decay, the number of decays is proportional to the number of parent species (N), the time interval (Δt), and the decay constant (λ) which is related to the half-life $T_{\frac{1}{2}}$ by (see Eqs. 42–4 and 42–8):

$$\Delta N = -\lambda N \, \Delta t = -\frac{\ln 2}{T_{\frac{1}{2}}} N \, \Delta t.$$

Thus, dealing only with magnitudes,

$$T_{\frac{1}{2}} = \frac{N}{\Delta N} \Delta t \ln 2$$

which tells us that, for $\Delta N < 1$ over the four-year trial:

$$T_{\frac{1}{2}} > N(4 \, \text{yr})(0.69)$$

where N is the number of protons in 3300 tons of water. To determine N, we note that each molecule of H_2O contains $(2 + 8 =)$ 10 protons. So one mole of water (18 g) contains $10 \times 6 \times 10^{23}$ protons in 18 g of water, or about 3×10^{26} protons per kilogram. One ton is 10^3 kg, so the chamber contains $(3.3 \times 10^6 \, \text{kg}) \cdot (3 \times 10^{26} \, \text{protons/kg}) \approx 1 \times 10^{33}$ protons. Then our very rough estimate for a lower limit on the proton half-life is $T_{\frac{1}{2}} > (10^{33})(4 \, \text{yr})(0.7) \approx 3 \times 10^{33}$ yr.

Connection
with
cosmology

An interesting prediction of unified theories relates to cosmology (see Chapter 45). It is thought that during the first 10^{-35} s after the theorized big bang that created the universe, the temperature was so extremely high that particles had energies corresponding to the unification scale. Baryon number would not have been conserved then, thus allowing an imbalance that might account for the observed predominance of matter ($B > 0$) over antimatter ($B < 0$) in the universe.

This last example is interesting, for it illustrates a deep connection between investigations at either end of the size scale: theories about the tiniest objects (elementary particles) have a strong bearing on the understanding of the universe as a whole. We will look at this more in the next chapter.

Even more ambitious than grand unified theories are attempts to also incorporate gravity, and thus unify all four forces in nature into a single theory. (Such theories are sometimes referred to misleadingly as **theories of everything**.) The only consistent theory so far that attempts to unify all four forces is called **string theory**, in which the elementary particles (Table 44–5) are imagined not as points but as one-dimensional strings about 10^{-35} m long.

String theory

Supersymmetry

A related idea is **supersymmetry**, which applied to strings is known as **superstring theory**. Supersymmetry predicts that interactions exist that would change fermions into bosons and vice versa, and that all known fermions have supersymmetric boson partners. Thus, for every quark there would be a *squark*, and for every lepton there would be a *slepton*. Likewise, for every known boson (photons and gluons, for example), there would be a supersymmetric fermion (*photinos* and *gluinos*). But why hasn't this "missing half" of the universe ever been detected? The best guess is that supersymmetric particles might be heavier than their conventional counterparts, perhaps too heavy to have been produced in today's accelerators. Until a supersymmetric particle is found, however, supersymmetry is just an elegant guess.

The world of elementary particles is opening new vistas. What happens in the near future is bound to be exciting.

Summary

Particle accelerators are used to accelerate charged particles, such as electrons and protons, to very high energy. High-energy particles have short wavelength and so can be used to probe the structure of matter at very small distances in great detail. High kinetic energy also allows the creation of new particles through collision (via $E = mc^2$).

Van De Graaff and linear accelerators use high voltage to accelerate particles along a line. Cyclotrons and synchrotrons use a magnetic field to keep the particles in a circular path and accelerate them at intervals by high voltage. **Colliding beams** allow higher interaction energy.

Particle **detectors** that can image particle tracks include photographic emulsions, cloud chambers, bubble chambers, and today wire drift chambers.

An **antiparticle** has the same mass as a particle but opposite charge. Certain other properties may also be opposite: for example, the antiproton has **baryon number** (nucleon number) opposite to that for the proton.

In all nuclear and particle reactions, the following conservation laws hold: momentum, angular momentum, mass-energy, electric charge, baryon number, and the three **lepton numbers**.

Certain particles have a property, called **strangeness**, which is conserved by the strong force but not by the weak force. The more recently noted properties, **charm**, **bottomness**, and **topness**, also are believed to be conserved by the strong force but not by the weak.

Just as the electromagnetic force can be said to be due to an exchange of photons, the strong nuclear force was first thought to be carried by *mesons* that have rest mass, or, according to more recent theory, by massless **gluons**. The W and Z particles carry the weak force. These fundamental force carriers (photon, W and Z, gluons) are called **gauge bosons**.

Other particles can be classified as either *leptons* or *hadrons*. **Leptons** participate in the weak and electromagnetic interactions. **Hadrons**, which today are considered to be made up of **quarks**, participate in the strong interaction as well. The hadrons can be classified as **mesons**, with baryon number zero, and **baryons**, with nonzero baryon number.

All particles, except for the photon, electron, neutrinos, and proton, decay with measurable half-lives varying from 10^{-25} s to 10^3 s. The half-life depends on which force is predominant in the decay. Weak decays usually have half-lives greater than about 10^{-13} s. Electromagnetic decays have half-lives on the order of 10^{-16} to 10^{-19} s. The shortest lived particles, called **resonances**, decay via the strong interaction and live typically for only about 10^{-23} s.

Today's standard model of elementary particles considers **quarks** as the basic building blocks of the hadrons. The six quarks are called **up**, **down**, **strange**, **charmed**, **bottom**, and **top** quarks. It is expected that there are the same number of quarks as leptons (six of each), and that quarks and leptons are the truly elementary particles

along with the gauge bosons (γ, W, Z, gluons). Quarks are said to have **color**, and, according to **quantum chromodynamics** (QCD), the strong color force acts between their color charges and is transmitted by **gluons**. **Electroweak theory** views the weak and electromagnetic forces as two aspects of a single underlying interaction. QCD plus the electroweak theory are referred to as the **Standard Model**.

Grand unified theories of forces suggest that at very short distances (10^{-30} m) and very high energy, the weak, electromagnetic, and strong forces appear as a single force, and the fundamental difference between quarks and leptons disappears.

Questions

1. Give a reaction between two nucleons, similar to Eq. 44–4, that could produce a π^-.

2. If a proton is moving at very high speed, so that its kinetic energy is much greater than its rest energy $(m_0 c^2)$, can it then decay via p → n + π^+?

3. What would an "antiatom," made up of the antiparticles to the constituents of normal atoms, consist of? What might happen if *antimatter*, made of such antiatoms, came in contact with our normal world of matter?

4. What particle in a decay signals the electromagnetic interaction?

5. Does the presence of a neutrino among the decay products of a particle necessarily mean that the decay occurs via the weak interaction? Do all decays via the weak interaction produce a neutrino?

6. Why is it that a neutron decays via the weak interaction even though the neutron and one of its decay products (proton) are strongly interacting?

7. Which of the four interactions (strong, electromagnetic, weak, gravitational) does an electron take part in? A neutrino? A proton?

8. Check that charge and baryon number are conserved in each of the decays in Table 44–2.

9. Which of the particle decays in Table 44–2 occur via the electromagnetic interaction?

10. Which of the particle decays in Table 44–2 occur by the weak interaction?

11. By what interaction, and why, does $\Sigma^\pm$ decay to Λ^0? What about Σ^0 decaying to Λ^0?

12. The Δ baryon has spin $\frac{3}{2}$, baryon number 1, and charge $Q = +2$, +1, 0, or −1. Why is there no charge state $Q = -2$?

13. Which of the particle decays in Table 44–4 occur via the electromagnetic interaction?

14. Which of the particle decays in Table 44–4 occur by the weak interaction?

15. Quarks have spin $\frac{1}{2}$. How do you account for the fact that baryons have spin $\frac{1}{2}$ or $\frac{3}{2}$, and mesons have spin 0 or 1?

16. Suppose there were a kind of "neutrinolet" that was massless, had no color charge or electrical charge, and did not feel the weak force. Could you say that this particle even exists?

Problems

Sections 44–1 and 44–2

1. (I) What is the total energy of a proton whose kinetic energy is 6.35 GeV?

2. (I) Calculate the wavelength of 35-GeV electrons.

3. (I) What strength of magnetic field is used in a cyclotron in which protons make 2.8×10^7 revolutions per second?

4. (I) What is the time for one complete revolution for a very high-energy proton in the 1.0-km-radius Fermilab accelerator?

5. (I) If α particles are accelerated by the cyclotron of Example 44–2, what must be the frequency of the voltage applied to the dees?

6. (II) (a) If the cyclotron of Example 44–2 accelerated α particles, what maximum energy could they attain? What would their speed be? (b) Repeat for deuterons (^2_1H). (c) In each case, what frequency of voltage is required?

7. (II) Which is better for picking out details of the nucleus: 30-MeV alpha particles or 30-MeV protons? Compare each of their wavelengths with the size of a nucleon in a nucleus.

8. (II) The voltage across the dees of a cyclotron is 55 kV. How many revolutions do protons make to reach a kinetic energy of 25 MeV?

9. (II) What is the wavelength, and maximum resolving power attainable, using 900-GeV protons at Fermilab?

10. (II) A cyclotron with a radius of 1.0 m is to accelerate deuterons (^2_1H) to an energy of 10 MeV. (a) What is the required magnetic field? (b) What frequency is needed for the voltage between the dees? (c) If the potential difference between the dees averages 22 kV, how many revolutions will the particles make before exiting? (d) How much time does it take for one deuteron to go from start to exit, and (e) how far does it travel during this time?

11. (II) Protons are injected into the 1.0-km-radius Fermilab Tevatron with an energy of 8.0 GeV. If they are accelerated by 2.5 MV each revolution, how far do they travel and approximately how long does it take for them to reach 900 GeV?

12. (II) The Fermilab Tevatron takes about 20 seconds to bring the energies of the stored protons from 150 GeV to 900 GeV. The acceleration is done once per turn. Estimate the energy given to the protons on each turn. (You can assume that the speed of the protons is essentially c the whole time.)

13. (II) Show that the energy of a particle (charge e) in a synchrotron, in the relativistic limit ($v \approx c$), is given by E(in eV) $= Brc$, where B is magnetic field strength and r the radius of the orbit (SI units).

14. (II) What magnetic field intensity is needed at the 1.0-km-radius Fermilab synchrotron for 900-GeV protons?

Sections 44–3 to 44–6

15. (I) How much energy is released in the decay
$$\pi^+ \to \mu^+ + \nu_\mu ?$$

16. (I) About how much energy is released when a Λ^0 decays to n $+ \pi^0$? (See Table 44–2.)

17. (I) How much energy is required to produce a neutron–antineutron pair?

18. (I) Estimate the range of the strong force if the mediating particle were the kaon instead of the pion.

19. (II) Two protons are heading toward each other with equal speeds. What minimum kinetic energy must each have if a π^0 meson is to be created in the process? (See Table 44–2.)

20. (II) Which of the following decays are possible? For those that are forbidden, explain which laws are violated.
(a) $\Xi^0 \to \Sigma^+ + \pi^-$ (b) $\Omega^- \to \Sigma^0 + \pi^- + \nu$
(c) $\Sigma^0 \to \Lambda^0 + \gamma + \gamma$.

21. (II) Estimate the range of the weak force using Eq. 44–3, given the masses of the W and Z particles as about 80 to 90 GeV/c^2.

22. (II) What are the wavelengths of the two photons produced when a proton and antiproton at rest annihilate?

23. (II) (a) Show, by conserving momentum and energy, that it is impossible for an isolated electron to radiate only a single photon. (b) With this result in mind, how can you defend the photon exchange diagram in Fig. 44–10?

24. (II) What would be the wavelengths of the two photons produced when an electron and a positron, each with 420 keV of kinetic energy, annihilate?

25. (II) In the rare decay $\pi^+ \to e^+ + \nu_e$, what is the kinetic energy of the positron? Assume the π^+ decays from rest.

26. (II) What minimum kinetic energy must a neutron and proton each have if they are traveling at the same speed toward each other, collide, and produce a K^+K^- pair in addition to themselves? (See Table 44–2.)

27. (II) Calculate the kinetic energy of each of the two products in the decay $\Xi^- \to \Lambda^0 + \pi^-$. Assume the Ξ^- decays from rest.

28. (III) Could a π^+ meson be produced if a 100-MeV proton struck a proton at rest? What minimum kinetic energy must the incoming proton have?

29. (III) Calculate the maximum kinetic energy of the electron in the decay $\mu^- \to e^- + \bar{\nu}_e + \nu_\mu$. [Hint: In what direction do the two neutrinos move relative to the electron in order to give the latter the maximum kinetic energy? Both energy and momentum are conserved; use relativistic formulas.]

Sections 44–7 to 44–11

30. (I) Use Fig. 44–14 to estimate the energy width and then the lifetime of the Δ resonance using the uncertainty principle.

31. (I) The measured width of the J/ψ meson is 88 keV. Estimate its lifetime.

32. (I) The measured width of the ψ(3685) meson is 277 keV. Estimate its lifetime.

33. (I) What is the energy width (or uncertainty) of (a) η^0, and (b) Σ^0?

34. (I) The B$^-$ meson is presumed to be a b$\bar{u}$ quark combination. (a) Show that this is consistent for all quantum numbers. (b) What are the quark combinations for B$^+$, B^0, $\bar{B}^0$?

35. (II) What are the quark combinations that can form (a) a neutron, (b) an antineutron, (c) a Λ^0, (d) a $\bar{\Sigma}^0$?

36. (II) What particles do the following quark combinations produce? (a) uud, (b) $\bar{u}\bar{u}\bar{s}$, (c) $\bar{u}s$, (d) d$\bar{u}$, (e) $\bar{c}$s.

37. (II) What is the quark combination needed to produce a D^0 meson ($Q = B = S = 0$, $C = +1$)?

38. (II) The D$_S^+$ meson has $S = C = +1$, $B = 0$. What quark combination would produce it?

39. (II) Draw possible Feynman diagrams using quarks (as in Fig. 44–16c) for the reactions (a) $\pi^- + p \to \pi^0 + n$, (b) $\bar{p} + p \to 2\pi^0$.

40. (II) Draw a possible quark Feynman diagram (see Fig. 44–16c) for reaction K$^-$ + p $\to$ K$^-$ + p.

General Problems

41. What is the total energy of a proton whose kinetic energy is 25 GeV? What is its wavelength?

42. Assume there are 5×10^{13} protons at 900 GeV stored in the 1.0-km-radius ring of the Tevatron. (a) How much current (amperes) is carried by this beam? (b) How fast would a 1500 kg car have to move to carry the same kinetic energy as this beam?

43. The 4.25-km-radius LEP tunnel will be reused to house the magnets for the Large Hadron Collider (LHC). If the design calls for proton beams of energy 7.0 TeV, what magnetic field will be required?

44. (a) How much energy is released when an electron and a positron annihilate each other? (b) How much energy is released when a proton and an antiproton annihilate each other?

45. Which of the following reactions are possible, and by what interaction could they occur? For those forbidden, explain why.
   (a) $\pi^- + p \rightarrow K^+ + \Sigma^-$
   (b) $\pi^+ + p \rightarrow K^+ + \Sigma^+$
   (c) $\pi^- + p \rightarrow \Lambda^0 + K^0 + \pi^0$
   (d) $\pi^+ + p \rightarrow \Sigma^0 + \pi^0$
   (e) $\pi^- + p \rightarrow p + e^- + \bar{\nu}_e$
   (f) $\pi^- + p \rightarrow K^0 + p + \pi^0$
   (g) $K^- + p \rightarrow \Lambda^0 + \pi^0$
   (h) $K^+ + n \rightarrow \Sigma^+ + \pi^0 + \gamma$
   (i) $K^+ \rightarrow \pi^0 + \pi^0 + \pi^+$
   (j) $\pi^+ \rightarrow e^+ + \nu_e$

46. For the decay $\Lambda^0 \rightarrow p + \pi^-$, calculate (a) the Q-value (energy released), and (b) the kinetic energy of the p and π^-, assuming the Λ^0 decays from rest. (Use relativistic formulas.)

47. Symmetry breaking occurs in the electroweak theory at about 10^{-18} m. Show that this corresponds to an energy that is on the order of the mass of the $W^\pm$.

48. The mass of a π^0 can be measured by observing the reaction $\pi^- + p \rightarrow \pi^0 + n$ at very low incident π^- kinetic energy (assume it is zero). The neutron is observed to be emitted with a kinetic energy of 0.60 MeV. Use conservation of energy and momentum to determine the π^0 mass.

49. Calculate the Q-value for each of the reactions, Eq. 44–4, for producing a pion.

50. Calculate the Q-value for the reaction $\pi^- + p \rightarrow \Lambda^0 + K^0$, when negative pions strike stationary protons. Estimate the minimum pion kinetic energy needed to produce this reaction.

51. How many fundamental fermions are there in a water molecule?

52. Determine the maximum kinetic energy of (a) the positron, and (b) the π^-, in the decay $K^0 \rightarrow \pi^- + e^+ + \nu_e$.

53. (a) Show that the so-called unification distance of 10^{-30} m in grand unified theory is equivalent to an energy of about 10^{14} GeV. Use either the uncertainty principle or de Broglie's wavelength formula, and explain how they apply. (b) Calculate what temperature this corresponds to.

54. For the reaction $p + p \rightarrow 3p + \bar{p}$, where one of the initial protons is at rest, use relativistic formulas to show that the threshold energy is $6m_p c^2$, equal to three times the Q-value of the reaction, where m_p is the proton mass.

55. The lifetimes listed in Table 44–2 are in terms of *proper time*, measured in a reference frame where the particle is at rest. If a tau-lepton is created with a kinetic energy of 450 MeV, how long would its track be as measured in the lab, on average, ignoring any collisions?

56. A particle at rest, with a rest-energy of $m_p c^2$, decays into two fragments with rest energies of $m_{D1} c^2$ and $m_{D2} c^2$. Show that the kinetic energy of fragment D1 is

$$K_{D1} = \frac{1}{2m_p c^2}\left[(m_p c^2 - m_{D1} c^2)^2 - (m_{D2} c^2)^2\right]$$

57. Use the quark model to describe the reaction

$$\bar{p} + n \rightarrow \pi^- + \pi^0$$

58. What fraction of the speed of light c is the speed of a 7.0 TeV proton?

Hubble Space Telescope photo of columns of gas and dust inside M16, the Eagle Nebula. The Eagle Nebula is thought to be a site of recent star formation. To discuss the nature of the universe as we understand it today, we first discuss stars and galaxies and how they form and evolve, including the role of nucleosynthesis. We briefly discuss Einstein's general theory of relativity, which deals with gravity and curvature of space. Finally we look at the evidence for the expansion of the universe, and the standard moded of the univese evolving from an initial Big Bang.

CHAPTER 45

Astrophysics and Cosmology

I n the previous chapter, we studied the tiniest objects in the universe—the elementary particles. Now we leap to the largest—stars and galaxies. These two extreme realms, elementary particles and the cosmos, are among the most intriguing and exciting subjects in science. And, surprising though it may seem, these distant realms are related in a fundamental way, as already hinted in Chapter 44.

Use of the techniques and ideas of physics to study the heavens is often referred to as **astrophysics**. At the base of our present theoretical understanding of the universe is Einstein's *general theory of relativity* and its theory of gravitation— for in the large-scale structure of the universe, gravity is the dominant force. General relativity serves also as the foundation for modern **cosmology**, which is the study of the universe as a whole. Cosmology deals especially with the search for a theoretical framework to understand the observed universe, its origin, and its future. The questions posed by cosmology are complex and difficult; the possible answers are often unimaginable. They are questions like "Has the universe always existed, or did it have a beginning in time?" Either alternative is difficult to imagine: time going back indefinitely into the past, or an actual moment when the universe began (but, then, what was there before?). And what about the size of the universe? Is it infinite in size? It is hard to imagine infinity. Or is it finite in size? This is also hard to imagine, for if the universe is finite, it does not make sense to ask what is beyond it, because the universe is all there is.

Our survey of astrophysics and cosmology will be necessarily brief and qualitative, but we will nonetheless touch on the major ideas. We begin with a look at what can be seen beyond the Earth.

45–1 Stars and Galaxies

According to the ancients, the stars, except for the few that seemed to move (the planets), were fixed on a sphere beyond the last planet. The universe was neatly self-contained, and we on Earth were at or near its center. But in the centuries following Galileo's first telescopic observations of the heavens in 1610, our view of the universe has changed dramatically. We no longer place ourselves at the center, and we view the universe as vastly larger. The distances involved are so great that we specify them in terms of the time it takes light to travel the given distance: for example, 1 light-second $= (3.0 \times 10^8 \text{ m/s})(1.0 \text{ s}) = 3.0 \times 10^8 \text{ m} = 3.0 \times 10^5 \text{ km}$; 1 light-minute $= 18 \times 10^6 \text{ km}$; and 1 **light-year** (ly) is

$$1 \text{ ly} = (2.998 \times 10^8 \text{ m/s})(3.156 \times 10^7 \text{ s/yr})$$
$$= 9.46 \times 10^{15} \text{ m} \approx 10^{13} \text{ km}.$$

Light-year

For specifying distances to the Sun and Moon, we usually use meters or kilometers, but we could specify them in terms of light. The Earth–Moon distance is 384,000 km, which is 1.28 light-seconds. The Earth–Sun distance is $1.50 \times 10^{11} \text{ m}$, or 150,000,000 km; this is equal to 8.3 light-minutes. The most distant planet in the solar system, Pluto, is about $6 \times 10^9 \text{ km}$ from the Sun, or $6 \times 10^{-4} \text{ ly}$. The nearest star to us, other than the Sun, is Proxima Centauri, about 4.3 ly away. (Note that the nearest star is 10,000 times farther from us than the farthest planet.)

On a clear moonless night, thousands of stars of varying degrees of brightness can be seen, as well as the elongated cloudy stripe known as the Milky Way (Fig. 45–1). It was Galileo who first observed, about 1610, that the Milky Way is comprised of countless individual stars. A century and a half later (about 1750), Thomas Wright suggested that the Milky Way was a flat disc of stars extending to great distances in a plane, which we call the **Galaxy** (Greek for "milky way").

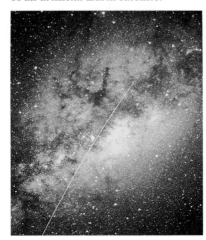

FIGURE 45–1 A section of the Milky Way. The thin line is the trail of an artificial Earth satellite.

FIGURE 45–2 Our Galaxy, as it would appear from the outside: (a) "end view," in the plane of the disc; (b) "top view," looking down on the disc. (If only we could see it like this— from the outside!) (c) Infrared photograph of the inner reaches of the Milky Way, showing the central bulge of our Galaxy. This very wide angle photo extends over 180° of sky, and to be viewed properly it should be wrapped in a semicircle with your eyes at the center. The white dots are nearby stars.

100,000 ly

Our Sun

(b)

Our Sun

2000 ly

(a)

(c)

Our Galaxy has a diameter of almost 100,000 light-years and a thickness of very roughly 2000 light-years. It has a bulging central nucleus and spiral arms (Fig. 45–2). Our Sun, which seems to be just another star, is located more than halfway from the center to the edge, about 28,000 ly from the center. Our Galaxy contains about 10^{11} stars. The Sun orbits the galactic center approximately once every 200 million years, so its speed is about 250 km/s relative to the center of the Galaxy. The total mass of all the stars in our Galaxy is about 3×10^{41} kg.

EXAMPLE 45–1 ESTIMATE Our Galaxy's mass. Estimate the total mass of our Galaxy using the orbital data of the Sun (including our solar system) about the center of the Galaxy. Assume that most of the mass of the Galaxy can be approximated as a uniform sphere of mass (the central bulge, Fig. 45–2a).

SOLUTION Our Sun and solar system orbit the center of the Galaxy, according to the best measurements as mentioned above, with a speed of about $v = 250$ km/s at a distance from the Galaxy center of about $r = 28,000$ ly. We use Newton's second law, $F = ma$, with a being the centripetal acceleration, $a = v^2/r$, and F being the universal law of gravitation (Chapter 6)

$$F = ma$$
$$G\frac{Mm}{r^2} = m\frac{v^2}{r}$$

where M is the mass of the Galaxy and m is the mass of our Sun and solar system. Solving this, we find

$$M = \frac{rv^2}{G} \approx \frac{(28,000\,\text{ly})(10^{16}\,\text{m/ly})(2.5 \times 10^5\,\text{m/s})^2}{6.67 \times 10^{-11}\,\text{N·m}^2/\text{kg}^2} \approx 3 \times 10^{41}\,\text{kg}.$$

In terms of *numbers* of stars, if they are like our Sun $(m = 2.0 \times 10^{30}\,\text{kg})$, there would be about $(3 \times 10^{41}\,\text{kg})/(2 \times 10^{30}\,\text{kg}) \approx 10^{11}$ or about 100 billion stars.

FIGURE 45–3 This globular star cluster is located in the constellation Hercules.

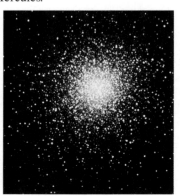

We can see by telescope, in addition to stars both within and outside the Milky Way, many faint cloudy patches in the sky which were all referred to once as "nebulae" (Latin for "clouds"). A few of these, such as those in the constellations Andromeda and Orion, can actually be discerned with the naked eye on a clear night. Some are **star clusters** (Fig. 45–3), groups of stars that are so numerous they appear to be a cloud. Others are glowing clouds of gas or dust (Fig. 45–4), and it is for these that we now mainly reserve the word **nebula**. Most fascinating are those that belong to a third category: they often have fairly regular elliptical shapes and seem to be a great distance beyond our Galaxy. Immanuel Kant (about 1755) seems to have been the first to suggest that these latter might be circular discs, but appear elliptical because we see them at an angle, and are faint because they are so distant. At first it was not universally accepted that these objects were **extragalactic**—that is, outside our Galaxy. The very large telescopes constructed in the twentieth century revealed that individual stars could be resolved within these extragalactic objects and that many contained spiral arms. Edwin Hubble (1889–1953), who did much of this observational work in the 1920s using the 2.5-m (100-inch) telescope[†] on Mt. Wilson near Los Angeles, California, was also able to demonstrate that these objects were indeed extragalactic because of their great distances. The distance to the Andromeda nebula (a galaxy), for example, is over 2 million light-years, a distance 20 times greater than the diameter of our Galaxy. Thus it was determined that these nebulae are **galaxies** similar to ours. Today, the largest telescopes can see about 10^{11} galaxies. See Fig. 45–5. (Note that it is usual to capitalize the word galaxy only when it refers to our own.)

Galaxies tend to be grouped in **galaxy clusters**, with anywhere from a few to many thousands of galaxies in each cluster. Furthermore, clusters themselves seem to be organized into even larger aggregates: clusters of clusters of galaxies, or **superclusters**. The galaxies nearest us are about 2 million light-years away. The farthest detectable galaxies are thousands of times farther away, on the order of 10^{10} ly. (See Table 45–1.)

[†] 2.5 m (= 100 inches) refers to the diameter of the curved objective mirror. The bigger the mirror, the more light it collects and the less diffraction there is, so more and fainter stars can be seen. See Chapters 34 and 36.

FIGURE 45–4 This gaseous nebula, found in the constellation Carina, is about 9000 light-years from us.

TABLE 45–1
Heavenly Distances

Object	Approx. Distance from Earth (ly)
Moon	4×10^{-8}
Sun	1.6×10^{-5}
Farthest planet in our solar system (Pluto)	6×10^{-4}
Nearest star (Proxima Centauri)	4.3
Center of our Galaxy	3×10^{4}
Nearest galaxy	2×10^{6}
Farthest galaxies	10^{10}

FIGURE 45–5 Photographs of galaxies. (a) Spiral galaxy in the constellation Hydra. (b) Two galaxies: the larger and more dramatic one is known as the Whirlpool galaxy. (c) The same galaxies as in (b), but this is a false-color infrared image which shows the arms of the spiral as being more regular than in the visible light photo (b); the different colors correspond to different light intensities.

(a)   (b)   (c)

CONCEPTUAL EXAMPLE 45–2 | **Looking back in time.** Astronomers often think of their telescopes as time machines, looking back toward the origin of the universe. How far back do they look?

RESPONSE The distance in light-years measures exactly how long in years the light has been traveling to reach us, so Table 45–1 tells us also how far back in time we are looking. For example, if we saw Proxima Centauri explode into a supernova today, then the event would have really occurred 4.3 years ago. The most distant objects, galaxies 10^{10} ly away, emitted the light we see now 10^{10} years ago, so what we see was how they were then, close to the beginning of the universe.

Besides the usual stars, clusters of stars, galaxies, and clusters and superclusters of galaxies, the universe contains a number of other interesting objects. Among these are stars known as *red giants*, *white dwarfs*, *neutron stars*, *black holes* (at least theoretically), and exploding stars called *novae* and *supernovae*. In addition there are *quasars* ("quasistellar radio sources"), which, if we judge their distance correctly, are galaxies thousands of times brighter than ordinary galaxies. Furthermore, there is radiation that reaches the Earth but does not emanate from the bright pointlike objects we call stars: it is a background radiation that seems to arrive uniformly from all directions in the universe. We discuss all these phenomena in this chapter.

How astronomical distances are measured

We have talked about the vast distances of objects in the universe. But how do we measure these distances? One basic technique employs simple geometry to measure the **parallax** of a star. By parallax we mean the apparent motion of a star, against the background of more distant stars, due to the Earth's motion about the Sun. As shown in Fig. 45–6, the sighting angle of a star relative to the plane of Earth's orbit (angle θ) is measured at different times of the year. Since we know the distance d from Earth to Sun, we can reconstruct the right triangles shown in the figure and can determine[†] the distance D to the star.

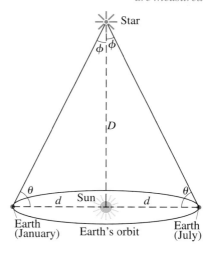

FIGURE 45–6 Distance to a star determined by parallax. (Not to scale.)

EXAMPLE 45–3 **ESTIMATE** | **Distance to a star using parallax.** Estimate the distance D to a star if the angle θ in Fig. 45–6 is measured to be $89.99994°$.

SOLUTION The angle ϕ in Figure 45–6 is $0.00006°$, or about 1.0×10^{-6} radians. Since $\tan \phi = d/D$, then the distance D to the star is

$$D = \frac{d}{\tan \phi} = \frac{d}{\phi} = \frac{1.5 \times 10^8 \text{ km}}{1.0 \times 10^{-6} \text{ rad}} = 1.5 \times 10^{14} \text{ km},$$

or about 15 ly. (We can use $\tan \phi \approx \phi$ since ϕ is very small.)

The distances to stars are often specified in terms of parallax angle given in seconds of arc: 1 second (1″) is $\frac{1}{60}$ of a minute (′) of arc, which is $\frac{1}{60}$ of a degree, so $1″ = \frac{1}{3600}$ of a degree. The distance is then specified in parsecs (meaning *par*allax angle in *sec*onds of arc), where the **parsec** (pc) is defined as $1/\phi$ and ϕ is given in seconds. In the Example we just did, $\phi = (6 \times 10^{-5})°(3600) = 0.22″$ of arc, so we would say the star is at a distance of $1/0.22″ = 4.5$ pc. It is easy to show that the parsec is given by

$$1 \text{ pc} = 3.26 \text{ ly}$$
$$= (3.26 \text{ ly})(9.46 \times 10^{15} \text{ m}) = 3.08 \times 10^{16} \text{ m}.$$

Parsec

Parallax can be used to determine the distance to stars as far away as

[†]This is essentially the way the heights of mountains are determined, by "triangulation." See Example 1–7.

100 light-years (≈30 parsecs). Beyond that distance, parallax angles are too small to measure. For greater distances, more subtle techniques must be employed. We can, for example, compare the apparent brightnesses of galaxies, and—using the inverse square law (intensity drops off as the square of the distance)—estimate their relative distances. This method cannot be very precise since we cannot expect all galaxies to have the same intrinsic brightness. A better estimate compares the brightest stars in galaxies or the brightest galaxies in galaxy clusters. It is reasonable to assume, for example, that the brightest stars in all galaxies are similar and have about the same intrinsic luminosity. Consequently, their *apparent* brightness would be a measure of how far away they are.

Another important technique for estimating distance is via the "redshift" in the line spectra of elements, which is related to the expansion of the universe, as will be discussed in Section 45–4.

As we look farther and farther away, the measurement techniques are less and less reliable, so there is more and more uncertainty in the measurements of large distances.

45–2 Stellar Evolution: The Birth and Death of Stars

The stars appear unchanging. Night after night the heavens reveal no significant variations. Indeed, on a human time scale, the vast majority of stars (except novae and supernovae) change very little. Although stars *seem* fixed in relation to each other, many move sufficiently for the motion to be detected. Indeed, the speeds of stars relative to neighboring stars can be hundreds of km/s, but at their great distance from us, this motion is detectable only by careful measurement. Furthermore, there is a great range of brightness among stars. The differences in brightness are due both to differences in the amount of light stars emit and to their distances from us.

A useful parameter for a star or galaxy is its **absolute luminosity**, L, by which we mean the total power radiated in watts. Also important is the **apparent brightness**, l, defined as the power crossing unit area perpendicular to the path of the light at the Earth. Given that energy is conserved, and ignoring any absorption in space, the total emitted power L at a distance d from the star will be spread over a sphere of surface area $4\pi d^2$. If we let d be the distance from the star to the Earth, then L must be equal to $4\pi d^2$ times l (power per unit area at Earth). That is,

Luminosity and brightness of stars

$$l = \frac{L}{4\pi d^2}. \qquad (45\text{–}1)$$

EXAMPLE 45–4 **Apparent brightness.** Suppose a particular star has absolute luminosity equal to that of the Sun but is 10 pc away from Earth. By what factor will it appear dimmer than the Sun?

SOLUTION The luminosity L is the same for both stars, so the apparent brightness depends only on their relative distances. We use the inverse square law as stated above in Eq. 45–1: the star appears dimmer by a factor

$$\frac{l_1}{l_2} = \frac{d_2^2}{d_1^2} = \frac{(1.5 \times 10^8 \text{ km})^2}{(10 \text{ pc})^2 (3.26 \text{ ly/pc})^2 (10^{13} \text{ km/ly})^2} \approx 2 \times 10^{-13},$$

where d_1 and d_2 are the distances from Earth to the star and to the Sun, respectively.

Careful study of nearby stars has shown that for most stars, the absolute luminosity depends on the mass:[†] *the more massive the star, the greater its luminosity.* Another important parameter of a star is its surface temperature, which can be determined from the spectrum of electromagnetic frequencies it emits, just as for a blackbody (Section 38–1). As we saw in Chapter 38, the spectrum of hotter and hotter bodies shifts from predominantly lower frequencies (and longer wavelengths, such as red) to higher frequencies (and shorter wavelengths such as blue). Quantitatively, the relation is given by Wien's law (Eq. 38–1): the peak wavelength λ_P in the spectrum of light emitted by a blackbody (and stars are fairly good approximations to blackbodies) is inversely proportional to its kelvin temperature T; that is, $\lambda_P T = 2.90 \times 10^{-3}$ m·K. The surface temperatures of stars typically range from about 3500 K (reddish) to perhaps 50,000 K (bluish).

An important astronomical discovery, made around 1900, was that for most stars, the color is related to the absolute luminosity and therefore to the mass. A useful way to present this relationship is by the so-called Hertzsprung–Russell (H–R) diagram. On the H–R diagram, the horizontal axis shows the temperature T whereas the vertical axis is the luminosity L, and each star is represented by a point on the diagram, Fig. 45–7. Most stars fall along the diagonal band termed the **main sequence**. Starting at the lower right we find the coolest stars, reddish in color; they are the least luminous and therefore of low mass. Farther up toward the left we find hotter and more luminous stars that are yellowish-white, like our Sun. Still farther up we find still more massive and more luminous stars, bluish in color. Stars that fall on this diagonal band are called *main-sequence stars.* There are also stars that fall outside the main sequence. Above and to the right we find extremely large stars, with high luminosities but with low (reddish) color temperature: these are called *red giants.* At the lower left, there are a few stars of low luminosity but with high temperature: these are the *white dwarfs.*

Luminosity increases with star's mass

H–R diagram

Main-sequence stars

Red giants

White dwarfs

[†] The mass of a star can be determined by observing its gravitational effects. Most stars are part of a cluster, the simplest being a binary star in which two stars orbit around each other, allowing their masses to be determined using Newtonian mechanics.

FIGURE 45–7
Hertzsprung–Russell (H–R) diagram.

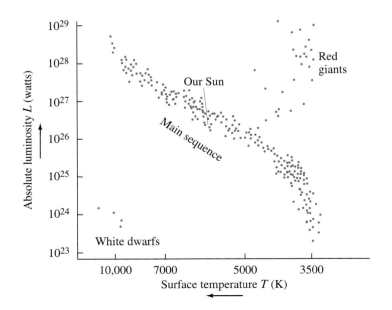

Before considering the different types of stars, let us first look at a couple of Examples that show how physical principles can be used to gain information about stars.

EXAMPLE 45–5 **Determining star temperatures and size.** Suppose that the distances to two nearby stars can be reasonably estimated and this data, together with their measured apparent brightnesses, suggests that the two stars have about the same absolute luminosity, L. The spectrum of one of the stars peaks at about 700 nm (so it is reddish). The spectrum of the other peaks at about 350 nm (bluish). Use Wien's law (Eq. 38–1) and the Stefan-Boltzmann equation (Section 19–10) to determine (a) the surface temperature of each star, and (b) how much larger one star is than the other.

SOLUTION (a) Wien's law states that $\lambda_P T = 2.90 \times 10^{-3}\, \text{m} \cdot \text{K}$. So the temperature of the reddish star is

$$T_r = \frac{2.90 \times 10^{-3}\, \text{m} \cdot \text{K}}{700 \times 10^{-9}\, \text{m}} = 4140\, \text{K}.$$

The temperature of the bluish star will be double this since its peak wavelength is half (350 nm vs. 700 nm); just to check:

$$T_b = \frac{2.90 \times 10^{-3}\, \text{m} \cdot \text{K}}{350 \times 10^{-9}\, \text{m}} = 8280\, \text{K}.$$

(b) The Stefan-Boltzmann equation, which we discussed in Chapter 19 (see Eq. 19–15), states that the power radiated *per unit area* of surface from a body is proportional to the fourth power of the kelvin temperature, T^4. Now the temperature of the bluish star is double that of the reddish star, so the bluish one must radiate $(2^4) = 16$ times as much energy per unit area. But we are given that they have the same luminosity (the same total power output), so the surface area of the blue star must be $\frac{1}{16}$ that of the red one. Since the surface area of a sphere is $4\pi r^2$, we conclude that the radius of the reddish star is $\sqrt{16} = 4$ times larger than the radius of the bluish star (or $4^3 = 64$ times the volume).

EXAMPLE 45–6 **ESTIMATE** **Distance to a star using H–R and color.** Suppose that detailed study of a certain star suggests that it most likely fits on the main sequence of an H–R diagram. Its measured apparent brightness is $l = 1.0 \times 10^{-12}\, \text{W/m}^2$, and the peak wavelength of its spectrum is $\lambda_P \approx 600\, \text{nm}$. Estimate its distance from us.

SOLUTION The star's temperature, from Wien's law (Eq. 38–1), is

$$T \approx \frac{2.90 \times 10^{-3}\, \text{m} \cdot \text{K}}{600 \times 10^{-9}\, \text{m}} \approx 4800\, \text{K}.$$

A star on the main sequence of an H–R diagram at this temperature has absolute luminosity of about $L \approx 1 \times 10^{26}\, \text{W}$, read off of Fig. 45–7. Then, from Eq. 45–1,

$$d = \sqrt{\frac{L}{4\pi l}} \approx \sqrt{\frac{1 \times 10^{26}\, \text{W}}{4(3.14)(1.0 \times 10^{-12}\, \text{W/m}^2)}} \approx 3 \times 10^{18}\, \text{m}.$$

Its distance from us in light-years and parsecs is

$$d = \frac{3 \times 10^{18}\, \text{m}}{10^{16}\, \text{m/ly}} \approx 300\, \text{ly} \approx \frac{300\, \text{ly}}{3.26\, \text{ly/pc}} \approx 90\, \text{pc}.$$

Why are there different types of stars, such as red giants and white dwarfs, as well as main-sequence stars? Were they all born this way, in the beginning? Or might each different type represent a different age in the life cycle of a star? Astronomers and astrophysicists today believe the latter is most likely the case. Note, however, that we cannot actually follow any but the tiniest part of the life cycle of any given star since they live for ages vastly greater than ours, on the order of millions or billions of years. Nonetheless, let us follow the process of **stellar evolution** from the birth to the death of a star, as astrophysicists have theoretically reconstructed it today.

Birth of a star

Contraction due to gravity

Stars are born, it is believed, when gaseous clouds (mostly hydrogen) contract due to the pull of gravity. A huge gas cloud might fragment into numerous contracting masses, each mass centered in an area where the density was only slightly greater than that at nearby points. Once such "globules" formed, gravity would cause each to contract in toward its center of mass. As the particles of such a *protostar* accelerate inward, their kinetic energy increases. When the kinetic energy is sufficiently high, the Coulomb repulsion that keeps the hydrogen nuclei apart can be overcome and nuclear fusion can take place. In a star like our Sun, the "burning" of hydrogen[†] occurs via the proton–proton cycle (Section 43–4, Eqs. 43–6), in which four protons fuse to form a $_2^4$He nucleus with the release of γ rays and neutrinos. These reactions require a temperature of about 10^7 K, corresponding to an average kinetic energy (kT) of about 1 keV. In more massive stars, the carbon cycle produces the same net effect: Four $_1^1$H produce a $_2^4$He—see Section 43–4. The fusion reactions take place primarily in the core of a star, where T is sufficiently high. (The surface temperature is, of course, much lower—on the order of a few thousand kelvins.) The tremendous release of energy in these fusion reactions produces a pressure sufficient to halt the gravitational contraction, and our protostar, now really a young *star*, stabilizes on the main sequence. Exactly where the star falls along the main sequence depends on its mass. The more massive the star, the farther up (and to the left) it falls on the diagram. To reach the main sequence requires perhaps 30 million years, if it is a star like our Sun, and (assuming our theory is right) it will remain there[‡] about 10 billion years (10^{10} yr). Although most stars are billions of years old, there is evidence that stars are actually being born at this moment.

Fusion begins when T (and $\overline{K}$) is large enough

Reaching the main sequence

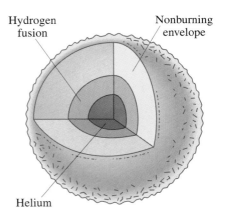

Hydrogen fusion

Nonburning envelope

Helium

FIGURE 45–8 A shell of "burning" hydrogen (fusing to become helium) surrounds the core where the newly formed helium gravitates.

As hydrogen "burns"—that is, fuses to form helium—the helium that is formed is denser and tends to accumulate in the central core where it was formed. As the core of helium grows, hydrogen continues to "burn" in a shell around it, Fig. 45–8. When much of the hydrogen within the core has been consumed, the production of energy decreases and is no longer sufficient to prevent the huge gravitational forces from once again causing the core to contract and heat up. The hydrogen in the shell around the core then "burns" even more fiercely because of this rise in temperature, causing the outer envelope of the star to expand and to cool. The surface temperature, thus reduced, produces a spectrum of light that peaks at longer wavelength. By this time the star has left the main sequence. It has become redder, and as it has grown in size, it has become more luminous. So it will have moved to the right and upward on the H–R diagram, as shown in Fig. 45–9. As it moves upward, it enters the **red giant** stage. Thus, theory explains the origin of red giants as a natural step in a star's evolution. Our Sun, for example, has been on the main sequence for about $4\frac{1}{2}$ billion years. It will probably remain there another 4 or 5 billion years. (We can take comfort in that!) When our Sun leaves the main sequence, it is expected to grow in size (as a red giant) until it occupies all the volume out to approximately the present orbit of Earth.

Red giants

[†] The word "burn" is put in quotation marks because these high-temperature fusion reactions occur via a *nuclear* process, and must not be confused with ordinary burning (of, say, paper, wood, or coal) in air, which is a chemical reaction at the *atomic* level (and at a much lower temperature).

[‡] More massive stars, since they are hotter and the Coulomb repulsion is more easily overcome, "burn" much more quickly, and so use up their fuel faster, resulting in shorter lives. A star 10 times more massive than our Sun, for example, will remain on the main sequence only for about 10^7 years. Stars less massive than our Sun live much longer than our Sun's 10^{10} yr.

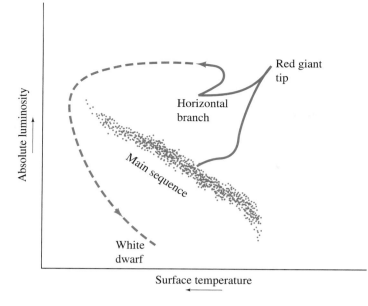

FIGURE 45–9 Evolution of a star like our Sun represented on an H–R diagram.

As a star's envelope expands, the core is shrinking and heating up. When the temperature reaches about 10^8 K, even helium nuclei, in spite of their greater charge and hence greater electrical repulsion, can then undergo fusion. The reactions are

$$\begin{aligned}
{}_{2}^{4}\text{He} + {}_{2}^{4}\text{He} &\longrightarrow {}_{4}^{8}\text{Be} + \gamma \quad (Q = -95\,\text{keV}) \\
{}_{2}^{4}\text{He} + {}_{4}^{8}\text{Be} &\longrightarrow {}_{6}^{12}\text{C} + \gamma. \quad (Q = 7.4\,\text{MeV})
\end{aligned}$$
(45–2) *Nucleosynthesis*

The first reaction is (slightly) endothermic, and the ${}_{4}^{8}\text{Be}$ formed is unstable and can decay quickly $(10^{-16}\,\text{s})$ back to two alpha particles. So the second reaction must occur very quickly after the first, and this is aided by the particularly large cross section for the second reaction. The net effect is

$$3\,{}_{2}^{4}\text{He} \longrightarrow {}_{6}^{12}\text{C}. \quad (Q = 7.3\,\text{MeV}).$$

The burning of helium causes a rapid and major change in the star, and the star moves rapidly to the "horizontal branch" on the H–R diagram (see Fig. 45–9). Further reactions are possible, creating elements of higher and higher Z, up to $Z = 10$ or 12. If the mass of the star is greater than about 0.7 solar masses, then still higher Z elements can be formed. The star can get even hotter and in the range $T = 2.5\text{–}5 \times 10^9\,\text{K}$, nuclei as heavy as ${}_{26}^{56}\text{Fe}$ and ${}_{28}^{56}\text{Ni}$ can be made. But here the process of **nucleosynthesis**, the formation of heavy nuclei from lighter ones by fusion, ends. As we saw in Fig. 42–1, the average binding energy per nucleon begins to decrease for A greater than about 60. Further fusions would *require* energy, rather than release it. (Elements heavier than Ni are probably formed mainly by neutron capture, in supernova explosions, as we shall discuss shortly.) *Production of heavy elements*

What happens next depends on the mass of the star. If the star has a residual mass less than about 1.4 solar masses[†] (known as the *Chandrasekhar limit*), no further fusion energy can be obtained and the star collapses under the action of gravity. As it shrinks, the star cools and typically follows the route shown in Fig. 45–9, according to theory, descending from the upper left downward, becoming a **white dwarf**. A white dwarf with a mass equal to that of the Sun would be about the size of the Earth. What determines its size is that it collapses to the point at which the electron clouds of the atoms start to overlap: at this point it collapses no further because, as the Pauli exclusion principle claims, no two electrons can be in the same quantum state. A white dwarf continues to lose internal energy, decreasing in temperature and becoming dimmer and dimmer until its light goes out. It has then become a *black dwarf*, a dark cold chunk of ash. *White dwarfs*

[†] This cutoff refers to the mass of the star remaining *after* the red giant stage.

FIGURE 45–10 These glowing, expanding gas rings, a light-year in diameter, were observed by the Hubble Space Telescope a decade after the explosion of supernova SN1987a.

Neutron stars

Supernovae

FIGURE 45–11 SN1987a, the supernova that appeared in February 1987, is the very bright star on the right. The bright nebulous object is the Tarantula Nebula.

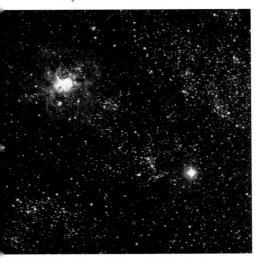

More massive stars (residual mass greater than 1.4 solar masses) are thought to follow a quite different scenario. A star with this great a mass can contract under gravity and heat up even further, reaching extremely high temperatures. The kinetic energy of the nuclei is then so high that fusion of elements heavier than iron is possible even though the reactions require energy input ($Q < 0$). Furthermore the high-energy collisions can also cause the breaking apart of iron and nickel nuclei into He nuclei, and eventually into protons and neutrons:

$$\substack{56\\26}\text{Fe} \rightarrow 13\,{}_2^4\text{He} + 4\text{n}$$
$${}_2^4\text{He} \rightarrow 2\text{p} + 2\text{n}.$$

These are energy-requiring (endoergic) reactions, but at such extremely high temperature and pressure there is plenty of energy available, enough even to force electrons and protons together to form neutrons in inverse β decay:

$$\text{e}^- + \text{p} \rightarrow \text{n} + \nu.$$

As the core contracts under the huge gravitational forces, the tremendous mass becomes essentially an enormous nucleus made up almost exclusively of neutrons. The size of the star is no longer limited by the exclusion principle applied to electrons, but rather applied to neutrons, and the star begins to contract rapidly toward forming an enormously dense **neutron star**. The contraction of the core would mean a great reduction in gravitational potential energy. Somehow this energy would have to be released. Indeed, it was suggested in the 1930s that the final core collapse to a neutron star may be accompanied by a catastrophic explosion whose tremendous energy could form virtually all elements of the periodic table and blow away the entire outer envelope of the star (Fig. 45–10), spreading its contents into interstellar space. Such explosions are believed to produce one type of observed **supernova** (there may be other mechanisms as well). Indeed, the presence of heavy elements on Earth and in our solar system suggests that our solar system may have formed from the debris from at least one supernova.

In a supernova explosion, a star's brightness is observed to suddenly increase billions of times in a period of just a few days and then fade away over the next few months. In February 1987 such a supernova occurred in an appendage to our Galaxy known as the Large Magellanic Cloud, only 170,000 ly away, and was visible to the naked eye in the southern hemisphere (Fig. 45–11). Supernovae are seen relatively often in distant galaxies, but this recent one, called SN1987a, is the first to have occurred close enough to be visible to the naked eye since the famous one of 1604 observed by Kepler and Galileo (and thus SN1987a is the first since the invention of the telescope). A supernova visible to the naked eye was observed by Chinese astronomers in A.D. 1054, and its remains are still visible in the sky (in the Crab Nebula), in the midst of which there is now a **pulsar**. Pulsars are astronomical objects that emit sharp pulses of radiation at regular intervals on the order of a second. They are now believed to be neutron stars which, because of conservation of angular momentum, increase greatly in rotational speed as their moment of inertia decreases during their contraction. The intense magnetic field of such a rapidly rotating star (perhaps as high as 10^6 to 10^{10} T) could trap and accelerate charged particles, which then give off radiation in a beam that rotates with the star. The discovery of pulsars in 1967 has lent support to the theory that neutron stars are the result of supernova explosions.

The core of a neutron star contracts to the point at which all neutrons are as close together as they are in a nucleus. That is, the density of a neutron star is on the order of 10^{14} times that of normal solids and liquids on Earth. A neutron star that has a mass 1.5 times that of our Sun would have a diameter of only about 10 km.

If the mass of a neutron star is less than about two or three solar masses, its subsequent evolution is thought to be similar to that of a white dwarf. If the mass is greater than this, the gravitational force may become so strong that the star contracts to an even smaller diameter and an even greater density, overcoming, in a sense, even the neutron exclusion principle. Gravity would then be so strong that even light emitted from it could not escape—it would be pulled back in by the force of gravity. In other words, the escape velocity from such a star is greater than c, the speed of light. Since no radiation could escape from such a star, we could not see it—it would be black. A body may pass by it and be deflected by its gravitational field, but if it came too close, it would be swallowed up, never to escape. This is a **black hole**. Black holes are predicted by theory to exist. Experimentally, evidence for their existence is strong but not everyone agrees it is sufficient to fully confirm the existence of black holes. One possibility is there may be a giant black hole at the center of our Galaxy—its mass is estimated at several million times that of the Sun—and it seems possible that many or all galaxies have black holes at their centers.

Black holes

45–3 | General Relativity: Gravity and the Curvature of Space

We have seen that the force of gravity plays a dominant role in the processes that occur in stars and, in general, in the evolution of the universe as a whole. The reasons gravity, and not one of the other of the four forces in nature, plays the dominant role in the universe are (1) it is long-range and (2) it is always attractive. The strong and weak nuclear forces act over very short distances only, on the order of the size of a nucleus; hence they do not act over astronomical distances (although of course they act between nuclei and nucleons in stars to produce nuclear reactions). The electromagnetic force, like gravity, acts over great distances; but it can be either attractive or repulsive. And since the universe does not seem to contain large areas of net electric charge, a large net force does not occur. But gravity acts as an attractive force between *all* masses, and there are large accumulations in the universe of only the one "sign" of mass (not + and − as with electric charge).

However, the force of gravity as Newton described it in his law of universal gravitation shows discrepancies on a cosmological scale. Einstein, in his general theory of relativity, developed a theory of gravity that resolves these problems and forms the basis of cosmological dynamics.

In the *special theory of relativity* (Chapter 37), Einstein concluded that there is no way for an observer to determine whether a given frame of reference is at rest or is moving at constant velocity in a straight line. Thus the laws of physics must be the same in different inertial reference frames. But what about the more general case of motion where reference frames can be *accelerating*?

It is in the **general theory of relativity** that Einstein tackled the problem of accelerating reference frames and developed a theory of gravity. The mathematics of general relativity is very complex, so our discussion will be mainly qualitative.

We begin with Einstein's famous **principle of equivalence** (briefly discussed in Chapter 6), which states that

> **No experiment can be performed that could distinguish between a uniform gravitational field and an equivalent uniform acceleration.**

If observers sensed that they were accelerating (as in a vehicle speeding around a sharp curve), they could not prove by any experiment that in fact they weren't simply experiencing the pull of a gravitational field. Conversely, we might think we are being pulled by gravity when in fact we are undergoing an "inertial" acceleration having nothing to do with gravity.

As a thought experiment, consider a person in a freely falling elevator near the Earth's surface. If our observer held out a book and let go of it, what would happen? Gravity would pull it downward toward the Earth, but at the same rate $(g = 9.8 \text{ m/s}^2)$ at which the person and elevator were falling. So the book would hover right next to the person's hand (Fig. 45–12). The effect is exactly the same as if this reference frame were at rest and *no* forces were acting. On the other hand, suppose the elevator were far out in space where there is no gravitational field. If the person released the book, it would float, just as it does in Fig. 45–12.

FIGURE 45–12 A freely falling elevator. The released book hovers next to the owner's hand.

(a)

(b)

If, instead, the elevator (out in space) were accelerating upward at an acceleration of $9.8 \, \text{m/s}^2$, the book as seen by our observer would fall to the floor with an acceleration of $9.8 \, \text{m/s}^2$, just as if it were falling because of gravity. According to the principle of equivalence, the observer could do no experiment to determine whether the book fell because the elevator was accelerating upward at $a = 9.8 \, \text{m/s}^2$ in the absence of gravity, or because a gravitational field with $g = 9.8 \, \text{m/s}^2$ was acting downward and he was at rest (say on the Earth). The two descriptions are equivalent.

Mass: inertial vs. gravitational

The principle of equivalence is related to the concept that there are two types of mass. Newton's second law, $F = ma$, uses **inertial mass**. We might say that inertial mass represents "resistance" to any type of force whatever. The second type of mass is **gravitational mass**. When one body attracts another by the gravitational force (Newton's law of universal gravitation, $F = Gm_1 m_2 / r^2$, Chapter 6), the strength of the force is proportional to the product of the gravitational masses of the two bodies. This is much like the electric force between two bodies which is proportional to the product of their electric charges. The electric charge of a body is not related to its inertial mass; so why should we expect that a body's gravitational mass (call it gravitational charge if you like) be related to its inertial mass? We have assumed they were the same. Why? Because no experiment—not even high-precision experiments—has been able to discern any measurable difference between inertial and gravitational mass. This is another way to state the equivalence principle: *gravitational mass is equivalent to inertial mass.*

Light bending

The principle of equivalence can be used to show that light ought to be deflected due to the gravitational force of a massive body. Consider another thought experiment, in which an elevator is in free space where no gravity acts. If a light beam enters a hole in the side of the elevator, the beam travels straight across the elevator and makes a spot on the opposite side if the elevator is at rest (Fig. 45–13a). If the elevator is accelerating upward as in Fig. 45–13b, the light beam still travels straight across in a reference frame at rest. In the upwardly accelerating elevator, however, the beam is observed to curve downward. Why? Because during the time the light travels from one side of the elevator to the other, the elevator is moving upward at ever-increasing speed. Now according to the equivalence principle, an upwardly accelerating reference frame is equivalent to a downward gravitational field. Hence, we can picture the curved light path in Fig. 45–13b as being the effect of a gravitational field. Thus we expect gravity to exert a force on a beam of light and to bend it out of a straight-line path.

That light is affected by gravity is an important prediction of Einstein's general theory of relativity. And it can be tested. The amount a light beam would be

FIGURE 45–13 (a) Light beam goes straight across an elevator not accelerating. (b) The light beam bends (exaggerated) in an elevator accelerating in an upward direction.

(a)

(b)

deflected from a straight-line path must be small even when passing a massive body. (For example, light near the Earth's surface after traveling 1 km is predicted to drop only about 10^{-10} m, which is equal to the diameter of a small atom and not detectable.) The most massive body near us is the Sun, and it was calculated that light from a distant star would be deflected by 1.75″ of arc (tiny but detectable) as it passed near the Sun (Fig. 45–14). However, such a measurement could be made only during a total eclipse of the Sun, so that the Sun's tremendous brightness would not overwhelm the starlight passing near its edge. An opportune eclipse occurred in 1919 and scientists journeyed to the South Atlantic to observe it. Their photos of stars around the Sun revealed shifts in accordance with Einstein's prediction. (See Fig. 45–15.)

A light beam must travel by the shortest, most direct, path between two points. If it didn't, some other object could travel between the two points in a shorter time and thus have a greater speed than the speed of light—a clear contradiction of the special theory of relativity. If a light beam can follow a curved path (as discussed above), then this curved path must be the shortest distance between the two points—which suggests that *space itself is curved* and that it is the gravitational field that causes the curvature. Indeed, the curvature of space—or rather, of four-dimensional space-time—is a basic aspect of general relativity.

What is meant by *curved space*? To understand, let us recall that our normal method of viewing the world is via Euclidean plane geometry. In Euclidean geometry, there are many axioms and theorems we take for granted, such as that the sum of the angles of any triangle is 180°. Other geometries, non-Euclidean which involve curved space, have also been imagined by mathematicians. Now it is hard enough to imagine three-dimensional curved space, much less curved four-dimensional space-time. So let us explain the idea of curved space by using two-dimensional surfaces.

Consider, for example, the two-dimensional surface of a sphere. It is clearly curved, Fig. 45–16, at least to us who view it from the outside—from our three-dimensional world. But how would hypothetical two-dimensional creatures determine whether their two-dimensional space were flat (a plane) or curved? One way would be to measure the sum of the angles of a triangle. If the surface is a plane, the sum of the angles is 180°, as we learn in plane geometry. But if the space is curved, and a sufficiently large triangle is constructed, the sum of the angles will *not* be 180°. To construct a triangle on a curved surface, say a sphere, we must use the equivalent of a straight line: that is, the shortest distance between two points, which is called a **geodesic**. On a sphere, a geodesic is an arc of a great circle (an arc contained in a plane passing through the center of the sphere) such as the Earth's equator and the Earth's longitude lines. Consider, for example, the large triangle of Fig. 45–16 whose sides are two "longitude" lines passing from the "north pole" to the equator, a part of which forms the third side. The two longitude lines make 90° angles with the equator (look at a world globe to see this more clearly). If they make, say, a 90° angle with each other at the north pole, the sum of these angles is 90° + 90° + 90° = 270°. This is clearly *not* a Euclidean space. Note, however, that if the triangle is small in comparison to the radius of the sphere, the angles will add up to nearly 180°, and the triangle (and space) will seem flat.

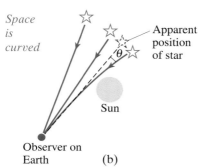

FIGURE 45–14 (a) Three stars in the sky. (b) If the light from one of these stars passes very near the Sun, whose gravity bends the rays, the star will appear higher than it actually is.

Geodesic

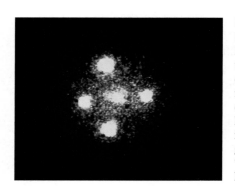

FIGURE 45–15 Photograph, taken by the Hubble Space Telescope, of the so-called "Einstein cross" (or Huchra's cross after its discoverer, John Huchra). It is thought to represent "gravitational lensing": The central spot is a relatively nearby galaxy, whereas the four other spots are thought to be images of a single quasar *behind* the galaxy, and the galaxy bends the light coming from the quasar to produce the four images. See also Fig. 45–14. (If the shape of the nearby galaxy were perfectly symmetric, we would expect the "image" of the distant quasar to be a circular ring or halo, instead of the four spots seen here.)

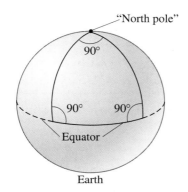

FIGURE 45–16 On a two-dimensional curved surface, the sum of the angles of a triangle may not be 180°.

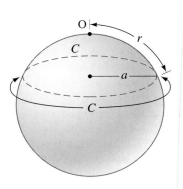

FIGURE 45–17 On a spherical surface, a circle of circumference C is drawn about point O as the center. The radius of the circle (not the sphere) is the distance r along the surface. (Note that in our three-dimensional view, we can tell that $2\pi a = C$; since $r > a$, then $2\pi r > C$.)

FIGURE 45–18 Example of a two-dimensional surface with negative curvature.

The universe: open or closed?

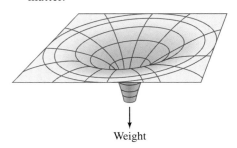

FIGURE 45–19 Rubber-sheet analogy for space-time curved by matter.

Weight

Another way to test the curvature of space is to measure the radius r and circumference C of a large circle. On a plane surface, $C = 2\pi r$. But on a two-dimensional spherical surface, C is *less* than $2\pi r$, as can be seen in Fig. 45–17. The proportionality constant between C and r is *less* than 2π. Such a surface is said to have *positive curvature*. On the saddlelike surface of Fig. 45–18, the circumference of a circle is greater than $2\pi r$, and the sum of the angles of a triangle is less than 180°. Such a surface is said to have a *negative curvature*.

Now, what about our universe? On a large scale (not just near a large mass), what is the overall curvature of the universe? Does it have positive curvature, negative curvature, or is it flat (zero curvature)? In the nineteenth century, Carl Friedrich Gauss (1777–1855) tried to determine whether our natural three-dimensional space deviated from Euclidean space by measuring the angles of a triangle formed by three mountain peaks using light rays as sides of the triangle. He was unable to detect any deviation from 180°, nor have experiments today detected any deviation.

Nonetheless, the question of the curvature of space is an important one in cosmology. If the universe has a positive curvature, then the universe would be *finite*, or *closed*. This does not mean that in such a universe the stars and galaxies would extend out to a certain boundary, beyond which there is empty space. Rather, galaxies would be spread throughout the space, and the space would fold back and "close on itself." There is no boundary or edge in such a universe. If a particle were to move in a straight line in a particular direction, it would eventually return to the starting point—albeit eons of time later. To ask "What is beyond such a closed universe?" is futile, since the space of the universe is all there is. If the curvature of space is zero or negative, on the other hand, the universe would be *open*. It would just go on and on and never fold back on itself. An open universe would be *infinite*. Whether the universe is open or closed depends, in part, on how much total mass there is in the universe, as we will discuss in Section 45–7. If the mass is great enough, it would bend space into a positively curved, closed, and finite space.

According to Einstein's theory, space-time is curved, especially near massive bodies. To visualize this, we might think of space as being like a thin rubber sheet; if a heavy weight is hung from it, it curves as shown in Fig. 45–19. The weight corresponds to a huge mass that causes space (space itself!) to curve. Thus, in Einstein's theory[†] we do not speak of the "force" of gravity acting on bodies. Instead we say that bodies and light rays move along geodesics (the equivalent of straight lines in plane geometry) in curved space-time. Thus, a body at rest or moving slowly near the great mass of Fig. 45–19 would follow a geodesic toward that body.

[†] Alexander Pope (1688–1744) wrote an epitaph for Newton:
  "Nature, and Nature's laws lay hid in night:
  God said, *Let Newton be!* and all was light."
Sir John Squire (1884–1958), perhaps uncomfortable with Einstein's profound thoughts, added:
  "It did not last: the Devil howling '*Ho!*
  Let Einstein be!' restored the status quo."

The extreme curvature of space-time shown in Fig. 45–19 could be produced by a **black hole**. A black hole, as we saw in the previous Section, is so dense that even light cannot escape from it. To become a black hole, a body of mass M must undergo **gravitational collapse**, contracting by gravitational self-attraction to within a radius called the **Schwarzschild radius**:

Black holes

$$R = \frac{2GM}{c^2}$$

where G is the gravitational constant and c the speed of light. If a body collapses to within this radius, it is predicted by general relativity to rapidly $(\approx 10^{-5}\,\text{s})$ collapse to a point at $r = 0$, forming an infinitely dense **singularity**. This prediction is uncertain, however, since in this realm we need to combine quantum mechanics with gravity, a unification of theories not yet fully achieved.

The Schwarzschild radius also represents the **event horizon** of a black hole. By event horizon we mean the surface beyond which no signals can ever reach us, and thus inform us of events that happen. As a star collapses toward a black hole, the light it emits is pulled harder and harder by gravity, but we can still see it. Once the matter passes within the event horizon the emitted light cannot escape, but is pulled back in by gravity.[†]

Event horizon

All we can know about a black hole is its mass, its angular momentum (there could be rotating black holes), and its electric charge. No other information, no details of its structure or the kind of matter it was formed of, can be known.

How might we observe black holes? We cannot see them because no light can escape from them. They would be black objects against a black sky. But they do exert a gravitational force on nearby bodies. The suspected black hole at the center of our Galaxy was discovered by examining the motion of matter in its vicinity. Another technique is to examine stars which appear to be rotating as if they were members of a *binary system* (two stars rotating about their common center of mass), although the companion is invisible. If the unseen star is a black hole, it might be expected to pull off gaseous material from its visible companion (Fig. 45–20). As this matter approached the black hole, it would be highly accelerated and should emit X-rays of a characteristic type. One of the candidates for a black hole is in the binary-star system Cygnus X-1.

[†] According to quantum mechanics, matter and radiation could escape from a black hole by tunneling, but the rate at which this could happen would be expected to be extremely low.

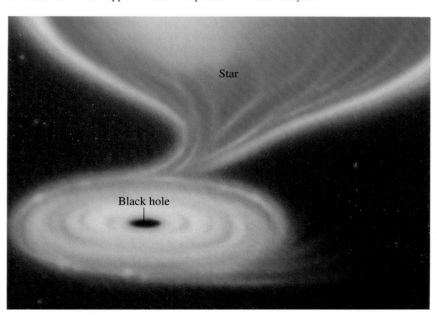

FIGURE 45–20 Artist's conception of how a black hole, which is one star of a binary pair, might pull matter from the other (more normal) star.

Star

Black hole

45–4 | The Expanding Universe

We discussed in Section 45–2 how individual stars evolve from their birth to their death as white dwarfs, neutron stars, and black holes. But what about the universe as a whole: is it static, or does it evolve? The evolution of stars suggests that the universe as a whole evolves. One of the most important scientific ideas of this century proposes that distant galaxies are racing away from us, and that the farther they are from us, the faster they are moving away. How astronomers arrived at this astonishing idea and what it means for the past history of the universe as well as its future, will occupy us for the remainder of the book.

Hubble and the redshift

The idea that the universe is expanding was first put forth by Edwin Hubble in 1929. It was based on observations of the Doppler shift of light emitted by stars. In Chapter 16 we discussed how the frequency and wavelength of sound are altered if the source is moving toward or away from an observer. If the source moves toward us, the frequency is higher and the wavelength is shorter. If the source moves away from us, the frequency is lower and the wavelength is longer. The **Doppler effect** occurs also for light, but the shifted wavelength or frequency is given by a formula slightly different from that for sound because for light (according to special relativity), we can make no distinction between motion of the source and motion of the observer. (Recall that sound travels in a medium, such as air, but light does not; according to relativity, there is no ether.) According to special relativity, the formula is (see Section 37–12)

$$\lambda = \lambda_0 \sqrt{\frac{1 + v/c}{1 - v/c}}, \qquad \left[\begin{array}{l}\text{source and observer moving} \\ \text{away from each other}\end{array}\right] \quad \textbf{(45–3)}$$

where λ_0 is the emitted wavelength as seen in a reference frame at rest with respect to the source, and λ is the wavelength measured in a frame moving with velocity v away from the source along the line of sight. Note that this relation depends only on the relative velocity v. (For relative motion toward each other, $v < 0$ in this formula.) Thus, when a source emits light of a particular wavelength and the source is moving away from us, the wavelength appears longer to us: the color of the light (if it is visible) is shifted toward the red end of the visible spec-

Redshift

trum, an effect known as a **redshift**. If the source moves toward us, the color shifts toward the blue (shorter wavelength) end of the spectrum. The amount of shift depends on the velocity of the source (Eq. 45–3). For speeds not too close to the speed of light, it is easy to show (Problem 30) that the fractional change in wavelength is proportional to the speed of the source to or away from us (as was the case for sound):

$$\frac{\Delta\lambda}{\lambda_0} = \frac{\lambda - \lambda_0}{\lambda_0} \approx \frac{v}{c}. \qquad [v \ll c] \quad \textbf{(45–4)}$$

In the spectra of stars and galaxies, lines are observed that correspond to lines in the known spectra of particular atoms (see Section 38–9). What Hubble found was that the lines seen in the spectra of galaxies were generally *redshifted*, and that the amount of shift seemed to be approximately proportional to the distance of the galaxy from us. That is, the velocity, v, of a galaxy moving away from us is proportional to its distance, d, from us:

Hubble's law

$$v = Hd. \qquad \textbf{(45–5)}$$

This is **Hubble's law**, one of the most important astronomical ideas of this century. The constant H is called the **Hubble parameter**. Hubble's law does not work well for nearby galaxies—in fact some are actually moving toward us ("blueshifted"); but this is believed to merely represent random motion of the galaxies. For more distant galaxies, the velocity of recession (Hubble's law) is much greater than that of random motion, and so is dominant.

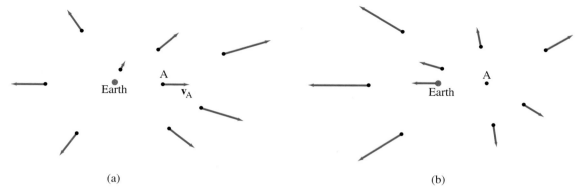

(a)   (b)

FIGURE 45–21 Expansion of the universe looks the same from any point in the universe.

The value of H is not known very precisely. It is now estimated to be about

$$H \approx 70\,\text{km/s/Mpc}$$

(that is, 70 km/s per megaparsec of distance) with an uncertainty of about 20 percent. If we use light-years for distance, then $H \approx 20$ km/s per million light-years of distance.

What does it mean that distant galaxies are all moving away from us, and with ever greater speed the farther they are from us? It is as if there had been a great explosion at some distant time in the past. And at first sight we seem to be in the middle of it all. But we aren't, necessarily. The expansion appears the same from any other point in the universe. To understand why, see Fig. 45–21. In Fig. 45–21a we have the view from Earth (or from our Galaxy). The velocities of surrounding galaxies are indicated by arrows, pointing away from us, and greater for galaxies more distant from us. Now, what if we were on the galaxy labeled A in Fig. 45–21a? From Earth, galaxy A appears to be moving to the right at a velocity, call it $\mathbf{v}_A$, represented by the arrow pointing to the right. If we were *on* galaxy A, Earth would appear to be moving to the left at velocity $-\mathbf{v}_A$. To determine the velocities of other galaxies relative to A, we vectorially add the velocity vector, $-\mathbf{v}_A$, to all the velocity arrows shown in Fig. 45–21a. This yields Fig. 45–21b, where we see clearly that the universe is expanding away from galaxy A as well; and the velocities of galaxies receding from A are proportional to their distance from A.

Thus the expansion of the universe can be stated as follows: All galaxies are racing away from *each other* at an average rate of about 70 km/s per megaparsec of distance between them. The ramifications of this idea are profound, and we discuss them in a moment.

There is, however, a class of objects called **quasars**, or "quasistellar radio sources," that do not seem to conform to Hubble's law. Quasars are as bright as nearby stars but display very large redshifts. If quasars are normal participants in the general expansion of the universe according to Hubble's law, their large redshifts suggest they are very distant galaxies. If they are so far away, they must be incredibly bright—sometimes thousands of times brighter than normal galaxies. On the other hand, it has been suggested that some quasars, at least, are not of abnormal brightness because they are nearer than their redshifts suggest. In the first case we would have an unresolved brightness problem, in the second case an unresolved redshift problem. An interesting piece of evidence is that the population density of quasars seems to increase with distance from us. If they are at great distances from us—as their redshifts suggest—this would merely mean they were more common in the early universe than they are now. But if instead they are much closer to us—as their brightness suggests—it would seem that *we* are in a special place in the universe, the place where quasars are least populous. Most astronomers are unwilling to accept the latter possibility, for it would violate a basic assumption of uniformity, the so-called *cosmological principle*, which we discuss in the next paragraph. A recent suggestion that could resolve the quasar question is that these mysterious objects may be cores of galaxies that emit enormous amounts of energy when they draw in nearby material, for example when galaxies "collide." If the center of a galaxy is a black hole, a large amount of matter accelerated toward it would give off a prodigious amount of energy. It seems possible, then, that quasars are powered by black holes.

Cosmological principle

A basic assumption in cosmology has been that on a large scale, the universe looks the same to observers at different places at the same time. In other words, the universe is both isotropic (looks the same in all directions) and homogeneous (would look the same if we were located elsewhere, say in another galaxy). This assumption is called the **cosmological principle**. On a local scale, say in our solar system or within our Galaxy, it clearly does not apply (the sky looks different in different directions). But it has long been thought to be valid if we look on a large enough scale, so that the average population density of galaxies and clusters of galaxies ought to be the same in different areas of the sky. The expansion of the universe (Fig. 45–21) is consistent with the cosmological principle; and the uniformity of the cosmic microwave background radiation (discussed in the next Section) supports it. But matter (i.e., galaxies), even on the largest scale, seems not to be homogeneous but tends to clump, as already discussed. Although astronomers are reluctant to give up the cosmological principle, there is now some doubt about its validity. One possible resolution might be that over 90 percent of the universe may be nonluminous dark matter, which might be uniformly distributed. In any case, the cosmological principle has aided us in treating the universe as a single evolving entity and not as a random collection of material bodies. Another way of stating the cosmological principle is this: There is nothing special about the Earth on a cosmological scale; our large-scale observations are no different from those that might be made elsewhere in the universe.

The expansion of the universe, as described by Hubble's law, strongly suggests that galaxies must have been closer together in the past than they are now. Further, Hubble's law is consistent with all galaxies having been quite close together at some time in the past. This is, in fact, the basis of the *Big Bang* theory of the origin of the universe (discussed in the next Section) which pictures the beginning of the universe as a great explosion. Let us see what can then be said about the age of the universe.

Age of the universe

We can estimate the age of the universe using the Hubble parameter. With $H \approx 20 \, \text{km/s}$ per million light-years, then the time required for the galaxies to arrive at their present separations would be approximately (starting with $v = d/t$ and using Eq. 45–5):

$$t = \frac{d}{v} = \frac{d}{Hd} = \frac{1}{H} \approx \frac{(10^6 \, \text{ly})(10^{13} \, \text{km/ly})}{(20 \, \text{km/s})(3 \times 10^7 \, \text{s/y})} \approx 15 \times 10^9 \, \text{yr},$$

or roughly 15 billion years. The age of the universe calculated in this way is called the *characteristic expansion time* or "Hubble age." It is not very precise, in part because we don't know the value of H precisely.

There are two other independent checks on the age of the universe. The first is determination of the age of the Earth (and solar system) from radioactivity, primarily using uranium, which places the age of the solar system at about $4\frac{1}{2}$ billion years. Second, by using the theory of stellar evolution, the ages of stars have been estimated to be about 10 to 15 billion years. These independent and unrelated determinations are consistent with a Big Bang occurring, according to our best estimates today, 10 to 15 billion years ago. The lower value determined from radioactivity is consistent since we would expect the origin of the Earth (and the solidification of rocks) to have occurred somewhat after the origin of the universe as a whole.

Before discussing the Big Bang theory in more detail, we briefly discuss one of the alternatives to the Big Bang—the **steady-state model**—which assumes that the universe is infinitely old and on the average looks the same now as it always has.[†] According to the steady-state model, no large-scale changes have taken place in the universe as a whole, particularly no Big Bang. To maintain this view in the face of the recession of galaxies away from each other, mass–energy conservation must be violated. That is, matter is assumed to be created continuously, keeping the density of the universe constant. The rate of mass creation required is very small—about one nucleon per cubic meter every 10^9 years.

Steady-state model

The steady-state model provided the Big Bang model with healthy competition a half century ago. However, the discovery of the cosmic microwave background radiation (next Section), as well as the observed expansion of the universe, has made the Big Bang model the almost universally accepted model.

45–5 The Big Bang and the Cosmic Microwave Background

The expansion of the universe seems to suggest, as we have seen, that the matter of the universe was once much closer together than it is now. This is the basis for the idea that the universe began about 15 billion years ago with a huge explosion known affectionately as the **Big Bang**.

The Big Bang

If there was a Big Bang, it must have occurred simultaneously at all points in the universe. If the universe is *finite*, the explosion would have taken place in a tiny volume approaching a point. However, this point of extremely dense matter is not to be thought of as a concentrated mass in the midst of a much larger space around it. Rather, the initial dense point *was* the universe—the entire universe. There wouldn't have been anything else. If, on the other hand, the universe is *infinite*, then the explosion would have occurred at *all* points in the universe since an infinite universe, even if smaller at an earlier time, would still have been infinite. In either case, when we say that the universe some time after the Big Bang was smaller than it is now, we mean that the average separation between galaxies was less. Thus, it is the *size of the universe itself* that has increased since the Big Bang.

What is the evidence supporting the Big Bang? First, the age of the universe as calculated from the Hubble expansion, from stellar evolution, and from radioactivity all point to a consistent time of origin for the universe, as we saw in the last Section. Another, and crucial, piece of evidence was the discovery in the 1960s of the **cosmic microwave background** radiation (or CMB), which came about as follows.

In 1964, Arno Penzias and Robert Wilson were experiencing difficulty with what they assumed to be background noise, or "static," in their radio telescope (a large antenna device for detecting radio waves from the heavens, Fig. 45–22).

FIGURE 45–22 Robert Wilson (left) and Arno Penzias, and behind them their "horn antenna."

[†]That the universe should be uniform in time as well as in space is essentially an extension of the cosmological principle and is sometimes referred to as the *perfect cosmological principle*.

Eventually, they became convinced that it was real and that it was coming from outside our Galaxy. They made precise measurements at a wavelength $\lambda = 7.35$ cm, which is in the microwave region of the electromagnetic spectrum. (See Fig. 32–12; this radiation is called "microwave" because the wavelength, though much greater than that of visible light, is somewhat smaller than wavelengths for ordinary radio waves, which are typically meters or hundred of meters.) The intensity of this radiation was found initially not to vary by day or night or time of year, nor to depend on direction. It came from all directions in the universe with equal intensity (within less than one part per thousand). It could only be concluded that this radiation came from beyond our Galaxy, from the universe as a whole. The remarkable uniformity of the cosmic microwave background radiation was in accordance with the cosmological principle. But theorists also felt that there needed to be some small inhomogeneities in the CMB that would have provided "seeds" around which galaxy formation could have started. Such small inhomogeneities, on the order of parts per million, were finally detected in 1992.

The intensity of this radiation as measured at $\lambda = 7.35$ cm corresponds to blackbody radiation (see Section 38–1) at a temperature of about 3 K, now measured more precisely to be 2.73 ± 0.01 K. When radiation at other wavelengths was measured, the intensities were found to fall on a blackbody curve as shown in Fig. 45–23, corresponding to a temperature of 2.73 K.

The discovery of the CMB at a temperature of 2.7 K ranks as one of the two most significant cosmological discoveries of the twentieth century. (The other was Hubble's expanding universe.) It is highly significant because it provides strong evidence in support of the Big Bang, and it gives us some idea of conditions in the very early universe. In fact, in the late 1940s, George Gamow and his collaborators calculated that a Big Bang origin of the universe should have generated just such a microwave background radiation.

To understand why, let us look at what a Big Bang might have been like. There must have been a tremendous release of energy. The temperature must have been extremely high, so high that there could not have been any atoms in the very early stages of the universe. Instead, the universe must have consisted solely of radiation (photons) and elementary particles. The universe would have been opaque—the photons in a sense "trapped," since as soon as they were emitted they would have been scattered or absorbed, primarily by electrons. Indeed, the microwave background radiation is strong evidence that matter and radiation were once in equilibrium at a very high temperature. As the universe expanded, the energy would have spread out over an increasingly larger volume and the temperature would have dropped. Only when the temperature had reached about 3000 K some 300,000 years later, could nuclei and electrons have stayed together as atoms.

FIGURE 45–23 Spectrum of cosmic microwave background radiation, showing blackbody curve and experimental measurements including that of Penzias and Wilson. (Thanks to G. F. Smoot and D. Scott, 3/00. The vertical bars represent the experimental uncertainty in a measurement.)

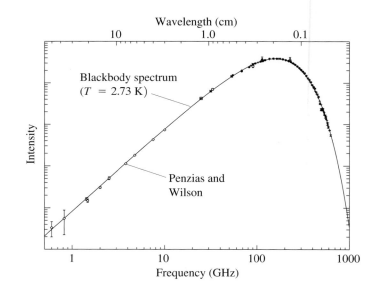

With the disappearance of free electrons, as they combined with nuclei to form atoms, the radiation would have been freed—"decoupled" from matter, we could say—to spread throughout the universe. As the universe expanded, so too the wavelengths of the radiation lengthened (you might think of standing waves, Section 15–9), thus redshifting to longer wavelengths that correspond to lower temperature (recall Wien's law, $\lambda_P T$ = constant, Section 38–1), until they would have reached the 2.7 K background radiation we observe today.

Photons "decoupled"

Although the total energy associated with the cosmic microwave background is much larger than that from other radiation sources (such as the light emitted by stars), it is small compared to the energy, and mass (remember $E = mc^2$), associated with matter. In fact today, radiation is believed to make up less than $\frac{1}{1000}$ of the energy of the universe: today the universe is *matter-dominated*. But it was not always so. The cosmic microwave background radiation strongly suggests that early in its history the universe was *radiation-dominated.*[†] But, as we shall see in the next Section, that period lasted less than $\frac{1}{10,000}$ of the history of the universe (thus far).

45–6 | The Standard Cosmological Model: the Early History of the Universe

It is now generally agreed that the evolution of the universe must have been determined in the first few moments of the Big Bang. In the last decade or two, a convincing theory of the origin and evolution of the universe has developed, now known as the **standard model**. Much of this theory is based on recent theoretical and experimental advances in elementary particle physics. Indeed, cosmology and elementary particle physics have cross-fertilized to a surprising extent.

Let us go back now to the earliest of times—as close as possible to the Big Bang—and follow a standard model scenario of events as the universe expanded and cooled after the Big Bang. Initially, we will be talking of extremely small time intervals, as well as extremely high temperatures, far beyond anything in the universe today. Figure 45–24 is a graphical representation of the events, and it may be helpful to consult it as we go along.

The standard-model "scenario" of the history of the universe after the Big Bang

We begin at a time only a minuscule fraction of a second after the Big Bang, 10^{-43} s. Although this is an unimaginably short time, predictions as early as this can be made based on present theory, albeit somewhat speculatively. Earlier than this instant, we can say nothing since we do not yet have a theory of quantum effects on gravity which would be needed for the incredibly high densities and temperatures then.

[†] If there had not been such intense radiation in the first few minutes of the universe, nuclear reactions might have produced a much larger percentage of heavy nuclei than we see. Instead, nearly 75 percent of visible matter is hydrogen, presumably because the intense radiation immediately blasted apart any heavy nuclei that formed into their constituent protons and neutrons.

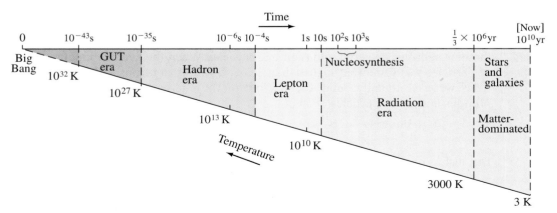

FIGURE 45–24 Graphical representation of the development of the universe after the Big Bang, according to the standard model.

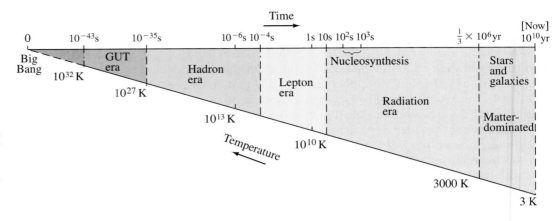

FIGURE 45–24 (repeated) Graphical representation of the development of the universe after the Big Bang, according to the standard model.

All four forces unified

It is imagined, however, that prior to 10^{-43} s, the four forces in nature were unified—there was only one force. The temperature would have been about 10^{32} K, corresponding to particles moving about every which way with an average kinetic energy K of 10^{19} GeV (see Eq. 18–4):

$$K \approx kT \approx \frac{(1.4 \times 10^{-23} \text{ J/K})(10^{32} \text{ K})}{1.6 \times 10^{-19} \text{ J/eV}} \approx 10^{28} \text{ eV} = 10^{19} \text{ GeV}.$$

Symmetry broken (gravity condensed out)

(Note that the factor $\frac{3}{2}$ in Eq. 18–4 is usually ignored in such calculations.) At $t = 10^{-43}$ s, a kind of "phase transition" is believed to have occurred during which the gravitational force, in effect, "condensed out" as a separate force. This, and subsequent phase transitions, are somewhat analogous to the phase transitions water undergoes as it cools from a gas, condenses into a liquid, and with further cooling freezes into ice. The symmetry of the four forces was broken, but the strong, weak, and electromagnetic forces were still unified. At this point, the universe entered the so-called **grand unified era** (after grand unified theory—see Chapter 44). There was no distinction between quarks and leptons; baryon and lepton numbers were not conserved. Very shortly thereafter, as the universe expanded considerably and the temperature had dropped to about 10^{27} K, there was another phase transition during which the strong force condensed out. This probably occurred about 10^{-35} s after the Big Bang. Now the universe was filled with a soup of leptons and quarks. The leptons included electrons, muons, taus, neutrinos, and all their antiparticles. The quarks were initially free (something we have not seen in our present universe), but soon they began to "condense" into more normal particles: nucleons and the other hadrons and their antiparticles.

Quark confinement (hadron era)

With this **confinement of quarks**, the universe entered the **hadron era**.

We can think of this "soup" as a grand mixture of particles and antiparticles, as well as photons—all in roughly equal numbers—colliding with one another frequently and exchanging energy.

By the time the universe was only about a microsecond (10^{-6} s) old, it had cooled to about 10^{13} K, corresponding to an average kinetic energy of 1 GeV, and the vast majority of hadrons disappeared. To see why, let us focus on the most familiar hadrons: nucleons and their antiparticles. When the average kinetic energy of particles was somewhat higher than 1 GeV, protons, neutrons, and their antiparticles were continually being created out of the energies of collisions involving photons and other particles, such as

Most hadrons disappear

$$\text{photons} \rightarrow \text{p} + \bar{\text{p}},$$
$$\rightarrow \text{n} + \bar{\text{n}}.$$

But just as quickly, particles and antiparticles would annihilate: for example

$$\text{p} + \bar{\text{p}} \rightarrow \text{photons or leptons.}$$

So the processes of creation and annihilation of nucleons were in equilibrium. The numbers of nucleons and antinucleons were high—roughly as many as there were electrons, positrons, or photons. But as the universe expanded and cooled, and the average kinetic energy of particles dropped below about 1 GeV, which is the minimum energy needed to create nucleons and antinucleons (about 940 MeV each) in a typical collision, the process of nucleon creation could not continue. The process of annihilation could continue, however, with antinucleons annihilating nucleons, until there were almost no nucleons left. But not quite zero. To explain our present world, which consists mainly of matter (nucleons and electrons) with very little antimatter in sight, we must suppose that earlier in the universe, perhaps around 10^{-35} s after the Big Bang, a slight excess of quarks over antiquarks was formed.[†] This would have resulted in a slight excess of nucleons over antinucleons. And it is these "leftover" nucleons that we are made of today. The excess of nucleons over antinucleons was about one part in 10^9. Earlier, during the hadron era, there should have been about as many nucleons as photons. After it ended, the "leftover" nucleons thus numbered only about one nucleon per 10^9 photons, and this ratio has persisted to this day. Protons, neutrons, and all other heavier particles were thus tremendously reduced in number by about 10^{-6} s after the Big Bang. The lightest hadrons, the pions, disappeared as the nucleons had; because they are the lightest mass hadrons (140 MeV), they were the last hadrons to go, around 10^{-4} s after the Big Bang. Lighter particles, including electrons, positrons, neutrinos, photons—in roughly equal numbers—dominated, and the universe entered the **lepton era**.

Why is there matter now?

Lepton era

By the time the first full second had passed (clearly the most eventful second in history!), the universe had cooled to about 10 billion degrees, 10^{10} K. The average kinetic energy was about 1 MeV. This was still sufficient energy to create electrons and positrons and balance their annihilation reactions, since their masses correspond to about 0.5 MeV. So there were about as many e$^+$ and e$^-$ as there were photons. But within a few more seconds, the temperature had dropped sufficiently so that e$^+$ and e$^-$ could no longer be formed. Annihilation (e$^+$ + e$^-$ → photons) continued. And, like nucleons before them, electrons and positrons all but disappeared from the universe—except for a slight excess of electrons over positrons (later to join with nuclei to form atoms). Thus, about $t = 10$ s after the Big Bang, the universe entered the **radiation era**. Its major constituents were now photons and neutrinos. But the neutrinos, partaking only in the weak force, rarely interacted. So the universe, until then experiencing an energy balance between matter and radiation, became **radiation-dominated**: much more energy was contained in radiation than in matter, a situation that would last hundreds of thousands of years (Fig. 45–24).

Radiation-dominated universe

Meanwhile, during the next few minutes, crucial events were taking place. Beginning about 2 or 3 minutes after the Big Bang, nuclear fusion began to occur. The temperature had dropped to about 10^9 K, corresponding to an average kinetic energy $\overline{K} \approx 100$ keV, where nucleons could strike each other and be able to fuse (Section 43–4), but now cool enough that newly formed nuclei would not be immediately broken apart by subsequent collisions. Deuterium, helium, and very tiny amounts of lithium nuclei were probably made. But the universe was cooling too quickly, and larger nuclei were not made. After only a few minutes, probably not even a quarter of an hour after the Big Bang, the temperature dropped far enough that nucleosynthesis stopped, not to start again for millions of years (in stars).

Making He nuclei

[†]An alternative possibility is that there was perfect symmetry between quarks and antiquarks, matter and antimatter, but that the universe somehow separated into domains, some containing only matter, others only antimatter. If this were true, we would expect antiparticles from such distant domains to reach us, at least occasionally, in cosmic rays; but none has ever been detected.

Thus, after the first hour or so of the universe, matter consisted mainly of bare nuclei of hydrogen (about 75 percent) and helium (about 25 percent)[†] and electrons. But radiation (photons) continued to dominate.

Our story is almost complete. The next important event is presumed to have occurred about 300,000 years later. The universe had expanded to about $\frac{1}{1000}$ of its present size, and the temperature had cooled to about 3000 K. The average kinetic energy of nuclei, electrons, and photons was less than an electron volt. Since ionization energies of atoms are on the order of eV, then as the temperature dropped below this point, electrons could orbit the bare nuclei and remain there (without being ejected by collisions), thus forming atoms. With the birth of atoms, the photons—which had been continually scattering from the free electrons—now became much freer to spread nearly unhindered throughout the universe. We say that the photons become decoupled from matter. The total energy contained in radiation had been decreasing (lengthening in wavelength as the universe expanded), until at this point it was about equal to the total energy contained in matter. As the universe continued to expand, the radiation cooled further (to 2.7 K today, forming the cosmic microwave background radiation we detect from everywhere in the universe), and lost energy. But the mass of material particles did not decrease, so beginning at about this time, the energy of the universe became increasingly concentrated in matter rather than in radiation: the universe became **matter-dominated**, as it remains today.[‡]

Birth of stable atoms

Matter-dominated universe

After the birth of atoms, stars and galaxies could begin to form—probably by self-gravitation around mass concentrations (inhomogeneities). This began perhaps a million years after the Big Bang. The universe continued to evolve (see Section 45–2) until today, some 15 billion years later.

This scenario is by no means "proven" in any sense. Nor does it answer all questions. But it does provide a tentative picture, for the first time, of how the universe may have begun and evolved. It does have problems, however, some of which have been resolved by a modification first proposed in the early 1980s known as the **inflationary scenario**. According to the inflationary scenario, at the earliest stages of cosmological evolution, around 10^{-35} s after the Big Bang, the universe underwent a very rapid exponential expansion associated with the phase transition (symmetry breaking) that separated the strong force from the electroweak (as discussed earlier in this Section). After the brief inflationary period, the scenario settles back to the standard expansion already discussed. The appeal of the inflationary scenario is that it provides natural explanations for a number of problems not resolved by the standard model, such as why the universe is as close to being as flat as it seems to be, and why the cosmic microwave background radiation is so uniform.

Inflation

There are, however, questions we have not yet treated such as what is the future of the universe, which we will look at next, in the final Section of this book.

[†] This standard model prediction of a 25 percent primordial production of helium is fully in accord with what we observe today—the universe *does* contain about 25 percent He—and it is strong evidence in support of the standard Big Bang model. Furthermore, the theory says that 25 percent He abundance is fully consistent with there being three neutrino types, which is the number we observe so far. And it sets an upper limit of four to the maximum number of possible neutrino types. Actually, the fourth could be another type of low-mass particle, a *photino* or a *gravitino*, for example. Here we have a situation where cosmology actually makes a specific prediction about fundamental physics.

[‡] Although today matter contains more of the energy of the universe than does radiation, there are many more photons (perhaps 10^9 times more) than atoms, nuclei, and electrons. But each photon (at $T \approx 2.7$ K) has very little energy.

According to the standard Big Bang model, the universe is evolving and changing. Individual stars are evolving and dying as white dwarfs, neutron stars, black holes. At the same time, the universe as a whole is expanding. One important question is whether the universe will continue to expand forever. This question is connected to the curvature of space-time (Section 45–3) and to whether the universe is open (and infinite) or closed (and finite). There are three possibilities as shown in Fig. 45–25. If the curvature of the universe is *negative*, the expansion of the universe will never stop, although the rate of expansion might be expected to decrease due to the gravitational attraction of its parts. Such a universe would be *open* and infinite. If the universe is *flat* (no curvature), it would still be open and infinite but its expansion would slowly approach a zero rate. Finally, if the universe has *positive* curvature, it would be *closed* and finite. The effect of gravity in this case would be strong enough so that the expansion would eventually stop and the universe would then begin to contract. All matter eventually would collapse back onto itself in a **big crunch**. If, in this last case, the maximum expansion of the universe corresponded to, say, an intergalactic separation twice what it is now, the maximum expansion would occur about 30 or 40 billion years from now. Then, as the universe began to contract, the big crunch would occur about 100 billion years after the Big Bang.

An open or closed universe?

A big crunch?

Whether we live in an open and continually expanding universe, or a closed one that eventually will contract, is connected also to the average mass density in the universe. If the average mass density is above a critical value known as the **critical density**, estimated to be about

$$\rho_c \approx 10^{-26} \, \text{kg/m}^3$$

Critical density of the universe

(i.e., a few nucleons/m^3 on average throughout the universe), then gravity will prevent expansion from continuing forever, and will eventually pull the universe back into a big crunch. To say it another way, if $\rho > \rho_c$, there would be sufficient mass that gravity would give space-time a positive curvature. If instead the actual density is equal to the critical density, $\rho = \rho_c$, the universe will be flat and open. If the actual density is less than the critical density, $\rho < \rho_c$, the universe will have negative curvature and be open, expanding forever.

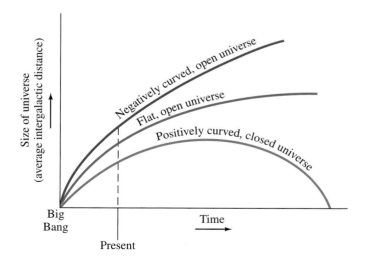

FIGURE 45–25 Three future possibilities for the universe.

EXAMPLE 45–7 **ESTIMATE** **Critical density of the universe.** Use energy conservation and the concept of escape velocity (Section 8–7) to estimate the critical density of the universe.

SOLUTION At the critical density, ρ_c, any given galaxy of mass m will just be able to "escape" away from our Galaxy. As we saw in Section 8–7, escape can just occur if the total energy E of the galaxy satisfies

$$E = K + U = \tfrac{1}{2}mv^2 - G\frac{mM}{R} = 0.$$

Here R is the distance of that galaxy m from our Galaxy, and we assume on the enormous scale of the universe that each galaxy is essentially a point. The total mass M that pulls inward on m is the total mass of all galaxies within a sphere of radius R (see Appendix C). If we assume the density of galaxies is roughly constant, then $M = \tfrac{4}{3}\pi\rho_c R^3$. Substituting this M into the above equation, and setting $v = HR$ (Hubble's law, Eq. 45–5), we obtain

$$\frac{GM}{R} = \tfrac{1}{2}v^2$$

or

$$\frac{G(\tfrac{4}{3}\pi\rho_c R^3)}{R} = \tfrac{1}{2}(HR)^2.$$

We solve for ρ_c:

$$\rho_c = \frac{3H^2}{8\pi G} \approx \frac{3\left[(70 \times 10^3\,\text{m/s/Mpc})(1\,\text{Mpc}/3.08 \times 10^{22}\,\text{m})\right]^2}{8(3.14)(6.67 \times 10^{-11}\,\text{N·m}^2/\text{kg}^2)}$$

$$\approx 10^{-26}\,\text{kg/m}^3.$$

Great efforts have gone into measuring the actual density of the universe. Estimates of the amount of visible matter in the universe put the actual density at between one and two orders of magnitude less than the critical density, thereby suggesting an open universe. However, there is evidence for a significant amount of nonluminous matter in the universe, often referred to as the **missing mass** or **dark matter**, enough to bring the density to almost exactly ρ_c. For example, observations of the rotation of galaxies suggest that they rotate as if they had considerably more mass than we can see. And observations of the motion of galaxies within clusters also suggest that they have considerably more mass than can be seen. Furthermore, the highly regarded inflationary theory offers a strong argument in favor of ρ being closely equal to ρ_c, and thus that space on a grand scale is Euclidean. If there is nonluminous matter in the universe, what might it be? One suggestion is that the missing mass consists of previously unknown weakly interacting massive particles (WIMPS), possibly supersymmetric particles known as "neutralinos," or perhaps small primordial black holes made in the early stages of the universe. Another possibility for the missing mass (or part of it) are MACHOS (massive compact halo objects), which would be chunks of matter in the form of large planets (like Jupiter) or stars that are too small to sustain fusion; they might glow faintly due to energy released from gravitational contraction (thus they are sometimes referred to as *brown dwarfs*), but too faintly to be seen. Evidence for MACHOS was found in 1993, inferred from their (assumed) gravitational effect on light passing by them from a distant star.

Another proposal for the missing mass is that neutrinos, once believed to be massless, may actually have nonzero rest mass. Since the universe probably contains roughly as many neutrinos of each type as it does photons (that is, about 10^9 times the number of nucleons, although this neutrino background has yet to be detected), neutrino masses of only a few eV could help to bring the actual density

Missing mass or dark matter

of the universe up to the critical density. The supernova of 1987 offered an opportunity to estimate the electron neutrino mass. If neutrinos have nonzero rest mass, then their velocities are less than the speed of light ($v < c$). High-energy neutrinos emitted from the supernova should in this case have arrived at Earth earlier than lower-energy (and therefore slower) neutrinos emitted at the same instant. Since they traveled a distance of about 170,000 ly from SN1987a, the time difference ought to be measurable. To get an idea of the size of the effect, let us consider the following Example.

EXAMPLE 45–8 ESTIMATE Neutrino mass estimate from a supernova.
Suppose two neutrinos from SN1987a were detected on Earth (via the reaction $\bar{\nu}_e + p \rightarrow n + e^+$) 10 seconds apart, with measured kinetic energies of about 20 MeV and 10 MeV. (a) Estimate the rest mass of the neutrino, m_ν, using this data, assuming both neutrinos were emitted at the same time. (b) Theoretical models of supernova explosions suggest that the neutrinos are emitted in a burst that lasts from a second or two up to perhaps 10 s. If we assume the neutrinos are not emitted simultaneously but rather at any time over a 10-s interval, what then can we say about the neutrino mass based on the data given above?

SOLUTION (a) We expect the neutrino mass to be less than 100 eV (from laboratory measurements). Since our neutrinos have kinetic energy of 20 MeV and 10 MeV, we can make the approximation $m_\nu c^2 \ll E$, where E (the total energy) is essentially equal to the kinetic energy. From Eq. 37–11, we have

$$E = \frac{m_\nu c^2}{\sqrt{1 - v^2/c^2}}.$$

We solve this for v, the velocity of a neutrino with energy E:

$$v = c\left(1 - \frac{m_\nu^2 c^4}{E^2}\right)^{\frac{1}{2}} = c\left(1 - \frac{m_\nu^2 c^4}{2E^2} + \cdots\right),$$

where we have used the binomial expansion $(1 + x)^{\frac{1}{2}} = 1 + \frac{1}{2}x + \cdots$, and we ignore higher-order terms since $m_\nu^2 c^4 \ll E^2$. The time t for a neutrino to travel a distance $d (= 170,000 \text{ ly})$ is

$$t = \frac{d}{v} = \frac{d}{c\left(1 - \frac{m_\nu^2 c^4}{2E^2}\right)} \approx \frac{d}{c}\left(1 + \frac{m_\nu^2 c^4}{2E^2}\right),$$

where again we used the binomial expansion $\left[(1 + x)^{-1} = 1 - x + \cdots\right]$. The difference in arrival times for our two neutrinos of energies $E_1 = 20$ MeV and $E_2 = 10$ MeV is

$$t_2 - t_1 = \frac{d}{c}\frac{m_\nu^2 c^4}{2}\left(\frac{1}{E_2^2} - \frac{1}{E_1^2}\right).$$

We solve this for $m_\nu c^2$ and set $t_2 - t_1 = 10$ s:

$$m_\nu c^2 = \left[\frac{2c(t_2 - t_1)}{d}\frac{E_1^2 E_2^2}{E_1^2 - E_2^2}\right]^{\frac{1}{2}}$$

$$= \left[\frac{2(3 \times 10^8 \text{ m/s})(10 \text{ s})}{(1.7 \times 10^5 \text{ ly})(10^{16} \text{ m/ly})}\frac{(400 \text{ MeV}^2)(100 \text{ MeV}^2)}{(400 \text{ MeV}^2 - 100 \text{ MeV}^2)}\right]^{\frac{1}{2}}$$

$$= 22 \times 10^{-6} \text{ MeV} = 22 \text{ eV}.$$

We thus estimate the mass of the neutrino to be 22 eV/c^2, but there would of course be experimental uncertainties.
(b) If the two neutrinos of energy $E_1 = 20$ MeV and $E_2 = 10$ MeV were emitted at unknown times over a 10-s interval, then the 10-s difference in their arrival times could be due to a 10-s difference in their emission time. In this case our data are consistent with zero rest mass and it puts an approximate *upper limit* on the neutrino mass of 22 eV/c^2.

In the actual experiments, the most sensitive detector consisted of several thousand tons of water in an underground chamber. It detected 11 events in 12 seconds, probably via the reaction $\bar{\nu}_e + p \rightarrow n + e^+$. There was not a clear correlation between energy and time of arrival. Nonetheless, a careful analysis of this experiment and others, has set a rough upper limit on the electron anti-neutrino mass of about 4 eV. The upper limits for the masses of the other neutrinos are much higher (see Table 44–2). Very recent experiments that investigate mass differences among neutrinos suggest that at least one of the neutrinos may have nonzero mass.

The scenarios we have been describing are based on some well-founded assumptions. As of this writing (year 2000) one of those basic assumptions—that gravity should be slowing down the expansion of the universe—has been called into question, and there are suggestions that there might be a previously undetected repulsive component of gravity.

Latest experiments

Finally, very recent experiments on the early CMB are consistent with the universe being flat, and fully in accord with the inflation theory.

The questions raised by cosmology can seem absurd at times, they are so removed from everyday "reality." We can always say, "the Sun is shining, it's going to burn on for an unimaginably long time, all is well." Nonetheless, the questions of cosmology are deep ones that fascinate the human intellect. One aspect that is especially intriguing is this: calculations on the formation and evolution of the universe have been performed that deliberately varied the values—just slightly—of certain fundamental physical constants. The result? A universe in which life as we know it could not exist. [For example, if the difference in mass between proton and neutron were zero, or small (less than the mass of the electron, $0.511 \text{ MeV}/c^2$), there would be no atoms: electrons would be captured by protons never to be freed again.] Such results have given rise to the so-called **Anthropic principle**, which says that if the universe were even a little different than it is, we could not be here. It seems as if the universe is exquisitely tuned, almost as if to accommodate us.

Summary

The night sky contains myriads of stars including those in the Milky Way, which is a "side view" of our **Galaxy** looking along the plane of the disc. Our Galaxy includes about 10^{11} stars. Beyond our Galaxy are billions of other galaxies.

Astronomical distances are measured in **light-years** ($1 \text{ ly} \approx 10^{13} \text{ km}$). The nearest star is about 4 ly away and the nearest other galaxy is 2 million ly away. Our galactic disc has a diameter of about 100,000 ly. Distances are often specified in **parsecs**, where 1 parsec = 3.26 ly.

Stars are believed to begin life as collapsing masses of hydrogen gas (protostars). As they contract, they heat up (potential energy is transformed to kinetic energy). When the temperature reaches 10 million degrees, nuclear fusion begins and forms heavier elements (**nucleosynthesis**), mainly helium at first. The energy released during these reactions balances the gravitational force, and the young star stabilizes as a **main-sequence** star. The tremendous luminosity of stars comes from the energy released during these thermonuclear reactions. After billions of years, as helium is collected in the core and hydrogen is used up, the core contracts and heats further. The envelope expands and cools, and the star becomes a **red giant** (larger diameter, redder color). The next stage of stellar evolution depends on the mass of the star. Stars of residual mass less than about 1.4 solar masses cool further and become **white dwarfs**, eventually fading and going out altogether. Heavier stars contract further due to their greater gravity: the density approaches nuclear density, the huge pressure forces electrons to combine with protons to form neutrons, and the star becomes essentially a huge nucleus of neutrons. This is a **neutron star**, and the energy released from its final core contraction is believed to produce **supernovae** explosions. If the star's residual mass is greater than two or three solar masses, it may contract even further and form a **black hole**, which is so dense that no matter or light can escape from it.

In the **general theory of relativity**, the **equivalence principle** states that an observer cannot distinguish acceleration from a gravitational field. Said another way, gravitational and inertial mass are the same. The theory predicts gravitational bending of light rays to a degree consistent with experiment. Gravity is treated as a curvature in space and time, the curvature being greater near massive bodies. The universe as a whole may be curved. If there is sufficient mass, the curvature of the universe is positive, and the universe is **closed** and **finite**; otherwise, it is **open** and **infinite**.

Distant galaxies display a **redshift** in their spectral lines, interpreted as a Doppler shift. The universe seems to be **expanding**, its galaxies racing away from each other at speeds (v) proportional to the distance (d) between them:

$$v = Hd,$$

which is known as **Hubble's law** (H is the **Hubble parameter**). This expansion of the universe suggests an explosive origin, the **Big Bang**, which probably occurred about 15 billion years ago.

Quasars are objects with a large redshift (suggesting great distance) and high luminosity (suggesting closeness or, more likely, extraordinary energy output).

The **cosmological principle** assumes that the universe, on a large scale, is homogeneous and isotropic.

Important evidence for the Big Bang model of the universe was the discovery of the **cosmic microwave background** radiation (CMB), which conforms to a blackbody radiation curve at a temperature of 2.7 K.

The **standard model** of the Big Bang provides a possible scenario as to how the universe developed as it expanded and cooled after the Big Bang. Starting at 10^{-43} seconds after the Big Bang, according to this model, there was a series of **phase transitions** during which previously unified forces of nature "condensed out" one by one. The **inflationary scenario** assumes that during one of these phase transitions, the universe underwent a brief but rapid exponential expansion. Until about 10^{-35} s, there was no distinction between quarks and leptons. Shortly thereafter, quarks were **confined** into hadrons (the **hadron era**). About 10^{-4} s after the Big Bang, the majority of hadrons disappeared, having combined with anti-hadrons, producing photons, leptons and energy, leaving mainly photons and leptons to freely move, thus introducing the **lepton era**. By the time the universe was about 10 s old, the electrons too had mostly disappeared, having combined with their antiparticles, and the universe became **radiation-dominated**. A couple of minutes later, nucleosynthesis began, but lasted only a few minutes. It then took several hundred thousand years before the universe was cool enough for electrons to combine with nuclei to form atoms. Also about this time, the background radiation had expanded and cooled so much that its total energy equaled the energy in matter. As the radiation cooled further, losing energy, the universe became **matter-dominated** (not in numbers, but in energy). Then stars and galaxies formed, producing a universe not much different than it is today—some 15 billion years later.

If the universe is open, it will continue to expand indefinitely. If it is closed, gravity is sufficiently strong to halt expansion and the universe will eventually begin to collapse back on itself, ending in a *big crunch*. Whether the universe is open or closed depends on whether its average mass density is above or below a critical density. The evidence today suggests that the universe is flat.

Questions

1. The Milky Way was once thought to be "cloudy," but no longer is considered so. Explain.

2. If you were measuring star parallaxes from the Moon instead of Earth, what corrections would you have to make? What changes would occur if you were measuring parallaxes from Mars?

3. Certain stars called *cepheid variables* change in luminosity with a period of typically several days. The period has been found to have a definite relationship with the absolute luminosity of the star. How could these stars be used to measure the distance to galaxies?

4. A star is in equilibrium when it radiates at its surface all the energy generated at its core. What happens when it begins to generate more energy than it radiates? Less energy? Explain.

5. Describe a red giant star. List some of its properties.

6. Select a point on the H–R diagram. Mark several directions away from this point. Now describe the changes that would take place in a star moving in each of these directions.

7. Does the H–R diagram reveal anything about the core of a star?

8. Why do some stars end up as white dwarfs, and others as neutron stars or black holes?

9. Can we tell, by looking at the population on the H–R diagram, that hotter suns have shorter lives? Explain.

10. What is a geodesic? What is its role in general relativity?

11. If it were discovered that the redshift of spectral lines of galaxies was due to something other than expansion, how might our view of the universe change? Would there be conflicting evidence? Discuss.

12. All galaxies appear to be moving away from us. Are we therefore at the center of the universe? Explain.

13. If you were located in a galaxy near the boundary of our observable universe, would galaxies in the direction of the Milky Way appear to be approaching you or receding from you? Explain.

14. What is the difference between the Hubble age of the universe and the actual age? Which is greater?

15. Compare an explosion on Earth to the Big Bang. Consider such questions as: Would the debris spread at a higher speed for more distant particles, as in the Big Bang? Would the debris come to rest? What type of universe would this correspond to, open or closed?

16. When the primordial nucleus exploded, thus creating the universe, into what did it expand? Discuss.

17. If nothing, not even light, escapes from a black hole, then how can we tell if one is there?

18. Explain what the 2.7 K cosmic microwave background radiation is. Where does it come from? Why is its temperature now so low?

19. The birth of atoms—that is, the combination of electrons with nuclei about which they orbit—occurred when the universe had cooled to about 3000 K and is generally called *recombination*. Why is this term misleading?

20. Why were atoms unable to exist until hundreds of thousands of years after the Big Bang?

21. Explain why today the universe is said to be matter-dominated, yet there are probably 10^9 times as many photons as there are massive particles.

22. If the universe is open, what will eventually happen to the cosmic microwave background radiation? If it is closed, what will happen to it?

23. Under what circumstances would the universe eventually collapse in on itself?

Problems

Sections 45–1 and 45–2

1. (I) Using the definitions of the parsec and the light-year, show that 1 pc = 3.26 ly.

2. (I) A star exhibits a parallax of 0.38 seconds of arc. How far away is it?

3. (I) The parallax angle of a star is 0.00019°. How far away is the star?

4. (I) A star is 36 parsecs away. What is its parallax angle? State (a) in seconds of arc, and (b) in degrees.

5. (I) What is the parallax angle for a star that is 55 light-years away? How many parsecs is this?

6. (I) A star is 35 parsecs away. How long does it take for its light to reach us?

7. (I) If one star is twice as far away from us as a second star, will the parallax angle of the first star be greater or less than that of the second star? By what factor?

8. (II) We saw earlier (Chapter 19) that the rate energy reaches the Earth from the Sun (the "solar constant") is about $1.3 \times 10^3 \, \text{W/m}^2$. What is (a) the apparent brightness l of the Sun, and (b) the absolute luminosity L of the Sun?

9. (II) What is the apparent brightness of the Sun as seen on Jupiter? (Jupiter is 5.2 times farther from the Sun than the Earth.)

10. (II) Estimate the angular width that our Galaxy would subtend if observed from the nearest galaxy to us (Table 45–1). Compare to the angular width of the Moon from Earth.

11. (II) When our Sun becomes a red giant, what will be its average density if it expands out to the orbit of Earth (1.5×10^{11} m from the Sun)?

12. (II) When our Sun becomes a white dwarf, it is expected to be about the size of the Moon. What angular width will it subtend from Earth?

13. (II) Calculate the density of a white dwarf whose mass is equal to the Sun's and whose radius is equal to the Earth's. How many times larger than Earth's density is this?

14. (II) A neutron star whose mass is 1.5 solar masses has a radius of about 11 km. Calculate its average density and compare to that for a white dwarf (Problem 13) and to that of nuclear matter.

15. (II) Calculate the Q-values for the He burning reactions of Eq. 45–2. (The mass of $^8_4$Be is 8.005305 u.)

16. (II) In the later stages of stellar evolution, a star (if massive enough) will begin fusing carbon nuclei to form, for example, magnesium:

$$^{12}_6\text{C} + ^{12}_6\text{C} \rightarrow ^{24}_{12}\text{Mg} + \gamma.$$

(a) How much energy is released in this reaction (see Appendix D). (b) How much kinetic energy must each carbon nucleus have (assume equal) in a head-on collision if they are just to touch (use Eq. 42–1) so that the strong force can come into play? (c) What temperature does this kinetic energy correspond to?

17. (II) Repeat Problem 16 for the reaction

$$^{16}_8\text{O} + ^{16}_8\text{O} \rightarrow ^{28}_{14}\text{Si} + ^4_2\text{He}.$$

18. (II) Suppose two stars of the same apparent brightness l are also believed to be the same size. The spectrum of one star peaks at 800 nm whereas that of the other peaks at 400 nm. Use Wien's law (Section 38–1) and the Stefan-Boltzmann equation (Eq. 19–15) to estimate their relative distances from us. [*Hint*: See Examples 45–5 and 45–6.]

19. (III) Stars located in a certain cluster are assumed to be about the same distance from us. Two such stars have spectra that peak at $\lambda_1 = 500$ nm and $\lambda_2 = 700$ nm, and the ratio of their apparent brightness is $l_1/l_2 = 0.091$. Estimate their relative sizes (give ratio of their diameters). [*Hint*: Use the Stefan-Boltzmann equation, Eq. 19–15.]

Section 45–3

20. (I) Describe a triangle, drawn on the surface of a sphere, for which the sum of the angles is (a) 360°, and (b) 180°.

21. (I) Show that the Schwarzschild radius for a star with mass equal to that (a) of our Sun is 2.95 km, and (b) of Earth is 8.9 mm.

22. (II) What is the Schwarzschild radius for a typical galaxy?

23. (II) What is the maximum sum-of-the-angles for a triangle on a sphere?

24. (II) Calculate the escape velocity, using Newtonian mechanics, from a body that has collapsed to its Schwarzschild radius.

Section 45–4

25. (I) If a galaxy is traveling away from us at 1.0 percent of the speed of light, roughly how far away is it?

26. (I) The redshift of a galaxy indicates a velocity of 3500 km/s. How far away is it?

27. (I) Estimate the speed of a galaxy (relative to us) that is near the "edge" of the universe, say 12 billion light-years away.

28. (II) Estimate the observed wavelength for the 656-nm line in the Balmer series of hydrogen in the spectrum of a galaxy whose distance from us is (a) 1.0×10^6 ly, (b) 1.0×10^8 ly, (c) 1.0×10^{10} ly.

29. (II) Estimate the speed of a galaxy, and its distance from us, if the wavelength for the hydrogen line at 434 nm is measured on Earth as being 610 nm.

30. (II) Starting from Eq. 45–3, show that the Doppler shift in wavelength is $\Delta\lambda/\lambda_0 \approx v/c$ (Eq. 45–4) for $v \ll c$. [Hint: Use the binomial expansion.]

Sections 45–5 to 45–7

31. (I) Calculate the wavelength at the peak of the blackbody radiation distribution at 2.7 K using the Wien law.

32. (II) The critical density for closure of the universe is $\rho_c \approx 10^{-26} \ \text{kg/m}^3$. State ρ_c in terms of the average number of nucleons per cubic meter.

33. (II) The size of the universe (the average distance between galaxies) at any one moment is believed to have been inversely proportional to the absolute temperature. Estimate the size of the universe, compared to today, at (a) $t = 10^6$ yr, (b) $t = 1$ s, (c) $t = 10^{-6}$ s, and (d) $t = 10^{-35}$ s.

34. (II) At approximately what time had the universe cooled below the threshold temperature for producing (a) kaons $(M \approx 500 \ \text{MeV}/c^2)$, (b) Υ $(M \approx 9500 \ \text{MeV}/c^2)$, and (c) muons $(M \approx 100 \ \text{MeV}/c^2)$?

General Problems

35. Suppose that three main-sequence stars could undergo the three changes represented by the three arrows, A, B, and C, in the H–R diagram of Fig. 45–26. For each case, describe the changes in temperature, luminosity, and size.

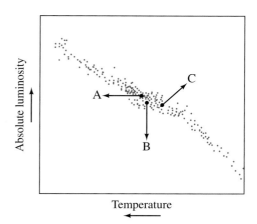

FIGURE 45–26 Problem 35.

36. Assume that the nearest stars to us have an intrinsic luminosity about the same as the Sun's. Their apparent brightness, however, is about 10^{11} times fainter than the Sun. From this, estimate the distance to the nearest stars. (Newton did this calculation, although he made a numerical error of a factor of 100.)

37. Use conservation of angular momentum to estimate the angular velocity of a neutron star which has collapsed to a diameter of 10 km, from a star whose radius was equal to that of our Sun $(7 \times 10^8 \ \text{m})$, of mass 1.5 times that of the Sun, and which rotated (like our Sun) about once a month.

38. By what factor does the rotational kinetic energy change when the star in Problem 37 collapses to a neutron star?

39. A certain pulsar, believed to be a neutron star of mass 1.5 times that of the Sun, with diameter 10 km, is observed to have a rotation speed of 1.0 rev/s. If it loses rotational kinetic energy at the rate of 1 part in 10^9 per day, which is all transformed into radiation, what is the power output of the star?

40. Estimate the rate of which hydrogen atoms would have to be created, according to the steady-state model, to maintain the present density of the universe of about $10^{-27} \ \text{kg/m}^3$, assuming the universe is expanding with Hubble constant $H = 70 \ \text{km/s/Mpc}$.

41. Estimate what neutrino rest mass (in eV) would provide the critical density to close the universe. Assume the neutrino density is, like photons, about 10^9 times that of nucleons, and that nucleons make up only (a) 2 percent of the mass needed, or (b) 5 percent of the mass needed.

42. Two stars, whose spectra peak at 600 nm and 400 nm, respectively, both lie on the main sequence. Use Wien's law, the Stefan-Boltzmann equation, and the H–R diagram (Fig. 45–7) to estimate the ratio of their diameters. [Hint: See Examples 45–5 and 45–6.]

43. The farthest we can measure with parallax is about 30 parsecs. What is our minimum angular resolution (in degrees), based on this information?

44. Through some coincidence, the Balmer lines from singly ionized helium in a distant star happen to overlap with the Balmer lines from hydrogen (Fig. 38–20) in the Sun. How fast is the star receding from us?

45. What is the temperature that corresponds to the 1.8 TeV collisions at the Fermilab collider? To what era in cosmological history does this correspond?

46. Astronomers have recently measured the rotation of gas around what might be a supermassive black hole of about 2 billion solar masses at the center of a galaxy. If the radius from the galactic center to the gas clouds is 60 light-years, what Doppler shift $\Delta\lambda/\lambda_0$ do you estimate they saw?

47. Astronomers use an *apparent magnitude* (m) scale to describe apparent brightness. It uses a logarithmic scale, where a higher number corresponds to a less bright star. (For example, the Sun has magnitude −27, whereas most stars have positive magnitudes.) On this scale, a change in apparent magnitude by +5 corresponds to a decrease in apparent brightness by a factor of 100. If Venus has an apparent magnitude of −4.4, whereas Sirius has an apparent magnitude of −1.4, which is brighter? What is the ratio of the apparent brightness of these two objects?

48. How large would the Sun be if its density equaled the critical density of the universe, $\rho_c \approx 10^{-26} \text{ kg/m}^3$? Express your answer in light-years and compare with the Earth–Sun distance and the size of our Galaxy.

49. Calculate the Schwarzschild radius using classical (Newtonian) gravitational theory, by calculating the minimum radius R for a sphere of mass M such that a photon (rest mass = 0) can escape from the surface. (General relativity gives $R = 2GM/c^2$.)

50. Estimate the radius of a white dwarf whose mass is equal to that of the Sun by the following method, assuming there are N nucleons and $\frac{1}{2}N$ electrons (why $\frac{1}{2}$?): (a) Use Fermi-Dirac statistics (Section 41–6) to show that the total energy of all the electrons is

$$E_e = \frac{3}{5}\left(\frac{1}{2}N\right)\frac{h^2}{8m_e}\left(\frac{3}{\pi}\frac{N}{2V}\right)^{\frac{2}{3}}.$$

[*Hint*: See Eqs. 41–12 and 41–13; we assume electrons fill energy levels from 0 up to the Fermi energy.] (b) The nucleons contribute to the total energy mainly via the gravitational force (note that the Fermi energy for nucleons is negligible compared to that for electrons—why?). Use a gravitational form of Gauss's law to show that the total gravitational potential energy of a uniform sphere of radius R is

$$-\frac{3}{5}\frac{GM^2}{R},$$

by considering the potential energy of a spherical shell of radius r due only to the mass inside it (why?) and integrate from $r = 0$ to $r = R$. (See also Appendix C.) (c) Write the total energy as a sum of these two terms, and set $dE/dR = 0$ to find the equilibrium radius, and evaluate it for a mass equal to the Sun's $(2.0 \times 10^{30}\text{ kg})$.

51. Determine the radius of a neutron star using the same argument as in Problem 50 but for N neutrons only. Show that the radius of a neutron star, of 1.5 solar masses, is about 11 km.

52. (a) Use special relativity and Newton's law of gravitation to show that a photon of mass $m = E/c^2$ just grazing the Sun will be deflected by an angle $\Delta\theta$ given by

$$\Delta\theta = \frac{2GM}{c^2R}$$

where G is the gravitational constant, R and M are the radius and mass of the Sun, and c is the speed of light. (b) Put in values and show $\Delta\theta = 0.87''$. (General relativity predicts an angle twice as large, $1.74''$.)

APPENDIX

Mathematical Formulas

A–1 Quadratic Formula

If
$$ax^2 + bx + c = 0$$
then
$$x = \frac{-b \pm \sqrt{b^2 - 4ac}}{2a}$$

A–2 Binomial Expansion

$$(1 \pm x)^n = 1 \pm nx + \frac{n(n-1)}{2!} x^2 \pm \frac{n(n-1)(n-2)}{3!} x^3 + \cdots$$

$$(x + y)^n = x^n \left(1 + \frac{y}{x}\right)^n = x^n \left(1 + n\frac{y}{x} + \frac{n(n-1)}{2!}\frac{y^2}{x^2} + \cdots\right)$$

A–3 Other Expansions

$$e^x = 1 + x + \frac{x^2}{2!} + \frac{x^3}{3!} + \cdots$$

$$\ln(1 + x) = x - \frac{x^2}{2} + \frac{x^3}{3} - \frac{x^4}{4} + \cdots$$

$$\sin \theta = \theta - \frac{\theta^3}{3!} + \frac{\theta^5}{5!} - \cdots$$

$$\cos \theta = 1 - \frac{\theta^2}{2!} + \frac{\theta^4}{4!} - \cdots$$

$$\tan \theta = \theta + \frac{\theta^3}{3} + \frac{2}{15}\theta^5 + \cdots \qquad |\theta| < \frac{\pi}{2}$$

In general:

$$f(x) = f(0) + \left(\frac{df}{dx}\right)_0 x + \left(\frac{d^2f}{dx^2}\right)_0 \frac{x^2}{2!} + \cdots$$

A–4 Areas and Volumes

Object	Surface area	Volume
Circle, radius r	πr^2	—
Sphere, radius r	$4\pi r^2$	$\frac{4}{3}\pi r^3$
Right circular cylinder, radius r, height h	$2\pi r^2 + 2\pi rh$	$\pi r^2 h$
Right circular cone, radius r, height h	$\pi r^2 + \pi r \sqrt{r^2 + h^2}$	$\frac{1}{3}\pi r^2 h$

A–5 Plane Geometry

1.

If line a_1 is parallel to line a_2, then $\theta_1 = \theta_2$.

FIGURE A–1

2.

If $a_1 \perp a_2$ and $b_1 \perp b_2$, then $\theta_1 = \theta_2$.

FIGURE A–2

3. The sum of the angles in any plane triangle is 180°.

4. *Pythagorean theorem:*

In any right triangle (one angle $= 90°$) of sides a, b, and c:

$$a^2 + b^2 = c^2$$

where c is the length of the hypotenuse (opposite the 90° angle).

FIGURE A–3

A–6 Trigonometric Functions and Identities

(See Fig. A–4.)

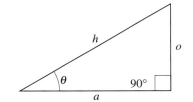

FIGURE A–4

$$\sin \theta = \frac{o}{h} \qquad \csc \theta = \frac{1}{\sin \theta} = \frac{h}{o}$$

$$\cos \theta = \frac{a}{h} \qquad \sec \theta = \frac{1}{\cos \theta} = \frac{h}{a}$$

$$\tan \theta = \frac{o}{a} = \frac{\sin \theta}{\cos \theta} \qquad \cot \theta = \frac{1}{\tan \theta} = \frac{a}{o}$$

$$a^2 + o^2 = h^2 \qquad \text{[Pythagorean theorem]}.$$

Figure A–5 shows the signs (+ or −) that cosine, sine, and tangent take on for angles θ in the four quadrants (0° to 360°). Note that angles are measured counterclockwise from the x axis as shown; negative angles are measured from *below* the x axis, clockwise: for example, $-30° = +330°$, and so on.

FIGURE A–5

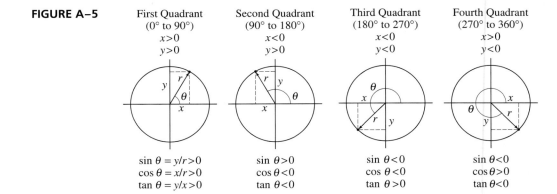

First Quadrant (0° to 90°) $x>0$ $y>0$	Second Quadrant (90° to 180°) $x<0$ $y>0$	Third Quadrant (180° to 270°) $x<0$ $y<0$	Fourth Quadrant (270° to 360°) $x>0$ $y<0$
$\sin \theta = y/r > 0$ $\cos \theta = x/r > 0$ $\tan \theta = y/x > 0$	$\sin \theta > 0$ $\cos \theta < 0$ $\tan \theta < 0$	$\sin \theta < 0$ $\cos \theta < 0$ $\tan \theta > 0$	$\sin \theta < 0$ $\cos \theta > 0$ $\tan \theta < 0$

The following are some useful identities among the trigonometric functions:

$$\sin^2\theta + \cos^2\theta = 1, \quad \sec^2\theta - \tan^2\theta = 1, \quad \csc^2\theta - \cot^2\theta = 1$$

$$\sin 2\theta = 2\sin\theta\cos\theta$$

$$\cos 2\theta = \cos^2\theta - \sin^2\theta = 2\cos^2\theta - 1 = 1 - 2\sin^2\theta$$

$$\tan 2\theta = \frac{2\tan\theta}{1 - \tan^2\theta}$$

$$\sin(A \pm B) = \sin A\cos B \pm \cos A\sin B$$

$$\cos(A \pm B) = \cos A\cos B \mp \sin A\sin B$$

$$\tan(A \pm B) = \frac{\tan A \pm \tan B}{1 \mp \tan A\tan B}$$

$$\sin(180° - \theta) = \sin\theta$$

$$\cos(180° - \theta) = -\cos\theta$$

$$\sin(90° - \theta) = \cos\theta$$

$$\cos(90° - \theta) = \sin\theta$$

$$\cos(-\theta) = \cos\theta$$

$$\sin(-\theta) = -\sin\theta$$

$$\tan(-\theta) = -\tan\theta$$

$$\sin\tfrac{1}{2}\theta = \sqrt{\frac{1 - \cos\theta}{2}}, \quad \cos\tfrac{1}{2}\theta = \sqrt{\frac{1 + \cos\theta}{2}}, \quad \tan\tfrac{1}{2}\theta = \sqrt{\frac{1 - \cos\theta}{1 + \cos\theta}}$$

$$\sin A \pm \sin B = 2\sin\left(\frac{A \pm B}{2}\right)\cos\left(\frac{A \mp B}{2}\right).$$

For any triangle (see Fig. A–6):

$$\frac{\sin\alpha}{a} = \frac{\sin\beta}{b} = \frac{\sin\gamma}{c} \qquad \text{[Law of sines]}$$

$$c^2 = a^2 + b^2 - 2ab\cos\gamma. \qquad \text{[Law of cosines]}$$

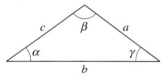

FIGURE A–6

A–7 Logarithms

The following identities apply to common logs (base 10), natural logs (base e) which are often abbreviated ln, or logs to any other base.

$$\log(ab) = \log a + \log b$$

$$\log\left(\frac{a}{b}\right) = \log a - \log b$$

$$\log a^n = n\log a.$$

A–8 Vectors

Vector addition is covered in Sections 3–2 to 3–5.
Vector multiplication is covered in Sections 3–3, 7–2 and 11–1.

APPENDIX B

Derivatives and Integrals

B–1 Derivatives: General Rules

(See also Section 2–3.)

$$\frac{dx}{dx} = 1$$

$$\frac{d}{dx}\left[af(x)\right] = a\frac{df}{dx} \qquad (a = \text{constant})$$

$$\frac{d}{dx}\left[f(x) + g(x)\right] = \frac{df}{dx} + \frac{dg}{dx}$$

$$\frac{d}{dx}\left[f(x)g(x)\right] = \frac{df}{dx}g + f\frac{dg}{dx}$$

$$\frac{d}{dx}\left[f(y)\right] = \frac{df}{dy}\frac{dy}{dx} \qquad \text{[chain rule]}$$

$$\frac{dx}{dy} = \frac{1}{\left(\dfrac{dy}{dx}\right)} \qquad \text{if } \frac{dy}{dx} \neq 0.$$

B–2 Derivatives: Particular Functions

$$\frac{da}{dx} = 0 \qquad (a = \text{constant})$$

$$\frac{d}{dx}x^n = nx^{n-1}$$

$$\frac{d}{dx}\sin ax = a\cos ax$$

$$\frac{d}{dx}\cos ax = -a\sin ax$$

$$\frac{d}{dx}\tan ax = a\sec^2 ax$$

$$\frac{d}{dx}\ln ax = \frac{1}{x}$$

$$\frac{d}{dx}e^{ax} = ae^{ax}$$

B–3 Indefinite Integrals: General Rules

(See also Section 7–3.)

$$\int dx = x$$

$$\int af(x)\,dx = a\int f(x)\,dx \qquad (a = \text{constant})$$

$$\int [f(x) + g(x)]\,dx = \int f(x)\,dx + \int g(x)\,dx$$

$$\int u\,dv = uv - \int v\,du \qquad [\text{integration by parts}]$$

B–4 Indefinite Integrals: Particular Functions

(An arbitrary constant can be added to the right side of each equation.)

$$\int a\,dx = ax \qquad (a = \text{constant})$$

$$\int x^m\,dx = \frac{1}{m+1}x^{m+1} \qquad (m \neq -1)$$

$$\int \sin ax\,dx = -\frac{1}{a}\cos ax$$

$$\int \cos ax\,dx = \frac{1}{a}\sin ax$$

$$\int \tan ax\,dx = \frac{1}{a}\ln|\sec ax|$$

$$\int \frac{1}{x}\,dx = \ln x$$

$$\int e^{ax}\,dx = \frac{1}{a}e^{ax}$$

$$\int \frac{dx}{x^2 + a^2} = \frac{1}{a}\tan^{-1}\frac{x}{a}$$

$$\int \frac{dx}{x^2 - a^2} = \frac{1}{2a}\ln\left(\frac{x-a}{x+a}\right) \qquad (x^2 > a^2)$$

$$= -\frac{1}{2a}\ln\left(\frac{a+x}{a-x}\right) \qquad (x^2 < a^2)$$

$$\int \frac{dx}{\sqrt{x^2 \pm a^2}} = \ln(x + \sqrt{x^2 \pm a^2})$$

$$\int \frac{dx}{(x^2 \pm a^2)^{\frac{3}{2}}} = \frac{\pm x}{a^2\sqrt{x^2 \pm a^2}}$$

$$\int \frac{x\,dx}{(x^2 \pm a^2)^{\frac{3}{2}}} = \frac{-1}{\sqrt{x^2 \pm a^2}}$$

$$\int \sin^2 ax\,dx = \frac{x}{2} - \frac{\sin 2ax}{4a}$$

$$\int xe^{-ax}\,dx = -\frac{e^{-ax}}{a^2}(ax + 1)$$

$$\int x^2 e^{-ax}\,dx = -\frac{e^{-ax}}{a^3}(a^2x^2 + 2ax + 2)$$

B–5 A few Definite Integrals

$$\int_0^\infty x^n e^{-ax}\,dx = \frac{n!}{a^{n+1}}$$

$$\int_0^\infty e^{-ax^2}\,dx = \sqrt{\frac{\pi}{4a}}$$

$$\int_0^\infty xe^{-ax^2}\,dx = \frac{1}{2a}$$

$$\int_0^\infty x^2 e^{-ax^2}\,dx = \sqrt{\frac{\pi}{16a^3}}$$

$$\int_0^\infty x^3 e^{-ax^2}\,dx = \frac{1}{2a^2}$$

$$\int_0^\infty x^{2n} e^{-ax^2}\,dx = \frac{1 \cdot 3 \cdot 5 \cdots (2n-1)}{2^{n+1}a^n}\sqrt{\frac{\pi}{a}}$$

Gravitational Force due to a Spherical Mass Distribution

In Chapter 6 (Section 6–1), we stated that the gravitational force exerted by or on a uniform sphere acts as if all the mass of the sphere were concentrated at its center, if the other mass is outside the sphere. In other words, the gravitational force that a uniform sphere exerts on a particle outside it is

$$F = G\frac{mM}{r^2}, \qquad \qquad [m \text{ outside sphere of mass } M]$$

where m is the mass of the particle, M the mass of the sphere, and r the distance of m from the center of the sphere. Now we will derive this result. We will use the concepts of infinitesimally small quantities and integration.

First we consider a very thin, uniform spherical shell (like a thin-walled basketball) of mass M whose thickness t is small compared to its radius R (Fig. C–1). The force on a particle of mass m at a distance r from the center of the shell can be calculated as the vector sum of the forces due to all the particles of the shell. We imagine the shell divided up into thin (infinitesimal) circular strips so that all points on a strip are equidistant from our particle m. One of these circular strips, labeled AB, is shown in Fig. C–1. It is $R d\theta$ wide, t thick, and has a radius $R \sin \theta$. The force on our particle m due to a tiny piece of the strip at point A is represented by the vector $\mathbf{F}_A$ shown. The force due to a tiny piece of the strip at point B, which is diametrically opposite A, is the force $\mathbf{F}_B$. We take the two pieces at A and B to be of equal mass, so $F_A = F_B$. The horizontal components of $\mathbf{F}_A$ and $\mathbf{F}_B$ are each equal to

$$F_A \cos \phi$$

and point toward the center of the shell. The vertical components of $\mathbf{F}_A$ and $\mathbf{F}_B$ are of equal magnitude and point in opposite directions, and so cancel. Since for every point on the strip there is a corresponding point diametrically opposite (as with A and B), we see that the net force due to the entire strip points toward the center of the shell. Its magnitude will be

$$dF = G\frac{m\,dM}{l^2}\cos\phi,$$

where dM is the mass of the entire circular strip and l is the distance from all

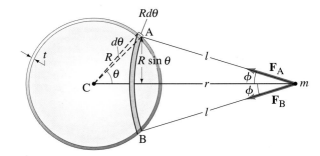

points on the strip to m, as shown. We write dM in terms of the density ρ; by density we mean the mass per unit volume (Section 13–1). Hence, $dM = \rho\, dV$, where dV is the volume of the strip and equals $(2\pi R \sin\theta)(t)(R\, d\theta)$. Then the force dF due to the circular strip shown is

$$dF = G\frac{m\rho 2\pi R^2 t \sin\theta\, d\theta}{l^2}\cos\phi. \qquad (C-1)$$

To get the total force F that the entire shell exerts on the particle m, we must integrate over all the circular strips: that is, from $\theta = 0°$ to $\theta = 180°$. But our expression for dF contains l and ϕ, which are functions of θ. From Fig. C–1 we can see that

$$l\cos\phi = r - R\cos\theta.$$

Furthermore, we can write the law of cosines for triangle CmA:

$$\cos\theta = \frac{r^2 + R^2 - l^2}{2rR}. \qquad (C-2)$$

With these two expressions we can reduce our three variables (l, θ, ϕ) to only one, which we take to be l. We do two things with Eq. C–2: (1) We put it into the equation for $l\cos\phi$ above:

$$\cos\phi = \frac{1}{l}(r - R\cos\theta) = \frac{r^2 + l^2 - R^2}{2rl};$$

and (2) we take the differential of both sides of Eq. C–2 (because $\sin\theta\, d\theta$ appears in the expression for dF, Eq. C–1):

$$-\sin\theta\, d\theta = -\frac{2l\, dl}{2rR} \qquad \text{or} \qquad \sin\theta\, d\theta = \frac{l\, dl}{rR},$$

since r and R are considered constants when summing over the strips. Now we insert these into Eq. C–1 for dF and find

$$dF = Gm\rho\pi t\,\frac{R}{r^2}\left(1 + \frac{r^2 - R^2}{l^2}\right)dl.$$

Now we integrate to get the net force on our thin shell of radius R. To integrate over all the strips $(\theta = 0°$ to $180°)$, we must go from $l = r - R$ to $l = r + R$ (see Fig. C–1). Thus,

$$F = Gm\rho\pi t\,\frac{R}{r^2}\left[l - \frac{r^2 - R^2}{l}\right]_{l=r-R}^{l=r+R}$$

$$= Gm\rho\pi t\,\frac{R}{r^2}\,(4R).$$

The volume V of the spherical shell is its area $(4\pi R^2)$ times the thickness t. Hence the mass $M = \rho V = \rho 4\pi R^2 t$, and finally

$$F = G\frac{mM}{r^2}. \qquad \left[\begin{array}{c}\text{particle of mass } m \text{ outside a} \\ \text{thin uniform spherical shell of mass } M\end{array}\right]$$

This result gives us the force a thin shell exerts on a particle of mass m a distance r from the center of the shell, and *outside* the shell. We see that the force is the same as that between m and a particle of mass M at the center of the shell. In other words, for purposes of calculating the gravitational force exerted on or by a uniform spherical shell, we can consider all its mass concentrated at its center.

What we have derived for a shell holds also for a solid sphere, since a solid sphere can be considered as made up of many concentric shells, from $R = 0$ to $R = R_0$, where R_0 is the radius of the solid sphere. Why? Because if each shell has mass dM, we write for each shell, $dF = Gm\,dM/r^2$, where r is the distance from the center C to mass m and is the same for all shells. Then the total force equals the sum or integral over dM, which gives the total mass M. Thus the result

$$F = G\frac{mM}{r^2} \qquad \left[\begin{array}{c}\text{particle of mass } m \text{ outside} \\ \text{solid sphere of mass } M\end{array}\right] \quad \text{(C-3)}$$

is valid for a solid sphere of mass M even if the density varies with distance from the center. (It is not valid if the density varies within each shell—that is, depends not only on R.) Thus the gravitational force exerted on or by spherical objects, including nearly spherical objects like the Earth, Sun, and Moon, can be considered to act as if the objects were point particles.

This result, Eq. C–3, is true only if the mass m is outside the sphere. Let us next consider a point mass m that is located inside the spherical shell of Fig. C–1. Here, r would be less than R, and the integration over l would be from $l = R - r$ to $l = R + r$, so

$$\left[l - \frac{r^2 - R^2}{l}\right]_{R-r}^{R+r} = 0.$$

Thus the force on any mass inside the shell would be zero. This result has particular importance for the electrostatic force, which is also an inverse square law. For the gravitational situation, we see that at points within a solid sphere, say 1000 km below the earth's surface, only the mass up to that radius contributes to the net force. The outer shells beyond the point in question contribute no net gravitational effect.

The results we have obtained here can also be reached using the gravitational analog of Gauss's law for electrostatics (Chapter 22).

Selected Isotopes

(1) Atomic Number Z	(2) Element	(3) Symbol	(4) Mass Number A	(5) Atomic Mass†	(6) % Abundance (or Radioactive Decay Mode)	(7) Half-life (if radioactive)
0	(Neutron)	n	1	1.008665	β^-	10.4 min
1	Hydrogen	H	1	1.007825	99.985%	
	Deuterium	D	2	2.014102	0.015%	
	Tritium	T	3	3.016049	β^-	12.33 yr
2	Helium	He	3	3.016029	0.000137%	
			4	4.002603	99.999863%	
3	Lithium	Li	6	6.015122	7.5%	
			7	7.016004	92.5%	
4	Beryllium	Be	7	7.016929	EC, γ	53.12 days
			9	9.012182	100%	
5	Boron	B	10	10.012937	19.9%	
			11	11.009305	80.1%	
6	Carbon	C	11	11.011434	β^+, EC	20.39 min
			12	12.000000	98.90%	
			13	13.003355	1.10%	
			14	14.003242	β^-	5730 yr
7	Nitrogen	N	13	13.005739	β^+	9.965 min
			14	14.003074	99.63%	
			15	15.000108	0.37%	
8	Oxygen	O	15	15.003065	β^+, EC	122.24 s
			16	15.994915	99.76%	
			18	17.999160	0.20%	
9	Fluorine	F	19	18.998403	100%	
10	Neon	Ne	20	19.992440	90.48%	
			22	21.991386	9.25%	
11	Sodium	Na	22	21.994437	β^+, EC, γ	2.6019 yr
			23	22.989770	100%	
			24	23.990963	β^-, γ	14.9590 h
12	Magnesium	Mg	24	23.985042	78.99%	
13	Aluminum	Al	27	26.981538	100%	

†The masses given in column (5) are those for the neutral atom, including the Z electrons.

(1) Atomic Number Z	(2) Element	(3) Symbol	(4) Mass Number A	(5) Atomic Mass[†]	(6) % Abundance (or Radioactive Decay Mode)	(7) Half-life (if radioactive)
14	Silicon	Si	28	27.976927	92.23%	
			31	30.975363	β^-, γ	157.3 min
15	Phosphorus	P	31	30.973762	100%	
			32	31.973907	β^-	14.262 days
16	Sulfur	S	32	31.972071	95.02%	
			35	34.969032	β^-	87.32 days
17	Chlorine	Cl	35	34.968853	75.77%	
			37	36.965903	24.23%	
18	Argon	Ar	40	39.962383	99.600%	
19	Potassium	K	39	38.963707	93.2581%	
			40	39.963999	0.0117%	
					β^-, EC, γ, β^+	1.28×10^9 yr
20	Calcium	Ca	40	39.962591	96.941%	
21	Scandium	Sc	45	44.955910	100%	
22	Titanium	Ti	48	47.947947	73.8%	
23	Vanadium	V	51	50.943964	99.750%	
24	Chromium	Cr	52	51.940512	83.79%	
25	Manganese	Mn	55	54.938049	100%	
26	Iron	Fe	56	55.934942	91.72%	
27	Cobalt	Co	59	58.933200	100%	
			60	59.933822	β^-, γ	5.2714 yr
28	Nickel	Ni	58	57.935348	68.077%	
			60	59.930791	26.233%	
29	Copper	Cu	63	62.929601	69.17%	
			65	64.927794	30.83%	
30	Zinc	Zn	64	63.929147	48.6%	
			66	65.926037	27.9%	
31	Gallium	Ga	69	68.925581	60.108%	
32	Germanium	Ge	72	71.922076	27.66%	
			74	73.921178	35.94%	
33	Arsenic	As	75	74.921596	100%	
34	Selenium	Se	80	79.916522	49.61%	
35	Bromine	Br	79	78.918338	50.69%	
36	Krypton	Kr	84	83.911507	57.0%	
37	Rubidium	Rb	85	84.911789	72.17%	
38	Strontium	Sr	86	85.909262	9.86%	
			88	87.905614	82.58%	
			90	89.907737	β^-	28.79 yr
39	Yttrium	Y	89	88.905848	100%	
40	Zirconium	Zr	90	89.904704	51.45%	
41	Niobium	Nb	93	92.906377	100%	
42	Molybdenum	Mo	98	97.905408	24.13%	

[†]The masses given in column (5) are those for the neutral atom, including the Z electrons.

(1) Atomic Number Z	(2) Element	(3) Symbol	(4) Mass Number A	(5) Atomic Mass[†]	(6) % Abundance (or Radioactive Decay Mode)	(7) Half-life (if radioactive)
43	Technetium	Tc	98	97.907216	β^-, γ	4.2×10^6 yr
44	Ruthenium	Ru	102	101.904349	31.6%	
45	Rhodium	Rh	103	102.905504	100%	
46	Palladium	Pd	106	105.903483	27.33%	
47	Silver	Ag	107	106.905093	51.839%	
			109	108.904756	48.161%	
48	Cadmium	Cd	114	113.903358	28.73%	
49	Indium	In	115	114.903878	95.7%; β^-, γ	4.41×10^{14} yr
50	Tin	Sn	120	119.902197	32.59%	
51	Antimony	Sb	121	120.903818	57.36%	
52	Tellurium	Te	130	129.906223	33.80%	7.9×10^{20} yr
53	Iodine	I	127	126.904468	100%	
			131	130.906124	β^-, γ	8.0207 days
54	Xenon	Xe	132	131.904154	26.9%	
			136	135.907220	8.9%	
55	Cesium	Cs	133	132.905446	100%	
56	Barium	Ba	137	136.905821	11.23%	
			138	137.905241	71.70%	
57	Lanthanum	La	139	138.906348	99.9098%	
58	Cerium	Ce	140	139.905434	88.48%	
59	Praseodymium	Pr	141	140.907647	100%	
60	Neodymium	Nd	142	141.907718	27.13%	
61	Promethium	Pm	145	144.912744	EC, γ, α	17.7 yr
62	Samarium	Sm	152	151.919728	26.7%	
63	Europium	Eu	153	152.921226	52.2%	
64	Gadolinium	Gd	158	157.924101	24.84%	
65	Terbium	Tb	159	158.925343	100%	
66	Dysprosium	Dy	164	163.929171	28.2%	
67	Holmium	Ho	165	164.930319	100%	
68	Erbium	Er	166	165.930290	33.6%	
69	Thulium	Tm	169	168.934211	100%	
70	Ytterbium	Yb	174	173.938858	31.8%	
71	Lutecium	Lu	175	174.940767	97.4%	
72	Hafnium	Hf	180	179.946549	35.100%	
73	Tantalum	Ta	181	180.947996	99.988%	
74	Tungsten (wolfram)	W	184	183.950933	30.67%	
75	Rhenium	Re	187	186.955751	62.60%; β^-	4.35×10^{10} yr
76	Osmium	Os	191	190.960927	β^-, γ	15.4 days
			192	191.961479	41.0%	
77	Iridium	Ir	191	190.960591	37.3%	
			193	192.962923	62.7%	
78	Platinum	Pt	195	194.964774	33.8%	

[†] The masses given in column (5) are those for the neutral atom, including the Z electrons.

(1) Atomic Number Z	(2) Element	(3) Symbol	(4) Mass Number A	(5) Atomic Mass†	(6) % Abundance (or Radioactive Decay Mode)	(7) Half-life (if radioactive)
79	Gold	Au	197	196.966551	100%	
80	Mercury	Hg	199	198.968262	16.87%	
			202	201.970625	29.86%	
81	Thallium	Tl	205	204.974412	70.476%	
82	Lead	Pb	206	205.974449	24.1%	
			207	206.975880	22.1%	
			208	207.976635	52.4%	
			210	209.984173	β^-, γ, α	22.3 yr
			211	210.988731	β^-, γ	36.1 min
			212	211.991887	β^-, γ	10.64 h
			214	213.999798	β^-, γ	26.8 min
83	Bismuth	Bi	209	208.980383	100%	
			211	210.987258	α, γ, β^-	2.14 min
84	Polonium	Po	210	209.982857	α, γ	138.376 days
			214	213.995185	α, γ	164.3 μs
85	Astatine	At	218	218.008681	α, β^-	1.5 s
86	Radon	Rn	222	222.017570	α, γ	3.8235 days
87	Francium	Fr	223	223.019731	β^-, γ, α	21.8 min
88	Radium	Ra	226	226.025402	α, γ	1600 yr
89	Actinium	Ac	227	227.027746	β^-, γ, α	21.773 yr
90	Thorium	Th	228	228.028731	α, γ	1.9116 yr
			232	232.038050	100%; α, γ	1.405×10^{10} yr
91	Protactinium	Pa	231	231.035878	α, γ	3.276×10^4 yr
92	Uranium	U	232	232.037146	α, γ	68.9 yr
			233	233.039628	α, γ	1.592×10^5 yr
			235	235.043923	0.720%, α, γ	7.038×10^8 yr
			236	236.045561	α, γ	2.342×10^7 yr
			238	238.050782	99.2745%; α, γ	4.468×10^9 yr
			239	239.054287	β^-, γ	23.45 min
93	Neptunium	Np	237	237.048166	α, γ	2.144×10^6 yr
			239	239.052931	β^-, γ	2.3565 d
94	Plutonium	Pu	239	239.052157	α, γ	24,110 yr
			244	244.064197	α	8.08×10^7 yr
95	Americium	Am	243	243.061373	α, γ	7370 yr
96	Curium	Cm	247	247.070346	α, γ	1.56×10^7 yr
97	Berkelium	Bk	247	247.070298	α, γ	1380 yr
98	Californium	Cf	251	251.079580	α, γ	898 yr
99	Einsteinium	Es	252	252.082972	α, EC, γ	472 d
100	Fermium	Fm	257	257.095099	α, γ	101 d
101	Mendelevium	Md	258	258.098425	α, γ	51.5 d
102	Nobelium	No	259	259.10102	α, EC	58 min
103	Lawrencium	Lr	262	262.10969	α, EC, fission	216 min

† The masses given in column (5) are those for the neutral atom, including the Z electrons.

(1) Atomic Number Z	(2) Element	(3) Symbol	(4) Mass Number A	(5) Atomic Mass†	(6) % Abundance (or Radioactive Decay Mode)	(7) Half-life (if radioactive)
104	Rutherfordium	Rf	261	261.10875	α	65 s
105	Dubnium	Db	262	262.11415	α, fission, EC	34 s
106	Seaborgium	Sg	266	266.12193	α, fission	21 s
107	Bohrium	Bh	264	264.12473	α	0.44 s
108	Hassium	Hs	269	269.13411	α	9 s
109	Meitnerium	Mt	268	268.13882	α	0.07 s
110			271	271.14608	α	0.06 s
111			272	272.15348	α	1.5 ms
112			277	277	α	0.24 ms
114			289	289	α	20 s
116			289	289	α	0.6 ms
118			293	293	α	0.1 ms

† The masses given in column (5) are those for the neutral atom, including the Z electrons.

Answers to Odd-Numbered Problems

CHAPTER 1

1. (a) 1×10^{10} yr; (b) 3×10^{17} s.

3. (a) 1.156×10^3; (b) 2.18×10^1;
(c) 6.8×10^{-3}; (d) 2.7635×10^1;
(e) 2.19×10^{-1}; (f) 2.2×10^1.

5. 7.7%.

7. (a) 4%; (b) 0.4%; (c) 0.07%.

9. 1.0×10^5 s.

11. 9%.

13. (a) 0.286 6 m; (b) 0.000 085 V;
(c) 0.000 760 kg;
(d) 0.000 000 000 060 0 s;
(e) 0.000 000 000 000 022 5 m;
(f) 2,500,000,000 volts.

15. 1.8 m.

17. (a) 0.111 yd²; (b) 10.76 ft².

19. (a) 3.9×10^{-9} in;
(b) 1.0×10^8 atoms.

21. (a) 0.621 mi/h;
(b) 1 m/s = 3.28 ft/s; (c) 0.278 m/s.

23. (a) 9.46×10^{15} m;
(b) 6.31×10^4 AU; (c) 7.20 AU/h.

25. (a) 10^3; (b) 10^4; (c) 10^{-2}; (d) 10^9.

27. ≈20%.

29. 1×10^5 cm³.

31. (a) ≈600 dentists.

33. ≈3×10^8 kg/yr.

35. 51 km.

37. $A = [L/T^4] = $ m/s⁴,
$B = [L/T^2] = $ m/s².

39. (a) 0.10 nm; (b) 1.0×10^5 fm;
(c) 1.0×10^{10} Å; (d) 9.5×10^{25} Å.

41. (a) 3.16×10^7 s; (b) 3.16×10^{16} ns;
(c) 3.17×10^{-8} yr.

43. (a) 1,000 drivers.

45. 1×10^{11} gal/yr.

47. 9 cm.

49. 4×10^5 t.

51. ≈4 yr.

53. 1.9×10^2 m.

55. (a) 3%, 3%; (b) 0.7%, 0.2%.

CHAPTER 2

1. 5.0 h.

3. 61 m.

5. 0.78 cm/s (toward $+x$).

7. ≈300 m/s.

9. (a) 10.1 m/s;
(b) +3.4 m/s, away from trainer.

11. (a) 0.28 m/s; (b) 1.2 m/s;
(c) 0.28 m/s; (d) 1.6 m/s;
(e) −1.0 m/s.

13. (a) 13.4 m/s;
(b) +4.5 m/s, away from master.

15. 24 s.

17. 55 km/h, 0.

19. 6.73 m/s.

21. 5.2 s

23. -7.0 m/s², 0.72.

25. (a) 4.7 m/s²; (b) 2.2 m/s²;
(c) 0.3 m/s²; (d) 1.6 m/s².

27. $v = (6.0 \text{ m/s}) + (17 \text{ m/s}^2)t$,
$a = 17$ m/s².

29. 1.5 m/s², 99 m.

31. 1.7×10^2 m.

33. 4.41 m/s², $t = 2.61$ s.

35. 55.0 m.

37. (a) 2.3×10^2 m; (b) 31 s;
(c) 15 m, 13 m.

39. (a) 103 m; (b) 64 m.

41. 31 m/s.

43. (b) 3.45 s.

45. 32 m/s (110 km/h).

47. 2.83 s.

49. (a) 8.81 s; (b) 86.3 m/s.

51. 1.44 s.

53. 15 m/s.

55. 5.44 s.

59. 0.035 s.

61. 1.8 m above the top of the window.

63. 52 m.

65. 19.8 m/s, 20.0 m.

67. (a) $v = (g/k)(1 - e^{-kt})$;
(b) $v_{\text{term}} = g/k$.

69. $6h_{\text{Earth}}$.

71. 1.3 m.

73. (b) $H_{50} = 9.8$ m; (c) $H_{100} = 39$ m.

75. (a) 1.3 m; (b) 6.1 m/s; (c) 1.2 s.

77. (a) 3.88 s; (b) 73.9 m; (c) 38.0 m/s,
48.4 m/s.

79. (a) 52 min; (b) 31 min.

81. (a) $v_0 = 26$ m/s; (b) 35 m; (c) 1.2 s;
(d) 4.1 s.

83. (a) 4.80 s; (b) 37.0 m/s; (c) 75.2 m.

85. She should decide to stop!

87. $\Delta v_{0\text{down}} = 0.8$ m/s, $\Delta v_{0\text{up}} = 0.9$ m/s.

89. 29.0 m.

CHAPTER 3

1. 263 km, 13° S of W.

3. $\mathbf{V}_{\text{wrong}} = \mathbf{V}_2 - \mathbf{V}_1$.

5. 13.6 m, 18° N of E,

7. (a)

(b) $V_x = -11.7$, $V_y = 8.16$;
(c) 14.3, 34.9° above $-x$-axis.

9. (a) $V_N = 476$ km/h,
$V_W = 421$ km/h;
(b) $d_N = 1.43 \times 10^3$ km,
$d_W = 1.26 \times 10^3$ km.

11. (a) 4.2, 45° below $+x$-axis;
(b) 5.1, 79° below $+x$-axis.

13. (a) 53.7, 1.40° above $-x$-axis;
(b) 53.7, 1.40° below $+x$-axis.

15. (a) 94.5, 11.8° below $-x$-axis;
(b) 150, 35.3° below $+x$-axis.

17. (a) $A_x = \pm 82.9$; (b) 166.6, 12.1°
above $-x$-axis.

19. $(7.60 \text{ m/s})\mathbf{i} - (4.00 \text{ m/s})\mathbf{k}$; 8.59 m/s.

21. (a) Unknown; (b) 4.11 m/s², 33.2°
north of east; (c) unknown.

23. (a) $\mathbf{v} = (4.0\,\text{m/s}^2)t\mathbf{i} + (3.0\,\text{m/s}^2)t\mathbf{j}$;
   (b) $(5.0\,\text{m/s}^2)t$;
   (c) $\mathbf{r} = (2.0\,\text{m/s}^2)t^2\mathbf{i} + (1.5\,\text{m/s}^2)t^2\mathbf{j}$;
   (d) $\mathbf{v} = (8.0\,\text{m/s})\mathbf{i} + (6.0\,\text{m/s})\mathbf{j}$,
      $|\mathbf{v}| = 10.0\,\text{m/s}$,
      $\mathbf{r} = (8.0\,\text{m})\mathbf{i} + (6.0\,\text{m})\mathbf{j}$.

25. (a) $-(18.0\,\text{m/s})\sin(3.0\,\text{s}^{-1})t\mathbf{i}$
      $+ (18.0\,\text{m/s})\cos(3.0\,\text{s}^{-1})t\mathbf{j}$;
   (b) $-(54.0\,\text{m/s}^2)\cos(3.0\,\text{s}^{-1})t\mathbf{i}$
      $- (54.0\,\text{m/s}^2)\sin(3.0\,\text{s}^{-1})t\mathbf{j}$;
   (c) circle; (d) $a = (9.0\,\text{s}^{-2})r$, $180°$.

27. 44 m, 6.3 m.

29. $38°$ and $52°$.

31. 1.95 s.

33. 22 m.

35.

37. 5.71 s.

39. (a) 65.7 m; (b) 7.32 s; (c) 267 m;
   (d) 42.2 m/s, $30.1°$ above the horizontal.

43. Unsuccessful, 34.7 m.

45. (a) $\mathbf{v}_0 = 3.42\,\text{m/s}$, $47.5°$ above the horizontal; (b) 5.32 m above the water; (c) $\mathbf{v}_f = 10.5\,\text{m/s}$, $77°$ below the horizontal.

47. $\theta = \tan^{-1}(gt/v_0)$ below the horizontal.

49. $\theta = \frac{1}{2}\tan^{-1}(-\cot\phi)$.

51. $7.29g$ up.

53. $5.9 \times 10^{-3}\,\text{m/s}^2$ toward the Sun.

55. $0.94g$.

59. 2.7 m/s, $22°$ from the river bank.

61. 23.1 s.

63. 1.41 m/s.

65. (a) 1.82 m/s; (b) 3.22 m/s.

67. (a) 60 m; (b) 75 s.

69. 58 km/h, $31°$, 58 km/h opposite to $\mathbf{v}_{12}$.

71. $0.0889\,\text{m/s}^2$.

73. $D_x = 60\,\text{m}$, $D_y = -35\,\text{m}$,
   $D_z = -12\,\text{m}$; 70 m; $\theta_h = 30°$ from the x-axis toward the $-y$-axis,
   $\theta_v = 9.8°$ below the horizontal.

75. 7.0 m/s.

77. $\pm28.5°$, ±25.2.

79. 170 km/h, $41.5°$ N of E.

81. $1.6\,\text{m/s}^2$.

83. 2.7 s, 1.9 m/s.

85. (a) $Dv/(v^2 - u^2)$;
   (b) $D/(v^2 - u^2)^{1/2}$.

87. $54.6°$ below the horizontal.

89. (a) 464 m/s; (b) 355 m/s.

91. Row at an angle of $23°$ upstream and run 243 m in a total time of 20.7 min.

93. 1.8×10^3 rev/day.

CHAPTER 4

3. 6.9×10^2 N.

5. (a) 5.7×10^2 N; (b) 99 N;
   (c) 2.1×10^2 N; (d) 0.

7. 107 N.

9. -9.3×10^5 N, 25% of the weight of the train.

11. $m > 1.9$ kg.

13. 2.1×10^2 N.

15. $-1.40\,\text{m/s}^2$ (down).

17. a (downward) $\geq 1.2\,\text{m/s}^2$.

19. $a_{max} = 0.557\,\text{m/s}^2$.

21. (a) $2.2\,\text{m/s}^2$; (b) 18 m/s; (c) 93 s.

23. 3.0×10^3 N downward.

25. (a) 1.4×10^2 N; (b) 14.5 m/s.

27. Southwesterly direction.

29.

(a)   (b)

31. (a) $1.13\,\text{m/s}^2$, $52.2°$ below $-x$-axis;
   (b) $0.814\,\text{m/s}^2$, $42.3°$ above $+x$-axis.

33. (a) $m_1g - F_T = m_1a$, $F_T - m_2g$
   $= m_2a$.

35. (a) lower bucket $= 34$ N, upper bucket $= 68$N; (b) lower bucket $= 40$ N, upper bucket $= 80$ N.

37. 1.4×10^3 N.

39. $F_B = 6890$ N, $\mathbf{F}_A + \mathbf{F}_B = 8860$ N.

41. (a) 2.2 m up the plane; (b) 2.2 s.

43. $\frac{5}{2}(F_0/m)t_0^2$.

47. (a)

(b) $a = m_2g/(m_1 + m_2)$,
   $F_T = m_1m_2g/(m_1 + m_2)$.

49. $a = [m_2 + m_C(\ell_2/\ell)]g/$
   $(m_1 + m_2 + m_C)$.

51. $1.74\,\text{m/s}^2$, $F_{T1} = 22.6$ N,
   $F_{T2} = 20.9$ N.

53. $(m + M)g\tan\theta$.

55. $F_{T1} = [4m_1m_2m_3/$
   $(m_1m_3 + m_2m_3 + 4m_1m_2)]g$,
   $F_{T3} = [8m_1m_2m_3/$
   $(m_1m_3 + m_2m_3 + 4m_1m_2)]g$,
   $a_1 = [(m_1m_3 - 3m_2m_3 + 4m_1m_2)/$
   $(m_1m_3 + m_2m_3 + 4m_1m_2)]g$,
   $a_2 = [(-3m_1m_3 + m_2m_3 + 4m_1m_2)/$
   $(m_1m_3 + m_2m_3 + 4m_1m_2)]g$,
   $a_3 = [(m_1m_3 + m_2m_3 - 4m_1m_2)/$
   $(m_1m_3 + m_2m_3 + 4m_1m_2)]g$.

57. $v = \{[2m_2\ell_2 + m_C(\ell_2^2/\ell)]g/$
   $(m_1 + m_2 + m_C)\}^{1/2}$.

59. 2.0×10^{-2} N.

61. 4.3 N.

63. 1.5×10^4 N.

65. 1.2 s, no change.

67. (a) $2.45\,\text{m/s}^2$ (up the incline);
   (b) 0.50 kg; (c) 7.35 N, 4.9 N.

69. 1.3×10^2 N.

71. $8.8°$.

73. 82 m/s (300 km/h).

75. (a) $F = \frac{1}{2}Mg$;
   (b) $F_{T1} = F_{T2} = \frac{1}{2}Mg$, $F_{T3} = \frac{3}{2}Mg$,
   $F_{T4} = Mg$.

77. -8.3×10^2 N.

79. (a) $0.606\,\text{m/s}^2$; (b) 150 kN.

CHAPTER 5

1. 35 N, no force.

3. (a)

(b)

(c)

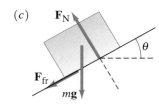

5. 0.20.

7. 69 N, $\mu_k = 0.54$.

9. 8.0 kg.

11. 1.3 m.

13. 1.3×10^3 N.

15. (a) 0.58; (b) 5.7 m/s; (c) 15 m/s.

17. (a) 1.4 m/s^2; (b) 5.4×10^2 N;
(c) 1.41 m/s^2, 2.1×10^2 N.

19. (a) 86 cm up the plane; (b) 1.5 s.

21. (a) 2.8 m/s^2; (b) 2.1 N.

23. $a = \{\sin\theta - [(\mu_1 m_1 + \mu_2 m_2)/ \\ (m_1 + m_2)]\cos\theta\}g,$
$F_T = [m_1 m_2(\mu_2 - \mu_1)/ \\ (m_1 + m_2)]g\cos\theta.$

25. (a) $\mu_k = (v_0^2/2gd\cos\theta) - \tan\theta$;
(b) $\mu_s \geq \tan\theta$.

27. (a) 2.0 m/s^2 up the plane;
(b) 5.4 m/s^2 up the plane.

29. $\mu_k = 0.40$.

31. (a) $c = 14$ kg/m; (b) 5.7×10^2 N.

33. $F_{min} = (m + M)g(\sin\theta + \mu\cos\theta)/ \\ (\cos\theta - \mu_s\sin\theta)$.

35. $v_{max} = 21$ m/s, independent of the mass.

37. (a) 0.25 m/s^2 toward the center;
(b) 6.3 N toward the center.

39. Yes, $v_{top, min} = (gR)^{1/2}$.

41. 0.34.

43. 2.1×10^2 N.

45. 5.91°, 14.3 N.

47. (a) 5.8×10^3 N; (b) 4.1×10^2 N;
(c) 31 m/s.

49. $F_T = 2\pi mRf^2$.

51. 66 km/h $< v <$ 123 km/h.

53. (a) $(1.6$ m/s$^2)$i;
(b) $(0.98$ m/s$^2)$i $- (1.7$ m/s$^2)$j;
(c) $-(4.9$ m/s$^2)$i $- (1.6$ m/s$^2)$j.

55. (a) 9.0 m/s^2; (b) 15 m/s^2.

57. $\tau = m/b$.

59. (a) $v = (mg/b)$
$+ [v_0 - (mg/b)]e^{-bt/m}$;
(b) $v = -(mg/b)$
$+ [v_0 + (mg/b)]e^{-bt/m}$,
$v \geq 0$.

61. (b) 1.8°.

63. 10 m.

65. $\mu_s = 0.41$.

67. 2.3.

69. 101 N, $\mu_k = 0.719$.

71. (b) Will slide.

73. Emerges with a speed of 13 m/s.

75. 27.6 m/s, 0.439 rev/s.

77. $\Sigma F_{tan} = 3.3 \times 10^3$ N,
$\Sigma F_R = 2.0 \times 10^3$ N.

79. (a) $F_{NC} > F_{NB} > F_{NA}$;
(b) heaviest at C, lightest at A;
(c) $v_{Amax} = (gR)^{1/2}$.

81. (a) 1.23 m/s; (b) 3.01 m/s.

83. $\phi = 31°$.

85. (a) $r = v^2/g\cos\theta$; (b) 92 m.

87. (a) 59 s; (b) greater normal force.

89. 29.2 m/s.

91. 302 m, 735 m.

93. $g(1 - \mu_s\tan\phi)/4\pi^2 f^2(\tan\phi + \mu_s)$
$< r < g(1 + \mu_s\tan\phi)/ \\ 4\pi^2 f^2(\tan\phi - \mu_s)$.

CHAPTER 6

1. 1.52×10^3 N.

3. 1.6 m/s^2.

5. $g_h = 0.91 g_{surface}$.

7. 1.9×10^{-8} N toward center of square.

9. $Gm^2\{(2/x_0^2) + [3x_0/(x_0^2 + y_0^2)^{3/2}]\}$i
$+ Gm^2\{[3y_0/(x_0^2 + y_0^2)^{3/2}] \\ + (4/y_0^2)\}$j.

11. 1.26.

13. 3.46×10^8 m from Earth's center.

15. (b) g decreases with an increase in height; (c) 9.493 m/s^2.

19. 7.56×10^3 m/s.

21. 2.0 h.

23. (a) 56 kg; (b) 56 kg; (c) 75 kg;
(d) 38 kg; (e) 0.

25. (a) 22 N (toward the Moon);
(b) -1.7×10^2 N (away from the Moon).

27. (a) Gravitational force provides required centripetal acceleration;
(b) 9.6×10^{26} kg.

29. 7.9×10^3 m/s.

31. $v = (Gm/L)^{1/2}$.

33. 0.0587 days (1.41 h).

35. 1.6×10^2 yr.

37. 2×10^8 yr.

39. $r_{Europa} = 6.71 \times 10^5$ km,
$r_{Ganymede} = 1.07 \times 10^6$ km,
$r_{Callisto} = 1.88 \times 10^6$ km.

41. 9.0 Earth-days.

43. (a) 2.1×10^2 A.U. $(3.1 \times 10^{13}$ m);
(b) 4.2×10^2 A.U.; (c) 4.2×10^2.

45. (a) 5.9×10^{-3} N/kg;
(b) not significant.

47. 2.7×10^3 km.

49. 6.7×10^{12} m/s^2.

51. 4.4×10^7 m/s^2.

53. $G' = 1 \times 10^{-4}$ N·m^2/kg$^2 \approx 10^6 G$.

55. 5 h 35 min, 19 h 50 min.

57. (a) 10 h; (b) 6.5 km;
(c) 4.2×10^{-3} m/s^2.

59. 5.4×10^{12} m, in the Solar System, Pluto.

61. $2.3 g_{Earth}$.

63. $m_P = g_P r^2/G$.

67. 7.9×10^3 m/s.

CHAPTER 7

1. 6.86×10^3 J.

3. 1.27×10^4 J.

5. 8.1×10^3 J.

7. 1 J $= 1 \times 10^7$ erg $= 0.738$ ft·lb.

9. 1.0×10^4 J.

13. (a) 3.6×10^2 N; (b) -1.3×10^3 J;
(c) -4.6×10^3 J; (d) 5.9×10^3 J;
(e) 0.

15. $W_{FN} = W_{mg} = 0$,
$W_{FP} = -W_{fr} = 2.0 \times 10^2$ J.

21. (a) -16.1; (b) -238; (c) -3.9.

23. $C = -1.3$i $+ 1.8$j.

25. $\theta_x = 42.7°, \theta_y = 63.8°, \theta_z = 121°$.

27. 95°, $-35°$ from x-axis.

31. 0.089 J.

33. 2.3×10^3 J.

35. 2.7×10^3 J.

37. $(kX^2/2) + (aX^4/4) + (bX^5/5)$.

39. (a) 5.0×10^{10} J.

41. (a) $\sqrt{3}$; (b) $\frac{1}{4}$.

43. -5.02×10^5 J.

45. 3.0×10^2 N in the direction of the motion of the ball.

47. 24 m/s (87 km/h or 54 mi/h), the mass cancels.

49. (a) 72 J; (b) -35 J; (c) 37 J.

51. 10.2 m/s.

53. $\mu_k = F/2mg$.

55. (a) 6.5×10^2 J; (b) -4.9×10^2 J; (c) 0; (d) 4.0 m/s.

57. (a) 1.66×10^5 J; (b) 21.0 m/s; (c) 2.13 m.

59. $v_p = 2.0 \times 10^7$ m/s, $v_{pc} = 2.0 \times 10^7$ m/s; $v_e = 2.9 \times 10^8$ m/s, $v_{ec} = 8.4 \times 10^8$ m/s.

61. 1.74×10^3 J.

63. (a) 15 J; (b) 4.2×10^2 J; (c) -1.8×10^2 J; (d) -2.5×10^2 J; (e) 0; (f) 10 J.

65. (a) 12 J; (b) 10 J; (c) -2.1 J.

67. 86 kJ, $\theta = 42°$.

69. $(A/k)e^{-(0.10\,\text{m})k}$.

71. 1.5 N.

73. 5.0×10^3 N/m.

75. (a) $6.6°$; (b) $10.3°$.

CHAPTER 8

1. 0.924 m.

3. 2.2×10^3 J.

5. (a) 51.7 J; (b) 15.1 J; (c) 51.7 J.

7. (a) Conservative; (b) $\frac{1}{2}kx^2 - \frac{1}{4}ax^4 - \frac{1}{5}bx^5 + $ constant.

9. (a) $\frac{1}{2}k(x^2 - x_0^2)$; (b) same.

11. 45.4 m/s.

13. 6.5 m/s.

15. (a) 1.0×10^2 N/m; (b) 22 m/s^2.

17. (a) 8.03 m/s; (b) 3.44 m.

19. (a) $v_{max} = [v_0^2 + (kx_0^2/m)]^{1/2}$; (b) $x_{max} = [x_0^2 + (mv_0^2/k)]^{1/2}$.

21. (a) 2.29 m/s; (b) 1.98 m/s; (c) 1.98 m/s; (d) $F_{Ta} = 0.87$ N, $F_{Tb} = 0.80$ N, $F_{Tc} = 0.80$ N; (e) $v_a = 2.59$ m/s, $v_b = 2.31$ m/s, $v_c = 2.31$ m/s.

23. $k = 12Mg/h$.

25. 4.5×10^6 J.

27. (a) 22 m/s; (b) 2.9×10^2 m.

29. 13 m/s.

31. 0.23.

33. 0.40.

35. (a) 0.13 m; (b) 0.77; (c) 0.46 m/s.

37. (a) $K = GM_E m_S/2r_S$; (b) $U = -GM_E m_S/r_S$; (c) $-\frac{1}{2}$.

39. (a) 6.2×10^5 m/s; (b) 4.2×10^4 m/s, $v_{esc}/v_{orbit} = \sqrt{2}$.

45. (a) 1.07×10^4 m/s; (b) 1.17×10^4 m/s; (c) 1.12×10^4 m/s.

47. (a) $dv_{esc}/dr = -\frac{1}{2}(2GM_E/r^3)^{1/2}$ $= -v_{esc}/2r$; (b) 1.09×10^4 m/s.

49. $GmM_E/12r_E$.

51. 1.1×10^4 m/s.

55. 5.4×10^2 N.

57. (a) 1.0×10^3 J; (b) 1.0×10^3 W.

59. 2.1×10^4 W, 28 hp.

61. 4.8×10^2 W.

63. 1.2×10^3 W.

65. 1.8×10^6 W.

67. (a) -25 W; (b) $+4.3 \times 10^3$ W; (c) $+1.5 \times 10^3$ W.

69. (a) 80 J; (b) 60 J; (c) 80 J; (d) 5.7 m/s at $x = 0$; (e) 32 m/s^2 at $x = \pm x_0$.

71. $a^2/4b$.

73. 8.0 m/s.

75. 32.5 hp.

77. (a) 28 m/s; (b) 1.2×10^2 m.

79. (a) $(2gL)^{1/2}$; (b) $(1.2gL)^{1/2}$.

81. (a) 1.1×10^6 J; (b) 60 W (0.081 hp); (c) 4.0×10^2 W (0.54 hp).

83. (a) 40 m/s; (b) 2.6×10^5 W.

87. (a) $29°$; (b) 6.4×10^2 N; (c) 9.2×10^2 N.

89. (a) $-\dfrac{U_0}{r}\left(\dfrac{r_0}{r} + 1\right)e^{-r/r_0}$; (b) 0.030; (c) $F(r) = -C/r^2$, 0.11.

91. 6.7 hp.

93. (a) 2.8 m; (b) 1.5 m; (c) 1.5 m.

95. 76 hp.

97. (a) 5.00×10^3 m/s; (b) 2.89×10^3 m/s.

CHAPTER 9

1. 6.0×10^7 N, up.

3. (a) 0.36 kg·m/s; (b) 0.12 kg·m/s.

5. $(26$ N·s$)\mathbf{i} - (28$ N·s$)\mathbf{j}$.

7. (a) $(8h/g)^{1/2}$; (b) $(2gh)^{1/2}$; (c) $-(8m^2 gh)^{1/2}$ (up); (d) mg (down), a surprising result.

9. 3.4×10^4 kg.

11. 4.4×10^3 m/s.

13. -0.667 m/s (opposite to the direction of the package).

15. 2, lesser kinetic energy has greater mass.

17. $\frac{3}{2}v_0\mathbf{i} - v_0\mathbf{j}$.

19. 1.1×10^{-22} kg·m/s, $36°$ from the direction opposite to the electron's.

21. (a) $(100$ m/s$)\mathbf{i} + (50$ m/s$)\mathbf{j}$; (b) 3.3×10^5 J.

23. 130 N, not large enough.

25. 1.1×10^3 N.

27. (a) $2mv/\Delta t$; (b) $2mv/t$.

29. (a)

(b) 1.2 N·s; (c) 1.2 N·s; (d) 3.9 g.

31. (a) $(0.84$ N$) + (1.2$ N/s$)t$; (b) 18.5 N; (c) $(0.12$ kg/s$)\{[(49$ m^2/s$^2) - (1.18$ m^2/s$^3)t]^{1/2} + (9.80$ m/s$^2)t\}$, 18.3 N.

33. $v_1' = -1.40$ m/s (rebound), $v_2' = 2.80$ m/s.

35. (a) 2.7 m/s; (b) 0.84 kg.

37. 3.2×10^3 m/s.

39. (a) 1.00; (b) 0.89; (c) 0.29; (d) 0.019.

41. (a) 0.32 m; (b) -3.1 m/s (rebound), 4.9 m/s; (c) Yes.

43. (a) $+M/(m + M)$; (b) 0.964.

45. $141°$.

47. (b) $e = (h'/h)^{1/2}$.

49. (a) $v_1' = v_2' = 1.9$ m/s; (b) $v_1' = -1.6$ m/s, $v_2' = 7.9$ m/s; (c) $v_1' = 0$, $v_2' = 5.2$ m/s; (d) $v_1' = 3.1$ m/s, $v_2' = 0$; (e) $v_1' = -4.0$ m/s, $v_2' = 12$ m/s; result for (c) is reasonable, result for (d) is not reasonable, result for (e) is not reasonable.

51. $61°$ from first eagles's direction, 6.8 m/s.

53. (a) $30°$; (b) $v_2' = v/\sqrt{3}$, $v_1' = v/\sqrt{3}$; (c) $\frac{2}{3}$.

55. $\theta_1' = 76°$, $v_n' = 5.1 \times 10^5$ m/s, $v_{He}' = 1.8 \times 10^5$ m/s.

59. 6.5×10^{-11} m from the carbon atom.

61. 0.030 nm above center of H triangle.

63. $x_{CM} = 1.10$ m (East), $y_{CM} = -1.10$ m (South).

65. $x_{CM} = 0$, $y_{CM} = 2r/\pi$.

67. $x_{CM} = 0$, $y_{CM} = 0$, $z_{CM} = 3h/4$ above the point.

69. (a) 4.66×10^6 m.

71. (a) $x_{CM} = 4.6$ m; (b) 4.3 m; (c) 4.6 m.

73. $mv/(M + m)$ up, balloon will also stop.

75. 55 m.

77. 0.899 hp.

79. (a) 2.3×10^3 N; (b) 2.8×10^4 N; (c) 1.1×10^4 hp.

81. A "scratch shot".

83. 1.4×10^4 N, 43.3°.

85. 5.1×10^2 m/s.

87. $m_2 = 4.00m$.

89. 50%.

91. (a) No; (b) $v_1/v_2 = -m_2/m_1$; (c) m_2/m_1; (d) does not move; (e) center of mass will move.

93. 8.29 m/s.

95. (a) 2.5×10^{-13} m/s; (b) 1.7×10^{-17}; (c) 0.19 J.

97. $m \leq M/3$.

99. 29.6 km/s.

101. (a) 2.3 N·s; (b) 4.5×10^2 N.

103. (a) Inelastic collision; (b) 0.10 s; (c) -1.4×10^5 N.

105. 0.28 m, 1.1 m.

CHAPTER 10

1. (a) $\pi/6$ rad = 0.524 rad; (b) $19\pi/60$ = 0.995 rad; (c) $\pi/2$ = 1.571 rad; (d) 2π = 6.283 rad; (e) $7\pi/3$ = 7.330 rad.

3. 2.3×10^3 m.

5. (a) 0.105 rad/s; (b) 1.75×10^{-3} rad/s; (c) 1.45×10^{-4} rad/s; (d) zero.

7. (a) 464 m/s; (b) 185 m/s; (c) 355 m/s.

9. (a) 262 rad/s; (b) 46 m/s, 1.2×10^4 m/s² radial.

11. 7.4 cm.

13. (a) 1.75×10^{-4} rad/s²; (b) $a_R = 1.17 \times 10^{-2}$ m/s², $a_{tan} = 7.44 \times 10^{-4}$ m/s².

15. (a) 0.58 rad/s2; (b) 12 s.

17. (a) $(1.67 \text{ rad/s}^4)t^3 - (1.75 \text{ rad/s}^3)t^2$; ($b$) $(0.418 \text{ rad/s}^4)t^4 - (0.583 \text{ rad/s}^3)t^3$; ($c$) 6.4 rad/s, 2.0 rad.

19. (a) ω_1 is in the $-x$-direction, ω_2 is in the $+z$-direction; (b) $\omega = 61.0$ rad/s, 35.0° above $-x$-axis; (c) $-(1.75 \times 10^3 \text{ rad/s}^2)\mathbf{j}$.

21. (a) 35 m·N; (b) 30 m·N.

23. 1.2 m·N (clockwise).

25. 3.5×10^2 N, 2.0×10^3 N.

27. 53 m·N.

29. (a) 3.5 kg·m²; (b) 0.024 m·N.

31. 2.25×10^3 kg·m², 8.8×10^3 m·N.

33. 9.5×10^4 m·N.

35. 10 m/s.

37. (a)

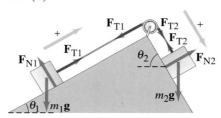

(b) $F_{T1} = 47$ N, $F_{T2} = 75$ N; (c) 7.0 m·N, 1.7 kg·m².

39. Thin hoop (through center): $k = R_0$;

Thin hoop (through diameter): $k = [(R_0^2/2) + (w^2/12)]^{1/2}$;

Solid cylinder (through center): $k = R/\sqrt{2}$;

Hollow cylinder (through center): $k = [(R_1^2 + R_2^2)/2]^{1/2}$;

Uniform sphere (through center): $k = (2r_0^2/5)^{1/2}$;

Rod (through center): $k = \ell/\sqrt{12}$;

Rod (through end): $k = \ell/\sqrt{3}$;

Plate (through center): $k = [(\ell^2 + w^2)/12]^{1/2}$.

41. (a) 4.18 rad/s²; (b) 8.37 m/s²; (c) 421 m/s²; (d) 3.07×10^3 N; (e) 1.14°.

43. (a) $I_a = Ms^2/12$; (b) $I_b = Ms^2/12$.

45. (a) $5.30MR_0^2$; (b) -15%.

47. (a) $9MR_0^2/16$; (b) $MR_0^2/4$; (c) $5MR_0^2/4$.

51. (b) $M\ell^2/12$, $Mw^2/12$.

53. 0.38 rev/s.

55. (a) As moment of inertia increases, angular velocity must decrease; (b) 1.6.

57. (a) 7.1×10^{33} kg·m²/s; (b) 2.7×10^{40} kg·m²/s.

59. 0.45 rad/s, 0.80 rad/s.

61. 2.33×10^4 J.

63. 5×10^9, loss of gravitational potential energy.

65. 1.4 m/s.

67. (a) 2.5 kg·m²; (b) 0.58 kg·m²; (c) 0.35 s; (d) -72 J; (e) rotating.

69. 12.4 m/s.

71. 1.4×10^2 J.

73. (a) 4.48 m/s; (b) 1.21 J; (c) $\mu_s \geq 0.197$.

75. $v = [10g(R_0 - r_0)/7]^{1/2}$.

77. (a) 4.5×10^5 J; (b) 0.18 (18%); (c) 1.71 m/s²; (d) 6.4%.

79. (a) $12v_0^2/49\mu_k g$; (b) $v = 5v_0/7$, $\omega = 5v_0/7R$.

81. (a) 4.5 m/s², 19 rad/s²; (b) 5.8 m/s; (c) 15.3 J; (d) 1.4 J; (e) $K = 16.7$ J, $\Delta E = 0$; (f) $a = 4.5$ m/s², $v = 5.8$ m/s, 14.1 J.

83. $\theta_{Sun} = 9.30 \times 10^{-3}$ rad (0.53°), $\theta_{Moon} = 9.06 \times 10^{-3}$ rad (0.52°).

85. $\omega_1/\omega_2 = R_2/R_1$.

87. $\ell/2$, $\ell/2$.

89. (a) $-(I_W/I_P)\omega_W$ (down); (b) $-(I_W/2I_P)\omega_W$ (down); (c) $(I_W/I_P)\omega_W$ (up); (d) 0.

91. (a) $\omega_R/\omega_F = N_F/N_R$; ($b$) 4.0; ($c$) 1.5.

93. (a) 1.5×10^2 rad/s²; (b) 1.2×10^3 N.

95. (a) 0.070 rad/s²; (b) 40 rpm.

97. 7.9 N.

99. (b) 2.2×10^3 rad/s; (c) 24 min.

101. (a) 2.9 m; (b) 3.6 m.

103. (a) 1.2 rad/s; (b) 2.0×10^3 J, 1.2×10^3 J, loss of 8.0×10^2 J, decrease of 40%.

105. (a) 1.7 m/s; (b) 0.84 m/s.

107. (a) $h_{min} = 2.7R_0$; (b) $h_{min} = 2.7R_0 - 1.7r_0$.

109. (a) 0.84 m/s; (b) 0.96.

CHAPTER 11

7. (a) $-7.0\mathbf{i} - 14.0\mathbf{j} + 19.3\mathbf{k}$; ($b$) 164°.

11. $-(30.3$ m·N$)\mathbf{k}$ (in $-z$-direction).

13. $(18$ m·kN$)\mathbf{i} \pm (14$ m·kN$)\mathbf{j}$ $\mp (19$ m·kN$)\mathbf{k}$.

19. $(55\mathbf{i} - 90\mathbf{j} + 42\mathbf{k})\,\text{kg}\cdot\text{m}^2/\text{s}$.

21. $(a)\; [(7m/9) + (M/6)]\ell^2\omega^2$;
$(b)\; [(14m/9) + (M/3)]\ell^2\omega$.

23. $2.30\,\text{m/s}^2$.

25. $(a)\; L = [R_0 M_1 + R_0 M_2 + (I/R_0)]v$;
$(b)\; a = M_2 g/[M_1 + M_2 + (I/R_0^2)]$.

27. Rod rotates at 7.8 rad/s about the center of mass, which moves with constant velocity of 0.21 m/s.

31. $F_1 = [(d + r\cos\phi)/2d]m_1 r\omega^2 \sin\phi$,
$F_2 = [(d - r\cos\phi)/2d]m_1 r\omega^2 \sin\phi$.

33. $16\,\text{N}, -7.5\,\text{N}$.

35. $3m^2 v^2/g(3m + 4M)(m + M)$.

37. $(1 - 4.7 \times 10^{-13})\omega_E$.

39. $(a)\; 14\,\text{m/s}$; $(b)\; 6.8\,\text{rad/s}$.

41. $1.02 \times 10^{-3}\,\text{kg}\cdot\text{m}^2$.

43. $2.2\,\text{rad/s}\;(0.35\,\text{rev/s})$.

45. $\tan^{-1}(r\omega^2/g)$.

47. $(a)\; g$, along a radial line; $(b)\; 0.998g$, $0.0988°$ south from a radial line; $(c)\; 0.997g$, along a radial line.

49. North or south direction.

51. (a) South; $(b)\; \omega D^2 \sin\lambda/v_0$; $(c)\; 0.46\,\text{m}$.

53. $(a)\; (-9.0\mathbf{i} + 12\mathbf{j} - 8.0\mathbf{k})\,\text{kg}\cdot\text{m}^2/\text{s}$; $(b)\; (9.0\mathbf{j} - 6.0\mathbf{k})\,\text{m}\cdot\text{N}$.

55. (a) Turn in the direction of the lean; $(b)\; \Delta L = 0.98\,\text{kg}\cdot\text{m}^2/\text{s}$, $\Delta L = 0.18 L_0$.

57. $(a)\; 1.8 \times 10^3\,\text{kg}\cdot\text{m}^2/\text{s}^2$; $(b)\; 1.8 \times 10^3\,\text{m}\cdot\text{N}$; $(c)\; 2.1 \times 10^3\,\text{W}$.

59. $v_{\text{CM}} = (3g\ell/4)^{1/2}$.

61. $(19\,\text{m/s})(1 - \cos\theta)^{1/2}$.

63. $(a)\; 2.3 \times 10^4\,\text{rev/s}$; $(b)\; 5.7 \times 10^3\,\text{rev/s}$.

CHAPTER 12

1. $379\,\text{N}, 141°$.

3. $1.6 \times 10^3\,\text{m}\cdot\text{N}$.

5. $6.52\,\text{kg}$.

7. $2.84\,\text{m}$ from the adult.

9. $0.32\,\text{m}$.

11. $F_{T1} = 3.4 \times 10^3\,\text{N}$,
$F_{T2} = 3.9 \times 10^3\,\text{N}$.

13. $F_1 = -2.94 \times 10^3\,\text{N}$ (down),
$F_2 = 1.47 \times 10^4\,\text{N}$.

15. Top hinge: $F_{Ax} = 55.2\,\text{N}$,
$F_{Ay} = 63.7\,\text{N}$; bottom hinge:
$F_{Bx} = -55.2\,\text{N}, F_{By} = 63.7\,\text{N}$.

17. (a)

$(b)\; 1.5 \times 10^4\,\text{N}$; $(c)\; 6.7 \times 10^3\,\text{N}$.

19. $F_T = 1.4 \times 10^3\,\text{N}$ (up),
$F_{\text{bone}} = 2.1 \times 10^3\,\text{N}$ (down).

21. $2.7 \times 10^3\,\text{N}$.

23. $89.5\,\text{cm}$ from the feet.

25. $F_1 = 5.8 \times 10^3\,\text{N}, F_2 = 5.6 \times 10^3\,\text{N}$.

27. $(a)\; 2.1 \times 10^2\,\text{N}$; $(b)\; 2.0 \times 10^3\,\text{N}$.

29. $7.1 \times 10^2\,\text{N}$.

31. $F_T = 2.5 \times 10^2\,\text{N}$,
$F_{AH} = 2.5 \times 10^2\,\text{N}$,
$F_{AV} = 2.0 \times 10^2\,\text{N}$.

33. $(a)\; 1.00\,\text{N}$; $(b)\; 1.25\,\text{N}$.

35. $\theta_{\max} = 40°$, same.

37. $(a)\; F_T = 182\,\text{N}$; $(b)\; F_{N1} = 352\,\text{N}$,
$F_{N2} = 236\,\text{N}$; $(c)\; F_B = 298\,\text{N}, 52.4°$.

39. $1.0 \times 10^2\,\text{N}$.

41. $(a)\; 1.2 \times 10^5\,\text{N/m}^2$; $(b)\; 2.4 \times 10^{-6}$.

43. $(a)\; 1.3 \times 10^5\,\text{N/m}^2$; $(b)\; 6.5 \times 10^{-7}$; $(c)\; 0.0062\,\text{mm}$.

45. $9.6 \times 10^6\,\text{N/m}^2$.

47. $9.0 \times 10^7\,\text{N/m}^2, 9.0 \times 10^2\,\text{atm}$.

49. $2.2 \times 10^7\,\text{N}$.

51. $(a)\; 1.1 \times 10^2\,\text{m}\cdot\text{N}$; (b) wall; (c) all three.

53. $3.9 \times 10^2\,\text{N}$, thicker strings, maximum strength is exceeded.

55. $(a)\; 4.4 \times 10^{-5}\,\text{m}^2$; $(b)\; 2.7\,\text{mm}$.

57. $1.2\,\text{cm}$.

61. $(a)\; F_T = 129\,\text{kN}$;
$F_A = 141\,\text{kN}, 23.5°$;
$(b)\; F_{DE} = 64.7\,\text{kN}$ (tension),
$F_{CE} = 32.3\,\text{kN}$ (compression),
$F_{CD} = 64.7\,\text{kN}$ (compression),
$F_{BD} = 64.7\,\text{kN}$ (tension),
$F_{BC} = 64.7\,\text{kN}$ (tension),
$F_{AC} = 97.0\,\text{kN}$ (compression),
$F_{AB} = 64.7\,\text{kN}$ (compression).

63. $(a)\; 4.8 \times 10^{-2}\,\text{m}^2$; $(b)\; 6.8 \times 10^{-2}\,\text{m}^2$.

65. $F_{AB} = 5.44 \times 10^4\,\text{N}$ (compression),
$F_{ACx} = 2.72 \times 10^4\,\text{N}$ (tension),
$F_{BC} = 5.44 \times 10^4\,\text{N}$ (tension),
$F_{BD} = 5.44 \times 10^4\,\text{N}$ (compression),
$F_{CD} = 5.44 \times 10^4\,\text{N}$ (tension),
$F_{CE} = 2.72 \times 10^4\,\text{N}$ (tension),
$F_{DE} = 5.44 \times 10^4\,\text{N}$ (compression).

67. $12\,\text{m}$.

69. $M_C = 0.191\,\text{kg}, M_D = 0.0544\,\text{kg}$,
$M_A = 0.245\,\text{kg}$.

71. $(a)\; Mg[h/(2R - h)]^{1/2}$;
$(b)\; Mg[h(2R - h)]^{1/2}/(R - h)$.

73. $\theta_{\max} = 29°$.

75. $6, 2.0\,\text{m}$ apart.

77. 3.8.

79. $5.0 \times 10^5\,\text{N}, 3.2\,\text{m}$.

81. $(a)\; 600\,\text{N}$; $(b)\; F_A = 0, F_B = 1200\,\text{N}$;
$(c)\; F_A = 150\,\text{N}, F_B = 1050\,\text{N}$;
$(d)\; F_A = 750\,\text{N}, F_B = 450\,\text{N}$.

83. $6.5 \times 10^2\,\text{N}$.

85. $0.67\,\text{m}$.

87. Right end is safe, left end is not safe, $0.10\,\text{m}$.

89. $(a)\; F_L = 3.3 \times 10^2\,\text{N}$ up,
$F_R = 2.3 \times 10^2\,\text{N}$ down;
$(b)\; 65\,\text{cm}$ from right hand;
$(c)\; 123\,\text{cm}$ from right hand.

91. $\theta \geq 40°$.

93. (b) beyond the table;
$(c)\; D = L \sum_{i=1}^{n} \dfrac{1}{2i}$; $(d)\; 32$ bricks.

95. $F_{TB} = 134\,\text{N}, F_{TA} = 300\,\text{N}$.

97. $2.6w, 31°$ above horizontal.

CHAPTER 13

1. $3 \times 10^{11}\,\text{kg}$.

3. $4.3 \times 10^2\,\text{kg}$.

5. 0.8477.

7. $(a)\; 3 \times 10^7\,\text{N/m}^2$;
$(b)\; 2 \times 10^5\,\text{N/m}^2$.

9. $1.1\,\text{m}$.

11. $8.28 \times 10^3\,\text{kg}$.

13. $1.2 \times 10^5\,\text{N/m}^2$,
$2.3 \times 10^7\,\text{N}$ (down),
$1.2 \times 10^5\,\text{N/m}^2$.

15. $6.54 \times 10^2\,\text{kg/m}^3$.

17. $3.36 \times 10^4\,\text{N/m}^2$ $(0.331\,\text{atm})$.

19. $(a)\; 1.41 \times 10^5\,\text{Pa}$; $(b)\; 9.8 \times 10^4\,\text{Pa}$.

21. $(a)\; 0.34\,\text{kg}$; $(b)\; 1.5 \times 10^4\,\text{N}$ (up).

23. $(c)\; \geq 0.38h$, no.

27. $4.70 \times 10^3\,\text{kg/m}^3$.

29. $8.5 \times 10^2\,\text{kg}$.

31. Copper.

33. $(a)\; 1.14 \times 10^6\,\text{N}$; $(b)\; 4.0 \times 10^5\,\text{N}$.

35. (b) Above the center of gravity.

37. 0.88.

39. $7.9 \times 10^2\,\text{kg}$.

43. $4.1\,\text{m/s}$.

45. $9.5\,\text{m/s}$.

47. $1.5 \times 10^5\,\text{N/m}^2 = 1.5\,\text{atm}$.

49. $4.11 \times 10^{-3}\,\text{m}^3/\text{s}$.

51. $1.7 \times 10^6\,\text{N}$.

59. (a) $2[h_1(h_2 - h_1)]^{1/2}$;
(b) $h_1' = h_2 - h_1$.

61. $0.072 \ \text{Pa} \cdot \text{s}$.

63. $4.0 \times 10^3 \ \text{Pa}$.

65. 11 cm.

67. (a) Laminar; (b) 3200, turbulent.

69. 1.9 m.

71. $9.1 \times 10^{-3} \ \text{N}$.

73. (a) $\gamma = F/4\pi r$; (b) 0.024 N/m.

75. (a) 0.88 m; (b) 0.55 m; (c) 0.24 m.

77. $1.5 \times 10^2 \ \text{N} \le F \le 2.2 \times 10^2 \ \text{N}$.

79. 0.051 atm.

81. 0.63 N.

83. 5 km.

85. $5.3 \times 10^{18} \ \text{kg}$.

87. 2.6 m.

89. 39 people.

91. 37 N, not float.

93. $d = D[v_0^2/(v_0^2 + 2gy)]^{1/4}$.

95. (a) 3.2 m/s; (b) 20 s.

97. $1.9 \times 10^2 \ \text{m/s}$.

CHAPTER 14

1. 0.60 m.

3. 1.19 Hz.

5. (a) 2.4 N/m; (b) 12 Hz.

7. (a) $0.866 \ x_{\text{max}}$; (b) $0.500 \ x_{\text{max}}$.

9. $0.866 \ A$.

11. $[(k_1 + k_2)/m]^{1/2}/2\pi$.

13. (a) 8/7 s, 0.875 Hz; (b) 3.3 m, -10.4 m/s; (c) $+18$ m/s, $-57 \ \text{m/s}^2$.

15. 3.6 Hz.

19. (a) $y = -(0.220 \ \text{m}) \sin[(37.1 \ \text{s}^{-1})t]$;
(b) maximum extensions at 0.0423 s, 0.211 s, 0.381 s, …; minimum extensions at 0.127 s, 0.296 s, 0.465 s, ….

21. $f = (3k/M)^{1/2}/2\pi$.

25. (a) $x = (12.0 \ \text{cm}) \cos[(25.6 \ \text{s}^{-1})t + 1.89 \ \text{rad}]$;
(b) $t_{\text{max}} = 0.294 \ \text{s}, 0.539 \ \text{s}, 0.784 \ \text{s}, \ldots$;
$t_{\text{min}} = 0.171 \ \text{s}, 0.416 \ \text{s}, 0.661 \ \text{s}, \ldots$;
(c) -3.77 cm; (d) $+13.1$ N (up);
(e) 3.07 m/s, 0.110 s.

27. (a) 0.650 m; (b) 1.34 Hz; (c) 29.8 J;
(d) $K = 25.0 \ \text{J}, U = 4.8 \ \text{J}$.

29. 9.37 m/s.

31. $A_1 = 2.24 A_2$.

33. (a) $4.2 \times 10^2 \ \text{N/m}$; (b) 3.3 kg.

35. 352.6 m/s.

39. 0.9929 m.

41. (a) 0.248 m; (b) 2.01 s.

43. (a) $-12°$; (b) $+2.1°$; (c) $-13°$.

45. $\frac{1}{3}$.

47. 1.08 s.

49. 0.31 g.

51. (a) 1.6 s.

53. 3.5 s.

55. (a) 0.727 s; (b) 0.0755;
(c) $x = (0.189 \ \text{m})e^{-(0.108/s)t}$
$\sin[(8.64 \ \text{s}^{-1})t]$.

57. (a) $8.3 \times 10^{-4}\%$; (b) 39 periods.

59. (a) 5.03 Hz; (b) $0.0634 \ \text{s}^{-1}$;
(c) 110 oscillations.

61.

F_0/k

65. (a) 198 s; (b) $8.7 \times 10^{-6} \ \text{W}$;
(c) $8.8 \times 10^{-4} \ \text{Hz}$ on either side of f_0.

69. (a) 0.63 Hz; (b) 0.65 m/s; (c) 0.077 J.

71. 151 N/m, 20.3 m.

73. 0.11 m.

75. 3.6 Hz.

77. (a) 1.1 Hz; (b) 13 J.

79. (a) 90 N/m; (b) 8.9 cm.

81. $k = \rho_{\text{water}} g A$.

83. Block will oscillate with SHM, $k = 2\rho g A$, the density and the cross section.

85. $T = 2\pi(ma/2k \ \Delta a)^{1/2}$.

87. (a) 1.64 s; (b) 0.67 m.

CHAPTER 15

1. 2.3 m/s.

3. 1.26 m.

5. 0.72 m.

7. 2.7 N.

9. (a) 75 m/s; (b) $7.8 \times 10^3 \ \text{N}$.

11. (a) $1.3 \times 10^3 \ \text{km}$;
(b) cannot be determined.

13. (a) 0.25; (b) 0.50.

17. (a) 0.295 W; (b) 0.25 cm.

19. $D = D_M \sin[2\pi(x/\lambda + t/T) + \phi]$.

21. (a) 41 m/s; (b) $6.4 \times 10^4 \ \text{m/s}^2$;
(c) 41 m/s, $8.2 \times 10^3 \ \text{m/s}^2$.

23. (a, c)

(b) $D = (0.45 \ \text{m}) \sin[(3.0 \ \text{m}^{-1})x - (6.0 \ \text{s}^{-1})t + 1.2]$.

(d) $D = (0.45 \ \text{m}) \sin[(3.0 \ \text{m}^{-1})x + (6.0 \ \text{s}^{-1})t + 1.2]$.

25. $D = -(0.020 \ \text{cm}) \cos[(8.01 \ \text{m}^{-1})x - (2.76 \times 10^3 \ \text{s}^{-1})t]$.

27. The function is a solution.

31. (a) $v_2/v_1 = (\mu_1/\mu_2)^{1/2}$;
(b) $\lambda_2/\lambda_1 = v_2/v_1 = (\mu_1/\mu_2)^{1/2}$;
(c) lighter cord.

33. (c) $A_T = [2k_1/(k_2 + k_1)]A = [2v_2/(v_1 + v_2)]A$.

35. (b) $2D_M \cos(\frac{1}{2}\phi)$, purely sinusoidal;
(d) $D = \sqrt{2} \ D_M \sin(kx - \omega t + \pi/4)$.

37. 440 Hz, 880 Hz, 1320 Hz, 1760 Hz.

39. $f_n = n(0.50 \ \text{Hz}), n = 1, 2, 3, \ldots$;
$T_n = (2.0 \ \text{s})/n, n = 1, 2, 3, \ldots$.

41. 70 Hz.

45. 4.

47. (a) $D_2 = (4.2 \ \text{cm}) \sin[(0.71 \ \text{cm}^{-1})x + (47 \ \text{s}^{-1})t + 2.1]$;
(b) $D_{\text{resultant}} = (8.4 \ \text{cm}) \sin[(0.71 \ \text{cm}^{-1})x + 2.1] \cos[(47 \ \text{s}^{-1})t]$.

49. 308 Hz.

51. (a)

(b)

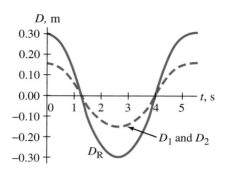

53. 5.4 km/s.

55. 29°.

57. 24°.

59. Speed will be greater in the less dense rod by a factor of $\sqrt{2}$.

61. (a) 0.050 m; (b) 2.3.

63. 0.99 m.

65. (a) solid curves,

(c) dashed curves;

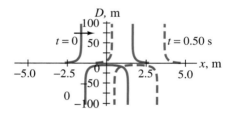

(b)
$D = (4.0 \text{ m}^3)/\{[x - (3.0 \text{ m/s})t]^2 - 2.0 \text{ m}^2\}$;

(d)

$D = (4.0 \text{ m}^3)/\{[x + (3.0 \text{ m/s})t]^2 - 2.0 \text{ m}^2\}$.

67. (a) 784 Hz, 1176 Hz, 880 Hz,

1320 Hz; (b) 1.26; (c) 1.12; (d) 0.794.

69. $\lambda_n = 4L/(2n - 1)$, $n = 1, 2, 3, \dots$.

71. $y = (3.5 \text{ cm}) \cos[(1.05 \text{ cm}^{-1})x$

$- (1.39 \text{ s}^{-1})t]$.

73.

$t = 0$

$t = 1.0 \text{ s}$

$t = 2.0 \text{ s}$

$t = 3.0 \text{ s}$

CHAPTER 16

1. 2.6×10^2 m.

3. 5.4×10^2 m.

5. 1200 m, 300 m.

7. (a) 1.1×10^{-8} m; (b) 1.1×10^{-10} m.

9. (a) $\Delta P = (4 \times 10^{-5} \text{ Pa})$
$\sin[(0.949 \text{ m}^{-1})x$
$- (315 \text{ s}^{-1})t]$;

(b) $\Delta P = (4 \times 10^{-3} \text{ Pa})$
$\sin[(94.9 \text{ m}^{-1})x$
$- (3.15 \times 10^4 \text{ s}^{-1})t]$.

11. (a) 49 dB; (b) 3.2×10^{-10} W/m².

13. 150 Hz to 20,000 Hz.

15. (a) 9; (b) 9.5 dB.

17. (a) Higher frequency is greater by a factor of 2; (b) 4.

19. (a) 5.0×10^{-13} W; (b) 6.3×10^4 yr.

21. 87 dB.

23. (a) 5.10×10^{-5} m; (b) 29.8 Pa.

25. (a) 1.5×10^3 W; (b) 3.4×10^2 m.

27. (b) 190 dB.

29. (a) 570 Hz; (b) 860 Hz.

31. 8.6 mm < L < 8.6 m.

33. (a) 110 Hz, 330 Hz, 550 Hz, 770 Hz;
(b) 220 Hz, 440 Hz, 660 Hz, 880 Hz.

35. (a) 0.656 m; (b) 262 Hz, 1.31 m;
(c) 1.31 m, 262 Hz.

37. −2.6%.

39. (a) 0.578 m; (b) 869 Hz.

41. 215 m/s.

43. 0.64, 0.20, −2 dB, −7 dB.

45. 28.5 kHz.

47. (a) 130.5 Hz, or 133.5 Hz;
(b) ±2.3%.

49. (a) 343 Hz; (b) 1030 Hz, 1715 Hz.

53. 346 Hz.

57. (a) 1690 Hz; (b) 1410 Hz.

59. 30,890 Hz.

61. 120 Hz.

63. 91 Hz.

65. 90 beats/min.

67. (a) 570 Hz; (b) 570 Hz; (c) 570 Hz;
(d) 570 Hz; (e) 594 Hz; (f) 595 Hz.

71. (a) 120; (b) 0.96°.

73. (a) 37°; (b) 1.7.

75. 0.278 s.

77. 55 m.

79. 410 km/h (255 mi/h).

81. 1, 0.444, 0.198, 0.0878, 0.0389.

83. 18.1 W.

85. 15 W.

87. 2.3 Hz.

89. $\Delta P_M/\Delta P_{M0} = D_M/D_{M0} = 10^6$.

91. 50 dB.

93. 17.5 m/s.

95. 2.3 kHz.

97. 550 Hz.

99. (a) 2.8×10^3 Hz; (b) 1.80 m;
(c) 0.12 m.

101. (a) 2.2×10^{-7} m; (b) 5.4×10^{-5} m.

CHAPTER 17

1. 0.548.

3. (a) 20°C; (b) ≈3300°F.

5. 104.0°F.

7. −40°F = −40°C.

9. $\Delta L_{Invar} = 2.0 \times 10^{-6}$ m,
$\Delta L_{steel} = 1.2 \times 10^{-4}$ m,
$\Delta L_{marble} = 2.5 \times 10^{-5}$ m.

11. −69°C.

13. 5.1 mL.

15. 0.06 cm³.

19. −40 min.

21. -2.8×10^{-3} (0.28%).

23. 3.5×10^7 N/m².

25. (a) 27°C; (b) 4.3×10^3 N.

27. −459.7°F.

29. 1.07 m³.

31. 1.43 kg/m³.

33. (a) 14.8 m³; (b) 1.81 atm.

35. 1.80×10^3 atm.

37. 37°C.

39. 3.43 atm.

41. 0.588 kg/m³, water vapor is not an ideal gas.

43. 2.69×10^{25} molecules/m³.

45. 4.9×10^{22} molecules.

47. 7.7×10^3 N.

49. (a) 71.2 torr; (b) 157°C.

51. (a) 0.19 K; (b) 0.051%.

53. (a) Low; (b) 0.017%.

55. 1/6.

57. 5.1×10^{27} molecules, 8.4×10^3 mol.

59. 11 L, not advisable.

61. (a) 9.3×10^2 kg; (b) 1.0×10^2 kg.

63. 1.1×10^{44} molecules.

65. 3.3×10^{-7} cm.

67. 1.1×10^3 m.

69. 15 h.

71. 0.66×10^3 kg/m³.

73. ± 0.11 C°.

77. 3.6 m.

CHAPTER 18

1. (a) 5.65×10^{-21} J; (b) 7.3×10^3 J.

3. 1.17.

5. (a) 4.5; (b) 5.2.

7. $\sqrt{2}$.

11. (a) 461 m/s; (b) 19 s⁻¹.

13. 1.00429.

17. Vapor.

19. (a) Gas, liquid, vapor;
(b) gas, liquid, solid, vapor.

21. 0.69 atm.

23. 11°C.

25. 1.96 atm.

27. 120°C.

29. (a) 5.3×10^6 Pa; (b) 5.7×10^6 Pa.

31. (b) $b = 4.28 \times 10^{-5}$ m³/mol,
$a = 0.365$ N·m⁴/mol².

33. (a) 10^{-7} atm; (b) 300 atm.

35. (a) 6.3 cm; (b) 0.58 cm.

37. 2×10^{-7} m.

39. (b) 4.7×10^7 s⁻¹.

43. 7.8 h.

45. (b) 4×10^{-11} mol/s; (c) 0.7 s.

47. 2.6×10^2 m/s,
4×10^{-17} N/m² $\approx 4 \times 10^{-22}$ atm.

49. (a) 2.9×10^2 m/s; (b) 12 m/s.

51. Reasonable, 70 cm.

53. $mgh = 4.3 \times 10^{-5}(\frac{1}{2}mv_{rms}^2)$, reasonable.

55. $P_2/P_1 = 1.43$, $T_2/T_1 = 1.20$.

57. 1.4×10^5 K.

59. (a) 1.7×10^3 Pa; (b) 7.0×10^2 Pa.

61. 2×10^{13} m.

CHAPTER 19

1. 1.0×10^7 J.

3. 1.8×10^2 J.

5. 2.1×10^2 kg/h.

7. 83 kcal.

9. 4.7×10^6 J.

11. 40 C°.

13. 186°C.

15. 7.1 min.

17. (b) $mc_0[(T_2 - T_1) + a(T_2^2 - T_1^2)/2]$;
(c) $c_{mean} = c_0[1 + \frac{1}{2}a(T_2 + T_1)]$.

19. 0.334 kg (0.334 L).

21. $\frac{2}{3}m$ steam and $\frac{4}{3}m$ water at 100°C.

23. 9.4 g.

25. 4.7×10^3 kcal.

27. 1.22×10^4 J/kg.

29. 360 m/s.

31. (a) 0; (b) 5.00×10^3 J.

33.

35. (a) 0; (b) −1300 kJ.

37. (a) 1.6×10^2 J; (b) $+1.6 \times 10^2$ J.

39. $W = 3.46 \times 10^3$ J, $\Delta U = 0$,
$Q = +3.46 \times 10^3$ J (into the gas).

41. +129 J.

45. (a) +25 J; (b) +63 J; (c) −95 J;
(d) −120 J; (e) −15 J.

47. $W = RT \ln\left(\dfrac{V_2 - b}{V_1 - b}\right) + a\left(\dfrac{1}{V_2} - \dfrac{1}{V_1}\right)$.

49. 22°C/h.

51. 4.98 cal/mol·K, 2.49 kcal/kg·K;
6.97 cal/mol·K, 3.48 kcal/kg·K.

53. 83.7 g/mol, krypton.

55. 46 C°.

57. (a) 2.08×10^3 J; (b) 8.32×10^2 J;
(c) 2.91×10^3 J.

59. 0.379 atm, −51°C.

61. 1.33×10^3 J.

63. (a) $T_1 = 317$ K, $T_2 = 153$ K;
(b) -1.59×10^4 J; (c) -1.59×10^4 J;
(d) $Q = 0$.

65. (a)

(b) 231 K;
(c) $Q_{1\to2} = 0$,
$\Delta U_{1\to2} = -2.01 \times 10^3$ J,
$W_{1\to2} = +2.01 \times 10^3$ J;
$W_{2\to3} = -1.31 \times 10^3$ J,
$\Delta U_{2\to3} = -1.97 \times 10^3$ J,
$Q_{2\to3} = -3.28 \times 10^3$ J,
$W_{3\to1} = 0$,
$\Delta U_{3\to1} = +3.98 \times 10^3$ J,
$Q_{3\to1} = +3.98 \times 10^3$ J;
(d) $W_{cycle} = +0.70 \times 10^3$ J,
$Q_{cycle} = +0.70 \times 10^3$ J,
$\Delta U_{cycle} = 0$.

67. (a) 64 W; (b) 22 W.

69. 4.8×10^2 W.

71. 31 h.

73. (a) 1.7×10^{17} W; (b) 278K (5°C).

75. (b) $\Delta Q/\Delta t = A(T_2 - T_1)/\Sigma(\ell_i/k_i)$.

77. 22%.

79. 4×10^{15} J.

81. 2.8 kcal/kg.

83. 30 C°.

85. 682 J.

87. 2.8 C°.

89. 2.58 cm, rod vaporizes.

91. 4.3 kg.

95. (a) 2.3 C°/s; (b) 84°C;
(c) convection, conduction, evaporation.

97. (a) $\rho = m/V = (mP/nR)/T$;
(b) $\rho = (m/nRT)P$.

99. (a) 1.9×10^5 J; (b) -1.4×10^5 J;

(c)

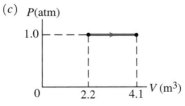

101. 3.2×10^5 s = 3.7 d.

103. 10 C°.

CHAPTER 20

1. 24%.

3. 816 MW.

5. 18%.

7. 13 km^3/day, 63 km^2.

9. 28.0%.

13. 1.2×10^{13} J/h.

15. 1.4×10^3 m/day.

17. 660°C.

19. 3.7×10^8 kg/h.

21. (a) $P_a = 5.15 \times 10^5$ Pa,
$P_b = 2.06 \times 10^5$ Pa;
(b) $V_c = 30.0$ L, $V_d = 12.0$ L;
(c) 2.83×10^3 J; (d) -2.14×10^3 J;
(e) 0.69×10^3 J; (f) 24%.

23. 5.7.

25. -21°C.

27. 2.9.

29. (a) 3.9×10^4 J; (b) 3.0 min.

31. 76 L.

33. 0.15 J/K.

35. $+11$ kcal/K.

37. $+0.0104$ cal/K·s.

39. 1.7×10^2 J/K.

43. (a) 0.312 kcal/K;
(b) > -0.312 kcal/K.

45. (a) Adiabatic process;
(b) $\Delta S_i = -nR \ln 2, \Delta S_a = 0$;
(c) $\Delta S_{surr,i} = nR \ln 2, \Delta S_{surr,a} = 0$.

47. (a) Entropy is a state function.

51. (a) 5/16; (b) 1/64.

53. (b), (a), (c), (d).

55. 69%.

57. 2.6×10^3 J/K.

59. (a) 17; (b) 5.9×10^7 J/h.

61. (a) 5.3 C°; (b) $+77$ J/kg·K.

63. $\Delta S = K/T$.

65. (a)

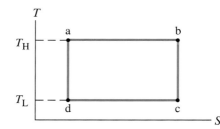

(b) area $= Q_{net} = W_{net}$.

67. $e_{Stirling} = (T_H - T_L) \ln(V_b/V_a)/$
$[T_H \ln(V_b/V_a)$
$+ \frac{3}{2}(T_H - T_L)]$,
$e_{Stirling} < e_{Carnot}$.

69. 0.091 hp.

71. (a) 1/379; (b) $1/1.59 \times 10^{11}$.

CHAPTER 21

1. 6.3×10^9 N.

3. 2.7×10^{-3} N.

5. 5.5×10^3 N.

7. 8.66 cm.

9. -5.4×10^7 C.

11. 83.8 N away from the center of the triangle.

13. 2.96×10^7 N toward the center of the square.

15. $\mathbf{F}_1 = (kQ^2/\ell^2)[(-2 + 3\sqrt{2}/4)\mathbf{i} + (4 - 3\sqrt{2}/4)\mathbf{j}], \mathbf{F}_2 = (kQ^2/\ell^2)[(2 + 2\sqrt{2})\mathbf{i} + (-6 + 2\sqrt{2})\mathbf{j}],$
$\mathbf{F}_3 = (kQ^2/\ell^2)[(-12 - 3\sqrt{2}/4)\mathbf{i} + (6 + 3\sqrt{2}/4)\mathbf{j}], \mathbf{F}_4 = (kQ^2/\ell^2)[(12 - 2\sqrt{2})\mathbf{i} + (-4 - 2\sqrt{2})\mathbf{j}].$

17. (a) $Q_1 = Q_2 = \frac{1}{2}Q_T$;
(b) Q_1 (or Q_2) $= 0$.

19. $0.402Q_0, 0.366\ell$ from Q_0.

21. 60.2×10^{-6} C, 29.8×10^{-6} C;
-16.8×10^{-6} C, 106.8×10^{-6} C.

23. $\mathbf{F} = -(1.90kQ^2/\ell^2)(\mathbf{i} + \mathbf{j} + \mathbf{k})$.

25. 2.18×10^{-16} N (west).

27. 7.43×10^6 N/C (up).

29. $(1.39 \times 10^2 \text{ N/C})\mathbf{j}$.

33.

35. 8.26×10^{-10} N/C (south).

37. 4.5×10^6 N/C up, 1.2×10^7 N/C, 56° above the horizontal.

39. 5.61×10^4 N/C away from the opposite corner.

41. $Q_1/Q_2 = \frac{1}{4}$.

43. (a) $2Qy/4\pi\epsilon_0(y^2 + \ell^2)^{3/2}\,\mathbf{j}$.

45. $\dfrac{Q}{4\pi\epsilon_0}\left[\dfrac{x\mathbf{i} - (2a/\pi)\mathbf{j}}{(x^2 + a^2)^{3/2}}\right]$.

49. $\dfrac{-2\lambda \sin\theta_0}{4\pi\epsilon_0 R}\,\mathbf{i}$.

51. (a) $\dfrac{\lambda}{4\pi\epsilon_0 x(x^2 + L^2)^{1/2}}$
$\{L\mathbf{i} + [x - (x^2 + L^2)^{1/2}]\mathbf{j}\}$.

53. $(\sigma/2\epsilon_0)\mathbf{k}$.

55. (a) $\mathbf{a} = -(3.5 \times 10^{15} \text{ m/s}^2)\,\mathbf{i}$
$-(1.41 \times 10^{16} \text{ m/s}^2)\,\mathbf{j}$; (b) $\theta = -104°$.

57. $\theta = -28°$.

59. (b) $2\pi(4\pi\epsilon_0 mR^3/qQ)^{1/2}$.

61. (a) 3.4×10^{-20} C; (b) No;
(c) 8.5×10^{-26} m·N;
(d) 2.5×10^{-26} J.

63. (a) $\theta \ll 1$;
(b) $(pE/I)^{1/2}/2\pi$.

65. (b) Direction of the dipole.

67. 6.8×10^3 C.

69. 5.7×10^{13} C.

71. $\mathbf{F}_1 = 0.30$ N, 265° from x-axis,
$\mathbf{F}_2 = 0.26$ N, 139° from x-axis,
$\mathbf{F}_3 = 0.26$ N, 30° from x-axis.

73. 4.2×10^5 N/C up.

75. $0.444Q_0, 0.333\ell$ from Q_0.

77. 5.60 m from the positive charge, and 3.60 m from the negative charge.

79. (a) In the direction of the velocity, to the right; (b) 2.1×10^2 N/C.

81. $\theta_0 = 18°$.

83. $(1.08 \times 10^7$ N/C)/
$\{3.00 - \cos[(12.5 \text{ s}^{-1})t]\}^2$, up.

85. $E_A = 4.2 \times 10^4$ N/C (right),
$E_B = -1.4 \times 10^4$ N/C (left),
$E_C = -2.8 \times 10^3$ N/C (left),
$E_D = -4.2 \times 10^4$ N/C (left).

87. $d(1 + \sqrt{2})$ from the negative charge, and $d(2 + \sqrt{2})$ from the positive charge.

CHAPTER 22

1. (a) 41 N·m^2/C; (b) 29 N·m^2/C;
(c) 0.

3. $\Phi_{net} = 0$,
$\Phi_{x=0} = -(6.50 \times 10^3 \text{ N/C})\ell^2$,
$\Phi_{x=\ell} = +(6.50 \times 10^3 \text{ N/C})\ell^2$,
$\Phi_{all \ others} = 0$.

5. 12.8 nC.

7. -1.2 μC.

9. -3.75×10^{-11} C.

11. (a) -1.0×10^4 N/C (toward wire);
(b) -2.5×10^4 N/C (toward wire).

13.

15. (a) 5.5×10^7 N/C (away from center);
(b) 0.

17. (a) -8.00 μC;
(b) $+1.00$ μC.

19. (a) 0; (b) σ/ϵ_0;
(c) unaffected.

21. (a) 0; (b) $\sigma_1 r_1^2/\epsilon_0 r^2$;
 (c) $(\sigma_1 r_1^2 + \sigma_2 r_2^2)/\epsilon_0 r^2$;
 (d) $\sigma_2/\sigma_1 = -(r_1/r_2)^2$; (e) $\sigma_1 = 0$.

23. (a) $q/4\pi\epsilon_0 r^2$;
 (b) $(1/4\pi\epsilon_0)[Q(r^3 - r_1^3)$
  $+ q(r_0^3 - r_1^3)]/(r_0^3 - r_1^3)r^2$;
 (c) $(q + Q)/4\pi\epsilon_0 r^2$.

25. (a) $q/4\pi\epsilon_0 r^2$; (b) $(q + Q)/4\pi\epsilon_0 r^2$;
 (c) $E(r < r_0) = Q/4\pi\epsilon_0 r^2$,
  $E(r > r_0) = 2Q/4\pi\epsilon_0 r^2$;
 (d) $E(r < r_0) = -Q/4\pi\epsilon_0 r^2$,
  $E(r > r_0) = 0$.

27. (a) $\sigma R_0/\epsilon_0 r$; (b) 0; (c) same.

29. (a) 0; (b) $Q/2\pi\epsilon_0 Lr$;
 (c) 0; (d) $eQ/4\pi\epsilon_0 L$.

31. (a) 0;
 (b) -2.3×10^5 N/C (toward the axis);
 (c) -1.8×10^4 N/C (toward the axis).

33. (a) $\rho_E r/2\epsilon_0$; (b) $\rho_E R_1^2/2\epsilon_0 r$;
 (c) $\rho_E(r^2 + R_1^2 - R_2^2)/2\epsilon_0 r$;
 (d) $\rho_E(R_3^2 + R_1^2 - R_2^2)/2\epsilon_0 r$;
 (e)

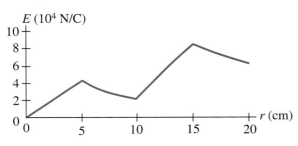

35. $Q/\epsilon_0\sqrt{2}$.

37. $\oint \mathbf{g} \cdot d\mathbf{A} = -4\pi GM$.

39. $Q_{\text{enclosed}} = \epsilon_0 b \ell^3$.

41. 3.95×10^2 N·m²/C,
 -1.69×10^2 N·m²/C.

43. (a) 0; (b) $Q/25\pi\epsilon_0 r_0^2 \le E \le Q/\pi\epsilon_0 r_0^2$;
 (c) not perpendicular;
 (d) not useful.

45. (a) $0.677e = +1.08 \times 10^{-19}$ C;
 (b) 3.5×10^{11} N/C.

47. (a) $\rho_E r_0/6\epsilon_0$ (right);
 (b) $-17\rho_E r_0/54\epsilon_0$ (left).

49. (a) 0; (b) 5.65×10^5 N/C (right);
 (c) 5.65×10^5 N/C (right);
 (d) -5.00×10^{-6} C/m²;
 (e) $+5.00 \times 10^{-6}$ C/m².

CHAPTER 23

1. -4.2×10^{-5} J (done by the field).

3. 3.4×10^{-15} J.

5. $V_a - V_b = +72.8$ V.

7. 7.04 V.

9. 0.8 μC.

11. (a) $V_{BA} = 0$; (b) $V_{CB} = -2100$ V;
 (c) $V_{CA} = -2100$ V.

13. (a) -9.6×10^8 V;
 (b) $V(\infty) = +9.6 \times 10^8$ V.

15. (a) The same;
 (b) $Q_2 = r_2 Q/(r_1 + r_2)$.

17. (a) $Q/4\pi\epsilon_0 r$;
 (b) $(Q/8\pi\epsilon_0 r_0)[3 - (r^2/r_0^2)]$;
 (c)

19. (a) $V_0 + (\sigma R_0/\epsilon_0)\ln(R_0/r)$;
 (b) $V = V_0$; (c) $V \ne 0$.

21. (a) 29 V;
 (b) -29 eV $(-4.6 \times 10^{-18}$ J).

23. $+0.19$ J.

25. 4.2 MV.

27. 2.33×10^7 m/s.

29. $V_{BA} = (1/2\pi\epsilon_0)q(2b - d)/b(d - b)$.

31. $\dfrac{\sigma}{2\epsilon_0}[(x^2 + R_2^2)^{1/2} - (x^2 + R_1^2)^{1/2}]$.

33. $\dfrac{Q}{8\pi\epsilon_0 L}\ln\left(\dfrac{x + L}{x - L}\right), x > L$.

35. $\dfrac{a}{6\epsilon_0}[(x^2 + R^2)^{1/2}(R^2 - 2x^2) + 2x^3]$.

37. 3.2 mm.

39. (a) 8.5×10^{-30} C·m; (b) zero.

41. (a) -0.088 V; (b) 1%.

43. (a) 5.2×10^{-20} C.

47. $\mathbf{E} = 2y(2z - 1)\mathbf{i} - 2(y + x - 2xz)\mathbf{j}$
 $+ (4xy)\mathbf{k}$.

49. (a) 9.6×10^4 eV; (b) 1.9×10^5 eV.

51. -2.4×10^4 V.

53. (a) $U = (1/4\pi\epsilon_0)(Q_1 Q_2/r_{12}$
  $+ Q_1 Q_3/r_{13} + Q_1 Q_4/r_{14} + Q_2 Q_3/r_{23}$
  $+ Q_2 Q_4/r_{24} + Q_3 Q_4/r_{34})$.
 (b) $U = (1/4\pi\epsilon_0)(Q_1 Q_2/r_{12}$
  $+ Q_1 Q_3/r_{13} + Q_1 Q_4/r_{14} + Q_1 Q_5/r_{15}$
  $+ Q_2 Q_3/r_{23} + Q_2 Q_4/r_{24} + Q_2 Q_5/r_{25}$
  $+ Q_3 Q_4/r_{34} + Q_3 Q_5/r_{35} + Q_4 Q_5/r_{45})$.

55. (a) 2.0 keV; (b) 42.8.

57. (a) $(-4 + \sqrt{2})Q^2/4\pi\epsilon_0 b$; (b) 0.

59. $3Q^2/20\pi\epsilon_0 r$.

61. 5.4×10^5 V/m.

63. 9×10^2 V.

65. (a) 1.1 MV; (b) 13 kg.

67. 7.2 MV.

69. 1.58×10^{12} electrons.

71. 1.7×10^{-12} V.

73. 1.03×10^6 m/s.

75. $V_a = -3.5\, Q/4\pi\epsilon_0 L$,
 $V_b = -5.2\, Q/4\pi\epsilon_0 L$,
 $V_c = -6.8\, Q/4\pi\epsilon_0 L$.

77. (a) 5.8×10^5 V; (b) 9.2×10^{-14} J.

79. $V_a - V_b = (\lambda/2\pi\epsilon_0)\ln(R_b/R_a)$.

83. (a) $\rho_E(r_2^3 - r_1^3)/3\epsilon_0 r$;
 (b) $(\rho_E/6\epsilon_0)[3r_2^2 - r^2 - (2r_1^3/r)]$;
 (c) $(\rho_E/2\epsilon_0)(r_2^2 - r_1^2)$, potential is
 continuous at r_1 and r_2.

CHAPTER 24

1. 2.6 μF.

3. 6.3 pF.

5. 0.80 μF.

7. 2.0 C.

9. 1.8×10^2 m².

11. 7.1×10^{-4} F.

13. 23 nC.

17. 4.5×10^4 V/m.

19. (a) $\epsilon_0 A/(d - \ell)$; (b) 3.

21. 2880 pF, yes.

23. (a) $(C_1 C_2 + C_1 C_3 + C_2 C_3)/$
  $(C_2 + C_3)$;
 (b) $Q_1 = 350$ μC, $Q_2 = 117$ μC.

25. (a) 3.71 μF; (b) $V_{ab} = V_1 = 26.0$ V,
 $V_2 = 14.9$ V, $V_3 = 11.1$ V.

27. 18 nF (parallel), 1.6 nF (series).

29. (a) $3C/5$; (b) $Q_1 = Q_2 = CV/5$,
 $Q_3 = 2CV/5$, $Q_4 = 3CV/5$,
 $V_1 = V_2 = V/5$, $V_3 = 2V/5$,
 $V_4 = 3V/5$.

31. $Q_1' = C_1 C_2 V_0/(C_1 + C_2)$,
 $Q_2' = C_2^2 V_0/(C_1 + C_2)$.

33. (a) $Q_1 = Q_3 = 30\ \mu$C;
$Q_2 = Q_4 = 60\ \mu$C;
(b) $V_1 = V_2 = V_3 = V_4 = 3.75$ V;
(c) 7.5 V.

35. 3.0 μF.

37. $C \approx (\epsilon_0 A/d)[1 - \frac{1}{2}(\theta \sqrt{A}/d)]$.

39. 2.0×10^{-3} J.

41. 2.3×10^3 J.

43. 1.65×10^{-7} J.

45. (a) 2.5×10^{-5} J; (b) 6.2×10^{-6} J;
(c) $Q_{par} = 4.2\ \mu$C, $Q_{ser} = 1.0\ \mu$C.

47. (a) 2.2×10^{-4} J; (b) 8.1×10^{-5} J;
(c) -1.4×10^{-4} J; (d) stored
potential energy is not conserved.

51. 1.5×10^{-10} F.

53. 0.46 μC.

55. 3.3×10^2 J.

57. $C = 2\epsilon_0 A K_1 K_2/d(K_1 + K_2)$.

59. (a) $0.40Q_0, 1.60Q_0$; (b) $0.40V_0$.

61. (a) 111 pF; (b) 1.66×10^{-8} C;
(c) 1.84×10^{-8} C;
(d) 1.17×10^5 V/m; (e) 3.34×10^4 V/m;
(f) 150 V;
(g) 172 pF; (h) 2.58×10^{-8} C.

63. 22%.

65. $Q = 4.41 \times 10^{-7}$ C,
$Q_{ind} = 3.65 \times 10^{-7}$ C,
$E_{air} = 2.69 \times 10^4$ V/m,
$E_{glass} = 4.64 \times 10^3$ V/m;

67. 11 μF.

69. (a) 4×; (b) 4×; (c) $\frac{1}{2}$×.

71. 10.9 V.

73. (b) 1.5×10^{-10} F/m.

75. $U_2/U_1 = 1/2K$, $E_2/E_1 = 1/K$.

77. (a) 19 J; (b) 0.19 MW.

79. (a) 0.10 MV; (b) voltage will
decrease exponentially.

81. 660 pF in parallel.

83. $Q_1 = 11\ \mu$C, $Q_2 = Q_3 = 13\ \mu$C,
$V_1 = 11$ V, $V_2 = 6.5$ V, $V_3 = 4.4$ V.

85. $Q^2 x/2A\epsilon_0$.

87. (a) 7.4 nF, 0.33 μC, 1.5×10^4 V/m,
7.5×10^{-6} J; (b) 27 nF, 1.2 μC,
1.5×10^4 V/m, 2.7×10^{-5} J.

89. (a) 66 pF; (b) 30 μC; (c) 7.0 mm; (d)
450 V.

CHAPTER 25

1. 9.38×10^{18} electron/s.

3. 2.1×10^{-11} A.

5. 7.5×10^2 V.

7. 2.1×10^{21} electron/min.

9. (a) 16 Ω; (b) 6.8×10^3 C.

11. 0.57 mm.

13. $R_{Al} = 1.2 R_{Cu}$.

15. 1/6 the length, 8.3 Ω, 1.7 Ω.

17. 58.3°C.

19. 1.8×10^3 °C.

21. $R_2 = \frac{1}{4}R_1$.

25. $R = (r_2 - r_1)/4\pi\sigma r_1 r_2$.

27. 3.2 W.

29. 37 V.

31. (a) 240 Ω, 0.50 A; (b) 96 Ω, 1.25 A.

33. 0.092 kWh, 22¢/month.

35. 1.1 kWh.

37. 3.

39. 5.3 kW.

41. 0.128 kg/s.

43. 0.094 A.

45. (a) Infinite; (b) 1.9×10^2 Ω.

47. 636 V, 5.66 A.

49. 1.5 kW, 3.0 kW, 0.

51. (a) 7.8×10^{-10} m/s; (b) 10.5 A/m^2
along the wire; (c) 1.8×10^{-7} V/m.

53. 2.7 A/m^2 north.

55. 12 h.

57. 6.67×10^{-2} S.

59. (a) 8.6 Ω, 1.1 W; (b) 4×.

61. (a) \$44; (b) 1.8×10^3 kg/yr.

63. (a) −19.5%; (b) percentage decrease
in the power output would be less.

65. (a) 1.44×10^3 W; (b) 17 W; (c) 11 W;
(d) 0.8¢/day.

67. (a) 1.5 kW; (b) 12.5 A.

69. 2.

71. 0.303 mm, 28.0 m.

73. (a) 1.2 kW; (b) 100 W.

75. 1.4×10^{12} protons.

77. $j_a = 2.8 \times 10^5$ A/m^2,
$j_b = 1.6 \times 10^5$ A/m^2.

CHAPTER 26

1. (a) 8.39 V; (b) 8.49 V.

3. 0.060 Ω.

5. 360 Ω, 23 Ω.

7. 25 Ω, 70 Ω, 95 Ω, 18 Ω.

9. Series connection.

11. 4.6 kΩ.

13. 310 Ω, 3.7%.

15. 960 Ω in parallel.

17. 105 Ω.

19. (a) V_1 and V_2 increase; V_3 and V_4
decrease; (b) I_1 ($= I$) and I_2
increase; I_3 and I_4 decrease;
(c) increases; (d) $I = I_1 = 0.300$ A,
$I_2 = 0, I_3 = I_4 = 0.150$ A; $I = 0.338$ A,
$I_2 = I_3 = I_4 = 0.113$ A.

21. 0.4 Ω.

23. 0.41 A.

25. $I_1 = 0.68$ A, $I_2 = -0.40$ A.

27. $I_1 = 0.18$ A right, $I_2 = 0.32$ A left,
$I_3 = 0.14$ A up.

29. $I_1 = 0.274$ A, $I_2 = 0.222$ A,
$I_3 = 0.266$ A, $I_4 = 0.229$ A,
$I_5 = 0.007$ A, $I = 0.496$ A.

31. 52 V, −28 V.
The negative value means the
battery is facing the other direction.

33. 70 V.

35. $I_1 = 0.783$ A.

39. (a) $R(3R + 5R')/8(R + R')$; (b) $R/2$.

41. (a) 3.7 nF; (b) 22 μs.

43. $t = 1.23\tau$.

45. (a) $\tau = R_1 R_2 C/(R_1 + R_2)$;
(b) $Q_{max} = \mathscr{E}R_2 C/(R_1 + R_2)$.

47. 2.1 μs.

49. 50 μA.

51. (a) 2.9×10^{-4} Ω in parallel;
(b) 35 kΩ in series.

53. 22 V, 17 V, 14% low.

55. 0.85 mA, 4.3 V.

57. 9.6 V.

61. 3.6×10^{-2} C°.

63. Two resistors in series.

65. 2.2 V, 116 V.

67. 0.19 MΩ.

69. (a) 0.10 A; (b) 0.10 A; (c) 53 mA.

71. 2.5 V.

73. 46.1 V, 0.71 Ω.

75. (a) 72.0 W; (b) 14.2 W; (c) 3.76 W.

77. (a) 40 kΩ; (b) between b and c.

79. 375 cells, 3.8 m × 0.090 m.

81. (a) 0.50 A; (b) 0.17 A; (c) 3.3 μC;
(d) 32 μs.

83. (a) + 6.8 V, 10.2 μC; (b) 28 μs.

CHAPTER 27

1. (a) 6.7 N/m; (b) 4.7 N/m.

3. 2.68 A.

5. 0.243 T.

7. (a) South pole; (b) 3.5×10^2 A;
(c) 5.22 N.

9. 5.5×10^3 A.

13. 1.05×10^{-13} N north.

15. (a) Down; (b) in; (c) right.

21. 1.6 m.

23. (a) 2.7 cm; (b) 3.8×10^{-7} s.

25. 1.034×10^8 m/s (west), gravity can be ignored.

27. $(6.4\mathbf{i} - 10.3\mathbf{j} - 0.24\mathbf{k})] \times 10^{-16}$ N.

29. 5.3×10^{-5} m, 3.3×10^{-4} m.

31. (a) $45°$; (b) 3.5×10^{-3} m.

33. (a) $2\mu B$; (b) 0.

35. (a) 4.33×10^{-5} m · N;
(b) north edge.

39. 29 μA.

41. 1.2×10^5 C/kg.

43. (a) 2.2×10^{-4} V/m;
(b) 2.7×10^{-4} m/s;
(c) 6.4×10^{28} electrons/m^3.

45. (a)Determine polarity of the emf;
(b) 0.43 m/s.

47. 1.53 mm, 0.76 mm.

51. 3.0 T up.

53. 1.1×10^{-6} m/s west.

55. 0.17 N, $68°$ above the horizontal toward the north.

57. 0.20 T, $26.6°$ from the vertical.

59. (c) 48 MeV.

61. Slower protons will deflect more, and faster protons will deflect less, $\theta = 12°$.

63. 2.0 A, down.

65. 7.3×10^{-3} T.

67. -2.1×10^{-20} J.

CHAPTER 28

1. 1.7×10^{-4} T, $3.1\times$.

3. 0.18 N attraction.

5.

7. 8.9×10^{-5} T, $70°$ above horizontal.

9. 4.0×10^{-5} T, $15°$ below the horizontal.

11. (a) $(2.0 \times 10^{-5}$ T/A$)(15$ A $- I)$ up;
(b) $(2.0 \times 10^{-5}$ T/A$)(15$ A $+ I)$ down.

13. 21 A down.

15. $[(\mu_0/4\pi)2I(d - 2x)/x(d - x)]\mathbf{j}$.

17. 4.12×10^{-5} T.

19. (b) $(\mu_0/4\pi)(2I/y)$.

21. 0.123 A.

23. (a) 6.4×10^{-3} T; (b) 3.8×10^{-3} T;
(c) 2.1×10^{-3} T.

25. (a) 51 cm; (b) 1.3×10^{-2} T.

27. (a) $(\mu_0 I_0/2\pi R_1^2)r$ circular CCW;
(b) $\mu_0 I_0/2\pi r$ circular CCW;
(c) $(\mu_0 I_0/2\pi r)(R_3^2 - r^2)/(R_3^2 - R_2^2)$ circular CCW; (d) 0.

29. 3.6×10^{-6} T.

31. $\mu_0 I/8R$ out of the page.

33. (a) $\mu_0 I(R_1 + R_2)/4R_1 R_2$ into the page;
(b) $\frac{1}{2}\pi I(R_1^2 + R_2^2)$ into the page.

35. (a) $\dfrac{Q\omega R^2}{4}\mathbf{i}$;

(b)

$$\frac{\mu_0 Q\omega}{2\pi R^2}\left[\frac{R^2 + 2x^2 - 2x\sqrt{R^2 + x^2}}{\sqrt{R^2 + x^2}}\right]\mathbf{i};$$
(c) yes.

37. (b) $B = \mu_0 IL/4\pi y(L^2 + y^2)^{1/2}$ circular.

39. (a) $(\mu_0 I_0/2\pi R)n \tan(\pi/n)$ into the page.

41. $B = \dfrac{\mu_0 I}{4\pi}\Bigg\{ \dfrac{(x^2 + y^2)^{1/2}}{xy}$

$+ \dfrac{[(b - x)^2 + y^2]^{1/2}}{y(b - x)}$

$+ \dfrac{[(a - y)^2 + (b - x)^2]^{1/2}}{(a - y)(b - x)}$

$+ \dfrac{[x^2 + (a - y)^2]^{1/2}}{x(a - y)}\Bigg\}$,

out of the page.

43. (a) 26 A · m^2; (b) 31 m · N.

45. 30 T.

47. $F_M/L = 5.84 \times 10^{-5}$ up,
$F_N/L = 3.37 \times 10^{-5}$ N/m $60°$ below the line toward P,
$F_P/L = 3.37 \times 10^{-5}$ N/m $60°$ below the line toward N.

49. 0.27 mm, 1.4 cm.

51. $B = \mu_0 jt/2$ parallel to the sheet, perpendicular to the current (opposite directions on the two sides).

53. Between long, thin and short, fat.

55. 3×10^9 A.

57. B will decrease.

59. 2.1×10^{-6} g.

61. $2\mu_0 I/L\pi$ (left).

63. 4×10^{-6} T, about 10% of the Earth's field.

CHAPTER 29

1. -3.8×10^2 V.

3. Counterclockwise.

5. 0.026 V.

7. (a) Counterclockwise; (b) clockwise;
(c) zero; (d) counterclockwise.

9. Counterclockwise.

11. (a) Clockwise; (b) 43 mV; (c) 17 mA.

13. 1.1×10^{-5} J.

15. 4.21 C.

17. (a) 5.2×10^{-2} A; (b) 0.32 mW.

19. 1.7×10^{-2} V.

21. $(\mu_0 Ia/2\pi) \ln 2$.

23. (a) 0.15 V; (b) 5.4×10^{-3} A;
(c) 4.5×10^{-4} N.

25. (a) Will move at constant speed;
(b) $v = v_0 e^{-B^2\ell^2 t/mR}$.

27. (a) $\dfrac{\mu_0 Iv}{2\pi}\ln\left(\dfrac{a + b}{b}\right)$ toward long wire;

(b) $\dfrac{\mu_0 Iv}{2\pi}\ln\left(\dfrac{a + b}{b}\right)$ away from long wire.

31. 0.33 kV, 120 rev/s.

33. 100 V.

35. 13 A.

37. 3.54×10^4 turns.

39. 0.18.

41. (a) 5.2 V;
(b) step-down transformer.

43. 549 V, 68.6 A.

45. 56.8 kW.

47. 0.188 V/m.

49. (b) Clockwise; (c) dB/d$t > 0$.

51. (a) IR/ℓ (constant); (b) $\dfrac{\mathcal{E}_0}{\ell}e^{-B^2\ell^2 t/mR}$.

53. 31 turns.

55. $v = 0.76$ m/s.

57. 184 kV.

59. 1.5×10^{17}.

61. (a) 23 A; (b) 90 V; (c) 6.9×10^2 W;
(d) 75%.

63. (a) 0.85 A; (b) 8.2.

65. $\frac{1}{2}B\omega L^2$ toward the center.

71. $B\omega r$, radially out from the axis.

CHAPTER 30

1. $M = \mu_0 N_1 N_2 A/\ell$.

3. $M/\ell = \mu_0 n_1 n_2 \pi r_2^2$.

5. $M = (\mu_0 w/2\pi) \ln(\ell_2/\ell_1)$.

7. 1.2 H.

9. 2.5×10^{-6} H.

11. $r_1 \geq 2.5$ mm.

13. 3.

15. (a) $L_1 + L_2$; (b) $= L_1 L_2 / (L_1 + L_2)$.

17. 15.9 J.

19. (a) $u_E = 4.4 \times 10^{-4}$ J/m^3,
  $u_B = 1.6 \times 10^6$ J/m^3, $u_B \gg u_E$;
  (b) $E = 6.0 \times 10^8$ V/m.

21. 4.4 J/m^3, 1.6×10^{-14} J/m^3.

23. $(\mu_0 I^2 / 4\pi) \ln(r_2/r_1)$.

25. $t/\tau = 4.6$.

27. $(dI/dt)_0 = V_0/L$.

29. (a) $(LV^2/2R^2)(1 - 2e^{-t/\tau} + e^{-2t/\tau})$;
  (b) $t/\tau = 5.3$.

31. (a) 213 pF; (b) 46.5 μH.

35. (a) $Q = Q_0/\sqrt{2}$; (b) $T/8$.

37. $R = 2.30 \ \Omega$.

41. Decrease, 1.15 kΩ.

43. 20 mH, 95 turns.

45. (a) 21 mH; (b) 45 mA;
  (c) 2.2×10^{-5} J.

47. 3.0×10^3 turns, 95 turns.

51. (b) Positioning one coil
  perpendicular to the other;
  (c) $L_1 L_2 / (L_1 + L_2)$;
  $(L_1 L_2 - M^2)/(L_1 + L_2 \mp 2M)$.

55. (a) $\frac{1}{2}(Q_0^2/C)e^{-Rt/L}$.

57.

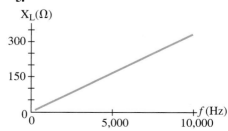

CHAPTER 31

1. (a) $3.7 \times 10^2 \ \Omega$; (b) $2.2 \times 10^{-2} \ \Omega$.

3. 9.90 Hz.

5.

7. 0.13 H.

11. (a) 5.0%; (b) 98%.

13. (a) 9.0 kΩ; (b) 10.2 kΩ.

15. (a) 18 mA; (b) $-29°$; (c) 1.8 W;
  (d) $V_R = 105$ V, $V_C = 58$ V.

17. (a) 0.38 A; (b) $-89°$; (c) 0.29 W.

19. 332 Ω.

21. (a) 0; (b) $\frac{2}{\pi} V_0$, $\overline{V}_{1/2} = \frac{2\sqrt{2}}{\pi} V_{rms}$.

23. 8.78 kΩ, $-7.66°$, 91.1 mA.

25. 265 Hz, 324 W.

27. 52.5 mA.

29. (b) $\omega'^2 = [(1/LC) - (R^2/2L^2)]$;
  (c) $k \leftrightarrow 1/C$, $m \leftrightarrow L$, $b \leftrightarrow R$.

31. (a) $V_0^2 R/[2R^2 + 2(\omega L - 1/\omega C)^2]$;
  (b) $\omega'^2 = 1/LC$; (c) $\Delta \omega = R/L$.

33. 4 Ω.

35. 9.76 nF.

37. 27.9 mH.

39. 1.6 kHz.

41. 14 Ω, 75 mH.

43. 2.2×10^3 Hz, 1.1×10^4 Hz.

45. (a) 23.6 kΩ, 10.8°; (b) 1.88×10^{-5} W;
  (c) 2.8×10^{-5} A, 0.66 V, 4.7×10^{-4} V,
  0.126 V.

49. $\{(R_1 + R_2)^2$
  $+ [\omega(L_1 + L_2) -$
  $(C_1 + C_2)/\omega C_1 C_2]^2\}^{1/2}$.

51. 19 Ω, 62 mH.

53. $I = \left(\dfrac{V_0}{Z}\right) \sin(\omega t + \phi)$,

$I_C = \left(\dfrac{V_0}{X_C}\right)\left[-\left(\dfrac{R}{Z}\right)\cos(\omega t + \phi) + \cos(\omega t)\right]$,

$I_L = \left(\dfrac{V_0}{X_L}\right)\left[\left(\dfrac{R}{Z}\right)\cos(\omega t + \phi) - \cos(\omega t)\right]$,

$Z = \sqrt{R^2 + \left(\dfrac{X_C X_L}{X_L - X_C}\right)^2}$,

$\tan \phi = \dfrac{X_C X_L}{(X_L - X_C)R}$.

CHAPTER 32

1. 9.2×10^4 V/m $\cdot$ s.

3. 7.9×10^{14} V/m $\cdot$ s.

7. $\oint \mathbf{B} \cdot d\mathbf{A} = \mu_0 Q_m$,
  $\oint \mathbf{E} \cdot d\boldsymbol{\ell} = \mu_0 \, dQ_m/dt - d\Phi_B/dt$.

9. 1.4×10^{-13} T.

11. (a) $B_0 = E_0/c$, $-y$-direction;
  (b) $-z$-direction.

13. (a) 1.08 cm; (b) 3.0×10^{18} Hz.

15. 314 nm, ultraviolet.

17. (a) 4.3 min; (b) 71 min.

19. 1.77×10^{-6} W/m^2.

21. 7.82×10^{-7} J/h.

23. 4.50×10^{-6} J.

25. 3.8×10^{26} W.

29. $r < 3 \times 10^{-7}$ m.

31. 302 pF.

33. 2.59 nH $\leq L \leq$ 3.89 nH.

35. (a) 441 m; (b) 2.979 m.

37. 5.56 m, 0.372 m.

39. (a) 1.28 s; (b) 4.3 min.

43. 469 V/m.

45. Person at the radio hears the voice
  0.14 s sooner.

47. (a) 0.40 W; (b) 12 V/m; (c) 12 V.

49. 1.5×10^{11} W.

51. (a) Parallel;
  (b) 8.9 pF $\leq C \leq$ 80 pF;
  (c) 1.05 mH $\leq L \leq$ 1.12 mH.

CHAPTER 33

1. (a) 2.21×10^8 m/s;
  (b) 1.99×10^8 m/s.

3. 8.33 min.

5. 3 m.

7. 3.4×10^3 rad/s.

9. I_3 is the desired image:

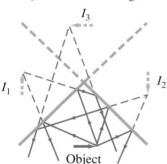

Depending on where you put your eye,
two other images may also be visible.

11. 5°.

15. 36.4 cm.

19. 4.5 m.

21. Convex, -20 m.

23. (a) Center of curvature; (b) real;
  (c) inverted; (d) -1.

29. (a) Convex mirror; (b) 22 cm behind
  surface; (c) -98 cm; (d) -196 cm.

33. 45.6°.

35. 24.9°.

37. 4.6 m.

43. 3.0%.

45. 0.22°.

47. 61.7°, lucite.

49. 93.5 cm.

51. $n_{\text{liquid}} \geq 1.5$.

55. 17.0 cm below the surface of the glass.

59. (a) 3.0 m, 4.0 m, 7.0 m; (b) toward you, away from you, toward you.

61. −3.80 m.

63. Chose different signs for the magnification; 13.3 cm, 26.7 cm.

65. $\geq 56.1°$.

69. 81 cm (inside the glass).

CHAPTER 34

1. (a)

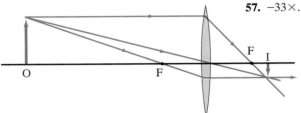

(b) 24.9 cm.

3. (a) 3.64 D, converging; (b) −16.0 cm, diverging.

5. (a) −0.26 mm; (b) −0.47 mm; (c) −1.9 mm.

7. (a) 81 mm; (b) 82 mm; (c) 87 mm; (d) 24 cm.

9. (a) Virtual; (b) converging lens; (c) 7.5 D.

11. (a) −10.5 cm (diverging), virtual; (b) +203 cm (converging).

13. 22.9 cm, 53.1 cm.

15. Real and upright.

17. Real, 21.3 cm beyond second lens, −0.708 (inverted).

19. (a) +7.14 cm; (b) −0.357 (inverted); (c)

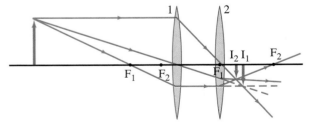

23. 1.54.

25. 8.1 cm.

27. −1.87 m (concave).

29. +1.15 D.

31. $f/2.3$.

35. 41 mm.

37. +2.3 D.

39. Glasses would be better.

41. (a) −1.33 D; (b) 38 cm.

43. −26.8 cm.

45. 17 cm, 100 cm.

47. 8.3 cm.

49. 6.3 cm from the lens, $3.9\times$.

51. (a) −49.4 cm; (b) $4.7\times$.

53. (a) $7.2\times$; (b) $2.2\times$.

55. 3.2 cm, 83 cm.

57. $-33\times$.

59. $12\times$.

61. $f_o = 4.0$ m, $r = 8.0$ m.

63. $7.5\times$.

65. 1.7 cm.

67. (a) 0.85 cm; (b) $230\times$.

69. (a) 14.4 cm; (b) $137\times$.

71. (a) 15.9 cm; (b) 14.3 cm; (c) 1.6 cm; (d) $r = 0.46$ cm.

73. $6.87\ \text{m} \leq d_o \leq \infty$.

75. 100 mm, 200 mm.

77. 79.4 cm, 75.5 cm.

79. 0.101 m, −2.7 m.

81. $0 < -d_o < -f$.

83. (c) $\Delta d = \sqrt{d_T^2 - 4d_T f}$,

$$m = \left(\frac{d_T + \sqrt{d_T^2 - 4d_T f}}{d_T - \sqrt{d_T^2 - 4d_T f}} \right)^2.$$

85. $1/f' = [(n/n') - 1][(1/R_1) + (1/R_2)]$; $(1/d_o) + (1/d_i) = 1/f'$, where $1/f' = [(n/n') - 1]/f(n - 1)$; $m = -d_i/d_o$.

87. +3.6 D.

89. $2.9\times, 4.1\times$.

91. (a) $-2.5\times$; (b) 5.0 D.

93. $-20\times$.

CHAPTER 35

3. 5.9 μm.

5. 3.9 cm.

7. 0.21 mm.

9. 613 nm.

11. 533 nm.

15. (a) $I_\theta/I_0 = (1 + 4\cos\delta + 4\cos^2\delta)/9$; (b) $\sin\theta_{\max} = m\lambda/d$, $m = 0, \pm1, \pm2, \dots$; $\sin\theta_{\min} = (m + \tfrac{1}{3}k)\lambda/d; k = 1, 2$; $m = 0, \pm1, \pm2, \dots$.

17. Orange-red.

19. 179 nm.

21. 9.1 μm.

23. 120 nm, 240 nm.

25. 1.26.

29. 0.47, 0.23.

31. 0.221 mm.

33. 0.289 mm.

35. (a) 17 lm/W; (b) 156.

37. (a) Constructive; (b) destructive.

39. 464 nm.

41. 646 nm.

43. (a) 81.5 nm; (b) 127 nm.

45. 0.5 cm.

47. $\theta = 63.3°$.

49. $\sin\theta_{\max} = (m + \tfrac{1}{2})\lambda/2S, m = 0, 1, 2, \dots$, $\sin\theta_{\min} = m\lambda/2S, m = 0, 1, 2, \dots$.

51. $I/I_0 = \cos^2(2\pi x/\lambda)$.

CHAPTER 36

1. 2.26°.

3. 2.4 m.

5. 5.8 cm.

7. 4.28 cm.

9. (b) Average intensity is $2\times$.

11. 10.7°.

13. $d = 4a$.

15. (a) 1.8 cm; (b) 11.0 cm.

17.

Center
E_0

Slit 1   Slit 2

First Minimum

Slit 2

$E_0 = 0$ Slit 1

$\sin \theta = \lambda/a$

Next maximum

E_0 Slit 2

Slit 1

Slit 1

$E_0 = 0$

Slit 2

19. 2.4×10^{-7} rad $= (1.4 \times 10^{-5})°$
  $= 0.050''$.

21. 820 lines/mm, 102 lines/mm.

23. $5.61°$.

25. Two full orders.

27. 497 nm, 612 nm, 637 nm, 754 nm.

29. 600 nm to 750 nm of second order overlaps with 400 nm to 500 nm of third order.

31. 621 nm, 909 nm.

35. (a) Two orders; (b) 6.44×10^{-5} rad $= 13.3''$, 7.36×10^{-5} rad $= 15.2''$, 2.52×10^{-4} rad $= 52.0''$.

37. (a) 1.60×10^4, 3.20×10^4; (b) 0.026 nm, 0.013 nm.

39. $\Delta f = f/mN$.

41. (a) $62.0°$; (b) 0.21 nm.

43. 0.033.

45. $57.3°$.

47. (a) $35°$; (b) $63°$.

49. $36.9°$, $53.1°$.

51. $I_0 / 32$.

55. $31°$ on either side of the normal.

57. 12,500 lines/cm.

59. $\sin \theta = \sin 20° - (m\lambda/a)$,
  $m = \pm 1, \pm 2, \ldots$.

61. Two orders.

63. $11.7°$.

65. (a) 16 km; (b) $0.42'$.

67. (a) 0; (b) $0.094 I_0$; (c) no light gets transmitted.

69. (a) $30°$; (b) $18°$; (c) $5.7°$.

73. 0.245 nm.

CHAPTER 37

1. (a) 1.00; (b) 0.99995; (c) 0.995; (d) 0.436; (e) 0.141; (f) 0.0447.

3. 2.07×10^{-6} s.

5. $0.773c$.

7. $0.141c$.

9. (a) 99.0 yr; (b) 27.7 yr; (c) 26.6 ly; (d) $0.960c$.

11. $0.89c = 2.7 \times 10^8$ m/s.

13. (a) (470 m, 20 m, 0); (b) (1820 m, 20 m, 0).

15. (a) $0.80c$; (b) $-0.80c$.

17. 60 m/s, $24°$.

19. 2.7×10^8 m/s, $43°$.

21. (a) $L_0 \sqrt{1 - (v/c)^2 \cos^2 \theta}$; (b) $\tan \theta' = \tan \theta / \sqrt{1 - (v/c)^2}$.

23. Not possible in the boy's frame.

25. $0.866c$.

27. (a) 0.5%; (b) 13%.

29. 5.36×10^{-13} kg.

31. 8.20×10^{-14} J, 0.511 MeV.

33. 9×10^2 kg.

37. (a) 11.2 GeV (1.79×10^{-9} J); (b) 6.45×10^{-18} kg · m/s.

39. 7.49×10^{-19} kg · m/s.

41. $0.941c$.

43. $M = 2m/\sqrt{1 - (v^2/c^2)}$,
  $K_{loss} = 2mc^2\{[1/\sqrt{1 - (v^2/c^2)}] - 1\}$.

45. $0.866c$, 4.73×10^{-22} kg · m/s.

47. 39 MeV (6.3×10^{-12} J), 1.5×10^{-19} kg · m/s, -6%, -4%.

49. $0.804c$.

51. 3.0 T.

57. (a) $0.80c$; (b) $2.0c$.

61. 3.8×10^{-5} s.

63. $\rho = \rho_0/[1 - (v^2/c^2)]$.

65. (a) 1.5 m/s less than c; (b) 30 cm.

67. 1.02 MeV (1.64×10^{-13} J).

69. 2.2 mm.

71. 0.78 MeV.

73. Electron.

75. $5.19 \times 10^{-13}\%$.

81. (a) $\alpha = \tan^{-1}[(c^2/v^2) - 1]^{1/2}$; (c) $\tan \theta = c/v$, $u = \sqrt{v^2 + c^2}$.

CHAPTER 38

1. 6.59×10^3 K.

3. 5.4×10^{-20} J, 0.34 eV.

5. (a) 114 J; (b) 228 J; (c) 342 J; (d) $114n$ J; (e) -456 J.

7. (b) $h = 6.63 \times 10^{-34}$ J · s.

9. 3.67×10^{-7} eV.

11. 2.4×10^{13} Hz, 1.2×10^{-5}m.

13. 400 nm.

15. (a) 2.18 eV; (b) 0.92 V.

17. 3.46 eV.

19. 1.88 eV, 43.3 kcal/mol.

21. (a) 2.43×10^{-12} m; (b) 1.32×10^{-15} m.

23. (a) 5.90×10^{-3}, 1.98×10^{-2}, 3.89×10^{-2}; (b) 60.8 eV, 204 eV, 401 eV.

27. 1.82 MeV.

29. 212 MeV, 5.85×10^{-15} m.

31. 3.2×10^{-32} m.

33. 19 V.

35. (a) 0.39 nm; (b) 0.12 nm; (c) 0.039 nm.

37. 1.84×10^3.

39. 3.3×10^{-38} m/s, no diffraction.

41. 0.026 nm.

43. 3.4 eV.

45. 122 eV.

49. 52.5 nm.

51.

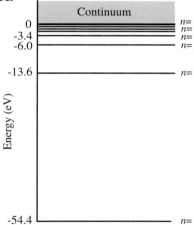

Continuum

0 — $n = \infty$, $n = 5$
-3.4 — $n = 4$, $n = 3$
-6.0

-13.6 — $n = 2$

Energy (eV)

-54.4 — $n = 1$

53. $U = -27.2$ eV, $K = +13.6$ eV.

55. Justified.

61. 3.28×10^{15} Hz.

65. 2.78×10^{21} photons/s · m².

67. 8.3×10^6 photons/s.

69. $\theta = 89.4°$.

71. 4.7×10^{-14} m.

73. 10.2 eV.

75. 4.4×10^{-40}, yes.

77. 653 nm, 102 nm, 122 nm.

79. 0.64 V.

83. 5×10^{-12} m.

85. 3.

CHAPTER 39

1. 1.8×10^{-7} m.
3. $\pm 1.3 \times 10^{-11}$ m.
5. 7.2×10^3 m/s.
7. 2.4×10^{-3} m, 1.4×10^{-32} m.
9. (a) 6.6×10^{-8} eV; (b) 6.5×10^{-9}; (c) 7.9×10^{-7} nm.
13. (a) $\psi = A \sin (3.5 \times 10^9 \text{ m}^{-1})x + B \cos (3.5 \times 10^9 \text{ m}^{-1})x$; (b) $\psi = A \sin (6.3 \times 10^{12} \text{ m}^{-1})x + B \cos (6.3 \times 10^{12} \text{ m}^{-1})x$.
17. 3.6×10^6 m/s.
19. (a) 52 nm; (b) 0.22 nm.
23. $E_1 = 0.094$ eV, $\psi_1 = (1.00 \text{ nm}^{-1/2}) \sin (1.57 \text{ nm}^{-1} x)$; $E_2 = 0.38$ eV, $\psi_2 = (1.00 \text{ nm}^{-1/2}) \sin (3.14 \text{ nm}^{-1} x)$; $E_3 = 0.85$ eV, $\psi_3 = (1.00 \text{ nm}^{-1/2}) \sin (4.71 \text{ nm}^{-1} x)$; $E_4 = 1.51$ eV, $\psi_4 = (1.00 \text{ nm}^{-1/2}) \sin (6.28 \text{ nm}^{-1} x)$.
25. (a) 4 GeV; (b) 2 MeV; (c) 2 MeV.
27. (a) 0.18; (b) 0.50; (c) 0.50.
29.

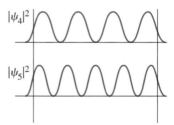

31. 0.03 nm.
33. 9.2 eV.
35. (a) Decreases by 8%; (b) decreases by 5%.
37. (a) 32 MeV; (b) 56 fm; (c) 8.8×10^{20} s^{-1}, 10^{10} yr.
39. 21 MeV.
41. 3.00×10^{-10} eV/c^2.
43. r_1, the Bohr radius.
45. 0.5 MeV, 5×10^6 m/s.
47. (b) 4 s.
49. 14% decrease.

CHAPTER 40

1. $\ell = 0, 1, 2, 3, 4, 5.$
3. 32 states, $(4,0,0,-\tfrac{1}{2})$, $(4,0,0,+\tfrac{1}{2})$, $(4,1,-1,-\tfrac{1}{2})$, $(4,1,-1,+\tfrac{1}{2})$, $(4,1,0,-\tfrac{1}{2})$, $(4,1,0,+\tfrac{1}{2})$, $(4,1,1,-\tfrac{1}{2})$, $(4,1,1,+\tfrac{1}{2})$, $(4,2,-2,-\tfrac{1}{2})$, $(4,2,-2,+\tfrac{1}{2})$, $(4,2,-1,-\tfrac{1}{2})$, $(4,2,-1,+\tfrac{1}{2})$, $(4,2,0,-\tfrac{1}{2})$, $(4,2,0,+\tfrac{1}{2})$, $(4,2,1,-\tfrac{1}{2})$, $(4,2,1,+\tfrac{1}{2})$, $(4,2,2,-\tfrac{1}{2})$, $(4,2,2,+\tfrac{1}{2})$, $(4,3,-3,-\tfrac{1}{2})$, $(4,3,-3,+\tfrac{1}{2})$, $(4,3,-2,-\tfrac{1}{2})$, $(4,3,-2,+\tfrac{1}{2})$, $(4,3,-1,-\tfrac{1}{2})$, $(4,3,-1,+\tfrac{1}{2})$, $(4,3,0,-\tfrac{1}{2})$, $(4,3,0,+\tfrac{1}{2})$, $(4,3,1,-\tfrac{1}{2})$, $(4,3,1,+\tfrac{1}{2})$, $(4,3,2,-\tfrac{1}{2})$, $(4,3,2,+\tfrac{1}{2})$, $(4,3,3,-\tfrac{1}{2})$, $(4,3,3,+\tfrac{1}{2})$.
5. $n \geq 5, m_\ell = -4, -3, -2, -1, 0, 1, 2, 3, 4$, $m_s = -\tfrac{1}{2}, +\tfrac{1}{2}.$
7. (a) 6; (b) -0.378 eV; (c) $\ell = 4$, $\sqrt{20}\hbar = 4{,}72 \times 10^{-34}$ kg$\cdot$m^2/s; (d) $m_\ell = -4, -3, -2, -1, 0, 1, 2, 3, 4$.
13. (a) $-[3/(32\pi r_0^3)^{1/2}] e^{-5/2}$; (b) $(9/32\pi r_0^3) e^{-5}$; (c) $(225/8r_0) e^{-5}$.
15. 1.85.
17. 17.3%.
19. (a) $1.34r_0$; (b) $2.7r_0$; (c) $4.2r_0$.
21. $\dfrac{r^4}{24 r_0^5} e^{-r/r_0}.$
27. (a) $\dfrac{4r^2}{27 r_0^3}\left(1 - \dfrac{2r}{3r_0} + \dfrac{2r^2}{27 r_0^2}\right)^2 e^{-2r/3r_0}$;

(b)

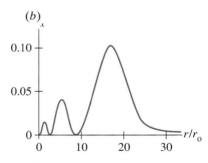

(c) $r = 13r_0$.
29. (a) $(1,0,0,-\tfrac{1}{2})$, $(1,0,0,+\tfrac{1}{2})$, $(2,0,0,-\tfrac{1}{2})$, $(2,0,0,+\tfrac{1}{2})$, $(2,1,-1,-\tfrac{1}{2})$, $(2,1,-1,+\tfrac{1}{2})$; (b) $(1,0,0,-\tfrac{1}{2})$, $(1,0,0,+\tfrac{1}{2})$, $(2,0,0,-\tfrac{1}{2})$, $(2,0,0,+\tfrac{1}{2})$, $(2,1,-1,-\tfrac{1}{2})$, $(2,1,-1,+\tfrac{1}{2})$, $(2,1,0,-\tfrac{1}{2})$, $(2,1,0,+\tfrac{1}{2})$, $(2,1,-1,-\tfrac{1}{2})$, $(2,1,-1,+\tfrac{1}{2})$, $(3,0,0,-\tfrac{1}{2})$, $(3,0,0,+\tfrac{1}{2})$.
31. (a) $1s^2 2s^2 2p^6 3s^2 3p^6 3d^{10} 4s^2 4p^4$; (b) $1s^2 2s^2 2p^6 3s^2 3p^6 3d^{10} 4s^2 4p^6 4d^{10} 4f^{14} 5s^2 5p^6 5d^{10} 6s^1$; (c) $1s^2 2s^2 2p^6 3s^2 3p^6 3d^{10} 4s^2 4p^6 4d^{10} 4f^{14} 5s^2 5p^6 5d^{10} 6s^2 6p^6 5f^3 6d^1 7s^2$.
33. 5.8×10^{-13} m, 0.115 MeV.
37. 0.041 nm, 1 nm.

41. 0.19 nm.
43. Chromium.
47. (a) 0.25 mm; (b) 0.13 mm.
49. (a) $\tfrac{1}{2}, \tfrac{3}{2}, \tfrac{1}{2}\sqrt{3}\ \hbar, \tfrac{1}{2}\sqrt{15}\hbar$; (b) $\tfrac{5}{2}, \tfrac{7}{2}, \tfrac{1}{2}\sqrt{35}\hbar, \tfrac{1}{2}\sqrt{63}\hbar$; (c) $\tfrac{3}{2}, \tfrac{5}{2}, \tfrac{1}{2}\sqrt{15}\hbar, \tfrac{1}{2}\sqrt{35}\hbar$.
51. (a) 0.4 T; (b) 0.4 T.
53. 5.6×10^{-4} rad, 1.7×10^2 m.
55. 3.7×10^4 K.
57. (a) 1.56; (b) 1.4×10^{-10} m.
59. (a) $1s^2 2s^2 2p^6 3s^2 3p^6 3d^7 4s^2$; (b) $1s^2 2s^2 2p^6 3s^2 3p^6 3d^{10} 4s^2 4p^6$; (c) $1s^2 2s^2 2p^6 3s^2 3p^6 3d^{10} 4s^2 4p^6 5s^2$.
61. (a) 2.5×10^{74}; (b) 5.0×10^{74}.
63. $r = 5.24r_0$.
65. (a) ϕ is unknown; (b) L_x and L_y are unknown.
67. (a) 1.2×10^{-4} eV; (b) 1.1 cm; (c) no difference.
69. (a) 3×10^{-171}, 7×10^{-203}; (b) 1.1×10^{-8}, 6.3×10^{-10}; (c) 6.6×10^{15}, 3.8×10^{14}; (d) 7×10^{23} photons/s, 4×10^{22} photons/s.
71. 182.
73. 2.25.

CHAPTER 41

1. 5.1 eV.
3. 0.7 eV.
7. (a) 13.941 u; (b) 7.0034 u; (c) 0.9801 u.
9. (a) 6.86 u; (b) 1.85×10^3 N/m.
11. (a) 1.79×10^{-4} eV; (b) 7.16×10^{-4} eV, 1.73 mm.
13. 2.36×10^{-10} m.
15. -7.9 eV.
17. 0.283 nm.
19. (b) -6.9 eV; (c) -10.8 eV; (d) 3%.
21. 1.8×10^{21}.
23. (a) 6.9 eV; (b) 6.8 eV.
25. 6.3%.
27. 3.2 eV, 1.05×10^6 m/s.
29. (a) 1.79×10^{29} m^{-3}; (b) 3.
33. (a) 0.021, reasonable; (b) 0.979; (c) 0.021.
37. A large energy is required to create a conduction electron by raising an electron from the valence band to the conduction band.
39. 1.1 μm.
41. 5×10^6.
43. 1.91 eV.

45. 13 mA.

47. (a) 2.1 mA; (b) 4.3 mA.

49. (a) 9.4 mA (smooth);
(b) 6.7 mA (rippled).

51. 4.0 kΩ.

53. 0.43 mA.

55. (a) 3.5×10^4 K; (b) 1.2×10^3 K.

57. (a) 0.9801 u; (b) 482 N/m, 88% of the constant for H_2.

59. States with higher values of L are less likely to be occupied, so less likely to absorb a photon; I will depend on L.

61. 2.8×10^4 J/kg.

63. 1.24 eV.

65. 1.09 μm, could be used.

67. (a) 0.094 eV; (b) 0.63 nm.

69. (a) 145 V $\leq V \leq$ 343 V;
(b) 3.34 kΩ $\leq R_{load} < \infty$.

CHAPTER 42

1. 3729 MeV/c^2.

3. 1.9×10^{-15} m.

5. (a) 2.3×10^{17} kg/m^3; (b) 184 m;
(c) 2.6×10^{-10} m.

7. 28 MeV.

9. $^{31}_{15}$P.

11. (a) 1.8×10^3 MeV;
(b) 7.3×10^2 MeV.

13. 7.48 MeV.

15. (a) 32.0 MeV, 5.33 MeV;
(b) 1636 MeV, 7.87 MeV.

17. 12.4 MeV, 7.0 MeV, neutron is more closely bound in ^{23}Na.

19. (b) Stable.

21. 0.782 MeV.

23. β^+ emitter, 1.82 MeV.

25. $^{228}_{90}$Th, 228.02883 u.

27. (a) $^{32}_{16}$S; (b) 31.97207 u.

29. 0.862 MeV.

31. 5.31 MeV.

33. (b) 0.961 MeV, 0.961 MeV to 0.

35. 3.0 h.

37. 1.2×10^9 decays/s.

39. 1.78×10^{20} nuclei.

41. 7 α particles, 4 β^- particles.

43. (a) 4.8×10^{16} nuclei;
(b) 3.2×10^{15} nuclei;
(c) 7.2×10^{13} decays/s; (d) 26 min.

45. 1.68×10^{-10} g.

47. 2.6 min.

49. 3.4 mg.

51. 8.6×10^{-7}.

53. $^{211}_{82}$Pb.

55. $N_D = N_0(1 - e^{-\lambda t})$.

57. (a) 0.99946; (b) 1.2×10^{-14};
(c) 2.3×10^{17} kg/m^3, $10^{14}\times$.

59. 28.6 eV.

61. (a) 7.2×10^{55}; (b) 1.2×10^{29} kg;
(c) 3.2×10^{11} m/s^2.

63. 6×10^3 yr.

65. 6.64 $T_{1/2}$.

69. Calcium, stored by body in bones,
193 yr, $^{90}_{38}$SR $\rightarrow$ $^{90}_{39}$Y + $^{0}_{-1}$e + $\bar{\nu}$,
$^{90}_{39}$Y is radioactive,
$^{90}_{39}$Y $\rightarrow$ $^{90}_{40}$Zr + $^{0}_{-1}$e + $\bar{\nu}$,
$^{90}_{40}$Zr is stable.

73. (a) 0.002603 u, 2.425 MeV/c^2; (b) 0;
(c) -0.094909 u, -88.41 MeV/c^2;
(d) 0.043924 u, 40.92 MeV/c^2;
(e) $\Delta \geq 0$ for $0 \leq Z \leq 8$
and $Z \geq 85$, $\Delta < 0$ for $9 \leq Z \leq 84$.

75. 0.083%.

77. $^{228}_{88}$RA, $^{228}_{89}$Ac, $^{228}_{90}$Th, $^{224}_{88}$Ra, $^{220}_{87}$Rn,
$^{231}_{90}$Th, $^{231}_{91}$Pa, $^{227}_{89}$Ac, $^{227}_{90}$Th, $^{223}_{88}$Ra.

79. (b) $\approx 10^{17}$ yr; (c) ≈ 60 yr; (d) 0.4.

CHAPTER 43

1. $^{28}_{13}$Al, β^- emitter, $^{28}_{14}$Si.

3. Possible.

5. 5.701 MeV is released.

7. (a) Can occur; (b) 19.85 MeV.

9. +4.730 MeV.

11. (a) $^{7}_{3}$Li; (b) neutron is stripped from the deuteron; (c) +5.025 MeV, exothermic.

13. (a) $^{31}_{15}$P(p, γ)$^{32}_{16}$S; (b) +8.864 MeV.

15. $\sigma = \pi(R_1 + R_2)^2$.

17. Rate at which incident particles pass through target without scattering.

19. (a) 0.7 μm; (b) 1 mm.

21. 173.2 MeV.

23. 0.116 g.

25. 25.

27. 0.11.

29. 1.3 keV.

33. 6.1×10^{23} MeV/g, 4.9×10^{23} MeV/g,
2.1×10^{24} MeV/g,
5.1×10^{23} MeV/g.

35. Not independent.

37. 3.23×10^9 J, $65\times$.

39. (a) $4.9\times$; (b) 1.5×10^9 K.

41. 400 rad.

43. 167 rad.

45. 1.7×10^2 counts/s.

49. 0.225 μg.

51. (a) 0.03 mrem $\approx$ 0.006% of allowed dose; (b) 0.3 mrem $\approx$ 0.06% of allowed dose.

53. (a) 1; (b) $1 \leq m \leq 2.7$.

55. (a) $^{12}_{6}$C; (b) +5.70 MeV.

57. 1.004.

59. 51 mrem/yr.

61. 4.6 m.

63. 18.000953 u.

65. 6.31×10^{14} J/kg, $\approx 10^7\times$ the heat of combustion of coal.

67. 1×10^{24} neutrinos/yr.

69. (a) 6.8 bn; (b) 3×10^{-14} m.

71. (a) 3.7×10^3 decays/s;
(b) 5.2×10^{-4} Sv/yr ≈ 0.15 background.

CHAPTER 44

1. 7.29 GeV.

3. 1.8 T.

5. 13 MHz.

7. $\lambda_\alpha = 2.6 \times 10^{-15}$ m $\approx$ size of nucleon, $\lambda_p = 5.2 \times 10^{-15}$ m $\approx$ 2(size of nucleon), α particle is better.

9. 1.4×10^{-18} m.

11. 2.2×10^6 km, 7.5 s.

15. 33.9 MeV.

17. 1.879 GeV.

19. 67.5 MeV.

21. 2.3×10^{-18} m.

23. (b) Uncertainty principle allows energy to not be conserved.

25. 69.3 MeV.

27. 8.6 MeV, 57.4 MeV.

29. 52.3 MeV.

31. 7.5×10^{-21} s.

33. (a) 1.3 keV; (b) 8.9 keV.

35. (a) n = d d u; (b) $\bar{n}$ = $\bar{d}$ $\bar{d}$ $\bar{u}$;
(c) Λ^0 = u d s; (d) Σ^0 = $\bar{u}$ $\bar{d}$ $\bar{s}$.

37. D^0 = c$\bar{u}$.

39. (*a*)

(*b*)

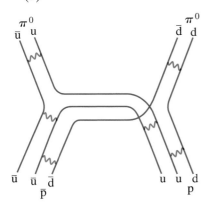

41. 26 GeV, 4.8×10^{-17} m.

43. 5.5 T.

45. (*a*) Possible, through the strong interaction; (*b*) possible, through the strong interaction; (*c*) possible, through the strong interaction; (*d*) forbidden, charge is not conserved; (*e*) possible, through the weak interaction; (*f*) forbidden, charge is not conserved; (*g*) possible, through the strong interaction; (*h*) forbidden, strangeness is not conserved; (*i*) possible, through the weak interaction; (*j*) possible, through the weak interaction.

49. -135.0 MeV, -140.9 MeV.

51. 64.

53. (*b*) 10^{27} K.

55. 6.58×10^{-5} m.

57. $\bar{u}\,\bar{u}\,\bar{d} + u\,d\,d \rightarrow \bar{u}\,d + d\,\bar{d}$.

CHAPTER 45

3. 4.8 ly.

5. $0.059''$, 17 pc.

7. Less, $\phi_1/\phi_2 = \frac{1}{2}$.

9. 48 W/m^2.

11. 1.4×10^{-4} kg/m^3.

13. 1.8×10^{9} kg/m^3, 3.3×10^{5}.

15. -92.2 keV, 7.366 MeV.

17. (*a*) 9.594 MeV is released; (*b*) 7.6 MeV; (*c*) 6×10^{10} K.

19. $d_2/d_1 = 6.5$.

23. 540°.

25. 1.4×10^{8} ly.

27. $0.86c$.

29. $0.328c$, 4.6×10^{9} ly.

31. 1.1 mm.

33. (*a*) 10^{-3}; (*b*) 10^{-10}; (*c*) 10^{-13}; (*d*) 10^{-27}.

35. (*a*) Temperature increases, luminosity is constant, size decreases; (*b*) temperature is constant, luminosity decreases, size decreases; (*c*) temperature decreases, luminosity increases, size increases.

37. 8×10^{3} rev/s.

39. 7×10^{24} W.

41. (*a*) 46 eV; (*b*) 18 eV.

43. $\approx (2 \times 10^{-5})°$.

45. 1.4×10^{16} K, hadron era.

47. Venus is brighter, $\ell_{\mathrm{V}}/\ell_{\mathrm{S}} = 16$.

49. $R \geq GM/c^2$.

51. $R = \dfrac{h^2}{16m_{\mathrm{n}}^{8/3}GM^{1/3}}\left(\dfrac{18}{\pi^2}\right)^{2/3}$.

Index

Note: The abbreviation *defn* means the page cited gives the definition of the term; *fn* means the reference is in a footnote; *pr* means it is found in a problem or question; *ff* means also the following pages. Page numbers such as A-3 refer to the appendices, after the main text.

A

Copier, electrostatic, 555
Core, of reactor, 1093
Coriolis force, 292–94
Cornea, 850
Corrective lenses, 850–52
Correspondence principle, 943, 971, 976 *pr*, 978
Cosmic microwave background radiation, 1159–61, 1164
Cosmic rays, 1102, 1114
Cosmological principle, 1157 *ff* "perfect," 1159 *fn*
Cosmology, 1136, 1140–69
Coulomb, Charles, 549–50
Coulomb (unit), 550, 712 operational definition of, 712
Coulomb barrier, 998, 1098, 1148
Coulomb's law, 549–53, 575–86, 966 vector form of, 553
Coulomb potential, 1034
Counter emf, 742–43
Counter torque, 743
Covalent bond, 1030–33, 1044
Crane, 304
Creativity in science, 3
Crick, F.H.C., 906
Critical damping, 376
Critical density, of universe, 1165–68
Critical mass, 1092–95
Critical point, 473
Critical temperature, 473
Crossed Polaroids, 908
Cross product, 279–80
Cross section, 1088–90
CRT, 605–6, 700, 805
Crystal lattice, 446–47, 1044
Crystallography, 906
CT scan, 1105–6
Curie, M. and P., 726, 1067
Curie (unit), 1101
Curie temperature, 722
Curie's law, 726
Current (*see* Electric current)
Current density, 647–49
Current gain, 1055
Current sensitivity, 674
Curvature of field, 859
Curvature of universe (space-time), 149, 1151–55, 1165
Cycle (*defn*), 363
Cyclic universe, 1168
Cyclotron, 707 *pr*, 1116–17
Cyclotron frequency, 694, 1116

D

Damped harmonic motion, 374–77
Dark matter, 1166
Dating, radioactive, 1078–79
Daughter nucleus (*defn*), 1068
Davisson, C. J., 960
dB (unit), 421
DC circuits, 658–77
DC generator, 741
DC motor, 698
de Broglie, Louis, 959–61, 971, 977–78
de Broglie wavelength, 959, 971–72
Debye's law, 514 *pr*
Decay:
    of elementary particles, 1124–34
    radioactive (*see* Radioactivity)
Decay constant, 1073–76
Decay series, 1077–78
Deceleration (*defn*), 24
Deceleration parameter, 1168
Decibel (db), 421
Declination, magnetic, 688
Decommissioning nuclear plant, 1094
Dee, 1116
Defects of the eye, 851
Defibrillator, 632 *pr*, 651 *fn*
Degrees of freedom, 500–501
Del, 603 *fn*
Delayed neutron, 1093
Delta particle, 1128
Demagnetization, 725
Density, 332–33
    and floating, 342
Density of occupied states, 1047
Density of states, 1045–48
Depth of field, 849
Derivatives, 21–22, A-4–A-5
    partial, 399
Derived quantities, 8
Descartes, R., 146
Destructive interference, 404, 430, 871, 1031
Detection of radiation, 1080
Determinism, 147, 984–85
Deuterium, 1086, 1092, 1095–1100
Deuteron, 1086
Dew point, 476
Diamagnetism, 725–26
Diamond, 826–27
Dielectric constant, 621–22
Dielectrics, 621–26
    molecular description of, 624–26

Dielectric strength, 622
Diesel engine, 540 *pr*
Differential cross section, 1089
Diffraction, 867, 887–911, 1020
    by circular opening, 896–97
    as distinguished from interference, 896
    in double slit experiment, 896
    of electrons, 960
    Fraunhofer, 888 *fn*
    Fresnel, 888 *fn*
    of light, 867, 887–911
    as limit to resolution, 896–97
    by single slit, 888–93
    of water waves, 410
    X-ray, 905–6
Diffraction grating, 900–905
    resolving power of, 903–5
Diffraction limit of lens resolution, 896–97
Diffraction patterns, 888 *ff*
    of circular opening, 896–97
    of single slit, 888–93
    X-ray, 905–6
Diffraction spot or disc, 896
Diffuse reflection, 813
Diffusion, 479–80
    Fick's law of, 480
Diffusion constant, 480
Diffusion equation, 480
Digital information, 749
Digital voltmeter, 675
Dimensional analysis, 12–13
Dimensions, 12–13
Diodes, 1052–54
Diopter, 838
Dip, angle of, 688
Dipole antenna, 793
Dipole bonds, 1036–37
Dipoles and dipole moments:
    of atoms, 1015–18
    electric, 565–67, 578, 601–3
    magnetic, 695–97, 721
Dirac, P. A. M., 1005, 1046 *fn*
Direct current (dc) (*see* Electric current)
Discharge tube, 963
"Discovery" in science, 700
Disintegration energy (*defn*), 1068–69
Disorder and order, 533–34
Dispersion, 402, 824–25
Displacement, 17–18, 46–47
    angular, 246–47

types of, in nature, 147–48, 545 *fn*, 1123, 1136
units of, 81
van der Waals, 1036–37
velocity-dependent, 122–23
viscous, 344, 350–51
weak, 1066, 1072, 1121–36
work done by, 156 *ff*
Forced convection, 505
Force diagram (*see* Free-body diagram)
Forced vibrations, 378–79
Force pump, 353
Fossil-fuel power plants, 741
Foucault, J., 868
Four-dimensional space-time, 932
Fourier integral, 401
Fourier's theorem, 401
Fovea, 850, 899
Fractal behavior, 446
Fracture, 312–15
Frames of reference, 17, 79, 291
Franklin, Benjamin, 546, 586
Fraunhofer diffraction, 888 *fn*
Free-body diagram, 88–96, 303
Free charge, 625
Free electrons, 548
Free-electron theory of metals, 1045–48
Free expansion, 498, 531, 537
Free fall, 31–33
Free particle, and Schrödinger equation, 989–94
Freezing (*see* Phase, changes of)
Frequency, 65, 242
  angular, 366
  of audible sound, 422
  beat, 429–32
  carrier, 804
  of circular motion, 65
  cyclotron, 694
  fundamental, 406, 424–26
  infrasonic, 419
  of light, 799, 869
  natural, 366, 378, 405
  resonant, 378–79, 405, 780
  of rotation, 242–43
  ultrasonic, 418, 437
  of vibration, 363, 366
  of wave, 390
Frequency modulation (FM), 804
Fresnel, A., 887
Fresnel diffraction, 888 *fn*
Friction, 78, 106–14, 118–20, 178, 184

coefficients of, 107, 112
  kinetic, 106 *ff*
  in rolling, 106, 263–64, 267
  static, 107 *ff*
Fringes, interference, 871
Frisch, O., 1090
f-stop, 848
Full-wave rectifier, 1053
Fundamental frequency, 406, 424–26
Fuse, 644–45
Fusion, heat of, 490–93
Fusion, nuclear, 1095–1100
  in stars, 1148–49, 1163
Fusion reactor, 1097

G

g-factor, 1017
Galaxies, 1141–45, 1155 *ff*
Galaxy clusters, 1143
Galilean-Newton relativity, 917–19, 932–36
Galilean telescope, 856
Galilean transformation, 932–36
Galileo, 2–3, 16, 31–32, 56–57, 78–79, 146, 339, 448, 811, 854–56, 854 *fn*, 917–18, 932, 1141, 1150
Galvani, Luigi, 635
Galvanometer, 674, 697, 755 *pr*
Gamma camera, 1104
Gamma decay, 1067, 1072–73
Gamma rays, 957, 1067, 1072–73
Gamow, G., 1160
Gas constant, 457
Gases, 332, 445–46, 454–81
  adiabatic expansion of, 502–3
  change of phase, 473–76, 490–93
  heat capacities of, 498–99
  ideal, 456–60, 466–72
  real, 473–74
  work done by, 496–98
Gas laws, 454–59
Gauge bosons, 1026–27, 1132, 1134
Gauge pressure, 337
Gauges, pressure, 338–40
Gauge theory, 1134
Gauss, C. F., 1154
Gauss (unit), 691
Gauss's law, 575–86, 713
  for magnetism, 791
Gay-Lussac, Joseph, 456
Gay-Lussac's law, 456

Geiger counter, 612 *pr*, 1080
Gell-Mann, M., 1130
General motion, 262–68, 283–84
General theory of relativity, 148, 149, 169, 922, 1140–41, 1151–55
Generator, electric, 740–42
Genetic damage, 1101
Geodesic, 1153–54
Geological dating, 1079
Geometric optics, 811 *ff*
Geophone, 759–60
Geosynchronous satellite, 140, 145
Germer, L. H., 960
Glaser, D., 1120
Glashow, S., 1134
Glasses, eye, 851–52
Glueballs, 1133 *fn*
Gluino, 1136
Gluons, 1123, 1133
Gradient:
  concentration, 480
  electric potential, 602
  pressure, 351
  velocity, 350
Gradient operator, 603 *fn*
Gram (unit), 7
Grand unified era, 1162
Grand unified theories, 148, 1134–36
Grating, diffraction (*see* Diffraction grating)
Gravitation, universal law of, 133 *ff*
Gravitational collapse, 1155
Gravitational constant (*G*), 135
Gravitational field, 146–47, 562, 1151–55
Gravitational force, 85–86, 133 *ff*
  due to spherical mass distribution, A-6–A-8
Gravitational mass, 148, 1152
Gravitational potential energy, 178–80, 192–94
Gravitino, 1164 *fn*
Graviton, 1122–23
Gravity, 32–33, 85, 133 *ff*, 1123, 1150–55
  acceleration of, 32–33, 85, 137–39
  center of, 223
  free fall under, 31–36
  specific, 333
Gravity anomalies, 138
Gray (unit), 1102
Greek alphabet, inside front cover
Grimaldi, F., 867, 872
Grounding, electrical, 651–52

Ground state, of atom, 967 *ff*
GUT, 148, 1134–36
Gyromagnetic ratio, 1017
Gyroscope, 289

H

h-bar (*h*), 1013
Hadron era, 1162
Hadrons, 1126–36, 1162
Hahn, O., 1090
Hair dryer, 646
Halflife, 1073–76
Half-wave rectification, 1053
Hall, E. A., 701
Hall effect, Hall emf, Hall field, Hall probe, 701
Halogen, 1013
Hard drive, 243–44
Harmonic motion:
  damped, 374–77
  forced, 378–79
  simple, 364–71, 390
Harmonics, 406, 424, 426–27
Harmonic wave, 398
Hazards of electricity, 651–52
Headlights, 643
Hearing, threshold of, 422
Heart, 353
  defibrillator, 632 *pr*, 651 *fn*
  pacemaker, 673, 758
Heat, 485–544
  compared to work, 493
  conduction, convection, radiation, 503–8
  distinguished from internal energy and temperature, 487
  as energy transfer, 189–90, 485 *ff*
  in first law of thermodynamics, 493–98
  as flow of energy, 485–86
  of fusion, 490–93
  latent, 490–93
  mechanical equivalent of, 486
  specific, 488 *ff*, 498–99
  of vaporization, 491
Heat capacity, 515 *pr* (*see also* Specific heat)
Heat conduction, 503–5
Heat death, 534–35
Heat engine, 517–25, 1093
  Carnot, 520–25
  efficiency of, 519, 522–23

internal combustion, 518–19
  steam, 518–19
Heating element, 642
Heat of fusion, 490–93
Heat pump, 527–28
Heat reservoir, 495
Heat transfer, 503–8
Heat of vaporization, 491
Heavy water, 1092
Heisenberg, W., 949, 978, 981–84
Heisenberg uncertainty principle, 981–84, 996, 1031
Helium:
  liquid, 474
  primordial production, 1164
Helium-neon laser, 1021
Helmholtz coils, 733 *pr*
Henry, Joseph, 734
Henry (unit), 757
Hertz, Heinrich, 798
Hertz (unit of frequency), 243, 363, 418
Hertzsprung-Russell diagram, 1146–49
Higgs boson, 1134
Higgs field, 1134
High energy accelerators, 1115–20
High energy physics, 1114–36
Highway curves, banked and unbanked, 118–20
Holes (in semiconductors), 1050–52
Holograms, 1023–24
Hooke, Robert, 309, 878 *fn*
Hooke's law, 162, 181, 309
Horsepower (unit), 195
H-R diagram, 1146–49
Hubble, E., 943, 1143, 1156–57
Hubble age, 1158
Hubble's law, 1156–58
Hubble parameter, 1156–58
Hubble Space Telescope, 897, 1140
Humidity, 476
Huygens, C., 867
Huygens' principle, 867–69
Hydraulic brakes, 338
Hydraulic lift, 338
Hydrodynamics, 343 *ff*
Hydroelectric power, 741
Hydrogen atoms, 604–5
  Bohr theory of, 965–71
  magnetic moment of, 697
  quantum mechanics of, 1004–10
  spectrum of, 968–70
Hydrogen bomb, 1098

Hydrogen bond, 1036
Hydrogen-like atoms, 966 *fn*
Hydrogen molecule, 1030–31, 1033, 1040–42
Hydrometer, 342
Hyperopia, 851
Hysteresis, 724–25
Hysteresis loop, 725

I

Ideal gas, 456–60, 466 *ff*
  internal energy of, 487–88
Ideal gas law, 456–60, 466 *ff*
Ideal gas temperature scale, 460
Ignition (fusion), 1099–1100
Illuminance, 882
Image:
  as tiny diffraction pattern, 896–97
  CAT scan, 1105–6
  fiber optic, 826–27
  formed by lens, 836 *ff*
  formed by plane mirror, 812–15
  formed by spherical mirror, 816–22
  NMR, 1064, 1107–8
  PET and SPET, 1106–7
  real, 814, 818, 839
  ultrasound, 437
  virtual, 814, 821, 840
Image distance, 814, 819
Imaging, medical, 437, 1105–9
Imbalance, rotational, 287–88
Impedance, 759, 774–75, 781–82
Impedance matching, 781–82
Impulse, 211–13
Impulsive forces, 212
Incidence, angle of, 403, 408, 813, 822
Inclines, motion on, 94–95, 262–64
Incoherent source of light, 873–74, 877
Indeterminacy principle (*see* Heisenberg uncertainty principle)
Index of refraction, 811–12
  in Snell's law, 822–24
Induced electric charge, 548–49, 625
Induced emf, 734–35, 739–40
  counter, 742–43
  in electric generator, 740–42
  in transformer, 744–47
Inductance, 756–60
  in ac circuits, 758–63, 772–83
  mutual, 756–58
  self-, 758–60

produces electric field and current, 747–49
of solenoid, 716–18
sources of, 709–26
of straight wire, 710, 720
of toroid, 708
Magnetic field lines, 687
Magnetic flux, 736, 791
Magnetic flux density (*see* Magnetic field)
Magnetic force, 686, 689–95
on electric current, 689–92
on moving electric charges, 692–95
Magnetic induction (*see* Magnetic field)
Magnetic lens, 961
Magnetic moment, 695–97, 1015–18, 1064
Magnetic permeability, 710, 724–25
Magnetic poles, 686–87
of Earth, 688
Magnetic quantum number, 1005–6, 1016
Magnetic resonance imaging, 1064, 1108–9
Magnetic tape and discs, 749
Magnetism, 686–726
Magnetization (vector), 726
Magnification:
angular, 853
of electron microscope, 961
lateral, 819, 841
of lens, 841–43
of lens combination, 843–45
longitudinal, 833 *pr*
of magnifying glass, 853–54
of microscope, 856–58, 898–99
of mirror, 819
sign conventions for, 819, 841
of telescope, 855, 898–99
useful, 899, 961
Magnifier, simple, 853–54
Magnifying glass, 836, 853–54
Magnifying power, 853, 855 (*see also* Magnification)
Magnitude (stars and galaxies), 1172 *pr*
Main-sequence, stars, 1146–49
Manhattan Project, 1095
Marconi, Guglielmo, 803
Manometer, 338
Mass, 7, 79, 82
atomic, 446
center of, 221–24

critical, 1092–95
gravitational vs. inertial, 148, 1152
inertial, 148, 1152
molecular, 446
in relativity theory, 936–38
rest, 938–40, 1063
standard of, 7
units of, 7, 79
variable, 227–29
Mass-energy transformation, 938–42
Mass excess, 1084 *pr*
Mass increase in relativity, 936–38
Mass spectrometer (spectrograph), 702
Matter, states of, 332, 446–47
Matter-dominated universe, 1161, 1164
Maxwell, James Clerk, 470, 787, 792, 797–98, 918
Maxwell distribution of molecular speeds, 470–72
Maxwell's equations, 787, 792–806, 918–19, 934
Mean free path, 478–79
Mean life, 1075, 1084 *pr*
Mean speed, 469
Measurement, 4–5, 981
Mechanical advantage, 93, 303
Mechanical energy, 182–89
Mechanical equivalent of heat, 486
Mechanical waves, 388–411
Mechanics (*defn*), 16
Medical imaging, 437, 1105–9
Meitner, L., 978, 1090
Melting point, 491 (*see also* Phase, changes of)
Mendeleev, D., 1012
Meson exchange, 1122
Mesons, 1122–36
Metallic bond, 1044
Metals, free electron theory of, 1045–48
Metastable state, 1019–20, 1073
Meter (unit), 6, 812 *fn*, 881
Meters, 674–76, 697
correction for resistance of, 676
Metric system, 6–8
MeV (million electron volts) (*see* Electron volt)
Mho (unit), 656 *pr*
Michelson, A., 811–12, 881, 919–22
Michelson interferometer, 881, 919–22

Michelson-Morley experiment, 919–23, 923 *fn*
Micrometer, 10
Microphones:
capacitor, 677
magnetic, 749
ribbon, 749
Microscope:
compound, 856–58
electron, 899, 961, 999
magnification of, 856–58, 898–99
resolving power of, 898–99
useful magnification, 899
Microscopic description of a system, 445
Microstate, 535–36
Microwaves, 798
Milky Way, 854, 1141
Millikan, R. A., 700, 953
Millikan oil-drop experiment, 700
Mirage, 869
Mirror equation, 819–22
Mirrors:
concave and convex, 819 *ff*
focal length of, 817, 821
plane, 812–15
spherical, 816–22
used in telescope, 856
Missing mass, 1166–68
MKS units (*see* SI units)
mm-Hg (unit), 339
Models, 3
Moderator, 1092–94
Modern physics (*defn*), 1, 916
Modulation, 804
Moduli of elasticity, 309–12
Molar specific heat, 498–99
Mole (*defn*), 456
volume of, for ideal gas, 457
Molecular mass, 446
Molecular rotation, 1038–40
Molecular spectra, 1037–43
Molecular speeds, 466–72
Molecular vibrations, 501, 1040–42
Molecules, 1030–43
bonding in, 1030–33
polar, 1032–33
P E diagrams for, 1033–36
spectra of, 1037–43
Moment arm, 247
Moment of a force, 247
Moment of inertia, 249–56, 374
Momentum, 206 *ff*, 936–38, 941
angular, 256–60, 281–89, 965

done by a gas, 496–98
in first law of thermodynamics, 494 *ff*
from heat engines, 517 *ff*
relation to energy, 164–69, 177–83, 190–91, 193, 195, 260–61
units of, 156
Work-energy principle, 165–68, 182, 190–91, 261
Work function, 953, 1048
Working substance (*defn*), 519
Wright, Thomas, 1141

X

X-ray crystallography, 906
X-ray diffraction, 905–6
X-rays, 799, 905–6, 1013–15, 1073, 1105–9

and atomic number, 1013–15
in electromagnetic spectrum, 799
spectra, 1013–15

Y

Young, Thomas, 870
Young's double-slit experiment, 870–73
Young's modulus, 309–10
Yukawa, H., 1121–22

Z

Z^0 particle, 1123, 1126–28, 1133
Zeeman effect, 708 *pr*, 1005, 1016, 1018, 1107
Zener diode, 1053

Zero, absolute, 455, 538
Zero-point energy, 992, 1041
Zeroth law of thermodynamics, 449–50
Zoom lens, 850
Zweig, G., 1130

Photo Credits

29-15 Tomas D.W. Friedmann/Photo Researchers, Inc. 29-20 Jon Feingersh/Comstock 29-28b National Earthquake Information Center, U.S.G.S. CO-30 Adam Hart-Davis/Science Photo Library/Photo Researchers, Inc. CO-31v1 Albert J. Copley/Visuals Unlimited CO-31v2 Mary Teresa Giancoli CO-32 Richard Megna/Fundamental Photographs 32-1 AIP Emilio Segre' Visual Archives 32-14 The Image Works CO-33 Douglas C. Giancoli 33-6 Douglas C. Giancoli 33-11a Mary Teresa Giancoli and Suzanne Saylor 33-11b Mary Teresa Giancoli 33-21 Mary Teresa Giancoli 33-25 David Parker/Science Photo Library/Photo Researchers, Inc. 33-28b Michael Giannechini/Photo Researchers, Inc. 33-34b S. Elleringmann/Bilderberg/Aurora & Quanta Productions 33-40 Douglas C. Giancoli 33-42 Mary Teresa Giancoli CO-34 Mary Teresa Giancoli 34-1c Douglas C. Giancoli 34-1d Douglas C. Giancoli 34-2 Douglas C. Giancoli and Howard Shugat 34-4 Douglas C. Giancoli 34-7a Douglas C. Giancoli 34-7b Douglas C. Giancoli 34-18 Mary Teresa Giancoli 34-19a Mary Teresa Giancoli 34-19b Mary Teresa Giancoli 34-26 Mary Teresa Giancoli 34-30a Franca Principe/Istituto e Museo di Storia della Scienza, Florence, Italy 34-32 Yerkes Observatory, University of Chicago 34-33c Palomar Observatory/California Institute of Technology 34-33d Joe McNally/Joe McNally Photography 34-35b Olympus America Inc. CO-35 Larry Mulvehill/Photo Researchers, Inc. 35-4a John M. Dunay IV/Fundamental Photographs 35-9a Bausch & Lomb Incorporated 35-16a Paul Silverman/ Fundamental Photographs 35-16b Richard Megna/Fundamental Photographs 35-16c Yoav Levy/Phototake NYC 35-18b Ken Kay/ Fundamental Photographs 35-20b Bausch & Lomb Incorporated 35-20c Bausch & Lomb Incorporated 35-22 Kristen Brochmann/ Fundamental Photographs CO-36 Richard Megna/Fundamental Photographs 36-2a Reprinted with permission from P.M. Rinard, American Journal of Physics, Vol. 44, #1, 1976, p. 70. Copyright 1976 American Association of Physics Teachers. 36-2b Ken Kay/ Fundamental Photographs 36-2c Ken Kay/Fundamental Photographs 36-11a Richard Megna/Fundamental Photographs 36-11b Richard Megna/Fundamental Photographs 36-12a and b Reproduced by permission from M. Cagnet, M. Francon, and J. Thrier, The Atlas of Optical Phenomena. Berlin: Springer-Verlag, 1962. 36-15 Space Telescope Science Institute 36-16 The Arecibo Observatory is part of the National Astronomy and Ionosphere Center which is operated by Cornell University under a cooperative agreement with the National Science Foundation. 36-21 Wabash Instrument Corp./Fundamental Photographs 36-26 Photo by W. Friedrich/Max von Laue. Burndy Library, Dibner Institute for the History of Science and Technology, Cambridge, Massachusetts. 36-29b Bausch & Lomb Incorporated 36-36 Diane Schiumo/Fundamental Photographs 36-39a Douglas C. Giancoli 36-39b Douglas C. Giancoli CO-37 Image of Albert Einstein licensed by Einstein Archives, Hebrew University, Jerusalem, represented by Roger Richman Agency, Beverly Hills, California 37-1 AIP Emilio Segre Visual Archives CO-38 Wabash Instrument Corp./Fundamental Photographs 38-10 Photo by S.A. Goudsmit, AIP Emilio Segre' Visual Archives 38-11 Education Development Center, Inc. 38-20b Richard Megna/Fundamental Photographs 38-21abc Wabash Instrument Corp./Fundamental Photographs CO-39 Institut International de Physique/American Institute of Physics/Emilio Segre Visual Archives 39-01 American Institute of Physics/Emilio Segre Visual Archives 39-02 F.D. Rosetti/American Institute of Physics/Emilio Segre Visual Archives 39-04 Advanced Research Laboratory/Hitachi Metals America, Ltd. 39-18 Driscoll, Youngquist, and Baldeschwieler, Caltech/Science Photo Library/Photo Researchers, Inc. CO-40 Patricia Peticolas/Fundamental Photographs 40-16 Paul Silverman/Fundamental Photographs 40-22 Yoav Levy/Phototake NYC 40-24b Paul Silverman/Fundamental Photographs CO-41 Charles O'Rear/Corbis CO-42 Reuters Newmedia Inc/Corbis 42-03 Chemical Heritage Foundation 42-07 University of Chicago, Courtesy of AIP Emilio Segre Visual Archives CO-43 AP/Wide World Photos 43-07 Gary Sheahan "Birth of the Atomic Age" Chicago (Illinois); 1957. Chicago Historical Society. 43-10 Liaison Agency, Inc. 43-11 LeRoy N. Sanchez/Los Alamos National Laboratory 43-12 Corbis 43-16a Lawrence Livermore National Laboratory/Science Source/Photo Researchers, Inc. 43-16b Gary Stone/Lawrence Livermore National Laboratory 43-22a Martin M. Rotker/Martin M. Rotker 43-22b Simon Fraser/Science Photo Library/Photo Researchers, Inc. 43-26b Southern Illinois University/Peter Arnold, Inc. 43-28 Mehau Kulyk/Science Photo Library/Photo Researchers, Inc. CO-44 Fermilab Visual Media Services 44-06 CERN/Science Photo Library/Photo Researchers, Inc. 44-08 Fermilab Visual Media Services CO-45 Jeff Hester and Paul Scowen, Arizona State University, and NASA 45-01 NASA Headquarters 45-02c NASA/Johnson Space Center 45-03 U.S. Naval Observatory Photo/NASA Headquarters 45-04 National Optical Astronomy Observatories 45-05a R.J. Dufour, Rice University 45-05b U.S. Naval Observatory 45-05c National Optical Astronomy Observatories 45-10 Space Telescope Science Institute/NASA Headquarters 45-11 National Optical Astronomy Observatories 45-15 European Space Agency/NASA Headquarters 45-22 Courtesy of Lucent Technologies/Bell Laboratories

Table of Contents Photos p. v (left) NOAA/Phil Degginger/Color-Pic, Inc. **p. v** (right) Mark Wagner/Tony Stone Images **p. vi** (left) Jess Stock/Tony Stone Images **p. vi** (right) Photograph by Dr. Harold E. Edgerton, © The Harold E. Edgerton 1992 Trust. Courtesy Palm Press **p. vii** (left) Ch. Russeil/Kipa/Sygma Photo News **p.vii** (right) Tibor Bognar/The Stock Market **p. viii** (left) Richard Megna/Fundamental Photographs **p. viii** (right) Douglas C. Giancoli **p. ix** (top) S. Feval/Le Matin de Lausanne/Sygma Photo News **p. ix** (bottom) Richard A. Cooke III/Tony Stone Images (right) David Woodfall/Tony Stone Images and AP/ Wide World Photos **p. x** (left) Fundamental Photographs **p. x** (right) Richard Kaylin/Tony Stone Images **p. xi** (left) Mahaux Photography/The Image Bank **p. xi** (right) Manfred Cage/Peter Arnold, Inc. **p. xii** Werner H. Muller/Peter Arnold, Inc. **p. xiii** (left) Mary Teresa Giancoli. **p. xiii** (right) Larry Mulvehill/Photo Researchers, Inc. **p. xiv** Image of Albert Einstein licensed by Einstein Archives, Hebrew University, Jerusalem, represented by Roger Richman Agency, Beverly Hills, California **p. xiv** (right) Donna McWilliam/AP/Wide World Photos **p. xv** (left) Charles O'Rear/Corbis (right) Fermilab **p. xvi** (left) Jeff Hester and Paul Scowen, Arizona State University, and NASA (right) R. J. Dufour, Rice University.

Periodic Table of the Elements[§]

Transition Elements

Key:

Symbol	Cl 17	— Atomic Number
Atomic Mass[§]	35.4527	
	$3p^5$	— Electron Configuration (outer shells only)

Element	Z	Atomic Mass	Configuration
Group I			
H	1	1.00794	$1s^1$
Li	3	6.941	$2s^1$
Na	11	22.989770	$3s^1$
K	19	39.0983	$4s^1$
Rb	37	85.4678	$5s^1$
Cs	55	132.90545	$6s^1$
Fr	87	(223)	$7s^1$
Group II			
Be	4	9.012182	$2s^2$
Mg	12	24.3050	$3s^2$
Ca	20	40.078	$4s^2$
Sr	38	87.62	$5s^2$
Ba	56	137.327	$6s^2$
Ra	88	(226)	$7s^2$
Group III (transition)			
Sc	21	44.955910	$3d^14s^2$
Y	39	88.90585	$4d^15s^2$
	57–71[†]		
	89–103[‡]		
Group IV (transition)			
Ti	22	47.867	$3d^24s^2$
Zr	40	91.224	$4d^25s^2$
Hf	72	178.49	$5d^26s^2$
Rf	104	(261)	$6d^27s^2$
Group V (transition)			
V	23	50.9415	$3d^34s^2$
Nb	41	92.90638	$4d^45s^1$
Ta	73	180.9479	$5d^36s^2$
Db	105	(262)	$6d^37s^2$
Group VI (transition)			
Cr	24	51.9961	$3d^54s^1$
Mo	42	95.94	$4d^55s^1$
W	74	183.84	$5d^46s^2$
Sg	106	(266)	$6d^47s^2$
Group VII (transition)			
Mn	25	54.938049	$3d^54s^2$
Tc	43	(98)	$4d^55s^2$
Re	75	186.207	$5d^56s^2$
Bh	107	(264)	
Group VIII (transition)			
Fe	26	55.845	$3d^64s^2$
Co	27	58.933200	$3d^74s^2$
Ni	28	58.6934	$3d^84s^2$
Ru	44	101.07	$4d^75s^1$
Rh	45	102.90550	$4d^85s^1$
Pd	46	106.42	$4d^{10}5s^0$
Os	76	190.23	$5d^66s^2$
Ir	77	192.217	$5d^76s^2$
Pt	78	195.078	$5d^96s^1$
Hs	108	(269)	
Mt	109	(268)	
110	110	(271)	
Group I (transition)			
Cu	29	63.546	$3d^{10}4s^1$
Ag	47	107.8682	$4d^{10}5s^1$
Au	79	196.96655	$5d^{10}6s^1$
111	111	(272)	
Group II (transition)			
Zn	30	65.39	$3d^{10}4s^2$
Cd	48	112.411	$4d^{10}5s^2$
Hg	80	200.59	$5d^{10}6s^2$
112	112	(277)	
Group III			
B	5	10.811	$2p^1$
Al	13	26.981538	$3p^1$
Ga	31	69.723	$4p^1$
In	49	114.818	$5p^1$
Tl	81	204.3833	$6p^1$
Group IV			
C	6	12.0107	$2p^2$
Si	14	28.0855	$3p^2$
Ge	32	72.61	$4p^2$
Sn	50	118.710	$5p^2$
Pb	82	207.2	$6p^2$
114	114	(289)	
Group V			
N	7	14.00674	$2p^3$
P	15	30.973761	$3p^3$
As	33	74.92160	$4p^3$
Sb	51	121.760	$5p^3$
Bi	83	208.98038	$6p^3$
Group VI			
O	8	15.9994	$2p^4$
S	16	32.066	$3p^4$
Se	34	78.96	$4p^4$
Te	52	127.60	$5p^4$
Po	84	(209)	$6p^4$
116	116	(289)	
Group VII			
F	9	18.9984032	$2p^5$
Cl	17	35.4527	$3p^5$
Br	35	79.904	$4p^5$
I	53	126.90447	$5p^5$
At	85	(210)	$6p^5$
Group VIII			
He	2	4.002602	$1s^2$
Ne	10	20.1797	$2p^6$
Ar	18	39.948	$3p^6$
Kr	36	83.80	$4p^6$
Xe	54	131.29	$5p^6$
Rn	86	(222)	$6p^6$
118	118	(293)	

†Lanthanide Series

Element	Z	Atomic Mass	Configuration
La	57	138.9055	$5d^16s^2$
Ce	58	140.115	$4f^15d^16s^2$
Pr	59	140.90765	$4f^35d^06s^2$
Nd	60	144.24	$4f^45d^06s^2$
Pm	61	(145)	$4f^55d^06s^2$
Sm	62	150.36	$4f^65d^06s^2$
Eu	63	151.964	$4f^75d^06s^2$
Gd	64	157.25	$4f^75d^16s^2$
Tb	65	158.92534	$4f^95d^06s^2$
Dy	66	162.50	$4f^{10}5d^06s^2$
Ho	67	164.93032	$4f^{11}5d^06s^2$
Er	68	167.26	$4f^{12}5d^06s^2$
Tm	69	168.93421	$4f^{13}5d^06s^2$
Yb	70	173.04	$4f^{14}5d^06s^2$
Lu	71	174.967	$4f^{14}5d^16s^2$

‡Actinide Series

Element	Z	Atomic Mass	Configuration
Ac	89	(227.02775)	$6d^17s^2$
Th	90	232.0381	$6d^27s^2$
Pa	91	231.03588	$5f^26d^17s^2$
U	92	238.0289	$5f^36d^17s^2$
Np	93	(237)	$5f^46d^17s^2$
Pu	94	(244)	$5f^66d^07s^2$
Am	95	(243)	$5f^76d^07s^2$
Cm	96	(247)	$5f^76d^17s^2$
Bk	97	(247)	$5f^96d^07s^2$
Cf	98	(251)	$5f^{10}6d^07s^2$
Es	99	(252)	$5f^{11}6d^07s^2$
Fm	100	(257)	$5f^{12}6d^07s^2$
Md	101	(258)	$5f^{13}6d^07s^2$
No	102	(259)	$5f^{13}6d^07s^2$
Lr	103	(262)	$5f^{14}6d^17s^2$

[§] Atomic mass values averaged over isotopes in percentages they occur on Earth's surface. For many unstable elements, mass of the longest-lived known isotope is given in parentheses. 1999 revisions. (See also Appendix D.)